D1252263

Biology of the Prokaryotes

Biology of the Prokaryotes

Edited by

Joseph W. Lengeler
University of Osnabrück

Gerhart Drews
University of Freiburg/Br.

Hans G. Schlegel
University of Göttingen

940 illustrations
150 tables

Thieme
Stuttgart
New York 1999

Blackwell
Science

Library of Congress Cataloging-in-Publication Data

Biology of the prokaryotes / edited by Joseph W. Lengeler, Gerhart Drews, Hans G. Schlegel.
p. cm
Includes bibliographical references and index.
ISBN 3-13-108411-1.
1. Prokaryotes. I. Lengeler, Joseph W. II. Drews, Gerhart III. Schlegel, Hans Günter, 1924-
[DNLM: 1. Prokaryotic Cells–physiology. 2. Bacteria. QW 51B615 1999]
QR41.2.B55 199
579.3–dc21
DNLM/DLC
for Library of Congress 98-36569
CIP

First published 1999

Distributed in the United Kingdom, Europe (outside Germany, Austria and Switzerland), the Middle East, Africa, the Indian Subcontinent and Asia by:
Blackwell Science Ltd
Osney Mead
Oxford OX2 0EL

Orders should be addressed to:
Marston Book Services Ltd
PO Box 269
Abingdon, Oxon OX14 4YN
Tel: 01235 465500
Fax: 01235 465555

Distributed in the USA by:
Blackwell Science, Inc.
Commerce Place
350 Main Street
Malden, MA 02148 5018
Tel: 800 759 6102
781 388 8250
Fax: 781 388 8255

Distributed in Australasia by:
Blackwell Science (Asia) Pty Ltd
54 University Street
Carlton, Victoria 3053
Tel: 3 9347 0300
Fax: 3 9347 5001

A catalogue record for this title is available from the British Library

Cover design by Ruth Hammelehle

ISBN- 978-0-632-05357-5

Preface

Microbiology is More Than the Biochemistry or Molecular Biology of Microorganisms

Bacteria were detected in the 17th century as minute unicellular organisms that lacked any detectable structure and occurred almost everywhere. Late in the 19th century, they were identified as a large group of organisms with distinct and specific physiological properties, such as the ability to ferment carbohydrates, to grow photoautotrophically, and to act as pathogens. The microbes were grouped systematically together with other small and allegedly primitive organisms, in particular unicellular algae, fungi, and protozoa. Today we distinguish between prokaryotic and eukaryotic microorganisms. In contrast to eukaryotic cells, prokaryotes lack a nuclear membrane (i.e., a nucleus), mitochondria and plastids, and mitosis and meiosis, but they contain particular cell wall and membrane components not found in eukaryotes. Prokaryotes are small, but neither simple nor primitive. At the morphological level, they are not "bags full of enzymes," but highly structured cells, able to grow and multiply at an astonishing speed, with cell divisions as accurate as in the eukaryotes, and with compartments that separate various metabolic activities. At the physiological level, however, the immense diversity of the prokaryotes has always been considered as their hallmark, together with their surprising adaptability to environmental changes.

Biology has been subdivided traditionally, according to the main types of organisms, into botany, zoology, and microbiology. The latter dealt with bacteria (prokaryotes), with eukaryotes of lower complexity, and with viruses. Modern biology, however, is subdivided into sections more defined by structures and organisms of increasing complexity, in particular from macromolecules and genes to the living cells, organisms, and populations. This classification facilitates the recognition of universal principles common to all living systems. In unicellular organisms, the cell is by definition also the organism. Their biology thus includes everything from the molecular structures of the cell and from cellular physiology to differentiation processes and their behavior as members of complex ecosystems.

Why a new Textbook on the Biology of the Prokaryotes?

Although many cellular components and universal biochemical mechanisms are present in all living organisms, the tremendous physiological diversity and adaptability of the prokaryotes, together with their fundamental role in environmental, biotechnological, and medical research and application, justify their separate treatment. Hence, this book is restricted to prokaryotic organisms, i.e., the true bacteria (eubacteria) and the archaea (archaebacteria), and their viruses (bacteriophages), which at the DNA- or RNA-level correspond to plasmids and not to true organisms.

Molecular biology has developed largely through studies with bacteria. This includes the rise of recombinant DNA or gene technology. It is safe to conclude that despite a shift of interest in recent times to eukaryotic organisms, the prokaryotes will continue to retain a central place both in fundamental and in applied biological research. This will, however, require new textbooks, such as this one, which presents an integrated view of the prokaryotic cell as an organism and of all prokaryotes as a large population in which all organisms communicate among themselves and with the rest of the environment.

Bacteria, although autonomous cells and complete organisms, cannot be fully understood if viewed as single cells, much as a sequenced gene cannot be understood unless its role in the biology of its organism is also considered. In this context, one of the most outstanding capacities of the prokaryotes is their extended horizontal gene transfer under natural conditions. A bacterium has access to any useful gene of any other strain and the sum of all the genes of all organisms of a community constitutes a large collective genome. Gene transfer, however, is optional and involves only a small percentage of the genes in a single transfer event. Of these, only the species-specific genes will recombine into the cellular chromosome of a cell. All others will be lost by curing unless under counter-selection. Life in temporary ecosystems of mixed populations with complementary metabolic and morphological capacities is the prokaryotic equivalent of multicellular life. Any bacterium with its cellular chromosome and variable autonomous genetic elements which is a member of an ecosystem thus resembles a differentiated cell in an eukaryotic multicellular organism. Furthermore, because no strict genetic isolation exists, speciation is not as pronounced in the prokaryotic world as in the eukaryotic world. This requires a new type of systematics. Viewed in this way, the lifestyle of the archaea (archaebacteria) resembles, despite important biochemical differences, the lifestyle of the bacteria (eubacteria) more than it resembles that of the eukaryotes.

How Is the Book Organized?

This book is based on a physiological and functional approach in which the diversity of the prokaryotic world is made visible by characteristic examples and in which up-and-coming developments are indicated. The book is divided into nine sections; the beginning sections provide the basic facts needed to understand the later sections. In this way, the book proceeds from the description of cellular structures through metabolic pathways and metabolic reactions to the genes and regulatory mechanisms. At a higher level of complexity, cell differentiation processes will be followed by a description of the diversity of prokaryotes and of their role in the biosphere. The book will end with a section on man and microbes, i.e., applied microbiology.

What Are the Aims and Scope of the Book?

The book is written for upper-level undergraduates, graduate and postgraduate students and for researchers working in fundamental research or using bacteria only as a tool, for example, in recombinant DNA technology, in biotechnology, and in medicine. Rather than presenting all the details known in biochemistry and in genetics and that can be found in such corresponding textbooks, this book concentrates on central concepts of the bacterial lifestyle and on the physiological significance that the various cellular structures, metabolic pathways, and regulatory networks have. Parts of the book, especially those dealing with the genetics of the prokaryotes and gene control, may appear "colicentric." This is because much more is understood at all levels about *Escherichia coli* than about any other bacterium, even after the complete sequenc-ing of several other bacterial chromosomes. Wherever similar phenomena are suspected to exist or have been analyzed in molecular detail and wherever new phenomena have been reported in other bacteria, e.g., sporulation and antibiotic biosynthesis, these have been used as examples. Moreover, it has been forgotten all too often that basic research in microbiology is the foundation on which applied microbiology rests. Most techniques dealing with or using prokaryotes in modern medicine, agriculture, industrial production, and en-vironmental processes profited vastly from progress in basic research. Wherever new developments and promising areas in applied microbiology can be antic-ipated, they have been pointed out.

Pedagogical Aids

Each section is preceded by a general introduction in which the subjects treated and the connections which link them are briefly described. Where possible, links to other sections are also indicated. This is especially conspicuous for Table 20.1, in which major global regulatory networks of prokaryotes are listed. These networks can best be used, as has been attempted in this book, to define the inherent logic of bacterial metabolism and to bring together seemingly unrelated phenomena that are parts of the same global network, e.g., bacterial taxes and carbon catabolite control, both involved in the quest for food; sporulation and antibiotic biosynthesis, both part of the same differentiation process; and cell surface and chromosomal rearrangements, both part of pathogen strategies in host infection. In all chapters, essential definitions are given and essential conclusions are highlighted in **shaded areas**. The corresponding pages are listed in the Index, and the sum constitutes a glossary. Historical and outstanding experiments, basic and new methods, or information for the "specialist" appear in **boxes**. All of the chapters offer **Further Reading** in which mostly recent papers and reviews are listed that can be used for further studies and research. Writing, editing, and coordination of the work was done by a team of individuals, each with expertise in the area that they covered. A list of their names and their contribution is given below. We hope that the general concept of the book and its content will increase the fascination of a broad readership for the world of the prokaryotes.

Acknowledgements

We would like to express our gratitude to all those who made this book possible. Mrs M. Hauff-Tischendorf from

Georg Thieme Verlag deserves the credit for having persuaded us to undertake the writing of a new type of textbook, the type of which she left open to us. We are indebted to the professionals who by some magic transformed our manuscripts and hand-drawn figures into a printed book. This is especially true for our copy editors Karen A. Brune and Lynn Rogers-Blaut.

The Editors

Addresses

Sankar Adhya
Laboratory of Molecular Biology
National Cancer Institute
Bethesda, Maryland 20892
USA

Carl-Alfred Alpert
INRA Institut National de la
Recherche Agronomique
Laboratoire Vivande
78352 Jouy-en-Josas Cedex
France

Ian R. Booth
University of Aberdeen
Dept. of Molecular + Cell Biology
Marishall College
Aberdeen AB9 1AS
Great Britain

Wolfgang Buckel
Fachbereich Biologie
Universität Marburg
Karl-von-Frisch-Straße
35043 Marburg
Germany

Arnold L. Demain
Department of Biology
Massachusetts Institute of Technology
77 Massachusetts Avenue
Cambridge, Mass. 02139
USA

D. N. Dowling
Dept. of Food Microbiology
University College Cork
Cork
Ireland

Gerhart Drews
Albert-Ludwigs Universität
Institut für Biologie II
Mikrobiologie
Schänzlestraße 1
79104 Freiburg/Br
Germany

Bärbel Friedrich
Humboldt-Universität
Mathemat.-Naturwissensch. Fakultät I
Institut für Biologie/Mikrobiologie
Chausseestraße 117
10115 Berlin
Germany

Georg Fuchs
Albert-Ludwigs-Universität
Inst.f.Biologie II – Mikrobiologie
Schänzlestr. 1
79104 Freiburg
Germany

Fergal O'Gara
Dept. of Food Microbiology
University College Cork
Cork
Ireland

Mike Goodfellow
Department of Microbiology
The Medical School
Framlington Place
Newcastle upon Tyne NE2 4HH,
Great Britain

Jörg Hacker
Institut für Molekulare
Infektionsbiologie
Röntgenring 11
97070 Würzburg
Germany

Wolfgang Hillen
Institut für Mikrobiologie
und Biochemie der Universität
Staudtstraße 5
91058 Erlangen
Germany

Gary R. Jacobson
Boston University
Department of Biology
2, Cummington Street
Boston, Mass. 02215
USA

Klaus Jann
Barbara Jann
Max-Planck-Institut für
Immunbiologie
Stübeweg 51
79108 Freiburg
Germany

Börries Kemper
Institut für Genetik der Universität zu
Köln
Zülpicher Str. 47
50674 Köln
Germany

Rolf Knippers
Fakultät für Biologie
Universität Konstanz
Universitätsstraße 10
78464 Konstanz
Germany

Werner Köhler
Adolf Reichwein Str. 26
07745 Jena
Germany

Reinhard Krämer
Institüt für Biochemie
der Universität Köln
Zülpicher Str. 47
50674 Köln
Germany

Achim Kröger
Institut für Mikrobiologie
Biozentrum Niederursel
Marie-Curie-Str. 9
60439 Frankfurt a.M.
Germany

J. Gijs Kuenen
Laboratory of Microbiology
Delft University of Technology
Julianalaan 67A
2628 BC Delft
The Netherlands

Giancarlo Lancini
Lepetit Research Center
Via R. Lepetit 34
21040 Gerenzano (Varese)
Italy

Erich Lanka
MPI für Molekulare Genetik
Ihnestraße 73
14195 Berlin
Germany

Joseph W. Lengeler
FB Biologie/Chemie
Universität Osnabrück
Postfach 4469
49076 Osnabrück
Germany

Edmond C.C. Lin
Harvard Medical School
Dept. of Microbiology and Molecular Genetics
Longwood Avenue
Boston, Mass. 02115
USA

Wolfgang Ludwig
Lehrstuhl für Mikrobiologie
Technische Universität München
Arcisstraße 21
802890 München
Germany

Mohamed A. Marahiel
Universität Marburg
Fachbereich Chemie
Hans-Meerwein-Straße
35043 Marburg/Lahn
Germany

Frank Mayer
Institut für Mikrobiologie
Georg-August-Universität
Grisebachstraße 8
37077 Göttingen
Germany

Walter Messer
MPI für Molekulare Genetik
Abteilung Trautner
Ihnestraße 73
14195 Berlin
Germany

Kurt Nordström
Department of Microbiology
Uppsala University
Biomedical Center
Box 581
75123 Uppsala
Sweden

M.P. Nuti
Dipt. di Biotechnologie Agrarie
Universita di Padova
Via Gradenigo 6
Padova
Italy

Werner Pansegrau
Institute for Molecular Plant Sciences
Clusius Laboratory
Leiden University
2333 AL Leiden
The Netherlands

Pieter W. Postma
Universiteit van Amsterdam
E.C. Slater Institute for Biochemical Research
Plantage Muidergracht 12
1018 TV Amsterdam
The Netherlands

Ursula B. Priefer
Institut für Botanik
Rheinisch-Westfälische Technische Hochschule
Worringer Weg
52056 Aachen
Germany

Alfred Pühler
Fakultät für Biologie VI (Genetik)
Universität Bielefeld
Postfach 10 01 31
33615 Bielefeld
Germany

Hermann Sahm
Institut für Biotechnologie I
Forschungszentrum Jülich GmbH
Postfach 1913
52428 Jülich
Germany

Bernhard Schink
Fakultät für Biologie
Universität Konstanz
Universitätsstraße 10
78464 Konstanz
Germany

Hans-Günter Schlegel
Institut für Mikrobiologie
Universität Göttingen
Grisebachstraße 8
37077 Göttingen
Germany

Erko Stackebrandt
DSM – Deutsche Sammlung von Mikroorganismen u. Zellkulturen
Mascheroder Weg 1b
38124 Braunschweig
Germany

Brian Tindall
DSM - Deutsche Sammlung von Mikroorganismen u. Zellkulturen
Mascheroder Weg 1b
38124 Braunschweig
Germany

Gottfried Unden
Institut für Mikrobiologie u. Weinforschung
FB21/Biologie
Joh. Gutenberg-Univ. Mainz
Becherweg 15
55099 Mainz
Germany

Peter Zuber
Department of Biochemistry and Molecular Biology
Louisiana State University
Shreveport, Louisiana
USA

Contents

Section I The Prokaryotic Cell

F. Mayer

Section II Basic Prerequisites for Cellular Life

A. Kröger, G. Fuchs

Section IV The Genetics of the Prokaryotes and Their Viruses

B. Friedrich, J.W. Lengeler

Section V Gene Expression and Regulatory Mechanisms

J.W. Lengeler

Section VI Cell Growth and Differentiation

G. Drews, J.W. Lengeler

Section VIII Prokaryotes in the Biosphere

B. Schink

Section IX Applied Microbiology

A. Pühler

Biochemical Nomenclature

Table **1. Biochemical units**

Physical quantity	SI unit		Expression in SI base units
	Name	Symbol	
Length	meter	m	
Mass	kilogram	kg	
Volume		m^3, l,	
Time	second	s	
Electric current	Ampere	A	
Density	ρ	g/cm^3	
Absorbance ("optical density")	A	dimensionless ($-\log T$)	
Transmittance	T	dimensionless (I/I_0)	
Temperature t	Celsius	°C	
Thermodynamic temperature T	Kelvin	K	
Amount of substance	mole	mol	
Force	Newton	N	$m\ kg\ s^{-2}$
Energy, work, heat	Joule	J	$Nm = m^2 kg\ s^{-2}$; 4.1854 J = 1 cal
Molecular mass	Dalton	Da	
Molar mass	M	g/mol	
Relative molecular mass	M_r	dimensionless	
Electric capacitance	farad	F	$CV^{-1} = m^{-2}\ kg^{-1}\ s^4\ A^2$
Electric charge	Coulomb	C	A s
Electric conductance	Siemens	S	$\Omega^{-1} = m^{-2}\ kg^{-1}\ s^3\ A^2$
Electric potential	Volt	V	$J\ C^{-1} = m^2\ kg\ s^{-3}\ A^{-1}$
Electric resistance	Ohm	Ω	$V\ A^{-1}\ s = m^2\ kg\ s^{-3}\ A^{-2}$
Frequency	Hertz	Hz	s^{-1}
Illuminance	lux	lx	$cd\ sr\ m^{-2}$
Power, radiant flux	Watt	W	$J\ s^{-1} = m^{-2}\ kg\ s^{-3}$
Pressure	Pascal	Pa	$N\ m^{-2}$; 1 bar = 10^5 Pa; 1 atm = 101325 Pa 1mmHg = 1 Torr = 133.2 Pa
Radioactive activity	Becquerel	Bq	s^{-1}
Gibbs standard Free-energy change		ΔG^0	at 1 M concentrations of reactants at 25°C
Free energy change		$\Delta G^{0\prime}$	at pH 7, 1 M conc. 25°C
Gas constant		R	8.31451 $J\ K^{-1}\ mol^{-1}$
Redox potential Oxidation-reduction potential		E_0	The potential of an electrode immersed in a solution of 1 M oxidant and 1 M reductant relative to a standard hydrogen electrode

SI prefixes

Submultiple	Prefix	Symbol	Multiple	Prefix	Symbol
10^{-1}	deci	d	10	deca	da
10^{-2}	centi	c	10^2	hecto	h
10^{-3}	milli	m	10^3	kilo	k
10^{-6}	micro	μ	10^6	mega	M
10^{-9}	nano	n	10^9	giga	G
10^{-12}	pico	p	10^{12}	tera	T
10^{-15}	femto	f	10^{15}	peta	P
10^{-18}	atto	a	10^{18}	exa	E

Table 2. **Standard symbols for chemical groups.** These symbols may be used without definition.

Symbol	Name
A, Ado	adenosine
Ac	acetyl
Ala, A	alanine
Arg, R	arginine
Asn, N	asparagine
Asp, D	aspartic acid
Asx, B	aspartic acid or asparagine
C, Cyd	cytidine
Cho	choline
Cya	cysteic acid
Cys, C	cysteine
Dol	dolichyl
Et	ethyl
Etn	ethanolamine
Fuc	fucose
Fru	fructose
Gal	galactose
GalNAc	2-deoxy-2-*N*-acetylamino-D-galactose
G, Guo	guanosine
Glc	glucose
GlcN	2-amino-2-deoxyglucose
GlcNAc	2-deoxy-2-*N*-acetylamino-D-glucose
Gln, Q	glutamine
Glu, E	glutamic acid
Glx, Z	glutamic acid or glutamine
Gly, G	glycine
Gly	glycerol
His, H	histidine
Hyl	hydroxylysine
Hyp	hydroxyproline
Ile, I	isoleucine
I, Ino	inosine
Ins	inositol
Leu, L	leucine
Lys, K	lysine
Man	mannose
Me	methyl
Met, M	methionine
Myr	myristoyl
N	an unspecified nucleoside
Neu	neuraminic acid
Ole	oleoyl
Orn	ornithine
P, p	phosphate
Pam	palmitoyl
Ph	phenyl
Phe, F	phenylalanine
Pr	propyl
Pro, P	proline
Ptd	phosphatidyl
R	an unspecified purine nucleoside
Rib	ribose
Ser, S	serine
T, Thd	ribosylthymine (not thymidine, which is dT or dThd)
Thr, T	threonine
Trp, W	tryptophan
Tyr, Y	tyrosine
U, Urd	uridine
Ψ, Ψrd	pseudouridine (5-ribosyluracil)
Val, V	valine
X, Xao	xanthosine
Xaa, X	an unspecified amino acid
Xyl	xylose
Y	an unspecified pyrimidine nucleoside

Table **3. Standard abbreviations for semi-systematic or trivial names.** These abbreviations may be used without definition.

Abbreviation	Name
ADP-Rib	adenosine (5′)diphospho(5)-β-D-ribose
AMP, ADP and ATP[a]	adenosine 5′-phosphate, 5′-diphosphate and 5′-triphosphate
bp	base pairs
cAMP and cyclic AMP	adenosine 3′,5′-phosphate
CD	circular dichroism
CMP, CDP and CTP[a]	cytidine 5′-phosphate, 5′-diphosphate and 5′-triphosphate
CoA (or CoASH)	coenzyme A
CoASAc	acetyl-coenzyme A
DEAE-cellulose	*O*-(diethylaminoethyl)-cellulose
DNA, cDNA, mtDNA and nDNA	deoxyribonucleic acid, complementary DNA, mitochondrial DNA and nuclear DNA
EDTA	ethylenediaminetetraacetate
EGTA	[ethylenebis(oxonitrilo)]tetraacetic acid
ELISA	enzyme-linked immunoabsorbant assay
EPR	electron paramagnetic resonance
ESR	electron spin resonance
FAD	flavin-adenine dinucleotide
FMN	riboflavin 5′-phosphate
FPLC	fast protein liquid chromatography
GC	gas chromatography
GLC	gas-liquid chromatography
GMP, GDP and GTP[a]	guanosine 5′-phosphate, 5′-diphosphate and 5′-triphosphate
Hb, HbCO and HbO_2	hemoglobin, carbon-monoxide hemoglobin and oxyhemoglobin
HPLC	high-performance liquid chromatography
HPTLC	high-performance thin-layer chromatography
IEF	isoelectric focussing
IgA, etc.	immunoglobulin A, etc.
IMP, IDP and ITP[a]	inosine 5′-phosphate, 5′-diphosphate and 5′-triphosphate
kb	10^3 bases
mAb	monoclonal antibody Mb, MbCO and MbO_2
MS	mass spectrometry
NAD, NAD^+ and NADH	nicotinamide-adenine dinucleotide and its oxidized and reduced forms
NADP, $NADP^+$ and NADPH	nicotinamide-adenine dinucleotide phosphate and its oxidized and reduced forms
NMN	nicotinamide mononucleotide
NMP, NDP and NTP[a]	unspecified nucleoside 5′-phosphate, 5′-diphosphate and 5′-triphosphate
NMR	nuclear magnetic resonance
NOE	nuclear Overhauser enhancement
ORD	optical rotatory dispersion
PAGE	polyacrylamide gel electrophoresis
P_i	inorganic phosphate
poly(A), etc.	(3′-5′)poly(adenylic acid), etc.
PP_i	inorganic pyrophosphate
RIA	radioimmunoassay
RNA, hnRNA, mRNA, mtRNA, nRNA, rRNA and tRNA	ribonucleic acid, heterogeneous nuclear RNA, messenger RNA, mitochondrial RNA, nuclear RNA, ribosomal RNA and transfer RNA
RNP	ribonucleoprotein
SDS	sodium dodecyl sulfate
TLC	thin-layer chromatography
TMP, TDP and TTP[a]	ribosylthymine 5′-phosphate, 5′-diphosphate and 5′-triphosphate[b]
U	unit
UDP-Glc	uridine(5′)diphospho(1)-α-D-glucose
UMP, UDP and UTP[a]	uridine 5′-phosphate, 5′-diphosphate and 5′-triphosphate
XMP, XDP and XTP[a]	xanthosine 5′-phosphate, 5′-diphosphate and 5′-triphosphate

[a] The d prefix may be used to represent the corresponding deoxynucleoside phosphates, e.g. dADP. The various isomers are written 2′NMP, 3′NMP or 5′NMP (in case of possible ambiguity).

[b] The thymidine derivatives, containing deoxyribose, are written dTMP, dTDP and dTTP, respectively.

Genetic Nomenclature

Bacteria. The key provision to describe phenotypes and genotypes of bacteria is that each observable property is given a three-letter code, usually the abbreviation of a mnemonic, according to Demerec et al. (1966). The commonly used symbols are given in Table 4.

(i) Phenotypic designations, e.g. His^+ for histidine biosynthesis, are not italicized and the first letter of the symbol is capitalized. They may be specified by lowercase superscripts which must be defined, e.g. His^+ for the wildtype, His^- for a negative phenotype, Str^S for streptomycin sensitivity, etc. Phenotypic designations can also be used to identify the protein product of a gene, e.g. the LacZ protein.

(ii) Genotypic designations in contrast are indicated in lowercase italics (e.g. *lac, his, mal*). They should correspond to the phenotypic symbol but may deviate for historical reasons. Thus to Str^S or Nal^S (for nalidixic acid sensitivity) correspond *rpsL* (for **r**ibosomal **pro**tein, **s**mall **L**) and *gyr*A (for **gyr**ase subunit **A**). If several loci govern related functions, these are distinguished by italized capital letters following the locus symbol (e.g. *lacZ, lacY, lacA*). Promoter, operator, terminator, and initiator sites are given as *lacZp* (reads promoter of gene **lac** *Z*), $lacZ_O$, $lacA_t$, and $araB_i$, for multiple sites in front of an operon as $galE_{p1}$ and $galE_{p2}$.

(iii) Wild type alleles are indicated with a superscript plus (lac^+ his^+ mal^+), but no superscript minus may be used to indicate a mutant locus. Thus, one refers to a *lac* mutant rather than to a lac^- strain. Individual mutation sites or alleles are designated by serial isolation numbers (e.g. *lacY1, lacZ4*). These cannot be assigned freely but are registered and assigned to laboratories on request. If only a single locus is known, or if it is not known in which of several related loci the mutation has occurred, a hyphen is used instead of a capital letter (e.g. *mtl-1, ara-23*).

(iv) No other superscript than + may be used for genotypes. Other mutations are indicated by phenotypic properties that follow the allele number, notably (Am) for **amber**, (Ts) for **temperature-sensitive**, (Con) for **constitutive**, (Cs) for **cold-sensitive**, (Hyb) for production of a **hybrid protein** [e.g. *ara-230(Am) hisD21(Ts)*]. **Deletions** are indicated by the symbol Δ placed before the deleted gene (e.g. Δ*trpA432*, Δ(*lac-pro*)...], fusions by a Φ [e.g. Φ((*ara-lac*)95], inversions by IN, and insertions by Ω, or in simple cases as e.g. *galT236::Tn5.*

(v) New genes, whose function have yet to be established, should be given the name of a homologous gene already identified in another organism or given a provisional name according to Demerec *et al.* (1966).

Bacteriophages, plasmids and transposable elements. The genetic nomenclature for these elements tends to differ from that for bacteria. Thus genetic symbols for phages may have from one to three letters (e.g. A11 int2 cI857 for mutations in genes A, int and cI of phage λ, respectively) and superscript plus and minus symboles may be used. To designate plasmid coded antibiotic resistances, two letter symbols are used (e.g. Ap^r for ampicillin and Sm^r for streptomycine resistance). The presence of a plasmid or an episome is indicated by parentheses [e.g. *E. coli* (F^+)], of an integrated episome as described for inserted elements, and exogenotes as e.g. W3110/F'8. The origin of a phage lysate is given as e.g. P1· *E. coli* K-12 for a P1 lysate grown on *E. coli* K-12. To designate, finally, transposon insertions at sites where there are no known loci, the chromosome of *E. coli* and related bacteria has been divided into 10 × 10 min sections marked as za., zb., zc. etc. for min 1 to 10, 11 to 20, 21 to 30 etc. Each section is divided into 1 min distances described as zaa, zab, zac, etc. for min 1, 2, 3, etc. Thus *zef-135::Tn5* designates a Tn5 insertion number 135 close to min 46 of the gene map.

For further information consult J. Bacteriol., issue 1 of each year.

Table **4. Genetic Markers**

Gene Symbol	Phene Symbol	Mnemonic	Phenotypic trait affected
Commonly used abbreviations for bacterial genetic markers			
ace	Ace	**Ace**tate	utilization
ade	Ade	**Ade**nine	biosynthesis
ala	Ala	L-**Ala**nine	biosynthesis
ara	Ara	L-**Ara**binose	utilization
arg	Arg	L-**Arg**inine	biosynthesis
aro	Aro	**Aro**matic	amino acid biosynthesis
asn	Asn	L-**As**paragi**n**e	biosynthesis
asp	Asp	L-**Asp**artate	biosynthesis
atp	Atp	ATP	ATP synthase (also *unc*)
att	Att	**Att**achement	sites for phages
bgl	Bgl	**β-Gl**ucosides	utilization
bio	Bio	**Bio**tin	biosynthesis
bla	Amp (Ap)	**β-La**ctamase	ampicillin/penicillin resistance
cat	Cam (Cm)	**C**hloramphenicol **a**cetyl-**t**ransferase	chloramphenicol resistance
che	Che	**Che**motaxis	response
chl	Chl	**Chl**orate	biosynthesis of molybdopterin
cps	Cps	**Caps**ular	polysaccharide biosynthesis
crp	Crp	**C**yclic AMP **r**eceptor **p**rotein	global regulator
crr	Crr	**C**atabolite **r**epression **r**esistance	IIA^{Glc} subunit of the glucose PTS
cya	Cya	**cy**clic **A**MP	adenylate cyclase
cys	Cys	L-**Cys**teine	biosynthesis
dna	Dna	**DNA**	biosynthesis
env	Env	**Env**elope	cell envelope
fim	Fim	**Fim**bria	morphopoiesis
flg	Flg	**Flag**ella	morphopoiesis
flh	Flh	**Flag**ella	morphopoiesis
fli	Fli	**Flag**ella	morphopoiesis
fnr	Fnr	**f**umarate-**n**itrate **r**eductase	global regulator
fru	Fru	D-**Fru**ctose	utilization
fts	Fts		cell division
gal	Gal	D-**Gal**actose	utilization
gln	Gln	L-**Gl**utami**n**e	biosynthesis
glp	Glp	sn-**Gl**ycerol 3-**p**hosphate	utilization
glt	Glt	L-**Gl**u**t**amate	utilization
gly	Gly	**Gly**cine	biosynthesis
gua	Gua	**Gua**nine	biosynthesis
gyr	Nal	**Nal**idixic acid resistance	DNA gyrase
his	His	L-**His**tidine	biosynthesis
hut	Hut	L-**Hist**idine	utilization
ile	Ile	L-Isol**eu**cine	biosynthesis
ilv	Ilv	L-Isol**eu**cine/L-**V**aline	biosynthesis
kan	Kan (Km)	**Kan**amycin	resistance
lac	Lac	**Lac**tose	utilization
leu	Leu	L-**Leu**cine	biosynthesis
liv	Liv	**L**eu-**I**le-**V**al	biosynthesis
lys	Lys	L-**Lys**ine	biosynthesis
mal	Mal	**Mal**tose	utilization
man	Man	D-**Man**nose	utilization
mel	Mel	**Mel**ibiose	uitilization
met	Met	L-**Met**hionine	biosynthesis
mot	Mot	**Mot**ility	flagellar motor

(*continued*)

Table **4.** ***(Continued)***

Gene Symbol	Phene Symbol	Mnemonic	Phenotypic trait affected
mtl	Mtl	D-**Mannitol**	utilization
mut	Mut	**Mut**ator	mutagenesis
nar	Nar	**N**itr**a**te **r**eductase	utilization
omp	Omp	**O**uter **m**embrane **p**rotein	
ori	Ori	**ori**gin	of replication
pfk	Pfk	**P**hospho**f**ructo**k**inase	
phe	Phe	L-**Phe**nylalanine	biosynthesis
pho	Pho	**Pho**sphate	utilization
pol	Pol	**Pol**ymerase	DNA polymerases
pro	Pro	L-**Pro**line	biosynthesis
pts	Pts	**P**hospho**t**ransferase **s**ystem	
pur	Pur	**Pur**ine	biosynthesis
pyr	Pyr	**Pyr**imidine	biosynthesis
rec	Rec	**Rec**ombination	enzymes
rfa	Rfa	Rough	lipopolysaccharide biosynthesis
rpl	Rpl	**R**ibosomal **p**rotein **l**arge	50s ribosomal proteins
rpo	Rpo	**R**NA **po**lymerase	subunits and sigma factors
rpoB	Rif	**Rif**ampicin resistance	RNA polymerase β subunit
rps	Rps	**R**isosomal **p**rotein **s**mall	30s ribosomal proteins
rpsL	Str^r	Streptomycin resistance	30s ribosomal protein S12
sec	Sec	**Sec**retion	of envelope proteins
ser	Ser	L-**Ser**ine	biosynthesis
suc	Suc	**Suc**cinate	utilization
sup	Sup	**Sup**pressor	mutations
tet	Tet(Tc)	**Tet**racycline	resistance
thi	Thi	**Thi**amine	biosynthesis
thr	Thr	L-**Thr**eonine	biosynthesis
thy	Thy	**Thy**mine	biosynthesis
ton	Ton	**T-one**	resistance to phage T1
tra	Tra	Gene **tra**nsfer	in conjugations
trp	Trp	L-**Tryp**tophan	biosynthesis
tsx	Tsx	**T-six**	resistance to phage T6
tyr	Tyr	L-**Tyr**osine	biosynthesis
uhp	Uhp	**U**ptake of **h**exose-**p**hosphates	
ura	Ura	**Ura**cil	biosynthesis
uvr	Uvr	**UV r**adiation	resistance
val	Val	L-**Val**ine	biosynthesis
xyl	Xyl	D-**Xyl**ose	utilization

1 Bacteriology Paved the Way to Cell Biology: a Historical Account

It was not before the last third of the nineteenth century that the bacteria were recognized as a group of organisms distinguished by a specific form, a characteristic type of cell propagation, and specific metabolic traits. Nevertheless, the bacteria became the model organisms for experimental research on the central questions of cellular biology. Thus, bacteriology paved the way to cell biology. The use of bacteria resulted in a considerable enhancement of biological research, and many "modern" concepts in today's biology developed in the nineteenth and in the first half of the twentieth century.

Among the events leading to new knowledge, the most worthy ones to remember are the beginnings. When, where, and by whom have the basic observations been made, the seminal ideas been conceived and the crucial experiments been performed? With these questions in mind, the reader will be guided through a few areas of biological research and will be able to follow their development.

1.1 New Concepts and Experimental Approaches Paved the Way for Progress

From the beginning, progress in biology profited from other sciences, such as mathematics, physics, and chemistry. The achievements of science were promoted both by hard experimental work and by new visions or concepts. A good theoretical background, phantasy, and intuition as well as practical reasoning and the skill required for performing experiments are not always united in a single person. Thus, there are many examples of brilliant ideas that have gone almost unnoticed and of otherwise outstanding scientists who had obtained the essential experimental data but failed to reach the enlightening conclusions.

Many of the most spectacular, surprising discoveries are accidental and often made during experimentation in unrelated areas. Eduard Buchner (1860–1917) discovered the cell-free fermentation by yeast press-juice when he added sugar to preserve the juice overnight for immunological studies. Alexander Fleming (1881–1955), while studying the colonial morphology of staphylococci, observed indications of bacterial lysis by a *Penicillium notatum* contamination. F. Griffith discovered DNA transformation while studying the epidemiology of *Pneumococcus*. "Serendipity" is the fashionable word for the old "faculty of making fortunate and unexpected discoveries", according to one dictionary.

Many discoveries were made independently at the same time by different individuals and a large number of researchers often contributed to a given discovery. In this overview, these will be united under one "roof." There are fewer laurels than there are merits! Furthermore, a few ingenious and influential individuals, such as Anton de Bary (1831–1888), Louis Pasteur (1822–1895), Ferdinand J. Cohn (1828–1898), Martinus W. Beijerinck (1851–1931), Sergius N. Winogradsky (1856–1953), Wilhelm Pfeffer (1845–1920), and Robert Koch (1843–1910), built with their discoveries and concepts the foundations for several routes of research, so they will be mentioned repeatedly. The present chapter ends with the concepts of the early nineteen sixties that lead to the experiments and the results described in the remainder of this book.

1.2 Observations and Speculation Lead to the First Concept of the Existence of Living Infectious Agents

In ancient times, epidemics were regarded as penalties of God. It was not before the end of the Middle Ages that **infectious agents** were considered to be involved in **epidemic diseases**. An outstanding figure of the time before bacteria had been made visible to the eye was Girolamo Fracastoro (1478–1553), a physician and poet in Verona. In his treatise "De contagionibus et contagiosis morbis et eorum curatione libri tres" ("Three books on Infections and Infectious Diseases and Their Treatment"), he came to the conclusion that infections are due to the transfer of a **contagium**, minute bodies which are alive and always produce the same disease. He regarded contagia to be involved in not only human and animal disease but also the decay of plants and fruits. Fracastoro is also known for his medical poem, entitled "Syphilis sive Morbus Gallicus" ("Syphilis or the French Disease"; 1530). Fracastoro's ideas, which were premature for his contemporaries, can only be appreciated in the light of bacteriological research and the epidemiological relationships that became known in the nineteenth century. Thus, the germ theory of disease may be attributed to a man who never even saw an infective microbe.

1.2.1 The Decisive Discovery in Experimental Microbiology: Bacteria Were Made Visible

It is easy to identify the beginning of bacteriology. The founder of the **concept of the bacterium** is Antonie van Leeuwenhoek (1632–1723). He arrived at this concept by making bacteria visible. Having seen, drawn, and described bacteria, he demonstrated the existence of organisms smaller than the known plants and animals by using simple microscopes, that had only biconvex lenses and did not exceed a 280-fold magnification. The discovery of bacteria can be followed in his correspondence (from 1673 on) with the Secretary of the Royal Society in London, Henry Oldenburg. From 1674 on, van Leeuwenhoek repeatedly reported on the "beesjes" (small beasts) and "cleijne Schepsels" (small creatures) that he saw in water, saliva, and dental tartar. The letters were translated into English and published by the Royal Society, and the terms "little/very little animalcules" were used. In his thirty-ninth letter, dated 17 September 1683, he included the now-famous drawing showing cocci, rods, and spirilla.

However, the effect of this great discovery was astonishingly small during the century that followed. Even the further **development of the microscope** was very slow. The first effective achromatic microscope was built in 1821 by Giovanni B. Amici (1784–1863), an outstanding Italian mathematician and physicist in Medina. Apochromatic objectives were introduced at the end of the century. The first microphotographs were produced by R. Koch in 1876/1877, using water-immersion lenses. The first oil-immersion lenses were introduced by Ernst Abbe (1840–1905) and Carl Zeiss (1816–1888) in Jena in 1878. The ultraviolet microscope, developed by J.E. Barnard in 1919, was followed by the development of the phase-contrast method by Frits Zernike (1888–1966) in Groningen and the phase-contrast microscope (1935), which became available from the firm Carl Zeiss, Jena as of 1940. The electron microscope was developed in 1934 in Belgium by L. Marton and in Germany by Ernst Ruska (1906–1988).

A. van Leeuwenhoek's observations are a milestone in the history of biology. The visualization of organisms smaller than worms and algae stimulated the imagination and the interest in experimentation. First, there was much speculation whether bacteria arise from nonliving matter de novo, by abiogenesis. This idea of **spontaneous generation** dates back to the ancient Greeks, who believed that fish, worms, and frogs arose from the mud of rivers and ponds. Later, the doctrine withdrew to smaller animals such as flies and their maggots and finally centered on the smallest of organisms, the bacteria. Second, the **transfer of diseases** by contact had been known since ancient times. The idea of a "contagium animatum" as the causative agent of infectious diseases appeared early, but experimental evidence was first presented nearly 200 years after van Leeuwenhoek's discovery. Third, it took a similarly long period to identify the bacteria as members of a new, large class of **independent, complete organisms** and to unite the germ theories of disease, fermentation, and putrefaction.

1.2.2 Attempts to Disprove the Doctrine of Spontaneous Generation Promoted Experimental Work

The reports of the experiments on spontaneous generation performed by Francesco Redi (1626–1697); Louis Joblot (1645–1723); John T. Needham (1713–1781); Georges Leclerc, Comte de Buffon (1707–1788); Lazzaro Spallanzani (1729–1799); Theodor Schwann (1810–1882); and others to confirm or disprove the doctrine of spontaneous generation already have been well compiled by Friedrich Loeffler (1852–1915) in his "Lectures on the historical development of bacteriology" (1887). The doctrine was widely known and confused the philosophers as well as the great experimental scientists of the time. The search for experimental evidence for or against the doctrine of spontaneous generation was the most efficient driving force for maintaining bacteriological research. Impressive evidence for rejecting the doctrine was presented by L. Pasteur in Paris. In 1858, F.A. Pouchet presented at the Paris Academy of Sciences a paper in which he claimed to have demonstrated spontaneous generation. In 1860, the Academy offered a prize with the guideline "To attempt by carefully conducted experiments, to throw new light on the question of the so-called spontaneous generations." Pasteur had just been studying yeast and bacterial fermentations, so he devoted much work to the problem. Previous investigators, Franz Schulze (1815–1873) and T. Schwann, had already shown (1836–1837) that a boiled infusion of decomposable substances would remain sterile if the air allowed to return to the headspace of the vessel had been sterilized by passage through a strong acid/alkali or through molten metal. The advocates of spontaneous generation claimed that this treatment had destroyed the "vital force" in the air. Although Pasteur's opponents could not repeat his spectacular experiments, the prize was granted to him in 1862. The reason for the contradictory results of Pouchet and Pasteur became obvious after F. Cohn discovered the heat-resistant endospores of the "Hay bacillus" in 1877: Pasteur had used a yeast infusion, which is easy to sterilize by boiling, and Pouchet had used a hay infusion, which cannot be sterilized by simple boiling due to the presence of bacterial spores. Pasteur's ingenuity was paired with fortune in this case.

1.3 Bacteria are Members of a New, Large Group of Independent Organisms

After small unicellular microorganisms had been made visible with the microscope, it became fashionable to observe various aquatic organisms, which were described by quite a number of "virtuosi", both scholars and amateurs. The first microscopist to bring order into the multitude of organisms described was Otto Friedrich Müller (1730–1784) in Copenhagen. His work, "Animalcula infusoria fluviatilia et marina" ("Animalcules of infusions, rivers, and the sea"), was published in 1786 and contains the descriptions of 379 different species.

One of the most spectacular bacterial phenomena was described at the beginning of the nineteenth century. It had preoccupied people for more than two thousand years in the form of "blood miracles" appearing on moist bread and other carbohydrate-rich foods. The phenomenon even had fatal consequences after the Catholic church started to use "hostia", the host, for communion in the 12th century. When kept in moist churches, the host often started to "bleed." This "miracle of the bleeding host" lead to many prosecutions until, in 1823, a commission of professors at the University of Padua was asked to explain this phenomenon. Bartolomeo Bizio, who described the "blood" droplets mistakenly as stalkless fungal bodies, gave them the name *Serratia marcescens* in honor of his physics teacher, Serrafino Serrati. Christian Gottfried Ehrenberg (1795–1876), who did not know of Bizio's publication, described the bacterium in 1848 under the name *Monas prodigiosa* (prodigium = miracle).

Pigmented microorganisms that appeared as pellicles, swarms, and clouds in aquatic habitats also raised the early attention of naturalists, for example, the giant phototrophic bacteria later known as *Chromatium okenii* and *Thiospirillum jenense* were observed by C.G. Ehrenberg during a walk in 1836 near Jena. The formal description, under the genus names *Monas* and *Ophidomonas*, respectively, was included in Ehrenberg's work "Die Infusionsthierchen als vollkommene Organismen" ("The infusion animalcules as complete organisms," 1838). *Gallionella ferruginea*, responsible for the formation of ocher, was also described by Ehrenberg (1838); its associate organism, the filamentous *Leptothrix ochracea*, was descirbed in 1843 by Friedrich Traugott Kützing (1807–1893).

During the following years, other pigmented bacteria were described. In "Zur Kenntnis kleinster Lebensformen" ("On the knowledge of the smallest forms of life," 1852), the Swiss botanist Maximilian Perty coined the genus name *Chromatium* and described *C. vinosum*, *C. weissei*, *C. violascens*, and *C. erubescens*. In 1873, R. Lankester (1847–1929) described the peach-red mats covering the mud of ponds, named the organism *Bacterium rubescens*, and assigned it to the bacteria. He called the characteristic pigment bacteriopurpurin. The term "purple bacteria" was coined by the Danish botanist, Eugen Warming (1841–1924).

Our modern knowledge of the forms of bacteria dates from the brilliant research of the plant physiologist F.J. Cohn in Breslau. The main results of his work were published between 1853 and 1877. When he repeated and extended experiments aiming at a better understanding of the bacteria, Cohn developed ideas and definitions which altered further research in bacteriology. He recognized that bacteria are members of a special group of organisms distinguished by a specific form, a characteristic type of cell propagation (binary fission), and specific metabolic traits. He assigned the bacteria to the plant kingdom and adopted the binary nomenclature introduced by Carl Linné (Linnaeus, 1707–1778). Cohn proposed the first bacteriological system that proved useful; it became compulsory for both medical and general bacteriologists. His main discovery was the observation of highly refractive spores in the "Hay bacillus," renamed *Bacillus subtilis*, isolated from hay decocts (1877). The description of *Bacillus anthracis* by R. Koch (1877) confirmed the value of the endospores as a differentiating characteristic in systematics.

How difficult it was at that time to accept the arguments in favor of the constancy of species of bacteria as issued by Cohn follows from the controversy of **pleomorphism** and **monomorphism** of bacteria. Around 1850, the description of rust and smut fungi and their morphological changes in the course of the "alteration of generations" had caused confusion among the early bacteriologists. They assumed that one and the same bacterium is able to appear in quite different forms (pleomorphic), to cause completely different diseases, and to form totally different metabolic products, only dependent on the growth conditions. This doctrine was promoted by great names of the time, such as T. Billroth, Carl W. Naegeli (1817–1891), E. Hallier, J. Lister, R. Lankester, and W. Zopf, before and after pure-culture methods had been achieved, and lasted until 1884. Others, such as L. Pasteur, F. Cohn, R. Koch, and O. Brefeld repeatedly argued for the necessity of pure-culture studies and emphasized the value of diligent work. Thus, the acceptance of the concept of monomorphism, i.e., that even bacteria are constant species, took a long time. Nothing is more difficult than to change one's mind. Understandably, the emphasis on the constancy of bacterial species caused confusion when variations and mutants were observed in pure cultures.

1.4 The Introduction of Solid, Defined Media and Pure-Culture Methods Marks a True Revolution

The first media used for the cultivation of bacteria were liquid. Even L. Pasteur grew the organisms in decocts. He had realized that yeasts, like all other plants, are composed of carbon, oxygen, hydrogen, nitrogen, and several of ash constituents (1858). Yeasts and bacteria were able to grow in a solution containing distilled water, ammonium tartrate, rock sugar, and yeast ash. The last component was replaced later by the respective mineral salts and the carbon source was variied. **Serial dilution** of bacteria in liquid media to achieve a pure culture was introduced in 1873 by Joseph Lister (1827–1912), and the strain of *Streptococcus lactis* that he isolated using this procedure was proved (1909) indeed to be a pure culture (Sigurd Orla-Jensen, 1870–1949). However, this procedure did not give pure cultures in the hands of every researcher. Thus, the development of appropriate solid media meant a true revolution. **Solid media**, natural substrates such as animal excrements, dung balls, or slices of carrots or potatoes, had been used for growing fungi for decades. In 1852, when transparent nutrient media were required for studying the life cycle of fungi gelatin was added (C. Vittadini). Oskar Brefeld (1839–1925) used this technique to study *Empusa muscae* (1868). The preparation of solid, transparent gelatin media and the design of the **pour plate method** (on glass plates) required the ingenuity of R. Koch (1881), whose familiarity with gelatin as a solidifying agent may be due to his practice of preparing photographic plates himself. The proposal to replace gelatin with **agar** as a solidifying agent was made by Fanny A. Hesse, whose husband was occupied with counting air-borne bacteria (1880); and Walther Hesse (1846–1911) brought this innovation into the laboratory of Koch (1882). Apparently, none of the researchers knew of the "Cookbook... for the advanced cuisine," which already recommended "agar-agar" for preparation of gelées in 1873. The traditional "glass plate" was later replaced by the **glass Petri dish** (1887).

1.5 The New Bacteriological Methods Proved that the Causative Agents of Infectious Diseases are Bacteria

In the middle of the eighteenth century, the physician Marcus Antonius von Plenciz (1705–1786) published in Vienna the book "Opera medico-physica" (1762) and postulated that a specific "seminium" (seed) is responsible for each disease. He even mentioned the putrefaction of apples and cherries and the transfer of sour dough (leaven) as examples. At the time, there was a real mania supposing minute worms to be everywhere. Thus, the ideas as to the involvement of the "contagium animatum" in diseases were made ridiculous; they were rejected and nearly forgotten. Medicine did not become the pacemaker of bacteriology until L. Pasteur and R. Koch entered the scene.

From 1850 on, L. Pasteur provided several examples for the concept that a specific fermentation is caused by a specific microorganism. He therefore assumed that the causative agents of diseases are also just as specific. Although he was not trained in veterinary or human medicine, Pasteur became a pioneer in elucidating infectious diseases. He not only accepted the task of studying the diseases of beer and wine but also investigated (from 1865 on) animal diseases, such as the silkworm disease and fowl cholera. In Pasteur's laboratory, C. Chamberland (1851–1908) introduced the **autoclave** for sterilization at 120°C and the method of pressing crude extracts through unglazed porcelain **filters** to remove bacteria.

Definitive proof of bacteria as causative agents of infectious diseases was obtained by R. Koch, while studying the anthrax disease of cattle. This disease was widely known; between 1850 and 1860, A. Pollender (1800–1879) and C.J. Davaine (1811–1882) reported that vast numbers of rod-like bodies were present in the blood of animals that had died of anthrax. Koch became familiar with anthrax when he was district physician in Wollstein, Silesia. In 1875, he was able to propagate from highly infectious material the easily visible rods, which were passed from rabbit to rabbit. By using the aqueous humor of rabbit or cattle eyes as culture media and paraffin-sealed slide cultures, Koch found that growth required oxygen and elevated temperatures (30–35°C). Finally, he observed the rods forming endospores. These experiments were done within about five weeks and demonstrated the consecutive steps later known as "Koch's postulates", which were formulated by E. Klebs (1877) and F. Loeffler (1884). Koch demonstrated his experiments regarding the life cycle of *Bacillus anthracis* to F. Cohn, J. Cohnheim, E. Eidam, and C. Weigert in Breslau and finished his manuscript, "The etiology of anthrax, based on the life cycle of *Bacillus anthracis*," which was submitted on May 27, 1876. The study provided proof for the concept that a specific disease is caused by a specific bacterium. Subsequent papers dealt with "Methods to study, conserve, and photograph bacteria" and "Studies on the etiology of wound infections." Koch had not yet developed pure-culture techniques when he published his paper on anthrax in 1877. The cultures used were "enrichment cultures" resulting from the selection of the dominating pathogenic bacterium by animal passage. By designing, standardizing, and diligently describing the fundamental pure-culture methods, Koch and his group of research workers provided the guidelines for discovering and for investigating bacteria. These methods, developed after Koch began work at the Kaiserliches Gesundheitsamt (Imperial Health Center) in Berlin, are still being used today in a nearly unaltered form, and **Koch's postulates** are still valid (see Chapter 33). The impetus given to microbiology was dramatic, marking the start of its golden age. The merits of Koch are not restricted to a few concepts and to the discovery of several species of bacteria; they consist of the creation of basic tools and principles, which served as amplifiers in this area of research.

The idea that bacteria can cause **plant diseases** was accepted only with great hesitation. Fungi as plant pathogens had been well known since A. de Bary (1853). In 1878, the American botanist T.J. Burril (1839–1916) described bacteria found in the tissues of pear trees diseased with fire blight and called the infective agent *Micrococcus amylovorus*. Ten years later, the American bacteriologist Erwin Frank Smith (1854–1927) described several bacteria and even dared to generalize that there are at least as many bacterial diseases of plants as there are of animals. E. Smith described many bacterial plant pathogens, among them the now famous *Agrobacterium tumefaciens* and its "gall" formation (1907). The name of the genus *Erwinia* has served to honor him since 1920.

1.6 Studies on Fermentation Founded Bacterial Physiology and Biochemistry

Fermented beverages and milk products were produced long before the causative agents were known. Simultaneously in 1837, three papers appeared by F.T. Kützing, Charles Cagniard-Latour (1777–1859), and T. Schwann, which claimed that living organisms were involved in the formation of alcohol. However, the "yeast theory" did not prevail among the chemists represented by Friedrich Wöhler (1800–1882), Jacob Berzelius (1779–1848), and Justus Liebig (1803–1873), who regarded fermentation to be caused by a chemical catalyst. The anonymous publication of Liebig in 1839, "The demystified secret of alcoholic fermentation" (Annalen der Pharmazie 29, 100–104), was written to make the vitalistic theory appear ridiculous and demonstrates the unrivaled arrogance of some leading scientists of the time. This publication tended to retard relevant research until L. Pasteur repeated Schwann's experiments and concluded that living yeasts cause fermentation. Pasteur observed (1857–1861) that different fermentations and the formation of vinegar are caused by distinct organisms. Pasteur recognized the existence of obligately **anaërobiontic (anaerobic) bacteria** and concluded that organisms of this type are also involved in putrefaction (1863). He referred to the causative agents of fermentation as "living cells" or "ferments," using the words "cell" and "ferment" as synonyms: in Pasteur's view, the cell was the ferment. Moritz Traube (1828–1894), a student of Liebig, did not deny the function of microorganisms in fermentation and putrefaction, but instead he attributed fermentation to the proteins present inside the cells (1858). The term for these proteins, the **enzymes**, was introduced in 1878 by W.F. Kühne (1837–1900).

Further substrate conversions in crude biological systems were described, e.g., in the laboratory of Ernst F. Hoppe-Seyler (1825–1895) in Strasbourg. When sugars or the calcium salts of organic acids such as formate, acetate, lactate, tartrate, and malate were added to sewage sludge and incubated under anoxic conditions, violent fermentations occurred. Hoppe-Seyler measured the products (including H_2 and CO_2) quantitatively and made many observations that could be understood only many decades later.

Escherichia coli, the most widely used bacterium in biochemical and genetic studies, was validly described in 1885 by Theodor Escherich (1857–1911) as *Bacterium coli commune*. Its fermentation products were first determined (1901) by Arthur Harden (1865–1940), whose general equation does not differ from that accepted today. Harden even correctly deduced that the hydrogen arises from formic acid and later recognized (1906) the *Aerobacter* (now *Klebsiella*) modification of the fermentation scheme.

1.7 Lithoautotrophy Is the Ability of Bacteria to Obtain Energy from the Oxidation of Inorganic Compounds and Carbon from Carbon Dioxide

Lithoautotrophy was first described in 1887 by the Russian plant physiologist S.N. Winogradsky, who was working in the laboratory of A. de Bary in Strasbourg at the time. When incubating the giant filamentous sulfurbacterium *Beggiatoa* in a moist chamber in the presence of hydrogen sulfide and air, he observed the bacterium microscopically and saw the deposition of sulfur globules in the cells, the disappearance of the globules in the absence of hydrogen sulfide, and the formation of sulfuric acid. Winogradsky's conclusion that *Beggiatoa* respired sulfur compounds instead of organic substances was immediately acknowledged as a new modus vivendi called **inorgoxidation** or **chemosynthesis**, now known as **lithotrophy**. Winogradsky also discovered the source of the carbon required for cell synthesis when he studied the nitrifying bacteria (1891). He succeeded in isolating nitrifiers from the soil and in growing them in a purely mineral medium. By determining the amounts of nitrite, nitrate, and cell carbon formed, he showed a constant stoichiometric ratio between the products of ammonia oxidation and the carbon assimilated. As no organic substrate had been added, the cell carbon could only have been derived from the air by **carbon dioxide assimilation**, a capability attributed until then only to green plants. The plant physiologist Wilhelm Pfeffer (1845–1920) in Leipzig introduced the term **autotrophy** to designate the ability to grow with carbon dioxide as the sole carbon source (1897). Lithoautotrophy enables a large metabolic group of bacteria to grow in mineral solutions with inorganic hydrogen donors, such as ammonia, nitrite, sulfur, hydrogen sulfide, thiosulfate, ferrous iron, molecular hydrogen, and carbon monoxide, and with carbon dioxide as the sole carbon source. Representative bacteria were isolated and studied after the turn of the century, and many lithoautotrophs that grow in very extreme environments were discovered quite recently. The new modus vivendi proved to be a concept of enormous importance.

Due to their modest but specific nutritional requirements, chemolithotrophic bacteria were the first physiological type of bacteria to be isolated by using selective culture methods, widely known as **enrichment culture techniques**. The method had already been applied by S.N. Winogradsky (1887) to cultivate colorless and purple sulfur bacteria in crude culture. Nowadays, practical courses for beginners often start with a demonstration of the "Winogradsky column." M.W. Beijerinck in Delft applied the enrichment culture methods to isolate a multitude of specialized bacteria. The majority of bacteria characterized by special nutritional requirements or environmental tolerances are isolated today according to the methods and principles introduced by Beijerinck and Winogradsky, even their latest and most advanced techniques using computerized screening devices. The same principles are also the basis of selection and counterselection techniques, which are applied to isolate mutants and recombinants in modern genetic approaches.

1.8 Light-Dependent Processes such as Phototaxis, Light-Induced Energy Transduction, and Photoassimilation of Carbon Dioxide Took a Long Time to be Understood

Anoxygenic phototrophic bacteria. As mentioned before, several phototrophic bacteria attracted the attention of naturalists long before their ability to use light energy for growth could be envisaged. These pigmented bacteria were described by C.G. Ehrenberg (1838), M. Perty (1852), R. Lankester (1873), E. Warming (1875), and F. Cohn (1875). Theodor Wilhelm Engelmann (1843–1910) was the first investigator to study the purple bacteria under physiological aspects (1883). He discovered various phenomena elicited by light, for example, the emission of oxygen from a lighted green algal thallus to which motile bacteria are attracted ("aerotaxis"). He had applied a technique known as "Engelmann's light trap" to study the **phototaxis** of motile purple bacteria and thus recognized the correlation between photosynthesis and phototaxis. He arrived at the conclusion that purple bacteria—since they are phototactic—must have a **photosynthetic metabolism**. In contrast, S.N. Winogradsky's studies were based on the aerobic sulfur bacterium *Beggiatoa* (1888), and he considered the metabolism of the sulfur purple bacteria as a type of chemosynthesis. He had recognized that the purple bacteria oxidize hydrogen sulfide under anaerobic conditions and had concluded that the necessary oxygen is provided by a light-dependent cleavage of water. Basically, he regarded the metabolism of the colorless and the purple sulfur bacteria as identical. When the Viennese botanist Hans Molisch (1856–1937) published his book on purple bacteria (1907), he criticized the theories of Engelmann and Winogradsky. Molisch concluded that purple bacteria assimilate organic substances in the light and do not produce oxygen. There is no doubt that Molisch had studied solely nonsulfur purple bacteria. In 1919, the Leipzig botanist Johannes Buder (1884–1966) explained, on the basis of his own studies, the reason for all the confusion by showing that Engelmann, Winogradsky, and Molisch had used quite different bacteria. To him, it was evident that there are two types of purple bacteria, sulfur-containing and nonsulfur purple bacteria. The former group shares with the colorless sulfur-oxidizing bacteria a lithoautotrophic metabolism; the latter group has an organotrophic metabolism like the heterotrophic bacteria. Buder recognized that the purple bacteria represent a new type of metabolism: the assimilation of CO_2 or organic substances in the light. Thus, he united the concepts of **photosynthesis** (from Engelmann) and **chemosynthesis** (from Winogradsky). A concept uniting the photosynthesis of green plants, sulfur purple bacteria, and nonsulfur purple bacteria was finally proposed by Cornelis B. van Niel (1897–1985) in 1931. He had worked with pure cultures of both groups of purple bacteria and had arrived at a general equation for photosynthesis:

$$CO_2 + 2\,H_2A \xrightarrow{\text{light}} [CH_2O] + H_2O + 2A,$$

in which A represents oxygen (in green plants), sulfur (in sulfur purple bacteria), or an oxidized organic compound (in nonsulfur purple bacteria). This includes the concept that purple bacteria perform an anoxygenic photosynthesis, i.e., no oxygen is generated by the primary photosynthetic processes.

Oxygenic phototrophic bacteria. The blue-green "algae" (now cyanobacteria), the oxygenic phototrophic bacteria that produce oxygen during photosynthesis, were definitely assigned to the prokaryotes in the nineteen sixties. Since the Middle Ages, the blue-green algae had been placed with the plants because of the color of their thalli and their aquatic or moist habitats. Many researchers had described the algae and the blue-green algae in the eighteenth and nineteenth centuries. By the middle of the nineteenth century, not only the morphology of the filaments and colonies but also the presence of **hormogonia**, **akinetes**, and **heterocysts** of many genera was known. F.T. Kützing (1843) named the accessory pigments **phycocyanin** and **phycoerythrin**, and the pigment system was called **phycochrome** (C. Naegeli, 1849). The dependence of the blue-green algae on light and carbon dioxide for growth suggested a plant-like metabolism. The decisive step to place the blue-green algae (i.e., the cyanobacteria) and the bacteria together was made by F.J. Cohn (1875). With respect to their small size and their characteristic mode of cell-division (called binary fission), Cohn adopted the term **Schizomycetes**, formerly used by Naegeli to designate the bacteria, and he called the united groups **Schizophytae**. In his combined system, he primarily considered morphological properties and only secondarily differentiated between colorless and phycochrome-containing forms. As a result of the delineation of the Schizophytae from the plants and animals, bacteria were no longer assigned to the worms, the protozoa or the unicellular algae. Furthermore, it prepared the way for the concept of a fundamental difference between the bacteria (including the cyanobacteria) and all other organisms. The patterns of cellular organization justified the combination of the classical bacteria with the cyanobacteria and the division of all living organisms into **prokaryotes** and **eukaryotes**, as done by Roger Y. Stanier (1916–1982) and C.B. van Niel in 1962.

The assignment of the cyanobacteria to the bacteria had its consequences. As long as the blue-green algae belonged to the **plants**, they were treated under the Botanical Code; as a result, living organisms were not maintained. When considered to be **bacteria**, they were subject to the Bacteriological Code, which prescribed the maintenance of viable cultures in an acknowledged culture collection of bacterial species, a procedure requiring pure cultures. The first pure cultures were obtained and used by Richard Harder (1888–1973) in 1917 for studying the developmental cycles of *Nostoc punctiforme*. Many more axenic cultures of cyanobacteria were isolated in Stanier's laboratory in Berkeley and in Paris.

1.9 Dinitrogen Fixation Is Unique to the Prokaryotes

The age-old experience of farmers that soil fertility is improved by leguminous plants was shown by J.B. Boussingault's (1802–1887) studies to be the result of nitrogen gain (1837). The **root nodules** of leguminous plants were regarded as root galls; J. Lachmann (1858) and M. Woronin (1866) showed that the nodules are filled with rods resembling bacteria. These "untypical" bacteria were named **bacterioids** (J. Brunchorst, 1885). Observations (1879) indicated that the nodules are formed due to infection from the soil [A.B. Frank (1839–1900)]. The association of root nodules with nitrogen fixation is the merit of H. Hellriegel (1831–1895) and H. Wilfarth. In very precise experiments (1884–1886), they showed that the nodules supply combined nitrogen to the plant. The causative bacterium was then isolated by M. Beijerinck (1888) and named *Bacterium radicicola*. The discovery of the first **free-living, nitrogen-fixing bacteria** is the merit of S.N. Winogradsky (1895). When he set up enrichment cultures containing sugar but no source of combined nitrogen, he isolated *Clostridium pasteurianum* and recognized its ability to fix nitrogen. Winogradsky failed to observe the pellicle that must have grown in the same enrichment culture, and thus he left the isolation of the bacterium that fixes nitrogen under oxic conditions, *Azotobacter chroococcum*, to M. Beijerinck and A. van Delden (1902). These discoveries gave rise to a well-based concept.

M. Beijerinck already had concluded in 1901 that cyanobacteria, such as *Anabaena* and *Nostoc*, can fix dinitrogen. Quantitative measurements by K. Drewes (1928) and G.E. Fogg (1942) confirmed this assumption. Because all hydrogenase-containing bacteria should be able to fix nitrogen, dinitrogen fixation in purple bacteria was predicted and detected by H. Gest (1950) using [$^{15}N_2$] in *Rhodospirillum* and *Chromatium*.

1.10 The Analysis of Anabolic and Catabolic Metabolism Lead to the Discovery of Substrates, Products, Apoenzymes, and Coenzymes, and, in the end, of Metabolic Pathways

L. Pasteur did not deny the chemical problems involved in fermentation, and he analyzed, for example, the excretion of enzymes by yeasts. M.P.E. Berthelot (1860) and J. Liebig (1870) isolated invertin from yeasts as did P. Miquel (1893) for urease from bacterial cultures. However, neither the fermentation enzyme "alcoholase" from living yeast cells nor the oxygen-evolving system from leaves could be isolated. The discovery of **cell-free fermentation** is an example of serendipity. In 1897 in Tübingen, E. Buchner was preparing yeast press juice for immunological studies; he attempted to stabilize and to conserve the juice by adding sugar. Within 20 minutes, foaming started. This was the first demonstration of a complex biochemical process outside the cell. After Buchner had observed the stimulating effect of orthophosphate on the evolution of CO_2 (1903), A. Harden and W.J. Young in Cambridge investigated the process in more detail and discovered the formation of fructose 1,6-bisphosphate (1906). E. Buchner and W. Antoni (1905) and Harden and Young (1906) discovered the coenzyme effect. In 1918, Otto F. Meyerhof (1884–1951) discovered the exchangeability of the coenzymes involved in fermentation by yeast press juice and in lactic acid fermentation by ground muscle tissue.

Further impulses were provided (1911–1912) by Carl Neuberg (1877–1956) in Berlin and S. Kostytschew and colleagues (1912–1913) in Moscow, who were investigating phytochemical reductions at the time. The use of substrate hydrogen by a large number of substrate-specific enzymes was discovered when methylene blue was applied as hydrogen acceptor. The reduction of methylene blue by tissues had already been discovered by Paul Ehrlich (1854–1915) in 1885. In 1912, Heinrich Wieland (1877–1957) proposed a theory of biological oxidation in which he attributed the main function in respiration to hydrogen. In contrast, Otto Warburg (1883–1970) attributed the main function in respiration to oxygen; he concluded that respiration is an iron catalysis taking place on surfaces and that the iron is bound to heme (1928). Heme pigments were described as early as 1886 by C.B. McMunn, and the name cytochrome was introduced by D. Keilin in 1925. O. Warburg already included bacteria such as *Acetobacter pasteurianum* and *Azotobacter* species in his studies (1932); since then, the function of cytochromes was mainly studied using aerobic bacteria.

The discoveries made by E. Buchner and A. Harden did not promote research on cell-free systems in bacteria immediately. However, many important insights were contributed by T. Thunberg (1916) and by A. Harden and J.H. Quastel of the Cambridge school (1924–1928), who used **resting cells** (i.e., washed suspensions) to study the dehydrogenases of bacteria. By 1920, pure cultures of the majority of fermentative bacteria were already available. S. Orla-Jensen (1919) studied a large number of various lactic acid bacteria and differentiated between homofermentative and heterofermentative species (1919). Harden determined the products of glucose fermentation by *Escherichia coli* and, by 1906, he recognized that fermentation by *Aerobacter* (i.e., *Klebsiella*) and *Serratia* differs from that of *E. coli*. Formation of acetoin was also discovered when bacilli and pseudomonads were used. Butyric acid fermentation, discovered in 1861 by L. Pasteur, was later studied by using *Granulobacter (Clostridium) butyricum* (Beijerinck, 1894) and *Clostridium pasteurianum* (Winogradsky, 1895).

Unity in biochemistry. Because of the lack of a unifying theory of the metabolism of microorganisms studied before 1925, Albert J. Kluyver (1884–1957), the successor of M.W. Beijerinck in Delft, decided "we shall endeavour to bring order into this chaos of phenomena." In 1925, Kluyver and H.J.L. Donker assigned the known sugar-fermenting microorganisms to a limited number of groups, for example, the alcohol yeasts, the true lactic acid bacteria, the true propionic acid bacteria, the coli bacteria in a general sense, and the butyric acid–butanol bacteria. They recognized **hydrogen transfer** as the common denominator for the dissimilatory processes of all fermentative bacteria and the involvement of only a few reactions: "From the elephant to the butyric acid bacterium—it is all the same." In a more detailed publication, "Die Einheit in der Biochemie" ("Unity in Biochemistry"), Kluyver and Donker (1926) developed the concept of reduction-oxidation processes occurring within an energetic gradient. This concept of the "unity in biochemistry" was immediately accepted.

Studies on the metabolism of bacteria provided many significant concepts for differentiating between catabolism and anabolism and for elucidating the biochemical reactions of intermediary metabolism, DNA replication and transcription, mRNA translation, membrane functions, and many other fundamental functions. For lack of space, only a few examples will be mentioned:

Heterotrophic CO_2 fixation was discovered (1935) when Harland G. Wood (1907–1991) worked on propionic acid fermentation in C.H. Werkman's laboratory in Ames, Iowa. More specifically, Wood studied the fermentation of glycerol by *Propionibacterium arabinosum* and concluded that CO_2 is not only the product of respiration and the substrate of autotrophic organisms but also a substrate for various reactions involved in intermediary metabolism.

Autotrophic CO_2 fixation was explored within 15 years after radioactive [^{14}C]-carbon had been discovered (S. Ruben and M. Kamen, 1940) and paper chromatography had been developed (A.J.P. Martin and R.L.M. Synge, 1944). The study of autotrophic bacteria began only after Melvin Calvin (1911–1997), A.A. Benson, and J.A. Bassham had elucidated the biochemical pathway of carbon in *Chlorella* and *Scenedesmus* in 1954.

Growth factors and vitamins were first identified in lactic acid bacteria. After S. Orla-Jensen (1936) reported on the riboflavin requirement of lactic-acid bacteria, E. Snell in Madison described (1936–1951) the requirements for pantothenic, nicotinic, and folic acid, pyridoxal derivatives, and biotin, and discovered the biotin-binding capability of avidin. These discoveries contributed to general biochemistry.

Antibiotics and antimetabolites. Molecular biology owes many impulses to antibiotic research. Observations on antibiotic effects date back to the last century. The discovery of penicillin by A. Fleming in 1928, followed by the systematic investigations by Howard W. Florey (1898–1968) and Ernst B. Chain (1906–1979) in the years 1939 and 1940, represent the breakthrough of antibiotic research. These studies, which centered on the action mechanisms of antibiotics on their cell targets, added a great deal to molecular biology, especially to the exploration of the mechanisms of DNA replication and transcription, mRNA translation, and membrane functions.

The concepts derived from the recognition of **competition effects** between metabolites and their structural analogues provided new tools for research and strategies for designing new agents with highly selective effects, such as chemotherapeutics and herbicides. From 1910 to 1914, L. Michaelis observed the competitive relationships between the products formed by carbohydrate-hydrolyzing enzymes and inhibitors. In 1927, J. Quastel and W.R. Wooldridge showed the competitive inhibition of succinate dehydrogenase by malonic acid, the **structural analogue of the substrate**. The decisive discovery was made when Gerhard Domagk (1885–1964) studied experimental pneumococcal infections and found the dramatic curative effect of prontosil red in vitro and in vivo (1935). The same year, J. Tréfouël explained this effect as due to the sulfanilamide part of the molecule. The mode of action of sulfanilic acid as a structural analogue or **antimetabolite** of 4-aminobenzoic acid was elucidated by D.D. Woods in Oxford (1940).

1.11 Studies on Inclusion Bodies and the Structures and Functions of Cell Envelopes Revealed the Organization of the Bacterial Cell

The interior of the bacterial cells remained obscure until the last decade of the nineteenth century. Before then, only C.G. Ehrenberg (1838) had given a highly prejudiced description of *Monas okenii*. Inclusion bodies and their chemical nature were first studied by botanists and chemists.

Polysaccharides, such as starch and glycogen, were described in spore formers by P. van Tieghem (1877) and M. Beijerinck (1893). The "**fat bodies**" described by the botanist Arthur Meyer (1899, 1912) were identified as **poly-β-hydroxybutyric acid** by the French chemist M. Lemoigne (1926). Granules described as **metachromatic** (A. Guilliermond, 1903) or **volutin granules** (A. Meyer, 1912) were identified as inorganic polyphosphate by J.M. Wiame (1947). The rhomboid, **parasporal inclusion bodies** were found when E. Berliner (1911) isolated *Bacillus thuringiensis* from deceased larvae of *Ephestia kuehniella*, and the idea to use the toxic protein for biological pest control came from E.A. Steinhaus (1951) in England. **Gas vacuoles** were first seen by F. Cohn and described by S.N. Winogradsky (1888) in *Lamprocystis roseopersicina*. With the "hammer, cork, and bottle" experiment, H. Klebahn (1895) showed that the vacuoles contain gas. The search for **bacterial nuclear equivalents** started after R. Feulgen (1924) had discovered that the nuclei of plant and animal cells can be stained with alkaline fuchsin dye. A. Rippel (1932), C.F. Robinow (1942, 1953), and others made many efforts. By enzymatic hydrolysis (with RNase, DNase), proof was obtained that the stained nucleoids consisted of DNA (D. Peters and R. Wiegand, 1953). Many questions raised by cytologists were finally answered by using electron-microscopical fine-structure studies (O. Maaløe and A. Birch-Andersen, 1956), genetic methods (J. Lederberg, 1947), and radioautography (J. Cairns, 1963).

The exploration of the **bacterial cell wall** made rapid progress after M.R.J. Salton and R.W. Horne (1951) had elaborated methods for the isolation of its basic components. The degradability of the cell wall by lysozyme (Salton, 1952) and the excretion of nucleotides (J.T. Park, 1952) lead to the concept of a saclike mural macromolecule, the murein sacculus (W. Weidel, 1964). Research on **lipopolysaccharides**, the somatic antigens responsible for the endotoxic properties of enterobacteria, began with the serodiagnostic differentiation of hundreds of *Salmonella* strains (B. White, 1931; F. Kauffmann, 1937) and lead to the characterization of the O-specific heteropolysaccharides by chemical methods (O. Westphal and O. Lüderitz, 1952).

The **motility** of bacteria already had been mentioned by A. van Leeuwenhoek in 1683. **Flagella** were first described by F. Cohn (1872) and stained by R. Koch (1877) and F. Loeffler (1889). Flagellar rotation was studied by J. Buder (1915) and P. Metzner (1920), and the first electron-microscopic pictures of flagella were provided by G. Piekarski and E. Ruska (1939). Further work on the flagellar structure came from W. van Iterson, C. Weibull, A. Pijper, E. Leifson, and A.L. Houwink. The basic concepts of **tactile responses** were provided by T.W. Engelmann and W. Pfeffer. **Aërotaxis** and **photophobotaxis** were discovered by Engelmann in 1881 and 1882, respectively. Chemotaxis, in a general sense the attracting or repelling action of various dissolved substances on spermatozoids of ferns and mosses, on bacteria, and on flagellates was described by Pfeffer (1865, 1888). He developed the glass-capillary and soft-agar methods, and he discussed stimulus perception, threshold phenomena, signal transduction, sensory adaptation, and the essential role of the cell membrane, demonstrating his modern conceptual understanding. Pfeffer assumed that the basic molecular mechanisms of sensory reception and signal transduction in bacteria, fungi, algae, higher plants, and even animals are the same (1904). With his experimentally founded theories on the cellular plasma membrane, substrate transport, sensory processes, enzyme regulation, and energy metabolism, Pfeffer was far ahead of his time. His seminal concepts were spread worldwide by his many students, doctoral candidates, and foreign guests and by his textbooks (1897, 1904); many of his ideas are valid to this day.

Thus, already long before 1950, it was shown that prokaryotes contain unique structural macromolecules and functional supercomplexes, are endowed with spectacular abilities, and are highly organized—in short, they are truly self-sufficient living organisms. These properties and the concepts derived thereof provided a firm basis for the development of molecular genetics and modern cell biology.

1.12 Bacterial Adaptation was Well Recognized Before the Genetic Approach Revealed the Basis of Molecular Mechanisms of Regulation

The exploration of the regulation of microbial metabolism began after pure-culture methods had become common practice. A few concepts of the "selfregulation" of metabolism could be derived from plant physiology. The phenomenon of substrate specificity already had been discovered by L. Pasteur in 1857, when he demonstrated that *Penicillium glaucum* utilized only the dextrorotatory tartaric acid but not the levorotatory acid when grown on the racemic mixture of both acids (*para*-tartaric acid). Further observations pointed to regulatory processes. For example, it became known that the secretion of depolymerases such as amylase during growth on starch is suppressed when the corresponding monomer (i.e., glucose) is also present. In his textbook on plant physiology (1904), W. Pfeffer asked the cardinal question, "What does the plant do if two or more compounds of carbon, nitrogen, or another element are simultaneously available?", which aimed at catabolite repression.

The phenomena observed after transfer of a bacterium from one substrate to another were considered to be **adaptations** to new environmental conditions. The vague definition of the term adaptation caused confusion for a long time. Adaptation in the sense of a transient change of enzyme activity (then called physiological adaptation) in response to the composition of the nutrient substrates was studied by H. Karström (1930) in Helsinki. Working with *Bacterium aerogenes*, he showed that xylose utilization is an adaptive property, but glucose utilization is a constitutive property. Karström made the generalization that enzymes for general substrates, such as glucose and fructose, or lactate, pyruvate, and succinate, are usually constitutive, whereas enzymes for rare sugars or rare organic acids are adaptive. Studies on enzyme adaptation were considerably facilitated by using resting cells (J. Quastel, 1925), and the dependence of the adaptation process on nitrogen source, temperature, and growth inhibitors (or even the response of UV-killed cells) was rapidly explored (Marjory Stephenson (1885–1948) and J. Yudkin, 1936). Many of the experiments were repeated after 1953, when the molecular processes were better understood.

The epoch-making investigations and **regulatory models** of Jacques Monod (1910–1976) began with very simple growth experiments, first with protozoa, then mainly with *Escherichia coli* and *Bacillus subtilis*. Cell growth was monitored photometrically by measuring the bacterial density. For the sequential growth curve characteristic for the utilization of two different substrates, Monod (1942) coined the word **diauxie**. He confirmed Karström's conclusions and those of H.M.R. Epps and E.F. Gale. J. Monod and M. Cohn introduced (1952) the terms "enzyme induction," "induced enzymes," "internal inducer," and "exogenous gratuitous inducer." G. Cohen-Bazire and H. Jolit isolated (1952) the first mutants to utilize lactose constitutively. These mutants made the work on enzyme regulation accessible to genetic analysis. A first model for the **induction** of catabolic enzymes (e.g., β-galactosidase) and the **repression** of anabolic enzymes in *Escherichia coli* was proposed (1952). The term "operon" was introduced by Francois Jacob and coworkers in 1960, with essential parts of the concept relying on data obtained in studies with the temperent bacteriophage λ (lambda).

The regulation of metabolism at the level of **enzyme activity** was discovered as late as 1956. The existence of enzymes in *E. coli* that do not function if a distinct end product is added could be concluded from experiments implying isotopic competition (R.B. Roberts and colleagues; 1954, 1955). These and other experiments (M.S. Brook, 1954) pointed to a specific effect of the end product of a pathway on the enzyme(s) involved. The mechanism of **feedback inhibition** was elucidated by R. Yates and A.B. Pardee (1956), when studying the synthesis of pyrimidine derivatives, and by H.E. Umbarger (1956), when studying isoleucine biosynthesis. Molecular studies lead to the concept of **allosteric enzymes** and to the theory of regulation of allosteric enzymes, which was formulated by J. Monod, J. Wyman, and J. Changeux (1965).

1.13 Studies on the Metabolic Types of Bacteria Revealed Their Functions in the Biosphere

A. Kircher already reported in 1671 on copious amounts of very small living beings in soil, water, and putrefying material. The question as to the role of microorganisms in these ecosystems arose much later. Even the leading plant physiologist J.B. Boussingault discussed only the mutual dependence of animals and plants without recognizing the **third segment of the cycle of matter**, mineralization or destruction. It is the merit of L. Pasteur and some of his contemporaries, such as F. Cohn, A. de Bary, and O. Brefeld, to have considered the role of the "infinitely small" and to have combined the mutual dependencies to the so-called cycle of matter. In fact, several concepts leading to the discovery of bacteria that represent new metabolic types originate from observations made in agriculture. The existence of bacteria involved in **nitrogen fixation**, **nitrification**, and **denitrification** was postulated on the basis of quantitative measurements of the fate of nitrogen in agricultural soil, as already described (Chapter 1.7).

The term **ecology** was coined by Ernst Haeckel (1834–1919) in Jena. Originally defined as "the total science of the relationships of the organism to its environment..." (1866), this definition has remained almost unchanged (R.E. Hungate, 1962). Terms such as ecosystem, habitat, ecological niche, community, biocoenosis, food chain, food web, and microenvironment had to be redefined for use in microbial ecology (T.D. Brock, 1966; M. Alexander, 1971). Some terms for interspecific relationships such as mutualistic and antagonistic symbiosis (A. de Bary, 1878), rhizosphere (L. Hiltner, 1904), and mycorrhiza (A.B. Frank, 1885) had to be modified, and new terms such as autochthonous (indigenous) and allochthonous (nonindigenous) bacteria (S.N. Winogradsky, 1925) and syntrophism and interspecies hydrogen transfer (M.J. Wolin, 1975) had to be added. Most of these terms imply essential concepts in microbial ecology.

While many soil bacteria were studied in pure culture soon after their discovery in order to examine their basic metabolism and metabolic capacities, S.N. Winogradsky and some of his students designed methods to study bacteria *in situ*, within their natural habitats. After his return from Strassburg and Zürich to St. Petersburg, Winogradsky continued to investigate different soils microscopically. N. Cholodny applied the growth plate technique, and B.V. Perfiliev used the capillary method. In his memoirs "Microbiologie du sol, problèmes et méthodes" (1949), Winogradsky emphasized that the conclusions drawn from laboratory studies on the behavior of bacteria in the natural habitat may be misleading because laboratory strains might only be "artefacts", i.e., selected by the chosen nutrient medium, temperature, aeration, and agitation. Consequently, he warned against using bacterial strains from culture collections as typical "wild-type" cells. Winogradsky's concepts on soil microbiology and on the differentiation between autochthonous and allochthonous populations in the soil are the basis of relevant, current research.

1.14 The Goals and Methods of the Classification of Bacteria Have Changed

Today, the tiny organisms portrayed by A. van Leeuwenhoek would be addressed as rod, coccus, vibrio, and spirillum. In C. Linné's (Linnaeus') system, these organisms were grouped as *species dubiosa* among the worms. C.G. Ehrenberg (1838) retained the designations *Monas* and *Vibrio* given by O.F. Müller (1789) and supplemented the names *Bacterium*, *Spirillum*, and *Spirochaeta*, whereas M. Perty (1852) assigned new names. Later, F. Cohn understood that bacteria are a special group of small organisms. He recognized the similarity between the cyanobacteria (schizophyceae, "fission algae") and bacteria (schizomycetes, "fission fungi") and combined them as **schizophytae** ("fission plants"). Thus, the concept of the bacteria as a separate group of organisms has to be credited to him.

S. Orla-Jensen presented in Copenhagen an essay on "Die Hauptlinien des natürlichen Bakteriensystems" ("The mainlines of the natural bacterial system," 1909). In this work, the lithoautotrophic bacteria were considered to be the most primitive bacteria and thus were placed at the beginning. Although no real phylogenetic tree of bacteria was developed, the work proved very instructive and stimulating. Again on the basis of morphological characters, A.J. Kluyver, C.B. van Niel, and R.Y. Stanier attempted to outline a **natural system** (1936, 1941). **Numerical taxonomy**, promoted since 1957, was a new concept (P.H.A. Sneath, 1963, 1973) based on the determination of a very large number of characters on alternate decisions, and on the computerized evaluation of combinations of characters. This method results in similarity matrices and dendrograms, which are similar to genealogical trees but bear no real relationship to phylogeny.

On the basis of the differentiating properties known up to 1961, R.Y. Stanier and C.B. van Niel (1962) arrived at "The concept of a bacterium." Due to the brilliant work of A. Ryter and E. Kellenberger (1958) and E.L. Wollman and F. Jacob (1959), the basic differences between the structure of the genetic material of bacteria and that of the other organisms had become obvious. The terms **prokaryotic** and **eukaryotic**, proposed by E. Chatton (1937), were used to designate the two patterns of cellular organization. The new concept appeared well-founded, clear, and easily comprehensible, and it was immediately accepted.

A completely different concept for creating a natural system of bacteria is that of Carl Woese. After chemists had succeeded in determining the amino acid sequence of proteins and the nucleotide sequence of nucleic acids, Woese made a new proposal for a **phylogenetic tree of bacteria** (1977). Evaluation of sequence data led to the division of this realm of organisms into archaebacteria (now the domain **Archaea**), eubacteria (**Bacteria**), and eukaryota (**Eukarya**). The phylogenetic tree of bacteria partially confirms long-known relationships and—most important—reveals completely new relationships. None of the proposed concepts is satisfactory, however, because they rely on too few properties (e.g., 16S RNA sequence) or they do not consider sufficiently the high horizontal gene-transfer rate between prokaryotes, which requires a new definition of species ("collective gene pool").

1.15 Bacterial Viruses (Bacteriophages) Were Detected as Lytic Principles

Bacteriology started with the visibilization of its subjects, the bacteria. In contrast, viruses remained unseen until the end of the nineteen thirties, when the tobacco mosaic virus and bacteriophages were made visible in the electron microscope. Credit for a first **concept of the virus** has to be assigned to F. Loeffler, P. Frosch, and M.W. Beijerinck, while modern phage research started with Max Delbrück (1906–1981).

The history of research on **bacterial viruses** began with reports on a "glassy transformation" of micrococcal colonies by a transferable principle (F.W. Twort, 1915) and on "an invisible microbial antagonist of dysentery bacilli" (Felix d'Herelle, 1917). d'Herelle continued his studies (1919–1926): he counted the holes (**plaques**) "eaten" by viruses in bacterial lawns, and concluded that the **bacteriophage** (i.e., eater of bacteria) is a discrete particle, in essence, a virus specific for bacteria. Further progress (1929–1936) was made by F.M. Burnet (1899–1985), who identified the phases of adsorption, propagation, and the sudden liberation of phages (lysis) and differentiated between bacterial immunity and resistance to phages. Max Schlesinger obtained the first physical data on phages; he saw the bacterial cells burst and the release of phages, which he was able to count by means of dark-field microscopy (1932, 1934). **Lysogeny** was simultaneously discovered by J. Bordet, M. Ciuca, and E. Gildemeister in 1921. A key question was whether each cell of a lysogenic bacterial strain has the genetic capacity to produce phages; this was proved by Burnet, who worked with a lysogenic strain of *Salmonella* (1929). While studying a lysogenic strain of *Bacillus megaterium* (1931), Den Dooren de Jong concluded that the phage in a lysogenic bacterium is in a **latent state** (or prophage phase). His experiments were confirmed and extended by Eugène and Elisabeth Wollman (1936) and, after 1945, by André Lwoff (1902–1994) and his followers, among them J. Monod and F. Jacob. In 1950, *E. coli* became the model bacterium for studying lysogeny again. Esther M. Lederberg discovered (1951) that *E. coli*, strain K12, which had been used for the first conjugation experiment (1946) by J. Lederberg and Edward L. Tatum (1909–1975), is lysogenic and carries a temperate phage, which she called "lambda" (λ). Most of today's laboratory strains of *E. coli* K12 are non-lysogenic (cured) derivatives of the original strain. K12 is still susceptible to λ. It is this strain which served the studies on zygotic induction, the integration of λ into the host chromosome, the lytic versus the lysogenic cycle, and episomes as performed by François Jacob and Elie L. Wollman. The studies on lysogeny provided immensely important concepts, for example, the repressor/operator and the operon concepts in bacteria, and the concept of oncogenic viruses in tumor biology.

E. coli K12 with its phages λ and P1 were also vital in the phenomenon of host-controlled modification and restriction of bacteriophages. The phenomenon as such had already been described by L. Bertani and J. Weigle (1953). Werner Arber and collaborators (1959–1968) discovered the sequence-specific, post-replicative methylation (modification) of the phage DNA in the host cell and the function of an endodeoxynuclease involved in DNA restriction. The possibilities that the discovery of specifically cutting restriction endonucleases provided for molecular genetics were recognized and discussed by Arber already in 1969.

New extrachromosomal DNA elements, the **plasmids**, were discovered when the distribution of antibiotic resistance in different bacteria was studied. In Japan, bacterial strains of *Shigella flexneri* that were resistant to several antibiotics had been described since 1952. T. Watanabe (1923–1972) and T. Fukasawa reported the involvement of plasmids, which carry resistance factors and transfer factors (1961). The term "plasmid" had been coined by J. Lederberg already in 1952, and phages had been recognized as a special class of plasmids. The plasmids remained undetected until 1961 when centrifugation experiments revealed the first plasmid bands ("satellite DNA"), and agarose gelelectrophoresis facilitated the recognition of plasmids. The discovery of various plasmids for metabolic, resistance, and virulence functions supported the concept that extrachromosomal elements play an enormous role in nature. They were joined shortly after by transposable elements (insertion sequences [IS] and transposons) detected originally by Barbara McClintock (1902–1992) as "mobile genetic elements in maize" (1937–1942).

1.16 Studies on Heredity in Bacteria Provided the Decisive Principles and Concepts for the Promotion of Modern Biology Including Gene Technology

Sexuality in plants has been known since the end of the seventeenth century, and chromosomal cleavage in the nuclei of plants and animals was recognized in 1885. The laws of heredity were discovered by Gregor J. Mendel (1822–1884). Mendel's concepts on the nature of the "hereditary markers" (genes) were premature for his contemporaries and remained unnoticed for nearly 35 years, until they were rediscovered and confirmed in 1900 by Hugo de Vries (1848–1935), Carl Erich Correns (1864–1933) and Erich von Tschermak (1871–1962). A few weeks after H. de Vries had given a lecture in Amsterdam (1900), in which he introduced the terms

mutation and **mutant**, M.W. Beijerinck lectured "On different forms of hereditary variation of microbes" and noted that some bacterial variants could also be called mutants. This lecture already contained the concept that microbes are useful for studying not only the laws of heredity and variability but also the principle of competition between species. By 1912, Beijerinck completely conformed with de Vries ("Die Mutationstheorie" 1901, 1903) but defined the term "gene," which was used in 1909 by Wilhelm Johannsen (1857–1927) in Copenhagen, as follows: "The genes or hereditary units are the carriers of the visible properties and can best be thought to be parts of the protoplasm or cell nucleus." Beijerinck's ideas on mutation, reversion (atavism), and adaptation were based on his studies in yeasts and in *Serratia marcescens*.

Although the character of a gene as the functional genetic unit ("cistron") and of a mutation became increasingly clearer through studies with higher organisms, some geneticists still regarded mutations as directed processes. For final clarification, Salvador E. Luria (1912–1991) and Max Delbrück performed the "fluctuation experiment" (1943). The statements were confirmed by the "spreading experiment" of H.B. Newcombe (1949) and the "indirect selection test" (1952) via replica plating of colonies by J. and E.M. Lederberg.

There are only few reports on early attempts to find indications for sexuality in bacteria. Copulating pairs of the giant purple bacterium *Chromatium okenii* and of *Rhodospirillum* were observed by F. Förster (1892) and H. Potthoff (1922). A mating experiment using two nutritional mutants of *Escherichia coli* was done in 1925 but published later (J.M. Sherman and H.U. Wing, 1937). The use of biochemical mutants of *Neurospora crassa* (1941) and the discovery of transformation in *Pneumococcus* (see below) prompted J. Lederberg to study the genetic structure of bacteria and to do crossing experiments with *E. coli*. The decisive mating experiment, which lead to the concept of bacterial conjugation, was then reported by E.L. Tatum (1909–1975) and J. Lederberg (1946), using two stable, double-auxotrophic mutants as partners.

The discovery of DNA transformation is credited to epidemiological studies on *Pneumococcus (Streptococcus) pneumoniae*. In 1928, F. Griffith discovered the transformation of rough (R) mutants by smooth (S) mutants *in vivo*. *In vitro* experiments by O.T. Avery, C.M. MacLeod, and M. McCarty (1944) revealed the transforming principle as DNA. The discovery of DNA as a carrier of genetic information meant a revolution. The transformation experiment provided a new challenge, which was first met by J. Lederberg and by Erwin Chargaff. Chargaff (1950) discovered by careful quantitative experiments that the members of various taxonomical groups differ with respect to the guanine + cytosine (G + C) content of their DNA and that there is the general rule: A = T, G = C, and A + G = C + T. From 1946 on, many new concepts were developed, among them the description of the molecular structure of DNA by James D. Watson and Francis H.C. Crick (1953). The concept of the bacterial chromosome as a circular structure and as a replicon is the result of the investigation of the mechanism of conjugation, of the recognition of the role of the F (fertility) plasmid, and of the Hfr strains i.e., with a high frequency of recombination (W. Hayes, 1953). The mechanism of conjugative transfer through cell-to-cell contact was elucidated by E.L. Wollman and F. Jacob (1955).

1.17 Epilogue

This review of the beginnings of bacteriological research and the concepts that emerged can by no means be complete. It is an attempt to record the main lines of development of bacteriology from its roots in medicine, in plant physiology, and in agricultural chemistry (i.e., as applied biology) and to emphasize the milestones. As a very important part of our culture, bacteriology exemplifies well how science was practiced down the ages and how strongly it was influenced by the spirit of each age. Historical analysis reveals these connections, which often remain hidden to contemporaries. Thus, it seems logical that this review ends with the concepts of the early nineteen sixties. In the remainder of the book, the reader will hopefully rediscover how much of the breathtaking progress and the diversification of modern bacteriology is the logical consequence of work and ideas and thought contributed by a great number of researchers during the past 250 years.

Table 1.1 **Milestones in the Development of Microbiology Until the Early 1960s (NP = Nobel Prize)**

Year(s)	Milestone
1676	A. van Leeuwenhoek first observes bacteria microscopically
1776	L. Spallanzani shows that microorganisms can be killed by boiling, and he disproves the hypothesis of spontaneous generation. His experiments lead to the preservation of food by boiling and sealing (i.e., canning; F. Appert, 1810)
1796	E. Jenner first vaccinates man with the cowpox virus (vaccinia)
1834–37	F.T. Kützing, C. Cagniard-Latour, and T. Schwann recognize independently that yeasts are living organisms, which multiply by budding and which ferment sugar to alcohol and carbon dioxide
1838	J.B. Boussingault shows that clover increases the nitrogen content of the soil
1838	C.G. Ehrenberg describes microorganisms in his book "Die Infusionsthierchen als vollkommene Organismen" and presents the species description of *Monas okenii* and *Ophidomonas jenensis*
1853	H. Schröder and T. von Dusch disprove the hypothesis of spontaneous generation and introduce culture-vessel stoppers of cotton, which allow air to pass but keep sterilized nutrient media free of contaminants
1854	A. de Bary recognizes that rust and smut of cereals are caused by parasitic fungi, thus promoting the development of mycology
1857–61	L. Pasteur describes the formation of ethanol by yeasts, of lactic acid by bacteria, of butyric acid by vibrios, and of acetic acid by *Mycoderma aceti*
1869	O. Brefeld cultivates the fungus *Empusa muscae*, a pathogen to insects, on nutrient media solidified with gelatin
1876–77	F.J. Cohn describes and classifies the bacteria known to date, describes the endospores of bacilli, and eliminates doubts as to the monomorphism of bacteria
1877	R. Koch (NP 1905) provides final proof that *Bacillus anthracis* is the causative agent of anthrax
1878	J. Lister isolates *Streptococcus lactis* in pure culture by the dilution method
1881	R. Koch reports on the pour plate technique and starts the "Golden Age" of medical microbiology
1882	T. Schloesing and A. Müntz show by soil percolation experiments that nitrification is a biological process
1883/89	T.W. Engelmann studies the phototactic behavior of purple bacteria and discusses their photosynthetic properties
1884	C. Gram discovers the differential staining of bacteria
1884	W. Hesse, on the advice his wife Fanny A. Hesse, introduces agar for solidifying nutrient media
1884/88	W. Pfeffer describes bacterial taxis as a sensory process
1885/86	H. Hellriegel and H. Wilfarth prove that nitrogen fixation in legumes is correlated to root-nodule formation, which depends on infection by soil bacteria
1887	J.R. Petri publishes a short note on the covered glass dish later named after him
1887	S.N. Winogradsky recognizes energy generation by respiration of hydrogen sulfide in *Beggiatoa* (lithotrophy)
1888	M.W. Beijerinck isolates in pure culture the nodule-forming bacteria from root nodules of leguminous plants
1891	S.N. Winogradsky obtains proof for autotrophic carbon dioxide assimilation by stoichiometric experiments on the oxidation of ammonia and on the assimilation of CO_2. He isolates *Nitrosomonas* and *Nitrobacter* in pure culture
1892	D.J. Ivanowsky describes the filterable agent (virus) causing the tobacco mosaic disease, and M.W. Beijerinck (1899) develops the concept of the "contagium vivum fluidum" ("fluid infectious principle"), the nature of the virus
1895	S.N. Winogradsky isolates the first free-living, dinitrogen-fixing bacteria (*Clostridium pasteurianum*), followed by M.W. Beijerinck (*Azotobacter chroococcum*; 1901)
1897	E. Buchner (NP 1907) discovers the fermentation of sugar by yeast press juice in the absence of living cells
1904/06	A. Harden (NP 1929) and W.J. Young study the effect of phosphate on cell-free fermentation by yeast press juice in detail and discover the accumulation of hexose 1,6-bisphosphate, marking the start of the study of cell-free systems
1909/19	S. Orla-Jensen describes many different lactic acid bacteria
1911	C. Neuberg begins his study of alcoholic fermentation: decarboxylation of pyruvate, formation of acetaldehyde, and application of trapping methods
1915/17	F.W. Twort and F. d'Herelle describe viruses of bacteria (bacteriophages)
1919	J. Buder perceives the purple bacteria as anaerobic photolithoautotrophic organisms, and C.B. van Niel (1931) formulates a unifying theory and the general equation of photosynthesis
1920/43	O. Warburg (NP 1931) discovers the function of respiratory enzymes in the cell and isolates NAD, NADP, FAD, and apoenzymes
1925/26	A.J. Kluyver and H.J.L. Donker conceive a unifying theory of reduction–oxidation reactions in tissue metabolism and microbial fermentation, the "unity of biochemistry", and a system of fermentations
1928	A. Fleming (NP 1945) discovers penicillin, and S. Waksman (NP 1952) discovers streptomycin (1943)
1928	J.H. Quastel and W.R. Wooldridge discover the inhibition of succinate dehydrogenase by malonate, thus paving the way for understanding the action of structural analogues and antimetabolites
1931/33	E. Ruska (NP 1986) constructs the first electron microscope

1933	F. Zernike (NP 1953) invents the phase-contrast microscope, which became available from the firm Carl Zeiss Jena (circa 1946)
1935	H.G. Wood and C.H. Werkman discover heterotrophic carbon dioxide fixation while studying glycerol fermentation by *Propionibacterium arabinosum*
1936	G. Domagk (NP 1939) discovers the antibacterial effect of prontosil (sulfonamide); its action as a structural analogue of 4-aminobenzoic acid is discovered by D.D. Woods in 1941
1936/50	E.E. Snell discovers the requirement of lactic acid bacteria and yeasts for several growth factors and vitamins (e.g., pantothenic acid, nicotinic acid, riboflavin, biotin (and avidin), folic acid, and pyridoxal derivatives
1937	H.A. Krebs (1900–1981; NP 1953) discovers the citric acid cycle
1941	G. Beadle (1903–1989) and E.L. Tatum (both NP 1958) isolate mutants of *Neurospora crassa* and use these mutants to explore the biosynthetic pathway of a metabolite ("one gene, one enzyme" concept)
1941	F.A. Lipmann (1899–1986; NP 1953) recognizes the role of the energy-rich phosphate bond in metabolism and discovers coenzyme A
1944	O. Avery, C. MacLeod, and M. McCarty discover DNA transformation, the transfer of genetic markers in *Pneumococcus* by uptake of DNA, and thus obtain evidence that DNA is the carrier of genetic information
1947	J. Lederberg (NP 1958) discovers the transfer of genetic markers by conjugation of *Escherichia coli* mutants
1948	J. Lederberg and B.D. Davis apply penicillin to select auxotrophic mutants and thus make possible the rapid exploration of metabolic pathways
1953	J. Watson and F. Crick (both NP 1962) recognize the structure of DNA as a helical double strand composed of two complementary single strands
1953	A.D. Hershey (NP 1969) and M.C. Chase show that viral DNA ("vegetative phage") is the necessary component for virus multiplication
1961	J. Monod and F. Jacob (both NP 1965) explain the genetic control of enzyme formation and present the operon model. Together with J.-P. Changeux, they describe (1963) the regulation of allosteric enzymes
1961	P.D. Mitchell (1920–1992; NP 1978) discovers the linkage of electron transport and proton translocation in the cytoplasmic membrane of a *Micrococcus* species and presents the chemiosmotic theory
1961	M.W. Nirenberg (NP 1968) and J.H. Matthäi perform the first experiments to synthesize polypeptides in a cell-free system, using natural or synthetic polyribonucleotides as matrix; the results lead eventually to the clarification of the genetic code
1962	W. Arber (NP 1978) discovers restriction and modification of DNA
1962	R.Y. Stanier and C.B. van Niel emphasize the essential differences between prokaryotic and eukaryotic cells, adopting the designations given by E. Chatton (1937)

Further Reading

Allen, P.W. (1932) The story of microbes. Knoxville: Bookmill

Brock, T.D. (1988) Robert Koch–A life in medicine and bacteriology. Madison: Science Tech Publishers

Bulloch, W. (1938) The history of bacteriology. Oxford: Oxford University Press

Collard, P. (1976) The development of microbiology. Cambridge: Cambridge University Press

Dobell, C. (1932) Antony van Leeuwenhoek and his "little animals." London: Constable

Dubos, R. (1950). Louis Pasteur, free lance of science. Boston: Little, Brown and Co

Iterson, G. van, Den Dooren de Jong, L.E., and Kluyver, A.J., eds. (1921–1940) Verzamelde Geschriften van M.W. Beijerinck, vols. 1–6. Delft: Martinus Nijhoff, 1940

Kamp, A.F., LaRivière, J.W.M., and Verhoeven, W. (1959) Albert Jan Kluyver, his life and work. Amsterdam: North-Holland Publishing

LeChevalier, H.A., and Solotorovsky, M. (1965) Three centuries of microbiology. New York: McGraw-Hill

Loeffler, F. (1887) Vorlesungen über die geschichtliche Entwicklung der Lehre von den Bakterien. Leipzig: Vogel Verlag

Mochmann, H., and Köhler, W. (1984) Meilensteine der Bakteriologie. Jena: Gustav Fischer Verlag

Pfeffer, W. (1897/1904) Pflanzenphysiologie. Ein Handbuch der Lehre vom Stoffwechsel und Kraftwechsel in der Pflanze, vol. 1 (1897), vol. 2 (1904). Leipzig: Wilhelm Engelmann Verlag

Sackmann, W. (1985) Biographische und bibliographische Materialien zur Geschichte der Mikrobiologie. Frankfurt Bern New York: Verlag Peter Lang

Stent, G.S. (1963) Molecular biology of bacterial viruses. San Francisco London: Freeman

Stephenson, M. (1950) Bacterial metabolism, 3rd edn. London: Longman

Thimann, K.V. (1955) The life of bacteria: their growth, metabolism, and relationships. New York: Macmillan.

Section I

The Prokaryotic Cell

2 Cellular and Subcellular Organization of Prokaryotes

The majority of prokaryotes are **unicellular microorganisms**. Typical prokaryotic cells range in size from around 0.5 μm to about 20 μm. At first sight, the prokaryotic cell does not exhibit the same degree of compartmentalization as a typical eukaryotic cell, which contains a variety of organelles and membrane-enclosed compartments. The multicellular eukaryotes comprise differentiated cells in specialized tissues to fulfill all of the tasks necessary for survival and multiplication. In general, prokaryotes are not differentiated in the same way. Instead, the single cell has a highly regulated and flexible metabolism. The functional entities engaged in this system are, in fact, located within the prokaryotic cell in **functional compartments** not necessarily enclosed by a bilayer membrane (see Chapter 24). Alternatively, cells with a special function (e.g., heterocysts) originate by cell differentiation from a vegetative cell.

2.1 Prokaryotes, Though Small, Contain all Structural Elements Necessary for Survival and Multiplication

The most prominent structural feature that distinguishes the prokaryotes from the eukaryotes (and which was taken as the basis for the term "pro"-karyote) is that prokaryotes lack a membrane-enclosed nucleus. Their DNA is located in the cytoplasm. Although prokaryotes are in most cases unicellular in organization, they exhibit various degrees of organization. This has been shown by electron microscopy. Using this technique, DNA organized in a nucleoid, ribosomes, the cytoplasmic membrane, and the cell wall were found to be the minimum structural equipment. Prokaryotes of higher complexity may exhibit additional structural constituents, such as external layers of the cell envelope exposed to the environment, pili and fimbriae, flagella, intracytoplasmic membranes, and inclusion bodies. As an example, Figure 2.**1** depicts whole cells of *Escherichia coli* with appendages. In Figure 2.**2** and Table 2.**1**, most of the structural elements of a bacterial cell are compiled. Not all of them occur at the same time or in all cells. Many of these components are present in both eubacteria and archaebacteria.

Prokaryotes commonly are thought to be simple organisms. Nevertheless, a list of the macromolecules comprising even one of the less complex bacteria, *Escherichia coli*, is extensive and impressive (Table 2.**1**). Synthesis of such a cell from glucose and a handful of ions involves distinct steps, leading first from precursor metabolites to building blocks, then to macromolecules, and finally to the structures or organelles proper. Highly regulated biosynthetic, polymerization, and assembly reactions take place; these add the new quality of "order", making the cell more than just the sum of its components. An intrinsic feature of cell structure is that a cell is not static. Rather, it is a dynamic system. Components are exchanged or degraded, and physiological activities are started or switched off. The resulting physiological state measured in the laboratory and the cell structure as observed under the electron microscope only represent momentary states of cultures or cells, respectively.

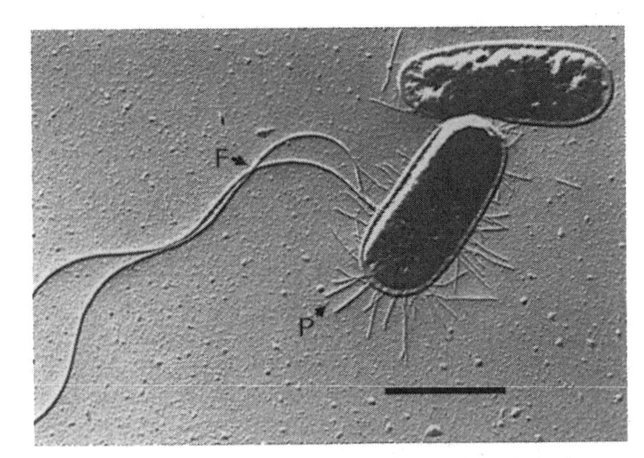

Fig. 2.**1** **Whole cells of *Escherichia coli*** (metal-shadowing preparation). One of the two cells exhibits two flagella (F) and peritrichous type 1 "pili" (P) (fimbriae). Bar = 0.5 μm. Courtesy of R. Kueper, K. Mlejnek, and S. Nakotte

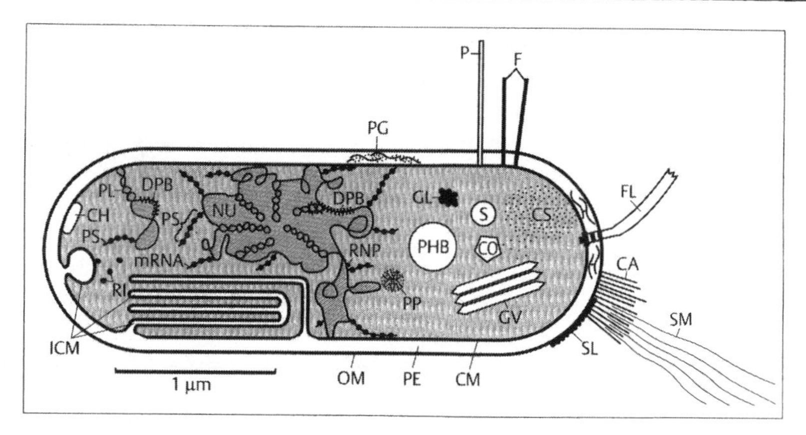

Fig. 2.2 **Diagrammatic view of an idealized bacterial cell** exhibiting basic structural features and facultative components (longitudinal section, not drawn to scale). The cell envelope is of the Gram-negative type. Only one copy each of the chromosome and a plasmid are drawn.

CA, capsule;
CH, chlorosome;
CM, cytoplasmic membrane;
CO, carboxysome;
CS, cytoplasm;
DBP, DNA-binding protein;
F, fimbriae;
FL, flagellum;
GL, glycogen granule;
GV, gas vesicle;
ICM, intracytoplasmic membrane;
mRNA, messenger RNA;
NU, nucleoid;
OM, outer membrane;
P, F pilus;
PE, periplasm in the periplasmic space;
PG, peptidoglycan ("murein");
PHB, polyhydroxybutyrate;
PL, plasmid;
PP, polyphosphate (volutin);
PS, polysome;
RI, ribosome;
RNP, RNA polymerase;
S, sulfur globule;
SL, surface layer (S-layer);
SM, slime layer

Table 2.1 **Macromolecular composition of structures of an *Escherichia coli* cell**. For abbreviations, see Fig. 2.2

Envelope and appendages				
Envelope	*Cell wall*	**CA**	Capsule:	1 complex polysaccharide
		SL	S-layer:	if present: > 5 000 units; 1 subunit; 4 or 6 subunits/unit
		OM	Outer membrane:	> 50 proteins; 4 abundant; 10^6 molecules/cell 5 phospholipids; ~ $5 \cdot 10^6$ molecules/cell 1 lipopolysaccharide (LPS); ~ $9 \cdot 10^6$ molecules/cell
		PG	Peptidoglycan:	1 molecule/cell
		PE	Periplasm:	~ 50 proteins; ~ 10^4 molecules/cell
	Cytoplasmic membrane	**CM**		> 200 proteins; ~ $2 \cdot 10^5$ molecules/cell
Flagella		**FL**		> 10 proteins; ~ $2 \cdot 10^4$ molecules/cell
Pili fimbriae		**F**		1 protein; $2 \cdot 10^4$ molecules/cell
Cytoplasm		**CS**		≫ 1 000 proteins; > 10^6 molecules/cell 60 tRNAs; ~ $2 \cdot 10^5$ molecules/cell
		GL	Glycogen:	variable
Ribosomes/mRNAs		**RI**	Ribosome:	20–100 · 10^3 ribosomes/cell in > 1 000 polysomes (PS) 55 proteins; 1 of each per 70S ribosome 3 rRNAs (5S, 16S, 23S; 1 of each per 70S ribosome) > 10^3 mRNAs; 1 per polysome (PS)
Nucleoid (NU) and plasmids				
			Chromosome:	haploid; ~ 1 molecule; circular; ~ 4 200 kb
		PL	Plasmid(s):	if present: 1 to many copies circular or linear small to very large (up to 1/3 of the chromosome)
		DBP	DNA-binding proteins (dsDNA, ssDNA) DNA replication and transcription machinery	

Prokaryotes are microscopically small cells. Although they normally lack structural compartments comparable to those of eukaryotic cells, they are equipped with functional compartments that contain all structural elements necessary for survival and multiplication in a highly organized way. An average prokaryotic cell comprises approximately 3500 structural and regulatory genes; about 50% of these code for enzymes and structural proteins, about 50% are used for regulation.

2.2 Cellular Structures Can Be Made Visible or Identified by Numerous Methods

The small size and low optical density and refractive index of most prokaryotic cells make them nearly invisible under the conventional brightfield light microscope even at high magnification. The first method to overcome this obstacle was staining of cells with cationic or anionic dyes. Application of the basic dye crystal violet was introduced by the Danish bacteriologist H.C.J. Gram in 1884 (**Gram staining**). The procedure begins with addition of the dissolved stain to fixed bacteria, followed by treatment with an iodine solution. Crystal violet and iodine form a complex moderately soluble in ethanol or acetone, and insoluble in water. "Differentiation" is performed by treatment with ethanol: the "Gram-positive" bacteria retain the dye–iodine complex and remain deep blue-purple. "Gram-negative" cells are destained; they are rendered visible by counterstaining with fuchsin. The dye–iodine complex appears to be localized on the protoplast. The positive Gram-staining reaction is caused by the wall of these cells being strongly resistant to extraction of the dye complex. The wall of Gram-negative bacteria lacks this property. The Gram-staining procedure is still in use. It is an important taxonomic feature that correlates with many other properties of bacteria.

A big step forward was the invention of **phase-contrast microscopy** by F. Zernike in 1934 and its refinement by A. Köhler and W. Loos. In conventional light microscopy, contrast is achieved by the different degrees of light absorption of the various structural elements in the sample. This kind of sample is called an absorption or amplitude object because it changes the amplitude of the incident light. Phase objects, such as many bacteria, do not absorb the light at all, or absorption is equal over the entire object. Only differences in thickness or refractive index exist, and therefore only the phase, not the amplitude, of the incident light wave is altered. The human eye cannot detect these phase differences. After transformation of phase differences into amplitude differences (i.e., differences in brightness), the structural details are rendered visible. This transformation is achieved by interference caused by the object proper (Fig. 2.**3a**). An alternative approach is **interference microscopy**. This kind of light microscopy makes use of interference between a reference beam passing not through the object, but only through the surrounding medium, and the beam traveling through the object. The difference between the two refractive indices results in a pseudo-three-dimensional image and can be measured with a compensator. Thus, the thickness of the object can be determined.

A very powerful technique with rapidly extending fields of application is **fluorescence epimicroscopy** or **transmission light microscopy**. Most methanogenic bacteria emit blue–green light when illuminated with light of the wavelength 420 nm (blue light) (Fig. 2.**3b**), caused by the naturally occurring coenzyme F_{420} in the cell. One of the two known classes of plant-pathogenic members of the genus *Pseudomonas* (*P. aeruginosa*) produces a characteristic blue pigment, pyocyanin, which belongs to the phenazine family. This pigment is very important in species identification. A second kind of approach of fluorescence light microscopy is based on the addition of fluorochromes. Application of the fluorochrome DAPI (4,6-diamidinophenylindole) results in the binding of this dye to double-stranded DNA. This technique allows localization of DNA within the cell. A variety of prokaryotes can be classified and identified using the fluorescent antibody technique (Fig. 2.**3c**). This technique uses fluorescine-conjugated antibodies directed specifically against cell constituents (e.g., cell appendages, envelope components, mRNA) of respective strains.

An extension of the power of light microscopy was achieved by the introduction of the **confocal laser scanning microscope**. A laser replaces the conventional light source. The laser beam is scanned over the specimen; the resulting image is composed of individual pixel intensity values. Based on the specific properties of

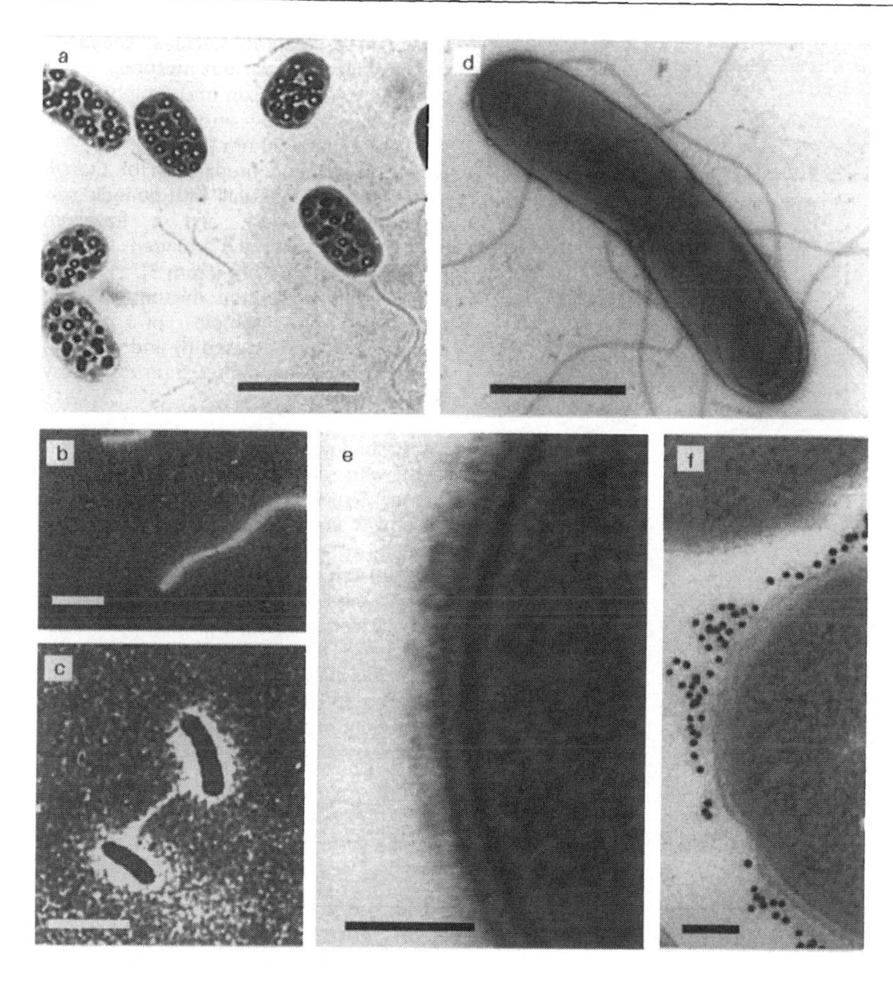

Fig. 2.**3a–f** **Methods of preparation and imaging of prokaryotes by light and electron microscopy.**
a Phase-contrast light micrograph of *Chromatium okenii*. The cell is filled with sulfur globules. Bar = 10 μm. [From 1].
b Fluorescence light micrograph of the methanogenic archaebacterium *Methanospirillum hungatei* excited at 420 nm. Bar = 10 μm. [From 2].
c Light microscopic visualization of the F pilus connecting two *E. coli* cells in an early state of conjugation, with fluorescence-labeled F-pilus-specific MS2 bacteriophages. Bar = 5 μm. [From 3].
d Transmission electron micrograph of a flagellated cell of *Alcaligenes eutrophus*. Bar = 1 μm. Courtesy of F. Mayer.
e, f Demonstration of the presence of adhesin on the cell surface of *Enterococcus faecalis* after induction by sex pheromones (ultrathin sections).
e Resin-embedded cell exhibiting a fringe of delicate fibers at its surface. Bar = 0.1 μm. Courtesy of R. Wirth.
f Labeling with anti-adhesin antiserum complex, via a second type of antibody, with colloidal gold. The adhesion is present as hair-like structures located between cell wall proper and gold label. Bar = 0.1 μm. Courtesy of R. Wirth

laser light (coherence), this technique allows generation of images of thin planes within the specimen and eliminates contaminating light from out-of-focus regions. Thus, the object is optically sectioned into "confocal" images. The approach allows computer-aided three-dimensional reconstruction of the object and a large range of quantitative evaluations.

The resolving power of light microscopy is limited to approximately 0.25 μm. **Electron microscopy** permits visualization of details at the level of macromolecules, i.e., in the nanometer range. However, conventional electron microscopy has a major drawback. Water has to be removed from the sample prior to insertion of the specimen into the microscope. Artifact formation is the inevitable consequence. Therefore, preparation techniques of biological samples for electron microscopy that eliminate or reduce this kind of artifact formation had to be developed. Often the specimen is too thick to allow the electron beam to pass it. Furthermore, a biological sample is composed of chemical elements that cause only a weak interaction with the imaging electron beam. The consequence is low contrast. Chemical fixation prior to removal of water reduces deformations and loss of material. Recently designed cryo preparation and imaging techniques, allowing the presence of frozen water in the sample, made occurrence of artifacts a much less severe problem to deal with. Reduction of sample thickness can be achieved by preparation of **ultrathin sections** from a resin-embedded, chemically fixed, and dehydrated sample, or from a frozen object. Both kinds of sectioning allow a prokaryotic cell of an average diameter do be cut into about 20 slices. Improved contrast can be gained by incubation of the sample in a solution containing heavy metal salt ions (resulting in "positive" staining of ultrathin sections; Fig. 2.**3e**), by metal "shadowing" of the dried sample mounted on a support film (Figs. 2.**1**, 2.**4b-e**), or by "negative" staining

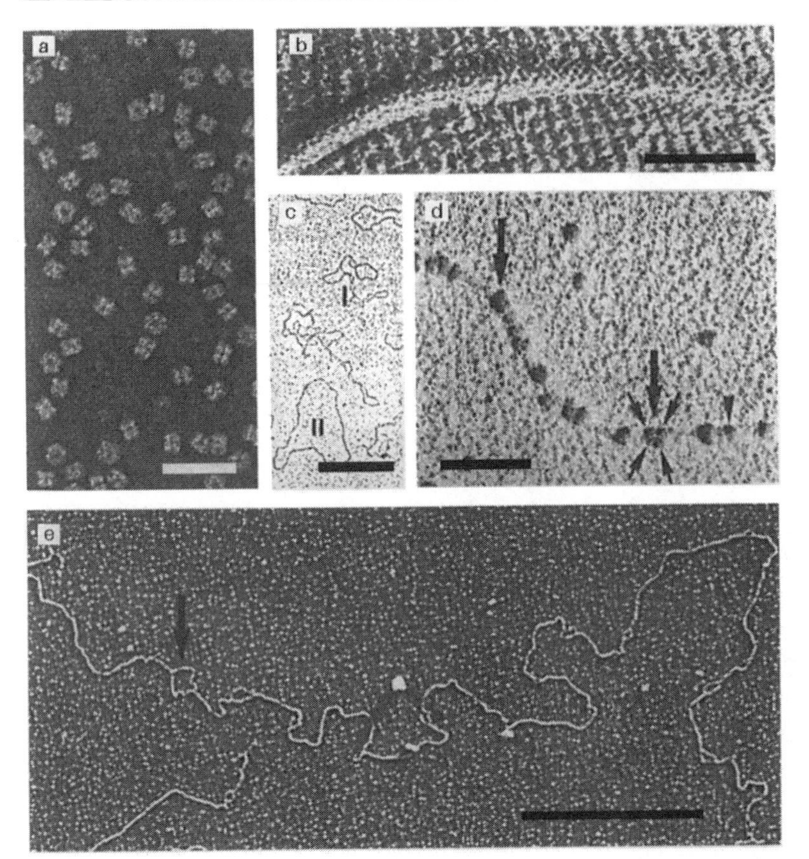

Fig. 2.4a–e **Macromolecules prepared and visualized by various methods.**
a Transmission electron micrograph of the negatively stained enzyme glutamine synthetase. Bar = 50 nm [From 4].
b Freeze-fracturing preparation of *Clostridium aceticum*, cell wall with periodic surface layer (S-layer), and a flagellum composed of helically arranged flagellin molecules. Bar = 50 nm [From 5].
c Transmission electron micrograph of a rotary-shadowed sample of plasmid pBR322; covalently closed (I) and open (II) forms can be seen. Bar = 0.5 μm. Courtesy of W. Johannssen.
d Transmission electron micrograph of rotary-shadowed plasmid pBR322 (linearized) with attached restriction endonuclease *Eco*RI molecules. Large arrows, tetrameric enzyme particles; small arrows, subunits of a tetrameric enzyme particle. Bar = 0.1 μm [From 6].
e Transmission electron micrograph of partially denatured, metal-shadowed, bacteriophage DNA (bacteriophage 16-6-12 from *Rhizobium lupini*) prepared by cytochrome c spreading. Arrow indicates strand separation; metal shadowing was from two directions. Bar = 0.5 μm. Courtesy of F. Mayer

(Figs. 2.**3d**, 2.**4a**) achieved by depositing heavy metal salt, dissolved in water, around the specimen. After blotting dry, the salt remaining around the object delineates the contours of the object. An indirect image of the object can be obtained by application of the replica technique, which provides structural information on outer surfaces of the sample. Inner surfaces can be depicted with this method by preparation of a replica of faces exposed by fracturing of the frozen sample ("**freeze fracturing**"; Figs. 2.**4b**, 2.**12d**, 2.**13b,c**). All contrasting procedures for electron microscopy depict the distribution of heavy metals rather than the structural elements of the sample proper. A recently developed approach, imaging of the sample in the "frozen hydrated" state, no longer requires introduction of heavy metals. The contrast between water and the biological sample is sufficient for imaging.

In analogy to the light-microscopic fluorescent antibody technique, similar approaches have been developed for electron microscopy. Instead of with a fluorescent dye, the antigen-specific antibodies are labeled with electron-dense markers, such as colloidal gold (Fig. 2.**3f**). The procedure is called "**electron microscopic immunocytochemistry**".

In addition to conventional transmission electron microscopy and its cryo variations, various **scanning modes of imaging** of samples at cellular and macromolecular levels have been developed. They necessitate modified sample preparation procedures.

Prokaryotic morphology and ultrastructure can be investigated by light **microscopy** and electron microscopy. Light microscopic resolution is limited to approximately 0.25 μm; electron microscopy reveals details at the cellular level and below and extends imaging into the nanometer range. Prior to imaging, biological specimens have to undergo various steps of preparation, which often cause artifacts.

2.3 Prokaryotes May Occur as Single Cells or as Cell Associations

2.3.1 Single Cells Have Different Size and Shape

Unicellular prokaryotes usually are too small to be viewed with the naked eye. Exceptions exist, such as *Epulopiscium fishelsoni*, which is about one million times larger in volume than a typical bacterium. Its unusual size may be related to its living symbiotically within the intestine of the Red Sea surgeon fish *Acanthurus nigrofuscus*. It is expected that there are many more unusual microorganisms. In addition to sphere-like prokaryotic cells ("cocci"), straight rods with rounded ends, bent or curved rods, helically bent rods, cells with buds and protuberances (prostheca), or even rectangular platelets are typical cell shapes (Fig. 2.**5**). Depending on growth conditions or state within a life cycle, both the size and the shape of prokaryotes, even of those belonging to a given species or strain, may vary within genetically defined borderlines. A look at individual cells all belonging to the genus *Clostridium* (Fig. 2.**5a–c**) illustrates this observation and demonstrates that size and shape of a prokaryotic cell alone are not valuable parameters for classification at the genus level.

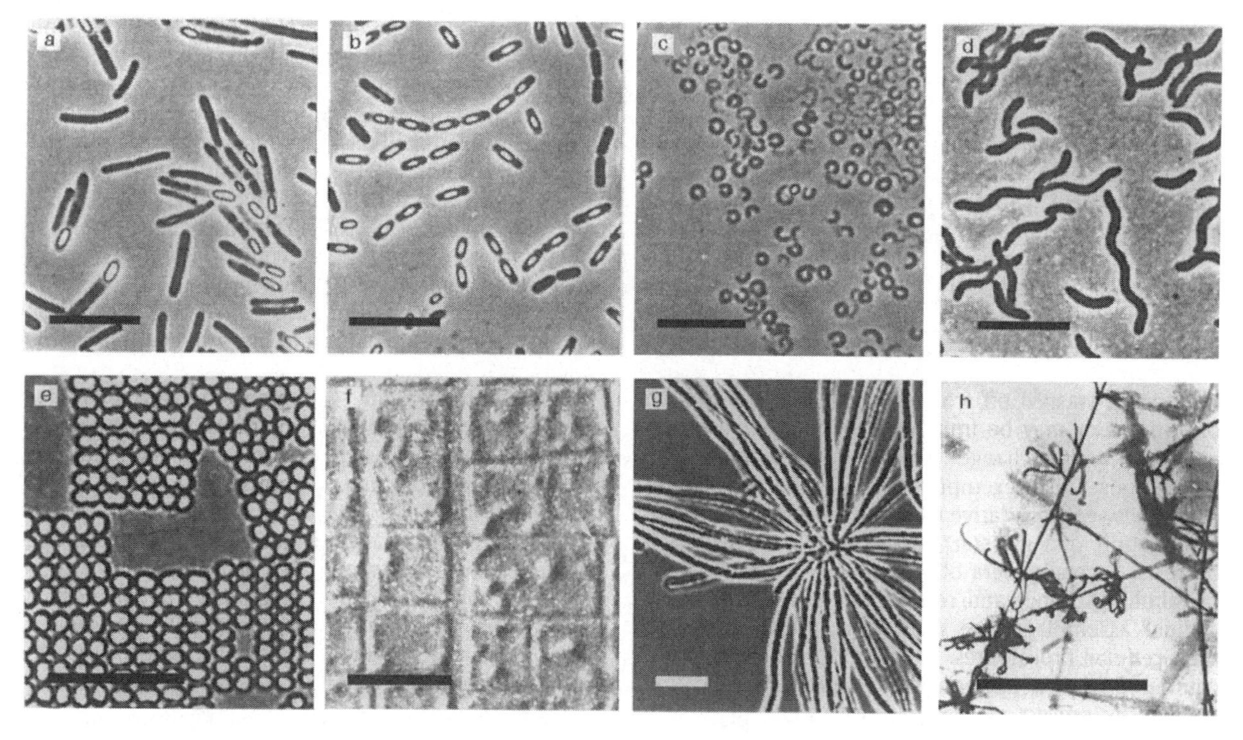

Fig. 2.**5a–h** **Light micrographs of prokaryotic cells and cell associations**, demonstrating the diversity of cell shapes, sizes, and organization of associations.
a–c Various cell shapes and endospore locations within the genus *Clostridium*. Bar = 10 μm [From 7].
a *Clostridium butyricum*; simple elongated rods, terminal spores, light scattering.
b *C. bifermentans*; pairs and chains of short rods, central spores.
c *C. cocleatum*; single bent rods.
d *Rhodospirillum rubrum*; shape of a corkscrew. Bar = 10 μm.
e *Thiopedia rosea*; highly ordered 2-D cell association. Bar = 10 μm.
[**d** and **e** from 1].
f Square bacteria depicted by interference light microscopy; regular association. Bar = 5 μm [from 8].
g *Leucothrix* spec.; filaments forming a rosette. Bar = 20 μm [from 9].
h Aerial mycelium of *Streptoverticillium* spec.; complex cell association; ends of spore chains sometimes hook-like. Bar = 0.1 mm [from 10].

Prokaryotic cells may be sphere-like, straight, bent, curved or helically bent rods, platelets, irregularly shaped, or may have prostheca or buds. The size and shape of a prokaryote can vary appreciably within genetically determined borderlines, depending on growth conditions.

2.3.2 Prokaryotic Cells Can Associate to Form Clusters

Prokaryotes are not only found as single cells. They may form cell associations of various complexity due to, for example, incomplete cell separation or division (**homologous associations**). Examples are *Nitrosolobus multiformis*, which forms cell clusters, or the trichomes formed by cyanobacteria. Prokaryotic cells may be kept "glued together" after division by external layers consisting of polysaccharides. These layers may form a slimy matrix (*Zoogloea ramigera*) or a sheath (*Sphaerotilus natans*). Other examples are *Lampropedia hyalina, Sarcina ventriculi*, and *Acetobacter aceti* var. *xylinum*. *A. aceti* excretes cellulose in the form of threads. This fibrillar material surrounds the cells, giving rise to the formation of the "mycoderma aceti", a leather-like skin. Associations may also be brought about by fimbriae and pili, i.e., cellular appendages that are proteinaceous in nature. Cell associations may reduce the probability of cells being washed off from sites of favorable growth conditions, or may be important for functional differentiation and exchange of metabolites and signal compounds. This is exemplified in cyanobacteria (interaction between vegetative cells and heterocysts), in cell conjugation, and in "star" formation observed in soil bacteria. Several genera of Cyanobacteria, which show metabolic exchange and cell differentiation within the cellular assemblage (see Fig. 2.**21e**), are examples of multicellular prokaryotes. A different situation is encountered in myxobacteria (see Chapter 25.3). At first glance, the process of aggregation that takes place in these systems leads to differentiation and "tissue" formation as they are known in eukaryotes. The process is a typical example of a prokaryotic differentiation. It comprises cooperative cellular morphogenesis, brought about by exchange of information coupled to specific cellular function and the formation of propagation units. The event is analogous to a generation cycle. It includes a vegetative cell cycle, sporulation, and germination of spores set free from fruiting body cysts or sporangioles. The result of spore germination closes the cycle with the formation of vegetative cells.

2.3.3 Cells Can Aggregate to Form Specific Consortia or Associations With a Host

A variety of prokaryotes may form associations of various complexity, either with other prokaryotes, or with eukaryotic cells or tissues. These **heterologous associations** include symbiotic and parasitic systems, commensalism, and neutralism. In a **symbiosis**, the association is advantageous to both partners, which usually are adapted to each other. When one partner is harmed in the association, the association is called **parasitism**. **Commensalism** is an association in which only one partner gains an advantage, but the other is not harmed. In many cases, the partners co-exist without any influence on each other; this is called **neutralism**.

In symbiotic systems, two modes of spatial relation can exist. In an ectosymbiosis, one of the partners exists outside the cells of the other; in an endosymbiosis, one partner lives inside the cells of the other (Figs. 34.**5**, 34.**12**). The larger of the two partners is usually referred to as the host. The associations can have nutritional advantages (fixation of nitrogen, supply of basic nutrients or accessory factors), serve recognition functions (luminescent bacteria in fish), or simply provide protection.

The green sulfur bacteria have the potential to form stable symbiotic associations (consortia) with colorless bacteria. Examples are "Chlorochromatium aggregatum", "Chlorochromatium glebulum", and "Cylindrogloea bacterifera". In addition to these green Chlorochromatium consortia, there are brown ones (Fig. 30.**24**).

"Chlorochromatium aggregatum" was first described as early as 1906. The center of this consortium is formed by a colorless motile bacterium of unknown taxonomic position and physiological activity. It is surrounded by 6–12 regularly arranged, non-motile green sulfur bacteria. A high degree of metabolic interdependence between the members of this consortium is obvious; cell growth and division occur synchronously. The non-motile *Chlorobium* symbionts, when irradiated, trigger a phototactic reaction in the central, motile bacterium to which they are connected. Thus, the consortium achieves the exposure of its photosynthetically active members to favorable light conditions (Chapter 30.7.7).

Other associations between bacteria often are clearly parasitic. An example is *Bdellovibrio bacteriovorus*. The genus *Bdellovibrio* comprises three species of small, Gram-negative, predatory bacteria. Two alternate forms of cells are observed. One is a motile, polarly flagellated, non-reproducing free form. The other is a non-motile reproductive form. *Bdellovibrio* is found in the periplasm of host cells, such as enteric bacteria, *Pseudomonas*, *Rhizobium*, *Rhodospirillum*, *Photobacte-*

rium, and *Chromatium*. These cell forms are, in fact, part of a life cycle, which starts with an initial, reversible stage after contact, followed by an irreversible stage. After penetration through the outer membrane of the host bacterium, the parasite sheds its flagellum and enters the periplasm. A septate filament is formed which grows at the expense of the host and fragments into flagellated cells. These cells lyse the host cell wall and swim away.

Chlamydia psittaci and other *Chlamydia* species are further examples of intracellular parasitic bacteria. Their hosts are animal cells. The life cycle of these bacteria is also divided into two states: extracellular and intracellular. The extracellular resting state, called "elementary body", is highly infectious. The "reticulate body" is the intracellular state; in this state, the cells grow and divide.

Prokaryotes may form associations of various complexity, such as trichomes, regularly or irregularly shaped aggregates, or stable and unstable associations with other cells, including eukaryotic cells and tissues. In these cases, symbiotic and parasitic systems may arise.

2.4 The Structural Components of Prokaryotic Cell Envelopes Are Organized as Barriers and Interfaces

2.4.1 Various Prokaryotic Cell Envelopes Have Common Properties

The **envelope** of a eubacterial cell is composed of the cell wall, the periplasm, and the cytoplasmic membrane. The **cell wall** comprises surface structures exposed to the environment, the outer membrane in Gram-negative eubacteria, and the peptidoglycan, which determines and preserves cell shape and integrity. The **periplasm** contains numerous proteins in the space between the outer membrane and the cytoplasmic membrane; the **cytoplasmic membrane** is the border of the cytoplasm.

Surface structures exposed to the environment are capsules, S-layers, fimbriae/pili, adhesins, and non-cytoplasmic appendages. These envelope components provide protection of the cell against unfavorable environmental conditions. They mediate various interactions, such as adhesion to surfaces of biotic or abiotic origin, or contact with partner cells, and contain receptors. The multiple functions are reflected by a diversity of chemical and structural features.

Capsules may be divided into three categories: macrocapsules, microcapsules, and slime layers. The presence of a macrocapsule can be demonstrated by light microscopic negative staining (Fig. 2.**6a**) with India ink (the capsule is revealed as a clear zone between the opaque medium and the cell wall), by fixation (Fig. 2.**6b**) or by the use of antibodies directed against the capsular antigens. Microcapsules cannot be detected by light microscopic negative staining because they are too thin. Their presence can be shown by serological techniques. Slime layers are secretions loosely adhering to the cell surface, without a clearly defined external border. Slime may segregate into the medium when the organism is grown in liquid culture. Swarm cells (e.g., from *Proteus mirabilis* and *Myxococcus xanthus*) are surrounded by slime that provides a matrix through which the cells can migrate.

In capsule-forming prokaryotic pathogens, good correlation exists between capsulation and pathogenicity (see Chapter 33). Prokaryotes that have lost their capsules commonly appear to be more susceptible to phagocytosis. Capsule formation may vary depending on the growth conditions even within cells of one species (e.g., *E. coli*). Many prokaryotes have totally lost their capsule during extended cultivation in the laboratory (e.g., *E. coli* K-12).

Specific modifications of capsules are "holdfasts" and "sheaths". Holdfasts are composed of amorphous material that is attached at one pole of the cell, such as at the end of the stalk of *Caulobacter* spec. and *Asticcacaulis* spec.; the holdfast mediates contact of the cell to biotic or abiotic surfaces by "polymer bridging". Sheaths are long hollow tubes (e.g., in *Sphaerotilus natans*) enclosing chains of bacterial cells.

Capsules in their various forms are not only found in eubacteria, but also in archaebacteria (e.g., in *Methanobacterium formicicum*). Sheaths have been observed surrounding square bacteria and *Methanospirillum hungatei*.

Water is the principal constituent of capsules. The most common organic components are polysaccharides. Polypeptides, mixtures of polysaccharides and polypeptides, or cellulose may occur. Capsular material may be connected to the cell wall by ionic bonding, but covalent linkages may also be involved in certain cases, such as in *Bacillus anthracis*, where capsular material is linked to

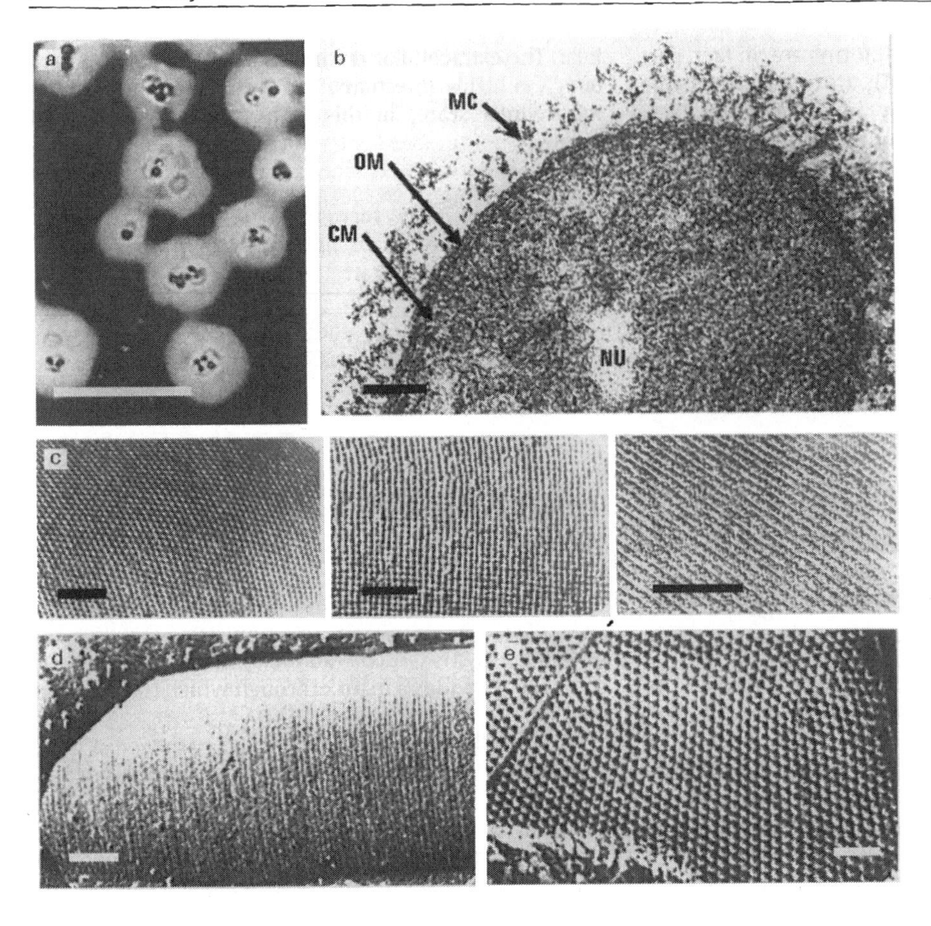

Fig. 2.6a–e **The components of the prokaryotic cell periphery.**
a Slime visualized by light microscopic negative staining with India ink (*Amoebobacter pendens*, a species belonging to the family Chromatiaceae). Bar = 10 μm [from 11].
b Macrocapsule of *Klebsiella pneumoniae* depicted, after chemical fixation using alcian blue lanthanum and tris(-1-aziridinyl) phosphine oxide-aldehyde-osmium, by transmission electron microscopy. MC, macrocapsule; CM, cytoplasmic membrane; OM, outer membrane; NU, nucleoid. Bar = 0.1 μm [from 12].
c Eubacterial surface layers (S-layers) of freeze-fractured preparations depicted by transmission electron microscopy, exhibiting various types of symmetry. (All three examples are Gram-positive bacteria.)
Bar = 0.1 μm. Left, *Clostridium thermohydrosulfuricum*; center, *Desulfotomaculum nigrificans*; right, *Bacillus stearothermophilus* [from 13].
d and **e** Freeze-fracture preparations of two different kinds of S-layer of methanogenic archaebacteria [from 14].
d Trichome of *Methanospirillum hungateii*; a prominent feature is the arrangement of the S-layer units in rows. Bar = 0.1 μm.
e *Methanogenium marisnigri* exhibiting a hexagonal order of the S-layer lattice and lattice disorders. Bar = 50 nm

peptidoglycan. By definition, the lipopolysaccharides of Gram-negative bacteria (see below) are not components of capsules, but of the "outer membrane" of these cells.

Prokaryotes may exhibit an ordered array of proteinaceous units. Usually, these units are organized as a monolayer covering the cell. This layer is referred to as **S-(surface) layer** (Fig. 2.**6c–e**). S-layers are found in nearly every taxonomic group, in Gram-positive and Gram-negative eubacteria, and in archaebacteria. S-layers exhibit structural features typical for two-dimensional crystals. In eubacteria, monolayer lattices with hexagonal (p6), square (p4), or oblique (p2) symmetry are observed. S-layers of archaea have predominantly hexagonal symmetry. Lattice distortions make sure that also the rounded ends of the cells are covered. Most S-layer structural units are composed of a single, homogeneous protein or glycoprotein species. The subunits of the structural units interact with each other and with the underlying cell envelope component, commonly through noncovalent forces. In extremely thermophilic archaebacteria, the S-layer is highly resistant to denaturants, possibly because of covalent linkage between the S-layer subunits.

Comparative studies on the distribution and uniformity of S-layers indicate that individual strains of a species can exhibit a remarkable degree of heterogeneity regarding the kind of symmetry and the molecular weight of the constituent subunits. After extended culturing of eubacteria, the S-layer may be lost without loss of viability.

2.4.2 The Cell Wall of Gram-negative Eubacteria Is Very Complex

A main constituent of the cell wall of Gram-negative eubacteria is the outer membrane (OM; Fig. 2.7), which appears as a wavy, membrane-like structure in electron microscopic preparations fixed by chemical treatment. Between the outer membrane and the cytoplasmic membrane (CM; Chapter 2.4.5), which delimits the protoplast, the murein or peptidoglycan network (PG) or sacculus is visible as a dark line in these preparations. The "periplasmic space", i.e., the volume enclosed by outer and cytoplasmic membrane, does not exhibit electron microscopically visible contents other than the peptidoglycan layer in ultrathin sections of cells fixed chemically by standard procedures. Application of cryopreparation techniques demonstrates that this is not the true architecture of the cell wall of Gram-negative eubacteria. The outer membrane is not wavy, but is smooth. The peptidoglycan is not a thin layer; it forms a very loosely organized three-dimensional network that seems to fill the entire periplasmic space (structure and biosynthesis of peptidoglycan, see Chapter 23.1). For the description of the state of the contents (the "periplasm") present in this space, the term "periplasmic gel" was coined, indicating that it is a highly hydrated gel. It contains numerous types of protein. These layers of the cell wall of Gram-negative eubacteria are interconnected by lipoproteins.

In contrast to the cytoplasmic membrane, the outer membrane is not a typical diffusion barrier. Its molecular architecture is asymmetric. The outer leaflet of the outer membrane contains the lipid A moiety of the lipopolysaccharides (see Chapter 23.2); its inner leaflet contains phospholipids. Pore proteins (porins) form channels through the outer membrane for the exit and entrance of small molecules (<600 molecular weight). The structure and function of porins are described in Chapter 5.6.1. Other proteins that mediate export of

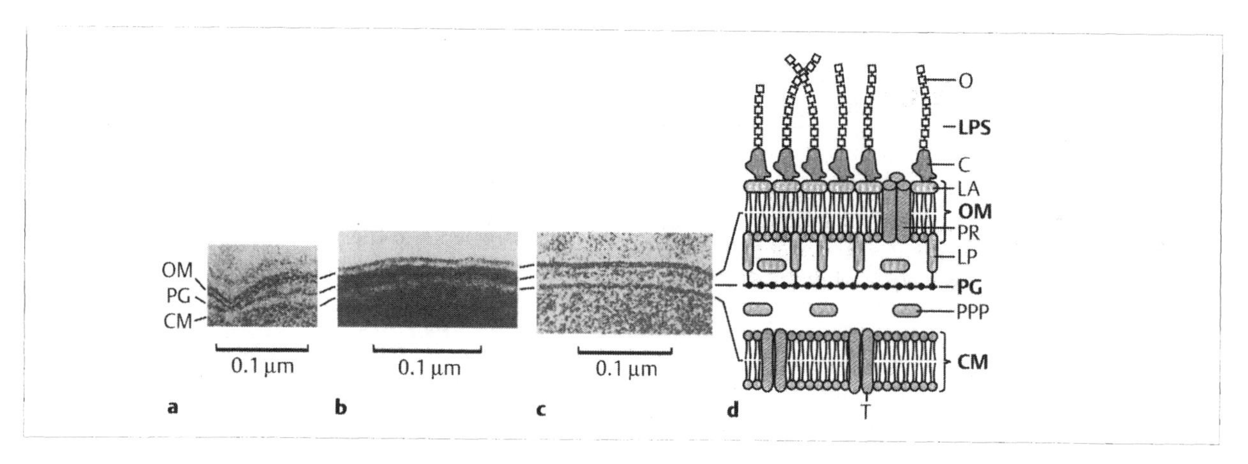

Fig. 2.7a–c **The cell envelope of Gram-negative eubacteria** (**a–c** ultrathin sections; bars = 0.1 µm).
a Conventionally chemically fixed, dehydrated, resin-embedded, and ultrathin-sectioned cell (*Acinetobacter* spec. strain MJT/F5/5); note the wavy appearance of the outer membrane, shrinkage by loss of material, and "empty" appearance of the space between the cytoplasmic and outer membrane, which seems to contain only the peptidoglycan layer [from 15].
b Cryosection of an *Alcaligenes eutrophus* cell (stabilization of the section with methyl cellulose and contrasting with uranyl acetate). Note the compact appearance of the cell envelope. The peptidoglycan present between the cytoplasmic and outer membrane is not discernible as a distinct layer. Courtesy of M. Rohde.
c Cryosection of a frozen-hydrated, not chemically fixed, unstained sample of *E. coli*. The layers of the cell envelope are clearly visible. In the space between the cytoplasmic and outer membrane, only the peptidoglycan can be discerned. Because of a lack of contrast, the other components present in the periplasmic space cannot be seen [from 4].
d Diagrammatic view of the macromolecular architecture; the lines indicate the respective layers visible in the ultrathin sections (**a–c**) [after 16]. C, core region of the lipopolysaccharide; CM, cytoplasmic membrane; LA, lipid A; LP, lipoprotein; LPS, lipopolysaccharide; O, O-specific side chains of lipopolysaccharide (sugar); OM, outer membrane; PR, porin; PG, peptidoglycan (murein); PPP, periplasmic protein; T, transmembrane protein

proteins or function as bacteriophage receptor and specific uptake channels are part of the outer membrane.

In cyanobacteria, a polysaccharide is covalently bound to peptidoglycan. Although cyanobacteria possess an outer membrane, the thick peptidoglycan network or sacculus and the polysaccharide may be responsible for the Gram-positive staining of these bacteria. In addition to porins, also carotenoids are components of the outer membrane of cyanobacteria.

The **cell wall** of typical **Gram-negative** eubacteria comprises a peptidoglycan network or sacculus covered on the outside by an outer membrane that contains porins and other proteins, lipopolysaccharides and phospholipids, and by the capsule. On the inside of the peptidoglycan sacculus, the cytoplasmic membrane (which is not part of the cell wall, but is part of the cell envelope) is located. Enclosed between the outer membrane and the cytoplasmic membrane is the periplasmic space. In addition to the peptidoglycan sacculus, it harbors soluble proteins. The peptidoglycan and the outer membrane are interconnected by lipoproteins.

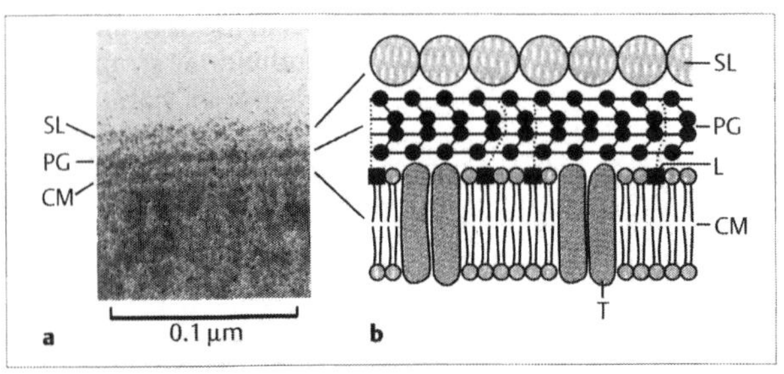

Fig. 2.**8a, b** **The cell envelope of Gram-positive eubacteria.** **a** Conventionally chemically fixed, dehydrated, resin-embedded, and ultrathin-sectioned cell (*Thermoanaerobacterium thermosulfurigenes* EM1) exhibiting a cytoplasmic membrane (CM), thick peptidoglycan layer (PG), and a surface layer (SL). Bar = 0.1 μm [from 17].
b Diagrammatic view of the macromolecular architecture; the lines indicate the respective layers visible in the ultrathin section (**a**). The lipoteichoic acid molecules (L) extend into the cell wall. T, transmembrane protein. Courtesy of F. Mayer

2.4.3 The Cell Wall of Gram-positive Eubacteria Exhibits a Sophisticated Organization That Is not Evident Under the Electron Microscope

Gram-positive eubacteria do not contain an outer membrane. A positive Gram reaction indicates that the cell wall of these bacteria (Fig. 2.**8**) is stable enough to retain the reaction product of the staining procedure, a crystal-violet-iodine complex, during the final step of the treatment. In eubacteria, this is due to the presence of a thick **peptidoglycan** sacculus in the envelope, located directly outside of the cytoplasmic membrane, and an additional polymer, the teichoic acids (Chapter 23.3). **Teichoic acids** consist of polyhydric alcohol phosphates, often carrying side-chains of oligosaccharide units and ester-linked D-alanine residues. Other additional components can be teichuronic acid (Chapter 23.4) and lipoteichoic acids (Chapter 23.5). In these compounds, a glycerol teichoic acid is linked to a glycolipid. The latter component enables the molecule to be attached with one end to the cytoplasmic membrane, with the hydrophilic part of the molecule reaching into the wall. Teichoic acids are covalently linked to peptidoglycan.

In mycobacteria, nocardia, and corynebacteria, 30% of the weight of the wall consists of **lipids**. Also wax, a tetrasaccharide of peptidoglycan joined to esterified arabinogalactan, has been found in several mycobacteria. Fragments of wax probably originating from partial autolysis, have importance as immunoadjuvants. Mycolic acids are further constituents of the wall of genera such as *Mycobacterium*, *Nocardia*, *Corynebacterium*, *Rhodococcus*, and *Caseobacter*. Mycolic acids are high-molecular-weight 3-hydroxy acids with a long alkyl branch at position 2 and constitute valuable chemotaxonomic markers for classification and identification of these taxa.

Components of the wall of Gram-positive eubacteria in addition to polysaccharides, which occasionally render the bacteria pathogenic, may be proteins. Examples are protein A, which is associated with the cell wall of *Staphylococcus aureus* and which interacts with human IgG antibodies in a pseudoimmune Fc reaction, and staphylococcal enterotoxins, which are transiently associated with the cell wall before being released extracellularly.

The major constituents of the **cell wall** of typical **Gram-positive** eubacteria are multiple peptidoglycan layers and teichoic acids linked to peptidoglycan, lipoteichoic acids, and teichuronic acids. Additional wall components can be lipids, wax and mycolic acids, and proteins. Neutral polysaccharides or a proteinaceous surface layer form the outermost wall layer. An outer membrane is absent.

2.4.4 Cell Walls of Archaea: the "old" Solution

The absence of a typical peptidoglycan is a feature characteristic of Archaea (Fig. 2.**9**). Five morphological types of cell wall have been found. Type 1, an electron-dense layer when viewed in ultrathin sections, consists mainly of pseudomurein, a sulfated or a nonsulfated acidic heteropolysaccharide (Fig. 7.**31b**). Gram staining of type 1 cells results in a positive reaction. A rare type, type 2, is also Gram-positive. The cell wall is rigid and consists of pseudomurein covered by a surface layer of protein (Fig. 2.**9**; see Chapter 23.10). Many Gram-negative Archaea, such as many halophiles and methanogens and all thermoacidophiles, exhibit type 3 cell walls. This type consists only of a surface monolayer composed of protein or glycoprotein subunits and lacks pseudomurein (see Chapter 23.10). Type 4 is rather complex; several cells are held together by a sheath consisting of protein fibrils, and the individual cell is surrounded by an electron-dense layer, which is probably also of proteinaceous nature. Type 5 lacks a cell wall; the cytoplasmic membrane is the only constituent of the cell envelope.

Cell walls of Archaea are different from those of Bacteria. Pseudomurein, polysaccharides, and proteins or glycoproteins organized as a surface layer or a sheath, may form the cell wall. Certain Archaea lack a cell wall.

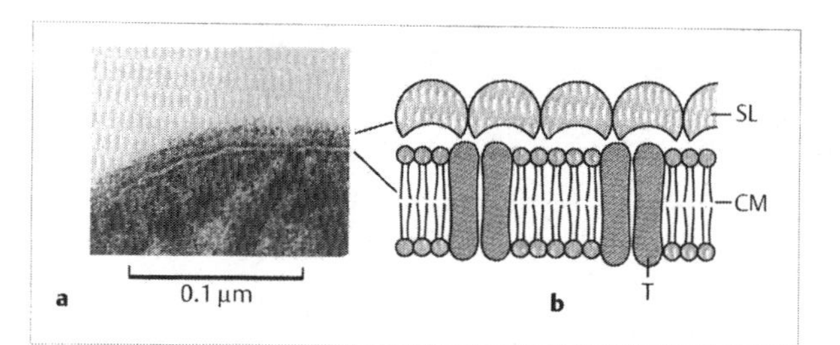

Fig. 2.**9a,b** **The cell envelope of an archaebacterium, *Methanogenium marisnigri*.**
a Conventionally chemically fixed, dehydrated, resin-embedded, and ultrathin-sectioned cell, exhibiting an envelope of very low structural complexity. Bar = 0.1 µm [from 18].
b Diagrammatic view of the macromolecular architecture. The cell envelope is composed of the cytoplasmic membrane (CM), which contains proteins (T), and a surface layer (SL). Note that this view is only one of the variations of the structural organization of the cell envelope of Archaea (see text). Courtesy of F. Mayer

2.4.5 Cytoplasmic and Intracytoplasmic Membranes Confine the Protoplast

The cytoplasmic membrane limits the cytoplasm and mediates transport of ions, solutes, metabolites, and macromolecules between the cytoplasm and the environment of the cell. The main functions are the specific import and export (selectivity), and the generation of an electrochemical gradient of ions (H^+, Na^+) across the membrane (see Chapters 4, 5 and 11).

In electron micrographs of conventionally prepared ultrathin sections (Fig. 2.**7a**), this membrane is visible as trilaminar layer approximately 5–6 nm thick, made up of a central bright sublayer with two adjacent electron-dense sublayers. This appearance reflects the bilayer composition of the membrane, which contains phospholipids orientated with their hydrophilic "heads" toward the outside and toward the cytoplasm of the cell. Numerous proteins (about 50–70% of the dry weight of the membrane), which bridge the lipid bilayer (integral membrane proteins) or are more or less loosely bound to one of the surfaces (peripheral proteins), serve in asymmetrically organized multimeric functional complexes of the cytoplasmic membrane. The lipid bilayer also contains redox carrier-like quinones and in most taxonomical groups, hopanoids (see Chapters 7.10 and 29.3.4). These triterpenoid derivatives act as stabilizers in bacterial membranes, which otherwise lack sterols (e.g., cholesterol), a major constituent of eukaryotic plasma membranes. The membrane components are in a fluid-crystalline state and may diffuse laterally rapidly, or very slowly by transverse diffusion (flip-flop). The occurrence of membrane-integrated proteins that span the membrane is obvious in freeze-fractured samples (Fig. 2.**10**). This preparation technique cleaves the frozen membrane between the two sublayers. Membrane-associated proteins, especially enzymes, can clearly be seen in negatively stained membrane fragments or vesicles because they are elevated above the plane of the membrane (Fig. 2.**10b,c**).

There are striking biochemical differences, although not visible under the electron microscope, between membranes of Bacteria and Archaea with respect to the linkage of residues in the lipid components. Fatty acids may be replaced, in some extreme halophiles, by ether-substituted isoprenoid chains; lipids containing glycerol-ether-linked residues are found in *Thermoplasma acidophilum*; in methanogens, the lipids are branched glycerol ethers. The cytoplasmic membrane of halobacteria contains patches of membrane-integrated bacteriorhodopsin molecules (Fig. 2.**10d,e**; "purple membrane"; see Chapter 13.4 and Fig. 13.**13**). Intracytoplasmic membranes in phototrophic bacteria increase the membrane surface and bear the

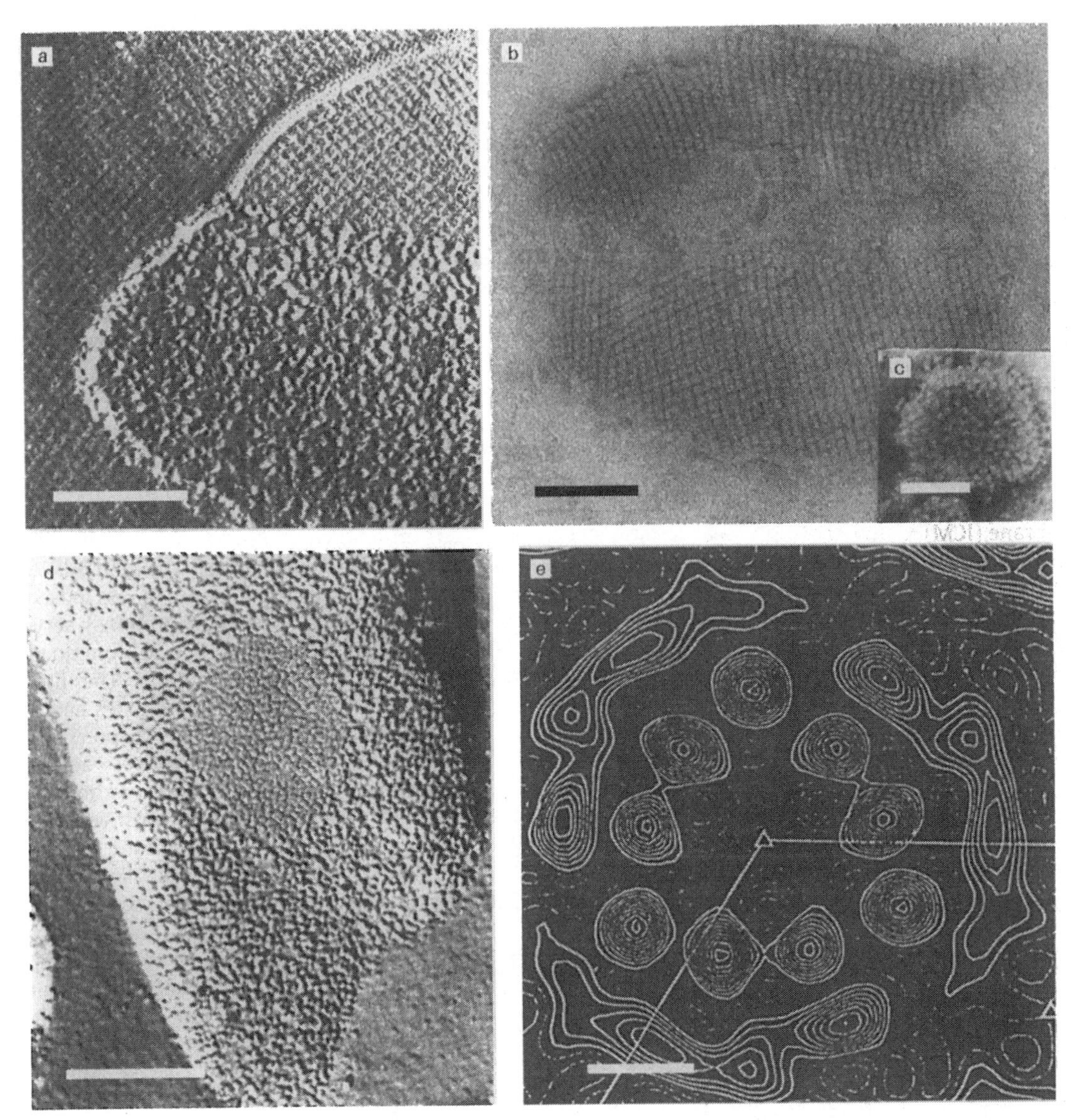

Fig. 2.**10a–e** **Structural aspects of prokaryotic cytoplasmic membranes.**
a Freeze-fracture preparation (compare with Fig. 2.**4b**) of *Clostridium aceticum*; below the S-layer, the fractured cytoplasmic membrane is exposed, exhibiting numerous membrane proteins. Bar = 0.1 μm [from 5].
b Inside-out membrane vesicle from *Nitrobacter hamburgensis* X14 (negative staining) showing nitrite oxidoreductase enzyme particles attached to the cytoplasmic membrane in a paracrystalline fashion. Note that the enzyme, under normal conditions, is exposed to the cytoplasm. Bar = 0.1 μm. Courtesy of E. Spieck.
c Inside-out membrane vesicle from the archaebacterium *Methanosarcina mazei* strain Göl (negative staining) exhibiting ATP synthase complexes. Normally, the enzyme is exposed toward the cytoplasm. Bar = 0.1 μm. Courtesy of F. Mayer.
d and **e** The purple membrane of *Halobacterium halobium* [from 19].
d Freeze-fracture replica of the cytoplasmic membrane containing a patch of purple membrane (center). Bar = 0.1 μm.
e Structure of the purple membrane projected onto the plane of the membrane. The figure is a reconstruction of the repeating unit from an electron micrograph. Part of a unit cell (containing three molecules of bacteriorhodopsin) is outlined in the bottom half of the figure. The triangle indicates a threefold axis of rotation. One molecule of bacteriorhodopsin comprises two adjacent and approximately parallel rows of α-helices; three of these double rows are grouped around the threefold axes at the corners of the unit cell. The inner row of each pair contains three α-helices, which are seen essentially end-on; the outer row contains four α-helices, which are inclined at increasing angles to the plane of the membrane from one end of the row to the other. The spaces around the bacteriorhodopsin molecules are presumably filled by lipid molecules in a bilayer arrangement. Bar = 1 nm

The **cytoplasmic membrane** is the innermost layer of the prokaryotic cell envelope and mediates the transport between cytoplasm and environment. In Bacteria, the cytoplasmic membrane has a bi-layer composition, made up of phospholipids. Proteins (most of them enzymes, sensors, and transport systems) are attached to the cytoplasmic membrane on its outside or inside, or they may be transmembrane proteins. Membranes of Archaea look similar to those of Bacteria under the electron microscope; however, they contain glycerol ethers with C_{20} or C_{40} isoprenoids in place of the fatty acid glycerol esters found in bacteria.

photosynthetic apparatus. Four basic types of intracytoplasmic membrane (ICM) in phototrophic bacteria are known (Fig. 2.**11**). A vesicle type of ICM, ICM consisting of tubuli, ICM in the form of flat thylakoid-like membranes organized in regular membrane stacks, and large thylakoids that are partially stacked and irregularly arranged have been observed (Fig. 2.**12**). All these types originate from the cytoplasmic membrane by invagination except perhaps for the thylakoids of cyanobacteria (see Chapter 24.2).

Diverse light-harvesting systems have developed in phototrophic bacteria. Light-harvesting systems serve to

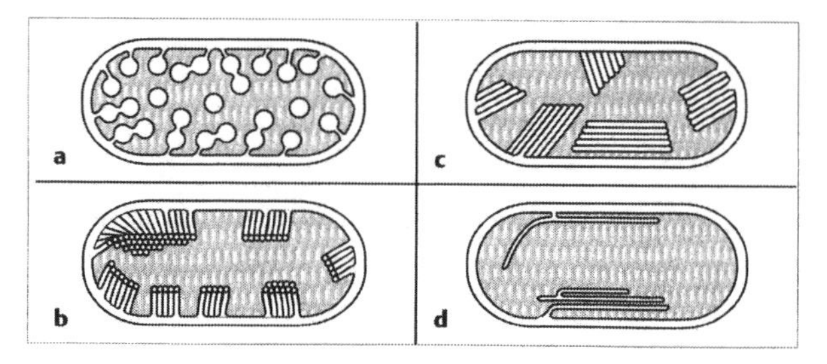

Fig. 2.**11a–d** **Scheme of the structure and arrangement of intracytoplasmic membranes (ICM) bearing the photosynthetic apparatus in phototrophic bacteria.**
a Vesicle type, found in *Rhodospirillum rubrum*, *Rhodobacter capsulatus*, *Chromatium vinosum*, *Thiocapsa roseopersicina*, and other species.
b ICM consisting of tubuli in *Thiocapsa pfennigii* and some other bacteria.
c ICM organized as flat, thylakoid-like membranes in regular membrane stacks, present in *Rhodospirillum molischianum*, *Ectothiorhodospira mobilis*, and other species.
d Large thylakoids, partially stacked and irregularly arranged, are present in *Rhodopseudomonas palustris* and *Rps. viridis* [from 20].

absorb photons and to funnel the excitation energy to the photochemical reaction center (see Chapters 13 and 24). In principle, three different types of light-harvesting systems exist:

1. Type 1, present in purple bacteria (Proteobacteria), are integral membrane proteins that bind bacteriochlorophyll (BChl) and carotenoids in stoichiometric amounts.
2. Type 2 are chlorosomes (Fig. 2.**13**), which are attached to the cytoplasmic side of the cytoplasmic membrane in chlorobiaceae and chloroflexaceae. Bacteriochlorophyll *c* aggregates are located within a closed lipid monolayer that forms a vesicle-like organelle and are organized into rod elements. The contact between the chlorosome and the cytoplasmic membrane is mediated by a crystalline baseplate that makes up part of the chlorosome; the site of attachment in the cytoplasmic membrane carries reaction centers.
3. Type 3 comprises the phycobilisomes of the cyanobacteria (Figs. 2.**14** and 13.**11**). The phycobilisomes are attached to the surface of the thylakoids of these organisms, often forming regular arrays. They contain the water-soluble accessory pigments allophycocyanin, phycocyanin, and phycoerythrin covalently bound to biliproteins, and exhibit a distinct macromolecular architecture. A triangular core is assembled from three stacks of disk-shaped subunits. Each stack contains two disks that are 12 nm in diameter. Radiating from two of the three sides of the triangular core are groups of three rods, each rod consisting of stacks of disk-shaped subunits. For certain strains showing chromatic adaptation, the number of disks per rod is dependent upon the wavelength of light to which the cells are exposed during growth. The orderly array of phycobilisomes on the thylakoid surface may reflect the orderly arrangement of the reaction centers (PS II) within these membranes. Cyanobacteria contain, in addition to phycobilisomes, integral chlorophyll-*a*-binding proteins.

Intracytoplasmic membranes in phototrophic bacteria bear the photosynthetic apparatus and enlarge the area of the membrane surface. Vesicular, tubular, and stacked membranes, and thylakoids are common, all of which originate from the cytoplasmic membrane by invagination, except for perhaps the thylakoids in cyanobacteria. Diverse light-harvesting systems have developed during evolution.

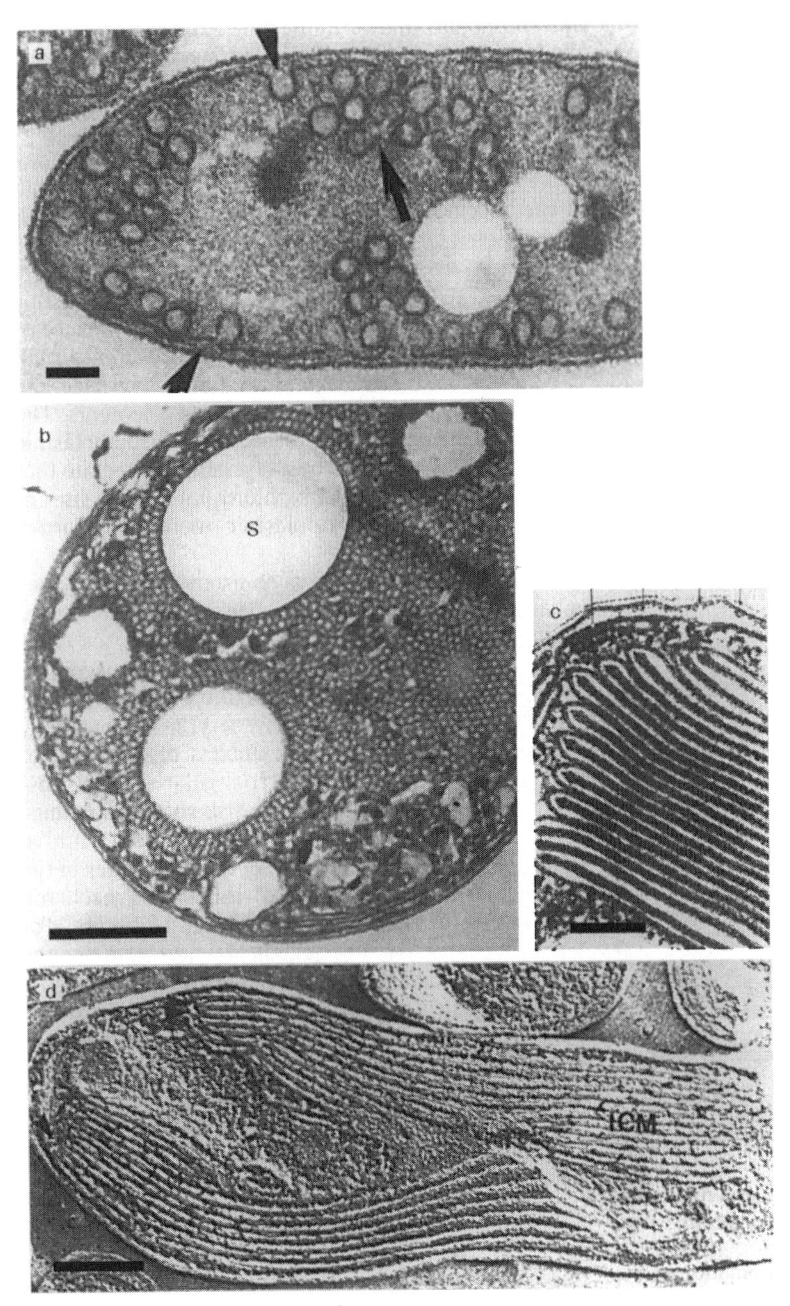

Fig. 2.**12a–d Electron micrographs of intracytoplasmic membranes in phototrophic bacteria.**
a Ultrathin section of *Rhodobacter sphaeroides*; note membrane invaginations. Bar = 0.2 μm.
b Ultrathin section of *Thiocystis violacea* showing packed chromatophores, sulfur globules (S), and lamellar to tubular membranes characteristic of old cells of this species. Bar = 0.5 μm.
[**a** and **b** from 21].
c Ultrathin section of *Ectothiorhodospira mobilis* with stacks of photosynthetic lamellae. Bar = 0.1 μm [from 22].
d Freeze-fracture preparation of *Rhodopseudomonas palustris* showing the organization of the photosynthetic membranes; the lamellae (ICM) are parallel to the cytoplasmic membrane. Bar = 0.2 μm [from 23].

Non-phototrophic prokaryotes may also form intracytoplasmic membranes. Extensive intracellular membrane systems in non-phototrophic eubacteria (Fig. 2.**15**) such as in nitrifying, nitrogen-fixing, and methane-utilizing bacteria, are derived from the cytoplasmic membrane by invagination as are those of phototrophic bacteria. The ratio of intracytoplasmic to cytoplasmic membrane surface is regulated. In *Azotobacter vinelandii*, intracytoplasmic and cytoplasmic membranes do not differ with respect to respiratory activities; both membranes represent differently located parts of an otherwise identical membrane system.

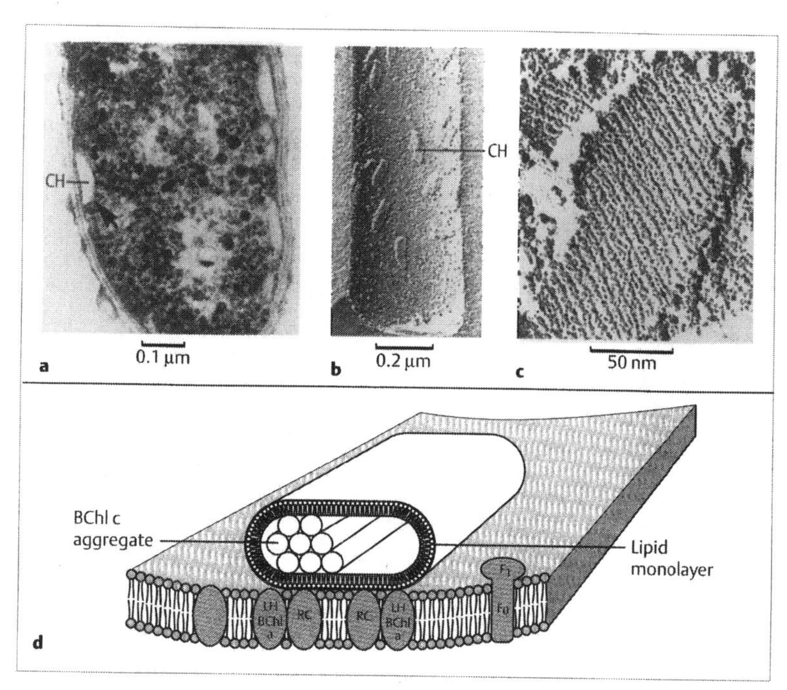

Fig. 2.**13** **Architecture of chlorosomes in the green bacteria *Chlorobium limicola* (a, c, d) and *Chloroflexus aurantiacus* (b).**
a Ultrathin section showing chlorosomes (CH) attached to the inner side of the cytoplasmic membrane. Bar = 0.1 µm [from 24].
b Freeze-fracture sample; the fracture plane goes through the cell periphery, exhibiting the sites of attachment of the chlorosomes (CH) to the cytoplasmic membrane. Nearly all chlorosomes in a cell have approximately the same orientation. Bar = 0.2 µm [from 25].
c Freeze-fracture sample showing the baseplate of a chlorosome with crystalline substructure. Bar = 50 nm. (Source, see **a**.)
d Model of the chlorosome and its associated cytoplasmic membrane. BChl, bacteriochlorophyll; F_0F_1, ATP synthase; LH, light harvesting complex; RC, reaction center. Courtesy of G. Drews

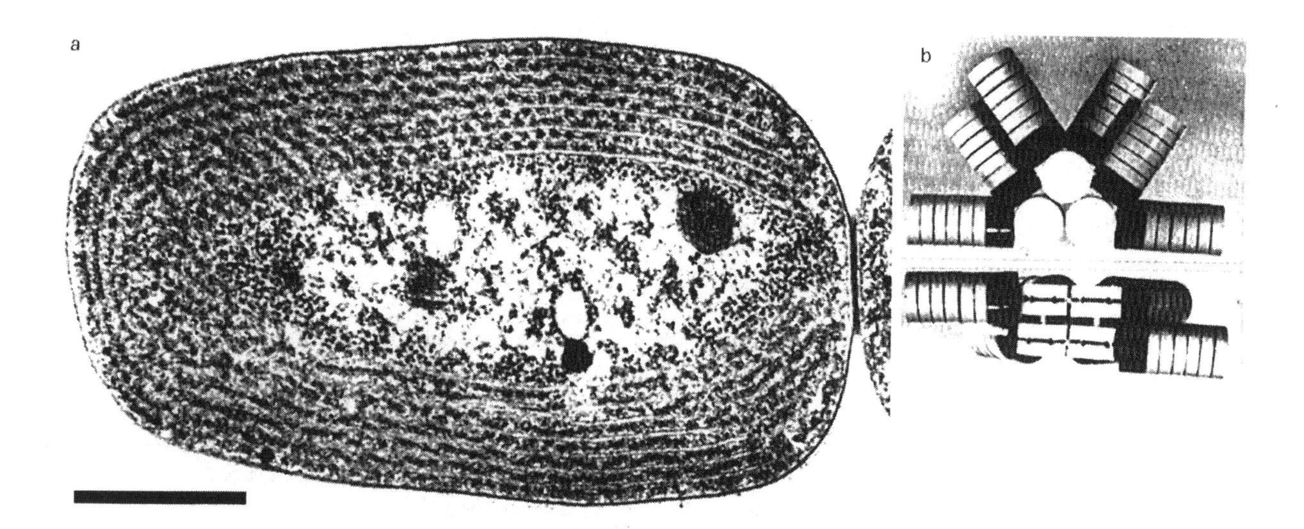

Fig. 2.**14a,b** **Structural organization of a cyanobacterium.**
a Ultrathin section through a cell of a filamentous cyanobacterium belonging to the *Lyngbya-Plectonema-Phormidium* group, with a cortical parallel array of thylakoids bearing phycobilisomes. Bar = 0.5 µm [from 26].
b Two views (top, face view; bottom, side view) of a scale model of the phycobilisome. White, black, and gray disks represent allophycocyanin, phycocyanin, and phycoerythrin, respectively. Single disks represent phycobiliprotein trimers $(\alpha\beta)_3$ with a diameter of 12 nm and a thickness of 3 nm [from 27]

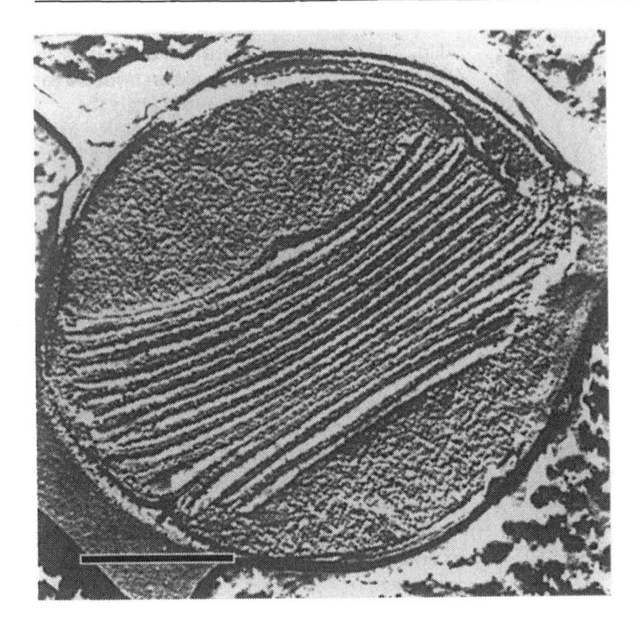

Fig. 2.15 **Intracytoplasmic membrane system of the non-phototrophic, nitrite-producing marine eubacterium *Nitrosococcus oceanus*.** A freeze-fractured cell is shown, which exhibits packets of membrane lamellae formed by flat vesicles arranged in parallel, some of which are connected to the cytoplasmic membrane. Bar = 0.5 μm [from 28]

Convoluted intracytoplasmic membrane systems seen on electron micrographs of *Methanobacterium thermoautotrophicum* were shown to be preparation artifacts. The same holds true for "mesosomes", i.e., convoluted membrane systems in eubacteria which appear to have been caused by unfavorable conditions of chemical fixation. The occurrence of mesosomes might indicate that the cytoplasmic membrane has patches with a composition different from that of other areas; this difference could make the specific sites of the membrane susceptible to deformation.

Nitrifying, nitrogen-fixing, and methane-utilizing eubacteria (i.e., non-phototrophic eubacteria) exhibit extensive intracellular membrane systems. By this differentiation, the membrane area is increased, giving more space for membrane-bound enzymes.

2.5 The Setup of the Intracellular Structures Reflects the High Degree of Organization in the Prokaryotic Cell

2.5.1 Ribosomes and Polysomes Are not the Only Constituents of the Cytoplasm

The cytoplasm contains a huge number of proteins (in the form of enzymes, enzyme complexes, and structural molecules), mRNA, tRNA, other macromolecules, smaller constituents, and solutes. Nevertheless, in electron micrographs of ultrathin sections through a bacterial cell, the cytoplasm is devoid of any visible structure. Also in cryo-preparations, no defined structures are visible. This finding does not imply that the cytoplasm is structureless. Possibly, existing structured organization cannot be depicted because it might be lost during preparation or its constituents are too small or lack contrast.

Clearly visible are the ribosomes (Fig. 2.16). In most cases, the ribosomes are organized as polysomes located in the cytoplasm or close to the cytoplasmic membrane. Ribosomes have the important function of translating the genetic information of messenger RNA (mRNA) into proteins. The prokaryotic ribosomes belong to the 70S type (S = Svedberg units, sedimentation constant in the ultracentrifuge). They are composed of 55 proteins and the 5S, 16S, and 23S r(ribosomal)RNA. The amount of ribosomes per cell (between 20 and $100 \cdot 10^3$) is under strict control (see Chapter 20.3).

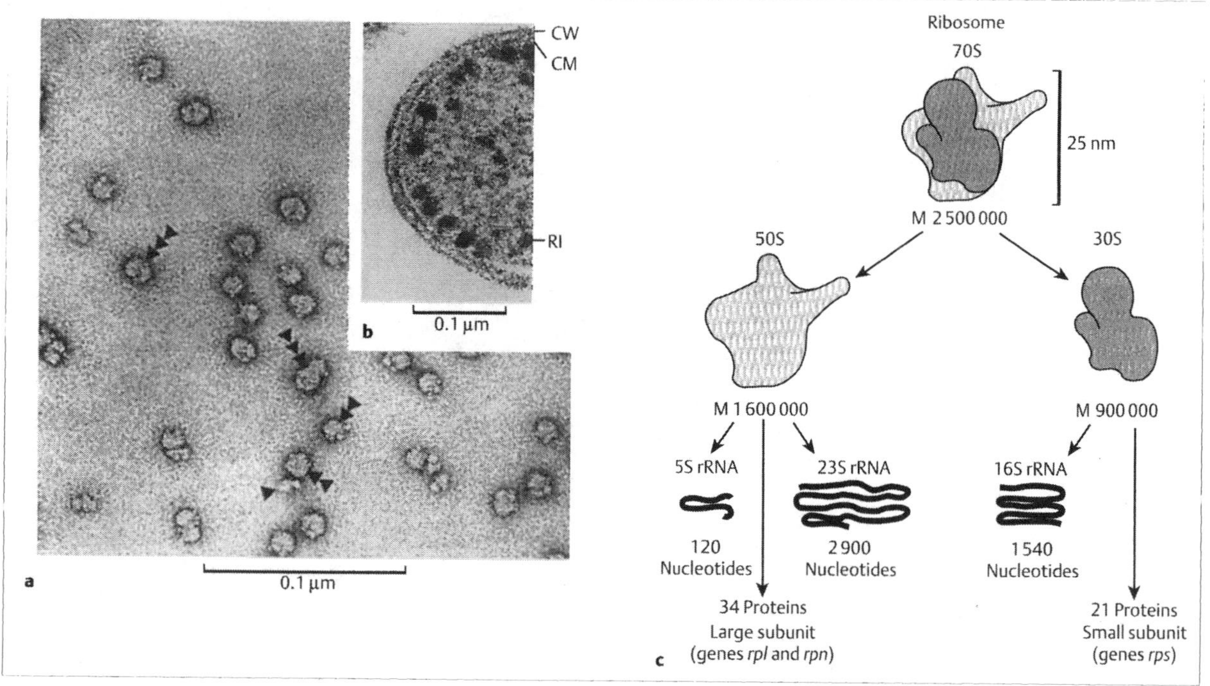

Fig. 2.**16a–c** **Structure of 70S ribosomes.** **a** Isolated ribosomes from *Escherichia coli*. Intact 70S ribosomes are marked with three arrowheads, 50S particles with two arrowheads, and 30S particles with single arrowheads. Negative staining. Bar = 0.1 µm [from 29].
b Ultrathin section through a cell of *Thermoanaerobacterium thermosulfurigenes* EM1 exhibiting ribosomes (RI) in the vicinity of the cytoplasmic membrane, arranged in a chain-like fashion (polysome). CW, cell wall; CM, cytoplasmic membrane. Courtesy of F. Mayer. **c** Model of the bacterial ribosome (consisting of a small and large subunit) [from 30]

In ultrathin sections, the prokaryotic **cytoplasm** appears to be devoid of any visible structure. Ribosomes, the organelles of protein synthesis, are visible as single particles, or they may form polysomes (i.e., chains of ribosomes arranged along mRNA).

2.5.2 The Organization of the Prokaryotic Nucleoid Provides the Prerequisites for Highly Regulated Processes

A distinct nucleoplasmic region not enclosed by a membrane is visible in conventional ultrathin sections of prokaryotes (Fig. 2.**17a**). This region is devoid of ribosomes and exhibits a fibrillar substructure. A less clear picture is obtained when cryo-preparation and imaging techniques are applied. There is only a very minor difference in contrast when the cytoplasm and the **nucleoid** region are compared. It may be assumed that a distinct visibility of the nucleoid is due to artificial changes in ion composition of the cell contents. Such a change may be caused by the destruction of the permeability barrier, which under physiological conditions is constituted by the cytoplasmic membrane. Additional reasons could be the disruption of the association of transcription-translation complexes with DNA, or the loss of DNA-binding proteins.

A bacterial chromosome released from the cell by osmotic shock shows only "naked" DNA when applied to a supporting film after spreading (Fig. 2.**17c**). However, when the cell is very gently lysed instead of lysed by osmotic shock treatment, the DNA can be seen to be organized together with bound proteins and RNAs, as necklace-like strands (Fig. 2.**17b**). Those strands are occasionally interwound to form higher aggregates. The occurrence of these beaded fibers supports the view that prokaryotic DNA-binding proteins and RNAs may

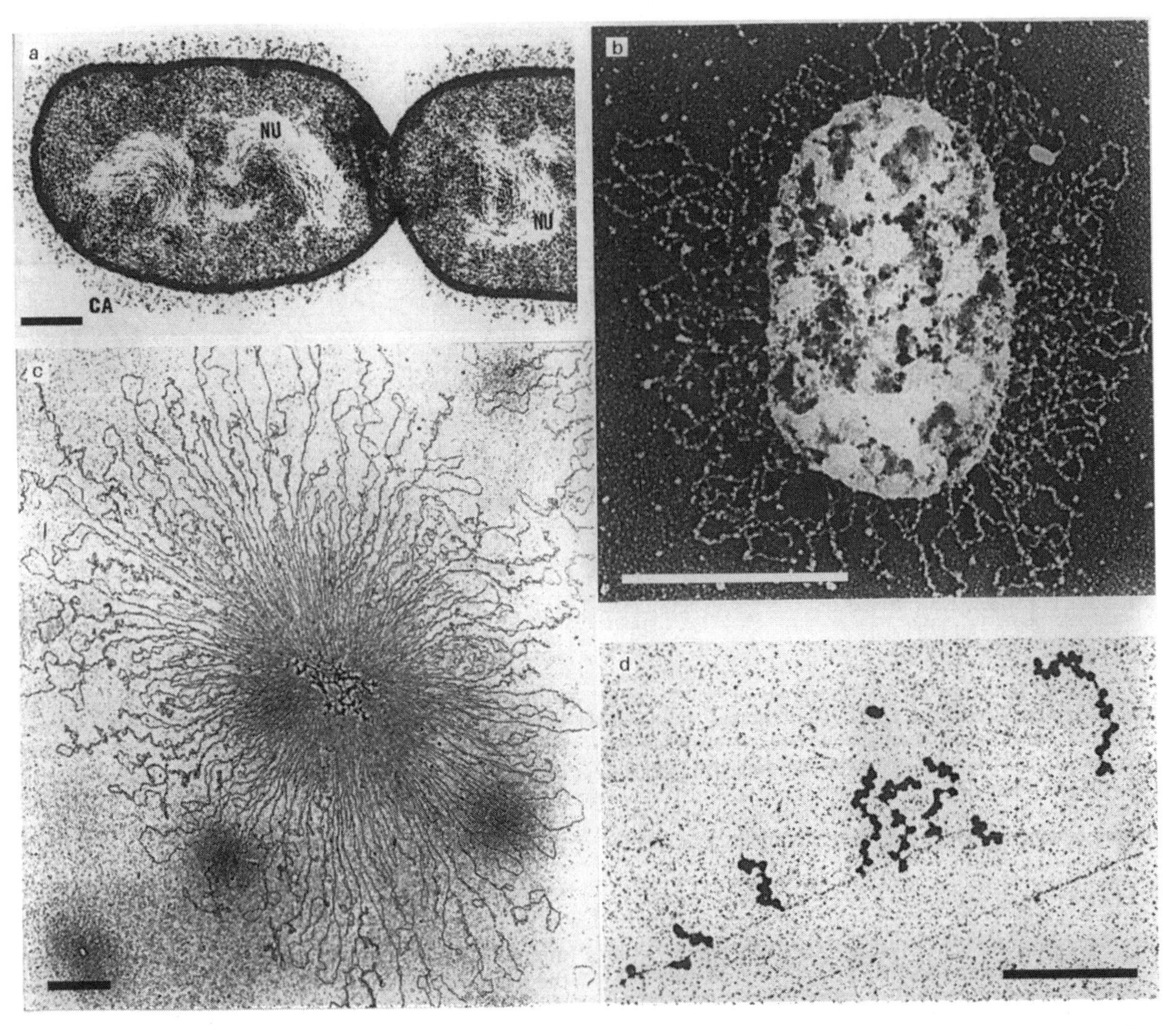

Fig. 2.**17a–d** **Structural organization of the prokaryotic nucleoid.**
a Ultrathin section through cells of *Klebsiella pneumoniae* exhibiting a fibrillar ultrastructure of the nucleoid (NU). CA, capsule. Bar = 0.1 μm [from 31].
b Metal-shadowing preparation of a cell of *Escherichia coli* disrupted on the electron microscopic supporting film. The DNA appears in regularly condensed chromatin-like fibers. Bar = 0.5 μm [from 32].
c Isolated nucleoid from *E. coli*. Cell-envelope-free spread DNA loops can be seen. Positive staining. Bar = 1 μm [from 33].
d Active region of the *E. coli* genome, showing an operon with presumptive initiation sites for transcription. Positive staining of a spread sample. Bar = 0.2 μm [from 34]

play an important role in condensation of the DNA and in regulation of processes taking place along the DNA. After all, at least 1.2 mm of double-stranded DNA have to be packed into a cell a few μm length, without loss of the potential for strict regulation of replication and transcription. It is assumed that in prokaryotes that live at extremely high temperatures, protein–DNA interactions reinforce DNA stability.

When bacterial nucleoids are isolated under low-salt conditions and without the addition of detergents, polyamines, or Mg^{2+}, the nucleoids are partially unfolded due to disruption of protein–nucleic acid complexes. The chromosomal DNA, which is a covalently closed circular macromolecule, is observed to be organized in independent negatively supertwisted domains restrained by remaining protein–nucleic acid interactions. The number of these domains is more than 50 (Fig. 2.**17c**).

After gentle preparation, nucleoids are found to be connected to the cytoplasmic membrane, which may contribute to the regulation of the process of DNA replication and nucleoid partition. Nucleoid and cell division are coordinated, but are not strictly correlated processes. A cell may contain more than one nucleoid.

Electron micrographs from gently lysed bacterial cells exhibit a further feature. Temporarily existing complexes are visible, comprising DNA-dependent mRNA polymerase (RNP) attached to the DNA, complexed with nascent mRNA to which ribosomes are attached; these are polysomes in the process of translation (Fig. 2.**17d**).

One should keep in mind that the nucleoid has to be a very dynamic structure; only sites along the DNA temporarily "exposed" to the cytoplasm can be transcribed. The components of the transcription–translation machinery do not have access to the interior of the nucleoid. Structure, replication of the nucleoid, and separation of the daughter chromosomes are described in Chapters 14.1, 14.2 and 22.3.

The prokaryotic **chromosome** is located within a nucleoplasmic region devoid of ribosomes. This "**nucleoid**" is not enclosed by a membrane. The chromosomal DNA, complexed with DNA-binding proteins, is a covalently closed circular molecule organized in independent supertwisted domains restrained by protein–nucleic acid interactions. Polysomes in the process of translation, attached to mRNA polymerase sitting on the DNA strand, have been observed.

2.5.3 Cell Inclusion Bodies May not Only Serve for the Storage of Polymers

Naturally occurring **inclusion bodies** in prokaryotic cells (Fig. 2.**18**) are characterized by a number of properties: they are products of cell metabolism, their contents belong to a variety of substance classes, often they make up a considerable portion of the cell dry weight, and usually their contents are not soluble. Often, they are visible under the light microscope due to light refraction. Their main function is storage of material produced by overflow metabolism; this stored material is used up under metabolic stress and during detoxification. Inclusion bodies are not essential for metabolism, but are of advantage in special growth phases or under special environmental conditions. A number of inclusion bodies are surrounded by an envelope, which is not a typical bilayer membrane and can be of various thicknesses and composition. Usually, the envelope contains proteins and may harbor enzymes that catalyze synthesis or degradation of the storage material. Table 2.**2** summarizes examples of prokaryotic inclusion bodies. It is evident that inclusion bodies, such as gas vesicles and magnetosomes, exist that fulfill other functions (e.g., that of orientation).

Other special inclusion bodies are intact and defective bacteriophages, which often have bacteriocin properties. This includes rhapidosomes and refractile bodies in bacteria, the so-called R bodies. Bacteria carrying these inclusion bodies exist as endosymbionts in Paramecia; also some free-living bacteria exhibit R bodies. Inclusion bodies containing polyhydroxy fatty acids are of interest in biotechnology as are recombinant proteins often deposited as inclusion bodies in bacterial cells that overproduce these proteins.

Inclusion bodies are common in prokaryotic cells. Usually, they contain a storage material in polymerized form, often enclosed by an envelope made up of proteins. Special cases are gas vesicles and magnetosomes, and intact and defective bacteriophages, which often are bacteriocins. Recombinant proteins produced in bacteria may be organized as artificial inclusion bodies.

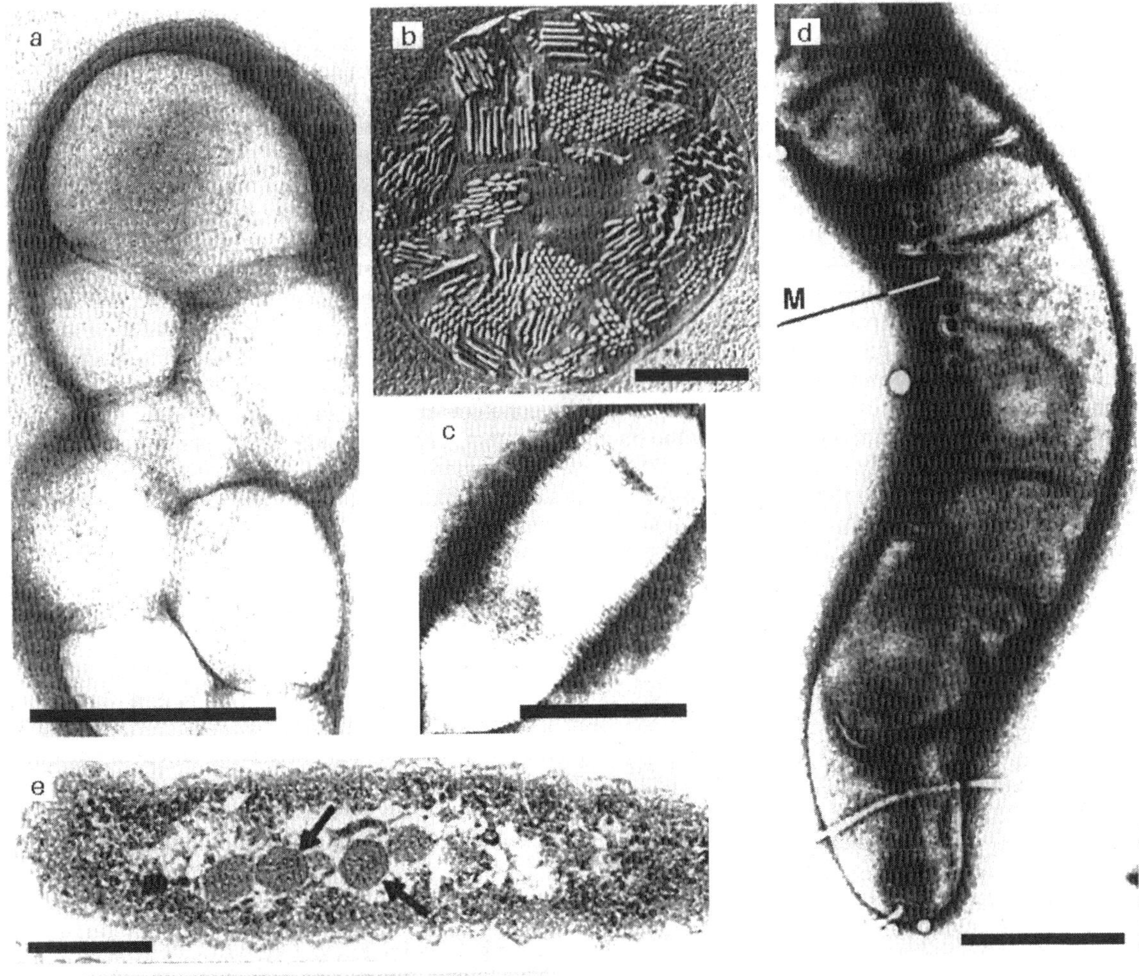

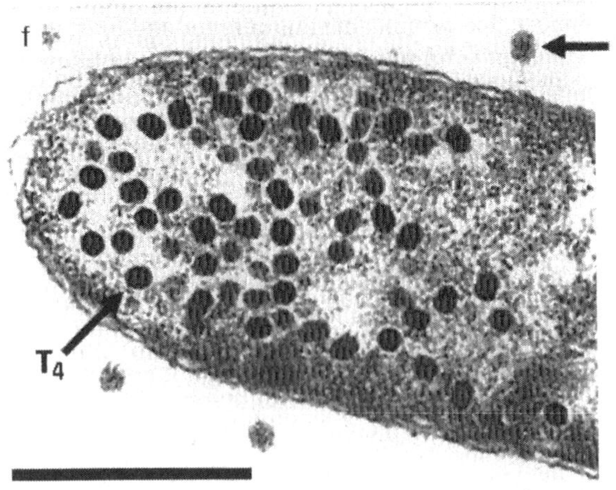

Fig. 2.**18a–f** **Bacterial inclusion bodies.**
a Ultrathin section showing polyhydroxybutyric acid (PHB) inclusion bodies in a cell of *Alcaligenes eutrophus*. Bar = 0.5 μm. Courtesy of B. Vogt.
b Freeze-fracture preparation of *Microcystis aeruginosa* showing groups of gas vesicles. Bar = 1 μm [from 35].
c Negatively stained isolated gas vesicle from *Anabaena flos-aquae*. Bar = 0.1 μm [from 36].
d Ultrastructure of a magnetotactic spirillum after negative staining. M, magnetosomes arranged in a chain-like fashion. Bar = 0.5 μm [from 37].
e Ultrathin section showing carboxysomes (arrows) in *Thiobacillus neapolitanus*. Bar = 0.2 μm [from 36].
f Ultrathin section of a cell of *E. coli* infected with bacteriophage T4; bacteriophages attached to the cytoplasmic membrane and within the cell can be seen. Bar = 0.5 μm [from 38]

Table 2.2 **Prokaryotic inclusion bodies**

Inclusion body	Major contents	Occurrence/examples
Carboxysomes	Ribulose-1,5-bisphosphate carboxylase	Bacteria with Calvin cycle; *Nitrosomonas, Thiobacillus*, many cyanobacteria
Polyhydroxy fatty acid globules	Polyesters of hydroxy fatty acid	Very common among *Pseudomonas; Bacillus, Alcaligenes eutrophus, Pseudomonas oleovorans*
Gas vesicles	Gas	Purple bacteria: *Lamprocystis, Amoebobacter, Thiodictyon* Green bacteria: *Pelodictyon* Many cyanobacteria: *Oscillatoria, Aphanizomenon, Microcystis* Non-pigmented eubacteria: *Pelonema, Peloploca* Archaea: *Halobacterium, Methanosarcina*
Magnetosomes	Fe_2O_3 (magnetite)	Rods, cocci, spirilla (*Aquaspirillum magnetotacticum*)
Sulfur globules	Sulfur (orthorhombic)	Aerobic H_2S-oxidizing bacteria: *Beggiatoa, Thiothrix, Achromatium, Thiovolum* Anoxygenic phototrophic sulfur bacteria: *Chromatium, Thiospirillum*, cyanobacteria; the sheathed bacterium *Sphaerotilus natans*
Polysaccharide granula	Starch-like material, glycogen	*Clostridium pasteurianum, C. butyricum, Acetobacter pasteurianus, Neisseria, Escherichia coli, Bacillus polymyxa, Micrococcus luteus, Arthrobacter*, enteric bacteria
Polyphosphate (volutin) granula	Polyphosphate, Mg^{2+}, K^+	*Spirillum volutans, Acinetobacter johnsonii, Klebsiella pneumoniae*
Cyanophycin	Protein (Asp/Arg)	Cyanobacteria
Protein crystals	Protein	Toxin: *Bacillus thuringiensis, B. laterosporus, B. medusa* RNA polymerase mutant: *Bacillus subtilis, Clostridium cochlearium, C. perfringens*
Calcium carbonate crystals	$CaCO_3$	*Achromatium oxaliferum, Macromonas mobilis*
Lipid	Neutral lipids, waxes	Very rare: *Pseudomonas aeruginosa* 44T1
Hydrocarbons	Hexadec-1-ene, nickel naphthenate	*Acinetobacter*
Intact and defective bacteriophages ("bacteriocins")	Protein +/− nucleic acid	Very common: *Escherichia coli, Serratia marcescens, Alcaligenes eutrophus*
Rhapidosomes	Protein	*Saprospira grandis, Aquaspirillum itersonii*
R-bodies	Protein	Endosymbiotic bacteria in *Paramecium* *Pseudomonas taeniospiralis* (not endosymbiotic)
Recombinant proteins	Protein	In bacteria that overproduce proteins in biotechnological processes

2.6 Cell Appendages Serve for Locomotion and Cell Recognition

2.6.1 Prokaryotes May Exhibit Various Modes of Locomotion

Most motile prokaryotes move with the aid of flagella (see below), several by gliding. Representative gliding organisms are filamentous cyanobacteria, such as *Oscillatoria* or *Anabaena*, and myxobacteria, such as *Myxococcus*. Although the mechanism of gliding is unknown, many genes essential for the gliding process have been identified. Slime seems to be necessary for this mode of locomotion.

2.6.2 Flagella Are Rotating Organelles Used for Cell Locomotion

A wealth of information is available on the ultrastructure, assembly, and molecular biology of flagella, and on their role in chemotaxis (see Chapters 20.5 and 24.1). The number and site of insertion of flagella in the prokaryotic cell, and parameters such as diameter, wavelength of these helically structured organelles, and amplitude of the helix are genetically controlled and highly specific and, hence, are of taxonomic value. There are two kinds of flagella in many species: polar or peritrichous, and medial flagella for swimming in liquid media, and lateral flagella for swarming on wet, solid surfaces (Fig. 2.**19a–c**). If both types of flagella can be present in one organism (such as in *Proteus* and *Serratia*), both types are rarely expressed simultaneously.

A typical flagellum of a Gram-negative eubacterium is composed of the filament, which is a helically wound, hollow cylinder with a diameter between 15 and 20 nm, made up of protein subunits ("flagellin"; Fig. 2.**19b**), a "hook", and a complex basal body, which mediates insertion of the flagellum in the cell envelope and its rotation relative to the cell body (Figs. 20.**21** and 24.**2**). A pair of outer rings, the L ring and the P ring, appear to function as bearings for the flagellum; they are connected to the outer membrane and the peptidoglycan, respectively, and are missing in Gram-positive bacteria. A complex consisting of an M/S ring, the components of the switch apparatus (responsible for direction of rotation), and the export apparatus (involved in flagellar growth), located in the plane of the cytoplasmic membrane, forms the inner part of the basal body. The basal body is surrounded by the membrane-integrated components of the Mot system and mediates, in an only partially understood way, the rotation of the flagellum relative to the static components ("stator") of the Mot system. The driving force for rotation is a proton gradient (or, in certain cases, a sodium gradient) between periplasm/outside and cytoplasm (see Chapter 20; the assembly of flagella is described in Chapters 23.9 and 24.1).

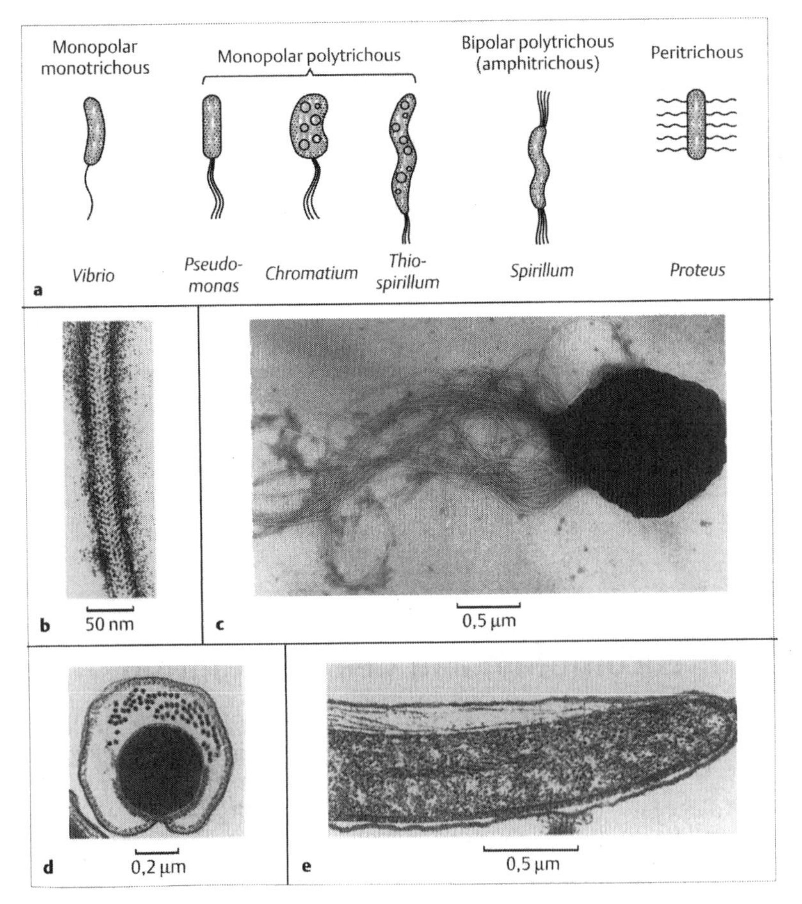

Fig. 2.**19a–e** **Flagellar systems of prokaryotes.**
a Modes of insertion of flagella [from 12].
b Negatively stained flagellar filament of *Pseudomonas facilis*. Note the individual flagellin molecules. Bar = 50 nm [from 4].
c Negatively stained flagellar fascicle of the methanogenic archaebacterium *Methanococcus jannaschii* isolated from a submarine hydrothermal vent. Bar = 0.5 µm [from 18].
d and **e** Periplasmic fibrils in spirochetes [from 39].
d Cross section.
e Longitudinal cell section.

In spirochetes, motility is brought about by filamentous structures wound around the helical protoplasmic cylinder (Fig. 2.**19 d,e**). Major structural aspects of these "periplasmic fibrils" are similar to flagella. The fibrils are not exposed to the environment, but are enclosed by a sheath covering the entire cell. Movement of spirochetes is not by swimming and tumbling as observed for other bacteria, but by contractions, undulations, flexions, wave propagations, and vibration of the highly flexible cell body caused by the action of the periplasmic fibrils. Archaea exhibit modes of flagellation comparable to those of typical Bacteria. In addition to only one or few flagella per cell, also tufts of flagella (Fig. 2.**19c**) have been observed. The major aspects of ultrastructural organization of the flagella of archaebacteria are similar to those of eubacterial flagella.

The **flagellum** is the most common organelle for the locomotion of prokaryotic cells. Mode of flagellation and structural parameters of flagella are genetically determined. A flagellum consists of a filament composed of protein subunits ("flagellin") forming a tube-like structure, a "hook", and a complex basal body mediating insertion of the flagellum into the cell envelope. The basal body functions as the flagellar "motor" similar to a "ship's screw". Motility in spirochetes is brought about by "periplasmic fibrils". Flagella of Archaea share many principal structural features with those of Bacteria.

2.6.3 Pili, Fimbriae, and Adhesins Function as Devices for Cell–Cell Recognition

Interactions of eubacterial cells with their environment, i.e., non-specific or specific attachment to surfaces and recognition reactions, may also be mediated by fimbriae and pili (Figs. 2.**1** and 2.**20**; see Chapter 23.8). Fimbriae and pili are filamentous cell appendages with diameters between 3 and 10 nm and of various length. They are composed of helically arranged protein subunits that form a tube-like organelle. The possession of such fibrillar appendages favors the attachment to surfaces, which carry a surface charge identical to that of the bacterial wall. Fimbriae and pili may be assumed to penetrate the "cloud" of surface charge, which otherwise would hinder attachment of bacterial cells by repulsion. Pili such as the F pilus of *Escherichia coli* mediate, during the initial step of conjugation, specific interaction with the partner cell, whereas fimbriae (including type 1 "pili" of *E. coli*) act less specifically. F pili, like other cell appendages, may also serve as receptors for bacteriophages (Figs. 2.**3c** and 2.**20a**).

After induction by sex pheromones, cells of *Enterococcus faecalis* form hair-like structures, adhesins, at certain areas of their surface. Adhesins provide highly specific tools that enable effective recognition of partner cells (Fig. 2.**3f**).

Fimbriae, **pili**, and **adhesins** may mediate cell attachment to surfaces and recognition reactions.

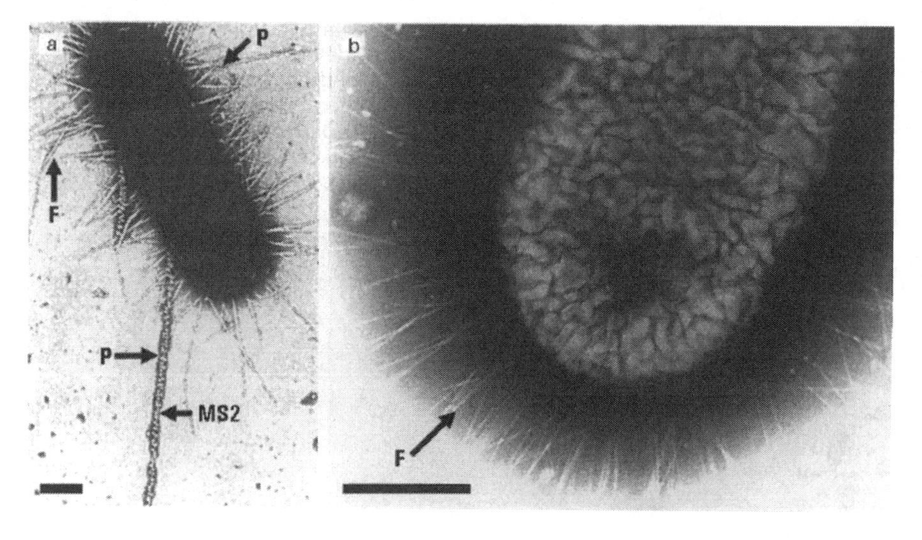

Fig. 2.**20a,b** **F pili and type 1 "pili".**
a Electron microscopic visualization (negative staining) of *Escherichia coli* carrying two F pili (P) with attached F pilus-specific bacteriophages (MS2) (compare with Fig. 2.**3c**). Type 1 "pili" (fimbriae; F) are also visible. Bar = 0.2 μm [from 40]. **b** *E. coli* cell with type 1 "pili" (fimbriae; F). Negative staining. Bar = 0.2 μm. Courtesy of B. Vogt

2.7 Bacteria May Form Spores and Other Resting Cells

Bacterial spores and other resting forms are products of cell differentiation designed to survive under unfavorable conditions. Different types of these differentiated cells with respect to structure, composition, and development have evolved (Fig. 2.**21**). These are forms referred to as endospores of Gram-positive bacteria (genera *Bacillus*, *Clostridium*, *Sporolactobacillus*, *Sporosarcina*, *Thermoactinomyces*) and the Gram-negative bacterium *Sporomusa*, the cysts of *Azotobacter arthrospora* and *Methylocystis*, the conidial spores of *Streptomyces*, the various kinds of resting cells of cyanobacteria (hormocysts of *Westiella*, baeocysts or endospores of the *Pleurocapsaspecies*, the akinetes of *Anabaena*, exospores of *Chamaesiphon*), and forms referred to as exospores of Gram-negative bacteria, among them the myxospores of myxobacteria (*Myxococcus*, *Stigmatella*). These persistent forms can be visualized and identified under the light microscope because of the optical properties which differ from those of vegetative cells (e.g., by high refractivity; Fig. 2.**5a,b**).

The modes of formation of the survival states are:

1. Differentiation of true endospores within vegetative cells of Gram-positive eubacteria (described in Chapter 25.1).
2. Transition of entire vegetative cells into resistant forms, primarily by wall thickening (cysts or arthrospores, myxospores, akinetes). *Methylosinus trichosporium* forms exospores by a process similar to budding of vegetative cells, combined with wall thickening; *Rhodomicrobium vannielii* produces clusters of unusual resting cells, somewhat similar to exospores formed by *Methylosinus*, at the tip of prosthecae situated at the tip of the cellular filament. Formation of spores in *Streptomyces* and *Myxobacterium* is described in Chapters 25.2 and 25.3.

Under favorable conditions, the persistent forms "germinate". Germination is preceded by water uptake and swelling, followed by a progressive disintegration of the exterior parts of the envelope by crack formation.

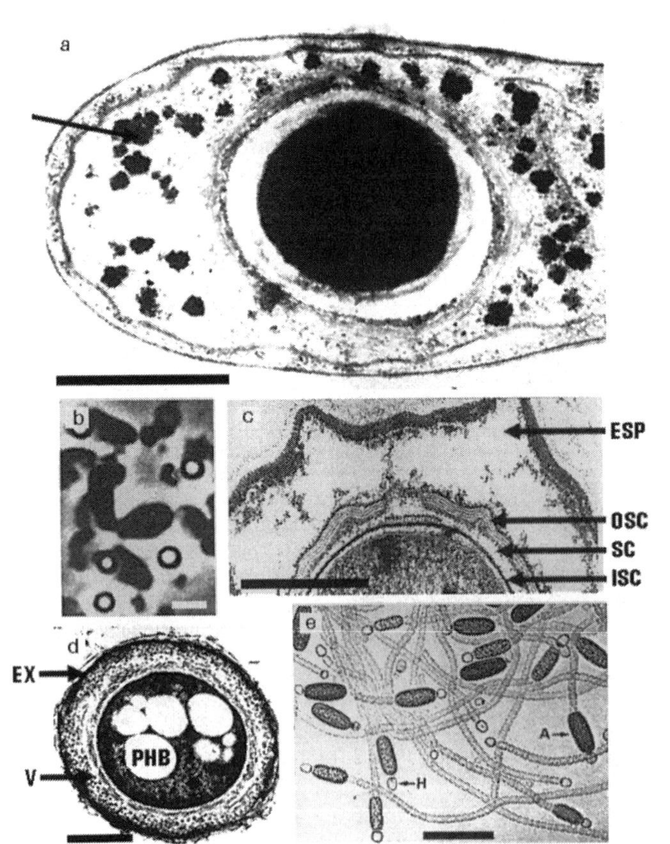

Fig. 2.**21a–e** **Bacterial survival forms.**
a Ultrathin section showing the late stage of spore formation in *Clostridium formicoaceticum*. Arrow, storage granules. Bar = 0.5 µm [from 4].
b Light micrograph of the Gram-negative eubacterium *Sporomusa* spec. showing vegetative and endospore-forming cells and free spores. Bar = 2 µm [from 41].
c Ultrathin section showing a germinating spore of *Bacillus polymyxa*. The spore coat (arrow) is cracked. Bar = 0.1 µm. ESP, exosporium; ISC, inner spore coat; OSC, outer spore coat; SC, spore cortex [from 42].
d Ultrathin section showing a cyst of *Azotobacter vinelandii*. Bar = 0.2 µm. EX, exine; PHB, polyhydroxybutyrate; V, vesicular intine [from 43].
e Light micrograph of the nostocacean cyanobacterium, *Cylindrospermum* spec., grown in the absence of a combined nitrogen source. Note the exclusively terminal heterocysts (H) and subterminal akinetes (A). Bar = 20 µm [from 26]

After rupture of the wall, a typical vegetative cell develops. Molecular details of sporulation will be discussed in Chapter 25.

Certain bacteria may form cell states that render them resistant to unfavorable environmental conditions. These forms are called endospores, cysts or arthrospores, akinetes, and exospores (including myxospores).

Further Reading

Methods

Aldrich, H.C., Todd, W.J., eds. (1986) Ultrastructure techniques for microorganisms. New York: Plenum Press

Glauert, A.M., ed. (1985) Practical methods in electron microscopy, vol. 10: Robards, A.W., and Sleytr, U.B. (eds.) Low temperature methods in biological electron microscopy. Amsterdam: Elsevier

Hancock, I., and Poxton, I. (1988) Bacterial cell surface techniques. Chichester: Wiley

Harris, I.R., and Horne, R.W. (1994) Negative staining: a brief assessment of current technical benefits, limitations and future possibilities. Micron 25: 5–13

Mayer, F., ed. (1988) Methods in microbiology, vol. 20. Electron microscopy in microbiology. London: Academic Press

Pawley, J., ed. (1995) Handbook of biological confocal microscopy, 2nd edn. New York: Plenum Press

White, J.G., Amos, W.B., and Fordham, M. (1987) An evaluation of confocal versus conventional imaging of structures by fluorescence light microscopy. J. Cell Biol 105: 41–48

Prokaryotic cell structure

Beveridge, T.J. (1995) The periplasmic space and the periplasm in Gram-positive and Gram-negative bacteria. ASM News 61: 125–130

Doyle, R.J., and Rosenberg, M., eds. (1990) Microbial cell surface hydrophobicity. Washington, DC: ASM Press

Fuller, R., and Lovelock, D.W., eds. (1976) Microbial ultrastructure. London: Academic Press

Kellenberger, E. (1990) Intracellular organization of the bacterial genome. In: Drlica, K., and Riley, M. (eds.) The bacterial chromatin. Washington, DC: ASM Press

Mayer, F. (1993) Principles of functional and structural organization in the bacterial cell: "compartments" and their enzymes. FEMS Microbiol Rev 104: 327–346

Mayer, F. (1993) Cell structure. In: Rehm, H.-J., and Reed, G. (eds.) Biotechnology, 2nd edn., vol. 1: Sahm, H. (ed.) Biological fundamentals. Weinheim: VCH; 5–46

Mayer, F. (1986) Cytology and morphogenesis of bacteria. Stuttgart: Bornträger

Ofek, I., and Doyle, R.J. (1994) Bacterial adhesion to cells and tissues. New York: Chapman & Hall

Sources of Figures

1 Trüper, H.G., and Pfennig, N. (1981) In: Starr, M.P., Stolp, H., Trüper, H.G., Balows, A., and Schlegel, H.G. (eds.) The prokaryotes, vol. 1. Berlin, Heidelberg, New York: Springer; 299–312

2 Doddema, H.J., and Vogels, G.D. (1978) Appl Environ Microbiol 36: 752–754

3 Jarchau, T. (1985) Untersuchungen zum Vorkommen und zur Häufigkeit von Proteinvorstufen bei der Bildung von 1- und F-Pili in *Escherichia coli.* Doctoral Thesis, University of Göttingen

4 Mayer, F., ed. (1986) Cytology and morphogenesis of bacteria. Berlin, Stuttgart: Gebrüder Bornträger

5 Braun, M., Mayer, F., and Gottschalk, G. (1981) Arch Microbiol 128: 288–293

6 Johannssen, W., Schuette, H., Mayer, H., and Mayer, F. (1984) Arch Microbiol 140: 265–270

7 Gottschalk, G., Andreesen, J.R., and Hippe, H. (1981) The genus *Clostridium* (nonmedical aspects). In: Starr, M.P., Stolp, H., Trüper, H.G., Balows, A., and Schlegel, H.G. (eds.) The prokaryotes, vol. 2. Berlin, Heidelberg, New York: Springer; 1767–1803

8 Kessel, M., and Cohen, Y. (1982) Bacteriol 150: 851–860

9 Brock, T.D. (1992) The genus *Leucothrix.* In: Balows, A., Trüper, H.G., Dworkin, M., Harder, W., Schleifer, K.-H. (eds.) The prokaryotes, 2nd edn., vol. 4. Berlin, Heidelberg, New York: Springer: 3247–3255

10 Kutzner, H.J. (1981) The family Streptomycetaceae. In: Starr, M.P., Stolp, H., Trüper, H.G., Balows, A., and Schlegel, H.G. (eds.) The prokaryotes, vol. 2. Berlin, Heidelberg, New York: Springer: 2028–2090

11 Andreesen, M., and Schlegel, H.G. (1974) Arch Microbiol 100: 351–361.

12 Cassone, A., and Garaci, E. (1977) Can J Microbiol 23: 684–689

13 Sleytr, U.B., and Messner, P. (1983) Annu Rev Microbiol 37: 311–339

14 Kandler, O. (1979) Naturwissenschaften 66: 95–105

15 Sleytr, U.B. (1978) Int Rev Cytol 53: 1–64

16 Kleinig, H., and Sitte, P. (1986) Zellbiologie. Stuttgart: Gustav Fischer

17 Antranikian, G., Herzberg, C., Mayer, F., and Gottschalk, G. (1987) FEMS Microbiol Lett 41: 193–197

18 Jones, W.J., Leigh, J.A., Mayer, F., Woese, C.R., and Wolfe, R.S. (1983) Arch Microbiol 136: 254–261

19 Unwin, P.N.T., and Henderson, R. (1975) J Mol Biol 94: 425–440

20 Drews, G., and Imhoff, J.F. (1991) Phototrophic purple bacteria. In: Shively, J.M., and Barton, L.L. (eds.) Variations in autotrophic life. London: Academic Press; 75

21 Remsen, C.C. (1978) Comparative subcellular architecture of photosynthetic bacteria. In: Clayton, R.K., and Sistrom, W.R. (eds.) The photosynthetic bacteria. New York, London: Plenum Press; 3160

22 Remsen, C.C., Watson, S.W., Waterbury, J.B., and Trüper, H.G. (1968) J Bacteriol 95: 2374–2392

23 Varga, A.R., and Staehelin, L.A. (1983) J Bacteriol 154: 1414–1430

24 Staehelin, L.A., Golecki, J.R., and Drews, G. (1980) Biochim Biophys Acta 5879: 30–45

25 Staehelin, L.A., Golecki, J.R., Fuller, R.C., and Drews, G. (1978) Arch Microbiol 119: 269–277

26 Stanier, R.Y., Pfennig, N., and Trüper, H.G. (1981) Introduction to the phototrophic bacteria. In: Starr, M.P., Stolp, H., Trüper, H.G., Balows, A., and Schlegel, H.G. (eds.) The prokaryotes, vol. 1. Berlin, Heidelberg, New York: Springer: 197–211

27 Bryant, D.A., Guglielmi, G., Tandeau de Marsac, N., Castets, A., and Cohen-Bazire, G. (1977) Arch Microbiol 123: 113–127
28 Remsen, C.C., Valois, F.W., and Watson, S.W. (1967) J Bacteriol 94: 422–433
29 Schlegel, H.G. (1986) General microbiology. 6th edn. Cambridge: Cambridge University Press
30 Alberts, B., Bray, D., Lewis, J., Raff, M., Roberts, K., Watson, J.D. (1987) Molekularbiologie der Zelle, 2nd edn. Weinheim: VCH
31 Cassone, A., and Garaci, E. (1977) Can J Microbiol 23: 684–689
32 Griffith, J.D. (1976) Proc Natl Acad Sci USA 73: 563–567
33 Rogers, H.J. (1983) Bacterial cell structure. Wokingham: Van Nostrand Reinhold
34 Hamkalo, B.A., and Miller Jr., O.L. (1973) Annu Rev Biochem 42: 379–396
35 Lehmann, H. (1979) Biologie in Unserer Zeit (BIUZ) 9: 129–134
36 Shively, J.M. (1974) Annu Rev Microbiol 28: 167–187
37 Balkwill, D.L., Maratea, D., and Blakemore, R.P. (1980) J Bacteriol 141: 1399–1408
38 Alberts, B., Bray, D., Lewis, J., Raff, M., Roberts, K., and Watson, J.D. (1983) Molecular biology of the cell. New York, London: Garland
39 Margulis, L., To, L.P., and Chase, D.G. (1981) The genera *Pillotina, Hollandina*, and *Diplocalyx*. In: Starr, M.P., Stolp, H., Trüper, H.G., Balows, A., and Schlegel, H.G. (eds.) The prokaryotes, vol. 1. Berlin, Heidelberg, New York: Springer: 548–554
40 Streyer, L. (1981) Biochemistry. San Francisco: Freeman
41 Müller, B., Ossmer, R., Howard, B.H., Gottschalk, G., and Hippe, H. (1984) Arch Microbiol 139: 388–396
42 Ghuysen, J.A., and Shockman, G.D. (1973) Biosynthesis of peptidoglycan. In: Leive, L. (ed.) Bacterial membranes and walls. New York: Dekker: 37–130
43 Lin, L.P., Pankratz, S., and Sadoff, H.L. (1978) J Bacteriol 135: 641–646

Section II

Basic Prerequisites for Cellular Life

Section I described how prokaryotic cells are organized. All cellular structures have been evolved to fulfill functions and can only be understood in that context. Section II now turns to metabolism. In contrast to eukaryotic metabolism with diverse organelle compartmentation, bacterial metabolism is characterized by a few compartments: cytoplasm including the chromosomal region or nucleoid, cytoplasmic membrane—in Gram negative bacteria—the periplasmic space, and the outer layers of the cell wall. Before forces and fluxes of cellular processes are discussed, how energy conservation and energy coupling work will be covered. The first two chapters will introduce the principles of catabolism, or how cells extract energy from their environment. In Chapter 3 substrate level phosphorylation, i.e., ATP synthesis via "energy-rich" intermediates, will be described; this process occurs in the cytoplasm. Chapter 4 will cover how the bacterial cytoplasmic membrane is energized in respiration and photosynthesis by generation of an electrochemical proton potential that is coupled with ATP synthesis. Chapter 5 will detail the multiple roles of these energized cell membranes, especially the specificity, dynamics, and mechanisms of transport processes. Metabolic energy and suitable substrates are a basis for growth. Principles of nutrition, kinetics of growth, and methods to measure growth are described in Chapter 6. Chapter 7 deals with the biosynthesis of building blocks from simple growth substrates which are the reservoir for the formation of cellular macromolecules. It will be basically assumed that the elements for growth are provided in an appropriate redox state and molecular form, e.g., glucose as carbon source, ammonia as nitrogen source, phosphate as phosphorus source, and sulfide as sulfur source. The principles of all these processes are similar in all organisms.

3 Substrate-Level Phosphorylation

The chemical reactions making up cellular metabolism can be divided into two categories (Fig. 3.**1**). One category, which comprises the processes designed for ATP synthesis from ADP and P_i, is termed **catabolism** or **energy metabolism**. The other category of processes is characterized by **ATP consumption**, which results in ATP hydrolysis to ADP and P_i or to AMP and pyrophosphate, which is subsequently hydrolized to P_i. In growing bacteria, most of the ATP is used in the synthesis of cell constituents (**biosynthesis** or **anabolism**) from anabolic substrates that provide carbon, nitrogen, sulfur, and other cellular constituents (Chapters 7 and 15). In addition, certain processes such as motion (Chapter 2.6), transport of compounds across the cytoplasmic membrane (Chapter 5), and regulatory processes (Chapter 20) may be driven at the expense of phosphoryl transfer from ATP or by other exergonic processes that are related to ATP synthesis. In this chapter, the basic principles of catabolism will be discussed using anaerobic glycolysis and aerobic glucose degradation to CO_2 as examples (see Chapter 9.11).

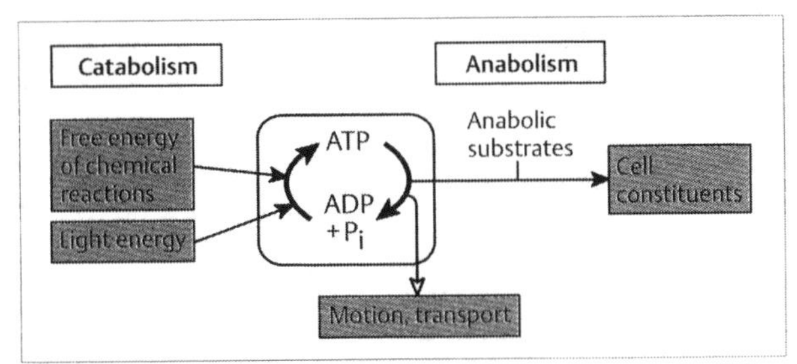

Fig. 3.**1** **The two categories of metabolic processes in living cells, catabolism and anabolism.** The catabolic processes lead to ATP formation, while anabolism and other processes are associated with ATP hydrolysis

3.1 ATP Synthesis Is Coupled to Exergonic Reactions

The main purpose of the catabolic reactions of bacterial metabolism is ATP synthesis from ADP and P_i (phosphorylation) with the maximum possible ATP yield. The endergonic phosphorylation reaction is driven by exergonic redox reactions in most cases. There are two types of mechanisms by which the redox reactions are coupled to phosphorylation: (1) **substrate-level phosphorylation (SLP)** and (2) **electron-transport-coupled phosphorylation (ETP**; Chapters 4 and 12). As a common principle, the phosphoryl group that is formed from P_i is finally transferred to ADP to give ATP. In Fig. 3.**2**, the oxidation of an aldehyde to the corresponding carboxylate (SLP[1]) and the reduction of oxygen to water (ETP[1]) are presented as examples of redox reactions that drive phosphorylation. In SLP, the phosphoryl group is bound in a "high-energy" compound. In ETP, the phosphoryl group is an intermediate formed at the active site of ATP synthase (Chapter 4).

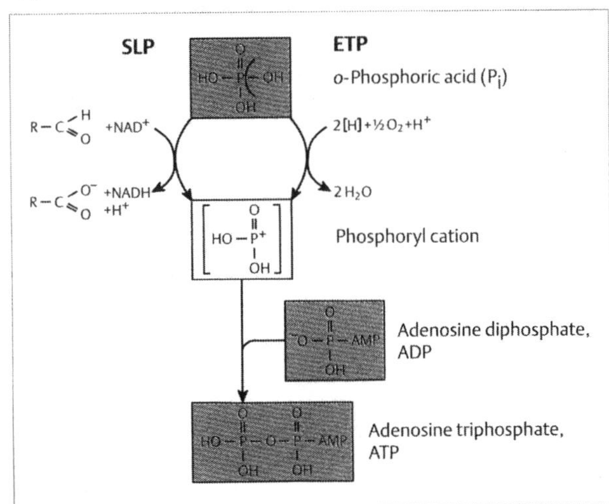

Fig. 3.**2** **Simplified mechanism of ATP formation by SLP and ETP.** Each of the two redox reactions indirectly causes the formation from phosphoric acid of a phosphoryl cation which is transferred to ADP. [H], reducing equivalents

[1] Abbreviations and symbols used in this chapter are listed in Appendix

3.2 The ATP Yield Is a Function of the Free Energy of the Driving Reaction

The ATP yield is a function mainly of the free-energy change available from the reaction that drives ATP synthesis (Box 3.1). This will be exemplified by a comparison of the ATP yield during aerobic and anaerobic glucose degradation.

> **Box 3.1** The processes occurring in living organisms obey the laws of energetics (thermodynamics). Consequently, living organisms can neither generate energy de novo nor can they dissipate energy to nothing (first law). Living organisms represent energy converters that transform one form of energy to another. However, living cells cannot convert thermal energy to other forms of free energy, but other forms of free energy can be transformed to thermal energy (second law). These rules apply to each of the chemical reactions involved in metabolism and help in understanding, in particular, coupled reactions (see Box 3.2).

Many bacteria drive ATP synthesis at the expense of glucose degradation. The initial steps of glucose degradation usually proceed according to the Embden-Meyerhof glycolytic pathway with pyruvate as an intermediate (Fig. 3.3). Pyruvate is reduced to lactate by some anaerobic bacteria (Fig. 3.**3a**). In aerobic bacteria, pyruvate is oxidized to CO_2 (Fig. 3.**3b**). With oxygen as substrate, the free-energy change of glucose degradation is:

$$\begin{aligned} &\text{Glucose} + 6\,O_2 \rightarrow 6\,CO_2 + 6\,H_2O \\ &\Delta G^{\circ\prime} = -2830 \text{ kJ/mol glucose} \end{aligned} \tag{3.1}$$

This is much larger than that of anaerobic glucose fermentation:

$$\begin{aligned} &\text{Glucose} \rightarrow 2\,\text{Lactate} + 2\,H^+ \\ &\Delta G^{\circ\prime} = -198 \text{ kJ/mol glucose} \end{aligned} \tag{3.2}$$

As a consequence, the ATP yield per mol glucose is much higher under oxic conditions than under anoxic conditions. The oxidation of glucose to pyruvate is coupled to SLP and yields 2 mol ATP/mol glucose. The reduction of oxygen is coupled to ETP and may yield 34 mol ATP/mol glucose, while the reduction of pyruvate to lactate is not coupled to phosphorylation.

The coupling mechanisms (SLP and ETP) are designed to conserve part of the change in free energy ($n_p \cdot \Delta G'_p$) of the driving reaction [ΔG, in J/mol substrate (S) consumed] in the phosphorylation reactions (Fig. 3.**4**), while the residual part is converted to thermal energy. The ratio of the free-energy change conserved to that of the driving reaction is called energetic efficiency (η; see Box 3.2):

$$\eta = n_p\,\Delta G'_p/\Delta G \tag{3.3}$$

> **Box 3.2** Theoretically, the change in free energy (ΔG) of the driving reaction ($S \rightarrow P$) could be fully conserved in that of the phosphorylation reaction ($n_p \cdot \Delta G'_p$) (Fig. 3.**4**). If the phosphorylation reaction were coupled to the driving reaction by an ideal energy converter, the energetic efficiency (η) would be 1 and the amount of ATP (n_p) formed would be equal to the ratio $\Delta G/\Delta G'_p$ (Eqn. 3.3). This would require the operation of the overall reaction close to equilibrium and at infinitely small velocity. However, metabolic reactions proceed at considerable velocities and often far from equilibrium. Therefore, part of the free-energy change of the driving reaction is lost as thermal energy and $\eta < 1$.

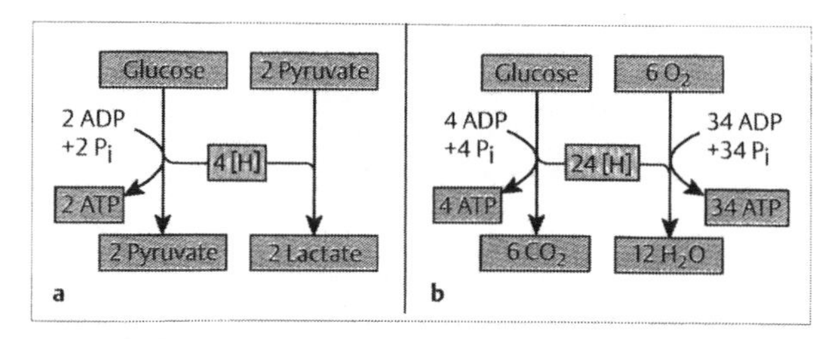

Fig. 3.3 **Schematic presentation of anaerobic glycolysis (a) and aerobic glucose oxidation (b).** The removal of reducing equivalents ([H]) from glucose is associated with SLP and the reduction of O_2 is associated with ETP

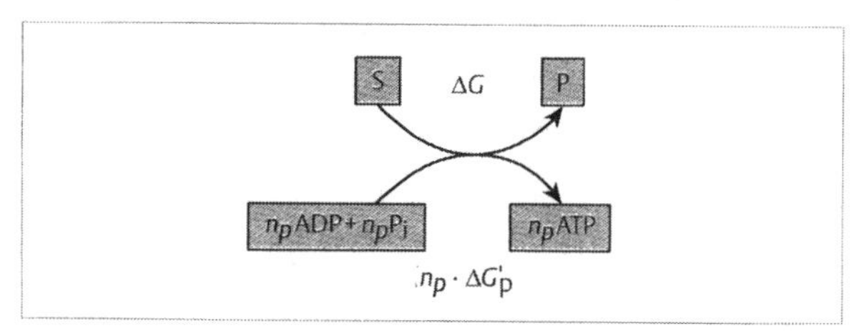

Fig. 3.4 **Coupling of an exergonic chemical reaction ($S \rightarrow P$) to the formation of ATP from ADP and P_i.** The free energy of the chemical reaction is ΔG (kJ/mol S) and that of ATP formation is $\Delta G'_p$ (kJ/mol ATP). n_p designates the molar amount of ATP formed per mol S

The amount of ATP formed is termed n_p (mol ATP/mol S consumed). $\Delta G'_p$ (kJ/mol ATP) represents the free-energy change required for ATP synthesis. The energetic efficiency may vary between 0 and 1. The average η of most catabolic processes is roughly 0.5. This allows estimation of the ATP yield (n_p) of a catabolic process from the change in free energy of the driving reaction (ΔG) and from that of phosphorylation ($\Delta G'_p$) according to Equation 3.3. As shown in the following, the energetic efficiency of ATP synthesis driven by the fermentation of glucose to lactate (Reaction 3.2) is close to 0.5.

Box 3.3 The free-energy change of a reaction ($S \rightarrow P$) is a function of the distance of the actual concentrations ([ ]) of the reactants from their equilibrium concentrations ($[\ ]_{eq}$):

$$\Delta G = RT\left(\ln\frac{[P]}{[S]} - \ln K_{eq}\right) \quad (3.4)$$

$$K_{eq} = \frac{[P]_{eq}}{[S]_{eq}} \quad (3.5)$$

The standard-free-energy change (ΔG°) of the reaction ($S \rightarrow P$) refers to 1 M concentrations of the reactants in aqueous solution at 25 °C. At 1 M reactant concentrations, Equation 3.4 yields Equation 3.6, which shows how ΔG° relates to K_{eq}, the equilibrium constant of the reaction ($S \rightarrow P$):

$$\Delta G^\circ = -RT \ln K_{eq} \quad (3.6)$$

The combination of Equations 3.4 and 3.6 results in Equation 3.7, which allows the calculation of ΔG from ΔG° or from the equilibrium constant (K_{eq}) and the actual concentrations of the reactants:

$$\Delta G = \Delta G^\circ + RT \ln\frac{[P]}{[S]} \quad (3.7)$$

Exergonic reactions exhibit negative values of ΔG. The standard-free-energy changes of reactions liberating or consuming protons usually refer to pH 7 (not to 1 M H^+) and to 1 M concentrations of the other reactants. These values are termed $\Delta G^{\circ\prime}$. The changes in free energy at pH 7 and at variable concentrations of the other reactants are designated $\Delta G'$.

As delineated in Box 3.3, the change in free energy of a reaction is a function of the reactant concentrations. The free-energy change of glucose fermentation (Reaction 3.2) is calculated according to Equation 3.8 for several reactant concentrations (Table 3.1):

$$\Delta G = \Delta G^\circ + RT \ln\frac{[\text{lactate}]^2\,[H^+]^2}{[\text{glucose}]} \quad (3.8)$$

The ΔG value almost doubles when the pH is shifted from standard (1 M H^+) to neutral conditions. A decrease in the concentrations of glucose and lactate by two orders of magnitude causes an increase of the (negative) ΔG value by only 5%.

The free-energy difference required for ATP synthesis according to Reaction 3.9 at pH 7 is called phosphorylation potential ($\Delta G'_p$), which is generally calculated from the ratio of the bacterial contents of ATP, ADP, and P_i according to Equation 3.10, assuming that the reactants are freely available in the bacterial cytoplasm:

$$ADP + P_i + H^+ \rightarrow ATP + H_2O \quad (3.9)$$

$$\Delta G'_p = \Delta G_p^{\circ\prime} + RT \ln\frac{[ATP]}{[ADP][P_i]} \quad (3.10)$$

Nearly equal amounts of ATP and ADP have been found in growing bacteria. With $\Delta G_p^{\circ\prime} = 32$ kJ/mol ATP and $[P_i] = 10$ mM, the difference in free energy required for the synthesis of 1 mol ATP ($\Delta G'_p$) is 44 kJ. This value has to be considered a rough estimate, since the concentrations of the free reactants within the bacterial cytoplasm are not known. Part of the reactants are probably bound to proteins, and part of the nucleotides form complexes with magnesium ions.

Division of the $\Delta G^{\circ\prime}$ of glucose fermentation (198 kJ/mol glucose; see Tab. 3.1) by $\Delta G'_p$ (44 kJ/mol ATP) results in the theoretical ATP yield of glucose fermentation to lactate, 4.5 mol ATP/mol glucose, which would be valid if the energy conversion operated at maximum energetic efficiency ($\eta = 1$). The actual yield of the process is known to be 2 mol ATP/mol glucose. Hence, η is 0.42, according to Equation 3.3.

Table 3.1 **Free-energy change of glucose fermentation to lactate as a function of the concentrations of the reactants**

Glucose (M)	Lactate (M)	pH	ΔG (kJ/mol glucose)
1	1	0	−118 (ΔG°)
1	1	7	−198 ($\Delta G^{\circ\prime}$)
10^{-2}	10^{-2}	7	−209

See Equation 3.2. ΔG is calculated according to Equation 3.8, with $R = 8.3 \cdot 10^{-3}$ kJ K^{-1} mol^{-1} and $T = 298$ K. ΔG° was calculated from the standard free energies of formation of the reactants, from Thauer et al. (1977; see Further Reading)

3.3 Coupling of ATP Synthesis to Glucose Degradation Requires C–C Cleavage and Subsequent Oxidation

Glycolysis leading to pyruvate is the pathway of glucose degradation that is common to the majority of the fermentative and aerobic bacteria (Fig. 3.**5**; see also Chapter 9.11). Glucose is converted to fructose 1,6-bisphosphate by the transfer of two phosphoryl groups from ATP. Fructose 1,6-bisphosphate is cleaved in the aldolase reaction to give two triose phosphates (glyceraldehyde 3-phosphate and dihydroxyacetone phosphate) that are interconverted by triosephosphate isomerase. In the subsequent oxidation of glyceraldehyde 3-phosphate, P_i is taken up and a phosphoacylanhydride (1,3-bisphosphoglycerate) is formed, the phosphoryl group of which is subsequently transferred to ADP. The resulting 3-phosphoglycerate is isomerized to 2-phosphoglycerate, from which phospho*enol*pyruvate is formed by dehydration. The final transfer of the phosphoryl group from phospho*enol*pyruvate to ADP is driven by the tautomerization of the resulting enol to pyruvate.

The energetic events associated with the individual glycolytic reactions are illustrated in Figure 3.**6**. On the

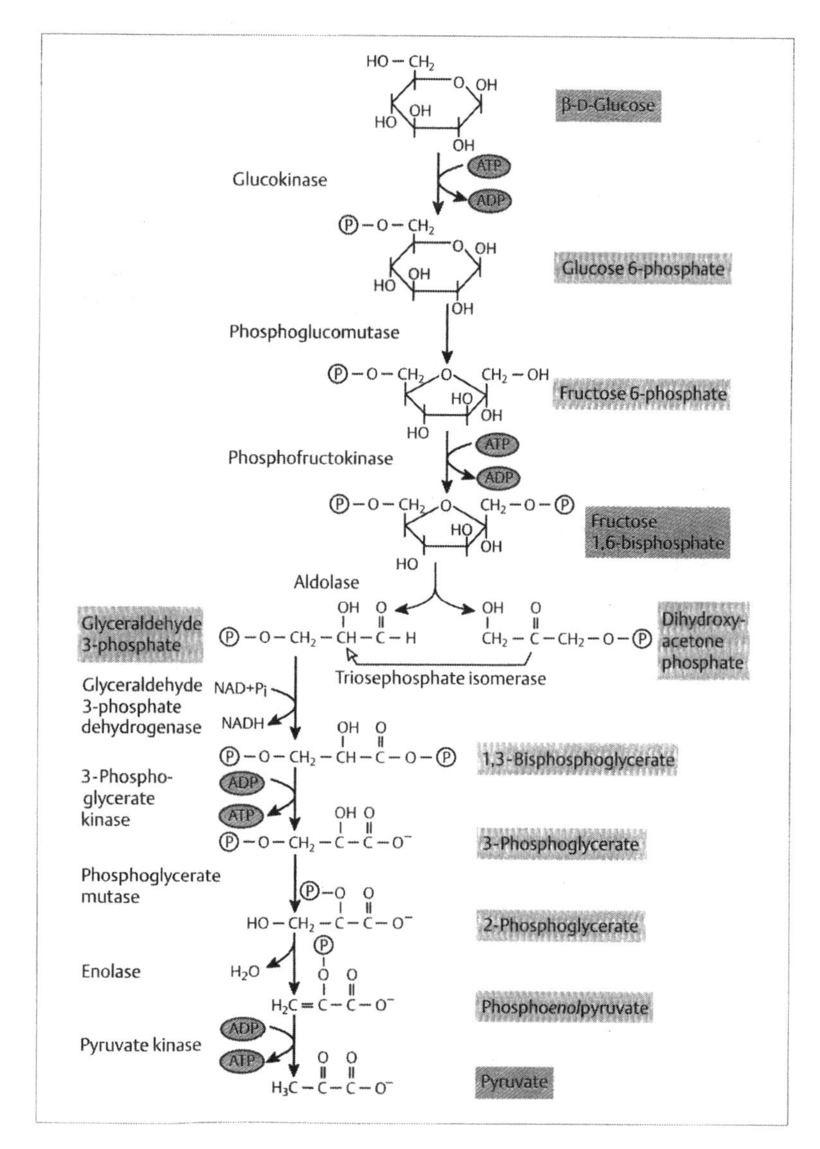

Fig. 3.**5** **Intermediates and enzymes of the Embden-Meyerhof-Parnas glycolytic pathway.** The phosphoryl residue is designated Ⓟ

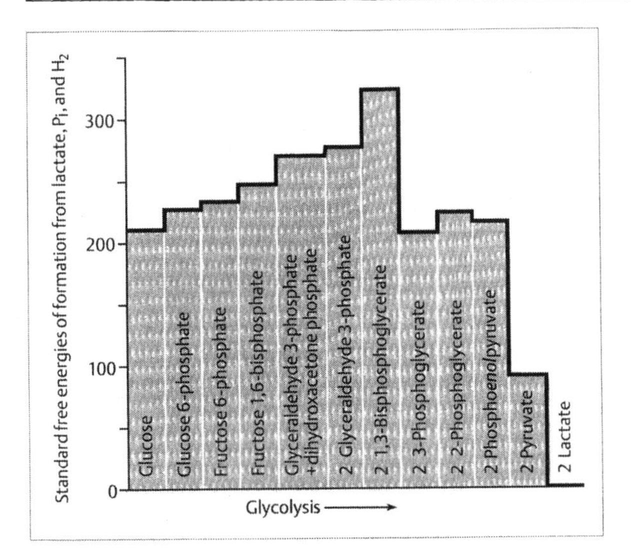

Fig. 3.**6** **Energy profile of the glycolytic pathway.** The standard free energies of formation of the intermediates from lactate, P_i, and H_2 are given on the ordinate. The free energies associated with ADP/ATP and NAD/NADH interconversion are not considered

ordinate, the free formation energies of the glycolytic intermediates from lactate and P_i are given. The free energy of interconversion of NAD, NADH, ADP, and ATP are not considered in Figure 3.**6**. Molecular hydrogen is assumed to be formed in the glyceraldehyde-dehydrogenase reaction and to be consumed in pyruvate reduction to lactate. Note that the free-energy values refer to standard conditions at pH 7, and not to the concentrations of the glycolytic intermediates in growing bacteria. The free-energy content of the intermediates gradually increases from glucose to glyceraldehyde 3-phosphate. The large increase in free energy, which is caused by the glyceraldehyde-3-phosphate-oxidation reaction, reflects the formation of the "high-energy" phosphate bond in 1,3-bisphosphoglycerate. Actually, this increase in free energy is balanced by the exergonic pyruvate reduction since the two reactions are coupled by hydrogen transfer via NAD. A dramatic decrease in free energy is associated with the reactions that are coupled to ATP synthesis (i.e., the phosphoglycerate kinase and pyruvate kinase reactions). Approximately 120 kJ are available from each of these reactions (under standard conditions) for the synthesis of one mol ATP.

One of the surprising properties of glycolysis is the expenditure of two mol ATP for hexose phosphorylation in a pathway for the purpose of ATP formation. The reason for the formation of fructose 1,6-bisphosphate prior to cleavage of the sugar molecule is an energetic one. The two negatively-charged phosphate groups of fructose 1,6-bisphosphate are in close proximity and thus favor the aldolase reaction, which is nevertheless endergonic under standard conditions. Therefore, the equilibrium concentrations of the products are distinctly lower than that of the substrate. The cleavage of fructose or fructose 6-phosphate in an alternative hypothetical pathway would be nearly 10 kJ more endergonic than the cleavage of fructose 1,6-bisphosphate, and the product concentrations at equilibrium would be approximately an order of magnitude lower than those of the aldolase reaction. As a consequence, the velocity of the reactions following C–C cleavage would probably be insufficient to fulfill the catabolic function of glycolysis.

3.4 A "High-Energy" Compound Is Formed in SLP

The reaction driving SLP has to be sufficiently exergonic. Since the phosphorylation potential in growing bacteria is nearly 50 kJ/mol ATP and the energetic efficiency of catabolism is about 50%, the change in free energy provided by the driving reaction for the synthesis of 1 mol ATP has to be at least 100 kJ. In the glycolytic pathway, the oxidation of an aldehyde (i.e., glyceraldehyde 3-phosphate) by NAD (Box 3.4) serves as the driving reaction; the mechanism will be described in the following (Fig. 3.**9**).

Box 3.4 Nicotinamide adenine dinucleotide (NAD) and nicotinamide adenine dinucleotide 3′-phosphate (NADP) are the main transport metabolites for the reducing equivalents liberated and consumed in metabolism (Fig. 3.**7**). The pyridinium ring of the oxidized pyridine nucleotides is reduced by hydride transfer from donor substrates. The reduced pyridine nucleotides (NADH and NADPH) are reoxidized by hydride transfer to certain acceptors. Both reactions are catalyzed by specific dehydrogenases.

Box 3.5 The oxidation of an aldehyde by NAD might be envisaged as a direct abstraction of a hydride from the aldehyde and transfer of the hydride to NAD (Fig. 3.**8**). This would result in the formation of an acyl cation as an intermediate at the active site of the enzyme. Subsequent transfer of the acyl cation to P_i would yield a "high-energy" acyl phosphate as product. Acyl phosphates are known to serve as phosphoryl donors for ATP formation from ADP (e.g., the 3-phosphoglycerate-kinase reaction).

A comparison of the hypothetical mechanism of aldehyde oxidation (Fig. 3.**8**) to the actual mechanism of glyceraldehyde-3-phosphate dehydrogenase (Fig. 3.**9**) demonstrates the strategy of energy conservation in SLP. The theoretical mechanism is not used by the dehydrogenase probably because the highly reactive acyl cation cannot be stabilized at the active site of the enzyme. The probable fate of the acyl cation would be hydrolysis, with the carboxylate as product (Fig. 3.**8**). Thus, the oxidation of the aldehyde to the carboxylate would occur without energy conservation. In contrast, the strategy of SLP consists of the formation of a "high-energy" intermediate that tends to split off an acyl or phosphoryl group but is sufficiently stable to resist hydrolysis or an attack by other nucleophiles.

The first step in glyceraldehyde 3-phosphate oxidation, as catalyzed by glyceraldehyde-3-phosphate dehydrogenase, is not a direct abstraction of a hydride from the aldehyde (see Box 3.5). Instead, a thiohemiacetal is formed from the aldehyde and the sulfhydryl group present at the active site of the enzyme (Fig. 3.**9**).

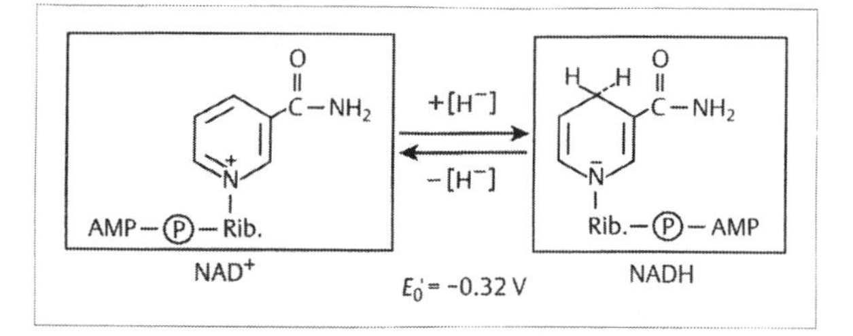

Fig. 3.**7** **Reduction of NAD and oxidation of NADH by hydride transfer.** [H⁻] is a hydride enzymatically abstracted from a donor substrate. Rib., 1,5-substituted ribose. NADP differs from NAD by a phosphate group in the 2′-position of the ribose residue

Subsequently, a hydride is transferred from the thiohemiacetal to NAD. This transfer results in the formation of a "high-energy" thioester in which most of the free energy of the oxidation reaction is conserved. In the final step of catalysis, the thioester is converted to an acyl phosphate (1,3-bisphosphoglycerate), which is released from the enzyme. Much of the change in free energy of the thioester is conserved in the acyl phosphate group of 1,3-bisphosphoglycerate, which represents a "high-energy" compound as well and serves as a phosphoryl donor for ATP formation from ADP in the subsequent 3-phosphoglycerate-kinase reaction (Figs. 3.**6** and 3.**9**).

Acyl phosphates such as 1,3-bisphosphoglycerate, thioesters such as acyl-CoA, and ATP are called "high-energy" compounds because their standard free energy of hydrolysis is more negative than −30 kJ/mol. The free energy of hydrolysis is a measure of the driving force for acyl or phosphoryl transfer to nucleophilic compounds such as ADP. The free energies of hydrolysis of various "high-energy" compounds are listed in Table 3.**2** (see also Box 3.6).

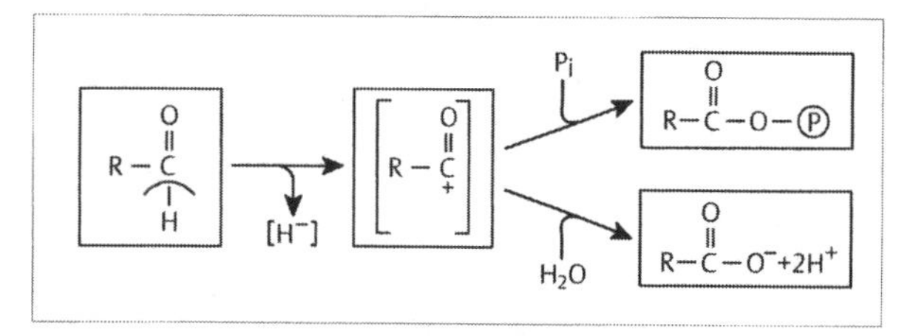

Fig. 3.**8** **Schematic abstraction of a hydride from an aldehyde and reaction of the resulting carbonium cation with P_i or H_2O**

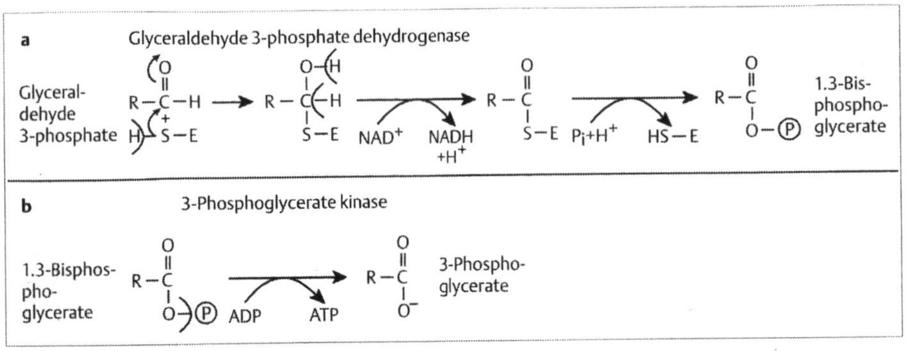

Fig. 3.9 **Mechanism of glyceraldehyde-3-phosphate dehydrogenase (a) and of 3-phosphoglycerate kinase (b).** H-S-E designates the sulfhydryl group in the active site of the dehydrogenase

Table 3.2 **"High-energy" compounds involved in substrate-level phosphorylation**

Type of compound	Compound	Hydrolysis products	$\Delta G^{\circ\prime}$ of hydrolysis (kJ/mol)
Phosphoacyl anhydride	1,3-Bisphosphoglycerate	3-Phosphoglycerate + P_i	−52
	Acetyl phosphate	Acetate + P_i	−45
	Carbamoyl phosphate	HCO_3^- + NH_4^+ + P_i	−39
Phosphoanhydride	ATP	ADP + P_i	−32
	ATP	AMP + PP_i	−42
	PP_i	2 P_i	−22
	ADP	AMP	−32
Phosphoenol ester	PEP	Pyruvate + P_i	−52
Acyl thioester	Acetyl-CoA	Acetate + CoA	−36
	Propionyl-CoA	Propionate + CoA	−36
	Butyryl-CoA	Butyrate + CoA	−36
	Succinyl-CoA	Succinate + CoA	−35

Box 3.6 The oxidation of glucose 6-phosphate by NADP, as catalyzed by glucose-6-phosphate dehydrogenase (Fig. 3.10; Chapter 9), results in the formation of an ester (6-phosphogluconolactone), which is not a "high-energy" intermediate and is not used for ATP synthesis, in contrast to thioesters. The hemiacetal linkage within glucose 6-phosphate is not converted to a thiohemiacetal or another suitable aldehyde derivative, the oxidation of which would allow formation of a "high-energy" bond. Oxidation of glucose at C1 without SLP is the reason that only 1 ATP/glucose results from pathways of glucose degradation involving glucose-6-phosphate oxidation [e.g., KDPG (2-keto-3-deoxy-6-phosphogluconate) pathway, heterofermentative glucose degradation by lactic acid bacteria; Chapters 9.11 and 12.2].

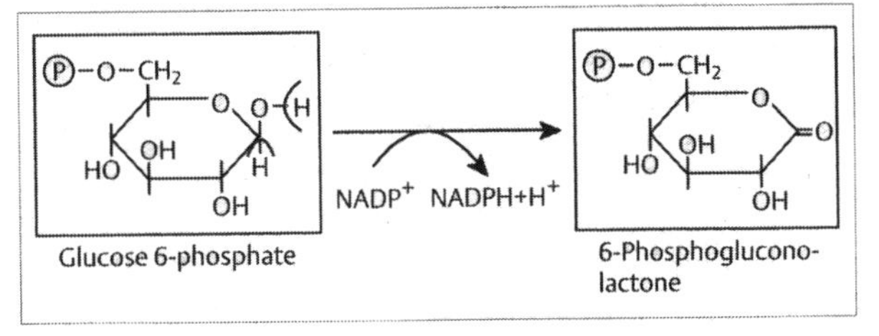

Fig. 3.10 **The glucose-6-phosphate dehydrogenase reaction**

3.5 Pyruvate Oxidation Is Coupled to Energy Conservation

Pyruvate formed from glucose is usually oxidized to CO_2 by aerobic bacteria. The first step in this pathway, the oxidation of pyruvate by NAD to acetyl-CoA, is catalyzed by pyruvate dehydrogenase:

$$\text{Pyruvate} + \text{CoA} + \text{NAD} \rightarrow \text{Acetyl-CoA} + CO_2 + \text{NADH} \quad (3.11)$$

Pyruvate dehydrogenase is a large enzyme complex harboring three prosthetic groups, thiamine diphosphate (TPP), lipoate, and FAD. The function of the prosthetic groups is illustrated in Figure 3.**11**. In the first step of catalysis, pyruvate is bound to TPP and the product is decarboxylated. The resulting TPP-bound acetaldehyde is transferred to oxidized lipoate. In this step, the bound aldehyde is oxidized to give the acetyl thioester of reduced lipoate. After transfer of the acetyl group to CoA, the reduced lipoate is reoxidized by NAD, with FAD acting as a redox mediator.

The catalysis resembles that of glyceraldehyde-3-phosphate dehydrogenase (Fig. 3.**9**): the oxidation of an aldehyde results in the formation of a "high-energy" compound with a thioester as intermediate. Again, the aldehyde is not oxidized directly by hydride abstraction. In pyruvate oxidation, the aldehyde is linked to TPP by a C–C-bond and the oxidation occurs simultaneously with the cleavage of this bond. In contrast, the aldehyde forms a thiohemiacetal, which is oxidized without cleavage at the active site of glyceraldehyde-3-phosphate dehydrogenase. While an acyl phosphate is the product of the latter enzyme, a thioester is formed in pyruvate oxidation. In many anaerobic bacteria, acetyl-CoA is converted to acetylphosphate (Reaction 3.12), which is subsequently used for ADP phosphorylation to ATP (Reaction 3.13). The reactions are catalyzed by phosphotransacetylase and acetate kinase:

$$\text{Acetyl-CoA} + P_i \rightarrow \text{Acetylphosphate} + \text{CoA} \quad (3.12)$$

$$\text{Acetylphosphate} + \text{ADP} \rightarrow \text{Acetate} + \text{ATP} \quad (3.13)$$

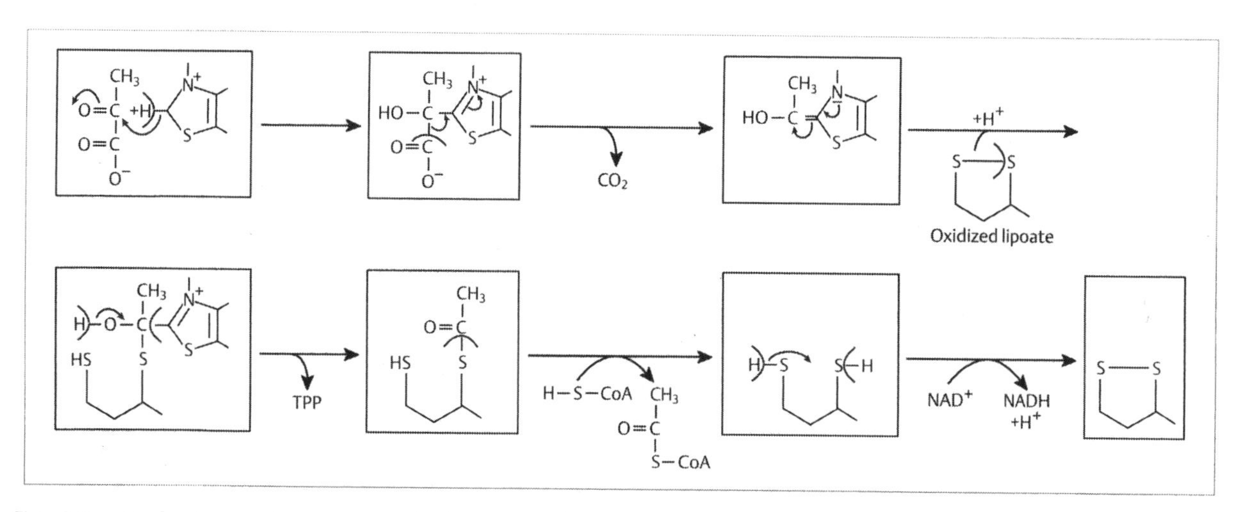

Fig. 3.**11** **Mechanisms of pyruvate dehydrogenase,** demonstrating the interaction of the prosthetic groups [thiamine diphosphate (TPP) and lipoate] and coenzymes (CoA and NAD^+)

3.6 The Catabolic Function of the Citrate Cycle Is to Provide Reducing Equivalents for Oxidative Phosphorylation

Of the total 24 reducing equivalents ([H]) derived from the conversion of glucose to CO_2 (Fig. 3.**3b**), 4[H] are liberated by glycolysis, 4[H] in the pyruvate-dehydrogenase reaction, and the remaining 16[H] in the citrate cycle (Table 3.**3**). The main purpose of [H] liberation is to provide reducing equivalents for oxidative phosphorylation at the most negative possible redox potential (Box 3.7; see also Chapter 7). The more negative the redox

Table 3.3 **The ATP yield and the amount of reducing equivalents liberated in the processes of aerobic glucose degradation**

Pathway	Substrate	Products	ATP or GTP (mol/mol glucose)	[H][1]
Glycolysis	Glucose	2 Pyruvate	2	4
Pyruvate dehydrogenase	2 Pyruvate	2 Acetyl-CoA + 2 CO_2	–	4
Citrate cycle	2 Acetyl-CoA	4 CO_2	2	16
Oxidative phosphorylation	24[H] + 6 O_2	12 H_2O	34	–

[1] [H], reducing equivalents liberated

potential of the reducing equivalents transferred to oxygen is, the higher the ATP yield of oxidative phosphorylation.

The overall reaction catalyzed by the enzymes of the citrate cycle consists of the oxidation of the acetyl group of acetyl-CoA to CO_2 and the concomittant reduction of NAD and ubiquinone (Q):

$$\text{Acetyl-CoA} \rightarrow 2\,CO_2 + \text{CoASH}; \quad 3\,\text{NAD} + \text{Q} \rightarrow 3\,\text{NADH} + QH_2 \qquad (3.14)$$

The cycle is initiated by the formation of citrate from oxaloacetate and acetyl-CoA, which is catalyzed by citrate synthase (Fig. 3.**12**). The reaction resembles the aldolase reaction. The nucleophilic attack on oxaloacetate requires an activated acetyl group; acetate itself would not be sufficiently reactive. The activation consists of proton mobilization from the methyl group, which is brought about by the thioester carbonyl group. This is the reason why the "high-energy" bond of acetyl-CoA is not used for ATP formation (see Reactions 3.12 and 3.13) in bacteria using the citrate cycle (Box 3.7).

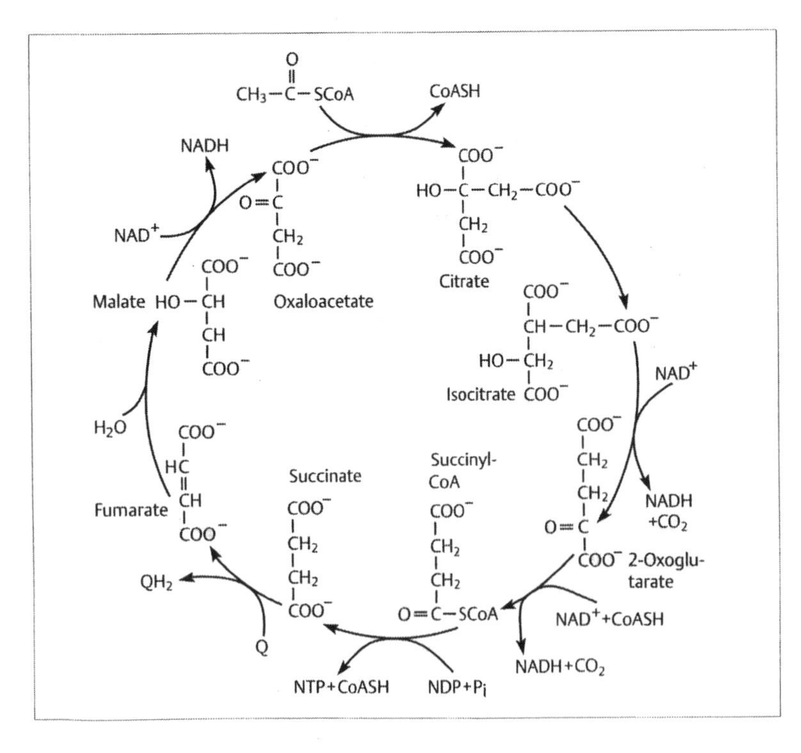

Fig. 3.**12** **The intermediates of the citrate cycle.** The enzymes are given in Table 3.**4**. NTP, ATP or GTP; NDP, ADP or GDP; $-CO_2^-$, carboxylate

Box 3.7 The direct oxidation of acetate to CO_2 in the presence of O_2 (without the intermediate formation of citrate) would provide half the amount of reducing equivalents for oxidative phosphorylation (Fig. 3.**13**), as compared to the citrate cycle. This is due to the requirements of the first step in that theoretical pathway. Acetate (or acetyl-CoA) cannot be oxidized to glycolate in a dehydrogenase reaction. Instead, acetate oxidation would require a hydroxylase which catalyzes [H] consumption and not [H] liberation. Therefore, two of the six [H] gained from glycolate oxidation to CO_2 would be consumed in the hydroxylation reaction and could not be provided for oxidative phosphorylation. This is probably the reason why the theoretical pathway (Fig. 3.**13**) does not occur as a substitute for the citrate cycle in living organisms.

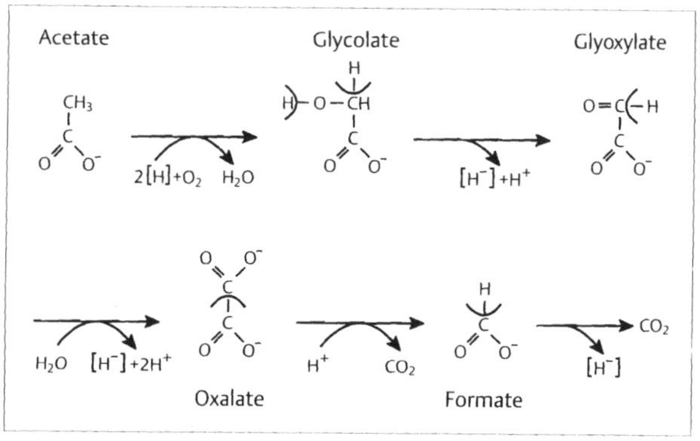

Fig. 3.**13** **Hypothetical pathway of acetate oxidation**

The citrate formed by citrate synthase is oxidized to oxaloacetate and 2 CO_2 in the residual part of the citrate cycle. To this end, the tertiary alcohol citrate is first converted by aconitase to the secondary alcohol isocitrate, which can be oxidized. The oxidation of isocitrate by NAD (or NADP) is catalyzed by isocitrate dehydrogenase and yields 2-oxoglutarate and CO_2. The oxidative decarboxylation of 2-oxoglutarate is catalyzed by an enzyme complex (2-oxoglutarate dehydrogenase), which is similar to pyruvate dehydrogenase with respect to composition, prosthetic groups, and mechanism of action. The free energy of the thioester bond of the succinyl-CoA resulting from the oxidation of 2-oxoglutarate is used for nucleotide triphosphate (ATP or GTP) formation from nucleoside diphosphate and P_i via SLP. The succinate thiokinase catalyzing this reaction combines the activities of both an acyl transferase and a kinase (see Reactions 3.12 and 3.13 for comparison). The resulting succinate is oxidized to fumarate, which, after hydration to malate (catalyzed by fumarase), is oxidized by NAD to yield oxaloacetate (catalyzed by malate dehydrogenase).

Succinate dehydrogenase, in contrast to the other enzymes of the citrate cycle, is integrated in the bacterial membrane. The reducing equivalents derived from succinate are transferred to ubiquinone, which is located in the membrane, while NAD serves the other dehydrogenases as the acceptor. The redox potential of the NAD/NADH couple ($E'_0 = -0.32$ V) is 0.35 V more electronegative than that of fumarate/succinate ($E'_0 = 0.03$ V), while the redox potential of Q/QH_2 ($E'_0 = 0.11$ V) is more electropositive. Thus, succinate oxidation is exergonic with Q as acceptor and would be extremely endergonic with NAD as acceptor. Succinate

Table 3.**4** **Standard-free-energy differences at pH 7 and equilibrium constants of the reactions of the citrate cycle**

Enzyme	Substrates	Products	$\Delta G^{\circ\prime}$ [1] (kJ/mol)	K_{eq} [2]
Citrate synthase	Acetyl-CoA + oxaloacetate	Citrate + CoA	−37.5	$3.8 \cdot 10^6$
Aconitase	Citrate	Isocitrate	+6.7	0.066
Isocitrate dehydrogenase	Isocitrate + NAD (NADP)	2-Oxoglutarate + NADH(NADPH) + HCO_3^-	−4.6	6.4 M
2-Oxoglutarate dehydrogenase	2-Oxoglutarate + NAD + CoA	Succinyl − CoA + NADH + HCO_3^-	−29.4	$1.4 \cdot 10^5$
Succinate thiokinase (Succinyl-CoA synthetase)	Succinyl − CoA + NDP + P_i	Succinate + NTP + CoA	−3.3	3.8
Succinate dehydrogenase	Succinate + Q	Fumarate + QH_2	−16.0	650
Fumarase	Fumarate + H_2O	Malate	−3.7	4.5
Malate dehydrogenase	Malate + NAD	Oxaloacetate + NADH	+28.8	$8.9 \cdot 10^{-6}$
			−59.0	

[1] $\Delta G^{\circ\prime}$, standard-free-energy difference at pH 7
[2] K_{eq}, equilibrium constant

dehydrogenase and ubiquinone may be regarded as part of the respiratory chain.

The catabolic function of the citrate cycle is best demonstrated by comparing the free-energy change of acetyl-CoA oxidation to CO_2, with NAD and ubiquinone as acceptors (Reaction 3.14), to the free energy available from reoxidation by O_2 of the NADH (Reaction 3.15) and ubiquinol (Reaction 3.16) formed in the cycle:

$$3\,NADH + \tfrac{3}{2}O_2 + 3\,H^+ \rightarrow 3\,NAD + 3\,H_2O \qquad \Delta G^{\circ\prime} = -677\ kJ/3\ mol\ NADH \tag{3.15}$$

$$QH_2 + \tfrac{1}{2}O_2 \rightarrow Q + H_2O \qquad \Delta G^{\circ\prime} = -137\ kJ/mol\ QH_2 \tag{3.16}$$

The free-energy change of acetyl-CoA oxidation (Reaction 3.14) is given by the sum of the $\Delta G^{\circ\prime}$ values of the individual reactions of the cycle (Table 3.**4**). This value (−59 kJ/mol acetyl-CoA) is more than ten times smaller than the change in free energy available in oxidative phosphorylation (814 kJ/mol acetyl-CoA). Only a small amount of the overall free-energy change available from acetyl-CoA oxidation by O_2 is used to drive the citrate cycle, while the major part is conserved as reducing equivalents at a relatively negative redox potential. This part is subsequently used for ATP synthesis by oxidative phosphorylation.

Further Reading

Gottschalk, G. (1985) Bacterial metabolism. New York: Springer-Verlag

Thauer, R.K., Jungermann, K., and Decker, K. (1977) Bacteriol Rev 41: 100–180

4 Electron-Transport-Coupled Phosphorylation

The function of electron-transport-coupled phosphorylation (ETP) is ATP synthesis. ATP synthesis is driven by a vast variety of redox reactions. Not only oxygen but a variety of other compounds may serve as terminal electron acceptors in bacterial electron transport (Table 4.1; Chapter 12). ETP[1] with oxygen as electron acceptor is called **oxidative phosphorylation** (Chapter 11). The coupling mechanism of ETP is different from that of SLP[1]. ETP is catalyzed by the combined action of a multienzyme **electron transport chain** and the **ATP synthase**, which are integrated in the bacterial cytoplasmic membrane (Fig. 4.1). Topologically, the ATP synthase is separated from the electron transport chain. However, the enzyme systems are functionally coupled by the **electrochemical proton potential** (**Δp**, also termed proton motive force) across the cytoplasmic membrane. The electron transport chain is designed to conserve the free energy of the electron transport reaction in the free energy of the Δp. This is done by coupling the electron-transport reaction to the consumption of protons from the cytoplasm and to the liberation of protons on the outside of the bacterial membrane. The ATP synthase couples the translocation of protons from the outside to the cytoplasm. This process is driven by Δp.

Table 4.1 **Standard redox potentials at pH 7 (E'_0) of various redox donors or acceptors of ETP**

Redox couples	E'_0 (V)
HCO_3^-/HCO_2^-	−0.41
H^+/H_2	−0.42
HCO_3^-/acetate	−0.35
HCO_3^-/CH_4	−0.33
NAD/NADH	−0.32
S^0/HS^-	−0.27
Acetaldehyde/ethanol	−0.20
Pyruvate/lactate	−0.19
Dihydroxyacetone-P/glycerol-P	−0.19
Oxaloacetate/malate	−0.17
HSO_3^-/HS^-	−0.12
SeO_4^{2-}/SeO_3^{2-}	+0.02
Fumarate/succinate	+0.03
$NO_2^-/HONH_3^+$	+0.06
TMAO/TMA[1]	+0.13
DMSO/DMS[2]	+0.16
NO_2^-/NH_4^+	+0.34
NO_2^-/NO	+0.35
NO_3^-/NO_2^-	+0.43
Fe^{3+}/Fe^{2+}	+0.77
O_2/H_2O	+0.82
NO/N_2O	+1.18
N_2O/N_2	+1.36

Redox donors (D_{red}) and acceptors (A_{ox}) in Fig. 4.1
[1] TMAO, trimethylamine oxide; TMA, trimethylamine
[2] DMSO, dimethylsulfoxide; DMS, dimethylsulfide

4.1 In ETP, the Amount of ATP Formed Corresponds With the Free Energy of the Driving Redox Reaction

In SLP, one mol ATP is formed per mol substrate of the reaction that drives phosphorylation. Therefore, the free energy difference of the driving reaction has to exceed −44 kJ/mol substrate, and less exergonic reactions cannot be coupled to phosphorylation. In contrast, the oxidation of one mol substrate in ETP may yield fractions of one mol ATP; phosphorylation can be driven by reactions that are less exergonic than −44 kJ/mol substrate. This is achieved by variation of the amount of protons shifted across the membrane per electron transported in the driving reaction. This energetic flexibility is probably the reason for the widespread occurrence of ETP in anaerobic bacteria, especially in pathways operating at small ΔG values that do not allow SLP (Chapter 12). A list of redox couples serving as

[1] Abbreviations and symbols used in this chapter are listed in Appendix **

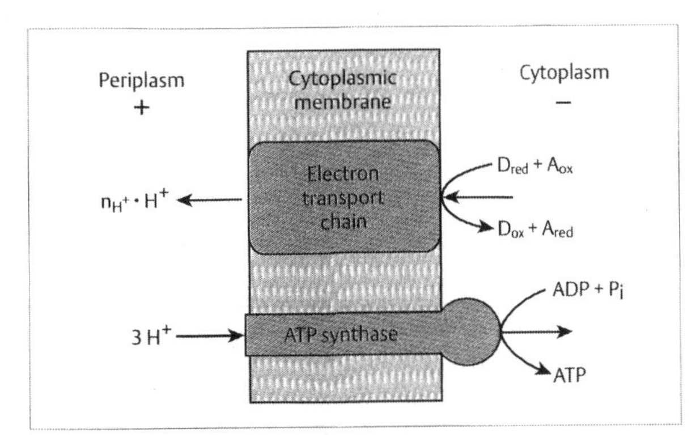

Fig. 4.1 **Coupling mechanism of ETP.** The electron transport chain and the ATP synthase are located in the bacterial cytoplasmic membrane. The oxidation of a donor substrate D_{red} (e.g., NADH or succinate) by an acceptor A_{ox} (e.g., O_2) is coupled to proton uptake from the inside and to proton liberation on the outside of the membrane. The substrate sites of the electron transport enzymes are assumed to face the bacterial cytoplasm. ATP synthesis is coupled to proton translocation from the outside to the cytoplasm

donors or acceptors (D_{red}, A_{ox} in Fig. **4.1**, respectively) in ETP is given in Table **4.1**.

In aerobic ETP, the ATP yield is usually presented as the "**P/O ratio**," which is the amount of ATP gained from the reduction of half an O_2 to H_2O and refers to the transport of two electrons. In the following, the ATP yield will be given as the ratio $\mathbf{n_p/n_e}$, which is the ATP yield per electron and also refers to ETP with electron acceptors other than O_2. The n_p/n_e ratio can be calculated from the free energies of the electron-transport and the phosphorylation reactions according to Equation 4.1 (delineated in Box 4.1):

$$n_p/n_e = \Delta E \cdot F \cdot \eta / \Delta G'_p \qquad (4.1)$$

n_p/n_e is shown as a function of ΔE in Fig. **4.3**, with $\eta = 1$. In aerobic glucose degradation, NADH (Reaction 4.2) and succinate (Reaction 4.3) are the donor substrates of oxidative phosphorylation:

$$\mathrm{NADH} + \tfrac{1}{2}\,O_2 + H^+ \rightarrow \mathrm{NAD}^+ + H_2O \qquad 3\ \mathrm{ADP} + 3\ P_i \rightarrow 3\mathrm{ATP} \qquad (4.2)$$

$$\mathrm{Succinate} + \tfrac{1}{2}\,O_2 \rightarrow \mathrm{fumarate} + H_2O \qquad 2\ \mathrm{ADP} + 2P_i \rightarrow 2\mathrm{ATP} \qquad (4.3)$$

The most accurate n_p/n_e ratios have been measured in mitochondria: 1.5 for NADH respiration (Reaction 4.2) and 1 for succinate respiration (Reaction 4.3). Note that 2 mol electrons are transported per mol NADH or succinate. These values are used for calculating the energetic efficiencies η of the two reactions. This is done by comparing the theoretical n_p/n_e ratios, calculated according to Equation 4.1 with $\eta = 1$, to the values actually measured (Table **4.2**). Thus, the energetic efficiency of oxidative phosphorylation with NADH and succinate in mitochondria is approximately 0.6.

Box 4.1 ETP functions as a free energy converter. Three forms of free energy can be discriminated in the process of ETP (Fig. **4.2**). (1) The amount of free energy of the redox reaction catalyzed by the electron transport chain is given by the redox potential difference (ΔE, in Volt) between the donor (D_{red}) and the acceptor substrate (A_{ox}) multiplied by the amount of redox equivalents transported ($n_e F$, in Ampere·s). The molar amount of electrons is designated n_e. F represents the Faraday constant (96.5 Ampere·s/mol electrons). (2) The amount of free energy conserved in the Δp is obtained by multiplying Δp (in Volt) with the electrical charge ($n_{H^+} F$, in Ampere·s) transported across the membrane by protons. The proton transport is driven by electron transport. (3) The third form of free energy is conserved in the phosphorylation reactions, which is driven by Δp. Its amount is determined by the phosphorylation potential ($\Delta G'_p = 44$ kJ/mol ATP) multiplied by the molar amount of ATP formed (n_p).

If ETP operated at 100% efficiency ($\eta = 1$), the free energy of electron transport would be fully conserved in the free energy of phosphorylation, and the amounts of free energy conserved in the three forms would be the same. In biological ETP, only a fraction η (energetic efficiency) of the free energy of the driving reaction is conserved in that of the driven reaction. Equations 4.1 and 4.4 are obtained from the expressions given in Fig. **4.2** after the introduction of η.

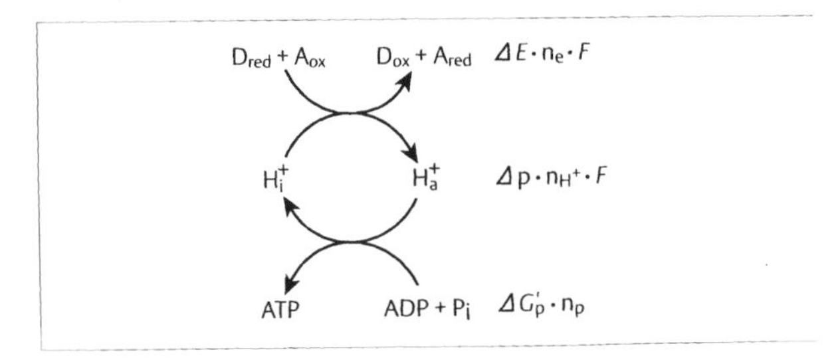

Fig. 4.2 **Forms of free energy in ETP and their amounts.** The difference in redox potentials of the substrates of electron transport (ΔE) and the electrochemical proton potential across the membrane (Δp) are in Volt. n_e, n_{H^+} and n_p designate molar amounts of electrons, of protons, and of ATP, respectively. $\Delta G'_p$ (44 kJ/mol ATP) is the phosphorylation potential

Table 4.2 **The ATP yield and the amount of protons translocated per electron in NADH and succinate respiration**

Donor	Acceptor	$\Delta E'_0$ (V)	Theoretical n_{H^+}/n_e [2]	Theoretical n_p/n_e [1]	Experimental n_{H^+}/n_e [2]	Experimental n_p/n_e [1]
NADH	O_2	−1.14	7.6	2.5	4.5	1.5
Succinate	O_2	−0.79	5.3	1.7	3	1

The theoretical values are calculated according to Equations 4.1 and 4.4, with $\eta = 1$, $\Delta G'_p = 44$ kJ/mol ATP, and $\Delta p = 0.15$ V. The amount of protons to be translocated for ATP synthesis (n_{H^+}/n_p) is assumed to be 3 (see Fig. 4.1). The $\Delta E'_0$ values are calculated from the E'_0 values given in Table 4.1

[1] n_p/n_e, ATP yield (mol ATP/mol electrons)

[2] n_{H^+}/n_e, amount of protons translocated per electron (mol H^+/ mol electrons)

The amount of protons shifted across the membrane per electron transported ($n_{H^+}n_e$) can be calculated from ΔE and Δp according to Equation 4.4, provided that the energetic efficiency η of the process is known:

$$n_{H^+}/n_e = \Delta E\ \eta/\Delta p \qquad (4.4)$$

Equation 4.4 is delineated in Box 4.1. n_{H^+}/n_e is a function of ΔE in Fig. 4.3, with $\eta = 1$. The straight line designates the maximum theoretical values of n_{H^+}/n_e with $\Delta p = 0.15$ V, which is the average proton potential measured with most ETP systems (Box 4.2).

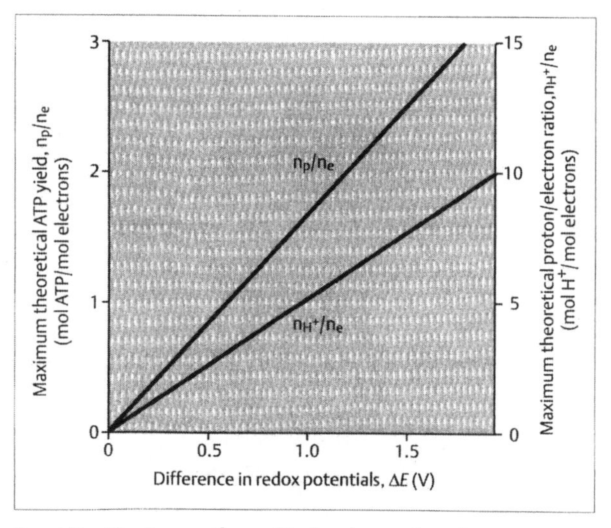

Fig. 4.3 **Maximum theoretical n_p/n_e and n_{H^+}/n_e ratios as a function of the difference in redox potentials between the donor and acceptor substrate in ETP.** The lines are drawn according to Equation 4.1 for n_p/n_e and Equation 4.4 for n_{H^+}/n_e, with the energetic efficiency $\eta = 1$, $\Delta p = 0.15$ V, and $\Delta G'_p =$ 44 kJ/mol ATP

Box 4.2 Proton transport yields an electrical potential across the membrane. The translocation of protons across the membrane creates both an electrical potential and a pH gradient (ΔpH). Therefore, the Δp (also termed proton motive force) is defined as being composed of an electrical term ($\Delta\psi$) and the proton diffusion potential:

$$\Delta p = \Delta\psi + \Delta pH\ RT\ 2.3/F \qquad (4.5)$$

Measurements in the steady state of electron transport indicate that most of Δp consists of $\Delta\psi$, while the proton diffusion potential is almost negligible in most bacteria. This observation can be explained by comparing the electrical capacity and the buffering capacity of the membrane.

The electrical capacity of biological membranes (C) has been determined to be 10^{-6} Ampere·s $V^{-1}\,cm^{-2}$. The electrical capacity is given by

$$\Delta H^+/\Delta\psi = C \cdot O \qquad (4.6)$$

where O ($2.6 \cdot 10^6\ cm^2$/g phospholipid) represents the area of the bilayer formed from 1 g phospholipid. The contribution of the membrane-integrated proteins to the membrane area is neglected. The buffering capacity of phospholipid (B) has been measured as 10^{-4} mol H^+/g phospholipid:

$$\Delta H^+/\Delta pH = B \qquad (4.7)$$

ΔH^+ designates the molar amount of protons added (or removed) to cause the pH change, ΔpH. Combination of Equations 4.6 and 4.7, by substitution of ΔH^+, results in Equation 4.8:

$$\Delta pH = C \cdot O \cdot \Delta\psi/B \qquad (4.8)$$

The ΔpH corresponding to the amount of the protons to be translocated for generating a $\Delta\psi$ of 0.15 V across the membrane is calculated according to Equation 4.8 to be 0.04. This is equivalent to a proton diffusion potential of less than 3 mV. In this calculation, it has to be considered that the electrical equivalent of 1 mol H^+ is equal to the Faraday constant F (96.5 Ampere·s).

Thus, the proton diffusion potential generated by proton translocation is expected to be very small, which is in agreement with the experimental results. If proton translocation were associated with appreciable counterflux of other cations or anion symport, part of $\Delta\psi$ would be converted to ΔpH. However, these processes appear to be much slower than ETP under the usual conditions of bacterial growth.

In Table 4.1, the maximum theoretical n_{H^+}/n_e values of NADH and succinate respiration are calculated according to Equation 4.4 with $\eta = 1$. These values are compared to those measured with mitochondria. The measured values amount to approximately 60% of the theoretical ones. Thus, nearly the same value of η is valid for proton translocation and for phosphorylation driven by the mitochondrial electron transport. This result reflects the general observation that free-energy dissipation in ETP occurs mainly in Δp generation by electron transport, while Δp-driven ATP synthesis operates close to equilibrium and at maximal energetic efficiency.

4.2 All ATP Synthases Operate According to the Same Mechanism

In contrast to the electron transport chains, which are made up of several enzyme complexes, the ATP synthase represents a single protein complex that can be isolated from the membrane. The basic structure and the amino acid sequences of the main subunits are conserved among the enzymes of prokaryotic and eukaryotic organisms. The enzyme is present in aerobic, anaerobic, and phototrophic bacteria and in mitochondria, chloroplasts, and even in bacteria such as *Streptococcus faecalis* that do not perform oxidative phosphorylation. In the latter case, the enzyme is used as a proton-translocating ATPase to maintain the proton potential required for transport processes.

After isolation and incorporation into liposomes, some ATP synthases have been demonstrated to catalyse ATP synthesis from ADP and P_i at the expense of an artificial Δp (Fig. 4.4). In the liposomes, the substrate site of the ATP synthase is exposed to the outside, but in bacteria it faces the cytoplasm. The liposomes contain a buffer at a low pH and are suspended in a medium of high pH. Proton export mediated by the ATP synthase is driven by the ΔpH across the membrane. The proton export is coupled to phosphorylation. Proton flux across the membrane can occur only if accompanied by cation flux in the opposite direction. In the experiment described in Fig. 4.4, the K^+ ionophore, valinomycin, is integrated in the liposomal membrane; K^+ is present in the external medium.

The ATP synthase consists of a hydrophilic domain (F_1), which is exposed to the bacterial cytoplasm, and a membrane-integrated domain (F_O; Fig. 4.5). The two domains are linked by a stalk. The F_1 domain is an approximate sphere (10 nm in diameter) that carries the catalytic substrate sites for ADP, ATP, and P_i. Disruption of the stalk releases the F_1 ATPase from the membrane. F_1 consists of five different polypeptides with the composition $\alpha_3\beta_3\gamma\delta\varepsilon$. The α- and β-subunits form a ring-shaped $\alpha_3\beta_3$ hexamer; in the central opening, two helices of the γ-subunits are situated, forming a somewhat eccentric axis. The hexamer has three catalytic sites that differ in conformation.

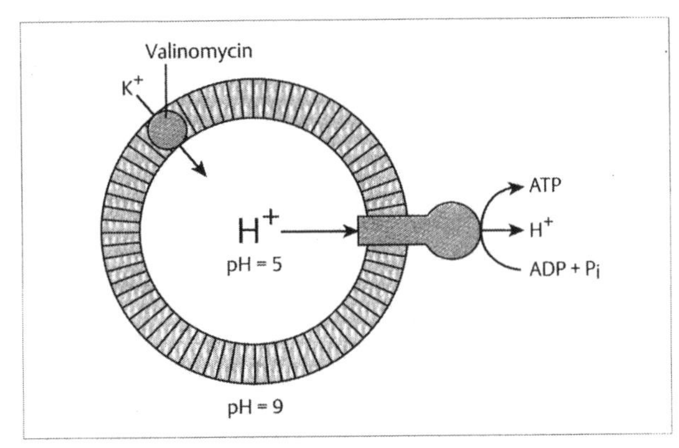

Fig. 4.4 **ATP synthesis catalyzed by ATP synthase incorporated into the membrane of liposomes.** ATP synthesis is driven by the ΔpH across the membrane. Valinomycin catalyzes the import of potassium ions, which is coupled to the export of protons and prevents polarization of the membrane

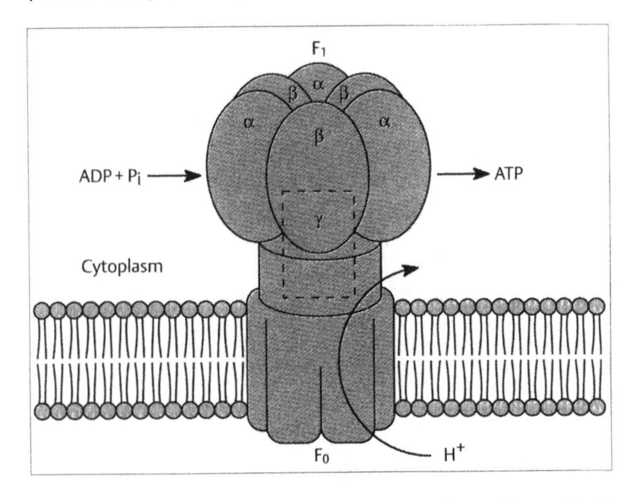

Fig. 4.5 **Schematic view of the ATP synthase.** The catalytic domain (F_1) is made up of five different polypeptides (α–ε). The membrane-integrated domain (F_O) of the *E. coli* enzyme consists of three different subunits (a, b, and c; not shown). F_1 and F_O are linked by subunit γ. Subunits δ and ε are not shown

Each catalytic site is thought to assume one of three conformations, and the transition from one conformation to the other is probably caused by rotation of the γ-subunit relative to the hexamer. The rotation may be driven by the proton translocation that is catalyzed by F_0. This view is in agreement with the hypothesis that ADP and P_i are first loosely bound at the active site, which is subsequently transformed to a conformation characterized by tight binding (Fig. 4.**6**). In this state, ATP is formed spontaneously since water has no access to the site and phosphorylation is exergonic under these conditions. In the third state, ATP is released from the site. The free energy of the phosphorylation reaction is reflected in the binding constants of the individual states. The free energy required for the transitions from one state to the other is provided by proton translocation in the direction of Δp ($H_{a^+} \longrightarrow H_{i^+}$). The conformational coupling mechanism of proton translocation to rotation is not known.

In some bacteria, ATP synthesis is driven by the transport of Na^+ across the membrane. The sodium-translocating ATP synthases that catalyze this reaction are structurally very similar to the proton-translocating ATP synthases. The similarity between the two enzymes has greatly influenced the view on the mechanism of ATP synthesis in ETP. It is now almost settled that protons (or Na^+) have to cross the membrane from the outer to the inner aqueous phase in order to drive phosphorylation. Furthermore, the protons crossing the membrane during phosphorylation appear to be not directly involved in the condensation reaction of P_i with ADP to ATP.

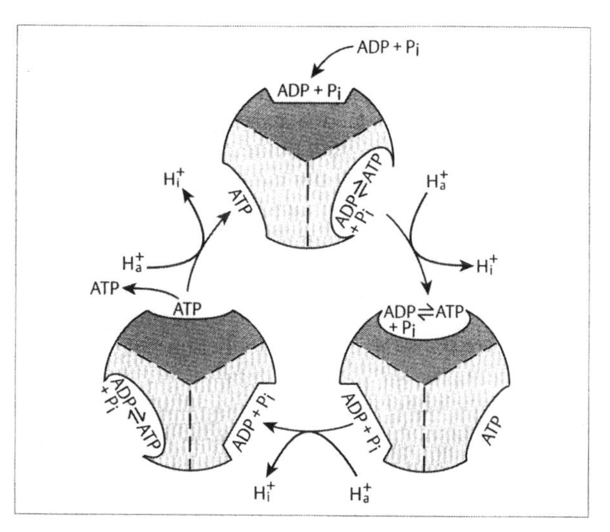

Fig. 4.**6** **Hypothetical mechanism of the ATP synthase.** Each of the three catalytic sites of the enzyme occurs in one of three different conformational states. The reaction cycle only of the upper catalytic site is considered. Proton translocation through F_0 is designated by $H_{a^+} \rightarrow H_{i^+}$

4.3 There Are Many Different Respiratory Chains

The electron transport chains catalyze the oxidation of certain donor substrates (D_{red} in Fig. 4.**1**; e.g., NADH) by certain acceptor substrates (A_{ox} in Fig. 4.**1**; e.g., O_2) and couple these reactions to proton translocation across the bacterial cytoplasmic membrane. The electron-transport chains do not represent protein complexes, but consist of at least two topologically separate enzymes, a dehydrogenase and a reductase (Fig. 4.**7**). These enzymes move independently within the phospholipid bilayer and usually differ in the amounts present in the membrane. The dehydrogenases and the reductases are functionally linked by a respiratory quinone that freely diffuses within the lipid phase of the membrane. The dehydrogenases catalyze the reduction of the quinone by the donor substrates. Reoxidation of the quinol by the acceptor substrates is catalyzed by reductases. The structures of the respiratory quinones are given in Fig. 4.**8**.

When *E. coli* grows aerobically on glucose, NADH and succinate are the donor substrates of the electron-transport chain (Fig. 4.**7**). Therefore, NADH dehydrogenase and succinate dehydrogenase represent the major constituents of the respiratory chain that catalyze the reduction of ubiquinone, the respiratory quinone responsible for aerobic respiration in *Escherichia coli*. The reoxidation of ubiquinol by O_2 is mainly catalyzed by cytochrome-*bo*-quinol oxidase.

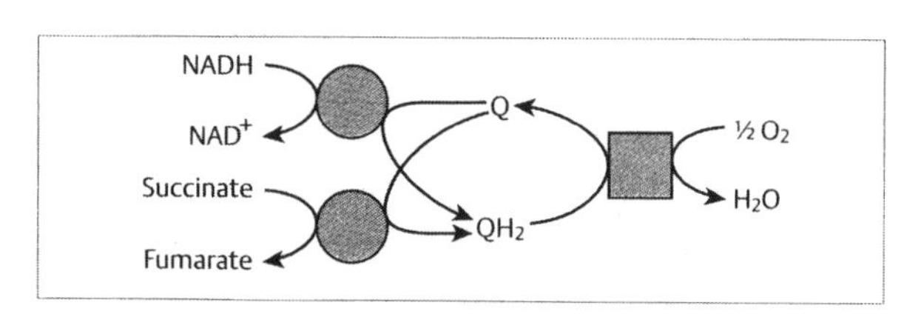

Fig. 4.**7** **Composition of the electron-transport chain of *E. coli* growing aerobically on glucose.** NADH dehydrogenase and succinate dehydrogenase are integrated in the membrane and catalyze the reduction of ubiquinone (Q). QH_2 is reoxidized by a membranous quinol oxidase

Bacteria may contain various dehydrogenases and reductases, dependent on the prevailing metabolic conditions. The enzymes that may constitute the respiratory chains of *E. coli* are given in Fig. **4.9**. The electrons derived from each of the donor substrates can be transported to each of the acceptor substrates by the respiratory quinones. The function of the respiratory quinones within the cytoplasmic membrane is similar to the function of NAD in the cytoplasm (Fig. **4.10**). Both redox mediators collect the reducing equivalents derived from donor substrates and transport them to enzymes that catalyze the reduction of acceptor substrates by the reduced mediators.

The respiratory enzymes may contain proteins harboring heme *a*, *b*, *c* or *d*, the structures of which are given in Fig. **4.11**. These proteins are termed cytochromes. Heme *c* is covalently linked to the apoprotein by two cysteinyl residues. Heme *a* occurs mainly in oxidases. Heme a_3 is a heme *a* that is directly involved in binding O_2 at the catalytic site of an oxidase, together with a copper ion (Chapter 11.)

While quinol oxidation by O_2 is catalyzed by a quinol oxidase in *E. coli*, other bacteria may contain a quinol–cytochrome-*c* oxidoreductase instead of or in addition to the quinol oxidase (Chapter 11). In these

Ubiquinone (UQ)
Menaquinone (MK)
Demethylmenaquinone (DMK)
Methylmenaquinone

Fig. 4.**8** **The structures of respiratory quinones.** The number of isoprenyl units (n) making up the side chains may vary from 6 to 10

bacteria, the reduction of O_2 is catalyzed by a cytochrome oxidase (Fig. **4.12**). The cytochrome *c* that represents the redox mediator between the quinol–cytochrome-*c* oxidodreductase and the oxidase is usually attached to the periplasmic surface of the membrane or, in some cases, may be integrated in the membrane with the heme site exposed to the periplasm. In some bacteria, the oxidation of certain donor

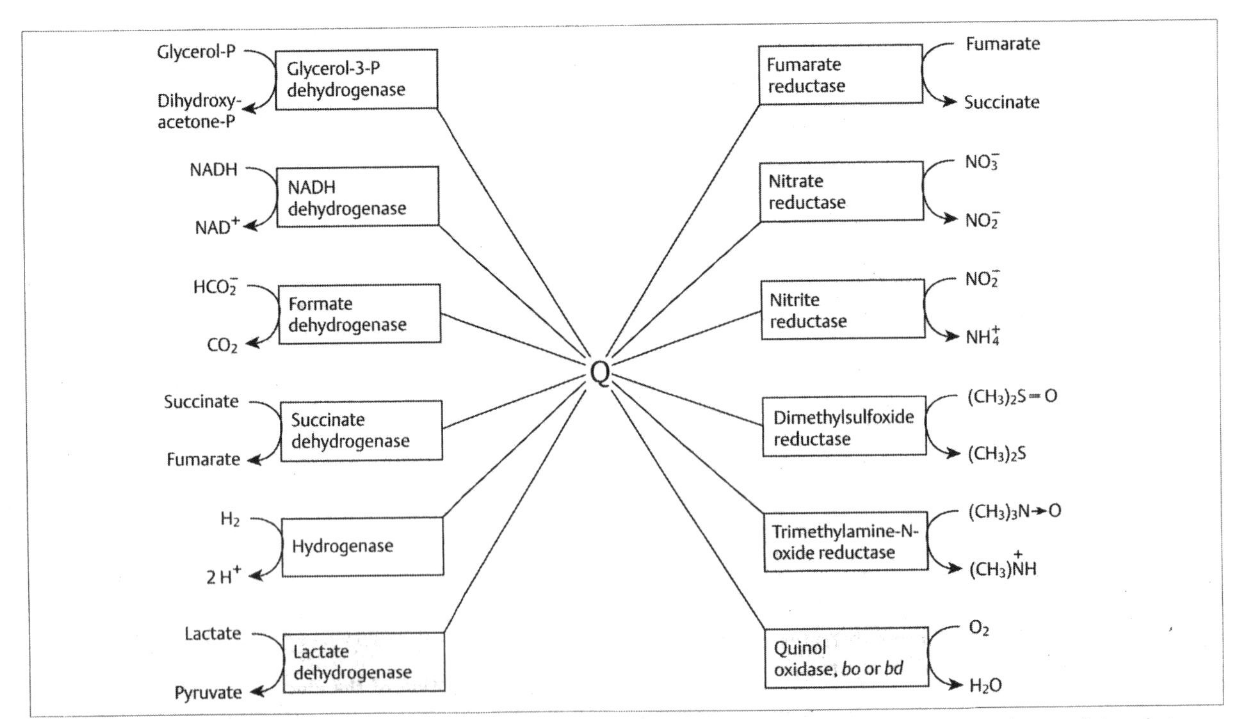

Fig. 4.**9** **Dehydrogenase and reductases (oxidases) that may be components of the electron-transport chains of *E. coli* and other bacteria.** Ubiquinone, menaquinone or demethylmenaquinone (Q) serve as redox mediators between the dehydrogenases and reductases in *E. coli*

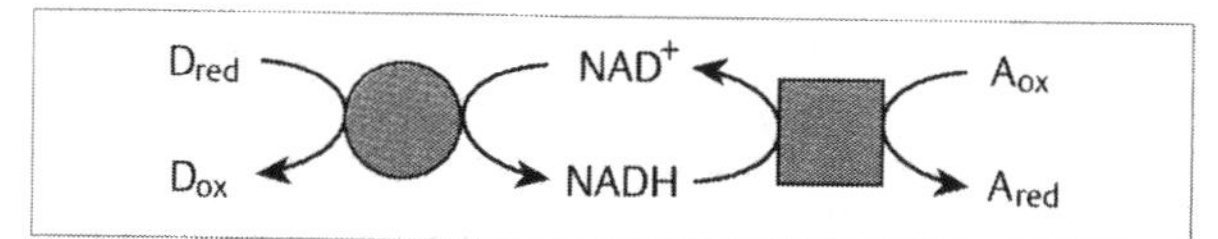

Fig. 4.10 **NAD/NADH as a metabolic intermediate collecting and distributing reducing equivalents in the cytoplasm.** The respiratory quinones serve a similar function within the bacterial membrane. The circle and the square represent dehydrogenases reacting with NAD/NADH. D, donor; A, acceptor

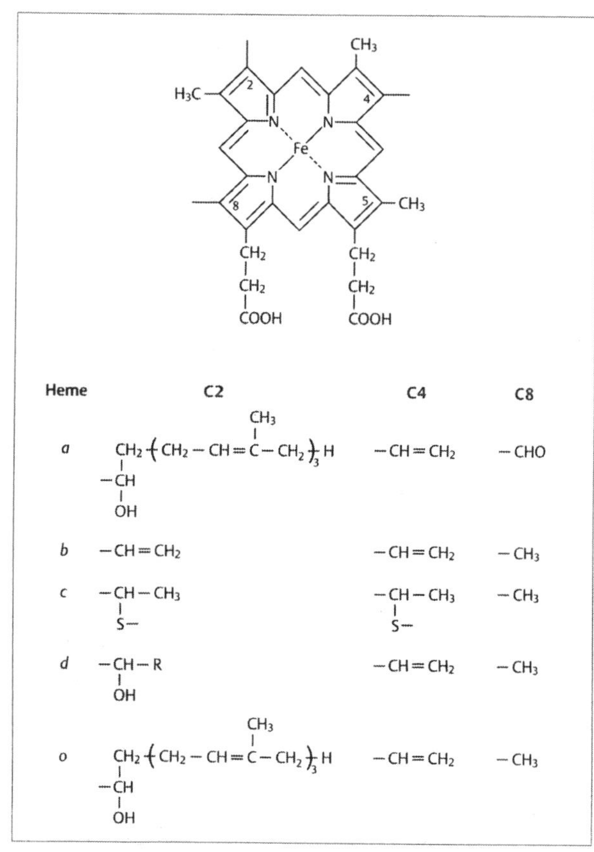

Fig. 4.11 **The structures of the various hemes found in bacterial cytochromes.** Note that heme *d* is an 8,9-dehydroporphyrin; heme *c* is covalently bound to two cysteinyl residues of the apoprotein

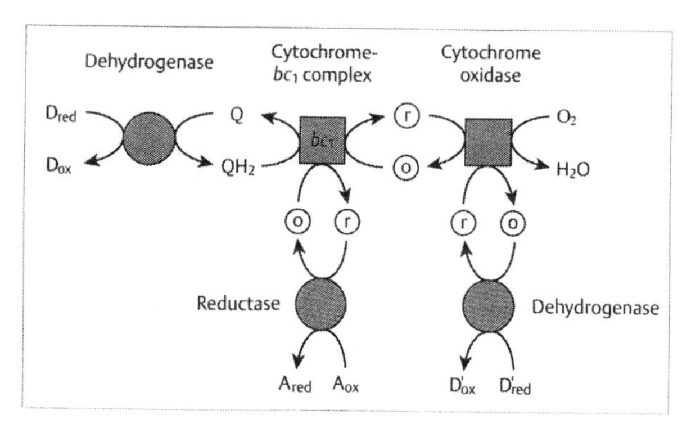

Fig. 4.12 **Composition of the electron-transport chains containing a cytochrome-bc_1 complex.** o/r designates a periplasmic cytochrome *c* or a redox mediator protein serving a similar function. D_{red}/D_{ox} and D'_{red}/D'_{ox} designate donor substrates and A_{ox}/A_{red}, acceptor substrates, Q, ubiquinone or menaquinone; QH_2, reduced Q

substrates (D'_{red} in Fig. 4.**12**; e.g., methanol) by O_2 is not mediated by a respiratory quinone. The corresponding dehydrogenase transfers the reducing equivalents to the periplasmic cytochrome *c* or another mediator, which is then reoxidized by cytochrome oxidase. In the absence of O_2, electrons may be transported to other acceptors (A_{ox} in Fig. 4.**12**; e.g., nitrate). The corresponding reductase may react with the periplasmic cytochrome *c* or another mediator that also accepts electrons from the quinol–cytochrome-*c* oxidoreductase.

The quinol–cytochrome-*c* oxidoreductase is also termed "cytochrome-bc_1 complex" or "complex III." These enzymes are made up of at least three different subunits and contain a 2 Fe–2S center, a heme *c*, and two heme *b* groups. The mechanism of the enzyme is described in Chapter 11.

4.4 There Are Many Different Mechanisms of Coupling Electron Transport to Proton Transport

Electron transport is coupled to proton liberation on the outside of the membrane and to proton consumption on the internal side of the membrane. These processes cause the formation of a Δp across the membrane, in which part of the free energy of the electron transport reaction is conserved. The coupling mechanisms connecting electron transport to proton liberation and consumption are as diverse as the composition of the electron transport chains in the various organisms. Three different principles of coupling mechanism are depicted in Figure 4.**13**.

Figure 4.**13a** refers to electron-transport-driven proton pumps. The electron transfer within certain electron-transport enzymes is directly coupled to proton translocation across the membrane. NADH dehydrogenase and cytochrome oxidase are typical examples of electron-transport-driven proton pumps. The coupling mechanisms of these or other electron-transport enzymes of this type are not known. Cytochrome oxidases (Chapter 11) catalyze the oxidation of reduced cytochrome *c* to oxidized cytochrome *c* ($[Fe^{2+}]_p$ to $[Fe^{3+}]_p$ in Reaction 4.9) by O_2 and the simultaneous translocation of 1 H^+ per electron from the cytoplasmic to the periplasmic side:

$$2\,[Fe^{2+}]_p + \tfrac{1}{2}\,O_2 + 2\,H_c^+ \longrightarrow 2\,[Fe^{3+}]_p + H_2O \qquad (4.9)$$
$$2\,H_c^+ \;\; 2\,H_p^+$$

The indices *p* (periplasmic side) and *c* (cytoplasmic side) refer to the location of cytochrome *c* and H^+. Electrical polarization of the membrane is caused not only by the proton translocation but also by the redox reaction (Reaction 4.9) since cytochrome *c* is oxidized on the external side of the membrane, while the protons consumed in O_2 reduction are taken up from the inside. Thus, per two electrons transported to O_2, four positive charges (two of which represent protons) are generated on the periplasmic side of the membrane and four protons disappear from the cytoplasmic side.

An alternative coupling mechanism is depicted in Figure 4.**13b**, where proton translocation is coupled to the redox reactions of ubiquinone or menaquinone (Q) that are localized in the lipid phase of the membrane. It is assumed that the transfer of electrons to the quinone, which is catalyzed by a dehydrogenase, is coupled to the uptake of protons from the cytoplasm to yield the corresponding quinol (QH_2). The subsequent oxidation of QH_2 to Q is thought to be coupled to proton liberation on the periplasmic side of the membrane. It is likely that the protons are guided by the structure of the enzymes that reduce Q and oxidize QH_2. While the redox reactions of a respiratory quinone may be coupled to proton translocation in some cases, proton uptake and release may occur at the same side of the membrane in others (not shown). In the latter case, the redox reactions of the quinone would not contribute to Δp generation.

In the mechanisms depicted in Figure 4.**13c**, Δp is generated without proton translocation. The substrate site of the dehydrogenase is exposed to the outside, and the substrate site of the reductase faces the cytoplasm. The electrons derived from the donor substrate (DH_2) are transported across the membrane from the substrate site of the dehydrogenase to that of the reductase. The protons released in DH_2 oxidation are liberated at the active site of the dehydrogenase to the outside of the membrane. Reduction of the acceptor substrate (A) by the electrons is associated with proton uptake from the cytoplasm.

In the mechanism of Figure 4.**13c**, the electron transfer from the dehydrogenase to the reductase is usually mediated by a quinone (see Fig. 4.**7**). In other electron-transport chains, the electron transfer from the dehydrogenase to the reductase is mediated by a cytochrome *c* (see Fig. 4.**12**).

The three coupling mechanisms presented in Figure 4.**13** may occur in combination in one electron-transport chain or even in a single electron-transport enzyme. Thus, the mechanism of cytochrome oxidase (Reaction 4.9) may be envisaged as a combination of Figure 4.**13a** and **c**, while mechanisms (a) and (b) apply to the cytochrome-bc_1 complex (Chapter 11).

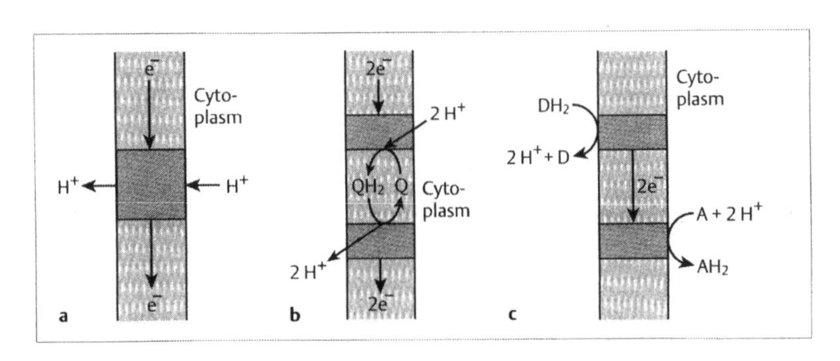

Fig. 4.13a–c **Mechanisms of Δp generation by electron transport.** The shaded boxes represent electron transport enzymes within the membrane. Q, ubiquinone or menaquinone, QH_2, reduced Q; DH_2/D donor substrate; A/AH_2, acceptor substrate

4.5 In Photophosphorylation, Electron Transport and Proton Translocation Are Driven by Light

Phototrophic eubacteria use the free energy of light to drive ATP synthesis from ADP and P_i (see Fig. 3.1). The process is catalyzed by the cytoplasmic membrane of phototrophic bacteria, and its mechanism resembles that of ETP (Fig. 4.14). The electron transport is driven by the energy of light and is coupled to the formation of Δp across the membrane, which drives ATP synthesis in the same way as in ETP. The composition and function of the chain that catalyzes light-driven electron transport is described in Chapter 13. The electron-transport chain involved in photophosphorylation differs from that of ETP mainly in the presence of a photochemical reaction center.

The energetics of photophosphorylation are depicted in Table 4.3. The free energy of a light quantum (E_{LQ}) is a function of the wavelength (λ) of the light:

$$E_{LQ} = h\, c\, N_A\, \lambda^{-1} \quad (4.10)$$

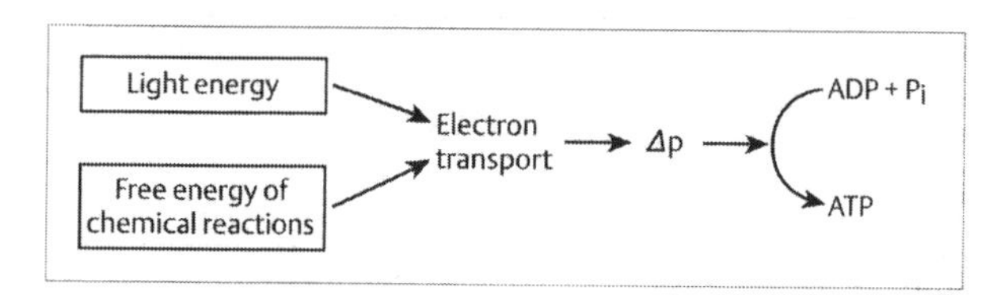

Fig. 4.14 **Photophosphorylation is mechanistically similar to ETP.** A Δp is generated by electron transport in both cases. The Δp is used to drive ATP synthesis by the ATP synthase (see Chapter 4.2). The electron transport reaction is driven by the energy of light in photophosphorylation

Table 4.3 **Free energy of light quanta as a function of the wavelength.**

Wavelength, λ (10^{-9} m)	Color	E_{LQ}[1] (kJ/mol quanta)
400	blue	300
600	red	200
800	near-infrared	150
1000	near-infrared	120

Calculated according to Equation 4.10, where h designates the Planck constant ($6.6 \cdot 10^{-34}$ J·s); c, the speed of light ($3 \cdot 10^{8}$ m·s^{-1}); and N_A, Avogadro's number ($6 \cdot 10^{23}$)
[1] E_{LQ}, free energy of light quanta

It decreases with increasing wavelength. Even the quanta of near-infrared light carry sufficient energy to allow the synthesis of an ATP per quantum at an energetic efficiency η of 0.3–0.4 (see Eqn. 3.3).

Further Reading

Nichols, D.G., and Ferguson, S.J. (1992) Bioenergetics 2. London: Academic Press
Harold, F.M. (1986) The vital force: a study of bioenergetics. New York: Freeman
Anthony, C. (1980) Bacterial energy transduction. London: Academic Press
Abrahams, J.P., Leslie, A.G.W., Lutter, R., and Walker, J.E. (1994) Nature 370: 621–628

5 Multiple Roles of Prokaryotic Cell Membranes

All cells are separated from the surrounding medium by membranes. Cell membranes have thus two basic functions: (1) Membranes separate the internal from the external compartment, thereby creating a separate "system" in which particular reactions can occur. These reactions may not occur otherwise because of, for example, unfavorable pH, inappropriate concentrations of reactants, instability of proteins, or presence of toxic compounds. (2) Membranes mediate the exchange of both matter and information between cells and their environment.

Thus, besides its role as a physical barrier encapsulating the cytoplasm, the plasma membrane incorporates essential metabolic, sensory, and reproductive functions. The variety of functions include the exclusion of toxic compounds or conditions, the maintenance and regulation of turgor pressure, the harboring of membrane-bound enzymes, the exchange and processing of information, energy transduction, cell motility, growth, differentiation and, last but not least, processes related to the uptake of nutrients, export of metabolic end products and building blocks for cell wall synthesis, and disposal of toxic compounds. Actually, the majority of cellular functions are somehow connected to and dependent on transport events. This is reflected by the conservation of many fundamental principles of membrane transport and sensory processes from prokaryotic to human cells.

5.1 Bacterial Membranes Function as Permeability Barriers

The cytoplasmic membrane of bacterial cells consists of a phospholipid bilayer, which functions as a permeability barrier for most solutes. Polar solutes (e.g., carbohydrates) and charged molecules (ions, carboxylic acids, amino acids) have a very low rate of passive flux across lipid bilayer membranes (Table 5.1). In addition to some small solutes and molecules, such as water, ethanol, ammonia, or oxygen, only apolar (hydrophobic) compounds (e.g., phenylalanine, glycerol, or fatty acids) are significantly permeable. Bacteria, however, need to transport solutes at high rates across the cell wall and the cytoplasmic membrane for growth and metabolism. The solute transfer across bacterial membranes is mediated by specific membrane proteins called **transporters, transport systems, carriers** or, in analogy to enzymes, **permeases**. By means of these proteins, the transfer rate across phospholipid membranes can be significantly increased (Table 5.1). The presence and activity of carrier systems in a relatively impermeable membrane is the reason for the observed concentration gradients of solutes across the cell membrane. These range from 10–30-fold (external/internal Na^+ and internal/external K^+) to more than 10000-fold (external/internal free Ca^{2+}), up to a 200000-fold accumulation of some solutes in the cytoplasm (e.g., maltose and particular amino acids).

Since lipid bilayers are permeable to water, the plasma membrane is also responsible for the osmotic behavior of the cell. The concentrations of solutes inside

Table 5.1 **Comparison of diffusion-controlled and carrier-mediated solute fluxes across bacterial plasma membranes**

Transported solute	Typical transfer rate		
	Diffusion-controlled at a concentration difference of		Carrier-mediated (V_{max})
	10 μM	10 mM	
K^+	0.00002	0.02	100
Glutamate	< 0.00005	< 0.05	25
Glucose	0.001	1	50
Isoleucine	0.0015	1.5	30
Phenylalanine	0.08	8	1
Urea	0.04	40	5

The transfer rates [in $\mu mol \cdot min^{-1} \cdot (g\ dry\ mass)^{-1}$] for diffusion-controlled processes are calculated on the basis of the known passive permeability of bacterial plasma membranes, whereas the rates of carrier-mediated processes represent examples of typical transport systems specific for the respective solutes at full saturation (V_{max})

the cell are in general higher than outside; thus, water has a constant tendency to enter the cell. Consequently, without a cell wall, the cell would burst. The difference in osmotic pressure between cytoplasm and medium, cell turgor, is caused by a combination of an active, selectively permeable plasma membrane and a rigid cell wall. A number of particular transport systems maintain the turgor essential for cell function at a relatively constant 1–5 (Gram-negative bacteria) to 15–20 (Gram-positive bacteria) atmospheres.

Integral membrane proteins (called transporters, carrier proteins, or permeases) and transport systems catalyze the transfer of solutes across the permeability barrier of the phospholipid bilayer.

5.2 The Structure and Function of Transport Systems Is Dictated by the Membrane

Solute transport systems (at least the essential components) consist of integral membrane proteins, which can be regarded as membrane-bound enzymes. Instead of catalyzing the conversion of substrate to product, as soluble enzymes do, they mediate the vectorial reaction of solute transfer from one compartment to the other, i.e., from "out to in" or from "in to out". Because of the different metabolic purposes of solute transport, the widely differing nature of transported solutes, and different membranes across which transport occurs, there is a large diversity in transport mechanisms and systems. This diversity is most extreme among the bacteria, most likely because these small unicellular organisms have to cope with frequently and drastically changing solute concentrations, and because bacteria use transport systems for other (e.g., sensory) functions as well. Some systems in fact only occur in prokaryotes, namely, binding-protein-dependent systems and phosphotransferase systems.

The observed enormous diversity in transport systems can be explained by the different functional and physiological requirements which have to be met by those transport systems. Among these are (1) the need for specificity with respect to the structural demands of particular substrates, (2) energetic requirements (e.g., high accumulation ratios) under which a particular solute is transported, and (3) the need for mechanisms to couple various kinds of energy input to

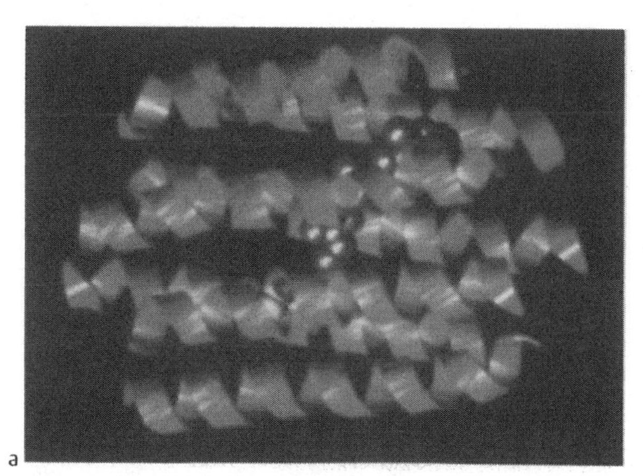

a

b

Fig. 5.1 **Schematic drawing of the three-dimensional structure of bacteriorhodopsin from *Halobacterium halobium*** as elucidated by electron diffraction as a side view **a** showing the membrane-embedded helices or as a top view **b** looking onto the plane of the membrane in the direction of the protein α-helix axes. The protein consists of seven transmembrane helices and bears a retinal molecule (shown in purple), which is responsible for the light absorption. The retinal is covalently bound to a lysine residue of the protein (yellow) via a Schiff base (blue). The protons to be translocated are accepted via an entrance channel (upper part), bound to the central retinal moiety, and finally released through an exit channel (lower part). Courtesy of J. Granzin; the structural data [from 1] obtained from the Protein Data Bank and the software "O" [from 2] were used.

the solute translocation. The latter requirement is frequently met with the help of additional domains (e.g., in binding protein-dependent or in phosphotransferase systems; see Chapter 5.6).

Only in 1980, the first primary structure of a transporter protein, the lactose permease of *Escherichia coli* was determined. At present, the amino acid sequences of a large number of prokaryotic and eukaryotic transporters are available, and in some cases, a model of their **transmembrane topology** is also available. Not a single transporter has been crystallized so far (due to their hydrophobic nature), and thus a truly three-dimensional structure has not yet been obtained. In order to give an impression of what a membrane-inserted transporter would probably look like, the three-dimensional structure of a primordial transporter, the light-driven proton pump bacteriorhodopsin, is shown in Figure 5.1. The basic arrangement of an ordered set of membrane-spanning α-helices around a substrate pathway is probably not far from the correct structure of a "typical" solute transporter, although the number of transmembrane segments differs from that of a typical transporter (see below).

Most, if not all, of these transporters, whether of prokaryotic or eukaryotic origin, seem to have a common structural design. Irrespective of the wide variety in function and irrespective of whether the membranous part of the transporter consists of a dimer (e.g., some binding protein-dependent systems and the Enzyme II of phosphotransferase systems), a monomer (e.g., secondary transporters), or even only a domain within a complex protein (e.g., some ATPases), the majority of protein structures responsible for membrane transport seem to be constructed of an array of approximately 12 (6 + 6) transmembrane segments (Fig. 5.2). This structural paradigm is best conserved in the large class of secondary transporters, irrespective of whether they function in a uniport, symport, or antiport process (see below). The finding of a common structural motif leads to two conclusions: (1) the common design of transporters is possibly related to a common evolutionary origin and (2) a general structural concept may be correlated to a general principle of function.

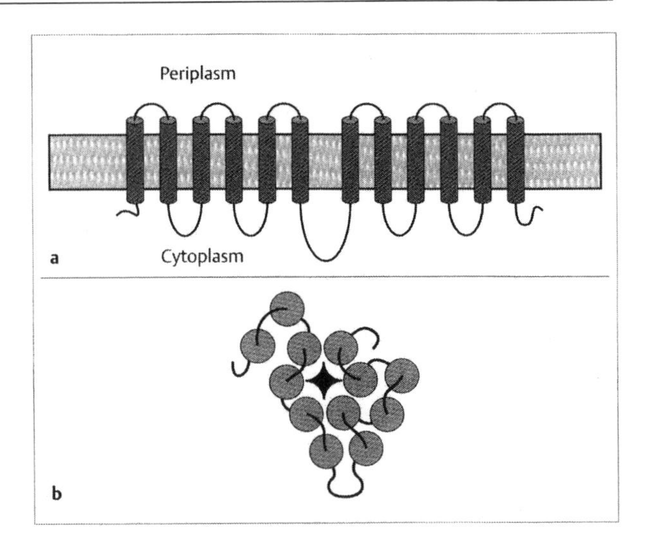

Fig. 5.2 **The common structural motif of transport proteins of the 12-transmembrane-helix family. a** The 12 transmembrane segments are connected by hydrophilic loops, which are in general larger on the cytoplasmic side. Frequently, the central loop is especially extended. This most likely indicates an ancient fusion between two originally separated monomers, thus emphasizing the "6+6 helix" motif of membrane transporters. **b** Model for a possible topological arrangement of a 12-transmembrane-segment transporter as viewed in the plane of the membrane. The star indicates a hypothetical solute pathway through the protein moiety

5.3 Formal Concepts of Transporter Function

A useful model describing the basic function of a transporter is shown in Figure 5.3. Although transport catalysis is not fully understood, such models are helpful in rationalizing the molecular events. The "essentials", i.e., the basic steps (binding, translocation, and release), and the occurrence of two different basic states (conformation "e" and "i") are properly described by the model. The **translocation** step as the central reaction involves a major conformational change of the transporter protein, thus representing the "switch" in the transport reaction. The simplest rationale for this conformational change is a model assuming an alternating (i.e., orientation toward one side of the membrane or the other) access of the solute to the binding site of the protein. An alternative model, the rotational movement of the transporter in the membrane, resulting in the exposure of the same binding site on either side, has been excluded for energetic reasons.

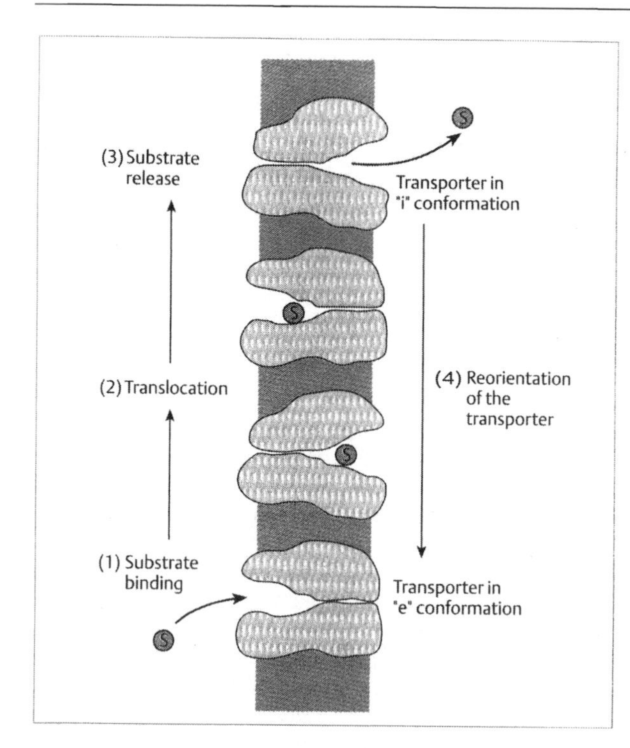

Fig. 5.3 **Model of conformational changes related to the transport reaction of a transporter.** The basic reactions are (1) binding of the substrate (S) to the transporter, which is in conformational state "e" (for "external"), on one side of the membrane, (2) translocation of the substrate involving major conformational changes of the carrier protein, which thereby switches to conformation "i" (for "internal"), and (3) release of the substrate on the opposite side of the membrane. (4) The transporter then switches back to state "e", depending on the type of mechanism (e.g., uniport or antiport), in the loaded or unloaded state

The function of a transporter is similar to that of an enzyme. The interaction with the transported solute comprises: (1) binding, (2) translocation (i.e., the essential "switch" of the transport reaction), and (3) release of the solute.

The function of a uniport system can be described by the simple kinetic scheme shown in Figure 5.**4**. Each form of the transporter represents an intermediate within the catalytic cycle of transport. The transporter may either exist as a free form (C) or as a form loaded with substrate (CS); furthermore, the binding site of the transporter may either face the "external" (C_e) or the "internal" space (C_i). Although such kinetic models are

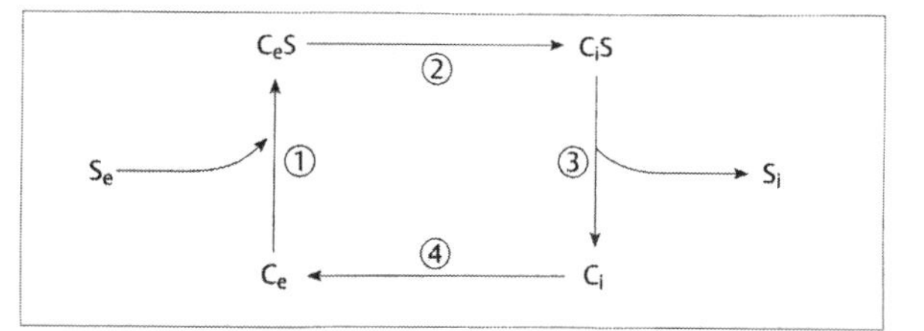

Fig. 5.**4** **Kinetic scheme of a uniporter.** The substrate (S) is bound on the external side to the unloaded carrier in the "e" conformation (C_e). The substrate-loaded form (C_eS) reorients to the "i" conformation (C_iS; translocation step). The substrate is released (S_i) into the internal compartment, and the unloaded carrier (C_i) reorients back to the "e" conformation

very useful for understanding transport mechanisms, one should keep in mind that the "movement" of a carrier (reactions 2 and 4 in Fig. 5.**4**) is only a formal one and actually indicates the conformational changes of the transporter protein during transport catalysis, as explained above.

A more detailed example of this type of kinetic interpretation is represented by the paradigm of a secondary transporter, the lactose/H^+ symporter (or LacY) from *Escherichia coli* (Fig. 5.**5**; see also Chapter 5.6.2). The so-called lactose permease transports lactose (L) or other β-galactopyranosides and protons in a symport mode from the outside (e, external) to the inside (i, internal). The transporter exists in different states with respect to the number and type of bound

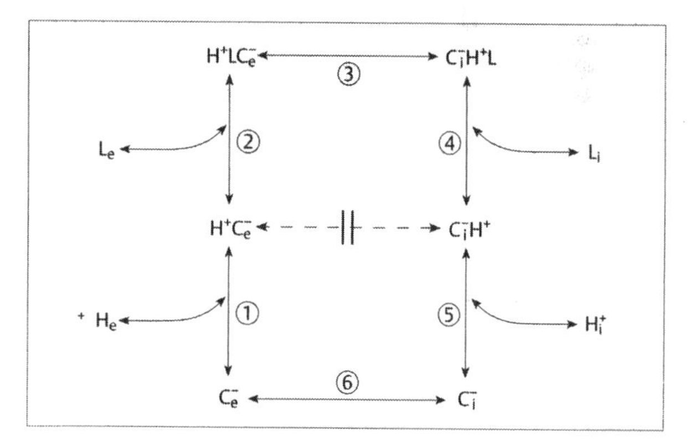

Fig. 5.**5** **Kinetic scheme of the lactose/H^+ symporter (lactose permease, LacY) in *E. coli*.** A minimum number of catalytic steps, labeled with consecutive numbers, is shown. The broken line indicates an energetically or kinetically "forbidden" transition. The left side represents the binding steps (L, lactose) on the outside (C_e, transporter in the "e" conformation), whereas the right side shows the catalytic steps with the inward-facing binding site (C_i, transporter in the "i" conformation)

ligands (C, CL, HC, HLC), and the conformational state of the protein (C_e and C_i). Solid arrows represent kinetic transitions between different states; the broken line indicates a transition that, although formally possible, does not occur with significant probability. The complete set of transitions and states represent a kinetically feasible description of the transport cycle, which can be used for mathematical analysis. It should be kept in mind, however, that a change in transporter conformation (C_e/C_i) does not mean a movement of the transporter protein across the membrane, as exemplified in Figure 5.**3**.

5.4 Studying Kinetics of Transport Is Useful for Identification and Characterization of Transport Processes

Whether a diffusion-controlled process or a carrier-mediated transport is involved in solute transfer across a phospholipid membrane can be determined by observing the transport rate of the substrate (Fig. 5.**6**). The rate of diffusion linearity increases with the increase in concentration of the molecule to be transported, whereas carrier-catalyzed transport reaches a maximum value, i.e., the transporter becomes saturated with substrate. There is a perfect analogy to enzyme kinetics, in that Figure 5.**6a** and **b** also describe the difference between a chemical reaction and an enzyme-catalyzed process with respect to their dependence on substrate concentration. The treatment of transport kinetics with the same formalism as established for enzyme kinetics is very useful. In fact, a transporter can be regarded as an enzyme, the reaction of which is not a change in the structure of the substrate, but of its location. Also, graphical analysis of the experimental data similar to that used in enzyme kinetics can be performed. In particular, this is true for inhibition kinetics. The well-known plots such as velocity versus substrate concentration (Michaelis–Menten diagram, Fig. 5.**6b**) and reciprocal plots (Lineweaver–Burk diagram, Fig. 5.**6d**) can be used to identify the type of inhibition (e.g., competitive, noncompetitive).

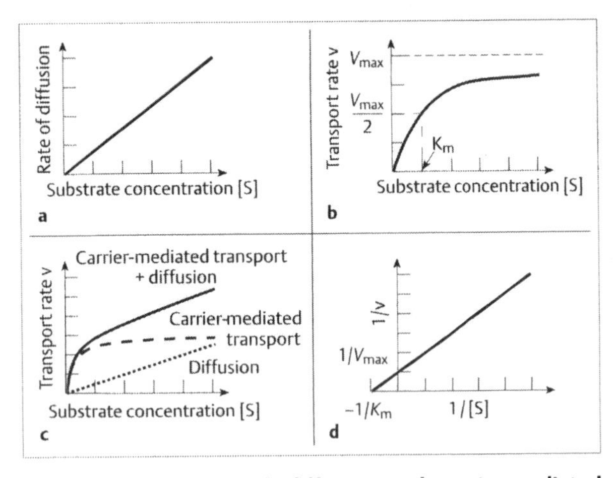

Fig. 5.**6a–d** **Kinetics of diffusion and carrier-mediated transport.** **a** Dependence of the solute transfer rate on substrate concentration in facilitated diffusion, **b** dependence of the solute transfer rate on substrate concentration in a typical carrier-mediated transport (Michaelis–Menten diagram), **c** Transport rate in the presence of both a diffusion component and a carrier-mediated transport, **d** reciprocal plot (Lineweaver–Burk plot) of the data of panel **b**

A solute can cross the permeability barrier of the membrane by the most simple process, **diffusion**, i.e. the movement of a solute without direct interaction with membrane proteins. The rate of unidirectional transfer (v_1) from one compartment (I) to the other (II) is directly proportional to the concentration of the solute (S) in compartment I (cf. Fig. 5.**6a**).

$$v_1 = PA[S]_I \tag{5.1}$$

P is the permeability coefficient, and A is the area of the membrane. A corresponding equation holds for the reverse flux from side II to side I. The net flux of a solute (S) from side I to side II is then described by

$$V = PA([S]_I - [S]_{II}) \tag{5.2}$$

Net transport by diffusion occurs only from the compartment of higher concentration to that of lower concentration and is not specific. For most (hydrophilic) solutes, passive diffusion across the cytoplasmic membrane is low and not of physiological relevance in prokaryotes.

> Passive, diffusion-controlled transmembrane solute movement depends on the concentration difference of the respective solute across the membrane. In contrast, carrier-mediated transport is characterized by: (1) saturation kinetics, (2) specificity, and (3) susceptibility to inhibition.

Carrier-mediated transport involves the participation of a proteinaceous component of the membrane, i.e., the transporter. It is not always that simple to prove unequivocally the involvement of a transport protein in a transport process. The major arguments indicating the presence of a transporter are mainly of a kinetic type. These are, similar to enzymes, as follows: (1) saturation kinetics, (2) substrate specificity, (3) inhibition by specific agents, and (4) the particular kinetic observation of counterflow of substrate molecules between the two compartments.

For a basic discrimination between carrier-mediated transport and diffusion, it is convenient to start with secondary uniport, also called **facilitated diffusion** (see Chapter 5.6). Facilitated diffusion is a passive process, but since it is carrier-mediated, the net flux is, in general, not simply proportional to the difference in solute concentration "inside" and "outside". Many transport systems, not only those of facilitated diffusion, can be described by Michaelis–Menten kinetics (Fig. 5.**6b**).

$$v = V_{max} \cdot [S]/(K_m + [S]) \qquad (5.3)$$

K_m, used by analogy with enzyme kinetics, means the solute concentration at which the transport rate v reaches half its maximum (V_{max}), and is also called K_t (for **t**ransport). Kinetic analysis often helps to find out whether more than one process is involved in the transport of a given solute. This is exemplified for a combination of carrier-mediated and diffusion-controlled transport in Fig. 5.**6c**.

5.5 Energetics of Carrier-mediated Transport: The Concept of Coupling

A process of diffusion (or facilitated diffusion) does not need an input of metabolic energy in order to take place since it occurs with a negative change in free energy. This means that it is not coupled to an exergonic reaction, but is merely driven by the concentration difference of the solute across the membrane. In biological systems, however, the transport reaction frequently leads to the accumulation of a given solute in the cytoplasm. This "uphill" transport, as an endergonic reaction, is characterized by a positive ΔG, which means that this reaction will never occur spontaneously. In order to take place, this reaction must be "coupled" to another reaction with a sufficiently negative ΔG, so that the overall change in free energy of the coupled reactions is negative. Figure 5.**7a** elucidates the basic principle of **coupling** a "driving" (exergonic) to a "driven" (endergonic) reaction. The endergonic reaction on the right side will only take place if it is coupled to the exergonic reaction on the left. Compounds that disrupt this interrelation are therefore called "uncouplers". The system will work from left to right only if the negative value of ΔG of the exergonic reaction is greater than the positive one of the endergonic reaction; otherwise, it will work in the opposite direction. If the positive ΔG of the reaction on the right side increases (e.g., by accumulation of product D) to a value balancing that of the reaction on the left side, the exergonic reaction will be stopped by the coupling mechanism. Many different concepts have been suggested to explain these principles in terms of transport coupling. Currently, only two main mechanisms are known: chemical coupling (Fig. 5.**7b**, primary

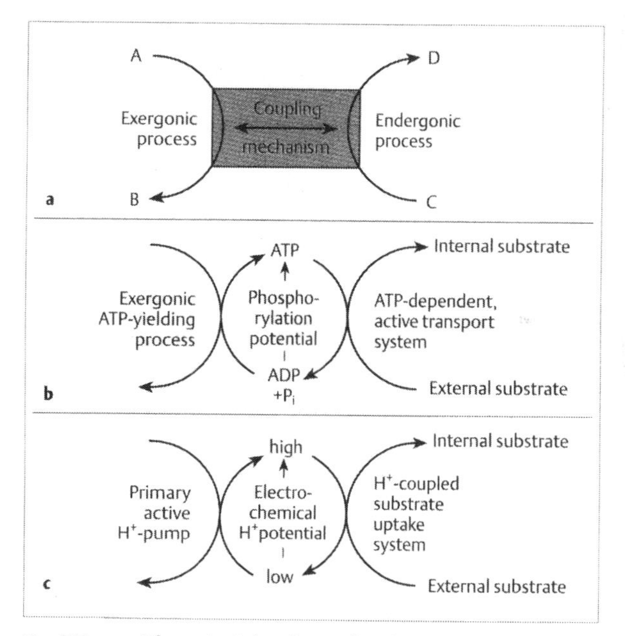

Fig. 5.**7a–c** **The principle of coupling in transport systems.** **a** An exergonic process ("driving" or "downhill" reaction) is mechanistically linked to an endergonic process ("driven" or "uphill" reaction) by a particular coupling mechanism. Panels **b** and **c** represent the two basic mechanisms of coupling. **b** In chemical coupling, the exergonic (driving) reaction provides a compound with high free energy (in general ATP or phospho-*enol* pyruvate), which is then used by the endergonic (driven) reaction. **c** In chemiosmotic coupling, the electrochemical energy of an ion gradient, provided by the action of a primary pump, is coupled to the flux of the driven solute, thus providing energy for its uphill flux.

transport) and coupling by ion currents (Fig. 5.**7c**, secondary transport). Consequently, these two mechanisms are associated with two different types of general "currency" of energy: ATP, phospho*enol*pyruvate, and a few other high-energy compounds on the one hand (chemical coupling), and electrochemical ion potentials on the other (coupling by ion currents).

The principle of coupling in transport reactions means the mechanistic connection between a driving (exergonic) and a driven (endergonic) reaction in such a way that one reaction cannot take place without the other.

5.6 There Are Many Different Transport Mechanisms in Prokaryotes

Transport processes can be classified according to different aspects, such as the structure of the transporter protein, the solute transported, or kinetic and energetic criteria. About 25 years ago, Peter Mitchell introduced the now widely accepted concept based on the utilization of different energy sources for transport. Transport processes can be divided into four classes (Fig. 5.**8**):

I. Some solutes pass the permeability barrier of the membrane by **diffusion**. The diffusion-controlled movement of solutes across otherwise impermeable membranes can be mediated by channel-type proteins (e.g., porins) in the outer membrane of Gram-negative bacteria (Chapter 5.6.1; Fig. 5.**8**, a, b).
II. The only driving force in **secondary transport** is the electrochemical potential of a given solute (Chapter 5.6.2). This energy can be utilized to drive the "uphill" transport of another solute, i.e. against its own concentration gradient. This is achieved either by cotransport (symport, Fig. 5.**8**, d) or by countertransport (antiport, Fig. 5.**8**, e) of the driving and the driven solute. Because of the similarity in mechanism and in the dependence on the presence of a solute gradient, carrier-mediated unidirectional transport driven simply by its own electrochemical gradient is also included in this class of mechanisms (uniport, also called facilitated diffusion, Fig. 5.**8**, c).
III. In **primary transport** the solute translocation is directly coupled to a chemical or photochemical reaction. Primary transport systems thus directly convert light or chemical energy into electrochemical energy, i.e., the electrochemical potential of a given solute (Chapter 5.6.3; Fig. 5.**8**, f).
IV. **Group translocation**, catalyzed by the phospho*enol*pyruvate: carbohydrate phosphotransferase systems (PTSs) in bacteria, differs from all other mechanisms in the modification (phosphorylation) of the transported solute during transport (Chapter 5.6.4; Fig. 5.**8**, g).

The transport systems of class III and IV and the symport and antiport systems of class II may lead to "uphill" transport of a given solute against its electro-

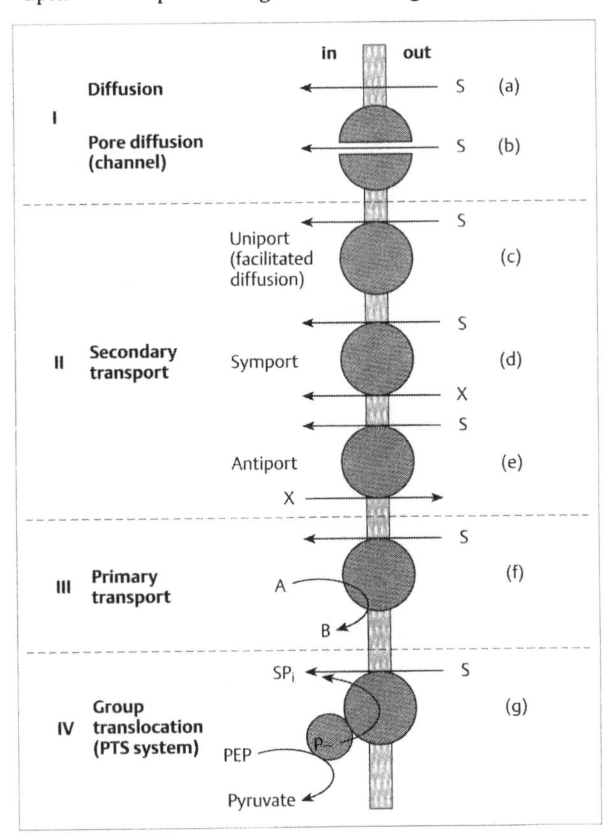

Fig. 5.**8** **I–IV** **Fundamental mechanisms of solute transfer across biological membranes.** Solute molecules (S) may cross the phospholipid membrane of bacteria by facilitated diffusion **I**(a) or channel-type mechanisms **I**(b) on the one hand, or by carrier-mediated processes on the other. The latter processes can be classified, due to the mechanism of energy utilization and the product as secondary transport processes **II**, primary transport **III** or group translocation **IV**. PEP, phospho*enol*-pyruvate; X, a second solute; A, energy source; B, energy product.

chemical gradient. To function properly, these systems need the input of metabolic energy and are thus also called primary or secondary **active** transport systems.

Transport processes are divided into four classes according to the energy sources used to drive the transport reaction: (facilitated) diffusion, secondary transport, primary transport, and group translocation.

5.6.1 Transport by Diffusion and Through Bacterial Porins

Besides gases such as O_2, CO_2, NH_3, and some small molecules such as water or ethanol, also hydrophobic (lipid-soluble) molecules can cross the membrane more or less freely (Fig. 5.**8**, a). This holds for aliphatic (e.g., butanol) or aromatic molecules (e.g., benzene). A number of solutes, (i.e., weak acids or bases), which are membrane-permeable in one state of protonation (in the uncharged form), are relatively impermeable in the other (charged) form. Typical examples are acetic acid (permeable if protonized) or ammonium (permeable if deprotonized).

These processes should not require proteins for facilitating permeation. However, even for solutes which in principle cross the membrane relatively freely, transport systems are frequently provided by the cell since, in the presence of a weak solute gradient, diffusion-mediated fluxes may prove insufficient for physiological processes (e.g., fast growth). A typical example is glycerol, which despite its high membrane permeability is taken up in *E. coli* and other bacteria by specific **facilitator** systems (see Chapter 5.6.2). Other examples are ammonium ions, aromatic amino acids, or fatty acids. In many cells, even channels specific for water molecules have been found.

Unlike the plasma membrane, the outer membrane of Gram-negative bacteria is relatively permeable to small solutes. This membrane basically consists of a lipid bilayer, asymmetrically constructed of lipopolysaccharides and phospholipids. It is thus, in principle, a barrier for hydrophilic molecules such as carbohydrates, amino acids, or simple ions, as well as for larger molecules such as proteins. Transport of these molecules across the outer membrane is achieved by channel-type proteins (Fig. 5.**8I**,b), the **porins**, and not by transporters. Channel-type proteins in the outer membrane can be divided into three classes: (1) non-specific channels, the so-called general porins, (2) more or less specific porins, and (3) active transport systems related to porins. The last class comprises uptake systems for iron chelators and vitamin B_{12}. They are energized by conformational coupling to the plasma membrane through protein components spanning the periplasm (the Tol/Ton system). Their mechanism, although poorly understood, differs from that of the porins.

Porins are channel-type proteins which mediate the passive transfer of solutes across the outer membrane of Gram-negative bacteria.

Members of classes (1) and (2) can be regarded as water-filled pores. Their solute specificity is due in some cases to the presence of ligand-binding sites in the channel. Thus, they also show some properties of transporters (e.g., substrate specificity and saturation kinetics) because of the saturability of the internal binding site. A typical example of a specific porin is the LamB protein in *E. coli*, an outer-membrane protein that increases the flux of maltose to the periplasm, where maltose is accepted by its specific binding protein-dependent uptake system (see Fig. 5.**15**). Further specific channels have been found for sucrose and for nucleosides. Examples of non-specific porins are the proteins encoded by the *ompF*, *ompC*, and *phoE* genes in *E. coli*. These porins are very abundant in the outer membrane; in an *E. coli* cell, for example, there are about 10^5 copies.

The three-dimensional structure of porins from *Rhodobacter capsulatus* and *E. coli* have been elucidated by X-ray crystallography (Fig. 5.**9**). The porins consist of trimeric complexes of identical subunits with molecular masses of approximately 35 kDa. The channel diameter of the general porins is roughly 1 nm and allows more or less free passage for molecules up to a mass of about 600 Da. In general, porins seem to form a barrel of antiparallel β-sheets, and thus they also differ from the common structure of transport proteins in this respect.

5.6.2 Secondary Transport Systems

Secondary transport systems (Fig. 5.**8II**) use pre-existing gradients, which are established by the action of primary systems, as driving forces. This is exemplified in Figure 5.**10** for the lactose/H^+ symporter (or LacY) of *E. coli*, which is discussed in more detail below. A primary active transport system, the respiratory chain in this case, pumps ions (in general, H^+ or Na^+) across the membrane, thus establishing an electrochemical ion potential. This potential is then used for driving the

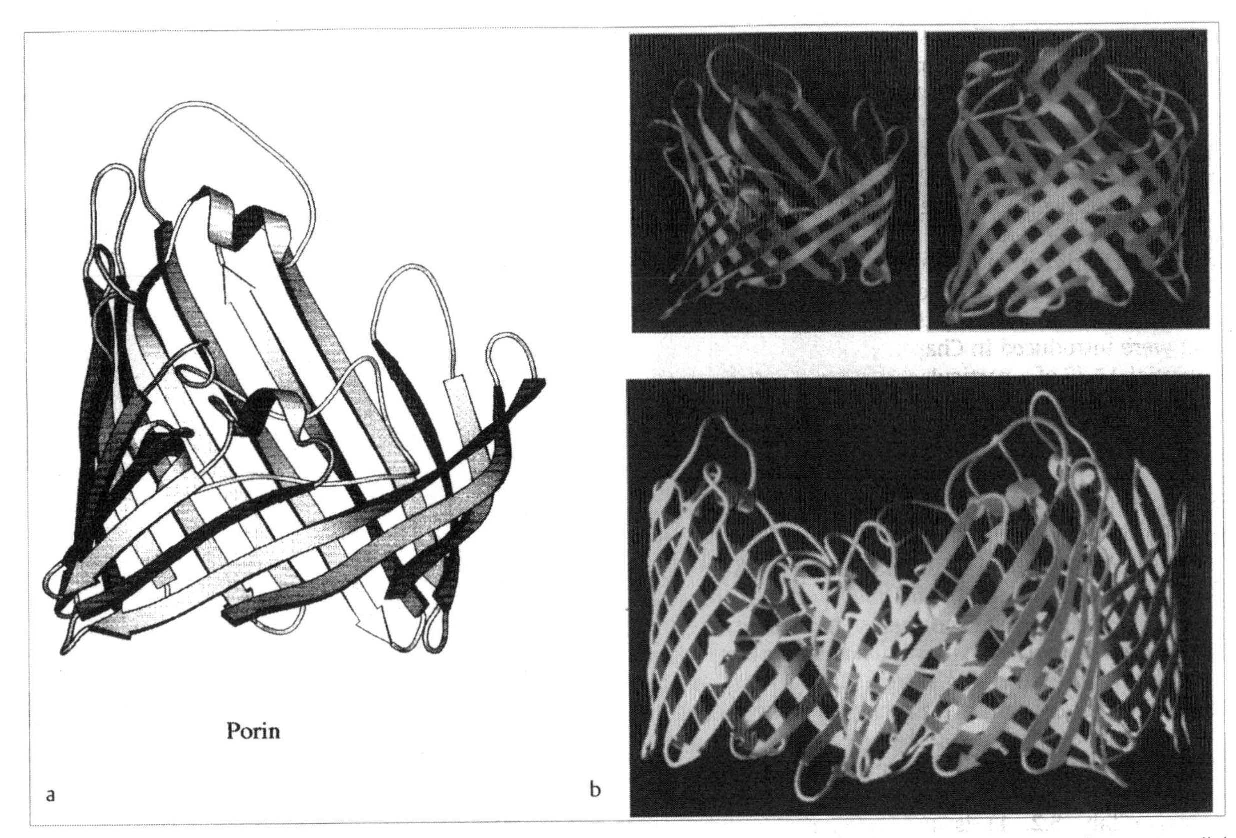

Fig. 5.9 **Three-dimensional structure of the outer membrane porin from *Rhodobacter capsulatus* as elucidated by X-ray diffraction.** Panel **a** shows one porin monomer; the functional trimer is given in panel **b**. Porins of the outer membrane of Gram-negative bacteria consist of 16 antiparallel β-sheet elements organized in a barrel-type structure. Three combined barrels form a functional porin channel. Courtesy of W. Welte

uptake of transport substrate (lactose) into the cell against its concentration gradient. The observation of accumulative uptake has led to the term "secondary active transport". The sum of fluxes, however, must be in a "downhill" direction because the overall change in free energy must be negative. This is essentially the principle of energetic coupling (see Chapter 5.5).

Secondary transport mechanisms are divided into uniport, symport, and antiport mechanisms, according to the number of solute molecules involved and the way of coupling of solute fluxes. The driving force in secondary transport is the sum of the electrochemical potentials of the solute molecules involved in the transport reaction.

In **uniport** mechanisms (Fig. 5.**8**, c), the permeation of a single solute is facilitated. The transport is driven merely by the concentration difference of the solute across the membrane, provided that the solute does not carry a net charge. In **symport** mechanisms (Fig. 5.**8**, d), two solutes (S and X) are transported simultaneously in

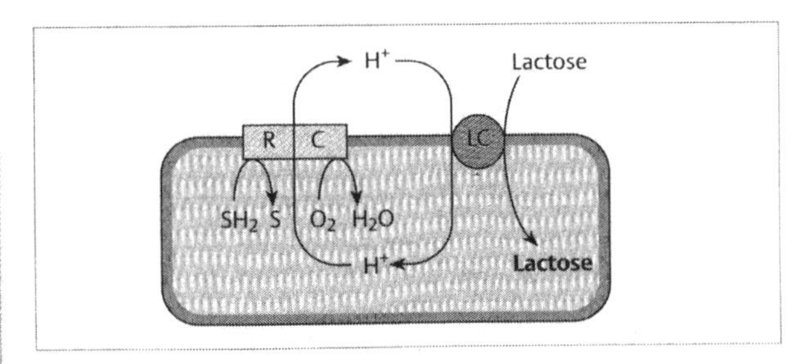

Fig. 5.10 **Chemiosmotic coupling of secondary transport systems in a bacterial cell.** The respiratory chain (RC, a primary pump) extrudes protons, thus creating an electrochemical proton gradient (proton motive force). This is then used by the secondary lactose permease (LC) for driving the cytoplasmic accumulation of lactose. S, substrate

the same direction. Two solutes are transported in opposite directions in the case of **antiport** mechanisms (Fig. 5.**8**, e). The principle of secondary transport is to couple the two fluxes [i.e., of the driving ("downhill") and the driven ("uphill") substrate] so that one flux cannot take place without the other. The coupling of a set of various transport systems by gradients across the bacterial plasma membrane is exemplified in Table 5.**2**, where the driving forces of the respective transport processes also are presented. The driving forces are quantitatively described by the chemiosmotic principles that were introduced in Chapter 4. The electrochemical potential $\Delta\tilde{\mu}_S/F$ of a particular ion S (present in the two compartments I and II) is given by

$$\Delta\tilde{\mu}_S/F = n\Delta\psi + RT/F \cdot \ln[S_{II}]/[S_I] \qquad (5.4)$$

When expressed for protons ($S = H^+$, $n = 1$), in general, Mitchell's term **"proton motive force" (Δp)** is used. At 25 °C, Δp becomes (in millivolts)

$$\Delta p = \Delta\psi - 59\ \Delta pH \qquad (5.5)$$

The energy stored in the electrochemical potential of protons or other coupling ions, such as Na^+, may be used in turn by secondary transport systems to drive the vectorial movement of a solute against its concentration gradient (Tab. 5.**2**).

The driving force of an electroneutral uniport process (Tab. 5.**2**, 1) is provided by the diffusion potential of the solute S^0, which may be calculated (in volts) from the internal $[S^0]_{in}$ and external $[S^0]_{ex}$ solute concentrations according to the Nernst equation. If the solute in a uniport process is charged, the electrical potential (membrane potential $\Delta\psi$) adds to the driving force (Tab. 5.**2**, 2). In this case, net transport vanishes when the diffusion potential balances with $\Delta\psi$. Under these conditions, a calculation of the steady-state concentration gradient becomes possible. At a membrane potential of 120 mV (positive outside), the internal concentration of S^+ (Tab. 5.2, 2) would be 100 times the external concentration. When uncharged lactose (S^0, Tab. 5.**2**, 3) is transported in symport with a proton, a charge is transferred across the membrane. In this electrogenic (i.e., charge-moving) transport process, the membrane potential ($\Delta\psi$), the diffusion potential of S^0, and the proton diffusion potential ($Z \cdot \Delta pH$) therefore have to be added. Na^+-linked proline uptake (Tab. 5.**2**, 4) differs from the former process (Tab. 5.**2**, 3), in that the diffusion potential of protons is substituted for that of Na^+. Since transport of two Na^+ is coupled to uptake of glutamate by the GltS system of *E. coli*, twice the sodium diffusion potential adds to the driving force (Tab. 5.**2**, 5). The contribution of the proton diffusion potential is threefold when citrate is transported in symport with three protons (Tab. 5.**2**, 6). As a net negative charge is transported in the malate/proton symport (Tab. 5.**2**, 7), the negative value of $\Delta\psi$ adds to the driving force. The driving forces of the two antiport processes (Tab. 5.**2**, 8 and 9) are obtained in the same way as those of the uniport and symport mechanisms. However, the direction of transport of the individual solutes has to be considered.

Uniport systems are common in eukaryotes (e.g., glucose transport in various cells including yeasts), but are rare in bacteria. The only well-known systems in bacteria are the **glucose facilitator** of *Zymomonas mobilis* (Tab. 5.**3**) and glycerol transport in a variety of bacteria.

Antiport systems are frequently found in bacteria. Well-known examples are the bacterial precursor/product antiport mechanisms. These mechanisms couple the uptake of a precursor substrate with the corresponding efflux of the product of the respective metabolic pathway, for example, malate/lactate antiport in the malolactic fermentation in *Lactococcus lactis* (Fig. 5.**11**), oxalate/formate exchange in *Oxalobacter formigenes*, and lactose/galactose and arginine/ornithine antiport in lactic acid bacteria. Antiport systems may exchange very similar substrates (e.g., in precursor/product antiport), and relatively different solutes. Examples of the latter are the phosphate/sugar-phosphate exchange system in *E. coli* (Tab. 5.**3**) and various detoxification mechanisms (e.g., Tab. 5.**2**, 9).

Solute uptake by symport mechanisms is especially widespread in bacteria and is, in general, coupled to the flux of H^+ or Na^+ ions, as is the case for many amino acids, monosaccharides, and disaccharides. The best characterized bacterial transporter is the **β-galactoside/H^+ symporter** from *E. coli*, also called **lactose permease (or LacY)**. It catalyzes symport of a variety of galactosides with one proton. Three different modes of transport can be observed (Fig. 5.**12**):

1. H^+-coupled lactose uptake against a lactose gradient. In this mode, lactose accumulation in the cytoplasm is coupled to the entry of protons "down" their electrical and/or concentration gradient.
2. H^+-coupled lactose efflux along the lactose gradient. This transport mode converts the chemical potential of lactose (lactose gradient) into an electrochemical potential of protons (proton motive force).
3. Electroneutral equilibrium exchange of external against internal lactose.

These different transport modes can be functionally explained by the kinetic model described above (cf. Fig. 5.**5**). The ternary complex, composed of the transport

Table 5.2 **Different driving forces in some representative secondary transport systems**

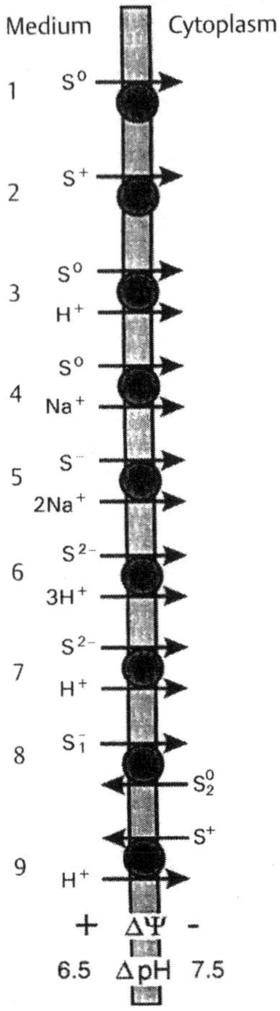

Mechanism (Medium → Cytoplasm)	Driving forces	Transport system (gene)
1 S^0 →	$Z\log\frac{[S^0_{ex}]}{[S^0_{in}]}$	Glucose (S^0) uptake in *Zymomonas mobilis (glf)*
2 S^+ →	$Z\log\frac{[S^+_{ex}]}{[S^+_{in}]}+\Delta\Psi$	Lysine (S^+) uptake in *Bacillus stearothermophilus*
3 S^0 →, H^+ →	$Z\log\frac{[S^0_{ex}]}{[S^0_{in}]}+\Delta\Psi+Z\Delta pH$	Lactose (S^0) uptake in *Escherichia coli (lacY)*
4 S^0 →, Na^+ →	$Z\log\frac{[S^0_{ex}]}{[S^0_{in}]}+\Delta\Psi+Z\log\frac{[Na^+_{ex}]}{[Na_{in}]}$	Proline (S^0) uptake in *Escherichia coli (putP)*
5 S^- →, $2Na^+$ →	$Z\log\frac{[S^-_{ex}]}{[S^-_{in}]}+\Delta\Psi+2Z\log\frac{[Na^+_{ex}]}{[Na^+_{in}]}$	Glutamate (S^-) uptake in *Escherichia coli (gltS)*
6 S^{2-} →, $3H^+$ →	$Z\log\frac{[S^{2-}_{ex}]}{[S^{2-}_{in}]}+\Delta\Psi+3Z\Delta pH$	Citrate ($HCit^{2-}=S^{2-}$) uptake in aerobic *Escherichia coli (citH)*
7 S^{2-} →, H^+ →	$Z\log\frac{[S^-_{ex}]}{[S^-_{in}]}-\Delta\Psi+Z\Delta pH$	Malate (S^{2-}) uptake in *Lactococcus lactis (malP)*
8 S^-_1 →, ← S^0_2	$Z\log\frac{[S^-_{ex}]}{[S^-_{in}]}-\Delta\Psi+Z\log\frac{[S^0_{2in}]}{[S^0_{2ex}]}$	Malate ($HMal^-=S^{-1}$)/lactate ($HLac=S^0_2$) antiport in *Lactococcus lactis (mlfP)*
9 ← S^+, H^+ →	$Z\log\frac{[S^+_{in}]}{[S^+_{ex}]}+Z\Delta pH$	Cation/tetracycline (S^+) efflux in *Escherichia coli (tetA)*
+ $\Delta\Psi$ − ; 6.5 ΔpH 7.5		

Mechanisms: 1,2 are uniporters, 3–7 are symporters, and 8,9 are antiporters
For reasons of simplification, the sign of the different driving forces was adjusted in the following way: positive sign, driving in the direction of the indicated transport system; negative sign, opposing force. S, substrate; Z, 2.3 *RT*/*F*; $\Delta\Psi$, membrane potential ΔpH, pH gradient.

protein, the substrates lactose and H^+, and the unloaded transporter, undergoes a conformational change in the translocation steps (3) and (6); these two forms of the transporter are therefore called the "mobile" forms. For both uptake (Fig. 5.**12**, mode a) and efflux (mode b), the complete catalytic cycle of the transporter (Fig. 5.**5**, steps 1–6) is involved, whereas for equilibrium exchange (mode c) only steps (2)–(4) are needed. The membrane potential affects only uptake and efflux (i.e., the two electric transport modes), but not exchange.

Many different techniques have been utilized to establish a well-defined secondary structure of the *E. coli* lactose permease, the topological arrangement of α-helices in the membrane with connecting loops at both the cytoplasmic and periplasmic sides (Fig. 5.**13**). As already shown in principle in Figure 5.**2**, the common array of 12 transmembrane segments (α-helices) and the prominent central loop between helix VI and VII can be detected. Site-directed mutations of the corresponding gene (*lacY*) have been obtained, and the functions of many of the mutant proteins have been studied. This has demonstrated the importance of particular amino acids of the transporter for catalyzing particular functions. Among these are substrate recognition and

Table 5.3 **Representative examples of secondary transport systems in bacteria**

Class	Example	Gene	Coupling ion	Organism	Major function
Uniport	Glucose carrier	*glf*	–	*Zymomonas mobilis*	Glucose uptake
Symport	Lactose permease	*lacY*	H^+	*Escherichia coli*	β-galactoside uptake
	Proline carrier	*putP*	Na^+	*Escherichia coli*	Proline uptake
	Glutamate carrier	*gltT*	$H^+ + Na^+$	*Bacillus stearothermophilus*	Glutamate uptake
	Galactose carrier	*galP*	H^+	*Escherichia coli*	Galactose uptake
	K^+ carrier	*trkAGH*	H^+	*Escherichia coli*	K^+ transport, osmoregulation
Antiport	Malate/lactate antiporter	*mlfP*	ap[1]	*Lactococcus lactis*	Prec./proc. antiport generation of Δp
	Oxalate/formate antiporter	*oxlT*	ap[1]	*Oxalobacter formigenes*	Prec./prod. antiport generation of Δp
	Arginine/ornithine antiporter	*arcD*	ap[1]	*Pseudomonas aeruginosa*	Precursor/product antiport
	Glucose 6-P/Pi antiporter	*uhpT*	ap[1]	*Escherichia coli*	Glucose 6-P and phosphate uptake

[1] In precursor/product antiport systems, the two substrates are both driving and driven ions. prec., precursor; prod., product

specificity, coupling of substrate and proton movement, and single steps in the catalytic cycle (Fig. 5.**5**). In some cases, alteration of a single amino acid can in fact change the protein from a coupled (electrogenic) transporter into a transporter that catalyzes the facilitated diffusion of a single substrate (lactose or H^+). In other words, these mutant proteins are "uncoupled". Hence, only after the two substrates occupy the binding site is the energy barrier to conformational change of the transporter lowered to the extent necessary for translocation.

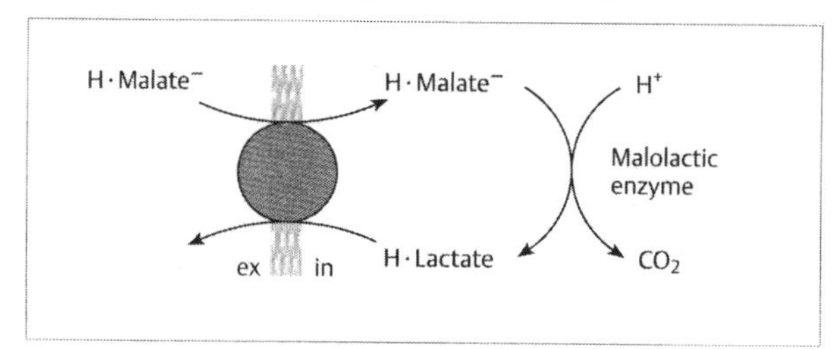

Fig. 5.**11** **Precursor/product antiport in malolactic fermentation in *Lactococcus lactis*.** The import of substrate in anionic form, export of an uncharged product, and consumption of a H^+ in the cytoplasm lead to formation of an electrochemical proton potential (outside positive), which is used as an energy source in this organism

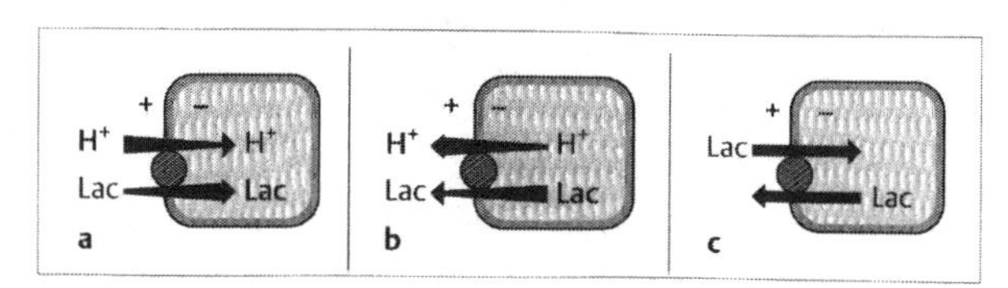

Fig. 5.**12a–c** **Three possible transport modes of the β-galactoside/H^+ symporter from *E. coli*** (Lac, lactose). **a** The "uphill" lactose uptake is driven by the proton motive force; **b** the "downhill" lactose efflux creates a proton potential; **c** electroneutral lactose counterflow

In general, the coupling ions in secondary transport are either H^+ or Na^+ (Tab. 5.**3**). In some cases, the choice of the appropriate coupling ion is obvious. Secondary solute uptake systems in bacteria that grow in alkaline environments, for example, "prefer" Na^+ because of the abundance of this cation in the surroundings. The same holds for bacteria which grow in extremely hot environments, where protons are unsuitable coupling ions because the membranes are highly pervious to protons at these temperatures. Other systems use both Na^+ and H^+ as coupling ions (e.g., the glutamate transporter of *Bacillus stearothermophilus*), and some systems use either Na^+ or H^+ as the coupling ions. The melibiose transporter, as an example, accepts either Na^+, Li^+, or H^+ as a coupling ion, depending on the type of substrate transported.

In principle, the systems used for uptake may also work in the direction of excretion. Hydrophobic metabolic end products, such as alcohols (butanol), acetone, and some organic acids (in undissociated form), may leave the cell by diffusion. Other compounds, such as amino acids and carboxylic acids, are excreted by carrier-mediated processes. An important example of secondary mechanisms of excretion for metabolites (i.e., precursor/product antiport), has already been mentioned. In *E. coli* and *Lactococcus lactis*, lactate produced

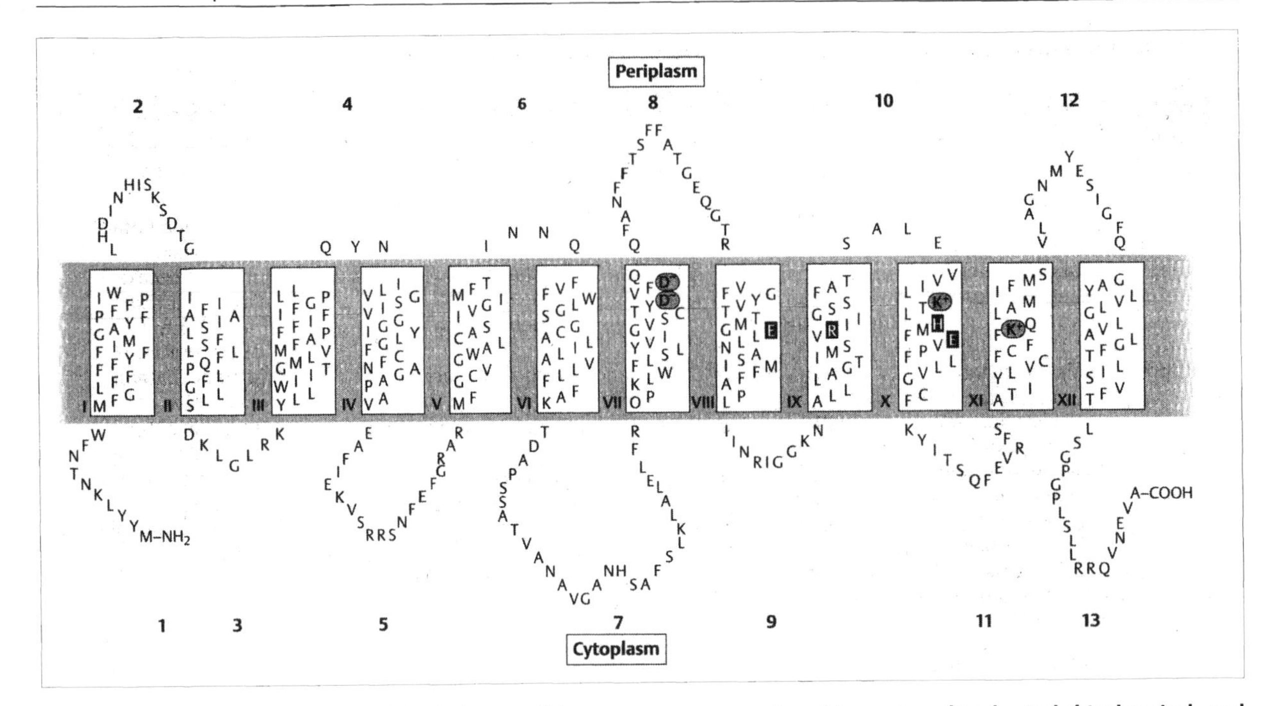

Fig. 5.13 **Secondary structure model of the *E. coli* lactose permease analyzed by various biophysical, biochemical, and genetic techniques.** The one-letter code for amino acids is used. The putative transmembrane helices are shown in Roman numerals and emphasized by boxes; the loops are emphasized by arabic numerals. Courtesy of H. Jung,

during glucose fermentation can be excreted in symport with protons, thus leading to generation of an electrochemical proton potential. A different class of solutes is excreted by a family of transport proteins that are collectively referred to as bacterial multidrug-resistance systems. The majority of these systems function as secondary transport systems and are responsible for active efflux of a large variety of antibiotics (e.g., tetracycline in Tab. 5.**2**.9) or other toxic compounds. This leads to increased resistance to these substances. Primary ATP-dependent excretion systems for a variety of toxic compounds, including heavy-metal ions, are also found in bacteria (see Chapter 5.6.3).

5.6.3 Primary Transport Systems

In **primary transport systems** (Fig. 5.**8III**), the solute transport is directly coupled to a chemical or photochemical reaction (A → B; Fig. 5.**8**, f). There are three classes of primary transport processes (Tab. 5.**4**). In class 1 systems, transport is driven by a redox reaction or by light; class 2 transport systems couple solute movement to ATP hydrolysis or synthesis. Class 3 systems couple Na^+ transport to a decarboxylation reaction. Figure 5.**14** and Table 5.**4** give some typical examples for these types of transport systems.

Class 1. The general principle of the systems of class 1, which couple redox or light reactions to H^+ or Na^+ translocation in respiration and photosynthesis, is described in Chapters 4 and 13. A possible primordial case of this type of transport systems, and presumably

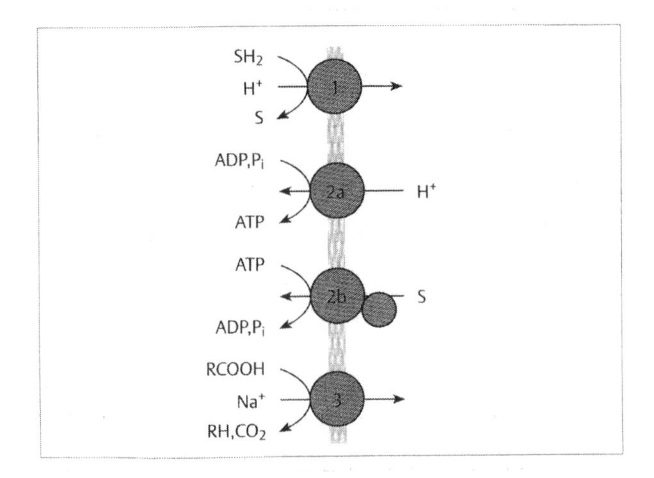

Fig. 5.**14** **Overview of bacterial primary transport systems.** H^+-pumping activity of the respiratory chain (member of class 1), F_1F_0 ATP synthase (class 2a), binding-protein-dependent systems (class 2b), decarboxylation-driven Na^+ extrusion (class 3). S, substrate

Table 5.4 **Representative examples for primary transport systems in bacteria**

Class		Example	Transported solute	Energy source	Organism (gene)	Function
1	Light-driven pumps	Bacteriorhodopsin	H^+	h · ν	Halobacteria	Generation of proton motive force
		bacterial photosystem	H^+	h · ν	phototrophic bacteria	
	Redox-driven pumps	Respiratory chain	H^+	Redox energy	Most bacteria	Generation of proton (sodium) motive force
2a	"ATPases"	F_1F_0-ATPase (F-type)	$H^+ (Na^+)$	$\Delta u\ (Na^+)$[1]	Most bacteria	ATP synthesis
		Kdp-ATPase (P-type)	K^+	ATP	*E. coli* and others	K^+ transport
		Cd-ATPase (P-type)	Cd^{2+}	ATP	*E. coli* and others	Cadmium resistance
2b	ABC[2] transporters	BPDS[3]: maltose transport	Maltose	ATP	*E. coli* (*malEFGK*)	Maltose uptake
		MDR[4]: hemolysin transport	Hemolysin	ATP	*E. coli* (*hlyBCL*)	Cytotoxicity
3	Decarboxylation	oxaloacetate decarboxylation	Na^+	Oxalo-acetate	*Klebsiella pneumoniae*	Generation of sodium motive force

[1] In a few bacteria (e.g., *Vibrio alginolyticus*), Na^+-transporting respiratory chains exist
[2] ABC, transport systems characterized by the structural motif of an **A**TP-**b**inding **c**assette (see text)
[3] BPDS, **b**inding **p**rotein-**d**ependent **s**ystem
[4] MDR, **m**ulti**d**rug **r**esistance system

for transporters in general, is H^+ and Cl^- translocation in Halobacteria, in which ejection of protons (via bacteriorhodopsin) or uptake of Cl^- (via halorhodopsin) is directly coupled to light absorption. These two light-driven ion pumps contain a retinal molecule as chromophore, covalently bound to a lysine residue by means of a Schiff's base (cf. Fig. 5.**1**). The retinal moiety absorbs photons, and its subsequent photoisomerization causes conformational changes in the protein. This then triggers the vectorial ion movement along identified amino acid residues in a pathway formed by the seven α-helices of the protein (Figs. 13.**13**, 13.**14**).

Class 2. This very heterogeneous class includes primary transport systems coupled to ATP synthesis or hydrolysis, which are known as **"ATPases"** or **ATP synthases.** For the majority of these systems, the common designation as "ATPase" does not seem to be appropriate since, in general, these enzymes are involved in either ion transport or ATP synthesis, but not in ATP hydrolysis. The reason for the expression "ATPase" is that the function is usually tested in the direction of ATP hydrolysis. ATPases are mainly involved in translocation of small monovalent (H^+, K^+, Na^+) or divalent cations (Ca^{2+}, Mg^{2+}). They convert free energy of ATP into the electrochemical gradient of the transported ion or, as in the case of respiration-coupled ATP synthases, the other way around. They are divided into at least three different subclasses (Tab. 5.**4**, class 2a):

1. **F-type ATP synthases** (e.g., F_1F_0-ATPase) are multicomponent systems involved in primary energy conversion, such as respiration- and photosynthesis-coupled movement of H^+ (and Na^+) in bacteria, mitochondria, and chloroplasts. Their function in ATP synthesis and their structure have already been discussed in Chapter 4. At least in facultative bacteria, the F-type ATPases operate in both directions. During oxidative phosphorylation, they catalyze ATP synthesis driven by the proton flux in the direction of the membrane potential. During fermentation, the electric potential across the plasma membrane is created by ATPase-mediated H^+ flux in the opposite direction at the expense of ATP hydrolysis.
2. **P-type ATPases** (e.g., eukaryotic Na^+/K^+-ATPase, K^+-ATPase in bacteria) are commonly found in the plasma membrane and consist of only one or two subunits. They form a phosphorylated high-energy intermediate within the reaction cycle. Bacterial P-type ATPases are widespread and transport K^+, Ca^{2+}, or Mg^{2+}.
3. **V-type ATPases** (not in Tab. 5.**4**) are present in the vacuolar membrane of eukaryotic cells and have not yet been found in prokaryotes.

Binding-protein-dependent transport systems catalyze another kind of primary transport directly coupled to ATP (Fig. 5.**15**). They are mainly found in Gram-negative bacteria and comprise an ensemble of protein

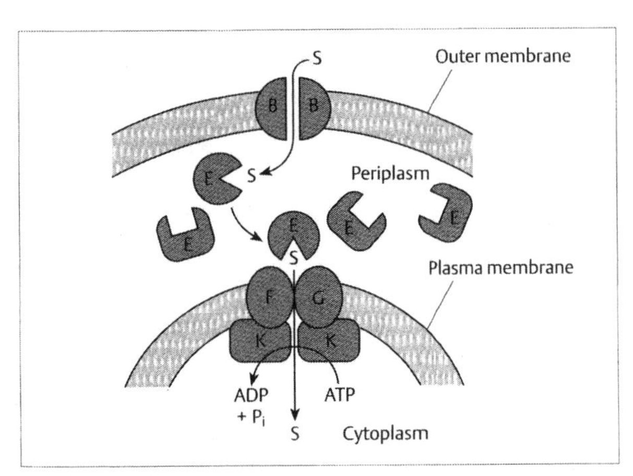

Fig. 5.**15** **Binding protein-dependent maltose uptake system of *E. coli*.** The functions of the different subunits located in the cytoplasmic membrane, in the periplasm and in the outer membrane are explained in the text. The abbreviations for the subunits are: B, LamB; E, MalE; F, MalF; G, MalG; K, MalK. LamB actually is not a member of the maltose transport system; however, it increases the flux of maltose across the outer membrane. S, substrate

components. In *E. coli*, more than 20 binding-protein-dependent systems are known so far. Very diverse substrates are transported, including amino acids, peptides, monosaccharides, disaccharides, organic anions, nucleotides, coenzymes, and inorganic ions such as sulfate or phosphate. Well-characterized examples are the uptake systems for maltose, histidine, and oligopeptides (in *E. coli*, *Salmonella typhimurium*). These systems are typically composed of four distinct subunits or domains (Fig. 5.**15**); two are highly hydrophobic and thus fully inserted into the membrane, whereas the two other domains are presumably only associated with the cytoplasmic face of the membrane. The characteristic feature is the presence of an additional soluble substrate-binding protein in the periplasm. Some systems, for example, that of maltose transport, furthermore involve a specific channel protein (porin) in the outer membrane, which mediates the passive permeation of the substrate. A basic functional scheme of the maltose transport system is shown in Figure 5.**15**. After passage through the maltose-specific porin (LamB), the substrate is tightly bound to the binding protein (MalE), which thereby dramatically changes its conformation, i.e., the bi-lobed, two-domain structure switches from an "open" to a "closed" state. Upon interaction of the substrate-loaded binding protein with the membrane subunits (MalF, MalG), the substrate is released and translocated to the cytoplasm. The translocation event is coupled to ATP hydrolysis by the two membrane-associated subunits (MalK).

Binding-protein-dependent uptake systems are essentially unidirectional; they are characterized by high substrate affinity in the micromolar range, but a relatively low maximal transport rate. A characteristic of these systems in Gram-negative bacteria is their sensitivity to osmotic shock, by which the binding proteins are released and transport activity is lost. In addition to their role as high-affinity acceptors for the solute to be transported, several binding proteins in the periplasm fulfill essential functions by signaling changes in the availability of nutrients to the cell. These mechanisms are described in Chapter 20 in connection with bacterial chemotaxis. Until recently, binding-protein-dependent transport systems were considered to be unique to Gram-negative bacteria, since Gram-positive bacteria do not have a periplasmic space. An increasing number of this type of transport system have now also been found in Gram-positive bacteria. In this case, the binding protein is anchored to the outer surface of the plasma membrane by a lipid moiety at the N-terminal end.

Bacterial binding-protein-dependent transport systems belong to a large superfamily of transporters present both in prokaryotes and eukaryotes. Because of a common sequence motif of ATP-binding sites, this class of transport ATPases is called the **ABC family** (**A**TP-**b**inding **c**assette). Members of this family are involved in a multitude of biological processes in bacteria (such as solute uptake, antibiotic resistance, cell development) and in eukaryotes [e.g., **multidrug resistance** (MDR) systems; peptide and Cl^- transport]. ABC transporters are related to each other in sequence and structural organization, and they probably have a common evolutionary origin. Many ABC transporters are involved in solute excretion. A very diverse group of substrates is exported, ranging from proteins (e.g., hemolysin) and polysaccharides (e.g., glucans and capsular material) to various toxic compounds. In addition, the extrusion of toxic heavy metal ions such as cadmium or arsenate is frequently mediated by so-called "export ATPases", which are related to the P-type ATPases.

Class 3. A unique class of primary transport mechanisms, namely, **decarboxylation-driven Na^+ transport**, was recently discovered in some bacteria (*Propiogenium modestum*, *Klebsiella pneumoniae* and *Salmonella typhimurium*). The extrusion of Na^+ ions is coupled to the decarboxylation of carboxylic acids (e.g., oxaloacetate or methylmalonyl CoA) by a membrane-bound, biotin-containing decarboxylase (see Fig. 5.**14**). The energetic coupling of this vectorial reaction to the ATPase by the electrochemical Na^+ potential has been discussed in Chapter 4.

Primary transport mechanisms couple: (1) light or redox reactions, (2) the chemical energy of ATP (e.g., ATPases, ABC transporters), or (3) the chemical energy of decarboxylation reactions to the vectorial process of solute translocation. They are based on direct coupling of a chemical or photochemical reaction to the vectorial reaction of solute transport.

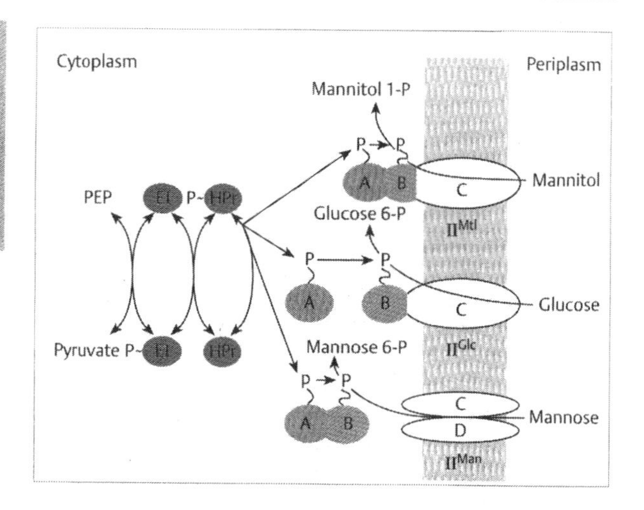

Fig. 5.16 **Various types of phosphotransferase systems (PTSs).** Enzyme I (EI) and Histidine Protein (HPr) are the general proteins for all PTSs. The Enzymes II (EII), specific for mannitol (II^{Mtl}) for glucose (II^{Glc}), and for mannose (II^{Man}), represent the variety of EIIs found in bacteria; all contain at least three autonomous domains, which may separate into independent proteins or unite into one large EII as indicated. The hydrophilic domain IIA (formerly Enzyme III or III) contains the first phosphorylation site (P ~ His); the hydrophilic domain IIB contains the second phosphorylation site (usually a P ~ Cys, rarely a P ~ His). The membrane-bound (hydrophobic) or transporter domain IIC may separate into two parts (IIC and IID). P ~ indicates the phosphorylated form of the soluble proteins/domains, while arrows indicate the direction of the biochemical steps [after 3]

5.6.4 Group Translocation

The mechanism of **group translocation** is unique because the transported solute is chemically modified (i.e., phosphorylated) during the process. Only the bacterial **phospho*enol*pyruvate (PEP)-dependent carbohydrate phosphotransferase systems (PTSs)** catalyze group translocation. PTSs, of which up to 20 different ones are found in a cell, mediate the uptake of many different carbohydrates. Their basic structure is similar in all bacteria and is shown schematically in Figure 5.**16**. During group translocation, a carbohydrate is translocated across the membrane and is phosphorylated while still bound to the transporter. The phosphoryl group is derived from PEP and is transferred sequentially via a protein kinase, called **Enzyme I** (EI), and several phospho-acceptor proteins, the Heat-stable or **Histidine Protein** (HPr) and **Enzyme II** (EII), to a particular carbohydrate. EI and HPr are common to all PTSs of a cell, while each substrate is recognized by one or more substrate-specific EIIs. The phosphoryl group is attached covalently to a histidine residue in EI and HPr.

Group translocation by the phospho*enol*pyruvate phosphotransferase systems (PTS) leads to phosphorylation of the transported sugar or sugar alcohol during solute uptake.

Studies of the amino acid sequence and structure of EIIs have revealed a modular structure. Each EII contains three (or rarely, four) domains: IIA, IIB, and IIC (IID, the fourth) (Fig. 5.**16**). These domains may be united to form a large protein, as in the case of the mannitol-specific EII (II^{Mtl}), or they form a unit of two to four proteins, of which at least one is membrane-bound, as, for example, the glucose-specific EII ($IICB^{Glc}$ IIA^{Glc}) and the mannose-specific EII ($IIAB^{Man}$ IIC^{Man} IID^{Man}) from enteric bacteria. In all cases, the phosphoryl group from P ~ HPr is transferred during group translocation to a His residue of the IIA domain and subsequently to a Cys (or rarely a His) residue in the IIB domain. Hence, PTS-dependent translocation and phosphorylation of a substrate comprise the following reactions (see also Fig. 5.**16**):

$$\text{PEP} + \text{EI} \rightarrow \text{P} \sim \text{His EI} + \text{pyruvate} \quad (5.6)$$

$$\text{P} \sim \text{EI} + \text{HPr} \rightarrow \text{P} \sim \text{His HPr} + \text{EI} \quad (5.7)$$

$$\text{P} \sim \text{HPr} + \text{IIA} \rightarrow \text{P} \sim \text{His IIA} + \text{HPr} \quad (5.8)$$

$$\text{P} \sim \text{IIA} + \text{IIB} \rightarrow \text{P} \sim \text{Cys IIB} + \text{IIA} \; (\text{or P} \sim \text{His IIB} + \text{IIA}) \quad (5.9)$$

$$\text{P} \sim \text{IIB} + \text{substrate}_{out} \xrightarrow{\text{IIC(IID)}} \text{substrate} - \text{P}_{in} + \text{IIB} \quad (5.10)$$

Because the free energy of a P ~ His (or P ~ Cys) is roughly equal to the energy of PEP ($\Delta G^{\circ\prime} = -61.5$ kJ/mol), Reactions 5.6 to 5.9 are fully reversible. In contrast, Reaction 5.10 is virtually irreversible under physiological conditions. The phosphoryl group is transferred from P ~ IIB to a substrate bound to the transporter that comprises domain(s) IIC (IID), and the EII is rephosphorylated via P~HPr. Carbohydrates can only be translocated efficiently when IIB is phosphorylated. According to present models, substrates bind in a first step with high affinity to a binding site of the transporter IIC, which is oriented toward the periplasmic side of the membrane. In an unphosphorylated EII, such tightly bound substrates are translocated very slowly, if at all, by a process that resembles facilitated diffusion and involves a reorientation of the binding site toward the inner face of the membrane. In contrast, low-affinity substrate analogues diffuse readily through unphosphorylated EIIs and even through truncated IIC variants that lack the IIA and IIB domains completely. High-affinity (i.e., normal) substrates apparently "lock" a IIC transporter binding site in an "occluded" state, thus preventing complete translocation of the substrate through the membrane and subsequent substrate phosphorylation. Translocation is increased drastically by a phosphorylation of IIB, perhaps because this "unlocks" the occluded state. Phosphorylation of the substrate occurs simultaneously with the "unlocking" or, subsequently, while the substrate is still bound to a low-affinity state of IIC at the inside of the cytoplasmic membrane. In any event, the dissociation of the phosphorylated substrate into the cytoplasm terminates uptake, while the rephosphorylation of all PTS proteins dephosphorylated during the process closes the PTS phosphorylation cycle. Under normal physiological conditions, (i.e., with high-affinity substrates and a phosphorylated EII) the transport and the phosphorylation process appears to be tightly coupled during group translocation. Mechanistically, however, the two partial reactions are sequential events that can be uncoupled readily either by mutations in the IIC transporter or by the use of carbohydrates having a low affinity for PTS.

At present, it cannot be decided whether EIIs constitute a unique class of transporters or whether the mechanisms involved in group translocation fit into a basic concept common to other transport systems as well. The answer is complicated further because the PTSs as a whole correspond to a complex chemosensor and signal transduction system through which bacteria survey their environment and control in a global way most functions related to the quest for food. These functions will be discussed in detail in Chapter 20.2.

5.7 Regulation and Diversity of Transport Systems

Transport systems take part in a variety of functions that are essential for the cell, including uptake of nutrients, export of metabolic end products and building blocks for cell wall synthesis, and signal processing. To maintain an optimal function, these processes have to be tightly regulated by the cell. This, for example, includes adaptation to different environmental conditions such as the presence (or lack) of nutrients in widely varying concentrations or the necessity of choosing between carbon sources of different metabolic quality. Furthermore, the activity of transport systems has to be adapted to the demand for maintaining homeostatic conditions between solutes such as K^+, Na^+, H^+, and Ca^{2+} in the cell, to name a few. Transport regulation is directly related to the physiological significance of the observed diversity in transport mechanisms (Tab. 5.**5**). Different mechanisms exist for catalyzing solute uptake, namely, passive and active systems as well as various types of coupling, such as antiport or symport. Furthermore, more than one type of uptake system for a certain solute frequently is found in a particular organism. In *E. coli*, for example, there are at least seven different uptake systems that accept D-glucose and D-galactose and at least five uptake systems for the amino acids glutamate and aspartate.

The observed diversity in mechanisms can be explained by the availability of a particular solute and by the energetic demands of metabolism and transport. One may ask why complex multisubunit primary transport systems such as binding-protein-dependent systems or phosphotransferase systems evolved in addition to simple single-subunit secondary transport systems. The answer: different systems are used for different purposes. Primary transport systems normally have a high substrate affinity and are often practically unidirectional, thus allowing very high accumulation ratios. In contrast, secondary systems are in general reversible; they may thus lead to "leakage" of the transport substrate in situations of low external substrate concentration or low energy. Furthermore, secondary uptake systems are coupled to membrane potential and/or ion gradients. Accumulation ratios (internal/external), therefore, in general do not exceed 10^2–10^3. The driving force of primary uptake systems, on the other hand, provides enough free energy for high accumulation ratios, which may be necessary if a given solute is present in the environment at a very low concentration. ATP-driven binding-protein-dependent uptake of maltose in *E. coli*, for example, causes up to a $2 \cdot 10^5$-fold accumulation of the substrate in the cell.

Table 5.5 **Some examples illustrating the diversity of transport systems**

Transport system (mechanism)	Gene(s)	Substrate(s)	Kinetic parameters K_m (μM)	V_{max} (nmol · min^{-1} · mg^{-1})	Function	Regulation
Uptake of branched-chain amino acids in *Escherichia coli*						
Binding protein-dependent system	*livGMHKLJ*	Leu, Ile, Val	0.01–0.2	?	Amino acid uptake	Regulated by leucine
Secondary system (Na^+-coupled)	*liv-II*	Leu, Ile, Val	2–4	?	Amino acid uptake	Constitutive
Proline uptake in *E. coli* and *Salmonella typhimurium*						
Secondary system (Na^+-coupled)	*putP*	Pro	2	?	Proline utilization	Induced by proline
Secondary system (H^+-coupled)	*proP*	Betaine/Pro	40/300	32	Osmotic stress response	Induced and activated by hypo-osmotic conditions
Binding protein-dependent system	*proUVW*	Betaine/Pro	1/35	?	Osmotic stress response	Induced and activated by hypo-osmotic conditions
Citrate uptake in *Klebsiella pneumoniae*						
Aerobic system	*citH*	H_2Cit^{2-} + 3 H^+	2000	?	Citrate uptake	Expressed under oxic conditions
Anaerobic system	*citS*	H_2Cit^{2-} + Na^+/2H^+	1000	?	Citrate uptake	Expressed under anoxic conditions
Glutamate/aspartate uptake in *Escherichia coli*						
Binding protein-dependent system	?	Glu + Asp	0.5	30	Amino acid uptake	
Secondary system (H^+-coupled)	*gltP*	Glu + Asp	5	68	Amino acid uptake	
Secondary system (Na^+-coupled)	*gltS*	Glu	1.5	50	Amino acid uptake	
Secondary system	?	Asp	1.5	5	Amino acid uptake	
Secondary system	*dct*	Dicarboxylates, Asp	30	?	Amino acid uptake	

Further examples of nature's selection of appropriate mechanisms for particular needs are precursor/product antiport systems. Since fermentative metabolism provides only relatively little useful energy for the cell, minimization of the energy required for substrate uptake is essential. Frequently, this problem has been solved by systems in which the same transporter takes up the precursor and removes the product of the fermentative pathway. A well-known example is the malate/lactate antiport system in *Lactococcus lactis*, which takes up malate and excretes lactate (Fig. 5.**11**) during malolactic fermentation. Thus, the internal consumption of malate leads to an inwardly directed malate gradient, and the production of lactate in the cytoplasm creates a corresponding driving force for the solute lactate to be excreted. Since malate carries one negative charge more than lactate at physiological pH, the coupled transport reaction establishes a membrane potential, negative inside. This is in fact the energy source that this organism uses for ATP synthesis via the ATP synthase. Other well-documented systems providing metabolic energy by excretion of metabolic end products are the oxalate/formate antiport in *Oxalobacter formigenes* (Tab. 5.**3**) and the histidine/histamine antiport in *Lactobacillus buchneri*.

Also, the frequent observation of a multitude of uptake systems for the same solute in bacteria can be explained in physiological terms. On the one hand, only one uptake system (or none) for rare and unusual substrates may be found or, on the other hand, a very simple uptake system if the substrate is constantly present in high concentrations. An example of the latter case is *Z. mobilis*, normally found in a habitat with a high sugar content, in fruit juice. This bacterium consequently has only one single sugar uptake system involving the simplest mechanism possible, i.e., facilitated diffusion (uniport). In general, multiple uptake systems are present when the respective solute is an important substrate for the cell in a given habitat and if it is present in widely varying concentrations. As a general rule, when several systems for a given substrate are present, there is at least one constitutive system with a comparably low affinity (high K_m), but high capacity (high V_{max}). Additional, inducible systems are thus, in general, of high affinity, but low capacity (cf. Tab. 5.**5**). The latter systems are thus provided for "cases of emergency" and are also called "**scavenger**" systems. Frequently, inducible transport systems are induced by their major substrate, which is reflected by a regulatory link between the genes for the transport system and the genes for respective catabolic enzymes of this substrate. In other words, these genes are often organized in an operon or regulon.

Typically, the diversity of transport systems for a particular solute comprises at least a high-capacity, low-affinity system, which in general is constitutively expressed, in addition to a low-capacity, high-affinity uptake system, which in general is expressed only in the case of the presence or the absence of a particular solute.

In addition to mechanisms controlling the level of expression, mechanisms on the activity level may be present, thus providing a meaningful regulation of transport systems. Transporters can be regulated by the energy status of the cell, by the intracellular and extracellular pH, or by the osmolarity of the medium. Further effectors are internal metabolites (e.g., cyclic AMP), intracellular regulatory proteins or components of other transport systems [e.g., the IIA component of PTS systems (see Chapter 5.6.4)]. Communication between transport systems and cytoplasmic enzymes can be very complex, according to the "preference" of the particular organism for particular substrates, thus providing a hierarchy of catabolic sequences. This means, for example, that in the presence of glucose and galactose, *E. coli* will first use glucose and then galactose ("glucose effect" or diauxic growth). The principles of this type of network are discussed in more detail in Chapter 20.1. Finally, a single component from a transport system may have an alternative function, hence requiring its separation from the other components. For example, the maltose-binding protein MalE serves as a chemoreceptor, and Enzyme IIA^{Glc} of the glucose PTS serves as a signal transducer.

5.8 Secretion of Macromolecules

Although mainly the transport of small molecules is discussed here, obviously the transport of macromolecules also is essential for growth and survival of the cell. The mechanisms are extremely diverse and include translocation of proteins, complex lipids (e.g., components of the outer membrane), carbohydrates (e.g., extracellular polysaccharides), and nucleic acids (e.g., in transformation). Because the mechanisms are intimately linked to the phenomena of cellular compartmentation (Chapter 19), cell membrane and cell wall assembly (Chapter 23), or gene transfer (Chapter 16), they will be treated in detail in the indicated chapters.

Further Reading

Ames, G.F.-L., Mimura, C.S., Holbrook, S.R., and Shyamala, V. (1992) Traffic ATPases; a superfamily of transport proteins operating from *Escherichia coli* to humans. Adv Enzymol 65: 1–47

Anthony, C., ed. (1988) Bacterial energy transduction. New York: Academic Press

Antonucci, T.K., and Oxender, D.L. (1986) The molecular biology of amino-acid transport in bacteria. Adv Microbiol Physiol 28: 146–180

Boos, W., and Lucht, J.M. (1996) Periplasmic binding protein-dependent ABC transporters. In: Neidhardt, F.C. (ed.). *Escherichia coli* and *Salmonella*. 2nd. edn. Washington DC: ASM Press

Dimroth, P. (1991) Na^+-coupled alternative to H^+-coupled primary transport systems in bacteria. Bioessays 13: 463–468
Harold, F. (1986) The vital force: a study of bioenergetics. New York: Freeman
Henderson, P.J.F. (1990) Proton-linked sugar transport systems in bacteria. J Bioenerg Biomemb 22: 525–569
Kaback, H.R. (1992) In and out and up and down with lac permease. Intern Rev Cytol 137A: 97–125
Kaback, H.R., Voss, J., and Wu, J.H. (1997) Helix packing in polytopic membrane proteins – the lactose permease of *Escherichia coli.* Curr Opin Struct Biol 7: 532–542
Konings, W.N., Kaback, H.R., and Lolkema, J.S. (1996) Transport processes in eukaryotic and prokaryotic organisms. Handbook of biological physics, vol 2. Amsterdam New York: Elsevier
Krämer, R. (1994) Functional principles of solute transport systems: concepts and perspectives. Biochim Biophys Acta 1185: 1–34
Krulwich, T.A., ed. (1990) Bacterial energetics. The bacteria, vol. 12. San Diego: Academic Press
Lengeler, J.W. (1993) Bacterial transport under environmental conditions, a black box? Leuwenhoek Antonie van 63: 275–288
Maloney, P.C. (1990) Anion exchange reactions in bacteria. J Bioenerg Biomemb 22: 509–523
Marger, M.D., and Saier, M.H. (1993) A major superfamily of transmembrane facilitators that catalyse uniport, symport and antiport. Trends Biochem Sci 18: 13–20
Mathies, R.A., Lin, S.W., Ames, J.B., and Pollard, T.W. (1991) From femtoseconds to biology: mechanisms of bacteriorhodopsin's light-driven proton pump. Annu Rev Biophys Chem 20: 491–518
Nikaido, H. (1992) Porins and specific channels of bacterial outer membranes. Mol Microbiol 6: 435–442
Nikaido, H., and Saier, M.H. (1992) Transport proteins in bacteria-common themes in their design. Science 258: 936–942
Poolman, B. (1990) Precursor product antiport in bacteria. Mol Microbiol 4: 1629–1636
Poolman, B., and Konings, W.N. (1993) Secondary solute transport in bacteria. Biochim Biophys Acta 1183; 5–39
Postma, P.W., Lengeler, J.W., and Jacobson, G.R. (1993) Phosphoenolpyruvate-carbohydrate phosphotransferase systems of bacteria. Microbiol Rev 57; 543–594
Pugsley, A.P. (1993) The complete general secretory pathway in Gram-negative bacteria. Microbiol Rev 57; 50–108
Senior, A.E. (1990) The proton-translocating ATPase of *Escherichia coli.* Annu Rev Biophys Chem 19: 7–41
Stein, W.D. (1990) Channels, carriers, and pumps. New York London: Academic Press
Weiss, M.S., Abele, U., Weckesser, J., Welte, W., Schiltz, E., and Schulz, G.E. (1991) Molecular architecture and electrostatic properties of a bacterial porin. Science 254: 1627–1630

Sources of Figures

1 Henderson, R., Baldwin, J.M., Ceska, T.A., Zemlin, F., Beckmann, E., and Downing, K.H. (1990) J Mol Biol 213: 899–929
2 Jones, T.A., Zou, J.-Y., Cowan, S.W., and Kjeldegaard, M. (1991) Acta Cryst A47: 110
3 Postma, P.W., Lengeler, J.W., and Jacobson, G.R. (1993) Microbiol Rev. 57: 543–594

6 Growth and Nutrition

Advances in microbiology depend on the knowledge of the growth characteristics and nutritional demands of the microorganism. Methodology of bacterial growth and pure-culture techniques began with the era of R. Koch; yet, after more than 100 years, probably only a few percent of the extant bacterial species have been isolated and studied. Molecular biological techniques allow the detection of even a single bacterial cell in complex samples and the establishment of its approximate phylogenetic relationship to other bacteria. DNA technologies allow the identification and use of individual genes and gene products derived therefrom without any knowledge of the species from which they come. Still, as long as a bacterium cannot be studied in laboratory culture, one can only intelligently guess what its properties and its role in nature may be. For instance, it may be impossible to determine whether it is a pathogen. This limitation is also true for the potential biotechnological use of more than 90% of all bacteria that are still "hidden". Nutrition implies much more than what an organism needs to grow. It gives an idea about the organism's role in nature and allows the organism to be distinguished from other organisms for taxonomic, ecological, physiological, and applied reasons. Growth characteristics reflect underlying physiological events in the individual cells. Imagine that genome sequencing has revealed many "orfan genes" of hypothetical or unknown function; again, screening for potential roles requires some growth response. Finally, for any practical purpose, manipulation of growth is the basis for success. These introductory remarks stress the importance of the study of the growth and nutrition of bacterial cultures. The growth characteristic of a batch culture will be discussed first, and then the physicochemical factors affecting bacterial growth will be considered.

6.1 Growth Characteristics of a Batch Culture Is a Reflection of Cell Physiology

Microbial growth involves an increase in the number of cells. When individual bacterial cells grow, the cell length increases and finally the cell divides into two daughter cells (**binary fission**). The increase in cell number can also result from budding (see Chapter 2). The daughter cells have the same cellular composition and size as the parent cell, including a copy of the chromosome. If the time for the doubling of the cell number is constant over a period of time, the cells are in the **exponential growth phase** (incorrectly also called the logarithmic, or log, growth phase, a name derived from plotting the growth on a semi-logarithmic scale). The time required for doubling the number of cells in the exponential phase is the **generation** or **doubling time, t_d**. Growth is **balanced** when the number is proportional to the cell mass or any cell constituent, such as protein, RNA, or DNA. The rate of formation of any of these quantities (N) will follow Equations 6.1 and 6.2,

$$\frac{dN}{dt} = \mu \cdot N \qquad (6.1)$$

and N at any time t will be

$$N_t = N_0 e^{\mu(t-t_0)} \qquad (6.2)$$

where N_t and N_0 are the amount of cell mass or constituent at the cultivation time t and at the beginning t_0, respectively. This exponential equation can be expressed in logarithmic terms:

$$\ln\frac{N_t}{N_0} = \mu(t - t_0) \qquad (6.3)$$

or

$$\ln N_t - \ln N_0 = \mu(t - t_0) \qquad (6.4)$$

or

$$\log N_t - \log N_0 = \frac{\mu}{2.303}(t - t_0) \qquad (6.5)$$

The meaning of μ becomes evident when one considers the doubling of N_0 to $2N_0$, which takes one generation time t_d. Then, with $N_t = 2N_0$ and $t - t_0 = t_d$, Equation 6.3 is converted to Equation 6.6, which gives the relation between μ and t_d:

$$\mu = \frac{\ln 2}{t_d} = \frac{0.693}{t_d} \qquad (6.6)$$

μ has the dimension 1/time unit (e.g., 1/h) and is called the **specific growth rate**.

Box 6.1 From an anthropocentric point of view, bacteria with a high specific growth rate appear to be highly evolved and adapted. However, fast growth is just one strategy. Other bacteria are adapted to low nutrient concentrations, which requires high affinity for the substrates. Or, high flexibility of substrate consumption may be useful under one condition, high specialization under another condition. Finally, high efficiency of energy extraction from a substrate, i.e., high growth yield, may be a valuable strategy.

During balanced growth, no matter what measure of growth is used (see below), the same μ value will be obtained, and μ will be constant over time. The size of μ depends on many parameters, such as temperature, pH, osmolarity, the types of major nutrients, the availability of building blocks in a rich medium, or the availability of an electron acceptor. Hence, μ depends on the enzymatic capacity of the bacterial species, on the medium composition, and on all the biophysical factors that affect the growth rate. The t_d for a given bacterium may vary from 15 min in the laboratory to days or weeks in nature, depending on the conditions. The shortest generation times of individual species vary from 15 min in fast-growing species to 1 day in slow-growing ones.

Consider one cell of *Escherichia coli* with an approximate size of 1 μm³, weighing 10^{-12} g (wet weight), and growing with a generation time of 1 h. After 48 h of growth, 2^{48} cells or $3 \cdot 10^{14}$ cells weighing 300 g, and after 72 h, $5 \cdot 10^{21}$ cells weighing 50 tons would be obtained. Clearly, in a **batch culture**, which means growth in a fixed volume, nutrients will be exhausted and products will have accumulated to a toxic level, resulting in a gradual cessation of growth.

Cells reach the **stationary phase** and finally die (**death phase**). Common causes for death are the depletion of cellular energy and the activity of autolytic (self-destructive) enzymes, which is accompanied by **cell lysis**. Some bacteria escape death in this situation by forming spores or cysts that survive harsh conditions. In the stationary phase, complex adaptations to starvation and survival take place (see Chapter 28.5). There may be cryptic growth of survivors on the remnants of lysed cells. Also, at the beginning, growth may start at a slower speed because the cells used to inoculate the medium may require time to adapt to the new situation; this phase is called the **lag phase**, which reflects the shock of rapid change in the culture environment. The length of the lag phase can depend on the size and age of inoculum, and changes in nutrient composition and concentration experienced by the cells. Time is required to reverse the adaptations to the stationary phase and to prepare cells for growth. The lag phase may be long when cells come from a stationary phase pre-culture or from a different, richer growth medium. Inversely, transfer of exponential-phase cells grown on the same medium results in immediate growth without a lag phase. A typical **growth curve** in a **semilogarithmic plot**, which shows these growth phases, is shown in Figure 6.**1**.

Adaptation and repair of damaged cells require the induction of new enzymes. This is also true when cultures are grown on mixtures of "good" and "poor"

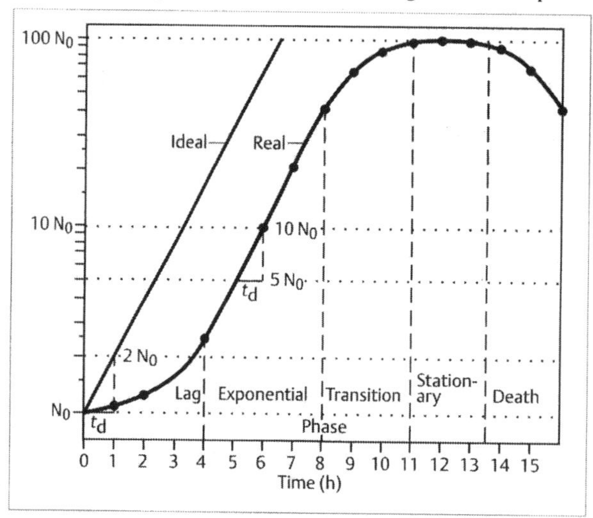

Fig. 6.**1** **Typical growth curve of a batch culture** in a semilogarithmic plot of time versus log N (1% inoculum). An ideal exponentially growing culture is shown for comparison. Note that a similar lag phase would result simply if only 10% of the initial cells N_0 (= inoculum) were viable. The slope of the line is equal to $\mu/2.303$; note that log N rather than ln N is plotted

substrates, of which the "good" substrates are preferentially used. After exhaustion of the "good" substrate, growth pauses and resumes only after derepression of the capacity to use the poorer second substrate. This phenomenon is called **diauxic** growth (see Chapter 20.2.1; catabolite repression). **Linear growth** occurs when the supply of a critical nutrient is constant, such as by drop-wise addition or by diffusion of an essential gas (e.g., O_2) into the medium; clearly, growth will respond in a linear manner. Another reason for linear growth could be that an essential enzyme cannot be made anymore, for example, because of the exhaustion of a cofactor, and the existing amount of enzyme allows only constant production of an intermediate (and therefore linear growth) until this enzyme becomes diluted too much by cell fissions. This is often the case during transition from exponential to stationary phase. Similar arguments can be used for linear hyphal growth and growth in pellets.

In many biotechnological processes, cell mass is used as a biocatalyst, and often desired products are only produced in the stationary phase. Optimal cell growth is achieved in the **trophophase**, and cells are then shifted to a productive non-growth condition (**idiophase**).

6.2 Physicochemical Factors Affect Growth and More

Physicochemical factors not only affect the growth rate, but also the cell yield, the viability, the metabolic pattern, differentiation, and—to some extent—even the chemical composition. Notably, the pH, temperature, and oxygen supply are critical with every bacterial culture and need to be controlled. Some bacteria can grow under dramatic circumstances, and no other group of living beings has evolved such remarkable tolerances to extreme environments (Fig. 6.**2**).

6.2.1 The pH Value of Culture Media Must Be Controlled

The pH value of a solution indicates the negative decadic logarithm of the proton activity in solution. Most growth media are dilute solutions in which the activity of the H^+ ion is nearly identical with its molar concentration.

Internal pH. Most species can grow over a wide pH range of ~4 units, with reasonably fast growth in a narrower range of 2 pH units. Yet they maintain their internal pH near a fixed optimal value, which can be quite different from the more alkaline or more acidic external pH. The internal pH is controlled by an unknown sensory system and the proton motive force. If the pH outside is more alkaline, the bacterium must transport protons into the cell and simultaneously create a membrane potential, $\Delta\psi$. The $\Delta\psi$ is exchanged for ΔpH (see Chapter 28.3).

pH range. Bacteria with a pH optimum for growth around pH 7.0, like *E. coli*, are **neutrophiles. Extremophiles** that grow optimally at very low pH are **acidophiles**; an example is *Thiobacillus ferrooxidans*, which grows between pH 2.0 and 8.0. **Alkaliphiles** depend on alkaline pH, such as *Bacillus alkalophilus* with a growth optimum around pH 10.5. The cellular pH of these three representatives is near 7.6, 6.5, and 9, respectively. Some bacteria can tolerate extremes of pH for some time, e.g., *Acetobacter* and *Lactobacillus* spec. As a rule, most fungi are adapted to a weakly acidic environment.

Occurrence. In Figure 6.**2**, the pH spectrum of some natural habitats and microorganisms growing in these habitats are shown. Alkaliphiles can tolerate pH values > 10, but growth in general is limited by a pH value of 10. Many bacteria consume and/or produce acidic or basic ions during growth. For instance, if NH_3 is assimilated from NH_4Cl as nitrogen source, HCl is left over.

Buffers. To avoid large changes in pH, growth media are normally buffered. The Henderson-Hasselbalch equation allows the calculation of the buffer composition for a given pH:

$$pH = pK + \log\frac{[Base^-]}{[Acid]} \qquad (6.7)$$

The buffer capacity is highest at the pK value of the buffering substance, where $[Base^-] = [Acid]$

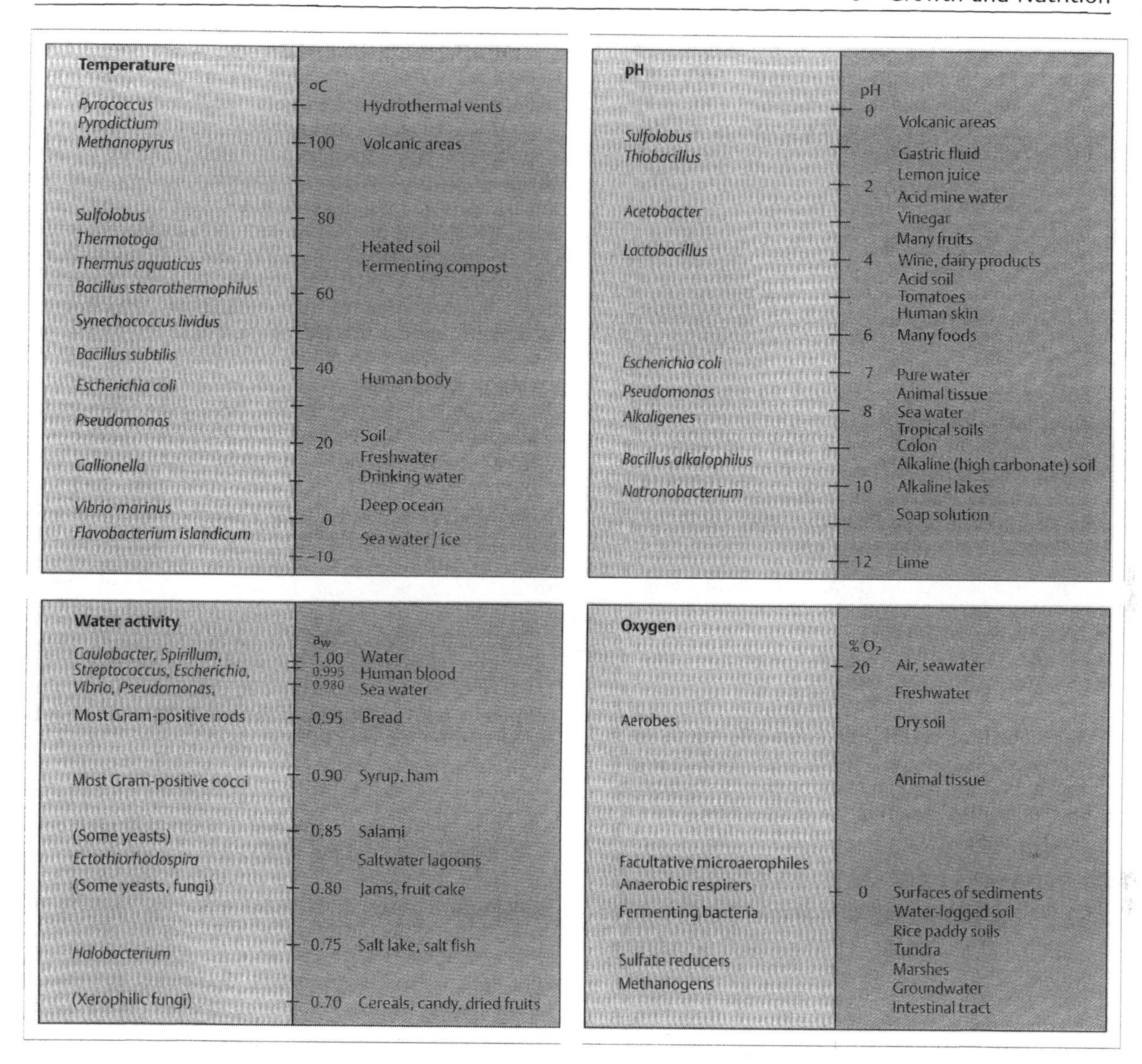

Fig. 6.2 **Physicochemical parameters of potential bacterial habitats and some representative organisms**

Box 6.2 The bicarbonate/dissolved CO_2 ($pK'_1 = 6.35$, 30 °C) and bicarbonate/carbonate ($pK'_2 = 10.3$, 30 °C) buffers are based on the reactions of gaseous CO_2 in aqueous solutions:

$$[\text{dissolved } CO_2 + H_2O \rightleftarrows H_2CO_3] \overset{pK'_1}{\rightleftarrows} HCO_3^- + H^+ \overset{pK'_2}{\rightleftarrows} CO_3^{2-} + 2\,H^+ \quad (6.8)$$

The equilibrium concentration of dissolved CO_2 and H_2CO_3 is approximately 1000 : 1; in calculations, the term $[H_2CO_3]$ can therefore be neglected. These buffers also meet the CO_2 requirement of many heterotrophic bacteria and are generally used for autotrophic bacteria. If solid media constituents are acceptable, the addition of finely ground chalk ($CaCO_3$) perfectly buffers even strong acid production; on agar plates, acid producers are recognized by clearing zones in which the white $CaCO_3$ (0.3%) becomes dissolved by the acid.

In the case of CO_2/bicarbonate, the Henderson-Hasselbalch equation is

$$pH = pK'_1 + \log\frac{[HCO_3^-]}{[\text{dissolved } CO_2]}$$

The solubility coefficient α for CO_2 varies with temperature, salinity, and the atmospheric partial pressure of CO_2 (gas). For many practical purposes, the following equation is sufficiently exact at normal pressure:

$$pH = pK'_1 + \log[HCO_3^-] - \log[\alpha \cdot (\text{vol\% } CO_2 \text{ gas}) \cdot (4.35 \cdot 10^{-4})] \quad (6.9)$$

with $\alpha = 0.64$ and $pK'_1 = 6.35$ at 30 °C. α indicates milliliters of CO_2 (reduced to 0 °C and 1 atm) that will dissolve in 1 ml water at a gas pressure of 1 atm at the stated temperature.

pH control. Many growth media contain pH indicators to indicate pH changes during growth, which may be used as a diagnostic criterium. pH indicators are strongly colored weak organic acids or bases that change color upon protonation/deprotonation. Examples are bromophenol blue (pK_a 4.0), methyl red (pK_a 5.2), and thymol blue (pK_a 8.9), which in the acid form turn yellow. When the pH of a culture must be controlled within a narrow range, an automatic pH control system is used. It consists of a steam-sterilizable pH electrode, a pH meter, a controller and controls for setting the desired pH limit, and pumps for the addition of acid and/or base.

pH effects on bacterial cells. What effects a exerted by unsuited pH values on bacterial cells? T cell interior is well buffered, with a buffering capacity > 0.1 M acid or alkali required to change the pH value 1 unit. The cell envelopes are directly affected changing pH; dissociation/protonation of their mac molecules (e.g., lipopolysaccharides, surface layer pr teins, cytoplasmic membrane) results in a different lo charge distribution, which results in changing morph logy, impaired cell division, changed adhesion, floccu tion, or dissolution of the cytoplasmic membra Metabolism is also affected. pH shifts are often used biotechnology to stop growth of a culture and indu excretion of a product (trophophase → idiophase, Cha ter 27). Solvent fermentations are shifted from ac $+ H_2$ production to alcohol production when the pH h reached a critical acidic pH value. Toxicity of ma compounds is a function of pH. Weak acids, such acetic, propionic, sorbic (each $pK_a = 4.7$), and benz acids ($pK_a = 3.7$) passively diffuse in the protonat form [AH] and are trapped inside the cell by dissociati to the impermeable anionic form $[A^-]$ and H^+ ("i trap"); the internal pH thereby drops. These compour only act as inhibitors if the external pH is acidic enou i.e., in the range of the pK value of the respective ac These acids, which are tolerated by the human body, a commonly used in food preservation. Adaptation to changes is described in Chapter 28.

6.2.2 The Temperature Range of Growth Extends Frc Below 0 °C to Above 100 °C

Temperature range. Most bacteria have the capacity grow over a temperature range of approximately 40 with a reasonable fast growth in a narrower range 20 °C (Fig. 6.3). The **minimum, optimum,** and **ma imum temperatures** (**cardinal temperatures**) a characteristic of each type of organism. The major of bacteria are **mesophilic**, with a temperatu optimum for growth between 20 and 42 °C. **Therm tolerant** species can tolerate temperatures up to 50 and **thermophiles** can grow up to 70 °C. These temp atures are reached in rotting compost and on su heated soil surface. **Extreme thermophiles** (also call **hyperthermophiles**) have temperature optima ov 70 °C, e.g., *Sulfolobus acidocaldarius* (80 °C) and *Pyrod tium occultum* (105 °C). These temperatures are reach in nature only in volcanic areas. Parts of the oceans w temperatures of approximately 5 °C and polar regic are habitats for **psychrophilic** bacteria, which ha optimal growth rates below 20 °C. Some grow well temperatures of −10 °C on the ice/seawater interpha in microscopic pockets of water; the bacteria seem to be limited by the availability of liquid water. Psychrophiles can be killed by a brief warming to room temperature. In contrast, most mesophiles are **psychrotolerant.**

Effects of temperature. The dependence of the growth rate on the temperature is basically similar in all bacteria; the following temperature values are exemplarily given for *E. coli* (Fig. 6.**3**). In a limited mid-range (21–37 °C), the dependence follows the Arrhenius equation, and there is no special temperature regulation during this "normal" temperature range. Thus, the logarithm of the growth rate is proportional to the reciprocal of absolute temperature, with a negative slope of the line. For most bacteria, the temperature coefficient (Q_{10}) for the growth rate in this limited range is approximately 2, as for many chemical reactions, i.e., the growth rate doubles with a 10 °C increase in temperature. In the normal range, the physiological state of bacteria is unaffected by the temperature. Cellular reactions remain coordinated simply by changes in enzyme activity. As will be discussed in Chapter 7, the fatty acid composition of the phospholipids in bacterial membranes varies with temperature

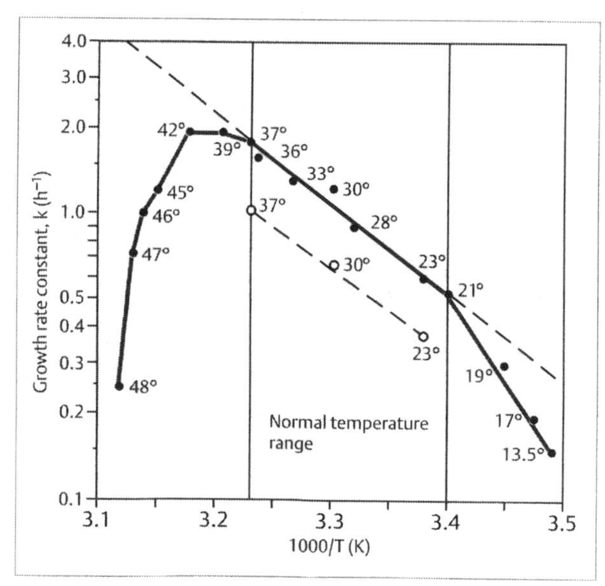

Fig. 6.3 **Growth rate of a mesophilic bacterium (*Escherichia coli*) as a function of temperature.** The log of the specific growth rate is plotted against 1/T, similar to the Arrhenius plot of a chemical reaction rate versus 1/T. k is equal to $1/t_d$ (h^{-1}), T is the absolute temperature (K = kelvin). For comparison, the °C temperatures are indicated. ●, in a rich medium; ○ in a glucose-minimal medium. The dashed extrapolation lines in the rich medium curve indicate the linear response of the growth rate, which is constant in response to 1/T, and highlight the linear response in the normal temperature range. [after 1]

and counteracts the changing membrane fluidity. The molecular adaptations of psychrophiles and thermophiles to extreme temperatures are not considered here (see Chapter 28.2).

Rapid shifting a culture from the normal range towards temperature extremes, e.g., sudden chilling to 4 °C or heating to 54 °C, can be deleterious. However, stationary-phase cells are much less sensitive to these changes. Susceptibility to killing by heat depends on the moisture content of the environment and also on the physiological state (e.g., low sensitivity in the stationary phase versus high sensitivity in the growth phase). The physiological state is also important in rapid chilling. These effects will be explained in Chapter 28.2.

6.2.3 Supply of Oxygen and Other Gases Can Easily Become Limiting

Obligately aerobic bacteria require O_2 for growth; **facultative aerobes** prefer aerobic growth, but can live without O_2. A large number of bacteria, the **microaerophiles,** are adapted to extremely low oxygen concentrations (e.g., 0.1–0.5%), a situation common in many natural habitats, and these bacteria often cannot yet be cultured in the laboratory, or if so, not in submersed culture, but rather in thin bacterial layers on floating particles, where they create their low-oxygen atmosphere by their own respiratory oxygen consumption.

Gas solubility. Oxygen is the terminal electron acceptor for aerobic bacteria, but other gases can serve for other functions, e.g., H_2, CH_4, and CO as electron donor and/or carbon source, N_2 as nitrogen source, H_2S as a reductant to keep off oxygen and as a sulfur or electron source, and CO_2 as an additional or sole carbon source. Oxygen, like many other gases, is relatively poorly soluble in water (see Chapter 30.1.3). The mass of a gas dissolved depends linearly on the partial pressure of that gas, and increasing temperature and osmolarity result in lower solubility.

O_2 supply. Dense cultures of aerobes require forced aeration. Many culture devices are designed to improve the oxygen mass transfer rate and to increase the gas/liquid interface. The opening of the culture vessel is covered with porous material (e.g., cotton) to allow maximum gas exchange. Examples of such culture vessels are shallow liquid volumes in flat-bottom vessels (Erlenmeyer or Fernbach flasks), flasks with indentations ("baffles") that are shaken, and stirred fermenters (volumes of cultures > 0.5 l) in which air or oxygen is sparged. The goal is to keep as many air bubbles as small and as long as possible in the culture broth to allow effective O_2 transfer into the liquid and to avoid excess foaming.

O_2 control. Dissolved oxygen is measured polarographically with commercial oxygen electrodes (see Fig. 11.**1**). Constant oxygen concentrations in solution are achieved by controlling the gas flow rate, the oxygen content in the gas supply, and the stirring speed. The effectiveness of a fermenter with respect to oxygen supply is expressed as the **oxygen absorption rate**, i.e., mmol O_2 transferred into 1 l oxygen-free culture volume per min. Bacteria growing in very thin layers on surfaces are normally well supplied with oxygen. Note, however, that oxygen consumption in a biofilm or in a bacterial colony on agar plates generally exceeds oxygen supply by diffusion, resulting in an oxygen-free zone in the interior of the colony and below the colony in the agar. This O_2 gradient also applies to biofilms or sediments. Hence, many bacteria shift to fermentation or anaerobic respiration under such conditions.

CO_2. Autotrophic bacteria require CO_2 as sole carbon source, which is often supplied as $NaHCO_3$ in closed vessels. Note that the concentration of dissolved CO_2 and HCO_3^- and therefore the distribution of total "CO_2" between the liquid and gas phase depend on the pH of the medium (see Chapter 6.2.1). Also heterotrophic bacteria require CO_2, mostly for biosynthetic carboxylations. Many pathogenic or symbiotic bacteria are adapted to the high CO_2 partial pressure of their hosts and often require 10% CO_2 in the gas phase. If CO_2 is rigorously absorbed by alkali, growth of most bacteria stops.

6.2.4 Water Activity and Osmotic Pressure Depend on the Number of Dissolved Osmotically Active Particles

Water is the solvent of biological systems. Water is a substrate in many enzyme reactions, hydration is responsible for the correct structure of macromolecules, and water maintains the cell turgor. Water availability depends on the water content of a microbial habitat, that is, how moist or dry a solid habitat such as soil or a piece of bread is. Furthermore, in aquatic habitats, dissolved solutes such as ions ($Na^+ + Cl^-$ from salt), sugars, or other substances compete for the solvent water and "fix" water, making it unavailable to organisms. In this sense, high salt or sugar content can be considered analogous to dryness. It is clear that these are two different situations that require different adaptations. Both types of "dryness" have in common that water diffuses from a region of high water concentration (wet; low solute concentration) to a region of lower water concentration (dry; high solute concentration) in the process of osmosis.

The availability of water to microorganisms is determined by the water activity (a_w) of the medium. This term is equal to the ratio of the vapor pressure of the solution or solid medium to that of pure water. In media containing dissolved solutes, the osmotic pressure (Π) of the medium and a_w are interrelated:

$$\Pi = \frac{R \cdot T}{V_w} \ln a_w \qquad (6.10)$$

where R is the gas constant, T the absolute temperature, and V_w the volume of 1 mol of water. Ideally, the osmotic pressure of a solution containing 1 mol osmotically active particles (such as Na^+ or Cl^- or glucose molecules) per liter amounts to 22 bar (2.2 MPa). This is equal to the gas pressure of 1 mol gas molecules in a liter. Values for a_w of some natural habitats of microorganisms are given in Figure 6.**2**.

Obviously, "dryness" can have different meanings and requires different molecular mechanisms of adaptation. Most common are habitats with high concentrations of salt; seawater contains about 3% NaCl plus small amounts of other minerals and elements. Marine organisms usually require Na^+ for growth and are **halophilic**. Other bacteria can tolerate salt (**halotolerant**), but do not require it. The spectrum of halophily reaches from low (1–6%) to moderate (6–15%) to extreme (15–30%). Most bacteria cannot survive at low water activity; this is the reason why salting, drying, or adding sugar preserves food. In fact, most bacteria prefer a water activity of 0.98–0.99. Organisms able to live in environments high in sugar are called **osmophiles**, and those able to grow in very dry environments (made dry by lack of water) are called **xerophiles**. The osmotic pressure of the medium normally is empirically adjusted by adding salt (NaCl, KCl, Na_2SO_4; artificial seawater). Note, however, that different ions at higher concentrations can have adverse effects. See Chapters 28.4 and 31.4 for adaptation to osmotic extremes.

6.2.5 Pressure Has Little Effect on Growth

Surprisingly, hydrostatic pressure has little or no effect on the growth of most bacteria in the range of 0–100 bar (0–10 MPa). However, **barophilic** bacteria exist in the deep sea that not only are adapted to, but require pressures similar to those encountered in their natural habitat (e.g., 600 bars or 60 MPa at 6000-m depth). The study of deep-sea microorganisms requires sophisticated equipment to maintain high pressure.

6.2.6 Anoxic Conditions Are Essential for Growth of Anaerobes

Occurrence. Many habitats are very low in dissolved oxygen or anoxic, or temporarily become anoxic (Fig. 6.2). Oxygen-free conditions are created by limited O_2 diffusion and O_2 consumption by microorganisms. In addition, many anaerobes produce reducing end products, such as H_2S, which may be able to reduce O_2 to water.

Anaerobes. Facultative anaerobes can grow under oxic or anoxic conditions. **Microaerophiles** are organisms that live in more-or-less stable oxygen gradients; they require O_2 for respiration, but tolerate only low oxygen concentrations. **Anaerobes** are unable to respire with O_2, but may grow in the presence of oxygen and tolerate it (**aerotolerant**). **Obligate** (**strict**) **anaerobes**, however, die upon exposure to O_2 or growth is at least severely inhibited. Their metabolism is adapted to a low redox potential, and some enzymes are inherently oxygen-sensitive. Strict anaerobes not only require a virtually oxygen-free medium, but in addition require reducing conditions. The medium has to be adjusted to a redox potential (E_0') more negative than 0 V. The reasons for oxygen toxicity and mechanisms for elimination of and protection against reactive oxygen species are discussed in Chapter 11.

> **Box 6.3** The redox potential of a redox system is described by the Nernst equation. E_0' is the standard redox potential at pH 7.0 versus the hydrogen electrode. If half of a reducing agent at pH 7.0 is in the reduced state and the other half in the oxidized state, the redox potential E' is equal to the standard value E_0 (which can be found in the literature), corrected for pH 7.0 (see Table 4.1).

Removal of oxygen. A culture medium is first boiled to render it low in oxygen. Residual oxygen is removed by gassing, e.g., with oxygen-free N_2 or N_2/CO_2 mixtures. Then a reducing agent is added, and the vessel is sealed. Reducing agents act in a manner analogous to that of buffers: some poise the redox potential, others the pH. Examples of such reducing agents are Na_2S, FeS, dithionite, organic thiols such as thioglycolate and cysteine, and titanium (III) citrate. Their standard redox potential at pH 7.0 varies from approximately −0.2 V (cysteine) to −0.5 V [Ti(III)]. Reducing agents reduce traces of O_2 to H_2O and create a redox buffer. Traces of oxygen in gases can be removed by absorption; alternatively, ~5% of H_2 is added, which with the aid of a palladium catalyst, reacts with O_2 to form H_2O. The appropriate low redox potential can be conveniently controlled by adding trace amounts of non-toxic, colored (in the oxidized state), autoxidizable redox dyes, which turn colorless in the reduced form. The standard potential at pH 7.0 (E_0') of the most common dyes varies from around 0 V (methylene blue, resazurin) to −0.25 V (phenosafranine). Hence, when resazurin is colorless at pH 7.0, the E' of the medium is below 0 V and suitable for the growth of strict anaerobes.

Handling of strict anaerobes (e.g., plating), is greatly facilitated by a flexible plastic anoxic glove box chamber (anaerobic chamber). Materials are taken in and out through an air-lock, which can be evacuated and refilled with N_2 gas containing 5% H_2 (see above). Less-sophisticated equipment also works. Typically, pre-reduced media are prepared in rubber-stoppered glass tubes or bottles that are flushed with oxygen-free N_2 gas and filled completely to the top. Additions and samplings are done using sterile syringes. Less-sensitive anaerobes can be grown even on agar plates in anaerobic jars in which oxygen is removed by H_2. H_2 is evolved from $NaBH_4$ at acid pH in a separate bag to which water is added. Then, the jar is closed and $2\,H_2$ react with O_2 to form $2\,H_2O$ due to a catalyst.

6.2.7 Light Can Serve as an Energy Source, but Can Also Be Deleterious

Light of the correct spectral region (wavelengths) and intensity is required as an energy source for the growth of **phototrophic bacteria**. For phototrophs and non-phototrophs, intense light can be inhibitory due to photochemical reactions. Hence, phototrophs harbor photopigments, which serve two functions, as light-energy converting systems and as protecting systems; the latter systems are found also in many non-phototrophs (see Chapters 13, 30.1, and 31.1).

6.3 Growth Media Provide all Essential Nutrients

Bacteria require for growth—in addition to the proper physicochemical conditions—an energy source and sources for all elements that occur in the living cell (see Chapter 7). **Specialists** have defined and narrow requirements, whereas **generalists** can use a wide range of nutrients.

Prototrophs can grow on simple medium containing one single compound as the source of carbon and energy and a few inorganic salts as the supply of the other elements. **Auxotrophs** depend on preformed cell constituents, be it trace amounts of vitamins as precursors of coenzymes and prosthetic groups, or larger amounts of amino acids and other organic growth factors that are constituents of cell polymers or essential organic solutes. Often the growth requirements are not known and are approximated by complex natural materials, such as blood serum, ruminal fluid, earth decoction, yeast extract, and digested protein (peptones). A few **obligate parasitic bacteria** or **symbiotic bacteria** cannot be grown yet outside the living host cells. Similarly, some anaerobic bacteria that ferment "low-energy substrates" can only be grown in the presence of a **syntrophic partner** organism that removes the fermentation products by way of a specialized metabolic pathway (see Chapters 30.7).

6.3.1 Energy, Electron, and Carbon Sources and Electron Acceptors Are the Quantitatively Most Important Nutrients

Energy source. Light is the primary energy source on earth, and **phototrophic bacteria** can use it as the sole energy source for growth. If light energy cannot be used, chemical compounds have to be provided that can be either oxidized or fermented (by **chemotrophs**). Oxidation requires an electron acceptor, commonly O_2; however, many bacteria are able to perform anaerobic respirations. Hence, either O_2 or another oxidant has to be provided as electron acceptor for respiring bacteria.

Electron source. An inorganic compound is oxidized by **lithotrophs** to serve as electron donor for respiration and biosynthesis, whereas an organic molecule serves these functions in **organotrophs**.

Carbon source. In **heterotrophs**, the organic substrate also serves as the source of cell carbon. CO_2 can be used by **autotrophic** bacteria as sole carbon source, provided that energy and reducing power are available. Therefore, if light energy cannot be used, organisms have to be provided with an organic or inorganic substrate that can be used as the source of electrons and energy. Respiring bacteria in addition need an electron acceptor (see Chapters 10–12).

6.3.2 Carbon, Nitrogen, Sulfur, and Phosphorus Are the Major Elements

Carbon, nitrogen, sulfur, and phosphorus are required by all bacteria for cell synthesis; together with H and O they form the polymers. The amounts needed can be estimated from the elemental composition of cells (see Chapter 7), which is approximated by $C_4H_7O_{1.5}N$ plus small amounts of P, S, Fe, alkali and earth alkali elements plus trace elements. Note that organic carbon compounds generally serve a dual purpose, as a source of energy and as a supply of **carbon** (50% of cell mass). As a rule, approximately half of the organic carbon compound is assimilated by aerobic organisms. Fermenting bacteria can assimilate much less, typically 10–20%; the rest is required as an energy source. The whole spectrum of naturally occurring organic substances must be considered as potential substrates, with glucose being the most common organic molecule and preferred carbon source of many bacteria.

Similarly, **nitrogen** (14% of cell mass) can be presented in many molecular forms; NH_4^+ is generally preferred. Nitrate, urea, amino sugars, and amino acids, among others, will satisfy the nitrogen requirement of many bacteria. Amino acids are mostly provided as inexpensive protein hydrolysates (peptones). N_2 fixation is restricted to few bacteria. The oxidized nitrogen sources are reduced to NH_3 inside the cell, and NH_3 is the form that is assimilated.

The **sulfur** ($\leqslant$1% of cell mass) source in nature is mostly organic sulfur, except for marine environments where sulfate occurs at a concentration of 28 mM. The preferred organic thiols are cysteine, cystine, or methionine (often in peptone). Rarely, elemental sulfur can be used. The oxidized sulfur sources are reduced to H_2S inside the cell, and H_2S is the form that is assimilated.

Phosphorus (3% of cell mass) is normally provided as phosphate or as phosphate esters in nature. There are no reduced phosphorus compounds that are stable. Phosphate does not need to be reduced since the cellular phosphorus is in the oxidation state of phosphate.

Oxygen and **hydrogen** are derived from water and/or from the organic carbon source. There are only few exceptions where O_2 is required for hydroxylation, i.e., the introduction of an OH group from molecular oxygen in the biosynthesis of cell constituents.

6.3.3 Minerals Are Minor, but Important Constituents

Cells are dependent on minor amounts of some alkali (Na^+, K^+) and earth alkali (Mg^{2+}, Ca^{2+}) metal ions as electrolytes and cocatalysts. In addition, trace elements are required, particularly Fe, and to a lesser extent Zn, Mn, Co, Mo, Cu, Ni, W, Se, V, B, and others (Table **6.1**). Not all of these elements are needed by all bacteria. The concentrations required vary from the millimolar range

Table 6.1 **Common trace elements and their cellular function in bacteria**

Element	Function and location
Iron (Fe)	Heme, siroheme, iron-sulfur centers in many oxidoreductases and some other classes of enzymes, electron carriers
Manganese (Mn)	Enzymes reacting with active oxygen species, such as superoxide dismutase, water-splitting enzyme of photosystem II
Cobalt (Co)	Cobalamine (B_{12}), required specially for carbon rearrangement reactions and CH_3-transfer
Nickel (Ni)	Ni-tetrapyrrole of methanogens, in urease, hydrogenase, CO dehydrogenase
Copper (Cu)	Many enzymes that react with O_2, such as cytochrome *c* oxidase, some superoxide dismutases, plastocyanin, oxygenases
Zinc (Zn)	Different enzyme classes, notably in carbonic anhydrase, alcohol dehydrogenase, RNA and DNA polymerases, some proteinases
Molybdenum (Mo)	Many oxidoreductases that catalyze the formal introduction of an OH group from water into the substrate or the reverse reaction (molybdopterin), most nitrogenases
Tungsten (W)	Substitute for Mo in some oxidoreductases that normally would contain Mo; cannot be replaced by Mo in these cases
Selenium (Se)	Incorporated in selenocysteine in some oxidoreductases; occurs in tRNA
Vanadium (V)	V-dependent nitrogenase, bromoperoxidase

(alkali, earth alkali metals) to a few micromolar (Fe) and less than 1 μM (10^{-6} to 10^{-8} M) for other trace elements. Trace elements are toxic at higher concentrations. Most divalent and trivalent metal cations tend to form insoluble hydroxides or phosphates at neutral to alkaline pH, making these elements unavailable for use to bacteria. Concentrated stock solutions of inorganic salts therefore are often kept anoxically at acidic pH, and minimal amounts of a complexing agent such as EDTA or nitrilotriacetate (NTA) are incorporated; some medium constituents [e.g., some amino acids (Cys, His), or carboxylic acids (malate, citrate)] can also form soluble metal complexes. Over-chelation similarly results in metal starvation, as does precipitation.

6.3.4 Vitamins Are Sometimes Required for Coenzyme Synthesis

Vitamins are required by some bacteria, which cannot synthesize them. Often, just one is required, such as vitamin B_{12} for growth of *E. coli* on certain substrates. Vitamins are organic precursors or constituents of coenzymes or prosthetic groups (Table 6.**2**) and are required in very small amounts (10^{-6} to 10^{-7} M). Vitamins occur in many habitats, due to decaying biological material, and since their synthesis requires many enzymes, bacteria often prefer to take them up by high-affinity transport systems rather than to make them. Yeast extract (0.1–0.5%) normally meets the minimal vitamin requirement, but synthetic stock

Table 6.**2** **Common vitamins and their cellular function in bacteria**

Vitamin	Function
Folic acid	Coenzyme of C_1-metabolism and C_1-transfer
p-Aminobenzoic acid	Precursor of folic acid
Biotin	Coenzyme of ATP-driven carboxylations of $-CH_2-$ next to a $-C{=}O$ group, e.g., in fatty acid synthesis
Nicotinic acid	Precursor of NAD(P)
Pantothenic acid	Precursor of CoA
Riboflavin	Precursor of FAD, FMN
Lipoic acid	Irreversible oxidative decarboxylation of 2-oxo acids
Thiamine	Decarboxylation of 2-oxo acids, transketolase
Pyridoxal, pyridoxamine	Precursor of pyridoxalphosphate, coenzyme in amino acid transformation reactions
Cobalamine (B_{12})	CH_3-transfer reactions; precursor of coenzyme B_{12} which is required in several carbon rearrangement reactions, ribonucleotide reductase

solutions are preferred. These solutions should be kept dark, cold, anoxic, sterile (by filtration), and at slightly acidic pH (pH 6) and should be added only after autoclaving the medium.

6.3.5 Amino Acids and Other Growth Factors Are Required by Auxotrophs

Amino acids and other growth factors are constituents or precursors of cellular macromolecules or of important solutes that are required in relatively high amounts (millimolar concentrations) by bacteria that cannot make them themselves. Examples of growth factors other than amino acids, are peptides, pyrimidines, purines, (unsaturated) fatty acids, cholesterol, mevalonic acid, choline, betaine, and polyamines. Unidentified growth factors can be present in serum or yeast extract.

6.3.6 Undefined Media Are Less Expensive and Easier To Work With

In the previous chapters, the rationale of medium composition based on pure individual chemical compounds was explained (**defined** or **synthetic media**) (Table 6.3). **Minimal media** contain only those compounds that are essential for growth. Such media are indispensable for studying the minimal growth requirements. In nature as well as in practice, **undefined** or **complex media** made up of inexpensive food stocks are common (see Table 6.3). Stark, malt, or molasses often serve as carbon and energy source. Enzymatic, acid, or alkali hydrolysates of inexpensive protein (peptones from casein, soy protein, meat, fish) or corn steep liquor are used mainly as sources of N, S, P, and amino acid building blocks. Yeast extract made by autolysis of baker's yeast is high in amino acids, peptides, water-soluble vitamins, and carbohydrates, and contains also essential trace elements. Even supplements such as tomato or fruit juice or skim milk mimic the milieu of natural habitats of certain bacteria. Pathogens may require serum, rumen bacteria rumen fluid. It is possible to satisfy growth requirements of most lactic bacteria by using one complex medium: milk. A defined culture medium for the same bacteria would require some 40–50 chemicals, including most amino acids, vitamins, and trace elements.

Table 6.3 **Example of a synthetic minimal growth medium.** This medium contains approximately 20 different chemicals. For comparison, a complex medium (LB) for growth of enterobacteriaceae contains only tryptone (10 g/l), yeast extract (5 g/l), NaCl (5 g/l), and NaOH to adjust the pH to 7.0. In both cases, agar medium contains in addition 15 g agar.

Compound (M_r)	g/l	Final concentration in the growth medium
Na_2HPO_4 (142)	6 }	
KH_2PO_4 (136)	3 }	
Glucose (180)	4	22 mM
NH_4Cl (53)	1	19 mM
$MgSO_4 \cdot 7\,H_2O$ (246)	0.2	0.8 mM
$CaCl_2 \cdot 2\,H_2O$ (147)	0.01	68 μM
$FeSO_4 \cdot 7\,H_2O$ (278)	0.01	36 μM
Trace element solution (100×)	(10 ml)	0.5–5 μM each
Water (pH 7.3)	(1 l)	

6.3.7 Other Additions, Such as Buffers and Agar, Are not Used as Nutrients

All media require a **buffer** to avoid strong pH shifts. Common buffers are phosphate ($pK_2 = 7$), bicarbonate/CO_2 ($pK'_1 = 6.3$), carboxylic acid (acetic, succinic, citric acid (pK values between 3.1 and 6.4), and protein buffers (average pK around 7.0). In media, the organic natural buffer used should not be consumed by the bacteria. A series of synthetic buffers such as Tris or "Good" buffers (zwitterionic organic sulfonic acids) are widely used in the laboratory. It has to be taken into account that many buffers are by no means inert, but may complex di- and trivalent metal ions, serve as a nutrient, or form insoluble precipitates with media components. Notably, many bacteria do not tolerate high phosphate concentrations.

Solid media are normally supplemented with **agar** as solidifying agent (0.8% for soft agar, 1.5–2% for normal solid agar). Agar is a complex polysaccharide from marine red algae that melts at 100 °C and, once molten, remains liquid down to 45 °C. Bacteria growing at extremely high temperatures, high salt concentrations, or at very acidic pH, or those that can degrade agar are grown on media solidified with **silica gel**. This solidifying inorganic agent is also useful for the demonstration of autotrophic growth.

6.4 Sterilization of Media and Equipment Are a Must for Maintaining Pure Cultures

Sterilization frees the object of any living organism. The sterilization methods and their uses are described below.

Autoclaving. Media and heat-resistant equipment are normally steam-sterilized or autoclaved for 15 min at 121 °C, which is sufficient to kill bacterial cells and spores (note, however, that some extremophiles can survive even this treatment). At this temperature, pure water steam has a pressure of 2 bar (0.2 MPa), which is obtained before the autoclave reaches 121 °C if air is present (which contributes to the total pressure when heated). Since autoclaves for the sake of safety are normally pressure-controlled rather than temperature-controlled, the air in the chamber must be expelled and replaced by steam.

Kinetics of death are nearly always exponential. The killing rate follows first-order kinetics; the same proportion of cells is killed per time unit. The efficiency of the sterilization process and the sensitivity of an organism is characterized by the decimal reduction time (D_{10}), i.e., the time required to kill 90% of the cells. Especially heat-resistant endospores of *Bacillus stearothermophilus* are used as indicators, with a D_{10} value of approximately 30 s at 121 °C (Fig. 6.**4**).

The practical consequences of the different heat sensitivity of cells and spores and of the different decimal reduction times, D_{10}, are important. D_{10} decreases essentially exponentially with increasing temperature, and vice versa. Any material with high numbers of endospores will require special attention; the time taken for a definite fraction of the cells to be killed is independent of the initial concentration.

Pasteurization. For many practical purposes, it is sufficient to reduce drastically the number of viable microbes by heating for a short period of time (minutes) at 65–75 °C. By this treatment, most pathogenic bacteria are killed, and the storage of heat-treated foods such as milk is improved. For example, milk is heated for 20 s at 71–74 °C, and canned food is normally heated to 80 °C for 20 min. Note that this treatment is not equal to sterilization since spores survive; however, the acidic pH of many sorts of canned food prevents germination and growth of the spores.

Dry heat. Bacterial endospores are quite resistant to dry heat. Equipment can be sterilized by incubation for 2 h at 160 °C or 30 min at 180 °C. Again, note that the equipment to be sterilized and not only the incubator must attain this temperature.

Filtration. Heat-labile or volatile water-soluble compounds are sterilized by filtration of their aqueous solutions. For small scales, membrane filters with pore sizes of 0.2 μm and 0.45 μm are available, and pre-sterilized filtration systems are almost foolproof. For large-scale filtration different filtration systems are used, based for example, on thick porous porcelain or kieselgur (diatomaceous earth) layers.

Radiation. X-rays, γ-rays, and other ionizing radiation methods are used only on an industrial scale. UV light is used on clean benches and laboratories to keep the number of viable bacteria low.

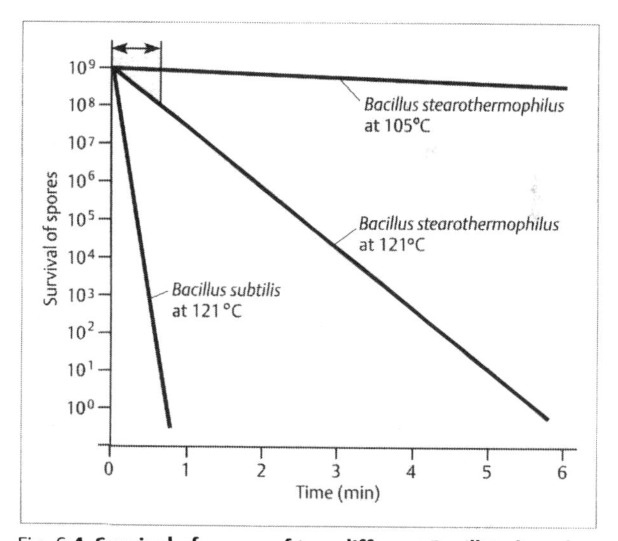

Fig. 6.**4** **Survival of spores of two different Bacilli in liquid at different temperatures.** (Note the logarithmic scale.) For example, the decimal reduction time of *B. stearothermophilus* at 121 °C is approximately 0.6 min, as indicated by the double-headed arrow

6.5 Growth Can Be Measured by Various Methods

Growth of bacteria is associated with cell division and, therefore, the increase in the total number of bacteria. One method of growth determination is therefore based on the count of individual cells. However, this method does not distinguish between dead and live cells, and sometimes even discrimination of small cells and other particles is difficult. Therefore, the increase in colony forming units may be a useful method under certain circumstances, e.g., in highly diluted samples. Increase in cell mass is the most common growth criterion that can easily be followed. For mycelial or filamentous growth or for cells that clump, this is even the method of choice. The increase in dry mass concentration is most conveniently determined by measuring turbidity. Finally, for many practical purposes, the consumption of a substrate (e.g., O_2) or the formation of products (e.g., CO_2, acid) may be proportional to bacterial growth.

Turbidimetry. Bacterial size is in the order of the wavelength of visible light, and bacteria scatter light quite well. They appear turbid, and turbidity can be measured by a spectrophotometer (Fig. 6.**5**). Such instruments measure how much of the intensity of light, which passes through the sample and reaches the photocell, was lost by scattering (and absorption). A constant path length d of the cuvette is required for comparison. Absorbance is related to the light intensity of incident (I_0) and transmitted (I) light, following $A = \log(I_0/I)$. Strictly speaking, Lambert-Beer's law, $A = \varepsilon \cdot c \cdot d$, applies to non-particulate solutions. However, there is a linear relationship between the "absorbance" A and the bacterial cell concentration N: $A \sim d \cdot N$, up to $A = 0.3$. Normally, light of wavelengths between 500 and 660 nm is used (the wavelength is given as a subscript of A); this is the spectral region in which most cell constituents do not absorb. The instrument reading is often referred to as "optical density" OD (1 cm cuvette) measured against a cuvette containing medium only, or as "Klett units" for the Klett-Summerson photometer. Dense cultures obviously need to be diluted before measurement.

Light scattering not only depends on the number of particles, but also on their size and form. However, most bacteria, independent of size, have nearly the same absorbance per unit dry mass concentration. Using a standard curve of the dry cell mass determination versus the OD reading, turbidometry therefore allows a very precise and convenient indirect measurement of cell mass. Note, however, that different absorbances are found per individual bacteria with different cell sizes.

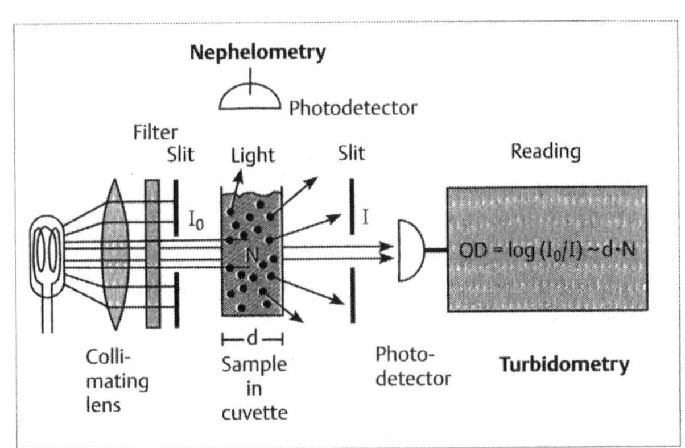

Fig. 6.**5** **Illustration of the measurement of turbidity caused by suspended bacteria.** The incident light is I_0 and the transmitted light is I. The instrument reading provides $\log(I_0/I) = OD$ (optical density). Similarly, scattered light is measured by nephelometry. d, path length; N, bacterial cell concentration

A less common method called **nephelometry** measures the intensity of scattered light at a 90° angle from the primary beam (note that most of the light is scattered, however, at a low angle range of 2–12°) (Fig. 6.**5**). This method is quite sensitive, but there can be false signals from particulate impurities in the medium. Special equipment is needed, and the instrument requires frequent calibration.

Biomass measurement. Turbidometry fails with bacteria that exhibit mycelial or surface growth or with those that form aggregates, long filaments, and clumps. Also, growth of mixed cultures with varying proportions of the different species cannot be followed adequately. Measuring the wet weight normally is too inaccurate and variable. Therefore, after centrifugation or filtration of cells and washing with water, the dry weight obtained after drying in an oven at 105 °C to a constant weight is determined. Alternatively, cells can be freeze-dried. The dry mass content is usually 10–20% of the wet mass. The biomass can be estimated indirectly by measuring the protein content of a culture. Cells are acid-precipitated, and protein is determined colorimetrically. The relationship between the two quantities biomass and protein is only constant when growth is balanced, i.e., the composition of the cell does not vary.

This is, however, not the case in most batch cultures at the end of exponential growth or when cells form huge amounts of storage material. Also, the C or N content can be measured.

Metabolic parameters proportional to growth. If growth is to be expected to be more-or-less balanced, and no uncoupling of metabolism from biosynthesis is to be feared, the consumption of substrates (O_2, carbon source) or formation of products (CO_2) can simply be followed. This can be done in the gas inlet and outlet, e.g., by mass spectrometry or infrared absorption spectroscopy. If the growth yield is known, the cell mass formed can be estimated.

Colony counts. Counting colonies is a valuable method for determining the number of viable cells that develop and divide under given conditions. However, the way in which the inoculum is exposed to the new environment can lead to irreproducible results. In addition, resistant stages of bacteria can be overlooked. Normally, bacteria need to be diluted (serial dilution) in, for example, phosphate-buffered saline. Since the approximate **viable count** is unknown ahead of time, several 10-fold or 100-fold dilutions of the sample usually have to be made. In contrast, many aquatic samples contain very few bacteria and need to be concentrated by filtration or—less conveniently—by centrifugation. Cells in 0.1 ml can be spread on the surface of solid media (**spread plates**) or 0.1–1 ml are suspended in the not-yet solidified medium and poured into plates (**pour plates**). It is assumed (and has to be checked) that after short exposure to agar medium at 45 °C the cells survive. Anaerobes are similarly grown in so-called **roll tubes**, which are stoppered tubes containing a small amount of liquified agar medium. Rolling the tubes horizontally in the cold results in a thin agar film on the wall. Another method, the **layered plate**, uses a solid supporting medium onto which the bacteria are spread. A second thin layer of soft agar is spread over this plate. Small, compact colonies develop in the subsurface. Similărity, in **thin-layer plates**, bacteria are suspended in a small amount of molten, but cool soft (0.75%) agar medium; the mixture is then poured onto hardened medium, and the thin overlay is allowed to harden. On the average, between 30 and 300 colonies formed per plate (or roll tube) are counted. Several hundred colonies need to be counted to keep the statistical error low. The standard deviation is equal to the square root of the counted colonies. Each of these methods requires that the **plating efficiency** is controlled by comparison with direct counts (see below) and that the efficiency is near 100%. It must be ensured that the cells do not form clumps. The result is normally presented as **colony forming units** (**CFU**).

The **membrane filter method** is the method of choice when counting bacteria in dilute samples (e.g., drinking water). The samples are filtered onto an appropriate membrane filter, which is then placed on an agar medium plate or even onto blotter pads soaked with liquid medium. The nutrients penetrate the filter and colonies develop on the filter surface. These methods may suffer from the sensitivity of certain bacteria to substances present in agar. Colony count methods can be combined with physiological tests by using indicator agar.

Most probable numbers (MPN). This method allows the rough estimation of viable cells by determining the fraction of multiple cultures that fail to show growth in a parallel series of dilution tubes containing a suitable medium. The tubes exhibiting no growth are assumed to have failed to receive even a single viable cell. The distribution of such cells must follow a **Poisson distribution**. The mean number m plated at this dilution can be calculated from the formula

$$m = -\ln P_0 \tag{6.11}$$

where P_0 is the ratio number of tubes with no growth to the total number of tubes at this dilution. The mean number is then multiplied by the dilution factor and corrected for the inoculum size to yield the viable count of the original sample. This sounds cumbersome compared to direct counts; however, this is the method of choice when a bacterium does not grow on solidified medium. It also detects bacteria that grow much slower than others in the same sample. Most importantly, if other organisms not of interest are present in the sample, the bacterium to be studied can still be identified, for example, by microscopic examination or by a characteristic product (e.g., sulfate reducers form H_2S, which in the presence of Fe(II) forms black FeS precipitates), or other bacteria can be excluded by selective media.

Direct counts. Bacteria can be counted directly in a simple counting chamber by microscopic observation. Such chambers consist of a special glass slide with a small, precisely defined depression (usually 0.02 mm). The depression is filled with the culture sample. A grid is marked on the surface of the chamber with squares of a small known area. When tightly covered with a rigid cover slip the bacteria are counted in a very small volume of known size. Even membrane filters after staining of the bacteria can be used for direct microscopic count. The membranes are dried and made transparent by immersion oil. This method is useful when samples contain less than 10^6 cells/ml. Alternatively, cells can be concentrated before by centrifugation. Electronic particle counting allows the determination of the number of bacteria and also the size distribution (using a **Coulter counter**).

Growth measurement in natural environments is a complicated task. Normally, quite indirect methods have to be used, such as incorporation of radioactivity into biomass from $^{14}CO_2$ or $[^3H]$thymidine, which may be taken as an indirect measurement of growth. Also DNA staining with fluorescent dyes (e.g., 4′,6-diamidinno-2-phenylindole, DAPI) or reaction with fluorescent DNA probes allows direct count by use of a fluorescence microscope.

6.6 Culture Preservation Maintains Bacteria Alive for a Long Time

Culture preservation aims at maintaining a bacterial strain alive, uncontaminated, and without variation or mutation, as was the original isolate. Success depends on the use of the proper medium and cultivation procedure and on the age of the culture at the time of preservation.

Short-term methods only partly fulfill these criteria. One method is the periodic serial transfer to fresh medium and subsequent growth at low temperature and intermediate storage in a refrigerator. Cultures of spore formers that contain mature spores can be kept dry in sterile soil, on sterile filter paper, or as cultures dried by silica gel. Other bacteria can survive in dried gelatin drops. For many purposes, cells can be kept viable for years when frozen at $-20\,^{\circ}C$ or better at $-70\,^{\circ}C$. In general, cells in the mid- to late-exponential phase of growth should be used, concentrated, and resuspended in a small amount of fresh growth medium. Freezing requires the presence in the medium of cryoprotective agents, such as glycerol (15–40 vol%) or dimethylsulfoxide (DMSO; 5–10 vol%) to prevent cellular damage during the freezing process, which should proceed at a cooling rate of approximately $1\,^{\circ}C$/min. In contrast, rapid thawing of frozen cultures is recommended.

Long-term methods are now widely used and use either freeze-drying or **ultrafreezing** in liquid nitrogen ($-196\,^{\circ}C$) or above the liquid nitrogen ($-150\,^{\circ}C$). Each of these methods has to be evaluated in terms of the percentage of survivors after a given storage time. The expected shelf life of bacteria preserved by both long-term methods is >40 years, as compared to 1–3 years in the case of deep-freezing. Routinely, glycerol (10 vol%) or DMSO (5 vol%) are added for ultrafreezing. Cryoprotective agents are water-soluble and cell-membrane-permeable, have a low melting point, and can substitute for water as a hydration shell, thus reducing the severity of dehydration upon freezing of water. Hence, freezing and formation of external ice crystals is depressed. As ice crystals outside the cell continue to form, the solute concentration increases and water begins to migrate out of the cell. Eventually, all of the free water crystallizes as pure ice, leaving a concentrated solute that finally freezes with only little formation of crystals that would damage the cell.

Freeze-drying involves the removal of water from frozen cells by sublimation in vacuo. The dried cells can be stored for long periods if kept away from oxygen, moisture, high temperature, and light. Often, skim milk (20%, w/v), horse serum (10%, v/v), sucrose (12%, w/v) or other cryoprotective chemical agents need to be added.

6.7 Selective Culture Methods, Enrichment, and Isolation Are Permanent Tasks of Microbiologists

So far, all the essential steps in growing and maintaining a given bacterial culture have been described. Many new scientific questions and applications, however, require a new look at nature and the isolation of new species with unprecedented or unknown properties. To find these, enrichment cultures are required, as introduced 100 years ago by S. N. Winogradsky and M. W. Beijerinck.

Principle. Most samples taken from nature contain hundreds of different bacterial species in one gram. Still, many thousands of bacterial species exist, and the one desired might be missing. Therefore, the right **inoculum**

first has to be selected from an appropriate habitat. For instance, extreme thermophiles are likely to be found in a hot spring, anaerobes in the sediment. Natural enrichments exist already in nature when a specific substrate is continuously supplied under more-or-less constant conditions. The goal of this technique then is to design a medium and a set of physicochemical conditions that are **selective** for the desired organism and are unfavorable for the others. Unfortunately, this method generally results in a dominance of rapidly growing species, which finally will overgrow the others. The slowly growing species will be discovered only if the samples are plated early in the enrichment. The samples can even be directly plated on selective solid media, and visible colonies might form with time under these conditions (**direct plate**).

Selectivity may be achieved by combining different energy, electron, and carbon sources, different sources of the other elements (N, P, S, etc.), different electron acceptors, the presence of inhibitory compounds, differences in light and oxic conditions, and differences in temperature, pH, water activity, salt, etc. The concentrations of chemicals can be varied, and the spectral region of light can be defined as appropriate.

Pure cultures derived from single cells (clone cultures) are obtained basically as described in the sections describing growth measurement using different plating methods (Chapter 6.5). The simplest way is to streak a sample of the enrichment culture on solid medium (**streak plate**), or to add a sample of the enrichment culture to molten agar medium and pour it into plates (**pour plates**) or tubes (**agar shake**). A second plate is streaked from cells from a well-isolated colony. If all the colonies on the second plate appear identical, a well-isolated colony can be used to establish a liquid culture. However, rapidly moving bacteria, notably spirochetes, can spread over the moist surface and contaminate the plate. Alternatively, serial 1 : 2 dilutions can be made; the most diluted tube that shows growth might have been inoculated with just one bacterial cell (**liquid dilution**). The dilution method can be used only to isolate the numerically predominant member of an enrichment culture. When neither plating nor dilution methods can be applied, the microscopically controlled isolation of a single bacterium using a fine capillary pipette and a micromanipulator is an alternative. These steps need to be repeated, and purity of a colony or culture need to be proven rigorously; the latter in many cases is all but trivial.

6.8 Continuous Cultures Are Valuable Tools

6.8.1 Chemostat

Bacteria can adjust μ according to the conditions imposed by the growth medium. The response of μ to variations in substrate concentration is nicely demonstrated by means of a chemostat (continuous) culture (Fig. 6.**6**). Continuous cultures differ from batch cultures by a device for continuous and regulated addition of growth medium in which one substrate is limiting growth. The volume of the culture is kept constant by an overflow. The culture is stirred to ensure instantaneous distribution of the added fresh medium.

Chemostat cultures exert two surprising properties. First, at a constant inflow of new medium, the culture maintains a constant cell density (Fig. 6.**7**). Secondly, the steady state cell density, N, remains almost unchanged even upon a large variation of the flow velocity, F (dimension: volume/time unit, e.g., l/h), of medium into the culture. At a higher flow velocity, however, the steady-state cell density drops within a relatively narrow range and reaches zero at a critical value, which is related to the maximal specific growth rate (μ_{max}) observed during the exponential growth phase with the

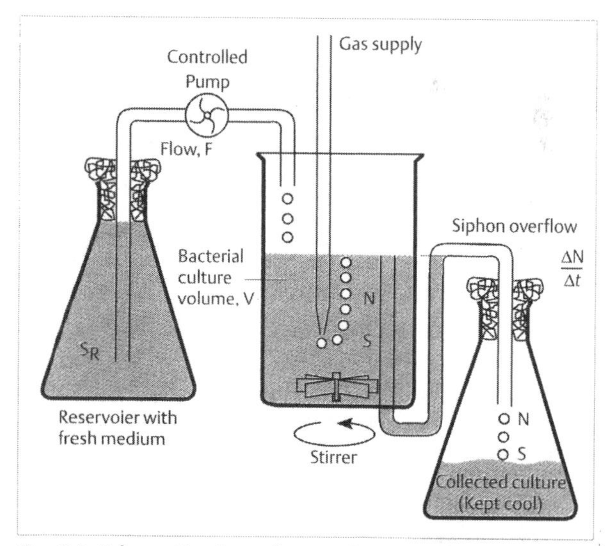

Fig. 6.**6** **Schematic view of a continuous culture device or chemostat.** The volume of fresh medium added equals that of the culture fluid leaving through the overflow. Thus, the culture volume (V) is kept constant. S_R, substrate concentration in the reservoir; F, flow velocity; N, cell density; S, substrate concentration in the culture (and overflow)

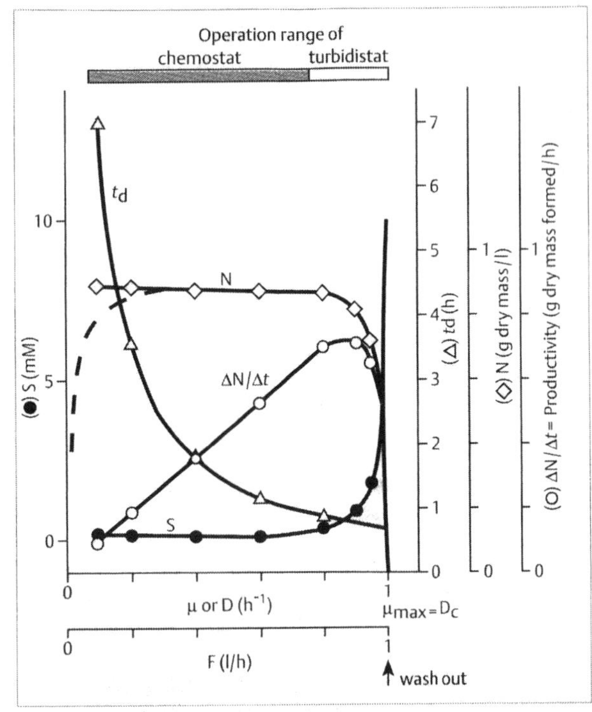

Fig. 6.7 **Cell density N (or cell number) and concentration of the growth limiting substrate S in the steady state of a chemostat as a function of flow velocity (F), dilution rate (D), or specific growth rate (μ).** The critical dilution rate D_c equals μ_{max}, the specific growth rate of the equivalent batch culture. t_d is the generation time; the productivity of the chemostat is $\Delta N/\Delta t$. The graph shows the ideal values for a 1-l culture with $\mu_{max} = 1\ h^{-1}$, Y = 0.5 g cells formed per g substrate consumed, a substrate (glucose) concentration in the reservoir of 10 mM (1.8 g/l), and a K_s of 0.1 mM glucose. The real values for N at low μ are indicated by a dashed line (for explanation, see the section on maintenance energy, Chapter 6.8.4). If one multiplies D with the cell mass of N in the chemostat, then the product D · N represents the rate of loss (output or productivity in g cells formed per time unit)

equivalent batch culture (see Fig. 6.**1**). The residual substrate concentration in the culture is usually very small, but increases hyperbolically with the flow velocity. Near the critical flow velocity, the residual substrate concentration rises about as sharply as the cell density drops.

The establishment of a constant cell density at a given flow velocity can be explained on the basis that in the **steady state**, the velocity of cell formation due to growth at a constant culture volume, V,

$$+\frac{dN}{dt} = \mu \cdot N \qquad \text{(see Eqn. 6.1)}$$

is equal to the velocity of cell removal through t overflow,

$$-\frac{dN}{dt} = \frac{F}{V} \cdot N \qquad (6.$$

Thus, Equation 6.13 results for the steady state:

$$\mu \cdot N = \frac{F}{V} \cdot N \qquad (6.$$

Substitution of F in Equation 6.13 using Equation 6 results in Equation 6.15, where the dilution rate, D, c be conceived as the frequency of exchange of the to culture volume per time unit

$$F/V = D \qquad (6.$$

$$\mu = D \qquad (6.$$

The change of the bacterial cell mass dN/dt is compos of two velocity terms

$$dN/dt = \mu \cdot N - D \cdot N,$$

and Equation 6.15 represents the steady-state conditi The stability of the steady state in a chemostat is bas on the limitation of the growth rate by one substrate. a constant dilution rate D, if the concentration substrate in the reservoir, S_R, is increased, there is o an increase in the steady-state value of N (i.e., cell m per volume unit), but no change in the steady-st growth rate, i.e., μ. Of course, cells will initially gr faster and the cell density will increase when increases. However, soon a new steady state will reached in which the increased number of cells use almost all of the available nutrient as it enters; t dilution rate again controls growth rate.

After changing the dilution rate, the system au regulates and reaches the new value of μ quite fast. T surprising property of bacteria can be understo considering the hyperbolic reaction of μ and substr concentration S (Fig. 6.**8**). The relation is described Equation 6.16, which is called the Monod equation:

$$\mu = \mu_{max} \cdot S/(K_s + S) \qquad (6.$$

K_s is the substrate affinity constant of the organism a is equal to the concentration of the growth-limiti substrate at which half-maximal growth rate (μ_{max}/ occurs. K_s values are typically in the micromolar ran

The bacterial growth rate is related to the substr concentration by the same function (Eqn. 6.16) as t activity of a single enzyme is related to the concent tion of its substrate. A simple explanation of t phenomenon would be that the growth rate is esse tially determined by the velocity of the slowest me bolic step, often the substrate transport through t bacterial membrane, which is catalyzed by a sin

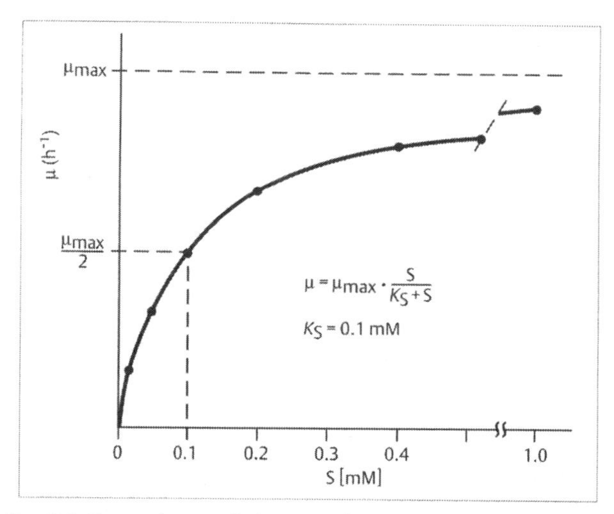

Fig. 6.8 **Dependence of the specific growth rate μ on the concentration of a limiting substrate S**, following the Monod equation (Eqn. 6.16)

A **chemostat**:

1. Guarantees a constant supply of fresh cells grown under exactly defined conditions at a defined specific growth rate; this is a prerequisite for precise studies of regulatory processes.
2. Allows the determination of critical growth parameters, such as K_s, μ_{max}, m_s, and of any parameters directly linked to μ, as well as physiological responses to nutrient limitation.
3. Imitates many natural growth situations quite well (see Chapters 30.3, 30.4).
4. Allows the study of the preferential use of mixed substrates or the competition or coexistence of two or three species under a predetermined condition.
5. Allows the feeding of toxic substrates at higher concentration.
6. Allows the selection of mutants that have a better K_s for the limiting substrate or a higher μ_{max}.
7. Allows the study of stable two-membered cultures, such as syntrophic associations, as well as protozoa that feed on bacteria.

carrier. Hence, the growth rate depends on the substrate supply following a saturation curve, in which S is the prevailing (residual) substrate concentration in the culture (not in the in-flowing medium). The substrate concentration present in the chemostat can be derived from Equation 6.17, which is obtained by rearrangement of Equation 6.16.

$$S = K_s \cdot \frac{\mu}{\mu_{max} - \mu} \tag{6.17}$$

Since, at equilibrium in the chemostat, $\mu = D$, Equation 6.17 becomes Equation 6.18.

$$S = \frac{K_s \cdot D}{\mu_{max} - D} \tag{6.18}$$

It is worth noting that the substrate concentration in the chemostat at equilibrium is independent of the substrate concentration in the reservoir S_R, but only is a function of K_s and D at a given μ_{max} of the culture. The substrate concentration in the reservoir S_R (and therefore in the influent) dictates only the prevailing cell density N (Eqn. 6.19).

$$N = Y(S_R - S) = S_R - \frac{K_s \cdot D}{\mu_{max} - D} \tag{6.19}$$

The more cells are formed by the substrate used up by the continuous culture ($S_R - S$), the more energy can be obtained from the substrate. Y is the growth yield, which will be explained in Chapter 6.8.3 (Eqn. 6.20).

6.8.2 Turbidostat

In a turbidostat, cell concentration (biomass) is continuously monitored and maintained at a constant level predetermined by the experimenter by adjusting the feed rate of fresh nutrients. The substrate needs not be limiting. Operation is most stable when the specific growth rate μ of the culture is near μ_{max}, i.e., when changes in substrate supply result in changes of the biomass content that is to be measured. This is the range in which the chemostat is least stable. In practice, the most difficult problem to be solved is the continuous monitoring of the biomass. Normally, parameters that are tightly linked to cell growth and metabolism, such as CO_2 in the exhaust, are measured.

6.8.3 Growth Yield

The mass of cells (N) produced per mol of energy substrate consumed (catabolic substrate S_c) is called the growth yield, Y (Eqn. 6.20).

$$\frac{dN}{dS_c} = Y \tag{6.20}$$

Usually Y is constant during exponential growth, provided that growth is limited by the catabolic substrate. (The growth yield may also be defined as cell mass formed in relation to the amount of a typical product formed). Y_{ATP} represents the cells mass formed per mol ATP, and is related to Y according to Equation 6.21, where n represents the molar amount of ATP formed per mol of the catabolic substrate S_c.

$$Y = n \cdot Y_{ATP} \tag{6.21}$$

Y can be used to estimate the molar amount (n) of ATP formed per mol S_c in the catabolism of the growing bacteria. Assuming that the cell composition and the ATP requirements of the biosynthetic pathways are the same, a common Y_{ATP} should apply for all bacteria using the same energy substrate. However, Y_{ATP} is expected to vary considerably with the carbon source. From the biosynthetic routes, it can be calculated that the amount of ATP required for cell synthesis is roughly inversely proportional to the number of carbon atoms of the carbon source (with similar redox state). Thus, with lactate or acetate, Y_{ATP} would be two or three times smaller than with glucose as carbon source.

Under the culture conditions usually applied, the substrate (e.g., glucose) is used both in catabolism and biosynthesis. For determining Y with respect to biomass formed per mol substrate used in energy metabolism (S_c), the proportion of substrate used as carbon source (S_B) has to be subtracted from the total amount of substrates consumed (Eqn. 6.22). The proportion of substrate used as carbon source (S_B/S) can be calculated from Equation 6.23, which is obtained by combining Equations 6.22, 6.24, and 6.25.

$$S = S_c + S_B \tag{6.22}$$

$$S_B/S = 1/(1 + 24\,a/n \cdot Y_{ATP}) \tag{6.23}$$

$$N = 24 \cdot a \cdot S_B \tag{6.24}$$

$$N/S_c = n \cdot Y_{ATP} \tag{6.25}$$

Equation 6.24 states that the cell mass (N) formed amounts to twice the carbon mass derived from the carbon source, because half of the cellular dry mass is carbon (the molar mass of carbon is 12 g/mol). The number of carbon atoms of the carbon source is designated a. Equation 6.25 results from the combination of Equations 6.20 and 6.21 and the substitution of dN/dS_c by N/S_c.

According to Equation 6.23, the proportion of substrate used as carbon source (S_B/S) is essentially a function of the ATP gain (n) since a/Y_{ATP} is nearly the same with, for example, glucose, malate, lactate, or acetate as carbon source. In Figure 6.**9**, S_B/S is drawn as a function of n with $a/Y_{ATP} = 0.6$ mol ATP/g cells. This ratio corresponds to a Y_{ATP} of 10 g cells/mol ATP, which is usually measured with glucose in batch culture. As seen from Figure 6.**9**, 12% (70%) of the growth substrate

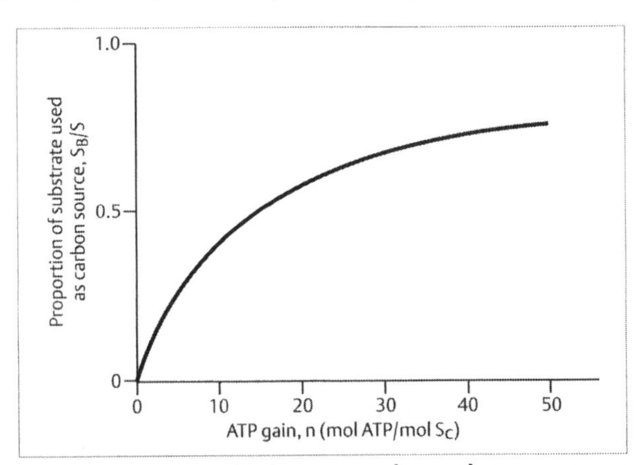

Fig. 6.**9** **Proportion of substrate used as carbon source as a function of the ATP gain of catabolism.** The substrate is used in catabolism and biosynthesis. The curve is calculated according to Equation 6.22 with $a/Y_{ATP} = 0.6$ mol ATP/g cells

is used as carbon source, if 2 (30) mol ATP is gained per mol of the catabolic substrate. These two numbers refer to two common types of glucose metabolism by, for example, *E. coli*: lactic or alcoholic fermentation (2 ATP) and complete oxidation of glucose to CO_2 (30 ATP).

6.8.4 Energetic Costs of Bacterial Maintenance

The values of Y_{ATP} calculated (Eqn. 6.21) from known values of n using growth yields measured with batch culture are considerably smaller than the theoretical Y_{ATP} derived from the biosynthetic pathways (see Chapter 7). The discrepancy is generally attributed to the consumption of ATP or energetic equivalents of ATP by processes that are not related to cell synthesis and are commonly termed bacterial **maintenance**. The proportion of ATP equivalents required for the maintenance of bacteria is expected to be smaller the faster the bacteria grow. This hypothesis can be tested experimentally using a chemostat by measuring Y at increasing growth rates. These measurements usually give a linear relationship between $1/Y$ and $1/\mu$ (Fig 6.**10**). The relationship can be explained on the basis of two assumptions:

1. Part of the catabolic substrate (S_c) is used for maintenance (S_M) and the rest (S_E) to drive the endergonic reactions of biosynthesis (Eqn. 6.26). Differentiation of Equation 6.26 gives Equation 6.27.

$$S_c = S_E + S_M \tag{6.26}$$

$$\frac{dS_c}{dt} = \frac{dS_E}{dt} + \frac{dS_M}{dt} \tag{6.27}$$

2. The velocity of S_M consumption is proportional to cell density (Eqn. 6.28);

$$\frac{dS_M}{dt} = m_s \cdot N \qquad (6.28)$$

m_s is termed the **maintenance coefficient**. Substitution of the other two terms of Equation 6.27 using Equations 6.1 and 6.20 yields the Pirt equation (Eqn. 6.29), where Y_{max} is the ratio of the amount of cells formed per mol S_E.

$$1/Y = 1/Y_{max} + m_s/\mu \qquad (6.29)$$

Equation 6.29 predicts a linear relation between $1/Y$ and $1/\mu$, in agreement with the experimental results (Fig. 6.**10**). The maximal growth yield corrected for maintenance (Y_{max}) is obtained by extrapolating the measured values of Y to infinite μ (or $1/\mu=0$). In several experiments, the Y_{max} determined by extrapolation was close to the values calculated according to Equation 6.21 from n and the theoretical Y_{ATP} derived from the ATP requirement of cell synthesis. The slope (m_s) of the line in Figure 6.**10** would represent the amount of catabolic substrate consumed per time unit and per cell mass that is used for the purpose of maintenance, and represents the maintenance coefficient. The term m_{ATP} refers to the amount of ATP used for maintenance (Table 6.4). Unfortunately, there is no second method for measuring m_s independently to confirm the validity of the assumptions leading to the Pirt equation. Some values for maintenance coefficients measured with various bacteria are given in Table 6.**4**.

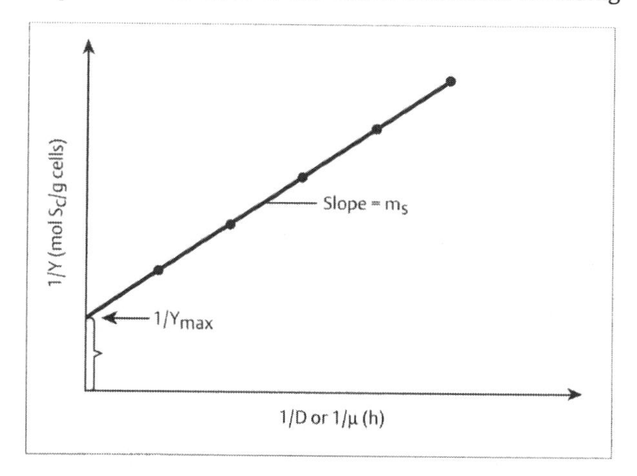

Fig. 6.**10** **Growth yield (Y) as a function of the dilution rate (D) in a hypothetical experiment with a chemostat.** The dots are obtained from measured values of N and S_c. See the Pirt equation (Eqn. 6.29)

6.8.5 Estimation of Enzyme Contents From μ and Y

Catabolic enzymes. The velocity of consumption of a catabolic substrate (dS_c/dt) in an exponentially growing culture is related to μ and Y according to Equation 6.30

$$\frac{dS_c}{dt} = (\mu/Y) \cdot N \qquad (6.30)$$

which is obtained by combining Equations 6.1 and 6.20. The ratio μ/Y is called the **metabolic quotient** and represents the specific activity of substrate consumption by the bacteria, which is proportional to that of the enzymes involved in the catabolic pathway. Comparison of the theoretical enzyme activities so obtained to the specific activities measured with the bacterial homogenate provides a clue to an understanding of the catabolic pathway used by the growing bacteria. With the assumption that the actual activities are close to V_{max}, the enzyme activity A (mol substrate consumed per time unit) per gram of dry cell mass N equals μ/Y, which necessarily has the same dimension (Eqn. 6.31),

$$\frac{A}{N} = \frac{\mu}{Y} \qquad (6.31)$$

Table 6.**4** **Maintenance coefficients measured with various bacteria**

Species	Cultivation conditions	m_s (μmol substrate g^{-1} h^{-1})	m_{ATP} [μmol ATP (g dry matter)$^{-1}$ h^{-1}]
Escherichia coli	Glucose, aerobic	0.2–0.4	4.0–8.0
Escherichia coli	Glucose-fermenting	2.3	6.9
Paracoccus denitrificans	Glucose, aerobic	0.08	2.7
Pseudomonas fluorescens	Acetate, aerobic	0.25–0.6	2.5–6.3
Klebsiella aerogenes	Glucose, aerobic	0.35	6.9
Desulfovibrio vulgaris	Ethanol + sulfate	4.3	4.3
Nitrosomonas europaea	Ammonia + O_2	1.5	1.5
Desulfobulbus propionicus	Ethanol + sulfate	0.9	0.9
Propionigenium modestum	Succinate-fermenting	0.87	0.29

If one knows the turnover number T (mol substrate converted per time unit) of the enzyme, the molar enzyme content E per g cells follows Equation 6.32.

$$v_{max} = E.T. \quad (6.32)$$

As the enzymes are not expected to operate at substrate saturation in growing bacteria, the activities usually measured with the cell homogenate (V_{max}) (Eqn. 6.31 and 6.32) should exceed the activities calculated according to Equation 6.30.

Anabolic enzymes. The velocity of synthesis of a cell constituent C (dC/d_t) in an exponentially growing culture can be estimated from μ and the cellular content of the constituent (in mol per gram cell mass) according to Equation 6.33.

$$\frac{dC}{dt} = \mu \cdot C \quad (6.33)$$

This equation is obtained from Equation 6.1 upon substitution of N by C, assuming that the composition of the bacteria does not change during growth. Equation 6.33 can be used for estimating the specific activities (based on bacterial cell mass) of the enzymes involved in the synthesis of a cell constituent. Comparison of this theoretical activity to that measured with the disrupted bacteria helps in elucidating the biosynthetic pathways of a bacterium.

6.8.6 Semi-continuous cultures

When a batch culture is continuously fed in an exponential manner with an essential nutrient, this is called a **fed batch culture**. It allows exponential growth, and one can periodically remove and harvest part of the culture (repeated fed batch). Fed batch is a valuable technique if substrates are toxic in higher concentrations, and it allows higher cell mass or product productivity than an ordinary batch operation with little additional equipment. By this means, **high-density batch cultures** with up to 150 g of dry mass of cells/l can be obtained simply by supplying essential nutrients in increasing amounts to meet, but not exceed the growth needs. If biomass is desirable as a biocatalyst, it may be fed back from continuous or semi-continuous cultures (**biomass feed-back**). If a product rather than the cells is of interest, **dialysis cultures** can be useful. The culture is contained in a vessel separated by a dialysis membrane from recirculating medium from which the product is continuously removed. The nutrients and products freely pass the membrane. Dialysis culture allows (1) the attainment of very high densities of cells, (2) relief of product inhibition and production of metabolites free from cells, (3) the study of pure cultures in natural systems, and (4) the study of interactions between two separate pure cultures.

6.9 Preservation Methods Are Based on the Inhibition of Microbial Growth

Physical factors. Any factor that affects the growth rate of bacteria is inhibitory if applied in the extreme range. Although there are extremophiles that can tolerate such conditions, in practice these conditions are suited to conserve any organic material, e.g., food, and prevent spoilage by pathogenic, food-spoiling, or toxin-producing microorganisms. Physical treatments include heating or freezing, acidification, and drying or salting or adding large amounts of sugar to decrease water availability. Also, exclusion of oxygen adds to the adverse effects of extreme temperature, pH, and osmolarity. The conventional methods of food preservation are based upon these effects. Filtration and ionizing radiation have been mentioned earlier (Chapter 6.4).

Chemicals. Some chemical compounds that in lower concentration are safe for humans have proven to be of value in growth inhibition of microorganisms. For instance, addition of SO_2 gas to water results in formation of sulfurous acid (H_2SO_3, $pK_1 = 2$, $pK_2 = 7$); the protonated form can penetrate the cell membrane and acts as reductant (being oxidized to H_2SO_4) and cleaves –S–S–bridges, forming $-S-SO_3^- + HS^-$. These effects are used in conserving acidic food products such as wine or dry fruits, provided that the pH is ≤ 3.5–4. Addition or microbial in situ formation of weak organic acids can prevent growth if the pH is kept low to maintain a substantial proportion of the acid in the membrane-permeable protonated from (AH) (see Chapter 6.2.1). This applies to lactic, acetic, formic, propionic, citric, tartaric, benzoic, and sorbic acid. Esters of benzoic acid are perfectly membrane permeable at neutral pH and are cleaved by cellular esterases; these esters therefore are growth inhibitory as an additive to non-acidic food. Food that is preserved by acid will be spoiled by yeast and fungi if air is not excluded. High alcohol content prevents microbial growth. Smoking adds antimicrobial agents to the smoked goods; preservatives in smoke include phenols, methylphenols (cresols), aldehydes, and other less-safe components.

Growth inhibitory compounds more powerful than those used in food preservation can be applied in other fields. They include heavy metals such as sublimate ($HgCl_2$), hypochloric acid ($HOCl$), bromine, iodine, detergents (soap), and organic solvents (alcohol), not to mention the huge variety of synthetic or microbial antibiotics.

Further Reading

Atlas, R. M. (1993) Handbook of microbiological media. Boca Raton: CRC Press

Demain, A. L., and Soloman, N. A. (eds.) (1986) Manual of industrial microbiology. Washington, DC: American Society for Microbiology

Gerhardt, P., Murray, R. G. E., Wood, W. A., and Krieg, N. R. (eds.) (1994) Methods for general and molecular bacteriology. Washington, DC: American Society for Microbiology

Harder, W., and Dijkhuizen, L. (1983) Physiological responses to nutrient limitation. Annu Rev Microbiol 37:1–23

Kirsop, B. E., and Doyle, A. (1991) Maintenance of microorganisms and cultured cells, 2nd edn. New York, London: Academic Press

Krieg, N. R., and Holt, J. G. (eds.) (1989) Bergey's manual of systematic bacteriology, vol. 1. Baltimore: Williams & Wilkins

Monod, J. (1949) The growth of bacterial cultures. Annu Rev Microbiol 3:371–394

Neidhardt, F. C., Ingraham, J. L., and Schaechter, M. (1990) Physiology of the bacterial cell. Sunderland: Sinauer Associates

Neidhardt, F. C., Curtiss, R., Ingraham, J. L., Lin, E. C. C., Low, K. B., Magasanik, B., Reznikoff, W., Riley, M., Schaechter, M., and Umbarger, H. E. (eds.) (1996) *Escherichia coli* and *Salmonella*: cellular and molecular biology. 2nd edn. Washington, DC: ASM Press

Pirt, S. J. (1975) Principles of microbe and cell cultivation. New York: Wiley

Veldkamp, H. (1976) Continuous culture in microbial physiology and ecology. Durham: Meadowfield Press

White, D. (1995) The physiology and biochemistry of prokaryotes. Oxford: Oxford University Press

Source of Figure

1 J. L. Ingram, and A. G. Marr (1996) Effect of temperature, pressure, pH, and osmotic stress on growth. In F. C. Neidhardt et al. eds. (1996) *Escherichia coli* and *Salmonella*: Cellular and molecular biology, vol. 2, 2nd edn. Washington, DC: ASM Press

7 Biosynthesis of Building Blocks

This chapter describes how prokaryotic organisms derive from their substrates all the main building blocks (small molecules, e.g., amino acids) from which they synthesize the different cell components: soluble compounds, large molecules, polymeric macromolecules, and complex structures composed of different macromolecules. In quantitative terms, the polymers protein, RNA, DNA, cell-wall peptidoglycan, reserve material, and membrane lipids are the most important end products (96% of the cell dry weight). Different stages in the synthesis of new cell material from substrates are illustrated in Figure 7.**1**.

Other chapters of this book deal with the synthesis of macromolecules, including the activation of building blocks, the polymerization of activated building blocks, and the assembly of complex structures from the respective polymers and complex monomers. Turnover, repair, and adaptation to changes in the environment all require a sophisticated interplay of biosynthetic and degradative mechanisms that are not always understood. Turnover of the rigid cell wall and chromosome replication during growth are examples of such complex processes.

Most information in this chapter was obtained from studies of the enteric bacteria *Escherichia coli* and *Salmonella typhimurium* (Proteobacteria). Less is known about biosynthesis in other prokaryotic groups; still much of the knowledge probably can be generalized. As a rule of thumb, the more remote a prokaryotic natural group (e.g., Archaea) is from the evolutionary branch of the Proteobacteria, the more likely are deviations from the biosynthetic pathways outlined here. Also, new structures with novel building blocks are to be expected. However, the reactions by which the building blocks for protein (amino acids) and for nucleic acids (nucleotides) are formed are very similar in all organisms ("universality of cellular biochemistry"). Other constituents (e.g., cell walls, lipids, or polysaccharides) show a considerable diversity of the structures and of the biosynthetic pathways involved in their synthesis.

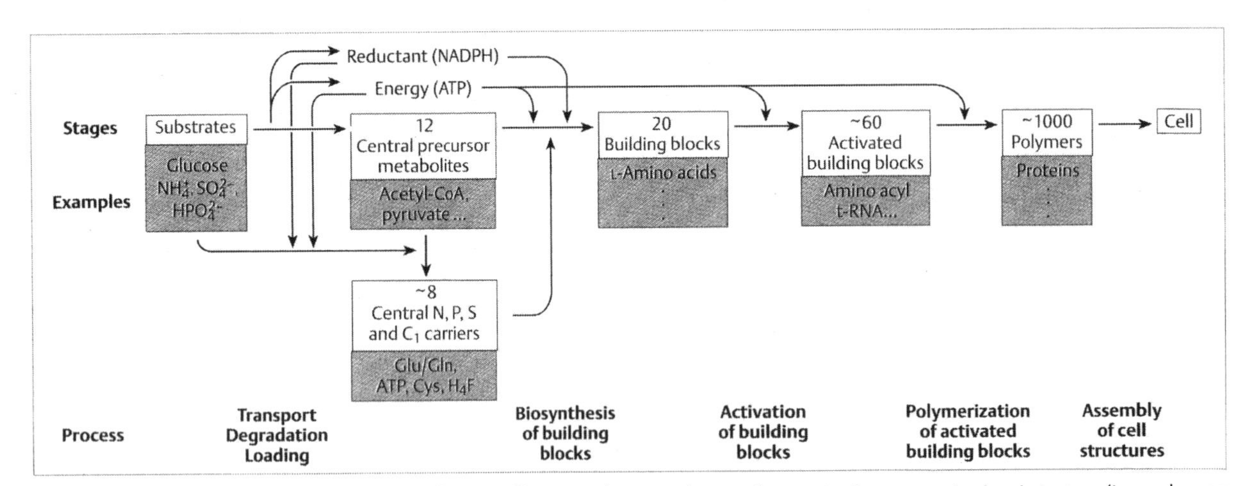

Fig. 7.1 **Different stages in the synthesis of new cell material from the substrate of a prokaryotic organism.** This scheme also applies to eukaryotic organisms. As an example, the synthesis of protein from standard substrates (i.e., glucose, ammonium, phosphate, and sulfate) is given. H_4F tetrahydrofolate, Glu glutamate, Gln glutamine, Cys cysteine

7.1 The Molecular Composition Reflects the Complexity of the Prokaryotic Cell

First, a few basic facts pertinent to biosynthesis in prokaryotic cells will be considered.

Protein accounts for about half of the cell dry weight. How many building blocks are required to build 1 gram of cells? The molecular composition of an average prokaryotic cell illustrates best its needs for biosynthesis (Table 7.1). Protein accounts for approximately half of the cell dry weight, whereas nucleic acids account for approximately one fifth. Stable RNA (i.e., rRNA and tRNA) by far exceeds DNA. About half of the RNA is 23S rRNA, 27% is 16S rRNA, 2% is 5S rRNA, 15% accounts for the 60-odd different species of tRNA, and 4% is composed of the mRNA transcribed from approximately 1000 operons and 4000 genes. The main monomeric building blocks of the macromolecules, 70 to 100 different compounds, are also listed in Table 7.1.

The cell composition varies to some extent with the growth rate since the relative amount of the protein-synthesizing apparatus, which can reach nearly half of the cell dry weight, increases with the growth rate. Since the ribosomes consist of nearly 50% RNA, the cellular RNA content varies from a few percent in slowly growing cells to nearly 20% in fast growing cells. Also, in Gram-positive bacteria, the (thick) peptidoglycan cell wall comprises many layers of peptidoglycans, whereas Gram-negative bacteria contain a cell wall that comprises a single peptidoglycan layer, an outer membrane, and layers of polysaccharides.

Building blocks are made from the central precursor metabolites. How many genes are required for biosynthesis? This number, reflecting the complexity of a prokaryotic cell, can be estimated because all genes and approximately half of the gene products of *E. coli* are known. Nearly one-fifth of them are involved in the synthesis of small molecules, the **building blocks** for polymers and cofactors. This figure assumes that the central intermediates of metabolism, from which the individual building blocks are made, are already available due to the operation of central metabolic pathways. Those intermediates of central metabolic pathways from which all building blocks are made are called **central precursor metabolites**. They must be made from any substrates a prokaryotic cell can use for growth.

Cells prefer preformed building blocks. What happens if preformed building blocks are available? Cells are economic, and they normally use preformed building blocks as substrates for macromolecular synthesis if these building blocks are available in the environment and if these compounds can be taken up into the cell (see Chapter 7.4). If this is the case, the synthesis of building blocks or central precursor metabolites from simple substrates is repressed.

Many bacteria are unable to synthesize all building blocks; they are then termed **auxotrophic** for given compounds (e.g., certain amino acids, unsaturated fatty acids, purines, or precursors of cofactors). These compounds, called **growth factors** or **vitamins**, may be available in their natural environment; in the laboratory, these compounds are supplied, for example, in the form of yeast extract, protein hydrolysates, "blood agar," or vitamin mixtures.

Finally, bacteria living on complex organic material, such as blood, milk, decaying plants, juices, or in the intestinal tract of animals, have often lost important parts of their biosynthetic potential. This is especially true of some prokaryotic symbionts and obligate cell parasites. On the other hand, prokaryotic symbionts often provide their hosts with essential amino acids (e.g., tryptophan) or vitamins that cannot be made by the host. The hosts (e.g., insects) are often nutritional specialists; their food rarely contains such building blocks.

Table 7.1 **Average composition of a prokaryotic cell (Gram-negative eubacterium *Escherichia coli*)**

	Individual compounds (% of dry weight)	μmol of building blocks for 1 g cells	Stoichiometric amounts per molecule	Building blocks	
Macromolecules (≈96% of cell dry weight)	Protein (50–60%)	5000	–	L-Amino acids (20 different) activated as aminoacyl- tRNAs (rare amino acids: Trp, Lys, His)	
	RNA (10–20%)	300–600	–	Nucleoside triphosphates (4 different) (ATP, GTP, CTP, UTP)	
	DNA (≈3%)	100	–	2′-Deoxynucleoside triphosphates (4 different) (dATP, dGTP, dCTP, dTTP), A and T, G and C, purines and pyrimidines each in equal amounts	
	Lipids (≈10%)	130	1	Glycerol 3-phosphate, later activated as CDP-diacylglycerol	
		260	2	Acyl carrier protein–fatty acids (e.g., 16 : 0, 16 : 1Δ9, 18 : 1Δ11)	
		130	1	Polar head groups (e.g., ethanolamine, glycerol, L-serine, phosphatidylglycerol)	
	Murein (3–10%)	30–60	1 1	UDP- *N*-acetylglucosamine UDP- *N*-acetylmuramic acid	Glycan
		90–180	3	D- and L-amino acids (e.g., L-alanine, D-glutamate, *meso*-diaminopimelic acid), activated by ATP	Peptide
			1	di-D-alanine, activated by ATP	
	Lipopolysaccharides (≈3–4%)	16	2	UDP-*N*-acetylglucosamine	Lipid A
		48	6	Acyl carrier protein—fatty acids (e.g., 14 : 0, R-3-OH 14 : 0, 16 : 0)	
		24	1	CDP-ethanolamine	

(*continued*)

Table 7.1 (*continued*)

	Individual compounds (% of dry weight)	μmol of building blocks for 1 g cells	Stoichiometric amounts per molecule	Building blocks	
			2	CDP-ethanolamine	
		24	3	CMP-2-deoxy-D-mannooctulosonate (KDO)	Core
		24	3	ADP-heptose	
		32	4	UDP- hexoses (e.g., glucose, galactose)	
		100–200	≈40	NDP-sugars (repeating units of 3–5 sugars)	O-specific side chain
	Glycogen (2.5–25%)	150–1500	–	ADP-glucose	
Soluble compounds: (≈4% of cell) dry weight)	Polyamines	60	–	Derived from basic amino acids	
	Dissociable coenzymes, cofactors, electron carriers			NAD, NADP, CoA, pyridoxal phosphate, tetrahydrofolate, ubiquinone and menaquinone, undecaprenol, thioredoxin, glutaredoxin, glutathione	
	Prosthetic groups			FAD, FMN, biotin, heme and porphyrins, lipoic acid, thiamine pyrophosphate, molybdopterin, cobalamine	
	Osmoprotectants			e.g., betaine	
	Pool of metabolic intermediates			Degradative, central metabolic, biosynthetic routes (see above)	
	Pool of soluble building blocks			e.g., amino acids	
	Inorganic ions			Na^+, K^+, Mg^{2+}, Ca^{2+}, Mn^{2+}, Cl^-, P_i, PP_i	

7.2 The Few Central Precursor Metabolites Are Intermediates of the Central Metabolic Pathways

Only approximately a dozen compounds, virtually the same in all organisms, belong to central precursor metabolites; they are the intermediates of three indispensable pathways: (1) the **citric acid cycle** and (2) the **glycolytic** (or other sugar-degrading mechanism)/ **gluconeogenic pathway** have been discussed in Chapter 3. The third pathway (3), the **pentose-phosphate cycle**, is mainly responsible for the interconversion of sugar phosphates (see Chapter 7.7.1).

Box 7.1 Despite the apparent homogeneity of biosynthetic reactions, the enzymes involved may differ substantially, even if they catalyze the same reaction (isoenzymes, divergent evolution). But also convergent evolution occurs; for example, there are different classes of fructose-bisphosphate aldolases, glyceraldehyde-3-phosphate dehydrogenases, or glutamine synthetases. Normally, isoenzymes of a cell are differently regulated. The regulatory properties of biosynthetic enzymes may vary in different bacteria. Also, the functional and genetic organization of the genes and the means adopted for regulation may differ. Separate, small enzyme entities are often found in prokaryotes, whereas polyfunctional, large proteins are more often found in eukaryotes. Many enzymes are made up of functional domains or modules that were combined independently during evolution to yield different enzymes.

The central metabolic pathways have a dual function. The structures of the central precursor metabolites and the amounts required for the synthesis of the building blocks for 1 gram of new cells are given in Figure 7.**2**. The building blocks of other organisms can be used by prokaryotes as organic substrates; in quantitative terms, plants constitute the most important food source. These substrate molecules are normally broken down into one of the central precursor metabolites, thus linking degradation with biosynthesis. Any degradative pathway has to fulfill a dual role, to degrade organic substrates and to provide all central precursor metabolites for biosynthesis. Even when CO_2 or one-carbon (C_1) compounds are the only carbon sources of an organism and when energy is derived from inorganic substrates, the carbon fixation mechanisms are connected to these central precursor metabolites. The biosynthetic routes leading to the building blocks are essentially the same in all living cells, with few notable exceptions. This is in remarkable contrast to the variety of degradation pathways.

Substrate molecules are normally broken down to one of the **central precursor metabolites**, which link degradation with biosynthesis. Any degradative pathway has to fulfill a dual role: to serve the degradation of the organic substrates and to provide all central precursor metabolites for biosynthesis.

Central precursor metabolites, for example, acetyl-CoA, pyruvate, oxaloacetate, 2-oxoglutarate, phospho*enol*pyruvate (PEP), hexose phosphates, and a few other metabolites are the starting material for the synthesis of approximately 70–100 building blocks (Fig. 7.**2**). Together with soluble compounds (e.g., coenzymes and polyamines), they amount to approximately 150 different low-molecular-weight compounds. The amounts of building blocks needed by growing cells is illustrated in Table 7.**1**. The most important building blocks are L-amino acids for the synthesis of proteins, nucleotides for nucleic acids, activated sugars for the cell wall and other structures, and fatty acids (or isoprane alcohols in Archaea) for lipids. The types and amounts of other building blocks vary with the bacterium. For example, lipopolysaccharides are required by Gram-negative bacteria (e.g., *E. coli*), whereas Gram-positive ones require far more peptidoglycan precursors in order to form a thick cell wall.

There is a **need for nitrogen, phosphorus and sulfur, as well as C_1 units.** The biosynthetic precursor metabolites contain only carbon, hydrogen, and oxygen atoms (plus some phosphate) because most organic growth substrates eventually lose nitrogen, phosphorus, and sulfur during degradation. In contrast, most cell constituents contain these elements. In order to form building blocks from these metabolites, the cell needs compounds which provide nitrogen, phosphorus, and sulfur, and C_1 units. Very few biosyntheses in aerobic prokaryotes also require molecular oxygen. Last but not least, energy (mostly in the form of the phosphoric acid anhydride bonds in ATP) and a reductant (in the form of the hydride of NADPH) are required. These prerequisites will be discussed first before turning to the biosynthesis of the individual building blocks.

Simple inorganic compounds and low-molecular-weight or macromolecular compounds can serve as

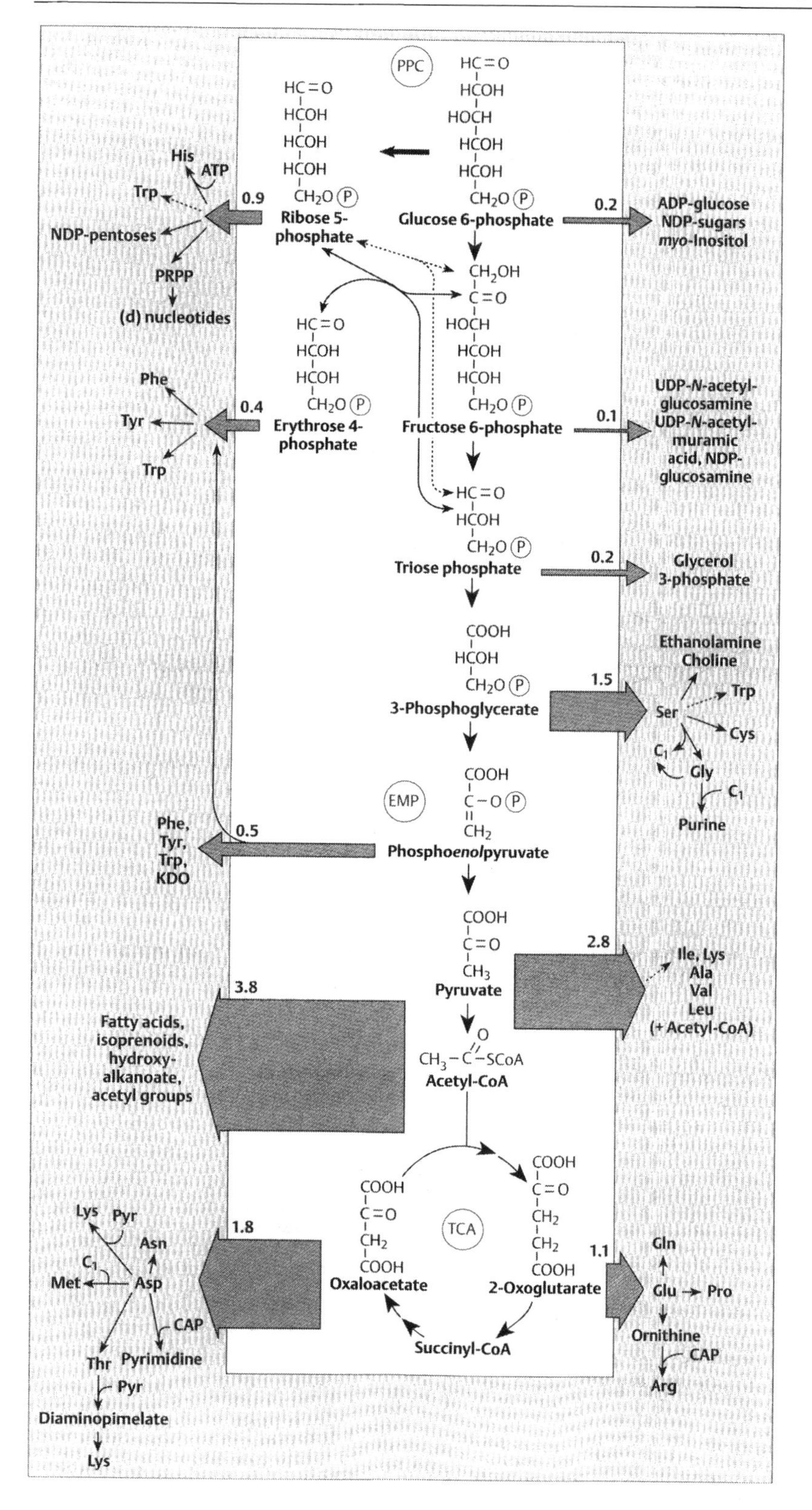

Fig. 7.2 **Formation of building blocks from central intermediates of carbon metabolism.** The figures indicate how much of the precursor (in mmol) is required to synthesize 1 gram of cells. **TCA**, **tri**carboxlic **a**cid cycle; **EMP**, **E**mbden- **M**eyerhof-**P**arnas pathway; **PPC**, **p**entose **p**hosphate **c**ycle; CAP, carbamoyl phosphate; Pyr, pyruvate; KDO, an octonic acid of lipopolysaccharides; NDP, nucleoside diphosphate; PRPP, 5-phosphoribosyl 1-pyrophosphate; C_1, C_1 compounds linked to tetrahydrofolate

substrates. They are taken up and channeled to the central metabolic pathways, which link degradation with biosynthesis and provide a pool of central precursor metabolites for the biosynthesis of building blocks.

A knowledge of the basic chemistry of amino acids, carbohydrates, aromatic and heterocyclic compounds, and of their functional groups is essential for this and subsequent chapters. The reader should make sure that the differences between various bonds are clear (e.g., ether, ester, glycosidic, anhydride, amide, and peptide bonds). Also, a knowledge of the basic stereochemical features and the correct nomenclature of compounds is required.

7.3 How Nitrogen, Phosphorus, Sulfur, One-carbon Units, and Molecular Oxygen Are Incorporated Into Cell Compounds

The prerequisites for the synthesis of 1 gram of cells are summarized in Table 7.2. How are these elements and the C_1 units supplied? A few central molecules mediate their incorporation into the building blocks; these molecules take over the respective element from the precursor form (e.g., NH_3) and donate it to the building blocks (e.g., amino acids). The loading and unloading of these "carriers" (not to be confused with "transport" carriers!) will now be discussed.

7.3.1 Most Bacteria Can Use Several Nitrogen Compounds as Nitrogen Source

The synthesis of 1 gram of cell mass requires about 140 mg of nitrogen. Nitrogen is available in different oxidation states (in parenthesis) and molecular substrate forms, for example, NH_4^+ (−3), NO_3^- (+5), N_2 (0), amino acids (−3), urea (−3), purines (−3), and pyrimidines (−3). The oxidation state of most forms of cellular nitrogen (amino groups, heterocyclic N) is equivalent to ammonia (−3). Nitrogen substrates therefore are reduced or hydrolyzed to NH_4^+ ($NH_3 + H^+ \rightleftharpoons NH_4^+$, $pK_b = 9.3$) and then incorporated by means of the four central nitrogen carriers (i.e., **glutamate**, **glutamine**, and to some extent secondarily via **aspartate** and **carbamoyl phosphate**; Fig. 7.3) through the three steps leading from substrates to cellular materials:

(1) **Transport and reduction steps.** Nitrogen sources first have to be transported into the cell. For example, NH_4^+ transport is one of the major energy-consuming transport processes of the cell. Inside the cell, ammonia is formed by reduction of nitrate or N_2 or by hydrolytic release from organic substrates that contain nitrogen.

(2) **Loading of nitrogen carriers. Glutamine** directly transfers about 15% of the nitrogen. This N-donor is

Table 7.2 **Requirements for the synthesis of monomeric building blocks for 1 gram of cells starting from the central precursor metabolites.** Activation of the building blocks requires additional ATP. In a complex (rich) medium, virtually all building blocks are present and may be taken up. This saves up to 100% of the NADPH and 85% of the ATP otherwise required

Requirement	Amount	Compound
Carbon skeleton (~40 mmol C)	14 mmol	Central precursor metabolites, some CO_2
NH_3	10 mmol	Glutamine, glutamate, aspartate, carbamoyl phosphate
Phosphate	1 mmol	ATP
H_2S	0.3 mmol	Cysteine
C_1 units	1.2 mmol	C_1-tetrahydrofolate, S-adenosyl-Met
Reductant	~19 mol	NADPH
Oxidant	~3.5 mmol	NAD^+
Energy	~19 mmol	ATP

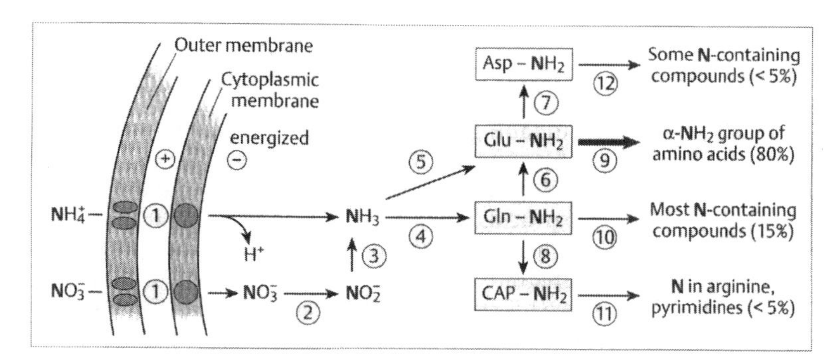

Fig. 7.3 **Assimilation of nitrogen.** The percent values indicate the relative amount of nitrogen carried by the individual compounds (e.g., 80% via the α-amino group of glutamate, abbreviated as Glu-NH_2 to indicate its NH_3-carrier function). 1, Transport of N sources (note the energized membrane); 2, nitrate reductase; 3, nitrite reductase; 4, glutamine synthetase; 5, glutamate dehydrogenase; 6, glutamate synthase; 7, glutamate: oxaloacetate transaminase; 8, carbamoyl-phosphate synthetase; 9, amino transferases; 10, amido transferases; 11, carbamoyl transferases; 12, two-step transaminations with aspartate

abbreviated here as Gln-NH_2. The only reaction for the assimilation of NH_4^+ into glutamine is the L-glutamine synthetase reaction (Fig. 7.**4**). It requires ATP to facilitate the amidation of the γ-carboxyl group of glutamate (i.e., replacement of –OH by –NH_2 requires dehydration, for mechanistic reasons), and to make ammonia fixation unidirectional (for thermodynamic reasons). Glutamine is the direct N donor for most nitrogen compounds, except for the α-amino group of amino acids for which glutamine is only an indirect donor (see below).

Glutamate is the universal donor for α-amino groups in amino acids, which account for about 80% of cell nitrogen. This N-donor is abbreviated in the figures as Glu-NH_2. Glutamate may form large intracellular pools (up to > 100 mM; e.g. during osmoadaptation). Glutamate can be formed from 2-oxoglutarate in two ways, depending on the availability of NH_3 and on the organism, which may possess one or both of the synthetic pathways.

Nitrogen excess. At high concentrations of NH_3 (generally > 1 mM), NH_4^+ is directly incorporated. L-Glutamate dehydrogenase catalyzes the reversible amination of 2-oxoglutarate to 2-iminoglutarate and the reduction with NADPH to 2-aminoglutarate (note that degradation of amino acids involves an

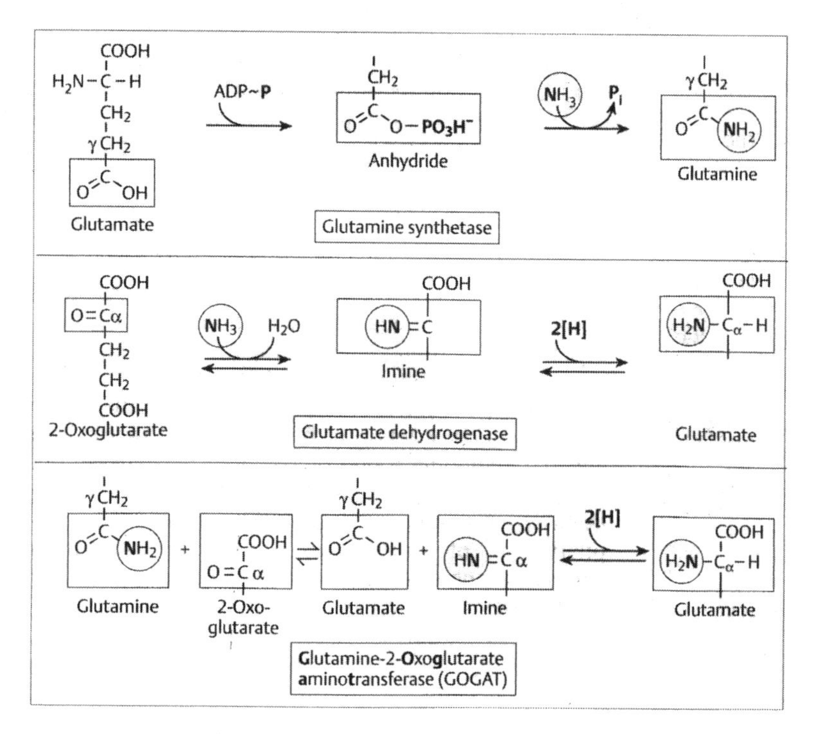

Fig. 7.4 **NH_3-assimilating reactions**

NAD^+-dependent enzyme functioning in the reverse direction). In some bacteria, alanine dehydrogenase (pyruvate + NADH + NH_4^+ → alanine + NAD^+ + H_2O) may replace glutamate dehydrogenase; glutamate is then formed by transamination to 2-oxoglutarate (alanine + 2-oxoglutarate ⇄ pyruvate + glutamate). However, the affinity of those dehydrogenases for NH_4^+ is even lower (K_m approximately 5 mM), and they only function at high cellular ammonium concentrations and under energy and carbon restriction.

Nitrogen limitation. Under most growth conditions, nitrogen is limiting and glutamate dehydrogenase is dispensible. At low NH_3 concentrations (generally < 0.5 mM), NH_4^+ is indirectly incorporated. Glutamate is then formed by the reductive amination of 2-oxoglutarate, with glutamine as NH_3 donor and NADPH as reductant (Fig. 7.**4**). Under these conditions, the enzyme glutamate synthase (also referred to as GOGAT = **g**lutamine : 2-**o**xo**g**lutarate **a**mino**t**ransferase, a flavin–iron–sulfur protein) is formed and the glutamine synthetase level is increased. The overall fixation of ammonia indirectly requires 1 mol ATP/mol NH_3; note that the fixation of NH_3 in the C5 amide group of glutamine requires ATP! What is the benefit of using the glutamine synthetase/glutamate synthase reactions? The apparent K_m of glutamine synthetase for NH_3 is rather low (≤ 0.5 mM), and the reaction is almost irreversible due to ATP cleavage. Thus, this route functions only when sufficient energy is available or when the ammonium or phosphate concentrations become low.

The assimilation of nitrogen is under the control of the nitrogen source (type and amount) and the energy state of the cell (see Chapters 19 and 20).

There is a relationship between the free-energy change of an enzymatic reaction and the kinetic constants of the enzyme (rate constants k, dissociation constants K_d). ATP cleavage therefore can be used to lower the K_d of the enzyme–substrate complex for a critical substrate or to increase the rate constant k of the forward reaction, or both. A good example is glutamine synthetase, which has a much lower K_m value for NH_4^+ than glutamate dehydrogenase has.

The secondary N carriers, **carbamoyl phosphate** (CAP) and **aspartate** ($Asp\text{-}NH_2$) are discussed below since their nitrogen is indirectly derived from glutamine or glutamate. How is nitrogen from the α-amino group of glutamate and from the amide group of glutamine transferred to other compounds?

(3) Unloading of nitrogen carriers (N transfer; Fig. 7.**5**). Aminotransferases (transaminases) transfer the α-amino group of **L-glutamate** ($Glu\text{-}NH_2$) to 2-oxo acids, which are the direct precursors of α-amino acids. Figure 7.**5a** illustrates the synthesis of aspartate from oxaloacetate. Most cells contain several transaminases; each one is responsible for a certain group of 2-oxo acids.

In a few biosynthetic reactions, the α-amino group of **aspartate** serves as the nitrogen donor in a two-step reaction (Fig. 7.**5c**), whereby it is transferred to an acceptor molecule with an oxo group (=C=O). A covalently linked aspartate-acceptor intermediate is formed, and fumarate is only released by a second enzyme. The product is the imine (=C=NH; see synthesis of L-arginine, of N1 of purines, and of the amino group of adenine). The accepting carbonyl group is activated by ATP to form a better "leaving group" for oxygen.

Amidotransferases (transamidases) transfer the amide group from **L-glutamine** ($Gln\text{-}NH_2$) in the biosynthesis of most other nitrogen-containing compounds (i.e., purines, pyrimidines, tryptophan, histidine, and glucosamine 6-phosphate). The acceptor molecule generally carries a hydroxyl-acceptor group which must be activated (Fig. 7.**5b**). A few glutamine-dependent aminations require no ATP because the acceptor is already activated (5-phosphoribosyl 1-pyrophosphate + Gln–NH_2 → 5-phosphoribosyl 1-amine + Glu) or because the product is isomerized [D-fructose 6-phosphate (ketose) + Gln – NH_2 → D-glucosamine 6-phosphate (aldose) + Glu]. Others are driven by ATP cleavage to ADP and P_i (for energetic reasons), which also helps to expulse the –OH substituent, a poor "leaving group" (for mechanistic reasons). The "nascent NH_3" formed in glutamine-dependent aminations can be replaced by (unphysiologically) high concentrations of free ammonia; therefore, glutamine-dependent amidotransferases also can use very high levels of NH_3.

Carbamoyl phosphate (CAP). This intermediate is the immediate N donor in the synthesis of pyrimidines and the guanidino group of arginine. This energy-rich compound is formed by carbamoyl-phosphate synthetase, which requires 2 mol ATP: one for the phosphorylation of bicarbonate, and one for amination. Carbamoyl phosphate is transferred by carbamoyl transferases to the amino group of aspartate (pyrimidines) or of ornithine (arginine) (Fig. 7.**5d**).

Control. Besides ammonia (and/or nitrate), many prokaryotes can use nearly two dozen organic nitrogen

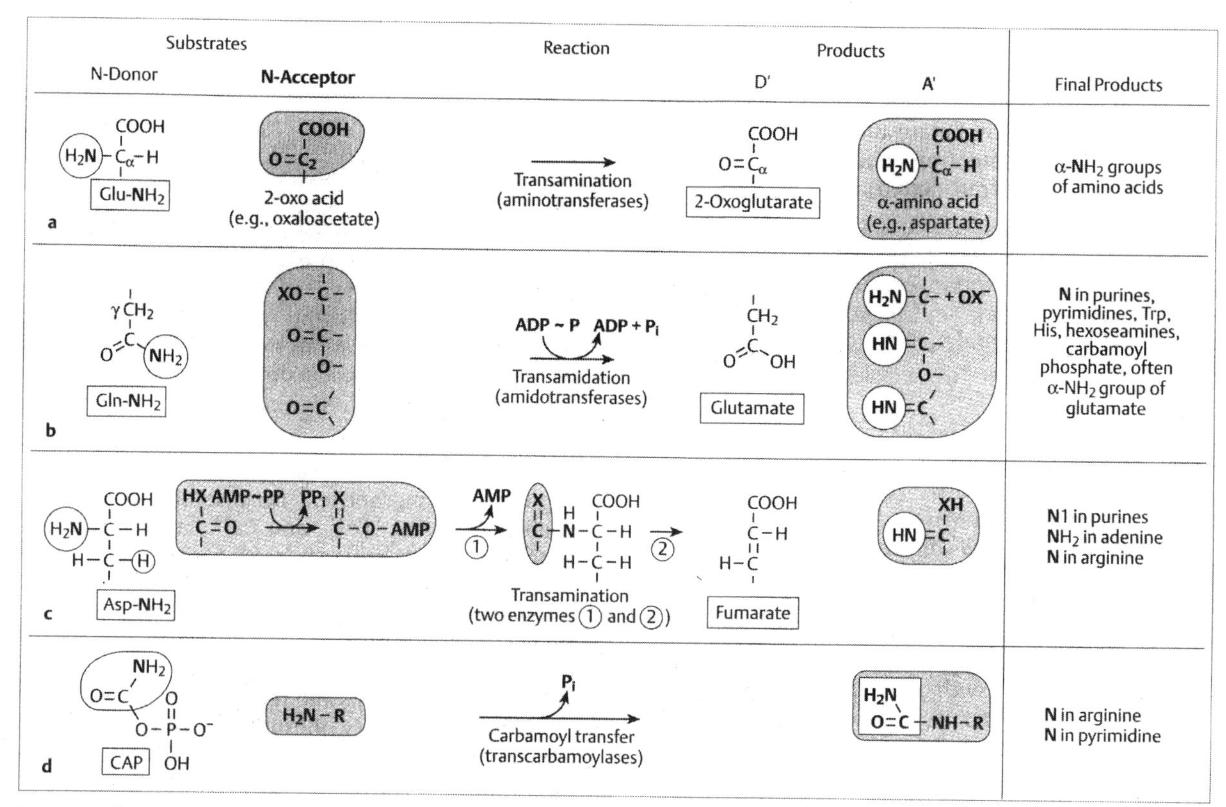

Fig. 7.5a–d **Four reactions for amino-group transfer.** Typical reactions for transfer from the N carriers
a glutamate,
b glutamine,
c aspartate, and
d carbamoyl phosphate

compounds as sole sources of nitrogen. The utilization of other N sources often requires inducible enzymes. NH_4^+ is the preferred nitrogen source, and growth with other N sources can be considered to be N-limited because the release of NH_3 is the rate-limiting step.

Glutamine synthetase, the most important enzyme in NH_3 assimilation, is strictly regulated by covalent modification and at the transcriptional level. The general signal for nitrogen availability is the ratio of the concentrations of 2-oxoglutarate (the main NH_4^+ acceptor) and glutamine (the main N donor under N limitation), but the complete system responds also to many nitrogenous compounds derived from glutamine. The different types of regulation of glutamine synthetase synthesis and activity in response to nitrogen (i.e., NH_4^+) availability will be discussed in Chapters 19.3 and 20.4.3.

Carbamoyl-phosphate synthetase activity is inhibited in many bacteria by the first pyrimidine nucleotide in the biosynthetic pathway (UMP), which is antagonized by ornithine (the precursor of arginine) and IMP (the first purine nucleotide in the biosynthetic pathway). Synthesis of the enzyme is cumulatively repressed by arginine and pyrimidine nucleotides. In this repression either one or two enzymes (one for arginine and one for pyrimidines) are involved.

7.3.2 Phosphate Is the Main Source of Phosphorus in Prokaryotes

One gram of cells contains about 30 mg of phosphorus, which is incorporated in **ATP**, the **P carrier**. The main phosphorus-containing products are nucleic acids, phospholipids, and coenzymes, which all contain this element as phosphate.

Phosphorus is easily oxidized by air; therefore, it exists almost completely as phosphate, which is the main inorganic phosphorus source of prokaryotes and does not need to be reduced. The hydrolysis products of phosphate-containing polymers are the main organic phosphorus sources available to bacteria. Various scavenging enzymes that cleave phosphate mono- or diesters and phosphate-anhydride bonds are located in the periplasmic space. Examples are the relatively unspecific alkaline phosphatase and the 5′-nucleotidase, which splits phosphate from nucleotides (which for good reasons are not transported).

Many processes that in Gram-negative eubacteria take place in the periplasmic space often are carried out in other bacteria by enzymes (lipoproteins) associated with the outer phase of the cytoplasmic membrane.

Transport. $H_2PO_4^-$ and HPO_4^{2-} are two phosphate species at a neutral pH ($pK_a = 7$) that are transported by several transport systems: (1) At high phosphate concentrations (> 0.1 mM), a constitutively expressed, relatively unspecific, **low-affinity system** [**P_i** (inorganic phosphate) **t**ransport, **Pit**] operates with a high capacity. It is driven by the proton motive force via a H^+ symport (equivalent to less than one ATP). (2) At low phosphate concentrations (< 10 μM), a well-regulated, **high-affinity system** operates with a low capacity (**p**hosphate-**s**pecific **t**ransport, **Pst**); its synthesis is repressed by extracellular phosphate. This binding-protein–dependent or ABC transport system is driven by the hydrolysis of ATP.

Loading and unloading of the phosphorus carrier (P transfer; Fig. 7.**6**). The P carrier, ATP, is formed from ADP and phosphate by ATP synthase, which is driven by the proton motive force created by electron transport and by substrate-level phosphorylation. ATP is also the universal energy carrier. The transfer of phosphorus proceeds via phosphoryl-group transfer.

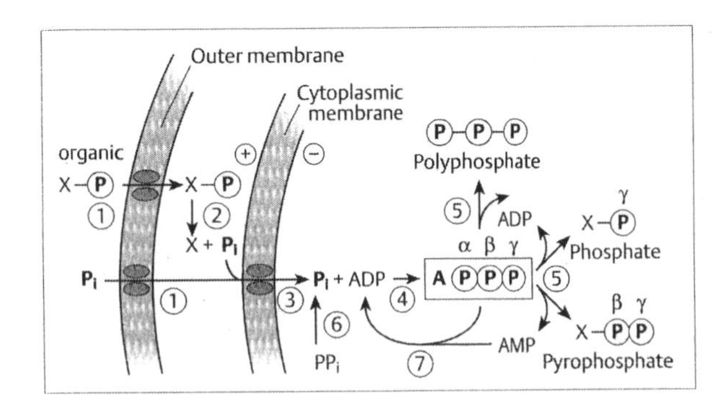

Fig. 7.**6 Assimilation of phosphorus.** 1 and 3, Inorganic and organic phosphate uptake systems respectively; 2, phosphatases acting on organic phosphates; 4, ATP formation from ADP by substrate-level and electron- transport phosphorylation; 5, phosphoryl or pyrophosphoryl transfer; 6, pyrophosphatase; 7, adenylate kinase. Note that ATP also serves to activate acids by forming an acyl-AMP intermediate whereby PP_i is released

The γ-phosphoryl group (PO_3^{2-}) of the phosphate anhydride, rather than the phosphate group ($HOPO_3^{2-}$), is transferred to various acceptor groups by kinases; acceptors may be –OH, –NH_2, –COOH, and other groups. Phosphoryl transfer often facilitates reactions, such as the expulsion of an –OH group during amination or reduction. The –OH is then released in the form of phosphate ($HOPO_3^{2-}$); ATP assists in the "dehydration." Therefore, ATP is also a universal co-substrate in many mechanistically difficult reactions. Many examples in which ATP has a dual (i.e., thermodynamic and mechanistic) function will be given in this chapter.

Control. Under many natural conditions, phosphate is a growth-limiting nutrient because of the very low solubility of its Al^{3+}, Fe^{3+}, and Ca^{2+} salts. Phosphate assimilation is efficiently regulated in most prokaryotes (see Chapter 20.1.1). Genes whose expression is regulated by phosphate concentration are part of a phosphate *pho* modulon, of which the *pho* regulon is a part (Chapter 20). The *pho* modulon overlaps with other global regulatory systems and regulates the induction of not only the high-affinity phosphate-uptake system but also the organic-phosphate–hydrolyzing enzymes when phosphate is limiting. With very few exceptions, organic phosphate compounds are not transported through the cytoplasmic membrane; rather, phosphatases or esterases at the cell surface or in the periplasmic space are responsible for phosphate release. These enzymes are often repressed by phosphate. Phosphonic acids ($R\text{-}PO_3H_2$, those with direct C–P bonds) are rare biological compounds; they are transported via a separate system and cleaved by cytoplasmic "C–P lyases." Polyphosphate, which serves as an energy and phosphorus reservoir, is formed by many bacteria when excess phosphate and energy are available but growth is limited by another factor (e.g., nitrogen or iron). Cellular pyrophosphate, PP_i, is cleaved by the ubiquitous cellular pyrophosphatase to two molecules of inorganic phosphate, P_i. Some anaerobes can use PP_i instead of ATP for phosphorylation of substrates, such as acetate.

7.3.3 Sulfate Is the Principal Sulfur Source for Bacteria

The synthesis of 1 gram of cells requires 3–10 mg of sulfur. **Cysteine** is the cellular **sulfur carrier** (Cys-SH) from which most (if not all) of the other sulfur-containing compounds obtain sulfur directly or indirectly (Fig. 7.7). Cellular sulfur is mainly at the oxidation level of H_2S, −2. In contrast, reduced sulfur compounds are easily oxidized by air at pH ≥7; consequently sulfate (SO_4^{2-}) and other sulfur oxo-compounds [e.g.

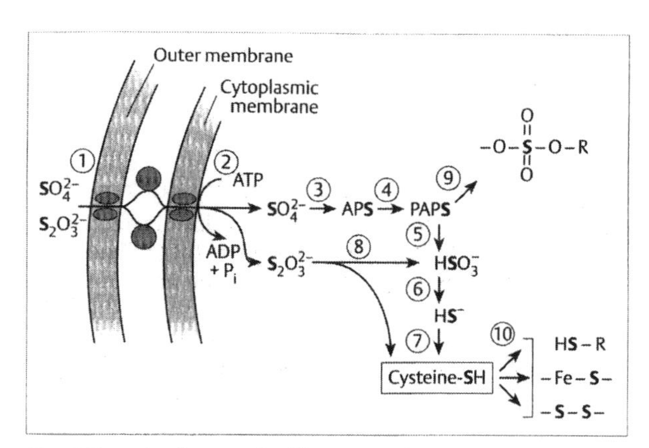

Fig. 7.7 **Assimilation of sulfur.** APS, adenosinephosphosulfate; PAPS,3′-phosphoadenosinephosphosulfate (see Chapter 8.7). 1 and 2, sulfate- and thiosulfate-transport components; 3, ATP sulfurylase; 4, APS kinase; 5, PAPS sulfotransferase/glutathion- or thioredoxin-dependent reduction; 6, sulfite reductase; 7, serine transacetylase and the *O*-acetylserine sulfhydrylases A and B; 8, *S*-sulfocysteine reduction; 9, sulfate-ester formation; and 10, sulfhydryl-transfer reactions

thiosulfate ($S_2O_3{}^{2-}$)] are the principal sulfur sources for aerobes. Hydrogen sulfide, H_2S, can also be released from sulfur-containing organic substrates.

Transport and reduction to H_2S. Most prokaryotes transport $SO_4{}^{2-}$ or $S_2O_3{}^{2-}$ and reduce them to H_2S by assimilatory sulfate reduction (Fig. 7.**7**). Sulfate is transported via an ATP-dependent, high-affinity ABC transport system (K_m 0.1 μM) similar to the high-affinity phosphate-specific transport system. The ABC system also transports thiosulfate and its toxic analogue, chromate ($CrO_4{}^{2-}$); negative mutants are chromate-resistant. In addition, a low-affinity transport system may be present. In sulfate-reducing bacteria (Chapter 12.1.8), which use sulfate as terminal electron acceptor, two symport systems are operative, which either co-transport 2 H^+ (at a high sulfate level) or 3 H^+ (at a low sulfate level). Na^+ may replace H^+ in marine strains.

Under anaerobic conditions, sulfate is reduced to H_2S by sulfate reducers in an anaerobic respiration. This H_2S is then available as a sulfur source for other bacteria (H_2S does not need to be transported, but because of its toxicity may not be present in too high a concentration). At the anaerobic/aerobic borderline, thiosulfate ($S_2O_3{}^{2-}$) and other sulfur oxides are present; these and H_2S are preferred sulfur substrates for many anaerobes. Elemental sulfur (S_8), which arises from microbial or air oxidation of H_2S, is poorly soluble. S_8 is used only by a few prokaryotes as electron acceptor in anaerobic respiration.

Loading of the sulfur carrier. H_2S, responsible for the "rotten-egg smell," is very toxic and volatile ($H_2S \rightleftarrows HS^- + H^+$, $pK_a = 7$), and therefore is immediately trapped in cysteine by *O*-acetylserine sulfhydrylase (Fig. 7.**8a**). The substrate *O*-acetylserine (derived from serine via acyl transfer from acetyl-CoA) provides a good acyl "leaving group," which facilitates the incorporation of even traces of H_2S. By a similar enzyme, thiosulfate is trapped as cysteine thiosulfonate, which is reduced to cysteine and sulfite ($SO_3{}^{2-}$), which can subsequently be reduced to H_2S.

Sulfate esters (R-O-$SO_3{}^-$) are probably formed from activated sulfate (PAPS; see Chapter 8). The rare organic sulfonates (R-$SO_3{}^-$) possibly are derived from sulfite.

Unloading of the sulfur carrier (S transfer). Probably all other sulfur compounds ultimately derive their sulfur from cysteine, for example, methionine, coenzyme A, lipoic acid, thiamine pyrophosphate, or glutathione (Fig. 7.**8b**). The synthesis of [Fe-S] centers, a major pool of sulfur in most prokaryotes, is poorly understood. It may be assumed that the acid-labile sulfur in these inorganic clusters is derived from cysteine via an enzyme-bound cysteinyl persulfide (R-S-S^-). Seleno-cysteine formation will be discussed in Chapter 8.8.2.

Sulfate transport and cysteine synthesis are controlled by the *cys* regulon. In nature, much of the sulfur is present in organic compounds. Most prokaryotes can use more than one sulfur source, but many

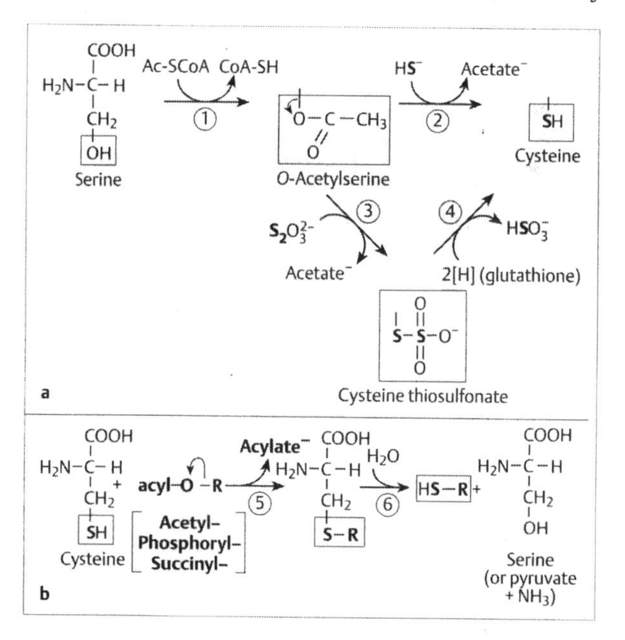

Fig. 7.**8a,b** **H_2S-assimilation and sulfhydryl-transfer reactions.**
a H_2S assimilation: 1, serine transacetylase; 2, *O*-acetylserine sulfhydrylase; 3, cysteine-thiosulfonate-forming enzyme; 4, thiosulfonate-reducing enzyme.
b Sulfhydryl transfer: 5, cysteine-adduct (e.g., cystathionine) synthase; and 6, cysteine-adduct lyase (see Fig. 7.**18**)

bacteria, such as *E. coli* prefer the compounds in the following order (of decreasing preference): the amino acid cysteine or its oxidized disulfide cystine, H_2S (in low concentrations!), HSO_3^-, $S_2O_3^{2-}$, SO_4^{2-}, the common tripeptide glutathione, and methionine, which is often used only as methionine rather than as general sulfur source. These preferences require a sophisticated regulation that is achieved by positive genetic control. Sulfate transport and cysteine synthesis are not linked, but both are controlled by a complex regulon (*cys* regulon). *O*-Acetylserine formation is inhibited by cysteine (via feedback inhibition), whereas the expression of the sulfate-reducing enzymes is induced by *O*-acetylserine. This regulation is intelligible because sulfate reduction costs (per mol sulfate) 3 mol ATP and 4 mol NADPH. On the other hand, organic sulfur compounds other than cysteine induce sulfur-releasing enzymes (e.g., sulfohydrolases), which hydrolyze sulfate esters.

7.3.4 One-Carbon Units Are Required for the Synthesis of Many Compounds

One-carbon (C_1) units are those carbon units that are separated from other carbon atoms by hetero atoms (i.e., O, N, or S). They are biosynthetically derived from C_1-carrying coenzymes in which the C_1 units are bound to an N atom or S atom. Their degradation often results in the reloading of C_1 carriers. **Biotin** is the carrier of activated bicarbonate.

Methyl, methenyl, or formyl groups linked to a hetero atom (N, S, O) do not occur in the biosynthetic precursor metabolites, but they are present in many building blocks (e.g., in methionine and purines). Further methylations of building blocks or cellular polymers also require C_1 donors.

Loading of C_1 carriers. The universal C_1 carriers are **S-adenosylmethionine** (SAM) and **tetrahydrofolate** (H_4F) or a similar tetrahydropteridine (Fig. 7.**9**). C_1 units are mostly derived from serine by serine hydroxymethylase, which catalyzes the formal reaction $HOCH_2–CHNH_2–COOH + H_2O \rightarrow HOCH_2OH + CH_2NH_2–COOH$.

The carrier of the formaldehyde hydrate, $HOCH_2OH$, is tetrahydrofolate (H_4F), in the form of methylene-H_4F. The product glycine may also serve as an additional C_1 donor in the reaction catalyzed by glycine decarboxylase. Methylene-H_4F is either oxidized to the formyl level or reduced to the methyl level. Many bacteria can also utilize formate as a C_1 source by using the reversible, ATP-consuming formyl-tetrahydrofolate-synthetase reaction.

Unloading of C_1 carriers (C_1 transfer). These C_1–H_4F compounds serve as C_1 precursors (e.g., in methionine synthesis, purine synthesis, or *N*-formyl methionine synthesis). Methyl groups often are indirectly derived from methyl-H_4F via formation of ***S*-adenosyl methionine (SAM)** (catalyzed by SAM synthetase) and a subsequent methyl transfer; the product is adenosylhomocysteine.

7.3.5 Virtually all Cellular Oxygen Is Derived From Water

Biosynthetic reactions normally do not require molecular oxygen. This seems plausible if one takes into account that molecular oxygen was available only after the evolution of oxygenic photosynthesis, hence, at a time when biosynthetic pathways were elaborated a billion years ago. Still, obligate aerobic bacteria make use of O_2 in a few **hydroxylation reactions** or in the

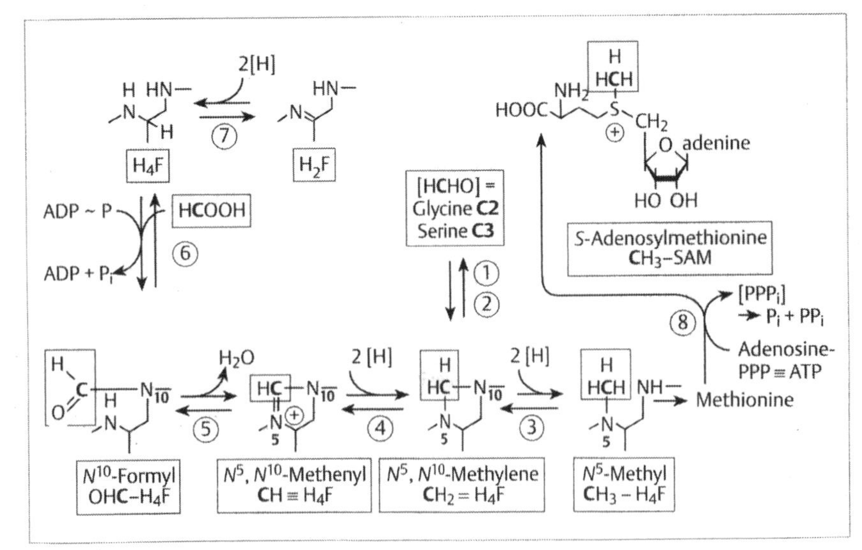

Fig. 7.**9 One-carbon units: coenzymes (C_1 carriers) and some enzymes involved in interconversion.** Only the catalytically most essential part of the coenzyme tetrahydrofolate (H_4F) is shown; H_2F denotes dihydrofolate. Enzymes 1 and 2, serine hydroxymethyl(transfer)ase and glycine decarboxylase or similar glycine-cleavage enzyme; 3, methylene-H_4F reductase; 4, methylene-H_4F dehydrogenase; 5, methenyl-H_4F cyclohydrolase; 6, formyl-H_4F synthetase; 7, dihydrofolate reductase; 8, S-adenosylmethionine synthetase

introduction of double bonds into saturated compounds. Examples are the synthesis of ubiquinone, of monounsaturated fatty acids in some bacteria (see Chapter 7.10.2) and of polyunsaturated fatty acids.

7.4 The Formation of Building Blocks From Central Precursor Metabolites, Polymerization, and the Final Assembly of Macromolecules Cost ATP and NADPH

Table 7.3 gives an account of the costs for the synthesis of the building blocks required to produce 1 gram of cells of a typical member of the Gram-negative bacteria, with glucose, ammonium, phosphate, and sulfate as the principal substrates.

7.4.1 How Much ATP and NADPH Are Needed?

Costs of building-block synthesis. As expected, the synthesis of proteins and nucleic acids requires the major portion of the ATP. NADPH is mostly required for protein and lipid biosynthesis. Most bacteria prefer preformed building blocks in complex (i.e., rich) media. This can save up to 85% of the ATP otherwise required for the synthesis of the building blocks from the central precursor metabolites (100% = 19 mmol ATP/g cells) and 100% of the NADPH (100% = also 19 mmol NADPH/g cells). In addition, the spared precursor metabolites can be used in energy metabolism.

Costs of polymerization. Polymerization of amino acids to protein is the most energy-consuming process of the prokaryotic cell. More energy is required for protein synthesis than for the synthesis of all the precursor building blocks: One gram of cells contains about 0.5 gram of protein, which is composed of approximately 5 mmol amino acids. Aminoacyl-tRNA formation requires two of the energy-rich, phosphate anhydride bonds of ATP; in addition, aminoacyl-tRNA binding and translocation require 4 energy-rich nucleotides (ATP, GTP) per building block. Considerable energy is required for mRNA turnover (An half-life of only 3 min allows approximately 20 translations per mRNA) and transport, but less for movement.

The total energy requirement may be twice as high when the cells grow on poor substrates, such as acetate and nitrate. This explains why many prokaryotes favor glucose and ammonium.

Table 7.3 **Estimated costs of energy and reductant for the synthesis of 1 g bacterial cells from central precursors metabolites.** It is assumed that *E. coli* is growing aerobically in a minimal medium with phosphate, ammonia, and sulfate as P, N, and S sources, respectively, and with glucose as carbon and energy source. The discrepancy between the theoretical and experimental values for growth yield may be due to (1) less-efficient energy conservation than presumed, (2) underestimation of energy requirement of synthesis and transport, and/or (3) energy requirement for other growth-associated processes. ca. means approximately (circa). ~P means ATP equivalent

Cellular component or parameter	Synthesis of building blocks	Activation of building blocks and polymerization	NADPH for synthesis of building blocks mmol
	mmol ~P required		
Protein	7.3	29	11.5
RNA	6.5	1.5	0.4
DNA	1.1	0.3	0.2
Lipid	2.6	0.3	5.3
Lipopolysaccharide	0.5	–	0.6
Murein	0.3	0.2	0.2
Glycogen	0.2	–	–
Total	ca. 19	ca. 32	ca. 19
Transport	ca. 6		
Total amount of mmol ATP/g cells	ca. 57		
Experimentally determined amount of mmol ATP/g cells	72		

Box 7.2 The role of ATP in biosynthesis is fivefold: (1) **Phosphoryl transfer** (of PO_3^{2-} rather than phosphate, $HOPO_3^{2-}$) is common in kinase and phosphatase reactions. (2) **Pyrophosphoryl transfer** occurs in the synthesis of 5-phosphoribosyl 1-pyrophosphate (PRPP), of guanosine 5′-diphosphate 3′-diphosphate (PPGPP), and also of phospho*enol*pyruvate (PEP) via PEP synthase and pyruvate phosphate dikinase. (3) **Nucleotidyl transfer** (e.g., of AMP) is common in carboxylate activation, synthesis of adenosine phosphodiester derivatives, sugar activation for polysaccharide synthesis, phospholipid synthesis, sulfate activation, and adenylation of enzymes for catalytic or regulatory purposes. The hydrolysis of the product, pyrophosphate (PP_i), drives the reaction. The abbreviations ADP~P, AMP~PP and AMP~ are used for these different roles of ATP. Phosphoryl and pyrophosphoryl groups are indicated by ~P and ~PP_i, respectively. (4) **Adenyl transfer** is important in adenosylcobalamine (coenzyme B_{12}) and *S*-adenosylmethionine formation. (5) Finally, in **histidine biosynthesis**, ATP is used as a donor of C_1 units and of nitrogen.

7.4.2 ATP Is Regenerated From ADP via Electron-Transport Phosphorylation or Substrate-Level Phosphorylation

ATP is formed from ADP mainly via ATP synthase, through a process formerly called "electron-transport phosphorylation" (see Chapter 4) because proton gradients generated during electron transport in respiration and photosynthesis drive the synthesis. Substrate-level phosphorylation (see Chapter 3) contributes only little to ATP formation, except in anaerobes. AMP, which is formed in the activation of acids (see Box 7.2), is converted to ADP by the ubiquitous, reversible adenylate kinase reaction:

$$AMP + ATP \overset{Mg^{2+}}{\rightleftharpoons} 2ADP$$

Nucleotides are important regulatory compounds of metabolism. The ratio

$$\frac{[ATP] + \frac{1}{2}[ADP]}{[AMP] + [ADP] + [ATP]},$$

called the **"energy charge,"** is kept near 9/10 in the living cell. As a rule, catabolic ATP-yielding reactions are stimulated by AMP or ADP, whereas biosynthetic, ATP-consuming reactions are blocked by AMP or ADP. AMP signals "hunger". The ATP turnover is so fast (half-life of approximately 10–100 ms) that this regulation ensures a very rapid response. If the synthesis of 1 gram of cells costs approximately 70 mmol of ATP ($M_r = 507$), this corresponds to 35 grams of ATP!

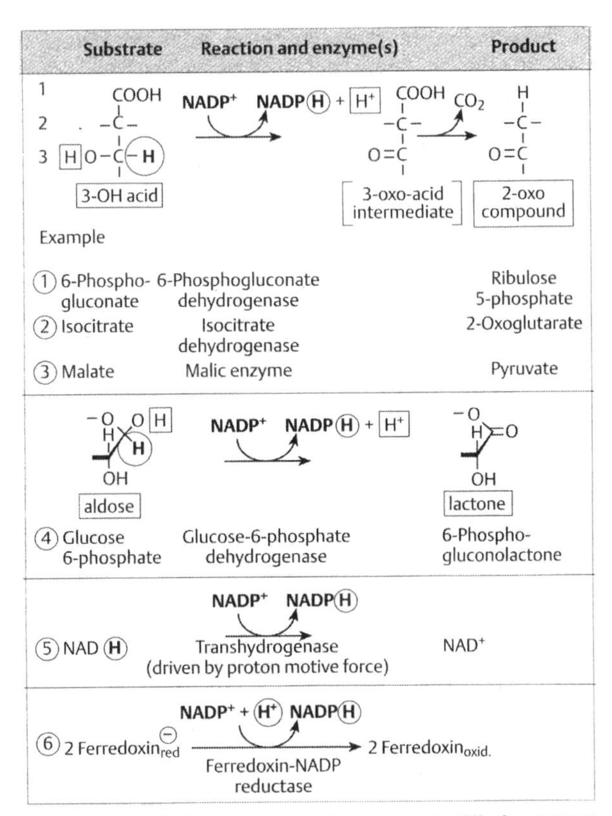

Fig. 7.10 **Metabolic reactions that serve to fill the NADPH pool.** These reactions involve either the **oxidation of a 3-hydroxy acid** (e.g., 6-phosphogluconate, isocitrate, or malate) to the corresponding labile 3-oxo acid in a reaction driven by the decarboxylation forward; or the **oxidation of the aldehyde group of an aldose** (e.g., glucose 6-phosphate) to form a lactone (i.e., an intramolecular ester). The most common NADPH source is the oxidative pentose phosphate pathway, by which glucose 6-phosphate is oxidized in three steps to ribulose 5-phosphate. This pathway is also the most common route for the synthesis of pentose-phosphate building blocks. One mol NADPH is formed in the glucose-6-phosphate-dehydrogenase reaction; the product, 6-phosphogluconolactone, is hydrolyzed by lactonase, and the resulting 6-phosphogluconic acid is oxidatively decarboxylated to ribulose 5-phosphate by 6-phosphogluconate dehydrogenase under NADPH formation

7.4.3 NADPH Is Regenerated in a Few Catabolic Reactions

Normally, reduction steps in biosynthesis are **NADPH-dependent**, whereas oxidation steps in metabolism are **NAD^+-dependent** (except for NADPH-loading reactions). NADPH is the cellular reductant, NAD^+ the cellular oxidant. As a rule, pyridine-nucleotide-dependent oxidoreductase reactions are fully reversible.

In order to drive reductive biosynthetic (anabolic) steps and oxidative catabolic steps in the right direction, the cell keeps the ratio $[NADPH]/[NADP^+]$ high and the ratio $[NADH]/[NAD^+]$ low.

$$\frac{[NADH]}{[NAD^+]+[NADH]}$$

is kept near 1/20 (**catabolic reduction charge**);

$$\frac{[NADPH]}{[NADP^+]+[NADPH]}$$

is kept near 1/2 (**anabolic reduction charge**).

Hence, pyridine nucleotides are important regulatory compounds (allosteric regulation) whose rapid turnover allows an extremely fast response.

How does the cell keep the reductant NADPH "loaded"? Some catabolic enzymes that catalyze rather exergonic oxidation reactions fill up the NADPH pool (Fig. 7.**10**).

Box 7.3 Normally, far more than 2 mol NADPH per pentose-phosphate building block are required for biosynthesis. More NADPH is generated in two ways: (1) The pentose phosphate pathway is closed to a cyclic process by which six molecules of pentose 5-phosphate are reconverted into five molecules of glucose 6-phosphate, which again can serve as an NADPH and pentose source (see Fig. 7.**20**). One turn of this cycle results in the net oxidation of one carbon atom per hexose to CO_2 and in the formation of 2 mol NADPH. It is called the oxidative pentose phosphate cycle because six turns of the cycle result in the net oxidation of one molecule of hexose phosphate via pentose phosphates (see Chapter 7.7). (2) Other NADPH-forming reactions are isocitrate dehydrogenase and, less often, malic enzyme. Various bacteria catalyze an energy (proton motive force)-driven $NADP^+$ reduction with NADH by a membrane-bound transhydrogenase. Anoxygenic phototrophic bacteria (with only an equivalent of photosystem II) and chemolithotrophs drive NADH formation and therefore indirectly NADPH formation by reversed electron transport from reduced quinones. Oxygenic photosynthetic organisms with a photosystem I form NADPH via ferredoxin : NADP oxidoreductase. Many anaerobes can reduce ferredoxin by the oxidation of 2-oxo acids (e.g., with pyruvate ferredoxin reductase), formate (via formate dehydrogenase), or H_2 (via hydrogenase). The redox potential of ferredoxin is between 70 mV and 130 mV more negative than that of $NADPH/NADP^+$.

7.5 Central Biosynthetic Precursor Metabolites From Which the Cellular Building Blocks Are Formed Derive From Three Main Metabolic Pathways

These pathways are: (1) the citric acid cycle, (2) glycolysis/glucogenesis, and (3) the pentose phosphate cycle (Table 7.**1**; Fig. 7.**2**). Even those bacteria that use alternatives to these pathways have them in an incomplete form (e.g., incomplete citric acid cycle lacking 2-oxoglutarate dehydrogenase) or the alternatives still provide the essential precursor metabolites (e.g., the Entner-Doudoroff pathway). The ambivalent (biosynthetic/degradative) features of these pathways requires a strict regulation at both the levels of enzyme activity and of enzyme synthesis. Figure 7.**11** indicates a few very common irreversible enzyme reactions of central metabolic pathways that are under allosteric (i.e., activity) control (see also Fig. 19.**7**). Note that fully reversible enzyme reactions are never regulated. The principles of regulation of complex biosynthetic pathways and networks are discussed in Chapters 18 to 21.

Half of the protein (10 out of the 20 amino acids) is derived from the citric acid cycle, and intermediates also are drained off for the synthesis of tetrapyrroles and other cell constituents. Therefore, the intermediates of the citric acid cycle have to be replenished; other-

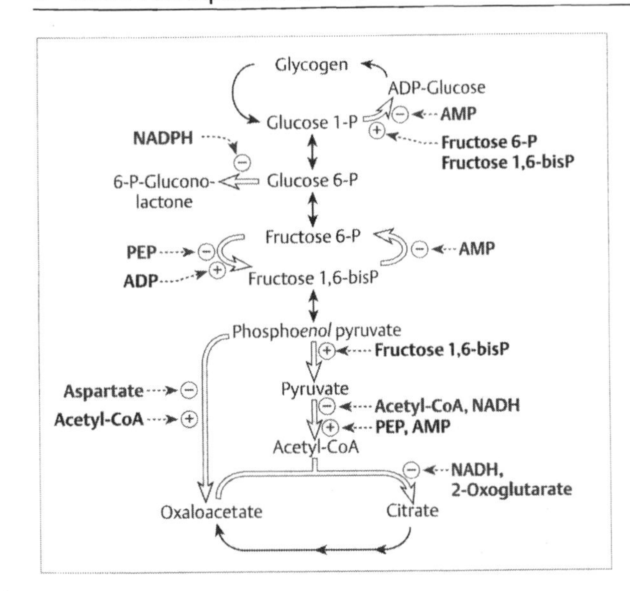

Fig. 7.11 **Allosteric activity control of irreversible enzyme reactions of the central metabolic pathways that provide biosynthetic precursor metabolites.** The scheme applies to *E. coli* and related organisms. +, activation; −, inhibition [after 1]

wise, the cyclic acetyl-CoA oxidation would come to a halt because C_4 dicarboxylic acid acceptors no longer would be available. These reactions, called **anaplerotic** (i.e., filling up), very much depend on the growth substrate (Fig. 7.**12**).

Anaplerotic reactions and heterotrophic CO_2 fixation fill up the oxaloacetic pool. The most common oxaloacetate-forming enzyme is PEP carboxylase. Oxaloacetate is formed from pyruvate by the biotin- and, therefore, ATP-dependent pyruvate carboxylase; the enzyme in most bacteria is allosterically activated by acetyl-CoA (as in mammals). Very rarely, PEP carboxytransphosphorylase catalyzes the reaction

$$\text{PEP} + CO_2 + P_i \rightarrow \text{oxaloacetate} + PP_i,$$

which replaces PEP carboxylase. In propionic acid bacteria, which use the succinate pathway of propionate formation (Chapters 12.2.2 and 12.2.3), the intermediate methylmalonyl-CoA acts as CO_2 donor in the reaction catalyzed by methylmalonyl-CoA pyruvate transcarboxylase:

$$\text{methylmalonyl-CoA} + \text{pyruvate} \rightleftarrows \text{oxaloacetate} + \text{propionyl-CoA}.$$

Although the PEP-carboxykinase reaction (Fig. 7.**12**) in principal is reversible, the enzyme normally functions in

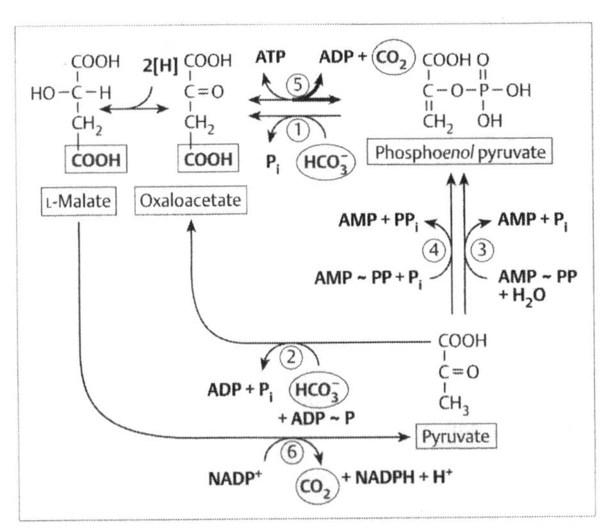

Fig. 7.**12** **Anaplerotic reactions and heterotrophic CO_2 fixation.** 1, PEP carboxylase; 2, pyruvate carboxylase (biotin enzyme); 3, pyruvate water dikinase (i.e., PEP synthetase); 4, pyruvate phosphate dikinase; 5, PEP carboxykinase; 6, malic enzyme

the direction of PEP formation during growth on C_4 and C_5 dicarboxylic acids. The same is true for malic enzyme, which forms pyruvate.

Variations of anaplerotic reactions result from different carbon sources. A different problem arises if C_3 compounds, such as lactate or pyruvate, are the carbon sources. Pyruvate kinase generates pyruvate from phosphoenolpyruvate and 1 mol ATP, which is an irreversible step; therefore, PEP formation from pyruvate requires different enzymes and the input of 2 mol ATP. Two enzymes form PEP directly from pyruvate, PEP synthetase (pyruvate water dikinase) and pyruvate phosphate dikinase (known also from C_4 plants) (Fig. 7.**12**). In both cases, pyrophosphorylated enzymes are intermediates; one energy-rich phosphate group is transferred to the substrate pyruvate and the other phosphate group to water or phosphate. Pyrophosphate is usually hydrolyzed by pyrophosphatase, which drives this reaction forward. Both PEP-forming routes therefore require 2 mol ATP/mol PEP. PEP may be formed indirectly via pyruvate carboxylase and PEP carboxykinase, which also requires two high-energy phosphoryl bonds.

The most diverse anaplerotic and pyruvate- or PEP-forming reactions occur when microorganisms grow either with C_2 compounds and substrates that are degraded via acetyl-CoA (e.g., fatty acids) or with C_1 compounds (e.g., methanol). These reactions are discussed in Chapters 8.2 and 8.3.

7.6 Amino Acids Derive From Few Central Precursor Metabolites

The building blocks of the major cell constituent, protein (50–60% of dry weight), are the 20 natural L-amino acids. Nitrogen-fixing cyanobacteria may form storage proteins. D- and L-amino acids are constituents of most cell-wall polymers, whereby the D-isomers are formed by racemases from the L-forms. L-amino acids, also the precursor molecules in nucleotide synthesis, all are derived from only a few central precursor metabolites (Fig. 7.**2**).

Pyridoxal phosphate is a common prosthetic group in many enzymes that transform amino acids.

Amino acids traditionally are grouped into five biosynthetic families, which are named after their common precursor, and histidine. The biosynthetic pathways are rather uniform, with a few exceptions. Since most reduction steps in biosynthesis are NADPH-dependent (+2[H]) and most oxidation steps are NAD^+-dependent (−2[H]), the pyridine nucleotides involved in redox reactions are not specified. Exceptions to this rule are the "NADPH-loading" oxidation reactions (see Chapter 7.4.3).

Box 7.4 The biosynthetic pathways and their discoveries illustrate well the way radioisotope studies, mutant methodology, and enzyme studies were often combined to decipher metabolic pathways in prokaryotes. These are still essential techniques today. Examples are: (1) Mutants often accumulate (or even excrete) those intermediates that cannot be processed because of a metabolic block. (2) Another method may allow the identification of the metabolic step affected by a mutation. Hypothetical intermediates are fed to mutants, whereby the compounds that occur after the biosynthetic step blocked by the mutation enable the microorganism to grow. (3) Cells are fed with specifically [^{14}C]-labeled precursors; this is followed by chemical degradation of the labeled products (accumulated/excreted by mutants or found in the wild-type cell) to determine the [^{14}C]-content of each C atom and thus to deduce its origin. With this information, it is often possible to trace a biosynthetic pathway retrospectively. This cumbersome method of determining the label of the individual carbon atoms has been replaced by another technique. (4) [^{13}C]-labeled precursors are fed to the bacteria, and the [^{13}C]-label distribution in the carbon skeleton of products (e.g., amino acids) can be precisely determined by [^{13}C]-**n**uclear **m**agnetic **r**esonance (**NMR**) spectroscopy.

7.6.1 The Amino Acids of the Glutamate Family Derive From 2-Oxoglutarate

The glutamate and the aspartate family have in common that all reactions take place at the side-chain carboxyl group, which requires ATP for chemical activation. There are two primary reactions, amidation or reduction. Note that glutamate and glutamine are also the main N carriers in biosynthesis.

The glutamate family comprises four amino acids (i.e., **glutamate**, **glutamine**, **proline**, and **arginine**), which contain a C_5 backbone (arginine $= C_5$ plus guanidino group) and are derived from 2-oxoglutarate via glutamate (Fig. 7.**13**). The formation of **glutamate** and **glutamine** has already been discussed in Chapter 7.3.1 since these amino acids are the main N carriers. Glutamine synthesis requires ATP-dependent glutamyl-phosphate formation in order to eliminate the -OH group and to drive the amidation reaction forward.

Similarly, the reduction of the C5 carboxyl group to an aldehyde in **proline** and **arginine** biosynthesis requires the ATP-dependent chemical activation of the group. The resulting aldehyde spontaneously forms a cyclic Schiff's base, the precursor of proline, and proline is subsequently formed by reduction.

In order to avoid cyclization in arginine synthesis, the α-NH_2 group is transiently protected by acylation, whereby the acyl group later is cleaved off. The aldehyde is transaminated; this finally yields the diamino acid, ornithine. Addition of a carbamoyl group yields L-citrulline; its oxo group is replaced by an imino group, and the NH_2-donor (aspartate) is converted to fumarate (see Fig. 7.**5c**). L-Ornithine often is a constituent of murein. Arginine synthesis is also important for polyamine synthesis (see Chapter 7.12.1).

Control. The regulation of glutamine synthesis is discussed in Chapters 19.3 and 20.4.3. In proline synthesis, the first enzyme (γ-glutamyl kinase) is regulated by feedback inhibition by proline. Proline formerly was thought to act as an osmoprotectant since it tended to accumulate from the medium under hyperosmotic stress. Betain compounds turned out to be the physiological osmoprotectants, and the rate of proline synthesis is not regulated in response to osmotic stress. Arginine exerts a feedback inhibition on *N*-acetylglutamate kinase, and the expression of all enzymes of the arginine pathway is repressed by arginine.

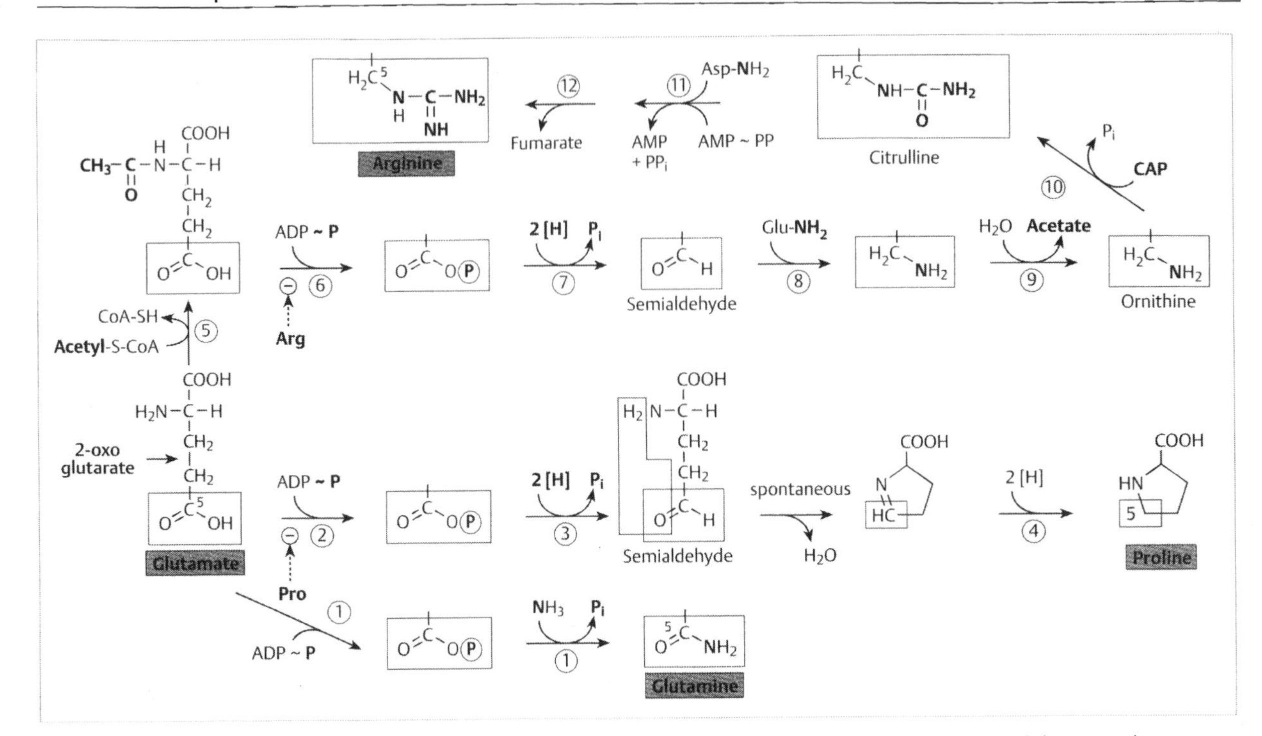

Fig. 7.13 **Synthesis of the amino acids of the glutamate family.** Glutamate (Glu): reductive amination of 2-oxoglutarate. Glutamine (Gln): 1, glutamine synthetase. Proline (Pro): 2, γ-glutamyl kinase; 3, glutamate-γ-semialdehyde dehydrogenase; 4, Δ^1-pyrroline-5-carboxylate reductase. Arginine (Arg): 5, *N*-acetylglutamate synthase; 6, *N*-acetylglutamate kinase; 7, *N*-acetylglutamyl-phosphate reductase; 8, *N*-acetylornithine-δ-transaminase; 9, *N*-acetylornithinase; 10, ornithine carbamoyltransferase (or isoenzymes); 11, argininosuccinate synthetase; 12, argininosuccinase. CAP = carbamoyl phosphate

7.6.2 Oxaloacetate and Pyruvate Are the Precursors of two Families of Amino Acids

The **aspartate family** typically comprises six amino acids (i.e., **aspartate**, **asparagine**, **threonine**, **methionine**, **lysine**, and **isoleucine**), which are derived from aspartate (Fig. 7.**14**). Also, ***meso*-diaminopimelate** (cell-wall amino acid), **dipicolinate** (an endospore constituent), and ***S*-adenosylmethionine** (universal C_1 carrier) are derived from this complex, multiply branched pathway. **Aspartate** is formed from oxaloacetate by transamination with glutamate. **Asparagine** is formed from aspartate by L-asparagine synthetase in a reaction analogous to that of glutamine synthetase. Aspartate, threonine, and methionine (C_4+C_1) have a C_4 backbone that is formed in a common pathway via the C4 semialdehyde. As in the glutamate family, the modification of the β-carboxyl group of aspartate also requires an ATP-dependent reduction to the aldehyde, followed by one of three possible steps: (1) Further reduction to the alcohol homoserine and then isomerization affords **threonine**; the shift of the -OH group is facilitated by phosphorylation of the C4 alcohol. (2) The alcoholic function is replaced by a thiol group. As in cysteine synthesis, a good "leaving group" is introduced (acetyl or succinyl group), and the -SH group donor is cysteine. The product, homocysteine, is methylated to **methionine** with CH_3-tetrahydrofolate as C_1 donor by an enzyme dependent on or independent of vitamin B_{12}. (3) In most prokaryotes, **lysine** and **isoleucine** are also formed from aspartate semialdehyde plus additional pyruvate (with exceptions). The first branching step towards lysine formation (Fig. 7.**15**) is the condensation of aspartate semialdehyde with pyruvate, and cyclization (Schiff's base formation) to a dihydropyridine (2,3-dihydrodipicolinate), which is reduced to the tetrahydro-form. **Dipicolinic acid** is easily formed by oxidation of the dihydro-form during sporulation. In endospores, it is a major soluble compound that confers heat stability. The subsequent strategy is known from proline and arginine synthesis. The $-N{=}C{=}$group can be considered a Schiff's base formed by addition of an $-NH_2$ to a $O{=}C{=}$group: *N*-acylation (by acetyl- or succinyl-CoA) forces the N=C bond to break and thereby the ring to open. Then the 2-oxo group is transaminated, and the acyl group is hydrolyzed when the possibility of cycliz-

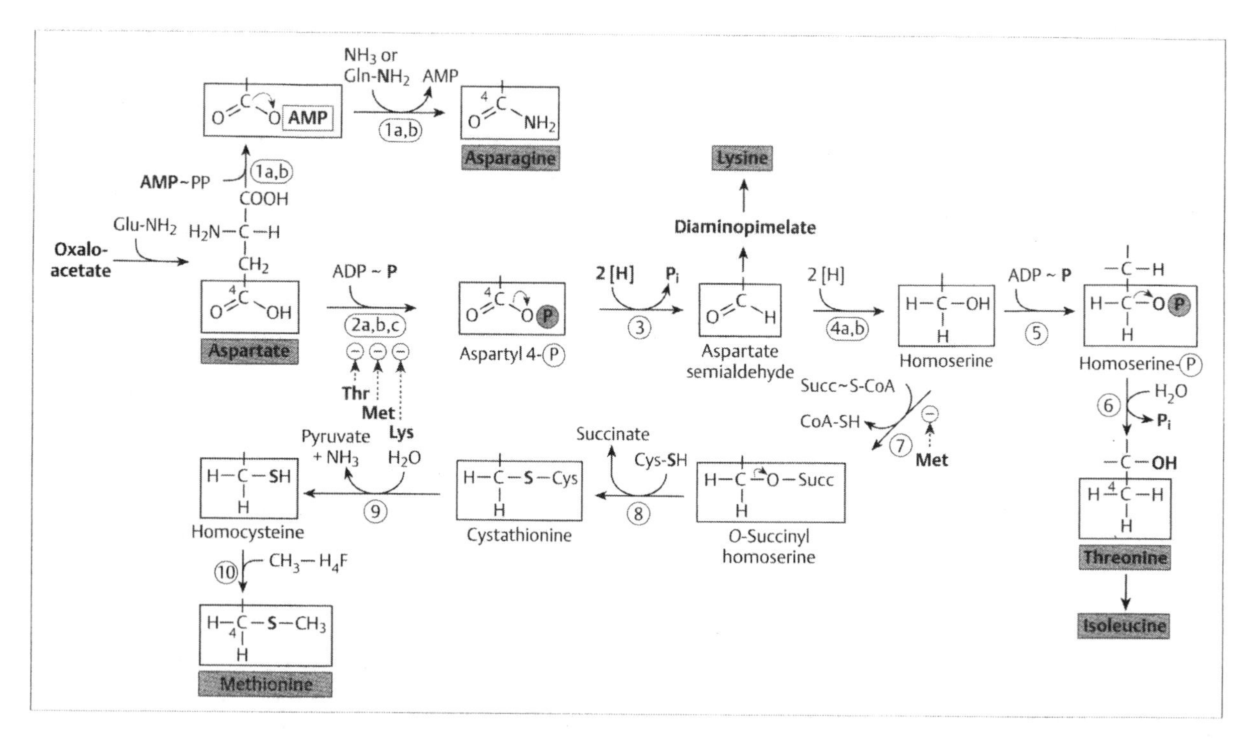

Fig. 7.14 **Synthesis of four amino acids of the aspartate family.** Aspartate (Asp): transamination of oxaloacetate with glutamate. Asparagine (Asn): 1a/1b, asparagine synthetases. Common pathway and threonine (Thr) synthesis: 2a/2b/2c, aspartate kinases I, II, and III; 3, aspartate semialdehyde dehydrogenase; 4a/4b, homoserine dehydrogenase I, II (i.e., aspartate kinases I, II; bifunctional enzymes); 5, homoserine kinase; 6, threonine synthase (pyridoxal-phosphate enzyme). Methionine (Met) branch: 7, homoserine succinyltransferase; 8, cystathionine γ-synthase (pyridoxal-phosphate enzyme); 9, cystathionine β-lyase (pyridoxal-phosphate enzyme); 10, homocysteine methylase (vitamin B_{12}–dependent and vitamin B_{12}–independent enzymes). The enzyme reactions numbered 2–4 are the common pathway to methionine, threonine, and isoleucine (Ile), and reactions 2–3 lead to lysine (Lys). Broken lines indicate feedback inhibition of the enzymes

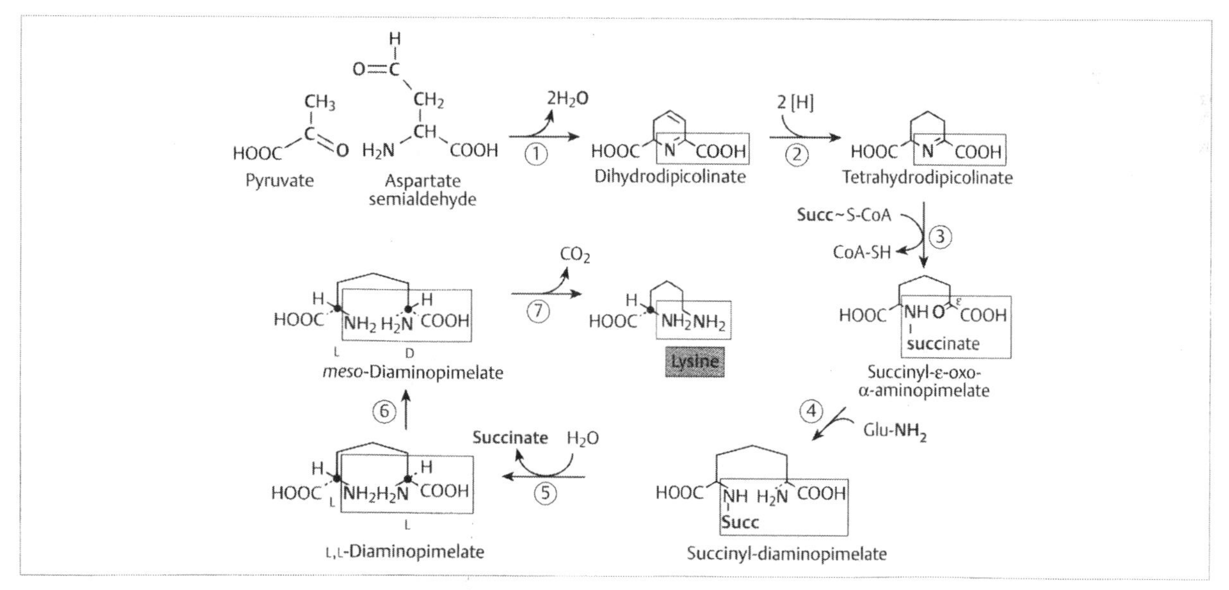

Fig. 7.15 **Synthesis of the amino acid lysine (aspartate family) and of diaminopimelate.** Lysine (Lys) branch: 1, dihydrodipicolinate synthase; 2, dihydropicolinate reductase; 3, tetrahydrodipicolinate succinylase; 4, succinyl diaminopimelate (DAP) aminotransferase; 5, succinyl diaminopimelate desuccinylase; 6, diaminopimelate epimerase; 7, diaminopimelate decarboxylase

ation has passed. The resulting L,L-α,ε-diaminopimelate is epimerized to the *meso* form, which is decarboxylated to L-lysine. This ***meso*-diaminopimelate** is also a common constituent of murein. Threonine can be deaminated to 2-oxobutyrate; addition of pyruvate and conversion to isoleucine are discussed below.

Box 7.5 There are variations of the pathways that are **representative for Proteobacteria** and outlined in this chapter. **Lysine.** In some Gram-positive prokaryotes with lysine instead of diaminopimelate as diamino acid in the cell wall, the enzymatic steps 3 through 6 of Fig. 7.**15** are replaced by a single enzyme, *meso*-diaminopimelate dehydrogenase. An alternative synthesis of lysine is via the α-amino-adipate pathway, whereby 2-oxoglutarate is condensed with acetyl-CoA and then converted to 2-oxoadipate (a C_6 dicarboxylic acid), very much the same as 2-oxoglutarate is formed in the citric acid cycle. Transamination of the 2-oxo group, reduction of the ε-carboxyl group to the aldehyde, and transamination of the product yield lysine. This pathway has been found in many members of the Eukarya, but also in Archaea. **Branched-chain amino acids.** The 2-oxo-acid precursors may be formed by anaerobes via the reductive carboxylation of CoA thioesters of the fermentation products propionate (leading to 2-oxobutyrate), 2-methylbutyrate (yielding isoleucine), and isobutyrate (resulting in valine). The compound 2-oxobutyrate may also be formed from methionine. Others use the so-called citramalate pathway for isoleucine synthesis, in which acetyl-CoA and pyruvate are condensed to citramalate. **Arginine.** The acyl group introduced from acetyl-CoA in the first step can be recycled by a glutamate acetyltransferase. **Glycine** and **serine.** In some C_1-utilizing bacteria, glycine plus the formaldehyde C_1 unit are the precursors of serine. Glycine can also be formed from threonine aldolase, which cleaves threonine into glycine and acetaldehyde. **Aromatic amino acids.** In some prokaryotes, a slightly different route of phenylalanine and tyrosine synthesis is used: prephenate is first transaminated to arogenate, and the subsequent steps are, in principal, as in the standard pathway. **Acylations.** The protecting or activating acyl groups introduced during synthesis may vary, for example, acetyl-CoA or succinyl-CoA.

Alanine, **valine** and **leucine**, which form the **pyruvate family**, all are derived from pyruvate. The grouping is somewhat arbitrary since pyruvate is also required for lysine and isoleucine synthesis. **Alanine** results directl from the transamination of pyruvate. There are tw alanine racemases, one for D-alanine degradation (D,L alanine-inducible) and one for the biosynthesis of D alanine for cell-wall synthesis (constitutive).

The synthesis of the branched-chain amino acid **isoleucine**, **valine**, and **leucine** is based on man common principles (Fig. 7.**16**). The following fou reactions are even catalyzed by the same set of enzyme in parallel: (1) A 2-oxoacid [e.g., pyruvate (2-oxopro pionate) and 2-oxobutyrate, derived from threonin deamination] and activated acetaldehyde from pyruvat are condensed in a thiamine-pyrophosphate-dependen reaction. The resulting α-acetyl group is reduced to th 2,3-dihydroxyl compound; water is eliminated, accom panied by an interesting shift of the alkyl group. Th resulting 2-oxo group is transaminated to valine an isoleucine, respectively. The immediate 2-oxoacid pre cursor of valine is condensed with acetyl-CoA to α isopropylmalate (this common reaction type is calle **Claisen condensation**; see e.g., citrate synthesis an Chapter 8.2). The elongated intermediate is converted t leucine.

Control of aspartate family (see also Chapter 19. and Fig. 7.**11**). The synthesis of the aspartate famil shows a high degree of variation. Therefore, its control i not uniform, and the regulatory pattern representativ for Enterobacteriaceae—as shown here—is quite differ ent (e.g., in *Bacillus*). **Common pathway**: There are tw asparagine synthetases; one uses glutamine as amin donor and functions under nitrogen limitation, and th other uses NH_3 and functions under nitrogen exces (compare with glutamate formation via glutamine- an NH_3-dependent enzymes!). Both enzymes hydrolyz ATP to AMP and PP_i. There are three aspartate kinases numbered I, II, and III. Aspartate kinases I and II als catalyze the homoserine-dehydrogenase reactior **Threonine branch**: The genes form an operon whos expression is repressed by threonine and isoleucin ("multivalent repression"). Aspartate kinase I is als inhibited by threonine (feedback inhibition). **Methio nine branch**: Methionine represses the expression c aspartate kinase II of the common pathway and affect all steps but one of the methionine branch, for example the first enzyme is allosterically controlled by methio nine and *S*-adenosylmethionine. **Lysine branch**: Aspar tate kinase III is inhibited, and its expression i repressed by lysine, which also affects steps of th lysine branch.

Control of branched-chain amino acids. Th branched-chain amino acid synthesis is well regulatec However, note that in Gram-positive bacteria als branched-chain fatty acids are derived from the 2-oxo acid precursors.

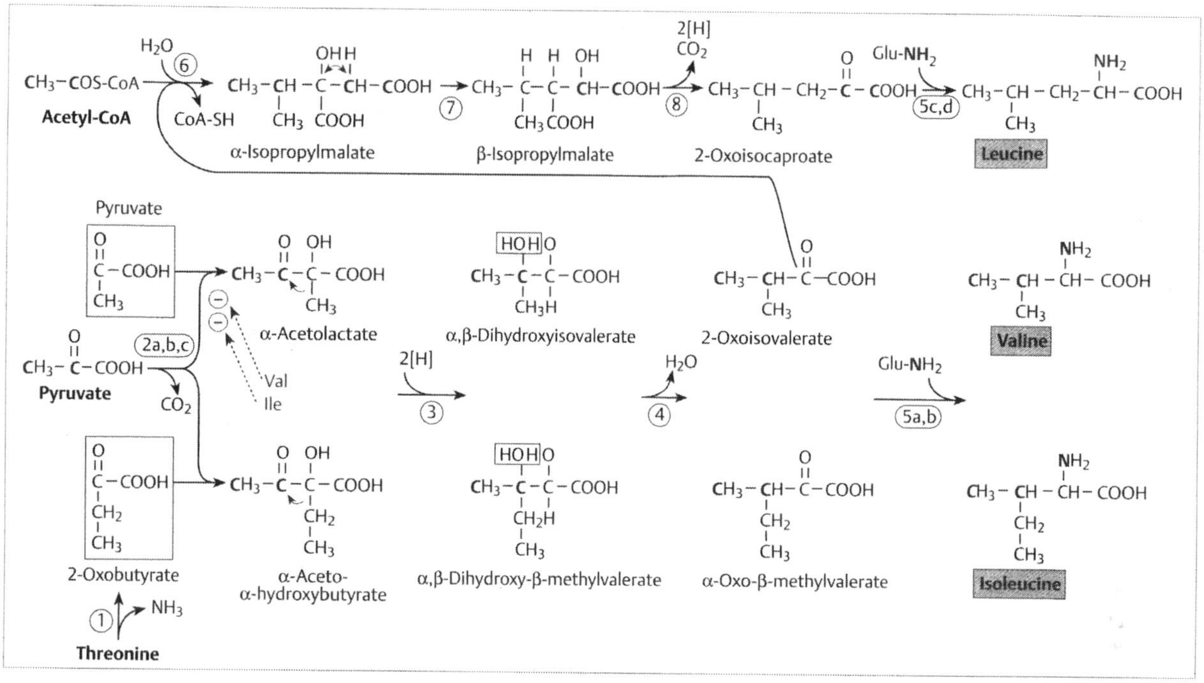

Fig. 7.16 **Synthesis of the branched-chain amino acids valine, isoleucine, and leucine.** Valine, Val; isoleucine, Ile; leucine, Leu. 1, Threonine deaminase; 2a/2b, end-product–inhibited and 2c, end-product–noninhibited acetohydroxyacid synthases; 3, acetohydroxyacid isomeroreductase; 4, dihydroxyacid dehydrase; 5a/5b, transaminases, 5c/5d, transaminases; 6, isopropylmalate synthase; 7, isopropylmalate isomerase; 8, β-isopropylmalate dehydrogenase

7.6.3 The Serine–Glycine Family (Including Serine, Cysteine, and Glycine) Normally Has Serine as the Common Precursor

Serine is derived from 3-phosphoglycerate in three steps (Fig. 7.**17**), oxidation of the free hydroxy group, transamination, and phosphate removal. Serine is the precursor of **cysteine** (the cellular S carrier; Fig. 7.**8**) and of **glycine**.

Control. The enzymes yielding serine are constitutive. The first enzyme of the pathway is inhibited by serine. Expression of serine hydroxymethyltransferase (serine hydroxymethylase) is repressed by several products of C_1 metabolism (i.e., serine, glycine, methionine, purines, and thymine). Although cysteine is derived from serine, its synthesis is regulated differently (see *cys* regulon). In many organisms, there are additional sources of C_1 units. Cysteine desulfurylation to alanine is the first step in cysteine degradation.

7.6.4 The Aromatic Amino Acids Are Formed From Erythrose 4-Phosphate and Phospho*enol*pyruvate

The aromatic ring of the three aromatic amino acids, **phenylalanine**, **tyrosine**, and **tryptophan**, is formed from erythrose 4-phosphate and phospho*enol*pyruvate; the C_3 side chain also comes from phospho*enol*pyruvate or (in tryptophan) from serine. The two carbon atoms of the heterocyclic ring of tryptophan are derived from C1 and C2 of 5-phosphoribosyl 1-pyrophosphate (Fig. 7.**18**). All three have a common route (the **shikimate pathway**), which begins with the condensation of the two precursor metabolites to a C_7 compound, 3-**d**eoxy-D-**a**rabino-**h**eptulosonate-7-**p**hosphate (DAHP) catalyzed by three isoenzymes, the DAHP synthases.

The identification of the various intermediates of aromatic amino acid synthesis made use of auxotrophic mutants. Shikimate was the only compound out of 55 aromatic and hydroaromatic compounds tested that could replace the requirement of certain aromatic auxotrophs of *Escherichia coli* for tyrosine, phenylalanine, and tryptophan. In the common pathway, other mutants accumulated shikimate and other intermediates, mostly in the dephosphorylated form.

The C_7 compound DAHP is cyclized to 5-dehydroquinate and converted via shikimate into chorismate. This is the branching point from which three indepen-

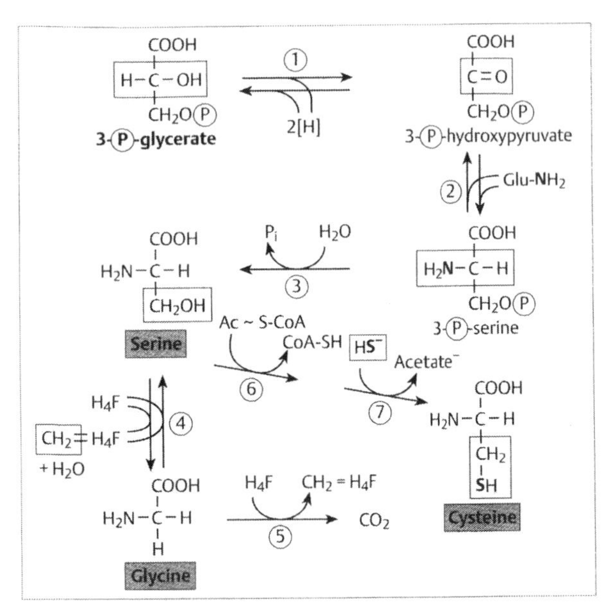

Fig. 7.**17** **Synthesis of the amino acids of the serine–glycine family.** Serine, Ser; glycine, Gly. 1, 3-phosphoglycerate dehydrogenase; 2, 3-phosphoserine aminotransferase; 3, 3-phosphoserine phosphatase; 4, serine hydroxymethyltransferase; 5, glycine cleavage enzymes; 6, serine transacetylase, and 7, *O*-acetylserine sulfhydrylase (see Fig. 7.**8**)

dent terminal pathways lead to the three amino acids. A mutase reaction (i.e., an intramolecular group transfer) yields prephenate, the precursor of **phenylalanine** and **tyrosine**. Anthranilate is the precursor of **tryptophan**. The 5-phosphoribosyl-*N*-glycoside formation with anthranilic acid is followed by cyclization (comparable with late steps in pyrimidine nucleotide synthesis or with early steps in purine nucleotide synthesis). The exchange of the C_3 chain at the indole ring is a β-replacement of the hydroxyl group of serine by indole, a reaction catalyzed by the pyridoxal-phosphate–dependent tryptophan synthase. This enzyme can also form tryptophan from serine and indole and catalyze the β-elimination of NH_3 from serine to give pyruvate.

There are four additional terminal pathways that all lead from chorismate to three important **aromatic coenzymes** (folate-containing **4-aminobenzoate**, derived from 4-hydroxybenzoate; **ubiquinone**, which contains benzoquinone; and **menaquinone**, which contains naphthoquinone) and to the iron chelator **enterochelin**, which contains 2,3-dihydroxybenzoate respectively (see Chapter 8.8).

Control (see also Chapter 19). Each one of the three DAHP synthases is inhibited by and its expression is repressed by one of the aromatic end products; the other enzymes of the common pathway are constitutive. The genetic organization of the genes of the terminal pathways (e.g., the *trp* operon for tryptophan biosynthesis or the *tyrR* regulon, which includes tyrosine transport) and their transcription and translation are described elsewhere in this book. In addition, the regulation of the aromatic biosynthetic pathway by feed-back control is described in Chapter 19.2. Again the genetic organization and regulation differ in various bacteria, reflecting evolutionary history and regulatory constraints.

7.6.5 Histidine Is Synthesized in a Completely Separate Pathway

The **histidine** pathway (Fig. 7.**19**) comprises nine or ten enzymes and eleven reactions. Histidine is composed of the carbon atoms from 5-phosphoribosyl 1-pyrophosphate (PRPP), which form the C_3 side chain and the two adjacent carbon atoms (–C=CH–) of the heterocyclic ring. The atoms of the -N=CH-group are derived from the purine base (N1 and C2) of ATP (Note that C2 of adenine itself is derived from C_1 metabolism and that the imidazole ring of adenine is not incorporated into the imidazole ring of histidine!). The -NH-group comes from glutamine, and α-amino group stems from glutamate. As will be seen, the side product (AICAR) itself is an intermediate in purine-nucleotide synthesis. This fact may help to remember this curious ATP-dependent synthesis.

Control. In *E. coli*, the genes for the enzymes are in the *his* operon. In other bacteria, the *his* genes may be organized in several unlinked clusters. The activity of the first enzyme is modulated by feedback inhibition by histidine.

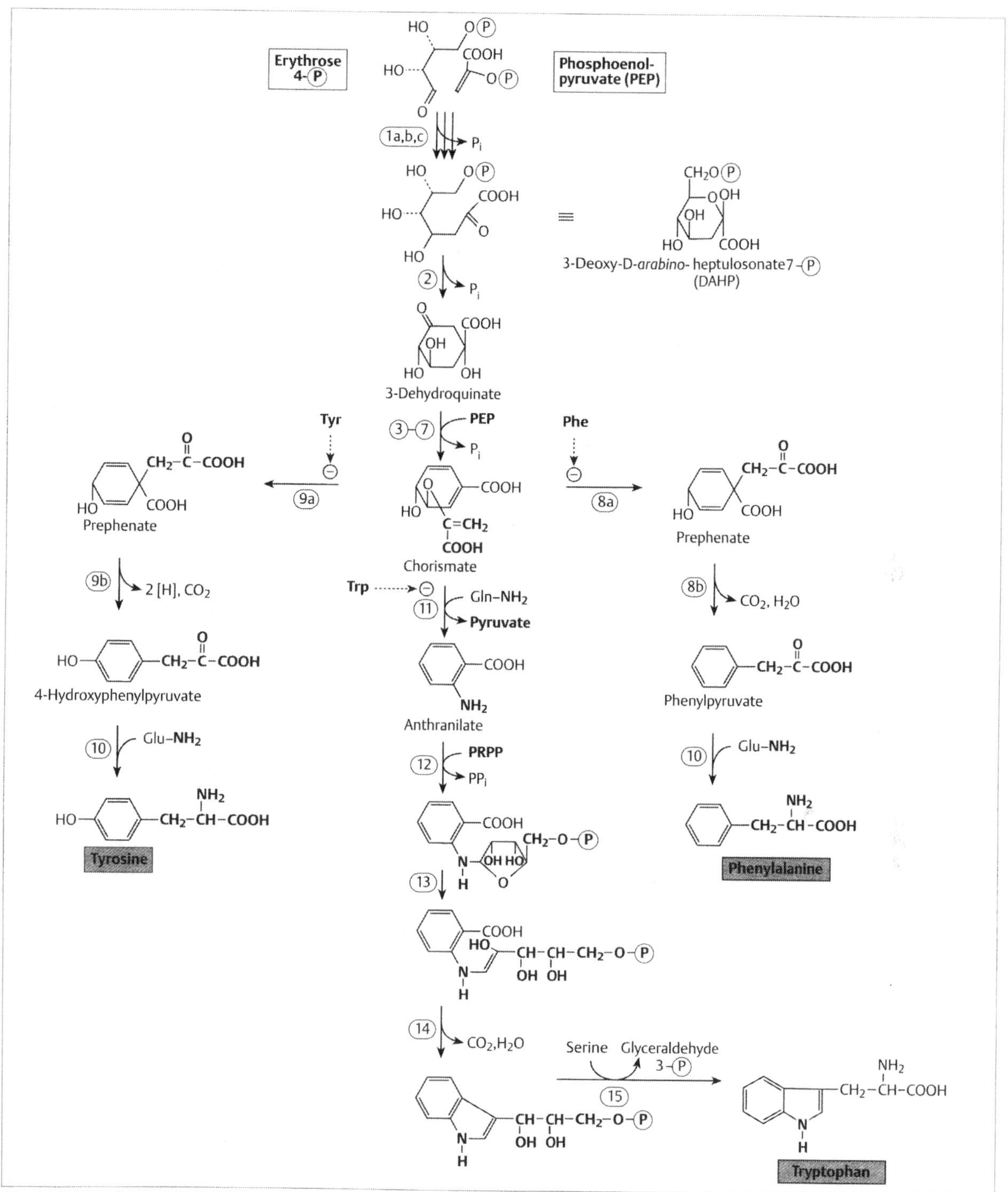

Fig. 7.18 **Synthesis of the aromatic amino acids phenylalanine, tyrosine, and tryptophan.** Common pathway: 1a/1b/1c, 3-deoxy-D-arabino-heptulosonate-7-phosphate (DAHP) synthases (phe, tyr, trp); 2, dehydroquinate synthase; 3, dehydroquinate dehydratase; 4, shikimate dehydrogenase; 5, shikimate kinases (the function of two shikimate kinases is unclear); 6, 5-enoylpyruvoylshikimate-3-phosphate synthase; 7, chorismate synthase. Phenylalanine (Phe) branch: 8a/8b, chorismate mutase/prephenate dehydratase (phe; bifunctional enzyme); 9, tyrosine aminotransferase. Tyrosine (Tyr) branch: 9a/9b, chorismate mutase/prephenate dehydrogenase (tyr; bifunctional enzyme); 10, tyrosine aminotransferase. Tryptophan (Trp) branch: 11, anthranilate synthase; 12, anthranilate-phosphoribosyl transferase; 13, phosphoribosyl-anthranilate isomerase; 14, indoleglycerolphosphate synthetase (enzymes 13 and 14 may be a single enzyme or two enzymes); 15, tryptophan synthase

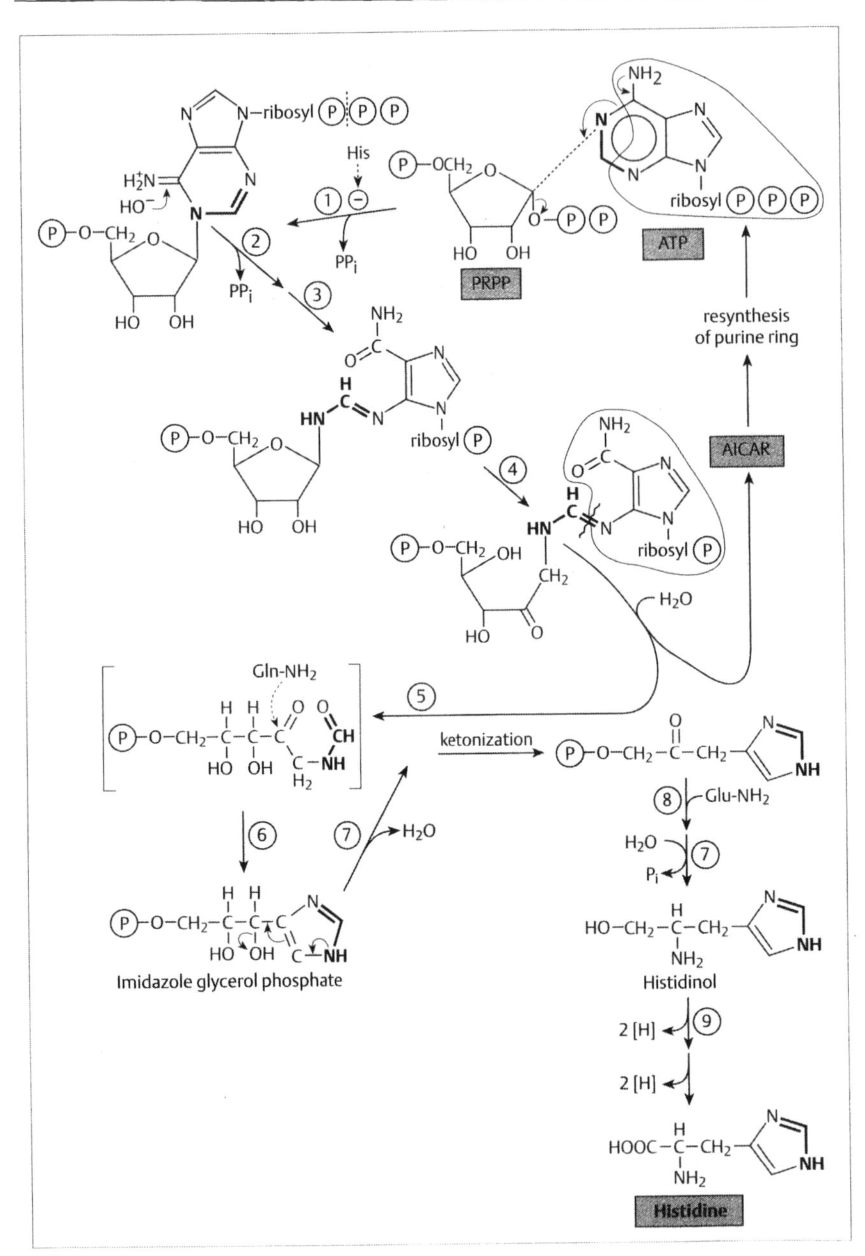

Fig. 7.19 **Synthesis of histidine** PRPP, 5-phosphoribosyl 1-pyrophosphate. 1, Phosphoribosyl-ATP pyrophosphorylase; histidine, His; 2 phosphoribosyl-ATP pyrophosphohydrolase; 3, phosphoribosyl-AMI cyclohydrolase; 4, phosphoribosylformimino-5-aminoimidazole carboxamide ribonucleotide isomerase 5, phosphoribosylformimino-5-aminoimidazole carboxamide ribonucleotide: glutamine aminotransferase; 6 cyclase; the name of the compoun released (encircled) is 5-**a**mino**i**midazole-4-**c**arbox**a**mide **r**ibonucleotid (AICAR); 7, imidazoleglycerolphosphate dehydratase; 8, histidinol phosphate transaminase; 7, histidinol-phosphate phosphatase (bifunctional enzyme 7 or two separat enzymes); 9, histidinol dehydrogenase (catalyzes the two-step dehydrogenation of histidinol to L-histidine)

7.7 Carbohydrate Phosphates Are Important Precursors of Many Building Blocks

Some reactions in amino acid biosynthesis require phosphorylated carbohydrates [erythrose 4-phosphate for aromatic amino acids, 5-phosphoribosyl 1-pyrophosphate (PRPP) in tryptophan and histidine synthesis]. Nucleotide and deoxynucleotide synthesis demand even more pentose phosphates. The synthesis of cell-wall constituents, for example, peptidoglycans, glycolipids, lipopolysaccharides, or surface polysaccharides, also requires activated (sugar) precursors.

Therefore, before turning to the next important building blocks, we should know how carbohydrate phosphates are interconverted, how they are activated for further interconversions, and how they function as building blocks of nucleotides and various glycans. Most naturally occurring carbohydrates are D sugars. We will first discuss the basic reactions by which most prokaryotes can form and transform sugar phosphates from central precursor metabolites. Then, the basic activation steps will be described.

The interconversion of the phosphate esters of C_3 and C_6 aldoses and ketoses by the glycolytic/glucogenic pathways has already been described (Chapter 3). How then are the other C_4 and C_5 sugar phosphates, which are considered central precursor metabolites, synthesized? Obviously, all cells need pathways for the synthesis of these compounds. How are all other sugars and carbohydrates derived from them?

There are four steps leading from the intermediates of glycolytic/glucogenic pathways to these sugar building blocks:

1. **The central pentose-phosphate cycle interconverts many sugar phosphates** (Fig. 7.**20**). Therefore, this pathway is indispensible for all organisms. The products are **central precursor metabolites** esterified with phosphate at the "terminal" carbon atom.
2. From the central precursor metabolites of Fig. 7.**1**, only a few primary sugar 1-phosphates are formed by simple reactions (Note that the phosphate group is glycosidically bound!). These compounds are indicated in Fig. 7.**21**.
3. In order to function as building blocks, sugar 1-phosphates are further activated with nucleoside triphosphates (NTP) via nucleotidyl transfer to C1 (with PP_i release; Figs. 7.**21**, 7.**22c**). Similarly, ribose 5-phosphate is activated at C1 by pyrophosphoryl transfer (with NMP release), and 5-phosphoribosyl 1-pyrophosphate (PRPP) is formed (Fig. 7.**22a**).
4. The activated pentose 5-phosphate (PRPP) acts directly as precursor for syntheses (Fig. 7.**22b**). The activated primary sugars [nucleoside diphosphate (NDP) sugars] function either as building blocks, in which they form a glycosidic bond between the activated C1 and various acceptor groups with retention or inversion of the configuration at C1 (Fig. 7.**22d**), or they are converted into other activated sugar building blocks (Chapter 7.9).

Box 7.6 The contribution of the individual carbohydrate degradation pathways can be determined experimentally. Specifically [^{14}C]-labeled hexose is fed to the organisms, and the amount of [^{14}C] in the CO_2 formed early is determined. Oxidation via the pentose phosphate pathway yields CO_2 from C_1; glycolysis and pyruvate oxidation yield CO_2 from C3 and C4. The Entner-Doudoroff pathway yields CO_2 from C1 and C4. The method is called **radiorespirometry** (i.e., the measurement of radioactivity in respiratory CO_2).

7.7.1 The Pentose Phosphate Cycle Is the Major Source of Central Sugar-Precursor Metabolites and NADPH

This process of sugar interconversion can be divided into three steps. The first step (1) involves the oxidation of hexose phosphates. The formation of ribulose 5-phosphate by oxidation of glucose 6-phosphate has been discussed in the section on NADPH formation (Chapter 7.4.3; oxidative pentose phosphate pathway). This irreversible reaction sequence is combined with other reactions to a cyclic process, the oxidative pentose phosphate cycle (also called "shunt" because it diverts glucose 6-phosphate from glycolysis when cells need NADPH; Fig. 7.**20**). This not only is the major way to fill up the NADPH pool but also allows for the synthesis of all central sugar-precursor metabolites.

The next two steps involve non-redox reactions that are fully reversible. (2) In the interconversion of pentose phosphates, the isomerization (i.e., aldose $\rightleftharpoons$ ketose) of ribulose 5-phosphate yields ribose 5-phosphate. Epimerization is the inversion of the configuration R_1–CHOH-R_2 at one of several centers of chirality, i.e., a special type of isomerization, which affords xylulose 5-phosphate from ribulose 5-phosphate, for example. This step is followed by (3), the interconversion of other sugar phosphates. Erythrose 4-phosphate is formed in

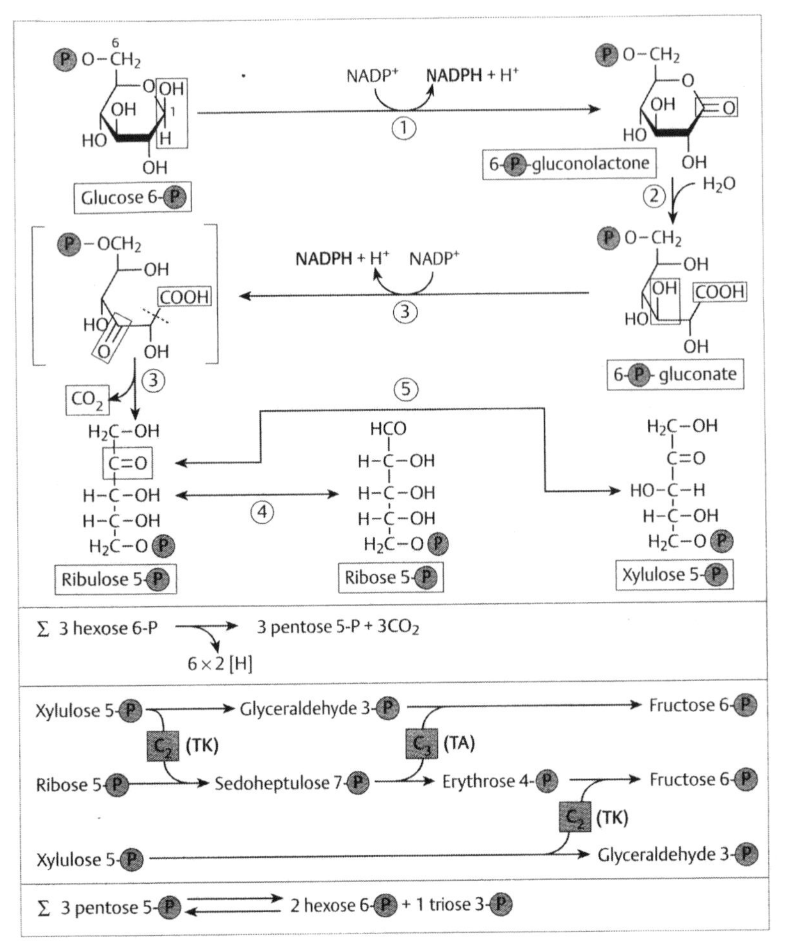

Fig. 7.**20** **Oxidative pentose phosphate cycle.** The entire cycle is formulated with three molecules of hexose. NADPH and pentose formation: 1, glucose-6-phosphate dehydrogenase; 2, lactonase; 3, 6-phosphogluconate dehydrogenase. Pentose interconversion: 4, ribose-5-phosphate isomerase; 5, ribulose-5-phosphate 3-epimerase. Reconversion by transaldolase (TA) and transketolase (TK) reactions

the course of cyclic regeneration of a hexose 6-phosphate (i.e., fructose 6-phosphate) from xylulose 5-phosphate and ribose 5-phosphate. These reactions are catalyzed by transaldolase and by the thiamine-pyrophosphate–dependent transketolase. The transketolase transfers the C_2 group glycolaldehyde (C1 and C2 of a ketose phosphate, the donor) to C1 of the acceptor, an aldose phosphate (see Chapter 8). Transaldolase transfers a C_3 unit (dihydroxyacetone group, C1 to C3 of a ketose phosphate as donor) to the acceptor, an aldose phosphate. Fig. 7.**20** schematically shows the conversion of three molecules of pentose 5-phosphate into two molecules of hexose 6-phosphate and one triose-3-phosphate molecule.

These reversible reactions allow the synthesis of C_4 and C_5 sugar phosphates also from fructose 6-phosphate and glyceraldehyde 3-phosphate when NADPH is synthesized in other reactions (see Chapter 7.4.3). Pentose phosphates are channeled into hexose 6-phosphates by the same mechanism in many bacteria. Most bacteria capable of growing on hexoses, however, use at the same time both the main carbohydrate degradation pathway (e.g., the Embden-Meyerhof-Parnas pathway and/or the Entner-Doudoroff pathway) and the oxidative pentose phosphate pathway for the formation of NADPH and pentose phosphates. The relative involvement of these pathways varies with the dependence on other NADPH-forming reactions and the need for pentose phosphates. This combination of reversible reactions allows a high degree of metabolic flexibility: the interconnected sugar phosphates form a common pool from which sugars can be drained off.

7.7.2 There Are Only a few Primary Sugar 1-Phosphates Involved in Metabolism

From the central precursor metabolite sugar phosphates (Fig. 7.**2**), sugar 1-phosphates are formed by simple

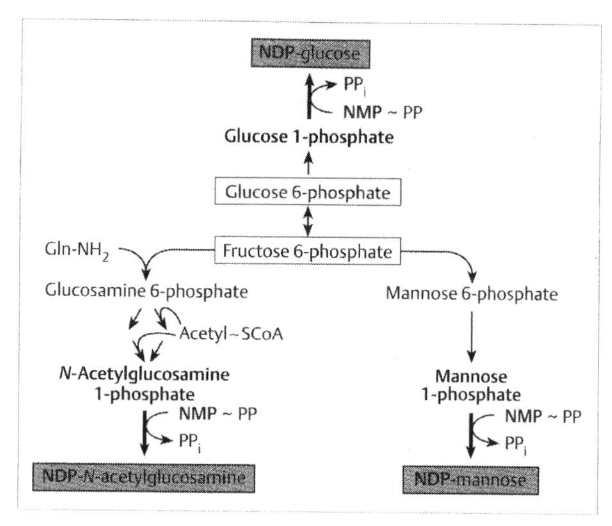

Fig. 7.**21** **Formation of primary sugar 1-phosphates from central precursor metabolites.** The few D-hexose 1-phosphates are activated by nucleotidyl (AMP, UMP, GMP, CMP, dTMP) transfer from the corresponding nucleoside triphosphate (NTP) to form nucleoside diphosphate (NDP)-sugars. As a rule, other sugar interconversions proceed on the NDP–hexose level. Galactose 1-phosphate may be formed directly from the substrate galactose, followed by UDP activation with UDP-glucose, yielding UDP-galactose and glucose1-P

reactions (Fig. 7.**21**). Isomerases interconvert aldose phosphates and ketose phosphates (e.g., fructose 6-phosphate and mannose 6-phosphate, glucose 6-phosphate and fructose 6-phosphate). Mutases convert 6-phosphohexoses (phosphate esters) into 1-phosphohexoses (e.g., glucose 6-phosphate to glucose 1-phosphate); the aldose-1-phosphate formation (in which phosphate is glycosidically linked) is energetically unfavorable. Still, these sugar interconversion reactions are, in principle, reversible and can be both catabolic and anabolic. Usually, very few primary sugar 1-phosphates (such as glucose, mannose, *N*-acetylglucosamine, and galactose) are formed from hexose 6-phosphates. Amino sugars are formed by transamination (with glutamine) of fructose 6-phosphate to glucosamine 6-phosphate. Acetyl donor for *N*-acetylated sugars is acetyl-CoA. Sugar 1-phosphates are converted in an irreversible step to sugar nucleotides, the activated sugars.

7.7.3 Sugar Phosphates Need to be Activated for Further Conversions and to Serve as Building Blocks

In order to be reactive enough for further conversions and to serve as building blocks for synthesis (i.e., to undergo glycosidic bonds), ribose 5-phosphate is provided with a good "leaving group" at C1; this is accomplished by transferring a pyrophosphoryl group from a **n**ucleoside **trip**hosphate (NTP), with NMP release (Fig. 7.**22a**). Likewise, sugar 1-phosphates are activated by transferring an NMP group from NTP (with PP$_i$ release; Fig. 7.**20**).

Pyrophosphoryl transfer. Pentoses are mostly derived from ribose 5-phosphate via the activated form, 5-phospho-α-ribose 1-pyrophosphate (PRPP), whereby PP$_i$ is the "leaving group". PRPP obtains its pyrophosphate group via pyrophosphoryl transfer from the β- and γ-phosphates of ATP by PRPP synthetase (Fig. 7.**22a**). PRPP is the common substrate of enzymes forming (β-) glycosidic bonds at the activated C1 of ribose (Fig. 7.**22b**).

Nucleotidyl transfer. Aldohexoses are activated by nucleoside diphosphate (NDP) activation of C1, which is catalyzed by NTP-hexose 1-phosphate nucleotidyl transferases; 1-phosphohexose is converted with NTP to **NDP-hexose** + PP$_i$ (with NDP as the "leaving group") (Fig. 7.**22c**). The equilibrium of the reaction is moved in the direction of NDP-sugar formation owing to the cleavage of the product PP$_i$ to two P$_i$. The free energy of the glycosidic linkage in the activated building blocks is about -33 kJ, which provides the driving force for the formation of glycosidic bonds in the synthesis of nucleotides and of sugar-containing polymers, (Fig. 7.**22d**). Compare the free energy of the glycosidic linkage of NDP-hexose with the free energy of *N*- or *O*-glycosidic linkage, approximately -20 kJ/mol or with that of phosphate esters, such as glucose 6-phosphate, approximately -14 kJ/mol.

The fourth step towards carbohydrate synthesis, the use of NDP-sugars in the biosynthesis of complex carbohydrates by glycosyl transfer (Fig. 7.**22d**) and in the interconversion into other sugar building blocks, is discussed in Chapter 7.9. The principal glycosyl-transfer reactions are shown in Fig. 7.**28**.

Control. All strongly exergonic steps in the synthesis of sugar building blocks are controlled. **Oxidative pentosephosphate cycle.** The first enzyme of the cycle, glucose 6-phosphate dehydrogenase is inhibited by NADPH. **PRPP synthetase.** Nearly twelve enzymes compete for PRPP (e.g., for the synthesis of purine nucleotides, pyrimidine nucleotides, histidine, tryptophan, and nicotinamide coenzymes). Therefore, PRPP synthetase is strictly regulated on two levels, that of enzyme synthesis and that of enzyme activity. NTP-hexose 1-phosphate nucleotidyl transferases (also called NDP–sugar pyrophosphorylases) are strictly controlled by feedback inhibition.

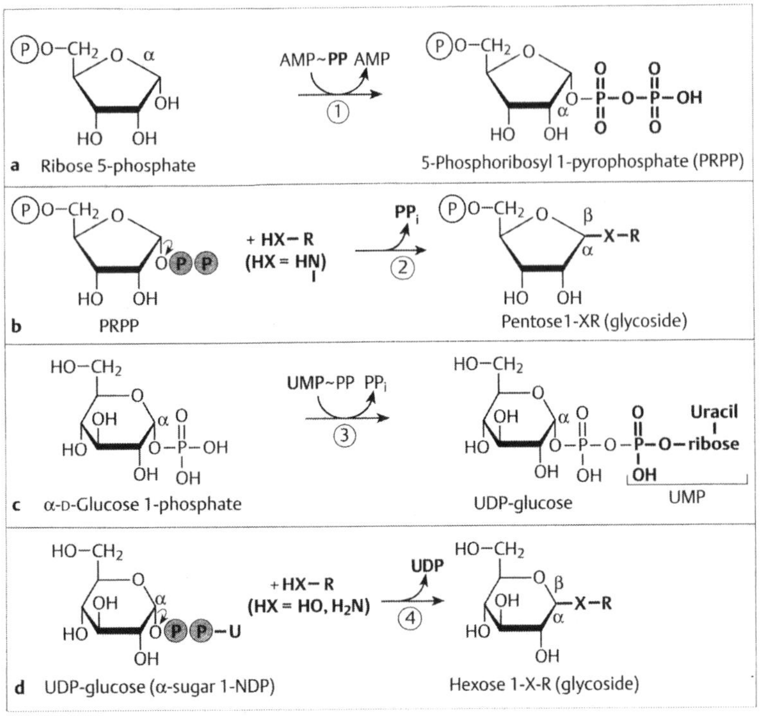

Fig. 7.**22a–d** **Activation of 5-phosphoribose by pyrophosphoryl transfer to C1 (a) and of sugar 1-phosphates by nucleotidyl transfer to C1 (c).** The activated sugars are transferred in glycoside formation reactions **b** and **d**.
a 5-phosphoribose activation by pyrophosphoryl transfer to C1; 1, 5-phosphoribosyl-1-pyrophosphate (PRPP) synthetase.
b Pentose glycoside formation from PRPP; 2, enzymes catalyzing this formation.
c Sugar-1-phosphate activation by nucleotidyl transfer to C1; 3, NTP-hexose-1-phosphate nucleotidyl transferases.
d Formation of hexose glycoside from UDP-glucose; 4, enzymes involved in this formation

7.8 The Building Blocks of Nucleic Acids Are Ribonucleotides and Deoxyribonucleotides

Nucleic acids (RNA, DNA) account for up to 20% of cell dry weight. The ribonucleotides and deoxyribonucleotides contain a purine base or a pyrimidine base, a ribose or a deoxyribose, and phosphate groups. The 2-deoxy-ribonucleotides are formed by reduction of ribonucleotides, which is catalyzed by ribonucleotide reductase. With the exception of CTP, all nucleoside triphosphates are synthesized from the corresponding nucleoside monophosphates. The pentose moiety is always derived from PRPP, but at a different stage in synthesis: In purine synthesis, different groups are added to PRPP to form the heterocyclic ring; in pyrimidine synthesis, PRPP is introduced only after completion of the heterocyclic ring.

Ribonucleotides are the building blocks of RNA and the precursors of the DNA building blocks. Since mRNAs are metabolically unstable, their rapid turnover results in a high demand for nucleoside triphosphates; these are regenerated mostly from the nucleoside monophosphates formed during mRNA degradation. The synthesis of stable RNA and DNA requires de novo synthesis. Nucleotides are also the constituents of many coenzymes.

Box 7.7 Ribonucleotides and deoxyribonucleotides consist of three components: (1) a purine or a pyrimidine base called **nucleobase**, whereby adenine and guanine are the purines, and cytosine, thymine, and uracil are the pyrimidines; (2) a ribose or 2-deoxyribose; nucleobases attached to C1 of (deoxy)pentose are called **nucleosides** (ribonucleosides or deoxyribonucleosides) and are symbolized by the letters A for adenosine, G for guanosine, C for cytidine, T for thymidine, U for uridine, and d for their respective deoxy derivatives (e.g., dA); (3) a monophosphate, diphosphate, or triphosphate group attached to C5′ of ribose (the prime indicates the position in the sugar as opposed to the base; the numbering of the atoms of the bases is shown in Fig. 7.**23**). All monophosphate, diphosphate, or triphosphate compounds (e.g., AMP, ADP, ATP and their respective deoxy forms) are called **nucleotides** (i.e., either **ribonucleotides** or **deoxyribonucleotides**), either purine or pyrimidine nucleotides. DNA differs chemically from RNA in two respects: it contains deoxyribose instead of ribose and thymidine nucleotide instead of uridine nucleotide.

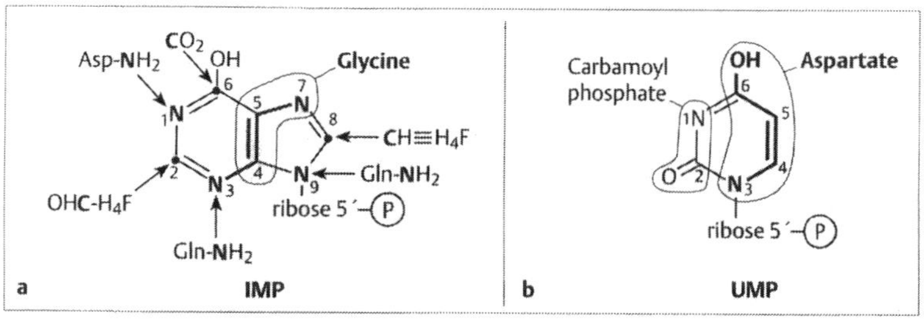

Fig. 7.**23a,b** **Origin of the atoms of IMP (purine) and UMP (pyrimidine) nucleotide precursors.**
a IMP.
b UMP

Nucleoside monophosphates and diphosphates (NMP, NDP) are formed at a very high rate due to the universal use of NTPs as energy-rich compounds and P carriers. Mononucleotides and dinucleotides can successively be phosphorylated with ATP by nucleotide-specific NMP kinases (e.g., GMP kinase) and by NDP kinases (e.g., GDP kinase); both types of kinases also act on dNMP and dNDP, respectively. In addition, adenine nucleotides are interconvertible by adenylate kinase (see Chapter 7.4.2). The synthesis of purine and pyrimidine nucleotides follows quite different strategies; the precursor metabolites are indicated in Fig. 7.**23**.

7.8.1 Purine Ribonucleotides Are Synthesized Stepwise From Activated Ribose 5-Phosphate

The purine ring is formed in an amazing sequence of ten enzyme reactions by stepwise addition of functional groups to the activated form of ribose 5-phosphate, 5-phosphoribosyl 1-pyrophosphate (PRPP), until the ring is completed (Fig. 7.**24**). The common intermediate in purine ribonucleotide synthesis is inosine 5′-monophosphate (**IMP**), with the nucleobase hypoxanthine (a 6-hydroxypurine). The functional groups added are glycine, amino groups from glutamine and aspartate, C_1 units at the methenyl or formyl level, and CO_2. In the first amination reaction, the C_1 of ribose is inverted from α-to β-configuration. First, the imidazole ring and then the pyrimidine ring is completed. The nucleotides **AMP** and **GMP** are made from IMP via separate branches, IMP is aminated at C6 to adenosine monophosphate (AMP) or oxidized at C2 to xanthosine monophosphate (XMP, nucleobase xanthine) and then aminated at C2 to guanosine monophosphate (GMP). Subsequent phosphorylations give ADP and GDP, then ATP and GTP, respectively. These nucleotides are also the precursors of effector molecules and alarmones, such as cAMP (cyclic AMP) or ppGpp (guanosine 5′-diphosphate 3′-diphosphate).

Control. The first reaction in purine nucleotide synthesis, PRPP amination, is controlled via feedback inhibition by AMP and GMP. Note that purine synthesis is linked to histidine synthesis.

7.8.2 Pyrimidine Ribonucleotides Are Formed From Aspartate and Carbamoyl Phosphate

Pyrimidine ribonucleotides are formed in three steps (Fig. 7.**25**): (1) A common pyrimidine, **orotate** (pyrimidine ring substituted with 2 hydroxyl and 1 carboxyl), is formed from aspartate and carbamoyl phosphate. (2) The intermediate is linked to 5-phosphoribosyl 1-pyrophosphate and decarboxylated to **uridine monophosphate (UMP)**. (3) UMP is phosphorylated to **UTP**; subsequent amination yields **CTP**. The biosynthetic pathway is shown in Fig. 7.**25**, and the origin of the individual atoms in UMP is indicated in Fig. 7.**23**.

Control (see Fig. 19.**3**). Pyrimidine biosynthesis is regulated at three strategic points: the enzymes carbamoyl-phosphate synthase, aspartate carbamoyltransferase, and CTP synthase all are under allosteric control. Aspartate transcarbamylase (also called carbamoyltransferase) is probably the best-studied regulatory enzyme. Ornithine carbamoyltransferase and aspartate carbamoyltransferases are evolutionary related.

7.8.3 The 2′-Deoxyribonucleotides Derive From Ribonucleotides by Reduction

The reduction of ribonucleotides to 2′-deoxyribonucleotides, the DNA precursors, takes place either at the diphosphate level or the triphosphate level. Ribonucleotide reductase catalyzes an intricate reduction, which proceeds via a radical mechanism involving a coenzyme B_{12}-, an iron–, or a non-iron-dependent enzyme (Fig. 7.**26**). At least five classes of this enzyme have evolved, all of which are strictly regulated in many ways. The electron donor, in general, is reduced thioredoxin, a small dithiol-containing peptide, which is reduced in turn by NADPH. The deoxynucleoside diphosphates are phosphorylated to the triphosphates with ATP. One reduction product, deoxyuridine triphosphate (dUTP), is not a DNA building block. Curiously,

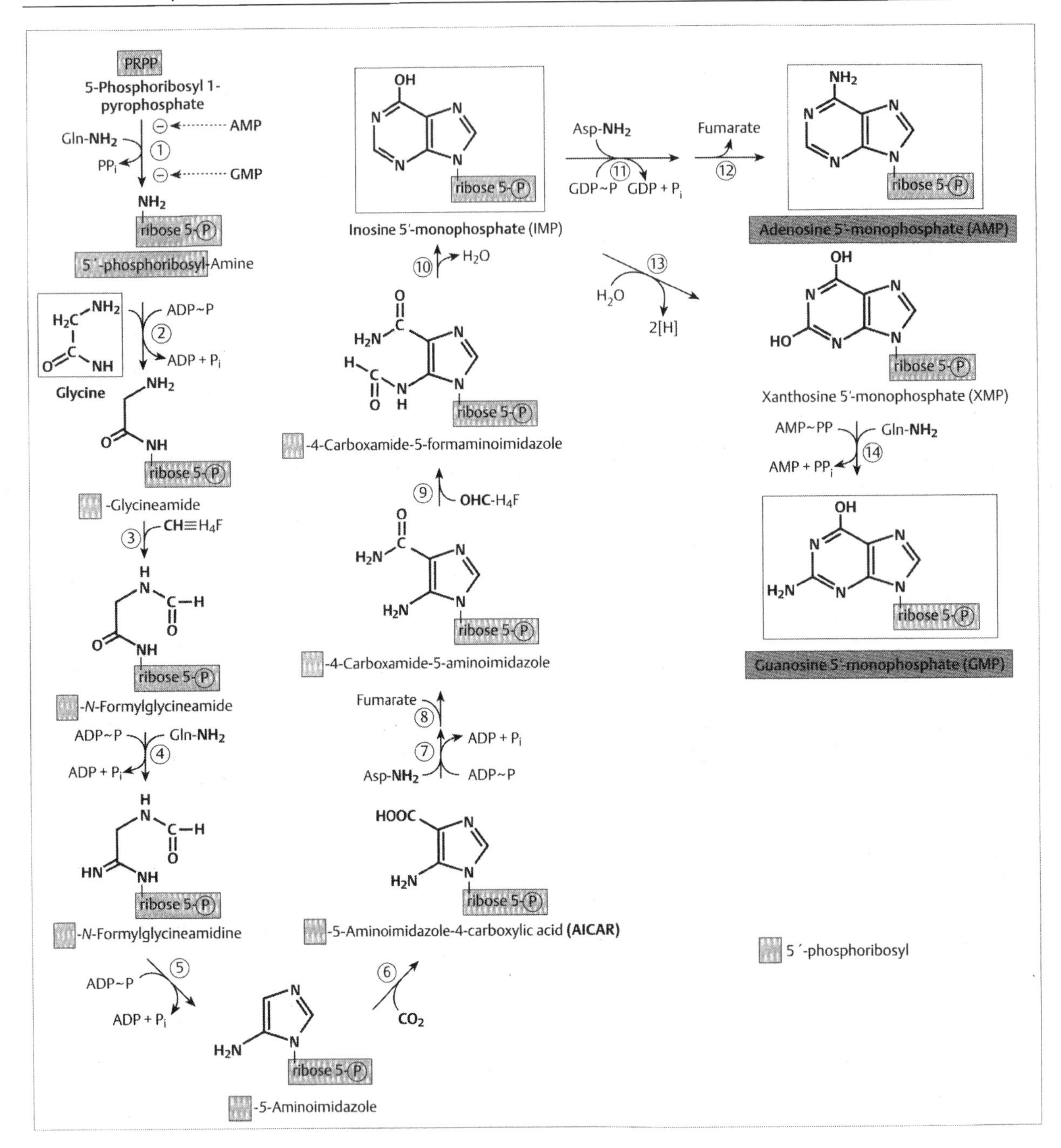

Fig. 7.**24** **Synthesis of the purine nucleotides AMP and GMP.** 1, PRPP amidotransferase; 2, phosphoribosyl-glycineamide synthetase; 3, phosphoribosylglycineamide formyltransferase; 4, phosphoribosyl-formylglycineamidine synthetase; 5, phosphoribosyl-aminoimidazole synthetase; 6, phosphoribosyl-aminoimidazole carboxylase (forming AICAR); 7, phosphoribosylaminoimidazole-succinocarboxamide synthetase; 8, adenylosuccinate lyase (bifunctional enzyme identical with 12); 9, phosphoribosylaminoimidazole-carboxamide formyltransferase; 10, IMP cyclohydrolase; 11, adenylosuccinate synthetase; 12, identical with 8; 13, IMP dehydrogenase; 14, GMP synthetase

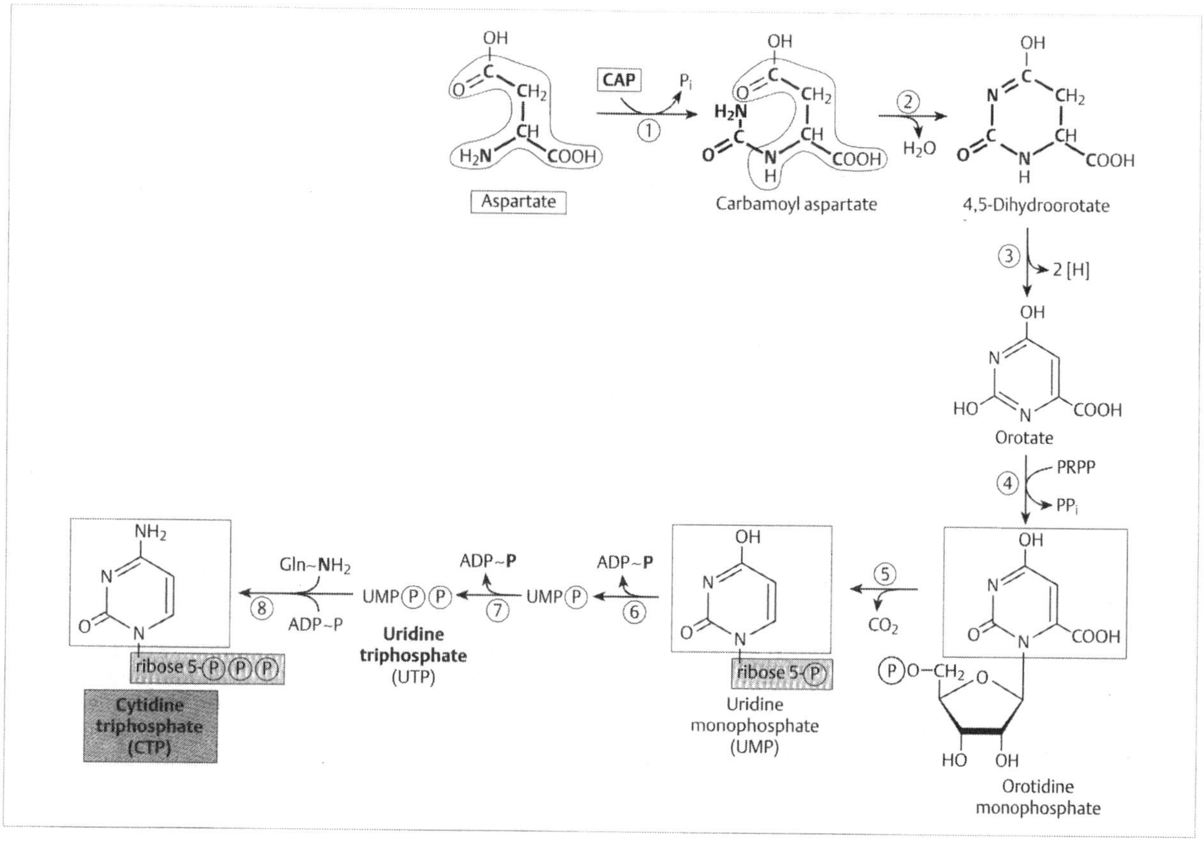

Fig. 7.**25** **Synthesis of the pyrimidine nucleotides.** 1, aspartate transcarbamoylase; 2, dihydroorotase; 3, dihydroorotate dehydrogenase; 4, orotate phosphoribosyltransferase; 5, orotidine 5-phosphate decarboxylase; 6, nucleoside-monophosphate kinase; 7, nucleoside-diphosphate kinase; 8, CTP synthase. Carbamoyl-phosphate synthesis (CAP) is described in Chapter 7.3.1

dUTP is turned into UMP by a pyrophosphatase, then dUMP is methylated to dTMP in the interesting thymidylate-synthetase reaction (Fig. 7.**26**). Methylene tetrahydrofolate (methylene H_4-folate), which actually carries bound formaldehyde, donates the methylene group and simultaneously reduces it to the methyl level by oxidizing the carrier H_4-folate to H_2-folate. The oxidized coenzyme H_2-folate subsequently has to be reduced to H_4-folate by dihydrofolate (H_2-folate) reductase. dTTP may also be formed differently.

7.8.4 Salvage Pathways Enable Utilization of Nucleosides and Nucleobases

Prokaryotes contain a collection of enzymes, the salvage enzymes, which have three functions: (1) They enable the cell to utilize preformed nucleobases or nucleosides for nucleotide synthesis when these are available in the medium. These compounds are actively transported into the cell. (2) The enzymes also make the pentose and the nucleobases available as sources of carbon, energy, and nitrogen. Nucleotides very rarely are taken up. Instead, they are dephosphorylated to nucleosides by periplasmic 5′-nucleotidases and phosphatases (see Chapter 7.3.2). (3) Another function of the salvage enzymes is to reutilize the nucleobases and nucleosides produced endogenously as a result of nucleotide turnover (also signal molecules such as cAMP or ppGpp). All three functions of salvage pathways are beneficial to the cell if the de novo synthesis decreases concomitantly. This decrease is accomplished by different regulatory mechanisms.

The salvage mechanisms (which include degradation steps of nucleic acid components) differ considerably in various bacteria. The most common reactions can be grouped as follows (Fig. 7.**27**):

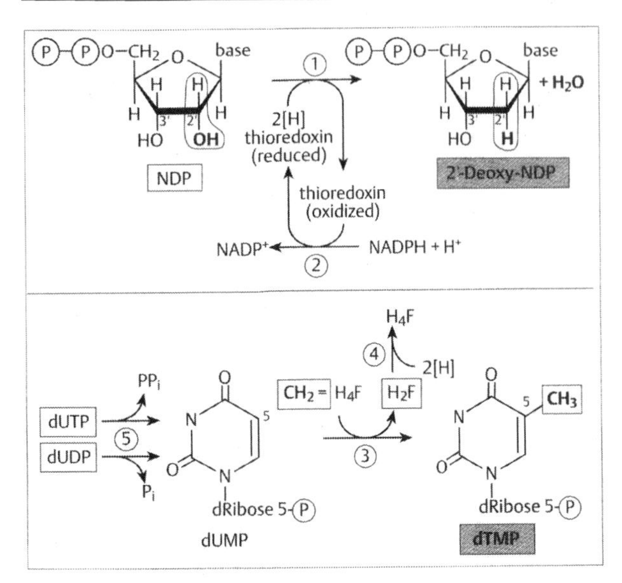

Fig. 7.**26** **Synthesis of deoxyribonucleotides.** 1, Ribonucleotide reductase; 2, thioredoxin reductase; 3, thymidylate synthase; 4, dihydrofolate reductase; 5, dUTP pyrophosphatase or dUTP phosphatase. H_4F tetrahydrofolate

(1) Phosphoribosyltransferases add the nucleobase to the C1-activated pentose 5-phosphate, PRPP. This mechanism is common in the utilization of nucleobases. Equilibrium is far on the nucleotide side.
(2) Nucleobases and (d)nucleosides can be interconverted by nucleoside phosphorylases, which replace the *N*-glycosidic linkage with a phosphate *O*-glycosidic linkage; the equilibrium constant is near unity.
(3) (d)Nucleosides can be phosphorylated to (d)mononucleotides by nucleoside kinases; equilibrium is far on the nucleotide side. Reactions (2) and (3) provide an alternative mechanism to (1) for nucleobase conversion to NMPs. Thymine or thymidine can experimentally be used to specifically label DNA; however, these precursors are not incorporated by all bacteria.
(4) NMP glycosylases hydrolyze the *N*-glycosidic bond of NMPs.
(5) The various specific kinases that phosphorylate NMPs and NDPs have already been mentioned.
(6) Although the last steps in nucleotide synthesis are not reversible, they can be bypassed by interconversions of the nucleobases (e.g., deamination of cytidine to uracil or of (d)cytosine to (d)uridine). Similar interconversions of adenine and guanine compounds also are known, for example, adenine (6-aminopurine) or the adenine base in (d)adenosine can be deaminated to the nucleobase hypoxanthine in IMP. GMP (with 2-hydroxy-6-aminopurine) may be deaminated and then reduced at C2 to IMP, the common precursor of purine nucleotides. These reactions are also involved in purine degradation (see Chapters 9.15 and 12.2.4).
(7) When histidine is synthesized, the ATP fragment AICAR (a precursor of purine nucleotide synthesis) can be recycled in a few steps into IMP (Fig. 7.**24**).

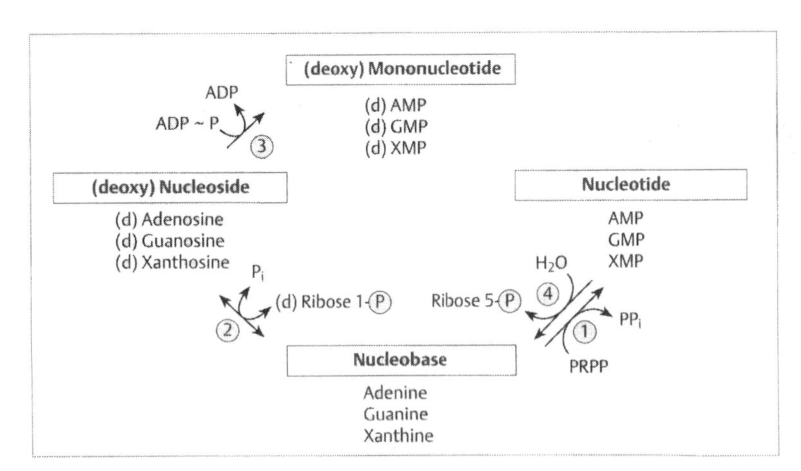

Fig. 7.**27** **Some reactions of salvage pathways and interconversions of constituents of nucleic acids.** 1, purine phosphoribosyltransferases; 2, nucleoside phosphorylases; 3, nucleoside kinases; 4, NMP glycosylases

7.9 Sugars and Sugar Nucleotides Are Important Building Blocks

Carbohydrates, sugars and amino sugars are not only the main food source for bacteria. They are also the major building blocks of many sugar-derived polymers and lipids that form the (generally non-proteinaceous) antigenic surface structures of prokaryotes. Examples are the constituents of the cell walls of most bacteria, archaeal cell-wall components, lipopolysaccharides of Gram-negative bacteria, teichoic and lipoteichoic acids of Gram-positive bacteria, glycoprotein surface layers, polysaccharides and capsular substances (which often contain sugar acids), a homopolymer of *N*-acetylneuraminic acid, and glycolipids. Surface glycans are often referred to as forming a glycocalix (see Chapter 23.5). Intracellularly, polyglucose is a common reserve material. The prokaryotic structural polysaccharides differ remarkably, and they are reliable phylogenetic markers, although they are composed of similar building blocks.

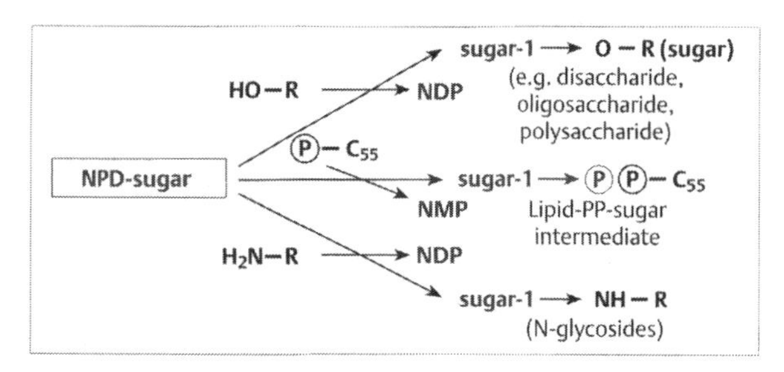

Fig. 7.**28** **Transfer of glycoside from NDP sugars to acceptors**

7.9.1 Metabolic Reactions for the Interconversion of Carbohydrates Appear Complex

These reactions can be divided into two groups, the interconversions of phosphorylated sugars and the reactions of carbohydrates activated as sugar nucleotides. Reactions of the first type and the activation of ribose 5-phosphate to PRPP and of sugar 1-phosphates to nucleoside-diphosphate (NDP) sugars have already been discussed (Chapter 7.7). Usually, very few primary sugar 1-phosphates (e.g., glucose, mannose, *N*-acetylglucosamine, galactose) are converted to sugar nucleotides (Fig. 7.**21**).

7.9.2 NDP Sugars Have Important Functions as Intermediates in Carbohydrate Synthesis

This section deals with the fate of NDP-activated sugars. Three types of reactions are required for polymerizations starting from sugar phosphates: (1) **Activation** of sugar 1-phosphates with NTP to NDP-sugars (catalyzed by pyrophosphorylases), (2) **Modification** of NDP-sugars by reactions such as epimerization or deoxysugar synthesis (catalyzed by epimerases and reductases), and (3) **Transfer** of NDP-sugars to appropriate acceptors (catalyzed by glycosyl transferases), the final step in polymer synthesis, as shown in Fig. 7.**28**.

In the process of complex carbohydrate polymer synthesis, nucleotide sugars have three major functions: (1) as intermediates during the formation of most carbohydrate building blocks in prokaryotes, (2) as precursors or activated building blocks (glycosyl donors) for carbohydrate-containing polymers, and (3) as regulatory mediators. The NTP-hexose-1-phosphate nucleotidyltransferases (NDP-sugar pyrophosphorylases), the first enzymes that form activated sugars by nucleotidyl transfer to sugar 1-phosphates, are specific not only with respect to the sugar moiety but also for the type of NTP involved (Fig. 7.**22c**). The type of nucleotide predestinates the sugar for different functions: ADP-glucose is used in glycogen synthesis in most bacteria, whereas UDP-glucose is the precursor of peptidoglycan and lipopolysaccharide. Hence, sugar nucleotides function as regulatory mediators that channel the sugars (even one species) into separate pathways leading ultimately to the many different sugar-containing polymers. This allows an independent metabolic control by allosteric regulation of the NDP-sugar–forming pyrophosphorylases (see Chapter 19.4.2).

Sugar interconversions. Some sugar interconversions occur on the sugar-phosphate level (Fig. 7.**21**). A much larger group of NDP-sugars, which are formed from a few primary NDP-sugars, undergoes various sugar interconversions (Fig. 7.**29**). Surprisingly, most reactions require NAD^+. This is understandable if C6 is oxidized in two steps to **uronic acids** by NDP-sugar dehydrogenases with NAD^+ as electron acceptor (Fig. 7.**29a**). However, intramolecular hydrogen transfer mediated by enzyme-bound NAD^+ prior to sugar transformations is also important in epimerization, in deoxy-sugar formation, or in the synthesis of branched sugars. First, a ketosugar is formed by oxidation, which may be rearranged in different ways, and becomes reduced either with inversion or retention of the -OH configuration.

Epimerization (i.e., inversion of the configuration of a chiral $R_1R_2HC\text{-}OH$ center) proceeds via intermediary oxidation to the keto group ($R_1R_2C{=}O$) and sub-

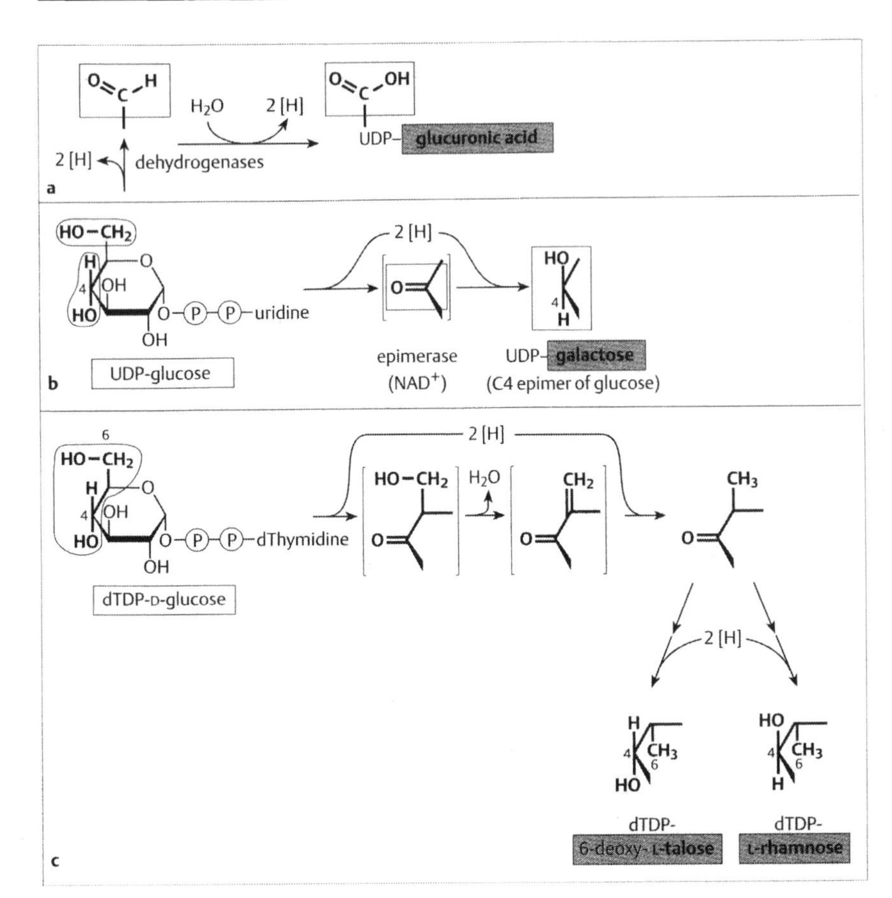

Fig. 7.29a,b,c **Frequent interconversions of NDP sugars.**
a Oxidation to uronic acids.
b Epimerization.
c Reduction to deoxy-sugars

sequent reduction with inversion (Fig. 7.**29b**). Examples are the actions of UDP-glucose 4-epimerase at C4 (UDP-glucose ⇄ UDP-galactose) or UDP-glucose 2-epimerase at C2 (UDP-glucose ⇄ UDP-mannose). **6-Deoxy-hexoses** are formed in three steps (Fig. 7.**29c**): (1) oxidation of C4 to a =CO group, then dehydration of C6 and reduction of the $=CH_2$ group, followed by (2) epimerization via an enediol intermediate, and finally (3) reduction to deoxyhexoses.

7.9.3 Activated Sugars Are Transferred Across the Membrane to Acceptors

Transfer to acceptors (Fig. 7.**28**). The acceptor to be glycosylated is often an HO group at C4 or C6 (less often at C3 or C2) of an acceptor aldose; in ketoses, the HO group at C2 of the ketal form is the acceptor. The acceptor X-R is a nucleophile, which attacks the C1 carbon of the activated sugar with concomitant fission of the C1–oxygen bond and often with inversion of the α-configuration to the β-configuration in the product. Examples are disaccharide, polysaccharide, and glycoprotein synthesis, or the mechanistically similar reaction in the 5-phosphoribosyl-1-(β)-amine formation (purine synthesis) from 5-phosphoribosyl 1-pyrophosphate (PRPP). The C1-hydroxyl group is a poor "leaving group," whereas the -OPP_i group is an excellent one (see Fig. 7.**22**), which also blocks the ring opening (possible in the hemiacetal structure).

NDP-activated glycans are bound to **lipid carriers** for transport across the membrane. All NDP-activated glycans (i.e., monosaccharides or oligosaccharides), which are constituents of extracellular polymers (e.g., peptidoglycans, lipopolysaccharides, or surface glycoproteins), are water-soluble. Still, these glycans have to be translocated through the hydrophobic lipid layer of the cytoplasmic membrane in an activated form in order to be transferred at the outer phase to an acceptor, be it a peptidoglycan, a chrondroitin-like glycan, or an S-layer protein. The lipid carriers are C_{55} (to C_{60}) polyisoprenoid

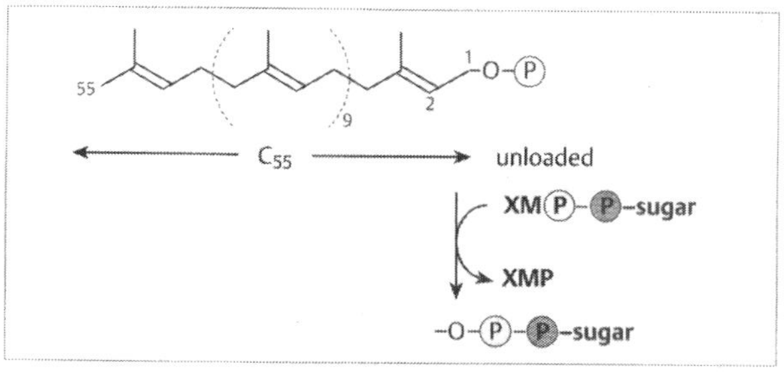

55 ... 9 ... 2 ... 1 O–(P)
C$_{55}$ unloaded
XM(P)–(P)–sugar
XMP
–O–(P)–(P)–sugar

Fig. 7.**30** **Structure of undecaprenol (bactoprenol) phosphate.** The C$_{55}$-isoprenoid membrane lipid carries oligosaccharides from XDP–sugars through the cytoplasmic membrane, a process which requires ABC transporters. Similar or identical lipids are involved in the synthesis of peptidoglycans, lipopolysaccharides, capsular polysaccharides, and teichuronic acids. Dolichol phosphate is a similar carrier, which transfers glycosides to extracellular glycoproteins; the red double bond is reduced

membrane lipids to which the oligosaccharides are glycosidically linked via a pyrophosphate group (Fig. 7.**30**). This membrane carrier serves an an intermediate prior to transfer to the final glycosyl receptor. The lipid carriers are **undecaprenyl phosphate** (i.e., 11 isoprene units), involved in the synthesis of peptidoglycan and glycans, and **dolichol phosphate** (the double bond of the first isoprene unit is reduced), which plays a role in the synthesis of glycoproteins. This is another sorting mechanism which helps to direct sugars in extracellular polymers to the right destination.

7.9.4 Hexosamines Are a Main Building Block of Structural Polysaccharides of the Cell Envelope

Peptidoglycan and lipopolysaccharide synthesis are detailed in Chapter 23. The glycan strand of murein (peptidoglycan) is synthesized from *N*-acetyl-D-glucosamine and *N*-acetylmuramic acid (Fig. 7.**31**). Fructose 6-phosphate is aminated at C2 with glutamine, and the product glucosamine 6-phosphate is *N*-acetylated and then mutated to the 1-phosphate (Fig. 7.**21**). *N*-acetyl-D-glucosamine 1-phosphate is subsequently activated with UTP to **UDP-*N*-acetyl-D-glucosamine** (compare with Fig. 7.**22c**). Thus, one building block of the glycan strand of peptidoglycans (murein, pseudomurein) and of lipid A of lipopolysaccharides is formed. The second (eubacterial) murein precursor, formed by C3-lactyl-ether formation with the C3-OH group of UDP-*N*-acetylglucosamine, is called **UDP-*N*-acetyl-muramic acid**; phospho*enol*pyruvate (with a reactive enol) is used for the ether linkage, and then reduced (Fig. 7.**31a**). L- and D-amino acids plus D-alanyl-D-alanine are sequentially added to the carboxyl group of the lactyl residue; each amide-bond formation requires ATP for activation of the growing oligopeptide, probably via terminal amino-acyl-phosphate formation. The final second building block is **UDP-*N*-acetylmuramic acid pentapeptide** (Fig. 7.**31c**) (for assembly, see Chapter 23.1).

Pseudomurein is the cell-wall component in *Methanobacteriales*. The cell walls of the Archaea differ from those of the bacteria, and there are differences within each domain. There are murein-like peptidoglycan structures (called pseudomurein) in Methanobacteriales (Fig. 7.**31b**) analogous to structures in the bacteria; however, pseudomurein synthesis starts from other nucleotide-activated compounds. There are several differences: (1) Pseudomurein comprises **only L-amino** acids. (2) *N*-acetyl muramic acid is replaced by ***N*-acetyl-L-talosamine-uronic acid**. (3) A **UDP-activated disaccharide** is formed in which UDP-*N*-acetylglucosamine (or the galactosamine) is linked in β-3 $\rightarrow$ 1 (rather than 1 $\rightarrow$ 4) to *N*-acetyltalosamineuronic acid. (4) To the UDP-disaccharide, L-amino acids are added. The amide bond formation between the first amino acid (L-glutamate) of the oligopeptide and the uronic acid requires **UDP-activated L-glutamate** (the phosphate is P–N linked to the α-NH_2 group!). The UDP-glutamate is subsequently built up to a UDP-pentapeptide. Addition of the next amino acids requires the ATP-dependent activation of the carboxyl group of the last amino acid of the growing UDP-oligopeptide.

7.9.5 Non-peptidoglycan-type Cell-wall Components Are Present in Several Archaea and in Some Bacteria

These components contain **glycosamine *O*-glycans.** These glycans are similar to the chondroitine sulfate (i.e., mucopolysaccharide) part of proteins of connecting tissues (i.e., proteoglycans); however, they lack the sulfate and the peptide moieties. Examples are methanochondroitin in *Methanosarcina* species and teichuronic acids in *Bacillus* species. The building blocks are UDP-activated *N*-acetylgalactosamine and UDP-glucuronic-acid residues, which are combined to UDP-activated oligosaccharides to form the repeating oligosaccharide units of the wall.

Many Archaea contain a **glycoprotein surface layer (S-layer)** as cell wall; the Bacteria may contain these layers in addition to the murein. The glycan is linked to the amino group of asparagine (via ***N*-glycosylation**) or to the hydroxyl group of serine, threonine, or tyrosine (via ***O*-glycosylation**) in a specific peptide surrounding (called sequon structure in *N*-glycosides). NDP-activated

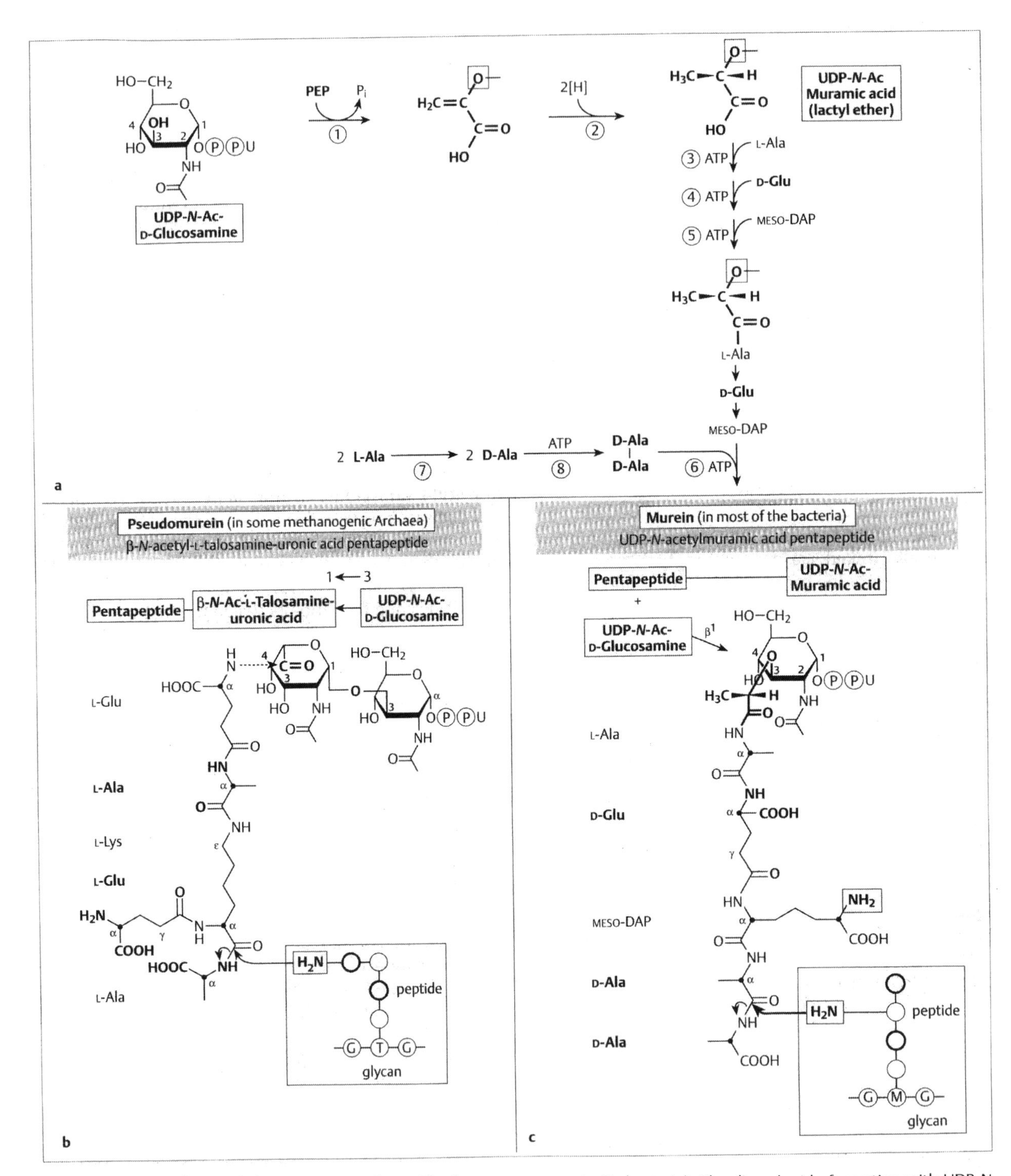

Fig. 7.31a–c **Synthesis of the precursors of peptidoglycans and a comparison of pseudomurein and murein.**
a Peptidoglycan precursor synthesis: 1, UDP-*N*-acetylglucosamine-3-enolpyruvylether synthase; 2, UDP-*N*-acetylenolpyruvylglucosamine reductase; 3–5, enzymes forming peptide bonds with individual amino acids; 6, enzyme adding D-alanine dipeptide; 7, alanine racemase; 8, D-alanyl-D-alanine synthetase.
b Comparison of pseudomurein (in some methanogenic Archaea) and **c** murein (in bacteria): The disaccharide formation with UDP-*N*-Ac-D-glucosamine proceeds on the lipid carrier. The inset shows schematically a neighboring peptidoglycan strand and the site of transpeptidation. MESO-DAP, *meso*-diaminopimelic acid; G, *N*-acetyl-D-glucosamine; M, *N*-acetylmuramic acid; T, *N*-acetyl-L-talosamineuronic acid. o represents an amino acid residue. See Chapter 24. The α-atoms of the amino acids are marked. The arrow in peptide linkages in **a** indicates the direction carboxyl group → amino group

oligosaccharides are formed from nucleoside-diphosphate-activated monosaccharides (e.g., *N*-acetyl glucosamine, mannose, galactose) followed by modification reactions (e.g., partial *O*-methylation or acetylation). NDP-activated oligosaccharides form the building blocks of the repeating units of the intact glycan.

Despite their different structures, surface polymers often fulfill similar ("analogous") functions: determination of the shape of the organism, molecular sieve, osmotic resistance, protection against enzymes or phagocytosis, adhesion and cell contact (important in pathogenesis and biofilms), formation of a periplasmic space, and absorption of ions.

7.10 The Biosynthesis of Lipids Is as Complex as Their Structure

The building blocks of membranes are lipids (about 10% of cell dry weight), which function together with proteins in transport, energy conservation, biosynthesis, excretion, and other cell processes. Prokaryotes form no triglycerides (i.e., glycerol esterified with three fatty acids) and only few other neutral lipids. Lipids contain hydrophobic hydrocarbon chains linked to glycerol (the core lipid), which form a bilayer, and polar head groups directed towards to water phase. The synthesis of eubacterial and archaeal phospholipids and glycolipids, among other compounds, will be discussed.

Biosynthetic precursors. The various groups of prokaryotes differ remarkably in the structure and synthesis of lipids, which serve as reliable systematic marker molecules in chemotaxonomy. The constituents of most lipids are as follows: (1) **Glycerol phosphate** is directly formed from dihydroxyacetone phosphate by reduction (see, however, Archaea). (2) Two compounds with a **hydrocarbon chain** are derived from acetyl-CoA. In the **Bacteria**, these are C_{14}–C_{18} fatty acids activated by thioesterification with a small **a**cyl **c**arrier **p**rotein (ACP) (Fig. 7.**33**). The fatty acids are esterified with glycerol. In the **Archaea**, these are C_{20}-isoprane alcohols linked to pyrophosphate. The isoprane alcohols form ether bonds with glycerol. (3) **A polar head group** is derived from various compounds (e.g., serine or sugar phosphates). In **phospholipids**, phosphate is bound to glycerol; in **glycolipids**, a sugar is glycosidically bound (see Fig. 7.**39**).

7.10.1 Lipids Are the Most Diverse Molecules in Prokaryotes

Besides the profound differences in lipids of the major phylogenetic groups, lipids may also vary with the growth temperature and other factors (see Chapter 29.2 and 29.3). Mycobacteria contain very complex lipids. Membrane lipids may contain small amounts of squalene (in Archaea), hopanoids (in bacteria), carotenoids, and other neutral lipids. Gram-negative bacteria, in addition, contain an outer membrane made up of two layers: the inner layer has a composition similar to that of the cytoplasmic membrane [i.e., phospholipids, approximately 2×10^7/cell], and the outer layer is composed of lipopolysaccharides (approximately 2×10^6/cell). Lipoproteins contain covalently linked lipids, which serve as anchors in the outer membrane or cytoplasmic membrane. Many prokaryotes contain a lipid-like reserve material, poly-3-hydroxy-fatty acids (polyhydroxyalkanoates; see Chapter 7.11).

7.10.2 Most Eubacterial Species Contain Only a Few Major Fatty Acids

Types of fatty acids. These are C_{14} to C_{18} fatty acids, including $C_{14:0}$, $C_{16:0}$, $C_{18:0}$, $C_{16:1\Delta9}$, $C_{18:1\Delta11}$, cyclopro-

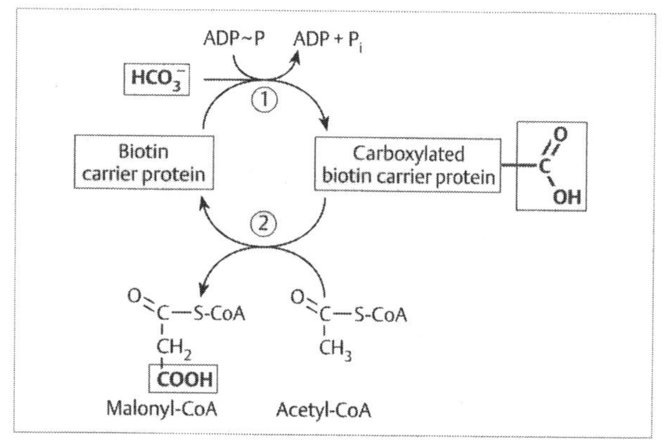

Fig. 7.**32** **Reactions catalyzed by acetyl-CoA carboxylase.** 1, biotin carboxylase subunit, which catalyzes the Mn^{2+}- and ATP-dependent carboxylation of biotin carboxyl carrier protein; 2, carboxyl transferase subunit, which transfers "activated bicarbonate" from carboxy-biotin protein to acetyl-CoA

pane, 10-methyl $C_{16:0}$, or 2- or 3-hydroxy acids. The fatty acids are saturated or unsaturated (i.e., with one *cis* double bond in the middle of the chain). Unsaturated fatty acids increase the fluidity of the membrane. In most Gram-positive bacteria, branched-chain or alicyclic fatty acids fulfill the same function (i.e., that of thermal adaptation). The higher the content of unsaturated or branched-chain fatty acids, the lower is the solid-to-liquid-phase transition temperature of lipids. **Polyunsaturated fatty acids** occur in cyanobacteria. From these few fatty acids, a few phospholipid (rarely glycolipid) species are made. Table 7.**4** shows the lipid composition of *E. coli* (representative of organisms with unsaturated fatty acids) at two different growth temperatures and of *Bacillus subtilis* (with branched-chain fatty acids).

Other destinations of fatty acids and phospholipids. Most (90%) of the **fatty acids** at the level of acyl-ACPs (see below) are integrated in the phospholipids with $C_{16:0}$ as the main component; 10% also serve as the precursor of lipid A in lipopolysaccharides with 3-OH $C_{14:0}$ as the main component. Traces are used for fatty acid chain-like moieties of the prosthetic groups lipoic acid and biotin. Small amounts of **phospholipids** are used for the synthesis of membrane-derived oligosaccharides and lipoproteins.

The enzymes of **lipid synthesis** are compartmented between the cytoplasma and the inner membrane. The enzymes that synthesize glycerol 3-phosphate and fatty acids are soluble; all other enzymes are membrane-bound.

Synthesis of glycerol 3-phosphate. Dihydroxyacetonephosphate is reduced by the biosynthetic glycerol-3-phosphate dehydrogenase to glycerol 3-phosphate. Catabolic isoenzymes are involved in glycerol utilization. Glycerol taken up from the medium can be phosphorylated, directly.

Synthesis of long-chain saturated fatty acids. Bacterial fatty acid synthesis resembles chloroplast fatty acid synthesis (type II or dissociated fatty acid synthase) and polyketide synthesis in bacteria. The individual reactions are carried out by separate proteins encoded by separate genes as opposed to the multifunctional protein complexes in animals, in which protein domains catalyze the individual reactions (type I synthesis).

Box 7.8 The nomenclature of fatty acids involves a system of abbreviations, exemplified as follows: $C_{16:1\Delta9\ cis}$ means the carbon chain length (C_{16}) with the number of double bonds (one). The Δ indicates a double bond, here between the carbon atoms C9 and C10, and *cis* denotes a double bond with *cis* configuration. There are many **variations in fatty acid synthesis**. Numerous isozymes carry out the same basic chemical reaction in fatty acid synthesis, but they differ in substrate specificity. These isozymes contribute to the regulation of the distribution of products from the pathway. Some fractions of straight-chain fatty acids must have *cis*-unsaturation to avoid rigid membrane structures (in many Gram-negative bacteria); on the other hand, branched or alicyclic fatty acids "disturb" the pseudocrystalline order of the alkyl chains (in many Gram-positive bacteria).

Table 7.**4** **Typical lipid composition at 37°C of a Gram-negative bacterium of the Proteobacteria (e.g., *E. coli*) and of a Gram-positive bacterium of the low G + C group (e.g., *B. subtilis*)**

	Escherichia coli [1]		*Bacillus subtilis*	
Fatty acids	Palmitic acid n-$C_{16:0}$	45%	Anteiso- $C_{15:0}$	40% [2]
	Palmitoleic acid n-$C_{16:1\Delta9}$	35%	Iso-$C_{17:0}$	20%
	cis-Vaccenic acid n-$C_{18:1\Delta11}$	18%	Iso-$C_{15:0}$	15%
	Myristic acid n-$C_{14:0}$	2%	n-$C_{16:0}$	5%
			n-$C_{16:1}$	7%
Polar head groups	-P-Ethanolamine	75%	-Glucose	
	-P-Glycerol	18%	-Glucose-*O*-glucose	
	-P- Phosphatidylglycerol	5%	-P-Ethanolamine	
	-P- Serine	trace	-P-Glycerol	

[1] When cells are grown at 25°C, the relative amounts of n-$C_{16:1\Delta9}$ and n-$C_{18:1\Delta11}$ are reversed. The phosphatidyl-glycerol level is higher in the exponential growth phase, whereas the diphosphatidyl-glycerol (i.e., cardiolipin) level rises in the stationary phase. [2] % By mass. Melting points of some fatty acids: $C_{12:0}$ 40°C, $C_{18:0}$ 70°C, $C_{18:1}$ 13°C, $C_{18:2}$ −5°C,

Acetyl-CoA carboxylase. Acetyl-CoA, the precursor of fatty acids (Fig. 7.**32**), is activated by the complex, biotin-containing, strictly regulated acetyl-CoA carboxylase to malonyl-CoA:

$$\text{acetyl-CoA} + \text{ATP} + HCO_3^- \longrightarrow \text{malonyl-CoA} + \text{ADP} + P_i.$$

Acyl carrier protein (ACP). Fatty acid synthesis proceeds via a heat-stable, very acidic protein cofactor (M_r approximately 9000), the acyl carrier protein (ACP; see Fig. 7.**33**). ACP carries the growing acyl chain from one enzyme to another enzyme and supplies precursors for the condensation reactions. The acyl intermediate is bound to the protein as a thioester attached to a prosthetic group, 4′-phosphopantetheine. This prosthetic group is the catalytically important part of CoA from which it is derived by phosphodiester linkage to a serine residue (Fig. 7.**33**). ACP is one of the most abundant proteins (0.25% of cell protein); however, the CoA pool in most prokaryotes is approximately eight times higher than the ACP pool.

Transacylation and initiation (Fig. 7.**34a**). Acetyl-CoA and malonyl-CoA are transacylated to ACP by acyltransferases. There are various ways to initiate fatty acid synthesis: by the formation of either acetyl- or acetoacetyl-ACP from malonyl-CoA and/or acetyl-CoA. Further condensation requires the transfer of the malonyl moiety to another ACP by malonyl-CoA ACP transacylase:

$$\text{malonyl-CoA} + \text{ACP} \rightleftarrows \text{malonyl-ACP} + \text{CoA}.$$

Elongation (Fig. 7.**34b**). The elongation includes four steps. (1) An acetyl unit of malonyl-ACP is condensed, releasing CO_2 and ACP. This condensing reaction is the only irreversible step in the elongation process and is catalyzed by several 3-ketoacyl-ACP synthase isoenzymes. (2) The 2-oxo group is reduced to the D-(−)-β-hydroxyl group (by the β-ketoacyl-ACP reductase reaction). (3) The hydroxyl group is dehydrated to the β-*trans*-unsaturated fatty acyl-AMP (by the dehydrase reaction). (4) The double bond is reduced (by enoyl-ACP reductase), and the resulting saturated acyl-ACP either serves as substrate for the next elongation cycle, or is withdrawn as a building block. Both reductases are NADPH-specific.

Synthesis of unsaturated fatty acids (Fig. 7.**34c**). A C_{10}-ACP intermediate is the precursor of unsaturated fatty acids ($C_{16:1\Delta 9\,cis}$, $C_{18:1\Delta 11\,cis}$). A specific dehydrase catalyzes not only dehydration but also isomerization of (enzyme-bound) *trans*-2-decenoyl-ACP to *cis*-3-decenoyl-ACP. The *cis* isomer is not recognized by the *trans*-specific reductase of step 4 in Fig. 7.**34b**, but is elongated by one of the synthases (I) to $C_{16:1\Delta 9\,cis}$ (step 1). Synthetase II is able to elongate $C_{16:1\Delta 9\,cis}$, its activity is high at low temperatures and low at high temperatures, a simple control mechanism for increased synthesis of $C_{18:1\Delta 11\,cis}$ at low temperatures to increase membrane fluidity. This so-called **anaerobic pathway** of synthesis of unsaturated fatty acids occurs in anaerobes and in many aerobes.

Most Gram-positive aerobic prokaryotes use an **aerobic pathway** (similar to that of Eukarya). With the help of molecular oxygen, a fatty acid desaturase introduces a double bond in the middle of a saturated C_{16} (palmityl) acyl-ACP; the product is palmitoleic ($C_{16:1\Delta 9\,cis}$) acyl-ACP. The formation of polyunsaturated fatty acids (in cyanobacteria and in chloroplasts) always requires this O_2-dependent mechanism.

Synthesis of iso-, anteiso-, and omega-alicyclic fatty acids. Some bacteria (*Bacillus* species and many other Gram-positive bacteria) have (1) **iso**-fatty acids (i.e., those with a methyl group at the penultimate carbon atom of the acyl chain), (2) **anteiso**-fatty acids (i.e., with the methyl group at the third-to-last carbon atom), and (3) **omega-alicyclic** fatty acids [i.e., those with a terminal saturated carbon C_5–C_7 ring, with or without substitution (i.e., unsaturation or hydroxylation)], shown in Fig. 7.**35**. The most common branched-chain fatty acids are the iso-16, iso-15/17, anteiso-15/17, and 10-methyl-17/18 types. The difference between the two synthetic pathways lies only in the respective primer acyl-ACPs and in the condensing enzymes (synthases), which prefer modified primer acyl chains for elongation instead of the "normal" acetoacetyl primer. These modified primers are made from three different precursors: (1) **Branched-chain 2-oxoacids** are the intermediates in the synthesis of the branched-chain amino acids leucine, valine, and isoleucine. The 2-

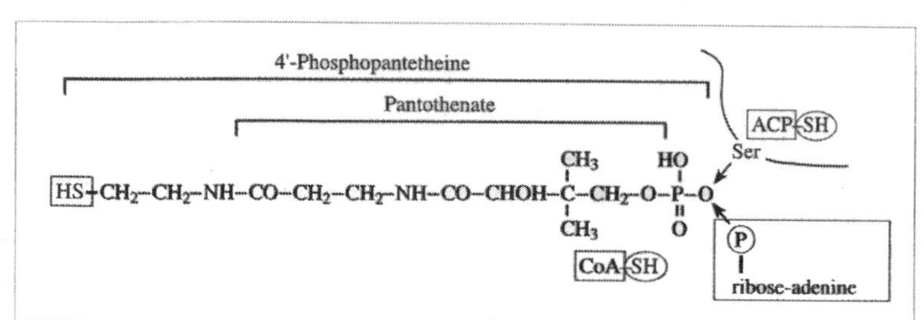

Fig. 7.33 **Common structure of the prosthetic group (4′-phosphopantetheine) of acyl carrier protein and coenzyme A.** ACP-SH, acyl carrier protein; CoA-SH, coenzyme A

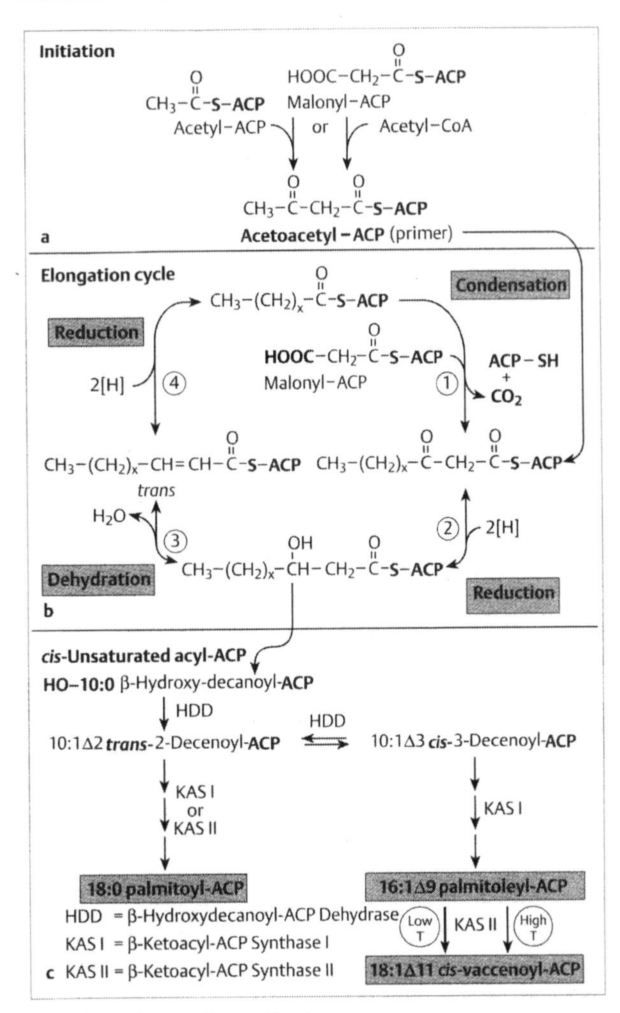

Fig. 7.**34** **Fatty acid synthesis.**
a Initiation.
b Elongation cycle: 1, condensation of malonyl-ACP with acyl-ACP by one of the *β*-ketoacyl-ACP synthases; 2, reduction by *β*-ketoacyl-ACP reductase (NADPH-dependent); 3, dehydration by *β*-hydroxyacyl-ACP dehydrase; 4, reduction by enoyl-ACP reductase (NADPH-dependent).
c Synthesis of unsaturated fatty acids and regulation of product distribution. A special *β*-**h**ydroxy**d**ecanoyl-ACP **d**ehydrase (HDD) catalyzes not only dehydration but also isomerization of the 2-*trans* to the 3-*cis* isomer, the key step in the synthesis of unsaturated fatty acids. A *β*-**k**eto**a**cyl-ACP **s**ynthase isoenzyme I (KAS I) is required for the elongation of these unsaturated acyl-ACPs. Isoenzyme KAS II is responsible for elongation to $C_{18:1}$

oxoacids are decarboxylated, and the aldehyde is added to malonyl-ACP; this condensation, yielding branched *β*-hydroxy-acyl-ACPs, is less clear. (2) **Branched, short-chain carboxylic acids** supplied in the medium are taken up, activated to the acyl-CoA, and transferred to ACP by special transacylase(s). This is common, for

iso– anteiso– ω-alicyclic- (e.g., cyclohexyl–) cis 10 9 Cyclopropane– 10-methyl–
a
Branched-chain amino acid
NH_2 Ⓡ-CH-COOH → R-C(=O)-COOH → (CO_2) [R-C(=O)-] Malonyl-ACP → Branched-chain primer-ACP
Ⓡ= CH_3-CH(-CH_3)- ; CH_3-CH(-CH_3)-CH_2- ; CH_3-CH(-CH_2-CH_3)-
b

Fig. 7.**35** **Structures of modified fatty acids.**
a Structures of branched-chain, omega-alicyclic, and modified fatty acids and
b structures of the precursors of branched-chain fatty acids

example, in bacteria living in the rumen, which depend on such fermentation products (e.g., isobutyrate, isovalerate, or 2-methylbutyrate). (3) Newly synthesized cyclohexyl and cycloheptyl carboxylic acids are similarly incorporated as in (2). Their synthesis, possibly from shikimate (cyclohexyl) or another compound (cycloheptyl), is not well understood.

Fatty acids in bacterial lipids are routinely analyzed by gas-liquid chromatography for diagnostic and identification purposes, and in systematic studies. The lipids are hydrolyzed under alkaline conditions, and the fatty acids are methylated to render them volatile.

Final steps in lipid synthesis: transfer of the acyl chains (Fig. 7.**36**). The final steps are associated with the inner membrane. The first step is the transfer of the acyl chains from acyl-ACP into the membrane phospholipids by glycerol phosphate acyltransferase. The acceptor is the 1-position of *sn*-glycerol 3-phosphate; another enzyme esterifies the 2-position of glycerol phosphate to give the so-called **phosphatidic acids** (i.e., diacylglycerol 3-phosphate). The different preferences of the two acyltransferases for certain acyl chain lengths and *cis*-unsaturated or branched-chain acids ensure asymmetric distribution of fatty acids between positions 1 and 2.

Diversification of polar head groups (Fig. 7.**36**). The diacylglycerol 3-phosphate is then converted to a **phospholipid** of one of the major phospholipid classes.

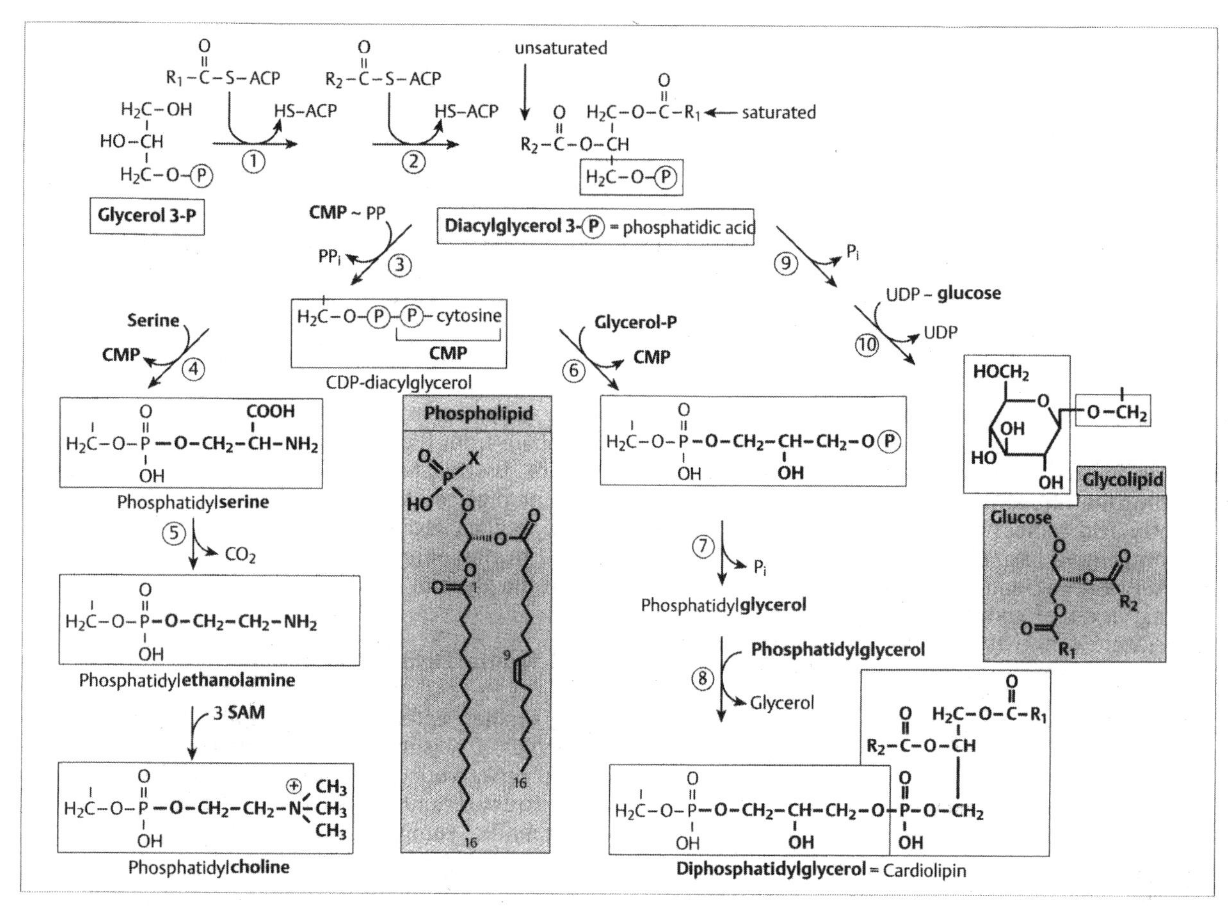

Fig. 7.**36 Synthesis of phospholipids and glycolipids from precursors.** The insets show the structures of typical phospholipids and glycolipids. 1–2, transfer to the membrane by acyl-transferring glycerol-3- phosphate acyltransferase(s); 3, cytidyl transfer by CDP-diglyceride synthase; 4, phosphatidylserine (PS) synthase; 5, PS decarboxylase; 6, phosphatidylglycerolphosphate (PGP) synthase; 7, PGP phosphatase; 8, cardiolipin (i.e., phosphatidylglycerolphosphatidyl) synthase; 9, phosphatidate phosphatase; 10, diacylglycerol glycosyltransferases (e.g., glucosyltransferase). Note that phosphatidylglycerol is located at a branching point that also leads to the synthesis of membrane-derived oligosaccharides (MDO) and lipoproteins. SAM, *S*-adenosylmethionine

In *E. coli*, there are only three types of phospholipids, **phosphatidylethanolamine**, **phosphatidylglycerol**, and **cardiolipine**, plus traces of a few others (Table 7.**4**). The diacylglycerol 3-phosphate is activated by nucleotidyl transfer from CTP, thus forming CDP-diacylglycerol; the reaction is catalyzed by CDP-diglyceride synthase. A polar head group (R-OH; e.g., **glycerol**, **serine**, or **phosphatidylglycerol**) is esterified by specific enzymes with the CMP-activated phosphoryl group, and CMP is released. The **serine** moiety can be decarboxylated to the **ethanolamine** moiety. Further methylations (rare in prokaryotes) serve to generate choline (i.e., phosphatidylcholine, also called lecithin). Other head groups of **phospholipids** not shown are *myo*-inositol (a six-membered ring with six OH groups), monosaccharides, disaccharides, or oligosaccharides. *myo*-Inositol 1-phosphate is formed by isomerization of D-glucose 6-phosphate. If the sugar is directly linked glycosidically to glycerol, the lipid is called a **glycolipid.**

Postsynthetic modifications. Some reactions occur at the membrane phospholipids. For example, cyclopropane fatty acids arise by addition of a methylene bridge (from SAM) to the double bond of unsaturated fatty acids.

Control of lipid synthesis. There are several aspects of lipid metabolism and function to be considered.

Response to external factors. Several external factors influence the lipid composition of phospholipids and (to some extent) lipopolysaccharides, which thereby affects the fluidity and other properties of membranes. The most important factors are temperature and the presence of organic, lipophilic compounds in the medium. The two groups of bacteria that contain either "unsaturated" or "branched" fatty acids obviously control membrane fluidity in different ways, (1) by varying the proportion of unsaturated fatty acids (essential for those with unsaturated fatty acids) or (2) by altering the proportion of methyl-branched and other fatty acids; unsaturated fatty acids are nonessential for those with methyl-branched fatty acids. Both groups control the relative proportions of unsaturated or branched-chain fatty acids and the chain length, but only to a limited extent by altering the composition of the head groups.

Content of unsaturated or branched-chain fatty acids, the chain length, and postsynthetic modifications of complete lipid molecules (e.g., methylation, cyclopropane-ring formation). Part of the regulation of the fatty acid composition can simply be explained as the competitive balance at the activity level between acyl-ACP–forming and acyl-ACP–using enzymes. This simplest level of control is based on differences in temperature sensitivity (in the activation energy), in kinetic constants for reactions with different acyl-ACP substrates (e.g., K_m or V_{max} values), and in the substrate specificity of these enzymes. Examples are (1) the different 3-ketoacyl-ACP synthetases (condensing enzymes), which elongate the saturated or unsaturated chain; (2) the β-ketoacyl-ACP reductase, which forms the 3-hydroxyacyl-ACP; and (3) the glycerol-phosphate acyltransferase and the corresponding enzymes in lipopolysaccharide synthesis, which withdraw the acyl products.

Total fatty acid content. This is regulated on the level of acetyl-CoA carboxylase and initiation of fatty acid synthesis by formation of acyl-ACP. The growth-rate control of lipid synthesis is not understood.

7.10.3 Lipid A, Membrane-Derived Oligosaccharides, and Lipoproteins Are Constituents of the Cell Envelope

Lipid A. Lipopolysaccharides are composed of three formal domains (i.e., **lipid A, core**, and **O-antigen**). Lipid A is well conserved in contrast to O-antigen. In *E. coli*, lipid A is composed of a β-1 $\rightarrow$ 6-linked glucosamine disaccharide acylated with four 3-hydroxy-$C_{14:0}$ fatty acids (3-hydroxymyristinic acid) and two fatty acids esterified with the 3-hydroxyl group. Here is its synthesis in brief: Glucosamine is converted to UDP-*N*-acetylglucosamine, which is acylated to UDP-2,3-diacylglucosamine; this product is formally similar to CDP-diacylglycerol. The first NDP-activated sugar lipid is then transferred to C6′ of a second lipid, 2,3-diacylglucosamine 1-phosphate (called lipid X), in which both acyl moieties are R-3-OH C_{14} acids. A kinase phosphorylates the C4′ of the former glucosamine unit. Next, the beginning core (i.e., two units of 2-keto-3-deoxy*manno*-octonic acid, KDO) is added to C6, and the free hydroxyl groups of the fatty acids are acylated with two additional, short-chain, saturated or unsaturated fatty acids (C_{12}–C_{16}). Details are presented in Chapter 23.2.

Membrane-derived oligosaccharides and lipoproteins. Phosphatidylglycerol (Fig. 7.**36**) is the intermediate in the synthesis of water-soluble, membrane-derived oligosaccharides (MDOs) in the periplasm, which play a role in osmoregulation. *sn*-Glycerol 1-phosphate from the polar head group of phosphatidylglycerol is transferred to the oligosaccharide; the remnant diglyceride is phosphorylated by diglyceride kinase to diacylglycerol 3-phosphate. Bacterial lipoproteins (in the outer membrane and those excreted) obtain the glycerol moiety in the same way.

7.10.4 Neutral Lipids Occur in all Bacteria

To a smaller extent, **neutral lipids** (e.g., squalene, hopanoids, C_{40} isoprenoids, carotenoids, and lipids of special prokaryotic groups) may be essential for membrane structure and function (Fig. 7.**37**). This large group of lipophilic compounds differing in structure and function is derived from acetyl-CoA via C_5-isoprene units, which are polymerized to **isoprenoids**. Examples of these compounds are (1) quinones, (2) undecaprenol

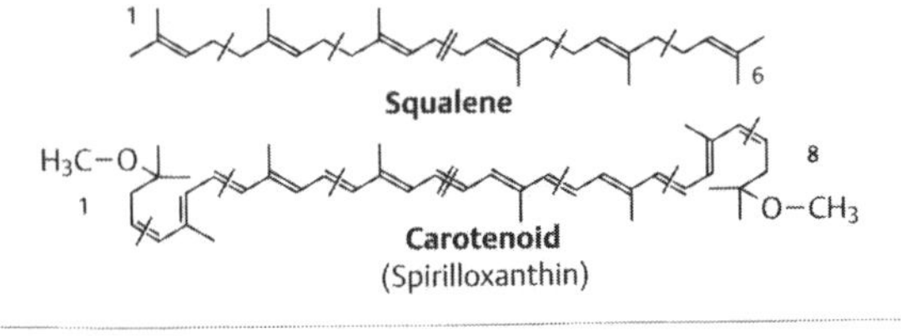

Hopanoid (Bacteriohopantetrol)

Squalene

Carotenoid (Spirilloxanthin)

Fig. 7.**37** **Structure of a some neutral lipids that play different roles.** Hopanoids are polyterpenoids found in many species of bacteria. Squalene is present in many Archaea. Carotenoids are pigments present in phototrophic as well as in many non-phototrophic prokaryotes

and dolichol, (3) carotenoids and retinal, and (4) squalene and hopanoids. These polymers function in (1) electron transport, (2) sugar and oligosaccharide transfer via the membrane, (3) light absorption for protection from light or to harvest light, and (4) membrane stability.

The first step in the synthesis of neutral lipids is the condensation of two acetyl-CoA molecules to an acetoacetyl-CoA molecule, followed by a (Claisen) condensation of a third molecule of acetyl-CoA to form 3-hydroxy-3-methyl-glutaryl-CoA (Fig. 7.**38**). The CoA-activated acid group is reduced to the alcohol group in mevalonic acid, to which a diphosphate group is added. Decarboxylation and water elimination—facilitated by phosphorylation of the -OH group—affords one **isoprenyl** unit, **isopentenyl-diphosphate**, which is isomerized to the second **prenyl** unit, dimethylallyl-diphosphate. There is another pathway in some eubacteria, green algae and chloroplasts of higher plants that starts from glyceraldehyde 3-phosphate and pyruvate forming 1-deoxyxylulose-5-phosphate. This intermediate is converted to isopentenyl-diphosphate. The diphosphate-activated prenyl and isoprenyl units undergo a polymerization chain reaction. The pyrophosphate "leaving group" pulls out the oxygen atom and thus converts the allyl compound into a reactive carbenium cation (head), which easily adds to the double bond of the electrophilic isopentenyl unit (tail). The product is then stabilized by H^+ elimination. The C_{55} isoprenoids and quinones are examples that this process can go on and on. Finally, two polyprenyl-diphosphate units may be condensed head-to-tail (as above), head-to-head, or tail-to-tail. The polyprenyl compounds contain two (terpenes), three (sesquiterpenes, one-and-a-half terpenes), four (diterpenes), or more isoprenoid units. Those with more than four units are mostly derived from two sesquiterpene (with a total of six units) or two diterpene units (with a total of eight units).

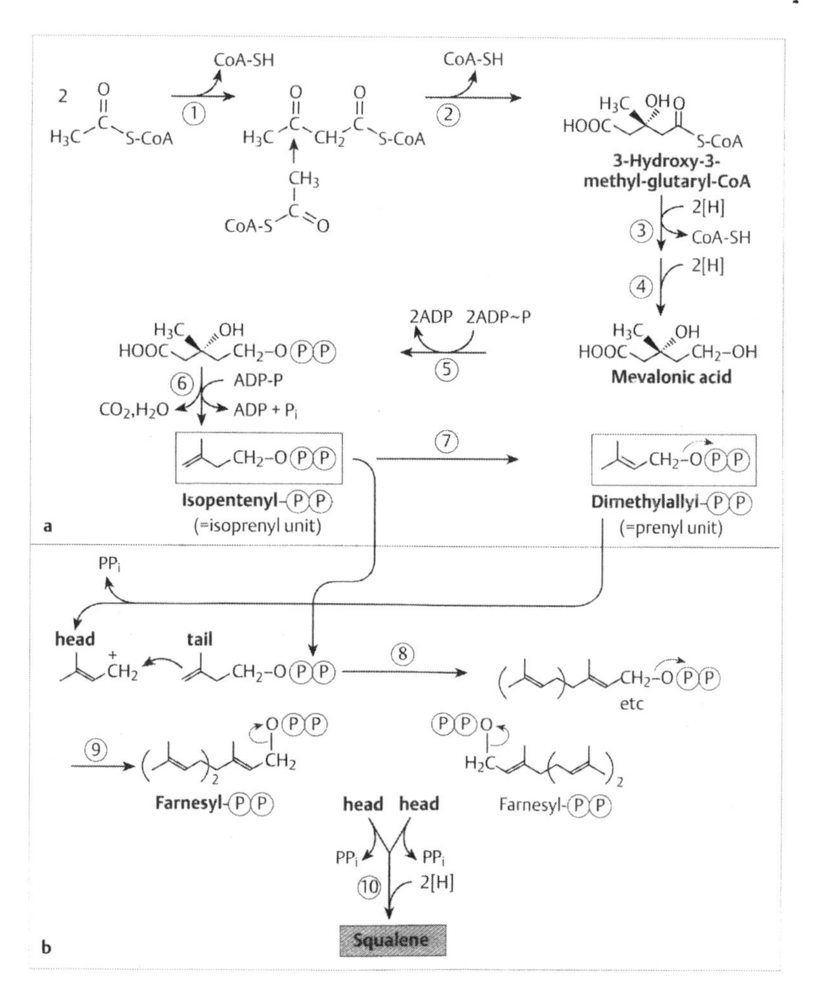

Fig. 7.**38** **Polyisoprenoid synthesis.**
a Formation of the activated isoprene units: 1, β-ketothiolase; 2, 3-hydroxy-3-methyl-glutaryl-CoA (HMG) synthase; 3/4, HMG reductase; 5, two kinase reactions; 6, decarboxylation and concomitant water elimination; 7, isomerization.
b Condensation of the activated isoprene units (prenyl transfer) and of higher terpenes. 8/9, head-to-tail condensations; 10, head-to-head condensation. A third possibility, tail-to-tail condensation, is not shown

Squalene (C_{30}), with six isoprenoid units, is formed by head-to-head condensation of two sesquiterpenes farnesyl-PP (C_{15}; see Fig. 7.**37**). Squalene can be cyclized in one reaction to hopanoids with four alicyclic rings analogous to the steroid ring system (which normally is not formed by prokaryotes). **Carotenoids** are similarly formed by condensation of the C_{20} precursor phytol-PP; in some carotenoids, aromatic rings are formed from alicyclic six-membered rings followed by oxidation (aromatization, the second way to form aromatic rings, with the third being via the so-called polyketide pathway). Reduction of part of the double bonds results in unsaturated isoprane alcohols, which form the alkyl chains of archaeal lipids.

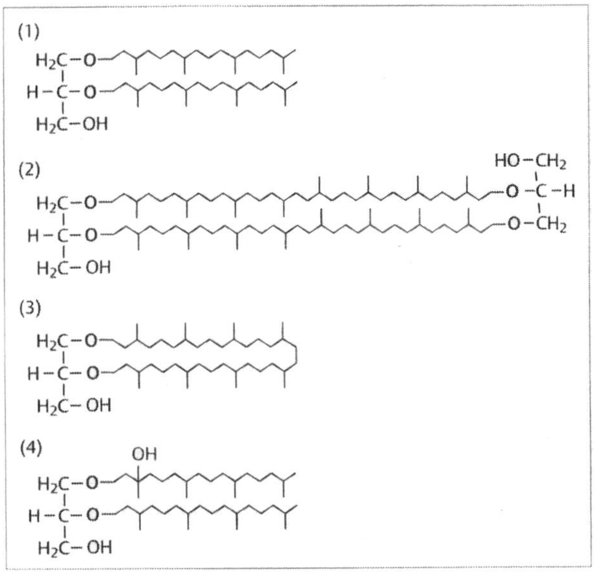

7.10.5 Archaeal Lipids Differ From Those of Bacteria and Eukarya

Membrane lipids in Bacteria (and Eukarya) differ from those of Archaea in three respects (Fig. 7.**39**): (1) in the nature of the linkage between glycerol and hydrocarbon chains (ester versus ether), (2) in the nature of the hydrocarbon chains (straight fatty-acyl chains versus highly methyl-branched, mostly saturated isopranyl chains), and (3) in the stereochemistry of the glycerol moiety (*sn*-1,2-di-O-R glycerol versus *sn*-2,3-di-O-R glycerol).

Archaea may also contain, in addition to the diether lipids, tetraether lipids in which two C_{40} isoprane dialcohols (formed by head-to-head condensation of C_{20} isoprenoid alcohols) are ether-linked at both sides to a glycerol phosphate (Fig. 7.**40**). Lipids of bacteria placed on the lowest branch of the domain Bacteria contain long-chain (e.g., C_{30}) dicarboxylic acids.

The differences between the lipids of Bacteria and those of Archaea reflect their differing origins rather than differing environmental adaptations. Ether lipids may still have advantages over ester lipids in survival at extremely high temperatures, but fatty acid ester lipids can easily be adapted to diverse and changeable environments.

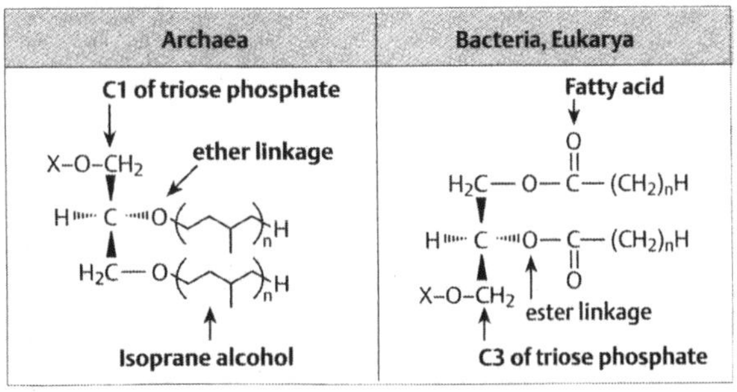

Fig. 7.39 **Major differences between lipids in Archaea and lipids in Bacteria or Eukarya**

Fig. 7.**40** **Some archaeal lipid structures.** (1) Diether archaeol; (2) tetraether caldarchaeol; (3) macrocyclic diether archaeol; (4) diether with a 3-hydroxy-archaeol. Cyclopropane rings in the phytanyl chains are common in thermoacidophilic Archaea [after 2]

The biosynthesis of archaeal lipids and its control are much less understood. A plausible biosynthetic pathway is given in Fig. 7.**41**. The polar head groups also vary considerably. The third hydroxyl group of glycerol is esterified with phosphate (in phospholipids), or glycosidically linked to a hexose C (in glycolipids).

The major steps in the biosynthesis of archaeal ether lipids are (1) the synthesis of the phytanyl chain, (2) the formation of ether bonds, and (3) the completion of the 2,3-di-O-R *sn*-glycerol structure. Tetraethers require additional biosynthetic steps; the exact precursors and the head-to-head condensation remain to be elucidated. The phytanyl chains (C_{20}) are formed from acetyl-CoA via four mevalonate units, resulting in the diterpenyl alcohol, geranylgeranyl pyrophosphate. This reaction sequence is similar to, but probably not identical with, isoprenoid synthesis via the mevalonate pathway in most Bacteria. The ether bond is formed via nucleophilic attack by the primary (*s*)-3-OH group of glycerol phosphate on geranylgeranyl pyrophosphate. The exact C_3 precursor and how the hydroxyl group at the *sn*-2 position of the glycerol-phosphate moiety is inverted are not known; there may be two independent pathways.

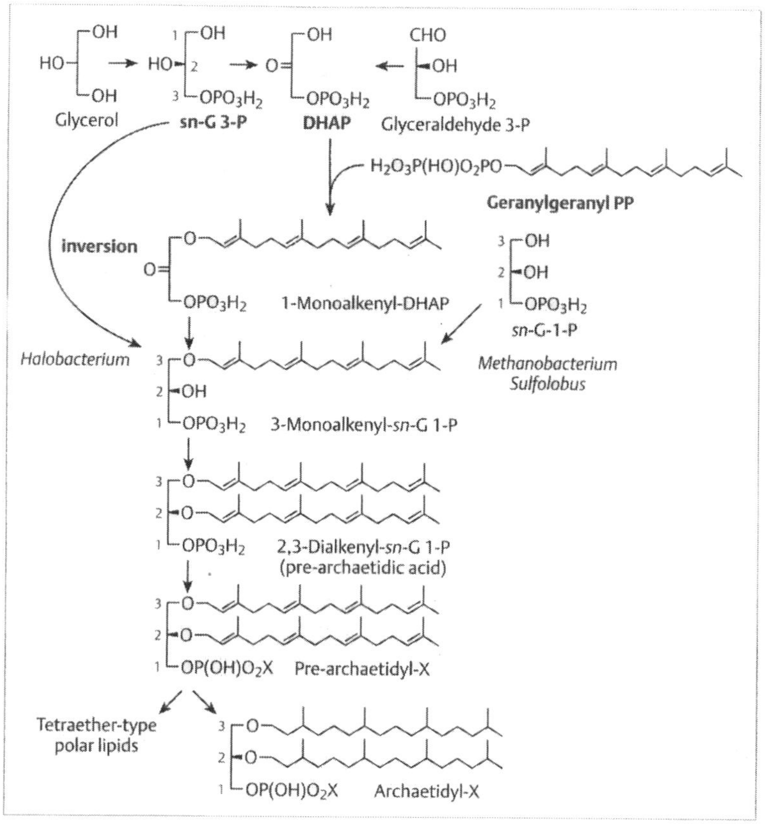

Fig. 7.**41** **Plausible biosynthetic pathway of archaeal ether lipids.** The first steps leading from dihydroxyacetone phosphate plus isoprenoid alcohol linked to pyrophosphate (e.g., geranylgeranyl pyrophosphate) to 3-monoalkenyl-*sn*-glycerol 1-phosphate may differ in various Archaea. The mechanism of inversion of *sn*-glycerol 3-phosphate to *sn*-glycerol 1-phosphate is at issue. The numbers 1, 2, and 3 near the glycerol skeleton indicate the *sn* numbers of the carbon atoms of the glycerophosphate moiety [after 2]. The incorporated isoprenoid alcohol is reduced to the saturated isoprenoid alcohol level. The polar head groups X are discussed in the text

7.11 All Bacteria Synthesize One or More Polymeric Storage Compounds

Most prokaryotes are able to deposit intracellularly one (or more) type(s) of homopolymeric reserve materials (Fig. 7.**42**). These materials form water-insoluble granules (“**inclusion bodies**”), which can be detected and differentiated by appropriate staining methods. Reserve materials are formed when ample energy is available but growth is limited because of shortage of an essential element (N, P, S, or Fe). The macromolecular reserve can be used for maintenance metabolism during starvation and to start growth when the limiting nutrient is available again. The regulation of storage-polymer metabolism is poorly understood.

Polyglucose (energy and carbon reserve) (Fig. 7.**42a**). **ADP-glucose** is normally the precursor of the highly branched polysaccharide **glycogen** (i.e., polyglucose) in which the glucose moieties are α-(1 → 4)- and α-(1 → 6)-linked. The enzyme that synthesizes the building block, ADP-glucose pyrophosphorylase, is allosterically inhibited: glycolytic intermediates act as activators, and AMP and ADP act as inhibitors. Glycogen synthase catalyzes the glycosidic linkage between the activated C1 of the ADP-glucose moiety and the non-reducing C4 end of the growing oligosaccharide chain (which functions as “primer”). This reaction is similar to the glycosidic bond formation in disaccharide synthesis between an NDP-activated sugar and another sugar (resulting in, e.g., lactose, maltose, sucrose, or trehalose; for structures, see Chapter 9.11 and Table 9.**1**). The formation of the α-(1 → 6) branches is catalyzed by **transglucosylation**. In this reaction, an oligosaccharide is transferred from the C4 position to the C6-OH group of a glucose moiety located several residues further along the helical sugar chain.

Polyhydroxyalkanoate (energy and carbon reserve) (Fig. 7.**42b**). Acetyl-CoA is the precursor of the polyester formed by condensation of 3-hydroxy-fatty acid CoA thioesters. The lipophilic polyhydroxyalkanoate (PHA) can amount to 90–95% of cell dry weight. The

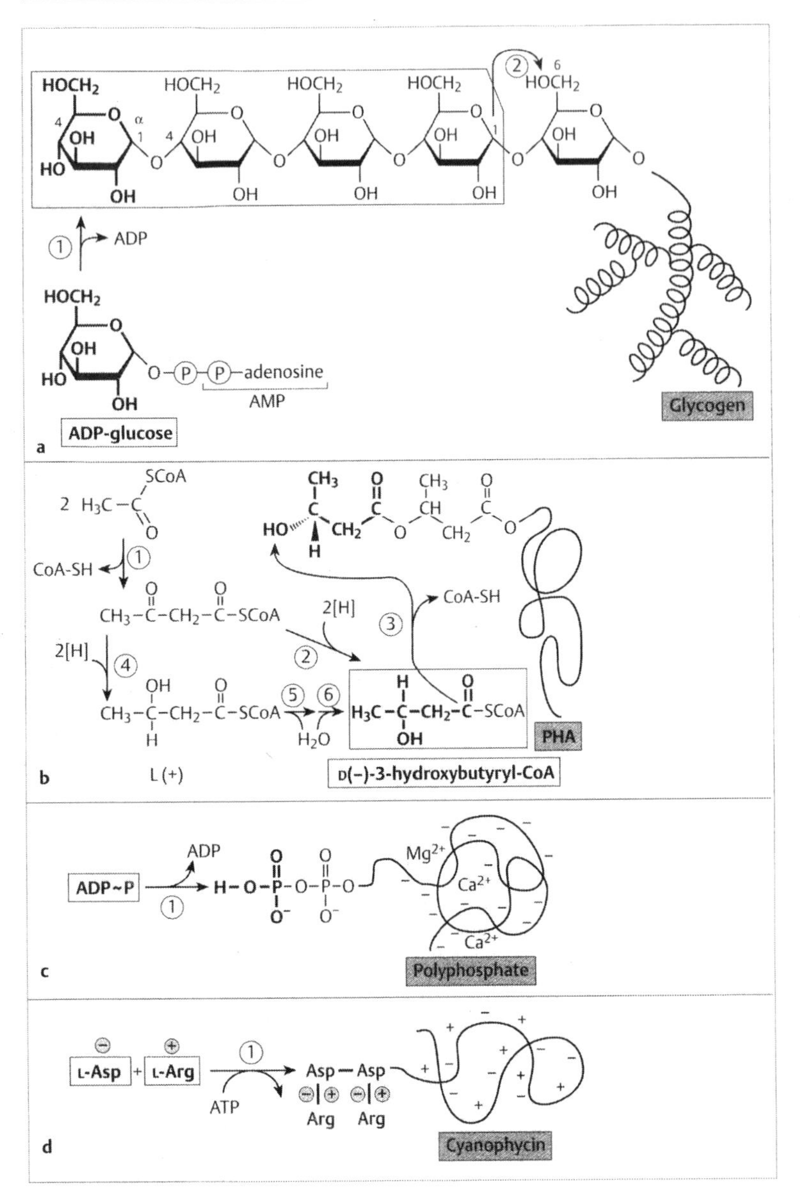

Fig. 7.**42** **Building blocks for the synthesis of reserve materials.**
a Glycogen: For ADP-glucose synthesis by AMP transfer to glucose 1-phosphate, see NDP-hexose-1-phosphate nucleotidyl-transferase reaction. 1, Glycogen synthase; 2, transglycosylase (branching enzyme).
b Polyhydroxyalkanoate (PHA): 1, β-Ketothiolase; 2, D-(−)-3-hydroxybutyryl-CoA dehydrogenase; 3, D-(−)-3-hydroxybutyryl-CoA polymerase. An alternative, enzyme 4, is required when enzyme 2 is L-specific. Then, two crotonases, 5 [L(+)-3-OH-specific] and 6 [D(−)-3-OH-specific], are combined.
c Polyphosphate: 1, polyphosphate kinase.
d Reserve protein cyanophycin: 1, enzyme that alternatively condenses aspartate and arginine

polymer consists of several hundred to several thousand monomeric building blocks. The carbon chain length of the substrate determines the range of monomer units incorporated into PHA. Under most conditions, it is 3-hydroxybutyrate. The chain may reach a C_{12} length when cells are grown on long-chain alkanes, alcohols, or fatty acids (i.e., alkanoates). The synthesis of the monomeric building block D-(−) 3-hydroxybutyryl-CoA from two molecules of acetyl-CoA corresponds to fatty acid synthesis. The polymerase that esterifies the CoA-activated building block with the free 3-hydroxyl group of the growing chain is granule-bound.

Polyphosphate (energy and phosphorus reserve) (Fig. 7.**42c**). Polyphosphate (**volutin**) is ubiquitous and is made from **ATP** when ample energy and phosphate are available. A polyphosphate kinase associated with the membrane transfers the γ-phosphate of ATP to the growing linear polyphosphate chain, which contains several hundred to several thousand phosphate groups. The reaction is freely reversible. The negative charges are compensated for by Mg^{2+}, Ca^{2+}, and other cations, which need to be co-transported into the cell by the low-affinity phosphate-transport system. Ca^{2+}-Polyphosphate and PHA are presently being discussed as membrane constituents that possibly confer the cell with competence for transformation. The genes for the polyphosphate kinase and an exopolyphosphatase constitute an operon and belong to the *pho* modulon in *E. coli*.

Sulfur ([H] reserve). Many phototrophic purple sulfur bacteria and aerobic sulfur oxidizers transiently accumulate intracellular sulfur globules when excess H_2S is available. Surplus H_2S not required as H donor for carbon fixation or respiration is incompletely oxidized to sulfur, whereby the chemical nature of this "sulfur" is not exactly known. Upon H_2S depletion, the sulfur is further oxidized to sulfate.

Reserve protein (energy, nitrogen, and carbon reserve) (Fig. 7.**42d**). Many cyanobacteria form a reserve protein called **cyanophycin**, which can represent up to 8% of cell dry weight. It is composed of poly-L-Asp-L-Arg. As expected, its synthesis is ribosome-independent. Cyanophycin is used as an energy, nitrogen, and carbon reserve.

7.12 Many Soluble Compounds Are Essential and Only Required in Small Amounts

Besides the soluble building blocks, cells contain many different soluble compounds that are essential (e.g., coenzymes) and only required in small amounts. The synthesis of these minor compounds therefore has less impact on the strategy of the cell in terms of the flow of energy, of reductants, and of elements. Only a few major groups of such essential, soluble compounds can be mentioned.

7.12.1 Polyamines Are Required for Optimal Growth

The properties of polyamines are due to their polybasic nature. Their function is not well understood. They may constitute a defense mechanism against transient cell acidity, neutralize the negative charges of nucleic acids, adjust the intracellular ionic strength, stabilize ribosomal structure, and perhaps indirectly be involved in protein synthesis. They are basically derived from the arginine and the lysine biosynthetic pathways; equal amounts of arginine and polyamines are required. Some structures are given in Fig. 7.**43**.

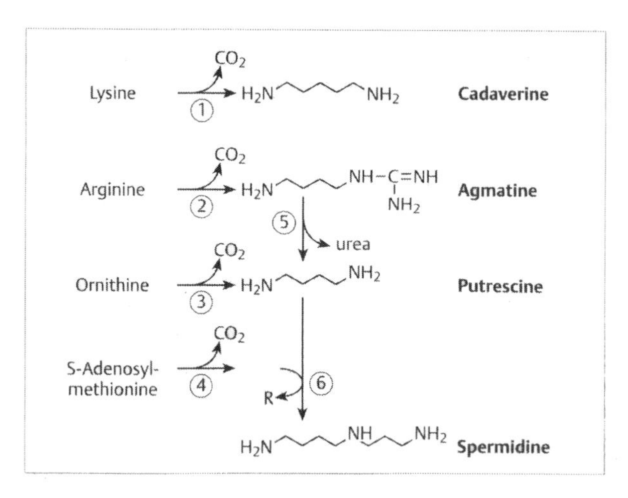

Fig. 7.**43** **Structure and synthesis of polyamines.** Enzymes 1–4, decarboxylases; 5, ureahydrolase; 6, spermidine synthetase (i.e., putrescine aminopropyltransferase); R, 5′-methylthioadenosine

Polyamines are formed by decarboxylation of arginine (to **agmatine**), lysine (to **cadaverine**), or ornithine (to **putrescine**) (Fig. 7.**43**). Agmatine can be hydrolyzed to putrescine and urea when arginine in the medium represses arginine biosynthesis. Putrescine and the decarboxylation product of *S*-adenosylmethionine give rise to **spermidine**. The complex regulation of the enzymes involved in polyamine biosynthesis supports the presupposition of an important biological function; for example, there are two ornithine decarboxylases, a degradative one and a biosynthetic one. The synthesis of the biosynthetic enzyme is highly regulated. The most striking regulatory mechanism seems to be an interaction of an inhibitory protein (antizyme) with this enzyme, which causes enzyme inhibition.

7.12.2 Tetrapyrroles Play Very Different Roles in Metabolism

Many pigments, prosthetic groups, and coenzymes contain four molecules of the five-membered, N-heterocyclic pyrrole ring, the tetrapyrroles. Linear tetrapyrroles (i.e., bile pigments) are linear chains of four pyrroles. Macrocyclic tetrapyrroles contain in addition a complexed metal ion, which may undergo a redox change during catalysis (Fig. 7.**44**). The number (eight to eleven) and the position of the conjugated

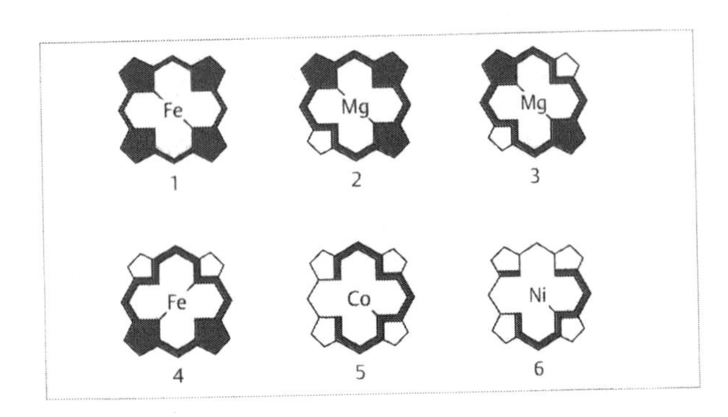

Fig. 7.**44** **Schematic structures of cyclic tetrapyrroles.** 1, Hemes; 2, chlorophylls and bacteriochlorophylls *c*, *d*, and *e*; 3, bacteriochlorophylls *a*, *b*, and *g*; 4, siroheme; 5, corrinoids; 6, coenzyme F_{430}. The scheme emphasizes the oxidation/reduction state; double bonds and systems of conjugated double bonds are shaded red [after 3]

double bonds cause the different colors and other properties of these compounds.

Tetrapyrroles and their functions are summarized in Table 7.**5**. The common precursor is δ-aminolevulinic acid (ALA). The overall synthesis comprises the following five steps (Fig. 7.**45**): (1) **Formation of ALA** proceeds along one of two ways. As a rule, synthesis begins with glutamate (C_5 pathway), via the "older" phylogenetic pathway. The synthesis of ALA from succinyl-CoA and glycine (C_4 pathway or Shemin pathway, named after its discoverer) appears to be confined basically to the α-branch of Proteobacteria (from which the mitochondria are derived; mitochondria have this pathway, in contrast to the chloroplasts). **C_5 pathway.** Interestingly, the α-carboxyl group of glutamate is activated by $tRNA^{Glu}$ amino acylation as in protein synthesis. This activation allows reduction to the semialdehyde. A pyridoxal-phosphate–dependent isomerase changes the position of the carbonyl oxygen atom and the amino group by a curious transamination, which yields ALA. **C_4 pathway.** ALA is formed in a single step by condensation of succinyl-CoA with glycine. (2) **Formation of the pyrrole ring**. Two ALA molecules are condensed such that the amino group of the first molecule forms a Schiff's base with the keto group of the second molecule, and the keto group of the first molecule condenses with the methylene C3 of the second molecule. The product is called porphobilinogen; the condensing enzyme is ALA dehydratase. (3) **Joining of the pyrrole rings, synthesis of uroporphyrinogen III.** In a common pathway, four pyrrole rings are condensed to the macrocyclic tetrapyrrole, uroporphyrinogen III, at a branching point in the pathway. For the steps that follow consult Further Reading (end of chapter); the sequence of events varies, depending on the type of tetrapyrrole. (4) Conversion by two separate routes to the secondary branching points, **protoporphyrin IX** and **dihydrosirohydrochlorin**. (5) **Metal insertion** by an ATP-dependent metal chelatase. (6) Possibly further **modifications** of the ring, its substituents, and introduction of side groups and ligands that may coordinate to the metal atom.

Table 7.**5** **Macrocyclic and linear tetrapyrroles: their function and occurrence**

Macrocyclic tetrapyrroles	Central atom	Function in	Occurrence
Hemes	$Fe^{2+/3+}$	Respiration, electron transport, energy conservation, hydroxylation	Most bacteria
Chlorophylls[1]	Mg^{2+}	Light absorption, energy conservation	Cyanobacteria
Bacteriochlorophylls[1]	Mg^{2+}	Light absorption, energy conservation	Phototrophic bacteria, except cyanobacteria and halobacteria
Siroheme	Fe^{2+}	HSO_3^- reduction to H_2S, NO_2^- reduction to NH_3	
Corrinoids	$Co^{1+/2+/3+}$	Diverse reactions, isomerization, methyl transfer	Most bacteria
Coenzyme F_{430}	Ni^{2+}	CH_3-Coenzyme M reduction to CH_4	Methanogens (Archaea)
Pheophytine	H^+	Electron transport in photosynthetic reaction center II	Phototrophs with photosystem II equivalent
Linear tetrapyrroles Phycobilines	—	Light absorption	Cyanobacteria

[1] See Fig. 13.1

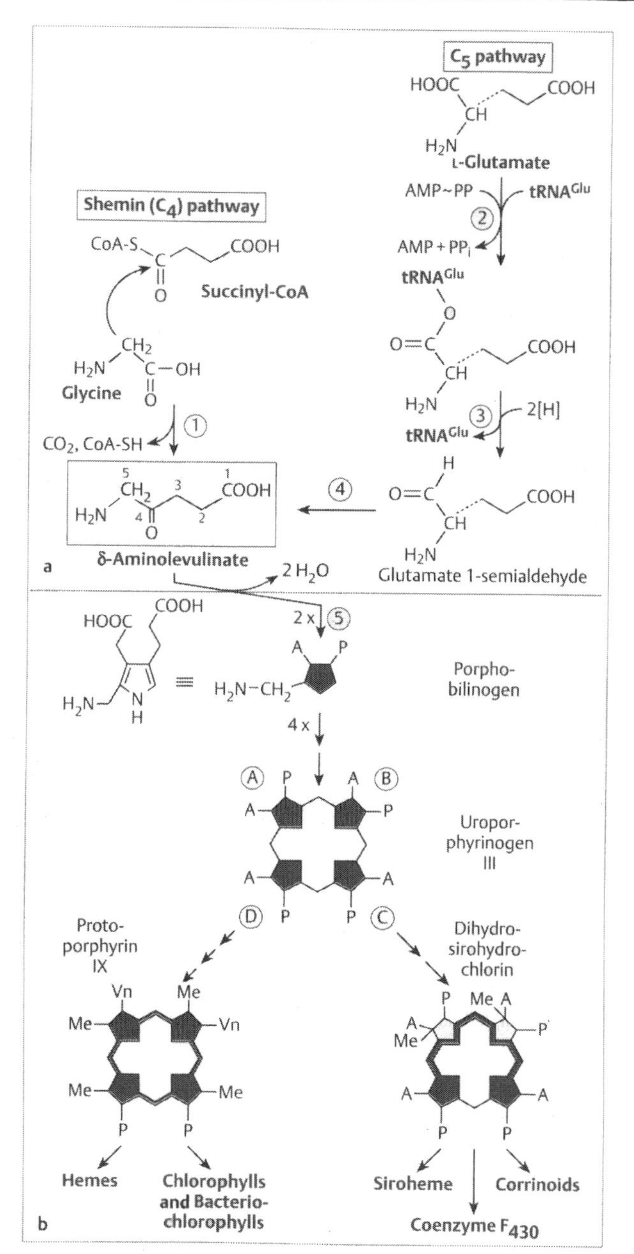

Fig. 7.45 **Synthesis of tetrapyrroles.**
a Synthesis of δ-**aminolevulinic acid** (ALA): 1, δ-aminolevulinic-acid synthase; 2, glutamyl-tRNA synthetase; 3, glutamyl-tRNA reductase; 4, glutamate-1-semialdehyde aminotransferase.
b Synthesis of macrocyclic tetrapyrroles from δ-ALA via uroporphyrinogen III and either protoporphyrin IX or dihydrosirohydrochlorin: 5, ALA dehydratase. For structures, see Fig. 7.**44**. A, acetyl (-CH_2-COOH); P, propionyl (-CH_2-CH_2-COOH); M, methyl (-CH_3); Vn, vinyl (-CH=CH_2); A, B, C, and D (encircled) denote rings A, B, C, and D, respectively

7.12.3 The Synthesis of Other Pigments, Cofactors, Prosthetic Groups, and Electron Carriers Requires a Considerable Portion of the Genetic Information of the Cell

There are many more small compounds that—despite the low amounts required for growth—fulfill important metabolic functions. Their synthesis is complex and requires a substantial amount of the genetic information of most prokaryotes. For details of their synthesis, consult the literature suggested below.

Further Reading

Boom, T. V. (1989) Genetics and regulation of bacterial lipid metabolism. Annu Rev Microbiol 43:317–343

Crawford, I. P. (1989) Evolution of a biosynthetic pathway: the tryptophan paradigm. Annu Rev Microbiol 43:567–600

Cunin, R., Glansdorff, N., Piérard, A., and Stalon, V. (1986) Biosynthesis and metabolism of arginine in bacteria. Microbiol Rev 50:314–352

DeRosa, M., Gambacorta, A., and Gliozzi, A. (1986) Structure, biosynthesis, and physicochemical properties of archaeal lipids. Microbiol Rev 50:70–80

Enzyme nomenclature (1992) Nomenclature committee of the International Union of Biochemistry and Molecular Biology. New York, London: Academic Press

Gerhard, P., Murray, R. G. E., Wood, W. A., and Krieg, N. R., eds. (1994) Methods for general and molecular bacteriology. Washington, DC: ASM Press

Gottschalk, G. (1986) Bacterial metabolism, 2nd edn. Berlin, New York, Heidelberg: Springer

Kaneda, T. (1991) Iso- and anteiso-fatty acids in bacteria: biosynthesis, function, and taxonomic significance. Microbiol Rev 55:288–302

Kates, M., Kushner, D. J., and Matheson, A. T., eds. (1993) The biochemistry of Archaea (Archea). Amsterdam, New York: Elsevier

Koga, Y., Masateru, N., Morii, H., and Akagawa-Matsushita, M. (1993) Ether polar lipids of methanogenic bacteria: structures, comparative aspects, and biosynthesis. Microbiol Rev 57:164–182

Lechner, J. (1989) Structure and biosynthesis of prokaryotic glycoproteins. Annu Rev Biochem 58:173–194

Lederberg, J., ed. (1992) Encyclopedia of microbiology, vols. 1–4. New York, London: Academic Press

Madigan, M. T., Martinko, J. M., and Parker, J. (1997) Biology of microorganisms, 8th edn. Englewood Cliffs, NJ: Prentice-Hall

Magnuson, K., Jackowski, S., Rock, C. O., and Cronan, J. E., Jr. (1993) Regulation of fatty acid biosynthesis in *Escherichia coli*. Microbiol Rev 57:522–542

Neidhardt, F. C., Curtiss, R., Ingraham, J. L., Lin, E. C. C., Low, K. B., Magasanik, B., Reznikoff, W., Riley, M., Schaechter, M., and Umbarger, H. E. (eds.) (1996) *Escherichia coli* and *Salmonella*: cellular and molecular biology. 2nd edn. Washington, DC: ASM Press

Neidhardt, F. C., Ingraham, J. L., and Schaechter, M. (1990) Physiology of the bacterial cell: a molecular approach. Sunderland, Mass.: Sinauer; 351–388

Raetz, C. R. H. (1990) Biochemistry of endotoxins. Annu Rev Biochem 59:129–170

Riley, M. (1993) Functions of the gene products of *Escherichia coli*. Microbiol Rev 57:862–952

Sahm, H., Rohmer, M., Bringer-Meyer, S., Sprenger, G. A., and Welle, R. (1993) Biochemistry and physiology of hopanoids in bacteria. Adv Microbiol Physiol 35:247–273

Schnaitman, C. A., and Klena, J. D. (1993) Genetics of lipopolysaccharide biosynthesis in enteric bacteria. Microbiol Rev 57:655–682

Sonenschein, A. L., Hoch, J. A., and Losick, R., eds. (1993) *Bacillus subtilis* and other Gram-positive bacteria. Washington, DC: ASM Press

Stanier, R. Y., Ingraham, J. L., Wheelis, M. L., and Painter, P. R. (1986) The microbial world, 5th edn. Englewood Cliffs, NJ: Prentice-Hall

Walsh, C. (1979) Enzymatic reaction mechanisms. San Francisco: Freeman

Woese, C. R. (1987) Bacterial evolution. Microbiol Rev 51:221–271

Sources of Figures

1 Gottschalk, G. (1986) Bacterial metabolism, 2nd edn. Berlin, Heidelberg, New York: Springer
2 Koga, Y., Masateru, N., Morii, H., and Akagawa-Matsushita, M. (1993) Microbiol Rev 57:164–182
3 Friedman, H. C., and Thauer, R. K. (1992) Macrocylic tetrapyrrole biosynthesis in bacteria. In: Lederberg, J. (ed.) Encyclopedia of microbiology, vol. 3. New York, London: Academic Press; 1–19

Section III

Diversity of Metabolic Pathways

In the chapters of Section II, the basic facts of bacterial energy metabolism (catabolism), including the fate of the most common substrate glucose and the different modes of ATP generation, were described. The multiple role of the cytoplasmic membrane in every conversion and other processes was discussed. The principles of growth and nutrition as well as general microbiological techniques were introduced. These aspects apply to virtually all prokaryotes. In Chapter 7, the biosynthesis of essential cellular building blocks from simple growth substrates was described. These biosynthetic pathways (anabolism) show some diversity in prokaryotes, yet there are more common traits than differences.

In the chapters of Section III, the impressive diversity of bacterial metabolism will be described. In Chapter 8, the simplistic view of the preceding chapter on biosynthesis will be widened. Other molecular forms of the main growth substrates, e.g., CO_2, N_2, or sulfate, will be taken into account. In order to form all building blocks from these precursors, prokaryotes need special pathways, many of which occur only in specialists. Prokaryotes possess even diverse pathways for the same metabolic process, for example, for autotrophic CO_2 fixation. The subsequent chapters will be devoted to the great variety encountered in prokaryotes in making their living by oxidizing all kinds of organic substrates (chemotrophy, Chapter 9) or inorganic substrates (lithotrophy, Chapter 10), In Chapter 11, the diversity in handling oxygen in respiration will be presented, as well as the other roles of oxygen in metabolism. Anaerobic bacteria can obtain energy by oxidizing substrates with an electron acceptor other than oxygen (anaerobic respiration) or by fermentation, i.e., oxidation of part of the substrate and transfer of electrons to metabolic intermediates, forming reduced end products (Chapter 12). Finally, Chapter 13 will deal with light as a source of energy for prokaryotes. Again, the metabolic diversification, originated by a long process of adaptation to special environments during evolution, is impressive. Still behind the diversity of metabolic pathways, there is a unity of basic mechanisms.

8 Assimilation of Macroelements and Microelements

In the previous chapter, the principal biosynthetic pathways starting from a dozen central precursor metabolites were discussed. It was presumed that the elements were provided in the growth medium in an easily usable form: carbon was assumed to be provided as an organic substrate (e.g., glucose serving as an energy and carbon source). Glucose degradation immediately yields the central precursor metabolites for syntheses (see Chapters 3, 4, and 7). N was taken as NH_4^+ (pK of NH_3 is 9.3), P as $H_2PO_4^-$ (pK_2 of H_3PO_4 is 7), and S as HS^- (pK_1 of H_2S is 7). The trace elements were provided in the appropriate oxidation state (e.g., iron as Fe^{2+}). The transport of these compounds into the cell was discussed in Chapter 5.

However, such conditions do not represent the situation prevailing in nature, where the elements often are provided in other molecular forms. The reasons are evident: many different bacteria may encounter environments where a suitable organic compound is lacking, and where CO_2 or C_1 compounds are the sole or main source of cell carbon; this means that the bacteria have to form first one of the central precursor metabolites and finally all of them from (inorganic) carbon sources. NH_3 is rapidly oxidized to nitrate by nitrifying bacteria. H_2S is air-oxidized or oxidized by bacteria (e.g., of the *Thiobacillus* type) to elemental sulfur and further to sulfate. Iron, the most important metal, exists under oxic conditions as Fe(III), which forms virtually insoluble oxohydroxy or carbonate compounds. Only phosphorus is normally present in the proper oxidation state of phosphate since phosphate reduction would require a reductant so strong that the reductant and reduced P compound would be unstable under environmental conditions. In this chapter, the mechanisms by which prokaryotes form their building materials from less suitable, mostly inorganic molecular forms, will be discussed.

The bacterial assimilatory pathways show some metabolic versatility. This is in contrast to the rather uniform anabolism of the autrophic green plants. Note, however, that their assimilatory potential is based on the metabolism of only one endosymbiotic prokaryotic group, that of the cyanobacteria. The prokaryotic endosymbiont enabled the host cell to grow at the expense of carbon dioxide, nitrate, phosphate, and sulfate as sole sources of carbon, nitrogen, phosphorus, and sulfur. It is not surprising that the plant assimilatory enzymes are mostly located in the chloroplast and resemble the prokaryotic systems.

8.1 Autotrophic Bacteria Use CO_2 as Sole Source of Carbon

A considerable number of bacteria are able to use CO_2 as sole source of cell carbon; CO_2 is also required for heterotrophic growth. Important reactions by which bacteria incorporate CO_2 into cell carbon are summarized in Table 8.**1**. Autotrophic (= self-nurishing) bacteria belong to quite different evolutionary branches. Fixation of CO_2 is characterized by Reaction 8.1, where $\langle CH_2O \rangle$ represents cell carbon at the average oxidation level zero of carbohydrates (equal to formaldehyde):

$$CO_2 + 4\,H + nATP \leftrightarrows \langle CH_2O \rangle + H_2O + nADP + nPi \qquad (8.1)$$

Four different mechanisms of CO_2 fixation are known. The Calvin-Benson-Bassham cycle (Calvin cycle), which is active in chloroplasts, is also the dominant extant mechanism of carbon fixation in prokaryotes. It is restricted to aerobic Eubacteria; in Archaea it appears to be absent. In several anaerobic Eubacteria and Archaea, two alternatives have been found, the reductive citric acid cycle and the reductive acetyl-CoA pathway, both of which bring about the synthesis of acetyl-CoA from 2 CO_2, although by different means. Finally, a fourth pathway, the 3-hydroxypopionate cycle, appears to operate in *Chloroflexus aurantiacus* of the phototrophic green non-sulfur bacteria and possibly in aerobic Archaea. The four mechanisms of CO_2 fixation are shown in Table 8.**2**.

Autotrophic modes of life. In order to fix one CO_2 (oxidation state +4) to the level of cell carbon (oxidation state about 0), four electrons and energy

Table 8.1 **Important CO_2-fixation reactions in bacteria**

Reaction	$\Delta G^{\circ\prime}$ (2[H] = H_2) (kJ/mol CO_2)	CO_2 acceptor	Enzyme
Ribulose 1,5-bisphosphate + CO_2 + H_2O → 2 3-phosphoglycerate		Ribulose 1,5-bisphosphate	Ribulose-1,5-bisphosphate carboxylase
$CO_2 + 2[H] \rightleftharpoons HCOO^- + H^+$	+ 3.4	Mo or W enzyme	Formate dehydrogenase
$CO_2 + 2[H] \rightleftharpoons [CO] + H_2O$	< + 20.1	Ni enzyme	CO dehydrogenase
CO_2 + Methanofuran ⇄ Carboxymethanofuran	?	Methanofuran	Formyl methanofuran dehydrogenase
$CH_3\text{–}CO\text{–}SCoA + CO_2 + 2[H] \rightleftharpoons CH_3\text{–}CO\text{–}COO^- + H^+ + CoASH$	+ 16.3	Acetyl-CoA	Pyruvate synthase
$^-OOC\text{–}CH_2\text{–}CH_2\text{–}CO\text{–}SCoA + CO_2 + 2[H] \rightleftharpoons {}^-OOC\text{–}CH_2\text{–}CH_2\text{–}CO\text{–}COO^- + H^+ + CoASH$	+ 14.8	Succinyl-CoA	2-Oxoglutarate synthase
$^-OOC\text{–}CH_2\text{–}CH_2\text{–}CO\text{–}COO^- + CO_2 + 2[H] \rightleftharpoons {}^-OOC\text{–}CH_2\text{–}CH(COO^-)\text{–}CHOH\text{–}COO^- + H^+$	− 9.6	2-Oxoglutarate	Isocitrate dehydrogenase
$CH_3\text{–}CO\text{–}COO^- + HCO_3^- + ATP \rightarrow {}^-OOC\text{–}CH_2\text{–}CO\text{–}COO^- + ADP + P_i$	+ 0.1	Pyruvate	Pyruvate carboxylase
$CH_2{=}C(O\text{–}PO_3^{2-})\text{–}COO^- + HCO_3^- \rightarrow {}^-OOC\text{–}CH_2\text{–}CO\text{–}COO^- + P_i$	− 19.7	Phospho*enol*pyruvate (PEP)	PEP carboxylase
$PEP + HCO_3^- + P_i \rightleftharpoons {}^-OOC\text{–}CH_2\text{–}CO\text{–}COO^- + PP_i$	+ 2.3	Phospho*enol*pyruvate (PEP)	PEP carboxykinase (PP)
Acetyl-CoA + HCO_3^- + ATP → Malonyl-CoA + ADP + P_i		Acetyl-CoA	Acetyl-CoA carboxylase
Propionyl-CoA + HCO_3^- + ATP → Methylmalonyl-CoA + ADP + P_i		Propionyl-CoA	Propionyl-CoA carboxylase

Table 8.2 **Comparison of the autotrophic CO_2-fixation pathways**

Pathway	Amount of ATP for synthesis of 1 triose phosphate	Reductants for synthesis of 1 triose phosphate	CO_2-fixing enzymes	CO_2-fixation product	Key enzymes
Calvin-Benson-Bassham cycle (reductive pentose phosphate cycle)	9 ATP	6 NAD(P)H	Ribulose-1,5-bis-phosphate carboxylase	3-Phosphoglycerate	Ribulose-1,5-bisphosphate carboxylase, Phosphoribulokinase
Reductive citric acid cycle	5 ATP	3 NAD(P)H 1 unknown donor 2 ferredoxin	2-Oxoglutarate synthase, Isocitrate dehydrogenase Pyruvate synthase	2-Oxoglutarate, Isocitrate, Pyruvate	2-Oxoglutarate synthase, ATP citrate lyase
Reductive acetyl-CoA pathway	4–5 ATP	3-4 NAD(P)H 2–3 ferredoxin 1 H_2 (in methanogens)	Acetyl-CoA synthase/CO dehydrogenase, Formate dehydrogenase, Pyruvate synthase	Enzyme-bound CO Formate, Pyruvate	Acetyl-CoA synthase/CO dehydrogenase, Enzymes reducing CO_2 to CH_3–H_4–pterin
3-Hydroxypropionate cycle	?	?	Acetyl-CoA carboxylase Propionyl-CoA carboxylase	Malonyl-CoA, Methylmalonyl-CoA	Enzymes reducing malonyl-CoA to propionyl-CoA, Malyl-CoA lyase

are required. **Obligate autotrophs** cannot utilize organic compounds to a substantial extent, probably because of the lack of appropriate transport systems. **Facultative autotrophs** shut down autotrophic carbon fixation when an appropriate organic substrate is available, whereas **mixotrophs** continue to fix CO_2 while simultaneously incorporating the organic carbon source. Some anoxygenic **phototrophic facultative autotrophs** prefer organic substrates; however, when the organic substrate (e.g., a fatty acid) is more reduced than the average cell carbon, they save the excess reducing equivalents by fixing CO_2. In these bacteria, the Calvin cycle plays an important function in maintaining the redox balance of the cell; even alternative means to fix CO_2 may come into play.

Box 8.1 The fixation of carbon dioxide is the second most important biochemical process in quantitative terms, next to ATP synthesis. ATP synthase has a higher catalytic activity than ribulose-bisphosphate carboxylase which is inherently inefficient (low turnover number). Therefore, the carboxylase is the most abundant enzyme protein in nature. In chemo- lithotrophs, it amounts to up to 20% of the total protein; in chloroplasts, the content is even higher. In some chemolithotrophs, the carboxylase forms crystalline inclusion bodies called **carboxysomes**. Their biological function is at issue. In view of the importance of CO_2 fixation, it is very likely that early life forms depended on autocatalytic carbon assimilation mechanisms. Autotrophy, therefore, may be considered a primitive trait. This possibility has been denied by many evolutionary biologists, who claim the assimilation of abiotically preformed organic compounds. The reserves of chemically fixed organic compounds must have been adequate for the emergence of life, but probably not for its further evolution. In any case, autotrophic bacteria existed long before the event of endosymbiosis between an autotrophic cyanobacterial-like prokaryote and a eukaryotic heterotrophic cell, which gave rise to the kingdom of plants. Before this event, which may have happened several times two billion years ago, bacteria were the only primary producers.

Role, habitats, and limits of autotrophic bacteria in nature. The contribution of bacterial CO_2 fixation, especially in aquatic systems, to the overall carbon fixation can hardly be estimated. In the past, limnologists often overlooked the role of microplankton (prokaryotes) in this process. The dominant aquatic autotrophic bacteria are the **oxygenic photolithotrophic cyanobacteria**; they obtain energy from light, and reducing power is derived from light-driven water cleavage (see Chapter 13). On a global scale, the cyanobacteria are the main CO_2-fixing bacteria. The **anoxygenic phototrophs** live at the oxic/anoxic borderline where light is still available as energy source; reduced inorganic compounds such as H_2S, Fe^{2+}, or H_2 diffuse from the anoxic zone and serve as electron donor.

Chemolithoautotrophic CO_2 assimilation has a low efficiency of growth. In **chemolithoautotrophic aerobic bacteria**, more than half of the energy is consumed for reversed electron transport to generate reducing power (an exception are the Knallgas bacteria); however, they can dominate in an ecological niche such as in ore deposits of copper sulfide. **Chemolithoautotrophic anaerobic bacteria** are a minority among bacteria. They are often limited by the supply of reducing power and persist in areas where H_2 or formate are being formed by fermentative bacteria and where electron acceptors such as CO_2 or sulfate are present.

There are special environments, however, in which autotrophs are abundant. The **anoxic/oxic borderline** (of soil or stratified water bodies) is the ecological niche where the **aerobic chemolithoautotrophs** and **anoxic photolithoautotrophs** dominate. Even more spectacular are the **hot vents** in the deep sea, where sulfide-rich hydrothermal fluid emanates from the ocean floor into the ambient, oxic sea water (chemolithoautotrophs). This ecosystem is fueled by true chemoautotrophic primary production.

8.1.1 The Calvin-Benson-Bassham Cycle Is the Major Pathway of CO_2 Fixation

The key enzymes of the Calvin cycle are the CO_2-fixing enzyme, ribulose-1,5-bisphosphate carboxylase, and ribulose-5-phosphate kinase. The entire process can be divided into three sections (Fig. **8.1**):

1. **Irreversible carboxylation** of ribulose 1,5-bisphosphate, yielding two molecules of 3-phosphoglyceric acid as the primary fixation products. One of the two carboxyl groups is derived from CO_2.
2. **Reduction of the two carboxyl groups to the aldehyde level**, catalyzed by the freely reversible glucogenic enzymes, 3-phosphoglycerate kinase and glyceraldehyde-3-phosphate dehydrogenase. The bacterial dehydrogenase is generally NADH-specific, whereas the plant enzyme uses NADPH. This reduction process requires activation of the carboxylate group (by formation of a carboxylic acid–phosphoric

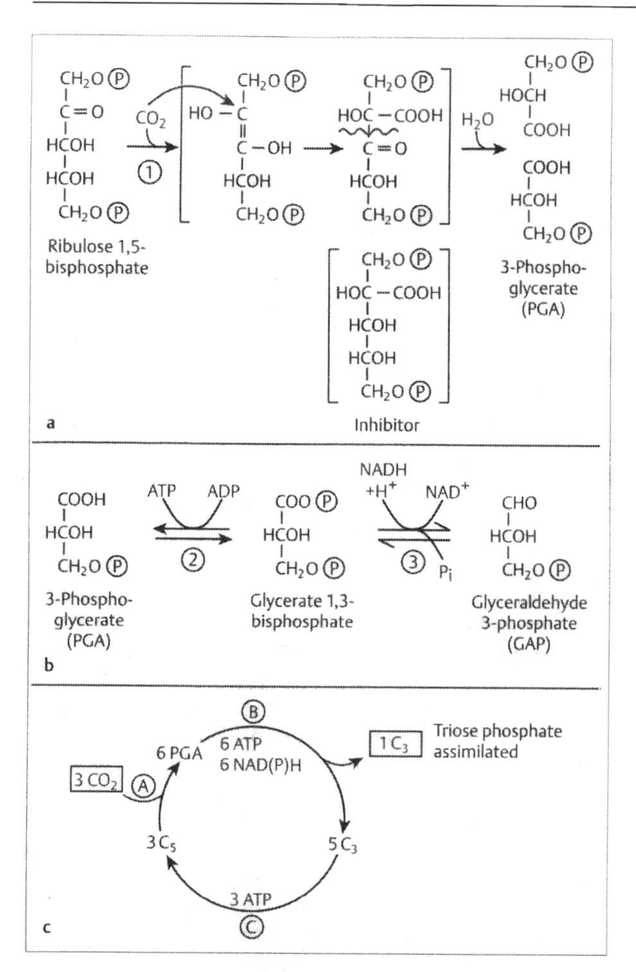

Fig. 8.1a–c Reactions of the Calvin cycle.
a Carboxylation catalyzed by (1) ribulose-1,5-bisphosphate carboxylase. The inhibitor, 2-carboxy-D-arabinitol-1,5-bisphosphate, is an analogue of the transition stae,
b Reduction phase catalyzed by (2) 3-phosphoglycerate kinase and (3) glyceraldehyde-3-phosphate dehydrogenase. These reactions are freely reversible.
c Scheme of the complete cycle, with phases A and B, and regeneration of the acceptor molecules in phase C. One triose phosphate is formed that can be assimilated into cell material

acid anhydride, 1,3-bisphosphoglycerate) that will be reduced, and consumes all reducing power and most of the energy required for the process.

3. **Regeneration of the acceptor.** Formation of fructose 6-phosphate, as in glucogenesis, is followed by unidirectional rearrangement reactions to yield pentose 5-phosphate(s). Another phosphorylation (at C1-OH) is required to regenerate the original CO_2 acceptor, ribulose 1,5-bisphosphate (Figure 8.2).

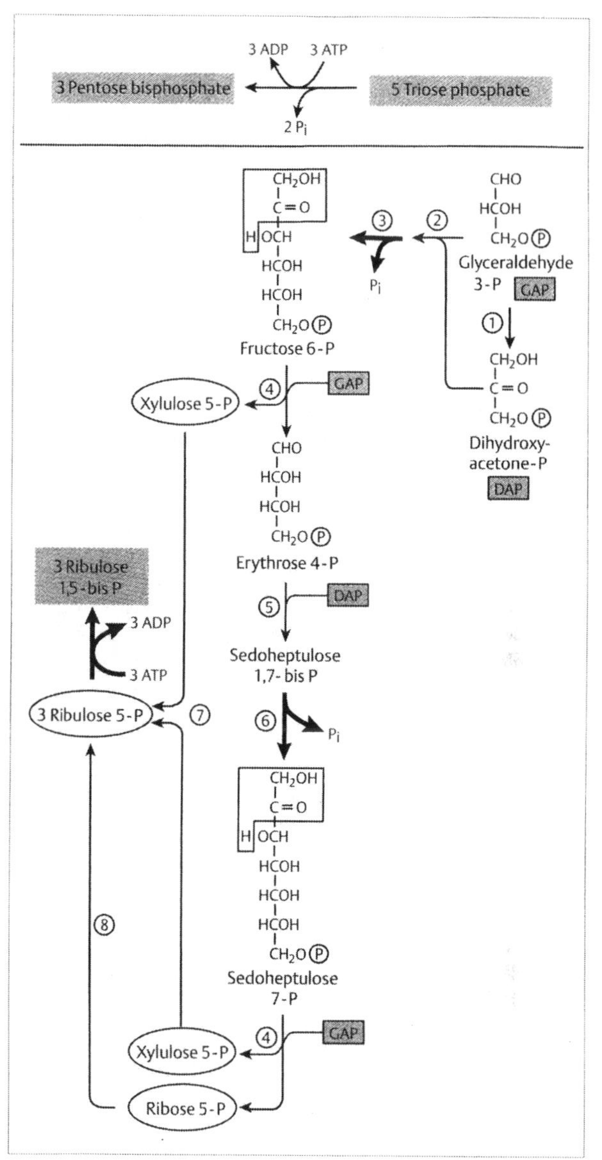

Fig. 8.2 Reactions involved in the regeneration of the CO_2 acceptor molecule ribulose 1,5-bisphosphate. The conversion of 5 C_3 to 3 C_5 is shown. All reactions are freely reversible except for those indicated by bold arrows; (1) triose-phosphate isomerase (2) fructose-bisphosphate aldolase, (3) fructose-bisphosphate phosphatase, (4) transketolase, (5) sedoheptulose-bisphosphate aldolase, (6) sedoheptulose-bisphosphate phosphatase, (7) phosphopentose-3-epimerase, (8) ribose-phosphate isomerase, and (9) phosphoribulokinase (ribulose-5-phosphate kinase)

The **carbon rearrangement** requires a few enzymes of glucogenesis and of the pentose phosphate cycle; these carbon rearrangement reactions and en-

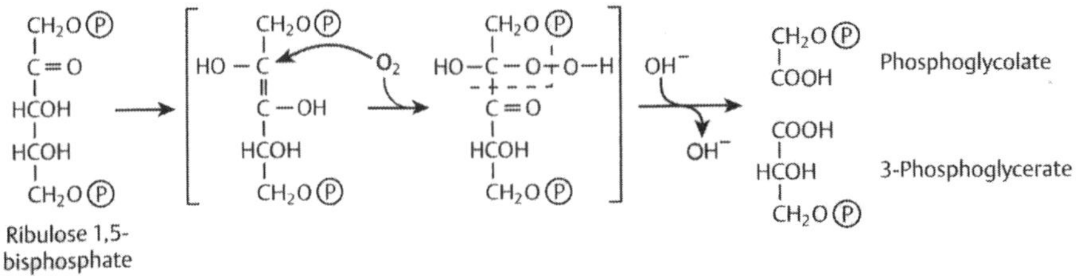

Ribulose 1,5-bisphosphate

Phosphoglycolate

3-Phosphoglycerate

Fig. 8.3 **Ribulose-1,5-bisphosphate oxygenase reaction catalysed by the carboxylase in the presence of molecular** oxygen. Note the reactivity of the enediol(ate) intermediate

zymes are essentially the same in heterotrophic bacteria (see Chapter 7). There is, however, a different way to form the heptose sugar sedoheptulose 7-phosphate (Fig. 8.**2**). In the conventional pentose phosphate cycle, a C_3 unit is transferred by transaldolase from fructose 6-phosphate to erythrose 4-phosphate yielding sedoheptulose 7-phosphate plus glyceraldehyde 3-phosphate. In the Calvin cycle, the C_3 unit is glyceraldehyde 3-phosphate, yielding the 1,7-bisphosphate of the heptose, which requires irreversible dephosphorylation at C_1 by a phosphatase. This hydrolysis of sugar–phosphate diesters renders the interconversion of sugar–phosphates unidirectionally towards pentose phosphate synthesis. The overall stoichiometry can be formulated as follows:

$$3\ CO_2 + 9\ ATP + 6\ NAD(P)H + 6\ H^+ \rightarrow 1\ \text{triose phosphate}_{\text{assimilated}} + 9\ ADP + 8\ P_i + 6\ NAD(P)^+ \quad (8.2)$$

The product of three turns of the cycle is one triose phosphate, which is ready to become assimilated (Fig. 8.**1**). This requires a net input of 3 ATP/CO_2. Two ATP are required for reduction of the carboxylic group of two molecules of 3-phosphoglycerate to the aldehyde level. Another ATP is required for C1-OH phosphorylation of the pentose 5-phosphate. Hence, only two additional enzymes are required to confer autotrophy to an organism, **ribulose-1,5-bisphosphate carboxylase** and **phosphoribulokinase**. These enzymes are called the **key enzymes of the Calvin cycle** and have been well studied. In facultative autotrophs, their activities are highest during autotrophic growth.

Ribulose-1,5-bisphosphate carboxylase. There are two forms of ribulose-1,5-bisphosphate carboxylase. The dominant form consists of eight large subunits (L, $M \sim 53$ kDa) each with a catalytic center, and eight small subunits (S, $M \sim 14$ kDa), the function of which is not known. A minor form, which occurs in *Rhodospirillum rubrum*, consists of only two large subunits, which share a high degree of similarity in functional regions with the large subunits of the other carboxylases. The original name of the enzyme, **carboxydismutase**, refers to the reduction of C2 of the acceptor pentose during carboxylation, and the oxidation of C3 (dismutation = internal redox reaction). The active carboxylase contains bound Mg^{2+} and CO_2 [note that CO_2 rather than bicarbonate (HCO_3^-) is the actual substrate]. It forms a highly reactive **enediol(ate) intermediate**, which reacts with CO_2 (carboxylase activity) and with O_2 (oxygenase activity) (see below). Loss of a proton from the 3-OH group forms the enolate anion needed for the carboxylation. The postulated reaction intermediate of the carboxylase, which undergoes a hydrolytic cleavage (Fig. 8.**1**), is mimicked by a transition state analogue, which acts as strong inhibitor.

In the absence of CO_2 and in the presence of oxygen, the enzyme catalyzes the oxygen-dependent cleavage of ribulose 1,5-bisphosphate; the products of this oxygenase reaction are **phosphoglycolate** and **phosphoglycerate** (Fig. 8.**3**). The ratio of the specificity constants (V_{max}/K_m values) for carboxylase:oxygenase activity is between 80:1 (in aerobic bacteria) and 10:1 (in anoxygenic phototrophs). The affinity for CO_2 is much higher than for O_2. Phosphoglycolate is dephosphorylated by a phosphatase, and glycolate is mostly excreted as a waste product. In plants, glycolate is oxidized at the expense of oxygen (photorespiration).

Other enzymes. The second key enzyme, **phosphoribulokinase**, consists of six or eight identical subunits, in contrast to the homodimeric plant enzyme. Most bacterial sedoheptulose-1,7-bisphosphate aldolases and sedoheptulose-1,7-bisphosphate phosphatases are two bifunctional enzymes; they each convert both fructose bisphosphates and sedoheptulose bisphosphates. Some species, however, each have two separate enzymes. Facultatively autotrophic strains contain isoenzymes of those enzymes of the cycle (e.g., aldolase and fructose-bisphosphate phosphatase) that are common to other pathways; isoenzymes are regulated differently. Those bacteria that carry autotrophic carbon-fixation genes (operons) both on the chromosome and on a plasmid, contain two similar sets of carbon-fixation enzymes.

Enzyme regulation by metabolites and control of enzyme synthesis. The key enzymes of the Calvin cycle and possibly other proteins and genes required for autotrophic carbon fixation are controlled by different mechanisms, both at the level of enzyme activity and at the level of transcription. Evidently, this is brought about by various signals and regulatory proteins. Regulation depends on whether the organisms are obligate or facultative autotrophs, oxygenic or anoxygenic phototrophs, chemolithotrophs, etc. Of primary importance are signals reflecting the energetic state and the availability of reducing power. However, also feedback control is important. For example, most bacterial carboxylases are inhibited by 6-phosphogluconate; this compound signals that the oxidative pentose phosphate cycle is operating at the same time, which inevitably results in a futile cycle.

The genes for carbon fixation and lithotrophic growth may be encoded on a (mega)plasmid and/or on the chromosome. The lithoautotrophic character often can be transferred by conjugation from the wild-type strain to mutants deficient in this trait, or even to heterotrophic bacteria from rather distantly related genera. In *Alcaligenes eutrophus*, homologous carbon-fixation operons on the chromosome and on the megaplasmid are simultaneously expressed; they form a regulon and are under the control of a common regulator protein, a transcriptional activator protein. The intracellular signal molecule that reflects the need for autotrophic CO_2 fixation, which acts as a positive or negative effector of the regulatory protein, is not yet known. This area has also been well studied in the autotrophic bacteria *Rhodobacter sphaeroides*, *Rhodospirillum rubrum*, and *Xanthobacter flavus*.

Box 8.2 The pathways by which CO_2 is incorporated into cellular compounds have been elucidated using various techniques:

1. **Short-term labeling** with $^{14}CO_2$ leads to the detection of the primary $^{14}CO_2$-fixation product. The ^{14}C-tracer technique was introduced in 1940 by Kamen and Ruben. Thick suspensions of autotrophically grown cells are supplied with $^{14}CO_2$. After various periods of time, the cells are quickly killed by pouring them into boiling or ice-cold alcohol; thereby, soluble intermediates are extracted. The soluble compounds are separated by TLC and the labeled fixation products are detected by autoradiography and identified. The kinetics of ^{14}C incorporation is indicative of whether a labeled product is an early or even the first intermediate. The first intermediate carries almost all label fixed into soluble products after a very short time. Late products derived from the primary fixation product become labeled only later. The absolute amount of ^{14}C in an individual compound after prolonged ^{14}C fixation is a measure of the pool size and of the carbon content of the compound.

2. **Long-term labeling**. Selectively ^{14}C- or ^{13}C-labeled putative intermediates in the carbon-fixation cycle are fed to growing cells; also compounds that are directly transformed into one of those intermediates can be used (e.g., acetate gives rise to acetyl-CoA). The majority of cell carbon still comes from CO_2; hence, the tracer molecule behaves like a spy who follows the autotrophic carbon flow into the major polymers of cell material. After several generations of balanced growth, the building blocks of cell polymers, such as sugars or amino acids, are isolated from the cell wall or protein, respectively. The label distribution is determined by chemical degradation, mass spectroscopy or ^{13}C-NMR spectroscopy. The biosynthetic pathways are retrospectively deduced from the labeling pattern of the individual carbon atoms in the carbon skeleton.

3. **Demonstration of key enzymes** that occur only in organisms that are able to fix CO_2 via the given pathway. Key enzymes for the Calvin cycle are ribulose-1,5-bisphosphate carboxylase and phosphoribulokinase; for the reductive citric acid cycle, ATP citrate lyase plus 2-oxoglutarate:ferredoxin oxidoreductase; for the reductive acetyl-CoA pathway, acetyl-CoA synthase/CO-dehydrogenase; and for the 3-hydroxypropionate cycle, the enzymes reducing malonyl-CoA to propionyl-CoA.

4. **Mutants** affected in the key enzymes are reduced in their capacity to grow autotrophically and indicate the role of the enzyme in the pathway.

5. **Detection of the genes** encoding the enzymes by hydridization with labeled gene probes. This does not necessarily prove that the enzymes encoded by these genes are functional. It also requires rather conserved regions in the corresponding genes; otherwise, false negative results can be obtained.

6. **Methods to study regulation** on the transcriptional or posttranscriptional level are discussed in Chapters 19–21.

8.1.2 Alternative Pathways Function in Strictly Anaerobic or Microaerophilic Bacteria and in Archaea

Two pathways, the reductive citric acid cycle (Figure 8.**4**) and the reductive acetyl-CoA pathway (Figure 8.**5**),

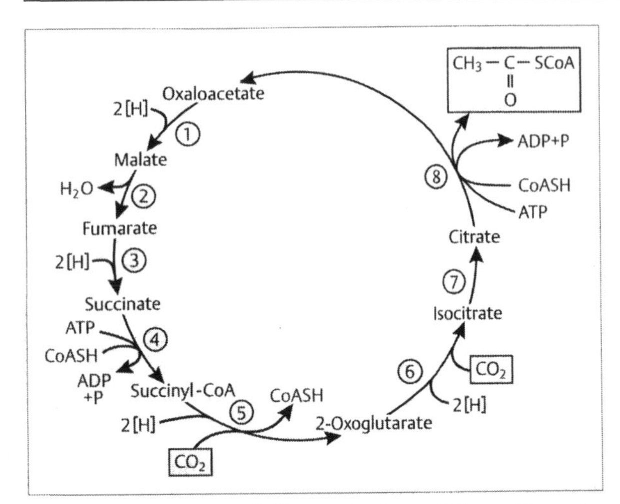

Fig. 8.4 **Reactions of the reductive citric acid cycle:** (1) malate dehydrogenase, (2) fumarate hydratase (fumarase), (3) fumarate reductase, (4) succinyl-CoA synthetase, (5) 2-oxoglutarate:ferredoxin oxidoreductase (2-oxoglutarate synthase), (6) isocitrate dehydrogenase, (7) aconitate hydratase (aconitase), and (8) ATP-citrate lyase

which have in common a synthesis of acetyl-CoA from 2 CO_2 according to Reaction 8.3, will be considered first.

$$2\ CO_2 + 8[H] + nATP + CoASH \rightarrow Acetyl\text{-}CoA + nADP + nP_i \quad (8.3)$$

Surprisingly, these pathways can be reversed under heterotrophic conditions and serve for acetyl-CoA oxidation to 2 CO_2 (see Chapter 12).

The 3-hydroxypropionate cycle will then be considered, followed by a comparison of the four different CO_2-fixation pathways.

Reductive citric acid cycle. In order to reverse the citric acid cycle (Fig. 8.**4**), three reactions which normally are irreversible, have to be catalyzed by enzymes different from those of the citric acid cycle. First, fumarate reductase reduces fumarate to succinate, whereas the reverse reaction is catalyzed by succinate dehydrogenase. Second, 2-oxoglutarate:ferredoxin oxidoreductase brings about the reversible reductive carboxylation of succinyl-CoA to 2-oxoglutarate. The conventional 2-oxoglutarate dehydrogenase catalyzes the irreversible oxidative decarboxylation of the 2-oxoacid. Third, citrate has to be cleaved in an ATP-dependent, fully reversible reaction catalyzed by ATP-citrate lyase (Reaction 8.4);

$$Citrate + CoASH + ATP \rightarrow Oxaloacetate + Acetyl\text{-}CoA + ADP + P_i \quad (8.4)$$

whereas citrate synthase forms citrate irreversibly. The synthesis of acetyl-CoA requires two ATP, one for the succinyl-CoA synthetase reaction, the other for the ATP-citrate lyase reaction.

This pathway, with several modifications, occurs in the phototrophic green sulfur eubacteria (*Chlorobium*), in some sulfate-reducing eubacteria (*Desulfobacter*), in the thermophilic microaerophilic Knallgas eubacteria (*Hydrogenobacter*), and in sulfur-dependent anaerobic archaebacteria (*Thermoproteus*).

Reductive acetyl-CoA pathway. In contrast to the previously discussed cyclic pathways, the reductive acetyl-CoA pathway (Fig. 8.**5**) is non-cyclic and brings

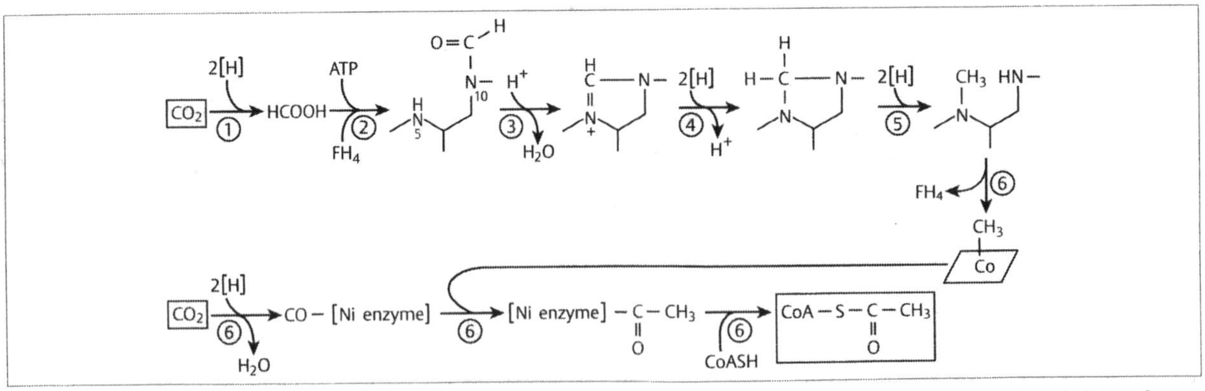

Fig. 8.5 **Reactions of the reductive acetyl-CoA pathway.** The variant that operates in acetogenic bacteria is shown; (1) formate dehydrogenase, (2) formyl-pterin-H_4 synthetase, (3) methenyl-pterin-H_4 cyclohydrolase, (4) methylene-pterin-H_4 dehydrogenase, (5) methylene-pterin-H_4 reductase, and (6) acetyl-CoA synthase complex consisting of methyl transferase, corrinoid protein, and CO-dehydrogenase/acetyl-CoA synthase. The pterin-H_4 can be tetrahydrofolate (FH_4) or other tetrahydropterins

about the synthesis of acetyl-CoA from 2 CO_2 via an enzyme-bound carbonyl (CO) group and a tetrahydropterin-bound methyl group. The final step corresponds to industrial synthesis of acetic acid from CO and CH_3OH. The two CO_2 molecules are reduced independently of each other by separate routes. One CO_2 is reduced to formate, which is then transferred to tetrahydrofolate by the ATP-requiring formyl-tetrahydrofolate synthetase. The coenzyme-bound formyl group is reduced stepwise to the methyl group; this route is common to biosynthetic C_1 transformations in non-autotrophs (Chapter 7).

The unique step is the formation of acetyl-CoA from CO_2 and methyltetrahydropterin; three components are required. A methyl transferase transfers the methyl group from methyltetrahydrofolate to a corrinoid protein, which acts as methyl carrier. The key enzyme of this pathway is the nickel enzyme acetyl-CoA synthase/CO dehydrogenase, which appears to be membrane-associated. The enzyme brings about the reduction of the other molecule of CO_2 to an enzyme-bound CO group (CO-dehydrogenase activity). This endergonic reaction is driven by the electrochemical potential of the energized membrane. Next, the enzyme accepts the methyl group from the corrinoid protein and combines it with the carbonyl group to form an enzyme-bound acetyl group, which is finally released by coenzyme A as acetyl-CoA (acetyl-CoA synthase activity). This pathway will be described in detail in Chapter 12.1.6. It functions not only in autotrophic carbon fixation, but also in total synthesis of acetate by acetogenic bacteria, and—in the reverse direction—in the cleavage of acetate into $CO_2 + CH_4$ by acetate-utilizing methanogens and in complete oxidation of acetate by various anaerobes.

The reductive acetyl-CoA pathway is the autotrophic pathway in some sulfate-reducing eubacteria (*Desulfobacterium*) and archaebacteria (*Archaeoglobus*), in most acetogenic eubacteria, and in methanogens (i.e., in strict anaerobes only). The characteristic enzymes are those that reduce CO_2 to the level of CH_3-tetrahydropterin and notably acetyl-CoA synthase/CO-dehydrogenase. Many variations of the pathway exist; formate may be a free intermediate or bound to a new coenzyme during CO_2 reduction (Table 8.1), the C_1-carrying coenzymes vary (e.g., tetrahydrofolate or tetrahydromethanopterin), and also the electron donors differ (e.g., ferredoxin, factor 420).

Assimilation of acetyl-CoA formed from 2 CO_2. Both the reductive citric acid cycle and the reductive acetyl-CoA pathway require additional enzymes for assimilation of the CO_2-fixation product acetyl-CoA. Acetyl-CoA is reductively and reversibly carboxylated to pyruvate by pyruvate:ferredoxin oxidoreductase (Reaction 8.5):

$$\text{Acetyl-CoA} + CO_2 + \text{Ferredoxin}_{red} + 2\,H^+ \rightarrow \text{Pyruvate} + \text{CoASH} + \text{Ferredoxin}_{ox} \quad (8.5)$$

Pyruvate assimilation proceeds via different mechanisms (see anaplerotic reactions, Chapter 7).

Box 8.3 This chapter mainly deals with the formation of C–C bonds. There are two prominent enzyme-catalyzed reactions that lead to the formation or cleavage of C–C bonds (Fig. 8.7), the aldol reaction and the Claisen condensation. The **aldol reaction**—when considered in the direction of C–C bond formation—involves the condensation of the α carbon of an aldehyde or ketone to the carbonyl carbon of another aldehyde or ketone. The hydrogen that is α to the carbonyl group in one molecule is abstracted in base to form the nucleophile enolate. The carbanion, which is stabilized, attacks the electron-deficient carbonyl carbon in the second molecule of aldehyde (or ketone); the new C–C bond is produced. The condensation product from two molecules of acetaldehyde is called aldol, which provides the generic name for the reaction. Aldol reactions are fully reversible. The **Claisen condensation** can involve condensation between two ester molecules, where one becomes acylated at the α-carbon, or it can involve acylation of the ester component by an aldehyde or ketone molecule. Again, the reactive species is the enolate of the ester, which attacks the electrophilic carbonyl carbon of its condensation partner. In biological systems, the nucleophilic component is the α-anion of acylthioesters of coenzyme A, rather than acyl oxygen esters. As in the aldol condensation, the first step is the abstraction of acidic α-hydrogen to yield the stabilized carbanion. The nucleophilic partner then condenses with another ester or a ketone as the electrophilic component. Claisen condensations normally are unidirectional towards C–C bond formation. Examples for aldol reactions are aldolases or transaldolases; for Claisen condensations, malate synthase, citrate synthase, ATP-citrate lyase, citrate lyase, isocitrate lyase, and thiolase. (For further information, see Further Reading: C. Walsh, Enzymatic reaction mechanisms.)

3-Hydroxypropionate cycle. The 3-hydroxypropionate cycle (Figure 8.6) is another cyclic pathway that

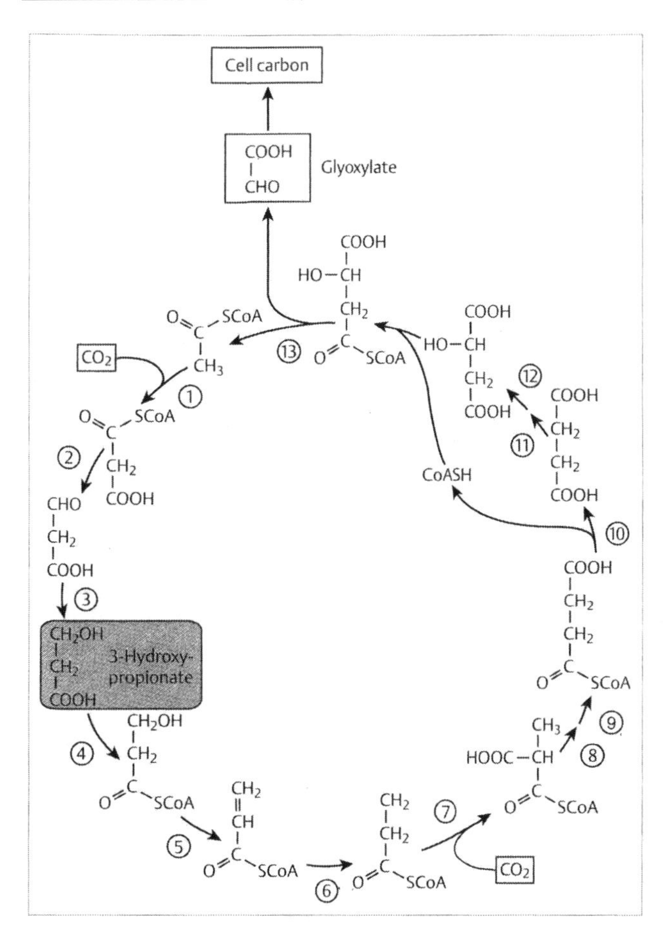

Fig. 8.6 **Reactions of the proposed 3-hydroxypropionate cycle:** (1) acetyl-CoA carboxylase, (2) malonate-semialdehyde dehydrogenase, (3) 3-hydroxypropionate dehydrogenase, (4) 3-hydroxypropionate-CoA ligase, (5) acryloyl-CoA hydratase, (6) acryloyl-CoA reductase, (7) propionyl-CoA carboxylase, (8) methylmalonyl-CoA epimerase, (9) methylmalonyl-CoA mutase, (10) succinyl-CoA:malate CoA-transferase, (11) succinate dehydrogenase, (12) fumarate hydratase, and (13) malyl-CoA lyase

brings about the synthesis of glyoxylate from two bicarbonate, according to Reaction 8.6:

$$2\ HCO_3^- + 2\ NADPH + 3\ H^+ + 3\ ATP \rightarrow CHO{-}COO^- + 2\ NADP^+ + 3\ ADP + 3\ P_i \qquad (8.6)$$

Glyoxylate is the precursor of cell carbon. The pathway occurs in the green phototrophic non-sulfur bacterium *Chloroflexus aurantiacus* and possibly in aerobic archaebacteria.

In essence, acetyl-CoA is carboxylated to malonyl-CoA by the biotin-dependent, ATP-dependent acetyl-CoA carboxylase, which is known from fatty acid synthesis. The CoA-activated carboxyl group is prepared to become reduced in two steps to the C3-OH group, 3-hydroxypropionate. This intermediate is activated to the CoA thioester, which consumes another ATP. From 3-hydroxypropionyl-CoA, water is eliminated, yielding acryloyl-CoA, which is reduced to propionyl-CoA. Propionyl-CoA acts as a second CO_2 acceptor and is carboxylated by propionyl-CoA carboxylase to methylmalonyl-CoA. This is converted via succinyl-CoA to malyl-CoA, which is cleaved into acetyl-CoA, the primary CO_2 acceptor, and glyoxylate, the assimilation product. The latter reaction is catalyzed by malyl-CoA lyase (Reaction 8.7):

$$\text{Malyl-CoA} \rightarrow \text{Acetyl-CoA} + \text{Glyoxylate} \qquad (8.7)$$

The pathway by which glyoxylate is assimilated into one of the central precursor metabolites remains to be elucidated.

Comparison of the autotrophic pathways. The four different CO_2-fixation pathways are compared in Table 8.**2**. The most important difference is the estimated amount of ATP required to form, for example, one molecule of triose phosphate from 3 CO_2. Another trait is the use of different reductants. The anaerobic pathways use reduced ferredoxin as electron donor, in addition to NAD(P)H, and the CO_2-fixation reactions are mostly reductive carboxylations. The most remarkable trait of the reductive citric acid cycle and the reductive acetyl-CoA pathway is the reversibility of the overall process; these pathways may serve two functions in one organism, depending on the substrate supply: autotrophic CO_2 fixation as well as end oxidation of acetyl-CoA.

8.2 Generation of Precursor Metabolites From C_2 Compounds as Carbon Source Requires Specific Reactions

When organisms grow with acetate as sole source of cell carbon, they are confronted with the problem of generating all biosynthetic precursor metabolites from acetyl-CoA alone. This also applies to growth with substrates like fatty acids, or when polyhydroxybutyrate storage material is mobilized; their degradation yields acetyl-CoA. Pyruvate and phospho*enol*pyruvate cannot be made directly from acetyl-CoA by reversal of the irreversible pyruvate dehydrogenase and pyruvate kinase reactions. Therefore, there must be a different solution. The enzymes of the citric acid cycle are supplemented by two enzymes that form the glyoxylate cycle.

Glyoxylate cycle. The key enzymes of the glyoxylate cycle (Fig. 8.**8**) are isocitrate lyase, which catalyzes Reaction 8.8:

$$\text{Isocitrate} \rightarrow \text{Glyoxylate} + \text{Succinate} \quad (8.8)$$

and malate synthase, which catalyzes Reaction 8.9:

$$\text{Glyoxylate} + \text{Acetyl-CoA} + H_2O \rightarrow \text{Malate} + \text{CoA-SH} \quad (8.9)$$

The malate synthase reaction is analogous to the citrate synthase reaction: both are Claisen condensations. The purpose of the cycle is to generate from two acetyl-CoA molecules (C_2) one additional molecule of oxaloacetate (C_4) to serve as the precursor for pyruvate and phospho*enol*pyruvate (PEP) synthesis. From PEP, all biosynthetic routes can be reached.

Acetyl-CoA assimilation by strict anaerobes. Strict anaerobes do not use the glyoxylate cycle for acetyl-CoA assimilation when they contain pyruvate:ferredoxin oxidoreductase. This reversible reaction can be reversed for the reductive carboxylation of acetyl-CoA to pyruvate (Reaction 8.5). Starting from pyruvate, there are several options for making oxaloacetate and PEP (see anaplerotic reactions, Chapter 7). Examples of anaerobic assimilation of acetyl-CoA are the alternative autotrophic CO_2-fixation pathways, which form acetyl-CoA from 2 CO_2 (reductive citric acid cycle, reductive acetyl-CoA pathway).

Assimilation of other C_2 compounds. When small molecules such as glyoxylate, glycolate, or oxalate are the only growth substrate, non-standard assimilatory pathways are required to form acetyl-CoA, pyruvate, and PEP. Some of these reactions will be discussed in Chapter 9.

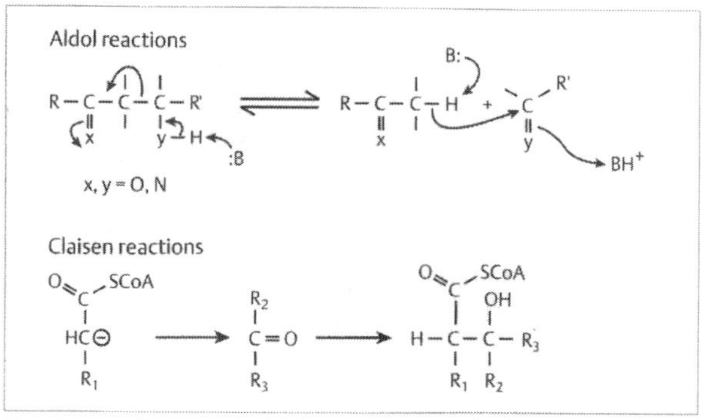

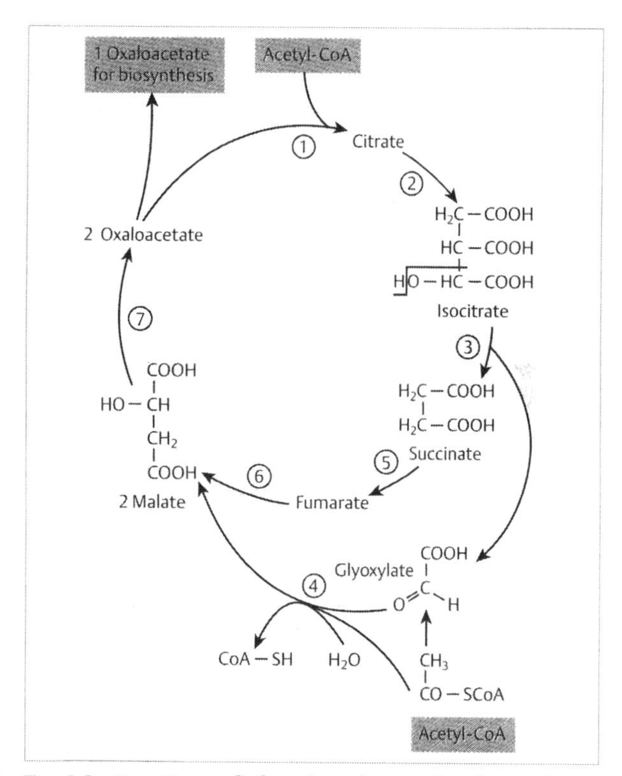

Fig. 8.8 **Reactions of the glyoxylate cycle of acetyl-CoA assimilation.** As a result of the glyoxylate cycle, two molecules of acetyl-CoA make one molecule of oxaloacetate, which serves as precursor for all central biosynthetic precursor metabolites (except for acetyl-CoA itself). Acetyl-CoA is oxidized via the citric acid cycle to generate energy; (1) citrate synthase, (2) aconitase, (3) isocitrate lyase, (4) malate synthase, (5) succinate dehydrogenase, (6) fumarate hydratase, and (7) malate dehydrogenase

Aldol reactions

x, y = O, N

Claisen reactions

Fig. 8.7 **Scheme of aldol reactions and Claisen condensations**

8.3 Methylotrophic Bacteria Use C_1 Compounds as the Only Carbon Source

Various bacteria can use CO, formate, formaldehyde, methanol, methylamine, methylmercaptane, or methane as sole source of cell carbon. Others are able to use one-carbon units linked via a hetero-atom (N, S, O) to the rest of an organic molecule; examples for this type of substrate are the very common aryl-*O*-methyl ether compounds in plants. Aerobic bacteria cleave methyl-ether compounds by monooxygenases, which transform the $R{-}O{-}CH_3$ group into an $R{-}O{-}CH_2OH$ group. This product decomposes to the corresponding alcohol (R–OH) and formaldehyde (CH_2O). Anaerobes catalyze a completely different reaction; by *O*-demethylation, the alcohol ($R{-}O^-$) is released, and the methyl group (CH_3^+) is directly transferred to a protein and from there to tetrahydrofolate. How do these different bacteria form one of the central precursor metabolites (Chapter 7) from one-carbon compounds?

8.3.1 Aerobic Bacteria Have Two Possible Ways to Assimilate C_1 Substrates

Aerobic bacteria can assimilate C_1 substrates by oxidizing the C_1 substrate to CO_2 and refixing CO_2 via the Calvin cycle. This possibility exists in non-specialist C_1 oxidizers; however, it wastes energy since the endergonic reduction of CO_2 to the level of cell carbon

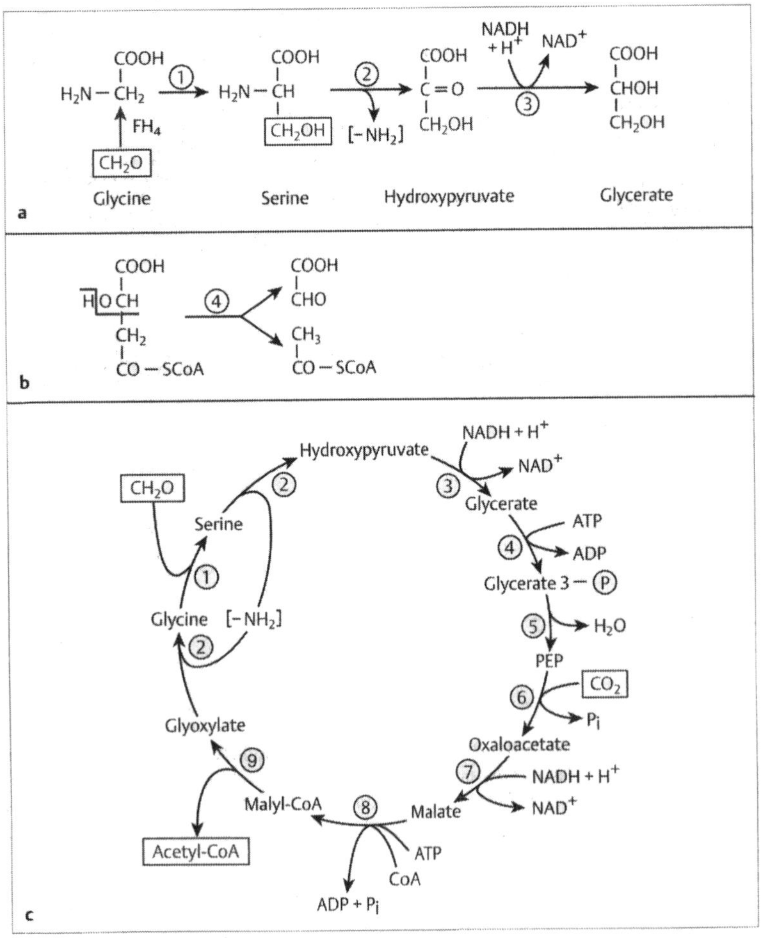

Fig. 8.9a–c **Serine pathway of formaldehyde fixation.**
a and **b** Key reactions: (1) serine hydroxymethylase, (2) transaminase, (3) hydroxypyruvate reductase and (4) malyl-CoA lyase.
c The complete cycle with the regeneration of glycine: (1) serine hydroxymethylase, (2) a transaminase that converts serine into hydroxypyruvate and glyoxylate into glycine, (3) hydroxypyruvate reductase, (4) glycerate kinase, (5) phosphoglycerate mutase and enolase, (6) PEP carboxylase, (7) malate dehydrogenase, (8) malyl-CoA synthetase, and (9) malyl-CoA lyase

(=formaldehyde) consumes energy. Aerobes can also assimilate C_1 substrates by directly assimilating formaldehyde, an intermediate at the oxidation state of carbohydrates, which is formed in the oxidation of methyl compounds. Three formaldehyde fixation cycles have been found, two of which occur in methylotrophic bacteria. They are named after the first product: serine pathway, hexulose-phosphate pathway (ribulose-monophosphate cycle), and dihydroxyacetone pathway (the latter occurs only in methylotrophic yeasts).

Serine pathway. As indicated in Chapter 7, the main route of generation of C_1 compounds for biosyntheses proceeds via conversion of serine to glycine. This reversible reaction is catalyzed by serine hydroxymethylase and is linked to the formation of N^5, N^{10}-methylene tetrahydrofolate (FH_4) from C3 of serine and tetrahydrofolate. The C3 group of serine is equivalent to formaldehyde, and methylene-FH_4 arises spontaneously when tetrahydrofolate and formaldehyde are mixed. The serine pathway (Fig. 8.**9**) uses the reverse reaction, the formation of serine from glycine, to fix formaldehyde. In order to regenerate the C_1 acceptor molecule glycine, a cyclic process operates in which serine is first converted to glycerate and then to phospho*enol*pyruvate (PEP). Carboxylation of PEP to oxaloacetate fixes another carbon atom from CO_2. Malyl-CoA is formed with malate in an ATP-consuming reaction catalyzed by malyl-CoA synthetase. Malyl-CoA is cleaved by malyl-CoA lyase into glyoxylate and the assimilation product acetyl-CoA. Note that acetyl-CoA contains carbon atoms from one CO_2 and one formaldehyde. Glyoxylate is transaminated with serine as $-NH_2$ donor to regenerate the formaldehyde acceptor glycine. Hence, four additional key enzymes are required for this cycle to function: transaminase, hydroxypyruvate reductase malyl-CoA synthetase, and malyl-CoA lyase.

Further assimilation of acetyl-CoA formed in the cycle requires the operation of the glyoxylate cycle characterized by the enzymes isocitrate lyase and malate synthase. This pathway has been discussed in Chapter 7. The entire C_1-assimilation cycle is shown in Figure 8.**9**. This pathway costs more ATP to form one molecule of triose phosphate than the ribulose monophosphate cycle (see below) because CO_2 is also fixed.

Hexulose-phosphate pathway. In this cycle (Figure 8.**10**), ribulose 5-phosphate acts as a formaldehyde acceptor, and formaldehyde is transferred to C1 by hexulose-6-phosphate synthase. The product of CH_2O addition, a hexulose 6-phosphate, is isomerized to fructose 6-phosphate. The acceptor, ribulose 5-phosphate, is regenerated unidirectionally as described for the Calvin cycle (see Fig. 8.**2**). The only difference is that part of fructose 6-phosphate is phosphorylated to the

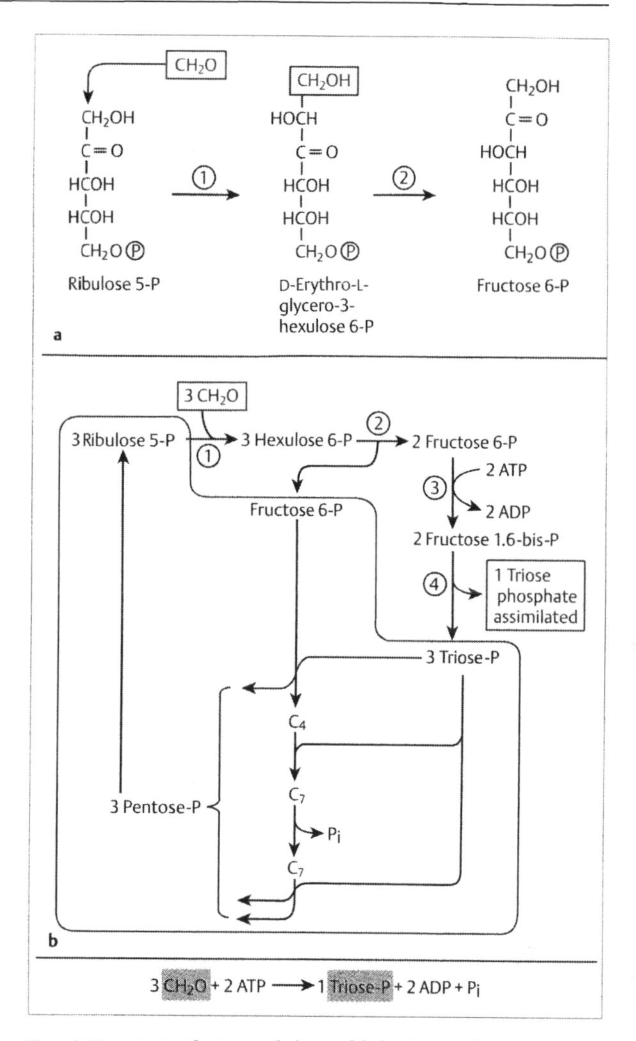

Fig. 8.**10** **Assimilation of formaldehyde in the hexulose-phosphate pathway.**
a Initial reactions: (1) hexulose-phosphate synthase, and (2) hexulose-phosphate isomerase.
b Regeneration of the C_1 acceptor. (1) and (2) are as in **a**, (3) phosphofructokinase, and (4) fructose-1,6-bisphosphate aldolase. The reactions in the box bring about the formation of 3 pentose phosphate from fructose 6-phosphate and 3 triose phosphate. This part of the pathway is virtually identical to the corresponding part in the Calvin cycle, as shown in Fig. 8.**2**

1,6-bisphosphate, and aldolase cleavage forms triose phosphates. The overall process is shown in Figure 8.**10**, with a simplified version of the regeneration of ribulose 5-phosphate from fructose 6-phosphate. In Table 8.**3** the main characteristic features of the C_1-assimilation pathways are summarized.

Variations in formaldehyde-fixation pathways. In the serine pathway, the assimilation of acetyl-CoA may

Table 8.3 **Comparison of the C_1-assimilation pathways**

Pathway	C_1-fixing enzyme	CO_2-fixing enzyme	C-assimilation products
Serine pathway	Serine-hydroxymethylase	PEP carboxylase	Serine, oxaloacetate
Hexulose-phosphate pathway	Hexulose-phosphate synthase	–	Hexulose phosphate
Acetyl-CoA pathway (anaerobes only)	Formyltetrahydrofolate synthetase, spontaneous reaction of formaldehyde with FH_4	CO dehydrogenase	Methylene tetrahydropterin, Enzyme-bound CO

not proceed via the glyoxylate cycle. Several bacteria lack isocitrate lyase. Similarly, malyl-CoA synthetase may be undetectable. The explanation for how these bacteria assimilate or form acetyl-CoA remains to be elucidated. In the ribulose-monophosphate cycle, the enzymes of the Entner-Doudoroff pathway may be active, instead of the enzymes of the Embden-Meyerhof pathway. In this case, pyruvate is the precursor of cell biomass, and KDPG aldolase and transaldolase are involved, instead of fructose-1,6-bisphosphate aldolase and sedoheptulose bisphosphatase.

8.3.2 Assimilation of C_1 Substrates by Anaerobic Bacteria Follows Another Pathway

In principle, the aerobic formaldehyde-assimilation pathways could also function in anaerobes. Most methylotrophic anaerobes, however, use a quite different pathway. The reductive-acetyl-CoA pathway of autotrophic CO_2 fixation can be used also for assimilation of various C_1 compounds (Fig. 8.**11**). Note that in this pathway coenzyme-bound or enzyme-bound C_1 units are formed from CO_2. These intermediates can also be formed from external C_1 compounds. Only a few additional peripheral enzymes are required. For instance, specific methyl transferases channel the methyl group from methanol or from aryl-*O*-methyl ether compounds into the pathway via methyltetrahydrofolate. The product of the *O*-demethylation reaction is the corresponding alcoholic (e.g., phenolic) compound, that is, the oxygen atom remains with the organic molecule (e.g., aromatic ring).

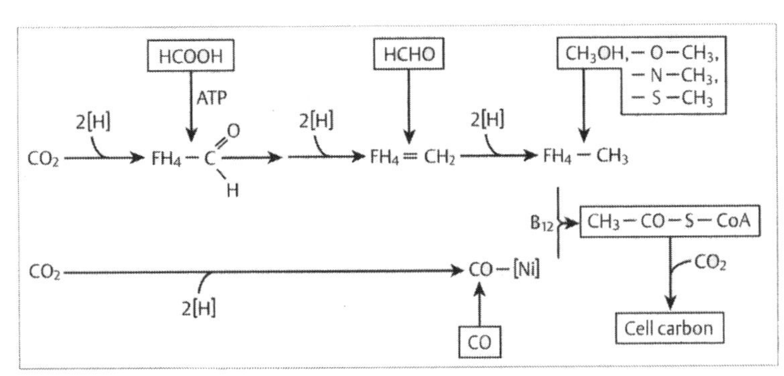

Fig. 8.**11** **Anaerobic assimilation of C_1 compounds via the reductive acetyl-CoA pathway.** For an explanation, see Fig. 8.**5** and the text

8.4 Ammonia Can Be Obtained From Various Nitrogen Sources

Ammonia is released by various mechanisms from organic compounds containing nitrogen (see Chapters 9 and 12). Common organic nitrogen sources are the NH_3 detoxification products of animals, urea and uric acid, and allantoin in plants. Release of NH_3 from urea (which can be considered the simplest organic compound) is catalyzed by the nickel enzyme urease according to Reaction 8.10:

$$H_2N{-}CO{-}NH_2 + H_2O \rightarrow 2\ NH_3 + 1\ CO_2 \qquad (8.10)$$

The degradation of uric acid will be discussed in Chapter 9.

Assimilatory nitrate reduction. Nitrate is the most common form of dissolved nitrogen, and many bacteria can reduce nitrate to NH_3. Nitrate and nitrite (which occurs in low concentrations only) are anions that need to be transported. Nitrate reduction proceeds in two steps; both reactions are strongly exergonic (Reactions 8.11 and 8.12):

$$HNO_3 + 2\,[H] \rightarrow HNO_2 + H_2O \qquad (8.11)$$

$$HNO_2 + 6\,[H] \rightarrow NH_3 + 2\,H_2O \qquad (8.12)$$

Nitrate reduction (Reaction 8.11) by two electrons to form nitrite is catalyzed by the molybdo-flavo-iron-sulfur enzyme nitrate reductase. Mo is present in the **Molybdo-cofactor** (see Chapter 12). The enzyme also reduces the analogous chlorate (ClO_3^-) to toxic chlorite (ClO_2^-); mutants devoid of this enzyme activity are resistant to chlorate. **Nitrite reduction** (Reaction 8.12) by six electrons to form ammonia is catalyzed by nitrite reductase. The most common type of nitrite reductase contains iron-sulfur centers and the iron porphyrin **siroheme** (Chapter 12.2.2); siroheme is also the prosthetic group of sulfite reductase, which catalyzes a similar reaction, the six-electron reduction of HSO_3^- to HS^-. **Assimilatory** nitrate and nitrite reductase are soluble enzymes, their synthesis is repressed by ammonia and induced by nitrate. This is in contrast to the **dissimilatory** systems, which are membrane-associated or periplasmic, and their synthesis is repressed by aerobiosis (Chapter 12). Electron donors for both assimilatory enzymes can be NAD(P)H or ferredoxin, depending on the species. Growth of *Escherichia coli* and some other bacteria with NO_3^- as N source is accomplished through the dissimilatory nitrate and nitrite reductases. These bacteria, as expected, cannot utilize nitrate during aerobic growth.

8.5 Reduction of N_2 to NH_3 Is Catalyzed by Nitrogenase

In many habitats, none of the nitrogen sources mentioned is available. These habitats can only be settled by bacteria that are able to fix atmospheric dinitrogen into cell material by reducing it to ammonia (**diazotrophy**). Eukaryotes are unable to fix nitrogen. Reduction of N_2 to 2 NH_3 is widespread in eubacteria and archaebacteria and is catalyzed by the enzyme **nitrogenase**. This reaction requires the hydrolysis of two ATP per electron; for every N_2 reduced (6 e^-) two protons in addition are reduced to H_2 (2 e^-). The overall stoichiometry is shown in Reaction 8.13:

$$N_2 + 8\,H^+ + 8\,e^- + 16\,ATP + 16H_2O \rightarrow 2\,NH_3 + H_2 + 16\,ADP + 16\,P_i \qquad (8.13)$$

Energy requirement and supply of reduced ferredoxin. N_2 reduction to 2 NH_3 with 3 H_2 as reductant is exergonic by 91 kJ/mol. Nevertheless N_2 reduction requires much ATP and reducing power, and therefore nitrogenase synthesis in free-living bacteria is repressed by NH_3. ATP is required to overcome the high activation energy required to reduce the extremely stabile $N{\equiv}N$ triple bond to the double bond at ambient temperatures. The industrial **Haber-Bosch process** must be carried out at high temperatures, and high pressures of hydrogen and nitrogen must be applied to overcome this energy barrier; efficient metal catalysts must also be used. The amount of ATP and reducing power required in biological nitrogen fixation is so high that growth yields and growth rates under N_2-fixing conditions are considerably lower than those during growth in the presence of NH_3. Note that 14% of cell dry matter is N and that 8 H required to reduce N_2 are equivalent to the oxidation of 1 acetate to 2 CO_2; almost one additional acetate has to be sacrificed to produce the required amount of ATP.

Nitrogen fixation depends on reduced **ferredoxin** as electron donor. **Flavodoxin**, a small electron-carrying flavoprotein, can substitute for ferredoxin when the iron supply is limiting. It acts, in place of ferredoxin, as the electron acceptor of various ferredoxin-reducing enzymes and as the electron donor in nitrogenase. The ferredoxin(s), which donates electron(s) to the nitrogenase, may be reduced in various ways (Fig. 8.**12**).

Oxygen sensitivity and control by oxygen. Dinitrogen reduction requires a strongly reducing active center, which is intrinsically easily destroyed by oxidation with O_2. Therefore, all nitrogenases are oxygen-sensitive, that is, they are inactivated in air, and their synthesis is therefore repressed by aerobiosis. This important constraint makes nitrogen fixation a strictly anaerobic process. Oxygen sensitivity is no problem for strict anaerobes, whereas it is a challenge for aerobes and facultative anaerobes. Those bacteria that can reduce N_2 at low levels of oxygen have developed mechanisms to avoid or get rid of oxygen:

1. Some aerobic soil bacteria, such as *Azotobacter* species, consume oxygen rapidly by respiration; this phenomenon is called **respiratory protection**. These bacteria have a very active branched respiratory chain (see Chapter 11.3.2). When fixing nitrogen,

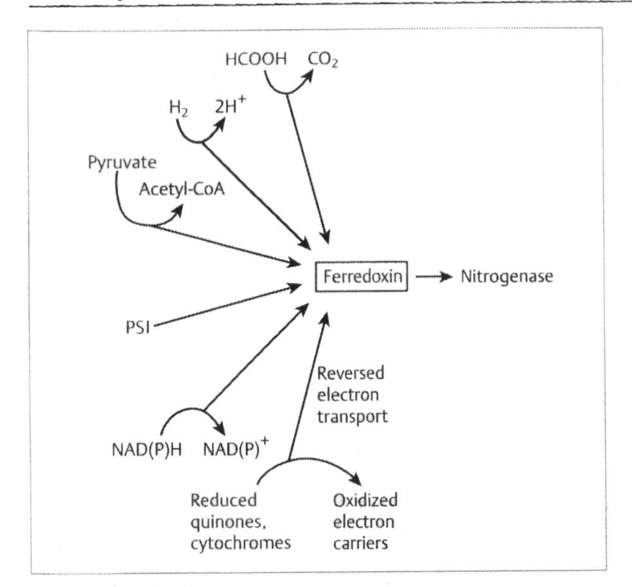

Fig. 8.**12** **Reactions by which ferredoxin is reduced to serve as reductant for the nitrogenase reaction.** This scheme applies also to the alternative electron carrier flavodoxin. In the oxygenic phototrophic cyanobacteria, ferredoxin is reduced by photosystem I in the light or by pyruvate:ferredoxin oxidoreductase in the dark. In anaerobic chemotrophs, it is reduced by ferredoxin-dependent oxidoreductases, such as pyruvate: ferredoxin oxidoreductase, hydrogenase, and formate dehydrogenase. In aerobic chemotrophs or in anoxygenic phototrophs, ferredoxin can be reduced by NAD(P)H through ferredoxin:$NAD(P)^+$ oxidoreductases. In aerobic chemolithotrophs, an energy-driven reverse electron transport from NAD(P)H or more positive reduced components of the electron transport chain to ferredoxin is mediated by membrane proteins . ATP is supplied either by respiration, photosynthesis, or fermentation

they use those branches that are coupled to only one phosphorylation site, but efficiently reduce O_2 to H_2O. [The cytochrome *d* terminal oxidase in *Azotobacter* induced in response to high O_2 concentration has a low affinity for O_2, but a high V_{max}. Note that cytochrome *d* normally has a high affinity and is synthesized in response to O_2 limitation (Chapter 10)]. This is a waste of NADH, but does protect nitrogenase against oxygen damage. Under other conditions, they use the other branch which—like in mitochondria—is coupled to three phosphorylation sites.

2. Other bacteria, such as facultative phototrophs, protect nitrogenase at low oxygen levels by a protecting protein; this phenomenon, which is still not understood well, is called **conformational protection**. Binding to the protecting protein is thought to change the conformation of the nitrogenase, thereby rendering the enzyme inactive, but oxygen-stable. This process can be reversed ("on-off switch").
3. Some bacteria avoid oxygen by **morphological and behavioral adaptation. Heterocyst formation** in cyanobacteria is a protection mechanism (see Chapters 24.3 and 34.2). **Leghemoglobin** in symbiotic nitrogen-fixing bacteria provides maintenance of a low oxygen partial pressure. Also, **slime production** may minimize convection and thus O_2 access. Furthermore, **negative aerotaxis** away from O_2, aggregation, or association with aerobic heterotrophs may contribute to the avoidance of oxygen damage in motile bacteria.

Proton reduction to H_2 is mechanistically connected to N_2 reduction; therefore, virtually all N_2-fixing bacteria possess a membrane-bound hydrogenase, whereas symbiotic Rhizobia do not contain hydrogenase. This nickel enzyme serves to scavenge H_2 and to reduce ferredoxin, possibly explaining why, in general, nitrogen fixation not only depends on the trace element molybdenum (or vanadium), but also on nickel. However, H_2-dependent respiration is so slow that it certainly cannot protect nitrogenase from oxygen damage.

Distribution of nitrogen fixation among prokaryotes. Dinitrogen fixation is restricted to prokaryotes. All eukaryotes depend on this capability, either indirectly or directly by endosymbiosis with nitrogen-fixing eubacteria. On a global scale, dinitrogen reduction is essential for maintenance of the N cycle; remember that N_2 is lost by bacterial anaerobic nitrate respiration. The ability to fix nitrogen is spread over the entire system of prokaryotes; it is found in all major taxonomic units, in anaerobic, aerobic, and phototrophic bacteria. It is believed that frequently the genes for nitrogen fixation have been acquired by lateral gene transfer from one species or genus to another. This possibility has experimentally been verified by conjugational transfer of the **ni**trogen **f**ixation (*nif*) genes from *Klebsiella* to *Escherichia*.

Nitrogenases. Three types of nitrogenases are known. The most common enzyme contains **molybdenum** complexed by homocitrate and an FeS-center (Fig. 8.**13**). Alternatively, a **vanadium**-dependent and an **iron**-dependent enzyme are also found. A bacterium may be able to form two or all three enzymes (e.g., *Azotobacter* sp.), depending on the availability of molybdenum or vanadium. All nitrogenases are composed of two different independent proteins. **Component 1**, sometimes referred to as **MoFe-protein** (or **dinitrogenase** or molybdoferredoxin), is the actual dinitrogenase that reacts with N_2. **Component 2** is

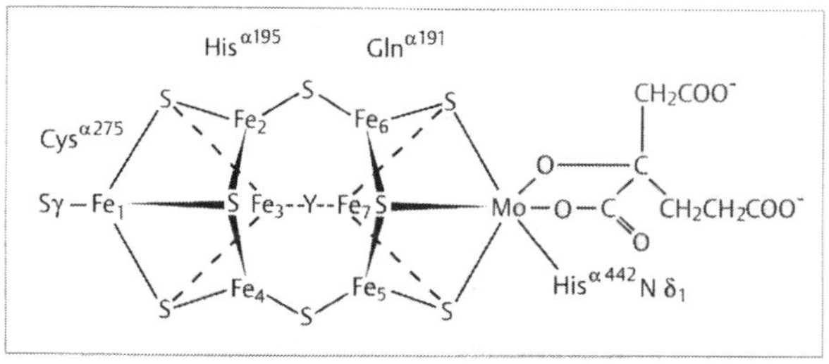

Fig. 8.13 **Structure of the reaction center of Mo nitrogenase.** The Fe-Mo-cofactor contains seven Fe(II) and one Mo, which are partly complexed with sulfur, partly with homocitrate. The ligand Y is possibly S [from 1]

dinitrogenase reductase, sometimes referred to as **Fe-protein** (or azoferredoxin). It is an iron-sulfur protein that accepts electrons from reduced ferredoxin or flavodoxin and transfers them in an ATP-dependent reaction to component 1.

Box 8.4 Nitrogenase catalyzes the ATP-dependent reduction of other compounds, which is probably of little significance in nature:

$$HC{\equiv}CH + 2\,H \rightarrow H_2C{=}CH_2 \quad (8.14)$$

$$HN_3 + 2\,H \rightarrow NH_3 + N_2 \quad (8.15)$$

$$HCN + 6\,H \rightarrow NH_3 + CH_4 \quad (8.16)$$

$$N_2O + 2\,H \rightarrow N_2 + H_2O \quad (8.17)$$

Acetylene reduction to ethylene is commonly used as an indirect measure of nitrogenase activity. The product ethylene can be detected by gas chromatography with high sensitivity. This convenient assay considerably promoted research on nitrogen fixation. The in vitro assay requires strictly anoxic conditions ($O_2 < 1$ ppm). The alternative enzymes reduce ethylene to ethane, which allows the differentiation between these forms even in living cells. Also, incorporation of ^{15}N from $^{15}N_2$ into cell material can be followed by mass spectroscopy. When N_2-dependent growth is tested in aerobic cultures, vials must not be shaken, in order to maintain microoxic conditions. Growth then occurs in a thin layer at the surface or a few millimeters below. Gene probes against conserved regions of nitrogenase genes may also be used to test for diazotrophs. A disadvantage of this method is that a negative result may arise due to an unsuitable gene probe; a positive response needs to be confirmed by other criteria. An advantage of this method is that nitrogenase genes can be detected even if they are silent under many growth conditions.

Structure of nitrogenase, role of ATP, and possible reaction mechanism. The average composition of dinitrogenase (MoFe-protein) is $\alpha_2\beta_2$ (M approx. 240 kDa), that of the reductase (Fe-protein) is α_2 (approx. 2 × 60 kDa). The **MoFe-protein** component 1 contains 30 Fe, 34 inorganic S, and 2 Mo atoms complexed in a unique way in two pairs of metal centers: two identical FeMo-cofactor units (paramagnetic M-centers) each contain 1 Mo:7 Fe:9 S and one homocitrate molecule (Fig. 8.**13**). The diamagnetic P cluster pairs contain two sulfur-bridged 4 Fe:4 S clusters. The **Fe-protein** component 2 is a dimer of identical subunits bridged by a single 4 Fe:4 S cluster, which undergoes a one-electron redox cycle. It contains two nucleotide binding sites per dimer and is intimately involved in the coupling of ATP hydrolysis to the electron transfer to component 1.

The reactions are described by the following model (Fig. 8.**14**). The Fe-protein must dock to the MoFe-protein to donate one electron at a time to the P-cluster pair, which is already highly reduced. The complex formation involves switching between alternate conformational states of the two components, driven by nucleotide binding and hydrolysis at the interface between the different subunits. Electrons are then transferred from the P cluster to the FeMo cluster, where all substrates are reduced. Reduction of dinitrogen can be considered to proceed in three two-electron steps via the enzyme-bound intermediates HN=NH (diimide) and $H_2N{-}NH_2$ (hydrazine). The **major barrier** should be the **two-electron reduction of dinitrogen to diimide**. However, none of these intermediates can be detected, and NH_3 is the only product formed.

Regulation. Nitrogenase synthesis and nitrogen metabolism are regulated either by a hierarchic cascade of transcriptional activators or by parallel regulatory networks (see Chapters 18–20). Nitrogenase is only formed when the concentration of cellular ammonia is very low. This is not the case when NH_4^+ or compounds that are easily transformed to NH_4^+ are the growth substrates. NH_4^+ represses transcription of nitrogen fixation (*nif*) genes. Furthermore, nitrogenase activity is inhibited by ADP and NH_4^+. The best-studied bacteria are *Klebsiella pneumoniae* and *Rhizobium* species. The direct regulator of *nif* genes expression, NifA, acts as an activator of an alternative sigma factor, σ^{54}. This sigma factor is often required for the expression of non-standard metabolic functions and turns on transcription of the *nif* operons. *nif* gene expression is turned off by the regulator protein NifL, which complexes NifA, when NH_3 supply is high or when oxygen is present. Expression of *nifL* and *nifA* genes is regulated by the general nitrogen two-component system NtrB–NtrC in *Klebsiella*. In *Rhizobium*, microaerobic regulation of the

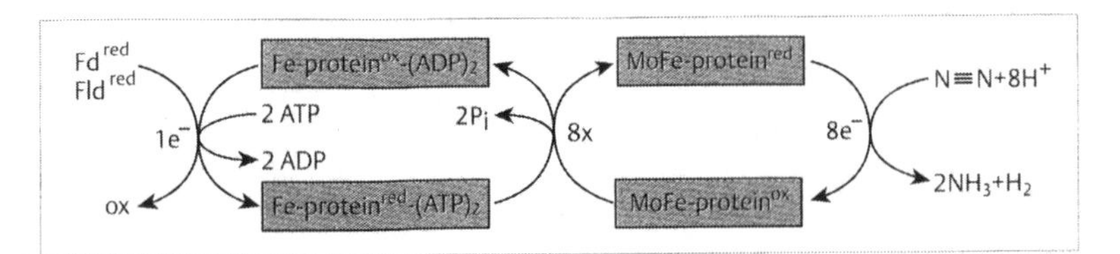

Fig. 8.14 **Model of the nitrogenase reaction.** Fd ferredoxin, Fld flavodoxin

nif genes is controlled by a separate two-component system, FixJ–FixL. The sensor, FixJ, phosphorylates the regulator, FixL, in response to anaerobiosis; the active regulator in turn activates transcription of the genes for the two activator proteins of *Rhizobium nif* genes, *nifA* and *fixK* (see Chapter 34).

Symbiotic nitrogen fixation. Many plants depend on symbiotic nitrogen-fixing bacteria that are harbored in special organs. The host plant supplies a carbon and energy source (e.g., a dicarboxylic acid) and all other nutrients and receives ammonia. Symbiotic nitrogen fixation can be as high as 100–300 kg per year per hectar. Nitrogen fixation by free-living bacteria accounts for probably only a few percent of this amount. Bacteria able to fix nitrogen in symbiosis with plants belong to various groups. The process of infecting the host plant and establishing symbiosis in special plant organs (nodules) as well as the exchange of low-molecular-weight nutrients and signals between the partner organisms are described in Chapter 34.2. Nitrogenase is protected from oxygen by respiration of the nodule tissue, and leghemoglobin in the nodules guarantees constant supply of oxygen for respiration of the strict aerobic symbiotic bacteria.

The most common symbiotic nitrogen-fixing bacteria belong to the α-subgroup of Proteobacteria. The genera are *Rhizobium*, *Bradyrhizobium*, and *Azorhizobium*. They form root nodules and stem nodules (*Azorhizobium*) with leguminoses. The *Rhizobium* system is the most common symbiosis that contributes most to N_2 fixation. In the β-subgroup of Proteobacteria, the genus *Azoarcus* and related genera are found as symbionts of the tropical Kallar grass. Gram-positive bacteria of the genus *Frankia*, members of the *Actinomyces* line, are the endosymbionts in the root nodules of trees such as alder (*Alnus*) and sea buckthorn (*Hippophaë*). Cyanobacteria (*Anabaena azollae*) are symbionts of the water fern *Azolla*, which contributes much to N_2 fixation in rice fields; *Azolla* grows at the surface of water-logged rice fields. Other cyanobacteria (*Nostoc*) live in symbiosis with liver moss as well as with a tropical plant, *Gunnera macrophylla*. Cyanobacteria are also associated with fungi in lichens. Lichens are the first organisms to settle completely inorganic habitats (e.g., rocks) and act as pioneers for other organisms.

In the rhizosphere of various plants living on poor soil, an accumulation of nitrogen-fixing bacteria has been observed. This relationship between prokaryotes and plants is not a symbiosis; however, it is believed that both partners have a mutual benefit from their association. To date there are no symbiotic nitrogen-fixing systems known in the marine environment. The role of free-living cyanobacteria in nitrogen fixation in the sea has not yet been assessed.

Nitrogen fixation by filamentous cyanobacteria. Free-living, N_2-fixing cyanobacteria may live as **single cells** (e.g., *Synechococcus*) or as **multicellular filaments**. Several filamentous nitrogen-fixing cyanobacteria (e.g., *Anabaena* or *Nostoc*), when grown in the absence of a nitrogen source, form differentiated larger cells with a thick cell wall, called *heterocysts*; their differentiation is described in Chapter 24.3. Heterocysts harbor the N_2-fixation system; they possess only photosystem I. This allows ATP synthesis via photosynthetic cyclic electron transport. Lack of photosystem II avoids the production of O_2, which would be deleterious for nitrogenase. Heterocysts obtain a carbon and electron source (e.g., maltose) from the surrounding cells in the filament, which have both photosystems and are able to fix CO_2 into maltose. Oxidation of maltose in the heterocysts proceeds via the oxidative pentose-phosphate cycle, yielding NADPH, which is used to reduce ferredoxin. Heterocysts supply the "multicellular" organism with NH_4^+. Not unexpectedly, single cellular cyanobacteria and those filamentous nitrogen-fixing species that do not form heterocysts fix nitrogen only at low partial pressures of oxygen.

The various components of any nitrogen-fixing (*nif*) system and the complex global regulatory networks make it difficult to establish this ability in plants by genetic engineering. Several prerequisites have to be fulfilled: supply of trace elements, sufficient supply of ATP and of reduced ferredoxin, protection against oxygen, and altered control of synthesis and activity of Nif proteins. Finally, the assembly and maturation of nitrogenase is complex and requires further enzymes and proteins, for example, for Mo uptake, FeMo-cofactor and homocitrate synthesis, and various processing steps.

8.6 Assimilation of Phosphorus Does Not Require Redox Reactions

Phosphorus exists in the +5 valence state under both oxic and anoxic conditions. Phosphorus is the limiting element under many environmental conditions, with concentrations in the micromolar range, whereas cellular phosphate is in the millimolar range. Inorganic orthophosphate (P_i) is the preferred phosphorus source and represses the transcription of other genes involved in the acquisition of alternative phosphorus sources (*pho* regulon). The alternative phosphorus sources are organic phosphate esters, diesters, and anhydrides. Some phosphorus is present as phosphonic acids in which phosphorus is directly linked to carbon (for a scheme of P_i assimilation, see Chapter 7).

Acquisition of external phosphates. Inorganic phosphate enters the cell through two transport systems (see Fig. 7.**7**), a low-affinity, constitutive permease (Pit) and a high-affinity, ATP-dependent transport system (Pst) ($K_d < 1\ \mu M\ P_i$), which is activated by P_i limitation. Organophosphate compounds may enter the cell in one of two ways: (1) Some common phosphorus-containing compounds are directly taken up via binding-protein-dependent transport systems; an example is glycerol 3-phosphate, which is released by lipases from food lipids. These compounds may serve as carbon, phosphorus, and energy sources. (2) Other, non-transportable organophosphates still are usable as phosphorus sources. They enter the periplasm, where they are hydrolyzed by periplasmic non-specific or relatively specific esterases (phosphomonoesterases or phosphodiesterases). The P_i released is actively transported. Similar systems operate in Gram-positive bacteria. Long-chain polyphosphates (volutin, inorganic polyphosphate) and possibly also oligonucleotides enter through specific pores and are hydrolyzed by periplasmic polyphosphatase and nucleases plus phosphatases, respectively.

Phosphonate utilization. Phosphonic acids contain a chemically very stable C–P bond ($R_3C–PO_3H^-$). Degradation of these compounds, therefore, requires specific, inducible enzymes and transport systems. The cleavage of the critical C–P bond proceeds via direct dephosphonation catalyzed by complex enzymes; the cleavage is hydrolytic if the substrate is activated, such as in phosphonoacetaldehyde, or occurs possibly by a radical mechanism in most other cases.

Regulation by P_i. In *Escherichia coli*, more than 20 promoters are regulated by P_i and the synthesis of almost 100 proteins is markedly enhanced during P_i-limited growth. These genes induced by **p**hosphate **st**arvation (*pst*) belong to different classes. One class belongs to the ***pho* regulon** for acquisition of alternative phosphorus sources at extracellular P_i levels in the micromolar range. Other genes belong to various control systems. For a discussion of regulon control, two-component regulatory systems, and the consensus boxes see Chapter 20.

8.7 Common Sources of Cell Sulfur Are Sulfate and Thiosulfate

Sulfate and thiosulfate have to be reduced to H_2S to become incorporated. Consequently, many bacteria and the green plants can use them as sulfur sources. Sulfate or thiosulfate anions need to be taken up by specific transport systems (e.g., by a permease) that transport sulfate, thiosulfate, and sulfite.

Sulfate. Assimilatory reduction of sulfate (see Fig. 7.**8**) proceeds via several steps. Sulfate is first reduced in a two-electron step to sulfite, which is reduced in a six-electron step to H_2S. However, sulfate reduction to sulfite is energetically unfavourable (SO_4^{2-}/SO_3^{2-}; $E_0' = -600$ mV). Therefore, the acid must be activated before it can be reduced. The reactions involved are:

$$SO_4^{2-} + ATP \rightarrow \text{Adenosine phosphosulfate} + PP_i \quad (8.18)$$

$$PP_i + H_2O \rightarrow 2\ P_i \quad (8.19)$$

$$\text{Adenosine phosphosulfate} + ATP \rightarrow 3'\text{-Phosphoadenosine 5' phosphosulfate} + ADP \quad (8.20)$$

$$3'\text{-Phosphoadenosine 5' phosphosulfate} + 2[H] \rightarrow 3'\text{-Phospho-AMP} + HSO_3^- \quad (8.21)$$

$$HSO_3^- + 6[H] \rightarrow HS^- + 3\ H_2O \quad (22)$$

ATP sulfurylase (Reaction 8.18) forms an energy-rich phosphoric acid–sulfuric acid anhydride, adenosine 5′-phosphosulfate (APS). However, this reaction is endergonic and is pulled by hydrolysis of pyrophosphate by pyrophosphatase (Reaction 8.19); hence, two energy-rich phosphate anhydride bonds are spent. To force the reaction further to the side of the activated sulfate, a third energy-rich phosphoryl group is spent to phosphorylate the 3′-OH of ribose, forming 3′-phosphoadenosine 5′-phosphosulfate (PAPS), catalyzed by APS kinase (Reaction 8.20). PAPS is reduced to sulfite by "PAPS reductase" which releases 3′-phosphoadenosine 5′-phosphate (Reaction 8.21); this reaction proceeds in two steps. First, sulfotransferase transfers the sulfonyl moiety of PAPS to a thiol acceptor to form an acceptor–$S{-}SO_3^-$ intermediate, for example, with the small protein thioredoxin. The enzyme–thiosulfate ester is similar to an organic thiol ester like acetyl-CoA (CoA–S–CO–CH_3). Two-electron reduction of the thiosulfate ester yields free sulfite and the disulfide form of the acceptor; the disulfide has to become reduced subsequently. Some bacteria, however, seem to reduce APS rather than PAPS, as do the dissimilatory sulfate reducers in anaerobic sulfate respiration. Sulfite is then reduced by the complex siroheme-containing sulfite reductase to H_2S (Reaction 8.22). The electron donor is normally NADPH, which reduces the siroheme via a flavoprotein. H_2S is immediately bound to *O*-acetylserine, giving rise to cysteine (Chapter 7).

All these **assimilatory** reactions are catalyzed by soluble enzymes that are normally under the control of cysteine (repression). Note that **dissimilatory** sulfite reduction is catalyzed by different, membrane-associated enzyme(s) (Chapter 12).

Thiosulfate. Thiosulfate can react with *O*-acetylserine to yield an *S*-sulfonate derivative, which is subsequently reduced to cysteine.

Sulfonic acids. Sulfonic acids ($R{-}SO_3H$) are not a very common natural sulfur source; an example is the sulfonolipid of chloroplasts, 6-sulfo-D-quinovose. Sulfonic acids can be used as a sulfur source by various bacteria, both aerobes and anaerobes; however, the C–S bond is relatively stable, explaining why utilization of sulfonic acids is rare. The aerobic attack on sulfonic acids proceeds via dioxygenases or monooxygenases, which release sulfite (see Chapter 9.17.2). The anaerobic degradation is not well studied. The release of the sulfur moiety may proceed via different mechanisms, either hydrolytically or reductively.

Regulation of sulfur assimilation. Feedback inhibition by cysteine of serine transacetylase (see Chapter 7) constitutes a major kinetic regulation mechanism of cysteine biosynthesis. This enzyme provides the acceptor form for sulfide, *O*-acetylserine. The sulfur assimilatory pathways are under transcriptional control. *O*-Acetylserine is also important as the co-activator of the regulator protein of the cysteine (*cys*) regulon, *cysB*. The cysteine regulon in *Escherichia coli* consists of those genes required for cysteine biosynthesis from sulfate/thiosulfate and serine. Presumably, cysteine or a derivative thereof represses the *cys* regulon. Poor sulfur sources (e.g., djenkolic acid or glutathione) result in low intracellular cysteine pools and derepression of the *cys* regulon.

8.8 Trace Elements and Electrolytes Are Taken Up By Specific Transport Systems

Every cell is made up of the **macroelements** (C, O, H, N, P, S), which form the polymers and soluble organic compounds, and of **essential alkali** (K, Na) and **earth alkali elements** (Mg, Ca), which act as electrolytes and enzyme cofactors. In addition, **trace elements** are required, which include the transition metals V, Mn, Fe, Co, Ni, Cu, Zn, Mo, W, and a few non-metal elements such as Se, B, Cl, and others. Not all of the trace elements are required by individual species. These inorganic compounds are normally present as simple cations or anions and are actively taken up by relatively specific transport systems. Some of the elements occur in the form of oxyanions, such as SeO_3^- (selenite), WO_4^{2-} (tungstate), or MoO_4^{2-} (molybdate). They undergo enzyme-catalyzed redox reactions in the cell to become transformed into the correct valence state. The uptake of iron is of special importance; its supply is often the growth-limiting factor.

8.8.1 Iron Is Taken Up as Chelate Complexes by Different Transport Systems

Iron is required in relatively high amounts. It is present in heme, iron-sulfur centers, and in the active site of many enzymes. Although it is the fourth most common element in the earth's crust (O > Si > Al > Fe), it is not as readily available as one might think. Under oxic conditions, iron is oxidized to Fe(III), which forms extremely unsoluble oxide hydrates FeO(OH), carbonate $Fe_2(CO_3)_3$, and magnetite Fe_3O_4 [iron(II/III) oxide]. These compounds become only soluble at very acidic pH. Therefore, most aerobic microorganisms produce and secrete Fe^{3+}-complexing organic compounds called **siderophores** (= iron carriers) and transport the iron complexes into the cell. Anaerobically, iron is not limiting; Fe^{2+} is the prevailing soluble form, which is transported as a cation by a simple transport system.

Cations like K^+ or Zn^{2+} are easily taken up by transport systems specific for these cations; uptake is often driven by the membrane potential (inside negative). The uptake of iron chelated by siderophores requires quite different transport systems because of the different size and charge of the iron complexes. Uptake and utilization require a sophisticated interplay of various processes that are strictly regulated: (1) synthesis of siderophores, (2) excretion of siderophores, (3) transport of iron (III) siderophores, (4) release of iron(III) and reduction to Fe(II), and (5) incorporation of Fe^{2+} in cofactors and prosthetic groups.

The formation and utilization of siderophores can be tested by various methods: (1) Iron is supplied to the agar medium in the form of colored Fe(III) complexes. Siderophores—which are colorless or less colored—compete for iron binding and cause bleaching of a zone around siderophore-producing colonies. (2) A siderophore is applied on a small filter disk, as in an antibiotic bioassay. Bacteria that can use this specific siderophore, will grow in a zone around the disk.

Siderophores. The uptake of Fe(III) is well-studied in *Escherichia coli*. Three types of siderophores (Fig. 8.**15**) are produced: (1) **Catechol-type siderophores**, such as enterobactin, contain three molecules of the complexing agent 2,3-dihydroxybenzoate; they are linked to a cyclic trilactone made up of three molecules of serine. Fe(III) enterobactin is positively charged. (2) **Hydroxamate-type** siderophores, such as ferrioxamine, make use of the well-known property of hydroxamic acids to produce colored stable complexes with Fe(III). Three molecules of hydroxamic acids form a cyclic compound, which in the Fe(III)-loaded form is uncharged. (3) **Dicarboxylic or tricarboxylic acids**, notably citrate, are good iron-complexing agents.

Ferrichromes are cyclic hexapeptide siderophores formed by fungi, but can also be used by bacteria, for example by *E. coli*. A comparison of the Fe^{3+} binding capacity of siderophores and of inorganic oxo-hydroxy complexes is shown in Table 8.**4**.

Transport systems for siderophores. The two types of pore-forming proteins (**porins**), the nonspecific (type I channels) and the more specific porins (type II channels) of the outer membrane of Gram-negative bacteria are not sufficient for permeation of siderophores. A third type of outer membrane proteins

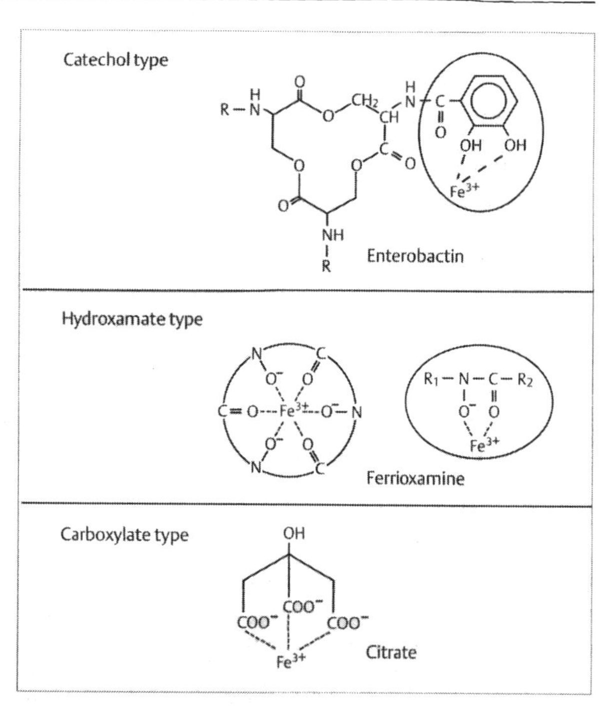

Fig. 8.15 **General structures of siderophores**

called receptors, is required; they bind larger molecules tightly, such as various siderophores and vitamin B_{12}. After binding to the multifunctional receptors, siderophores are passed through the outer membrane in an energy-consuming process and then bind to binding proteins in the periplasm. Uptake of Fe(III) siderophores into the cell proceeds via various ATP-dependent transport systems (Fig. 8.**16**).

Box 8.5 Certain phages (e.g., phage T1), antibiotics (e.g., albomycin), and colicins bind to multifunctional receptors or are even transported by iron-transport systems. The multifunctional properties of the receptors provide opportunities for using several ligands to unravel their modes of action. For instance, certain mutants deficient in iron uptake are resistant to infection by phage T1 (**T-one**). The corresponding gene product, the **Ton** protein is an essential component of siderophore transport (see below). Siderophores are also mimicked by various analogous antibiotics that misuse their transport systems (Fig. 8.**16**).

Energy-dependent transport across the outer membrane of Gram-negative bacteria. Iron in the form of ferric [Fe(III)]–siderophore complexes and

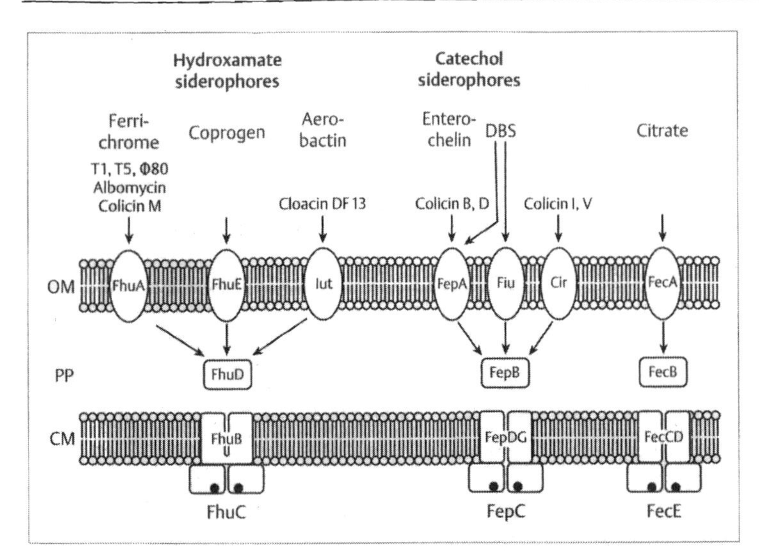

Fig. 8.16 **Scheme of the iron-transport systems of *Escherichia coli*.** The receptor proteins contained in the outer membrane (OM) specifically bind ferric hydroxamates, ferric catecholates, and ferric citrate, which are then transported into the periplasm (PP) where they interact with binding proteins that donate the substrates to the hydrophobic transport proteins in the cytoplasmic membrane (CM). Transport through CM is catalyzed by proteins at the expense of ATP hydrolysed by the proteins associated with the inner face of the CM T1, T5, ϕ80 are phages. DBS = dihydroxybenzoic acid [from 2]

vitamin B_{12} are transported through the outer membrane of Gram-negative bacteria by a mechanism that consumes energy. There is no known energy source in the outer membrane or in the adjacent periplasmic space; therefore, energy has to be provided by the electrochemical potential across the cytoplasmic membrane. How are the receptors in the outer membrane energized?

A transmembrane protein complex (TonB complex) in the cytoplasmic membrane gets into contact—via a long periplasmic polypeptide chain—with the receptors in the outer membrane. The closed channel of the receptor is thought to be forced by conformational energy to open a lid when the TonB complex changes its conformation due to the electrochemical potential of the cytoplasmic membrane.

Release of Fe from siderophores. The siderophores release iron either by reduction to Fe^{2+}, (e.g. for ferrichrome, which does not bind Fe^{2+}), or by enzymic hydrolytic destruction of the siderophore (e.g. for enterobactin), which necessitates its de novo synthesis. In the case of Fe(III) citrate, only Fe(III) enters the cell.

Regulation. Because iron-uptake systems are expensive to make, they are only synthesized and excreted when needed, that is, when iron limits growth. Iron regulation involves more than two dozen proteins, some of which are components of the transport systems. Iron-regulated genes are controlled by the **Fur** (**f**erric **u**ptake **r**egulation) protein, which acts as a **repressor** of most genes, when $\mathbf{Fe^{2+}}$ is present as the **co-repressor**. The TonB complex is required for induction of the transport system by Fe(III) citrate; citrate does not have to enter the cell for induction. Note that there are a great variety of siderophores and transport mechanisms, for example, the greenish fluorescent pigments of *Pseudomonas* sp. function as siderophores.

Box 8.6 Higher organisms keep iron in highly insoluble complexed form by iron-binding and iron-storage proteins (transferrin, lactoferrin, ferritin) (Tab. 8.**4**). These proteins bind iron so strongly in tissues, blood, milk, and tears that microbial growth is iron-limited. Pathogens cope with this problem by (see Chapter 33.3.5) (1) producing **siderophores** that have such an incredibly high affinity for iron that they actually remove iron from the animal iron-binding proteins. Iron transport is often plasmid-encoded and contributes to pathogenicity; (2) lysing red blood cells to reach iron in hemoglobin (**hemolysins**); (3) binding the iron-transporting proteins on their surface and stealing its Fe(III); or (4) not requiring iron for catabolism (such as lactic acid bacteria, which grow under extreme iron-shortage conditions). Also note that excess of free iron ions is deleterious due to the iron-catalyzed Fenton's reaction, which generates OH˙ radical (see Chapter 9).

Table 8.**4** **Apparent stability constants of Fe(III) complexes at pH 7.** For the reaction between metal (M) and ligand (L), M + L⇌ML, the stability constant is defined by $K = [ML]/[M]\cdot[L]$. The stability constant of Fe(III) hydroxide at pH7 is approximately 10^{38}

Compound	Stability constant at pH 7
Fe(III) enterobactin	10^{52}
Fe(III) Ferrioxamine E	10^{32}
Fe(III) Ferrichrome A	10^{29}
Fe(III) Transferrin A	10^{24}
Fe(III) EDTA	10^{25}

8.8.2 Considerable Efforts Are Required to Provide the Cell With Other Essential Trace Elements Such as Selenium

In order to avoid growth limitation because of the lack of a specific trace element, bacteria often have developed different enzyme systems with different trace elements in the active site; a good example is nitrogenase. Selenium is the non-metal trace element whose biological role is best studied; Se assimilation will serve as a representative example.

Selenium-containing cell compounds. It has long been known that *Escherichia coli*, when grown anaerobically, requires traces (10^{-8} M) of molybdate (MoO_4^{2-}) and selenite ($HSeO_3^-$) in the growth medium to oxidize formate; during fermentation of glucose, formate is produced from pyruvate by pyruvate:formate lyase. Three types of selenium compounds are known: (1) Se has been found in the active site of oxidoreductases, for example, in formate dehydrogenases (which also contain molybdenum, explaining the requirement of *E. coli* for Se and Mo mentioned above), hydrogenases, and glycine reductase. These selenoproteins contain a **selenocysteine**, that is, a cysteine in which S is replaced by Se. (2) A few molybdoenzymes, such as xanthine dehydrogenases and nicotinic-acid dehydrogenases contain Se in a **molybdenum cofactor** in which S is replaced by Se. (3) Some tRNAs contain a **modified uracil** base in the anticodon, in which C=O at position 2 of the pyrimidine ring is replaced by C=S. This sulfur atom can be substituted in Se-containing tRNA molecules by Se.

What is the advantage of using selenium? Se belongs to group VI elements of the periodic table; it follows after sulfur. The chemical properties of sulfur and selenium are similar, yet there are important differences: sulfur exists under oxic conditions mostly in the +6 oxidation state (sulfate, SO_4^{2-}), selenium in the +4 oxidation state (selenite, $HSeO_3^-$). The form used in biological macromolecules is the selenol (–SeH, –2 oxidation state). **Selenide** (**HSe^-**) and **selenols** (**RSe^-**) are more powerful nucleophiles at neutral pH than the S counterparts, and nature uses this property for catalysis.

How is selenocysteine formed and incorporated? Selenocysteine incorporation is determined by an in-frame TGA stop codon in the gene, and selenocysteine incorporation into selenoproteins occurs by a **cotranslational process** on the ribosome, rather than via posttranscriptional modification of precursor proteins (Fig. 8.**17**). Before this, selenite must be transported and reduced to selenide by specific systems.

There are several requirements for incorporation of selenide, which are fulfilled by the expression of four genes involved in the synthesis of selenoproteins.

1. An unusual $tRNA_{UCA}$ species ($tRNA^{Sec}$) exists, which is able to decode the stop codon UGA on the mRNA. This tRNA becomes loaded with serine by seryl-tRNA synthetase to give seryl-$tRNA^{Sec}$.

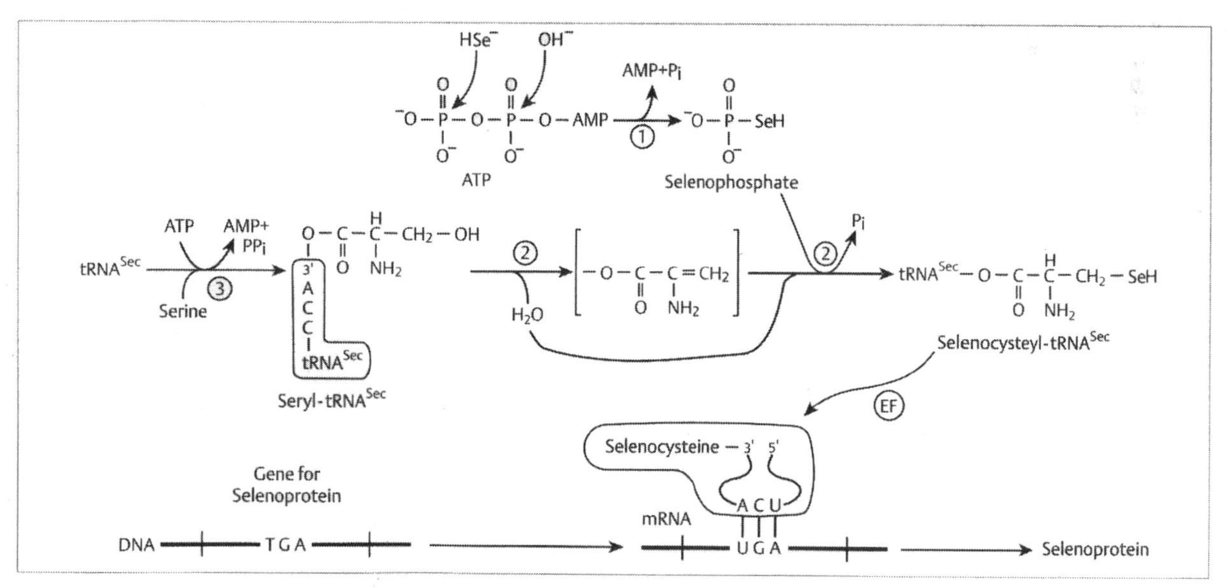

Figure 8.**17** **Scheme of incorporation of selenium into selenoproteins containing selenocysteine in the active site:** (1) selenophosphate synthetase, (2) selenocysteine synthase, and (3) seryl-tRNA synthetase. Further specific components are $tRNA^{Sec}$ and a specific elongation factor for selenocysteyl-$tRNA^{Sec}$

2. A specific biosynthetic pathway must provide selenocysteyl-charged, UGA-decoding tRNA. This is accomplished in two steps. First, selenide is activated by selenophosphate synthetase to **phosphoroselenoate**, $HSePO_3^{2-}$ ("**selenophosphate**"), according to Reaction 8.23:

$$ATP + HSe^- \rightarrow \text{Phosphoroselenoate} + AMP + P_i \quad (8.23)$$

This activated selenide is then used to form **selenocysteine**, catalyzed by pyridoxalphosphate-containing selenocysteine synthase (Reaction 8.24):

$$\text{Seryl-tRNA}^{Sec} + \text{Phosphoroselenoate} \rightarrow \text{Selenocysteyl-tRNA}^{Sec} + P_i \quad (8.24)$$

Thereby, water is eliminated from tRNA-bound serine, and selenide is added to the aminoacryloyl-tRNA intermediate.
3. The UGA condon directing insertion of selenocysteine residues must be differentiated from a real termination codon. A unique elongation factor is required since $tRNA^{Sec}$ does not bind to the normal elongation factor EF-Tu. This modified elongation factor also discriminates between $tRNA^{Sec}$ and other tRNA species.
4. Termination of translation at the UGA codon that determines selenocysteine incorporation must be prevented. This does not require an extra gene product, but the special sequence surrounding the selenocysteine codon on the selenoprotein mRNA determines that the UGA codon is not mistaken as stop codon. The required recognition factor is the special elongation factor.

The incorporation of selenium into the modified Se-containing tRNA proceeds also via selenophosphate, but details are not known.

Further Reading

Braun, V. (1995) Energy-coupled transport and signal transduction through the Gram-negative outer membrane via TonB-ExbB-ExbD-dependent receptor proteins. FEMS Microbiol Rev 16: 295–307

Cole, J.A., and Ferguson, S.J. (1988) The nitrogen and sulphur cycles. Cambridge: Cambridge University Press

Heider, J., and Böck, A. (1993) Selenium metabolism in microorganisms. In: Rose, A.H., ed. Advances in microbial physiology 35: 71–109

Large, P.J. (1983) Methylotrophy and methanogenesis. Washington, DC: ASM Press

Neidhardt, F. C., Curtiss, R., Ingraham, J. L., Lin, E. C. C., Low, K. B., Magasanik, B., Reznikoff, W., Riley, M., Schaechter, M., and Umbarger, H. E. (eds.) (1996) *Escherichia coli* and *Salmonella*: cellular and molecular biology. 2nd edn. Washington, DC: ASM Press

Postgate, J.R. (1982) The fundamentals of nitrogen fixation. Cambridge: Cambridge University Press

Schlegel, H.G., and Bowien, B. (1989) Autotrophic bacteria. Berlin Heidelberg New York: Springer

Shively, J.M., and Barton, L.L. (1991) Variations in autotrophic life. London: Academic Press

Walsh, C. (1979) Enzymatic reaction mechanisms. San Francisco: Freeman

Sources of Figures

1 Karlson, P., Doenecke, D., Koolman, I. (1994) Biochemie, 14th edn. Stuttgart: Thieme

2 Braun, V. (1995) FEMS Microbiol Rev 16:295–307

9 Oxidation of Organic Compounds

As discussed thus far, many—though not all—aerobic bacteria grow on the most common substrate, glucose. Most bacteria can utilize a vast number of organic compounds, some *Pseudomonas* sp, for example, can utilize up to 200 compounds. Often peripheral parts of metabolic pathways are plasmid-encoded, which makes prokaryotes even more metabolically flexible. Yet, there are specialists, such as methane-oxidizing bacteria, which can utilize only a few related compounds.

The most important **organic substrates** are the main constituents (biopolymers) of plants, which supply the most substrates for animals, fungi, and bacteria. Their bodies are remineralized to inorganic compounds by prokaryotic microorganisms, together with eukaryotic microorganisms (fungi, yeasts, algae, protozoa) and small animals like nematodes, worms, insects and their larvae, other arthropods, and mollusks. Fungi contribute half of the biomass of soil. Note that anaerobic or microaerophilic bacteria, and to some extent protozoa and fungi, are also essential to food digestion in the intestinal tract of animals and man.

Hence, the primary organic substrates are macromolecules that cannot penetrate the cell membrane; therefore, bacteria excrete enzymes as scavengers that degrade the macromolecules to soluble oligomers and monomers. These can be transported to provide low-molecular-weight nutrients for the host bacterium. Excreted enzymes, mostly hydrolases, are either set free or are bound to the cell surface. The monomeric or oligomeric soluble substrates, after transport into the cell, are degraded into a few central intermediates. These are completely oxidized to CO_2 when oxygen or an alternative electron acceptor is available.

Under anoxic conditions, substrates are incompletely degraded to fermentation products that are excreted; these serve as food for aerobes when oxygen becomes available ("nutrient chain") ("food chain" means that different organisms serve as food for others). In the complete absence of oxygen, the end products of the anaerobic nutrient chain are methane and CO_2 (biogas), i.e., true "waste products". Anaerobic degradation is a domain of prokaryotes; only few anaerobic fungi and protozoa have been described to be involved in this process. Note that even aerobic degradation of biomass by animals indirectly depends on the action of anaerobic bacteria in their intestinal tracts; these microorganisms are responsible for the cleavage of most of the fiber material (lignocellulose, chitin). In nature, bacteria are always associated with other bacterial species, fungi, or yeasts. Complete degradation of biopolymers requires a **complex interplay of microorganisms**.

Box 9.1 Investigations of the metabolism of microorganisms use various approaches: (1) **Isotope studies.** ^{14}C- or ^{13}C-labeled compounds are fed to the culture and labeled products are detected by tracer or NMR techniques, either in the culture supernatant or in the cells. When short periods of incubation are chosen, this approach is called "pulse-labeling", and allows the detection of early intermediates. (2) **Detection and identification of intermediates.** Intermediates in the growth medium are detected especially when metabolic inhibitors are added. All modern analytical techniques can be useful tools, but even simple ones, such as thin-layer chromatography with detection of the compounds by spraying the plates with indicators may work. (3) **Simultaneous adaptation.** Suspensions of cells grown with a given substrate may immediately consume the intermediates in the pathway, whereas degradation of compounds that are not in the degradation pathway, may take from 20 min to several hours of induction of degradative enzymes. (4) **Mutant studies.** Mutants that are blocked in the consumption of a given substrate may accumulate metabolic intermediates that occur before the site blocked. The mutants grow on substrates that are found downstream of the affected enzyme in the pathway. First clues may be obtained from the substrate spectrum of the wild-type strain, which should grow with postulated intermediates.

The degradative routes can be divided schematically into an oxidative and a reductive part (Fig. 9.1). The **oxidative part** comprises all steps required to oxidize the organic substrate completely to CO_2. O_2 is not required in these oxidations; rather, O in CO_2 comes from water, and the electrons are transferred to coenzymes. Exceptions are inert substrates with stable chemical bonds, the degradation of which requires molecular oxygen as co-substrate for oxygenases. In the **reductive part**, the reduced coenzymes reduce an exogenous electron acceptor; this part is associated with the main energy conservation (see Chapter 4 for aerobic electron transport and Chapter 12 for anaerobic electron transport).

Degradation of organic substrates to the level of the common precursor metabolites is independent of the terminal electron acceptor (for exceptions, oxygen-dependent pathways, see below). The organic growth substrates normally are the sole precursors of cellular building blocks. Therefore, the degradative pathways also have to supply all central biosynthetic precursor metabolites (see Chapter 7).

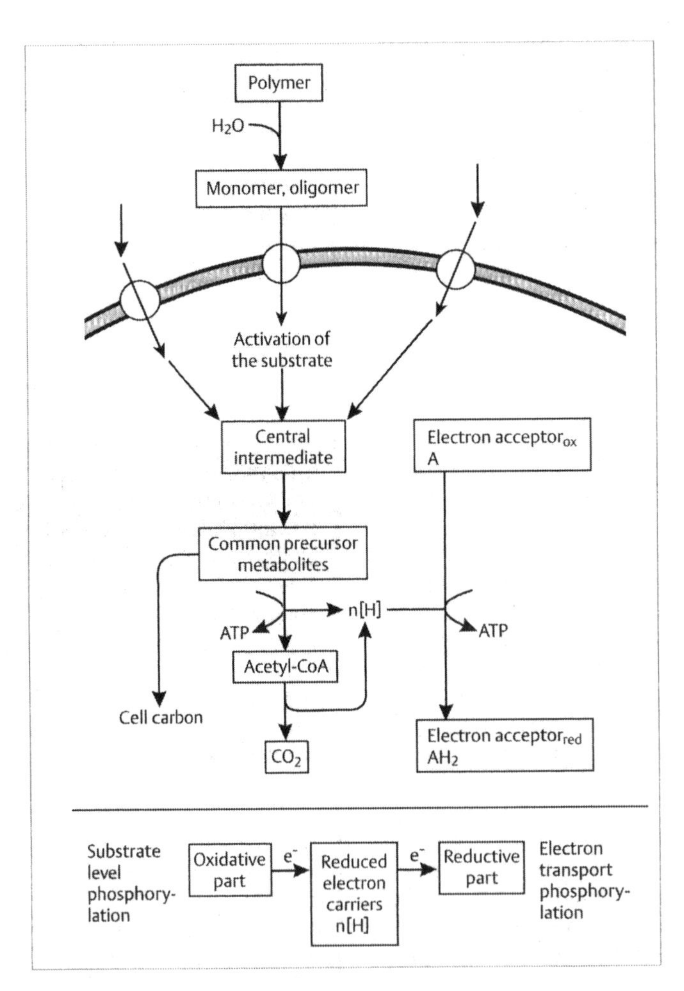

Fig. 9.1 **Oxidative and reductive parts of degradative routes.** The individual branches may be divided into several steps. Oxidative part: (1) polymer degradation, transport of soluble products, activation of substrates for degradation, channeling of the many different compounds of a single class (e.g., sugars or aromatic compounds) into a few central intermediates (e.g., glucose 6-phosphate or catechol). These pathways are peripheral; (2) conversion of the central degradation intermediates to the common (precursor) metabolites (e.g., pyruvate; see Chapter 7); this is the central pathway. The common (precursor) metabolites serve as precursors for the synthesis of all cellular compounds; (3) conversion of the common precursor metabolites to acetyl-CoA, and end oxidation of acetyl-CoA to CO_2. Reductive part: (1) oxidation of reduced electron acceptors; (2) electron transport chain; (3) reduction of terminal electron acceptor

9.1 Utilization of Polymeric Organic Substrates Depends on Extracellular Steps of Degradation

Most potential substrates for microbes are primarily organic polymers that are often water-insoluble. Before dealing with the individual classes of substrates, some common aspects of the utilization of polymers will be discussed.

General aspects of polymer degradation. Microbial extracellular enzymes are responsible for much of the cycling of polymeric organic matter. The diversity of substrates, environments, and organisms has led to the evolution of a prolific variety of enzymes. These hydrolytic exoenzymes are the most commonly used enzymes in industry. Polymer degradation is considered the rate-limiting step of degradation under many different conditions. Degradation is hampered by association of polymers with chemically resistant structures, crystallinity of polymers, and low water solubility.

There is one remarkable exception to the rule that polymers are hydrolytically cleaved (for a minor exception of β-elimination of sugars, see lyases Chapter 9.5.5 and Fig. 9.**9**). Lignin, a major polymeric substrate, cannot be hydrolyzed, but degradation requires H_2O_2. Exoenzymes for lignin degradation therefore require an oxidizable co-substrate and O_2 to form H_2O_2. Lignin degradation is rare in prokaryotes, but is a domain of fungi. Oxygen plays a singular role also in the degradation of other inert compounds, such as hydrocarbons or ethers, which normally cannot be attacked by water (see, however, anaerobic degradation Chapters 9.13.4 and 9.13.5).

Natural structures like wood are composed of various complex polymers with various monomers. In order to degrade them a complete set of enzymes is required, which is not produced by a single organism. Generally, complete degradation of polymeric substrates requires the interaction of different microorganisms. Poor substrates are often only degraded (**cometabolized**) concomitantly with "good" substrates (co-substrates); alone the poor substrates would not allow cell growth. An example is the degradation of wood, a process in which cellulose degradation provides energy and electrons that generate H_2O_2 which is required to attack lignin.

Exoenzymes are among the most important **virulence factors** of pathogenic microorganisms. Exoenzymes allow the invader to penetrate the cell border and to live at the expense of the host's cellular structures.

Which strategy is used to prevent dilution of the polymer-degrading enzymes? How does the cell prevent competing organisms from taking over the hydrolyzed food? How are these exoenzymes transported? How does the cell know that a polymeric substrate is available? How is the synthesis of the exoenzymes regulated? These questions will be addressed here.

Biopolymers are the principal substrates for microbial metabolism. Degradation of polymeric organic substances often requires interaction of different microorganisms. Polymer degradation is the rate-limiting step initiated by the action of exoenzymes. Further degradation of released soluble products can be divided into an oxidative part and a reductive part.

9.2 Extracellular Enzymes Are Exported Into the Surrounding Medium or to the Cell Surface

Extracellular enzymes are exported across the cytoplasmic membrane and excreted into the surrounding medium. Membrane-bound enzymes can also be included in this definition. The localization of extracellular proteins in the prokaryotic cell depends largely on the cell structure.

Synthesis and localization of extracellular enzymes. In Gram-positive cells with a thick peptidoglycan wall surrounding the cytoplasmic membrane, enzymes are localized on both the inner and outer surfaces of the membrane and traverse it. Those that are released from the outer surface of the membrane pass through the wall to become truly extracellular. Others, like β-lactamase, are bound to the membrane in the form of a lipoprotein. The lipid portion of this protein is similar to that found in the lipoproteins of the outer envelope of Gram-negative bacteria. Gram-negative bacteria, surrounded by two hydrophobic barriers, the cytoplasmic and outer membranes, offer a variety of locations for enzymes. Proteins are located on and in the cytoplasmic membrane, in the periplasm, and on or in the outer membrane. Some enzymes show variable distributions—cytoplasmic, membrane-bound, periplasmic, and extracellular—depending on culture age and conditions. Examples of well-studied extracellular enzymes are pullulanase, amylase, laevansucrase, alkaline

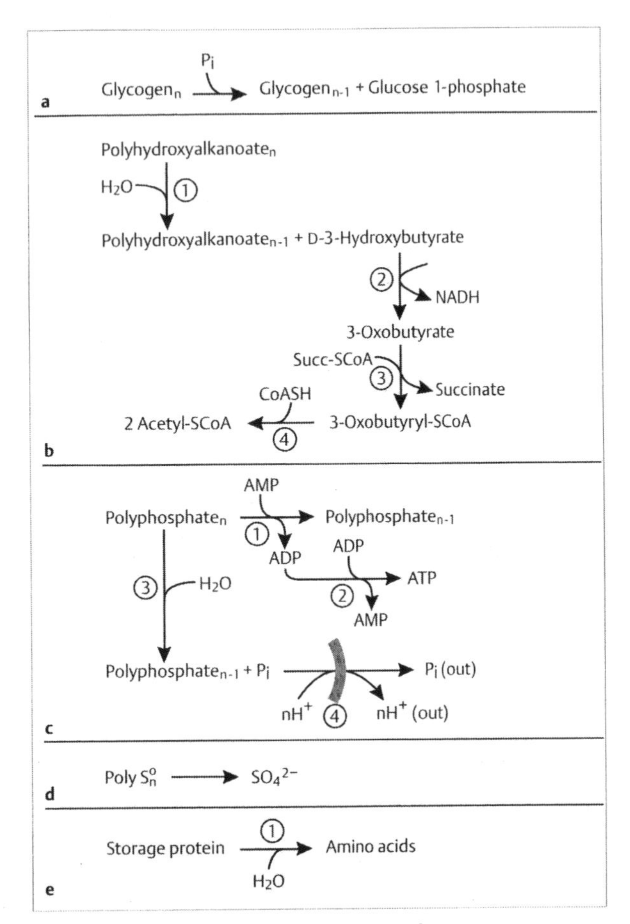

Fig. 9.**2a–e** **Degradation of storage polymers.**
a Glycogen, (1) glycogen phosphorylase;
b polyhydroxy alkanoate, (1) depolymerase, (2) NAD-dependent D-3-hydroxybutyrate dehydrogenase, (3) succinyl-CoA CoA-transferase, (4) ketothiolase.
c polyphosphate, (1) polyphosphate-AMP-phosphotransferase, (2) adenylate kinase, (3) polyphosphatase, (4) H^+-symport (export)
d sulfur, oxidation to sulfate;
e storage protein, (1) proteinase

phosphatase, β-lactamases, proteinases, cellulases, and lipases (see Chapter 19).

It has often been suggested that the membrane- and wall-associated enzymes of a Gram-positive bacterium are equivalent to the Gram-negative periplasmic enzymes, and, indeed, there are many similarities between the types of enzymes found in these two locations. Analysis of enzyme localization has been assisted greatly in recent years by the development of gene fusions containing "reporter" genes for secreted proteins. These genes encode proteins that are enzymatically active only when transported out of the cytoplasm.

Some of the hydrolytic enzymes are localized on the surface of the cell in large discrete structures that have a high affinity for both the bacterial cell and the substrate and are thus responsible for binding the cell to it; an example is the cellulosome. Some bacteria have even adopted the alternative approach of assimilating the macromolecule, rather like the eukaryotic lysosome. Intact cells bind, for example, starch through a maltose-inducible surface protein, and presumably starch is accumulated in the periplasm and hydrolyzed by periplasmic amylase and pullulanase, and the maltodextrins are transported into the cytoplasm for further hydrolysis by glucosidase.

Adhesion and chemotaxis. The binding of exoenzymes to cell surfaces, the attachment to the substrate, and possibly the uptake of polymeric substrate into the periplasmic space prevents dilution of the enzyme and a large loss of hydrolyzed food to competing organisms. Adhesion may allow the cell to continuously monitor the nutrient status of its environment; chemotaxis to polymer hydrolysis products may be inducible.

Secretion. The process of secretion seems to have been conserved throughout evolution and retains much in common between bacteria and higher eukaryotes. Secretion pathways and processing of proteins by the export machinery will be discussed in Chapters 5, 16, 19, 24, and 33. Secretion of an enzyme into the environment has several drawbacks. First, the conditions may not be conducive to enzymic activity, because of extremes of pH or the presence of inhibitory or inactivating ions or molecules. Second, if the conditions are conducive to enzymic activity, the products from the degradation of the macromolecular substrate will be available for competing microorganisms. For these reasons, sophisticated regulatory mechanisms might be expected to have evolved to provide fine control of extracellular enzyme synthesis.

9.3 Formation and Secretion of Extracellular Enzymes Are Regulated

Extracellular enzymes are generally secreted at a basal level and, if conditions are favorable, the enzymes hydrolyze the substrate in the environment. The low-molecular-weight products are transported into the cell and, assuming that other conditions are satisfied, such as the absence of catabolite repression, the products effect induction of the enzyme up to several dozen-fold. Some enzymes are constitutive; although the levels of synthesis vary with the carbon source, this is generally due to a relief of catabolite repression. The control of extracellular enzyme synthesis is sophisticated. Synthesis of most enzymes responds to the familiar stimuli of induction and catabolite repression. In addition, regulation may be superimposed by global regulation of stationary-phase metabolism, as well as sporulation or even motility, and competence for DNA-mediated transformation.

Catabolite repression. Extracellular enzymes are commonly controlled by catabolite repression. That is, in the presence of a rapidly metabolized carbon source, generally glucose, but also other "good" sources, there is little or no synthesis of the extracellular enzymes. This makes sense for energy efficiency since it is wasteful for the bacterium to synthesize enzymes of peripheral carbon metabolism (even in the presence of inducer) when a preferred carbon source is present. Thus, catabolite repression overrides induction of metabolic enzymes. Despite these uniform regulatory features, there is variation in the molecular mechanisms involved (see Chapters 18–20).

Temporal regulation. In some bacteria (e.g., bacilli), synthesis of many inducible and constitutive extracellular enzymes (e.g., proteases, amylase) is repressed during exponential growth and derepressed during stationary phase. Toward the end of exponential growth, scavenger enzymes are secreted that provide nutrients for continued vegetative growth in the absence of easily assimilable low-molecular-weight compounds.

Signal transducing systems. Bacteria constantly sense their environment and respond to it by modulating expression of sets of operons, often termed regulons. Many bacteria use common systems to respond to these environmental stimuli. Two-component signal-transducing systems play crucial roles in the regulation of, for example, sporulation and extracellular enzyme synthesis (see Chapter 20).

Synthesis and excretion of exoenzymes are regulated by sophisticated mechanisms. Several mechanisms have evolved to prevent dilution of the enzymes and loss of the soluble low-molecular-weight degradation products. Examples are binding of exoenzymes to cell surfaces or even uptake of polymeric substrates.

9.4 Intracellular Storage Polymers Are Degraded by Cellular Enzymes

Virtually all prokaryotes form intracellular storage polymers that are degraded by intracellular enzymes when energy (and/or phosphate in the case of polyphosphate) becomes limiting (Fig. 9.**2**). Their mobilization requires intracellular (rather than extracellular) enzymes, and the intracellular degradation of polymers often differs from the extracellular degradation of food biopolymers. These processes will be considered before turning to extracellular biopolymer degradation because most microorganisms can carry out synthesis and remobilization of storage polymers.

Polyglucose. Glycogen is degraded from the nonreducing end by glycogen phosphorylase, which forms glucose 1-phosphate. The α-1,4-glycosidic linkage is phosphorolytically cleaved. The reaction is reversible, but is not involved in polyglucose synthesis. Glucose 1-phosphate is rapidly isomerized to glucose 6-phosphate, which is degraded. At branching points, the α-1,6-linkages are hydrolyzed by 1,6-glucosidase. This phosphorolytic degradation is different from the hydrolytic degradation of food polyglucose. In addition, degradation of storage polyglucose results in rapid formation of glucose 1-phosphate, while the polymer remains intact, whereas extracellular hydrolytic degradation results in more rapid destruction of the polymer.

Polyhydroxyalkanoate. A granule-associated depolymerase forms D(-)-3-hydroxybutyrate (or related 3-hydroxyacid). Oxidation occurs by NAD-specific D(-)-3-hydroxybutyrate dehydrogenase to 3-oxobutyrate (acetoacetate). CoA transfer from succinyl-CoA forms acetoacetyl-CoA, which is CoA-thiolyzed to two acetyl-CoA.

Polyphosphate. Polyphosphate is degraded when phosphate and/or energy is limited, e.g., under oxygen limitation. Polyphosphate-AMP-phosphotransferase phosphorylates AMP to ADP. Two ADP are converted via adenylate kinase to ATP and AMP (Fig. 9.**2**). This reaction sequence is freely reversible; no intermediates of polyphosphate are detectable. Alternatively, polyphosphatase hydrolytically releases *ortho*-phosphate, which is probably exported by symport with one or two protons; ATP is conserved via H^+-ATP synthase; small polyphosphate granules are formed.

Sulfur. Polysulfur $[S^0]_n$ is oxidized to sulfate when ATP and carbon source (e.g., CO_2) are plentiful, but reductant is limiting. Rather little is known about degradation of storage sulfur. For sulfur oxidation to sulfate, see Chapter 10.

Storage protein. The storage proteins of cyanobacteria are mobilized by proteinases that are induced under starvation and nitrogen-limitation conditions.

9.5 Degradation of Major Polymeric Substrates Into Soluble Products Is Catalyzed by Exoenzymes

The fate of polymeric carbohydrates that form the most important food source for bacteria will be considered first in Chapters 9.5.1–9.5.5.

9.5.1 Cellulose Is the Most Abundant Organic Substrate in Nature

Cellulose is the major carbohydrate synthesized by plants and occurs in the cell wall as lignocellulose, together with other compounds (hemicelluloses, lignin). (Note that some bacteria also produce cellulose, e.g., *Acetobacter xylinum*). Cellulose is the most common organic substrate in nature. Cellulose is a linear polymer made of 100-10 000 glucose units linked by β-1,4 bonds. As in all structural polysaccharides, the β-linkage confers high tensile strength. Each glucose residue is rotated by 180° relative to its neighbors, so that the basic repeating unit is cellobiose (Fig. 9.**3**). The cellulose chains form numerous intramolecular and intermolecular hydrogen bonds, which make the fibrils rigid and insoluble. The fibrils form crystalline regions (with various crystalline cellulose forms), which are highly resistant to enzymic hydrolysis, and amorphous regions, which are attacked more easily. Cellulose microfibrils can form thick fibrils embedded in a matrix of hemicellulose and lignin (Fig. 9.**4**). β-1,4-Mannans or β-1,3-xylans are the structural analogues of cellulose in some algae. **Hemicellulose** is composed of complex carbohydrate polymers, with xylans and glucomannans as the main components. **Xylans** contain a xylose backbone carrying glucuronic acid and arabinose side chains esterified to aromatic (phenolic) acids, which in turn are ether-linked to lignin. Hence, there is a lignin-hemicellulose cross-link of ether bridges. **Glucomannans** contain β-1,4-linked D-glucose and D-mannose. **Lignin**, a highly branched, random, aromatic polymer, is highly resistant to biodegradation and protects cellulose and hemicellulose against enzymic hydrolysis.

Cellulose degradation is studied with various standardized cellulose forms. These forms differ in crystallinity, available surface area, pore size, solubility, degree of substitution of sugar OH groups by carboxymethyl groups, and other properties. Examples of amorphous, easily degradable forms are acid-swollen cellulose and soluble carboxymethyl cellulose. Chromogenic artificial substrates contain a colored compound β-glycosidically linked to the anomeric C1 carbon of cellobiose. A measure of catalytic activity of cellulases is, for example, the release of soluble reducing sugars (glucose equivalents) or color release due to hydrolysis of chromogenic substrates.

Cellulolytic and related enzymes. Cellulases are exoenzymes composed of different components that physically and synergistically interact with each other, with the fibrils, and with the surface of the cell. The components often possess a substrate-binding site independent from the catalytic site, or they are multifunctional proteins composed of distinct domains of similar sequences and biochemical properties, which can be arranged in various combinations. The nature of the cellulolytic enzymes varies according to the source. Many bacteria, unlike fungi erode the cellulose surface using cell-bound enzymes. This may be advantageous in

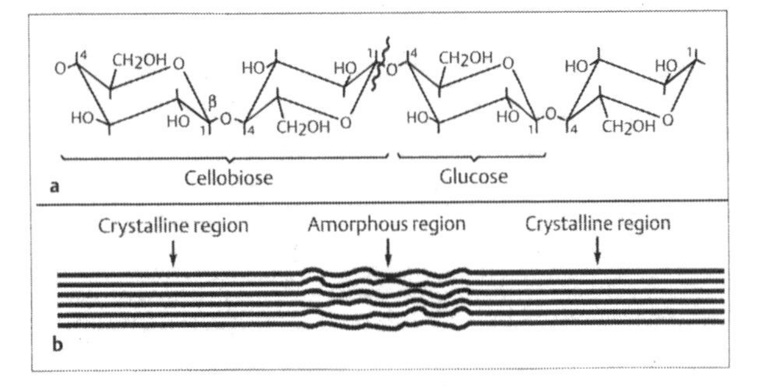

Fig. 9.**3a,b** **Structure of cellulose.**
a β-Glucosidic bonds;
b schematic structure of a fibril. [From 1]

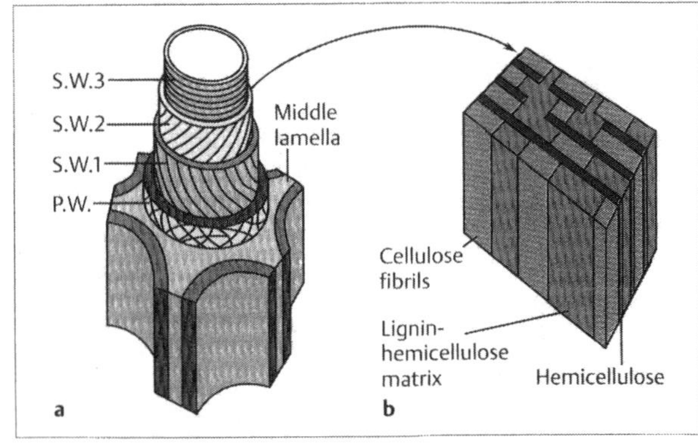

Fig. 9.**4a,b** **Plant cell wall, an important food source for microorganisms.**
a Cutaway view showing the organization of the cell wall layers composing woody fibers.
b Probable relationship of lignin and hemicelluose to the cellulose microfibrils in the secondary walls. The diameter of each cell is approximately 25 μm. P.W., primary cell wall; S.W. 1–S.W. 3, secondary cell walls. [From 1]

aquatic or rumen environment and when exposed to predatory protozoa. Each cellulolytic organism possesses a battery of enzymes that have different specificity with respect to *endo/exo* mode of action (see below), activity towards amorphous or crystalline regions, or preference for substrates of different chain length. This diversity is needed to cope with (1) the physical heterogeneity of the substrate, and (2) the changes due to degradation, requiring different enzymes at different times. The degradation of the complex hemicellulose is performed by a similarly complex set of enzymes that hydrolyze the different backbone structures and side chains. The final products of the cellulolytic enzymes are cellobiose (which forms the structural units of cellulose; bi = two glucose units) and glucose. Cellobiose is hydrolyzed by cellobiase to form two glucose.

Classification. Sequence comparison of hundreds of cellulases and xylanases has led to a detailed classification of these enzymes and their domains. Conventionally, the enzymes are loosely defined as exoglucanases (end-wise acting cellobiohydrolases splitting off cellobiose from the non-reducing end of the cellulose chain), endoglucanases and β-glucosidases acting on soluble disaccharides or oligosaccharides. This specificity is reflected in the geometry of their active sites, as revealed by X-ray diffraction of crystalline enzymes. In endoglucanases and xylanases, the substrate lies in an open cleft, which can straddle the polyglucose chain anywhere along the strand. In exoglucanases the active site forms a hole into which the non-reducing end fits.

The reaction mechanism of glycosidases may be a similar acid-base mechanism. First, the oxygen of the glycosidic link is protonated by one amino acid residue, followed by nucleophilic attack of a second residue, promoting OH^- formation. Hydrolysis may proceed with or without inversion of configuration (see Fig. 9.**8**).

The cellulolytic and xylanolytic enzymes are divided into several families of glycosyl hydrolases. These enzymes have spread to some extent by horizontal gene transfer, either as complete genes or as modules of different domains. As a rule, a non-conserved catalytic domain is linked by a flexible hinge region with a conserved binding domain; these components are integrated into the proteins in different orders.

The cellulosome concept. The case for stable multienzyme cellulase complexes is most firmly established for several anaerobic microorganisms, such as *Clostridium thermocellum*. A variety of enzymes capable of attacking the various forms, notably crystalline forms, of cellulose as well as other carbohydrates present in cell walls, form a physically integral complex of > 2 MDa. This **cellulosome** (Fig. 9.**5**) consists of more than a dozen peptides and glycoprotein components; some are present in a much greater proportion than others. The majority has cellulase, xylanase, or lichenase activity. One cellulose chain at a time appears to be processed to cello-oligosaccharides. A central protein is responsible for cellulose binding and binding of the catalytic components (this is called "scaffolding"); the central protein consists of a cellulose-binding domain and reiterated anchoring sites to which the various catalytic peptides attach with their conserved binding segments.

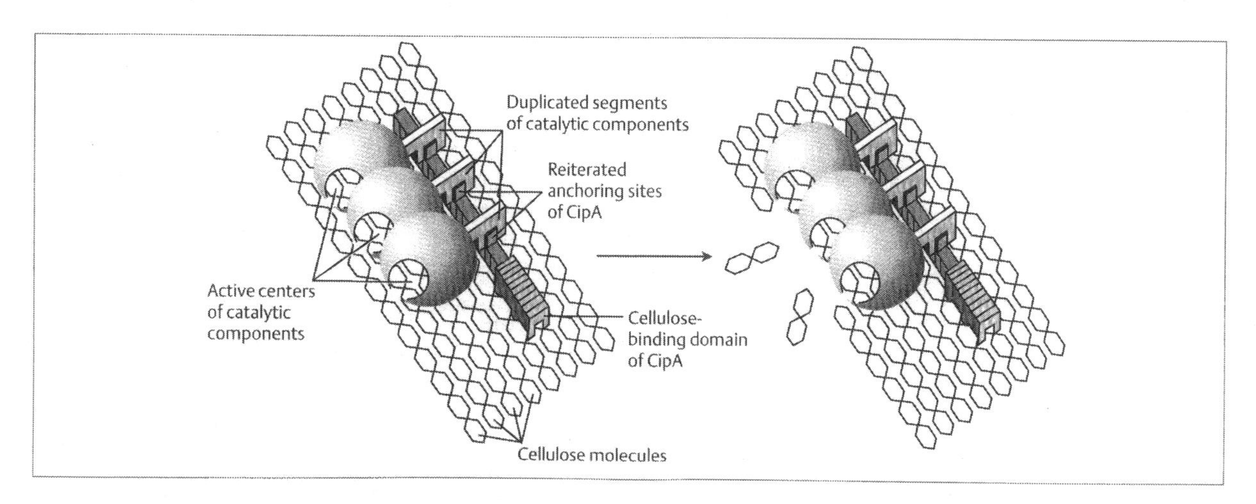

Fig. 9.**5** **Hypothetical model of the topological organization of subunits within the cellulosome.** The diagram shows a portion of the cellulosome with the cellulose-binding domains of the cellulose-binding proteins and three of its domains responsible for binding the duplicated segment borne by the catalytic components. Glucose residues are not drawn to the same scale as proteins. See text for details. [From 1]

Attachment of cellulolytic organisms to the substrate. Close association between cellulolytic microorganisms and their substrate is very common and is of obvious advantage to optimize the recovery of soluble hydrolysis products. Cellulosomes may be embedded in the surface layer as large clusters visible as protuberances by electron microscopy, may form polycellulosomes covering all of the residual cellulose, or are only loosely associated with the cell surface.

Cellulose degradation in the intestinal tract of herbivores. Herbivores are able to live essentially from the digestion of cellulose, xylanes, fructosans, and other polymeric carbohydrates. In ruminants, the rumen harbors a symbiotic community of anaerobic bacteria and protozoa (and rarely fungi) that operate as a semicontinuous culture of microorganisms. (See Chapter 31.5).

Control of the cellulase systems. Cellulase systems are repressed in the presence of low-molecular-weight carbon sources that are more easily metabolized than cellulose (catabolite repression). In fact, many cellulolytic bacteria cannot grow with glucose alone. In addition, the systems may be induced in the presence of cellulose or its oligomeric soluble degradation products (cellobiose, cellodextrins). The cellulase system is probably always present in small amounts, allowing cellulose first to undergo limited hydrolysis. The soluble hydrolysis products are transported and cause induction of cellulase synthesis. The true inducers are not known; possibly cellobiose (β-1,4) is transglycosylated to another disaccharide (β-1,2). Most of the genes are monocistronic, and transcription of different genes is not necessarily strictly coordinated.

Application. Cellulases and hemicellulases are biotechnologically important for the modification of cellulose and hemicellulose by partial hydrolysis, such as for breaking up straw as a substrate for mushrooms, to clear fruit juices, in the brewing industry, in future silage techniques, in textile processing, in the processing of paper pulp, and in the treatment of cellulosic wastes. Complete hydrolysis allows cellulose to be utilized as an inexpensive source of glucose.

Cellulolytic organisms. Cellulolytic organisms and enzymes are found in many prokaryotic genera, in fungi (*Trichoderma* or the anaerobic rumen fungus *Neocallimastix*), and in Protozoa (*Diplodinium, Eudiplodinium, Entodinium*). Examples of cellulolytic prokaryotes are the aerobic genera *Acetivibrio, Actinomyces, Alcaligenes, Bacillus, Cellulomonas, Erwinia, Microbispora, Pseudomonas, Streptomyces, Thermomonospora*, and the anaerobic genera *Bacteroides, Butyrivibrio, Clostridium, Fibrobacter, Thermoanaerobacter*, and *Ruminococcus*. Only a few produce a complete set of enzymes capable of degrading native cellulose efficiently. Others produce incomplete systems; they are probably associated with other prokaryotes.

9.5.2 Starch and α-Glucan Polymers Are Storage Materials That Are Favored Microbial Substrates

Starch is a branched α-D-glucose polymer composed of an α-1,4-linked backbone (**amylose**), with varying amounts of α-1,6-linked branches (**amylopectin**). Starch is the most common storage material of plants. One chain contains a few hundred to a few thousand glucose units, which form a helical structure in which iodine can be trapped. This results in the blue color of starch after iodine staining. **Glycogen** is more highly branched, which results in a brown color upon iodine staining. The polymers differ considerably regarding the degree of polymerization, branching, content of Ca^{2+} and phosphate, and other properties. **Pullulan**, a linear polymer of maltotriose units linked at the α-1,6 position, is produced by the yeast-like fungus *Aureobasidium pullulans*. **Cyclodextrins** are α-1,4 glycosidically linked glucose units (6 units = α-cyclodextrin, 7 = β, 8 = γ), which form a cyclic structure by reaction of the terminal C4 with the C1 of the first glucose residue (with no reducing end).

Amylosaccharidases cleave α-1,4 and α-1,6 linkages of starch and related polymers to form a mixture of soluble smaller glycosides ranging from glucose to maltodextrins with ten glucose units on avarage (Fig. 9.**6**). The enzymes are classified according to the substrate specificity and whether they display *endo*- or *exo*-hydrolytic action. Further criteria are the type of end product(s) formed (glucose, maltose, maltotriose, maltodextrin, cyclodextrins), the α or β configuration of the anomeric carbon of the products (see α- or β-amylase), and the bond that is cleaved (α-1,4, α-1,6, or both). All amylases possess similar conserved amino acid sequences in three to four regions that are essential for substrate binding and for general acid–base hydrolysis mechanism, but have low overall amino acid homology.

Enzymes. Endo-acting enzymes that produce a mixture of malto-oligosaccharides are α-amylases, pullulanases, and cyclodextrin glucosyl transferases. These enzymes rapidly destroy the polymeric structure. The best known example is **α-amylase**. **Exo**-acting enzymes cleave α-1,4 linkages from the non-reducing end, producing maltose; the best known example is **β-amylase**. Amyloglucosidases similarly release glucose units. Exo-acting enzymes leave the polymeric structure

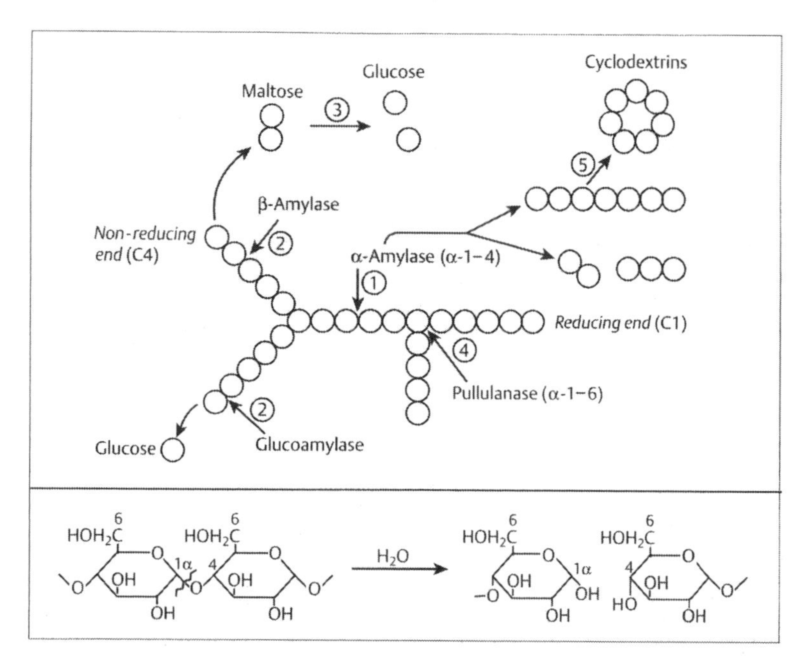

Fig. 9.6 **Enzymes involved in the degradation of starch.** (1) Endo-acting enzymes, (2) exo-acting enzymes, (3) glucosidases, (4) pullulanase (is also endo-acting); and (5) cyclodextrin glucosyl transferase. For hydrolytic glycosyl-oxygen bond breakage, see Fig. 9.9

intact for a longer time and produce instantaneously maltose or glucose. **Glucosidases** cleave short-chain dextrins like maltose into glucose. **Pullulanase** cleaves α-1,6 linkages of starch and pullulan. **Cyclodextrin glucosyl transferases** have α-amylase activity and, in addition, cyclize the maltosaccharides of 6–8 glucose units formed. This transglycosylation is of considerable technical interest (see below). The cyclic products (cyclodextrins) form inclusion complexes, which offer many applications. Isoamylases debranch glycogen to limit dextrins (i.e., short dextrins with many 1,6-glycosidic bonds). Some plants produce **fructans** (also called levans) instead of or in addition to glucans. They are similarly hydrolyzed by α-fructanase. Maltose and lower sugar oligomers and glucose are transported by specific transport systems. Maltose can also be hydrolyzed by the exoenzyme maltase into two glucose units.

Control of starch degradation. The starch and α-glycan polymer-degrading enzymes are normally excreted; occasionally they are cell-bound, for example, to S-layers. The factors that control the cellular localization are unknown. The synthesis may be growth-associated and constitutive at low levels. Low-molecular-weight degradation products (maltodextrins, maltose) act as inducers. Glucose acts as catabolite repressor, the mechanisms of repression being different in, for example, Enterobacteria, *Pseudomonas*, and *Bacillus*.

Application. The enzymes are used for the production of sugars from starch, such as various dextrins, maltose, glucose plus maltose syrup, glucose, or cyclodextrins. Besides the classical enzymes, there are new enzymes with interesting intermediary properties. The thermostable enzymes from mesophilic microorganisms and from extreme thermophiles are of special practical interest since they can be heated up to 100°C for considerable periods of time without loss of activity.

Organisms. Representative starch degraders are found in the genera *Bacillus*, *Bacteroides*, *Clostridium*, *Klebsiella*, *Pseudomonas*, *Pyrococcus*, *Streptomyces*, *Thermoanaerobacter*, *Thermotoga*, *Thermus*, *Thermomonospora*, and *Thermoactinomyces*.

9.5.3 Hemicelluloses Are Associated With Cellulose in Plant Cell Walls

Softwood (gymnosperms), hardwood (angiosperms), and grasses (graminaceous plants) have evolved separately, and they contain different lignin and hemicellulose constituents.

Hemicelluloses are found in association with cellulose in the secondary cell walls, and less in the primary cells walls of plants; they are relatively easily extracted and probably represent the second most important source of carbohydrates in nature. Hemicelluloses are homopolymers or heteropolymers consisting largely of β-glycosidically (β-1,4, β-1,6, β-1,3, etc) linked xylose, mannose, glucose, and galactose (all D-pyranose forms), with a number of substituents (acetyl ester, methyl ether, uronic acids, other pentoses), types of linkage of sugar residues, and degree of branching.

Hemicelluloses are only slightly branched. Most often the backbone consists of β-1,4-linked xylose or mannose. Those rich in D-xylose, D-mannose, or galactose are called xylans, mannans, and galactans, respectively. The xylan chain can be considered a cellulose counterpart: β-D-xylose formally derives from β-D-glucose just by replacement of the CH_2–OH group (C6) by a H atom. Yet, the degree of polymerization is much lower, the chain is branched, and it contains additional arabinose, glucose, galactose, and glucuronic acid.

Xylans and other constituents of hemicellulose are more easily degraded than cellulose, and the capacity to degrade hemicelluloses is more widespread. Many cellulose degraders also produce xylanases. As in the case of cellulose, xylan degradation at acid pH is a domain of fungi, and at neutral or alkaline pH of bacteria.

Enzymes. According to the different structures of hemicelluloses, an entire set of enzymes is required for complete degradation of hemicelluloses. Hemicellulases are usually characterized by their action on defined substrates. Hemicellulases (xylanases, etc) involved in main-chain breakdown consist of endo-1,4-β-D-xylanase, which catalyzes endohydrolysis, and β-xylosidase, which removes succesive D-xylose residues from the non-reducing ends (exoglycosidase). Hemicellulases are most commonly assayed by measuring the rate of reducing-group formation. Products are, for example xylose and xylobiose.

Accessory enzymes. Accessory enzymes release the substituents from the main carbohydrate chains. For example deacetylation makes native hemicellulose less water-soluble, but increases its susceptibility to enzyme attack.

Regulation. Xylanase may be produced constitutively by a specialist (e.g., *Clostridium* sp.) or is induced by xylans.

9.5.4 Chitin and Chitosan (Glucosaminoglycans) Are the Cellulose Counterpart in Many Invertebrate Animals and in Fungi

Chitin consists of (1–4)-β-linked *N*-acetyl-D-glucosamine (Fig. 9.**7**). The stability of chitin resides in the many hydrogen bridges due to the *N*-acetyl side groups, which are responsible for the strength. As a supporting material, chitin is a cellulose counterpart. Chitin is produced in enormous amounts in the sea (zooplankton) mainly by animals (tough structural component in most invertebrates forming the exoskeleton), in the soil by fungi (main component of cell wall), by the many insects, and also by protista. Prokaryotes and plants do not contain chitin; however, eubacteria contain murein,

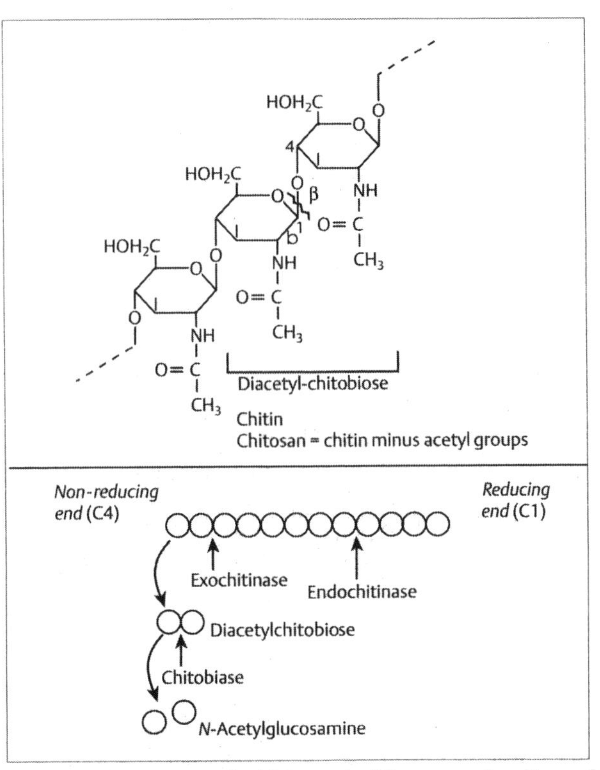

Fig. 9.7 **Structures of chitin and chitosan and action of chitin-degrading enzymes**

which is structurally related (see Figs. 7.**32** and 23.**1**). Chitin is considered to be the second or third most frequent carbohydrate. The crystalline form contains mostly antiparallel chains. As with other structural polymers, chitin is crosslinked to other components, such as proteins or β-glucans. Mineralization (calcification) and sclerotization with phenolic and lipid molecules are common. **Chitosan**, the deactylated form (Fig. 9.**7**), is formed by deacetylase in some yeasts and fungi.

Enzymes and pathways of chitin and chitosan degradation. Typically, a chitinolytic microorganism—either free-living or in association with animal guts—will produce several **chitinases** and ***N*-acetylglucosaminidases** with different substrate specificities. Chitin degradation can be compared to cellulose degradation. **Exochitinase** cleaves diacetylchitobiose units from the non-reducing end; **endochitinase** cleaves glycosidic linkages randomly along the chain, eventually yielding di- or triacetylchitobiose. **β-N-Acetylglucosaminidase** (chitobiase) hydrolyzes the dimer chitobiose to *N*-acetylglucosamine. Alternatively, chitin is deacetylated to chitosan, which is hydrolyzed by chitosanase to chitobiose, which in turn is hydrolyzed by glucosaminidase to glucosamine. Assays are based on consumption of macromolecular chitin forms (clearing zones on agar plates) or production of soluble oligomers. The intracellular enzymes play an important morphogenetic role in chitin-containing organisms. Chitinases are (glyco)proteins that are classified as a special group of the glycosyl hydrolase family.

Attachment to the substrate and control. Often a direct contact between the organism and chitin fibrils exists, as in cellulolytic prokaryotes (e.g., in chitinolytic invertebrate pathogens that attack the chitinous exoskeleton or the peritrophic gut membrane). Chitinase production is inducible by chitin oligomers and low levels of *N*-acetylglucosamine.

Applications. Chitinases (and their transglycosylase activity) and chitosanases play a role in the biotechnological use of chitin waste as a renewable resource and in protoplast formation of fungi. Chitosan is used in large scale mostly because of its gluey consistency.

Organisms. Chitin-degrading bacteria and fungi are very common in soil and sediment. Chitin is degraded by both aerobic and anaerobic bacteria. Actinomycetes can easily be enriched with chitin. Examples of chitin-degrading species are found in the genera *Alteromonas, Cytophaga, Chromobacterium, Curtobacterium, Vibrio, Pseudomonas, Aeromonas, Streptomyces, Clostridium, Photobacterium, Bacillus, Haloanaerobacter, Serratia, Flavobacterium, Lysobacter*, and *Arthrobacter*.

9.5.5 Pectins and Other Carbohydrate Containing Polymers Are Plant Cell Wall Matrix Polysaccharides

Pectins and hydrolytic enzymes. Pectins are plant cell wall matrix polysaccharides that also occur in some plant juices. They are polyanionic polymers of α-(1-4)-linked sugar acids, such as galacturonic acid (galacturonans). Some of the carboxyl groups are methyl esters, and pentoses may be present. Pectinase is a mixture of methylesterase, which produces methanol, and of endo- and exoenzymes (analogous to amylases and cellulases). Alginic acids are the functional equivalents of galacturonans in algae.

Pectin lyases. There are many other polymers composed of carbohydrates that are hydrolyzed by bacteria in the various kingdoms of organisms. In addition to hydrolytic enzymes, there are pectin lyases, which break the hexose-1,4-α- or β-uronic acid C–O bond by **β-elimination**, leading to unsaturated products (Fig. 9.**8**).

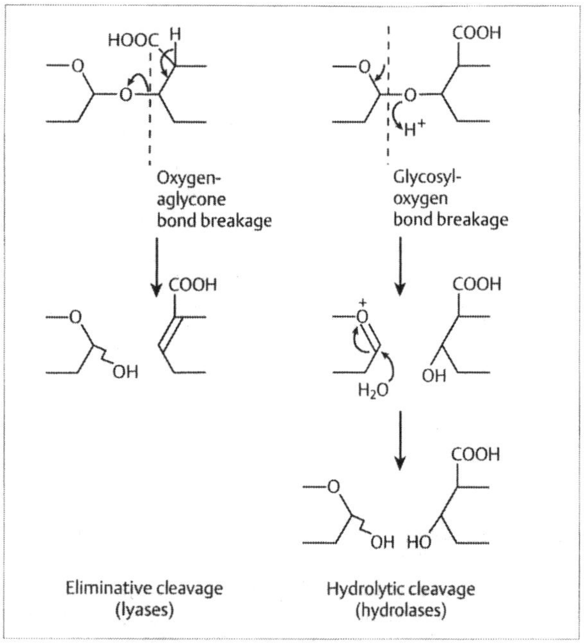

Fig. 9.**8** **The mechanism of β-elimination observed through the action of polysaccharide lyases compared with the hydrolytic action of glycanhydrolases.** [From 2]

Other glycosidases. There are many other glycosidases that hydrolyze *O*-glycosyl compounds such as inulin (1,2-β-D-fructan) or dextran (1,6-α-D-glucan). Sialidase (neuraminidase) hydrolyzes α-2,3-, α-2,6-, and α-2,8-glycosidic linkages of terminal sialic acid residues in, for example, oligosaccharides and glycoproteins. Hyaluronate is a polymer in which *N*-acetyl-β-D-glucosamine and D-glucuronate residues are β-1,5-linked (hyaluronase). Chondroitin, chondroitin 4- and 6-sulfates, and dermatan consist of 1,4-β-glycosidic-linked *N*-acetyl-galactosamine or *N*-acetylgalactosamine sulfate and glucuronic acid. Levans are 2,6-β-linked D-fructose. Agarose consists of 1,3-β-linked galactose (agarase). Eubacterial cell walls are composed of peptidoglycan. Lysozyme hydrolyzes the 1,4-β-linkages between *N*-acetylmuramic acid and *N*-acetyl-D-glucosamine residues. It is one of the best-studied enzymes. Xanthan, from *Xanthomonas campestris*, is an exopolysaccharide consisting of a complex repeating unit of sugars, which is hydrolyzed by xanthanase.

9.6 Lignin Is a Heteropolymer of Phenylpropane Units

Lignin is one of the few natural polymers that can be degraded only with the help of molecular oxygen by secreted oxidative enzymes. It is the most abundant aromatic polymer in nature, and is contained in the cell walls of higher (vascular) plants. Lignin is composed of phenylpropane building blocks (*p*-coumaryl, coniferyl, and sinapyl alcohol) (Fig. 9.**9a**), the monomeric units varying with plant species and location in the cell wall. Lignin surrounds cellulose and hemicellulose (lignocellulose) in the plant cell wall, forming a three-dimensional matrix, which is itself resistant to degradation. Wood contains 20–30% lignin and has a much greater C/N ratio than many other foods; hence, under such nitrogen-limiting conditions, excess carbon has to be disposed of by oxidation or production of secondary metabolites, while nitrogen is taken up into the growing cells.

Lignocellulose is quantitatively the most important organic substrate for microorganisms and is rapidly degraded in decaying wood. However, it is only very slowly and incompletely degraded in soil, instead giving rise to humic acids and other recalcitrant insoluble compounds, which account for most of the carbon content of soil. These compounds are important for soil fertility. Enzymic attack on lignin involves radical oxidation by peroxidases. Peroxidases oxidase a substrate with the help of H_2O_2, which acts as a co-substrate that is formed in separate reactions. This most likely explains why lignin degradation as a rule is found in organisms with hyphal growth. This allows them to penetrate inert substances such as wood, to come into close contact with the substrate, and to produce in situ oxidizable co-substrates and exoenzymes that support extracellular free-radical attack on the lignin polymer (see Fig. 9.**10**).

Despite its simple building blocks, the structure of lignin is extremely complex as a result of its biosynthesis via random nonstereospecific polymerization mechanisms. Phenylpropanoid precursors (Fig. 9.**9a**) are oxidized by peroxidases with H_2O_2 or by laccases with O_2 to yield free aromatic alcohol radicals, which polymerize randomly and spontaneously in various coupling reactions ("dehydrogenative polymerization"). Ether and inert carbon–carbon linkages are formed, most of which are non-hydrolyzable. The most abundant structure is the β-aryl ether between propyl side chains and aromatic nuclei (**β-*O*-4 ether**), which accounts for about half of the interphenylpropane linkages (Fig. 9.**9b**). The next most abundant are **carbon–carbon bonds**, found primarily between aromatic nuclei and propyl side chains. Ester linkages occur between the free carboxyl group of uronic acids in

Box 9.3 Peroxidases are donor:hydrogen peroxide oxidoreductases; they are **hemoproteins** containing one molecule of iron-protoporphyrin IX per enzyme molecule. During the catalytic cycle, the peroxidases are oxidized to form higher-valency reactive oxidation states, while H_2O_2 is reduced to water. Peroxidases also play an important role in the oxidation of xenobiotics. Lignin peroxidases or ligninases are able to remove one electron from aromatic substrates, yielding an unstable aryl cation radical, which undergoes C–C bond cleavage and a variety of other degradative reactions. **Manganese peroxidases** oxidize free Mn(II) to Mn(III), which in turn diffuses away from the enzyme and oxidizes substrates with free phenolic groups [note that Mn(III) is a relatively weak oxidant, and that non-phenolic benzyl ether C–O–C bonds are extremely stable, whereas phenolic benzyl ether bonds are more easily hydrolyzed, e.g., by alkali treatment]. Lignin peroxidases, but not manganese peroxidases, are able to oxidize the principle non-phenolic structures of lignin. Yet, some lignin degraders do not produce lignin peroxidase, and Mn(III) cannot perform the entire job because non-phenolic structures are hardly attacked. Other ligninolytic mechanisms must therefore exist.

Box 9.2 The complexity of the structure of lignin complicates the use of lignin in biochemical studies of its degradation. Elucidation of degradation pathways has depended heavily on simpler **synthetic dimeric model compounds** that contain an intermonomeric bond representative of lignin. A model for the β-aryl ether (β-O-4 ether) linkage is shown in Fig. 9.**9b**. This substructure has hydroxyl groups in the α- and γ-position which—like free phenolic hydroxyl groups—may be linked to carbohydrates. Commonly, biodegradation of lignin is followed by trapping the $^{14}CO_2$ respired from active cultures growing on [^{14}C]lignin (biologically or chemically formed from specifically labeled precursors or building blocks). A significant fraction of the lignin may be mineralized, but some is generally converted into insoluble polymeric products.

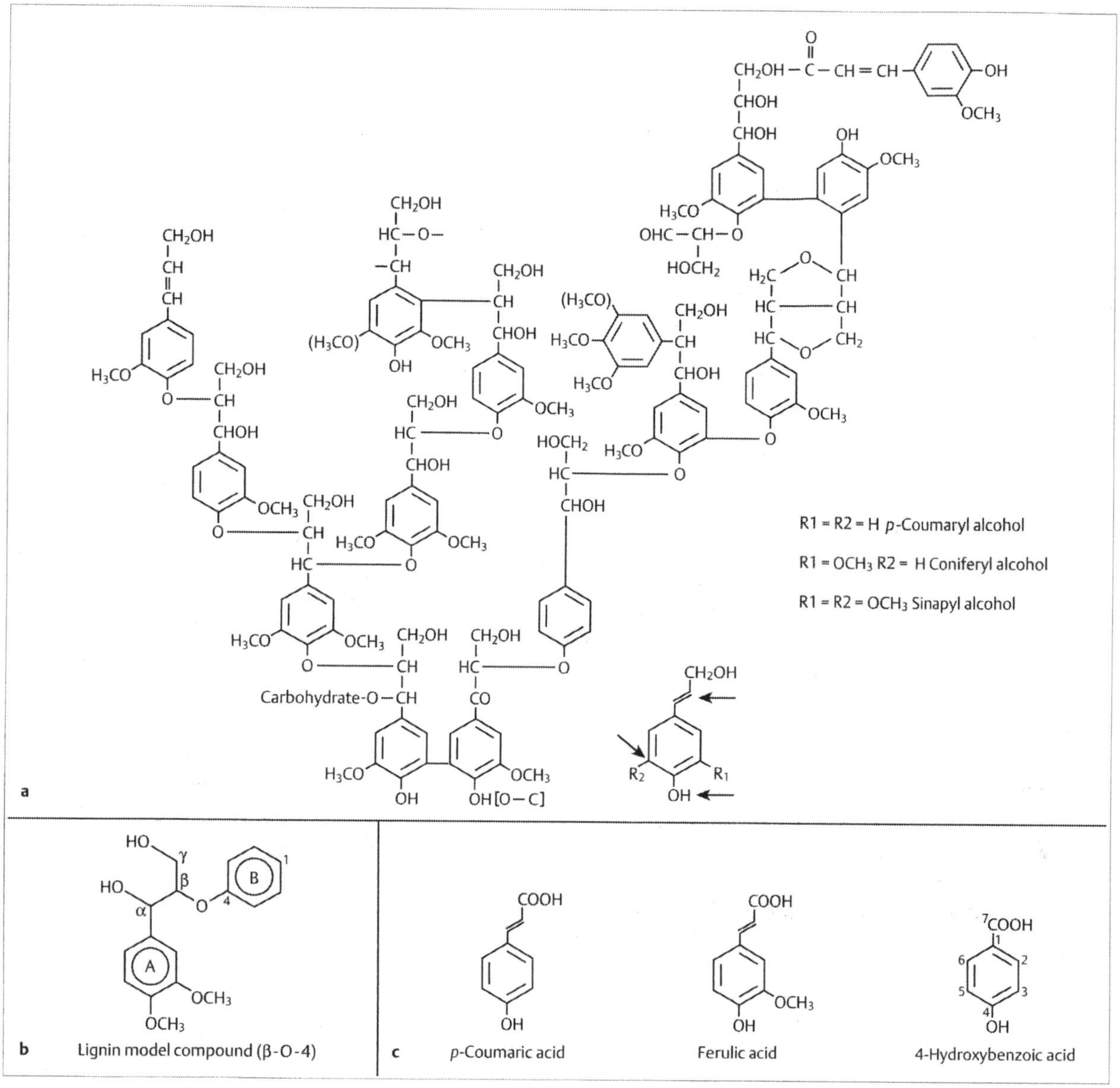

Fig. 9.**9a,b,c** **Lignin structure and nomenclature of carbon units.**
a The three precursor alcohols of lignin. One-electron oxidation of these lignin building blocks and subsequent polymerization reactions produces the three-dimensional lignin. The arrows indicate the position of intermonomeric bonds in lignin, including arylether, biphenyl, and diphenylether bonds.
b Lignin model compound with *β*-*O*-4 ether linkage.
c Aromatic acids found in lignin-carbohydrate bands.

hemicellulose and the propyl side chain of lignin. The multiplicity of intermonomeric bonds and the irregularity in their arrangement can be deduced from Fig. 9.**9a**. Figure 9.**9c** shows the structures of some common low-molecular-weight aromatic constituents of plant cell walls.

Organisms. Despite the importance of microbial ligninolysis, little is known about bacterial lignin degradation. There is no conclusive evidence for lignin degradation under anoxic conditions except for removal of hydrolyzable residues and O-methyl groups and modification of side chains. Most of the studies have

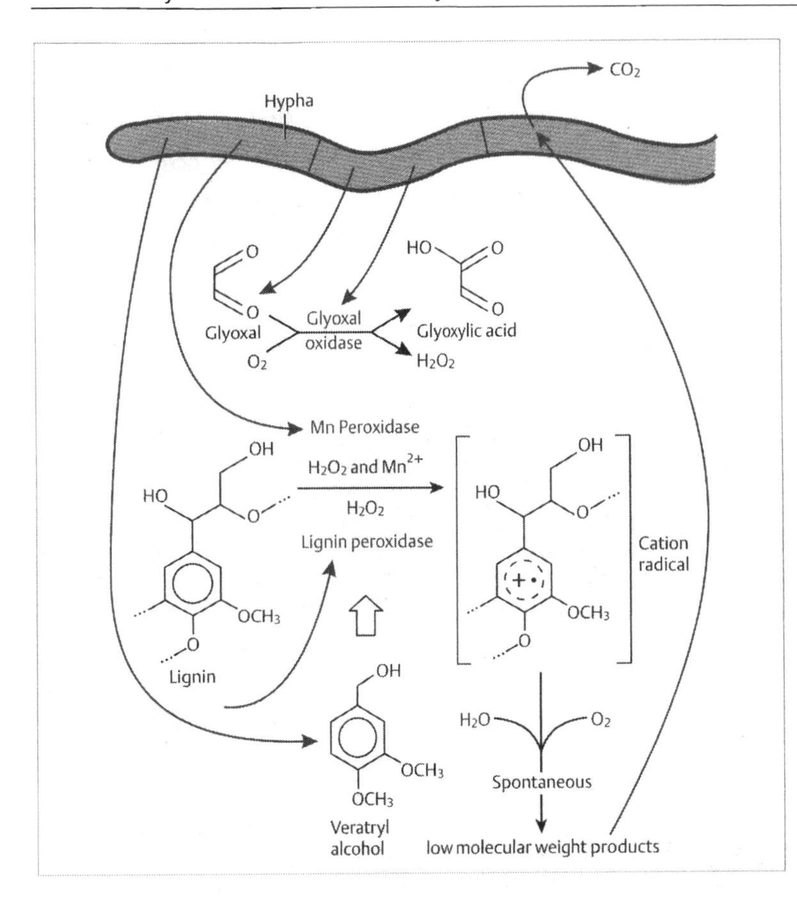

Fig. 9.**10** **The ligninolytic system of the white-rot fungus *Phanerochaete chrysosporium.*** Glyoxal (or methylglyoxal) is produced by the fungus; glycolaldehyde originates from the *β*- and *γ*-carbon of the phenylpropane units during lignin degradation. Oxidation of these substrates by glyoxal oxidase generates H_2O_2, and at the same time the chelators glyoxylate and finally oxalate are formed. These chelators complex Mn(III) formed by manganese peroxidase. In addition, aromatic alcohols such as veratryl alcohol are excreted or formed in the course of degradation; they are substrates for lignin peroxidase and promote ligninolysis by this enzyme via mechanisms that remain to be defined. Mn(III) chelates oxidize the more reactive phenolic structures that make up approximately 10% of lignin. This may facilitate later attack by the bulkier, but more powerful oxidant lignin peroxidase. The aryl cation radical formed by these peroxidases spontaneously decomposes to low-molecular-weight (LMW) products as well as to lesser amounts of residual polymeric products. [After 3]

dealt with low-molecular-weight lignin model compounds that can be taken up and metabolized intracellularly, whereas ligninolysis is a necessarily extracellular process. It appears that lignin degradation is mostly the domain of filamentous **white-rot fungi**, such as the basidiomycete *Phanerochaete chrysosporium* (Fig. 9.**10**), which degrade the (brown) lignin in wood in order to gain access to the readily metabolized (white) cellulose and hemicellulose. White-rot fungi appear to have adopted this strategy because lignin itself cannot support cell growth. Most white-rot fungi are saprophytes, but a few (e.g., *Armillaria mellea*) are pathogens on living plants. A variety of lower fungi and bacteria are also found in decaying wood, but their role in ligninolysis remains unclear. **Brown-rot** and **soft-rot fungi** attack polysaccharides predominantly, and do not degrade lignin extensively.

Although bacteria may not mineralize lignin efficiently, some evidently can solubilize it. This ability is found in genera *Streptomyces*, *Thermomonospora*, *Actinomadura*, *Pseudomonas*, and *Xanthomonas*, which can grow on natural lignocellulose as a carbon source. By coverting lignin to a water-soluble form, these bacteria presumably facilitate its removal so that they can utilize the cellulose and hemicellulose. Given the lack of information on mechanisms that bacteria might use to degrade lignin, it is perhaps most productive to review the important and relatively well-characterized fungal system (Fig. 9.**10**).

Enzymes. Fungal attack on lignin requires a battery of different enzymes, co-substrates, metals, and molecular oxygen. The process is located mainly at the tip of the hyphae, where exoenzymes are excreted by exocytosis. In white-rot fungi, the hemoproteins lignin peroxidase and manganese-dependent peroxidase, as well as the copper-containing laccase, are families of enzymes that have been implicated in the oxygen-dependent degradation of lignin. They are present as multiple, functionally similar isozymes. They have in common that they are one-electron oxidants, producing substrate-free radicals that undergo a variety of post-enzymic degradative reactions.

The enzyme principally responsible for ligninolysis in many fungi is thought to be lignin peroxidase, which oxidizes lignin structures by one electron, yielding **aromatic cation radical intermediates** that undergo spontaneous fission reactions (Fig. 9.**10**). For example the predominant arylglycerol *β*-aryl ether (*β*-O-4) substructure of lignin (Fig. 9.**9b**) is cleaved between C_α and C_β of its propyl side chain. The products of this fission reaction are a C_α-linked benzaldehyde, a phenol, and glycolaldehyde. A variety of strategies are employed to produce the H_2O_2 needed for lignin peroxidase turnover. Some secrete aldehydes, such as glyoxal, and utilize an extracellular glyoxal oxidase to oxidize it; other aldehydes, such as glycolaldehyde, are derived from lignin. By oxidation of these aldehydes, O_2 is reduced to H_2O_2. Other fungi secrete various aromatic alcohols along with H_2O_2-producing aryl alcohol oxidases.

Limitation of the proposed mechanisms. Exoenzymes such as lignin peroxidase appear to be too large to penetrate sound wood. Some investigators, therefore, believe that these enzymes are involved chiefly in the later stages of fungal ligninolysis. The initial depolymerization reactions instead involve diffusible low-molecular-weight factors that can react to generate oxyradicals within the wood cell wall, thus initiating nonspecific free-radical cleavage reactions in the lignin. The well-known **Fenton system**, which uses Fe^{2+} and H_2O_2 to form the hydroxyl radical, $OH^\cdot$, is one example of such a system.

The strong oxidizing power of a mixture of H_2O_2 and Fe^{2+} was recognized almost 100 years ago by Fenton. The actual oxidant generated is the hydroxyl radical, $OH^\cdot$ (Fenton reaction, Reaction 9.1):

$$Fe^{2+} + H_2O_2 \rightarrow Fe^{3+} + OH^\cdot + OH^- \qquad (9.1)$$

The hydroxyl radical has an extremely high redox potential for one-electron reduction to form water (Reaction 9.2):

$$OH^\cdot + H^+ + e^- \rightarrow H_2O, E_0' = +2\,180\text{ mV} \qquad (9.2)$$

The hydroxyl radical is the most potent unselective oxidizing agent in biological systems and has been proposed to play a role in lignin degradation. A problem with the Fenton system is its extremely low selectivity—$OH^\cdot$ reacts with virtually all biomolecules and would therefore have to be produced in large quantities at a distance from the fungal hyphae to carry out ligninolysis. No mechanism for such a process is yet known.

Applications. The potential use of the process in biopulping and bleaching, improvement of lignocellulosic feed, clean up of toxic waste sites, and effluent treatment is evident.

9.7 Protein Is a Substrate for Many Microorganisms

Protein is a major constituent of living organisms and therefore represents an important food source for microorganisms. Protein is built up by the 20 natural L-amino acids linked by peptide bonds; these bonds are hydrolyzed by proteinases (hydrolases). Protein degradation has to be initiated outside the cell. Products are oligopeptides and/or amino acids that are finally transported into the cell by more or less specific transport systems. Many structural proteins are highly insoluble, which renders them difficult to attack enzymically. Examples are keratin, elastin, and collagen. Proteins can be connected with polyphenolic compounds (tanning); removal of non-proteinaceous compounds is rate-limiting. Often, degradation of proteins is enhanced by denaturation (i.e. loss of biological activity due to unfolding) of the proteins by extremes of pH, temperature, or other factors.

Enzymes. Numerous proteinases are produced by microorganisms—depending on the individual strain—which may produce several enzymes under various culture conditions. Proteinases are often synthesized in an inactive pre-form that becomes active by processing, either during transport or by autocatalytic activation. Endopeptidases cleave the polypeptide chain, followed at the extracellular or intracellular site by further hydrolysis by exopeptidases that split off the carboxy- or amino-terminal amino acid residue (Fig. 9.**11**) Proteinases are the best-studied enzymes of all; their structures and catalytic mechanisms are mostly well understood. They are assayed using chromophoric synthetic substrates that mimic the natural amino acids and peptide links hydrolyzed.

Classification. Proteinases are generally classified by the nature of their active center (rather than by their substrates). These classes—serine, cysteine, metal, and

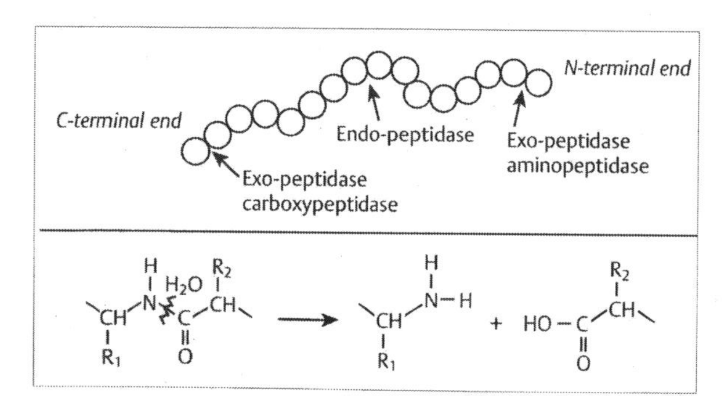

Fig. 9.**11 Sites of cleavage of polypeptide chain by proteinases**

aspartic proteinases—can be distinguished by their sensitivity toward specific inhibitors. Serine -OH groups are covalently blocked by diisopropylphosphofluoridate or organophosphorus derivatives (e.g., subtilisin produced by *Bacillus subtilis*). Cysteine -SH groups are inhibited by metal ions, and alkylating or oxidizing agents (e.g., trypsin-like enzymes produced by many bacteria). Metals are removed by complexing agents or by dialysis (e.g., thermolysin). The aspartate β-carboxyl group can be inactivated by certain alkylating agents (e.g., pepsin-like enzymes, which are rare in bacteria, but common in fungi and yeasts). These four classes are further distinguished by the side-chain specificity, i.e., the amino acids at the site that is cleaved.

Applications. A large quantity of microbial proteinases are produced by fermentation for industry. The reasons are obvious: (1) microorganisms produce a huge variety of proteinases; the pH range can be from 0.1 M HCl to 0.1 M NaOH, the temperature range from 4 °C to 110 °C, and up to 8 M urea may be tolerated; (2) inexpensive raw materials can be used as substrates; (3) mass production is easy; some strains produce 10 g proteinase or more per liter; (4) strains can be improved by mutation and selection after genetic engineering; the screening methods and protein engineering depend on the purpose of the proteinase wanted, e.g., the side-chain specificity, pH and temperature range and stability, behavior with detergents, salts, urea, and inhibitors, catalytic activity, and other properties; and (5) there are many applications of proteinases in food industry (e.g., milk-clotting), leather industry, chemical industry (notably in the production of household detergents for laundrying), and medicine (e.g., processing of protein hormones produced by engineered microorganisms). Proteinases can also be used as catalysts in low-water environments to catalyze peptide synthesis, i.e., using their reverse reaction.

Cellular functions and control. Many cellular (cytoplasmic, membrane-associated, periplasmic, outer membrane) proteinases play a key role in the regulation of cellular processes, rather than in food digestion. Limited proteolysis cleavages are essential in maturational processing of many proteins, inactivation of certain regulatory proteins, transport, or development (spore formation) (see ATP-dependent proteaosome Chapter 10). Their cellular action needs to be highly controlled, a process not fully understood in most cases.

Proteinases are also important virulence factors of pathogens, which often live at the expense of the host's protein. Examples are subtilisin, streptokinase, hemolysins, and elastase, which have special targets and cleave cellular structural proteins (see Chapter 33).

Organisms. Protein degradation occurs in most groups of bacteria, both aerobically and anaerobically. Many proteolytic bacteria are able to hydrolyze gelatin. This can easily be recognized with gelatin containing Chinese ink, which is set free upon degradation of the protein matrix. Colonies on solid media form a clearing zone in which gelatin can no longer be precipitated by acid.

9.8 Ribonucleic and Deoxyribonucleic Acids Are Ubiquitous Biopolymers That Are Easily Degraded by Extracellular Hydrolases

The catabolic phosphodiesterases act on nucleic acids in a sequence-independent way. The products are various low-molecular-weight transportable compounds, mononucleotides, or oligonucleotides. Nucleotides are normally dephosphorylated by periplasmic phosphatases, and phosphate and nucleoside are taken up independently. Ribose, deoxyribose, nucleotides, nucleosides, and nucleobases can be used by many microorganisms. Intracellular enzymes with more site-specific action are responsible for essential cellular processes and will not be considered here.

Basically, RNase and DNase enzymes are classified as those that: (1) act on DNA, RNA, or DNA-RNA hybrids, (2) act on a double- or single-stranded substrate, (3) act at the 5′- or 3′-termini (exo-acting); enzymes that cleave the 3′-adjacent phosphodiester bond yield nucleoside 3′-phosphate and those that cleave the 5′-adjacent phosphodiester bond yield nucleoside 5′-phosphate, (4) cleave internal phosphodiester bonds (endo-acting), (5) recognize particular sequences, and (6) act nonspecifically (nucleases). Nucleases, very much like proteinases, play important roles in cellular metabolism, which requires sophisticated control mechanisms of cellular enzymes and protection of the cellular nucleic acids.

Applications. The nonspecific extracellular enzymes are used in small scale for processing of nucleic acids or nucleotides. The highly sequence-specific intracellular restriction enzymes are important tools for molecular genetic techniques.

9.9 Lipids Are Membrane Constituents of all Organisms

Although lipids are not polymers in a strict sense, many rules of polymer degradation apply to these poorly water-soluble compounds. Lipids are membrane constituents of all organisms. In addition, in eukaryotes, oils and fats are triacylglycerol storage materials, which are liquid or solid at ambient temperatures; they contain low amounts ($<5\%$) of phospholipids or glycolipids, carotenoids, and steroids, whereas membrane lipids contain no triacylglycerol. Carotenoids and steroids are nonsaponifiable. Saponification of the other components yields fatty acids, glycerol, and polar head groups. Fatty acids are oxidized (see Chapter 9.14.1) and/or assimilated, either indirectly via acetyl-CoA, or directly after modification (elongation, desaturation). For glycerol assimilation, see Chapter 7. Many microorganisms produce **surfactants**, mostly glycolipids, for microdroplet formation when grown with lipids.

Lipases. Acylhydrolases that catalyze the hydrolysis of triglycerides into diglycerides, monoglycerides, glycerol, and fatty acids are called lipases. These enzymes normally hydrolyze tri-, di-, and monoacylglycerols equally well. In solvents of low water activity, acylhydrolases catalyze the reverse reaction, i.e., esterification of glycerol with fatty acids, and other reactions (Fig. 9.**12a**). Also, some lipoproteins (egg yolk) are good substrates (opacity formation on egg yolk agar). Some lipases have also phospholipase activity. Lipases are exoenzymes or periplasmic enzymes or are bound to the cell surface. They can handle both water-soluble and water-insoluble substrates, but require for action the lipid–water interphase of droplets or reverse-phased micelles. The requirement of emulsified lipids presents a difficulty for various in vitro assays, e.g., using colored or fluorescent substrates. Alternatively, titration of fatty acids released is measured.

Lipases have different specificity (1) Some catalyze a nonspecific reversible hydrolysis of triacylglycerols, with preference for long-chain fatty acids (the most common specificity) (or ester synthesis in alcohols with $<1\%$ water), (2) some have a specificity for the acyl substituent, and (3) some have a specificity for the position at which hydrolysis occurs (Fig. 9.**12b**). Their function requires association at an oil-water interface, in contrast to esterases. A lipophilic α-helix loop of the polypeptide chain acts as a "lid". It opens up the catalytic site of the enzyme (a serine hydrolase) to receive the substrate only when the enzyme is bound to the lipid–water interphase, a phenomenon called "**interfacial activation**". This property causes unusual kinetic behavior. The substrate-binding domains of lipases show high similarity; the remainder of the enzyme molecules show little similarity.

Control. Lipases are either constitutively formed or are under complex control. Synthesis and excretion are influenced, for example, by iron, carbon source, or the presence of non-metabolizable polysaccharides. The stimulation of exolipase yield by some unrelated compounds that are not growth substrates may be caused by competitive detachment of lipases from the cell surface.

Phospholipases. These are mostly intracellular, often membrane-associated enzymes that hydrolyze phospholipids at different sites (Fig. 9.**13**). The products, lysophospholipids, are excellent surfactants that readily lyse biological membranes. Therefore, phospholipases need to be strictly controlled. Furthermore, lysophosphatidic acids (e.g., 1-acylglycerol-3-phosphate) are powerful intracellular signalling molecules in eukaryotes. Their release by pathogenic phospholipid degraders is harmful, and some bacterial toxins are phospholipases; hence, lipases are pathogenicity factors involved in colonization and persistence of pathogenic bacteria on the skin and cell surfaces. The four types of

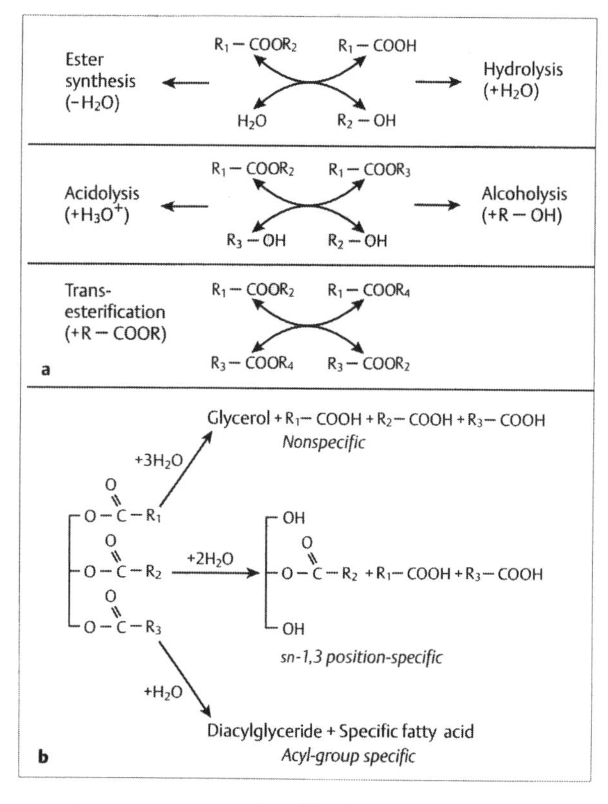

Fig. 9.**12** **Reactions catalyzed by lipases (a) and details of hydrolytic reactions (b).** R_1, R_2 = alkyl chains

phospholipases that attack the substrate at distinct sites are indicated in Fig. 9.13. Only phospholipase C (see below) is found extracellularly; its product diacylglycerol diffuses through the membrane. There is no enzyme B; the removal of the last acyl group is carried out by lysophospholipases L_1 (at C1) and L_2 (at C2). A_1 and A_2 also catalyze acyl transfers; D catalyzes the transfer and exchange of the polar head group. These enzymes function well in organic solvents.

Applications. Various reactions catalyzed by lipases are used technically: hydrolysis or glycerolysis of fats and oils, esterification, transesterification, and acylation of various HO-containing acceptor molecules (sugar alcohols, glycerol, monosaccharides), and formation of lactones. Lipases are required in food industry and in the production of soap, detergents, surfactants (monoacylglycerol, sugar monoacylesters), toiletries, plastics, and paints. Notably alkaliphilic and thermostable enzymes, and those with unusual catalytic properties are of practical interest. Some lipases are extraordinarily heat-stable, which can lead to the deterioration of sterilized food.

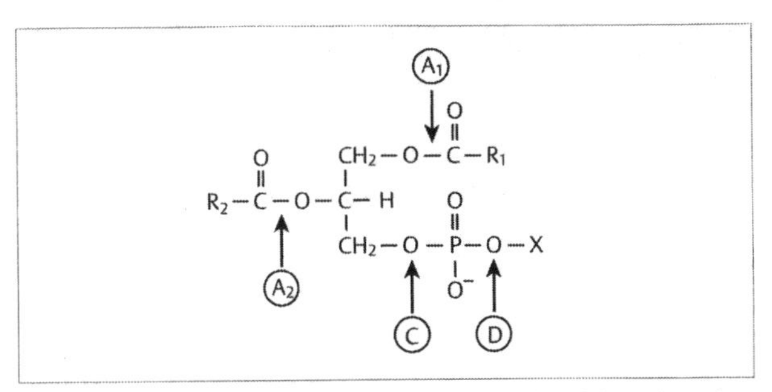

Fig. 9.13 **Sites of cleavage of a phospholipid by phospholipases A_1, A_2, C, and D.** It is uncertain which bond, C–O or P–O, is cleaved by enzymes C and D

Organisms. It would be more unusual to find bacterial genera without lipases than to find lipase-producing ones. Well-studied examples are species of the genera *Staphylococcus*, *Pseudomonas*, *Bacillus*, *Streptomyces*, *Aeromonas*, *Xenorhabdus*, *Moraxella*, *Propionibacterium*, *Chromobacterium*, and *Serratia*.

9.10 There Are Variants of the Central Pathways Into Which Products of Biopolymer Degradation Are Channeled

The release of monomeric or oligomeric building blocks from polymeric substrates has been discussed. These building blocks are then oxidized mostly to pyruvate, acetyl-CoA, or intermediates of the citric acid cycle. Before turning to the huge variety of catabolic pathways, a few special aspects of the central metabolic pathways that have not been considered in the previous chapters should be mentioned. These aspects concern facultative or even strict anaerobes that in the presence of an inorganic electron acceptor such as nitrate or sulfate may be able to oxidize organic compounds completely to CO_2. Anaerobic oxidizers have many peculiarities in their central metabolic pathways.

Pyruvate and acetyl-CoA oxidation. Pyruvate is not always oxidized in the conventional way. There are five enzymes known to degrade pyruvate (see Chapter 12.2.2). In aerobes the pyruvate dehydrogenase complex mediates oxidative decarboxylation. Other enzymes come into play in facultative and strict anaerobes; these bacteria are able to oxidize organic compounds to CO_2 completely (see Chapter 12). Variations also occur in the end-oxidation pathways of acetyl-CoA. In aerobes, the citric acid cycle is the end-oxidation pathway. In some strict anaerobic complete oxidizers, either modifications of the citric acid cycle are found or an alternative route, the oxidative acetyl-CoA/CO dehydrogenase pathway, is used. For details, see Chapter 12.

Gluconeogenesis versus carbohydrate degradation. It appears that so far all prokaryotes form hexose phosphates from C_3 compounds via glucogenesis, i.e., a reversal of the Embden-Meyerhof-Parnas glycolytic pathway. The enzymes for glucose oxidation are normally constitutive. As a rule, all other proteins and enzymes required for transport and degradation of other sugars and glycosides are inducible and are catabolite repressed. In contrast to the seemingly uniform biosynthetic routes, the degradative routes vary depending on the organism and the substrate.

9.11 Several Pathways Serve for the Oxidation of Carbohydrates, the Most Common Substrates

How oligosaccharides are handled and how the different monomeric sugars are channeled into the common intermediates of degradative pathways will now be discussed. Figure 9.**14** shows the principal structures of common carbohydrate building blocks and derivatives.

9.11.1 Peripheral Reactions Channel Hexoses and Pentoses Into the Central Degradation Pathways

The channeling of the great variety of low-molecular-weight carbohydrates into the central degradation pathways requires a set of interconversion reactions; this part of metabolism is referred to as peripheral. The peripheral enzyme outfit is substrate-induced, whereas the central degradation routes normally are constitutive. The mechanisms of sugar interconversions include phosphorylation, phosphorolysis of glycosides, *keto-enol* isomerization, oxidation or reduction, and aldol cleavage. Additional reactions remove substituents such as $-NH_2$, $-OCH_3$, or $-OOC-CH_3$ groups. The regulation of the uptake and metabolism of various sugars is described in Chapters 18 and 20.

Hexose | Pentose

D-Glucose (Aldose) | D-Fructose (Ketose) | D-Xylose (Aldose) | D-Xylulose (Ketose)

Hexuronide | Hexuronate | Hexonate

β-D-Glucuronide | D-Glucuronate | D-Gluconate

Glucoside | Amino sugar | Hexitol (polyhydric alcohol)

α-D-Glucoside | β-D-Glucosamine | D-Glucitol

Fig. 9.**14** **Structures of hexoses, pentoses (aldose, ketose), glucosides, amino sugars, hexuronides, hexuronates, hexonates, and polyalcohols**

Conversion of oligosaccharides and disaccharides into monosaccharides. The most common disaccharide or trisaccharide substrates are summarized in Table 9.**1**. Many bacteria can transport common disaccharides like sucrose, cellobiose, or maltose by permeases (Fig. 9.**15**). **Phosphorolytic cleavage** of the glycosidic bond of the disaccharide (e.g., maltose) by specific phosphorylases results in **glucose 1-phosphate** and the second sugar residue. In addition, glucose can be released from maltodextrins by amylomaltase, which catalyzes the reversible Reaction 9.3:

$$(\text{Glucose})_n + (\text{Glucose})_m \leftrightarrow (\text{Glucose})_{n+m-1} + \text{Glucose} \quad (9.3)$$

Others, like lactose (β-galactoside) and melibiose (α-galactoside), are **hydrolyzed** by specific galactosidases to the free sugars. Alternatively, D-glucosides like trehalose (α-glycoside) are **transported** by the phos-

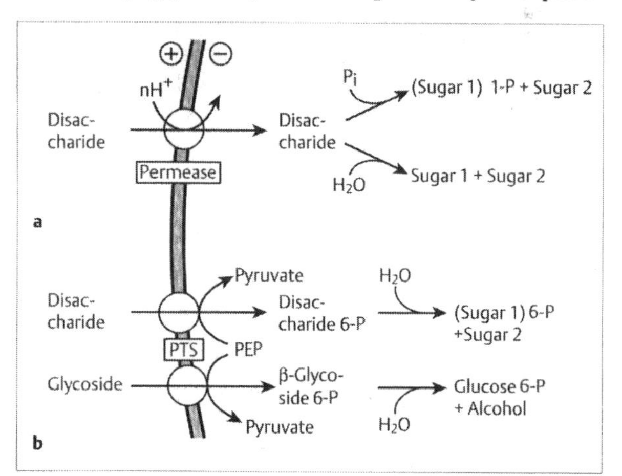

Fig. 9.**15a,b** **Uptake of disaccharides (sugar 1 – sugar 2) and glycosides and conversion to monosaccharides (and alcohol).** PTS, phosphotransferase system; PEP, phospho*enol*-pyruvate. Examples for disaccharides in (b) are trehalose, lactose or sucrose; a well-studied glucoside is arbutin (the glycoside of glucose and hydroquinone)

Table 9.1. **Some common disaccharides and trisaccharides**

	Sugar 1 (pyranose)	Glycosidic linkage	Sugar 2 (pyranose, unless indicated otherwise)
Disaccharide			
Cellobiose	D-Glucose	β 1-4	D-Glucose
Maltose	D-Glucose	α 1-4	D-Glucose
Isomaltose	D-Glucose	α 1-6	D-Glucose
Trehalose	D-Glucose	α 1-1 α	D-Glucose
Sucrose	D-Glucose	α 1-2 β	D-Fructose (furanose)
Lactose	D-Galactose	β 1-4	D-Glucose
Melibiose	D-Galactose	α 1-6	D-Glucose
Gentiobiose	D-Glucose	β 1-6	D-Glucose
Chitobiose	D-Glucosamine	β 1-4	D-Glucosamine
Chondrosine	D-Glucosamine	β 1-3	D-Galactose
Hyalobiuronic acid	D-Glucuronic acid	β 1-3	D-Glucosamine
Xylobiose	D-Xylose	β 1-4	D-Xylose
Trisaccharide			
Raffinose	D-Galactose	α 1-6	Sucrose (C6 of glucose unit)

photransferase system; the resulting disaccharide 6-phosphate is hydrolyzed to **glucose 6-phosphate** and the second sugar residue. Similarly, β-glycosides are transformed into 6-phosphoglycosides and hydrolyzed to glucose 6-phosphate and alcohol (Fig. 9.**15**).

Channeling sugars into intermediates of degradation pathways. Glucose is converted to glucose 6-phosphate by the phosphotransferase system (PTS) (Chapters 5, 20), while being taken up. Similarly, other hexose 6-phosphates are formed during PTS-dependent transport. Glucose 1-phosphate is formed from storage glycogen by phosphorylase; it is subsequently isomerized to the 6-phosphate and further metabolized. What about other hexoses, aldoses, ketoses, amino hexoses, and other compounds that, for example, are formed during disaccharide metabolism? They require additional inducible enzymes. Free **D-glucose** formed from disaccharides by glycosidases is converted by glucokinase to glucose 6-phosphate (Reactio 9.4).

$$\text{Glucose} + \text{ATP} \rightarrow \text{Glucose 6-phosphate} + \text{ADP} \tag{9.4}$$

Free glucose is formed for example, from lactose (by β-galactosidase) or melibiose (by α-galactosidase). **D-Galactose** (e.g., formed by β-galactosidase from lactose or when taken up by permease) requires three additional enzymes for conversion to glucose 1-phosphate (Fig. 9.**16**). **D-Fructose, D-mannose, and L-sorbose** are taken up by the PTS and are thereby converted to fructose 1-phosphate, mannose 6-phosphate, and glucitol 6-phosphate. Their conversion to D-fructose 1,6-bisphosphate is shown in Fig. 9.**17**.

Amino sugars such as *N*-acetyl-D-glucosamine and D-glucosamine are the main sugars derived from structural components such as chitin, chitosan, or bacterial cell walls. They are transported via the phosphotransferase system, which results in phosphorylation of the C6 OH group. The phosphorylated form is deacetylated and/or deaminated to fructose 6-phosphate. Note that D-glucosamine 6-phosphate is synthesized from fructose 6-phosphate with glutamine as NH_2-donor (Chapter 7).

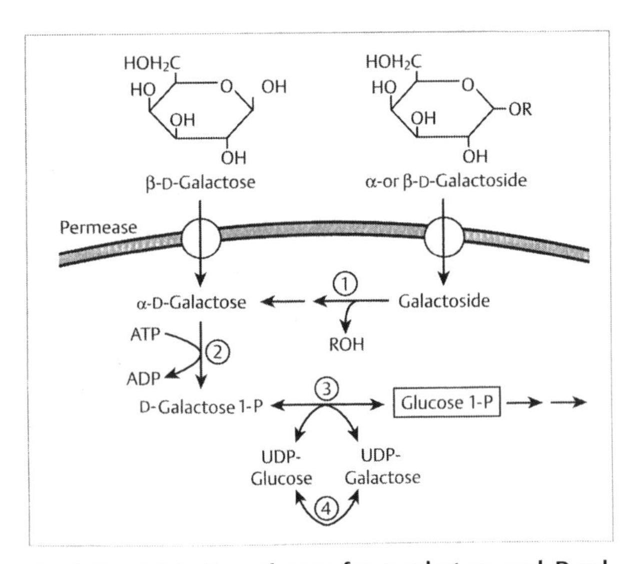

Fig. 9.**16** **Catabolic pathways for D-galactose and D-galactosides.** (1) α- and β-Galactosidases (2) kinase, (3) UDP-transferase, and (4) epimerase. In case of β-galactose and β-galactosides a rotamutase converts β-galactose into α-galactose

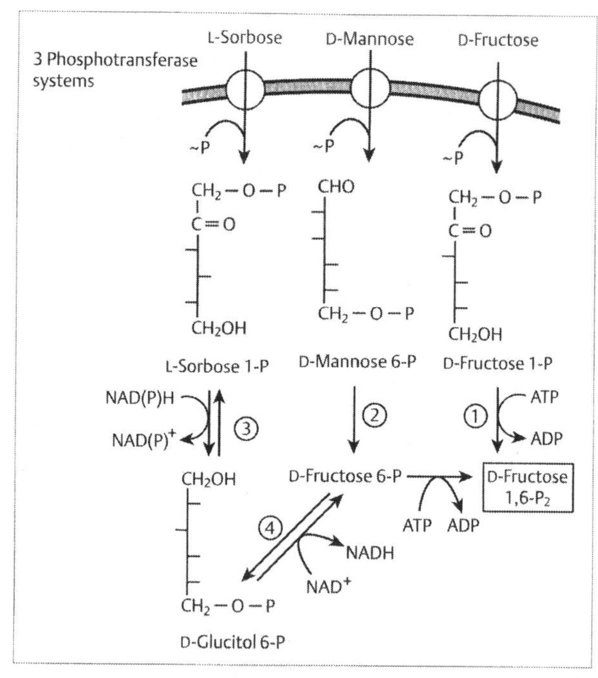

Fig. 9.17 **Catabolic pathways for D-fructose, D-mannose, and L-sorbose.** (1) kinase, (2) isomerase, (3) reductase, and (4) dehydrogenase

Channeling pentoses into intermediates of degradation pathways. Pentoses such as D-xylose, D-ribose, or L-arabinose are very common substrates (see hemicelluloses, Chapter 9.5.3 and nucleic acids, Chapter 9.8). They are actively taken up by permeases. **Pentitols** are first oxidized to the ketopentoses. Pentoses and ketopentoses are converted to D-xylulose 5-phosphate. There are several options for the further metabolism of pentose phosphates. Normally, D-xylulose 5-phosphate and ribose 5-phosphate are interconverted to fructose 6-phosphate and glyceraldehyde 3-phosphate by transaldolase and transketolase of the pentose phosphate cycle. Further degradation is via the Embden-Meyerhof-Parnas pathway. However, there are other options.

Methylpentoses, such as L-rhamnose and L-fucose, are derived from hexoses by reduction of the $-CH_2OH$ group at C6 to a $-CH_3$ group. They are transported via permeases, converted to the 1-phosphoketulose form, and cleaved by specific aldolases to dihydroxyacetone phosphate and L-lactaldehyde; the latter is oxidized via L-lactate to pyruvate.

Pentoses from nucleosides and deoxynucleosides are released as follows. Nucleic acid degradation results in nucleotides and deoxynucleotides, which are dephosphorylated before the (deoxy)nucleosides and phosphate are transported. The nucleobases of some (deoxy)nucleosides [(deoxy)cytidine, (deoxy)adenosine] are deaminated by specific nucleoside deaminases [to (deoxy)uridine and (deoxy)inosine]. The (deoxy)ribose is then extracted from the (deoxy)nucleosides as shown in Fig. 9.**18**. The *N*-glycosidic bond is phosphorolytically cleaved by phosphorylase to nucleobase and (deoxy)ribose 1-phosphate (note that this reaction is reversible). A mutase forms the 5-phosphopentose. Deoxyribose 5-phosphate is cleaved by a class I aldolase (Schiff's base enzyme) into glyceraldehyde 3-phosphate and acetaldehyde. Ribose 5-phosphate is further metabolized, for example, by the non-oxidative pentose phosphate pathway.

Channeling other related compounds into intermediates of degradation pathways. Hexuronides (e.g., glycosides of glucuronic acid), **hexuronates** (e.g., glucuronic acid), and **hexonates** (e.g., gluconate) are transported by permeases and converted in several steps into 2-keto-3-deoxy-gluconate 6-phosphate, the characteristic intermediate of the Entner-Doudoroff pathway of hexose catabolism. **Polyols** (sugar alcohols, such as glucitol = sorbitol, mannitol, or galactitol = dulcitol) are transported by the PTS system and are thereby converted into their 6-phosphates. Oxidation leads to fructose 6-phosphate or the analogous tagatose 6-phosphate (from galactitol). Phosphorylation leads to the ketose 1,6-bisphosphates, which are cleaved by specific fructose bisphosphate aldolase and tagatose bisphosphate aldolase, respectively. D-Arabinitol, ribitol, and xylitol are oxidized by NAD^+-dependent dehydrogenases to the corresponding pentulose and then activated by ATP-dependent kinases.

After channeling the different carbohydrates into a few central intermediates, these intermediates are further oxidized through the central degradation pathways. A summary of the most common pathways of sugar degradation is given in Fig. 9.**19**. As will be seen, many variants of the main pathway occur. This is especially true for the Archaea, in which virtually only modifications of these pathways occur. Transport of

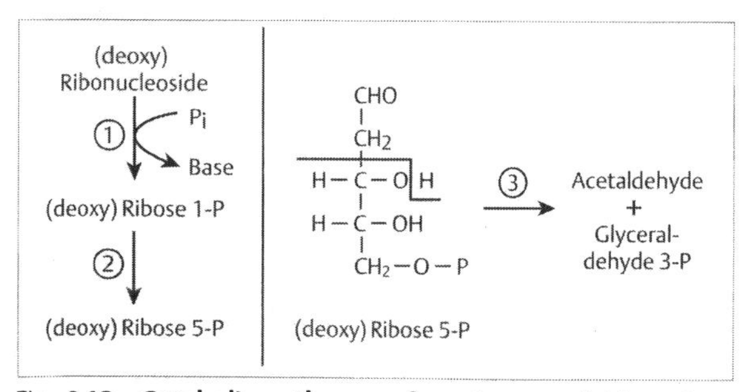

Fig. 9.18 **Catabolic pathways of pentoses and (deoxy)nucleosides from nucleic acids.** (1) Phosphorylases, (2) mutase, and (3) deoxyriboaldolase

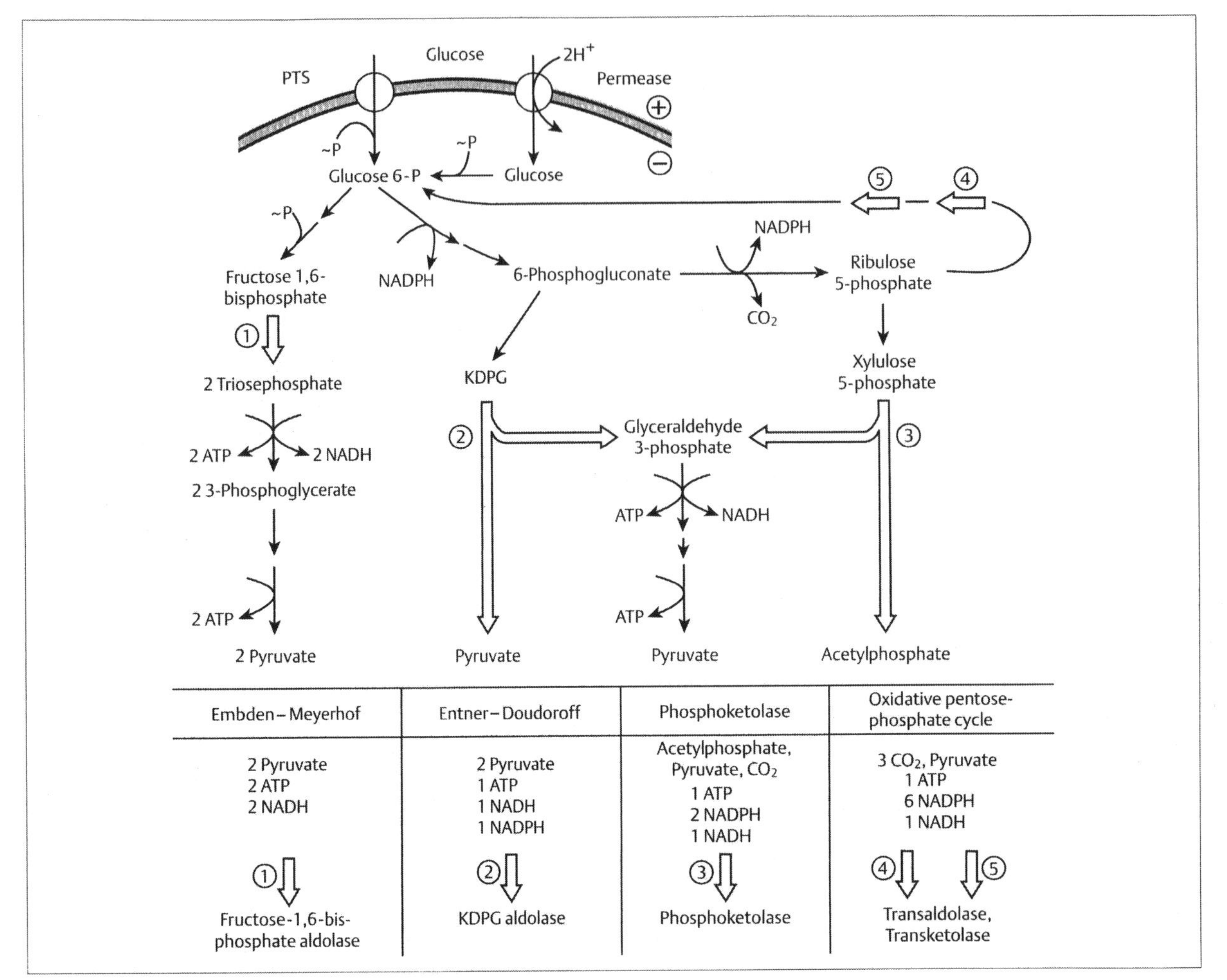

Fig. 9.**19** **Overview of the most common glucose-degradation pathways.** The figure summarizes the pathways and important intermediates, products, and key enzymes that catalyze C–C bond cleavage. (1) Fructose-1,6-bisphosphate aldolase, (2) KDPG aldolase, (3) phosphoketolase, and (4), (5) transaldolase, transketolase. KDPG, 2-keto-3-deoxygluconate 6-phosphate; PTS, phosphotransferase system

sugars is discussed in Chapter 5, regulation of sugar degradation will be described in Chapters 18–20, glycolysis in Chapter 3, and the Entner-Doudoroff pathway in Chapter 12.

9.11.2 There Are Variations of the Most Common Route of Hexosephosphate Degradation, the Embden-Meyerhof-Parnas Pathway

Hexoses are most commonly oxidized via the **Embden-Meyerhof-Parnas** or **glycolytic pathway** to 2 pyruvate and 2 NADH (see Fig. 3.**5**). However, there are important variations of the standard pathway discussed in Chapter 3. Some bacteria contain phosphofructokinases, which use pyrophosphate instead of ATP as phosphoryl donor. In sulfur-reducing anaerobic archaebacteria (Thermococcales, such as *Pyrococcus* sp.), an interesting modification was detected in which the ATP-dependent hexokinase and phosphofructokinase (forming ADP) are ADP-dependent enzymes forming AMP (Fig. 9.**20**). In this pathway, all oxidizing enzymes are ferredoxin-dependent. Glyceraldehyde 3-phosphate is oxidized by a tungsten enzyme, the ferredoxin-dependent glyceraldehyde-3-phosphate oxidoreductase; pyruvate is oxidized by pyruvate:ferredoxin oxidoreductase. Energy is obtained by these anaerobes mainly by Reaction 9.5):

$$\text{Acetyl-CoA} + \text{ADP} + \text{P}_i \rightarrow \text{Acetate} + \text{CoASH} + \text{ATP} \quad (9.5)$$

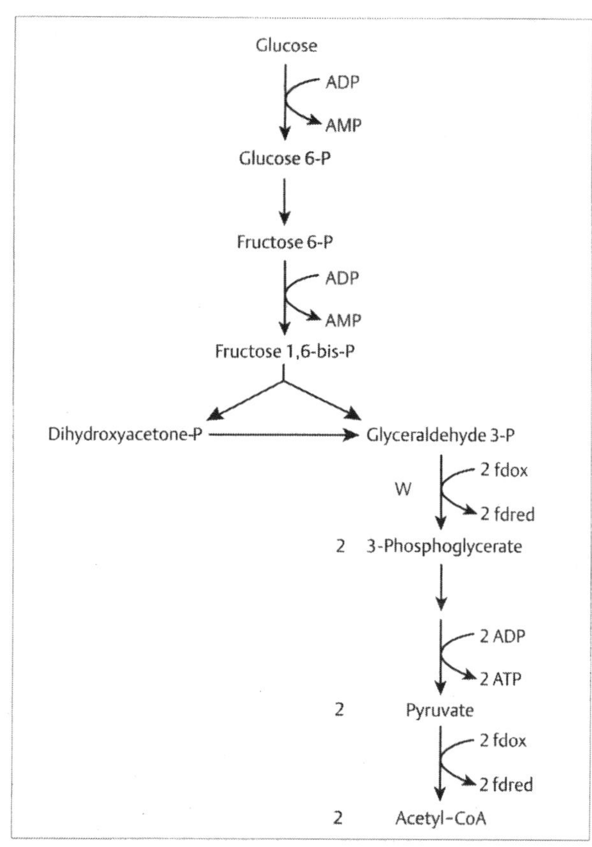

Fig. 9.**20** **Variation of the glycolytic pathway in *Pyrococcus furiosus*, an anaerobic archaebacterium**. fd Ferredoxin, W tungsten

In many bacteria (e.g., *E. coli*), not only the glycolytic enzymes are found, but also enzymes of other sugar degradation pathways (Fig. 9.**21**).

9.11.3 The Entner-Doudoroff Pathway and Variations May Represent an Ancient Way of Sugar Degradation

In many bacteria, the glycolytic pathway does not function, whereas enzymes of glucogenesis are present. Phosphofructokinase, a key enzyme of glycolysis, is absent. These bacteria use a second pathway for carbohydrate breakdown, which is named after their discoverers Entner and Doudoroff. This pathway is of great importance in bacteria and may represent an ancient way of sugar degradation (see Chapter 12; schematic presentation in Fig. 9.**21**). Different sugars may be degraded via different pathways; for instance, in *Halococcus*, glucose is metabolized via a modified Entner-Doudoroff pathway, whereas fructose is degraded via the Embden-Meyerhof-Parnas pathway.

Outline of the Entner-Doudoroff pathway. Glucose 6-phosphate is oxidized via 6-phosphogluconate to pyruvate and glyceraldehyde 3-phosphate, with the concomitant reduction of NADP. By conventional glyceraldehyde 3-phosphate oxidation to pyruvate, one NAD is reduced and a net one ATP is formed (note that PEP is required for glucose transport). There are two main differences between this and the glycolytic pathway: one NADPH is formed, and only one ATP results from substrate level phosphorylation. The lower ATP yield may explain why this pathway is normally not used by fermenting bacteria that have low ATP yields. The pathway has only two reactions in addition to those of the glycolytic and oxidative pentose phosphate pathway. As in the pentose phosphate, glucose 6-phosphate is oxidized to 6-phosphogluconate. **6-Phosphogluconate dehydratase** forms 2-keto-3-deoxygluconate 6-phosphate (2-**k**eto-3-**d**eoxy-6-**p**hospho**g**luconate = **KDPG**; the correct designation of KDPG is 2-dehydro-...). A second enzyme, **KDPG aldolase**, catalyzes aldol cleavage of KDPG, yielding pyruvate and glyceraldehyde 3-phosphate.

The Entner-Doudoroff pathway has two functions. It is found to be the main pathway for carbohydrate breakdown in an increasing number of mostly Gram-negative bacteria. In many others, such as *E. coli*, glycolysis and the pentose phosphate pathway are the two central—and constitutive—routes of intermediary carbon metabolism; the Entner-Doudoroff pathway is a third optional and inducible route (Fig. 9.**21**). It is induced for metabolism of gluconate, which is converted to gluconate 6-phosphate by gluconokinase. Also hexuronic acids are degraded via KDPG aldolase cleavage (see below).

Oxidation of glucose via gluconate and oxidation of hexonates. The role of the Entner-Doudoroff pathway in hexose metabolism in bacteria with a functioning glycolytic pathway may be questionable; however, in the presence of **p**yrrolo-**q**uinoline **q**uinone (PQQ) (Fig. 9.**42b**), a cofactor of glucose oxidase, some bacteria may oxidize all or part of glucose to gluconate using this enzyme. Gluconate is phosphorylated to 6-phosphogluconate, which is further metabolized via the Entner-Doudoroff pathway. Alternatively, gluconate is further oxidized to 2-ketogluconate by **gluconate dehydrogenase**, then phosphorylated and reduced again to 6-phosphogluconate. Both dehydrogenases are periplasmic and feed electrons directly into the respiratory chain. The sugar acids are transported. Other sugars, such as fructose, are metabolized via the conventional Entner-Doudoroff pathway. This route of glucose channeling into 6-phosphogluconate appears to be useful for bacteria that are specialized in gluconic and 2-ketogluconic acid transport and degradation;

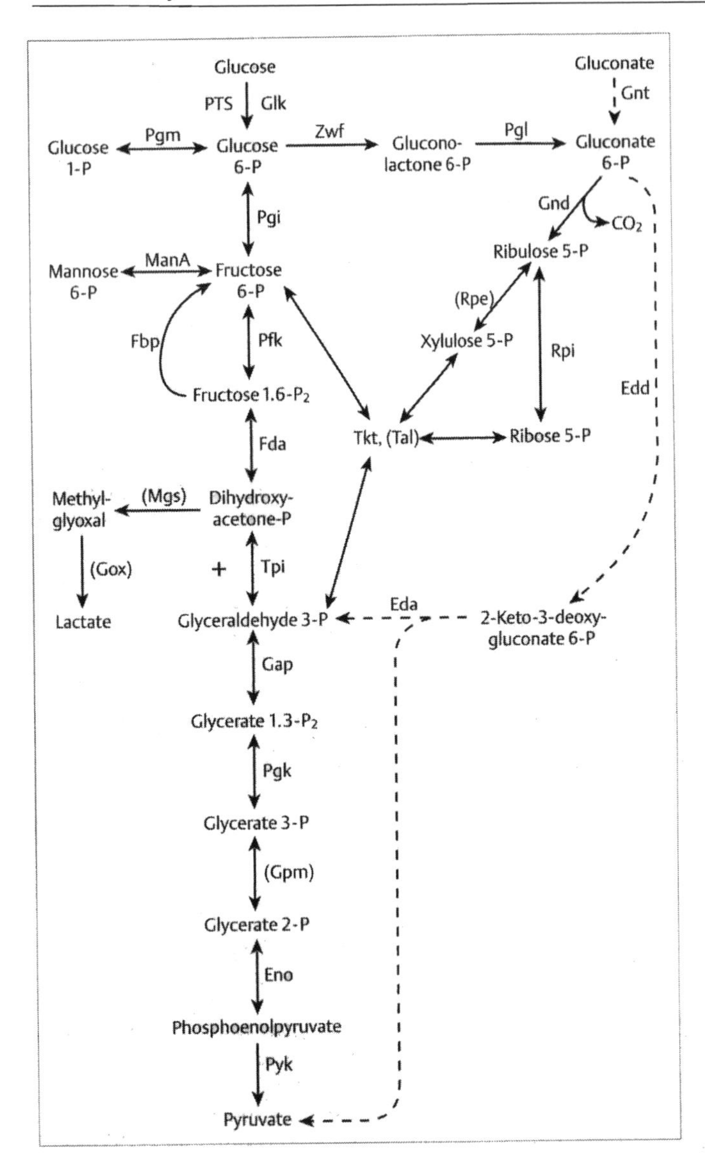

Fig. 9.**21** **Glycolysis, the pentose phosphate pathway, and the Entner-Doudoroff pathway.** The reactions are schematized, and cofactors and co-substrates (ADP, P_i, NAD, etc.) are not shown. Gene product symbols are those used (or, in parentheses, suggested) for *E. coli enzymes*

for these bacteria glucose is a less common natural substrate.

The various hexuronic acids (hexoses oxidized at C6 = **hexuronates**) are transformed into 2-keto-3-deoxy-gluconate in three steps, followed by phosphorylation. The glycosides of uronic acids (= **hexuronides**) are hydrolyzed.

Modified Entner-Doudoroff pathways. Two modifications of this pathway are found: (1) In halophilic archaebacteria and a few eubacteria, glucose is oxidized to gluconate and then dehydrated to 2-keto-3-deoxy-gluconate, which only then is phosphorylated to 2-keto-3-deoxy-6-phosphogluconate (Fig. 9.**22**): aldol cleavage generates pyruvate and glyceraldehyde 3-phosphate. (2) In the archaebacterial genera *Thermoplasma* and *Sulfolobus*, a pathway functions in which none of the intermediates are phosphorylated (**non-phosphorylated Entner-Doudoroff pathway**; (Fig. 9.**22**). Therefore, aldol cleavage of 2-keto-3-deoxygluconate affords pyruvate and glyceraldehyde. Glyceraldehyde is oxidized to glycerate and then phosphorylated to 2-phosphoglycerate, which is converted via phosphoenolpyruvate to pyruvate. The ATP gained in the pyruvate kinase reaction is spent in glycerate activation.

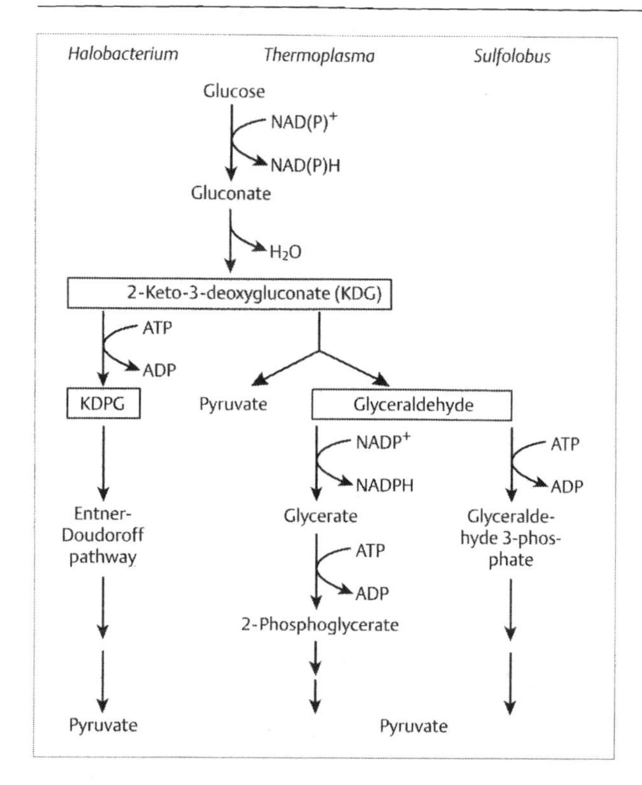

Fig. 9.**22** **Variations of the Entner-Doudoroff pathway in aerobic archaebacteria**

Hence, there is no net yield of ATP during the oxidation of glucose to pyruvate. The dehydrogenases in this pathway are NAD(P)$^+$-linked.

9.11.4 The Oxidative Pentose Phosphate Cycle Allows Complete Sugar Oxidation

The pentose phosphate pathway is normally used in a non-cyclic way for pentose and NADPH formation by oxidizing C1 of hexoses to CO_2; this is common in many bacteria. The interconversion of different sugar phosphates has a major catabolic role in pentose degradation via conversion into hexoses, and in gluconate metabolism. These sugar interconversion reactions are called the non-oxidative branch of the pentose phosphate cycle.

When the oxidative branch (hexose phosphate oxidation to pentose phosphate $+ CO_2$) is combined with the non-oxidative branch (pentosephosphate conversion to hexose phosphate), the complete cyclic operation results in oxidation of hexose 6-phosphate to CO_2 and triose phosphate and reduction of NADP (Fig. 9.**23**. Hence, glucose can be oxidized by the **oxidative pentose phosphate cycle** without participation of the Embden-Meyerhof-Parnas pathway or the Entner-Doudoroff pathway. This cycle occurs in many cyanobacteria, which may completely oxidize glucose to CO_2 via this pathway; their citric acid cycle is incomplete (2-oxoglutarate dehydrogenase is lacking) and, therefore, they cannot oxidize acetyl-CoA. Biosynthesis starts from triose phosphate. For similar reasons, some acetic acid bacteria (*Gluconobacter* sp.) oxidize triose phosphate further to acetate, which is excreted. In some other bacteria (*Thiobacillus novellus, Brucella abortus*), this cycle replaces the Embden-Meyerhof-Parnas pathway or the Entner-Doudoroff pathway.

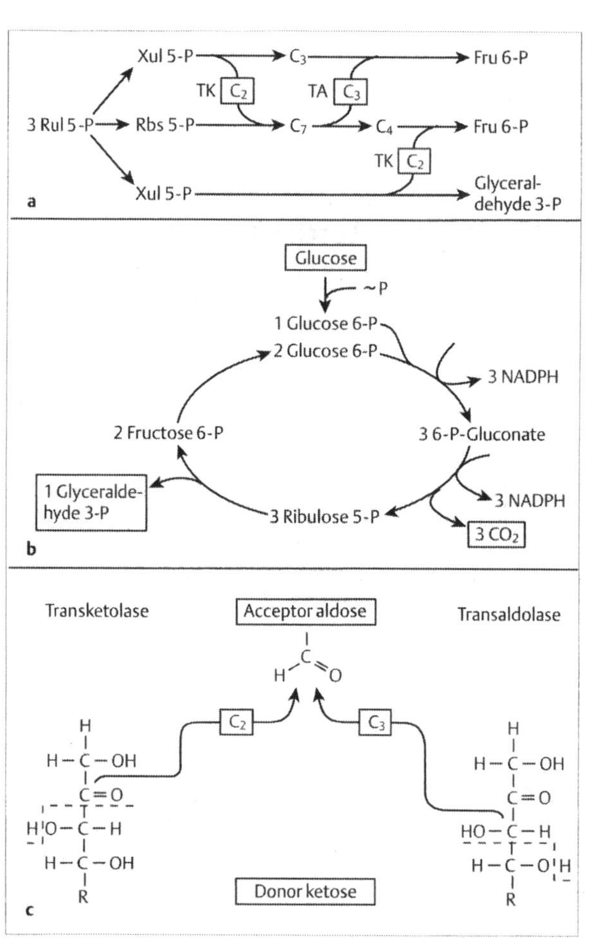

Fig. 9.**23a–c** **Oxidative pentose phosphate cycle.**
a Reactions leading from pentose phosphate back to hexose phosphate. Rul, ribulose; Rbs, ribose; Xul, xylulose; Fru, fructose; TK, transketolase, and TA, transaldolase.
b oxidation of 1 glucose to 3 CO_2 and 1 triose phosphate through the cycle, with the production of 6 NADPH.
c Reactions catalyzed by transaldolase and transketolase

9.11.5 The Methylglyoxal Bypass Is Active at Low Phosphate Concentrations

When low concentrations of phosphate limit the oxidation of glyceraldehyde 3-phosphate by glyceraldehyde-3-phosphate dehydrogenase, *E. coli* and some other bacteria oxidize dihydroxyacetone phosphate to pyruvate via another three-step route. First, a lyase removes phosphate, affording methylglyoxal. Hydration gives D-lactate which is oxidized by a flavin-linked membrane-bound D-lactate oxidase to give pyruvate (Fig. 9.**24**). Hence, the oxidation of triose phosphate to acetyl-CoA is possible without extra phosphate. In presence of phosphate, the bypass is shut off since the first enzyme is inhibited by phosphate.

In anaerobes, more sugar degradation pathways are found. For instance, some bacteria specialized in pentose degradation convert hexoses and pentoses into xylulose 5-phosphate, which is cleaved by phosphoketolase (this enzyme, like transketolase and 2-oxoacid dehydrogenases, contains thiamine pyrophosphate and catalyzes an internal redox reaction, forming the energy-rich intermediate acetyl phosphate; see Chapter 12).

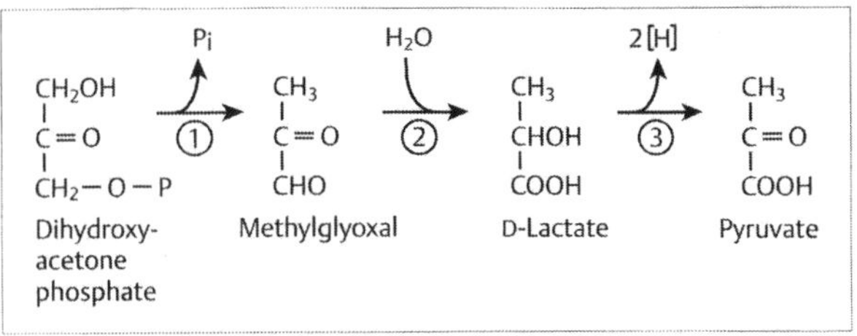

$CH_2OH{-}C{=}O{-}CH_2{-}O{-}P$ (Dihydroxyacetone phosphate) $\xrightarrow[\text{(1)}]{-P_i}$ $CH_3{-}C{=}O{-}CHO$ (Methylglyoxal) $\xrightarrow[\text{(2)}]{+H_2O}$ $CH_3{-}CHOH{-}COOH$ (D-Lactate) $\xrightarrow[\text{(3)}]{-2[H]}$ $CH_3{-}C{=}O{-}COOH$ (Pyruvate)

Fig. 9.**24** **Methylglyoxal bypass.** (1) Methylglyoxal synthase, (2) glyoxalase I and II, and (3) D-lactate oxidase (flavin-linked)

9.12 Amino Acids Are the Second Most Common Substrates

Next to sugars, amino acids are common growth substrates for microorganisms. Under many conditions, proteins, oligopeptides, and mixtures of amino acids constitute the sources of cell carbon, nitrogen, and energy for bacterial growth. Inexpensive sources of amino acids, such as peptones, tryptones, and casamino acids, are used in laboratory growth media. How much of the individual amino acids are degraded depends on the availability of other energy sources and on the C/N ratio. Before discussing some degradation pathways, the principal reactions in preparing amino acids for degradation will be discussed.

9.12.1 The Initial Steps of Degradation Are Deamination or Decarboxylation

Most naturally occurring amino acids are L-amino acids. Still, D-amino acids are constituents of, for example, murein. Amino acid racemases catalyze the equilibration of the D- and L-stereoisomers. The initial steps in the degradation of these stereoisomers are deamination or decarboxylation. Generally, the first step in the metabolism of amino acids is the removal of the α-amino group, this removal proceeds in different ways (for reductive substitution, see Chapter 12.2.3):

1. **Oxidative deamination.** The α-carbon is oxidatively deaminated by different dehydrogenases, which oxidize the amino to the imino group; the imino intermediate is hydrolyzed to NH_3 and the α-oxo compound. **D- or L-amino acid oxidases** linked with cytochrome feed directly to cytochromes of the respiratory chain. These flavoenzymes are rather nonspecific and may act on several amino acids. **Amino acid dehydrogenases** catalyze the endergonic oxidation linked with NAD^+, for example of glutamate to 2-oxoglutarate + NH_3 (glutamate dehydrogenase) and alanine to pyruvate + NH_3 (alanine dehydrogenase). Note that the biosynthetic enzyme functioning in glutamate formation from 2-oxoglutarate and NH_3 is NADPH-dependent (Chapter 7).
2. **Transamination.** Transamination has been described in Chapter 7; by transamination, the α-amino groups of amino acids are derived from glutamate. Similarly, **oxidation via transamination** requires two steps (Fig 9.**25**). In the first step, the α-amino acid donates its NH_2-group to 2-oxoglutarate or pyruvate; the corresponding 2-oxoacid is formed. Subsequent oxidative deamination of the products glutamate or alanine regenerates the NH_2-acceptor form (Fig. 9.**25**). In summary, the amino acid is oxidized by NAD^+ to the corresponding 2-oxoacid and NH_3 (Reactions 9.6–9.8):

$$\text{Amino acid} + \text{2-Oxoglutarate} \rightarrow \text{2-Oxoacid} + \text{L-Glutamate} \quad (9.6)$$

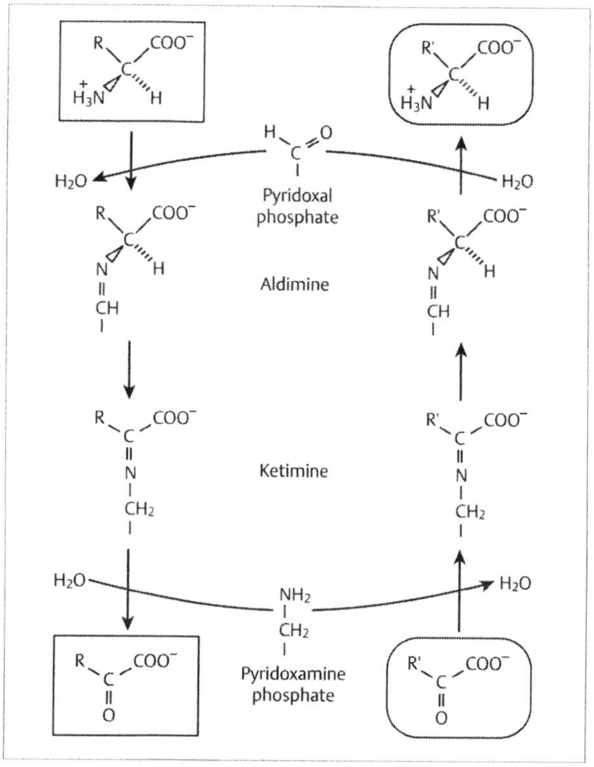

Fig. 9.**25** **Mechanism of amino transferases with pyridoxal 5′-phosphate.** The prosthetic group is bound via a Schiff base to a lysine residue of the enzyme (this is not shown). Addition of the amino acid substrate yields a new Schiff base, which undergoes an intramolecular redox reaction to pyridoxamine 5′-phosphate and the corresponding 2-oxoacid. The prosthetic group then reacts in a similar manner with the amino acceptor 2-oxoglutarate to form glutamate, regenerating pyridoxal 5′-phosphate. The first product (2-oxoacid) leaves before the second substrate enters. This is called a ping-pong mechanism

$$\text{L-Glutamate} + NAD^+ + H_2O \rightarrow NH_3 + NADH + H^+ + \text{2-Oxoglutarate} \quad (9.7)$$

$$\text{Amino acid} + NAD^+ + H_2O \rightarrow NH_3 + \text{2-Oxoacid} + NADH + H^+ \quad (9.8)$$

3. **Deamination by β-elimination** (Fig. 9.**26**). Electron withdrawing -OH or -SH groups in the β- or γ-position relative to the carboxyl group facilitate elimination of the α-amino group by lyases. This mechanism is used whenever possible. **Dehydratases** ($-H_2O$) act on serine or threonine, desulfhydrase ($-H_2S$) acts on cysteine. They catalyze the pyridoxal-phosphate-dependent formation of an α, β C=C bond (enamine) and subsequent conversion (tautomerization) to the corresponding imine C=N (ketimine). The imine is hydrolyzed to NH_3 and the 2-oxo compounds pyruvate (from serine), 2-oxobutyrate (from threonine), and pyruvate (from cysteine).

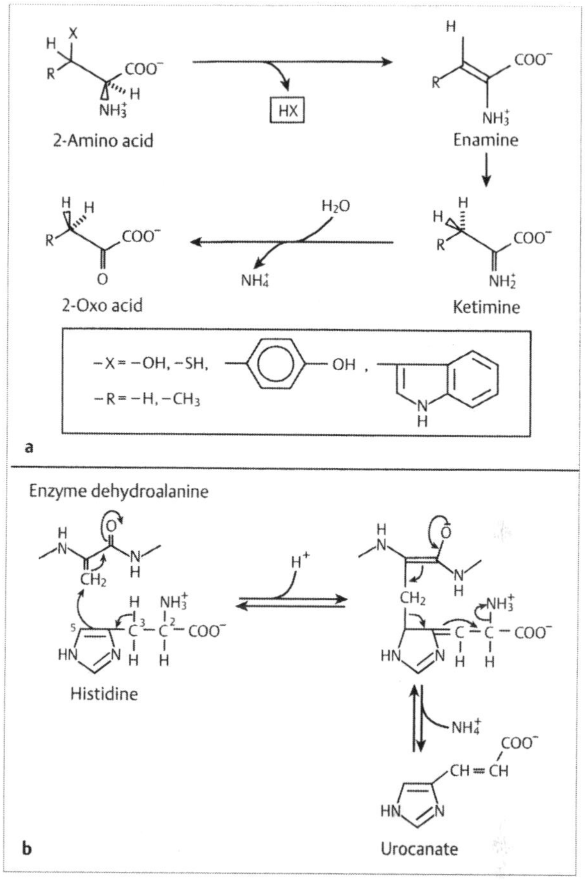

Fig. 9.**26a,b** **β- and γ-elimination of amino acids.**
a General scheme for serine, threonine, methionine, cysteine, tyrosine, and tryptophan. With the exception of L-serine, the substrate and the intermediates are bound to pyridoxal 5′-phosphate during the reaction. NH_3 instead of HX is directly β-eliminated when X=COOH [aspartate, (2*S*, 3*S*)-3-methylaspartate], imidazole (histidine), or phenyl (phenylalanine), yielding the α-, β-unsaturated acids, i.e., fumarate or methylfumarate (mesaconate), urocanate, and cinnamate, respectively. It is still uncertain whether bacteria contain phenylalanine ammonia-lyase, like plants do. Formation of a Schiff base of the α-amino group with the electron-withdrawing pyridoxal 5′-phosphate decreases the pK_a of the α-H and thus facilitates β- or γ-elimination. In the case of L-serine, a [4Fe-4S] cluster, like in aconitase, coordinates the hydroxyl group, which facilitates its removal.
b β-Elimination of NH_3 by histidine ammonia-lyase with a dehydroalanine residue. The C5 of histidine is attacked, which thereby activates the hydrogen at the conjugated C3. This facilitates the α-, β-elimination of ammonia. Dehydroalanine is derived by dehydration of a serine residue in the active center

Amino acid ammonia lyases acting on aspartate, histidine, and phenylalanine catalyze a direct elimination of NH_3 with concomitant formation of a *trans* double bond. These amino acids have in common an electron-withdrawing substituent at the β-carbon atom with delocalized electrons ($-COO^-$, imidazole, or phenyl residue). This facilitates NH_3 elimination as well as subsequent stabilization of the resulting double bond. The conjugated olefinic products are fumaric acid (from aspartate), urocanic acid (from histidine), and cinnamic acid (from phenylalanine). Finally, **γ-elimination** of tyrosine and tryptophan gives rise to phenol and indol.

Many bacteria growing with peptides or amino acids can **decarboxylate α-amino acids** to give the corresponding primary amines. These can be oxidized to the corresponding fatty acids.

9.12.2 There Are Special Pathways for the Oxidative Degradation of Amino Acids

In nature, single amino acids are rarely encountered. Rather, protein degradation results in a mixture of amino acids, dipeptides and oligopeptides, which are consumed concomitantly. The further degradation of a few representative amino acids and their corresponding 2-oxoacids will be discussed here. The degradation of the aromatic amino acids will be discussed in Chapter 9.13.2 (aromatic compounds) and in Chapter 9.15 (*N*-heterocyclic compounds). For anaerobic metabolism, notably fermentations, see Chapter 12.

Glutamate family. Glutamate is the center of the amino acid degradation network; it is oxidatively deaminated to 2-oxoglutarate. The amide group of **glutamine** (or **asparagine**) is often hydrolyzed by glutaminase (asparaginase) to give glutamate (aspartate) and ammonia. **Proline** is oxidized to δ-1-pyrroline-5-carboxylate and the N=C double bond between C5 and the α-N is hydrolyzed to give glutamate semialdehyde (for reactions, see biosynthesis of proline, Chapter 7.6.1); the semialdehyde is oxidized to glutamate. **Arginine** degradation is initiated by four different enzymes. Two of the pathways proceed via ornithine, which is oxidized to glutamate. The other two routes lead to 4-aminobutyrate, which is oxidized to succinate. Note that carbamoyl phosphate released from citrulline in the arginine deiminiase pathway allows also for the synthesis of ATP (Chapter 12). Some bacteria possess more than one of these pathways. Arginine with its four nitrogen atoms is a valuable N source under nitrogen limitation. Transamination or deamination of glutamate, aspartate, and alanine result in formation of 2-oxoglutarate, oxaloacetate, and pyruvate, which are central metabolic intermediates.

Aspartate family. Asparagine is normally hydrolyzed by asparaginase to aspartate and ammonia. **Threonine** is first dehydrated to give the en-amine, which is rearranged to the α-imino acid and subsequently hydrolyzed to 2-oxobutyrate and ammonia.

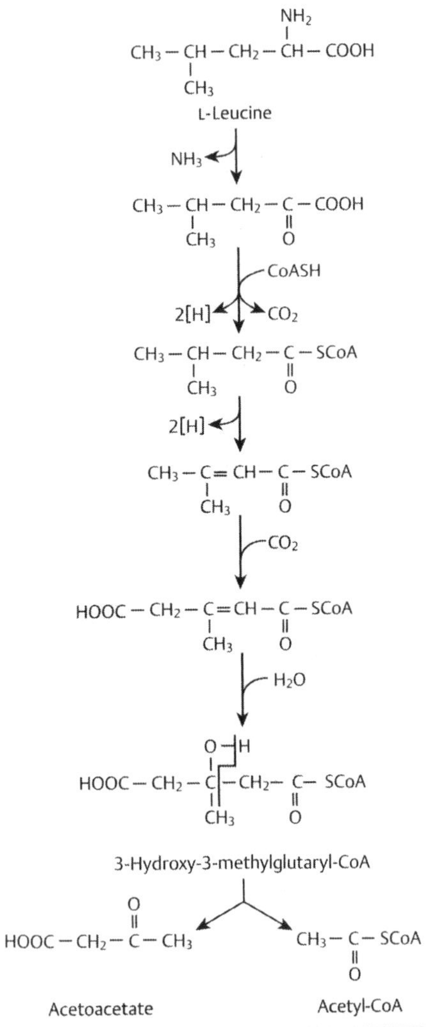

Fig. 9.27 **Degradation of a representative of the branched-chain amino acids, L-leucine.** This pathway shows that utilization of a growth substrate and, therefore, also growth may depend on the presence of CO_2 because of heterotrophic CO_2 fixation. Note that the intermediate 3-hydroxy-3-methylglutaryl-CoA is formed from three acetyl-CoA in the biosynthesis of isoprenoid compounds

Alternatively, it is cleaved by threonine aldolase into glycine and acetaldehyde. **Methionine** can be degraded by different routes.

Branched-chain amino acids. The branched-chain amino acids valine and leucine are oxidized finally to acetyl-CoA and propionyl-CoA. The degradation of leucine is shown in Fig. 9.**27**.

Serine family, aromatic amino acids, histidine. Serine and **cysteine** are degraded to pyruvate (Fig. 9.**26a**). Deamination of **glycine** forms glyoxylate, which can be converted to C_3 or C_4 compounds (see Chapter 9.16). Aspects that are specific for anaerobic bacteria are discussed in Chapter 12. **Histidine** is broken down to glutamate and formamide as shown in Fig. 9.**28**. For the degradative pathways of the other compounds, consult biochemistry textbooks. Note, however, that the anoxic degradation of aromatics presents some problems, which will be addressed next.

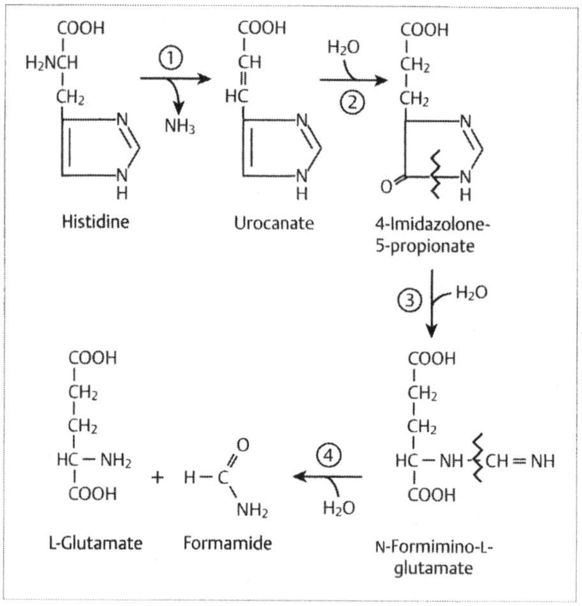

Fig. 9.**28** **Degradation of histidine.** (1) Histidase, (2) urocanase, (3) imidazolone propionase, and (4) formiminoglutamate hydrolase

9.13 Aerobic Degradation of Aromatic Compounds Requires Molecular Oxygen; Anaerobic Pathways Are Quite Different

Aromatic compounds are formed in large amounts by all organisms. A few examples are aromatic amino acids, quinones, and phenolic compounds. The most prolific producers of aromatic compounds are plants; the most predominant compound is lignin. The aerobic degradation of these compounds requires a variety of enzymes, but only a few general principles are involved in the attack on the aromatic structure. The initial steps in degradation require molecular oxygen for (1) introducing hydroxyl groups and (2) cleavage of the aromatic ring. The anoxic degradation of aromatic compounds is different (Fig. 9.**29**).

Degradation can be divided into three steps: (1) modification and conversion of the many different compounds into a few central aromatic intermediates (ring-fission substrates) that can undergo (2) oxidative ring cleavage by dioxygenases, and (3) further degradation of the non-cyclic, non-aromatic ring-fission products to intermediates of central metabolic pathways. Step (1) is referred to as peripheral and involves considerable modification of the ring or perhaps elimination of a substituent group. The most common central intermediates are **catechol** (1,2-dihydroxybenzene) and **protocatechuate** (3,4-dihydroxybenzoate) into which most of the aromatic compounds are transformed; some aromatic compounds are degraded via **gentisate** (2,5-dihydroxybenzoate). Figure 9.**30** shows exemplarily how a small selection of compounds are chanelled into this ring fission substrate (= **peripheral pathways**). As a general rule, 1,2-disubstituted and many mono-substituted aromatic compounds are transformed to catechol; examples are salicylate or phenol. Most unsubstituted aromatic hydrocarbons (benzene, naphthalene) are degraded via 1,2-diphenolic intermediates and therefore also lead to catechol. Compounds with two disubstitutions in the 1,3 and 1,4 positions and polysubstituted compounds are metabolized via protocatechuate; examples are 3-hydroxybenzoate, 4-hydroxybenzoate, and vanillate.

The peripheral pathways are characterized by introduction of hydroxyl groups at the ring and often by removal of other substituents; both reactions are catalyzed by oxygenases. Chloro- nitro-, and sulfonate-groups are replaced by hydroxyl groups. Aliphatic side chains may be shortened in different ways, or they remain intact. The oxygenolytic cleavage of the aromatic ring occurs only with a few central aromatic intermediates, either by *ortho*- or by *meta*-cleavage.

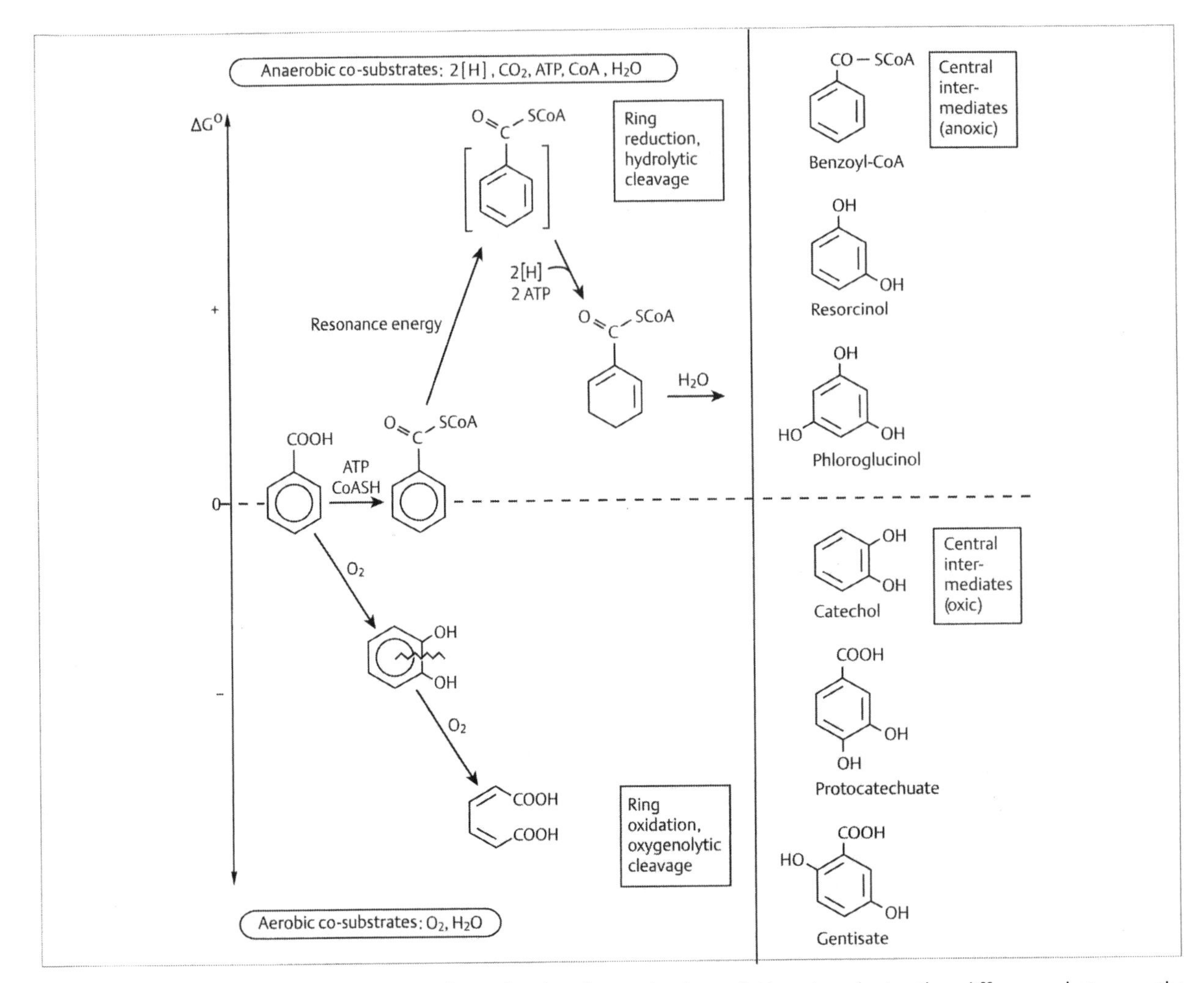

Fig. 9.**29** **Comparison of aerobic (oxygen-dependent) and anaerobic (anoxic) metabolism of aromatic compounds, the co-substrates used, and the central intermediates formed in the peripheral metabolism.** The metabolism of benzoate is shown schematically. Under oxic conditions, benzoate is converted, for example to catechol, and the aromatic ring subsequently is oxygenolytically cleaved. Under anoxic conditions, the high resonance energy does not allow direct reduction of the ring (note the difference between the delocalized π-electron system and the hypothetical cyclohexatriene structure). Rather, the CoA-thioester is formed, which requires ATP. Then, the aromatic ring is reduced at the cost of ATP, and finally the ring is opened hydrolytically. CO_2 is used for the carboxylation of some phenolic compounds (not shown, see Fig. 9.**35**)

9.13.1 The *ortho*-Cleavage Pathway of Catechol and Protocatechuate Leads to 3-Oxoadipate

Figure 9.**31** shows the *ortho*-pathway of catechol and protocatechuate degradation via the common intermediate 3-oxoadipate to succinate and acetyl-CoA. The two branches are catalyzed by two sets of enzymes. Catechol 1,2-dioxygenase and protocatechuate 3,4-dioxygenase are the ring-cleaving enzymes that incorporate both atoms of oxygen into the product *cis,cis*-muconate and the β-carboxy analogue. The products are cycloisomerized to the five-membered muconolactone and its 4-carboxy analogue, respectively. The double bond is shifted by an isomerase to produce an instable enol-lactone, which is decarboxylated in the case of the 4-carboxymuconolactone. The enol-lactone is easily hydrolyzed and yields the characteristic intermediate 3-oxoadipate (the pathway therefore is referred to as **3-oxoadipate pathway**). One of the final products, succinyl-CoA, is used to activate this C_6 dicarboxylic

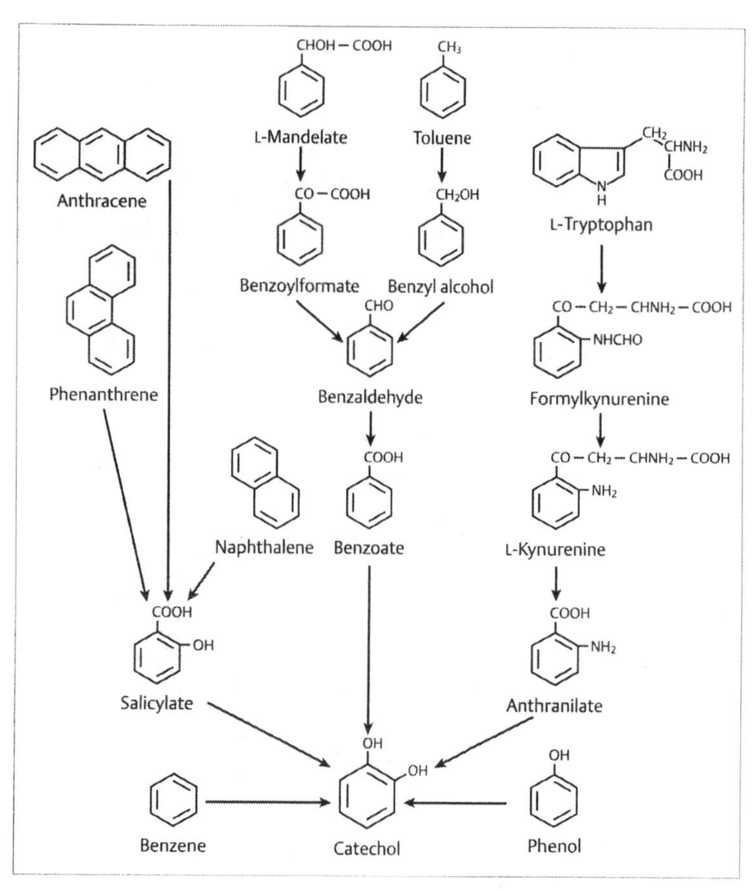

Fig. 9.**30** **Aromatic compounds that can be converted to catechol.** To reach these compounds from the huge variety of aromatic compounds, substituents have to be modified, often by oxygenases. The ring-fission substrates catechol and protocatechuate can be cleaved by two types of dioxygenases, by **ortho**-cleavage (between the two hydroxyl groups) or by the **meta**-cleavage (adjacent to the hydroxyl groups) [from 4]

acid by CoA transfer. The 3-oxoacyl-CoA compound is thiolytically cleaved by a thiolase, as in β-oxidation of fatty acids. Cleavage products are succinyl-CoA and acetyl-CoA.

9.13.2 Homogentisate Can Similarly Be Ortho-Cleaved

The aerobic degradation of the aromatic amino acids **phenylalanine** and **tyrosine** are outlined in Fig. 9.**32**. The key intermediate in this case is **homogentisate**, which is oxidatively cleaved to maleyl acetoacetate and metabolized further to fumarate and acetoacetate. The introduction of hydroxyl groups and ring cleavage proceed via oxygenases. In a similar way, several other aromatic compounds are channeled into gentisate, which is cleaved to fumaryl pyruvate; pyruvate and fumarate are the products.

Tryptophan is degraded in several different ways (Fig. 9.**33**). One possibility is to cleave off by β-elimination the C_3 side chain and to use pyruvate and indole separately (see below). Indole is oxidized via 2,3-dihydoxyindole and anthranilate. The most common way uses tryptophan 2,3-dioxygenase for oxygenolytic cleavage of the five-membered ring. The reaction product *N*-formylkynurenine is hydrolyzed to kynurenine and formate. At this point, the pathway diverges. Some bacteria form anthranilate, which is degraded normally via catechol through the "aromatic pathway". Others form the *N*-heterocyclic kynurenic acid, which is degraded similarly as the heterocyclic quinoline (see Chapter 9.15).

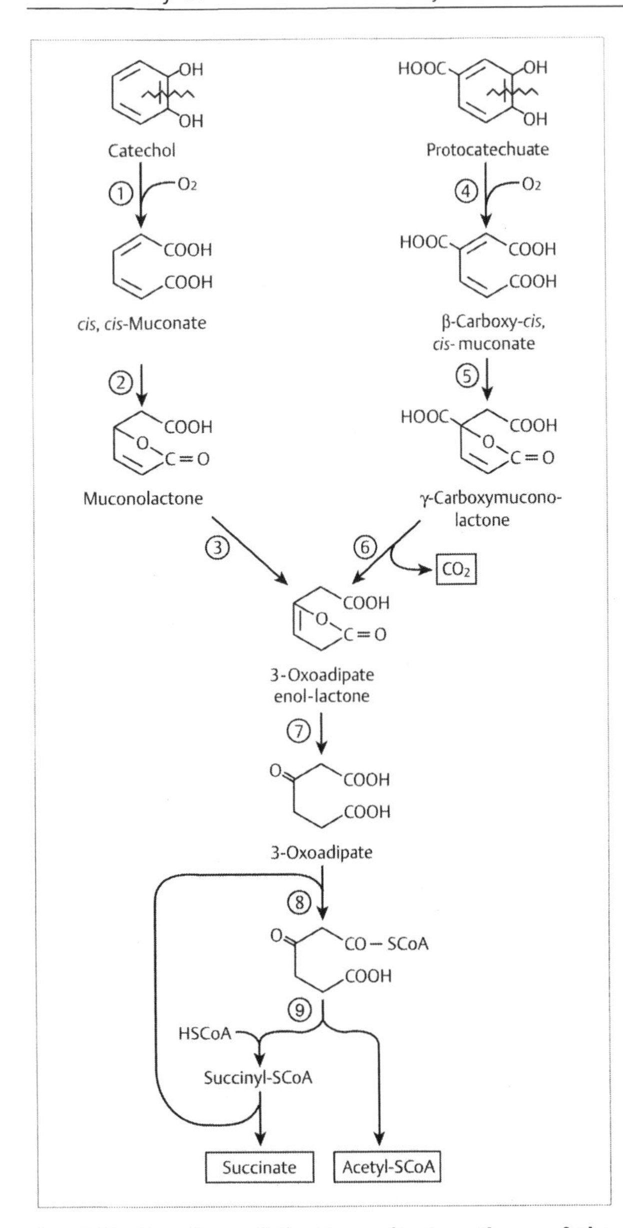

Fig. 9.**31** **Reactions of the 3-oxoadipate pathway of the *ortho*-cleavage pathway.** (1) Catechol 1,2-dioxygenase, (2) muconate-lactonizing enzyme, (3) muconolactone isomerase, (4) protocatechuate 3,4-dioxygenase, (5) β-carboxymuconate-lactonizing enzyme, (6) γ-carboxymuconolactone decarboxylase, (7) 3-oxoadipate *enol*-lactone hydrolase, (8) 3-oxoadipate succinyl-CoA transferase and (9) 3-oxoadipate-CoA thiolase

Box 9.4 Oxygenases are oxidoreductases that incorporate into the substrate(s) oxygen from O_2: one atom of oxygen (monooxygenases, often called hydroxylases) or both atoms (dioxygenases). **Dioxygenases** are responsible for the oxygenolytic ring cleavage of dihydroxylated aromatic compounds (catechol, protocatechuate, gentisate) and the cleavage of β-carotene to two molecules of retinal; they often require Fe^{2+} or Fe^{3+}. When dioxygenases act on an aliphatic double bound (e.g. in unsaturated fatty acids, carotene), they may form hydroperoxy [-C(OO)C-] bonds; examples are lipoxygenases. A few dioxygenases oxidize α-amino or α-hydroxy acids, thereby oxidatively decarboxylating them (internal monooxygenases). A special case are dioxygenases that require two oxidizable substrates, and oxygen from O_2 is incorporated into one or both of the substrates; examples are enzymes that oxidize 2-oxoglutarate as co-substrate to succinate, thereby incorporating one oxygen atom; the other is incoporated into the actual substrate to be hydroxylated. Other dioxygenases require NAD(P)H and O_2. Examples are dioxygenases that act on aromatic hydrocarbons forming *cis*-diol compounds, or dioxygenases that form dihydroxylated aromatic compounds and in addition remove (carboxy-, chloro-, sulfo-, amino-) substituents from the ring. The **monooxygenases** require an electron donor such as NAD(P)H, a reduced flavin or flavoprotein, a reduced iron-sulfur protein, or a reduced pteridine; these enzymes are numerous. A few others concomitantly oxidize a second organic substrate instead of a reduced coenzyme.

9.13.3 Meta Cleavage Is the Oxygenolytic Cleavage of the Aromatic Ring Adjacent to the Hydroxyl Groups

Less common is the oxygenolytic opening by dioxygenases of the aromatic ring adjacent to the hydroxyl groups, i.e., *meta*-cleavage (Fig. 9.**34**). The product of catechol 2,3-dioxygenase is 2-hydroxymuconic semialdehyde, and of protocatechuate 4,5-dioxygenase, the corresponding 4-carboxy derivative. Metabolic products are pyruvate, formate, and acetaldehyde. The two types of ring opening may be carried out by the same strain, depending on whether the growth substrate is benzoate or phenol, respectively, and the genes coding for one set of the cleavage enzymes may be located on plasmids.

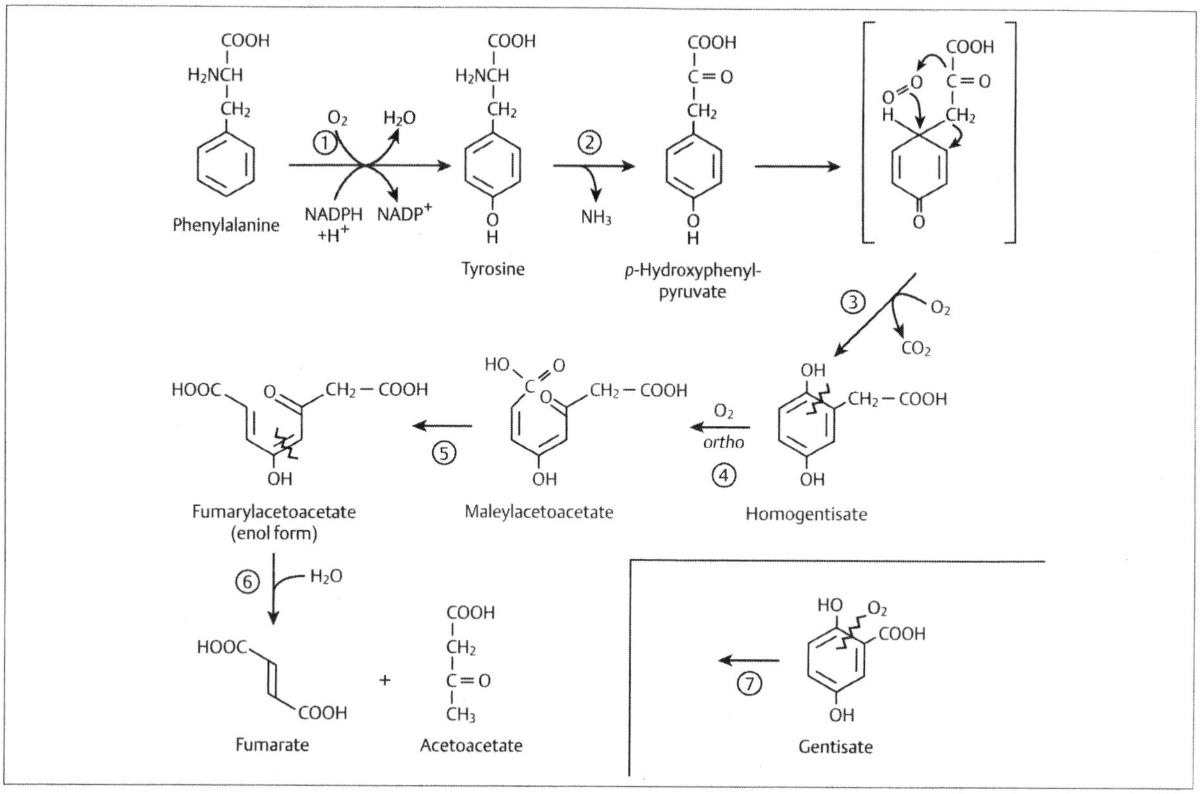

Fig. 9.32 **Degradation of phenylalanine and tyrosine (homogentisate pathway).** (1) Phenylalanine hydroxylase, (2) transaminase, (3) *p*-hydroxyphenylpyruvate oxidase; note that hydroxylation at C1 requires the carboxyl as H-donor, which is oxidatively decarboxylated; hydroxylation at C1 forces the side chain to migrate, (4) homogentisate-1,2-dioxygenase, (5) maleylacetoacetate isomerase (glutathione-dependent), (6) fumarylacetoacetate hydrolase. The metabolism of gentisate by (7) gentisate-1,2-dioxygenase is analogous (gentisate pathway)

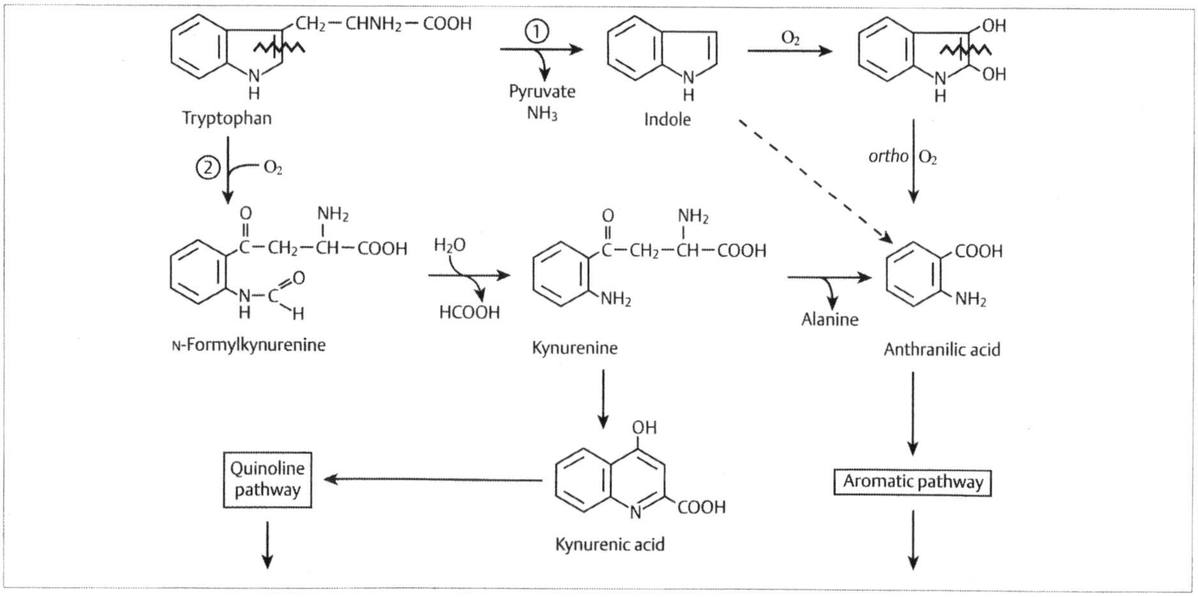

Fig. 9.33 **Degradation of tryptophan via different routes.** (1) Tryptophan indole lyase and (2) tryptophan 2,3-dioxygenase

9.13.4 Anaerobic Oxidation of Aromatic Compounds Requires a Different Strategy

Facultative anaerobes and anaerobes face the problem that molecular oxygen, the "magic bullet" of aerobes, is not available for activation of the ring by hydroxylation nor for its oxygenolytic cleavage. They depend on "poor men's" co-substrates: water, CO_2, ATP, coenzyme A thioesters, and reduced and oxidized coenzymes.

Catechol
2-Hydroxymuconic semialdehyde
HCOOH
2-Oxopent-4-enoate
4-Hydroxy-2-oxovalerate
Pyruvate + Acetaldehyde

Protocatechuate
2-Hydroxy-4-carboxymuconic semialdehyde
HCOOH
2-Oxo-4-carboxypent-4-enoate
4-Hydroxy-4-carboxy-2-oxovalerate
2 Pyruvate

Fig. 9.**34** **Dissimilation of catechol and protocatechuate by the pathways involving *meta*-cleavage.** (1) Catechol 2,3-dioxygenase, (2) 2-hydroxymuconic semialdehyde hydrolase, (3) 2-oxopent-4-enoic acid hydrolase, (4) 4-hydroxy-2-oxovalerate aldolase, (5) protocatechuate 4,5-dioxygenase, (6) 2-hydroxy-4-carboxymuconic semialdehyde hydrolase, (7) 2-oxo-4-carboxypent-4-enoic acid hydrolase, and (8) 4-hydroxy-4-carboxy-2-oxovalerate aldolase

The essence of their completely different strategy is that the aromatic ring is reduced and the alicyclic ring formed is opened hydrolytically. The central intermediates are **benzoyl-CoA**, **phloroglucinol** (1,3,5-trihy-

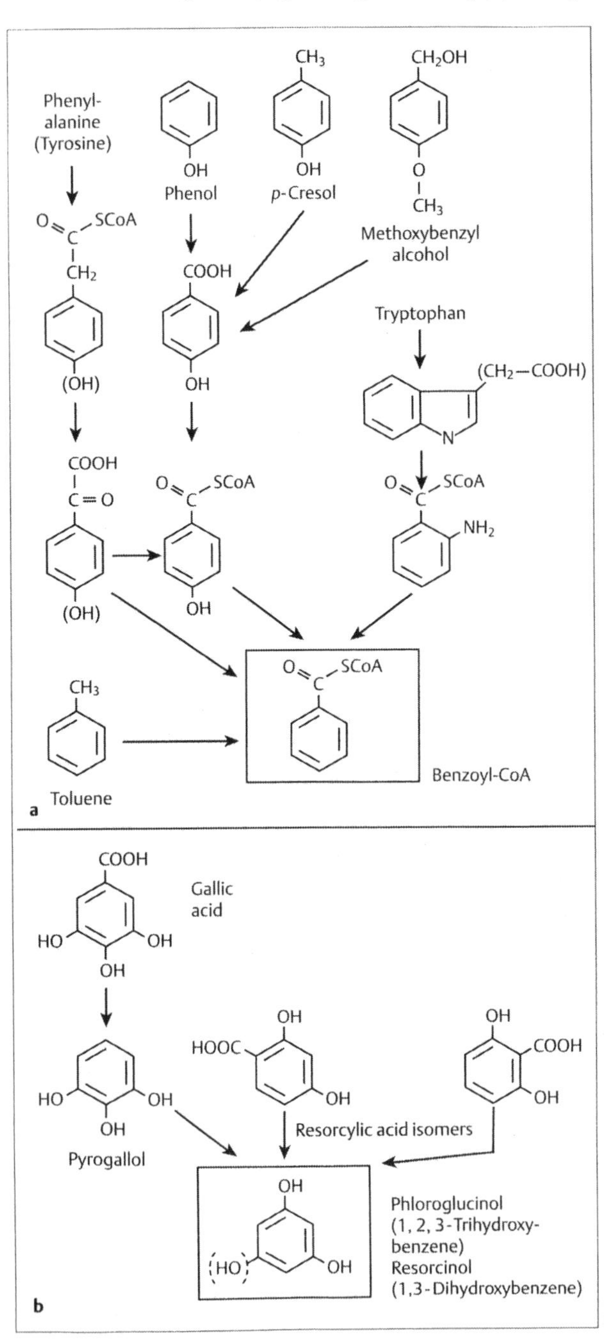

Fig. 9.**35** **Some aromatic compounds that are transformed anoxically**
(a) to benzoyl-CoA and
(b) to resorcinol and phloroglucinol

droxybenzene), **resorcinol** (1,3-dihydroxybenzene), and possibly others. It is intelligible that the central intermediates of anaerobic pathways must be different from those of the aerobic pathways (catechol, protocatechuate, gentisate) since the intermediates have to be reduced rather than to react with O_2. Thus, a set of different enzyme reactions comes into play: carboxylations of phenolic compounds, reductive removal of substituents (–OH, $-NH_2$, –halogen), *O*-methyl ether cleavage, transhydroxylations, addition of fumarate to $-CH_3$ groups, α-oxidation of $-CH_2-COOH$ groups, and ring reduction (Fig. 9.**35**). Facultative anaerobes, therefore, have two sets of enzymes, one for aerobic degradation and a second one for anaerobic degradation.

The anaerobic pathways may be divided into three parts: (1) **peripheral reactions** channeling the variety of substrates into a few central intermediates and thereby activating the substrates for ring reduction, (2) **ring reduction**, formation of 1,3-diketone structure, and hydrolysis of the diketone, and (3) **β-oxidation** to one of the central metabolites (acetyl-CoA). The central intermediates phloroglucinol and resorcinol are formed from phenolic compounds that contain already three or two phenolic –OH groups in *meta* position to each other.

Figure 9.**35a** shows a set of enzyme reactions required to transform a selection of aromatic compounds into **benzoyl-CoA** (channeling reactions). Figure 9.**35b** shows how **phloroglucinol** and **resorcinol** are formed. These examples illustrate the function of the principal anaerobic alternative reactions. The aromatic ring is reduced in a two-electron step to afford a diene. In the case of benzoyl-CoA the reaction requires 2 ATP to overcome the high activation energy associated with overcoming the resonance energy; this is reminiscent of the nitrogenase reaction in which the stable N≡N triple bond has to be split (Fig. 9.**36a**). The diene is then converted to the β-keto structure. In case of resorcinol and phloroglucinol (with much less aromatic character), ring reduction is ATP-independent, and the resulting endiol is already completely in the β-diketo form (Fig. 9.**36b**). The subsequent reactions are common to β-oxidation and oxidation of dicarboxylic acids. The end product is acetyl-CoA, which is a central metabolite.

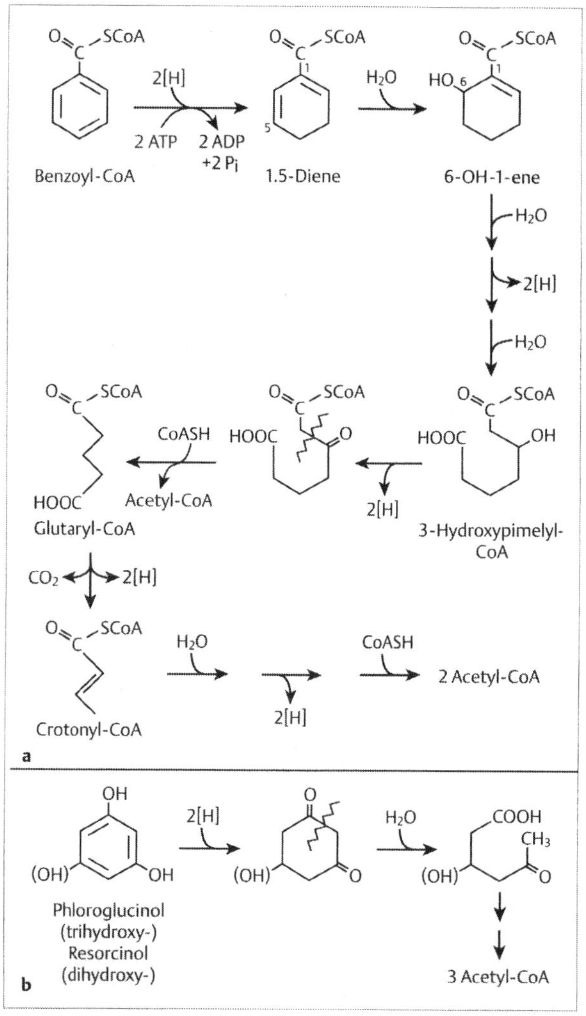

Fig. 9.**36** **Ring reduction and further metabolism of (a) benzoyl-CoA and (b) phloroglucinol and resorcinol under anoxic conditions**

9.13.5 There Are Common Principles of Anoxic Oxidation of Inert Compounds That Normally Require Oxygen for Attack (Aromatic Compounds, Hydrocarbons, Ether Compounds)

Degradation of aromatic compounds and, as will be shown, of other chemically inert compounds, requires molecular oxygen to attack these structures oxidatively to make them degradable. These include cyclic, linear, and branched aliphatic hydrocarbons, terpenoids, steroids, and ether compounds. In the absence of oxygen, different, oxygen-independent reactions have to be applied by phototrophic, fermenting, or anaerobically respiring bacteria (e.g., the aromatic ring is reduced). Introduction of hydroxyl groups proceeds via dehydrogenation and addition of water. Carboxylations may introduce polar carboxyl groups. Ether bonds are cleaved using strong nucleophiles such as Co(I) in B_{12}-dependent proteins, which form reactive Co(III)-alkyl groups.

9.14 Fatty Acids, Waxes, Hydrocarbons, Sterols, and C_1 Compounds Can Be Used by Specialists

Lipids and phospholipids cannot be transported and, therefore, have to be hydrolyzed by exoenzymes to fatty acids, glycerol, and polar head groups. Waxes, which are esters of long-chain fatty acids and alcohols, are hydrolyzed by esterases; the alcohols are oxidized to the fatty acids.

9.14.1 Fatty Acids Have to Be Transported and Activated Before Degradation

Long-chain (C_9–C_{18}) **fatty acids** are transported (Fig. 9.**37**). This requires a receptor protein in the outer membrane of Gram-negative bacteria. The passage through the cytoplasmic membrane is less clear. Possibly, acyl-CoA synthetase (which has broad chain-length specificity) is loosely associated with the inner side of the cytoplasmic membrane and immediately activates entering fatty acids to avoid the unwanted detergent effect (Reaction 9.9),

$$\text{R–COOH} + \text{CoA–SH} + \text{ATP} \rightarrow \text{R–CO–SCoA} + \text{AMP} + \text{PP}_i \quad (9.9)$$

The pyrophosphatase reaction (Reaction 9.10)

$$\text{PP}_i + H_2O \rightarrow 2\ \text{P}_i \quad (9.10)$$

renders the reversible reaction (Reaction 9.9) irreversible; hence, two high-energy phosphate equivalents are required. **Medium-chain** (C_7–C_{11}) **fatty acids**, which are more toxic, may diffuse and then become activated. The degradation of CoA-activated fatty acids is called **β-oxidation** because the carbon β to the activated carbon is prepared to be oxidized.

The cyclic process of β-oxidation of **saturated fatty acids** releases one **acetyl-CoA** per turn and forms one $FADH_2$ and one NADH (Fig. 9.**37**). For acetyl-CoA assimilation, see Chapter 7. **Odd-numbered fatty acids** also afford **propionyl-CoA**, which is normally carboxylated to methylmalonyl-CoA and isomerized to succinyl-CoA. For possible other propionate oxidation pathways, see Chapter 9.16 (organic acids).

β-Oxidation of **mono-unsaturated fatty acids** (e.g., 18:1Δ9) eventually results in a 3-*cis*-unsaturated acyl-CoA, which has the double bond in the wrong position (it should be at position 2) and in the wrong configuration (it should be *trans*, not *cis*). *cis-trans* isomerase activity of 2,3-enoyl-CoA hydratase (crotonase)/3-hydroxyacyl-CoA dehydrogenase (see legend to Fig. 9.**37**) forms the correct 2-*trans* isomer, catalyzing 3-*cis*-acyl-CoA→2-*trans* acyl-CoA. This product then continues in the β-oxidation cycle. With **polyunsaturated fatty acids** (e.g., 18:2Δ9, Δ11) a further inducible enzyme, 2,4-dienoyl-CoA reductase, is required to convert the 2-*trans*-, 4-*cis*-dieneoyl-CoA to the 3-*trans*-enoyl-CoA. The *cis-trans* isomerase activity acts again and forms the 2-enoyl-CoA isomer. Degradation of short-chain fatty acids (C_4–C_6) and acetoacetate in addition requires another thiolase isoenzyme and acetoacetyl-CoA transferase (Fig. 9.**38**).

The CoA ligase and the enzymes of the subsequent β-oxidation system are induced by growth on oils, fatty acids, or alkanes and are catabolite-repressed by glucose. Long-chain fatty acids act as inducer of the β-oxidation system; medium-chain fatty acids cannot act as an inducer, even when they can serve as substrate. Acetoacetate induces the short-chain oxidation system. The long-chain and medium-chain fatty acid degradation system and the short-chain oxidation system are under the control of different regulatory proteins.

Long-chain dicarboxylic acids are transported, then activated by a special dicarboxylate:CoA ligase to the mono-CoA thioester, followed by β-oxidation. The product of even-numbered dicarboxylic acids is succinyl-CoA. Odd-numbered dicarboxylic acids result in glutaryl-CoA (C_5 dicarboxylic acid) formation. Glutaryl-CoA dehydrogenase oxidizes glutaryl-CoA to glutaconyl-CoA, which decarboxylates spontaneously to crotonyl-CoA (see Chapter 12). There are numerous other routes of fatty acid oxidation, many of which involve monooxygenases.

Oxidation of glycerol. Glycerol and glycerol 3-phosphate are the other products of lipid metabolism. They are aerobically oxidized via 3-phosphoglycerate (Fig. 9.**39**). Anaerobically, glycerol is fermented (Chapter 12).

9.14.2 Hydrocarbons Are Present in Natural Waxes, Resins, and Other Biological Materials

Besides being found in these natural sources, hydrocarbons are also introduced into the biosphere by mineral oil and technical products. Furthermore, under anoxic conditions, when no alternative electron accepts like sulfate or nitrate are present, half of the organic carbon is fermented to methane. Hydrocarbons are extremely inert (i.e., paraffin). Still, microorganisms can use them as a sole organic substrate. As a rule, the first

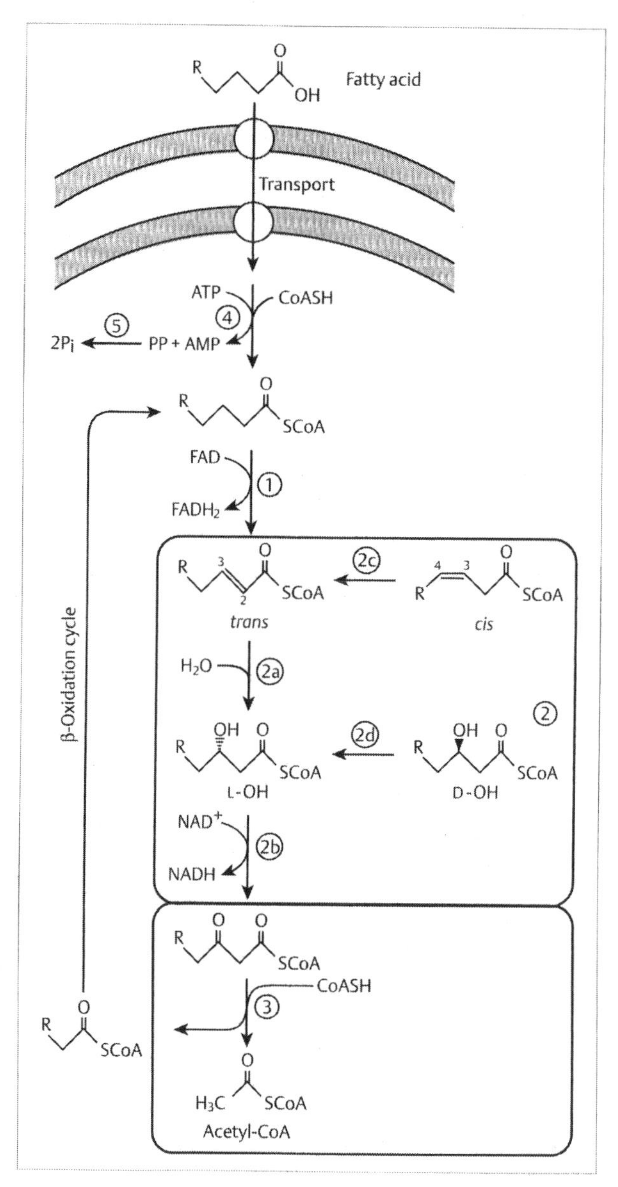

Fig. 9.37 **β-Oxidation of long-chain and medium-chain fatty acids.** The process can be divided into four steps: (1) Fatty acyl-CoA dehydrogenase (1), a flavoenzyme, forms an α-β *trans* double bond (*trans*-isomer of 2,3-enoyl-CoA) and reduces FAD. The redox potential of the dehydrogenation E_0 is near zero; therefore, the reaction cannot be coupled to NAD^+ reduction. $FADH_2$ feeds into the electron transport chain on the quinone level. The remaining reactions are catalyzed by two enzymes in a multifunctional enzyme complex (approximately 260 kDa), keeping the pool of metabolic intermediates low. The first of these two enzymes (2) has four activities [reactions (2a–2d)]. (ii) 2,3-Enoyl-CoA hydratase (crotonase) (2a) forms the L-3-hydroxy isomer (in contrast to the biosynthetic D-isomer). (iii) L-3-Hydroxyacyl-CoA dehydrogenase (2b), specific for the L-isomer and NAD^+, forms the 3-oxo compound. When *cis*-unsaturated fatty acids are degraded, two additional activities are also carried out by this protein: *cis-trans* isomerization is catalyzed by isomerase (2c) and D-3-hydroxy fatty acyl-CoA is epimerized to the L-isomer by epimerase (2d). (iv) The second enzyme of the complex, 3-oxoacyl-CoA thiolase (thiolase) (3), thiolytically cleaves off acetyl-CoA, leaving an acyl ester that is now two carbon atoms shorter than the initial acyl chain. The shortened fatty acyl-CoA thioester then repeats the sequence, the end product of even-numbered fatty acids is acetyl-CoA. (4) Acyl-CoA synthetase, (5) pyrophosphatase

enzymic attack of hydrocarbons, and, therefore, growth with hydrocarbons depends on molecular oxygen. The anaerobic degradation of some hydrocarbon compounds is rather slow. Although alkanes have high heats of combustion, growth yields are comparatively low, probably because the carbon supply is limiting. The utilization of methane and other C_1 compounds is restricted to a few specialists.

Requirement for biosurfactants. Hydrocarbons are virtually water-insoluble, which makes their uptake and enzymic attack a difficult task. (There is no life in water-free oil!) Rapid degradation requires a large oil–water interphase in which these bacteria grow. For that purpose, most alkane-degrading microorganisms excrete low-molecular, extracellular and cell-bound compounds that function as biosurfactants for emulsifying the substrate. The micelles or reversed micelles facilitate uptake. The amphiphilic emulsifiers have various structures and are composed of a hydrophilic and a lipophilic moiety. Examples are glycolipids, acylpolyols, lipopeptides, fatty acids, phospholipids, and neutral lipids.

Uptake of hydrocarbons. Uptake is considered the rate-limiting step. Only low-molecular-weight hydro-

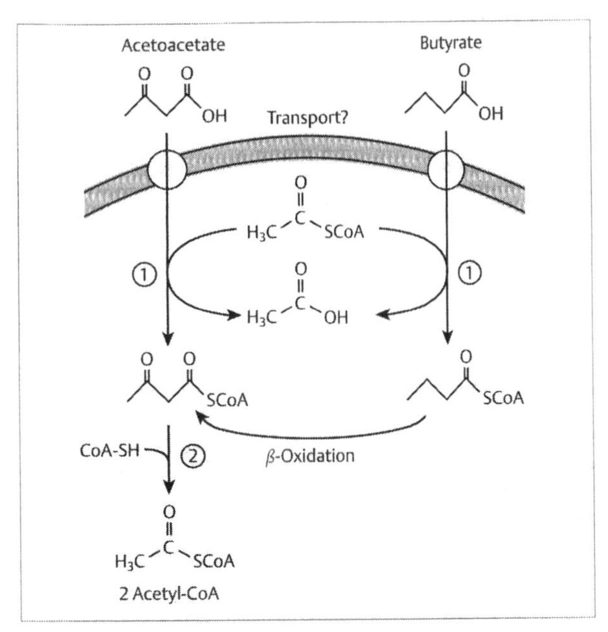

Fig. 9.**38** **Short-chain fatty acid degradation in *Escherichia coli*.** (1) Acetoacetyl-CoA transferase, (2) thiolase isoenzyme

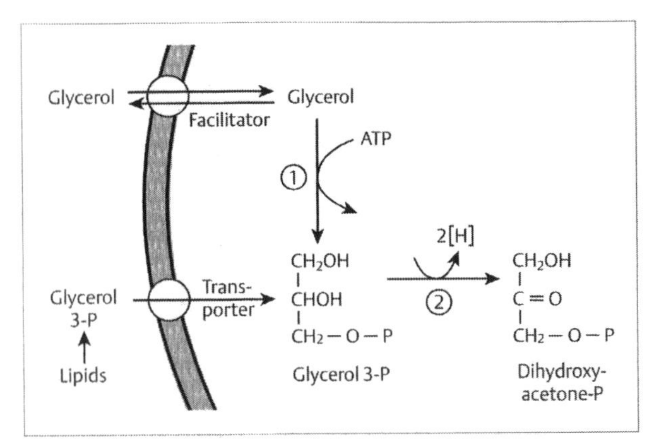

Fig. 9.**39** **Metabolism of glycerol and glycerol 3-phosphate in Enterobacteriaceae.** (1) Kinase and (2) aerobic and anaerobic dehydrogenases

carbons up to C_8 are soluble or volatile enough that diffusion (even from the vaporous phase) may be sufficient for substrate supply. High-molecular-weight compounds are extremely poorly and slowly soluble and therefore are taken up as microdroplets. They are frequently observed within the cell; their uptake is not understood. In addition, many hydrocarbon-degrading microorganisms have hydrophobic cell surfaces, either constitutive or induced by the hydrocarbon substrate; these surfaces allow the cells to bind to the water-insoluble substrate. Hydrocarbon droplets may even be encapsulated in small membrane vesicles, which are then taken into the cell by fusion.

Aerobic degradation of hydrocarbons. The most common substrates are the aliphatic hydrocarbons. Long-chain hydrocarbons (C_{10}–C_{18}) can be used frequently and rapidly by, for example, many high G+C Gram-positive bacteria (*Arthrobacter, Corynebacterium, Nocardia, Mycobacterium, Acinetobacter*) and *Pseudomonas* strains. Only a few bacteria can oxidize C_2–C_8 hydrocarbons. Degradation of these *n*-alkanes requires activation of the inert substrate by molecular oxygen with the help of oxygenases by three possible ways (Fig. 9.**40**):

1. The most common initial attack is at the end, and the monooxygenase produces the corresponding alkan-1-ol (Reaction 9.11):

$$R{-}CH_3 + O_2 + NAD(P)H + H^+ \rightarrow R{-}CH_2OH + NAD(P)^+ + H_2O \quad (9.11)$$

2. Attack by dioxygenase produces the hydroperoxides, which are reduced to yield also alkan-1-ol (Reaction 9.12):

$$R{-}CH_3 + O_2 \rightarrow R{-}CH_2OOH + NAD(P)H + H^+ \rightarrow R{-}CH_2OH + NAD(P)^+ + H_2O \quad (9.12)$$

3. Rarely, subterminal oxidation at C2 by monooxygenase yields secondary alcohols.

These hydroxylations and subsequent oxidations proceed usually in the membrane. The electron donor for the monooxygenase is reduced rubredoxin, a small protein in which Fe^{2+}/Fe^{3+} is bound to cysteine. Rubredoxin is reduced in turn by a soluble NADH-dependent enzyme. Similarly, reduced cytochrome P-450 may act as co-substrate; it is reduced by NADH via a flavo-iron-sulfur protein. Subsequent metabolism may follow a number of pathways. Normally, *n*-alkyl alcohols are oxidized to the fatty acids. Secondary alcohols are oxidized to the ketones. An oxygen atom is inserted next to the keto group by a monooxygenase, forming the acetyl ester. The ester is hydrolyzed to alcohol (which is oxidized to the fatty acid) and acetate (Fig. 9.**40**). The properties of alcohol dehydrogenase and aldehyde dehydrogenases vary. They may be soluble or membrane-bound, require NAD^+ or $NADP^+$ as cofactor, and may be inducible or constitutive.

Branched alkanes are more slowly degraded, normally via α-oxidation at both ends. **Alkenes** are attacked either on the double bond or by the same mechanism employed in the *n*-alkane mechanism: (1) oxygenase converts the terminal methyl to the alken-1-ol, (2) subterminal oxygenase attack produces an alkenol with a secondary alcohol function, (3) oxidation across the double bond gives an epoxide, which is subsequently hydrolyzed to yield a dihydroxy alkane,

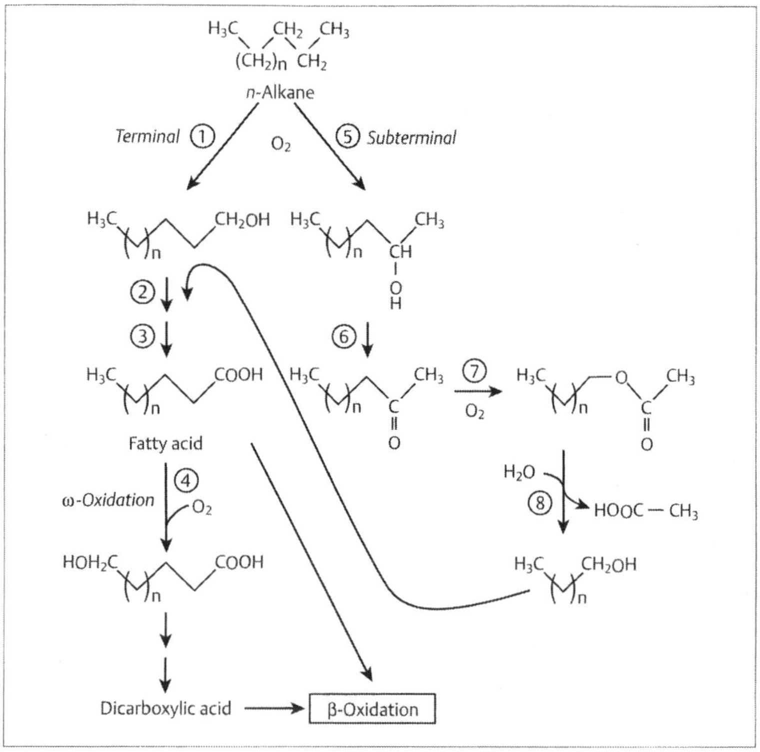

Fig. 9.**40** **Basic metabolic pathways involved in the metabolism of *n*-alkanes, with terminal oxidation being the most common.** (1) *n*-Alkane monooxygenase, (2) alcohol dehydrogenase, (3) aldehyde dehydrogenase, (4), (5), (7) monooxygenases, (6) secondary alcohol dehydrogenase and (8) acetylesterase

and (4) oxidation across the double bond produces a diol.

Of the many hydrocarbon and related **cycloaliphatic compounds**, only the degradation of **cyclohexane** is shown in Fig. 9.**41**. The opening of the ring is brought about by a monooxygenase that copies the chemical Baeyer-Villiger oxidation of such cyclic oxo-compounds. There are other pathways involving the cyclohexane-1,2-diol.

Anaerobic degradation of aliphatic hydrocarbons. Aliphatic hydrocarbons were considered to be completely metabolically inert in the absence of molecular oxygen. However, sulfate- and nitrate-reducing bacteria were discovered to metabolize these compounds slowly and use them as sole organic substrate. The initial attack under anoxic conditions is either chain elongation by one carbon (i.e., addition of an activated C_1 unit) yielding the C_{n+1} compound (further oxidation proceeds via the C_{n+1} fatty acid) or the terminal $-CH_3$ is oxidized to a $-COOH$ group in the corresponding C_n fatty acid, with water as source of oxygen. Alkenes are attacked by adding water to the double bond to give the secondary alcohol. The elucidation of these pathways is just emerging.

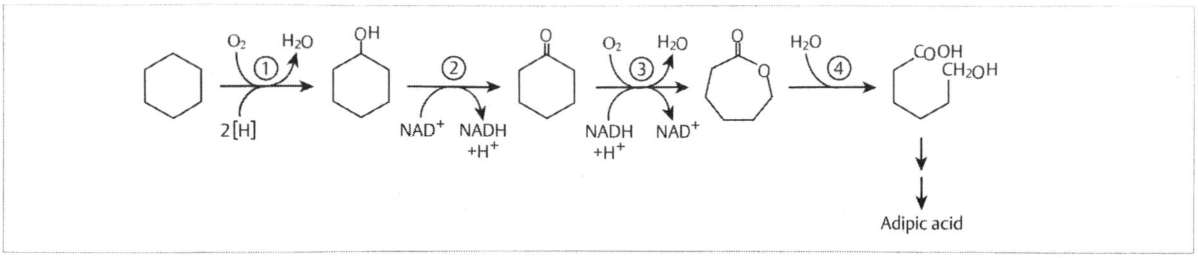

Fig. 9.**41** **Metabolic pathway involved in the metabolism of cyclohexane *(Nocardia, Pseudomonas)*.** (1) Cyclohexane monooxygenase, (2) cyclohexanol dehydrogenase, (3) cyclohexanone monooxygenase, and (4) ε-caprolactone hydrolase

9.14.3 Sterols, Hopanoids, Carotenoids, Monoterpenoids, and Other Isoprenoids Are Common Poorly Water-soluble Compounds

Monoterpenoids (C_{10}) are major components of plant oils and are synthesized from two isoprene units. They form acyclic, monocyclic, or bicyclic structures. There is a bewildering set of combinations of reactions involved in their degradation. Some of the principal reactions have been discussed in the previous chapter. Because of the methyl branching, the final products are acetyl-CoA and propionyl-CoA.

9.14.4 C_1 Compounds Are Organic Compounds With One Carbon Atom

Functional C_1 groups, in which a one-carbon unit is separated from the carbon unit chain by a heteroatom (O, N, S) such as in phenolic $-O-CH_3$ groups, are degraded via C_1 compounds. The most prominent C_1 compound is CH_4 which is formed in anaerobic fermentation of organic matter to biogas; however, also methanol (CH_3OH), formaldehyde (CH_2O), formate ($HCOOH$), methylamine (CH_3-NH_2), methylmercaptan (CH_3-SH), and others play an important role in nature. Evidently, these compounds cannot be oxidized through the citric acid cycle, they require special treatment. Two groups of microorganisms can be differentiated: (1) obligate methylotrophs that can only grow with C_1 compounds; this group includes all bacteria that can oxidize methane (methanotrophs), and (2) facultative methylotrophs that can utilize, in addition to other substrates, CH_3OH or CH_3NH_3, for example, but not CH_4; however, there are exceptions.

Obligate methylotrophs. A large number of bacteria are found in the borderline of the oxic and anoxic zones of sediments and wet soil that can oxidize CH_4 to CO_2 with the help of molecular oxygen. These methylotrophic bacteria contain internal membrane stacks or membranes lying round the cell periphery. These membranes contain the enzyme that oxidizes methane to methanol, the electron transport chain, and terminal oxidases. Cells grown on methanol have no membrane system.

Methane is oxidized in four two-electron oxidation steps (Fig. 9.**42**). The first step catalyzed by the membrane-bound methane monooxygenase requires molecular oxygen to attack the chemically inert CH_4 and split the stable C–H bond (bond dissociation energy of 435 kJ/mol) (paraffin means nonreactive). The nonspecific enzyme, which can hydroxylate various other compounds, contains copper. Under some conditions (e.g., copper deficiency) a soluble methane monooxygenase is used that contains a hydroxo-bridged dinuclear iron center. NADH or reduced cytochrome *c* are used for hydroxylation.

The product **methanol** cannot be oxidized to formaldehyde ($E_0' = -182$ mV) with NAD^+ ($E_0' = -320$ mV); instead a special quinone, methoxatin (= PQQ, **p**yrrolo**q**uinoline **q**uinone; $E_0' = +120$ mV) serves as electron acceptor (Fig 9.**42b**) (this coenzyme is used in other oxidations of alcoholic groups, e.g., with glucose oxidase). Methanol dehydroxygenase is a periplasmic enzyme. Formaldehyde dehydrogenase and formate dehydrogenase are NAD^+-dependent. NADH is

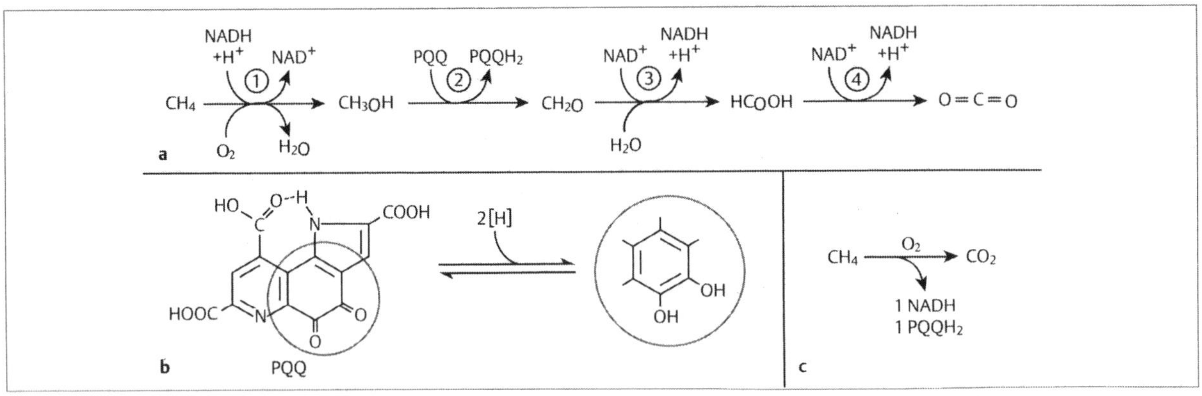

Fig. 9.**42a–c** **Oxidation of methane.**
a Reactions: (1) methane monooxygenase, (2) methanol dehydrogenase, (3) formaldehyde dehydrogenase, and (4) formate dehydrogenase. Methane monooxygenases oxidize other hydrocarbons; also, NH_3 which is similarly stable, is oxidized to NH_2OH. Ammonia monooxygenase and methane monooxygenase have many common features. This nonspecificity reflects the strong oxidizing reagent generated in the active site.
b structure of the coenzyme PQQ and its redox reaction. **c** Net reaction of methane oxidation

fed into the respiratory chain at the level of quinone, PQQ is fed at the level of cytochrome *c*. **Formaldehyde** adds to glutathione and is oxidized as adduct (*S*-hydroxymethylglutathione) to *S*-formylglutathione, which becomes hydrolyzed. **Formate** is oxidized to CO_2 by formate dehydrogenase.

Methanol can be used as an inexpensive organic substrate for biomass production using methylotrophic bacteria ("**single-cell protein** production"). N- and S-containing compounds require additional enzymes. For instance, **trimethylamine** is oxidized by a flavoenzyme according to Reaction 9.13:

$$(CH_3)_3N + H_2O + [FAD] \rightarrow (CH_3)_2)NH + CH_2O + [FADH_2] \quad (9.13)$$

Alternatively, trimethylamine is hydroxylated by a monooxygenase to trimethylamine *N*-oxide, which is converted to dimethylamine and formaldehyde by an aldolase. Several mechanisms have been described for **methylamine** oxidation.

Facultative methylotrophs. A large number of bacteria can utilize C_1 compounds other than methane, but prefer other organic substrates when available. Electrons are fed into the quinone pool. Some of these bacteria use a dissimilatory hexulose phosphate cycle for formaldehyde oxidation (see Chapter 8). In this pathway, formaldehyde adds to ribulose 5-phosphate, and the resulting hexulose phosphate is oxidized via fructose 6-phosphate to ribulose 5-phosphate + CO_2.

C_1 compounds as carbon source. Methylotrophic bacteria can utilize CH_4 and other C_1 compounds also as sole source of cell carbon. This requires special assimilation pathways, and formaldehyde is the intermediate that is assimilated. In a few cases of non-specialist C_1 utilizers, C_1 compounds are first completely oxidized, and CO_2 is incorporated via the Calvin cycle (see Chapter 8).

Anaerobic pathways. Methane probably cannot be oxidized under anoxic conditions; the anaerobic metabolism of other C_1 compounds by acetogenic and methanogenic bacteria as well as their assimilation into cell material are discussed in Chapter 12 and Chapter 8, respectively.

9.15 The Nitrogen-containing Heterocyclic Compounds Are a Large and Heterogeneous Group of Biologically Important Substances

Nucleic acids contain purines and pyrimidines, proteins contain tryptophan, proline, and histidine, and various natural metabolites and man-made xenobiotics are *N*-heterocyclic. Most *N*-heterocyclic compounds are π-electron-deficient aromatic compounds, yet the electronegative nitrogen makes the C2 or C4 amenable to nucleophilic attack. Other systems have excessive electrons and thus are susceptible to electrophilic substitution.

Introduction of hydroxyl groups is either brought about by specific monooxygenases (Reaction 9.14):

$$N\text{-Heterocyclic compound} + 2[H] + O_2 \rightarrow \text{Hydroxylated } N\text{-heterocyclic compound} + H_2O \quad (9.14)$$

or molybdo-enzyme dehydrogenases introduce oxygen from water according to Reaction 9.15:

$$N\text{-Heterocyclic compound} + H_2O + \text{electron acceptor} \rightarrow \text{Hydroxylated } N\text{-heterocyclic compound} + \text{Reduced electron acceptor } [2H] \quad (9.15)$$

9.15.1 Purine and Pyrimidine Bases Are Released During DNA, RNA, and Nucleotide Degradation

Purines. Purine bases are aerobically oxidized to glyoxylate and urea (Fig. 9.**43**); (for anaerobic metabolism, see Chapter 12). The different bases are transformed to hypoxanthine and xanthine, both of which are oxidized to uric acid by xanthine oxidase. Uric acid is oxidized further to allantoin, which is hydrolyzed to yield allantoate. Different microorganisms then convert it to urea and glycine by different branches. Urea is hydrolyzed to ammonia and CO_2 by the nickel enzyme urease (Reaction 9.16):

$$H_2N{-}CO{-}NH_2 + H_2O \rightarrow 2\ NH_3 + CO_2 \quad (9.16)$$

Pyrimidines. Pyrimidine bases are aerobically oxidized to β-alanine (Fig. 9.**44**). Cytosine is deaminated to uracil. Degradation of uracil is almost a reversal of its synthesis (Chapter 7). The ring is hydrogenated and hydrolyzed between N3 and C4. Since the C1 carboxyl of aspartate is lost during synthesis of UMP, the product of degradation of uracil is β-alanine rather than aspartate. β-Alanine can be oxidatively deaminated to malonic semialdehyde and oxidized to malonyl-CoA. Thymine is degraded

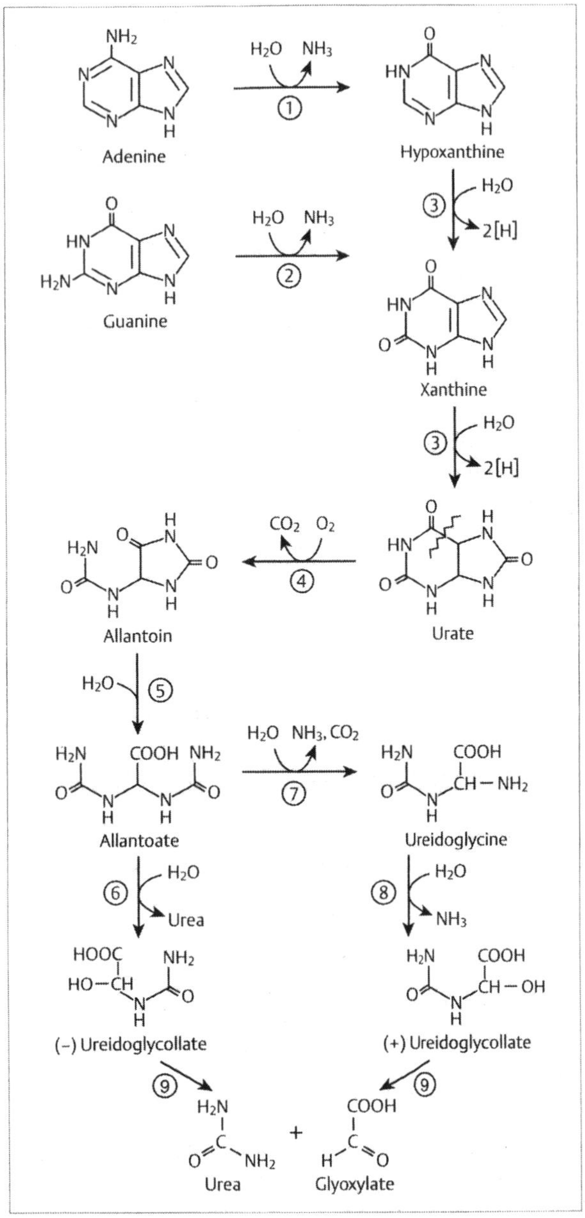

Fig. 9.**43** **Aerobic degradation of purine bases.** (1) Adenine deaminase, (2) guanine deaminase, (3) xanthine oxidase, (4) uricase, (5) allantoinase, (6) allantoicase, (7) allantoate amidohydrolase, (8) ureidoglycine aminohydrolase, and (9) ureidoglycolase

analogously, but β-aminoisobutyrate is formed. The latter can be oxidatively converted to methylmalonate (Reaction 9.17),

$$H_2N-CH_2-CH(CH_3)-COOH \rightarrow\rightarrow HOOC-CH(CH_3)-COOH \quad (9.17)$$

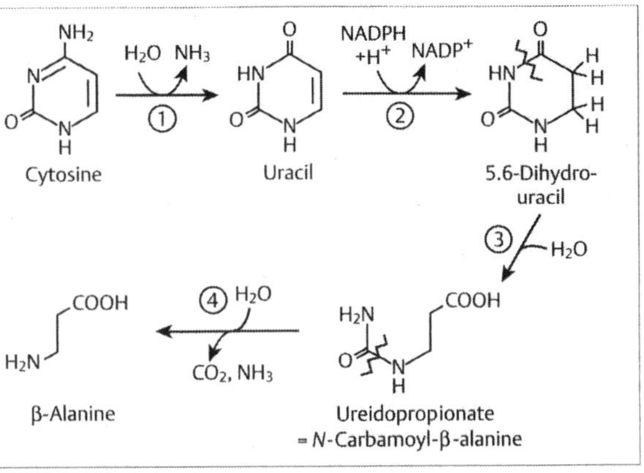

Fig. 9.**44** **Aerobic degradation of pyrimidine bases (cytosine and uracil).** (1) Cytosine deaminase, (2) 5,6-dihydrouracil dehydrogenase, (3) 5,6-dihydrouracil hydrolase, and (4) β-ureidopropionase

which can enter the methylmalonyl-CoA pathway to succinate.

9.15.2 *N*-Heterocyclic Compounds Containing One Nitrogen Are Degraded via Introduction of Hydroxyl Groups From Water or From O_2

Pyridine is first reduced to 1,4-dihydropyridine. This compound either undergoes a hydrolytic ring cleavage following deamination to glutaric dialdehyde or the ring is oxygenolytically cleaved, and succinate semialdehyde and formate are hydrolytically released. In the case of hydroxylated pyridines, a second hydroxyl group is introduced, either from water via a dehydrogenase-type of enzyme, or mediated by a monooxygenase. The resulting diols are cleaved by diol dioxygenases.

Similarly, **pyridoxine**, **pyridoxamine** and **pyridoxal** (the three forms of vitamin B_6) and pyridine mono- or dicarboxylic acids are readily degraded. To illustrate the variety of degradation mechanisms and metabolic products, the degradation of nicotinic acid (pyridine 3-carboxylic acid) is mentioned. In the aerobic pathways, oxygen from water is introduced by a molybdo enzyme at C6; 6-hydroxynicotinic acid is then oxidatively decarboxylated yielding the 2,3-diol, and the pyridine ring is cleaved by insertion of two atoms of O_2. Alternatively, another hydroxyl is introduced from water at position 2 by a molybdo enzyme. Anaerobically, 6-hydroxynicotinic acid is reduced and converted to α-methylene-glutarate.

A similar interplay of enzymes that introduce hydroxyl groups from O_2 (monooxygenases) or water (dehydrogenases, mostly Mo-dependent) is found in the metabolism of other *N*-heterocyclic compounds such as **quinoline** or **isoquinoline** and derivatives. **Indole** and **tryptophan** degradation was discussed in Chapter 9.13.2.

9.16 Organic Acids Are Common Natural Substrates That Are Also Formed Secondarily by Fermentations

Organic acids can be metabolized by many bacteria. Their uptake requires more or less specific transport systems.

Acetate can be activated to acetyl-CoA in four different ways:

$$\text{Acetate} + \text{HSCoA} + \text{ATP} \rightarrow \text{Acetyl-SCoA} + \text{AMP} + \text{PP}_i \quad (9.18)$$

$$\text{Acetate} + \text{ATP} \rightarrow \text{Acetyl-phosphate} + \text{ADP} \quad (9.19)$$

$$\text{Acetyl-phosphate} + \text{HSCoA} \rightarrow \text{Acetyl-CoA} + \text{P}_i \quad (9.20)$$

$$\text{Acetate} + \text{Succinyl-CoA} \rightarrow \text{Acetyl-CoA} + \text{Succinate} \quad (9.21)$$

$$\text{Acetate} + \text{ATP} + \text{HSCoA} \rightarrow \text{Acetyl-CoA} + \text{ADP} + \text{P}_i \quad (9.22)$$

Acetyl-CoA is then oxidized by aerobes via the citric acid cycle. If acetate serves as sole carbon source, the reactions of the glyoxylate cycle have to supply C_4 and C_3 (phospho*enol*pyruvate) precursors for biosynthesis. In anaerobes, acetyl-CoA can completely be oxidized via the citric acid cycle or its variations, or via the acetyl-CoA pathway.

Propionate results from the degradation of odd-numbered fatty acids. It is activated to propionyl-CoA and is generally converted into succinyl-CoA. The reactions in principal are the reversal of propionate fermentation via succinyl-CoA (see Chapter 12). First, propionyl-CoA is carboxylated at the carbon next to the carbonyl in an ATP-dependent reaction catalyzed by the biotin enzyme propionyl-CoA carboxylase, which resembles pyruvate carboxylation. Methylmalonyl-CoA undergoes a rearrangement reaction leading to succinyl-CoA, catalyzed by a vitamin B_{12}-dependent mutase. Succinyl-CoA then is oxidized in the citric acid cycle. There might be different pathways of propionate oxidation operating in bacteria, such as oxidation via 3-hydroxypropionate or oxidation of propionyl-CoA via condensation with oxaloacetate to methylcitrate. Methylcitrate is isomerized to 2-methylisocitrate which is cleaved by a lyase to pyruvate and succinate, similar to the glyoxylate cycle. For **butyrate** and longer-chain fatty acids, see Chapter 9.14.1, β-oxidation.

Hydroxyacids and **keto acids**, such as malate or pyruvate, that are intermediates of central pathways are easily metabolized. Growth with malate requires formation of pyruvate catalyzed by malic enzyme (Reaction 9.23)

$$\text{L-malate} + \text{NAD}^+ \rightarrow \text{Pyruvate} + \text{NADH} + \text{H}^+ + \text{CO}_2 \quad (9.23)$$

An $NADP^+$-dependent isoenzyme provides NADPH for biosynthesis. Phospho*enol*pyruvate may be formed from oxaloacetate or directly from pyruvate (see Chapter 7).

Glyoxylate is a product of purine and glycine degradation and is also formed by oxidation of **glycolate**. The conversion of this oxidized substrate into an intermediate of central metabolism requires additional enzymes of the **glycerate pathway** (Fig. 9.**45a**). The product is 3-phosphoglycerate, an intermediate of the Embden-Meyerhof-Parnas pathway. Glyoxylate also serves to replenish the intermediates of the citric acid cycle: it is condensed with acetyl-CoA to form malate, catalyzed by malate synthase (Fig. 9.**46**). An alternative pathway leading to oxaloacetate, described in Fig. 9.**45b**, consists of three steps: transamination to glycine, condensation of glycine with another molecule of

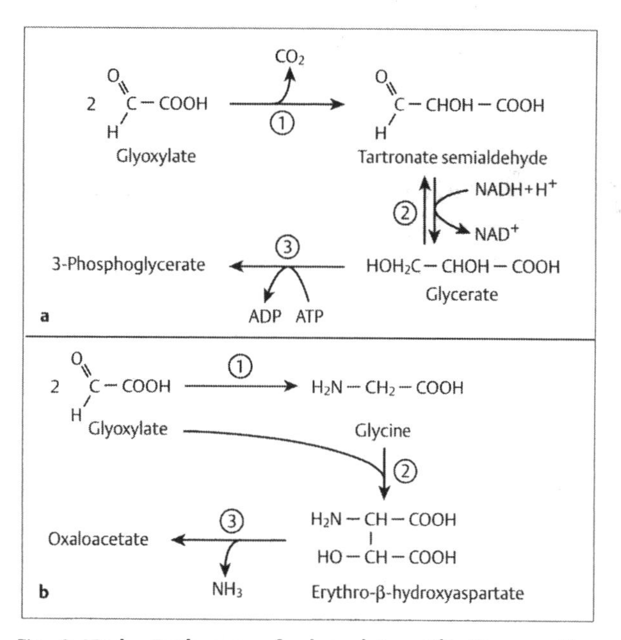

Fig. 9.**45a,b** **Pathways of glyoxylate utilization. a** The glycerate pathway. (1) glyoxylate carboligase, (2) tartronate semialdehyde reductase, (3) glycerate kinase. **b** The β-hydroxyaspartate pathway. (1) Transaminase, (2) erythro-β-hydroxyaspartate aldolase, (3) erythro-β-hydroxyaspartate dehydratase

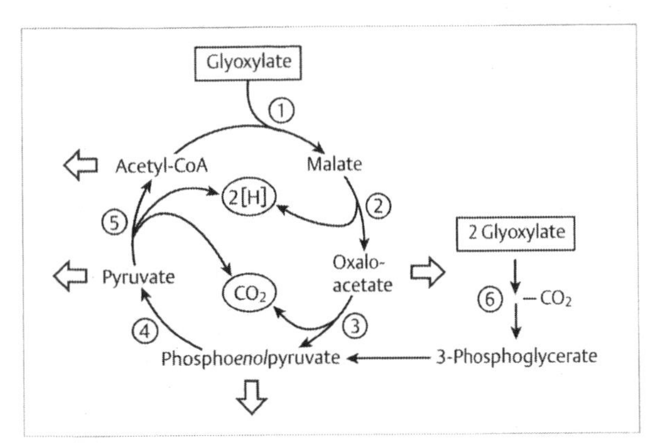

Fig. 9.**46** **Assimilation of glyoxylate by a dicarboxylic acid cycle when the glycerate pathway is functioning.** (1) Malate synthesis B, (2) malate dehydrogenase, (3) phospho*enol*-pyruvate carboxykinase, (4) pyruvate kinase, (5) pyruvate dehydrogenase, (6) glyoxylate carboligase. Thick arrows represent fluxes to biosynthetic pathways

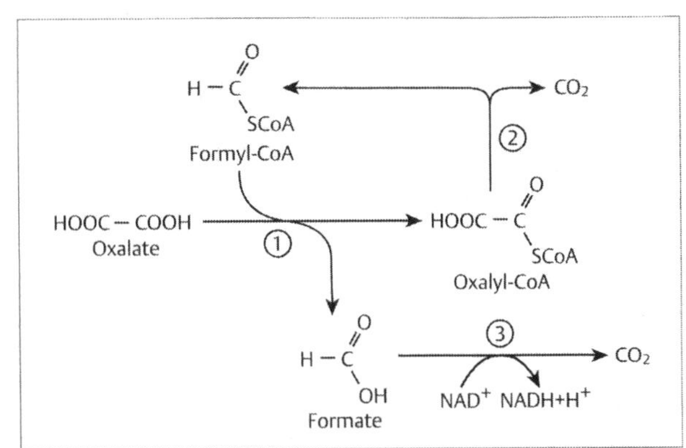

Fig. 9.**47** **Oxidation of oxalate by *Pseudomonas oxalaticus*.** (1) CoA transferase, (2) oxalyl-CoA decarboxylase, (3) formate dehydrogenase

glyoxylate to form β-hydroxyaspartate, and elimination of ammonia.

Oxalate, the most highly oxidized C_2 compound, is oxidized to CO_2 in three steps via formyl-CoA (Fig. 9.**47**). First oxalyl-CoA is formed by CoA transfer from formyl-CoA, then oxalyl-CoA is decarboxylated to yield formyl-CoA, which yields formate upon CoA transfer, and formate is oxidized to CO_2. Cell compounds are synthesised from glyoxylate (via the glycerate pathway), which is formed by reduction of the activated oxalate (Reaction 9.24):

$$\text{Oxalyl-SCoA} + \text{NADH} + \text{H}^+ \rightarrow \text{Glyoxylate} + \text{NADP}^+ + \text{CoA-SH} \quad (9.24)$$

Ethylene glycol (CH_2OH-CH_2OH) is oxidized—via glycolaldehyde and glycolate—to glyoxylate, which is then assimilated and oxidized.

9.17 Xenobiotic Compounds Are Man-made Recalcitrant Pollutants That Normally Do Not Occur in Nature

This definition for xenobiotic compounds also applies to natural toxic (e.g., phenol) or recalcitrant (e.g., hydrocarbons) compounds, the natural concentrations of which are orders of magnitude lower than that prevailing in man-made environments; xenobiotics are harmful to man, animals, and plants. Their recalcitrance can often be predicted on chemical grounds, i.e, from the stability of the chemical bonds. However, nature produces an unimaginable variety of compounds; in fact, an increasing number of xenobiotic compounds or closely related compounds are found in unpolluted ecosystems. Microorganisms, therefore, are prepared during evolution to handle many of these compounds and use them as substrates. Yet, the rate of xenobiotic degradation is often extremely slow. For example, halogenated organic compounds, a large group of man-made environmental toxic pollutants, are produced in large amounts in natural abiotic and biotic processes. Examples of producers are many marine algae. Almost 1000 natural halogenated compounds are known so far.

9.17.1 Dehalogenation Is Often the First Step in the Degradation of Halogenated Compounds

As a general rule, recalcitrance increases with substituents in the order I, Br, Cl, F and with the number of halogen substituents. Reductive dehalogenations are a domain of anaerobes.

The principal enzymic dehalogenation mechanisms are described in Fig. 9.**48**. In addition, spontaneous dehalogenation reactions may occur as a result of chemical decomposition of unstable primary products of an unassociated enzyme reaction. This dehalogenation is called "spontaneous". "Fortuitous" dehalogena-

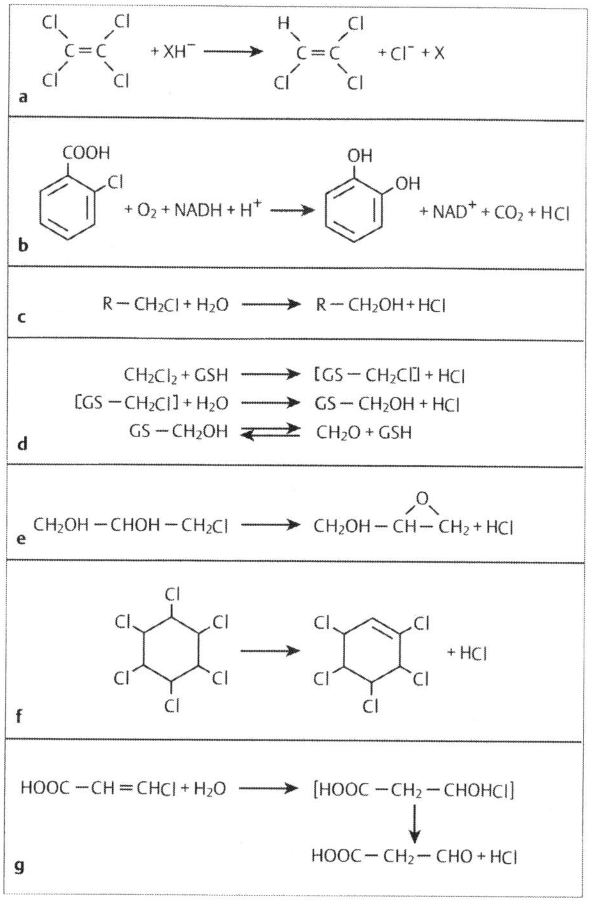

a
$$C_2Cl_4 + XH^- \longrightarrow C_2HCl_3 + Cl^- + X$$

b
$$\text{2-chlorobenzoate} + O_2 + NADH + H^+ \longrightarrow \text{catechol} + NAD^+ + CO_2 + HCl$$

c
$$R-CH_2Cl + H_2O \longrightarrow R-CH_2OH + HCl$$

d
$$CH_2Cl_2 + GSH \longrightarrow [GS-CH_2Cl] + HCl$$
$$[GS-CH_2Cl] + H_2O \longrightarrow GS-CH_2OH + HCl$$
$$GS-CH_2OH \rightleftharpoons CH_2O + GSH$$

e
$$CH_2OH-CHOH-CH_2Cl \longrightarrow CH_2OH-CH(O)CH_2 + HCl$$

f
$$C_6H_6Cl_6 \longrightarrow C_6H_5Cl_5 + HCl$$

g
$$HOOC-CH=CHCl + H_2O \longrightarrow [HOOC-CH_2-CHOHCl] \longrightarrow HOOC-CH_2-CHO + HCl$$

Fig. 9.**48a–g** **Dehalogenation mechanisms in the enzymic cleavage of the carbon–halogen bond.**
a Reductive dehalogenation catalyzed by reductases; the halogen substituent is replaced by hydrogen.
b Oxygenolytic dehalogenation catalyzed by mono- or dioxygenases; the halogen substituent is replaced by oxygen from O_2.
c Hydrolytic dehalogenation catalyzed by halidohydrolases; the halogen substituent is replaced by a hydroxy-group derived from water.
d "Thiolytic" dehalogenation catalyzed by enzymes forming a glutathione conjugate (glutathione S-transferases), which is hydrolyzed.
e Intramolecular substitution of vicinal haloalcohols by -OH group (nucleophilic displacement), yielding epoxides.
f Dehydrohalogenation by HCl elimination, leading to the formation of a double bond.
g Hydration of vinylic halogenated compounds, which spontaneously eliminate HCl [from 5]

tions are brought about by enzymes with broad substrate specificity, but there are transitions between the two processes. Xenobiotics are often co-metabolized, degradation therefore requires an additional growth substrate.

9.17.2 Organic Sulfonic Acids Can Serve as Carbon and/or Sulfur Source

Substantial amounts of sulfur in the biosphere are present as organic sulfonic acids ($R-SO_3^-$), especially in humus. There are low-molecular-weight sulfonates such as coenzymes M (Fig. 12.**10**), the chloroplast lipids contain 6-sulfoquinovose as polar head group. In addition, sulfonates play an important role as detergents and dyes and are xenobiotics. They are metabolized notably under starvation conditions.

The C–S bond in sulfonates is rather stable. Still, there are several ways to remove the sulfonate group as sulfite oxygenolytically. In the case of aromatic sulfonic acids, desulfonation may occur before or during cleavage by oxygenases, or after ring cleavage by hydrolysis. In "oxygenolytic" desulfonation, the substrate is dioxygenated and spontaneously loses sulfite. The hydroxyl group causes the $C-SO_3^-$ bond to be labile and makes sulfite a good leaving group. The dioxygenation of the natural substrate with the stable intermediate is shown in Figure 9.**48** for comparison. Sulfite can also be eliminated by monooxygenase reactions to serve as sulfur source. Desulfonation occurs even under anoxic conditions; the mechanisms are unknown.

9.18 Some Bacteria Oxidize Simple Organic Compounds Incompletely

Complete degradation of complex organic substrates requires the cooperation of different microorganisms. However, some bacteria degrade even simple organic compounds incompletely and excrete partially oxidized compounds, such as organic acids. This acid production (e.g., acetic acid from ethanol in vinegar by acetic acid bacteria), lowers the pH of the environment and prevents other bacteria from growing. Incomplete oxidations are also referred to as "oxidative fermentations". Products excreted are acetic, gluconic, fumaric, citric, and other organic acids, oxoacids, ketones, and even amino acids. These processes are of great biotechnological interest. The transformation of organic compounds that are not associated with growth (co-metabolism) will not be considered here.

> **Box 9.5 Incomplete oxidizers** of simple organic compounds are strictly aerobic bacteria. They can be considered specialists that rapidly grow by partial oxidation of a common low-molecular-weight substrate, which in the organism's habitat temporarily accumulates in very high concentrations. This applies to ethanol formed by rapid fermentation of sugars in plant juices by acid-tolerant yeasts; acetic acid bacteria live together with these yeasts and oxidize ethanol to acetic acid. The high concentrations of ethanol and of the organic acid produced (acetic acid) as well as the low pH can be tolerated (acid- and solvent-tolerant), but keep away competitors that are not adapted to these extremes. When ethanol is completely taken up and consumed, most of these bacteria slowly grow at the expense of complete acetate oxidation to CO_2. Succinate dehydrogenase in the citric acid cycle is lacking in the ethanol-oxidation phase, it becomes induced in the acetate-oxidation phase. Similar arguments hold true for glucose oxidation to gluconic acid and other products. If the organic acids are reutilized, the bacteria belong to the "**peroxydans**-group"; if the organic acids are not reutilized, the bacteria belong to the "**suboxydans**-group". There are transitions between these two extremes.

Acetic acid bacteria, e.g., *Acetobacter* sp. and *Gluconobacter* sp., are closely related Gram-negative bacteria that can form organic acids from excess alcohol or sugars. They oxidize ethanol to acetate in two steps; both alcohol dehydrogenase and aldehyde dehydrogenase are NAD^+ independent enzymes that reduce the quinone PQQ (see Chapter 9.11.3, 9.14.4 and Fig. 9.**42**). PQQ feeds electrons into the respiratory chain at the level of cytochrome *c*. ATP solely is produced by electron transport phosphorylation. *Acetobacter aceti* is the well-known vinegar producer. Growth on C_2 compounds requires anaplerotic reactions to produce C_3 and C_4 compounds for biosynthesis (see glyoxylate cycle and C_3–C_4 interconventions, Chapter 7). Note that the incomplete citric acid cycle only operates in biosynthesis.

Gluconobacter oxidizes glucose via the oxidative pentosephosphate cycle to glyceraldehyde 3-phosphate, which is oxidized via pyruvate and acetyl-CoA to acetate. Glucose normally is first oxidized by the PQQ-containing glucose oxidase to gluconate, which may be taken up. Another biotechnologically useful process performed by acetic acid bacteria is incomplete oxidation of D-sorbitol to L-sorbose (compare to gluconate oxidation to 5-oxogluconate). D-Sorbitol (=glucitol) is formed by certain bacteria by reduction of glucose when excess glucose is available. L-Sorbose is oxidatively converted in a chemical process to L-ascorbate.

Alcohol dehydrogenases and aldehyde dehydrogenases are not specific for ethanol and acetaldehyde in incomplete oxidizers. Therefore, secondary alcohol formed in fermentation are transformed to ketones (e.g., isopropanol→acetone; glycerol→dihydroxyacetone; gluconate→5-oxogluconate). Sugar alcohols are oxidized to aldoses or ketoses. *Acetobacter* and *Gluconobacter* contain the PQQ enzymes alcohol oxidase, aldehyde oxidase, and glucose oxidase in the periplasmic space, which react with cytochrome *c* located at the periplasmic site of the membrane. The substrates and products do not penetrate the cell. This is one reason for the tolerance against the toxically high concentrations of substrates and/or products.

Bacilli, when growing on excess carbohydrates, incompletely oxidize glucose to acetate, but also pyruvate, acetoin, and butanediol are formed (for structural formulas, see Chapter 12.2.2). These latter reactions are normally found in fermenting bacteria. When other nutrients, such as nitrogen or phosphorus, become limiting, Bacilli sporulate and use the excreted products as energy source; these products are completely oxidized. As in acetic acid bacteria, the citric acid cycle is incomplete during the first rapid growth phase. Acetoin can be oxidized by an enzyme complex very similar to the 2-oxoacid dehydrogenase complexes; thiamine pyrophosphate is essential for catalysis (see Chapter 3.5).

Further Reading

Gottschalk, G. (1986) Bacterial metabolism, 2nd ed. New York: Springer

Neidhardt, F. C., Curtiss, R., Ingraham, J. L., Lin, E. C. C., Low, K. B., Magasanik, B., Reznikoff, W., Riley, M., Schaechter, M., and Umbarger, H. E. (eds.) (1996) *Escherichia coli* and *Salmonella*: cellular and molecular biology. 2nd edn. Washington, DC: ASM Press

Ratledge, C., ed. (1993) Biochemistry of microbial degradation. Dordrecht: Kluwer

Sources of Figures

1 Beguin P, and Aubert J.-P. (1994) FEMS Microbiol Rev 13:25–58
2 Sutherland, I. W. (1995) FEMS Microbiol Rev 16:323-347
3 De Jong, E., Field, J. A., and de Bont J. A. M. (1994) FEMS Microbiol Rev 13:153–188
4 Gottschalk G. (1986) Bacterial metabolism. Berlin Heidelberg New York: Springer
5 Fetzner, S. and Lingens F. (1994) Microbiol Rev 58:641–685

10 Oxidation of Inorganic Compounds by Chemolithotrophs

10.1 Chemolithotrophs Derive Energy From the Oxidation of Inorganic Compounds

Nature harbors a large variety of bacteria that can derive metabolically useful energy from the oxidation of inorganic compounds (see Fig. 10.1), such as hydrogen (H_2), carbon monoxide (CO), inorganic (reduced) sulfur and nitrogen compounds (S^0 etc., NH_4^+), and from divalent ions [e.g., ferrous (Fe^{++}) and manganous (Mn^{++}) ions]. These bacteria are called chemolithotrophs. Such oxidations can only proceed if a suitable electron acceptor is available. Many chemolithotrophs use oxygen as the terminal electron acceptor. The facultatively anaerobic chemolithotrophs can also use nitrate or nitrite as electron acceptor. A variety of obligately anaerobic bacteria can use sulfate or bicarbonate as electron acceptor for growth on, for example, hydrogen. The latter two groups are discussed in Chapter 12. Many of the chemolithotrophs are also dependent on carbon dioxide as their (main) carbon source; these bacteria are called chemolithoautotrophs. In nature, a wide variety of (partially) reduced inorganic compounds are found that can serve as the energy source for chemolithotrophic growth. The most important energy-generating reactions involving oxygen or nitrate as electron acceptor are listed in Table 10.1.

Although formally CH_4 can be regarded as an "organic" compound, the methane-oxidizing bacteria, the methanotrophs, share many basic properties with the

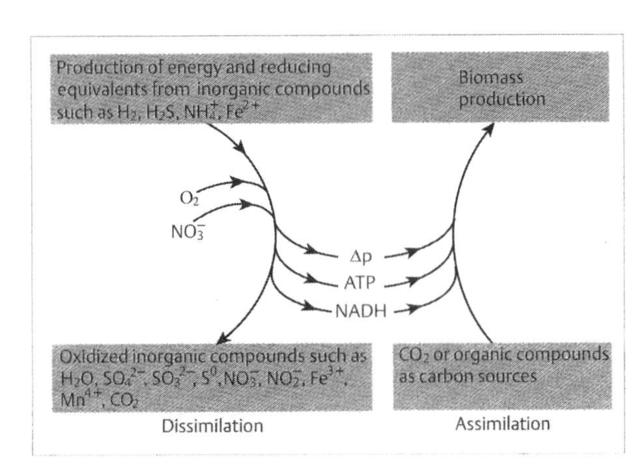

Fig. 10.1 **Chemolithoautotrophy.** This mode of living is a combination of energy conversion from the oxidation of inorganic compounds and of autotrophic carbon dioxide fixation. Δp, proton motive force or proton potential

Table 10.1 **Examples of energy-generating reactions for chemolithotrophs**

Example of genus or species	Reaction	$\Delta G_R^{0'}$
Nitrosomonas europaea	$2\ NH_4^+ + 3\ O_2 \rightarrow 2\ NO_2^- + 2\ H_2O + 4\ H^+$	– 551.2
Mixed cultures	$NH_4^+ + NO_2^- \rightarrow N_2 + 2\ H_2O$	– 357.0
Nitrobacter winogradskyi	$2\ NO_2^- + O_2 \rightarrow 2\ NO_3^-$	– 74.3
Alcaligenes eutrophus	$2\ H_2 + O_2 \rightarrow 2\ H_2O$	– 472.5
Paracoccus denitrificans	$5\ H_2 + 2\ NO_3^- + 2\ H^+ \rightarrow 6\ H_2O + N_2$	– 958.8
Desulfovibrio desulfuricans	$4\ H_2 + SO_4^{2-} + 2\ H^+ \rightarrow H_2S + 4\ H_2O$	– 154.4
Methanobacterium thermoautotrophicum[1]	$4\ H_2 + HCO_3^- + H^+ \rightarrow CH_4 + 3\ H_2O$	– 138.6
Pseudomonas carboxydovorans	$2\ CO + O_2 \rightarrow 2\ CO_2$	– 504.9
Methylomonas spec.[2]	$CH_4 + 2\ O_2 \rightarrow HCO_3^- + H^+ + H_2O$	– 814.5
Thiobacillus thioparus and *Beggiatoa* spec.	$HS^- + 2\ O_2 \rightarrow SO_4^{2-} + H^+$	– 798.2
Thiobacillus thiooxidans	$2\ S^0 + 3\ O_2 + 2\ H_2O \rightarrow 2\ SO_4^{2-} + 4\ H^+$	– 588.2
Thiomicrospira denitrificans	$5\ S_2O_3^{2-} + 8\ NO_3^- + H_2O \rightarrow 10\ SO_4^{2-} + 2\ H^+ + 4\ N_2$	– 3925.6
Desulfovibrio sulfodismutans[1]	$S_2O_3^{2-} + H_2O \rightarrow HS^- + SO_4^{2-} + H^+$	– 18.9
Thiobacillus ferrooxidans	$4\ Fe^{2+} + O_2 + 4\ H^+ \rightarrow 4\ Fe^{3+} + 2\ H_2O$	– 17.7
Leptothrix spec.	$2\ Mn^{2+} + O_2 + 2\ H_2O \rightarrow 2\ MnO_2 + 4\ H^+$	– 77.6
Stibiobacter spec.	$2\ Sb^{3+} + O_2 + 4\ H^+ \rightarrow 2\ Sb^{5+} + 2\ H_2O$	n.a.[3]

[1] For further details, see Chapter 12 (anaerobic respiration and fermentation).
[2] For further details on the methanotrophs and methylotrophs, see Chapter 9.
[3] Data not available.

chemolithotrophs, and hence CH_4 oxidation is mentioned here. For a discussion of methane oxidation, see Chapter 9.

10.1.1 Major Sources of Reduced Inorganic Compounds Are Intermediates of the Bio(geo)chemical Cycles

Most reduced inorganic compounds are produced biologically under anoxic conditions by anaerobic respiration or fermentation. For example, H_2 is a by-product of many (bacterial) fermentation reactions. Ammonium is released during the (an)aerobic breakdown of organic nitrogen compounds, as well as by anaerobic reduction of nitrate to ammonia in various processes. Hydrogen sulfide and carbon dioxide are the main end products of the anaerobic breakdown of organic compounds by sulfate-reducing bacteria (Fig. 10.**1**). Hydrogen sulfide can be reoxidized to sulfate by the chemolithotrophs in the presence of oxygen or nitrate. In this way, the chemolithotrophs play an essential role in the biogeochemical sulfur cycle and, likewise, in the nitrogen cycle, the iron cycle, and many other cycles.

It is important to note that the driving force behind these cycles is sunlight, which allows the primary production of organic compounds from carbon dioxide by photosynthesis. Subsequently, organic compounds provide the reducing power for the reductive part of the element cycles. Some important aspects of the element cycles in which chemolithotrophs play a role are mentioned in Chapter 10.5, while an extensive discussion can be found in Chapters 30–32.

Some reduced inorganic compounds may also be of volcanic origin, such as hydrogen sulfide and ferrous iron, or may originate from geological deposits. Human activities, including agriculture, the oil and chemical industries, and combustion processes, also represent substantial sources of inorganic reduced compounds.

The concept of chemolithoautotrophic growth was formulated by S. N. Winogradsky, who studied the sulfide-oxidizing *Beggiatoa* species and showed that these bacteria can oxidize hydrogen sulfide to elemental sulfur and then further to sulfate using oxygen as the oxidant (see Chapter 1). That *Beggiatoa* could use the energy from sulfide oxidation to assimilate CO_2 as its main carbon source was difficult to prove because of the instability of sulfide in the presence of oxygen. The ability to fix carbon dioxide and to grow autotrophically was experimentally proven when Winogradsky studied ammonia-oxidizing bacteria. On the basis of both studies, the concept of chemolithoautotrophy was created.

Details of the oxidation reactions of relevant inorganic compounds are summarized in Table 10.**1**. In order to proceed, an oxidation-reduction reaction requires a suitable electron donor and an electron acceptor. Oxygen is very suitable, having a redox potential of +816 mV. Other electron acceptors are listed in Table 10.**2**. Note that the values given have been corrected to the standard potential at pH 7.0 (E_0').

Box 10.1 From the data shown in Table 10.**2**, it can be calculated that the oxidation of molecular hydrogen with O_2, NO_3^-, SO_4^{2-}, or HCO_3^- as terminal electron acceptor has a negative Gibbs free energy (ΔG) value of 237, 224, 38, and 34 kJ/mol H_2, respectively. Hence, it is not surprising that these reactions occur in this order, or hierarchy, in nature: it is normally observed that as long as oxygen is available, this is the preferred electron acceptor, and only when the concentration of oxygen becomes low is nitrate consumed. Both sulfate and HCO_3^- reduction (methanogenesis) normally only occur under anoxic conditions, when both oxygen and nitrate are depleted. As long as the SO_4^{2-} concentration is sufficiently high, sulfate reduction is preferred over methanogenesis from bicarbonate. It should be noted, however, that there are exceptions to the normal pattern. Furthermore, physical, (bio)chemical, and biological possibilities under any given set of conditions may cause deviations from the general pattern.

The terms "**oxic**" and "**anoxic**" literally mean with and without oxygen, respectively. Their equivalents "**aerobic**" and "**anaerobic**" literally mean life with and without air. In the microbiological literature, there is no rule as to how to use these terms. Sometimes the terms (an)oxic are used to indicate environments, localities, layers, conditions, etc., while the terms (an)aerobic are used to indicate the way of life of a microorganism. Thus, one could state that an anaerobic bacterium must grow under anoxic conditions. In the engineering literature, however, the term "anaerobic" is often used for conditions without oxygen, nitrate, or nitrite, while the term "anoxic" is used for conditions without oxygen. Hence, in the latter terminology, anaerobic and anoxic refer to the average redox potential of the environment.

Table 10.2 **Redox half reactions of electron-donating and electron-accepting couples of the (chemo)lithotrophs and of some relevant redox half reactions of coenzymes and redox carriers of the electron transport chain** at pH 7.0 and a temperature of 25°C. The standard Gibbs free energy (or $\Delta G_R^{o\prime}$ at pH 7) of a redox reaction is given by: $\Delta G_R^{o\prime} = -nF\ [E_0'$ (acceptor couple) $-E_0'$ (donor couple)], where n = number of electron transferred in the reaction and F = 96.5 kJ/V mol. It must be noted that both electron-donating and electron-accepting reactions are written as electron-donating reactions. Depending on the growth conditions, some reactions may act either as donating or as accepting couples. Other reactions, marked with an asterisk, act exclusively as electron-accepting reactions

Redox half reactions	E_0' (potential of electrons removed) (mV)
1. Hydrogen	
$H_2 \rightarrow 2\ H^+ + 2\ e^-$	−418 0 (pH = 0)
2. Reduced one-carbon compounds	
$CH_4 + 2\ H_2O \rightarrow CO_2 + 8\ H^+ + 8\ e^-$	−239
$CO + H_2O \rightarrow CO_2 + 2\ H^+ + 2\ e^-$	−492
3. Reduced sulfur compounds	
$HS^- + 4\ H_2O \rightarrow SO_4^{2-} + 9\ H^+ + 8\ e^-$	−218
$S^0 + 4\ H_2O \rightarrow SO_4^{2-} + 8\ H^+ + 6\ e^-$	−200
$HS^- \rightarrow S^0 + H^+ + 2\ e^-$	−278
$SO_3^{2-} + H_2O \rightarrow SO_4^{2-} + 2\ H^+ + 2\ e^-$	−526
$S_2O_3^{2-} + 8\ H^+ + 8\ e^- \rightarrow 2\ HS^- + 3\ H_2O$	−193
$S_2O_3^{2-} + 5\ H_2O \rightarrow 2\ SO_4^{2-} + 10\ H^+ + 8\ e^-$	−242
$SO_3^{2-} + 7\ H^+ + 6\ e^- \rightarrow HS^- + 3\ H_2O$	−116
4. Reduced nitrogen compounds	
$NH_4^+ + 2\ H_2O \rightarrow NO_2^- + 8\ H^+ + 6\ e^-$	+340
$NO_2^- + H_2O \rightarrow NO_3^- + 2\ H^+ + 2\ e^-$	+431
$2\ NO_3^- + 12\ H^+ + 8\ e^- \rightarrow N_2 + 6\ H_2O$	+824
$2\ NO_2^- + 8\ H^+ + 6\ e^- \rightarrow N_2 + 4\ H_2O$	+956
$N_2O + 2\ H^+ + 2\ e^- \rightarrow N_2 + H_2O$	+1354
$2\ NH_4^+ \rightarrow N_2 + 8\ H^+ + 6\ e^-$	−277
5. Reduced metals	
$Fe^{2+} \rightarrow Fe^{3+} + e^-$	+770
$Mn^{2+} + 2\ H_2O \rightarrow MnO_2 + 4\ H^+ + 2\ e^-$	+615
6. Pyridine nucleotides	
$NADH + H^+ \rightarrow NAD^+ + 2\ H^+ + 2\ e^-$	−320
$NADPH + H^+ \rightarrow NADP^+ + 2\ H^+ + 2\ e^-$	−324
7. Quinones	
$QH_2 \rightarrow$ coenzyme Q $+ 2\ H^+ + 2\ e^-$	+100
8. Cytochromes	
2 cytochrome $b_{k(red)} \rightarrow$ 2 cytochrome $b_{k(ox)} + 2\ e^-$	+30
2 cytochrome $c_{red} \rightarrow$ 2 cytochrome $c_{ox} + 2\ e^-$	+254
2 cytochrome $o_{red} \rightarrow$ 2 cytochrome $o_{ox} + 2\ e^-$	+270
2 cytochrome $a_{3(red)} \rightarrow$ 2 cytochrome $a_{3(ox)} + 2\ e^-$	+385
9. Oxygen	
$O_2 + 4\ H^+ + 4\ e^- \rightarrow H_2O$	+816 +1054 (pH = 3) +1233 (pH = 0)

10.1.2 The Gibbs Free Energy Changes of the Reactions Are Dependent on the Concentrations of Reactants and Products

As pointed out before, corrections for the actual concentrations of reactants and products must be made in order to obtain a reasonable estimate of the feasibility and magnitude of a reaction. For example, for the reaction

$$aA + bB \Rightarrow cC + dD \quad (10.1)$$

the correction becomes

$$\Delta G_R = \Delta G_R^\circ + RT \ln \frac{[C]^c[D]^d}{[A]^a[B]^b} \quad (10.2)$$

in which ΔG_R is the Gibbs free energy of the reaction, ΔG_R° is the standard Gibbs free energy of the reaction, R is the universal gas constant ($= 8.314\ J\ mol^{-1}\ K^{-1}$), and T is the temperature in K. All physical/chemical tables give ΔG_R° values that are defined at concentrations of 1 mol/l of all reactants and products, including $[H^+]$ (i.e., pH = 0). As this is not relevant for biological systems, biological tables often give the $\Delta G_R^{\circ\prime}$, which is corrected to pH 7 (i.e., $[H^+]$ is 10^{-7} according to Eqn. 10.2). Alternatively, the redox potential of the redox couple can be used to make an estimate of the feasibility and magnitude of a particular redox half-reaction:

$$E' = E_0' + \frac{RT}{nF} \ln \frac{C_{ox}}{C_{red}} \quad (10.3)$$

E' and E_0' are the midpoint potentials for the redox half-reactions at pH 7. F is the Faraday constant ($96.5\ kJ\ V^{-1}\ mol^{-1}$), and n is the number of electrons transferred.

When a donor couple and an acceptor couple combine, the following is obtained:

$$\Delta E_0{}' = E_0'(\text{acceptor}) - E_0'(\text{donor}) \quad (10.4)$$

Since $\Delta G_R^{\circ\prime} = -nF\Delta E_0'$, it is easy to convert the values for redox reactions into Gibbs free energy changes. Corrections for the true concentrations (i.e., lower than the standard of 1 mol/l) are particularly important when reactions yielding a small Gibbs free energy change are considered. Indeed, many oxidations of inorganic compounds have a relatively small Gibbs free energy change, and hence, low concentrations of either the substrate (the reactant) or its products may strongly influence the magnitude and the sign (+ or −) of the ΔG, as the following three examples show:

1. Substrate concentration much lower than standard. The oxidation of hydrogen with bicarbonate under standard conditions ($pH_2 = 1$, pH = 7.0)

   $$4\ H_2 + H^+ + HCO_3{}^- \longrightarrow CH_4 + 3\ H_2O \quad (10.5)$$

   has a $\Delta G_R^{\circ\prime} = -138.2$ kJ/reaction. However, when the partial pressure of the hydrogen (pH_2) reaches 10^{-6}, $\Delta G_R' = 3.8$ kJ/reaction, which is positive, implying that the reaction no longer proceeds in the original direction, but can run in reverse. This example is relevant because in natural environments the concentrations of hydrogen are usually extremely low.
2. Hydrogen ion concentration deviating from standard.

   $$4\ Fe^{++} + O_2 + 4\ H^+ \longrightarrow 4\ Fe^{+++} \downarrow + 2\ H_2O \quad (10.6)$$

   This reaction has a ΔG_R of only −17.7 kJ/reaction at pH 7.0, but at pH 3.0, ΔG_R becomes −91.0 kJ/reaction due to proton uptake by the half reaction of oxygen. This example relates to the existence of iron-oxidizing bacteria able to grow at neutral or extremely low pH values.
3. Product concentration much lower than standard because of a poorly soluble product. This is the reality of Reaction 10.6 because many ferric salts (e.g., the hydroxides and carbonates) are very insoluble (symbol ↓). Thus, the reaction is "pulled" to the right by the removal of the product. If the solubility product of $Fe(OH)_3$ is taken to be $4 \cdot 10^{-38}$, at pH 7, the iron concentration would be $4 \cdot 10^{-17}$ M. At that concentration, the $\Delta G_R'$ would become −373.88 kJ/reaction. Another example of a considerable change in the ΔG due to low end-product concentration occurs if the end product is metabolized by other organisms. Interspecies hydrogen transfer is a good example of this (see Chapter 12).

Even if a reaction is thermodynamically possible, this does not necessarily guarantee that it will yield biologically useful energy, even if an organism is able to carry it out. Not only must the organism possess the appropriate enzymes but it must also have a means of coupling the reaction to the conservation of the chemical energy. This takes the form of an electrochemical gradient (often the **p**roton **m**otive **f**orce or proton potential, Δp) or a high energy compound such as ATP. In the chemolithotrophs, this is primarily the Δp (see Chapter 4). It should be noted that the pumping of one proton from the inside to the outside of the membrane, at a Δp of −200 mV, would require 19.3 kJ per "mol" of H^+, indicating that the oxidation/reduction reaction must yield a minimum amount of energy (approximately 20 kJ) in order to be biologically useful.

All energy-generating reactions of the chemolithotrophs are oxidation-reduction reactions. Some of these reactions only involve small Gibbs free energy changes. In order to predict whether or not a reaction will run under realistic conditions, it is, therefore, important to correct the standard values of the $\Delta G_R^{0\prime}(=\Delta G_R^0$ at pH 7) for the actual concentrations of reactants and products.

10.2 There Are Numerous Metabolic Types Among the Chemolithotrophs

Winogradsky's concept of the "inorgoxidants" originally referred exclusively to the **obligate chemolithoautotrophs**, that is, organisms that are able to derive energy only from inorganic compounds and carbon only from carbon dioxide. However, in addition to the obligate chemolithoautotrophs, many genera of bacteria also include species of **facultative chemolithoautotrophs** and **chemolithoheterotrophs**. This spectrum can be found among the hydrogen bacteria, the nitrifying bacteria and the colorless sulfur bacteria. The latter are a good example, and their typical characteristics are therefore listed in Table 10.**3**.

The **obligate chemolithotrophs** are highly specialized organisms (i.e., "specialists"), whereas the **facultative chemolithotrophs** are the most versatile of the group as they can grow autotrophically, using a sulfur compound as energy source, and heterotrophically. The facultative chemolithoautotrophs are sometimes called "mixotrophs" since they are able to grow on mixtures of substrates under certain conditions. **Mixotrophic growth** involves the simultaneous use of separate metabolic pathways (e.g., gaining energy from sulfide and/or an organic compound; using carbon dioxide and/or an organic compound as a source of carbon). Mixotrophy usually occurs only at low concentrations of the mixed substrates. At high substrate concentrations, **diauxie** (i.e., sequential use of the two substrates) can be observed. The **chemolithoheterotrophs** can use sulfur compounds as an additional energy source, but they cannot fix carbon dioxide; thus, they require an organic carbon source under all conditions.

A last important group in this spectrum comprises chemoheterotrophic organisms that can oxidize sulfur compounds, but are unable to derive metabolically useful energy from the reaction. These heterotrophs use an organic compound as their normal carbon and energy source. The reasons why these organisms oxidize reduced sulfur compounds are not generally understood. Sulfide oxidation may be a detoxification reaction. Alternatively, the sulfide may be used to detoxify hydrogen peroxide generated during metabolism. The incidental oxidation of inorganic compounds is not restricted to sulfur oxidizers. Among the ammonium oxidizers, it is known as "heterotrophic nitrification" whereby ammonium (or other reduced nitrogen compounds) is converted to nitrite. A special case of

Table 10.**3** **Physiological types among the nonphototrophic sulfur-oxidizing bacteria.** Note that chemoorganoheterotrophs are able to oxidize sulfur compounds, but are unable to derive metabolically useful energy from the oxidation of sulfur compounds

	Energy source (electron donor)		Carbon source	
Physiological type	Inorganic sulfur compound	Organic compound	CO_2	Organic compound
Obligate chemolithoautotroph	+	–	+	–
Facultative chemolithoautotroph	+	+	+	+
Chemolithoheterotroph	+	+	–	+
Chemoorganoheterotroph	–	+	–	+

incidental oxidation is the aerobic oxidation of ammonium by methane-oxidizing bacteria. In these bacteria, the key enzyme for methane oxidation, a monooxygenase, has a low affinity for ammonia, which is an analogue of methane, and hence the enzyme oxidizes the ammonia "by accident". This type of fortuitous metabolism is known as **"co-metabolism"**.

10.3 Mechanisms of Energy Conservation in Chemolithotrophs Are Principally the Same as in Chemoheterotrophs

Table 10.**2** clearly shows that hydrogen and carbon monoxide are the compounds that have the lowest redox potential and consequently are the most energy-rich substrates for chemolithotrophic growth. Their low redox potential also implies that they can be oxidized with sulfate or bicarbonate as electron acceptors in contrast to, for example, ammonium or ferrous iron. Although the systematics of the table are evident, it should be realized that not all feasible half reactions are listed, and that unexpected reactions sometimes are thermodynamically possible. One such reaction is the disproportionation of thiosulfate to sulfide and sulfate:

$$S_2O_3{}^{2-} + H_2O \longrightarrow HS^- + SO_4{}^{2-} + H^+ \qquad (10.7)$$

$\Delta G_R^{o\prime} = 21.9$ kJ/reaction. This reaction has been proven to provide energy for chemolithotrophic growth of a sulfate-reducing species (Table 10.**1**). Another apparently unlikely, but feasible reaction is the oxidation of ammonium with nitrite or nitrate:

$$NH_4{}^+ + NO_2{}^- \longrightarrow N_2 + 2\,H_2O \qquad (10.8)$$

$\Delta G^{o\prime} = -360$ kJ/reaction. This reaction supports (autotrophic) growth of a community of organisms, the identity of which has not yet been established.

10.3.1 Negative $\Delta G^{o\prime}$ Values and Coupling Mechanisms Are Essential for Energy Conservation

As was pointed out in the beginning of this chapter, a negative free energy change is a boundary condition for a reaction, but it does not automatically follow that the reaction can run biochemically or biologically, yielding metabolically useful energy. It must generate an electrochemical gradient (Δp), or directly yield a high-energy compound such as ATP. It is important to recall the three mechanisms used to build up a proton motive force (Δp, see Chapters 3 and 4; or Figs. 4.**1**–4.**7**): 1) the operation of "conformational" proton pumps in the electron transport chain (for example cytochrome oxidase) leading to the transport of (usually) protons from the inside to the outside of the membrane, 2) the transfer of protons from the inside to the outside through the Q cycle, and 3) the formation of a Δp, without proton translocation by the operation of redox loops in which the active sites for H^+ consumption and H^+ production by the redox reactions (often termed scalar H^+ consumption/production) are on different sides of the cytoplasmic membrane. Two examples will demonstrate the third mechanism for the chemolithotrophs.

The bacterium *Thiobacillus ferrooxidans* lives at a permanently low pH because of the oxidation of sulfur compounds to sulfuric acid. As its cytoplasm has a (normal) near-neutral pH, a permanent proton gradient exists over its membrane. This can be used for ATP synthesis provided that the protons that enter the cytoplasm can be neutralized by a chemical reaction. This is accomplished by ferrous iron oxidation:

$$4\,Fe^{++} \longrightarrow 4\,Fe^{+++} + 4\,e^- \quad \text{(outside)} \qquad (10.9)$$

$$O_2 + 4\,H^+ + 4\,e^- \longrightarrow 2\,H_2O \quad \text{(inside)} \qquad (10.10)$$

The ferrous iron is oxidized to ferric iron, and the electrons removed are transported through the membrane (via electron carriers) to oxygen, whereby protons are consumed. Thus, by means of the operation of a redox loop, the scalar consumption of protons on the inside of the cytoplasmic membrane allows ATP production in a permanently acid environment. For further details, see Figure 10.**13**.

A second example is the oxidation of hydrogen by the sulfate reducers, as shown below. Hydrogen is oxidized to protons on the outside of the membrane. The liberated electrons are transported via the electron transport chain to the inside of the membrane, where they are consumed in the reaction whereby sulfate is reduced and protons are consumed. Even if proton pumping did not occur, this oxidation reduction reaction would result in the build-up of a proton motive force:

$$4\,H_2 \longrightarrow 8\,H^+ + 8\,e^- \quad \text{(outside)} \qquad (10.11)$$

$$SO_4{}^{2-} + 8\,H^+ + 8\,e^- \longrightarrow S^{2-} + 4\,H_2O \quad \text{(inside)} \qquad (10.12)$$

$$\text{net result:} \quad 4\,H_2 + SO_4{}^{2-} + 8\,H^+ \;\text{(inside)} \longrightarrow H_2S + 4\,H_2O + 8\,H^+ \;\text{(outside)} \qquad (10.13)$$

In many oxidation/reduction reactions, the proton pumping and the scalar consumption/production will occur simultaneously. When calculating the net effect of such a reaction, both mechanisms must therefore be taken into account.

10.3.2 Substrate Level Phosphorylation Is a Rare Energy-conserving Mechanism in Chemolithotrophs

Another mechanism that may rarely be responsible for the production of biochemically/biologically useful energy in the chemolithotrophs is **substrate phosphorylation**, as discussed in Chapter 3. There is only one known example of this mechanism occurring in chemolithotrophic energy generation, namely in the oxidation of sulfite by a few sulfur-oxidizing bacteria (e.g., *Thiobacillus thioparus* and *T. denitrificans*), as shown in Eqn. 10.14–10.16. Adenosine phosphosulfate (APS) is the energy-rich intermediate formed from AMP during the strongly exergonic oxidation of sulfite. APS is then converted to ADP. Two molecules of ADP can be used to produce 1 ATP.

$$SO_3{}^{2-} + AMP \xrightarrow{\text{APS reductase}} APS + 2\,e^- \quad (10.14)$$

$$APS + P_i \xrightarrow{\text{ADP sulfurylase}} ADP + SO_4{}^{2-} \quad (10.15)$$

$$2\,ADP \xrightarrow{\text{Adenylate kinase}} ATP + AMP \quad (10.16)$$

10.3.3 The Chemolithotrophs Operate Electron Transport Chains Very Similar to Those of the Chemoorganoheterotrophs

Few cytochrome chains of chemolithotrophs have been investigated in depth. Although the details may differ, the principles underlying these systems can be demonstrated by considering the respiratory cytochrome chain (Fig. 10.**2**) of a well-investigated facultative chemolithoautotroph, *Paracoccus denitrificans*, which is able to grow chemolithoautotrophically on both hydrogen and reduced inorganic sulfur compounds. Similar cytochrome chains have been found in other bacteria, including the closely related *Thiosphaera pantotropha*, as shown in Figure 10.**2**.

When organic compounds serve as the energy source, NADH is generated during their oxidation. NADH delivers its electrons via the relatively low redox potential flavoproteins to quinones (Table 10.**2**). Subsequently, the electrons flow down the chain via a number of proton-translocating loops (indicated by the vertical arrows in Fig. 10.**2**) to the terminal electron acceptor, in this case nitrate, nitrite, or oxygen. If hydrogen is the electron donor, however, the organism does not oxidize it with an NAD-dependent oxidoreductase, but with a hydrogenase that delivers its electrons directly at the level of the quinone, or cytochrome *b*. This demonstrates that, although in theory the oxidation of H_2 yields more energy than that of NADH (see Table 10.**2**), hydrogen actually yields less biologically useful energy, in terms of a build-up of a Δp. There are two reasons for this. One is the kinetics of the reaction: in order to produce energy at a sufficient rate, the organism sacrifices one proton-translocating loop. Secondly, if one takes into account the natural partial pressure of hydrogen gas (10^{-4}–10^{-5} atm or 1–10 Pa), the free energy change ($pH_2 = 10^{-4}$) becomes considerably lower, as shown in Reactions 10.17–10.20. The result is that less energy is conserved, but at the same time, the free energy change of the total reaction is more negative so that Reaction 10.19 (and even Reaction 10.20) can run faster than Reaction 10.18.

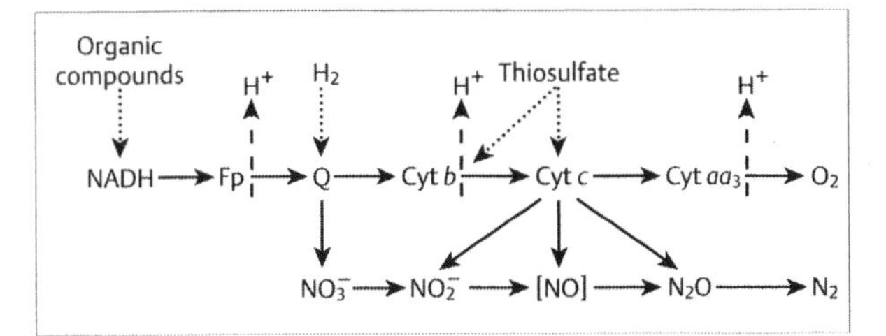

Fig. 10.2 **Schematic representation of electron flow to oxygen and the denitrification pathway in *Paracoccus denitrificans*** and related bacteria. Solid arrows show electron flow through the cytochrome chain; broken arrows show electrons entering the cytochrome chain and proton translocation. Fp, flavoprotein; Cyt, cytochrome; Q, quinone

$$NADH + H^+ + \tfrac{1}{2}\,O_2 + 3\,ADP + 3\,P_i \longrightarrow NAD^+ + H_2O + 3\,ATP \quad (10.17)$$
$$\Delta G^{\circ\prime} = -131.2\ \text{kJ/reaction}$$

$$H_2 + \tfrac{1}{2}\,O_2 + 3\,ADP + 3\,P_i \longrightarrow H_2O + 3\,ATP \quad (10.18)$$
$$\Delta G^{\circ\prime} = -149.4\ \text{kJ/reaction}$$

$$H_2 + \tfrac{1}{2}\,O_2 + 2\,ADP + 2\,P_i \longrightarrow H_2O + 2\,ATP \quad (10.19)$$
$$\Delta G^{\circ\prime} = -178.8\ \text{kJ/reaction}$$

$$H_2\,(10^{-4}\ \text{atm; 10 Pa}) + \tfrac{1}{2}\,O_2 + 2\,ADP + P_i \longrightarrow H_2 + 2\,ATP \quad (10.20)$$
$$\Delta G' = -156.2\ \text{kJ/reaction.}$$

Once the electrons from hydrogen oxidation are delivered to the level of the quinone or cytochrome *b*, as shown in Figure 10.**2**, their redox potential is too high to reduce NAD^+. In some hydrogen bacteria (e.g., *Alcaligenes eutrophus*, see Chapter 10.8), a second hydrogenase, a soluble NAD^+-dependent oxidoreductase is present. However, in most of the known hydrogen bacteria and in virtually all other chemolithotrophs, NAD^+ must be reduced via an energy-requiring pathway because the inorganic energy source/electron donor has a redox potential too high to deliver electrons at the level of NAD^+. In many cases, the redox couple of the substrate (e.g., thiosulfate) is such that the electrons from its oxidation can only be delivered at the level of cytochrome *c*. In a few cases, delivery to the cytochrome *b* level might be possible. This means that the electrons will pass only one or, sometimes, two proton-translocating loops on their way to oxygen. In many sulfur-oxidizing bacteria, the electrons are delivered at the cytochrome *c* level, probably again for kinetic reasons. Clearly in the denitrifying bacteria, some electrons must be delivered at the cytochrome *b* level in order to reduce nitrate to nitrite. Nitrite, in turn, can be reduced by cytochrome *c*, as shown in Figure 10.**2**. (Note that most strains of *Paracoccus denitrificans* and *Thiosphaera pantotropha*, while denitrifying, can grow on hydrogen, but only very slowly or not at all on thiosulfate.)

"**Reversed electron transport**" is used to reduce NAD^+ to NADH when the substrate (e.g., thiosulfate) is a compound with a midpoint potential far more positive than that of the NAD^+/NADH couple. Electrons derived from the oxidation of thiosulfate can go in two directions: either down the electron transport chain to oxygen, generating the proton motive force, or up the electron transport chain, against the thermodynamic gradient, at the expense of the proton motive force. In other words, the reactions in the electron transport chain can be considered as at least partially reversible reactions that can carry electrons from a moderately negative or even positive redox potential to the highly negative potential of NADH. It is estimated that the reversed transport of two electrons from the cytochrome *c* level to NADH requires the equivalent of at least 2 ATP. Measurements of growth (biomass) yields of sulfur oxidizers, grown on thiosulfate as energy/electron donor and carbon dioxide as carbon source show that between 10–20% of the electrons from thiosulfate serve as reducing power, while the rest is used for energy generation. Obviously, this percentage may vary with the energy source/electron donor and the metabolic details of the organism studied.

In autotrophic, CO_2-fixing chemolithotrophs, the reduced inorganic electron donor serves two purposes: It donates electrons to the electron transport chain for (1) energy conversion (Δp build-up) and (2) production of reducing equivalents (NADH) at the expense of the Δp. The NADH is subsequently used for CO_2 reduction.

10.4 The Carbon Metabolism of Chemolithotrophs Is not Different From That of Heterotrophs and Phototrophs

There are many autotrophic chemolithotrophs among the eubacteria and the archaebacteria. In most respects, their general carbon metabolism does not appear to be different from that of common heterotrophs. However, the autotrophs must obviously possess a pathway for CO_2 fixation. In most eubacteria, this pathway is the Calvin-Benson-Bassham cycle. (Calvin cycle); this pathway is discussed in detail in Chapter 8.1.1, together with the three other autotrophic CO_2-fixation pathways. In the methanogenic, acetogenic, and some sulfate-reducing bacteria, the acetyl-CoA pathway operates. In a third group of organisms, CO_2 fixation proceeds via a reversed tricarboxylic acid (TCA) cycle. More recently another pathway, the 3-hydroxypropionate cycle, has been described.

The majority of the chemolithoautotrophs employ the Calvin cycle. It is therefore important to discuss some aspects of carbon and energy metabolism related to the operation of this cycle in chemolithotrophic bacteria. The net costs of the Calvin cycle are high. It appears that for each CO_2 fixed in the Calvin cycle, 3 ATP and 2 NADH are required to produce organic compounds at the level of hexoses. When the cost of NADH production is calculated in terms of ATP, it is obvious that the expense of CO_2 fixation will depend on the lithotrophic energy source: for some hydrogen oxidizers, NADH can be produced directly from H_2 and does not then require additional energy. In this case, the net cost is 3 ATP/1 CO_2 fixed. However, if it is assumed that reversed electron transport will cost a minimum of 1 ATP per NADH formed if electrons are delivered at the level of cytochrome *b*, and 2 ATP if delivered at the level of cytochrome *c*, it is clear that the fixation of 1 CO_2 would require $(3+2)$ or $(3+4)$ ATP. Per hexose formed, this would be 30 or even 42 ATP. Biomass yields of the chemolithotrophs are roughly in line with the redox level of the energy source/electron donor for autotrophic growth, as shown in Table 10.**2**, but may vary considerably with the specific organism involved. For iron oxidation at low pH, the biomass yield ranges from 0.4– 0.7 g/2 e^-, when the yield is expressed per 2 (mol) electrons transferred to oxygen. For thiosulfate oxidation, the range is 0.7– 2.0 g/2 e^-, and for ammonium and hydrogen, 0.6 g/2e^- and 2.0 g/2e^-, respectively.

"Anaplerotic carboxylation reactions". As in common heterotrophs, other essential carboxylation reactions, known as "anaplerotic carboxylation reactions", must occur in autotrophs. One such reaction is the carboxylation of phospho*enol*pyruvate to oxaloacetate, which is essential for the operation of the enzymes of the tricarboxylic acid cycle (TCA). However, during chemolithoautotrophic growth, the enzymes of the TCA cycle are only used for biosynthetic purposes and not for energy generation. The pathway operates not as a cycle, but rather as a "horseshoe" (Fig. 10.**3**), due either to the absence of α-ketoglutarate dehydrogenase in the obligate chemolithoautotrophs or to the complete repression of this enzyme during autotrophic growth of the facultative autotrophs. Thus, oxaloacetate is not only used for the production of α-ketoglutarate via the normal route, but also, in the reverse direction, for the synthesis of the malate and succinate required. A compound such as acetate can only enter the right hand part of the horseshoe. Since the complete glyoxylate cycle is absent in these organisms, acetate can only serve as a carbon source for the synthesis of the glutamate family of amino acids and for providing carbon to other biosynthetic reactions involving acetyl-CoA.

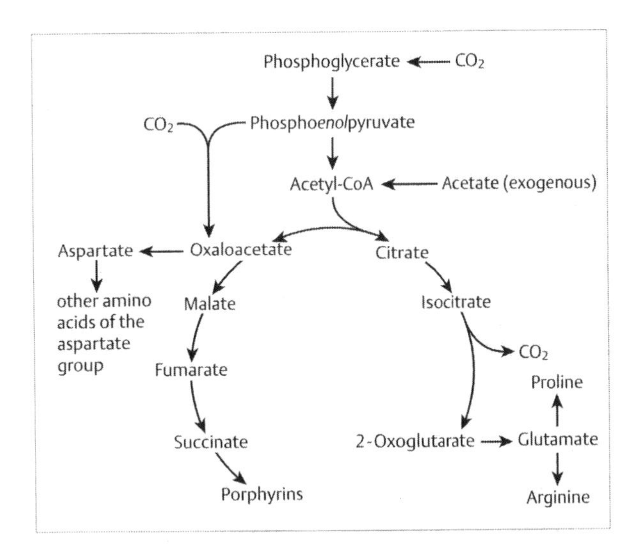

Fig. 10.3 **The "horseshoe", an incomplete tricarboxylic acid cycle operating in obligate chemolithoautotrophs.** The biosynthetic roles of reactions normally associated with the operation of the tricarboxylic acid cycle in organisms that cannot convert α-ketoglutarate to succinate. Note that carbon from exogenous acetate can enter the amino acids of the glutamate family via citrate and α-ketoglutarate, but cannot enter those of the aspartate family via succinate and oxaloacetate, as it does in organisms with a functional tricarboxylic acid cycle

The biochemical basis of obligate chemolithoautotrophy is the transport problem. As the obligate autotrophs do not appear to be able to grow on organic compounds, extensive studies have been made to discover the reasons for this inability. It is particularly puzzling because these organisms can oxidize internal reserves of polyglucose by means of the pentose phosphate cycle, and thus generate energy for maintenance purposes. It is now clear that the obligate chemolithoautotrophs lack the carrier proteins for the transport of sugars into the cells. Even if transport carriers are available to the cells, as is the case for some amino acids, their maximum capacity appears to be too low to support heterotrophic growth. Indeed, when ^{14}C-labeled acetate and amino acids are supplied to these organisms, they contribute to not more than 20–30% of the total biomass carbon, implying that CO_2 remains the main carbon source under all growth conditions. Finally, it has been shown that the NADH-oxidizing capacity of the obligate chemolithotrophs remains very low, even in the presence of exogenous organic compounds. This makes heterotrophic growth very unlikely. In summary, it appears that the obligate nature of autotrophic metabolism is an extreme specialization in which all necessary metabolic pathways have become constitutive and very few inducible enzymes exist. A

The obligate chemolithoautotrophs pay a very high price for their specialization because of the energetic constraints and the difficulty of providing sufficient reducing equivalents for CO_2 reduction. To compete, they all have very high oxidation capacities for their special substrates, usually 2–3 times more than those of facultatively chemolithoautotrophic organisms. It may well be that the presence of these high concentrations of membrane-bound respiratory enzyme/cytochrome systems simply excludes the "economic" maintenance of carrier proteins for organic compounds. At the same time, their high rate of substrate oxidation allows these obligate chemolithoautotrophs to compete successfully for inorganic compounds with the facultative autotrophs when only low concentrations of organic compounds are available. Hence, "obligate" autotrophy may also be considered as an adaptation to growth in environments poor in organic nutrients.

notable exception is the GS-GOGAT pathway used for nitrogen assimilation under nitrogen limitation (see Chapter 7).

The inability of obligate chemolithoautotrophs to grow on organic media renders these organisms much less accessible for modern molecular genetic recombinant DNA techniques because metabolic lesions caused by mutations in the assimilation and dissimilation pathways tend to be lethal. Therefore, the facultative chemolithoautotrophs are the preferred choice for most investigations of the chemolithoautotrophic pathways and their genetic regulation. However, there are ways around this problem, for example, by cloning genes from obligate chemolithotrophs into suitable (facultatively chemolithotrophic) host organisms. A practical point relevant to the problems encountered in using cloning techniques is that many obligate chemolithoautotrophs grow very poorly, or not at all, as colonies on plates, hindering the screening of colonies for desirable properties.

10.5 Chemolithotrophs Are Adapted to Specific and Very Often Extreme Environments Deficient in Organic Matter

The chemolithotrophic bacteria, like any other form of life, are dependent on suitable physicochemical conditions for growth and survival, for example, with respect to pH, temperature, redox potential, and salt concentration. Chemolithotrophs are known to be able to live even at the extremes of these limits. Some extremely thermophilic hydrogen utilizers are able to grow at 110 °C, some acidophilic sulfur oxidizers can still grow at pH values below 2, and other bacteria are able to tolerate high concentrations of heavy metal ions. The first obvious need for a chemolithotroph is the availability of a suitable energy source/electron donor and an electron acceptor.

The habitats of chemolithotrophs are rich in sources of reduced inorganic compounds. Figure 10.**4** gives an overview of biological, geological, and anthropogenic inorganic energy sources for chemolithotrophs. Estimates of the quantitative contribution of the various global sources and sinks are given in Table 10.**4**.

The chemolithotrophic bacteria play an essential role in the maintenance of the element cycles. These cycles will be discussed in detail in Chapter 32. Figure 10.**1** showed how photosynthesis (primary production) provides the organic compounds that serve as the source of energy for all other forms of life. Many of these compounds will be reoxidized under oxic conditions, whereby the oxygen produced during photosynthesis is used and recycled to water (not shown in Fig. 10.**1**). Once oxygen is depleted, fermentation processes can occur, or nitrate, ferric ion, sulfate, or bicarbonate (among others) can be used as terminal electron acceptors for anaerobic respiration with hydrogen, dinitrogen, ammonia, ferrous ion, hydrogen sulfide, and methane as examples of end products. With the exception of dinitrogen, these reduced inorganic end products can subsequently be reoxidized (recycled) by the chemolithotrophs using oxygen or nitrate as the electron acceptor. A typical habitat for most physiological types of chemolithotrophs is near the interface between oxic/anoxic conditions, for example near the surface of sediments or in stratified water bodies, where oxidized and reduced inorganic compounds meet.

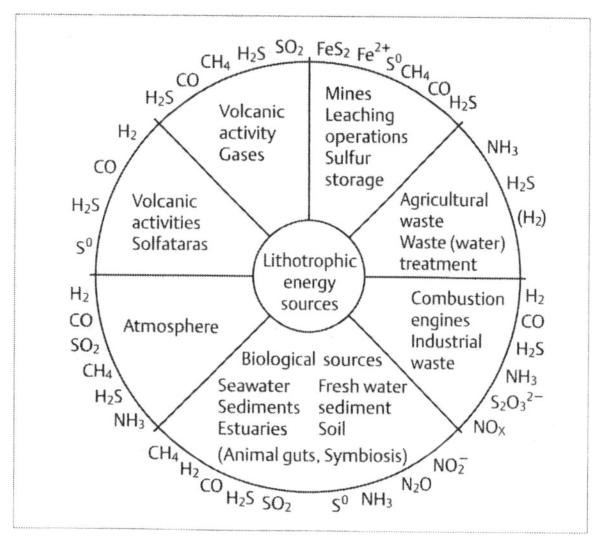

Fig. 10.4 **Geological, biological, and anthropogenic sources of reduced inorganic compounds supporting chemolithotrophic life.** Biological sources represent the vast majority of the total

Table 10.4 **Estimates (or ranges) of the rates of production/emission (= sources) and consumption/disappearance (= sinks) of atmospheric gases on earth.** The ranges given are compilations from various authors, to indicate the order of magnitude of the various global fluxes. Tg = 10^{12} g

Compound	Tg/year[1]	
	Production	Consumption
CO	3,315 ± 1,700	2,500 ± 770
CH_4	170–854	?
H_2	89	89
H_2S	(43)	?[3]
SO_2	(86)	?[3]
NO_x	39–48.4	24–43
N_2O	15.3 ± 6.7[2]	14 ± 3.5[2]
NH_3	54	49

[1] After Warneck, P. (1988) Chemistry of the natural atmosphere. San Diego: Academic Press.
[2] After Davidson, E. A. (1991) Fluxes of nitrous oxide and nitric oxide from terrestrial ecosystems. In: Rogers, J. E., and Whitman, W. B. (eds.) Microbial production of greenhouse gases: methane, nitrogen oxides and halomethanes. Washington DC: ASM Press; 219–236
[3] Budgets for individual S compounds are difficult on a global scale because of multiple interconversions in the atmosphere

In summary, a large variety of reduced inorganic compounds are available as electron donors under conditions suitable for the growth and survival of chemolithotrophs. Evolution has created an impressive diversity of chemolithotrophs with an equally impressive physiological and metabolic diversity. Many of the metabolic types that have evolved may have very similar physiological properties, but as judged from their 16S rRNA sequences, they have only a remote taxonomic (phylogenetic) relationship. This implies that chemolithotrophic capabilities may have evolved independently among different evolutionary lines, and hence are the result of evolutionary convergence. That so many different bacteria with similar properties can

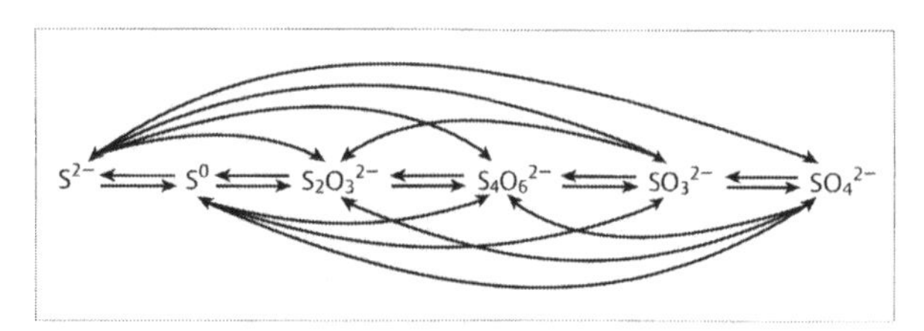

Fig. 10.5 **"Mini" cycles in the sulfur cycle**

Box 10.2 The **geochemical cycles**: 1) **Molecular hydrogen** is produced as an end product of many fermentative pathways and is recovered by oxidation to its most oxidized state, H_2O. Obviously, most of the hydrogen will be produced under anoxic conditions, and hence much of it will be oxidized by sulfate-reducing bacteria, methanogens, or acetogens. If hydrogen reaches a zone containing nitrate or oxygen, it can be oxidized with these electron acceptors. 2) **Sulfur compounds.** Hydrogen sulfide is produced during the (an)aerobic hydrolytic breakdown of organic sulfur compounds. Major amounts of hydrogen sulfide originate from sulfate reduction. Sulfide can be oxidized by chemolithotrophs (and also by phototrophs, see Chapter 13) via various intermediates to sulfate, as shown in Figure 10.**5**. Since all these intermediates can also be used as electron acceptors by the sulfate-reducing bacteria, "mini" cycles may be found, depending on the availability of suitable electron donors for the sulfate reducers or suitable electron acceptors for the sulfur-compound-oxidizing bacteria. 3) **Metals.** Ferrous-iron- or manganese (Mn^{++})-oxidizing bacteria may also be found near the interface, as these two reduced ions are potential end products of the anaerobic respiration of (in)organic compounds with ferric iron or MnO_2 as terminal electron acceptors. Indeed, the iron cycle (Fe^{++}/Fe^{+++}) with intermediate formation of the insoluble ferrous sulfide, ferric phosphate, and ferric hydroxide, and the manganese cycle with the highly insoluble MnO_2, are thought to be important electron-carrying cycles at relatively high redox potentials in anoxic sediments. 4) **Nitrogen compounds.** Ammonium is produced from the anaerobic or aerobic hydrolysis of organic nitrogen compounds. This is referred to as "ammonification". The ammonia can be oxidized under oxic conditions by the nitrifying bacteria to nitrite and, subsequently, to nitrate in the process known as "nitrification". As sediments are a rich source of ammonium, one of their important habitats is the interface. Ammonium is also produced by dissimilatory nitrate reduction by some heterotrophic anaerobes during the anaerobic oxidation of organic compounds. Thus, such organisms can create a "mini" cycle at the oxic/anoxic interface, whereby an aerobic nitrifier would produce nitrite, which could immediately be reduced to ammonium, and then reoxidized to nitrite again.

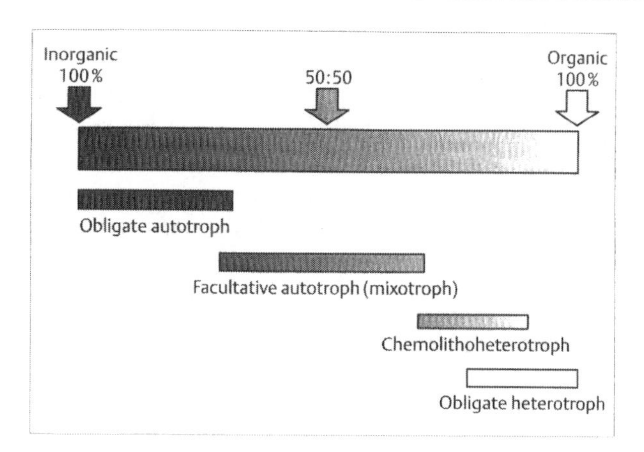

Fig. 10.6 **Occurrence of different metabolic types of sulfur-oxidizing bacteria in fresh water during energy-limiting growth conditions.** The distribution is a function of the relative turnover rate of reduced inorganic sulfur and organic substrates available

live, and even coexist, in one and the same environment is because of the large number of ecological niches available, as determined not only by the abiotic conditions, but certainly also by biotic factors. Hence, when the spectrum of chemolithotrophs is regarded, their habitats and also the many possible microbial interactions, such as mutualism, predation, and competition for limiting resources, all of which play a crucial role in the growth and survival of these organisms, must be considered. In the following sections, the various groups of chemolithotrophs will be discussed, according to their respective substrates, together with some important applications of these organisms in wastewater treatment and in "microbial leaching" (see also Chapter 32).

Box 10.3 The physiological spectrum of metabolic types among the chemolithotrophs is related to their ecological niches. It is often asked how the spectrum of metabolic types shown in Table 10.**3** can survive, and even coexist, in one and the same habitat. The current most likely explanation is that the "niche" of each metabolic type is linked to the relative turnover rates of inorganic versus organic compounds in a particular environment, as shown in Figure 10.**6**. This concept starts with the reasonable assumption that in most environments the carbon and energy sources limit the rate of growth of the population. Under these conditions, mixotrophy can, in principle, take place. When the supply (influx) of inorganic compounds is high relative to that of organic compounds, the specialists have an advantage, while the mixotrophic (facultative) organisms are favored when the influx (and hence the turnover) of inorganic and organic compounds is of the same order of magnitude. Chemolithoheterotrophs are favored when the ratio of organic to inorganic compounds available to the organisms is high. Although this mechanism of competition has been proven for the sulfur-oxidizing bacteria in model systems in the laboratory (energy-limited continuous cultures), it must be realized that in nature many other factors and mechanisms determine the ecological niche of an organism. For example, the diurnal cycle may cause strong fluctuations in the supply of the mixtures of nutrients, while at the same time the pH and oxygen concentration may change and obviously, other nutrients (e.g., a trace element) may be limiting. Clearly such variables may change the outcome of competition for substrates. The success of an organism therefore depends on all relevant parameters and not just on one or two.

10.6 The Sulfur-oxidizing Bacteria Are a Heterogeneous Group

Hydrogen sulfide, metal sulfides, polysulfides, elemental sulfur, thiosulfate, polythionates, and sulfite can all be oxidized by an extremely heterogeneous group of bacteria, commonly known as the **colorless sulfur bacteria**. The term "colorless" discriminates these organisms from the pigmented phototrophic sulfur-oxidizing bacteria (Chapter 13). Important genera of the colorless sulfur bacteria are listed in Table 10.**5**. The table shows a very diverse group of organisms with, in many cases, no taxonomic relationships, emphasizing the point made in the previous paragraph. Even within one genus, such as *Thiobacillus*, organisms are found that have very little taxonomic relationship despite very similar physiological properties. This is demonstrated by the phylogenetic tree shown in Figure 10.**7**.

Most of the known sulfur-oxidizing bacteria have been assigned to the subdivision of the Proteobacteria, which belongs to the kingdom of the Bacteria. This is

Table 10.5 **Genera of the colorless bacteria traditionally recognized as being capable of growth on reduced sulfur compounds and their environmental parameters.** The letters (given in brackets behind the genus name): A, Archaea; B, Bacteria; P. class (division) of the Proteobacteria within B; U, unknown give the taxonomic position based on 16S RNA analysis; S, symbionts known in genus; N, neutrophilic; Ac, acidophilic; M, mesophilic; T, thermophilic; D, denitrifier; EA, electron acceptor S^0 or Fe^{3+}. After Robertson, L. A. and Kuenen J. G. (1992) The colorless sulfur bacteria. In: Balows, A., Trüper, H. G., Dworkin, M., Harder, and Schleifer, K. H. (eds.) The Prokaryotes. Berlin, Heidelberg, New York. Springer; 385–413

Genus	pH requirement	Thermal requirement	Anaerobic growth
Traditional colorless sulfur bacteria			
Thiobacillus (P)	N,Ac	M,T	D
Thiomicrospira (P) (S)[1]	N	M	D
Thiosphaera (P)	N	M	D
Sulfolobus (A)	Ac	T	
Acidianus (A)	Ac	T	EA
Thermothrix (U)	N	T	D
Thiovulum (P)	N	M	
Beggiatoa (P)	N	M	EA
Thiothrix (P)	N	M	
Thioploca[2] (P)	N	M	(D)[3]
Thiobacterium (U)	N	M	EA
Macromonas (U)			
Achromatium[2] (P)	N	M	EA
Other bacteria capable of growth on reduced sulfur compounds			
Paracoccus (P)	N	M	D
Hyphomicrobium (B)	N	M	
Alcaligenes (P)	N	M	D
Pseudomonas (P)	N	M	D
Hydrogenobacter (B)	N	T	

[1] 16S rRNA analysis indicates a possible relationship
[2] Axenic cultures are not available
[3] End product of nitrate reduction not yet known

presently the only recognized kingdom in the "domain" of the Bacteria. Both *Sulfolobus* and *Acidianus* belong to the domain of the Archaea. Many of the colorless sulfur bacteria are very large (up to 200 μm in diameter), and may contain visible sulfur globules, making them easily recognized under the microscope. Many of these conspicuous bacteria were therefore described and named even before pure cultures were obtained.

Most of the sulfur bacteria can oxidize the reduced inorganic sulfur compounds to sulfate; a few can only carry out partial oxidations. Elemental sulfur (S°), thiosulfate ($S_2O_3{}^{2-}$) and tetrathionate ($S_4O_6{}^{2-}$) are common intermediates (see Table 10.2). The role of sulfur compound oxidation in most of the colorless sulfur bacteria appears to be the generation of energy.

A list of some of the well-studied genera and species is given in Table 10.**6**, showing aerobic and denitrifying species, obligate and facultative autotrophs, chemolithoheterotrophs, neutrophiles, acidophiles, and thermophilic organisms. Much of our present knowledge on the sulfur and carbon metabolism of the sulfur-oxidizing bacteria is from research on these organisms.

All colorless sulfur bacteria are able to oxidize reduced inorganic sulfur compounds in the dark, but not all organisms capable of carrying out this oxidation are categorized under this name. For example, a number of the phototrophic sulfur bacteria, such as *Chromatium* species, can grow chemolithoautotrophically in the dark under oxic conditions. Furthermore, it must be emphasized that some organisms, for example *Paracoccus denitrificans*, were classified before it was discovered that they could grow chemolithoautotrophically on thiosulfate. Had *P. denitrificans* been originally enriched on thiosulfate, it would undoubtedly have been classified as a "*Thiobacillus*" species.

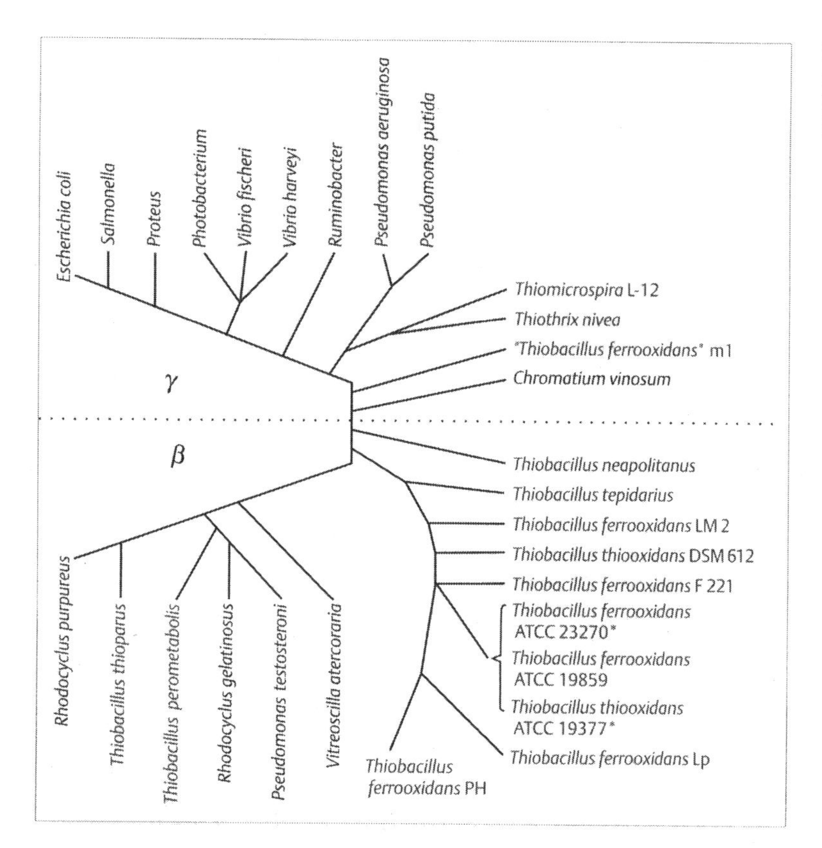

Fig. 10.7 **The 16S rRNA relationships of a number of *Thiobacillus* isolates within the β and γ subdivisions of the Proteobacteria.** * indicates a type strain [after 1]

Colorless sulfur bacteria are non-phototrophic bacteria capable of deriving metabolically useful energy from the oxidation of inorganic sulfur compounds. Virtually all known colorless sulfur bacteria can use oxygen as their terminal electron acceptor. Some can denitrify, using nitrate, nitrite, or nitrous oxide as the electron acceptor. A few of these bacteria can use ferric iron as an alternative electron acceptor. Colorless sulfur bacteria can be found in many environments with extreme physicochemical conditions: there are acidophiles (pH 1–5), alkaliphiles (pH 8–11), thermophiles (up to 95 °C), and halophiles (NaCl concentrations up to 150 g/l).

10.6.1 The Biochemical Pathway for Sulfur Compound Oxidation Has not yet Been Completely Resolved

This is due in part to the reactivity of sulfur compounds, which leads to spontaneous oxidation products, and to the lack of sufficiently sensitive analytical techniques for the detection and determination of these compounds. In addition, it appears to be very difficult to obtain cell-free extracts that have retained the full capacity to oxidize the diverse sulfur compounds in vitro. Although some biochemical "unity" may be found, different pathways apparently exist.

The energy source of the colorless sulfur bacteria, inorganic sulfur, occurs in many forms and oxidation states [$H_2S(-2)$, $S_8(0)$, $SO_3^{2-}(+4)$, and $SO_4^{2-}(+6)$] and in many more complex forms, such as $H{-}S_4{-}H$ polysulfanes, $R{-}S_n{-}H$ mono-organylsulfanes, $R{-}S_n{-}R$ biorganylsulfanes, $^-O_3S{-}S_n{-}SO_3^-$ polythionates, and $^-S{-}S_n{-}S^-$ polysulfides. Well-known examples of (poly)thionates are thiosulfate, $^-O_3S{-}SO_3^-$, and tetrathionate, $^-O_3S{-}S{-}S{-}SO_3^-$. Water-soluble sulfides, such as Na_2S, dissolve in and react with water, forming a protonated sulfide, i.e., HS^- and H_2S. The sulfide ion, S^{2-}, only exists at very high pH values (pH>14) because of the extremely alkaline pKa value of HS^- (14.9). Solutions containing HS^- dissolve elemental sulfur as a polysulfide with the number of S atoms varying from 2 to 7. Elemental sulfur has a low, but discrete solubility in water (5 μg/l water).

The acidophilic and neutrophilic sulfur-oxidizing bacteria are metabolically diverse. Schemes illustrating this diversity are shown in Figure 10.**8**. Obvious differences in these pathways concern the fate of the polythionates (tetrathionate, trithionate) and thiosul-

Table 10.6 **Some well-studied unicellular, non-filamentous species among the sulfur-oxidizing bacteria** with their physiological type. All can use oxygen as electron acceptor

	pH for growth	Inorganic electron donor
Thiobacillus thiooxidans (o)	2–5	S^{2-}, S, $S_2O_3{}^{2-}$
Thiobacillus ferrooxidans (o)	2–6	S^{2-}, S, $S_2O_3{}^{2-}$, $S_4O_6{}^{2-}$ Fe^{2+}, FeS_2, sulfidic minerals
Thiobacillus acidophilus (f)	3–6	S^{2-}, $S_2O_3{}^{2-}$, S^0, $S_3O_6{}^{2-}$, $S_4O_6{}^{2-}$
Thiobacillus thioparus (o)	5–8	S^{2-}, S^0, $S_2O_3{}^{2-}$, SCN^-, $S_4O_6{}^{2-}$ CS_2, $(CH_3)_2S$
Thiobacillus neapolitanus (o)	5–7	S^{2-}, S^0, $S_2O_3{}^{2-}$, $S_4O_6{}^{2-}$
Thiobacillus tepidarius (o)	6–8[2]	S^{2-}, S^0, $S_2O_3{}^{2-}$, $S_4O_6{}^{2-}$
Thiobacillus denitrificans (o)	6–8	S^{2-}, S^0, $S_2O_3{}^{2-}$, $S_4O_6{}^{2-}$
Thiobacillus intermedius (f)	4–7	S^{2-}, S^0, $S_2O_3{}^{2-}$, $S_4O_6{}^{2-}$
Thiobacillus novellus (f)	6–8.5	$S_2O_3{}^{2-}$, $S_4O_6{}^{2-}$
Thiobacillus versutus (f)	6–8.5	S^{2-}, $S_2O_3{}^{2-}$
Thiomicrospira pelophila (o)	6–8.5	S^{2-}, S^0, $S_2O_3{}^{2-}$, $S_4O_6{}^{2-}$
Thiosphaera pantotropha (f)	6–8.5	S^{2-}, S^0, $S_2O_3{}^{2-}$
Sulfolobus acidocaldarius (f)	2–3[3]	S^0, (Fe^{2+}, sulfidic minerals)

[1] o, obligate chemolithotroph; f, facultative chemolithoautotroph (can grow on various organic compounds, depending on species)
[2] Optimum growth temperature 43–44 °C
[3] Optimum growth temperature 66–74 °C

fate. In at least some of the acidophilic organisms (*Thiobacillus ferrooxidans, Thiobacillus acidophilus*), the metabolism occurs via tetrathionate as an intermediate. However, in some of the neutrophilic organisms the pathway may proceed via thiosulfate and its subsequent hydrolysis to sulfur (in the zero valence) and sulfite. Whether elemental sulfur is a true intermediate or occurs bound to an organic compound (R) or as a polysulfide of the form $R-(S)_nS^-$ remains a matter of speculation. In some thiosulfate-oxidizing bacteria (for example *Thiobacillus versutus*), a thiosulfate-oxidizing multi-enzyme complex appears to operate. This complex contains at least two enzymes and cytochromes and converts thiosulfate directly to sulfate without detectable intermediates. The electrons liberated in these reactions by oxidoreductase-types of enzymes ("reductases") would be fed into the electron transport chain. This is thought to occur at the cytochrome *c* level in most organisms, but in the denitrifying bacteria, such as *Thiobacillus denitrificans* (*T. denitrificans*) and *Thiomicrospira denitrificans*, the electrons may feed into cytochrome *b* (see Fig. 10.**2**). This would explain their ability to reduce nitrate while using reduced sulfur compounds and their relatively high yields during aerobic growth on sulfur compounds.

Elemental sulfur, which occurs primarily in the form of S_8, is one of the naturally occurring substrates of the sulfur bacteria. Although it is not entirely insoluble, it is possible that elemental sulfur is also metabolized through the initial formation of soluble or cell-bound (organic) polysulfide. It is a matter of debate whether zero-state sulfur (S^0) is oxidized by an oxidoreductase, which would give off its electrons to the electron transport chain, or by an oxygenase-type of enzyme (a mono- or di-oxygenase), which has been detected in several sulfur-oxidizing bacteria. Oxygenases bind O_2 directly to the substrate, and sulfite ($SO_3{}^{2-}$) would be the first product. There is evidence that in *Thiobacillus denitrificans*, sulfide is oxidized by a series of reactions catalyzed by a sulfide/sulfite oxidoreductase similar to that found in some sulfate-reducing bacteria. The oxidation of elemental sulfur by *T. denitrificans* might be preceded by its initial reduction to sulfide, followed by oxidation to sulfite.

Thiobacillus thioparus and *Thiobacillus denitrificans* appear to have the possibility for substrate-level phosphorylation in the oxidation of sulfite to sulfate (Fig. 10.**8b**), as mentioned in Chapter 10.3.

The sulfur-oxidizing enzymes are constitutively expressed in the obligate chemolithoautotrophs *Thiobacillus* and *Thiomicrospira* spec., but their maximum capacity for sulfur-compound oxidation can change by a factor of three to four. In the facultative *Thiobacillus* species, the sulfur-oxidizing (as well as the CO_2-fixing) capacity is often inducible and subject to (catabolite) repression.

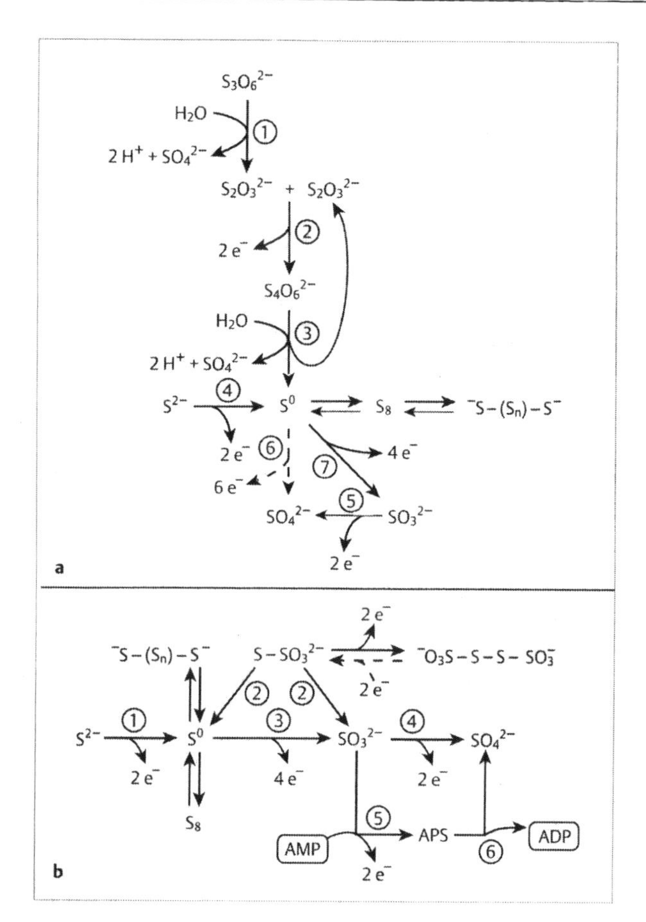

Fig. 10.8a **Tentative scheme for sulfur-compound oxidation by the acidophilic sulfur oxidizers, *Thiobacillus ferrooxidans* and *Thiobacillus acidophilus*.** Enzymes: 1, trithionate hydrolase; 2, thiosulfate dehydrogenase; 3, tetrathionate hydrolase; 5, sulfite-oxidizing enzyme; 4, 6, and 7, not yet characterized. $S^{-}-(S_n)S-^{-}$ represents polysulfides.
b The most important reactions in the oxidation of sulfur compounds by neutrophilic sulfur-oxidizing bacteria such as *T. thioparus*. Enzymes: 1, sulfide oxidoreductase; 2, thiosulfate-cleaving enzymes (rhodanese); 3, sulfur-oxidizing enzyme (sulfur dioxygenase or sulfur oxidoreductase); 4, sulfite oxidoreductase; 5, adenosine phosphosulfate reductase; 6, ADP sulfurylase

Since most of the sulfur-oxidizing bacteria produce sulfuric acid, many of them are tolerant of low pH values (pKa of HSO_4^- is 1.9). Many of the neutrophilic thiobacilli can survive pH values down to 3–4. The specialized acidophilic sulfur oxidizers are extremely pH tolerant. For example, *Thiobacillus thiooxidans* has an optimum for growth between pH 3 and 4. *Thiobacillus ferrooxidans* may still grow at pH 1.8 and survives at pH values below 1.0. Some of the enzymes of sulfur-compound oxidation are located in, or at least facing, the periplasm. Their exposure to acid conditions is reflected in the very low pH optima of some of these enzymes (e.g., enzymes 1, 2, and 3 in Fig. 10.**8**).

10.6.2. The facultative sulfur oxidizers can grow mixotrophically

In *Thiobacillus* and *Thiomicrospira* species the Calvin cycle is constitutive, but in the facultative organisms, it is completely repressed during heterotrophic growth. During simultaneous limitation by organic compounds and a sulfur compound, the facultative sulfur oxidizers can grow mixotrophically (see Chapter 10.2). Special cases of mixolithotrophic growth can be observed in both obligately and facultatively chemolithoautotrophic organisms capable of growing on different inorganic compounds, with the restriction that these mixtures be presented under energy-limited conditions. *Thiobacillus ferrooxidans* can grow on a mixture of tetrathionate and ferrous iron and on pyrite (FeS_2), which is an "intrinsic" mixed substrate for this organism.

In some strains of *T. ferrooxidans*, the sulfur-oxidizing enzymes are constitutive, but during prolonged growth on sulfur, the organism represses its iron-oxidizing capacity. The moderately thermophilic hydrogen bacteria can grow on a mixture of hydrogen and sulfur compounds. *Thiosphaera pantotropha* and *Paracoccus denitrificans* can grow on a combination of hydrogen and thiosulfate under energy limitation (sometimes called mixolithotrophy). In chemolithoheterotrophic thiobacilli, growth is obligately dependent on the presence of an organic compound as the carbon (and energy) source, while sulfur compounds can serve as a supplementary energy source and increase the growth yield.

Sulfolobus and *Acidianus* species are examples of thermophilic, acidophilic colorless Archaea. The known members of both genera are facultatively chemolithotrophic. Relatively little is known about their sulfur metabolism. The oxidation of elemental sulfur may proceed either by a dioxygenase, producing sulfite via Reaction 10.21 or via a reductive/oxidative step whereby sulfide and sulfite are formed (Reaction 10.22), initially followed by reoxidation of the sulfide to elemental sulfur (Reaction 10.23).

$$S^0 + H_2O + O_2 \longrightarrow HSO_3^- + H^+ \quad (10.21)$$

$$3\,S^0 + 3\,H_2O \longrightarrow HSO_3^- + 2\,HS^- + 3\,H^+ \quad (10.22)$$

$$2\,HS^- + 2\,H^+ + O_2 \longrightarrow 2\,S^0 + 2\,H_2O \quad (10.23)$$

In Reactions 10.22 and 10.23, sulfite would be the product, as in Reaction 10.21, the difference being, however, that electrons from Reactions 10.22 and 10.23 might be channeled into the electron transport chain, allowing the build-up of a proton motive force. As mentioned before, the autotrophic metabolism of *Sulfolobus* probably proceeds through a reductive TCA pathway, but little is known about its metabolic regulation.

As the sulfur-oxidizing bacteria often live at the interface between oxic and anoxic layers, many of them are equipped with survival mechanisms for anaerobic energy metabolism. Intracellular polyglucose can be fermented via a heterolactic fermentation to lactate and ethanol, (*Thiobacillus neapolitanus*) or intracellularly stored elemental sulfur can be used as the electron acceptor for the oxidation of organic compounds (*Beggiatoa*). Some sulfur oxidizers can grow anaerobically on elemental sulfur as the energy source in the presence of ferric ions as electron acceptor (*Thiobacillus ferrooxidans*).

The best-studied group among the conspicuous, large (>5 μm), colorless sulfur bacteria are *Beggiatoa* species, among which even the smallest are 5–10 μm wide, and the largest have a diameter of 100 μm or more. It is likely that among the *Beggiatoa* species, the full metabolic spectrum as shown in Table 10.**3** exists. Most of the described *Beggiatoa* species are either chemolithoheterotrophs or belong to the group that does not gain energy from sulfide oxidation, but benefits from it in other ways (chemoorganoheterotrophs, Table 10.**3**). A few marine *Beggiatoa* species are capable of chemolithoautotrophic growth on H_2S and CO_2 using the Calvin cycle for CO_2 fixation. These organisms can only be cultivated in the laboratory at the sulfide/oxygen interface of agar gradient cultures (Fig. 10.**9**). The typical habitat of *Beggiatoa* species is the sulfide/oxygen interface of sediments. Using their ability to glide on solid surfaces, these organisms can follow the sulfide oxygen interface as it moves during diurnal or tidal cycles. The above-mentioned anaerobic metabolism would thus represent an adaptation crucial for life at the oxic/anoxic interface. A *Thiothrix* species can grow chemolithoautotrophically on thiosulfate, CS_2 and certain thiophenes.

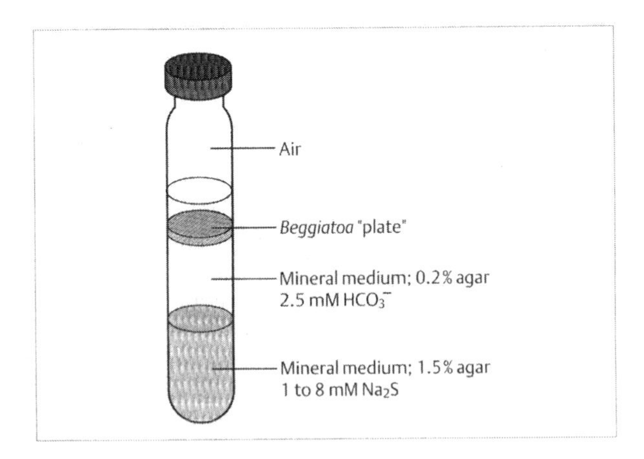

Fig. 10.**9** **Initial geometry of marine gradient medium for growing *Beggiatoa* spec. immediately after construction.** The air space contains sufficient O_2 for complete oxidation of H_2S to sulfate. HCO_3^- serves as carbon source. The red region shows where *Beggiatoa* proliferate as a plate following inoculation [after 2]

In many heterotrophic *Beggiatoa* species, as well as in similar metabolic types such as *Thiobacterium* and *Macromonas* species, the ability to oxidize sulfide may be linked to the lack of catalase. During heterotrophic growth, hydrogen peroxide is produced. In the absence of catalase, H_2O_2 can be detoxified by a spontaneous reaction with sulfide, whereby elemental sulfur is produced and deposited intracellularly.

Recent work has identified the giant, filamentous *Thioploca* species as major constituents of dense microbial mats in sediments along the coast of Chile. These bacteria form filaments up to 7 cm long, consisting of chains of single cells 15–40 μm wide and 60 μm long. They can be observed living together in bundles of filaments in a sheath of polymeric, slimy material, measuring 1.5 × 100–150 mm. Due to the very high seasonal photosynthetic production of organic material in this coastal environment, very active bacterial sulfate reduction takes place in the sediment. The resulting sulfide is effectively reoxidized to sulfate by the *Thioploca* mat, using nitrate that, in contrast to oxygen, is abundantly available in the overlying water. The *Thioploca* cells actively move out of their sheath into the water and accumulate high concentrations (500 mmol/l) of nitrate in large vacuoles. They subsequently glide back down their sheath, into the sulfide-containing sediment. This behavior explains the abundant occurrence of thousands of square kilometers of *Thioploca* mats in this coastal area.

Symbiotic sulfur bacteria live together with tube worms, mussels, and other invertebrates at hydrothermal vents (Chapter 31.1.4). These vents are very special habitats for free-living and symbiotic sulfide-utilizing bacteria; they are described in Chapter 31.1.7. Their main inhabitants are members of the genera *Beggiatoa*, *Thiomicrospira*, and *Thiobacillus* and some hydrogen- and methane-oxidizing bacteria. Mussels and other invertebrates feed on the rich bacterial production, while 1.5 m long tube worms feed directly on the sulfide which is transferred to symbiotic autotrophic sulfide-oxidizing bacteria that live within the eukaryotic cells.

This type of symbiosis is not limited to hydrothermal vents but has also been found in many other invertebrates, including much smaller mussels, sea urchins, clams, and other invertebrates living at the oxic/anoxic interface at, for example, tidal mud flats.

The colorless sulfur-oxidizing bacteria are used in waste-water treatment and in microbial leaching. Sulfide-oxidizing bacteria are a crucial subpopulation of nearly all waste-water treatment sludges, in which they are responsible for the oxidation of malodorous and toxic hydrogen sulfide to relatively harmless sulfate. In most situations, the oxidation is carried out under oxic conditions. An obvious disadvantage of this procedure is that the sulfide-containing waste water must be aerated. Hydrogen sulfide may be stripped from the system and become a nuisance or even danger to the environment. As a remedy, aeration tanks can be covered and the off-gas treated, but an attractive alternative is to oxidize the sulfide to sulfate under anoxic conditions with nitrate as the oxidant. In these two procedures, sulfate is the end product. However, release of quantities of sulfate into freshwater environments may boost naturally occurring sulfate reduction processes, and the output of sulfate should preferably be limited. A novel process using obligately chemolithoautotrophic *Thiobacillus* species involves the partial oxidation of sulfide under oxygen-limiting conditions, producing insoluble elemental sulfur. This sulfur can be recycled for further use. Boosting of the sulfur cycle by human activity can thus be avoided.

Box 10.4 In aerated waste-water treatment with a relatively low sulfide supply, filamentous *Thiothrix* species may develop within bacterial flocs. This makes the flocs less easily separated from the water phase, and hence is highly undesirable. It is one of the causes for poor settling of the sludge known as "bulking".

The sulfur-oxidizing bacteria are also responsible for environmental problems and economic damage due to corrosion. This is linked to the sulfuric acid produced by the oxidation of sulfur compounds, which may lead to dramatic acidification of soil-containing insoluble sulfides or sulfur. For example, if land is reclaimed from the sea, the oxidation of pyrite (FeS_2) present in the soil may lead to pH levels around 1, thereby making the soil unsuitable for agriculture. In such cases, calcium carbonate is applied to neutralize the soil. Another example is the oxidation of sulfur in concrete sewage pipes, leading to extensive damage by dissolution of the carbonates in the concrete by sulfuric acid.

10.7 The Nitrifying Bacteria Oxidize Ammonium and Nitrite

Ammonium [in the form of ammonia at higher pH (pK = 9.3)], hydroxylamine, and nitrite are among the inorganic nitrogen compounds that can be used by various nitrifying chemolitho(auto)trophic bacteria. Ammonium is oxidized to nitrate in two discrete steps. The nitrosobacteria oxidize the ammonium to the level of nitrite, while the nitrobacteria oxidize the nitrite further to nitrate. Table 10.7 shows the most important genera of nitrifying bacteria. The genera shown share the capability to grow chemolithoautotrophically. Most of the nitrifying bacteria described thus far are obligate chemolithoautotrophs, with a few facultatively chemolithoautotrophic species, and there are no known chemolithoheterotrophs. However, a wide variety of

Table 10.7 **Nitrifying bacteria**

Reactions	Organisms
Ammonium-oxidizing (nitroso-) $NH_4^+ + 1\frac{1}{2}\ O_2 \rightarrow NO_2^- + 2\ H^+ + H_2O$	*Nitrosococcus oceanus, Nitrosolobus multiformis, Nitrosomonas europaea, Nitrosospira briensis, Nitrosovibrio tenuis*
Nitrite-oxidizing (nitro-) $NO_2^- + \frac{1}{2}\ O_2 \rightarrow NO_3^-$	*Nitrobacter hamburgensis, Nitrobacter vulgaris, Nitrobacter winogradskyi, Nitrococcus mobilis, Nitrospina gracilis, Nitrospira marina*

heterotrophic organisms are able to co-metabolize (co-oxidize) ammonium to nitrite, and nitrite to nitrate, a phenomenon known as "heterotrophic nitrification".

Elaborate intracellular membrane structures harbor the enzymes for ammonium or nitrite oxidation. Most of what is known of the ammonium and nitrite oxidation pathways and the carbon metabolism of the nitrifiers has been derived from the study of only a few species. Experimental difficulties are linked with a relatively low rate of growth, a low biomass yield, and the toxicity of millimolar concentrations of nitrite. Hence, a co-culture of the two types of nitrifiers is much easier to cultivate than axenic cultures.

The nitrosobacteria oxidize ammonia to nitrite. The biochemistry of the oxidation of ammonium to nitrite has been investigated in detail for *Nitrosomonas europaea*. The overall oxidation reaction of ammonium with oxygen is acid producing:

$$2\,NH_3 + 4\,O_2 \longrightarrow 2\,NO_3^- + 2\,H_2O + 2\,H^+ \qquad (10.24)$$

The reaction proceeds via hydroxylamine and an (NOH) intermediate and gives nitrite as the end product.

$$NH_3 \longrightarrow NH_2OH \longrightarrow (NOH) \longrightarrow NO_2^- \qquad (10.25)$$

The first step is catalyzed by ammonia monooxygenase. In this reaction, the oxygen atom of hydroxylamine originates from molecular oxygen. The reaction requires the input of reducing power:

$$NH_3 + \text{"NAD(P)H"} + H^+ + O_2 \longrightarrow NH_2OH + H_2O + NAD(P)^+ \qquad (10.26)$$

In the absence of oxygen, the oxidation of ammonia to hydroxylamine has a positive Gibbs free energy change. By the sacrifice of NADH, this reaction is changed into a reaction with a negative Gibbs free energy change. Although NAD(P)H may act as the electron donor for this reaction in cell-free extracts in the test tube, the physiological donor is likely to be a compound of the electron transport chain (Fig. 10.**10**). Hydroxylamine dehydrogenase, a periplasmic enzyme that reduces periplasmic cytochrome *c*, catalyzes:

$$NH_2OH + H_2O \longrightarrow HNO_2 + 4\,[H] \qquad (10.27)$$

Since the electrons from hydroxylamine oxidation are likely to be delivered at the level of the quinones, it is assumed that reduced quinones donate the reducing power for the monooxygenase reaction in vivo. Hence, for each ammonium oxidized to hydroxylamine, the cell must sacrifice two energy-rich electrons, which otherwise would have been available for energy generation. The delivery of the electrons from hydroxylamine at the level of quinones also implies that reversed electron transport is required for the generation of NADH for CO_2 fixation. Given the low energy yield of ammonium oxidation and electron transport being membrane-bound, the presence of large membrane systems appears to be a necessary adaptation to run these complex reactions at a sufficiently high rate.

Ammonia monooxygenase does not appear to be very specific for ammonia. This explains why these organisms are capable of oxidizing a variety of other compounds which, to various extents, can be considered to be analogues of NH_4^+. For example, methane can be oxidized to methanol, although at a much lower rate (100 times less) than ammonia. Other substrates that can be oxidized very slowly include carbon monoxide, ethylene, propylene, cyclohexane, benzyl alcohol, and phenol. As the observed turnover rates of these compounds in nature are often extremely small, nitrifying populations may contribute significantly to their breakdown.

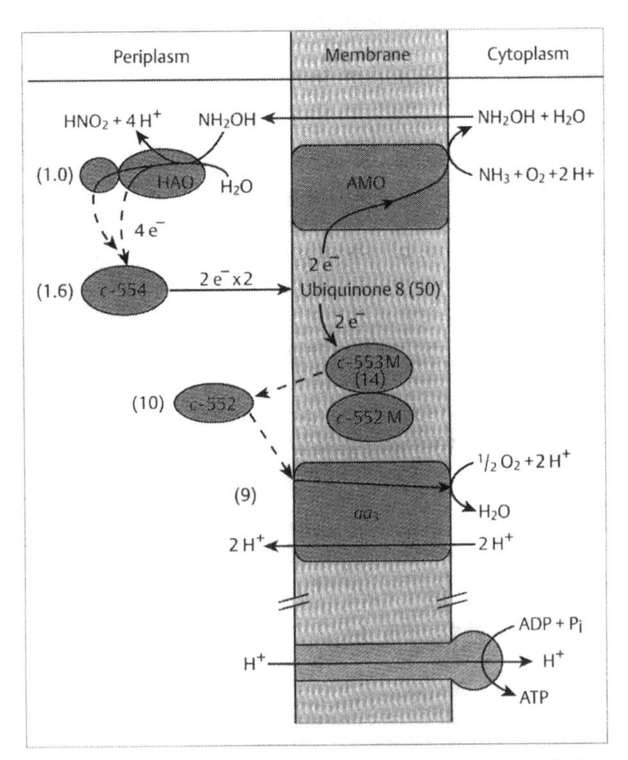

Fig. 10.**10** **Arrangement of enzymes of nitrogen oxidation and electron transport in membranes of *Nitrosomonas*.** Numbers in parentheses indicate molar ratios of components relative to hydroxylamine oxidoreductase (HAO). The site of the ammonia monooxygenase (AMO) reaction is hypothetical. ATP synthase is shown at the bottom. *a* and *c* indicate cytochromes *a* and *c* [after 3]

The nitrobacteria oxidize nitrite to nitrate. The oxidation of nitrite to nitrate by *Nitrobacter* species proceeds according to the following two half reactions:

$$NO_2^- + H_2O \longrightarrow NO_3^- + 2\,H^+ + 2\,e^- \quad (10.28)$$

$$\tfrac{1}{2}\,O_2 + 2\,H^+ + 2\,e^- \longrightarrow H_2O \quad (10.29)$$

The nitrite/nitrate oxidoreductase is located at the inside of the cytoplasmic membrane, and a proton motive force is built up by a proton-pumping cytochrome oxidase. Of course, *Nitrobacter* must also possess a complete electron carrier system to allow production of NAD(P)H by reversed electron transport. Although heterotrophic nitrification is a widespread property of many genera of prokaryotes and fungi, the biochemistry of these reactions is very diverse. In some organisms, the inorganic pathway as shown above is present, while in other organisms, nitrite is produced with organic nitrogen compounds as intermediates. It is characteristic of heterotrophic nitrification that the specific activity is 100- to 10,000-fold lower than that of the chemolithoautotrophs. Heterotrophic nitrifiers do not gain energy from this reaction. An interesting observation is that many heterotrophic nitrifiers may convert part of the nitrite produced into gaseous compounds such as N_2O or N_2, even under fully oxic conditions.

Many of the known chemolithotrophic ammonium oxidizers and nitrite oxidizers are typical obligate chemolithoautotrophs. All the currently known chemolithotrophic nitrifiers use the Calvin cycle to fix CO_2. Although the *Nitrosomonas* species can be considered as classical examples of obligate chemolithoautotrophs, at least one strain appears to be able to obtain energy, albeit very slowly, by anaerobic denitrification with pyruvate as electron donor. Indeed, its ability to denitrify may well be a survival mechanism under anoxic conditions. While facultative chemolithotrophy is rare among the ammonium-oxidizing bacteria, it is relatively common among the nitrite oxidizers. It had been known for many years that the "obligately chemolithotrophic" *Nitrobacter agilis* could grow very slowly on acetate, and it now appears that some species of *Nitrobacter* (e.g., *Nitrobacter hamburgensis*) are facultative chemolithoautotrophs. Other strains can grow heterotrophically under denitrifying conditions.

Habitats of nitrifiers. The nitrifiers can be found in nature whenever ammonium is liberated and oxygen is available, in soil and aquatic environments. Long-term use of ammonium-based fertilizers has resulted in a massive enrichment of autotrophic nitrifiers in arable fields. Like so many other chemolithotrophs, the nitrifiers are particularly active at the oxic/anoxic interface of sediments and water bodies. The contribution of heterotrophic nitrification to the total oxidation of ammonium is negligibly small as long as the ratio of the turnover of organic compounds (C) versus ammonium (N) is relatively low. However, when the C/N ratio is more than 10, the heterotrophic population becomes so dominant that heterotrophic nitrification may contribute significantly. It has also been observed that at the oxic/anoxic interface of water bodies, where significant methane-oxidizing populations are present, ammonia is oxidized by the methane monooxygenase and then further to nitrite by heterotrophic nitrifiers. In this case, co-metabolic ammonia oxidation may compete effectively with autotrophic nitrification.

Nitrifiers in waste-water treatment. The chemolithoautotrophic ammonium- and nitrite-oxidizing bacteria play a vital role in modern waste-water treatment for nitrogen removal. Ammonium produced by degradation of organic compounds is oxidized by nitrifying bacteria to nitrite and nitrate. The nitrate is subsequently reduced by denitrifying bacteria to dinitrogen gas. Figure 10.**11** gives a typical example of the set-up of a treatment system. In order to ensure that nitrate is completely reduced by the available organic compounds, nitrate is fed into the first anaerobic step. The remainder of the organic material emerging from the first step is aerobically oxidized in the second stage with a relatively short residence time. In a final step, with a much larger residence time, and often using biomass retention systems (either by immobilization or by sedimentation), the nitrifying bacteria carry out the oxidation of ammonia to nitrate, which is then recycled to the first stage. This third stage, separate from the second, ensures that the slow-growing nitrifying bacteria cannot be overgrown by heterotrophs, and that excess sludge produced in the second step can be removed without removing nitrifying capacity. Because of the legal limits for nitrogen release into the environment, nearly all modern waste-water treatment plants possess a nitrification/denitrification system.

The role of nitrifiers in soil and agriculture. Ammonium, when used as a fertilizer in agriculture, is

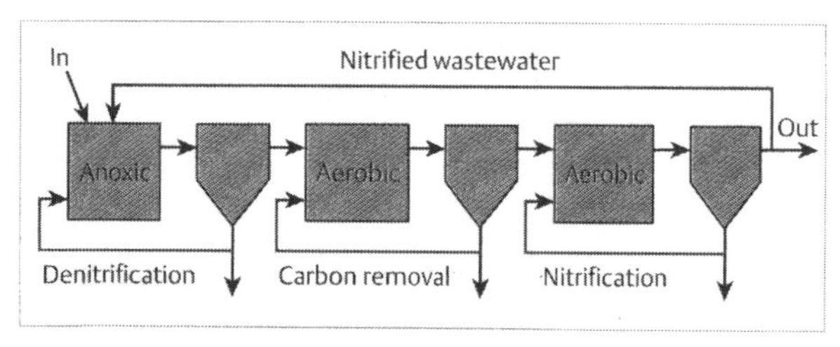

Fig. 10.**11** **Simple reactor combination for the treatment of waste water**

bound firmly to the negatively charged soil minerals and humic acids. Thus, during the growth of plants, ammonium is released slowly and oxidized by the nitrifying bacteria to nitrate. Nitrate is more readily taken up than ammonium by most plants. Nitrite or nitrate, however, as negatively charged ions, are easily washed out by rain. In addition, water-logging during heavy rainfall can also temporarily cause anoxic conditions, allowing denitrification to occur. Indeed, it has been estimated that as much as 55% of the ammonium nitrate used as fertilizer in corn (maize) fields is lost by run off or denitrification.

Degradation of limestone and soil acidification. In agricultural (e.g., pig sites, intensive chicken and cattle-rearing houses) and other anthropogenic activities, ammonia is volatilized and reabsorbed into soil, water, and even the porous stones of buildings. In addition to undesired fertilization, volatilized ammonia has a corrosive effect on limestone because of the presence of nitrifying bacteria growing a few millimeters or even centimeters deep into the porous stone. Here the ammonia is oxidized by these bacteria, whereby acid is produced (Reaction 10.27). The acid combines with the carbonate of the limestone which is dissolved releasing CO_2. In this way, old limestone buildings, and sculptures are deteriorating. The nitrifiers are also involved in the acidification of poorly buffered soils when the nitrifiers oxidize ammonia from atmospheric pollution. The low pH mobilizes trace metals, which then can be leached out by rainfall. Lack of essential trace metals has been suggested as at least one cause of the "Waldsterben" (or death) of the forests. One of the oldest sources of saltpeter for gunpowder manufacture was animal and human urine. Ammonia liberated from the urine would seep through the porous stone walls of manure pits and be oxidized by the nitrifiers to nitrate. Water slowly evaporated and the nitrate became visible as "blooms" of crystalline saltpeter on the wall. This was scraped off, processed, and used.

The nitrifying chemolithotrophs are most important in the removal of ammonia (ammonium) from waste water. They oxidize ammonia to nitrite and nitrate, which is subsequently reduced to dinitrogen gas by denitrifying bacteria. The nitrifiers represent one of the bottle-necks in the nitrogen cycle because of their relatively low rate of growth. Nitrifiers are held partially responsible for the erosion of limestone, leading to the deterioration of buildings and statues.

10.8 The "Knallgas" Bacteria Gain Energy From Hydrogen Oxidation

Hydrogen can be used as a substrate by a wide variety of bacteria, both aerobic and (facultatively) anaerobic. The electron acceptors for the oxidation also vary widely: oxygen, ferric iron, nitrate, nitrite, or nitrous oxide. Most of the aerobic hydrogen bacteria (often referred to as the "Knallgas" bacteria) are facultative chemolithoautotrophs. Only a few known species are obligate chemolithotrophs or chemolithoheterotrophs. Among the obligate anaerobes described in Chapter 12, many methanogens are, in fact obligate chemolithoautotrophs. In this section, the (facultatively autotrophic) aerobic hydrogen- and carbon-monoxide-oxidizing bacteria will be discussed. Table 10.**8** gives an overview of most of the genera and species of these hydrogen bacteria (see also Table 10.**9**). Obviously, there is no taxonomic relationship between the members of this physiological group. Among the few described obligately chemolithoautotrophic species, the acidophilic *Thiobacillus ferrooxidans*, which can grow on hydrogen, ferric iron, or (reduced) inorganic sulfur compounds, and the thermophilic *Hydrogenobacter thermophilus*, able to grow on hydrogen and inorganic sulfur compounds, are found. Both organisms can grow mixolithotrophically, i.e., they can utilize mixtures of these compounds simultaneously during autotrophic growth. Among the phototrophic bacteria, one can also find species that can grow as Knallgas bacteria. A significant number of the hydrogen bacteria can use carbon monoxide as their energy source (see below). The oxidation of carbon monoxide yields approximately the same Gibbs free energy change as hydrogen (−514 kJ/mol and −474 kJ/mol, respectively).

10.8.1 The Oxidation of Hydrogen Can Be Catalyzed by Two Types of Hydrogenase

The most common hydrogenase is membrane-bound and delivers its electrons at the cytochrome *b* level. Only

Table 10.8 **Genera and species of Knallgas bacteria**

Gram-negative Bacteria:	*Acidovorax facilis, Alcaligenes eutrophus, Alcaligenes eutrophus* CH34, *Alcaligenes hydrogenophilus, Alcaligenes latus, Alcaligenes paradoxus, Alcaligenes ruhlandii, Aquaspirillum autotrophicum, Azospirillum lipoferum, Calderobacterium hydrogenophilum, Derxia gummosa, Flavobacterium autothermophilum, Hydrogenophaga flava, Hydrogenophaga palleronii, Hydrogenophaga pseudoflava, Hydrogenobacter thermophilus, Microcyclus aquaticus, Paracoccus denitrificans, Pseudomonas hydrogenothermophila, Pseudomonas hydrogenovora, Pseudomonas saccharophila, Pseudomonas thermophila, Renobacter vacuolatum, Rhizobium japonicum, Xanthobacter autotrophicus, Xanthobacter flavus*, most carboxidobacteria
Gram-positive Bacteria:	*Arthrobacter* strain 11/X, *Bacillus schlegelii, Bacillus tusciae, Mycobacterium gordonae, Nocardia autotrophica, Rhodococcus opacus*
Archaea:	*Aquifex pyrophilus*

a few Knallgas bacteria also contain a cytoplasmic ("soluble") NAD^+-dependent hydrogenase (Fig. 10.**12**). In principle, the electrons from these two hydrogenases can be used for both energy generation and the production of reducing power, but under physiological conditions the NAD-dependent enzyme is primarily used for the latter function. This is a rare example where an inorganic electron donor can directly be used for the reduction of NAD. Organisms that only have the membrane-bound hydrogenase must generate reducing power via reversed electron transport. Bacteria with both types of hydrogenase have a higher biomass yield per hydrogen than those with only the membrane-bound enzyme. Referring to Figures 10.**2** and 10.**12**, and given that the membrane-bound hydrogenases operate on the outside of the cytoplasmic membrane, the build-up of a proton motive force during hydrogen oxidation is due to a combination of proton pumping and scalar production/consumption of protons on the outside/inside of the membrane. *Alcaligenes eutrophus* has both hydrogenases and has been studied in detail (Fig. 10.**12**). It contains a membrane-bound hydrogenase facing the periplasmic space. This enzyme consists of large and small subunits coupled to a cytochrome-*b*-like anchor protein. The active site of the enzyme is located in the large (67 kDa) subunit and contains a bimetallic center consisting of nickel and iron. The small (35 kDa) subunit, which coordinates several [Fe–S] clusters, is the polypeptide transferring the electrons to the electron transport chain via the cytochrome-*b*-receptor. The soluble (i.e., cytoplasmic) hydrogenase (SH) also is a [Ni–Fe] hydrogenase (Fig. 10.**12**). It consists of two functional dimeric moieties, a hydrogenase and an NADH oxidoreductase (diaphorase). The hydrogenase moiety is related to the membrane-bound hydrogenase and the diaphorase part shows homology with the NADH : ubiquinone oxidoreductase (complex I) of respiratory chains.

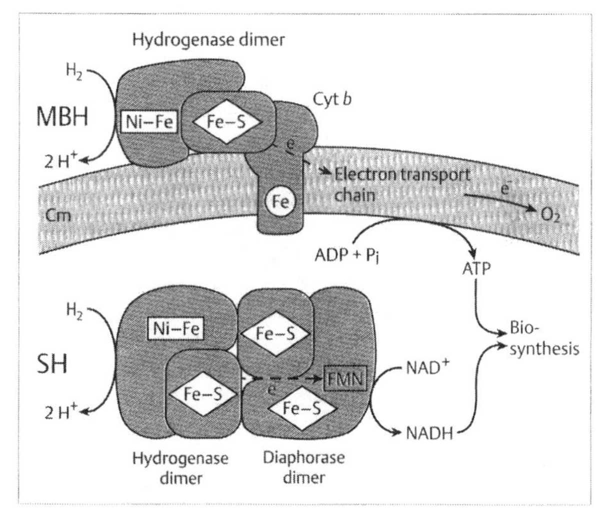

Fig. 10.**12** **Structural model and function of the two NiFe-hydrogenases present in *Alcaligenes eutrophus*.** MBH, membrane-bound; SH, soluble hydrogenase. Courtesy of B. Friedrich

The genes of the two hydrogenases in *A. eutrophus* strain H16 are located on a 450-kb megaplasmid. This plasmid harbors, in addition to the structural genes for the soluble (SH) and membrane-bound hydrogenases (MBH), sets of accessory genes required for nickel uptake, hydrogenase maturation, electron transport functions, and transcriptional regulation. In *A. eutrophus* strain H16 the synthesis of both hydrogenases is coordinately regulated in response to environmental conditions, such as the energy status (i.e., the presence of excess or limiting substrates). On "rich" substrates, hydrogenase synthesis is repressed, but during substrate limitation, hydrogenase is derepressed, irrespective of the presence of hydrogen as inducer. In other hydrogen bacteria, the expression of hydrogenase is strictly dependent on the presence of hydrogen. Considering the low concentration of hydrogen and other substrates in the environment, it must be expected that most of the facultative hydrogen bacteria will be typical "mixotrophs" (see below and also Box 10.5).

10.8.2 Nearly All of the Hydrogen Bacteria Use the Calvin Cycle for CO_2 Fixation

Notable exceptions are *Hydrogenobacter thermophilus* and *Aquifex pyrophilus*, which utilize a reversed tricarboxylic acid cycle (see Chapter 10.4, Fig. 8.**4**). Like the obligately autotrophic sulfur and ammonium oxidizers, *H. thermophilus* can assimilate small amounts of organic compounds, but CO_2 remains the main carbon source for growth under all conditions tested. The facultatively chemolithotrophic Knallgas bacteria are typical heterotrophs that often have a wide metabolic potential and diversity. Many of these organisms show diauxic growth when presented with high concentrations of substrate, but can grow mixotrophically on mixtures of organic, or organic and inorganic compounds under substrate limitation. It is therefore not surprising that their hydrogen metabolism and CO_2-fixation pathways are rather strictly regulated. Although the mode of regulation may differ, the general pattern is that the autotrophic potential is repressed if there is sufficient organic carbon present. Similarly, hydrogenase may be induced/repressed in the presence/absence of hydrogen, but in some bacteria, the enzyme is constitutive.

All nitrogen-fixing bacteria use nitrogenase for the conversion of molecular nitrogen to ammonia, with ferredoxin-type reducing equivalents. This enzyme also has hydrogenase activity, which results in the inevitable production of significant quantities of molecular hydrogen (about the same amount as when nitrogen is fixed). In order to "recover" this hydrogen, most of the nitrogen-fixing bacteria rely on uptake hydrogenases to reoxidize it. Hydrogen utilization is a trait primarily used to deal with endogenous hydrogen production. However, a few, such as *Rhizobium japonicum*, *Azospirillum lipoferum*, and a number of *Xanthobacter* species, are capable of chemolithoautotrophic growth on hydrogen with molecular nitrogen as the only nitrogen source.

Box 10.5 The genes coding for hydrogenase, the ability to grow autotrophically, nitrogenase, and the denitrification enzymes are frequently located or harbored on plasmids. The same is true for the ability to grow on carbon monoxide. This implies that these properties may be laterally transmitted through a microbial population, as has in fact, been shown to happen in a few cases. This may explain, at least in part, why these unique properties can be found in such a diverse range of taxonomically nonrelated microorganisms.

10.9 The Carbon Monoxide-oxidizing Bacteria Are Facultative Chemolithoautotrophs

Many of the carbon-monoxide-utilizing bacteria can also grow on hydrogen (Table 10.**9**). Several of the bacteria listed, for example *Rhizobium* species, can even grow on carbon monoxide while denitrifying. All known isolates are facultative chemolitho(auto)trophs. For *Hydrogenophaga pseudoflava*, mixotrophic growth on H_2 or CO and pyruvate has been demonstrated. The biomass yields for the hydrogen bacteria range from 0.10–0.22 mol CO_2 fixed/mol hydrogen, depending on the hydrogenase involved. In *Pseudomonas carboxydovorans*, which like all other known CO-utilizing bacteria employs the Calvin cycle for CO_2 fixation, the yield is 0.18 mol CO_2 fixed/mol CO oxidized.

CO dehydrogenase. The enzymology of the CO-oxidizing bacteria has been studied in detail with *Pseudomonas carboxydovorans* as the model organism. The CO dehydrogenase is a relatively large (230–300 kDa) enzyme containing a molybdenum-carrying pterin (bactopterin) as cofactor. This pterin (or a similar molybdenum-carrying pterin cofactor) is present in all molybdo-enzymes (oxidoreductases/dehydrogenases) except for nitrogenase. The CO dehydrogenase of *P. carboxydovorans* is located at the inside of the membrane, where it delivers its electrons to a *b*-type cytochrome. Thus, for NADH production, reversed electron transport is required. If CO is the substrate, the electrons are transferred to oxygen via a CO-insensitive electron transport chain involving a cytochrome b^{563}, and allowing a charge separation of 4 H^+ per 2 electrons transferred. If hydrogen is the electron donor, a membrane-bound dehydrogenase can donate its electrons to a more efficient cytochrome chain with a cytochrome *c* and a terminal cytochrome *a* type, whereby a charge separation of 6 H^+ per 2 electrons is possible.

Table 10.9 **Characteristics of selected carboxydotrophic bacteria** [after Meyer, O. (1989) Aerobic carbon monoxide-oxidizing bacteria. In: Schlegel, H.G., and Bowien, B. (eds) Autotrophic bacteria. Madison, Wisc.: Science Tech; 331–350.] All these carboxydotrophic bacteria use oxygen as electron acceptor, a few can use nitrate as alternative. Some strains grow on $H_2 + CO_2(H_2)$; (NO_3^-) can denitrify; (N_2), dinitrogen can be used as N-source; (M), mesophile; (T), thermophile

Species
Arthrobacter spec. (H_2, M)
Hydrogenophaga pseudoflava (H_2, M)
Pseudomonas carboxydohydrogena (H_2, M)
Pseudomonas carboxydovorans (H_2, M)
Pseudomonas compransoris (H_2, M)
Pseudomonas gazotropha (H_2, M)
Bacillus schlegelii (H_2, T)
Pseudomonas thermocarboxydovorans (T)
Streptomyces G26 (T)
Alcaligenes carboxydus (M)
Acinetobacter spec. (M)
Azomonas spec. (M)
Azotobacter spec. (H_2, N_2, M)
Rhizobium japonicum (NO_3^-, H_2, N_2, M)
Isolates S17 and A305 (H_2, N_2, M)

Industrial applications. The use of hydrogen for the synthesis of poly-β-hydroxy butyrate (PHB) and other polyhydroxyalkanoates for the production of biodegradable plastics has been considered. An interesting potential application is the use of denitrifying hydrogen bacteria for the removal of nitrate from drinking water. The attractiveness of this method lies in the non-toxic nature of hydrogen: hydrogen is bubbled through an anoxic cylindrical reactor with immobilized hydrogen bacteria, which convert the nitrate to dinitrogen gas. The effluent liquid is then aerated in a second reactor, where traces of nitrite are reoxidized to nitrate and traces of organic material are remineralized. Dissolved excess H_2 is also oxidized. Finally, the effluent is filtered through a conventional sand bed to remove the last traces of soluble and particulate material. Mixed cultures of carbon-monoxide- and hydrogen-oxidizing bacteria also play a role in the effective treatment of polluted air and industrial waste gas in compost and other gas/air biofilters that are used on a very large scale in various industries.

Nearly all known "Knallgas" bacteria and CO-oxidizing bacteria are facultative chemolithoautotrophs, implying that these organisms can also use organic compounds as the sole or as a supplementary energy and/or carbon sources. When hydrogen and/or CO_2 are used simultaneously with an organic substrate, the bacteria grow mixotrophically.

10.10 The Iron- and Manganese-Oxidizing Bacteria

Although iron is one of the major elements on earth, relatively few bacteria are known to be able to derive energy from the aerobic oxidation of iron. This is because of a number of factors. Ferrous iron is not often present at high concentrations, while insoluble compounds containing ferrous iron [e.g., as (hydr)oxides or sulfides] are much more abundant. Ferrous iron can be generated from insoluble ferrous compounds by anaerobic respiration processes, but the concentration of free ferric insoluble iron at neutral pH is extremely low; therefore, ferrous iron production is also limited. As explained in paragraph 10.1.2, example 3, the Gibbs free energy change of the oxidation reaction is considerable, but at neutral pH, the bacteria have to compete with the spontaneous oxidation of ferrous iron with molecular oxygen. However, at lower pH values, the solubilities of iron (II) compounds are much higher, and below pH 4, ferrous iron is chemically stable in the presence of oxygen. At a pH of 3, the Gibbs free energy of the reaction (see Reaction 10.33) is −91.0 kJ/mol. Hence, it is not surprising that many of the known chemolithotrophic iron-oxidizing bacteria are acidophilic. The acidophilic species are listed in Table 10.**10**. Many of these species belong to the sulfur-oxidizing bacteria shown in Table 10.**5**. *Leptospirillum* species have only been shown to oxidize iron, even though they can grow on pyrite, because the ferric iron produced in the reaction acts as a chemical oxidant for the pyrite, according to the following reaction:

$$FeS_2 + 14\ Fe^{3+} + 8\ H_2O \longrightarrow 15\ Fe^{2+} + 2\ SO_4^{2-} + 16\ H^+ \quad (10.30)$$

It has been suggested that the same reaction would also explain the growth of *Thiobacillus ferrooxidans* on pyrite, but there is ample evidence that this organism pos-

Table 10.10 **Acidophilic iron-oxidizing bacteria**

Mesophiles	Moderate thermophiles	Extreme thermophiles
Leptospirillum ferrooxidans	*Sulfobacillus thermosulfidooxidans*	*Acidianus brierleyi*
Thiobacillus ferrooxidans	Unnamed *Thiobacillus*-like isolates (strains m-1, TH1, TH3, ALV, BC and K)	*Metallosphaera sedula*
		Sulfolobus acidocaldarius
		Sulfolobus solfataricus

sesses sulfur-oxidizing enzymes, and that it derives more energy from the electrons from sulfur than from those from iron.

Among the acidophilic iron oxidizers, obligate and facultative chemolithoautotrophs are found, as well as "incidental" oxidizers, such as *Acidiphilium cryptum*. At present, it is not known whether chemolithoheterotrophic iron oxidizers exist. The acidophilic iron-oxidizers can be found in large numbers in geological deposits of sulfidic minerals, such as pyrite and other metal sulfides and, if water and oxygen are available, they may cause dramatic acidification of their environment. This occurs, for example, in coal mines. If acid mine water reaches natural waters with a neutral pH, precipitates of basic ferric sulfates (called jarosites) may be formed. These compounds can block drainage pipes and smother vegetation, and the low pH may cause the death of most other forms of life.

Gallionella ferruginosa and *Leptothrix* species are among the best-known neutrophilic, iron oxidizers (Table 10.**11**). *Leptothrix* species are filamentous, sheathed bacteria which, in the presence of oxidizable ferrous iron, deposit large quantities of ferric iron precipitates around the cells. Recently, the chemolithoautotrophic nature of *Gallionella ferruginosa* has been established with the finding that it can use ferrous iron oxidation to drive CO_2 fixation via the Calvin cycle. Natural blooms of the neutrophilic, iron-oxidizing bacteria are found in drainage pipes where ferrous iron leached from the surroundings reaches the oxic environment. Recently, common rod- and vibrio-shaped chemolithotrophs have been isolated from such blooms, using Fe^{2+}–O_2 gradient cultures similar to Fig. 10.**9**.

In *Thiobacillus ferrooxidans*, ferrous iron is oxidized by a periplasmic enzyme system. Nearly all biochemical details known about microbial iron oxidation stems from work with *T. ferrooxidans*. Ferrous iron is oxidized by a periplasmic enzyme system, whereby ferric iron is produced outside the cells (Fig. 10.**13**). Electrons from this oxidation are transferred by a cytochrome-containing iron-oxidizing enzyme to a redox mediator (rusticyanin), which channels the electrons to a terminal cytochrome oxidase. This oxidase delivers its electrons, probably at the cytoplasmic side of the membrane, to oxygen and protons, whereby water is formed. Protons are thus consumed inside the cell. Since *T. ferrooxidans* lives in a very acid environment, which is kept acid by sulfuric acid formation, and the pH of the cytoplasm of the organism is near 6, a permanent proton gradient exists over the membrane. This Δp can produce ATP by means of ATP synthase. Recent investigations show that the pathway of iron oxidation by acidophiles is not universal. There is a large diversity of mechanisms involving different soluble and membrane-bound electron carriers.

Acidophilic iron-oxidizing bacteria are used in the leaching of ores. Mixed populations of acidophilic, iron-oxidizing bacteria play a major role in the leaching of

Table 10.11 **Neutrophilic iron bacteria associated with the deposition of ferric iron.** After Jones, J. G. (1986) Adv Microbial Ecol 9: 149–185

	Genus or group	
Morphological type	In axenic culture	Not yet isolated
Filamentous	*Leptothrix, Sphaerotilus*	*Clonothrix, Crenothrix, Lieskeella, Toxothrix*
Prosthecate/appendaged	*Hyphomicrobium, Gallionella, Pedomicrobium, Planctomyces*	*Metallogenium, Seliberia?*
Encapsulated/coccoid		*Naumanniella, Ochrobium, Siderocapsa (Arthrobacter), Siderococcus*
Rods, vibrios	γ-Proteobacteria	

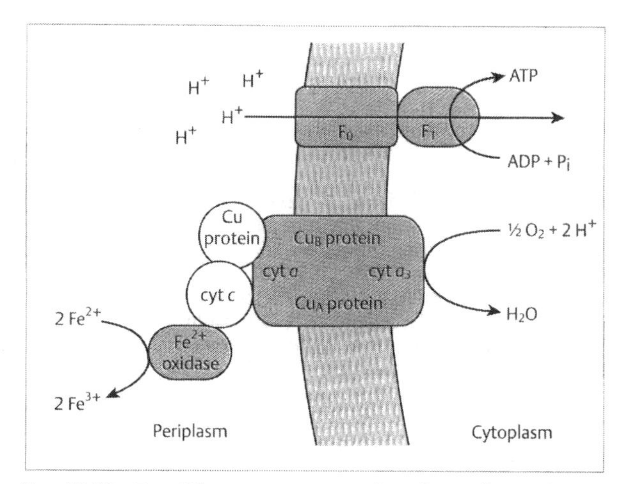

Fig. 10.**13** **Possible arrangement for the Fe^{2+}-oxidation electron transport system of *Thiobacillus ferrooxidans*.** Fe^{2+} oxidase, Fe(II)-cytochrome c_{552}; cy + c, soluble cytochrome c_{552}; Cu protein, rusticyanin; Cytochrome *a* (cyt *a*), cytochrome a_3 (cyt a_3), Cu_A protein and Cu_B protein are likely components of a terminal oxidase. F_0 and F_1 are the membrane-integral and membrane-associated portions, respectively of ATP synthase [after 4]

sulfidic minerals for the recovery of copper and other metals from ores too poor for normal metallurgical extraction methods. Heaps of ore or tailings from the metallurgical extraction process are sprinkled with acid water, which subsequently percolates the ore. There, the iron-oxidizing bacteria oxidize the sulfidic minerals to sulfuric acid and ferric iron. During this process, the pH may become as low as 1.0–1.5, in which range the pH is buffered by HSO_4^{2-}/SO_4^{2-} (pK 1.9). The soluble ferric iron may also act as a chemical oxidant of the minerals. Table 10.**12** shows some of the reactions involved. The acid dissolves insoluble copper in the form of divalent copper. The leachate from the heap is pumped up, and the copper is removed from it by metallurgical techniques. The ferrous iron is reoxidized to ferric iron by bacteria in an aerated reactor, and the resulting "soup" is sprinkled onto the ore heaps again. Bacterial leaching is an economically important operation, and is responsible for 15–20% of all copper produced in the United States.

Manganese-oxidizing bacteria can be found among many of the genera shown in Table 10.**13**. The Gibbs free energy of the oxidation of divalent manganese to MnO_2 is negative; its magnitude is increased by the high insolubility of the end product. There is strong evidence that at least some of the manganese-oxidizing bacteria are chemolithoautotrophs. It has been suggested that the manganese-oxidizing bacteria may play a role at, or near, the oxic/anoxic interface of sediments and water

Table 10.**12** **Various reactions involved in the bacterial leaching process.** These reactions are carried out by a number of acidophilic bacteria, such as *Thiobacillus ferrooxidans*. Pyrite (FeS_2) is oxidized by the bacteria to sulfate and ferrous ions, which are subsequently oxidized to ferric ions. The ferric ions can chemically oxidize insoluble sulfides, such as pyrite or copper sulfide, whereby sulfur (S^o) and ferrous ions are produced. The ferrous ions and sulfur are biologically oxidized to sulfate and ferric ions, respectively. When the chemical oxidation step is involved, the oxidation of sulfide is said to occur via the "indirect" mechanism. Some sulfides can be directly oxidized by the bacteria. This is called the "direct" mechanism

Reactions
$2\,FeS_2 + 2\,H_2O + 7\,O_2 \rightarrow 2\,Fe^{2+} + 4\,SO_4^{2-} + 4\,H^+$
$4\,Fe^{2+} + O_2 + 4\,H^+ \rightarrow 4\,Fe^{3+} + 2\,H_2O$
$2\,Fe^{3+} + FeS_2 \rightarrow 3\,Fe^{3+} + 2\,S^0$
$2\,S^0 + 3\,O_2 + 2\,H_2O \rightarrow 2\,SO_4^{2-} + 4\,H^+$
$4\,Fe^{3+} + CuS \rightarrow 4\,Fe^{2+} + 2\,Cu^{2+} + S^0$

Table 10.**13** **Bacterial genera with manganese-oxidizing representatives.** After Nealson, K. H. (1983) The microbial manganese cycle. In: Krumbein, W. E. (ed.) Microbial geochemistry. Oxford: Blackwell; 191–221

Aeromonas	*Leptothrix*
Arthrobacter	*Metallogenium*
Bacillus	*Micrococcus*
Caulococcus	*Nocardia*
Clonothrix	*Oceanospirillum*
Flavobacterium	*Pedomicrobium*
Hyphomicrobium	*Pseudomonas*
Kuznetsovia	*Streptomyces*

bodies, where they are involved in the intricate cycles that shuttle electrons from sulfide via the manganese and iron cycle to oxygen (see Chapters 30–32). They may also be associated with the manganese nodules found on the sea bed.

Iron-oxidizing bacteria oxidize ferrous iron (Fe^{2+}) to ferric iron (Fe^{3+}). As ferric iron is a strong oxidizing compound, it can act as the end oxidant for other metals in their environment. The acidophilic iron oxidizers, which can often oxidize sulfur compounds as well, are used in microbial leaching, whereby insoluble metal compounds such as CuS can be dissolved and extracted from metal ores. The iron-oxidizing bacteria are also (partially) responsible for various forms of pollution, such as the acidification of soil and the production of ferric iron which, especially at elevated pH, leads to heavy precipitates of iron hydroxides and basic ferric sulfates (known as jarosites).

Further Reading

Balows, A., Trüper, H. G., Dworkin, M., Harder, W., and Schleifer, K. H. (eds.) (1992) The prokaryotes, 2nd edn. Berlin, Heidelberg, New York: Springer; Chapters 14 Kelly, D. P. (chemolithotrophs); 15 Aragno, M., and Schlegel, H. G. (Knallgas bacteria); 16 Robertson, L. A., and Kuenen, J. G. (colorless sulfur bacteria); 17 Bock, E., Koop, H. P., Ahlers, B., and Harms, H. (nitrifiers); 18 Lidstrom, M. E. (methylotrophs); 113 Bock, E., and Koops, H. P. (*Nitrobacter*); 114 Nealson, K. H. (manganese oxidizers); 115 Van Verseveld, H. W., and Stouthamer, A. H. (*Paracoccus*); 137 Koops, H. P., and Möller, U. C. (manganese oxidizers); 138 Kuenen, J. G., Robertson, L. A., and Tuovinen, O. H. (*Thiobacillus*, *Thiomicrospira*, and *Thiosphaera*); 166 Nelson, D. C. (*Beggiatoa*); 215 Felbeck, H., and Distel, D. L. (symbiotic lithotrophs); 217 Aragno, M. (thermophilic Knallgas bacteria); 218 la Rivière, J. W. M., and Schmidt, K. (conspicuous sulfur-oxidizing eubacteria); 236 Hanert, K. (iron and manganese oxidizers)

De Zwart, J. M. M., and Kuenen, J. G. (1992) C_1 cycle of sulfur compounds. Biodegradation 3: 37–59

Friedrich, B., and Schwartz, E. (1993) Molecular biology of hydrogen utilization in aerobic chemolithotrophs. Annu Rev Microbiol 47: 351–383

Kelly, D. P., and Kuenen, J. G. (1984) Ecology of the colorless sulphur bacteria. In: Codd G. A. (ed.) Aspects of microbial metabolism and ecology. London: Academic Press; 211–240

Kuenen, J. G., Robertson, L. A., and Van Gemerden, H. (1985) Microbial interactions among aerobic and anaerobic sulfur oxidizing bacteria. Adv Microbial Ecol 8: 1–59

Kuenen, J. G., Pronk, J. T., Hazeu, W., Meulenberg, R., and Bos, P. (1993) A review of bioenergetics and enzymology of sulfur compound oxidation by acidophilic thiobacilli. In: Torma, A. E., Apel, M. L., and Brierley, C. L. (eds.) Biohydrometallurgical technologies, vol. 2. Warrendale, Penn.: Minerals, Metals & Materials Society; 487–494

Kuenen, J. G., and Robertson, L. A. (1994) Combined nitrification-denitrification processes. FEMS Microbiol Rev 15: 109–117

Nealson, K. H., and Saffarini, D. (1994) Iron and manganese in anaerobic respiration: Environmental significance, physiology, and regulation. Annu Rev Microbiol 48: 311–343

Pronk, J. T., and Johnson, D. B. (1992) Oxidation and reduction of iron by acidophilic bacteria. Geomicrobiol J 10: 153–171

Rawlings, D. E., and Kusano, T. (1994) Molecular genetics of *Thiobacillus ferrooxidans*. Microbiol Rev 58: 39–55

Schlegel, H. G., and Bowien, B. (eds.) (1989) Autotrophic bacteria. Madison, Wisc.: Science Tech

Sources of Figures

1 Lane, D. J., Harrison, A. P., Stahl, D., Pace, B., Giovannoni, S. J., Olsen, G. J., and Pace, N. R. (1992) J Bacteriol 174:269–278

2 Nelson, D. C. (1989) Physiology and biochemistry of filamentous sulfur bacteria. In: Schlegel, H. G., and Bowien, B. (eds.) Autotrophic bacteria. Madison, Wisc.: Science Tech: 219–238

3 Hooper, A. H. (1989) Biochemistry of the nitrifying lithoautotrophic bacteria. In: Schlegel, H. G., and Bowien, B. Autotrophic bacteria. Madison, Wisc.: Science Tech: 239–265

4 Rawlings, D. E., and Kusano, T. (1994) Microbiol Rev 58: 39–55

11 Aerobic Respiration and Regulation of Aerobic/Anaerobic Metabolism

This chapter deals with the roles of oxygen in the life of aerobic, microaerobic, and facultatively anaerobic prokaryotes. Due to its wide-spread occurrence and favorable redox potential, molecular oxygen is the most important and preferred oxidant in metabolism. After a short consideration of some important chemical and physical properties of molecular oxygen, its function in catabolism, anabolism, and as a toxic agent will be discussed.

11.1.1 Properties of Molecular Oxygen

The atmosphere consists of 21% molecular oxygen. **Air-saturated water** contains rather low concentrations of dissolved oxygen (0.276 mmol/l at 20 °C) (Table 11.**1**). This amount of dissolved oxygen is sufficient for the oxidation of only 0.046 mmol/l of glucose to CO_2 and water. At increased temperature or high ionic strength, the solubility of O_2 in water is even lower. The O_2 supply in water is limited by the small diffusion coefficient of O_2 in water, which is four orders of magnitude lower than that in air. Therefore, under conditions of rapid O_2 consumption, the oxygen supply in water solution becomes limiting. To the cell interior, O_2 is supplied by diffusion through the cytoplasmic membrane. Phospholipid membranes do not represent diffusion barriers to O_2 and similar gases. Due to the small size of microorganisms, O_2-carrier proteins like hemoglobin apparently are not required in microorganisms. For growth of aerobic bacteria, the O_2 content in the medium is an important growth parameter. To study the effect of O_2 on bacterial growth, the bacteria can be grown at defined O_2 levels in an **oxystat**. In an oxystat, defined and constant O_2 levels are maintained by measuring the dissolved O_2 continuously with an O_2 electrode (Fig. 11.**1**) and by regulation of the O_2 supply.

Molecular oxygen largely consists of "**triplet oxygen**," which is in the ground state, that is, not activated. This form of molecular oxygen represents the substrate for aerobic bacteria. In aerobic metabolism, reactive and hence toxic oxygen species are produced, such as the **superoxide radical anion** ($O_2^{-}\cdot$), **hydrogen peroxide** (H_2O_2), or the **hydroxyl radical** ($\cdot OH$) by enzymic or non-enzymic redox reactions in the cell (Table 11.**2**). Normally, the concentrations of these compounds in cells are kept low in aerobic bacteria by detoxifying enzymes that degrade H_2O_2 and $O_2^{-}\cdot$. H_2O_2 is relatively stable and of limited toxicity, whereas $O_2^{-}\cdot$ has severe toxic effects. The most toxic product is the hydroxyl radical, which has a half-life of microseconds and reacts with most organic molecules, resulting in their destruction. In lipid peroxidation by $\cdot OH$ and $O_2^{-}\cdot$, peroxy radicals ($ROO\cdot$) are formed, which are able to destroy many other cell components and to disturb various cell functions.

Table 11.**1** **O_2 content under air saturation and diffusion in various media**

Medium		O_2 content	Diffusion coefficient ($cm^2\ s^{-1}$)
Air	(20 °C)	0.21 atm[a]	$2 \cdot 10^{-1}$
Water, air-saturated	(20 °C)	0.276 mmol/l	$2.3 \cdot 10^{-5}$
	(35 °C)	0.22 mmol/l	–
Ethanol	(20 °C)	1.35 mmol/l	–
Petroleum	(20 °C)	3.8 mmol/l	–
Phospholipid membrane	(20 °C)	–	$\approx 2 \cdot 10^{-5}$

[a] 1 atmosphere (atm) ≅ 1 bar = 705 Torr (mm Hg) = 10^5 Pa

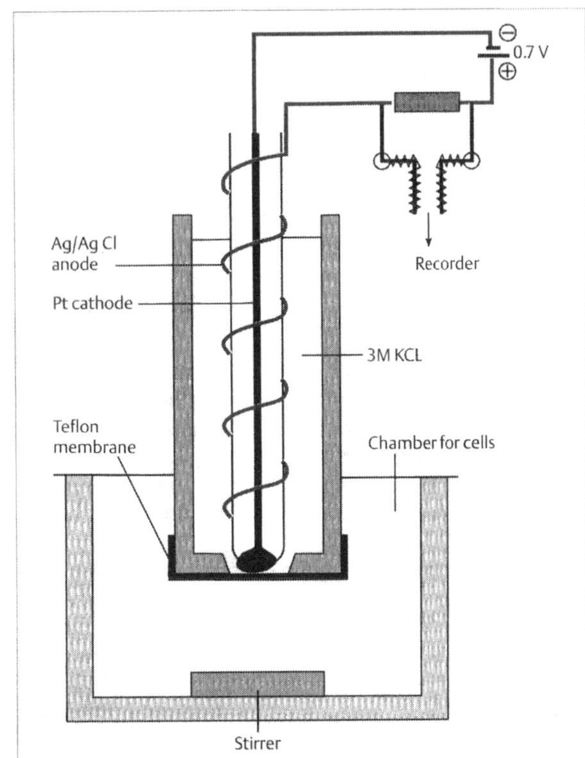

Fig. 11.1 **Clark oxygen electrode for measuring dissolved O_2.** The electrode consists of a platinum (Pt) cathode and a silver/silver chloride (Ag/AgCl) reference electrode (anode), which are polarized by applying 0.7 V. At the Pt cathode, O_2 is reduced to H_2O by electrons supplied by Ag oxidation. The current is proportional to the O_2 concentration at the Pt electrode and can be recorded. The electrode space is separated from the reaction chamber by a thin, O_2-permeable teflon membrane. Equal concentrations of O_2 in the reaction chamber and at the electrode are ensured by short diffusion distances and high diffusion rates through the membrane. The electrode is calibrated with air-saturated buffer and buffer made anoxic by addition of dithionite.

Anode reaction: $4\,Ag + 4\,Cl^- \longrightarrow 4\,AgCl + 4\,e^-$

Cathode reaction: $4\,H^+ + 4\,e^- + O_2 \longrightarrow 2\,H_2O$

Sum: $4\,Ag + 4\,H^+ + 4\,Cl^- + O_2 \longrightarrow 4\,AgCl + 2\,H_2O$

11.1.2 Oxygen Played a Major Role During the Evolution of Life

The early earth atmosphere was anoxic, and molecular oxygen is of biological origin. The major O_2-generating process is photosynthesis, which presumably arose about $2.7 \cdot 10^9$ years ago. After oxidation of reduced compounds, notably ferrous iron, and about 500 million years later, free O_2 appeared in the atmosphere. Since then, the O_2 content in the atmosphere increased slowly. Only since about 400 million years ago, atmospheric O_2 reached the present level, which is fairly constant. The appearance of O_2 made the **evolution of aerobic metabolism** possible. The efficiency of aerobic metabolism is assumed to have altered evolution drastically. Eukaryotic cells, for example, probably developed only after the appearence of aerobic metabolism. Nowadays, most of the O_2 turnover on earth is caused by aerobic life. O_2 generation by oxygenic photosynthesis (Reaction 11.1; $[CH_2O]$ stands for carbohydrates) and O_2 consumption by aerobic respiration (Reaction 11.2) are quantitatively the most important processes in O_2 turnover and are close to balance (Table 11.**3**). Other biotic and abiotic processes are quantitatively less important.

Table 11.**2** **Biologically important oxygen species and their generation by one-electron reduction steps**

Reaction	Species	E'_m (V)
$O_2 + e^- \rightarrow O_2^{-}\cdot$	Superoxide anion	−0.33
$O_2^{-}\cdot + e^- + 2\,H^+ \rightarrow H_2O_2$	Hydrogen peroxide	0.89
$H_2O_2 + e^- + H^+ \rightarrow \cdot OH + H_2O$	Hydroxy radical	0.38
$\cdot OH + e^- + H^+ \rightarrow H_2O$	Water	2.33
$O_2 + 4\,e^- + 4\,H^+ \rightarrow 2\,H_2O$		0.82

$$2\,H_2\mathbf{O} + CO_2 \longrightarrow [CH_2O] + H_2O + \mathbf{O_2} \quad (11.1)$$

$$[CH_2O] + \mathbf{O_2} \longrightarrow H_2\mathbf{O} + C\mathbf{O}_2 \quad (11.2)$$

Table 11.**3** **O_2-generating and O_2-consuming processes involved in O_2 turnover**

Processes	Amount (10^{12} mol O_2/year)
O_2 generation	
Oxygenic photosynthesis	15,000
Dissociation of H_2O (abiotic)	0.007
	≈15,000
O_2 consumption	
Respiration and aerobic metabolism	14,940
Methane oxidation (abiotic)	50
Weathering (C-, S-, FeO-containing minerals)	10
Burning (wood, coal, oil)	350
	≈15,350

11.2 Oxygen Fulfills Diverse Functions in Metabolism

Molecular oxygen serves many different functions in the metabolism of microorganisms. In aerobic or microaerobic bacteria, O_2 is used as the preferred electron acceptor in respiration. Aerobic respiration is coupled to the generation of an electrochemical membrane potential and to the phosphorylation of ADP. In some bacteria, such as O_2-tolerant anaerobes, O_2 is also utilized as an electron sink for the oxidation of NADH or other reduced compounds without direct energy conservation. A rection of this type is catalyzed by **lactate oxidase** (Reaction 11.3):

$$\text{L-Lactate} + O_2 \longrightarrow \text{Pyruvate} + H_2O_2 \qquad (11.3)$$

The enzyme is found in some lactic acid bacteria. Aerobes use O_2 as a co-substrate for many catabolic and anabolic reactions. The degradation of chemically inert substances such as alkanes and aromatic compounds under oxic conditions requires activation of the substrates with O_2 by monooxygenases or dioxygenases. In aerobes, some hydroxyl groups in cell constituents are derived from O_2 by hydroxylation reactions in biosynthesis. Molecular oxygen in the cells, however, is also the origin for toxic reaction products. Due to the multiple and adverse effects of O_2, bacteria respond by many reactions to O_2. Hence, molecular oxygen is an important signal for transcriptional regulation of catabolism and anabolism. Many bacteria also respond to favorable oxygen concentrations by positive aerotaxis, and to disadvantageous concentrations by negative aerotaxis (see Chapter 20.5.7).

11.2.1 A Great Variety of Enzymes and Cofactors React With Oxygen

Various types of enzymes catalyze redox reactions with molecular oxygen as a substrate or a product (Table 11.**4**). The enzymes contain prosthetic groups like metal ions, such as iron, copper, and manganese, heme groups, or flavins at the active site for the interaction with O_2. These compounds can react (slowly) also non-enzymically with O_2 if they are not bound to enzymes. The metal ions and heme groups are bound to the enzymes in a way that leaves one ligand of the metal ion free for binding and reaction with the O_2 molecule.

Enzymes catalyzing the transfer of electrons to O_2 are termed **oxidases** (Table 11.**4**). The enzymes are membrane-bound, such as the oxidases of aerobic respiratory chains, or soluble, such as glucose oxidase. H_2O or H_2O_2 can be the products. At the active sites of the respiratory oxidases, O_2 is bound by a binuclear heme–heme or a heme–Cu center and reduced to H_2O. **Oxygenases** catalyze the incorporation of O_2 into the substrates. **Dioxygenases** catalyze the cleavage of molecular oxygen with subsequent incorporation of both oxygen atoms into organic substrates. Important

Table 11.**4** **Types of bacterial enzymes and cofactors acting on molecular oxygen (substrate or product)**

Enzyme	Cofactor for interaction with O_2	Reaction
Oxidases		
Cytochrome *c* oxidase aa_3	Heme a_3, Cu_B	$4\ \text{Cyt}\ c_{red} + O_2 + 4\ H^+ \rightarrow 2\ H_2O + 4\ \text{Cyt}\ c_{ox}$
Quinol oxidase bo_3	Heme o_3, Cu_B	$2\ QH_2 + O_2 \rightarrow 2\ H_2O$
Glucose oxidase	FAD^+	$\text{Glucose} + O_2 \rightarrow \text{Gluconate} + H_2O_2$
Oxygenases		
Catechol dioxygenase	Fe^{3+}	Catechol + O_2 → *cis*, *cis*-Muconate
Methane monooxygenase	Binuclear Fe–Fe	$CH_4 + NADH + H^+ + O_2 \rightarrow CH_3{-}OH + NAD^+ + H_2O$
Luciferase (*Vibrio fischeri*)	FMN	$R{-}CHO + NAD(P)H + H^+ + O_2 \rightarrow NAD(P)^+ + H_2O + R{-}COOH + h\nu$
Miscellaneous (O_2 production)		
Catalase	Heme *b*	$H_2O_2 + H_2O_2 \rightarrow 2\ H_2O + O_2$
Superoxide dismutase	Fe^{3+} or Mn^{3+}	$O_2^{-}\cdot + O_2^{-}\cdot + 2\ H^+ \rightarrow H_2O_2 + O_2$
Water-splitting enzyme (Photosystem II)	Mn complex (4 Mn ions)	$2\ H_2O + 4\ h\nu \rightarrow O_2 + 4\ H^+ + 4\ e^-$

R, alkyl residue; Cyt, cytochrome; Q, ubiquinone

dioxygenases of aerobic bacteria are required for the cleavage of aromatic rings. **Monooxygenases** (hydroxylases) insert only one oxygen atom of O_2 into the substrate, the other is reduced to H_2O. Hydroxylases often contain heme *b* or non-heme iron at the active site. The O_2 is bound and reduced by the cofactor to generate a reactive oxygen species, which is required for the hydroxylation of the substrate. Many monooxygenases use NADH as the reductant. The monooxygenase luciferase from Gram-negative luminous bacteria contains FMN as a covalently bound cofactor. The reaction is coupled to light emission.

The enzymes that produce O_2 from various reactive oxygen species such as peroxides (H_2O_2, R-OOH) or $O_2^{-}\cdot$ are important detoxifying enzymes in many prokaryotes. The enzymes use mostly the same cofactors as the O_2-consuming enzymes. **Catalases** contain heme *b* as prosthetic group and catalyze the reduction of H_2O_2 to water with H_2O_2 as the electron donor. **Superoxide dismutase** dismutates the $O_2^{-}\cdot$ anion and contains ferric iron or Mn^{3+} as a cofactor. Mn ions play also an essential role in O_2 production from H_2O by the water-splitting enzyme of photosystem II in oxygenic photosynthesis.

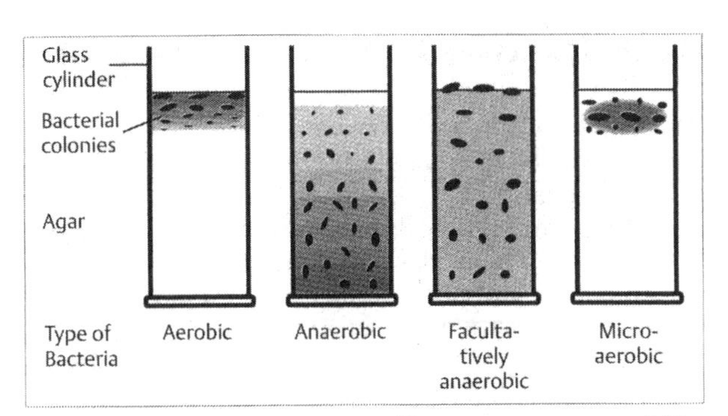

Fig. 11.2 **Different responses of bacteria to oxygen.** Different types of bacteria are suspended in molten soft agar containing appropriate growth substrates. After solidification of the agar and incubation, aerobic, anaerobic, facultative anaerobic, and microaerophilic bacteria grow in the agar zones as shown

11.2.2 The Availability of O_2 Is Important for Bacterial Ecology and Physiology

The significance of O_2 as a substrate on one hand and its toxic effects on the other determine the relationship of the bacteria towards O_2 (Fig. 11.**2**). The relationship to O_2 determines which biotopes are accessible to the various microorganisms. This can be demonstrated by suspending bacteria in a column of soft agar. Due to the slow diffusion of O_2, only the upper layers are oxic. **Aerobic** bacteria grow and form colonies only on or near the surface, whereas **obligate anaerobic** bacteria only grow in deeper layers. **Facultatively anaerobic** bacteria form colonies all over the agar column. **Microaerobic** bacteria are found in a distinct layer of appropriate O_2 tension within the agar.

Aerobic bacteria require O_2 for growth and metabolism (i.e., respiration), whereas anaerobic bacteria grow in the absence of O_2. Strictly anaerobic bacteria cannot use O_2 as a substrate for metabolism and are mostly inhibited or killed by O_2. However there are also **O_2-tolerant anaerobic** bacteria, such as lactic acid bacteria, which are able to tolerate O_2 at normal levels. Facultatively anaerobic bacteria are able to grow aerobically as well as anaerobically and show either an aerobic or anaerobic type of metabolism depending on the growth conditions. In many biotopes such as soil, microaerobic bacteria make up a considerable part of the bacterial population. They require O_2 for respiration, but can cope only with reduced O_2 tensions. This might be due to toxic effects of O_2 at higher O_2 tensions.

11.2.3 Bacteria Cope With Limiting and Changing O_2 Supply

Many bacteria have to cope with the situation that the supply of O_2 is limiting or that O_2 is not permanently supplied in their habitates.

Many environments are microoxic. In many biotopes, the available O_2 concentration is low because of the slow supply by diffusion from the atmosphere. Habitats with O_2 concentrations below 5% of air saturation, corresponding to a concentration of approx. 10 μM dissolved O_2 in water, are regarded as microoxic. A characteristic **microoxic habitat** exists in root nodules of legumes. Microoxic conditions of ≤ 1 μM O_2 are maintained by the plant to enable bacterial growth as well as N_2 fixation by symbiotic bacteria (*Rhizobium* or *Bradyrhizobium*).

Biotopes with **changing O_2 content** are found in great number. They exist at the interfaces between oxic and anoxic areas in aquatic habitats, in sediments, or in solid matter such as soil. In soil particles and other comparable biotopes, steep oxygen gradients exist. A few millimeters below an oxic surface, anoxic conditions can exist. The oxic/anoxic interfaces move because of changes in oxygen supply, such as reduced diffusion due to wetness, or because of microbial growth, which results in rapid O_2 consumption. Thus, even in oxic environments, microbial populations of high cell densities such as biofilms, microbial mats, or colonies, cannot be supplied sufficently with O_2 and become anoxic. The O_2 concentrations in and around a surface colony of bacteria can be measured with microelectrodes and presented in a contour map (Fig. 11.**3**). At already a few micrometers below the colony surface, anoxic conditions are found, and in the deeper zones anoxic or fermentative metabolism prevails. This can be demonstrated by the inclusion of pH indicator dyes in the agar, which indicate the production of acidic fermentation products. In aquatic habitats, another reason for changing O_2 supply is the alternating O_2 generation by oxygenic photosynthesis during the day-night cycle. In such rapidly changing microenvironments, facultatively anaerobic bacteria profit from their flexible metabolism, which allows growth under oxic, anoxic, or microoxic conditions.

Microaerobic bacteria are adapted to low O_2 tensions. Although high O_2 concentrations are inhibitory to microaerobic bacteria, these bacteria require O_2 for growth by aerobic respiration. The terminal oxidases of microaerobic bacteria have low apparent K_m values for O_2, which can be well below 1 µM. In many "facultatively" microaerobic bacteria, which can grow at normal and reduced O_2 levels, specific respiratory pathways with high-affinity terminal oxidases are expressed in microaerobic growth. Thus, *Rhizobium* and *Bradyrhizobium* strains express respiratory pathways during symbiotic growth in nodules different than during growth in soil. *Escherichia coli* also is able to express a high-affinity terminal oxidase (quinol oxidase *bd*) during growth at low O_2 tension. The microaerobic quinol oxidase *bd* is able to scavenge traces of O_2.

Microaerobic bacteria form a large and important physiological group that are difficult to grow because of their critical demands on the prevailing O_2 concentrations. One distinctive group of microaerobic bacteria is represented by the magnetotactic bacteria, such as *Aquaspirillum magnetotacticum*. These bacteria are capable of directed movement in a magnetic field ("**magnetotaxis**"), which has been suggested to be advantageous in the directed search of and migration toward microoxic zones.

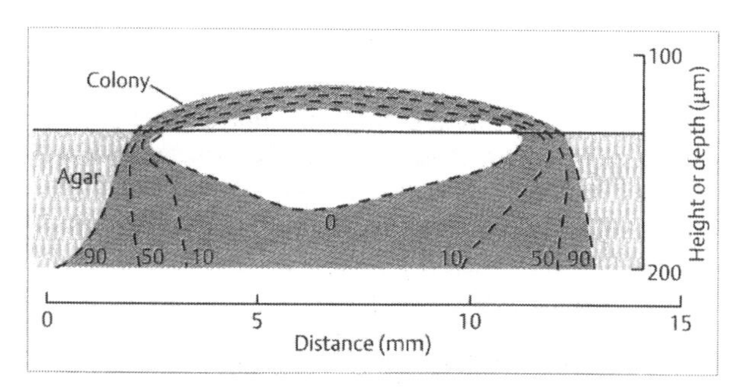

Fig. 11.**3** **Contour map of O_2 concentration in and around a surface colony of *Bacillus cereus* after growth on complex agar.** The O_2 concentrations were determined with an oxygen microelectrode. The numbers on the contour lines give the O_2 concentrations in %; 100% corresponds to air-saturated medium

Facultatively anaerobic bacteria are able to grow with or without O_2. Facultatively anaerobic bacteria conserve energy in anoxic growth either by fermentation, anaerobic respiration, anoxygenic photorespiration, or by combinations of these metabolic types. In most cases, a regulatory switch triggered by O_2 takes place, ensuring that only the preferred catabolic pathway is expressed. In aerobic and anaerobic growth, largely different metabolic pathways are observed. The switch affects expression of catabolic as well as anabolic pathways. In anaerobically growing *E. coli*, most citrate-cycle enzymes are repressed to 5–20% of the aerobic activities. Due to a nearly complete lack of oxoglutarate dehydrogenase, the citrate cycle in anaerobically grown *E. coli* is separated into an oxidative and a reductive branch terminating at oxoglutarate and succinyl-CoA (Fig. 11.**4**). Oxoglutarate and succinyl-CoA have anabolic functions and are required as precursors for glutamate and for some other syntheses. The reductive branch, starting from phospho*enol*pyruvate, serves also catabolic functions by supplying fumarate as an electron acceptor for fumarate reduction. Concurrent to the citrate cycle, synthesis of pyruvate dehydrogenase is repressed and that of pyruvate-formate lyase increased. By the repression of the citrate cycle, complete oxidation of acetyl-CoA to CO_2 is no longer possible, and acetate, ethanol, or related products have to be excreted.

In many denitrifying bacteria on the other hand, such as *Pseudomonas*, *Paracoccus*, and *Alcaligenes*, the citrate cycle is not repressed by the absence of oxygen and functions when nitrate is present. Here, quinol and NADH are reoxidized by nitrate respiration.

11.2.4 O_2 Is a Dominant Regulatory Signal

In facultatively anaerobic bacteria, the switch from aerobic to anaerobic metabolism requires a large change in catabolic and anabolic enzyme equipment. The shift mainly occurs at the transcriptional level by increasing or reducing the synthesis of enzymes of aerobic or anaerobic metabolism. In *E. coli* more than 150 genes that are positively or negatively regulated by O_2 are known. This corresponds to approximately 5% of the

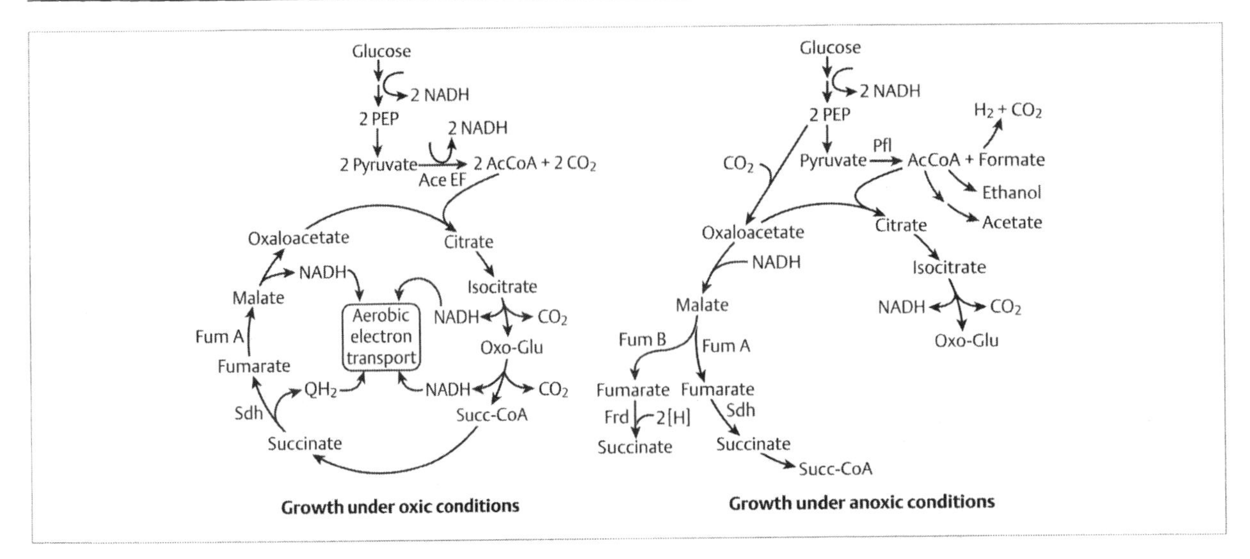

Fig. 11.**4** **Operation of the citrate cycle of *Escherichia coli* and of related reactions during growth of the bacteria under oxic or anoxic conditions.** The major pathways under oxic or anoxic conditions are given in red. Reactions catalyzed by isoenzymes under oxic and anoxic growth conditions are indicated: FumA and FumB, fumarases; Sdh, succinate dehydrogenase; Frd, fumarate reductase; AceEF, pyruvate dehydrogenase; Pfl, pyruvate : formate lyase

chromosomal genes. In facultatively anaerobic bacteria, therefore, O_2 is one of the most important regulatory signals.

In facultatively anaerobic bacteria, pathways with high ATP yield are preferred. Certain facultatively anaerobic bacteria can grow with various electron acceptors in addition to O_2, such as nitrate, nitrite, and fumarate. The synthesis of the enzymes of the corresponding catabolic pathways is usually regulated by the presence or absence of the electron acceptors according to a hierarchical principle.

In *E. coli*, for example, the selection of the pathways is determined by the electron acceptors, and acceptors of higher priority repress pathways of lower priority. Thus, O_2 represses all anaerobic pathways, and nitrate represses fumarate respiration and fermentation. The hierarchical rank is related to the $\Delta G^{\circ\prime}$ values of the respective pathways and the ATP yields. O_2 often also regulates the synthesis of the pertinent cofactors or coenzymes and of biosynthetic enzymes that require O_2 as a cosubstrate, such as oxygenases (Table 11.**5**). In some bacteria, such as *Thiosphaera*, however, electron acceptors such as O_2 and nitrate do not repress other catabolic processes, and different electron acceptors can be used concomitantly. Thus, nitrate respiration may proceed in the presence of O_2, and fumarate respiration in the presence of nitrate.

Bacteria contain various regulatory systems to sense O_2. The transcriptional regulation of metabolism in response to O_2 is effected in *E. coli* by three O_2-responsive regulators. The Arc (**a**erobic **r**espiratory **c**ontrol) regulatory system is mainly responsible for the regulation of the genes required for aerobic metabolism. The expression of anaerobic respiration, on the other hand, requires the transcriptional activator Fnr (**f**umarate **n**itrate reductase **r**egulation). Synthesis of some enzymes of fermentation, mainly of formate/hydrogen metabolism are regulated by FhlA (**f**ormate **h**ydrogen **l**yase regulation). Each of the regulatory systems senses O_2 in a different way. The operation and interaction of these regulatory systems is described in Chapter 21.

Bacteria show tactic responses in O_2 gradients. Many flagellated bacteria swim to areas of optimal O_2 concentrations, when they are exposed to local O_2 gradients. This response is called **aerotaxis** (see also Chapter 20.5.7). Positive aerotaxis is a movement toward areas of higher O_2 concentration, negative aerotaxis to areas of lower concentration. Aerobic and facultatively anaerobic bacteria show positive aerotaxis; anaerobic bacteria show negative aerotaxis. In areas of hyper-optimal O_2, aerobic bacteria may also respond negatively.

Apparently, the aerotactic response of facultative (and aerobic) bacteria is the primary response to maintain the preferred oxic conditions and metabolism. Only if this strategy is not successful, metabolic adaptations take place by a switch to anaerobic metabolism. Microaerobic bacteria are able, by positive and negative aerotactic response, to move to regions

Table 11.5 **Metabolic pathways and enzymes that are regulated by the presence or absence of O_2 in facultatively anaerobic bacteria**

Reaction/pathway	Regulatory effect of O_2 (negative or positive)	Bacterium (example)
Catabolism		
Aerobic respiration	Positive	*Escherichia coli*
Citrate cycle	Positive	*Escherichia coli*
Pyruvate dehydrogenase	Positive	*Escherichia coli*
Oxygenase (aromate degradation)	Positive	*Pseudomonas*
Anaerobic respiration	Negative	*Escherichia coli*
Fermentation	Negative	*Escherichia coli*
Photosynthesis	Negative	*Rhodobacter*
Pyruvate formate lyase	Negative	*Escherichia coli*
Anabolism/intermediary metabolism		
Glyoxylate cycle	Positive	*Escherichia coli*
N_2 fixation	Negative	*Klebsiella*

with optimal O_2 content and to concentrate in narrow zones. Upon changes in the O_2 gradient, the bacterial layers or clouds change their position. The signal transduction system and the mechanisms, which regulate the rotation of the flagellum motor in aerotaxis involve elements not found in the signal transduction pathway of chemotaxis.

11.3 Oxygen Makes High Energy Yields Possible

11.3.1 Oxygen Is the Superior Electron Acceptor

The significance of O_2 in the mineralization of organic matter is due to its widespread occurrence and to the high amount of free energy that can be conserved under oxic conditions (see Chapter 4). Because of the advantages of aerobic metabolism, probably more than 90% of organic matter is mineralized aerobically.

The high ATP gain of aerobic catabolism is mainly due to the electropositive nature of O_2 as an electron acceptor (see Table **4.1**), but also due to the more complete oxidation of substrates in the presence of O_2. In the aerobic metabolism of many bacteria, glucose or other substrates are oxidized to carbon dioxide and water. In the absence of O_2, more reduced end products are formed, sometimes even in the presence of suitable acceptors such as nitrate (Fig. 11.**5**). As a consequence, fewer electrons can be fed into the respiratory pathway, which results in a lower ATP yield. For example, in *E. coli*, 24 [H] are fed into aerobic respiration from glucose oxidation to CO_2. During nitrate respiration, glucose is incompletely oxidized to 2 acetyl-CoA and 2 CO_2, and only 8 [H] are available for nitrate respiration. In *Pseudomonas* strains and other denitrifying bacteria, which oxidize glucose to 6 CO_2 with nitrate as the acceptor, 24 [H] are transferred into the nitrate pathway with a concomitantly higher $\Delta G^{o\prime}$ value.

The $\Delta G^{o\prime}$ value of the redox reaction defines the maximal amount of charge or H^+ that can be transferred across the membrane in respiratory electron transport. Based on the calculations given in Chapter 4, maximally 12 $H^+/2\ e^-$ can be translocated in the electron transfer from NADH to O_2. In mitochondria and some bacteria, $H^+/2\ e^-$ ratios up to 10 have been measured. Many bacteria, however, contain electron transfer chains in which aerobic respiration is coupled to translocation of less H^+. As ATP synthesis requires 3 or 4 H^+/ATP, the amount of ATP that can be gained in respiration is directly related to the H^+/e^- ratios.

11.3.2 Different Types of Aerobic Respiratory Chains Are Known in Bacteria

The aerobic electron-transfer chains of bacteria can be classified into two types, one containing and the other lacking quinol : cytochrome *c* oxidoreductase (Fig. 11.**6**).

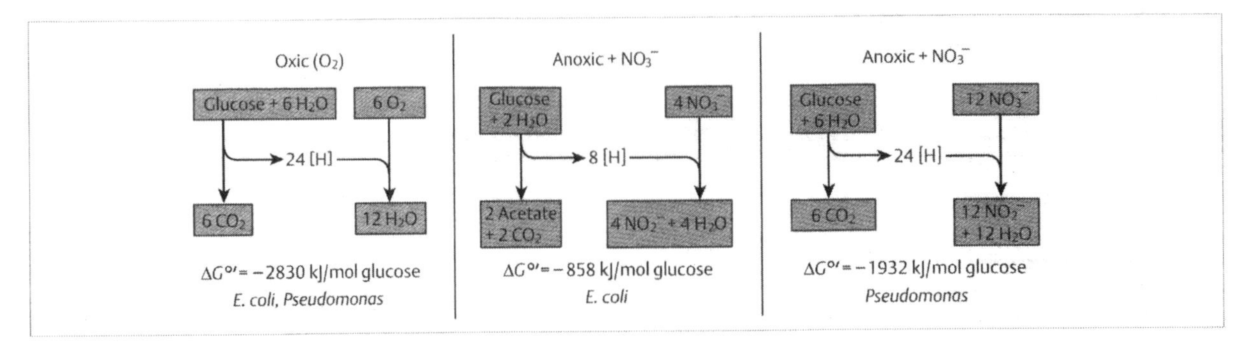

Fig. 11.**5** **Glucose oxidation during aerobic or nitrate respiration in *Escherichia coli* and *Pseudomonas*.** *E. coli* oxidizes glucose completely to CO_2 and water only during aerobic growth. The $G^{o\prime}$ values refer to the overall metabolism. For simplicity, it is assumed that nitrate is reduced only to nitrite

Both types branch at the level of the quinones. The former consists of three oxidoreductases that are linked by two redox mediators, quinone and cytochrome *c*. Cytochrome *c* is membrane-associated and links the oxidoreductases by diffusion. Unlike the mitochondrial cytochrome *c*, the protein from some bacteria contains a lipophilic membrane anchor. The terminal reductase is cytochrome *c* oxidase. In the alternative chains, quinol is oxidized by a quinol oxidase. Many bacteria contain more than one oxidase, which often are expressed under different growth conditions. The oxidases differ with respect to their affinity for O_2, sensitivity to inhibitors, and proton-pumping activity. Oxidases with high affinity for O_2 are synthesized under conditions of low O_2 concentrations.

The two types of aerobic respiratory chains can be differentiated by means of the **oxidase test** that has been used for long time as an empirical taxonomic criterion of aerobic bacteria. In the oxidase test, the oxidation of *N,N*-dimethyl-*p*-phenylenediamine ("TMPD") or dichlorophenol-indophenol by the bacteria is tested. The dyes are colorless in the reduced state and become colored upon oxidation. Dye oxidation depends on the presence of a cytochrome *c* that is sufficiently electropositive to serve as acceptor and that is accessible to the dyes because of the periplasmic location. In the respiratory chain, these cytochromes *c* represent the donor substrates of bacterial cytochrome *c* oxidases.

Bacteria can feed electrons from a wide variety of donor substrates into the respiratory chain by means of specific dehydrogenases (Fig. 11.**6**). Some dehydrogenases occur as **isoenzymes**, that is, there are two or more enzymes that catalyze the same reaction, but comprise different subunits and regulatory properties and may use different cofactors. Thus, *sn*-glycerol 3-phosphate is oxidized in the facultatively anaerobic *E. coli* by either of two dehydrogenases, which are encoded by different genes. One enzyme (*glpD* gene product) operates during aerobic growth, feeding electrons to the aerobic respiratory chain. The other enzyme is encoded by the *glpABC* genes. It is produced during growth under

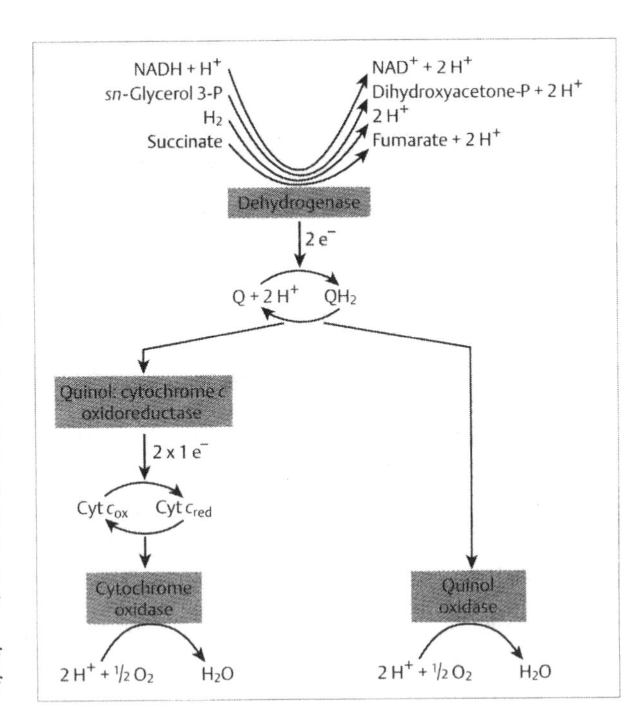

Fig. 11.**6** **Composition of bacterial aerobic respiratory chains with and without a quinol : cytochrome c oxidoreductase.** Various substrates for primary dehydrogenases are given. The quinones and cytochrome *c* collect the electrons from the various sources and channel them to the oxidases. Q, ubiquinone; Cyt, cytochrome

anoxic conditions and takes over the same function in anaerobic respiration. Both enzymes are membrane-bound and contain FAD as a prosthetic group.

For the **oxidoreductases** of the respiratory chains, different designations are used, which refer either to the donor or the acceptor substrate. The enzymes oxidizing the primary donor mostly are termed **dehydrogenases**, the enzymes reducing the electron acceptor are called (terminal) **reductases**, or **oxidases** if O_2 is the acceptor. Hence, Reaction 11.4 is catalyzed by succinate:Q oxidoreductase (systematic name) or succinate dehydrogenase (recommended name), but can also be regarded as a quinone reductase. The enzyme of Reaction 11.5 is quinol oxidase.

$$\text{Succinate} + Q \longrightarrow QH_2 + \text{fumarate} \qquad (11.4)$$
$$QH_2 + 0.5\,O_2 \longrightarrow Q + H_2O \qquad (11.5)$$

Quinol oxidases and cytochrome *c* oxidases are functionally related. Three major types of oxidases are known (Fig. 11.7):

1. The **cytochrome *c* oxidases.** The best-characterized enzyme of this type stems from *Paracoccus denitrificans*. The enzyme functions very similarly as the mitochondrial enzyme. It consists of two catalytic subunits (I and II). Subunit I contains two heme *a* and one copper ion and operates as a proton pump with $2\,H^+/2\,e^-$. In addition, 2 negative charges are transferred in the opposite direction, resulting in an overall ratio of 4 charges (equivalent to $4\,H^+$)/$2\,e^-$. Subunit II carries two copper ions and accepts electrons from cytochrome *c*.
2. The copper-containing **quinol oxidases.** The prototype of this group is the quinol oxidase bo_3 of *E. coli*. The enzyme contains catalytic subunits I and II, homologous to cytochrome *c* oxidases. Two heme groups (heme *b*, heme *o*) and one copper ion are found in subunit I. O_2 reduction and proton pumping are very similar to that of cytochrome *c* oxidase. Subunit II carries no copper, but is required for the electron transfer from the quinol to subunit I. By the reaction of quinol oxidase bo_3, a net translocation of $4\,H^+/2\,e^-$ takes place. $2\,H^+/2\,e^-$ are achieved by the operation of the enzyme as a proton pump. Additionally, $2\,H^+/2\,e^-$ are due to the orientation of the substrate sites for quinol oxidation and H_2 formation. The cytochrome *c* oxidases and the copper-containing quinol oxidases are termed heme copper respiratory oxidases.
3. The quinol oxidase *bd* from *E. coli* represents a different type of enzyme. It carries no copper and consists of two subunits with few sequence similarities to the other enzymes. The electrons from quinol are delivered via one heme *b* to a binuclear heme *b*/heme *d* center with the active site for O_2 reduction. The enzyme does not operate as a proton pump. However, a net H^+ translocation of $2\,H^+/2\,e^-$ results

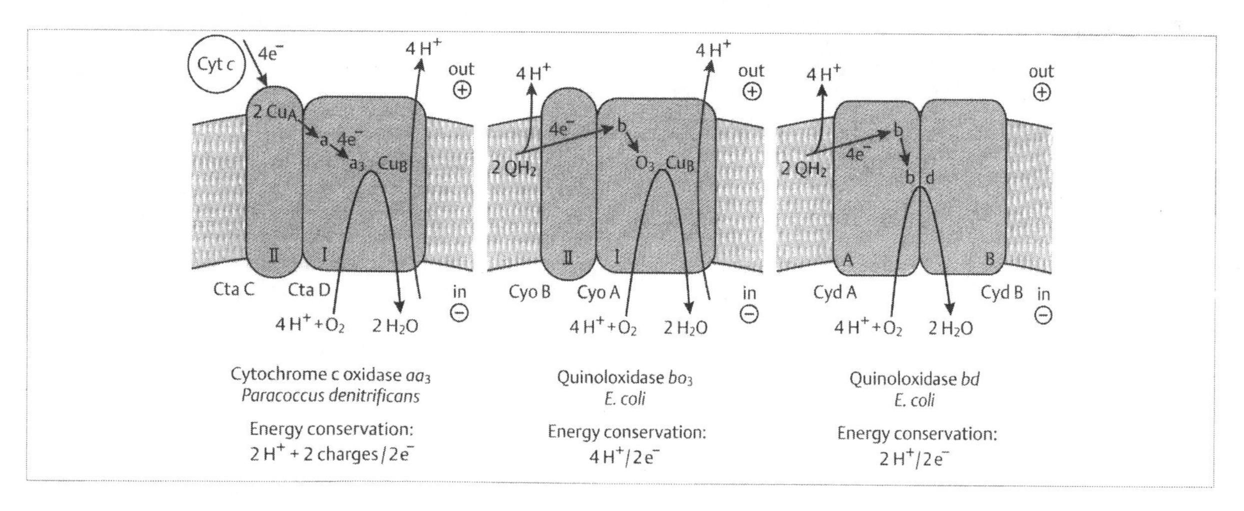

Fig. 11.7 **Types of bacterial cytochrome *c* oxidases and quinol oxidases.** Only the catalytic subunits (I and II, A and B) and the corresponding gene products in *Paracoccus denitrificans* or *Escherichia coli* are given. Subunits I and II of cytochrome *c* oxidase and of quinol oxidase bo_3 show strong sequence similarities, a, b, o, and d represent heme *a*, *b*, *o*, and *d*. The heme group interacting with O_2 is marked by subscript "3". Cytochrome *c* oxidase and quinol oxidase *bo* are heme-copper oxidases. Q, Ubiquinone; Cyt, cytochrome

from the location of quinol oxidation and O_2 reduction on opposite sides of the membrane.

Box 11.1 Three basic mechanisms are used for the generation of a proton motive force (Δp) by aerobic respiration. The number of protons that are translocated in aerobic respiration depends on the enzymes constituting the respiratory chains. Three mechanisms can be involved (see Chapter 4): (1) redox loops, i.e., redox chains, in which the sites for H^+ uptake and release are on opposite sides of the cytoplasmic membrane, (2) conformational H^+ pumps (Fig. 11.**8**), and (3) Q cycles (Fig. 11.**9**). In the respiratory chain of *Paracoccus denitrificans* that catalyzes NADH oxidation by O_2, the topology of the enzymes and the mechanism of Δp generation are probably very similar to those of mitochondria (Fig. 11.**10**). In respiratory chains of this type, $6\,H^+/2\,e^-$ are contributed from the H^+-pumping NADH dehydrogenase and cytochrome *c* oxidase. The rest ($4\,H^+/2\,e^-$) stems from H^+ translocation by the Q cycle associated with the quinol : cytochrome *c* oxidoreductase.

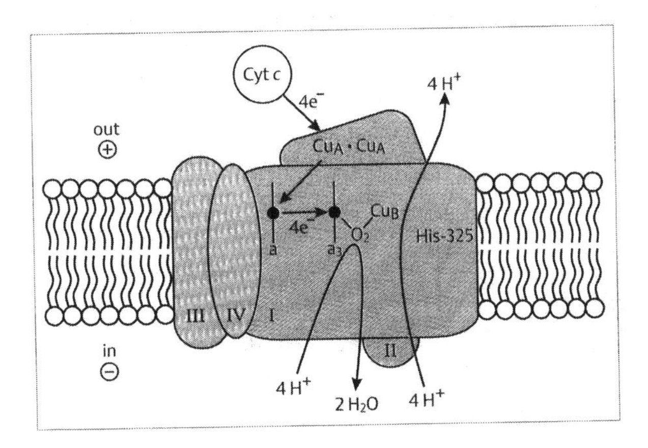

Fig. 11.**8** **Function of cytochrome *c* oxidase from *Paracoccus denitrificans* as a redox-driven proton pump: pathways of electrons and protons.** The enzyme has been well-characterized by X-ray crystallography and biochemical analyses. It is composed of four subunits (I– IV); only subunits I and II are catalytically active. The binuclear center heme a_3–Cu_B is the active site for O_2 reduction. Cu_B, heme *a*, heme a_3 are located in the periplasmic half of subunit I. The binding site for cytochrome *c* is close to the binuclear Cu_A center in subunit II at the periplasmic side of the membrane. The pathway of the electrons is $Cu_A \cdot Cu_A \rightarrow$ heme $a \rightarrow$ heme a_3. The chemical protons required for H_2O formation and the pumped protons ($4\,H^+/O_2$) are delivered via different pathways in the protein. A histidine residue close to Cu_B promotes transport of the pumped protons to the periplasmic part of the pathway in a manner directly linked to O_2 reduction

11.3.3 *Paracoccus denitrificans* Contains Cytochrome *c* Oxidases and Quinol Oxidases

P. denitrificans contains aerobic respiratory chains with and without quinol : cytochrome *c* oxidoreductase (Fig. 11.**11**). The former is very similar to the mitochondrial respiratory chain. The similarity refers to the enzyme equipment and the catalytic mechanism of the enzymes. The main enzymes, namely NADH dehydrogenase, quinol : cytochrome *c* oxidoreductase, and cytochrome oxidase aa_3 show the same composition of prosthetic groups and catalytic subunits and the same mechanism of H^+-translocation as the enzymes from mitochondria. The overall H^+/e^- stoichiometry amounts to $10\,H^+/2\,e^-$. The enzymes from *Paracoccus*, however, are composed only of the catalytically essential subunits, whereas the mitochondrial enzymes contain a large number of additional subunits of mostly unknown function. Thus NADH dehydrogenase, quinol : cytochrome *c* oxidoreductase, and cytochrome oxidase

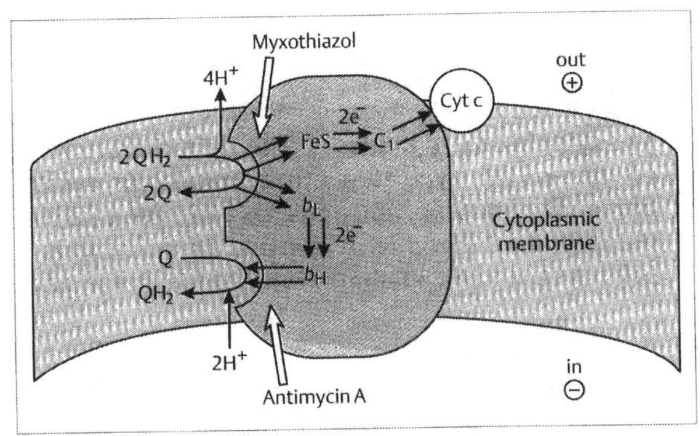

Fig. 11.**9** **Proton translocation by a quinone (Q) cycle as catalyzed by quinol : cytochrome *c* oxidoreductases.** The Q cycle was suggested for quinol : cytochrome *c* oxidoreductase from mitochondria and most likely operates in a similar way in the enzyme of *Paracoccus denitrificans* and other bacteria. The electron flow is depicted in red. It starts with $4\,e^-$ from 2 quinols (QH_2) and branches to transfer $2\,e^-$ to cytochrome *c* and $2\,e^-$ to another Q, which becomes reduced to QH_2. Two different active sites for QH_2 oxidation and Q reduction are present. Oxidation of $2\,QH_2$ results in the release of $4\,H^+$ to the periplasmic side, oxidation of Q in the uptake of $2\,H^+$ from the cytoplasmic side. The electron transfer in the enzyme is effected by 2 heme *b* groups with low (b_L) and high (b_H) midpoint potential. Reaction of the quinones at the two active sites can be specifically inhibited by myxothiazol or antimycin A, respectively. The partial and overall reactions are:

$$2\,QH_2 + 2\,\text{Cyt}\,c_{ox} \longrightarrow 2\,Q + 2\,\text{Cyt}\,c_{red} + 4\,H^+ + 2\,e^-$$
$$Q + 2\,e^- + 2\,H^+ \longrightarrow QH_2$$
$$QH_2 + 2\,H^+ + 2\,\text{Cyt}\,c_{ox} \longrightarrow Q + 2\,\text{Cyt}\,c_{red} + 4\,H^+$$

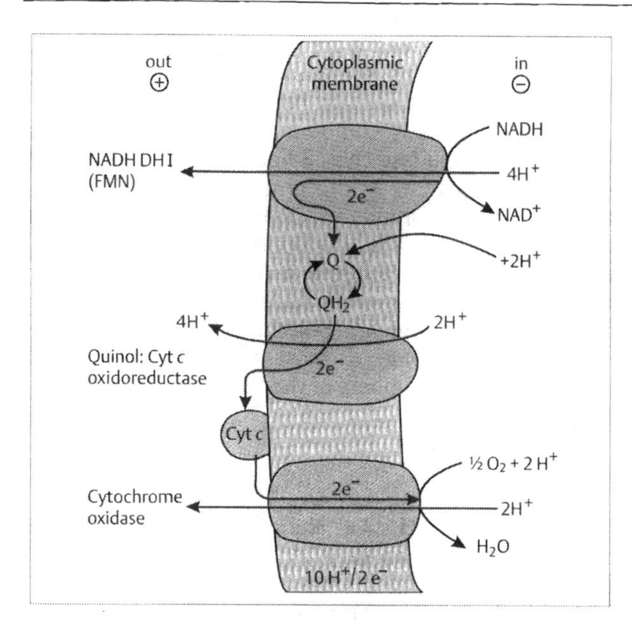

Fig. 11.**10** **Topology of aerobic respiratory enzymes of mitochondria and reactions contributing to the generation of a proton motive force (Δp).** The organization and function of aerobic respiration of *Paracoccus denitrificans* is assumed to correspond to that from mitochondria. Cyt, cytochrome; DHI, dehydrogenase I; Q, ubiquinone

aa_3 are composed of 14, 4, and 3 subunits in *P. denitrificans*, but of 39, 11, and 13 subunits, respectively, in the mitochrondrial enzymes. The NADH dehydrogenase contains FMN as prosthetic group and couples NADH oxidation to H^+ translocation. This contrasts to many bacterial NADH dehydrogenases, which carry FAD and do not operate as a H^+ pump. The cytochrome oxidase aa_3 from *P. denitrificans* was the first respiratory proton-pumping enzyme from which the three-dimensional structure was resolved by X-ray crystallography at high resolution. Detailed information on the coupling of the redox reaction and proton pumping is available (Fig. 11.**8**).

The close similarity of the *Paracoccus* and mitochondrial electron transport chains supports the endosymbiont hypothesis. This hypothesis assumes that an ancestor, which belongs to the same subclass of proteobacteria as *P. denitrificans*, invaded a eukaryotic cell as an endosymbiont and represents the progenitor of mitochondria. This hypothesis is also supported by close phylogenetic relationships between mitochondria and *Paracoccus* based on 16S rRNA sequences.

Two more oxidases are found in the bacterium which also operate as proton pumps. One is a cytochrome c oxidase, the other a quinol oxidase that is expressed under low O_2 supply. Methanol ($E_m' \approx -0.2$ V) or methylamine can also be used as electron donors for respiration (Fig. 11.**11**). The corresponding dehydrogenases are located in the periplasm and supply electrons to the chain at the level of cytochrome *c*.

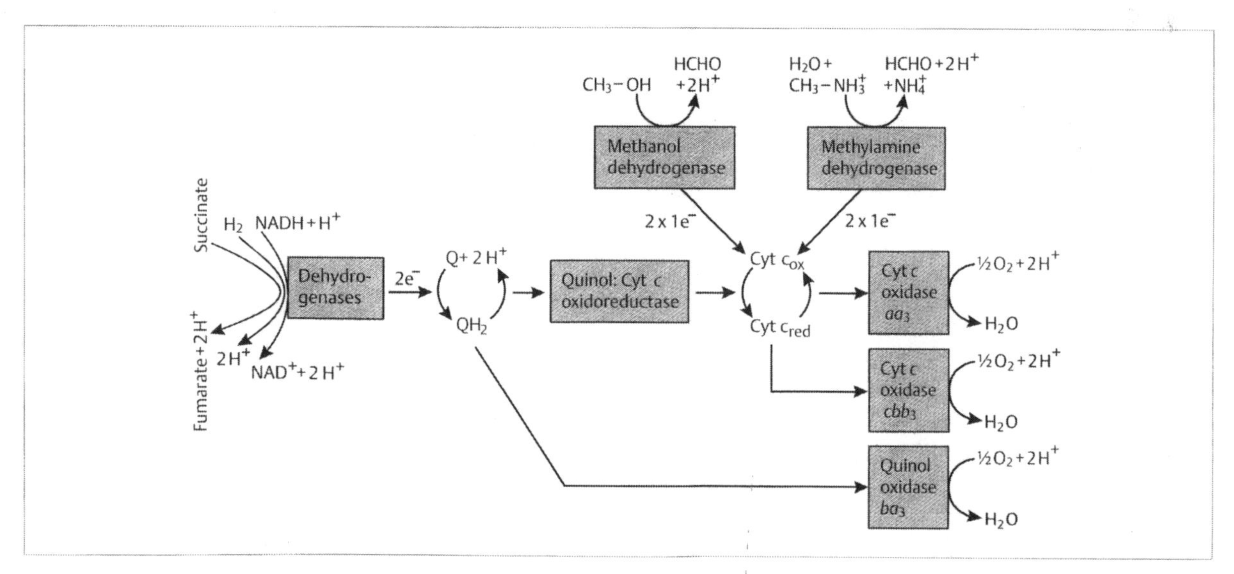

Fig. 11.**11** **Aerobic electron transport chains and enzymes of *Paracoccus denitrificans*.** Electrons from NADH, H_2, and succinate are delivered by the dehydrogenases at the level of ubiquinone (Q), from methanol and methylamine at the level of cytochrome c (Cyt c). Methanol dehydrogenase, methylamine dehydrogenase and cytochrome *c* are located in the periplasm, the other enzymes in the cytoplasmic membrane

11.3.4 *Escherichia coli* Contains Only Quinol Oxidases

In *E. coli* grown under oxic conditions, no *c*-type cytochromes are present. The electrons are transferred from ubiquinol to O_2 by two quinol oxidases, quinol oxidase bo_3 and *bd* (Figs. 11.7 and 11.**12**). In cells grown with high aeration, quinol oxidase bo_3 dominates, and is replaced by quinol oxidase *bd* during growth with limiting O_2 supply. Quinol oxidase *bd* has a distinctly higher affinity for O_2 than quinol oxidase bo_3. The adaptation enables *E. coli* to scavenge trace amounts of O_2 under microoxic conditions, when O_2 becomes limiting.

NADH, formate, succinate, *sn*-glycerol 3-phosphate, lactate, and pyruvate can be used as electron donors by *E. coli*. Each of the dehydrogenases acting on one of the substrates transfers the electrons to the universal acceptor, ubiquinone (Figs. 4.**9** and 11.**12**). Most of the substrates are also used as electron donors for anaerobic respiration; however, different dehydrogenases may be expressed under anoxic conditions. The two NADH dehydrogenases of *E. coli* differ in energy conservation. NADH dehydrogenase I is composed of 14 subunits, contains FMN, and is similar to the H^+-translocating enzyme of mitochondria and of *Paracoccus denitrificans* in composition and function. It couples NADH oxidation to H^+ translocation (4 H^+/2 e^-). NADH dehydrogenase II is made up of one subunit, contains FAD, and does not operate as a H^+ pump.

11.3.5 The Aerobic Respiratory Chains of Many Bacteria Are Diverse

The aerobic respiratory chains of most aerobic bacteria consist of more than one terminal oxidase. Quinol oxidases or cytochrome *c* oxidases are present in various combinations and with a great variability with respect to the heme groups present (Table 11.**6**). Thus, the quinol oxidases can contain heme *a*, heme *b*, heme *o*, and heme *d* in many combinations. In some enzymes such as quinol oxidase cbb_3, a heme *c* group within the enzyme replaces Cu_A as the primary electron acceptor from the quinol. A similar variability with respect to the types of heme groups is also found for the cytochrome *c* oxidases. Also in this enzyme, heme *c* can be found as a third heme group. However, here, heme *c* is present in addition to Cu_A and accepts the electrons from the diffusible cytochrome *c*. Most of the enzymes are homologous and members of the family of heme copper oxidases.

11.3.6 In Bacteria, H^+/2 e^- Ratios of Aerobic Respiration Can Vary

In *E. coli*, the H^+/2 e^- ratio of aerobic respiration with NADH can vary greatly depending on the enzymes involved (Fig. 11.**13**). In aerobically grown *E. coli*, a maximum ratio of 8 H^+/2 e^- can be achieved. An amount of 6 H^+/2 e^- is delivered by NADH dehydrogenase I and quinol oxidase bo_3. In addition, 2 H^+/2 e^- are translocated because of the different orientation of the active sites for Q reduction and oxidation (redox loop). Under microoxic conditions, quinol oxidase bo_3 is replaced by quinol oxidase *bd* with a lower H^+/2 e^- ratio. Thus, the ability to grow with trace amounts of O_2 is payed for with lower H^+/e^- ratios. If NADH dehydrogenase II is used under these conditions, the overall H^+/2 e^- ratio can be as low as 2 H^+/2 e^-. *E. coli* and other bacteria, therefore, are able to adapt the efficiency of energy conservation by the selection of enzymes. Hence, the amount of ATP gained in respiration not only depends on the $\Delta G^{\circ\prime}$ value of the redox

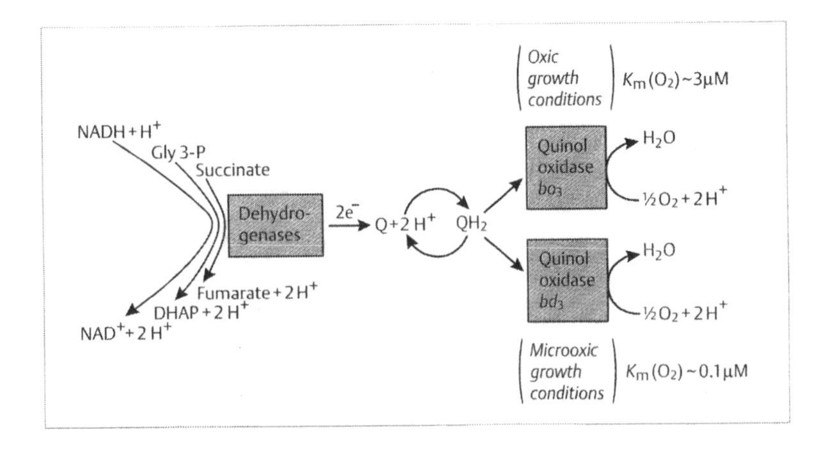

Fig. 11.**12** **Schematic representation of oxidases in the aerobic electron transport of *Escherichia coli*.** Gly 3-P, *sn*-glycerol 3-phosphate; DHAP, dihydroxyacetone phosphate; Q, ubiquinone

Table 11.6 **Diversity of aerobic respiratory oxidases in various bacteria.** The quinol and cytochrome *c* oxidases present in various bacteria are shown. The enzymes are designated according to the heme groups present. Cyt, cytochrome; Q, ubiquinone; MK, menaquinone

	Quinol oxidases	Cyt c oxidases	Quinones
Escherichia coli	Cyt bo_3, Cyt bd_3	Absent	Q, MK
Paracoccus denitrificans	Cyt ba_3	Cyt aa_3, Cyt cbb_3	Q
Bacillus subtilis	Cyt aa_3	Cyt caa_3	MK
Bradyrhizobium japonicum and *Rhodobacter sphaeroides*	Cyt bb_3	Cyt aa_3, Cyt cbb_3	Q

reaction, but also on the mechanisms that are available for energy conservation.

The aerobic respiratory chains of bacteria can be of two types, the quinol oxidase pathway and the quinol : cytochrome *c* oxidoreductase plus cytochrome *c* oxidase pathway. The latter is similar to the mitochondrial respiratory chain. Cytochrome *c* oxidase and many quinol oxidases are related heme-copper enzymes. Many bacteria contain more than one oxidase, which can be of different types. The $H^+/2\,e^-$ ratio in different pathways varies and depends on the enzyme composition.

11.3.7 Nitrite or Fe^{2+} as Electron Donors Require Reverse Electron Transport for NAD Reduction

Nitrobacter and *Thiobacillus ferrooxidans* use nitrite and Fe^{2+}, respectively, as electron donors for aerobic respiration. The midpoint potentials of nitrite/nitrate ($E_m{}' = +0.42$ V) and Fe^{2+}/Fe^{3+} ($E_m' = +0.78$ V at pH 2, and $+0.2$ V at pH 6) are positive compared to most other donors. As a consequence, the corresponding $H^+/2\,e^-$ ratios and ATP yields (<1 ATP/nitrite) are

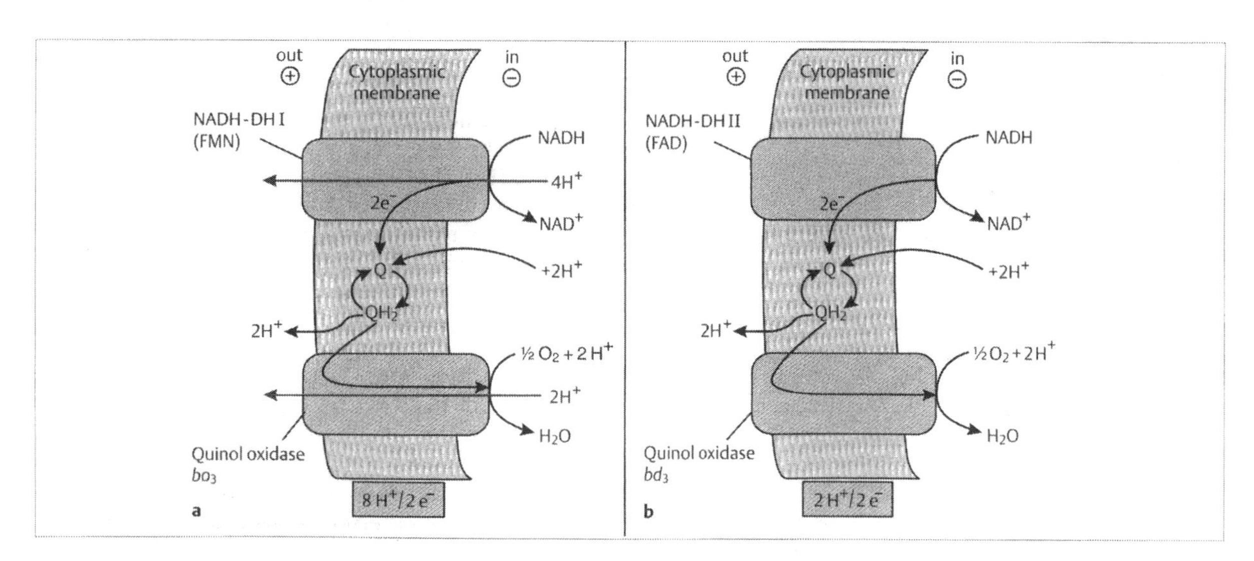

Fig. 11.**13a, b** **Variation of H^+/e^- ratio in *Escherichia coli* by the use of alternative enzymes.** The topology of aerobic electron transport enzymes from NADH to O_2 and the generation of a proton motive force (Δp) is shown. Under oxic growth conditions, mainly quinol oxidase bo_3 is present, under microoxic conditions mainly quinol oxidase *bd*. The alternative NADH dehydrogenases are present under either condition. NADH dehydrogenase I contains FMN as a cofactor and operates as a proton pump ($4\,H^+/2\,e^-$). NADH dehydrogenase II is not a proton pump and contains FAD.
a If electron transport from NADH to O_2 is exerted by the H^+-translocating NADH dehydrogenase I (DHI) and quinol oxidase bo_3, the $H^+/2\,e^-$ ratio is 8.
b Use of NADH dehydrogenase II (DHII) and quinol oxidase *bd* result in a $H^+/2\,e^-$ ratio of 2. Under most conditions, both of the alternative enzymes will be used to varying extents, resulting in intermediate $H^+/2\,e^-$ ratios. Q, Ubiquinone

low. *Nitrobacter* grows at the expense of the following reaction (Reaction 11.6):

$$NO_2^- + 1/2\ O_2 \longrightarrow NO_3^-$$
$$\Delta G^{\circ\prime} = -82\ \text{kJ/mol nitrite} \qquad (11.6)$$

The oxidation of nitrite by O_2 is catalyzed by an electron transport chain consisting of nitrite oxidase and cytochrome *c* oxidase which are presumably linked by a diffusible cytochrome *c* ($E_m' = +0.2$ V). Nitrite oxidase is similar to the nitrate reductase known from respiratory nitrate reduction. However, nitrite oxidase uses cytochrome *c* as electron acceptor, whereas nitrate reductase accepts electrons from quinols.

Nitrobacter grows autotrophically and assimilates CO_2 by means of the Calvin–Benson–Bassham cycle. The electrons for NAD(P) reduction, which is required for CO_2 fixation and biosyntheses, stem from nitrite. This reaction is extremely endergonic since the donor nitrite is 0.74 V more positive than the acceptor, NAD(P). The reaction is driven by the Δp across the membrane (Reaction 11.7):

$$NO_2^- + NAD(P) + H_2O \xrightarrow{H_o^+ \rightarrow H_i^+} NO_3^- + NAD(P)H + H^+ \qquad (11.7)$$

Assuming a Δp of 0.18 V, at least 5 H^+/e^- are required to drive the reduction of NAD(P) with NO_2^- as the electron donor. This process is called **reversed electron transport**. The reaction is catalyzed by a respiratory chain that comprises nitrite oxidase, cytochrome *c*, quinol:-cytochrome *c* oxidoreductase, ubiquinone, and a H^+-pumping NADH dehydrogenase. The reversed electron transport relies on the reversibility of the redox and H^+-translocation reactions of the enzymes and their coupling. Thus, in *Nitrobacter*, the Δp generated by aerobic respiration with nitrite has to drive ATP synthesis as well as NAD(P) reduction. Both processes are inhibited by uncouplers. The same requirement for reversed electron transport exists for other autotrophic bacteria that use electropositive electron donors, such as Fe^{2+} or Mn^{2+} (see Chapter 10).

11.4 O_2 Can Serve as a Co-substrate for Metabolism

Molecular oxygen is required as a co-substrate for many catabolic or anabolic reactions in aerobically growing bacteria. Anaerobes either have developed O_2-independent mechanisms or are incapable of accomplishing such reactions (see Chapter 9.13).

11.4.1 Some Catabolic Processes Require O_2 as Co-substrate

Hydrocarbons cannot serve directly as electron donors for dehydrogenases. They have to be activated by monooxygenases to yield alcohols, which can be further degraded. In the absence of O_2, therefore, alkanes, alkenes, and aromatic compounds are not degraded by most microorganisms, even if an alternative electron acceptor such as nitrate is available. This inability is due to the requirement for O_2 in the hydroxylation reactions, which are the first steps in the degradation of these substrates (Fig. 11.**14**). A similar situation is observed in nitrification (oxidation of ammonia) by *Nitrosomonas*. Here, NH_3 has to be hydroxylated by ammonia monooxygenase to yield hydroxylamine. Hydroxylamine is oxidized by a dehydrogenase of the aerobic respiratory pathway (see Chapter 10).

The chemical stability of aromatic compounds also requires activation with O_2. In aerobic metabolism, the introduction of hydroxyl residues by monooxygenases

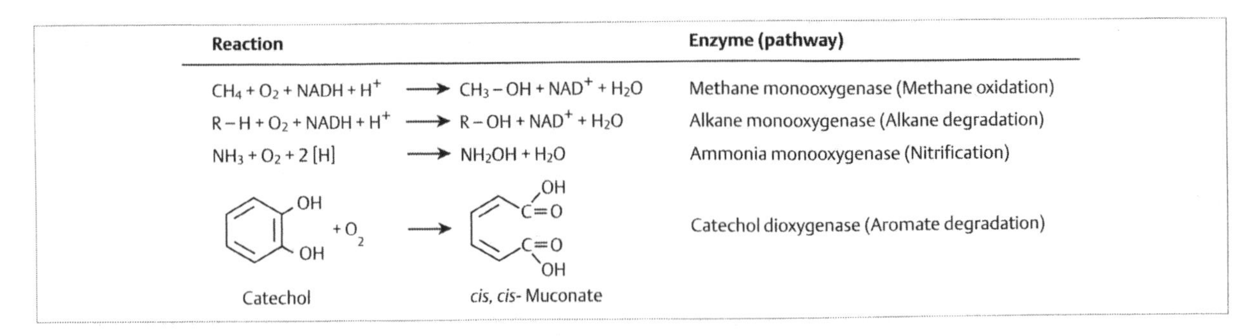

Reaction	Enzyme (pathway)
$CH_4 + O_2 + NADH + H^+ \longrightarrow CH_3-OH + NAD^+ + H_2O$	Methane monooxygenase (Methane oxidation)
$R-H + O_2 + NADH + H^+ \longrightarrow R-OH + NAD^+ + H_2O$	Alkane monooxygenase (Alkane degradation)
$NH_3 + O_2 + 2\,[H] \longrightarrow NH_2OH + H_2O$	Ammonia monooxygenase (Nitrification)
Catechol $+ O_2 \longrightarrow$ *cis, cis*-Muconate	Catechol dioxygenase (Aromate degradation)

Fig. 11.14 **Examples of catabolic reactions that require O_2 as a co-substrate**

or dioxygenases destabilizes the aromatic ring for cleavage by a dioxygenase. In anaerobic bacteria, ring cleavage is effected by a different, reductive pathway. Degradation of lignin, which represents an abundant polymeric aromatic compound, can be accomplished only with molecular oxygen as a co-substrate (see Chapter 9.13).

11.4.2 O_2 Is a Co-substrate for Biosynthetic Reactions

In aerobic microorganisms, the biosynthesis of tetrapyrroles, ubiquinones, pyrimidines, sterols, and unsaturated fatty acids involves O_2-dependent reactions. O_2 either serves as a co-substrate or as an oxidant. In anaerobic bacteria, these reactions are replaced by O_2-independent reactions or pathways. In ubiquinone biosynthesis by *E. coli*, three hydroxyl groups are introduced into the molecule by hydroxylases, the oxygen atoms being derived from O_2. In synthesis under anoxic conditions, the oxygen atom is derived from water and incorporated by a hydratase reaction. Heme and pyrimidine biosynthesis includes oxidation reactions in which O_2 serves as electron acceptor under oxic conditions. In anaerobic growth, other oxidants such as fumarate replace O_2 as the electron acceptor. In enzymes like ribonucleotide reductase, molecular oxygen is used for transforming the enzyme to the active state by the introduction of a radical into the enzyme. In anaerobic bacteria, such reactions are catalyzed by different enzymes and mechanisms (see Chapter 7).

11.5 Bacteria Respond to the Effect of Toxic Oxygen Species

Aerobic metabolism inevitably generates reactive oxygen species, such as the superoxide anion ($O_2^{-\bullet}$), hydrogen peroxide (H_2O_2), and the hydroxyl radical ($\bullet OH$) by one-electron reduction steps (Table 11.**2** and Fig. 11.**15**). The reaction sequence from O_2 to $\bullet OH$ can occur by non-enzymic reactions with quinones and Fe^{2+} as the electron donors. $O_2^{-\bullet}$ is also generated from O_2 by nonspecific oxidations of reduced flavoproteins.

11.5.1 O_2 Can Be Excluded From Sensitive Bacteria

As a protective mechanism, bacteria often hinder O_2 from unrestricted access to the cell. This principle is verified in aerobic N_2-fixing bacteria, for which O_2 is an essential, but toxic substrate. Nitrogenase is extremely O_2-sensitive, but O_2 is required for phosphorylative electron transport. In aerobic N_2-fixing bacteria, thick cell walls or capsules provide a diffusion barrier to O_2. Examples are the heterocysts of cyanobacteria or the capsules and slime layers formed from extracellular polysaccharides (e.g., by *Azotobacter*). In the heterocysts, glycolipid layers appear to represent the major diffusion barrier to O_2. In root nodules of legumes that harbor the N_2-fixing species of *Bradyrhizobium* or *Rhizobium*, the nodule cortex serves the same function by providing a diffusion barrier to maintain a microoxic environment for the bacteroids. A different mechanism protecting nitrogenase may be "respiratory protection," as suggested for *Azotobacter*, which fixes N_2 during aerobic growth. In such bacteria, high respiratory rates may decrease the diffusion of O_2 to the intracellular space and to nitrogenase (see Chapter 8.5).

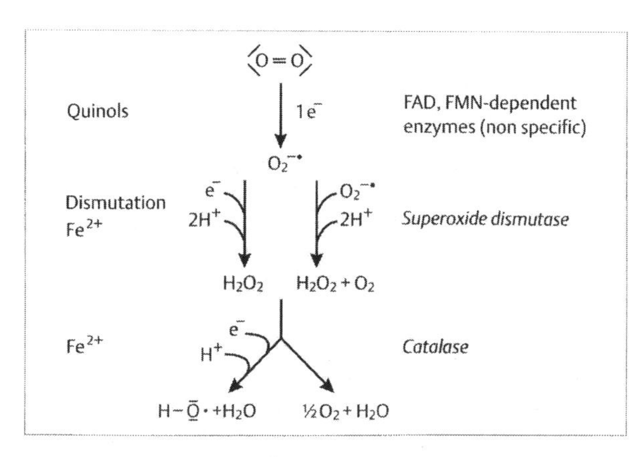

Fig. 11.**15** **Generation of reactive oxygen species.** The left part shows the nonenzymic production of the intermediates; the right part shows reactions catalyzed by enzymes

11.6.2 Reactive Oxygen Species Are Detoxified in Bacteria

In aerobic bacteria, reactive oxygen species can be inactivated by enzymic and non-enzymic reactions. Non-enzymic detoxification is effected by glutathione, which is present in high concentrations (up to 10 mM) in many bacteria. Reduction of $O_2^{-\bullet}$, $^\bullet OH$, and H_2O_2 with glutathione yields H_2O as the final product. The main detoxifying activity, however, is due to catalases, peroxidases, and superoxide dismutases. **Superoxide dismutase** (SOD) converts $O_2^{-\bullet}$ to the less toxic H_2O_2

(Reaction 11.8), **catalase** dismutases H_2O_2 to non-toxic products (Reaction 11.9 and Fig. 11.**15**).

$$2\,O_2^{-}\cdot + - H^+ \longrightarrow H_2O_2 + O_2 \quad (11.8)$$

$$2\,H_2O_2 \longrightarrow 2\,H_2O + O_2 \quad (11.9)$$

Superoxide dismutase is present in nearly all aerobic bacteria. In prokaryotes, two types of superoxide dismutases are known, a Fe^{3+}- and a Mn^{3+}-containing enzyme. Whereas most bacteria contain only one type, *E. coli* contains both types. The Fe-containing SOD is found in aerobic as well as in anaerobic cells. The Mn-containing SOD, on the other hand, is induced during aerobic growth. Catalases are generally present in aerobic and facultatively anaerobic bacteria. In *E. coli*, the synthesis of the detoxifying enzymes is regulated by two distinct "oxidative stress" modulons, which respond to the presence of H_2O_2 and $O_2^{-}\cdot$.

Oxidatively damaged cell constituents can only partially be regenerated. Alkylhydroperoxides are converted to alcohols by **peroxidases**, which catalyze a reaction similar to catalase (Reaction 11.10). Repair of oxidative damages to DNA is achieved by the SOS-DNA repair system, which is induced by oxidative stress.

$$R{-}OOH + 2\,[H] \longrightarrow R{-}OH + H_2O \quad (11.10)$$

11.6 Bioluminescent Systems Require Oxygen for Light Emission

Bioluminescence is widely distributed among living organisms, including bacteria, dinoflagellates, fungi, insects, squids, and fish. The various light-emitting systems show significant differences in the bioluminescence reactions, enzymes, and substrates. All systems require O_2 for the bioluminescence reaction. The luminous bacteria are found in marine, fresh-water, and terrestrial environments as free-living species, saprophytes, or as symbionts in the digestive tracts or light organs in the teleost fishes and in squid. Most of the luminous bacteria are members of the genera *Photobacterium* and *Vibrio*, both of which represent Gram-negative facultatively anaerobic bacteria with polarly inserted flagella.

11.6.1 Biochemistry of Bioluminescence

Two enzymes are essential for bacterial bioluminescence, **luciferase** and a fatty acid reductase complex. The light emission is effected by the luciferase reaction, a monooxygenase. The enzyme couples the oxidation of NAD(P)H and of an aldehyde (R-CHO, decanal) by O_2 to the emission of blue-green light (Reaction 11.11):

$$NAD(P)H + H^+ + R{-}CHO + O_2 \longrightarrow$$
$$NAD(P)^+ + H_2O + R{-}COOH + h\nu \quad (490\ nm) \quad (11.11)$$

The free energy of the redox reaction is used for light emission, but not for ATP regeneration, as in aerobic respiration. Accordingly, the reactions take place within the cytoplasm. The turnover of the luciferase is very slow (about 1–0.1 s^{-1}). The enzyme contains bound FMN, which is reduced by NAD(P)H. $FMNH_2$ reacts with O_2 to form peroxyflavin and a stable intermediate with the aldehyde. After oxidation of the substrates, the complex slowly decays, releasing the products. This step is coupled to the emission of light (E = enzyme; Reaction 11.12):

$$E{-}FMNH_2 \xrightarrow{O_2} E{-}FMNH(OOH) \xrightarrow{R{-}CHO \;\to\; R{-}COOH} E + FMN + H_2O + h\nu \quad (11.12)$$

The production of the aldehyde from the corresponding fatty acid is catalyzed by the fatty acid reductase complex consisting of a reductase, a transferase, and a synthetase (Reaction 11.13):

$$R{-}COOH + ATP + NADPH \longrightarrow$$
$$NADP + AMP + PP_i + R{-}CHO \quad (11.13)$$

The synthetase activates the fatty acid and forms fatty-acyl-AMP. The acyl group is transferred to the reductase (reductase ~ acyl), where it is reduced by NADPH and released. The intensity of light emission is directly related to the amount of NAD(P)H or ATP consumed in the reaction. The reaction, therefore, can be used for a very sensitive coupled assay to determine low amounts of NAD(P)H or of ATP with the bacterial or fly (*Photinus pyralis*) luciferase system, respectively.

The enzymic functions required for light production in *Vibrio* are encoded by the *luxCDABEGH* genes. The *luxAB* genes encode the subunits of the luciferase, *luxCDE* the components of the fatty acid reductase complex. The *luxGH* gene products are required for flavin reduction.

11.6.2 Light Production Is Regulated by Cell Density

Light production in most luminous bacteria is strongly regulated in response to the density of the cell culture. Light emission per cell can be 1,000-fold higher in dense cultures than in dilute cultures. Cell-density-dependent regulation ("quorum sensing") is at the transcriptional level. It has been studied with the light organ symbiont *Vibrio fischeri*. The bacteria synthesize continuously a small extracellular signal molecule, an autoinducer (Fig. 11.**16**), which accumulates in the growth medium. The level of autoinducer appears to be proportional to growth and cell densities. The autoinducer is able to diffuse freely across the cell membrane. At high cell densities, critical levels of autoinducer are achieved, which induce the expression of luminescence ("autoinduction"). Under these conditions, a transcriptional regulator (LuxR protein), which functions as an autoinducer-receptor, activates the expression of the *luxCDABEGH* genes. For symbiotic luminous bacteria, the cell-density-dependent luminescence regulation appears to be reasonable. Luminescence is developed only under the high cell densities found in the light organ of the fish. However, the same type of regulation is found also in free-living luminous bacteria, which normally do not reach cell densities sufficient for induction of bioluminescence (for further details, see Chapter 20.1.4).

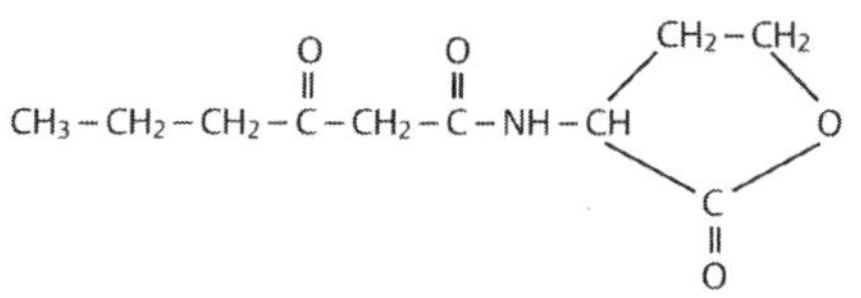

$CH_3-CH_2-CH_2-C(=O)-CH_2-C(=O)-NH-CH$ (ring: $CH-CH_2-CH_2-O-C(=O)-$)

Fig. 11.**16** **Autoinducer of bioluminescence from *Vibrio fischeri*.** The autoinducer (β-ketocaproyl homoserine lactone, or *N*-(3-oxohexanoyl)-L-homoserine lactone) is one of a larger family of homoserine-lactone derivatives operating in signal transduction

Further Reading

Babcock, G. T., and Wikström, M. (1992) Oxygen activation and the conservation energy in cell respiration. Nature 356: 301–309

Friedrich, T., Steinmüller, K., and Weiss, H. (1991) The proton-pumping respiratory complex I of bacteria and mitochondria and its homologue in chloroplasts. FEBS Lett 367: 107–111

Garcia-Horsman, J. A., Barquera, B., Rumbley, J., Ma, J., and Gennis, R. B. (1994) The superfamily of heme-copper respiratory oxidases. J Bacteriol 176: 5587–5600

Iwata, S., Ostermeier, C., Ludwig, B., and Michel, H. (1995) Structure at 2.8 Å resolution of cytochrome *c* oxidase from *Paracoccus denitrificans*. Nature 376: 660–669

Karlson, P., Doenecke, D., and Koolman, J. (1994) Biochemie, 14th edn. Stuttgart: Thieme

Meighen, E. A. (1991) Molecular biology of bacterial bioluminescence. Microbiol Rev 55: 123–142

Miles, J. S., and Guest, J. R. (1987) Molecular genetic aspects of the citric acid cycle of *Escherichia coli*. Biochem Soc Symp 54: 45–65

Nicholls, D. G., and Ferguson, S. J. (1992) Bioenergetics 2. London: Academic Press

Unden, G., and Bongaerts, J. (1997) Alternative respiratory pathways of *Escherichia coli*: Energetics and transcriptional regulation. Biochim Biophys Acta 1320: 217–234

Voet, D., and Voet, J. G. (1995) Biochemistry, 2nd edn. New York: Wiley

Wimpenny, J. W. T. (1982) Responses of microorganisms to physical and chemical gradients. Philos Trans R Soc Lond Biol 297: 497–515

Zannoni, D. (1995) Aerobic and anaerobic electron transport chains in anoxigenic phototrophic bacteria. In: Anoxigenic Bacteria, Blankenship, R. E., Madigan, M. T., and Bauer, C. E., eds. pp. 943–971 Dordrecht: Kluwer

12 Anaerobic Energy Metabolism

12.1 Anaerobic Electron Transport Phosphorylation Uses Electron Acceptors Other Than Oxygen

Most energy-conserving reactions in living organisms are redox reactions in which one substrate is oxidized and another substrate is reduced. In aerobic heterotrophic organisms, the electron donor is usually an organic compound such as glucose and the electron acceptor is molecular oxygen. In the oxidative branch of metabolism, energy for growth is almost exclusively conserved via substrate level phosphorylation, whereas in the reductive branch, ATP synthesis is coupled to the electron transport with H^+ or, in a few cases, with Na^+ as coupling ion (**e**lectron-**t**ransport-coupled **p**hosphorylation, ETP; Fig. 12.**1** and Chapter 4). Many bacteria, however, are able to live under anoxic conditions and to use organic or inorganic electron acceptors instead of molecular oxygen. All such anaerobic energy-conserving processes in which electron transport is coupled to phosphorylation are called **anaerobic respiration**. In contrast, anaerobic energy-conserving redox processes with substrate-level phosphorylation but no electron-transport-coupled phosphorylation are called **fermentations**. Whereas the reactions of fermentations usually occur in the cytoplasm, respirations involve the ATP synthase and, for at least one step, an ion pump or an electron transport that is integrated in the membranes.

In Table 4.**1** all important biological electron acceptors are listed together with their redox potentials. Anaerobic respirations can be divided into two groups, (1) those with positive redox potentials (i.e., the reductions of nitrate, nitrite, ferric ion, fumarate, dimethyl sulfoxide and *N*-oxides) and (2) those with negative redox potentials (i.e., the reductions of sulfur, sulfate, and carbonate). The "high-potential anaerobic respirations," often referred to as "anoxic" processes, mainly occur in facultative anaerobes. Thus, these respirations have to compete with oxygen, which usually suppresses the anaerobic pathways, whereas the "low-potential anaerobic respirations" are found only in strictly anaerobic bacteria that do not grow under aerobic conditions.

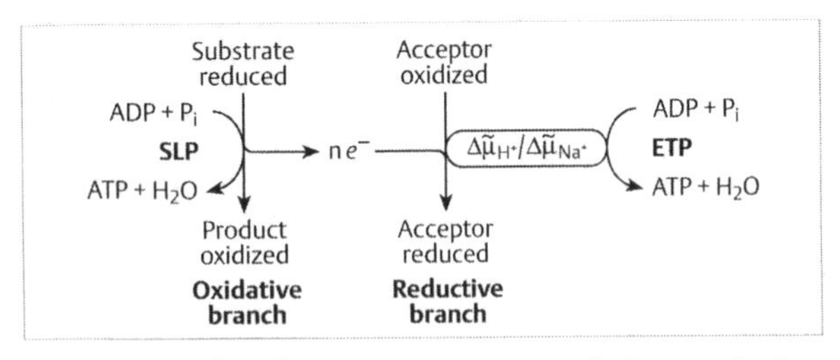

Fig. 12.**1** **Modes of energy conservation during respiration.** Substrate, organic substrate or molecular hydrogen; Product, usually $H_2O + CO_2$; Acceptor, electron acceptor; $\Delta\mu_{H^+}/\Delta\mu_{Na^+}$, electrochemical potential difference for protons and/or sodium ions, respectively, across the membrane; SLP, substrate-level phosphorylation; ETP, electron-transport phosphorylation. SLP is not necessarily involved in respiration

In anaerobic respirations, organic substrates or hydrogen are oxidized, and an electron acceptor other than oxygen is reduced. The acceptor can be nitrate, nitrite, fumarate, sulfur, ferric ion, dimethyl sulfoxide, trimethylamine *N*-oxide, carbon dioxide, sulfate, or organic chlorocompounds. Energy is conserved via an electrochemical proton and/or sodium-ion gradient; this process is also called electron-transport phosphorylation.

12.1.1 Two Pathways for Nitrate and Nitrite Respiration

Many organisms are able to use nitrate (NO_3^-) as electron acceptor, which is reduced to nitrite (NO_2^-, Reaction 12.1) and further reduced either via nitric oxide (NO, Reaction 12.2) and nitrous oxide (N_2O, Reaction 12.3) to molecular nitrogen (N_2, Reaction 12.4, "**denitrification**") or more directly reduced with NAD(P)H to ammonia (NH_3/NH_4^+, Reaction 12.5, "**ammonification**," Fig. 12.**2**):

$$NO_3^- + 2\,e^- + 2\,H^+ \rightarrow NO_2^- + H_2O \quad \text{(nitrate reductase)} \quad (12.1)$$

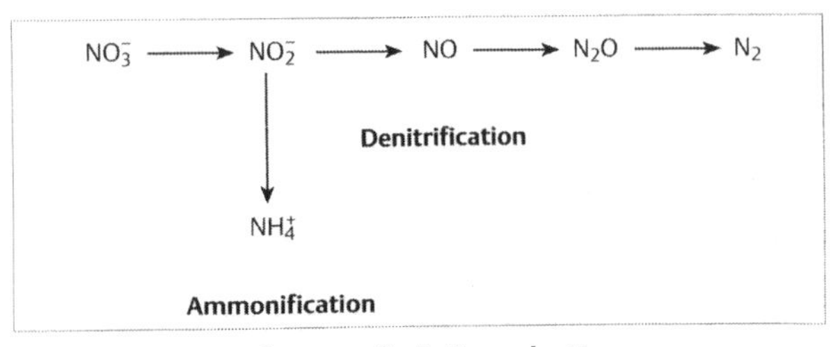

Fig. 12.**2** **Two pathways of nitrite reduction**

$$NO_2^- + e^- + 2\,H^+ \rightarrow NO + H_2O$$
(nitrite reductase, NO-forming) (12.2)

$$2\,NO + 2\,e^- + 2\,H^+ \rightarrow N_2O + H_2O$$
(nitric-oxide reductase) (12.3)

$$N_2O + 2\,e^- + 2\,H^+ \rightarrow N_2 + H_2O$$
(nitrous-oxide reductase) (12.4)

$$NO_2^- + 3\,NAD(P)H + 5\,H^+ \rightarrow NH_4^+ + 3\,NAD(P)^+ + 2\,H_2O$$
(nitrite reductase, ammonia-forming) (12.5)

There are two pathways leading from nitrate to ammonia; the respiratory pathway is called **dissimilatory ammonification**, whereas **assimilatory ammonification** denotes the reduction of nitrate to ammonia for the biosynthesis of nitrogenous compounds (Chapter 8). These two pathways differ: the enzymes of the respiratory pathways are integrated in cytoplasmic membranes or located in the periplasm, and their synthesis is repressed by oxygen, whereas the biosynthetic pathways use soluble enzymes, the synthesis of which is repressed by ammonia. Among higher **Eukarya**, only plants and fungi are capable of assimilatory ammonification. Bacteria, however, can use both pathways as well as denitrification, which is only involved in energy-conserving processes.

The biologically important inorganic nitrogen compounds are listed in Table 12.1; the most stable compounds are dinitrogen, ammonia, nitrate, and nitrous oxide (in order of decreasing stability). Although nitrite is stable at pH 7, in acidic solutions it slowly reacts with ammonia to dinitrogen (non-enzymatic denitrification):

$$NH_4^+ + NO_2^- \rightarrow N_2 + 2H_2O \quad (12.6)$$

Nitric oxide (NO) and nitrogen dioxide (NO_2) are odd molecules, since they each contain an unpaired electron which accounts for their high reactivity. NO_2 is generated from NO in the presence of molecular oxygen:

$$2\,NO + O_2 \rightarrow 2\,NO_2 \quad (12.7)$$

However, the concentration of biologically formed NO in the atmosphere is very low. Therefore, the rate of Reaction 12.7, which is proportional to $[NO]^2$ and requires a simultaneous collision of three atoms, is negligible [i.e., below 10 parts per million (ppm) NO]. In the higher atmosphere, however, NO is oxidized more efficiently to NO_2 by ozone (O_3) or by OH radicals. In water, NO_2 disproportionates to nitric acid and nitrous acid:

$$2\,NO_2 + H_2O \rightarrow NO_3^- + NO_2^- + 2\,H^+ \quad (12.8)$$

In animals, NO acts as a neurotransmitter and is formed from arginine and molecular oxygen in concentrations even exceeding 10 ppm:

$$2\,Arginine^+ + 3\,O_2 \rightarrow 2\,citrulline + 2\,NO + H_2O_2 + 2\,H^+ \quad (12.9)$$

Hydroxylamine is an enzyme-bound intermediate in ammonification and in **nitrification**, the oxidation of ammonia by molecular oxygen (Chapter 10.7). In addition, hydroxylamine (NH_2OH) occurs in the form of hydroxamic acids in certain siderophores (Chapter 8). Finally, the very reactive diimine (NH=NH) and the relatively stable hydrazine (NH_2–NH_2) are enzyme-bound intermediates of **nitrogen fixation** (Chapter 8).

Respiratory ammonification is used for energy conservation by most enterobacteria and staphylococci in the presence of nitrate and in the absence of oxygen (Reactions 12.10 to 12.12).

$$Glucose + 12\,NO_3^- \rightarrow 6\,CO_2 + 12\,NO_2^- + 6\,H_2O$$
$$\Delta G^{\circ\prime} = -1926\ kJ/mol\ glucose \quad (12.10)$$

$$Glucose + 4\,NO_2^- + 8\,H^+ \rightarrow 6\,CO_2 + 4\,NH_4^+ + 2\,H_2O$$
$$\Delta G^{\circ\prime} = -1713\ kJ/mol\ glucose \quad (12.11)$$

$$Glucose + 3\,NO_3^- + 6\,H^+ \rightarrow 6\,CO_2 + 3\,NH_4^+ + 3\,H_2O$$
$$\Delta G^{\circ\prime} = -1767\ kJ/mol\ glucose \quad (12.12)$$

$$NO_2^- + 3\,NAD(P)H + 5\,H^+ \rightarrow NH_4^+ + 3\,NAD(P)^+ + 2\,H_2O \quad (12.13)$$

Table 12.1 **Oxidation states of biologically important nitrogen compounds**

Oxidation state	Compound/function
+5	NO_3^-, nitrate, electron acceptor
+4	NO_2, nitrogen dioxide,* see text
+3	NO_2^-, nitrite, electron acceptor
+2	NO, nitrogen oxide,* denitrification
+1	N_2O, dinitrogen oxide, denitrification
±0	N_2, dinitrogen
−1	NH_2OH, hydroxylamine, ammonification
−1	NH=NH, diimine, nitrogen fixation
−2	NH_2-NH_2, hydrazine, nitrogen fixation
−3	NH_3, ammonia

* reactive compounds with unpaired electrons

The NADH generated by the oxidation of glucose via the Emben-Meyerhof pathway reduces nitrate to nitrite via a membrane-bound electron transport system, whereby energy is conserved (Reaction 12.10; see denitrification). The subsequent six-electron reduction of nitrite to ammonia serves to conserve nitrate because less nitrate is required as electron acceptor (Reaction 12.11). This process, however, does not contribute to energy conservation, if the nitrite reductase involved is the soluble enzyme that uses NAD(P)H as electron donor (Reaction 12.13). Hence, incomplete reduction of nitrate to nitrite may lead to a higher ATP yield per mol glucose than that resulting from the complete reduction of nitrate to ammonia.

In *Wolinella succinogenes*, however, nitrite reduction to ammonia with molecular hydrogen or formate is involved in energy conservation. The nitrite reductase is integrated in the membrane and requires reduced menaquinone (Fig. 12.3) as electron donor.

In strictly anaerobic organisms such as *Veillonella parvula*, *Clostridium perfringens*, and species of *Propionibacterium*, dissimilatory nitrate reduction to ammonia merely is used as an electron sink in order to save the organic substrate for oxidations in which ATP is conserved via substrate-level phosphorylation. Therefore, this mode of energy conservation may be called fermentative nitrate reduction.

Denitrification (i.e., nitrate reduction to N_2) is widespread. Denitrification is found in over 40 genera of the Bacteria (formerly eubacteria), whereas among the Archaea (archaebacteria) only a group of extreme halophiles has been recognized as capable of denitrification. This mode of energy conservation is mainly found among the Gram-negative proteobacteria such as *Pseudomonas*, *Thauera*, *Alcaligenes*, *Paracoccus*, *Hyphomicrobium*, and *Thiobacillus*, but also in some Gram-positive *Bacillus* species. The respiration involves the complete oxidation of a suitable organic substrate to CO_2 and water:

$$\text{Glucose} + 8\ NO_2^- + 8\ H^+ \rightarrow 6\ CO_2 + 4\ N_2 + 10\ H_2O \quad \Delta G^{\circ\prime} = -3144\ \text{kJ/mol glucose} \qquad (12.14)$$

$$5\ \text{Glucose} + 24\ NO_3^- + 24\ H^+ \rightarrow 30\ CO_2 + N_2 + 42\ H_2O \quad \Delta G^{\circ\prime} = -2657\ \text{kJ/mol glucose} \qquad (12.15)$$

The pathways used for oxidation of organic substrates by anaerobic respirations usually are the same as those found under oxic conditions. Exceptions are pathways involving substrates whose oxidation requires molecular oxygen as a reactive agent, for example, aromatic compounds and hydrocarbons. These substrates are transformed to common metabolites by oxygen-independent pathways (e.g., reductive dearomatization of benzoyl-CoA by *Thauera aromatica*; Reaction 12.16) or they are not degraded at all under denitrifying conditions (e.g., lignin).

$$\text{Benzoyl} - \text{CoA} + 2\ e^- + 2\ H^+ + 2\ \text{ATP} + 2\ H_2O \rightarrow \text{Cyclohexa-1,3-diene-2-carboxyl-CoA} + 2\ \text{ADP} + 2\ P_i \qquad (12.16)$$

In most denitrifying organisms, the NADH formed by the oxidation of glucose is dehydrogenated by an NADH : ubiquinone oxidoreductase; this enzyme funnels the electrons into the chain leading from nitrate to dinitrogen (Reactions 12.1–12.4). In contrast to NADH : ubiquinone oxidoreductase, the terminal electron-accepting enzymes are not proton pumps. The respiratory nitrate reductase, however, is able to generate an electrochemical proton gradient by an alternative mechanism linking the movements of two electrons from the outside (positive side) to the inside (negative side) of the membrane with the release of two protons at the outside and uptake at the inside (Fig. 12.4). It is not yet known whether nitrite reductase also generates a proton gradient. Although nitrate reduction is associated with a smaller change in free energy than nitrite reduction (compare Reactions 12.10 and 12.15), apparently more ATP is formed in the former, kinetically superior reaction. As shown with several organisms only nitrate is reduced in the early growth phase and nitrite accumulation is transient. The nitrite is reduced to nitrogen only after the nitrate has been consumed.

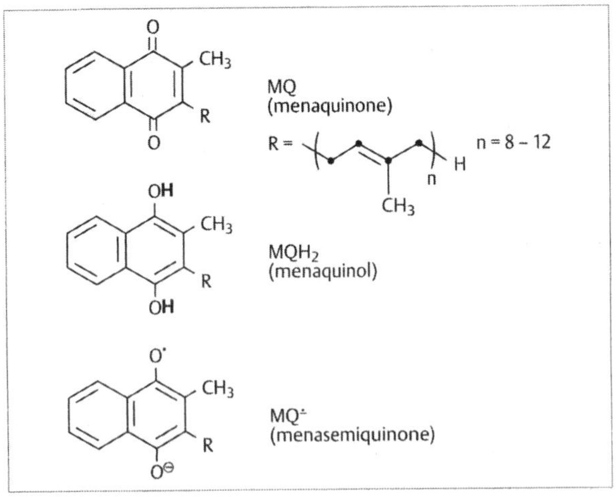

Fig. 12.3 **Menaquinone**. Menaquinone is the short form for **me**thyl**na**phtho**quinone**. Reduction by one electron yields the menasemiquinone anion, which upon further reduction by the second electron takes two protons, thus forming the menahydroquinone (also called menaquinol)

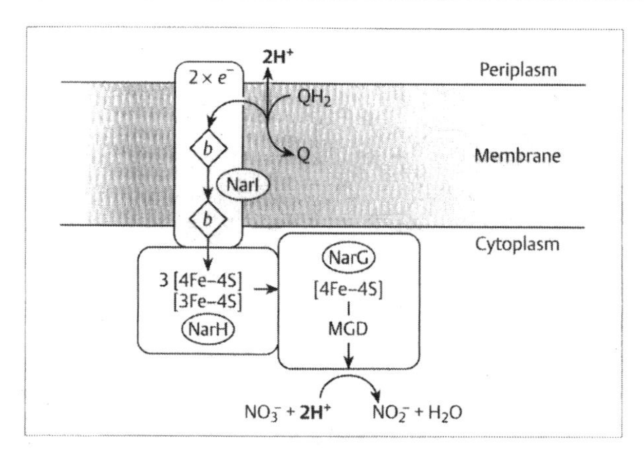

Fig. 12.**4** **Structure and function of nitrate reductase.** For abbreviations see text; in *E. coli* ubiquinone (Q) is replaced by menaquinone (Fig. 12.**3**) [after 1]. NarG, H, and I specify the three different subunits of nitrate reductase; *b*, cytochrome *b*; MGD, molybdopterin guanine dinucleotide (see Fig. 12.**5**)

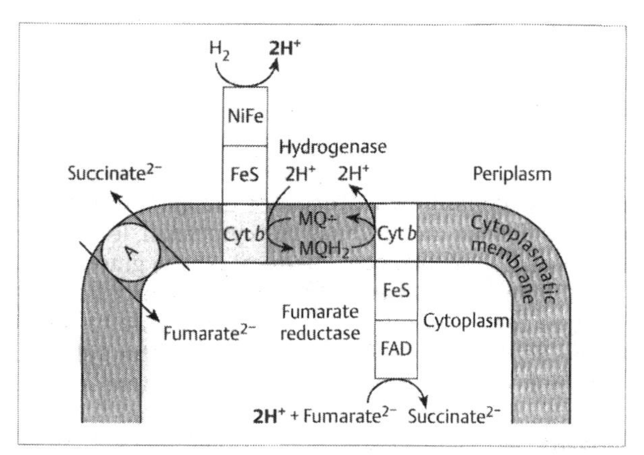

Fig. 12.**5** **Generation of an electrochemical proton gradient by electron transport in *Wolinella succinogenes*.** NiFe, FeS, Cyt *b*, FAD denote the prosthetic groups of the subunits of hydrogenase and fumarate reductase. Nitrate or sulfur reductases may replace fumarate reductase

Box 12.1 An alternative mechanism to generate a proton motive force has been developed in some anaerobes, for example *Wolinella succinogenes* (Fig. 12.**5**). Oxidation of hydrogen or formate at the outer side of the membrane releases $2\,H^+$, whereas reduction of different electron acceptors (nitrate, fumarate, nitrite, sulfur, dimethyl sulfoxide) at the inner side consumes $2\,H^+$. This allows the synthesis of 2/3 mol ATP, assuming the requirement of $3\,H^+$/ATP in the ATP-synthase reaction (see Chapter 4). This calculation agrees well with the growth yield of 6–7 g dry cells/mol formate. Since the usual growth yield (Y_{ATP}) is 10 g dry cells/mol ATP, 2/3 mol ATP is a very reasonable value. From a thermodynamic point of view, however, Reaction 12.17 would allow the formation of approximately 2 mol ATP, since a $\Delta G^{\circ\prime}$ of −70 to −80 kJ has been shown to be sufficient for the generation of 1 mol ATP in many respirations and fermentations.

$$HCOO^- + H^+ + NO_3^- \rightarrow NO_2^- + CO_2 + H_2O \quad \Delta G^{\circ\prime} = -167\ \text{kJ/mol} \qquad (12.17)$$

Enzymes involved in denitrification and ammonification contain molybdopterins. The nitrate reductases of the fermentative anaerobes and those with assimilatory functions are soluble enzymes using NADH or NADPH as electron donor (Reaction 12.18). The enzymes are flavoproteins (FAD or FMN) and contain molybdopterin dinucleotide bound to the protein (Fig. 12.**6**).

$$NAD(P)H + NO_3^- + H^+ \rightarrow NAD(P)^+ + NO_2^- + H_2O \quad \Delta G^{\circ\prime} = -143\ \text{kJ/mol} \qquad (12.18)$$

The **dissimilatory nitrate reductases** involved in anaerobic respirations are membrane associated and composed of three different subunits, which are designated by the genetic abbreviations NarI, NarH, and NarG (Nar = nitrate reductase, Fig. 12.**4**). NarI is a cytochrome *b* composed of five membrane-spanning helices between which two hemes are imbedded. The hemes conduct the electrons released from ubiquinol (QH_2) or menaquinol to the two peripheral membrane

Box 12.2 **Molybdoenzymes**, with the exception of the nitrogen-reducing enzyme nitrogenase, contain molybdopterin (Eukarya) or molybdopterin dinucleotide (Bacteria and Archaea, Fig. 12.**6**). In most cases, the base in the dinucleotide is guanine; it may be replaced by adenine, uracil, or cytosine. Molybdopterin-containing enzymes generally catalyze the reversible introduction of an oxygen atom from water into the substrate. Further examples are formate dehydrogenase, formylmethanofuran dehydrogenase (in methanogenesis, Fig. 12.**9**), aldehyde dehydrogenase, sulfite oxidase, xanthine dehydrogenase (Chapter 12.2.4), carbon-monoxide oxidase, molybdenum carbon-monoxide dehydrogenase (but not nickel-containing acetyl-CoA synthase, Fig. 12.**8**) as well as trimethylamine *N*-oxide and dimethyl sulfoxide reductases. In addition, there are some isoenzymes in which molybdenum (Mo) is replaced by tungsten (W).

proteins, NarH and NarG. The electrons are transported via the one [3 Fe–4 S] and the three [3 Fe–4 S] clusters of NarH to the [4 Fe–4 S] cluster of NarG and finally to molybdopterin guanosine dinucleotide (MGD), where nitrate is reduced to nitrite. Two protons are released into the periplasm by oxidation of the quinol, whereas two protons are consumed in the cytoplasm, leading to the formation of an electrochemical proton gradient.

Nitrite reductases involved in denitrification catalyze the one-electron reduction of nitrite to nitric oxide (Reaction 12.2). There are two different types of enzymes: one contains cytochrome cd_1 (*Pseudomonas aeruginosa*), the other contains copper (other pseudomonads and *Alcaligenes*). Cytochrome d_1 has been identified as a protein containing 1,3-porphyrindione (Fig. 12.**7a**). Many nitrite reductases also have an oxidase activity, i.e. they catalyze the reduction of oxygen to water. The NO-reductase has been characterized as a novel cytochrome *bc* complex. It is the only enzyme known to catalyze the formation of an N–N

In nitrate respiration, nitrate (oxidation state of N is +5) is reduced to nitrite (oxidation state: +3), a process catalyzed by a cytochrome *b*, iron–sulfur centers, and a molybdopterin-containing membrane enzyme. Energy is conserved via an electrochemical proton gradient. Nitrite is further reduced to ammonia (oxidation state: −3) in a six-electron step catalyzed by a siroheme-containing enzyme (ammonification) or to dinitrogen, N_2 (oxidation state: 0) in three consecutive steps via nitric oxide (NO, oxidation state: +2) and nitrous oxide (N_2O, oxidation state: +1). The latter three steps are catalyzed by 1,3-porphyrinedione and copper containing enzymes (denitrification).

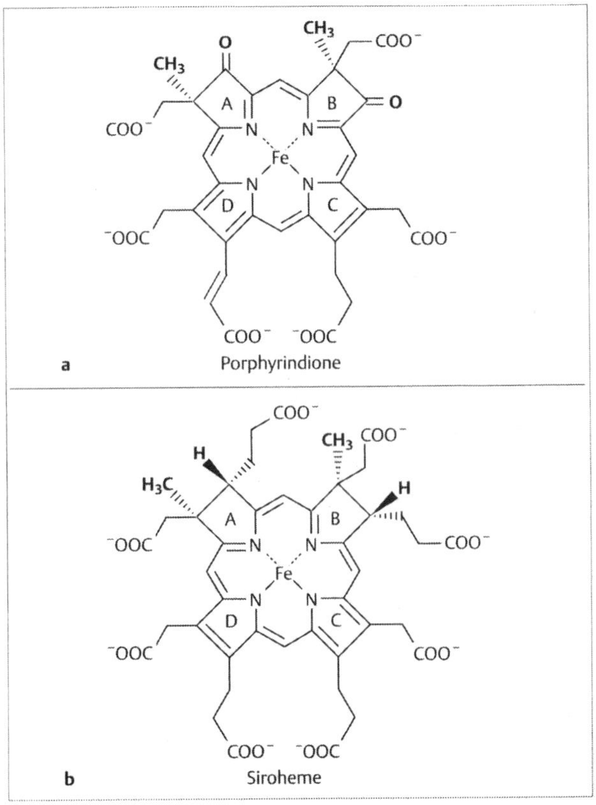

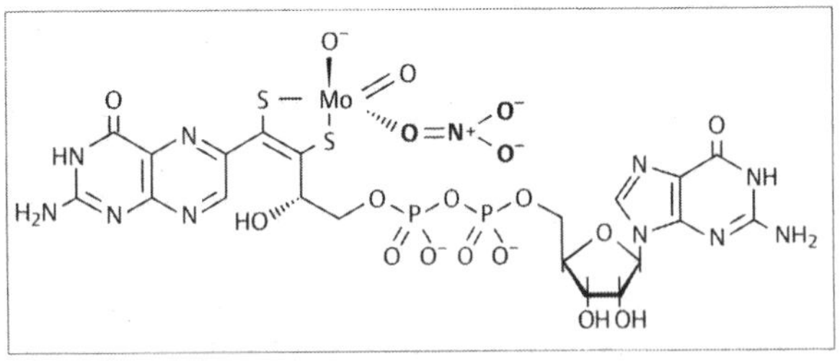

Fig. 12.6 **Molybdopterin guanine dinucleotide.** In bacterial and archaeal molybdopterins the nucleotide base can also be cytidine, adenine, or uracil, instead of guanine. GMP is not found in the molybdopterins of eukaryotes. The molybdopterins are prosthetic groups of many dehydrogenases or reductases. During reduction the substrate (e.g., nitrate) is probably coordinated to Mo

Fig. 12.7 **Prosthetic groups of nitrite reductases.**
(**a**) The green colored porphyrindione (also called incorrectly heme d_1) is the prosthetic group of the nitrite reductases involved in denitrification (e.g., in *Paracoccus denitrificans*). A second class of these enzymes contains Cu (e.g., in *Alcaligenes eutrophus*).
(**b**) Siroheme is the prosthetic group of the nitrite reductases involved in ammonification (e.g., in *Escherichia coli*). The molecule still contains all the carboxylated substituents present in the precursor uroporphyrin III, from which also heme is synthesized. In contrast to heme, porphyrindione and siroheme are methylated (rings A and B). The biosynthesis of porphyrindione has not been elucidated yet

bond (see Table 12.**1**). The N_2O-reductase also is a copper enzyme with a binuclear Cu center of the same type (Cu_A) as found in the mitochondrial cytochrome c oxidase. All these enzymes are more or less firmly attached to or integrated in the cytoplasmic membrane.

The nitrite reductase that catalyzes the reaction involved in ammonification (Reaction 12.14) is a soluble enzyme containing siroheme (Fig. 12.**7b**) and a [4 Fe–4 S] cluster as prosthetic groups. The likewise six-electron reduction of hydrogen sulfite (HSO_3^-, oxidation state of sulfur +4) to sulfide (oxidation state of −2) is catalyzed by a very similar enzyme, the dissimilatory hydrogen sulfite reductase (see Chapter 12.1.8, Reaction 12.63).

12.1.2 Fumarate Reduction Is Coupled to Electron Transport

The reduction of α,β-unsaturated monocarboxylic acids (either in the free form or as CoA esters) to the corresponding saturated fatty acids is catalyzed by soluble cytoplasmic enzymes that use reduced pyridine nucleotides as electron donors. These reactions are not coupled to energy conservation, possibly because of their low redox potentials ($E_0'= -30\,\text{mV}$). In contrast, fumarate reduction ($E_0'= +30\,\text{mV}$) with a variety of electron donors (e.g., NADH, H_2, formate) is catalyzed by membrane-integrated electron-transport chains. These chains are made up of dehydrogenases and fumarate reductase, which are connected by menaquinone (Fig. 12.**3**). The fumarate reductases are similar to the nitrate reductases, except that the molybdopterin-containing subunit is replaced by a subunit in which FAD is covalently bound to a specific histidine residue. The mode of forming an electrochemical proton gradient is essentially identical in both types of enzyme.

Interestingly, *Escherichia coli* is able to synthesize two related enzymes that catalyze the succinate/fumarate interconversion. Under oxic conditions the organism contains succinate dehydrogenase, with ubiquinone ($E_0'= +45\,\text{mV}$) as electron acceptor. Succinate dehydrogenase is replaced under anoxic conditions by fumarate reductase and menaquinol serves as electron donor ($E_0'= -100\,\text{mV}$).

Although fumarate is not abundant in natural habitats, it may be generated from sugars, malate, citrate, and especially aspartate and asparagine, which are frequent amino acids in proteins. Most of the fumarate-reducing bacteria contain fumarase, aspartate ammonia lyase and asparaginase. The synthesis of all these enzymes is controlled by the oxygen-sensing regulator Fnr (see Chapters 11 and 21). Enterobacteria and *Helicobacter* are known to convert asparagine and aspartate to succinate. Aspartate is also formed by the fermentation of the pyrimidine precursor orotate in *Clostridium oroticum* (Reaction 12.19). In propionibacteria and *Veilonella parvula* fumarate is formed from lactate via pyruvate and oxaloacetate. Propionibacteria convert succinate further to propionate via methylmalonyl-CoA (Fig. 12.**25**).

$$\text{Orotate}^- + 2\,H_2O + 3\,H^+ + 2\,e^- \rightarrow NH_4^+ + CO_2 + \text{L-aspartate} \quad (12.19)$$

> The reduction of fumarate to succinate (fumarate respiration) is catalyzed by an enzyme similar to nitrate reductase, in which molybdopterin is replaced by FAD. The electron donor is menaquinol. In aerobes, the reverse reaction is catalyzed by the related but different succinate dehydrogenase, with ubiquinone as primary electron acceptor.

12.1.3 Some Bacteria and Archaea Use Sulfur Respiration

Sulfur reducers live in various habitats ranging from mesophilic and alkaline to extremely thermophilic and acid conditions. Some of these organisms oxidize simple organic compounds like acetate, formate, and ethanol. Some strains require either sugars or peptides for growth. Several extremely thermophilic Archaea ferment these organic substrates in the absence of elemental sulfur. But addition of sulfur increases the cell density due to the removal of molecular hydrogen (see Chapter 12.2 for a general discussion on the importance of H_2 formation).

Elemental sulfur usually occurs in rings of eight sulfur atoms (S_8). It is practically insoluble in water at 25 °C; therefore, it is unlikely to be directly reduced to sulfide by bacteria. **Polysulfide** ($^-S\text{-}S_n^-$) was shown to be the soluble compound that serves as the intermediate of sulfur reduction in *Wolinella succinogenes*. However, it is not known whether this also pertains to other bacteria. Polysulfide (mainly S_4^{2-} and S_5^{2-}) is formed abiotically from elemental sulfur and sulfide (Reactions 12.20 and 12.21).

$$S_8 + HS^- \rightarrow {}^-S - S_8^- + H^+ \quad (12.20)$$

$$^-S - S_8^- + HS^- \rightarrow S_4^{2-},\ S_5^{2-} \text{ and other polysulfides} \quad (12.21)$$

The amount of S^0 that can be dissolved in a sulfide solution at pH 8 and 37 °C is nearly equivalent to the sulfide content. With decreasing pH, the stability of polysulfide drops so drastically that it becomes an unlikely substrate for sulfur reducers at pH below 5.

The mechanism of polysulfide reduction has been investigated in greater detail only in *W. succinogenes*. The electron-transport chain that catalyzes the reduction of polysulfide with formate or H_2 consists of polysulfide reductase and formate dehydrogenase or hydrogenase. The three enzymes are integrated in the cytoplasmic membrane with the substrate sites facing the periplasm. A quinone is apparently not involved in the electron transport. Electron transfer from a dehydrogenase to polysulfide reductase appears to require diffusion and collision of the enzyme complexes within the membrane. Polysulfide reductase consists of two hydrophilic and one hydrophobic subunit. The larger of the two hydrophilic subunits contains the substrate site together with molybdopterin guanine dinucleotide (Fig. 12.**6**). The smaller hydrophilic subunit is an iron-sulfur protein. The hydrophobic subunit anchors the enzyme in the membrane. Growth with H_2 or formate (electron donor) and polysulfide ([S], acceptor) demonstrates that the electron transport must be coupled to phosphorylation:

$$H_2 + [S] \rightarrow HS^- + H^+ \qquad (12.22)$$

The difference in redox potentials between the donor (H_2, $E_0' = -414\,mV$) and the acceptor ([S], $E_0' = -270\,mV$) is very small. This means that the H^+/e^- ratio should be less than one. Assuming the H^+/e^- ratio as 0.5, then 1/3 mol ATP per mol H_2 would be synthesized.

The reduction of sulfur to H_2S ($E_0' = -270\,mV$) requires hydrogen or formate as electron donor. Since elemental sulfur (S_8) is insoluble, the actual substrate for the molybdopterin dinucleotide containing sulfur reductase is polysulfide (S_n^{2-}, n = 4, 5).

12.1.4 Reduction of the Ferric Ion Has a High Redox Potential

The reduction of ferric ion (Fe^{3+}) to ferrous ion (Fe^{2+}) has a very high redox potential ($E_0' = +770\,mV$) similar to that of oxygen ($E_0' = +820\,mV$, Table 4.**1**). Hence, during the dissimilatory oxidation of 1 mol acetate by ferric ions up to 11 mol ATP could be conserved (Reaction 12.23). This is almost the same amount to be expected from the oxidation of acetate by oxygen in mitochondria.

$$\text{Acetate}^- + 8\,Fe^{3+} + 2\,H_2O \rightarrow 2\,CO_2 + 8\,Fe^{2+} + 7\,H^+$$
$$\Delta G^{\circ\prime} = -815\ \text{kJ/mol acetate.} \qquad (12.23)$$

However, because $Fe(OH)_3$ has an extremely low solubility product ($[Fe^{3+}][OH^-]^3 = 10^{-39}$), the concentration of ferric ion at pH 7 is 10^{-18} M. Only below pH 4 does $[Fe^{3+}]$ reach the micromolar range. Therefore at pH 7 the free energy of Reaction 12.23 ($\Delta G' = -34$ kJ/mol acetate) is not sufficient to allow the synthesis of 1 mol ATP. Since the naturally occurring ferric minerals in the soil [e.g., goethit (α-FeOOH), hematit (α-Fe_2O_3), ferrihydrit ($5\,Fe_2O_3 \cdot 9\,H_2O$), lepidokrokit (γ-FeOOH), and maghemit (γ-Fe_2O_3)] have even lower solubility products, nothing is known about the exact amount of energy conserved in reaction 12.23.

Members of the genera *Bacillus*, *Clostridium*, *Escherichia*, *Pseudomonas*, and *Serratia*, among others, are able to reduce ferric to ferrous ions. In most cases it has not yet been shown whether this process represents a dissimilatory reduction coupled to energy conservation or whether it has an assimilatory function. In addition the reduction of ferric ion serves solely as an electron sink in fermentative organisms and is used to remove reducing equivalents. In *Geobacter metallireducens* and *Shewanella putrefaciens*, growth is clearly dependent on ferric-ion reduction. Remarkably, *G. metallireducens* is able to oxidize toluene with ferric ions under anoxic conditions. The biochemistry of the enzymes and the electron-transport chain involved in dissimilatory ferric-ion reduction has not been elucidated yet. It is to be expected that the ferric-ion reductase protrudes from the membrane in order to come into contact with the insoluble ferrihydrit (FeOOH), especially in Gram-negative organisms. The readily soluble ferrous ion is therefore formed outside the cell. In contrast, assimilatory ferric-ion reduction is well understood (see Chapter 8).

Recently an anaerobic phototrophic eubacterium was isolated that uses ferrous ion as electron donor for carbon-dioxide reduction. Hence, in the Precambrian biosphere, the formation of ferric iron would have been possible prior to the generation of oxygen. Therefore ferric iron could have been the first inorganic electron acceptor used by bacteria before oxygen, nitrate, and sulfate became available. Today, ferric-ion reduction continues to be an important process in the oxidation of organic matter in many modern sedimentary environments. There is evidence that other metals in higher oxidation also serve as dissimilatory electron acceptors, for example, the reduction of manganese(IV) in the insoluble MnO_2 to the soluble Mn(II) ion. The biochemistry and geochemical importance of MnO_2 reduction apparently are similar to the same aspects of ferric-ion reduction.

Although ferric ion (Fe^{3+}) has a redox potential similar to oxygen, bacteria using this ion as electron acceptor have to cope with its extreme insolubility at a neutral pH. Nevertheless, ferric-ion reduction is an important process for oxidation of organic matter in many sedimentary environments.

12.1.5 Dimethyl Sulfoxide and Trimethylamine *N*-Oxide Respirations Lead to Malodorous Products

Like fumarate or nitrate, dimethyl sulfoxide and trimethylamine *N*-oxide act as electron acceptors in anaerobic respirations, mainly in enterobacteria and in *Wolinella succinogenes*. Both compounds are reduced by two electrons each to dimethyl sulfide (Reaction 12.24) and trimethylamine (Reaction 12.25), catalyzed by the molybdopterin-containing dimethyl sulfoxide reductase and trimethylamine *N*-oxide reductase, respectively.

$$\text{Dimethyl sulfoxide} + 2\,H^+ + 2\,e^- \rightarrow \text{dimethyl sulfide} + H_2O$$
$$CH_3 - SO - CH_3 \qquad\qquad CH_3 - S - CH_3 \qquad (12.24)$$

$$\text{Trimethylamine } N\text{-oxide} + 2\,H^+ + 2\,e^- \rightarrow \text{trimethylamine} + H_2O$$
$$(CH_3)_3N^+ - O^- \qquad\qquad (CH_3)_3N \qquad (12.25)$$

The enzymes are integral membrane proteins with ubihydroquinol or menahydroquinol as electron donors. The modes of electron transport and energy conservation are similar to those of the respiration of fumarate or nitrate in the corresponding organisms.

Like fumarate and nitrate, dimethyl sulfoxide and trimethylamine *N*-oxide are used as electron acceptors in facultative anaerobes. The process is coupled to energy transduction in an electron-transport chain.

Box 12.3 Dimethyl sulfide is a degradation product of dimethylsulfopropionate, $(CH_3)_2S^+-CH_2-CH_2-COO^-$, an important osmolyte of marine plants. The volatile dimethyl sulfide is the most abundant biogenic sulfur compound that enters the atmosphere from the sea where it is oxidized in part to dimethyl sulfoxide. Phototrophic bacteria also produce dimethyl sulfoxide by using dimethyl sulfide as an electron donor for the fixation of carbon dioxide. In industry, dimethyl sulfoxide is generated during the production of cellulose and is used as an excellent solvent. Its application in medicine as a transporter of drugs through the skin, however, is no longer permitted due to its toxicity. Trimethylamine *N*-oxide is formed in rotting fish by the oxidation of trimethylamine, which is a degradation product of trimethylammonium compounds (e.g., choline).

12.1.6 Exergonic Acetate Synthesis From Two One-Carbon Units

Acetogens, strictly anaerobic members of the Bacteria, form a taxonomically very heterogenic group of bacteria from the Gram-positive phylum. The following genera comprise the acetogens: *Peptostreptococcus*, *Butyrobacterium*, *Eubacterium*, *Clostridium*, *Acetobacterium*, *Sporomusa*, and *Syntrophococcus*. Acetogens synthesize acetate from two C_1 units, most commonly carbon dioxide, which is reduced by molecular hydrogen (Reaction 12.26). Other precursors are methanol (Reaction 12.27) and carbon monoxide (Reaction 12.28).

$$2\,CO_2 + 4\,H_2 \rightarrow \text{acetate}^- + 2\,H_2O + H^+ \quad \Delta G^{\circ\prime} = -95 \text{ kJ/mol acetate} \qquad (12.26)$$

$$4\,\text{Methanol} + 2\,CO_2 \rightarrow 3\,\text{acetate}^- + 3\,H^+ + 2\,H_2O \quad \Delta G^{\circ\prime} = -70 \text{ kJ/mol acetate} \qquad (12.27)$$

$$4\,CO + 2\,H_2O \rightarrow \text{acetate}^- + 2\,CO_2 + H^+ \quad \Delta G^{\circ\prime} = -175 \text{ kJ/mol acetate} \qquad (12.28)$$

In addition, most acetogens are able to attack molecules containing more than one carbon atom and to hydrolyze or to oxidize them to one-carbon precursors of acetogenesis. A well-known example is the degradation of one mol glucose to three mol acetate by *Clostridium aceticum* (Reaction 12.29). Acetic acid, $C_2H_4O_2$, may thus be regarded as a simple carbohydrate with the general formula $C_nH_{2n}O_n$; n = 2 for acetic acid, whereas n = 6 for glucose.

$$\text{Glucose} \rightarrow 3\,\text{acetate}^- + 3\,H^+ \quad \Delta G^{\circ\prime} = -104 \text{ kJ/mol acetate} \qquad (12.29)$$

Initially, glucose is fermented via the Embden-Meyerhof pathway to pyruvate (2 mol/mol glucose), which is further oxidized to 2 mol acetate and 2 mol CO_2, where-by 4 mol ATP are formed. The resulting reducing equivalents are used to reduce the 2 mol CO_2 to the third mol acetate. Many acetogens are able to cleave ethers such as methylated phenolic compounds, which are found among the degradation products of lignin. The liberated methyl group is converted via methyl tetrahydrofolate to the methyl group of acetate, whereas the carboxyl group is derived from carbon dioxide; for example, *Clostridium formicoaceticum* is able to convert vanillin [(3-methoxy)-4-hydroxybenzaldehyde] and carbon dioxide to acetate and protocatechuate (3,4-dihydroxybenzoate). In this example, the aldehyde also provides the necessary reducing equivalents (Reactions 12.30–12.31).

$$\text{Vanillin} + 2\,H_2O \rightarrow \text{protocatechuate}^- + [CH_3OH] + 2\,[H] + H^+ \quad (12.30)$$

$$[CH_3OH] + 2\,[H] + CO_2 \rightarrow CH_3COO^- + H^+ + H_2O \quad (12.31)$$

$$\text{Vanillin} + H_2O + CO_2 \rightarrow \text{acetate}^- + \text{protocatechuate}^- + 2\,H^+ \quad (12.32)$$

Equation 12.31 describes almost exactly the final reaction catalyzed by acetyl-CoA synthase, the key enzyme of acetogenesis (also called CO dehydrogenase) (i.e., the formation of acetyl-CoA from CO_2 and a methyl group attached to tetrahydrofolate; Reaction 12.32).

$$N^5\text{-Methyl tetrahydrofolate} + 2\,[H] + CO_2 + \text{CoASH} \rightarrow \text{acetyl-S-CoA} + \text{tetrahydrofolate} + H_2O \quad (12.32)$$

The extremely oxygen-sensitive acetyl-CoA synthase is a very complex enzyme (Fig. 12.**8**). It has four functions, (1) a cobalamine-dependent methyl-transferase (methylform of vitamin B_{12} similar to the cobamide in Fig. 12.**10**) which accepts the methyl group from tetrahydrofolate; (2) a nickel and [4 Fe–4 S] containing CO dehydrogenase, which provides CO by reduction of CO_2; (3) a nickel and [4 Fe–4 S] containing carbonylase,

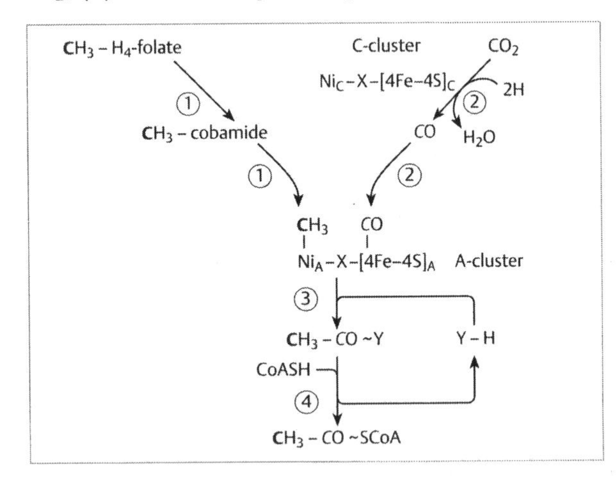

Fig. 12.**8** **Acetyl-CoA synthase (carbon monoxide dehydrogenase).** The enzyme complex has the following partial activities: 1, methyl-transferase; 2, CO dehydrogenase; 3, carbonylase; 4, acetyl-transferase; cobamide, a derivative of vitamin B_{12} (Fig. 12.**10**); [4 Fe–4 S], four iron–four sulfur cluster; H_4-folate, tetrahydrofolate (THF); ~ "energy-rich" bond; X, probably sulfur; Y, acetyl carrier, probably a cysteine residue of the enzyme

which joins the methyl group with CO to yield the acetyl group; and finally (4) an acetyl-transferase, by which the acetyl-CoA is formed. The chemical precedent of step (3) of acetyl-CoA synthase is the technical Monsanto process for acetic acid production (Reaction 12.33), in which CO is inserted into the C–O bond of methanol.

$$CO + CH_3OH \rightarrow CH_3\text{-}COOH \quad (12.33)$$

The formation of the methyl group from CO_2 requires three two-electron reductions involving the intermediates formate and formaldehyde. Due to its high reactivity, formaldehyde is toxic and is therefore handled in the cell normally as N^5,N^{10}-methylene-tetrahydrofolate; in fact, this intermediate is spontaneously formed from formaldehyde and tetrahydrofolate. Five enzymes are necessary to convert CO_2 to N^5-methyl tetrahydrofolate:

$$CO_2 + 2\,H \rightarrow HCOO^- + H^+ \quad \text{(formate dehydrogenase).} \quad (12.34)$$

$$HCOO^- + ATP^{4-} + \text{tetrahydrofolate} \rightarrow N^{10}\text{-formyltetrahydrofolate} + ADP^{3-} + HPO_4^{2-} \quad \text{(formyltetrahydrofolate synthetase)} \quad (12.35)$$

$$N^{10}\text{-Formyltetrahydrofolate} \rightarrow N^5,N^{10}\text{-methylenetetrahydrofolate} + H_2O \quad \text{(formyltetrahydrofolate cyclohydrolase)} \quad (12.36)$$

$$N^5,N^{10}\text{-Methylenetetrahydrofolate} + NAD(P)H + H^+ \rightarrow N^5,N^{20}\text{-methylene-tetrahydrofolate} + NAD(P)^+ \quad \text{(methylene-tetrahydrofolate dehydrogenase)} \quad (12.37)$$

$$N^5,N^{10}\text{-Methylenetetrahydrofolate} + NADPH + H^+ \rightarrow N^5\text{-methyl-tetrahydrofolate} + NADP^+ \quad \text{(methyl-tetrahydrofolate : NADP}^+\text{ oxidoreductase)} \quad (12.38)$$

NAD(P)H and the unknown electron donor of formate dehydrogenase are kept in the reduced state by molecular hydrogen in reactions catalyzed by hydrogenases (Reaction 12.39).

$$H_2 \rightarrow 2\,H^+ + 2\,e^- \quad \text{(hydrogenase, see Box 12.4).} \quad (12.39)$$

Energetic considerations. Under standard conditions the formation of acetate from CO_2 and H_2 provides sufficient energy to produce 1 mol ATP/mol acetate (Reaction 12.26). However, in nature, acetogenic bacteria have to cope with much lower H_2 concentrations. Since 4 mol H_2 are required per mol acetate, the free energy change decreases dramatically with decreasing H_2 concentrations:

$$\Delta G' = \Delta G^{\circ\prime} + RT \ln([\text{acetate}^-][CO_2]^{-2}[H_2]^{-4}) \quad (12.40)$$

Thus under the conditions $[H_2] = 100$ Pa (equivalent to 1 mM), $[CO_2] = 10^5$ Pa (equivalent to 1 M) and [acetate] = 1 M, then $\Delta G' = -95 + 5.7 \lg 10^{+12} = -27$ kJ/mol. This amount is not enough to generate 1 mol ATP/mol acetate, for which at least −70 kJ are required. This calculation shows that substrate-level phosphorylation appears to be impossible. Moreover, the ATP generated from acetyl-CoA, which is formed in the acetyl-CoA synthase reaction (Reaction 12.32), is consumed in the activation of formate (Reaction 12.35).

Hence, with CO_2 as the substrate, electron-transport phosphorylation (i.e., anaerobic respiration) should be the only way for energy conservation. If preformed methyl groups are available, however, additional substrate-level phosphorylation from acetyl-CoA via phosphate acetyltransferase (Reaction 12.41) and acetate kinase (Reaction 12.42) is possible.

$$\text{Acetyl-CoA} + HPO_4^{2-} \rightarrow \text{acetyl phosphate}^{2-} + \text{CoASH} \quad \text{(phosphate acetyltransferase)} \quad (12.41)$$

$$\text{Acetyl phosphate}^{2-} + ADP^{3-} \rightarrow ATP^{4-} + \text{acetate}^- \quad \text{(acetate kinase)} \quad (12.42)$$

The acetogenic clostridia *C. thermoaceticum* and *C. formicoaceticum* contain cytochrome *b* and they use H^+ as coupling ion for electron-transport phosphorylation. Na^+ is the coupler in the cytochrome *b*-deficient *Acetobacterium woodii*, in which an Na^+-dependent ATP synthase was discovered. The site of the generation of the electrochemical gradient has not been established in all cases.

In the presence of nitrate, some acetogenic bacteria repress their capability to reduce carbon dioxide. They reduce nitrate to ammonia under these conditions, whereby over six times more energy is released; compare Reaction 12.26 with Reaction 12.43.

$$NO_3^- + 4\,H_2 + 2\,H^+ \rightarrow NH_4^+ + 3\,H_2O \quad \Delta G^{\circ\prime} = -600 \text{ kJ/mol} \quad (12.43)$$

Acetogenic bacteria are able to form acetate by reduction of carbon dioxide with low-potential electron donors (e.g., hydrogen, reduced ferredoxin from 2-oxo acids, or aldehydes). Methoxy groups may be converted directly to the methyl group of acetate. Acetyl-CoA is synthesized from methyl tetrahydrofolate, CO_2, and reducing equivalents catalyzed by acetyl-CoA synthase, a complex enzyme containing the three transition metals iron, cobalt, and nickel. Energy is conserved via an electrochemical proton or sodium-ion gradient.

12.1.7 Methanogenesis Is Restricted to Archaea

Methanogenic bacteria produce methane under strictly anoxic conditions by cleavage of acetate to CO_2 and methane (CH_4; biogas) and by reduction of C_1 compounds. The C_1-reducing methanogens produce methane mainly from CO_2/H_2, but also from carbon monoxide, formate, methanol, methanol/H_2, methylamines, methyl sulfides, and CO_2/alcohols. Methanogens are found only in the Archaean domain Euryarchaeota. They comprise 17 genera with 50 different species. Some of the genera are only as distantly related to each other as Gram-positive bacteria are to Proteobacteria, for example, the extreme thermophilic *Methanopyrus* [guanine plus cytosine content of DNA, $(G+C) = 60$ mol%] that grows only on CO_2/H_2 at $\leqslant 110\,°C$ and the mesophilic *Methanosarcina* ($G+C = 40$ mol%) at 37 °C that has the broadest substrate spectrum among the methanogens.

The **methanogenic reactions** may be divided into three groups, (1) reductions of C_1 compounds with molecular hydrogen or alcohols with two or more carbon atoms as electron donors (Reactions 12.44–12.46), (2) disproportionations (i.e., internal redox reactions) in which C_1 compounds serve as electron acceptor and electron donor (Reactions 12.47–12.50), or (3) the apparent decarboxylation of acetate, which is actually a disproportionation (Reaction 12.51).

Reductions of C_1 compounds:

$$CO_2 + 4\,H_2 \rightarrow CH_4 + 2\,H_2O \quad \Delta G^{\circ\prime} = -131 \text{ kJ/mol } CH_4 \quad (12.44)$$

$$CH_3OH + H_2 \rightarrow CH_4 + H_2O \quad \Delta G^{\circ\prime} = -112 \text{ kJ/mol } CH_4 \quad (12.45)$$

$$CO_2 + 2\,CH_3CH_2OH \rightarrow CH_4 + 2\,CH_3COO^- + 2\,H^+ \quad \Delta G^{\circ\prime} = -111 \text{ kJ/mol } CH_4 \quad (12.46)$$

Disproportionations of C_1 compounds:

$$4\ HCOO^- + 4\ H^+ \rightarrow CH_4 + 3\ CO_2 + 2\ H_2O \quad (12.47)$$
$$\Delta G^{\circ\prime} = -144\ \text{kJ/mol}\ CH_4$$

$$4\ CO + 2\ H_2O \rightarrow CH_4 + 3\ CO_2 \quad (12.48)$$
$$\Delta G^{\circ\prime} = -448\ \text{kJ/mol}\ CH_4$$

$$4\ CH_3OH \rightarrow 3\ CH_4 + CO_2 + 2\ H_2O \quad (12.49)$$
$$\Delta G^{\circ\prime} = -319\ \text{kJ/mol}\ CH_4$$

$$4\ CH_3NH_3^+ + 2\ H_2O \rightarrow 3\ CH_4 + CO_2 + 4\ NH_4^+ \quad (12.50)$$
$$\Delta G^{\circ\prime} = -235\ \text{kJ/mol}\ CH_4$$

Disproportionation of acetate:

$$CH_3COO^- + H^+ \rightarrow CH_4 + CO_2 \quad (12.51)$$
$$\Delta G^{\circ\prime} = -36\ \text{kJ/mol}\ CH_4$$

The pathway of the reduction of CO_2 to methane via molecular hydrogen has been elucidated in great detail and hence will be described as an example for all the others (Reaction 12.44). The fate of the carbon comprises seven steps, in addition to the reduction of the oxidized coenzymes by hydrogenases (Fig. 12.**9**). Seven new coenzymes unique to the Archaea are involved in this pathway (Fig. 12.**10**): methanofuran, tetrahydromethanopterin, coenzyme F_{420}, coenzyme M, coenzyme B, 5-hydroxybenzimidazolyl-hydroxycobamide, and coenzyme F_{430}. Only tetrahydromethanopterin (H_4MPT), coenzyme F_{420}, and 5-hydroxybenzimidazolyl-hydroxycobamide resemble coenzymes of the Bacteria and the Eukarya.

Tetrahydromethanopterin is related to tetrahydrofolate (H_4F), and coenzyme F_{420} is a 5-deazaflavin derivative with the properties of NAD^+. As shown in Fig. 12.**10**, a part of the structure of the reduced coenzyme is identical to that of NADH. The redox potential of coenzyme F_{420} (oxidized/reduced) is, however, 40 mV more negative than that of the NAD^+/NADH couple (−320 mV). Finally, 5-hydroxybenzimidazolyl-hydroxycobamide is very similar to the hydroxycobalamin (hydroxy form of vitamin B_{12}) present in the acetogenic Bacteria; the only difference is the presence of a 5-hydroxyimidazole base (instead of dimethylbenzimidazole) as axial ligand to the cobalt.

Tetrahydromethanopterin has the same task in methanogenic Archaea as tetrahydrofolate has in the Bacteria and the Eukarya, that of carrying the C_1 compounds (as the formyl to the methyl residue) and thus protecting the organism from the reactive free formaldehyde (Table 12.**2**). Formyl residues are attached to N^5 of tetrahydromethanopterin (unlike to N^{10} in tetrahydrofolate) and dehydrated to N^5,N^{10}-methenyl

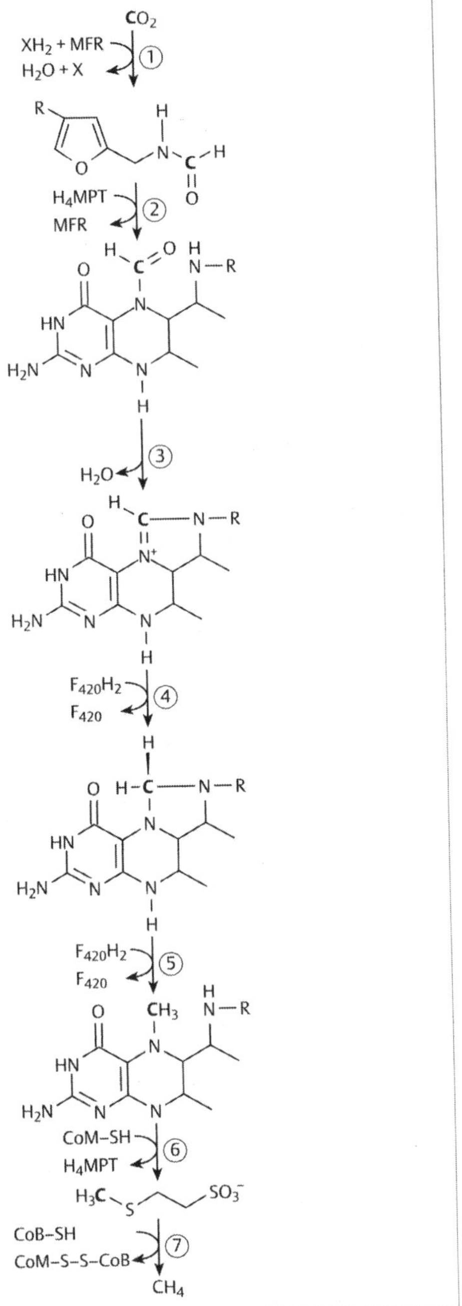

Fig. 12.**9** **Fate of carbon in methanogenesis.** 1, Formylmethanofuran dehydrogenase; X, unknown hydrogen donor; 2, formyltransferase; 3, cyclohydrolase; 4, F_{420}-dependent N^5,N^{10}-methenylmethanopterin dehydrogenase; 5, F_{420}-dependent N^5,N^{10}-methylenemethanopterin dehydrogenase; 6, methyltransferase; 7, methyl-S-CoM reductase [after 2]. For further abbreviations, see Fig. 12.**10**

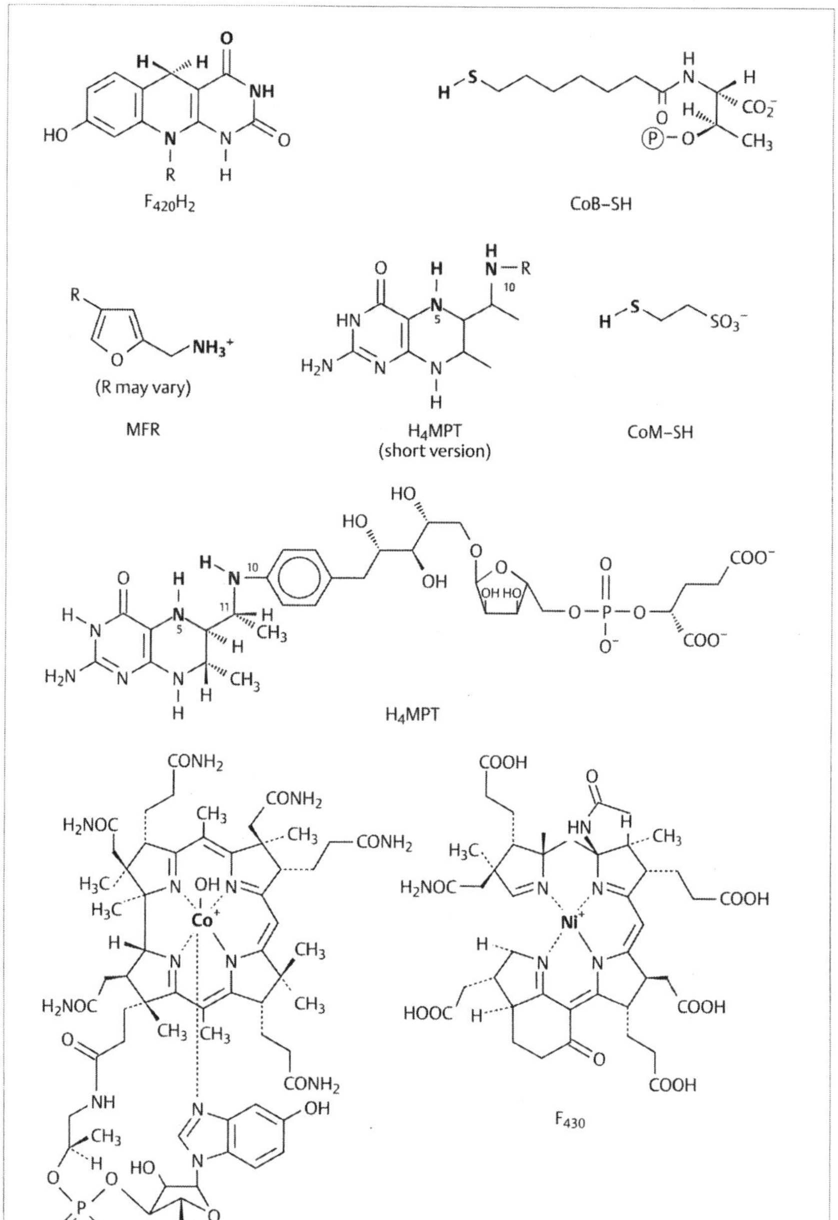

Fig. 12.**10** **New coenzymes in methanogenesis.** F_{420}, a fluorescing factor, is a deazaflavin with an absorbance maximum at 420 nm, part of its structure resembles that of NAD (in red); CoB-SH or HS-HTP, 7-mercaptoheptanoyl threonine phosphate; CoM-SH, mercaptoethanesulfonate; MFR, methanofuran; H_4MPT, tetrahydromethanopterine; F_{430} is a nickel tetrapyrrole with an absorbance maximum at 430 nm

Table 12.2 **Biologically important oxidation states of C_1 compounds**

Oxidation state	Compound
+4	Carbon dioxide, CO_2; bicarbonate, HCO_3^-; carbonate, CO_3^{2-}
	Carbamate, R-NH-COO$^-$
	Urea, NH_2-CO-NH_2
	Guanidinium, R-NH-C(=NH_2^+)-NH_2
	Cyanide, C≡N$^-$
+2	Carbon monoxide, CO
	Formate, HCOO$^-$
	Formamide, NH_2-CHO; *N*-formyl, R_1R_2N-CHO
	N,N′-Methenyl, R_1R_2N-CH=N-$R_3R_4^+$
±0	Carbon, C
	Hydrated formaldehyde, HO-CH_2-OH
	N,N′-Methylene, R_1R_2N-CH_2-NR_3R_4
−2	Methanol, CH_3OH
	Methyl, R_1R_2N-CH_3 or R-O-CH_3 or R-S-CH_3
−4	Methane, CH_4

tetrahydromethanopterin. Reduction to the formaldehyde level (oxidation state 0) by molecular hydrogen either directly or indirectly via reduced coenzyme F_{420} yields N^5,N^{10}-methylene-tetrahydromethanopterin. This compound is further reduced to N^5-methyl-tetrahydromethanopterin, from which the methyl group is transferred to the SH group of coenzyme M, yielding methyl-S-CoM, the actual substrate for the final reduction to methane. The reductant is HS-HTP (7-mercapto-**h**eptanoyl**t**hreonine **p**hosphate) which is also called coenzyme B (Reaction 12.52). This last step is catalyzed by methyl-CoM reductase, the prosthetic group of which is the tetrapyrrole coenzyme F_{430} with nickel as the central atom (Fig. 12.10).

$$\text{Methyl-SCoM} + \text{HSHTP} \rightarrow CH_4 + \text{CoMS-SHTP} \quad \text{(methyl-CoM reductase)} \quad \Delta G^{\circ\prime} = -45\ \text{kJ/mol methane} \tag{12.52}$$

A methyl group attached to the central nickel atom has been postulated as an intermediate in methane formation. Besides methane, a heterodisulfide of coenzyme M and coenzyme B, CoM-S-S-CoB, is generated. The reduction of the heterodisulfide by molecular hydrogen ($\Delta G^{\circ\prime} = -40$ kJ/mol) is catalyzed by a membrane-bound enzyme complex composed of a hydrogenase and a heterodisulfide reductase. The complex, which contains several iron-sulfur clusters and FAD, is thought to act as an electrogenic proton pump by which the major amount of the energy released in methanogenesis is conserved. The first step of this pathway, however, the reduction of CO_2 to the oxidation state of formate in formylmethanofuran, requires energy ($\Delta G^{\circ\prime} = +16$ kJ/mol; this value even increases at the low hydrogen concentrations found in nature, which are much lower than 10^5 Pa, equivalent to 1 M). The enzyme catalyzing this step contains a molybdopterin dinucleotide able to act as a ligand for either Mo or W. These so-called molybdopterins or tungstopterins are not unique to methanogenesis since they are also found in the Bacteria (Fig. 12.**6**). The source of energy for CO_2 activation is not known yet; it could be either ATP or the sodium-ion gradient generated by the membrane enzyme methyl-transferase containing 5-hydroxybenzimidazolylcob(I)amide. In this reaction the free energy ($\Delta G^{\circ\prime} \approx -30$ kJ/mol) released by the transfer of a methyl group from N to S is converted into an electrochemical sodium-ion gradient. A chemically similar methyl transfer is the *S*-methylation of homocysteine by N^5-methyltetrahydrofolate catalyzed by methionine synthase. This biosynthetic enzyme, however, is soluble and does not generate a sodium-ion gradient.

Hence, the energy-conserving reactions of CO_2 reduction to methane (Reaction 12.44) involve only electrochemical ion gradients but no substrate-level phosphorylation. This enables the organisms to adjust their energy yield to varying substrate concentrations in nature. Whereas standard conditions (10^5 Pa H_2; $\Delta G^{\circ\prime} = -131$ kJ/mol CH_4) would allow the production of more than 1 mol ATP/mol CH_4, in most methanogenic habitats the H_2 concentrations range from 1–10 Pa, leading to $\Delta G'$ values of −22 to −45 kJ/mol. This large dependence on the H_2 concentration is due to the requirement of 4 mol H_2/mol CH_4 (Reaction 12.44 and Equation 12.53).

$$\Delta G' = \Delta G^{\circ\prime} + RT\ \ln([CH_4][CO_2]^{-1}[H_2]^{-4}) \tag{12.53}$$

$$\text{with } [CH_4] = [CO_2] \text{ and } [H_2] = 10^{-5}\ \text{M}\ (1\ \text{Pa}/10^5\ \text{Pa})$$

$$\Delta G' = -131 + 5.7\ \lg[10^{-5}]^{-4}$$

$$\Delta G' = -131 + 114 = -17\ \text{kJ/mol}\ CH_4$$

In the disproportionation of methanol to methane and carbon dioxide (Reaction 12.49), all reactions from CO_2 to methyl-H_4MPT are reversible. The only irreversible reactions in the catabolism of methanogens are the final steps leading from methyl-H_4MPT to methane. Although there is geochemical evidence for the anaerobic consumption of methane, it has not been established whether this occurs by reversal of methanogenesis or by a novel C–H activating mechanism.

Among all methanogens, only *Methanosarcina* and *Methanothrix* are able to cleave acetate to methane and

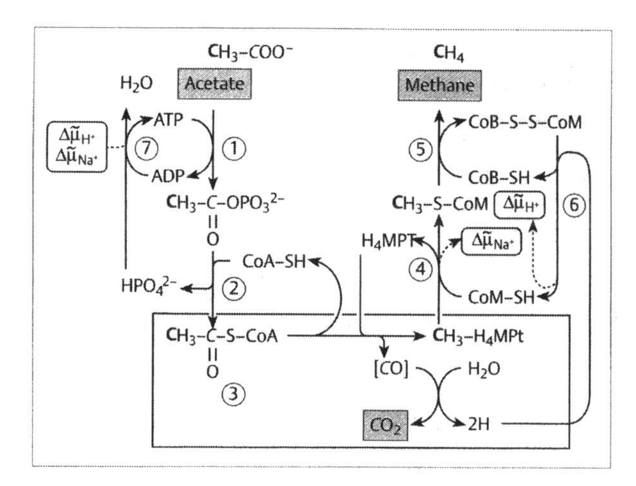

Fig. 12.**11** **Methanogenesis from acetate.** 1, Acetate kinase; 2, phosphate acetyltransferase; 3 (in the large box), acetyl-CoA synthase, (CO dehydrogenase, see Fig. 12.**8**); 4, methyltransferase generating $\Delta\mu_{Na+}$; 5, methyl-CoM reductase; 6, heterodisulfide reductase generating $\Delta\mu_{H^+}$; 7, ATP synthase using $\Delta\mu_{H^+}$ or $\Delta\mu_{Na^+}$ as driving force. For abbreviations of the coenzymes, see Fig. 12.**10**

carbon dioxide, which is the predominant methane-forming reaction in nature. As stated in the introduction to Chapter 12.1.7, the overall decarboxylation of acetate to CO_2 and methane is an internal redox process (Fig. 12.**11**). Initially, acetate is activated to acetyl-CoA by the combined action of acetate kinase (Reaction 12.41) and phosphate acetyltransferase (Reaction 12.42) or it is activated directly by an acetyl-CoA synthetase (Reaction 12.57).

$$\text{Acetate}^- + \text{ATP}^{4-} + \text{CoASH} \rightarrow \text{acetyl-CoA} + \text{AMP}^{2-} + \text{diphosphate}^{3-} \qquad (12.57)$$

Box 12.4 There are at least **three different hydrogenases** present in the methanogenic Archaea: coenzyme-F_{420}-reducing hydrogenases (Reaction 12.54), coenzyme F_{420}-nonreducing hydrogenases (Reaction 12.55), and the hydrogen-forming N^5,N^{10}-methylenetetrahydromethanopterin dehydrogenase (Reaction 12.56).

$$H_2 + F_{420} \rightarrow H_2F_{420} \qquad (12.54)$$

(coenzyme-F_{420}-reducing hydrogenase).

The enzyme contains nickel. The reduced coenzyme F_{420} is required for the methenyl- and methylene-H_4MPT reductases.

$$H_2 + 2A \rightarrow 2H^+ + 2A^- \qquad (12.55)$$

(coenzyme-F_{420}-nonreducing hydrogenase).

The enzyme also contains nickel, A = acceptor.

$$H_2 + \text{methenyl-}H_4\text{MPT} \rightarrow H^+ + \text{methylene-}H_4\text{MPT} \qquad (12.56)$$

(hydrogen-forming N^5,N^{10}-methylene-tetrahydromethanopterin dehydrogenase)

At high H_2 concentrations, a direct reduction of methenyl-H_4MPT by H_2 is possible (Reaction 12.56). The reaction is catalyzed by an enzyme devoid of transition metals. This is unusual, because it is generally accepted by chemists that the activation of hydrogen is only possible in the presence of metals such as Ni and/or Fe. In this enzyme, however, a direct attack of H_2 at the methine carbon of methenyl-H_4MPT is assumed, yielding an intermediate with a 5-coordinated C atom. The other hydrogenases of methanogens are "normal", i.e., they contain a Ni-Fe center and [Fe–S] clusters. In addition, in some of these enzymes, one ligand of Ni is selenocysteine, which leads to specific activities (up to $10^5\,s^{-1}$) much higher than the corresponding hydrogenases that contain only sulfur. In the F_{420}-reducing hydrogenase, the hydrogen is bound to the Ni-Fe center, where the electrons are passed to a chain of [Fe–S] clusters which deliver the electrons via FMN to F_{420}. The protons generated at the Ni-Fe center, deeply embedded in the protein, are transported to the surface of the enzyme via a histidine-rich channel. The F_{420}-nonreducing hydrogenases are less well characterized: they are part of the two energy-transducing enzyme complexes, the formylmethanofuran dehydrogenase and the heterodisulfide reductase. In some methanogens such as *Methanosarcina barkeri*, the latter complex contains a cytochrome *b*.

Consequently, acetyl-CoA is decarbonylated to methyl-H_4MPT and CO, which is oxidized to CO_2. The resulting reducing equivalents are used to generate methane by the pathway shown in Fig. 12.**9**. The CO dehydrogenase (or acetyl-CoA synthase), which catalyzes the decarbonylation of acetyl-CoA to methyl-H_4MPT, CoASH, and CO and the subsequent oxidation of CO, is a nickel-cobalt-iron enzyme similar to the acetyl-CoA synthase of the acetogenic Bacteria. The amount of energy conserved by the decarboxylation of acetate is not very large since most of the ATP synthesized by the H^+-translocating ATP-synthase is used for the activation of acetate. The sodium-ion gradient formed by the methyl-residue transfer also contributes to ATP synthesis via a Na^+/H^+ antiporter. Overall, approximately 1/3 mol ATP may be conserved per mol CH_4 formed, which agrees well with the overall change in free energy ($\Delta G^{\circ\prime} = -30$ kJ/mol).

All methanogenic Archaea are able to reduce CO_2 to methane with molecular hydrogen. Most of the methane, however, is produced by only two genera, *Methanosarcina* and *Methanotrix*, through disproportionation (apparent decarboxylation) of acetate to CO_2 and CH_4. Methanogenesis from CO_2 involves seven unique coenzymes found neither in Bacteria nor in Eukarya: methanofuran, tetrahydromethanopterin, coenzyme F_{420}, coenzyme M, coenzyme B, 5-hydroxybenzimidazolylhydroxycobamide, and coenzyme F_{430}. Energy is conserved via an electrochemical proton gradient *and* sodium-ion gradient.

12.1.8 Dissimilatory Sulfate Reduction to H_2S Is Widespread in Nature

Sulfate-reducing bacteria and Archaea are able to use sulfate or thiosulfate as electron acceptors for anaerobic respiration (Table 12.**3**), which is called dissimilatory sulfate reduction. For the formation of the final product

Table 12.**3** **Biologically important inorganic sulfur compounds**

Oxidation state	Compound
+6	Sulfur trioxide, SO_3
	Sulfate, SO_4^{2-}
	Hydrogen sulfate or bisulfate, HSO_4^-; $pK_2 = 1.9$
+5	Dithionate, $^-O_3S\text{-}SO_3^-$
+10/(**x**+2)	Polythionate, $^-O_3S\text{-}S_x\text{-}SO_3^-$
+4	Sulfur dioxide, SO_2
	Sulfite, SO_3^{2-}
	Bisulfite, $^-HSO_3$ or $HOSO_2^-$; $pK_2 = 6.9$
	Disulfite, $^-O_3S\text{-}SO_2^{-*}$
+3	Dithionite, $^-O_2S\text{-}SO_2^-$
+2	Thiosulfate $^-S\text{-}SO_3^-$
±0	Sulfur, S_8 or S_x
−1	Disulfide, $^-S\text{-}S^-$
−2/(**x**+2)	Polysulfide, $^-S\text{-}S_x\text{-}S^-$
−2	Sulfide, S^{2-}
	Bisulfide, HS^-; $pK_2 = 12$
	Hydrogen sulfide, H_2S; $pK_1 = 7$

Usually, $\mathbf{x} < 6$, but species with $\mathbf{x} > 8$ also are known.
* as solid, hydrolyzes in water: $S_2O_5^{2-} + H_2O \rightarrow 2^- HSO_3$

Table 12.**4** **Electron donors of sulfate reducers**

Common substrates	Less common substrates
Hydrogen	Alanine, glutamate
Formate, acetate	Choline, glycerol
Propionate, butyrate (C_4) to arachinate (C_{20})	Fructose
Lactate	Nicotinate
Fumarate	Indol
Succinate	Hydrocarbons, preferentially C_{16}, but not CH_4
Methanol	Starch and peptides (only *Archaeoglobus*)
Ethanol, propanol	Benzoate, phenylacetate, phenol

hydrogen sulfide, eight electrons/mol sulfate are consumed. The electron donor may be molecular hydrogen (Reactions 12.58 and 12.59) in addition to a vast variety of organic compounds (Table 12.**4**). Hence, sulfate-reducing organisms have a much broader substrate specificity than methanogens.

$$SO_4^{2-} + H^+ + 4\,H_2 \rightarrow HS^- + 4\,H_2O \quad \Delta G^{\circ\prime} = -152\,\text{kJ/mol} \tag{12.58}$$

$$SSO_3^{2-} + H^+ + 4\,H_2 \rightarrow 2\,HS^- + 3\,H_2O \quad \Delta G^{\circ\prime} = -174\,\text{kJ/mol} \tag{12.59}$$

Four unrelated bacterial groups are capable of dissimilatory sulfate reduction. The largest group, which belongs to the δ-subdivision of the Proteobacteria comprises the genera *Desulfo*..., including *Desulfovibrio*, *Desulfobulbus*, *Desulfomicrobium*, *Desulfobacter*, *Desulfobacterium*, *Desulfosarcina*, *Desulfococcus*, *Desulfonemia*, among others. The other groups include the Gram-positive genus *Desulfotomaculum* (related to clostridia), the Gram-negative genus *Thermodesulfobacterium* (belonging to a separate bacterial phylum), and the archaean genus *Archaeoglobus*.

The pathway of the **dissimilatory sulfate reduction** is depicted in Fig. 12.**12**. This process can be divided in two steps: (1) endergonic reduction to sulfite catalyzed by soluble enzymes and (2) exergonic sulfite reduction to sulfide, probably catalyzed by membrane-bound enzymes. Since the first step, the reduction of sulfate to sulfite ($E_0' = -516$ mV), requires a reductant more negative than H_2 ($E_0' = -414$ mV), the substrate is activated by ATP, which raises the redox potential to $E_0' = -60$ mV.

$$SO_4^{2-} + ATP^{4-} + H^+ \rightarrow APS^{2-} + HP_2O_7^{3-} \quad \text{(ATP sulfurylase)} \quad \Delta G^{\circ\prime} = +46\,\text{kJ/mol} \tag{12.60}$$

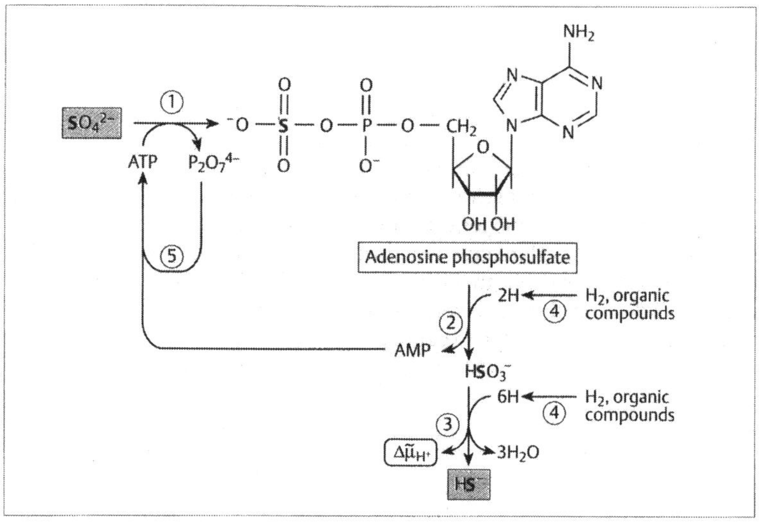

Fig. 12.**12** **Dissimilatory sulfate reduction.** 1, ATP sulfurylase; 2, adenosylphosphosulfate(APS) reductase; 3, sulfite reductase; 4, hydrogenases or dehydrogenases; 5, hypothetical H^+-transporting ATP diphosphatase

$$APS^{2-} + 2\,H \rightarrow AMP^{2-} + HSO_3^- + H^+ \quad \text{(APS-reductase)} \qquad (12.61)$$

$$HP_2O_7{}^{3-} + AMP^{2-} \rightarrow ATP + H_2O \quad \Delta G^{\circ\prime} \sim +35\ \text{kJ/mol} \quad \text{(ATP diphosphatase, driven by } \Delta\mu_{H^+}\text{)} \qquad (12.62)$$

The endergonic activation of sulfate (Reaction 12.60) is driven by the subsequent reduction of APS (adenosylphosphosulfate) to sulfite (Reaction 12.61) which shifts the reaction to the right side. The formation of ATP from AMP and diphosphate may conserve the energy-rich bond of the latter (Reaction 12.62), but the reaction has not been established yet. The activation of sulfate is similar to the activation of acetate to acetyl-CoA as found in some Archaea and in Eukarya (Reaction 12.57). The two-electron reduction of APS to sulfite is catalyzed by a flavoprotein containing [Fe–S] clusters. The natural electron donor is unknown. This reduction differs from the **assimilatory sulfate reduction** occurring in many organisms that do not reduce sulfate, (e.g., in plants; see Chapter 8).

There are four different dissimilatory sulfite reductases in sulfate-reducing organisms: desulfoviridin, desulforuberin, desulfofuscidin, and P-582 (similar to cytochrome P). They all contain siroheme (Fig. 12.**7b**) and [4 Fe–4 S] clusters, but they differ in their spectroscopic properties as indicated by their names (Latin, viridis = green, rubrum = red, fuscus = dark brown). The intermediates of this six-electron reduction are possibly trithionate ($^-O_3SSSO_3^-$) and thiosulfate (SSO_3^-) (Reaction 12.63).

In the first step, only one molecule of sulfite is reduced to form [S^{2+}], which is sandwiched between two sulfite molecules. A further two-electron reduction of this sulfur atom to [S^0] yields thiosulfate, which is finally reduced to sulfide. In both step 2 and step 3, a carrier sulfite is regenerated. Again, the natural electron donor remains to be identified.

$$(1)\ 3\,HSO_3^- + 2\,e^- + 3\,H^+ \rightarrow {}^-O_3S{-}S{-}SO_3^- + 3\,H_2O$$

$$(2)\ {}^-O_3S{-}S{-}SO_3^- + 2\,e^- + H^+ \rightarrow {}^-S{-}SO_3^- + HSO_3^-$$

$$(3)\ {}^-S{-}SO_3^- + 2\,e^- + 3\,H^+ \rightarrow HSO_3^- + H_2S$$

$$\text{Sum: } HSO_3^- + 6\,e^- + 7\,H^+ \rightarrow H_2S + 3\,H_2O \quad \text{(sulfite reductase)} \qquad (12.63)$$

The mode of energy conservation in sulfate-reducing organisms is an elusive problem. Since the activation of sulfate requires 1 mol ATP/mol SO_4, the reduction of sulfite to H_2S has to generate more than 1 mol ATP/mol SO_4. The ubiquitous occurrence of cytochromes and menaquinones in sulfate reducers suggests that sulfite reduction involves ATP synthesis via a proton pump with membrane-bound protein complexes similar to the mitochondrial complexes I, III, and IV as well as an F_1F_0-ATP synthase (Fig. 12.**12**).

Sulfate reducers are versatile in carbon metabolism. In 1886 Hoppe-Seyler demonstrated a complete oxidation of cellulose in anoxic enrichments with mud in the presence of gypsum ($CaSO_4$), which was reduced to H_2S. Despite this early discovery, it was assumed until the late nineteen seventies that sulfate reducers were able to oxidize organic compounds only as far as acetate, for example, *Desulfovibrio vulgaris*, which grows on lactate (Reaction 12.64).

$$2\ \text{Lactate}^- + SO_4^{2-} + 2\,H^+ \rightarrow 2\ \text{acetate}^- + 2\,CO_2 + H_2S + 2\,H_2O \quad \Delta G^{\circ\prime} = -160\ \text{kJ/mol} \qquad (12.64)$$

During the nineteen eighties many sulfate reducers that are able to oxidize organic compounds completely to CO_2 were discovered. It now appears that the majority of these organisms are complete oxidizers; the remaining ones are called "incomplete oxidizers." The incomplete oxidizers are easier to isolate because they grow much faster than the complete oxidizers. The complete oxidizers may be divided into two groups: those using modified versions of the Krebs cycle (see below) and those using the reverse acetogenic pathway, which involves the CO-dehydrogenase complex (Fig. 12.**8**). Although *Archaeoglobus* belongs to the latter group, it contains the archaean version of this pathway in which tetrahydromethanopterin (H_4MPT, Fig. 12.**10**) is used instead of tetrahydrofolate (H_4F, see Chapter 9)

Whereas in *Desulfurella acetivorans* the Krebs cycle operates similarly to that in mitochondria, *Desulfobacter* and *Desulfuromonas* contain a succinyl-CoA : acetate CoA-transferase instead of a succinate thiokinase (Reaction 12.65).

$$\text{Succinyl-CoA}^- + \text{acetate}^- = \text{succinate}^{2-} + \text{acetyl-CoA} \quad \text{(succinate CoA-transferase)} \quad \Delta G^{\circ\prime} \approx 0\,\text{kJ/mol} \tag{12.65}$$

Thus, the substrate acetate is directly activated by succinyl-CoA to acetyl-CoA without involving ATP. Most sulfate reducers use citrate synthase in the Krebs cycle (Reaction 12.66); the only known exceptions being *Desulfobacter* species, which contain ATP-citrate lyase (Reaction 12.67). Hence, one mol ATP may be conserved by substrate-level phosphorylation.

$$\text{Acetyl-CoA} + \text{oxaloacetate}^{2-} + H_2O \rightarrow \text{citrate}^{3-} + \text{CoASH} + H^+ \quad \text{(citrate synthase)} \quad \Delta G^{\circ\prime} = -33\,\text{kJ/mol} \tag{12.66}$$

$$\text{Acetyl-CoA} + \text{oxaloacetate}^{2-} + ADP^{3-} + HPO_4^{2-} \rightleftharpoons \text{citrate}^{3-} + \text{CoASH} + ATP^{4-} \quad \text{(ATP citrate lyase)} \quad \Delta G^{\circ\prime} \approx 0\,\text{kJ/mol} \tag{12.67}$$

Sulfate reducers and methanogens compete with each other for H_2 and other substrates. Sulfate reducers are commonly found in marine sediments, whereas methanogens are present in freshwater sediments, rice fields, and rumen of cattle. This distribution reflects the different sulfate concentrations, which are as high as 28 mM in seawater, while freshwater lakes and ponds only contain ≥0.2 mM sulfate. Furthermore, high sulfate concentrations inhibit methane formation. The higher energy yield enables the sulfate reducers ($\Delta G^{\circ\prime} = -152$ kJ/mol sulfate, Reaction 12.58, as compared to methanogenesis, $\Delta G^{\circ\prime} = -131$ kJ/mol CH_4, Reaction 12.44) to grow at lower hydrogen concentrations (see Reaction 12.39 and the following discussion); in addition they can utilize a much broader range of substrates and therefore outcompete the methanogens. Although methanogenesis and sulfate reduction are extremely oxygen-sensitive processes, both groups of organisms are able to survive under oxic conditions. Some sulfate reducers even contain terminal oxidases that use molecular oxygen as electron acceptor for respirations.

Sulfate reduction is a reversible process, so sulfate-reducing organisms should also be able to disproportionate intermediates of this pathway to sulfate and sulfide. Thermodynamic calculations show indeed that, for the disproportionations of sulfite and thiosulfate, $\Delta G^{\circ\prime} < 0$ (Reactions 12.68 and 12.69), and that they may be involved in energy conservation.

$$S_2O_3^{2-} + H_2O \rightarrow SO_4^{2-} + HS^- + H^+ \quad \Delta G^{\circ\prime} = -22\,\text{kJ/mol thiosulfate} \tag{12.68}$$

$$4\,SO_3^{2-} + H^+ \rightarrow 3\,SO_4^{2-} + HS^- \quad \Delta G^{\circ\prime} = -59\,\text{kJ/mol sulfite} \tag{12.69}$$

The recently isolated sulfate reducer *Thiocapsa thiozymogenes* is able to grow either on thiosulfate or sulfite as an energy source. In addition this organism is able to disproportionate elemental sulfur to sulfide and sulfate (Reaction 12.70).

$$4\,S^0 + 4\,H_2O \rightarrow SO_4^{2-} + 3\,HS^- + 5\,H^+ \quad \Delta G^{\circ\prime} = +10\,\text{kJ/mol}\ S^0 \tag{12.70}$$

Dissimilatory sulfate reduction occurs in two steps, endergonic activation of sulfate and reduction to sulfite, followed by exergonic reduction of sulfite to hydrogen sulfide, whereby energy is conserved via an electrochemical proton gradient. In contrast to methanogens, which are restricted to C_1 compounds and acetate, sulfate reducers are able to use not only molecular hydrogen but also a vast variety of organic compounds as electron donors. According to their carbon metabolism, sulfate reducers are differentiated by their final products: those forming CO_2 are **complete oxidizers**, whereas those excreting acetate are called **incomplete oxidizers**. The complete oxidations are performed either via modifications of the citric acid cycle or more frequently via reverse acetogenesis. In contrast to the sulfate reducers of the Bacteria, *Archaeoglobus* uses some coenzymes of methanogenesis for complete oxidations.

Due to the unfavorable thermodynamics, growth is not possible. Addition of ferric hydroxide, however, allows slow growth. Ferric hydroxide is reduced to ferrous sulfide and elemental sulfur is regenerated (reaction 12.71).

$$3\ HS^- + 2\ FeOOH + 3\ H^+ \rightarrow S^0 + 2\ FeS + 4\ H_2O \quad \Delta G^{\circ\prime} = -144\ kJ/mol\ S^0 \qquad (12.71)$$

The sum of Reactions 12.70 and 12.71 yields a complete **inorganic fermentation** (Reaction 12.72).

$$3\ S^0 + 2\ FeOOH \rightarrow SO_4^{2-} + 2\ FeS + 2\ H^+ \quad \Delta G^{\circ\prime} = -34\ kJ/mol\ S^0 \qquad (12.72)$$

12.1.9 Energy Conservation Coupled to Reductive Dechlorination

The sulfate reducer *Desulfomonile tiedjei* is able to use 3-chlorobenzoate and related *meta*-chloroaromatic compounds as electron acceptors (Fig. 12.**13**). Chlorobenzoate is thereby reduced to benzoate (Reaction 12.73).

$$\text{3-Chlorobenzoate}^- + 2\ e^- = H^+ \rightarrow \text{benzoate}^- + Cl^- \quad \text{(3-chlorobenzoate dehalogenase)} \qquad (12.73)$$

$$\text{Formate}^- \rightarrow CO_2 + 2\ e^- + H^+ \quad \text{(formate dehydrogenase)} \qquad (12.74)$$

This reaction is coupled to the oxidation of formate (Reaction 12.74) or molecular hydrogen (hydrogenase), whereby energy is conserved via an electrochemical proton gradient. Competition experiments indicate that the electron carriers involved in the reduction of sulfite to hydrogen sulfide are also used for the dehalogenation.

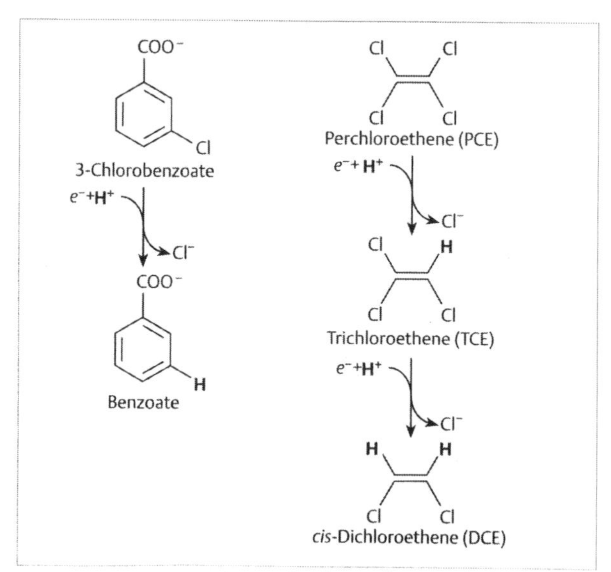

Fig. 12.**13** **Compounds involved in reductive dechlorination**

Recently, it was shown that *Dehalospirillum multivorans* (ε-group of the Proteobacteria) is able to grow on molecular hydrogen and the industrial solvent tetrachloroethene (perchloroethene, PCE) as electron acceptor (Reaction 12.75), whereby chloride, transiently trichloroethene (TCE), and finally *cis*-dichloroethene (DCE) are formed (Fig. 12.**13**).

$$\text{Perchloroethene} + 2\ H_2 \rightarrow \textit{cis}\text{-dichloroethene} + 2\ H^+ + 2\ Cl^- \quad \Delta G^{\circ\prime} = -376\ kJ/2\ mol\ Cl^- \qquad (12.75)$$

A hydrogenase (Box 12.4) and a PCE/TCE-dehalogenase (Reaction 12.75) have been detected in the organism. The latter was purified and shown to contain corrinoids (derivatives of vitamin B_{12}) and, probably, two different [4 Fe–4 S] clusters.

Some anaerobic bacteria are able to reduce C–Cl bonds to C–H bonds, whereby energy is conserved via an electrochemical proton gradient. The mechanism of this process, called **reductive dechlorination**, has not been elucidated yet.

12.1.10 In Most Cases the Redox Potential Determines the Selection of the Electron Acceptor

In nature, a variety of electron acceptors are usually available for the organisms. Hence, the question arises whether the electron acceptors are used randomly or in a specific order. For a growing organism, it would be more economical to produce only one type of electron-transport chain at a time rather than to be prepared to use all available electron acceptors simultaneously. In sediments, it has always been observed that electron acceptors are used according to the energy yield. Therefore, those oxidants available in high concentrations and with the most positive redox potential have priority. This may be the reason why facultative anaerobes "prefer" oxygen over all other acceptors. Only if oxygen has been consumed, do these organisms use nitrate, ferric ion, fumarate, trimethylamine *N*-oxide, and dimethyl sulfoxide in this order. The more negative electron acceptors such as sulfate or carbon dioxide can only be used by specialized anaerobes which generally are unable to use oxygen. Finally, if no electron acceptor is left, fermentation is used for energy conservation. *Wolinella* however, prefers sulfur, with the lowest redox potential, over nitrate, with a much higher E_0' (see Table 4.1). The specific order of electron acceptors used by a certain organism is genetically determined and the regulatory mechanisms involved are very complex. There are not only thermodynamic but also kinetic and ecological reasons why such a specific order has developed in a certain organism during evolution.

For respirations, facultative anaerobes generally use electron acceptors according to their redox potential, preferably the most positive ones. Strict anaerobes usually use only electron acceptors with low redox potentials.

12.2 Fermentation Is an Anaerobic Redox Process

Many anaerobic bacteria are able to live from organic substances in the absence of suitable inorganic electron acceptors. These organisms occur in all places where organic matter anaerobically decomposes, mainly in the soil, in marine and freshwater sediments, in anoxic sewage sludge, and also in the intestinal tract and other anoxic niches of the animal and human body. The mode of energy metabolism (catabolism) of these organisms is called fermentation.

A fermentation is an anaerobic redox process, in which ATP is generated via substrate-level phosphorylation coupled to the oxidation of the substrate (oxidative branch of metabolism), whereas the electron transport to an acceptor molecule (reductive branch of metabolism) is usually not coupled to energy conservation (Fig. 12.**14**). The electron donors and acceptors are derived from organic molecules of medium redox states such as sugars, organic acids, amino acids, and heterocyclic compounds. Highly oxidized compounds (e.g., carbon dioxide) or highly reduced compounds (e.g., hydrocarbons or fatty acids) are not suitable for fermentations. Their anaerobic transformations involve respirations that require inorganic electron donors (e.g., molecular hydrogen) or acceptors (e.g., nitrate, sulfate, or carbon dioxide; see Chapter 12.1).

Whereas in most respirations the organic substrate is converted to water and CO_2, in fermentations a variety of compounds are formed (for example Table 12.**5**). Hence, it may become necessary to check whether all the products have been quantitatively recovered. Therefore two values are important, the carbon recovery, and the oxidation/reduction balance, which should be close to 100% and 1.0, respectively. The latter is derived by dividing the amount (in mol) of the oxidized compounds (products – substrates) by the amount (in mol) of the reduced compounds (products – substrates, Table 12.**5**).

In most fermentations a single compound that serves both as electron donor and as electron acceptor is fermented. Only in some amino acid fermentations does the one substrate (S_1), which is transformed to the actual electron donor, differ from the other one (S_2), yielding the acceptor (Stickland reaction, Fig. 12.**14a**; see also Fig. 12.**31**). In the fermentation of single substrates, oxidative steps are generally followed by reductive steps (Fig. 12.**14b**) or the pathways are branched in which a series of steps lead to an intermediate (pyruvate, acetyl-CoA, crotonyl-CoA) capable of being oxidized as well as reduced (Fig. 12.**14c**); sometimes this redox reaction

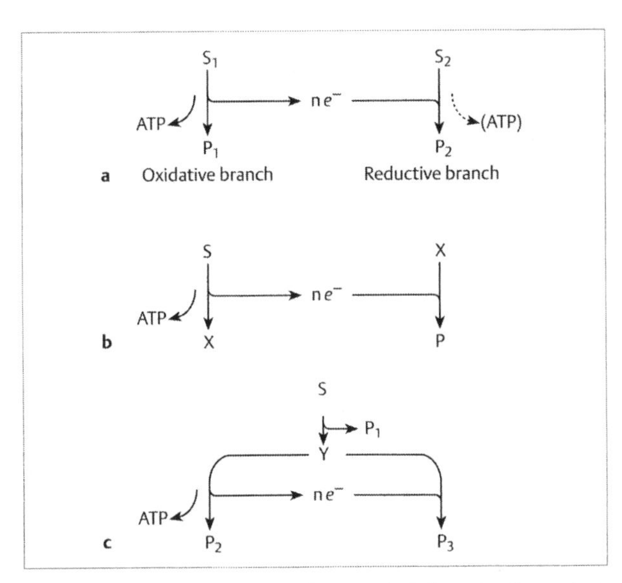

Fig. 12.**14** **Fermentation pathways.** S, Substrate; P, product; X, Y intermediates; n, number of electrons (e^-) transferred from the oxidative branch to the reductive branch.
a, Fermentation of two substrates (e.g., Stickland reaction): S_1, alanine; S_2, 2 glycine; n = 4; P_1, acetate, CO_2, and ammonia; P_2, 2 acetate and 2 ammonia.
b linear fermentations of single substrates (e.g., homolactate fermentation): S, glucose; X, 2 pyruvate; n = 4 (2 NADH); P, 2 lactate (see Fig. 12.**18**).
c Branched fermentation of a single substrate (e.g., glutamate fermentation): S, glutamate; Y, crotonyl-CoA, n = ≧2; P_1, CO_2 and ammonia; P_2, acetate; P_3, butyrate and H_2

Table 12.5 **Calculations of carbon recovery and oxidation/reduction balance for the mixed acid fermentation of glucose by *E. coli*; for the pathway see Fig. 12.22.**

Substrate or products	(mol)	Summation formula	Carbon (mol)	Oxidation state	Excess reduced(−) (mol)	or oxidized(+)
Glucose	100	$C_6H_{12}O_6$	600	±0	0	0
2,3-Butanediol	0.3	$C_4H_{10}O_2$	1.2	−6	−1.8	
Acetoin	0.06	$C_4H_8O_2$	0.2	−4	−0.2	
Glycerol	1.4	$C_3H_8O_3$	4.2	−2	−2.8	
Ethanol	50.0	C_2H_6O	100.0	−4	−200.0	
Formic acid	2.4	CH_2O_2	2.4	+2		+4.8
Acetic acid	36.5	$C_2H_4O_2$	73.0	0	0	
Lactic acid	79.5	$C_3H_6O_2$	238.5	0	0	
Succinic acid	10.7	$C_4H_6O_2$	42.8	+2		+21.4
CO_2	88.0	CO_2	88.0	+4		+352.0
H_2	75.0	H_2	0	−2	−150.0	
Total (Σ)			550.3		−354.8	+378.2

The oxidation state is calculated from the summation formula, with C = 0, H = −1, O = +2, N = +3, and S = +2. The excess reduced or oxidized is calculated by multiplying the oxidation state with the amount of substrate or product.

$$\text{C recovered} = \frac{\Sigma \text{ C products}}{\Sigma \text{ C substrate fermented}} = \frac{550.3}{600} \cdot 100 = 91.7(\%)$$

Σ reduced (products − substrate) = 354.8; Σ oxidized (products − substrate) = 378.2

$$\text{oxidation/reduction (O/R) balance} = \frac{\Sigma \text{ oxidized (products − substrate)}}{\Sigma \text{ reduced (products − substrate)}} = \frac{378.2}{354.8} = 1.07$$

occurs intramolecularly as in xylulose 5-phosphate (Fig. 12.**16**, phosphoketolase pathway). The oxidative steps or branches transform the substrate or intermediate to an energy-rich organic phosphate able to phosphorylate ADP. The reductive steps or branches are required solely as electron sinks, since the reductions do not contribute to substrate-level phosphorylation. A notable exception are some clostridia and related Gram-positive bacteria, which are capable of reducing glycine to acetyl phosphate (Fig. 12.**32**), a process followed by phosphorylation of ADP. This is the only example of a substrate-level phosphorylation coupled to a reduction.

The products of the fermentations are carbon dioxide, molecular hydrogen, formate, acetate, and other short-chain fatty acids. In addition, amino acids give rise to ammonia, hydrogen sulfide, methyl mercaptan, branched-chain fatty acids, and aromatic acids. Some of these products may be converted further also by fermentative organisms. However, the complete anaerobic transformation of these products to carbon dioxide and methane requires a consortium of syntrophic bacteria (Chapter 12.2.5), acetogens (Chapter 12.1.6), and methanogens (Chapter 12.1.7).

The formation of molecular hydrogen deserves some attention, since it is an exception to the generalizations stated above. Molecular hydrogen is generated via the reduction of a proton by a hydride or via the reduction of two protons by two electrons catalyzed by the enzyme hydrogenase (Box 12.4). Hence, the electron acceptor is a proton rather than the organic substrate of the fermentation. Therefore, the formation of hydrogen conserves substrates for oxidation, leading to an increase in the yield of ATP per mol of substrate consumed. However, the removal of the reducing equivalents as molecular hydrogen is thermodynamically less favorable ($E_0' = -420\,mV$) than the reduction of most substrates which have a more positive redox potential. Apparently, many fermentative bacteria produce as much hydrogen as thermodynamically possible which results in optimization of their growth yield.

An intriguing question regarding the fermentations is the amount of ATP conserved per mol of substrate fermented. A rough estimate can be obtained from the free energy ($\Delta G'$) of the overall process. Since fermentations are irreversible processes, the energy required for the formation of 1 mol ATP is ≥70–80 kJ/mol rather than 40–50 kJ/mol under equilibrium conditions. In many cases, the maximum amount of conserved ATP is simply calculated by dividing the $\Delta G'$ of the overall process by −80. The error may be substantial, however, since $\Delta G'$ is usually equated to $\Delta G^{\circ\prime}$ for practical reasons. In addition, sometimes it is mechanistically impossible to conserve ATP even if $\Delta G^{\circ\prime}$ of the reaction is more negative than −80 kJ/mol.

There is no strict **nomenclature for fermentations**. They are named after the substrates or after the products, depending on which is more characteristic for the reaction. For instance, the fermentations of glucose to lactate by lactic acid bacteria, to the solvents butanol and acetone by *Clostridium acetobutylicum*, or to ethanol and carbon dioxide by yeasts and *Zymomonas* are called lactic acid fermentation, solvent fermentation, or ethanol fermentation, respectively. On the other hand, the fermentations of amino acids to ammonia, carbon dioxide, short-chain fatty acids, and molecular hydrogen are specified by the individual amino acid substrate rather than by the less characteristic products, for example, the fermentation of glycine to acetate and ammonia is called glycine fermentation, and the fermentation of glutamate to ammonia, CO_2, acetate, butyrate, and H_2 commonly is called glutamate fermentation, but also the name butyrate fermentation has been applied to this process. Sometimes the term "amino acid fermentation" is misused to describe general biotechnological processes, for example, the aerobic conversion of sugars to certain amino acids, such as the partial oxidation of glucose to glutamate by *Corynebacterium glutamicum*.

A fermentation is an anaerobic redox process in which ATP is generated via substrate-level phosphorylation coupled to the oxidation of the substrate (oxidative branch of metabolism), whereas the electron transport to an acceptor molecule (reductive branch of metabolism) is normally not coupled to energy conservation. The electron donors and acceptors are derived from organic molecules of medium redox states such as sugars, organic acids, amino acids, and heterocyclic compounds, which are converted to carbon dioxide, molecular hydrogen, formate, acetate, and other short-chain fatty acids. In addition, amino acids give rise to ammonia, hydrogen sulfide, methyl mercaptane, branched-chain fatty acids, and aromatic acids. Further conversion of these products requires respiration. A rough estimate of the amount of ATP (mol/mol substrate) synthesized during a fermentation is obtained by dividing the overall $\Delta G'$ (kJ) of the reaction by −80 kJ.

12.2.1 Four Pathways of Sugar-Phosphate Degradation Are Known

The most common substrates for fermentative microorganisms are hexoses and pentoses. In addition, bacteria are able to ferment degradation products of these sugars, for example, sugar acids, glycerol, citrate, malate, succinate, pyruvate, lactate, ethanol, and acetate. Prior to their cleavage to smaller compound units, hexoses and pentoses are phosphorylated. However, it has been discovered recently that some members of the Archaea degrade sugars without initial phosphorylation. There are at least four pathways to convert sugar phosphates to either pyruvate alone or to additional acetyl phosphate via: (1) fructose-1,6-bisphosphate aldolase (Embden-Meyerhof pathway and archaen variants of it), (2) 2-dehydro-3-deoxy-6-phosphogluconate aldolase (KDPG-aldolase, Entner-Doudoroff pathway), (3) xylulose-5-phosphate phosphoketolase (phosphoketolase pathway), or (4) a combination of transaldolase and ketolase reactions (*Bifidobacterium bifidum* pathway). Pyruvate is exclusively produced by the latter two pathways, which contain aldolases as key enzymes, whereas in phosphoketolase reactions also acetyl phosphate is formed. The term key enzyme defines an enzyme specific for the pathway.

Oxidation steps determine the Embden-Meyerhof pathway. The Embden-Meyerhof pathway is widespread in nature and occurs in animals, plants, fungi and yeasts, and in many bacteria such as *Escherichia coli*, clostridia, and homofermentative lactic acid bacteria. Its reactions are described in Chapter 3. The term "Embden-Meyerhof" only applies to the **oxidative part** of the pathway leading from 1 mol glucose to 2 mol pyruvate whereby 2 mol ATP are produced from ADP and inorganic phosphate. Other hexoses are also degraded by this route. Differences among organisms are due to differences in the further fate of pyruvate, which provides electron acceptors for the NADH generated by the oxidation of glyceraldehyde 3-phosphate to 1,3-bisphosphoglycerate (3-phosphoglyceroyl-phosphate). The formation of the initial glucose 6-phosphate and fructose 6-phosphate is also subject to variations that depend on the kind of phosphorylation of glucose, either by ATP or by PEP (phosphotransferase system, see Chapter 3) or the mode of introducing other hexoses into the pathway. In some Archaea, the initial phosphorylation reactions are catalyzed by ADP-dependent enzymes that produce AMP (see Chapter 9).

In the **Entner-Doudoroff pathway** only 1 ATP is formed from glucose. Hence, this pathway is commonly found in aerobic organisms or bacteria capable of anaerobic respiration coupled to electron-transport phosphorylation in order to provide additional ATP.

The only fermentative bacteria that use this pathway for glucose degradation are *Zymomonas* species; these bacteria are adapted to high sugar concentrations. In addition, this pathway is used by *Escherichia coli*, for the fermentation of gluconate. In the Entner-Doudoroff pathway (Fig. 12.**15**) glucose is phosphorylated to glucose 6-phosphate, which is followed by an $NADP^+$- or NAD^+-dependent oxidation to glucono-1,5-lactone 6-phosphate. In the next step, the lactone hydrolyzes to 6-phosphogluconate. In the latter two reactions, an aldehyde is oxidized to a carboxylic acid but, contrary to the 3-phosphoglyceraldehyde-dehydrogenase reaction, no energy-rich acyl phosphate is formed, which explains the lower ATP yield. The two consecutive steps are catalyzed by the key enzymes of the pathway, phosphogluconate dehydratase and 2-dehydro-3-deoxy-6-phosphogluconate aldolase. The 2,3-dehydration of 6-phosphogluconate yields an enol, which spontaneously and almost irreversibly tautomerizes to 2-dehydro-3-deoxy-6-phosphogluconate. Frequently, this compound is incorrectly called 2-keto-3-deoxy-6-phosphogluconate (KDPG). In the consecutive aldolase reaction, a Schiff's base is formed between a specific lysine residue of the enzyme and the carbonyl group of the substrate, which facilitates the cleavage of the 3,4-carbon bond and yields pyruvate and glyceraldehyde 3-phosphate.

In the **phosphoketolase pathway** glucose 6-phosphate is oxidized to carbon dioxide and a pentose 5-phosphate (see also pentose phosphate pathway, Chapter 9.11.4) which is cleaved by inorganic phosphate to glyceraldehyde 3-phosphate and acetyl phosphate. Glyceraldehyde 3-phosphate is then converted to pyruvate as in the Embden-Meyerhof pathway. Pentoses are channeled into the pathway by direct phosphorylation without oxidation (Fig. 12.**16**). Glucose 6-phosphate

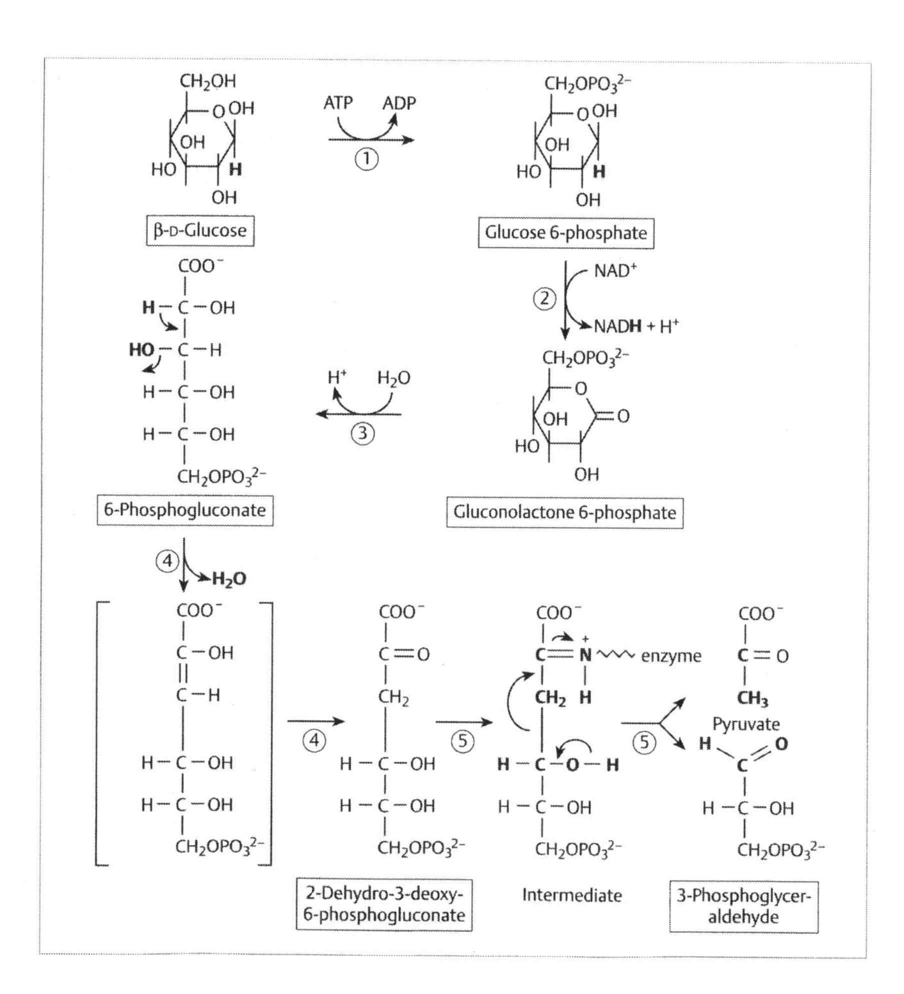

Fig. 12.**15** **Entner-Doudoroff pathway.** 1, Glucokinase; 2, glucose-6-phosphate dehydrogenase; 3, lactonase; 4, 6-phosphogluconate dehydratase; 5, 2-dehydro-3-deoxy-6-phosphogluconate aldolase

Box 12.5 Thiamine diphosphate is required by enzymes that cleave the bond adjacent to the carbonyl group (1,2-cleavage) as in phosphoketolase, transketolase, or pyruvate decarboxylase. **Aldolases**, which cleave the bond between carbon atoms 2 and 3, require a lysine residue for the formation of a Schiff's base, which after protonation attracts electrons better than the carbonyl group alone. A second group of aldolases uses an electron-deficient metal ion as a Lewis acid (Mg^{2+}, Mn^{2+}, Zn^{2+}) for this purpose. Note that the structures of the intermediates in **ketolases** and in aldolases that form a Schiff's base are almost identical (Figs. 12.**15** and 12.**17**; red structures).

is oxidized as in the Entner-Doudoroff or pentose phosphate pathways to 6-phosphogluconate (Fig. 12.**15**). A second oxidation by NAD^+ at carbon atom 3 yields a β-ketoacid, which readily decarboxylates to ribulose 5-phosphate, which epimerizes at carbon atom 3 to xylulose 5-phosphate (Fig. 12.**16**). The next step is catalyzed by the thiamine diphosphate-dependent key enzyme of the pathway, phosphoketolase (Fig. 12.**17**). The proton at the 2′-carbon atom of thiamine diphosphate readily dissociates (pK≈18) and the resulting anion attacks the carbonyl group of xylulose 5-phosphate. After protonation, the adduct is fragmented as shown in Fig. 12.**17**, whereby the positively charged nitrogen atom attracts the electrons, and glyceraldehyde 3-phosphate is released. Dehydration of the remaining residue followed by tautomerization yields acetyl-thiamine diphosphate, which is cleaved by inorganic phosphate to acetyl phosphate and the coenzyme (prosthetic group) is regenerated.

In the ***Bifidobacterium bifidum*** **pathway** 2 mol glucose 6-phosphate are cleaved and oxidized to 2 mol pyruvate and 3 mol acetyl phosphate without formation of carbon dioxide ($2\,C_6 \rightarrow 2\,C_3 + 3\,C_2$) (Fig. 12.**18**). After

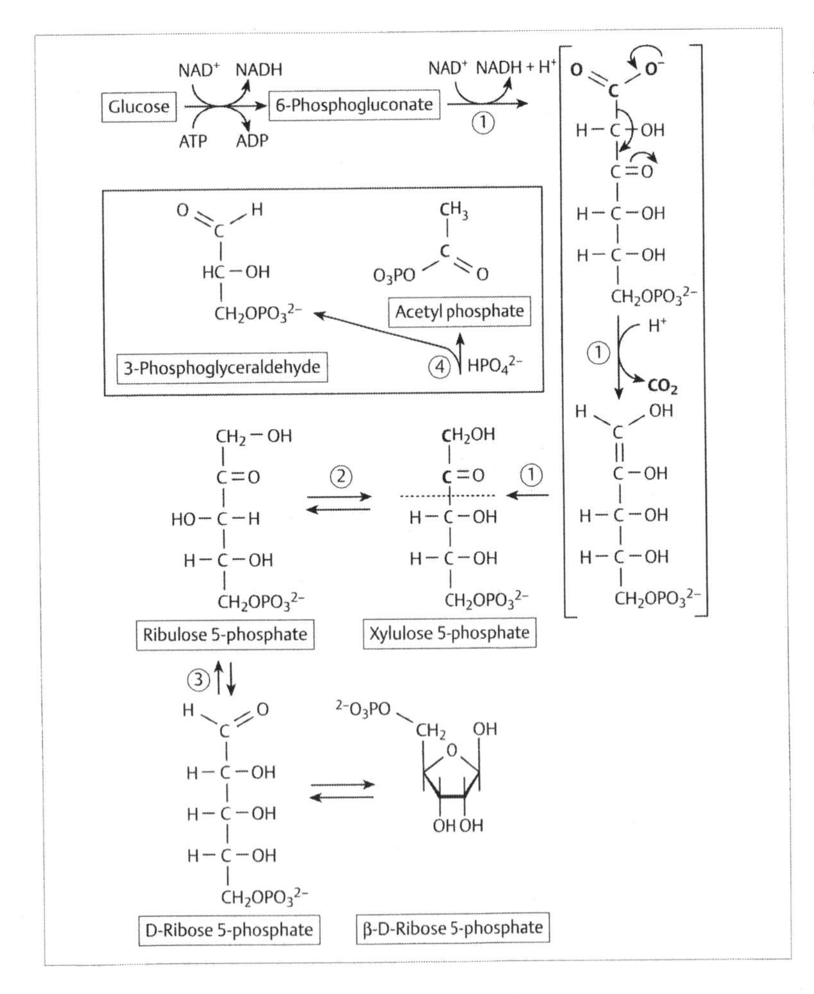

Fig. 12.**16** **Phosphoketolase pathway.** For the conversion of glucose to 6-phosphogluconate see Fig. 12.**15**. 1, 6-Phosphogluconate dehydrogenase; 2, ribulose-5-phosphate epimerase; 3, ribose-5-phosphate isomerase; 4, phosphoketolase-2 with thiamine diphosphate as prosthetic group

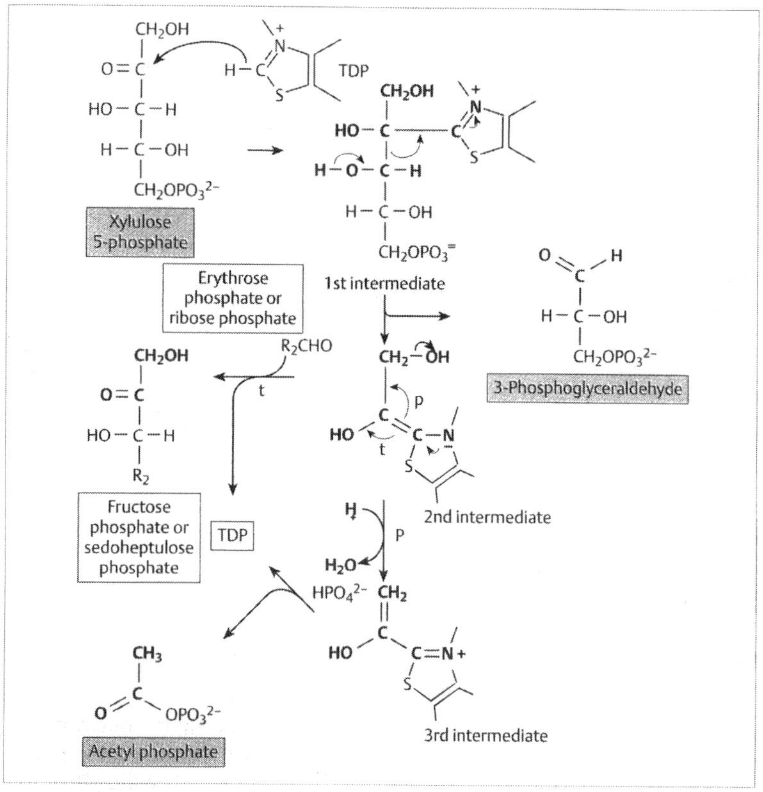

Fig. 12.17 **Mechanism of transketolase and phosphoketolase-2.** TDP, essential part of thiamine diphosphate; p, phosphoketolase-2; t, transketolase. Compare the first intermediate with that of 2-dehydro-3-deoxy-6-phosphogluconate aldolase (Fig. 12.14)

isomerization of 2 mol glucose 6-phosphate to 2 mol fructose 6-phosphate, one mol fructose 6-phosphate (plus inorganic phosphate) is cleaved to acetyl phosphate and erythrose 4-phosphate. The second mol fructose-6-phosphate undergoes transaldolation (Schiff's base or metal ion) with the erythrose 4-phosphate to form glyceraldehyde 3-phosphate and sedoheptulose 7-phosphate ($C_6 + C_4 \rightarrow C_3 + C_7$). In the following transketolase reaction (thiamine diphosphate as intermediate carrier) both products interchange a C_2 unit to yield xylulose 5-phosphate and ribose 5-phosphate ($C_3 + C_7 \rightarrow 2\,C_5$). The latter pentose is isomerized to ribulose 5-phosphate followed by epimerization at carbon atom 3 to xylulose 5-phosphate. Finally, xylulose 5-phosphate (2 mol) is cleaved (together with inorganic phosphate) to 2 mol acetyl phosphate and 2 mol glyceraldehyde 3-phosphate, which is oxidized to 2 mol pyruvate. This reaction is followed by reduction to 2 mol L-lactate in order to regenerate the NAD.

$$\text{Sum: } 2\,\text{Glucose} \rightarrow 2\,\text{L-lactate}^- + 3\,\text{acetate}^- + 5\text{H}^+$$
$$\Delta G^{\circ\prime} = -510\ \text{kJ/2 mol glucose}$$
$$(102\ \text{kJ/mol ATP}). \quad (12.76)$$

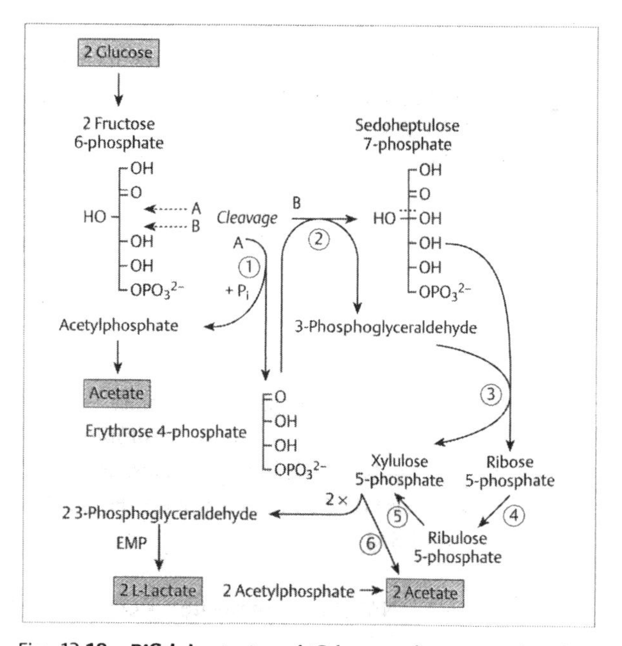

Fig. 12.18 ***Bifidobacterium bifidum* pathway.** 1, Phosphoketolase-1; 2, transaldolase; 3, transketolase; 4, ribose-5-phosphate isomerase; 5, ribulose-5-phosphate 3-epimerase; 6, phosphoketolase-2. EMP, Embden-Meyerhof pathway. For other enzymes and pathways, see previous figures and text

There are at least four pathways to oxidize sugar phosphates to either pyruvate alone or to additional acetyl phosphate via the following key enzymes: (1) fructose 1,6-bisphosphate aldolase (Embden-Meyerhof pathway and its archaean variant, (2) 2-dehydro-3-deoxy-6-phosphogluconate aldolase (KDPG-aldolase, Entner-Doudoroff pathway), (3) xylulose 5-phosphate phosphoketolase (phosphoketolase pathway), or (4) a combination of transaldolase and ketolase reactions (*Bifidobacteriumm bifidum* pathway).

12.2.2 Pyruvate Is a Very Versatile Intermediate

Phospho*enol*pyruvate^{3-} + H^+ + ADP^{3-} → pyruvate$^-$ + ATP^{4-}
(pyruvate kinase, occurs in many organisms)
$\Delta G^{\circ\prime} = -31.4$ kJ/mol (12.77)

2-Dehydro-3-deoxygluconate 6-phosphate^{3-} → pyruvate$^-$ + glyceraldehyde 3-phosphate^{2-}
(KDGP aldolase, see above)
$\Delta G^{\circ\prime} = +17$ kJ/mol (12.78)

Oxaloacetate^{2-} + H^+ → pyruvate$^-$ + CO_2
(oxaloacetate decarboxylase, occurs in citrate fermenting bacteria, Fig. 12.**27**)
$\Delta G^{\circ\prime} = -31$ kJ/mol (12.79)

(*S*)-Citramalate^{2-} → pyruvate$^-$ + acetate$^-$
(citramalate lyase, occurs in clostridia fermenting glutamate via 3-methylaspartate, Fig.12.**37**)
$\Delta G^{\circ\prime} = +3.6$ kJ/mol (12.80)

(2*R*,3*S*)-2, 3-Dimethylmalate^{2-} → pyruvate$^-$ + propionate$^-$
(dimethylmalate lyase, occurs in *Clostridium barkeri* fermenting nicotinate, Fig.12.**43**)
$\Delta G^{\circ\prime} = +1.8$ kJ/mol (12.81)

(*S*)-Malate^{2-} + NAD(P)$^+$ → pyruvate$^-$ + CO_2 + NAD(P)H
(malic enzyme, widespread)
$\Delta G^{\circ\prime} = -3$ kJ/mol (12.82)

L-Alanine + 2-oxoglutarate^{2-} → pyruvate$^-$ + L-glutamate$^-$
(alanine aminotransferase, widespread)
$\Delta G^{\circ\prime} = +1$ kJ/mol (12.83)

L-Alanine + NAD^+ → pyruvate$^-$ + NH_4^+ + NADH + H^+
(alanine dehydrogenase in some *Bacillus* sp.)
$\Delta G^{\circ\prime} = +29$ kJ/mol (12.84)

L-Serine or D-serine → pyruvate$^-$ + NH_4^+
(L-serine dehydratase, widespread; D-serine dehydratase in *E. coli*)
$\Delta G^{\circ\prime} = -43$ kJ/mol (irreversible)

L-Cysteine + H_2O → H_2S + pyruvate$^-$ + NH_4^+
(cysteine desulfurylase, widespread)
$\Delta G^{\circ\prime} = -11$ kJ/mol. (12.86)

As shown in the previous paragraph, pyruvate is commonly formed in the pyruvate-kinase reaction, the last step of the Embden-Meyerhof pathway (Reaction 12.77). Another source is the 2-dehydro-3-deoxygluconate-6-phosphate-aldolase reaction (Reaction 12.78). In addition, cleavage of citrate yields acetate and oxaloacetate (citrate lyase), which decarboxylates to pyruvate (Reaction 12.79), whereas the analogous cleavages of (*S*)-citramalate (2-methylmalate) or (2*R*,3*S*)-2,3-dimethylmalate directly lead to pyruvate and acetate (Reaction 12.80) or propionate (Reaction 12.81), respectively. The malic enzymes catalyze the NAD(P)$^+$-dependent oxidative decarboxylation of malate to pyruvate (Reaction 12.82). In the fermentation of alanine (Fig. 12.**36**) pyruvate is the product of an aminotransferase reaction involving 2-oxoglutarate (Reaction 12.83) or, less frequently, of the direct oxidation by NAD^+ (Reaction 12.84). Finally β-eliminations of serine and cysteine afford pyruvate (Reactions 12.85 and 12.86).

The further fate of pyruvate is described by the following reactions:

Pyruvate$^-$ + NAD^+ + CoASH → acetyl-SCoA + CO_2 + NADH
(pyruvate dehydrogenase, common in almost all aerobic organisms)
$\Delta G^{\circ\prime} = -40$ kJ/mol (irreversible) (12.87)

Pyruvate$^-$ + 2 ferredoxin + CoASH → acetyl-SCoA + 2 reduced ferredoxin$^-$ + CO_2 + H^+
(pyruvate ferredoxin oxidoreductase also called pyruvate synthase, common in Archaea, anaerobic Gram-positive bacteria, and anaerobic unicellular eukaryotic parasites)
$\Delta G^{\circ\prime} = -20$ kJ/mol (reversible) (12.88)

Pyruvate$^-$ + CoASH → acetyl-CoA + formate$^-$
(pyruvate formate lyase, in enterobacteria, lactobacilli and some clostridia only under strict anoxic conditions, Fig. 12.**24**)
$\Delta G^{\circ\prime} = -17$ kJ/mol (reversible) (12.89)

Pyruvate$^-$ + H^+ → acetaldehyde + CO_2
(pyruvate decarboxylase, *Zymomonas* and yeast)
$\Delta G^{\circ\prime} = -20$ kJ/mol (irreversible) (12.90)

2 Pyruvate$^-$ + H^+ → 2-acetolactate$^-$ + CO_2
(2-acetolactate synthase, enterobacteria; *Bacillus*, and all organisms able to synthesize valine; see below). (12.91)

Homolactic-fermenting organisms produce exclusively D- or L-lactate from hexoses. The fermentation follows the Embden-Meyerhof pathway to pyruvate, which is reduced to lactate (Reaction 12.92) by NAD^+-dependent lactate dehydrogenases (Reaction 12.93).

Glucose → 2 lactate + 2 H^+
$\Delta G^{\circ\prime} = -198$ kJ/mol glucose (12.92)

Pyruvate$^-$ + NADH + H^+ → D- or L-lactate$^-$ + NAD^+
(D- or L-lactate dehydrogenase)
$\Delta G^{\circ\prime} = -32$ kJ/mol. (12.93)

Thus, 2 mol ATP are produced per mol glucose (Fig. 12.**19**), and 99 kJ are released to allow the formation of 1 mol ATP. Additional ATP (up to 2/3 mol) may be formed in *Streptococcus* by the export of 2 mol lactate together with four protons; two protons drive the ATP synthesis catalyzed by the ATP synthase. The exclusive formation of lactate only occurs in the presence of an excess of substrate (e.g., lactose, as in milk). When hexoses become growth-limiting some *Streptococcus* and *Lactobacillus* species are able to switch in order to increase the yield of ATP. Pyruvate is then cleaved by pyruvate formate lyase (Reaction 12.89) to acetyl-CoA

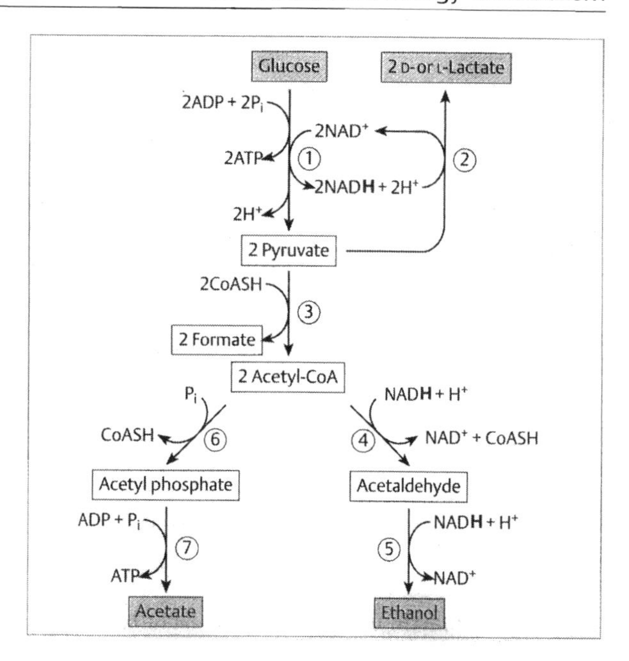

Fig. 12.**19** **Homolactate fermentation.** 1, Embden-Meyerhof pathway; 2, D- or L-lactate dehydrogenase; 3, pyruvate formate lyase; 4, acetaldehyde dehydrogenase, CoASH acetylating; 5, alcohol dehydrogenase; 6, phosphate acetyltransferase (also known as phosphotransacetylase); 7, acetate kinase. In *Escherichia coli* the activities of enzyme 4 and enzyme 5 as well as the inactivase of pyruvate formate lyase (Fig. 12.**24**) are all located on one protein (AdhE)

and formate. Consequently, acetyl-CoA serves as electron acceptor and is reduced by 2 mol NADH/mol acetyl-CoA via acetaldehyde to ethanol (Reactions 12.94 and 12.95). The other acetyl-CoA is used for substrate-level phosphorylation in order to provide a third mol of ATP (Reactions 12.41 and 12.42).

Acetyl-CoA + NADH + H^+ → acetaldehyde + NAD^+ + CoASH
(acetaldehyde dehydrogenase, CoASH acetylating) (12.94)

Acetaldehyde + NADH + H^+ → ethanol + NAD^+
(alcohol dehydrogenase) (12.95)

Sum : glucose + H_2O → acetate$^-$ + 2 formate$^-$ + ethanol + 3 H^+
$\Delta G^{\circ\prime} = -219$ kJ/mol. (12.96)

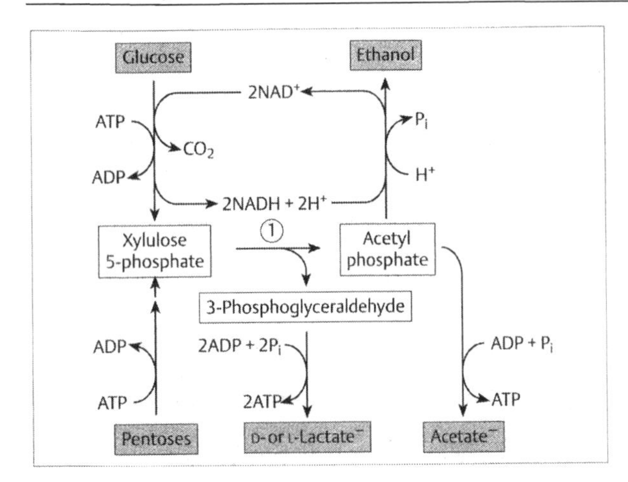

Fig. 12.**20** **Heterolactate fermentation.** 1, Phosphoketolase-2 (Figs. 12.**16** and 12.**17**). For other enzymes and pathways, see Figs. 12.**18** and 12.**21** and text

During **heterolactic fermentation** to lactate, CO_2 and ethanol or acetate, hexoses are oxidized to pentose 5-phosphates, whereas pentoses require no oxidation. All sugars are converted to xylulose 5-phosphate which is cleaved to pyruvate and acetyl phosphate. If a hexose serves as substrate, the excess NADH is used to reduce acetyl phosphate via acetyl-CoA to ethanol (Reactions 12.94 and 12.95). Hence, pentoses give rise to additional ATP. Heterolactate-fermenting bacteria that grow on hexoses are easily recognized by the formation of CO_2, which readily evolves as visible bubbles (Fig. 12.**20**) due to the acidic pH of the medium.

$$\text{Glucose} \rightarrow CO_2 + \text{lactate}^- + H^+ + \text{ethanol}$$
$$\Delta G^{\circ\prime} = -211\,\text{kJ/mol (1 ATP)} \qquad (12.96)$$

$$\text{Ribose} \rightarrow \text{lactate}^- + \text{acetate}^- + 2\,H^+$$
$$\Delta G^{\circ\prime} = -210\,\text{kJ/mol (2 ATP)} \qquad (12.97)$$

Ethanol fermentation without the formation of acid (Reaction 12.98) involves the thiamine diphosphate-containing pyruvate decarboxylase (Reaction 12.90). The product acetaldehyde is reduced to ethanol (Reaction 12.95).

$$\text{Sum: Glucose} \rightarrow 2\,CO_2 + 2\,\text{ethanol}$$
$$\Delta G^{\circ\prime} = -236\,\text{kJ/mol glucose} \qquad (12.98)$$

Hence, acetaldehyde may be generated in two ways: (1) by reduction of acetyl-CoA as shown above (lactic acid bacteria, clostridia, and enterobacteria) or (2) by decarboxylation of pyruvate via acetaldehyde (in yeasts and in *Zymomonas*). For pyruvate production, yeasts such as *Saccharomyces cerevisiae* use the Embden-Meyerhof pathway (2 ATP), whereas *Zymononas* ferments glucose via the Entner-Doudoroff pathway (1 ATP). Since the lower ATP yield results in the production of less cell mass to be disposed of, ethanol fermentation by *Zymomonas* is of technical interest.

Butyrate is generated by *Clostridium pasteurianum* (clostridial cluster 1) via the Embden-Meyerhof pathway. Pyruvate is oxidized to acetyl-CoA and CO_2 by pyruvate: ferredoxin oxidoreductase. Due to its low redox potential ($E_0' = -420\,\text{mV}$), the reduced ferredoxin is able to reduce protons; this leads to molecular hydrogen. Acetyl-CoA serves as electron acceptor for the NADH formed in the Embden-Meyerhof pathway. This reaction yields butyryl-CoA, which undergoes substrate-level phosphorylation directly to butyryl phosphate or indirectly via CoA-transfer to acetate (Reaction 12.99). Thus 3 mol ATP/mol glucose are formed (Fig. 12.**21**).

$$\text{Sum: Glucose} \rightarrow 2\,CO_2 + 2\,H_2 + \text{butyrate}^- + H^+$$
$$\Delta G^{\circ\prime} = -255\,\text{kJ/mol glucose (85 kJ/mol ATP)} \qquad (12.99)$$

Alternatively, a portion of the NADH serves to reduce ferredoxin, which in turn reduces protons to H_2, catalyzed by a hydrogenase. Hence, 4 mol ATP/mol glucose are formed by substrate-level phosphorylation, since 2 mol acetyl-CoA rather than 1 mol butyryl-CoA are generated (Reaction 12.100). This requires, however, an energy-dependent reverse electron transport from NADH ($E_0' = -320\,\text{mV}$) $\rightarrow$ H_2 ($E_0' = -420\,\text{mV}$) via an electrochemical proton gradient ($\Delta\tilde{\mu}_{H^+}$) by which the extra ATP is consumed again. In reality, when *C. pasteurianum* grows in batch culture, it ferments glucose by Reaction 12.101 leading to 3.3 mol ATP/mol glucose (Fig. 12.**27**). In the presence of organisms thriving on hydrogen, which are able to rise the E' of H_2 to more than $-320\,\text{mV}$, a reverse electron transport is not necessary (see Chapter 12.2.5).

$$\text{Glucose} \rightarrow 2\,CO_2 + 4\,H_2 + 2\,\text{acetate}^- + 2\,H^+$$
$$\Delta G^{\circ\prime} = -206\,\text{kJ/mol glucose (52 kJ/mol ATP)} \qquad (12.100)$$

$$\text{Glucose} \rightarrow 2\,CO_2 + 2.6\,H_2 + 0.7\,\text{butyrate}^- + 0.6\,\text{acetate}^- + 1.3\,H^+$$
$$\Delta G^{\circ\prime} = -240\,\text{kJ/mol glucose (71 kJ/mol ATP)} \qquad (12.101)$$

Hence, 0.6 mol H_2 is derived from NADH for which about 0.3 mol ATP is consumed. This extra hydrogen production, which is of no energetic benefit for the organism, presumably serves to save glucose as outlined earlier in this chapter.

Clostridium acetobutylicum (cluster 1) **produces solvents from acids.** In the initial growth phase at pH 7, glucose is fermented to the same products as in *C. pasteurianum*. In a later growth phase, prior to sporulation, when the formed acids have lowered the pH to 5, the fermentation is shifted to three branching pathways that serve to form neutral products (ethanol, acetone, 2-propanol, and 1-butanol) rather than acids, which would kill the organism (Fig. 12.**21**). The regulation of the branching-point enzymes and the genetics of this pathway are subjects of intensive investigations, since 1-butanol is a valuable technical solvent.

Mixed acid and butane-2,3-diol fermentation is characteristic of enteric bacteria and bacilli. The pyruvate, derived from the Embden-Meyerhof pathway, is in part reduced to lactate and in part cleaved by pyruvate formate lyase to acetyl-CoA and formate (Reaction 12.89). Formate may be excreted or oxidized to CO_2 concomitantly with hydrogen formation (formate hydrogen lyase). Acetyl-CoA is converted to acetate, whereby ATP is formed, or reduced to ethanol. Another path for removing reducing equivalents leads from phospho*enol*pyruvate (PEP) to succinate. ATP is formed by electron-transport phosphorylation (ETP) coupled to fumarate reduction (Chapter 12.1.2). Finally, 2 mol pyruvate may form CO_2 and (*R*)-2-acetolactate,

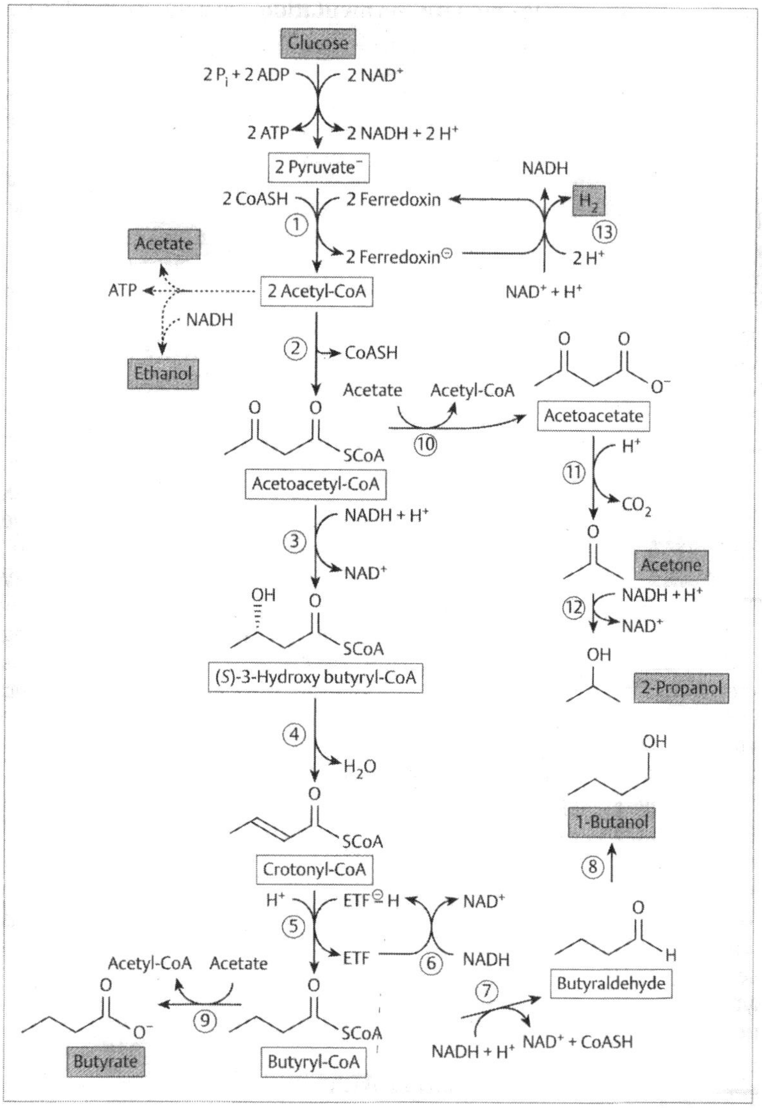

Fig. 12.**21** **Fermentation of glucose by Clostridia.** Glucose is degraded to pyruvate via the Embden-Meyerhof pathway. 1, Pyruvate ferredoxin oxidoreductase (also called pyruvate synthase); 2, acetyl-CoA acetyltransferase or thiolase; 3, 3-hydroxybutyryl-CoA dehydrogenase; 4, crotonyl-CoA hydratase or crotonase;
5, butyryl-CoA dehydrogenase, ETF = electron-transferring flavoprotein; 6, diaphorase; 7, butyraldehyde dehydrogenase, CoASH-acylating; 8, 1-butanol dehydrogenase; 9, butyrate CoA-transferase; 10, acetoacetate CoA-transferase; 11, acetoacetate decarboxylase; 12, 2-propanol dehydrogenase; 13, NADH : ferredoxin oxidoreductase + hydrogenase. For other reactions of acetyl-CoA, see Fig. 12.**19**

Box 12.6 The interconversion and mechanism of pyruvate formate lyase has been elucidated. Pyruvate formate lyase (Reaction 12.89) from *E. coli* is synthesized as an inactive metal- and coenzyme-free homodimer (170 kDa). The enzyme is post-translationally modified by *S*-adenosylmethionine and a reduced flavodoxin in a reaction catalyzed by an iron-containing activase. A hydrogen atom is abstracted from a specific glycine residue, yielding methionine and 5′-deoxyadenosine. The formed free radical (HS-enzyme, Fig. 12.**24**), which is stable under anoxic conditions, is involved in a two-step catalysis.

$$\text{Pyruvate} + \text{HS-enzyme} \rightarrow \text{acetyl-S-enzyme} + \text{formate}$$

$$\text{Acetyl-S-enzyme} + \text{CoASH} \rightarrow \text{acetyl-SCoA} + \text{HS-enzyme}$$

Hence, pyruvate is cleaved homolytically rather than heterolytically. When *E. coli* is shifted from anoxic to oxic conditions, the radical is removed by NAD^+ in a reaction catalyzed by a trifunctional deactivase with alcohol dehydrogenase and acetaldehyde dehydrogenase activities. Pyruvate formate lyase is inactivated and pyruvate is converted to acetyl-CoA by the aerobic enzyme pyruvate dehydrogenase (Reaction 12.86).

Box 12.7 Acetolactate synthase catalyzes a thiamine diphosphate-dependent reaction in which pyruvate is decarboxylated to hydroxyethylthiamine diphosphate (Fig. 12.**23**). Reaction with a second molecule of pyruvate liberates the β-keto-acid acetolactate, and the coenzyme is regenerated. The enzyme also is called **"pH 6 enzyme"** indicating its activity optimum at a slightly acidic pH. Like in the clostridial solvent fermentation, the formation of butanediol is a physiological response to acidic conditions by avoiding the production of acids. In the pathway of valine biosynthesis, there is a different anabolic acetolactate synthase with an activity optimum at pH 8. The latter enzyme, which is subject to feedback inhibition by valine, contains thiamine diphosphate and FAD.

which is decarboxylated to (*R*)-acetoin; this is reduced to (*R,R*)-butane-2,3-diol or oxidized to diacetyl (Figs. 12.**22** and **23**).

Sum (non stoichiometric):

$$\text{Glucose} \rightarrow \text{succinate}^{2-} + \text{L-lactate}^- + \text{acetate}^- + \text{formate}^- + \text{ethanol} + (R,R)\text{-butane-2,3-diol} + \text{diacetyl} + CO_2 + H_2 + H^+ \quad (12.102)$$

Table 12.**6** shows the amounts of these products formed by *E. coli*, *Serratia marcescens*, and *Bacillus subtilis*.

Propionate fermentation, mainly from glucose and lactate via two different pathways, is carried out by a variety of organisms. It is also produced from glycerol, succinate, and several amino acids such as alanine, serine, cysteine, threonine, methionine, and glutamate. The two different pathways can be distinguished by the use of [2- or 3-^{13}C]lactate. In bacteria belonging to the genus *Propionibacterium* and in most other propionate producers, the pathway (Fig. 12.**25**) involves a transcarboxylation of pyruvate to oxaloacetate. Propionate formation proceeds via the symmetrical intermediate succinate, causing a random distribution of the label between C-2 and C-3. Hence both labeled lactate species give rise to [2,3-^{13}C]propionate, which is easily analyzed by ^{13}C NMR spectroscopy. The other pathway (see Fig. 12.**36**) found in *Clostridium propionicum* and in *Megasphera elsdenii* proceeds via the asymmetrical intermediate acryloyl-CoA. Therefore, the labeling in propionate is identical to that of the original lactate. These pathways are called "random" and "non-random", respectively. In the random pathway (Fig. 12.**25**), the reduction of fumarate to succinate is coupled to electron-transport phosphorylation yielding an additional 2/3 mol ATP. The non-random pathway is also used in the fermentation of alanine and is described in Chapter 2.2.3 (Fig. 12.**36**). A third possibility for propionyl-CoA formation involves oxidative decarboxylation of 2-oxybutyrate, which is derived either by β-elimination of water from L-threonine or by γ-elimination of methylmercaptan (CH_3-SH) from L-methionine (see Chapter 12.2.3).

Propionate fermentations used by propionibacteria via succinate are as follows (random pathway, Reactions 12.103–12.105):

$$3\,\text{Lactate}^- \rightarrow \text{acetate}^- + CO_2 + 2\,\text{propionate}^- \quad \Delta G^{\circ\prime} = -169\ \text{kJ/mol acetate (2.3 ATP; 72kJ/mol ATP)} \quad (12.103)$$

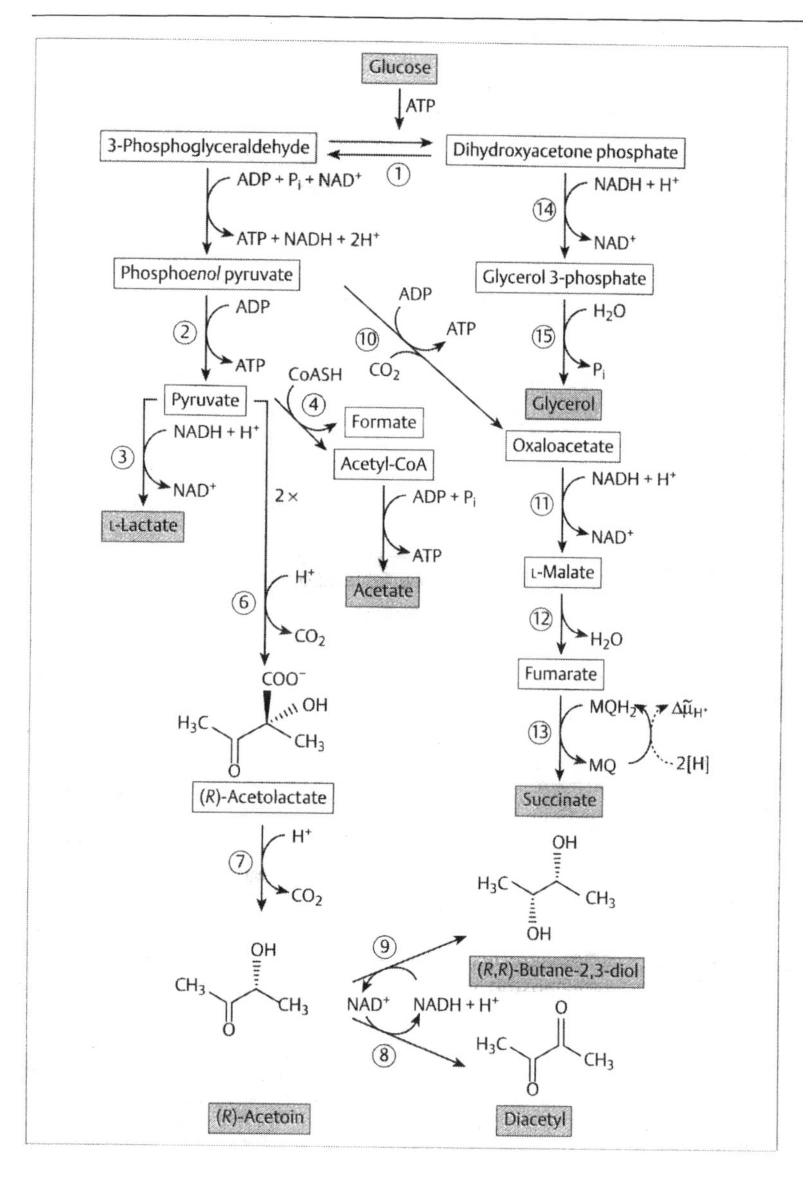

Fig. 12.**22** **Mixed acid fermentation.** 1, Triosephosphate isomerase; 2, pyruvate kinase; 3, L-lactate dehydrogenase; 4, pyruvate formate lyase; 5, formate dehydrogenase and hydrogenase; 6, acetolactate synthase; 7, acetolactate decarboxylase; 8, acetoin dehydrogenase; 9, 2,3-butandiol dehydrogenase; 10, phosphoenolpyruvate carboxykinase; 11, malate dehydrogenase; 12, fumarase; 13, fumarate reductase (Fig. 12.**4**). Contrary to the mammalian fumarase, the two isoenzymes from *E. coli* contain iron-sulfur clusters as does the aconitase; this results not only in a higher turnover number but also in oxygen sensitivity. Hence, under oxygen stress, the organism induces a third fumarase similar to the mammalian one. 14, Glycerol-3-phosphate dehydrogenase; 15, glycerol 3-phosphatase

$$3\,\text{Glucose} \rightarrow 2\,\text{acetate}^- + 2\,CO_2 + 4\,\text{propionate}^- + 6\,H^+$$
$$\Delta G^{\circ\prime} = -934\,\text{kJ/2 mol acetate (10.7ATP; 87 kJ/mol ATP)} \quad (12.104)$$

$$\text{Glycerol} \rightarrow \text{propionate}^- + H^+ + H_2O$$
$$\Delta G^{\circ\prime} = -149\,\text{kJ/mol; (1.7 ATP; 87.6 kJ/mol ATP).} \quad (12.105)$$

Enterobacteria ferment glycerol to propane-1,3-diol, acetate, and formate by the following Reaction (12.106).

$$3\,\text{Glycerol} \rightarrow 2\,\text{propane} - 1,3\text{-diol} + \text{acetate}^- + \text{formate}^- + 2\,H^+ + H_2O \quad (12.106)$$

Glycerol is oxidized via glycerol 3-phosphate to acetate, formate, and other products, as has been shown for the mixed acid and butane-2,3-diol fermentations in *E. coli*, whereby ATP is formed (Fig. 12.**26**). The key enzyme of this pathway is a coenzyme B_{12}-dependent glycerol dehydratase, which catalyses the dehydration of glycerol to 3-hydroxypropionaldehyde (see Fig. 12.**35**), which is reduced by NADH to the commercially interesting propane-1,3-diol. 3-Hydroxypropionaldehyde may also dehydrate non-enzymatically to the tear gas acrolein (propenal), which caused problems in whisky making.

Citrate is readily fermented by enterobacteria, lactic acid bacteria, *Veillonella*, and clostridia. Photo-

Table 12.6 **Mixed acid fermentations**

Products	*Escherichia coli*	*Serratia marcescens* (mol/100 mol glucose)	*Bacillus subtilis*
2,3-Butanediol	0.3	64.0	54.6
Acetoin	0.06	1.9	1.6
Glycerol	1.4	1.3	56.8
Ethanol	50.0	46.0	7.7
Formate	2.4	48.2	1.3
Acetate	36.5	3.8	0.2
Lactate	79.5	10.1	17.6
Succinate	10.7	8.2	1.1
CO_2	88.0	116.8	117.8
H_2	75.0	0.0	0.16
C recovered (%)	91.7	102.5	98.0
O/R balance	1.07	1.01	1.00

Theoretical values: for C recovered, 100%; for O/R (oxidation/reduction) balance, 1.00, see Table 12.**5**

trophic Proteobacteria (*Rhodopseudomonas gelatinosa*) use citrate as carbon source under anoxic conditions. A notable source of citrate is milk. Citrate is cleaved by citrate lyase to acetate and oxaloacetate. Under fermentative conditions, oxaloacetate is decarboxylated to pyruvate, which is further cleaved to formate and acetyl-CoA (Reaction 12.107; Figs. 12.**27** and 12.**28**).

$$\text{Citrate}^{3-} + H_2O \rightarrow \text{formate}^- + 2\,\text{acetate}^- + CO_2 \quad \Delta G^{\circ\prime} = -79\,\text{kJ/mol citrate (1 ATP/citrate)} \qquad (12.107)$$

Enterobacteria may convert oxaloacetate and pyruvate also to other products. One possibility involving respiration is the reduction of some oxaloacetate to succinate by formate (see mixed-acid fermentation). The lactic acid bacteria *Streptococcus lactis* subsp. *diacetylactis* and *Leuconostoc cremoris* use the pyruvate derived from citrate in milk for the synthesis of acetoin and diacetyl,

Fig. 12.**23** **Mechanism of acetolactate synthesis from pyruvate catalyzed by acetolactate synthase.** 1, acetolactate synthase

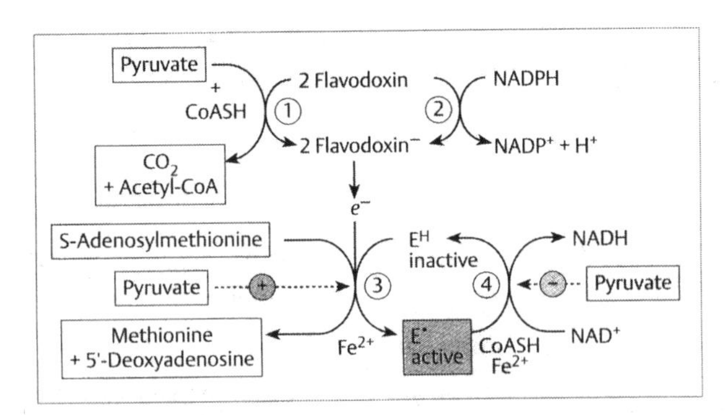

Fig. 12.**24** **Interconversion of pyruvate formate lyase between radical (E•) and nonradical forms (E^H).** Fld, Flavodoxin (a small flavoprotein with a redox potential as negative as ferredoxin); 1, pyruvate : flavodoxin oxidoreductase; 2, NADPH : flavodoxin oxidoreductase; 3, activase; 4, inactivase (AdhE, see legend to Fig. 12.**19**). The signs + and − indicate the positive or negative effects, respectively, of pyruvate on the reactions

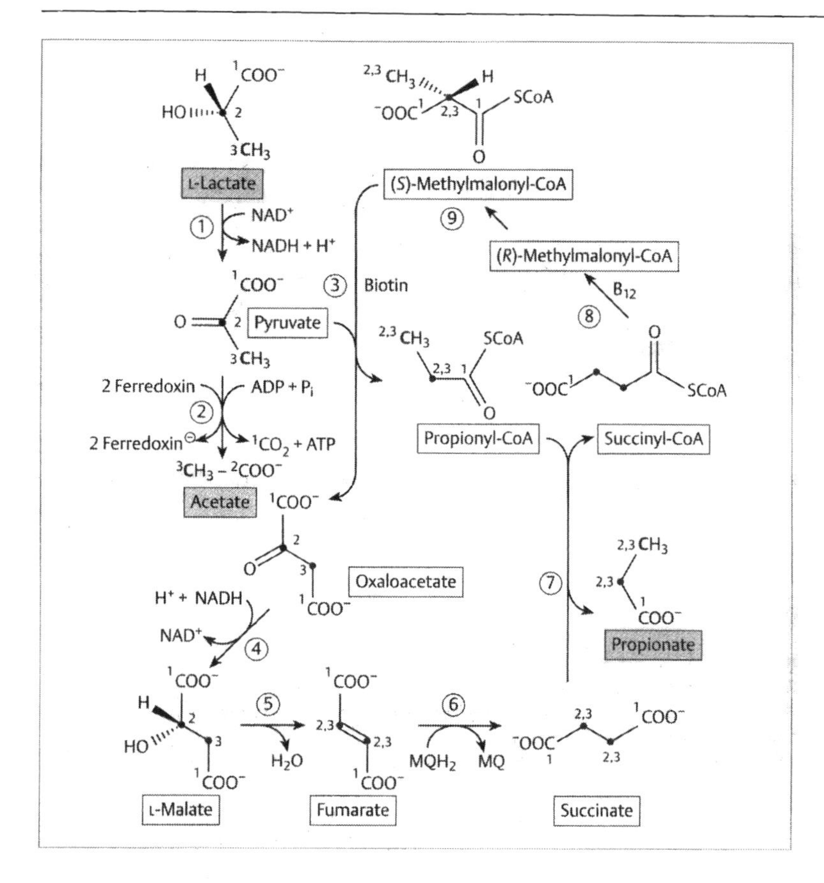

Fig. 12.**25** **Propionate fermentation in *Propioni-bacterium* (randomizing pathway).** 1, L-Lactate dehydrogenase; 2, pyruvate: ferredoxin oxidoreductase, phosphate acetyl-transferase (phosphotransacetylase) and acetate kinase; 3, transcarboxylase (contains biotin); 4, malate dehydrogenase; 5, fumarase; 6, fumarate reductase (see Fig. 12.**5**); 7, propionate CoA-transferase; 8, methylmalonyl-CoA mutase (B_{12}, coenzyme B_{12} or adenosylcobalamin; see Fig. 12.**10** and Chapter 12.2.2); 9, methylmalonyl-CoA epimerase

which give butter its flavor (Fig. 12.**22**). *E. coli* contains citrate lyase but lacks oxaloacetate decarboxylase. Therefore, the organism is able to grow anaerobically on citrate, if oxaloacetate is reduced to succinate by an additional reductant such as glycerol or glucose (Reaction 12.108).

$$\text{Citrate}^{3-} + 4\,[\text{H}] \rightarrow \text{acetate}^- + \text{succinate}^{2-} + \text{H}_2\text{O} \quad (12.108)$$

The **fermentation of ethanol** by *Clostridium kluyveri* (cluster 1) to fatty acids proceeds via acetyl-CoA derived from the oxidation of ethanol and yields 1 mol ATP/mol ethanol (Reaction 12.109).

$$\text{Ethanol} + \text{H}_2\text{O} \rightarrow \text{acetate}^- + \text{H}^+ + 2\,\text{H}_2 \qquad \Delta G^{\circ\prime} = +9.6\,\text{kJ/mol ethanol} \quad (12.109)$$

Since the free energy of Reaction 12.109 is positive, it has to be coupled to the highly exergonic syntheses of butyrate and caproate, which yield no additional ATP (Reactions 12.110 and 12.111). A typical fermentation is represented by Reaction 12.112.

$$\text{Ethanol} + \text{acetate}^- \rightarrow \text{butyrate}^- + \text{H}_2\text{O} \qquad \Delta G^{\circ\prime} = -39\,\text{kJ/mol} \quad (12.110)$$

$$2\,\text{Ethanol} + \text{acetate}^- \rightarrow \text{caproate}^- + 2\,\text{H}_2\text{O} \qquad \Delta G^{\circ\prime} = -77\,\text{kJ/mol} \quad (12.111)$$

$$6\,\text{Ethanol} + 3\,\text{acetate}^- \rightarrow 3\,\text{butyrate}^- + \text{caproate}^- + \text{H}^+ + 2\,\text{H}_2 + 4\,\text{H}_2\text{O} \qquad \Delta G^{\circ\prime} = -183\,\text{kJ/mol H}^+\ (1\ \text{ATP}) \quad (12.112)$$

Experiments with the protonophor 3,5,3′,5′-tetrachlorosalicylanilide suggest that the coupling agent is probably the proton motive force. However, the mechanism of its generation and the mode of coupling has not been established yet. The coupling is apparently not very efficient, since − 183 kJ are required for a yield of 1 mol ATP.

Another way to make the anaerobic oxidation of ethanol exergonic is the use of succinate rather than

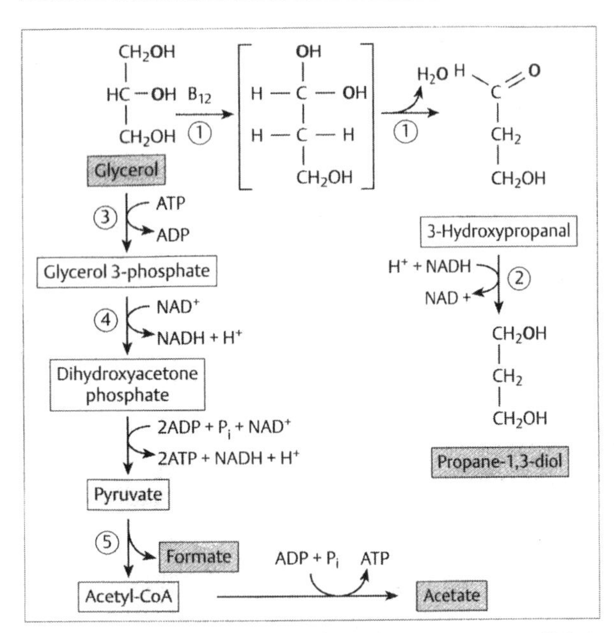

Fig. 12.**26** **Fermentation of glycerol to 1,3-propanediol.** 1, Glycerol dehydratase, coenzyme B_{12}-dependent; 2, propane-1,3-diol dehydrogenase; 3, glycerol kinase; 4, glycerol-3-phosphate dehydrogenase; 5, pyruvate formate lyase. For the further fate of dihydroxyacetone phosphate, see mixed-acid fermentation (Fig. 12.**21**)

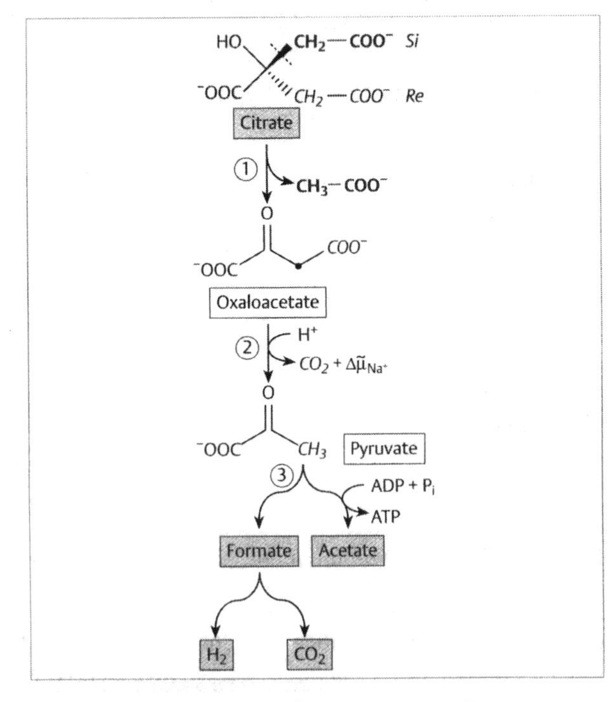

Fig. 12.**27** **Fermentation of citrate.** 1, Citrate lyase (Citrate Si-lyase; the stereospecific enzyme converts the Si-carboxymethyl residue of citrate to acetate); 2, oxaloacetate decarboxylase; 3, pyruvate formate lyase

Fig. 12.**28** **Mechanism of citrate lyase.** The enzyme (1) consists of 6 × 3 different subunits: transferase, lyase, and acyl carrier. The acetyl or citryl residue is bound as a thiol ester to phosphoribosyl-dephospho-CoASH, which is covalently connected to a serine residue of the acyl carrier. The acetyl enzyme may hydrolyze (3) to the inactive HS-enzyme, which can be spontaneously reactivated by acetylation with acetic anhydride or enzymatically with ATP and acetate catalyzed by an activase (2). In *Rhodocyclus gelatinosus*, reaction 3 is catalyzed by an inactivase, which is inhibited by glutamate. Thus, citrate cleavage can only proceed if sufficient glutamate and, hence, nitrogen for growth is available. The amino acid is synthesized by the NAD(P)H-dependent reduction of ammonia and 2-oxoglutarate obtained, whose source is citrate via isocitrate

protons as electron acceptor (Reaction 12.113, Fig. 12.**29**).

$$2\,\text{Succinate}^{2-} + 3\,\text{ethanol} \rightarrow 2\,\text{butyrate}^- + 3\,\text{acetate}^- + H^+ + H_2O \qquad \Delta G^{\circ\prime} = -165\,\text{kJ/mol}\,H^+ \qquad (12.113)$$

Although 3 mol ethanol are oxidized to 3 mol acetyl-CoA, only 1 mol ATP can be formed, since 2 mol succinate have to be activated to the thiol ester in order to be reduced by NADPH to succinate semialdehyde and further to 4-hydroxybutyrate by NADH. A specific CoA-transferase (Reaction 12.114, Chapter 12.2.2) catalyzes the formation of 4-hydroxybutyryl-CoA which subsequently is dehydrated to crotonyl-CoA (Fig. 12.**30**). The final products butyrate and acetate are generated as shown in Fig. 12.**21**.

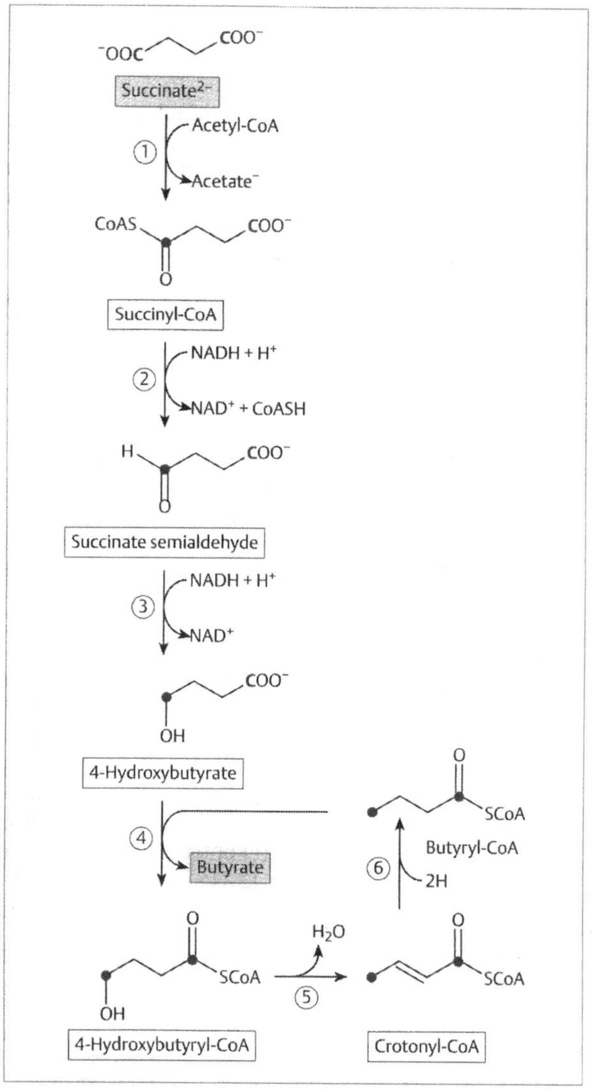

Fig. 12.29 **Reduction of succinate to butyrate by *Clostridium kluyveri*.** 1, Succinate CoA-transferase; 2, succinate semialdehyde dehydrogenase, CoASH-succinylating; 3, 4-hydroxybutyrate dehydrogenase; 4, 4-hydroxybutyrate CoA-transferase; 5, 4-hydroxybutyryl-CoA dehydratase; 6, butyryl-CoA dehydrogenase

The great diversity of carbohydrate fermentation stems—besides the four different pathways for glucose—mainly from the different modes of pyruvate reduction. Thus frequent products are D- and L-lactate, CO_2, H_2, formate, acetate, propionate, butyrate, valerate, caproate, ethanol, acetone, isopropanol, 1-butanol, acetolactate, acetoin, butane-2,3-diol, diacetyl, and propane-1,3-diol.

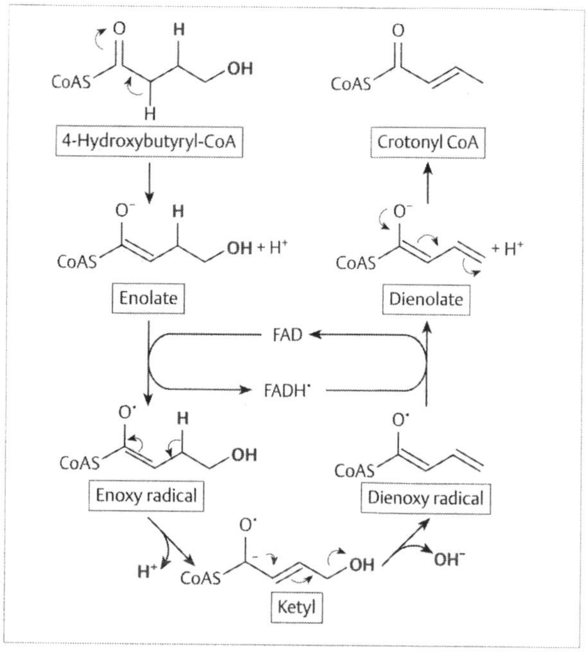

Fig. 12.30 **Mechanism of dehydration of 4-hydroxybutyryl-CoA.** In order to facilitate the removal of the unactivated β-hydrogen as a proton, the substrate is oxidized by H•-abstraction ($H^+ + e^-$). The resulting enoxy radical can be deprotonated at the β-position to the resonance-stabilized ketyl (radical anion), which readily eliminates the OH^- group. Finally, H• is donated back yielding the product. The mechanism is related to that of coenzyme B_{12}-dependent rearrangements (Fig. 12.**35**)

12.2.3 Amino Acids Are Suitable Substrates for Fermentative Organisms

The 20 proteinogenous amino acids have an average redox state similar to that of sugars; therefore, they are suitable for fermentative redox reactions. Although anaerobic organisms capable of degrading amino acids were discovered a century ago, the first stoichiometric fermentation balances were described by L.H. Stickland in the nineteen thirties. He showed that pairs of amino acids were fermented by *Clostridium sporogenes* (cluster 1); one amino acid (for example, alanine) served as electron donor, whereas a different amino acid (for example, glycine or proline) was used as electron acceptor (Fig. 12.**31**). Later, in the nineteen forties, H.A. Barker and others isolated clostridia and related non-sporulating anaerobes that fermented single amino acids. Thus, *C. propionicum* (cluster 14b) uses alanine as an electron acceptor and as an electron donor. A more recent isolate, *Eubacterium acidaminophilum* (cluster 15), is able to ferment glycine. Almost all known organisms capable of fermenting amino acids are

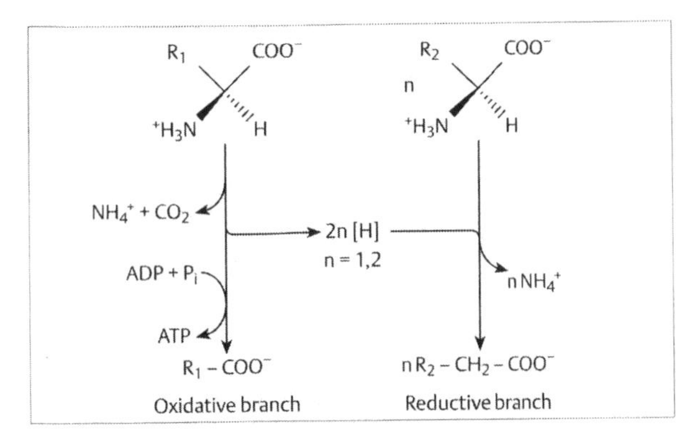

Fig. 12.**31** **General scheme of amino acid fermentations.** The amino acids are fermented pairwise; one is oxidized, and the other is reduced. The pair may be formed from different or identical amino acids. The following amino acids are used in both the oxidative and the reductive branch: alanine, cysteine, glycine, methionine, leucine, phenylalanine, serine, threonine, tryptophan, and tyrosine. Isoleucine and valine are used exclusively as electron donors, whereas proline only serves as electron acceptor. With the exception of glycine fermentations, ATP is formed only in the oxidative branch as indicated. In addition, some of the amino acids mentioned above and glutamate, glutamine, histidine, arginine, aspartate, asparagine, and lysine use special pathways. *Clostridium acetireducens* (Clostridia group I) requires, in addition to certain amino acids, which only are oxidized, acetate for growth, since the synthesis of butyrate from two molecules of acetate is used as the electron acceptor

clostridia and related organisms (clostridial clusters 1–29). This relationship has been established by a comparison of the sequences of their 16 S ribosomal RNAs.

Amino acid fermentations are anaerobic redox reactions, in which the oxidative branch of fermentation is always coupled to substrate-level phosphorylation, which is necessary for growth of the organism (Fig. 12.**31**). Most commonly, an acyl-CoA is formed; this is converted to acetyl-CoA by a CoA-transferase (Reaction 12.114), followed by phosphate acetyltransferase (also called phosphotransacetylase, Reaction 12.41) and acetate kinase (Reaction 12.42).

$$\text{R-CO-SCoA} + \text{acetate}^- \rightarrow \text{R-COO}^- + \text{acetyl-SCoA}$$
$$\Delta G^{\circ\prime} \approx 0,\ \text{depending on R} \qquad (12.114)$$

Generally, in the reductive branch, no ATP is formed. Exceptions are the fermentations of glycine, which is the only example of a reductive substrate-level phosphorylation (Fig. 12.**32**), and of aspartate, which is converted to fumarate, a substrate for anaerobic respiration (see Chapter 12.1.2). The products of amino acid fermentations are ammonia, CO_2, hydrogen, alcohols, and short-chain fatty acids. The fatty acids produced in the reductive branch (Fig. 12.**30**) often have the same carbon skeletons as the amino acid substrate, for example, leucine can be reduced to 4-methylvalerate, or tryptophan can be reduced to indolepropionate. The oxidation of an amino acid via the corresponding 2-oxo intermediate leads to a fatty acid that is shorter by one carbon atom. The most important exceptions to this generalization are the fermentations of glutamate (Figs. 12.**37** and 12.**38**) and lysine (Fig. 12.**39**).

In amino acid fermentations performed almost exclusively by clostridia and related organisms, one amino acid serves as electron donor, and the same or another amino acid serves as electron acceptor. The main products are ammonia, carbon dioxide, hydrogen, and short-chain fatty acids, which in many cases retain the carbon skeleton of the original amino acid. Energy is conserved via substrate-level phosphorylation.

There are three general strategies for the **removal of amino groups from amino acids**. Anaerobic amino acid-fermenting organisms remove amino groups by oxidation, elimination, or reduction. Whereas aerobic organisms also use oxidation and elimination for this purpose (see amino acid metabolism in aerobic organisms, Chapter 9), the reductive substitution of the amino groups of glycine and proline are unique to anaerobes. The reduction of glycine to acetyl phosphate is catalyzed by glycine reductase (Reaction 12.115), which is a three-enzyme complex (Fig. 12.**32**).

$$\text{Glycine} + \text{NADH} + 2\,\text{H}^+ + \text{HPO}_4^{2-}$$
$$\rightarrow \text{acetyl phosphate}^{2-} + \text{NH}_4^+ + \text{H}_2\text{O} + \text{NAD}^+$$
$$\Delta G^{\circ\prime} = -23\ \text{kJ/mol glycine} \qquad (12.115)$$

D-Proline is reduced without an acylphosphate as intermediate to 5-aminovalerate (Reaction 12.116). Hence, no energy is conserved in this reaction. The naturally occurring L-proline is converted to the D-enantiomer by a racemase (Reaction 12.117).

$$\text{D-Proline} + 2\,\text{H}^+ + 2\,e^- \rightarrow \text{5-aminovalerate} \qquad (12.116)$$

$$\text{L-Proline} = \text{D-Proline} \qquad (12.117)$$

The 2-oxoacids generated by deamination of amino acids—either by oxidation, transamination or elimination—are often further oxidized under decarboxylation to the CoA esters of the corresponding fatty acids. Whereas aerobic organisms generally use NAD^+ as oxidant for this purpose (for example pyruvate dehydro-

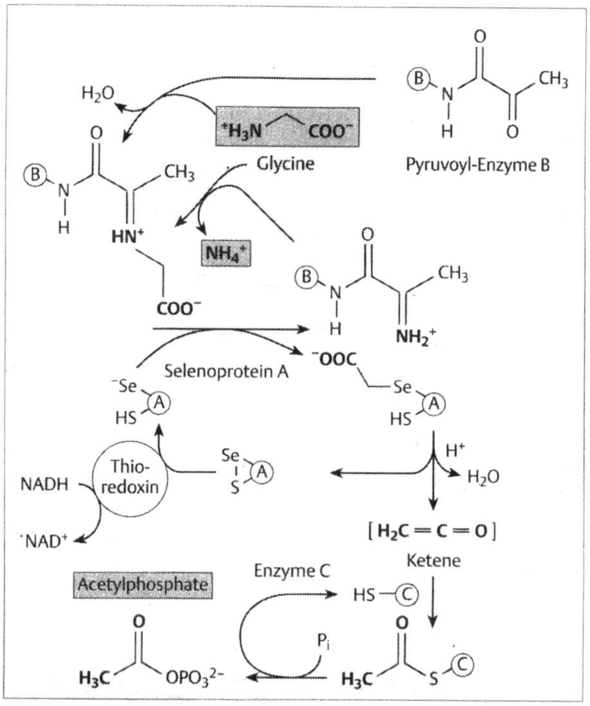

Fig. 12.**32** **Mechanism of glycine reductase.** Protein Ⓐ (12 kDa) contains selenocysteine. The pyruvoyl residue of enzyme Ⓑ forms a Schiff's base with glycine. The α-carbon of the bound glycine is attacked by the protein Ⓐ-Se$^-$-anion to yield carboxymethylselenocysteine linked to protein Ⓐ and the iminopyruvoyl residue is formed; ammonia is released in the next turnover or by hydrolysis. Elimination of ketene with the assistance of the adjacent cysteine yields the oxidized protein Ⓐ-Se-S intermediate, which is reduced by thioredoxin, a small protein containing two adjacent SH-groups forming an internal disulfide bridge upon oxidation. Reduction of thioredoxin is catalyzed by thioredoxin oxidoreductase with NADH or another electron donor. The hypothetical ketene intermediate adds to the cysteine residue of protein Ⓒ, whereby S-acetylcysteine is formed, which is cleaved by P_i to form acetyl phosphate. Proline reductase also contains a pyruvoyl residue but no Se

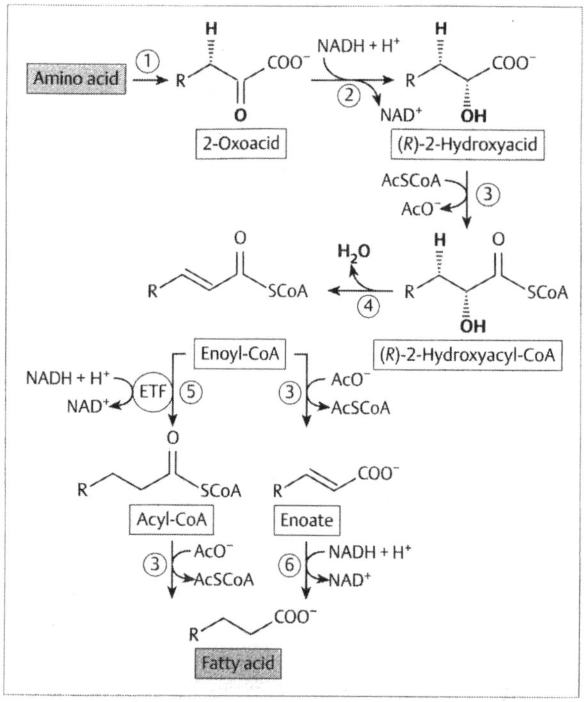

Fig. 12.**33** **The 2-hydroxy acid pathway.** 1, Initially, the amino acid is converted to the corresponding 2-oxo acid either by oxidation, transamination or elimination; 2, (*R*)-2-hydroxy acid dehydrogenase; 3, CoA-transferases; 4, (*R*)-2-hydroxyacyl-CoA dehydratase; 5, butyryl-CoA dehydrogenase (ETF, electron transferring protein); 6, enoate reductase

genase, Reaction 12.87), strictly anaerobic organisms require ferredoxin as electron acceptor. Ferredoxins (fd) are small acidic proteins (< 10 kDa), which contain either [4 Fe–4S] or [2 Fe–2 S] clusters and accept only one electron per cluster (fd^-). The oxidative decarboxylations are catalyzed by 2-oxoacid ferredoxin oxidoreductases (Reaction 12.118; for example pyruvate ferredoxin oxido-reductase, Reaction 12.88), which contain thiamine diphosphate and iron-sulfur clusters as prosthetic groups. In contrast to the NAD^+-dependent 2-oxo acid dehydrogenases, these reactions are reversible because of the more negative redox potential of ferredoxin ($E_0' = -420\,mV$) as compared to that of NAD^+ ($E_0' = -320\,mV$).

$$\text{R-CO-COO}^- + 2\text{fd} + \text{CoASH} \rightarrow \text{R-CO-SCoA} + CO_2 + 2\text{fd}^- + H^+$$
(2-oxo acid ferredoxin oxido-reductase)
$$\Delta G^{\circ\prime} \approx -20\,\text{kJ/mol} \qquad (12.118)$$

The **2-hydroxyacid pathway**, a widespread reductive pathway, involves the reduction of the 2-oxoacids to the corresponding (*R*)-2-hydroxyacids followed by activation to the CoA ester and *syn*-dehydration to enoyl-CoA (Fig. 12.**33**). A second reduction catalyzed by acyl-CoA dehydrogenases or enoyl-CoA reductases yields the saturated CoA esters, from which the free acid is liberated by CoA-transferase. A variant to this scheme is the CoA transfer from the enoyl-CoA to the 2-hydroxyacid, liberating a free enoate that is reduced by enoate reductase to the saturated acid (*Clostridium sporogenes*). The pathway is involved in the reduction of alanine, cysteine, glutamate, glutamine, histidine, leucine, methionine, phenylalanine, serine, threonine,

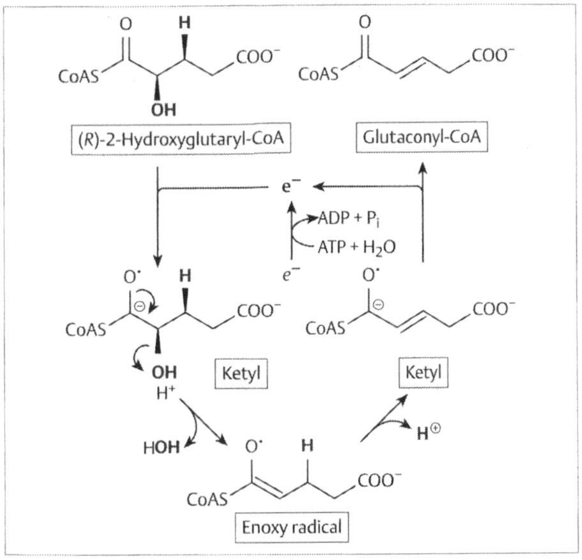

Fig. 12.**34** **Mechanism of dehydration of (*R*)-2-hydroxyglutaryl-CoA to glutaconyl-CoA.** Initially, the thiolester carbonyl group, which can be regarded as a ketone, is reduced by one electron to an anion radical, the ketyl radical. The electron donor may be either the reduced flavin, which is oxidized to a semiquinone, or the iron-sulfur cluster. The ketyl radical is able to eliminate the neighboring hydroxyl group. The resulting enoxy radical may be deprotonated to yield the ketyl radical of glutaconyl-CoA, which is reoxidized to form the product. The high energy electron, which is able to reduce a thiol ester, is funneled into the cycle by a reductant and is energized by hydrolysis of ATP. The electron remains in this high-energy state for many turnovers, since only catalytical amounts of ATP are required

tryptophan, and tyrosine by certain organisms; other organisms use different pathways, for example, the fermentation of glutamate (Figs. 12.**37** and 12.**38**).

The reversible dehydration of (*R*)-2-hydroxyacyl-CoA to enoyl-CoA is an intriguing catalytic reaction since the β-hydrogen atom to be eliminated is not activated by the thiol ester. The pK of the activated α-hydrogen atom is around 20, while that of the non-activated β-hydrogen is >30; the latter cannot be abstracted by a base from the enzyme. A well-studied example is the dehydration of (*R*)-2-hydroxyglutaryl-CoA to glutaconyl-CoA (Fig. 12.**34**) in the strict anaerobe *Acidaminococcus fermentans* (cluster 9). The extremely oxygen-sensitive enzyme, 2-hydroxyglutaryl-CoA dehydratase (100 kDa), contains [Fe–S] clusters, reduced riboflavin, and $FMNH_2$. The activation of the dehydratase, which requires a reducing agent and catalytic amounts of ATP and Mg^{2+}, is catalyzed by an activator containing an [Fe–S] cluster. *In vivo* NADH and an additional enzyme serve as reducing agent, whereas Ti(III) citrate can be used *in vitro*. A novel mechanism involving thiol ester-derived ketyls (radical anions) has been proposed for these dehydrations (Fig. 12.**34**). Remarkably, the benzoyl-CoA reductases from Proteobacteria (*Thauera*, *Azoarcus*, and *Rhodopseudomonas*, see Reaction 12.16 and Chapter 9) are related to this dehydratase. In both enzymes, hydrolysis of ATP increases the reductive power of an electron (i.e., E_0' becomes more negative).

Crotonyl-CoA (2-butenoyl-CoA) is an intermediate in certain pathways of the fermentation of glutamate, lysine, threonine, and methionine. Similarly, the fermentation of 5-aminovalerate leads to homocrotonyl-CoA (2-pentenoyl-CoA, Fig. 12.**42**). Crotonyl-CoA is a versatile intermediate since it may be oxidized to acetate via substrate-level phosphorylation or may be reduced to butyrate (Fig. 12.**21**). The reduction, however, is not coupled to respiration as with fumarate (Chapter 12.1.2.).

General strategies of amino acid fermentations include:

(1) Direct reductive elimination involves the amino group in glycine and proline. The reduction of glycine leads to acetyl phosphate, which serves to phosphorylate ADP. This is the only known reductive-substrate-level phosphorylation.
(2) Oxidation, transamination, or β-elimination leads to 2-oxoacids, which are oxidatively decarboxylated to acyl-CoA and reduced to (*R*)-2-hydroxy acids. Activation of this acid to the CoA derivative, followed by a mechanistically difficult inverse β-elimination of water, yields enoyl-CoA which is further reduced by acyl-CoA (hydroxy acid pathway, HAP).
(3) Glutamate and lysine can be rearranged to β-amino acids, from which the amino groups are easily removed by β-elimination.
(4) Crotonyl-CoA is a versatile intermediate in many amino acid fermentations.
(5) Aspartate and asparagine are deaminated to fumarate which serves as electron acceptor in a respiration (Chapter 12.1.2).

Enzymes containing coenzyme B_{12} (adenosylcobalamin) are involved in the fermentations of lysine, of glutamate via 3-methylaspartate, and the heterocyclic compound nicotinate. In all these coenzyme B_{12}-dependent rearrangements a hydrogen atom migrates *inter*molecularly to the adjacent carbon, whereas a group X migrates intramolecularly in the reverse direction (Fig. 12.**35**). Although the conversion of α-lysine to β-lysine (Fig. 12.**39**) resembles a coenzyme B_{12}-dependent reaction, the enzyme lysine 2,3-aminomutase from *C. subterminale* (cluster 1) does not contain or require adenosylcobalamin. The enzyme contains iron-sulfur clusters and pyridoxal 5′-phosphate. Furthermore, the reaction requires *S*-adenosylmethionine; its 5′-deoxyadenosine residue is involved in the intermolecular hydrogen transfer from the β- to the α-carbon, as is the case in adenosylcobalamine-dependent rearrangements. Therefore, the much simpler *S*-adenosylmethionine has been called by H.A. Barker "a poor man's adenosylcobalamin". Remarkably, the consecutive migration of the ε-amino group of β-lysine to the δ-position in *C. subterminale* (Fig. 12.**39**) requires an enzyme which in addition to pyridoxal-5′-phosphate, contains coenzyme B_{12}!

Glycine is fermented by *Eubacterium acidaminophilum* to acetate, NH_4, and CO_2 as follows:

$$\text{Oxidative branch: Glycine} + 2\,H_2O \rightarrow 2\,CO_2 + NH_4^+ + 5\,H^+ + 6\,e^- \quad (12.119)$$

$$\text{Reductive branch: 3 Glycine} + 6\,H^+ + 6\,e^- \rightarrow 3NH_4^+ + 3\,\text{acetate}^- \quad (12.120)$$

$$\text{Sum: 4 Glycine} + 2\,H_2O + H^+ \rightarrow 4\,NH_4^+ + 2\,CO_2 + 3\,\text{acetate}^-$$
$$\Delta G^{\circ\prime} = -217\,\text{kJ/4 mol glycine} \quad (12.121)$$

The pathway of the oxidative branch (Reaction 12.119) is also found in aerobic organisms (Chapter 9), whereas the reductive branch (Reaction 12.120) is unique to clostridia (Fig. 12.**32**). The free energy ($\Delta G^{\circ\prime} = -217$ kJ/4 mol glycine) is just sufficient for the formation of 3 mol ATP (72 kJ/mol ATP). In theory, 4 mol ATP should be formed by substrate-level phosphorylation, three mol via the glycine-reductase reaction and one mol in the oxidative branch. However, in this case, the fermentation (Reaction 12.121) is in equilibrium with ATP formation from ADP and inorganic phosphate. Therefore, part of the transiently generated acetyl phosphate is probably hydrolyzed in order to drive the fermentation to completion.

In the **fermentation of L-alanine** by *C. propionicum* (Reaction 12.122) the amino acid is transaminated to pyruvate (Reaction 12.83), which is oxidized to acetyl-CoA. ATP formation via acetyl phosphate follows (Fig. 12.**36**).

$$3\,\text{L-Alanine} + 2\,H_2O \rightarrow 3\,NH_4^+ + CO_2 + \text{acetate}^- + 2\,\text{propionate}^-$$
$$\Delta G^{\circ\prime} = -135\,\text{kJ/mol acetate} \quad (12.122)$$

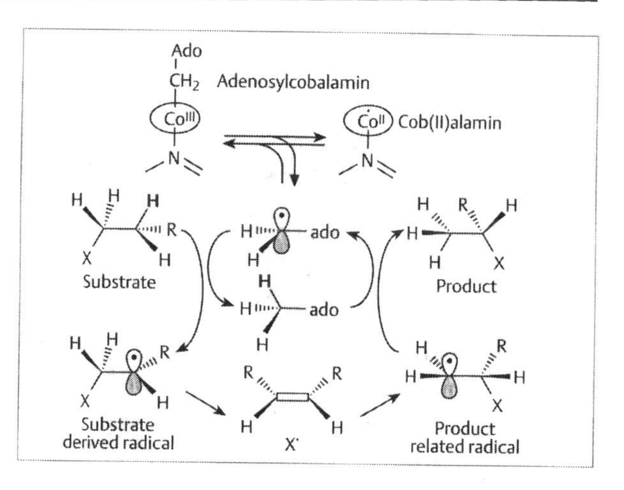

Fig. 12.35 **Mechanism of coenzyme B_{12}-dependent rearrangements.** A 5′-deoxyadenosine radical (Ado-$CH_2^•$) is generated by homolytic cleavage of the carbon-cobalt bond of adenosylcobalamin (coenzyme B_{12}). The cleavage is induced by the enzyme together with the substrate. The 5′-deoxyadenosine radical abstracts an H• atom from carbon 1 of the substrate to yield 5′-deoxyadenosine (Ado-CH_3) and a substrate-derived radical, from which X• is eliminated. Addition of X• to carbon 1 leads to the product-related radical to which one of the three H-atoms of 5′-deoxyadenosine is redonated, yielding the product and regenerating the 5′-deoxyadenosine radical for the next turnover. Hence, the group X migrates *intra*molecularly, whereas the H-atom migrates *inter*molecularly via the methyl group of 5′-deoxyadenosine. The group X is variable depending on the enzyme reaction. In the carbon skeleton-rearranging enzymes methylmalonyl-CoA mutase (Fig. 12.**25**), glutamate mutase (Fig. 12.**38**), and 2-methyleneglutarate mutase (Fig. 12.**43**), X represents formylyl-CoA, 2-glycyl, or 2-acrylyl, respectively. The three reactions have in common that the substrate-derived radicals fragment into X• and acrylate; (R = COO^-). The stereochemistry shown in the figure above is that observed with glutamate mutase. In β-lysine 5,6-mutase (Fig. 12.**39**) X stands for an amino group attached to pyridoxal 5′-phosphate. In the mechanism of glycerol dehydratase (Fig. 12.**26**) the hydrogen atom is abstracted from C-1 of glycerol and exchanged by the OH-group from C-2

In the reductive branch, pyruvate is reduced to propionate via the hydroxy acid pathway (Fig. 12.**33**). In contrast to crotonyl-CoA reduction, it is assumed that acryloyl-CoA reduction like fumarate reduction is coupled to electron-transport phosphorylation (see Chapter 12.1.2) resulting in 1/3 mol ATP/2e^-. Hence, almost two (5/3) mol ATP should be formed in Reaction 12.122, which agrees well with the overall free energy of the fermentation (81 kJ/mol ATP). The rumen bacterium *Megasphera elsdenii* (cluster 9) also uses this pathway, but butyrate and valerate are formed additionally by reductive condensation of acetyl-CoA with a second molecule of acetyl-CoA or propionyl-CoA, respectively.

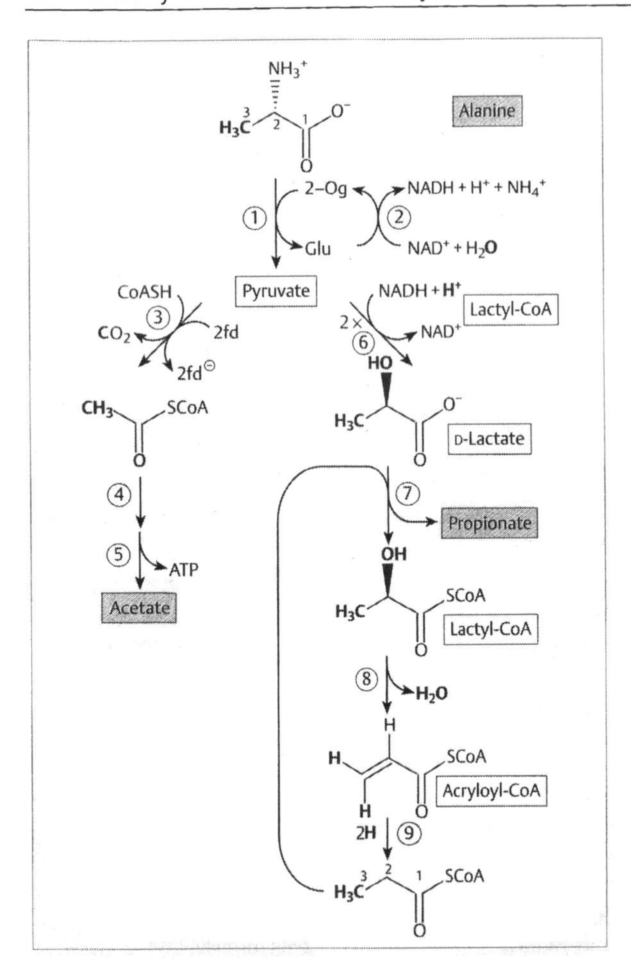

Fig. 12.**36** **Fermentation of L-alanine by *Clostridium propionicum*.** 1, Alanine aminotransferase (2-Og, 2-oxoglutarate; Glu, glutamate); 2, glutamate dehydrogenase; 3, pyruvate ferredoxin oxidoreductase (fd, ferredoxin); 4, phosphate acetyltransferase; 5, acetate kinase; 6, D-lactate dehydrogenase; 7, lactate CoA-transferase; 8, D-lactoyl-CoA dehydratase; 9, propionyl-CoA dehydrogenase

The fermentation of L-serine and cysteine in *C. propionicum* also involves the same kind of pyruvate disproportionation as shown above for alanine (Reactions 12.123 and 12.124), although the stoichiometry differs due to the higher oxidation state of these substrates.

$$3\ \text{L-Serine} + H_2O \rightarrow 3\ NH_4^+ + 2\ CO_2 + 2\ \text{acetate}^- + \text{propionate}^- \quad (12.123)$$

$$3\ \text{L-Cysteine} + 4\ H_2O \rightarrow 3\ NH_4^+ + 3\ H_2S + 2\ CO_2 + 2\ \text{acetate}^- + \text{propionate} \quad (12.124)$$

The fermentation of L-serine is initiated by a [4 Fe–4 S] cluster located in L-serine dehydratase (Reaction 12.85), which catalyzes the overall deamination of L-serine to pyruvate. Water is eliminated in the first step, which is facilitated by coordination of the hydroxy group to an iron atom of the cluster. The product, dehydroalanine, tautomerizes non-enzymatically to iminopyruvate, which finally hydrolyzes to ammonia and pyruvate. Interestingly, D-serine and threonine are deaminated by enzymes which contain pyridoxal phosphate and are devoid of any iron-sulfur clusters (Chapter 9). The pyridoxal 5′-phosphate-containing cysteine desulfurylase (Reaction 12.86) catalyzes the initial elimination of H_2S to dehydroalanine followed by hydrolysis to pyruvate.

Threonine fermentation can be carried out by at least three different pathways, depending on the organism.

(1) $$3\ \text{L-Threonine} + H_2O \rightarrow 3\ NH_4^+ + +2\ CO_2 + 2\ \text{propionate}^- + \text{butyrate}^- \quad (12.125)$$

In *C. propionicum* threonine is deaminated just as serine (but pyridoxal 5′-phosphate-dependent) to 2-oxobutyrate which disproportionates to propionate and butyrate (Reaction 12.125) via reactions analogous to those described for serine via the hydroxyacid pathway.

(2) $$\text{L-Threonine} + H_2O \rightarrow NH_4^+ + 2\ \text{acetate}^- + H^+ \quad (12.126)$$

$$\text{L-Threonine} \rightarrow \text{glycine} + \text{acetaldehyde} \quad \text{(threonine aldolase).} \quad (12.127)$$

In *C. pasteurianum* (cluster 1), threonine is fermented to ammonia and acetate (Reaction 12.126). Initially, the amino acid is cleaved to glycine and acetaldehyde, a reaction catalyzed by a pyridoxal 5′-phosphate-dependent threonine aldolase (Reaction 12.127). The resulting aldehyde is oxidized to acetyl-CoA (catalyzed by acetaldehyde dehydrogenase) (Reaction 12.94), whereas glycine is reduced to acetyl phosphate and ammonia. Hence, 2 mol ATP should be formed from 1 mol threonine.

(3) $$\text{L-Threonine} + NAD^+ \rightarrow \text{L-2-amino-3-oxobutyrate} + NADH + H^+ \quad (12.128)$$

$$\text{L-2-Amino-3-oxobutyrate} + \text{CoASH} \rightarrow \text{glycine} + \text{acetyl-CoA} \quad \text{(glycine } C\text{-acetyltransferase).} \quad (12.129)$$

$$\text{Acetoacetyl-CoA} + \text{CoASH} \rightarrow 2\ \text{acetyl-CoA} \quad \text{(thiolase).} \quad (12.130)$$

In an alternative route found in *C. sticklandii* (cluster 11), which also leads to acetate and ammonia, threonine is oxidized to L-2-amino-3-oxobutyrate (2-*C*-acetylglycine, Reaction 12.128). The subsequent cleavage to glycine and acetyl-CoA (Reaction 12.129) is similar to the well-known thiolase reaction (Reaction 12.130, acetyl-CoA *C*-acetyltransferase; see also Fig. 12.**21**). In contrast, glycine *C*-acetyltransferase contains pyridoxal 5′-phosphate. Glycine is metabolized further according to Reaction 12.121.

Glutamate, one of the most abundant amino acids, **is fermented** via at least four pathways. The two most important pathways proceed either via **3-methylaspartate** or via **2-hydroxyglutarate** and lead to identical products, and are approximated as follows:

$$5\,\text{Glutamate}^- + 6\,H_2O + 2\,H^+ \rightarrow 5\,NH_4^+ + 5\,CO_2 + 6\,\text{acetate}^- + 2\,\text{butyrate}^- + H_2$$
$$(0.6\,\text{ATP/mol glutamate})$$
$$\Delta G^{\circ\prime} = -59\,\text{kJ/mol glutamate}\ (98\,\text{kJ mol ATP}) \qquad (12.131)$$

The fermentation of glutamate via (*R*)-2-hydroxyglutarate ('hydroxyglutarate pathway') initially follows the 2-hydroxy acid pathway (Fig. 12.**33**). Glutaconyl-CoA, the product of the dehydration of (*R*)-2-hydroxyglutaryl-CoA, is decarboxylated to crotonyl-CoA (Fig. 12.**37**), whereby the free energy of decarboxylation is conserved as an electrochemical Na^+ gradient (Chapter 12.3). Disproportionation of crotonyl-CoA leads to the formation of acetate, butyrate, hydrogen, and ATP (via substrate-level phosphorylation). This pathway via 2-hydroxyglutarate has been found in certain clostridia and related organisms, which live in anoxic niches of humans, mammals, or birds: *Acidaminococcus fermentans* (cluster 9), *Fusobacterium nucleatum* (19), *Clostridium symbiosum* (14a), *C. aminophilum* (14a), *C. sporosphaeroides* (4), and *Peptostreptococcus asaccharolyticus* (13).

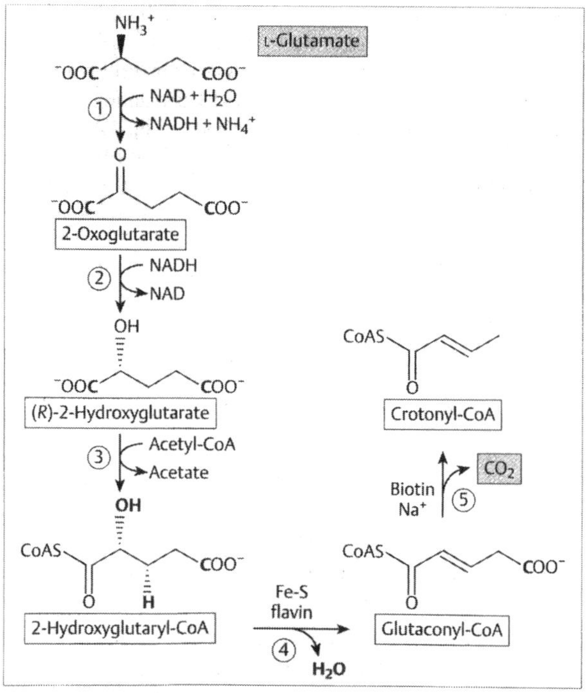

Fig. 12.**37** **Conversion of L-glutamate to crotonyl-CoA.** 1, Glutamate dehydrogenase; 2, (*R*)-2-hydroxyglutarate dehydrogenase; 3, glutaconate or 2-hydroxyglutarate CoA-transferase; 4, (*R*)-2-hydroxyglutaryl-CoA dehydratase; 5, glutaconyl-CoA decarboxylase

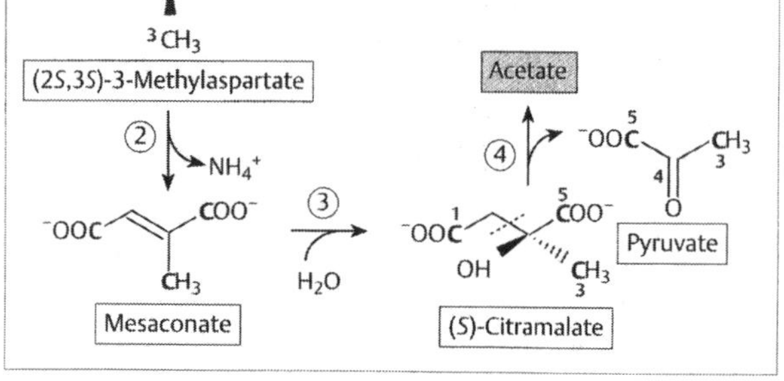

The methylaspartate pathway is an alternative pathway in which the linear carbon skeleton of glutamate is rearranged to that of the branched-chain amino acid (2*S*,3*S*)-3-methylaspartate, catalyzed by the coenzyme B_{12}-dependent glutamate mutase (Fig. 12.**38**). The carbon-skeleton rearrangement facilitates the subsequent elimination of ammonia by β-methylaspartase, since the β-hydrogen is then activated by the adjacent carboxyl group. In the third step, the mesaconate formed is hydrated to (*S*)-citramalate, which–analogous to citrate (Fig. 12.**28**)–is cleaved to acetate and pyruvate (Reaction 12.80). Oxidative decarboxylation of pyruvate gives rise to acetyl-CoA and reduced ferredoxin, which is reoxidized during the synthesis of butyryl-CoA from two mol acetyl-CoA. In contrast to the hydroxyglutarate pathway, the 3-methylaspartate pathway is found in a closely related group of clostridia occurring in the soil: *C. tetani* (cluster 1), *C. tetanomorphum* (1), *C. cochlearium* (1), *C. malenominatum* (1), and

Fig. 12.**38** **Conversion of L-glutamate to pyruvate.** 1, Glutamate mutase; 2, methylaspartase; 3, mesaconase; 4, citramalate lyase

C. limosum (2). It may be mentioned that these bacteria use either pathway to ferment also glutamine and D-glutamate, since they all contain glutaminase and glutamate racemase.

The γ-aminobutyrate pathway of glutamate fermentation requires two organisms. One is able to decarboxylate glutamate (Reaction 12.132) and the other is able to ferment the resulting 4-aminobutyrate (γ-aminobutyrate, GABA; Reaction 12.133).

$$\text{L-Glutamate}^- + H^+ \rightarrow \text{4-aminobutyrate} + CO_2 \quad \text{(glutamate decarboxylase).} \qquad (12.132)$$

$$2 \times \text{4-Aminobutyrate} + 2\,H_2O \rightarrow 2\,NH_4^+ + 2\,\text{acetate}^- + \text{butyrate}^- + H^+ \qquad (12.133)$$

The 4-aminobutyrate is formed by many Eukarya and Bacteria (Reaction 12.132). In animals, it serves as neurotransmitter (GABA), whereas many bacteria excrete this amino acid. In some bacteria, the free energy of decarboxylations of amino acids is conserved in a very simple manner, as shown in Chapter 12.3. *Clostridium aminobutyricum* (cluster 11) is the only organism known to ferment 4-aminobutyrate (Reaction 12.133). The substrate is first transaminated to succinate semialdehyde, which in turn is reduced by NADH to 4-hydroxybutyrate. Hence, the amino group is replaced by a hydroxy group. The further fate of 4-hydroxybutyrate is similar to that in *C. kluyveri* (Figs. 12.**29** and 12.**30**).

The propionate pathway of glutamate fermentation (Reaction 12.134) is found in *Selenomonas acidaminophila* and *Barkera propionica* (cluster 9). The pathway is assumed to be a combination of the methylaspartate pathway (Fig. 12.**38**, from glutamate to acetate and pyruvate) and the pathway in which pyruvate disproportionates to propionate and additional acetate (as in *Propionibacterium*, Fig. 12.**25**). Hence, this pathway should contain two consecutive coenzyme B_{12}-dependent enzymes, glutamate mutase and methylmalonyl-CoA mutase, whereby 2 mol ATP should be conserved from 3 glutamate. An additional 2/3 mol ATP should be obtained by electron-transport phosphorylation in the fumarate-reductase reaction (Chapter 12.1.2).

$$3\,\text{Glutamate}^- + 4\,H_2O \rightarrow 3\,NH_4^+ + 2\,CO_2 + 5\,\text{acetate}^- + \text{propionate}^- \quad \Delta G^{\circ\prime} = -187\ \text{kJ/3 mol glutamate} \ (70\ \text{kJ/mol ATP}) \qquad (12.134)$$

Histidine is fermented by the same organisms that are able to use glutamate as substrate.

$$\text{L-Histidine} + 3\,H_2O \rightarrow NH_4^+ + \text{formamide} + \text{L-glutamate}^- \qquad (12.135)$$

$$\text{Formamide} + H_2O \rightarrow NH_4^+ + HCOO^- \quad \text{(formamidase)} \qquad (12.136)$$

L-Lysine is fermented to acetate and butyrate by *Clostridium subterminale* (cluster 1) as well as by *Fusobacterium nucleatum* (cluster 19) (Reaction 12.137; Fig. 12.**39**).

$$\text{L-Lysine}^+ + 2\,H_2O \rightarrow 2\,NH_4^+ + \text{acetate}^- + \text{butyrate}^- + H^+ \qquad (12.137)$$

L-Lysine [(2*S*)-2,6-diaminohexanoate] is first converted to β-lysine [(3*S*)-3,6-diaminohexanoate)] by an *S*-adenosylmethionine-dependent but coenzyme B_{12}-independent intramolecular migration of the α-amino group to the β-position (see general strategies). In the next step, a coenzyme B_{12}-dependent intramolecular migration of the 6-amino group to the 5-position yields (3*S*,5*S*)-3,5-diaminohexanoate, which is oxidized by NAD^+ to ammonia and (5*S*)-3-oxo-5-aminohexanoate. An unusual reaction with acetyl-CoA leads to (*S*)-3-aminobutyryl-CoA and acetoacetate (Fig. 12.**40**). The 3-aminobutyryl-CoA is deaminated by β-elimination to crotonyl-CoA which is reduced to butyryl-CoA. CoA-transfer to acetoacetate yields butyrate and acetoacetyl-CoA, which is cleaved into 2 mol acetyl-CoA. Finally, 1 mol ATP/mol lysine is generated by the catalysis of phosphate acetyl transferase and acetate kinase.

L-Arginine is hydrolyzed to ammonia and L-citrulline (Reaction 12.138; Fig. 12.**41**). This is followed by phosphorolysis of citrulline to L-ornithine and carbamoyl phosphate (Reaction 12.139), the phosphoryl group of which is transferred to ADP (Reaction 12.140).

$$\text{L-Arginine}^+ + H_2O \rightarrow \text{L-citrulline} + NH_4^+ \quad \text{(arginine deiminase)} \quad \Delta G^{\circ\prime} = -38\ \text{kJ/mol} \qquad (12.138)$$

$$\text{L-Citrulline} + HPO_4^{2-} + H^+ \rightarrow \text{carbamoylphosphate}^{2-} + \text{L-ornithine}^+ \quad \text{(ornithine carbamoyltransferase)} \quad \Delta G^{\circ\prime} = +29\ \text{kJ/mol} \qquad (12.139)$$

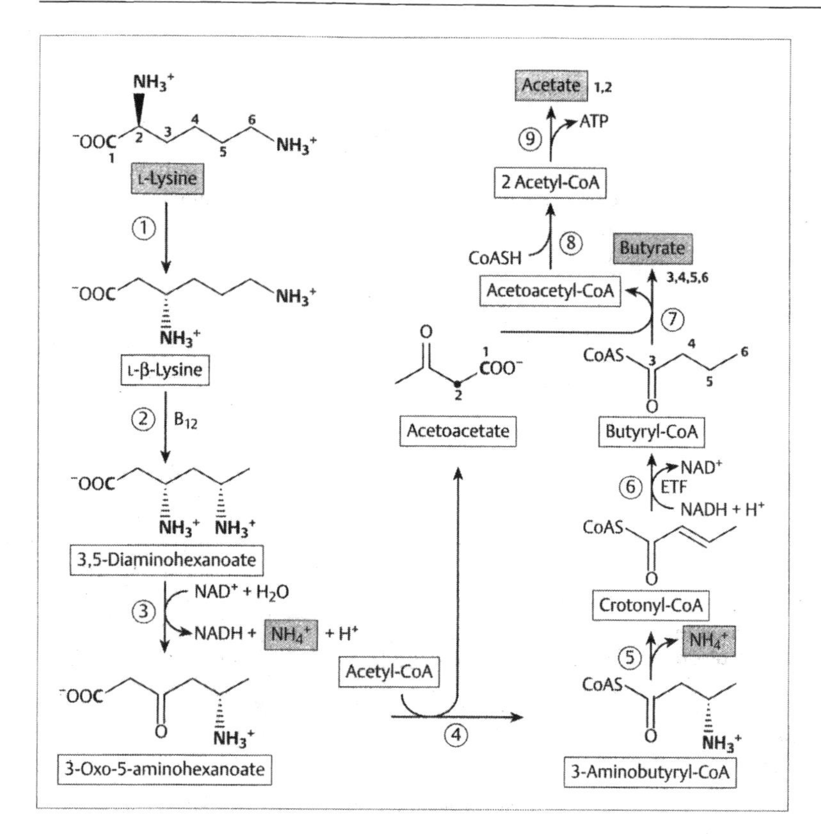

Fig. 12.**39** **Fermentation of L-lysine by *Clostridium subterminale* and *Fusobacterium nucleatum*.** 1, Lysine 2,3-aminomutase; 2, β-lysine 5,6-aminomutase; 3, 3,5-diaminohexanoate dehydrogenase; 4, 3-oxo-5-aminohexanoate acetyltransferase; 5, 3-aminobutyryl-CoA ammonia lyase; 6, butyryl-CoA dehydrogenase; 7, acetoacetate CoA-transferase; 8, thiolase; 9, phosphate acetyltransferase and acetate kinase

$$\text{Carbamoyl phosphate}^{2-} + ADP^{3-}H_2O + 2\,H^+ \rightarrow CO_2 + NH_4^+ + ATP^{4-}$$

(carbamate kinase)

$$\Delta G^{\circ\prime} = -29\ \text{kJ/mol} \quad (12.140)$$

$$\text{Sum: L-Arginine}^+ + ADP^{3-} + 2\,H_2O + HPO_4^{2-} + 3\,H^+ \rightarrow \text{L-ornithine}^+ + CO_2 + 2\,NH_4^+ + ATP^{4-}$$

$$\Delta G^{\circ\prime} = -38\ \text{kJ/mol} \quad (12.141)$$

Although the formation of carbamoyl phosphate is endergonic, it is obviously driven by the preceding irreversible hydrolysis of arginine. Thus, citrulline accumulates in the cell and reaches a high concentration which pushes the consecutive cleavage involving phosphate. The pathway from arginine to ornithine is widespread among Bacteria and not restricted to Gram-positive species. This pathway even has been found in Archaea, for example in *Halobacterium halobium*, which under anaerobic conditions in the dark uses arginine as energy source.

Clostridium viride **(cluster 3) contains three green enzymes.**

$$2\ \text{5-Aminovalerate} + 2\,H_2O \rightarrow 2\,NH_4^+ + \text{acetate}^- + \text{propionate}^- = \text{valerate}^- + H^+ \quad (12.142)$$

In fermentative organisms, proline is exclusively reduced to 5-aminovalerate. Hence, 5-aminovalerate would accumulate under anaerobic conditions unless an organism would be present that degrades this compound. The only known organism fermenting 5-aminovalerate is *C. viride* (cluster 3), formerly called *C. aminovalericum* sp. By way of amino transfer to 2-oxoglutarate and an NADH-dependent reduction of the glutarsemialdehyde, the amino group of 5-aminovalerate is replaced by a hydroxy group (Fig. 12.**42**). The resulting 5-hydroxyvalerate is activated to the thiol ester by CoA-transfer (Reaction 12.114) and oxidatively dehydrated to 2,4-pentadienoyl-CoA. The 1,4-reduction of the conjugated double bonds leads to 3-pentenoyl-CoA, which isomerizes to 2-pentenoyl-CoA. Finally,

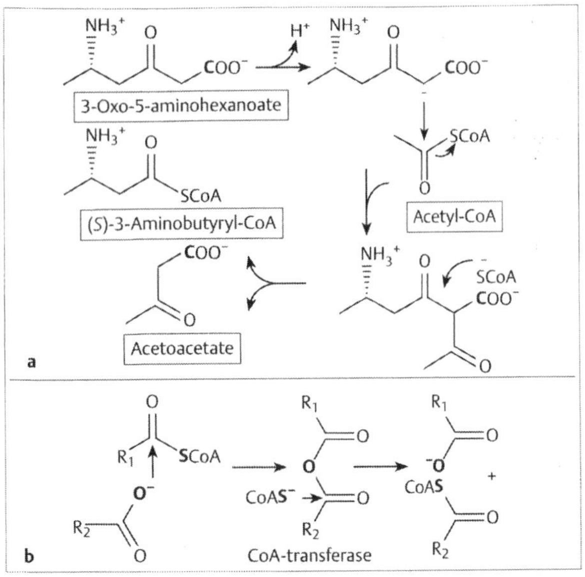

Fig. 12.40 **Mechanism of the cleavage of 3-oxo-5-aminohexanoate by acetyl-CoA and comparison to that of CoA-transferases.**
a The acidic α-hydrogen of 3-oxo-5-aminohexanoate is removed by a base. The resulting carbanion condenses with acetyl-CoA to a branched 1,3-dioxo-C_8-intermediate. The liberated $CoAS^-$ takes part in the cleavage of the 1,3-dioxoamino acid into acetoacetate and (*S*)-3-aminobutyryl-CoA.
b In the CoA-transferase reaction a γ-glutamyl anion of the enzyme (R_2-COO^-) rather than a carbanion attacks the thiol ester to form a mixed anhydride. Hence, an enzyme-CoA-ester is formed, which reacts in the same manner with the carboxylate anion of the second substrate to yield the CoA-ester product

disproportionation of 2-pentenoyl-CoA yields acetate, propionate, and valerate whereby 0.5 ATP/5-amino valerate is conserved. The FAD-containing enzyme hydroxyvaleryl-CoA dehydratase, 2,4-pentadienoyl-CoA reductase, and valeryl-CoA dehydrogenase have a green color caused by a blue charge-transfer absorption band at 700 nm in addition to the yellow flavin absorption at 450 nm.

In **amino acid fermentations** almost all proteinogenous amino acids (17 out of 20) can act both as electron donors and as electron acceptors, whereas isoleucine and valine are only used as electron donors, and proline appears to serve exclusively as electron acceptor (Table 12.**7**). Interestingly, some amino acids are fermented by different pathways that lead to the same products. Glutamate is fermented to butyrate by a coenzyme B_{12}-dependent pathway via 3-methylaspartate and by an independent pathway via 2-hydroxyglutarate. The same is found with lactate (from alanine), which can be reduced to propionate via the random pathway (coenzyme B_{12}-dependent, Fig. 12.**24**) and the nonrandom pathway (Fig. 12.**35**), respectively.

12.2.4 *N*-Heterocyclic Compounds Are Fermented Like Amino Acids

The most common *N*-heterocyclic compounds in nature are the purine and pyrimidine bases of the nucleic acids. Other heterocyclic compounds are creatinine, from vertebrate muscles, and nicotinic acid, a constituent of NAD^+. The average oxidation state of the carbons in

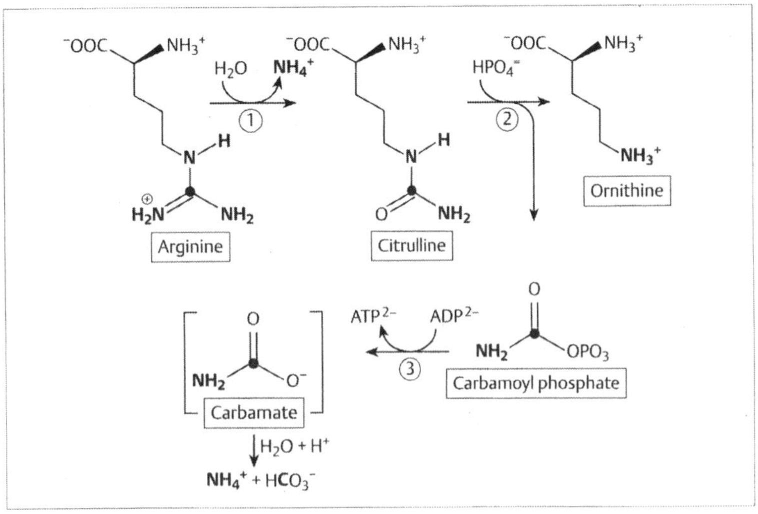

Fig. 12.41 **Fermentation of arginine to ornithine.** 1, Arginine deiminase; 2, ornithine transcarbamoylase; 3, carbamate kinase

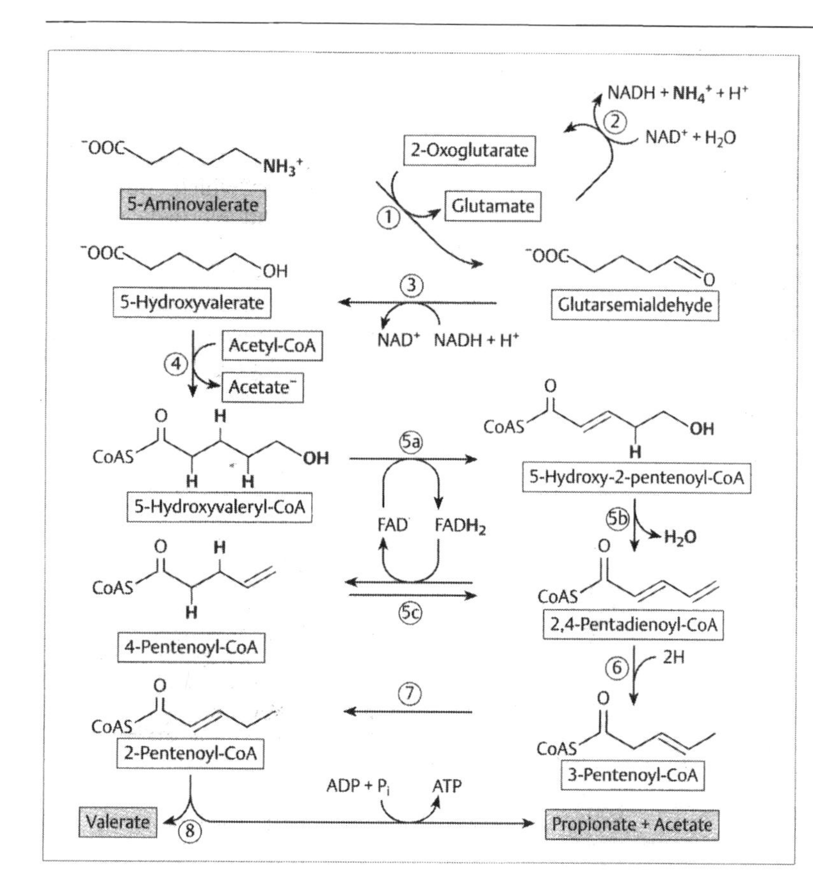

Fig. 12.**42** **Fermentation of 5-aminovalerate by *Clostridium viride*.** 1, 5-Aminovalerate aminotransferase; 2, glutamate dehydrogenase; 3, 5-hydroxyvalerate dehydrogenase; 4, 5-hydroxyvalerate CoA-transferase; 5(a,b,c), 5-hydroxyvaleryl-CoA dehydratase; 6, pentadienoyl-CoA reductase; 7, 3-pentenoyl-CoA Δ-isomerase; 8, the disproportionation of 2-pentenoyl-CoA to valerate, propionate, and acetate is analogous to that of crotonyl-CoA to butyrate and acetate. 4-Pentenoyl-CoA is only formed with the isolated 5-hydroxyvaleryl-CoA dehydratase (5c) in the absence of an external electron acceptor in order to regenerate FAD

these heterocyclic compounds is +2. Hence, these compounds are suitable for fermentations. No fermentation by a single bacterial species has been reported for indole, indoleacetate, and indolepropionate, which are formed from tryptophan. These compounds are too reduced for this type of energy conservation.

Purines are readily fermented to ammonia, carbon dioxide, formate, and acetate by *Clostridium acidiurici* (cluster 12), *C. purinolyticum* (cluster 12), and *Peptostreptococcus asaccharolyticus* (cluster 13) (Reactions 12.143–12.145).

$$\text{Guanine} + 7\,H_2O + 4\,H^+ \rightarrow 5\,NH_4^+ + 3\,CO_2 + \text{acetate}^- \quad (12.143)$$

$$\text{Adenine} + 8\,H_2O + 3\,H^+ \rightarrow 5\,NH_4^+ + 2\,CO_2 + \text{formate}^- + \text{acetate}^- \quad (12.144)$$

$$2\,\text{Urate}^- + 12\,H_2O + 7\,H^+ \rightarrow 8\,NH_4^+ + 6\,CO_2 + 2\,\text{formate}^- + \text{acetate}^- \quad (12.145)$$

The main intermediate of purine degradation is xanthine, to which all other purines have to be converted like the reactions in aerobic bacteria (Chapter 9).

Pyrimidines are not fermented as readily as purines. The only well-documented conversions are the reductions of uracil and orotate. In a complex medium, *C. uracilium* (unclassified) reduces uracil to dihydrouracil, which is hydrolyzed to CO_2, ammonia, and β-alanine (Reaction 12.146).

$$\text{Uracil} + 2\,H_2O + 3\,H^+ + 2\,e^- \rightarrow NH_4^+ + CO_2 + \beta\text{-alanine} \quad (12.146)$$

In a simple β-elimination reaction, β-alanine may be converted via β-alanyl-CoA to acryloyl-CoA, which is reduced to propionate via propionyl-CoA (see Fig. 12.**36**). Orotate is reduced by *C. oroticum* (cluster 14a) to CO_2, ammonia, acetate and succinate via aspartate (Reaction 12.147).

$$\text{Orotate}^- + 2\,H_2O + 3\,H^+ + 2\,e^- \rightarrow NH_4^+ + CO_2 + \text{L-aspartate}^- \quad (12.147)$$

The only known organism able to **ferment nicotinic acid** is *C. barkeri* (cluster 15, Reaction 12.148). In addition, sulfate reducers are able to completely oxidize nicotinate under anoxic conditions by an unknown pathway.

$$\text{Nicotinate}^- + 4\,H_2O \rightarrow NH_4^+ + CO_2 + \text{acetate}^- + \text{propionate}^- \quad (12.148)$$

Nicotinate (pyridine-3-carboxylate) is oxidized by $NADP^+$ to 6-hydroxynicotinate followed by reduction with reduced ferredoxin to 4,5-dihydro-6-hydroxynicotinate (Fig. 12.**43**). The next established intermediate is 2-methyleneglutarate, which might be derived from the latter compound by hydrolysis of the amide. This results in an enamine which is tautomerized and hydrolyzed to 2-formylglutarate. After reduction of 2-formylglutarate by NAD(P)H to 2-hydroxymethylglutarate, this branched 3-hydroxy acid should easily eliminate water to yield the unsaturated 2-methyleneglutarate. The fermentation of 2-methyleneglutarate is well established. In a coenzyme B_{12}-dependent reaction, 2-methyleneglutarate is rear-

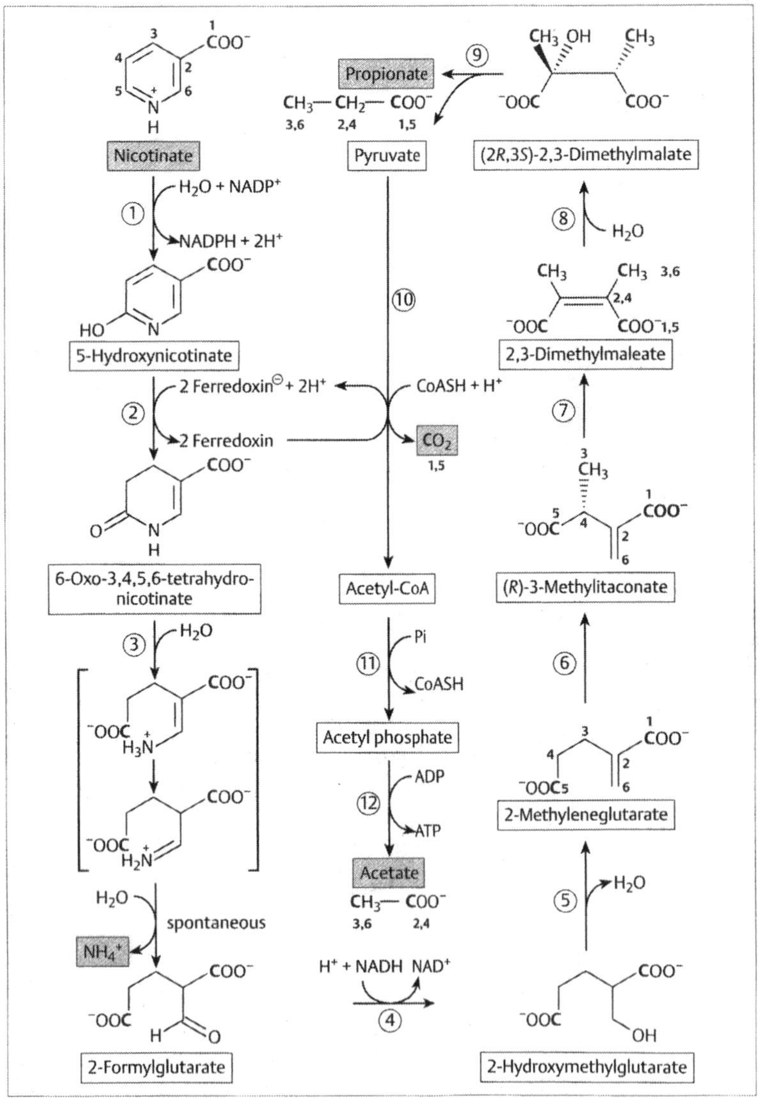

Fig. 12.**43** **Fermentation of nicotinate by *Clostridium barkeri*.** 1, Nicotinate hydroylase; 2, 5-hydroxynicotinate reductase; 3, 5-oxo-3,4,5,6-tetrahydronicotinate hydrolase; 4, 2-hydroxymethylglutarate dehydrogenase; 5, 2-hydroxymethylglutarate dehydratase; 6, 2-methyleneglutarate mutase (co-enzyme B_{12}); 7, 3-methylitaconate Δ-isomerase; 8, 2,3-dimethylmalate dehydratase; 9, 2,3-dimethylmalate lyase; 10, pyruvate ferredoxin oxidoreductase; 11, phosphate acetyltransferase; 12, acetate kinase

ranged to (*R*)-3-methylitaconate (2-methylene-3-methylsuccinate) followed by isomerization to 2,3-dimethylmaleate (Fig. 12.**35**). Addition of water leads to (2*R*,3*S*)-dimethylmalate, which is cleaved to pyruvate and propionate (Reaction 12.81). Pyruvate ferredoxin oxidoreductase (Reaction 12.88) catalyzes the formation of acetyl-CoA and reduced ferredoxin, which is required for the second step of the nicotinate pathway. One mol ATP is formed per mol nicotinate.

Box 12.8 Unusual enzymes of nicotinate fermentation by *Clostridium barkeri* include nicotinate hydroxylase, 2-methyleneglutarate mutase, and dimethylmalate lyase. The $NADP^+$-dependent hydroxylase resembles xanthine dehydrogenase (Chapter 9). It contains flavin, iron-sulfur clusters, molybdopterin (Fig. 12.**6**) and even selenium but not as selenocysteine. The coenzyme B_{12}-dependent 2-methyleneglutarate mutase is explained in Fig. 12.**35**. (2*R*,3*S*)-Dimethylmalate is cleaved to propionate and pyruvate by an enzyme that is mechanistically related to isocitrate lyase. Both enzymes contain a reactive sulfhydryl group but not the acetyl-CoA-like prosthetic groups, which have been detected in citrate and citramalate lyases (Fig. 12.**28**).

The fermentation of *N*-heterocyclic compounds is limited to only a few clostridia and related organisms that are able to ferment purines, pyrimidines, and nicotinate. All purines are converted to xanthine, which is degraded to ammonia, carbon dioxide, formate, and glycine. Among the pyrimidines, only uracil and orotate fermentations have been reported in the literature. These compounds are cleaved to β-alanine and aspartate, respectively. The fermentation of nicotinate to ammonia, carbon dioxide, acetate, and propionate requires very special enzymes.

12.2.5 Syntrophic Bacteria Link Fermentation and Methanogenesis

Among the various products formed by the fermentative and acetogenic organisms, only acetate, H_2, CO_2, and C_1 compounds serve as substrates for methanogenic Archaea, whereas fatty acids larger than acetate (e.g., propionate and butyrate), alcohols larger than ethanol, and aromatic acids are not attacked (Chapter 12.1.7). In the absence of any suitable electron acceptor such as nitrate or sulfate (Chapter 12.1), these compounds would persist in nature under anoxic conditions, unless they would not be oxidized to acetate and H_2 by syntrophic bacteria (obligate H^+-reducers). These bacteria are difficult to grow in axenic cultures and therefore they are not well characterized. The oxidation of straight-chain fatty acids, e.g., butyrate, by *Syntrophobacter wolinii*, *Syntrophomonas wolfei*, or *Syntrophospora bryantii* (all in cluster 8) is shown by reaction 12.149.

$$\text{Butyrate}^- + 2\,H_2O \rightarrow 2\,\text{acetate}^- + H^+ + 2\,H_2 \qquad \Delta G^{\circ\prime} = +48.1\ \text{kJ/mol butyrate} \tag{12.149}$$

The free energy required for this process clearly shows that this reaction cannot proceed under standard conditions. A simple calculation similar to that of Equation 12.53, however, indicates that at a partial pressure of about 10 Pa (10^{-4} bar) H_2, the reaction is in equilibrium ($\Delta G' = 0$); below 1 Pa H_2, ATP formation becomes possible. In order to maintain such low H_2 pressures, methanogenic Archaea are required. Hence, the syntrophs can only exist in combination with methanogens. This is expressed by the greek name **syntroph**, which means 'eating together'. Since the methanogens can thrive at a partial H_2 pressure above 0.1 Pa, less than 20 kJ are available for the syntrophs, which is hardly sufficient to form 1/3 mol ATP. Therefore, electron-transport phosphorylation via $\Delta\tilde{\mu}_{H^+}$ or $\Delta\tilde{\mu}_{Na^+}$ rather than substrate-level phosphorylation via acetyl phosphate appears feasible. The mechanism of the generation of this electrochemical ion gradient has not yet been elucidated. The hydrogen transferred from the syntroph to the methanogen does not occur free in solution; instead it is rather **transferred directly** from species to species which need to be in close contact with each other. Hence, this process is called "**interspecies hydrogen transfer**". In some cases, it has been observed that the hydrogen is transferred together with CO_2 in the form of formate. This is catalyzed by formate hydrogen lyase (Reaction 12.157).

$$H_2 + CO_2 = HCOO^- + H^+ \qquad \Delta G^{\circ\prime} = +3.5\,\text{kJ/mol} \tag{12.150}$$

For the conversion of propionate to acetate, two pathways have been detected using a consortium of syntrophs and methanogens (Reactions 12.151 and 12.152).

$$\text{Propionate}^- + H_2 + CO_2 \rightarrow 2\,\text{acetate}^- + H^+ \qquad \Delta G^{\circ\prime} = -23.7\ \text{kJ/mol propionate} \tag{12.151}$$

$$\text{Propionate}^- + 2\,H_2O \rightarrow \text{acetate}^- + 3\,H_2 + CO_2 \qquad \Delta G^{\circ\prime} = +71.3\ \text{kJ/mol propionate} \tag{12.152}$$

In the first pathway (Reaction 12.151), for which no methanogen is required, propionyl-CoA is reductively carboxylated to 2-oxobutyrate. This process is analogous to the conversion of acetyl-CoA to pyruvate catalyzed by pyruvate ferredoxin oxidoreductase (Reaction 12.88). The further reduction and dehydration of 2-oxobutyrate to crotonyl-CoA follows the 2-hydroxy acid pathway (Fig. 12.**33**). Crotonyl-CoA is then oxidized to 2 molecules of acetate. In the second pathway (Reaction 12.152) propionyl-CoA is converted via methylmalonyl-CoA to succinyl-CoA (Fig. 12.**25**) which enters the Krebs cycle.

Little is known about the fate of the aromatic acids. It is likely, however, that these compounds are converted to benzoyl-CoA. This reaction is followed by the reduction of the aromatic ring (Reaction 12.16). Subsequent hydrations and oxidations lead to acetyl-CoA, crotonyl-CoA, and CO_2 as shown for the denitrifying proteobacterium *Thauera aromatica*.

Fermentations of polymeric carbohydrates, proteins, and nucleic acids finally lead, via the monomers, to ammonia, H_2S, acetate, CO_2, and H_2. In the absence of electron acceptors such as nitrate, or sulfate, CO_2 serves as electron acceptor for acetogenesis and methanogenesis. In the last step, acetate is cleaved by certain methanogens to CH_4 and CO_2, called "biogas", the end product of the anaerobic food chain. As soon as CH_4 escapes from the anoxic niches, most of it is oxidized by methanotrophic bacteria to CO_2, which is ready to enter a new round of the carbon cycle via the photosynthetic organisms.

Syntrophic bacteria use reduced compounds such as fatty acids as electron donors ($E_0' \geq -320$ mV), and protons ($E_0' = -420$ mV) as electron acceptors. Hence, the fermentations are endergonic, unless a hydrogen-consuming organism (usually a methanogen) is also present. The hydrogen is transferred directly from the syntroph to the methanogen without entering in the medium. This process is called "interspecies hydrogen transfer".

12.3 Energy Can Be Conserved by Non-Redox Reactions

In aerobic chemotrophic organisms, energy conversions are coupled to oxidations in which oxygen serves as the final electron acceptor. It is the aim of Chapter 12.1 to show that energy is also conserved in anaerobic organisms by reactions involved in electron transport. The only difference is the replacement of oxygen by a variety of other inorganic and organic electron acceptors. In most cases, the redox potentials of these acceptors are much less positive than that of oxygen, which leads to energy differences that allow the synthesis of much less ATP. The amount of ATP generated is sometimes even less than 1 mol ATP/mol substrate, for example, in acetogenesis and methanogenesis. Hence, the substrate turnover in these organisms must be much higher than in aerobic organisms.

In addition the anaerobes conserve energy even from reactions, which are neglected by aerobes. These reactions, which are neither respirations nor fermentations, comprise a third kind of energy conservation and are summarized here in order to show that dehydrogenation and electron transport are very frequent but not absolute requirements for energy conservation and in order to demonstrate again the versatility of the anaerobes.

During degradation of arginine to ornithine the **energy-rich carbamoyl phosphate** is formed and is able to phosphorylate ADP (Fig. 12.**41**). This mode of energy conservation is widespread among the Archaea, the Bacteria, and some of the Eukarya living under anoxic conditions.

Biotin-dependent membrane-bound **decarboxylases** generate $\Delta\tilde{\mu}_{Na^+}$, for example, the oxaloacetate decarboxylase involved in citrate fermentation (Fig. 12.**27**), the methylmalonyl-CoA decarboxylase involved in the decarboxylation of succinate to propionate by *Propionigenium modestum* (clostridial cluster 19; the pathway is related to propionate fermentation, Fig. 12.**25**), and the glutaconyl-CoA decarboxylase involved in glutamate fermentation via 2-hydroxyglutarate (Fig. 12.**37**). The energy change between the enzyme-bound negatively-charged carboxybiotin and the uncharged biotin may generate $\Delta\tilde{\mu}_{Na^+}$.

Decarboxylation of oxalate and amino acids consumes one proton inside the cell, leading to a product with one negative charge less than the substrate. Product and substrate are exchanged by an antiporter. Hence, the generation of $\Delta\tilde{\mu}_{H^+}$ of 1/3 mol ATP/mol substrate is possible. *Oxalobacter formigenes*

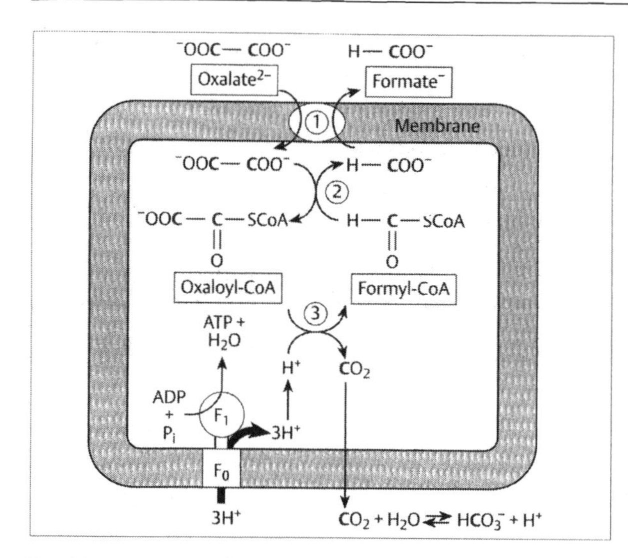

Fig. 12.**44** **ATP-synthesis coupled to oxalate decarboxylation in *Oxalobacter formigenes*.** 1, Oxalate^{2-}/formate^{-} antiporter; 2, oxalate CoA-transferase; 3, oxaloyl-CoA decarboxylase (thiamindiphosphate); F_1/F_0, ATP-synthase

thrives solely on the decarboxylation of oxalate to formate, via oxalyl-CoA and formyl-CoA (Fig. 12.**44**).

Export of lactate^{-} together with $2\,H^+$ generates $\Delta\tilde{\mu}_{H^+}$. This mode of energy conservation has been proposed as an additional energy source for lactic acid bacteria, which ferment the uncharged glucose to 2 mol lactate^{-} $+2\ H^+$ per mol glucose.

Methyl-group transfer from an N atom to an SH group in methanogenic bacteria ($\Delta G^{\circ\prime} \approx -30$ kJ/mol) generates $\Delta\tilde{\mu}_{Na^+}$. Besides the electron-transport chain with the heterodisulfide as terminal acceptor, the methyltransferase (Figs. 12.**09** and 12.**11**) conserves the energy released in methanogenesis via an electrochemical Na^+ gradient. During methyl-group transfer the cobamide coenzyme changes from the methylated, hexacoordinated Co(III)-form to the unmethylated, tetracoordinated Co(I)-form. Hence, the 5th ligand to the cobalt, an axial histidine of the protein, moves towards the Co-atom during methylation ('base on') and away from it during demethylation ('base off'). That this movement may drive the Na^+-pump, is an attractive hypothesis.

Dehydrogenation or electron transport is the most frequent but not an absolute requirement for the generation of energy-rich compounds for substrate-level phosphorylation, or of electrochemical H^+- and Na^+-gradients for electron-transport phosphorylation. In anaerobes, also non-redox reactions are used for these processes.

These examples demonstrate that non-redox reactions play an important role in energy conservation in anaerobic bacteria.

Further Reading

Andreesen, J. R. (1994) Glycine metabolism in anaerobes. Antonie van Leeuwenhoek 66: 223-327

Barker, H. A. (1961) Fermentations of nitrogenous organic compounds. In Gunsalus, I. C., and Stanier, R. Y. (eds.). The bacteria, vol. 2. New York: Academic Press; pp 151-207

Barton, L. L., ed. (1995) Sulfate-reducing bacteria. New York, London: Plenum Press

Buckel, W. (1990) Amino acid fermentation: coenzyme B_{12}-dependent and -independent pathways. In: The molecular basis of bacterial metabolism. 41st Colloquium - Mosbach 1990 (Hauska, G., and Thauer, R., eds.). Berlin, Heidelberg, New York: Springer; pp 21-30

Buckel, W. (1996) Unusual dehydrations in anaerobic bacteria: considering ketyls (radical anions) as reactive intermediates in enzymatic reactions. FEBS Lett 389: 20-24

Buckel, W., and Barker, H. A. (1974) Two pathways of glutamate fermentation by anaerobic bacteria. J Bacteriol 117: 1248-1260

Buckel, W., and Bobi, A. (1976) The enzyme complex citramalate lyase from *Clostridium tetanomorphum*. Eur J Biochem 64: 255-262

Buckel, W., and Golding, B. T. (1996) Glutamate and 2-methyleneglutarate mutase: from microbial curiosities to paradigms for coenzyme B_{12}-dependent enzymes. Chem Soc Rev 25: 329-337

Collins, M. D., Lawson, P. A., Willems, A., Cordoba, J. J., Fernandez-Garayzabal, J., Garcia, P., Cai, J., Hippe, H., and Farrow, J. A. (1994) The phylogeny of the genus *Clostridium*: proposal of five new genera and eleven new species combinations. Int J Syst Bacteriol 44: 812-826

Dimroth, R., and Schink, B. (1998) Energy conservation in the decarboxylation of dicarboxylic acids by fermenting bacteria. Arch Microbiol (in press)

Drake, H. L., ed. (1994) Acetogenesis. Vols 1-3. New York, London: Chapman & Hall.

Ermler, U., Grabarse, W., Shima, S., Goubeaud, M., and Thauer, R. K. (1997) Crystal structure of methyl-coenzyme M reductase: The key enzyme of biological methane formation. Science 278: 1457-1462

Fontecillacamps, J. C., Frey, M., Garcin, E., Hatchikian, C., Montet, Y., Piras, C., Vernede, X., and Volbeda, A. (1997) Hydrogenase—a hydrogen-metabolizing enzyme. What do the crystal structures tell us about its mode of action? Biochimie 79: 661-666

Kisker, C., Schindelin, H., and Rees, D. C. (1997) Molybdenum-cofactor-containing enzymes—structure and mechanism. Annu Rev Biochem 66: 233-267

Kröger, A., Geisler, V., Lemma, E., Theis, F., and Lenger, R. (1992) Bacterial fumarate respiration. Arch Microbiol 158: 311-314

Lovley, D. R., Giovannoni, S. J., White, D. C., Champine, J. E., Philipps, E. J., Gorby, Y. A., and Goodwin, S. (1993) *Geobacter metallireducens* gen. nov. sp. nov., a microorganism capable of coupling the complete oxidation of organic compounds to the reduction of iron and other metals. Arch Microbiol 159: 336–344

Schauder, R., and Kröger, A. (1993) Bacterial sulphur respiration. Arch Microbiol 159: 491–497

Schink, B. (1997) Energetics of syntrophic cooperation in methanogenic degradation. Microbiol Mol Biol Rev 61: 262–280

Thauer, R. K., Jungermann, K., and Decker, K. (1977) Energy conservation in chemotrophic anaerobic bacteria. Bacteriol Rev 41: 100–180

Wagner, A. F. V., Frey, M., Neugebauer, F. A., Schäfer, W., and Knappe, J. (1992) The free radical in pyruvate formate-lyase is located on glycine-734. Proc Natl Acad Sci USA 89: 996–1000

Widdel, F., Schnell, S., Heising, S., Ehrenreich, A., Assmus, B., and Schink, B. (1993) Ferrous iron oxidation by anoxygenic phototrophic bacteria. Nature 362: 834–835

Zumft, W. G. (1997) Cell biology and molecular basis of denitrification. Microbiol Mol Biol Rev 61: 533–616

Sources of Figures

1 Berks, B. C., Page, M. D., Richardson, D. J., Reilly, A., Cavill, A., Outen, F., and Ferguson, S. J. (1995) Mol Microbiol 15: 319–331

2 Weiss, D. S., and Thauer, R. K. (1993) Cell 72: 819–822

13 Utilization of Light by Prokaryotes

The free energy consumed by biological systems derives ultimately from solar energy. The conversion of light energy into chemical energy is one of the most important biological processes which originated on earth very early during the evolution of life. The mechanism of light-energy transduction in reaction centers follows the same principle in higher plants and in algae as in cyanobacteria, phototrophic purple and green bacteria. The pigments, which absorb photons, and the proteins of the photosynthetic membrane, which bind the pigments and redox components, are basically similar. One major difference between the groups of photosynthetically active organisms is the formation of oxygen in oxygenic photosynthesis or the absence of oxygen production in anoxygenic photosynthesis. Oxygenic photosynthesis is present in all photosynthetic eukaryotes (plants and algae) and in the prokaryotic cyanobacteria and Prochlorophyta, while anoxygenic photosynthesis is restricted to purple and green bacteria and the heliobacteria (Table 13.**1**). Generally speaking, the only product of anoxygenic photosynthesis is the formation of an electrochemical gradient of protons across the photosynthetic membrane; this gradient can be used for ATP synthesis and other energy-consuming processes (Chapter 4). Anoxygenic phototrophic bacteria possess only one type of photochemical reaction center. In contrast, oxygenic photosynthetic organisms have two types of reaction centers, photosystems I and II, which work in series. Oxygenic photosynthetic organisms generate an electrochemical gradient of protons and reducing equivalents in the form of NADPH for CO_2-fixation. The higher number of similarities compared to the dissimilarities in organization and function of the photosynthetic apparatuses suggests that the photosynthetic apparatus originated only once during evolution. The eukaryotic, oxygenic photosynthetic apparatus located in chloroplasts seems to have evolved in the cyanobacteria (i.e., it is of prokaryotic origin).

Table 13.**1** **Groups of anoxygenic photosynthetic bacteria.** Nos. 1–6, groups of a few representative genera, based on 16S rRNA cataloging; alpha, beta, and gamma refer to Proteobacteria; motility: (+), mobile by flagellum; (+/–), some species mobile, others not; glid., mobile by gliding; localization of antenna: (+/–), in intracytoplasmic membranes (ICM), cytoplasmic membranes (CM), or chlorosomes; PS, photosystem; RC, reaction center of type I or II, see text; other enzymes: CO_2 fixation via Calvin cycle (+); by other mechanisms (–); photoautotroph: (+), obligate photoautotroph or facultative but mainly photoautotroph

No.	Genera	16S rRNA cataloging	Motility	Cell wall type (Gram type)	Localization of antenna	BChl	Type of PS (RC)	Calvin enzymes cycle	Other enzymes	H_2S or S^0 as only e⁻ donor	Sulfur globules inside the cell	Facultative chemotroph (aerobic, dark)	Photoautotroph obligatory or main path
1	*Chromatium*	gamma	+	–	intracytoplasmic membrane (ICM)	*a* or *b*	II	+	(+)	+	+	+/–	+
	Thiocapsa		–	–		*a* or *b*	II	+	(+)	+	+	+/–	+
	Thiocystis		+	–		*a*	II	+	(+)	+	+	+/–	+
	Amoebobacter		–	–		*a*	II	+	(+)	+	+	+/–	+
2	*Ectothiorhodospira*		+	–		*a* or *b*	II	+	(+)	+	–	–/+	+
3	*Rhodobacter*	alpha	+/–	–		*a*	II	+	(+)	–	–	+	
	Rhodopseudomonas		+/–	–		*a* or *b*	II	+	(+)	–	–	+/–	
	Rhodospirillum		+	–		*a*	II	+	(+)	–	–	+	
	Rhodocyclus	beta	+/–	–	+/–	*a*	II	+	(+)	–	–	+	
4	*Heliobacillus*	Gram	+	+	CM	*g*	I		+	–	–	–	–
	Heliobacterium		glid.	+		*g*			+	–	–	–	–
5	*Chlorobium*	green sulfur bact.	–	–	chlorosomes +CM	*a* + *c*	I	–	+	+	–	–	+
	Prosthecochloris		–	–		*a* + *c*, *d*, *e*	I	–	+	+	–	–	+
6	*Chloroflexus*	green	glid.	–		*a* + *c*	II	–	+	–	–	+	–

13.1 Three Groups of Pigments Are Active in Photosynthesis

The photochemical process of light-energy transduction in the reaction centers is catalyzed by chlorophylls. Absorption of photons and the formation and migration of excited states is mediated by light-harvesting or antenna pigments (i.e., carotenoids, chlorophylls, and phycobilins; Figs. 13.**1**–13.**4**). **Chlorophylls** (Chls) in oxygenic photosynthetic organisms and **bacteriochlorophylls** (BChls) in anoxygenic photosynthetic bacteria are substituted cyclic tetrapyrroles with a fifth, isocyclic ring that is derived biosynthetically from the C13 propionic acid side chain of protoporphyrin (Fig. 13.**1**, structure 2). The four N atoms of the pyrroles are coordinated to a Mg atom. Chls and BChls are light absorbers with a very high absorption cross section owing to their network of alternating single and double bonds (Fig. 13.**1**, structure 2). On ring IV, a long-chain hydrophobic alcohol, which is phytol in most Chls and BChls, is esterified (Fig. 13.**1**). Chls and BChls have strong absorption bands in the visible, the near-UV, and the near-infrared regions of the spectrum. In vivo, they are non-covalently bound to proteins. A highly conserved histidine in the hydrophobic, α-helical region of the pigment-binding protein most likely is the fifth ligand to the central Mg atom. Carbonyl groups of the tetrapyrrole ring (e.g., R_1 in BChl *a* and *b*; Table in Fig. 13.**1**) form hydrogen bonds to respective groups of the polypeptides. Hydrophobic interactions between the BChl moiety and regions of the protein contribute also to localization and orientation of the tetrapyrrole ring in the protein scaffold. Pigment–protein and pigment–pigment interactions shift the lowest singlet absorption bands to the longer-wavelength region of the spectrum (Fig. 13.**3**); for example, the near-infrared absorption band of BChl *a* is at 770 nm in methanol–acetone but at 800–880 nm when bound to proteins. Six BChls (*a*, *b*, *c*, *d*, *e*, *g*) and two Chls (*a*, *b*) are known (Fig. 13.**1**). Chl *a* and BChls *a*, *b*, and *g* are the photochemically active tetrapyrroles in reaction centers. They are also present in light-harvesting complexes in addition to the other BChls (see Table 13.**1**, Fig. 13.**1**, and text).

Carotenoids are polyenes, usually tetraterpenoids (Fig. 13.**2**). A large number of substitutions have been found. Due to their double-bond systems, carotenoids function in photosynthetic organisms as light-harvesting molecules and they prevent harmful photo-oxidative reactions related to the presence of singlet oxygen.

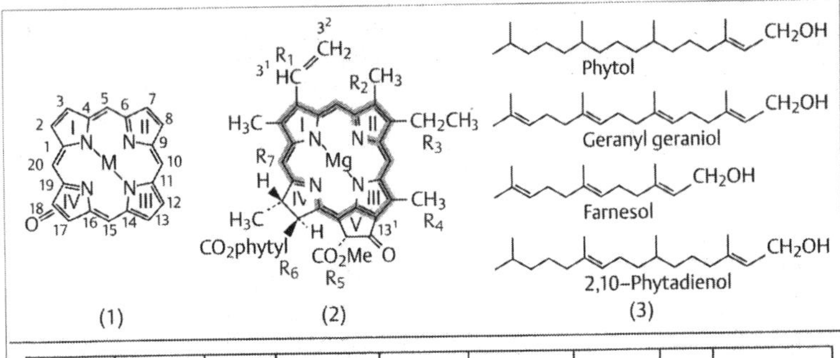

Pigment	R_1	R_2	R_3	R_4	R_5	R_6	R_7	Absorbance (nm)
BChl *a*	$-CO-CH_3$	$-CH_3$	$-CH_2CH_3$	$-CH_3$	$-CO-OCH_3$	Phytyl or geranyl geraniol	-H	800–880
BChl *b*	$-CO-CH_3$	$-CH_3$	$=CH-CH_3$	$-CH_3$	$-CO-OCH_3$	Phytyl or phytadienyl	-H	1020
BChl *c*	-CHOH $-CH_3$	$-CH_3$*	$-CH_2H_5$ to $-C_4H_9$	$-CH_3$ to $-C_2H_5$	-H	Farnesyl or stearyl	$-CH_3$	750
BChl *d*	-CHOH $-CH_3$	-CHO*	$-CH_2H_5$ to $-C_5CH_{11}$	$-CH_3$ to $-C_2H_5$	-H	Farnesyl	-H	725–745
BChl *e*	-CHOH $-CH_3$	-CHO*	$-C_2H_5$ to $-C_4CH_9$	$-C_2H_5$	-H	Farnesyl	$-CH_3$	715–725
BChl *g*	$-CH=CH_3$	$-CH_3$*	$=CH-CH_3$	CH_3	$-CO-OCH_3$	Farnesyl or geranyl geraniol	-H	788

Fig. 13.**1** **The tetrapyrrolic ring system of chlorophylls and bacteriochlorophylls.** Structure 1, numbering scheme for C atoms and pyrrole rings of porphyrin; M, metal ligand. Structure 2, numbering scheme for chlorophyll a; R_1 to R_7 (red), substituents of chlorophyll a that vary in different forms of bacteriochlorophyll (BChl, shown in the table); Me, methyl group; red background, the conjugated system of alternating single and double bonds. (3), structures of the esterifying alcohols in position R_6. Bonds in the tetrapyrrole ring between C7 and C8 are unsaturated in BChls *c*, *d*, and *e* and in Chls *a* and *b* and saturated in BChls *a* and *b*. In the table, stars (under R_2) denote unsaturated bonds between C7 and C8. Absorbance (nm): in-vivo, near-infrared absorbance bands of the protein-bound pigments

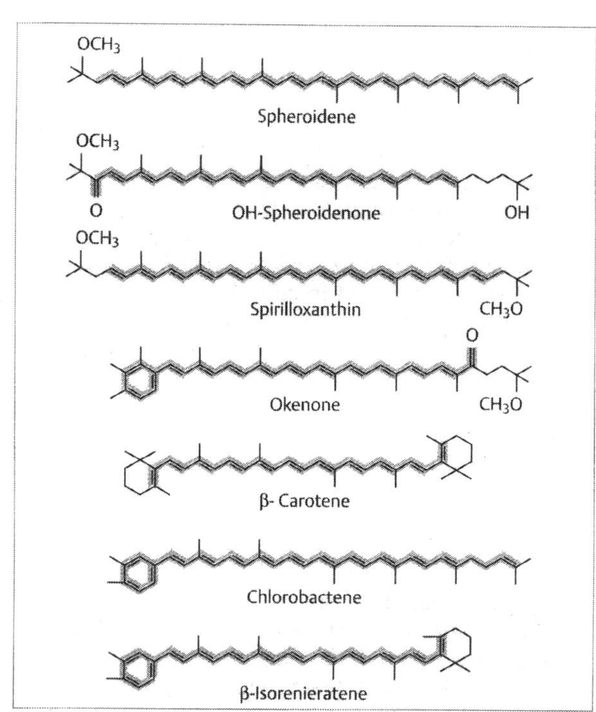

Fig. 13.2 **Structure of several carotenoids present in purple and green bacteria.** The conjugated double bonds (marked in red) are important for spectral features and functional properties

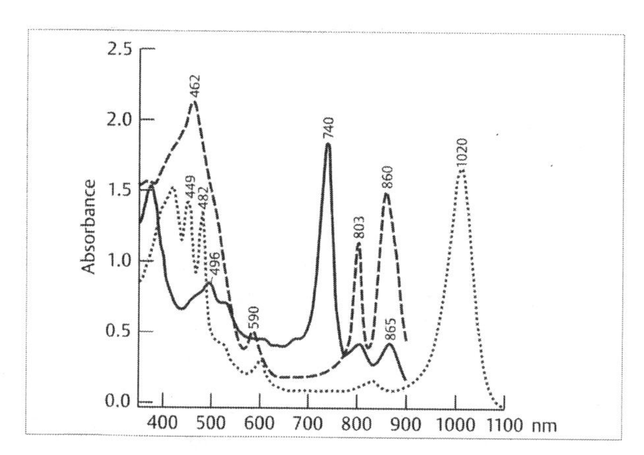

Fig. 13.3 **Absorbance spectra of pigment-proteins bound to membranes** isolated from *Chloroflexus aurantiacus*, a gliding, green bacterium containing BChl *c* (absorbance at 740 nm) as dominating species in chlorosomes, BChl *a* (865 nm), and carotenoids (496 nm) —— (solid line); *Rhodopseudomonas palustris*, containing BChl *a* (370, 590, 803, and 860 nm) and carotenoids (462 nm) - - - - - (broken line); *Rhodopseudomonas viridis*, containing BChl *b* (370, 600, and 1020 nm) and carotenoids (449 and 482 nm) ········· (dotted line). *Rps. palustris* and *Rps. viridis* have lamellar intracytoplasmic membranes

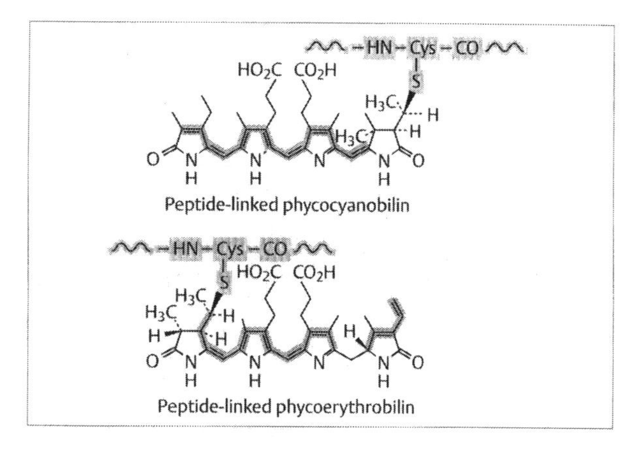

Fig. 13.4 **Formulas of the open-chain tetrapyrroles phycocyanobilin and phycoerythrobilin.** Phycobiliproteins are accessory light-harvesting systems for the operation of photosystem II in cyanobacteria. They consist of proteins and the prosthetic groups and are derived biosynthetically from porphyrins by loss of one carbon atom as carbon monoxide. They are covalently bound via cysteinyl (–Cys–) residues to proteins. They are, in addition to other phycobilins, important light-harvesting pigments in cyanobacteria (see Fig. 13.**11**). The in vivo absorption maxima of phycobiliproteins are between 565 and 670 nm (Fig. 13.**5**); they fill the gap of the chlorophyll-*a* and carotenoid maxima in the green-yellow-orange region of the visible light spectrum, which is important for the ecophysiology of cyanobacteria. The position of the absorbance maxima depends on various factors, including the state of aggregation. The phycobiliproteins are organized with linker proteins in phycobilisomes (Figs. 2.**14** and 13.**11**)

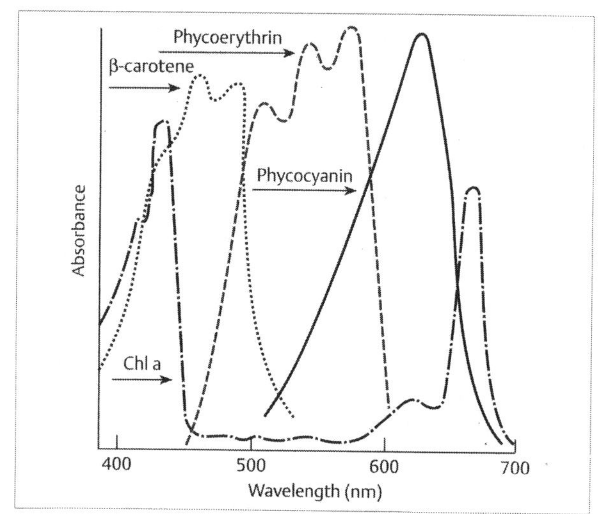

Fig. 13.5 **Absorption spectra of pigment-proteins from cyanobacteria.** — · —· chlorophyll *a* (Chl *a*), - - - - - phycoerythrin, ········· *β*-carotene, —— phycocyanin

Carotenoid-free mutants of purple bacteria are killed within seconds when exposed to oxygen in the presence of light.

Phycobilins are open-chain tetrapyrroles that are covalently bound by thioether bonds to cysteine residues of phycobiliproteins (Fig. 13.**4**). Phycobilins are light-harvesting molecules in cyanobacteria (Fig. 13.**11**) and red algae. They absorb photons in the range from 480 to 650 nm within the visible spectrum (Fig. 13.**5**).

Chlorophyll *a*; the bacteriochlorophylls *a, b, c, d, e,* and g, several phycobilins, and numerous carotenoids are pigments with conjugated systems (alternating single and double bonds). These polyene structures strongly absorb light in the visible, the near-UV, and the near-infrared regions of the spectrum. They are non-covalently or covalently bound to proteins of the photosynthetic apparatus of photosynthetic prokaryotes. They are involved in absorption and transduction of light energy.

13.2 The Photosynthesis of Green Bacteria, Purple Bacteria, and Heliobacteria Is Anoxygenic

13.2.1 Light-harvesting Systems of Purple Bacteria Are Membrane-bound Pigment-Proteins

Light-harvesting (LH) or antenna systems serve to absorb photons and to convert them to a singlet excited state of BChl. The carotenoid and BChl molecules are non-covalently bound to integral membrane proteins so that the tetrapyrrole rings are in a fixed position, a set distance from each other and to the plane of the membrane. The pigment-proteins form oligomeric structures which are organized in arrays to optimize energy transfer to the reaction center. Most light-harvesting systems of purple bacteria are localized in intracytoplasmic membranes (Figs. 2.**11**, 2.**12**, 13.**6**). The orientation of BChl molecules can be demonstrated by absorption spectroscopy with linearly polarized light. Charged groups in the protein-binding sites and electric coupling between the chromophores affect the absorption spectra in vivo. The absorption cross section of the antenna molecules per reaction center, which can be quantified by the average number of BChl molecules per reaction center (size of the photosynthetic unit), is variable within genetically fixed limits. This cross section can be modified by growth conditions, especially by the incident radiance (see Chapter 24.2). One of the electrons in the chromophore occupying a molecular orbital of the lowest available energy is promoted, following absorption of a photon, to an orbital not occupied in the ground state. The energy of the photon is thus used to boost the electron to a higher energy level resulting in the **excited state of the molecule**. The lifetime of a single BChl*a* site is about 0.1 to 0.2 ps ($1\ ps = 10^{-12}\ s$). The transfer of excitation energy between the BChl molecules of one LH complex is ultrafast (0.1 to 0.2 ps). Most photosynthetic systems are heterogeneous in pigment-protein composition; which leads to multiple absorption bands in the near-infrared absorption region of the spectrum. The various absorption bands correlate with the occurrence of specific pools of pigment-protein complexes, which are organized in such a way that pigments absorbing at a relatively high energy level (i.e. at shorter wavelengths) are at the periphery of the antenna system, whereas those absorbing at a lower energy level (i.e., at longer wavelengths) surround the photochemical reaction center (RC). Equilibration of the excitation density within a homogeneous population occurs rapidly. Excitation transfer over the antennae to the RC is not a linear process, but depends on many factors: influx of light energy into the system, size and organization of the antennae, distances between pigment molecules, overlapping of emission bands of donor molecules and absorption bands of acceptor molecules, temperature, and the state of the RC. In an "open" state, the RC is ready to receive excitation energy; in a "closed" state, the RC is in action, i.e., the primary donor is oxidized. The excitation energy is transferred between weakly coupled pigments by a dipol–dipol inductive resonance transfer (Förster) mechanism over long distances (≈ 2 nm) with a time of 0.1 to 1.0 ps required for a single step. There are two other mechanisms of energy transfer between closely interacting pigments (≈ 0.5 nm). Excitation energy migrates over the different populations of antenna complexes to the RC, where the energy is trapped within about 50 ps (open RCs) or >200 ps (closed RCs). Excitation energy migrates from the shorter to the longer wavelength (i.e., higher to lower free energy of quanta) of overlapping absorption bands of pigments (Crt, carotenoids; B..., antenna absorption bands; RC, reaction center), for example:

$$\text{Crt} \Rightarrow \text{(BChl)B800} \Rightarrow \text{B850} \Rightarrow \text{B870} \Rightarrow \text{RC}$$

$$\Downarrow \quad \Uparrow \quad \Uparrow \quad \Uparrow \quad \Uparrow$$

Excitation energy that cannot be trapped by RCs dissipates without radiation of light or as fluorescence. The rate-limiting step of the overall light-harvesting process is the transfer of the excitations from the core antenna to the RC.

The **core antenna of purple bacteria**, named **LH I (BChl *a* B870** in *Rhodobacter capsulatus* and **BChl *b* B1020** in *Rhodopseudomonas viridis*), surrounds the reaction center. It is synthesized in a stoichiometric ratio of about 25 to 30 mol BChl per mol RC (Figs. 13.**6** and 13.**7**).

Most purple bacteria have an additional, **variable antenna (LH II) complex** with a major BChl *a* absorption band between 820 nm and 860 nm and a second absorption band at 800 nm (Fig. 13.**3**). The complex LH II likewise consists of an α,β-heterodimer (M_r 5000–7000) to which 3 mol of BChl *a* and 1–2 mol of carotenoids per mol complex are noncovalently bound. This subunit forms oligomers up to the nonamers. The active unit of the complex LH II of *Rhodopseudomonas acidophila* consists of two concentric cylinders of helical protein subunits. Eighteen BChl-*a* molecules sandwiched be-

Box 13.1 The LH-I core antenna subunit is a heterodimer of the polypeptides (M_r about 5000–7000) α and β, to which BChl *a* and carotenoids are bound in a 1 : 1 stoichiometry. Resonance Raman spectroscopy and high-resolution X-ray spectroscopy have revealed the specific binding of the BChl molecules in their proteinaceous surroundings. The highly conserved histidines, which are located in the hydrophobic transmembrane α-helix in the periplasmic side of the double lipid layer of the membrane, are the fifth ligands of the Mg central atom of the BChls. There are additional hydrogen bonds between the carbonyl and keto side-chains of the tetrapyrrole ring and the charged groups of amino acids in the surroundings of the tetrapyrrole moiety. The phytol side-chain of BChl interacts with hydrophobic, transmembrane stretches of the polypeptides. The monomeric subunits absorb at 820 nm in vivo. They form oligomeric LH-I complexes (Fig. 24.**6a**) and have a near-infrared absorption band at 870–880 nm.

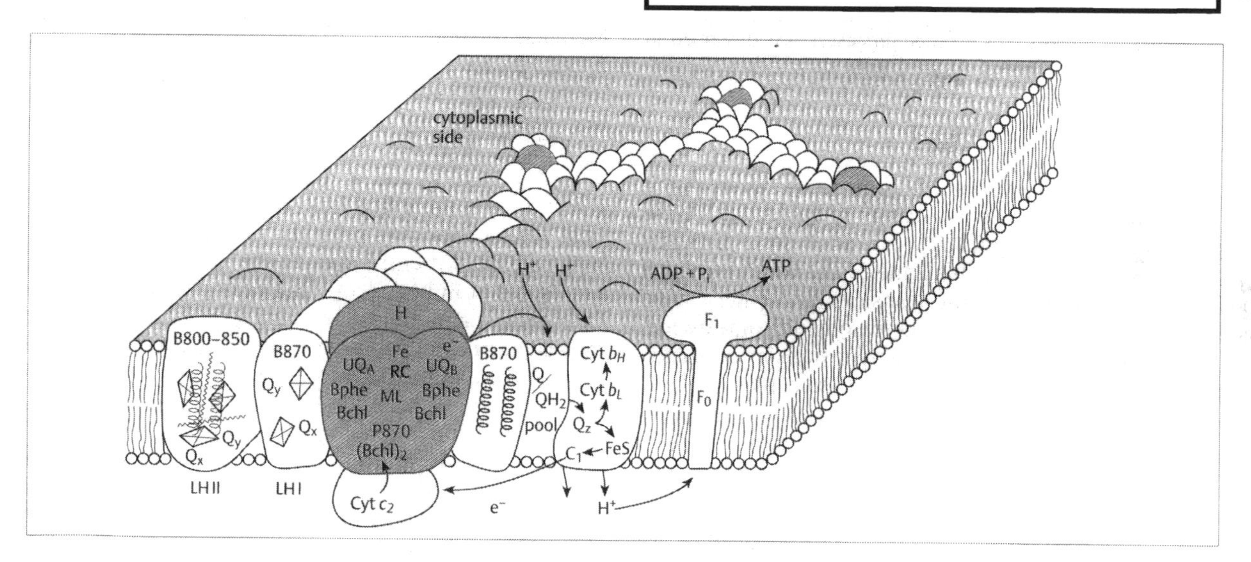

Fig. 13.**6** **Schematic view of an intracytoplasmic membrane of the purple bacterium *Rhodobacter capsulatus* and related species.** The photochemical reaction center (RC, red background) consists of the protein subunits M, L (ML), and H. The special pair of BChl [$(BChl)_2$], two monomeric BChls, two bacteriopheophytins (Bphe), the ubiquinones A and B (UQ_A, UQ_B), and one Fe atom are bound to the subunits M and L (ML). The H subunit does not bind pigments. The RC is surrounded by the core antenna B870 (LH I), the immediate donor of excitation energy to the RC. The two helices symbolize the low-molecular-weight proteins which span the membrane and bind two BChl molecules. The Q_Y (870 nm; ring I–III) and the Q_X (590 nm; ring II–IV) transitions, that is, the orientation of the tetrapyrrole ring relative to the plane of the membrane, can be determined spectroscopically by using linearly polarized light. Under low light intensities, the peripheral antenna B800–850 (LH II) is the dominating light-harvesting complex, which connects the different photochemical units to form an efficient excitation-energy-distributing system. Two polypeptides bind three molecules of BChl, of which one BChl differs in its orientation from the other two. The pool of quinones in the membrane (Q/QH_2) connects the RC with the cytochrome-b/c_1 complex containing the cytochromes c_1 (C_1), high-potential cytochrome *b* (Cyt b_H), and low-potential cytochrome *b* (Cyt b_L). Cytochrome c_2 in the periplasmic space shuttles electrons between the b/c_1 complex and RC. The proton : ATP synthase (F_1F_0) uses the proton gradient across the membrane to generate ATP from ADP and Pi

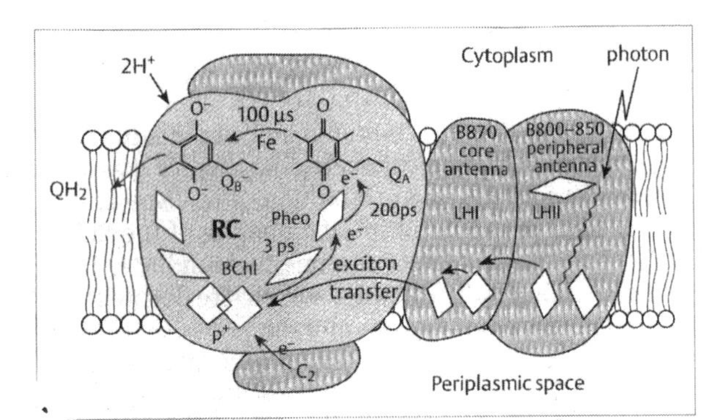

Fig. 13.7 **Schematic view of the primary processes of photosynthesis in anoxygenic photosynthesis with a reaction center of type II.** Light energy absorbed by peripheral antenna systems B800–850 (LH II) is transformed into excitation energy, which migrates within the antenna system to the core antenna B870 (LH I) and finally to the RC (red), where primary photochemistry starts. The special pair of BChl (P) is excited (P^*), and one electron is transferred within 3 ps by formation of the radical cation $[BChl_2](P^+)$ via a BChl monomer (BChl) to a bacteriopheophytin (Pheo) then within 200 ps to ubiquinone A (Q_A). By formation of P^+Q^-, a charge separation across the membrane dielectricum and a redox potential difference is formed. The further processes are described in the text. QH_2, ubiquinol; C_2, cytochrome c_2

tween the helices form a continuous overlapping system of BChl-*a* light-harvesting molecules. Nine additional BChl-*a* molecules are positioned between the outer helices, with the plane of the bacteriochlorin rings perpendicular to the transmembrane-helix axis. The BChl phytol chains and the carotenoids intertwine. This organization allows an efficient transfer of excitation energy between the pigment molecules. A third non-pigment-binding protein can be present and serves to stabilize the complex. The relative amounts of LH II to the complexes LH I of the RC depend strongly on the growth conditions, especially the light intensity. The complex LH II becomes the dominating pigment-protein complex if the cells are cultivated under low light intensity and the size of the photosynthetic unit (i.e., the molar ratio of antenna BChl per mol RC) increases. As a result, the light-harvesting area of the antenna is enlarged. Some species such as *Rhodopseudomonas palustris*, *Rhodopseudomonas acidophila*, and *Chromatium vinosum* have the ability to develop multiple LH-II systems with different proteins in response to changes in temperature or light intensity. The complexes LH II probably interconnect the RC LH-I core units (Fig. 13.**6**). The efficiency of energy transfer from the peripheral antenna to the RC can be up to 95%.

The light-harvesting or antenna complexes of purple bacteria are oligomeric structures of pigment-proteins located in the intracytoplasmic membrane. They absorb light energy and convert them to singlet excited states, which migrate to other pigments or decay by emission of light (fluorescence) or heat. Finally, excitation energy is trapped by photochemical reaction centers where the excitation energy is transduced into a redox potential difference across the membrane.

13.2.2 Antenna Systems of Green Bacteria and Heliobacteria Are Chlorosomes and Integral Membrane Complexes

Green bacteria are divided into the obligately anaerobic non-motile, phototrophic green or brown sulfur bacteria (e.g., ***Chlorobium***) and the filamentous, gliding, anaerobic or facultatively aerobic, primarily photoheterotrophic, yellow-colored *Chloroflexus* (Table 13.**1**). Members of both groups contain **chlorosomes** (Figs 2.**13**, 13.**8**). These are oblong, lipid-rich vesicles that are located in the cytoplasm and attached to the cytoplasmic membrane by a baseplate. The baseplate is a crystalline structure, which possibly contains complexes of BChl *a* and a light-harvesting protein. The chlorosome is separated from the cytoplasm by a single lipid layer containing proteins. Chlorosomes are filled with rod-like structures, which are presumably stacked aggregates of BChl *c*, *d*, or *e* molecules (Figs. 13.**1**, 13.**8**). Carotenoids and lipids are also present. The BChl-*a* molecules in chlorosomes are not protein-bound. There are approximately 1000 mol BChl *c*, *d*, or *e* per mol RC. The BChl *a* molecules localized in the baseplate or in the cytoplasmic membrane mediate the transfer of excited states from the chlorosomes to the RC. The overall energy transfer processes take place within several tens of ps (the numbers following BChl are the absorption maxima in the near-infrared; P840 is the primary donor in the reaction center):

[BChl *c* ⇒ BChl *a* 795] ⇒ [BChl *a* 809] ⇒
Chlorosome **Baseplate**
[BChl *a* 813–836 ⇒ P840
Membrane

Green bacteria have no intracytoplasmic membranes.

Heliobacterium species (Table 13.**1**) do not have chlorosomes. The photochemical RCs are associated with 30–40 light-harvesting BChl *g* (Fig. 13.**1**) molecules

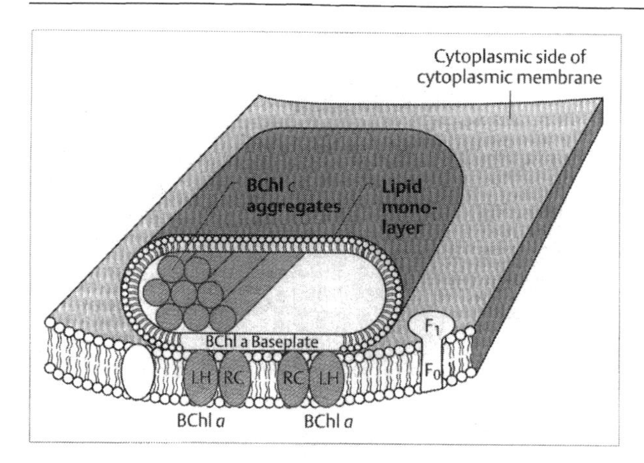

Fig. 13.8 **Chlorosome organization.** The scheme is deduced from electron-microscopic and biochemical data. Chlorosomes are attached by a proteinaceous baseplate to the cytoplasmic side of the cytoplasmic membrane; they absorb light via the linearly arrayed BChl-*c*, -*d*, or -*e* molecules (rod-shaped elements in chlorosomes). The baseplate harbors presumably BChl-*a*-containing, light-harvesting complexes. Below the baseplate are the reaction centers (RC) in the cytoplasmic membrane. F_1F_0, ATP synthase; LH, light-harvesting complex; see text for further details

Box 13.2 Three different spectral species of BChl *g* (i.e., B778, B793, and B808) are present. The excitations in the antenna rapidly accumulate in the long-wavelength complex BChl *g* B808. The lifetime for excited antenna BChl *g* is less than 1 ps for excitations of BChl *g* B778, whereas excited BChl *g* B793 decays in about 2 ps. The excitations that have accumulated in BChl *g* B808 decay with time constants of approximately 4 ps, 20 ps, and 100 ps. The main relaxation process of about 4 ps is probably associated with energy transfer to the reaction center.

Chlorosomes of green sulfur and non-sulfur bacteria are oblong, lipid-rich vesicles attached to the cytoplasmic membrane. Chlorosomes, which contain the major portion of the light-harvesting bacteriochlorophyll *c*, *d*, or *e* molecules as stacked aggregates, are structures analogous to the phycobilisomes of cyanobacteria. Light-harvesting pigments of heliobacteria are located in the cytoplasmic membrane (BChl *g*).

13.2.3 In Photochemical Reaction Centers, Excitation Energy Is Transduced into Charge Separation

Two types of reaction centers (RCs) have been detected in green plants, algae, and photosynthetic bacteria. Type-II RCs (photosystem II or PS II in plants) are present in the sulfur and non-sulfur purple bacteria (e.g., *Rhodopseudomonas viridis*, *Rhodobacter sphaeroides*, and *Chromatium vinosum*), the green non-sulfur bacterium *Chloroflexus*, cyanobacteria, and green eukaryotes. The atomic structure of the RCs of *Rhodopseudomonas viridis* and *Rhodobacter sphaeroides* has been determined. Structure, composition, and function of this type of RC is similar to the core of the PS II in plants and is representative for all purple bacteria and *Chloroflexus* (Figs. 13.**7**, 13.**9**). Type-I RCs are present in green sulfur bacteria (e.g., *Chlorobium limicola* and *Prosthecochloris aestuarii*), in the Gram-positive bacteria *Heliobacterium chlorum* and *Heliobacillus mobilis*, in cyanobacteria (e.g., *Anacystis nidulans* and *Anabaena variabilis*), and in algae and higher plants (Figs. 13.**10**, 13.**11**).

Type-II RC (quinone type): The primary electron donor in the type-II RC of purple bacteria is a special pair of BChl *a* or *b* molecules ($[BChl_2]$, P), which is excited by excitation energy received from the antennae or directly by absorption of a photon; an excited state P^* is then formed. The electronically excited primary donor P^*, a very strong reductant, donates an electron to a series of electron acceptors in an electron-transfer chain. Within 3.5 ps, the electron travels from the special pair to the accessory $BChl_A$. A very short-lived, transiently populated, radical-pair state $P^+B_A^-$ (0.9 ps) is formed before bacteriopheophytin (BPh) is reduced (P^+BPh^- state, Fig. 13.**7**). P^+ has a redox potential of about +0.4 V. The electron is transferred within 200 ps from BPh^- to a quinone A (Q_A), which is located in the cytoplasmic side of the membrane, approximately 2.5–3.0 nm from $[BChl]_2$. This step stabilizes the charge separation across the lipid double layer of the membrane ($P^+ Q_A^-$). The subsequent electron transport from Q_A^- to quinone B (Q_B) is much slower (50–200 μs). By a second turnover of the RC, which includes a second excitation of P and transfer of the electron to Q_B, ubiquinol (QH_2) is formed by the uptake of two H^+. The RC is "closed" for uptake of new excitation energy as long as the primary acceptor is reduced. The ubiquinol B is exchanged with a ubiquinone from the quinone pool of the membrane. P^+ is reduced directly by electrons from the soluble, periplasmic cytochrome c_2 (in *Rhodobacter sphaeroides*; Fig. 13.**9**) or indirectly via a four-cytochrome subunit bound to most RCs (e.g., in *Rhodopseudomonas viridis* and *Chromatium vinosum*). These RCs are "open," that is, ready for uptake of

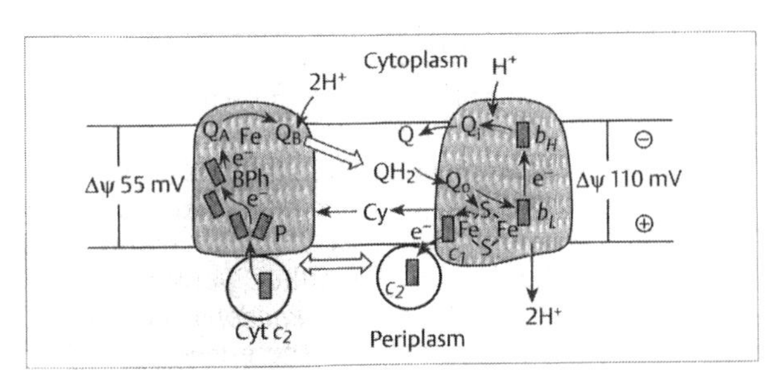

Fig. 13.**9** **The cyclic electron transport system of anoxygenic purple bacteria.** 1. Charge separation and formation of redox potential difference in reaction center (P^+Q^-) creates a membrane potential ($\Delta\psi$) of about 55 mV. 2. In a second turnover of the RC (gray structure, left), ubiquinol (QH_2) is formed, which interchanges with the ubiquinol/ubiquinone (Q) pool in the membrane. Ubiquinol is transferred by diffusion in the lipid double layer of the membrane to the quinol acceptor site Q_0 of the b/c_1 complex (gray structure, right). 3. Proton translocation from the cytoplasmic to the periplasmic space, reduction of cytochrome c_2 (c_2) and increase of the membrane potential to about 110 mV by the sequential redox processes of the Q cycle in the b/c_1 complex. b_L and b_H, low- and high-potential cytochrome *b*, respectively; Cy, c_1, c_2; cytochromes *c*; BPh, bacteriopheophytin; Q_A, ubiquinone A; Q_B, ubiquinone B; Q_i, quinone reduction site; Q_o, quinone oxidation site; Fe, iron atom; FeS, iron–sulfur center

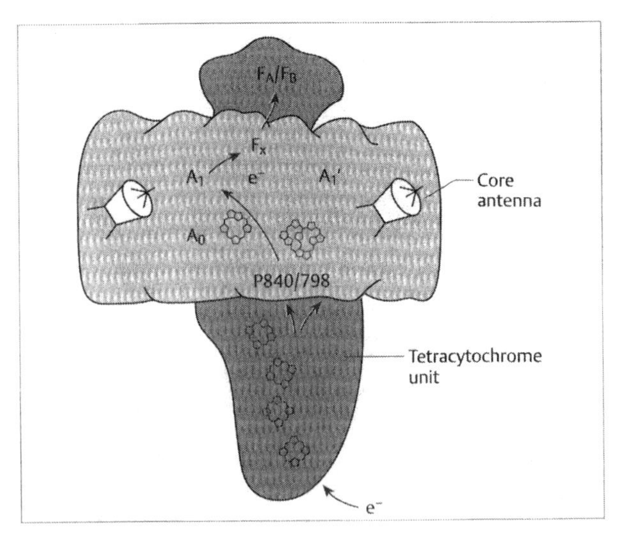

Fig. 13.**10** **Photochemical reaction center type I in green sulfur bacteria of the genus *Chlorobium* and *Heliobacterium*.** The primary electron donor P798 in *Heliobacterium* is presumably a BChl-*g* dimer; P840 is a BChl-*a* dimer in green sulfur bacteria. A_0, the primary acceptor in *Heliobacterium*, is presumably an 8-hydroxychlorophyll *a* (P670). A_1 is a quinone and F_X, a 4-Fe-4-S cluster. F_A/F_B are also iron–sulfur clusters. Electrons (e^-) are passed to the oxidized primary donor via a tetracytochrome subunit (four rings)

excitation energy. Electron flow is generally confined to the L branch of the RC (Fig. 13.**7**), although a second branch, the M branch, is present. In RCs, the excitation energy obtained from antenna pigments is transduced into a charge separation (P^+ Q^-; $\Delta\Psi$) and a redox potential difference (ΔE_h) across the membrane. This process is very efficient because the back reactions are several orders of magnitude slower. The high electron-transfer rates in the first steps of charge separation (from $[BChl]_2$ to pheophytin) are determined by the proximity of the donor and acceptor, the ΔG^0 values of the participants, the proteinaceous surrounding of the reactants, and the reorganization energy.

The primary electron transfer events of the **type-I RC (iron–sulfur**; Fig. 13.**10**) are similar to those of the type-II RC. A special pair of BChls (P) acts as the primary electron donor: P840, a dimer of BChl *a* in green sulfur bacteria, or P798, a dimer of BChl g in heliobacteria. The midpoint potential is at +240 mV, about 150 mV lower than the midpoint potentials of the primary donors in either purple bacteria or PS I of cyanobacteria. The first acceptor (A_0) is thought to be the monomeric BChl B663, which is a Chl-*a* isomer present in green sulfur bacteria, heliobacteria, and PS I of cyanobacteria. A_0 donates the electron to the secondary electron acceptor A_1, which is presumably a quinone. The electron is transferred from the bound quinone A_1 to an iron–sulfur cluster Fx[4 Fe–4 S]. This iron-sulfur cluster occupies the same relative position as the non-heme iron in the RC of purple bacteria halfway between Q_A and Q_B (Fig. 13.**9**). F_X interacts with the bound iron–sulfur clusters F_A and F_B, and undergoes redox reactions (Fig. 13.**10**). The redox centers of green sulfur bacteria are bound to a homodimer of an 82-kDa hydrophobic protein, each of which was predicted to span the membrane by 11 transmembrane helices. About 2 carotenoids per P840 are in the RC. Green sulfur bacteria are capable of direct photoreduction of $NAD(P)^+$ via ferredoxin.

The primary electron donor of heliobacteria is P798 (Fig. 13.**10**), which has a redox potential of 225–250 mV. P798 is a dimer of BChl *g* or its 13^2-epimer (Fig. 13.**1**). As in green sulfur bacteria, antenna BChls are bound to the RC complex. The homodimer of the 68-kDa polypeptide to which all RC and antenna pigments are bound, has 40% sequence similarity in the C-terminal region to the polypeptides A and B of PS I. The N-terminal half of the protein has a significant similarity to the CP47 and CP43 antenna proteins (**c**hlorophyll-binding **p**rotein; molecular mass in kDa) of PS II, particularly in the hydrophobic regions. The primary acceptor A_0 in the RC of Heliobacteria is a 8^1-hydroxychlorophyll *a* (P670). Bleaching of the 670-nm transition, which is accompanied by changes in absorbance around 785 nm, occurs with a time constant of 10 ps. P798 is reduced by cytochrome

c_{553}. Menaquinone possibly functions as a secondary electron acceptor analogous to phylloquinone in PS I. The position and orientation of the primary donor and the three Fe–S centers are as in PS I. A_1 in PS I is a quinone-type acceptor which has a position similar to that in bacterial RCs of green sulfur bacteria. The heliobacterial reaction center functionally resembles PS I. In view of this homology and the similarity of electron-transport processes, a model has been developed in which PS I and PS II both originated from a common homodimeric ancestral reaction center, which may have similarities with the RC of heliobacteria.

In reaction centers, excitation energy is transduced into charge separation (membrane potential $\Delta\Psi$) and a redox potential difference across the membrane (ΔE_h). The charge separation is between the primary electron donor, which is a dimer of chlorophyll or bacteriochlorophyll localized on the extracytoplasmic side of the membrane, and the secondary electron acceptors, which are quinones in type-II reaction centers or iron–sulfur clusters in type-I reaction centers, localized on the cytoplasmic side of the membrane. The similarity in the amino acid sequence of pigment-binding proteins and in the organization and function of all reaction-center cores supports the idea that all reaction centers have the same origin in evolution.

13.2.4 The Photosynthetic Electron Transport Contributes to the Formation of an Electrochemical Gradient of Protons

The facultative photosynthetic purple bacteria and **Chloroflexaceae** generate an electrochemical proton gradient across the cytoplasmic or intracytoplasmic membrane either by a cyclic, anaerobic, light-driven electron transport or a linear, respiratory-type electron transport. In the cyclic electron transport, ubiquinol (provided by the RC) is the electron donor for the ubiquinone : cytochrome b/c_1 oxidoreductase (see Chapter 11, Fig. 11.**9**). The cytochrome b/c_1 complex is a proton pump that contributes to generation of a protonmotive force (electrochemical gradient of protons, $\Delta\tilde{\mu}_{H^+}$) (Fig. 13.**9**). The enzyme oxidizes ubiquinol, reduces cytochrome c_2, and uses the electrochemical potential generated by the redox reactions of the so-called Q cycle to translocate protons (H^+) from the cytoplasm to the periplasm.

Box 13.3 Ubiquinol (QH_2), reduced ubiquinone, binds at the Q_o-site of the cytochrome-b/c_1 complex. Oxidation of QH_2 involves transfer of one electron to FeS and cytochrome c_1 and the other electron to cytochrome b_L (low-potential cytochrome *b*; b_L in Fig. 13.**9**) and release of two H^+ into the periplasm. The electron in cytochrome b_L is transferred rapidly to cytochrome b_H (=high-potential cytochrome *b*). Quinone from the pool in the membrane binds at the Q_i-site and is reduced to a semiquinone by uptake of an electron and a H^+ from the cytoplasm. A second QH_2 binds at the Q_o site and is oxidized as described before. In this way, a second electron passes into the *b*-type cytochrome chain. The second electron reduces the semiquinone with the uptake of a H^+ from the cytoplasm. Thus, four mol protons are pumped to the periplasm per two mol QH_2 oxidized.

The reduced cytochrome c_2 (Cyt c_2) shuttles back to the RC and reduces the oxidized primary donor P^+. In mutants defective in cytochrome c_2, cytochrome *y* (Cy in Fig. 13.**9**) has the function of an electron shuttle vector. After two turnovers of the RC, one mol QH_2 is released into the quinone pool of the membrane.

In the respiratory chain, cytochrome c_2 donates electrons to the cytochrome oxidase. In purple bacteria, the protonmotive force is generated by a cyclic electron transport from the RC via the quinone pool in the membrane, the cytochrome b/c_1 complex, and cytochrome c_2 back to the RC (Fig. 13.**9**). CO_2 reduction is not directly coupled to the photosynthetic electron transport as in plants, but the protonmotive force can drive a reverse electron flow from the cytochrome-b/c_1 complex via NADH dehydrogenase to reduce NAD^+. In green sulfur bacteria, redox components with higher electronegativity are generated, and electrons from the iron–sulfur center of the RC are possibly transferred to ferredoxin, which is used directly or indirectly as electron source for CO_2 fixation (Chapter 8.1).

The major function of anoxygenic photosynthesis is the generation of an electrochemical gradient of protons (membrane potential and proton gradient) across the photosynthetic membrane. The photochemical reaction center, the cytochrome-b/c_1 complex, the quinone pool in the membrane, and a cytochrome *c* shuttle system are the components of the cyclic electron transport. Production of reducing equivalents is generally not coupled.

The electrochemical gradient of protons generated by the light-driven cyclic electron transport or the respiratory, linear electron-transport chain can be used to produce ATP from ADP–Mg and inorganic phosphate, which is catalyzed by the ATP synthase (see Chapter 4.2 and Fig. 4.**5**).

13.2.5 Light-driven and Respiratory Electron Transport Compete With Each Other

Some members of the anoxygenic photosynthetic bacteria such as *Rhodobacter capsulatus*, *Rhodobacter sphaeroides*, and *Rhodospirillum rubrum* can grow both under anoxic conditions in the light (phototrophically) and under oxic conditions in the dark by respiratory metabolism (chemotrophically). Both the photosynthetic and the respiratory apparatuses are membrane-bound, and they transduce light energy and respiratory redox energy, respectively, into an electrochemical gradient of protons and a membrane potential across the cytoplasmic and intracytoplasmic membranes. Both electron transport pathways use the same membrane-bound redox systems and thus interact and compete with each other for electrons. In the natural habitat, the facultatively phototrophic bacteria have to adapt to continuous changes in light intensity and oxygen partial pressure to which they are exposed.

Light inhibits oxygen uptake by respiration, possibly owing to competition between cyclic and non-cyclic electron transport. The limiting redox carrier in the electron transport system is presumably cytochrome c_2. Reduced cytochrome c_2 is consumed for reduction of the oxidized P^+ in the RC. Electrons are therefore not available for electron transport to cytochrome oxidase. This hypothesis does not explain why the quinol oxidase, which is not dependent on cytochrome c_2 and is present in many purple bacteria, is also inhibited by light. In any case, cells of facultatively photosynthetic bacteria respond to transient changes of oxygen tension and light intensity by switching off one of the energy-transducing systems. The characteristics of some anoxygenic bacteria are summarized in Table 13.**1**.

13.3 The Photosynthesis of Cyanobacteria Is Oxygenic

Cyanobacteria are unicellular or multicellular, often filamentous prokaryotes and have no membrane-enclosed compartments such as chloroplasts or mitochondria. Most species are obligate photoautotrophs. The photosynthetic apparatus of Cyanobacteria is very similar to that of eukaryotic plants. The major differences between both types are as follows:

1. The photosynthetic apparatus is localized on thylakoids, i.e., sac-like closed membranes that occupy the peripheral part of the cytoplasm, which are not enclosed by a double membrane as are the chloroplasts in eukaryotes. The membranes are arranged in parallel layers like onion skins and branched or folded in complex patterns, but they are not stacked like grana membranes in chloroplasts (Fig. 2.**14**), and are apparently not connected regularly with the cytoplasmic membrane.
2. Cyanobacteria contain phycobiliproteins, which are organized in phycobilisomes attached to the surface of the thylakoid membrane (Fig. 2.**14**, and 13.**11**). Phycobilisomes are the major antenna systems of Cyanobacteria; they are also present in red algae. The membrane-bound, Chl-containing antennae are much smaller in Cyanobacteria than in plant thylakoids. Cyanobacteria contain Chl *a* but not Chl *b*.
3. At least parts of a respiratory system are localized on thylakoids (NADH dehydrogenase, cytochrome-b_6/f complex, cytochrome-*c* oxidase).

The light-harvesting system of cyanobacteria consists of phycobilisomes and a small Chl-*a* antenna bound to the RC core.

13.3.1 The Photosystem II of Cyanobacteria Is a Water : Plastoquinone Oxidoreductase

Photosystem II (PS II) is a protein complex embedded in the thylakoid membrane (Fig. 13.**11**). It consists of more than twenty different protein subunits and acts as a water : plastoquinone oxidoreductase. The functional activity of charge separation is located on a heterodimer consisting of the proteins D1 and D2. A dimer of Chl *a* (P680, primary donor; $E_m = +600–800$ mV), pheophytin *a* (Pheo), β-carotene, an additional Chl *a*, and presumably two quinones (Q_A, Q_B) and one Fe-molecule between Q_A and Q_B are bound to proteins D1 and D2, which are equivalent to the L and M subunits in the RCs of purple bacteria (Fig. 13.**11**). On the lumen side of the thylakoid membrane there are proteins attached to PS II, which, together with the proteins D1 and D2, are the

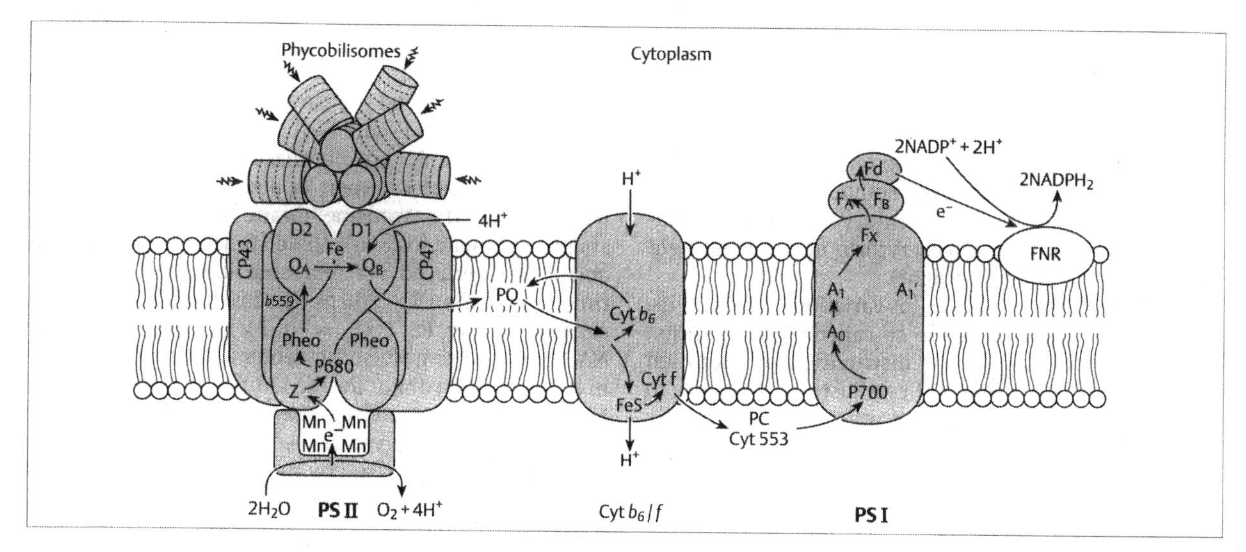

Fig. 13.**11** **The oxygenic photosynthetic apparatus of cyanobacteria with two photosystems.** Photosystem II (PS II) obtains light energy via phycobilisomes, which are attached on the cytoplasmic side of the thylakoid membrane, and small Chl *a*-containing antennae (CP43, CP47). The "water-splitting" system of PS II (see also text) is on the lumen side of the thylakoid membrane. The plastoquinone pool (PQ) of the membrane connects PS II and the cytochrome-b_6/f complex, which functions as a proton pump similar to the cytochrome b/c_1 complex. Plastocyanin (PC) or cytochrome 553 (cyt 553) connect the b_6/f complex with the photosystem I (PS I), a plastocyanin : ferredoxin oxidoreductase. P700 is the primary donor (Chl *a*) and A_0, the primary acceptor; A_1, the intermediate acceptor, is a quinone. F_X, iron–sulfur cluster; FA and FB, the terminal acceptors, iron-sulfur clusters. PS I is similar to the RCs of green sulfur bacteria and *Heliobacterium*. Fd, ferredoxin; FNR, ferredoxin : NADP oxidoreductase; Z, acceptor of electrons from the "water-splitting" system (Mn-cluster); *b*559, cytochrome *b* absorbing at 559 nm. Other symbols are explained in the text of Chapter 13.3

manganese binding sites of the oxygen-evolving system. In contrast to PS I and the RCs of purple bacteria, the primary donor of PS II is a powerful oxidant, which is required for extracting electrons from water. The electrons from the water-oxidizing machinery of PS II are used to reduce the oxidized primary donor. Protons and O_2 are produced by this reaction. *Oscillatoria limnetica* is capable of both oxygenic (oxygen-evolving) and anoxygenic (sulfide-dependent) photosynthesis. The anoxygenic reaction in *O. limnetica* is induced by sulfide under anoxic conditions. Following charge separation in PS II, reduced plastoquinone (PQ) is produced, which is fed into the plastoquinone pool of the thylakoid membrane. Instead of a cytochrome-b/c_1 complex, cyanobacteria have a plastoquinone–cytochrome-b_6/f complex which contributes to the proton gradient across the membrane. The cytochrome-b_6/f complex is equivalent to the cytochrome-b/c_1 complex of bacteria. At the acceptor site of the cytochrome-b_6/f complex, the electrons are passed to plastocyanin (a copper protein) or cytochrome c_{553}, which are the electron donors for PS I.

Light energy is absorbed by the **phycobilisomes** (Fig. 13.**11**), which are arranged perpendicularly on the external side of the thylakoid. Two PS II complexes are generally associated with one phycobilisome. Phycobilisomes are composed of the core and radially arranged rods. The core cylinders lie antiparallel and consist of discs of allophycocyanin trimers with an absorption band at 620–670 nm. The rods of phycobilisomes consist of discs which are connected by linker proteins. A hexamer of phycocyanin (absorption maximum 615–620 nm) is located near the core to which it is attached. The following hexameric discs are composed of phycocyanin, phycoerythrocyanin (568, 590 nm), or phycoerythrin (565, 568 nm; Fig. 13.**5**). Excitation energy flows from the shorter to the longer absorption bands of phycobilines and then via the light-harvesting Chl-*a* components (CP) in the membrane to the PS II. In response to modifications of light parameters in the environment, the amount of light-harvesting pigment varies inversely with photon flux rate. Upon transfer of cells from high to low light intensities, the phycobilisomes increase in size by elongation of the rods, followed by an increase in the number of phycobilisomes per unit thylakoid area.

Cyanobacteria adapt not only to the quantity but also to the quality of incident light. An increase in the light absorbed by PS II (470–660 nm, absorbed mainly by phycobilisomes) causes an increase in the relative amount of PS I. Upon transfer to red light (above 660 nm) or blue light (approximately 450 nm), absorbed preferentially by Chl *a* of PS I, the relative amount of PS I decreases. In some strains, the composition of phycobilisomes can change (complementary chromatic adaptation, e.g., formation of phycoerythrin under green light or phycocyanin under red light).

A short-term response to variations in light intensity or light quality is the so-called state transition process. In this process, the distribution of excitation energy between PS II and PS I is adjusted in response to excitation of either one. There are no morphogenetic processes involved.

13.3.2 Photosystem I of Cyanobacteria Is a Plastocyanin : Ferredoxin Oxidoreductase

The PS-I reaction center is a membrane-bound complex that functions as a light-driven plastocyanin : ferredoxin oxidoreductase (Fig. 13.**11**). The primary donor P700, a dimer of chlorophyll molecules, is excited directly by light or by excitation energy from the antenna systems (50–100 Chl *a* molecules). P700* reduces within 10 ps the primary acceptor A_0, most likely a monomeric Chl-*a* molecule. The primary photochemical event is the charge separation between P700 and A_0. The $P700^+$–A_0^- radical pair has a short lifetime, and the electron is stabilized by transfer via the intermediate acceptors A_1, a phylloquinone, and F_X (iron–sulfur cluster) to the terminal acceptors F_A and F_B (both iron–sulfur clusters). P700, A_0, and F_X are localized on the PsaA/PsaB proteinaceous heterodimer; F_A and F_B are on a 8.9-kDa polypeptide. The transmembrane distance A_0–A_1 is small in comparison with the distance P700–A_1. F_X, F_A and F_B are reduced via multistep kinetics within the time range from 15 to 200 ns. The low-potential 4 Fe–4 S centers of the PS I ($E_m = -0.7$ V) reduce NAD^+ or $NADP^+$ via other iron–sulfur centers. Reduced ferredoxin is the electron donor for the reduction of NADP by the ferredoxin : NADP oxidoreductase (FNR) or for reduction of nitrate. As in anoxygenic photosynthesis, the photophosphorylation is driven by the proton gradient across the membrane and the membrane potential.

The oxygenic photosynthetic apparatus of cyanobacteria contains two photosystems, the water : plastoquinone oxidoreductase (PS II) and the plastocyanin : ferredoxin oxidoreductase (PS I), which are coupled by a linear electron-transport system. The electrochemical proton gradient, NADPH, and O_2 are the products of oxygenic photosynthesis. Phycobilisomes serve as the major antenna system. Cyanobacteria contain only chlorophyll *a*.

13.4 Bacteriorhodopsin I Is a Light-driven Proton Pump of *Halobacterium halobium*

All photosynthetic apparatuses described previously in Chapter 13.1–3 are organized and function according to the same basic principle; they depend on chlorophyll or bacteriochlorophyll. Photosynthetic apparatuses are present in several groups of eukaryotic and prokaryotic microorganisms but have never been found in Archaea (archaebacteria). The extremely halophilic archaebacterium *Halobacterium halobium* is able to synthesize a light-driven proton pump. Under low oxygen tension, crystalline patches of **bacteriorhodopsin (BR)** are inserted into the cytoplasmic membrane, which then appears purple in color. This purple membrane contains (by weight) approximately 25% lipids and 75% of a single protein, bacteriorhodopsin, which consists of **retinal** bound by a protonated Schiff's base to lysine 216 (K216) of the apoprotein (M_r 27 000; Fig. 13.**12**). BR spans the membrane with seven α-helical structures, which enclose a transmembrane space that is occupied by retinal and protonatable amino acid residues. The Schiff's base of retinal is about midway, and it divides the interhelical space into an extracellular "channel" containing the charged residues aspartic acid at positions 85 and 212 (D85, D212) and arginine 82 (R82), bound water, and a cytoplasmic "channel" containing many hydrophobic residues, i.e., the important amino acid residues threonine 46 (T46), aspartic acid 96 (D96) and arginine 227 (R227), and bound water (Fig. 13.**13**). The extracellular channel functions as the proton release domain; the cytoplasmic channel serves as the proton uptake domain. Aspartic acid 96 is the internal proton donor to the Schiff's base; aspartic acid 85 is the proton acceptor. The entire domain structure on the cytoplasmic and extracellular side of the transmembrane space is responsible for proton translocation

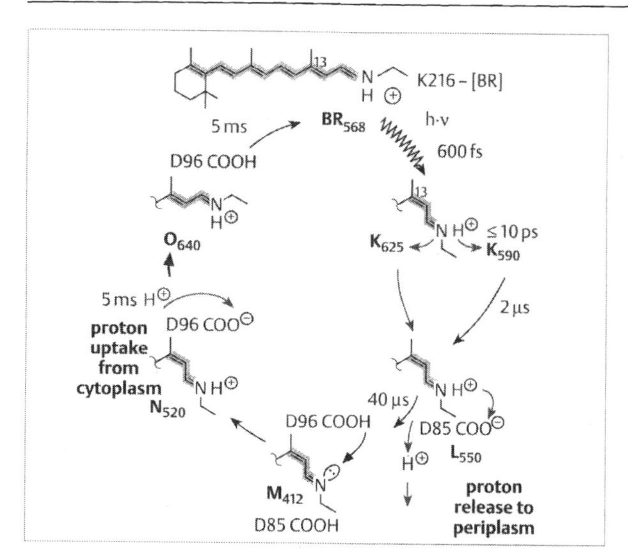

Fig. 13.12 **Bacteriorhodopsin in the purple membrane of the archaebacterium *Halobacterium halobium*–photocycle of the pigment retinal.** Light-dependent *cis–trans* isomerization of retinal and transient deprotonation of the Schiff's base causes a proton translocation across the membrane. Ground state and intermediates (e.g., K_{625}) are indicated by letters (K, L, M, N, O) with subscripts denoting the respective absorbance maximum in nm. The zigzag line with an arrow indicates the only light-dependent step. The key aspartic acid residues D85 and D96 are essential for proton translocation. K216 is a lysine residue forming a Schiff's base. The reaction times near the arrows indicate that the first steps of the photocycle until proton release are very fast (in the picosecond range). Proton uptake and regeneration of bacteriorhodopsin 568 (BR_{568}) are processes taking place in the millisecond range. Proton transfer steps are shown in red

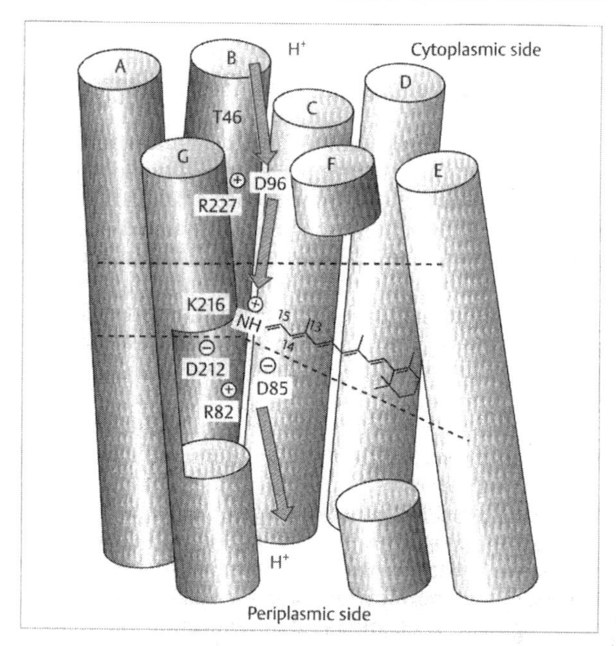

Fig. 13.13 **The protein chain of bacteriorhodopsin is folded and inserted into the cytoplasmic membrane** with seven hydrophobic, transmembrane α-helices (A–G). Retinal is bound by a Schiff's base to lysine 216. Aspartate 96 (D96) participates in the proton uptake on the cytoplasmic side of the membrane and aspartate 85 (D85), in the proton release site on the periplasmic side of the membrane. The outward (periplasmic) proton-release channel and the inward (cytoplasmic) proton-uptake channel (red arrows) are open only transiently during the cycle. The Schiff's-base region acts as a light-driven switch between the two channels and translocates a proton from the cytoplasmic to the periplasmic side of the membrane [from 1]

Absorption of photons effects a *cis–trans* isomerization of retinal and a transient deprotonation of the Schiff's base (Fig. 13.**12**). The free energy retained in the K-state (after absorption of a photon) is close to 50 kJ/mol. The vectorial force generated by the photocycle of BR and the resulting protonation–deprotonation translocates protons from the cytoplasm through a proton channel (D85, D96) to the periplasmic space (Fig. 13.**13**). The electrochemical proton gradient of 300 mV can be used for ATP formation at the proton: F_1F_0 ATP synthase. Light-mediated ATP production in *Halobacterium halobium* supports slow growth under anoxic conditions and shortage of organic substrates. The proton gradient generated by the light-stimulated proton pump is also used to pump Na^+ out of the cell by action of a Na^+/H^+ antiporter, for the uptake of ions by symport mechanisms (see Chapter 5.**6**), and to drive the flagellar motor.

Absorption of light by bacteriorhodopsin effects a *cis–trans* isomerization of retinal and a transient deprotonation of the Schiff's base. The vectorial force generated by the photocycle of retinal and the resulting protonation–deprotonation drives protons from the cytoplasm through a proton channel to the periplasm. The electrochemical proton gradient is used for ATP formation or transport of solutes.

13.5 Spectroscopical, Biochemical, and Genetic Methods Are Used to Study Bacterial Photosynthesis

The kinetics of light-induced photochemical reactions has been elucidated by the powerful technique of time-resolved absorption and fluorescence emission spectroscopy. The biological preparations (isolated reaction centers or photosynthetic membranes) are excited by a short, non-saturating flash, which, with present advanced techniques, can be ultrashort (i.e., in the femtosecond range, 10^{-15} s). By the use of sensitive absorption measurements, short-lived states have been detected. By using repetitive flashes and averaging small signals, a good signal-to-noise ratio can be obtained. Kinetics of protein–pigment interaction during primary processes of photosynthesis can be recorded by Fourier-transformed infrared Raman spectroscopy (measurements of vibrational properties of the entire molecules and resonance enhancement). Another approach is the isolation and purification of membrane-bound complexes by detergent-mediated solubilization and various chromatographic techniques. In a few cases, crystallization of the complexes (e.g., the reaction centers of *Rhodopseudomonas viridis* and *Rhodobacter sphaeroides* and the light-harvesting complex of *Rhodopseudomonas acidophila* and *Rhodospirillum molischianum* has been achieved. The resolution of the structure at an atomic level in combination with time-resolved spectroscopy yielded important results regarding the steps of charge separation and electron transfer in RCs.

In a third group of methods, molecular-genetic techniques are used to clone and to determine the sequence of proteins involved in photosynthesis and to exchange single or groups of amino acids by site-directed mutagenesis. The exchange of highly-conserved amino acid positions resulted in a modification of structural and functional properties and paved the way for identification of functional aminoacyl residues. For example, the exchange of the conserved histidine residue that binds coordinately the central magnesium atom of BChl to alanine in RCs resulted in a BChl-free, photoinactive cell.

Further Reading

Barber, J. and Andersson, B. (1994) Revealing the blueprint of photosynthesis. Nature 370; 31–34

Blankenship, R.E., Madigan, M.T., and Bauer, C.E., eds. (1995) Anoxygenic photosynthetic bacteria. Dordrecht, Netherlands: Kluwer Academic Publishers

Bryant D.A., ed. (1994) The molecular biology of cyanobacteria. Dordrecht, Netherlands: Kluwer Academic Publishers

Cramer, W.A. and Knaff, D.B. (1990) Energy transduction in biological membranes. Berlin: Springer-Verlag

Dutton, P.L. and Mosser, C.C. (1994) Quantum biomechanics of long-range electron transfer in protein: hydrogen bonds and reorganization energies. Proc Nat Acad Sci USA, 91: 10247–10250

Haupts, V., Tittar, J., Bamberg, E. and Oesterhelt, D. (1997) General concept for ion translocation by halobacterial retinal proteins. Biochemistry 36: 2–7

van Grondelle, R., Dekker, J.P., Gillbro, T., and Sundström, V. (1994) Energy transfer and trapping in photosynthesis. Biochim Biophys Acta 1187: 1–65

Source of Figure

1 Lanyi, J. K. (1993) Biochim Biophys Acta 1183: 241–261

Section IV

The Genetics of the Prokaryotes and Their Viruses

The lack of a nucleus, of chromosomes and of sexuality was long considered to represent major differences between the bacteria and the "true", i.e., the eukaryotic cells and organisms. It thus came as a surprise when in 1961 a book by F. Jacob and E. L. Wollman appeared with the provocative title "Sexuality and the genetics of bacteria." This book was followed in 1964 by another classic, W. Hayes' "The genetics of bacteria and their viruses." Both not only summarized a wealth of information accumulated during the last 30 years on the genetics of the bacteria and their viruses, but also made clear that bacteria possess chromosomes and show a highly developed sexuality involving "donor" and "recipient" cells. The number of concepts described or explained for the first time during this period which are considered as central in modern biology is truly amazing. This includes, as described in Chapter 14, DNA as a carrier of the genetic information and its semi-conservative replication by replisomes. These complex organelles catalyze the coordinated replication of both the "leading" and the "lagging" DNA strands of the cellular chromosome. The latter, although not covered by histones and not surrounded by a nuclear envelope, is not naked either. Instead it is highly organized in a "nucleoid" and complexed with a multitude of RNAs and proteins. The replisomes bind the replicating DNA (chromosome) to the membrane and ensure a tight coupling of the replication of the chromosome with cell division. By definition, the chromosome with its bi-directional origin of replication (*ori*) constitutes the largest replicon of a cell. As a general rule, the cellular chromosome carries essential ("house-keeping") and species-specific genes, i.e., a central part of, but not the complete genome. Normally, the genome of a cell contains additional genes for facultative properties, e.g., genes allowing horizontal gene transfer between donor and recipient cells, genes conferring resistance to toxic agents and new metabolic capacities, genes for symbiosis and pathogenicity. The genes for the facultative determinants are located on transposable genetic elements (**i**nsertion **s**equences or **IS**; **tra**nsposable elements or **Tn**; **c**onjugative **tran**sposable elements or **CTn**; "**pa**thogenicity **is**lands" or **PAIS**) and on plasmids. Plasmids which include, strictly speaking, the bacterial viruses or bacteriophages are autonomously replicating DNA elements or replicons which may occur in one to many (up to 50) copies per cell, may be circular or linear, conjugative or non-conjugative, and may range from small cryptic plasmids without a visible phenotype (10 kb, 3 genes) to large megaplasmids (>500 kb) coming close to the size of small bacterial chromosomes (ca. 700 kb).

As described in Chapter 15 the genes from prokaryotes (typically 1 kb in size) are often organized in transcriptional units (operons). Transcription of gene groups from one promoter and direct translation of the corresponding "polycistronic" mRNA by several ribosomes ("polysome") facilitates their coordinated regulation. High growth rates, if the possibility is given, and rapid adaptation to changing environmental conditions are the outstanding hallmarks for prokaryotic cells. Rapid adaptation involves, except for the direct coupling of transcription and translation, the presence of several RNA polymerases per cell which have a core part in common. Each type of polymerase contains one specific sigma subunit which determines the type of promoter to be recognized. As expected, various environmental conditions require different sigma subunits and hence assure the transcription of specific genes. Another strategy followed by the prokaryotes to ensure high adaptation is an extensive horizontal gene transfer through transformation, transduction and conjugation as described in Chapter 16. The three processes allow the efficient exchange of parts of the genome ("merogenotes or merodiploids") and DNA recombination. If the phenomenon of sexuality as found in the eukaryotes is freed from secondary elements (male and female; fecundation; meiosis) and reduced to what is essential (donor and recipient; gene exchange and recombination), then the prokaryotes show sexuality in a form which is perfectly adapted to their way of living, to their cellular organization.

The last chapter of this section (Chapter 17) describes basic principles of recombinant DNA technology also known as gene technology. Because gene technology was developed from bacterial genetics and bacteriophage studies, because essential tools (e.g., restriction enzymes, vectors, and genes) derive from prokaryotic organisms and because even most eukaryotic projects involve intermediate steps in *E. coli*, it is appropriate to include the chapter here. However, because the development in recombinant DNA technology is very fast, a great multitude of variants have been generated, and not more than basic principles can be given here. For details, the reader is referred to the numerous examples described in sections V to IX and to the Further Reading given below and in the corresponding chapters.

Further Reading

Beckwith, J., and Silhavy, T.J. (1992). The power of bacterial genetics. Cold Spring Harbor, New York: Cold Spring Harbor Press

Birge, E.A. (1994). Bacterial and bacteriophage genetics. 3rd edn. New York: Springer

Hayes, W. (1964). The genetics of bacteria and their viruses. New York: Wiley

Jacob, F., and Wollman, E.L. (1961). Sexuality and the genetics of bacteria. New York: Academic Press

Joset, F., and Guespin-Michel, J. (1993). Prokaryotic genetics. Genome organization, transfer and plasticity. London: Blackwell Scientific

Miller, J.H. (1992). A short course in bacterial genetics. A laboratory manual and handbook for *Escherichia coli* and related bacteria. Cold Spring Harbor, New York: Cold Spring Harbor Press

Steips, U.N., and Yasbin, R.E., eds. (1991). Modern microbial genetics. New York: Wiley-Liss

Snyder, L., and Champness, W. (1997). Molecular genetics of bacteria. Washington, DC: ASM Press

14 DNA, Chromosomes, and Plasmids

More than a century ago, Gregor Mendel discovered that heritable properties of organisms are separate units, and that the responsible genes can be more or less freely combined. In the course of the development of genetic concepts, it was realized that genes are arranged linearly on chromosomes. The finding that bacteria also have chromosomes and genes was a big step forward because large numbers of bacteria can be grown and handled easily. In addition to chromosomes, genetic information is stored in smaller units, such as plasmids, viruses, and transposable elements. The accurate duplication and the vertical transmission of this information to daughter cells is the key process in the cell cycle. In this chapter, the structure and the properties of the different genetic elements and the biochemical rules for their replication will be discussed.

Genetic information is arranged linearly on **chromosomes**. In bacteria, these comprise double-stranded DNA (deoxyribonucleic acid) molecules, RNA and proteins, which are organized in a cytoplasmic structure called the **nucleoid**.

14.1 DNA Is the Carrier of Genetic Information in Bacteria

The genetic information that is passed vertically from generation to generation is encoded in **DNA** (**deoxyribonucleic acid**), in most cases double-stranded (ds) DNA. Exceptions to this rule are only some viruses of prokaryotes (bacteriophages, see Chapter 26), which contain RNA as their genetic material. The original erroneous assumption that DNA was not complex enough to be the genetic material was overcome by the concept that the genetic information is **encoded in the sequence** of the bases of DNA. The essential experimental evidence for DNA being the genetic material comes from two experiments with bacteria and with bacteriophages:

i) In 1944, O. T. Avery, C. M. MacLeod, and M. McCarty transformed an apathogenic (rough) strain of *Pneumococcus* (*Streptococcus*) *pneumoniae* in the test tube with an extract from a pathogenic (smooth) strain. The transformants were then able to cause disease in mice. The authors demonstrated that the **transforming principle** was DNA.
ii) A. D. Hershey and M. Chase demonstrated in 1952 that upon infection of *Escherichia coli* with the bacteriophage T2, only the DNA of the phages enters the host cells and must, therefore, be responsible for the information that governs phage replication, the structure of its proteins, and its morphogenesis (see chapter 26).

With these pieces of evidence that the genetic material of bacteria and (most of) their viruses is DNA, the notion that DNA is the universal carrier of genetic information became generally accepted, although the direct demonstration for eukaryotic cells came later. As will be seen below, it is the unique structure of DNA that makes it especially well suited to carry genetic information because the structure is independent of the particular sequence.

14.1.1 DNA Is a Polynucleotide Containing Deoxynucleotides

Components of DNA. The monomeric building blocks of DNA are the nucleotides. Nucleotides contain a base that is linked via an N-glycosidic bond to position 1 (called 1′ in nucleotides) of 2-deoxyribose. This base plus sugar is called a nucleoside. Nucleosides, more precisely **2′-deoxynucleosides** in the case of DNA, can carry one, two, or three phosphate groups at the 3′ or 5′ positions of the sugar, and are then called **deoxynucleotides**. In the polynucleotide DNA, the 3′ position of one deoxyribose ring is connected via a phosphodiester linkage to the 5′ position of the next pentose. The sugar-phosphate backbone thus consists of an alternating series of sugar and phosphate residues. This gives the DNA strand a polarity with a 5′-phosphate end and a 3′-hydroxyl end.

The bases found in DNA are the **purines, adenine** (6-aminopurine) and **guanine** (2-amino-6-oxypurine), and the **pyrimidines, cytosine** (2-oxy-4-aminopyrimidine) and **thymine** (2,4-dioxy-5-methylpyrimidine). Usually only their initial letters A, G, C, and T (correctly dA, dG, dC, and dT) are used (see Chapter 7.8). In some cases, a few bases in DNA are modified subsequent to replication to, for example, 4- or 5-methylcytosine or 6-methyladenine (see Chapter 15.6). The DNA of *E. coli* phages T2, T4, and T6 contains 5-hydroxymethyl-deoxycytosine instead of deoxycytosine, that can, in addition, be glycosylated (see Chapter 26).

E. Chargaff found in the early 1950s by chemical analysis of a large number of DNA molecules from different organisms that the frequency of A corresponds to the frequency of T, and that of G to C, and that the ratio of A+T to G+C was characteristic for a given organism. This is because bases are able to form hydrogen bonds to other bases: A is able to pair with T, forming two hydrogen bonds, and G forms three hydrogen bonds with C (Fig. 14.**1**).

Antiparallel strands and the double helix. In 1953, J. Watson and F. C. Crick combined X-ray diffraction data of M. H. F. Wilkins and R. Franklin with the Chargaff rule of the correspondence of A to T and G to C to form a convincing and elegant model of DNA. In this model, two DNA strands of opposite polarity (antiparallel; Fig. 14.**2**) are wound around each other in a right-handed helix with the hydrophobic bases facing inside. Since small pyrimidine bases (T or C) pair with large purine bases (A or G, respectively) the double helix has a constant diameter of 2.0 nm, in agreement with the postulate that the overall structure of DNA is independent of the actual sequence. The bases lie flat like the steps of a staircase, and each base pair is rotated 34.6 °C around the helix axis relative to the previous one. Therefore, 10.4 base pairs are in one complete turn, with a helical pitch of 3.4 nm. This model describes what is now called **B-form DNA** (Fig. 14.**3**). The sugar-phosphate backbones on the outside of the double helix leave a major and a minor groove, in which the sequence of base pairs can be sensed by proteins. Since the information content of the major groove is higher, most regulatory proteins bind there.

B-form DNA contains two complementary strands arranged in an antiparallel way. The DNA has a diameter of 2 nm, a helical pitch of 3.4 nm, and contains 10.4 base pairs per turn.

Because of the base pairing, the two antiparallel DNA strands in the double helix contain equivalent information. Each strand can be used as a template for the complementary strand. This is the basic feature for the perpetuation of the genetic information during replication and repair.

Most of the DNA in cells or in solution is present in B-form. However, other configurations are possible. **A-form DNA** is less hydrated than B-DNA, and the structure is more compact. The bases are slightly tilted relative to the helix axis, there are 11 base pairs per helical turn, and the helix diameter is 2.3 nm. **Z-form DNA** is found in vitro in polynucleotides with alternat-

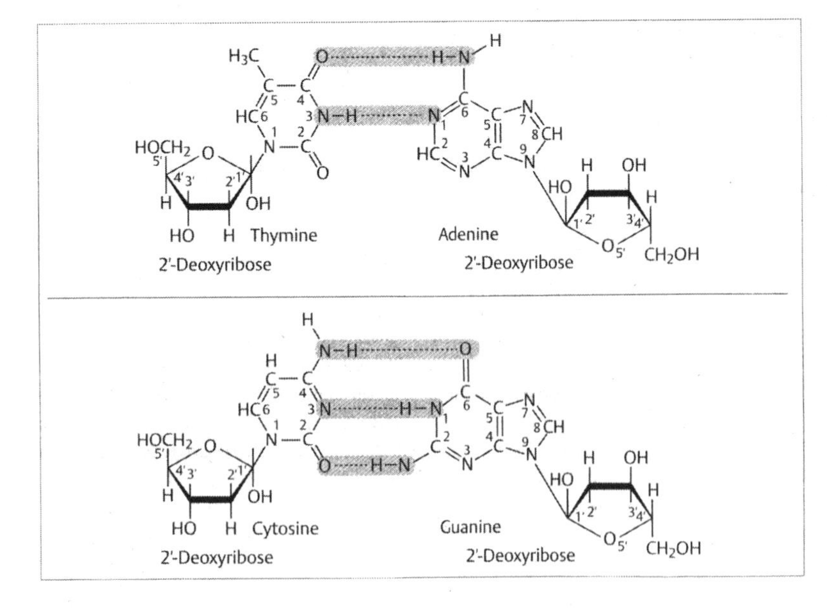

Fig. 14.**1** **Purine and pyrimidine base pairs (deoxynucleosides) found in DNA.** The adenine–thymine (A–T) pair forms two hydrogen bonds, and the guanine–cytosine (G–C) pair forms three hydrogen bonds. (Hydrogen bonds are shown in red.) Note the presence of 2-deoxy-ribose and the numbering of the bases (1–9) and the sugars (1′–5′)

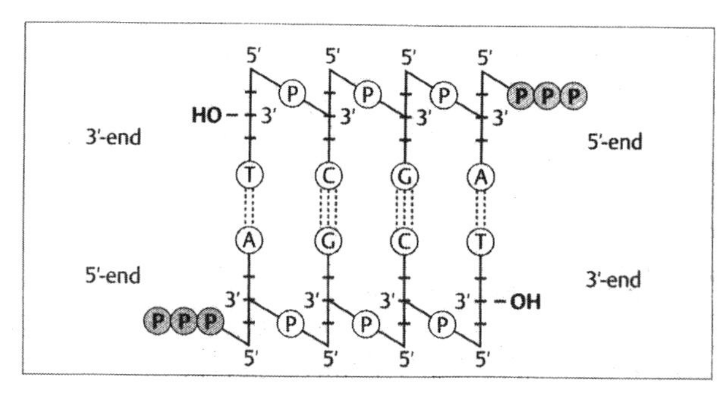

Fig. 14.2 **Complementary and antiparallel strands in double-stranded DNA.** By definition the phosphate-carrying ends (red) are called the 5′ ends and the hydroxyl-carrying ends (red) are called the 3′-ends. Note the antiparallel arrangement of the two complementary strands

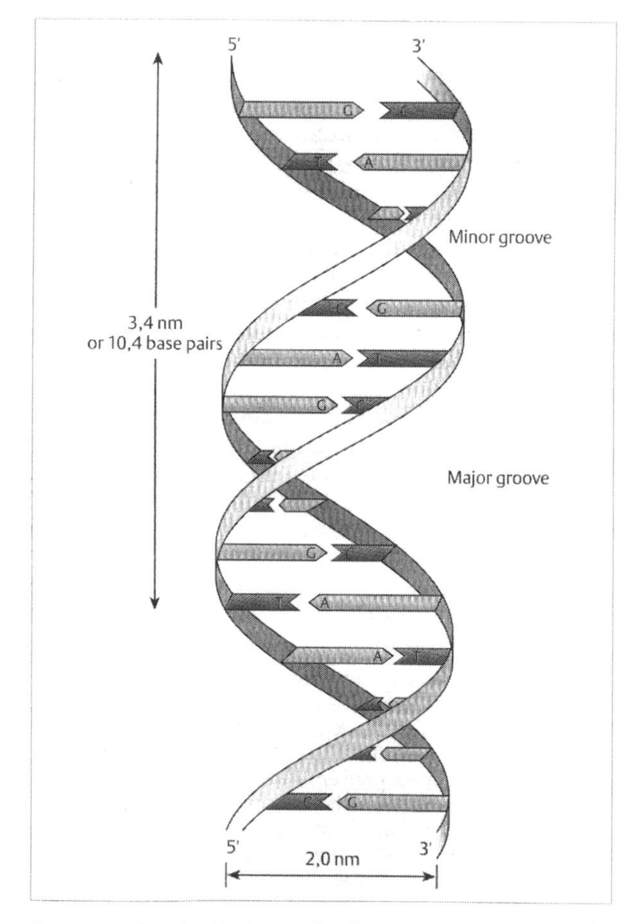

Fig. 14.3 **The double-helix of B-form DNA.** Two antiparallel DNA strands are wound in a right-handed double-helix, forming a major and a minor groove in which DNA-binding proteins can recognize the base pairs. (Pyrimidine bases are shown in red)

ing purines and pyrimidines at high salt concentrations. It is the only left-handed helix described for DNA, and the name was given because the sugar-phosphate backbone follows a zigzag path along the helix. Whether Z-DNA has a physiological role is not known.

DNA can have different conformations. DNA can be a linear double-stranded molecule. However, most DNA molecules in prokaryotes are present as **c**ovalently **c**losed **c**ircles (ccc DNA). If such a DNA ring can be placed flat in a plane, it is called relaxed. Supercoils are introduced when DNA is twisted around its axis or wound around a protein core, and then the ends are covalently closed to a circle. Each one of such twists results in one superhelical turn, which can only be released by breaking the covalent bonds of the DNA backbone. Molecules with the same sequence, but a different topology, such as relaxed and supercoiled rings, are topoisomers (Figs. 14.**4** and 14.**5**), and consequently, enzymes that catalyze such conversions are called **topoisomerases**.

Negative supercoils are introduced by twisting DNA in a direction opposite to the clockwise turns of the double helix. Such DNA is underwound, and the torsional stress of the molecules can be relieved by partial unwinding (Figs. 14.**4** and 14.**5**).

Within the structure of B-form DNA, small variations are possible. The angle of rotation of the base pairs relative to each other can vary slightly, which results in a range of 10–10.5 base pairs per helical turn. Similarly, the angle of base pairs with the helical axis may deviate somewhat from the normal 90 °C. Such variations among others are dependent on the primary sequence, but represent information that is not immediately obvious when reading the sequence. DNA thus is not necessarily a straight rod-shaped structure, but can be

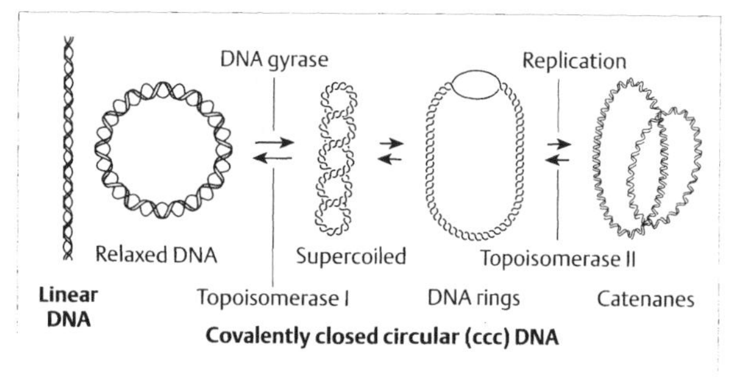

Fig. 14.4 **Different topological conformations (topoisomers) of DNA molecules.** Otherwise identical DNA molecules can assume various topological conformations (topoisomers). Special DNA nucleases, ligases, and topoisomerases (helicases, gyrases) actively convert, for example, linear DNA into covalently closed circular DNA, or supercoiled and catenane rings into relaxed forms

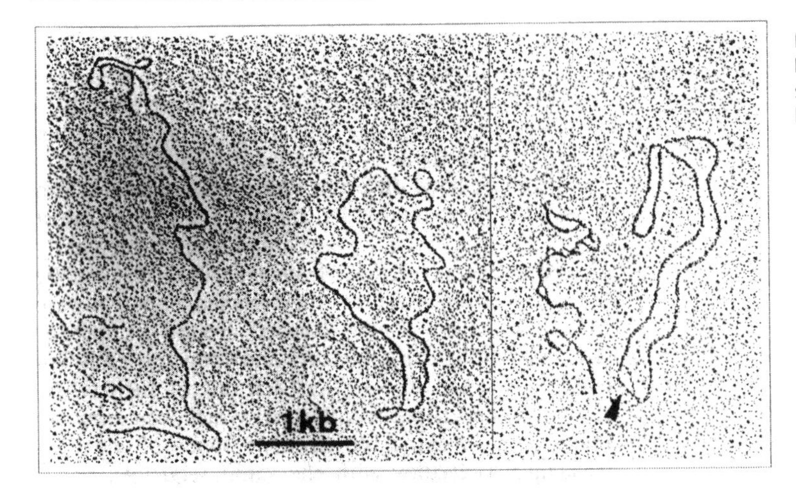

Fig. 14.5 **Topoisomers of plasmid (RFS1010) DNA.** From left to right: linear, relaxed, circular, supercoiled, and partially unwound (see arrowhead) supercoiled DNA. Courtesy of R. Lurz

bent or curved. Bends are deviations from the straight axis between adjacent base pairs, curves are the sum of bends over a distance of several base pairs. Bends and curves affect the interaction with DNA binding proteins (see Chapter 18.8).

14.1.2 The Biochemistry of DNA Synthesis Is Universal

DNA replication is semiconservative. The double-stranded structure of DNA immediately suggests a possible mode of replication. The two strands, containing equivalent information, are separated, and each is used as a template for the synthesis of a complementary strand ("zipper model"). In each of the resulting daughter molecules, half of the parental material is conserved. This is called semiconservative replication (Fig. 14.**6**).

Repair replication (better repair DNA synthesis) occurs in small patches in which parental DNA in one strand is replaced by newly synthesized DNA. A similar local DNA synthesis, also using the complementary strand as a template is found during homologous recombination. It is therefore also called **dispersive** DNA synthesis and will be dealt with in Chapters 15.6 and 16.4.

DNA-dependent DNA polymerases synthesize DNA. These enzymes or enzyme complexes synthesize DNA from single nucleotides using an existing DNA strand as template. *E. coli* (and as far as is known, other bacteria as well) has three DNA polymerases. (Tab. 14.**1**). **DNA polymerase I** (gene *polA*) is a single polypeptide with three enzymatic functions located on separate domains: the polymerizing activity; a 3′-5′ exonuclease activity, the proofreading function; and a 5′-3′ exonuclease that is able to remove RNA from RNA–DNA hybrids. The main function of DNA polymerase I is in repair DNA synthesis, but at least the 5′-3′ exonuclease activity is also used in replication to remove RNA primers from the lagging strand (see below). The 5′-3′ exonuclease domain can be proteolytically removed. The remainder is called the "Klenow fragment" of polymerase I, and is widely used for the in vitro manipulation of DNA.

All DNA polymerases synthesize DNA by the same basic mechanism. A nucleoside 5′-triphosphate is added

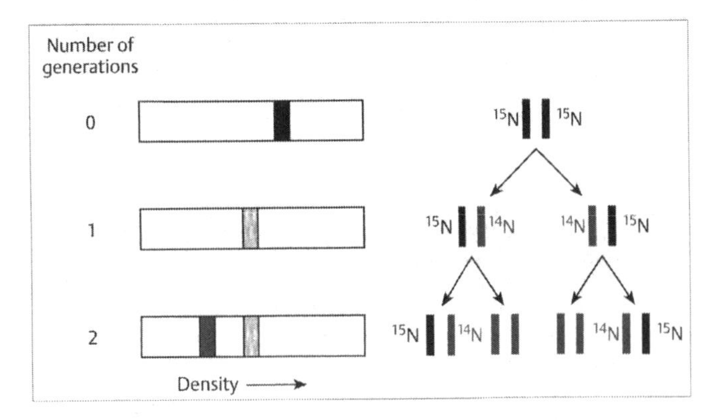

Fig. 14.**6** **Semiconservative replication (Meselson-Stahl experiment).** Heavy (^{15}N) parental DNA (black), hybrid ($^{15}N/^{14}N$) first generation DNA (gray), and light ($^{14}N/^{14}N$) progeny DNA (red) as separated in a CsCl equilibrium density gradient (left) are shown schematically on the right. Shifts in the density of chromosomal DNA fragments (ca 30 kbp) from *E. coli* during a transition from a "heavy" to a "light" medium were analysed after 0, 1, and 2 generations, respectively

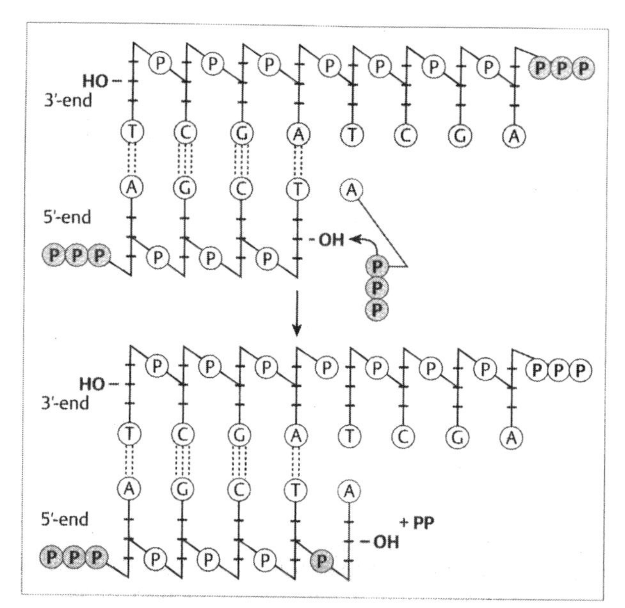

Fig. 14.7 **Biochemistry of DNA polymerization (synthesis).** The template and the growing DNA chain are shown. A 5'-deoxynucleotide triphosphate (red) is added to the 3'-OH end (red) of the growing chain according to the complementary base pair rule, and pyrophosphate (PP) is released

with its α-phosphate group to the 3'-OH end of a growing DNA chain. The energy for the process comes from cleaving the phosphate bond and the release of pyrophosphate. DNA polymerases select the correct nucleotide that is able to base pair with the base on the complementary strand (Fig. 14.7). A nascent DNA strand grows at the 3'-end, and the direction of synthesis is 5'→ 3'.

In the synthesis of double-stranded DNA, one strand (the **leading strand**) can therefore be synthesized continuously. The other strand is synthesized in small pieces (the Okazaki fragments) that start with an **RNA primer**, which provides the 3'-OH end. These primers are synthesized by **primases**, which are RNA polymerases used exclusively for the synthesis of short primers. The 5' → 3' synthesis direction for Okazaki fragments is opposite to the direction of the movement of the replication fork. Therefore, this strand is called the **lagging strand** (Fig. 14.8). How the synthesis of the two strands is coordinated will be described below.

DNA synthesis is semiconservative and semidiscontinuous. Individual strands are synthesized invariably in the 5' → 3' direction. DNA polymerases require a template and a primer with a free 3'-OH end. Primers are normally synthesized by primases.

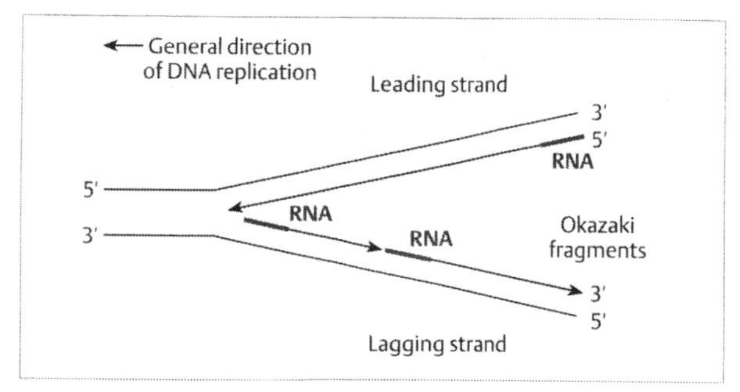

Fig. 14.8 **Schematic representation of a DNA replication fork.** After an initial RNA-primer-dependent start, the leading strand is copied continuously (5' → 3'). The lagging strand, however, is synthesized discontinuously; each Okazaki fragment begins with a new RNA primer (5' → 3'). The general direction of DNA replication in the figure is from right to left; RNA primers are shown in red

DNA polymerases not only select the correct nucleotide to be added opposite the next free base on the template, they also check whether the previous position is correctly base paired. If not, the erroneous nucleotide is removed by the 3'-5' exonuclease activity, the proofreading function, and the enzyme retracts by one base pair. In the case of DNA polymerase I, proofreading is done by a domain on the same polypeptide; in the case of DNA polymerase III, a separate subunit is responsible for proofreading. The error frequencies of DNA polymerases with proofreading activity, such as polymerases I and III, are about 10^{-8}, whereas enzymes without this activity have error rates between 10^{-4} and 10^{-6}.

DNA polymerase III holoenzyme is the enzyme complex that is responsible for replication of chromosomes, most plasmids, and phages. It is much more complex and consists of ten different subunits (Tab. 14.1), which form an asymmetric dimer, one half for the leading and one half for the lagging strand synthesis (Fig. 14.9). The coordination of the different parts of DNA polymerase III will be discussed below.

DNA polymerases have, in addition to the polymerizing activity, a proofreading function, mediated by a 3'-5' exonuclease. An exception is, for example, the thermoresistant Taq DNA polymerase.

Other proteins involved in DNA synthesis. Many other proteins besides DNA polymerases are also involved in DNA synthesis in bacteria (Table 14.2). Prominent among them are **DNA helicases** and **primases**. Some of the other proteins will be discussed in other chapters in the context of repair (Chapter 15.6) and recombination (Chapter 16.4) processes.

Table 14.1 **DNA-dependent DNA polymerases of *Escherichia coli***

	Polymerase I	Polymerase II	Polymerase III
Size (kDa)	103	90	918
Structure	Monomer	Monomer	Heteromultimer
Molecules per cell	300	Unknown	10
Enzymatic activities			
Polymerization	+	+	+
3′-5′ Exonuclease (proofreading)	+	+	+
5′-3′ Exonuclease	+	–	–
Function	Repair and replication	Repair	Replication
Genes (subunits)	*polA*	*polB*	*dnaE* (α), *dnaN* (β), *dnaQ* (ε), *dnaX* (τ+γ), *holA* (δ), *holB* (δ'), *holC* (χ), *holD* (ψ), *holE* (θ)

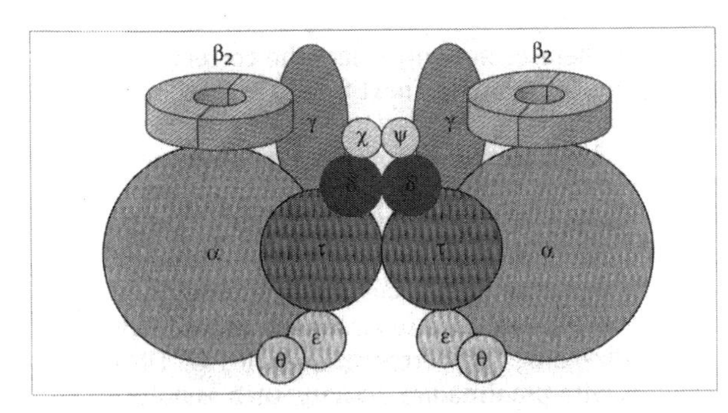

Fig. 14.**9** *Escherichia coli* **DNA polymerase III (holoenzyme).** A schematic structure of the asymmetric dimers coupled by protein τ is shown. Both dimers contain the core subunits (α, ε, θ) and a sliding or β-clamp, all of which are necessary for fast polymerization (approx. 10^3 nucleotides s^{-1}). The γ-complex (subunits γ, δ, δ', χ, ψ) is for clamp loading. The volumes of the subunits are drawn approximately to scale

A very important enzyme for the sealing of breaks in DNA as they occur after the synthesis of Okazaki fragments and after repair replication is **DNA ligase**. This enzyme covalently closes nicks with a 5′-phosphate and a 3′-OH group.

Histone-like proteins. HU, IHF (**i**ntegration **h**ost **f**actor), FIS (**f**actor for **i**nversion **s**timulation), and H-NS (**h**istone-like **n**ucleoid **s**tructuring protein) have an architectural (morphopoietic) role in shaping the initiation complex. This is why they are considered with initiation proteins. In a more general sense, especially FIS and IHF play a role wherever strong bends need to be introduced into DNA, for example, during transcription initiation (see Chapter 18.8) and phage excision/integration (see Chapter 26.4).

Four **topoisomerases** have been identified in *E. coli*: topoisomerase I (gene *topA*), DNA gyrase (topoisomerase II, genes *gyrA,B*), topoisomerase III (gene *topB*), and topoisomerase IV (genes *parC,E*). Only gyrase introduces supercoils in vitro. Apparently, it is also the major source of supercoiling in vivo since inhibitors of gyrase (e.g., nalidixic acid or coumermycin) block the introduction of negative supercoils and cause a loss of supercoils from the chromosome and from plasmids. *topA* mutants, deficient in topoisomerase I, have elevated levels of supercoiling, but they accumulate compensatory mutations in one of the gyrase genes very quickly. Thus gyrase and topoisomerase I act antagonistically also in vivo. The main function of DNA gyrase, in addition to the maintenance of superhelicity, is the release of torsional stress ahead of the replication fork. Gyrase and topoisomerases III and IV are involved in the decatenation of the interlinked end products ("concatemers") of replication of circular chromosomes and plasmids. All topoisomerases introduce a transient break into DNA strands by formation of a covalent protein–DNA intermediate. In this way, the enzymes allow the passage of DNA strands through one another.

The replicative **helicase** in *E. coli* is the product of the *dnaB* gene (see below). It moves, propelled by ATP hydrolysis, in the 5′ → 3′ direction along one DNA strand and unwinds the double-strand ahead of it.

Topoisomerases introduce transient single-strand breaks (type I enzymes: topoisomerases I and III) or double-strand breaks (type II enzymes: gyrase, topoisomerase IV), and allow movement of DNA strands through one another. Gyrase introduces negative supercoils, topoisomerase I relaxes them. Some antibiotics (e.g., **nalidixic acid** and **coumermycin**) act by inhibiting DNA gyrase activity. Helicases are enzymes that unwind double-stranded DNA. The replicative helicase for *E. coli* is the *dnaB* gene product.

Table 14.2 **Proteins involved in replication of the *Escherichia coli* chromosome**

Enzyme	Size (kDa)	Molecules per cell[1]
Initiation		
DnaA	52	1,000
HU-$\alpha\beta$	19	40,000
IHF-$\alpha\beta$	21.8	3,000
FIS (dimer)	22.4	30,000
H-NS	15.5	20,000
RNA-polymerase ($\alpha_2\beta\beta'\sigma$)	450	
Initiation and elongation		
Helicase ($DnaB_6$)	$6 \times 50 = 300$	20
DnaC ($DnaC_6$)	$6 \times 29 = 174$	20
Primase	60	50–100
DNA gyrase (A_2+B_2)	$2 \times 97 + 2 \times 90 = 374$	500
SSB (single-strand binding protein)	$4 \times 18.9 = 75.6$	800
Elongation		
DNA polymerase III $\alpha_2\varepsilon_2\theta_2\tau_2\gamma_2\delta\delta'\chi\psi\beta_4$	918	10
DNA polymerase I	103	300
DNA ligase	75	3
Topoisomerase I	100	1,000
Topoisomerase III	74	10
Topoisomerase IV ($ParC_2$+$ParE_2$)	$(2 \times 84)+(2 \times 70) = 308$	500

[1] Approximate number in cells growing with 1–2 doublings per hour

A protein that is involved in virtually all reactions of DNA is single-strand DNA binding protein, **SSB** (gene *ssb*). It stabilizes and protects single-stranded DNA as it is exposed during replication, recombination, and repair processes by cooperative binding to the single-stranded regions.

14.1.3 Different Mechanisms of the Initiation of DNA Replication Define Where and How Replication Starts

In their **replicon model**, F. Jacob, S. Brenner, and F. Cuzin suggested in 1963 that for each DNA able to replicate autonomously, which they called a **replicon**, an **initiator** protein acts upon the replicator, a DNA site in *cis* (i.e., on the replicon), which is now called the **replication origin**. Although the model at that time was highly speculative, it was seminal for subsequent experiments. Replicons vary in size. A bacterial chromosome is a replicon, normally the largest replicon in a cell. However, also phage DNAs and plasmids are replicons that have the ability to replicate autonomously.

Replicons are autonomous units of replication with initiators (proteins) that act upon replication origins (replicators). The replication origin is located on the DNA (in *cis*) that is replicated.

Initiation is the first step in DNA replication at which the major replication control mechanisms act, typically through an initiator. This is not surprising since, in general, cells adapt to changing conditions by adjusting the rate with which new chains of macromolecules are initiated and not by varying the velocity of polymerization. Replication initiation faces three problems: 1) finding the correct start site, the replication origin, 2) providing a primer molecule for DNA synthesis, and 3) loading DNA helicase onto the template. In some systems, these steps are intimately connected; in others, they are separated in time and in space. Before the mechanisms of origin recognition and the various DNA replication steps are explained, the various priming mechanisms and the primosomes central to initiation will be described.

Priming mechanisms using RNA (primers) were analyzed using single-stranded phage models. DNA polymerases cannot start new chains and must rely on a priming system for the start of DNA replication. **Primosomes** fulfill two functions in this process. They provide the free 3′-OH end that is required by DNA polymerases, and they mediate the loading of DNA helicase that is required to unwind and replicate a double-stranded template.

In most cases, the primers are short RNA molecules. Since their polymerization must occur without proofreading, RNA is a convenient label for the primers as potentially erroneous molecules that are subject to future removal. Three different phages of *E. coli* use different strategies to synthesize a complementary strand on the infecting viral strand, also called (+) strand, that is covered by single-strand DNA binding protein (SSB). In phage M13, the (+) strand carries a secondary structure that is recognized by RNA polymerase (i.e., the regular RNA polymerase that operates in transcription), which synthesizes a short RNA primer. In phage G4, a different hairpin structure is recognized by a different RNA polymerase, primase, the product of the *E. coli dnaG* gene. The most complicated priming mechanism is found for phage ΦX174. A hairpin structure on the (+) strand is recognized by the PriA protein, followed by the assembly of the preprimosomal complex consisting, in addition to PriA, of PriB, PriC, and DnaT proteins. This complex recruits DnaB helicase with the assistance of the DnaC protein, and the addition of DnaG primase forms the complete (ΦX174-type) primosome. In both G4 and ΦX174, primase synthesizes an RNA primer that is extended by DNA polymerase III to the (−) strand, which together with the template (+) strand forms a double-stranded molecule called RFII (**r**eplicative **f**orm II). Removal of RNA primers by DNA polymerase I and ring closure by DNA ligase converts it to the covalently closed RFI form.

Primosomes are responsible for the loading of helicase. In *E. coli* there are two basic types of primosomes. The ΦX174-type primosome discussed above is the primosome typical for single-stranded DNA. The PriA protein recognizes a hairpin structure called *pas* (for **p**rimosome **a**ssembly **s**ite), which can only form if the DNA has been made partially single-stranded by another mechanism, e.g., by the formation of an R-loop or a D-loop. Therefore, this is the primosome for plasmids such as ColE1 and for backup replication systems, such as stable DNA replication (SDR), which will be discussed later.

Primosomes are protein complexes that load helicase onto the template. Since primase binds to helicase, primosomes also provide RNA primers for DNA synthesis. There are two basic types of primosomes in *E. coli*: 1) the ΦX174-type primosome, consisting of PriA, PriB, PriC, and DnaT proteins, and DnaB/DnaC and primase; and 2) the DnaA primosome, consisting of DnaA protein and DnaB/DnaC and primase.

The primosome that operates in chromosomal replication is the **DnaA primosome**. In this primosome, the preprimosomal proteins PriA, PriB, PriC, and DnaT are replaced by DnaA. DnaA protein binds to double-stranded DNA and interacts with the hexameric DnaB–DnaC complex to load DnaB helicase. Primase enters the complex by binding to helicase.

The recognition of replication origins is different for different priming mechanisms. Most prokaryotic replicons are covalently closed double-stranded circles (ccc DNAs). There are two basic modes for the replication of circular templates. These modes are named according to the similarity of the DNA structures involved with letters of the Greek alphabet: the θ bidirectional type of replication and the σ or rolling circle type.

There are two possibilities for origin recognition in the **θ-type replication**. The most frequently used initiation mechanism (**oriC-type**), typical for chromosomal and most plasmid and phage origins, is the binding of several subunits of an initiator protein to repeated (iterated) recognition sites (or iterons) present within the origin, resulting in a specialized nucleoprotein structure. The next step is a local unwinding in a nearby AT-rich region, presumably because of the structural distortions in the initiation complex (Fig. 14.**10**). With the help of a primosome, one or two primer RNA molecules are synthesized. Such an R-loop is the initial unit for the assembly of replication forks. If a single leading strand primer is deposited, replication will be unidirectional. Two primers, and two primo-

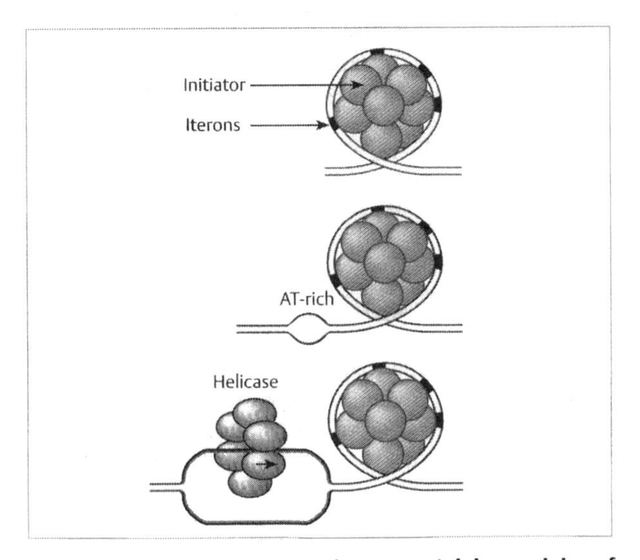

Fig. 14.**10** **Initiation at an iteron-containing origin of replication (*oriC*-type initiation).** Several subunits of the initiator protein bind to iterated (repeated) binding sites of an *ori* and help to unwind a nearby AT-rich region. This allows the binding of one primosome (one RNA primer) for a unidirectional DNA replication or of two primosomes (two RNA primers) for a bidirectional DNA replication. DNA helicases (here DnaB) help in replication initiation

somes, are required for bidirectional replication. Regulation of this type of initiation is primarily through the availability of the initiator protein (see Chapter 22).

R-loops are triple-stranded regions ("bubbles") with a DNA–RNA hybrid and a corresponding single-stranded DNA.

An alternative way for creating an R-loop with an RNA primer is used by **ColE1-type** plasmids. Here, the RNA primer is synthesized by the normal DNA-dependent RNA polymerase, starting at a regular promoter, and is subsequently processed by RNaseH, an enzyme that cuts RNA in RNA–DNA hybrids. Promoter recognition and recognition of the processing site constitute together the recognition of the origin. ColE1-type initiation and its regulation is described in detail in Chapter 22.

Initiation of θ-type replication of circular templates occurs: 1) by the *oriC* mode involving a specialized nucleoprotein structure made from the iteron-containing replication origin and the initiator protein binding to the iterons; or 2) by the ColE1 mode using an RNA primer synthesized by RNA polymerase and processed by RNaseH.

A completely different type of initiation and origin recognition on circular templates operates in **rolling circle replication** (or σ replication; Fig. 14.**11**). Here priming is not mediated by RNA, but by a free 3′-OH end of DNA. This is the preferred mode of replication for plasmids of Gram-positive bacteria, for transfer replication during conjugation, and for some (small) phages. Double-stranded RFI form of ΦX174 and similar phages also replicates by this mechanism. The initiator protein introduces a specific cleavage at a site called *dso* (for **d**ouble-**s**trand **o**rigin). The resulting 3′-end is used as a primer for leading strand DNA synthesis, using the circular strand as a template and peeling off the free (+) DNA strand. Once replication passes a site called *sso* (for **s**ingle-**s**trand or lagging-strand **o**rigin), this site becomes activated on the (+) strand and lagging-strand synthesis can be primed with the help of a primosome. DNA synthesis can continue around the circular template several times. Rolling circle replication can therefore result in a long **concatemer** that contains several genomes, which are subsequently processed into monomers as in the case of phage λ (see Chapter 26.3). Plasmids normally process and circularize the monomeric (+) strand.

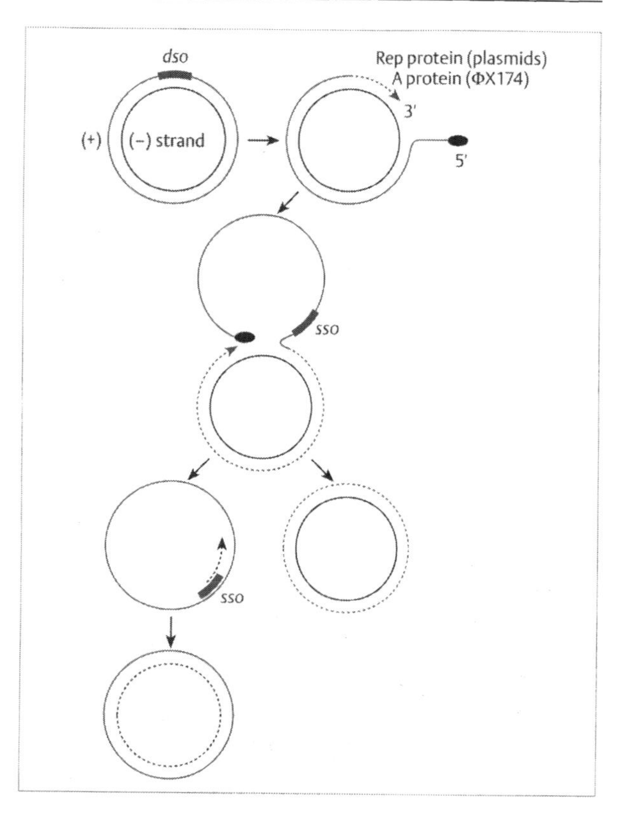

Fig. 14.**11** **The rolling-circle or sigma-type replication of plasmids.** Many double-stranded DNA plasmids (e.g., pT181) and phages (e.g., ΦX174) are cleaved by an initiator protein at a site *dso* (double-strand origin). The resulting 3′-end is used as the primer for leading strand synthesis, while the free 5′-strand (+strand) becomes the template for the lagging strand once a single-strand origin (*sso*) is in the single-stranded form and binds a primosome. Replication of the circular (−strand) can continue several times, resulting in long repeats or concatemers which are processed to monomeric form, e.g., during phage head packaging

Bacteria use all three modes of replication and initiation. Chromosomal replication occurs via the *oriC* type. This involves origin recognition by DnaA initiator protein, formation of a complex between *oriC* and DnaA that results in local unwinding and loading of DnaB helicase with the help of the DnaA primosome. *E. coli*, and probably other bacteria as well, have backup systems for replication called **stable DNA replication** (**SDR**) because the initiation of this type of replication is independent of the otherwise required de novo protein synthesis. **Constitutive stable DNA replication** (**cSDR**) is observed in mutants deficient in RNaseH. Stable **R-loops** are formed at several sites around the chromosome. **Induced stable DNA replication** (**iSDR**) is found upon induction of the SOS response (see Chapter 15.7).

Primers, in this case DNA primers, are provided by recombination intermediates. A homologous single-stranded DNA can invade a double-stranded DNA at a transiently formed bubble, mediated by RecA protein (see Chapter 15). The result is a **D-loop**, a triple-stranded DNA structure with a free 3′-OH end that can be used as a primer and be extended by DNA polymerase. A primosome is required to allow the replication of extended stretches of DNA. Only a few sites on the chromosome fulfill the structural requirements, such as closeness to a primosome loading site, to result in productive replication. Initiation via recombination intermediates is also used by some phages (e.g., phage T4 of *E. coli*).

Besides priming by RNA and by DNA, there is still a third possibility, priming by proteins. The *B. subtilis* phage ϕ29 is a paradigm for organisms with permanently linear DNAs. Its synthesis is primed by a protein at the end of the DNA molecules, to which the first nucleotide of the newly synthesized DNA strand is covalently attached.

14.1.4 In the Chromosomal Replication Fork, Several Proteins Cooperate To Synthesize DNA

As discussed previously, there is a basic difference between replication of the two strands in double-stranded DNA because of the polarity of DNA strands. The leading strand is synthesized continuously, whereas the lagging strand must be synthesized discontinuously by synthesis and joining of Okazaki fragments (see Fig. 14.**8**). The machine that is responsible for this synthesis, DNA polymerase III holoenzyme (Fig. 14.**9**) consists of a polymerase and proofreading part (**PolIII core**), a "**sliding clamp**" (β subunit), and "brace" or "**clamp loading**" proteins (the γ complex, $\gamma\delta\delta'\chi\psi$ subunits). The brace proteins recognize the 3′-end of the primer in an R-loop or D-loop, and load the sliding clamp. This can be the initial primer at the origin of leading strand synthesis, or RNA primers for Okazaki fragments in lagging strand synthesis. The sliding clamp is the subunit that is responsible for processivity and forms a ring around the DNA. This ring has a sixfold symmetry made up of two subunits in the case of *E. coli* DNA polymerase III β-subunit.

DNA polymerase III consists of the PolIII core for polymerizing and proofreading functions, the sliding clamp (β-subunit) for processivity, and the clamp loading proteins (γ-complex). It is an asymmetric dimer that simultaneously synthesizes the leading and lagging DNA strands (Fig. 14.**9**).

The brace and the sliding clamp load polymerase III core, and in the presence of dNTPs, the complex can synthesize long stretches of greater than 50,000 nucleotides at more than 500 nucleotides s^{-1}. The lagging strand is synthesized by the same enzyme complex. Once the synthesis of an Okazaki fragment is finished, the sliding clamp stays attached to the DNA, whereas the polymerase III core dissociates and reattaches to elongate the next Okazaki primer. Both leading and lagging strand DNA replication occur coordinately because of the dimerization of the two polymerase complexes. For this coordinate synthesis, the lagging strand must fold back in order to allow movement of both polymerase complexes in the same direction of the progressing fork (Fig. 14.**12**).

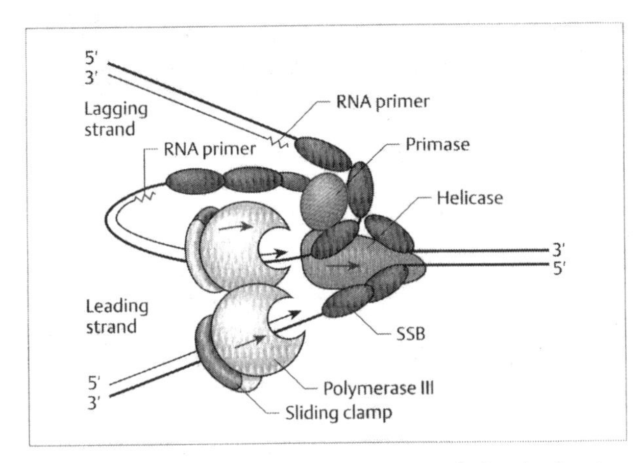

Fig. 14.**12** **The chromosomal replication fork.** The lagging strand is folded back, which allows the asymmetric DNA polymerase III dimer to perform a coordinated DNA synthesis on both strands. Symbols are like those in Fig. 14.**9** and in the text. SSB, single-stranded binding protein

14.2 Most Bacteria Have One Circular Chromosome

In most prokaryotes, the "house-keeping" genes, i.e., the essential and species-specific genes, are arranged in a linear sequence on a single covalently closed circular DNA called the **chromosome**. Although the genes are clustered on this macromolecule, the cytological term chromosome is not synonymous to the genetic term **genome**, which designates the sum of all genes of a cell whether located on chromosome(s), plasmids, or bacteriophages. Biochemically speaking, the chromosome corresponds to a large replicon.

14.2.1 How Genetic Maps of Bacteria Correlate With Chromosomes as Physical Elements

The first genetic map of a microorganism was the map of *E. coli*. It was obtained by conjugation and transduction analysis. Since longer distances were mapped by interrupted mating, which required about 100 min for the transfer of a full chromosome, the *E. coli* genetic map is still calibrated in (centi-)minutes. The genetic map is circular. In 1963, J. Cairns published an autoradiograph of a [^{3}H]-thymidine-labeled circular *E. coli* chromosome. The contour length of this circle was 1.3 mm, about 500 times the length of an *E. coli* cell. This proved for *E. coli* that the chromosome is circular, and that there is a single chromosome per cell. Only recently was it discovered that the large chromosomes of *Streptomyces* (7.8 Mb) and *Borrelia* (9.5 Mb) are linear, and that, for example, the genome of *Burkholderia cepacia* consists of three circular chromosomes (3.6, 3.2, 1.1 Mb).

Considerable effort was invested to correlate the physical and the genetic maps. However, with increasing sequence information, much of this information will soon be of merely historical interest. At the moment, the chromosomes of *Mycoplasma genitalium* (580 kb), *Mycoplasma pneumoniae* (816 kb), *Helicobacter pylori* (1.8 Mb), *Haemophilus influenzae* (1.83 Mb), and *Methanococcus janaschii* (1.67 Mb), *Bacillus subtilis* (4.2 Mb), *Enterococcus faecalis* (3.0 Mb) and *Escherichia coli* K-12 (4.7 Mb) are completely sequenced.

"Standard" chromosomes, like the ones of *E. coli* or *B. subtilis*, start replication from a unique **origin** (*oriC*) and replicate bidirectionally to the terminus. The **terminus region** is bordered by termination sites, which constitute binding sites for a termination protein. They are polar and allow replication forks to pass into the region, but not to leave it. More details about termination and the regulation of replication in the cell cycle can be found in Chapter 22. *Streptomyces* chromosomes also have a unique origin, but little more is known about the replication of these linear chromosomes.

14.2.2 Nucleoids and the Intracellular Organization of the Bacterial Chromosome

Bacterial chromosomes differ from their eukaryotic counterparts by their relatively low degree of protein-mediated condensation. Bacterial DNA is not organized into nucleosomes and is not surrounded by a membrane, but is located in a distinct area called the **nucleoid**. Bacterial DNA is more loosely covered with histone-like proteins, especially HU and H-NS (histone-like nucleoid structuring protein). Nucleoids can be isolated from cells by lysing the cells at high salt

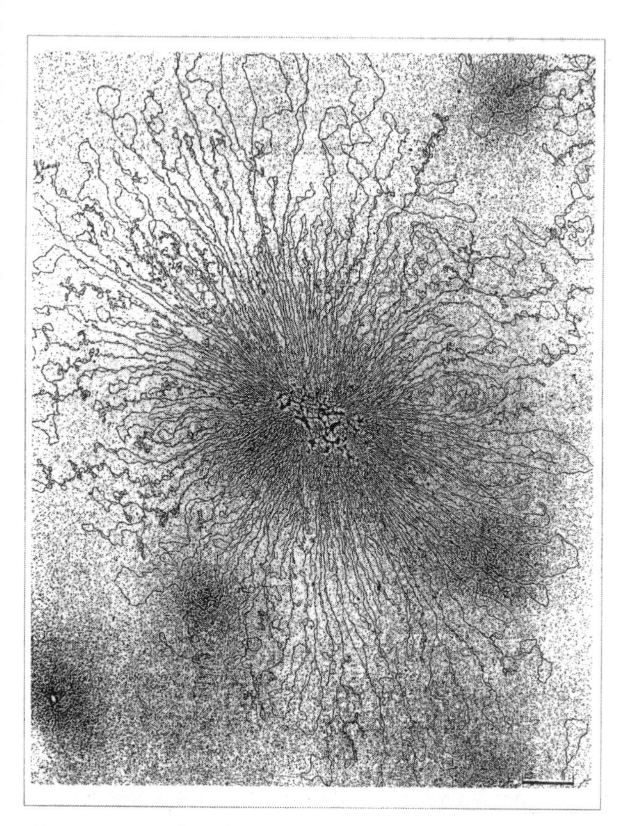

Fig. 14.**13** **Isolated nucleoid of *Escherichia coli*.** After isolation from the cells, nucleoids show loops of chromosomal domains, many of which are supercoiled. The nucleoid here was treated with detergent and spread carefully using cytochrome c. (Bar ≈1 μm) [from 1]

concentrations. Upon expansion of the DNA, a structure can be seen in electron micrographs with loops emerging from a central mass. The loops are mostly supercoiled, and it is assumed that the loops constitute separate **supercoil domains** of the chromosome (Fig. 14.**13**).

Within the cells, nucleoids are those cytoplasmic regions that are (relatively) free of ribosomes. Due to the size of the chromosome relative to the bacterial cell (1.3 mm to 3 μm, e.g., for *E. coli*), DNA in nucleoids is densely packed and is not easily penetrated by enzymes such as RNA polymerase or by ribosomes. Therefore, transcriptional activity is probably not associated with the bulk of the DNA in the nucleoids, but is restricted to its surface. We can imagine that the supercoil loops protrude into the region containing RNA polymerase and ribosomes, and that the entire structure is very dynamic.

Nucleoids are regions within bacterial cells in which DNA is densely packed. They are not surrounded by a nuclear membrane, and can be isolated by lysis at high salt concentrations.

The lack of a nuclear membrane around nucleoids has far-reaching consequences. Newly synthesized mRNA does not have to pass through a membrane. While it is synthesized, it is captured by ribosomes to form polysomes and is translated. This direct coupling of transcription and translation makes gene expression fundamentally different in prokaryotes and eukaryotes. The correlation between replication, nucleoid segregation, and cell division is described in Chapter 22.

14.3 Plasmids Are Independent Replicons

Naturally occurring plasmids range in size from a little more than 1 kb up to several hundred kilobasepairs. They are independent replicons; the vast majority of them are circular. The exceptions again are found in *Streptomyces* and *Borrelia* and in *Rhodococcus* species that contain linear plasmids.

Plasmids contain genes that are essential for plasmid maintenance, such as initiation and control of replication. They often code for facultative traits that give their hosts a selective advantage in special environments. Sometimes it may be difficult to discriminate large plasmids from chromosomes. By agreement, if such a large plasmid contains any housekeeping genes, it is called a chromosome, with the exception of the so-called F′ or R′ plasmids. They are the result of incorrect excision of **episomes** (plasmids that can integrate into the chromosome) from their chromosomal location and will be discussed in Chapter 16. Strictly speaking, the DNA of bacteriophages also corresponds to plasmids. For mainly historical reasons, they will, however, be treated separately in Chapter 26. Table 14.**3** lists some common plasmids of Gram-negative bacteria.

The nomenclature of plasmids is not uniform. Frequently, "p" for plasmid is followed by the initials of the person who constructed it and a number (e.g., pBR322) or by letters describing the function (e.g., pEX for an expression vector). For historical reasons, names for plasmids that encode genes for antibiotic resistance and fertility factors or colicins (e.g., R100, RSF1010, F, mini-F, ColE1) are used without the "p".

Plasmids are independent genetic units that carry facultative genes required in specific environments. Like the chromosome, plasmids are able to replicate autonomously. **Episomes** are plasmids that, in addition to autonomous replication, can exist integrated into the chromosome, the only lifestyle for transposable elements.

14.3.1 Plasmids Control Their Replication and Copy Number

Different plasmids use all the different initiation strategies that were already discussed. Independent of their mode of replication, they must control their copy number such that it is adjusted to the rate of chromosome replication: the copy number must be kept constant. For different plasmids, copy numbers can be very different, ranging from 1 per chromosome to more than 100, although for a given plasmid this number is relatively stable. All plasmids studied so far control their replication, and hence their copy numbers, by a negative feedback loop that operates at the initiation of replication. Three *trans*-acting elements that can act alone or in concert have been identified for negative feedback. One is a small countertranscript, an antisense RNA, that either directly inhibits initiation via interaction with the primer RNA, as in the case of ColE1-type plasmids, or that inhibits the translation of the initiator protein, usually called Rep (not to be confused with Rep

Table 14.3 **Common plasmids of Gram-negative bacteria**

Plasmid	Size (kb)	Copy number per chromosome	Characteristic markers[1]
ColE1	6	10–20	*cea*$^+$ (colicin E1)
pSC101	9	6	*tet*$^+$ (Tc)
R6K	38	10–20	*bla*$^+$ (Ap), *str*$^+$ (Sm), *tra*$^+$
PR4, RK2	60	5	*bla*$^+$(Ap), *aphA*$^+$ (Km), *tet*$^+$ (Tc), *tra*$^+$
R100	90	1	*aadA*$^+$ (Sm), *cat*$^+$ (Cm), *sul*$^+$ (Su), *tet*$^+$ (Tc), *mer*$^+$ (Hg), *tra*$^+$
F	90	1	*tra*$^+$

[1] Note that phenotypes of plasmid-encoded resistance markers are given in a two-letter code

helicase). The second element is a repressor of the *rep* gene. The third element is composed of a set of iterated short sequences, the same ones as in the origin, that bind Rep protein and thereby prevent its interaction with the origin (see Chapter 22 for additional information and figures).

Negative feedback control of replication initiation is the reason for plasmid **incompatibility**, the inability of two plasmids to be propagated stably in the same cell line. Plasmids with the same replication control or the same partition system are incompatible, whereas plasmids with a different initiation control are normally compatible. Additional genes coded by the plasmid may change, but incompatibility is a property of the basic plasmid replicon. Therefore it is usually used for classification: plasmids are classified into incompatibility groups, such as IncP (RP4, RK2), IncFI (F, ColV), IncFII (R1), and IncY (phage P1). To test for an incompatibility group, cells containing one plasmid of a known incompatibility group are transformed with the plasmid to be tested. After growth with selection for the second plasmid, the presence of the first plasmid is assayed. If it is still present, the two plasmids are said to be compatible. If they are in the same incompatibility group, the first plasmid is rapidly lost.

Plasmids with the same replication control are **incompatible**. They can only be maintained together in one cell with selection for both plasmids.

Partition and addiction systems. Plasmids ensure their maintenance in the population by special partition and addiction systems. Multiple genetic systems ensure that plasmids survive in bacterial populations. For multicopy plasmids, such a specialized system may not be required, but it is mandatory for low- or single-copy-number plasmids. Typically, plasmids, even plasmids with a copy number of one, are maintained in bacterial populations at a loss rate of $< 10^{-7}$ per cell division. Mechanisms that ensure this extraordinary stable inheritance of plasmids in bacterial populations act at three levels. Plasmids frequently carry several of these maintenance systems.

The products of θ-type replication are catenated rings. Dimeric or multimeric molecules can also be the products of recombination. All these multimers must be resolved into individual units that can be segregated individually to daughter cells. DNA topoisomerases of the host can exert this function. However, many plasmids and phages have specialized **site-specific recombination** systems for the efficient resolution of multimers.

True partition systems ensure that there is at least one plasmid at either side of the septum at cell division. Systems that have so far been analyzed comprise two *trans*-acting proteins and a *cis*-acting site on the DNA. For plasmid F (see Fig. 16.**3**), these are *sopA*, *sopB*, and *incD*; for phage P1, *parA*, *parB*, and *incB*; for R1, *stbA*, *stbB*, and *incA*. The *cis*-acting *inc* sites are the functional analogues of centromeres of eukaryotic chromosomes. How the *trans*-acting proteins act in equipartition is not known in detail. They probably mediate attachment of the *cis*-acting site on the plasmid to a special site on the cell membrane (see Chapter 22).

Addiction systems operate on the basis that a bacterium carrying a plasmid survives only as long as it retains the plasmid. In order to ensure that a cell dies if it is cured of a plasmid, plasmids produce a "poison" and an "antidote". The "poison" is relatively stable, whereas the "antidote" is a labile substance that blocks the action of the poison. When the plasmid is lost, the antidote decays and the remaining poison kills the cell. The biochemical basis for killing and for antidote actions can be different. In the case of plasmid F, both are proteins. In the case of plasmid R1, the antidote is a small antisense RNA that prevents expression of the mRNA for a toxic protein.

Plasmids can have very different host ranges. Some plasmids with narrow host range are confined to a single species. Other plasmids have a wide host range. RP4 can, for example (see Fig. 17.**1**), be transferred into practically all Gram-negative bacteria. Plasmids with narrow host range rely on the replication machinery of the host, whereas wide-host-range plasmids usually encode most replication proteins themselves. Genes, sites, and mechanisms responsible for horizontal distribution of plasmids by conjugation and mobilization are discussed in Chapter 16.

14.3.2 A Large Variety of Functions Are Encoded by Plasmids

Plasmids are made up of **modules**. All plasmids must have at least the **basic replicon**, consisting of a replication origin and functions that ensure and control replication. Many plasmids have one or more **partition systems**. Many of the larger (conjugative) plasmids code for **transfer functions** that allow horizontal transfer of the plasmid to other bacteria or even other organisms (see Chapter 16). These three modules are responsible for plasmid maintenance and propagation. There are natural plasmids, called "cryptic," that consist only of the basic replicon, and thus do not confer a visible phenotype to their hosts. Most plasmids, however, code for functions that allow their hosts to survive in a special environment. This property of plasmids contributes extensively to the physiological variability of bacteria.

Plasmids are made of modules. The module encompassing the basic replicon is always present. In addition, plasmids may contain partition modules, transfer modules, and modules coding for diverse biochemical phenotypes.

The unwise use of antibiotics in medicine and farming selected bacteria that are resistant to many (up to ten) antibiotics. This **antibiotic resistance** is often transmissible within a bacterial population and is caused by genes carried on resistance or **R-plasmids**. These resistance genes are often carried on **transposons**, which are mobile genetic elements that can be exchanged between different plasmids or between plasmids and the chromosome (see Chapter 14.4). A map of R100, a typical R-plasmid, is shown in Figure 14.**14**.

The mechanisms of resistance are very widespread. Genes such as *aphA* (*kan*, aminoglycoside phosphotransferase, Km, kanamycin resistance), *cat* (chloramphenicol acetyl transferase, Cm, chloramphenicol resistance), and *aadA* (*str*, aminoglycoside adenylyltransferase, Sm, streptomycin resistance) code for enzymes that modify the antibiotics by phosphorylation, acetylation, or adenylation. *bla* (Ap) codes for the enzyme β-lactamase, which degrades penicillins; Tc, coded by *tet*, alters the permeability of the cell membrane for tetracycline. Genes for Hg^{++} resistance code for an NADP(H)-dependent reductase that reduces Hg^{++} to the insoluble metal. In order to discriminate plasmid-encoded antibiotic resistance determinants from chromosomally encoded ones, a two-letter code is conventionally used for the phenotypes of plasmid determinants.

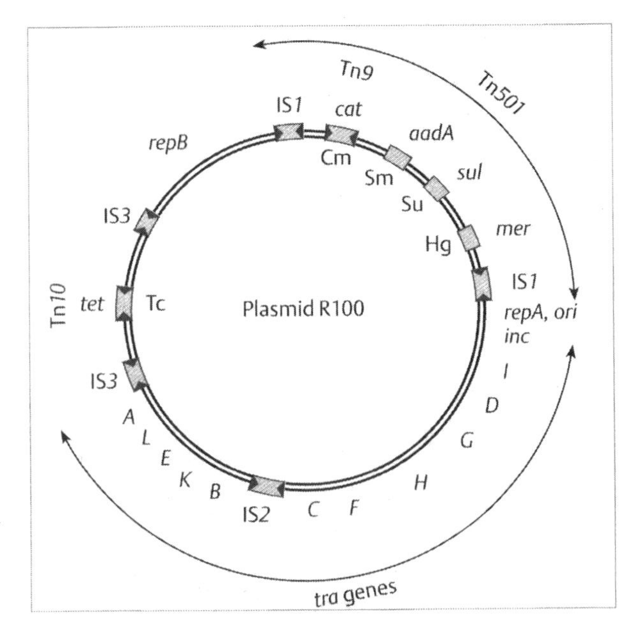

Fig. 14.**14** **Map of the resistance plasmid R100.** Genes are indicated on the outside and the phenotypes conferred by resistance genes (two-letter code) are given on the inside. For transposable elements (in red), arrows indicating inverted repeats are also given

Other plasmids of medical importance carry genes for **virulence** and **toxins**. *E. coli* can contain enterotoxins while other bacteria produce pili (fimbriae) as a virulence factor that mediate adhesion to host cells and are encoded by plasmids. These are just a few examples how plasmids modify medically important properties of their host bacteria (see Chapter 33 for details).

Many plasmids encode genes whose products are directed against other bacteria. **Bacteriocins** are proteins that are exported by some plasmid-carrying bacteria and kill the same or related bacteria. Bacteria that carry the bacteriocin-encoding ("bacteriocinogenic") plasmid are immune. Best known among bacteriocins are the colicins, produced by and directed against *E. coli*; the colicin-producing plasmid ColE1 is the ancestor of most plasmids used in gene technology. Different colicins act quite differently. Colicin E1 and others permealize the cytoplasmic membrane, colicin E2 leads to degradation of DNA, and colicin E3 leads to the degradation of ribosomal RNA. **Antibiotic production** is another way to exclude other bacteria from their own environment. Especially in *Streptomyces*, plasmids are found that code for the production of antibiotics. Host-controlled **restriction/modification systems** safeguard a cell against infection by phages or against the uptake of foreign DNA. Consequently, many plasmids encode restriction endonucleases and the cognate methyltransferases.

Degradative plasmids, frequently found in bacteria, govern the metabolism of a diverse group of substances, e.g., in *Pseudomonas* for aliphatic (octane, decane, etc.) and aromatic hydrocarbons (xylenes, toluene, naphthalene, etc.), including the products of their oxidative metabolism (salicylate, benzoate), and for terpenes and chlorinated hydrocarbons, including pesticides (see Chapters 9 and 34). Other bacteria carry plasmids that allow the degradation of more conventional carbon sources (e.g., sucrose, raffinose, and lactose in enteric bacteria). Such plasmids may encode a complete degradative pathway or only part of it, such that different microorganisms have to cooperate for complete degradation. The pathways are normally inducible, and the genes are organized in operons and regulons.

There are two kinds of interactions of plasmid-carrying bacteria with plants. *Agrobacterium tumefaciens* is a true parasite. It contains a large plasmid of about 500 kb, the **Ti-plasmid** (**t**umor-**i**nducing plasmid). Upon infection, about 10% of it (the T-DNA for **t**umor DNA) is transferred to the plant cells and induces them to produce amino-acid-like substances and to grow as undifferentiated callus ("tumor") cells. Since only the ends of this segment are required in *cis* for transfer (see Chapter 34), this plasmid is the basis for the transfer of foreign DNA into plants in gene technology.

Biological **nitrogen fixation**, mostly through symbiotic bacteria, is responsible for the conversion of about 10^8 tons of atmospheric nitrogen to ammonia per year. Twenty percent of it is contributed by *Rhizobium* species in the root nodules of legumes. The *nod/vir* genes for nodulation and the *nif* genes for the fixation of nitrogen are usually localized on a large plasmid, e.g., in the Rhizobiaceae (see Chapter 34).

Plasmids may have genes for antibiotic resistance, virulence, toxins, bacteriocins, production of antibiotics, restriction/modification systems, degradative pathways, tumor induction in plants, and nitrogen fixation.

Naturally occurring plasmids thus are responsible for a large part of the physiological variety found in bacteria.

14.4 Transposons Are Mobile Genetic Elements

Bacterial genomes, like other genomes, evolve by mutation and selection, by acquiring DNA from other bacteria (horizontal transfer) or by internal DNA rearrangements. The exchange of DNA between different bacterial cells will be discussed in Chapter 16. Another major cause for variation are **transposable** (or mobile) **DNA elements** which are able to move directly from one site to another site within the chromosome or onto extrachromosomal DNA within the same cell. A relationship between sequences at the donor and recipient sites is usually not required (**"illegitimate recombination"**). Some transposons, called **conjugative transposons** or CTn can move during the transposition process via a conjugation-like process to another cell (see Chapter 16). Transposable elements can have direct effects on the genome of their host by integrating into or close to genes, and thus interrupting the genes or placing them in a different regulatory environment. Recombination between two copies of the same type of transposable element can result in **deletions**, **duplications**, **amplifications**, **inversions**, and **transpositions** of segments of the genome (Fig. 14.**15**) or in the **fusion** of independent replicons. Transposable elements are the basis of many chromosomal mutations and are responsible for the high frequency of such mutation. Transposable elements can have different complexities ranging from simple IS elements to composite transposons.

Transposons are mobile DNA sequence elements that move directly from one site to another within the genome. Simple transposons (IS elements) have inverted sequences at their ends and code for an enzyme (transposase) that mediates transposition. Composite transposons have invertedly repeated ends, some in the form of IS elements and code for a transposase and additional genes.

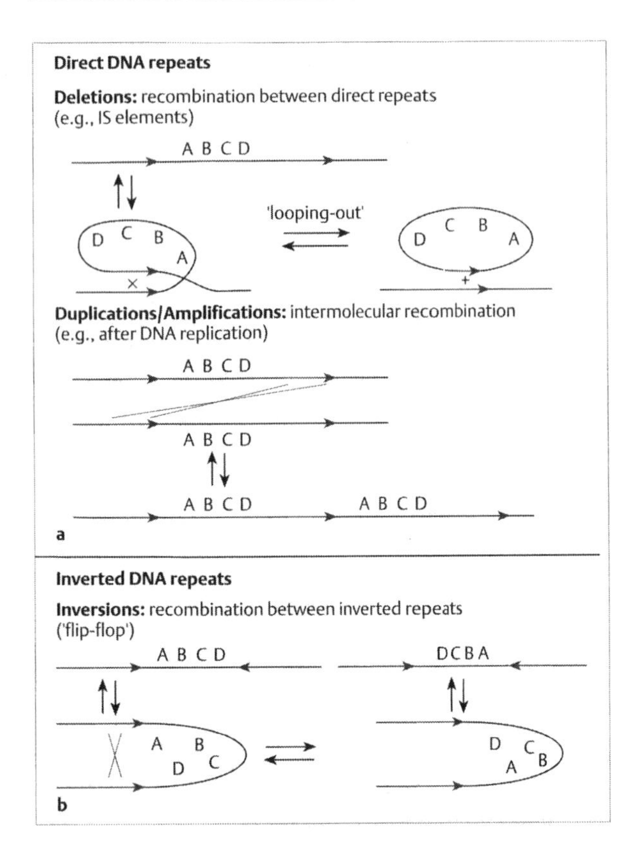

Fig. 14.**15a, b** **The genesis of chromosomal mutations and the role of transposable elements.**
a If by transposition (e.g., of IS elements) direct repeats (DR) of homologous DNA are generated, recombination can cause "looping-out" (deletion) or duplications, and through repeated recombination (e.g., with looped-out DNA rings or after DNA replication), gene amplification.
b In the presence of inverted repeats (IR), recombination causes the inversion (flip-flop) of DNA sequences located between the IRs

14.4.1 Insertion Sequences (IS Elements) Are the Simplest Transposable Elements

Insertion sequences are ubiquitous in prokaryotic cell and in their phages and plasmids. IS elements wer discovered around 1970 when P. Starlinger, H. Saedle and J. A. Shapiro analyzed an unusual type of mutatio (strong polar) in the *gal* operon of *E. coli*. There ar different IS elements, and bacteria normally contai several copies of the various IS elements in thei chromosome (Tab. 14.**4**). The transposition of suc elements to new sites is a rare event, and if the targe is within a gene, of course, this gene is inactivated. I elements range in size from approximately 750 t approximately 1550 base pairs.

The two ends of the elements carry nearly identica sequences in inverted orientation (inverted repeats o IR) that are required in *cis* for transposition. All I elements contain in their central part one open readin frame (*tnp*), except IS*1*, which contains two. The produc of this gene is called the **transposase**. It interacts wit the inverted ends during transposition. Transposon generate a short duplication (direct repeat) of a fev nucleotides at their target sites during transpositior The number of duplicated nucleotides is characteristi for each IS element; the actual sequence varies due t the (nearly) random selection of target sites (Fig. 14.**16** How this duplication is created will be discussed i Chapter 14.4.3.

14.4.2 Composite Transposons Carry Antibiotic Resistance or Other Markers in Addition to the Functions Required for Transposition

Composite transposons are named **Tn**, usually followe by a number, e.g., Tn*1*, Tn*5*, and Tn*10*. Transposons wit antibiotic resistance genes are commonly found in R plasmids, (e.g. R100 in Fig. 14.**14**) which may contai

Table 14.**4** **Common insertion sequence elements (IS) of *Escherichia coli***

IS element	Length (bp)	Inverted terminal repeats (bp)	Direct target repeats (bp)	Target selection	Copies per *E. coli* genome
IS*1*	768	23	9	random	5–10
IS*2*	1327	41	5	hot spots	4–8
IS*3*	1258	40	3	hot spots	5
IS*5*	1195	16	4	hot spots	10–12
IS*10*	1329	22	9	NGCTNAGCN	2
IS*50*	1531	9	9	hot spots	0

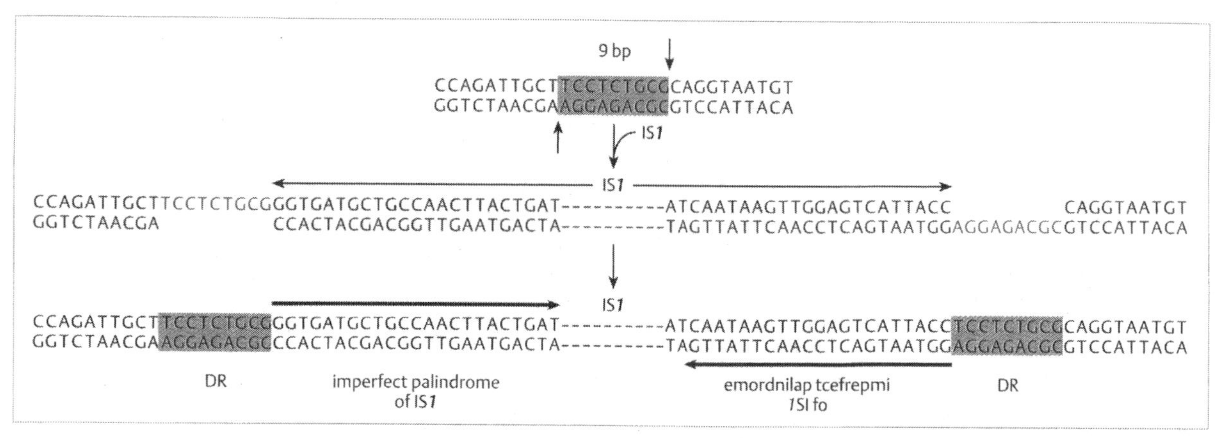

Fig. 14.16 **IS*1* insertion site within gene *galT* of *Escherichia coli*.** The IS*1*-encoded transposase (TnpA) causes at any target DNA, a staggered nick (9 bp), which is torn apart. The IS*1* element integrates between the free ends, and the single-strand stretches are filled by new synthesis, thus causing direct repeats (DR, in red) flanking the inserted IS*1*. Note the (imperfect) palindrome at the end of IS*1*; this palindrome is involved in integration/excision (see text)

several different transposons. The transposons are one reason for the fast acquisition of multiple drug resistance by medically important bacteria.

Composite transposons are categorized in two classes. **Class I transposons** have insertion elements at their ends. Two closely located IS elements can transpose any marker located between them. If such a marker is a drug resistance gene, selection for such a composite element can be high. The two IS elements can have either a direct or an inverted orientation (Fig. 14.**15** and Tab. 14.**5**). In many transposons the IS elements are not completely identical, and only one of them may code for a functional transposase (Tab. 14.**5**). This helps to control the frequency of transposition. Class I transposons use the nonreplicative mechanism for transposition.

The frequency of transposition has to be carefully controlled and depends on the availability of transposase. Different transposons use different strategies. In the case of **Tn*5***, both IS*50* elements are transcribed from promoters that are close to the outside ends. In IS *50*L (the left insertion element), a GAA codon within the potential reading frame is changed to a TAA codon. Therefore no transposase is synthesized from the left element; instead, the promoter is used to drive the kanamycin resistance gene, *aphA*, which encodes aminoglycoside phosphotransferase. Two proteins are expressed from IS*50*R (the right insertion element) from the same reading frame. One protein is 55 amino acids shorter at the N-terminus and inhibits transposition, presumably by forming a complex with the longer transposase and the outside end of Tn*5*. Tn*5* is also subject to regulation of transposition by Dam methylation similar to the way that Tn*10* is regulated.

Tn*10* (Fig. 14.**17**) is subject to multiple controls. IS*10*R is the active module. There are two promoters close to the outside end of IS*10*R. The promoter transcribing inward, P_{in}, is responsible for the expression of transposase. The outward promoter P_{out} is much more active. There are about 30 times more P_{out}

Table 14.5 **Examples of composite bacterial transposons**

Transposon	Length (bp)	Terminal structure	IS orientation	IS relationship	Functional IS	Resistance gene for:	Transposition mechanism
Class I							
Tn*5*	5820	IS*50* R/IS*50* L	Inverted	1-bp change	IS*50* R	Kanamycin	Nonreplicative
Tn*9*	2650	IS*1*	Direct	Identical	Both	Chloramphenicol	Nonreplicative
Tn*10*	9300	IS*10* R/IS*10* L	Inverted	Differ 2.5%	IS*10* R	Tetracyclin	Nonreplicative
Tn*903*	3094	IS*903*	Inverted	Identical	Both	Kanamycin	Nonreplicative
Class II							
Tn*3*	4957	38-bp inverted repeats				Ampicillin	Replicative
Tn*1000* ($=\gamma\delta$)	5981	38-bp inverted repeats				None	Replicative

transcripts than P_{in} transcripts, and the two transcripts overlap by 36 base pairs, resulting in antisense RNA regulation. The transposase acts preferentially in *cis*. This *cis*-preference, a common feature of transposases coded by IS elements, limits the frequency of transposition to the set value if more copies of, for example, Tn*10* are present in the cell. Dam methylation is an important regulation for Tn*10* transposition. Dam methyltransferase methylates the A in the recognition sequence GATC. Dam methylation of *oriC* is also required for the initiation of replication of Enterobacteriaceae and plays an important role in mismatch repair (see Chapter 15.6.1). Shortly after replication, DNA is hemimethylated. Hemimethylation activates the transposase (P_{in}) promoter and binding of transposase to IS*10* ends. Once these sites are completely methylated, transposition is blocked. Dam methylation thus provides a window for transposition. The window is open shortly after replication. For a transposon that transposes nonreplicatively by cutting the transposon out of the donor sequence, such a mechanism ensures that a complete daughter strand remains for survival of the cell.

Class II transposons also have inverted sequences at their ends, but these are not IS elements. As an example of transposons of the Tn*A* family, Tn*3* has genes for three proteins: *bla*, the gene for ampicillin resistance; *tnpA*, coding for transposase; and *tnpR*. *tnpA* and *tnpR* are divergently transcribed. Between them are the promoters for both genes and a site called *res*. The product of *tnpR* has a dual function. By binding to *res*, it represses both the transposase and its own gene. Class II transposons are transposed using the replicative pathway. Therefore, transposition products have to be resolved (see below). This is done by TnpR at the *res* site using site-specific recombination. TnpR is therefore also called resolvase.

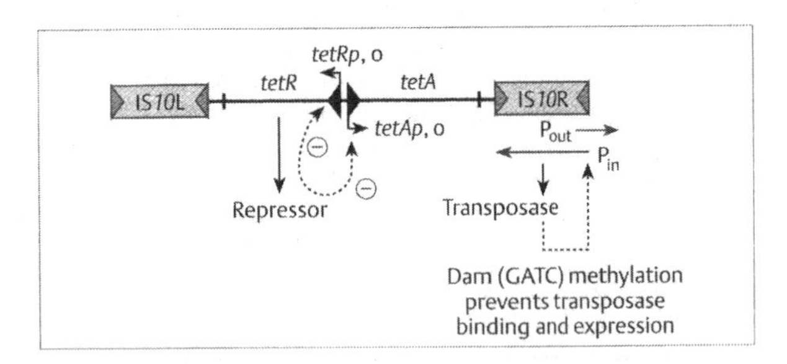

Fig. 14.17 **Map of transposon Tn*10*.** Tn*10* is flanked by the two IS elements IS*10* L (left) and IS*10* R (right); the transposase is encoded solely by IS*10* R. Expression of the transposase gene is regulated by two promoters (P_{in} and P_{out}) and antisense RNAs in a complex way. Transposase can bind only to hemimethylated IS*10* DNA, i.e., immediately after replication. The repressor TetR (gene *tetR*) controls the expression of resistance gene *tetA* (encodes an efflux protein) and auto-controls its expression

Class I transposons have complete IS sequences at their ends. They transpose nonreplicatively by a cut-and-paste mechanism. **Class II transposons** have short (38-bp) inverted repeats. The transposase is encoded in the internal sequence. They transpose by the replicative mechanism with a cointegrate as an intermediate.

14.4.3 The Mechanism of Transposition Can Be Replicative or Nonreplicative

Phage **Mu** is a transposon that can be packaged into a phage coat. Mu can be used to illustrate the two modes of transposition for transposable elements, nonreplicative and replicative. The transposition mechanisms for other transposons are most likely similar, using either the replicative or the nonreplicative pathway. The initial reactions are similar, and both modes share a common intermediate (Fig. 14.**18**).

Upon infection, Mu integrates into the chromosome by nonreplicative transposition. During the lytic cycle, the number of Mu molecules is amplified by replicative transposition. The common reaction is initiated by four single-strand cleavages by transposase. Two site-specific cleavages occur at the ends of the transposon in the donor molecule. The target sequence is cleaved at sites staggered by a small number of bases, five in the case of Mu. This results in the small duplication of target sequences characteristic for all transposition events. The 5′-ends of the target sequence are ligated to the 3′-ends of the transposon, resulting in the crossover-shaped structure shown in Figure 14.**18**. This structure can be resolved by two possible modes.

In the nonreplicative mode, the single strands are cut at the positions indicated by the small arrows in Figure 14.**18**, and the small gaps are filled by repair synthesis. The result of the nonreplicative transposition mechanism is a target DNA molecule containing the transposon and the small duplication at the target site, and a donor molecule containing a gap at the original position of the transposon, i.e., cut-and-paste transposition. In the presence of a complete sister strand, the gap in the donor DNA can be repaired.

In the replicative transposition mode, the 3′-ends of the target DNA in the crossover-shaped structure are used as primers for DNA synthesis. DNA synthesis proceeds through the transposon, and the new DNA is ligated to the 5′-end of the donor DNA. The result is a **cointegrate** (Fig. 14.**16**) that must be resolved, either by

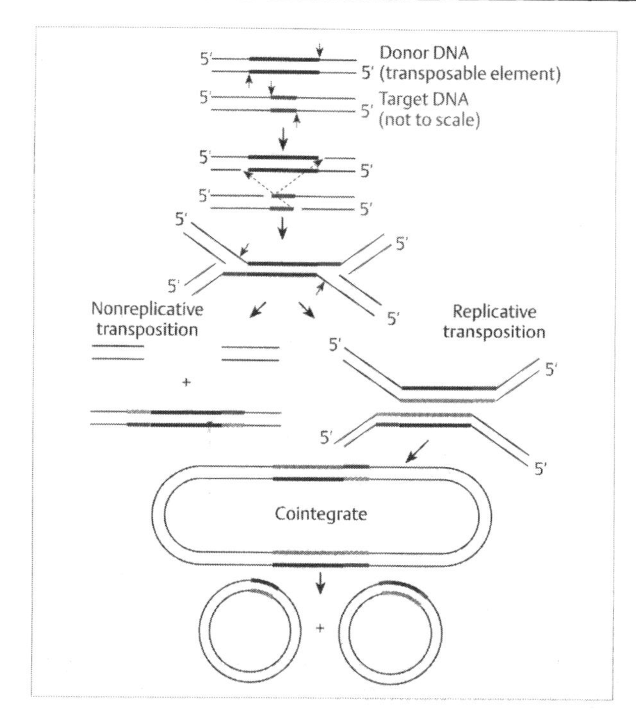

Fig. 14.**18** **Mechanism of nonreplicative and replicative transposition.** The model is that of phage Mu. Transposition begins with staggered nicks on both sides of the transposable element (donor DNA) and of the target DNA (see Fig. 14.**16**). Through crossing over, religation, and filling in of the single-stranded direct repeats in the target DNA, nonreplicative transposition is achieved. If transposition is accompanied by a DNA replication step, a cointegrate is formed. This cointegrate is normally resolved by a transposon-encoded resolvase at a *res* site, thus leaving donor and target DNA with a transposable element, the latter flanked by direct repeats. Since during nonreplicative transposition the cutting out of the transposable element from the donor DNA is within the direct repeat sequences, the duplication is reverted and leaves intact donor DNA

the transposon-encoded resolvase or, less efficiently, by homologous recombination by host enzymes.

IS elements provide regions of homology all over the *E. coli* chromosome that can be used by, for example, the F plasmid for integration (see Chapter 16). IS elements and transposons provide a flexibility to genomes; this is a very important feature in the evolution of chromosomes and of plasmids. They are instrumental for maintenance of the "**collective genome**" of bacterial species, i.e., the sum of all genes available to a species. The collective genome comprises the genes present in an individual cell, and all genes that can be made available by horizontal gene transfer. A cell of *E. coli*, for example, carries only about 80% of the genes found in the various strains of this species. The collective genome, hence, is larger than the individual genome and defines the complete species at the genetic level.

Further Reading

Berg, D. E., and Howe, M. M., eds. (1989) Mobile DNA. Washington, DC: ASM Press

Drlica, K., and Riley, M. (1990) The bacterial chromosome. Washington, DC: ASM Press

Holloway, B. W. (1993) Genetics for all bacteria. Annu Rev Microbiol 47: 659–684

Kelman, Z., and O'Donnell, M. (1995) DNA polymerase III holoenzyme: structure and function of a chromosomal replicating machine. Annu Rev Biochem 64; 171–200

Kittell, B. L., and Helinski, D. R. (1993) Plasmid incompatibility and replication control. In: Clewell, D. B. (ed) Bacterial conjugation. New York: Plenum Press

Kleckner, N. (1990) Regulation of transposition in bacteria. Annu Rev Cell Biol 6: 297–327

Kornberg, A., and Baker, T. A. (1992) DNA replication. New York: Freeman

Marians, K. J. (1992) Prokaryotic DNA replication. Annu Rev Biochem 61: 673–719

Neidhardt, F. C., Curtiss III, R., Ingraham, J., Lin, E. C. C., Low, K. B., Magasanik, B., Reznikoff, W. S., Riley, M., Schaechter, M., and Umbarger, H. E., eds. (1996) *Escherichia coli* and *Salmonella*, 2nd edn. Washington, DC: ASM Press

Skarstad, K., and Boye, E. (1994) The initiator protein DnaA: evolution, properties and function. Biochim Biophys Acta 1217: 111–130

Sonenshein, A. L., Hoch, J. A., and Losick, R., eds. (1993) *Bacillus subtilis* and other gram-positive bacteria: biochemistry, physiology, and molecular genetics. Washington, DC: ASM Press

Source of Figure

1 R. Kavenoff and B. C. Bowen (1976) Chromosoma 59: 89–101

15 The Genetic Information

Genetic information is stored in the DNA sequence of the chromosome or in extrachromosomal elements such as plasmids. Some viruses have genomes made from RNA. Early concepts of molecular biology were based on a clear separation between accumulation, storage, and reproduction of genetic information mediated by nucleic acids, and the utilization of genetic information mediated by proteins. The synthesis of functional proteins programmed by nucleic acid sequences is called **gene expression**. A unidirectional flow of information from DNA to RNA to protein was proposed as the central dogma of molecular biology after the role of DNA was recognized. Since then, the underlying molecular mechanisms of many genetic phenomena have been unraveled. A process called reverse transcription, first discovered in retroviruses, was recently also found in myxobacteria and *Escherichia coli*, and it was established that the flow of genetic information is not unidirectional from DNA to RNA. Instead, RNA can be copied into complementary DNA. Furthermore, self-splicing introns in eukaryotic mRNA provided the first demonstration that a purified RNA together with metal ions can catalyze a chemical reaction, a function that was previously only assigned to proteins. Thus, for RNA, there is no clear distinction between molecules carrying genetic information and those assuming a function.

Chromosome denotes a high-molecular-weight DNA. The term originates from cytological observations in eukaryotic cells. Bacterial cells contain generally one chromosome. **Genome** refers to the collection of all genes of an organism, regardless of whether they are located on the chromosome or a plasmid.

The organization and utilization of genetic information is fundamentally different in the three cellular kingdoms: Eukarya, Bacteria, and Archaea (see Table 29.**5**). The genetic information of Eukarya consists of mostly genes in which coding sequences called exons are interrupted by non-coding sequences called introns. Transcriptional units contain typically only a single gene and are produced in the nucleus, where they are assembled in spliceosomes. These are multi-component ribonucleoprotein complexes, in which the introns are removed, the exons are ligated together, and the processed mRNA is transported into the cytoplasm. Additional genetic variation is possible during these processes by alternative and *trans*-splicing as well as RNA editing in some organisms. Translation of the mRNA occurs then in the cytoplasm, from where the proteins are transported to their destination compartment.

In Bacteria, many transcriptional units encode several genes, which typically do not contain introns. Transcription and translation are coupled processes that take place in the cytoplasm (Fig. 15.**1**). Correct translation of a downstream gene in an operon depends often on successful translation of the upstream gene(s). Expression of genetic information in the Archaea contains, as far as it has been studied to date, mixed features from Eukarya and Bacteria. In this chapter, the

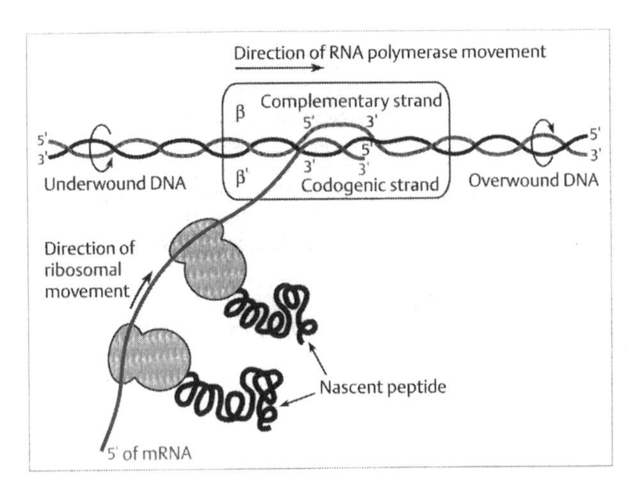

Fig. 15.1 **Transcription and translation are coupled in bacteria.** The DNA (thin black lines) is melted by RNA polymerase (large box) at the site of transcription. Movement of the RNA-polymerase-induced transcription bubble along the DNA leads to overwinding in front and underwinding behind the site of transcription. The nascent RNA (red line) is occupied by translating ribosomes, which expose the growing peptide chains (thick black lines). As a result of this coupling, free RNA occurs only when non-translated RNA (mostly rRNA and tRNA) is synthesized

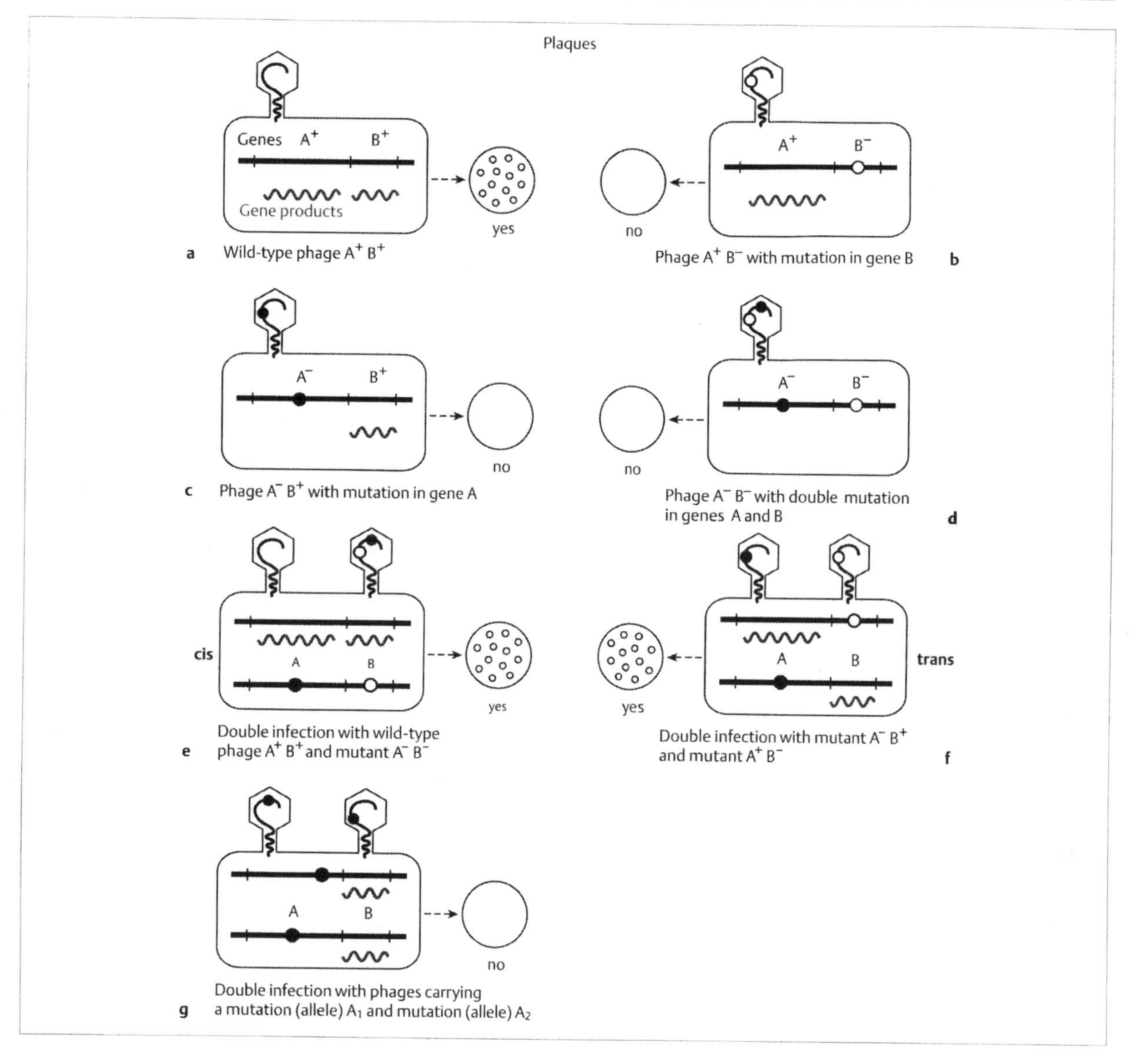

Fig. 15.2a–g **The *cis-trans* complementation test.** A schematic description of the outcome of various single (at a multiplicity of infection <1) and double infections with wild-type and mutant phages are shown. Assume that products of genes A and B are essential for lysis of the host.

a The wild-type (A^+B^+) phage lyses the host cell as indicated by the plaques on the plate.

b A phage with a mutation in gene B (A^+B^-) cannot lyse the host.

c A phage with a mutation in gene A (A^-B^+) cannot lyse the host.

d A phage (A^-B^-) with two mutations, one in A and the other in B cannot lyse the host.

e A double infection with wild-type and double mutants A^-B^- yields lysis of the host because functional products of genes A and B are present in the host cell. The phages in the plaques are of the parental strains, but only about 50% of them are wild-type and would be able to lyse a host in a following single infection.

f A double infection of the host with phages A^-B^+ and A^+B^- leads to lysis of the host. Progeny phages are either A^-B^+ or A^+B^-, none of them would be able to lyse a host in a following single infection.

g A double infection with phages A1 and A2 carrying different mutations (=alleles) in gene A. This combination is unable to lyse the host unless intragenic recombination in gene A occurs focus will be on the organization and expression of genes in Bacteria.

Transcription and **translation** are coupled processes in bacteria. Complete transcription often requires concomitant translation. Some regulatory mechanisms of transcription make use of this feature.

To understand the genetic information stored in nucleic acids, the organization of genes and operons will be discussed first, followed by a description of the details of gene expression and finally a discussion of the stability of genetic information resulting from the counterbalance of spontaneous mutations and repair.

15.1 The Organization of the Genetic Material in Bacteria

15.1.1 The Genetic Information Is Organized in Genes

The basic observations that led to modern genetics were already made by Gregor Mendel in 1865 and were based on the quantitation of the inheritance of properties after plant crossings. Detailed understanding of the processes underlying these observations was facilitated after the powerful techniques of bacterial and phage crossings were developed (see Chapter 26).

Some phages (see Chapter 26) display plaque morphologies that differed from that of the wild-type phage. These phages were **mutants**, and plaque morphologies were studied after cotransfecting independently isolated mutants into the same host cell. Some of these combinations gave wild-type plaques with a yield of 100%, whereas others yielded wild-type plaques only at a very low frequency. When the phenotype of the progeny of the individual plaques was analyzed, the progeny of those that yielded only wild-type plaques in the combinations retained their mutant phenotypes, whereas progeny of the wild-type phages occurring at a low frequency retained their wild-type phenotypes. Thus, the latter must have undergone sequence changes in their DNA, called **recombination**, whereas the former have not. The ability to nevertheless give a wild-type phenotype is called **complementation**. Successful complementation indicates in most cases that the mutations in the two genomes are located in independently expressed genetic units. However, intragenic complementation is also possible when the encoded protein forms oligomers. On the other hand, all mutations failing to yield a wild-type phenotype must be in the same genetic unit. This experiment became known as the ***cis-trans* test** (Fig. 15.2) and defines independent genetic units called **cistrons** or **genes**. The term gene includes **regulatory sequences** such as operators, which are typical *cis*-acting genetic elements. They exert their effect only on the gene(s) located next to them (*cis*-dominance), but may appear as *trans*-acting elements when they titrate the corresponding *trans*-acting protein. The difference between a **gene** and a **cistron**, also sometimes termed a structural gene, is that a cistron always encodes a diffusible product that can act in *trans*, whereas the term gene also covers genetic elements that act only in *cis*.

The *cis–trans* test defines **genes** by complementation. Structural genes encode diffusible products and are called **cistrons**. A gene also includes regulatory sequences able to titrate regulatory proteins.

15.1.2 Cistrons Can Be Cotranscribed in Operons

It is a characteristic feature of bacteria that several cistrons can be transcribed (see Chapter 15.2) into the same mRNA. This organization is called an **operon**, and such an mRNA is **polycistronic** (Fig. 15.3). Polar effects often occur because expression of the products encoded by downstream cistrons is dependent on the expression of cistrons located upstream on the mRNA.

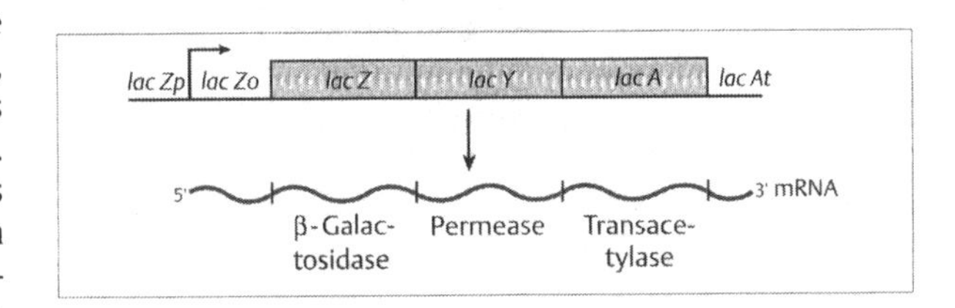

Fig. 15.3 **Genetic structure of the *lac* operon from *E. coli*.** The DNA with the open boxes indicating functional sequences (not drawn to scale) is indicated. *lacZp* denotes the promoter where initiation of transcription takes place; *lacZo* marks the operator where lactose-dependent regulation is exerted; the *lacI* gene encoding the LacI repressor is located upstream of the *lac* operon (not shown). Three genes, *lacZ*, *lacY*, and *lacA*, comprising the operon, encode the enzymes *β*-galactosidase, *β*-galactoside permease, and thio-*β*-galactoside-transacetylase, respectively. *lacAt* denotes the terminator of transcription defining the 3′-end of the mRNA (wavy line). Note that promoter, operator, and terminator act only in *cis*

Polarity describes the observation that a mutation in one cistron can affect the expression of another cistron located polar, i.e., downstream, in the direction of transcription on the same mRNA.

15.1.3 Most Cistrons Encode Proteins and a few Encode an RNA as the Functional Product

The expression of genetic information occurs in two reactions (Fig. 15.**1**): **transcription** yields an RNA complementary to the coding strand (also called codogenic or matrix strand) of the DNA. The produced RNA can either be the final product (e.g., rRNA, tRNA, or antisense RNA) or it can serve as an mRNA template for protein biosynthesis. **Translation** uses mRNAs to convert the information of a nucleic acid sequence into the linear amino acid sequence of the encoded protein. The rule for this conversion is called the **genetic code** (see Chapter 15.3). Nucleic acid triplets specify amino acids and the signals to start and stop protein synthesis. In order to carry out a function, the linear primary amino acid sequence of the protein has to be converted to the active tertiary or quaternary structure (see Chapter 19). The information for folding into the active structure is implicated in the sequence of amino acids.

The higher the molecular weight of a protein, the smaller is the chance that it folds into the active structure all by itself.

Box 15.1 The term *cis* dominance describes the property of a genetic element to act only on the loci to which it is connected. A typical example for a *cis*-dominant element is the operator of the *lac* operon needed for repression of *lacZYA* expression by the LacI repressor (*lacI*). Analysis of this property requires **merodiploid** strains in which two alleles for example, of *lacZ*, are present. Such strains can be constructed using plasmids. Assume that a mutation *lacZ1* located on the *E. coli* chromosome expresses a truncated, non-functional β-galactosidase, which can be detected by β-galactosidase-specific antibodies. In combination with a *lacZo*-(Con), i.e., operator constitutive, mutation, this truncated protein is constitutively expressed, (detectable by immunological methods) but the cell is deficient in β-galactosidase activity. The mutation defines a *lac* operator that cannot be recognized by the LacI repressor, which is itself constitutively expressed from *lacI*. Conjugation of an F′-plasmid carrying the wild-type *lacZo*$^+$ *lacZ*$^+$ alleles into this cell yields constitutive expression of *lacZ1* and inducible expression of *lacZ*$^+$. Thus, the Lac repressor is active in *trans* on the *lacZo*$^+$ *lacZ*$^+$ alleles on F′, but the *lacZo*(Con) operator is *cis* dominant, i.e., *lacZo* does not affect *lacZo*(Con) and vice versa.

In another merodiploid situation, *lacZo* may appear as *trans*-acting. When *lacZo* is present on a multi-copy plasmid such as pBR322, it competes with the chromosomally located *lacZo* for LacI repressor binding. Since the LacI repressor, like many other regulatory proteins, is only expressed at a low level, there are not enough LacI repressor molecules in the cell to saturate all *lacZo* sites. Thus, *lacZo* on a multi-copy plasmid is able to titrate LacI repressor away from the chromosomal *lacZo*, leading to depression of β-galactosidase. The resulting phenotype could be misinterpreted as if *lacZo* were acting in *trans*, but the phenotype results from the multi-copy artifact.

15.2 Transcription: the Synthesis of an RNA Complementary to the Codogenic DNA Strand

All bacterial cells contain at least three classes of RNAs. The most abundant class (up to 80%) contains the **ribosomal RNAs (rRNA)**, which are constituents of ribosomes (see Chapter 15.3.1), and the RNA present in other ribonucleoprotein complexes, such as the 4.5S RNA, which contributes to protein export in *E. coli* and *Bacillus subtilis*. Another as of yet only rarely found RNA species is present in some myxobacteria and *E. coli* strains and consists of a DNA–RNA hybrid made by reverse transcriptase. The function of this hybrid is unknown. The second class of RNA contains the **transfer RNAs (tRNA)**, which link the codons to their respective amino acids (see Chapter 15.3). RNA from both classes is usually processed and chemically modified after transcription.

The third class is the most diverse and contains the **messenger RNAs (mRNAs)**. mRNA is a transiently required template that provides the link between the genetic information contained in DNA and protein biosynthesis.

All RNAs in the bacterial cell are synthesized by **transcription**. In most cases, portions of the DNA are transcribed separately and independently of each other. Some phages can, however, produce a single transcript from their entire genome. Transcriptional units can comprise one or, in operons, several cistrons. Transcription is the DNA-template-directed polymerization of ribonucleoside triphosphates to a polymeric ribonucleic acid, whereby a pyrophosphate is released for every phosphodiester formed. Like any other synthesis of a biological macromolecule, transcription proceeds in three functionally distinct steps: initiation, elongation, and termination. Transcription is catalyzed by DNA-dependent RNA polymerase (RNP). As the nascent mRNA emerges from the synthesizing protein–nucleic acid complex, the signals for protein biosynthesis are exposed and are recognized by the translation machinery, and protein biosynthesis takes place on the nascent mRNA. This direct coupling of transcription and translation is typical for gene expression in bacteria. Thus, the growing mRNA is incorporated into **polysomes** and becomes part of a very complex multicomponent ultrastructure that moves with respect to the DNA template (Figs. 15.**1** and 15.**4**). Severe topological consequences for the usually supercoiled template DNA result from the limited rotational freedom of the nascent mRNA in that complex: the template is overwound in front of, and underwound behind the locus of transcription. As a result, the overall efficiency of transcription depends also on the superhelical status of the template and on the ability of the cell to remove the introduced superhelicity efficiently from the template.

15.2.1 Initiation of Transcription Is Linked to Promoter Structure and Function

Initiation of transcription takes place at sequence elements in the DNA called **promoters**, which are recognized by DNA-dependent RNA polymerases. These enzymes can be quite different, depending on whether they originate from eukaryotic, archaebacterial, or bacterial cells, or their respective phages. In this chapter, only the bacterial RNA polymerases will be described. They have been studied in a number of organisms and consist of a **core RNA polymerase** containing four polypeptides, two α subunits, one β, and one β' subunit, resulting in the $\alpha_2\beta\beta'$ quarternary structure. To initiate transcription, the core RNA polymerase must contain an additional subunit, called **sigma**. This subunit confers the ability to recognize specifically and to bind the initiation site of transcription, the promoters. The $\alpha_2\beta\beta'\sigma$ form is termed the **holo-RNA polymerase**. Bacteria contain a number of different σ-factors and, hence, an equivalent number of unique promoter consensus sequences. Groups of genes preceded by the same promoter can be co-regulated via expression or activation of their corresponding σ-factors, constituting an important global regulatory mechanism (see Chapter 20). This phenomenon is well studied in the sporulating bacterium *Bacillus subtilis*, where at least ten different σ-factors have been identified (see Chapter 25). A number of different σ-factors are also active in *E. coli* for example, the stress-induced genes are transcribed by RNA polymerase containing a 32 kDa σ-factor, called σ^{32} (RpoH), genes expressed in the stationary phase are transcribed by RpoS-containing RNA polymerase, and genes involved in nitrogen assimilation are transcribed by RNA polymerase with a σ-factor of 54 kDa mol. mass called σ^{54} (RpoN).

Sigma subunits confer sequence specificity to bacterial DNA-dependent RNA polymerase. They recognize promoters in front of genes needed to be expressed for a global task. For example, σ^{70} (RpoD) is needed for vegetative gene expression, σ^{54} (RpoN) is needed for genes encoding proteins for nitrogen assimilation and other special traits, σ^{32} (RpoH) is needed for transcription of heat shock genes, σ^{S} (RpoS) is needed for transcription of stationary phase genes.

σ^{70} (RpoD) is the most abundant σ-factor in *E. coli* where it is responsible for recognition of the most commonly found "vegetative" promoter sequence defined by the consensus elements TTGACA at −35 and TATAAT at −10 base pairs upstream of the transcriptional start site, the **−35 and −10 segments**. Sequence conservation is also observed for a region upstream of the −10 segment, called the extended −10 region, and a stretch of generally AT-rich sequence around the −43 position, called the **upstream activator region** (Fig 15.**4**).

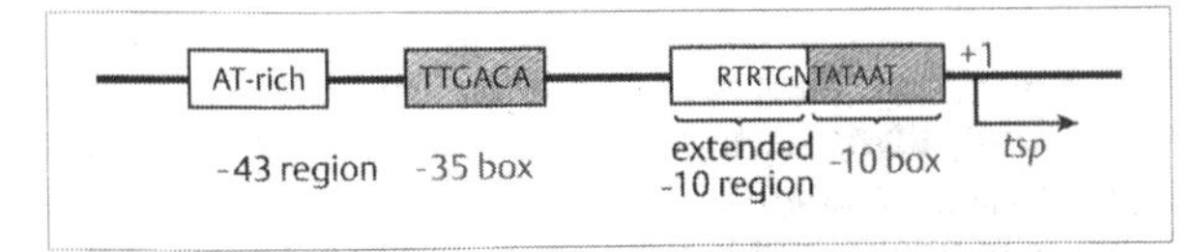

Fig. 15.**4** **Sequence features of a vegetative *E. coli* promoter.** The transcription start (*tsp*) is denoted +1 (in red) Conserved nucleotides (= consensus sequences) in many *E coli* and other bacterial promoters are indicated together with their designations (see also Fig. 15.**5**)

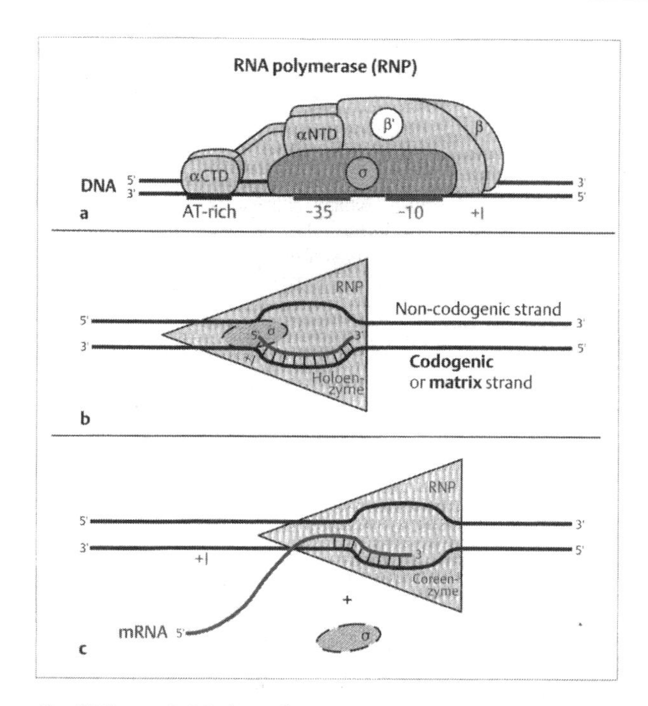

Fig. 15.**5a–c** **Initiation of transcription.**
a Schematic depiction of RNA polymerase (RNP) interacting with the promoter. Subunits β and β' are at the forward edge and σ recognizes the -10 and -35 promoter elements. The α-subunits are shown with their two functional domains. The N-terminal domains (αNTD) contact the other subunits, while the C-terminal domains (αCTD) are in contact with the AT-rich sequence element near base -43, which often occurs in strong promoters.
b Schematic drawing of the open complex, in which the DNA around the start site is melted and the RNA polymerase is able to synthesize a short piece of RNA, up to ten nucleotides long.
c Once RNA synthesis has proceeded beyond about 12 nucleotides, RNA polymerase moves along the DNA in an inch-worm-like fashion. This is called promoter clearance in which the σ-subunit leaves the complex

Initiation of transcription proceeds in three biochemically well-defined steps (Fig. 15.**5**). The first step is the thermodynamically controlled recognition and binding of the promoter sequence by RNA polymerase holoenzyme to form the so-called **closed** complex. Interaction of RNA polymerase with the AT-rich upstream activator region contributes to binding. The second step is the kinetically controlled **isomerization** to the **open** complex, which is characterized by melting of about 10 bp around the transcription start nucleotide (*tsp* or +1), followed by the formation of the first phosphodiester bond, resulting in an RNA dinucleotide that serves as the template for RNA chain elongation. This "open complex" is characterized by the presence of a single-stranded DNA bubble. In the presence of ribotriphosphates, the open complex catalyzes the synthesis of a short RNA of about 10 nucleotides. If this dissociates from the complex, one round of **abortive initiation** is completed. The third step, called **promoter clearance**, represents the transition from initiation to elongation of transcription.

During transcription initiation, σ^{70} RNA polymerase goes through **binding** to the promoter, **isomerization** to the open complex, followed by idling in the open complex until **promoter clearance** leads to elongation. It is not yet possible to predict the strength of a promoter on the basis of sequence conservation in the -10 and -35 elements.

15.2.2 Elongation and Termination of Transcription

Once an approximately 12-nucleotide long mRNA is synthesized, the σ-factor leaves the initiation complex, and RNA polymerase core enzyme enters the elongation mode. It is thought to move along the mRNA in an inch-worm-like fashion, in which a synthetically active mode of the enzyme that elongates the nascent mRNA progressively by about 10 nucleotides, alternates with a translocation mode, in which RNA polymerase moves about 10 bp along the template. This model calls for interactions of RNA polymerase with the nascent RNA over a length of at least 10 nucleotides.

The ability to synthesize RNA molecules several thousand residues long requires that transcription is a highly processive reaction. Nevertheless, RNA synthesis does not occur at a constant pace along any given template. Instead, synthesis can be paused at many sites, mostly at palindromic sequences, which have the ability to form hairpin structures in the nascent RNA. At each pausing site, RNA polymerase can either resume elongation or prematurely terminate RNA synthesis. Termination of RNA synthesis at the 3′-end is achieved by two mechanisms:

1. **Type I terminators** are factor-independent termination sites, in which a GC-rich stem–loop structure of the nascent RNA causes RNA polymerase to pause, and the following T-rich sequence of the template yields a thermodynamically unstable RNA–DNA hybrid, which facilitates dissociation of the RNA. This type of termination signals occurs often at the end of operons, e.g., the *E. coli trp* operon (Fig. 15.**6**). It only leads to termination when it is not translated. Thus, termination signals of this type are usually preceded by one or more stop

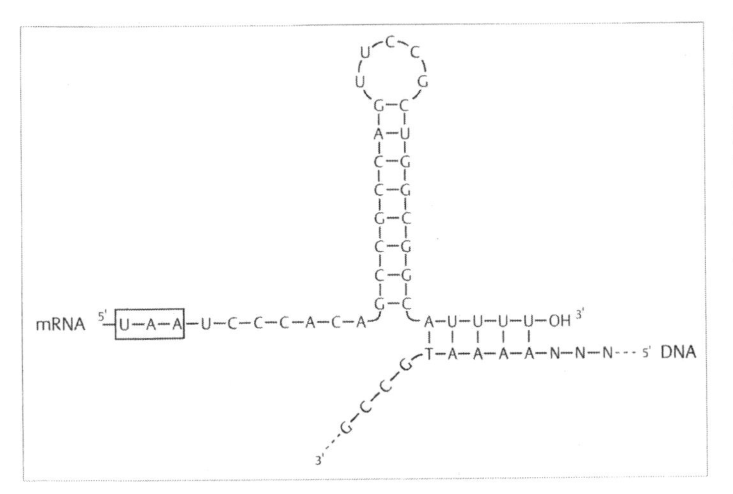

Fig. 15.6 **RNA sequence of the transcriptional terminator of the *E. coli trp* operon.** The 3′-end of the *trp* mRNA is shown in the assumed secondary structure. The stop codon of the last structural gene is boxed. Since the terminator sequence is not translated, it is able to fold into the indicated hairpin structure, which causes RNA polymerase to pause transcription. The following U-rich sequence results in an increased probability of dissociation of the ribo-U-desoxy-A hybrid

codons that terminate an open reading frame (compare below).

2. **Type II terminators** require a transcription termination protein, called **Rho**. Rho is a hexameric RNA-binding protein, which is proposed to wrap up the nascent RNA and track behind RNA polymerase. Type II terminators are less well defined in their primary structure. Important for in vitro termination is a region of about 200 bp upstream from the site of termination, called *rut* (Rho utilization). *rut* contains two sequence motifs: CACA$(Y)_3$ and C$(N)_3$CC. A pausing RNA polymerase can be terminated when Rho reaches the enzyme (see Chapter 26).

Terminators of transcription are sequences at which transcription stops and the synthesized RNA is liberated from RNA polymerase. Factor-independent terminators require no termination protein, whereas factor-dependent terminators require the additional protein factor Rho for function. Transcription of untranslated rRNA and tRNA requires a special form of RNA polymerase containing anti-termination factors.

Termination and pausing of transcription constitute a problem for the contiguous synthesis of RNAs that are never translated, but are highly structured, such as rRNA and tRNA. In order to be able to transcribe the respective genes, RNA polymerase needs additional proteins, called **antitermination** or **processivity factors**, such as the Nus (N-utilization, named after its first-characterized function in phage λ biology) proteins and the ribosomal protein S10 in *E. coli*. At the beginning of rRNA genes, a sequence element called the anti-terminator causes the RNA polymerase to bind additional proteins, which prevent it from pausing. This basic phenomenon of anti-termination is exploited for regulation of many genes and operons in bacteria and their phages (see Chapter 18.10).

Since bacterial and eukaryotic transcription proteins differ, the bacterial RNA synthesis can be specifically inhibited by the antibiotic rifampicin, which prevents RNA polymerase from initiating transcription by binding to the β-subunit.

15.2.3 Half-lives of Bacterial mRNAs Vary Dramatically

The steady-state levels of mRNAs depend on the rates of synthesis and degradation. Efficient degradation of mRNAs is an essential part of gene expression because it is often required for switching expression off. This is highlighted by the non-viability of *E. coli* mutants that lack the corresponding RNases. The half-lives of mRNAs in *E. coli* can vary between less than 0.5 min and 50 min, and are typically between 0.5 min and about 2 min. Degradation occurs either exonucleolytically from the 3′-end, or is in many cases initiated by endonucleolytical cleavages primarily in the 5′-portion of the mRNA. It should be kept in mind that generalizations about mRNA lifetimes are impossible because mRNA species are individuals owing to their enormous sequence and structure variation. However, it is assumed that many mRNA species are stabilized by translation, presumably because binding of ribosomes masks RNase cleavage sites. Introduction of stop codons, for instance, in a promoter-proximal gene causes polarity on distal genes by accelerated mRNA degradation. A specialized ribonuclease, called RNaseE, seems to perform the initial rate-limiting endonucleolytic cleavage for many mRNAs, as they are considerably stabilized in strains in which RNaseE activity is knocked out. While it is clear that mRNAs have different half-lives, which can be affected to different extents by the physiology of the bacterial cell, a biochemical description of a regulatory mechanism involving accelerated or delayed mRNA degradation is pending.

mRNA in *E. coli* is degraded by a specific endo-RNase, RNaseE (gene *rne*), and by several exo-RNases. The stabilities of different mRNAs vary to a great extent. mRNA stability may be regulated and is increased by translation.

15.3 Translation Is the Synthesis of Proteins Directed by RNA Templates

Translation is the process in which the genetic information contained in the linear sequence of a nucleic acid is converted into the linear amino acid sequence of a protein (Fig. 15.**7**). mRNA can contain one or more **open reading frames** (*orf*), each specifying a polypeptide (Orf) and carrying sequence signals for initiation, elongation, and termination of protein biosynthesis. They are utilized by the ribosomes, which read out segments of three nucleotides ("**triplets**", also called **codons**) and incorporate the corresponding encoded amino acid into the growing polypeptide chain. Decoding of the message is achieved within the ribosomes by tRNA–protein complexes. tRNAs are short, highly modified ribonucleic acids. All tRNAs have a clover-leaf secondary structure, which is folded into an L-shaped tertiary structure. The common structure of tRNAs is depicted in Figure 15.**8**. One end of the molecule contains the **anticodon**, a triplet complementary to the recognized codon, and the other end of the molecule contains the 3′-end of tRNA, at which the cognate amino acid is covalently bound as an activated *cis*-diol-ester. Charging of the tRNA with the cognate amino acid is catalyzed by amino acid **tRNA synthetase**, which uses ATP to activate the respective amino acid, recognize and select the proper tRNA out of a pool of structurally similar but non-cognate tRNA variants, and catalyze the formation of a chemical bond between the tRNA and the amino acid specified by the respective anticodon. Thus, aminoacyl-tRNA synthetases achieve the transition from nucleic acid to amino acid sequences. A tRNA loaded with the wrong amino acid will lead to misincorporation of this residue into the polypeptide.

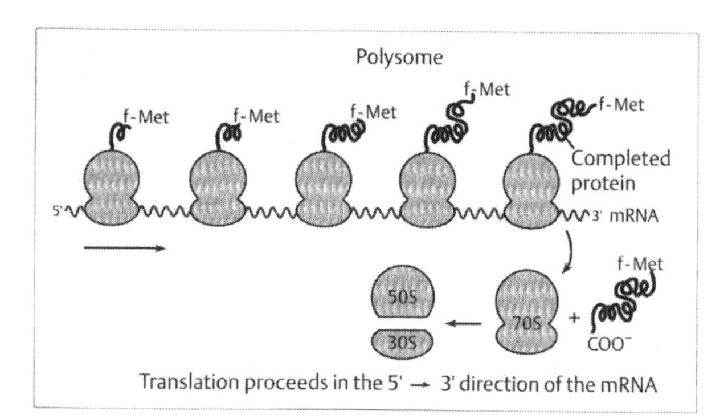

Fig. 15.7 **mRNA is simultaneously translated by many ribosomes.** The mRNA (red wavy line) is depicted with a few (of up to 500) ribosomes proceeding in the 5′ to 3′ direction (arrow) with polypeptides growing in the amino- to carboxy-terminal direction. Hence, the name **polysome** for this structure. The initiating amino acid is *N*-formyl-methionine (f-Met). Upon termination of translation, the 70S ribosome may dissociate from the mRNA followed by dissociation of the 70S ribosome into the 50S and 30S subunits

15.3.1 Ribosomes Have a Complex Structure

Ribosomes are large ribonucleoprotein complexes that serve to mediate recognition between codons in mRNAs and anticodons in charged tRNAs and that catalyze the formation of peptide bonds after binding of the correct tRNAs. Bacterial 70S ribosomes consist of two subunits, called the 30S (or small) and the 50S (or large subunit), which are formed in vitro at low Mg^{2+} concentrations. Denaturation of the 30S subunits leads to the 16S rRNA and about 20 different proteins (genes *rps*, **r**ibosomal **p**rotein **s**mall), the exact number depending on the harshness of the preparation, while the 50S subunit consists of 5S and 23S rRNAs and about 30 proteins (genes *rpl*, **r**ibosomal **p**rotein **l**arge) (see Fig. 2.**16**). The protein content differs according to the preparation procedure because a number of accessory protein factors are more or less loosely bound to the ribosome during different stages of translation. It has not been unambiguously possible to assign all of the ribosomal functions to the proteins. In fact, it has been discussed

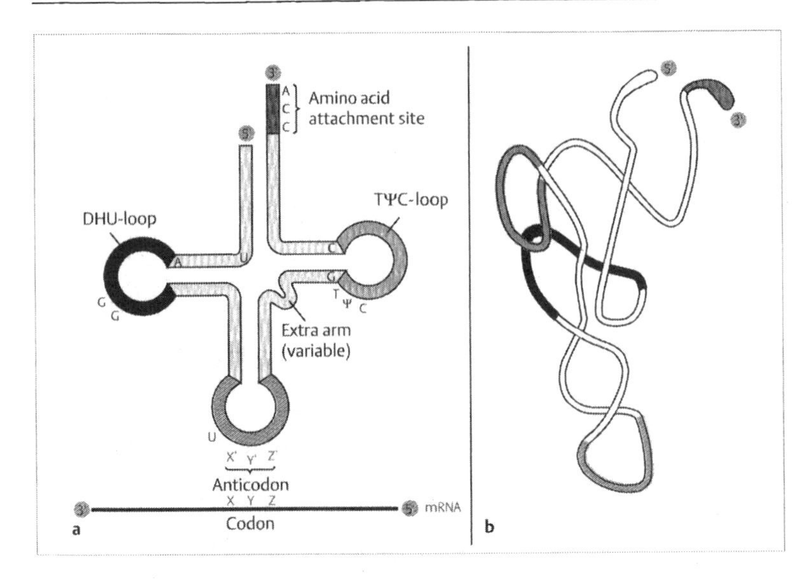

Fig. 15.**8a, b** **Conserved primary, secondary, and tertiary structures of transfer RNAs.**
a The clover leaf secondary structure of tRNAs. Conserved sequences are indicated. DHU, dihydrouridine; ψ, pseudouridine; and X′Y′Z′ the three nucleotides of the anticodon complementary and antiparallel to the codon XYZ.
b Conserved fold into the tertiary structure (L-form)

recently that at least some of the enzymatic activities (e.g., the "peptidyl transferase" in Fig. 15.**12**) may be carried out by the rRNAs rather than by the proteins in the ribosomes.

15.3.2 Initiation of Translation Begins at the 30S Subunit

The translation initiation complex is formed by the 30S subunit, the region around the start codon of mRNA, the initiator tRNA loaded with the initiator amino acid formylmethionine, and several initiation factors (IF) as summarized in Figure 15.**9**. The most common and preferred initiation codon in *E. coli* is AUG. In addition, GUG serves that purpose in about 7% and UUG in about 1% of the genes, albeit usually with a lower efficiency than AUG, for instance, to express regulatory proteins at a low level. Since AUG also encodes methionine in intragenic codons, additional mechanisms must exist that distinguish initiation from elongation codons. Extensive sequence comparisons of regions around start codons in many genes have revealed a conserved **ribosome binding sequence** (**RBS**), located on the 5′-side of the start codon. This sequence (called previously Shine-Dalgarno sequence) is complementary to the 3′-end of the 16S rRNA in the 30S subunit, which is available for base pairing. The recognition between these sequence elements is probably the first step in translation initiation. Figure 15.**10** shows the RBS and the complementary part of the 16S rRNA sequence for the A-protein mRNA from coliphage R17. The efficiency of initiation depends on the degree of complementarity to the 16S rRNA and the spacing between RBS and the start codon. The efficiency can also be affected by the nature of the second codon in the reading frame and possibly by other sequence elements, as is assumed based on the conservation of the respective nucleotides in many mRNAs.

Initiation is the rate-limiting step of translation in many genes and is also subject to regulation in several cases. For instance, translation of ribosomal proteins is product-inhibited because the proteins bind to their respective recognition sequences present in the 5′-region of their mRNA to mask the initiation site (see Chapter 18).

In addition to the sequence, the secondary structure of mRNA is important, as sequestering of RBS sequences or start codons in stem–loop structures interferes with translation initiation. The initiation of translation of proteins from polycistronic mRNA is often coupled in that the ribosome that finished translation of the first reading frame does not dissociate from the mRNA, but finds the next start codon by linear diffusion along the nucleic acid. It is assumed that this mechanism contributes to the coordinated synthesis of the encoded proteins. One of the most prominent examples of translational coupling is found in the late genes of phage lambda (genes A–W) encoding envelope proteins, which are needed at the same time in a defined ratio to build up the phage particle. These genes are organized such that the stop codon of the previous gene overlaps with the start codon of the following reading frame (see Fig. 15.**11**). It is hypothesized that the translating ribosome terminates and reinitiates protein biosynthesis at this site, leading to a coordinated, similar level of expression of the encoded proteins.

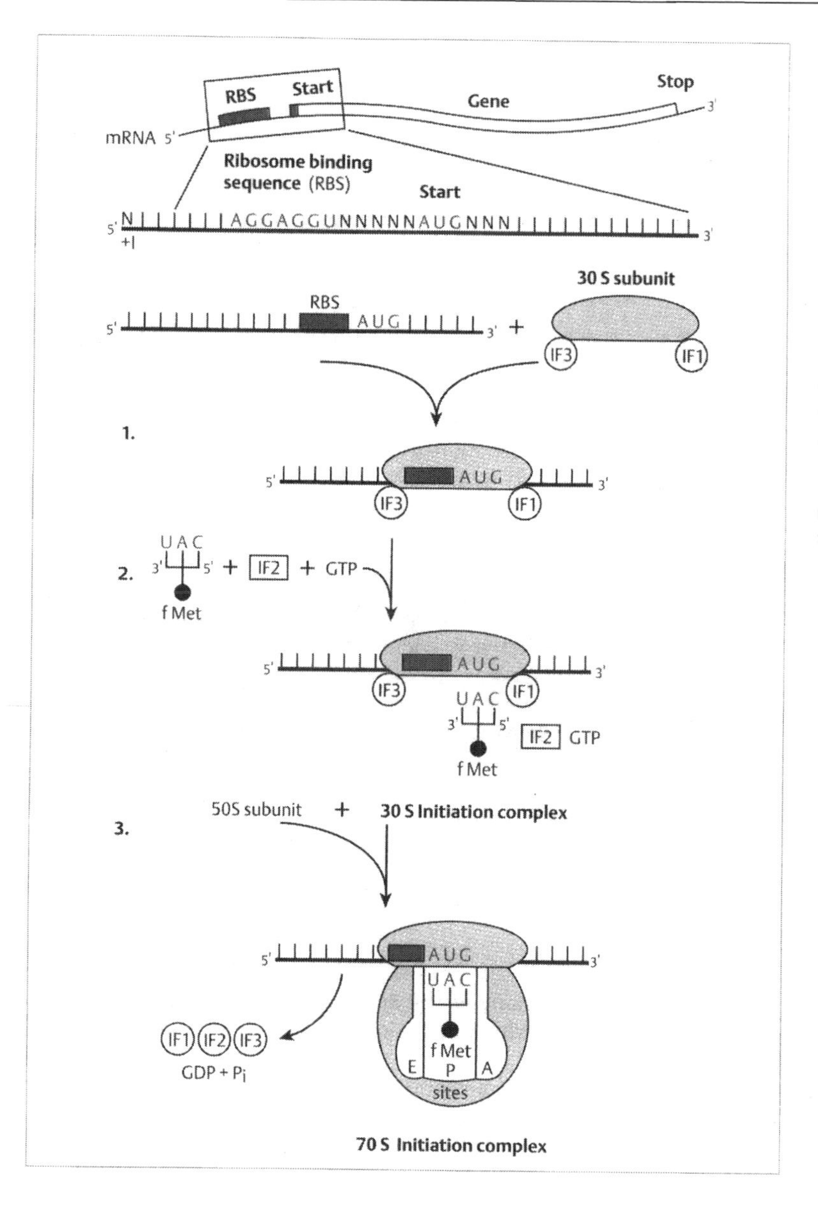

Fig. 15.**9** **Translational initiation in bacterial protein synthesis.** Initiation of translation begins with the formation of an initiation complex. (1) A 30S ribosomal subunit kept separated from 50S subunits by initiation factors IF1 and IF3 binds to a ribosomal binding (previously called "Shine-Dalgarno") sequence (RBS, red rectangle) of the mRNA. The RBS is located about 5 nucleotides upstream of the start codon AUG of a gene. (2) A charged fMet-tRNA^{fMet} activated by initiation factor IF2 and GTP binds to the initiation codon. (3) Upon dissociation of IF1, IF2 and IF3, and hydrolysis of GTP, a 50S ribosomal subunit binds to the 30S initiation complex. In the newly formed 70S initiation complex, the charged tRNA^{fMet} is located in the P (or peptide) site, which leaves the A (or acceptor) site open for the second charged tRNA (see Fig. 15.**12**) and the E (or exit) site empty

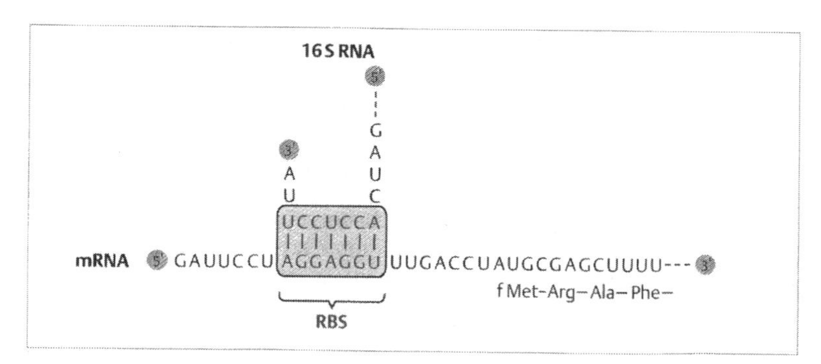

Fig. 15.**10** **Arrangement of a translational initiation site in *E. coli*.** The sequence of mRNA encoding protein A from phage R17 is shown with the start codon and the ribosome binding sequence (RBS, red rectangle). The proposed interaction of that site with the 3′-end of the 16S rRNA is indicated

```
2601                                                              2661
5' GATTACGCCCGTGCCTTATCCGGAGAGGATGAATGACGCGACAGGAAGAACTTGCCGCTGC 3'
     L   R   P   C   L   I   R   R   G
------------gene A-------------------|
                                     X   N   D   A   T   G   R   T   C   R   C
                                    |-------------------gene W--------------------
```

Fig. 15.11 **Translational coupling in late genes of bacteriophage lambda (λ).** The stop codon is underlined and the start codon is given in red print. The A-protein is a component of the terminase, which cleaves at *cos* sites; the W-protein is needed for head completion (see Chapter 26)

Initiation of translation requires a start codon preceded by a ribosome binding sequence in the proper distance and the lack of secondary structure involving any of these elements. The initiation is probably the rate-limiting step for translation of many genes.

15.3.3 Elongation and Termination of Translation Determine the Correct Sequence and Length of the Protein

The elongation cycle of protein biosynthesis (summarized in Fig. 15.**12**) ensures the correct incorporation of amino acids, determined by the respective codons in mRNA (**colinearity**), into the growing polypeptide. After initiation, the second codon in the reading frame is exposed for codon–anticodon interaction in the elongation complex. This so-called **A-site** (aminoacyl-tRNA site) is occupied by the ternary complex consisting of elongation factor Tu (EFTu), charged cognate tRNA, and GTP; the complex places the α-amino group of the respective amino acid in a position to form a peptide bond with the activated carboxyl function of the amino acid (or growing peptide) at the **P-site** (peptidyl-tRNA site). After peptide bond formation, another elongation factor, EFG, mediates the movement (**translocation**) of the complex along the mRNA and the subsequent release of the uncharged tRNA. The latter may occur simultaneously with binding of the new charged tRNA into the A-site, now exposing the next codon of the mRNA. The elongation factors are typical **G-proteins** in that they exhibit different conformations depending on the bound cofactor, GTP or GDP. It is thus assumed that conformational changes of the elongation complex are triggered by elongation-factor-bound GTP hydrolysis. The ribosome performs elongation with the remarkable speed of about 15–20 amino acids per second and the even more remarkable accuracy of no more than one error in over 50 000 incorporations. The antibiotic streptomycin decreases this accuracy drastically, causing the synthesis of faulty proteins. The detailed mechanisms underlying these properties are still under investigation.

Some proteins, for example, the polypeptide release factor-2 (RF-2) in *E. coli*, require a translational frameshift of the elongating ribosome for expression (Fig. 15.**13**). It is hypothesized that frame-shifting is possible at sites where elongation pauses, giving the ribosome an opportunity for slippage on the mRNA. Frame-shifting occurs at respective sites at a low frequency, thus ensuring low expression of the affected reading frame.

The end of a reading frame on mRNA is determined by the first stop codon (Fig. 15.**14**). There are no tRNAs complementary to the three stop codons. Thus, translation is first paused at a stop codon, followed by binding of a **release factor** (Rf), which triggers the dissociation of the completed polypeptide from the ribosome. The now idling ribosome can either dissociate from the mRNA or may find a new start codon in the vicinity of the stop codon, where reinitiation of protein biosynthesis may take place.

15.3.4 Many of the Best-Known Antibiotics Interfere With Translation in Bacteria

Bacterial translation differs enough from eukaryotic translation that many inhibitors show specificity for the bacterial 70S ribosome (see Figs. 15.**9** and 15.**12**). To some extent, these antibiotics are also active in inhibiting mitochondrial (and chloroplast) translation, reenforcing the concept of the close relationship of these organelles to the bacteria. For instance, **tetracyclines** inhibit the elongating ribosomes, presumably by blocking the A-site. **Chloramphenicol** prevents peptidyl transferase activity, and **streptomycin**, an aminoglycoside antibiotic, interferes with the onset of elongation after initiation of protein synthesis and with codon reading accuracy. **Erythromycin**, the prototype of the macrolide antibiotics, interferes with elongation by inhibiting the movement of the peptidyl-tRNA from the A-site to the P-site (translocation). See Chapter 27 for more details on antibiotics.

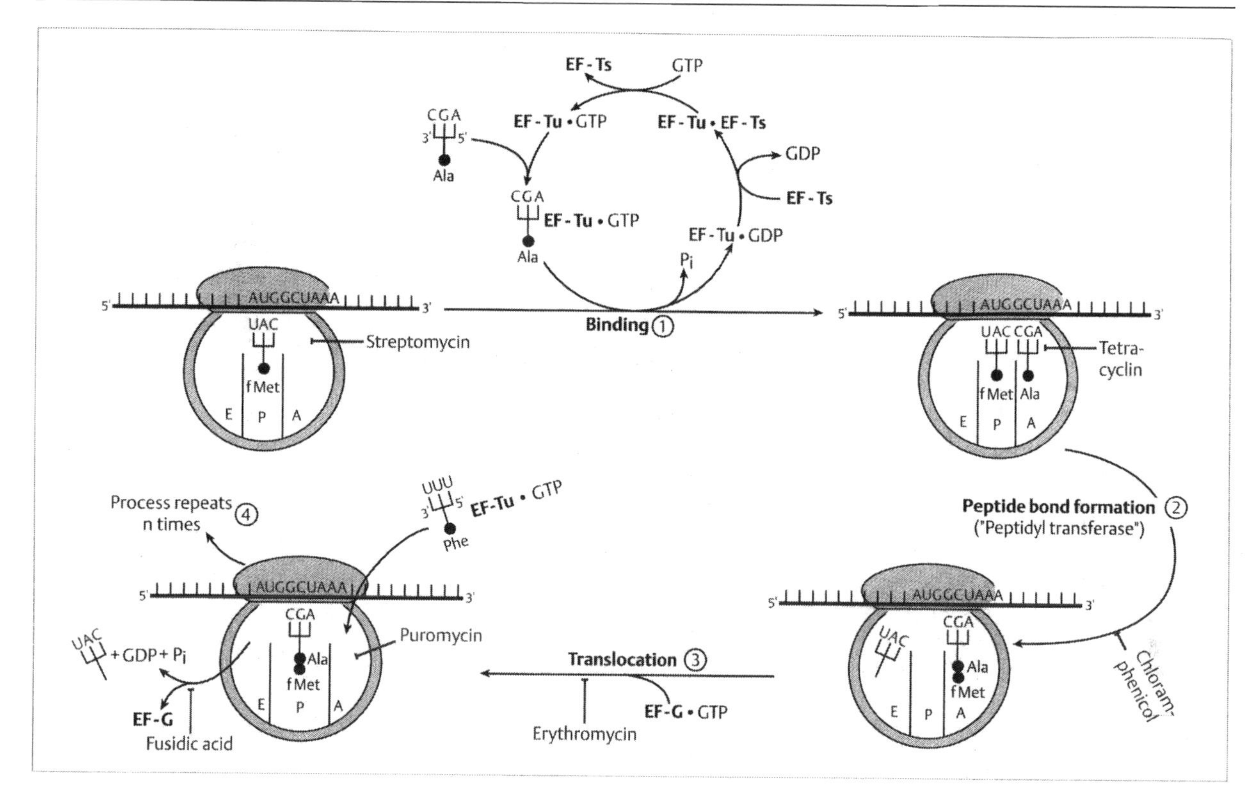

Fig. 15.**12** **Elongation cycles involved in protein translation.** (1) The translational elongation cycle begins by the binding of the activated (through elongation factor EF-Tu and GTP) and charged tRNA to the codon in the A-site (in the figure GCU for alanine). The process requires GTP hydrolysis and generates EF-Tu · GDP+P_i. Regeneration of EF-Tu · GTP requires the elongation factor EF-Ts and GTP as indicated. The incoming tRNA is tested for accurate pairing with codon 2, a process that requires the ribosomal protein S12 (gene *rpsL*). The antibiotic streptomycin interferes with this process, thus increasing translational misreading. (2) The α-amino group of the amino acid in the A-site forms a peptide bond with the activated carboxyl group of the amino acid (or growing peptide) at the P-site. Peptide formation is not catalyzed by an enzyme, but by an endogenous "peptidyl transferase" activity of the 50S subunit, a process inhibited by the antibiotic chloramphenicol. (3) After peptide bond formation, elongation factor EF-G mediates translocation of the complex along the mRNA; translocation requires hydrolysis of one GTP per triplet. Concomitantly, the uncharged tRNA is released through the E-site together with EF-G, and the A-site becomes free for the next cycle. If at this stage the antibiotic puromycin, an analog of the 3′-end of tRNAs, enters the A-site, a peptidyl-puromycin is synthesized, thus interrupting peptide elongation. Other antibiotics inhibit where indicated. (4) Under normal conditions, the elongation cycle repeats until a stop codon is encountered on the mRNA

Thr – Glu – Arg – Ser – Asp – Val – Leu – Arg – Gly – Tyr – Leu ||

mRNA 5′ AUG ··· ACG GAA CGC UCC GAC GUU CUU AGG GGG UAU CUU **UGA** C UAC GA ··· 3′

Leu – Asp – Tyr ···

Fig. 15.**13** **Translational frame shifting is required for expression of RF-2 in *E. coli*.** The figure shows the relevant sequence of the RF-2-encoding mRNA. The amino acids encoded by the reading frame on the left side are shown above the mRNA sequence. An internal RBS is indicated in red. The stop codon UGA is given in bold print. This stop codon is an inefficient stop signal in this sequence context when the RF-2 concentration is low. In that case, the translating ribosome slips at the three uracil residues (underlined) to cause a +1 frame shift and continues RF-2 synthesis in the new frame specifying the amino acid sequence given below the mRNA

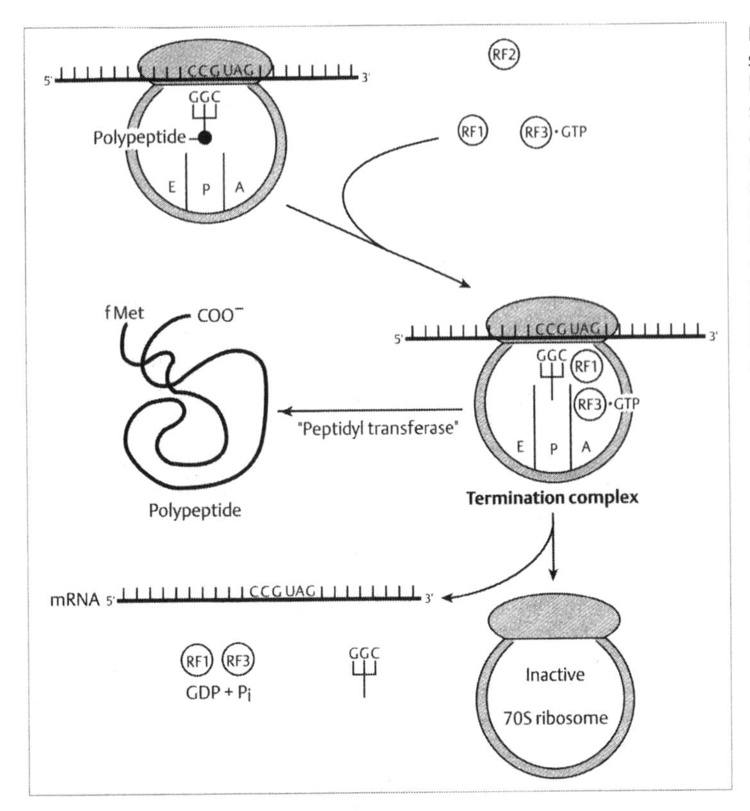

Fig. 15.**14** **Translational termination in protein synthesis.** Upon reaching a stop codon (UAG here), no corresponding tRNA is available (but see mutated suppressor tRNAs in Chapter 15.5.2), and release factors RF1 (for UGA and UAA) or RF2 (for UAA and UGA) and RF3 bind at the empty A-site to form the termination complex. The endogenous "peptidyl transferase" activity of the 50S subunit releases the growing polypeptide and causes dissociation of mRNA, release factors, tRNA, and an inactive 70S ribosome, which hydrolyzes GTP in the process. The 70S complex is dissociated by IF1 and IF3 into its 50S and 30S subunits, thus terminating translation

Translation is the process in which the genetic information contained in the linear sequence of a nucleic acid is converted into the linear amino acid sequence of a peptide. The process requires 30S and 50S ribosomal subunits, IF, EF, and RF factors, GTP, and charged tRNA molecules. Several antibiotics (**tetracyclines**, **chloramphenicol**, **aminoglycoside antibiotics**, and **erythromycin**) interfere with bacterial, but not with eukaryotic protein synthesis.

15.4 The Genetic Code Is a Degenerate and Universal Triplet Code

The genetic code specifies the conversion of the genetic information contained in a nucleic acid sequence into an amino acid sequence. Since most functions of a cell are either carried out or catalyzed by enzymes and proteins, the genetic code must be viewed as the key link between storage and reproduction of genetic information on the one hand and its conversion to biological functions on the other hand. Information is stored in the form of DNA sequences of genes and their organization into transcriptional units on the chromosomes. Each transcriptional unit contains regulatory information defining when in the life cycle of a cell and to what extent the encoded product is synthesized, and structural information specifying the primary sequence of the product to be synthesized, either an RNA or, in most cases, a polypeptide.

15.4.1 The Properties of the Genetic Code and Codon Usage Are Linked

Proteins usually consist of 20 different L-amino acids. The minimum number of base pairs needed to unambiguously define each of the 20 amino acids is 3, allowing a variance of $4^3 = 64$ combinations. The evolution of the genetic code has indeed minimized the number of base pairs used to specify an amino acid to 3 (**"triplet code"**) and makes use of the greater than required variance by using a **degenerate code**, meaning that most of the amino acids are specified by more than one triplet of base pairs. There are only two amino acids specified by a single codon, namely tryptophan by UGG and methionine by AUG, which is also used as the initiator codon as described before. The genetic code is depicted in Figure 15.**15**. Three of the possible codons do not specify an amino acid, but encode the end of the protein to be synthesized. Hence, they are called **stop codons**. These are sometimes falsely termed "nonsense" codons to distinguish them from the amino-acid-encoding "sense" codons. The logic of the genetic code is **universal** in all known types of cells. Few exceptions exist, in which one of the stop codons is used to encode an amino acid. The protein encoding **structural gene** consists of a reading frame beginning with the start codon and ending with a stop codon. The order of triplets in between specifies the primary structure of the encoded protein. A small number of proteins, such as formyl hydrogen-lyase (gene *fhl*) of *E. coli*, contains selenocysteine. This amino acid is encoded by the stop codon UAG, which, in the given sequence context, is not recognized as a stop codon, but by a specific selenocysteinyl-tRNA. Thus, **selenocysteine** constitutes the 21st biogenic amino acid (see Chapter 8.8.2).

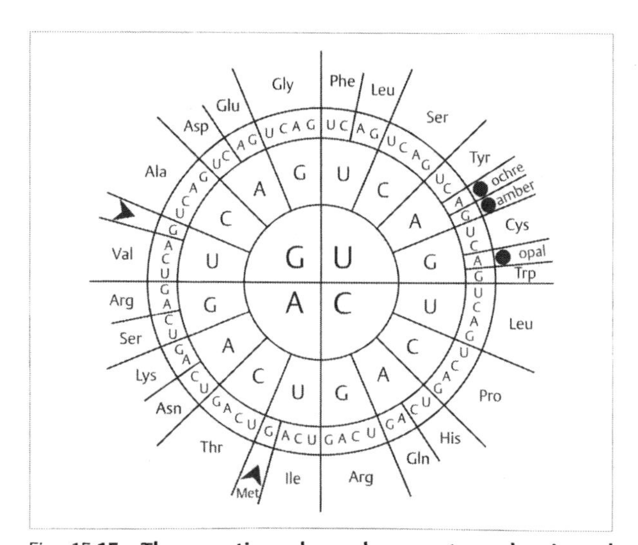

Fig. 15.**15** **The genetic code, a degenerate and universal triplet code.** All possible 64 codons consisting of three nucleotides are shown. The 5′ nucleotide is shown in the center, the middle nucleotide in the next circle, and the 3′ nucleotide on the outside. The encoded amino acid is given for each codon. The filled circles (●) denote stop codons, and the filled triangles (▲) start codons. The historical designations of the stop codons (ochre, amber, opal) are also given

> The link between nucleic acid and amino acid sequences is the **genetic code**. It consists of successive nucleotide triplets or codons that specify the amino acid sequence between the start and stop codons. The genetic code is degenerate, with 61 codons for 20 amino acids, and is universal for all known cell types.

The codons are recognized during translation by tRNAs. tRNAs contain the anticodon, which is complementary to the codon specifying that amino acid. There are more tRNAs than the 20 amino acids. Many amino acids can be linked to any one of the several **isoaccepting tRNAs**. To account for the decoding of the same codons by different tRNAs, the base-pairing rules for codon–anticodon interaction are somewhat relaxed as compared to the rules for replication and transcription. The **"wobble" hypothesis** calls for strict recognition of the first two bases in a codon, while the third base is subject to relaxed base-pairing constraints with the anticodon (e.g., a G residue can either pair with a C or a U residue). In this way, each of the 61 codons can be recognized specifically by one or more tRNA species. The wobble mechanism, together with two or more triplets encoding the same amino acid, are the reasons underlying the degeneracy of the genetic code.

While the logic of the genetic code is universal among all known organisms, the usage of different codons is not universal. When more than one codon specifies the same amino acid, different organisms may prefer the use of one or the other in their genes. Bacteria having a high G+C content in their DNA generally tend to make preferred use of the more G+C-rich codons and those having a low G+C content tend to make preferred use of the more A+T-rich codons. It has been hypothesized for some time that genes expressed at low levels tend to have more rare codons than highly expressed genes. In most cases, however, the initiation of translation rather than elongation appears to be the rate-limiting step for protein biosynthesis. The **codon usage** may, however, be an important consideration for the expression of genes from different organisms in a given host because there seems to be a clear correlation between the preferred codons and the abundancy of their cognate tRNAs.

15.4.2 tRNA-aminoacyl Synthetase Recognition Is a Specific Process

Decoding of the mRNA sequence by tRNAs via codon–anticodon interaction on the ribosome follows the basic complementarity rules for nucleic acid base pairing (A=U, G=C), albeit with a somewhat relaxed specificity. Thus, the crucial recognition of a nucleic acid and the subsequent specification of the encoded amino acid occurs in the assignment of amino acids to tRNAs. There appears to be no way of correcting an error made in charging the tRNA with a non-cognate amino acid during ribosomal protein biosynthesis. The obvious choice for recognition, the complete anticodon, contributes only in some cases to synthetase interaction. At least one nucleotide of the anticodon contributes to synthetase recognition in 17 out of the 20 *E. coli* isoaccepting tRNA groups. In many tRNAs, nucleotides in the acceptor stem and at position 73 are important in addition to the anticodon (see Fig. 15.**8**). Other specificity determinants are located in the variable ("synthetase") loop of tRNA.

The charging of a tRNA with the correct amino acid is the crucial information link between nucleic acid and amino acid sequences. It is carried out by aminoacyl-tRNA synthetases, which recognize different features of tRNA.

15.5 Mutations and Mutant Selection in Bacteria Are Essential Genetic Methods

Mutations are defined as inheritable changes in the genotype that are **alterations of the DNA sequence**. Most mutations, especially the ones usually studied for gene mapping purposes, cause altered phenotypes, i.e., the mutant strain has properties different than the wild-type strain from which it is derived. This definition implies that **wild-type** and **mutant** (i.e., an organism carrying a mutation) are relative terms. In the first part of this section, the various DNA sequence alterations will be classified and their potential effects on the encoded phenotypes will be discussed. The second part contains a description of external factors that increase the frequency of mutagenesis.

15.5.1 Mutations Can Be Classified Into Point and Chromosomal Mutations

The simplest mutations are changes of a single base pair in the DNA. These are called **point mutations** and can be a substitution of a base pair, an insertion of a base pair, or a deletion of a base pair. Substitutions that maintain the purine-pyrimidine order of the sequence are called **transitions** and those that change this order are called **transversions**. Thus, transitions are either T↔C or A↔G substitutions, and transversions are C or T↔A or G substitutions.

The consequences of mutations for the encoded protein vary greatly. No change in the phenotype is usually observed when a substitution changes the nucleotide in a wobble position in a codon (UUC→UUU, both triplets encode Phe) since wild-type and mutant DNA encode the same amino acid. This is often referred to as a **silent** or **neutral** mutation and usually is not noticed. Most substitutions lead to changes of the encoded amino acid (CUU→UUU leads to replacement of Leu by Phe) and are called **missense mutations**. Missense mutations lead to an altered phenotype if an amino acid essential for folding or activity of the encoded protein is changed; however, many missense mutations may not lead to an altered phenotype when non-essential positions of the protein are affected. Severe phenotype alterations usually result from mutations that convert a codon to a stop codon. These are called **nonsense mutations**. Translation of all downstream codons of the reading frame is abolished, and a truncated protein is synthesized. In addition, nonsense mutations may show **polar effects** because transcription of a usually translated RNA is uncoupled from translation (see Chapter 15.1.2).

Additions or deletions of single base pairs (**frameshift mutations**) disrupt the affected reading frame and, therefore, change not only the codon they occur in, but also the informational content of all downstream codons. In most cases, the new reading frame will terminate after several codons. Thus, insertions or deletions in genes cause usually clearly altered phenotypes.

Mutations involving multiple base pairs (**chromosomal mutations**) are the result of homologous or specific recombinations. These mutations can be classified as **duplications**, in which at least one, but sometimes more repetitions of the same sequence occur, while **deletions** and **insertions** describe the loss or addition of two or more base pairs, respectively (see Chapter 14 for details). **Inversions** contain DNA sequences that are reversed with respect to the wild-type orientation and are usually the result of flip-flop recombination events, for example, the antigenic variation of *Salmonella* and G inversion of phage Mu (see Chapter 18). **Translocation**, finally, is called the positioning of a DNA at a new place. Examples are the integration of plasmids at various locations in the host chromosome, or the exchange of homologous sequences between plasmid and chromosome or chromosome and chromosome after DNA duplication and cell division, or gene transfer in **diplogenotes**.

Mutations are inheritable changes in the genome. They can be point mutations, which cause alteration, deletion, or insertion of a single base pair. The change in phenotype depends on the nature of the mutation and the importance of the site. Chromosomal mutations involve more than one base pair.

15.5.2 Mutant Phenotypes Can Be Restored by Reversion and Suppression ("Second-site Mutations")

Most mutants can undergo a second mutation that restores the original phenotype. If the mutated base pair is converted back to the original base pair, the process is called a **reversion**. If a mutation at a different site restores the original phenotype, it is called a **suppression** or sometimes a **second-site reversion**. Point mutations revert at a high frequency, whereas deletions cannot simply revert. Suppression can occur within a codon, intragenically, or extragenically. A suppression within a codon occurs when the first mutation changes the codon to encode an amino acid incompatible with the function of the protein, and the suppression mutation leads to incorporation of a different amino acid compatible with the function of the protein. An **intragenic suppression** may occur at a different codon, leading to an amino acid which, for example, restores the structure of the encoded protein impaired by the original mutation. This type of second-site reversion is often used to study interactions of residues within a protein (e.g., between two domains). The most prominent example of an **extragenic suppression** is a mutation in a tRNA gene that enables the encoded tRNA to decode the stop codon introduced by the original mutation. This type of suppression is sometimes referred to as "**informational suppression**". Restoration of the original phenotype due to functional substitution by a different enzyme or by activation of an entirely new pathway is also possible (see Chapter 19.1).

15.5.3 Spontaneous Mutations Are Vital Events That Drive Evolution

Spontaneous mutations occur in all organisms. The mutations are called **spontaneous** because they arise in the absence of any obvious external factor known to promote mutagenesis. Spontaneous point mutations can be caused by the basic infidelity in DNA replication or by DNA-damaging activities of mutagenic chemicals and cosmic radiation. Mutations do not occur randomly in the DNA sequence, but rather at hot spots (Fig. 15.**16**). One reason for a mutational hot spot is the presence of a methylated guanosine, which can be converted to the

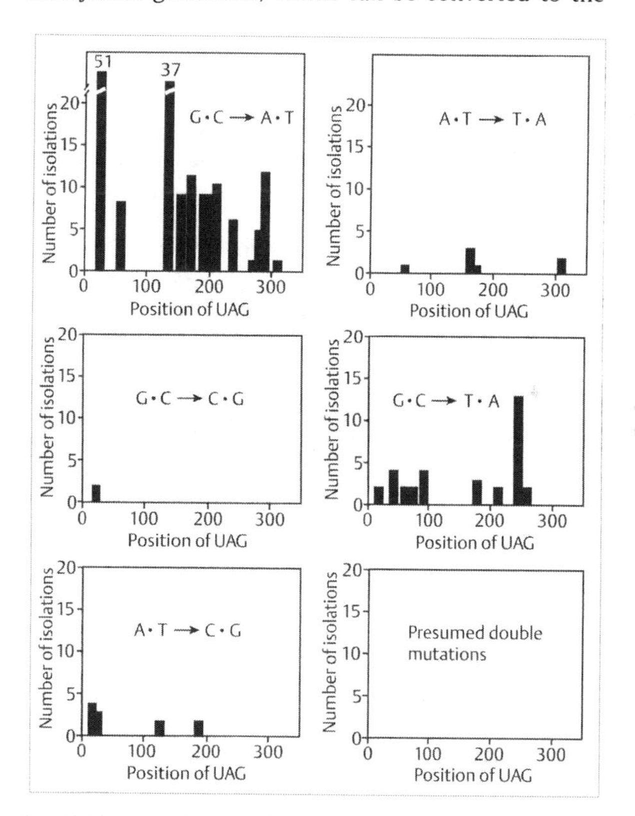

Fig. 15.**16** **Distribution of spontaneous mutations in gene *lacI* of *E. coli*.** There are 37 sites in *lacI* at which an amber (UAG) mutation can be generated by a single base pair exchange. Transitions (GC → AT; AT → GC) and transversions (all others) occur, but transitions predominate. These mutations occur at the highest frequency (hot spots) at positions of C-methylation and are probably caused by unrepaired deamination of 5-methyl cytosines

naturally occurring thymidine by deamination. The sequence T–T is photosensitive (see Chapter 15.6.3) and constitutes another hot spot for mutations. The frequency of spontaneous mutations can vary to a great extent. Spontaneous mutations occur at an average rate of about 10^{-7} per gene and generation in bacteria.

Spontaneous mutations **occur randomly** in many genes in bacteria. The environment provides **selective pressure** to support increased survival of advantageous mutations. Recently, an increased frequency of advantageous mutations in starving colonies of bacteria was reported; this stirred up some controversy because the mutations seemed to affect genes needed for survival at a higher frequency than they affected silent genes. The mechanisms underlying these so-called "adaptive mutations" are currently being investigated. It remains to be seen whether this observation identifies unknown genetic properties of bacteria (e.g., inducible mutation rates).

A unique class of spontaneous mutations arises from transposition events, which are sometimes accompanied by large deletions at either their site of insertion or excision, especially in the case of cut-and-paste transposition (see Chapter 14). Transposons have been changed by man into a versatile tool for mutagenesis experiments because the transposons can link the site of mutagenesis to an indicator gene incorporated into the transposable element. This strategy has become increasingly important in the study of bacteria for which no gene cloning system is available because many transposons have a broad host range.

Spontaneous mutations result mostly from inherent infidelity of replication and chemical instability of the bases in the cell. Mutations are needed for evolution. Mutations occur at different frequencies in all genetic loci.

15.5.4 Mutagenesis Is the Experimental Increase of the Spontaneous Mutation Rate

The spontaneous mutation rate can be increased substantially by chemical or physical treatment of bacteria and in so-called mutator strains (see Chapter 15.6). The nature of the most frequently occurring mutations depends to a large extent on the procedure used to induce it (Fig. 15.**17**). For instance, base analogs such as 5-bromo-uracil are incorporated into DNA where they exhibit relaxed base-pairing properties compared to uracil, leading to mutations in the progeny. It takes one round of replication between modification of the DNA and manifestation of the mutation caused by a base analog. Another set of mutagens (e.g., *N*-methyl-*N*′-nitro-*N*-nitrosoguanidine, ethyl methanesulfonate, hydroxylamine, nitrous acid) leads to chemical modification of DNA, mostly in a base-specific manner, and affects the fidelity of replication. The modified nucleotide often has different base-pairing properties owing to the chemical alteration. Both classes of mutagens act directly ("direct mutagenesis"). Polycyclic aromatic compounds (e.g., acridine derivatives) can intercalate between the base pairs of DNA and mimic the presence of an extra base pair during replication, leading to frame-shift mutations. The most common physical treatment of bacteria for mutagenesis is the exposure to UV irradiation. DNA absorbs UV light most efficiently at a wavelength of 260 nm, which leads to increased chemical activity of the bases. The most common reaction of DNA after UV irradiation is thymine dimerization, which leads to a cyclobutane ring structure between neighboring thymine residues in the same strand of DNA. Since DNA with this damage cannot be replicated, the cell can only divide after repair of the damage (compare below). Since UV repair involves an error-prone DNA synthesis with DNA polymerase I (PolI, gene polA), mutations are introduced in the vicinity of the originally damaged site, a process called **indirect mutagenesis** (see Chapter 15.7).

All chemicals produced by man are routinely checked for their mutagenic potential using a set of tests developed by B. Ames, who introduced several well-characterized point mutations in the operon encoding histidine biosynthesis in *Salmonella typhimurium*. Additional mutations in the test strains increase the uptake of chemicals and reduce mismatch repair, which together lead to an increased sensitivity to the mutagen.

In the so-called **Ames test**, the reversion frequencies of these various mutations induced by treatment with the tested compound are determined (Fig. 15.**18**). This system is very powerful because only revertants are able to grow on minimal medium lacking histidine ("positive selection"). Thus, even a low mutation frequency can be conveniently detected. Furthermore, different types of mutations (e.g., transversions, transitions, and frame-shifts) in the histidine biosynthesis operon can be included in this reversion analysis. The different reversion frequencies induced by the investigated compound or mixture of substances reveals, therefore, hints about the type of mutation induced by the agent. Some chemicals are **pro-mutagenic**, i.e., they are not mutagenic by themselves, but are converted to mutagens when chemically modified by detoxification reactions in the host liver. This potential

Mutagen	Action	Result
Base analog		
5-Bromo-uracil 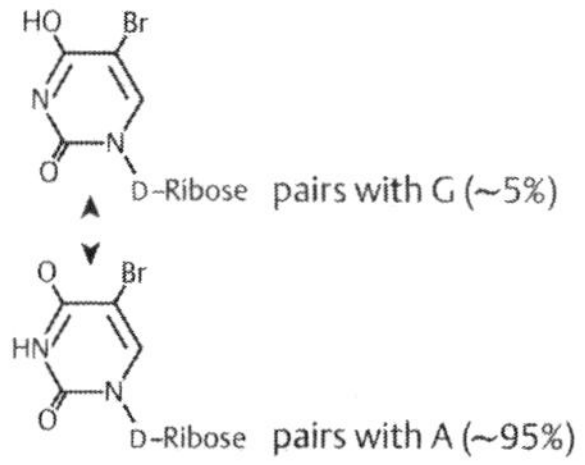	Incorporation instead of T, occasional mispairing with G	AT → GC (~5%)
D-Ribose pairs with G (~5%)		
D-Ribose pairs with A (~95%)		
Chemicals reacting with DNA		
Nitrous acid (HNO_2)	Oxidative deamination of A and C	AT → GC and GC → AT
Hydroxylamine (NH_2OH)	Changes $-NH_2$ at position 4 of C into $-NHOH$ or $-NHO-CH_3$	GC → AT
Alkylating agents		
Monofunctional		
e.g., ethylmethanesulfonate 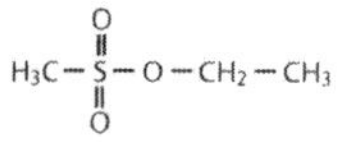 $H_3C-SO_2-O-CH_2-CH_3$	Methylation of G, mispairing with T	GC → AT
Bifunctional		
e.g., *N*-nitro-N'-nitrosoguanidine 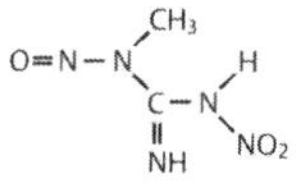	Alkylates and crosslinks DNA strands, modified region excised by DNase, misreading by PolI causes mutations	Point mutations and deletions
Intercalation		
Polycyclic aromatic compounds	Insert between two base pairs	Insertions and deletions
e.g., ethidium bromide 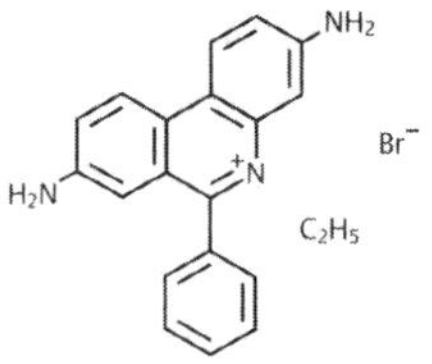		
Radiation		
UV	Pyrimidine dimer formation, very short patch excision, misreading by PolI causes mutations ("indirect mutagenesis")	Base pair exchange or deletion

Fig. 15.**17** **Action of various mutagens.** The left column shows some mutagens, the middle column their proposed effect on DNA, and the right column the resulting mutation. 5-Bromo-uracil, which upon incorporation instead of T in DNA, forms in 5% of the cases a base pair with G instead of A. Thus, transitions from AT to GC occur as the result. Nitrous acid and hydroxylamine undergo reactions with exocyclic amino groups in the bases, altering their H-bonding potential, which leads to mispairing. Alkylating reagents can exert their effect either by direct mispairing of the modified G or by leading to error-prone repair synthesis ("indirect mutagenesis") by DNA polymerase I (PolI). The latter effect is also caused by bifunctional reagents and pyrimidine dimer formation

is routinely investigated by performing the Ames test after incubation of the respective agent with rat liver extracts. The Ames test is probably the most widely used application of bacterial mutagenesis to date.

Induced mutations are the result of treatment with mutagenic chemicals or radiation. The mutagenic potential of compounds is evaluated by the Ames test using bacteria.

15.5.5 Use of Mutations and Mutants in Bacterial Genetics

The basic concept of gene expression, the chemical link between information and function in life, was analyzed by G. W. Beadle and E. L. Tatum, who studied mutations in the fungus *Neurospora crassa*. Mutations were used to identify DNA as the genetic material and to analyze the genetic code, led to the detection of the three naturally occurring DNA exchange mechanisms in bacteria, namely conjugation, transduction, and transformation, and allowed the mapping of genes (see Chapter 16). Using recently developed methods, nearly any desired single or multiple mutation can be introduced into any place of interest in many bacterial chromosomes (“**localized mutagenesis**”), helping to understand the role of a given genetic locus in physiology, or the role of a given amino acid in the function of the encoded protein. Furthermore, mutations provide one basis for cloning genes using complementation (see Chapter 17). Sequences of entire bacterial chromosomes are being determined. Most *orfs* encode proteins of presently unknown functions. It will undoubtly be through mutagenesis that the yet unknown functional roles played by many of the products of those genes in the life cycles of the various bacteria will be learned.

15.5.6 Isolation of Mutants Can Be Trivial or an Art

Mutations can generally be isolated for any scorable phenotype of bacteria, albeit with much different experimental efforts. Mutations can either lead to a growth advantage for the mutant or cause a deficiency. The former class of mutants can, in most cases, easily be isolated by **positive selection**. Examples for a positive selection are antibiotic or phage resistance, where a mutation prevents either uptake of the drug or the interaction of the antibiotic with the target, and recognition of the host by a phage. Resistant mutants will grow in the presence of the antibiotic or phage, whereas the wild-type will not. Since about 10^6 to 10^8

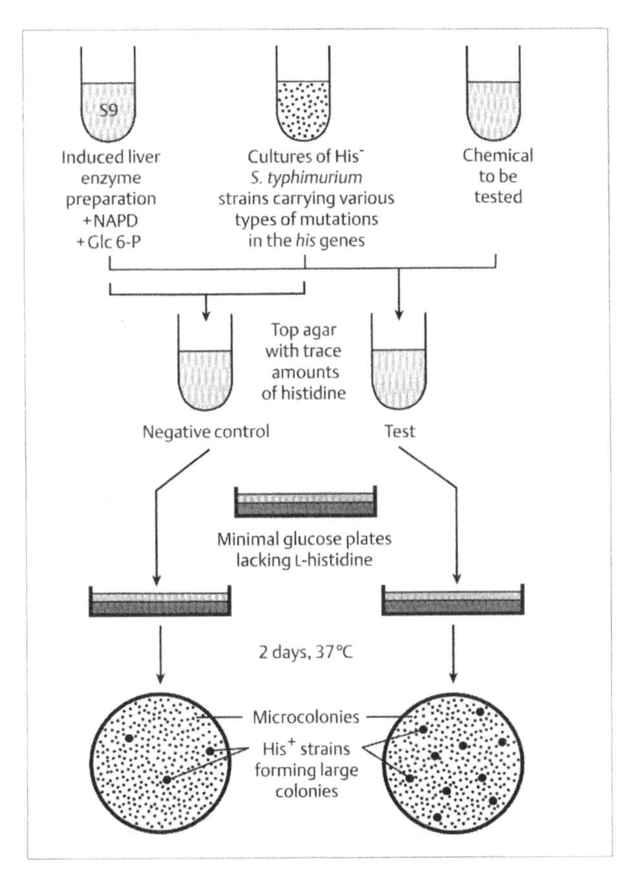

Fig. 15.18 **Schematic description of the Ames test.** The components needed for the Ames test to quantify the mutagenic potential of the chemical to be tested (top right) are shown at the top. An overnight culture of *Salmonella typhimurium* with a defined mutation in the histidine biosynthesis operon, an *rfa* mutation causing increased permeability of the cell wall, and a *uvrB* deletion causing increased mutation frequencies, is incubated with soluble protein extract from induced rat liver supplied with NADP and glucose 6-phosphate and with (test) or without (negative control) the chemical in top agar (soft agar) containing trace amounts of histidine to promote limited growth. The top agar is then plated out and incubated at 37°C for two days. Revertants of the different *his* mutants used appear as colonies on these plates, as the histidine supply allows the mutants to grow only as microcolonies. The difference between the test strains and the negative control reflects the mutagenic potential of the chemical for the type of mutations (transition, transversion, deletion) that caused the original histidine auxotrophy

cells can be spread on a single plate, it is usually easy to isolate mutants.

Mutants requiring a nutrient for growth are called **auxotrophs**, whereas the wild-type is a **prototroph**. Since auxotrophs lack a property, they cannot be easily selected, but must be screened. **Screening** is mostly done by replica plating, where colonies are replicated from a plate allowing growth of wild-type and mutant to a plate allowing growth of only the wild-type. The lack of a colony on the latter plate indicates the desired negative phenotype. Given the low frequency of mutations with special phenotypes and that only about 100 colonies per plate can be screened by replica plating, this procedure is too laborious for routine screening. In these cases, a **counter** or **negative selection** is often applied to enrich a bacterial population for defective mutants. Counter selection for antibiotic resistance makes use of the fact that only growing cells are killed by many antibiotics, e.g., penicillin, nalidixic acid, or streptozotocin (Fig. 15.**19**). Growth conditions are chosen that allow the wild-type to grow, while the auxotrophic mutant cannot grow when the culture is treated with the antibiotic. The antibiotic will kill the growing wild-type cells and leave the resting mutant cells unaffected. To achieve a sufficient enrichment of the mutants, this procedure has to be applied at least twice. The resulting mixture of cells should be enriched for auxotrophic mutants, which can subsequently be identified and isolated by replica plating.

Novel methods for mutant isolation makes use of combined selection and screening protocols that depend on preconstructed genetic tools to acquire the desired mutations. Transposons Tn*5* and Tn*10* (see Chapter 14) have a broad host range, only little target sequence preference, and encode kanamycin or tetracycline resistance, respectively. The collection of mutants obtained by Tn*5* or Tn*10* transposition can be selected by resistance. When the desired mutation has been isolated, the affected genetic locus is tagged by the Tn sequence, greatly facilitating cloning of the mutated gene.

Mutations in vital genes are lethal and cannot be obtained by the procedures described so far in this chapter. For these loci, **conditional lethal mutants** are needed. In most cases, **temperature-sensitive mutants** are isolated. These encode products active at reduced (permissive) temperature and inactive at elevated temperature. The respective mutants are identified by replica plating at different temperatures. The underlying amino acid exchanges usually reduce the stability of the encoded protein. Alternatively, suppressor (or *sup*) strains carrying mutated tRNA genes are used. These permissive hosts allow synthesis of proteins containing an artificial stop codon (see Chapter 15.5.2). In wild-type (*sup*$^{+}$) strains, these proteins are prematurely terminated.

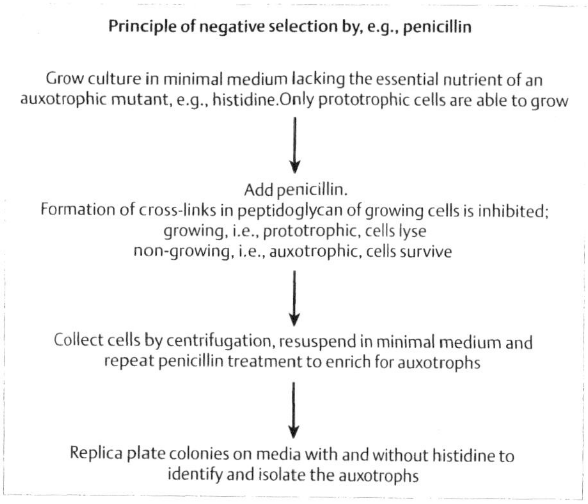

Principle of negative selection by, e.g., penicillin

Grow culture in minimal medium lacking the essential nutrient of an auxotrophic mutant, e.g., histidine.Only prototrophic cells are able to grow

↓

Add penicillin.
Formation of cross-links in peptidoglycan of growing cells is inhibited;
growing, i.e., prototrophic, cells lyse
non-growing, i.e., auxotrophic, cells survive

↓

Collect cells by centrifugation, resuspend in minimal medium and repeat penicillin treatment to enrich for auxotrophs

↓

Replica plate colonies on media with and without histidine to identify and isolate the auxotrophs

Fig. 15.**19** **Principle of negative mutant selection by, e.g., penicillin.** The strategy of penicillin selection is outlined. This protocol can be used to enrich auxotrophic mutants in a large number of prototrophs using various bactericidal agents

Mutants are isolated by selection or screening. Screening is often preceded by negative selection against the wild-type. Transposon mutagenesis tags the affected locus with the known nucleotide sequence of the transposon, and the pool of the mutants can be selected from the wild-type by resistance to the respective antibiotic, encoded by the transposon.

15.6 The Stability of the Bacterial Genome Is Increased by DNA Repair

Faithful reproduction of species requires stability of their genomes, whereas evolution is based on changes in genomes by mutations as discussed above. To counterbalance potential destruction by extensive mutagenesis, all organisms have developed DNA repair systems, which keep modifications and destructions of genetic information in check. To ensure some conservation in the genetic information of species, each frequently occurring mutagenic event is counterbalanced by a specialized repair system, the more important of which will be described in greater detail in the following paragraphs.

Repair counterbalances mutations and protects the genome from excessive alterations. Repair processes remove mismatches present after replication or recombination and damaged DNA.

15.6.1 Dam-Methylation-Directed (Long Patch) Mismatch Repair Removes Replication Errors

Mismatches may arise by replication errors, formation of heteroduplices during recombination, and deamination of 5-methylcytosine to thymidine. Repair of replication errors is more than just their enzymatic removal because it requires the distinction of the right from the wrong nucleotide in the mismatch. The best-understood repair system is the methyl-directed mismatch repair in *E. coli*, the Dam (**D**NA **a**denine **m**ethylase) system, which methylates GATC sites at the exocyclic 6-amino group of adenine (Fig. 15.**20**). Dam mutants of *E. coli* that lack the methylase (proteins MutH, MutL and MutS) show an increased frequency of spontaneous mutations. Direct evidence for the involvement of Dam-dependent methylation in repair was obtained from the analysis of the direction of mismatch repair in heteroduplices hemimethylated at GATC sites. The results indicated that the methylated strand serves as a template for the repair reaction, leading to removal of the nucleotide in the non-methylated strand. This result proves that the repair mechanism is able to faithfully remove replication errors because the newly synthesized DNA in semiconservative replication is hemimethylated for a short period of time, during which a replication error can be corrected. Gap repair synthesis followed by ligation of the remaining nick

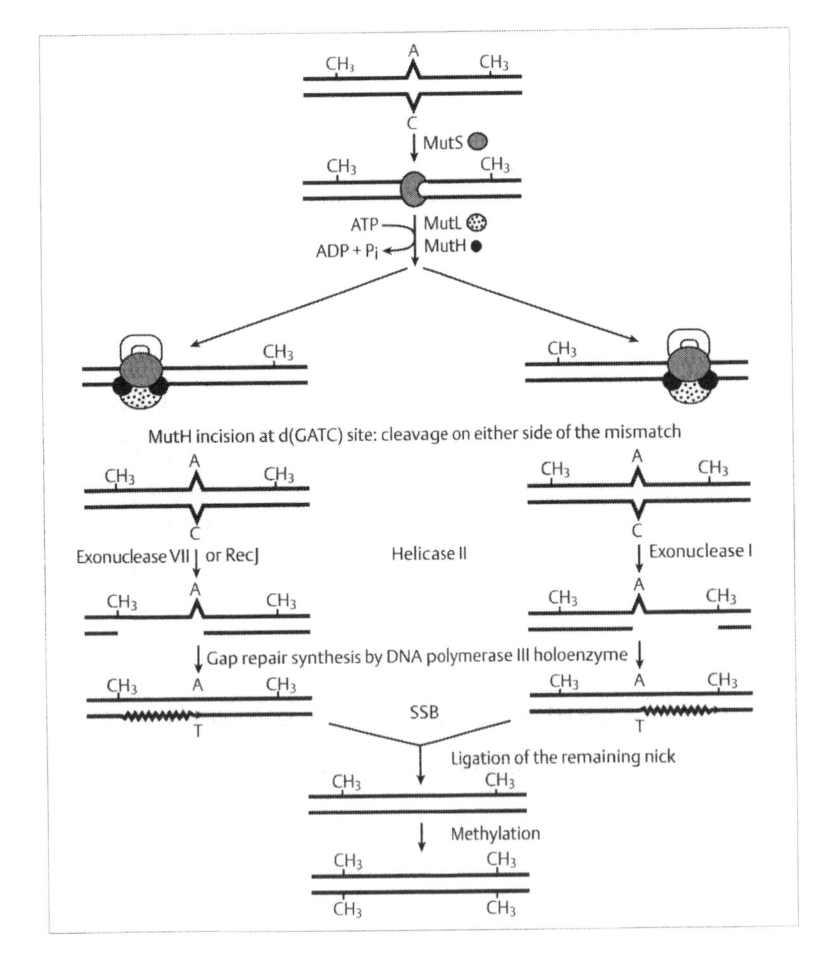

Fig. 15.**20** **Mechanism of DAM-directed mismatch repair.** A hemimethylated (CH_3) DNA double strand with an A–C mismatch is shown at the top. The MutS protein recognizes and binds the mismatched site, and MutL and MutH form a quarternary structure involving the hemimethylated sites at the expense of ATP hydrolysis. MutH cleaves the non-methylated strand on either side of the mismatch. The action of various exonucleases and DNA polymerase III, followed by ligation and methylation yields repaired DNA (long-patch mismatch repair)

leads to the correctly paired double strand. Since the transiently formed gap can extend over several hundred base pairs, this process is also called **long-patch mismatch repair**. Similar systems may contribute to replication fidelity in many, if not all organisms.

15.6.2 Very Short Patch Mismatch Repair Corrects Hot Spot Mutations

Short patch mismatch repair is distinguished from long patch repair by the limited size of the excision, usually about ten bases. The best-characterized example is the very short patch repair in *E. coli*, where **Dcm**-(**D**NA **c**ytosine **m**ethylase)-dependent methylation creates two inner 5-methylcytosines at CCTAGG sequences. These are hot spots for mutations because spontaneous deamination of 5-methyl-C yields T, generating a G·T mismatch. In the CCTAGG sequence context, this mismatch can be repaired in fully methylated DNA to a GC base pair by the action of MutS and MutL from the *DAM*-directed repair system and Vsr, a strand-specific mismatch endonuclease. Thus, this repair mechanism may be considered as an example of a defense system against a mutational hot spot.

Another mismatch-specific very short patch repair in *E. coli* is performed by the MutY protein, which recognizes and repairs A·G, G·A, and A·C mismatches. MutY is an adenine-DNA-glycosylase that removes the adenine residues from these mispairs. The concept of short patch repair represents another repair strategy that is used to increase genetic stability in many organisms.

15.6.3 Photo Repair Is a Specialized Repair System

All organisms exposed to sunlight are subject to DNA damage by UV irradiation. Therefore, a specialized repair mechanism has evolved, called **photo repair**, which is able to remove photo-induced pyrimidine dimers from double-stranded DNA. The underlying experimental result is that short exposure to UV light yields a higher mutation rate than extended exposures. Thus, continued exposure to light is anti-mutagenic, because a repair system is induced. The wavelength dependencies for damage and repair are different. In contrast to mismatch repair described above, photo repair does not lead to excision of the damaged nucleotides from the DNA. Instead, long wavelength light (>300 nm) is absorbed by an enzyme called **photolyase** (*phr*), which then recognizes pyrimidine dimers in double-stranded DNA and uses the energy of the absorbed light to reverse the cyclobutane structure directly, yielding the undamaged nucleotides. Enzymes with this activity are widely distributed among many species, probably as the result of the tremendous evolutionary pressure to protect genetic information from the frequently occurring light-induced damage.

15.6.4 Repair of DNA Alkylation in *E. coli* Leads to Modification of the Repairing Enzymes

E. coli shows an adaptive response to mutagenesis by alkylating reagents (e.g., dimethyl sulfate, *N*-ethyl-*N*-nitrosourea): The mutation frequency reaches a plateau value after initially increasing with the time of exposure. After about 60 min of exposure, the number of mutants obtained by treatment with methylating agents, e.g., *N*-methyl-*N*′-nitro-*N*-nitrosoguanidine or ethyl methanesulfonate, does not increase further. This phenomenon is not due to base excision repair, but depends on induction of synthesis of O^6-methylguanine-DNA-methyltransferase I, an enzyme that removes the methyl group from O^6 of guanine by transferring it to a cysteine residue within the enzyme. This type of repair reaction has also been demonstrated for O^4-alkylthymine and phosphotriesters in DNA. *E. coli* contains two distinct proteins with this enzymatic activity. Similar enzymes have been found in many other bacteria, and also in *Saccharomyces cerevisiae* and *Drosophila melanogaster*.

Another way of removing alkylated nucleotides in *E. coli* is provided by the action of DNA-glycosylases, some of which exhibit substrate specificity for chemically modified bases in DNA. The most prominent enzyme of this class is uracil-DNA-glycosylase (*ung*), which removes uracil residues by hydrolysis of their *N*-glycosidic bond. Other substrates include 3-methyladenosine, hypoxanthine, 5,6-dihydroxydihydropyrimidine, and pyrimidine dimers. The site from which the base was removed is subject to strand cleavage by an endonuclease, followed by regular repair synthesis.

Specialized repair mechanisms have evolved to counterbalance the most frequently occurring damaging reactions of DNA.

15.7 The SOS Response in *E. coli* Is Induced by Single-Stranded DNA

When DNA damage proceeds to the extent that single-stranded DNA appears in the cytoplasm of *E. coli*, a whole set (regulon) of about 20 genes is induced. The function of the induced proteins is to help the cell to cope with large lesions in the DNA. This phenomenon is called the **SOS response**. It is closely linked to recombination since the major recombination protein, RecA, functions as the sensor for single-stranded DNA. The RecA–ssDNA complex supports autoproteolysis of the LexA repressor, rendering it unable to repress the SOS genes (called *din* genes for **d**amage-**in**duced). Their encoded proteins catalyze SOS repair. Thus, distressing conditions lead to an increased repair capacity. Figure 15.**21** illustrates the regulatory network leading to the SOS response.

The SOS response is also involved in recombination since *recA* expression is controlled by the LexA repressor. It is hypothesized that gaps or double-stranded breaks, which may result from illegitimate recombination or DNA damage, can be repaired by the SOS system. DNA synthesis conducted during SOS response is an error-prone synthesis, which leads to increased numbers of mutations in the "repaired" DNA ("indirect mutagenesis"). This is consistent with the observation that UV light produces more mutations in *E. coli* that are wild-type for their SOS response, than in *E. coli* cells mutated in the SOS-controlled *umuCD* genes. This result suggests that the proteins encoded by these genes actually increase the mutation frequency. Mutants unable to induce the SOS response are hypersensitive to UV irradiation. Thus, the *umuCD* locus is involved in error-prone repair. The current hypothesis is that repair replication of DNA containing various lesions is supported by the UmuCD proteins at the expense of fidelity. It is speculated that an error-prone repair in combination with an increased recombination frequency may contribute to survival under growth conditions that promote large DNA lesions, by increasing the chance of favorable mutations in the population.

A general repair response is provided by the **SOS system** in *E. coli*. The SOS system responds to damages of DNA, independent of the nature of the primary lesion. It involves an error-prone repair synthesis that generates an increased number of mutations in response to DNA damage ("indirect mutagenesis").

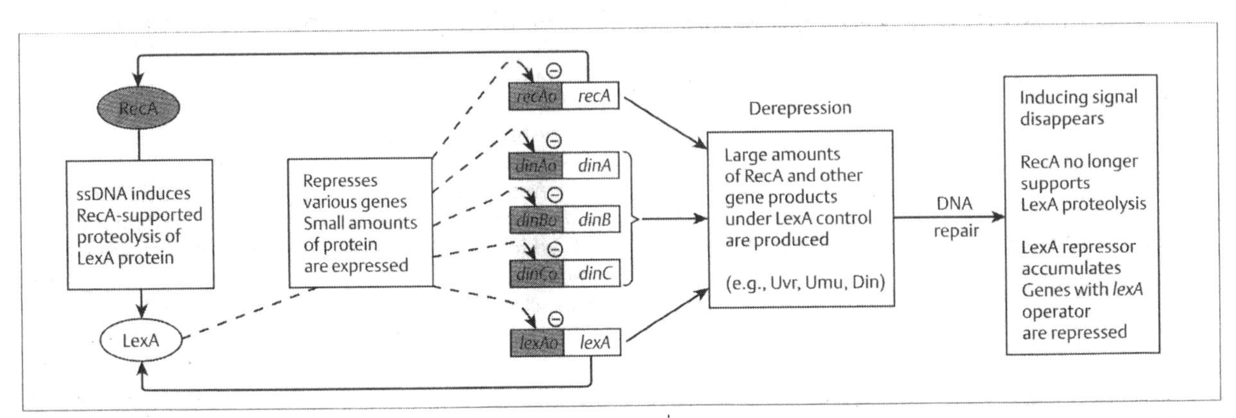

Fig. 15.**21** **The regulatory network controlling SOS repair in *E. coli*.** The occurrence of single-stranded DNA is sensed by RecA as an indication of damaged DNA. RecA activated by ssDNA catalyzes autoproteolysis of the general SOS repressor LexA, thereby inducing the SOS regulon. This regulon consists of a large number of genes (**d**amage-**in**duced or *din*) and operons all controlled through the binding of LexA to an operator with a common consensus sequence preceding these genes. After SOS repair has taken place, LexA accumulates and leads to repression of the genes involved in SOS response

Further Reading

Busby, S., and Ebright, R. H. (1994) Promoter structure, promoter recognition, and transcription activation in prokaryotes. Cell 79: 743–746

Dale, J. W. (1994) Molecular genetics of bacteria. New York: Wiley

Farabaugh, P. J. (1996) Programmed translational frameshifting. Microbiol Rev 60: 103–134

Friedberg, E. C., Walker, G. C., and Siede, W. (1995) DNA repair and mutagenesis. Washington, DC: ASM Press

Hopwood, D. A., and Chater, K. E. (1989) Genetics of bacterial diversity. New York, London: Academic Press

Knippers, R. (1997) Molekulare Genetik 7th edn. Stuttgart, New York: Thieme

Krohn, M., Pardon, B., and Wagner, R. (1992) Effects of template topology on RNA polymerase pausing during in vitro transcription of the *Escherichia coli rrnB* leader region. Mol Microbiol 6: 581–589

Kucherlapati, R., and Smith, G. R. (1988) Genetic recombination. Washington, DC: ASM Press

Moses, R. E., and Summers, W. C. (1988) DNA replication and mutagenesis. Washington, DC: ASM Press

Neidhardt, F. C., Curtiss, R., Ingraham, J. L., Lin, E. C. C., Low, K. B., Magasanik, B., Reznikoff, W., Riley, M., Schaechter, M., and Umbarger, H. E. (eds.) (1996) *Escherichia coli* and *Salmonella*: cellular and molecular biology. 2nd edn. Washington, DC: ASM Press

Nierhaus, K. H. (1993) Solution to the ribosome riddle: how the ribosome selects the correct aminoacyl-tRNA out of 41 similar contestants. Mol Microbiol 9: 661–669

Scaife, J., Leach, D., and Galizzi, A. (1985) Genetics of bacteria. New York, London: Academic Press

Singer, M., and Berg, P. (1991) Genes and genomes. Oxford, London: Blackwell

Snyder, L., and Champness, W. (1997). Molecular genetics of bacteria. Washington DC: ASM Press

Sonenshein, A. L., Hoch, J. A., and Losick, R. (1993) *Bacillus subtilis* and other Gram-positive bacteria. Washington, DC: ASM Press

Zubay, G. (1987) Genetics. Menlo Park, Calif.: Benjamin/Cummings

16 Genetic Exchange Between Microorganisms

Prokaryotes are unique in their ability to respond to environmental challenges by fast acquisition of appropriate genetic traits. Rapidly increasing problems caused by the emergence of antibiotic-resistant pathogens are a dramatic example of such an adaptation of bacteria to changed conditions. The genetic flexibility of prokaryotes is due to the natural existence of efficient means for **horizontal gene transfer** among bacteria, which are not parent and progeny, and between bacteria and other organisms. Genetic exchange, therefore, plays a prominent role in the evolution of prokaryotes. The mechanisms for DNA transfer include: (1) uptake of free DNA from the environment, a process known as **transformation**, (2) transfer of DNA from a bacterial donor to a recipient cell by **conjugation**, and (3) spread of bacterial genes by temperate bacteriophages, a process designated as **transduction**.

The fate of the transferred DNA in the recipient cell depends on the nature of the DNA molecule itself. If the DNA does not contain an active vegetative origin, the DNA has to be integrated into the host genome to become stably inherited. Most prokaryotic cells bear several mechanisms of homologous and illegitimate recombination involved in this integration.

Plasmids and bacteriophages replicate autonomously. However, as all foreign DNA, their genomes are subject to the restriction–modification systems of the host cell. Restriction of foreign DNA is thought to safeguard the bacterial cell against killing by bacteriophages and to ensure that the viability of a bacterial population is not impaired by too frequent changes in the genomes of individual cells. To overcome the restriction barrier, some plasmids and phages have acquired anti-restriction systems that protect imported DNA from degradation.

16.1 Uptake of Free DNA Can Result in Genetic Transformation of Bacteria

Bacteria become genetically transformed by taking up DNA from the environment and recombining it into their genome. Transformation has been studied intensively for more than three decades in a few bacterial species. The best-characterized systems include those in: *Streptococcus pneumoniae*, *Bacillus subtilis*, and *Haemophilus influenzae*. Transformation was discovered in *S. pneumoniae*. The observation that non-pathogenic mutants of *S. pneumoniae* can be rendered pathogenic by concomitant injection of virulent, but killed cells into mice revealed that DNA released from the cells acts as the transforming principle. Transformation frequencies can be very high, depending on the fraction of competent cells in the population.

Genetic competence is defined as the ability of bacteria to become transformed by the uptake of exogenous DNA.

Competence can be achieved by treatment of bacteria with physical or chemical agents to permit the uptake of transforming DNA. However, this "artificial" competence is different from "natural" competence, which is a genetically and physiologically determined property of a particular strain. Natural competence can be inducible or constitutive. The latter is observed in *Neisseria gonorrhoeae*. Induction of competence in *S. pneumoniae* can be triggered by diffusible factors—peptides that are excreted into the medium. The transformability of *B. subtilis* is growth-phase-dependent; under growth-rate-limiting conditions, about 10% of the cells can become competent. The transformation process can be subdivided into three stages: adsorption of the DNA to the cell envelope, penetration, and recombination. Uptake of the DNA requires its conversion to a single-stranded, nuclease-resistant form, and is followed by recombination with the chromosomal DNA of the host cell.

16.1.1 Genetic Competence in *Bacillus subtilis* Is Growth-phase-dependent

Emergence of competence. Competence of *B. subtilis* occurs under growth-limiting conditions, e.g., in the stationary phase of a culture. Development of competence is stimulated by glucose. Glutamate has an inhibitory effect, suggesting that nitrogen supply may be crucial for the competence state.

Binding of DNA. The first step in the transformation of *B. subtilis* consists in the wash-resistant association of exogenous DNA with the surface of competent cells (Fig. 16.**1**). DNA adsorption takes place at specialized structures proposed to traverse the cell wall and the cytoplasmic membrane. Assuming that one binding site can adsorb one molecule of DNA, it is estimated that competent cells have about 50 such binding sites. The association between the bacteria and the DNA is noncovalent since DNA can be recovered by treatment with detergents and phenol. DNA binding is not sequence-specific; however, glycosylated DNA and dsRNA are poor substrates, indicating that dsDNA is the natural substrate for binding.

Fragmentation. The bound DNA is subject to double-strand cleavage. Fragmentation occurs at random points, but with a defined size distribution. In the bound state, the DNA is still susceptible to added DNase, suggesting that the DNA forms an extended structure at the cell surface. Experiments involving genomic DNA of phage T7 (approximately 40 kb) reveal a mean size of 19 kb of the resulting fragments.

Uptake of transforming DNA. Following fragmentation, the recipient-attached DNA becomes completely inaccessible to DNase within 1–2 min. Current models interpret this observation either by transport of the DNA across the membrane or by penetration in a site at which DNA cannot be attacked by added nucleases. Experiments measuring the time of entry of distinct genetic markers indicate that DNA is taken up as a linear molecule. Uptake rates of 50–200 bp/s at 28°C are reported.

At this stage, the DNA can be recovered from the cells in the form of single strands. In a concerted process with the appearance of DNase-resistant transformants, acid-soluble DNA fragments can be detected in the supernatant of the cultures, suggesting that one strand of the transforming DNA is nucleolytically digested during uptake. Preliminary results indicated no preference for the chemical polarity with which the DNA single strand enters the cell. This would require a process strictly different from other known DNA transfer systems, such as conjugative DNA transfer or DNA injection by ssDNA phages. Therefore, the existence of two different DNA uptake systems in *B. subtilis* has been discussed. Alternatively, DNA could be taken up as a double strand, followed by intracellular digestion of one single strand. Since both alternatives seem rather unlikely to occur in the cell, more experimental data are required to determine the polarity of DNA uptake in *B. subtilis*.

Integration and resolution. Integration of transforming DNA proceeds by recombination in a RecE-dependent process. The *recE* gene of *B. subtilis* is an analogue of the *Escherichia coli recA* gene, which is essential for homologous recombination (see Chapter 16.4). About 75% of the ssDNA mass is eventually integrated. The product of recombination is a paired (three-stranded) heteroduplex of donor and recipient molecules. Integrated DNA segments are clustered, confirming the finding that donor DNA molecules are fragmented on the surface of the cell. If fragmentation is followed by efficient uptake and integration, this should result in a number of clustered, but independent integration events from each molecule that was initially bound.

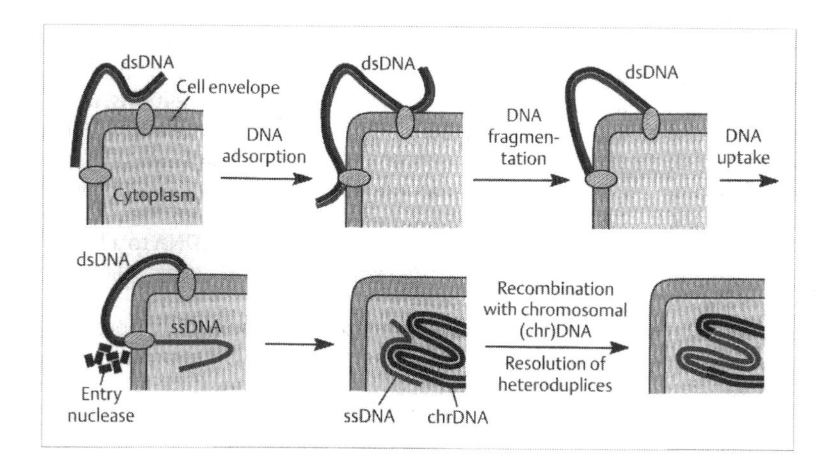

Fig. 16.1 **Model for transformation of *Bacillus subtilis*.** Double-stranded DNA (dsDNA) is bound by special protein complexes at the cell surface. Following DNA fragmentation at the DNA attachment sites, a DNA single strand (ssDNA) is taken up, while the complementary strand is degraded by an entry nuclease. The imported ssDNA is incorporated into the bacterial chromosome (chrDNA) by homologous recombination (see Chapter 16.4)

Genetics of natural competence. Based on the mode of expression and on the proposed function within the transformation process, genes involved in natural competence of *B. subtilis* have been classified into two types: regulatory ("early") genes, and "late" *com* genes (for **com**petence). The latter are expressed only transiently and are proposed to be directly involved in binding and processing of DNA or in the proper assembly of the competence machinery (see Chapter 25.1).

Mutational analyses of competence genes revealed that three units are transcribed during postexponential growth; mutations in these units (*comC*, *comDE*, and *comG*) can result in a competence-deficient phenotype. However, the Com$^-$ phenotypes generated so far are of only limited suitability in assigning functions to certain gene products. Proposed roles are based on sequence similarities found, for example, between *comG* gene products and proteins involved in the passage of macromolecules through bacterial cell membranes and to type 4 pilins of Gram-negative bacteria. An endopeptidase activity of the *comC* gene product has been demonstrated in vitro. Since certain mutations in *comDE* still allow binding of DNA to the cell surface, a role in DNA uptake of the corresponding gene products has been proposed. Two further genes seem to code a sensor (ComP) and a response-regulator (ComA) of a two-component system (see Fig. 25.**2**) involved in sensing starvation and triggering competence.

16.1.2 Mechanistic Variations in Natural Competence of Other Species Are Found

Haemophilus influenzae. The mechanism of transformation in *H. influenzae* seems to be different from that of other species, such as *S. pneumoniae* and *B. subtilis*. Upon entry into the stationary growth phase, vesicle-like structures form at the cell surface, and the lipid and protein composition of the cell wall changes. Parallel to the changes in the cell wall, the recombination frequency increases, and single-stranded gaps and "tails" appear in the chromosomal DNA, preparing the cell for incorporation of exogenous DNA into the genome. Virtually all cells of a culture can become competent. DNA uptake is highly specific for homologous DNA. Discrimination between heterologous and homologous DNA is possible because of the presence of 11-bp uptake sites (5′-AAGTGCGGTCA-3′) in the *H. influenzae* DNA, occurring in intervals of roughly 4 kb. The vesicle-like membrane extensions, designated as **transformosomes**, are involved in DNA recognition, DNA uptake, and protection of the DNA against nucleases. Transformosomes are located preferably at fusion points between the cytoplasmic and outer membrane ("adhesion sites," see Chapter 2). Addition of homologous DNA leads to efficient internalization of the transformosomes, and the DNA becomes insusceptible to nucleases. Transport into the cytoplasm is followed by release of the DNA and its incorporation into the genome by homologous recombination.

Streptococcus pneumoniae. Emergence of competence in *S. pneumoniae* is triggered by the accumulation of a diffusible **competence factor** (**CF**) in the culture medium. CF is a basic protein that apparently interacts with receptors in the cell envelope. The transformation process itself in general appears to be similar to that reported for *B. subtilis*. DNA uptake involves nucleolytic digestion of one of the incoming DNA strands. A membrane-located nuclease required for DNA uptake has been identified. This nuclease functions endonucleolytically when in solution, but apparently is an exonuclease in situ, which probably reflects its topology in the membrane. For *S. pneumoniae*, a 3′→5′ polarity of DNA uptake and a requirement for ATP have been indicated.

16.1.3 Efficient Transformation of *Escherichia coli* Requires Reversible Modification to the Cell Envelope by Chemical or Physical Means

In coliform bacteria, a high-efficiency competence state comparable to those described above does not exist. Transformation obviously is not an important mechanism for genetic exchange in these organisms. *E. coli*, however, always has had immense importance as a model organism for genetic studies. Therefore, the discovery that treatment of *E. coli* cells in the cold with $CaCl_2$ makes the cells competent for uptake of exogenous DNA during a short "heat pulse" was an important step in the development of genetic engineering.

Modifying the procedure by replacing $CaCl_2$ with a mixture of salts containing RbCl as the major component and adding dimethylsulfoxide improved the yield of transformants so that under optimal conditions, up to 20% of the viable cells can be transformed. The mechanism of action of $CaCl_2$ or other salts on the cells is still poorly understood. In contrast to the efficient transformation mechanisms that exist in nature and that involve conversion of the exogenous DNA to a linear single strand, artificially competent cells take up circular double-stranded DNA.

An alternative method, originally developed for transformation of eukaryotic cells, is the introduction of DNA into bacterial cells by **electroporation**. The cells are subjected to short pulses of a high electric field, which allows the DNA to pass the cell membranes. The main advantage of this method is that, in contrast to the $CaCl_2$ procedure, it can be applied to most bacterial species without a substantial change in the protocol. The yield of transformands obtained with this procedure generally is comparable to or higher than the yield obtained with the $CaCl_2$ procedure.

16.2 Conjugative DNA Transfer Requires Intimate Cell–Cell Contact

Conjugation allows the unidirectional transfer of genetic material from a donor to a recipient organism via direct cell–cell contact. Since this mode of DNA transfer permits exchange of genes not only between different bacterial species, but also the transfer of DNA from prokaryotes to eukaryotes (Fig. 16.**2**), conjugative DNA transfer is considered as a major route of genetic exchange in nature.

The transfer machinery required for this complex process usually is encoded by plasmids or transposons. Plasmids can be either self-transmissible or mobilizable (see Chapter 14). Conjugative transfer of mobilizable plasmids requires the presence of a co-resident self-transmissible plasmid that provides missing transfer functions.

Transfer functions are subdivided into **mating pair formation functions (Mpf)** required for establishing the cell–cell contact and **DNA transfer and replication functions (Dtr)** involved in DNA processing during bacterial conjugation.

The discovery of the phenomenon of bacterial conjugation by J. Lederberg and E. L. Tatum in the 1940s provides one of the cornerstones of prokaryote genetics. Mobilization of *E. coli* chromosomal genes

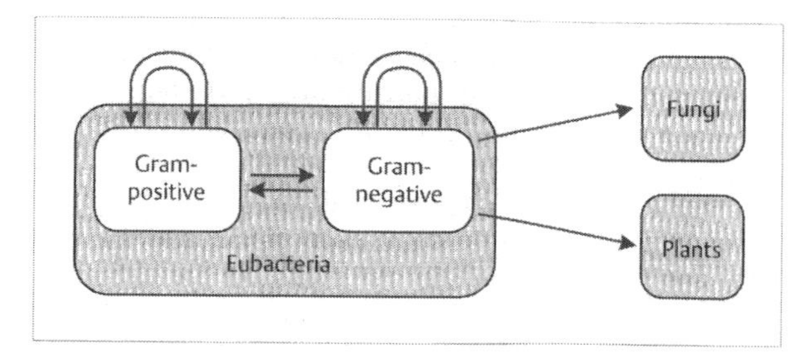

Fig. 16.**2** **Known interspecies and *trans*-kingdom gene transfer routes mediated by bacterial conjugation**

mediated by the "fertility factor" F, led to an initial understanding of the circular bacterial chromosome and of the process of homologous recombination. However, the F factor, later recognized by W. Hayes as a conjugative plasmid (renamed **F plasmid**) turned out to be quite atypical among the family of related "F-like" plasmids: expression of the F transfer (*tra*) genes is permanently derepressed. This results from the disruption of a co-repressor gene, *finO*, by insertion of an IS3 element. In other natural isolates of F-like plasmids, the *finO* gene product acts together with an antisense RNA, the transcriptional product of *finP*, to prevent expression of *traJ*, which encodes an activator of other transfer genes. F carries an entire collection of insertion elements that allow the plasmid to integrate into the *E. coli* chromosome at many sites. Integration occurs by reciprocal recombination with homologous IS elements already present in the *E. coli* genome. Chromosomal integration of F, a process known as **Hfr formation**, allows mobilization of adjacent host genes at a high frequency (see Chapter 16.2.1).

Without this serendipitous coincidence of unlikely details, the frequency of exchange of chromosomal markers between *E. coli* F^+ and F^- cells would have been far below the limits of detectability. The discovery of F in *E. coli* led to similar studies in other bacterial genera, resulting in the discovery of the FP factor in *Pseudomonas* and of a transmissible colicinogenic factor in *Vibrio*. Colicinogenic factors transmissible among enterobacteria formed the first major group of conjugative plasmids to be described.

A second even larger and more important group of conjugative plasmids, the resistance transfer factors, was discovered in the late 1950s and early 1960s in Japan. R factors (or **R plasmids** as they were later called) are responsible for the rapid spread of antibiotic resistance genes among enterobacteria, particularly in hospitals. The large number of plasmid isolates soon required a uniform taxonomic system. Classification was achieved via incompatibility grouping, which relies on the interrelationship between the regulatory systems controlling plasmid replication.

Box 16.1 Essential definitions related to bacterial conjugation:

- **Recipient.** Any cell that can accept DNA through horizontal transfer from a donor bacterium.
- **Donor.** Bacterium that contains a conjugative plasmid or a conjugative transposon and is able to transfer DNA through horizontal gene transfer to a recipient. The transfer is usually unilateral from the donor to the recipient.
- **Conjugative plasmid.** Self-transmissible plasmid that carries *tra* genes for proteins that promote cell–cell contact and DNA transfer to a recipient. Examples are the **F plasmid** (for **f**ertility) from enteric bacteria and **R plasmids** (for antibiotic **r**esistance) from various bacteria.
- **F^+** cell. Donor cell, for example, of *E. coli* carrying the F plasmid in an autonomous replicating form (replication from *oriV* of the plasmid).
- **Sex-duction.** Historic term for transfer of a conjugative plasmid to a recipient. If the F plasmid is involved, its transfer from an F^+ donor to an **F^-** recipient is also called **F-duction** (or conjugation). The process converts the original F^- to a secondary F^+ donor.
- **Hfr** (**H**igh **f**requency of **r**ecombination) strain. Donor in which the conjugative plasmid is integrated through insertion-sequence-(IS)-promoted homologous recombination into one of several (in *E. coli* about 12) chromosomal sites. The plasmid is replicated from *oriV* of the chromosome, and is hence strictly coordinated, while its *tra* genes are expressed throughout.
- **Episome.** Plasmid that replicates either as an autonomous unit (through an *oriV*) or is integrated into the host's chromosome (e.g., the F plasmid).
- **F′ plasmid** (also **R′ plasmid**). Plasmid derived from an Hfr strain, which through illegitimate recombination, picked up chromosomal genes from the host chromosome. The respective genes located on an F′ are indicated as F′*lac*$^+$, F′*gal*$^+$, F′*his*$^+$, etc.
- **Conjugation.** Transfer of (chromosomal) DNA by direct cell–cell contact from a donor bacterium to a recipient cell. Stable inheritance of the transferred DNA usually requires recombination into the chromosome.
- **Merogenote** or **diplogenote.** Since in bacterial conjugation only few genes become diploid (e.g., those located on F′ plasmids), the terms merogenote or diplogenote (mero stands for partial, i.e., not all genes are diploid) are used to describe the exconjugants. F′ plasmids are used in complementation and in *cis-trans* dominance tests as well as in gene mapping.
- **Interrupted mating.** During Hfr conjugations, transfer of the leading region of, for example, the F plasmid is followed by transfer of chromosomal genes. The process runs at a constant speed and requires about 100 min to transfer the entire chromosome of *E. coli*. If interrupted artificially (e.g., by vigorous shaking), the time required to transfer a given gene can be determined and can be used in combination with all Hfr strains and chromosomal mutants available to construct a **genetic map**.

Conjugative plasmids, common in bacteria, determine an unexpected diversity of horizontal gene transfer systems and play an important role in the practical aspect of antibiotic resistance transfer and also in the process of bacterial evolution. The latter is illustrated especially by the recent discovery of *trans*-kingdom gene transfer mediated by bacterial conjugation.

16.2.1 Plasmids Mediate Conjugative DNA Transfer in Gram-negative Bacteria

F and F-like plasmids. The prototype of plasmids of incompatibility group FI is the F plasmid, a plasmid of about 100 kb (Fig. 16.**3**). F is a low-copy-number plasmid (1–2 copies per chromosome equivalent). Many plasmids have been isolated that, as shown by hybridization studies, share extended regions of homology to F. Since these plasmids may be compatible to F or to each other, they were therefore designated as F-like plasmids and classified as belonging to incompatibility groups FII–FVII. The replication machinery of F is complex. The plasmid possesses at least two complete (basic) replicons (RepFIA and RepFIB) and part of a third (RepFIC). Several functions have been identified that are devoted to plasmid stabilization, partitioning, and coupling of the host's cell division to its own replication cycle (see Chapter 14).

The F plasmid carries at least three types of insertion elements: one copy each of IS2 and Tn*1000* (γ–δ) and two copies of IS3. All elements are clustered within a region of 20 kb. Since most naturally occurring strains of *E. coli* or *Salmonella typhimurium* carry multiple copies of IS2, IS3, and/or Tn*1000* (see Chapter 14.4), the presence of these IS elements on F allows the

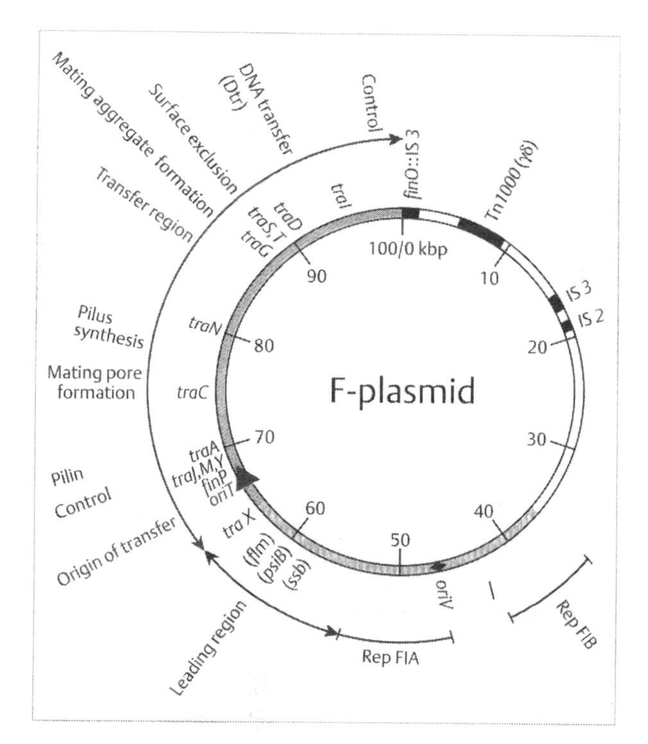

Fig. 16.3 **Structure of the F plasmid.** The F plasmid is a covalently closed circular DNA molecule (100 kb) with a transfer region (*tra* genes) and an origin of transfer (*oriT*, red arrowhead) at which DNA transfer during conjugation begins, continuing toward the leading region. (The leading region is transferred first.) Indicated are the transposable elements IS2, IS3, and Tn*1000* ($\gamma-\delta$) (black boxes), the extents of the leading region and the transfer region, and the two replication regions RepFIA and RepFIB, with the bidirectional vegetative region *oriV*. Numbers within the map indicate kilobase coordinates. The gene *finO* (encoding a repressor) is inactivated by an IS3 insertion, thus allowing synthesis of the regulator RNA *finP* and a constitutive expression of the large *tra* regulon (>30 kb, ~26 genes, ~34 gene products) [after 1]

plasmid to integrate at numerous sites into a host chromosome. Integration takes place via reciprocal recombination between homologous IS elements; therefore, the integrated plasmid is flanked by directly repeated IS sequences. Plasmid-encoded transfer functions are active also in the integrated state. When unidirectional plasmid transfer begins, it will not stop at the boundary between plasmid and chromosome; instead, it will result in conjugative transfer of chromosomal markers. Bacterial strains carrying a chromosomally integrated F plasmid have been isolated. These strains transfer chromosomal markers at a high frequency when mated with F^{-} strains, and therefore they have been called **Hfr** (**h**igh **f**requency of **r**ecombination) strains. Hfr strains are widely used for mapping chromosomal markers. DNA transfer is unidirectional, and in a particular Hfr strain, transfer initiates at a unique site within the chromosomal DNA. Therefore, the relative position of genetic markers can be deduced from their time-dependent order of entry into a recipient cell. The order of entry can be measured by interrupting the mating process (e.g., by vigorous shaking of the culture) at various times and determining the time-dependence of transfer frequencies of individual markers.

Chromosomal genes may become incorporated into F if the integrated plasmid is excised imprecisely by an illegitimate recombination event. The plasmids that are formed in this way are called **F′** (**F prime**) plasmids (or **R′** if the original plasmid was an R plasmid). The transfer (*tra*) genes of F are located in a continuous region of approximately 34 kb. Details of the F transfer apparatus will be described in Chapter 16.2.2.

IncP plasmids. IncP plasmids are characterized by their broad host range, which allows their stable inheritance in virtually all Gram-negative bacteria. For this reason, the IncP replicon has become an important tool for genetic studies in Gram-negative bacteria other than *E. coli*. IncP plasmids have a copy number of 6–8 per chromosome equivalent in *E. coli*. The minimal IncP replicon consists of only two elements: the vegetative replication of origin *oriV* and the product of the *trfA* gene (***tr****ans*-acting **r**eplication **f**unction), which binds and activates *oriV* for recognition by the host-encoded replication machinery. However, a replicon consisting of only these two components is easily lost and needs to be stabilized by a partitioning system.

Depending on their ability to exist stably in a variety of potential host organisms, plasmids are classified as having a broad or a narrow host range. However, the range of potential recipients of an associated conjugative DNA transfer apparatus usually is considerably greater. This is demonstrated impressively by DNA-transfer to yeasts directed by a narrow-host-range plasmid like F.

A further characteristic of IncP plasmids is the existence of a so-called ***kil–kor*** **regulon**, a network consisting of at least five repressor genes (*korA–G*) that regulate a set of functions, some of which are potentially lethal to the host when their expression is not regulated. Although almost nothing is known about the functions of these *kil* genes, *kor* gene products have been shown to be involved in the regulation of plasmid replication and maintenance and transfer functions. It is speculated that KorB mediates cross-talk between plasmid maintenance and transfer functions, allowing coordination of gene expression in response to the current state of the host cell.

Box 16.2 Based on DNA hybridization studies, IncP plasmids were classified as belonging to two subgroups: IncPα and IncPβ. However, since the differences exist mainly in the transfer regions, all these plasmids express P-type incompatibility, either symmetrically or asymmetrically. The **IncPα** subgroup contains the so-called Birmingham plasmids, which were isolated in the Burns Unit of the Birmingham Accident Hospital in 1969. These plasmids became noticeable since they encode resistance to three clinically important classes of antibiotics: β-lactam antibiotics (ampicillin, carbenicillin), aminoglycoside antibiotics (kanamycin, neomycin), and tetracyclines. Moreover, IncPα plasmids carry inducible cryptic tellurite-resistance genes that are expressed only when certain mutations activate the corresponding genes or gene products. Other IncPα plasmids are based on the Birmingham plasmids, but differ mainly in the various patterns of IS elements and transposons. The **IncPβ** subgroup is more heterogeneous, showing considerable divergence in the nucleotide sequences for plasmid maintenance and transfer functions. The best-characterized IncPβ plasmid, R751 (53 kb) confers resistance to trimethoprim to its host; the corresponding determinant specifies a dihydrofolate reductase and is located on the Tn*402* ele-ment.

Transfer genes of IncP plasmids are located in two separate regions occupying about 25 kb. Although the gene organization of transfer regions is very similar in IncPα and β plasmids, some specificity determinants cannot be exchanged between the two systems. This observation provided a clue for the identification of gene products that interact directly with the origin of transfer (*oriT*). Details of the transfer apparatus will be discussed in Chapter 16.2.2.

Plasmid mobilization. A number of small plasmids in Gram-negative bacteria are not self-transmissible by conjugation. However, virtually all naturally occurring plasmids can be mobilized efficiently if a co-resident conjugative plasmid provides **helper functions**. Typical representatives of this class of plasmids are ColE1 (6.6 kb), RSF1010 (8.7 kb), and pSC101 (9.3 kb).

Mobilization of these plasmids requires in addition to the helper plasmid a **transfer origin** (*oriT*) as a *cis*-acting element and a set of gene products encoded by the respective mobilizable plasmid that interacts specifically with this transfer origin. These are designated as mobilization or **Mob functions**. The nucleoprotein complex that forms at *oriT*—also called relaxosome (see Chapter 16.2.2)—is used by the transfer apparatus of the helper plasmid when transfer takes place. The efficiency of this interaction can vary: not every small plasmid can be mobilized by a certain conjugative plasmid. This observation might indicate different degrees of adaptation between relaxosomes of various mobilizable plasmids and the mating pore formation (Mpf) systems of the respective conjugative plasmids.

Plasmids can be either self-transmissible (conjugative) or mobilizable (non-selftransmissible). Whereas the former group encodes a complete conjugative DNA transfer apparatus (**Tra functions**), the latter group usually bears only the functions required for initiation of its own transfer DNA replication (**Mob functions**).

Interactions between transfer systems. Coexisting conjugative plasmids can mutually impair their transfer functions. For example, the fertility inhibition functions on IncFI and IncP plasmids are directed against IncP and IncW plasmids, respectively. IncFI plasmids, such as F, encode a protein, PifC, which normally is involved in replication control and regulation of the *pif* (**p**hage **i**nhibition by **F** plasmid) operon since the corresponding gene products provide exclusion of certain phages such as T7. The IncP plasmid RP4 (see Fig. 17.**1**) appears to have at least one binding site for PifC since the transfer rate of RP4 decreases by a factor of 10^2 to 10^3 in the presence of F (PifC$^+$). Accordingly, in the presence of RP4, expression of the *pif* genes is increased. The same effect is observed in the presence of the corresponding *pifO* operator sequence. The presence of additional *pifC* binding sites in the cell obviously lowers the concentration of free PifC protein, resulting in increased expression of *pif* operon genes. Inspection of the complete IncP sequence revealed a segment corresponding to one half of the symmetrical *pifO* operator sequence. The sequence is located within the IncP *traI* gene, which codes for the relaxase function involved in initiation of conjugative DNA replication (see below). It is likely that this sequence acts as a target for PifC, leading to down-regulation of IncP transfer gene products.

IncP plasmids, on the other hand, profoundly inhibit the transfer of coresident IncW plasmids, such as R388. This effect was assigned to two separate regions called *ftwA* and *ftwB* (**f**ertility **i**nhibition of

IncW plasmids). Both interactions are unidirectional: transfer of F is not influenced by the presence of RP4, and R388 does not inhibit transfer of RP4. Another example of the interaction of DNA transfer systems is the inhibition of agrobacterial tumorigenicity by IncW plasmids, most likely caused by repression of virulence function(s) of the Ti plasmid.

Fertility inhibition provides an evolutionary advantage for a plasmid capable of impairing the transfer functions of other conjugative plasmids coexisting in the same host cell.

16.2.2 A Model of Conjugative Plasmid DNA Transfer Between Gram-negative Bacteria

Two conjugative plasmids have been studied in great detail: the F plasmid, which served as a paradigm for bacterial conjugation for almost 50 years, and the IncPα plasmid RP4, which is mentioned here as representative of the virtually identical broad-range plasmids RP1, R18, RK2, and R68. This section will focus on the description of these two systems. The transfer regions of both plasmids have been sequenced completely. Moreover, the complete circular sequence of RP4 (60 099 bp) (see Fig. 17.**1**) and the IncPβ plasmid R751 (53 339 bp) have been compiled.

The current model of conjugative plasmid DNA transfer in Gram-negative bacteria involves six steps (Fig. 16.**4**): (1) mating aggregate formation and stabilization, (2) site-specific cleavage at the transfer origin (*oriT*) of a single strand in the DNA molecule destined for transfer, (3) unwinding of the DNA and transfer of the cleaved DNA strand with the 5′-end leading to the recipient, and concomitant DNA replacement synthesis in the donor cell by a rolling-circle-type mechanism, (4) recircularization of the transferred single strand, (5) complementary strand synthesis in the recipient and restoration of plasmid superhelicity, and (6) separation of the mating partners. Steps 4–6 are substituted in a conjugation involving chromosomal genes by their homologous recombination into the recipient chromosome, as will be described in Chapter 16.4.

Organization of transfer regions. The transfer regions of F and RP4 are characterized by the existence of very long polycistronic operons. The longest transcript in the F system (Fig. 16.**5a**) comprises more than 30 kb, encoding at least 34 gene products. In RP4 (Fig. 16.**5b**), the transfer functions are located in two regions

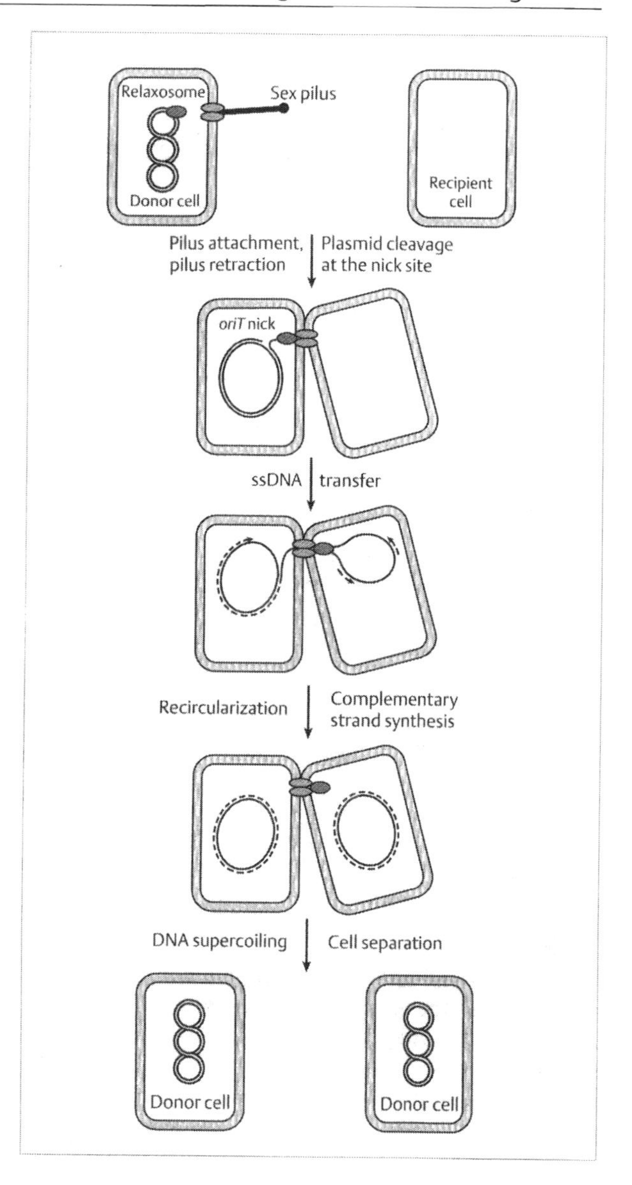

Fig. 16.**4** **Model of plasmid DNA transfer during conjugation.** A donor carrying a conjugative plasmid contacts a plasmid-free recipient cell (chromosomes are not shown) and starts mating through cell–cell contact as indicated. During the process called conjugation ("sex-duction"), a single strand of plasmid DNA is transferred to the recipient. This and the strand remaining in the donor are replicated and recircularized, leaving the old donor after cell separation with a plasmid and converting the previous recipient into a secondary donor. In true conjugation, chromosomal DNA is also transferred from the donor to the recipient, where it is integrated by homologous recombination into the host's chromosome. The recipient remains a recipient cell. For further explanations, see text

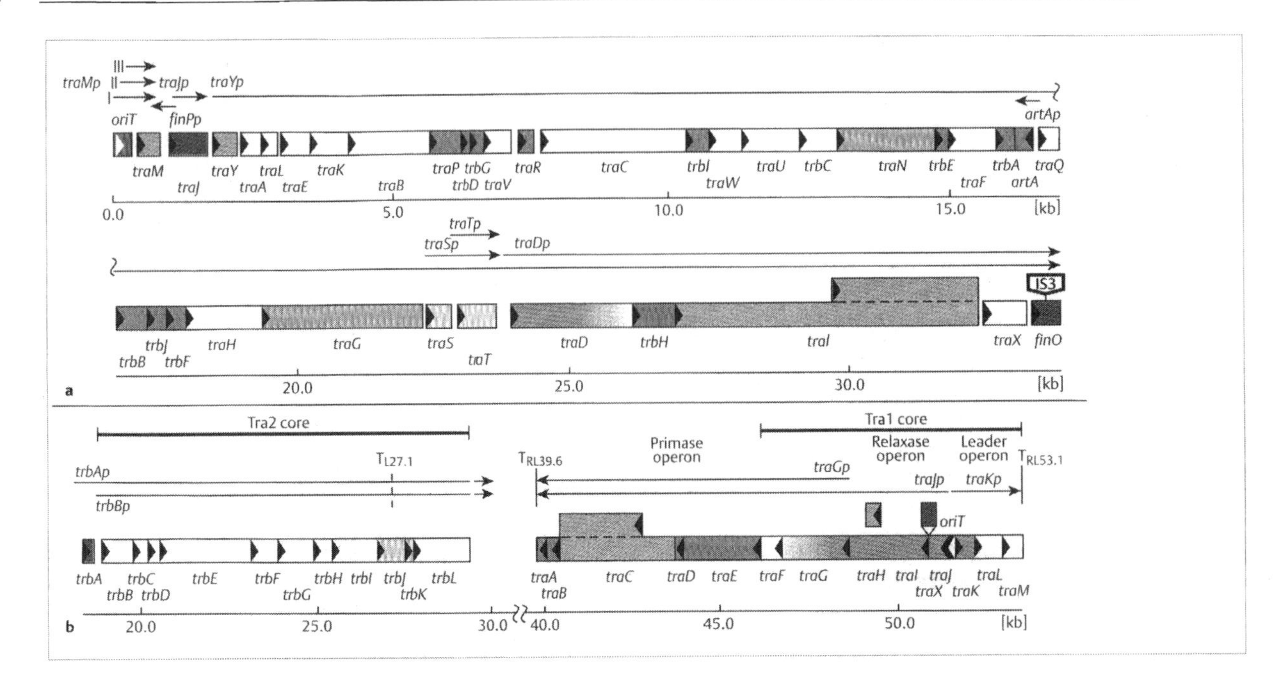

Fig. 16.**5a,b** **Organization of transfer (*Tra*) regions.** Transfer genes (*tra*) are drawn as bars; triangles within bars mark the 5′-ends. Functional classification of genes is indicated by the shading of bars: light red, Dtr system; white, Mpf system and pilus synthesis and assembly; light gray, entry exclusion; gray, mating pair stabilization; red, gene regulation; open, function not known. Transfer origins have white arrowheads that indicate the direction of DNA transfer. Arrows above the gene organization depict the extension of transcripts. Promoter sequences are designated *p*; rho-independent transcriptional terminators, t.
a Organization of the F Tra region.
b Organization of the RP4 Tra regions (see also Fig. 17.**1**)

designated Tra1 and Tra2, separated by the insertion element IS*8*, the kanamycin resistance gene, and the *mrs/par* locus, which is involved in plasmid segregation and stability.

Unlike those of F, the transfer genes of RP4 are functionally highly clustered: the Tra2 region is exclusively devoted to mating-pair formation and surface extension (Fig. 16.**5b**). Tra1 encodes all essential DNA processing functions, such as relaxosome components and DNA transport functions. In both systems, the transfer origins (*oriT*) are located within an intergenic region at one extreme of the Tra region. Plasmid transfer is polar in such a way that the bulk of *tra* genes enters a recipient cell last. Consequently, since in an Hfr conjugation, the DNA breaks before the entire chromosome has been transferred, the *tra* genes never enter the recipient, which does not become a donor as in conjugation involving an F^+ or F′ donor.

Step 1: mating aggregate formation and stabilization; conjugative or sex pili. Transfer of DNA by bacterial conjugation requires intimate cell–cell contact. To circumvent the repelling force caused by ζ-like potentials of bacterial cells in an ionic environment, conjugative plasmids of Gram-negative bacteria encode extracellular filaments that extend from the cell surface. These so-called **sex pili** mediate the first contact to a potential recipient cell (Figs. 16.**4** and 16.**6**). Pili encoded by plasmids of different incompatibility groups can be distinguished morphologically. *E. coli* cells carrying the F plasmid have 1–3 long, flexible pili (Figs. 2.**21** and 16.**6a**), those carrying RP4 have short, rigid pili (Fig. 16.**6b**). Only 1 out of 20 cells carrying RP4 appears to have a pilus when examined by electron microscopy. Pili morphology correlates with the ability of the cells to mate in liquid medium or on a solid surface: F is transferred better in liquid, whereas efficient RP4 transfer requires a solid surface.

Sex pili are hair-like appendices of the bacterial cell surface, the proteins of which are encoded by genes on conjugative plasmids of Gram-negative bacteria. Sex pili mediate the initial contact to a potential recipient cell.

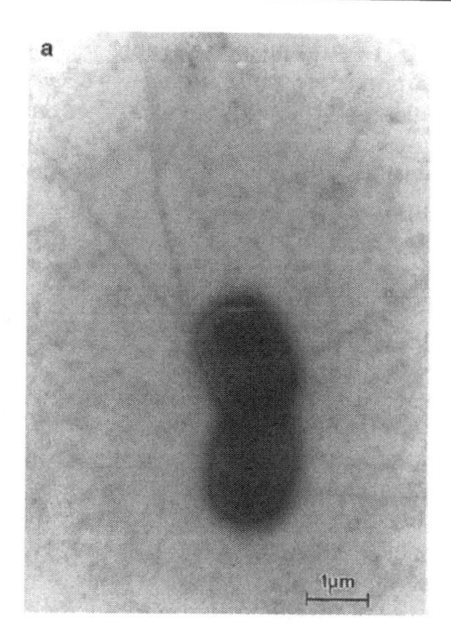

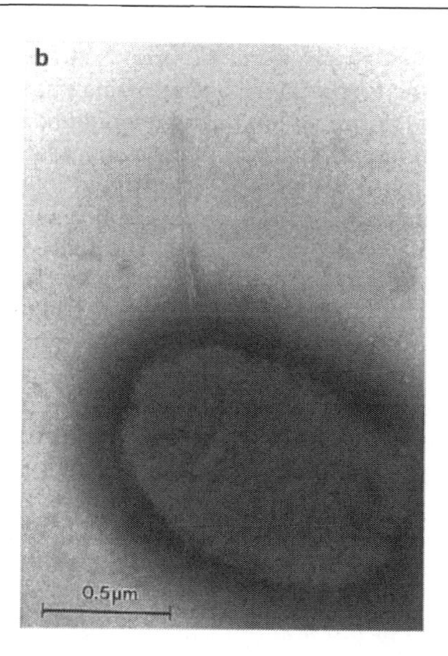

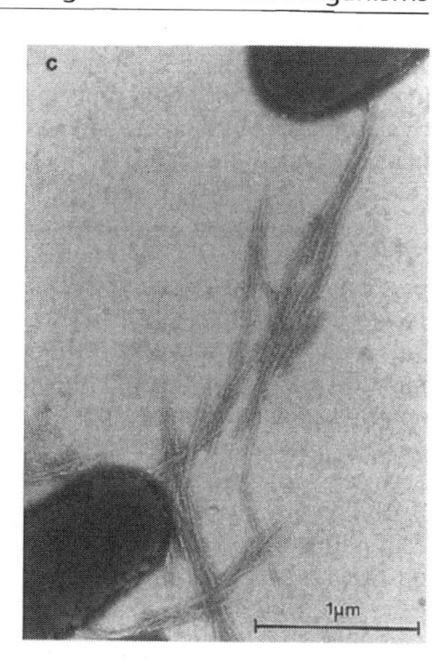

Fig. 16.**6a–c** **Piliated *Escherichia coli* cells.** *E. coli* JE2571 served as host for conjugative or recombinant plasmids. This strain does not encode common pili and fimbriael or flagella (Pil$^-$, Fla$^-$); therefore, plasmid-encoded pili can be identified unambiguously. **a** *E. coli* JE2571 (F), **b** *E. coli* JE2571 (RP4), **c** *E. coli* JE2571 [pML123 (Tra2), pWP471 (*traF*)]. The latter strain overexpresses RP4-encoded filaments. Cells were negative-stained with uranyl acetate. Courtesy of R. Lurz

Synthesis and assembly of functional F pili (see Chapter 23) requires 13 genes of the F *tra* region (Fig. 16.**5**). The pilus itself consists largely if not entirely of multiple subunits formed by the processed product of the *traA* gene. The *traA* gene specifies a propilin (121 amino acid residues, 13.2 kDa) from which a 51-amino-acid-residue segment is cleaved off to yield the mature pilin (7 kDa). Propilin processing requires the genes *traQ* and *traH*. Further maturation steps involve *N*-acetylation mediated by the non-essential *traX* gene product and possibly phosphorylation and glycosylation processes. The remaining ten pilus-specific genes are required for assembly of the pilus and formation of a putative transmembrane complex in the cell envelope that might establish a physical link to the DNA to be transferred. However, the individual functions of the corresponding gene products that are all located in the cell envelope remain to be elucidated. The **F pilus filament** consists of multiple pilin subunits arranged with fivefold rotational symmetry around the pilus axis, forming a hollow cylinder of 8-nm diameter with an axial hole of 2 nm (Fig. 2.**20**). Although this cylinder could allow the passage of a naked DNA single strand, DNA transfer through the pili has not been demonstrated.

The **initial mating contact** is formed between the pilus tip and a receptor site on the recipient cell surface. However, these initial mating aggregates are unstable in the presence of shear forces and ionic detergents. Stabilization involves retraction by depolymerization of the pilus, bringing the cells in close contact. Estimations indicate that F$^+$ cells contain about 10^5 copies of the pilin subunit, but of these, only 10–15% are assembled into pili. Therefore, a flow equilibrium of pilus assembly and disassembly is suggested that could explain why donor and recipient cells are pulled together after the initial contact. **Mating aggregate stabilization** requires the genes *traG* and *traN* of F and, from the recipient, the major outer membrane protein OmpA together with appropriate lipopolysaccharide membrane structures. Mating aggregate formation is thought to evoke some kind of trigger signal that is transmitted to the plasmid-encoded Dtr system, resulting in initiation of transfer DNA replication.

RP4-specified (P-type) pili are extremely sensitive to shear forces. In fact, observation of the pili by electron microscopy seems to be only possible if the cells are grown on solid medium. Consistent with this observation is that efficient transfer of RP4 requires donor and recipient cells to be incubated on a solid surface. This adaptation of the RP4 conjugative apparatus to DNA transfer on surfaces or in biofilms is in accordance with the hypothesis that IncP plasmids originated in soil bacteria (e.g., *Pseudomonas* spec.). Synthesis and assembly of P-type pili require the genes *trbB–trbL* of the RP4 Tra2 region—the so-called Tra2 core—plus the *traF* gene product, which is located in Tra1 (Fig. 16.**5**). The RP4 *trbC* gene product shares significant sequence similarity with the F TraA propilin and is therefore a likely candidate for being processed to the P-type pilin. Overexpression of the Tra2 core genes and of *traF* results in overproduction of pili (Fig. 16.**6c**), demonstrating that this set of genes is sufficient for P-type pilus assembly.

Propagation of donor-specific phages. Bacterial surface structures encoded by conjugative plasmids, in particular the pili, can serve as specific receptor sites for certain bacteriophages. Since these bacteriophages infect exclusively bacteria carrying their cognate plasmid, the associated plasmid phenotype is designated as donor-specific phage sensitivity (Dps). Donor-specific phages can serve as tool to elucidate relationships between Mpf systems of plasmids of different incompatibility groups. For example, phage PRD1, a lipid-containing phage with a linear double-stranded genome of 14.5 kb, infects cells carrying plasmids of incompatibility groups N, P, and W, but not F. Indeed, DNA sequence analysis indicates a close relationship of Mpf gene products of N-, P-, and W-type plasmids, but not of N- and F-type, P- and F-type, or W- and F-type plasmids (see Chapter 16.2.5). The existence of donor-specific phages probably is a major reason why pilus synthesis and hence conjugative DNA transfer are repressed on most conjugative plasmids under normal conditions.

Surface (or **Entry**) **exclusion.** Most conjugative plasmids, including F and RP4, drastically reduce the ability of their host cells to serve as recipients in matings with cells carrying identical or closely related plasmids. This phenomenon has been designated **surface** (or **entry**) **exclusion** (Sfx).

Surface exclusion has to be distinguished from **plasmid incompatibility**, which rather acts on the level of replication and plasmid partitioning during cell division. Surface exclusion, in contrast, prevents the formation of stable mating aggregates and/or the transport of the DNA into the recipient cell.

Cells carrying the F plasmid typically have a 100–300-fold reduction in their ability to act as recipients in $F^+ \times F^+$ matings relative to an F^- cell. Two plasmid-borne genes, *traS* and *traT*, are responsible for the Sfx phenotype. Although their contributions are synergistic, their gene products operate through quite unique mechanisms. The product of *traS* (16.9 kDa) is an inner membrane protein that is thought to act by inhibiting the triggering of conjugative DNA replication. The *traT* gene product (23.8 kDa), a lipoprotein that is located at an exposed site in the outer membrane, blocks conjugation at an earlier stage, before the cells form stable mating aggregates. Two models of TraT action are presently discussed: TraT could block a specific site on the major outer membrane protein OmpA that otherwise would be recognized by the pilus tip of a potential donor to initiate the mating process. Alternatively, TraT itself could interact with the pilus tip, thereby preventing the normal mating contact.

The entry exclusion function (Eex) of RP4 is encoded by Tra2. In contrast to the F plasmid, only one gene, *trbK*, is required for the reduced transmission of IncP plasmids into strains with a resident IncP-type plasmid. The phenotype is expressed independently of whether *trbK* is provided in cis or in trans. The mature TrbK appears to be an inner membrane lipoprotein of 47 amino acid residues which carries the lipid moiety linked to the N-terminal cysteine. The function of the protein depends on its presence in the recipient, but not in the donor cell. TrbK excludes plasmids of homologous systems of the IncP complex, but it is inert towards the IncI system. The likely target for TrbK action is the mating pair formation system because DNA or any of the components of the relaxosome have been excluded as possible targets.

Step 2: site-specific cleavage at *oriT* of a single strand in the DNA molecule destined for transfer; transfer origins. In the general model of conjugation, *oriT* represents the site on a conjugative or mobilizable plasmid where the transfer process is initiated by a specific single-stranded cleavage event. Common features of transfer origins are direct and inverted repeats thought to function as targets for protein-DNA recognition. The AT content of *oriT* usually is higher than that of the flanking regions, a feature that is also found in vegetative replication origins (see Chapter 14). In addition, *oriT* may include promoters responsible for transcription of *tra* or *mob* genes. Essential genes are transcribed divergently from promoters in the *oriT* region of IncP plasmids. Products of these genes specifically interact with the corresponding *oriT* DNA (Fig. 16.**7**).

Deletion analysis has shown that the F transfer origin is extensive, with domains that contribute individually to nicking and to transfer and to the overall structure. Major nicking determinants are inferred to lie in a region from about 22 bp to the right to 70 bp to the left of the nick site (Fig. 16.**8**). This domain includes an IHF binding site (consensus sequence YAANNNNT-TGATA/T). IHF binding sites enhance intrinsic bending in the F *oriT* DNA and may function in forming higher-order protein–DNA structures. IHF-dependent transfer has not been demonstrated for plasmids other than F-type plasmids.

Step 3: initiation of conjugative DNA transfer requires the assembly of a complex nucleoprotein

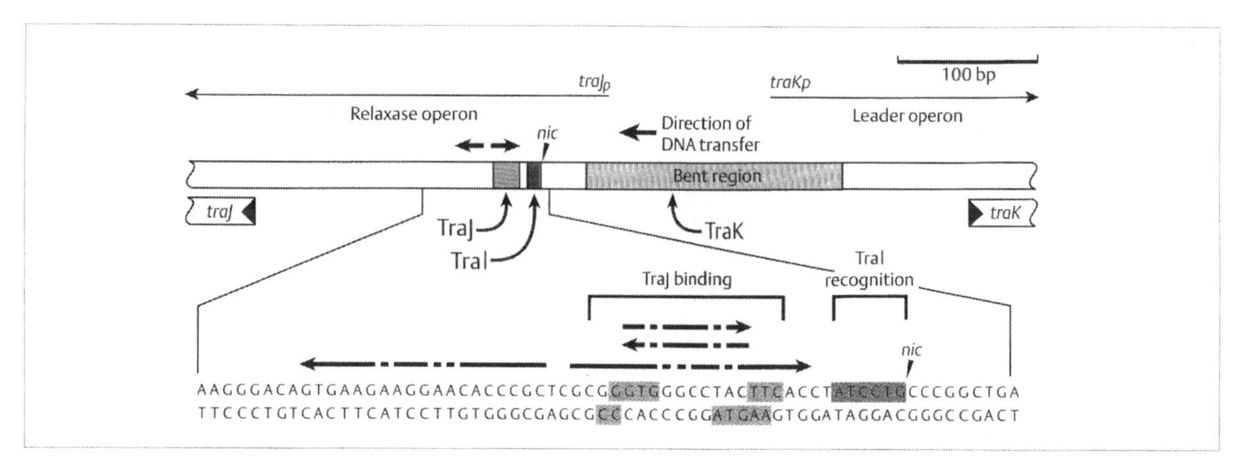

Fig. 16.7 **Modular structure of the RP4 transfer origin.** The RP4 transfer origin is represented by a bar. Binding sites for the proteins TraI and TraJ are drawn as dark and light red boxes, respectively. A bent DNA region that contains the TraK binding region is shown in dark gray. Transcription of the relaxase and leader operons initiates at two divergent promoters, *traJp* and *traKp*, respectively. The 5′-ends of the structural genes for TraJ and TraK are drawn as white bars. The DNA sequence of the region adjacent to the nick site is shown below. Inverted repeat sequences are represented by arrows, deviations from the symmetry are marked by dots. Nucleotide positions that are recognized by TraI and TraJ are shown with a red and a light red background, respectively. The nick site (*nic*) is indicated by an arrowhead. The upper DNA strand is the strand that is cleaved and transferred with 5′→3′ polarity to the recipient cell [after 2]

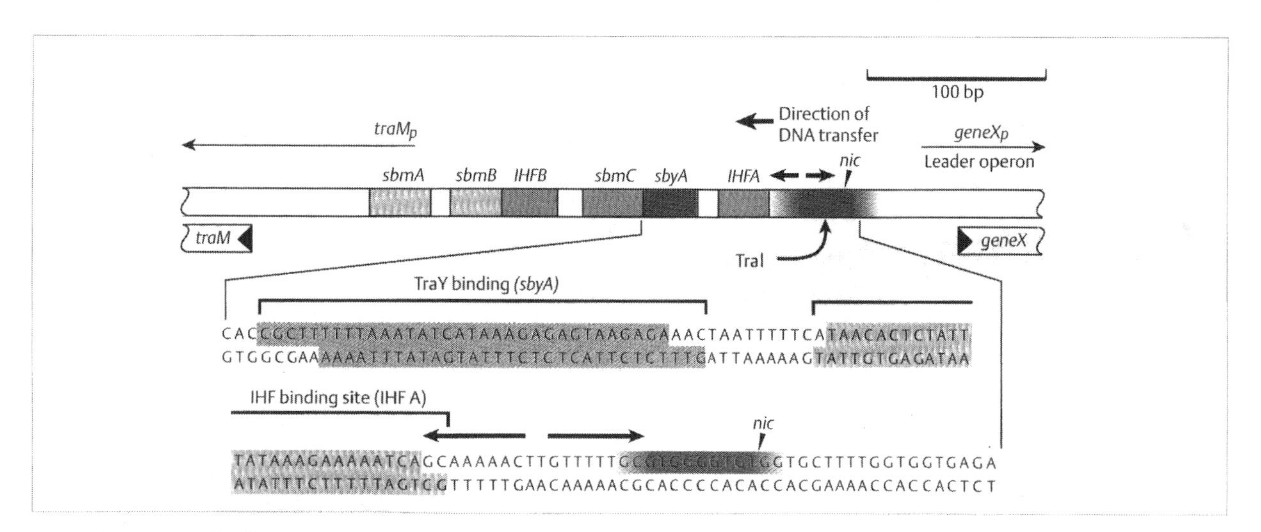

Fig. 16.8 **Protein binding sites within the F transfer origin.** Symbols correspond to those used in Fig. 16.7. Specific binding sites of TraM (*sbmABC*) and TraY (*sbyA*) are drawn in light gray and light red, respectively. Binding sites for the host-encoded histone-like protein IHF are shown in dark gray. The region recognized by the relaxase/helicase TraI, encoded by F, is shaded in dark red. General features are similar to those of the RP4 transfer origin, except that there are no known protein binding sites in the leading region of the F transfer origin

structure at the transfer origin—the relaxosome. Relaxosomes were first described for the non-self-transmissible, but mobilizable plasmid ColE1, where association of the transfer origin with three proteins was demonstrated. The molecular masses of these proteins agree quite well with those later deduced from the nucleotide sequence of mobilization genes mapping adjacent to *oriT*. Treatment of ColE1 relaxosomes with ionic detergents disrupts the interactions between the relaxase components and the DNA. Following this treatment, one protein (60 kDa, MbeA) remains tightly—most likely covalently—attached to plasmids that are captured in the nicked state.

Relaxosomes (Fig. 16.**9**) are specialized nucleoprotein complexes consisting of superhelical (form I) plasmid DNA and the relaxase components specifically attached to the transfer origin. Relaxases possess site- and strand-specific cleaving–joining activity. DNA in the cleaved state is constrained by covalent and non-covalent interactions with the relaxase, maintaining the superhelical form of the plasmid.

Genetic and biochemical studies on F and related plasmids, in particular R100 (Fig. 14.**14**), revealed a complex network of protein–protein and protein–DNA interactions. Proteins TraJ, FinO, and TraY are involved in regulation of expression of the *tra* regulon. Other proteins that bind at *oriT*, such as the histone-like integration host factor (IHF) and TraM, contribute to the overall structure of the F relaxosome (Fig. 16.**8**). Finally, the TraI protein, which has relaxase/helicase activity, catalyzes the *oriT*-specific cleavage reaction. The DNA helicase activity of TraI has been demonstrated to be required for conjugative DNA transfer. Moreover, physical linkage between the relaxase and helicase domains is also essential, suggesting that the TraI helicase

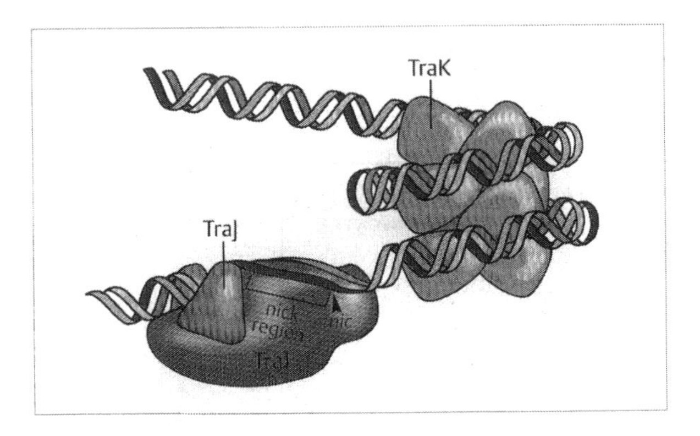

Fig. 16.**9** **Model of the IncP relaxosome.** DNA strands are drawn as ribbons, protein molecules as globular bodies. TraJ is the specificity determinant that directs the relaxase TraI to the nick region. Binding of TraJ and TraI to the DNA results in local strand separation; the nick region is thereby exposed as a single strand. Local unwinding is promoted by wrapping of a 180-bp region of the transfer origin around a core of TraK protein subunits, a histone-like transfer protein involved in relaxosome formation. The TraI–TraJ–*oriT* complex can be stabilized by TraH, which acts as a clamp between TraI and TraJ but does not bind by itself to the DNA (not shown) [from 2]

activity is not only responsible for separation of the DNA strands but might provide also the motive force for DNA translocation between donor and recipient. Interestingly, mobilization of small plasmids such as ColE1 takes place also in the absence of TraI, suggesting that either a host-encoded DNA helicase is recruited or that the strand separation/DNA translocation mechanism is fundamentally different from that of the F plasmid.

A model of the chemical reactions that take place during cleaving/joining of the DNA backbone is illustrated in Fig. 16.**10**. The aromatic hydroxyl group of Tyr22 is activated by proton abstraction mediated by His116. This permits a nucleophilic attack of Tyr22 on the phosphodiester at the nick site, leading to a *trans*-esterification reaction that couples TraI covalently to the 5′-terminal cytidyl residue of the DNA strand to be transferred. The 3′-terminal hydroxyl group is restored through the donation of a proton by His118. The joining reaction works similarly: His118 activates the 3′-terminal hydroxyl group, resulting in a strong nucleophile that attacks the phosphodiester between TraI and the 5′-end of the DNA. The resulting transesterification restores the phosphodiester at the nick site and yields the free TraI protein. Experiments with synthetic oligonucleotides have shown that TraI also catalyzes the sequence-specific cleavage and joining of single-stranded DNA. In contrast to the reactions with double-stranded DNA, cleaving/joining of a DNA single strand requires only Mg^{2+} and no auxiliary proteins. This observation suggests that the function of relaxosome auxiliary proteins required for cleaving/joining of double-stranded DNA consists of local unwinding of the nick region, permitting access of TraI to its target sequence. Local unwinding of the nick region is promoted by another accessory protein, TraK, which wraps an intrinsically bent *oriT* region of about 180 bp around a core of protein (Figs. 16.**7** and 16.**9**). TraK (14.7 kDa) is a basic protein that forms tetramers in solution. In the presence of TraK, the cleavage/joining equilibrium of RP4 relaxosomes is shifted toward the cleaved products. However, TraK is not essential for the cleaving/joining reaction on double-stranded DNA; it is essential for DNA transfer in vivo, suggesting that TraK fulfills additional functions during conjugation.

Conjugative DNA synthesis. Whereas IncP relaxosomes can be isolated from plasmid-containing cells independently of the presence or absence of potential recipient cells, initiation of transfer DNA replication in donor cells takes place only in the presence of a mating partner. A comparable observation was made in F-containing cells: transfer DNA replication of the F plasmid proceeds only after formation of mating aggregates. Therefore, a **"mating signal"** was postulated

Fig. 16.**10** **Proposed model of *oriT*-specific DNA cleaving/joining by RP4 relaxase TraI.** The amino acids and nucleotides involved in the reaction are drawn in black and red, respectively. B symbolizes an unidentified basic function [from 2]

to be transmitted to the relaxosome upon formation of mating aggregates that triggers irreversible cleavage at *oriT* and initiates transfer DNA replication. The transducer of the mating signal was suggested to be the F TraM protein. TraM is an essential transfer factor that binds specifically at *oriT* (Fig. 16.**8**), but is not required for relaxosome formation, pilus synthesis, or mating aggregate stabilization.

Cleavage at the nick site of *oriT* results in both F and RP4 systems in the formation of a free 3′-OH terminus, which is likely to be used as a primer for chain elongation by the host DNA replication machinery (i.e., DNA polymerase III holoenzyme).

Step 4: recirculation of the DNA strands. Elongation at the 3′-terminus results in displacement synthesis via a rolling-circle-type replication mechanism (Fig. 16.**11**). However, in case of the F plasmid, there is indirect evidence that additional priming events are required to initiate complementary strand synthesis in the donor cell. During transfer of F, DNA replication in the donor can be prevented when primosome activity is suppressed in a thermosensitive *dnaB* mutant strain under non-permissive conditions and RNA polymerase is inhibited by rifampicin. Since the function of host-encoded DNA polymerases should not be disturbed under these conditions, it is postulated that the 3′-terminus at *nic* of F is not accessible and therefore cannot serve as primer. Alternatively, after initial elongation of the 3′-OH terminus, replacement strand synthesis could be discontinuous and therefore dependent on primosome activity as observed in vivo for the viral strand synthesis of bacteriophage ϕX174.

The 5′-terminus with the covalently attached relaxase (TraI) is thought to be transported via a channel formed by the Mpf system across the cell membranes. Interaction between the relaxase and the Mpf system could be mediated by the proteins TraG and TraD of RP4 and F, respectively. The amino acid sequences of TraG and TraD are similar and contain NTP binding motifs. Therefore, TraG and TraD may also be involved in strand separation and energizing of the DNA transport process. Because both proteins are likely to be associated with the inner cell membrane, it is conceivable that they function as specialized helicases; strand separation therefore provides also the motive force that drives the DNA to the recipient cell (Fig. 16.**11**).

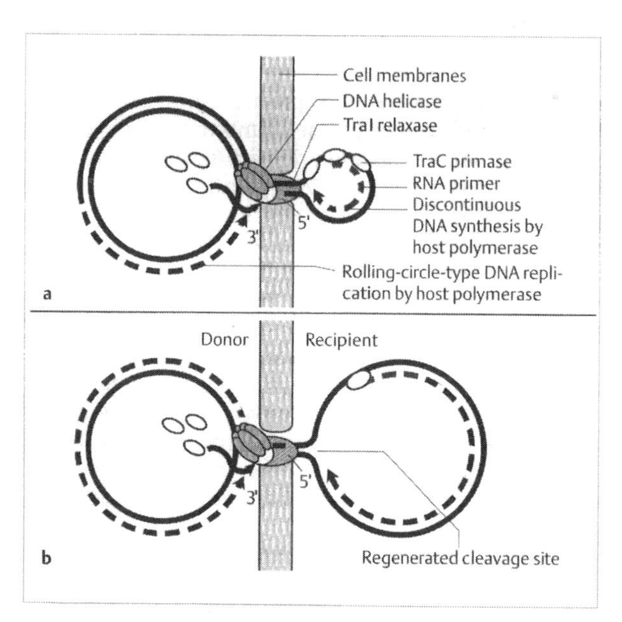

Fig. 16.**11a,b** **Rolling circle model of transfer DNA replication. a** Early stage of the transfer process following initiation. **b** Advanced stage prior to termination by a second cleavage at the regenerated cleavage site [after 3]

Step 5: complementary strand synthesis in the recipient and restoration of plasmid superhelicity. The DNA enters the recipient cell as a single strand; however, in the case of RP4, the *traC1* gene product is also transported to the recipient during conjugative DNA transfer. TraC1 is highly abundant in RP4-containing cells (>10^3 copies per cell); therefore, TraC1 most likely coats the DNA single strand during transfer and protects it from the action of nucleases. TraC1 is one of two gene products of the RP4 *traC* gene that results from two in-frame translational initiation sites, which give rise to products of 116.7 (TraC1) and 81.6 kDa (TraC2) (Fig. 16.**5**). Only the large form (TraC1) is transmitted during conjugation. Both products possess primase activity, meaning that they catalyze the template-dependent synthesis of short oligonucleotide primers on single-stranded DNA. The primers could be used by DNA polymerases of the recipient to initiate discontinuous complementary strand synthesis (Fig. 16.**11**). However, in most interspecific and intraspecific matings, the *traC* gene products are not required, indicating that other types of **single-stranded DNA binding proteins** (**SSBs**) can replace TraC1 during conjugation and that host-encoded priming mechanisms can also initiate complementary strand synthesis. The F plasmid does not encode a primase, but it specifies a single-stranded DNA binding protein that is similar to the host-encoded SSB. It is thought that either SSB can protect single-stranded DNA regions from the action of nucleases, but it has not been found that SSB coats the DNA during transfer. Complementary strand synthesis in the recipient must rely on primers synthesized either by RNA polymerase or by the primosome since neither rifampicin nor a *dnaB* mutation under non-permissive conditions alone prevents DNA synthesis. Assembly of primosomes requires the presence of a *pas* site on the single-stranded DNA, a sequence that frequently is associated with a characteristic secondary structure that can be recognized by the host-encoded *priA* gene product (also called n′). Binding of PriA initiates primosome assembly (see Chapter 14). Surprisingly, neither a promoter nor a *pas* site has been found in the first 130 nucleotides of the F leading region—the region that during DNA transfer enters the recipient cells first—and, therefore, location of the initiation sites of complementary strand synthesis awaits further biochemical studies.

Step 6: termination of DNA transfer and separation of the mating partners. Following transfer of a unit length of single-stranded DNA, a termination reaction must occur to ensure restoration of the plasmid. Termination is thought to take place by a second cleavage event at the regenerated nick site that concomitantly leads to recircularization of the transferred DNA. Again, the relaxase must play a central role in recognizing the nick region and in catalyzing the transesterification reactions that restore the phosphodiester bond at the nick site. To accomplish this function, it is conceivable that the relaxase remains attached at or near the pore that allows the DNA to transverse the cell envelope, scanning constantly the immigrant DNA for the occurrence of the single-stranded *oriT* nick region (Fig. 16.**11**). Experiments with RP4 TraI and single-stranded oligonucleotides have shown that TraI catalyzes site-specific recombination of two differently sized oligonucleotides at the nick site. However, this reaction occurs only if an oligonucleotide is offered that ends at the 3′-terminal nucleotide at the nick site. Therefore, either two TraI molecules are involved in the second cleavage/recircularization reaction or, as it was proposed for the F system, the 3′-OH terminus at *nic* is not elongated during transfer and, therefore, remains available for the TraI-catalyzed joining reaction that is also observed in vitro.

16.2.3 Leading Region Genes Provide Installation Functions

The leading region is the first segment to enter the recipient cell during conjugation, and it may carry determinants called installation genes that promote establishment of the immigrant plasmid in the new host cell. The leading region of F is defined as the first 13 kb between *oriT* and the primary replication region RepFIA (see Fig. 16.**3**). Genes identified in this region include *flm* of the *hok* gene family, which specifies a maintenance function; *psiB*, which determines inhibition of the cellular SOS response; and *ssb*, which encodes a single-stranded DNA binding protein (SSB) of the type that binds cooperatively to DNA with no sequence specificity.

The CoIIb leading region also contains an antirestriction gene designated *ard* (**a**lleviation of **r**estriction of **D**NA). The gene is located between *oriT* and *psiB* and, like other known genes in the leading region, its expression is increased by derepression of the IncI1 transfer system. Ard alleviates restriction of DNA by the three known families of type I restriction systems, including *Eco*K. Genetic studies have indicated that the alleviation process requires that *ard* is expressed in the recipient from the transferred plasmid. This raises the timing problem of how Ard accumulates before the immigrant plasmid is destroyed by the restriction enzyme. Possibly the route of DNA entry during conjugation allows Ard production before the plasmid becomes accessible to a type I restriction enzyme. However, the other possibility, that transferred leading region genes can be transcribed before they are converted into duplex form by complementary strand synthesis, should also be considered.

Box 16.3 Plasmid *ssb* genes are homologous to *E. coli ssb* and complement known defects of bacteria carrying a temperature-sensitive *ssb* mutation, indicating shared functions. The biological significance of plasmid SSBs remains unclear. *ssb* mutants of F-like or IncI plasmids show no obvious defect in conjugation; however, the regulation of plasmid *ssb* genes by the fertility inhibition system in IncFII and IncI1 plasmids implies a function in conjugative DNA metabolism. Furthermore, there is no evidence that SSBs coat the transferred DNA during its passage to the recipient cell. Possibly, plasmid *ssb* genes function to prevent depletion of SSB reserves in conjugating cells, and there may be a more stringent requirement for the plasmid *ssb* gene products in bacteria other than *E. coli* or under conditions that differ from those routinely used in the laboratory.

The *psiB* locus (**p**lasmid **S**OS **i**nhibition) specifies a polypeptide of 12 kDa that inhibits the bacterial SOS response as a function of its intracellular concentration. SOS induction requires activation of the co-protease function of RecA protein by a signal inferred to be single-stranded DNA (see Chapter 15.7). PsiB protein is thought to exert its inhibitory effect by direct interaction with activated RecA. A gene designated *psiA* is located immediately downstream of *psiB*; *psiA* is inessential for the Psi function, and its role has yet to be determined. In conjugation, expression of *psiB* on F and ColIb is strongly, but transiently induced in the recipient cell, giving levels of expression sufficient to cause the Psi phenotype. The implication is that *psiB* genes facilitate installation of the plasmid in the new host by preventing expression of the SOS response, which otherwise would be induced by progressive transfer of single-stranded DNA.

The leading region of RP4 has no known similarity to the equivalent part of plasmids in the set that carries related *ssb* and *psiB* genes. Moreover, unlike plasmids F and ColIb, there are one essential (*traK*) and two accessory *tra* genes (*traL, traM*) in the leading region of RP4 (see Fig. 17.**1**). However, RP4 encodes a gene product that, when overproduced, suppresses *ssb* mutations in *E. coli* and exhibits significant similarity to other SSBs. This *ssb* gene maps adjacent to the gene of the initiator protein of vegetative replication, *trfA*, and therefore apparently is part of the RP4 Rep system. Moreover, RP4 might possess a possibly defective antirestriction system consisting of the genes *klcA*, *klcB*, and *korC*. KlcA shows significant similarity to the ArdB protein of the IncN plasmid pKM101. ArdB specifically inhibits type I restriction, but does not influence type I modification. In IncPα plasmids such as RP4, the *klcB* reading frame is disrupted by the insertion of the ampicillin-resistance transposon Tn*1*. In IncP plasmids lacking Tn*1*, the *klcAB–korC* operon is autogenously regulated by *korC*. Thus, also IncP plasmids seem to possess genes that can be considered as installation genes; however, instead of being clustered in the leading region, these functions are spread over the plasmid genome.

16.2.4 Conjugative DNA Transfer System of Gram-positive Bacteria

***Staphylococcus* spec.** One of the prototype conjugative replicons in *Staphylococcus* spec. is the plasmid pGO1 (52 kb) which confers resistance to trimethoprim, gentamycin, and quaternary ammonium compounds. Two separated regions, one of less than 14.5 kb and a second of approximately 2 kb, identified by transposon mutagenesis, are sufficient for conjugative transfer to occur. The transfer origin has been identified recently, and the nick site has been determined. The DNA sequence of the nick region is identical to that of IncQ plasmids.

Mobilization of small non-self-transmissible plasmids by pGO1 requires the *mobA* and *mobB* gene products encoded by non-self-transmissible plasmids. MobA and MobB can introduce a site- and strand-specific nick located within an intergene region of the plasmids proposed to include the transfer origin. Moreover, the amino acid sequence of one of the gene products contains three motifs identified in relaxases of numerous transmissible plasmids of Gram-negative bacteria that are involved in DNA recognition and specific cleaving/joining (see also Chapters 14 and 15). One motif contains the tyrosine residue that covalently attaches to the 5′-end of the cleaved DNA strand, allowing the protein to "pilot" the DNA, with the 5′-end leading toward the recipient cell. However, this has been shown directly only for the related relaxases TraI (RP4) and VirD2 (pTi) of Gram-negative bacteria; a similar mechanism involving unidirectional transfer of a specific strand seems very likely.

Virtually nothing is known about the mechanism of mating-pair formation. Productive matings require a solid surface, and neither pili nor other structures on the cell surface that might be related to conjugation have been observed by electron microscopy in cells containing this class of plasmids.

Enterococcus faecalis. An outstanding feature of many *E. faecalis* plasmids is that their conjugative transfer is subject to a mating response mediated by **sex pheromones**. Plasmid-free cells of *E. faecalis* secrete multiple (at least five and probably many more) plasmid-specific sex pheromones. Contact of plasmid-containing cells with the corresponding pheromone results in the induction of a series of biosynthetic events that lead to induction of the conjugative DNA processing machinery and to cell aggregation. Acquisition of a plasmid results in the cessation of the synthesis of the cognate pheromone. Some plasmid-containing cells secrete specific competitive inhibitors of the mating response. These are thought to prevent induction by causing pheromone levels that are too low to result in production of mating aggregates or to prevent self-induction due to leakiness in the repression of pheromone production (see Chapter 20.1.4)

Box 16.4 The best-characterized model systems in *E. faecalis* conjugation are the plasmid pAD1 (60 kb), which encodes the production of hemolysin/bacteriocin, and pCF10 (54 kb), a plasmid that mediates resistance to tetracycline through the presence of a Tn*925* element. The pAD1-specific pheromone (**cAD1**) is a hydrophobic octapeptide with the sequence Leu-Pro-Ser-Leu-Val-Leu-Ala-Gly. Exposure of pAD1-containing cells to cAD1 results in cell aggregation caused by accumulation of a microfibrillar "**aggregation substance**" (**AS**) on the donor surface. The structural gene of the pAD1-encoded aggregation substance (*asa-1*) encodes a 140-kDa protein with a protein export signal sequence at the N-terminus. Immunological studies have shown that the gene product is present in the microfibrillar structures on the surface of induced cells. The amino acid sequence of the *asa-1* gene product contains several known motifs: an anchor segment in the C- terminal region that is found in surface proteins of several Gram-positive bacteria (e.g., the M-protein of *Streptococcus pyogenes*), a proline-rich segment probably associated with the cell wall, and two motifs proposed to be involved in adherence to integrin family proteins on the surface of eukaryotic cells. The pAD1 transfer region occupies about half of the genome; *oriT* maps at one end. Genes for pheromone response and regulatory functions
(approximately 15 kb) separate a region entirely devoted to DNA transfer (approximately 11 kb).

Conjugative transposons (CTn). This type of genetic element was first identified in *Enterococcus* and *Streptococcus* spec. The host range of conjugative transposons generally is very extended, including the Gram-negative bacteria, for which three conjugative transposons have been described, including one in enteric bacteria. Representatives examined so far seem to fall in two classes: those similar to Tn*916* (16.4 kb) and those similar to Tn*5252* (47.5 kb). Tn*916* and Tn*5252* confer resistance to tetracycline and chloramphenicol to their hosts, respectively. In fact, conjugative transposons probably are more responsible for the dissemination of antibiotic resistance in streptococci than are plasmids.

The current model of intercellular transfer of Tn*916*-like elements involves three steps (Fig. 16.**12**): (1) excision of the element from a donor replicon and formation of a circular intermediate not capable of autonomous replication, (2) conjugative transfer of the transposon to the recipient cell, conceivably via a single-stranded intermediate, and (3) transposition of the reconstituted element to a recipient replicon. Excision and integration take place by reciprocal recombination mechanisms resembling the Xis/Int system of lambdoid phages. Indeed, Tn*916* encodes two gene products, Xis-Tn and Int-Tn which share sequence similarity with the *xis* gene product of phage P22 and to a lesser extent of λ and of ϕ80, and with the Int-related family of recombinases, respectively.

Excision seems to be the rate-limiting key step in the transfer process since genetic markers present in the donor replicon and located adjacent to either side of the transposon are not mobilized at a detectable frequency. On the other hand, co-transfer of distinguishable, but functionally analogous Tn*916* derivatives takes place with great efficiency (e.g., 50%) even in the absence of selective pressure, suggesting that the movement of one transposon can trigger *trans*-activation of excision and transfer of a second transposon. Excision takes place by staggered cutting (6–7 bp overhang) adjacent to the transposon ends, followed by reciprocal joining of the overlapping ends of transposon and host DNA (Fig. 16.**12**). In contrast to other transposition events, integration of Tn*916* into a target molecule does not result in the generation of short direct repeats. Therefore, joining of the staggered ends usually leads to formation of a 6–7-bp heteroduplex core region, which is resolved only during subsequent DNA transfer or, in the case of the recircularized donor replicon, during replication.

Virtually nothing is known about the mechanism of intercellular DNA transfer. A region of about 9 kb of Tn*916* is required for DNA transfer to occur. However, neither the operon structure of this region is known nor has an origin of transfer been mapped. Integration of Tn*916*-like elements requires two 20-bp stretches of the

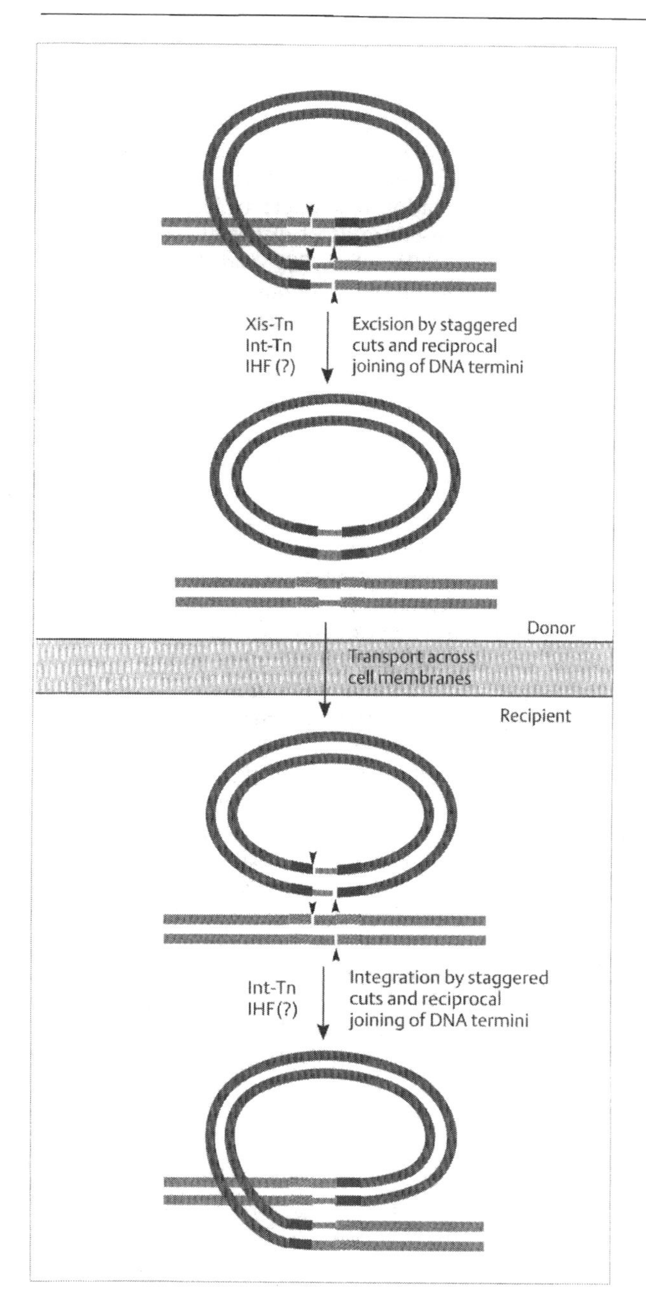

Fig. 16.12 **Model of conjugative transposition by Tn*916*.** DNA strands of the conjugative transposon are drawn as black lines, those of donor and recipient target replicons as gray lines. Transposon ends are marked in red; target sequences with limited similarity to the transposon ends are drawn in light red. Heteroduplex formation within the variable core region is indicated by different line thickness. Resolution of heteroduplices takes place during conjugative transfer or subsequent replication of integration/excision products (not shown). Cleavage positions are indicated by arrowheads

target molecule, with limited similarity to the transposon ends, separated by a 6-bp variable core region. Staggered cuts at the ends of the circular transposon and near the 6-bp variable core region initiate the integration event. Reciprocal joining of the overlapping ends completes the transposition. Again, heteroduplices of 6–7 bp can be formed that are resolved only during subsequent rounds of replication. Several potential binding sites for integration host factor (IHF) are found clustered near the ends of Tn*916*, which suggests that IHF might play a role in the integration/excision reaction of conjugative transposons similar to the role played by IHF in lambdoid bacteriophages.

Conjugative plasmids of *Streptomyces*. *Streptomyces* can harbor a wide variety of plasmids, ranging in size from less than 4 kb to more than 100 kb. Most plasmids are covalently closed circles, but also double-stranded linear plasmids have been described. pIJ101 is considered as a representative of small (8.8 kb) self-transmissible double-stranded circular plasmids. pIJ101 exhibits a high copy number with 300 copies per chromosome equivalent and was originally isolated from *S. lividans*. However, its host range is broad among mycelia-forming bacteria: stable inheritance has been demonstrated in several *Streptomyces* species, in *Micromonospora*, *Thermomonospora*, *Saccharopolyspora*, and in *Amycolatopsis*. pIJ101 replicates via a single-stranded intermediate, most likely by a rolling circle mechanism. Vegetative replication requires only the plasmid-encoded Rep protein (50 kDa) and a non-coding region upstream of the *rep* gene, most probably containing the origin of vegetative replication.

Although pIJ101 does not code for any selectable marker, conjugative transfer of the plasmid can easily be followed by a phenomenon called "**pock formation**." Pocks are circular zones of retarded growth in a confluent lawn of *Streptomyces* mycelia that form where conjugative transer of pIJ101 causes a burst of plasmid gene expression in the recipients. This impairs growth and sometimes leads to a change in the physiology of the recipient. The latter is most clearly reflected by the precocious production of pigmented antibiotics in the pocks of *S. coelicolor* and *S. lividans*. These antibiotics normally are produced only after vegetative growth ends.

Conjugative transfer of pIJ101 requires very little genetic information: sex pili have not been observed, and only one *tra* gene is required for DNA transfer to occur. *tra* specifies a polypeptide of 66 kDa and is the first gene of an operon containing three additional genes: *spdA*, *spdB*, and a "66-aa ORF." The products of these genes may be membrane-associated since the proteins have strong hydrophobic domains as deduced from the nucleotide sequence of the genes. Although *spdA* and *spdB* are not essential for DNA transfer, mutants in these genes exhibit reduced pock size (*spd*: "**spr**ea**d**"). Hence, these genes are proposed to be involved in the migration of the plasmid within the hyphae of the recipient.

Expression of the *tra* operon is negatively controlled by the product of *korA*, an autoregulated gene arranged divergently from the *tra* operon, but transcribed from a promoter located in the intergene region from which transcription of the *tra* operon initiates.

In those conjugative systems of Gram-positive bacteria that have been analyzed more extensively, there is good evidence for the existence of a single-stranded DNA transfer intermediate. This functional similarity to the Gram-negative systems is paralleled by sequence similarities in transfer gene products and nick regions of *Staphylococcus aureus* plasmids and conjugative transposons, suggesting a common ancestry of Gram-positive and Gram-negative conjugation systems. An important difference, however, is the lack of a conjugative pilus involved in mating pair formation. Apparently, in Gram-positive organisms, aggregation substances are a functional substitute for pili.

16.2.5 DNA Transfer Systems Are Phylogenetically Related

The various conjugative DNA transfer systems in Gram-negative bacteria are thought to share mechanistic properties: (1) the initial physical contact between a donor and recipient cell during mating pair formation is mediated by a conjugative pilus, and (2) DNA transmitted to the recipient as a single strand. Th mechanistic properties common to the known system may have an evolutionary basis because a number distinct sequence similarities have been found. Th mating pair formation machinery of IncP plasmids ha two analogues in plasmids of different incompatibilit groups: IncW (PilW of R388) and IncN (pKM101 Interestingly, the RP4 Tra2 region also shares simila ities with the Ti plasmid *virB* operon (Fig. 16.**13**). Th *virB* operon is involved in facilitating the *Agrobacteriu* spp.–plant cell contact, possibly via a pilus-like stru ture, and it is thought that VirB components, most like together with the *virD4* gene product of the pTi *vir* region (see below), mediate DNA transport. Experimen tal evidence for this hypothesis is provided by th finding that the IncQ plasmid RSF1010 (see Chapt 16.2.1) can be mobilized by the VirB region in th presence of VirD4, not only between *Agrobacteriu* cells, but also to plants. This demonstrates that the pT encoded *vir* genes can function as "conjugative transf genes" and that the *virD1*/*virD2* gene products are n necessarily required for DNA transfer to plants relaxosome components of a heterologous DNA tran port system are provided.

Extensive similarity has also been found betwee Tra2, *virB*, and the *ptl* operon of the human pathoge *Bordetella pertussis*, which apparently is involved i pertussis toxin export. These data suggest that th operons belong to an evolutionarily related superfamil of protein export machineries. These operons sha components (Fig. 16.**13**) the products of which migh energize the export processes by hydrolysis of NTPs an are likely to be parts of transport systems an components responsible for DNA transfer and replic tion. Also, the loci for DNA relaxases and accesso proteins involved in the initiation and termination

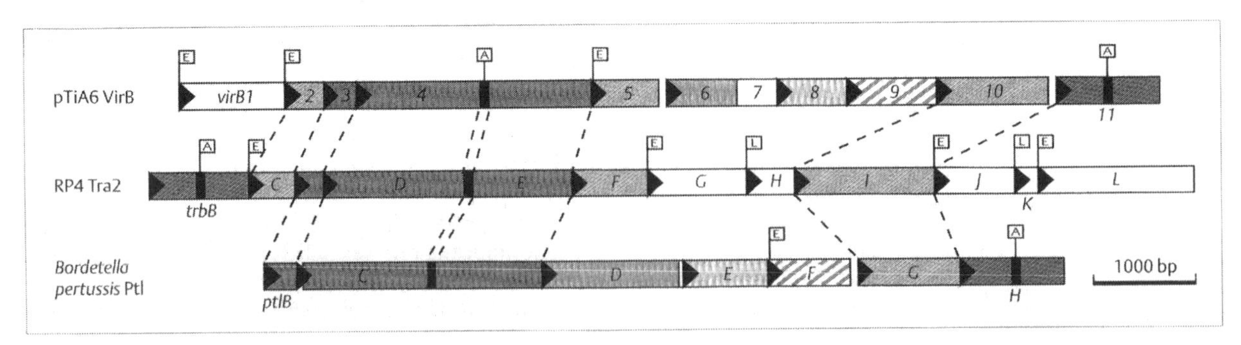

Fig. 16.**13** **Common gene organization of specialized macromolecule export systems.** Genes, the products of which exhib significant sequence similarity, are shown in the same color or are connected by dotted lines. Tags labeled with A, E, and L ma sequence motifs proposed to be involved in binding of NTPs, protein export, and lipid attachment, respectively (see also Fig. 16. [after 4]

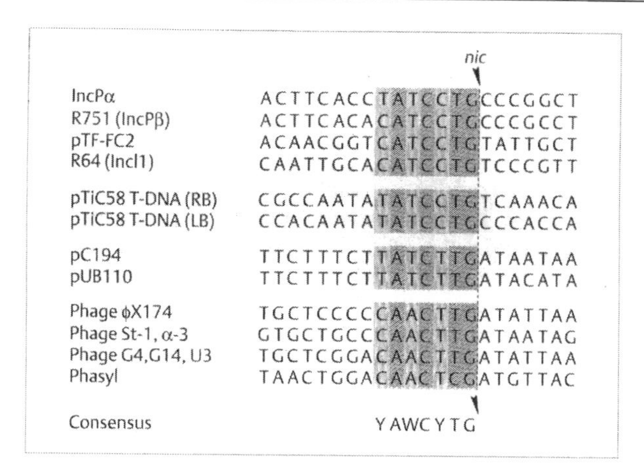

nic
IncPα ACTTCACCTATCCTGCCCGGCT
R751 (IncPβ) ACTTCACACATCCTGCCCGCCT
pTF-FC2 ACAACGGTCATCCTGTATTGCT
R64 (IncI1) CAATTGCACATCCTGTCCCGTT
pTiC58 T-DNA (RB) CGCCAATATATCCTGTCAAACA
pTiC58 T-DNA (LB) CCACAATATATCCTGCCCACCA
pC194 TTCTTTCTTATCTTGATAATAA
pUB110 TTCTTTCTTATCTTGATACATA
Phage ϕX174 TGCTCCCCCAACTTGATATTAA
Phage St-1, α-3 GTGCTGCCCAACTTGATAATAG
Phage G4,G14, U3 TGCTCGGACAACTTGATATTAA
Phasyl TAACTGGACAACTCGATGTTAC
Consensus YAWCYTG

Fig. 16.**14** **Conservation of *oriT* nick regions in rolling-circle-type vegetative origins.** Cleavage sites (*nic*) are marked by an arrowhead. Stringently conserved nucleotide positions are marked in red, and positions where pyrimidine residues (Y) or A/T residues (W) are conserved are marked in gray [after 5]

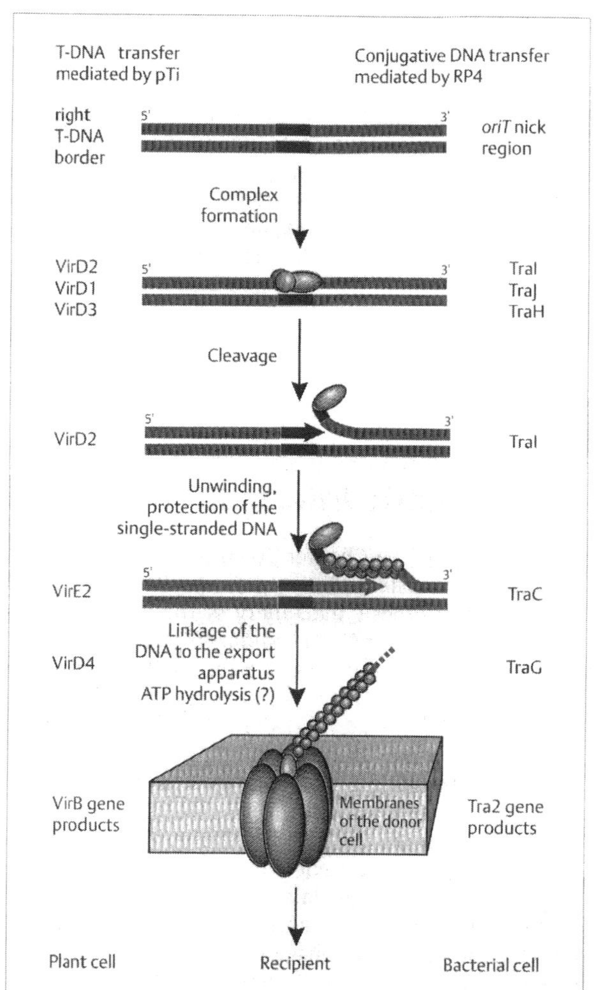

Fig. 16.**15** **Mechanistic analogies between bacterial conjugation and T-DNA transfer to plant cells.** See text and Chapter 16.2.2 for explanations [from 4]

DNA transfer are organized in comparable arrangements.

Functional analogy has been demonstrated for the VirD2 protein, which, like TraI, specifically cleaves single-stranded DNA containing either the pTi or the RP4 nick region in vitro. A more rudimentary consensus sequence is found when additional rolling circle systems are included in the comparison, such as the origins for asymmetric rolling circle replication in plasmids or the (+)-strand origin of certain single-stranded DNA phages (Fig. 16.**14**).

Extensive similarities between amino acid sequences of the pTi *virB* and *virD* regions and the RP4 Tra2 and Tra1 regions and functional similarities between certain gene products of these regions allow the postulation of a common mechanism for T-DNA transfer to plant cells and bacterial conjugation (Fig. 16.**15**). The RP4 *oriT* nick region and pTi border sequences share a core sequence of 7–8 bp, which serves as a recognition signal for the relaxases TraI and VirD2, and accessory proteins (TraHJ/VirD1D3) respectively. Cleavage at the nick site results in the covalent attachment of the relaxase to the DNA 5′-terminus. Replacement strand synthesis by host DNA polymerase leads to generation of the T-strand destined for transfer to the recipient. The ssDNA is coated by specialized single-stranded DNA binding proteins (TraC1/VirE2) that protect the DNA from degradation by nucleases and probably help to initiate complementary strand synthesis in the recipient bacteria. The T-complex is thought to interact with products of the Tra2/VirB region that are proposed to form some kind of pore in the bacterial membrane that facilitate the active transport of the T-complex with the relaxase (TraI/VirD2) heading the DNA at the 5′-terminus. Proteins VirE2 and VirD2 contain **n**uclear **l**ocalization **s**ignals (**NLS**) that might help to direct the T-complex to the plant nucleus. The remarkable mechanistic similarity of both processes strongly suggests that the systems evolved from a common ancestor and that T-DNA transfer to plants is a special case of bacterial conjugation adapted to the requirements of interkingdom DNA transfer.

Relationships described here do not include F or F-like plasmids because no extensive similarities are observed. In particular, the *oriT* nick regions of the F complex do not fit into the alignments shown in Figure 16.**14**. Thus, it seems that IncP-like systems have evolved almost independently from the F system. On the other hand, certain systems, R388 (IncW) and pkN101 (IncN), exist that apparently combine components of F-like and P-like transfer systems. These examples might illustrate how complex systems with new properties can emerge from accumulation and/or exchange of independently evolved components ("modular evolution").

Bacterial conjugation and T-DNA transfer to plant cells are mechanistically and genetically related processes.

16.3 Transduction Is Mediated by Bacteriophages Carrying Non-Viral Genetic Information

Bacteriophages (see Chapter 26) may be considered as "molecular parasites" that rely on the DNA replication and protein synthesis machinery of the host to reproduce themselves. However, similar to plasmids, bacteriophages occasionally function as vectors for DNA originating in their host cells. Therefore, phage infection can provide a means for the exchange of genetic material between bacteria, called transduction, and hence an evolutionary advantage for a bacterial population.

Depending on the type of life cycle in different phage species two different kinds of transduction can be distinguished: (1) **specialized transduction**, during which a narrow set of host genes, clustered at the attachment site of the phage on the bacterial chromosome is transferred, and (2) **generalized transduction**, which describes the possibility of transferring a wide variety of host genes that can be scattered over the entire genome.

Stable inheritance of transduced genes requires their incorporation into the genome. Usually, this takes place by homologous recombination (see Chapter 16.4). If a gene cannot be integrated, it may stably exist as a single copy in a single cell of a bacterial population. This kind of gene transfer is designated as "**abortive transduction**". During growth on a solid medium under conditions that select for the presence of the transduced gene, abortive transduction results in the formation of microcolonies that consist of approximately 10^4 cells. Microcolonies form because daughter cells of the recipient still contain the product of the abortive transduced gene and, therefore, survive under selective conditions. Further cell division and turnover of the gene product results in its depletion, and the cells stop growth after a few generations.

Genetic **transduction** is the virus-mediated transfer of non-viral genetic information from a donor to a recipient cell. Depending on the nature and the amount of DNA transferred, the process is called **specialized** or **general(ized)** transduction.

16.3.1 Specialized Transduction Is Mediated by Temperate Bacteriophages Capable of Existing in a Chromosomally Integrated State

Bacteriophage λ. Phage λ has been a paradigm for the phenomenon of lysogeny since its discovery by E. M. Lederberg in 1951. If a susceptible *E. coli* cell is infected by λ the phage can propagate in two different ways:

1. In the lytic pathway, the phage genome exists as an autonomous DNA molecule that replicates extensively and directs the synthesis of a large set of phage-specific proteins. The lytic pathway results in the production of several hundred phage particles per infected cell; these phage particles are released by cell lysis (see also Chapter 26).
2. The lysogenic pathway leads to the integration of the phage genome into the bacterial chromosome. In this state, most of the phage genes are repressed by the action of a central repressor protein (cI), and the phage DNA is replicated passively together with the chromosomal DNA.

The integration of phage λ into the bacterial chromosome is a **site-specific recombination** event that is catalyzed by the phage-encoded Int protein (Fig. 16.**16a**). The host-encoded histone-like protein IHF plays an accessory function in that it bends the DNA in a

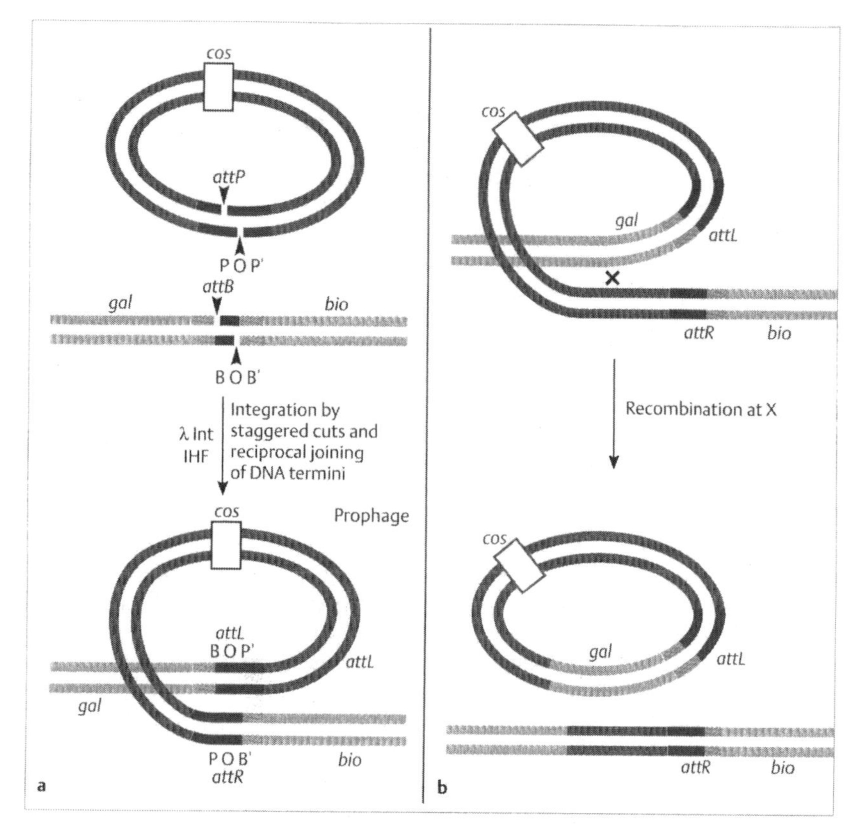

Fig. 16.**16a,b** **Model of specialized transduction.** **a** λ Int-mediated integration of the bacteriophage λ genome (black) into the *Escherichia coli* chromosome (gray). Attachment sites *attP* (POP′) and *attB* (BOB′) are drawn in red and light red, respectively. The core region (O), which is present in both sites, is drawn in red. **b** Emergence of specialized transducing phages by imprecise excision of the λ prophage. In the example shown here, illegitimate recombination between prophage and chromosomal DNA yields a phage genome carrying the *E. coli gal* operon. The presence of the λ *cos* site allows the DNA to be packaged efficiently during lytic growth

specific way. Specific recombination takes place at the so-called attachment sites, the loci *attP* (located on the phage genome) and *attB* (on the bacterial chromosome; also termed *att*λ). Both *attP* and *attB* consist of three parts, P, O, P′ and B, O, B′, respectively, with P, P′, B and B′ designated as arms and O designated as the core of the attachment sites. Integration of λ leads to formation of the chimeric sites *attL* and *attR*, which have the structures BOP′ and POB′, respectively. The normal excision reaction of the λ prophage that takes place when the phage enters the lytic pathway is the reverse of the integration reaction. The phage-encoded Xis protein plays an accessory role in this reaction since it inhibits integration, but is required for excision (see Chapter 26.4).

Specialized transducing phages result from imprecise excision, usually by illegitimate recombination (Fig. 16.**16b**). Since *attB* is flanked by the operons *galKTE* (galactose metabolism) and *bioABFCD* (biotin synthesis), transducing λ phages usually carry a portion of the chromosome containing either *gal* (λ*gal*$^+$) or *bio* (λ*bio*$^+$) markers. Normally, transducing λ phages are defective (λd) since they lack parts of their own genome essential for packaging or replication. However, these phages can be propagated if an intact helper phage, which provides the missing functions, is used for coinfection of susceptible cells. With this technique, cell lysates with high titers of identical specialized transducing phages (**HFT lysates**) can be obtained.

Specialized transduction occurs when temperate phages integrate at their attachment sites, but through imprecise recombination incorporate flanking chromosomal genes. The transducing phages either are still complete and form phages or they lose genes (d for defective phages) and can only duplicate by means of a complementing helper phage.

16.3.2 Erroneous Recognition of Bacterial DNA Sequences by the Phage Packaging Machinery Leads to Generalized Transduction

Bacteriophages P1 and P22. The life cycle of the temperate phage P1 of *E. coli* differs from that of λ in that its genome exists also in the lysogenic state as an autonomous replicon as a low-copy-number plasmid.

Phage P22 of *S. typhimurium* resembles phage λ more since it also integrates into the host genome. When phage P1 or phage P22 enters the lytic pathway, a special replication origin is activated, initiating rolling circle replication of the phage genome. The resulting concatemers, which contain several unit lengths of the phage genome, serve as substrates for the packaging reaction that yields the infectious phage particles. Packaging initiates at so-called *pac* sites. The amount of DNA that enters a phage head ($\sim$100 kb for P1 and 41.8 kb for P22) is determined by the head size ("**headful packaging**"); for phage P22, this is approximately 2% more than one genome unit length. This mechanism results in a limited **circular permutation** of the phage DNA and in **terminal redundancies** (1.7 kb) at the ends of the linear DNA molecules in the phage head.

The only requirement for the formation of generalized transducing phages particles is the presence of a *pac*-like sequence in the chromosomal DNA of the host. If such a sequence is recognized by the packaging machinery of the phage, the bacterial chromosome is treated like a phage DNA concatemer yielding a series of phage particles that contain only bacterial DNA and phage lysates in which in principle any chromosomal gene is contained (hence **general transduction**). In contrast to phage P1, which does not integrate into the host genome, phage P22 can also execute specialized transduction. However, this results from a mechanism that is analogous to that described for phage λ.

The gene transfer agent (GTA) of *Rhodobacter capsulatus*. A special gene transduction system exists in *R. capsulatus*, a Gram-negative photosynthetic bacterium. Most natural isolates of *R. capsulatus* produce phage-like particles, termed the **g**ene **t**ransfer **a**gent (GTA). GTA particles resemble tailed phages with a head 30 nm in diameter with short spikes and a tail of variable length (30–50 nm) with a diameter of 6 nm. Tail fibers and a collar are also present. GTA is thus smaller than any morphologically similar virus. Analysis of the DNA content revealed that GTA particles contain only chromosomal DNA (approximately 4.5 kb) from the donor bacterium. The DNA of GTA particles can be taken up by the recipient cells and recombines with the chromosome of the recipient, thereby replacing some of the genetic information of the recipient with a portion of the genome of the donor. Virtually any part of the donor's chromosome can be packaged; therefore, the GTA system has been used for genetic manipulation in *R. capsulatus*. Because GTA carries a relatively small fragment of the donor genome, it is well suited for the resolution of mutant phenotypes that involve more than one mutation since even markers that lie fairly close to one another are frequently separated upon transfer.

It is likely that GTA evolved from a chromosomally integrated temperate phage that lost the ability to package specifically its own genome. The small size of GTA particles and the amount of DNA that these particles contain exclude the possibility that the complete genome of a hypothetical GTA progenitor could fit into a single GTA particle. Therefore, also the morphology of GTA seems to be adapted to its function as a gene exchange system of the host rather than its function as a bacterial virus. The reason why this defective phage is conserved among various strains of *R. capsulatus* might be that it provides its host with an effective gene transfer system and therefore enhances its viability. The finding that a system like GTA could evolve and survive indicates that phage-mediated transduction plays a significant role in gene exchange in a natural environment.

Generalized transduction is used extensively as a tool for bacterial genetics to map the distance of gene markers and in strain construction to introduce or replace specific genes.

16.4 Homologous Recombination Is an Important Pathway for Genomic Incorporation of Exogenous DNA

Homologous recombination is a complex process during which a part of the cellular DNA is exchanged against genetic material that is either similar (homologous) or identical. The DNA that is used in the recombination process can be exogenous or it can originate in the cell itself. Whereas in the latter case, recombination plays an important role in maintaining the integrity of the cellular genetic information (see Chapter 15.5), recombination with exogenous DNA is a prerequisite for the stable inheritance of DNA that has entered the cell by horizontal gene transfer, i.e., by transformation, by transduction or, in the case of matings involving Hfr and F′ strains, by conjugation.

Multiple pathways of homologous recombination. In *E. coli*, a great number of mutants exhibiting different recombination deficiency (Rec$^-$) phenotypes

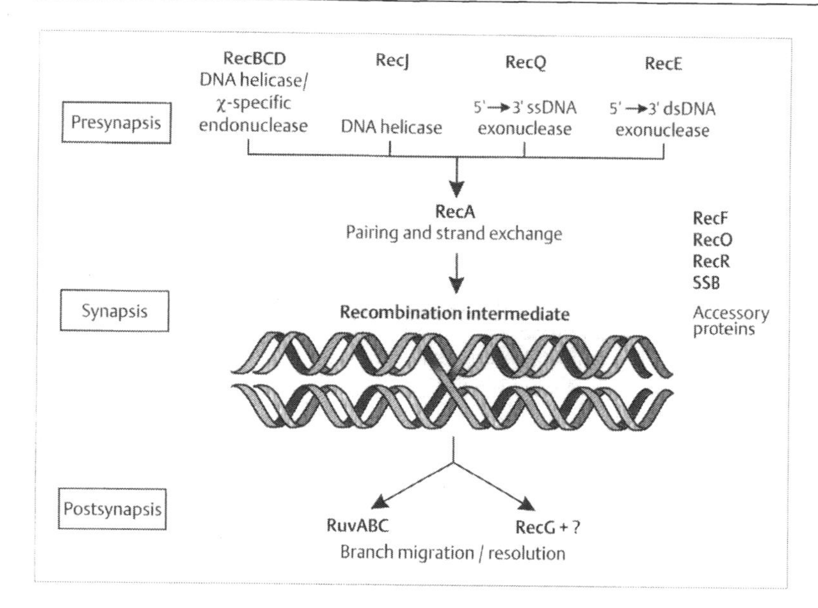

Fig. 16.**17** **Recombination pathways in *Escherichia coli*.** Four different but not independent pathways for homologous recombination and the steps they catalyze are indicated. For details, refer to the text [after 6]

have been isolated. Whereas some of the mutations lead only to the loss of the ability to use certain types of DNA substrates for the recombination reaction (i.e., ssDNA, dsDNA, linear or circular DNAs), other mutations lead to the complete loss of recombination potential. This finding suggests the existence of different, although not independent, pathways of homologous recombination (Fig. 16.**17**). The complicated situation was resolved only recently when powerful methods for the manipulation of poorly expressed genes became available. These techniques allowed the purification of significant quantities of the gene products involved and permitted the reconstitution of the homologous recombination reaction pathways step by step in vitro.

Homologous recombination reactions are subdivided into three major steps: During initiation (also designated **presynapsis**), a single-stranded DNA substrate is prepared that is used to initiate the homologous pairing and strand-exchange reaction (**synapsis**). Synapsis results in the formation of intermediates, so-called **Holliday junctions** (see below), which are resolved by specific cutting reactions (**postsynapsis**) (Fig. 16.**17**).

16.4.1 Presynapsis Involves the Generation of Single-Stranded DNA Required for Initiation of Recombination

Initiation of recombination in *E. coli* (and possibly in all organisms) requires the formation of single-stranded DNA. Single strands can be generated by the combined helicase–nuclease activities of the RecBCD multienzyme complex or by alternative pathways involving RecQ, RecJ, RecE, UvrD, or HelD. Any or all of these proteins might be expected to function as initiators of recombination (Fig. 16.**17** and Table 16.**1**).

The RecBCD pathway. The RecBCD enzyme is an essential component of the main pathway of homologous recombination in *E. coli*. In vitro, RecBCD enzyme degrades dsDNA by means of its ATP-dependent exonuclease activity and ssDNA by its ATP-stimulated endonuclease activity. RecBCD is also a helicase that can processively unwind large tracts of dsDNA (> 30 kb, without dissociation from the DNA) at rates of up to 1000 bp/s. RecBCD is an unusual helicase: linear duplex DNAs containing 3′- or 5′-tails less than 25 nucleotides in length are substrates for RecBCD helicase. During dsDNA unwinding, the ATP-dependent degradation of DNA is asymmetric, with the 3′-terminal DNA strand at the entry site for RecBCD being degraded more rapidly than the 5′-terminal strand. The unwinding reaction results in the formation of a loop at the 3′-end with respect to the end used for entry. Initiation of homologous recombination by RecBCD is stimulated by the presence of χ sites in the DNA. The χ recombination hotspots are composed of the DNA sequences 5′-GCTGGTGG-3′. Stimulation of recombination by χ occurs primarily at the 5′-side of the χ site and extends, with decreasing magnitude, more than 10 kb from χ.

The enzymatic activities of the RecBCD complex suggest the following model for initiation of homologous recombination in *E. coli* (Fig. 16.**18**): RecBCD invades a linear dsDNA at one end and moves rapidly along the DNA. The 3′-terminal DNA strand is subject to

Table 16.1 **Classification of *Escherichia coli* recombination genes and proteins**

Gene	Protein	Size of gene product (kDa)	Activities
Initiators (presynapsis)			
recB	RecBCD	134	ATPase, 5′→3′ dsDNA exonuclease, DNA helicase, χ-specific endonuclease
recC	RecBCD	129	ATPase, 5′→3′ dsDNA exonuclease, DNA helicase, χ-specific endonuclease
recD	RecBCD	67	ATPase, 5′→3′ dsDNA exonuclease, DNA helicase, χ-specific endonuclease
recE	RecE	96	5′→3′ dsDNA exonuclease
recJ	RecJ	63	5′→3′ ssDNA exonuclease
recQ	RecQ	68	3′→5′ DNA helicase
Homologous pairing and strand exchange (synapsis)			
recA	RecA	38	Formation of helical filaments on DNA. ATPase, catalyzes homologous pairing and strand exchange, co-protease when stimulated by ssDNA
recF	RecF	40.5	ssDNA binding protein
recO	RecO	27	Stimulates binding of RecA to ssDNA (with RecR)
recR	RecR	22	Stimulates binding of RecA to ssDNA (with RecO)
ssb	SSB	18	ssDNA binding protein
Branch migration/resolution (postsynapsis)			
ruvA	RuvA	22	Binds specifically to Holliday junctions, targets RuvB to DNA
ruvB	RuvB	37	ATPase, promotes branch migration (with RuvA)
ruvC	RuvC	19	Endonuclease, specifically resolves Holliday junctions
recG	RecG	76	ATPase, specifically binds to Holliday junctions, promotes branch migration

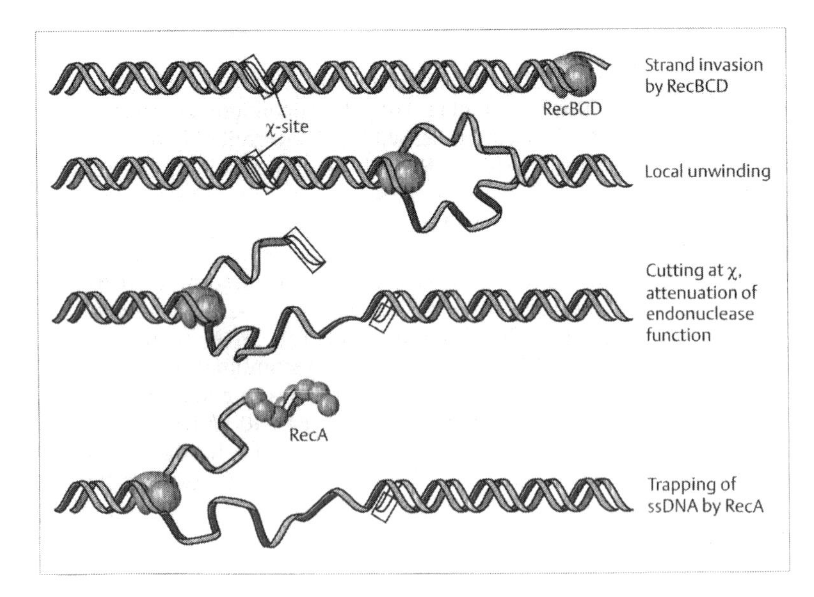

Fig. 16.**18** **Initiation of homologous recombination by RecBCD.** The model summarizes the steps involved using enzymes listed in Figure 16.**17** and Table 16.**1**

limited degradation (not shown in the figure). In the absence of SSB protein, the ssDNA produced by the RecBCD helicase activity reanneals behind the enzyme. Upon encountering a χ sequence in the appropriate orientation, the enzyme cuts the DNA at χ and continues to unwind the DNA. Following cleavage at χ, the endonuclease function is attenuated until RecBC(D) exits the DNA molecule. The DNA single strand that is stripped off following nicking at χ is trapped by RecA protein and used for the strand exchange reaction.

16.4.2 Synapsis Consists of Pairing and Exchange of Homologous DNA Strands

Following the production of single-stranded DNA, RecA protein promotes homologous pairing and strand exchange, leading to the formation of recombination intermediates (**Holliday junctions**). RecA plays a pivotal role in recombination as the central catalyst of homologous pairing and strand exchange. The protein binds to DNA to form long polymeric filaments within which DNA–DNA pairing takes place (Fig. 16.**19**). On ssDNA, filament assembly proceeds with $5' \rightarrow 3'$ polarity. The nucleoprotein filament is a helical structure with a pitch of 9.5 nm and 6.2 RecA monomers per turn. The DNA within the filament is stretched from 10.5 bp/turn (B-form DNA) to 18.6 bp/turn as RecA imposes its own helicity on the DNA. The importance of the structure is evident from the finding that it has been conserved throughout the prokaryotic kingdom, in yeasts, and most likely also in mammalians. Recent studies suggest that RecF, RecO, RecR, and SSB proteins are RecA accessory proteins (Table 16.**1**).

A second homologous DNA molecule is paired in a single-stranded region of the DNA within the nucleoprotein filament and strand exchange ensues. The initial pairing of two DNAs may involve the formation of a novel DNA triplex structure (Fig. 16.**19**). Strand exchange is unidirectional ($5' \rightarrow 3'$ with respect to the single strand within the filament) and relatively slow (3–10 bp/s). A key property of the RecA-mediated three-strand exchange reaction is its capacity to bypass structural barriers in one or both DNA substrates, including heterologous inserts of 50–100 bp. Four-strand exchange reactions are invariably initiated as a three-strand reaction in the single-stranded gap of the first DNA. As strand exchange proceeds into a double-stranded region of the first DNA molecule, a Holliday intermediate (Fig. 16.**20**) is formed.

Holliday junctions are intermediates of recombination in which two duplex DNAs are connected by partial exchange of homologous DNA strands (Fig. 16.**20**).

The RecA nucleoprotein filament has an intrinsic capacity to bind at least three strands of DNA and to promote strand exchange between them. ATP is not required for this activity, but ATP hydrolysis alters the properties of the strand exchange reaction fundamentally: it is slower, taking 5–10 min to proceed 3000 bp; it is unidirectional; it readily bypasses substantial structural barriers; it accommodates four DNA strands. These observations suggest that ATP hydrolysis indeed is coupled to DNA strand exchange and that it alters the characteristics of the reaction in such a way that it becomes useful for recombinatorial DNA repair. How-

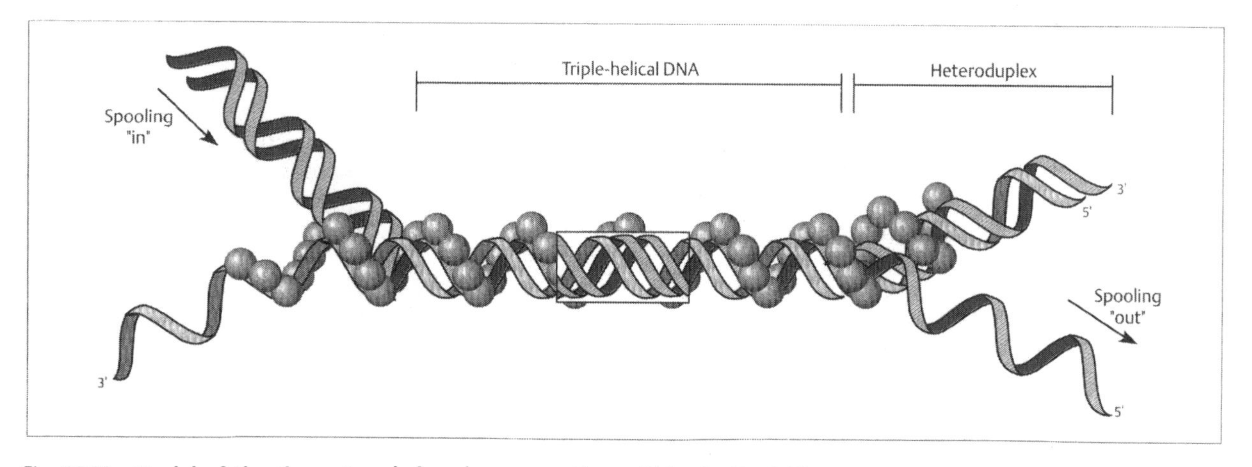

Fig. 16.**19** **Model of the three-stranded exchange reaction within the RecA filament.** Protein molecules (spheres) are not drawn to scale [after 7]

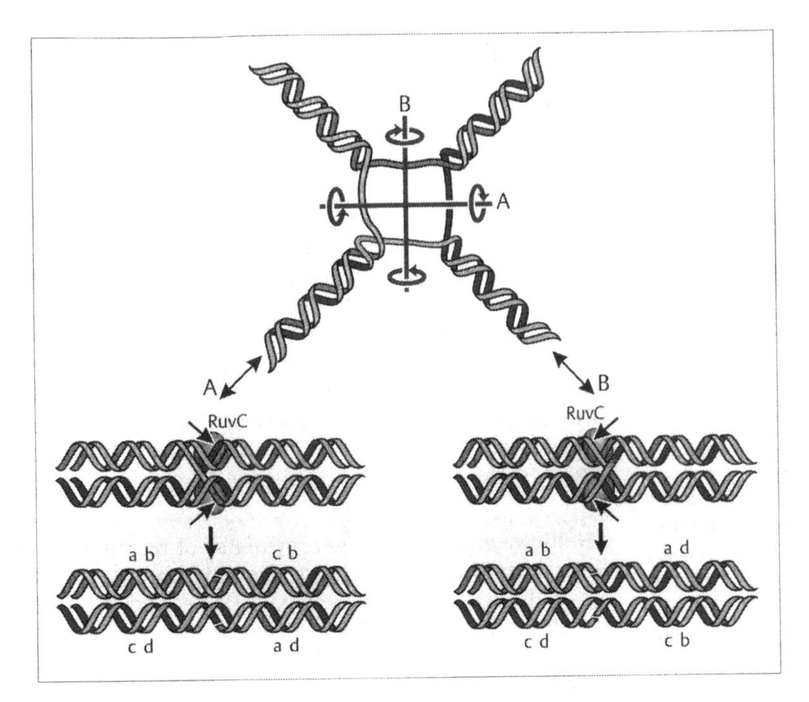

Fig. 16.**20** **Cleavage of Holliday junctions by RuvC results in two possible products.** Prior to resolution, Holliday junctions have to take on a conformation that is a substrate for RuvC. This conformation is achieved either by twisting the junction around axis A (left pathway) or around axis B (right pathway). Resolution by cleavage (thick arrows) results either in products abcb and cdad or products abad and cdcb

ever, the molecular basis of this coupling has not yet been established.

In addition to its strand exchange functions, RecA protein plays a key role in inducing the cellular SOS response and a set of operons, the expression of which is induced when the cellular DNA becomes damaged. DNA lesions result in single-stranded regions that are trapped by RecA (see Chapter 15.5.1).

16.4.3 Postsynapsis Resolves Recombinational Intermediates

Branch migration. In addition to the unidirectional branch migration that occurs in RecA filaments in the presence of ATP, at least two additional activities that mediate the translocation of Holliday junctions exist in *E. coli*:

1. The *E. coli* RuvA and RuvB proteins act together to promote branch migration of Holliday junctions. Addition of purified RuvA (binding) and RuvB (ATP-dependent branch migration) to a RecA-mediated recombination reaction stimulates the rate of strand exchange and the formation of heteroduplex DNA.
2. The second branch migration activity in *E. coli* is provided by RecG. A *recG* or *ruv* single mutation reduces conjugational recombination by approximately 3-fold; a *recG ruv* double mutant reduces recombination by more than 500-fold. The double mutants are also extremely sensitive to ultraviolet light, much more so than the single mutants. This synergism demonstrates the functional overlap between the *ruv* and *recG* products. The 76-kDa RecG protein is a DNA-dependent ATPase, like RuvB, and it binds to Holliday junctions as does RuvA.

Resolution and cleavage of Holliday structures. Completion of the recombination event requires the resolution of the Holliday junction to restore the DNA to two discrete molecules. One activity that is capable of performing Holliday junction resolution has been identified in *E. coli* cell-free extracts. This activity has been assigned to the RuvC protein. The purified protein in vitro specifically binds to Holliday junctions and resolves them by endonucleolytic cuts at the crossover point (Fig. 16.**20**). Depending on the conformation of the DNA at the crossover point, cutting can result in one of two different products.

16.5 DNA Restriction and Modification Protects the Cell Against Foreign DNA

Several mechanisms exist in prokaryotic cells that safeguard the genetic information content against physical or chemical damage (see Chapter 15.5) and also against invasion of the cell by foreign DNA. The latter mechanism provides the cell with immunity against many types of bacteriophages, but it is also a barrier for plasmid transfer. The plating efficiency of many bacteriophages on *E. coli* cells varies dramatically depending not only on the type of the strain infected, but also on that of the previous host. For example, phages grown on *E. coli* K-12 plate with equal efficiency on *E. coli* K-12 and *E. coli* C, but the converse is not true: phages grown on *E. coli* C plate poorly on *E. coli* K-12, but with high efficiency on *E. coli* C. Moreover, a single cycle growth on *E. coli* C removes the ability of the phages to grow well on *E. coli* K-12.

The molecular interpretation of this phenomenon is that *E. coli* K-12 degrades foreign DNA by a sequence-specific endonuclease that is not present in *E. coli* C. The DNA of a phage grown on *E. coli* K-12 is modified by a methyltransferase that recognizes the sites of the endonuclease, rendering the DNA resistant against endonucleolytic degradation. This combination of enzymatic activities has been designated **host-controlled restriction–modification (R–M) system**. Based on differences in cofactor requirements and subunit structure, restriction–modification systems have been classified into three groups (Table 16.2). Members of all three

Table 16.2 **Classification of restriction–modification systems**

Characteristic	Type I	Type II	Type III
Occurrence:	Enterobacteriaceae	All prokaryotes	*Escherichia coli, Haemophilus influenzae*
Examples:	*EcoK, EcoB*	*EcoRI, HindIII*	*EcoP1, HinfIII*
Subunits for restriction:	Heteropentamer ($HsdR_2{\cdot}HsdM_2{\cdot}HsdS$)	Heterodimer	Homodimer
Subunits for modification:	Heteropentamer ($HsdR_2{\cdot}HsdM_2{\cdot}HsdS$) or Heterodimer (HsdM·HsdS)	Heterodimer or Monomer	Monomer
Cofactors for restriction:			
Essential	AdoMet, ATP, Mg^{2+}	Mg^{2+}	ATP, Mg^{2+}
Stimulatory			AdoMet
Cofactors for modification:			
Essential:	AdoMet	AdoMet	AdoMet, Mg^{2+}
Stimulatory:	ATP, Mg^{2+}		
Recognition site:	Bipartite and asymmetric	4–8-bp sequence, often palindromic	5–7 bp, asymmetric
Cleavage site:	Random, >1000 bp from the recognition site	Within or close to the recognition sequence	25–30 bp downtream from recognition site
Enzymatic turnover:	No	Yes	Yes
Coordination of restriction/modification	Mutually exclusive	Separate reactions	Competing reactions
DNA translocation:	Yes	No	No
ATPase:	Yes	No	No

classes are found in *E. coli*. On the other hand, strains, such as *E. coli* C, can be found that have no detectable R–M system. The biological significance of R–M systems in limiting plasmid transfer and phage infection is demonstrated by the finding that many conjugative plasmids and phages encode anti-restriction systems (see Chapter 16.2.4).

16.5.1 Type II Restriction–Modification Systems

Type II systems are ubiquitous in prokaryotes. About 30% of the bacterial strains examined so far possess at least one type II system. R–M functions are mediated separately by two types of enzymes. The type II restriction endonucleases usually exist as homodimers and recognize a 4–6-bp sequence (in a few cases also 8 bp) that often is a palindrome. The corresponding modification methylases, although they recognize the same target sequence, are not similar in their amino acid sequence and act as monomers. The two components of type II systems most likely emerged by convergent evolution and not by gene duplication and subsequent adaptation to the different activities.

Palindromic target sequences may exist in three states: (1) nonmethylated, (2) hemimethylated, and (3) fully methylated on both strands. **Nonmethylated** DNA is the natural substrate for both the endonuclease and the methylase. However, it is much more likely that a nonmethylated DNA is endonucleolytically broken down than that it survives by gaining the modification pattern of its new host. The **hemimethylated** DNA species exists transiently when the DNA is replicated. It is not recognized by the restriction endonuclease, but is converted to **fully methylated** DNA by the modification methylase. Finally, the fully methylated DNA is not a substrate for the endonuclease or the methylase.

The modification **methylases** catalyze the transfer of methyl groups from *S*-adenosylmethionine (AdoMet) to specific residues within the target sequence. Depending on the R–M system, the methylated residue is either *N*6-methyladenosine or 5-methylcytosine. In the case of *Hha*I methyltransferase—the enzyme recognizes the sequence GCGC and the first cytidyl residue is modified—the crystal structure of a chemically trapped reaction intermediate has been determined. Interestingly, it was found that the cytidyl residue to be methylated swings completely out of the DNA double helix to be positioned in the active site of the enzyme. The enzyme stabilizes the complementary "orphan" guanosyl residue by hydrogen bonds in nearly the same conformation that it has in the undisturbed B-form DNA, preventing collapse of the local helix structure. Type II restriction endonucleases have a great importance in gene technology because they can be used to create specific DNA fragments. Many restriction endonucleases generate staggered cuts with an overhang of 2–4 bases. The resulting cohesive fragment ends can be ligated specifically and efficiently by DNA ligases (see Chapter 17.1).

In **host-controlled restriction–modification** a sequence-specific endonuclease ("restriction enzyme") cleaves unprotected (foreign) DNA at a consensus sequence site. The producer organism also contains a modification enzyme that binds at the same consensus sequence and modifies (methylation, glycosylation) a nucleotide, thus preventing cleavage (restriction) by the endonuclease.

16.5.2 Type I Restriction–Modification Systems

In contrast to the type II systems, type I enzymes function as complex multimers that can catalyze both the endonuclease and the modification methylase reactions. Typical representatives are the Hsd systems (**h**ost **s**pecificity of **D**NA) *Eco*K and *Eco*B of *E. coli*. *Eco*K contains one subunit of HsdS (55 kDa), which confers the specificity of the reactions, and two copies of each of HsdR (135 kDa) and HsdM (62 kDa), which mediate restriction and methylation activities, respectively. The target sequences recognized by all type I enzymes are asymmetric and consist of two defined components, one 3 bp and another 4 or 5 bp, separated by a non-specific spacer of fixed length (e.g., AAC$(N)_6$GTGC for *Eco*K and TGA$(N)_8$TGCT for *Eco*B). Consistent with this, the specificity polypeptides (HsdS) are known to contain two DNA recognition domains, each specifying recognition of one of the two defined components of the target sequence.

The restriction endonuclease activity of type I enzymes is completely dependent on AdoMet and ATP and requires Mg^{2+} (Table 16.**2**). AdoMet and ATP serve as cofactors and allosteric effectors. AdoMet binds to the HsdM subunit and allosterically changes the conformation of HsdS, allowing DNA binding. Reaction with ATP initiates either methylation or restriction, depending on the methylation state of the DNA. At an unmethylated site, AdoMet is released and restriction commences. DNA cleavage occurs after the enzyme has translocated on the DNA in an ATP-dependent reaction. During translocation, the enzyme remains attached to its recognition site, and a loop is formed. The DNA is cut at random sites located more than 1000 bp away from the recognition sequence. The restriction reaction leads to irreversible inactivation since there is no enzymatic turnover. The methylase reaction requires only AdoMet as cofactor and takes place on hemimethylated DNA. If the enzyme is bound to fully methylated DNA, ATP is required to release it from the DNA.

16.5.3 Type III Restriction–Modification Systems

Three type III systems have been characterized in greater detail: the *Eco*P1 and *Eco*P15 enzymes specified by *E. coli* phages P1 and P15, respectively, and *Hin*fIII, discovered in *Haemophilus influenzae*. Type III R–M enzymes consist of two subunits; one subunit (HsdMS, 75 kDa) is involved in recognition and methylation of the target sequence and the other subunit (HsdR, 108 kDa) provides the restriction endonuclease activity. Binding of the enzyme to the DNA requires ATP, but does not involve ATP hydrolysis (Table 16.**2**). Restriction and modification activities are expressed simultaneously, competing for reaction with the DNA. The modification reaction consists of the methylation of an adenine residue within the recognition sequence and requires AdoMet as donor for methyl groups. Restriction cleavage takes place 25–30 bases downstream of the recognition sequence, probably reflecting the subunit structure of type II enzymes with recognition and endonuclease functions in different subunits that contact different sites on the DNA. Cleavage leads to staggered cuts with an overhang of 2–4 bp. Surprisingly, only one strand of the recognition sites of *Eco*P1 and *Eco*P15 contains adenine residues; therefore, only this strand can be methylated (AGACC for *Eco*P1 and CAGCAG for *Eco*P15).

It is not known how the methylation state of the DNA is perpetuated during replication because one replica would be completely unmethylated. Therefore, it would be susceptible for endonucleolytic breakdown by the restriction enzyme.

Further Reading

Birge, E. A. (1994) Bacterial and bacteriophage genetics, 3rd edn. Berlin, Heidelberg, New York: Springer

Calendar, R., ed. (1988) The bacteriophages, vol. 1 and 2. New York, London: Plenum Press

Clewell, D. B., ed. (1993) Bacterial conjugation. New York: Plenum Press

Dubnau, D. (1991) Genetic competence in *Bacillus subtilis*. Microbiol Rev 55: 395–424

Holloway, B. W. (1993) Genetics for all bacteria. Annu Rev Microbiol 47: 659–684

Joset, F., and Guespin-Michel, J. (1993) Prokaryotic genetics. Genome organization, transfer and plasticity. Oxford: Blackwell

Kornberg, A., and Baker, T. (1992) DNA replication, 2nd edn. New York: Freeman

Mazodier, P., and Davies, J. (1991) Gene transfer between distantly related bacteria. Annu Rev Genet 25: 147–171

Neidhardt, F. C., Curtiss, R., Ingraham, J. L., Lin, E. C. C., Low, K. B., Magasanik, B., Reznikoff, W., Riley, M., Schaechter, M., and Umbarger, H. E. (eds.) (1996) *Escherichia coli* and *Salmonella*: cellular and molecular biology. 2nd edn. Washington, DC: ASM Press

Pansegrau, W., and Lanka, E. (1996) Enzymology of DNA transfer by conjugative mechanisms. Prog Nucl Acid Res Mol Biol 54: 197–251

Salyers, A. A., Shoemaker, N. B., Stevens, A. M., and Li, L.-Y. (1995) Conjugative transposons: an unusual and diverse set of integrated gene transfer elements. Microbiol Rev 59: 579–590

Scott, J., and Churchward, G. G. (1995) Conjugative transposition. Annu Rev Microbiol 49: 367–397

Smith, G. R. (1988) Homologous recombination in prokaryotes. Microbiol Rev 52: 1–28

Snyder, L., and Champness, W. (1997) Molecular genetics of bacteria. Washington, DC: ASM Press

Sonenshein, A. L., ed. (1993) *Bacillus subtilis* and other Gram-positive bacteria. Washington, DC: ASM Press

Spaink, H. P., Kondorosi, A., and Hoyykaas, P. F. F. eds. (1998) Kluwer Academic Publishers

West, S. C. (1992) Enzymes and molecular mechanisms of genetic recombination. Annu Rev Biochem 61: 603–640

Wilson, G. G., and Murray, N. E. (1991) Restriction and modification systems. Annu Rev Genet 25: 585–627

Sources of Figures

1 Neidhardt, F. C., Curtiss, R., Ingraham, J. L., Lin, E. C. C., Low, K. B., Magasanik, B., Reznikoff, W., Riley, M., Schaechter, M., and Umbarger, H. E. (eds.) (1996) *Escherichia coli* and *Salmonella*: cellular and molecular biology. 2nd edn. Washington, DC: ASM Press

2 Pansegrau, W., and Lanka, E. (1996) Prog Nucleic Acid Res Mol Biol 54: 197–251

3 Wilkins, B., and Lanka, E. (1993) In: Clewell, D. B. (ed). Bacterial conjugation. New York: Plenum; 105–136

4 Lessl, M., and Lanka, E. (1994) Cell 77: 321–324

5 Pansegrau, W., Lanka, E., Barth, P. T., Figurski, D. F., Guiney, D. G., Haas, D., Helinski, D. R., Schwab, H., Stanisch, V. A., and Thomas, C. M. (1994) J Mol Biol 239: 623–663

6 West, S. C. (1994) Cell 76: 9–15

7 West, S. C. (1992) Annu Rev Biochem 61: 603–640

17 Recombinant DNA Technology

Until about 1975, research in molecular biology was limited to a few bacteria and their viruses, the bacteriophages. Through a clever combination of genetic, biochemical, and molecular biological methods, scientists gained essential information about the structure and function of genes. The situation changed dramatically when a set of new methods became available, for which the term **gene technology** or **recombinant DNA technology** was coined. These methods influenced the research fields discussed here and are currently revolutionizing the entire field of biology from biophysics and biochemistry to systematics and ecology, from basic research to biotechnology, medicine, and agriculture.

New methods develop very fast since they profit from the vast experience accumulated from handling bacteria and bacteriophages. The basic techniques are often simple; the details of many experimental setups, however, become more complex since they are continuously improved and adapted from the initial more general strategy to specific variations for specific purposes. The basic principles of recombinant DNA technology as used in prokaryotic organisms will be discussed here. Numerous practical examples are given throughout Sections IV–XI.

Recombinant DNA technology involves four basic steps:

1. Production of defined fragments from naturally occurring long DNA molecules ("DNA restriction").
2. Separation of the various DNA fragments and establishment of a genomic library.
3. Isolation and propagation of a desired DNA fragment, followed by molecular biological analysis, i.e., determination of the DNA sequence and investigation of the encoded proteins.
4. Modification of DNA and heterologous gene expression.

17.1 Genomic Libraries Contain Defined DNA Fragments

The first step, the cutting of the native DNA molecules into defined fragments requires **restriction endonucleases**. These enzymes recognize short specific nucleotide sequences and cut a DNA strand (see Chapter 16.5). Two examples will be considered:

1. The enzyme *Eco*RI cuts DNA at locations with the nucleotide sequence GAATTC, independent of the source of the DNA. If such sequences are statistically distributed, the DNA will be cut into pieces of average lengths of 4000–5000 base pairs. *Eco*RI will therefore cut a bacterial chromosome of, for example, $5 \cdot 10^6$ base pairs, into $5 \cdot 10^6/5 \cdot 10^3$ or 1000 fragments.
2. The enzyme *Alu*I cuts DNA at the tetranucleotide sequence AGCT, which with naturally occurring DNA leads to fragments of 250–350 base pairs.

The statistical frequency can be estimated as follows. Consider DNA in which all four nucleotides are present in equal amounts. If a recognition sequence contains all four bases, such as AGCT for *Alu*I, cutting sites will theoretically appear in distances of $4^4 = 256$ nucleotides. Accordingly, the distance of hexanucleotide sequences will be $4^6 = 4096$ nucleotides. However, naturally occurring DNA molecules and many recognition sequences do not fit these simple rules because the distribution of specific restriction sites normally deviates considerably from statistics. Therefore, these calculations can only give an estimation of the average fragment lengths expected. Conversely, the frequency and distribution of cutting sites for a specific restriction endonuclease are characteristic ("fingerprint"), for example, for a given chromosomal DNA.

The important point is that an identical collection of DNA fragments will always be produced, independent of when and where a specific DNA is treated with a specific restriction endonuclease. This collection of fragments is a very complex mixture of molecules. The different constituents of this mixture must be separated accurately from one another. This is usually done by transferring the DNA fragments into competent bacteria (see Chapter 16). However, nonreplicating restriction fragments would be quickly digested and lost within transformed host cells. The fragments must therefore be integrated into DNA carriers or **cloning vectors**, as they are called in the terminology of gene technology, which allow the replication of the cloned DNA.

Commonly used vectors are plasmids and bacteriophages or a combination of both. Currently there are literally hundreds of different vectors available. As an introduction to this large field, only a few typical and often used examples of prokaryotic vectors will be described.

17.1.1 Many Vectors Are Derived From Plasmids

The most popular plasmid vectors are derived from R (resistance) plasmids (see Chapter 16). Figure 17.1 depicts the naturally occurring **RP4 plasmid** from the IncP family with transfer genes, with some genes that confer resistance to antibiotics, with an origin for bidirectional replication (*oriV*), and with a second origin (*oriT*), which is engaged in the unidirectional transfer replication during transfer of the plasmid from the donor to recipient bacterium.

Cloning vectors (or just **vectors**) are small DNA molecules (usually 3–11 kb) that can take up cloned DNA into specific restriction sites and multiply this DNA in a host cell. For reasons of safety, vectors normally lack an *oriT*, as well as *tra* and *mob* genes and therefore cannot be transferred into other cells by conjugation.

A natural R plasmid is inappropriate for the purposes of gene technology because 1) it is present in only one or two copies per bacterial cell, 2) it is easily damaged by shearing during biochemical procedures owing to its usually large size (up to 100 kb), and 3) most importantly, most natural R plasmids can be transferred from one bacterial cell to another. The latter characteristic would hamper the intended efficient separation of the restriction fragments. Therefore, derived plasmids are

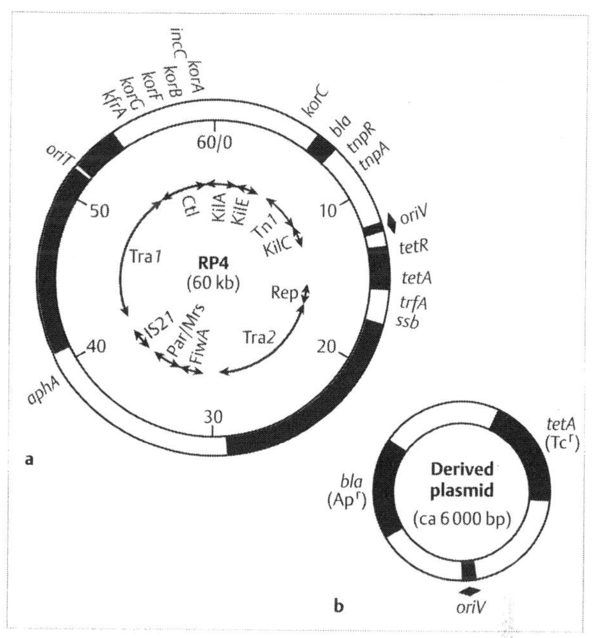

Fig. 17.**1a, b** **Natural R plasmid and derived vector plasmid.** **a** Natural plasmids are built in a modular way as shown here for the conjugative resistance plasmid RP4 of the IncPα family. RP4 contains functionally independent modules, such as the *IS21* element and transposon Tn*1*, and modules that constitute an integral part of the plasmid's maintenance (genes *inc*, *kfr*, *kor*, *ssb*, and *trf*, and gene products Ctl, Kil, Pat/Mrs, FiwA, and Rep) and conjugative transfer functions (see also Chapter 16). The latter are clustered in two regions, Tra1 and Tra2 (shown in red), for DNA processing and mating pair formation, respectively. Integral modules coevolved to become functionally adapted, and this evolutionary drift resulted in non-interchangeability of functionally analogous genes, e.g., the *tra* genes from RP4 and from the F plasmid IncF1.
b In derived vectors, only the origin of replication (*oriV*) and the antibiotic resistance genes *bla* and *tet* for ampicillin (Ap^r) and tetracycline (Tc^r) resistance, respectively, are retained to minimize unwanted gene transfer

normally used that only can multiply within bacterial cells to higher copy numbers by bidirectional replication and which cannot be transferred to other bacteria. A simple derived plasmid that was frequently used in the beginning of the gene technology era is **pBR322**. It contains an *oriV*, and *bla* and *tet* genes, each of which derives from a different plasmid and which allow replication in *Escherichia coli* and confer resistance to the antibiotics ampicillin (Ap^r) and tetracycline (Tc^r), respectively (Fig. 17.**2**). As is customary in the nomenclature of plasmid vectors, the lower case p indicates that the vector is a plasmid, and **BR** represents the initials of the scientists that constructed the vector (F. **B**olivar and R. L. **R**odrigues, in 1977), followed by a

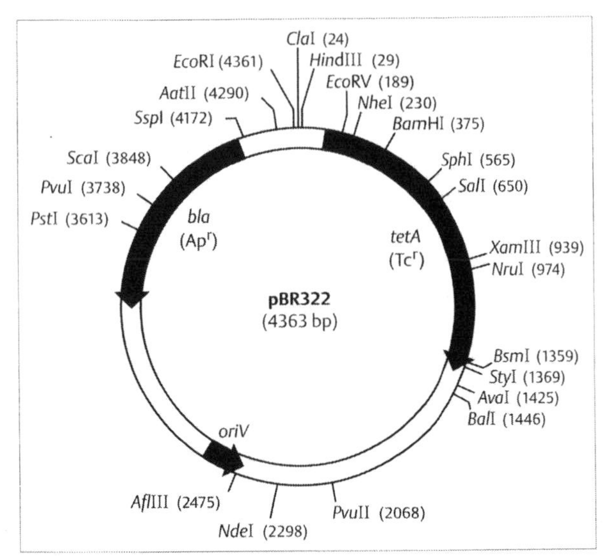

Fig. 17.2 **Plasmid pBR322.** Numbering starts at the unique *Eco*RI site and continues clockwise. There are several other unique restriction sites (positions in parentheses) that facilitate precise cloning, in addition to an ampicillin (Ap^r) and a tetracycline (Tc^r) resistance locus for easy selection, and an *oriV* for DNA replication. Note the absence of genes for conjugation and mobilization

laboratory number to discriminate between similar, but not identical plasmids. Furthermore, the plasmid-encoded phenotypes are given in a two-letter code (Ap^r, Tc^r) rather than a three-letter code (Amp^r, Tet^r) as is usual for chromosomal traits.

During cloning, restriction fragments are integrated into a vector. To facilitate this process, the same restriction endonuclease is used for cutting the foreign DNA and the vector DNA at a site that occurs only once in the vector. Consider the opening ("linearization") of pBR322 (Fig. 17.**3**) at the unique *Bam*HI site in the *tet* gene. Restriction fragments of the foreign DNA are added to the linearized vector DNA. The two DNA molecules are covalently linked to one another by DNA ligase. The resulting products are DNA rings with portions derived from the vector and from the foreign DNA.

When performing cloning experiments, it is important to prevent the religation of the ends of the cut vector DNA and to promote the insertion of the foreign DNA. This is possible when the foreign DNA is in excess of the vector DNA, or even better, if the vector DNA after linearization is treated with a phosphatase, which removes the 5′-phosphates from the ends of the vector. The presence of 5′-phosphates is a prerequisite for the covalent linkage of DNA ends by DNA ligase. Since only

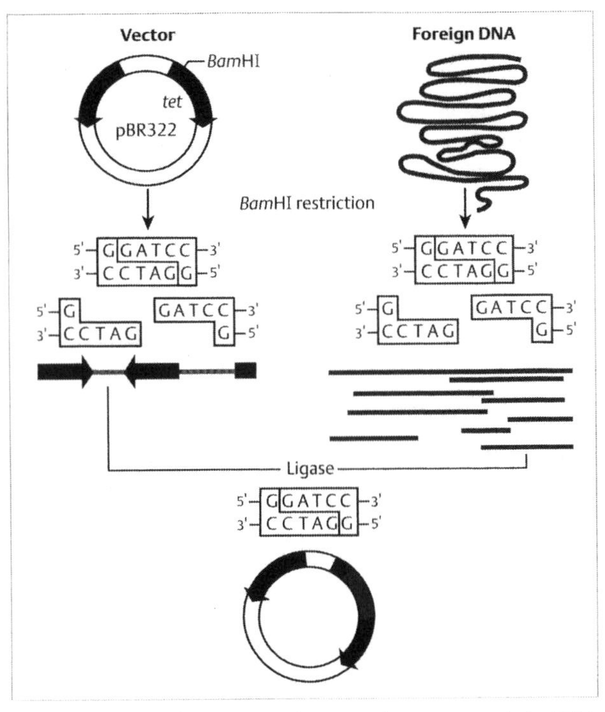

Fig. 17.**3** **Principles of DNA cloning.** The vector and the DNA to be cloned are cut with a restriction endonuclease, here *Bam*HI, which has a unique cutting site in the vector. The site shown here is located within the *tetA* gene of pBR322 (see Fig. 17.**2**). After insertion, the cointegrate thus loses its Tc^r phenotype, but retains the Ap^r resistance for easy selection. To prevent simple religation, the cut vector DNA is normally treated by a phosphatase. To ensure a unique orientation of the cloned DNA within the vector, DNA fragments and vectors can be cut by two distinct restriction endonucleases (e.g., *Hin*dIII and *Bam*HI) instead of just one

the foreign DNA provides 5′-phosphates, only foreign DNA fragments will be ligated to phosphatase-treated vector DNA. Hybrid molecules result in which one strand of the foreign DNA at each end is ligated to the vector. The opposite strands are not covalently linked in this reaction, but they will be linked by resident ligases after transformation into host cells.

Subsequent to these biochemical reactions, the separation of the DNA molecules occurs in bacterial cells. After treatment with calcium ions, *E. coli* cells become competent (see Chapter 16) and able to take up DNA. During transformation, only a single DNA molecule should be taken up by one bacterial cell. This can be accomplished using a number of bacterial cells in large excess over DNA molecules. The transformed bacteria are finally spread on agar plates so that they can develop separate colonies. In the example shown in Figure 17.**3**, the plates should contain the antibiotic ampicillin. Bacteria without plasmid will be killed by the antibiotic and only cells that have taken up pBR322 can continue to grow to form colonies. Bacteria that have taken up plasmids with foreign DNA can be identified by an additional screening method: bacteria with pBR322 form colonies on plates with both ampicillin and tetracycline; bacteria with pBR322 and inserted foreign DNA form colonies only on plates with ampicillin because the *tet* gene has been inactivated by the insertion of the foreign DNA. The separation of the mixture of restriction fragments of the foreign DNA is automatically ensured under these conditions because each individual colony consists of bacteria that contain an identical piece of foreign DNA inserted into the plasmid vector. Genetically identical descendants of a parental cell are traditionally called a **clone** in biology. Accordingly, the described procedure is frequently called **DNA cloning** or **gene cloning**, the restriction fragment contained in a colony is called a **DNA clone**, and the corresponding DNA has been **cloned**.

If all steps are successfully performed, a complete chromosome that has been separated into smaller fragments is available for further investigation. This set of vectors containing cloned DNA is frequently called a **genomic (DNA) library** (**"genotheque"**). Similar to checking out a book from a library, any DNA fragment of interest can be taken from the genomic library and investigated in more detail. Bacteria can easily be grown in large quantities so that almost any amount of the foreign DNA fragment can be made available in a short time. A prerequisite is that the plasmid DNA can be isolated easily from the bacteria and that the cloned DNA fragment can be released from the plasmid by restriction.

Because the correct identification of the inserted DNA in pBR322 is laborious, other plasmid vectors have become more popular, such as the frequently used plasmid pUC19, a representative of the large **pUC plasmid family** (Fig. 17.4). Similar to other plasmid vectors, pUC19 contains an *oriV*, the *bla* gene which confers ampicillin resistance and, in addition, a special DNA fragment with many different recognition sites for restriction endonucleases. This **multiple cloning site** (MCS) sequence is located in the 5′-part of the *lacZ* gene, which codes for the amino-terminal part of the β-galactosidase (LacZ) from *E. coli*. Suitable bacterial strains used in this selection system must contain the 3′-part of the *lacZ* gene encoding an intact carboxy-terminal part of the β-galactosidase. Bacteria containing pUC19 are ampicillin resistant and encode a functional β-galactosidase, which is assembled from the plasmid-encoded and the chromosome-encoded parts. These

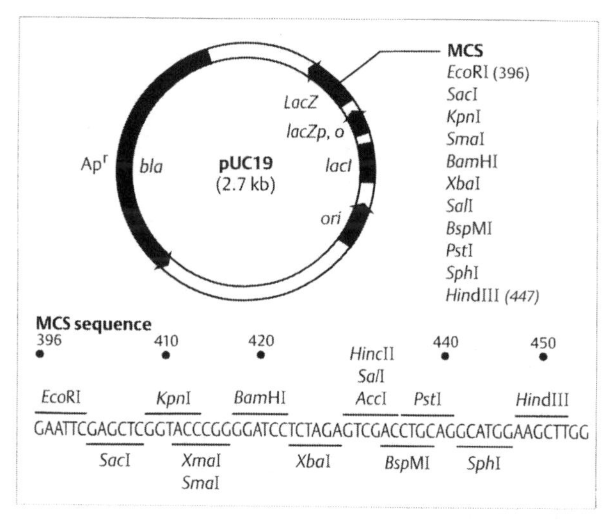

Fig. 17.4 **Vector pUC19 allows "blue-white" selection.** pUC19 represents a typical expression vector with an *oriV*, the *bla* gene for easy Ap^r selection, a truncated *lacz′* with its *lacZo,p* promoter/operator, and *lacI* for control and easy induction by means of the inducer IPTG. Gene *lacz′* contains a multiple cloning site (MCS) for several restriction endonucleases and codes for the amino-terminal part of LacZ, which allows "blue-white" selection (see text) in the corresponding selection strains

bacteria form blue colonies in the presence of the chromogenic substrate 5-bromo-4-chloro-3-indolyl-β-D-galactopyranoside (X-Gal). However, the insertion of foreign DNA into the MCS interrupts the reading frame of the *lacZ* gene. The bacteria containing a vector with cloned DNA will therefore remain colorless in the presence of X-Gal. In other words, colonies on ampicillin/X-Gal agar will be blue when the bacteria contain pUC19, but they will remain white when the bacteria carry foreign DNA inserted into pUC19. This **blue-white selection** is the basic principle for many recombinant DNA techniques.

17.1.2 Derivatives of Phage λ and Cosmids Are Also Often Used as Vectors

Cloning into plasmids is technically simple, but has the disadvantage that only relatively short DNA fragments (up to a few thousand base pairs) can be inserted. Longer inserted DNA fragments limit more and more severely the replication of plasmids in bacteria. Functionally related genomic sections, such as operons, may thus be separated during cloning in plasmids; this is often an unwanted effect. Therefore, vectors that allow the insertion of longer fragments of foreign DNA are frequently preferred.

A simple and popular system is derived from bacteriophage λ. A large part of the λ genome can be dispensed with and replaced by foreign DNA, leaving only a lytic cycle of infection: the *b2* region, the integration region with the genes *xis* and *int*, and the *red* genes for recombination enzymes (see Fig. 26.**13**). However, native λ DNA does not contain any suitable restriction sites that would allow the easy deletion of the expendable region. Therefore, a series of λ derivatives has been designed. The vector **EMBL3** will be discussed as an example (Fig. 17.**5**).

In EMBL3, the center region of the phage DNA is replaced by an optional DNA fragment, which is flanked on either side by cutting sites for various restriction enzymes. The procedure for cloning into λ vectors includes the following steps:

1. The DNA of interest is **partially** cut by restriction endonucleases in order to obtain long fragments.
2. The "arms" of the λ vector are isolated, mixed with the cut DNA, and covalently connected with ligase to form long DNA concatemers.
3. The concatemeric DNA is packaged in vitro into phage particles. The assembly of the phage structure from single components can be achieved in the test tube if the supplied DNA is present in concatemeric form and contains *cos* sites at the ends (see Chapter 26). DNA fragments of 10–20 kb can be inserted into λ vectors. The recombinant phages are separated by plaque formation on a lawn of bacteria. A single plaque contains more than a million identical phages (phage clones), which can be amplified to even larger numbers in order to allow an investigation of the inserted foreign DNA by molecular biological methods.

Cosmids are plasmids that carry the **cohesive ends** or ***cos* sites** of the λ DNA in addition to, for example, the *bla* gene for ampicillin resistance. Cosmids combine the advantages of plasmids (simple technical handling) with the advantages of λ vectors (insertion of larger DNA fragments). Foreign DNA of up to 50 kb can be inserted into cut cosmids. The concatemeric DNA can be packaged into phage particles and used for infection. The recipient bacteria acquire the potential to form colonies on agar plates containing ampicillin (Fig. 17.**6**).

17.1.3 cDNA and Genomic Libraries Are Often Used to Identify Genes

In many instances, RNA can be used as an alternative starting material for cloning purposes. If, for example, a regulated gene is to be cloned whose transcription can be induced to high levels, there will be many more

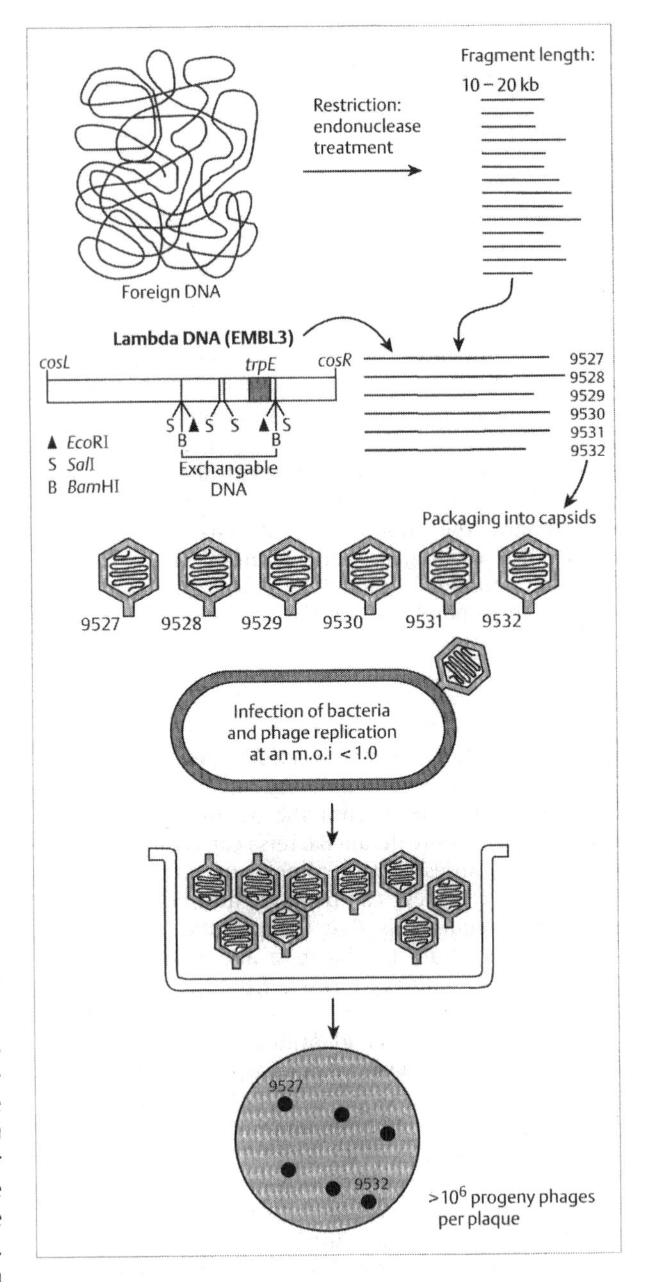

Fig. 17.**5** **Cloning using phage λ vectors.** The foreign DNA is cut by **partial** restriction into fragments of 10–20 kb. After cutting and separation from the dispensable DNA of the λ vector (here EMBL3), the "arms" of the vector are ligated to the DNA fragments to form long concatemers and are packed in vitro into capsids. The phages are plated on a lawn of sensitive host cells and allowed to form plaques. From each plaque, a clone containing $>10^6$ progeny phages, each with a different DNA fragment (here numbered 9527 to 9532) can be isolated and analyzed

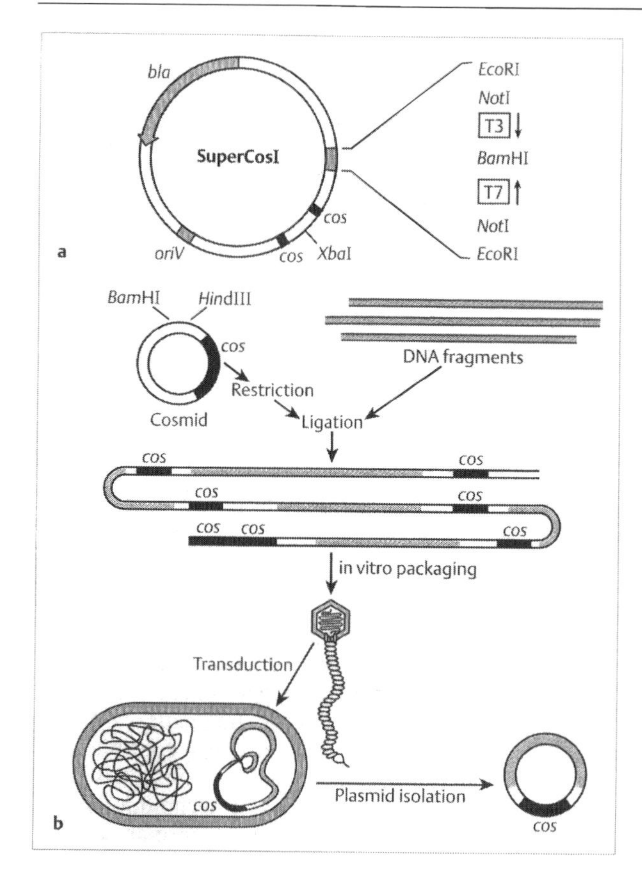

Fig. 17.6a, b **Cosmids as cloning vectors.**
a A cosmid vector contains an *oriV* of replication, the *bla* gene for Ap^r selection, and the *cos* sites of phage λ. The SuperCos1 vector shown here in a simplified version contains in addition a multiple cloning site as well as phage T7 and T3 promoters. These can be used as primer binding sites to sequence both ends of an insert for easy identification or for the in vitro production of strand-specific and end-specific RNA probes, such as for the identification of overlapping clones by hybridization methods. The cosmid is linearized at the *XbaI* site, followed by insertion of the foreign DNA into the MCS. After ligation, the DNAs are packaged in vitro into phage λ capsids; the dual *cos* sites allow a high efficiency of cloning and packaging. After transduction, Ap^r host cells containing the cosmids can be easily isolated and analyzed.
b When cosmids contain a single *cos* site, in vitro ligation of the DNA cloned into the cosmids produces concatemeric DNA by association of the natural *cos* sites. These can be packaged in vitro if the *cos* sites are about 37–52 kb apart. Packaging requires, in addition to the recombinant DNA, high concentrations of phage head precursor, and packaging and tail proteins. After injection into a host cell, the cosmid replicates as a plasmid and can be isolated from Ap^r host cells

copies of its mRNA present than of the single chromosomal DNA copy. Also, the higher relative amount of the mRNA as compared to that of other, nontranscribed genes, will be favorable for its preferred cloning. On the other hand, most bacterial mRNAs have a short lifetime. Therefore, the cDNA made from the mRNA may be short because of the presence of partially degraded mRNAs.

After isolation of total RNA from deregulated or induced bacterial cells, the RNA is converted to DNA by using the enzyme reverse transcriptase, which requires RNA as the template for synthesis of **copy** (complementary) **DNA** or **cDNA**. This cDNA can then be used like any other DNA for cloning, sequencing, or gel analysis.

The essential four steps in the synthesis of cDNA are depicted in Figure 17.7 and are described as follows:

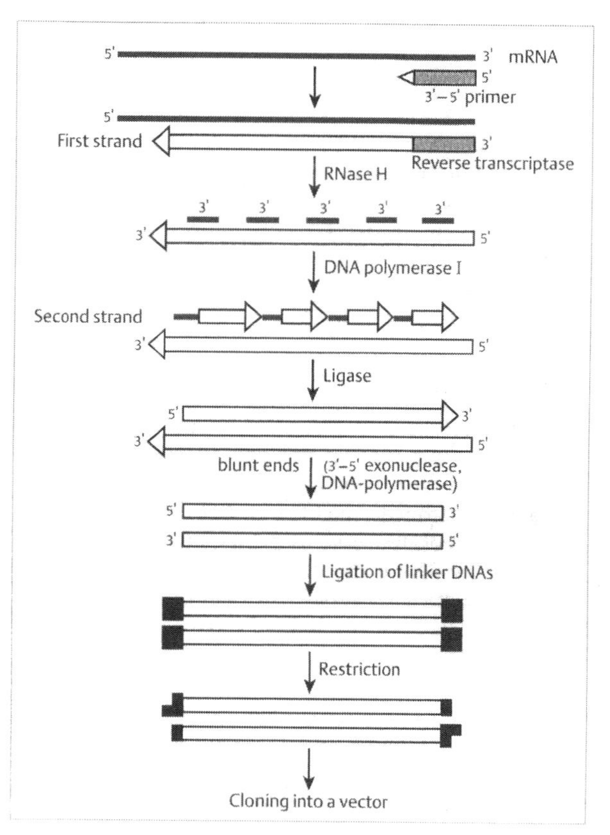

Fig. 17.7 **Generation of a cDNA (library) from RNA.** This method, sometimes used in less-well-analyzed prokaryotes, and regularly in eukaryotes, starts with a specific RNA as found, for example, only under certain growth conditions. First, reverse transcriptase is used to synthesize the first complementary DNA strand, and then one of a number of DNA polymerases is used to synthesize the second DNA strand, to obtain a complementary DNA or cDNA. For easy manipulation, artificial restriction sites are introduced on both ends of the cDNA before it is cloned into a vector and amplified. All RNAs of an organism can be isolated and cloned at a time, thus generating a cDNA library

1. **First strand synthesis** from deoxynucleotide triphosphates (dNTPs) by the enzyme reverse transcriptase is initiated either at short oligonucleotides with random sequences that hybridize randomly to any RNA molecule with complementary sequence or at oligonucleotides with specific sequences, which only hybridize to one specific RNA molecule. Hybrid DNA–RNA molecules are the result of this reaction. Frequently, the cDNA synthesis is interrupted at this point. If specific oligonucleotides that hybridize at an appropriate distance from the 5′-end of an mRNA molecule are used for the starting points of the synthesis, the reverse transcriptase will extend the primers to the very end of the RNA template. If this product is run alongside the products of a sequencing reaction that used the same primer, but used DNA from the organism as a template, the starting point of the transcription of the respective mRNA can be determined. This technique is called **primer extension analysis**.
2. **Second strand synthesis** requires that the RNA strand is interconverted into a DNA strand, and a double-stranded DNA molecule is then obtained. This can be accomplished by several methods. A popular method is the use of the enzyme **RNaseH**, which attacks and partially digests the RNA moiety of an RNA–DNA hybrid molecule. The reaction is interrupted before the RNA strand has been completely removed. The short remaining RNA oligonucleotides that are still bound to the first cDNA strand are good starting points for the DNA polymerase I from *E. coli*. The 3′-OH ends of the RNA oligonucleotides serve as primers for the synthesis, while the RNA is digested at the same time by the 5′–3′ exonuclease activity of DNA polymerase I.
3. The various consecutive DNA fragments can then be **covalently linked** by the enzyme DNA ligase. The cDNA synthesis is essentially completed. Note that it is inherent to the method that if random oligonucleotides are used in first strand synthesis, the extreme 5′-end of the RNA template most probably will not be copied.
4. The last step is the preparation of the cDNA for the **insertion** into cloning vectors. Protruding single-stranded DNA sections are digested by appropriate nucleases; frequently the 3′–5′ exonuclease activity of DNA polymerases is used. This is followed by the adaptation of the cDNA ends to the vector by the addition of short synthetic oligonucleotides, so-called **linkers**, to the ends of the cDNA by DNA ligase. These linkers are selected to carry appropriate restriction endonuclease sites that produce ends compatible with the ends in the vector.

Box 17.1 Genomic (DNA) libraries must be complete and easily accessible. The quality of a specific genomic library depends on whether the library is complete and overlapping or not. Completeness is a statistical problem, which can be assessed with the following equation:

$$N = \frac{\ln(1-P)}{\ln(1-f)} \qquad (17.1)$$

where P is the probability that a specific DNA fragment (a gene) will be contained in the library, f is the relation of the average length of the inserted DNA to the size of the whole DNA length, and N is the number of plasmid or phage clones.

If a probability of 0.99 is desired, λ libraries with average insert sizes of $2{\cdot}10^4$ bp will have to contain 228 clones for *Mycoplasma* (genome size about 10^6 bp) or 1103 clones for *E. coli* (genome size $4.8{\cdot}10^6$ bp).

A library should contain overlapping fragments because this facilitates or renders possible the analysis of contiguous DNA sections of the chromosome. If the end of one DNA fragment can be identified at the start of another one, a continuous section of the chromosome can be assembled, as illustrated in Figure 17.**8**. Similar techniques are used to assemble the DNA sequence information for complete chromosomes ("chromosome walking"), such as the recently finished sequencing of the entire chromosomes of *Bacillus subtilis, Enterococcus faecalis, Escherichia coli* K-12, *Haemophilus influenzae, Helicobacter pylori, Mycobacterium genitalium,* and *Methanococcus jannaschii.*

17.1.4 Some Vectors Allow Specific Expression of Cloned Genes

Frequently, researchers want to analyze the gene product(s) encoded by a specific cloned DNA fragment. There are now numerous plasmids available that have been constructed as **expression vectors** and that allow the control of the expression of the cloned genes. Each vector has been adapted to specific experimental requirements.

Two examples from the multitude of expression vectors will be described (Figs. 17.**9**, 17.**10**). They contain the usual vector elements on the plasmid, such as an *oriV* and a *bla* gene for antibiotic selection. However, pRSET (Fig. 17.**9**) features some additional elements:

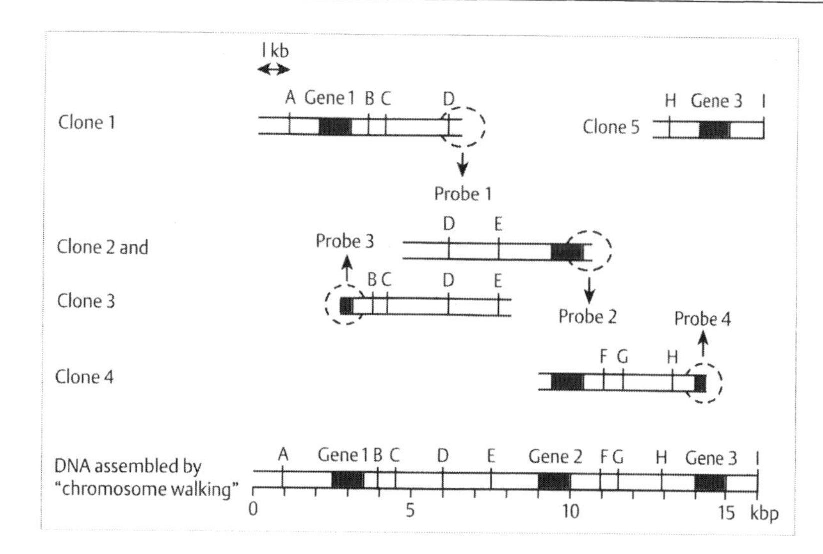

Fig. 17.**8** **Cloning contiguous DNA fragments by "chromosome walking."** Once a gene (DNA) has been cloned (Clone 1) and physically mapped or even sequenced, a DNA probe (Probe 1) corresponding to a distal end can be obtained. This probe can be used to identify by hybridization other DNA fragments that overlap the end of the first fragment. These DNA fragments will either extend further (Clone 2) and thus identify the flanking DNA, or cover parts of Clone 1 (Clone 3), thus confirming its physical mapping by using restriction endonucleases A–E. By repeating this procedure with the termini of Clones 2, 3, and 4 (and other clones) as probes 2, 3, and 4, a contiguous DNA can be assembled that covers several genes and even larger parts (restriction sites A–I) of a chromosome ("chromosome walking")

1. A promoter/operator sequence (*lacZp,o*), which is followed by a LacI repressor sequence (see also Fig. 17.**4**),
2. A ribosome binding site (*rbs*) and a short open reading frame (ORF) encoding the LacZ α fragment, which contains a synthetic MCS for the insertion of the foreign DNA fragment. If the inserted DNA does not carry its own *rbs*, the ORF encoded by the newly introduced DNA fragment can form together with the vector-derived ORF a single continuous ORF, which will result in the expression of a **fusion protein**.

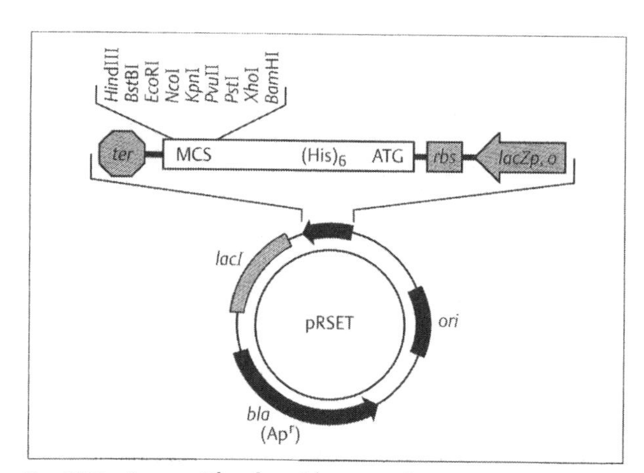

Fig. 17.**9** **A versatile plasmid expression vector.** Expression vectors, here pRSET (Invitrogen, San Diego, Calif., USA), contain the usual *oriV* and antibiotic resistance gene (*bla*) for easy replication and selection. pRSET contains in addition the *lacZp,o* promoter/operator, a ribosome binding site (*rbs*), an initiation codon ATG, a His-tag $(His)_6$, a multiple cloning site (MCS), and a terminator (*ter*). This system is under the control of the LacI repressor and allows the expression of transcription and of translation protein fusions that carry a His-tag for easy purification of any peptide whose gene is fused into the multiple cloning site

This vector is mute when the transcription of the inserted DNA is repressed by the LacI repressor, but transcription can easily be induced by adding the nonmetabolizable inducer IPTG (see Chapter 18.1.2). The bacteria will then begin to synthesize large amounts of the encoded (fusion) protein. Very often, these proteins are insoluble and form precipitates within the bacteria, which are called "**inclusion bodies**."

Expression vectors allow transcription and translation of a given gene product, often under the control of the *lac* operon regulatory genes and proteins (LacI repressor; *lacZp,o* promoter/operator; *rbs*). Other vectors allow cloning of any gene linked to the major phage λ promoters (λP_L or λP_R) under the control of the corresponding operators and CI repressor, the latter in its temperature-sensitive form (allele *cI857*) to allow easy induction. Still other vectors use a phage T7 promoter, which requires T7 RNA polymerase for transcription. These vectors are used for overexpression of lethal gene products that require tight control in the non-induced state.

With pRSET, the production of a fusion protein can be exploited for the purification of the protein by the addition of a poly-histidine peptide to the target protein. Consecutive histidines bind tightly to the free coordination sites of Ni chelates attached to solid chromatographic support materials. Proteins can be efficiently isolated from bacterial extracts by passing them through an appropriate column with nickel-charged agarose resin. The "**His-tag**" labelled proteins will be retarded, while the other material can be washed off. The bound protein can then be eluted by competition with imidazole or histidine or by lowering the pH. This method allows the recovery of nearly homogeneous proteins in a single chromatographic step. Although the vectors frequently provide recognition sequences for specific proteases in order to facilitate the removal of the His-tag once the protein has been recovered from the column, in many instances the His-tag has no influence on the activity of the protein under investigation. The His-tag system even works under denaturing conditions, such as when detergents are used for the solubilization of membrane proteins or when chaotropic agents such as urea for the solubilization of inclusion bodies are used. Purified proteins can be used for structural and functional analysis or for the production of antibodies.

Many proteins, especially membrane-bound proteins, are highly toxic if overproduced from multi-copy expression vectors that lack a tight control. This includes all *lacZp,o–lacI*-based vectors. To avoid such problems, many cloning vectors use phage-specific promoters that flank the MCS (Fig. 17.**10**). Probably the most frequently used phage promoter is the f10 promoter of phage T7, a 23-bp sequence. The T7 RNA polymerase is specific for T7 promoters and will not (or only poorly) recognize bacterial promoters. Only genes that are preceded by a T7 promoter will be transcribed by this polymerase if, in addition, the bacterial RNA polymerase is selectively inhibited by the antibiotic rifampicin. Since the system is very efficient, the T7 RNA polymerase expression itself has to be tightly controlled to prevent, among other things, premature cell death caused by depletion of ribonucleoside triphosphates. Special plasmids and/or bacterial strains have been constructed that allow precise control of the synthesis of the T7 RNA polymerase. Normally, the corresponding gene is cloned and expressed from the λP_L promoter, which is under the control of a temperature-sensitive repressor allele (*cI857*; see Chapter 26). The depicted vector also features another interesting element: the f1 origin, i.e., the origin of replication of the filamentous phage f1, a close relative of phage M13. The f1 origin allows the recovery of single-stranded plasmid DNA when the host strain containing the plasmid is co-infected with helper phage. This can be exploited for the preparation of single-stranded DNA used for DNA sequencing or site-directed mutagenesis.

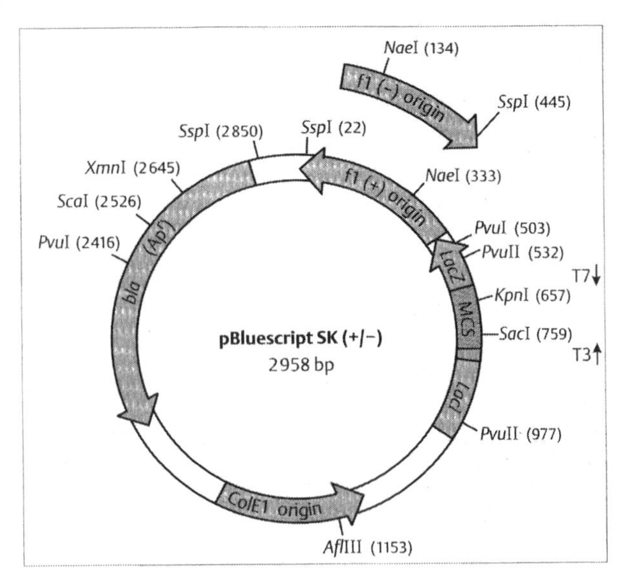

Fig. 17.**10** **Four versions of a versatile phagemid vector.** The four pBluescript vectors (Stratagene, La Jolla, Calif., USA) are phage-plasmid hybrids ("phagemids"). Due to the presence of a pColEI origin of replication, they correspond to multicopy vectors; due to a phage f1 origin, they can produce, in the presence of a helper phage, single-stranded DNA, either the sense (+) or the anti-sense (−) strand, for sequencing (see Fig. 17.**13**) or for localized mutagenesis (see Fig. 17.**16**). The *bla* gene allows Ap^r selection, and the *lacI*, *lacZp,o*, and *lacZ* genes allow a blue-white selection (see Fig. 17.**4**), as well as inducible synthesis of fusion proteins and screening with LacZ antibodies. The vectors contain a polylinker (MCS) in the two opposite orientations (SK and KS) with 21 unique restriction sites for easy production of nested deletions using, for example, exonuclease III (see Fig. 17.**16**), and within the MCS a phage T3 and a phage T7 promoter. These allow efficient in vitro synthesis of mRNA with purified phage RNA polymerase, or they can be used in vivo for efficient protein overexpression. This is usually done in host cells that contain a chromosomally encoded cassette in which the T7 polymerase is expressed from a tightly controlled promoter, e.g., *lacZp,o* and the LacI repressor

17.2 Genomic (DNA) Libraries Are Often Used as Starting Points for the Analysis of Genes and Their Products

One aim of the cloning procedure as described is the separation of a complex mixture of DNA fragments into single components by amplifying only one fragment via a plasmid clone or a phage clone. How a genomic library can be used to construct a physical gene map from overlapping DNA fragments has already been described (see Box 17.**1**). However, scientists are normally interested in a specific gene and its product, which can be isolated by two frequently used procedures. These procedures involve either the determination of short amino acid sequences and the synthesis of the corresponding DNA probe or the preparation of antibodies against a purified protein. These antibodies and probes can be used for screening.

Often it is sufficient to separate the protein from other proteins by polyacrylamide gel electrophoresis, followed by microsequencing of fragments obtained by proteolytic digestion. By referring to the genetic code, the nucleic acid sequence is determined from the amino acid sequence of the fragment and the appropriate oligonucleotide can be synthesized by chemical methods and labelled appropriately (Fig. 17.**11**). Since the genetic code is degenerate, there will frequently be ambiguities that can be accounted for by the synthesis of the various predicted sequences. These oligonucleotides constitute **DNA probes** that allow the identification of the corresponding gene by Southern hybridization. Specific antibodies are alternative **probes** that allow the **screening** of genomic libraries.

Both methods are used basically in the same way. Initially, a copy of an original agar plate with bacterial colonies or with bacteriophage plaques is blotted onto a nitrocellulose filter. The filter accepts—similar to the colony replication technique (see Chapter 15.5.6)—the original pattern of the colonies or plaques. The hybridization of the DNA bound to the filter (after lysis of the cells) with the labelled DNA oligonucleotide probes or the detection of bound proteins by antibodies is performed on the filter. Reactions will occur only in those places with "positive" colonies or "positive" plaques (Fig. 17.**11**). The position on the filter corresponds to the position on the original plate, which allows the identification of the few positive clones among the hundreds or even thousands of negative ones.

Positive colonies or plaques can be isolated and propagated by conventional microbiological methods. Bacterial colonies are suspended in growth media and cultivated to high cell densities prior to the isolation of plasmid or cosmid DNA. Bacteriophage clones can be used to infect large quantities of bacterial cells until enough DNA for further investigations is available.

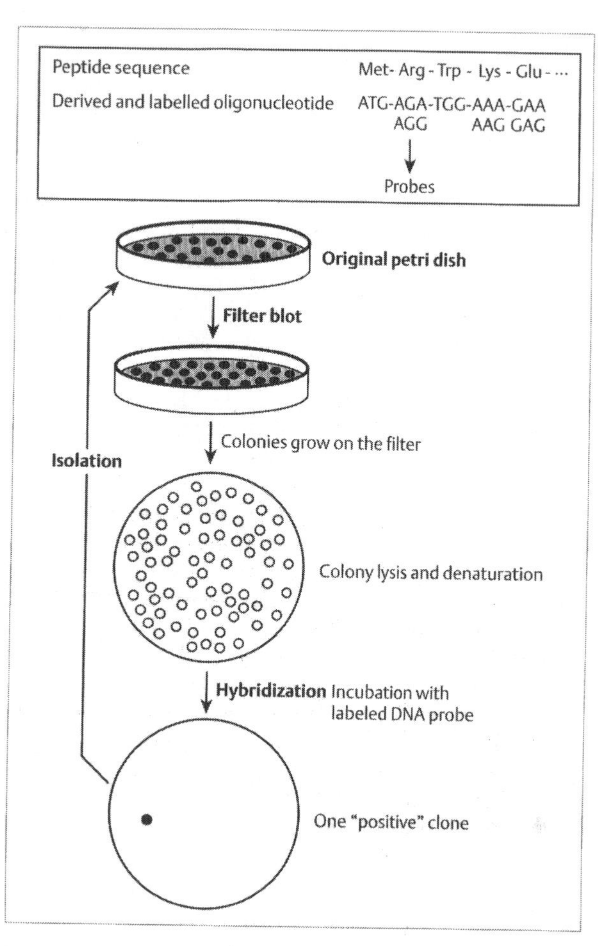

Fig. 17.**11** **Screening of a genomic (or a cDNA) library with oligonucleotide probes.** From a short peptide sequence (often the amino-terminus or a highly conserved motif), the corresponding oligonucleotide probe can be derived and synthesized. Due to the degeneracy of the genetic code, this can cause problems. Bacteria containing the genomic library cloned on plasmids, cosmids, or phages develop on the original petri dish, are then blotted onto a nitrocellulose filter, and grown further on the filter. The colonies are lysed while still on the filter, and liberated DNA is denatured using NaOH. After fixation to the filter, bound DNA and a labelled probe are hybridized, and positive clones containing DNA complementary to the probe are made visible. The corresponding colonies can be isolated from the original master plate for further analysis

17.3 DNA Sequencing Is an Important Technique in Biological Research

Further investigations frequently include the determination of the DNA sequence of the desired gene. The overview of the basic methods of gene technology would be incomplete without at least a short description of the most popular sequencing method. There are two basic methods used today for the determination of DNA sequences: the chemical method developed by A. M. Maxam and W. Gilbert in 1977 and the biochemical dideoxy or chain-termination method developed by F. Sanger in 1977. Although the chain-termination method is perhaps more laborious, most investigators prefer it because of its reliability and precision for longer sequences. More than 95% of all DNA sequence data stored in the international databanks have been obtained by this method. Only the biological part of the Sanger method will be presented here. A description of its chemical part and of the Maxam-Gilbert technique can be found in any biochemistry textbook.

Cloning of DNA into the M13 replicative form (RF) DNA in order to obtain single-stranded DNA rings is a standard procedure (Fig. 17.**12**). Molecular biologists use derivatives of the M13 phage for cloning purposes that have properties not unlike those of other plasmids, such as those of the pUC series (see Fig. 17.**4**). The intergenic region of M13 carries a part of the *lacZ* gene for blue-white selection and a multiple cloning site. This region allows the linearization of the M13 RF DNA for the insertion of foreign DNA. Hybrid constructs of RF and foreign DNA are then introduced into bacteria and multiplied. During the infection cycle, single-stranded

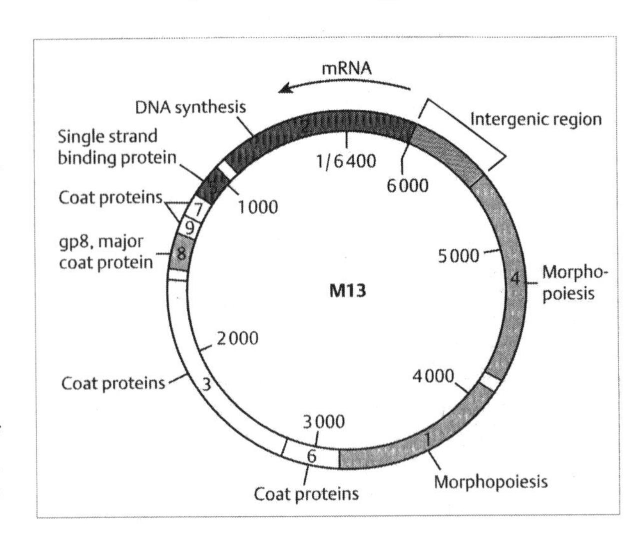

Fig. 17.12 **Genome of the filamentous phage M13.** The genome of phage M13 in capsids is a single-stranded DNA ring encoding ten genes for DNA replication, coat proteins, and assembly (morphopoiesis) of the mature phage. The infecting single strand is rapidly converted in the host cell into a double-stranded replicative form (RF) with an intergenic region into which foreign DNA can be inserted without interfering with phage replication and assembly. The single-stranded DNA present in the phage capsid can be used as a template in DNA sequencing using the dideoxy chain-termination method and in localized mutagenesis

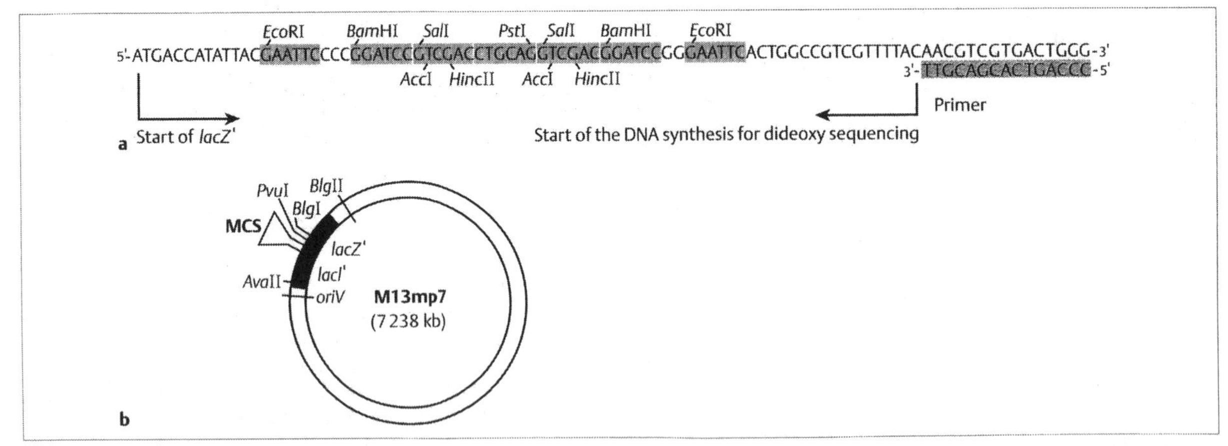

Fig. 17.13 **Phage M13 as a cloning and sequencing vector.** Phage M13mp7 is a derivative of the single-stranded DNA phage M13. Phage M13mp7 contains a multiple cloning site (MCS) shown in **a**. The primer used in the dideoxy chain-termination sequencing reaction is also indicated. **b** In M13mp7, here shown in its double-stranded replicative form (RF), the natural intergenic region (see Fig. 17.**12**) has been enlarged by genes *lacI* and *lacZ'* for a blue-white selection and by an MCS

phages that also carry the foreign DNA are produced. These single-stranded rings are the templates for in vitro DNA synthesis (Fig. 17.**13**).

Sequencing according to the dideoxy chain-termination method involves the following steps (Fig. 17.**14**): 1) a DNA primer (oligonucleotide) hybridizes to the region next to the MCS, and 2) the Klenow DNA polymerase I or another polymerase, which does not possess 5′–3′ exonuclease activity, is used to extend the 3′-OH end of the primer during synthesis of the complementary strand.

The trick of the sequencing reaction is the targeted, but statistically distributed disruption of the complementary strand synthesis. This is accomplished by the addition of **dideoxynucleotides** (ddNTPs) to the mixture of regular dNTPs. Dideoxynucleotides lack the 3′-OH group (see Chapter 14); this causes the interruption of the synthesis whenever a dideoxynucleotide is inserted into the growing DNA strand. Several hundred nucleotides of a DNA sequence can be "read" in a single experiment. In order to determine longer DNA sequences, the initial DNA fragment has to be divided into smaller units. Each fragment is then cloned in M13 vectors and sequenced separately. The necessity to produce smaller clones of large inserts is obviated by the possibility of synthesizing specific oligonucleotide primers, which hybridize to the ends of the sequences that were determined in a previous sequencing round.

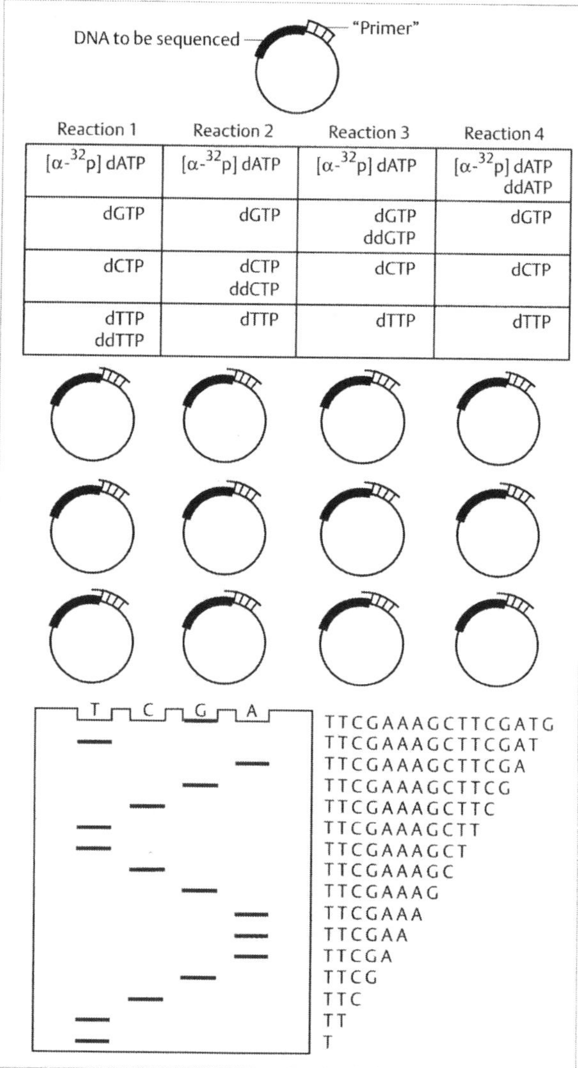

Fig. 17.**14** **Schematic presentation of the sequencing reactions using the dideoxy or chain-termination method.** Four reaction mixtures are prepared in parallel. Each reaction mixture contains a radioactively or otherwise labelled nucleotide (here α-[^{32}P]dATP is used) and the three other nonlabelled deoxynucleotides. In addition, each reaction contains a dideoxynucleotide, either ddTTP, ddCTP, ddGTP, or ddATP as shown. Depending upon the experimental situation, a ratio of 1/50, 1/100, or 1/200 of ddNTP to dNTP is chosen. The synthesis of the complementary strand is initiated by the addition of DNA polymerase. Synthesis is interrupted when a dideoxynucleotide is inserted into the active center of the DNA polymerase. In Reaction 1, this will happen when the sequence of the template strand calls for the insertion of a thymine nucleotide. Hence, in the first reaction, a collection of DNA fragments will accumulate, and their lengths indicate the positions of adenosine residues in the template strand. Similarly, Reaction 2 maps guanosine, Reaction 3 cytosine, and Reaction 4 thymidine residues in the template strand. The important experimental step is the exact separation of the synthesis products. The synthesized DNA strands are separated from one another by denaturation and analyzed on special polyacrylamide gels. The labelling of the synthesis products allows their detection by autoradiography or an appropriate staining procedure. The scheme of a gel electrophoresis run is shown at the bottom of the figure. The smallest fragment, which has migrated furthest toward the bottom of the gel, indicates the first nucleotide in the sequence, while the next, larger fragment indicates the second nucleotide, and so on

Box 17.2 **Bacteriophage M13** and its relatives (e.g., phages fd and f1) are **filamentous phages** of *E. coli* (see Fig. 26.**4**). Their capsid DNA consists of a single-stranded DNA [the (–) strand] of approximately 6400 nucleotides which is converted into its double-stranded or **replicative form** (**RF DNA**) in the host cell. The phage genome codes for ten genes with functions in DNA replication, for phage coat proteins, and for the assembly (morphopoiesis) of the phage particle (see Chapter 26). The intergenic region of about 500 nucleotides between genes 2 and 4 can be much larger than normal without affecting phage propagation. The capsid proteins enclose the single-stranded DNA, and the resulting phage progeny are extruded continuously from the infected host cell without lysis. The phage particles are simply extended by additional gp8 molecules in the protein tube to accommodate the longer DNA strand. This property and the ability of phage M13 to generate single-stranded DNA make it one of the most useful tools in molecular biology (e.g, for DNA sequencing).

The new specific primers can be used in another sequencing round as starting points for the determination of additional sequences from the insert DNA. By repeating these steps as often as necessary, the complete sequence of a DNA fragment can be determined without the need to produce any subclones. This method is called the **primer walking strategy**.

Most research groups involved in DNA sequencing either use automated DNA sequencing equipment or have the sequencing done by commercial companies. Although modern equipment can run sequence gels automatically and can give computer-aided printouts of the determined sequence in the double-stranded form, including the derived amino acid sequences, a scientist still can produce rarely more than 1000 base pairs per day, counting the cloning steps. However, the sequencing of entire prokaryotic genomes ranging from 700 to 7000 kb is now possible within months or a year and has become a routine procedure in modern biology.

17.4 Localized Mutagenesis Helps To Identify Specific Functions

Significant knowledge about the structure and function of genes is gained by the analysis of mutants. The usual procedure in genetics is the description of the effects that spontaneous or induced mutations have on the phenotype of the respective mutant. This is then traditionally followed by the isolation of the gene and the determination of its structure by DNA sequencing. Gene technology, however, allows the reverse procedure ("reverse genetics"). Targeted mutations are introduced by biochemical means (i.e., in vitro) into an isolated gene, which is then reintroduced into the organism to investigate the phenotypic changes.

Again, as in the other parts of this chapter, only the basic principle of the procedure will be described. Methods to transfer DNA into an organism by transformation will be described elsewhere (see Chapter 16; for more detailed descriptions, see Further Reading).

17.4.1 Single Nucleotides Can Be Exchanged

The targeted exchange of a single nucleotide is started using a DNA fragment of known sequence cloned into an M13 vector (Fig. 17.**15**). A complementary, usually synthetic, oligonucleotide, which contains one or several mismatches is hybridized to the DNA insert. The oligonucleotide is extended from the 3′-OH end by a DNA polymerase to form the complete complementary DNA strand. The resulting RF DNA is then introduced into bacteria. During the infection process, progeny will result from either the unchanged original DNA strand or the complementary strand containing the nucleotide exchange. The mutated DNA has to be selected or identified in another step.

Variations of this basic scheme help to detect mutants. The original M13 strand in the example in Figure 17.**15** results from an infection of *E. coli* Dut^- Ung^- host cells (see Chapter 15.6). The bacteria with an altered *dut* gene lack the enzyme dUTPase. dUTP is a regular by-product of deoxynucleotide synthesis. Usually, dUTP is metabolized by dUTPase to dUMP. However, in *dut* mutants, dUTP remains and can be incorporated into DNA instead of dTTP. When the enzyme uracil-DNA-glycosylase is missing due to an *ung* mutation, the uracil residues will be preserved in the DNA. Accordingly, the M13 DNA derived from infected *dut ung* mutants will contain several uracil residues. This DNA can be used without problems as a template for the in vitro synthesis of the complementary DNA strand since uracil pairs with adenine. The resulting double-stranded DNA is introduced into Ung^+ bacteria. The uracil-DNA-

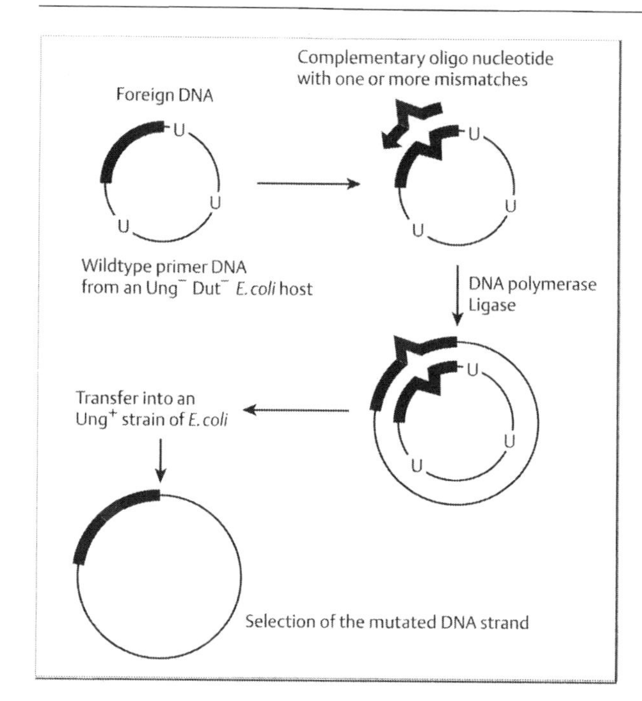

Fig. 17.**15** **in vitro (localized) mutagenesis.** The procedure begins with DNA cloned into M13, which contains several uracils due to replication in an Ung^- Dut^- host cell. A localized mutagenesis of the cloned DNA requires a complementary primer with one or more base exchanges. This primer is then elongated by DNA synthesis, resulting in a double-stranded or RF molecule. When this RF DNA is transfected into an Ung^+ host, the uracil-containing strand will be inactivated, while the mutated complementary strand survives and multiplies (see text for details)

glycosylase in these cells will attack the template DNA and produce apurinic sites, which result in strand breaks. Hence, the template DNA is lost with a high probability, while the mutated complementary strand, which lacks the uracil residues, survives.

17.4.2 Deletions and Insertions Help to Identify Functional Domains in Proteins

Cloned DNA fragments can also be mutated by introducing deletions or insertions. It is technically simple to introduce **short deletions** at the cutting sites of restriction endonucleases. The protruding ends are removed by single-strand-specific nucleases (e.g., the S1 nuclease from *Aspergillus oryzae*; see Chapter 15). The DNA is then religated by DNA ligase. This results in the deletion of a few base pairs, whose sequence depends upon the specific restriction enzyme used for cutting.

Site-directed or **localized mutagenesis** is an important procedure in modern biology. It allows the introduction of defined mutations into any DNA whose sequence is known. Single nucleotides can be exchanged with any other nucleotide and deletions of a known size and DNA insertions can be introduced using vectors specifically designed for this task. To inactivate a specific gene in the chromosome of an organism, DNA cassettes carrying antibiotic resistance markers are often used as insertions. These allow the selection of homogenotes, which by **homogenotization** have recombined the mutated allele into their chromosome.

More complex is the introduction of deletions with varying lengths, i.e., the production of a set of **nested deletions**. A popular tool is the enzyme **ExoIII** from *E. coli*, a 3′–5′ exonuclease that removes 5′-nucleotides from the 3′-OH end of double-stranded DNA. Since 3′-overhanging ends are not a substrate for the enzyme, it is possible to produce unidirectional deletions in a DNA molecule, such as in a plasmid carrying foreign DNA. The DNA is cut with two different restriction enzymes, one producing a protected 3′-overhanging end and one either a 5′-overhanging end or a blunt end as substrate for the ExoIII digestion. The exonucleolytic attack of the DNA by ExoIII for various lengths of time produces a set of DNA molecules with varying lengths of 5′-overhanging single-stranded DNA. In the next step, these 5′-overhangs as well as the protecting 3′-overhang are removed by a single-strand-specific nuclease (S1 or mung bean nuclease). The now blunt-ended molecules are religated with DNA ligase to again form circular plasmids that can be transferred into bacterial hosts by transformation. The effect which the deletions might have upon the function of the cloned foreign DNA fragment can be (phenotypically) analyzed; or the plasmids can be amplified and recovered as templates for DNA sequencing (Fig. 17.**16**). Special vectors for this purpose have been developed that feature MCSs with recognition sites for enzymes that produce protecting 3′-overhangs at their outer ends, while recognition sequences for either blunt-end- or 5′-overhang-producing enzymes are located at the center (see Fig. 17.**10**). Primer binding sites in both flanking regions complete the advantageous arrangement, which allows the introduction of nested deletions into a cloned DNA fragment from either end.

A useful method for the introduction of **insertions** starts out with the treatment of a DNA preparation with formic acid at a concentration that introduces on the average one apurinic site per molecule. This is followed by digestion with ExoIII, which is also an apurinic-site-specific endonuclease. The procedure results in one cut per molecule and produces a collection of linear DNA molecules that have been opened at different sites. Oligonucleotides are ligated to the ends of this DNA prior to renewed ring closure with DNA ligase. The result is a collection of DNA molecules with insertions at various sites. Other procedures use mutagenized insert DNA or cassettes with a selectable marker to generate insertions and to introduce these mutated copies by homogenotization into chromosomal markers (Fig. 17.**17**).

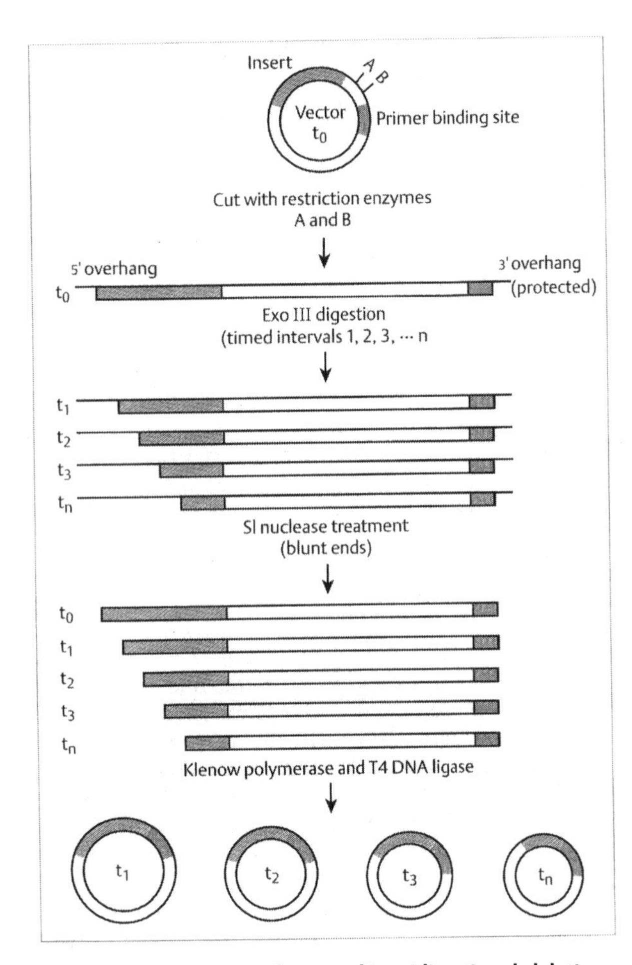

Fig. 17.16 **Generation of "nested" unidirectional deletions by ExoIII treatment.** The vector with cloned DNA (t_0) is cut with the restriction enzymes A and B, which generate a 3′-overhang next to the sequencing primer site and a 5′-overhang next to the inserted DNA. Digestion with exonuclease III for various times generates unidirectional deletions from the 5′-overhang end (t_1–t_n), while the 3′ side is protected (e.g., by α-thio-dNTPs). Treatment with S1 nuclease and Klenow polymerase creates blunt ends on both sides of the DNA. After ligation and transformation into host cells, clones from each time point can be isolated and the deletions identified by gel electrophoresis and DNA sequencing

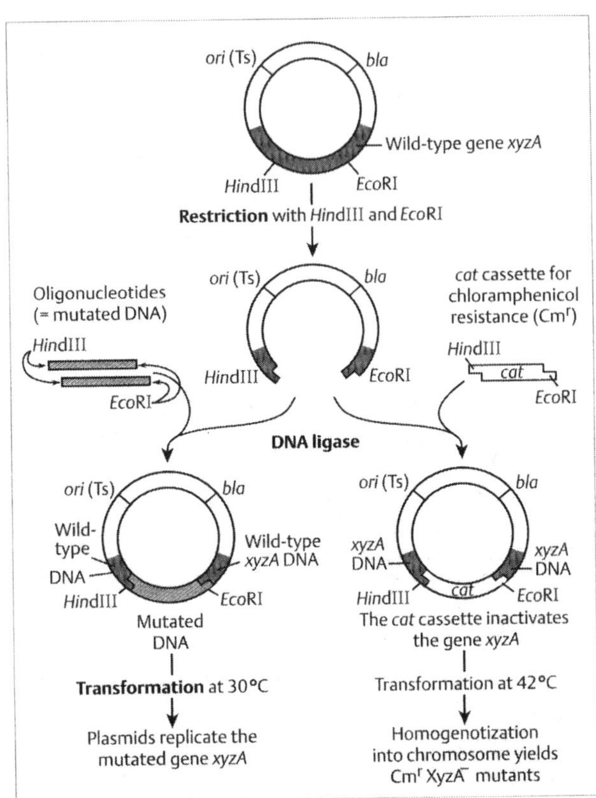

Fig. 17.**17** **Mutagenesis by DNA insertion ("cassette mutagenesis").** From a cloned DNA (here the hypothetical gene *xyzA*), a fragment is excised by two restriction endonucleases, e.g., *Hin*dIII and *Eco*RI. The fragment is replaced either by a mutated form of the same DNA (left) or by a foreign DNA cassette (right). A mutated double-stranded DNA or a set of mutated oligonucleotides can be used to generate mutations in *xyzA* which may also contain insertions and deletions. Alternatively, a DNA cassette with a selectable marker (here the *cat* gene for chloramphenicol resistance, Cm^r) is inserted into a cloned copy of *xyzA*. Homologous recombination between the chromosomal allele and the *xyzA* DNA flanking the cassette can occur (homogenotization). If the vector is unable to replicate in the host cell [here at 42 °C due to its temperature-sensitive origin *ori* (Ts)], selection of Cm^r cells at 42 °C will generate such chromosomal copies with the inserted cassette, and hence XyzA$^-$ mutants

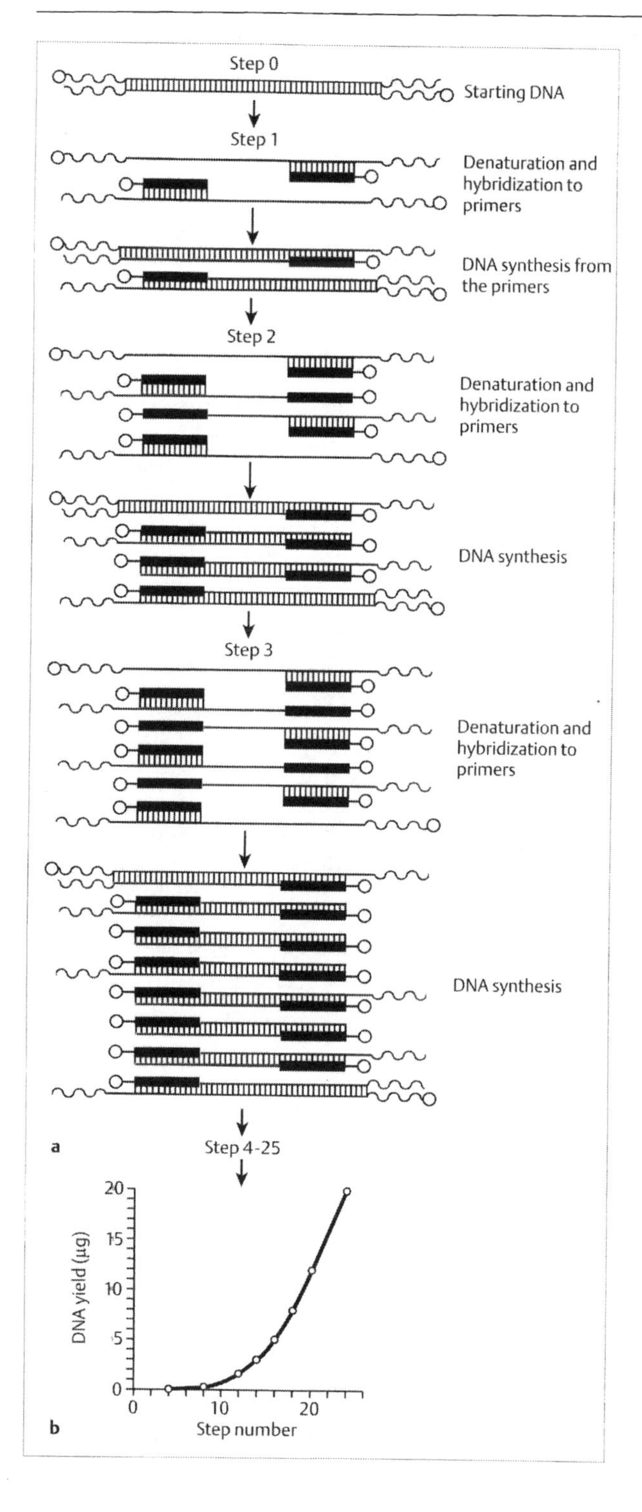

Fig. 17.**18a, b** **The polymerase chain reaction (PCR).**
a A sense and an antisense primer that flank both sides of the DNA to be replicated must be available. These are hybridized in Step 1 to the denatured DNA strands and are used by the thermostable Taq DNA polymerase as 3′-primers to start synthesis of both complementary DNA strands. In Step 2, the DNAs are denatured again to allow hybridization of the primers, and a second round of DNA synthesis. The first two molecules flanked by the primers appear. In Step 3 and all following steps, the denaturation, primer hybridization, and DNA synthesis steps are repeated again and again until
b after about 25 cycles, $>10^6$ copies of the DNA molecules flanked by the primers are found in microgram amounts. Thanks to the thermostable DNA polymerase, the cycles can be automated to run at a rate of 30–60 cycles per h

17.4.3 The Polymerase Chain Reaction Is a Versatile Method

Although it is per se not a recombinant DNA method, the **Polymerase Chain Reaction (PCR)** will be discussed here because of its extreme importance in modern biology, and because it is frequently used to complement other methods in gene technology. The method allows the in vitro amplification of minute amounts (theoretically just one molecule) of DNA, to produce quantities sufficient, for example, for sequencing, cloning, or gel analysis and "fingerprinting" of uncultured bacteria.

The basic principle is the enzymatic amplification/replication of a DNA fragment that is flanked by two oligonucleotide primers that hybridize to the opposite DNA strands (Fig. 17.**18**). A prerequisite for a targeted PCR analysis is thus the prior knowledge of the sequence flanking the DNA to be amplified and the availability of primers. The oligonucleotide primers are added in excess to the template DNA under hybridizing conditions (40–60 °C, depending upon the nature of the primer). After raising the temperature (to 72 °C), a thermostable DNA polymerase extends the primers by adding nucleotides at their 3′ OH ends, thereby synthesizing the complementary DNA sequences. The resulting DNA double strands are then denatured by raising the temperature to 94 °C. By lowering the temperature again to the hybridizing temperature, the oligonucleotide primers anneal to the original template strands as well as to the newly synthesized DNA strands, and a new round of in vitro replication can begin by raising the temperature and extending the primers. These cycles can be repeated many times (usually 20–50 times), and the DNA will be amplified exponentially. The multiplication of the initial amount of DNA can reach 10^6-fold or more depending upon the number of cycles.

When PCR was invented by K. B. Mullis in 1986, the Klenow fragment of the *E. coli* bacterial DNA polymerase was used. Because this enzyme is denatured at the high temperatures necessary to melt the DNA double strands, new enzyme had to be added after each round of denaturation and hybridization of the primers for the new round of synthesis. Currently, thermostable DNA polymerases from thermophilic bacteria or archaea are used. This allowed the automation of the method and furthered its widespread use. PCR machines are nowadays regarded as basic equipment in any molecular biology laboratory. The method is used routinely in the isolation of genes from cDNA and genomic libraries (see Chapter 17.2), in the isolation of RNAs from cells (RNA-PCR), in localized mutagenesis (see Chapter 17.4), as well as in modern bacterial systematics (see Chapter 29) and in applied microbiology (e.g., rapid clinical diagnosis; see Chapter 33).

The development of PCR-related techniques is very rapid and many new applications are described. One such interesting variant, for example, combines the cDNA technique of primer extension with thermal cycling. The template RNA used in conventional primer extension experiments is purified in a time-consuming process from the investigated organism, and the primer extension proper is performed once in a separate reaction. The new "in situ" primer extension takes advantage of the reverse transcriptase activity of the thermostable DNA polymerase from *Thermus thermophilus* in the presence of $MnCl_2$. The RNA to be investigated is released from the bacteria by heat treatment for 30 s at 94 °C, a process that concomitantly denatures all nucleic acids. Using a primer complementary to the 3′-end of this RNA, many copies of cDNA can be produced by repeated thermal cycling. As in any other PCR protocol, the process cycles between denaturation, primer annealing, and primer extension. The amplification of the reaction product, however, is only linear instead of exponential since priming takes place on the single-stranded RNA, which is not amplified during the process. Therefore, this technique requires higher numbers of cycles, usually between 70 and 100. On the other hand, it is not necessary to isolate the RNA, which can rather be used in "statu nascendi" since the cell extracts are used without delay after harvesting. Degradation of mRNA and artifacts caused by conventional RNA isolation procedures are thereby avoided, and only minute amounts of cell samples are needed.

Further Reading

Ausubel, F. M., Brent, R., Kingston, R. E., Moore, D. D., Seidman, J. G., Smith, J. A., and Struhl, K. (1989) Current protocols in molecular cloning. New York: Wiley

Brown, T. A. (1995) Gene cloning, 3rd edn. London: Chapman & Hall

Gassen, H., and Minol, K., eds. (1997) Gentechnik. Einführung in Prinzipien und Methoden, 4th edn. Stuttgart: Fischer

Glick, B. R., and Pasternak, J. J. (1994) Molecular biotechnology. Principles and applications of recombinant DNA. Washington, DC: ASM Press

Glover, D. M. (1985) Gene cloning: the mechanics of DNA manipulation. London: Chapman & Hall

Henikoff, S. (1987) Unidirectional digestion with exonucleases III in DNA sequence analysis. Meth Enzymol 155: 156–165

Kaspar, P., Zadrazil, S., and Fabry, M. (1989) An improved double-stranded DNA sequencing method using gene 32-protein. Nucleic Acid Res 17: 3316

Martinez, E., Bartholome, B., and de la Cruz, F. (1988) pACYC184-derived cloning vectors containing the multiple cloning site and *lacZ*, a reporter gene of pUC8/9 and pUC18/19 plasmids. Gene 68: 159–162

Maxam, A. M., and Gilbert, W. (1977) A new method for sequencing DNA. Proc Natl Acad Sci USA. 74: 560–564

Old, R. W., and Primrose, S. B. (1994) Principles of gene manipulation. An introduction to genetic engineering. London: Blackwell Scientific

Prentki, P., and Krisch, H. M. (1984) In vitro insertional mutagenesis with a selectable DNA fragment. Gene 29: 303–313

Sambrook, J., Fritsch, E. F., and Maniatis, T. (1989) Molecular cloning. A laboratory manual. Cold Spring Harbor: Cold Spring Harbor Laboratory Press

Sanger, F., Nicklen, S., and Coulson, A. R. (1977) DNA sequencing and chain terminating inhibitors. Proc Natl Acad Sci USA 74: 5463–5467

Tabor, S., and Richardson, C. C. (1985) A bacteriophage T7 RNA polymerase/promoter system for controlled exclusive expression of specific genes. Proc Natl Acad Sci USA 82: 1074–1078

Watson, J. D., Gilman, M., Witkowski, J., and Zoller, M. (1992) Recombinant DNA, 2nd edn. New York: Freeman

Winnacker, E. L. (1987) From genes to clones. Introduction to gene technology. Weinheim: VCH

Yanisch-Perron, C., Vieira, J., and Messing, J. (1985) Improved M13 phage cloning vectors and host strains: nucleotide sequences of the M13mp18 and pUC19 vectors. Gene 33: 103–119

Section V
Gene Expression and Regulatory Mechanisms

Similar to other living organisms, every prokaryotic cell is the product of its genes, of the prehistory of its ancestor cells, and of the actual environmental conditions. According to this view, the genome, i.e., the sum of all genes of an individual cell plus all the genes of its collective (species) genome, defines the potential of a cell that can be expressed in the cellular structures, in the various metabolic pathways, and in the different gene-exchange mechanisms, which were described in the previous sections (I–IV) of this book. Prehistory and the actual environment, however, determine which potential programs of the genome are realized.

This section deals with the molecular mechanisms by which prokaryotes sense changes in their environment and control gene expression as an adaptation to these changes. When unicellular organisms are considered, "outside" normally refers to the non-cellular (physical) surrounding. This contrasts with multicellular (eukaryotic) organisms in which a cell is normally surrounded by other cells of the same clone and in which highly specialized cells sense changes in the physical outside world. Despite appearance, this fundamental difference is more apparent than real at the molecular and cellular level. Regardless of whether a cell changes gene expression as the consequence of a sugar in the medium or of a hormone bound to its surface, which was secreted by a neighboring cell, both processes involve highly similar sensory transduction and gene control mechanisms. This is not as surprising as it may seem because prokaryotes show a sophisticated intercellular communication at the population level. A clonal bacterial population thus seems to be the functional analogue of a multicellular organism and of a population comprising such organisms. The underlying universal biochemistry would be consequently explained as the product of a joint and a common evolution.

In a modern view, prokaryotic chromosomes, plasmids, and prokaryotic viruses (bacteriophages) are equivalent as carriers of genetic information, i.e., they carry the same type of genes whose expression is controlled by the same types of regulatory mechanisms. The rules formulated specifically for one of these entities are thus valid for all three. Phages in particular correspond at the DNA level ("vegetative" and prophage) strictly to other autonomously replicating genetic elements, i.e., to plasmids. Like plasmids, they are no longer considered as living organisms, a term restricted to cellular organisms, but are considered as parts of living organisms. The specific examples chosen to illustrate relevant gene regulation and signal transduction mechanisms thus originate from the three genetic entities.

Historically, the **operon** model as formulated originally by F. Jacob and J. Monod in the late 1950s marks the beginning of an understanding of gene expression in molecular terms. The basic model is still valid today, although the number of variations found in, for example, induction, repression, attenuation, antitermination, or positive and negative control, is astonishing. Consequently one chapter (Chapter 18) will present the various mechanisms involved in operon control at the transcriptional and translational level (i.e., in enzyme synthesis). This includes **regulons**, i.e., groups of operons regulated by one common, but specific regulator. Similar to repressors, activators, attenuators, antiterminators, etc. of operons, these regulators normally respond to specific molecules, often the substrate of a catabolic pathway or the end product of an anabolic pathway. Since essential regulations also occur at the level of enzyme activity, a second chapter (Chapter 19) will deal with posttranslational control and protein modifications, including compartmentalization. This type of control is almost characteristic for the regulatory proteins themselves, which often are switched reversibly from an active to an inactive form during regulation.

Recently, it has become clear that cellular control is hierarchical, i.e., there are global control networks that are epistatic (superimposed) over the specific control systems mentioned thus far. As a general rule, such global systems are coupled to complex signal transduction systems, such as the two-component systems and the PEP-dependent carbohydrate : phosphotransferase systems (PTSs). These sense (drastic) changes in general states of a cell, such as starvation, oxic or anoxic conditions, and stress, that require more fundamental cellular adaptations than simply activation or deactivation of a single or a few operons. Groups of operons (up to 50 or more) controlled together by such an epistatic and global regulator are called a **modulon**. A **stimulon**, however, designates the sum of all genes that respond to the same stimulus. Consequently, the third chapter (Chapter 20) will treat such global control networks and their signal transduction pathways. Typical examples include the *crp* modulon involved in the quest of food (comprising the second messenger or alarmone cAMP, diauxie, carbon catabolite control, inducer exclusion) with its sensory system (PTS), the stringent response with its alarmone (indicator molecule) ppGpp, several two-component systems, and in particular, the well-understood chemotactic response.

Many physiological phenomena in the prokaryotic world must now be viewed as transiently acting adaptive mechanisms controlled by global regulatory networks. These include the switch of facultative anaerobes from oxic to anoxic growth conditions (and vice versa). The corresponding biochemical pathways and membrane-bound protein complexes (e.g., respiratory chains, ATP synthetases) have been described in previous chapters (see Chapters 4, 5, 9–12), but their global control will be given in some detail in the fourth chapter (Chapter 21) of this section of the book.

There are numerous additional examples of specific and global gene control beyond those discussed in Section V; some are listed in Table 20.**1**. These include many examples found in Sections VI and IX, especially those related to cell division, cellular differentiation processes such as sporulation and the "secondary" metabolism of the streptomycetes, phage morphogenesis, stress response, symbiosis, cellular parasitism, and molecular pathology. The relevance of the new concepts for such traditional microbial areas, such as systematics and "ecology," or for biotechnology also becomes more and more clear. The modular construction and the repeated use of similar biochemical and genetic modules at the different hierarchical levels, such as promoters, operators, repressors, and activators, which reappear in global control and even in intercellular communication mechanisms of the various prokaryotes (and of multicellular eukaryotes) argue for a new view of the mechanisms involved in prokaryotic ecosystems and a new definition of the prokaryotic species that is based on the gene pool. The intelligent use of these natural modules could pave the way to a new "semi-synthetic" biotechnology.

Further Reading

Lengeler, J. (1993) Carbohydrate transport in bacteria under environmental conditions, a black box? Antonie van Leeuwenhoek 63: 275–288

Neidhardt, F. C., Ingraham, J. L., and Schaechter, M. (1990) Physiology of the bacterial cell. A molecular approach. Sunderland, Mass.: Sinauer

Neidhardt, F. C., and Savageau, M. A. (1996) Regulation beyond the operon. In: Neidhardt, F. C., Curtiss, R., Ingraham, J. L., Lin, E. C. C., Low, K. B., Magasanik, B., Reznikoff, W., Riley, M., Schaechter, M., and Umbarger, H. E. (eds.), *Escherichia coli* and *Salmonella*, 2nd edn. Washington, DC; ASM Press; 1310–1324

Magasanik, B. (1988) Research on bacteria in the mainstream of biology. Science 240: 1435–1439

Shapiro, L., Kaiser, D., and Losick, R. (1993) Development and behavior in bacteria. Cell 73: 835–836 (A summary of the following six reviews in the issue)

Sonea, S. (1989) A new look at bacteria. ASM News 55: 584–585

18 Regulation of Gene Expression: Operons and Regulons

Thousands of organisms that constitute the prokaryotic kingdom have found the ecological niche where they can grow optimally. Prokaryotic cells are in direct contact with an environment that can vary from being ideal for growth to being adverse even for survival of the cell, but most, if not all, prokaryotic cell types are capable of adjusting to a wide range of environmental conditions. Fast and efficient adjustments to a rapidly changing environment are achieved in two ways: an instant response involving alterations in activities of critical metabolic enzymes, and a delayed but long-term response involving positive or negative regulation of gene activity in a coordinated fashion. In addition, cells may need only a few molecules of one type of protein, but many thousands of molecules of another type of protein. Thus, some genes need to be transcribed every second as opposed to others that are transcribed only once per generation. For efficient adaptation and harmony and economical use of cellular resources, it is essential that the cell regulates the expression of its genes to produce the optimal amount of each gene product and only when needed. The development of very powerful microbial genetic methods and the use of strong intuitive logic helped to make the conceptual breakthrough in our understanding of gene regulation in bacteria and provided the **operon model** of gene expression. The invention of modern gene technology and the development of sensitive biochemical methods have revealed fascinating features of gene regulation not envisioned before.

The operon concept originally described the regulation of transcription initiation, and its basic features are followed in all prokaryotic organisms. Regulation of protein synthesis also occurs at the levels of transcription elongation and termination, mRNA metabolism, translation, and protein stability. Posttranscriptional regulation may appear to be energetically wasteful; however, it allows fine-tuning and is found more often than expected, particularly among bacteriophages. Genes are also turned on and off purposefully by DNA recombination. This chapter will review the structural aspects of operons that are involved in the regulation of the operons in bacteria and bacteriophages, define the participating components and their function, and describe the molecular mechanisms of gene expression at the known levels. In developing integrated views of universal principles and specialized processes, examples will be used to explain the underlying concepts and added complexities, such as regulons. The integrated picture of the physiological significance of regulation by operons and regulons is discussed in Chapter 20. Models of prokaryotic gene regulation also serve as models of gene regulation in eukaryotes and have explained various developmental processes, genetic and infectious diseases, viral life cycles, and transformations.

18.1 Operons and Regulons Are Units of Transcription

In its simplest form, an **operon** is a unit of transcription (on the chromosome) with signals for RNA polymerase to start and stop. The unit may encode one (monocistronic) or more genes (polycistronic). Genes become subject to joint regulation through a polycistronic operon. The physical linkage makes the genes transcribed as a single RNA molecule, whose amount increases or decreases through the action of various intraoperonic and extraoperonic control elements.

A set of operons scattered throughout the chromosome, but subject to regulation by the same regulator, is called a **regulon** (see Fig. 20.1). The purpose of regulation through a regulon mode is to extend the same coordinated control to more than one operon that encode similar functions. In a **modulon**, a regulatory protein simultaneously modulates the expression of seemingly unrelated operons that are bound by a common global purpose (see Fig. 20.2). In addition to having their individualized local regulators, operons are regulated by a global regulator for inclusion in a modulon. Operons and regulons make global controls through modulons feasible (see Chapter 20.1).

Operons encode metabolic enzymes, structural components, housekeeping machinery, or proteins necessary for cell development and division. Construction and regulation of operons, regulons, and modulons allow a logical, i.e., purposeful and economical, use of metabolic pathways and other cellular processes.

Operons and **regulons** serve in gene control in many ways and influence the following activities: coordinated expression of functionally related genes, similar control of functionally related genes, integration of gene transcription and translation, synthesis of component gene products in the correct proportions, and conservation of transcriptional and translational energy.

18.1.1 Operons Contain Several Components

The structure of an operon may vary in interesting ways, but an operon usually has the following basic components (Fig. 18.**1**): promoter, transcription start point, region encoding the leader RNA, cistron(s), ribosome binding site, DNA control element, and transcription termination signal.

The **promoter (*p*)** is the DNA sequence that defines RNA polymerase binding and determines the rate at which transcription initiates at a fixed point. The promoter is most often expanded to include DNA control elements that regulate its activity. Promoter mutations are *cis*-dominant. The **transcription start point (*tsp*)** is the site of transcription initiation. The **leader RNA (*l*)** defines the region of mRNA of variable size encoded between *tsp* and the first cistron. *l* may contain *cis*-acting regulatory structural features and an open reading frame (*orf*) encoding a small polypeptide called the **leader peptide (L)**. A **cistron** is defined as a structural gene that encodes a protein. Each cistron begins with a start codon encoding methionine (usually ATG) and ends with a stop or nonsense codon: TAA (ochre), TAG (amber), or TGA (opal). In a polycistronic operon, the open reading frames (*orf*) of two successive cistrons may overlap. An overlap can be very small (e.g., ...ATGA...). Extensive overlaps are observed in compact bacteriophage chromosomes (see Chapter 26). A **ribosome binding site (*rbs*)** is a purine-rich stretch of five nucleotides. Each start codon is preceded by an *rbs* segment. A **DNA control element** is a *cis*-acting DNA sequence that either enhances (positive control) or lowers (negative control) the promoter activity by binding to a *trans*-acting regulatory protein. The regulator could be an **activator**, which induces positive control by binding to a DNA **activation site(s)** (***a***, sometimes designated as ***i***), or a **repressor**, which brings about negative control by binding to a DNA site called an **operator (*o*)**. The activity of the regulatory proteins may be modulated by small molecules called **effectors** (**inducer** or **corepressor**). The location of *a* or *o* can vary considerably. *a* and *o* can interpenetrate with the promoter or can be located far upstream; *o* can also be located even within structural genes. The DNA control elements can also be multipartite in nature. Whereas mutations in a DNA control element are *cis*-dominant, mutations in the *trans*-acting regulatory proteins are usually recessive, but occasionally *trans*-dominant (negatively complementing).

A **transcription termination signal (*t*)** signifies the end of the transcription unit in the operon. Transcription termination by RNA polymerase at this site may require a protein factor. Latent termination signals are found within an operon that is expressed under special situations (see Chapter 18.10.1). There can also be intraoperonic sites that may use specific proteins, called **anti-termination** factors, to suppress transcription termination at downstream regions.

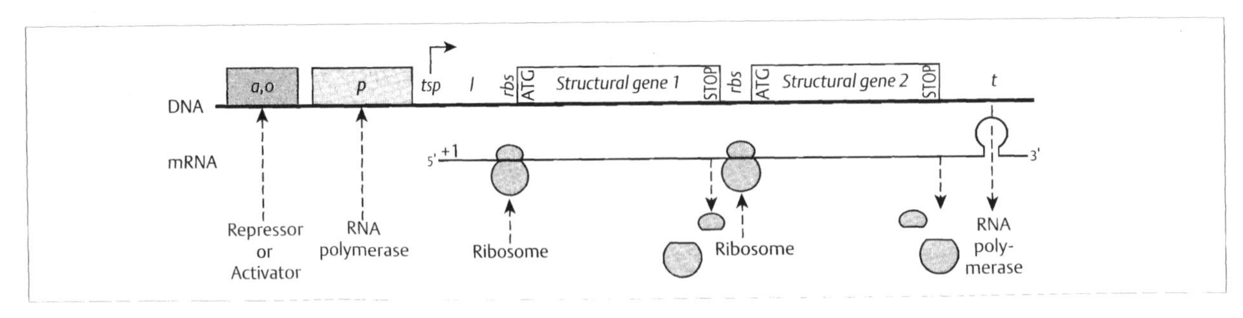

Fig. 18.**1** **Elements of an operon.** *a* (activation site), binding site of activator protein for transcription activation; *o* (operator), binding site of repressor protein for transcription repression; *p* (promoter), site of RNA polymerase binding and determinant of the intrinsic rate of transcription initiation; *tsp* (arrow), transcription start site also referred to as (RNA base) + 1; *l* (leader RNA), the region encoding the mRNA between *tsp* and the transcription attenuation site; the first structural gene or cistron 1; the second structural gene or cistron 2; *t*, the transcription termination site at the end of the operon

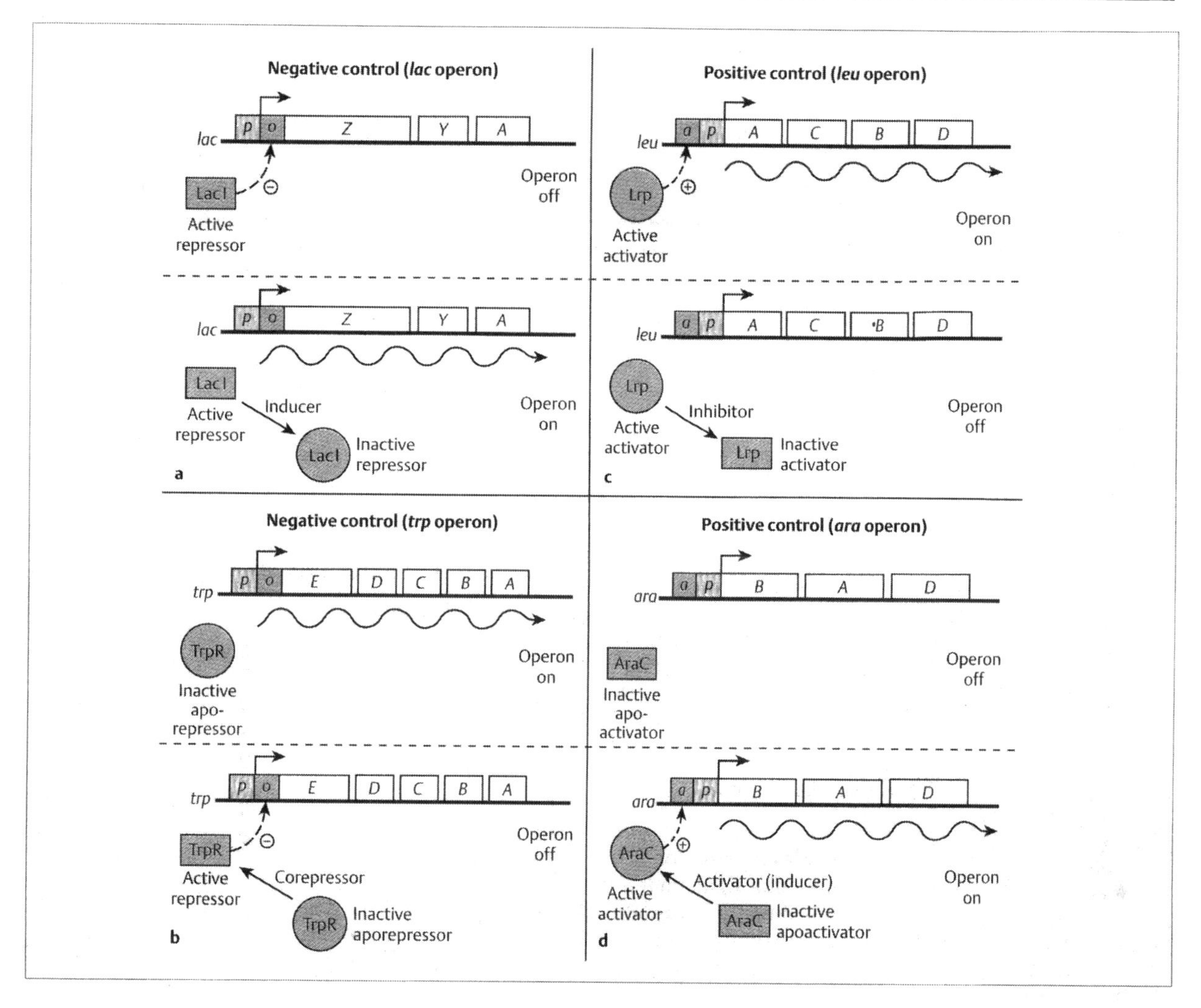

Fig. 18.**2a–d** **Strategies of regulation of transcription initiation.** The principles are explained using the paradigm examples from *Escherichia coli.*

a Negative control of the catabolic lactose (*lac*) operon by the interaction of the Lac repressor (LacI) with the operator *lacZo* (*o*). Active LacI is inactivated by the inducer. *p* is the promoter *lacZp*. *Z*, *Y*, and *A* are the structural genes.

b Negative control of the anabolic tryptophan (*trp*) operon by the interaction of the Trp aporepressor (TrpR) with the operator *trpEo* (*o*). Normally inactive, TrpR is activated by the corepressor tryptophan. *p* is the promoter *trpEp*. *E*, *D*, *C*, *B*, and *A* are the structural genes.

c Positive control of the anabolic leucine (*leu*) operon by the interaction of the activator leucine responsive protein (Lrp) with the activation site *leuAa* (*a*). Lrp is inactivated by the inhibitor leucine. *p* is the promoter *leuAp*. *A*, *C*, *B*, and *D* are the structural genes.

d Positive control of the catabolic arabinose (*ara*) operon by the interaction of the apoactivator AraC with the activation site *araBa* (*a*). *a* is now known to be multipartite (see Fig. 18.**5c**). AraC is activated by the coactivator (also called inducer) arabinose. *p* is the *araBp* promoter. *B*, *A*, and *D* are the structural genes. The systems are described here in a simple form (for the nomenclature, see Box 18.1)

Some operons do not encode any proteins (e.g., those for rRNA, tRNAs, and other small regulatory RNA molecules). These RNA products differ from most other mRNA products in two other respects; they are metabolically stable and are usually transcribed as a single large RNA chain which is then cleaved (processed) to generate the mature products (see Chapter 18.12).

18.1.2 Strategies of Regulation of Transcription Initiation Vary

Interactions of repressors and activators with effectors have been used by nature in generating both negative and positive control. All four strategies in prokaryotes, shown in Table 18.**1**, inducible and repressible, have

Table 18.1 **Strategies of regulation of transcription initiation.** Examples of strategies a–d are given in Figure 18.**2a–d**, respectively

Regulatory strategy	Function regulated	DNA control element	Regulatory protein	Effector	Operon transcription	
					No	Yes
a. Negative (inducible)	Catabolic	Operator (*o*)	Repressor	Inducer	Repressor	Repressor + inducer
b. Negative (repressible)	Anabolic	Operator (*o*)	Aporepressor	Corepressor	Aporepressor + corepressor	Aporepressor
c. Positive (repressible)	Anabolic	Activation site (*a*)	Activator	Inhibitor	Activator + inhibitor	Activator
d. Positive (inducible)	Catabolic	Activation site (*a*)	Apoactivator	Coactivator (inducer)	Apoactivator	Apoactivator + coactivator (inducer)

evolved to regulate transcription initiation of operons and of regulons in response to effectors. Depending on the system, the effector may be called an **inducer**, a **corepressor**, an **inhibitor**, or a **coactivator**. The four strategies are shown in the simplest forms in Figure 18.**2** using paradigm examples. For operons and regulons encoding enzymes of biochemical pathways, the effector is usually a substrate, an intermediate, or an end product of the metabolic pathway. If a non-metabolizable analogue of the natural effector is to perform the same function, it is called a **gratuitous effector**. In modulons, the effector is an environmental signal that is perceived by the regulatory protein directly or by an intermediary sensor (see Chapter 20). Although there is no strict rule, the strategies shown in Figure 18.**2 a** and **d** (repressor–inducer and apoactivator–coactivator, respectively) are generally used for the regulation of catabolic enzyme synthesis. The repressor–inducer strategy (Fig. 18.**2a**) allows the use of an energy source other than glucose by de novo synthesis of enzymes (e.g., utilization of sugar as a carbon source). In this way, cells make machinery only when it is beneficial. On the other hand, the aporepressor–corepressor (Fig. 18.**2b**) and activator–inhibitor (Fig. 18.**2c**) strategies are commonly associated with the synthesis of enzymes for anabolic processes (e.g., amino acid and vitamin synthesis). They help prevent unnecessary flux through a pathway when the end products are plentiful.

Allosteric effectors. Gene transcription can be controlled in an environmentally responsive way by small molecule effectors. An operon can be turned off or on by the presence or absence of the effector, and the amount of transcription of the operon when expressed is proportional to the concentration of the effector molecule present. The effector molecule interacts with

Box 18.1 The **operon** is a unit of transcription characteristic for the prokaryotes (Figs. 18.**1** and 18.**2**). The mnemonic name of an operon, e.g., *lac*, *his*, and *rps*), is written in lowercase letters and in italics, and is usually derived from the function (e.g., **lac**tose degradation, **his**tidine biosynthesis, and **r**ibosomal **p**rotein **s**mall). Most operons contain several structural genes or cistrons, each designated by a capital letter (in italics) and written in the sequence of their location behind the promoter (e.g., *lacZYA* or *trpEDCBA*). The **promoter** is named according to the first structural gene (e.g., *lacZp* or *trpEp*), and the **terminator** is named according to the last gene (e.g., *lacAt* or *trpAt*). Other essential elements are the binding sites or **operators** for **repressors** (e.g., *lacZo*) and the activator binding sites (e.g., *araBa*, sometimes *araBi* for **initiator**), the transcription initiation site (*tsp*; often shown as an arrow) at base pair +1, the **leader** sequence (e.g., *hisGl*), and the **ribosome binding sites** (*rbs*; called Rbs at the mRNA level). **Open reading frames** (e.g., *orf-1*, *orf-2*) for unknown gene products (Orf) are labelled in analogy to alleles (e.g., *lac-1*, *lac-4*). In figures describing complex operons, abbreviations are often used (e.g., P_1, P_2, P_3) to designate the various promoters of an operon. Operons regulated by the same specific regulator (i.e., those that form a **regulon**) use mnemonic names and nomenclature similar to those used for single operons. Groups of genes also controlled by a global regulator and organized in **modulons** (Fig. 20.**2**) (e.g., the *crp* modulon, the *lrp* modulon), however, use different mnemonic names.

the regulatory protein to change the behavior of the protein allosterically. A regulatory protein can be functional and inactivated by an effector, or nonfunctional and activated by an effector. The allosteric changes make the protein deficient (e.g., LacI) or proficient (e.g., TrpR) in specific DNA binding or even alter the pattern of their binding (e.g., LysR, see Chapter 18.6). Although the effect of allosteric modifiers on gene expression is slower than the instantaneous metabolic control, the association is usually noncovalent and rapidly reversible, permitting a continual response to environmental changes. Occasionally, a covalent modification of a regulatory protein is needed (see, for example, the global regulator NtrC in Chapter 20).

The strategies of regulation of transcription initiation (Tab. **18.1**) can vary in several ways. For example, a regulatory protein may not have an effector. Also, a regulatory protein may have a dual role, acting as an activator at one promoter and as a repressor at another. In addition, a regulatory protein may act as a repressor, which becomes an activator when bound to an effector molecule (see Chapter 18.6). If the operon is also part of a modulon (see Chapter 20), its transcription is regulated by two regulators: an operon (or regulon)-specific local regulator and a modulon-specific global regulator.

18.2 A Regulator Often Controls its own Synthesis: Autoregulation

Frequently, a regulatory protein needs to be maintained at a critical level in response to environmental changes. This is achieved by autoregulation, which uses negative or positive control. In negative control, the regulatory protein acts as a repressor by binding to an operator of its own operon. A decrease in the concentration of the protein decreases its binding to the operator and proportionately releases the repression and vice versa. Autoregulation of an activator protein by positive control alone, however, is expected to result in exponential synthesis of the activator protein and therefore should not exist. Thus, an autoregulatory system in which the regulatory protein acts as an activator must also be tied to a negative control. Indeed, autoregulation by a combination of activation and repression occurs in many cases (see Chapter 18.6).

By **autoregulation**, high cellular concentrations of the regulatory protein lead to the decrease in the transcription from its own promoter, whereas low cellular concentrations of the regulatory protein lead to the increase in transcription from its own promoter.

18.3 The Bacterial RNA Polymerase Is a Multisubunit Enzyme That Initiates Transcription at the Promoter

Two essential components of transcription initiation are the *trans*-acting RNA polymerase and the *cis*-acting promoter, one or both of which could be the target of regulation. The bacterial RNA polymerase is a multi-subunit enzyme that consists of a four-subunit ($\beta\beta'\alpha_2$) core enzyme and a σ-subunit (or σ-factor), which finds the promoter. There are several different σ-subunits in cells. The details of promoter structure and of RNA polymerase and its subunits are described in Chapter 15.

Operons frequently have more than one promoter. Promoters of the same or of different operons may be arranged in tandem or may even overlap. They are known to be separated by as little as 5 bp to as much as 600 bp. Tandem promoters for different operons may be convergent or divergent. There are several regulatory consequences of promoters being arranged in tandem and so close to each other: 1) a regulatory protein binding to a common DNA site can modulate two tandem promoters that drive the same operon or two divergent operons simultaneously and coordinately, but not necessarily in the same direction (i.e., negatively or positively). In the case of divergent tandem promoters, one of the promoters frequently drives the regulatory gene itself and thus is autoregulated (see Chapter 18.6); 2) a stronger promoter can act as an "antenna" to collect RNA polymerase molecules and shift them to the weaker promoter; 3) transcription from a strong upstream promoter through a downstream promoter can occlude the activity of the latter, and 4) in extreme examples, one of the promoters present within the operon can allow transcription of only the cistrons located downstream of the internal promoter and not of genes upstream of it.

18.4 Negative Control Decreases Transcription

The original concept of negative control postulated a repressor that inhibits transcription initiation by sterically hindering RNA polymerase binding to the promoter. However, the ability of repressors to inhibit transcription initiation from the variety of locations of the operators and the discovery of operons with multipartite operators, suggest that there should be differences in the exact level at which different repressors act. Several models of action are feasible:

1. In model A, the repressor interferes with RNA polymerase binding as originally conceived. Such a mechanism is used if the operator and the promoter loci overlap.
2. In model B, the repressor interferes with RNA polymerase activity. If the operator does not overlap with the promoter, the repressor clearly does not hinder RNA polymerase binding. Although such systems are sometimes characterized by multipartite operators, the repressor inhibits a step beyond RNA polymerase binding.
3. In model C, the repressor acts by changing the DNA structure, which inhibits promoter function by disabling it.
4. In model D, the repressor interferes with the activity of an activator (mixed control).

18.4.1 Two Paradigms of Negative Control Are the *lac* and *trp* Operons

One of the two systems that brought about the concept of negative control is the *lac* operon, which encodes lactose-catabolizing proteins in *Escherichia coli*. The operon follows the repressor–inducer strategy (Fig. 18.**2a**). The LacI repressor binds to the operator to decrease transcription initiation at the *lac* promoter by more than 1000-fold. Inactivation of LacI by the inducer allo-lactose (α-D-galactosyl-β1,6-D-glucopyranoside), a metabolic product of lactose (α-D-galactosyl-β1,4-D-glucopyranoside), or isopropyl-thio-β-D-galactoside (IPTG), a nonmetabolizable analogue, increases the transcription frequency to the derepressed level. LacI binding to an operator (*lacZo1*) centered at position +7 sterically hinders RNA polymerase binding, generating negative control (model A). The molecular mechanism by which LacI acts, nevertheless, is complicated by three operator loci: the primary operator *lacZo1* at position +7, a downstream operator *lacZo2* at position +401, and an upstream operator *lacZo3* at position −82. The affinities of *lacZo2* and *lacZo3* for the repressor are much weaker than that of *lacZo1*. It is believed that the former two loci enhance the repressor binding to the primary operator by cooperativity (Chapter 18.4.2).

The *trp* operon, which encodes the anabolic enzymes for the synthesis of the amino acid tryptophan, is the paradigm example of the other strategy of negative control (i.e., corepressor–aporepressor; Fig. 18.**2b**). An aporepressor, TrpR, requires the presence of a tryptophan molecule, the corepressor, to bind to a unique operator (*trpEo*) at position −3 and inhibit RNA polymerase binding (model A). Thus, when the tryptophan concentration in the cell is high, the corepressor–aporepressor complex decreases transcription initiation from the promoter *trpEp*. At low cellular tryptophan concentrations, the aporepressor becomes free and unable to bind to the operator.

18.4.2 The Simple Theme of Negative Control Often Deviates

The concept that a repressor is a unique protein that inhibits transcription initiation from a promoter by binding to a specific operator, as portrayed in Figure 18.**2**, is not a firm rule; deviations commonly occur.

One operator can control two promoters of a regulon. A single operator locus is sufficient to inhibit two divergent operons. The transcription start sites (*tsp*) of the two convergent promoters in the divergent *bio* operons, for example, are separated by only 9 bp, with a unique operator centered at positions +11 of the *bioAp* (P_A) promoter and −21 of the *bioBp* (P_B) promoter. The biotinyl-5′-adenylate·BirA repressor binding at this site inhibits transcription initiation from both promoters which hence constitutes the **bio regulon** (Fig. 18.**3**).

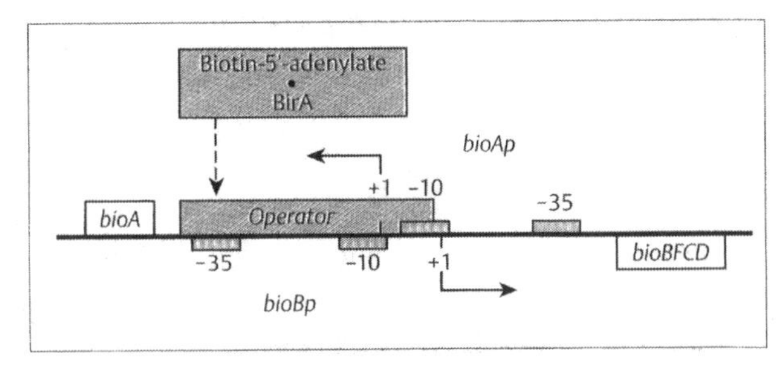

Fig. 18.**3** **Divergent *bio* operons in *Escherichia coli*.** The two overlapping promoters *bioAp* (P_A) and *bioBp* (P_B) guide two divergent *bio* operons. The corresponding transcription start sites (*tsp*; indicated by +1) are 9 bp apart. The 40-bp unique operator locus (shaded red) overlaps with both *bioAp* and *bioBp*. The structural genes are *bioA* and *bioBFCD*. BirA, the repressor, is also a synthetase

Repressor occupation at this position, which overlaps with both promoters, uses model A of negative control.

Enzymes can act as repressors. The divergent anabolic *bio* operons in *E. coli*, which encode enzymes for the synthesis of the vitamin biotin, are negatively regulated by an aporepressor, BirA. However, BirA has other functions. It acts as the enzyme biotin holoenzyme synthetase and synthesizes its own corepressor, biotinyl-5′-adenylate. The biotinyl-5′-adenylate·BirA co-complex either binds to the operator as an active repressor of the *bio* operons or transfers the biotinyl moiety to a lysine residue of the apoenzyme of acetyl-CoA carboxylase in *E. coli*. BirA is also involved in retention of biotin in cells.

One of the genes, *putA*, of the divergent *put* operons, which encodes enzymes for proline utilization in *Salmonella typhimurium*, encodes a bifunctional protein. In the presence of intracellular proline, PutA becomes membrane bound and catalyzes both enzymatic steps required to catabolize proline to glutamate. In the absence of proline, PutA becomes cytoplasmic and acts as a repressor of the two divergent *put* operons which

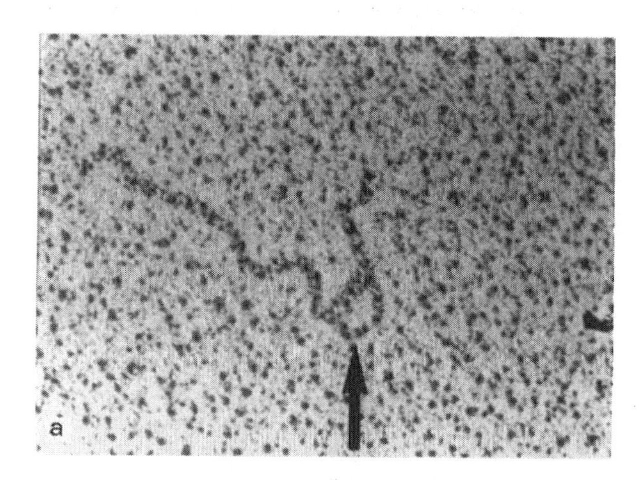

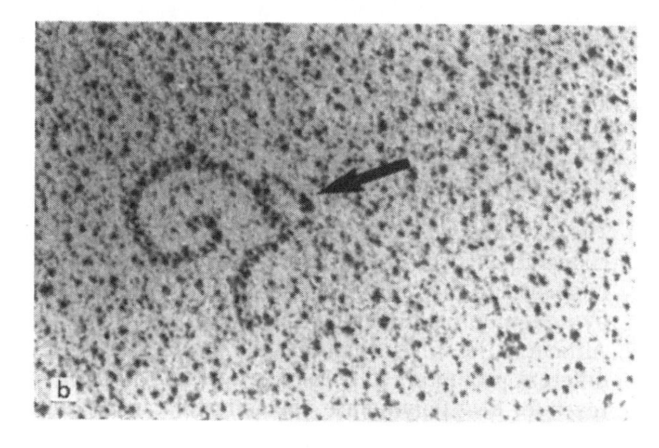

Fig. 18.**4a,b** **DNA looping.** Electron micrographs of Lac repressor (arrows) bound to two engineered *lac* operators separated by 113 bp in the *gal* operon of *Escherichia coli* [from 1]

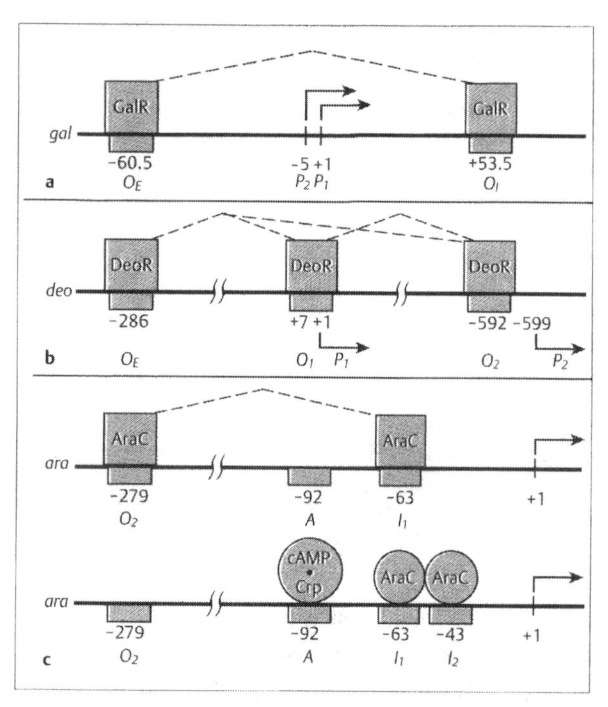

Fig. 18.**5a–c** **Multipartite operators in *Escherichia coli*.**
a The *galETK* operon contains two promoters *galEp1* (P_1) and *galEp2* (P_2), which are modulated differently by cAMP·Crp. cAMP·Crp stimulates *galEp1* (P_1) by direct contact with the RNA polymerase, whereas it inhibits *galEp2* (P_2) by inhibiting RNA polymerase binding. cAMP·Crp does this by binding at position −41.5 from the start site of P_1 (not shown). Superimposed on this control is the negative control of both *galEp1* (P_1), and *galEp2* (P_2) by the Gal repressor, which acts by binding to two operators *galEoE* (O_E) (at −60.5) and *galEoI* (O_I) (at +53.5). *galEoI* (O_I) is located within the first structural gene *galE*. The broken lines indicate loop formation by interaction of the indicated operator-bound repressors.
b The *deoCABD* operon contains two promoters *deoCp1* (P_1; at +1) and *deoCp2* (P_2; at +599). The DeoR repressor exerts repression by binding to three *deoCo* operator loci, *deoCo2*, (O_1; at +7), *deoCo2* (O_2; at +592), and *deoCoE* (O_E; at −286). The broken lines indicate the alternate looping between the operator-bound repressors.
c The *araBAD* operon. Negative control of the *araBAD* operon occurs by DNA loop formation between AraC protein bound to two operators *araBi* (I_1) and *araBo* (O_2). Addition of arabinose changes the AraC protein, which now binds to the tandem sites (*araBi1* (I_1) and *araBi2* (I_2) and becomes an activator of the promoter (Chapter 18.5). For activation, binding of cAMP·Crp to an activation site *araBa* (*A*) at −92 is also required

again constitute a regulon. Thus, BirA and PutA exemplify the use of enzymes also as gene regulatory proteins.

Multipartite operators act by DNA looping. Two or more spatially separated operator loci are needed to exert negative control in several operons. Such multipartite operators encompass the promoters they repress. In these systems, two operator-bound repressors contact each other, as if they were adjacent, but in the process cause the intervening (promoter-containing) region to form **DNA loops**, which can be observed by electron microscopy (Fig. 18.**4**). DNA looping in three operons will be described here.

The galETKM operon (Fig. 18.**5a**). The GalR repressor mediates loop formation by the bipartite operators *galEoE* (O_E) and *galEoI* (O_I) (113 bp apart) in the *galETKM* operon, which encodes enzymes of galactose metabolism in *E. coli*. The loop formation inhibits transcription initiation from the two tandem promoters *galEp1* (P_1) and *galEp2* (P_2) of the operon. DNA looping disables the promoters by altering DNA structure (model C).

The deo operon (Fig. 18.**5b**). In the *deo* operon, which encodes nucleoside- and deoxynucleoside-catabolizing enzymes of *E. coli*, the two promoters *deoCp1* (P_1) and *deoCp2* (P_2) (600 bp apart), are repressed by the DeoR repressor through three operators, *deoCo1*, *deoCo2*, and *deoCoE*, (O_1, O_2 and O_E, respectively). A DNA loop between *deoCo1* next to *deoCp1* and *deoCo2* next to *deoCp2* is required for repression of the promoters; any instability of this loop is rectified by *deoCoE*, which helps repression by forming an insurance loop with *deoCo1* or *deoCo2*. DNA looping enhances the affinity of DeoR toward the operators because of a cooperative binding rule. Because *deoCp1* overlaps with *deoCo1* and *deoCp2* with *deoCo2* (model A) steric hindrance is the likely cause of repression.

The araBAD operon (Fig. 18.**5c**). As opposed to the initial view about the strategies of regulation of the *ara* operon of *E. coli* as described in Figure 18.**2**, the apoactivator AraC also acts as a repressor by binding to two sites. In the absence of the effector (inducer) L-arabinose, AraC simultaneously binds to two spatially separated DNA sites *araBi1* (I_1) and *araBo2* (O_2) generating a repression loop. Addition of arabinose breaks up the repression loop and makes AraC bind to the adjacent sites *araBi1 araBi2* (I_1 I_2) as an activator. Thus, AraC performs a dual role: it acts both as a repressor and as an activator—a combination of the repressor–inducer and the apoactivator–coactivator strategies (Figs. 18.**2a** and **d**). For more examples of regulatory proteins with a dual role, see Chapter 18.6.

Negative control provides a fail-safe mechanism and may have originated from a system that once functioned constitutively. Negative control evolved to give the cells selective advantage because such a control imparts increased economy and efficiency.

18.5 Positive Control Activates Transcription

Positive control activates promoters that are normally inactive. Although promoter deficiencies can be optimized by changing the surrounding sequences, many promoters in bacteria are rate limited in one or more of the steps of initiation. In **positive control**, a specific rate-limiting step is overcome by an activator. The portrayal of the steps of initiation (Chapter 15) has been useful for biochemical analysis of the site of activator action. Regulation by activators can occur by an increase of closed or open complex formation or through the aid of promoter clearance. This is found in the control of operons/regulons and in the global control of modulons.

18.5.1 Activation Sites Can Act From Short Distances

An activation site (a) for positive control can be located near or far from the promoter. The usual mode of action by which an activator acts by binding to the activation site within 100 bp of the promoter is the following. The DNA-bound activator contacts and stimulates the activity of RNA polymerase. This principle is best illustrated, however, by the well-studied activation of *lacZp* by the global activator cAMP receptor protein (Crp) when complexed with cAMP using the apoactivator–coactivator strategy (see Chapter 18.1.2). Five significant lines of evidence show that cAMP·Crp contacts the RNA polymerase (Fig. 18.**6**):

1. The activators and RNA polymerase bind to DNA cooperatively. The cooperative binding of two protein molecules to the same face of DNA is easily explained by protein–protein contact (Fig. 18.**6b**).
2. Although the DNA binding sites for cAMP·Crp in different promoters in *E. coli* are located within easy reach for RNA polymerase, the activator can act only if these sites are separated from *tsp* by an integral number of DNA helix turns. The activator must be on the same face of the DNA helix as RNA polymerase, making the activator–RNA polymerase contact geometrically feasible. If the activator binding face is engineered on the opposite face of the promoter by

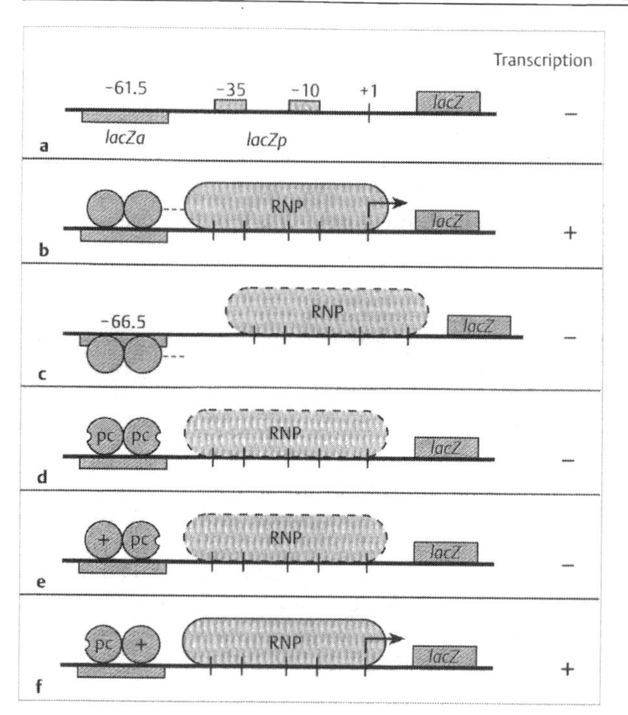

Fig. 18.6 **Transcription activation of the *Escherichia coli lacZ* promoter (a) by Crp.** When complexed with cAMP, Crp binds to the upstream activator *lacZa*, a palindromic DNA sequence centered at −61.5 from *tsp* (+1). Crp acts as a dimer, each monomer occupying a half-site on DNA.
b,c Crp makes contact with the α-subunit(s) of RNA polymerase (RNP) if placed on the same side of the DNA as the RNA polymerase, through a surface defined by several *pc* (positive control) mutations.
d The *pc* mutations do not activate transcription.
e,f Crp makes the useful contact with RNA polymerase only through one subunit of Crp. A Crp heterodimer of one wild-type subunit and one mutant subunit can activate transcription only if the wild-type subunit is facing the promoter

inserting 5 bp in between the activator binding face and the promoter, activation does not occur (compare Figs. 18.**6b, c**).

3. Genetic evidence to establish protein–protein contact is based on the idea that contact-defective mutant proteins would bind DNA normally, but would be unable to activate transcription (Fig. 18.**6d**). Such mutations (*pc*, for positive control) identify patches on the surface of Crp. The contribution of a patch to transcription activation depends on the location of the activation site in the promoters. In addition, *pc* may be an "acidic" patch (common among eukaryotic activators), such as in the cI regulator of phage λ (Chapter 18.6). The corresponding contact points in RNA polymerase reside in more than one of its subunits. Mutations in RNA polymerase that do not respond to cAMP·Crp and several other activators are located in the C-terminal region of the α-subunit, indicating that these activators contact the α-subunit. In contrast, RNA polymerase mutations that suppress the *pc* defect in the cI activator are located in the σ-subunit. Presumably, the mutations in the σ-subunit restore the contact defect in cI. Contact points for activators in β and β′ subunits of RNA polymerase are quite conceivable.
4. Since Crp is a dimer, when a heterodimer of Crp, consisting of one wild-type subunit and one *pc* mutant subunit, is bound to DNA, the wild-type subunit is located in one case promoter distal (Fig. 18.**6e**) and in the other case promoter proximal (Fig. 18.**6f**). Transcription is stimulated only in the latter case.
5. Protein–protein cross linking experiments have shown that the activation patch on the surface of Crp is in direct physical proximity to the C-terminal region of the α-subunit when bound. Examples of biochemical steps at which different activators work are shown in Table 18.**2**.

Table 18.2 **Steps at which activators work.** K_B corresponds to the initial binding of RNA polymerase to the promoter and k_f to the subsequent isomerization step (Chapter 15). *a*, activator binding site; *Enh*, enhancer; kb, *Enh* is located kilobases away from the promoter; +, increase in open complex formation at *glnAp* in the presence of phosphorylated NtrC; gp, gene product

Promoter	Activator	Location of *a*/*Enh*	Activation (-fold)	
			K_B	k_f
P_{RM} (phage λ)	cI	−42	None	11
lacZp (*Escherichia coli*)	cAMP·Crp	−61.5	20	None
P_{RE} (phage λ)	cII	−33	15	40
P_I (phage λ)	cII	−33	100	100
glnAp (*Salmonella typhimurium*)	NtrC∼P	kb	None	+
P_{late} (phage T4)	gp44, gp62, gp45	kb	None	8

Double positive controls and multipartite activation sites. In superimposed controls (see also Chapter 20), operons can have two activators—one specific and one global. Both activators must be functional for the promoter to be transcribed. An example of such superimposed positive controls is the activation of the *malEp* (P_E) and *malKp* (P_K) promoters of the two divergent *malB* operons in *E. coli*. These two promoters, inducible by maltotriose, are dependent upon the local apoactivator MalT, which binds to maltotriose and the global activator cAMP·Crp (Fig. 18.**7**). The 271-bp segment between the two divergent *tsp* contains an array of activator sites—five for maltotriose·MalT (T_1–T_5) and four for cAMP·Crp (C_1–C_4). cAMP·Crp binds to C_1, C_2, and C_3 strongly and to C_4 poorly. Binding to all three high-affinity sites is required for transcription from both *malEp* and *malKp*, and a mutation in any one of the sites reduces their activation. T_1 and T_2 are required for the proximal *malEp* promoter, and sites T_3, T_4, and T_5 are required for the proximal *malKp* promoter closer to the latter sites. Activation of *malEp* and *malKp* by two activators is a complex process and involves the formation of a DNA–multiprotein complex of higher order structure. For example, although the intrinsic binding to T_3, T_4, and T_5 is high, maltotriose·MalT binding repositions in the presence of the global regulator complex cAMP·Crp, to a second set of three binding sites, T_3', T_4', and T_5', which are closer to the −10 element of *malKp*. The second set of binding sites partially overlaps the first set. It is the DNA–protein complex with the repositioned MalT sites that triggers transcription from *malKp*. The active structure of the DNA–protein complex involves interaction between MalT bound to T_1–T_2 and T_3'–T_5'. The role of cAMP·Crp molecules bound to intervening segment facilitates the correct protein–protein and DNA–protein contacts by DNA bending. Interestingly, MalT action requires ATP hydrolysis.

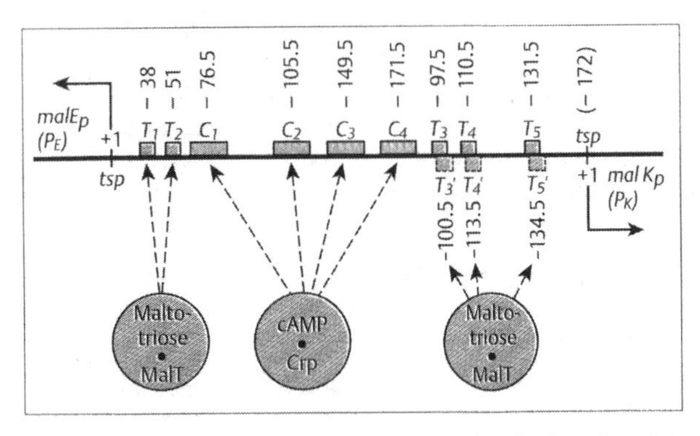

Fig. 18.7 **The divergent promoters *malEp* (P_E) and *malKp* (P_K) in the *malB* locus of *Escherichia coli.*** The multipartite activation sites for the two activators, MalT complexed with inducer maltotriose (maltotriose·MalT) and Crp complexed with cAMP (cAMP·Crp) are named T_1–T_5 (red boxes) and C_1–C_4 (black boxes), which are located in the order as shown. The three sites closer to *malKp* (T_3, T_4, T_5) reposition (dotted boxes; T_3', T_4', T_5') closer to *malKp* when cAMP·Crp binds to its own sites (C_1–C_4). The two *tsp* sites are separated by 271 bp. The location of the activation sites is shown with respect to *tsp* of *malEp*

Promoters requiring two activators, one of which is Crp, reveal differences in the mechanisms by which the global activator Crp acts.

A suicidal enzyme as an activator. The activators do not have to be an exclusive group of proteins. A metabolic enzyme can also be an activator. The DNA repair protein Ada removes mutagenic alkyl groups from purines in DNA (see Chapter 15.6.4). During repair, the alkyl groups are transferred from the DNA to the protein itself. Although methylation of Ada is suicidal for DNA repair (i.e., methylated Ada (meAda) is enzymatically dead, making the cell deficient in the repair activity), the condition is alleviated by derepression of the synthesis of fresh Ada protein because meAda, and not Ada, acts as a transcriptional activator of its own gene (*ada*). meAda acts by binding to an activator locus at the *ada* promoter.

18.5.2 Activation from Distant Sites and the Role of Enhancers

Activation sites located far away from the promoter, commonly found in eukaryotic organisms, are known as **enhancers**. The enhancer functions efficiently at large distances (many kb) both upstream and downstream from the promoter in either orientation. Bacterial DNA enhancer sequences work by binding activators. A family of prokaryotic enhancer binding activators has been found in Eubacteria (e.g. in the Proteobacteria and in the Gram-positive bacteria). These activators function in conjunction with promoters recognized by σ^{54}–RNA polymerase (see Chapters 15 and 28). A well-characterized global activator in this group is NtrC (also called NRII), which binds to two enhancer sites at the promoter of the *glnA* gene for transcription activation in *E. coli* and *Salmonella typhimurium* (Fig. 18.**8** and Chapter 20.4). Although more than one enhancer site is associated with a promoter, one appears to activate transcription quite well. Unmodified NtrC acts as an

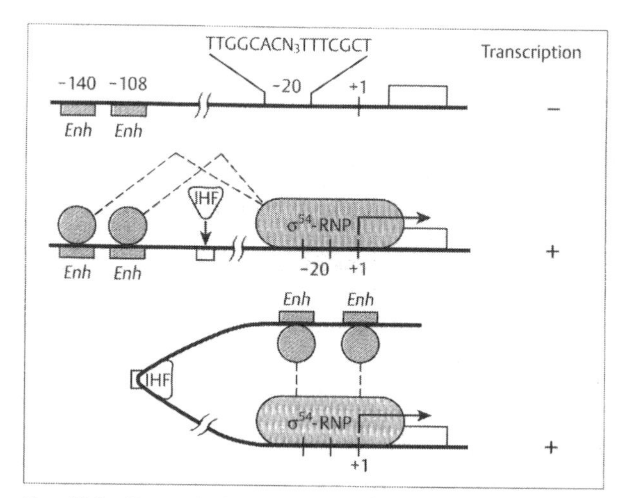

Fig. 18.8 **Transcription activation by activator binding to enhancer elements (*Enh*).** The discontinuity in the DNA (black lines) represents that the enhancer element could be located kilobases from the promoter. The DNA looping formed by contacts (indicated by broken lines) of enhancer-bound proteins (red) and the σ^{54}-RNA polymerase (gray) bound to the promoter (at −20 position) is shown at the bottom. DNA looping for some activators is aided by binding of a DNA bending protein IHF (integration host factor) at an intermediate position as shown

apoactivator and binds to the enhancer with high affinity; the apoactivator is phosphorylated to become an activator of transcription. The phosphorylated protein has a very short half-life. The system is a two-component system that signals nitrogen limitation in the cell (see Chapter 20.4). The enhancer-bound phosphorylated NtrC physically contacts the promoter-bound RNA polymerase by forming a DNA loop, which can be observed by electron microscopy.

Loop formation by the homologous NifA activator in enhancing transcription at the *nifH* promoter in *Klebsiella pneumoniae* is aided by the bending of the intervening segment by a DNA bending protein (Fig. 18.**8**). Interestingly, when an enhancer site and the promoter are present on different DNA circles (in *trans*), but are linked by a concatenate structure, the enhancer is still able to function, suggesting that a protein–protein contact between the enhancer-bound activator and the promoter-bound RNA polymerase is sufficient for transcription activation. The enhancer-binding activators act by stimulating isomerization of pre-bound RNA polymerase at the promoter, i.e., at the level of isomerization of the closed-to-open complex (Tab. 18.**2**). Like MalT, the enhancer-binding activator requires ATP hydrolysis to function. The bacterial enhancers serve as excellent model systems for the study of their eukaryotic counterparts.

The family of proteins that act in conjugation with RNA polymerase to activate transcription by binding to enhancers are modular in nature and have different functional domains: **DNA binding, transcription activation,** and **dimerization.** In many instances, combining a domain of one member with another domain from a second member produces a chimeric protein that demonstrates the expected functional properties of each parent.

18.5.3 Mobile Enhancers Are Typically Found in Bacteriophages

An intriguing mechanism of transcription activation occurs in bacteriophage T4 (see Chapter 26). Late promoters of T4 are transcribed by RNA polymerase containing a phage-encoded σ-factor (σ^{55}). T4 late transcription, which is coupled to DNA replication in vivo requires the transcription factor gp33 (gene product 33) and several T4 DNA replication proteins (gp44, gp62, and gp45) in vitro. Whereas gp33 along

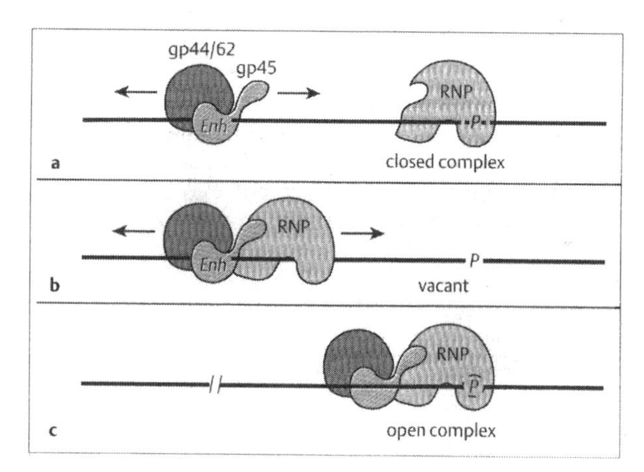

Fig. 18.**9a–c** **Models of DNA tracking by mobile enhancers.**
a,b Two possible modes of protein DNA tracking that result in the enhancement of bacteriophage T4 late transcription are shown:
a The movement of (all or some of) the DNA polymerase accessory proteins [gp44, gp62 (black), and gp45 (red)] from the enhancer (Enh) to the closed promoter–RNP (gray) complex; and
b The movement of an activated RNP from the enhancer to a vacant late promoter (*p*). Both models draw attention to the bidirectional (arrows) and orientation-dependent functions of the T4 late enhancer.
c The enhancer–promoter complex is arbitrarily drawn to contain all three DNA polymerase accessory proteins associated with RNP, although this is not an intrinsic requirement of the model [from 2]

with σ^{55}–RNA polymerase binds to the promoter, the replication proteins activate transcription in the absence of replication in vitro by recognizing, not a specific sequence of DNA, but a single-stranded DNA nick that may be present at varying distances upstream or downstream of the promoters. Transcription activation is only effective if the nick in the fork is present in the nontranscribed strand, regardless of the location (upstream or downstream) and the distance of the nick from the promoter. Because gp44, gp62, and gp45 are normally present in the replication fork that provides a nicked DNA and moves along the DNA, the fork serves as a mobile enhancer to activate transcription. Thus, unlike the looping mechanism of an enhancer, the mobile enhancer-mediated activation involves DNA tracking (Fig. 18.**9**). The mobile enhancer increases closed-to-open complex formation of the σ^{55}–RNA polymerase. The use of replication forks as a mobile enhancer is an intriguing way to couple replication and transcription, which is a common phenomenon among many bacteriophage late promoters.

18.6 Regulatory Proteins Often Have a Dual Role

Although Table 18.**1** simplifies a regulatory protein as being either an activator or a repressor, it is now clear that many regulatory proteins perform both roles. These examples are usually, but not necessarily, found among tandem divergent promoters, one of which is the promoter of the gene encoding the regulator itself.

Such regulators simultaneously act as repressors at one promoter and activators at another without the involvement of an effector, as exemplified by several bacteriophage regulators: cI of *E. coli* phage λ and p4 of *Bacillus subtilis* phage ϕ29.

An elegant, but very simple mechanism by which the same protein acts as repressor and activator occurs in the regulation of the divergent P_{RM} (**r**ight **m**aintenance) and P_R (**r**ight) promoters in phage λ (Fig. 18.**10**). In a prophage, the cI regulator is made from P_{RM} and acts as a repressor of P_R, which is a promoter for lytic growth of the phage (see Fig. 26.**15**). cI autoregulates to maintain a constant level for preserving the prophage state. At the concentrations of cI present in a lysogen, the protein, while repressing P_R, activates P_{RM}. Since cI autoregulates by positive control, the activation must be quenched by a negative control. Thus, at high concentrations, cI acts as a repressor at P_{RM}, turning down its own synthesis. This entire regulation is achieved by binding of the cI protein to three similar and contiguous operator ("right") loci (O_{R1}, O_{R2}, and O_{R3}) and based on three properties of cI: 1) the intrinsic affinities of cI for O_{R1}, O_{R2}, and O_{R3} are different–the affinity for O_{R1} is the highest and that for O_{R3} is the lowest; 2) two cI

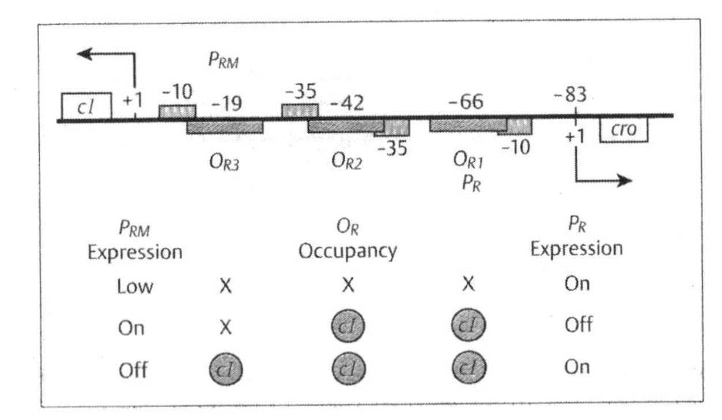

Fig. 18.**10** **Dual role and autoregulation of cI protein of phage λ.** Arrangement of the tripartite operators O_{R1}, O_{R2}, and O_{R3} (red) at the divergent promoters P_{RM} (**r**ight **m**aintenance) and P_R (right). Whereas the occupation of O_{R1} and O_{R2} by cI (red) turns off transcription at P_R, transcription at P_{RM} is turned on by occupation of O_{R2} and turned off by occupation of O_{R3}. X denotes that the corresponding DNA site is not occupied by cI. cI stimulates k_f (isomerization step after binding of RNA polymerase to the promoter) at P_{RM} (Table 18.**2**)

Box 18.2 At lower repressor concentrations, cI binds to O_{R1} and automatically binds to O_{R2} because of cooperativity, leaving O_{R3} vacant. Occupation of O_{R1} sterically hinders RNA polymerase binding to the overlapping P_R; thus, cI plays a repressor role. At the same time, cI occupying O_{R2} interacts with RNA polymerase at P_{RM} (see Chapter 18.5.1). This interaction stimulates transcription from P_{RM}, making cI an activator and autoregulator (Tab. 18.**2**). At higher repressor concentrations, cI also occupies O_{R3}. Since the O_{R3} sequence overlaps with P_{RM}, cI represses transcription initiation from P_{RM} also by steric hindrance. When the cI concentration falls to a lower level, it is lifted from O_{R3}, resuming stimulation of P_{RM}. This genetic switch merely responding to the concentrations of the cI protein and not influenced by any effector molecule has profound physiological consequences in the life cycle of λ (see Chapter 26.4).

Table 18.3 **Multiple regulatory roles of the leucine responsive promoter protein Lrp in *Escherichia coli*.** From Newman, E. B., D'Ari, R., and Lin, R. T. (1992) Cell 68: 617–619

Role of Lrp	Role of effector leucine	Operons regulated and their function	
Activator	Antagonist	*ilvIH*	Ile, Val, Leu biosynthesis
		serA	Ser biosynthesis
		gltBDF	Glu biosynthesis
		leuABCD	Leu biosynthesis
		Cp4	Unknown
		Cp5	Unknown
	Essential	*Cp52*	Unknown
	None	*geu*	Gly cleavage
Repressor	Antagonist	*sdaA*	L-Ser deaminase
		tdh/kbl	Thr dehydrogenase
		oppABCD	Oligopeptide uptake
		lysU	Lys tRNA synthetase
		Cp60	Unknown
		Cp61	Unknown
	Essential	*ilvJK*	Ile, Val uptake
		Cp36	Unknown
	None	*lrp*	Regulator

molecules interact with each other when occupying adjacent sites, resulting in cooperative binding; and 3) a cI molecule occupying O_{R2} interacts with RNA polymerase at P_{RM} for transcription activation.

Some regulators act as repressors, but in the presence of effectors become activators. Among them are the LysR, SorC and AraC proteins. Although the exact topography and mechanism can vary in detail, in general these regulators in the unliganded state act as repressors for both of the cognate divergent promoters. When bound to effectors, the regulators become activators of one of the two promoters. The bifunctional role of LysR, the activator that regulates lysine biosynthesis in *E. coli*, is shown in Figure 18.**11**. Binding of the effector lysine to the apoactivator LysR reduces the affinity of LysR to the autoregulatory operator, but allows the interaction with a new region of DNA that acts as an activation site for the divergent promoter of the *lysA* gene.

Regulators and their effectors in various combinations play a more diversified role. The global regulator Lrp (**l**eucine **r**esponsive promoter **p**rotein) and its effector leucine regulate using all six strategies (Tab. 18.**3**).

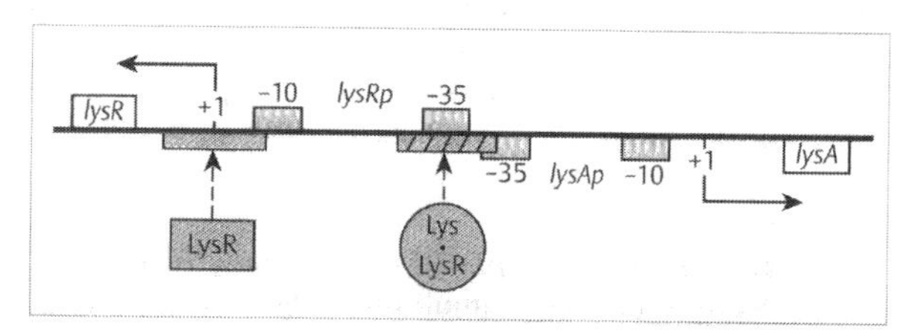

Fig. 18.**11** **Dual control in *Escherichia coli* by LysR of the two divergent promoters.** Transcription of the *lysR* gene is from the leftward promoter *lysRp*. Transcription of the structural gene *lysA* which encodes an enzyme for lysine biosynthesis is from the rightward promoter *lysAp*. The operator (red box) binds LysR (red rectangle), which represses transcription from the leftward promoter, and the second site (red hatched box) binds the Lys·LysR complex (red circle) and activates transcription from the rightward promoter

18.7 Mixed Control: Repressor Inhibits Activator

At least seven promoters in *E. coli* are negatively regulated by a local repressor (CytR) and the global cAMP·Crp regulator acting collectively, but cAMP·Crp, when present alone, acts as an activator. The *deoCp* (P_2) promoter of the *deoC* operon is an example of such a promoter (see Chapter 18.4.2) and is inducible by cytidine (Fig. 18.**12**). cAMP·Crp has two binding sites, A_1 and A_2. Binding at site A_1 is sufficient for activation,

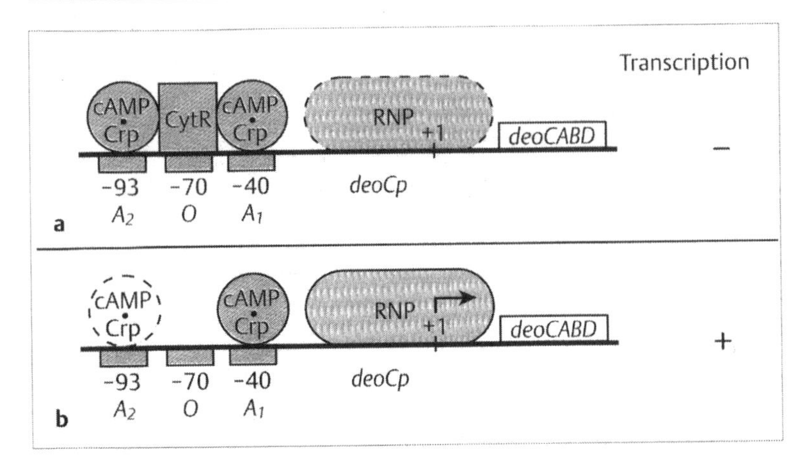

Fig. 18.**12a,b** **The *deoCp(P₂)* promoter of *Escherichia coli*.**
a CytR repressor (red square) exerts negative control by binding to an operator (*O*) located between two cAMP·Crp binding sites (A_1 and A_2).
b In the absence of CytR repressor, cAMP·Crp (red circle) binding to A_1 helps to activate transcription from the promoter. Binding to A_2 is not needed for positive control, but is essential for CytR repressor-mediated negative control. Similar combinatorial control also occurs at the promoters of the *cdd*, *tsx*, and *cytR* operons

but full activation requires binding at site A_2. Repression of *deoCp* requires binding of cAMP·Crp at the two spatially separated cAMP·Crp binding sites and binding of CytR at the CytR binding site in between. CytR physically interacts with the two DNA-bound cAMP·Crp complexes for repression. Mutated Crp proteins that do not contact CytR are unable to repress transcription from *deoCp*. Binding of CytR to DNA is not essential for repression, but provides stability to the repression loop complex. The negative control by CytR occurs by model D, that is, by inhibiting the activity of an activator.

18.8 DNA Structure Also Regulates Transcription Initiation

DNA structure can contribute to the regulation of transcription initiation at two levels: global and local.

18.8.1 Changing DNA Superhelicity Globally Affects Many Promoters Simultaneously

Regulation of transcription by changing DNA superhelicity is an example of the effect of DNA structure at a global level. DNA superhelicity can stimulate, inhibit, or have no effect on the activity of promoters. In general, σ^{70}-RNA polymerase initiates transcription from a large number of promoters more efficiently from supercoiled DNA templates than from relaxed templates by enhancing RNA polymerase binding and/or isomerization. Supercoiled templates may need less free energy for protein binding and strand separation. Addition of some drugs or a mutation that interferes with the activity of enzymes that affect supercoiling changes the activity of many promoters. Two major enzymes that affect DNA supercoiling are DNA gyrase (encoded by the *gyr* genes), which introduces negative supercoils, and topoisomerase I (encoded by *topA*), which removes (relaxes) negative supercoils (see Chapter 14.1.2). Inactivation of DNA gyrase decreases and inactivation of topoisomerase I increases transcription initiation from these promoters. Promoters are also known whose activities are increased by DNA gyrase inactivation and decreased by topoisomerase I inactivation (see Chapter 28.1.3). The physiological significance of DNA supercoiling controlling transcription is underscored by the increase in transcription from the promoter of the gene encoding the relaxing enzyme topoisomerase I (*topA*) with increased DNA superhelicity and the decrease in transcription from the promoters of the genes encoding DNA gyrase (*gyrA* and *gyrB*) under similar conditions. The autoregulation of DNA gyrase and topoisomerase I provides a homeostatic mechanism for maintaining a constant level of supercoiling in bacterial DNA (see Chapter 14.1.2).

18.8.2 Bending DNA Locally Affects Nearby Promoters

Most gene regulatory proteins that bind DNA bend the DNA at points within the region of interaction. Bending is detected by a reduction in the electrophoretic mobility of the bent DNA fragment in native gels compared with the mobility expected for a straight DNA fragment. DNA bending has the potential of influencing the structure of a nearby promoter by either facilitating or hindering the formation of a promoter–RNA polymerase complex of optimal structure (caging). Bending can also help or hinder dispersed DNA-bound proteins to make additional protein–protein or DNA–

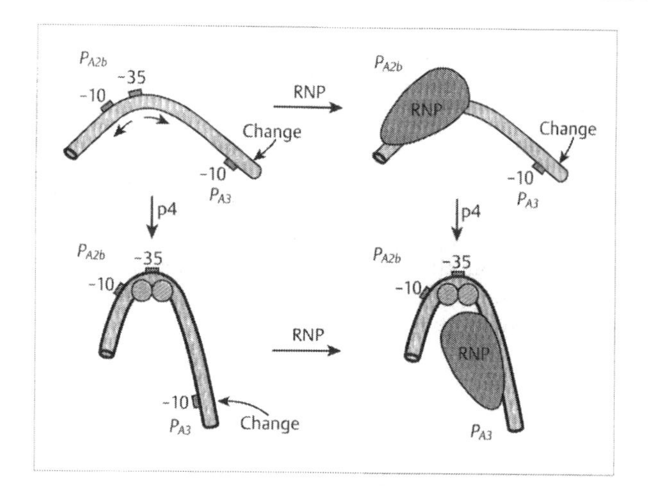

Fig. 18.**13** **Transcriptional regulation by DNA bending.** P_{A2b} and P_{A3} are two divergent promoters in phage ϕ29 of *Bacillus subtilis*. σ^A-RNA polymerase (RNP) recognizes P_{A3} or P_{A2b}, depending on the binding of the regulatory protein p4 (dimer in red). Note the orientation of RNA polymerase at the two promoters. P_{A3} does not have any sequence that resembles the −35 consensus sequence

protein contacts. An activator or a repressor can simply act by inducing a DNA bend. This is illustrated by two divergent promoters: P_{A2b} (an early promoter) and P_{A3} (a late promoter) transcribed by σ^A-RNA polymerase in *Bacillus subtilis* phage ϕ29. A phage-encoded protein p4, with dual behavior, represses P_{A3} and activates P_{A2b} by binding to a single site centered at −79 bp from P_{A3} *tsp* and −43 bp from P_{A2b} *tsp* (Fig. 18.**13**). p4 occupies the same face of DNA as RNA polymerase at P_{A3} and the opposite face of DNA as RNA polymerase at P_{A2b}.

If the promoter DNA must cage RNA polymerase for efficient transcription initiation, protein binding and bending of DNA helps the optimal wrapping for activation, but hinders it for repression.

18.8.3 Methylating DNA Can Locally Modulate Transcription From Nearby Promoters

Methylation of the DNA sequence close to the promoter can regulate transcription initiation. Some promoters contain the GATC sequence, which is methylated at the A residue by the enzyme DNA-adenine methyltransferase, the *dam* gene product in *E. coli* (see Chapter 15.6.1). In transposon Tn*10*, the promoter for the DNA transposase gene which contains GATC sequences, is activated tenfold when nonmethylated or methylated only on one strand (hemimethylated) and repressed when methylated on both strands (fully methylated). The promoter is transcribed in dam^+ cells primarily during the brief period of the cell cycle when the promoter is transiently hemimethylated after passage of the DNA replication fork through the region. In contrast, trans-cription from the *E. coli dnaA* gene promoter, which contains several GATC sites, is repressed when unmethylated or hemimethylated and activated when fully methylated. In this case, the promoter is transiently repressed by the passage of the replication fork. Full methylation of DNA near the promoter of *dnaA* signals transcription initiation in both directions and provides a mechanism of coupling gene regulation to, for example, the replication phase of the cell.

18.9 Regulation Occurs by Altering RNA Polymerase Structure and Function

18.9.1 RNA Polymerases Are Also Encoded by Bacteriophages

For temporal regulation of gene transcription, bacteriophages normally use the major host RNA polymerase holoenzyme to transcribe the early operons (see Chapter 26). An exception is the DNA bacteriophage N4 of *E. coli*, which brings a specialized RNA polymerase with its virion to transcribe the phage DNA. The delayed early and late genes of phages are transcribed with the help of the product of early gene(s). Switching to transcription of the later genes at appropriate times can occur by production of new RNA polymerases. Several bacteriophages (e.g., T3 and T7 of *E. coli*, SP6 of *S. typhimurium*, and SP01 of *B. subtilis*) make late-promoter-specific new RNA polymerases. Interestingly, these RNA polymerases are single polypeptide enzymes of less than 100 000 molecular weight. Their existence suggests that the apparatus required for RNA synthesis can be much smaller than that of the large multisubunit bacterial holoenzymes. Unlike the major holoenzyme, these RNA polymerases recognize a contiguous DNA sequence in the promoter and synthesize RNA faster (~200 nucleotides/s), consistent with the rapid life cycles of the bacteriophages.

18.9.2 Replacement of the σ-subunit Alters RNA Polymerase Specificity

The presence of multiple σ-subunits are common in prokaryotes (see Tab. 25.**1** and 28.**3**). The σ-subunits interact with the common core enzyme to generate a family of RNA polymerases. Because σ-factors are specific for DNA binding, the resulting holoenzymes are promoter-specific. The holoenzyme containing the major or vegetative σ-factor (σ^{70} in *E. coli* and σ^{43} in *B. subtilis*) transcribes the vast majority of the genes in the organism. The holoenzymes with the minor σ-factors transcribe discrete sets of operons and regulons in response to specific physiological requirements. The σ-factors, although of different sizes, have strong sequence similarity, suggesting a common ancestor. An exception is the σ^{54} of enterobacteria, which has no similarity to the other σ-subunits, suggesting that some structural aspect rather than homology is important for the contact to the common core enzyme. The minor σ-factors recognize specific −35 and −10 promoter elements. By changing the σ-subunit, which is an effective way to generate different RNA polymerases, the cell regulates two types of operons. One type encodes proteins that allow the cell to tackle specific stress situations or environmental changes (e.g., heat or chemical shock and nutritional starvation; see Tabs. 20.**1** and 28.**3**). The other type encodes proteins needed for developmental (differentiation) functions [e.g., in bacterial organelle (flagella) or phage development and bacterial sporulation; see Chapters 22–28 for numerous examples].

18.9.3 Covalent Modifications of RNA Polymerase Switch Transcription From Host to Phage Genes

RNA polymerase of host bacteria can be modified after bacteriophage infection to switch the cell away from expression of the host genes to expression of the phage genes. Phage T4 infection destabilizes the attachment of σ^{70} to the core enzyme by phage-encoded proteins that ADP-ribosylate the α-subunit and phosphorylate the β- and β′-subunits to modify or turn off host RNA synthesis at an appropriate stage of the phage life cycle.

18.9.4 Alteration of RNA Polymerase by ppGpp

Under conditions of amino acid starvation, accumulation of ppGpp redirects the cellular resources from the wasteful transcription of rRNA, tRNA, and ribosomal-protein operons to those encoding the enzymes of amino acid biosynthesis (see Chapter 20.4).

Modifying the RNA polymerase allows genes to be temporally regulated in an orderly fashion with a high degree of promoter specificity and efficiency. Covalent- or effector-induced modification turns off unneeded promoters and concentrates on useful transcription.

18.10 Transcription Elongation and Termination Are Also Regulated

It was initially assumed that RNA polymerase, once leaving the promoter as an elongating complex, proceeds uninterrupted and stops only at the termination signal at the end of a transcription unit (operon). However, the path of RNA polymerase is not smooth and is punctuated by pauses and premature termination at discrete sites. Both pausing and termination of the elongating complex can be influenced, for regulatory purposes, by other proteins whose target of action is RNA, not DNA.

Regulation of transcription termination (see Chapter 15.2.2) serves several purposes: 1) to delineate trans-cription units (operons), 2) to modulate transcription of an operon in response to excess gene products (**attenuation**), 3) to decrease the levels of transcription of successive genes within one transcription unit (**natural polarity**), 4) to terminate transcription if translation is prematurely stopped (**mutational polarity**), 5) to bring separate operons, some of which may be promoterless, into one transcription unit (**antitermination**), and 6) to protect the transcription of relatively long untranslated RNA from premature termination.

18.10.1 Intraoperonic Transcription Termination and Polar Effects

Nonsense and frameshift mutations characteristically have **polar effects** in operon expression. Such mutations inactivate the cistrons in which they occur and reduce or abolish the expression of the cistrons promoter-distal to the site of the mutation in polycistronic operons. This phenomenon is termed **mutational polarity**. Polarity results from transcription termination due to the

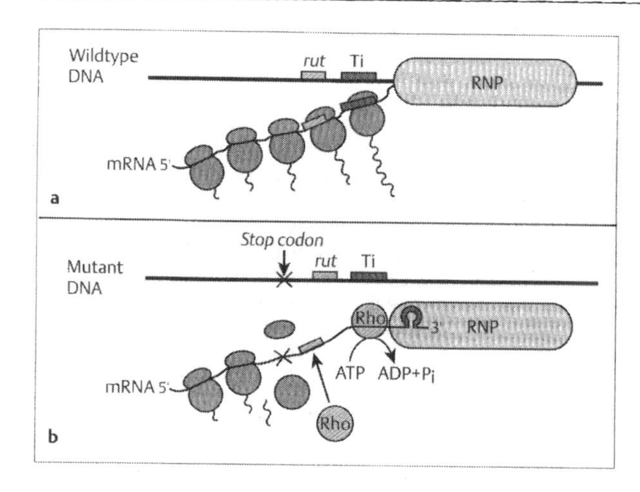

Fig. 18.**14a,b** **A model for polarity.**
a Coupled transcription and translation.
b Release of ribosomes (shaded dark gray) from an mRNA due to the presence of a nonsense mutation (stop codon; x) allows Rho (red circle) to bind to the nascent RNA using a *rut* site (red box). Upon ATPase-mediated tracking along the unmasked mRNA, Rho encounters RNA polymerase (RNP) paused at an intragenic transcription terminator (Ti) hairpin in RNA, causing premature termination and preventing transcription of distal genes of the same operon. Such premature termination does not occur if there is a mutation that makes Rho defective. Other proteins called NusA or NusG may play a role in transcription termination by influencing the amount of pausing of an elongating RNP at discrete positions

presence of *rut* sites and Rho-dependent t_i signals within the operon. A t_i signal is latent and becomes active when translation of the region is prematurely halted at a stop codon and uncoupled from transcription. This unmasks the nascent RNA (*rut* segment) preceding a t_i site and permits access of Rho to act on the elongating RNA polymerase (Fig. 18.**14**). Mutational polarity in transcription is released by mutations in the *rho* gene. A premature stopping of translation is not essential for activation of the latent intraoperonic termination sites. If ribosome loading occurs at a low frequency, it results in an extension of the spacing between the elongating RNA polymerase and the trailing ribosome and thus allows Rho access to RNA for causing termination. The process is reinforced by degradation of the unmasked mRNA. This phenomenon is **natural polarity** (see also Chapter 15.5).

18.10.2 Regulation During Promoter Clearance by RNA Polymerase Stuttering

A novel regulatory mechanism has been discovered in two bacterial operons, *pyrBI* and *galETKM* in *E. coli*. The expression of these operons is regulated by intracellular UTP concentrations. UTP reduces transcription from the *pyrBI* promoter and the *galEp2* promoter by causing the stuttering of RNA polymerase during promoter clearance. The initial RNA sequence at *tsp* of these two promoters is $ppp(A)_{1-2}(U)_3 \ldots$ At high UTP concentrations in vitro, RNA polymerase stutters at the uridine clusters and ends up making no real transcript, but a large amount of nonproductive pseudotemplated RNA products of the sequence $ppp(A)_{1-2}(U)_n$, where $n \geq 20$. At low UTP concentrations, transcription elongates in a regular fashion without stuttering. The importance of a "negative" control by UTP at *pyrBI*, which encodes enzymes for the synthesis of UTP precursors, is obvious; in *galETKM*, the negative control is perhaps correlated with the involvement of uridine residues as intermediates in the metabolism of galactose (e.g., UDP-galactose). Regulation of gene expression by changing the concentration of nucleotide pools may be more common than is known.

18.10.3 Regulation by Transcription Attenuation Controls Elongation and Termination

Regulation by attenuation of transcription at a specific termination site located between the leader RNA and the first structural gene occurs in both anabolic and catabolic operons. Both type I and type II terminators (Chapter 15.2.2) participate in attenuation control. Trans-cription of the entire operon occurs only if the attenuation signal is overcome.

Amino acid operons. Although amino acid operons vary in details, many use attenuation control mechanisms. The general concept is shown in Figure 18.**15** using the paradigm *trp* operon of *E. coli*. The leader transcript can form mutually exclusive RNA hairpin structures, one of which (3 : 4 pairing) is the type I terminator. Although the formation of the 1 : 2 hairpin allows the formation of the 3 : 4 hairpin, the formation of the 2 : 3 hairpin precludes 3 : 4 formation. The leader peptide contains "control codons" that specify the amino acid end product of the enzymes encoded in the operon (e.g., Trp codons in the leader peptide of the *trp* operon). When the supply of the amino acid and thus of the cognate aminoacyl-tRNA is deficient, the ribosome stalls at the control codons. The leader RNA then forms the pre-emptor 2 : 3 hairpin, allowing transcription through the attenuation signal. If the aminoacyl-tRNA is abundant, the ribosome stops at the end of the leader peptide, thereby prohibiting formation of the 2 : 3 hairpin and allowing formation of the 3 : 4 hairpin and thus attenuation. As expected, mutations in the genes

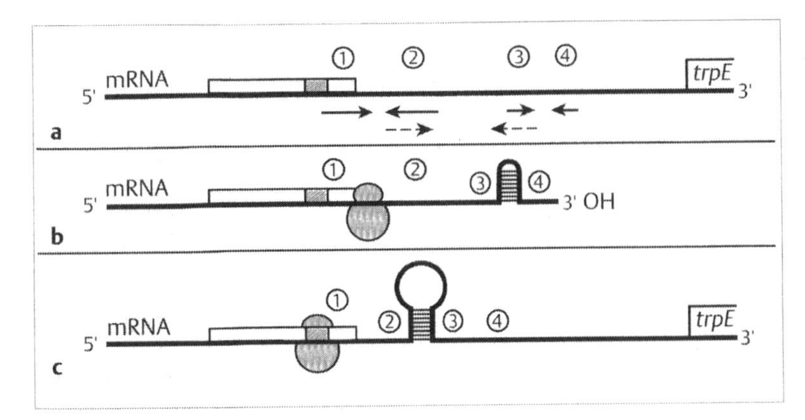

Fig. 18.**15a–c** **Regulation by transcription attenuation in *Escherichia coli*.**
a The 5′-end of the *trp* mRNA containing the leader region is shown. Open bar, leader peptide; red bar, control codons. The arrows indicate the regions (1–4) that form various potential RNA hairpin structures.
b If ribosomes (shaded) complete the translation of leader peptide (i.e., there is sufficient charged $tRNA^{Trp}$), they allow 3 : 4 terminator hairpin formation.
c If ribosomes stall at the control codons because of the lack of charged $tRNA^{Trp}$, they permit 2 : 3 hairpin formation, which precludes attenuation

for aminoacyl-tRNA synthetases and tRNAs specific for the control codons result in increased transcription through the attenuation signals, and mutations that affect leader-peptide synthesis cause increased attenuation. Base change or deletion mutations at key positions that alter the relative stability of the hairpins also derange the attenuation control in the expected manner. The control codons in the leader peptides of various amino acid operons are shown in Table 18.**4**.

Replacement of the set of control codons in a given amino acid operon by a set of another amino acid codons removes regulation in response to the starvation of the original amino acid. The new leader regions may even show regulation in response to starvation of the amino acid corresponding to the new control codons. Note that the leader peptide of an operon that encodes enzymes for the synthesis of more than one amino acid (by a branch pathway, e.g., *thr* and *ilvGMEDA*) contains control codons for all of them. Among the operons subject to attenuation control, some (e.g., *trp*) are still subject to transcription initiation control by a repressor; others (e.g., *his*, encoding histidine biosynthesis) are not.

Antitermination relies on the secondary structure of the leader RNA acting in *cis*, which is manifested by the act of translation of the leader peptide present in these operons and not by the peptide product. The movement of a translating ribosome determines the formation of alternative RNA hairpin structures that variously promote or preclude transcription termination at the attenuation site.

Pyrimidine biosynthesis. Attenuation control also occurs in bacterial pyrimidine biosynthetic operons. In the *pyrBI* operon in *E. coli*, the extent of coupling between transcription and translation in the leader

Table 18.**4** **Control codons in leader peptides of amino acid biosynthetic operons in *Escherichia coli***

Operon	Amino acid sequence of leader peptides	Controlling amino acids
trp	Met Lys Ala Ile Phe Val Leu Lys Gly Trp Trp Arg Thr Ser	Trp
thr	Met Lys Arg Ile Ser Thr Thr Ile Thr Thr Thr Ile Thr Ile Thr Thr Gly Asn Gly Ala Gly	Thr, Ile
his	Met Thr Arg Val Gln Phe Lys His His His His His His His Pro Asp	His
ilvGDMEA	Met Thr Ala Leu Leu Arg Val Ile Ser Leu Val Val Ile Ser Val Val Val Ile Ile Ile ... ... Pro Pro Cys Gly Ala Ala Leu Gly Arg Gly Lys Ala	Ile, Val, Leu
leu	Met Ser His Ile Val Arg Phe Thr Gly Leu Leu Leu Leu Asn Ala Phe Ile Val Arg Gly Arg Pro ... ... Val Gly Gly Ile Gin His	Leu
PheA	Met Lys His Ile Pro Phe Phe Phe Ala Phe Phe Phe Thr Phe Pro	Phe
ilvBN	Met Thr Thr Ser Met Leu Asn Ala Lys Leu Leu Pro Thr Ala Pro Ser Ala Ala Val Val Val ... ... Val Arg Val Val Val Val Val Gly Asn Ala Pro	Val, Leu

region determines whether the attenuation hairpin is formed. The *pyrBI* leader RNA does not have any other hairpin structures, but it encodes a leader peptide consisting of 44 amino acids and contains a stretch of 8 essential uridine residues positioned 20 nucleotides upstream of the termination site. At low UTP concentrations, RNA polymerase pauses at the region of uridine clusters. This provides sufficient time for a ribosome to initiate synthesis of the leader peptide and to catch up to the stalled RNA polymerase, thereby establishing coupling. The coupled translation–transcription helps to disrupt the terminator hairpin during eventual escape of RNA polymerase from the uridine clusters, and allows transcription through the terminator. In contrast, at high UTP concentrations, RNA polymerase transcribes through the uridine stretch without a pause, thus allowing the RNA polymerase to escape from coupling and to form the terminator hairpin. Ribosomal gene mutations that reduce the rate of translational elongation proportionately reduce expression of *pyrBI* even under conditions of UTP limitation. Interestingly, the leader peptide in the *pyrE* operon in *E. coli* is 238 amino acids long and is functionally the tRNA processing enzyme RNase pH. In effect, this operon is bicistronic, with the attenuation signal located between the two cistrons, the first also serving the leader peptide role.

Along with stuttering control (Chapter 18.10.2), attenuation control appropriately ensures the expression of the pyrimidine biosynthetic genes in amounts commensurate with the intracellular UTP concentrations.

Aminoacyl tRNA synthetases. When cells are starved for a specific amino acid, the corresponding synthetase level increases. This permits more efficient use of the residual amount of the limiting amino acid so that the cell can quickly recover by derepressing genes encoding the required amino acid biosynthetic enzymes. Although the synthetase genes in *E. coli* and its close relatives are regulated by a variety of mechanisms, including attenuation control as in amino acid operons, most synthetase genes in *B. subtilis* and very likely in *Lactobacillus casei* are regulated by a common attenuation mechanism in which extensive secondary structures in the leader RNA participate in the decision to terminate or not. Despite considerable variation in the primary sequences of the different synthetase leader RNA, the region exhibits striking conservations in the secondary structures. Such structures are illustrated in Figure 18.**16** using the example of tyrosinyl-tRNA synthetase. The signal responsible for responding to the cognate amino acid limitation is its codon present (specifier sequence) at an identical position in each leader structure. The regulatory mechanism requires the cognate tRNA as an effector, although the "control" codon is not translated. When the amino acid level is low, the level of the corresponding free tRNA is high. The free tRNA interacts with the nascent leader RNA by a codon–anticodon interaction at the specifier sequence (and probably at other sites), resulting in a structure that favors antitermination at the attenuator signal. When the amino acid level is high, tRNA is mostly charged and unable to interact appropriately with the leader, allowing the formation of the terminator structure. Changing the control codon to that of a different amino acid changes the specificity of the response, and the introduction of a nonsense codon at the position results in a noninducible phenotype.

The attenuation mechanism illustrates the use of codon–anticodon interaction for a purpose other than deciphering the genetic code.

Tryptophan degradation. Attenuation control also occurs in catabolic operons and in this case, simply follows the principle of mutational polarity. The transcription of the catabolic *tnaAB* operon in *E. coli*, encoding enzymes of tryptophan degradation, is induced by tryptophan. The attenuation site in *tna* is a Rho-dependent type II terminator signal with no strong hairpin structure in the leader RNA. A 319-nucleotide leader RNA contains a leader peptide consisting of 24 amino acids with a single codon for tryptophan at a critical position. In the absence of tryptophan, ribosomes stall at this position, allowing attachment of Rho to RNA and subsequent termination. Excess tryptophan induces *tnaAB* expression 100-fold by preventing Rho access and termination.

The mechanisms by which transcription elongation is regulated by stuttering and attenuation are simple and originate entirely from the DNA primary structure. Only the normal machinery of transcription and translation is used and no special regulatory protein is required.

18.10.4 Antitermination of Transcription That Requires Regulatory Proteins

Regulation of transcription by antitermination at many discrete termination sites located within large operons uses regulatory proteins. Factor-dependent antitermination operates in operons, such as in those encoding

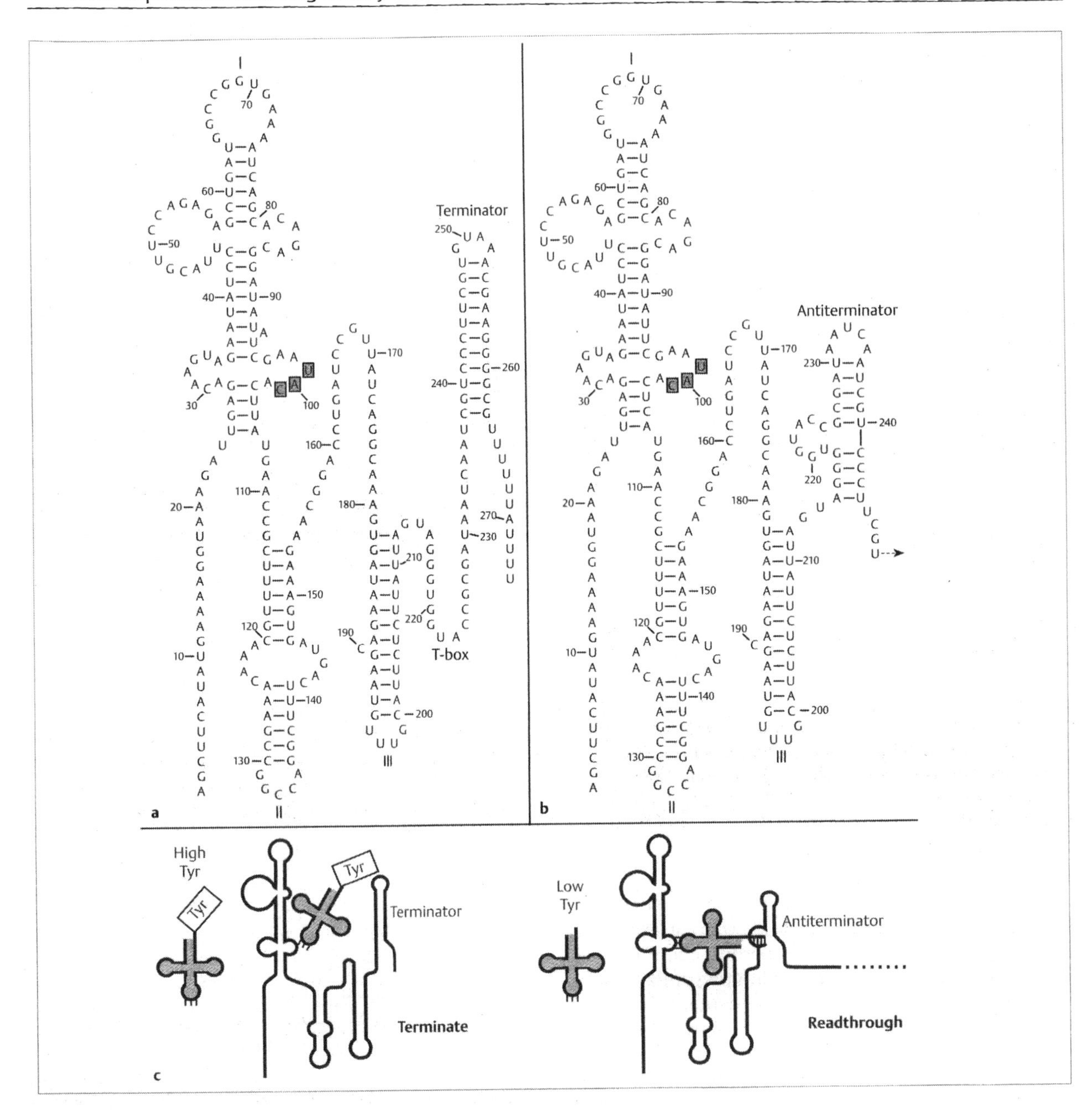

Fig. 18.**16a–c Structure of the tyrS leader region of *Bacillus subtilis*.**
a The terminator and
b the antiterminator forms are shown. Numbers designate positions relative to the *tsp*. Roman numerals indicate stem–loop regions I, II, and III. Between stem loops II and III is a perfectly conserved CGUUA sequence (positions 167 to 171 in *tyrS*) that is often found in a region of potential secondary structure. There is a highly conserved CAGAGA sequence at positions 52 to 57 of *tyrS*. The boxed residues in red indicate the specifier sequence. The translation start point is at position +298, downstream of the transcription terminator.
c Model for induction of *tyrS* by uncharged tyrosyl-tRNA. The red cloverleaf structure represents tyrosyl-tRNA uncharged or charged with tyrosine (box). The last stem–loop structure indicates the terminator structure (top) or the antiterminator structure (bottom). When tyrosine levels are high, most of the tyrosyl-tRNA is charged and unable to interact appropriately with the leader mRNA; therefore, the terminator structure forms and transcription terminates. When tyrosine levels decrease, the level of uncharged tyrosyl-tRNA increases. Uncharged tyrosyl-tRNA interacts with the nascent transcript by codon–anticodon interaction at the specifier sequence. This interaction results in formation (or stabilization) of the antiterminator structure, which results in readthrough of the terminator [after 3]

enzymes of β-glucoside degradation and purine biosynthesis in *E. coli* and of sucrose utilization in *B. subtilis*. In lambdoid bacteriophages, sophisticated antitermination mechanisms have evolved to handle such processes (see Chapter 26).

Every stage of transcription after the first phosphodiester bond formation can be regulated, having biological consequences. In such regulation, transcription elongation is coupled to different metabolic or genetic signals.

18.11 Genes Can Also Be Turned on and off by DNA Recombination

Use of site-specific reciprocal recombination to rearrange DNA primary structure for regulation of gene expression is another example of nature's unlimited ways to achieve developmental goals. Whereas a mutation that occurs at a frequency of about 10^{-6} or less can permanently switch genes on or off, DNA rearrangements that occur at frequencies of 10^{-5} to 10^{-2} can turn genes on and off and serve to preadapt the organisms to changes in their environment or to respond to long-term developmental changes in their life cycles.

18.11.1 Inversion of a Promoter: Phase Variation in *Salmonella*

Flagella responsible for bacterial motility normally change between two antigenic types in *Salmonella*, a process called **phase variation** (see Chapter 33.4.1). The variation of antigenicity allows *Salmonella* to evade the host immune response. The genes encoding these two mutually exclusive major flagellin antigenic proteins, HagA and HagB (known by various names in different strains of *Salmonella*) are located on different regions of the chromosome.

Phase variation occurs, depending on the *Salmonella* strain, at a frequency of 10^{-5} to 10^{-3} per bacterium per generation. HagR is a *trans*-acting repressor of the *hagA* gene. Its own gene *hagR* forms an operon with the *hagB* gene (Fig. 18.**17**). Thus, the coordinated synthesis of HagB and HagR results in turning off *hagA*. In the alternate phase, the controlling element in the *hagB* locus turns off the expression of *hagB* and *hagR*, derepressing the synthesis of HagA flagellin from the *hagA* gene. The "controlling element" of the *hagB* locus is an invertible DNA segment that contains the promoter for the *hagBR* operon and an independently expressed separate gene (*hin*) whose product Hin catalyzes the inversion of the segment. In one orientation, the segment fuses the promoter to *hagBR*, and in the opposite case, it orients the promoter in the opposite direction, explaining phase variation.

Hin-protein-mediated inversion occurs by a site-specific recombination between two inverted repeat sequences, *hixL* and *hixR*. The recombination is enhanced by binding of a protein (Fis) to a recombination enhancer site. Fis bends the DNA and helps to form the DNA–multiprotein recombination complex with appropriate DNA–DNA, DNA–protein, and protein–protein alignments. Mutations in *hin*, *hixL*, or *hixR* lock the flagellin expression in the HagA or HagB mode.

Like Hag, type I pili expression from the *fimA* gene in *E. coli* is controlled by an invertible promoter to switch cells between fimbriates and nonfimbriates. Type I pili are responsible for making virulent strains that attach to epithelial cells, erythrocytes, and leukocytes. However, unlike the *hin* system, the recombination events leading to the on state and off state are catalyzed by different proteins, FimE and FimB, respectively (see Chapter 33.4.1)

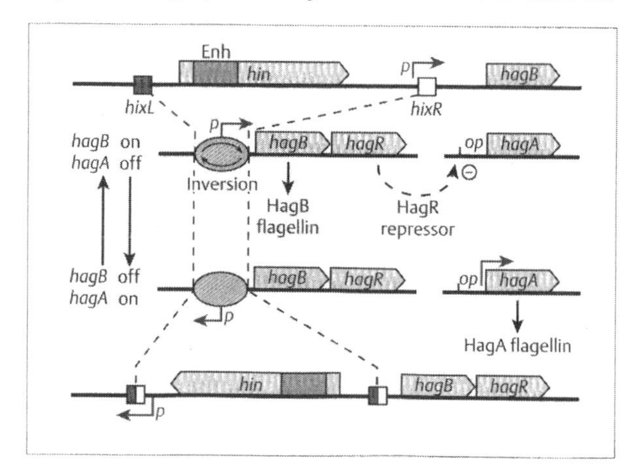

Fig. 18.**17** **Turning genes on and off by DNA rearrangement.** During phase variation in *Salmonella*, expression of the *hag* genes is controlled by recombinational inversion (switch) of a promoter (see text for details; Enh, enhancer; *p*, promoter; *o*, operator)

18.11.2 Host Range Alteration in Bacteriophages by Inversion of Structural Genes

To increase the chances of survival, bacteriophages Mu, D108, and P1 have homologous invertible DNA regions that control their host range and help the organisms to

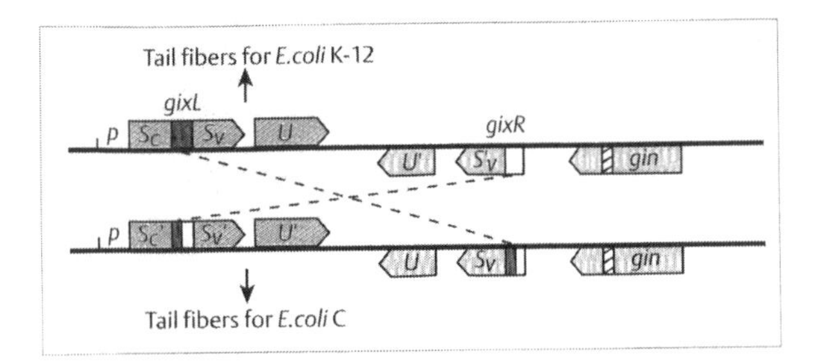

Fig. 18.**18** **Host range changes in bacteriophage Mu.** By recombinational inversion of the structural genes (see text for details), the host range of phage Mu is changed

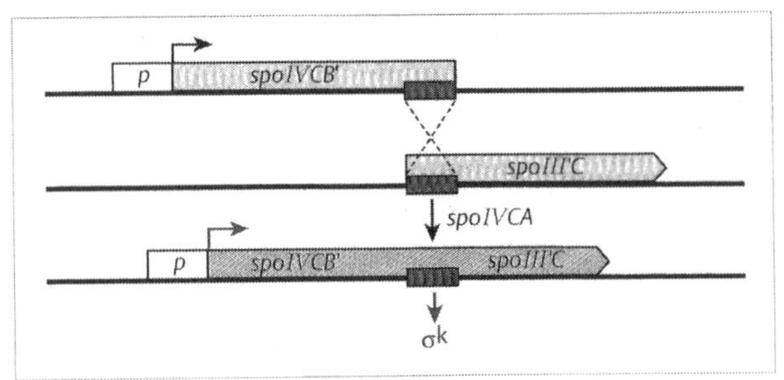

Fig. 18.**19** **Sigma subunit synthesis in sporulating *Bacillus subtilis*.** Sigma synthesis requires a recombinational (dotted "X") creation of a full structural gene from two split halves catalysed by SpoIVCA. The fused *spoIVCB'–spoIII'C* DNA segment encodes σ^K (see text for details)

preadapt part of the phage population to changes in the environment. The inversion is carried out by a recombinase. For example, phage Mu in one phase adsorbs to *E. coli* K-12 using lipopolysaccharide receptors; it no longer infects *E. coli* K-12 in the other phase, but infects such organisms as *E. coli* C, *Citrobacter*, *Shigella*, and *Serratia*. In **altering host range** by DNA inversion, the invertible segment does not contain the promoter, but rather two sets of host range tail fiber genes. The two orientations of the invertible DNA next to the promoter determine which set of genes is transcribed. In one orientation, the tail genes of phage Mu that help adsorb *E. coli* K-12 are expressed; in the other orientation, the genes specific for the other enteric bacteria are made (Fig. 18.**18**).

Interestingly, the recombinase Gin in phage Mu, Pin in phage P1, and Hin in *Salmonella* can cross-complement and have considerable similarities in amino acid sequences.

18.11.3 Sigma Subunit Synthesis During Sporulation in *B. subtilis* by Joining of Split Genes

Part of the sporulation-specific genes are transcribed in the mother cell of sporulating *B. subtilis* cells by σ^K–RNA polymerase (see Chapter 25.1.3). The σ^K gene is split in the bacterial chromosome into two separate segments, *spoIVCB* and *spoIIIC* (Fig. 18.**19**). The two split-gene segments are 10 kb apart and juxtaposed together in frame by a site-specific recombination between two directly repeated sequences at an intermediate stage of sporulation. The process is catalyzed by SpoIVCA. The recombinational event cannot be reciprocal or reversible, as the mother cell and its chromosome are discarded at the end of sporulation. σ^K is involved in the synthesis of cortex and coat structures of the prespore. Because protective layers of coats and cortex effectively seal off the prespore from the mother cell, the timing of σ^K synthesis is crucial, and the cell cannot afford leaky σ^K synthesis—thus, the split gene arrangement.

Several microbial developmental systems in terminal cells, e.g., fruiting body formation in myxobacteria, aerial mycelium formation in streptomycetes and heterocyst formation in *Anabaena* (see Chapters 24 and 25), are likely to employ recombinational control of gene expression. Some genetic systems concerning bacterial virulence and pathogenicity also use recombination to control gene expression (see Chapter 33.4.1).

Programmed **site-specific recombinational events** are used to turn genes on or off by either splitting up or bringing together promoters and structural genes or even different segments of the same structural gene. The mechanism is characteristic of irreversible processes that often occur in cellular differentiation.

18.12 Gene Expression Is Regulated by RNA Metabolism

One of the original models for regulation of induced enzyme synthesis involves constitutively produced, but rapidly degraded, mRNA. The presence of inducer prevents mRNA degradation, allowing its translation. The predominant control in prokaryotes occurs, nevertheless, not at the level of mRNA degradation, but at the level of mRNA synthesis (see Chapter 15.2.3).

Although mRNAs are degraded to prevent unnecessary translation after transcription has stopped, nature has utilized specific types of RNA degradation (**processing**) to regulate gene expression.

Prokaryotic mRNAs are translated mostly in the same form as they are made, with a few interesting exceptions that occur in the transcripts of some bacteriophage genes and of bacterial stable RNA genes in which processing of the RNAs by a ribonuclease (RNase III) is critical for their final usable form. The RNase III cleavage sites lie in an unpaired bubble contained within a hairpin structure.

18.12.1 Processing Is Needed to Make rRNA and tRNA From Larger Precursor RNAs

The stable RNAs in bacteria are transcribed as larger units. For example, a single RNA molecule carrying 23S, 16S, and 5S rRNA as well as more than one tRNA sequence is transcribed by antitermination control from the *rrn* operons (Fig. 18.**20**). The larger molecule is cleaved by RNase III to generate the individual species, which mature by further processing to rRNAs and tRNAs.

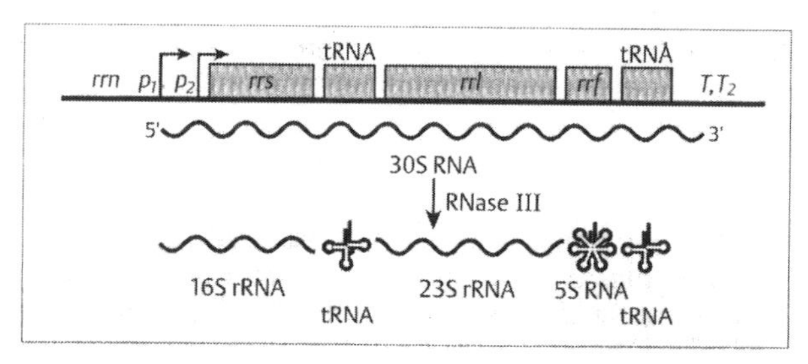

Fig. 18.**20** **RNA processing.** There are seven stable RNA operons (*rrn*) in *Escherichia coli*. Each operon contains genes for 23S rRNA (*rrl*), 165rRNA (*rrs*), 5SRNA (*rrf*) and tRNAs. The functional cistron-like RNA sequences are separated from each other by spacer regions and are transcribed as one RNA, which is processed by RNase III by cleaving at specific sites to produce mature functional rRNA, tRNA, and other molecules. Promoters (P_1, P_2) and terminators (T_1, T_2) are indicated

18.12.2 Processing of mRNA in Bacteriophages

Transcription of the early genes of bacteriophage T7 by host RNA polymerase results in a polycistronic mRNA encoding the phage early genes. Before mRNA can be translated, it is processed into several smaller species by RNase III. In contrast, RNase III processing destroys the *int* gene mRNA in phage λ (see Chapter 26.4.3). mRNA degradation can regulate gene expression in opposite directions by releasing mRNA from the clamp of secondary structures.

18.12.3 mRNA Splicing Is Rare in Prokaryotes

Introns are DNA sequences within a cistron that are missing in mRNA that is translated. Although splicing to remove introns in the primary RNA product is usual in eukaryotes, introns are not found in eubacteria, with a few rare exceptions. The genome of bacteriophage T4 has several introns that are removed from the mRNA by splicing during the phage growth. A G residue is added to the 5′-end of the intron mRNA during splicing in vitro. A single intron interrupts the T4 thymidylate synthetase (*td*) gene. The intron is removed from *td* mRNA by a self-splicing reaction in vitro. It is likely that splicing reactions require *trans*-acting protein factors of bacterial or phage origin in vivo.

18.13 Translation Is Often Regulated

Translation in prokaryotic systems is regulated. Translational control can occur at the initiation, elongation, or termination level. Whereas in operon and regulon arrangements, transcriptional control usually helps coordinate expression of a group of structural genes, translational control adds further tuning by making a difference in the amount of individual gene products. Translational control sometimes selects the reading frame of translation.

18.13.1 Translation Can be Repressed by Proteins or Antisense RNA

Translational repression occurs by either a protein or an RNA ("antisense RNA") molecule binding to the mRNA; this binding hinders the binding of the ribosome to the ribosome binding site (Rbs) either sterically or by inducing a conformational change in the RNA. Table 18.**5** shows examples of translational repressor proteins.

Table 18.5 **Translation repression**

Translation repressor proteins	Target	Site of binding
R17 coat	R17 replicase gene	Ribosome binding site (Rbs)
T4 DNA polymerase	T4 DNA polymerase	Ribosome binding site (Rbs)
T4 RegA	Early T4 genes	Initiating ATG
Ribosomal (R) proteins	R-protein genes	Initiating ATG
T4 gp32	T4 gene 32	5′ leader RNA, including the initiating ATG

Ribosomal (R)-protein operons. R-protein genes (*rps* and *rpl*) are contained in several operons, each of which has genes with assorted functions (Fig. 18.**21**). In each case, the regulatory R-protein binds to its own mRNA at a specific sequence, which is contiguous with the Rbs of the first R-protein gene of the regulated subset. The sequences of the specific sites in mRNA are similar to their corresponding binding sites in rRNA. In this way, the autoregulation of the synthesis of R-proteins is linked to cell growth control using rRNA synthesis as an intermediate. When rRNAs are made commensurate with a high growth rate, the regulatory R-proteins associate preferentially with the rRNAs because the affinity of the regulatory R-proteins for rRNA is higher than that for their binding sites in mRNA, leaving no free pool of R-proteins. When all the rRNA has been assembled into ribosomes, R-proteins accumulate, allowing the regulatory R-proteins to bind to the mRNA to repress translation further. By tying up the level of rRNA to the regulation by R-protein at the level of translation, the cell maintains a needed harmony. The exact molecular mechanism by which the synthesis of a subset of R-proteins, but not the synthesis of other proteins present in the same operon, is repressed by the autoregulatory mechanism is unknown. This strategy of translational feedback makes at least the nonribosomal proteins encoded in the same operons free from the constraint of growth control, but subject to other types of regulation (Fig. 18.**2**).

A feature common to the R-protein operons is **autoregulation** of a subset of R-protein genes within the same operon by one of the R-proteins in that operon.

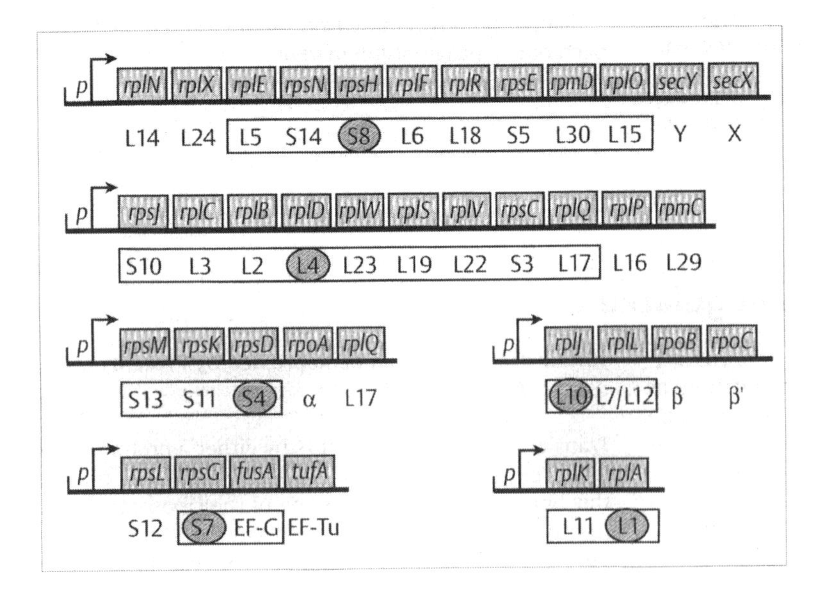

Fig. 18.**21** **Operons containing ribosomal (R)-protein genes in *Escherichia coli.*** A specific R-protein product (red) in each operon autoregulates the synthesis of itself and several contiguous products (boxed) at the level of translation

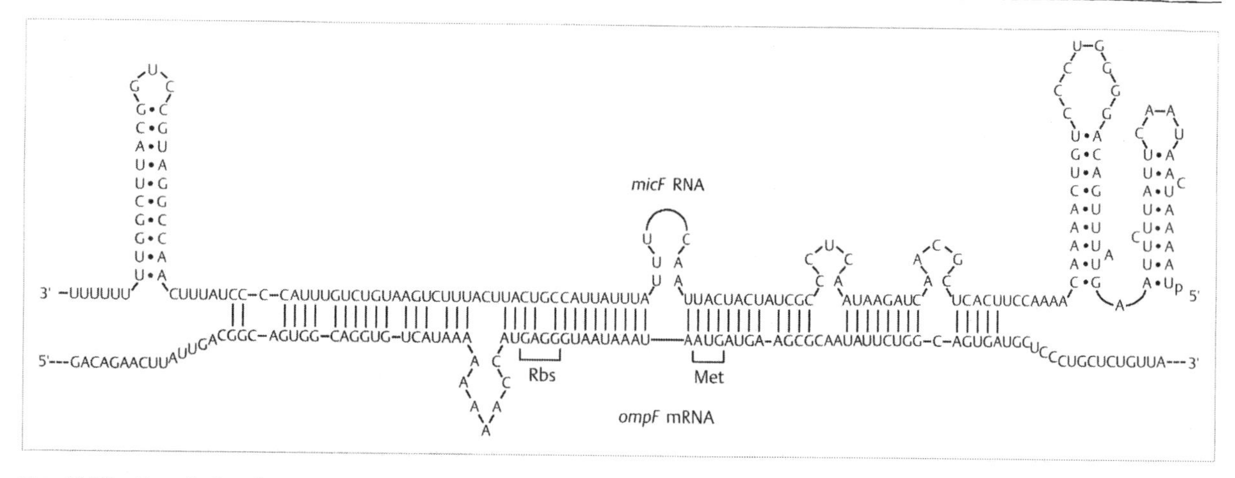

Fig. 18.**22** **Translational repression by antisense RNA in *Escherichia coli*.** The *micF* RNA (translational repressor, shown in red) is strongly complementary to the ribosome binding site–AUG region of *ompF* mRNA. The hybrid prevents translation of *ompF* mRNA

T4 proteins. Some translational repressors are of bacteriophage origin. T4 phage protein gp32 is a single-stranded DNA binding protein that binds to its own mRNA specifically rather than to other cellular mRNA and inhibits (autoregulates) translation of gp32 mRNA. This is an example of translational autoregulation. Normally, most gp32 is bound to ssDNA in cell. When ssDNA is not available, gp32 binds to the next structure in the hierarchy, i.e., its own mRNA, and represses its synthesis because the need is no longer there.

Antisense RNAs. Small RNA molecules, e.g., *micF* RNA, also can act as translational repressors. Like repressor proteins, the repressor or "**antisense RNA**" molecules are products of independent genes and act by binding to the 5′-end of mRNA using the complementary sequences. The area of binding encompasses Rbs, making the Rbs inaccessible to ribosomes. Such a mechanism is observed in the translational control of the *ompF* gene of *E. coli* (Fig. 18.**22**). When the translational repressor, called *micF* RNA, is not pairing with mRNA, the *micF* RNA assumes a secondary structure that triggers its cleavage. Following the principle of *micF* RNA, complementary RNA, made in the cell from artificial genes, has been used to turn off genes in eukaryotic cells.

18.13.2 Translational Attenuation Is Involved in Antibiotic Resistance

Control of gene expression by what may be termed translational attenuation has been observed in the induction of resistance to the macrolide, lincosamide, and streptogramin type B (MLS) antibiotics in Gram-positive pathogens, such as *Staphylococcus aureus*, *Streptococcus faecalis*, and *Streptomyces*. These organisms develop resistance to the MLS antibiotics by modification of the 23S rRNA, which reduces the critical affinity between the antibiotics and the 50S ribosome. In *S. aureus*, resistance to the antibiotic erythromycin is conferred by induced synthesis of an RNA methylase from a plasmid-borne gene (*erm*) in the presence of low concentrations of erythromycin. RNA methylase makes the cells resistant to very high concentrations of erythromycin by *N6*-dimethylation of the 23S rRNA.

The leader mRNA of the *erm* operon encodes a leader peptide of 19 amino acids and also contains four

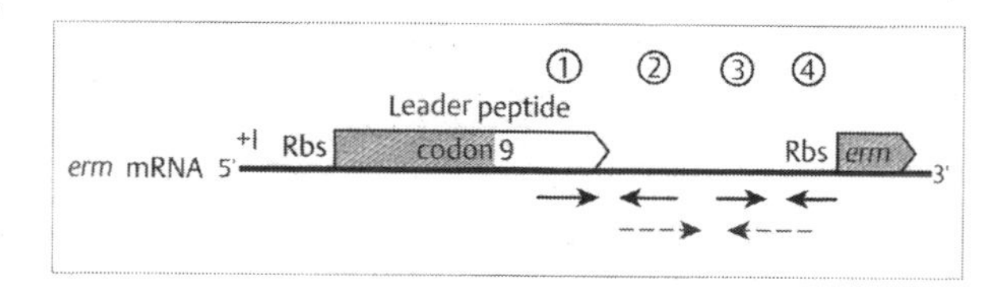

Fig. 18.**23** **A model for translational attenuation in *Staphylococcus aureus*.** The leader RNA segment for the gene (*erm*) for inducible erythromycin resistance in *Staphylococcus aureus* is shown with respect to the positioning of the leader peptide and segments 1, 2, 3, and 4, which form alternate RNA hairpins, indicated by black (non-induced) and red broken (induced) arrows. Formation of the 3 : 4 hairpin occludes *erm* translation. In the presence of erythromycin, the antibiotic–ribosome complex stalls at codon 9, permitting the pairing of segments 2 and 3, and allowing translation (Rbs, ribosome binding site)

complementary sequences 1, 2, 3, and 4, in which 1 is complementary to 2, 2 to 3, and 3 to 4 (Fig. 18.**23**). In the absence of the inducer erythromycin, ribosomes translating the full-length leader peptide cover both segments 1 and 2, allowing the formation of 3 : 4 hairpin. The 3 : 4 pairing sequesters the Rbs and the initiating ATG codon, making the mRNA unavailable for translation initiation. In the presence of inducing concentrations of erythromycin, the antibiotic–ribosome complex stalls at a specific sequence at codon 9 of the peptide, permitting the pairing of segments 2 and 3. This pairing makes the Rbs and the ATG codon free and the mRNA translationally active. The mechanism resembles the transcriptional attenuation by alternate RNA hairpin formation in amino acid operons (see Chapter 18.10.3).

The induction of RNA methylase occurs by an induced conformational change in the leader region of the RNA methylase mRNA, which converts the mRNA from an inactive to an active form for translation of the methylase.

18.13.3 Programmed Ribosome Frameshift and Jumping Controls Elongation

The non-overlapping triplet genetic code is read in a sequential manner from a fixed point on mRNA. Although translation of an mRNA by switching the reading frame is more common in eukaryotic viruses, such examples are now known to occur in *E. coli* and its phages and perhaps in other prokaryotes.

mRNA-sequence-dependent frameshift and jumping of ribosomes during translation is programmed and has genetic consequences.

Ribosome frameshift. A single gene, *dnaX*, encodes both the τ and the γ subunits of the DNA polymerase III holoenzyme in *E. coli* (Fig. 18.**24**). The τ-subunit contains 643 amino acids and is read in the 0 frame of the *dnaX* mRNA. The stretch of codons 428–442 of τ has six adenine residues in a row, followed by a UGA stop codon in the -1 frame encompassing codons 431 and 432. For the synthesis of the 431-amino-acid-long γ-subunit,

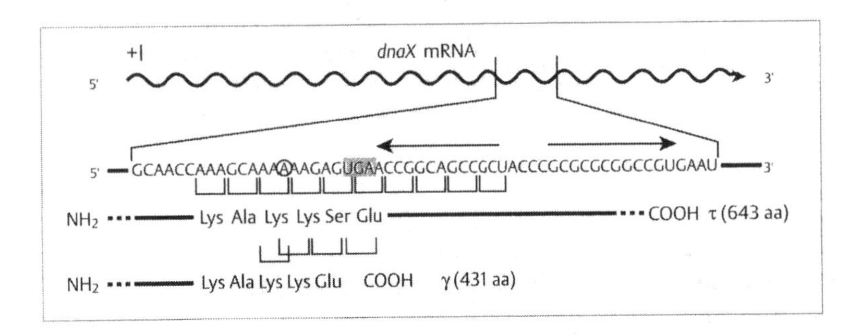

Fig. 18.**24** **Translation elongation control in the *dnaX* gene in *Escherichia coli* by ribosome frameshift.** The region of the mRNA where frameshifting occurs is expanded. The divergent arrows indicate the potential mRNA hairpin structure in the region. The τ-subunit is read in the 0 frame. The hairpin is essential for the γ-subunit production by frameshift of ribosome in the -1 frame. The UGA stop codon in the -1 frame is shaded red

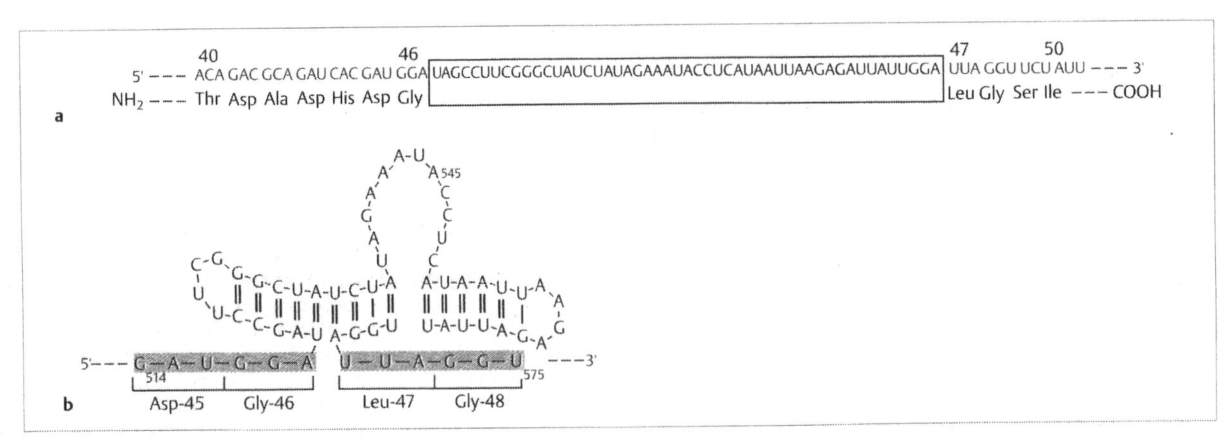

Fig. 18.**25a,b** **Translation elongation control in gene 60 of phage T4 by ribosome frameshift and jump.** The jumped RNA sequence between codons 46 and 47 is boxed in **a** and is shown in a potential secondary structure in **b**

which has the same amino acid sequence as the N-terminal portion of τ, ribosomes change the reading frame from 0 to -1 and terminate at the UGA codon. Frameshift occurs in the stretch of adenines at one of the two adjacent lysine codons at positions 429 and 430. A hairpin structure following the UGA codon enhances the frameshift by an unknown mechanism. The γ- and the τ-subunits are made in the ratio of 1 to 4. Similar slippery sequences are found in the frameshift sites in retroviral RNA.

Ribosome jumping. Frameshift may be accompanied by "jumping down" the mRNA. Phage T4 protein gp60 is a subunit of a topoisomerase. In the phage gene 60 mRNA, a stretch of 50 nucleotides separates the codons for amino acids 46 and 47 (Fig. 18.**25**). This segment is bypassed by the translating ribosome to make gp60, as if the ribosome "takes off" after translating up to codon 46 in the 0 frame and "lands" at codon 47 and resumes translation in the -1 frame. How the translational machinery conspires with the mRNA so that ribosome "jumps" and shifts the reading frame is not known. A likely model involves secondary structure formation of the intervening RNA. Why the cell uses frameshift to decide on the reading frame of a protein at the stage of translation elongation remains an intriguing question.

18.13.4 Protein Splicing, Inteins and Exteins

Protein-level splicing involving interrupted genes occurs when the spacer regions, called **inteins**, are removed from a protein precursor. The product of the splicing of the two end segments of the precursor is called the **extein**. Genetically specified protein splicing, as opposed to DNA excision, mRNA splicing, or ribosome jumping, has been occasionally found in prokaryotes. The *recA* gene in *Mycobacterium tuberculosis* and in *M. leprae* contains an intein. Genes containing two inteins also have been found. Extein splicing is autocatalytic.

Inteins can have a site-specific DNA endonuclease activity that participates in gene transposition at the DNA level. The excised intein promotes a unidirectional gene conversion with the movement of the intein segment from the intein-containing DNA into an empty site. The phenomenon is called **homing**, which is initiated by the endonuclease activity of the free intein. Free intein recognizes and cleaves the DNA sequence specifically at the vacant site of the copy of the gene lacking an intein. In *M. tuberculosis*, the free intein of *recA* also possesses similar site-specific endonuclease activity, suggesting that homing may occur. Intein amino acid sequences show conserved signature sequences and have regions homologous to DNA endonuclease, which initiates mating-type gene conversion in yeast.

18.13.5 Bacteria Make Use of Restrictions in Codon Usage

The genetic code is very degenerate. An essential factor in the degeneracy is that usually a variety of tRNA molecules (isoacceptors, with different anticodons) are used by a given amino acid. The other factor of degeneracy is the ability of some isoacceptors to pair with two or more synonymous codons of the corresponding amino acids ("wobbling"). A strong correlation also exists between the use of a particular codon of an amino acid (codon usage) in the primary structure of genes and the amount of the corresponding isoacceptor tRNA. In this way, the overall amino acid usage in proteins constrains the total level of tRNAs for the particular amino acid (see Chapter 15.4.1).

Organisms can use different codons for the same amino acid very differently, but synonymous codon usage patterns in different genes of the same prokaryotic organism are usually similar to each other, regardless of the gene functions. Bacteria are known to make use of restrictions in codon usage advantageously by excluding certain synonymous codons of a given amino acid.

In *Streptomyces coelicolor* and *Streptomyces lividans*, the gene for the transcriptional activator of the biosynthetic regulon of the antibiotic actinorhodin contains the leucine UUA codon, which is extremely rare in *Streptomyces*. Mutations inactivating the corresponding isoacceptor $tRNA^{Leu}$ gene block the production of the antibiotic in these organisms. Furthermore, changing UUA to another codon in the activator gene restores the normal synthesis of actinorhodin. Similarly, the arginine AGA codon is rare in *E. coli* and is decoded by a minor $tRNA^{Arg}$. A G $\rightarrow$ A mutation in the first nucleotide of the mature tRNA makes the tRNA ineffective and the mutant organism temperature-sensitive for growth. The mutation limits the expression of inducible genes with multiple AGA codons and of overproduced AGA-rich

Restriction in codon usage has evolutionary reasons. It arose from the need of nucleotides in RNA to fulfill pairing requirements for maintaining a particular secondary and tertiary structure important in translational regulation as well as in mRNA processing. Secondary structures are also important in packaging of RNA molecules in bacteriophages.

eukaryotic proteins, but has no effects on genes that lack AGA.

The restriction in codon usage in small RNA bacteriophages occurs more often than might be expected on a chance basis. In the coat protein gene of phage MS2, isoleucine uses two of its three codons, AUU and AUC, four times each, but never the third, AUA, in the 129-amino-acid-long protein. Similarly, of the two tyrosine codons, UAC is used four times for tyrosine, but UAU is not used at all.

18.14 Proteolysis Can Determine Levels of Gene Products

Increased gene transcription as demanded by environmental changes can occur by bringing transient stability of a critical protein that is normally unstable. Control of gene transcription by proteolysis is discussed in Chapter 19.3.4.

18.15 DNA–Protein Interactions Play Key Roles in Gene Regulation

DNA–protein interactions play key roles in gene regulation and in DNA replication, recombination, modification, and condensation. The regulatory proteins recognize a sequence of 8–20 bp in a background of millions of base pairs. The proteins are distinct from the other type of DNA binding proteins that have no sequence specificity. The prerequisite for the action of activators and repressors is their binding to DNA with high specificity. The DNA sites *a*, *i*, and *o* in Fig. 18.**2** and Fig. 18.**6**, are placed usually at strategically important locations. The purpose of sequence-specific binding of gene regulatory proteins is threefold: 1) to guide the proteins to their respective area of performance without disturbing the programmed gene expression, 2) to increase the local concentration of the regulatory proteins near the target of their actions, facilitating further protein–protein and DNA–protein interactions, and 3) to bring about any required structural changes in the interacting partners. A protein-induced DNA bending may be a required component in the process of transcription activation; alternatively, a DNA-induced conformational alteration in the activator protein may make it proficient for interaction with RNA polymerase.

Sequence-specific binding of specialized proteins to DNA is a fundamental step in DNA transcription, replication, recombination, modification, and condensation. Although a relatively small number of specific DNA–protein complexes have been studied, there seems to be no code in amino-acid-base interactions. In contrast, four kinds of structural motifs in proteins that interact in sequence-specific binding have been recognized: **helix-turn-helix**, **β-fold**, **leucine zipper**, and **zinc finger**.

Nonspecific DNA–protein interactions are frequently formed by hydrogen bonds between phosphate oxygen atoms of DNA and main chain NH-groups of the protein. Specific interactions are 1000-fold or more stronger and use hydrogen, hydrophobic, and ionic bonds. Most of these bonds involve amino acid side chains with base edges in DNA grooves available for interactions.

18.15.1 The Helix–Turn–Helix Motif Is the Most Common DNA Binding Motif in Prokaryotes

Many bacterial gene regulatory proteins are dimers. The problem of DNA recognition by these proteins is

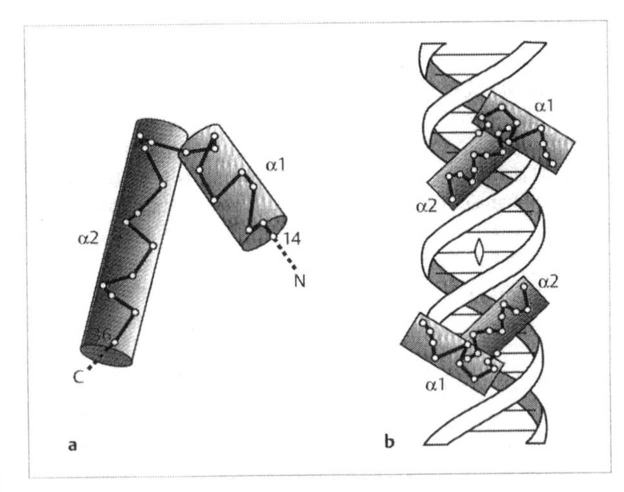

Fig. 18.**26a,b** **The helix–turn–helix motif and sequence-specific DNA binding.**
a Schematic diagram of the helix–turn–helix motif of the λ Cro repressor. The recognition helix is in red.
b The helix–turn–helix motif of Cro bound to operator DNA with the two recognition helices (in red) of the Cro dimer sitting in the major groove of the DNA [after 4]

simplified by the use of molecular symmetry. Unique stretches of 16–20 bp long DNA sequences with dyad symmetry bind dimeric proteins with symmetrically spaced DNA binding motifs. The DNA binding motif of this group of proteins is composed of two stretches of α-helices connected by a tight β-turn with an interhelical angle of about 90° (Fig. 18.**26a**). The two helices in one subunit are related by a twofold symmetry to their partners in the other subunit in the dimer. The two motifs, called **helix–turn–helix motifs**, in the dimer are 34 Å apart, equivalent to one turn of the B-DNA duplex. The second helix of the helix–turn–helix motif, called the **recognition helix**, lies in the major groove of a DNA half-symmetry (Fig. 18.**26b**). The recognition helix of the other subunit related to the first by twofold rotational symmetry similarly occupies the major groove in the other half of the DNA palindrome. The first helix in each subunit makes contact with backbone phosphates by ionic interactions through basic amino acid side chains. Several gene regulatory proteins with helix–turn–helix motifs can form tetramers. Tetramers can simultaneously bind to two DNA sites, each with a dyad symmetry. This property makes such proteins able to form DNA loops (Chapter 18.4). In all of the DNA–protein complexes that have been analyzed, the structure of the DNA is bent. The implication of DNA bending in gene regulation is described in Chapter 18.8.

The structure and amino acid sequence of the helix–turn–helix motifs. The three-dimensional structure of the helix–turn–helix motifs is preserved by having amino acids with special properties at crucial positions (Table 18.**6**). The motif is usually 20 amino

Table 18.**6** **Characteristics of amino acids at various positions of several helix–turn–helix DNA-binding motifs.** The sequence alignment of the helix–turn–helix motifs in prokaryotic gene regulatory proteins are shown. The positions subject to constraints are shown in gray, and those (in helix 2) that interact with DNA are in red. The proteins are from *Escherichia coli*, unless indicated otherwise. Proteins from bacteriophages λ, 434, and P22 are indicated as such. Subscripts denote proteins from other bacteria: Ec, *Escherichia coli*; Kp, *Klebsiella pneumoniae*; Pa, *Pseudomonas aeruginosa*; Rm, *Rhizobium meliloti*; St, *Salmonella typhimurium*

	← Helix 1 →							← Turn →				← Helix 2 →								
Proteins	1	2	3	4	5	6	7	8	9	10	11	12	13	14	15	16	17	18	19	20
AraC	A	K	L	L	L	S	T	T	R	M	P	I	A	T	V	G	R	N	V	G
Crp	R	Q	E	I	G	Q	I	V	G	C	S	R	E	T	V	G	R	I	L	K
CysB	V	S	S	T	A	E	G	L	Y	T	S	Q	P	G	I	S	K	Q	V	R
CytR	M	K	D	V	A	L	K	A	K	V	S	T	A	T	V	S	R	A	L	M
EbgR	L	K	D	I	A	I	E	A	G	V	S	L	A	T	V	S	R	V	L	N
434cI	Q	A	E	L	A	Q	K	V	G	T	T	Q	Q	S	I	E	Q	L	E	N
434Cro	Q	T	E	L	A	T	K	A	G	V	K	Q	Q	S	I	Q	L	I	E	A
Fnr	R	G	D	I	G	N	Y	L	G	L	T	V	E	T	I	S	R	L	L	G
FruR	L	D	E	I	A	R	L	A	G	V	S	R	T	T	A	S	Y	V	I	N
GalR	I	K	D	V	A	R	L	A	G	V	S	V	A	T	V	S	R	V	I	N
GalS	I	R	D	V	A	R	Q	A	G	V	S	V	A	T	V	S	R	V	L	N
Hin_{St}	R	Q	Q	L	A	I	I	F	G	I	G	V	S	T	L	Y	R	Y	F	P
IlvY	F	G	R	S	A	R	A	M	H	V	S	P	S	T	L	S	R	Q	F	Q
$LacI_{Ec}$	L	Y	D	V	A	E	Y	A	G	V	S	Y	Q	T	V	S	R	V	V	N
$LacI_{Kp}$	L	E	D	V	A	R	R	G	R	V	P	A	D	G	L	R	R	V	L	N
λcI	Q	E	S	V	A	D	K	M	G	M	G	Q	S	G	V	G	A	L	F	N
λcII	T	E	K	T	A	E	A	V	G	V	D	K	S	Q	I	S	R	W	K	R
λCro	Q	T	K	T	A	K	D	L	G	V	Y	Q	S	A	I	N	K	A	I	H
LexR	R	A	E	I	A	Q	R	L	G	F	R	S	P	N	A	A	E	E	H	L
LysR	L	T	E	A	A	H	L	L	H	T	S	Q	P	T	V	S	R	E	L	A
MalI	I	H	D	V	A	L	A	A	G	V	S	V	S	T	V	S	L	V	L	S
MetR	L	A	A	A	A	A	V	L	H	Q	T	Q	S	A	L	S	H	Q	F	S
$NodD_{Rm}$	L	T	A	A	A	R	S	I	N	L	S	Q	P	A	M	S	A	A	I	Q
OxyR	F	R	R	A	A	D	S	C	H	V	S	Q	P	T	L	S	G	Q	I	R
P22C2	Q	A	A	L	G	K	M	V	G	V	S	N	V	A	I	S	Q	W	Q	R
P22Cro	Q	R	A	V	A	K	A	L	G	I	S	D	A	A	V	S	Q	W	K	E
PurR	I	K	D	V	A	K	R	A	N	V	S	T	T	T	V	S	H	V	I	N
$ScrR_{Kp}$	I	K	D	I	A	E	L	A	G	V	S	K	A	T	A	S	L	V	L	N
TnpR	E	A	K	L	K	G	I	K	F	G	R	R	R	T	V	D	R	N	V	V
$TrpI_{Pa}$	I	S	L	A	A	E	E	L	H	V	T	H	G	A	V	S	R	Q	V	R
TrpR	Q	R	E	L	K	N	E	L	G	A	G	I	A	T	I	T	R	G	S	N

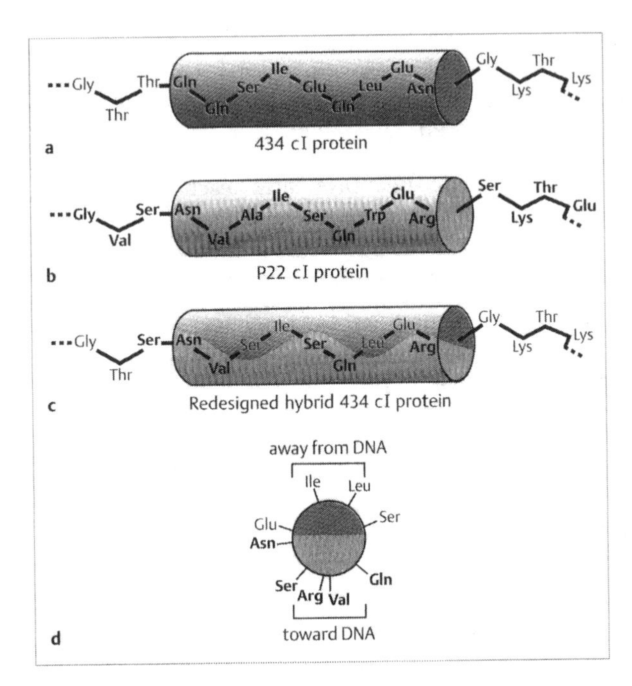

Fig. 18.**27a–d** **The DNA binding surface of the recognition helix of the *Escherichia coli* bacteriophage 434 cI repressor protein** was altered genetically to that of the *Salmonella typhimurium* bacteriophage P22 cI protein, changing six amino acid residues that participate in interaction with DNA. The amino acid sequences of the recognition helix of
a 434 cI (red),
b P22 cI (gray), and
c the redesigned 434 cI (red and gray) are shown. The view of **c** along the helix is shown in
d The redesigned 434 cI acquired all the DNA binding properties of P22 cI. Similar "helix swap" experiments have been performed between LacI and GalR, FruR and ScrR, and between Crp and Fnr [after 5]

acids long with 7 amino acids in the first helix, 4 in the turn, and 9 in the second helix. The alignment of amino acid sequences of several bacterial proteins with the helix–turn–helix binding motif shows the nature of the conserved amino acids and their positions, which provide stereochemical constraints to the structure: a conserved glycine at 9 in the turn to ensure against steric interference of the β-carbon with the main chain, hydrophobic or weakly polar and no charged amino acids at positions 4, 8, 10, and 15 to hold to the rest of the protein, no proline at helical residues 3–7 and 15–20 to prevent helix break, and no amino acid with branched side chain at 5 to maintain the relative alignment of the two helices. The solvent-exposed side chains that participate in interactions with DNA are some or all of the following residues: 11–13, 16–17, and 20.

Because of the nature of the structure of the helix–turn–helix motifs, it is sometimes possible to swap the recognition amino acids between two DNA binding proteins that are closely related in structure. Such a trade enables them to swap their respective DNA binding specificities (Fig. 18.**27**)

Family of proteins with helix–turn–helix motifs. The DNA binding proteins (both specific and global regulators) with helix–turn–helix motifs in bacteria and bacteriophages constitute several families, depending on amino acid sequences or structural similarities. Note that the helix–turn–helix motif is present sometimes in the NH_2-domain (e.g., LacI family) and sometimes in the C-domain (e.g., LysR family) of the proteins. Usually, there are a large number of specific regulators in an organism because they are needed for specific operons and regulons. Since the global regulators handle a large number of operons, there are only few of them in each organism. The compilation of Table 18.**6** suggests a wide-spread conservation of structural motifs among gene regulatory proteins in bacteria and has strong implications for the further study of peptide motifs in domain function and in protein evolution.

18.15.2 β-fold Motifs Are Rarely Used in Prokaryotes

The two-stranded antiparallel β-sheet as a DNA binding motif has been found in two cases among specific gene regulatory proteins of known structure: *E. coli* MetJ repressor and *Salmonella* phage P22 Arc repressor. The β-folds of both MetJ and Arc dimers recognize a simple major groove of DNA. MetJ aporepressor uses S-adenosylmethionine (SAM) as a corepressor in controlling the *met* regulon, which encodes enzymes of methionine biosynthesis. In the MetJ dimer, one β-strand of each subunit forms an antiparallel sheet protruding from the surface and binds to the major groove in the center of an 8-bp palindromic *met* operator sequence in the B-DNA form. The side chains of the β-strands interact tightly within the operator sequence, resulting in a kink that narrows the major groove (Fig. 18.**28**).

Genetic, biochemical, and physical studies have made available the intimate details of the structure of several gene regulatory proteins and their DNA complexes and have provided explanations for the conformational changes that affect gene regulation. How various macromolecular interactions transpire into regulation of biochemical reactions of RNA polymerase remains to be discovered.

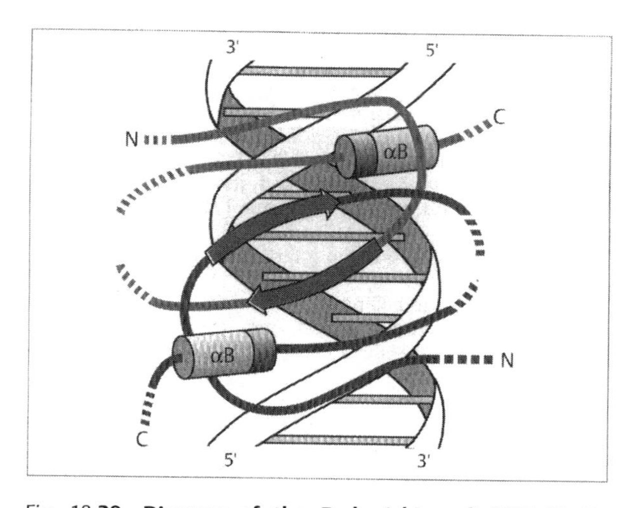

Fig. 18.**28** **Diagram of the *Escherichia coli* SAM·MetR–operator complex** comprising a β-fold motif. The complex illustrates the regions of the protein dimer that contact DNA. Two β-strands, one from each subunit (red arrows) bind to one major groove, where they form the sequence-specific interactions. The N-terminal flexible loops and the N-ends of the α-B helix interact nonspecifically with the phosphate backbone [after 4]

Two other DNA binding motifs, the leucine zipper and the zinc finger, used in specific DNA binding in eukaryotes, have not been observed among any of the prokaryotic gene regulatory proteins.

18.15.3 Allosteric Effectors Modulate the DNA Binding Capacity of Regulatory Proteins

The specific DNA binding capacity of many gene regulatory proteins is modulated allosterically by small molecules (see Chapters 18.1 and 19). Genetic, biochemical, and structural studies have indicated that these proteins usually have two domains connected by a hinge in each subunit: one domain binds to DNA and the other domain binds to the effector. Binding a ligand to one domain causes a conformational change in the protein that is transmitted to the other domain through the hinge. The effector-induced change either sets up the DNA binding motif for proper major groove contacts or alters it in such a way as to prohibit the necessary contacts.

18.16 Summary

Genes are arranged in the prokaryotic chromosome as **operons** (mostly intron-free) and **regulons** for the convenience of their regulation in response to extracellular and intracellular changes. There are two countenances in the mechanics of gene expression by the operon mode. First, the primary and secondary structural features of DNA and RNA that set up intrinsic strength of all aspects of transcription and translation define mRNA processing and even decide the reading frame and coupling of transcription and translation. Second, *trans*-acting regulatory proteins (sometimes RNA) and small effectors change the intrinsic rates and efficiencies of different steps. Diverse control mechanisms have evolved to regulate different steps of the gene expression mechanics. An operon is usually regulated at more than one level for finer tuning and to join the operon to a regulon. For these reasons, each operon has been customized. Superimposed global controls provide additional dimensions in gene regulation (**modulons**) by permitting the sensing of a wider range of signals (discussed in Chapter 20). These controls also give coherency rather than chaos, as would be expected if each system, however perfect, functioned independently. With the advent of modern molecular biological technology, the researcher's ability to do an experiment both in vivo and in vitro and to alter each component of the regulatory system at will has been invaluable in generating and verifying various regulatory hypotheses.

Further Reading

Adhya, S. (1989) Multipartite genetic control elements: communication by DNA loop. Annu Rev Genet 23: 227–250
Beckwith, J., and Silhavy, T. J. (1992) The power of bacterial genetics. Cold Spring Harbor, N.Y.: Cold Spring Harbor Laboratory Press
Busby, S., and Ebright, R. H. (1997) Transcription activation at class II CAP-dependent promoters. *Mol Microbiol* 23: 853–859
Calvo, J. M., and Matthews, R. G. (1994) The leucine-responsive regulatory protein, a global regulator of metabolism in *Escherichia coli*. Microbiol Rev 58: 461–490
Condon, C., Squires, C., and Squires, C. L. (1995) Control of rRNA transcription in *Escherichia coli*. Microbiol Rev 59: 623–645
Court, D. (1993) RNA processing and degradation by RNase III. In: Brawerman, G., and Beleaso, J. (eds.) Control of mRNA stability, vol. 5. San Diego: Academic Press; 71–116
Dorman, C. J. (1995) DNA topology and the global control of bacterial gene expression: implications for the regulation of virulence gene expression. Microbiology 141: 1271–1280

Farabaugh, P. J. (1996) Programmed translational frameshifting. Microbiol Rev 60: 103–134
Gallegos, M. T., Schleif, R., Bairoch, A., Hofmann, K., and Ramos, J. L. (1997) AraC/XylS family of transcriptional regulators. *Microbiol Mol Biol Rev* 61: 393–410
Gold, L. (1988) Posttranscriptional regulatory mechanisms in *Escherichia coli*. Annu Rev Biochem 57: 199–233
Ishihama, A. (1993) Protein–protein communication within the transcription apparatus. J Bacteriol 175: 2483–2489
Kolb, A., Busby, S., Buc, H., Garges, S., and Adhya, S. (1993) Transcriptional regulation by cAMP and its receptor protein. Annu Rev Biochem 62: 749–795
Lambowitz, A. M., and Belfort, M. (1993) Introns as mobile genetic elements. Annu Rev Biochem 62: 587–622
Lin, E. C. C., and Lynch, A. S., eds. (1996) Regulation of gene expression in *Escherichia coli*. Austin, Tex: Landes
Neidhardt, F. C., Curtiss, R., Ingraham, J. L., Lin, E. C. C., Low, K. B., Magasanik, B., Reznikoff, W., Riley, M., Schaechter, M., and Umbarger, H. E. (eds.) (1996) *Escherichia coli* and *Salmonella*: cellular and molecular biology. 2nd edn. Washington, DC: ASM Press
North, A. K., Klose, K. E., Stedman, K. M., and Kustu, S. (1993) Prokaryotic enhancer binding proteins reflect eukaryotic-like modularity: the puzzle of nitrogen regulatory protein C. J Bacteriol 175: 4267–4273
Ptashne, M. (1992) A genetic switch. Cambridge, Mass: Cell Press and Blackwell
Richet, E., Vidal-Ingigliardi, D., and Raibaud, O. (1991) A new mechanism for coactivation of transcription initiation: repositioning of an activator triggered by the binding of a second activator. Cell 66: 1185–1195
Schell, M. A. (1993) Molecular biology of the LysR family of transcriptional regulators. Annu Rev Microbiol 47: 597–626
Sonenschein, A. L., ed. (1993) *Bacillus subtilis* and other Gram-positive bacteria. Washington, DC: ASM Press
Summers, A. O. (1992) Untwist and shout: a heavy metal-responsive transcriptional regulator. J Bacteriol 174: 3097–3101
Travers, A. (1993) DNA-protein interactions. London: Chapman and Hall
Wagner, E. G. H., and Simons, R. W. (1994) Antisense RNA control in bacteria, phages and plasmids. Annu Rev Microbiol 48: 655–686

Sources of Figures

1 Mandal, N., Su, W., Haber, R., Adhya, S., and Echols, H. (1990) Genes Dev 4: 410–418
2 Herendren, D. R., Kassavetis, G. A., and Geiduschek, E. P. (1992) Science 256: 1298–1303
3 Grundy, F. J., and Henkin, T. M. (1993) Cell 74: 475–482
4 Branden, C., and Tooze, J. (1991) Introduction to protein structure. New York: Garland
5 Wharton, R., and Ptashne, M. (1985) Nature 316: 602

19 Posttranslational Control and Modifications of Proteins

As also described in other chapters in this section, bacterial cells must be able to respond rapidly to environmental changes, including changes in the availability of nutrients, in order to compete effectively with other organisms. Regulation of gene expression is one mechanism by which this is accomplished. However, this type of regulation is more "long-term" since there can be significant lag times in the induction of certain genes, and repression mechanisms may take several generations for the levels of pre-existing proteins and their activities to be diluted out. It is, therefore, also advantageous for bacteria to be able to adjust their metabolic activities instantaneously in response to changes in the medium. Bacterial cells use posttranslational control of proteins, which also includes cellular compartmentalization, for this purpose.

19.1 Posttranslational Control Allows Bacteria to Adapt Rapidly

19.1.1 Posttranslational Control and Modification of Proteins Are Involved in Many Processes in the Cell

Allosteric regulation and, in some cases, **covalent modification**, of metabolic enzymes are important posttranslational regulatory mechanisms. Posttranslational modifications (predominantly phosphorylations) are involved in the regulation of gene expression, bacterial behavioral responses such as chemotaxis, and global regulatory networks (see Chapter 20). In addition, proteolytic modifications are involved in the regulation of protein function, in protein turnover, and in protein targeting (**compartmentalization**) in bacterial cells. In this chapter, some well-studied examples of these mechanisms and their overall importance to bacterial physiology and adaptation will be explored.

Posttranslation control allows prokaryotes to adapt their metabolism rapidly (within seconds) to changes in the environment. It involves non-covalent modulation of enzyme activity, covalent protein processing, and protein compartmentalization.

19.1.2 Posttranslational Control Involves Both Non-covalent Modulation of Enzyme Activity and Covalent Protein Processing

Allosteric enzymes are those whose activities are modulated (activated or inhibited) by the non-covalent binding of metabolites to regulatory sites on the enzyme. This direct regulation serves to shut down rapidly the activity of an enzyme that is not needed (e.g., feedback inhibition of a biosynthetic enzyme by its end product), to increase the activity of such an enzyme

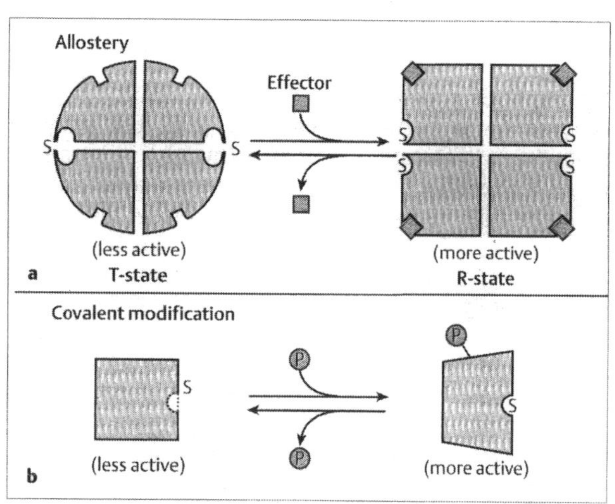

Fig. 19.1**a,b** **Control of enzyme activity by allostery and covalent modification.**
a In allostery, an effector (red square), which binds at a site distinct from the active site for the substrate (S), shifts the equilibrium between the T (less active) and R (more active) conformational states of the enzyme. In the hypothetical case shown, the effector is an allosteric activator.
b In covalent modification, the enzyme is reversibly modified, for example by phosphorylation (P), which may either increase (as shown) or decrease the activity of the enzyme. The state of covalent modification of the enzyme usually responds to environmental and/or metabolic factors

when there is an immediate need for a downstream end product, and also to modulate the activities of catabolic enzymes to ensure that they work neither too rapidly nor too slowly for the needs of the cell.

In general, allosteric regulation involves a conformational change in the enzyme induced by the reversible binding of an inhibitor or activator to a site on the protein that is distinct from the active site. In other cases, the activity of an enzyme may alternatively, or in addition, be modulated by a reversible covalent modification, which may be effected by enzymes that are themselves allosterically or covalently regulated (Fig. 19.**1**). In the next section, specific examples of both of these mechanisms will be discussed.

Covalent posttranslational protein modifications in bacteria include phosphorylation, methylation, acetylation, adenylylation, uridylylation, fatty acylation, and proteolysis (both specific and non-specific). A wide variety of cellular processes are carried out and controlled by these types of modifications, as will be discussed. Many such modifications, especially those affecting enzyme activity, gene expression, and behavior, are fully reversible with specific enzymes catalyzing both the modification and de-modification reactions. In other cases, such as proteolysis, the modifications are irreversible, reflecting processes in cells that do not need to be reversed (e.g., protein degradation as a source of amino acids, protein targeting, and completion of irreversible phases during generation cycles).

19.2 Allosteric Control of Enzyme Activity Is Widespread Among Prokaryotes

Both anabolic and catabolic pathways are, in general, regulated by allosteric mechanisms involving small-molecule metabolites acting as inhibitors or activators.

19.2.1 There Are Multiple Strategies of Allosteric Regulation

For **anabolic** pathways, the first "committed" (unique) enzyme for a particular pathway (e.g., the enzyme following a bifurcation of two pathways) is usually a target for **feedback inhibition** by an end product of the pathway (Fig. 19.**2a**). A buildup of a biosynthetic end product is a signal that it is being synthesized (or is available from the medium) in amounts that exceed the cell's capacity to use that end product for biosynthesis. For example, very commonly, biosynthetic pathways for amino acids, nucleotides, and vitamins are regulated in this manner. In branched biosynthetic pathways, in which a precursor molecule gives rise to more than one end product, a pathway may be regulated by so-called **sequential** feedback inhibition, in which each end product inhibits its first unique enzyme (Fig. 19.**2b**); by **cumulative** feedback inhibition, in which each end product partially inhibits an enzyme early in the pathway (Fig. 19.**2c**); or by a variation of cumulative feedback inhibition, called **concerted** feedback inhibition, in which each end product alone has no effect, but two or more together inhibit the enzyme. In addition, some bacterial biosynthetic pathways use **isozymes**, each of which catalyzes the same reaction at the beginning of a pathway, but each of which is regulated by a different end product (Fig. 19.**2d**).

An anabolic pathway may also be regulated by allosteric **activation** of an enzyme in the pathway. Most commonly, the activator molecule is a metabolite that is not part of the pathway being regulated, but whose presence in the cell is a signal that more of the end product of the pathway needs to be synthesized.

Catabolic pathways may also be regulated by feedback inhibition and allosteric activation, including "feedforward" regulation. For feedback inhibition in these cases, one or more intermediates (rather than end products) of the pathway most commonly feedback inhibit an enzyme earlier in the pathway. Activators of enzymes in catabolic pathways can be metabolites that are either part of the same pathway or in different pathways. In some cases, **"feedforward"** regulation also occurs, in which a metabolic intermediate modulates an enzyme further along in the pathway. In addition, the interconnected pathways of glycolysis, gluconeogenesis, and the TCA cycle, which are the central catabolic pathways in most heterotrophic bacteria, show a complex and interrelated regulatory pattern.

Allosteric control is widespread among anabolic and catabolic pathways of prokaryotes. It is used in the control of crucial enzymes and involves small-molecule metabolites acting as inhibitors or activators. Characteristic are various **feedback inhibitions** (sequential, cumulative, concerted), caused by the end product of a pathway, and **feedforward activation**, caused by metabolic intermediates.

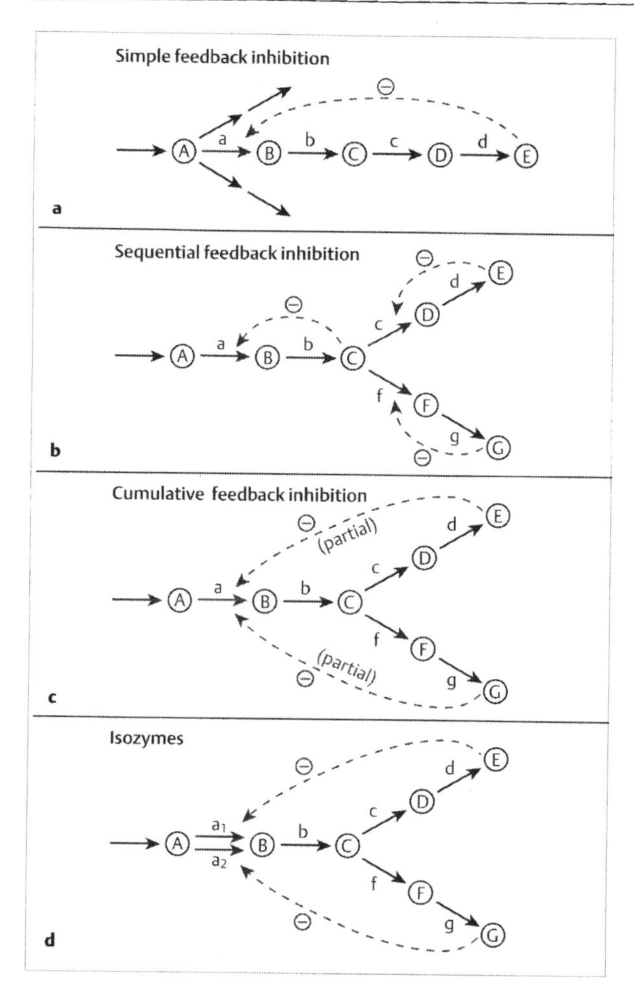

Fig. 19.**2a–d** **Mechanisms of allosteric regulation in anabolic pathways.** Hypothetical pathways are shown with intermediates in upper case letters and enzymes in lower case letters.

a Feedback inhibition of the first unique step of an anabolic pathway. The only fate of intermediate B is to form end product E. Therefore, E feedback inhibits the enzyme (a), which converts intermediate A to B.

b Sequential feedback inhibition in branched pathways. Each end product inhibits the enzyme catalyzing the first unique step in its biosynthesis. If both products are in excess, intermediate C builds up and inhibits the first enzyme unique to its biosynthesis (a), shutting down the entire pathway.

c Cumulative feedback inhibition. In a pathway giving rise to several end products, each end product inhibits the first enzyme (a) only partially. These effects are additive, so that enzyme (a) can only be completely inhibited when all end products are in excess.

d Differentially regulated isozymes (a1, a2). For some branched pathways, the first step is catalyzed by several isozymes, each of which is differentially inhibited by a different end product

19.2.2 Allosteric Regulation in Anabolic Pathways Is Highly Variable

In Table 19.**1** are listed a few of the anabolic pathways that are subject to allosteric regulation in bacteria, the enzyme targets, and their effectors. In most cases, these pathways and the allosteric enzymes have been the most thoroughly studied in *Escherichia coli* and *Salmonella typhimurium*. The biochemical details related to these pathways and their enzymatic steps are described in Chapters 7–12.

Example: aspartate carbamoyltransferase. Aspartate carbamoyltransferase (also called aspartate transcarbamylase or **ATCase**) catalyzes the first step unique to pyrimidine nucleotide biosynthesis, the condensation of carbamoyl phosphate and aspartate to yield carbamoylaspartate and inorganic phosphate (see Chapter 7.8.2 and Fig. 7.**25**). Native *E. coli* ATCase is composed of two trimeric catalytic (C) subunits ($M_r = 100000$ per subunit) and three dimeric regulatory (R) subunits ($M_r = 34000$ per subunit), which dissociate in the presence of certain sulfhydryl reagents, such as organic mercurial compounds. The isolated C subunits are sufficient to catalyze the ATCase reaction, while the R subunits possess no known catalytic activity.

The activity of native ATCase is inhibited by relatively low concentrations of CTP (an end product of the pyrimidine pathway), and is activated by ATP (an end product of the separate purine nucleotide biosynthetic pathway; see Chapter 7.8.1). This allosteric regulation results in a relatively low flux through the pathway when end products such as CTP are present in excess. On the other hand, the activating effect of ATP on ATCase activity presumably is a mechanism to increase flux through the pyrimidine pathway when purine nucleotides are in excess; therefore, more or less equal amounts of pyrimidine and purine nucleotides needed for nucleic acid biosynthesis can be maintained (Fig. 19.**3**).

The activity vs. aspartate concentration curve (Fig. 19.**4**) of native ATCase is **sigmoidal** and becomes more so in the presence of CTP and less so in the presence of ATP. For the isolated C subunit, however, this curve is hyperbolic, and the addition of either CTP or ATP has little effect on its shape. Moreover, isolated R subunits bind CTP and ATP at the same site. Thus, for the native enzyme, the binding of effectors on the R subunits influences the activity of the C subunits by either decreasing (for CTP) or increasing (for ATP) the apparent affinity of one of the substrates, aspartate. At the concentrations of aspartate normally present in the cell, this therefore leads to either inhibition or activation, respectively, of the enzyme.

Table 19.1 **Examples of anabolic pathways subject to allosteric regulation in *Escherichia coli* and/or *Salmonell[a] typhimurium*.** + Activator, − inhibitor

Pathway	Enzyme target	Effector(s)
Pyrimidine	Aspartate transcarbamylase	CTP (−) ATP (+)
Arginine	N-acetyl glutamic acid synthetase	Arg
Pyrimidine/arginine	Carbamoyl phosphate (CP) synthetase	UMP (−) IMP (+) Ornithine (+)
NH_3 assimilation	Glutamine synthetase	AMP, CTP, carbamoyl-P, glucosamine 6-P, Ala, Gly, His, Trp (all −)
Amino acids synthesized from L-aspartate	Aspartokinase I	Thr (−)
	Aspartokinase III	Lys (−)

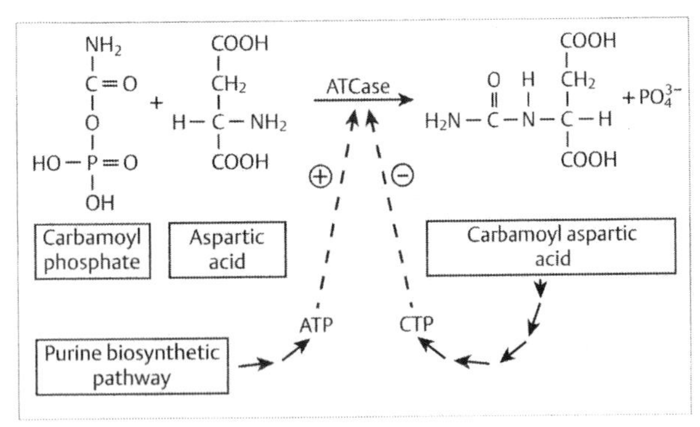

Fig. 19.3 **Regulation of the pyrimidine biosynthetic pathway.** CTP (an end product) allosterically inhibits aspartate carbamoyltransferase (ATCase) (−), while ATP (a purine nucleotide and the product of a separate biosynthetic pathway) activates the enzyme (+)

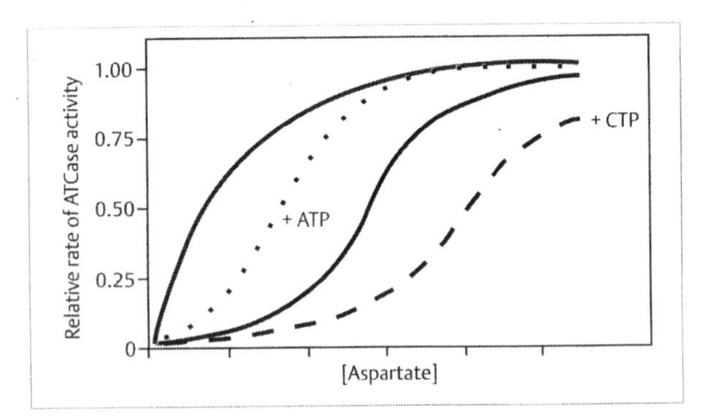

Fig. 19.4 **Rate vs. [aspartate] curves of ATCase.** This curve is sigmoidal for native ATCase (solid black), and becomes more sigmoidal in the presence of CTP (dashed black) and less so in the presence of ATP (dotted black). The curve for the isolated C subunit is hyperbolic (solid red), indicating a non-cooperative interaction of aspartate with the C subunit alone

In general, allosteric enzymes often show suc[h] sigmoidal activity curves for at least one substrate. Thi[s] type of activity vs. concentration of substrate curve i[s] indicative of a **cooperative** interaction of the substrat[e] with the enzyme. At low substrate concentrations, th[e] apparent affinity of the enzyme for the substrate i[s] lower than at high substrate concentrations. Thi[s] change in apparent affinity is presumably accomplishe[d] by a conformational change in the protein, which i[s] favored by binding of one or more molecules of th[e] substrate. Thus, in the case of ATCase, the binding o[f] aspartate to less than the full number (6) of catalyti[c] sites in the native enzyme increases the affinity of all o[f] the sites, including the unoccupied ones, for aspartate.

For many allosteric enzymes, at least two con[-]formational states have, therefore, been postulated: th[e] lower-affinity **T** (tight)-**state**, predominating at zero o[r] low substrate concentrations and in the presence of a[n] allosteric inhibitor, and the higher-affinity **R** (relaxed) **state** predominating at high substrate concentration[s] and in the presence of an activator. Intermediate state[s] are also possible. For ATCase, the structures of the T- an[d] R-states have been directly determined by X-ray crystal[-]lography and provide an explanation for how th[e] binding of aspartate at one site can influence the affinit[y] at the other sites, and also how the binding o[f] nucleotides can influence the equilibrium betwee[n] these two conformational states (Fig. 19.**5**).

Example: Carbamoyl phosphate synthetase an[d] the arginine and pyrimidine biosynthetic pathways Carbamoyl phosphate synthetase (**CP synthetase**) cat[-]alyzes the formation of carbamoyl phosphate from CO_2, ATP, and glutamine. Carbamoyl phosphate is, in turn, [a] substrate for ATCase in pyrimidine biosynthesis (se[e] Chapter 7.8.2) and a substrate for ornithine carbamoyl transferase (see Chapter 7.6.1), an enzyme in th[e]

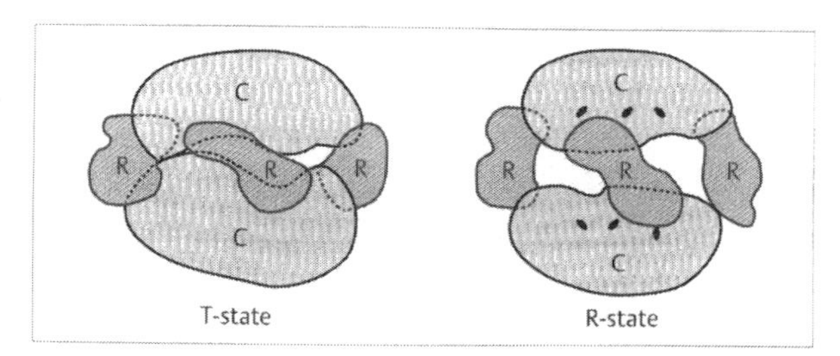

Fig. 19.5 **Schematic illustration of the structures of the T- and R-states of ATCase.** The two trimeric C subunits (black) form a "sandwich" around the three dimeric R subunits (red). In the T-state (determined without bound substrates), this "sandwich" is very compact. In the R-state (obtained in the presence of a bound substrate/product analogue shown as solid ovals), the sandwich has a much more "open" conformation. (Based on structures determined by W. N. Lipscomb and collaborators)

Allosteric enzymes comprise several (di-, tetra-, polymer) subunits that are identical (homomers) or different (heteromer). They contain, besides a substrate-binding site, one or several binding sites for **effector** molecules whose binding/dissociation modulates the enzymatic activity. Due to **cooperative** interactions between the subunits, allosteric enzymes often show **sigmoidal** activity curves, with a **threshold** substrate concentration below which almost no activity is seen, an optimal range within which small changes in substrate concentration show drastic activity changes, and a saturation range. Effectors either inhibit (often end products) or activate (often intermediates) the activity.

pathway for the biosynthesis of arginine from glutamate (Fig. 19.**6**). CP synthetase of *E. coli* consists of two subunits with molecular weights of 130000 and 42000, and shows an interesting allosteric regulation pattern indicative of its importance in these two pathways. CP synthetase is feedback inhibited by UMP, an intermediate in pyrimidine nucleotide biosynthesis, but this inhibition is antagonized by both IMP (an intermediate in the purine nucleotide pathway) and ornithine (Fig. 19.**6**).

Thus, in the presence of excess pyrimidines, carbamoyl phosphate synthesis is inhibited. However, if flux through the first half of the arginine pathway is high, ornithine will accumulate (due to the relative lack of carbamoyl phosphate). High levels of ornithine will partially overcome the inhibition of CP synthetase by UMP so that arginine can be synthesized. However, if both arginine and UMP are in excess, CP synthetase

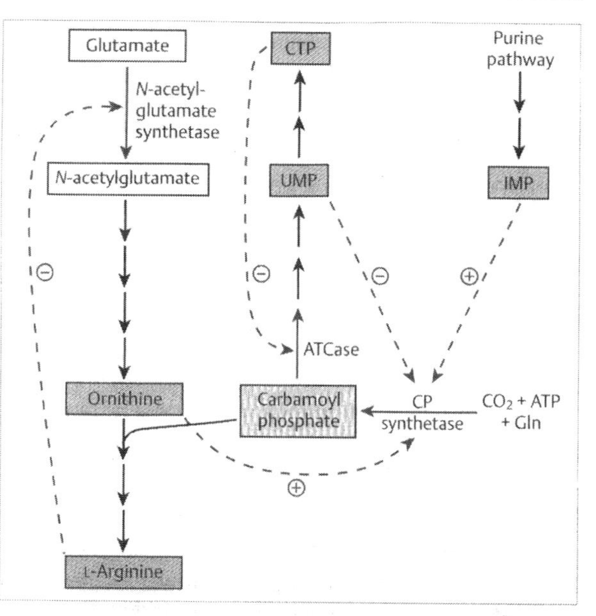

Fig. 19.**6** **The role of carbamoyl phosphate in both arginine and pyrimidine biosynthesis.** Because carbamoyl phosphate feeds into both of these pathways, carbamoyl phosphate (CP) synthetase is subject to complex allosteric regulation. It is feedback (dashed arrows) inhibited (−) by UMP (a pyrimidine nucleotide), but this effect is antagonized (+, for activation) by both IMP (a purine nucleotide) and ornithine (an intermediate in the arginine biosynthetic pathway). See text for further details

activity is virtually completely shut off because arginine feedback inhibits the first enzyme of its biosynthetic pathway, *N*-acetylglutamate synthetase, and ornithine therefore does not accumulate. The activating effect of IMP on CP synthetase presumably has an analogous purpose to the activating effect of ATP on ATCase, that is, in the coordination of the purine and pyrimidine biosynthetic pathways.

This example of the regulation of CP synthetase illustrates one of many similar strategies used by prokaryotes to ensure coordinate, and appropriate, regulation of interrelated biosynthetic pathways in response to the metabolic needs of cells.

Example: allosteric control of glutamine synthetase. Glutamine synthetase (**GlnS**) of *E. coli*, a dodecamer of identical 55-kDa subunits, catalyzes the synthesis of L-glutamine from ammonia and L-glutamate with the concomitant hydrolysis of ATP (see Chapter 7.6.1 and Fig. 7.**13**). Glutamine synthetase is important for the formation of glutamine used both for protein synthesis and as an amido-group donor for the biosynthesis of many nitrogen-containing compounds. Moreover, the enzyme is a primary means of ammonia assimilation under ammonia-limited conditions. Both the biosynthesis (see Chapter 18.5.2) and activity (see Chapter 19.3.1) of GlnS are regulated in a complex fashion by the availability of nitrogen, as reflected by the intracellular ratio of glutamine to α-ketoglutarate. Moreover, the activity of one form of GlnS (see Fig. 19.**10**) is subject to cumulative feedback inhibition by a variety of metabolites, most of which are products of glutamine metabolism, including: AMP, CTP, carbamoyl phosphate, glucosamine 6-phosphate, L-alanine, glycine, L-histidine, and L-tryptophan. Each of these compounds appears to have a separate binding site on GlnS and inhibits GlnS activity only partially. All of these compounds together, however, are capable of inhibiting GlnS completely.

Example: aspartokinase regulation involves isoenzymes. L-Aspartic acid is a biosynthetic precursor of the amino acids L-lysine, L-methionine, L-threonine, and L-isoleucine (see Chapter 7.6.2 and Figs. 7.**14** and 7.**15**). In *E. coli*, this branched biosynthetic pathway is regulated both by sequential feedback inhibition and by the use of three isofunctional aspartokinases (**AspKI**, **AspKII**, and **AspKIII**), each of which catalyzes the first step in the aspartate pathway, the synthesis of aspartyl phosphate from L-aspartate and ATP (Fig. 19.**7**). AspKI and AspKII also catalyze the third step in this pathway, the formation of homoserine, and therefore also are homoserine dehydrogenases. As shown in Table 19.**2**, the aspartokinases are regulated differently. Threonine is an allosteric inhibitor of AspKI, lysine is an inhibitor of AspKIII, and AspKII is not subject to allosteric control. However, the synthesis of all three enzymes is also differentially regulated by end products of the pathway that act as corepressors. This complex control ensures that an excess of one amino acid end product of the pathway will not cause starvation for another end product.

Anabolic regulatory strategies vary within and between bacterial species. In *E. coli*, examples of all of the various types of allosteric regulatory strategies shown in Figure 19.**2** have been given. However, the same biosynthetic pathways may also be regulated differently in other bacterial species. For example, while *E. coli* has three isofunctional aspartokinases, some species of *Pseudomonas* have only a single AspK that is allosterically inhibited by concerted feedback inhibition; that is, inhibition is observed only in the presence of all end products of the pathway. For 3-deoxy-arabinoheptulosonic acid-7-phosphate synthetase (the enzyme catalyzing the first step of the aromatic amino acid pathway), there are isofunctional enzymes in *E. coli* and other enteric bacteria. There is only one type of this enzyme in *Bacillus*, however, and it is regulated by sequential feedback inhibition. The enzyme in *Pseudomonas* is regulated either by concerted or cumulative feedback inhibition, depending on the species (see also Chapter 35.4.1 and Fig. 35.**5**).

Table 19.**2** **Feedback regulation and repression of the three aspartokinase (AspK) isozymes of *Escherichia coli***

Isozyme	Allosteric inhibitor	Repressed by
AspK—homoserine dehydrogenase I	Thr	Thr+Ile
AspK—homoserine dehydrogenase II	None	Met
AspK III	Lys	Lys

19.2.3 Allosteric Control Is Also Found in Catabolic Pathways

Enzymes of catabolic pathways are also subject to allosteric inhibition and activation. The allosteric reg-

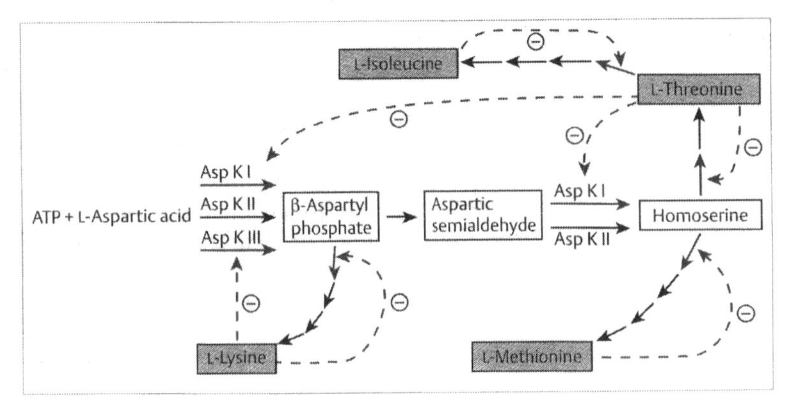

Fig. 19.**7** **Biosynthetic pathway for the aspartate family of amino acids.** The first step is catalyzed in *Escherichia coli* by three different aspartokinase (AspKI, II, III) isozymes. AspKI and AspKII also catalyze the third step in this pathway and are, therefore, also homoserine dehydrogenases. Feedback inhibition (red dashed arrows) through end products (–) is indicated

ulation of glycolysis via the Emden-Meyerhof-Parnas pathway, of gluconeogenesis and of the tricarboxylic acid (TCA) cycle in *E. coli* is illustrated in Figure 19.**8**. It should be noted that, unlike in biosynthetic pathways, allosteric regulation in these pathways is predominantly effected by intermediates, rather than by end products, of the particular pathway. This is due, in part, to the interrelation of the Embden-Meyerhof-Parnas, pentose phosphate, and TCA cycle pathways at the level of intermediates of the Embden-Meyerhof-Parnas pathway, and therefore, the amounts of these intermediates in the cell must be individually controlled. Moreover, the effector intermediates may either be inhibitors, activators, or both (affecting different enzymes). For example, accumulation of fructose 1,6-bisphosphate (signifying that its synthesis is occurring faster than its conversion to triose phosphates) serves to activate both ADP-glucose pyrophosphorylase (Gly C, involved in glycogen synthesis) and pyruvate kinase ("feedforward" regulation). On the other hand, an accumulation of phospho*enol*pyruvate (PEP) near the end of the Emden-Meyerhof-Parnas pathway serves to inhibit phosphofructokinase (PfkA), but activates pyruvate dehydrogenase, which converts pyruvate to acetyl-CoA for entry into the TCA cycle.

Finally, it should be noted that both adenine nucleotides and NADH are also allosteric effectors of various enzymes in these pathways. The relative

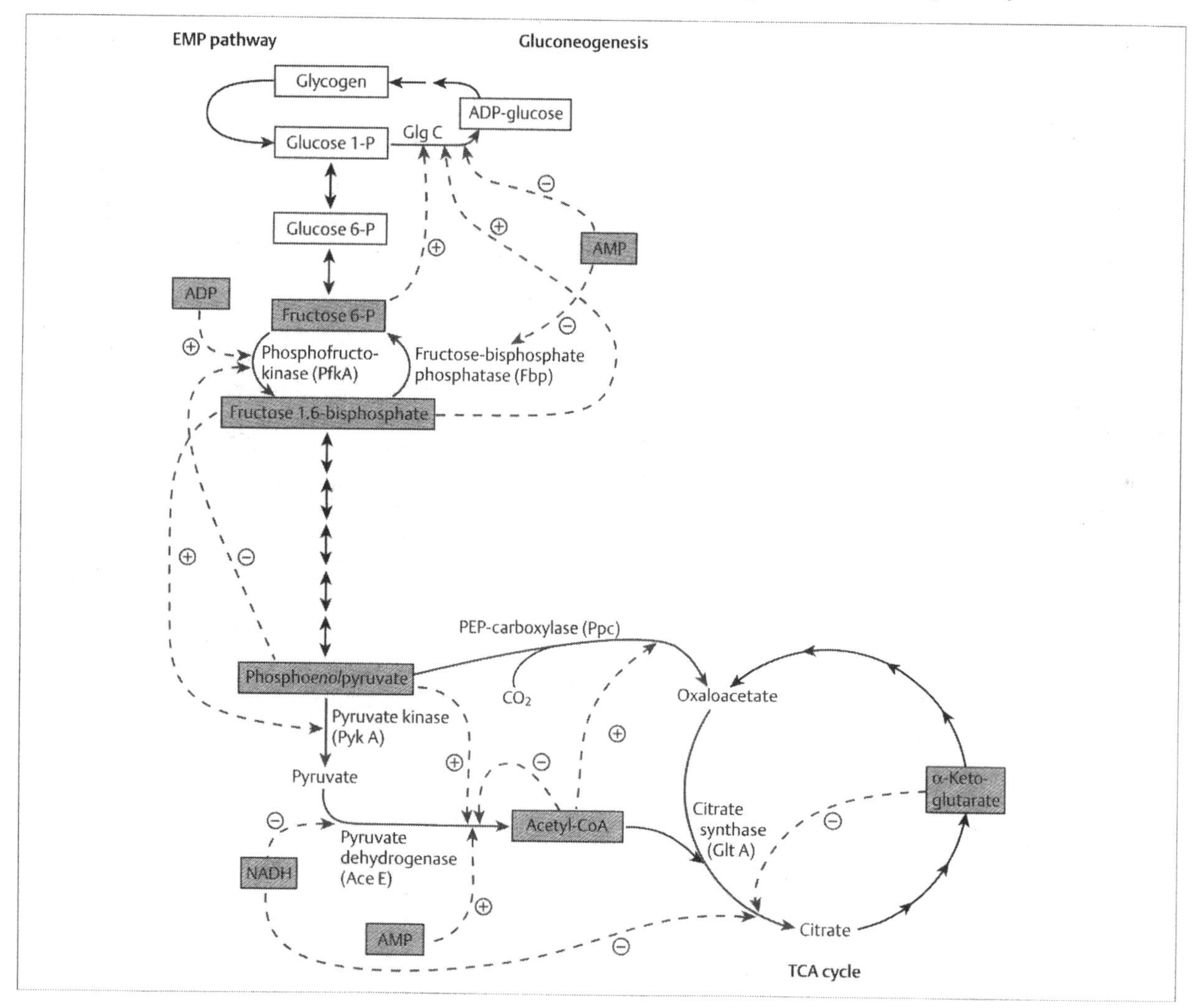

Fig. 19.**8** **Feedback regulation of catabolism: Emden-Meyerhof-Parnas (EMP) pathway, gluconeogenesis, and the tricarboxylic acid (TCA) cycle.** Dashed red lines denote negative feedback regulation (–) or positive allosteric regulation (+) by the indicated compounds (for further details, see text and Fig. 7.**11**). GlgC, ADP-glucose pyrophosphorylase

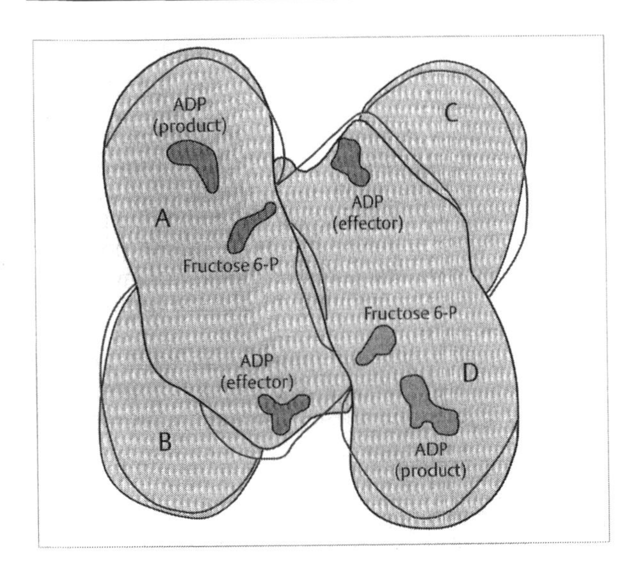

Fig. 19.**9** **Overall structural differences in phosphofructokinase from *Bacillus stearothermophilus* between the T- and R-states.** The four identical subunits (A–D) in the R-state (red) are rotated by about 7° compared to the subunits in the T-state (black). The binding sites for the substrate fructose 6-phosphate, and for ADP (product) and ADP (effector) are also shown [after 1]

amounts of the different adenine nucleotides define the energy state of the cell (the so-called "energy charge"), while the ratio of NADH to NAD^+ defines the overall "reducing power" in the cell (see Chapter 7.4.2). Since the catabolic pathways depicted are "fueling" reactions (produce ATP and NADH), it makes sense that enzymes of these pathways should also be responsive to these states of the cell.

Example: the structural basis for the allosteric control of phosphofructokinase. As indicated in Fig. 19.**8**, phosphofructokinase is allosterically inhibited by phospho*enol*pyruvate, and is activated by ADP. The structural basis for this regulation has been elucidated for phosphofructokinase from *Bacillus stearothermophilus* by determination of its crystal structure in the presence of substrates and effectors. *B. stearothermophilus* phosphofructokinase is a tetramer of identical subunits, each consisting of 319 amino acids, named A through D. Each subunit contains separate binding sites for the substrates (ATP and fructose 6-phosphate), products (fructose 1,6-bisphosphate and ADP), and the allosteric effectors. The binding of fructose 6-phosphate is highly cooperative, and phosphofructokinase has been presumed to undergo a more-or-less concerted transition from the T-state to the R-state upon binding of this substrate to any one of the four binding sites.

As shown in Fig. 19.**9**, the overall structure of the R-state (determined with bound fructose 6-phosphate and ADP) differs from that in the T-state (determined with bound 2-phosphoglycolate, a phospho*enol*pyruvate analogue) by an approximately 7° rotation of the four subunits with respect to one another. Although this structural difference appears rather slight in the overall structure, it profoundly affects the binding sites for fructose 6-phosphate, which are at the subunit–subunit interfaces. In the R-state, the phospho-group of bound fructose 6-phosphate is hydrogen-bonded to Arg-162 of phosphofructokinase. In the T-state, however, the side-chain of Arg-162 is rotated nearly 180° with respect to its orientation in the R-state, so that it no longer can interact with the substrate, and its position is replaced by the side-chain of the negatively charged Glu-161. These charge and hydrogen-bonding differences in the active sites of phosphofructokinase probably largely explain why the affinity of the enzyme for fructose 6-phosphate is more than three orders of magnitude higher in the R-state than in the T-state.

Allosteric regulation is typical in metabolic pathways that contain intermediates common to different anabolic or catabolic pathways, such as in amino acid and pyrimidine biosynthesis (e.g., aspartate carbamoyltransferase, carbamoylphosphate synthetase) and glycolysis/gluconeogenesis (e.g., phosphofructokinase), or for pathways that contain differently regulated isoenzymes (e.g., aspartokinases, 3-deoxy-arabinoheptulosonic acid-7-phosphate synthetase).

19.3 Posttranslational Covalent Modifications of Proteins Control a Wide Variety of Processes in the Prokaryotes

A wide variety of processes in prokaryotes are controlled by covalent modifications of proteins. These include the regulation of transcription, of enzyme activities, of a variety of signal transduction pathways, and of subcellular protein localization. The phosphorylation (and in some cases methylation) of proteins in the so-called two-component signal transduction pathways will be considered in more detail in Chapter 20, as will the catabolic control by phosphorylation/dephosphorylation of some phosphotransferase system (PTS) proteins. In this section, a few selected examples of how covalent modifications can affect the activities of certain bacterial proteins and the roles of proteolytic enzymes in bacteria will be discussed.

Covalent modification of prokaryotic proteins involves **phosphorylation** at various residues, **uridylylation, adenylylation, methylation, fatty acylation**, and **proteolysis**. While proteolysis constitutes an irreversible step, the former are usually reversible processes that often require two enzymes, such as a protein kinase and a phosphatase or a methyltransferase and a demethylase.

19.3.1 Uridylylation and Adenylylation Control Glutamine Synthetase Activity in *E. coli*

Glutamine synthetase (**GlnS**) is involved in ammonia assimilation at low concentrations of ammonia and also provides for the synthesis of glutamine for protein synthesis and as an amido-group donor. The synthesis of GlnS and its activity are subject to complex controls that respond to the glutamine/α-ketoglutarate ratio in the cell as an indication of nitrogen availability. Preexisting GlnS activity is primarily controlled by covalent modification and cumulative feedback inhibition (Fig. 19.**10**).

Each of the 12 subunits of GlnS can be adenylylated at a specific tyrosine residue by the transfer of an AMP group from ATP catalyzed by a specific **adenylyltransferase** (**ATase**), which also catalyzes the deadenylylation reaction. Fully adenylated GlnS is inactive, and partially modified GlnS has an activity that is decreased in proportion to the number of subunits modified. It is the partially adenylylated form that is apparently most sensitive to cumulative feedback inhibition (see Chapter 19.2.2).

The adenylylation of GlnS is controlled by two other proteins, $\mathbf{P_{II}}$ (a regulatory protein) and a uridylyltransferase (*UTase*). The non-uridylylated form of P_{II} interacts with ATase allosterically, stimulating adenylylation of GlnS and inhibiting its activity. The uridylylated form of P_{II} (also on a tyrosine residue) stimulates the deadenylylation activity of ATase, leading to unmodified, active GlnS (Fig. 19.**10**). It is the activity of the UTase that is subject to allosteric control by glutamine and α-ketoglutarate. Glutamine stimulates the deuridylylation activity of UTase, leading to unmodified P_{II} and adenylylation of GlnS, while α-ketoglutarate is an allosteric activator of the uridylyltransferase activity of UTase, which leads to modified P_{II} and deadenylylation

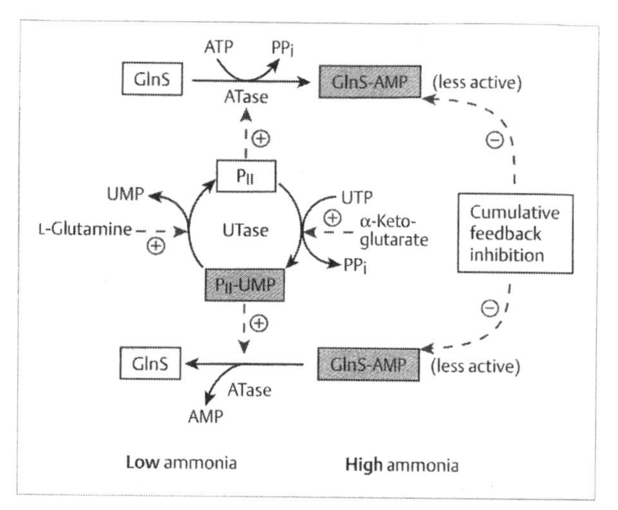

Fig. 19.**10** **Control of glutamine synthetase activity by covalent modification.** Glutamine synthetase (GlnS) is inhibited by adenylylation of a tyrosine residue catalyzed by an adenylyltransferase (ATase). The partially adenylylated GlnS form is most sensitive to the cumulative feedback inhibition (see Chapter 19.2.2) exerted by AMP, CTP, carbamoyl phosphate, GlcN6P, Ala, Gly, His, and Trp. ATase activity is in turn controlled by a regulatory protein, P_{II}, the regulatory activity of which is controlled by a uridylylation reaction catalyzed by a uridylyl-transferase (UTase). UTase activity is subject to allosteric control by glutamine and α-ketoglutarate. See text for further details [after 2]

(activation) of GlnS. Thus, at high glutamine/α-ketoglutarate ratios, signifying an **abundance of ammonia**, GlnS activity is inhibited and highly sensitive to feedback regulation, while at low ratios of glutamine to α-ketoglutarate, reflecting **limiting ammonia**, GlnS activity is high. These controls, along with these metabolites also controlling the synthesis of GlnS (Chapter 20), prevent the wasteful utilization of ATP for glutamine synthesis by GlnS when it is not needed by the cell.

As one of the most central enzymes involved in nitrogen metabolism, **glutamine synthetase** (GlnS) is regulated intricately at the activity level by uridylylation/adenylylation and in response to changes in ammonia concentration. At high Gln/α-ketoglutarate ratios (reflecting high ammonia), GlnS is inhibited; at low ratios (reflecting limiting ammonia), activity is high. Synthesis of the enzyme is controlled by a global regulatory system (Ntr, see Chapter 20.4) which responds to the same Gln/α-ketoglutarate ratio.

19.3.2 Phosphorylation of Proteins Controls Many Cellular Functions in Bacteria

Although the phosphorylation of proteins in bacteria, apart from the phosphotransferase system (PTS; see Chapters 5 and 20), has only been recognized over the past 20 years or so as being important for cellular functions, there has been a recent explosion in the identification of prokaryotic phospho-proteins and the kinases that phosphorylate them. A selective list of these phospho-proteins and kinases, along with their roles in sensing, signal transduction, and enzyme activity control, is given in Table 19.**3**. Phosphorylation of histidine (sensor proteins of two-component systems, PTS proteins), aspartate (regulator proteins of two-component systems), cysteine (PTS), tyrosine, serine, and threonine have all been demonstrated in a variety of bacteria. The protein kinases, traditionally classified according to their acceptor amino acids, transfer a phosphate group from a donor (generally the γ phosphate of ATP or GTP) to a specific residue on the target protein. The phosphate group either dissociates spontaneously or has to be removed by specific phospho-protein phosphatases. Except for the two-component systems and the PTS (see Chapter 20), relatively more is known about the protein kinases catalyzing these phosphorylations than about the protein targets and/or the physiological roles of these modifications.

In two-component systems (see Chapter 20.4), the sensor protein is phosphorylated at a histidine residue by ATP, usually autocatalytically in the dimeric form (autokinase), and this phosphorylation is modulated by an environmental signal. The phospho-sensor protein can then phosphorylate a response-regulator protein at an aspartate residue (receiver); this produces a response in the cell (e.g., modulation of flagellar rotation in chemotaxis, transcriptional regulation in osmotic control and in sporulation in *Bacillus*). In the PTS, which is responsible for carbohydrate transport and phosphorylation in many bacteria (Chapter 5), a phosphotransfer cascade from phospho*enol*pyruvate involves several phospho-histidyl proteins, and one phospho-cysteinyl protein (at least in the case of mannitol and glucose). Two of the intermediate phosphotransfer proteins of the PTS, IIA^{Glc} and HPr, appear also to be involved in the regulation of several cellular processes depending on their phosphorylation states (see Chapter 20).

In the remainder of this section, several examples of protein phosphorylation in bacteria that have been elucidated in some detail will be briefly discussed.

Covalent modification by **protein phosphorylation** is probably the most frequent control mechanism in prokaryotic posttranslational control and involves sequence-specific protein kinases and phosphoprotein-phosphatases specific for **histidine**, **cysteine**, **tyrosine**, **aspartate**, **threonine**, or **serine** residues. These enzymes have a central role in signal transduction (two-component systems), global gene control (phosphotransferase systems), and cell differentiation processes (e.g., sporulation).

Example: phosphorylation of isocitrate dehydrogenase in the control of the TCA cycle and the glyoxylate bypass. Isocitrate dehydrogenase (IDH) catalyzes the

Table 19.**3** **Some examples of phospho-proteins and protein kinases in bacteria**

Phospho-protein(s)	Residue(s)	Kinase(s)	Role(s)
Phosphotransferase system (PTS) proteins	His, Cys	PTS proteins	Carbohydrate uptake
IIA^{Glc} and HPr of the PTS	His	Enzyme I of the PTS	Regulation of nutrient assimilation
Sensor kinases (two-component systems)	His	Autokinase (transmitter)	A variety of sensory adaptations (see Table 20.**1**)
Response regulators (two-component systems)	Asp	Sensor kinase (receiver)	A variety of sensory adaptations (see Table 20.**1**)
HPr of the PTS	Ser	HPr kinase	Regulation of nutrient assimilation in Gram-positive bacteria
Isocitrate dehydrogenase (IDH)	Ser	IDH kinase	Regulation of IDH activity
SpoIIAA	Ser	SpoIIAB	Sporulation in *Bacillus*
Numerous	Ser/Thr	PKN1	Development in *Myxococcus*, etc.

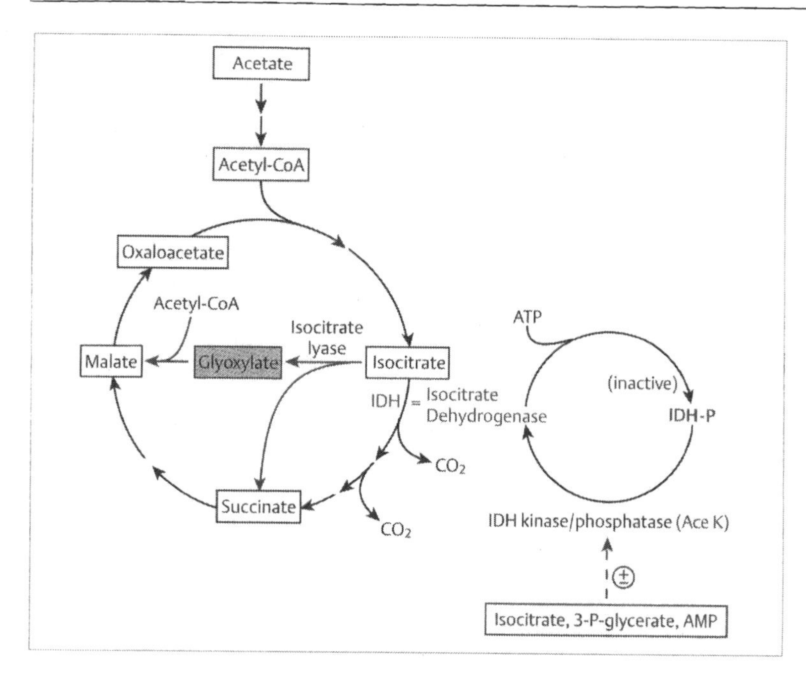

Fig. 19.**11** **The glyoxylate bypass, TCA cycle, and isocitrate dehydrogenase.** The glyoxylate bypass (solid red arrows) is necessary in *Escherichia coli* for growth on acetate or fatty acids. Isocitrate dehydrogenase (IDH) of the tricarboxylic acid (TCA) cycle is accordingly inhibited by a phosphorylation reaction under these conditions, diverting isocitrate through the glyoxylate bypass. For further explanations, see text

NADP-linked oxidation of isocitrate to α-ketoglutarate in the tricarboxylic acid (TCA) cycle (see Chapter 11.3.6). However, for growth on acetate or fatty acids, bacteria such as *E. coli* require the functioning of the glyoxylate bypass, the first step of which is the conversion of isocitrate to glyoxylate via isocitrate lyase (Fig. 19.**11**). Otherwise, the carbon entering the TCA cycle as acetyl-CoA, formed from acetate or fatty acids, would be quantitatively converted to CO_2, and there could be no net conversion of this carbon into metabolic intermediates.

During growth on acetate, isocitrate dehydrogenase in *E. coli* is inhibited by approximately 75% compared to during growth on most other carbon sources, thus helping to shunt isocitrate toward the glyoxylate bypass. This inhibition is caused by the phosphorylation of a single serine residue of isocitrate dehydrogenase (Ser-113), which completely inactivates the enzyme (Fig. 19.**11**). Phosphorylation of isocitrate dehydrogenase is carried out by an ATP-dependent IDH kinase/phosphatase (encoded by the *aceK* gene in *E. coli*), which also catalyzes dephosphorylation of phospho-isocitrate dehydrogenase. The importance of this regulatory control for growth on acetate is illustrated by the lack of growth of *aceK* mutants on acetate as the sole carbon source.

The isocitrate dehydrogenase phosphorylation cycle appears to be controlled by a number of metabolites acting as allosteric effectors of IDH kinase/phosphatase. Both isocitrate and 3-phosphoglycerate activate the phosphatase activity and inhibit the kinase activity of the enzyme. The levels of these compounds reflect the overall levels of metabolic intermediates, and therefore, when these levels are low (as, for example, when cells are shifted to an acetate-containing medium), the kinase activity is higher than the phosphatase activity, forcing more isocitrate through the glyoxylate bypass. AMP has also been shown to activate the isocitrate dehydrogenase phosphatase and inhibit the kinase. High AMP levels signify a general energy depletion in the cell and thus would tend to increase the amount of isocitrate going through the complete TCA cycle.

Example: serine phosphorylation in the PTS and nutrient acquisition in Gram-positive bacteria. The PTS is responsible for the uptake and phosphorylation of a variety of carbohydrates in both Gram-negative and Gram-positive bacteria (see Chapter 5.6.4). The PTS is a well-known regulator of catabolic systems in Gram-negative bacteria (see Chapter 20.2). In Gram-positive bacteria, the PTS also appears to be involved in catabolic regulation, but by different mechanisms (see Chapter 20.2.4). At least one of these mechanisms involves an ATP-dependent protein kinase, one of the first such kinases to be discovered in Gram-positive bacteria. This HPr kinase phosphorylates a serine residue (Ser-46) of the general PTS phosphotransfer protein, HPr, apparently only in Gram-positive species.

Phosphorylation of HPr by HPr kinase makes it a much poorer substrate for phosphorylation at His-15 by phospho-Enzyme I of the PTS, a phosphotransfer reaction necessary for PTS carbohydrate phosphorylation (Fig. 19.**12**). The HPr kinase is allosterically stimulated by such glycolytic intermediates as fructose

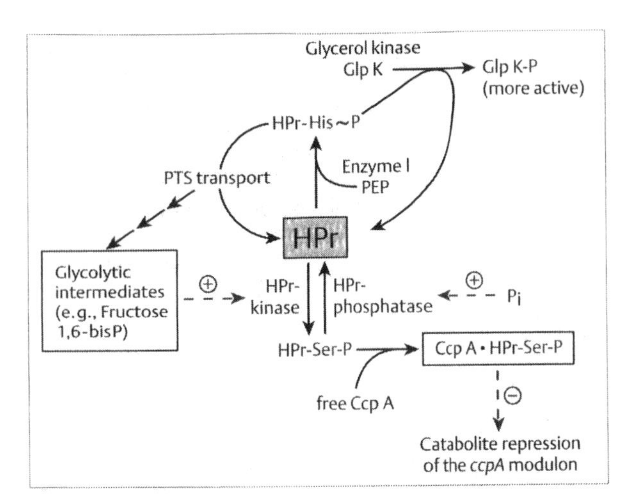

Fig. 19.**12** **Regulation of HPr of the PTS in Gram-positive bacteria by ATP-dependent phosphorylation.** Phosphorylation of HPr on a serine residue by an HPr kinase inhibits its ability to carry out its role in phosphotransfer (via a His residue) in PTS-substrate phosphorylation. This may act as a mechanism of feedback control in at least some Gram-positive organisms. Moreover, phospho-(Ser)-HPr is believed to be involved in repression of some catabolic genes in Gram-positive bacteria by virtue of its ability to form a complex with the catabolite control protein, CcpA. In addition, phospho-(His)-HPr phosphorylates other catabolic enzymes, such as glycerol kinase (GlpK), to regulate their activities

1,6-bisphosphate, while a separate protein, HPr phosphatase, is most active at high concentrations of inorganic phosphate. This has led to the suggestion that ATP-dependent phosphorylation of HPr may be a feedback regulatory mechanism on PTS-carbohydrate uptake, HPr activity being inhibited by ATP-dependent phosphorylation under conditions in which glycolytic intermediates are being formed faster than they can be utilized. While there is evidence for this in some Gram-positive organisms, this feedback mechanism has yet to be proven in any species.

Recent evidence does suggest a role for phospho-(seryl)-HPr in catabolite repression in *B. subtilis*. A number of catabolic genes in this organism are controlled via a *cis*-acting **c**atabolite **r**esponsive **e**lement (*cre*). Mutations in this sequence have been shown to lead to release certain genes (e.g., *amyE* encoding α-amylase) from catabolite repression by glucose and other PTS substrates. A *trans*-acting regulatory protein called **CcpA** (for **c**atabolite **c**ontrol **p**rotein) is believed to interact with various *cre* sequences and act as a repressor under certain conditions. Moreover, a direct protein–protein interaction between CcpA and phospho-(seryl)-HPr, but not other forms of this protein, has been demonstrated.

Phospho-(seryl)-HPr is likely to be a predominant form of this protein during uptake of PTS substrates, but probably exists at only low levels in the absence of these compounds. Therefore, it has been proposed that phospho-(seryl)-HPr may act as a corepressor along with CcpA in mediating at least some types of catabolite repression in *Bacillus* (Fig. 19.**12**). Consistent with this hypothesis is the observation that when a Ser-46→Ala mutant is expressed as the sole HPr species in *Bacillus subtilis*, a number of catabolic activities encoded by genes under the control of *cre* sequences and CcpA are relieved from catabolite repression by glucose.

Finally, yet another mechanism by which HPr may regulate catabolism in Gram-positive bacteria via covalent phosphorylation has been described. Glycerol (dihydroxyacetone) kinase of *Enterobacter faecalis* is phosphorylated at a histidine residue by phospho-(histidyl)-HPr. This phosphorylation of glycerol kinase increases its activity about ten-fold. This observation, coupled with the inhibition of glycerol uptake in *E. faecalis* by PTS substrates, suggests that phospho-(histidyl)-HPr (which predominates over free HPr in the absence of PTS substrates) is responsible for the higher glycerol kinase activity under these conditions (Fig. 19.**12**).

Protein phosphorylation is central to the control of carbohydrate transport and metabolism in the Gram-positive bacteria. It involves a histidine- and a serine-protein kinase/phosphatase system, the former promoting uptake and metabolism, the latter inhibiting and repressing ("**catabolite repression**") genes and enzymes for carbohydrate metabolism.

Example: serine (threonine) protein kinases may regulate development in diverse bacteria. Sporulation in *B. subtilis* is a complex and irreversible developmental process leading ultimately to the lysis of the "mother" cell and the release of a highly resistant endospore. A complex differentiation pathway controlled, in part, by the *spo* genes is involved in directing this process, and protein phosphorylation is used as a means of signal transduction during sporulation (also see Chapters 20 and 25). One sporulation-specific protein, Spo0A, is a typical response regulator of the two-component system class (see Fig. 25.**2**), and its phosphorylated form is a regulator (activator or repressor) of genes involved in the developmental program. There is evidence, however, that serine phosphorylation may also play a role in sporulation in *B. subtilis*, and in particular that a protein serine kinase called SpoIIAB may be involved.

The Gram-negative gliding bacterium, *Myxococcus xanthus*, undergoes a complex developmental process in response to nutrient deprivation in which the individual cells exhibit social behavior that leads to the formation of a multicellular fruiting body containing relatively resistant myxospores (see Chapter 25.3.3). Recent evidence implicates protein serine and/or threonine kinases in this process. The *pkn1* gene of *M. xanthus* has been cloned and sequenced. The deduced sequence of the protein product (PKN1, 693 residues) exhibits high similarity in its N-terminal half to eukaryotic protein serine/threonine kinases, most notably the rat Ca^{2+}/calmodulin-dependent protein kinase II. Like the eukaryotic kinases, PKN1 is autophosphorylated by ATP on serine and threonine residues. The protein substrate(s) of this kinase, however, remain to be identified. Interestingly, Southern blot analysis of a genomic library from *M. xanthus* has shown that there may be as many as 26 *pkn*-like genes in this organism, while attempts to find similar genes in *E. coli*, which of course does not have an extended developmental cycle, have been unsuccessful.

19.3.3 Methylation and Acetylation Are Sometimes Used in Protein Modification

Protein carboxymethylation is best known in bacteria as an adaptation mechanism in bacterial chemotaxis mediated by the so-called methyl-accepting chemotaxis proteins (MCPs), which are integral-membrane chemotaxis receptors (see Chapter 20.5.4). Methylation of the MCPs is carried out by the CheR methyltransferase using *S*-adenosylmethionine as the methyl group donor. Depending on the MCP, four to six glutamate residues in the carboxyl-terminal, cytoplasmic domain of these proteins can be methylated by CheR. Methylation of the MCPs appears to uncouple their receptor and cytoplasmic transmitter domains. This may, at least in part, allow them to respond to changes in the levels of chemo-effectors over a wide concentration range. Demethylation of the MCPs is catalyzed by the CheB methylesterase, the activity of which is stimulated by the phosphorylation of an aspartate residue of CheB by CheA, a sensor histidine kinase of the two-component regulatory system family. The significance of this for chemotaxis will be considered further in Chapter 20.

Although the methylation of other bacterial proteins is known (e.g., protein synthesis elongation factor EF-Tu is methylated on a lysine residue), the physiological significance is in general unknown. Likewise, a number of bacterial proteins have been shown to be acetylated, but again there is little evidence bearing on the physiological function, if any. The most notable example is ribosomal protein L7 in *E. coli*, in which the N-terminal residue is acetylated, but which is otherwise identical to ribosomal protein L12. L7/L12 (RplL) are the only proteins of the large subunit of the ribosome that appear to be present in multiple (four) copies. Again, however, the physiological significance of the acetylation of L7 on ribosomal function is unclear.

19.3.4 Proteolytic Processing as a Mechanism of Posttranslational Control

A remarkable number of endoproteases, well over 20, have been identified in *E. coli* cells (listed in Table 19.**4**), but the exact physiological role of many is not known. Many of these proteases undoubtedly have a "housekeeping" function, that is, the removal of mistranslated, damaged, improperly folded, or "foreign" proteins. Others have a regulatory role, degrading certain proteins "constitutively" soon after their synthesis to maintain a low steady-state level of such proteins, degrading "scaffold" proteins that have a role in the synthesis of complex structures such as flagella or fimbriae, or degrading proteins only in response to a specific signal, for example, starvation. Most of the proteolytic degradation, at least in *E. coli*, appears to require metabolic energy in the form of ATP. The number of different proteases in *E. coli* and the large amount of energy devoted to protein degradation suggest that this type of posttranslational modification is extremely important in the life of a bacterial cell.

It has been estimated based on pulse-labeling experiments in *E. coli* that as much as 20% of the protein synthesized during the pulse is degraded within one generation. However, different proteins clearly have different half-lives in these cells. For example, approximately 40% of the bulk protein has a half-life of greater than 3 days in exponentially growing cells, while 15% has a half-life of less than 5 h. Among the latter proteins are abnormal or mutant proteins, which may have very short half-lives of less than 1 h. Moreover, certain *E. coli* proteins having a regulatory function or roles in DNA replication or cell division are naturally very susceptible to proteolysis and have half-lives of less than 2 min, for example, the SulA protein or the heat-shock sigma subunit, σ^{32} (also see below). The general environmental conditions can also drastically affect the susceptibility of *E. coli* proteins to proteolytic degradation. For example, in contrast to exponentially growing cells, over 50% of the bulk protein is degraded with a half-life of less than 5 h in starved *E. coli* cells.

Table 19.4 **Endoproteases in *Escherichia coli*** (After Maurizi, M. R. (1992) Experientia 48: 178–201

Protease	Gene	Substrates in vivo (in vitro)	Protease type	Remarks
Cytoplasmic proteases				
Lon protease (protease La)	*lon*	SulA; λ N protein; RcsA; Tn*903* transposase	Serine	ATP-dependent; degrades λN in vitro
Clp protease (protease Ti)	*clpP* *clpA*	pro-ClpP; ClpA; LacZ fusions	Serine	ATP-dependent activation by Clp A ATPase activity
RecA/LexA	*recA* *lexA*	(λ cI repressor; UmuD) LexA	Serine	ATP-dependent; autolysis of substrates
Hfl A	*hflK* *hflC*	λ cII repressor	Serine	Low activity against cII in vitro
Proteases Do, Re, Fa, So		(Casein)	Serine	
Protease Ci		(Insulin)	Metallo	
ISP-L-Eco		(Z-AAL-PNA)	Serine	
Protease II		(BAEE)	Serine	Very low activity against proteins
(Alp-related)	(*alpA*)	SulA; RcsA		Alp affects expression of ATP-dependent protease
Periplasmic or membrane-associated proteases				
Protease Mi		(Casein)	Serine	
Protease Pi (protease III)	*ptr*	(Insulin)	Metallo	Insulinase superfamily
Signal peptidase I	*lep*	Precursors of exported proteins		Inner and outer membranes
Signal peptidase II	*lsp*	Precursors of lipoproteins		Outer membrane
Deg P (HtrA)	*degP*	PhoA fusions	Serine	σ^E heat shock protein
Protease I		(NAPNE)	Serine	Very low activity against proteins
Protease IV	*sppA*	Signal peptides after processing	Serine	Signal peptide peptidase
Protease IVa				
Protease V		(Z-Phe-ONP)	Serine	Inner and outer membranes
Protease VI		Membrane proteins	Serine	
Protease VII	*ompT*	Ferric enterobactin receptor protein	Serine	Cleaves between basic residues
Protease peri7		(Casein)	Metallo	Activated by ATP
Protease peri8		(Casein)	Serine	

Two ATP-dependent proteases, Lon and Clp, account for up to 60% of the total protein degradation in growing *E. coli* cells. These proteins, as well as the activation of a proteolytic function ascribed to the RecA protein, will be described briefly in this section as examples of activities that are responsible for both "housekeeping" and more specific regulatory functions of bacterial proteases.

The Lon protease is both a general and a specific protease. Protease Lon (also called protease La) is the product of the *lon* gene in *E. coli*, and is an ATP-dependent serine protease that degrades both specific regulatory proteins and abnormal proteins. Lon is an apparent tetramer of 87-kDa subunits, each consisting of 783 amino acids. ATP is an allosteric activator of the enzyme and is hydrolyzed during turnover of the enzyme, especially in the degradation of larger protein substrates. Lon possesses endoproteolytic activity, and cleaves its substrates to yield short oligopeptides, 5–20 residues in length.

A number of studies have implicated Lon protease in the general degradation of abnormal proteins. In *lon* mutants, the half-lives of naturally unstable proteins, temperature-sensitive proteins, and other abnormal proteins (such as those containing amino acid analogues or cloned foreign proteins), increase compared to such proteins in wild-type cells. Moreover, in vitro, large proteins are better substrates for Lon after they have been partially or completely denatured, suggesting that Lon recognizes for cleavage more extended or flexible regions of a polypeptide chain.

However, it is also clear that Lon is responsible for a number of more specific proteolytic cleavages of regulatory significance in vivo. *E. coli lon* mutants show an increased sensitivity to ultraviolet light, the result of an increased half-life of the SulA protein, a Lon substrate that is induced during the SOS response. SulA is an inhibitor of septum formation during cell division and has a transient role during DNA repair resulting from the SOS functions (for details, see Chapters 15.7 and 22.4.3). In Lon$^-$ cells, SulA accumulated during the transient SOS response is not rapidly degraded, and the cells form long filaments and die. A second phenotype of *lon* mutants is that these cells become mucoid as a result of an increase in capsular polysaccharide synthesis. This is due to an increase in the level of RcsA, the transcriptional activator of the *cps* genes (necessary for the synthesis of **c**apsular **p**oly**s**accharide). Under many environmental conditions, the level of RcsA is kept low, in part because of its rapid degradation by Lon. Thus, in these examples, Lon acts specifically either to eliminate a protein (SulA) rapidly that is needed only transiently after its induction, or to maintain under most conditions only a low, steady-state level of a protein (RcsA) needed in larger quantities only under special environmental conditions.

Interestingly, an apparent homologue of Lon has been found in *M. xanthus*. Cells of this organism carrying a mutation in the *bsgA* gene are defective in development and sporulation (see Chapter 25.3.2). The BsgA protein sequence is similar to that of Lon, and purified BgsA is also an ATP-dependent protease. Thus, this protease may play an important role in regulating the developmental pathway in *M. xanthus*. It is to be anticipated that homologues of Lon will be found in many bacteria and may play diverse regulatory functions in addition to the general degradation of abnormal proteins.

The ATP-dependent Clp protease of *E. coli* is conserved in a wide variety of organisms. A second major ATP-dependent protease of *E. coli* was identified in *lon* mutants as an activity that could degrade casein in vitro. This activity was purified and shown to be a complex of two different proteins, ClpA and ClpP, encoded by the *clpA* and *clpP* genes, respectively. Native Clp protease (also called protease Ti) has a molecular mass of 750 kDa and probably is a complex consisting of a dodecamer of 21.5-kDa ClpP subunits and a hexamer of 85-kDa ClpA subunits. Isolated ClpA has ATPase activity that is activated by ClpP and protein substrates and contains two consensus ATP-binding sequences. ClpP is the actual protease and appears to be a classical protease of the serine class. ClpP can cleave small peptides and denatured larger proteins in the absence of ClpA. However, ClpA and ATP together greatly increase the proteolytic activity of ClpP, especially toward larger proteins. ATP appears to have two roles for the activity of Clp: ATP is necessary for the association of ClpA with ClpP, and the hydrolysis of ATP is essential for the proteolysis of at least large proteins.

The physiological role of Clp is probably at least the degradation of abnormal proteins. Cells bearing mutations either in *clpA* or *clpP* are partially defective in this process, an effect that is much more apparent in a *lon* mutant background. Interestingly, ClpP is synthesized with an amino-terminal 14-amino-acid extension that is rapidly removed after synthesis. This removal appears to be an activity of ClpP itself, and is not dependent on ClpA. The role of this processing reaction, however, is unknown.

ClpP and ClpA levels increase in cells during the late exponential phase of growth, suggesting a growth-phase-dependent role of Clp protease. However, Clp is not essential to the cell. Mutants lacking Clp show no drastic defects in growth under a variety of conditions.

Both ClpA and ClpP are apparently highly conserved proteins in the biological kingdoms. Similar genes and/or proteins have been found in other bacteria, other microorganisms, plants, and animals. In most cases, the exact functions of these proteins are not known. The degree of conservation, however, suggests that these similar proteins also are ATP-dependent proteases in these organisms.

LexA repressor is inactivated by autoproteolysis. Several *E. coli* proteins are capable of a specific autoproteolysis reaction that is important for regulation of their activities. One such protein is the LexA repressor, which controls a number of genes that are expressed in response to DNA damage, including *sulA*, the gene discussed earlier as being involved in regulating cell septum formation (see Chapter 15.7). Upon damage to DNA, the LexA repressor is inactivated by a single proteolysis event at a specific Gly–Ala bond. This cleavage is dependent on RecA protein and ATP. Apparently, the RecA–ATP complex, possibly also in association with single-stranded DNA fragments, acts as an allosteric activator of a serine protease domain of LexA that cleaves, and inactivates, itself. The cleavage event also apparently requires ATP, and therefore LexA can be thought of as a highly specific, ATP-dependent autoprotease, but only in the presence of RecA. Interestingly, two other such autoproteases, phage λ CI repressor and the UmuD protein (under the control of the LexA repressor and involved in SOS-induced mutagenesis) are also autoproteolyzed in the presence of the RecA protein. Cleavage of the λ CI repressor results in its inactivation and hence prophage induction (see Chapter 26.4.3), while cleavage of the UmuD activates the protein.

Posttranslational control through **proteolysis** seems to be characteristic for **irreversible** metabolic steps and for steps committing a cell to irreversible differentiation processes, such as prophage induction, cell division, and sporulation. There are general proteases (e.g., Lon of *E. coli*) that degrade several regulatory proteins, but also damaged and abnormal proteins, and there are specific proteases (e.g., LexA, which degrades and inactivates itself).

19.4 Posttranslational Control Includes Enzymes and Protein Compartmentalization Within the Unicellular Prokaryotes

Unlike eukaryotic cells, most bacteria lack specialized intracellular organelles. This may explain, to a large extent, the intricate regulation of catabolic and anabolic pathways and macromolecular biosyntheses in bacteria because most of these processes take place in a continuous cytoplasm. To cope with this problem, anabolic and catabolic pathways for the same compound must be differentially regulated to avoid "futile cycles," the directions of bidirectional and branched pathways must be able to respond in different ways to growth conditions, and in some cases, gene expression must be regulated in a temporal fashion, i.e., consecutively. These types of regulation allow an efficient compartmentalization of metabolism in the absence of physical compartmentalization.

However, some bacterial proteins are subject to physical compartmentalization after their synthesis. Examples include proteins of the photosynthetic apparatus of photosynthetic bacteria (see Chapter 24.4.2), membrane-bound proteins, periplasmic proteins in Gram-negative bacteria, and proteins excreted into the extracellular medium. In this section, some of the mechanisms of compartmentalization will be briefly reviewed, with an emphasis on the physical compartmentalization of proteins. However, the reader is also referred to other chapters in which some of these processes in specific organisms are discussed in more detail.

Although the prokaryotes are unicellular organisms, they use posttranslational control by **compartmentalization** efficiently. Enzymatic steps running in opposite directions (e.g., biosynthesis/degradation of an amino acid) are separated at the **kinetic level** (different affinities and equilibrium), are expressed (induced) consecutively (**temporal** compartmentalization), or are arranged in separate enzyme complexes. The most efficient control, however, is through **physical compartmentalization.**

19.4.1 Protein Targeting to the Membrane or Beyond and Physical Compartmentalization

In Gram-negative bacteria, specific proteins are found in the inner membrane, the periplasmic space, the outer membrane, and in some cases are excreted completely from the cell. Gram-positive organisms lack an outer membrane, but also must target proteins to the cytoplasmic membrane and for secretion. Examples of periplasmic and excreted proteins include many hydrolases with a scavenging function, proteins conferring antibiotic resistances, exotoxins, enzymes involved in cell wall biosynthesis, and solute-binding proteins. Integral inner membrane proteins include certain enzymes, cytochromes, the F1/F0 ATP synthase, and specific transport proteins, as well as the porins and other types of transporters found in the outer membranes of Gram-negative bacteria.

Protein targeting for transport through the cytoplasmic membrane is usually accomplished via a signal-sequence-dependent mechanism, while targeting completely through both membranes in Gram-negative bacteria is carried out, at least in part, by different types of mechanisms. These are also common in the synthesis of flagella, pili, fimbriae, and other cell adhesive organelles. If the protein remains with the cell after targeting (e.g. in the periplasmic space or outer membrane of Gram-negative bacteria), the process is called **secretion**, but if the process results in release of the protein into the medium, it is called **excretion**. Integral membrane proteins are most likely targeted to the membrane largely because of their general hydrophobic nature, but at least in some cases, a membrane secretory apparatus may be involved. In a few other cases, association with the membrane is accomplished via a posttranslational fatty-acylation of the protein (see below).

There are several **physical compartments** in a prokaryotic cell: cytoplasm, cytoplasmic or inner membrane, periplasmic space, outer membrane, outer cell surface, and "out." Each protein is targeted specifically to its target locus. Translocation through the inner membrane is called **secretion**, while release into the medium ("out") is called **excretion**. The latter process is the normal process in Gram-positive bacteria.

Protein secretion through bacterial membranes frequently uses the Sec pathway. In many cases, protein secretion through the inner membrane of bacteria occurs by a mechanism dependent on an amino-terminal **signal sequence** (Fig. 19.**13**) that targets the secreted proteins to an inner-membrane-localized secretory apparatus. This appears to be the case for the vast majority of periplasmic and outer membrane proteins of Gram-negative bacteria, as well as for proteins that are excreted into the medium by Gram-positive organisms. Proteins that are secreted through both membranes by Gram-negative bacteria may traverse the inner membrane by either a signal-sequence-dependent or -independent mechanism. Genetic and more recent biochemical studies, primarily in *E. coli*, have identified a number of genes (collectively called the *sec* genes) and their products that mediate this process (Table 19.**5**).

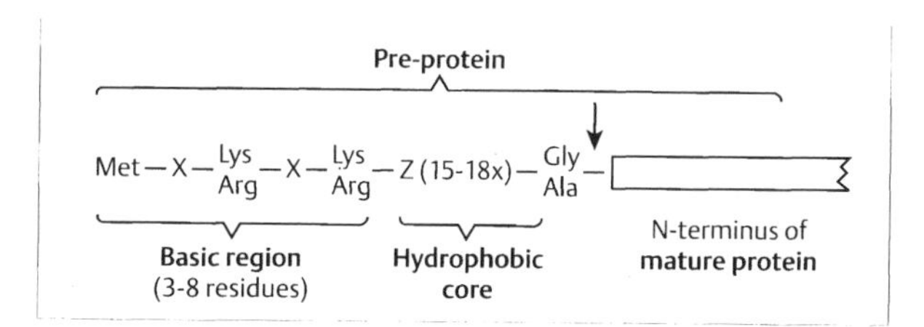

Fig. 19.**13** **Consensus signal sequence for the secretion of bacterial proteins through the cytoplasmic membrane.** In the newly synthesized pre-protein, a positively charged amino-terminus is followed by a hydrophobic "core" of (15–18) amino acids, terminating in a Gly or Ala residue, which denotes the site of cleavage (red arrow) by a signal peptidase. Cleavage releases the processed mature protein. X, hydrophilic or hydrophobic residue; Z, predominantly hydrophobic residues

A schematic illustration of the roles of the *sec* gene products in protein secretion across the cytoplasmic membrane is shown in Figure 19.**14**. The SecB protein binds to many pre-secretory-proteins (or to short **pre-proteins**) primarily through determinants on the "mature" region of the polypeptide. SecB is an example of a specific molecular chaperone. In general, **chaperones** are proteins that stabilize a particular conformational state of a protein either during, or more commonly after, its synthesis. This stabilization, in turn, assists in the subsequent folding ("morphopoiesis") of the protein and/or its assembly into a more complex structure.

In the case of Sec protein-dependent secretion, SecB is believed to keep the pre-protein (protein still containing its signal sequence) in a "translocation-competent" conformation. Secretion of some proteins is, however, SecB-independent. In most of these cases, other more general molecular chaperone proteins (e.g., GroEL) seem to play a role similar to that of SecB. The [pre-protein·SecB] complex binds to the SecA protein, which is partly free and partly peripherally associated with the inner surface of the cytoplasmic membrane by virtue of its binding to the integral-membrane SecY/E/G complex. SecA binds to determinants on SecB and to the signal sequence and other regions of the pre-protein. SecA possesses an ATPase activity that is activated when the [SecB·pre-protein·SecA] complex binds to SecY/E/G. ATP hydrolysis may, in part, serve to release the pre-protein from SecA and deliver it to the SecY/E/G "**translocase.**"

Table 19.**5** **Roles of the products of the *sec* genes of *Escherichia coli***

Protein	Polypeptide size (kDa)	Role
SecB	17 (tetramer)	Binds pre-proteins and keeps them in a "translocation-competent" conformation
SecA	102 (dimer)	"ATPase" that binds the SecB pre-protein complex, delivers it to SecY/E/G, and participates in translocation
SecY	49	Integral membrane protein; presumed "translocase"
SecE	14	Integral membrane protein that complexes with SecY; may be involved directly or indirectly in translocation
SecG	11	Integral membrane protein that may work in concert with SecA during secretion
SecD, SecF	67, 35	Membrane proteins that may have roles in the late steps of secretion

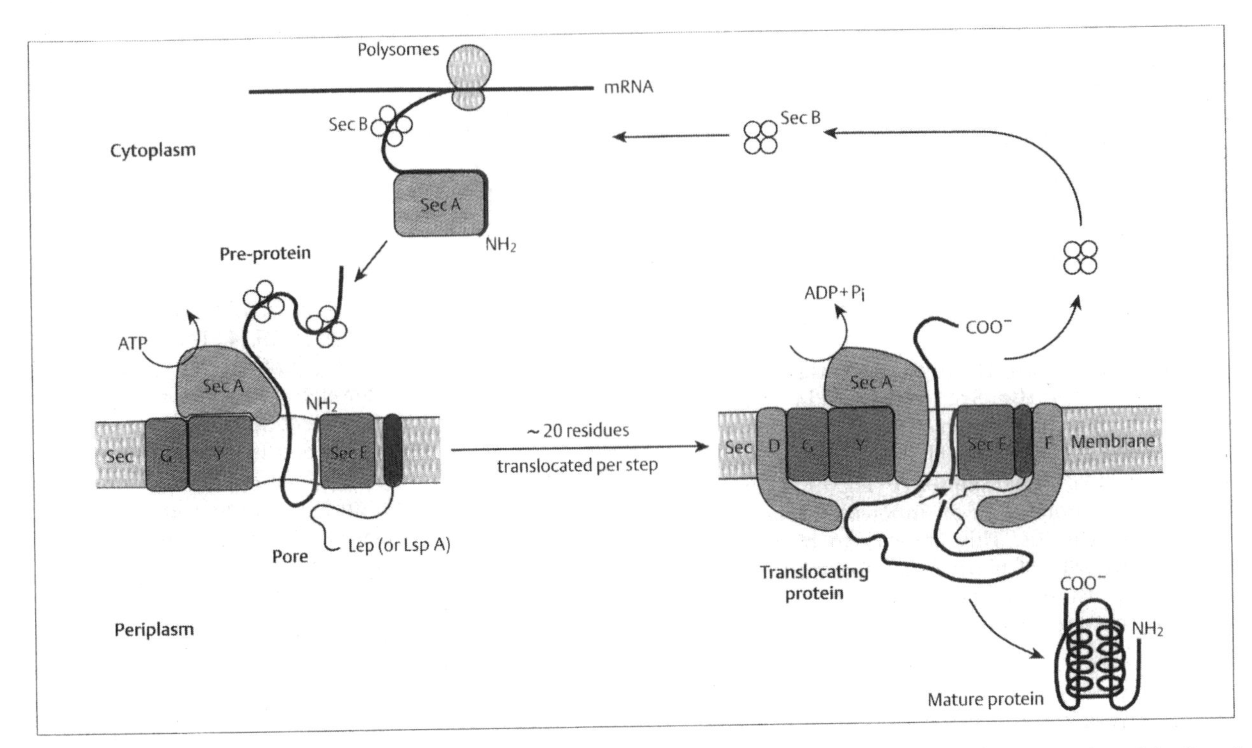

Fig. 19.**14** **Schematic illustration of protein secretion dependent on the sec gene products.** SecB is a molecular chaperone that keeps the pre-protein in its "translocation competent" state. SecA is a protein involved in delivering the [SecB·pre-protein] complex to the translocation complex (SecY/E/G) in the membrane and which also participates in the translocation process with SecG. The exact roles of SecD and SecF remain to be determined. The signal peptidase (Lep) is believed to act at the periplasmic surface of the cytoplasmic membrane. The signal peptide is shown in red. See text for further details [after 3]

Protein secretion through the inner membrane requires a complex secretion apparatus (encoded by the *sec* genes) that contains proteins (SecD, E, F, G, Y) and the integral membrane protein SecA, which hydrolyzes ATP to drive peptide translocation through the pore. Only peptides carrying a signal peptide will bind with the help of a chaperone (usually SecB) to SecA and be secreted. A signal peptidase at the periplasmic side processes the secreted protein into the mature form. The further targeting steps are largely unknown.

The detailed mechanism of secretion of the protein by Sec-protein machinery is unknown. However, recent evidence has shown that during translocation of the pre-protein, a portion of SecA inserts into the membrane such that parts of this protein become transiently exposed to the periplasmic surface. This is accompanied by a major change in the orientation of the SecG protein. Several insertion-deinsertion cycles of SecA may be necessary to translocate a protein, and this cycling process also requires ATP. This has led to the proposal that SecA, perhaps also in conjunction with SecG, acts as a sort of molecular "sewing machine" during protein translocation, each step transporting about 20 amino acid residues. SecY and SecE may help form some sort of pore or channel through which the polypeptide chain passes. Following secretion of most of the pre-protein, the signal sequence is cleaved by a "signal peptidase" (Table. 19.**4**), which is present on the outside surface of the cytoplasmic membrane, and the mature protein is released on this side of the membrane.

Protein secretion by this mechanism is absolutely dependent on the signal sequence. Deletion of, or certain mutations within, this sequence prevent secretion. Both the positively charged residues near the amino-terminus of a typical signal sequence and the hydrophobic nature of most of the rest of this sequence are important for efficient secretion. It is believed that the extreme amino-terminal, positively charged residues remain on the inside surface of the membrane during secretion, possibly in association with negatively charged phospholipids on the inner leaflet. Extrusion is then hypothesized to be initiated by a "looping" of the following hydrophobic stretch of the signal sequence through the membrane, followed by the rest of the protein, mediated by SecA and the SecY/E/G complex (Fig. 19.**14**).

Two additional proteins, SecD and SecF, also appear to play a role in this type of secretion mechanism (Tab. 19.**5** and Fig. 19.**14**). These proteins are associated with the cytoplasmic membrane, but with a considerable portion of their polypeptide chains exposed to the periplasm. The exact functions of SecD and SecF are unknown; however, they are believed to have roles during the late stages of the secretion process.

By this signal-sequence-dependent mechanism, most periplasmic and outer membrane proteins of Gram-negative bacteria and excreted proteins of Gram-positive bacteria traverse the cytoplasmic membrane. In the case of outer membrane proteins of Gram-negative bacteria, such as the porins, secretion is followed by insertion of the protein into the outer membrane by a largely unknown mechanism.

In a few cases, proteins that are secreted through the inner membrane by the Sec-dependent pathway have been shown to be **fatty-acylated**, with subsequent association of the protein with either the outer or inner membrane through the fatty-acyl chains. The best-studied example of this is the major lipoprotein ("Braun's lipoprotein") in *E. coli*, which is associated with both the peptidoglycan layer and the outer membrane. After synthesis, the eventual N-terminal residue of this protein (a cysteine) is modified by formation of a glyceryl thioether, followed by fatty acylation of the glycerol hydroxyl groups and of the amino group of the Cys residue (Fig. 19.**15**). Cleavage of the signal sequence by a specific enzyme, signal peptidase II (Tab. 19.**4**) then follows, concomitant with translocation. Fatty acylation of the major lipoprotein anchors it in the outer membrane, and furthermore, a portion of these molecules becomes covalently linked to the peptidoglycan layer by reaction of the C-terminal lysine residue with a diaminopimelic acid residue via an amide linkage (Fig. 19.**15**).

Lipoproteins with similar modifications have also been found in the inner membrane of *E. coli* (e.g., lipoprotein-28) and especially in Gram-positive bacteria (e.g., homologs of the periplasmic binding proteins found in Gram-negative bacteria). Interestingly, whether such a protein sorts to the outer or inner membrane in Gram-negative bacteria depends on the residue adjacent to the modified cysteine. If this residue is Asp (as in lipoprotein-28), the protein is anchored in

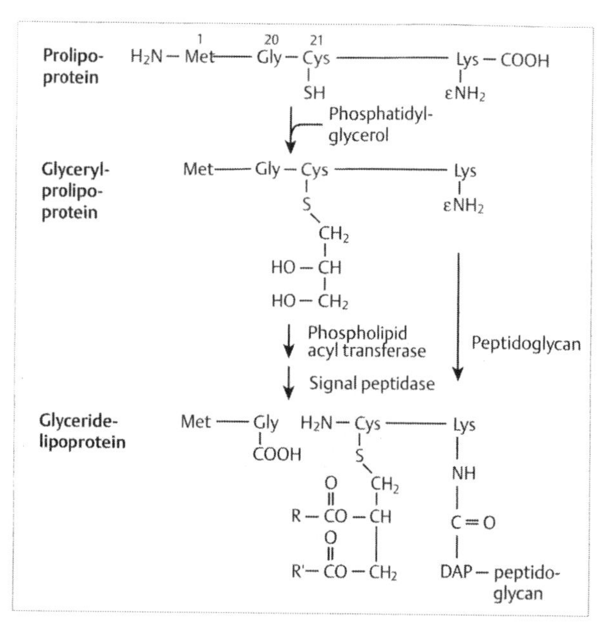

Fig. 19.**15** **Fatty acylation reactions involved in targeting of the major lipoprotein of *Escherichia coli*.** The eventual N-terminal cysteine residue in red of the major lipoprotein is modified to form a glyceryl thioether, followed by fatty acylation of the glyceryl hydroxyl groups. The fatty acyl groups anchor the protein in the outer membrane before it is processed. Some of these protein molecules are also covalently linked to *meso*-diaminopimelic acid (DAP) residues of the peptidoglycan through their C-terminal lysine residues

the inner membrane. If it is not Asp (as in the major lipoprotein), targeting is to the outer membrane.

Some proteins are secreted independently of Sec or require additional proteins. A number of proteins are excreted through both membranes of Gram-negative bacteria, and thus are true extracellular proteins. There are different mechanisms for this process, depending on the protein. Some proteins (e.g., pullulanase, an amylolytic enzyme in *Klebsiella*) traverse the inner membrane by the Sec-dependent pathway and the outer membrane by a different secretion process involving the products of as many as 14 genes. Other extracellular proteins, often involved in pathogenicity (see Chapter 33), including a number of bacterial toxins (e.g. α-hemolysin and colicin V from *E. coli*, leukotoxin from *Pasteurella haemolytica*, adenylate cyclase from *Bordetella pertussis*) and metalloproteases (in *Erwinia*, *Serratia*, and *Pseudomonas*) are secreted in a Sec-independent process without a periplasmic intermediate. In these cases, two inner membrane proteins and one outer membrane protein, at least, are required for secretion. One of the inner membrane proteins (e.g., HlyB for hemolysin and

PrtD for proteases of *Erwinia*) belongs to a superfamily of ATP-binding proteins (ABC cassette proteins, see Chapter 5) that include not only other bacterial proteins involved in transport, but also some eukaryotic transporters such as the mammalian multi-drug resistance protein (P-glycoprotein) and the cystic fibrosis transmembrane conductance regulator protein (CFTR). Morphopoiesis of more complex cell surface appendages (e.g., flagella, pili, and fimbriae; see Chapters 23, 24, and 33) requires, as a general rule, organelle-specific and ABC-type protein export systems.

Many proteins are **excreted** through both membranes of Gram-negative bacteria. Some use the Sec pathway to cross the inner membrane and another pathway for the outer membrane, others depend on their own specific pathway to cross both membranes. The latter systems are typical for proteins involved in the morphopoiesis of flagella, pili, fimbriae, and other **cell surface appendages**, bacterial toxins (e.g., hemolysins, leukotoxins), bacteriocins, and exo-proteins used by pathogenic bacteria to invade their host.

Finally, homologs to one protein subunit and a 7S RNA component of the eukaryotic **signal recognition particle** (SRP) have been identified in *E. coli*. In higher cells, SRP is an important player in the secretion of proteins into the endoplasmic reticulum. In *E. coli*, defects in the Ffh protein, which is similar to a 54-kDa subunit of SRP, appear to have variable effects on the secretion of proteins through the inner membrane. Thus, it is still unclear exactly what role Ffh and its associated 4.5S RNA plays in protein targeting in bacteria.

Targeting of integral cytoplasmic membrane proteins. Most cytoplasmic membrane proteins in bacteria are synthesized without a classical signal sequence. While, in a few cases, efficient membrane insertion of these proteins may be dependent on one or more Sec proteins, in other cases this process appears to be Sec-independent. A role for Ffh protein and its associated RNA has also been proposed in the process of integral membrane protein insertion in bacteria; however, definitive evidence for this is lacking.

Most likely, insertion of integral membrane proteins depends largely on their amino acid sequences and the high percentage of hydrophobic residues in most of these proteins. A topographical role for the distribution of positively charged residues in these proteins has also been proposed. Short cytoplasmic loops (approximately 10 residues) between putative transmembrane α-helices in these proteins are enriched in Lys and Arg compare to the periplasmic (or extracellular) loops and lon intracellular (>80 residues) loops. It is thought tha these positively charged regions help to anchor th cytoplasmic loops, possibly by interacting with nega tively charged head groups of phospholipids comprisin the inner leaflet of the membrane. A role for th electrical potential (negative inside) across the cyto plasmic membrane in this positional anchoring has als been proposed.

Most members of one class of cytoplasmic mem brane proteins, the Enzymes II of the PTS (see Chapter 5.4 and 20.2), possess a sequence of amino acids, ofte but not always at the amino-terminus, that has th potential to form a highly amphipathic α-helix, remi niscent of targeting sequences of eukaryotic mitochon drial membrane proteins. There is some evidence tha these sequences may be important for the efficien membrane targeting of the Enzymes II. However, suc sequences appear to be limited to this class o membrane proteins; therefore, this potential mechan ism of targeting is probably not a general mechanism i bacteria.

19.4.2 Other Mechanisms of "Compartmentalization"

Catabolic and anabolic pathways that share commo intermediates must be exquisitely regulated in bacteri to avoid so-called **"futile cycles"** in which, for example, compound synthesized by an anabolic reaction i degraded by a catabolic enzyme catalyzing the revers reaction, often with the wasteful expenditure of energ in the form of ATP. Examples of this include th phosphofructokinase (PfkA) and fructose 1,6-bisphos phatase (Fbp) reactions in glycolysis and gluconeogen esis, respectively (see Fig. 19.**8**), histidine metabolis via the catabolic *hut* system and the anabolic *his* syste (see Chapters 7 and 9), and the metabolism of amin sugars, in which the anabolic pathway leads to th biosynthesis of peptidoglycan and lipopolysacchari and the catabolic pathway leads to glycolysis (Fig 19.**16**).

In these cases, "compartmentalization" (whethe catabolism or anabolism is operative) occurs at th levels of the regulation of gene expression, allosteri control of the catabolic and anabolic enzymes, and i many cases, the kinetic and equilibrium properties o these enzymes. For example, for amino sugar metabo lism (Fig. 19.**16**) the catabolic enzyme D-glucosamine 6-phosphate-deaminase (NagB) is induced by *N*-acetyl glucosamine 6-phosphate as the physiological induce and has a low affinity for its substrate fructose 6 phosphate, while the biosynthetic enzyme, D-glucos

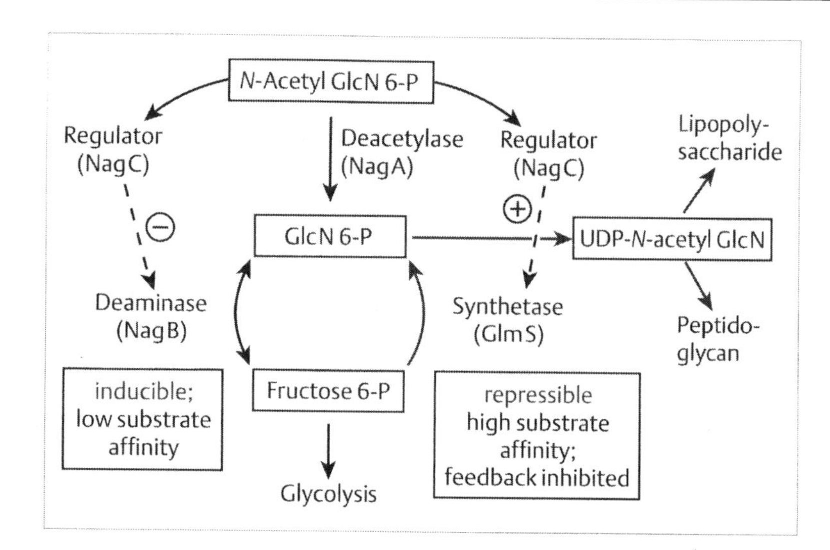

Fig. 19.16 **Catabolic and anabolic fate of D-glucosamine 6-phosphate (GlcN-6-P).** The synthesis and activity of the reversible catabolic enzyme (GlcN-6-P deaminase; NagB) is differentially regulated compared to the anabolic enzyme GlcN-6-P synthetase (GlmS) as shown. The regulator NagC represses the synthesis of NagB and activates the synthesis of GlmS. In the presence of the inducer N-acetyl-GlcN6-P, the synthesis of NagB is induced while that of GlmS is repressed. Under the same conditions, the activity of the GlmS synthetase is feedback inhibited [after 4]

amine-6-phosphate-synthetase (GlmS), has a high affinity for its substrates. GlmS is repressed under high N-acetyl-glucosamine concentrations and apparently subject to feedback inhibition. These types of control ensure that futile cycles, and the resultant dissipation of cellular energy, do not occur under normal conditions in bacteria.

Finally, separation of function in bacteria can also be accomplished by temporal "compartmentalization," i.e., the sequential induction of gene expression. The most obvious examples of this occur in bacteria that undergo developmental cycles often triggered by starvation, such as sporulation in *Bacillus*, fruiting body formation in *Myxococcus*, and the development of swarmer cells in *Caulobacter* (see Chapters 24 and 25). Sequential induction of genes in these cases ensures a smooth progression through the developmental cycle in response to the appropriate environmental signal(s). In some of these cases, for example sporulation in *Bacillus* (see Chapter 25 and the earlier discussion in this chapter), spatial compartmentalization (e.g., the developing spore vs. the mother cell) is also involved.

Further Reading

Freestone, P., Grant, S., Toth, I., and Norris, V. (1995) Identification of phosphoproteins in *Escherichia coli*. Mol Microbiol 15: 573–580

Gottesman, S., and Maurizi, M. R. (1992) Regulation by proteolysis: energy-dependent proteases and their targets. Microbiol Rev 56: 592–621

Hueck, C. J., and Hillen, W. (1995) Catabolite repression in *Bacillus subtilis*: a global regulatory mechanism for the Gram-positive bacteria? Mol Microbiol 15: 395–401

Kontrowitz, E. R., and Lipscomb, W. N. (1990) *Escherichia coli* aspartate transcarbamoylase: the molecular basis for a concerted allosteric transition. Trends Biochem Sci 15: 53–59

Miller, C. G. (1996) Protein degradation and proteolytic modifications. In: Neidhardt, F. C., Curtiss, R., Ingraham, J. L., Lin, E. C. C., Low, K. B., Magasanik, B., Reznikoff, W., Riley, M., Schaechter, M., and Umbarger, H. E. (eds.) *Escherichia coli* and *Salmonella*: cellular and molecular biology. 2nd edn. Washington, DC: ASM Press; 938–954

Min, K.-T., Hilditch, C. M., Diederich, J. E., and Yudkin, M. D. (1993) σ^F, the first compartment-specific transcription factor of *B. subtilis*, is regulated by an anti-σ factor that is also a protein kinase. Cell 74: 735–742

Murphy, C. K., and Beckwith, J. (1996) Export of proteins to the cell envelope in *Escherichia coli*. In: Neidhardt, F. C., Curtiss, R., Ingraham, J. L., Lin, E. C. C., Low, K. B., Magasanik, B., Reznikoff, W., Riley, M., Schaechter, M., and Umbarger, H. E. (eds.) *Escherichia coli* and *Salmonella*: cellular and molecular biology. 2nd edn. Washington, DC: ASM Press; 967–978

Neidhardt, F. C., Ingraham, J. L., and Schaechter, M. (1990) Physiology of the bacterial cell: a molecular approach. Sunderland, Mass.: Sinauer Associates; 302–319

Pugsley, A. P. (1992) Protein traffic in bacteria. In: Lederberg, J. (ed.) Encyclopedia of microbiology, vol. 3. San Diego: Academic Press; 466–479

Reitzer, L. J. (1996) Ammonia assimilation and the biosynthesis of glutamine, glutamate, aspartate, asparagine, L-alanine, and D-alanine. In: Neidhardt, F. C., Curtiss, R., Ingraham, J. L., Lin, E. C. C., Low, K. B., Magasanik, B., Reznikoff, W., Riley, M., Schaechter, M., and Umbarger, H. E. (eds.) *Escherichia coli* and *Salmonella*: cellular and molecular biology. 2nd edn. Washington, DC: ASM Press; 391–407

Saier, M. H., Jr. (1993) Introduction: protein phosphorylation and signal transduction in bacteria. J Cell Biochem 51: 1–6; and subsequent papers in this issue

Schirmer, T., and Evans, P. R. (1990) Structural basis of the allosteric behaviour of phosphofructokinase. Nature 343: 140–145

Wandersman, C. (1996) Secretion across the bacterial outer membrane. In: Neidhardt, F. C., Curtiss, R., Ingraham, J. L., Lin, E. C. C., Low, K. B., Magasanik, B., Reznikoff, W., Riley, M., Schaechter, M., and Umbarger, H. E. (eds.) *Escherichia coli* and *Salmonella*: cellular and molecular biology. 2nd edn. Washington, DC: ASM Press; 955–966

Wickner, W., Driessen, A. J. M., and Hartl, F.-U. (1991) The enzymology of protein translocation across the *Escherichia coli* plasma membrane. Annu Rev Biochem 60: 101–124

Sources of Figures

1 Schirmer, T. and Evans, P. R. (1990) Nature 343: 140–145
2 Reitzer, L. J. (1996) In: Neidhardt, F. C., Curtiss, R., Ingraham, J. L., Lin, E. C. C., Low, K. B., Magasanik, B., Reznikoff, W., Riley, M., Schaechter, M., and Umbarger, H. E. (eds.) *Escherichia coli* and *Salmonella*: cellular and molecular biology. 2nd edn. Washington, DC: ASM Press; 391–407
3 Wickner, W., Driessen, A. J. M., and Hartl, F.-U. (1991) Annu Rev Biochem 60: 101–124
4 Vogler, A. P., Trentmann, S., and Lengeler, J. W. (1989) J Bacteriol 171: 6586–6592

20 Global Regulatory Networks and Signal Transduction Pathways

Unicellular organisms such as the prokaryotes must be able to detect changes in their environment and to adapt their metabolism rapidly to external fluctuations. Immediate adaptation to such changes is as much a hallmark of the prokaryotes as is their rapid growth. Adaptations are in general transient, i.e., they last only a little longer than a given environmental condition persists. Prokaryotes monitor their surroundings through membrane-bound and intracellular sensors and through their transport systems. Complex signal transduction pathways and global regulatory networks are linked to the sensors. These networks lead eventually to the adaptation of the cell to the changed conditions. It has only recently become clear that all prokaryotes, the free-living and commensal, as well as the symbiotic and pathogenic bacteria, use similar adaptation strategies. Signal transduction pathways are also involved in global control of metabolic networks, cellular differentiation processes, and the behavior of bacterial populations.

20.1 Prokaryotes Contain Integrated Metabolic Networks and Signal Transduction Pathways

In the last two chapters, the regulation of specific metabolic pathways at the level of single operons and regulons, i.e., enzyme synthesis (Chapter 18), as well as the control of enzyme activity (Chapter 19) were described. In an **operon**, genes are grouped as a single transcriptional unit, thus allowing coordinated expression of genes coding for related functions (Fig. 20.**1**). Complex anabolic and catabolic pathways contain many genes, perhaps not easily accomodated in a single operon. Such a group of operons controlled by a common regulatory protein is called a **regulon**. Regulons seem to be typical for metabolic pathways used to degrade chemically related compounds that have to be active under different growth conditions. For example, the *glp* regulon of *Escherichia coli* (Fig. 20.**1**) is involved in the catabolism of glycerol and *sn*-glycerolphosphate. These compounds require two different transport systems and different enzymes for aerobic and anaerobic growth. Glycerokinase (GlpK) is only used in the degradation of glycerol. All *glp* operons, however, are induced by *sn*-glycerolphosphate and controlled through a single repressor, GlpR.

A different type of regulation, which modulates the expression of seemingly unrelated operons that are bound by a common global purpose, will now be discussed. This global control often causes rather drastic changes in cellular metabolism and is superimposed over the control by individualized local regulators.

20.1.1 Operons Bound by a Common Global Purpose Are Grouped in Modulons

Complex bacterial activities require coordination of whole blocs or networks of metabolic pathways. The corresponding genes are encoded in dozens of operons or regulons, and a higher level of organization is required. The term **modulon** designates a group of operons and regulons in which each member is not only regulated by its individual regulatory protein, but is in addition under the control of a **global regulator** common to all members of the modulon. The new regulatory system is **epistatic**, i.e., superimposed over the individual regulators. Mutations in the regulator cause pleiotropic phenotypes, i.e., they affect the expression of all members of the modulon. As an example, the peripheral catabolic operons and regulons from enteric bacteria involved in the quest for food (Fig. 20.**2**) are subject to cAMP · Crp-dependent catabolite repression (for molecular details, see Chapter 20.2) by the *crp* modulon, named after the structural gene *crp*, which encodes the pleiotropic regulatory protein Crp (also CRP, mnemonic for **c**AMP-**r**eceptor **p**rotein or **c**atabolite **r**epression **p**rotein). This protein was originally called CAP for **cA**MP-binding **p**rotein.

Typical examples of other metabolic blocs controlled by a superimposed global regulatory system are

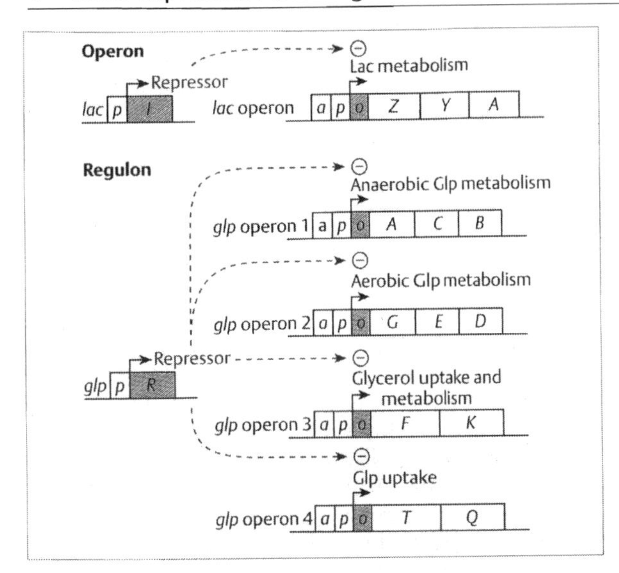

Fig. 20.1 **Schematic view of the *lac* operon and the *glp* regulon.** The *lac* operon for lactose (Lac) metabolism and the *glp* regulon for *sn*-glycerolphosphate (Glp) metabolism in *E. coli* are shown. Promoter (*p*) sequences are indicated with the structural genes (capital letters); the symbol for negative control (⊖) by the corresponding repressors at the operator (*o*) is indicated. Solid arrows indicate the direction of transcription. Symbol *a* designates a binding site for global regulators

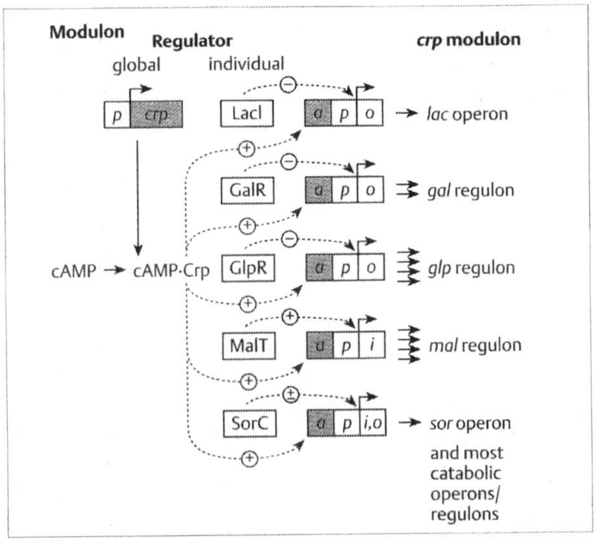

Fig. 20.2 **The *crp* modulon.** In the presence of cyclic adenosine 3′-5′-monophosphate (cAMP), the cAMP repressor protein Crp is activated and binds to a consensus sequence (a), where it acts as an activator (positive control, ⊕) for all operons (single arrow) and regulons (multiple arrows) containing this *crp* motif in their promoter. Not every promoter of a regulon carries the *crp* motif; those that lack the *crp* motif are not activated by Crp. The pleiotropic *crp* activation system is superimposed over the regulation of operons by their individual regulator at the operator (⊖ negative control, repressor) or initiator (*i*, positive control, activator)

the genes and enzymes involved in the synthesis of amino acids, nucleotides, or vitamins, in the translation machinery, or in sporulation, to name a few. Estimates for the number of such global systems range from 50 in a single bacterial cell to several hundred in the prokaryotic world. A necessarily incomplete list of typical and important examples, many treated in this book, is given in Table 20.1. The examples listed are grouped roughly into four classes: (1) modulons involved in nutrient and energy supply, (2) modulons involved in stress response, (3) modulons involved in cellular differentiation processes, and (4) modulons involved in cell aggregation and in cell-to-cell contact phenomena.

Modulons are groups or networks of operons and regulons that are controlled not only by individual regulators, but also by a common and epistatic, i.e., superimposed, global regulatory system. This system responds usually to environmental changes that are sensed through sensory receptor systems. Genes organized in a modulon have similar tasks, for example, in carbon, nitrogen, or phosphate supply, or common cellular activities, for example, in sporulation, symbiosis, or host cell infection.

20.1.2 Transiently Acting Signal Transduction Systems Are Essential in Global Regulation

Adaptation of bacterial populations to changes in the environment is, in general, transient. Apparent exceptions are differentiation processes like sporulation, prophage induction, or longer-lasting gene rearrangements, which are irreversible at the level of a single cell. Even these, however, are reversible adaptation processes at the level of the population. The initial state is restored after completion of a full generation cycle (see Chapters 22 and 24).

Sensory reception systems. Bacteria monitor their surroundings by an impressive array of extracellular and

Table 20.1 **Global regulatory networks of prokaryotes**

System or response	Stimulus	Sensor/regulator and genes	Regulated genes and systems	Details in Chapter
Nutrient supply				
Catabolite repression, Inducer exclusion	Carbon source level	PTSs, IIA^{Glc}, and cAMP·Crp or PTSs, P-Ser HPr, and CcpA	Most peripheral catabolic operons and carbohydrate transport systems; chemotaxis proteins	20
Stringent/relaxed response	tRNA pool	Ribosomes; RelA, SpoT, and ppGpp	Translational machinery, most anabolic enzymes	20
Nitrogen network	Nitrogen source level	Glutamine synthetase; NtrB/NtrC; NifR2/NifR1 Rpo N (σ^{54})	Most operons involved in nitrogen utilization and fixation; chemotaxis proteins	8, 19, 20
Phosphate network	Phosphate source level	PhoP,Q/PhoB,P	Several operons of phosphate (inorganic and organic) utilization	8, 20
Aerobic respiration	Oxygen level	ArcB/ArcA	Many genes for aerobic growth; aerotaxis	11, 21
Anaerobic respiration	Level of electron acceptors (other than oxygen)	NarQ,X/NarL; Nir; Fnr	Many genes for anaerobic respiration	11, 21
Fermentation	Absence of electron acceptors	Unknown	Many genes for fermentation pathways	11, 21
Lithoautotrophic metabolism	Hydrogen oxidation	HoxA; RpoN (σ^{54})	Many genes for lithoautotrophic growth in *Alcaligenes*	10, 20
Methanogenesis	Unknown	Unknown	Many genes	12, 20
Exoprotein synthesis	Starvation	DesS/DegU	Many genes for exoprotein synthesis and polymer degradation	9, 20
Stress response				
SOS response	DNA damages	LexA, RecA	About 20 **d**amage **in**ducible genes (*din*) for repair of DNA damage	15, 20
Alkylation response	DNA alkylation	Ada	Genes for removal of alkylated bases	15, 20
Oxidation response	H_2O_2 or other oxidants	OxyR, SoxR/SoxS	Several genes for oxidation protection	11, 28
Heat shock	Shift to high temperature	Ribosomes; RpoH (σ^{32})	Many genes for macromolecule processing, folding/unfolding (chaperones), degradation	20, 28
Cold shock	Shift to low temperature	Unknown	Genes and proteins similar to heat shock	28
Osmotic response	Shift to high or low osmolarity (turgor)	EnvZ/OmpR; KpdD/KpdE	Genes for outer membrane proteins, porins, MDO, potassium transporters	28
General starvation response	Cessation of cell growth	CreC/CreB; RpoS (σ^{38})	Many starvation-induced operons for survival in stationary phase	20, 28
Differentiation				
Stationary phase response	Growth cessation	Unknown, probably PilB/PilA; RcsC/RcsB; RpoS (σ^{38})	Genes for colony formation, surface adhesion, chromosomal rearrangements, increased mutagenesis	23, 24
Growth rate control	Unknown	Unknown	Unknown, but many related to translational machinery and cell division	22
Cell division	Unknown	Unknown, probably Sfx	Hundreds of genes for DNA replication, septum formation, and coordinated cell division	22

Table 20.**1** ***(continued)***

System or response	Stimulus	Sensor/regulator and genes	Regulated genes and systems	Details in Chapter
Sporulation (1)	(Energy) starvation	KinA/SpoOF(B,A); DivJ/DivK; KinB, PleC/SpoOF(B,A), several sigma factors in Bacilli	Many genes for endospore formation	25
Sporulation (2)	(Energy) starvation	Unknown in Streptomycetes	Many genes for exospore formation	25
Competence	(Energy) starvation	ComP/ComA	Genes for DNA uptake	16, 25
Secondary metabolism	Unknown	Unknown, probably Factor A; VanS/VanR; RteA	Many genes for antibiotic synthesis, secondary metabolites	27
Photosynthesic apparatus	Light, oxygen	RegB/RegA; PrrB/PrrA	Many genes for photosynthetic apparatus, phototaxis	24
Virulence (1) Animal cells	Shift in temperature, osmosis, pH, free iron	*Vibrio* (Tox); *Bordetella* (BvgS, LemA); many bacteria (Fur); *Yersinia* (Yop); *Shigella, Salmonella; Staphylococcus* (Agr, Exp); *Pseudomonas* (Alg)	Many chromosomally, plasmid-, and phage-encoded genes that control infection, colonization, toxin production, antihost defence, etc.	33
Virulence (2) Plant cells	Carbohydrates, wound exudate	*Agrobacterium tumefaciens*, (VirA/VirG); *Rhizobium* (FixL/FixJ); *Bradyrhizobium* (NodV/NodW)	Same as above	34
Intercellular contact				
Motility and development	Starvation	Systems in many bacteria for chemotaxis (CheA/CheY,B), phototaxis; sporulation in Myxobacteria (FrzE/FrzZ,G)	Many genes involved in motility, cell aggregation, sporophyte formation	20, 25
Conjugation	Cells of opposite sex	Two-component sensor (*cpx*) in enteric bacteria; pheromones in Gram-positive bacteria	Many *tra* genes for conjugation and coordination of chromosome and plasmid replication	16
Quorum sensing	Cell density	LuxN,Q/LuxO	Many genes	20

intracellular sensors. Sensors convert, as indicated schematically in Figure 20.**3**, a stimulus from the environment into a signal. Signals are communicated by transducers to individual or to pleiotropically acting regulators, such as those for modulons. Regulators cause a cellular response through the modulation of gene and enzyme activities. They also trigger adaptation, i.e., the process that ends the response despite the continuous presence of the stimulus.

The formal view of bacterial adaptive and differentiation processes as mediated by transiently acting sensory systems can improve our understanding of these complex phenomena. Some of the better understood systems, for example, the chemotaxis machinery of enteric bacteria (see Chapter 20.5) or the two-component systems (see Chapter 20.4), closely resemble or are similar to sensory systems of the eukaryotes. Just as the modulon has probably evolved from simpler operons and regulons, the sensory part of its regulatory system was not without simpler precedents. This can be exemplified by the response of the *lac* operon to the addition of lactose to a growing culture of *E. coli* (Fig. 20.**4**). Lactose in the medium (stimulus) is sensed by the permease (sensor). It is transported through the membrane and converted by β-galactosidase into allolactose, the internal inducer (signal) that binds to the lactose repressor. This regulator controls the synthesis of the lactose-degrading enzymes, which catalyze the first steps in lactose fermentation (response). If fermented at a high rate, sufficient to cause carbon catabolite repression and inducer exclusion, a new equilibrium is reached through adaptation. The lactose system may thus be considered as a precursor for systems that sense environmental changes indirectly in the form of changed rates of transport or changed pools of intracellular metabolic intermediates (here cAMP).

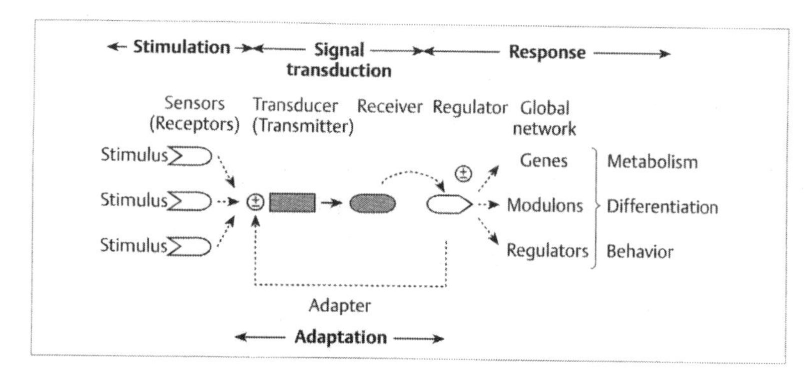

Fig. 20.3 **General scheme for sensory reception systems with a transient response to stimuli.** Central are the transducer, which converts a stimulus perceived by a sensor into a signal; the receiver, which accepts this signal and modulates accordingly the activity of a response regulator positively or negatively (⊕) and the adapter, which ends stimulation

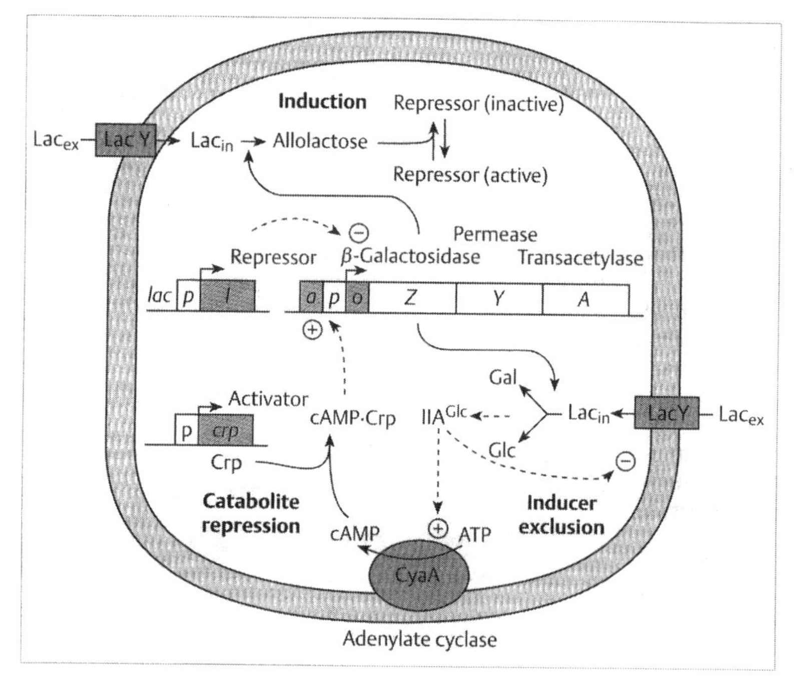

Fig. 20.4 **Schematic view of the *lac* operon and of lactose (Lac) metabolism as elements of a primitive sensory system.** In this model, induction corresponds to sensory input, hydrolysis of lactose corresponds to the response (output), and catabolite repression together with inducer exclusion corresponds to adaptation

20.1.3 Several Methods to Identify Global Regulatory Networks Are Available

The discovery and analysis of highly complex systems, such as modulons and sensory reception systems, frequently started with the analysis of mutants. Thus, the first mutants of enteric bacteria affected in catabolite repression and inducer exclusion were isolated as cells resistant to the various "glucose effects," years before the system was recognized as a modulon and before any biochemical details could be given. The same is true for *rel* mutants and the "stringent-relaxed" modulon, for *spo* mutants of *B. subtilis* and sporulation, and for many other examples. Molecular details, however, can only be elucidated by biochemical methods. A brilliant example of such a study is the analysis of the bacterial chemotaxis system discussed in Chapter 20.5. Signal transduction pathways rest on a few highly conserved proteins with similar functions and conserved amino acid sequences. A search for such conserved sequences can thus help to identify new members of the group.

Genetic methods. A genetic analysis usually starts by isolating mutants affected specifically in the different components (Fig. 20.**3**) of a sensory system or in the pleiotropic regulator of a modulon. Mutations in the regulator alter the expression of all member operons and thus define the extent of the modulon. Mutations in the sensor, in contrast, help to identify the corresponding stimulus.

A second genetic method uses gene fusions. A reporter gene, very often *lacZ* from *E. coli*, the activity of which can be followed easily by a β-galactosidase assay, is fused to various host promoters. The indicator gene allows identification of those promoters that respond to the same stimulus and to the same pleiotropic regulator. The simplest technique to introduce *lacZ* fusions involves a defective derivative of phage Mu. This derivative, called Mudlac (Fig. 20.**5**), contains (1) both Mu ends, which are essential for the integration of the prophage into the host chromosome, (2) the *lac* operon without a promoter, and (3) the *bla* gene encoding a β-lactamase (infected cells become resistant to ampicillin, allowing easy selection). In the presence of a Mu helper phage, Mudlac will integrate at various places of the bacterial chromosome. If strains carrying a deletion of the *lac* operon are infected, Lac^+ mutants arise only when the integration is such that the *lac* genes on the phage are expressed from an upstream host promoter. Consequently, conditions that regulate the activity of this promoter will control β-galactosidase expression. Strains carrying fusions to promoters belonging to the same modulon can be detected as colonies expressing the *lac* genes in response to the stimulus of that modulon, such as the stimuli listed in Table 20.**1**.

A third genetic method, called conformational suppression, is especially helpful to identify the members of complex signal transduction pathways. Mutations that interrupt the signal flux in such pathways can be isolated. They lead to alterations in the structure of one protein, which thus becomes unable to communicate with its immediate neighbor (Fig. 20.**6**). Secondary or suppressor mutations that restore a positive phenotype can easily be isolated, preferentially after cloning the genes on plasmids to allow extensive mutagenesis and easy recovery. These mutations frequently alter the structure or conformation of the neighbor protein in such a way that communication with the first mutated protein is again possible. Hence, the name **"conformational suppression"** is used to distinguish it from informational suppressor mutations acting through tRNA changes (see Chapter 15).

Conformational suppressor mutations are highly allele-specific. They identify proteins and even protein domains that function in complexes or in direct physical contact to other proteins, as is characteristic for signal transduction proteins.

Biochemical methods. To analyze global systems by biochemical methods, the two-dimensional gel electrophoresis of all proteins of a cell is used to monitor changes in the pattern of synthesis before and after an environmental change. Usually, a portion of an undisturbed growing culture (control) and a portion immediately after the shift to new growth conditions are labeled with $[^{35}S]$ L-methionine. The protein patterns are compared after gel electrophoresis of the cell extracts, and altered protein spots are identified on an

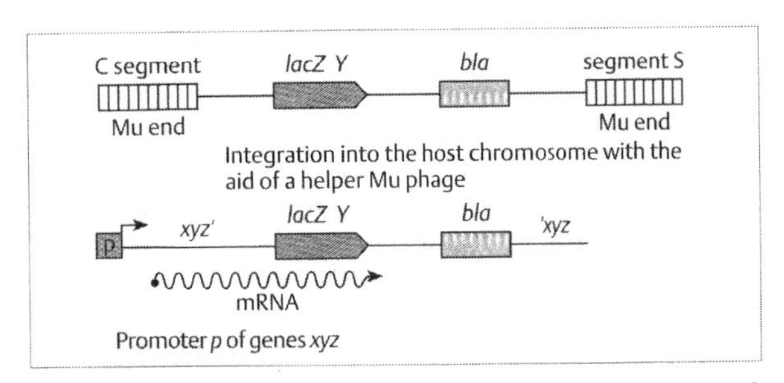

Fig. 20.5 **Construction of gene fusions with the help of bacteriophage Mudlac.** In the presence of a helper phage, a defective (d) Mu phage carrying the *lac* genes without a promoter, and a β-lactamase gene (*bla*), conferring ampicillin resistance to infected cells for easy selection, will integrate into many places on the chromosome of a host strain having a deletion of the *lac* genes. Lac^+ colonies will be generated if the phage *lac* genes are fused, as the consequence of a chromosomal rearrangement, to a promoter (*p*) and will become visible under conditions where this promoter is active. Genes *xyz* will be split by the insertion into a front half *xyz'* and a distal half *'xyz*

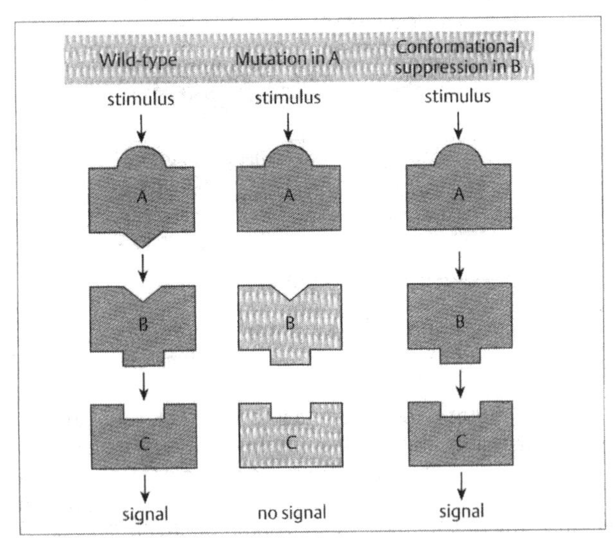

Fig. 20.6 **The use of conformational suppression to identify the nearest neighbors in a signal transduction pathway.** The effect of a mutation in protein A, which communicates with protein B through physical contact, is compensated for by a secondary or suppressor mutation in B

autoradiogram (Fig. 20.7). If available, the protein pattern from wildtype and from mutants affected in the regulation of a modulon may be used similarly. An as yet incomplete polypeptide map of *E. coli* has been compiled and can be used to identify components of various global systems. No such compilation is available for other bacteria. The proteins have thus to be eluted from the gel and used to obtain a partial or complete amino acid sequence. An oligonucleotide probe complementary to (parts of) the corresponding gene can be derived thereof and used in Southern hybridization tests to identify in a gene bank DNA fragments containing the gene that encodes the protein isolated from the gel. Details of this procedure, often erroneously called **"reverse genetics,"** have been described in Chapter 17.

A new technique for studying globally regulated genetic systems in *E. coli* combines detection, cloning, and physical mapping of groups of coregulated genes in one step. In this approach, total RNA is isolated from control and experimental cultures and converted by the reverse transcription technique into [^{32}P]-labeled cDNA. Complete genomic encyclopedias made from the chromosome of *E. coli* K-12 in bacteriophages λ and M13 can be used to prepare in the next step DNA dot blots in which the various labeled cDNAs are hybridized with chromosomal DNA fragments of the genomic library. Levels of mRNA transcribed from a cloned region in a control and in a test culture exposed to, for example, heat shock, osmotic shock, starvation of a nutrient, or changes in the levels of mRNA when bacterial cultures switch from exponential growth to stationary phase, from aerobic to anaerobic growth, or from life in a test tube to growth in the gnotobiotic mouse gut, can be compared to detect induced and repressed genes. Alternatively, the effects of pleiotropic mutations in *crp, fnr, rpoS, rpoH, rpoN, himA,* and in other putative global regulator genes on the expression of different genes can be analyzed. Once a clone of interest is found, its precise physical location on the gene map can be determined, its gene product or products can be identified, and its function can be ascertained by "reverse genetics." The advantage of this "one-step" method over the methods mentioned before is that it can be used for almost any organism and any (non-cryptic and non-silent) gene, whether essential or non-essential, provided a gene library is available.

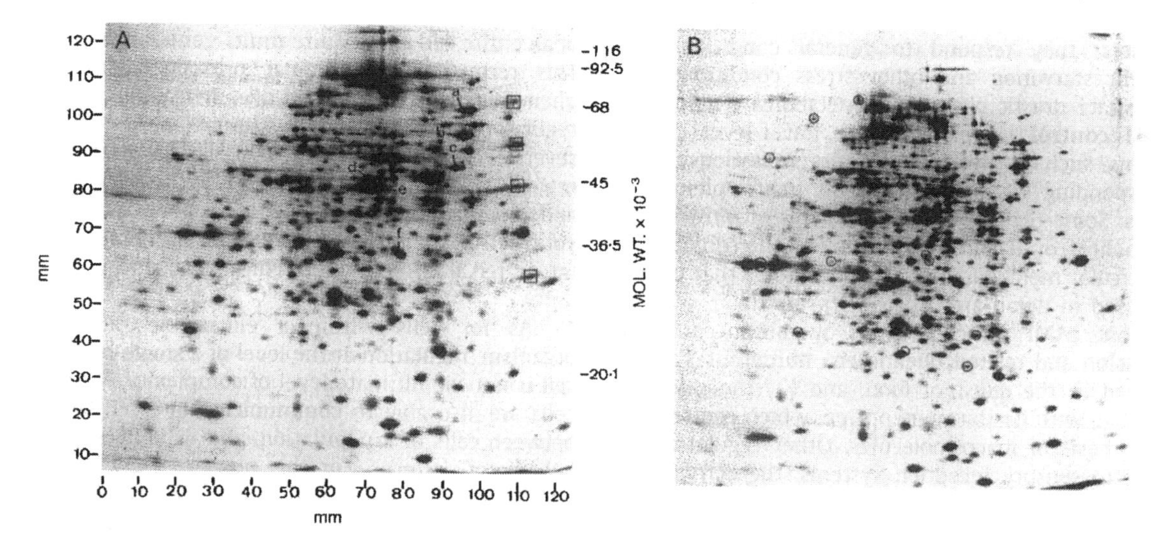

Fig. 20.7**A,B** **Two-dimensional electrophoresis analysis of *E. coli* proteins.** The standard two-dimensional polypeptide map of aerobically grown *E. coli* KL-19 used for the assignment of polypeptide coordinates is shown in
A. The identified polypeptides are ribosomal protein S1 (a), *groE* gene product (b), β-subunit of ATP synthase (c), enolase (d), and elongation factors Tu (e), and Ts (f). In the first dimension, polypeptides were separated by isoelectric focusing, in the second dimension by sodium dodecyl sulfate polyacrylamide gel electrophoresis according to a procedure first described by O'Farrell. Peptides induced under oxic conditions during growth on glucose are marked (□).
In panel **B,** peptides from a culture grown on glucose with formate or nitrate, but in the absence of oxygen, are shown, and the peptides induced under anoxic conditions are indicated (○) [from 1]

20.1.4 Hierarchical Regulatory Systems Require a New Nomenclature

The **operon**, as defined originally by F. Jacob and J. Monod, is a transcriptional unit expressed coordinately from a single promoter (forming a "polycistronic mRNA") and regulated by a common operator. The **operator,** in conjunction with the promoter, controls transcription initiation by binding or releasing allosteric regulatory proteins (repressors and activators). These respond to specific effectors that are usually substrates for the enzymes of the structural genes encoded in a particular operon or are end products of such enzymes. A set of operons, the locations of which are scattered throughout the chromosome, but that is regulated by the same specific repressor or activator, is called a **regulon.** Such operons usually encode genes for different parts of the same metabolic pathway (e.g., arginine biosynthesis in *E. coli*) or genes for different metabolic pathways for the same and closely related substrates (e.g., glycerol and *sn*-glycerolphosphate degradation; Fig. 20.**1**). Control of operons and regulons thus is at the lowest level, and is in fact identical.

In a **modulon,** in contrast, sets of operons and regulons that are bound by a common global goal, but that are otherwise seemingly unrelated, are controlled by a common regulator. These regulators do not respond to specific molecules, such as substrates and end products; they respond to general conditions like nutrient starvation and other stress conditions that necessitate drastic changes in metabolism. Since this **global control** is epistatic over lower-level control systems, such as specific regulators, mutations in the corresponding regulatory systems have pleiotropic effects. Some regulators measure pools of intracellular metabolites, or **alarmones,** that reflect the condition of a particular metabolic bloc. Two examples that will be discussed in detail are: (1) the *crp* modulon, with its alarmone cAMP, which controls by carbon catabolite repression and related phenomena numerous systems involved in the quest of food, and (2) the *relA/spoT* modulon, with its alarmone ppGpp, which controls the biosynthesis of macromolecules. Other regulators are linked to sensory reception systems, the activities of which reflect directly a particular environmental condition, such as the *pho* modulon, which reacts to changes in external concentrations of phosphate; the *envZ/ompR* modulon, which reacts to changes in osmotic conditions; the *ntr* modulon, which controls nitrogen metabolism; and the *che* system involved in the control of cellular motility. Several examples will be discussed in various sections of this book as listed in Table 20.**1**. The regulators involved in global control usually modulate RNA polymerase activity at the promoters of all operons and regulons constituting a modulon. Either they bind themselves as activators or repressors to a consensus sequence common to all members of a modulon, or they modulate the activity of such regulators, for instance, in the form of alternative sigma factors that allow the binding of RNA polymerase specifically to the promoters of a particular modulon.

A single stimulus (e.g., the addition of glucose to enteric bacteria, severe and long-lasting nutrient starvation, cold shocks or heat shocks, the damaging of DNA, or host cell infection) can trigger a cellular response that affects members of different modulons. The term **stimulon** has been proposed to designate such large groups of operons that respond together to an environmental stimulus, regardless of how many genes and regulatory elements are involved. The term is primarily of heuristic value, i.e., useful at the beginning of an analysis when it is not known which regulatory processes are involved and which responding operons comprise a modulon. It does not necessarily imply the presence of a global regulatory network that regulates all responding systems by a common mechanism.

Many cellular networks that are united by a common purpose often exceed the complexity of a modulon. Well-known examples are cell division, the formation of endospores and exospores during sporulation, and the bacterial cell surface changes upon host cell infection. Such processes approach the complexity of an entire cell and require **multi-gene family control.** This term is also used at present solely at the phenomenological level and does not imply a common regulatory system, such as a system superimposed over several modulons. The existence of such third-level regulatory systems epistatic over the control of operons and modulons, however, seems probable as a prokaryotic cell appears to be the sum of all metabolic steps integrated at different hierarchical levels to a purposeful unity.

As for cells within a eukaryotic multicellular organism, regulation at the level of a single prokaryotic cell is not the ultimate level of complexity. Prokaryotic cells are also able to communicate intercellularly, i.e., between cells of a population, and exhibit cooperative behavioral patterns. For this purpose, they use **auto-induction,** an environmental-sensing system that allows bacteria to monitor their own population density ("**quorum sensing**"). During growth, the cells excrete an autoinducer that accumulates in the environment. Only when its concentration passes a threshold value, usually at higher cell densities, will it cause autoinduction and activation of defined genes. In Gram-negative bacteria, the autoinducers belong to the chemical group of *N*-acyl-homoserinelactones. These amphiphilic molecules diffuse easily through membranes and bind to intracellular regulator proteins that form the so-called LuxR family and that control genes involved in autoinducer synthesis and in social function (see Chapter 11.6.2). Autoinducers control such diverse processes as: (1) **Bioluminescence** and **swarming motility** in species of the genera *Vibrio, Proteus,* and *Serratia. V. fischeri* is the symbiont in the light organs of certain marine organisms. Within the light organ, it reaches high cell titers (10^{10} to 10^{11} cells per ml), and luminescence is autoinduced, while isolated cells in free seawater remain repressed. Similar phenomena control the swarming motility on solid surfaces, which differs from normal swimming in free liquid media, as is characteristic for *P. mirabilis* and *S. marcescens.* (2) **Virulence** in *Erwinia carotovora* and *Pseudomonas aeruginosa.* Many plant or animal pathogenic bacteria require high cell densities ($>10^6$ cells per ml) to produce the high local amounts of exoenzymes (pectatelyases, polygalacturonases, cellulases, proteases) needed to lyse protective cell walls or to overcome host cell defense mechanisms. The corresponding genes, which often include genes for a special ("swarming") motility apparatus for social swimming along solid surfaces and gathering of the cells, are controlled by autoinducers found in most Gram-negative bacteria. (3) **Conjugation** in *Agrobacterium tumefaciens* (see Chapter 34.2). Other possible examples include the conversion of normal cells of *Rhizobium* spec. to bacterioids (see Chapter 34.3).

A further example, but with a different biochemistry, is found in *Enterococcus faecalis.* In this organism, donor and recipient cells exchange **pheromones** or diffusible mating signals that trigger cell aggregation as the prerequisite to conjugation. Similarly, sporulating myxococcal cells must produce extracellular signals during cell aggregation and fruit body formation (see Chapter 20.5.8), perhaps cAMP. Mutants are available that aggregate only when mixed with a second type of mutant, itself unable to aggregate. The first mutant excretes signaling molecules, but cannot detect them, while the second type of mutant detects the molecules, but cannot produce them. Together they complement each other and form fruiting bodies. In this respect, it is perhaps interesting to note that *E. coli* cells excrete large amounts of cAMP. Such pulses may signal the presence of nutrients to other members of the population. Finally, the central role of butyrolactone derivatives (e.g., factor A) in the sporulation of the actinomycetes (see Chapters 25.2 and 27.7) may indicate the presence of further intercellular communication systems in, for example, the generation and stable maintenance of complex microbial eco-systems.

Box 20.1 Regulatory systems that control gene expression in prokaryotes can be described either at the level of molecular mechanisms or at the phenomenological level. The terms **regulon** and **modulon,** defined previously, refer to groups of operons controlled by a single regulatory system (regulon) or by two different systems, one of which is dominant over the other (modulon). The term **stimulon,** in contrast, refers to all operons responding coordinately to an environmental stimulus. The term is of purely heuristic value, but is useful since the initial observation of a cellular response is often the coordinated expression of genes, without light being shed on the mechanisms involved. Hence, it is conceivable that some coordinately expressed operons do not share a regulatory protein or controlling elements, and that other stimulons are synonyms for modulons. The general terms **global control system** and **multi-gene family control** similarly do not imply any common regulatory elements, but solely indicate groups of operons that contribute to the achievement of one common goal in cellular metabolism. **Auto-induction,** finally, designates an environmental-sensing and control system by which bacteria communicate at an intercellular level and that allows the monitoring of their own population density ("**quorum sensing**"). Autoinduction is often the reason why populations at high cell densities differ strikingly from populations at low cell densities, e.g., in bioluminescence, swarming motility, sporulation, and other cooperative behavioral patterns, or in fermenters (clumping, flocculation).

20.2 Carbon Catabolite Repression Is Intimately Linked to Other Phenomena

In the previous section, some general concepts important in global regulation were described. Phenomena collectively called **carbon catabolite repression** that occur in many different microorganisms will now be discussed. Various processes can be subject to catabolite repression, including the synthesis of carbohydrate transporters and catabolic enzymes (see below), the synthesis of flagella or fimbriae, enzymes involved in

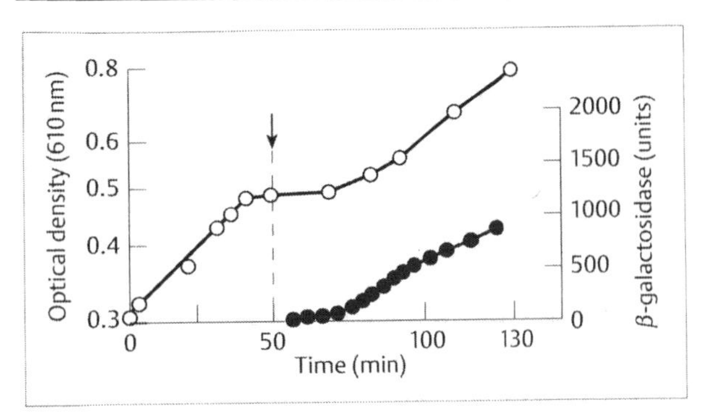

Fig. 20.**8** **Diauxic growth of *E. coli* on a mixture of glucose (0.04%) and lactose (0.2%).** Growth (as measured by the optical density at 610; ○) and synthesis of β-galactosidase [nmol (mg protein)$^{-1}$ ●] are shown as a function of time [after 3]

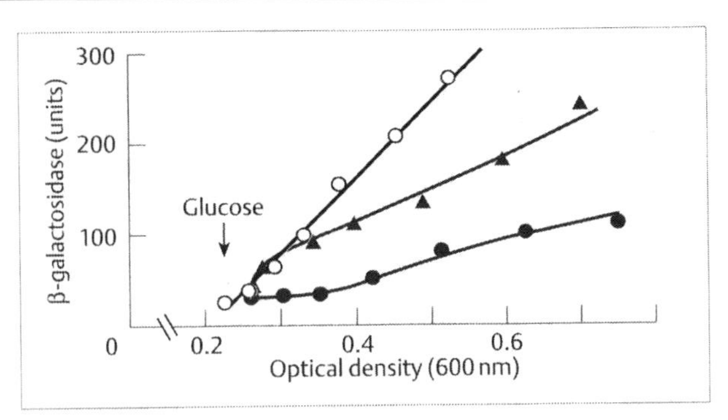

Fig. 20.**9** **Effects of the addition of extracellular cAMP on permanent and transient catabolite repression in *E. coli*.** *E. coli* K12 cells were grown in a rich Luria Broth medium containing 0.5 mM isopropyl β-D-thiogalactoside (IPTG) as an inducer of the *lac* operon. At the point indicated by the arrow, the following compounds were added: none, ○; glucose (0.8%), ●; glucose (0.8%) plus 5 mM cAMP, ▲. One unit of β-galactosidase corresponds to 1 nmol of o-nitrophenol produced per min at 28°C per ml [after 3]

glycogen synthesis, proteins involved in sporulation, and extracellular enzymes involved in the degradation of polymers. Focus will first be placed on the repression of enzymes involved in central carbon metabolism.

When bacteria are exposed to high concentrations (>1 mM) of two different carbon sources, usually one is used preferentially. Growth occurs in two distinct phases, separated by a lag period (Fig. 20.**8**). This lag can last up to one generation time, during which the cells adapt to the second carbon source. This phenomenon is called **diauxic growth** or **diauxie** and was studied extensively by J. Monod. Although most studies have been performed in enteric bacteria, particularly in *E. coli*, similar phenomena have been observed in many bacteria and with a multitude of nutrients. Since in many studies glucose has been used as the repressing carbon source, often the loose term **glucose effect** (glucose repression) is used rather than the more correct term carbon catabolite repression. It should be emphasized, however, that (1) many other carbon sources can elicit a response similar to that of glucose and (2) in non-enteric bacteria, sometimes the opposite is found, i.e., glucose utilization is repressed by other compounds. Later studies, in particular with enteric bacteria, have shown that carbon catabolite repression often occurs when there is an imbalance between catabolism and anabolism, i.e., when the rate of catabolism exceeds that of anabolism.

20.2.1 Carbon Catabolite Repression Is a Complex Phenomenon Involving Several Mechanisms in Enteric Bacteria

At least three different phenomena can be discriminated in the overall process that is called carbon catabolite repression: permanent repression, transient repression, and inducer exclusion. **Permanent repression** is generally weak (two- to threefold lower activities) and lasts as long as the repressing compound is present. In contrast, **transient repression** is strong, i.e., it causes more than 90% inhibition, but lasts only for 0.1–1 doubling time (Fig. 20.**9**). In the case of permanent repression, the repressing compound must be metabolized. Transient repression can also be elicited by non-metabolizable analogues. Although the molecular basis of the phenomena is not the same, both are connected to the intracellular second messenger cAMP and to the regulation of enzyme synthesis. **Inducer exclusion**, in contrast, is a process in which the activity of certain enzymes is inhibited rather than their synthesis (e.g., the entry of an inducer into the cell is prevented). A specific example is the exclusion of lactose, an inducer of the *E. coli lac* operon, by glucose. Each of these phenomena will be discussed in more detail, and how the control of cAMP synthesis and inducer entry regulates cellular metabolism will be described.

20.2.2 The *crp* Modulon Is Part of Carbon Catabolite Repression in Enteric Bacteria

Many bacteria, but not all, can synthesize cAMP (Fig. 20.**14**). Its presence was shown early on in *E. coli*, and the intracellular cAMP concentration was found to be regulated by synthesis, breakdown, and secretion. cAMP is synthesized from ATP by the enzyme adenylate cyclase, encoded in *E. coli* by the *cyaA* gene. The intracellular concentration of cAMP is in the order of 1–10 μM. It can be hydrolyzed via a phosphodiesterase (encoded by the *cpd* gene) or exported to the medium. The latter process is ill-understood, but large quantities of cAMP are found in the medium under certain growth conditions. In enteric bacteria, cAMP is essential for growth on a large number of carbon sources and can be considered as an **alarmone** for carbon starvation.

Alarmones are molecules that signal stress conditions. They are often derivatives of nucleotides, such as cAMP in the *crp* modulon for catabolic enzymes or guanosine tetraphosphate (ppGpp) for anabolic enzymes. Alarmones seem to be characteristic for global regulatory systems that control central metabolic activities and sense changes in the growth conditions through alterations of metabolic fluxes by means of intracellular sensors.

This became apparent when *cyaA* mutants were isolated that were defective in adenylate cyclase. Whereas growth on a complex medium is still possible, i.e., the mutation is not lethal, growth on many carbon sources (e.g., lactose, maltose, glycerol, mannitol, and Krebs cycle intermediates) in a minimal medium is prevented. Addition of high concentrations (>1 mM) of cAMP to the extracellular medium restores the growth of *cyaA* mutants. The corresponding operons (e.g., *lac, mtl*) or regulons (e.g., *mal, glp*) are members of the *crp* modulon (Fig. 20.**2**). Transcription of these operons and regulons requires cAMP together with the cAMP receptor protein Crp because their promoters are usually "closed," i.e., unable to bind RNA polymerase. Efficient RNA polymerase binding requires the simultaneous binding of the cAMP · Crp complex at a consensus sequence located upstream of the −35 motif (see Chapter 18 for details). Mutants that lack the Crp protein exhibit the pleiotropic carbohydrate-negative phenotype of *cyaA* mutants that are unable to synthesize cAMP. In contrast to the latter class of mutants, growth of *crp* mutants cannot be stimulated by the addition of cAMP.

The connection between carbon catabolite repression and cAMP became evident when it was shown that permanent and transient repression can be relieved by the addition of cAMP (Fig. 20.**9**). Furthermore, it could be shown that there is a correlation between the synthesis of, for instance, β-galactosidase and the intracellular cAMP concentration (Fig. 20.**10**). Lower cAMP concentrations, generated by using "strong catabolite repressors" like glucose, gluconate, or glucose 6-phosphate, correlated with a low β-galactosidase synthesis, whereas higher cAMP concentrations, using "weak catabolite repressors" like glycerol or lactate, resulted in three- to fivefold higher β-galactosidase activity. In earlier experiments, it had been found that synthesis of β-galactosidase was low in media with carbon sources that allowed fast growth (e.g., glucose or gluconate) and high in media that allowed slow growth (e.g., succinate or lactate) (Fig. 20.**10**). In other words, starvation conditions are signaled to the cells by high levels of the alarmone cAMP, and feast conditions are signaled by low levels of cAMP.

20.2.3 The PEP-dependent: Carbohydrate Phosphotransferase System is Central in Catabolite Repression and Inducer Exclusion

Explanations of the phenomena that comprise carbon catabolite repression should include an understanding of the regulation of the intracellular cAMP concentration and of its binding protein, Crp, as well as the control of inducer entry. In enteric bacteria, a major role in carbon catabolite repression is played by a carbohydrate-transport system, the phospho*enol*pyruvate (PEP)-

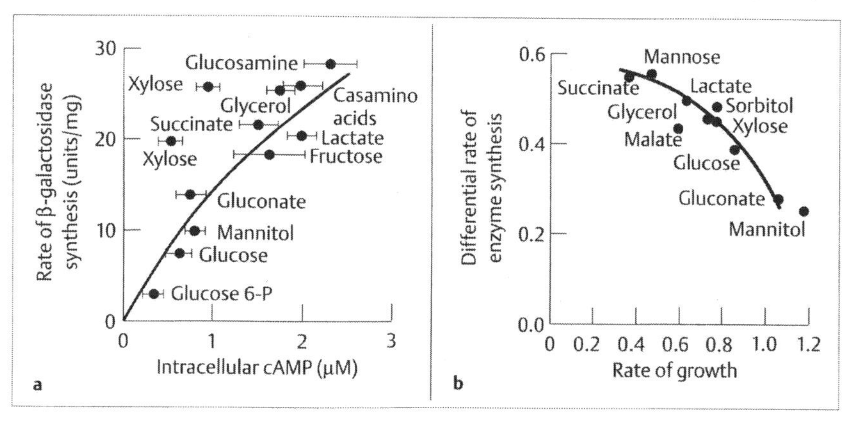

Fig. 20.**10a,b Correlation between growth rate, β-galactosidase synthesis, and intracellular cAMP concentration.**
a Relationship between the rate of β-galactosidase synthesis and the intracellular cAMP concentration in *E. coli* K-12 [after 4]
b Relationship between growth rate and β-galactosidase synthesis in *E. coli* K-12. Growth was on the substrates (20 mM) as indicated [after 5]

dependent phosphotransferase system (PTS). The PTS, as described in Chapter 5.6.4 is a major transport system in many different eubacteria. This exceptional system, first described by S. Roseman and coworkers, accumulates its substrates as the corresponding phosphate esters at the expense of PEP. PTSs are found in Gram-negative and Gram-positive bacteria, in obligate and facultative anaerobes and, although less frequently, in obligate aerobes. Their basic structure is similar in all bacteria and has been shown schematically in Fig. 5.**16**. In essence, the PTS transports and phosphorylates numerous carbohydrates by transferring the phosphogroup of PEP sequentially via the PTS proteins Enzyme I and HPr and the carbohydrate-specific Enzyme II complexes to the carbohydrates.

The phosphotransferase system, adenylate cyclase, and cAMP. The scheme presented in Figure 20.**11** predicts that mutants impaired in the general PTS proteins, EI (encoded by the *ptsI* gene) and HPr (encoded by the *ptsH* gene), would be unable to grow on any PTS carbohydrate, whereas mutants lacking a specific EII, such as II^{Mtl}, would be unable to grow on and metabolize mannitol, but would show normal growth on all other PTS carbohydrates. These predictions are supported by the mutant phenotype. However, *E. coli* and *S. typhimurium ptsH,I* mutants show an interesting additional phenotype: these mutants do not grow on a number of non-PTS carbon sources, i.e., they have a pleiotropic negative phenotype.

This pleiotropic phenotype of *ptsH,I* mutants is reminiscent of that of *cyaA* and *crp* mutants discussed above in that all three mutants are defective in growth on, for example, lactose, melibiose, maltose, glycerol, and Krebs cycle intermediates. The similarity is strengthened by the following observations: (1) growth of *ptsH,I* mutants on non-PTS carbon sources (but not on PTS carbon sources) is restored by the addition of extracellular cAMP, (2) the synthesis of cAMP in intact cells and the adenylate cyclase activity of toluene-treated cells is inhibited in the presence of a PTS carbohydrate such as glucose or its non-metabolizable analogue methyl α-glucoside, (3) adenylate cyclase activity is much lower in mutants that lack EI and/or HPr, than in cells with an intact PTS and (4) suppressor mutations that restore the growth of *ptsH,I* mutants on the non-PTS carbon sources mentioned above comprise mutations in the Crp protein. These *crp** [or *crp*(In) for **In**dependent] mutations result in a Crp protein that has become independent of cAMP. This class of mutations can be understood in the light of the effect of added cAMP on the phenotype of *ptsH,I* mutants: the mutant Crp protein binds to and activates promoters in the

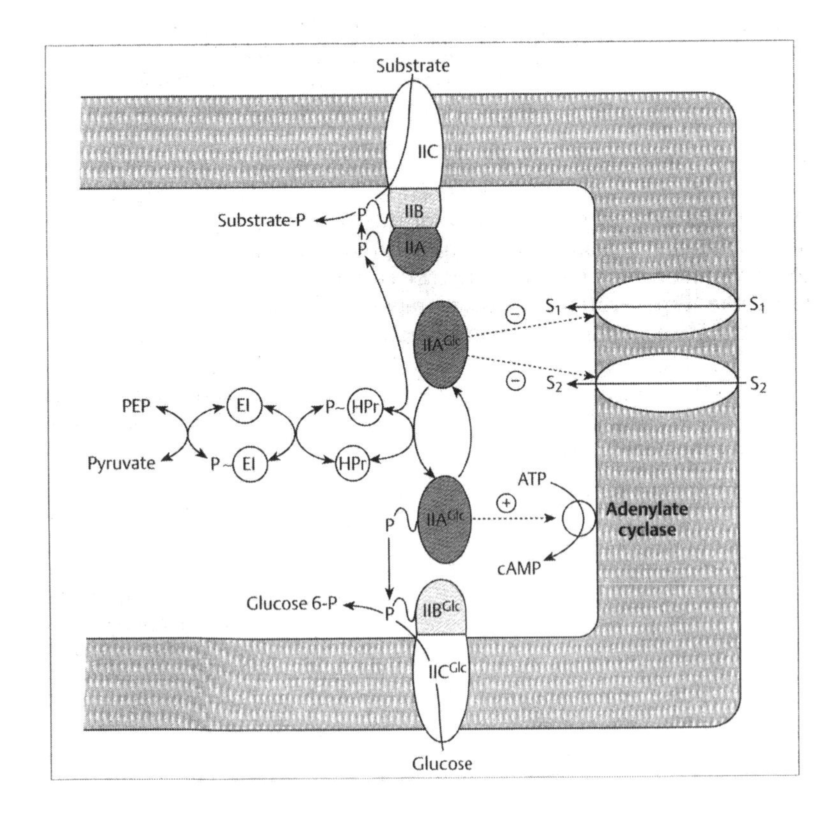

Fig. 20.**11** **A model for PTS-mediated carbon catabolite regulation in enteric bacteria.** The scheme shows the general enzymes of the PTS: Enzyme I (EI), HPr, and two Enzymes II [IIA-C and $IICB^{Glc}/IIA^{Glc}$ (formerly called III^{Glc})]. ~P indicates phosphorylation of the various PTS enzymes. Activation (+) of adenylate cyclase by phosphorylated IIA^{Glc} (P~IIA^{Glc}) and inhibition (−) of two non-PTS uptake systems, S_1 and S_2 (representing those involved in lactose, melibiose, maltose, or glycerol transport and metabolism) are also indicated [after 6]

absence of cAMP. It was inferred from the experiments that the dephosphorylation of one or more of the PTS proteins upon addition of a PTS carbohydrate to the cell (or upon introduction of a mutation) inhibit the adenylate cyclase activity. Alternatively, one or more of the phosphorylated PTS proteins could be involved in the activation of adenylate cyclase.

The phosphotransferase system and inducer exclusion. A second class of suppressor mutations in the *crr* gene (**c**arbohydrate **r**epression **r**esistance) has been isolated from enteric bacteria. These suppressor mutations restore growth of *ptsH,I* mutants on the non-PTS carbon sources lactose, maltose, melibiose, and glycerol. The *crr* gene encodes a component of the glucose PTS, IIA^{Glc} (Fig. 20.**11**). In parallel experiments, it has been shown that addition of any PTS carbohydrate or its non-metabolizable analogue inhibits the uptake of these non-PTS carbon sources in a matter of seconds. In *crr* strains, this inhibition is not observed. Thus, these effects are not at the level of enzyme synthesis, but concern the activity of the uptake system and involve IIA^{Glc}. The process has been called **inducer exclusion,** or more correct, **PTS-mediated inducer exclusion.** Genetic experiments have indicated that IIA^{Glc} interacts most likely with specific components of the various uptake systems since mutations that abolish PTS-mediated inducer exclusion have been isolated in the *lacY* (lactose permease), *melB* (melibiose permease), *malK* (component of the maltose permease), and *glpK* (glycerol kinase) gene. Biochemical experiments have shown that the purified non-phosphorylated IIA^{Glc} protein binds to and inhibits the proteins encoded by these genes, whereas phosphorylated IIA^{Glc} does not bind to these

For enteric bacteria, rapid growth on glucose ("glucose effect"), diauxic growth, and growth conditions that cause an excess of catabolism over anabolism cause **carbon catabolite repression.** These conditions are sensed by the PTS, which signals the information through $P{\sim}IIA^{Glc}$ to the enzyme adenylate cyclase and the cAMP-dependent *crp* modulon with its global regulator Crp, thus causing **transient** and **permanent catabolite repression**. The system controls, in parallel through IIA^{Glc}, the activity of carbohydrate transport systems and metabolic enzymes (**"inducer exclusion"**) and hence the uptake and synthesis of inducers. There are non-PTS-dependent elements in this complex control for which no mechanisms are known. In other than enteric bacteria, other carbon sources elicit carbon catabolite control (e.g., carboxylic acids in pseudomonads).

proteins. Recently, the crystal structure of one of these complexes, the glycerol kinase/IIA^{Glc} complex has been solved. Most likely, the phosphorylation of IIA^{Glc} on the His-90 residue prevents its binding to glycerol kinase due to direct electrostatic and steric repulsion. In Figure 20.**11**, the various processes are combined in a model of PTS-mediated regulation in enteric bacteria in which the glucose-specific IIA^{Glc} plays a central role and for which experimental support is available.

Quantitative aspects of PTS-mediated regulation. As indicated in Figure 20.**11**, IIA^{Glc} plays two important roles: (1) the non-phosphorylated form is an inhibitor of various non-PTS uptake systems, and (2) phosphorylated IIA^{Glc} is an activator of adenylate cyclase. It cannot be excluded at present, however, that additional proteins are involved in this activation process since the direct interaction of both proteins has not been shown.

In cells growing on non-PTS carbon sources, the PTS proteins will be fully phosphorylated because of the high free energy of hydrolysis of the phospho-group of PEP. In cells that encounter both a PTS carbohydrate and a non-PTS carbon source, uptake of the PTS carbohydrate results in dephosphorylation of the PTS proteins and their rephosphorylation via PEP. Depending on the rates of (de)phosphorylation, the steady-state level of phosphorylation of IIA^{Glc} can decrease, as has been shown under a limited set of conditions. An increase in the amount of non-phosphorylated IIA^{Glc} has two consequences. First, adenylate cyclase activity decreases since the concentration of the activator, P-IIA^{Glc}, decreases. As a consequence, transcription of cAMP-dependent operons and regulons will decrease, resulting in a lower level of the corresponding transport systems and metabolic enzymes. Second, remaining uptake systems will be inhibited by IIA^{Glc}. As a consequence, the entry of inducer will be reduced, resulting in an additional lowering of transcription.

When interactions between proteins like IIA^{Glc} and its various target proteins or between proteins like Crp and its specific target sequence are considered, stoichiometric relationships can become important. In cells in which either a target protein of IIA^{Glc}, for instance the lactose transporter or glycerol kinase, is expressed at very high levels or the number of IIA^{Glc} molecules is lowered below its normal level (approximately 20 000 copies per cell), inducer exclusion disappears or becomes less severe. Some target proteins remain free and active, resulting in escape from inducer exclusion. Cells also prevent the non-productive and wasteful binding of IIA^{Glc} to target proteins if their substrates are not available. Binding of IIA^{Glc} to, for example, the lactose transporter or glycerol kinase requires a substrate of that protein (β-galactoside or glycerol, respectively).

This implies that conformational changes have to occur before IIA^{Glc} can bind.

Similar stoichiometric considerations may apply to Crp. It has been shown that under certain conditions, such as growth in the presence of glucose, permanent repression cannot be relieved by the addition of extracellular cAMP. A possible reason is the lowered number of Crp molecules in these cells since in cells with increased Crp levels (using a plasmid that allows controlled synthesis of Crp), permanent repression is completely overcome by added cAMP. It is unclear at present how repressing carbon sources can lower the Crp concentration. It was mentioned in Chapter 20.2.1 that carbon sources that allow fast growth, but are not transported and phosphorylated via the PTS, such as gluconate and glucose 6-phosphate, can also induce strong catabolite repression. Possibly, these carbon sources exert their effect via modulation of the Crp level, as has been proposed for glucose.

Analysis of the scheme of PTS-mediated regulation shows an example of a signal transduction pathway in which the activity of an intracellular regulatory protein, IIA^{Glc}, is modulated by covalent modification in response to changes in the extracellular medium. Since all reactions of the PTS proteins are reversible, any PTS carbohydrate can act as the signal: P-IIA^{Glc} can be dephosphorylated **directly** via its associated $IICB^{Glc}$ (glucose) or **indirectly** via HPr, which can be dephosphorylated by each of the other EIIs (all PTS carbohydrates except glucose).

20.2.4 Carbon Catabolite Repression in Non-enteric Bacteria Also Involves the PTS

Carbon catabolite repression in most of the non-enteric bacteria is probably caused by different mechanisms. First, in a number of bacteria that are subject to catabolite repression, for example, *B. subtilis*, no intracellular cAMP is found. Second, there is no evidence that IIA^{Glc}, even if present, plays a role like that described for enteric bacteria. At present, insight to carbon catabolite repression in non-enteric bacteria is limited to a few examples.

In several Gram-positive organisms, mutations have been isolated that allow the expression of particular genes that are normally repressed by a number of carbon sources. *cis*-acting (promoter) sequences, called **c**atabolite **r**esponsive **e**lements *(cre)*, that mediate this repression have been detected. A consensus sequence has been proposed. The *cre* sequence sometimes overlap with the promoters of the genes that are sensitive to repression; however, in other cases, the *cre* sequences are part of the structural gene. In *Bacillus*, these *cis*-

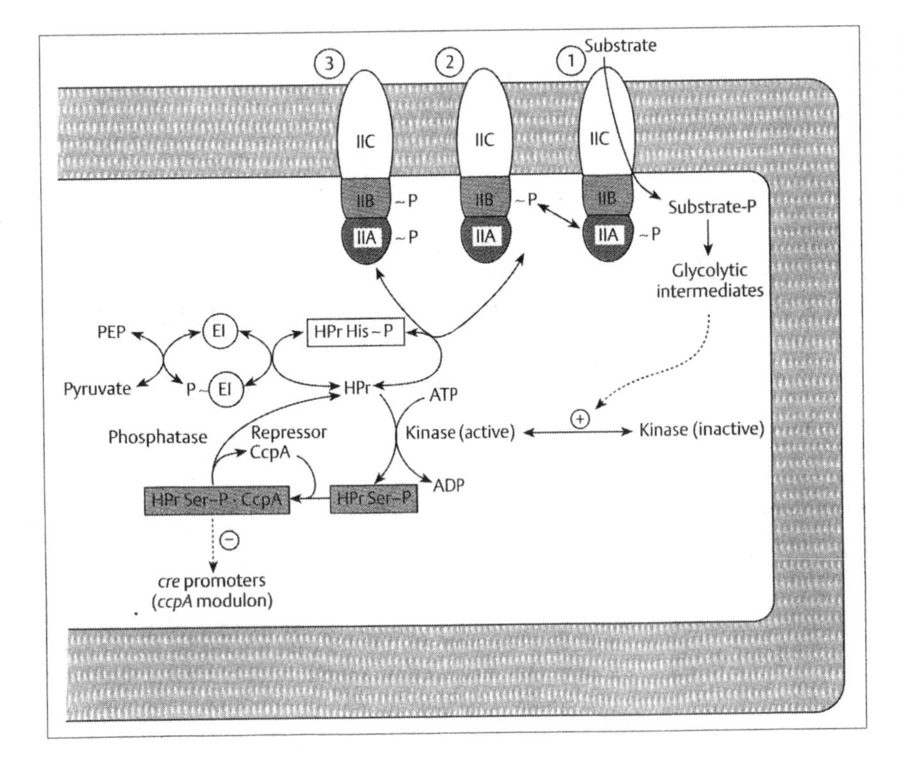

Fig. 20.**12** **A model for PTS-mediated carbon catabolite regulation in Gram-positive bacteria.** The PTS is drawn in analogy to Figure 20.**11**, with a single large Enzyme II. Uptake of a substrate dephosphorylates the various PTS proteins, in particular HPr, and generates glycolytic intermediates. These activate (⊕) an ATP-dependent protein kinase, which phosphorylates free HPr at Ser-46 (in *B. subtilis*). HPr-Ser-46-P is thought to activate the global repressor CcpA, which binds in this form to all promoters containing a *cre* binding site. These promoters and the genes they control constitute the *ccpA* modulon. A protein-phosphate phosphatase completes the cycle

acting sequences have been discovered, for instance, in the *amyE* gene, encoding α-amylase, and in operons involved in gluconate *(gnt)*, xylose *(xyl)*, glucan *(bgl)*, and histidine *(hut)* metabolism. Mutations that abolish *cre*-mediated catabolite repression occur in the *ccpA* gene, encoding the CcpA protein (**c**atabolite **c**ontrol **p**rotein) that belongs to the LacI family of transcriptional regulators and that may bind to the *cre* sequence. A connection between *cre*-mediated catabolite repression, CcpA, and the PTS exists in *B. subtilis* and other Gram-positive bacteria. In Chapter 5.6.4 it was described how HPr is phosphorylated by P-EI on a histidine residue. In Gram-positive bacteria, HPr can be phosphorylated on a second residue, a serine, by a protein kinase that utilizes ATP as a phospho-donor. The phospho-group can be removed by a phospho-protein phosphatase. Fructose 1,6-bisphosphate and other glycolytic intermediates stimulate the protein kinase, and P-(Ser)-HPr is present in cells with a high glycolytic activity. Recent data suggest that possibly P-(Ser)-HPr interacts with CcpA and promotes the binding of CcpA to the *cre* sequences, thus eliciting repression (Fig. 20.**12**). Absence of a readily usable carbon source favors the formation of HPr in its non-phosphorylated state, and thus, the relief of catabolite repression.

As in enteric bacteria, the phosphorylation state of a PTS protein, here HPr, is important in controlling gene expression of a modulon by determining the activity of the protein CcpA, which interacts with certain DNA elements (*cre* promoters). In both systems, the state of the peripheral catabolism is sensed by a signal transduction system, the PTS, and is signaled to a global regulator.

20.3 The RelA/SpoT Modulon Controls Anabolic Pathways and Macromolecule Biosynthesis

Depending on the growth conditions, the doubling time of a bacterial culture can vary from 20 min to 1 week. A characteristic rate of macromolecule biosynthesis and a typical cell size, i.e., a specific physiological state, corresponds to each growth rate. In general, fast-growing cells contain more DNA, RNA, proteins, phospholipids, and cell wall material, and increase in size. The relative composition of a cell may also change. Rapidly growing cultures contain larger numbers of ribosomes per cell, which increase, for instance in *E. coli*, from about 2 000 in poor media (≥180-min generation time) to 20 000 in rich media (20-min generation time). Under normal growth conditions, ribosomes catalyze protein synthesis close to the maximal speed (about 16 polymerization steps s^{-1}); in other words, the velocity of protein synthesis can only be increased by increasing the number of ribosomes. Changes in macromolecule composition following an abrupt change in the medium can be dramatic because up to half of a bacterial cell mass is protein and ribosomes. Changes in doubling times reveal sophisticated modulons and regulatory networks that control the synthesis of anabolic enzymes. One of these, the RelA/SpoT modulon of the enteric bacteria, is the equivalent of the *crp* modulon for

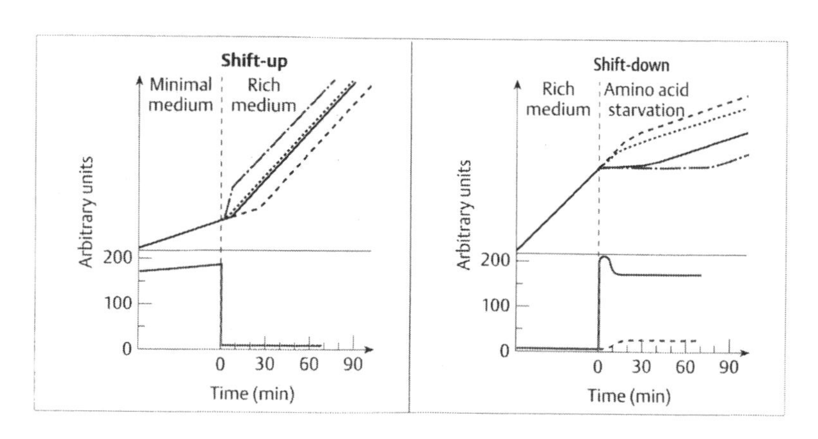

Fig. 20.**13** **Response of *E. coli* to shift up/shift-down under various growth conditions.** Indicated are the changes in RNA synthesis rates (–··–), DNA synthesis rates (...), optical density as a measure of cell mass (–), and cell number (- - -) after a shift in cell growth rate (vertical line). Also indicated in the lower part of the figure are the changes in ppGpp levels (in mM) for a Rel^+ (—) strain and a Rel^- (- - -) mutant

catabolic systems. It will be discussed in some detail next.

20.3.1 Changes in Doubling Time Trigger a Stringent Response

When a culture of *E. coli* growing at steady-state is enriched by the addition of nutrients (a "**shift-up**"), the rate of new cell mass formation, measured, for example, as an increase in optical density, increases after a short lag (Fig. 20.**13**), a pattern followed closely by the rate of DNA synthesis. RNA synthesis, however, increases immediately after the shift-up, and is even above average for a short time. Meanwhile, the number of cells increases at the slower preshift rate, and the cells become longer. As a consequence of these rearrangements, larger cells with more ribosomes and more DNA are generated. Finally, when a new macromolecular composition that corresponds to the new physiological state of the cells is reached, the cells begin to divide with a shortened doubling time.

Cells growing exponentially in rich medium can, on the other hand, be transferred to a minimal medium or into conditions that limit abruptly the availability of one or more amino acids. Such a "**shift-down**" can also be caused by exhausting the cells' primary carbon source. Growth measured as the increase in cell mass stops more or less completely under these conditions, while overall DNA synthesis and the increase in cell number continue temporarily at the preshift rate (Fig. 20.**13**). RNA synthesis, in contrast, stops immediately and for an extended period of time, until the macromolecular composition corresponds to the new growth rate. The cells become shorter and lighter and switch finally to an increased doubling time.

A closer look reveals that, during a shift-down, not only RNA and protein synthesis are inhibited temporarily. At the same time, the initiation of new DNA replication cycles is inhibited, as are the biosynthesis of carbohydrates, phospholipids, and cell wall constituents and the uptake of the low-molecular-weight precursors of these macromolecules. In contrast, other processes, in particular biosynthesis of amino acids and control of translational fidelity, are activated (Tab. 20.**2**). This set of responses, with its characteristic tight coupling between growth rate, ribosomal synthesis, and cell size, is referred to as the **stringent control** (or stringent response). Mutants are available in which the stringent control of RNA synthesis is **relaxed,** hence the mnemonic *rel* for the corresponding genes (Fig. 20.**13**). RelA$^-$ mutants in particular are affected in most of the processes listed in Table 20.**2**. These pleiotropic effects of a single mutation in *relA* suggested the existence of a global regulatory system. An efficient way to control all members of a modulon under stress conditions is the production of a single molecule (**alarmone**), which accumulates during the stress. Stringent control, which similar to catabolite repression affects central metabolic activities, uses ppGpp (Fig. 20.**14**), a derivative of guanosine, as an alarmone.

20.3.2 Synthesis of the Alarmone ppGpp Requires Stalled Ribosomes and Protein RelA

When starved for an amino acid, many bacterial species rapidly accumulate millimolar concentrations of unusual nucleotides, called originally Magic Spot I and II. These were later identified as the GTP derivatives guanosine 5′-diphosphate, 3′-diphosphate (ppGpp or guanosine tetra-

Table 20.**2** **Direct and indirect effects of the alarmone ppGpp**

Depletion of aminoacylated tRNAs (stalled ribosomes) or an excess of anabolism over catabolism cause the accumulation of ppGpp

Direct effects (Modulon)	Secondary effects (Stimulon)
Inhibition of RNA polymerase	**Inhibition** of ribosomal protein synthesis
rRNA and tRNA synthesis	tRNA processing
protein chain elongation	general protein synthesis
	phospholipid synthesis
	nucleotide synthesis
	cell wall constituents synthesis
	DNA replication (initiation)
	cell division
Activation of transcription	**Increase** of translation fidelity
for biosynthetic operons	chaperone synthesis
for catabolic operons	stress protein synthesis
	protein degradation
	catabolic pathways

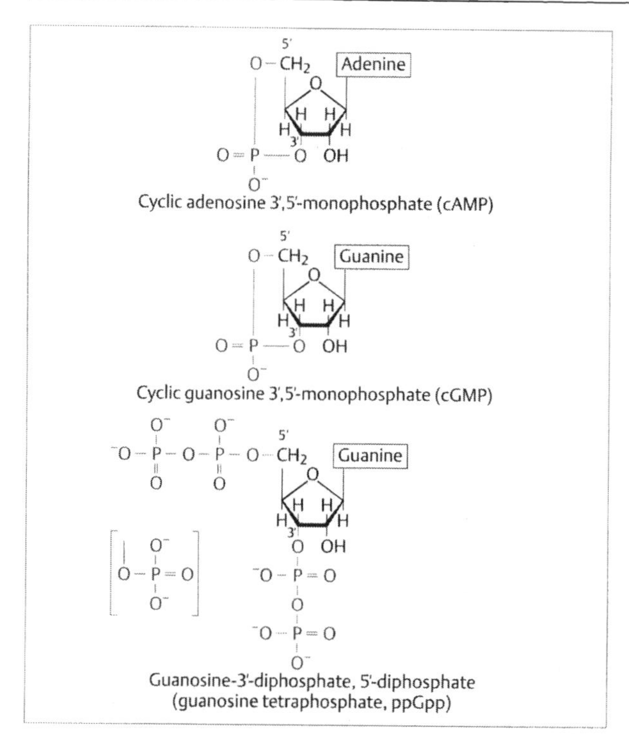

Fig. 20.**14** **Rare nucleotide derivatives (alarmones) involved in global cellular control.** Guanosine pentaphosphate (pppGpp) is guanosine 3′-diphosphate, 5′-triphosphate

phosphate) and guanosine 5′-triphosphate, 3′-diphosphate (pppGpp or guanosine pentaphosphate), respectively (Fig. 20.**14**). Synthesis of ppGpp is triggered by the binding of uncharged tRNAs, as they arise during amino acid starvation, to the acceptor (or A) site of a ribosome (Fig. 20.**15**). Such ribosomes, stalled in the process of peptide elongation, activate (p)ppGpp synthetase I. This enzyme is the product of the *relA* gene and appears to be bound to the large (50S) ribosomal subunit through peptide L11 (gene *rplK*). Activated synthetase I (or RelA) catalyzes a pyrophosphoryl group transfer of the β,γ-phosphates from ATP to the 3′-hydroxyl groups of GTP (and rarely of GDP). A stalled ribosome apparently undergoes cycles of uncharged tRNA binding, tRNA release, and (p)ppGpp synthesis, utilizing the GTP that is normally involved in peptide elongation. Deletion of the *relA* gene not only changes the stringent response of wild-type cells into a relaxed response, but also reduces ppGpp levels drastically, although not completely (Fig. 20.**13**). Hence, a second and RelA-independent biosynthetic pathway must be present. This alternative synthetase II is operative under different stress conditions.

The intracellular increase of ppGpp parallels a stringent response, and a decrease parallels relaxation. This rare nucleotide thus has all the properties of an alarmone. As outlined in Figure 20.**15**, not only biosynthesis, but also the degradation and, therefore, the steady-state levels of ppGpp are controlled. In addition to synthetase I (RelA), pppGpp-5′-phosphohydrolase (Gpp, gene *gpp*), nucleoside diphosphokinase (Ndk, gene *ndk*), and the protein SpoT (gene *spoT*) are essential. Mutations in any of these genes also cause a Rel^- phenotype.

20.3.3 An Alternative and RelA-Independent Pathway for ppGpp Synthesis Is Activated During Carbon Starvation

Tight *relA* mutants retain a residual ppGpp pool and a nearly normal stringent response towards certain types of starvation, such as that induced by carbon starvation. Both activities are lost if a second gene *spoT* (for magic "Spot") is also eliminated. A strain deleted in both *relA* and *spoT* is viable, but is severely impaired in its ability

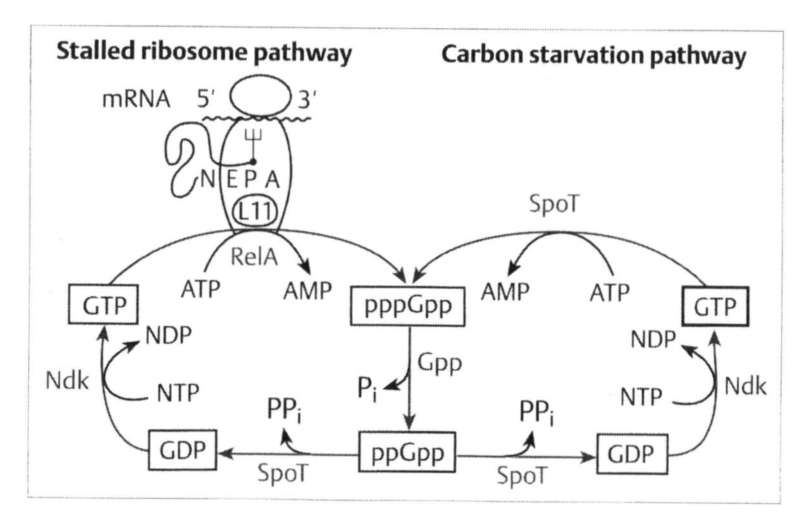

Fig. 20.**15** **Metabolism of the alarmone ppGpp in enteric bacteria.** In the absence of a charged tRNA in the A site of a ribosome, the ribosome stalls and triggers together with the RelA protein (ppGpp synthetase I) and L11 the synthesis of pppGpp. This synthesis can also be triggered through an unknown mechanism that involves the protein SpoT (ppGpp synthetase II activity) by severe carbon starvation. pppGpp is converted by a pppGpp 5′-phosphohydrolase (Gpp) to the alarmone ppGpp. Also involved in the complete cycle are SpoT [(p)ppGpp 3′-pyrophosphohydrolase], which now hydrolyzes ppGpp to GDP, and the nucleoside 5′-diphosphate kinase (Ndk)

to survive prolonged amino acid or carbon starvation. Gene *spoT* encodes the alternative (p)ppGpp synthetase II. This enzyme does not require stalled ribosomes for activation, but is activated through carbon starvation (Fig. 20.**15**). The SpoT protein is also involved in the degradation of ppGpp to GDP, even for the stalled ribosome pathway. This dual role has complicated the understanding of its precise role.

In the absence of charged tRNAs, ribosomes stall and trigger together with RelA the synthesis of the alarmone ppGpp. Long-lasting carbon starvation triggers ppGpp synthesis through an alternative pathway that involves SpoT.

20.3.4 Some Effects of the Alarmone ppGpp in Stringent Control Are Direct

The large number of adjustments following amino acid or carbon starvation makes it often difficult to distinguish between cause and effect. Thus, upon a sudden shift-down from rich to minimal medium, the shortage of charged aminoacyl tRNAs will stall almost immediately many ribosomes and reduce protein synthesis. Temporarily, no new DNA replication forks will be initiated, and no new ribosomal proteins will be synthesized. This will cause an ever-increasing feedback, first on total DNA and protein synthesis and then on the biosynthesis of their low-molecular-weight precursors. In the absence of cell growth, cell wall and phospholipid synthesis must also be stopped to avoid an unnecessary waste of energy. In the end, about half of the cellular proteins are affected negatively or positively during the stringent response. Because this response must be integrated with several other global regulatory systems, many of the observed effects are probably not a direct, but are an indirect consequence of the stringent control. A massive accumulation of partly finished peptides, for instance, activates chaperones and other stress proteins, and it triggers an increased protein degradation. Both processes help to overcome amino acid starvation quickly.

Strains deleted for genes *relA* and *spoT*, which lack consequently (p)ppGpp synthetases I and II, have been used to manipulate the intracellular concentration of ppGpp in the absence of amino acid or carbon source starvation. A truncated *relA* gene that encodes a smaller, but catalytically active peptide with ppGpp synthetase activity has been isolated. The truncated gene was cloned behind a *tac* promoter controlled by *lacI*^{q}. Induction with IPTG causes synthesis of the RelA peptide and a drastic increase in ppGpp levels. This gratuitous induction of ppGpp inhibits within minutes growth, ribosomal RNA transcription, and protein synthesis up to 90%. Based on this result and on previous in vitro studies, it is now generally accepted that an increase in ppGpp levels causes these inhibitions. The effect of ppGpp levels on transcription rates from various promoters has been tested in vitro. The results indicate that the promoters for tRNA and rRNA genes are the direct targets for stringent control and more precisely, an "upstream activator sequence," a ppGpp-dependent discriminator, and ppGpp-dependent pausing sites. Similar results have been obtained from in vivo studies in which these promoters were fused to a *lacZ* indicator gene. Many suppressor mutations that cause a relaxed phenotype, i.e., that eliminate stringent response, map in *rpoB*, the structural gene for the β-subunit of RNA polymerase. It is conceivable that in the presence of ppGpp, RNA polymerase cannot transcribe promoters of the RelA modulon. In line with this argument, the suppressed *rpoB* mutations would generate ppGpp-resistant RNA polymerase molecules.

The alarmone ppGpp inhibits transcription of the genes for stable RNAs; however, it appears to activate transcription of other genes. Consider again a cell deprived abruptly of a single or a few amino acids. Shutting down macromolecular biosynthesis, including general protein synthesis, and increasing protein degradation is reasonable only if the low amino acid supply is used to synthesize preferentially the biosynthetic enzymes for the missing component. There is evidence, although not very direct, that some operons involved in amino acid biosynthesis, in nucleotide biosynthesis, and in carbon source degradation are transcribed at an increased rate in the presence of ppGpp. A global increase in the synthetic capacity for all macromolecule precursors and in the degradative capacity for all carbon sources is reasonable under stringent-response conditions. All operons for which the end products are not limiting or for which no inducers are present in the medium will remain repressed, but the enzymes of the pathway for the limiting compound can be synthesized at an above-normal rate. In accordance with this model, it was found that the *relA-spoT* double mutants become auxotrophic for several amino acids and are unable to grow on certain carbon sources, namely lactose and succinate. The expression of the corresponding operons thus seems to depend, at least in part, on the presence of the alarmone ppGpp.

20.3.5 Many Effects Observed During Stringent Response Are Not Caused Directly by ppGpp

Two phenomena that are generally observed under stringent control are: (1) an inhibition of general protein synthesis and (2) an increase in translational fidelity. They are not caused directly by ppGpp, but are a consequence of the inhibition of stable RNA synthesis. (1) Inhibition of stable RNA synthesis causes the accumulation of free ribosomal proteins. These free proteins, as discussed in detail in Chapter 18, inhibit their own synthesis at the translational level by binding to the mRNAs that encode these proteins. This causes general inhibition of translation without any direct effect of ppGpp on the process. (2) When cells are deprived of an amino acid, net protein synthesis stops, but a residual synthesis continues. This synthesis depends on the amino acids generated through the increased protein degradation mentioned before as typical for stringent response. The residual synthesis is highly inaccurate in *relA* mutants, but not in *rel*$^+$ wild-type strains. One current model assumes that ppGpp slows down peptide elongation during residual protein synthesis. This then would increase the time needed for proofreading by a ribosome and hence translational accuracy.

Many other effects, listed in Table 20.**2**, are connected even less directly to stringent control. Inhibition of DNA synthesis may be caused through inhibition of the synthesis of initiator proteins. Phospholipid and cell wall material biosynthesis also require protein synthesis and protein excretion. None of these processes appear to be affected directly by ppGpp-inhibition of stable RNA bio-synthesis and activation of transcription of the operons, as mentioned before.

Upon a severe imbalance between anabolism and catabolism, cells readapt their macromolecule biosynthetic capacity through the stringent response. The system reacts directly through a modification of RNA polymerase specificity by ppGpp and modulation of the *relA/spoT* modulon. Also involved is a stimulon that comprises a plethora of functions controlled indirectly during stringent response.

20.3.6 Stringent Response Is Caused by an Imbalance Between Anabolism and Catabolism

Stringent response can also be described in terms of a transiently acting sensory system that globally controls anabolism and macromolecule biosynthesis. The system is poised to detect severe imbalances between anabolism and catabolism and to re-establish a new equilibrium quickly. The stimulus sensed by its sensors is a change in the amount of uncharged aminocyl tRNAs. The change is sensed by stalling ribosomes for RelA (synthetase I) and by an unknown mechanism for carbon-starvation-activated synthetase II. The signal is a change in the ppGpp-alarmone pool, which probably modulates the RNA polymerase itself as the response regulator. The immediate response to an increase in ppGpp levels is a decreased transcription of stable RNA operons and an enhanced transcription of specific operons involved in amino acid and carbon source supply. A large series of consequent effects (listed in Tab. 20.**2**) all concur subsequently to increase survival and recovery of the cells. Adaptation to the new growth conditions, finally, is by enzymes (listed in Fig. 20.**15**) that reset (p)ppGpp pools to a new level.

A stringent response as described for *E. coli* is found throughout eubacterial and archaebacterial species. It is probably the major system responsible for maintaining the correct relative level of amino acid pools in relation to normal, undisturbed cell growth. The stringent control with its alarmone ppGpp thus is the equivalent for anabolic operons of the *crp* modulon for catabolic operons. The Crp sensory system also senses changes in overall catabolism relative to anabolism through an alarmone, cAMP. Both systems react indirectly to changes in the environment. This differentiates them from global control systems that involve two-component systems. These often react directly to a specific external stimulus, as will be discussed next.

20.4 Signal Transduction Via Two-component Regulatory Systems Is Linked to Global Regulatory Systems

Many adaptive responses in bacteria are now known to be handled by systems of proteins that have been called **two-component regulatory systems.** Originally, it was thought that this family consisted of two associated proteins: a **sensor** (S) and a **response regulator** (RR). This classification was based on shared structural properties among the sensors (histidine protein kinase) and among the response regulators. The common feature is that each sensor autophosphorylates a conserved histidine residue (His) and subsequently transfers this phospho-group to an aspartate residue (Asp) of the associated response regulator. The following reactions are common to all two-component systems:

$$\text{ATP} + (\text{His})\text{-S} \rightarrow \text{ADP} + \text{P} \sim (\text{His})\text{-S}$$
$$\text{P} \sim (\text{His})\text{-S} + (\text{Asp})\text{-RR} \rightarrow (\text{His})\text{-S} + \text{P-(Asp)-RR}$$
$$\text{P-(Asp)-RR} + H_2O \rightarrow (\text{Asp})\text{-RR} + \text{P}_i$$

Two-component systems are signaling systems that have two protein components: a **sensor** and a **response regulator.** Essential in these systems are four domains: a variable **receptor** and a conserved **transmitter** (protein kinase) in the sensor component and a conserved **receiver** and a variable **regulator** in the response regulator component (Fig. 20.**16**). Information flows from the transmitter to the receiver by phosphorylation, involving histidine and aspartate residues.

It has become evident that many bacteria contain these sensory systems. The diversity of the processes that are dependent on these regulator proteins is impressive. They may mediate taxis towards many different stimuli or responses to redox changes, osmolarity, starvation, nitrogen fixation, or utilization of carbon sources, for example. It has been estimated that, for instance, in *E. coli*, 40–50 such systems may exist (some are indicated in Tab. 20.**1**). We will describe the basic principles of these sensory systems and describe some systems in more detail.

20.4.1 The General Structure of Two-Component Systems Reflects Their Function

The general structure and function of two-component systems is shown in Figure 20.**16**. The sensor is composed of an amino-terminal receptor domain and a carboxyterminal transmitter domain. Whereas the receptor domains are all different and specific for the different stimuli, the transmitter domains have several conserved features, including the histidine residue that becomes phosphorylated. The second component of these sensory systems, the response regulator, consists of a conserved amino-terminal receiver domain and a variable regulator domain that, like the receptor domain, is specific for the process to be regulated. In most cases, a sensor and its associated response regulator are separate polypeptides. Examples are known, however, in which the different domains are clearly recognizable, but in which the sensor and response regulator components are fused in a single polypeptide. In the remainder, sensors consisting of a receptor and a transmitter and response regulators consisting of a receiver and a regulator will be described. In Figure 20.**17**, a summary is given of a number of two-component systems. It illustrates that all systems are composed basically of the same modules, but it shows also that variations exist, both in the cellular localization of the domains and the number of polypeptides involved.

As mentioned above, sensors have autokinase activity. In a number of cases, it has been shown that the sensor has also phosphatase activity, releasing the phosphogroup from the phosphorylated response regulator. Some stimuli result in a higher kinase activity, others in a higher phosphatase activity. The different phosphorylated response regulators have different half-lives, which may range from seconds to many minutes. It should be evident that the half-life is an important characteristic of the system since it determines the duration of the response of the system to changing conditions. Biochemical studies have shown that sensors usually form dimers and that ATP binds to one of the subunits and phosphorylates the other, i.e., trans-phosphorylation occurs. Similarly, response regulators may dimerize upon phosphorylation and thus bind more strongly to DNA.

As all transmitters and receivers each share conserved features, one can ask whether a sensor belonging to one two-component system can phosphorylate the response regulator of another. Genetic and biochemical

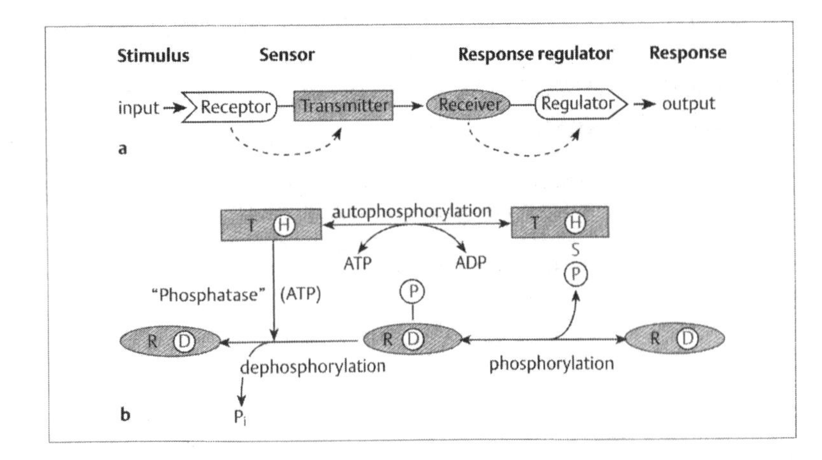

Fig. 20.**16a,b** **Sequence of domains (modules) and of phosphorylation activities involved in two-component systems.** Sensor and response regulator constitute the two components (**a**) that perceive a stimulus ("input") through a receptor domain and convert it to a signal. The signal corresponds to the modulation of a transmitter (T, red square), a protein kinase, which autophosphorylates at a histidine (H) residue and phosphorylates a receiver (R, red oval) at an aspartyl (D) residue (**b**). A spontaneous or phosphatase-catalyzed dephosphorylation completes the phosphorylation cycle. The receiver finally modulates a regulator domain, which triggers a response of the organism ("output") [after 7]

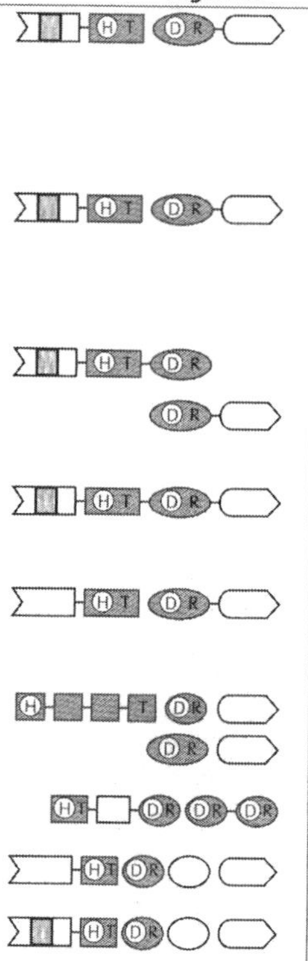

System	Organism	Sensor	Response regulator
Catabolite control	Eco	CreC	CreB
Osmolarity	Eco	EnvZ	OmpR
Turgor pressure	Eco	KdpD	KdpE
Phosphate	var	PhoQ,R	PhoB,P
Pilus synthesis		PilB	PilA
Antibiotics	Efa	VanS	VanR
Gene transfer	Bsu	ComP	ComA
Symbiosis	Rme	FixL	FixJ
Redox change	Eco	NarQ,X	NarL
Symbiosis	Bja	NodV	NodW
Carbon	Eco	UhpB	UhpA
Oxygen sensing	Eco	ArcB	ArcA
Quorum sensing	Vfi	LuxN,Q	LuxO
Capsule synthesis	Eco	RcsC	RcsB
Virulence	Atu	VirA	VirG
Virulence	Bpe	BrgS	
		LemA	
Antibiotics	Bac	RteA	
Starvation	Bsu	DegS	DegU
Nitrogen	var	NifR2	NifR1
Nitrogen	Sty	NtrB	NtrC
Chemotaxis	var	CheA	CheY
			FliG,M,N
			CheB
Sporulation	Mxa	FrzE	FrzZ,G
Sporulation	Bsu	KinA	Spo0F(B,A)
Division	var	DivJ	DivK
Sporulation	Bsu	KinB	Spo0F(B,A)
Sporulation	Bsu	PleC	Spo0F(B,A)

Fig. 20.**17** **Prokaryotic signaling proteins and communication modules.** Sensor and response regulators with a histidine (H) as phospho-donor and an aspartyl (D) residue as phospho-acceptor (symbols as in Fig. 20.**16**) are arranged in various combinations, with thin lines indicating peptide linkers and vertical bars indicating transmembrane helices in the receptors. Organisms are: Atu, *Agrobacterium tumefaciens;* Bac, *Bacteroides* spec.; Bja, *Bradyrhizobium japonicum;* Bpe, *Bordetella pertussis;* Bsu, *Bacillus subtilis;* Eco, *Escherichia coli;* Efa, *Enterococcus faecium;* Mxa, *Myxococcus xanthus;* Rme, *Rhizobium meliloti;* Sty, *Salmonella typhimurium;* var, various Gram-positive and Gram-negative bacteria; Vfi, *Vibrio fischeri* (see also Tab. 20.**1**)

studies have shown that this **cross-regulation** occurs under certain conditions.

Another pathway was discovered that results in the phosphorylation of certain response regulators in the absence of their associated sensor. Small phosphorylated compounds such as acetyl phosphate, carbamoyl phosphate, or phosphoramidate can directly phosphorylate response regulators like CheY or NtrC, while ATP or PEP cannot. Although this sensor-independent phosphorylation of a response regulator was discovered first in in vitro studies with the purified proteins, it may contribute to response regulator phosphorylation in intact cells.

Cross-regulation is the control of the response regulator by a signal that does not involve phosphorylation by its partner sensor. Its function might be the coordination of different, but directly linked systems, with the purpose of coordinating cell growth and metabolism.

Both the sensor and the response regulator contain domains, the receptor and the regulator, that are specific for a particular process, e.g., osmoregulation or nitrogen regulation. Receptors for external stimuli cross the cytoplasmic membrane at least twice. Part of the protein faces the periplasmic space, and this periplasmic-sensing domain recognizes the stimuli from the medium. The cytoplasmic transmitter domain is connected via a transmembrane domain to the receptor domain, but it is not yet clear how the external stimulus is converted into an internal signal. However, ligand-induced conformational changes have been observed, for instance, in the case of the receptors involved in chemotactic sensing (see Chapter 20.5). Mutations have been isolated in the transmembrane domains that lock the sensor in one of two conformations. Finally, hybrid sensors have been constructed from a receptor A and a transmitter B of another system. Stimuli of receptor A can now elicit the response in transmitter B, which normally is brought about by its own sensor domain.

Two examples of two-component systems will be discussed in detail. In one, the sensor/response regulator pair is located in the cytoplasmic membrane and the cytoplasm, respectively, and responds to the medium osmolarity. In a second example, both components are localized in the cytoplasm, but respond to intracellular stimuli. In Chapter 20.5, a third system, chemotaxis, will be discussed in which the response regulator, rather than acting as a transcriptional activator, controls the switching of the flagellar motor.

Two-component systems were discovered in bacteria and are now known to be involved in signal transduction in many species. As described in this section, the system uses the same basic building blocks; sometimes these domains are in a single polypeptide, and in other cases, they are separate proteins (Fig. 20.**17**). It was recently found that the same system is utilized by eukaryotes. In *Arabidopsis* and in several yeasts, proteins have been discovered that are involved in signal transduction and that show similarity with the bacterial two-component proteins. In *Arabidopsis* the *ETR1* gene product is involved in ethylene response. In yeast, the *SLN1* gene product also shows similarity with the transmitter and receiver domains of bacterial two-component systems, but its function is still unknown.

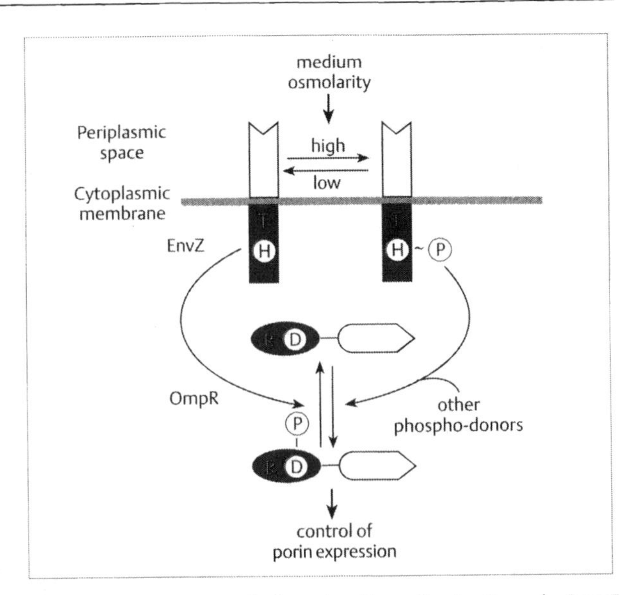

Fig. 20.**18** **Osmoregulation in *E. coli*.** EnvZ and OmpR respond to changes in the medium osmolarity. Autophosphorylation of EnvZ and dephosphorylation of P-EnvZ affect the phosphorylation state of OmpR, which in turn determines the expression of *ompC* and *ompF*, which encode two different porins. The scheme also indicates that the response regulator OmpR can be phosphorylated by other phospho-donors, either other phosphorylated sensors or low-molecular-weight compounds such as acetyl-phosphate (see text; symbols as in Fig. 20.**16**) [after 7]

20.4.2 The EnvZ/OmpR System Controls Osmoregulation as a Result of External Stimuli

Cells of *E. coli* react to changes in osmolarity by synthesizing different amounts of two outer membrane ***porins***, OmpC and OmpF. Both pores allow the movement of small molecules (molecular weight up to several hundred), but the diameter of OmpF is somewhat larger. Interestingly, an increase in the number of one type of pore is compensated by a decrease in the other. A two-component system, consisting of the **sensor** EnvZ and the **response regulator** OmpR, regulates the synthesis of both pores (Fig. 20.**18**). EnvZ is located in the cytoplasmic membrane and consists of a periplasmic domain, a cytoplasmic domain, and two membrane-spanning regions. It can sense, in an unknown way, the osmolarity in the extracellular medium and becomes phosphorylated upon an increase in osmolarity. The phosphoryl group is transferred to the cytoplasmic response regulator, OmpR. At low osmolarity, P-OmpR is dephosphorylated by the unphosphorylated EnvZ. The rate of synthesis of OmpC and OmpF is determined by the phosphorylation state of OmpR since it can act as a transcriptional activator at both the *ompC* and *ompF* loci, which contain OmpR binding sites. Transcription from the *ompC* promoter is stimulated by P-OmpR, whereas OmpR activates transcription from the *ompF* promoter and inhibits that from the *ompC* promoter. Thus, at high osmolarity, the cell contains relatively more OmpC pores, whereas at low osmolarity, OmpF pores predominate. The biological significance of the regulation is not clear.

20.4.3 The NtrB/NtrC System Controls Nitrogen Assimilation as a Result of Intracellular Stimuli

The synthesis of glutamine from glutamate and ammonia, catalyzed by glutamine synthetase, is an important pathway in nitrogen assimilation, ammonia being the preferred nitrogen source by many bacteria. If ammonia becomes scarce, bacteria respond by an increased transcription of a number of operons. In *E. coli*, the major one is the *glnALG* operon. The *glnA* gene encodes glutamine synthetase, and the *glnL,G* genes encode the components of a two-component system, GlnL (also called **NtrB** or NR_{II}) and GlnG (**NtrC** or NR_I), respectively. In the remainder, the terms NtrB and NtrC (mnemonic for **nit**rogen **r**egulation) will be used to avoid confusion. Upon lowering the intracellular glutamine concentration, the transcription of the *glnA* gene is increased. In addition, glutamine also influences the activity of glutamine synthetase. Although formally only the transcription activation is part of the two-component system, the regulation of enzyme activity will also be discussed since one of the proteins involved in this process affects the two-component system.

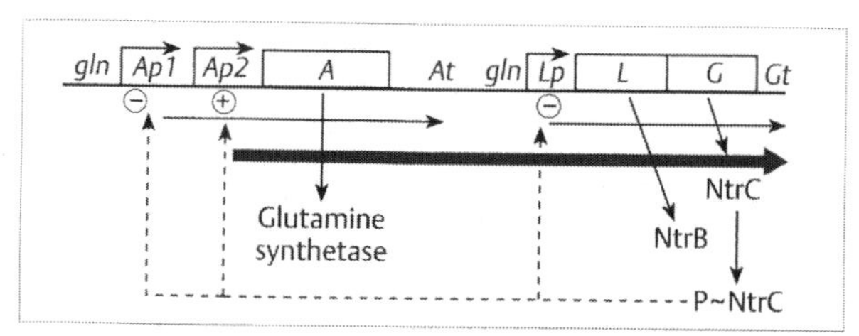

Fig. 20.**19** **Regulation of *glnA* and *glnL, G* expression**, *glnA* encodes glutamine synthetase and *glnL,G* encode the two-component system NtrB (NR_{II}) and NtrC (NR_I), respectively. Under conditions of nitrogen excess, transcription starts from the weak *glnAp1* and *glnLp* promoters, utilizing the σ^{70} sigma factor, and ends at *glnAt* and at *gnlGt*. Under conditions of nitrogen limitation, σ^{54}-dependent transcription starts from the strong *glnAp2* promoter (red arrow) and is activated by P-NtrC. Under these conditions, the *glnAp1* and *glnLp* promoters are shut off by P-NtrC (P-NR_I), and transcription terminates only at *glnGt*

The response regulator NtrC can act as a transcriptional activator at one of the *glnA* promoters (see Fig. 20.**19**). Under conditions of nitrogen excess, transcription starts from the *glnAp1* and *glnLp* promoters and proceeds at a low rate, resulting in a low level of glutamine synthetase synthesis as well as of the NtrB/NtrC components. Transcription from these promoters utilizes RNA polymerase containing the major sigma factor, σ^{70} or RpoD (encoded by the *rpoD* gene). Upon nitrogen starvation, transcription starts from a second promoter, *glnAp2*, and requires a minor sigma factor, σ^{54} or RpoN (product of the *rpoN* gene), as well as the response regulator NtrC. Under these conditions, the synthesis of both glutamine synthetase and NtrB/NtrC increases. The phosphorylated form of NtrC, P-NtrC, is formed via the transfer of the phospho-group from P~NtrB, the phosphorylated sensor. P-NtrC binds upstream of *glnAp2*, thus activating this promoter ("autoregulation") and inhibiting at the same time transcription from *glnAp1*. It also prevents σ^{70}-dependent transcription from *glnLp*.

Apart from the *glnALG* genes, other genes involved in nitrogen metabolism are under the same global control. In *Klebsiella pneumoniae*, for example, the *nif* operon, including the *nifHDK* genes involved in the synthesis of nitrogenase, is also activated by the phosphorylated NtrC response regulator. Upon nitrogen limitation, P-NtrC activates the synthesis of the NifA transcription factor, which in turn allows the synthesis of nitrogenase. The synthesis of the latter enzyme allows the cell to synthesize NH_3 from molecular nitrogen, N_2 (see Chapter 34). Other operons involved in nitrogen assimilation and controlled by NtrB/NtrC are, for instance, the *hut* operon for histidine utilization and the *put* operon for proline utilization.

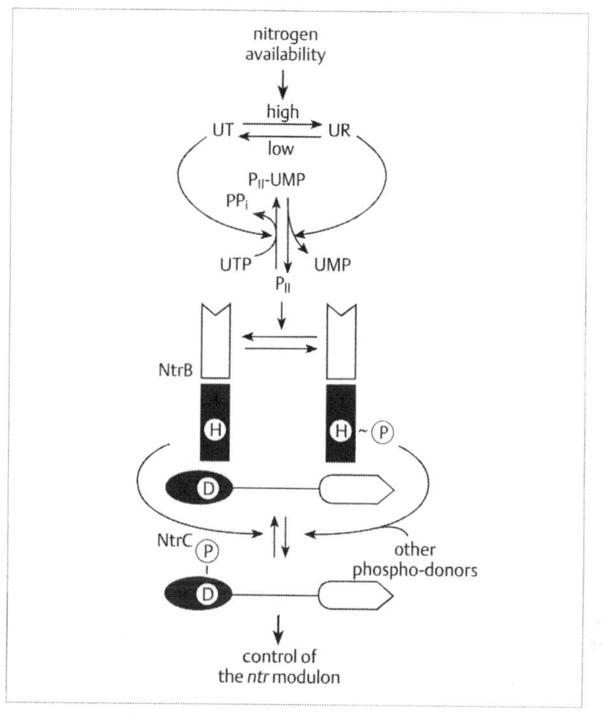

Fig. 20.**20** **Nitrogen regulation in *E. coli.*** The two-component system NtrC-NtrB (symbols as in Fig. 20.**16**) determines the expression of glutamine synthetase by the phosphorylation state of the response regulator NtrC. Phosphorylation of NtrC is controlled by the sensor NtrB, which has autokinase and phosphatase activity. The balance between these two activities is determined in turn by the uridylylation state of the enzyme P_{II} (see text) [after 7]

Two factors are crucial in determining the phosphorylation state of NtrC (see Fig. 20.**20**): (1) the rate by which it is phosphorylated by P-NtrB and (2) the rate by which P-NtrC is dephosphorylated by both NtrB and a second protein, P_{II}, which can stimulate NtrB. Figure 20.**20** shows that a second cascade of covalent protein modification operates. A bifunctional enzyme (UT/UR) can catalyze the uridylylation and de-uridylylation of tyrosine residues in P_{II}. The uridylyltransferase (UT) activity and the opposing uridylyl-removing (UR) activity are regulated by the 2-oxo-glutarate and glutamine concentrations (see also Chapter 19.3.1). UT activity is stimulated by nitrogen limitation, i.e., low levels of glutamine and high levels of 2-oxoglutarate. NtrC-dependent transcription from the *glnAp2* promoter can be summarized as follows: When the ratio of 2-oxoglutarate is low, UT activity will be predominant and convert P_{II} into its uridylylated form, UMP-P_{II}. Under these conditions, NtrB will mainly act as a protein kinase and phosphorylate NtrC. As a result, transcription will be stimulated. When there is excess ammonia, UR activity will predominate and result in P_{II}, which interacts with NtrB and increases its phosphatase activity, thus resulting in dephosphorylated NtrC and low transcription rates. In addition, NtrB in its non-phosphorylated form negatively regulates the *glnAp2* promoter. Thus, P_{II} shifts NtrB from a conformation in which it is a positive regulator (activator) to one in which it acts as a negative regulator (repressor) of transcription. The positive and negative regulatory functions can be separated genetically.

There is an extra twist to the regulation of nitrogen assimilation. In the previous paragraphs, the regulation of transcription and protein synthesis was discussed. As is the case with carbon catabolite repression, which was discussed in Chapter 20.2.3, regulation can occur at two levels: at the level of the synthesis of enzymes, which is long-term adaptation, and at the level of the regulation of enzyme activity, which is short-term adaptation.

Apart from its synthesis, the activity of glutamine synthetase is also controlled by covalent modification, and the same P_{II} enzyme is involved. Cells of *E. coli* contain adenylyltransferase (AT), another bifunctional enzyme that can transfer an adenylyl group to and remove this group from a tyrosine residue of glutamine synthetase (see Chapter 19.3.1). The adenylylated form of glutamine synthetase is inactive, and its formation is stimulated by P_{II}. Thus, P_{II} in its non-uridylylated form prevents the synthesis of glutamine synthetase and inhibits any remaining activity. Conversely, if the need for glutamine synthetase arises, inactive glutamine synthetase can easily be activated in the short term and more can be synthesized in the long term.

20.5 Bacterial Chemotaxis Is Another Paradigm for Signal Transduction Through a Two-component System

Many prokaryotes are motile and able to move in an ordered way (**taxis**). They can swim toward more favorable locations (positive taxis) and away from unfavorable conditions (negative taxis). Not all species of bacteria can move actively, but all motile species are most likely capable of reacting by taxis.

Taxis (plural **taxes**) is the oriented movement of freely motile organisms toward an attractant (positive taxis), or away from repellents (negative taxis). The terms **phobotaxis** ("avoidance reaction") and **topotaxis** (steered reaction) describe the mode of swimming, phobotaxis being predominant for prokaryotes. Depending on the stimulus, the terms **chemotaxis** (to soluble substances), **aerotaxis** (to oxygen), **osmotaxis** (to osmolarity), **phototaxis** (to light), **thermotaxis** (to temperature changes), **thigmotaxis** (to mechanical stimuli), **galvanotaxis** (to an electric current), or **magnetotaxis** (to a magnetic field) are used. **Kinesis** describes a response by which organisms change their swimming velocity as a result of stimulation.

20.5.1 There Are Various Mechanisms of Bacterial Movement

Prokaryotes move by two fundamentally different locomotory mechanisms: movement by flagella or by a gliding mechanism.

Movement by flagella. The majority of motile species, Gram-positive and Gram-negative eubacteria as well as the archaebacteria, swim by means of **flagella** (singular **flagellum**), all of which have a similar basic design (see Chapter 2, Fig. 20.**21** and Fig. 24.**2a**). A long filament, usually wound in a semi-rigid helix, is linked through a flexible hook to a basal body. This basal body comprises a rod that connects the filament to the motor,

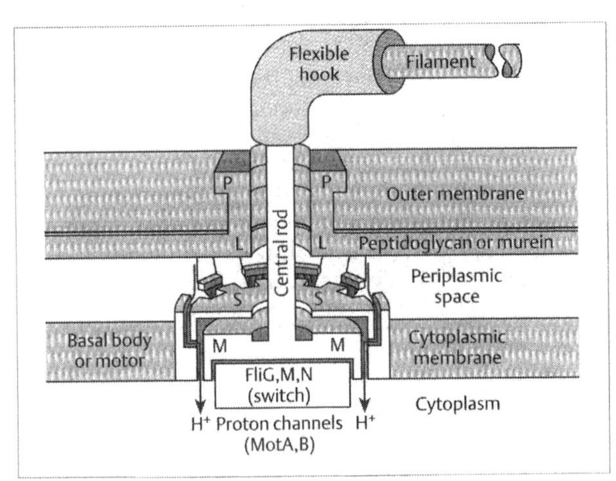

Fig. 20.**21** **Structure of a eubacterial flagellum.** The basal body or motor of a flagellum consists of a mobile disk (M) in the inner membrane and a stator (S) immobilized through covalent linkage to the peptidoglycan. The mobile disk is coupled through a central rod and a flexible hook to the filament, a long (~10 μm) flagellin polymer that ends with a capping protein. The mobile disk is surrounded by a set of proton channels (MotA, MotB) and complexed to the switch proteins (FliG, M, N). A proton flow sets the mobile disk in motion relative to the stator. Neither this process nor the switching in the direction of the rotor movement is understood. P, L, stabilizing disks

several disks that stabilize the rotating rod in the cell wall, and the motor proper. Essential parts of the motor are a mobile disk inserted into the inner or cytoplasmic membrane, and a stator immobilized through connection to the peptidoglycan. Proton channels (8 to 16), are located at the interface between the mobile ring and the surrounding stator. The movement of protons down the transmembrane gradient through these channels causes the rotation of the mobile disk relative to its stator, and thus the rotation of the connected filament. The mechanisms involved in the energization of the motor are not understood in detail.

Several bacteria have developed flagella specialized for swimming in specific media. Bacteria living in the rhizosphere, for example, can swim at speeds of up to 200 μm s^{-1} (40–50 body lengths). These fast-swimming bacteria use more rigid **complex flagella** adapted specifically to surroundings with increased viscosity. Others, for example species of *Proteus, Vibrio,* and *Serratia,* switch their morphology periodically from short and sparsely flagellated cells to elongated and highly flagellated cells. In liquid media, *Vibrio* swims by a single polar flagellum. When its motility is reduced, for example during growth in highly viscous media or on solid surfaces, lateral flagella are produced that are spread randomly over the surface (peritrichous). The new type of flagellation allows the cells to swarm in groups over solid surfaces, such as the moist surface of an agar Petri dish. This behavior, known as **swarming** (see Chapter 20.1.4) is the equivalent of other periodic differentiation processes such as cell surface rearrangements or **phase variation** in *Salmonella* species (Chapter 18). Spirochetes propel themselves by means of a screw-like movement, even through semi-solid media. Essential in this movement is an axial filament that runs along the length of the cell. It is apparently the equivalent of rotating flagella attached to opposite cell poles. The flagella are wound around the cell cylinder, but are inside the outer membrane. As flagella turn, they produce the helical waves that move the cell.

Gliding movement. Members of the Gram-negative myxobacteriales and cytophagales, as well as genera of phototrophic cyanobacteria move by a **gliding** mechanism when they are in direct contact with a solid surface. Such cells glide steadily back and forth along their long axis and sometimes roll or bend. Extracellular slime is deposited in tracks on surfaces over which cells move, and these tracks are followed preferentially by other cells. This behavior is reminiscent of swarming cells, which also produce large quantities of extracellular polysaccharide slime. Surface attachment along the cells and active changes in surface tension seem essential in gliding movements. The mechanism(s) underlying this form of bacterial movement, which requires a proton motive force as for the flagella, still are a mystery.

Motile bacteria move either by means of a rotating flagellum or a bundle made of several flagella that act like a ship's screw-propeller. Other forms of movement are swarming and gliding on solid surfaces.

20.5.2 The Normal Movement of Flagellated Enteric Bacteria Is a Random Walk

Bacterial flagella can be viewed as equivalent to a ship's screw-propeller (the filament) attached through a universal joint (the hook) to a motor. The flagella motor will turn at full speed (e.g., 60 Hz or revolutions per s for *E. coli*), even when attached to empty cell envelopes, provided the outside pH is adjusted to about four units less than that inside. In *E. coli,* about 1 000 protons move across the force-generating units of the motor per revolution. The efficiency of this energy-generating process is high (about 50%); the energy expenditure is low. If necessary, other ions can be used to drive the motor (e.g., a flow of sodium ions in alkaliphilic bacteria). An essential property of the motor is that it can rotate clockwise (CW) and counter-clockwise (CCW) (viewed along the filament toward the cell), and that it changes or switches the direction of rotation periodically.

Enteric bacteria and other peritrichously flagellated prokaryotes typically swim smoothly in nearly a straight line for about 1 s (Fig. 20.**22**). During such a **run**, a cell travels about 30 μm (or 10 body lengths). On the average of once per second, a cell **tumbles** briefly (about 0.1 s). Tumbling abruptly alters the course of swimming, on the average about 60°. When all filaments return to counter-clockwise rotation, the cell swims smoothly again, usually in a new direction. The net result of this sequence of run and tumble movements is an unbiased three-dimensional random walk.

The mechanisms for changing direction differ in various bacteria. Organisms with flagella at only one end of the cell, such as many pseudomonads, change direction simply by reversing the direction of flagellar rotation. Complex flagella have a stiffer filament that can only rotate in one direction. Organisms with complex flagella consequently swim by runs and transient stops. The same is true for bacteria that have a single flagellum inserted medially (e.g., *Rhodobacter* spec.). In bipolar bacteria, which have flagella at both ends, such as *Spirillum,* the ends must be synchronized to avoid permanent tumbling. This includes the spirochetes with their axial filament. Rotation of the axial

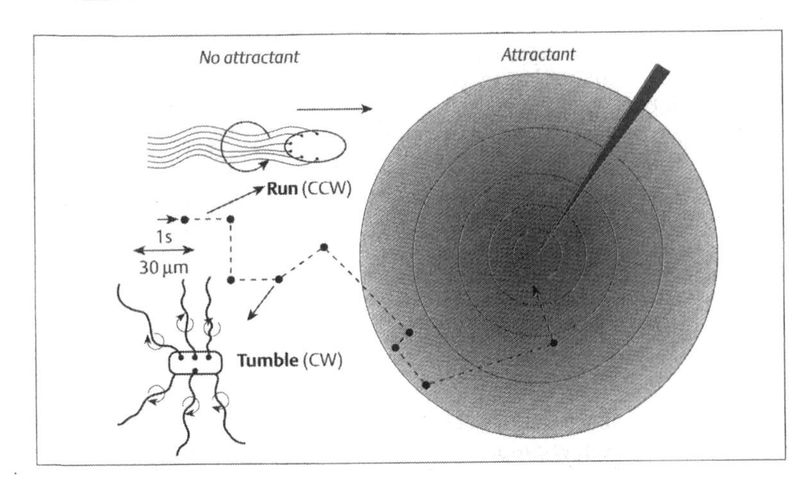

Fig. 20.**22** **How enteric bacteria swim.** In the absence of a stimulus, a cell swims in a straight line (a "run") for about 1 s; all of its flagella turn counter-clockwise (CCW) and form a bundle ("ship's screw"). A run is interrupted periodically by a tumble movement (clockwise or CW rotation of the flagella), which redirects the cell in a statistically oriented new direction. When a cell hits a gradient of an attractant, tumbling is suppressed for a longer period when it swims up the gradient, and tumbling increases when it swims down the gradient (phobotaxis). The sequence of steps eventually causes a positive chemotaxis at the mouth of a Pfeffer capillary tube

fibrils in one direction or in opposite directions seems essential in the rolling, bending, and screw-like movement of these bacteria.

20.5.3 Bacterial Taxis Is the Result of a Biased Random-Walk by Means of Phobotaxis

Prokaryotes respond by taxis to a variety of stimuli, as has already been shown in the 19th century by W. Pfeffer, T. W. Engelmann, E. Stahl, and others. Pfeffer demonstrated in a seminal experiment that motile bacteria are attracted to the mouth of a capillary tube containing an **attractant** ("Lockstoff"), such as carbon sources, amino acids, and ions, while they move away when the tube contains a **repellent** ("Schreckstoff"), such as weak organic acids and other noxious substances. He proposed the name **chemotaxis** for this response of freely motile organisms to chemical substances to distinguish it from responses to other stimuli as mentioned before. All taxes (including the chemotactic response to pH changes and to electron acceptors), except magnetotaxis, for flagellated cells depend on the run and tumble (or stop) movements described previously. Since no active steering is involved in these taxes, but merely the avoidance of the wrong direction, Pfeffer also called the process "Schreckreaktion" or **phobotaxis** ("avoiding reaction"). Magnetotactic bacteria, in contrast, seem to be oriented passively within a magnetic field through intracellular microcrystals of pure magnetite, which they synthesize (Fig. 2.**18d**). Taxis using oriented movements or active steering was called **topotaxis** by Pfeffer to distinguish it mechanistically from phobotaxis.

In the 1960s, J. Adler restarted the study of bacterial chemotaxis. He used *E. coli* and other enteric bacteria as model organisms and combined genetic, biochemical, and physiological approaches. Looking through a normal light microscope at a bacterium swimming in liquid medium, individual flagella cannot be seen, but the periodic switch between run an tumble movements, which causes a random zigzag path, can be observed (Fig. 20.**22**). When a bacterium moves toward an increasing concentration of an attractant (i.e., up the gradient), tumbling becomes less frequent. In contrast, when it moves away from an attractant, tumbling becomes more frequent. Repellents have opposite effects, i.e., addition causes tumbling, and removal prevents tumbling. As a consequence, cells swim for a prolonged time in favorable directions, and tumbling, not steering, serves to reorient a misdirected bacterium. The regulation of tumbling frequency thus is central to this form (phobotaxis) of chemotaxis. When a suspension of non-stimulated bacteria is rapidly mixed with a solution containing an attractant, tumbling is suppressed immediately, although no spatial gradient is present. The bacteria then swim for up to 5 min in straight lines (a run). This experiment demonstrates that the cells "remember" the absence of attractant before mixing and compare it with the higher concentration after mixing. They behave as if swimming up a gradient, i.e., they possess a temporal sensing mechanism or memory. The cell measures, stores this information (remembers), measures again, compares, and thus senses whether it is moving in a favorable or an unfavorable direction.

Similar to most biological sensory systems, bacteria do not respond to absolute concentrations; they respond to changes in concentrations. They must consequently adapt constantly to higher concentrations in order to detect still higher ones. This **adaptation** causes the cells in our mixing experiment to return after a certain time to their unbiased periodic run/tumble behavior as is characteristic for non-stimulated cells, although the stimulus (higher attractant concentration) is still present. When such adapted cells are diluted rapidly into a medium devoid of attractant (the equivalent of swimming down the gradient), they increase their tumble frequency temporarily until they are adapted to the new conditions.

The swimming of bacteria has been recorded in the form of continuous motility tracks and used to reconstruct with computer-aided programs the swimming path of stimulated and non-stimulated cells. An ingenious microscope follows automatically the motion of a swimming cell in the three dimensions and its locations at 80-ms intervals. Using darkfield micrographs and a very high intensity light source, the flagella bundle can be observed directly. During a run, the six to eight flagella of *E. coli* and *S. typhimurium* turn synchronously and counter-clockwise (CCW). They fuse into a single bundle that propels the cell smoothly through the medium. During a tumble, however, the flagella turn clockwise (CW), the bundle flies apart, and each flagellum pulls the cell in a different direction. Cells can also be stripped of the flagella by forcing them repeatedly through narrow-gauged needles. This treatment reduces the flagella filaments to short stumps, which may be used to tether a cell to, for example, a glass slide. When tethered through a single filament stump to a glass slide, the flagella motor rotates the entire cell slowly. This rotation can easily be followed in a light microscope. Such tethered cells again proved that a positive stimulus (increase of an attractant or decrease of a repellent) suppresses tumble movements, and that a negative stimulus (decrease of an attractant or increase of a repellent) induces tumbling. It could even be shown in this way that the cells integrate positive and negative stimuli when given simultaneously.

20.5.4 Complex and Multiple Signal Transduction Pathways and Protein Phosphorylation Steps Are Involved in Chemotaxis

Similar to other sensory reception systems that respond transiently to stimuli, the bacterial chemotaxis apparatus can be described as outlined in Fig. 20.**3**. There are at least three distinct signal transduction pathways and specific chemosensors in enteric bacteria: (1) four sensors known as **methyl-accepting chemotaxis proteins** (MCPs), (2) the Enzymes II and the general proteins Enzyme I and HPr of the PTS, and (3) components of the respiratory and electron-acceptor chains (e.g., cytochromes and cytochrome oxidases). First, the MCP-dependent chemotaxis in enteric bacteria will be described, and then other sensors not found in the enteric bacteria will be mentioned.

Swarming plates can be used in a simple, but ingenious way to select mutants with an altered chemotactic behavior (Fig. 20.**23**). A small aliquot of a motile culture is inoculated into the soft agar of a Petri dish, which contains an attracting carbon source (in micromolar concentrations). Cells able to utilize the attractant will use it up at the inoculation site, thus creating a gradient. All motile cells able to detect the substrate will move up the gradient and thus outward. The cells multiply and remain trapped within an ever-increasing concentric ring. Mutants, however, unable to move or unable to detect the gradient, are enriched at the inoculation site. In addition to mutants with

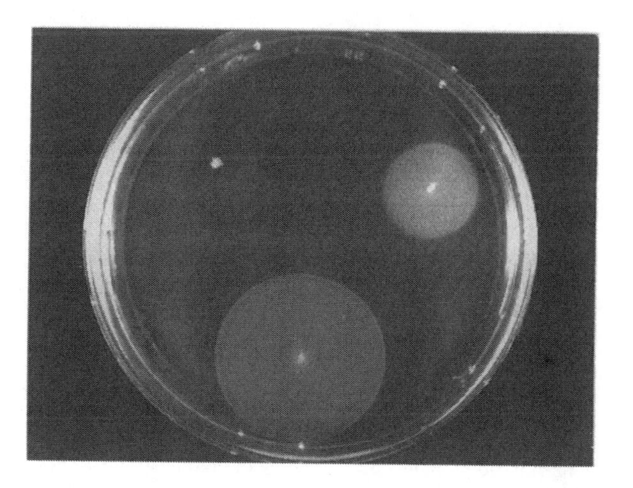

Fig. 20.**23** **The use of swarm plates in bacterial chemotaxis.** A petri dish is filled with soft agar (0.25%) that contains a mixture of 100 μM D-mannitol and 100 μM D-glucitol as carbon source. If inoculated with about $5 \cdot 10^7$ bacteria and incubated at 30°C, non-motile cells stay in place, forming a sharp spot (upper left), while motile cells, which are unable to sense the presence of the carbohydrates, swarm out statistically and grow in the form of a disk with "fuzzy edges" (upper right). Wild-type cells (below), in contrast, sense first the presence of the better substrate D-mannitol and use it up during growth, thus creating a gradient that they follow in the form of a concentric sharp ("outer") ring. Other wild-type cells "missed the train." These form a second population that senses and grows on the less favorable substrate D-glucitol, forming a second concentric ("inner") ring. Remaining in the middle are non-motile mutants and mutants that do not sense any of the two substates. A population of isogenic bacterial cells is divided during the process into two physiologically different, but stable subpopulations (**"maintenance-effect"**). [Courtesy of R. Lux]

impaired flagella (Fla$^-$) or motility (Mot$^-$), three other classes of chemotaxis mutants have been characterized: (1) Mutants that have rotating flagella, but do not respond to any chemotactic stimuli are classified as generally non-chemotactic (Che$^-$) mutants. They are permanently biased to CW or CCW rotation and rarely or never switch rotational direction. The mutants of this class are defective either in the motor switch or in a component of the central regulator system that is common to the various chemosensors. This central part comprises the products of four *che* genes, named accordingly CheA, CheW, CheY, and CheZ. (2) A second class of mutants are defective in sensing a limited range of stimuli. These defects affect one of the known four chemosensors or MCPs, named Tsr (taxis for serine and repellents), Tar (taxis for aspartate and repellents), Trg (taxis for ribose and galactose), and Tap (taxis for oligopeptides). (3) Specifically non-chemotactic mutants are defective in a single chemoreceptor, such as the maltose- or the galactose-binding protein.

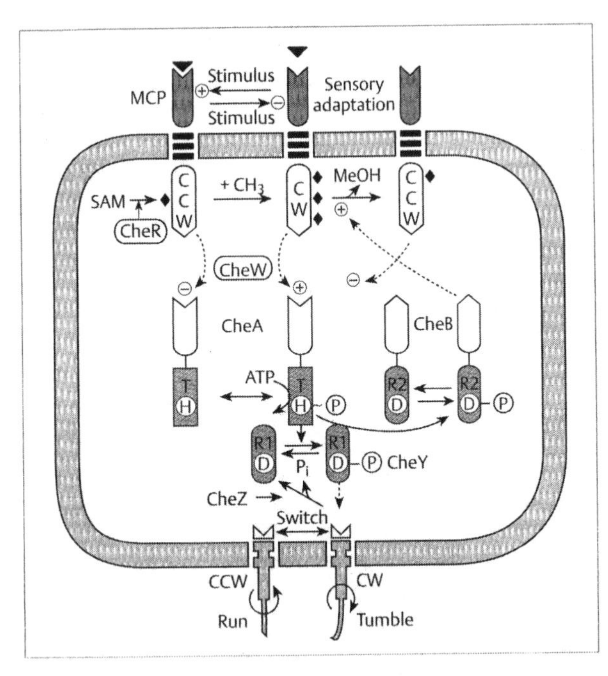

Fig. 20.**24** **MCP-dependent chemotaxis in enteric bacteria.** Upon stimulation by an attractant (upper left), a methyl-accepting chemotaxis protein (MCP) or sensor signals through the membrane and locks its intracellular signaler domain in the CCW configuration. This state inhibits its interaction through CheW with the transmitter CheA and allows a slow S-adenosine-methionine (SAM)-dependent methylation through the methyltransferase CheR at four glutamate residues (♦). If methylation is complete (adaptation), if the attractant dissociates from its MCP, or if a repellent binds to an MCP, the signaler domain switches to a second CW configuration and forms a complex with CheW and CheA. This triggers an autophosphorylation of CheA and the phosphorylation of the first receiver, CheY. P-CheY stabilizes the flagellar rotor in the CW ("tumble") state until it is hydrolyzed through the phosphatase CheZ. The freed rotor returns to the ground state (CCW) and causes a cell to run again. In parallel to these fast events (~100 ms), P~CheA phosphorylates the second receiver, CheB, which controls a methylesterase activity. Once activated, P-CheB demethylates all MCPs of a cell in the CW state, a process which may last up to 10 min. Full demethylation causes the signaler domains to return to the ground state. The cell is adapted. Periodic tumble generation of an unstimulated or an adapted cell probably relies on a slow cycling of the same process as involved in a stimulus response. The dissociation of a repellent corresponds to a positive stimulus. (Symbols as in Fig. 20.**16**)

MCP-dependent chemotaxis begins with the binding of chemicals to an MCP sensor, also called signal transducer (Fig. 20.**24**). Neither transport nor metabolism is necessary to trigger a response through the MCPs. Each transducer contains two membrane-spanning sequences (probably α-helices) that are linked through a periplasmic hydrophilic loop. This loop comprises the binding sites for the various attractants and repellents and thus confers specificity on the transducer. The binding of a ligand to the periplasmic receptor part of the transducer appears to cause a conformational change in the transmembrane helices and in the cytoplasmic part of the transducer. This carboxy-terminal cytoplasmic half of the MCPs is highly conserved among the four transducers; it interacts with CheA and CheW. The ligand-binding periplasmic parts of the MCPs thus correspond to variable input domains, while the cytoplasmic parts correspond to a conserved output domain. This output domain has two signaling states, one causing eventually CW rotation, the other CCW rotation of the flagellar motors. Mutations have been isolated that lock individual MCPs in one or the other state. They help to identify the corresponding signaling domains.

20.5.5. A Central Regulatory Unit Contains Elements of a Two-component System and Controls the Flagellar Motor Switch

In *E. coli* mutants lacking all Che proteins, the flagellar motor turns exclusively in the CCW direction. To restore tumbling, at least CheY must be supplied, while restoration of the periodic sequence of run/tumble movements requires in addition CheA, CheW, and CheZ.

CheA acts as the central processing unit in chemotactic signaling. Similar to other transmitter domains of two-component systems, it is autophosphorylated in an ATP-dependent reaction at a histidine residue. Phosphorylated CheA transfers its phospho-group to an aspartate residue of its associated receiver CheY. The presence of phospho-CheY increases CW rotation of the flagellar motor, probably because it locks the switch in the CW configuration. Neither the molecular details of the switching of the flagellar motor between the CCW and CW configuration, nor the modulation of this process through phosphorylated CheY is known in any detail. The proteins FliG, FliM, and FliN, however, appear to be essential.

The autokinase activity of CheA is modulated by the output (or signaler) domains of the MCPs, a process that also requires CheW (Fig. 20.**24**). Binding of an attractant to an MCP (or any positive stimulus) shifts the signaler domain of this MCP to the CCW state. This form prevents or lowers drastically the autophosphorylation of CheA. Binding of a repellent to an MCP (or any negative stimulus) in contrast, shifts the signaler domain to its CW state. This causes CheW and CheA to form a complex with a high autokinase activity and generates first phospho-CheA and then phospho-CheY. The phosphorylated forms of the effectors are short-lived, especially in the presence of CheZ. The phosphatase CheZ specifically dephosphorylates CheY molecules that are bound to the switch proteins of the flagellar motor.

The model as outlined for MCP-dependent signal transduction pathways postulates that the transmitter CheA and the receiver CheY correspond to the central unit of a two-component system, each MCP corresponds to an ancillary protein that couples the receptor MCP to the transmitter CheA. Such sophisticated receptors allow the differentiation between attractants and repellents as well as the integration of positive and of negative stimuli.

20.5.6 Adaptation to MCP-dependent Stimuli Requires Methylation/Demethylation and a Bacterial "Memory"

MCPs contain several glutamate residues that can be reversibly methylated. These residues are located close to the linker that couples the periplasmic receptor or input domain and the cytoplasmic transmitter or output domain. The methylation sites are recognized by two MCP-specific modifying enzymes. CheR, a slow, but permanently active methyltransferase, transfers methyl groups from S-adenosylmethionine (SAM) to the carboxyl side chain of glutamates, forming a glutamyl methyl ester. CheB, a methylesterase, hydrolyzes the methyl groups and generates free methanol in the process.

The relative activities of CheR and CheB determine the methylation of each MCP. In the absence of stimuli, about half of the sites are methylated. After positive stimulation, most sites are methylated, whereas after negative stimulation, few sites are methylated. An attractant-occupied MCP is a poor substrate for CheB, but a (good) substrate for CheR. Since CheR activity remains constant, only changes in CheB activity produce the transient fluctuations in the methylation level of all MCPs. CheB contains a receiver domain that is analogous to CheY, i.e., it can be phosphorylated at an aspartic residue by phospho-CheA. The esterase activity of CheB is activated upon phosphorylation. As indicated schematically in Fig. 20.**24**, positive stimuli inhibit the phosphorylation of CheA, and hence of CheB. The short-lived phosphorylated forms of CheA and CheB decay and the methylesterase activity is inactivated, while CheR continues to methylate the glutamate residues slowly. Negative stimuli, in contrast, cause an increase in CheA phosphorylation and, therefore, an activation of CheB. Activated CheB surpasses the CheR activity, thus causing a net demethylation of all MCPs.

Changes in the MCP methylation state lead to sensory adaptation, but play no role in triggering motor responses. Thus, mutants defective in CheB or CheR can initiate flagellar responses upon stimulation, but they continue to respond until the stimulating chemical is removed. Wild-type cells, in contrast, stop responding upon reaching an appropriate MCP methylation level. Present models assume that changes in receptor occupancy shift the MCP transmitter domains into the CCW (positive signal) or CW (negative signal) mode, and changes in the methylation state shift the equilibrium in the opposite direction to cancel the response, e.g., by uncoupling the receptor and the signaling domains. As expected according to the model, mutants lacking CheR or CheB have a Che$^-$ phenotype and run or tumble persistently.

In the chemotaxis system, sensory adaptation is a continuous process that enables the cells to make temporal comparisons as they swim about. Adaptation cancels recent stimulus responses, and the cell is poised to respond to any new change in chemoeffector concentration. Coupling rapid signaling with a slower canceling reaction thus is the biochemical base of the primitive bacterial **memory**. Its duration, up to 15 min or three-fourths of a generation in rapidly growing cells of *E. coli*, however, is well adapted to the bacterial life span.

20.5.7 Enteric Bacteria Use Several Signaling Pathways in Chemotaxis

The MCP-dependent signal transduction pathway described thus far is not the only pathway used by enteric bacteria in chemotaxis (Fig. 20.**25**). Mutants lacking the MCPs or CheR and CheB retain chemotaxis toward PTS carbohydrates and electron acceptors such as oxygen. Elements of the transport systems and metabolic pathways for the stimulating chemoeffectors are used as the sensors. Characteristic for such MCP-independent receptors seems to be that they sense stimuli that cover basic requirements for bacterial growth, i.e., the presence of energy sources, of oxygen and of other electron acceptors, or light for photosynthetic bacteria. In PTS-dependent chemotaxis of enteric bacteria, positive stimulation corresponds to uptake and phosphorylation of a substrate through an Enzyme II. Binding of a substrate alone is not sufficient to trigger chemotaxis; the intracellular accumulation of carbohydrate phosphates or extensive metabolism of the stimulating substrate is also not required for chemotaxis. In contrast, mutants lacking Enzyme I and HPr do not show a chemotactic response to any PTS carbohydrate, even if Enzymes II are present that can bind the substrates. All available data indicate that the stimulus in PTS-dependent chemotaxis is the rapid decrease in the phosphorylation level of Enzyme I as the consequence of substrate transport through an Enzyme II. Apparently, accumulating dephosphorylated Enzyme I inhibits the kinase CheA and hence decreases the level of phospho-CheY. This in turn causes cells to run and generates a positive chemotactic response.

Electron-transport-chain-dependent taxis, which includes aerotaxis to oxygen, is even less well understood. For most free-living bacteria, the responses to electron acceptors or compounds that alter the rate of electron transport may be important because they help to maintain the organism in environments where the proton motive force (Δp) can be sustained. In this type of taxis, it is often difficult to distinguish direct effects on the motor from effects on constituents of the signaling pathway. A newly discovered MCP called Aer (for **a**erotaxis and **e**nergy **r**esponses) and the MCP Tsr are involved in electron-transport-dependent chemotaxis. The sensory part must include essential elements of the respiratory chain. Thus, active electron transport is essential for a sensory response to oxygen. As will be discussed in Chapter 21, bacteria synthesize under different growth conditions different terminal cytochromes and reductases, e.g., those for oxygen, nitrate, and fumarate. The various pathways share common components, which facilitate the integration of different sensory signals, measured, for example, as a general change in redox potential of an electron transport component. The actual sensor could be a protein similar to other proteins that sense redox, energy and proton motive force changes.

Although the molecular details of these chemotactic responses are unknown, the physiological relevance is clear. When growing anaerobically with nitrate as an electron acceptor, *E. coli* is attracted by nitrate and also by (low concentrations of) oxygen. When growing aerobically, it loses its response to nitrate, but still retains aerotaxis to oxygen. A pulse of oxygen serves as an attractant for *Rhodobacter sphaeroides* growing aerobically. The same pulse acts as a repellent in photosynthetically growing cells. Many obligate anaerobes are repelled by oxygen, whereas most facultative anaerobes and strict aerobes are repelled by high concentrations of oxygen. The consistent pattern that emerges is that electron-transport-dependent taxes help a large variety of bacteria to find environments in which energization is optimal and to keep them within such environments.

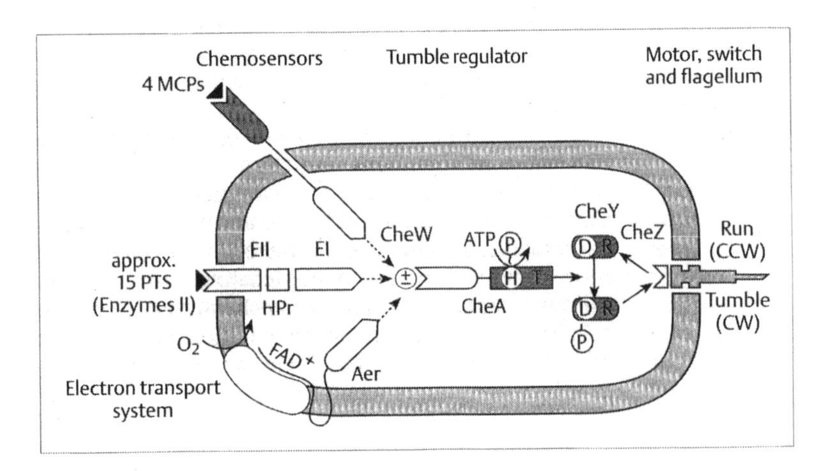

Fig. 20.**25** **Enteric bacteria contain several chemotactic signaling pathways.** During uptake and phosphorylation of many carbohydrates through an Enzyme II (EII), unphosphorylated HPr and Enzyme I (EI) accumulate. Dephosphorylated EI molecules inhibit autophosphorylation of the transmitter CheA and tumbling. Hence, PTS carbohydrates are attractants and cause a positive chemotaxis. The reaction to oxygen ("aerotaxis") and to electron acceptors, in contrast, apparently requires elements of the respiratory chains in addition to the old MCP Tsr and the new MCP Aer (for **a**erotaxis and **e**nergy **r**esponses) as well as the central elements CheA, CheW, and CheY

20.5.8 Other Bacteria May Have Different Chemotactic Components

Despite the astonishing progress in our understanding of MCP-dependent chemotaxis in enteric bacteria, our understanding of the physiological relevance and fantastic variety of motility and tactic behaviors in the prokaryotes within their natural environment is rather poor. Thus, in the Gram-positive *Bacillus subtilis*, the general MCP-dependent chemotactic machinery includes the equivalent of CheA and CheY as the central unit, and resembles the system of enteric bacteria. The systems deviate, however, in many details. Two major differences are the inversion of the tumble-generating process and a more complex adaptation mechanism. In *B. subtilis*, positive stimuli do not inhibit, but activate CheA to form phospho-CheY; phospho-CheY locks the motor switch in the CCW, not in the CW, rotary direction, and thus causes runs. Adaptation to positive stimuli involves the transfer of methyl groups from MCPs to an unidentified regulator protein and possibly to the motor proteins. A CheZ-like protein, in contrast, seems to be absent in *B. subtilis*. The purple non-sulfur bacterium *Rhodobacter sphaeroides* with its single flagellum inserted medially swims by a stop-and-go mechanism. It contains MCPs and apparently even the equivalent of CheA, CheB, and two CheYs. Furthermore, a step-up in the concentration of any limiting metabolite increases the speed of flagellar rotation and hence the swimming speed up to 20%. This behavior is called **chemokinesis.** It causes an increased spreading out of a population of bacteria in the presence of a metabolite. All metabolites are attractants, while non-metabolizable compounds, even if transported into the cell, cause no response. The faster swimming cells still respond by chemotaxis to gradients of the metabolites. They utilize in chemotaxis a sensory system that seems related to other bacterial two-component systems.

Proteins with essentially the same design as MCPs and the Che proteins have been found in *B. subtilis*, *Caulobacter crescentus* (Chapter 24.1), and *Rhizobium meliloti*, bacteria that resemble enteric bacteria in size and mode of locomotion. *Myxococcus xanthus* is a long bacterium that moves by gliding on solid surfaces and has a generation cycle. During vegetative growth on rich media, it moves in the form of single cells (or in small groups of cells) utilizing motility system A (for "Adventurous"). For the movement in large groups, typical during fruiting-body formation, system S (for "Social") is necessary. The cells move slowly (2–4 μm per min) and reverse their direction of gliding every 7–8 min. Frz (for "frizzy" colony morphology) mutants have been isolated that either rarely or with an increased frequency change their gliding direction. They correspond to *(che)* mutants of *E. coli* that either run or tumble. The mutated genes and their products closely resemble MCPs, the adaptory proteins CheB and CheR, and the proteins CheW and CheA. CheA is fused in this organism through a flexible Pro-Ala linker to CheY. Reversible protein phosphorylation and methylation catalyzed by the signaling proteins has also been found. Obviously, the system processes sensory information about nutrient conditions (starvation induces fruiting-body formation) and cell density (see Chapter 20.1.4) to regulate cell motility during both growth phases. It has thus evolved in these social and "multicellular" prokaryotes from a purely chemotactic sensory system to the precursor of a system involved in morphopoietic and differentiation processes with highly ordered cell movements (or taxes), as found regularly in multicellular eukaryotic organisms (see Chapter 25).

20.5.9 Phototaxis and Chemotaxis Are Integrated Sensory Processes in Archaebacteria

Many photoautotrophic bacteria, whether eubacterial or archaebacterial species, respond to light, a behavior termed **phototaxis.** The classical photoresponse in eubacteria is the phobotactic (avoiding) response in which cells change their direction of swimming when entering a region in which the light is of low intensity or of the wrong wavelength. Since they reverse their direction when leaving the lighted area, they accumulate in regions of optimal light intensity and wavelength. In most studies on bacterial phototaxis, large rod-shaped or spiral photosynthetic species have been used that swim by polar flagella (e.g., *Rhodobacter*) or by gliding motility (*Anabaena, Phormidium*, and other cyanobacteria). The results indicate that bacteriochlorophylls may act as the photoreceptors and the electron transport chains may act as signal transducers. Molecular details, however, are not yet available.

Phototaxis in the archaebacterium *Halobacterium salinarium*, however, has gained considerable interest recently because light and chemical stimuli are integrated, and a photosensor unique in the prokaryotic world is used. The photobiochemistry in halobacteria is based upon four retinal-containing rhodopsins: **bacteriorhodopsin** (BR), a light-driven proton pump that generates the proton motive force necessary to drive phosphorylation by the ATP synthase (Chapter 13.4), **halorhodopsin** (HR), a light-driven chloride pump, and two light sensors, **sensory rhodopsin I** and **II** (SRI, SRII). The primary photochemical process in the four rhodopsins is the isomerization of the all-*trans* retinal to its 13-*cis* state, followed by a conformational change in the protein. Retinal thus acts as a light-triggered switch between two signaling states (on/off) in the sensory rhodopsins.

When growing exponentially, *H. salinarium* cells have a monopolarly inserted flagella bundle that, upon CW rotation, pushes the cell into a forward motion and, during CCW rotation, pulls the cell backwards. In the absence of stimuli, the cells switch periodically between CW and CCW rotation and swim consequently backwards and forwards. Changes in direction seem to be passive, caused by small obstructions in the medium, and result in a random walk of the cells. When grown aerobically, *H. salinarium* synthesizes SRII at a low and constitutive level (400 copies per cell). No SRI is present under these conditions. In its ground state, SRII (abbreviated as $SRII_{480}$) absorbs blue and UV light and undergoes a simple photocycle (Fig. 20.**26**). The flux through the cycle corresponds to a negative signal, which is picked up by an as yet unidentified transducer. The system serves to prevent damage by blue and UV light by causing a classical avoidance reaction (phobotaxis) in the cells.

Halobacteria are facultatively anaerobic bacteria. Under anoxic conditions, they grow photosynthetically if no alternative carbon source is available. This capacity is vital since the oxygen tension can become very low under strong sunlight in the natural habitats of brines and salt ponds. A decrease in oxygen tension increases the numbers of SRI molecules from a few to about 4000 copies per cell. SRI is responsible for keeping cells in favorable regions of orange light and again out of blue light. The photosensor in its ground state (SRI_{578}) absorbs orange light and switches in <1 ms to a new signaling state (SRI_{373}). This metastable intermediate causes a negative response to near-UV light. It returns to the ground state either by slow thermal decay or photochemically through an alternative path by absorption of a new photon. The slow decay of SRI_{378} to SRI_{578} generates a positive response, i.e., suppresses motor reversals.

SRI is an integral membrane-bound protein (239 amino acids) that has minimal cytoplasmic loops. Analysis of a phototaxis mutant revealed a protein, HtrI (**h**alobacterial **t**ransducer for sensory **r**hodopsin I), that has many of the properties of MCPs in enteric bacteria. HtrI-deficient strains of *H. salinarium* no longer respond to positive light stimuli and lack a characteristic photo-induced demethylation. The deduced nucleotide sequence of HtrI predicts a larger molecular weight (56675) than is usual for other MCPs, two transmembrane helices near the amino-terminus, and a large cytoplasmic domain (490 amino acids). The 270 carboxy-terminal residues of this domain resemble the signaling domain of other eubacterial MCPs (30% identical residues). Similar to eubacterial MCPs, the signaling domain is flanked by characteristic glutamic acid residues as methylation sites. Unlike other MCPs, however, HtrI lacks an extended extramembranous receptor domain. One could postulate that the transmembrane helices and perhaps the 220 extra cytoplasmic residues located betwen these helices and the signaling domain sense photochemically induced changes in SRI and communicate the information to the signal domain of HtrI.

H. salinarium uses MCPs for phototaxis and for chemotaxis towards glucose, histidine, leucine, and phenol. Some MCPs appear to be involved in both chemotaxis and phototaxis. A role of the methylation system in tactic adaption seems likely, but no molecular details are yet available.

Photoreceptor signals are not propagated as electrical impulses along the membrane because the stimuli act only locally during excitation. Possible signals are Ca^{2+}, cGMP, a G-protein, and above all, fumarate. A

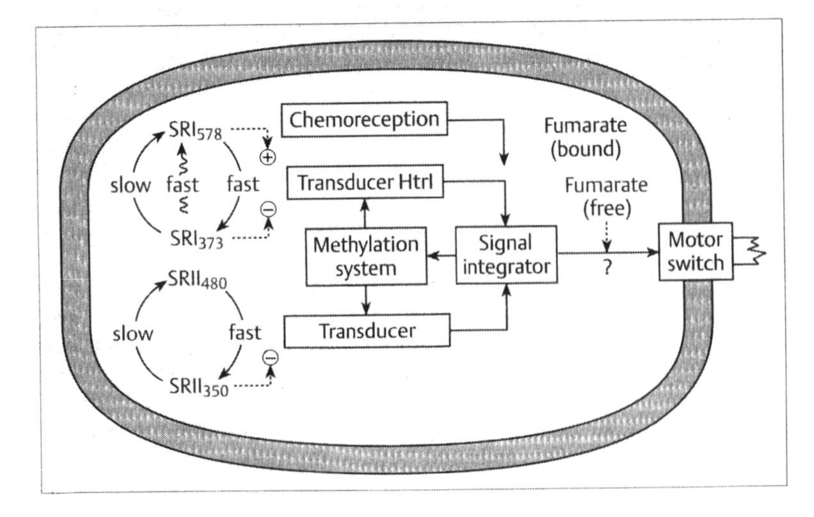

Fig. 20.**26** **Phototaxis in *Halobacterium salinarium*.** Photocycles from sensory rhodopsin I (SRI, orange light) and from sensory rhodopsin II (SRII, blue light) send attractant (⊕) or repellent (⊖) signals to transducers. These are HtrI for SRI and an unidentified transducer for SRII. The former, perhaps also involved in chemotaxis, and the latter are linked to a signal integrator (CheA, CheY), which apparently modulates the release of intracellular fumarate, a putative motor switch regulator. Adaptation seems to involve a methylation/demethylation system similar to the one from enteric bacteria

permanent smooth swimmer mutant of *H. salinarium* indeed was found to lack a switch factor that accumulates rapidly in wild-type cells after negative photostimulation. A switch factor restores flagellar motor reversal in the mutant and was identified as fumaric acid. Upon negative stimulation of halobacteria, fumaric acid is released rapidly from a cellular pool of (perhaps membrane-) bound molecules, and in large amounts (about 70 000 molecules per cell and 370 per physiological quantum of blue light). Release requires the presence of SRI or SRII and is not related to fumarate turnover through the citric acid cycle. These results indicate that the cytoplasmic concentration of free fumaric acid may control the flagellar motor switch. If its concentration reflects the sum of various external phototactic and chemotactic stimuli, then its intracellular pool would be the putative integrated signal of Fig. 20.**26**. If its concentration oscillates periodically, then the regular switch in swimming behavior would also be clear. The molecular analysis of this first example of an archaebacterial sensory system thus indicates elements of eubacterial signal transduction pathways. The utilization of fumaric acid as a diffusible second messenger, however, resembles eukaryotic signal transduction pathways in which Ca^{2+}, cGMP, cAMP, inositol 3-phosphate, and other small molecules are used.

Further Reading

Alex, L. A., and Simon, M. I. (1994) Protein histidine kinases and signal transduction in prokaryotes and eukaryotes. Trends Genet 10: 133–138

Armitage J. P., and Lackie, J. M., eds. (1990) Biology of the chemotactic response. Society for General Microbiology Symposium, vol. 46. Cambridge, New York: Cambridge University Press

Bremer, H., and Ehrenberg, M. (1995) Guanosine tetraphosphate as a global regulator of bacterial RNA synthesis: a model involving RNA polymerase pausing and queuing. Biochim Biophys Acta 1262: 15–36

Chuang, S.-E., Daniels, D. L., and Blattner, F. R. (1993) Global regulation of gene expression in *Escherichia coli*. J Bacteriol 175: 2026–2036

Hoch, J. A., and Silhavy, T. J., eds. (1995) Two-component signal transduction. Washington, DC: ASM Press

Hueck, C. J., and Hillen, W. (1995) Catabolite repression in *Bacillus subtilis:* a global regulatory mechanism for the Gram-positive bacteria? Mol Microbiol 15: 395–401

Kjellberg, S. (1993) Starvation in bacteria. New York: Plenum Press

Kolb, A., Busby, S., Buc, H., Garges, S., and Adhya, S. (1993) Transcriptional regulation by cAMP and its receptor protein. Annu Rev Biochem 62: 749–795

Kurjan, J., and Taylor, B. L., eds. (1995) Signal transduction. San Diego: Academic Press

Lengeler, J. W., and Jahreis, K. (1996) Phosphotransferase system or PTSs as carbohydrate transport and as signal transduction systems. In: Konings, W. N., Kaback, H. R., and Lolkema, J. S. (eds.). Handbook of biological physics, vol 2. Amsterdam: Elsevier; 573–598

Lin, E. C. C., and Lynch, A. S., eds. (1995) Regulation of gene expression in *Escherichia coli*. Austin, Tex: Landes

Manson, M. D., Armitage, J. P., Hoch, J. A., and Macnab, R. M. (1998) Bacterial locomotion and signal transduction. J Bacteriol 180: 1009–1022

Neidhardt, F. C., Ingraham, J. L., and Schaechter, M. (1990) Physiology of the bacterial cell: a molecular approach. Sunderland, Mass.: Sinauer Associates; 351–388

Parkinson, J. S., and Kofoid, E. C. (1992) Communication modules in bacterial signalling. Annu Rev Genet 26: 71–112

Piggot, P. J., Moran Jr., C. P., and Youngman, P., eds. (1994) Regulation of bacterial differentiation. Washington, DC: ASM Press

Postma, P. W., Lengeler, J. W., and Jacobson, G. R. (1993) Phosphoenolpyruvate: carbohydrate phosphotransferase systems of bacteria. Microbiol Rev 57: 543–594

Salmond, G. P. C., Bycroft, B. W., Stewart, G. S. A. B., and Williams, P. (1995) The bacterial "enigma": cracking the code of cell-cell communication. Mol Microbiol 16: 615–624

Schuster, S. C., and Khan, S. (1994) The bacterial flagellar motor. Annu Rev Biophys Biomol Struct 23: 509–539

Sonenshein, A. L., Hoch, J. A., and Losick, R., eds. (1993) *Bacillus subtilis* and other Gram-positive bacteria. Washington, DC: ASM Press

Spudich, J. (1993) Color sensing in the Archaea: a eukaryotic-like receptor coupled to a prokaryotic transducer. J Bacteriol 175: 7755–7761

Svitil, A. L., Cashel, M., and Zyskind, J. W. (1993) Guanosine tetraphosphate inhibits protein synthesis in vivo: a possible protective mechanism for starvation stress in *Escherichia coli.* J Biol Chem 268: 2307–2311

Torriani-Gorini, A., Yagil, E., and Silver, S., eds. (1994) Phosphate in microorganisms: cellular and molecular biology. Washington DC: ASM Press

Wagner, R. (1994) The regulation of ribosomal RNA synthesis and bacterial cell growth. Arch Microbiol 161: 100–109

Ward, M. J., and Zusman, D. R. (1997) Regulation of directed motility in *Myxococcus xanthus*. Mol Microbiol 24: 885–893

Sources of Figures

1 Sawers, R. G., Zehelein, E., and Böck, A. (1988) Arch Microbiol 149: 240–244

2 Epstein, W., Naono, S., and Gros, F. (1966) Biochem Biophys Res Commun 24: 588–592

3 Ishizuka, H., Hanamura, A., Kunimura, T., and Aiba, H. (1993) Mol Microbiol 10: 341–350

4 Epstein, W., Rothman-Denes, L. B., and Hesse, J. (1975) Proc Natl Acad Sci USA 72: 2300–2304

5 Okinaka, R. T., and Dobrogosz, W. J. (1967) J Bacteriol 93: 1644–1650

6 Postma, P. W., Lengeler, J. W., and Jacobson, G. R. (1993) Microbiol Rev 57: 543–594

7 Parkinson, J. (1993) Cell 73: 857–871

21 Regulation of Fermentation and Respiration

In previous chapters (see Chapters 4 and 9–12), the essential roles of oxygen in various metabolic processes and the reactive oxygen species as toxic agents have been described. Emphasis was given to the biochemical processes and on the mechanisms of these reactions. In this chapter, the sensory and regulatory networks involved in the control of the corresponding genes will be described. The description, however, is mostly restricted to the enteric bacteria because little is known about the networks in other bacteria and there is no indication that radical differences exist among them (see Chapter 34).

21.1 There Are Three Basic Modes of Metabolic Energy Generation

Metabolic energy generation may be regarded as biochemical reactions in which electrons flow stepwise from a positive to a negative pole in a harnessed manner. Such flow may involve solely dismutation reactions in the cytoplasm (as in glycolysis) or depend on exogenous electron acceptors accompanied by vectorial reactions across the plasma membrane (as in respiration). Bacterial physiologists often classify energy transduction processes according to levels of efficiency and probable order of evolutionary appearance: **fermentation, anaerobic respiration**, and **aerobic respiration**. It should be emphasized, however, that when a preferred pathway is inadequate as an electron sink, other pathways become available as auxiliaries or alternatives. The important principle is that the cell is endowed with elaborate means to take the best advantage of the energetic situation. Translating this concept into metabolic terms, oxidation by an exogenous agent takes precedence over internal dismutation, and, when several exogenous oxidants are available, the one with the highest redox potential is preferentially used (see Tab. 4.1). The regulatory mechanisms for energy metabolism in *Escherichia coli* will be largely used as the paradigm. It must be kept in mind, however, that evolution is extraordinarily opportunistic and capitalizes on almost every conceivable kind of regulatory mechanism. On the other hand, despite all the variations, it would be surprising if specific genetic regulatory mechanisms for fermentation and respiration were not coordinated by global controlling elements analogous to those operating in *E. coli.*

21.1.1 Energy Is Generated by Substrate-level Phosphorylation When Cells Grow Anaerobically (Fermentation)

Fermentation of a sugar such as glucose generates energy by substrate-level phosphorylation (see Chapter 3), while the redox balance is sustained by dismutation reactions. The pivotal metabolite is phospho*enol*pyruvate (PEP), which separates the reactions that generate reducing equivalents (i.e., H^+–e^- pairs) from those that dispose of them. Among the three modes of energy metabolism, fermentation is likely to be primordial for several reasons: (1) the phosphoylated intermediates and coenzymes (e.g., the triose phosphates, NADH/NAD^+, and ADP/ATP) also participate in numerous other pathways and therefore were probably founding units of life, (2) glucose likely emerged as the universal sugar because of its high stability in ring structure, which limits potentially deleterious noncatalyzed glycosylation of proteins, and (3) glycolysis is ubiquitous in the biological world, and many other carbon and energy sources are funneled into the process.

In the fermentation of one molecule of glucose to PEP, two molecules of ATP are generated from two molecules of ADP by substrate-level phosphorylation and four reducing equivalents (or $4H^+ + e^-$ pairs) are formed. The reducing equivalents are then disposed of by several terminal routes emanating from PEP. The relative flow through these routes varies according to the growth conditions.

21.1.2 Anaerobic Respiration Is Membrane-associated and Requires Electron Acceptors Other Than Oxygen

The establishment of a membrane-associated electron transport chain opened a novel means of energy generation since the series of redox reactions is coupled to the ejection of H^+ from the cytoplasm (see Chapter 4), thereby contributing to the proton potential (or proton motive force, also referred to as Δp). For this kind of energy source to become available, flavins, quinones, and hemes had to be added to the metabolic repertoire (see Chapters 4, 11, and 12).

Although ejection of protons is principally attributable to the respiratory process, it must not be ignored that metabolic events during fermentation can also contribute in an equivalent manner. For instance, the proton potential can be increased by any reaction that causes alkalinization of the cytoplasm. The reaction catalyzed by a hydrogenase is one example:

$$2\,H^+ + 2\,e^- \rightarrow H_2$$

Another example is the efflux of lactate together with a proton through the lactate/H^+ symport system.

21.1.3 The Most Efficient Energy Generation by Aerobic Respiration Appeared Late During Evolution

Evolution of aerobic respiration probably began shortly after the advent of oxygenic photosynthesis, which produced O_2. Effectiveness of aerobic respiration is manifested not only by enhanced growth rates, but also by amplified growth yields. It would appear that this efficient energy-generating power made certain ancient aerobes highly desirable endosymbionts of eukaryotes and predestined these eubacteria to become mitochondria. It is worth noting that fermentation, anaerobic respiration, and aerobic respiration can be viewed as different metabolic fates of the central metabolic pyruvate and the electron carrier NADH.

PEP and pyruvate are principal participants of different pathways for the generation of metabolic energy. The likely evolutionary sequel of energy metabolism is **fermentation, anaerobic respiration,** and **aerobic respiration.** The hypothesis is in accord with the "Horowitz principle" in that pathways closer to the core metabolism should appear before those that are more remote and with the fact that the number of coenzymes must increase as more efficient processes emerge.

21.2 Fermentation Involves Both Global and Specific Regulation of Gene Expression

In *E. coli*, which carries out mixed-acid fermentation, the end products arising from the pivotal intermediates PEP and pyruvate are succinate, D-lactate, acetate, ethanol, formate, CO_2, and H_2 (Fig. 21.**1**). Much remains to be discovered about how the proportion of the end products vary according to nutritional and environmental factors and how the pathways are balanced by gene expression.

21.2.1 The Global Regulator Fnr Is Essential During Growth Under Anoxic Conditions

Fnr (gene *fnr*) is an acronym of **f**umarate **n**itrate **r**eduction. The Fnr protein plays an overarching role as a global transcriptional regulator to ensure effective use of pathways for fermentation and/or anaerobic respiration, in particular those organized in the ***fnr* modulon.** This role is fulfilled by activating numerous operons which function under anoxic conditions, while repressing certain operons with aerobic function (Tab. 21.**1**). Except for the presence of a cysteine-rich N-terminal extension, Fnr is highly similar to Crp (for **c**yclic AMP **r**eceptor **p**rotein). Even the DNA consensus sequences of target promoters recognized by the two proteins are highly similar: the "Fnr box" contains a 5-bp inverted repeat TTGAT- - - -ATCAA, whereas the "Crp box" contains TGTGA- - - -TCACA (see Chapter 18.5).

It seems likely that at first the *crp* gene evolved to ensure the preferential use of the most rewarding carbon and energy source (abundant and energy-rich) when the biological world was still anoxic. With the advent of aerobiosis, *fnr* then evolved from a duplicate *crp* gene to ensure the preferential use of the most rewarding energy source.

Fnr senses the redox condition of the environment through the iron atom bound to a cluster of four cysteine residues that are highly conserved among all bacterial Fnr proteins. When the iron is in the Fe^{2+} state, the protein is functionally active as a transcriptional regulator (Fig. 21.**2**); when the iron is oxidized to the

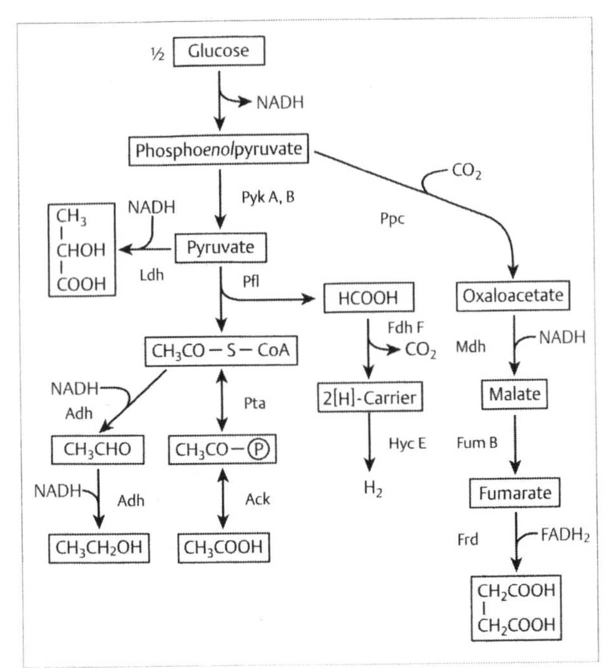

Fig. 21.1 **Scheme of fermentation pathways of *Escherichia coli* showing the balance of reducing equivalents (NADH and $FADH_2$).** Ack, ATP : acetate phosphotransferase or acetate kinase encoded by *ackA*; Adh, alcohol dehydrogenase or ethanol : NAD^+ oxidoreductase encoded by *adhE*; Fdh_H formate dehydrogenase F encoded by *fdhF* at min 92.6; the protein is part of the formate hydrogen-lyase complex, which also includes the HycB, HycC, HycD, HycE, HycF, and HycG proteins encoded by the *hyc* operon at min 58.8; Frd, fumarate reductase encoded by *frdABCD*; FumB, fumarate hydratase B or fumarase B, the anaerobic isozyme, encoded by *fumB*; HycE, hydrogenase 3 encoded by *hycE*; Ldh, D-lactate dehydrogenase or D-lactate : NAD^+ oxidoreductase encoded by *ldhA*; Mdh, malate hydrogenase or malate : NAD^+ oxidoreductase encoded by *mdh*; Pfl, pyruvate–formate lyase encoded by *pfl*; Ppc phospho*enol*pyruvate carboxylase encoded by *ppc*; Pta, phosphotransacetylase encoded by *pta*; PykA,B, pyruvate kinase A and B encoded by *pykA,B*

Fe^{3+} state, the protein is altered in conformation and becomes nonfunctional. O_2 is directly responsible for the oxidation of the metal cofactor, but the physiological reductant is still unknown.

Fnr is a transcriptional global regulator that senses the redox conditions of the environment. The protein is functionally active in the Fe^{2+} state and, under anoxic conditions activates numerous operons that constitute the ***fnr* modulon**.

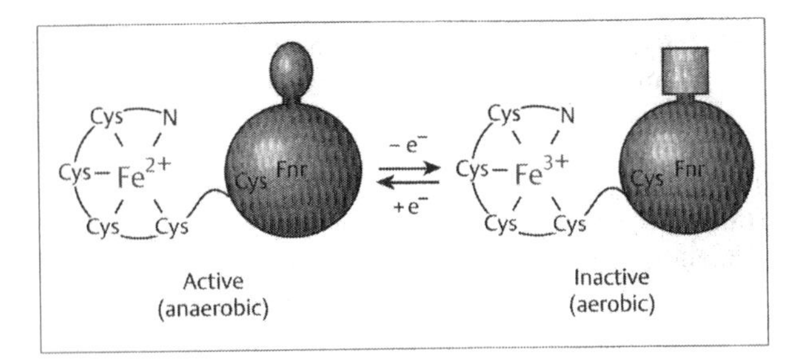

Fig. 21.2 **Schematic representation of active and inactive forms of Fnr.** The transcriptionally active Fnr protein is associated with Fe^{2+} under anoxic conditions. The protein binds to target promoters as a dimer. When the environment causes oxidization of Fe^{2+} to Fe^{3+}, the protein becomes nonfunctional. Fe^{3+} may remain associated with Fnr or may be captured by another chelator(s)

21.2.2 D-Lactate Formation From Pyruvate Is Elevated During Fermentation

The NAD^+-linked D-lactate oxidoreductase (encoded by *ldhA*) catalyzes the reduction of pyruvate to D-lactate with NADH as the electron donor (see Chapter 12.2). The activity level of this enzyme is most elevated during fermentative growth at low pH. The mechanisms of the regulation, however, are still obscure. Interestingly, no defect in growth has been discovered in mutants lacking this enzyme. Thus, the cell seems to have considerable flexibility in opting for its terminal pathways of fermentation, a point that will come up again.

21.2.3 Reactions Involved in the Cleavage of Pyruvate Are Highly Regulated

The main gateway for fermentation is controlled by pyruvate–formate lyase (formate C-acetyltransferase), which cleaves pyruvate to acetyl-CoA and formate (see Fig. 21.1 and Chapter 12.2). The enzyme is encoded by the *pfl* operon whose expression is controlled by seven promoters, allowing the synthesis of the enzyme to be adjusted some 15-fold between conditions favouring aerobic respiration to anaerobic fermentation. Transcription is activated by Fnr, in concert with the global regulator ArcA (in the phosphorylated form, see Chapter 21.4). Expression of *pfl* under anoxic conditions can be increased further by exogenous pyruvate. Transcription of the operon is repressible by nitrate, whose effect is mediated by NarL (in the phosphorylated form NarL-phosphate, see Chapter 21.3). Intriguingly, despite the importance of the pathway of pyruvate cleavage and the strong control of the *pfl* operon by Fnr, a mutant lacking

Table 21.1 **Representatives of the fnr and arcA modulons** +, postive regulation; –, negative regulation; 0, no effect; ND, not determined. After Guest, J. R., Green, J., Irvine, A. S., and Spiro, S. (1996) The *fnr* modulon and Fnr-regulated gene expression (317–337) and Lynch, A.S., and Lin, E. C. C. (1996) Responses to molecular oxygen (361–373). In: Lin, E. C. C., and Lynch, A. S. (eds.) Regulation of gene expression in *Escherichia coli*. Austin, Texas: Landes

Target operons	Enzyme/Function	Regulatory effects of	
		Fnr	ArcA-phosphate
aceBAK	Malate synthase, isocitrate lyase, and isocitrate dehydrogenase kinase/phosphatase	ND	–
acn	Aconitase	ND	–
arcA	ArcA (response regulator)	+	+
cydAB	Cytochrome *d* oxidase	–	+
cyoABDCE	Cytochrome *o* oxidase	–	–
dmsABC	Dimethylsulfoxide reductase	+	0
dcuB–fumB	Dicarboxylate transport and Fumerase B	+	ND
fdnGHI	Formate dehydrogenase N	+	–
fnr	Fnr (global regulator)	–	ND
fumA	Fumarase A (aerobic)	0	–
fumB	Fumarase B (anaerobic)	+	ND
frdABCD	Fumarate reductase	+	0
glpACB	*sn*-Glycerol-3-phosphate dehydrogenase (anaerobic)	+	0
glpD	*sn*-Glycerol-3-phosphate dehydrogenase (aerobic)	0	–
gltA	Citrate synthase	ND	–
hyaA-F	Hydrogenase 1	0	+
hypBCDE-fhlA	Hydrogenase activities and formate regulation	–	ND
icd	Isocitrate dehydrogenase	ND	–
lld (lctD)	L-Lactate dehydrogenase	0	–
mdh	Malate dehydrogenase	ND	–
narGHJI	Nitrate reductase	+	0
narX	NarX (sensor protein)	–	ND
ndh	NADH dehydrogenase II (aerobic)	–	ND
nirBDC	NADH-dependent nitrite reductase	+	ND
nrfA-G	Formate-linked nitrite reductase	+	ND
pdhR–aceEF–lpd	Pyruvate dehydrogenase complex and regulator	–	–
pfl	Pyruvate–formate lyase	+	+
sdhCDAB	Succinate dehydrogenase	–	–
sodA	Mn–Superoxide dismutase	–	–
sucAB	α-Ketoglutarate dehydrogenase	ND	–
sucCD	Succinate thiolkinase	ND	–

this global regulator can still grow fermentatively on glucose, albeit at a slower rate. This poor growth is greatly relieved by exogenous acetate.

Pyruvate–formate lyase is regulated transcriptionally by Fnr, NarL, and ArcA; at the enzyme level, its activity is controlled by activation through an activase Act and by deactivation through AdhE, i.e., ethanol dehydrogenase.

Pyruvate–formate lyase is posttranslationally interconvertible between active (anaerobic) and inactive forms (aerobic). The activase of the lyase is encoded by the *act* gene. The activated form of the lyase is deactivated by physical combination with ethanol dehydrogenase encoded by *adhE*. The biological significance of the deactivation phenomenon is not clear. It is possible that excess dehydrogenase acts as a negative feedback control of the lyase. Alternatively, by this process, both the lyase and the dehydrogenase are held in storage during aerobiosis.

21.2.4 Consistent with its Central Role, Acetyl-CoA Metabolism Is Regulated on Various Levels

Acetyl-CoA can have two fates: ending up as ethanol or as acetate. Ethanol is formed by two consecutive reductions at the expense of two NADH molecules. Typically, these reactions are catalyzed by two separate enzymes, but in *E. coli*, a large hybrid enzyme reversibly catalyzes:

$$\text{Acetyl-CoA} + 2\,\text{NADH} \longleftrightarrow 2\,\text{Ethanol} + 2\,\text{NAD}^{+} \quad (21.1)$$

This enzyme (encoded by *adhE*) is often called ethanol dehydrogenase or alcohol dehydrogenase, although considering its physiological function, the name acetyl-CoA reductase is more appropriate. Catalysis of two tandem reactions by a single enzyme protein probably minimizes the accumulation of free acetaldehyde, which is chemically very active and thus toxic (see Chapter 12).

Expression of *adhE* is transcriptionally controlled. The activity level of the enzyme can increase tenfold or more during anaerobic growth. Transcription seems to be regulated by two different mechanisms of almost equal importance: (1) enhancement of transcription is associated with a high NADH/NAD^{+} ratio, which may be sensed by an unidentified regulator protein, and (2) repression of transcription is exerted by the NarX/NarL system (see Chapter 21.3.1). For reasons still unclear, the expression of *adhE* is also under translational control. Synthesis of the enzyme requires the cleavage of a stem–loop structure of the *adhE* messenger RNA by ribonuclease III. In addition to translational control, the enzyme protein is rapidly degraded during aerobic metabolism. The Fe^{2+} at the catalytic site(s) serves as a self-destruct mechanism by catalyzing the formation of highly reactive oxygen radicals from hydrogen peroxide (see Chapter 21.5). The shortened aerobic half-life of the enzyme helps to direct acetyl-CoA to the tricarboxylic acid (TCA) cycle, and thus prevents wasteful dissipation of carbon and energy in the form of ethanol.

Acetyl-CoA ends up fermentatively as ethanol or as acetate depending on the growth conditions. Ethanol dehydrogenase and the enzymes of acetyl-phosphate metabolism play central roles. The switch between the two fates of acetyl-CoA involves various control mechanisms at the enzyme synthesis and activity level, and its complexity reflects the central role of this intermediate in catabolism.

Under anoxic conditions, an alternative pathway for acetyl-CoA converts it to acetyl phosphate with the help of phosphotransacetylase (encoded by *pta*). The phosphoryl group of acetyl phosphate can generate ATP from ADP with the help of acetyl kinase (encoded by *ackA*). The released acetate then exits the cell, probably via a H^{+}-symport system, thereby augmenting the proton potential (see Chapter 5). The activity levels of the transacetylase and the kinase are not significantly affected by the respiratory condition, possibly for the functional reason that the pathway can be partly employed for aerobic utilization of exogenous acetate. Perhaps one of the small prices for this lack of control is the significant leakage of acetate during aerobic growth on glucose.

21.2.5 The Regulation of Formate Metabolism Involves Intracellular Formate and Selenocysteine

Formate can also have two fates: ending up as CO_2 plus H_2 or simply being excreted. The splitting reaction is catalyzed by formate–hydrogen lyase. This enzyme is a protein complex comprising formate dehydrogenase H (encoded by *fdhF*) and hydrogenase 3 (encoded by *hycE*); together they convert formate to CO_2 and H_2. The other fate of formate, to be excreted, occurs perhaps in a manner similar to that of acetate.

Formate–hydrogen lyase is a multicomponent membrane-associated complex, the exact components of which have not yet been determined (see Chapter 12). About a dozen genes, distributed in the divergent *hpc* and *hyp* operons, have been uncovered. The genes encode subunits of the complex and proteins involved in its maturation. In addition, the catalytic activity of formate–hydrogen lyase requires a molybdenum cofactor, Ni and Fe. A highly interesting feature of formate dehydrogenase H is the presence of a **selenocysteine** (the S of cysteine is replaced by Se) as a novel amino acid (at position 140 from the N-terminal end and encoded by the codon UGA). This internal "termination" codon is subverted for the incorporation of selenocysteine with the help of a $tRNA^{Sec}$ and several translation factors. A genetically engineered enzyme in which the selenocysteine is replaced by cysteine has a K_{cat} that is two orders of magnitude lower than that of the wild-type enzyme. At pH7, the Se of selenocysteine exists mostly in the form of an anion, in contrast to the S of cysteine, which is in the protonated form. The exact biochemical function of selenocysteine, however, has not yet been elucidated.

The activity levels of the lyase are strongly diminished during aerobic growth. The *fdhF, hyc*, and *hyp* operons are activated during anaerobic growth by the FhlA regulatory protein. The FhlA protein shows partial homology to the response regulators of the two-component system and responds to formate as a signal.

Variation in the levels of induction under various anoxic growth conditions appears to reflect primarily cellular concentrations of formate, but how formate is sensed remains to be clarified. The discovery of *fhlB*, which encodes another regulatory protein, further complicates the picture. As investigations proceed rapidly, a more complex picture of transcriptional regulation is emerging. For instance, an Fnr-binding site (within *hypA*) is responsible for activating *hypBCDE*. More intriguingly, transcription of the operons *fdhF*, *hyc*, and *hyp* is absolutely dependent on the alternative sigma subunit σ^{54} (encoded by *rpoN*). This would suggest that regulation of fermentation is coordinated with nitrogen metabolism. It is conceivable that fermentative and nitrogen metabolism and their regulatory mechanisms converged during the evolution of the nitrogen fixation (reduction of N_2 to ammonia) under strictly anoxic conditions. The enzyme responsible, nitrogenase, is extremely sensitive to inactivation by O_2.

Formate-hydrogenase lyase activity is controlled at the level of synthesis by a system that apparently senses formate, but also involves Fnr, RpoN (σ^{54}), and other global regulators. The lyase contains the unusual selenocysteine.

21.2.6 Succinate Formation From PEP Consumes Reducing Equivalents Under Anoxic Conditions

For disposal of reducing equivalents, the most effective pathways are the formation of succinate or ethanol from pyruvate: in each case, the four H atoms (reducing equivalents) disposed of can regenerate 2 NAD^+ from the 2 NADH that arise during the conversion of glucose to pyruvate, permitting continued consumption of the sugar. The four-carbon skeleton of succinate is built by condensation of the three-carbon precursor PEP with CO_2, a reaction catalyzed by phospho*enol*pyruvate carboxylase (pyrophosphate : oxaloacetate carboxylase) encoded by *ppc*. The product, oxaloacetate, is reduced to malate by an NADH-dependent oxidoreductase. (It is not yet certain whether it is the same enzyme encoded by *mdh*). Malate is then converted to fumarate by fumarate hydratase B or fumarase B (encoded by *fumB* and activated by Fnr). Fumarate is reduced to succinate by fumarate reductase (encoded by *frdABCD*), thereby disposing of two more reducing equivalents (see Chapter 21.3.2).

Box. 21.1 *E. coli* belongs to the class of mixed-acid fermenters. Growth on carbohydrates results in the excretion of succinate, D-lactate, acetate, formate, ethanol, CO_2, and H_2 (see Chapter 12). The concentration patterns of these compounds vary according to the source of nutrition and the physical environments. The flexibility of the pathways would suggest the existence of several specific regulators in addition to the control by Fnr. *Lactobacillus lactis*, in contrast, can ferment glucose solely to lactic acid. Although the conversion of one molecule of the sugar to two molecules of lactic acid can balance the oxidation and reduction reactions in the fermentation pathway, all six carbons in glucose end up being excreted. For this reason, this bacterium must be provided with additional nutrients for anabolism and cannot grow fermentatively on glucose alone.

21.2.7 There Are Many More Fermentative Products of Other Bacteria

Many well-known compounds not mentioned here are excreted by other bacteria. A few examples are diacetyl and acetoin excreted by *Klebsiella pneumoniae*, butyrate and butanol by *Clostridium kluyveri*, and propionate by a variety of bacteria (see Chapter 12). Most mechanisms of regulation of gene expression still need to be worked out.

21.3 Anaerobic Respiration and its Control Require a Complex Regulatory Network

During anaerobic respiration, pyruvate is metabolized both by the pyruvate-formate lyase and the pyruvate dehydrogenase routes. A number of compounds can be exploited as exogenous electron acceptors. These include fumarate, dimethyl sulfoxide (DMSO), trimethylamine *N*-oxide (TMAO), nitrite, and nitrate. When presented as a mixture, most of these compounds can be utilized simultaneously. However, nitrate, the acceptor with the most positive midpoint potential (Tab. 4.1), is highly preferred over the other compounds. Under such a condition, the lyase-initiated pathway yields acetate as the major product, thus circumventing the necessity of wasteful excretion of D-lactate and ethanol. The cooperative action of several global regulators is necessary to ensure the choice of anaerobic respiration over fermentative pathways and the dominance of anaerobic nitrate respiration over the other anaerobic respiratory pathways.

21.3.1 Control of Anaerobic Respiration Involves a Pair of Two-Component Systems That Respond to Nitrate and Nitrite

With few exceptions, expression of genes involved in anaerobic respiration requires **Fnr** for activation. When nitrate and nitrite are available as exogenous electron acceptors, they recruit the participation of the **Nar transcriptional regulatory proteins**. Nar was proposed as a mnemonic term for **n**itr**a**te **r**eduction. The Nar proteins preside over two parallel signal transduction pathways: NarX (membrane sensor)/NarL (cytoplasmic regulator) and NarQ (membrane sensor)/NarP (cytoplasmic regulator). Interestingly, whereas **nitrate** stimulates both sensors NarX and NarQ to autophosphorylate and become kinases for NarP and NarL, **nitrite** stimulates NarQ-mediated phosphorylation of NarP, but promotes NarX-mediated dephosphorylation of Nar-phosphate (Fig. 21.**3**). As is true for other two-component systems, only phosphorylated forms of NarL and NarP (i.e., NarL-phosphate and NarP-phosphate) are functional as transcriptional regulators (Tab. 21.**2**). The effect of each phosphorylated regulator on the target promoters depends on their nucleotide sequence. The DNA consensus sequence TACYNKT (where Y=C or T, and K=A or C) has been proposed as the site of recognition by NarL-phosphate.

Although the detailed mechanisms of transcriptional regulation by the Nar proteins are still to be elucidated, one can surmise the underlying regulatory strategy. On one hand, nitrate should be utilized in preference to nitrite because nitrate has a greater potential capacity as an electron sink. On the the other hand, nitrite should not be allowed to accumulate above a certain concentration because of toxicity. Finally, the availability of either nitrate or nitrite makes it energetically inefficient for the cell to utilize other electron acceptors, such as fumarate (see Tab. 4.**1**, Fig. 21.**3**, and sections below).

In addition to nitrate and nitrite, molybdate is required as a co-effector for functional mobilization of the Nar system. To understand the rationale of this extra requirement, it is sufficient to realize that activation of operons encoding nitrate reductases coupled to repression of alternative anaerobic respiratory pathways would be self-defeating when there is an insufficient supply of a **molybdenum cofactor** for catalytic activity.

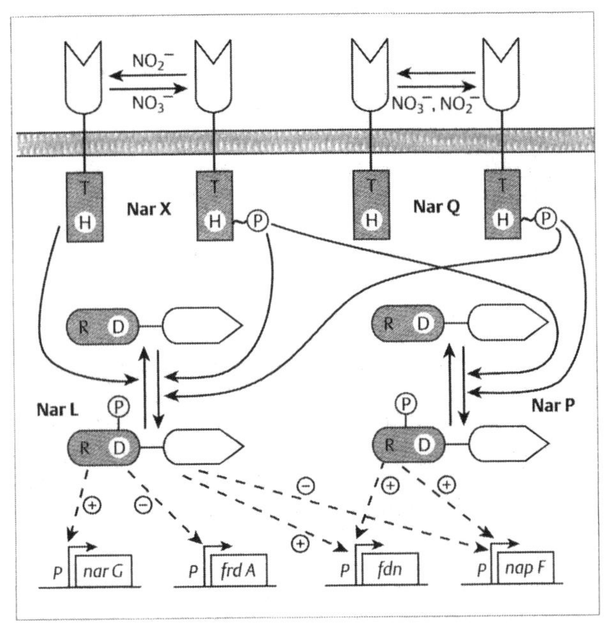

Fig. 21.**3** **A simplified model for Nar-controlled signalling processes.** NarX and NarQ are twin sensor kinases working with their corresponding response regulators NarL and NarP. The phosphorylated forms of NarL and NarP (NarL-phosphate and NarP-phosphate), formed at the expense of ATP, act to stimulate or inhibit transcription of target operons: the *fdn* operon (encoding the formate dehydrogenase N complex), the *frdA* operon (encoding fumarate reductase), the *napF* operon (encoding periplasmic nitrate reductase complex), and the *narG* operon (encoding the major nitrate reductase complex). Solid arrows indicate the effect of nitrate or nitrite on the phosphorylation state of the sensor kinases. Dashed arrows indicate the transcriptional effect of NarL-phosphate or NarP-phosphate on the promoters (*p*) of target operons (⊕, activation; ⊖, repression). Nitrate stimulates each sensor kinase to phosphorylate both NarL and NarP. In contrast, nitrite stimulates NarQ to phosphorylate both NarL and NarP, but promotes the dephosphorylation of NarL-phosphate by NarX. NarP-phosphate has no effect on the expression of the *narG* and *frdA* operons, but antagonizes the repression of the *napF* operon by NarL-phosphate. (Other symbols are as in Fig. 20.**16**)

21.3.2 Reductases of Fumarate, DMSO, and TMAO Are Expressed According to the Redox Potential and the Substrates Available

The evolutionary transition from fermentation to anaerobic respiration may be marked by the appearance of the **fumarate reductase** (encoded by *frdABCD*), a membrane enzyme complex that couples the reductive reaction with the ejection of protons into the environ-

Table 21.2 **Representatives of the *NarL* and *NarP* modulons.** +, postive regulation; −, negative regulation; 0, no effect; ND, not determined. NarL-phosphate activates the *nrfABCDEFG* operon in response to nitrite, but represses the operon in response to nitrate. After Darwin, A. J., and Stewart, V. (1996) In: Lin, E. C. C., and Lynch, A. S. (eds.) Regulation of gene expression in *Escherichia coli*. Austin, Texas: Landes; 343–359

Target operons	Enzyme/Function	Regulatory effects of	
		NarL-phosphate	NarP-phosphate
adhE	Alcohol (ethanol) dehydrogenase	−	ND
dmsABC	Dimethyl sulfoxide reductase	−	ND
fdnGHI	Formate dehydrogenase N	+	+
frdABCD	Fumarate reductase	−	0
modABCD	Molybdate uptake	+	ND
nap–ccm	Periplasmic nitrate reductase	−	+
narGHJI	Nitrate reductase	+	0
narK	Nitrite export	+	0
NarXL	Nitrate/Nitrite regulation	+	+
nirBDC	NADH–nitrite reductase	+	+
nrfABCDEFG	Formate–nitrite reductase	+/−	+
nuoA–N	NADH dehydrogenase I	+	+
pfl	Pyruvate–formate lyase	−	−

ment (the generation of a proton motive force). It is not yet clear whether the same dicarboxylic acid permease for the import of fumarate as an electron sink also provides the route of exit of the product succinate (see Chapter 12). Although no rigorous test has been carried out to see if the expression of *frdABCD* is endogenously induced, it is clear that the transcription is activated by Fnr and is strongly curtailed when nitrate is available as a superior electron acceptor. This curtailment is attributable to NarL-phosphate but not NarP-phosphate (Fig. 21.**3**).

DMSO (dimethyl sulfoxide) reductase, another enzyme requiring the molybdenum cofactor, is encoded by *dmsABC*. Anaerobic expression of the operon does not require substrate induction, but depends on Fnr stimulation. In the presence of nitrate, the expression of the operon is repressed by the Nar system.

TMAO (trimethylamine-*N*-oxide) reductase is encoded by *torABC*. TMAO is abundant in animals such as fish. Expression of the operon is inducible by TMAO, DMSO, tetrahydrothiophene 1-oxide, and pyridine *N*-oxide under anoxic conditions, all of which are substrates. The broad induction and substrate specificity, the independence from Fnr of the expression of the operon under anoxic conditions and the location of the enzyme mainly in the periplasm led to the speculation that this reductase is for detoxification rather than respiration. Consistent with this view, the operon is not subject to repression by the Nar system.

There is no evidence for interference of expression among *frdABCD, torABC*, and *dmsABC*. When their respective substrates are presented together, however, their utilization is in the order: DMSO, TMAO, fumarate. It is thought that both redox potential and substrate affinities influence this order of utilization.

21.3.3 The Membrane-bound and the Cytoplasmic Nitrite Reductases Have Different Functions

There are two pathways for the six-electron reduction of nitrite to ammonia: the major Nir pathway initiated by the **cytoplasmic** NADH-dependent enzyme (encoded by the *nirBDC* operon) and the minor Nrf pathway initiated by the **membrane-associated** and formate-dependent nitrite reductase (encoded by the *nrf* gene cluster; see Chapter 12). The cytoplasmic enzyme detoxifies nitrite and regenerates NAD^+, whereas the membrane enzyme, in addition, generates a proton potential. The *nirBDC* operon is activated by both NarL-phosphate and NarP-phosphate in the presence of nitrate or nitrite (see Tab. 21.**2**). In contrast, the *nrfA* operon is induced by nitrite (effected by both NarL-phosphate and NarP-phosphate), but repressed by nitrate (attributed primarily to the action of NarL-phosphate). Expression of both operons depends on Fnr.

21.3.4 The Expression of Nitrate Reductases Depends on Several Global Regulators

There are also two nitrate reductases. The major enzyme is encoded by *narGHJI*. The expression of *narGHJI*, as well as *fdnGHI* (encoding formate dehydrogenase N, which participates in nitrate respiration), is induced by either nitrate or nitrite (see Fig. 21.**3**). Whereas *narGHJI* is strongly activated by NarL-phosphate but not NarP-phosphate, *fdnGHI* is activated by both of the regulators. Curiously, the nitrate-induced expression of *narGHJI*, but not *fdnGHI*, requires the integration host factor (IHF). The **formate–nitrate pathway** is energetically the most efficient way of carrying out anaerobic respiration. In this pathway, formate dehydrogenase N pre-empts the role of the fermentative formate dehydrogenase F involved in fermentation. The minor nitrate reductase is encoded by *narZYWV*. Its expression does not require the Nar regulators, although the expression of both *narGHJI* and *narZYWV* requires Fnr.

To complete the Nar regulatory circuitry, it should be recalled that the *pfl* operon encoding pyruvate–formate lyase is repressed by both NarL-phosphate and NarP-phosphate. Since pyruvate–formate lyase is the bottle-neck of several routes of fermentative metabolism, the repression is an effective mechanism of curtailing fermentation during nitrate respiration.

The advent of nitrate reductase prepared the evolution of aerobic respiration. Probably during this transition to aerobiosis, the function of the menaquinones as intermediate electron carriers was gradually taken over by ubiquinone (with a higher mid-point potential, see Tab. 4.**1**).

Among the various electron acceptors that can be utilized, **nitrate** has the highest midpoint potential. The ability of the nitrate-stimulated **NarL-phosphate** to activate the operon encoding the major nitrate reductase and to repress operons encoding other reductases (e.g., fumarate reductase) is the critical molecular strategy for directing the electrons to nitrate. **Fnr** is involved in the activation of almost all of the operons concerned with anaerobic respiration. The regulatory role of this protein is seen also in certain operons involved in fermentation.

21.4 Control of Aerobic Respiration Involves a Two-component System

With a mid-point potential of +0.82 V, O_2 is the energetically most rewarding electron acceptor. Under oxic and weak catabolite repressive conditions, pyruvate derived from various carbon sources is rapidly fed into the tricarboxylic acid (TCA) cycle (Fig. 21.**4**). The flow rate is controlled by the pyruvate dehydrogenase complex, whose synthesis is repressed under anoxic conditions and whose activity is inhibited by NADH.

21.4.1 The Sensor Kinase ArcB and the Response-regulator ArcA Are Global Regulators for Aerobic Respiration

Like the transcriptional regulation of nitrate respiration, the central control of aerobic respiration also involves a two-component system, in which the ArcB protein acts as the sensor kinase and the ArcA protein as the regulator (Fig. 21.**5**). The term **Arc** is an acronym of **a**erobic **r**espiration **c**ontrol. There are several noteworthy structural differences between the Nar sensor kinases and the Arc sensor kinase (see also Fig. 20.**17**). Whereas the two Nar sensor proteins possess a significant periplasm domain (thought to be receptor sites for nitrate or nitrite), there is only a stretch of seven amino acid residues that connects the two transmembrane segments in ArcB. Another distinctive feature of ArcB is that on the C-terminal side of the classical transmitter domain, there is a receiver domain, and more distally, a second transmitter domain. The receiver region of ArcB shows similarity to that of the response regulator ArcA, but lacks a helix-turn-helix motif associated with DNA binding; the C-terminal secondary transmitter domain is only slightly similar to the primary transmitter domain. In contrast to ArcB, the structure of ArcA is structurally typical of all the known response regulators.

21.4.2 Stimuli and Effectors for the Sensor ArcB Seem To Be Metabolites

Stimulus. Like other sensor kinases, ArcB autophosphorylates itself at the expense of ATP when stimulated. Mutant analysis strongly indicates that O_2 itself is not the stimulus for ArcB. Instead, the data indicate that the sensor responds to diminished oxidative power within the cell caused by anoxia. There are two prinicpal terminal oxidase complexes in *E. coli* that reduce O_2: one with a low apparent affinity, but high V_{max} (encoded by *cyoABCDE*), and the other with a high apparent affinity, but low V_{max} (encoded by *cydAB*). In a double cytochrome-deletion mutant, the Arc system is desensitized to the presence or absence of O_2. Thus, ArcB may

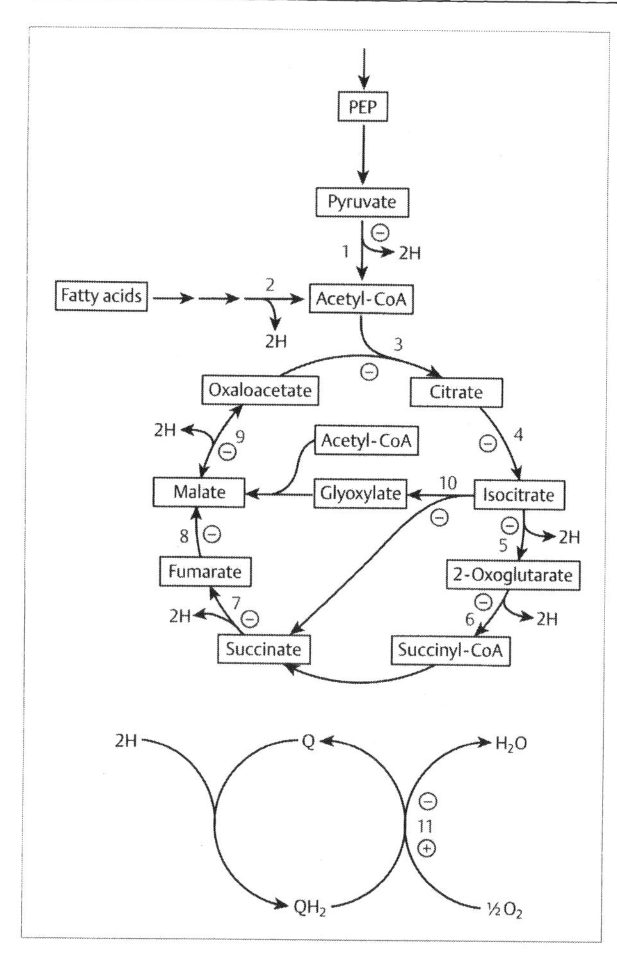

Fig. 21.4 **Three major aerobic pathways under Arc control.** Represented are enzymes of the TCA cycle, the glyoxylate bypass, and terminal electron transport to molecular oxygen. 1, Pyruvate dehydrogenase (*pdhR aceEF*); 2, 3-hydroxyacetyl coenzyme-A dehydrogenase (NAD^+-linked); 3, citrate synthase (*gltA*); 4, aconitase (*acn*); 5, isocitrate dehydrogenase (*icd*); 6, α-ketoglutarate dehydrogenase (*sucAB*); and succinate thiokinase (*sucCD*) (encoded by the same operon); 7, succinate dehydrogenase (*sdhCDAB*); 8, fumarase A (*fumA*); 9, malate dehydrogenase (*mdh*); 10, isocitrate lyase and malate synthase (*aceBAK*) (both enzymes are encoded by the same operon, which also specifies isocitrate dehydrogenase kinase/phosphatase, (see Chapter 19.3.2); 11, cytochrome *o* (*cyoABDCE*) and cytochrome *d* (*cydAB*) (terminal oxidases of ubiquinol-1). H, reducing equivalent; QH_2 (ubiquinol-1); Q, ubiquinone; ⊖, repression by the global regulator ArcA; ⊕, activation of cytochrome *d* by ArcA; numbers, enzymes repressed (or activated) by ArcA

respond to the accumulation of a metabolite in its reduced form, resulting from inadequate rate of electron drainage (see Chapter 20.5.7).

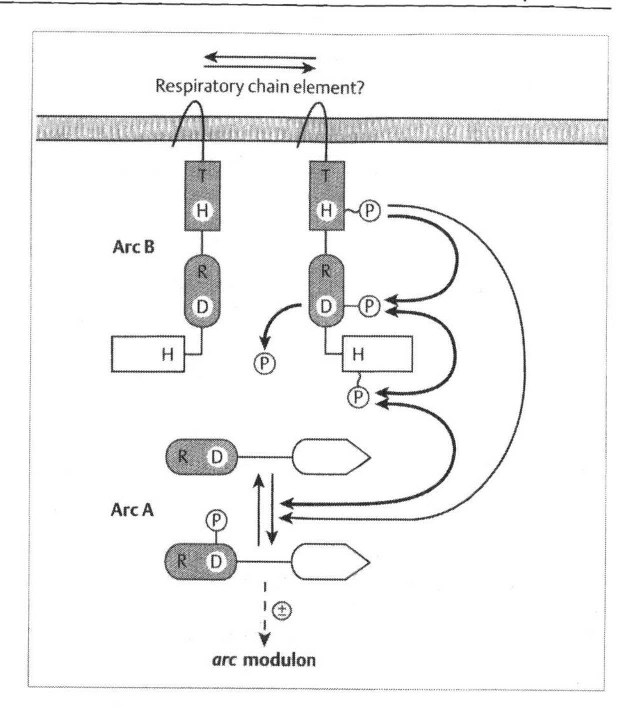

Fig. 21.5 **A model of the Arc signalling process.** The ArcB membrane sensor kinase has 778 amino acid residues: two transmembrane regions on the N-terminal side of the protein, the first transmitter (T) domain with a conserved His-292 residue, the receiver (R) domain with a conserved Asp-576 residue, and a second transmitter domain with a conserved His-717 residue. The ArcA response regulator has 238 amino acid residues: the receiver domain with a conserved Asp-54, and the helix–turn–helix domain for promoter recognition. Solid arrows indicate the direction of phosphoryl group (P) transfer, demonstrated (thick arrows) or postulated (thin arrows). The phosphoryl group at each receptor site may undergo hydrolysis catalyzed by a phosphatase domain(s) not yet demonstrated. (Other symbols are as in Fig. 20.**16**)

Signal decay. After the stimulation is over, those ArcB and ArcA molecules that are still in phosphorylated forms must undergo dephosphorylation so that the system is prepared to respond to another stimulus. There is evidence that a domain of ArcB also catalyzes dephosphorylation reactions. A purified version of ArcB (lacking the two transmembrane segments on the N-terminal side) undergoes rapid in vitro phosphorylation in the presence of ATP without stimulus. After exhaustion of ATP, the protein undergoes progressive dephosphorylation. In contrast, ArcA-phosphate is intrinsically stable once phosphorylated by ArcB. By analogy with certain other two-component systems, the sensor kinase ArcB catalyzes the dephosphorylation of the response regulator ArcA-phosphate.

Effectors. The rate of phosphorylation of the N-terminal truncated version of ArcB can be modified by certain metabolites. For instance, D-lactate and pyruvate, at well within intracellular concentrations, promote phosphorylation of the protein. Since the cellular levels of D-lactate and pyruvate are known to increase with anoxia, it appears that accumulation of these metabolites reinforces the source of signal through allosteric action on ArcB. The enhancing phenomena were confirmed with everted (turned inside out) membrane vesicles containing the wild-type ArcB protein.

21.4.3 A Working Model for Signal Transduction by the Arc System Can Be Proposed

Because the ArcB possesses three highly conserved sites as phosphoryl group receptors (His-292, Asp-576, and His-717), there should be several phosphorylated forms of this protein. Indeed, after phosphorylation of the protein by ATP—with its γ phosphoryl group labeled with [^{32}P]—the radioactivity is associated with histidyl and aspartyl residues. Furthermore, both His-292 and His-717 can be phosphorylated.

A plausible model for the signal transduction by ArcB can be proposed (Fig. 21.**5**). Data derived from both in vivo and in vitro experiments show that the Arc two-component system transmits the signal by a His-Asp-His-Asp phospho-relay reaction. Upon stimulation of ArcB at the membrane, the conserved His-292 of the primary (orthodox) transmitter domain undergoes autophosphorylation at the expense of ATP. The phosphoryl group then migrates to the conserved Asp-576 of the receiver domain. There, the phosphoryl group may undergo hydrolysis while associated with this domain or may migrate to the conserved His-717 of the secondary transmitter domain. From this secondary transmitter, the phosphoryl group may be transferred to the conserved Asp-54 of the ArcA protein, which thereupon acquires the specific ability to recognize the target promoters. The decay of the signal also involves a phospho-relay reaction: the phosphoryl group migrates from Asp-54 of ArcA to His-717 of ArcB and then to Asp-576, where hydrolysis may occur (see Fig. 21.**5**).

According to the proposed model, metabolites such as D-lactate and pyruvate would amplify the effect of the membrane-derived stimulus by enhancing the rates of phosphorylation of ArcB. This would result in elevated levels of ArcA-phosphate concentration and thus extend the control range of target promoters to include those that do not have strong binding constants. The DNA sequence (A/T)GTTAATTA(A/T) has been proposed as the site of recognition.

21.4.4 The *arc* Modulon Comprises Genes for Several Metabolic Pathways

The ***arc* modulon**, or the family of Arc-target operons, is known to comprise numerous flavo-dehydrogenases and enzymes of the TCA cycle, the glyoxylate bypass, fatty acid degradation pathway, and electron transport (see Tab. 21.**1** and Fig. 21.**4**). Two proteins of the last group are cytochrome *o* and cytochrome *d*. In the vast majority of cases, ArcA-phosphate functions as a transcriptional repressor.

Among the intensively studied target operons is *cydAB* (encoding cytochrome *d*). For the promoter of this operon, ArcA-phosphate serves as an activator and Fnr serves as a repressor. The complexity of transcriptional control is related to the enzyme being the most useful when the O_2 tension diminishes to a point that requires scavenging. Maximal *cydAB* expression occurs, therefore, under microoxic conditions (in the vicinity of 1% of the O_2 concentration in the atmosphere). Although the detailed mechanism of transcriptional regulation of this operon still has to be defined, preliminary data indicate that the net effect on gene expression depends on the association constants of ArcA-phosphate and Fnr with their respective DNA sites and the number and location of those sites.

The **Arc system**—and to a lesser extent the **Nar systems**—modulate the expression of genes involved in pathways of **respiration**. These pathways include enzymes that initiate the removal of electrons from donor substrates and those that catalyze the transfer of electrons to the final acceptors. For instance, synthesis of both primary dehydrogenases of the flavoprotein type and the terminal electron transporters, cytochrome *o* and cytochrome *d*, are progressively restricted as anoxia sets in. Equally important is the control of pathways of central metabolism, such as the TCA cycle, by the Arc system.

21.5 Aerobic Metabolism Is Associated With Formation of Harmful Oxygen Species (Oxygen Stress)

The high energy yield harvested by aerobic respiration is associated with a disadvantage that becomes even more serious under hyperbaric conditions (near the occurrence of oxygenic photosynthesis). During the course of reduction of molecular oxygen to water: toxic "reactive oxygen species" are unavoidably generated. Molecular oxygen may accept on electron at a time, as described below.

$$O_2 \xrightarrow{+e^-} O_2^{\bullet -} \xrightarrow[2H^+]{+e^-} H_2O_2 \xrightarrow[H^+]{+e^-} H_2O + OH^\bullet \xrightarrow[H^+]{+e^-} H_2O \quad (21.2)$$

H_2O_2 is routinely formed as a product by certain flavin enzymes; $O_2^{\bullet -}$ (superoxide) and $OH^\bullet$ (hydroxyl radical) arise as catalytic accidents. Aerobic respiratory electron transport chains are mostly responsible for generating these undesirable oxygen species (see Fig. 11.**15**).

21.5.1 There Are Several Defense Mechanisms Against Oxidative Damage (Detoxifying Enzymes)

The cell has evolved numerous mechanisms to protect the targets (see Chapter 11.5.2). For instance, $O_2^{\bullet -}$ is disarmed by superoxide dismutase; hydrogen peroxide, also harmful, is in turn destroyed by catalase.

No enzymes, however, have evolved to protect against $OH^\bullet$ because it is chemically so reactive; innumerable susceptible target groups are far more likely to be hit before the radical would encounter a catalytic site of a hypothetical detoxifying enzyme. $OH^\bullet$ radicals are generated when H_2O_2 reacts with Fe^{2+} (mostly bound on proteins):

$$H_2O_2 + Fe^{3+} \xrightarrow{\text{spontaneous}} OH^\bullet + Fe^{3+} \quad (21.3)$$

A likely target of the $OH^\bullet$ is an amino acid side chain near the site of the reaction, thereby irreversibly disabling the protein.

Turning disadvantage into advantage. Ironically, nature can also turn a destructive reaction to the advantage of the cell. For instance, oxidoreductases that catalyze NADH-dependent reduction of a carbonyl group to an alcohol group in fermentative pathways typically use the Fe^{2+} at the active site. When the cell switches to aerobic metabolism, such enzymes become a liability because they drain potentially valuable metabolites as excretion products. In these cases, the role of the Fe^{2+} as a self-destructing agent, in the presence of H_2O_2, greatly shortens the half-life of the protein. This "control" augments transcriptional regulation to achieve more rapid adaptation to aerobic metabolism. Ethanol oxidoreductase (see Chapter 21.2.4) is a concrete example. It is useful to note that in housekeeping NAD^+-coupled enzymes, needed both aerobically and anaerobically, Fe^{2+} is replaced by Zn^{2+}, which is not reactive with H_2O_2. It may be that when Fe^{2+} was abundant (before photosynthesis started to produce O_2), all NAD^+- and $NADP^+$-coupled oxidoreductases utilized this metal; with the accumulation of environmental O_2, Zn^{2+} oxidoreductase began to emerge. Such enzymes would not be rapidly destroyed during aerobic metabolism.

21.5.2 Chemical Neutralization of Toxic Oxygen Species Is Strongly Enhanced by Repair Enzymes

There are other kinds of defence against oxidative damages. One strategy is to annihilate the damaging oxidants stoichiometrically by metabolites, such as NADPH, NADH, thioredoxin, and glutathione (see Chapter 11.5.2). The other strategy is to repair the damaged

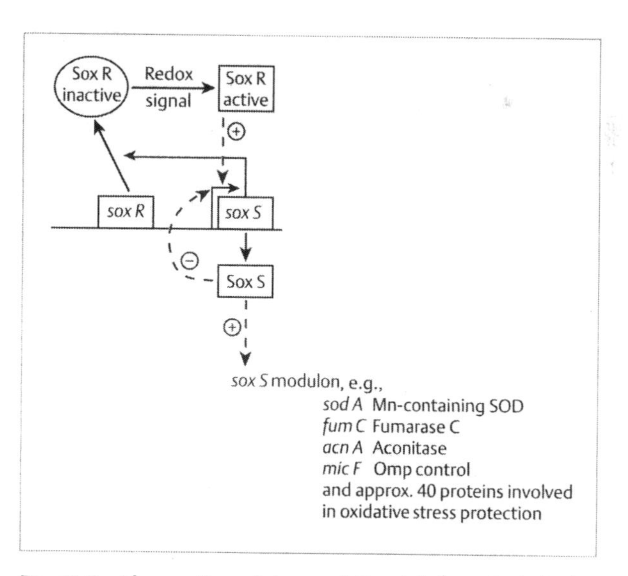

Fig. 21.**6** **The soxS modulon and its global control.** Redox signals activate SoxR, an FeS-containing activator, which controls transcription of *soxS*. SoxS represses its own synthesis (autoregulation) and activates most of the genes of the *soxS* modulon, a group of approximately 40 proteins whose function is protection against oxidative stress

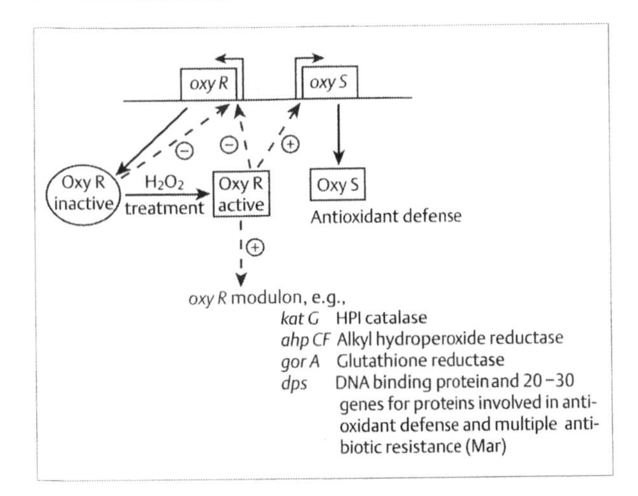

Fig. 21.7 **The *oxyR* modulon and its global control**. The global regulator OxyR is activated by H_2O_2, itself generated, for example, from superoxide ($O_2^{\bullet -}$) by SodA, a member of the *soxS* modulon (see Fig. 21.**6**). Upon activation, OxyR enhances transcription of the *oxyR* modulon, which controls approximately 20–30 proteins, and of OxyS, all involved in antioxidant defense and in multiple antibiotic resistance (Mar)

macromolecules: the repair of DNA is particularly successful (see Chapter 15.6).

Although our cognizance of the different ways by which proteins can protect against oxidative damage is still in an early phase, the importance of protection by various proteins is revealed by the existence of two sets of global regulatory genes that control the synthesis of repair and protecting enzymes.

The *soxRS* and *oxyRS* regulatory genes and their modulons. The *soxRS* genes are divergently transcribed and encode regulator proteins (Fig. 21.6). Redox-cycling agents (e.g., menadione), which mediate one-electron reduction of O_2 by diverting electrons from compounds like NADH or NADPH, activate SoxR (an FeS-containing protein). Activated SoxR in turn promotes the transcription of *soxS*, which encodes the global regulator SoxS SoxR can be activated by $O_2^{\bullet -}$ and also by $NO^{\bullet}$ (expecte to be a bactericidal agent of macrophages). The *so* modulon encodes about 40 proteins that protect agains oxidative stress and whose syntheses are generall under positive control by SoxR. One target operon i *sodA* (encoding the Mn-superoxide dismutase). Th importance of controlling the transcriptional activit of this operon is indicated by the involvement of at leas six regulators.

Another system that protects the cell agains oxidative stress is controlled by the *oxyRS* genes (Fig 21.**7**). These genes are also divergently transcribed. Th *oxyR* gene encodes a regulator protein that is activate by a H_2O_2-dependent signal or the compound itself. Th *sox* modulon encodes about 20–30 protective protein whose syntheses are generally under positive transcrip tional control. One target operon is *katG* (encoding HP catalase). The regulatory gene product of *oxyS* is surprisingly, a small untranslated mRNA. The role o these mRNAs in protection against oxidative stress is a active subject of study (see also Chapter 28.6).

The development of aerobic respiration also creates a problem for the cell. Aerobic metabolism unavoidably generates **reactive oxygen species**, regularly or accidentally. These agents can cause serious damage to nucleic acids, proteins, and lipids. Several kinds of defense mechanisms are employed by the cell, including stoichiometric neutralizing of the toxic-compounds, enzymatic destruction of those agents, and repairing the damages. Detailed molecular studies concerning oxidative damage is likely to be of importance to medicine because of the long and widely held theory that human aging is at least partially caused by progressive accumulation of damaged macromolecules that eluded breakdown or repair.

21.6 Overview

It is becoming increasingly apparent that many proteins are multifunctional and their possession of multiple domains allows them to interact with various other macromolecules. This is an ingenious way of integrating the molecular economy of the cell. It seems certain that much of our future efforts to deepen the understanding of genetic and metabolic regulation will depend on taking into consideration the informational linkage among different biochemical circuits. Interactions of several regulators may produce surprisingly comple responses.

There also remain the important questions of wha is the signal for energy sufficiency that stops the cel from gratuitous consumption and how the cell budget the major organic nutrients for energy and buildin blocks. It would seem very unlikely that such importan decisions are simply dictated by the laws of thermo dynamics.

Finally, it will be a challenge to evolutionists to dissect the archeological layers of regulatory networks for the energy-transducing pathways in the bacterial world. Perhaps this can be achieved through correlating geological changes with time and the homology of DNA, of RNA, and of proteins. It is staggering to think that in this long march of evolution, global regulators must have undergone innumerable and multi-faceted adaptations in concert with specific regulators and target operons, in order to achieve the efficient control circuits now in existence.

Further Reading

Aristarkhov, A., Mikulskis, A., Belasco, J. G., and Lin, E. C. C. (1996) Translation of the *adhE* transcript to produce ethanol dehydrogenase requires Rnase III cleavage in *Escherichia coli*. J Bacteriol 178: 4327–4332

Clark, D. P. (1989) The fermentation pathways of *Escherichia coli*. FEMS Microbiol Lett 63: 223–224.

Georgellis, D., Lynch, A. S., and Lin, E. C. C. (1997) In vitro phosphorylation study of the Arc two-component signal transduction system of *Escherichia coli*. J Bacteriol 179: 5429–5435

Gottschalk, G. (1985) Bacterial Metabolism, 2nd edn. Berlin Heidelberg New York: Springer

Harold, F. M. (1986) The vital force: a study of bioenergetics, Chapter 3. New York: Freeman

Lin, E. C. C., and Lynch, A. S. Eds. (1996) Regulation of gene expression in *Escherichia coli*. Austin, Texas: Landes

Neidhardt, F. C., Curtiss, R., Ingraham, J. L., Lin, E. C. C., Low, K. B., Magasanik, B., Reznikoff, W., Riley, M., Schaechter, M., and Umbarger, H. E. (eds.) (1996) *Escherichia coli* and *Salmonella*: cellular and molecular biology. 2nd edn. Washington DC: ASM Press

Parkinson, J. S., and Kofoid, E. C. (1992) Communication modules in bacterial signaling proteins. Annu Rev Genet 26: 71–112

Rossmann, R., Sawers, G., and Böck, A. (1991) Mechanism of regulation of the formate-hydrogen lyase pathway by oxygen, nitrate, and pH: definition of the formate regulon. Mol Microbiol 5: 2807–2814

Sonenshein, A. L., ed. (1993) *Bacillus subtilis* and other Gram-positive bacteria. Biochemistry, physiology and molecular genetics. Washington DC: ASM Press

Spiro, S., and Guest, J. R. (1990) FNR and its role in oxygen-regulated gene expression in *Escherichia coli*. FEMS Microbiol Rev 75: 399–428

Stewart, V. (1993) Nitrate regulation of anaerobic respiratory gene expression in *Escherichia coli*. Mol Microbiol 9: 425–434

Unden, G., and Bongaerts, J. (1997) Alternative respiratory pathways of *Escherichia coli*: energetics and transcriptional regulation in response to electron acceptors. Biochim Biophys Acta 1320: 217–234

Section VI
Cell Growth and Differentiation

Prokaryotes are directly exposed to the continuously altering physical and chemical environment. During evolution they have developed various strategies to optimize fitness in competition with other organisms and to use the resources of their ecosystem optimally. Some organisms have adapted to extreme, but constant conditions, others have developed an enormous flexibility that allows them to live in varying environments (see, for example, Chapter 28). In any case, the costs for structural and functional adaptation should not outweigh the benefits.

All organisms have developed **response mechanisms** that always function readily. These include, for example, housekeeping functions, use of carriers for export or import of ions in response to a transient salt stress, and the so-called state–transition process in oxygenic photosynthesis (distribution of excitonic energy between photo-systems I and II under different illumination, see Chapter 13.4.1).

Differentiation is the process by which a cell changes its functional and structural patterns under variable growth conditions. New functions and structures are established by regulated expression of a hierarchy of genes or gene clusters and by morphogenetic events that usually require protein synthesis. Differentiation is an irreversible process; the cell can only return to the previous state by a new process of differentiation. The term differentiation includes variations in response mechanisms, in gene expression, in the activity of enzymes, and in growth and morphogenesis. Differentiation can be induced by internal signals, as for example, by phases of the cell cycle (see Chapter 24.1.1, *Caulobacter*), but most processes are triggered by environmental stimuli (e.g., a shift from oxic to anoxic conditions, or from a poor to a rich growth medium).

Morphogenesis is the process by which the shape and structure of an organism, of a cell, or of a compartment of the organism is modified through regulated growth and cell differentiation. In contrast, the assembling of a virus, a flagellum, a pilus, and other cellular organelles has been named **morphopoiesis** because these structures are assembled from their subunits with the contribution of helper proteins (e.g., chaperones), even in the absence of growth.

Cell differentiation as described in this section has been intensively studied at the molecular level. Although many details are known, no process of cell differentiation has been completely elucidated. Instead, a limited number of general principles for signal recognition and amplification, for global control and gene expression, and for formation of cellular structures and organization seem to be the basis of differentiation for eubacteria, archaea, and eukaryotes. Mechanisms through which proteins find their target site and assemble to form complex structures have been described in Chapter 19.4. The apparently continuous membrane system of prokaryotic cells, which is not separated into typical membrane compartments like that of a eukaryotic cell, is in fact highly organized and clearly involved in differentiation. Unfortunately, only a small number of prokaryotes have been studied thoroughly as models for cell differentiation. Future research promises to broaden our insight into this fascinating field and should elucidate the general principles of cell differentiation.

In Chapter 22, the bacterial cell cycle is introduced, which directs the events of chromosome replication and cell division and their correlation. In Chapter 23, the formation of the cellular surface layers is described, which are important structures in the interaction between the environment and the cell. In Chapter 24, three examples of cellular differentiation and their regulation are discussed; the formation of swarmer and stalked cells during the cell cycle of *Caulobacter*, membrane differentiation in phototrophic bacteria, and heterocyst formation in cyanobacteria. Chapter 25 describes the events and regulation of sporulation in bacilli and streptomycetes and the multicellular development in *Myxococcus*. Chapter 26 discusses the biology and morphopoiesis of bacteriophages. In Chapter 27, the biosynthesis of antibiotics, secondary metabolites, and their regulation is described. Finally, Chapter 28 deals with the adaptation of bacteria to extreme environments.

22 The Bacterial Cell Cycle

Cells of all organisms grow by increasing their mass, and they eventually divide to generate two identical daughter cells. This process is called the **vegetative cell cycle.** Several discrete events take place during this cycle: duplication of the genetic material, separation of the two sets of chromosomes, and division of the cells. These processes are coordinated with each other and with the increase in cell mass. There are many similarities between the cell cycles of prokaryotic and of eukaryotic cells, but bacteria also show some cell-cycle characteristics that they do not share with eukaryotic cells. For example, the time it takes to replicate the genetic material can be much longer than the cell cycle; consequently, two or three chromosome replication cycles may take place simultaneously in a cell during rapid growth. This chapter is mainly concerned with the vegetative cell cycle of *Escherichia coli.*

22.1 Key Events Determine the Different Periods of the Vegetative Cell Cycle

A schematic representation of the cell cycles of prokaryotic and of eukaryotic cells is shown in Table 22.1, and the bacterial cell cycle is presented in more detail in Figure 22.1. The key events are as follows:

1. In eukaryotic cells, there is an initial phase, Gap1 or G1, during which the cells grow in mass; at a specific time in G1 termed START, the cells become committed to enter the cell cycle. In bacteria, the corresponding phase is called the **B period**. The B period is not seen at moderate and at high growth rates (see below and Fig. 22.**4**).
2. During the next phase (**C period** and DNA synthesis or S phase), the genetic material is duplicated. In eukaryotes, another period of growth, Gap2 or G2, follows the completion of the S phase.
3. The completed daughter chromosomes are separated from each other and moved to positions in the cell to ensure that each daughter cell receives one complete genome. This forms the first part of the **D period** in prokaryotes and the mitosis (M) phase in eukaryotes. The mitotic process in eukaryotes involves an elaborate structure, the spindle, which sorts and distributes the daughter chromosomes. The cells divide to form two identical daughter cells during the last part of the D period of bacteria and during the cytokinesis phase of eukaryotes.

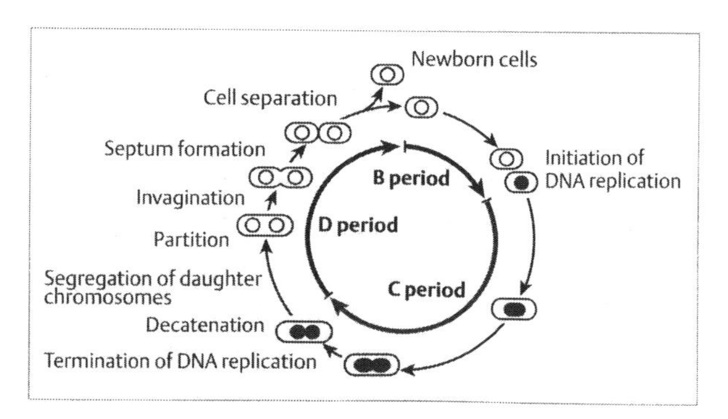

Fig. 22.**1**. **The vegetative bacterial cell cycle.** A schematic representation of the vegetative bacterial cell cycle (at a fairly slow growth rate). The inner circle indicates the B, C, and D periods. The key processes and the shape of the cells are shown in the outer circle. Nonreplicating or replicating nucleoids are shown in the center of the cells as white or red areas, respectively

During the cell cycle of a vegetative prokaryotic cell, the genetic material is duplicated, the newly formed chromosomes are separated, and the cell is divided.

The vegetative cell cycle consists of three phases, the B, C, and D periods. Replication of the chromosome occurs during the C period. Separation of the daughter nucleoids and cell division occur during the D period.

Table 22.1 **Cell-cycle events in prokaryotic and eukaryotic organisms**

Event	Prokaryotes	Eukaryotes
Growth in mass	B period[1]	Gap1 or G1 phase[2]
DNA synthesis	C period	DNA synthesis or S phase
Growth in mass	–	G2 phase
Separation of daughter chromosomes	D period	Mitosis or M phase[3]
Separation of cells	D period	Cytokinesis

1) At rapid growth, there is no separate B period, but replication cycles overlap.
2) There is a control point in the G1 phase, START, at which cells are committed to proceed into the cell cycle.
3) The M phase is often regarded as completed when the daughter nuclei have separated

22.1.1 Methods to Study the Timing and Duration of Cell-cycle Events

In an exponentially growing population, the cells are in different stages of the cell cycle; this gives a characteristic age distribution of the cells with twice as many newborn as dividing cells. In cell-cycle studies, it is sometimes advantageous if all cells are in the same stage of the cell cycle, i.e., the population is synchronized.

The timing and length of the various key processes during the cell cycle can be studied in exponentially growing populations with different methods: 1) experiments with the **membrane-elution technique**, 2) analysis of cells sorted according to size, and 3) flow cytometry.

C. E. Helmstetter and S. Cooper introduced the membrane-elution technique in cell-cycle studies. The principle of the experiment is shown in Figure 22.**2**. Exponentially growing populations of bacteria are collected on a nitrocellulose filter, and the newborn cells formed from the cells growing on the filter are collected. Hence, analysis of the newborn cells formed at different times gives information about the sequence of events during the cell cycle. The method was used to measure the time it takes to replicate the entire chromosome (the C period) and to finish cell division after completion of replication (the D period). It was also used to determine the length of the B period in slow-growing populations. By analysis of the label present in plasmid DNA, the method has been extended to determine at which times in the cell cycle plasmid replication occurs.

Bacteria grow during the cell cycle and can be sorted according to size (age) by centrifugation, for example, in sucrose gradients. Smaller cells migrate less rapidly than larger cells, and the collection of fractions after gradient centrifugation gives subpopulations of cells of different age. Exponentially growing populations can be pulse-labeled (Fig. 22.**2b**) and then separated by size for analysis, for example, of the degree of DN replication in the various fractions. This allows discrimination between periods of the cell cycle withou DNA replication and those during which replicatio occurs and the determination of the number of replica tion forks in operation.

Flow cytometry is a third powerful tool for study ing the cell cycle. It can be used to determine the cell size and DNA-content distributions of populations o bacteria. Figure 22.**3** describes the principle of th method and shows results obtained with *E. coli* grow exponentially in two different media. At a slow growt rate (Fig. 22.**3a**), a clear peak with one chromosome pe cell is obtained, which corresponds to cells in the period. At rapid growth (Fig. 22.**3b**), there are no peak with integer numbers of chromosomes because o overlapping replication cycles (see Fig. 22.**4**), but th results can be compared with distributions obtained b computer simulations using different values for th length of the C and D periods. In this way, the lengths o the B, C, and D periods have been calculated for a wid range of growth rates.

The length of the C period can also be calculate from a determination of the rate of movement of th replisome (i.e., the speed of the replication machinery approximately 800 nucleotides per second at 37°C for *E coli*) and the size of the chromosome (4.8 Mbp for *E coli*).

The methods described above are used to analyz exponentially growing populations. The cell cycle ma also be studied by using **synchronized populations** o cells, i.e., populations in which all cells are in the sam stage of the cell cycle. This can be achieved by collectin the newborn cells at 0°C with the membrane-elutio technique (Fig. 22.**2**) or by isolating the smallest cell from sucrose gradients and then initiating cell growt by the addition of prewarmed, fresh, growth medium. I this way, populations can be obtained in which all cell go through the cell cycle synchronously and divid simultaneously. Populations can also be synchronize

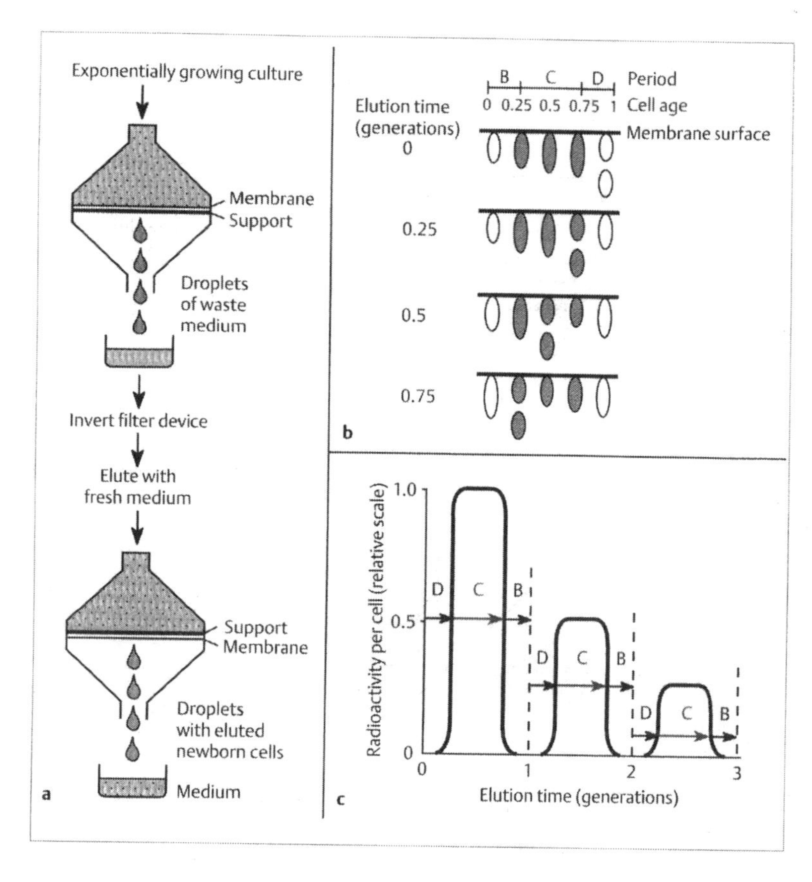

Fig. 22.**2a–c The membrane-elution technique.**
a An exponentially growing population of bacteria is rapidly chilled and then passed through a nitrocellulose filter. When a sufficient number of cells have been collected on the filter, the filter device is inverted, and prewarmed fresh culture medium is passed through the filter. The bacteria remain attached to the filter and grow on its surface. When a cell divides, one of the daughter cells (newborn cell) can be eluted and either collected and chilled as starting material for a synchronous culture or collected over time in fractions for analysis.
b The exponential population consists of cells of different ages. Newborn cells are of age 0, and cells that are finishing cell division are of age 1. These are represented on an enlarged cross section of the filter to which the cells are attached; they are shown in the order of increasing age, with the newborn cell on the left. The daughter cell formed from the oldest cell (right) is eluted first; with increasing elution time, daughter cells are released by cells that had been younger (going from right to left, in the order of decreasing age) at the beginning of the elution process. The exponential population in the figure was pulse-labeled with [^{3}H]-thymidine to mark the DNA that was being replicated during the pulse; these cells are shaded in red in the figure. The first daughter cells are from cells that were in the D period when collected on the membrane; they are unlabeled. Later, newborn cells from the C period or the B period emerge; they are labeled or unlabeled, respectively.
c The radioactivity is measured in fractions collected over the elution period. The graph shows the cyclic variations in the relative labeling of the newborn cells and allows an estimate of the length of each period (B, C, or D)

by amino acid starvation. However, this exposes the populations to stress, and the physiology of the cells may be altered.

22.1.2 Timing and Duration of Cell-Cycle Events Are Regulated by and Dependent on Growth Conditions

The methods, particularly the membrane-elution technique, presented in the previous section have been used to study the timing of key cell-cycle events. Initiation of chromosome replication was found to take place at a given time during the cell cycle for an exponentially growing population. However, this time varies with the growth rate (Fig. 22.**4**). The lengths of the C and D periods were found to be nearly the same at moderate and high growth rates (C and D are approximately 40 min and 20 min, respectively, for *E. coli* at 37°C). Hence, the length of the B period varies dramatically with the growth rate. The correlation between the cell-cycle events and the generation time is shown in Figure

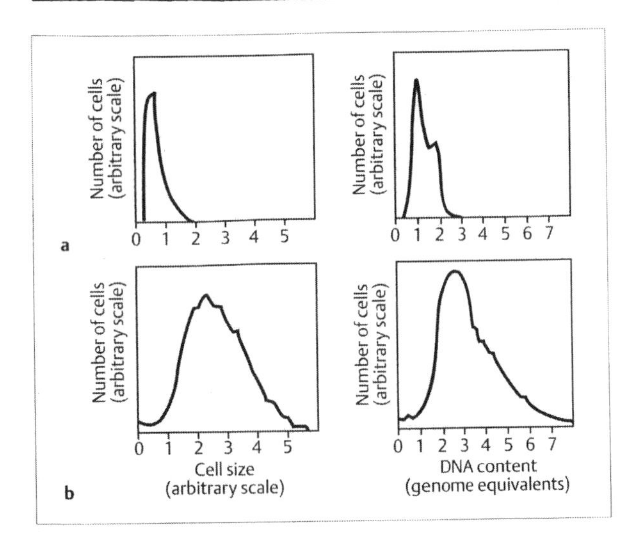

Fig. 22.**3a–b Flow cytometry.** A suspension of cells is passed through a measuring optical device. When light is passed through the suspension, particles scatter light, and the amount of light scattered is (approximately) proportional to the size of the particles. Similarly, if the DNA (or any other cell component) is allowed to react with a substance that fluoresces when illuminated by light of the appropriate wave length, each particle will give a fluorescence signal proportional to its content of DNA. The flow cytometer is able to measure the light-scattering and the fluorescence signal of each of a large number of particles that pass through the measuring device. The figure shows results obtained with *E. coli* at two different growth rates: **a** populations grown in minimal-acetate medium (slow growth) and **b** in rich medium (rapid growth). The results are presented as two-dimensional plots showing the cell-size and DNA-content distributions of the populations

22.**4**. When the generation time (τ) decreases, the length of the B period becomes shorter and shorter; when τ equals 60 min, the B period is zero, and at lower τ values, the B period becomes negative. Hence, initiation of replication takes place in the mother cells at $\tau = 60$ min and in the grandmother cells at $\tau = 30$ min. Consequently, there may be two or three overlapping DNA replication cycles, which progress simultaneously in the same cell. This appears to be typical of bacteria and has not been reported for eukaryotic cells.

The length of the C and the D periods of the cell cycle are essentially independent of the growth rate, whereas the length of the B period increases with increasing generation time.

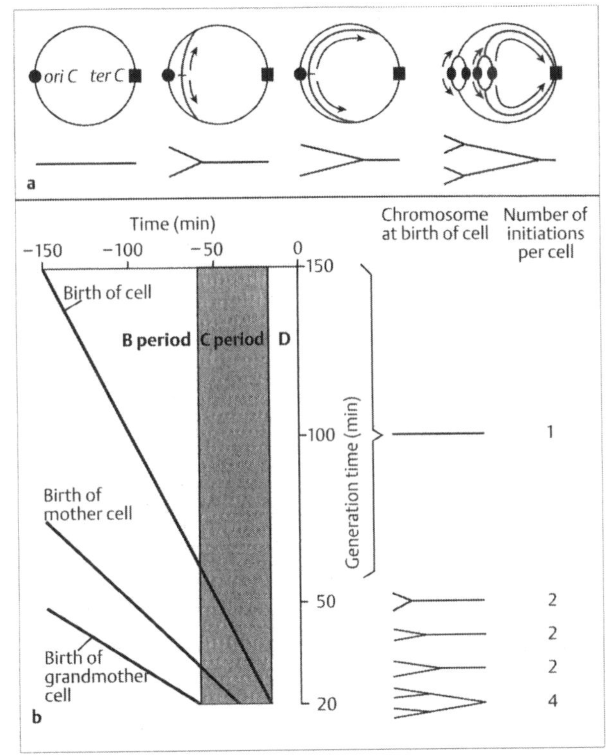

Fig. 22.**4a–b Overlapping replication cycles at rapid growth**
a Replication of the circular bacterial chromosome starts at a unique site, *oriC*, and proceeds bidirectionally to the terminus *terC*. The picture shows the chromosome at four different stages of replication. The circles viewed from the side appear as lines without or with branches.
b The generation time of bacteria can be varied by using different growth media. The C periods and D periods of *E. coli* are essentially independent of the growth rate and are approximately 40 min and 20 min, respectively, at 37°C. Time "−100" means 100 min until cell division. At time 0, cell division takes place. The figure shows the time of initiation of the cell cycle at different growth rates. At generation time beyond the C+D periods (60 min), there is a period without DNA replication (B period) in the young cells. The cells are born with a non-replicating chromosome. At generation time between 30 and 60 min, DNA replication starts in the mother cell, and the cells are born with a partly replicated chromosome; hence, two origins are present in these cells, and two initiations occur in each cell during one cell cycle. Finally, at generation times between 20 and 30 min, DNA replication starts in the grandmother cell, and the cells are born with chromosomes in which there are two consecutive replication cycles, and initiation of replication takes place at four origins during each cell cycle

22.2 Strategies of DNA Replication for Chromosomes and Plasmids

In most prokaryotes, the essential and species-specific genetic material forms one **c**ovalently-**c**losed **c**ircular (ccc) and **d**ouble-**s**tranded (ds) DNA molecule, the **chromosome**, which is approximately 5 Mbp in well-studied bacteria, such as *E. coli* and *Bacillus subtilis*, and may be significantly smaller (approximately 1 Mbp) in some bacteria (e.g., *Mycoplasma*). Other bacteria (e.g., *Streptomyces* species) seem to have one or several (up to seven chromosomes in the Rhizobiaceae) **linear** chromosome(s) (see Chapter 14.2).

It is the rule rather than the exception that bacteria contain additional DNA molecules, plasmids, which confer useful but nonessential properties to the host bacteria. Plasmids range in size from 1.5 kbp to more than 1 Mbp. It has been suggested that a megaplasmid can be distinguished from a chromosome by the presence of housekeeping genes, found only on the latter. Based on this criterion, bacteria with several chromosomes exist. Plasmids are discussed in Chapter 14.3.

At rapid growth, there are two or three overlapping replication cycles, which proceed simultaneously in the cells.

22.2.1 Methods for Studying Chromosome and Plasmid Replication

Meselson and Stahl's density-shift experiments initially were used to demonstrate that DNA replication is semiconservative (see Chapter 14.12). However, these experiments can also measure the time interval between consecutive replication events. The method is described in Figure 22.**5**. In a population that is growing exponentially in "heavy" medium, the DNA is radioactively labeled for a few minutes, and the time required for a second replication of the labeled DNA is determined (Fig. 22.**5a,b**). If DNA replication is cell-cycle specific, DNA of hybrid density would not appear until exactly one doubling time in cell mass (i.e., one generation time) after the density shift (Fig. 22.**5c**). This is the result observed for wild-type *E. coli* strains. Thus, each chromosomal origin initiates only once during each cell generation. If replication is extended in time, DNA of hybrid density appears much earlier (Fig. 22.**5c**).

The density-shift experiments have also been used to analyze the replication pattern of plasmids and minichromosomes (i.e., artificial plasmids that replicate from the same origin, *oriC*, as the chromosome. They are

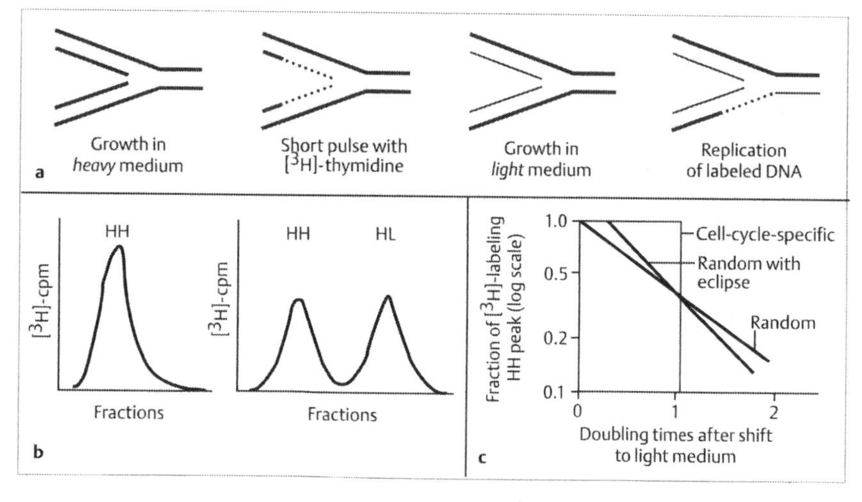

Fig. 22.**5a–c** **Meselson-Stahl density-shift experiments.**
a The bacteria are grown exponentially in "heavy" medium (e.g., containing [^{13}C]-glucose + [^{15}N]H_4Cl). A short pulse of [^{3}H]-thymidine is given immediately and is followed by a shift to growth in "light" (e.g., [^{12}C]-glucose + [^{14}N]H_4Cl) medium for two generation times or more. The scheme shows the pattern of heavy and light DNA in the replication fork at different times.
b Samples are taken at intervals and DNA is separated in CsCl gradients according to different densities. The radioactivity in the various fractions is measured. The graph shows the pattern obtained immediately after the [^{3}H] pulse (left) and at a time when approximately half of the [^{3}H]-labeled DNA has replicated a second time (right).
c The graph shows the results when replication is cell-cycle-specific, random, or random with an eclipse period between consecutive replication events. Symbols: ▬, heavy (H) DNA strand; —, light (L) DNA strand; ·····, pulse-labeled, heavy DNA strand; HH, both DNA strands heavy; HL, one heavy and one light DNA strand

useful models in studies of chromosome replication; see Chapter 14.1.3). The experiment is performed as described in the preceding paragraph but the content of radioactively labeled plasmid in the fractions is measured by hybridization. Plasmids have been found to replicate randomly with respect to both the choice of copies to be replicated and the timing of replication during the cell cycle: that is labeled plasmid DNA of intermediate density appears early after the shift, and the amount of this DNA decays exponentially (Fig. 22.**5c**). However, there is an eclipse period during which a newly replicated plasmid cannot participate in a second replication event (Fig. 22.**5c**). Replication of minichromosomes, on the other hand, follows the same pattern as that of the normal chromosomes.

It should be pointed out that the density-shift experiment determines the (inter-replication) time between two replication cycles of a replicon but does not give information about when replication is initiated during the cell cycle. Furthermore, if more than one initiation occurs in each cell during one cell generation, an exponential curve indicates random choice of copies for replication; this is the result observed for plasmids.

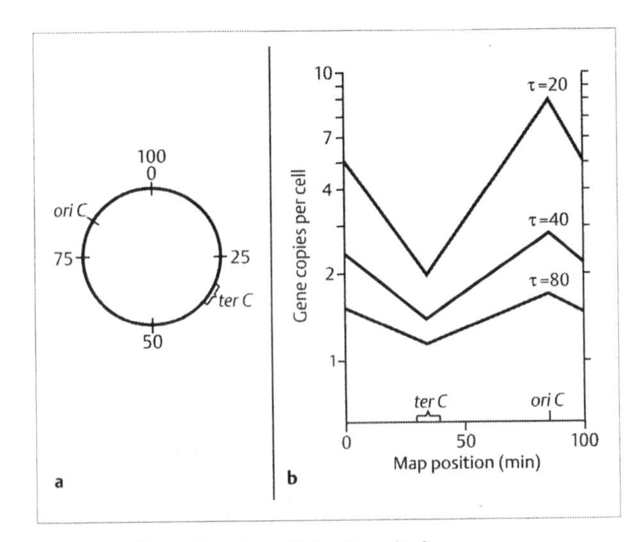

Fig. 22.**6a–b Replication of the *E. coli* chromosome.**
a The genetic map of *E. coli* showing the location of *oriC* and *terC*. The figures along the circle are the map coordinates (given in minutes).
b Gradient of gene dosages in exponentially growing populations. The lengths of the C and D periods are 40 min and 20 min, respectively. Curves are given for three generation times (τ) as indicated in the graph

22.2.2 Chromosome Replication Consists of Three Processes

These processes are (1) initiation, (2) elongation, and (3) termination.

(1) **Initiation.** Bacterial circular chromosomes replicate bidirectionally from a unique origin (*oriC*); hence, the two replisomes meet in the terminus (*terC*) region (Fig. 22.**6a**). The biochemistry of these processes is discussed in Chapter 14.1.2. In growing populations, genes close to the origin of replication are present in a higher number of copies than genes closer to the terminus. The average number of copies (F) of a gene per cell in an exponentially-growing population is

$$F = 2^{[C(1-x)+D]/\tau},$$

where C and D are the respective lengths of the C and D periods, τ is the generation time, and x is the relative distance of the gene from *oriC*; the x value is 0 for *oriC* and 1 for *terC*. This gives a gradient of gene dosages from *oriC* to *terC* (Fig. 22.**6b**). The equation also provides a way to determine the length of the C period by using hybridization with DNA probes directed against *oriC* and *terC* or genes in the close vicinity of these regions:

$$F_{oriC}/F_{terC} = 2^{C/\tau}$$

Hence, this ratio is independent of the length of the D period. Multiple initiations of replication during rapid growth are coordinated.

(2) **Elongation** of replication (the actual DNA synthesis) does not require de novo protein synthesis. Hence, inhibition of protein synthesis by the addition of chloramphenicol or of RNA synthesis by the addition of rifampicin leads to runout DNA synthesis: ongoing replications proceed to completion (Fig. 22.**7a**), the so-called runout replication. This process takes about 40 min in *E. coli* at 37°C, which is more clearly shown in the pulse-chase experiments in Figure 22.**7a**. The amount of DNA in the culture therefore increases during the runout synthesis; this relative increase is called the **increment** and can be calculated as follows:
The average number of genomes per cell (G) is

$$G = (\tau/C \cdot \ln 2) \cdot [2^{(C+D)/\tau} - 2^{D/\tau}]$$

The increment in the amount of DNA after a rifampicin runout is

$$\text{Increment} + 1 = F_{oriC}/G = C \cdot \ln 2/[\tau(1 - 2^{-C/\tau})]$$

Hence, the increment increases with increasing growth rate (Fig. 22.**7b**).

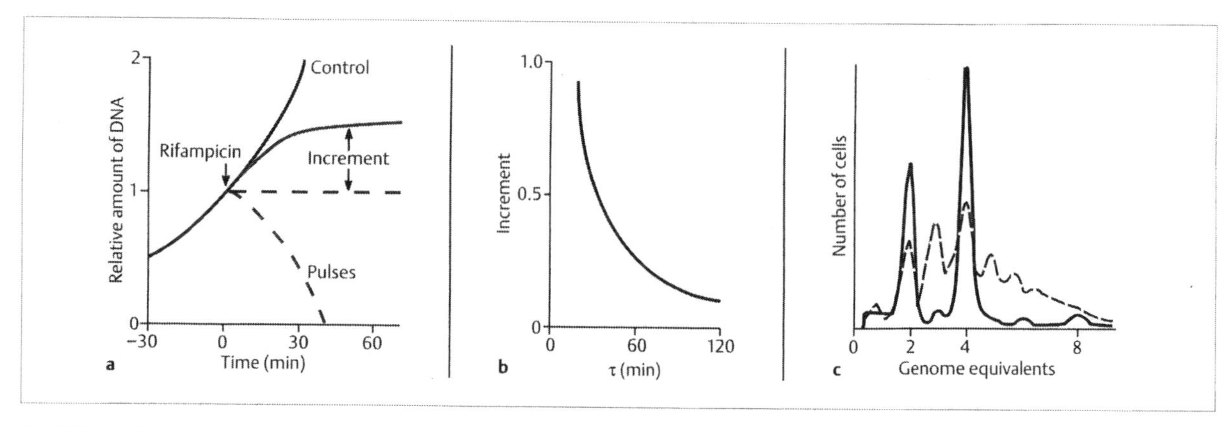

Fig. 22.7a–c **Runout replication.**
a At zero time, rifampicin is added to an exponentially growing population of bacteria. Cephalexin is also added in order to inhibit cell division during the runout experiments. These drugs do not prevent the ongoing replications from being completed. The graph shows total DNA (solid lines) and rate of DNA synthesis measured by pulse-labeling (broken line). The increment is the relative increase in DNA content during the runout experiment.
b The increment as a function of the generation time, τ.
c Flow cytograms. After runout of DNA replication, all cells contain integer numbers of genomes, and the number will equal the number of origins present in the cells at the time of addition of rifampicin. If all copies of *oriC* in each cell initiate simultaneously, the runout experiment gives cells with one, two, four, or eight genome equivalents (the synchronous phenotype, solid line). If, however, initiation of replication extends over a longer time period, rifampicin may block initiation of replication in some cells before all initiations have occurred. This results in cells with three, five, six, or seven genome equivalents (the asynchronous phenotype, broken line). In the absence of rifampicin, flow cytograms as shown in Figure 22.**3** are obtained

At the end of the runout experiment, the cells contain only complete chromosomes, and the number of those is the same as the number of origins present in the cells at the time of addition of rifampicin. This is demonstrated in the flow cytogram shown in Fig. 22.**7c**; the result of the control without rifampicin treatment is shown in Fig. 22.**3**.

Flow cytometry can also be used to determine the coordination of multiple origins in the same cell. If all copies of *oriC* in each cell initiate simultaneously, the runout experiment will give cells with one, two, four, or eight genome equivalents (Fig. 22.**7c**); this is called the synchronous phenotype. If, however, the initiation of replication is extended in time, rifampicin may block the initiation of replication in some cells before all initiations have occurred. In this case, also cells with three, five, six, or seven genome equivalents appear in the flow cytograms – the asynchronous phenotype (Fig. 22.**7c**). The longer the time over which replication is extended, the more abundant these aberrant numbers of genome equivalents will be. It should be noted, however, that such a result would also be obtained if initiation of replication is totally coordinated, but if fewer than all origins initiate in some cells. Likewise, at least in principle, four or eight genome equivalents could be produced by a mechanism that chooses origins at random for initiation, but always ensures a precise doubling of the number of origins during some short period during the cell cycle.

In *E. coli*, the A in GATC sequences is methylated by the Dam (**D**NA **a**denine **m**ethylase) enzyme. After replication, one of the strands will be nonmethylated (i.e., the DNA is hemimethylated). There are twelve GATC sites in *oriC* (see Chapter 14.1.3), and hemimethylated *oriC* DNA cannot participate in a new initiation event. This seems to be part of a mechanism that prohibits consecutive initiations at the same origin. Mutations that lead to the formation of an inactive Dam enzyme cause loss of coordination of multiple initiations of replication, i.e., lead to the asynchronous phenotype (Fig. 22.**7c**; see also Chapter 14.1). Asynchronous initiation of replication can also be obtained by replacing *oriC* with a plasmid replicon (***intR1*** strains; integrative suppression). Asynchronous initiation is not lethal for the cells but leads to a slight decrease in the exponential growth rate.

(3) **Termination.** The terminus (*terC*) region is flanked by *ter* sequences that block the progression of replisomes; the block requires that the Tus (**T**erminus **u**tilizing **s**ubstance) protein is bound to the *ter* sequences. The flanking sequences have the oppo-

site polarity. Therefore, the two replisomes meet in the *terC* region. The biological significance of the *terC* region is unknown, since it can be deleted. However, deletion of only one of the flanking regions is deleterious to the cells, presumably because the halting of a replisome at a *ter* sequence is difficult to overcome. Inactivation of the *tus* gene relieves some of the negative effects of unidirectional replication. The *terC* region might be important for maintaining an optimal gene dosage, or it may be required for proper segregation of the completed chromosomes. The *terC* region is a hot spot for recombination.

22.2.3 Plasmid Replication Is Autonomous

Plasmids are extrachromosomal genetic entities. They are autonomous and control their own replication (see Chapter 14.3).

Plasmid replication is independent of specific events in chromosome replication and of stages in the cell cycle – plasmids replicate randomly throughout the cell cycle. Hence, in a plasmid population, some plasmids do not replicate at all, some replicate once and some replicate two or more times during one cell generation. This has been demonstrated by measurement of DNA replication in cells separated by size (=age), by Meselson–Stahl density-shift experiments and by experiments using the membrane-elution technique.

Randomness in replication also means that there is random selection for replication among the plasmid copies present in the same cell. If a cell contains two differently marked derivatives of the same plasmid replicon, the randomization causes distortions in the proportion between the number of copies of the two derivatives, distortions that cannot be corrected later. This is one cause of plasmid incompatibility (see Chapter 14.3); eventually, pure lines appear that contain only one of the two plasmid derivatives.

22.3 The Nucleoid of Prokaryotes Is the Equivalent of the Nucleus of Eukaryotes

The **nucleoid** (Figs. 2.**17c** and 14.**13**) consists of the chromosome and all proteinaceous structures necessary for transcription, replication, and nucleoid processing. Chromosome replication and separation of the daughter chromosomes are coordinated but not strictly correlated with cell division. The process that separates the daughter chromosomes/nucleoids and moves them to the center of what will become the two daughter cells is called **partition**.

The process that separates the daugther chromosomes/nucleoids is called partition.

22.3.1 Nucleoid Processing Follows the Chromosome Replication

The bacterial chromosome, in some way linked to the cytoplasmic membrane, can be isolated from bacterial cells as a large complex of DNA–RNA–membrane, the so-called **folded chromosome**, which sediments rapidly (>2 000 S) in sucrose-gradient centrifugation. In most microscopic studies of the nucleoids, the cells are treated with chloramphenicol, which detaches the DNA from the membrane, and fixed before staining of the DNA. This leads to condensation of the DNA and, as a result, the nucleoid appears to occupy only a relatively small portion of the cell. However, if the DNA is stained without chloramphenicol treatment and without fixation of the cells ("vital" staining), it is evident that the chromosome occupies a large portion of the cell volume. This conclusion is further supported by electron microscopic studies. However, DNA replication and separation include steps of condensation and of relaxation of the chromosomes.

After completion of chromosome replication, the two daughter chromosomes are often entangled and interlocked. Hence, they have to be decatenated before they can be physically separated. At a later stage, the daughter chromosomes are moved to the center of what will become the two daughter cells. Conceptually, this corresponds to mitosis in eukaryotes. Recent data suggest that partition is a gradual rather than an abrupt process.

A class of mutants of the Par$^-$ phenotype was initially believed to be affected in the partitioning of the nucleoids. These mutants are characterized by aberrant nucleoid morphology and difficulty in separating the daughter nucleoids. Their cell division behavior resembles that of Min mutants since they go through polar divisions that yield cells without DNA (see below). Later they were found to be mutated in the genes for DNA gyrase and other topoisomerases and to be affected in the progression of DNA replication. Hence, these mutants did not really reveal the functioning of the partition process.

Mutants that seem to have lost the partition function have been isolated. In these mutants, a significant fraction (up to 15%) of the cells do not receive any chromosomal DNA at cell division. The mutated genes are called *mukA* and *mukB*. The MukA protein is identical to a previously described outer-membrane protein, TolC, whereas the MukB protein is a very large (molecular weight approximately 180 000) new protein. The latter protein contains several domains: a DNA-binding domain, an ATPase domain, and a region that resembles force-generating proteins. It might be part of a system analogous to the spindle apparatus in eukaryotes. However, a centromer analogue has not yet been found in bacteria.

Study of the partition process is difficult because bacteria are small and morphological studies do not reveal much detail.

After replication of the chromosomal DNA, the two daughter chromosomes are separated and moved to the center of the prospective daughter cells.

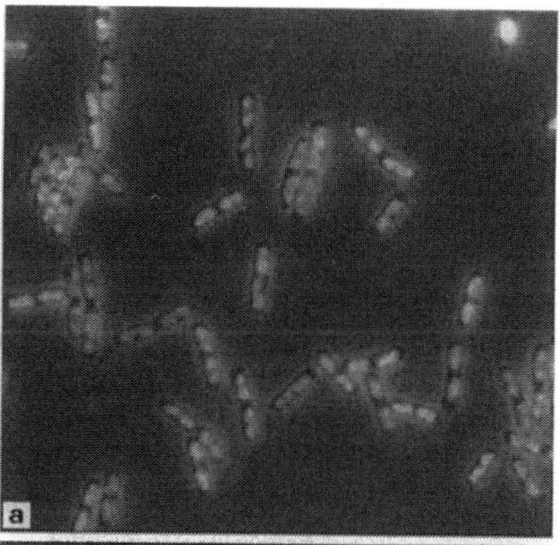

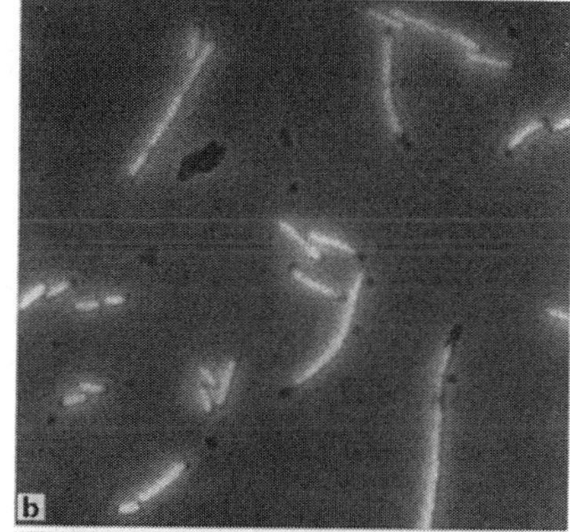

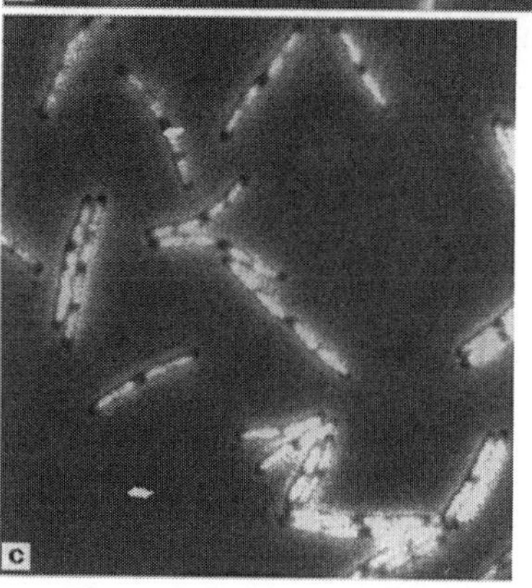

Fig. 22.**8a–c** **Three types of cell division.** The figure shows phase contrast–fluorescence micrographs of *E. coli* cells stained with DAPI (4′,6-diamidino-2-phenylindole) for visualization of the DNA.
a Normal wild-type cells with separated nucleoids in dividing cells.
b Cells of a *minB* mutant with nucleoids of aberrant morphology and divisions that give rise to DNA-less minicells.
c Wild-type cells after incubation in the presence of cephalexin, which inhibits cell division. The cells have formed filaments, but the nucleoids are of normal morphology and clearly separated

22.3.2 Cell Division Is the Final Stage in the Vegetative Cell Cycle

Cell division and separation occur during the last part of the D period (Fig. 22.**1**) and involve a directional change in the growth of the peptidoglycan layer, in which the cell envelope (cell wall and cytoplasmic membrane) invaginates in the middle of the cell (see Figs. 2.**17a**). This requires activation of one of the penicillin-binding proteins, **Pbp3**, which is involved in the synthesis of the peptidoglycan of the cross walls. A specific structure has been found in the middle of the cell, the **periseptal annulus**, which might define a compartment/organelle embracing the region where invagination of the cell envelope takes place. A complex cross-wall-forming and cell-separating structure has been described in *Staphylococcus aureus*.

As in chromosome replication, there is a time after which cell division proceeds even in the absence of protein synthesis. This time coincides approximately with the termination of chromosome replication.

A large number of mutants have been isolated in which cell division is affected. Since cell division is an essential process, most genes involved in this process have been discovered by analyzing conditional lethal mutants, in particular those that are viable at 30°C but unable to form colonies at 42°C. Several gene defects disturb cell division to such a degree that the morphogenesis and shape of the bacteria are affected, for example, the normally rod-shaped cells of *E. coli* may become round or irregular in shape (*envB* mutants). The phenotypes of the mutants have made it possible to identify at least two steps, septum formation and cell separation (see Fig. 22.**1**).

Many mutants belong to the ***fts*** (filamentation temperature-sensitive) group. In these conditional mutants, cell division is inhibited at the nonpermissive temperature. One special class of mutants (*ftsA* and *sefA*) is able to initiate but not to complete septum formation; these mutants form long cells with invaginations.

Two genes (*envA* and *cha*) are specifically involved in cell separation. The mutant cells form long chains. In these mutants, the cytoplasms are completely separated but the rest of the envelope is not – the peptidoglycan layers of the daughter cells are covalently bound to each other. The mutants have a very low activity of *N*-acetylmuramyl-L-alanine amidase, the enzyme that cleaves bonds in the peptidoglycan layers. The mutants are very sensitive to the entry of many different drugs.

One of the proteins specifically involved in an early stage of cell division is ***FtsZ***. This protein forms a ring in the middle of the cell just before cell division starts. The *ftsZ* gene is transcribed in a cell-cycle-specific manner.

Cell division normally takes place in the middle of the cells. However, there are mutants (gene locus *minB*), in which this ordered location of cell division is disturbed (see Fig. 22.**8**); at a significant frequency, cel division instead takes place near the cell poles, whicl results in the formation of a spherical, DNA-les: **minicell** and an elongated cell containing all the DN/ from the mother cell. One model proposes that the thre(gene products of the *minB* operon, the MinC, MinD, an(MinE proteins, function as follows: MinC and MinD inhibit cell division in the center of the cells and at th(potential division sites that are postulated to be close t(the cell poles and that are remnants from the previou: cell division. MinC acts directly on FtsZ (see Chapte 22.4.2). The MinE protein prevents the other two Mi proteins from acting at the cell center; hence, in th(wild type, division only takes place in the middle o each cell. An alternative model for the action of th(*minB* operon (discussed in Chapter 22.4.3) implies tha the MinB system is involved in DNA or nucleoi(processing and that the polar divisions occur becaus(DNA separation is delayed in the absence of MinB. Th(DNA occupies the cell center and thereby prohibits cel division at the center of the cell.

Cell division is initiated by a ring-like invagination of the cytoplasmic membrane in the middle of the cell. A large number of proteins participate in septum formation and in separation of the peptidoglycan layers that belong to the daughter cells. FtsZ is one protein involved in an early stage of cell division.

22.4 The Cell Cycle Consists of Controlled and Coordinated Processes

The three main processes in the cell cycle, (1) chromosome replication, (2) nucleoid processing, and (3) cell division, have to be carefully controlled to match the increase in cell mass and carefully coordinated with each other. Cells lacking DNA are rarely found in wild-type populations.

22.4.1 Several Models for Control Have Been Proposed

Present models related to the control of cell division are in general fairly speculative, and their biochemical basis is still unclear. Some models attempt to link the cell cycle to physical parameters, such as cell mass/size, cell length, and cell turgor, rather than to chemical parameters. More recently, Ca^{2+} and ppGpp have been proposed to play an important role in regulating th(bacterial cell cycle.

In 1963, F. Jacob, S. Brenner, and F. Cuzin propose(the classical **replicon model** (Fig. 22.**9a**). Presumabl for the first time, this model links the bacterial cell cycl(to a (putative) specific structure in the cytoplasmi(membrane. The model also couples replication an(partition. The model suggests that initiation occur when an initiator protein interacts with a fixed activato site at the DNA (equivalent to *oriC*) on each replicon. Th(replicating DNA is assumed to be linked to a structure i the membrane. At initiation of replication, this structur(is duplicated, and each daughter chromosome is boun(to one of the newly formed structures. These structure move apart because membrane material is synthesize(between them. Finally, this leads to a total separation o the daughter chromosomes. In support of this model

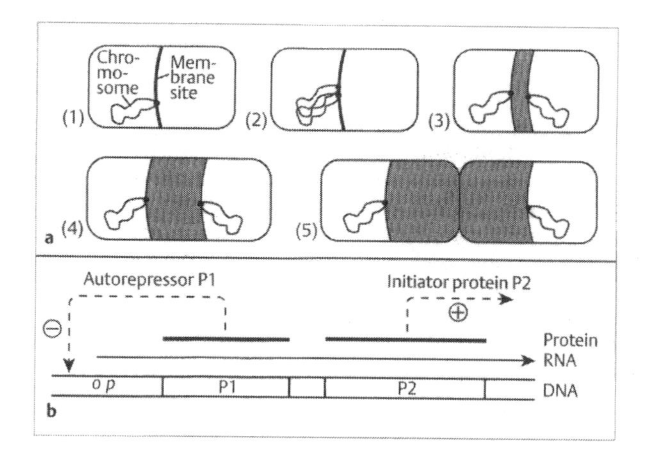

Fig. 22.9 **Replication models.**
a The replicon model as formulated by F. Jacob, S. Brenner, and F. Cuzin in 1963: (1) The bacterial chromosome (not to scale) is assumed to be attached by its origin of replication (*oriC*) to a specific site on the cytoplasmic membrane. (2) At a particular time during the cell cycle, a signal initiates replication. The two daughter chromosomes are attached side by side. (3) The bacterial membrane is assumed to grow between the daughter chromosomes; the newly synthesized membrane is shown as the shaded regions. (4) Membrane growth moves the daughter chromosomes apart. (5) Cell division is initiated in the middle, between the cells. The daughter chromosomes are now located in the middle of the daughter cells [after 1].
b The autorepressor model as formulated by L. Sompayrac and O. Maaløe in 1973: There are two genes (P1 and P2) in the regulatory circuit; they are parts of the same operon. The gene product P1 acts as a transcriptional repressor (⊖) of the operon (*o*, operator; *p*, promoter). The other protein (P2) is an initiator protein that is involved in initiation of replication (⊕) [after 2]

regions close to *oriC* have been found to bind specifically to outer-membrane fractions and to a specific protein. Unfortunately, these latter studies have not been completed. The replicon model does not explain how the duplication of the membrane structure that binds DNA is controlled nor does it address the question of the properties of the control circuitry. The model may be more important in explaining partition rather than control of initiation of replication.

R. H. Pritchard proposed the so-called **inhibitor-dilution model**. This is a negative-control model and consists of three components: (1) the origin of replication, (2) an initiator protein, and (3) a repressor that acts on the origin. There is a burst of synthesis of the repressor when replication is initiated, which doubles the concentration of the repressor. By interacting either with the origin or the initiator protein, the repressor inhibits further initiations until the cell volume has doubled. Thereby initiation of replication becomes linked to the growth of the cells. Plasmid replication has been shown to be negatively controlled, but not by interaction between an inhibitor and the origin of replication; instead, the negative control acts on the synthesis of a rate-limiting replication protein or on the transition of a preprimer to a primer of replication (see Chapter 14.1). There is no indication of the existence of a repressor of initiation of replication from *oriC*.

L. Sompayrac and O. Maaløe proposed the so-called **autorepressor model** (Fig. 22.**9b**), which was devised to explain how a defined number of copies of an initiator molecule could be formed during one cell cycle. Synthesis of the DnaA protein, a protein involved at a very early step in initiation of chromosome replication (Chapter 14.1.3), is autoregulated. This protein has been suggested to control initiation of chromosome replication. It has been proposed that the concentration of free DnaA molecules fluctuates during the cell cycle because it binds to many DnaA boxes that are spread over the entire chromosome. Not until all these boxes are saturated does free DnaA appear (in a burst) and bind to *oriC*. However, it is also possible that there is no control in the same sense as plasmid replication is controlled (i.e., by measurement of replicon concentration; see also Chapter 14.3), but rather that bacterial chromosomes behave as eukaryotic origins and replicate exactly once per cell cycle, irrespective of the number of origins. This conclusion is based on studies of cells containing minichromosomes; even the presence of many minichromosome copies does not affect the timing of replication in *E. coli*. However, there is an incompatibility between minichromosomes and the chromosome in, for example, *B. subtilis*.

W. D. Donachie made the two important observations: (1) that initiation of replication of the *E. coli* chromosome is initiated at a constant ratio, cell mass/*oriC* (the **initiation mass**), and (2) that cell division is initiated at a constant cell length regardless of the growth rate. The effector molecules that link cell mass and cell length, respectively, to these processes are not known (however, see the discussion of the DnaA protein above). Recent experiments, however, have demonstrated that there is some variation in the initiation mass. The fact that the presence of *oriC* plasmids does not affect the timing of chromosome replication in *Escherichia coli* may also reduce the significance of the concept of a critical mass.

Although several models for control of the bacterial cell cycle have been proposed, the process cannot yet be explained on a molecular basis.

22.4.2 There Are Presumably Several Control Points for the Steps of the Cell Cycle

The synthesis of the macromolecules DNA, RNA, and protein is controlled primarily at the level of initiation. Once started, the processes run to completion. The same is true for cell division. Initiation of chromosome replication and of cell division requires protein synthesis. Both processes proceed to completion even if protein synthesis has been inhibited after their initiation (see Chapter 22.2.2).

The cell cycle is not a true cycle in the sense that all later events require the completion of all previous ones and that a new cycle cannot be initiated before completion of all events in the previous cycle. Cell division can be inhibited (e.g., by cephalexin), but the cells continue to grow, the chromosome replicates, and the daughter nucleoids undergo partition, which leads to the formation of long filaments with two, four, eight, sixteen, or more clearly separated nucleoids (Fig. 22.**8c**). The same is the case with the fission yeast, *Schizosaccharomyces pombe*, and in many special situations involving eukaryotic cells. In eukaryotic cells, the M phase requires completion of the S phase, and vice versa. Since there are overlapping replication cycles in rapidly growing bacteria, completion of chromosome partition before chromosome replication is initiated is not required; within one cell cycle, partition and initiation of replication can take place in any order.

If the cell cycle were a true cycle, it might be sufficient to control only one event (e.g., initiation of replication); the other events could simply be triggered by replication. However, there are data that suggest that cell division has its own control. Whether nucleoid processing and partition are also separately controlled is not clear at present. In conclusion, it is possible that there are at least two (replication and division) or perhaps three parallel cell-cycle processes (Fig. 22.**10**). In eukaryotes, apparently two or three parallel processes are controlled at the same time period in the cell cycle, with one starting the S phase and the other leading (see Tab. 22.**1**) eventually to mitosis; in *Saccharomyces cerevisiae*, cell division (budding) also is initiated at the same time as the S and the M phases. The recent report that *ftsZ* expression reaches a maximum at the time of initiation of replication might also indicate that, in prokaryotic cells, replication and cell division are initiated at the same time in the cell cycle. FtsZ, a protein with some properties similar to the eukaryotic tubulins, performs an early step in cell division; this protein is abundant and found throughout the cell, but at the time of initiation of cell division, it forms a ring-like structure in the middle of the cell. In sporulating *Bacillus subtilis* cells, the ring is formed near both poles, but only the ring at which the forespore is being formed is retained. The assembly of the FtsZ ring may be supported by the FtsW protein. The assembly is assumed to be part of septum initiation. The protein may be involved not only in initiation of cell division but also in its control; overexpression of the *ftsZ* gene may result in cell division taking place in cells smaller than normal.

In eukaryotic cells, cyclic protein phosphorylation directs the initiation of the S and the M phases. No such cyclic processes (apart from the cyclic variation in *ftsZ* expression) have so far been found in bacteria. A distinctly different control could be based on structural elements. When the concentration and location of (several) key elements reaches a threshold, new structures may be created. This is the case for a special type of cell division, sporulation in *B. subtilis* (Chapter 25). The progression of the sporulation process involves a series of σ factors that consecutively turn on transcription of global regulons. Also, in *Caulobacter* (Chapter 24), each cell division results in two different daughter cells, a swarmer cell and a stalked cell. Hence, the search for a single control event may prove futile.

22.4.3 The Cell Cycle Is Under the Control of a Global Regulatory Network

The bacterial cell cycle consists of a number of processes that are carefully regulated to match the increase in mass (Fig. 22.**10**). Veto systems are used for their

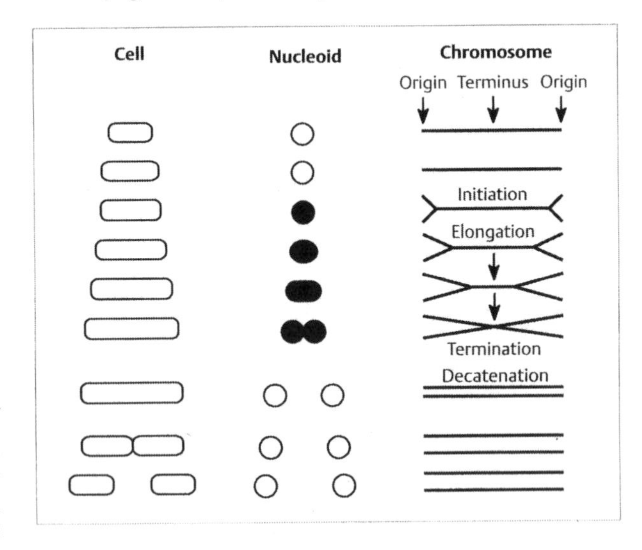

Fig. 22.**10** **The bacterial cell cycle can be viewed as involving three parallel processes.** Nonreplicating and replicating nucleoids are shown as filled or open structures, respectively. For clarity, slow-growing bacteria are shown; at higher growth rates, successive replication cycles may overlap

coordination. This can be studied by the use of conditional (e.g., temperature-sensitive, Ts) mutants and by the addition of toxic substances. The effects of such treatments will be discussed briefly.

Shifting a *dnaA*(Ts) mutant to a nonpermissive temperature results in runout of chromosome replication. Such cells can undergo division, which results in the formation of DNA-less cells of varying sizes. Inhibition of initiation of replication by rifampicin or chloramphenicol, for example, does not allow cell division because cell division requires protein synthesis at the time of termination of chromosome replication.

Shifting a *dna*(Ts)-mutant (in which elongation is affected) inhibits cell division and gives rise to filaments with one nucleoid.

Mutations [*gyrB*(Ts); DNA gyrase is affected] or treatments that cause the rate of DNA replication to slow down lead to disturbances in cell division. If the time required to complete replication is too long, the DNA becomes entangled, and the nucleoids do not separate well. This results in cell division at the poles and in the center of the cell, and DNA-less minicells are formed. This behavior resembles the phenotype of *minB* mutants: the nucleoids do not separate well in these mutants (Fig. 22.**8b**). This indicates that the MinB system is involved in some aspect of DNA or nucleoid processing (see also Chapter 22.3.2). C. L. Woldringh and coworkers have proposed the so-called **nucleoid occlusion model**, which states that cell division cannot occur where DNA is present. In fact, there is a high degree of similarity between the MinB protein and several proteins involved in partition or replication of plasmids, including the ATPase activity. **Over-replication** (i.e., an increased origin/mass ratio) of chromosomal DNA gives the same result: decreased rate of movement of the replisome, abnormal nucleoid morphology, and aberrant cell division. Hence, cell division does not occur in regions of the cell where there is DNA. **Under-replication** (i.e., a decreased origin/mass ratio) causes filamentation.

In yeast, *cut* mutations cause the cells to divide such that the genome is split between the daughter cells; there is one report of a similar class of mutants of *B. subtilis*. These mutants indicate that normally there is cross-talk between replication and cell division.

Lack of some proteins (e.g., FtsZ, MinC, MinD, and MinE) or their overproduction often causes difficulties in separation of the nucleoids, and, presumably as a consequence, cell division may become aberrant (omitted, misplaced, or the cells may even become branched). This suggests that the coordination of many processes and the maintenance of a proper balance between the concentrations of many proteins are essential for normal cell division.

Damage to DNA induces the so-called SOS response, which causes inhibition of cell division. This inhibition is mediated by the SulA(SfiA) protein, which interacts with FtsZ and thereby inhibits cell division. Under normal conditions, the expression of the *sulA* gene is inhibited by the LexA protein (see Chapter 15.7). This latter protein is cleaved (inactivated) by the RecA protein, which is induced by DNA damage. Inhibition of cell division by the SOS response is reversible because SulA is an unstable protein.

Blockage of DNA replication inhibits cell division, also in the absence of the SOS response. One protein, FtsA, has been implicated in this control loop.

Apparently, there also is a link to stress responses because inactivation of the *dnaK* gene, which encodes one of the major heat-shock proteins, blocks cell division.

The prokaryotic **cell cycle** is a delicately controlled process that involves several different events. These are carefully controlled and coordinated to match the increase in cell mass. Veto systems prohibit cell division if chromosome replication and nucleoid partition have not been completed. Damage to DNA or inhibition of DNA synthesis causes loss of the ability of the cells to divide. Deficiency in separation of daughter chromosomes may cause cell division to take place near the poles rather than in the middle of the cell. Neither the molecular basis for control of the various events nor its coupling to an increase in cell mass are yet understood. Cell division can be omitted and/or inhibited without disturbing the progression of other cell-cycle events.

22.5 Concluding Remarks

Prokaryotic and eukaryotic cells show many differences, and these have been of interest for a very long time. Since there are similarities between these two cell types, it is worthwhile to compare the cell cycles of prokaryotic and eukaryotic cells. These similarities have directed the interest to finding analogous systems in both types of cells.

A perhaps fundamental difference between the prokaryotic and the eukaryotic cell cycles is that genes involved in control of the progress of the cycle have only been found in eukaryotic systems. All prokaryotic genes found so far are concerned with the execution of the key events rather than with their control. For example, the DnaA protein acts at the first step in initiation of chromosome replication and, according to some microbiologists, this protein is also involved in control of the process. This also suggests that it is not always easy to discriminate between control and execution of a process.

This chapter has essentially dealt with the vegetative cell cycle of *E. coli*. This cell cycle results in the formation of two identical daughter cells (Fig. 22.**1**). However, there are many other types of cell division that participate in differentiation processes. Cell division in *Caulobacter crescentus* gives two different daughter cells, a free-living swarm cell and a stalked cell that adheres to surfaces (Chapter 24.1). Cell division in *Bacillus* may under certain circumstances result in sporulation; in this process, one daughter cell becomes the prespore and the other becomes the mother cell (Chapter 25.1).

Further Reading

Cooper, S. (1991) Bacterial growth and division: biochemistry and regulation of prokaryotic and eukaryotic division cycles. New York London: Academic Press

Donachi, W. D. (1993) The cell cycle of *Escherichia coli*. Annu Rev Microbiol 47: 199–230

Donachi, W. D., and Robinson, A. C. (1987) Cell division: parameter values and the process. In: Neidhardt, C., Ingraham, J. L., Brooks Low, K., Magasanik, B., Schaechter, M., and Umbarger, H. E. (eds.) *Escherichia coli* and *Salmonella typhimurium*: Cellular and molecular biology, 1st edn. Washington, DC: ASM Press; 1578–1593

Helmstetter, C. E. (1996) Timing of synthetic activities in the cell cycle. In: Neidhardt, C., Curtiss III, R., Ingraham, J. L., Lin, E. C. C., Brooks Low, K., Magasanik, B., Reznikoff, W. S., Schaechter, M., and Umbarger, H. E. (eds.) *Escherichia coli* and *Salmonella*: Cellular and molecular biology, 2nd edn. Washington DC: ASM Press; 1627–1639

Hiraga, S. (1992) Chromosome and plasmid partitioning in *Escherichia coli*. Annu Rev Microbiol 61: 283–306

Hunt, A., and Murray, T. (1993) The bacterial cell cycle, Chapter 11. In: Hunt, A., and Murray, T. The cell cycle. San Francisco: Freeman

Ishihama, A., and Yoshikawa, H., eds. (1991) Control of cell growth and division. Tokyo: Japan Scientific Societies Press

Jacob, F., Brenner, S., and Cuzin, F. (1963) On the regulation of DNA replication in bacteria. Cold Spring Harbor Symp Quant Biol 28: 329–348

Lutkenhaus, J., and Mukerjee, A. (1996) Cell division. In: Neidhardt, C., Curtiss III, R., Ingraham, J. L., Lin, E. C. C., Brooks Low, K., Magasanik, B., Reznikoff, W. S., Schaechter, M., and Umbarger, H. E. (eds.) *Escherichia coli* and *Salmonella*: cellular and molecular biology, 2nd edn. Washington DC: ASM Press; 1615–1626

Messer, W., and Weigel, C. (1996) Initiation of chromosome replication. In: Neidhardt, C., Curtiss III, R., Ingraham, J. L., Lin, E. C. C., Brooks Low, K., Magasanik, B., Reznikoff, W. S., Schaechter, M., and Umbarger, H. E. (eds.) *Escherichia coli* and *Salmonella*: cellular and molecular biology, 2nd edn. Washington DC: ASM Press; 1579–1601

Nordström, K., and Austin, S. J. (1989) Cell-cycle-specific initiation of replication. Mol Microbiol 10: 457–463

Nordström, K., Bernander, R., and Dasgupta, S. (1991) The *Escherichia coli* cell cycle: one cycle or multiple independent processes that are co-ordinated? Mol Microbiol 5: 769–774

Schmid, M. B., and von Freiesleben, U. (1996) Nucleoid Segregation. In: Neidhardt, C., Curtiss III, R., Ingraham, J. L., Lin, E. C. C., Brooks Low, K., Magasanik, B., Reznikoff, W. S., Schaechter, M., and Umbarger, H. E. (eds.) *Escherichia coli* and *Salmonella*: cellular and molecular biology, 2nd edn. Washington DC: ASM Press; 1662–1671

Skarstad, K., and Boyce, E. (1994) The initiator protein DnaA: evolution, properties and function. Biochim Biophys Acta 1217: 111–130

Sompayrak, L., and Maaløe, O. (1973) Autorepressor model for control of DNA replication. Nature New Biol. 241: 133–135

Vinella, D., and D'Ari, R. (1995) Overview of controls in the *Escherichia coli* cell cycle. Bioessays 17: 527–536

Woldringh, C. L., Mulder, E., Huls, P. G., and Vischer, N. (1991) Toporegulation of bacterial division according to the nucleoid occlusion model. Res Microbiol 142: 309–320

Sources of Figures

1 Jacob, F., Brenner, S., and Cuzin, F. (1963) Cold Spring Harbor Symp Quant Biol 28: 329–348

2 Sompayrak, L., and Maaløe, O. (1973) Nature New Biol 241: 133–135

23 Assembly of Cellular Surface Structures

Membrane biogenesis, i.e., the assembly of a membrane from its components, is one of the fundamental processes in biology. However, the mechanism of this process is poorly understood. It is however clear that the components of the outer bacterial membrane assemble on a template provided by the peptidoglycan and by the preexisting array of membrane lipids and proteins. Thus, **bacterial cell-surface assembly is not a de novo formation** but rather an extension of a preexisting and flexible arrangement by a concerted insertion of components from within the cell. Today, it is generally assumed that the ports of exit for membrane components are specialized (probably transient) sites of association between the inner and the outer membranes. For turnover of membrane components, these points are distributed over the whole cell surface (membrane junctions); for lateral growth prior to cell division, these sites are located parallel to the division plane (periseptal annuli).

The formation of bacterial cell-wall components (i.e., peptidoglycan, lipopolysaccharides, teichoic acids, proteins, and lipoproteins) and extracellular structures (i.e., capsules, fimbriae, and flagella) begins in the cytoplasm with the biosynthesis of activated precursors, continues with their assembly and translocation across the cytoplasmic membrane, and terminates with their transport to the cell surface and insertion into the cell-surface structures. Since the energy for assembly cannot be provided by energy sources, such as ATP, outside the cytoplasmic membrane, the precursors to be assembled outside have to be in an activated state. This is especially the case for complex carbohydrates, such as peptidoglycan, lipopolysaccharides (LPSs), teichoic acids, and capsular polysaccharides. A characteristic feature of the last components mentioned is the participation of undecaprenol phosphate as a cofactor. In archaebacteria, undecaprenol phosphate is replaced by dolichol phosphate, which otherwise occurs only in eukaryotic cells.

In the following, the assembly of the most important cell-surface components of Gram-positive and Gram-negative eubacteria and of archaebacteria are described, and the realization of specific mechanisms is discussed.

23.1 Peptidoglycan, the Rigid Layer of Cell Walls, Is Assembled From Activated Precursors

Peptidoglycan, also known as the murein sacculus, is the rigid layer of bacterial cell walls (Figs. 2.**2**, 2.**11**, 2.**13**) that renders the walls physically stable. It is multilayered in Gram-positive bacteria (e.g., *Staphylococcus*), but it consists of only one layer in Gram-negative bacteria (e.g., *Escherichia coli*). The peptidoglycan network is made up of glycan strands consisting of *N*-acetyl-glucosamine (GlcNAc) and 3-lactyl-GlcNAc (muramic acid, MurNAc) in an alternating sequence of β-1,4 linkages. The glycan strands are cross-linked by peptide chains of differing complexity, which are attached amidically to the carboxyl group of the lactic acid residue of muramic acid; hence, this explains the name peptidoglycan.

23.1.1 Assembly of the Structural Unit

The biosynthesis of peptidoglycan starts in the cytoplasm with the formation of UDPMurNAc from UDPGlc-NAc and phospho*enol*pyruvate (PEP), followed by sequential addition of L-Ala (alanine), D-Glu (glutamic acid) a di-amino acid [e.g., *m*-diamino-pimelic acid or L-Lys (lysine)], and two D-Ala residues from their tRNA-activated forms. The diamino acid residue in position 3 of the peptide (R3) is always linked with an L-center (L-R3). The last two D-Ala residues are transferred together as D-alanyl-D-alanine, which is formed from L-Ala by the action of alanine racemase and the ATP-dependent D-Ala-D-Ala ligase (see Chapter 7.9 and Figs. 7.**31a-c**).

The resulting muramic acid–pentapeptide is transferred as the 1-phosphate form from UDP to undecaprenol phosphate (undec-P) in the Mg^{2+}-dependent, reversible exchange reaction:

UDP-MurNAc + undec-P ⇌	undec-P-P-MurNac + UMP
L-Ala	L-Ala
D-Glu	D-Glu
L-R3	L-R3
D-Ala	D-Ala
D-Ala	D-Ala

This isoenergetic exchange reaction binds the disaccharide-pentapeptide to the inner side of the cytoplasmic membrane. Transfer of GlcNAc from UDPGlcNAc to C4 of the muramic acid residue completes the formation of the building block of the peptidoglycan. This is shown schematically on the right of Figure 23.**1**.

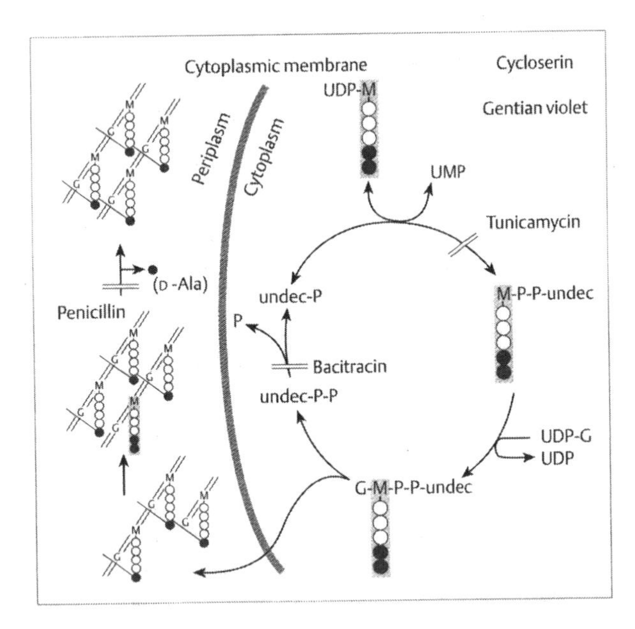

Fig. 23.**1** **Biosynthesis and assembly of peptidoglycan in *Escherichia coli*.** The synthesis of the building block is shown on the right; its incorporation into the peptidoglycan network is shown on the left. For details, see text. G, *N*-acetylglucosamine; M, muramic acid (3-*O*-lactyl-*N*-acetylglucosamine); undec, undecaprenol. The reactions occur as indicated in the cytoplasm or in the periplasm. Reactions inhibited by the antibiotics mentioned in the text are also indicated.

The disaccharide-pentapeptide is translocated across the cytoplasmic membrane and brought into position at the periplasmic side of the membrane in preparation for integration into the growing glycan strands.

23.1.2 Extension of Glycan Strands and Cross-linking

Integration of the undecaprenol diphosphate-bound disaccharide-pentapeptide building blocks into the peptidoglycan occurs by transfer of the disaccharide peptide from undec-P-P to the GlcNAc terminus of a free glycan strand in the peptidoglycan. Several peptidoglycan subunits can also be polymerized while bound to undecaprenol diphosphate and then transferred en bloc to the peptidoglycan strands (see Chapter 7.9.4)

Cross-linking of the glycan strands with the peptide substituents is achieved with the enzyme transamidase. In *E. coli*, this enzyme forms a direct cross-link from the subterminal D-Ala of one building block to the amino group of the L-R3 residue of an adjacent building block. The enzyme binds to the subterminal D-Ala unit with release of the terminal D-Ala moiety. The energy necessary for the formation of a cross-linking peptide bond is furnished by this concomitant cleavage of the terminal D-Ala–D-Ala linkage of the pentapeptide. Extension and cross-linking of peptidoglycan is shown schematically on the left of Figure 23.**1**. In Gram-positive bacteria, the free amino acid of the dibasic amino acid L-R3 (e.g., lysine in *Staphylococcus aureus*) is first substituted with several amino acids (e.g., glycine in *S. aureus*). Subsequent cross-linking between alanine of one strand and the amino-terminal glycine in the peptide side chain of an adjacent strand results in the looser peptidoglycan network of Gram-positive bacteria (as opposed to Gram-negative bacteria).

23.1.3 The Peptidoglycan Assembly Involves a Highly Coordinated and Regulated Teamwork Between Autolytic and Synthetic Enzymes

The synthesis of peptidoglycan occurs preferentially at the septum, the site where daughter cells separate during cell division. Peptidoglycan growth is a regulated process in which local disruptions of glycan strands and peptide substituents by autolytic enzymes are coordinated with extension of the loose ends thus formed. Peptidoglycan synthesis can be inhibited by a number of agents. **Penicillin** inactivates the enzyme transamidase by covalently binding to this protein (one of the penicillin-binding proteins, PBP), due to the partial steric identity of penicillin with the alanyl-alanine acceptor site on the peptidoglycan strands. Therefore, penicillin acts only on growing bacteria, which actively synthesize peptidoglycan. **Tunicamycin** inhibits the transfer of peptidyl–muramic acid from UDP to undecaprenol phosphate. **Bacitracin** inhibits the dephosphorylation of undecaprenol diphosphate by a Mg^{2+}-dependent complex formation with the undec-P-P phosphatase in the cytoplasmic membrane. This interrupts the recycling of the cofactor, undecaprenol phosphate, an inhibition also observed in other systems (lipopolysaccharides, capsules, and teichoic acids). Further inhibitors (and their target reactions) are **phosphonomycin** (formation of UDP–muramic acid), **gentian violet** (addition of L-Ala to UDP–muramic acid), and **cycloserine** (isomerization of L-Ala and formation of D-Ala–D-Ala).

The **assembly of peptidoglycan** involves the following steps: Undecaprenol diphosphomuramic acid, substituted with a pentapeptide, (undec-P-P-Mur-NAc–pentapeptide) is formed at the cytoplasmic side of the cytoplasmic membrane. Addition of GlcNAc results in the formation of the lipid-linked disaccharide-pentapeptide, undec-P-P-Mur–pentapeptide–GlcNAc, which is then translocated across the cytoplasmic membrane to the periplasm. The activated repeating unit is polymerized to give pentapeptide-substituted glycan strands, which are first integrated into preexisting peptidoglycan and then cross-linked with the peptide side chains with release of a terminal alanine. Since acceptor sites for new material are created by hydrolytic reactions on the peptidoglycan network, the extension of the net is a result of a well-regulated balance between lytic and synthetic processes (Fig. 23.**1**).

23.2 Lipopolysaccharides, Essential Components of the Gram-negative Cell Wall, Are Synthesized From Building Blocks at the Cytoplasmic Membrane

Lipopolysaccharides (LPSs) are characteristic components of Gram-negative bacteria (Fig. 2.**7**); they constitute the outer half of the outer membrane. These molecules exclude cell-damaging compounds, such as the bile salts of the intestine. LPSs are the major antigenic determinants of Gram-negative bacteria used for serological typing. They also serve as bacteriophage receptors or, together with outer membrane proteins, as complex bacteriophage receptors. Bacteriophage-receptor properties are the basis for bacteriophage typing of Gram-negative bacteria (especially well developed with species of *Salmonella*). The lipid part of LPS (lipid A, see Chapter 23.2.4) exerts a number of physiological effects in a host infected with Gram-negative bacteria. These effects are primarily due to the elicitation of regulatory factors, such as tumor necrosis factor (TNF), interleukins (IL-1, IL-6, IL-8), prostaglandins, and leukotrienes. For these reasons, lipid A is termed endotoxin (as opposed to the bacterial exotoxins, which are secreted bacterial proteins).

LPSs consist of three parts: lipid A, the core oligosaccharide, and the O-specific polysaccharide. Gram-negative bacteria may undergo a mutation that affects the formation of the O-specific polysaccharide. On agar, the wild-type bacteria have colonies with smooth, glossy surfaces and are therefore termed "S forms" (for smooth). The mutants with LPS lacking the O-specific polysaccharide grow in brittle colonies with a rough appearance and are therefore termed "R mutants". The corresponding LPSs are called S-LPSs and R-LPSs, respectively. It is noteworthy that R mutants are practically avirulent, even when derived from the virulent S forms.

The three regions of the LPS molecules are synthesized separately at the cytoplasmic membrane, then ligated, transported to the outer membrane, and integrated there. This is schematically shown in Figure 23.**2** (see Chapter 7.10.3).

In the outer membrane, the LPS molecules are absolutely restricted to the outer side of the outer membrane and are tightly associated with outer-membrane proteins. In many R mutants, the protein content of the outer membrane is lower than in S forms, and the corresponding space is filled with phospholipid. Thus, the outer membrane of R mutants contains areas with phospholipid bilayers, resulting in the greater sensitivity of these mutants to certain hydrophobic agents and also to bile salts.

23.2.1 Lipid A, the Anchor of LPS in the Outer Membrane

Lipid A synthesis starts from UDPGlcNAc, which first undergoes a 3-*O*-substitution and then an *N*-substitution with β-hydroxymyristic acid. The diacyl derivative

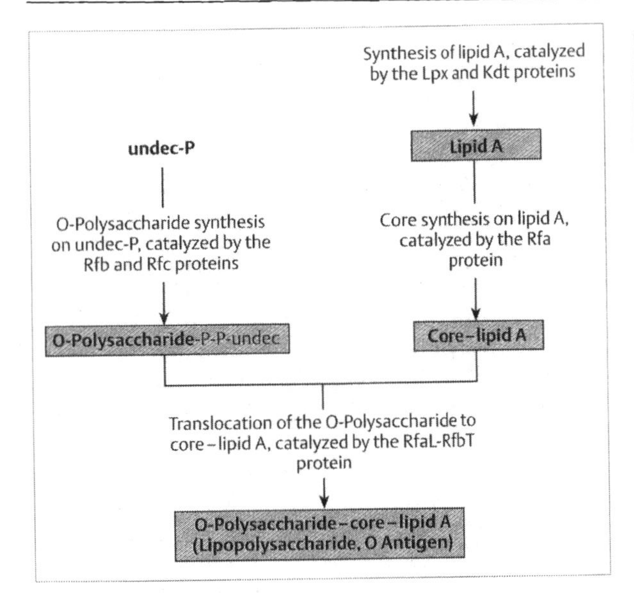

Fig. 23.2 **General pathway of lipopolysaccharide assembly.** The O-specific polysaccharide and the core–lipid A moieties of lipopolysaccharide (LPS) are synthesized separately and then joined together. The carrier molecule for the assembly of the O-polysaccharide is undecaprenol phosphate. The core oligosaccharide is assembled on lipid A. The genes directing LPS synthesis are scattered over the bacterial genome, with hot spots for polysaccharide (*rfb, rfc*) and core (*rfa*) syntheses

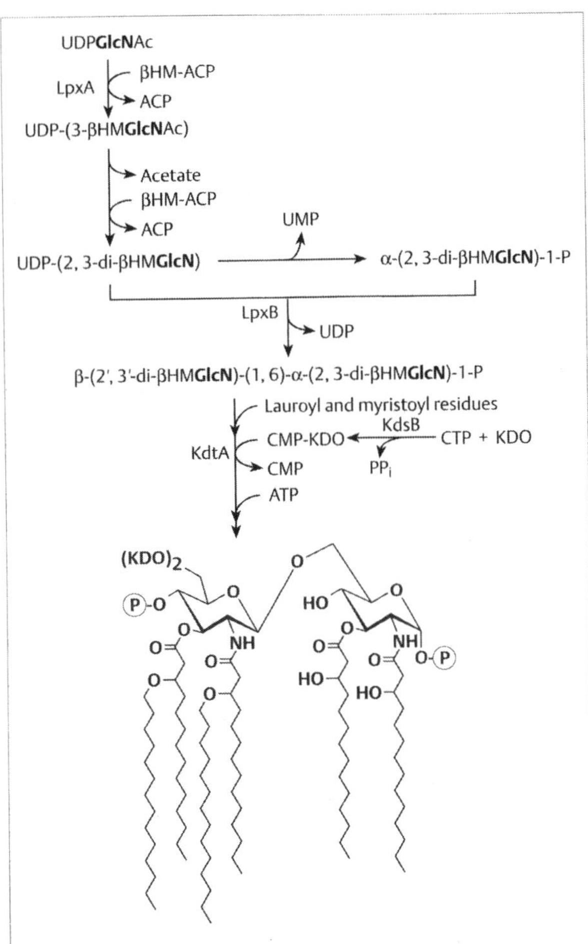

Fig. 23.3 **Synthesis of lipid A in *Escherichia coli*.** *β*-HM, *β*-hydroxymyristic acid; GlcNAc, *N*-acetylglucosamine; ACP, acyl carrier protein. *kdsB* is the gene determining the synthesis of CMP-KDO (CMP–2-keto-3-deoxyoctonic acid). The known gene products operative in the synthesis of lipid A are indicated [after 1]

is dimerized, and UMP is released. The resulting 1-phospho-dimer is substituted with 2-keto-3-deoxy-manno-octonic acid (KDO) from CMP–KDO, esterified with fatty acids at the hydroxyl groups of *β*-hydroxy-myristic acid, and finally phosphorylated at C-4′ (i.e., at the nonreducing glucosamine residue). The reaction scheme elucidated in a rough (R) mutant of *E. coli* is shown in Figure 23.**3**.

In *E. coli*, the acyl substituents are transferred directly from the acyl carrier protein of the fatty acid synthetase complex and not from the acyl-CoA derivatives. KDO substitution is a prerequisite for the esterification reaction at positions other than C-3 and C-3′. The KDO residues are transferred from CMP–KDO by only one KDO transferase. Cerulenin, an inhibitor of fatty acid synthesis, reduces the synthesis of lipid A and thus that of LPS (see Chapter 7.10.2–3).

23.2.2 The LPS Core Oligosaccharide Serves as Spacer Between Lipid A and the O-specific Polysaccharide

The biosynthesis of the core oligosaccharide part of LPS is directed by the *rfa* genes (*rf* stands for rough because the core oligosaccharide, which is determined by the *rfa* genes, is the characteristic and only carbohydrate in R-LPS). The core is formed on lipid A–$(KDO)_2$, which serves as an acceptor. The synthesis is initiated by the transfer of three units of L-glycero-D-manno-heptose (Hep, from ADPHep) to the lipid-proximal KDO and followed by their substitution with ethanolamine phosphate ($EtNH_2P$) and phosphate. The assembly of the core is completed by sequential transfer of Glc, Gal, and GlcNAc from their respective UDP derivatives.

Glucosyl transferase I and galactosyl transferase I have been isolated from bacterial homogenates. The activity of these enzymes is dependent on (i.e., stimulated by) phospholipids. A quarternary complex of phosphatidylethanolamine, glucosyl transferase I, galactosyl transferase I, and incomplete LPS as acceptor has been isolated by density-gradient centrifugation. Addition of UDPGlc and UDPGal to this complex resulted in a coupled transfer of both sugars to the incomplete LPS core.

In R mutants, LPS synthesis is terminated at this stage with the formation of core–lipid A, the R-LPS. In wild-type (smooth, S) bacteria, the LPS (S-LPS) is completed by the addition of the O-specific polysaccharide moiety (see below).

Lipid A is assembled from UDPGlcNAc and fatty acids, which are transferred from acyl carrier protein. The transfer of KDO from CMP-KDO is essential for the completion of lipid A by esterification with long-chain fatty acids. The **core** is attached to the KDO moiety of lipid A by sequential transfers of sugar units from their nucleotide precursors.

23.2.3 The O-Specific Polysaccharide Moiety of LPS Consists of Oligosaccharide Repeating Units

The polysaccharide moiety of LPS, a major antigen of Gram-negative bacteria, determines the serological O-specificity; it is therefore termed the O-specific polysaccharide. O-specific polysaccharides have repetitive structures consisting of oligosaccharide repeating units. These structures therefore can be represented by the respective repeating unit and the linkage which joins them. The biosynthesis of these polysaccharides consists of the synthesis of the oligosaccharides, which is followed by their polymerization (assembly), as depicted in general terms in Figure 23.**4**.

This pathway has been elucidated mainly in *Salmonella anatum* and in *S. typhimurium*. The O-specific polysaccharide of *S. anatum* has the structure [-(6)-β-D-Man-(1,4)-α-L-Rha-(1,3)-α-D-Gal-(1,]n, in which Man represents mannose, Rha is rhamnose, and Gal is galactose.

The pathway starts in this example with the reversible Mg^{2+}-dependent transfer of galactose 1-phosphate from UDPGal to undecaprenol phosphate (step 1). This reaction is directed by the *rfbP* gene. Subsequent transfer of the other sugar residues, rhamnose (R_2) and mannose (R_3), completes the formation of the undecaprenol diphosphate-bound oligosaccharide repeating unit (step 2). This part of the synthesis is directed by the *rfbN* and the *rfbU* genes. Subsequent polymerization of the undecaprenol-bound oligosaccharides (step 3; trisaccharide in *S. anatum*), with liberation of one molecule of undecaprenol diphosphate

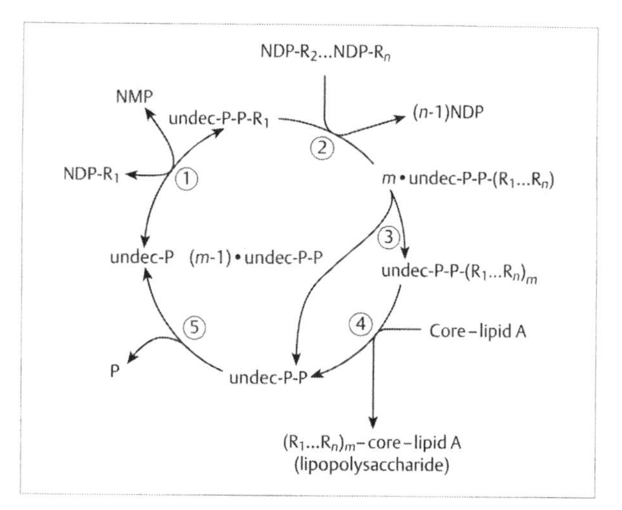

Figure 23.**4** **Synthesis of O Polysaccharide in *Salmonella* and its ligation to core–lipid A**. This is a general formulation; for details, see text. NDP, nucleotide diphosphate; NMP, nucleotide monophosphate; undec, undecaprenol. R_1 to R_n are sugar residues that are transferred from their nucleotide-diphosphate derivatives to the oligosaccharide, which is bound to the acceptor by a pyrophosphate linkage. The encircled numbers are the reaction steps. In *S. anatum*, R_1 is galactose (Gal; activated as UDPGal). R_2 is rhamnose (Rha; activated as TDPRha), and R_3 is mannose (Man; activated as GDPMan). Since the O-specific polysaccharide of *S. anatum* has a trisaccharide repeating unit, in this example n is three (step 1: transfer of a sugar 1-phosphate; step 2: transfer of sugars). Since the LPSs are heterogeneous with respect to the chain length of their polysaccharide moieties, m ranges from approximately 15 to 60, with an average of approximately 40. The assembly of the lipid-bound polysaccharide chain from m lipid-bound oligosaccharides (step 3) liberates (m-1)undec-P-P molecules. The last molecule of undec-P-P is liberated during the translocation of the finished polysaccharide from undec-P-P to core–lipid A (by RfaL/RfbT translocase; step 4). All undec-P-P molecules are recycled by the membrane-bound undec-P-P phosphatase (step 5)

per polymerization step, assembles the polysaccharide moiety of LPS bound to undecaprenoldiphosphate. In this process, the lipid-bound chain is transferred to the nonreducing end of a newly formed, lipid-bound oligosaccharide by a polymerase (Rfc). In some bacteria, the corresponding gene is part of the *rfb* gene cluster; in others, it is outside of the *rfb* locus. (The *S. anatum* polysaccharide is *O*-acetylated at C6 of each galactose unit after polymerization and prior to ligation).

The polymerization was reported to occur on the periplasmic side of the cytoplasmic membrane. Therefore, the lipid-linked oligosaccharides must first be translocated across the cytoplasmic membrane.

Box 23.1 Finished O-specific polysaccharides are sometimes substituted with a monosaccharide (e.g., glucose in *S. typhimurium*) to form a one-unit branch. Such monosaccharide transfers involve undecaprenol monophosphate sugars (as donors) and the concomitant liberation of undec-P. While this compound recycles directly into the synthetic pathway, the undec-P-P liberated during polymerization of the O-polysaccharide must first be dephosphorylated by a specific phosphatase (step 5). This phosphatase is extremely hydrophobic and an integral part of the cytoplasmic membrane.

In *E. coli* strains O8 and O9, which both have mannans as O-specific polysaccharide moieties, a different mechanism of O-polysaccharide biosynthesis was found. The reaction starts with the reversible transfer of GlcNAc from UDPGlcNAc (NDP-R_1) to undec-P, with the formation of undec-P-P-GlcNAc. This reaction is directed by the *rfe* gene. The polysaccharide grows while attached to this lipid and is extended at the non-reducing end by stepwise addition of mannose units directly from GDPMan without the participation of undec-P. In these strains, the polymerization of the O-specific polysaccharide occurs at the inner (cytoplasmic) side of the cytoplasmic membrane. This mechanism of polysaccharide synthesis is also discussed for certain O-specific polysaccharides of *Klebsiella*, which consist of mannose or galactose and are thus homopolysaccharides. The mechanism is independent of the *rfc* gene.

There are **two mechanisms for the formation of O-specific polysaccharides**:

1. In *Salmonella anatum*, the formation of the repeating units starts with the reversible transfer of Gal-1-P from UDPGal to undecaprenol phosphate (undec-P), with the formation of undec-P-P-Gal. The oligosaccharide repeating units are completed by transfer of the other sugars from their nucleotide forms. They are then assembled by transfer of the lipid-bound chain to the nonreducing end of the lipid-bound repeating unit. Thus, the **chain grows at the reducing end** and terminates with P-P-undec.
2. In *Escherichia coli* strain O9, the first reaction is the transfer of GlcNAc-1-P to undec-P. The polysaccharide chain is assembled on the product of undec-P-P-GlcNac, the previous reaction, by stepwise addition of the mannose units without further participation of undec-P. In this mechanism, the **chain grows at the nonreducing end** and terminates with P-P-undec.

23.2.4 Ligation of Core–Lipid A With O-Polysaccharide and Surface Expression of Lipopolysaccharide

Core–lipid A and the undec-P-P–linked O-polysaccharide are joined by a ligase (step 4 of Fig. 23.**4**) consisting of two subunits, RfaL and RfbT.

In *S. typhimurium* and *S. anatum*, the undec-P-P-linked polysaccharide is located at the periplasmic side of the cytoplasmic membrane. Therefore, the ligation must occur at this membrane face after the core–lipid A moiety also has been translocated.

Ligation of core-lipid A with the O-specific polysaccharide is coupled with the export of newly formed LPS. With a UDPGal-4-epimerase–negative mutant of *S. typhimurium*, which lacks the galactose present in the core and in the O-polysaccharide of the wild-type, newly formed LPS appears at discrete sites of the bacterial surface following galactose addition to the medium. These are regions where outer and inner membranes come into close apposition (**membrane adhesion sites**). Such sites may play a role in the transport of many macromolecules across the membranes, from inside to out (e.g., newly synthesized material) or from outside to in (e.g., bacteriophage DNA during phage infection).

In *E. coli* O9 and in *Klebsiella*, the finished polysaccharide is translocated across the cytoplasmic membrane with a membrane-located transport system of the ABC (**A**TP **b**inding **c**asette) transporter type. It consists of two copies of a transmembrane (channel) protein and, at the cytoplasmic side of the membrane, of two copies of a protein that energizes the transport by ATP hydrolysis. This transport system, which was also detected in the membrane translocation of capsular polysaccharides, is now termed the ABC-2 translocator. The membrane translocation processes postulated for the translocation and surface exposition of LPS in *S. typhimurium* and *E. coli* O9 are shown in Figures 23.**5** and 23.**6** (see also Chapter 19), respectively.

The **assembly of an O-specific polysaccharide** that grows at the reducing end occurs at the outer face of the cytoplasmic membrane. The oligosaccharide repeating units are first translocated and then assembled. An O-specific polysaccharide that grows at the non-reducing end is first assembled at the inner face of the cytoplasmic membrane and is then translocated with an ABC-2 transporter.

23.2.5 Regulation of LPS Biosynthesis

Regulation of LPS biosynthesis is mediated through the reversibility of the *rfbP* or *rfe*-directed glycosyl-1-

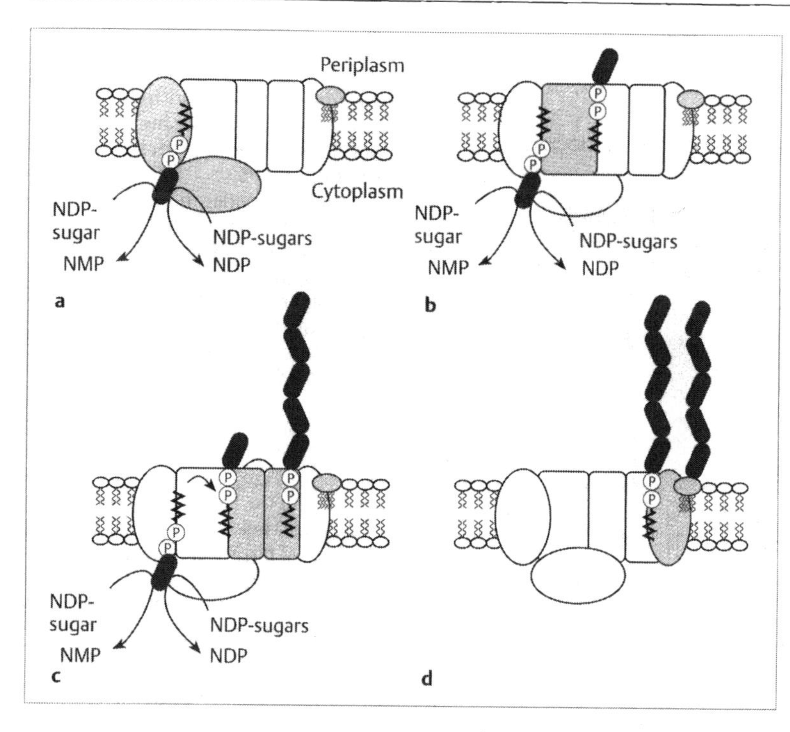

Fig. 23.5a–d **Topographic model of the formation of LPS in *S. typhimurium* by growth at the reducing end: synthesis of the O-specific repeating unit and its membrane translocation, assembly to the O-specific polysaccharide, and ligation to core–lipid A with the formation of LPS.**
a Formation of undecaprenoldiphospho-sugar from NDP (UDPGlcNAc or UDPGal) and undecP at the inner side of the cytoplasmic membrane, as the starting reaction (catalyzed by Rfe or RfbP);
b formation of the repeating unit (small dark red oval) at the inner side of the cytoplasmic membrane (catalyzed by *rfb*-encoded transferases, large light red ovals) and its translocation across the cytoplasmic membrane (probably by a protein termed Rfbx, light red rectangle);
c polymerization of the repeating units at the outer side of the cytoplasmic membrane cytoplasmic membrane (shown in red; catalyzed by the Rfc polymerase, in light red); catalyzed by the *rfc*-encoded polymerase;
d translocation of the finished polysaccharide to core–lipid A, with the formation of the O-specific LPS at the outer side of the cytoplasmic membrane (catalyzed by the RfaL and RfbT translocase/ligase). The mechanism for the transport of the complete LPS to the outer side of the outer membrane and its integration in the membrane is not known. NDP, nucleotide diphosphate [from 2]

phosphate transfer to undecaprenol phosphate (step 1 in Figure 23.**4**). This modulates the extent of O-specific polysaccharide expression in wild-type strains. The length of the individual polysaccharides is regulated by the product of the *rol* (**r**egulation **o**f **l**ength) gene. The Rol protein coordinates the activities of the Rfc polymerase and the RfaL/RfbT translocase. As a result, LPSs with different chain lengths are produced. A signal that modifies Rol activity is not known. The synthesis of LPS is not influenced by the growth temperature.

The involvement of undecaprenol associates LPS biosynthesis with that of other cell-wall components (e.g., peptidoglycan and capsules) of Gram-negative bacteria, which also need this lipid cofactor. An interdependence of these pathways can be demonstrated in vitro.

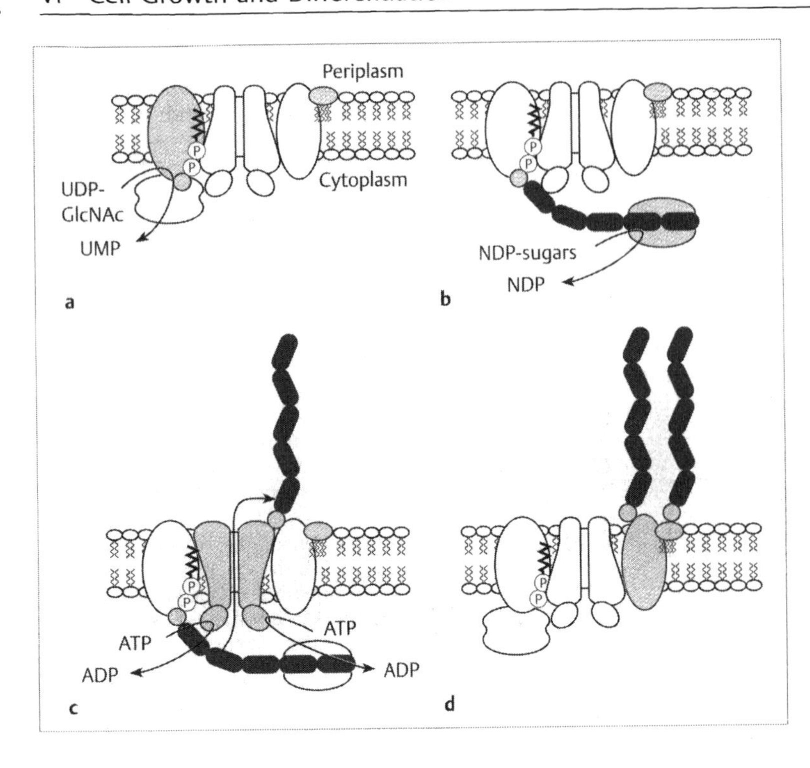

Fig. 23.**6a–d Topographic model of the formation of LPS in *E. coli* 09 by growth at the nonreducing end: synthesis of the polysaccharide, its membrane translocation, and its ligation to core–lipid A with the formation of LPS.**
a Formation of undecaprenoldiphospho-GlcNAc from UDPGlcNAc and undec-P at the inner side of the cytoplasmic membrane as starting reaction [catalyzed by an enzyme termed Rfe (light red oval)].
b Formation of the polysaccharide chain at the inner side of the cytoplasmatic membrane without the participation of undec-P [catalyzed by the Rfb sugar transferases (light red oval at the end of the polymer of red repeating units); there is no participation of an Rfc polymerase];
c translocation of the finished polysaccharide across the cytoplasmic membrane [catalyzed by the Rfb ABC-2 transporter, which consists of a transmembrane channel formed by a transmembrane protein (long, light red structures) and an ATP-binding protein as energizer of the translocation process (light red circle)];
d translocation of the O-specific polysaccharide to core–lipid A at the outer side of the cytoplasmic membrane, with the formation of LPS. The mechanisms for the transport of the complete LPS to the outer side of the outer membrane and the integration of LPS are not known. NDP, nucleotide diphosphate [from 2]

23.3 Teichoic Acids and Lipoteichoic Acids Are Polyol-containing Cell-Wall Components of Gram-positive Bacteria

Both teichoic acids and lipoteichoic acids are highly charged polymers. They concentrate cations at the cell wall and form complexes with proteins functional in cell division. They are therefore considered as regulators of cell-wall turnover, probably regulating enzymes of peptidoglycan autolysis and resynthesis. Teichoic acid, an important antigen of Gram-positive bacteria, is taxonomically relevant. Lipoteichoic acid of *S. pyogenes* is associated with the M protein of the cell wall, and a complex of both components is important in adhesion of the bacteria to target cells of the host and in resistance of the bacteria to phagocytosis.

Teichoic acids, the cell-wall antigens of Gram-positive bacteria (Fig. 2.**14**), are linked covalently to petidoglycan. They are polymers consisting of glycerol phosphate (Glp) or ribitol phosphate (RtlP), which may be substituted with sugars and/or amino acids (mostly alanine).

The synthesis of teichoic acid in *Staphylococcus aureus* starts with the reversible transfer of *N*-acetylglucosamine 1-phosphate to undecaprenol phosphate (see Fig. 7.**30** and step 1 in Fig. 23.**4**). The synthesis continues with the transfer of three glycerol phosphate residues from CMPGlp (corresponding to NDP-R_2 in step 2 of Fig. 23.**4**). The reaction product is substituted with a chain of about 40 ribitol-phosphate (Rtl-P) units, which are transferred from CDPRte.

The finished teichoic acids are transferred from undecaprenol diphosphate to C6 of a muramic acid residue in the peptidoglycan, resulting in the formation of a phosphodiester bond between C6 of muramic acid and C1 of the terminal *N*-acetylglucosamine of the

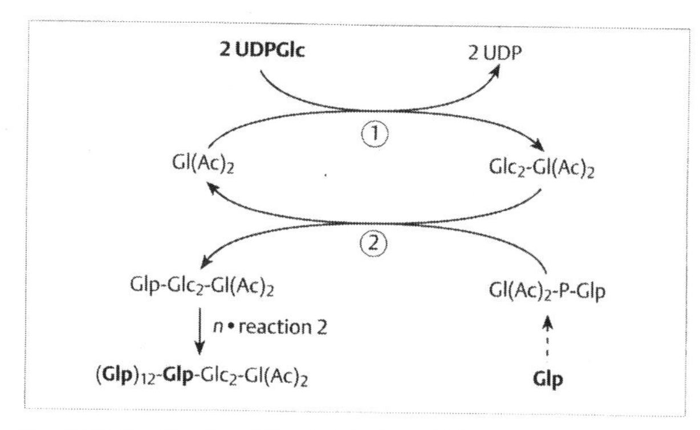

Fig. 23.7 **Synthesis of lipoteichoic acid in streptococci.** The acceptor for the polyglycerophosphate chain [(Glp)n] in the lipoteichoic acid synthesis is diglucosyldiacylglycerol [$(Glc)_2$-$Gl(AC)_2$. It is formed by glucosylation of diacylglycerol [$Gl(Ac)_2$]. During growth of the chain, the glycerol-phosphate donor is phosphatidylglycerol [$Glp(Ac)_2$-P-Glp]. The latter is obtained from *sn*-glycerol 3-phosphate (Glp), a starting compound for lipid synthesis. Glp, glycerol phosphate. Reactions 1 and 2 are denoted by encircled numbers (see text)

teichoic acid. The liberated undecaprenol phosphate is directly recycled into the synthetic pathway. It is obvious that the synthesis of teichoic acids has a mechanism that is in many respects similar to that of LPS synthesis.

In many teichoic acids, the hydroxyl groups of ribitol are substituted with monosaccharides or oligosaccharides. This modification is catalyzed by specific glycosyl transferases, probably after the polymerization. A participation of undecaprenol monophosphate has not been demonstrated. Alanyl substitution occurs by transfer of the alanine unit from an alanyl carrier protein.

Many Gram-positive bacteria produce, in addition to the teichoic acids, lipoteichoic acids (also known as membrane teichoic acids or glycerol teichoic acids). In contrast to the teichoic acids, they are not covalently attached to the peptidoglycan but associated with the cytoplasmic membrane through hydrophobic interaction. Lipoteichoic acids (Chapter 7.10.3) consist of poly (1-3-glycerol phosphate) linked to diglucosyl-1-3-diacylglycerol as the hydrophobic moiety. The polyglycerol phosphate chain is long enough to penetrate the peptidoglycan layer and reach the surface of the bacterium. The biosynthesis of the lipoteichoic acids differs significantly from that of the teichoic acids. The hydrophobic moiety is synthesized by transfer of two glucose units from UDPGlc to diacyl glycerol (Fig. 23.**7**, reaction 1). This acceptor is then substituted with glycerol phosphate (Glp), whereby phosphatidylglycerol is the donor (Fig. 23.**7**, reaction 2). Repetition of reaction 2 results in growth of the polymer chain.

Unlike teichoic acid synthesis, that of lipoteichoic acids is closely related to synthesis and turnover of membrane phospholipids.

> The synthesis of (ribitol) **teichoic acids** starts with the formation of undec-P-P-GlcNAc, to which glycerol phosphate (Glp) is transferred from CMP-glycerol. This core region is extended with ribitol phosphate (Rtl-P) from CMP-Rtl.
>
> The synthesis of (glycerol) **lipoteichoic acids** starts with the lipid glucosyl-diacylglycerol, which is elongated with Glp from phosphatidylglycerol. The precursors of teichoic acids are thus phospholipids.

23.4 Teichuronic Acids Are Hexuronic-acid–containing Surface Components of Gram-positive Bacteria

The teichuronic acids of Gram-positive bacteria contain hexuronic acid instead of a polyol phosphate. *Micrococcus luteus* teichuronic acid contains *N*-acetylmannosaminuronic acid (ManNAcA) and glucose; that of *B. licheniformis* contains glucuronic acid and *N*-acetylgalactosamine. In *M. luteus*, the starting reaction of teichuronic acid biosynthesis is the transfer of GlcNAc-1-P from UDPGlcNAc to undecaprenol phosphate (corresponding to step 1 in Fig. 23.**4**). This is followed by transfer of two units of ManNAcA and by alternating transfers of Glc and ManNAcA directly from their UDP-activated forms (step 2 of Fig. 23.**4**). The chain is finally transferred to C6 of muramic acid in the peptidoglycan with formation of a (-ManNAcA-ManNAcA-GlcNAc-1-P-MurNAc-) linkage region.

The synthesis of the teichuronic acid in *B. licheniformis* starts with the transfer of GalNAc-1-P from UDPGalNAc to undec-P, followed by sequential transfer of Glc and GalNAc from their UDP-activated forms. There is no linkage region between the peptidoglycan and the teichuronic acid chain.

Bacilli have the capacity to produce either teichoic acids or teichuronic acids. The pathways leading to one or the other of these surface components are regulated by the phosphate content of the growth medium: teichoic acid is favored at a high phosphate concentration and teichuronic acid at a low phosphate concentration.

23.5 The Capsules Surrounding Many Gram-positive and Gram-negative Bacteria as a Protective Layer Consist of Acidic Polysaccharides

Many Gram-positive and Gram-negative bacteria are encapsulated (Figs. 2.**6a,b**). The capsules protect the bacteria against adverse influences of the surroundings. Capsules are important virulence factors of pathogenic bacteria because they counteract the host defense mechanisms. In particular, capsules prevent the lytic and/or opsonic action of the complement system, and they protect the bacteria against phagocytosis. It has occasionally been observed that capsules mediate adhesion of bacteria to surfaces in vivo and in vitro. Although capsular material, like all the other bacterial surface components, is in part released into the medium, capsules should not be confused with bacterial slime, which is not anchored to the cell surface.

The capsules consist generally of **acidic polysaccharides**. Biosynthesis and surface expression have been studied with the capsular polysaccharides of *E. coli, Neisseria meningitidis, Haemophilus influenzae, Klebsiella*, and group A *Streptococcus*. The capsular polysaccharides of *E. coli* are divided into group I (*Klebsiella*-like) and group II (*Neisseria*- and *Haemophilus*-like).

E. coli (group I) and *Klebsiella* capsular polysaccharides are synthesized by enzymes encoded in genes that correspond to the *rfb* genes of LPS biosynthesis. Their polymerization may or may not include the intermediary formation of undecaprenol diphosphate-bound oligosaccharides. The final acceptor of some group II capsular polysaccharides in *E. coli, H. influenzae*, and *N. meningitidis* is phosphatidic acid. This substituent presumably mediates surface association of the capsules through hydrophobic interactions. In *E. coli* (group II) and in *N. meningitidis*, substitution of the polysaccharide with phosphatidic acid is coupled with its translocation across the cytoplasmic membrane. The export sites of the capsular polysaccharides of Gram-negative bacteria are membrane adhesion sites.

23.5.1 Synthesis and Surface Translocation of Capsular Polysaccharides

The synthesis of *E. coli* group II, *N. meningitidis*, and *H. influenzae* capsular polysaccharides is directed from the comparable gene clusters *kps* (*E. coli*), *cps* (*H. influenzae*), and *ctr* (*N. meningitidis*). The *kps* gene cluster is organized into three regions. A central gene region, which directs the polymerization of the capsular polysaccharide, is type-specific. It is flanked by two conserved gene regions that direct the surface expression of the capsular polysaccharides. The *cps* and *ctr* loci are similarly organized. These capsular polysaccharides are polymerized at the inner side of the cytoplasmic membrane. The translocation of the synthesized polysaccharide is mediated by an ABC-2 translocator in a reaction comparable to that shown in Figure 23.**6** for LPS. Thus, the translocation of the O-specific polysaccharide to core–lipid A and that of group II capsular polysaccharide to phosphatidyl-KDO seem to be comparable reactions.

Group A *Streptococcus* produces a capsule consisting of **hyaluronic acid**, a glycosaminoglycan with the structure $[3)$-β-GlcA(1,3)-β-GlcNAc-$(1]_n$. Synthesis and export of hyaluronic acid are combined processes. The polymerization proceeds by alternating transfer of glucuronic acid (GlcA) and GlcNAc from the UDP-activated sugars without the participation of undecaprenol phosphate as cofactor.

Since the Gram-positive streptococci have no outer membrane, the surface translocation of the capsular hyaluronic acid is a simpler process. It is assumed that the polysaccharide grows at the reducing end and is pushed out of the cell during the elongation process. A membrane-bound complex of hyaluronate synthetase (i.e., a protein with transferase activities for both GlcA and GlcNAc) and UDPGlc dehydrogenase synthesizes hyaluronic acid and concomitantly translocates it to the cell surface.

23.5.2 Regulation of Capsule Synthesis

The biosynthesis of *E. coli* (group I) and *Klebsiella* capsules is controlled by a two-component regulatory system (Chapter 20.4); *rcs* for **r**egulation of **c**apsule **s**ynthesis), which was first described for the synthesis of the slime or mucus (M) polysaccharide of *E. coli*. This regulatory system comprises the sensor RcsC and the response regulator RcsB. Both components function together with RcsA and the *E. coli* protease LonB. The stimulus for the membrane-bound *rcsC* product is not known. The regulating genes are located at different chromosomal sites outside the *rfb*-linked gene cluster for capsule synthesis.

The *Rcs* system does not regulate *E. coli* group II capsules. Instead, the expression of group II capsule genes is stably regulated by the *rfaH* gene in concert with a DNA sequence upstream of the *kps* genes. This DNA sequence was found **j**ust **u**pstream of **m**any **p**olysaccharide expression genes and was therefore termed "JUMP" start sequence. In this system, the RfaH (18-kDa basic protein) recognizes the 39-bp "JUMP" start sequence and then acts as an antiterminator for transcription of the capsule genes. In contrast to group I capsules, group II capsules are temperature-regulated. This regulation is exerted by some hitherto uncharacterized gene(s) upstream of the *kps* genes.

Regulation of capsule expression in the other genera mentioned is not known.

The group II *E. coli*, *N. meningitidis* and *H. influenzae* b **capsular polysaccharides** are synthesized at the inner side of the cytoplasmic membrane by elongation at the non-reducing end while attached to a primary acceptor (presumably undecaprenol phosphate). The finished polysaccharides are then translocated across the cytoplasmic membrane by an ABC-2 type transporter and transferred to the secondary acceptor, KDO–phosphatidic acid.

23.6 Polysaccharide Assembly Is Shunted to the Undecaprenol Phosphate Cycle

Undecaprenol phosphate, synthesized in bacteria from mevalonic acid with participation of ATP, is formed as a diphosphate monoester (see Chapter 7.9). Before it can enter the biosynthetic pathway of polysaccharides, it has to be dephosphorylated to its monophosphate by the bacitracin-sensitive phosphatase of the cytoplasmic membrane. In the membrane, there is a balanced equilibrium between undecaprenol (undec), undec-P, and undec-P-P determined by the involvement of the lipid in various synthetic systems (e.g., for peptidoglycan, polysaccharides, teichoic acids, and teichuronic acids) and by the rate of its synthesis. As indicated in Figure 23.**8**, oligosaccharides are formed and polymerized/transferred during the conversion of undec-P to undec-P-P.

Whereas undec-P-P always seems to be involved in the transfer of oligosaccharides, monosaccharides are transferred from undec-P. Thus, the undec-P liberated during glycosylations involving monosaccharides can directly reenter the cycle, whereas the undec-P-P liberated during oligosaccharide transfer first has to be dephosphorylated.

The same lipid cycle is operative in archaebacteria during the synthesis of oligosaccharides and their transfer to protein. This reaction, which is otherwise only encountered in eukaryotic cells, is mediated by a cycle that utilizes dolichol and its phosphorylated forms instead of undecaprenol and its phosphorylated forms (Fig. 7.**30**).

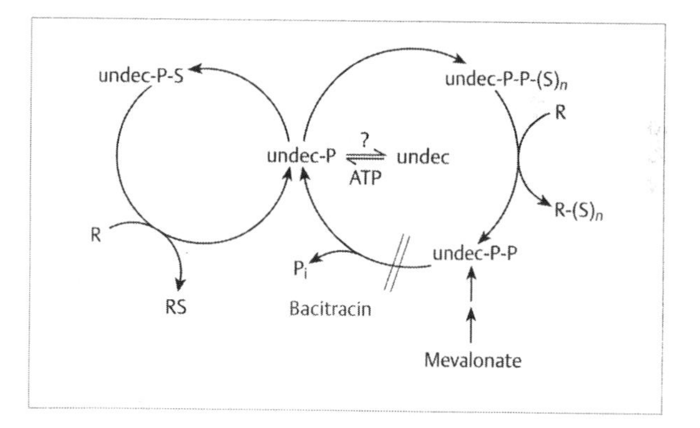

Fig. 23.**8** **The lipid cycle in polysaccharide assembly.** The central compound is undecaprenol phosphate (undec-P), which is synthesized from mevalonic acid. It is a cofactor for the transfer of oligosaccharides or polysaccharides [(S)n] and of monosaccharides (S). In the transfer of oligosaccharides or polysaccharides to an acceptor R, the linkage is a pyrophosphodiester originating from the original transfer of a sugar 1-phosphate. After transfer, the lipid is undec-P-P and has to be dephosphorylated for recycling (cycle on the right). In the transfer of a monosaccharide, the linkage is a monophosphodiester. After transfer, the lipid is undec-P and can be directly recycled (left cycle). Undecaprenol phosphate can also be dephosphorylated to undecaprenol and rephosphorylated with ATP

An essential lipid constituent of bacterial membranes is **undecaprenol** (undec), which occurs in two interchangeable forms, as a monophosphate or a diphosphate (i.e., undec-P or undec-P-P). Since undec-P is an acceptor in polysaccharide synthesis and is released during glycosyl transfer as undec-P-P, the turnover of these lipid forms is interrelated with polysaccharide (glycoside) synthesis.

23.7 Outer Membrane Proteins of Gram-negative Bacteria Are Either Pore-forming Proteins (Porins) or Structural Proteins

Outer membranes contain a number of proteins, of which the porins, such as OmpC, OmpF, or PhoE in *E. coli*, oligomerize to form diffusion channels (pores) for the transport of ions and small molecules. Similar pore-forming proteins have been described in *Salmonella*, *Neisseria*, *Haemophilus*, *Pseudomonas*, and *Rhodobacter capsulatus*. Other proteins, such as OmpA and lipoprotein of *E. coli*, are structural proteins that function in the assembly and maintenance of the outer membrane or as receptors for bacteriophages and bacteriocins.

The synthesis of **porins** is correlated with that of lipopolysaccharides (LPSs). Porins are synthesized as preproteins and are translocated across the cytoplasmic membrane in a Sec-dependent system; the signal peptide is then removed by signal peptidase I. Porins are integrated into the outer membrane in close association with LPS and form trimeric pores. LPS seems to be essential for porin formation and function.

23.7.1 Porins

The porins are synthesized on polysomes at the inner side of the cytoplasmic membrane as preproteins. These are translocated across the cytoplasmic membrane by the secretory (*Sec*) system, with participation of SecB as a chaperone, and then converted by signal peptidase I to mature porins. Their proper orientation in the outer membrane is determined by internal stop-transfer signals. This results in multiple looping of the protein across the outer membrane with several separate β strands that form β barrels (Fig. 5.**9**).

The pores of the outer membrane function in the uptake of specific solutes, such as phosphate (*E. coli* PhoE, *Pseudomonas* PhoP), glucose (*Pseudomonas* PhoD), or maltose (*E. coli* LamB), or in the less specific uptake of small molecules (*E. coli* OmpC and OmpF). The former are termed specific porins, and the latter are termed general porins. The expression of general porins is regulated by a two-component signaling (*osm*) system sensitive to the osmolarity of the medium. The expression of OmpC and OmpF can be greatly reduced by **cerulenin**, which inhibits fatty acid synthesis and thus reduces the synthesis of LPS. Thus, the synthesis of OmpC and OmpF is correlated with and dependent on the synthesis of LPS (see Chapter 20.4.2).

23.7.2 Lipoprotein of *E. coli*

The lipoprotein of *E. coli* exists in two forms: one is bound to the peptidoglycan, and the other is free in the outer membrane. Both forms are important in the maintenance of the outer membrane mainly by hydrophobic interaction with LPS and outer-membrane proteins. The peptidoglycan-bound form is also important for the association of the outer membrane with the peptidoglycan layer of the cell wall.

Lipoprotein is synthesized as preprotein with a signal peptide at the N terminus. The first amino acid after the signal peptide is cysteine. While the lipoprotein is still attached to its signal peptide, the cysteine moiety is substituted with glycerol from phosphatidylglycerol in a thioether linkage. After acylation of both hydroxyl groups of the glycerol moiety by fatty acid exchange from phospholipids, the preprotein is translocated across the inner membrane with removal of the signal peptide by signal peptidase II. The amino group of the N-terminal cysteine moiety is then acylated, resulting in the lipid terminus of the lipoprotein

```
                R–CO–O
                     |
R–CO–O–CH2–CHO–CH2–S
                         |
                        CH2
                         |
           R–CO–HN–CH–CO–......
```

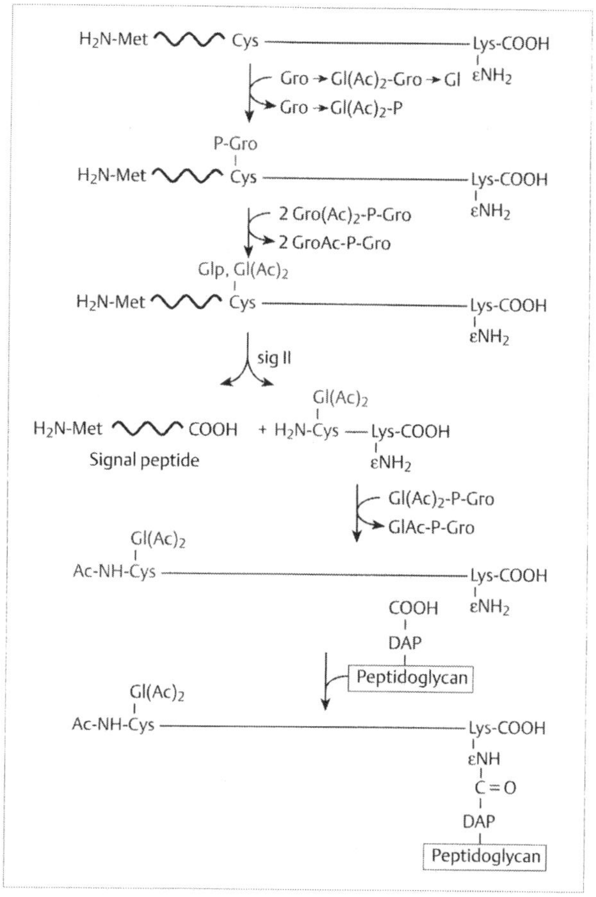

H_2N-Met ∿∿ Cys ——— Lys-COOH (εNH₂)

Gro → Gl(Ac)₂-Gro → Gl
Gro → Gl(Ac)₂-P

P-Gro
H_2N-Met ∿∿ Cys ——— Lys-COOH (εNH₂)

2 Gro(Ac)₂-P-Gro
2 GroAc-P-Gro

Glp, Gl(Ac)₂
H_2N-Met ∿∿ Cys ——— Lys-COOH (εNH₂)

sig II

H_2N-Met ∿∿ COOH + H_2N-Cys(Gl(Ac)₂) — Lys-COOH (εNH₂)
Signal peptide

Gl(Ac)₂-P-Gro
GlAc-P-Gro

Ac-NH-Cys(Gl(Ac)₂) ——— Lys-COOH (εNH₂)
COOH
DAP
Peptidoglycan

Ac-NH-Cys(Gl(Ac)₂) ——— Lys-COOH
εNH
C=O
DAP
Peptidoglycan

Fig. 23.9 **Synthesis of lipoprotein in *E. coli*.** The wavy line represents the signal peptide, and the solid line represents the actual protein part. Modification starts with the substitution of glycerol for the SH group of cysteine (with glycerol from phosphatidyl glycerol, resulting in the formation of a phosphatidic acid). The substituent is thioglycerol. The hydroxyl groups of the (thio)glycerol part are substituted by fatty acids (from two units of phosphatidylglycerol, resulting in the formation of two units of lysophosphatidylglycerol). At this stage, the signal peptide is removed by signal peptidase II (sig II), and the free amino group formed is also substituted by an acyl group (from phosphatidyl glycerol, resulting in the formation of lysophosphatidylglycerol). The finished lipoprotein either remains free in the outer membrane or is transferred to peptidoglycan, resulting in the formation of a peptide bond between the C-terminal lysine and the carboxyl group of the L-R3 residue in the peptide moiety of the peptidoglycan subunit (see Chapter 23.1.1). (Symbols/abbreviations as in Fig. 23.7)

Bound lipoprotein originates from the free form by ligation of the C-terminal lysine with the diaminopimelic acid of the *E. coli* peptidoglycan. The synthesis of free and bound lipoprotein is shown in Figure 23.**9**.

Lipoprotein, which contains three fatty acids, is synthesized as a preprotein, substituted with glycerol, and acylated at the glycerol moiety with two long-chain fatty acids, which are transferred from phospholipids. The acylated preprotein is translocated, processed by the specific signal peptidase II, and completed at the cysteine residue by acylation with the third fatty acid.

The free form of lipoprotein in the outer membrane is the precursor for the peptidoglycan-linked form.

23.8 The Filamentous Fimbriae Interact With Receptors of Cell Surfaces

Fimbriae, often termed pili in American publications, are hair-like appendages of bacterial cells allowing adhesion of the bacteria to surfaces (see Chapters 2.6.3, 3.5, and 33.3; Fig. 2.**20**). Their interaction with mammalian cells often initiates bacterial infections. Most fimbriae recognize complex carbohydrates on the surface of target cells. Fimbriae of *E. coli* exert different specificities: α-mannose for type I fimbriae, α-Gal-(1,4)β-Gal for P fimbriae, and α-NeuNAc for S fimbriae. Fimbriae are made up of the structural subunits (major peptides) and the adhesins (minor peptides), which are located at the tip.

The biogenesis of fimbriae is directed by complex gene clusters encoding genes for the synthesis of the adhesin and the fimbrial subunits and of a number of accessory proteins. The *pap* genes for the P fimbriae of uropathogenic *E. coli* are shown as an example in Figure 23.**10**.

The adhesin and the fimbrial subunits are synthesized as preproteins and translocated across the cytoplasmic membrane. With the P fimbriae, one of the accessory proteins, the PapD protein, functions as a shuttle protein and periplasmic chaperone for the adhesive minor peptides (i.e., the sugar-specific PapG

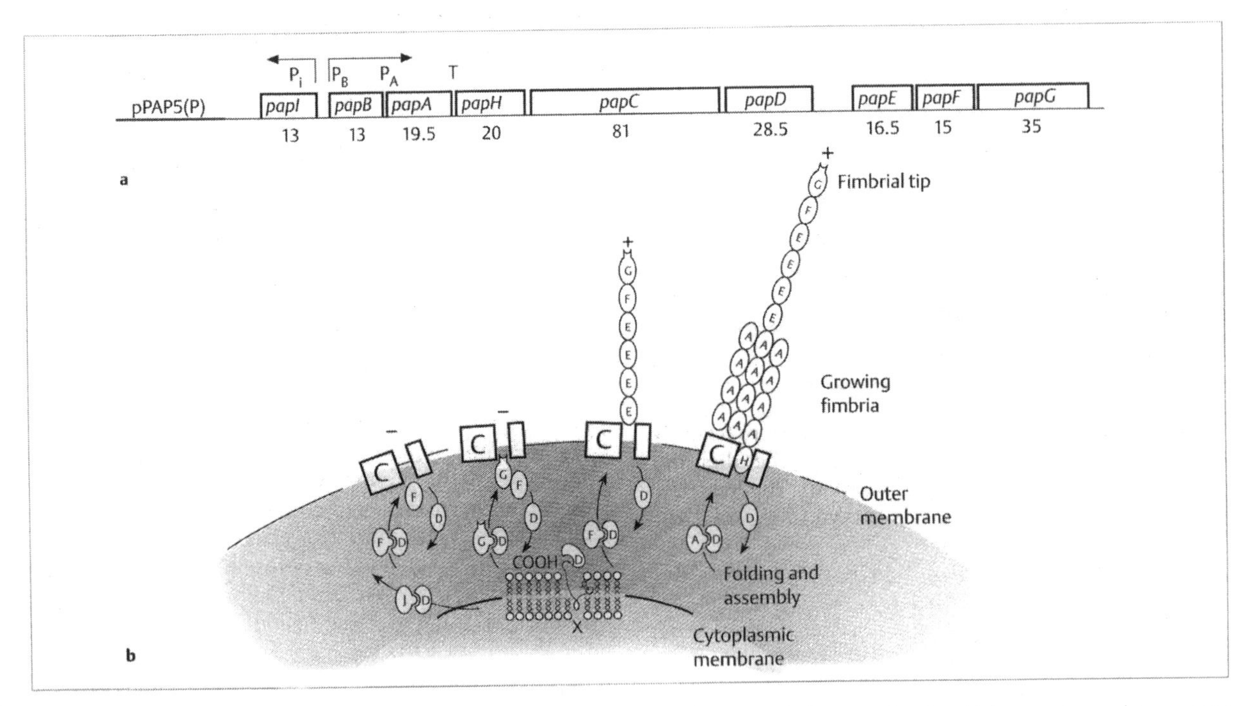

Fig. 23.**10a,b Genetics and assembly of P fimbriae of *E. coli*.** **a** The genes in the cluster determining the biogenesis of fimbriae: *papI* and *papB* are genes for regulatory proteins; the other genes are described in the text (p, promotor). The numbers below the genes represent the molecular masses of the proteins in kDa.
b The assembly process: The adhesive subunits (PapG = G, which recognize carbohydrates of the host cell surface) are transported across the cytoplasmic membrane and complexed with the chaperone (PapD = D). The complex is transported to the outer membrane. Interaction with the export site (PapC = C) in the outer membrane retains PapG in the membrane and liberates PapD which recycles and transports the other subunits (PapE and PapF, here E and F, respectively, minor adhesins recognizing components of the glycocalix of target cells; PapA, here A, subunit of the supporting fimbrial rod) in the same fashion. In this way, the fimbria (pilus) grows from the outer membrane and is pushed away from the membrane in the process. Growth of the fimbria is terminated by transport of PapH (H) and its insertion into the membrane as the final (membrane-proximal) fimbrial subunit. The components of the system are drawn out of proportion. T, terminator [after 3]

protein and the PapE and PapF proteins specific for components of host proteins) and for the major fimbrial subunit (PapA protein). The subunits dissociate from the PapD protein while interacting with an outer-membrane protein (PapH). The adhesin (PapG) is first shuttled to the outer membrane. It is pushed out of the membrane by several PapE and PapF proteins, which together with the PapG protein form the tip of the subsequently growing fimbrium. This process is schematized in Figure 23.**10**.

Fimbriae, which consist of specific recognition proteins on supportive fimbrial structures, are made up of subunits, which are synthesized as preproteins, individually translocated across the cytoplasmic membrane, and transported through the periplasmic space as complexes with chaperone proteins. The fimbriae grow by extrusion through the outer membrane; the recognition protein is extruded first.

23.9 The Flagella, Organelles of Motility, Are Assembled by Two Mechanisms

Bacterial flagella (see Chapters 2.3.8, 20.5; Figs. 2.**1**, 2.**19**, 20.**21** and 24.**2**) are very complex structures consisting of intracellular, membrane-associated, periplasmic regions and extracellular regions. The genetics, biosynthesis, and regulation of flagella are closely linked to those of chemotaxis proteins, in keeping with the functional association of these systems (see Chapter 20.5).

The **biogenesis of flagella** is a strictly regulated, sequential process. The precursors, synthesized intracellularly, are assembled without the aid of signal peptides. The assembly of the basal body, which is located in the cytoplasmic and the outer membranes, is followed by the formation of the hook, a region between the seal of the outer membrane (L ring) and the flagellar filament. The peptides that form the parts of the flagellum located in the cell wall (from M- to L-ring) are translocated across the membrane by the general bacterial export pathway involving signal peptides, the protein secretion (Sec) system, and signal peptidase. In the final stage, the filament is generated from subunits (flagellin) without signal peptides and protein maturation. The subunits travel through the hollow cylinder of the growing flagellum and assemble at the tip. This is different from the formation of fimbriae, which grow in the membrane-proximal region. The N-termini of the flagellar subunits contain a signal region, which enables the peptides to be recognized and to be transported by the flagellum-specific pathway. There is no control system for the flagellar length, which seems to be limited solely by the increasing resistance due to friction as new subunits travel within the central channel. The flagellar gene expression of *Caulobacter crescentus* is described in Chapter 24.1.3.

The subunits of the membrane-associated part (hook, basal body) are not synthesized as preproteins (i.e., without a signal sequence).

The subunits of the filament travel through the hollow center of the growing flagellum and are integrated at the tip.

There are at least two protein export systems operative in the flagellar synthesis.

23.10 The Surface and the Flagella of Archaebacteria Consist of Glycoproteins

The surface of several archaebacteria (Archaea), such as *Halobacterium*, consists of a closely and symmetrically packed array of highly negatively charged glycoproteins, the **S layer**. Flagella of these bacteria also consist of glycoproteins, and the carbohydrate moiety of the flagella has generally the same structure as that of the surface (S-layer) glycoproteins. Thus, the same carbohydrate intermediates are used in the biosynthesis of the cell surface and of the flagella (see Chapter 7.9.5)

There are two major types of carbohydrate substituents in archaeal glycoproteins, as analyzed for

Halobacterium halobium and *Halobacterium salinarum*, a glycosaminoglycan-like chain consisting of approximately 10–15 pentasaccharide repeating units

```
GlcNAc-(1,4)-GalA-(1,3)-GalNAc
   6            3
   1            1
3-O-Me-GalA   Gal
```

and the trisaccharide Glc-[-4]-HexA-$(1,]_{2-3}$. In the latter formulation, the hexuronic acid (HexA) may be either glucuronic acid (GlcA, as major component) or iduronic acid (IduA, as minor component). Both carbohydrates are substituted with two or three sulfate groups per oligosaccharide.

For the biosynthesis of the glycosaminoglycan chain, the pentasaccharide repeating unit is assembled on **dolichol diphosphate**, sulfated, and polymerized, with liberation of dolichol diphosphate. The complete chain is then transferred to an Asn-Ala-Ser consensus sequence of the protein in a membrane-associated, postpolymerization modification.

The trisaccharide also is synthesized on dolichol monophosphate, sulfated, and transferred to a similar asparagine-containing sequence.

During the formation of the glycosaminoglycan substituent and its transfer to the protein, dolichol diphosphate is liberated and then dephosphorylated before it re-enters the synthetic cycle. Substitution with the trisaccharide liberates dolichol monophosphate, which directly enters the cycle.

Similar to the assembly of peptidoglycan and the polymerization of lipopolysaccharides, the dolichol-P-P-bound oligosaccharides are first translocated across the cytoplasmic membrane and then transferred to the final acceptor (a protein).

The carbohydrate moieties of archaebacterial glycoproteins are synthesized on the C_{55}-polyprenoldolichol phosphate and then transferred to Asn-Ala-Ser consensus sequences of the protein. This mechanism is the same as that for the biosynthesis of eukaryotic glycoproteins. Characteristic of archaeal glycoproteins is the presence of sulfate groups in the carbohydrate moiety. Dolichol is generally found in eukaryotic cells but not in eubacteria.

Further Reading

Benz R. (1988) Structure and function of porins from Grar negative bacteria. Annu Rev Microbiol 42: 359–393

Fischer, W. (1988) Physiology of lipoteichoic acids in bacter Adv Microb Physiol 29: 234–302

Gottersman, S., and Stout, V. (1991) Regulation of capsul polysaccharide synthesis in *Escherichia coli* K12. M Microbiol 5: 1599–1606

Hayashi, S. and Wu, H. (1990) Lipoproteins in bacteria. Bioenerg Biomembr 22: 451–471

Höltje, J. V. (1995) From growth to autolysis: the mure hydrolyses in *E. coli*. Arch Microbiol 164: 243–254

Hultgren, S. J., and Normak, S. (1991) Chaperone-assist assembly and molecular architecture of adhesive pili. Anı Rev Microbiol 45: 383–415

Jann, K., and Jann, B. (1984) Structure and biosynthesis of antigens. In: Proctor, R. A. (ed.) Handbook of endotoxin, v 1, Rietschel, E. T. (ed.) Chemistry of endotoxin. Amsterda New York: Elsevier; 138–186

Jann, K., and Jann, B., eds. (1990) Bacterial capsules. Curr T Microbiol Immunol, vol 150. Berlin Heidelberg New Yor Springer

Jann, K., and Jann, B (1997) Capsules of *Escherichia coli*. In: I Sussman (ed.). *Escherichia coli*: mechanisms of virulenc Cambridge UK: Cambridge University Press; 113–143

Macnab, R. M. (1992) Genetics and biogenesis of bacteri flagella. Annu Rev Genet 26: 313–358

Mirelman, D. (1979) Biosynthesis and assembly of cell w peptidoglycan. In: Inouye, M. (ed.) Bacterial outer mer branes, biogenesis and functions. New York: Wiley; 11 166

Raetz, C. R. H. (1993) Bacterial endotoxins: extraordinary lipi that activate eucaryotic signal transduction. J Bacteriol 17 5745–5753

Raetz C. R. H. (1990) Biochemistry of endotoxins. Annu R Biochem 59: 129–170

Reizer, J., Reizer, A., and Saier, M. H., Jr (1992) A new subfami of bacterial ABC-type transport systems catalyzing expc of drugs and carbohydrates. Protein Sci 1: 1326–1332

Saier, M. H., Jr. (1994) Computer-aided analyses of transpc protein sequences: gleaning evidence concerning functio structure, biogenesis, and evolution. Microbiol Rev 5 71–93

Schnaitman, C. A., and Klena, J. D. (1993) Genetics lipopolysaccharide biosynthesis in enteric bacteria. Micr biol Rev 57: 655–682

Sleytr, U. B., Messner, P., Pum, D., and Sara, M. (1993) Crystalli bacterial surface layers. Mol Microbiol 10: 911–916

Sumper, M. (1987) Halobacterial glycoprotein biosynthes Biochem Biophys Acta 906: 69–79

Whitfield, C. (1995) Biosynthesis of lipopolysaccharide antigens. Trends Microbiol 3: 178–185

Whitfield, C., and Valvano, M. A. (1993) Biosynthesis aı expression of cell-surface polysaccharides in Gram-neg tive bacteria. Adv Microbial Pathophys 35: 135–246

Sources of Figures

1 Raetz, C. R. H. (1990) Annu Rev Biochem 59: 129–170
2 Whitfield, C. (1995) Trends Microbiol 3: 178–185
3 Hultgren, S. J., and Normak, S. (1991) Annu Rev Microbi 45: 383–415

24 Processes of Cellular Differentiation

In this chapter three examples of cell differentiation are discussed in detail: (1) cell-cycle-directed polar differentiation in *Caulobacter*, (2) oxygen-tension-regulated differentiation of the photosynthetic membranes, and (3) heterocyst formation in cyanobacteria upon deprivation of fixed nitrogen. Other examples are mentioned.

24.1 In *Caulobacter*, the Formation of Swarmer and Stalked Cells Is Regulated by the Cell Cycle

24.1.1 The Cell Cycle Is Coordinated With the Polar Differentiation

Caulobacter crescentus belongs to the group of Gram-negative, polarly flagellated, and stalked bacteria (Fig. 24.**1**). The **stalk** (prostheca, 0.5–3.0 μm in length) is continuous with the cell cytoplasm, but does not contain ribosomes or DNA. It is enveloped by the cytoplasmic membrane and the cell wall. The stalk is formed by an outgrowth at the former flagellar pole. At the tip of the stalk, the holdfast, a small mass of adhesive material, serves for the attachment of the cells on surfaces in the environment. The opposite pole of the cell bears a single **flagellum**, several pili, receptors for bacteriophage ΦCbK, and chemosensors for chemotactic response. The flagellum consists of three major components: (1) the basal body, and (2) the flexible hook, which is followed by (3) the flagellar filament (Fig. 24.**2**). The **basal body** consists of a series of rings localized within the outer membrane (L-ring), the peptidoglycan layer (P-ring), the periplasmic space (E-ring), and the cytoplasmic membrane (MS-ring) that anchors the flagellum in the cytoplasmic membrane and acts as a rotor. The rings are attached to each other and to the hook by the rod (see also Fig. 20.**21**). The **filament** is composed of three different flagellin proteins.

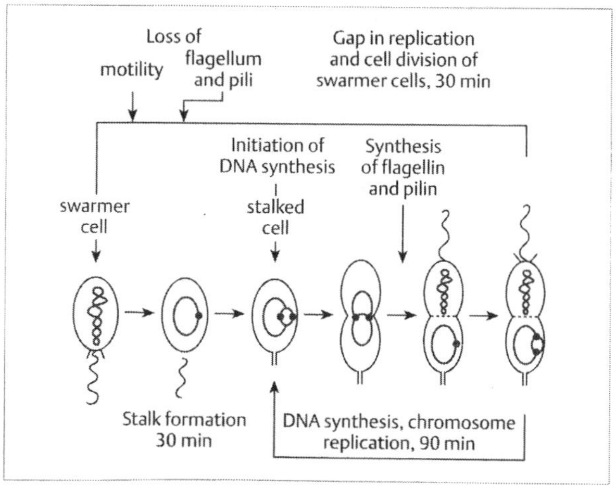

Fig. 24.**1** ***Caulobacter crescentus*: cell-cycle-dependent differentiation in swarmer and stalked cells.** After cell division, the stalked cell can immediately start with a new cell cycle, whereas swarmer cells have a gap in replication. The condensed and the relaxed genome structures and the DNA replication are symbolized in the cells. See text for details [after 1]

The **pili** are filaments about 4.0 nm in diameter and 1–4 μm in length. They consist of pilin, an 8.0-kDa hydrophobic protein that is accumulated in the cytoplasm and stored in the swarmer cell pole until the pili are assembled shortly before cell division.

During vegetative growth, *C. crescentus* divides asymmetrically by binary fission to produce two different cell types: a motile swarmer cell and a nonmotile stalked cell (Fig. 24.**1**). The stalked cell divides repeatedly to produce new swarmer cells from the stalk-distal pole. The swarmer cell, with its polar appendages at the old pole, cannot undergo DNA replication and cell division. It is motile for a short time until it matures into a stalked cell, which loses cell motility, flagellum, and pili. After the start of stalk formation at the former site of flagellum insertion, DNA synthesis is initiated, and the cell enters the S-phase (or C-phase) of the cell cycle (see Chapter 22.1). Cell division is initiated during chromosome replication and lasts about one-half of a cell cycle; cell division does not occur by rapid septation after completion of chromosome replication, as is usual, for example, for enteric bacteria. How the pole site is selected is not known. It has been proposed that an organizing center, coupled to the temporarily controlled expression of polar proteins, is positioned and activated when cell division starts.

Studies on cell-cycle-dependent differentiation are facilitated by the presence of the polar structures as markers of cell differentiation and the ease with which synchronous cultures can be prepared. *Caulobacter* is an excellent model system for cell-cycle-determined cell differentiation.

Caulobacter crescentus divides asymmetrically to produce a motile swarmer cell and a nonmotile stalked cell. Only the stalked cell undergoes DNA replication and cell division. The swarmer cell loses flagellum and pili and starts stalk formation before DNA synthesis is initiated.

24.1.2 The Chromosomal Replication Cycle Is a Cellular Clock

The biogenesis of polar structures depends on the initiation of a new round of DNA replication and on DNA synthesis, but not on the completion of cell division. The progeny swarmer cell contains a fast-sedimenting, compact nucleoid of more than 6000 S (S = Svedberg sedimentation constant), which is converted to a slow-sedimenting, normal nucleoid (3000 S) when the swarmer cell differentiates into a stalked cell later in the cell cycle (Fig. 24.**1**). The compact nucleoid of the swarmer cell may be compared to eukaryotic heterochromatin in which gene expression is silenced by higher-order packaging of DNA. The predivisional cell contains one compact nucleoid destined for the progeny swarmer cell and one normal nucleoid that is inherited by the stalked cell. The transition from the compact to the normal nucleoid is abrupt. The differential control of DNA synthesis in the sister swarmer and stalked cells is one of the most striking signs of developmental control in *C. crescentus*. The stalked cell begins DNA synthesis immediately after cell separation. The sister swarmer cell, in contrast, enters the S-phase after the stalk formation initiates (Fig. 24.**1**) and then undergoes the same S-(C-) and gap or G2(D)-period (Chapter 22.1).

Chromosome replication seems to act as a cell cycle clock. The *C. crescentus* chromosome utilizes one specific origin of chromosomal DNA replication (*oriC*), which cannot replicate in *E. coli*. It contains a potential binding motif for DnaA binding and a strong *hemE* (see below) promoter as essential elements. The initiation of chromosome replication has been analyzed by determining the replication of low-copy-number plasmids and by measuring the time of expression of corresponding genes. OriC-driven plasmid replication occurs specifically during the swarmer-to-stalked cell transition period, coincidently with the initiation of chromosomal replication. Chromosome replication and segregation require sequential expression of genes, the products of which are necessary for both processes. Expression of the genes *gyrB* (encoding the gyrase B subunit) and *orf-1* is under temporal control. Both genes are silenced in the swarmer pole of the predivisional cell, and they are specifically transcribed from the chromosome when the swarmer cell differentiates into a stalked cell. The transcription of *gyrB* and *orf-1* and other genes involved in DNA replication occurs from the replication-competent chromosome in stalked and predivisional cells. It has been speculated that *hemE* promoter activity also controls replication initiation in the predivisional cell. The genomes destined for the future swarmer and stalk cells are in some way programed during the G2(D)-period in the predivisional cell.

In cells of *C. crescentus*, a DNA methyltransferase is expressed during the cell cycle immediately prior to cell division. Chromosomal consensus sites of the type GANTC are fully methylated in swarmer cells; the recognition sequences are hemimethylated upon DNA replication in stalked cells and become remethylated just prior to cell division. When the gene of the methyltransferase is placed under the control of a constitutive promoter, the DNA sites are fully methylated throughout the cell cycle. In cells of this mutant population, a high proportion of morphologically aberrant cells and cells having an additional chromosome are observed. The temporal control of the methyltransferase apparently contributes to the exact control of DNA replication during cell cycle and cell-cycle-dependent differentiation.

PleC, a histidine kinase, and **DivK**, a response regulator, are members of a signal transduction pathway that couples motility and stalk formation to completion of a late cell-division-cycle event. DivK functions also in a signal transduction pathway required for cell division (see Fig. 20.**17**).

Chromosomal replication seems to act as a cell cycle clock and initiates unequal segregation of mRNA to the progeny cells. Selective silencing of groups of genes on the chromosome located in the swarmer or stalked cell poles of the predivisional cell may contribute to the differential gene expression. Temporal control of DNA methylation affects DNA replication and cell differentiation.

24.1.3 Flagellar Gene Expression Is Under Temporary Control

The formation of the flagellar apparatus is the best-understood process of cell differentiation. The multistep process is particularly regulated at the transcriptional level. Approximately 50 genes are required for assembly ("morphopoiesis") and function of the *C. crescentus* flagellum and for chemotaxis. These flagellar genes are arranged in a regulatory hierarchy consisting of four levels of classes of genes that are under strict temporal control. The genes are expressed sequentially in the order corresponding to the sequence of protein assembly into the flagellum (basal body → hook → filament; see Fig. 24.**2b**). The expression of genes at each level depends upon gene products at levels above them in the hierarchy. Class 1 (*pleC, divJK?*) genes are presumed to encode products that respond directly to cell cycle signals and are required for expression of class II genes. Class II genes encode proteins of the flagellar basal body, switch proteins, proteins involved in transport of flagellar proteins (FliQ and FliR), and sigma factor σ^{54} (gene product of *rpoN*) of RNA polymerase. The consensus sequences in the promoter region of class II flagellar genes are clustered mainly in the −35 region (Fig. 24.**2c**). Proteins of the P- and L-rings, hook, and rod are encoded by Class III genes. Genes of class IV encode three flagellins (FliJ, K, L) (Fig. 24.**2a,b**).

Genes at the same level in the hierarchy (of the same class), even if organized in different transcription units, are transcribed at about the same time. The timing of transcription is controlled by a cascade of

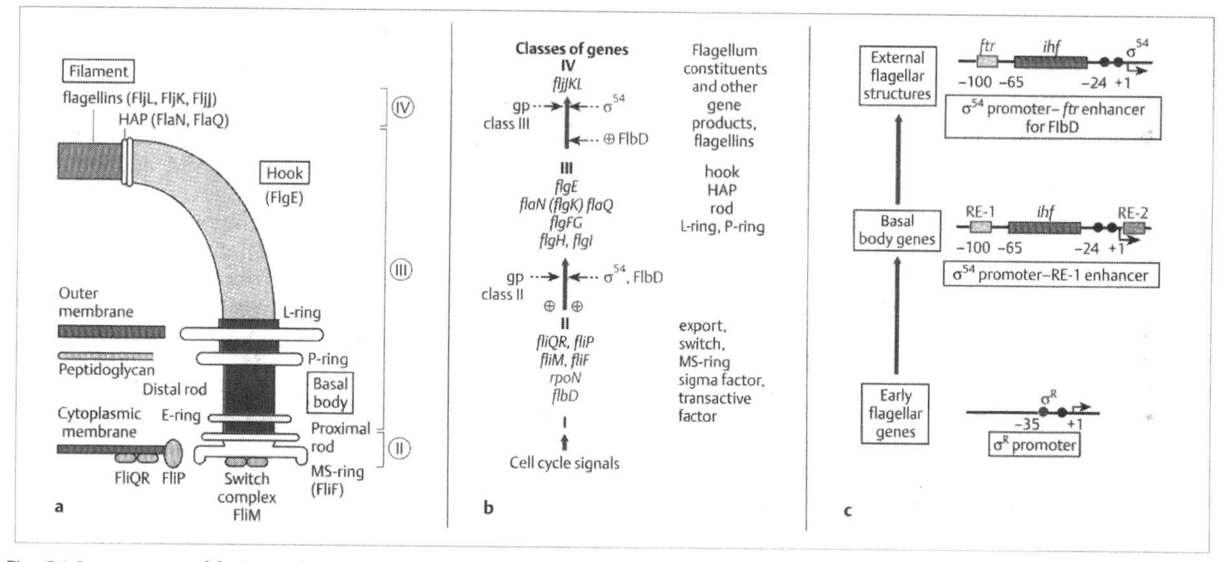

Fig. 24.**2a–c** **Assembly (morphopoiesis) of the *Caulobacter crescentus* flagellum.** Model of the regulatory hierarchy of expression of flagellar genes.
a Scheme of the flagellum structure showing the basal body, hook, and filament subunits and the proteins that are constituents of the flagellum or helper proteins. (HAP = **h**ook-**a**ssociated **p**roteins)
b Genes required for flagellum bio-synthesis are organized in four classes. Class I genes are reserved for genes that couple cell cycle signals to expression of class II genes. Each class of genes (II–IV) consists of several operons. The promoter regions of these operons are characterized by class-specific consensus sequences for sigma factors and *trans*-acting DNA-binding proteins (see Fig. 24.**2c**). The gene products of class II genes are required in addition to FlbD for activation of the promoters of the class III genes and together with phosphorylated FlbD for negative autoregulation of the *fliF* promoter. All genes in the class III transcription units are required for expression of class IV genes. The first morphological checkpoint, the completion of the MS-ring-switch complex, which requires all class II gene products, may be responsible for activation (phosphorylation) of the FlbD protein. The class II–IV genes and the function of their gene products are listed. The thick, red arrows indicate the hierarchy and the time course of gene expression. ⊕ at thin broken arrows indicates activation. Several gene products are essential for the assembly process (scaffold or morphopoietic proteins) or for regulation of gene expression.
c Different promoter classes operate at each level of the flagellar gene expression. Only the major flagellar promoter types are shown here. Several early class II genes are transcribed from a novel promoter sequence recognized by σ^R (regulatory sigma factor) of the RNA polymerase. Late flagellar genes possess σ^{54} promoters and IHF (similar to **I**ntegrated **H**ost **F**actor of *E. coli*)-binding sites. Differential regulation is achieved through the use of unique enhancer elements, such as RE-1 enhancer for the basal body, and RE-2 enhancer for the *flgF* operon. Genes encoding structural proteins of the hook and some flagellins have an *ftr* enhancer, the binding site for the FlbD transcriptional activator. The arrow at +1 indicates the transcription start site of the gene [after 1]

trans-acting regulatory proteins. Class II genes are transcribed by a σ^R-type RNA polymerase; genes of class III and IV are transcribed by a σ^{54}-type RNA polymerase. Upon a cell cycle signal, the response regulator FlbD (class II gene product) is likely phosphorylated by a kinase and binds in this form to DNA sequences, such as *ftr* elements, which control the timing of transcription of the σ^{54}-dependent class III and IV genes.

The different promoter classes that operate at each level of the flagellar gene expression are shown in Figure 24.**2c**. Differential regulation is achieved through unique enhancer elements. Promoters for the genes of the basal body contain an upstream RE-1 enhancer element; the *flgF* operon contains a downstream RE-2 enhancer element. The FlbD transcription activator binds to the *ftr* enhancer. Some of the genes in the hierarchy are negatively regulated. FlbD contains a sequence-specific DNA-binding activity within its carboxy-terminus; it exerts its effect as a positive and negative regulator by binding to *ftr* sequences to activate transcription of class III and IV genes and to repress transcription from a class II promoter. FlbD in its activated, phosphorylated form represses its own transcription and activates late flagellar genes; it is present at the swarmer cell pole. A protein equivalent to IHF (integration host factor) of *E. coli* accumulates in predivisional cells and binds to several class III and IV flagellar promoters, the transcription of which is specifically restricted to the chromosome in the swarmer cell pole. There is some evidence that a step in the DNA synthesis pathway is required for expression of some *fla* genes. Flagellar genes at or near the top of the hierarchy may be controlled, in part, by a unique transcription factor. The initiation of transcription of early genes in the flagellar regulatory hierarchy responds to cell cycle signals. Transcription of these genes is inhibited if DNA replication is interrupted. A model of the regulatory hierarchy for activation of genes of the flagellum apparatus is depicted in Figure 24.**2b,c**.

24.1.4 Gene Expression and Localization of Gene Products Are Compartment-specific

One mechanism for localization of gene products is the compartment-specific gene expression in the swarmer pole of the predivisional cell (for example, for *fliK* and *flaNQ*). Late flagellar genes continue to be expressed after septum formation. FlbD is present in both poles of the predivisional cell, but it seems to be phosphorylated in the swarmer pole, resulting in compartment-specific transcription of *ftr*-containing σ^{54} promoters (Fig. 24.**2c**).

The *C. crescentus* **pili** of strain CB15 are assembled on the swarmer cell pole from the 8-kDa pilin protein shortly after cell division and then retracted by the cell at about the time of stalk formation. The pilin subunits are synthesized continuously in stalked and predivisional cells and accumulate in the cytoplasm. The timing of the posttranslational temporal control of pili assembly is coupled to cell separation. As mentioned earlier, PleC and DivJ may contribute to regulation of polar morphogenesis and cell division.

Box 24.1. In the swarmer pole of the predivisional cell, late flagellar genes are transcribed and early flagellar genes are repressed; in the stalked pole, the early flagellar genes are transcribed. After cell division, the late flagellum structures are assembled in the swarmer cell. The early flagellum structures are assembled in the stalked cell. Components of the **chemotaxis machinery** (see Chapter 20) are localized to the swarmer pole of the predivisional cell. The chemotactic proteins methyltransferase (CheR), methylesterase (CheB), and methylreceptor proteins (MCPs) are synthesized just before cell division at the time of hook protein and flagellin synthesis. They segregate with the flagellar apparatus to swarmer cells at the time of cell division; they are, however, not exclusively targeted to a restricted membrane compartment, as are flagellins.

24.1.5 The Sec Functions Are Required for Cell Division and Stalk Formation

SecA protein mediates translocation of many proteins across the cytoplasmic membrane (see Chapter 19.4). *C. crescentus* mutants lacking SecA are able to release the polar flagellum, to degrade chemoreceptors, and to initiate DNA replication in swarmer-to-stalked cell transition, but they are unable to form a stalk, to complete DNA replication, or to carry out cell division. Hence, it has been concluded that SecA functions are required for cell division and stalk formation.

The genes for flagellum, pili, and chemotaxis are expressed in a regulatory hierarchy in which each level depends on the expression of the preceding gene level. The timing is controlled by *trans*-acting regulatory factors, sigma factors, and *cis* acting DNA elements. The global regulatory network responsible for cell-cycle-specific differentiation of cell poles seems to be stimulated by signals obtained from the DNA replication apparatus. Gene expression and localization of gene products is compartment-specific.

24.1.6 Polar Differentiation Has Been Observed in Other Bacteria

There are other examples of bacterial polar organization: formation of *Bradyrhizobium japonicum* surface-attachment proteins, actin assembly at the old cell poles of *Listeria monocytogenes*, and secretion of iron hydroxide at the concave site of the kidney-shaped cell of *Gallionella*.

24.2 Membrane Differentiation in Facultatively Phototrophic Bacteria Is Regulated by Shifts of Oxygen Tension and Light Intensity

Light is the energy source for photosynthesis (Chapter 13). Light is also an external stimulus for cell differentiation and phototaxis in phototrophic bacteria, as is oxygen. Oxygen is used as final electron acceptor in the respiratory chain of facultatively phototrophic bacteria under chemotrophic dark conditions. A change of oxygen partial pressure is the major external stimulus that causes changes in development of the photosynthetic apparatus. At low oxygen partial pressure, the formation of the photosynthetic apparatus is induced. Gradients in intensity and quality of light and concentrations of oxygen are stimuli for phototactic and chemotactic responses (Chapter 20.5).

24.2.1. The Syntheses of Pigments, Proteins, and the Photosynthetic Membrane Are Coordinately Regulated

The role of oxygen tension. During strict aerobic growth, *Rhodospirillum rubrum*, *Rhodobacter sphaeroides* and other facultatively phototrophic bacteria do not synthesize the photochemical reaction center (RC) and the light-harvesting (LH) pigment–protein complexes of the photosynthetic apparatus (see Chapter 13). The redox components, which are shared by the photosynthetic apparatus and the respiratory apparatus, are always present (constitutive) in the cytoplasmic membrane (CM) or in the periplasm, but their concentrations are modulated by light intensity and oxygen tension. Lowering of oxygen partial pressure (pO_2) in the dark below a species-specific threshold value induces a coordinated synthesis of bacteriochlorophyll (BChl), carotenoids, and pigment-binding proteins. In contrast, biosynthesis of NADH dehydrogenase and cytochrome oxidase and possibly of other components of the respiratory chain is induced when the pO_2 rises.

Concomitantly with formation of RC and LH complexes upon lowering of the pO_2, the CM invaginates and species-specific **intracytoplasmic membrane (ICM)** structures are formed (see Figs. 2.**11** and 24.**3**). The species-specific ICM structures are vesicles in *Rhodospirillum rubrum* and *Rhodobacter sphaeroides* and stacks of flat membrane sacs in *Rhodopseudomonas viridis* (see Figs. 2.**12** and 24.**3**). The formation of ICM is accompanied by the biosynthesis and incorporation of phospholipids and other membrane constituents into the growing membrane. **Phospholipids** are incorporated during cell division, while photosynthetic proteins are inserted into the membrane during the entire cell cycle. Under low pO_2, the ICM becomes the dominating membrane structure in the cell (Fig. 24.**3b**).

All ICM and CM structures are connected to each other by thin, tube-like membrane extensions and form a continuous membrane system. Cell homogenization with ultrasonic or French pressure treatment disrupts the delicate connections between ICM vesicles and the CM. Membrane fractions obtained after disruption of cells and separation of fragments by sucrose density gradient centrifugation differ in buoyant density, pigment and protein content, and enzymatic activities. These results and results of short-term labeling experiments with radioactive tracers have shown that the constituents of the photosynthetic apparatus and the respiratory chain and other activities are site-specifically incorporated and are not equilibrated in the membrane continuum by lateral diffusion. Although photosynthetic and respiratory apparatuses are localized on the same membrane continuum, the photosynthetic apparatus is enriched in the ICM system, and the respiratory apparatus dominates in the CM.

Low oxygen partial pressure is the major external stimulus that triggers biosynthesis of pigments and pigment-binding proteins of the **photosynthetic apparatus (PSA)** and of intracytoplasmic membranes (ICM). The constituents of the PSA are incorporated at specific sites into the cytoplasmic membrane (CM) and later, after invagination of the CM, preferentially into the ICM.

Light intensity and possibly light quality determine the amount and composition of the photosynthetic apparatus under anoxic or microoxic conditions. Lowering of light intensity causes an

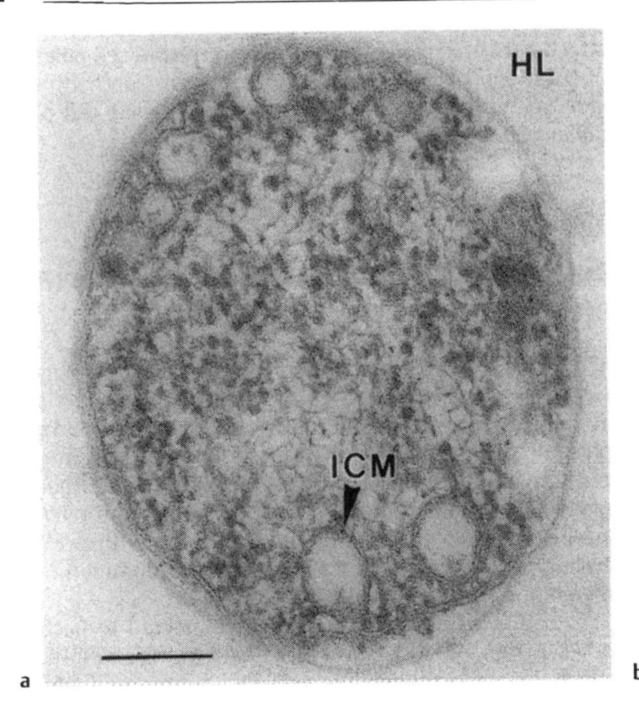

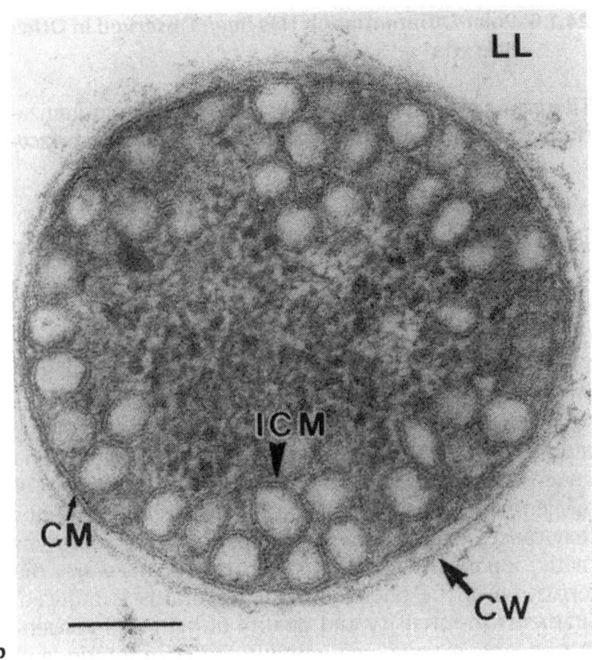

Fig. 24.3a,b **Cross sections of the photosynthetic bacterium *Rhodobacter capsulatus*** taken from cultures grown anaerobically under high light intensity (HL) and low light intensity (LL). The ICM (intracytoplasmic membrane) system fills most of the cell volume at low light intensity, but forms only a peripheral layer at high light intensity. The adaptations from low to high light intensity takes about two cell generations. The adaptation from high to low light, however, takes three to four generations because the growth is energy limited at the beginning. Bar equals 100 nm. Courtesy of J. R. Golecki

increase in the amount of ICM (Fig. 24.**3**) and an increase in the number of photosynthetic units per cell. In species such as *Rhodobacter capsulatus*, which contain a peripheral antenna in addition to the core antenna (see Fig. 13.**6**), the absorption cross section of the photosynthetic unit (i.e., the number of antenna BChl molecules per RC) increases when the light intensity is lowered. At high light intensities, the ratio of LH-BChl to RC-BChl decreases, i.e. the peripheral antenna (LHII, B800-850) become smaller, and the relative concentrations of ubiquinone and cytochromes *b* and *c* increase. The rate of photophosphorylation per RC also increases. The increase or reduction in the size of the antenna and the variation in the number of photosynthetic units in response to light intensity is a general principle of all photosynthetic organisms that allows them to optimize the light-harvesting process, i.e. to adapt the area and number of photon-absorbing pigments relative to the RC and to coordinate the light harvesting with the primary processes of photosynthesis in the RC (see Chapter 13). The structural organization and the mechanisms of regulation differ much more in light-harvesting systems than in reaction centers. It should be mentioned again that antenna systems are not only used to absorb photons and to deliver the excitation energy to RCs—they are also active in dissipation of superfluous excitation energy by fluorescence emission or radiationless decay.

Light intensity modulates the formation and composition of the photosynthetic apparatus under anoxic conditions. Lowering of light intensity causes an increase in the amount of photosynthetic units and in the ratio of antenna to reaction center complexes, and the area of intracytoplasmic membranes is enlarged. After an increase in light intensity, the concentration of electron carriers (cytochromes and quinones) and the rate of photophosphorylation per reaction center increase.

24.2.2 The Photosynthetic Genes Are Regulated by Regulatory Networks

The coordinated synthesis of BChl, carotenoids, pigment-binding proteins, quinones, cytochromes, and phospholipids, and their ordered assembly into a growing membrane implies that regulatory networks that coordinate all processes from signal transduction to membrane formation are active. The following description is based on our knowledge about *Rhodobacter capsulatus*, which belongs to the α-branch of the Proteobacteria. Comparative studies with other species of facultatively phototrophic purple bacteria have shown that they have similar gene organization and mechanisms of regulation. However, not much is known about the gene organization and regulation in purple sulfur bacteria (*Chromatiaceae*) and green sulfur bacteria (*Chlorobium* and relatives).

Most of the operons for photosynthetic genes are clustered. The genes encoding enzymes for bacteriochlorophyll (BChl) synthesis and the late steps of carotenoid synthesis, genes encoding proteins of the reaction center (RC) and the core antenna (LHI) complex, and genes encoding regulatory proteins are clustered in a 46-kb region of the *Rhodobacter capsulatus* chromosome, comprising about 1.5% of the entire genome. The operon encoding *puc* genes for the peripheral antenna complex LHII is located outside of this cluster (Fig. 24.**4**). The promoter regions of some operons are located within the upstream operons. For example, the promoters of the *puf* operon lie in the *bchCXYZ* gene cluster. Transcription of genes can start from different promoters, as indicated by the arrows above the genes in Figure 24.**4**. The strong promoter of the *puf* operon, which is activated under anoxic conditions, is located within the *bchZ* gene. The rate of transcription from this promoter can increase 20–50-fold. Under medium oxygen tensions, a readthrough coupling of transcription of the more weakly expressed *crt* and *bch* genes with the transcription of the *puf* and *puh* operons has been observed (see long and thin arrows in Fig. 24.**4**). This so-called "**superoperonal organization**" allows readthrough transcription of several operons and is largely responsible for a balanced low-level gene expression under oxic and semioxic growth conditions; it also allows coregulation under induced conditions.

Genes encoding enzymes for bacteriochlorophyll synthesis and carotenoid synthesis, pigment-binding proteins, and regulatory elements are organized in a large DNA cluster. A balanced low-level gene expression starts from weak promoters and comprises **superoperonal DNA structures**. Anoxic or microoxic conditions lead to coordinated transcription of several operons in different magnitudes.

Regulatory circuits for control of photosynthetic genes are stimulated by variations of oxygen tension or light intensity. The stimulus of lowering the oxygen tension below a species-specific threshold value in dark cultures of *Rhodobacter capsulatus* is presumably sensed by a redox system. It is not known how the stimulus of low pO_2 is sensed and converted to a signal for the cell. Upstream of the *puf* and *puc* operons, regulatory regions are present that are sensitive to cis- and trans-acting elements under the influence of changes in oxygen and light concentrations. With one stimulus, the various photosynthetic genes are expressed at quite different levels (3–100-fold). The putative *puf*, *puh*, and *puc* promoter regions share no similarity to known eubacterial promoter elements. Regions of dyad symmetry at positions up to −100 relative to the start site of transcription are putative binding sites for trans-acting regulatory proteins.

The two-component regulatory system of RegA and RegB of *Rhodobacter capsulatus* is involved in transactivation of gene expression for RC and LH proteins

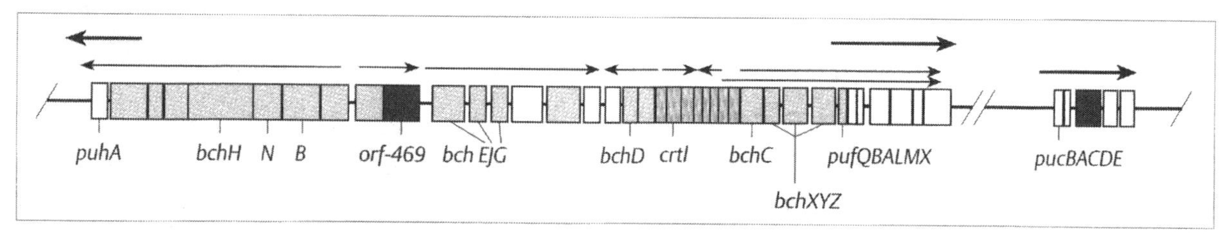

Fig. 24.**4** **The photosynthetic gene cluster of *Rhodobacter capsulatus* (46 kb) and its expression.** *bch*, genes encoding enzymes of bacteriochlorophyll synthesis (light red): *crt*, genes for enzymes of carotenoid synthesis (gray); *puf* operon, genes encoding the L and M subunits of the reaction center and the α- and β-proteins of the light-harvesting complex (B870) (white), PufQ and PufX are regulatory proteins of unknown function; *puhA*, encodes the H subunit of the reaction center; *puc* operon, located outside of the photosynthetic gene cluster on the chromosome of *Rhodobacter capsulatus*, encodes the pigment-binding proteins α, β, and γ (*pucBAE*) of the light-harvesting complex II (B800–850) and the regulatory proteins C and D (*pucCD*). The arrows above the gene boxes indicate the transcripts, some of which are readthrough transcripts of several operons. *orf-469* encodes a DNA-binding protein that acts as a repressor of promoters for Bchl, Crt, and light-harvesting II gene expression under oxic conditions. The heavy arrows indicate the transcripts that start from strong promoters. Red boxes indicate regulatory proteins

induced by low oxygen tension. RegB comprises a membrane-spanning sensor kinase that is autophosphorylated under anoxic conditions, and RegA, a cytoplasmic response regulator that communicates through a phosphorylation cascade. Under anoxic conditions, RegA-P positively controls the promoter activity of the *puf*, *puh*, and *puc* operons, similarly to other two-component systems (see Chapter 20).

In addition to molecular oxygen, light controls the photosynthetic genes (Fig. 24.**5**). The light effect is, however, overlapped by the strong influence of oxygen tension. On the genome of *Rhodobacter capsulatus*, an *hvrA* locus is located close to the *regB*, *regA* region. HvrA binds to a region 90–100 bp upstream of the transcription start of the *puf* and *puh* operons. The sensor for light is unknown; it is speculated to be a blue-light-absorbing pigment because strong blue-light inhibits photopigment expression. Although LHII polypeptide synthesis is inversely regulated by light intensity, the level of *puc* mRNA is not reduced at higher light intensity. The synthesis of LHII polypeptides might be regulated posttranscriptionally. A third regulatory system (Orf-469; Fig. 24.**5**) acts under high oxygen tension as a repressor of promoters for bacteriochlorophyll, carotenoid, and LHII gene expression.

There are more elements in the global regulatory network responsible for an optimal adaptation of the bacterial photosynthetic apparatus to variation in light intensity, oxygen tension, and presumably to stimuli of metabolism, the effects of which have been observed, but the regulatory circuits have not been resolved. The existence of a complex regulatory network for the differentiation of the photosynthetic apparatus has been indicated mainly by genetic methods (see Chapter 20.1.3).

Stimulus	Sensor	Regulator		Operons
$\Delta\Phi$ low light intensity	→ ?	→ HvrA activator	(+)	→ *puhA* → *pufQBALMX*
ΔpO_2 (high oxygen tension)	→ ?	→ Orf-469 repressor	(–)	→ *bchCXYZ* → *bchH* → *bchD* → *crtI* → *pucBACDE*
(low oxygen tension)	→ RegB	→ RegA	(+)	→ *pufQBALMX* *puhA* *pucBACDE*

Fig. 24.**5** **Regulatory circuits that control the transcription of photosynthetic genes** under the influence of changes in light intensity or oxygen partial pressure. The operon structures are depicted in Figure 24.**4**. The functions of the gene products are described in the text. The regulatory interaction of the modulons drawn in this figure is unknown.

24.2.3 mRNA Stability Is Posttranscriptionally Regulated

The gene products of an operon are required in very different amounts. Although *pufQBALMX* is a transcriptional unit, the gene products are synthesized in quite different ratios (e.g., PufBA/PufLM15/1 and the amount of PufQX synthesized is much less than that of PufLM). The synthesis of such very different amounts of proteins from the *puf* operon results mainly from different stabilities of the mRNAs: the half-life of the 2.7-kb *pufBALMX* mRNA is 5 min, and the half-life of the 0.5-kb *pufBA* mRNA is 20 min. A high pO_2 reduces and a low pO_2 increases the half-lives of these mRNA molecules. The half-life of mRNA is determined by individual segments of the polycistronic mRNA that have different stabilities. There are mRNA-destabilizing elements that serve to target initial endonucleolytic cleavage within the region coding for RC proteins. Secondary structures in regions coding for antenna proteins are stabilizing elements that prolong the lifetime of mRNA.

Other posttranscriptional processes that affect the biosynthetic rates of photosynthetic components are the efficiency of translation and the regulation of enzyme activities.

Oxygen tension and **light intensity** are stimuli that regulate the expression of the photosynthetic genes via a global regulatory system discordantly. Few of the *trans*-acting proteins that repress or activate the transcription of genes have been analyzed. The amount of gene products also depends on several posttranscriptional processes, such as mRNA stability, regulation of translation, and enzyme activity.

24.2.4 Bacteriochlorophyll Synthesis Is Regulated at Transcriptional and Posttranscriptional Levels

BChl, carotenoids, and pigment-binding proteins are assembled into RC and LH complexes in stoichiometric amounts, which are specific for each complex. The pools of free BChl and proteins are very small. The mechanism of the fine-tuned co-regulation is unknown. Pigment synthesis is regulated at the transcriptional and posttranscriptional levels. Transcription of *bch* and *crt* genes increases 2–12-fold in response to lowering of the oxygen tension. Furthermore, BChl synthesis is regulated, as are many anabolic pathways, by feed-back inhibition of aminolevulinate synthase, the first enzyme of the tetrapyrrol synthetic pathway, and at the branching point of heme and BChl synthesis at protoporphyrin IX, i.e., Mg-chelatase and methyltransferase. The end products, which affect feed-back inhibition at different key enzymes, have not been identified in vivo. In vitro, several porphyrins and hemes are active. Another element that might contribute to co-regulation is the *pufQ* gene product. PufQ is possibly a BChl-accepting protein that transfers the pigment moiety to the proteins of the antenna or RC complexes.

Changes in oxygen partial pressure and light intensity regulate bacteriochlorophyll and carotenoid synthesis at the transcriptional level (gene activation) and at the post-transcriptional level (feed-back regulation of key enzymes), transfer of BChl to their accepting proteins, and other unknown mechanisms.

24.2.5 The Assembly of the Functional Complexes of the Photosynthetic Apparatus Is a Multi-step Process

In the previous section it was mentioned that the constituents of the pigment–protein complexes of the photosynthetic apparatus are synthesized in stoichiometric amounts. Pigments or their precursors and proteins are imported from the cytoplasm into the membrane and assembled to form highly organized RC and LHI core complexes and the peripheral LHII antenna. The nascent pigment-binding, hydrophobic proteins are protected from misfolding after translation by binding to chaperones such as DnaK and GroEL.

Targeting of the proteins and insertion into the membrane is supported by membrane-associated helper proteins. After insertion, the antenna proteins are oriented with their N-terminal domains toward the cytoplasm and with the C-termini toward the periplasm (Fig. 24.**6a**). The N-termini of the α- and β-proteins interact with each other and stabilize the complex (Fig. 24.**6**). As the fifth ligand of the Mg atom in the tetrapyrrol ring of BChl the BChl molecules are bound to a highly conserved histidine residue in the hydrophobic transmembrane helix and by non-covalent bonds to amino acids in other positions. BChl–BChl and BChl–protein interactions stabilize the assembled complex and cause the spectral shift in the absorption from 770 nm (monomeric BChl in acetone) to 820 nm (pigment–protein monomer) to 870 nm (pigment–protein oligomer). The monomeric subunits of the antenna complexes aggregate to form ring-like oligomeric structures of defined size that optimize exciton transfer and stability of the complexes (Fig. 24.**6b**). After a shift to non-inducing conditions (e.g., phototrophic to chemotrophic conditions), the synthesis of the pigment–protein complexes is stopped; the complexes are not degraded, but are diluted out from the growing membrane.

The ring-like oligomeric structure of LH complexes has been verified by electron microscopy and image processing of the crystallized LHI of *Rhodospirillum rubrum* (Fig. 24.**6b**) and by high-resolution X-ray spectroscopy of the crystallized LHII complex of *Rhodopseudomonas acidophila*. This LHII complex is a nonamer of the subunit. Two concentric cylinders of 9 α-proteins (inner cylinder) and 9 β-proteins (outer cylinder, with a radius of 34 Å) sandwich 18 BChl molecules between the cylinders. The tetrapyrrol rings are located parallel to the cylinder axis (membrane normal). Nine additional BChl *a* molecules are packed between the β-subunits with their tetrapyrrol ring parallel to the membrane surface. The overlapping and interacting BChl molecules of one LH complex function like a storage ring and facilitate the delocalization of the excited states and equilibration of excitation energy within the LH complex and over the antenna system.

The **assembly** of the light-harvesting pigment–protein complexes is a multistep process, beginning with the synthesis of pigment and proteins. Translation, transport and targeting of proteins to the membrane are supported by helper proteins. Targeting is proposed to be a site-specific process, which is followed by insertion of the proteins into the membrane and translocation across the lipid double layer, a step that depends on the proton motive force. Pigments are bound non-covalently and stabilize the complex. The binding of BChl to protein and BChl–BChl interactions cause a shift of the near-infrared absorption peak to the red side of the spectrum, an orientation of the tetrapyrrol ring relative to the plane, and an optimal distance between the pigment molecules in order to optimize transfer of exitation energy.

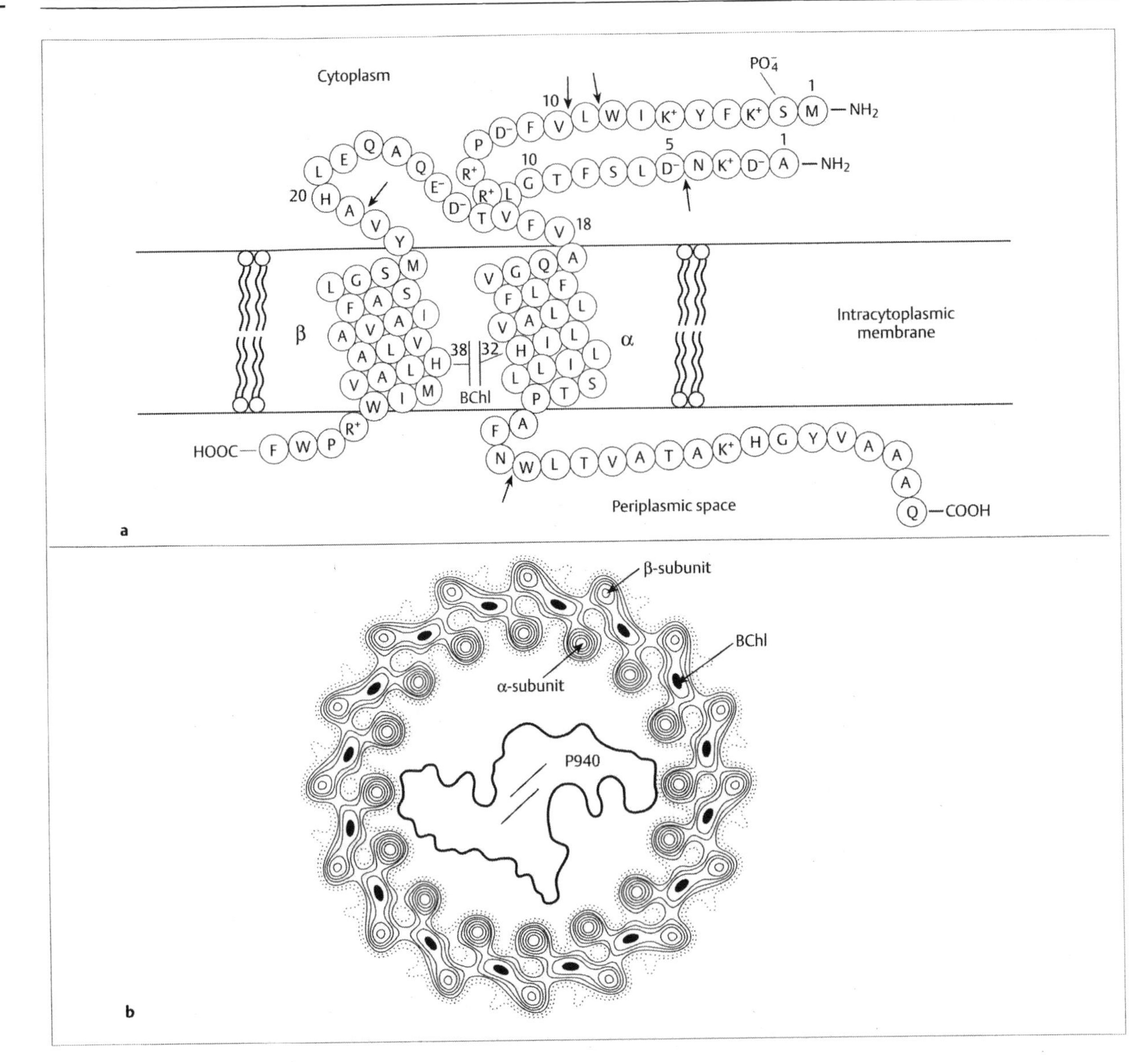

Fig. 24.6**a** **Scheme of the topology of the LHI α and β polypeptides in the membrane.** The three-dimensional structure of the oligomeric light-harvesting complexes is described in the text. The amino acid sequence of *R. capsulatus* LHI is given in the one-letter code. The NH_2-terminal domains of these and other LH proteins are exposed on the cytoplasmic surface and the COOH-terminal domains point toward the periplasmic side of the membrane. Bacteriochlorophyll is bound coordinately by its Mg ion to His-32 and His-38 in the transmembrane domain of the polypeptides. Arrows indicate where the NH_2-terminal and COOH-terminal peptides are split off by proteases. The central domains of the polypeptides are composed of hydrophobic amino acid residues and form an α-helical structure that traverses the lipid double layer of the membrane. Serine 2α is phosphorylated during insertion.

b The ring-like oligomeric structure of the light-harvesting complex I (LHI, B880). The projection map of two-dimensional crystals of the LHI complex from *Rhodospirillum rubrum* shows 16 pairs of the α- and β-subunits arranged in a 116-Å-diameter cylinder with a 68-Å hole in the center. The α-helices of the α- and β-subunits traverse the membrane; a top view is shown. It is hypothesized that the α-helices of the outer ring are more tilted relative to the membrane plane than the α-helices of the inner ring. The pigments bacteriochlorophyll *a* (BChl *a*) and spirilloxanthin are presumably located between the outer and the inner ring. The outline of the reaction center (RC) of *Rhodopseudomonas viridis* with the special pair of BChl *b* molecules (P940) is drawn in the ring to show how the core complex (RC + LHI) is presumably organized. [from 2]

24.3 Heterocysts Are Cells in Cyanobacteria Specialized for Fixation of Dinitrogen

Cyanobacteria are oxygenic photosynthetic prokaryotes (Chapter 13). Some genera of filamentous cyanobacteria can convert vegetative cells into heterocysts, akinetes (spores), or **hormogonia** (short, motile filaments, not sheathed, with a cell shape distinguishable from the vegetative cell) by cell differentiation. **Akinetes** are thick-walled resting cells produced by certain genera (e.g. *Cylindrospermum, Anabaena*) of heterocystous cyanobacteria in response to limitation in nutrients or energy supply. They are perennial structures that enable strains to survive periods of cold and dry conditions ("exospores").

Heterocysts are specialized cells with the primary function of dinitrogen fixation. They develop from vegetative cells of *Anabaena, Nostoc, Cylindrospermum*, and other genera of filamentous cyanobacteria upon depletion of combined nitrogen. Heterocysts are the only new cell type that develops under these conditions, within about 24 h. Heterocysts are the only cells that fix dinitrogen; they occur at regular intervals along each filament. The principal organism of most studies is *Anabaena*.

24.3.1 Heterocysts Differ from Vegetative Cells in Structure and Function

Mature heterocysts, slightly larger than vegetative cells, are surrounded by three thickened extra layers external

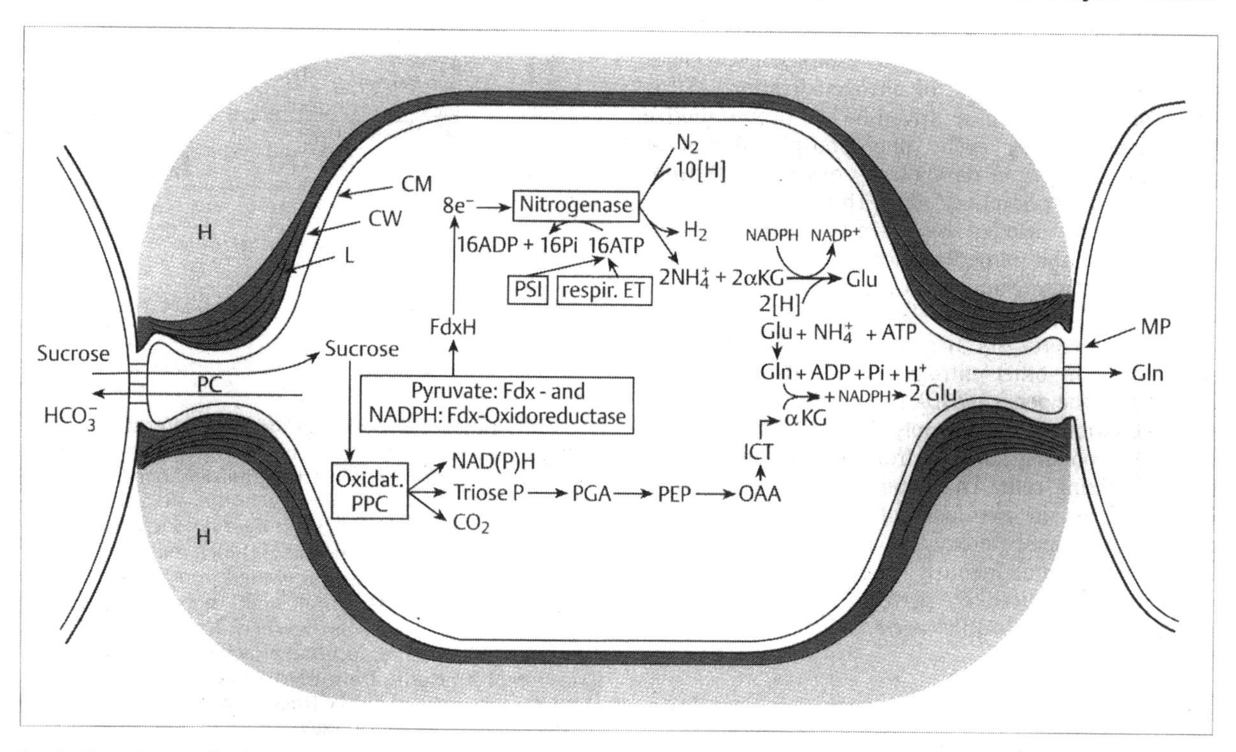

Fig. 24.7 **Scheme of a heterocyst and adjoining vegetative cells in a filamentous cyanobacterium (*Anabaena*).** Principal structural elements and metabolic interactions between heterocysts and vegetative cells. The heterocyst is enveloped by the cell wall (CW), the laminated glycolipid layer (L), and the polysaccharide layer (H). Microplasmodesmata (MP) join the cytoplasmic membranes (CM) of the neighboring cells at the end of the pore channel (PC) of the heterocyst. The heterocyst is a sink for carbohydrate, possibly sucrose, and exports glutamine to the neighbor cells.

Abbreviations:
ICT, isocitrate
Fdx, ferredoxin
Gln, glutamine
Glu, glutamate
αKG, α-ketoglutarate
OAA, oxaloacetate
oxidat. PPC, oxidative pentose phosphate cycle
P_i, inorganic phosphate
PEP, phospho*enol*pyruvate
PGA, 3-phosphoglycerate
PSI, photosystem I
respir. ET, respiratory electron transport
triose P, triose phosphate
[after 3]

to the vegetative cell wall (Fig. 24.**7**). The innermost laminated glycolipid layer (L) is followed by an intermediary homogenous layer and an outermost fibrous polysaccharide layer (H). The glycolipid and the polysaccharide layers of the heterocyst envelope are barriers for diffusion of gas into the heterocysts. The glycolipid layer thickens when the O_2 partial pressure increases. The exchange of metabolites between the heterocyst and the neighboring vegetative cell is facilitated by microplasmodesmata (MP), small pore channels in the cross wall that join the plasma membranes of both cells at the end of the pore channel of the heterocysts (Fig. 24.**7**).

In mature heterocysts, only photosystem I (Chapter 13.3) is active. The O_2-evolving photosystem II components, phycobilisomes, phosphoribulokinase, and ribulose bisphosphate carboxylase, are lost early during heterocyst differentiation. Thylakoids are present, but they are significantly modified at the final stage of heterocyst maturation. Increased respiration scavenges residual O_2 from the site of nitrogen fixation. ATP is supplied by ATP synthase driven by the proton motive force generated by cyclic photophosphorylation at photosystem I and in darkness by oxidative electron-transport-chain phosphorylation. The heterocyst-specific ferredoxin, encoded by the gene *fdxH*, is in its reduced form the immediate donor of electrons to the nitrogenase. Energy and reducing equivalents are conserved for fixation of dinitrogen by nitrogenase. Nitrogenase in heterocysts has a high similarity to its counterpart in other nitrogen-fixing bacteria and is encoded by the genes *nifHDK* (Chapter 8.5). N_2 fixation and respiration require a supply of reductant in the form of a sugar (possibly sucrose) from the adjacent vegetative, CO_2-fixing cells. The heterocysts in turn release fixed nitrogen in the form of glutamine, which is exported from the heterocyst to cells along the filament while being simultaneously metabolized. Ferredoxin seems to be reduced by pyruvate : ferredoxin oxidoreductase or via NADPH : ferredoxin oxidoreductase (Fig. 24.**7**).

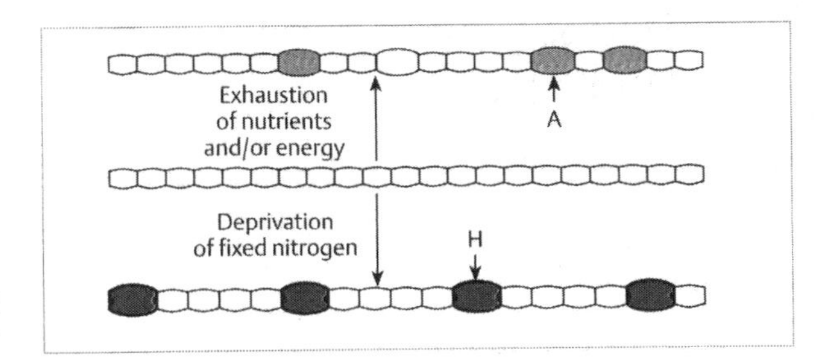

Fig. 24.**8** **Schematic diagram illustrating differentiation of vegetative cells of *Anabaena* into heterocysts (H) and akinetes (A, spores) under stress conditions.** Akinetes can germinate and give rise to new filaments. Heterocysts develop with a precise spatial relationship (approx. every tenth cell)

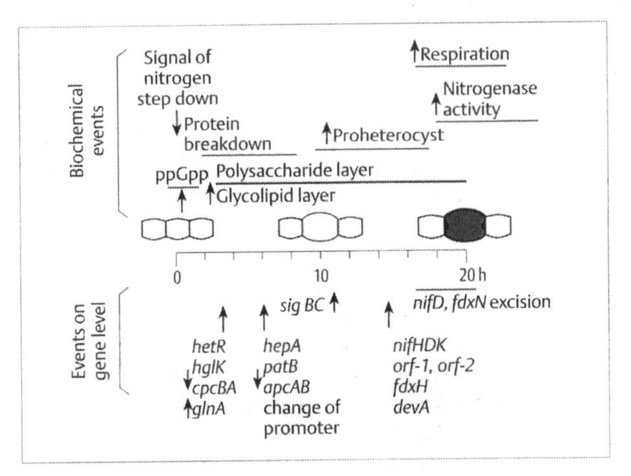

Fig. 24.**9** **Time course of heterocyst differentiation.** The regulatory circuits and the components of the global regulatory network (stimulon) that determine the development of the heterocyst are largely unknown. Few events are depicted. The external stimulus is a step-down in fixed nitrogen. *nifHDK* are the genes encoding the nitrogenase. During rearrangement of the genome, DNA elements are excised from the genes *nifD* and *fdxN*. Genes: *fdx*, ferredoxin; *sigBC*, sigma factor; *cpcBA*, phycocyanins; *apcAB*, allophycocyanins. Upward arrows indicate activation of genes, increase of activity of enzymes, and formation of structures. Downward arrows indicate repression of genes or decrease of fixed nitrogen. Arrows also indicate the start of the respective processes

24.3.2 The Differentiation of Heterocysts Comprises Extensive Changes in Structure and Function

Heterocysts arise in filaments when exogenous fixed nitrogen (NH_4^+, NO_3^-) is depleted and dinitrogen (N_2) is the only nitrogen source (Fig. 24.**8**). Complete differentiation of a vegetative cell into a mature, nitrogen-fixing heterocyst requires about 20 h. In an early step of differentiation, the external layers of polysaccharide are formed, followed by synthesis of the glycolipid layer. Within hours, two proteases are activated or synthesized. One breaks down phycobiliproteins (see Chapter 13.3 and Fig. 24.**9**); the other, a Ca^{2+}-dependent protease, degrades the enzymes ribulose-5-phosphate kinase, ribulose bisphosphate carboxylase of the Calvin cycle (Chapter 8.1.1), and other proteins.

The synthesis of a number of proteins characteristic of heterocysts appears to be initiated in vegetative cells and in presumptive heterocysts, but is continued only in developing heterocysts. A morphological distinction between presumptive heterocysts and all other cells can be observed after 5–10 h. This is the time of commitment to the completion of differentiation. At this point, proheterocysts do not revert to vegetative cells; however, isolated proheterocysts can undergo such a reversion. The thylakoids are reorganized to contain photosystem 1 and some components of photosystem II (such as the D1 protein), but they do not contain any O_2-evolving activity or electron transport to photosystem I. The thylakoids become contorted. Later, in proheterocysts, the fibrous and the homogeneous polysaccharide layers are completed, and the junctions to adjacent cells decrease in diameter. A plug of the storage copolymer **cyanophycin** (aspartate-arginine) can be deposited in the pore channel (PC in Fig. 24.7). The appearance of nitrogenase activity is a late step. Under oxic conditions, it lags behind appearance of mature heterocysts. Under anaerobiosis, deprivation of combined nitrogen and inhibition of O_2 evolution, nitrogenase synthesis is induced earlier. Admission of O_2 to induced, but undifferentiated cells results in rapid inactivation of nitrogenase and disappearance of the respective mRNA. In addition to nitrogenase, hydrogenase and enzymes of the oxidative pentose pathway, glutamine synthetase and elements of the transport systems for glutamine, and carbohydrates are synthesized. Heterocysts of *Anabaena* are formed at regular intervals, approximately after every tenth vegetative cell. The vegetative cell that differentiates is placed mid-way between two existing heterocysts. The spacing pattern can be altered by mutation and by environmental manipulation, such as reducing the length of the filament by mechanical treatment, changes in light intensity, and molybdenum starvation. Other cyanobacteria develop a single heterocyst at the end of a filament (e.g. *Cylindrospermum*).

24.3.3 The Regulatory Network for Heterocyst Formation Is Stimulated by Deprivation of Combined Nitrogen

Differentiation of vegetative cells into heterocysts is a multi-step process regulated by a cascade of two-component sensor–regulator systems (see Chapter 20), of which only few parts have been identified. The molecular mechanism of how pattern formation in a filament of *Anabaena* is initiated is unknown. A general signal for amino acid starvation, such as synthesis of polyphosphorylated nucleotides (ppGpp, "alarmones"), has been reported (within 30 min after nitrogen step-down), but a classical stringent response has not been observed. The differentiation of heterocysts may be initiated randomly. Inhibition of heterocyst formation then propagates bidirectionally along a filament from the proheterocyst.

Box 24.2 Two gradients are present in filament of *Anabaena* grown with N_2 as sole nitrogen source. From heterocysts, glutamine is exported to the vegetative neighboring cells, which is gradually consumed during the linear diffusion pathway. In the opposite direction, carbohydrate is produced in vegetative cells and consumed in heterocysts. An alternative model proposes transport systems that are activated by deprivation of nitrogen and that drain nitrogen from adjacent cells. Consequent deprivation of nitrogen from the adjacent cells activates transporters that drain nitrogen from their distal neighboring cells. The final cell, unable to drain any nitrogen other than N_2, is induced to become a heterocyst.

Box 24.3 *hetR* is expressed in all cells when fixed nitrogen is in the medium. The amount of HetR is controlled by positive autoregulation. Inactivation of HetR prevents differentiation of heterocysts. Supernumerary copies of ***hetR*** lead to the formation of heterocysts in the presence of combined nitrogen and to the formation of multiple contiguous heterocysts in the absence of fixed nitrogen. **HepA** is also required for heterocyst maturation and is induced within 6 h after removal of combined nitrogen from the medium. HepA is an ATP-binding transport protein at the cytoplasmic membrane. Presumably, HepA is involved in translocation of polysaccharides across the cytoplasmic and outer membranes. Heterocysts of *hepA* mutants lacking a polysaccharide layer are more O_2-permeable. The ***hglK*** gene is induced 3 h after nitrogen deprivation, and its product seems to be involved in the synthesis of the glycolipid layer. ***hetN*** may encode an enzyme that mediates intercellular interactions that regulate development of heterocysts in *Anabaena*.

PatA is likely a member of a pair of environment-sensing response regulator proteins that controls some steps of heterocyst differentiation. *patA* is transcribed at a low constitutive level in the presence and absence of combined nitrogen. The *patB* gene product contains in the C-terminal region a helix–turn–helix motif indicative for DNA-binding and in the N-terminal region a domain characteristic of [4Fe-4S] ferredoxins. ***patB*** is transcribed 3–6 h following nitrogen step-down. PatB appears to be required for repression of heterocyst differentiation in response to the redox potential of cells.

Transcriptional control seems to be an important mode of regulation during heterocyst differentiation. Patterns of gene expression from strong promoters have been made visible through transcriptional fusion of cyanobacterial genes to bacterial luciferase genes. Using this method, a spatial pattern of *hepA* expression, visualized by spaced luminescing cells, was demonstrated many hours before differentiation became discernible morphologically. Several genes, for example, *glnA* (encoding glutamine synthetase), *atpBE* (ATP synthase), *psbB* (a PSII protein), *sigA* (a sigma factor), and *hetR*, are transcribed from different promoters in heterocysts and in vegetative cells. Promoters of vegetative cells contain typical *E. coli*-like −10 sequences. Promoters of heterocysts are recognized by an RNA polymerase with a different sigma factor (encoded by *sigB* and *sigC*). The promoter for *nifHDK* is expressed only in heterocysts, and the promoter for *rbcLS* (gene for ribulose bisphosphate carboxylase) is expressed only in vegetative cells.

Within 2 h of nitrogen deprivation, the expression of the gene *hetR* increases strongly and remains active in mature heterocysts. HetR lacks any known DNA-binding motif, but is required for heterocyst maturation and is believed to act as a master switch of heterocyst differentiation.

24.3.4 The Genome of Heterocysts Is Rearranged During Differentiation

Vegetative cells of *Anabaena* contain two DNA elements that interrupt operons for nitrogenase and assembly proteins. During heterocyst differentiation, an 11-kbp DNA element imbedded in the *nifD* gene is excised by recombination between 11-bp repeats at the ends of the elements. The recombination is catalyzed by XisA, which is expressed during heterocyst development. A second element of 55 bp, which interrupts the locus ***fdxN*** [encoding genes for Fe–S centers and Mo cofactor of nitrogenase (see Chapter 8.5)], is also excised during heterocyst differentiation by site-specific recombination catalyzed by YisF, which is located within the element. The *trans*-acting factor NtcA (BifA) is present in vegetative cells and in heterocysts. NtcA binds to the upstream regions of *glnA*, *rbcL*, *nifH*, and *xisA*. Interruption of *ntcA* inhibits dinitrogen fixation and formation of heterocysts.

24.3.5 The DNA-dependent RNA Polymerase Uses Different Sigma Factors During Differentiation

The major form of RNA polymerase in vegetative cells of cyanobacteria binds to promoters with −10 and −35 consensus sequences that resemble the corresponding sequences from *E. coli*. Genes with housekeeping functions and those coding for proteins of the photosynthetic apparatus also have such promoter regions. Genes expressed only in heterocysts have no apparent promoter consensus sequences. The genes encoding sigma factors SigB and SigC are activated after starvation of combined nitrogen. The occurrence of multiple-factors has been observed also in the non-differentiating cyanobacterium *Synechococcus* and is not restricted to cell differentiation. More than 1 000 genes are differentially expressed in heterocysts. Hence, they constitute a stimulon (see Chapter 20). Few of these genes have been structurally and functionally identified. Several of their products have pleiotropic functions, and a few are related to well-known families of bacterial regulatory proteins. Many more must be identified in order to understand the complete cascade of regulatory events resulting in heterocyst formation.

Heterocysts are specialized cells in filamentous cyanobacteria with the primary function of N_2 fixation. Differentiation of vegetative cells into heterocysts and formation of a regular heterocyst pattern along the filament is a multi-step process, which is initiated by a step-down of bound nitrogen and regulated by a cascade of sensor–regulator systems. Wall thickening and synthesis of a glycolipid layer, loss of photosystem II activity, degradation of phycobiliproteins, loss of CO_2-fixation activity, rearrangement of DNA, increase of respiratory activity, and induction of nitrogenase formation are the major events upon deprivation of fixed nitrogen source in the medium. Heterocysts export glutamine into the neighboring vegetative cells and import carbohydrate from them.

Further Reading

Adams, D. G. (1992) Multicellularity in cyanobacteria. Symp Soc Gen Microbiol 47: 341–384

Blankenship, R. E., Madigan, M. T., and Bauer, C. E., eds. (1995) Anoxygenic photosynthetic bacteria. Dordrecht: Kluwer

Bryant, D. A., ed. (1994) The molecular biology of cyanobacteria. Dordrecht: Kluwer

Buikema, W. J., and Haselkorn R. (1993) Cyanobacterial development. Annu Rev Plant Mol Biol 44: 33–52

Drews, G., and Golecki, J. R. (1995) Structure, molecular organization, and biosynthesis of membranes of purple bacteria. In: Blankenship, R. E., Madigan, M. T., and Bauer, C. E. (eds.) Anoxygenic photosynthetic bacteria. Dordrecht: Kluwer; 231–257

Gober, J. W., and Marques, M. V. (1995) Regulation of cellular differentiation in *Caulobacter crescentus*. Microbiol Rev 59: 31–47

Herdman, M. (1987) Akinetes: structure and function. In: Fay, P., and van Baalen, C. (eds.) The cyanobacteria. New York: Elsevier; 227–250

Marczynski, G. T., and Shapiro, L. (1995) The control of asymmetric gene expression during *Caulobacter* cell differentiation. Arch Microbiol 163: 313–321

McDermott, G., Prince, S. M., Freer, A. A., Hawthornthwaite-Lawless, A. M., Papiz, M. Z., Cogdell, R. J., and Isaacs, N. W. (1995) Crystal structure of an integral membrane light-harvesting complex from photosynthetic bacteria. Nature 374: 517–521

Newton, A., and Ohta, N. (1990) Regulation of the cell division cycle and differentiation in bacteria. Annu Rev Microbiol 44: 689–719

Tandeau de Marsac, N., and Houmard, J. (1993) Adaptation of cyanobacteria to environmental stimuli: new steps towards molecular mechanisms. FEMS Microbiol Rev 104: 119–190

Whitton, B. A. (1987) Survival and dormancy of blue-green algae. In: Henis, Y. (ed.) Survival and dormancy of microorganisms. New York: Wiley: 109–167

Wolk, C. P. (1982) Heterocysts. In: Carr, N. G., and Whitton, B. A. (eds.) The biology of cyanobacteria. Oxford: Blackwell Scientific; 359–386

Wolk, C. P., Ernst, A., and Elhai, J. (1994) Heterocyst metabolism and development. In: Bryant, D. A. (ed.) The molecular biology of cyanobacteria. Dordrecht: Kluwer; 769–821

Wolk, C. P. (1996) Heterocyst formation. Annu Rev Genet 30: 59–78

Wu, J., and Newton, A. (1997) Regulation of the *Caulobacter* flagellar gene hierarchy; not just for motility. Molec Microbiol 24: 233–239

Sources of Figures

1 Gober, J. W., and Margues, M. V. (1995) Microbiol Rev 59: 31–47

2 Karrasch, S., Bullough, P. A., and Ghosh, R. (1995) EMBO J 14: 631–638

3 Wolk, C. P. (1982) Heterocysts. In: Carr, N. G., and Whitton, B. A. (eds.) The biology of cyanobacteria. Oxford: Blackwell Scientific; 359–386

25 Sporulation and Cell Differentiation

Bacteria possess the capacity to endure long periods of stress, surviving for years in nutritionally poor environments and withstanding encounters with bacteriocidal agents, both chemical and physical in nature. Responses to stress have evolved into elaborate schemes that include irreversible cellular differentiation processes as well as patterns of development that sometimes blur the distinction between unicellular and multicellular organisms. Often, differentiation results in the formation of a cell type that has remarkable longevity and resistance properties. The classic example is the bacterial spore, which is characteristic of *Bacillus*, *Clostridium*, *Streptomyces*, and the myxobacteria (see Figs. 2.**5** and 2.**21**). The bacterial spore can be thought of as fulfilling two ecological objectives. First, it serves as the last resort among a myriad of processes that a bacterium activates in response to a harsh environment, and it can survive long periods of stress. Secondly, it provides a means by which a species can be disseminated throughout the environment, enabling the species to colonize diverse habitats. Spores can be carried by air currents, animals, and rain run-off for great distances. Hence, spore formers are ubiquitous in nature. The soil bacterium, *Bacillus subtilis*, has been isolated from a variety of ecosystems, including ocean sediments. For these reasons, the ecological importance of the spore formers cannot be overstated.

Spore formers offer fertile ground for the study of cellular differentiation and multicellular development. Fundamental issues of developmental biology can be addressed utilizing sophisticated techniques practiced by modern molecular biologists. Among the questions asked by those who ponder the sporulation process are: (1) How is the pattern of gene expression altered in response to stress so that the functions needed for cellular differentiation are mobilized?, (2) How do cells communicate with one another so that they act in concert or interact appropriately to carry out complex multicellular development?, and (3) how is gene expression compartmentalized to create two cell types with two different developmental fates from a single cell?

25.1 Sporulation in *Bacillus subtilis* Is a Multi-step Process That Can Be Dissected Genetically; The Formation of Endospores

A large body of information about the formation and composition of bacterial spores has emerged from the study of *Bacillus subtilis*, *Streptomyces*, and *Myxococcus xanthus* through biochemical and microscopic investigation.

The **morphological events** that transpire during the sporulation of *B. subtilis* have been particularly well documented. A depiction of the sporulation process is shown in Figure 25.**1**. Under laboratory conditions, *B. subtilis* cells grow and divide rapidly, undergoing binary fission following the formation of a centrally located cell division septum. Upon starvation and at high cell density, the first morphological manifestation of the sporulation process is observed: the asymmetrically positioned cell division septum or the sporulation septum. As with the cell division septum, the sporulation septum is composed of two membranes with a thin layer of peptidoglycan in between. Prior to this event, the cell is said to be at **stage 0** of sporulation. When the sporulation septum has been formed, **stage II** begins. (Stage I originally described the formation of an axial filament composed of the fused daughter nucleoids; however, it is not known if this is an event specific to sporulation.) At stage II, the bacterium is divided into a large cellular compartment, which develops into the mother cell, and a smaller compartment, which will become the spore. As stage II proceeds, the peptidoglycan of the sporulation septum is degraded, and the membranes appear to fuse. The ends of the mother cell membrane appear to move towards the proximal pole, thereby engulfing the smaller of the two compartments, and a double membrane surrounding the prespore is formed. When engulfment is complete, the **prespore** (or **forespore**) protoplast is formed, marking the beginning of **stage III**, during which the double membrane structure (or the germ cell wall) appears. At this point,

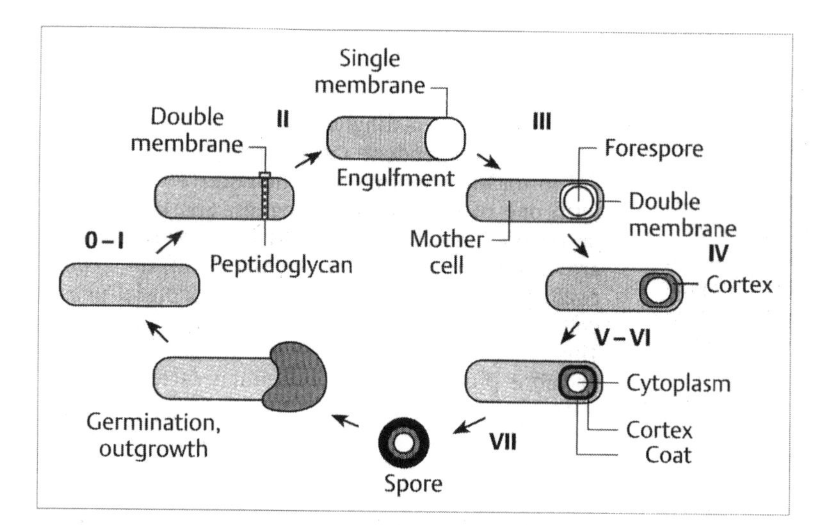

Fig. 25.1 **The developmental cycle of *Bacillus subtilis*.** The vegetative cell is shown at the left, labeled 0 to denote stage 0. The next cell in the clockwise direction is at stage II of sporulation. The double membrane and peptidoglycan of the sporulation septum are indicated. The single membrane of the engulfment step of stage II is shown, followed by the **forespore** of the stage III cell. The stage IV cell contains a forespore undergoing cortex formation. The stage V–VI cell shows a spore coat and completed cortex. This is followed by the release of the mature spore (stage VII). Germination and outgrowth involve rupturing of the spore and release of a vegetatively growing cell

stage IV begins, and peptidoglycan is deposited between the two membranes to form the **cortex** of the spore. Some time after cortex formation begins, the proteinaceous spore coat starts to assemble around the developing spore (**stage V**). The final stage of sporulation is **stage VI**, the maturation stage, when the spore becomes highly resistant to heat and chemicals (see Fig. 2.**21a,b**), becomes dormant, and has acquired the capacity to undergo germination (see Fig. 2.**21c**). Lysis (**stage VII**) of the mother cell follows, resulting in the release of a highly dehydrated and refractile (Fig. 25.**1**) spore.

A **genetic approach** to the study of spore formation in *B. subtilis* has involved the isolation of mutants that suffer a block in the sporulation process. The premise was that the mutations would identify genes that encode structural components of the endospore or important regulatory factors that govern the expression of genes that function specifically in differentiation. The mutations have been categorized according to the stage at which the sporulation process stops in the mutant cells. Thus, mutations of the *spo0* group (*spo0A*, *spo0B*, etc.) block the onset of sporulation, or before the sporulation septum is formed. The *spoII* mutant cells form the sporulation septum, but proceed no further. Cortex formation is blocked in the *spoIII* mutants, although the prespore protoplast is observed. Mutations that block post-engulfment events are not as easily classified. In general, the *spoIV* mutants contain an endospore with an incomplete cortex and no spore coat. *spoV* mutants have an incomplete spore coat, although the cortex appears normal. The *spoVI* mutants appear to have completed endospore formation, but lack the characteristic resistance properties of the mature spore (i.e., they are still sensitive to heat, chloroform, and/or lysozyme).

The formation of endospore involves events that occur in two separate cellular compartments. In the compartment that will become the spore, a set of prespore-specific genes is expressed (see Chapter 25.1.2). For example, the *ssp* genes, which encode the small acid-soluble proteins that bind to the spore chromosomal DNA, are expressed exclusively in the developing prespore. The *cot* genes, which encode the spore coat proteins that are deposited around the outside of the spore, are expressed in the mother cell. However, the influence of some prespore-specific and mother-cell-specific genes is not confined to events that occur within a single cellular compartment. There are genes that serve a regulatory function by exerting their control across the prespore/mother cell boundary. Thus, intercompartmental communication is an essential part of the sporulation process, just as intercellular signals serve to regulate development in higher organisms.

Sporulation is a simple form of differentiation that begins with an asymmetric cell division, yielding two cellular compartments that have different developmental fates: a prespore, which develops into the endospore, and a mother cell, which lyses, thereby releasing the endospore.

25.1.1 Initiation of Sporulation in *Bacillus subtilis* Is Activated by a Multitude of Cellular and Extracellular Signals

Two aspects of initiation of sporulation must be considered. One is the regulatory factors that function in activating sporulation-specific gene expression, and the other is the signals to which the regulatory factors respond. The **stimuli** are poorly defined, but appear to arise as a result of three conditions:

1. **Nutritional depletion.** Depletion of a readily metabolized carbon or nitrogen source is necessary for the activation of sporulation. A possible sporulation-activating signal that is generated under nutrient deprivation is a reduction in the level of **GTP**. When cultures growing exponentially under normally sporulation-repressing conditions are treated with purine analogues that inhibit GTP synthesis, the cells undergo sporulation despite the high concentrations of preferred carbon and nitrogen sources (e.g., glucose and glutamine) present in the growth medium. It was proposed that there is a positive regulatory factor that becomes activated or an inhibitor that becomes inactivated when GTP levels drop. Several proteins have been identified that are guanylylated as cells enter the stationary phase of growth. These include Enzyme I and HPr of the carbohydrate : phosphotransferase system (see Chapter 20.2.4). How this might relate to the induction of sporulation is not known.
2. **Cell density.** Interestingly, induction of sporulation is cell-density dependent. Cells of low density cultures fail to undergo sporulation when starved or treated with one of the purine analogues. Sporulation can be restored, however, when the cells at low density are resuspended in the cell-free filtrate ("preconditioned" medium) of late-exponential-phase cultures in which cells had begun to sporulate. This effect appears to be due to the presence of one or more proteinaceous extracellular factors, which are thought to serve as signals to ensure that differentiation is initiated at the appropriate cell density (quorum sensing, see Chapter 20.1.4).
3. **DNA synthesis** is required for the formation of the mother cell and spore chromosomes. Inhibition of DNA synthesis in late-exponential-phase cultures, in addition to blocking the initiation of the sporulation

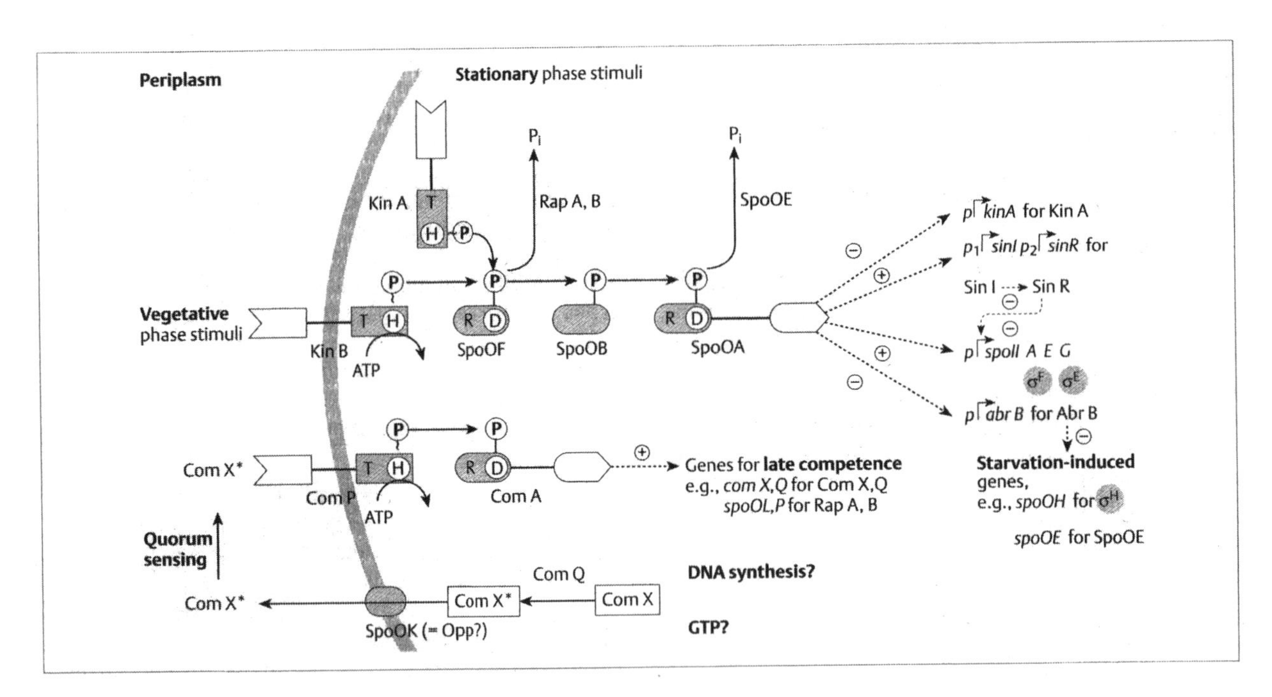

Fig. 25.2 **Hypothetical scheme of the Spo0 and Com phospho-relay pathways in *Bacillus subtilis*.** The KinA–Spo0A and the ComP–ComA two-component systems (symbols are as in Fig. 20.**16**), their multiple roles in the regulation of competence (see Chapter 16) and sporulation, and the links between both systems are shown. The proteins involved in the relays, the stimuli to which they respond, and the genes controlled by the response regulators Spo0A and ComA are indicated and described in the text. The phosphorylated form of a His (H) or an Asp (D) residue is indicated by a P added to the protein symbol. RapA,B and Spo0E represent protein phosphatases. Essential regulatory proteins controlled by the systems, e.g., sigma factors (σ) E, F, and H, or repressors SinR and AbrB are also shown. ComX represents a peptide (gene *comX*) processed by ComQ to a derivative ComX* which, after export through Spo0K ($=$Opp), probably is used in cell density (or quorum) sensing and in triggering ComP autophosphorylation activity. Broken arrows indicate positive ($\oplus$) or negative ($\ominus$) regulation at the gene or protein levels; solid arrows define a biochemical reaction

process, abolishes the expression of early-induced sporulation genes. The exact signal that arises from DNA synthesis and activates gene expression is not known.

The putative signals generated by the three conditions described above are integrated into a pathway that is composed, in part, of several of the *spo0* gene products. This signal transduction pathway, known as the **Spo0-phospho-relay**, is made up of the Spo0F, Spo0B, and Spo0A proteins (Fig. 25.**2**). Spo0F and Spo0A are members of the response regulator class of the two-component regulatory protein family (see Chapter 20.4). Spo0F is the target of histidine protein kinases, which are the gene products of *kinA* (*spoIIJ*) and *kinB*. According to the model, Spo0F becomes phosphorylated and passes the phosphate to Spo0B, a protein phosphotransferase, which then transfers the phosphate to Spo0A. Spo0A, thus activated by a phosphorylation at an aspartate residue, is a transcriptional regulator that acts both positively and negatively to control gene expression (see below). The multi-step nature of the phospho-relay and the presence of two kinases is thought to facilitate the integration of signals resulting from high cell density, DNA synthesis, and nutritional stress. Thus, the activity of a histidine protein kinase may be affected by a signal arising from carbon-source deprivation. There is evidence that the cell-density-dependent signal may be processed by the *spo0K* gene products. The *spo0K* operon encodes a complex that is a peptide permease with similarity to the oligopeptide transporter (Opp) complex of *Salmonella typhimurium*. It has been proposed that the low-molecular-weight extracellular factors (small oligo-peptides?) that accumulate in late-exponential-phase cultures regulate sporulation initiation and control the activity of histidine protein kinases and/or Spo0 products. DNA synthesis may exert its influence on the activity of either Spo0B or Spo0F, as evidenced by the stimulatory effects of these proteins on expression of sporulation gene transcription under conditions in which DNA synthesis is inhibited. What the exact signal is that arises from DNA synthesis and affects Spo0F and Spo0B activity is not known.

25.1.2 Phosphorylated Spo0A Is Central to Sporulation Initiation in *Bacillus subtilis*

The cumulative effect of the signals arising from conditions that promote sporulation is ultimately an increase in the **level of phosphorylated Spo0A** protein (Spo0A–P, Fig. 25.**2**). That Spo0A–P formation is the end result of the phospho-relay was determined in part by studies using mutations in *spo0A* that suppress null mutations in *spo0B*, *spo0F*, *spo0K*, and the *kin* genes. These mutations result in Spo0-independent phosphorylation of Spo0A or render Spo0A active in a non-phosphorylated state. The latter class of Spo0A mutations cause sporulation gene transcription to be independent of the conditions that otherwise promote sporulation. Thus, the signals that arise from these conditions contribute to the same outcome, the increase in Spo0A–P concentration.

Spo0A–P negatively regulates the transcription of genes that tend to suppress sporulation functions, including ***abrB***, which encodes a repressor of a number of genes that are induced by starvation, including sporulation genes. AbrB represses the transcription of ***spo0H***, which encodes an RNA polymerase sigma factor (σ^H) required for the transcription of a number of genes induced under sporulation-promoting conditions. $\boldsymbol{\sigma^H}$ is also required for transcription initiation from one of the two promoters of *spo0A*. Thus, by inhibiting *abrB* transcription, Spo0A–P stimulates production of Spo0H, thereby accelerating transcription from the *spo0A* promoter region and enhancing its own production (autoregulation).

Spo0A–P also directly activates the transcription of several sporulation genes, such as those of the spoIIA and spoIIG operons, which are involved in establishing the compartment-specific gene expression of the prespore and the mother cell. Spo0A–P performs its function by binding to sequences in or near the promoters of the genes it regulates. A sequence, the "Spo0A box," has been shown by footprinting analysis to be the site of Spo0A binding. In this way, Spo0A–P can either prevent RNA polymerase/promoter interaction or associate directly with RNA polymerase, thereby stimulating transcription initiation. Spo0A–P can also stimulate SpoIIA indirectly by positively controlling the transcription of *sinI*, which encodes a protein that binds and inhibits the repressor SinR (Fig. 25.**2**).

Initiation of sporulation in *B. subtilis* is facilitated by the integration of multiple developmental signals that activate histidine protein kinases, which then function to induce the phospho-relay pathway.

25.1.3 Compartmentalized Gene Expression Is Established During Sporulation in *Bacillus subtilis*

The Spo0-phospho-relay signal transduction system is also required for the asymmetric placement of the sporulation septum. The cell division proteins FtsZ and FtsA (see Chapter 22) are required for septum placement in vegetatively growing cells and in cells beginning to undergo sporulation. It is unclear why the sites of normal cell division septum placement are not used. The temporal program of gene expression during the sporulation process is summarized in Figure 25.**3** and the regulatory pathways governing prespore and mother cell gene expression are summarized in Figure 25.**4**. Table 25.**1** lists the various sporulation-specific sigma factors of *B. subtilis*.

Prior to formation of the septum, the operons *spoIIA* and *spoIIG* are transcriptionally activated through the action of Spo0A–P as described before. The ***spoIIA*** operon is composed of three genes, *spoIIAA*, *spoIIAB*, and *spoIIAC*. The last gene of the operon, *spoIIAC*, encodes the **sigma factor** σ^F, which is responsible for transcription initiation at promoters of genes expressed in the prespore compartment. The products of *spoIIAA* and *spoIIAB* serve to regulate SpoIIAC activity. The ***spoIIG*** operon contains two genes: *spoIIGA*, which encodes a membrane-bound protease, and *spoIIGB*, which encodes the **precursor** form of σ^E. The precursor form of σ^E is produced prior to the formation of the sporulation septum. According to the model of spoIIG function, pro-σ^E is processed to its active form by the *spoIIGA* product after the sporulation septum is formed. The mature σ^E form of RNA polymerase is probably active only in the mother cell.

Processing of pro-σ^E to active σ^E and compartmentalization of σ-factors. Interestingly, the processing of pro-σ^E to active σ^E requires the function of σ^F, which directs transcription of the genes expressed in the prespore (Fig. 25.**4**). Hence, two questions concerning the establishment of compartment-specific gene transcription remain unanswered: (1) How are the activities of σ^E and σ^F confined to different cellular compartments when the genes are transcribed prior to sporulation septum formation?, and (2) How is intercompartment communication implemented? It is thought that SpoIIAB acts as an inhibitor of σ^F and this prevents the expression of prespore gene transcription prior to septum formation. The inhibition is reversed by the interaction with SpoIIAA, an antagonist of SpoIIAB. Interactions between SpoIIAA and SpoIIAB, and SpoIIAB and σ^F have been demonstrated in vitro by cross-linking and by co-precipitation using antibodies against one of the SpoIIA proteins. Although it is not known how the control of σ^F activity relates to septum formation, there is in vitro evidence that differences in the energy charge between the prespore and mother cell may affect SpoIIAB activity. SpoIIAB can bind both ATP and ADP reversibly. SpoIIAB–ATP preferentially interacts with σ^F and inhibits transcription initiation catalyzed by σ^F RNA polymerase in vitro. SpoIIAB–ADP preferentially interacts with SpoIIAA. There may be a lower ATP/ADP ratio in the prespore compartment, than in the mother cell; these conditions might favor SpoIIAA–SpoIIAB interaction and high σ^F RNA polymerase activity.

It is also not clear how σ^E activity is confined to the mother cell. Either processing of pro-σ^E occurs only in the mother cell or processed σ^E is present in both compartments and active only in the mother cell. In either case, the formation of active σ^E depends on events that occur in the prespore compartment since σ^F activity is required. There may be a prespore-specific gene (transcribed by the σ^F RNA polymerase) that activates the SpoIIGA protease, presumably located in the sporulation septum membrane, so that pre-σ^E processing occurs.

Control of σ^G synthesis and activity. One of the genes transcribed by the σ^F-form of RNA polymerase is *SpoIIIG*, which encodes the prespore-specific **sigma factor σ^G**. The control of *spoIIIG* is interesting for two reasons. First, the activity of SpoIIIG (σ^G) is dependent on σ^E, the mother-cell-specific sigma factor. This activation is posttranslational and indirect, perhaps through "Y" (see Fig. 25.**5**). Again, this is indicative of another form of intercompartmental communication that co-

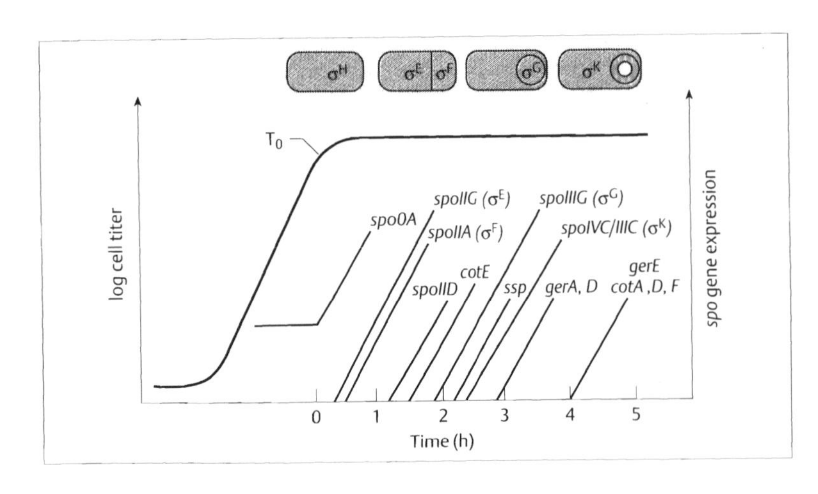

Fig. 25.**3** **Temporal pattern of sporulation gene expression of a *Bacillus subtilis* culture.** A typical growth curve (red line) of a *B. subtilis* culture is shown. T_0 denotes the point in the growth curve when exponential growth ends and stationary phase begins. Above the growth curve is a representation of the sporulation pathway and the cellular location of the stage-specific sporulation sigma factors. The lines below the growth curve indicate the timing and increase in *spo* gene expression. The gene designations are described in the text

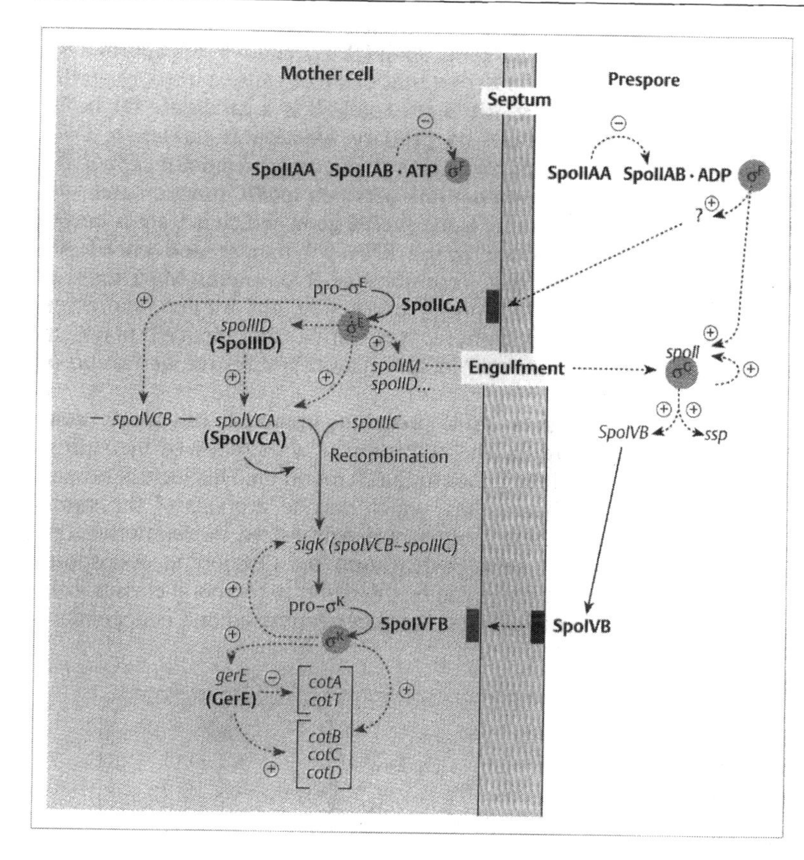

Fig. 25.4 **Control of sporulation gene expression and intercompartmental communication.** The diagram shows the pathways of *spo* gene regulation in the mother cell and in developing prespore. After sporulation septum formation, SpoIIAB inhibits SpoIIAC (σ^F) in the mother cell, while in the prespore compartment, SpoIIAA inhibits SpoIIAB, thereby activating σ^F. SpoIIGB (pro-σ^E) is processed proteolytically by a SpoIIGA-mechanism in the mother cell, a process dependent on σ^F activity in the prespore compartment. The resulting active SpoIIGB protein (σ^E) directs the transcription of genes required for engulfment (*spoIIM* and *spoIID*), *spoIIID* (a positive activator of some σ^E-dependent genes), and *spoIVCA* and *spoIVCB*. SpoIVCA is a site-specific recombinase that catalyzes the joining of the *spoIIIC* and *spoIVCB* sequences to form the *sigK* gene, whose product is the precursor form of the sigma factor σ^K. Completion of engulfment results in the activation of *spoIIIG* expression, which is also dependent on the prespore sigma factor σ^F. *spoIIIG* encodes the prespore sigma factor σ^G, which stimulates transcription of its own gene as well as genes encoding the acid-soluble proteins (Ssp) and the *spoIVB* gene. The activity of SpoIVB is necessary for the processing of pro-σ^K to σ^K, which involves the putative membrane-associated protease, SpoIVF. Active σ^K directs the transcription of *sigK*, the *cot* genes, which encode the spore coat proteins, and *gerE*, whose product regulates *cot* gene transcription. Broken arrows indicate positive (⊕) or negative (⊖) regulation at the gene or protein levels; solid arrows define a biochemical reaction

ordinates mother cell and prespore development. Secondly, *spoIIIG* transcription is not immediately followed by the appearance of σ^G activity, which is delayed by 30 min. This indicates control at the post-transcriptional level, possibly affecting σ^G activity through an interaction with the regulator of σ^F, SpoIIAB. Among the genes transcribed by the σ^G-form of RNA polymerase are the *ssp* genes that encode the acid-soluble proteins that bind to the prespore chromosome and in doing so, confer UV-irradiation resistance to the spore. The influence of σ^G extends beyond the prespore since σ^G is also required for mother cell development as described below.

The multiple roles of σ^E. σ^E-RNA polymerase is believed to transcribe (a) gene(s) that function(s) in the prevention of septum formation. Mutations in *spoIIG*

Table 25.1 ***Bacillus subtilis* sporulation-specific sigma (σ) factors**

Sigma factor	Gene(s)	(alternative name)	Function
σ^E	*spoIIGB*	(*sigE*)	Early mother cell gene expression
σ^F	*spoIIAC*	(*sigF*)	Early forespore gene expression
σ^G	*spoIIIG*	(*sigG*)	Late forespore gene expression
σ^K	*spoIVCB:spoIIIC*	(*sigK*)	Late mother cell gene expression

and certain other *spoII* genes result in a disporic phenotype, defined as the formation of two sporulation septa, one at each end of the cell. Although this suggests the presence of genes that function in the inhibition of septum placement, these have yet to be identified. Soon after active σ^E is produced, it directs the transcription of genes that function in engulfment (*spoIID*) and *spoIIM*) and in later stages of spore morphogenesis. However, σ^E-dependent genes are also required for the proper expression of genes in the prespore.

Another σ^E-regulated gene, *spoIIID*, encodes a protein that functions as a switch from one temporal class of σ^E-regulated genes to another. SpoIIID is a positive regulator of transcription initiation that acts at the regulatory regions of some genes transcribed by σ^E RNA polymerase. SpoIIID regulates its own synthesis by stimulating transcription from the *spoIIID* promoter. When high levels of SpoIIID are attained, the second temporal class of σ^E-regulated genes is activated.

Synthesis of σ^K requires site-specific recombination and processing. SpoIIID activates the σ^E-dependent transcription of the *spoIVCB* genes and is required for the function of the *spoIVCA* gene. *spoIVCA* and *spoIVCB* are convergently transcribed genes. The expression of *spoIVCA* results in an unusual form of regulation involving site-specific recombination (see Chapter 18.11.3). *spoIVCB* encodes the amino-terminal end of the **sigma factor σ^K**. *spoIVCA* encodes a site-specific recombinase. The carboxy-terminal end of σ^K is encoded by the *spoIIIC* gene, located 42 kbp downstream from the *spoIVCB* gene. The SpoIVCA recombinase catalyzes excision of the intervening 42 kbp of DNA, thereby joining the two *spoIVCB* and *spoIIIC* halves of the **sigma factor gene, sigK**. Once joined, *sigK* encodes the precursor form of the mother-cell-specific sigma factor σ^K. Thus, expression of *sigK* is under both transcriptional and recombinational control, but the production of active σ^K requires a third level of control that originates in the developing endospore.

The pro-σ^K contains a 20-amino-acid sequence at its amino terminus that must be removed to generate active σ^K. This processing event does not occur in cells bearing a mutation in *spoIIIG* (encoding σ^G). It has been postulated that there is a factor(s) encoded by a gene, transcribed in the prespore by σ^G RNA polymerase, that functions in transmitting a signal to the mother cell. This signal results in the activation of a protease that cleaves the pro-σ^K. Attempts to identify the genes that function in pro-σ^K processing involved the isolation of mutations that permit σ^K-dependent transcription in the absence of prespore gene expression. Such mutations, called *bof* (**b**ypass **of** **f**orespore), are located in the *spoIVF* operon. The first gene of the operon *spoIVFA* is the site of the mutations that include nonsense alleles. The second gene, *spoIVFB*, is required for sporulation and is the site of mutations that cause a block in pro-σ^K processing. Although SpoIVFB is a candidate for being the protease, its primary function is unknown. Conceivably, SpoIVFA acts as a negative regulator of SpoIVFB activity. Another link between *spoIIIG* function and pro-σ^K processing is the *spoIVB* gene, which is transcribed in the prespore by σ^G RNA polymerase and encodes a product that resembles a lipoprotein. Mutations in *spoIVB* block pro-σ^K processing, but are not suppressed by *bof* mutations, suggesting that SpoIVB plays an additional role in sporulation besides the activation of σ^K.

Genes and proteins involved in spore coat formation. The activation of σ^K is followed by expression of the genes in the σ^K modulon. This includes many of the *cot* genes, which encode proteins of the **spore coat**. These proteins are synthesized in the mother cell and are deposited around the developing endospore. The *cot* genes can be divided into temporal classes. *cotE* is one of the first *cot* genes expressed and is dependent

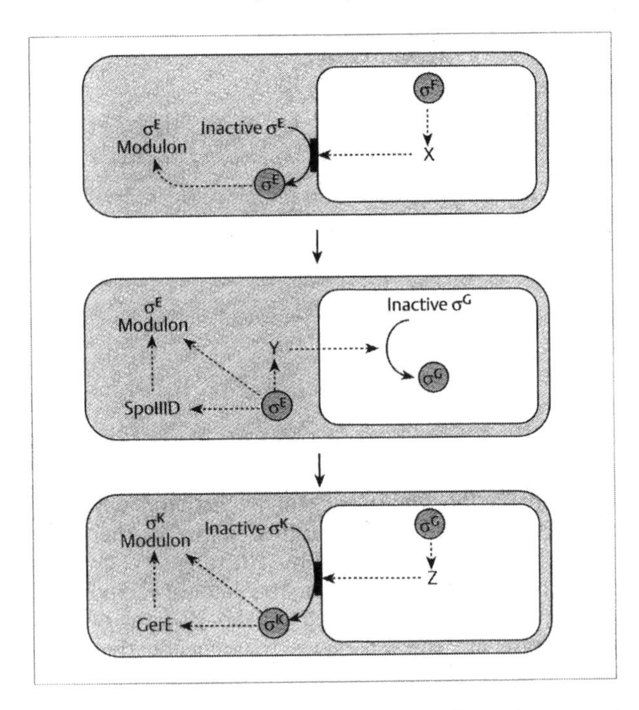

Fig. 25.5 **Diagram summarizing essential regulatory and signal transduction steps found in *Bacillus subtilis* sporulation.** Alternative sigma factors that function in the mother cell (σ^E and σ^K) and in the forespore (σ^F and σ^G) are shown. Hypothetical factors (X, Y, and Z) that function in intercompartment communication are activated at specific stages of sporulation. Sigma-specific modulons (σ^E and σ^K) and the temporal classes within each modulon determined by the transcriptional regulators SpoIIID and GerE are activated

on the σ^E-form of RNA polymerase and SpoIIID. CotE serves a morphogenetic function as it is necessary for the early steps of coat protein deposition and must be in place in order for coat assembly to proceed normally. Surprisingly, mutations in the other *cot* genes of the σ^K modulon have little or no effect on sporulation or germination, but their products are components of the spore coat. Another gene of the σ^K modulon is *gerE*. Originally identified as the site of mutations affecting germination, its product is a positive and negative regulator of *cot* gene transcription. Like SpoIIID, GerE is a small DNA-binding protein that stimulates the σ^K-dependent transcription of several *cot* genes (*cotB, C, D*), but it exerts a negative effect on the transcription of *cotT* and *cotA* genes. Transcription of *cotA* and *cotT* occurs first before GerE accumulates. Then, GerE shuts down the expression of *cotT* and *cotA* and activates the transcription of *cotB, C*, and *D*. In this way, the temporal classes of the *cot* genes are established.

Although there are many details that have yet to be clarified, the important regulatory strategies in spore development are evident. Both temporal and spatial control of sporulation gene expression ensures that spore-specific products are made at the appropriate time and morphological stage during development (summarized in Fig. 25.**5**). Functions necessary for completion of a structural component of the spore are also needed for turning on the next temporal class of sporulation genes. Thus, sporulation septum formation must be completed in order for the σ^E- and σ^F-specific modulons to be activated. Likewise, the completion of engulfment is a perquisite for the activation of the σ^G modulon. Posttranscriptional control plays an important role in regulating sigma factor activity. For example, σ^E and σ^K are first synthesized as precursors, which are processed once a morphological stage of sporulation is completed. The operons dependent on sigma factor activity and the regulators they encode can be divided into temporal classes. SpoIIID mediates the switch from one class of σ^E-transcribed genes to another, whereas GerE plays both a positive and a negative role in governing the temporal control of the *cot* genes. Such strategies of gene control are likely to appear in other examples of complex morphological differentiation in prokaryotes and are clearly evident in the developmental programs of eukaryotic organisms.

Generation of different cell types during sporulation in *B. subtilis* depends on the differential expression of genes in the forespore and mother cell compartments. This programmed gene expression is governed by the successive appearance of five developmental sigma factors.

25.1.4 Methods Used to Analyze Sporulation in *Bacillus subtilis*

Mutants of *B. subtilis* defective in sporulation are identified on solid medium by the absence of a characteristic brown pigment, which is dependent on the expression of the *cotA* gene. Traditional bacterial genetic techniques are used in isolating and characterizing the mutants, such as mutagenesis, generalized transduction mediated by the phage PBS1, and transformation of competent *B. subtilis* cells. More recently, transposable elements have been utilized to a great advantage in the identification and isolation of sporulation genes. The most widely used transposon is Tn*917*. Derivatives have been constructed that allow easy isolation of DNA flanking the site of transposon insertion and the construction of transcriptional fusions between the gene interrupted by the transposon and the *Escherichia coli lacZ* gene. A method for detecting compartment-specific gene expression involves the use of *lacZ* gene fusions and an immuno-gold conjugate made of anti-β-galactosidase antibody. The antibody is used on thin sections of sporulating cells prepared for electron microscopy. Spots of electron-dense immuno-gold-β-galactosidase complexes can be observed in the cell compartment where the gene fusion is expressed.

25.2 Sporulation in Streptomycetes Produces Exospores

The genus *Streptomyces* belongs to the actinomycetes, a heterogeneous group of Gram-positive bacteria characterized by an unusual branched mycelial growth (Fig. 25.**6**) and a high G + C (73%) content of their genomic DNA. *Streptomyces* species are abundant soil bacteria that show an unusual and complex **multicellular differentiation** (equivalent to a generation cycle) consisting of two distinct phases: vegetative mycelial growth under optimal conditions and an aerial, reproductive mycelial growth under starvation conditions. Growth of the mycelium culminates in the formation of chains of spores at the tips of the aerial hyphae ("sporangium"). A mature colony growing on a solid substrate may contain both types of differentiation stages: actively growing substrate mycelium at the periphery that penetrates and solubilizes the solid substrate, and aerial hyphae growth and spore development at the colony center where, due to nutrient limitation, long chains containing 50 or more spores are formed by specialized aerial hyphae that extend away

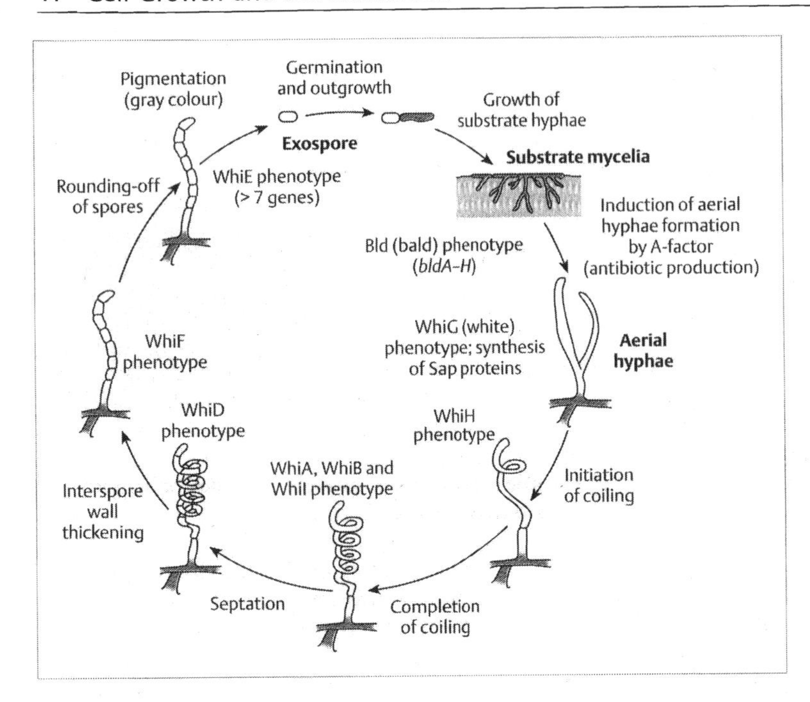

Fig. 25.6 **Diagram showing the generation cycle of *Streptomyces coelicolor*** [after 1]

from the substrate. The aerial hyphae are mainly developed through turnover of vegetative mycelium material.

Differentiation at the tips of aerial hyphae that result in formation of spore chains involves the creation of regularly spaced compartments containing single genomes. These compartments undergo changes in shape and wall thickness, ultimately giving rise to mature spores that have a sculptured hydrophobic surface and are similar in size to other bacterial spores. While these **exospores** are relatively resistant to desiccation, they are sensitive to high temperature.

During periods of limited resources when cells undergo complex cellular differentiation, many *Streptomyces* species produce a wide variety of so-called **secondary metabolites**, which include most of the known antibiotics (see Chapter 27). This developmental cascade of morphological (sporulation) and metabolic (secondary metabolite production) differentiation raises questions about the metabolic signals and the possible common genetic controls that govern these processes. Some of these issues are addressed below.

25.2.1 A Complex Network Couples the Physiology and Genetics of Differentiation in Streptomycetes

Metabolic signals. How are nutrient limitation and environmental stimuli that trigger aerial mycelium and spore formation sensed by *Streptomyces*? In *Bacillus subtilis*, there is good evidence that a reduction in the guanine nucleotide (GTP, GDP) pool is the crucial signal that triggers spore formation (see Chapter 25.1.1). However, it is unknown how this signal is transmitted to the apparatus that controls transcription of sporulation genes. In *Streptomyces* species, such a role for guanine nucleotides in triggering sporulation has been proposed. This is based on the observations that a sharp decrease in the **GTP** levels and an increase in **ppGpp** concentration occurs at the onset of aerial mycelia formation. An increase in ppGpp concentration after nitrogen limitation is also coupled to a sharp specific fall in the GTP pool in *S. griseus* cells of submerged, sporulating populations. Furthermore, **decoyinine**, an inhibitor of GMP-synthetase, induces formation of aerial mycelia and spores. Certain Rel^--like mutants of *S. griseus* and *S. antibioticus*, which show a phenotype like that of *E. coli* Rel^- mutants (reduction in total RNA synthesis and ppGpp accumulation, see Chapter 20.3), exhibit a delay in the formation of aerial mycelia. Decoyinine overcomes this delay and induces efficient sporulation, but ppGpp is not accumulated. This suggests that the level of the GTP pool has a more direct influence on the initiation of sporulation and that the effect of ppGpp concentration is more indirect since ppGpp inhibits IMP-dehydrogenase, which result in a reduction in the GTP pool.

Autoregulatory factors. Some *Streptomyces* species respond to endogenously produced **diffusible factors** known to induce sporulation at extremely low concentrations. The prototype of this group of low-molecular-weight substances, sometimes called "microbial hormones" or **pheromones**, is the **A-factor** (see Chapter 27.6.4).

This extracellular factor, which induces sporulation and streptomycin (Str) production in a sporulation- and Str-negative mutant strain, was discovered in a liquid culture of sporulating *S. griseus* cells. The chemical structure of A-factor (**2-isocapryloyl-3*R*-hydroxy-methyl γ-butyrolactone**) is shown in Figure 25.7. Chemically synthesized A-factor restores the phenotype of A-factor-deficient mutants at nanomolar concentrations. In *S. griseus*, A-factor production is induced immediately prior to streptomycin production and then rapidly disappears before streptomycin production reaches a maximum. Since A-factor is normally quite stable, its disappearance is believed to be a result of enzymatic inactivation.

A large family of **A-factor-like compounds** has been found in culture media of various *Streptomyces* spec. prompting the assumption that A-factor analogues may also act as chemical signals in other species (see Fig. 27.**21**). Most differ from A-factor at position 6 by having a β-hydroxy group instead of the usual β-keto group. There is also variation in the branching of the acyl side chains at position 2. Producer strains can often discriminate between A-factor and its analogues even though they are strikingly similar in structure; this discrimination is typical for pheromones.

Recent biochemical studies have resulted in the construction of a model for the biosynthetic pathway of one class of autoregulatory compounds, a **γ-butyrolactone** type produced by *S. antibioticus*. These compounds are assembled from two acetate, one isovalerate, and one glycerol molecule in a reaction resembling polyketide elongation. In this condensation reaction, the length and branching of the side chains at position 2 is dependent on the number of malonyl CoAs (derived from acetate) used in the reaction and the nature of the starter molecule. The precursors, usually β-keto acids, react with glycerol derivatives to form the corresponding γ-butyrolactone. The enzyme that catalyzes the condensation reaction for the synthesis of A-factor in *S. griseus* and *S. coelicolor* is believed to be encoded by the ***afsA*** gene. In the streptomycin producers *S. griseus* and *S. bikiniensis*, *afsA* is encoded on a plasmid, but it is chromosomally located in *S. coelicolor*. In *S. coelicolor*, the expression of *afsA* is controlled by ***afsR***, whose gene product is phosphorylated by AfsK. Like Spo0A and SpoIIJ in *B. subtilis*, AfsK and AfsR are members of the two-component regulatory protein family (see Chapter 20.4) and act as a sensor and a response regulator, respectively. AfsR in *S. coelicolor* is an independent regulator of the synthesis of A-factor and the antibiotics actinorhodin, undecyloprodigiosin, methylenomycin, and the calcium-dependent antibiotic CDA.

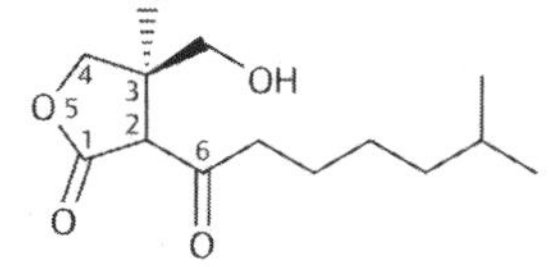

Fig. 25.**7** **Chemical structure of A-factor from *Streptomyces griseus***

In *S. griseus*, identification of A-factor-deficient mutants that are able to sporulate and produce streptomycin in the absence of A-factor led to the discovery of an **A-factor binding protein** that acts as a repressor of gene expression. Revertants of these mutants simultaneously lose their ability to produce streptomycin and to sporulate. All the data obtained suggest that the A-factor removes the A-factor binding protein from the DNA, where the binding protein acts as a repressor of the sporulation and streptomycin biosynthesis genes (see Fig. 27.**22**).

In *S. virginiae*, the producer of virginiae butanolide, a receptor protein may act as a repressor of both differentiation and secondary metabolite production. In contrast, morphological differentiation and secondary metabolite production in *S. coelicolor* appear to be independent of A-factor, suggesting the involvement of a different regulatory mechanism for differentiation.

Autoregulatory compounds of a chemical structure different from that of A-factor were also isolated from cultures of *Streptomyces* spec. **B-factor**, for example, a butyl ester of 3′-AMP, restores rifamycin B production in some mutant strains of *Nocardia* spec. The synthesis of B-factor from 2′,3′-cyclic AMP and butanol can be catalyzed by a crude extract of the producer strain. The structural similarity of B-factor to 3′,5′-cyclic AMP, an important regulator present in a wide variety of organisms, has led to the proposal of a potentially significant role in the regulation of differentiation. Another aerial mycelium activator, **panamycin**, produced by *S. alboniger*, exhibits antibacterial and antifungal activities. Panamycins represent a mixture of at least eight homologues, all having a 16-membered macrodiolide ring, that are able to induce aerial mycelium formation at low concentration and to inhibit growth of substrate mycelia at higher concentrations. Two other specific inhibitors of aerial mycelia formation in *Streptoverticillium* spec. are carbazomycinal and its 6-methoxy derivative. Further genetic and biochemical studies of these autoregulators are required to clarify their role in morphogenesis of the producer organism.

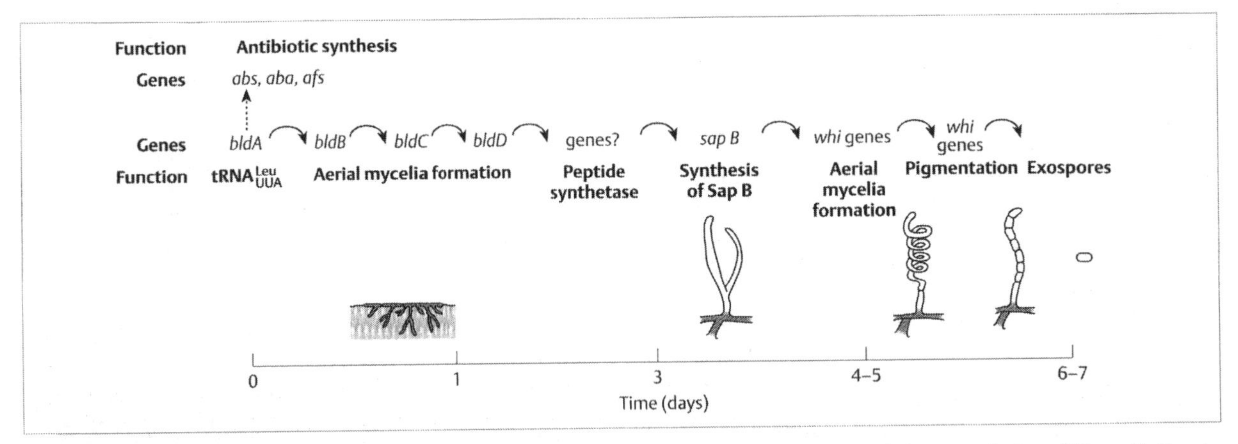

Fig. 25.**8** **Proposed gene cascade for the regulation of SapB synthesis and morphological changes during differentiation in *Streptomyces coelicolor***

SapB. Sap proteins (**s**mall **a**erial mycelium **p**roteins) are associated with the surface of **aerial mycelia** and **spores** of *S. coelicolor.* One of them, a small protein designated as SapB, acts as an extracellular inducer of aerial mycelium formation. SapB has a molecular weight of 1982, calculated from its overall composition: Ala(1), Arg(2), Asp(3), Gly(2), Ile(1), Leu(5), Ser(1), Thr(3). SapB reacts with Schiff's reagent, suggesting the presence of vicinal hydroxy groups, normally associated with sugar residues of glycoproteins. In addition, SapB synthesis in the aerial mycelium is insensitive to chloramphenicol, a finding that suggests a non-ribosomal mode of synthesis (Fig. 25.**8**). Using anti-SapB antibodies, it has been shown that the protein is present in a zone around colonies undergoing aerial hyphae formation. All *bld* mutants, which are blocked in aerial mycelium formation (see below), are impaired in SapB production.

Extracellular stimulation during differentiation in *Streptomyces* involves metabolic compounds, endogenously produced diffusible factors (pheromones), and the synthesis of morphogenic protein(s) that direct(s) the morphopoiesis of aerial hyphae. Among the pheromones, A-factor (2-isocapryloyl-3*R*-hydroxymethyl γ-butyrolactone) and derivatives are reminiscent of other autoinducers involved in quorum sensing (see Chapter 20.1).

However, the mutants produce transient aerial mycelium when grown next to a wild-type, SapB producer. Moreover, transient aerial mycelium formation is induced in *bld* mutants when the purified protein is provided. Thus, SapB may act as a diffusible, signal molecule (like A-factor) for aerial mycelium development, and also as a morphopoietic protein that enables the hyphae to grow into the air. This hypothesis is supported by the correlation of SapB temporal and spatial appearance with the formation of aerial mycelia.

25.2.2 Genes Involved in Sporulation and Aerial Mycelium Formation in *Streptomyces coelicolor*

The versatile genetic tools available for the sporulating *S. coelicolor* allow the investigation of complex morphological and physiological differentiation and their relationship to each other. Two general types of morphological mutants have been found: ***bld*** (bald), which lack an obvious aerial mycelium and do not sporulate, and ***whi*** (white) mutants, which develop aerial mycelia, but do not produce the characteristic gray-pigmented spores observed in a wild-type strain.

In *S. coelicolor,* at least seven different classes of *bld* mutants (*bldA–H*) defective in aerial mycelia formation (but not in vegetative growth) have been described (Fig. 25.**6** and **8**). Remarkably, *bldA, D, G, H,* and *E* mutants exhibit a conditional defect in aerial hyphae formation. The defect is observed when the mutants are grown on glucose-containing medium, but the mutants appear morphologically normal when glucose is replaced by another carbon source (for example, mannitol). With the exception of *bldC,* all mutants have partial or complete deficiencies in antibiotic production, which indicates a role for common elements in genetic regulation. The best characterized *bld* gene, ***bldA***, encodes a **leucyl-tRNA** bearing the anticodon UAA and, therefore, is capable of translating the rare leucine codon UUA. The **UUA codon** has been found primarily in genes involved in development, including several antibiotic-resistance and regulatory genes, and genes for aerial mycelia formation and antibiotic production. However, not all developmentally regulated genes contain the UUA codon. In contrast, it seems that all vegetative genes in *S. coelicolor* are devoid of it, as one would predict from the phenotype of *bldA* mutants. In general, UUA codons are very rare in *Streptomyces*, as expected from an organism with a high G + C content. Studies with the *bldA* promoter fused to an indicator gene suggest that *bldA* is transcribed at a relatively late stage when aerial mycelia formation starts. *bldA*-like DNA sequences seem to be widespread among other *Streptomyces* strains. Therefore, *bldA* could be a master gene that occupies a position near the top of a regulatory hierarchy that governs differentiation and secondary metabolite production. Recently isolated antibiotic regulatory mutants of *S. coelicolor* (*abs*, *aba*, *afs*), which are antibiotic deficient, but sporulation proficient, however suggest that there is another regulatory pathway for the induction of antibiotic synthesis that is distinct from sporulation control (Fig. 26.**8**).

In the morphological differentiation cascade, ***bldC*** is probably positioned downstream of *bldA*, as it appears to exert its control on morphogenesis and does not influence metabolic differentiation (in particular antibiotic production). However, *bldC* is located upstream of *whi* genes in the cascade since Whi^- mutants show no defect in aerial mycelia formation and antibiotic production, but are unable to form spores. At least nine *whi* genes have been identified by mutations (*whiA–I*) that block the formation of gray-pigmented spores in aerial mycelia (Fig. 25.**6**). Mutants defective in *whiG, H, A*, or *B* generally make no sporulation septa, but the aerial mycelia of *whiI* mutants show septation, albeit at a level lower than that of the wild-type strain. The *whiD*, *whiF*, and *whiE* products are involved in later stages of spore formation. Whereas mutations in *whiD* cause thin-walled spores, *whiE* mutants make apparently normal unpigmented spores. Microscopic examination of the development of aerial hyphae in *whi* mutants suggested the following order of expression: *whiG*, *whiH*, *whiA*, *whiB*, *whiI*, *whiD*, *whiF*, and *whiE* (Fig. 25.**6**).

The *whiE*, *whiB*, and *whiG* genes have been cloned recently. Interestingly, WhiG, is an RNA polymerase sigma factor that is dispensable for growth, but is essential for spore formation in aerial hyphae. WhiG is similar (38% identical residues) to σ^D, which is required for the transcription of motility genes of *B. subtilis*, but which does not function in spore formation. WhiG shows a lower degree of similarity to σ^H, which is also known to be essential for initiation of spore formation in *B. subtilis*. Disruption of the *whiG* gene inhibits the development of aerial mycelia to spores, but has no effect on growth. Two pieces of evidence suggest that the amount of WhiG determines not only the location, but also the level of sporulation. *whiG* expressed from a high-copy-number plasmid causes inappropriate sporulation in vegetative hyphae that are normally destined to undergo lysis during differentiation. On the other hand, introducing many copies of a *B. subtilis* σ^D-dependent promoter into *S. coelicolor* reduces sporulation, suggesting a partial sequestering of WhiG. Hence, it has been proposed that the *whiG* gene product plays a crucial role in the developmental fate of aerial hyphae.

Sequence analysis of the ***whiB*** gene revealed a gene product of 87 amino acids that is a putative transcription factor. The ***whiE*** locus, which is needed for immature white spores to acquire the gray color, contains at least seven genes encoding polyketide-like synthases, which are likely responsible for spore pigment formation.

The genetic complexity of morphological differentiation in *Streptomyces* is generally a function of the diversity of the integrated physiological processes. Intercellular communication by metabolic signals, autoregulatory factors, and extracellular signaling proteins contribute to the ordered biochemical events that culminate in aerial hyphae and spore formation.

The initiation of the 3- to 7-day program of multicellular development can proceed along alternative pathways as indicated by the ability of many *bld* mutants to produce spores when mannitol replaces glucose as the only carbon source, and by the observed SapB-induced aerial mycelium formation in rich medium.

25.2.3 Genetic Methods to Analyze Sporulation in the Streptomycetes

S. coelicolor mutants blocked in aerial mycelium (*bld*) or spore formation (*whi*) have been isolated by treating spores with UV or NTG (*N*-methyl-*N*′-nitro-*N*-nitrosoguanidine). The mutants were genetically analyzed using conjugation and transduction, and by transformation of protoplasts obtained by lysozyme treatment. Transformed and regenerated protoplasts of *S. coelicolor* are detected by "pock" formation (see Chapter 16.2.4). Pocks (i.e., circular zones of retarded growth in a confluent lawn of mycelia) are formed on solid media when a plasmid-containing colony develops in a confluent lawn of mycelia from another strain (usually *S. lividans*) that lacks the transformed plasmid or a closely related plasmid. A pock-forming transformant can be easily detected on a lawn of 10^9 plasmid-free regenerated protoplasts on a single agar plate.

Cloning systems for *S. coelicolor* using plasmids and the temperate phage ϕC31 have been recently developed. With some effort, these systems have also been adopted for use in other streptomycetes. Phage ϕC31 forms plaques on and lysogenizes many streptomycetes. A series of ϕC31 derivatives has been constructed and used for mutagenesis, chromosomal integration by homologous recombination, and "mutational cloning," when the vector is integrated via homologous recombination between a cloned fragment containing an internal region of a gene and the corresponding chromosomal DNA ("homogenotization"),

25.3 Multicellular Development in *Myxococcus xanthus*, a Social Organism

Myxococcus is a Gram-negative soil bacterium that exhibits gliding motility (see Chapter 20.5.1) and feeds by degrading insoluble organic material using an abundance of extracellular hydrolytic enzymes. This cooperative feeding occurs in dense populations containing thousands of cells, evidently to ensure a high local concentration of degradative enzymes for efficient hydrolysis of the organic material. Generally, the rate of growth of myxobacteria on a protein substrate increases with cell density, which is indicative of the cooperative feeding behavior. This cell-to-cell interaction extends to **fruiting body** formation, a complex and unique form of multicellular development that has perhaps evolved as a result of the selective advantage afforded by the intercellular cooperativity observed under optimal growth conditions.

25.3.1 Fruiting Body Development Is Initiated by Nutrient Deprivation

Nutrient deprivation, particularly amino acid depletion, coupled with high cell density on a solid surface, promotes the activation of a series of developmental events in myxobacteria (Fig. 25.**9**). This culminates in the formation of a multicellular structure called a **fruiting body** ("sporangium"), which can have a variety of shapes depending on the myxobacterial species examined. A fruiting body is about 0.2 mm high and comprises about 10^5 cells. In the early stages, when growth has slowed and cell density is high, cell aggregation commences and is completed in 4 h. Early in this aggregation state, ridge-like accumulations of cells move coordinately and rhythmically like ripples on the surface of a liquid. After about 12 h, the asymmetric aggregations of cells form elliptical or circular mounds. Through this movement and concomitant cell-to-cell interaction, the cell number and density in the developing centers increase, and a mature fruiting body is formed. At about 24 h of development, some of the cells inside the mound differentiate into spores, while others lyse. During these ordered morphological events, more than 30 new proteins are synthesized and new cell-surface antigens can be detected (Fig. 25.**9**).

25.3.2 Intercellular Signaling Precedes Aggregation and Fruiting Body Formation

The importance of cell-to-cell interaction and intercellular signaling for fruiting body development was demonstrated by the identification of four different complementing classes (A–D) of non-autonomous developmental mutants. All mutants can be rescued either by the addition of wild-type cells or by the addition of mutant cells of a different complementation group. These studies provided evidence for the existence of **extracellular factors** that are required in myxobacterial development. Genetic mapping studies using Tn5 insertions indicate that the four groups represent mutations in four different chromosomal loci, each responsible for a different signal(s). Mutations of **group A** are called *asg* (for **A-sig**nal) and those of **groups B**, **C**, and **D** are designated *bsg* (**B-sig**nal), *csg* (**C-sig**nal), and *dsg* (**D-sig**nal), respectively. Each group of mutations arrests development at a different stage or time, as judged from the pattern of developmentally regulated gene expression. The disruption of gene expression in *asg*, *bsg*, and *dsg* mutants, as monitored by measuring β-galactosidase activity of *lacZ* gene fusions to developmentally regulated promoters, is rescued by codevelopment with wild-type cells. The exact nature of the intercellular signals defined by *bsg* and *dsg* is unknown; however, the *bsgA* gene encodes an ATP-dependent protease and DsgA is homologous to *E. coli* initiation factor 3 (IF3).

The best-studied class of mutants are those belonging to the *asg* and *csg* classes. *asg* and *csg* act within the

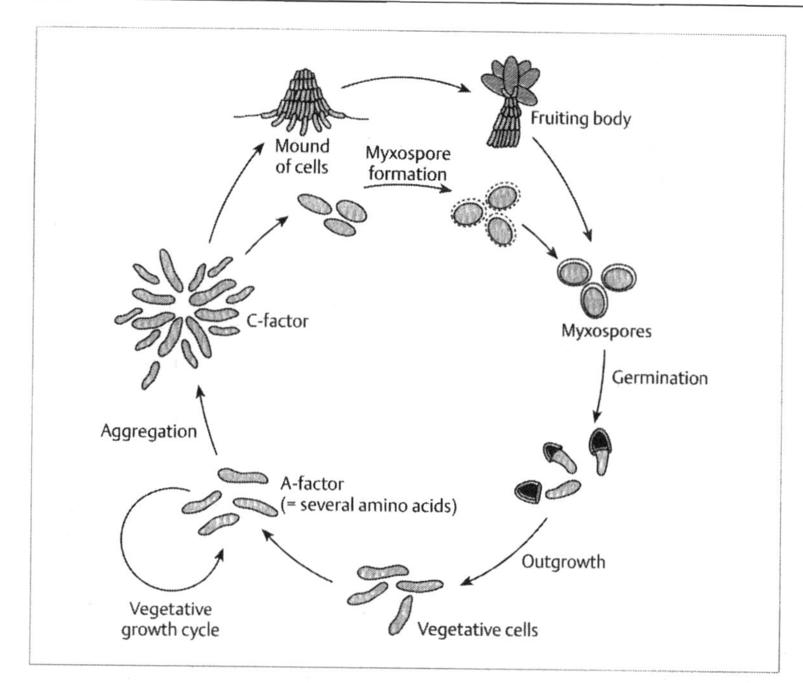

Fig. 25.9 **Developmental cycle of *Myxococcus xanthus*.** Vegetative growth is favored as long as adequate nutrients are present. Upon exhaustion of amino acids, vegetative cells aggregate to form a fruiting body that differentiates into stalk and head. Cells accumulating in the fruiting body head undergo morphogenesis to resting cells called myxospores. Under favorable conditions, myxospores germinate to yield vegetative rods

same regulatory pathway, and the A-signal (*asg*-dependent) is transmitted earlier than the C-signal (*csg*-dependent) during development. Three genetic loci, *asgA*, *asgB*, and *asgC* are involved in A-signaling, whereas only one locus, *csg* functions in C-signaling.

A-factor. Biochemical and genetic studies on the *asg* mutants led to the identification of a set of substances, collectively called **A-factor** or **A-signal** molecules (different from the A-factor described Chapter 25.2.1). The *asg* mutants release less than 5% of the wild-type level of A-factor molecules. A-factor restores the capacity of *asgB* and *asgC* mutants to aggregate and sporulate and induces aggregation in the *asgA* mutant. The major activity of the A-factor is associated with two fractions: a heat-labile fraction and a heat-stable fraction. The **heat-stable fraction** contains amino acids released by wild-type cells (proline, tyrosine, phenylalanine, tryptophan, leucine, isoleucine, and alanine) with a mixture of polypeptides that mainly consist of these amino acids. This fraction induces gene expression in the *asg* mutants at the micromolar level. The **heat-labile fraction** comprises at least two different proteases with a molecular mass of 10 and 27 kDa. The proteases purified from crude A-factor preparations have two distinct substrate specificities, suggesting that the proteases do not signal by cleaving a specific bond, but rather by producing extracellular amino acids. Indeed, several proteases, including trypsin and proteinase K, have heat-labile A-factor activity. Amino acids and small peptides are the primary A-factor signal molecules, which are missing in the *asg* mutants. *Myxococcus* apparently senses a threshold of amino acid concentration, above which developmental gene expression is repressed and below which development is activated. There is, however, a minimum extracellular concentration of amino acids below which development will not proceed; this is the defect conferred by the *asg* mutations. Thus, amino acid concentration is used both as an indicator of the nutritional environment and as a measure of cell density ("quorum sensing"), the latter being provided in starved cells through the activity of the *asg* genes.

The putative signal transduction system involved in mediating the cell's response to A-factor is uncharacterized, but a suppressor mutation of *asgB*, called *sasA*, may identify one of its components. Mutations in the *sasA* locus are recessive and occur at a frequency expected of those that cause loss of function, suggesting that it functions as an A-signal-sensitive repressor of certain developmental genes.

C-factor. The *csg* gene encodes a 17-kDa protein. Although no significant sequence similarity with other proteins was observed, C-factor does possess an amino terminal sequence that resembles the signal sequence required for protein translocation and secretion in enteric bacteria. Nanomolar concentrations of purified C-factor from fruiting bodies restores aggregation and sporulation in the *csgA* mutants. Furthermore, C-factor is tightly associated with the producer cells and is released only after detergent treatment and not by extensive salt treatment. How such a tightly bound protein can act as a signaling factor was elucidated from the observation that non-motile mutants of *M. xanthus* cannot complete fruiting body formation when combined with wild-type cells. However, purified C-factor promotes sporulation when added to the non-motile mutant. Motility is apparently required for the formation of tight packs of parallel-aligned cells, an arrangement that facilitates C-factor signal transmission. If non-motile mutant cells together with wild-type cells are artificially arranged in this way, then C-factor signal transmission is successful. It remains to be established whether an attached or physically transferred C-factor is responsible for the intercellular signaling. The cell response to C-factor is quite complex. C-factor obviously is a global regulator because it modulates the activity of over 50 genes, whose products have important functions in cell mobility, cohesion, aggregation, and morphogenesis.

Regulation of multicellular aggregation and spore formation in *M. xanthus* depends on cell-to-cell interaction and involves a number of intracellular and extracellular signaling systems. Morphological changes and developmentally regulated genes are controlled by at least four distinct chromosomal loci.

25.3.3 Kinases May Regulate Development in *M. xanthus*

A signal transduction cascade possibly using protein threonine/serine kinases that resembles those found in eukaryotes has been recently identified in *M. xanthus*. The prototype of this group in *M. xanthus* is encoded by the *pkn-1* gene, which encodes a protein of 693 residues. Like eukaryotic kinases, Pkn-1 is autophosphorylated by ATP on serine and threonine residues. To investigate its function during development, a *pkn-1*-deletion mutant was constructed. The mutant cells grow normally in rich media, but are defective in fruiting body formation. Cells aggregated earlier, and smaller, less compact fruiting bodies with a greatly reduced spore yield (about 35% of that of the wild-type) form. However, although these findings argue for an important role of a protein threonine/serine kinase in proper spore formation and timing of early developmental events in *M. xanthus*, many intriguing questions concerning the in vivo protein substrate(s) of this kinase remain to be answered.

25.3.4 Multicellularity and Sporulation in Various Organisms Utilize Common Strategies

Sporulation in myxobacteria differs from that of *Bacillus subtilis* and *Streptomyces*, as does the structure of the spores. *Myxococcus* vegetative cells respond to two distinct signals: the A-signal, which is needed prior to aggregation, and the C-signal, which is needed to complete aggregation and to initiate sporulation. Sporulation does not proceed through the formation of an asymmetric septation that gives rise to two cellular compartments, as in *Bacillus* sporulation, but instead involves complete differentiation of the entire vegetative cell. Within the fruiting body, rod-shaped vegetative cells of *Myxococcus* turn into almost spherical spores that contain specific spore proteins and show a high degree of cross-linking in their peptidoglycan. The resulting myxospores are also resistant to desiccation, heat, and detergent treatment, and can survive in soil for long periods with little metabolic activity. Although there are conspicuous differences in spore morphogenesis, *Bacillus*, *Streptomyces*, and *Myxococcus* utilize common strategies of sensing, signaling, and cell-to-cell interaction to control differentiation.

25.3.5 Methods to Analyze Sporulation in *M. xanthus*

To identify the signals passed between *Myxococcus* cells that coordinate fruiting body development and spore formation, mutants conditionally defective in development were isolated by conventional mutagenesis. Extracellular complementation groups were established based on pairwise testing and complementation by wild-type cells to produce spores of the mutant genotype. Signal factors were identified by using fractionated protein extracts of wild-type cells to restore development in arrested mutants. Developmentally regulated transcriptional units were identified by Tn5 mutagenesis. Tn*5lac*, a transposon that fuses the promoterless "reporter" *lacZ* gene to exogenous promoters has been used. The expression of β-galactosidase from developmentally regulated promoters was used as an assay to follow the purification of signaling factors (C-factor) from the wild-type cells. Generalized transducing myxophages (Mx4, Mx8, Mx9) are also available for transferring genes between strains of *M. xanthus* to construct mutants and to map loci by transduction and to form lysogens by chromosomal integration. By using the kanamycin-resistance of an inserted Tn5 for selection, it was possible for nearby genes to be cloned into an *E. coli* plasmid. The plasmid and its insert can be transduced back from *E. coli* to *M. xanthus* by bacteriophage P1 or by transformation. Alternatively, independent Tn*5* insertions were subjected to generalized transduction using the myxophage Mx8. Phage particles containing Tn*5* and adjacent *M. xanthus* genes were used to complement fruiting-body-negative mutants.

Further Reading

Champness, W. C., and Chater, K. F. (1994) Regulation and integration of antibiotic production and morphological differentiation in *Streptomyces* spp. In: Piggot, P. J., Moran Jr., C. P., and Youngman, P. (eds.) Regulation of bacterial differentiation. Washington, DC: ASM Press; 61–93

Chater, K. F. (1989) Sporulation in *Streptomyces*. In: Smith, I., Slepecky, R. A., and Setlow, P. (eds.) Regulation of procaryotic development: structural and functional analysis of bacterial sporulation and germination. Washington, DC: ASM Press; 277–299

Chater, K. F. (1993) Genetics of differentiation in *Streptomyces*. Annu Rev Microbiol 47: 685–713

Errington, G. (1993) *Bacillus subtilis* sporulation: regulation of gene expression and control of morphogenesis. Microbiol Rev 57: 1–33

Gill, E. R., and Shimkets, L. (1993) Genetic approach for analysis of myxobacterial behavior. In: Dworkin, M., and Kaiser, D. (eds.) Myxobacteria, Vol. 2. Washington, DC: ASM Press; 129–155

Grossmann, A. (1995) Genetic networks controlling the initiation of sporulation and the development of genetic competence in *Bacillus subtilis*. Annu Rev Genetics 29: 477–508

Haldenwag, B. (1995) The sigma factors of *Bacillus subtilis*. Microbiol Rev 59: 1–30

Hopwood, D. A., Bibb, M. J., Chater, K. F., Kieser, T., Bruton, C. J., Kieser, H. M., Lydiate, D. J., Smith, C. P., Ward, J. M., and Schrempf, H., eds. (1985) Genetic manipulation of *Streptomyces*. A laboratory manual. Norwich: John Innes Foundation

Horinouchi, S., and Beppu, T. (1992) Autoregulatory factors and communication in *Actinomycetes*. Annu Rev Microbiol 46: 377–398

Kaiser, D. (1989) Multicellular development in Myxobacteria. In: Hopwood, D. A., and Chater, K. F. (eds.) Genetics of bacterial diversity. New York London: Academic Press; 243–263

Kaiser, D., and Losick, R. (1993) How and why bacteria talk to each other. Cell 73: 873–885

Kim, S. K. (1991) Intracellular signaling in *Myxococcus* development: the role of C-factor. Trends Genet 7: 361–365

Losick, R., Youngman, P., and Piggot, P. J. (1986) Genetics of endospore formation in *Bacillus subtilis*. Annu Rev Genetics 20: 625–669

Mohan, S., Dow, C., and Cole, J. A., eds. (1992) Prokaryotic structure and function: a new perspective. Society for General Microbiology Symposium, vol 47. Cambridge: Cambridge University Press

Moran, C. P. Jr. (1993) RNA polymerase and transcription factors. In: Sonenshein, A., Hoch, J. A., and Losick, R. (eds.) *Bacillus subtilis* and other Gram-positive bacteria: biochemistry, physiology and molecular genetics. Washington, DC: ASM Press; 653–667

Pigott, P. J., Moran, C. P. Jr., and Youngman, P., eds. (1994) Regulation of bacterial differentiation. Washington, DC; ASM Press

Reichenbach, H. (1993) Biology of Myxobacteria: ecology and taxonomy. In: Dworkin, M., and Kaiser, D. (eds.) Myxobacteria, vol 2. Washington, DC: ASM Press; 13–62

Setlow, P. (1981) Biochemistry of bacterial forespore development and spore germination. In: Levinson, H. S., Sonenschein, A. L., and Tipper, D. J. (eds.) Sporulation and germination. Washington, DC: ASM Press; 13–38

Seung, K. K., and Kaiser, D. (1992) Control of cell density and pattern by intracellular signaling in *Myxococcus* development. Annu Rev Microbiol 46: 117–139

Shapiro, L., and Losick, R. (1997) Protein localization and cell fate in bacteria. Science 276: 712–718

Shimkets, L. J. (1990) Social and developmental biology of Myxobacteria. Microbiol Rev 54: 473–501

Youngman, P. (1990) Use of transposon and integrational vectors for mutagenesis and construction of gene fusion in *Bacillus* species. In: Harwood, C. R., and Cutting, S. M. (eds.) Molecular biological methods for *Bacillus*. New York: Wiley

Source of Figure

1 Chater, K. F., and Merrick, M.J. (1979) Streptomycetes. In Parish, J. H. (ed.) Developmental biology of the prokaryotes. London: Blackwell; 93–114

26 Bacteriophages as Models for Differentiation

26.1 Introduction and Experimental Techniques

26.1.1 Phages Are Parasites on the Molecular Level

Bacteriophages, usually abbreviated as **phages**, can be considered as obligatory "parasites on the molecular level" (S.E. Luria). They are viruses that prey on their bacterial hosts and "eat bacteria," a fact that gave them their name. Phages are found wherever bacteria live: in soil, fresh water, and salt water, and even under extreme conditions, such as in hot springs. Phages are definitely not organisms in a strict sense since they are not alive on their own. They depend completely on the life of their hosts and take advantage of them. Like all parasites, phages are selfish and use many of their host's proteins, enzymes, and chemical compounds for their own propagation (i.e., multiplication of their genetic material and generation of many progenies). The final release of the progeny, is usually accompanied by the death of the host cell or at least the extreme weakening of the host. Phages have many properties of other autonomous genetic elements and are best described at the DNA level as plasmids having acquired additional skills for survival outside of their hosts.

Since 1940, phages have become the most famous paradigms of molecular biology. Studies of phages and their hosts culminated in the detailed understanding of basic processes such as DNA replication, recombination, transcription, and gene regulation. Phages also opened the way to the development of many powerful techniques for a future-oriented gene technology.

26.1.2 Experimental Techniques for Handling Phages Are Simple

One of the reasons why phages became the best-studied subjects in molecular biology is the ease with which they and their bacterial hosts can be handled in the laboratory. Infections of bacteria can be performed at any time under controlled conditions on solidified agar plates or in liquid culture (Fig. 26.**1a**). Infected cells, which are mixed with an excess of uninfected cells in soft agar, after spreading and hardening the soft agar mixture on the surface of an agar plate, give rise to **plaques** ("cleared spots") after a few hours of incubation at 37°C (Fig. 26.**1b**). Plaques are areas where phages from the lysis of the first infected cell have infected uninfected growing cells in their immediate vicinity. The viscosity of the agar slows diffusion of progeny phages progressing from the site of lysis. If done under controlled conditions, characteristic plaque morphologies can be seen (Fig. 26.**2**). Since each plaque derives under these conditions from a single infected cell, the plaque comprises a population of genetically identical progeny, i.e., a **clone**. The phages from a plaque can be isolated either by simply cutting out a piece of agar containing the plaque and resuspending it in liquid medium (Fig. 26.**3**) or by picking some material from the surface of the plaque with a sterile toothpick or glass rod.

Infected cells can be produced in large quantities in Erlenmeyer flasks or in fermenters, thereby providing the researcher with ample material for biochemical analyses. The life cycle of most phages is short, and the time between infection and lysis (**latency period**) is correlated to the generation time of their hosts. It can be as short as 25 min.

The average number of phages produced by infected cells, called the **burst size**, varies somewhat among species, but usually several hundred progeny phages are released at the time of lysis. Usually, there is an equal amount of material of incomplete phages released at the same time.

The number of infected cells in a culture depends on the number of phages added at the time of infection.

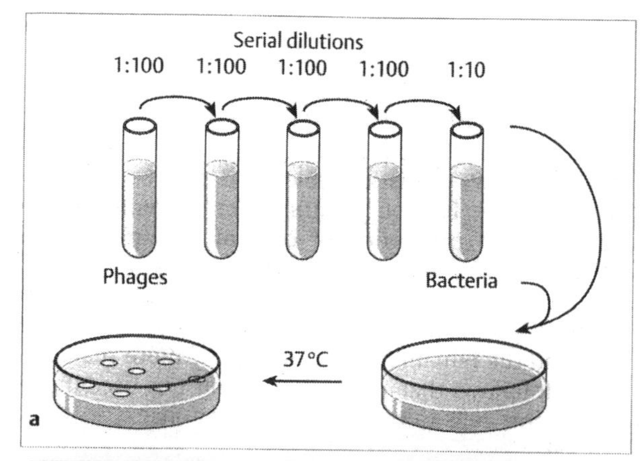

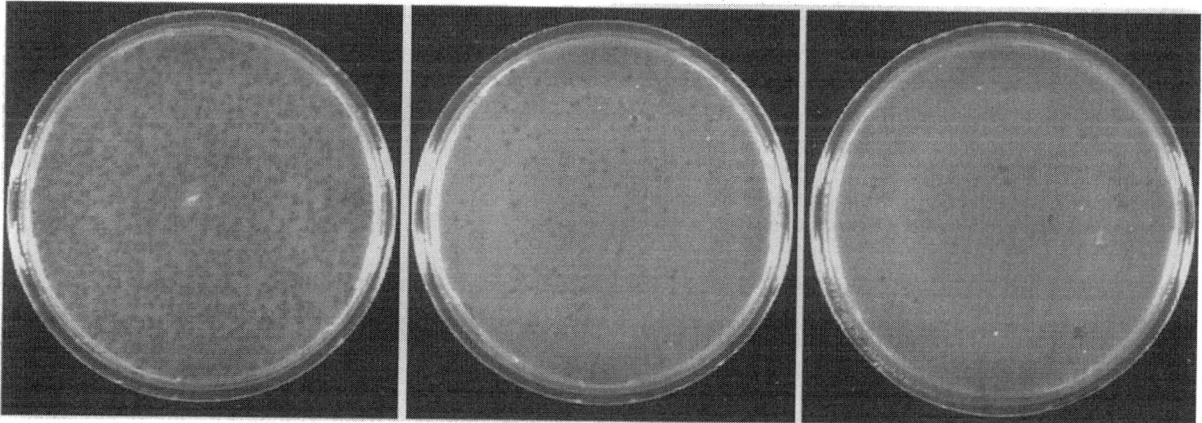

Fig. 26.**1a, b Titration of phages.**
a The diagram shows the principle of serial dilutions made in steps of 100 or 10 in a sterile dilution buffer containing only saline and gelatin for stabilization. An aliquot of each dilution is mixed with a small amount of freshly grown indicator bacteria in molten soft agar. The mixture is quickly poured onto a fresh agar plate and evenly distributed before it solidifies. After incubation at 37°C, each phage gives rise to a plaque in the layer of the bacterial lawn.
b Plates seeded with serial dilutions of 10^{-6}, 10^{-7}, and 10^{-8} phage particles made from a phage solution containing approximately 10^9 T4 particles per 1 ml are shown after incubation. By counting the number of plaques and multiplying with the dilution value, the phage lysate titer, i.e., the number of plaque-forming units (PFU) or active phages per volume can be calculated

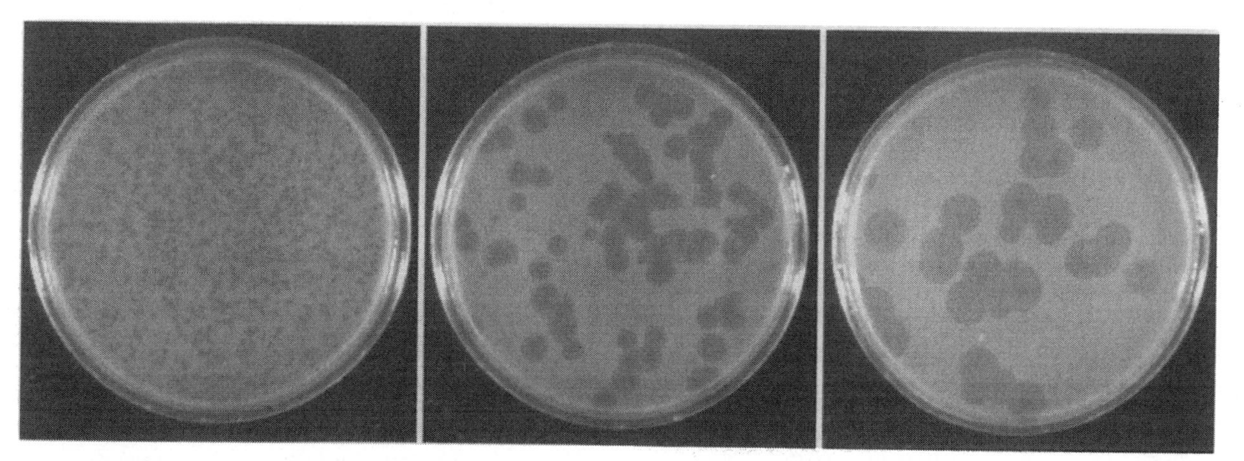

Fig. 26.**2 Plaque morphology is characteristic for each phage.** Examples of plaque morphologies of phages T4, T3, and T7 are shown. Note that in the boundaries of the large plaques, small colonies of resistant bacteria grew. The plaque morphology is typical for a given phage and can vary considerably between certain mutants from the same phage. This property is extensively used as a marker in studies of genetic recombination

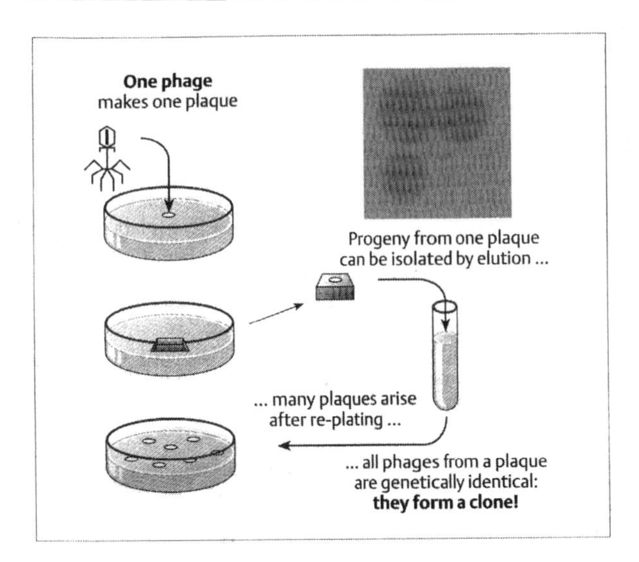
One phage
makes one plaque
Progeny from one plaque
can be isolated by elution ...
... many plaques arise
after re-plating ...
... all phages from a plaque
are genetically identical:
they form a clone!

Fig. 26.3 **All phages in a plaque form a clone.** One phage usually gives rise to many progeny phages concentrated in one plaque. Phages from one plaque are therefore genetically identical and constitute a clone. Phages can be isolated from a single plaque by extraction and then propagated by replating for further analyses

This number is called the **multiplicity of infection** (**m.o.i.**). Since infections depend on random collisions between a phage and cells, some cells may be hit more than once by different particles, while others are not hit at all. The proportion of multiply infected cells in a culture follows statistical rules and can be approximated by the Poisson equation. For example, if one wants a high proportion of cells of a culture to be infected by just one phage, the m.o.i. should be <1. According to the Poisson distribution, an m.o.i. of 1.0 yields 36% uninfected cells, 36% infected by one phage only, and the remaining 28% infected by more than one phage. It follows that an m.o.i. of >1 must be used if one wants most cells to be infected at least once.

26.2 Composition and Plaque Morphology Are Used for Classification of Bacteriophages

26.2.1 Bacteriophages Can Be Classified Into Various Groups (Families)

Virus classification is primarily based on chemical and physical similarities, such as the size and shape of the virion, the nature of the genomic nucleic acid, the number and function of component proteins, the presence of lipids and of additional structural features, such as envelopes, and serological interrelationships. Phages consist of RNA or DNA as their genetic material, surrounded by a protecting "shell" or "head," the **capsid**. In most cases, these protective envelopes are made from pure proteins, but a few examples are known in which the capsids contain also lipids and membraneous materials. Most phages have only one double- or single-stranded, linear or circular chromosome. However, a few examples with more than one chromosome have also been described. Some characteristics of phages are compiled in Table 26.1.

26.2.2 Phages Cause Distinct Phenotypes

The capsid of phages serves a dual function. It protects the genetic material from an unpleasant world outside the host, and it provides the adsorption device that is essential for entering another host cell. For more complex phages, an elaborate adsorption device, the **tail** and **base plate**, is formed, which is distinct from the capsid (head) proper. The complexity of the coat varies considerably among phages. Despite the many differences in the envelope described for different phages, only two basic shapes have been distinguished.

Filamentous phages such as fd and M13 have circular single-stranded chromosomes wrapped up in an elongated shell made from hundreds and thousands of subunits of the same protein. The tube-like structure is very flexible, and its length measures half the circumference of the circular DNA that it protects (Fig. 26.**4a**). The length of the chromosome determines the length of the phages since the capsid is assembled around the DNA until the entire chromosome is wrapped up. Hence, relatively large DNA fragments can be inserted into their chromosomes before pack-aging becomes problematic; this is a property of these phages used extensively in gene technology. A few copies of a special protein attached to one end of the phage provides an adsorption device.

Table 26.1 **General characteristics of some phages.** ds, double-stranded, ss, single-stranded; lin, linear; tr, terminal redundancies; cp, circular permuted genetic map; circ, circular

Genome structure and characteristics	Phage particle morphology	Examples: phage	Examples: hosts	Remarks
DNA ds, lin, tr, cp	Polyhedral oblong	T2, T4, T6	*Escherichia coli*	Virulent; "even" T-phage; T4 has two genes with introns
DNA, ds, lin, tr, cp	Icosahedral	P1, P7, PBS1, PBS2	*Escherichia coli/ Salmonella* *Bacillus subtilis*	Temperate; P1 resides in cells as free plasmid
DNA ds, lin, unique, sticky ends	Icosahedral	P2, P4	*Escherichia coli, Shigella, Serratia*	Temperate; P4 requires help of P2 for development
DNA ds, lin, unique, host DNA termini	Icosahedral	Mu (also termed μ)	*Escherichia coli* K-12, C, enteric bacteria	Temperate; "Mutator phage", integrates at random
DNA ds, lin, sticky ends	Icosahedral	λ, ϕ80, ϕ82, ϕ434, ϕ424	*Escherichia coli* K-12	Temperate
DNA ds, lin, unique	Icosahedral	ϕ29, X15, GA1, M2Y	*Bacillus subtilis*	Virulent
DNA ds, lin, unique tr	Octahedral	T1, T3, T5, T7, SPO1	*Escherichia coli* *Bacillus subtilis*	Virulent; "uneven" T-phage
DNA ds, lin,. cp, tr	Icosahedral	P22	*Salmonella, Escherichia coli*	Many similarities with λ
DNA, ss, circ	Icosahedral	ϕX174, S13, ϕR, GE	*Escherichia coli*	
RNA ss, lin, unique	Icosahedral	R17, f2, Qβ, MS2, M12, GA, SP, F1	*Escherichia coli*	Specific for male enterobacteria
DNA, ss, lin	Filamentous	M13, fd, Ff, IKe	*Escherichia coli*	Release from cells without lysis
DNA, ds, lin		PM2	*Alteromonas espejiano*	Marine, contains 10% lipids
RNA, ds, lin		ϕ6	*Pseudomonas phaseolicola*	Marine; three chromosomes; membraneous envelope; 20% lipids
DNA, ds, lin,		SM-1	Cyanobacteria (specific for *Synechococcus elongatus*)	First isolated cyanophage
DNA, ds, lin, cp		ϕH	*Halobacterium halobium*	Best-studied halobacteriophage

Large **polyhedral phages** have composite head structures of mostly an octahedral or icosahedral shape (Fig. 26.**4b**). A single chromosome is enclosed in a rather rigid capsid, which provides just enough room for a tightly packaged chromosome of unit size. These phages have developed elaborate packaging devices for threading the new chromosomes into pre-formed heads (see below). Tail structures of very different complexity are found attached to the base of the heads and provide the essential adsorption devices (Fig. 26.**5a,b**).

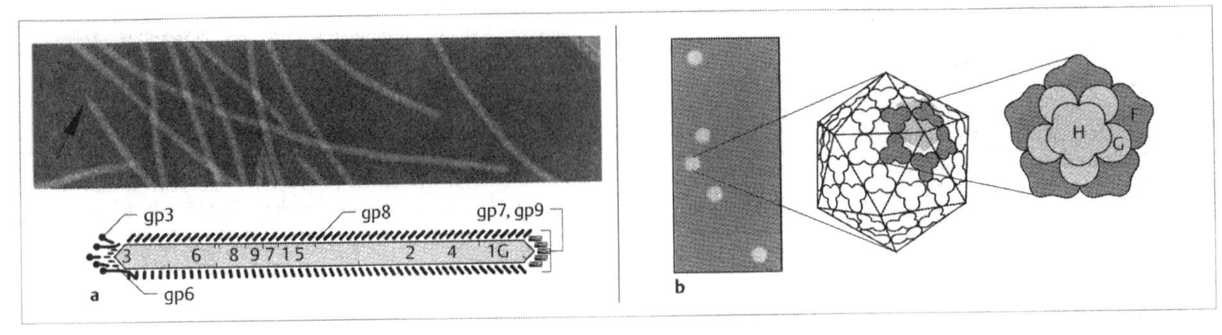

Fig. 26.**4a,b** **Structural composition of phages MI3 and ϕX174 and their visualization by electron microscopy.** The electron micrographs and sketches shown describe various phage types. Structural components are marked by the respective gene numbers or letters, preceded by gp (for **g**ene **p**roduct).
a The filamentous phage M13. The arrow in the electron micrograph points to a knob-like structure at one end. The sketch shows how the single-stranded circular chromosome is oriented within the protein shell. The DNA for specific genes (indicated by numbers) is placed at distinct locations relative to each other in the virion with the intergenic region (IG) always located at the end opposite from the adsorption complex (left). The DNA in the intercistronic region is rich in secondary structures with a large morphogenetic hairpin serving as a packaging signal (not shown). The single-stranded chromosome of 6 407 nucleotides codes for nine proteins. Structural proteins forming the coat are indicated. Phage M13 has a length of 895 nm and a width of 6 nm.
b Icosahedral, tail-less phage ϕX174. The electron micrograph shows an enlargement of the phage. The sketch shows the arrangement of subunits of the capsid. The capsid (25 nm in diameter) has 12 vertices of fivefold symmetry and is composed of 60 molecules of gpF. Each vertex consists of one molecule of gpH and five molecules of gpG which build together a knob-like structure or spike. The capsid holds the single-stranded chromosome of 5 386 nucleotides, which codes for 11 proteins [after 1]

Despite a great multitude of phages found among the bacteria and the archaea, their general morphology is based on a few general principles. As a general rule, linear double-stranded DNA is found in their capsids; more rarely, covalently closed circular and single-stranded DNA or even RNA is the carrier of genetic information. The capsids are polyhedral heads with or without tails and base plates as adsorption devices and contain a finite amount of DNA. The rare filamentous phages, however, assemble around their DNA; the length of the DNA determines the length of the mature phage. All of these properties are used to morphologically classify bacterial phages (see Table 26.**1**).

Regardless of their different lifestyles, all phages face the same basic five developmental problems: (1) adsorption, (2) penetration, (3) replication of DNA, (4) synthesis of new envelope components and phage assembly (morphopoiesis), and (5) release of progeny phages. **Virulent** phages multiply only through this lytic cycle. Some phages have, in addition to a lytic development, the choice of a **lysogenic** development. Instead of lysing their hosts after infection, the phages can also multiply their DNA in a peaceful coexistence with their host genomes, and are hence called **temperate** phages. Such coexistence in the form of a **prophage** can occur in the form of a free plasmid-like state (e.g., phage P1) or after integration of the phage chromosome into the host genome. The integration events occur either at many sites selected at random, as for phage Mu, or are restricted to special **attachment sites** (*att*), as for phage λ.

The lysogenic state of the host cell can be maintained for a long time before the dormant lytic functions of the prophage are revived again, leading to a normal development ending with cell lysis. This switch can be triggered from the outside, for example, by UV irradiation or starvation (Chapter 26.4.4).

The complexity (but not necessarily developmental efficiency) of the life style of a phage can be read from the number of genes it has on its chromosome. For example, phage ϕX174, with one of the smallest single-stranded phage genomes of 5 386 nucleotides, has only 10 genes (Fig. 26.**6**) and has a rather simple development. On the other hand, phage T4, with approximately 169 000 base pairs, stores its genetic information in at least 130 genes (Fig. 26.**7**). Both phages, however, produce almost the same number of progeny in nearly the same amount of time. Consequently, in comparison to phage ϕX174, phage T4 has an extremely complex morphopoietic development (see Chapter 26.3.5). Phage T4 uses the surplus of genes for many back-up mechanisms at all levels of its development and for

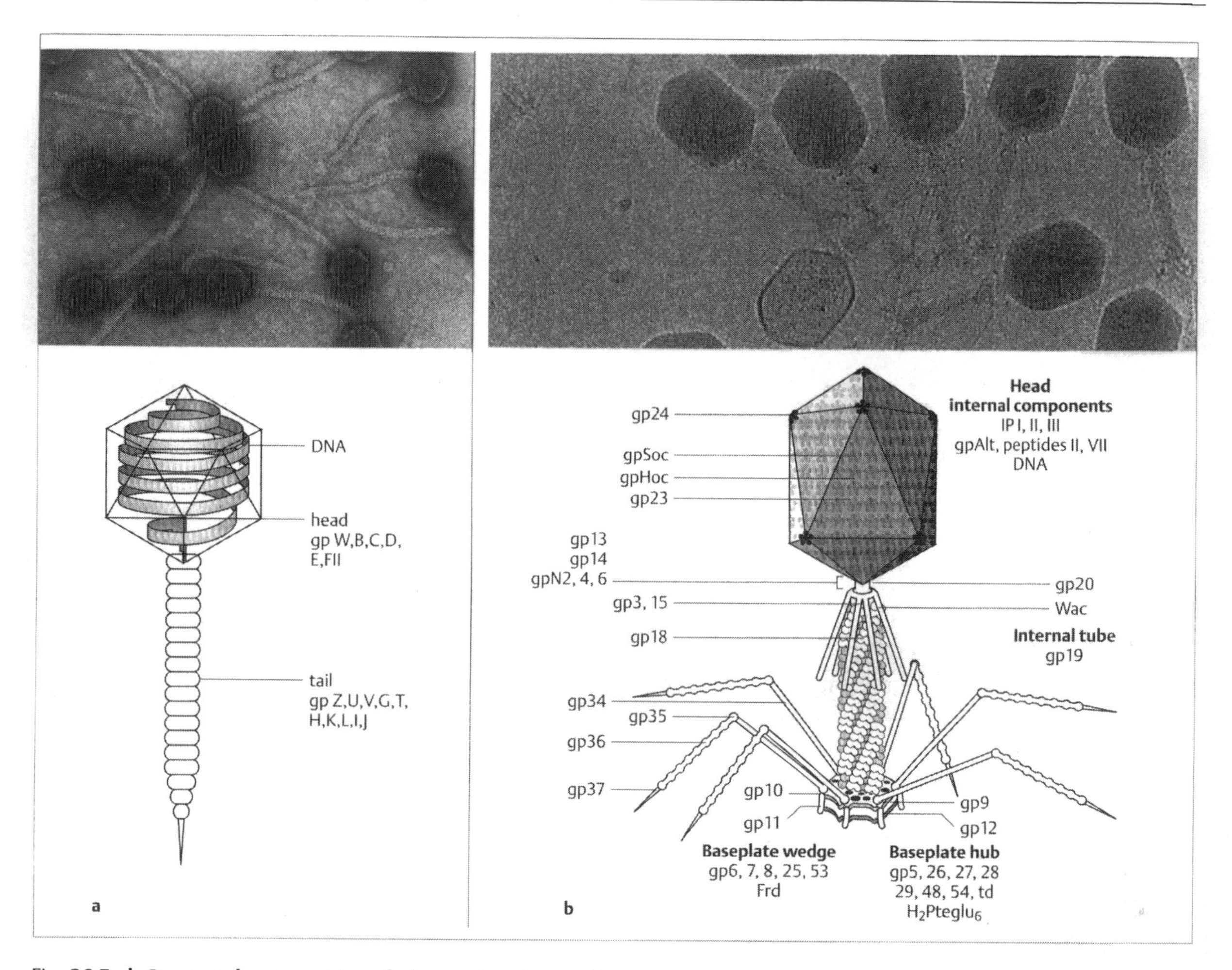

Fig. 26.**5a,b Structural composition of phages λ and T4 and their visualization by electron microscopy.**
a The icosahedral tailed phage λ. The electron micrograph shows a magnification of purified λ phages. The sketch shows the capsid and the tail composed of 6 and 11 proteins, respectively, marked by their letter code. The icosahedral capsid measures 50 nm in diameter. The tail is 135 nm long and ends in a small conical part with a single terminal tail fiber. The double-stranded DNA containing approximately 48 000 base pairs (about 50 genes) is linear and has cohesive ends. The left end, as determined by the genetic map, is packaged last and protrudes from the head into the upper part of the tail tube. This end is the first ejected [after2]
b The elongated icosahedral, long-tailed phage T4. The electron micrograph shows purified phages. The head measures 10 nm in length and 75 nm in width. The sketch indicates the locations of known major and minor proteins of the head and the tail. The vertices of the head are made from proteolytically cleaved gp24, the exterior between the vertices consists of gpSoc, gpHoc, and gp23. The double-stranded linear chromosome of approximately 166 000 base pairs for one chromosomal unit harbors about 130 known genes and nearly 100 functionally unidentified open reading frames. Each packaged genome carries a 5% redundancy at one end, which brings the complete chromosome to a total of 171 000 base pairs. The genomic map is circular permuted. Both characteristics are explained by a continuous packaging mechanism that operates on overlength, concatenated precursor DNA [after 3]

many fanciful structural additions. Visualization of these phages in the electron microscope reflects some of the structural differences (Figs. 26.**4b** and 26.**5b**). As another example, the genome of phage λ consists of 50 000 base pairs and is about tenfold larger than that of ϕX174 and one-third the size of the genome of phage T4. Phage λ has the choice between a lytic and a lysogenic pathway, and it uses many of its genes for fine tuning the control mechanisms required for making the decision between these pathways (see Chapter 26.4.2).

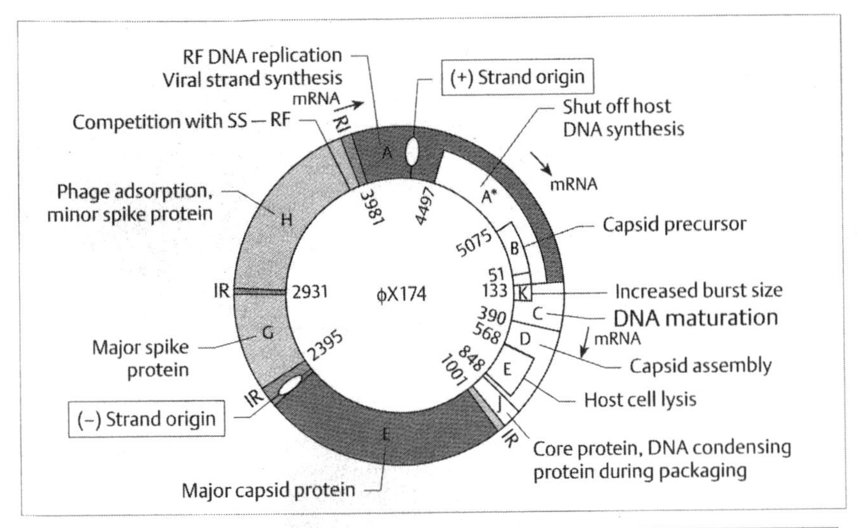

Fig. 26.**6 Genetic map of phage ϕX174.** The circular map of ϕX174 shows 10 genes (*A–H, J, K*) for making 11 proteins and 2 intercistronic regions (IR), which are not transcribed. One IR contains the origin of replication for (–) strand DNA synthesis and is organized in an extensively folded secondary structure. A few genes overlap each other saving space on the genome. Suggested functions of the gene products are indicated. The arrows indicate promotors and direction of transcription.

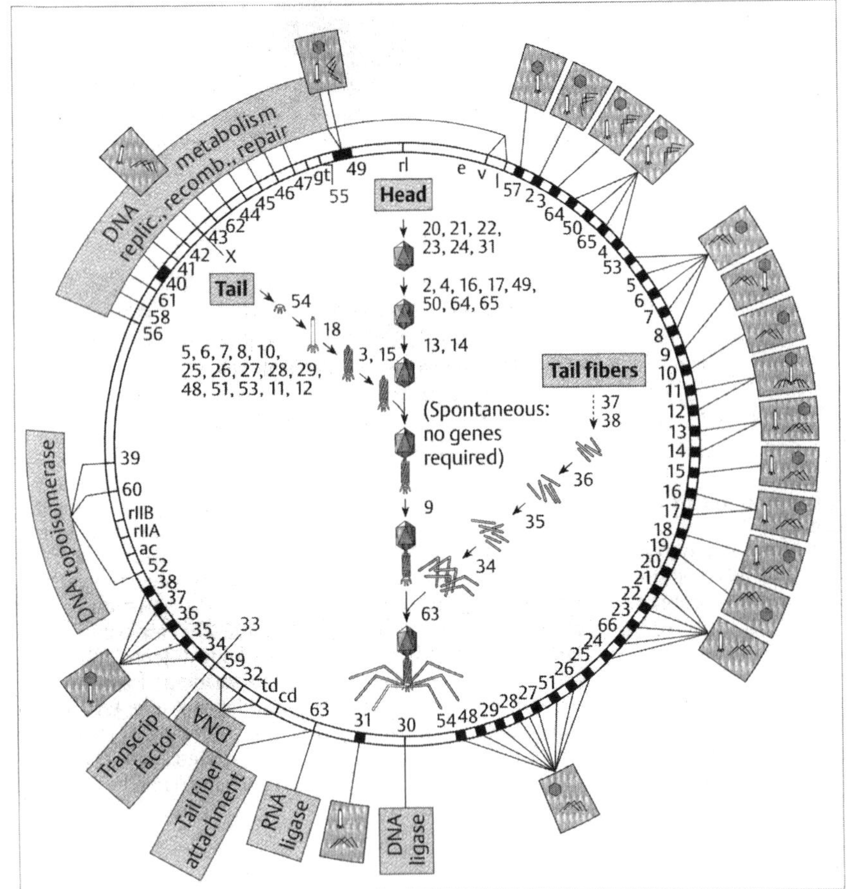

Fig. 26.**7 Genetic map of phage T4**. The circular map of phage T4 shows a selection of 76 genes, marked by numbers or letters. Morphogenetic genes, which belong to the class of late functions, are characterized by pictograms that show faulty products produced by phage mutants with mutations in the respective genes. Most of these genes are clustered on the right side of the map, which is transcribed clockwise. Within the circle, the suggested pathway of phage assembly is shown with three independently controlled branches for tail, head, and tail fiber assembly. The genes responsible for the steps in morphopoiesis are indicated by their numbers located beside the arrows connecting two developmental stages. Gene *63* is an interesting example with dual functions in nucleic acid metabolism and morphopoiesis since it catalyzes tail fiber attachment and can function on other occasions as an RNA ligase. Genes for DNA metabolism, such as DNA synthesis, recombination, repair, and transcription control map mostly on the left side of the circle, which is transcribed in a counterclockwise direction. In addition to the large number of genes, empty-appearing regions with a surplus of DNA are seen. These regions may be functionally neutral or act as spacers with unknown function, or they might contain genes that have not yet been discovered. Extensive sequencing of the genome has revealed a considerable number of open reading frames in these regions

26.3 The Lytic Infection Cycle Is Characteristic for Virulent Phages

26.3.1 The Growth of Virulent Phages Always Causes Cell Lysis

The first description of a lytic life cycle of phages was given by E. L. Ellis and M. Delbrück in 1939 when they performed their famous "one-step growth experiment" with phage T4 and *Escherichia coli* (Fig. 26.**8**). They allowed 10 phages per cell (m.o.i. of 10) to adsorb to *E. coli* in the presence of KCN, a metabolism-blocking agent that was initially given to the uninfected cells to prevent premature development. Synchronous development was started by diluting the infected cells into fresh medium without KCN. Aliquots from the culture were collected at intervals and the number of free phages were titrated after removing intact cells by centrifugation. Free phages were not seen before 22 min after infection, i.e., the latency period lasted for 22 min.

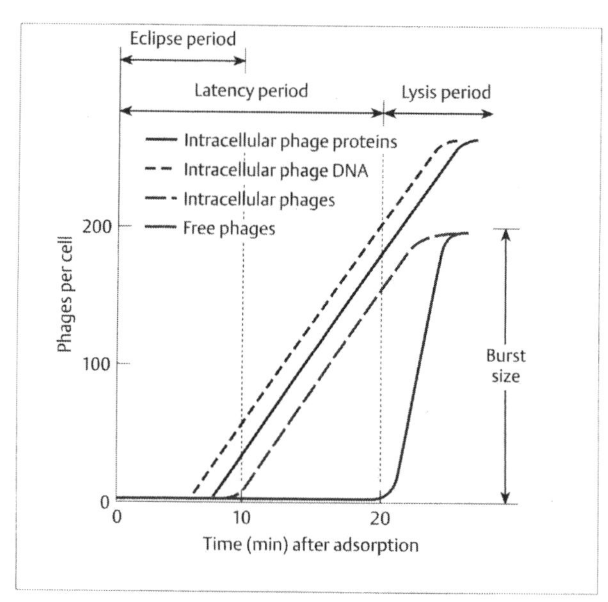

Fig. 26.**8** **"One-step growth curve" of phage T4.** The development of three intracellular components (DNA, envelope proteins, and viable phages) was followed over time and compared with the appearance of free phages in a synchronously developing cell culture. Three periods are distinguishable. The 10-min time interval between adsorption and the appearance of intracellular progeny is called the **eclipse period** and is contained within the **latency period**, which occurs from 0 to about 22 min, when free progeny phages become detectable in the medium at the beginning of the **lysis period**. Lysis is complete when the entire culture clears at about 30 min after infection. The average number of phages liberated per infected cell constitutes the **burst size**, in contrast to a **"single burst,"** which describes the progeny phages generated from a single cell

When A. H. Doerman repeated this experiment in 1952, he added an important new aspect and searched for viable phages inside the cells before natural lysis had occurred by lysing the cells prematurely with chloroform and titrating these lysates. Newly made phages appeared as early as 12 min after infection. The period from 0 to 12 min, called the **eclipse period**, was free from viable phages, but obviously contained phage DNA ("**vegetative phage**"). The time from 0 min until spontaneous lysis, when the first extracellular phages occur, named the **latency period**, is followed by the **lysis period** when all cells become disrupted and the culture finally clears completely (Fig. 26.**9**).

Many detailed investigations performed later revealed that the eclipse period is, despite its name (literally meaning "empty"), the most active period in phage development. This is the time when sophisticated regulatory programs are started and the metabolism of the hosts is brought under control of the incoming phage genome. Programmed decisions, such as what to take from the host and what to express from the phage's own genome, are quickly made. The phage functions are classified according to their time of requirement as **early** and **late**. The onset of DNA replication is taken as the logical border between the two classes (Fig. 26.**9**).

Early genes of phage T4 have normal promoters, which correspond to promoters from *E. coli* and are transcribed by the host RNA polymerase holoenzyme (containing the σ^{70} subunit; see Chapter 18.9). The early genes comprise a few genes whose gene products are responsible for modifying the subunits of *E. coli* RNA polymerase by ADP-ribosylation (genes *mod* and *alt*) or by adding new σ-factors (genes *33* for σ^{33} and *55* for σ^{55}), which make the RNA polymerase specific for certain early and all late promoters (Fig. 26.**7**). After modification, host genes are no longer transcribed by the altered RNA polymerase.

All functions involved in DNA metabolism, such as synthesis of nucleotide precursors and DNA replication, are found among the early genes. Late functions are mainly reserved for production of morphogenetic components and assembly of capsids. Interestingly, most genes are also organized as early and late in two separate sections on the genetic map, and they are transcribed in opposite orientation from different strands (Fig. 26.**7**).

Phage functions are classified traditionally as **early** and **late**. Early genes correspond roughly to those involved in overcoming the host cell defense and redirecting its metabolism towards phage production and to phage DNA replication. Late genes are those involved in production of phage envelope proteins and phage assembly, followed by cell lysis (see Fig. 26.**9**).

The time between adsorption and lysis varies considerably among phages. Phages T1 and T5 finish their growth cycle within 12 min after infection, while phage T4 requires 30 min and phage λ needs up to 65 min. The filamentous phages are exceptional in that they do not lyse their hosts cells, but rather produce progeny continuously by channeling them through the cell envelope. It takes hours before the host cells metabolism finally breaks down and the weakened cells can no longer produce phages.

26.3.2 The Growth of Phages Begins After Adsorption to the Cell Surface and Injection of Their DNA

The first contact between a phage and its host happens by random collision. If the cell carries specific **receptors** on its surface, stable binding (i.e., **adsorption**) of the phage can follow. The contact is usually made between the receptor proteins of the host and specific phage proteins located at the tip of the tail, tail fibers or, as in case of filamentous phages, at one of its ends. Phage T4 recognizes its receptors by means of protein gp37. A dimer of this polypeptide is located at the distal part of the long tail fibers. For some phages (λ and T4), stable adsorption is preceded by a stage of transient and reversible adsorption.

Many components of the envelope of bacteria can serve as phage receptors. These can be constituents of the outer or inner membrane or of the cell wall. Phage T4 attaches to a lipopolysaccharide (LPS) of the outer membrane. For adsorption, phage λ requires the membrane protein LamB, which is the maltodextrin porin of *E. coli* in the outer membrane. Filamentous DNA phages attach to the tip of sex pili of male bacteria, while RNA phages attach to the sides of the same pili. If a receptor gene is mutated, the cell may become resistant to the respective phages. **Resistance** can be evaded by phages if the phage genes for the receptor-recognizing proteins mutate (conformational suppression). These are called **host-range** mutations.

Phages Mu and P1 have developed a sophisticated genetic system that enables them to switch hosts. A piece of the gene coding for an adsorption protein is inverted by a site specific recombination event and transcribed in the opposite direction (see Fig. 18.**18**). Switching occurs frequently such that about 50% of a burst represents phages of one host range and 50% of the other (see Chapter 18.11.2).

Some phages require cofactors for efficient adsorption. For example, T4 needs L-tryptophan for a complete unfolding of its tail fibers prior to adsorption. Other phages, such as λ and P1 require addition of Ca^{++} and Mg^{++}, respectively.

Injection of DNA follows immediately after a phage has stably adsorbed to the cell surface. Although it is not yet clear how the injection process works in detail, it is generally assumed that repulsive forces between negatively charged side groups of the highly condensed DNA contribute to the energy required for its translocation. Phage T4 carries a few molecules of lysozyme attached to the base plate of the tail, perhaps for local hydrolysis of the peptidoglycan of the cell wall. Phage λ, however, requires the presence of a carbohydrate transport system, the mannose-specific PTS (genes *manXYZ*), for DNA injection. Translocation of phage DNA over the cytoplasmic membrane does not appear to require energetization [by the proton motive force (Δp) or ATP] and most likely is through pores or channels. In phage T4, the Δp appears essential to keep the pore (channel) open or to form an active attachment site.

With the exception of some phages such as T5, the injection process is very fast. Interestingly, phage T5 injects its DNA in two steps. Eight percent of the proximal end of the chromosome (FST DNA) is injected first; the distal portion (ss DNA) follows after a pause of about 4 min. During the pause, genes located at the proximal end of the genome are expressed and proteins are synthesized that break down the host genome, inactivate the restriction system of the host, and finally allow the injection process to proceed further. As expected, the proximal sequence of the chromosome is free from restriction sites sensitive to the host restriction system.

In a historical experiment performed by A. D. Hershey and M. Chase, it was shown that only DNA and no protein is injected into the cell. Protein and DNA of phage T2 were differentially labeled [S^{35}] and [P^{32}], respectively. Among the progeny released from an infection with doubly labeled parental phages, some daughter phages were found that were labeled with [P^{32}], but none with [S^{35}] were found (Fig. 26.**11**). Today a few exceptions are known in which a few proteins are injected together with the chromosomes of some phages. For example, phage T4 carries copies of gp2 bound to both ends of its DNA. The 25-kDa protein

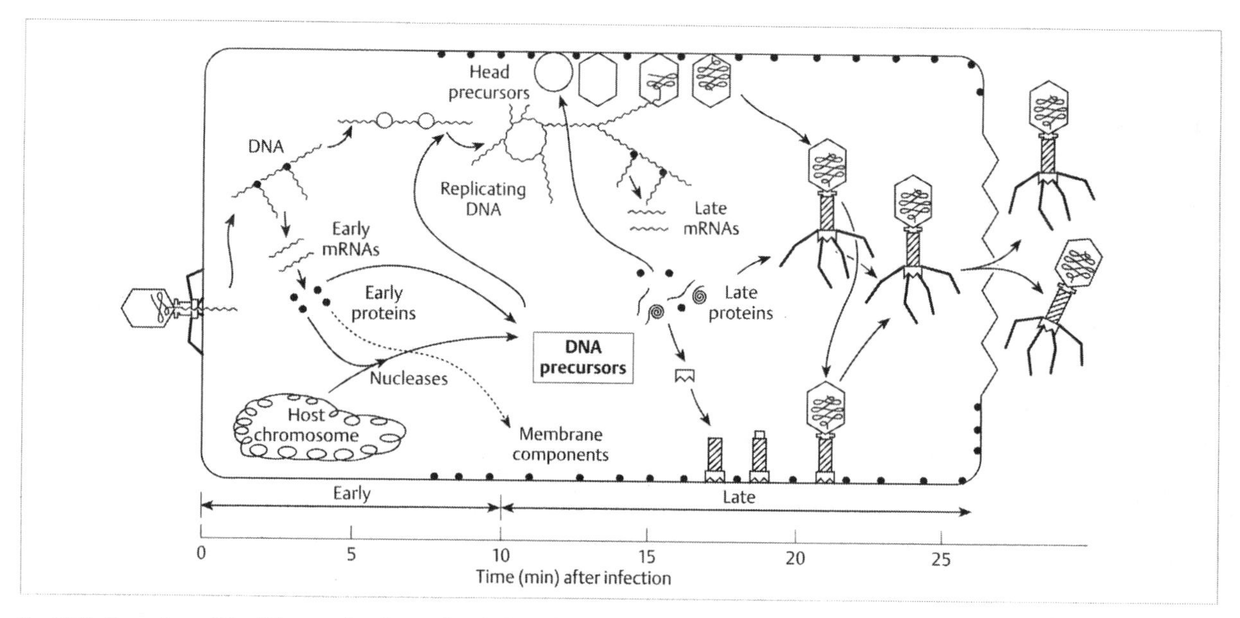

Fig. 26.9 **Overview of the T4 reproductive cycle.** The lytic development of phage T4 between infection and lysis of the host cell is drawn on a time scale, reading from left to right. Immediately after infection, the production of early proteins, translated from early mRNAs, initiate the production of DNA precursors and progeny DNA. Late proteins are translated from late mRNAs at 10 min after infection until lysis. Structural components such as heads and tails are attached to the cell membrane before they are assembled to complete viable virions.

protects the DNA efficiently against an early fragmentation by exonuclease V, a function of the RecBCD recombination and repair protein of the host. Consequently, phages that lack gp2 can only grow if the host has no RecBCD exonuclease function.

The experiment by Hershey and Chase would not have worked as it did if the authors had chosen filamentous phages instead of T2. For example, M13 is taken up as a whole by its host. The DNA is stripped from the coat protein during its transfer through the membrane. Later, the old coat protein is recycled and used for covering new progeny chromosomes on their way through the cell membrane to the outside.

26.3.3 Very Early Events in the Host Cell Redirect its Metabolism Toward Phage Production

Phage T4 produces about 600 copies of its chromosome per infection cycle. About 300 copies are packaged completely prior to the moment of lysis, and comprise the infectious particles of a burst. The remaining DNA, another 300 copies, is not successfully packaged and is lost after lysis. Overproduction seems to be the price for a rapid development. During the 30 min of its active life cycle, T4 DNA polymerase synthesizes about 1×10^8 base pairs ($600 \times 1.7 \times 10^5$) in the form of high molecular weight DNA. For comparison, in about the same time, *E. coli* replicates its chromosome only once, which is the equivalent of about 2×10^6 base pairs. Hence, T4 replicates its DNA nearly 100 times faster than *E. coli*. This requires fast enzymes, which T4 provides, and a sufficient supply of dNTPs. T4, being a "parasite", has solved this problem by using its host's chromosome as a nucleotide reservoir.

Observing the infection cycle of phage T4 more closely shows that dramatic changes occur in the host's chromosome, which is naturally folded into a compact structure, the **nucleoid**. Immediately after infection, the organization of the nucleoid is strongly affected, and three major changes can be seen. The nucleoid becomes unfolded, then disrupted, and finally degraded. The events are independent of each other and mutations are known that affect one event without impairing the other.

The chromosome of *E. coli*, which is normally folded into numerous negatively supercoiled domains, is gradually unfolded. This depends on the function of the T4 gene *unf/alc* gene product, which has a dual function and is also required for blocking transcription from cytosine-containing host DNA. The nuclear dis-

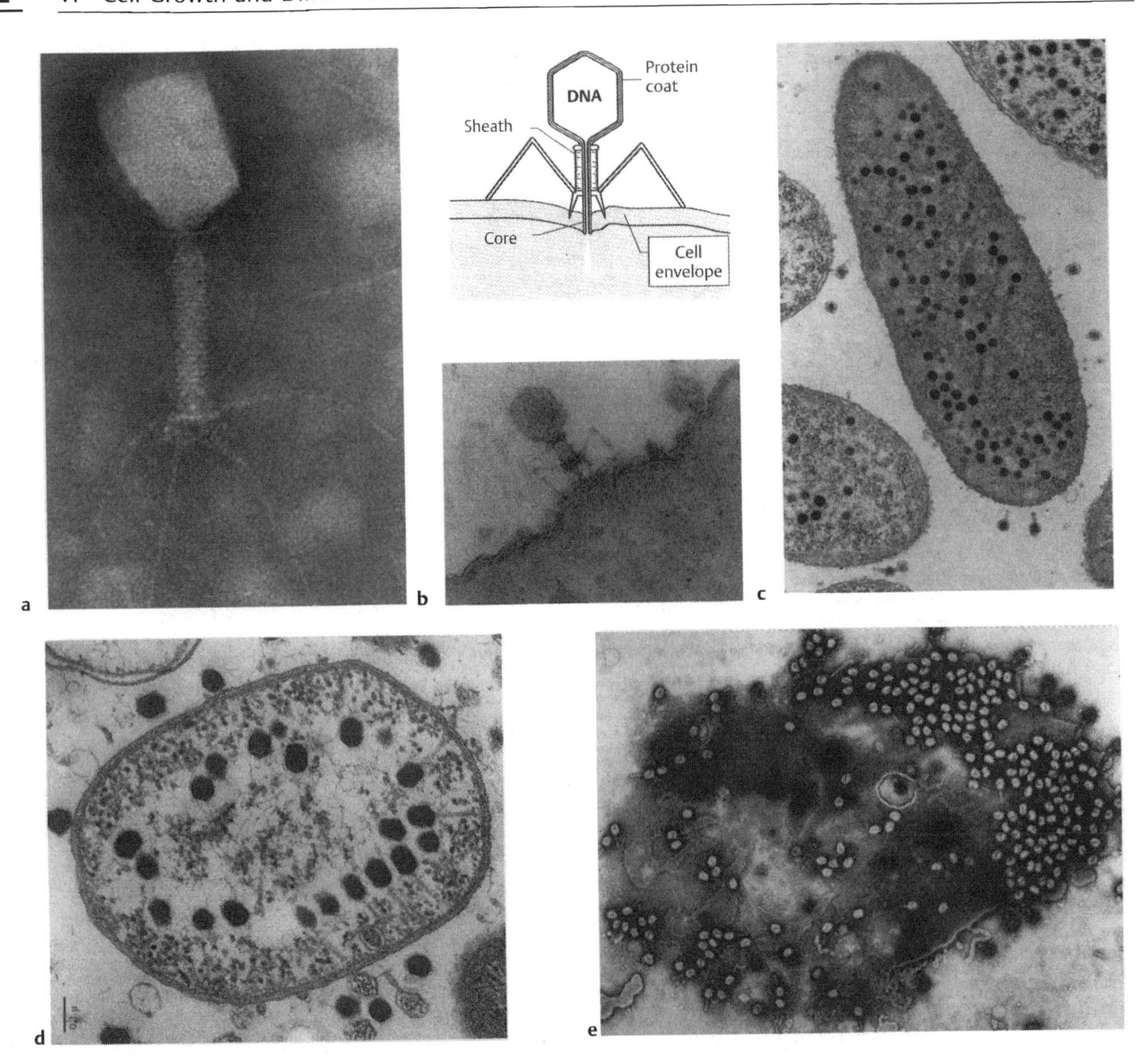

Fig. 26.10 **a–e Electron micrographs taken from different stages of phage T4 development**. Free phage, **b** phage adsorbed to the surface of *Escherichia coli* and injecting its DNA (inset shows an explanatory sketch), **c** cell with adsorbed phages and intracellular new phage, **d** cell with adsorbed phage, and membrane attached intracellular phages, and **e** lysed cell. The phage T4 head measures 100 nm in length and 75 nm in width [after 4]

ruption gene *ndd* is then responsible for moving the DNA from the central region of the cell toward its periphery, where it becomes attached to the inner membrane. The product of the gene, a 15-kDa protein, is synthesized very early after infection within 3–6 min. The protein is made in stoichiometric amounts, comprising the remarkable number of about 4000 molecules per infected cell by the end of this period.

Degradation of the host DNA follows and is initiated by two phage-encoded endonucleases, endonuclease II (gene denA = **d**eficient for **end**onuclease **A**) and endonuclease IV (gene denB = **d**eficient for **end**onuclease **B**). Both

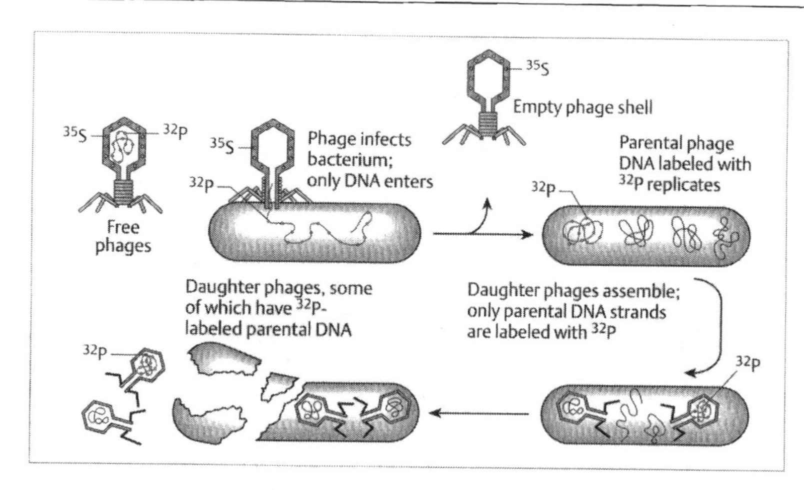

Fig. 26.11 **The historical experiment of Hershey and Chase.** Phage T2 protein and DNA were radioactively labeled with [^{35}S] and [^{32}P], respectively to demonstrate that DNA and not protein is the genetic material of this phage [after 6]

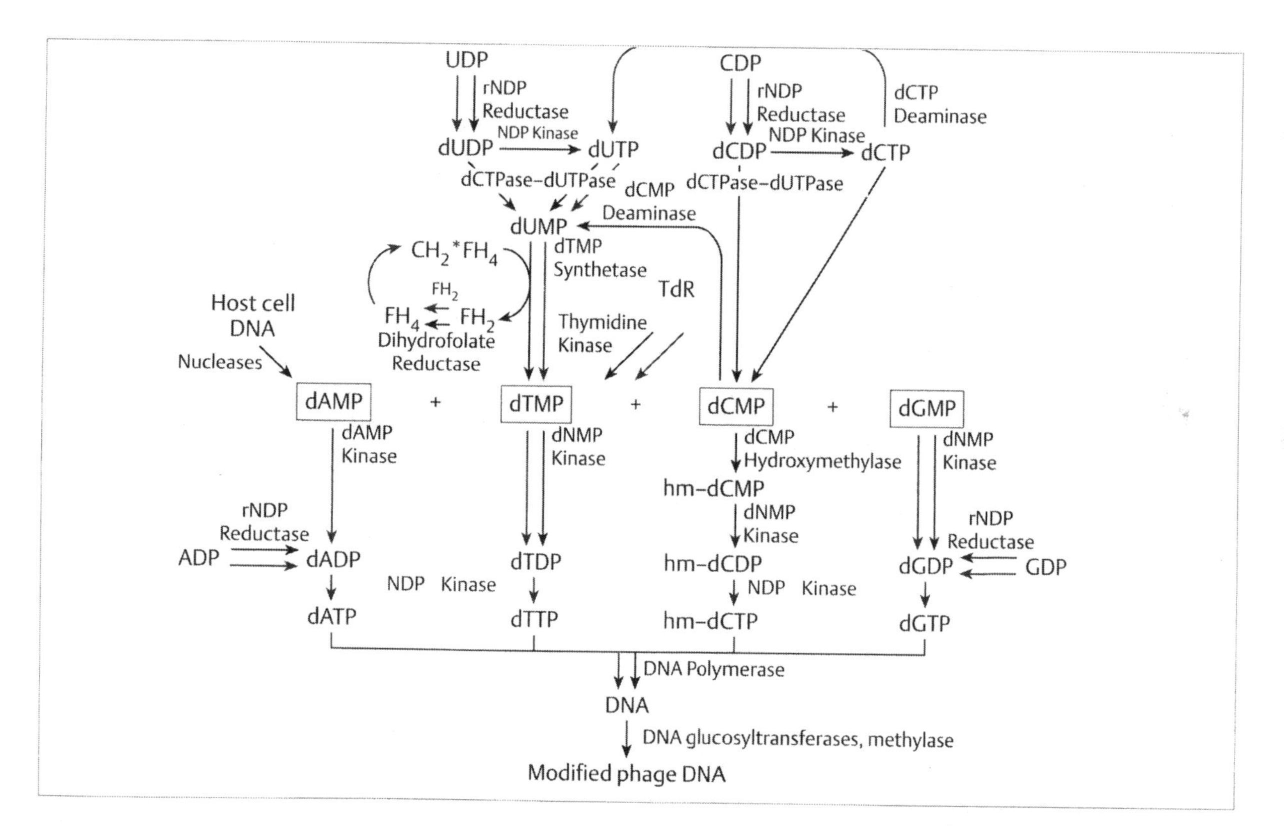

Fig. 26.12 **DNA precursor synthesis in T4-infected Escherichia coli.** The DNA of phage T4 contains glucosylated hydroxymethyl deoxycytosine (dHMC) instead of deoxycytosine (dC). To ensure complete replacement of dC by dHMC in progeny DNA, 14 gene products, 12 encoded by phage T4, are required. Glucosylated dHMC also makes T4 DNA resistant to almost all restriction enzymes and many regular nucleases. The central pool of dXMPs is fed from de novo synthesis and from phage-controlled nucleolytic degradation of host DNA. Several of the reactions are performed by host enzymes (red arrows), but often become reinforced by phage-encoded enzymes with similar functions (black arrows). The production of dHMC, which is a specialty of T4, depends exlusively on phage-borne genes. Glucosylation of cytosine residues, (starting from UDPGlc) occurs after DNA synthesis and is controlled entirely by phage-encoded genes [after 7]

enzymes degrade only cytosine-containing (i.e., host) DNA and do not degrade the modified phage DNA with fully glucosylated and hydroxymethylated cytosine. The degradation of host DNA is initiated by an extensive endonucleolytic cleavage by these enzymes and finally ended by production of mononucleotides by a membrane-bound exonuclease (presumably gp46 and gp47).

Phage T4 uses the nucleotides derived from host DNA degradation for its own development; this is sufficient for producing about 20 phage genomes. For the proper processing of these recycled nucleotides, phage T4 has evolved an elaborate system with 14 enzymes, 12 of which are encoded by T4 (Fig. 26.**12**).

26.3.4 Morphopoiesis (Assembly) of Phages Is a Late Process

The late phase of the life cycle of phages is dominated by morphopoietic assembly events. The constituents of the progeny are continuously assembled into new virions until lysis terminates further production. Two different modes of assembly have been observed among phages:

1. RNA and filamentous ssDNA phages use a process that is best described as **chromosome coating**. Newly made DNA still covered with single-stranded DNA binding proteins originating from replication is wrapped up with coat proteins at the moment when the virion is transported across the cell wall to the outside. In an exchange reaction, the single-stranded DNA binding protein is stripped from the DNA and is replaced by a coat protein that was stored in the cell membrane. The freed single-stranded DNA binding protein is recycled and used again to protect newly replicated DNA. One of the best-studied examples of this process is phage M13 (Fig. 26.**13**).
2. Large dsDNA phages and ssDNA phages with icosahedral heads (see Table 26.**1**) use an assembly process that involves the interaction of preassembled capsids (pro-heads) with the DNA (Fig. 26.**12**). The main constituents of a phage, namely **heads, tails**, and **tail fibers**, are synthesized through several genetically controlled steps via independent pathways. A defect in anyone of the genes along the assembly lines causes arrest of phage production and accumulation of unmatured precursors (Fig. 26.**7**). The DNA is actively packaged into a preformed protein container, the **prohead**, which has little room in excess of what is required for one tightly packaged chromosome in a highly condensed state ("**head-full**"). The finished elements are finally assembled with the help of additional proteins, giving viable phages waiting inside the cell for lysis to occur. Phages T4 and λ are the best-studied examples of this type of assembly. Phage T4 spends about 40% of its coding capacity for morphopoietic functions. The pathways of the assembly of subunits were worked out by R. Epstein and co-workers by careful studies of the phenotypes of mutants with mutations in the maturation genes.

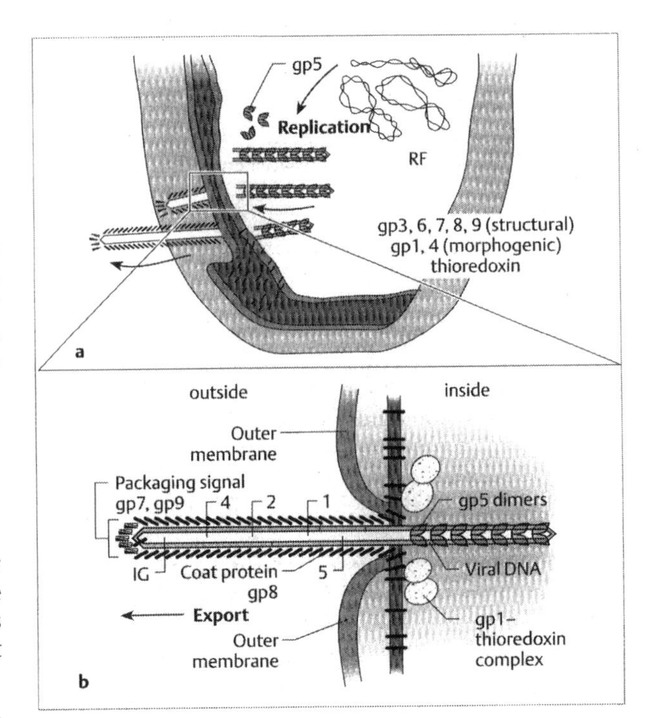

Fig. 26.**13 Assembly of filamentous phage M13 at the cell surface.**
a Progeny chromosomes of phage M13 emerge from replication as single-stranded circles, which are shielded against nucleases by a protective cover made from dimers of the single-stranded DNA binding protein gp5. Proteins gp7 and gp9 bind to a large hairpin structure in an intergenic region of the DNA, which serves as packaging signal. In the mature particle, these proteins are located at one end.
b The chromosomes are transferred to the inner cell membrane, where gp5 is removed from the particle and replaced by coat protein gp8. This occurs concomitantly with the extrusion of particles to the outside of the cell. gp8 was stored beforehand via its hydrophobic domain in the inner cell membrane. Proteins gp3 and gp6 mark the other end of the finished particle [after 8]

26.3.5 Head Assembly and DNA Packaging of Phage T4

The assembly of phage heads occurs regardless of whether DNA is present for packaging or not. Packaging of DNA, however, is a prerequisite for final fusions of tails to heads. Without DNA, empty appearing heads ("**ghosts**") accumulate in the cell. In phage T4, initiation of DNA packaging requires functioning proteins gp16 and gp17, while the progression of packaging depends largely on an intact gp49.

Prohead processing and maturation of DNA are shown in Figure 26.**14** in parallel pathways. Packaging of DNA begins when gp16 and gp17 mediate the attachment of processed DNA to competent heads. After filling, the head is stabilized and sealed by the addition of gpHoc and gpSoc to the outer capsid and neck proteins gp13 and gp14 to the base of the head. Filled heads are then ready for spontaneous attachment of separately assembled tails, as depicted in Fig. 26.**7**. Head structures are assembled as proheads attached to the inner membrane of the cell. The structure, held by a connector, is made from two major precursor proteins (gp23 and gp24) layered around a scaffold assembled from several major and minor proteins (gp21, gp22, gp67, gp68, IPI, IPII, IPIII, and gpAlt).

Proteolytic cleavage of prohead proteins and the release of the prohead from the membrane deliver head structures that are ready for DNA uptake. The DNA to be packaged is replicated in large, oversized, concatenated molecules (called 200S DNA because of its high sedimentation value; for comparison, unit size T4 DNA sediments with 63S). Packaging requires ATP, processed proheads, gp16 and gp17. The DNA is taken up until the head is filled to completeness, adding approximately 5% more than a unit genome length would require. Only then is the chromosome cleaved from its precursor catenate. The newly generated end is used further as a new packaging initiation site for filling one of the next heads, which again will take up about 105% of a complete genome, and so on. This procedure explains why the genetic map of phage T4 is **circularly permuted** with a **terminal redundancy** of 5%, although an individual chromosome shows physical linearity.

During the filling process, heads expand and gpHoc and gpSoc are added to the protein lattice. This head is competent for the next major step in phage assembly, the addition of tails. Mutations in genes *16*, *17*, or *49* severely affect the event of DNA packaging and head maturation. gp16 and gp17 are initiator proteins required for the attachment of DNA ends to processed proheads and initiation of the filling process. A defect in either of these genes causes the accumulation of mostly

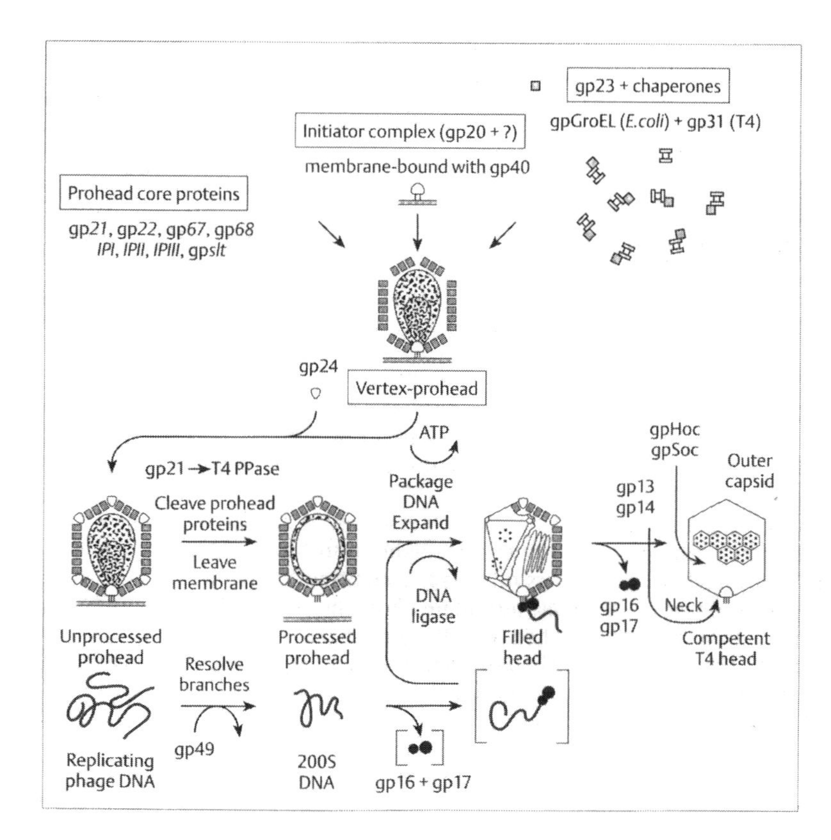

Fig. 26.**14** **T4 head assembly pathway.** Head formation starts with the assembly of prohead core proteins onto membrane-bound initiator complexes containing gp20. GroEl from *Escherichia coli* and gp31 from phage T4 work together as chaperones, providing gp23 with a molecular structure that allows its ordered assembly. Addition of gp23, the major capsid protein, followed by gp24, completes prohead formation. Proteolytic cleavage of the capsid proteins gp23, gp 24 and the core proteins by gp21 is responsible for the release of processed proheads from the membrane. These proheads are ready to take up DNA from concatenated molecules processed by gp 49 (i.e., removal of obstructing branches) before or during ongoing packaging. Uptake of DNA consumes ATP and requires gp16 and gp17 for the first contact between DNA and the entrance of the head with gp20. During the packaging process, heads expand and are stabilized by incorporation of the additional head proteins gpHoc and gpSoc. Packaging is terminated by cutting of the packaged portion of the DNA by gp17 from the precursor DNA pool and adding neck proteins gp13 and gp14. Refer to text for further details [after 9]

empty (ghost-like) heads and unpacked precursor DNA. gp49 is an endonuclease responsible for removal of branches and other structural deviations (e.g., branched three- and four-way junctions in newly replicated DNA, mismatches, gaps, nicks, single-stranded overhangs, and kinks) in precursor DNA that may interfere with packaging by steric hindrance. The enzyme functions as a guarding system examining the incoming DNA, marking structural failures by nicking, and thereby preparing them for further DNA repair events. Mutants with a defective gene *49* are still able to initiate DNA packaging, but do not proceed to completion and end the filling process prematurely. As a consequence, partially filled heads and highly branched unpacked precursor DNAs are seen in these cells (Fig. 26.**14**).

26.3.6 Phage λ DNA Packaging, a Paradigm for DNA "Head-full" Packaging

The chromosome of phage λ is double-stranded and comprises 48 502 bp with 5′-terminal single-stranded **cohesive ends** (or ***cos*** sites) of 12 bases. After infection of cells, the ends of the λ chromosome become covalently joined and many copies, arranged in a continuous oversized molecule (**concatenates**) are made by a rolling-circle-type of replication. Packaging of head-full loads of DNA proceeds unidirectionally, beginning with the left end on the genetic map towards the right end. The process requires a phage-coded terminase gpA, a host factor IHF, and two *cos* sites, which must flank the DNA to be packaged. A *cos* site consists of a nicking site within 16 bp, including the cohesive end of 12 bp, and further binding sites for gpA and IHF somewhere further away and located within the next 200 bp on both sides of the *cos* site. Packaging of the DNA can be initiated from any *cos* site along the concatenated DNA. To achieve this, the terminase binds simultaneously to the DNA of the *cos* site and to the connector protein gpNu1, which in turn binds to the base of the capsid. The sandwich-formation brings the empty capsid in close proximity to the packaging initiation site. The process requires ATP hydrolysis, proceeds via "looping" the DNA into the capsids, and is terminated at the next *cos* site again by gpA, which creates the sticky ends by introducing 12-bp staggered nicks. The end packaged last remains close to the entrance of the head and reaches into the tail structure after it is attached. This end is injected first after the next infection.

The amount of DNA packaged is determined by the distance between two *cos* sites and the volume of the capsid. Under native conditions, this amount corresponds exactly to one genome of phage λ. The DNA between two *cos* sites does not have to originate from phage λ, but can be from any other source as long as its length does not markedly exceed a head-full load. Head-full-type mechanisms of DNA packaging are also used by other icosahedral phages.

26.4 Lysogenic Pathways, an Alternative Life for Temperate Phages

26.4.1 Temperate Phages Can Integrate Into the Chromosome of Host Cells and Delay Entry Into the Lytic Pathway

Phages like λ, P22, P1, P2, and Mu have a choice between two developmental pathways. They can either propagate their progeny via a lytic pathway, which ends with lysis of the host cell and liberation of new phages as described above, or they can follow a **lysogenic** pathway, in which they seem to disappear for a while in the host, like a stowaway on a ship (Fig. 26.**15**). This is called the **prophage** state. This disappearance in the prophage state is based on two molecular mechanisms: (1) some phages (e.g., λ) have highly specialized tools for the integration of their DNA into the chromosome of the host, where they are passively copied with each round of host chromosomal replication; they can survive there for an indefinite period, unnoticed by the host system and without producing virions; and (2) other phages (e.g., the *E. coli* phage P1) change into low-copy-number plasmids that are replicated at a modest pace and stay under strict control of the host. Since the phages do not harm the cells in either of these lysogenic states, the phages are called **temperate** phages. This distinguishes them from **virulent** phages, which inescapably kill their hosts.

The status of lysogeny is quite stable in a cell culture since it is further transferred from one lysogenic cell to its offspring. Sometimes, however, the lysogenic pathway terminates and the lytic pathway is induced ("switched on"). Progeny phages are then made, and the life of the host cell is ended by a normal lysis with the liberation of new virions.

Phage P1 is an example of the second mode of lysogenization of *E. coli*. Phage P1 converts into a single copy plasmid (one copy per chromosome) that repli-

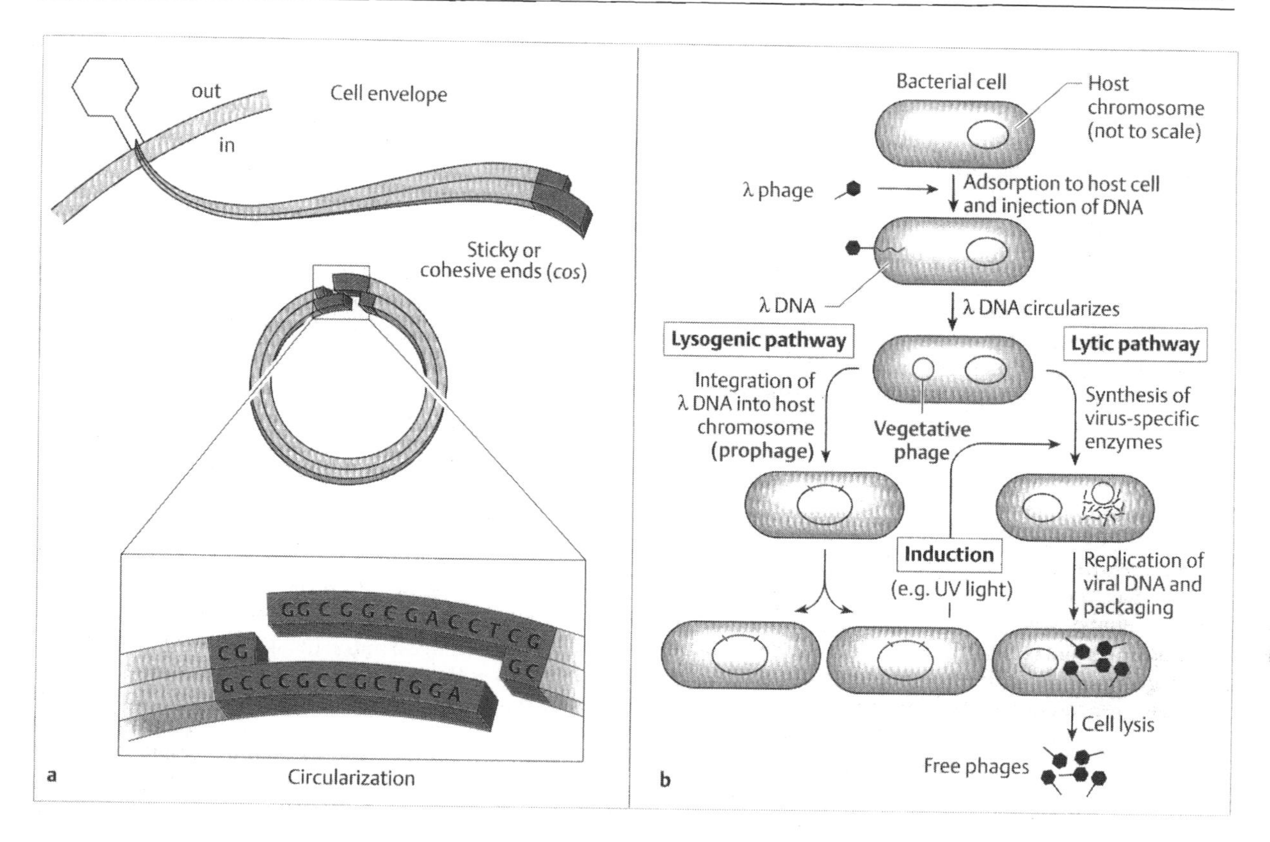

Fig. 26.**15 a,b Phage λ has the choice between lytic or lysogenic development.**
a Upon injection into the cell, the linear chromosome of phage is circularized (**vegetative phage** or "Hershey circle") and propagated further by either a lysogenic or a lytic pathway.
b Lysogeny begins with repression of lytic functions and integration of the phage DNA into the host chromosome. There it remains as a **prophage** until events from the outside (such as UV irradiation causing DNA damage or starvation) trigger its release from repression (**induction**), liberating the chromosome by excision from the host genome and initiating phage development towards lytic growth [after 6]

cates synchronously with the host chromosome. The linear P1 chromosome is circularized upon entry by the phage-encoded protein gpCre. This enzyme mediates site-specific recombination between two *loxP* sites, one located among genes of the linear map and another one in the region of terminal redundancy. The same *lox–cre*-dependent recombination is also responsible for resolving monomeric genomes from oversized concatenates shortly before cell division. In rare cases, P1 uses also the *lox-cre* recombination to integrate its genome into the genome of *E. coli*. The site-specific recombination event occurs between a phage-borne *loxP* site and a host-borne *loxB* site, which have a similar sequence.

Phage Mu is peculiar in that it corresponds to a transposable element that integrates its DNA at practically any place into the host chromosome by using a homology-independent ("**illegitimate recombination**") **transposition mechanism**. The lack of sequence selectivity causes frequent integrations into intact genes, which are thereby inactivated. This property gave the phage its name Mu, mnemonic for **mu**tator properties.

26.4.2 Bacteriophage λ, the Paradigm of a Temperate Phage

Lysogeny was first described in modern terms by A. Lwoff at the Pasteur Institute in Paris during the late

1940s. Its molecular nature was disclosed by work on phage λ and the *lac* operon in *E coli*. *trans*-recessive mutations in genes *cI*, *cII*, and *cIII* of λ (for **c**lear plaque, compared to the turbid plaques of a wild-type temperate phage) were recognized as causing a deficiency in three regulatory proteins. In contrast, the *cis*-dominant mutations in a λ*vir* phage (for **vir**ulent because it invariably lyses host cells, thus causing clear plaques) were interpreted as being promoter/operator mutations. Phage λ has remained popular as one of the best-understood prokaryotic model systems for the analysis of differentiation processes, such as the irreversible switch between the lytic and the lysogenic pathway. Phage λ is furthermore one of the preferred vectors used in gene technology.

Phage λ consists of about half protein and half DNA. Each particle contains one linear double-stranded DNA molecule (48 502 base pairs) encapsulated in an icosahedral head, which is connected to a long tail (Fig. 26.**5a**). The ends of the λ chromosome have a peculiar shape with 12-nt-long 3′-protruding single-stranded overhangs, which are called "sticky" because they can adhere to each other propagating ring closure. They are generated late after infection during packaging when a terminase cuts the DNA from oversized concatenates by introducing 12-bp staggered nicks (see Fig. 26.**15**). The sticky or **cohesive ends** (**cos**) are

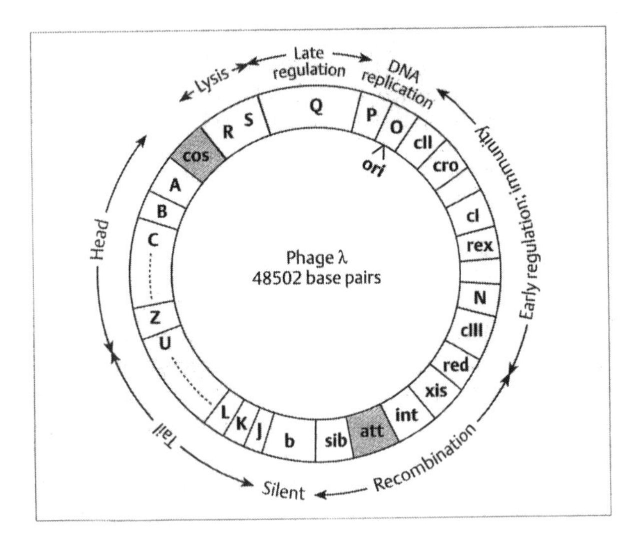

Fig. 26.**16 Simplified genetic map of phage λ.** A simplified genetic map of λ in its vegetative, i.e., circular form, is shown. The major functional groups are indicated (early functions in black, late functions in red), including the cohesive ends (*cos*) and the attachment site (*att*) (both shaded in red). Gene symbols and the function of their gene products are explained in the text. The origin of DNA replication (*ori*) is located within gene *O*.

essential for circularization of the chromosome shortly after injection inside the host. Circularization of the chromosome is essential for both prophage formation (see Chapter 16.3.1) and replication of the vegetative phage during virulent growth.

There are three ways to draw the genetic map of λ: the linear DNA in the phage head delimited by the two *cos* sequences, the circular vegetative phage covalently closed at the *cos* sites by the ligase, and the linearly integrated prophage. A simplified version of the genetic map of the vegetative phage is shown in Figure 26.**16**. As for other plasmids and phages, there is a distinct clustering of genetic functions, which roughly reflects the life cycle of the phage, i.e., the early and late functions. **Early functions** include the regulatory genes essential in the host's immunity and the decision between the lytic and the lysogenic cycles, and the genes involved in recombination (integration/excision) and in DNA replication. A typical origin of replication is located within gene *O*. It is used first in a slow bidirectional (or θ) replication, but switches later with the help of proteins O and P to a fast rolling-circle (or σ) replication, which produces the concatemeric DNA needed for packaging (see Chapter 26.3.6). **Late functions** are clustered in genes *SR* for cell lysis and *A–J* for head and tail protein synthesis and for phage assembly. Although the late genes appear on a linear map (as found, for example, in the head) as split at the *cos* sites, they are actually continuous in the cccDNA and are co-transcribed in a single polycistronic mRNA from promoter P_R (see Chapter 26.4.4). There are many more genes than given in Figure 26.**16**. The *red* locus, for example, corresponds to three genes involved in recombination, the Q locus corresponds to nine genes. Because many of these are facultative, needed for instance only in some host cells, they will not be discussed here. Other areas labeled **silent** appear to lack any genes and are probably only needed to ensure a full head during packaging. Silent DNA can be exchanged artificially for other genes, giving highly versatile vectors for gene technology (see Chapter 17.1.2).

26.4.3 Integration and Excision of the Genome of Bacteriophage λ Into and Out of the Bacterial Chromosome

If phage λ follows the lysogenic pathway, its entire DNA becomes integrated into the chromosome of *E. coli* (Fig. 26.**17**). The reaction requires phage-encoded integration protein gpInt, a monomer of 40 kDa, and the host-encoded integration factor IHF, a heterodimer of similar-sized subunits encoded by genes *himA* (protein 11.2 kDa) and *himB* (protein 10.6 kDa). The recombina-

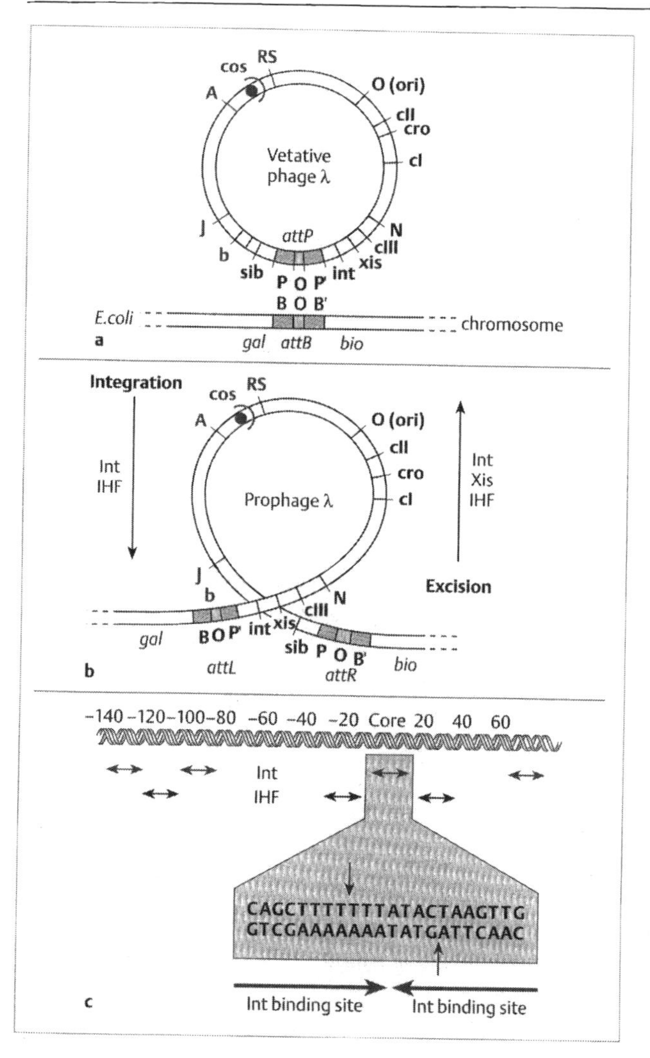

Fig. 26.**17a-c** **Integration and excision of phage λ from the host chromosome requires localized recombination.**
a The circularized vegetative λ carries an attachment site (*attP*) with similarity to an attachment site (*attB*) on the chromosome of *Escherichia coli* located between the *gal* and *bio* loci. During lysogenization, the phage-encoded integrase (gpInt), together with the host integration factor (IHF, genes *himA,B*) catalyze a localized recombination that splits and fuses ("over-cross") the 15 identical core base pairs of the *att* sites. As a consequence, prophage λ is integrated linearly in the chromosome of the host and is flanked by *gal* and *bio*.
b Upon induction, the prophage is excised by a process that resembles integration, but is not simply its reversal. The hybrid attachment sites (*attL*, *attR*) differ from *attP* and *attB* and the process requires Int, IHF, and the excision nuclease Xis (gpXis). If excision involves an illegitimate recombination, defective phages (λd) may be generated. These defective phages normally have genes *b* and *J* substituted by *gal* ($\lambda dgal^+$) and need a complementing helper phage for proliferation.
c Details of *attP* are shown. The common core of 15 bp is asymmetrically located within 240 bp. For integration, Int binds to four sites within *attP* (marked by red double-sided arrows) and IHF binds to three sites (marked by black double-sided arrows) that alternate with the Int binding sites. IHF together with Int organize the 240 bp of *attP* and the 23 bp of *attB* (not shown) such that both DNAs are wound around the protein complex, named the intasome. Their core sequences become perfectly aligned, and the exchange reaction involves 5′-located 7-bp staggered nicks in the core sequences of *attP* and *attB* (marked by arrows in the enlargement of the Int binding site). (For further details, see text) [after 10]

tion occurs between the phage attachment site *attP* and the bacterial attachment site *attB*. The attachment sites are unequal with 240 bp in λ versus 23 bp in *E. coli*. They share, however, a homologous core sequence of 15 bp. After binding several copies of gpInt and IHF at defined locations within *attP* and *attB*, an **intasome**, a structure of higher order, is formed, which permits the correct alignment of the two attachment sites. Pairing, exchange, and religation of the four strands is mediated by the gpInt protein within the intasome. The small protein IHF is responsible for bending the DNA during formation of the intasome. Integration is a two-step reaction that requires 7-nt staggered nicks within the core regions of the attachment sites. Exchange of strands begins at the left break point and proceeds through the right break point. The resulting recombinant attachment sites are called *attR* and *attL*, respectively.

Excision of the phage genome from the host chromosome after induction of the prophage is not simply the reverse reaction of the integration reaction since the recombinant attachment sites *attL* and *attR* differ considerably from the incoming attachment sites *attP* and *attB* (see Fig. 26.**17b,c**). The reverse direction is dependent on a special protein, gpXis (8.6 kDa), which is encoded by the phage. Xis depends also on gpInt and IHF, which help during the reaction. Under suboptimal

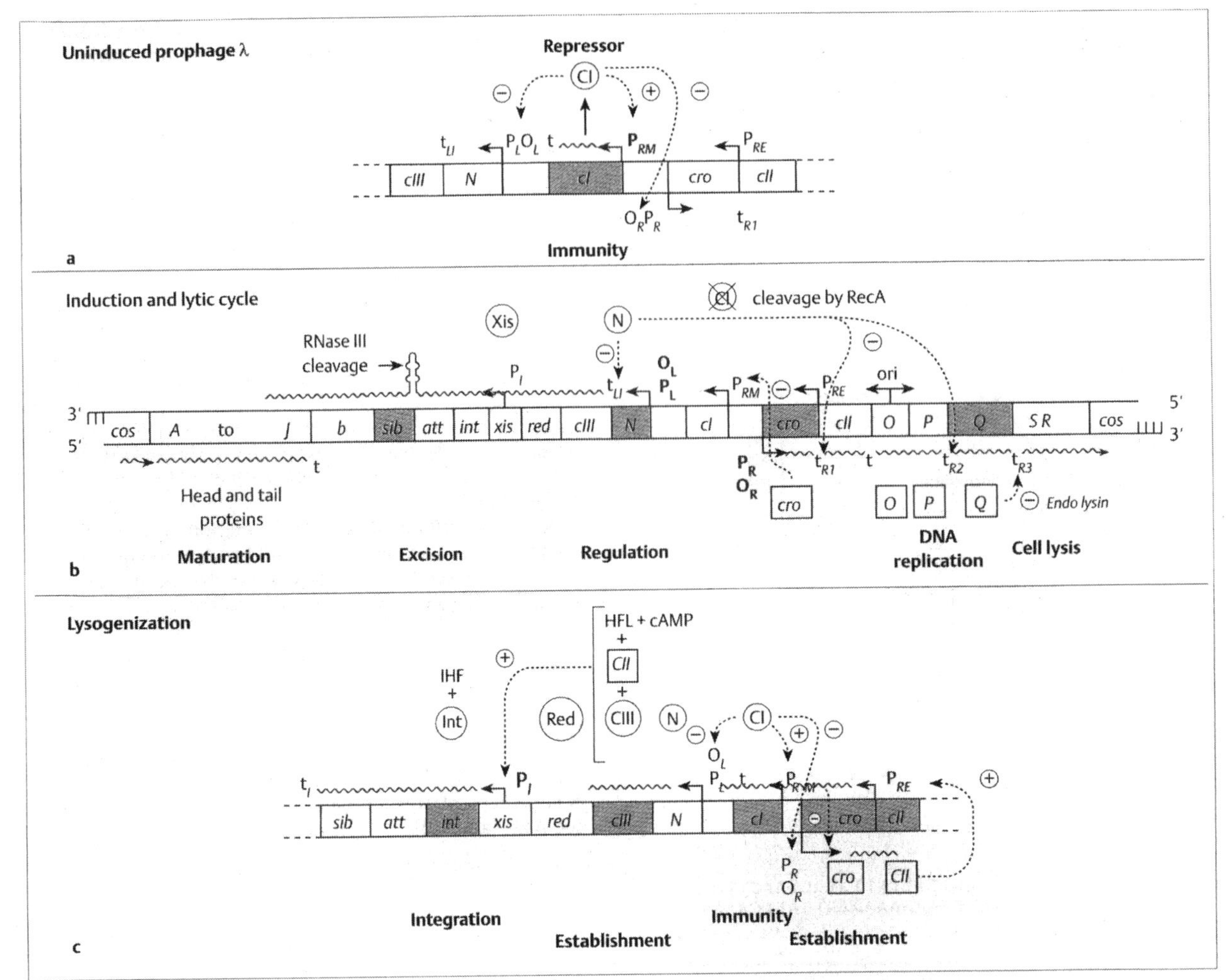

Fig. 26.**18 The three modes of phage λ.**

a In a lysogenic host cell, λ expresses the ressor CI at a low level. This ensures full repression of all other phage genes and causes host immunity against superinfecting λ or lambdoid phages.

b If, as a consequence of DNA damage, CI is cleaved (induction), transcription begins simultaneously at the leftward promoter P_L and the rightward promoter P_R. Transcription yields first the immediate-early proteins N and Cro and with their help the delayed-early proteins down to Xis and Q. This causes excision of the prophage from the chromosome and allows synthesis of all late proteins involved in phage assembly and cell lysis

c Immediately upon infection of a cell and during the decision (early) period after induction, the immediate-early and delayed-early proteins are synthesized. At a high m.o.i. ($\geqslant$5.0) and in starved cells (high cAMP), the activator CII is stabilized and causes first a very rapid synthesis of repressor CI and then integration of the phage through integrase Int into the chromosome. CII is eventually inactivated because CI represses further synthesis, the CI level drops to the maintenance level, and the prophage is stably integrated. In well-fed cells, the cAMP concentration is low, CII never reaches increased levels because of an inherent instability, and Cro, not CI, becomes dominant, driving the phage into the lytic cycle. Broken arrows indicate repression (−) and activation (⊕), promoters in red are active during a specific phase, and essential regulatory genes are shaded in red. For further explanations, see text

concentrations of gpXis, another ancillary *E. coli* protein, FIS, can be of additional help.

Prophage integration and excision involve a special type of recombination, the localized or "Campbell" recombination, named after A. Campbell, who first proposed the correct mechanism.

26.4.4 Lysogenic Host Cells Are Immune and Can Be Induced in a Process Involving Complex Regulatory Circuits

Integration of the λ chromosome into the chromosome of its host ensures its passive multiplication with each round of replication of the host's chromosome. The prophage expresses only one gene using the host's RNA polymerase, namely gene *cI*, which encodes the **repressor** (gpCI or CI). Transcription is from a single promoter P_{RM} (for **p**romoter **r**epressor **m**aintenance) and stops at a terminator (t) located behind the *cI* operon. CI has a dual role: it represses the two major promoters P_L and P_R of phage λ and activates P_{RM} in an autoregulatory process and hence its own synthesis (Figs. 18.**10** and 26.**18a**). There are about 200 molecules of CI repressor per lysogenic cell. If another phage λ superinfects a lysogenic cell, it would immediately be repressed by free-floating CI. This status is called **immunity** (or correctly homo-immunity) of lysogenic cells. It can also affect other λ-related or **lambdoid** phages as long as they react to the CI of λ ("hetero-immunity"). λ mutants lacking CI cannot lysogenize and cannot form **turbid** plaques, but, similar to virulent phages, form **clear** plaques (hence λ_{cI}). Upon superinfection of a lysogenic cell carrying a wild-type phage, *E. coli* (λ^+), the excess repressor represses the incoming λ_{cI} mutant, thus explaining the *trans*-recessive phenotype of λ_{cI} compared to λ^+ phages.

Several influences from the outside can induce prophage λ. DNA-damaging agents such as UV irradiation, carcinogens and mutagens, which cause DNA nicks, activate host-specific proteases in an SOS-repair dependent process (see Chapter 15.7 and 19.3.4). These proteases hydrolyze DNA-bound proteins, including CI, and hence cause **induction**. The early, delayed-early, and late lytic functions of λ are expressed basically from two promoters called P_L (for **l**eft) and P_R (for **r**ight), which transcribe divergently (Fig. 26.**18b**). After hydrolysis of CI, the complex repression at operators $O_{L1,2,3}$ and $O_{R1,2,3}$ (see Chapter 18.6) is relieved. Similarly, the sudden dilution occurring during the mating of a lysogenic Hfr donor with a non-lysogenic recipient lacking CI triggers a process called **zygotic induction**, which normally kills the recipient cells.

Lytic development of phages is dependent upon the temporal expression of very large operons. Their transcription is sure to encounter premature intraoperonic terminator (T_I) signals that naturally occur for the purpose of this temporal and modular expression of the genes. Consequently, anti-termination mechanisms are required for optimal transcription of the enclosed cistrons. One mechanism utilizes phage-specific RNA polymerases that do not recognize T_I. In lambdoid phages, however, alternative mechanisms are used. Thus, both the P_L and P_R operons contain several terminators (T_L/T_R) placed at discrete sites. A phage-encoded coded protein, N, made as the product of the first cistron transcribed from P_L helps the host RNA polymerase override the terminators T_{L1}, and T_{R1}/T_{R2} in both operons and extend transcription into downstream genes. Because transcription stops at T_{L1} and T_{R1} in the absence of N, this **anti-terminator** plays a major role in temporal regulation of phage gene expression. A *cis*-acting DNA site called *nut*, the host proteins NusA, NusB, RpsJ (synonym for S10 or NusE) and the anti-terminator Rho (NusD) are essential, as summarized in Figure 26.**19**.

Immediate-early genes *N* and *cro* are transcribed from P_L and P_R, respectively, as are the delayed-early genes. Cro can be considered as an **anti-(CI) repressor**. Similar to CI (see Chapter 18.6), it binds to operators O_L and O_R, but with opposite affinity, i.e., its affinity is highest for O_{R3} (O_{L3}) and lowest for O_{R1} (O_{L1}). Cro, especially in higher concentrations, thus directly prevents the binding of the few CI molecules present in an

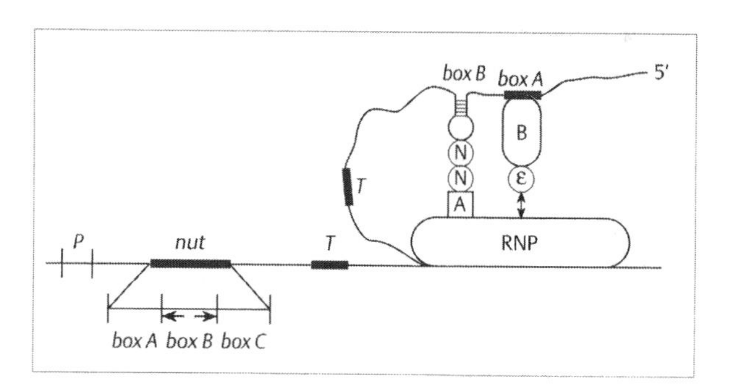

Fig. 26.**19 Model of transcription anti-termination by the N protein of λ.** The diagram indicates specific protein–protein and RNA–protein interactions and the nature of the anti-termination transcription apparatus. The *nut* site has three components: box A, box B, and box C. Box B in the RNA has an essential hairpin structure with a specific sequence in the loop. The phage protein N together with *Nus* A, B, and E (= ribosomal protein S10) help the host RNA polymerase (RNP) to override the termination signal (T) by binding at the RNA level to *nut* boxes A and B as indicated. The role of Rho (NusD) in anti-termination, although defined genetically, has not been established clearly [after 11]

induced cell to O_R and O_L and indirectly prevents the positive autoregulation of CI synthesis from P_{RM} by CI.

Once N has overridden T_{L1}, the leftward transcription generates a polycistronic mRNA that includes the gene *int*. Since transcription from P_L is not susceptible to normal termination signals, it proceeds beyond a region called *sib*. However, Int is not translated because *sib* in the P_L mRNA constitutes an RNase III cleavage site. Cleavage exposes a 3′-end for degradation of the *int* mRNA by other RNases. Because the expression of *int* is regulated by a site located downstream, the process is called **retroregulation**. The delayed-early gene product Xis, however, promotes excision of the prophage (see Chapter 26.4.3).

Once N has overridden T_{R1}, and T_{R2}, the rightward genes *O*, *P*, and *Q* are transcribed. This activates the origin of replication for λ (*ori*), which upon binding proteins O and P together with the host proteins also involved in normal replication (see Chapter 14.1), begins a slow bidirectional DNA replication. After several phage copies have been synthesized, the replication mode switches through Red-catalyzed recombination processes to a fast and unidirectional rolling-circle-type replication, which generates the concatemeric DNA needed for packaging (see Chapter 26.3.5). Phage λ has only one late promoter from which the entire set of 28 late genes (Figs. 26.**16** and 26.**18b**) is transcribed. First, however, transcription stops at terminator T_{R3}. A second anti-termination protein, Q, a delayed-early product of the P_R operon, helps override T_{R3} and all subsequent termination signals in the entire late operon. Q together with NusA act through binding at a DNA site *qut* and direct modification of the RNA polymerase. The late proteins SR form an endolysin involved in a controlled host cell lysis at the end of the latent phase while genes *A–J* encode the phage capsid proteins and its assembly or morphopoiesis into the mature progeny phages (see Chapter 26.3.6).

26.4.5 Genetic Control of the Integration of the λ Chromosome Into the Host's Chromosome

Immediately after infection, the λ chromosome forms the covalently closed Hershey-circle, and the "immediate-early" transcription of genes *N* and *cro* starts using the host RNA polymerase. Once N is present and overrides terminators T_{R1} and T_{L1}, delayed-early transcription of gene *cIII* from P_L and of genes *cII*, *O*, *P* and *Q* from P_R begins. It is only now that the decision between lysogenic and lytic pathways is made. This is even true for induced prophages. The switch enabling the choice between the two pathways comprises all the molecules mentioned thus far, in addition to CI, Int, and Xis. As expected, because success and failure of phage propagation depend intimately on the host's energy level, the switch is coupled through host-specific factors to the host's physiology and through it to enviromental conditions. Basically, the switch works by allowing CI to be made during lysogenization when Cro and Xis are not synthesized, and by allowing Cro (and Xis) to be made during the lytic pathway when CI is turned off (Fig. 26.**18c**). Furthermore, a high m.o.i. and starvation (i.e., a high cAMP level) favor lysogenization. Each position of the switch reflects a series of balanced interactions between activating and repressing regulatory proteins, between phage-specific and host ("environment")-specific conditions.

During the decision phase, several proteins (CII, CIII, host factors IHF and Hfl) as well as promoters P_R (for **i**ntegration) and P_{RE} (for **r**epressor **e**stablishment) become essential. As indicated, CII activates P_{RE} and transcription to a terminator located immediately downstream of the *cI* locus. This promotes the rapid synthesis of CI and hence repression at P_L/P_R, and then synthesis of an RNA complementary to gene *cro*. This antisense RNA helps repression of P_R through CI and more precisely the synthesis of the lytic proteins Cro, O, P, Q, and all late proteins. The activator CII, however, is a labile protein that is rapidly degraded by host-specific proteases (see Chapter 19.3.4) with the help of HflA (**h**igh **f**requency of **l**ysogenization by phage λ). A complex of CII and CIII seems to be stable against HflA-dependent degradation, and high intracellular concentrations of cAMP (as found during carbon starvation, see Chapter 20.2) also stabilize CII levels. Because P_{RE} is a very efficient promoter (sevenfold more efficient than P_{RM}), CI concentrations rise rapidly in the presence of CII, causing an increasing repression of Cro, CII, and CIII synthesis. Because CII is inherently labile, it disappears eventually and P_{RE} becomes inactive. Meanwhile, CI concentrations have, however, become sufficiently high to activate P_{RM}, the promoter for repressor maintenance (Fig. 26.**18a**). During these events, P_I is also activated by CII. Because P_I overlaps with gene *xis*, no Xis is synthesized from the corresponding mRNA. The P_I transcript terminates at terminator t_I, but it does not expose the RNase III cleavage site characteristic for the P_L transcript (see Chapter 26.4.4). Hence, Int is synthesized efficiently, but transiently from the P_I transcript and can promote the integration of the prophage into the host's chromosome with the help of another host protein, IHF (for **i**ntegration **h**ost **f**actor), as described in Chapter 26.4.3.

26.5 Phages Are Important Genetic Tools

26.5.1 Phages Can Transport Genes of Their Hosts

Most phages are capable of transducing foreign DNA from one host to another (see Chapter 16). As described above, the phages package their DNA from long, oversized concatenates starting from a phage-specific *pac* sequence. If the host genome contains similar or identical sequences, packaging can sometimes start mistakenly from one of these sequences, and a foreign or host piece of DNA becomes wrapped up. The event is not very frequent (approximagely 10^{-4}), and usually only a few phages in a population carry host DNA. These phages are called **transducing phages** since they transduce DNA from their former host to another cell. The ability of temperate phages P22 and P1 to **transduce** almost any sequence from *Salmonella typhimurium* and *Escherichia coli*, respectively (**general transduction**) has been studied extensively. However, virulent phages such as T1 and T4 are also capable of general transduction if appropriate selective conditions are applied. Phage λ became a paradigm for specialized transducing phages since λ can sometimes carry a piece of DNA originating from the immediate vicinity of the integration site *attB* (**specialized transduction**). If λ resides in its host as a prophage, it can excise itself from the chromosome upon induction (see Chapter 26.4.4). The excision process is not always precise and the DNA can sometimes be cut distant from the integration site. Since the phage capsid cannot accommodate more DNA than that equivalent to one phage genome, the false cut causes an equal loss of phage DNA and genes at one end of the genome. This explains why transducing λ phages are usually deficient and need a **helper phage** for propagation in the cell that they infect next. λ has its regular integration site located near the *gal* genes in *E. coli* (Fig. 26.**17**), and most transducing λ phages carry parts of or all *gal* genes. This specifies λ as specialized transducing phage. However, if the normal *attB* site is mutated (e.g., deleted), λ can integrate at second-choice attachment sites, creating specialized transducing phages for other genes.

Transduction is easily detectable if genetic markers are used that are transferred from a donor cell to a recipient cell, where they become stably incorporated into the genome by homologous genetic recombination.

26.5.2 Phages Create new Genotypes by Genetic Recombination

Exchange of genetic material between chromosomes of the same and sometimes even different species (**horizontal transfer**) are common features of the living world. Numerous processes describing different genetic recombination mechanisms have been studied in many systems. With the discovery of phages, systems became available that were experimentally easily accessible, allowing molecular studies of genetic recombination at all levels.

Exchange of genetic material may occur between intact homologous chromosomes or pieces thereof or between longer (>200 bp) homologous sequences on the same chromosome. These events are called intermolecular and intramolecular **homologous recombination** or **general recombination**. The latter name was coined to distinguish the events from **site-specific recombination** (as described in this chapter for phage λ), which is restricted to defined short sequences or locations. These sequences still share short regions of homology, which distinguishes them from sites of **illegitimate recombination** and **transposition**. The latter events can proceed without any homology and, therefore, occur at practically any site within a chromosome, as described for phage Mu.

Many phages have developed one or more complete recombination systems, which fall into one or more of these categories. Some phages, such as T4 or λ (in its prophage state), even depend on recombination steps as part of their regular developmental cycle. On the other hand, some phages, such as Mu, have highly specialized site-specific recombination systems that are not required for regular growth, and mutations in this pathway do not impair phage multiplication.

At the two ends of a hypothetical scale for complexity of recombination, phages that produce all enzymes necessary for recombination by themselves and phages that do not produce any enzymes for recombination and depend completely on available host functions for recombination are found. As a rule of thumb, one can expect that large phages with room for "luxury genes" in their genomes are equipped with more sophisticated recombination mechanisms than small phages. For example, phage T4, with a genome of 165 kb, has a complete system for homologous recombination. A series of proteins, UvsX (a RecA-like strand-transfer protein), UvsY (supporting factor of UvsX), gp32 (single-stranded DNA binding protein), gp46 and gp47 (control of exonuclease function), gp30 (DNA ligase), gp43 (DNA polymerase), gp49 (Holliday structure resolvase), and a series of other proteins, help to perform efficient DNA exchange reactions required for the development of phage T4. The speedy replication of phage T4 rests on a timely and continuous supply of additional start sites, called secondary origins of replication. These sites are provided through recombination between homologous sequences. After strand-transfer mediated by UvsX and its helper functions, DNA synthesis can be primed from 3′-OH terminated ends in displacement loops. Continuously ongoing rounds of recombination and DNA replication result in highly branched, concatenated replication intermediates that serve as a substrate for packaging head-full pieces of DNA after removal of all branches as described in Chapter 26.3.5 (Fig. 26.**14**).

Phage λ with a chromosome of 50 kb, encodes two recombination systems: (1) the Int pathway, for integration/excision recombination, with the components described in Chapter 26.4.3, and (2) the *red* pathway, for homologous recombination system, which requires proteins gpGam (protection protein against the RecBCD exonuclease from *E. coli*), gpExo (5′ → 3′ dsDNA-specific exonuclease) and gpb (promotes renaturation of complementary ssDNA and binds to gpExo). Both recom-bination systems also depend on additional host functions.

Phage T7, with a genome of 40 kb, has an almost complete set of proteins involved in general recombination. Unlike phage T4, however, this is not an essential part of the early replication machinery. Nevertheless, the system is required to ensure complete replication and packaging by providing concatenated DNA. The recombination system consists of an endonucleolytic Holliday-structure-resolving activity (protein gp3), and an ATP-dependent 5′-specific dsDNA exonuclease (gp6). A strand-transfer protein is not known, and phage T7 depends in this respect entirely on the RecA protein from *E. coli*.

Phage Mu, with a chromosome of 37 kb, has a site-specific inversion system (see Chapter 18.11.2) and also two mechanisms for integrative recombination that rest on illegitimate recombination mechanisms. They allow Mu either to integrate its genome without replication at any site in the host genome or to replicate its genome like a transposon, spreading genomes by copying them from the original, which remains at the site of integration.

The inversion system of phage Mu depends on the phage-encoded enzyme Gin and is used to invert a piece of DNA in front of a promoter that controls the expression of a tail fiber protein gene. The inversion event works precisely without loss or gain of any nucleotide. The message for the protein is read in either a forward or a backward direction from a promoter located outside of the invertable segment. Each protein thus formed determines a different host range. A similar inversion system is also found in phage P1. Here, the inversion event depends on the P1 encoded protein gpCin. The reaction also requires host proteins HU and Fis from *E. coli* for maximal activity. Under optimal conditions, switching during phage development can result in populations with nearly 50% specific for one host range and 50% specific for another host range in the same burst.

Phages M13 and ϕX174 with just a few thousand nucleotides and only a few genes do not express any recombination proteins. They depend completely on the host recombination functions.

26.5.3 Phages Provide Powerful Tools for Genetic Engineering

The detailed investigations of the development of phages were not only important for understanding life at the molecular level, but in the long run also provided new powerful techniques in genetic engineering. The initial discovery of the restriction and modification phenomenon by W. Arber and co-workers, for example, was made while studying phages P1 and λ. This was the beginning of the discovery of hundreds of restriction enzymes, which today are the most widely used tools in experimental molecular biology and gene technology. These enzymes, which are generally used for digesting DNA into defined pieces, also opened the way to recombine new DNA sequences from different sources in vitro, thereby allowing the creation of man-made constructs. It should be emphasized, however, that enzymological techniques used in gene technology ultimately originate from nature itself, where they are required for the same basic reactions.

The development of a powerful system for cloning genes from any source was based on the natural packaging mechanisms of phage DNA. Two of the cloning systems used most today originated from phages M13 and λ (see Chapter 17.1). A piece of foreign DNA is simply spliced into a predetermined site in the prepared DNA of either phage. In the case of λ, the spliced chromosome is then packaged in vitro to yield infectious and viable phages. Upon infection of their hosts, progeny phages are synthesized. With each round of replication the foreign piece of DNA fragment is also copied and can then be isolated in large quantities from purified DNA. The amount of foreign DNA that can be cloned in this system is limited by two major factors: (1) a capsid cannot hold more DNA than one λ chromosome equivalent, and (2) the amount of dispensable DNA that is replaced by foreign DNA without affecting vital functions is limited to about 20–30% of the phage genome, i.e., 10–15 kb.

The unique feature of filamentous phage M13 is its ability to package almost any size of DNA (Fig. 26.**13**), which allows the incorporation of large pieces (up to several kb) of foreign DNA. The system was recently used for the development of an advanced cloning vehicle, which under certain conditions produces unlimited amounts of functional eukaryotic antibodies in *E. coli*. The system rests on the availability of functional modules such as an authentic M13 packing start sequence in the DNA, a phage-specific replication origin from which a rolling circle mode of DNA synthesis can be initiated, and two phage-specific proteins, the coat protein gp*8*, and the adsorption protein gp3. The modules can derive from different species of filamentous phages and can be combined with other useful modules derived, for example, from plasmids. Because of this combinatorial freedom, the systems containing mixed features are called **phagemids** indicating the origin of their components from phages and plasmids. Details of such a vector are shown in Figure 26.**20**. It is worth noting that the coat protein gp*8* is not part of the vector itself, but is delivered from a helper phage added some time later during the development of phagemid-containing cells. The principle of the system is rather simple. The variable regions of a heavy chain and the corresponding light chain of an antibody are isolated together from the same specific antibody-producing cell (for example, hybridoma cells or spleen cells from mice). The heavy and light chains become linked by a short artificial peptide such that both antibody-derived peptides in conjunction can form their individual antigenic domain structure. This construct is then cloned behind an inducible promoter and in front of gene *3* of the vector. After induction, the linked antibody sequences are fused to and translated together with gp3. After packaging the vector DNA into phage particles, the fusion protein is transferred to the tip of the phages where gp3 finds its natural place as the adsorption protein. Packaging of DNA is initiated after helper phages are added to deliver gp*8* to the system. Each phage carries an antigenic-domain-forming fusion protein at its tip. Functional domains can be selected by

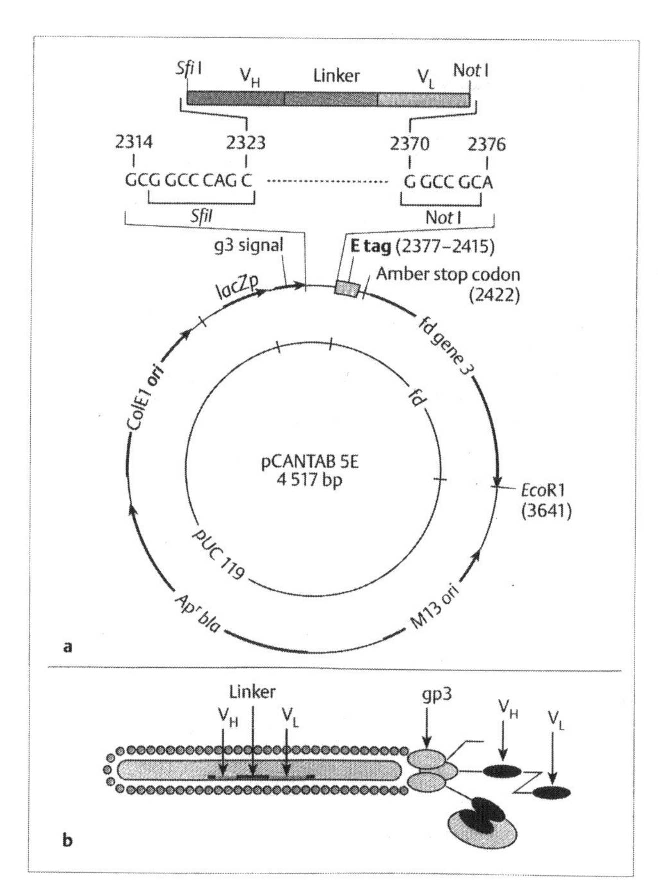

Fig. 26.**20** **a,b Phagemid-based antibody expression system.** Fragments of the variable regions of heavy (V_H) and light (V_L) chains from immunoglobulins can be expressed from an inducible (*lacZp*) promoter in *Escherichia coli* after cloning the corresponding gene in a phagemid.
a Functional modules of the phagemid pCANTAB 5E and their genealogy. Two replication origins, one from plasmid ColE1 for double-stranded DNA replication and one from phage M13 for the production of single-stranded progeny, flank the ampicillin-resistance gene *bla*, which is required for selection of phagemid-containing cells after transformation. This section of the phagemid and an inducible *lac* promoter (*lacZp*), which is under the control of $LacI^q$ repressor, derive from vector pUC119. Antibody variable region genes can be cloned between the leader sequence and the main body of the fd gene *3*, which is identical to the M13 gene *3*. After induction, a fusion protein is made that retains the function of both parent proteins. The gp3 leader sequence leads the protein to the inner membrane of *E. coli* where it is attached to the tip of the assembling phage. The fusion protein also contains a short sequence (E tag) for a peptide with an amber stop codon at its 5′-side. In suppressor-containing hosts, the stop signal is overcome and the complete fusion protein is made and finally translocated to the tip of progeny phage. In suppressor-negative hosts, however, translation stops at the amber codon, and only the antibody proteins with the leader sequence at their amino-terminus and the tag peptide at the carboxy-terminus are made. This protein is then transported into the periplasm. Since it will not be attached to phage progeny, it will be stored there and can be purified from cell wall preparations.
b A schematic representation of a complete phage after assembly of phagemid DNA with cloned immunoglobulin gene fragments V_H and V_L. The two gene fragments are spaced by a linker sequence (linker). The chromosome is covered by gp8 (circles) from phage M13 and carries gene *3* protein (gp3) from phage fd fused to immunoglobulin V_H and V_L peptides, which can form an antigenic domain

exposure of phage to the antigen that was originally used for immunization of the eukaryote. Negative phages with nonreactive antibodies are lost upon purification. Positive phages with reactive antibodies can be used for growth of unlimited amounts of phage-like particles. Even pure antibody proteins can be made, if a different host is used that prevents readthrough into gene *3* and makes no fusion protein. The antibody protein can finally be purified from crude cell extracts of *E. coli* using standard biochemical procedures.

Further Reading

Birge, E. A. (1994) Bacterial and Bacteriophage Genetics, 3rd edn. Berlin Heidelberg New York: Springer

Black, L. W. (1995) DNA packaging and cutting by phage terminases: control in phage T4 by a synaptic mechanism. *Bioessays* 17: 1025–1030

Calendar, R. (1988) The bacteriophage, vols. 1, 2. New York London: Plenum Press

Campbell, A. M. (1996) Bacteriophages. In: Neidhardt, F. C., Curtiss, R., Ingraham, J. L., Lin, E. C. C., Low, K. B., Magasanik, B., Reznikoff, W., Riley, M., Schaechter, M., and Umbarger, H. E. (eds.) *Escherichia coli* and *Salmonella*: cellular and molecular biology. 2nd edn. Washington, DC: ASM Press; 2325–2338

Casjens, S., and King, J. (1975) Virus assembly. *Annu Rev Biochem* 44: 555–611

Denhardt, D. T., Dressler, U., and Ray, D.S. (1978) (eds.) The single-stranded DNA phages. Cold Spring Harbor: Cold Spring Harbor Press

Fraenkel-Conrat, H. (1985) The viruses. New York London: Plenum Press

Hendrix, R. W., Roberts, J. W., Stahl, F. W., and Weisberg, R. A. (1983) Lambda II. Cold Spring Harbor: Cold Spring Harbor Press

Joset, F., and Guespin-Michel, J. (1993) Prokaryotic genetics. Genome organization, transfer and plasticity. London: Blackwell

Karam, J. D. (1994) Molecular biology of bacteriophage T4, 2nd edn. Washington, DC: ASM Press

Kornberg, A., and Baker, T. A. (1992) DNA replication, 2nd edn. New York: Freeman

Lewin, B. (1994) Genes V, 5th edn. Oxford: Oxford University Press

Linn, S. M., Lloyd, R. S., and Roberts, R. L. (1993) (eds.) Nucleases, 2nd edn. Cold Spring Harbor: Cold Spring Harbor Press

Luria, S. E., Darnell, J. E. Jr., Baltimore, D., and Campbell, A. (1978) General virology, 3rd edn. New York: Wiley

Mosig, G. (1994) Homologous recombination. In: Molecular biology of bacteriophage T4, 2nd edn. Washington, DC: ASM Press; 54–82

Ptashne, M. (1992) A genetic switch. Phage lambda and higher organisms, 2nd edn. Oxford: Cell Press and Blackwell

Stent, G. S., and Calendar, R. (1978) Molecular genetics, 2nd edn. New York: Freeman

Tsurushita, N., Fu, H., and Warren, C. (1996) Phage display vectors for in vivo recombination of immunoglobulin heavy and light chain genes to make large combinatorial libraries. Gene 172: 59–63

Yonesaki, T. (1995) Recombination apparatus of *T4* phage, Adv Biophys 31: 3–22

Yeo, A., and Feiss, M. (1995) Specific interaction of terminase, the DNA packaging enzyme of bacteriophage lambda, with the portal protein of the prohead. J Mol Biol 245: 141–150

Sources of Figures

1 Kornberg, A., and Baker, T. A. (1992) DNA replication, 2nd edn. New York: Freeman; 573

2 Ptashne, M. (1992) A genetic switch, 2nd edn. Oxford: Cell Press and Blackwell; 14

3 Karam, J. D., ed. (1994) Molecular biology of bacteriophage T4, 2nd edn. Washington DC: ASM Press; 209, 252

4 Levine, A. L. (1992) Viruses. New York: Freeman; 25, 21, 20

5 Mathews, C. K., Kutter, E. M., Mosig, G., and Berget, P. B., eds. (1983) Bacteriophage T4, 1st edn. Washington DC: ASM Press; 9

6 Watson, J. D., Gilman, M., Witkowski, J., and Zoller, M. (1992) Recombined DNA, 2nd edn. New York: Freeman

7 Mathews, C. K., Kutter, E. M., Mosig, G., and Berget, P. B., eds. (1993) Bacteriophage T4. Washington DC: ASM Press; 59–70

8 Kornberg, A., and Baker, T. A. (1992) DNA replication, 2nd New York: Freeman; 571

9 Mathews, C. K., Kutter, E. M., Mosig, G., and Berget, P. B., eds. (1983) Bacteriophage T4, 1st edn. Washington DC: ASM Press; 219–245

10 Lewine, B. (1994) Genes V, 5th edn. Oxford: Oxford University Press; 994

11 Das, A. (1993) Annu Rev Biochem 62: 893-930

27 Secondary Metabolism in Bacteria: Antibiotic Pathways, Regulation, and Function

Differentiation in microorganisms comprises **morphological differentiation** (morphogenesis, sporulation, germination) and **chemical differentiation** (secondary metabolism). Often these two aspects of differentiation share genes involved in regulation. Microbial secondary metabolites ("idiolites") are the low-molecular-weight products of secondary metabolism. Secondary metabolites are relatively small molecules, each produced by one microbial strain only or by a limited number of microbial strains, that, in contrast to primary metabolites, appear to have no obvious function in growth processes. In fact, producer strains that have lost by mutation the production ability, exhibit normal growth rates and characteristics under laboratory conditions.

Secondary metabolism may be viewed as a chemical form of differentiation. However, in contrast to morphological differentiation, which is typical of a species or of a genus, expression of secondary metabolism is strain specific since different strains belonging to the same species may produce different metabolites. Secondary metabolites have essential and multiple functions.

Secondary metabolites are involved in mechanisms of survival in nature, as antibiotics, pigments, pheromones, toxins, and as effectors of ecological competition and symbiosis. Some secondary metabolites are enzyme inhibitors; others are immunomodulating agents, receptor antagonists and agonists, pesticides, antitumor agents, and growth promotants of animals and plants. The majority of those isolated so far (> 7000 from cultures of prokaryotes) are **antibiotics**, i.e., substances produced by a living organism that inhibit at low concentrations the growth of different species of microorganisms. Their use exerts a major effect on the health, nutrition, and economics of our society. Most secondary metabolites have unusual structures. Their synthesis is dependent on nutrients and growth rates, and is regulated by feedback control and enzyme inactivation, by unique low-molecular-weight compounds, by tRNA, by sigma factors, and by gene products usually formed during post-exponential development. Enzymes involved in secondary metabolism are normally encoded by genes clustered on the chromosome but infrequently on plasmids. The biosynthetic pathways of a vast number of antibiotics have been elucidated, both as the result of academic interest and for the practical outcome of such research, such as in increasing production yields of antibiotics of commercial interest or to obtain molecules modified in their biological activity. In this chapter, the biosynthesis of small and large antibiotics, the regulation of their pathways, and the functions of these special metabolites will be discussed.

27.1 Most Secondary Metabolites Are Made by a Few Biosynthetic Pathways

In contrast to the huge variety of chemical structures presented by antibiotic molecules, the sequences of biological reactions by which they are made can be grouped into a relatively small number of biosynthetic pathways.

The key steps in the biosynthesis of a number of antibiotics are polymerization reactions, by which several similar units are linked together to form the backbone of a larger molecule. On the biochemical basis and, according to recent studies, on the basis of the similarity of the genes involved, three types of polymerization reactions are recognized.

1. **Condensation of acetate-malonate units.** Acetate-malonate units (sometimes propionate-methylmalonate units) are condensed by a mechanism denoted as polyketide synthesis, where chains are formed in which keto groups and methylene groups alternate. When methylmalonate substitutes for malonate, the chain becomes branched with methyl groups. Often

some of the keto groups are totally or partially reduced.

2. **Condensation of amino acids to oligopeptides.** Typical of secondary metabolism is a mechanism of formation of polypeptides (the thiotemplate mechanism) that is similar in several aspects to polyketide synthesis. There are, however, a few examples of larger peptide antibiotics synthesized through ribosomal protein synthesis. Often in the thiotemplate mechanism, hydroxy acids substitute for amino acids as building units; the resulting molecule is a depsipeptide, i.e., a chain in which amide and ester bonds alternate.
3. **Condensation of carbohydrate units** (often amino sugars) **to oligosaccharides.** In a number of important antibiotics, an aminocyclitol moiety is included, instead of a sugar, resulting in the formation of a pseudosaccharide.

These mechanisms of polymerization are similar to those of primary metabolism involved in building molecules that are components of cell envelopes. Polyketide synthesis and, to some extent, thiotemplate-directed systems, appear to be derived from the mechanism of membrane fatty acid biosynthesis (see Chapter 7.10), whereas the biosynthesis of sugar-derived antibiotics is similar to that of the polysaccharides constituting the O-antigens of Gram-negative organisms (see Chapter 23).

In addition to the antibiotics derived from polymerization, a number of antibiotic molecules are derived from modifications of one primary metabolite or by the condensation of a few modified metabolites. Their biosynthetic pathways can be conveniently classified according to the pathways of primary (intermediate) metabolism to which they are related. There are antibiotics, or antibiotic moieties, whose biosynthesis is tied to amino acid synthesis or catabolism, nucleoside metabolism, or coenzyme synthesis.

A high number of genes governing secondary metabolite biosynthesis have been recently isolated and characterized. Nucleotide sequence analysis demonstrated in all cases a high degree of similarity with the genes coding for enzymes that catalyze similar reactions in primary metabolism, e.g., fatty acid or O-antigen and cell wall biosynthesis. We can thus conclude that secondary metabolism evolved from primary metabolism.

In the following sections, examples of the biosynthesis of these small molecules are given, and the mechanisms of polymerization are analyzed in some detail.

27.2 Small Antibiotic Molecules Are Biosynthesized Through Reactions Related to Primary Intermediary Metabolism

27.2.1 Biosynthetic Pathways Related to the Metabolism of Amino Acids

Several antibiotics or moieties constituting antibiotic molecules derive from reaction sequences similar to those involved in the synthesis of amino acids. A typical example is **chloramphenicol**, produced by *Streptomyces venezuelae*. The carbon skeleton of this molecule is synthesized by the general pathway of aromatic amino acid biosynthesis (see Chapter 7.6.4). The path diverges from that of primary metabolism after the formation of chorismic acid. In chloramphenicol biosynthesis, chorismic acid is converted into *p*-aminophenylpyruvic acid rather than into *p*-hydroxyphenylpyruvic acid as in the synthesis of tyrosine (Fig. 27.**1**). It is interesting to note that the nitro-group, a functional group rarely observed in natural products, is formed by oxidation of an amino group. This appears to be a general rule in the formation of oxidized nitrogen functions, such as nitroso-, hydroxylamino-, and nitro-groups. However, the mechanism of the involved enzymatic reactions is not known. Another unusual reaction is the formation of dichloroacetic acid. Halogens are introduced into secondary metabolites by reactions catalyzed, in most cases, by peroxidases. The enzyme catalyzing the chlorination reaction in chloramphenicol biosynthesis has been identified as a dimeric heme protein with a high catalase activity and a low peroxidase activity.

A pathway of tyrosine metabolism that eventually produces melanins, is initiated by hydroxylation of the aromatic amino acid ring to give dihydroxyphenylalanine, followed by the formation of a pyrrole ring by cyclization of the amino acidic chain. A similar sequence of reactions leads to the alkylproline moieties of **lincomycin** (a product of *Streptomyces lincolnensis*) and of the **anthramycin** family of antibiotics (Fig. 27.**2**). In both cases, the pyrrolidine (or pyrroline) ring derives from cyclization of the alanyl chain of tyrosine,

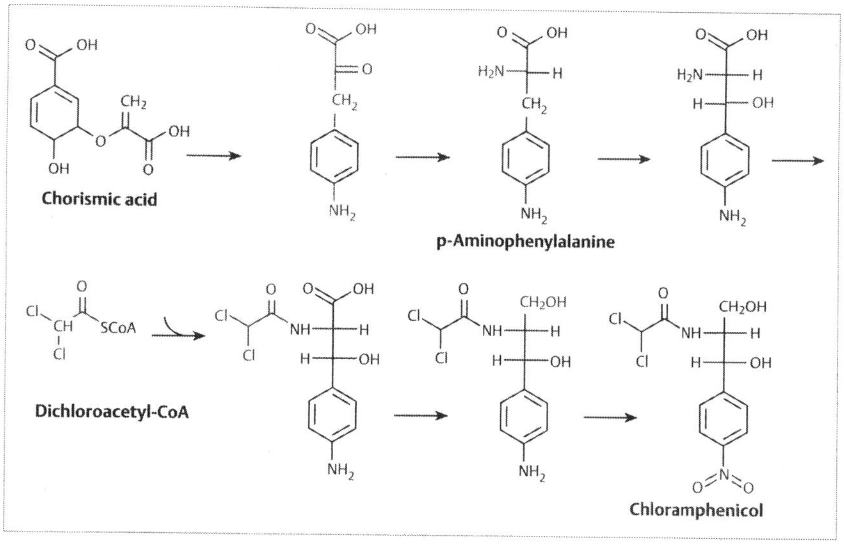

Chorismic acid

p-Aminophenylalanine

Dichloroacetyl-CoA

Chloramphenicol

Fig. 27.1 **Biosynthesis of chloramphenicol in *Streptomyces venezuelae*.** Chorismic acid (see Fig. 7.**18**) is converted by a series of multi-step modifications (in red) to chloramphenicol

whereas degradation of the aromatic ring provides the carbons of the aliphatic chain. The anthranilate moiety of anthramycin is derived from tryptophan through another well-known sequence of reactions, the kynurenine pathway (see Fig. 27.**2**).

A typical reaction encountered in the biosynthesis of several antibiotics is the **β-hydroxylation** of amino acids. This reaction is unknown in primary metabolism (however, the *β*-hydroxylation of aliphatic acids is very common), but is frequently the first reaction in pathways of secondary metabolism. It can be surmised that its function is to divert the amino acid from primary to secondary metabolism. We have already seen that *β*-hydroxylation of *p*-aminophenylalanine is one of the first steps in chloramphenicol biosynthesis. Another interesting example is the biosynthesis of the streptolidine moiety of **streptothricin F**. This heterocyclic compound derives from arginine by *β*-hydroxylation, followed by oxidation to *β*-keto-arginine and cyclization (Fig. 27.**3**). In this, as in several other cases, *β*-hydroxylation results in the formation of the erythro diastreoisomer of the hydroxy amino acid, with retention of the *S* configuration at the *α*-carbon and loss of the *α*-hydrogen. It has been proposed that the hydroxylation occurs on an intermediate product of condensation of pyridoxal phosphate on the amino group.

The biosynthesis of several secondary metabolites involves the transformation of *α*-amino acids into *β*-amino acids, e.g., of lysine (Fig. 27.**3**). The reaction proceeds through intramolecular migration of nitrogen from C-2 to C-3 and intermolecular migration of the pro-*R*-hydrogen from C-3 to C-2, with both steps resulting in inversion at the respective centers.

Several antibiotics derive from precursors of regular amino acids, e.g., chloramphenicol from the aromatic pathway and lincomycin from tyrosine; others require special reactions such as *β*-hydroxylation of amino acids or *β*-amino acids. Some of these intermediates may derive from "overflow" reactions when, at the end of the trophophase (the growth phase), the anabolic capacity far exceeds the catabolic supply in the cell.

27.2.2 Antibiotics Synthesized by Pathways Related to Nucleoside Metabolism

Over 150 antibiotics with a nucleoside structure are known. The majority are produced by actinomycetes, although many have been isolated from filamentous fungi. Their most frequent pattern of biosynthesis is the modification of a preformed purine or pyrimidine nucleotide (see Chapter 7.8). However, there are examples of antibiotics that originate by a modification of nucleotide synthesis.

An interesting example of an antibiotic whose biosynthesis parallels "de novo" synthesis of pyrimidine nucleotides is the antibiotic **PA 399** (5,6-dihydroazathymidine) produced by *Streptomyces platensis* (Fig. 27.**4**). The biosynthetic sequence is initiated by the condensation of the carbamoyl group with glyoxylurea, rather than with aspartic acid as in uridine biosynthesis. The subsequent reactions, cyclization, ribosylation, and decarboxylation, are analogous to those of pyrimidine nucleotide biosynthesis. Two steps are analogous to those of thymidine biosynthesis: ribose is converted to 2-deoxyribose, and the ring is methylated.

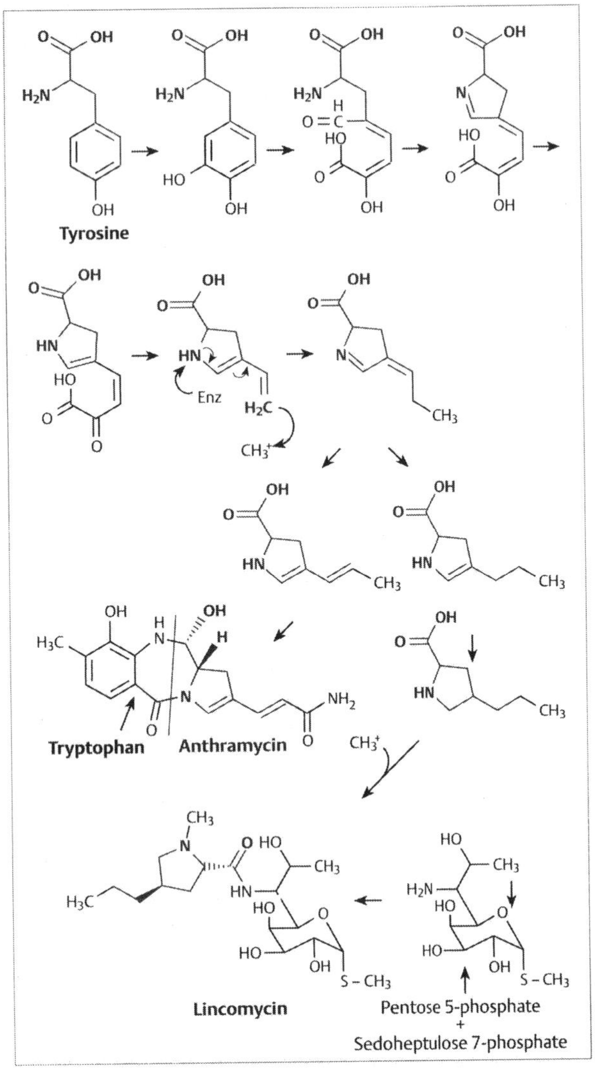

Fig. 27.2 **Biosynthesis of the alkyl-prolyl moieties of lincomycin** (in *Streptomyces lincolnensis*) **and anthramycin** (in *Streptomyces refuineus*). Red indicates the atoms of tyrosine that are incorporated into the final molecules

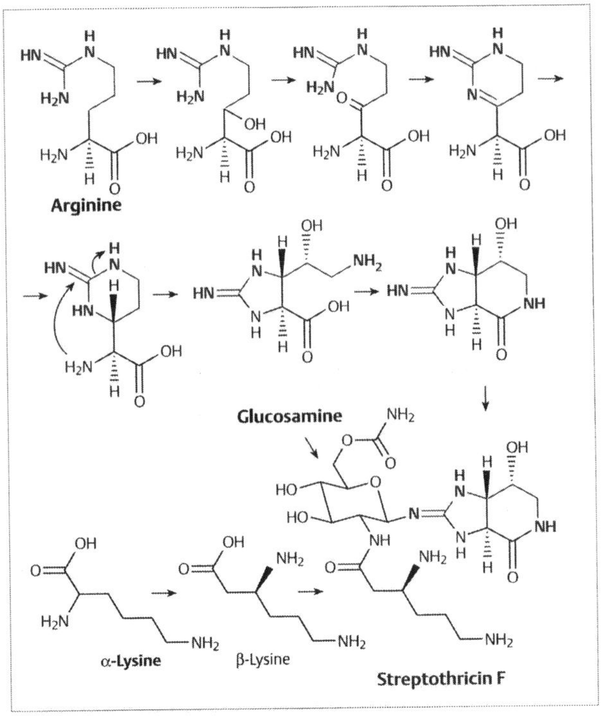

Fig. 27.3 **Biosynthesis of the streptolidine moiety o streptothricin F in *Streptomyces lavendulae*.** Red indicate the contribution of the amino acid chain to ring formation. Th synthesis of β-lysine from lysine implies an intramolecula migration of the amino group, with inversion of the configura tion

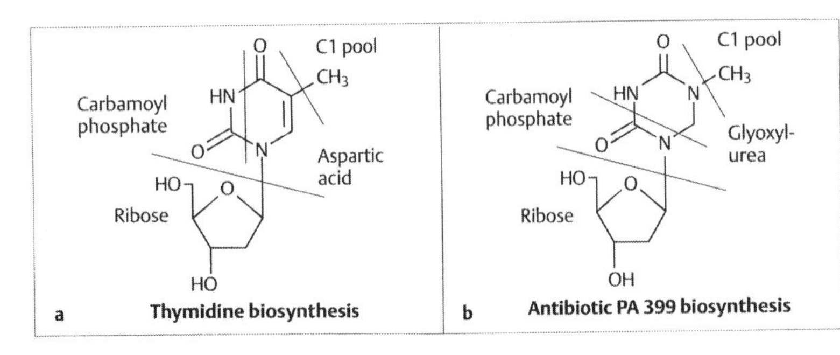

Fig. 27.4 **Biosynthesis of antibiotic PA 399** Comparison of the origin of the atoms com prising thymidine
(**a**) and the antibiotic PA 399,
(**b**) a product of *Streptomyces platensis* meta bolism

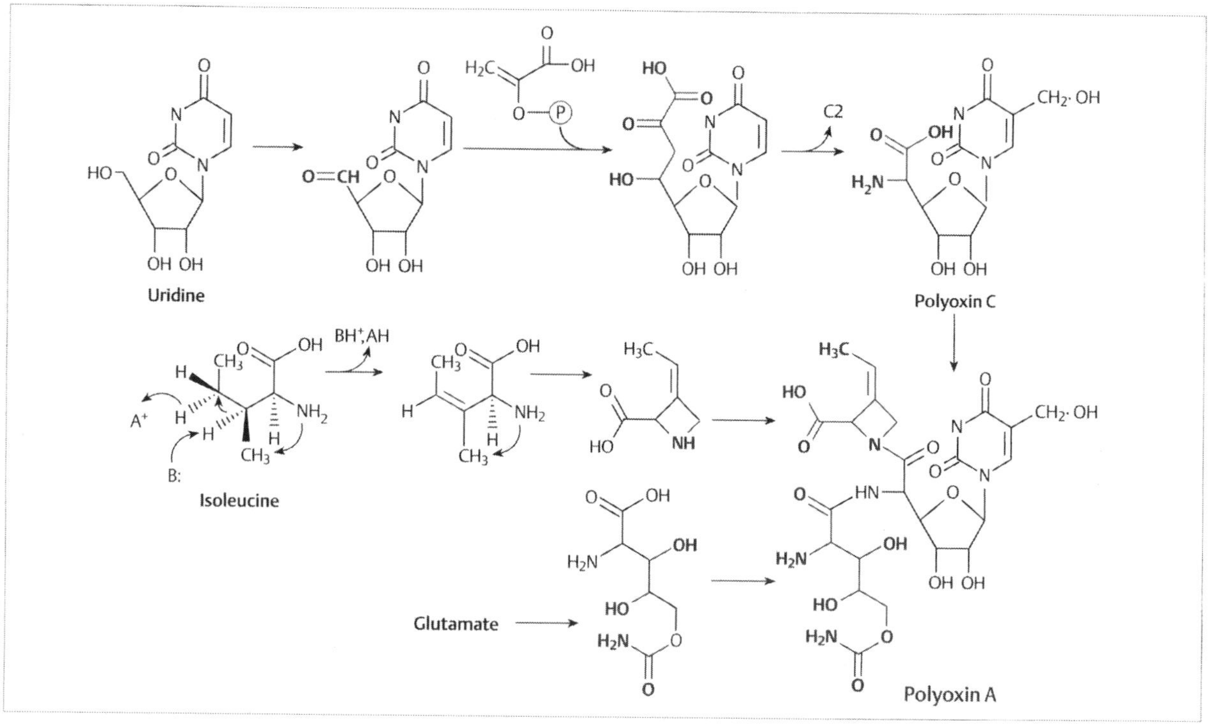

Fig. 27.5 **Biosynthesis of polyoxin A in *Streptomyces cacaoi*.** Synthesis starts from uridine, isoleucine, and glutamate and leads via multistep modifications (in red) to polyoxin A

As previously stated, several antibiotics derive from modifications of nucleotide or nucleoside molecules. Adenosine is the starting metabolite from which the antiviral agent vidarabine is synthesized by *Streptomyces antibioticus*. **Polyoxins** are a family of antibiotics, produced by *Streptomyces cacaoi*, used as antifungal agents in agriculture. All derive from uridine by the biosynthetic sequence shown for polyoxin in A (Fig. 27.**5**). The key reactions are the oxidation of the uridine 5′-hydroxyl group to an aldehyde and the condensation of the resulting intermediate with phospho*enol*pyruvate. Noteworthy is the formation of the unusual azetidine ring from isoleucine.

Many antibiotics derive from nucleosides either by modification of purine and pyrimidine nucleotides (e.g., vidarabine, polyoxins) or by a modification of the normal biosynthesis (e.g., PA 399). They often contain rare and unusual chemical groups.

27.2.3 Biosynthetic Pathways Related to Coenzyme Biosynthesis

Two different patterns are known for the biosynthesis of **nicotinic acid**, an intermediate in the synthesis of nicotinamide adeninedinucleotide. In mammals and in *Neurospora*, nicotinic acid originates from tryptophan, with the intermediate formation of 3-hydroxyanthran-ilic acid. In most bacteria and plants, however, it is synthesized by condensation of aspartic acid and a three-carbon unit, such as glycerol or a closely related metabolite.

As expected, coenzymes and their precursors can also be used as the starting molecule for essential antibiotics. Examples are actinomycins and anthramycin from kynurenine, an intermediate of nicotinic acid synthesis. Some pathways contain a curious mixture of prokaryotic and eukaryotic enzymes, perhaps indicative of an old horizontal gene transfer.

The aromatic core of **actinomycins**, produced by several *Streptomyces* species, derives from the condensation of two units of 3-hydroxy-4-methylanthranilic acid. The same anthranilic acid derivative is a precursor of the antitumor antibiotic anthramycin, a product of *Streptomyces refuineus* metabolism. This intermediate derives from tryptophan through 3-hydroxykynurenine, which is also an intermediate of nicotinic acid biosynthesis in *Neurospora* (Fig. 27.**6**). Thus, there is a sequence of reactions belonging to both primary metabolism of

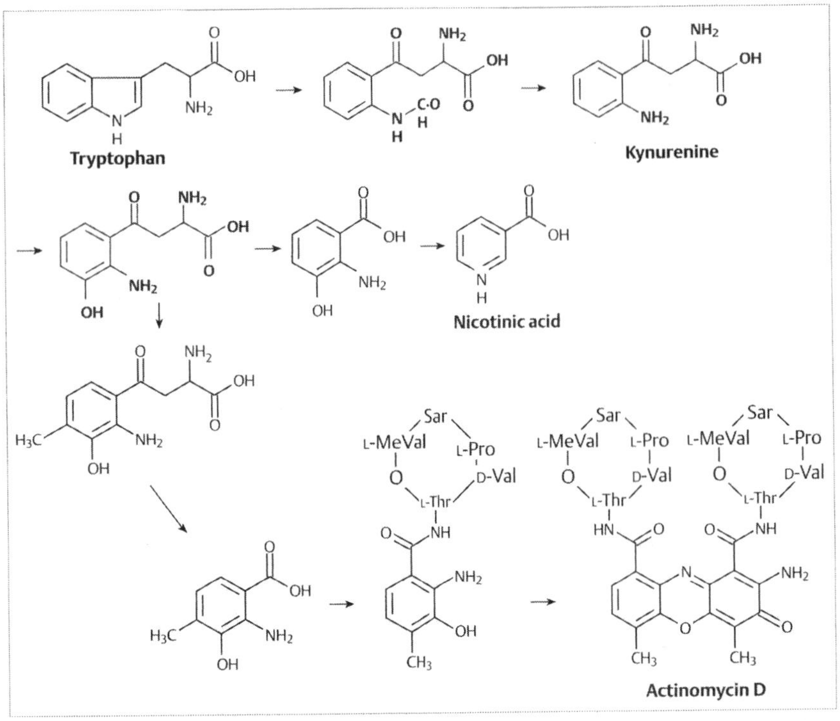

Fig. 27.6 **Origin of the aromatic core of actinomycin D in *Streptomyces antibioticus* and correlation with nicotinic acid synthesis in *Neurospora*.** The intermediates belonging to both the primary metabolism of *Neurospora* and the secondary metabolism of *Streptomyces antibioticus* are shown in red. The actinomycin peptide chain is built through the thiotemplate mechanism, as illustrated in Figure 27.**12** for the synthesis of gramicidin S. The 4-methyl-3-hydroxyanthranilic acid is the initiator molecule of the polymerization

eukaryotes and secondary metabolism of *Streptomyces* strains. It should be noted here that gene transfer from eukaryotic to prokaryotic microorganisms can explain the sequence similarity of the gene encoding the catalase involved in chloramphenicol biosynthesis with that encoding a eukaryotic catalase. Likewise, it has been hypothesized that the genes governing β-lactam biosynthesis have been transferred from *Streptomyces clavuligerus* to *Cephalosporium acremonium*.

27.3 Several Classes of Secondary Metabolites Are Derived From Polyketomethylene Chains

A large number of antibiotics derive from the polymerization of small acidic units by a mechanism denoted as polyketomethylene (or polyketide) synthesis. Among prokaryotes, the Actinomycetes are by far the most abundant and versatile producers; however, several metabolites having this origin have been isolated from myxobacteria, and to a lesser extent from pseudomonads. The mechanisms of polymerization present a close analogy with the process of fatty acid biosynthesis in the biochemical reactions (Box 27.1 and Figs. 7.**33**–7.**35**) and, to some extent, in the primary structure of the enzymes.

In microorganisms, a few variations are found in the general scheme of fatty acid synthesis (Box 27.1): in bacilli and actinomycetes, for instance, the initiator molecule is often *iso*-butyric, *iso*-valeric, or 2-methylbutyric acid rather than acetic acid, and branched aliphatic acids are formed. In other cases, one reduction step is omitted and a monounsaturated acid is produced. In the biosynthesis of secondary metabolites, the polymerization process leading to polyketomethylene chains is also catalyzed by two types of synthases: type I polyketomethylene synthase (**type I PKS**), a multifunctional enzyme, and **type II PKS**, a multienzyme complex. However, in contrast to the fatty acid synthases (FAS), either type may be found in strains belonging to the same genus, such as in *Streptomyces*.

Box 27.1 In most bacteria (and in plants), the biosynthesis of fatty acids is catalyzed by a multienzyme complex, denoted as type II fatty acid synthase (type II FAS) (see Chapter 7.10). Type I fatty acid synthase is present in vertebrates and consists of a single multifunctional enzyme. The basic steps constituting the biochemical process are:

Initiation. Acetate, activated as acetyl-CoA, and malonate, activated as malonyl-CoA, are bound to the synthase as thioesters. Acetate, the initiator molecule, binds to a domain denoted as the condensing enzyme, and malonate binds to the acyl carrier protein (ACP).

Elongation. (1) Acetate, through its carboxyl carbon, is condensed with the methylene carbon of malonate. At the same time, the free carboxyl group of malonate is eliminated as CO_2. The result of the reaction is formation of acetoacetate bound as a thioester to ACP. (2) The β-keto group of acetoacetate is reduced to methylene through a sequence of three reactions: reduction to a hydroxyl group, dehydration, and saturation of the formed double bond. (3) The growing chain is transferred to the condensing enzyme, another malonate molecule is linked to ACP, and the condensation and reduction steps are performed. The process is repeated until the proper length is reached.

Termination. The fatty acid (normally palmitic acid) is released by a thioesterase.

27.3.1 Aromatic Polyketides Include the Essential Tetracyclines

The basic structure of many microbial metabolites denoted as **polyketides** is constructed by a process similar to that of fatty acid synthesis, in which, however, the reduction steps are totally omitted. In prokaryotes, this is generally catalyzed by a type II PKS. The result is a chain in which keto groups and methylene groups alternate. Such a chain is very reactive and therefore tends to fold and form rings by aldol condensation and elimination of water between keto groups and methylene groups. According to the length of the chain and the nature of the enzymes involved, many different structures can be formed; these tend to be aromatic because of steric and energetic factors. A classical example of this process is the biosynthesis of **tetracycline** by *Streptomyces aureofaciens* (Fig. 27.**7**). It should be noted that in this case the initiator molecule is the monoamide of malonic acid.

27.3.2 Many Essential Antibiotics Are Derived From Complex Polyketide Chains

Several classes of secondary metabolites, each comprising important antibiotics, are biosynthesized by a process similar in its general outline to that of aromatic polyketides, but in which relevant modifications may be included.

One variant found frequently is the utilization of alkylmalonates, rather than malonate, as extender units. Since the condensation of the extender unit with the

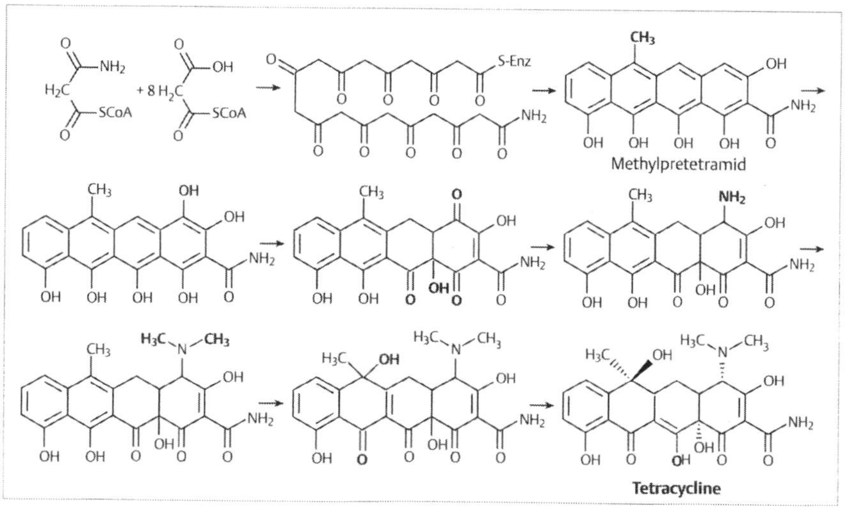

Fig. 27.**7** **Biosynthetic pathway of aromatic polyketides.** The biosynthesis of tetracycline in *Streptomyces aureofaciens* is shown. The polyketide chain is presumptive. For explanations see Box 27.1 and text. The sequential modifcations introduced during biosynthesis are indicated in red

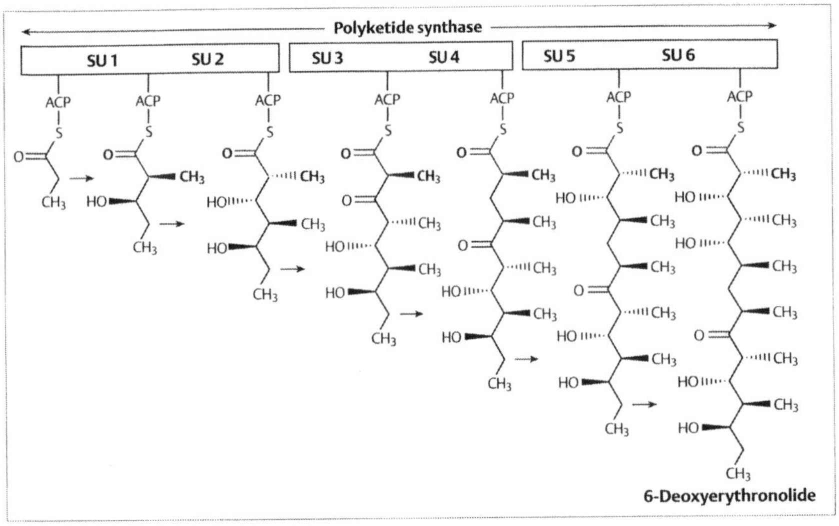

Fig. 27.8 **Biosynthetic pathwa' of complex polyketides.** Buildin of the chain constituting 6-deoxy erythronolide B in *Saccharopoly spora erythraea* by a polyketid synthase is shown. (SU, synthas unit; ACP, acyl-carrier protein). Th propionate units added at eacl elongation step are shown in red

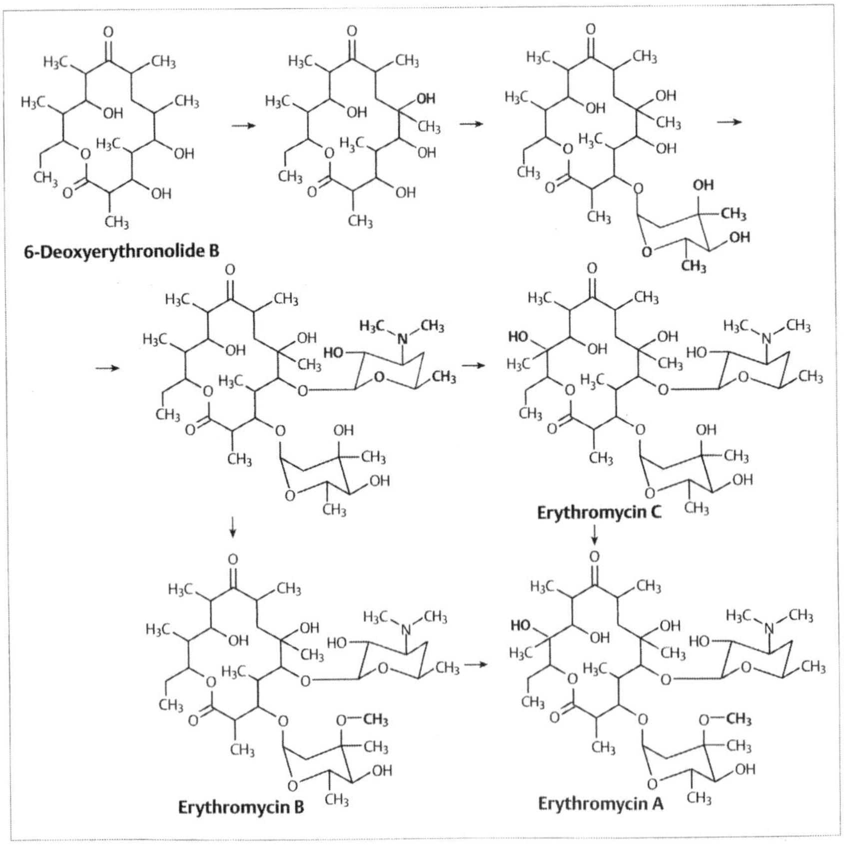

Fig. 27.9 **Pathway of the conver sion of 6-deoxyerythronolide I into erythromycins** (newly adde groups are shown in red)

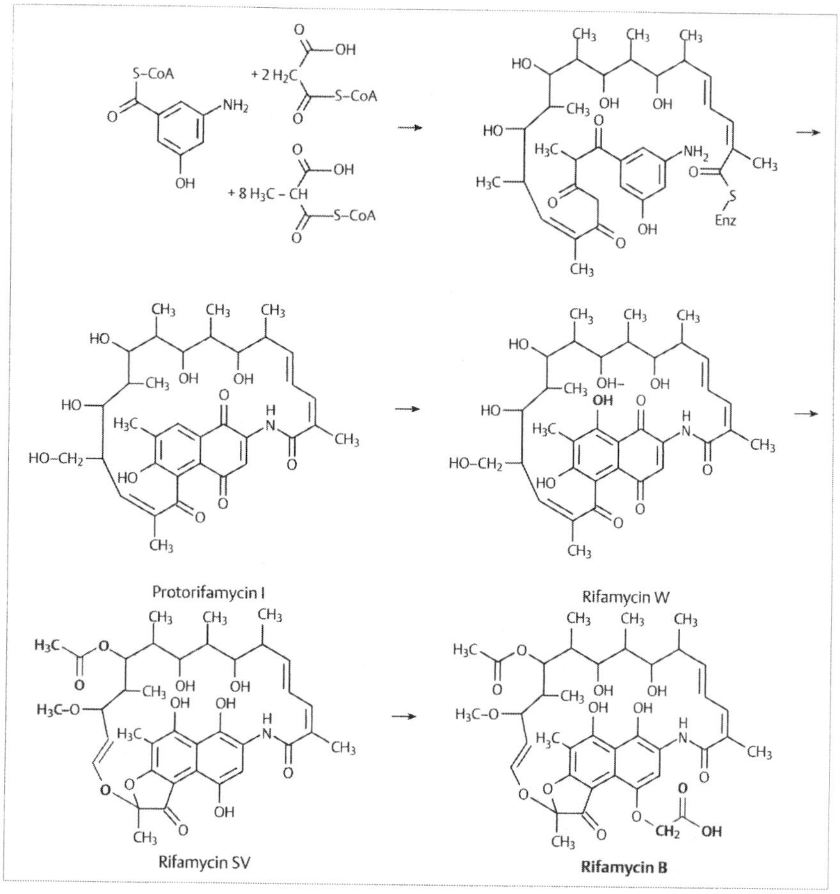

Fig. 27.**10** **Biosynthetic pathway of complex polyketides.** Biosynthesis of rifamycins in *Amycolatopsis mediterranei* is shown. The branched, reduced polyketide chain is hypothetical. For further explanations, see text

growing chain always occurs at the carbon adjacent to the carboxyl group, the result is a branched chain bearing methyl or ethyl groups as substituents, depending on whether methylmalonate or ethylmalonate is involved.

Another important source of variation is the partial or total reduction of most of the keto groups that are formed in the chain extension steps. The chain can then bear keto or hydroxy functions, and include double bonds and adjacent methylene groups.

As a consequence of these variations in its structure, the chain cannot be converted by the aldol reaction into aromatic rings, and either linear molecules or macrocyclic rings are formed. This is the origin, for instance, of the basic structures of the macrolide antibiotics, the **ansamycins**, and the **ionophoric polyethers**.

Many antibiotics are polymerized to polyketides in a process that resembles fatty acid synthesis. There are two types of polyketomethylene synthases (complexes): type I and type II PKS. Type I PKSs use a variety of acids as a starter molecule and alkylmalonates as extender units, omit some reduction steps, and produce linear or macrocyclic molecules. Type II PKSs use malonate as an extender unit, omit all the reduction steps of FAS, and produce aromatic molecules. These variations allowm the synthesis of a huge number of polyketide antibiotics, including tetracyclines, macrolides, erythromycins, rifamycins, and polyether antibiotics.

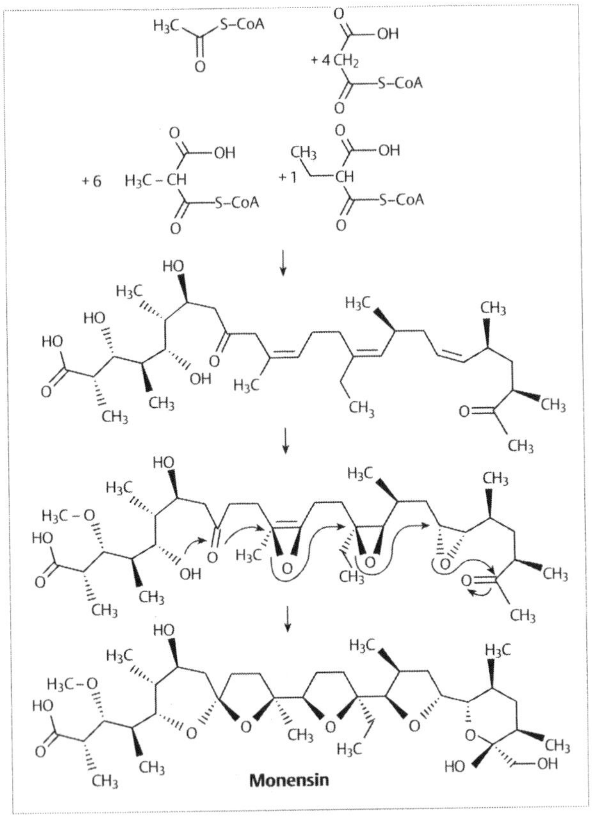

Monensin

Fig. 27.11 **Biosynthetic pathway of complex polyketides.** A presumptive pathway of the biosynthesis of monensin, a polyether antibiotic in *Streptomyces cinnamonensis*, is shown

It should be noted that in the synthesis of fatty acids and aromatic polyketides, almost all of the elongation cycles are identical. In the biosynthesis of these other classes of antibiotics, the chain assembly must be accurately programmed in order to insert the right extender unit with the appropriate level of reduction in each position of the molecule. In the few cases carefully studied, this programmed synthesis is performed by type I PKSs. This was elucidated for the first time in genetic studies on erythromycin biosynthesis.

Erythromycin, a product of *Saccharopolyspora erythraea*, is composed of a macrocyclic lactone of 14 atoms, bearing keto, hydroxy, and methyl groups and two sugars. The macrocycle is made by the polyketide process, from one propionate, the initiator unit, and six methylmalonate extender units (see Fig. 27.**8**). The synthase complex, as deduced from the gene organization and sequences, is composed of three multifunctional proteins, each comprising two FAS-like functional units. Thus, there are a total of six functional units, each responsible for one of the six elongation cycles needed to construct the chain. Each cycle includes the condensation of one methylmalonate with the growing chain and the appropriate total, partial, or non-reduction of the β-keto group. The growing chain is transferred from one unit to the adjacent one, so that the order in which the functional units are arranged determines the order of the different chemical groups in the final molecule (Fig. 27.**8**). The linear chain thus produced is then closed, forming the macrocyclic lactone of deoxyerythronolide B, the earliest identified intermediate of the biosynthetic pathway. The subsequent reactions, by which it is transformed into the antimicrobial erythromycins, are represented in Figure 27.**9**.

The antibiotics of the ansamycin family are synthesized by a similar process. A product of *Amycolatopsis mediterranei* is **rifamycin B**, the only ansamycin produced industrially. The building units of this metabolite are 3-amino-5-hydroxybenzoic acid (the initiator molecule), two malonates, and eight methylmalonates. The putative chain formed by the polymerization process is shown in Figure 27.**10**. The chain is then closed by an amide bond to form protorifamycin I, the earliest intermediate of the biosynthetic pathway isolated to date. It should be noted that the aliphatic chain folds and participates in the formation of the naphthalene ring. A series of reactions, including the insertion of an oxygen atom interrupting the all-carbon chain, give rise to the final microbiologically active molecules, rifamycin SV and rifamycin B.

Polyether antibiotics are a large class comprising not less than 80 members. Three products, monensin, lasalocid, and salinomycin, have been carefully studied because of their importance as coccidiostats and growth-promoting agents. All have polyketide derived linear structures, characterized by the presence of pyran or furan rings. The biosynthetic pathway leading to **monensin A** (produced by *Streptomyces cinnamonensis*) is an example of the process by which they are synthesized. The building blocks of monensin A are one acetate (the initiator unit), four malonates, seven methylmalonates, and one ethylmalonate. These are condensed to give a chain (whose putative structure is depicted in Figure 27.**11**) that is then converted into the final product through a series of intermediates not yet completely identified.

27.4 Two Polymerization Mechanisms Are Used for the Synthesis of Polypeptide Antibiotics

27.4.1 Ribosomal Synthesis of Antibiotics Uses Conventional Mechanisms

A small group of larger peptide antibiotics have been described that are made by the general transcription and translation system of protein synthesis. These are known as **lantibiotics**, because of the presence in their molecules of several lanthionine moieties derived from sulfide bridge formation between cysteine and serine (or threonine) residues. A few, such as nisin or subtilin, are produced by Gram-positive eubacteria; others, such as ancovenin and actagardine, are made by actinomycetes. Studies on their biosynthesis have demonstrated that the final antibiotic molecules derive from larger ribosomally synthesized peptides called prelantibiotics. These comprise a leader peptide and an amino acid sequence that give rise to the antibiotic structure through extensive posttranslational processing. The last biosynthetic step is the cleavage of the leader peptide.

27.4.2 The Thiotemplate Mechanism Is Unusual and Solely Catalyzed by Enzymes

The vast majority of peptide antibiotics are synthesized by the thiotemplate mechanism. The name reflects the main characteristic of the system, in which the sequence of the amino acids in the antibiotic molecule is determined by the order in which the precursors are linked, as thioesters, to one or more multifunctional enzymes that function as templates.

The process involves the activation of the amino acids as adenylates, the condensation of the carboxyl group of the amino acids to thiol groups of the enzymes to form thioesters, and step-wise polymerization. The polymerization initiates with the formation of a peptide bond between the carboxyl group of the first amino acid and the amino group of the second amino acid (the energy is provided by the breaking of the thioester bond). Then the thioester bond linking the formed dipeptide to the enzyme breaks, and the carboxyl group forms a second peptide bond with the third amino acid. The condensation is repeated until the completion of the chain, which is then released by a thioesterase (see Fig. 27.**12**).

The enzymatic complex that catalyzes the entire process can comprise up to four multifunctional enzymes. Several sites are present in a multifunctional enzyme, each comprising specific domains able to catalyze the activation of one amino acid, its esterification to the thiol group of a pantetheine moiety, and the formation of one peptide bond. Sometimes, in a site, domains are also present that catalyze the isomerization (from L to D) of an amino acid or the methylation of the nitrogen of a newly formed amide. The system presents several similarities with the polyketide synthesis catalyzed by type I PKS, as exemplified by the biosynthesis of the polyketide chain of erythromycin (see Fig. 27.**9**). The analogy is strengthened by the observation that the transfer of the growing peptide chain (either within a multifunctional enzyme, or from one multienzyme to the next) is mediated by the pantetheine-containing moieties present in each site, analogous to the acyl-carrier protein.

One of the best-studied peptide antibiotics, from the biosynthetic point of view, is **gramicidin S**. This is a cyclic peptide composed of two identical chains fused head to tail, each including five amino acid residues. The synthesis is catalyzed by two multifunctional enzymes, GS1, with a molecular weight of 130 000, and GS2 whose relative weight is about 500 000. The function of GS1 is to catalyze the activation of phenylalanine, its racemization, and the binding as thioester of the D-isomer, which is the initiator molecule. The GS2 enzyme activates the other four precursors and performs all the subsequent reactions. When the polymerization process is completed, two pentapeptide molecules thus formed are fused head to tail by peptide bonds, completing the antibiotic structure (Fig. 27.**12**).

27.4.3 β-Lactam Antibiotics

The classical ***β*-lactam antibiotics**, penicillins and cephalosporins, were first isolated from fungal strains, and for many years it was believed that prokaryotes were unable to synthesize structures of this type. It was later found that some actinomycetes produce penicillin N, cephalosporin C, and various cephamycins (7-methoxy-cephalosporins). Further cephalosporin-like molecules were subsequently isolated from different species of soil-inhabiting Gram-negative bacteria.

The first reaction in the biosynthetic pathway leading to **penicillins**, **cephalosporins** and **cephamycins** is the assembly of three amino acids, L-α-amino-adipic acid, L-cysteine, and L-valine, into a tripeptide, L-δ-aminodipyl-L-cysteinyl-D-valine (ACV). The oligomerization is performed, according to the thiotemplate mechanism, by a multifunctional enzyme, ACV

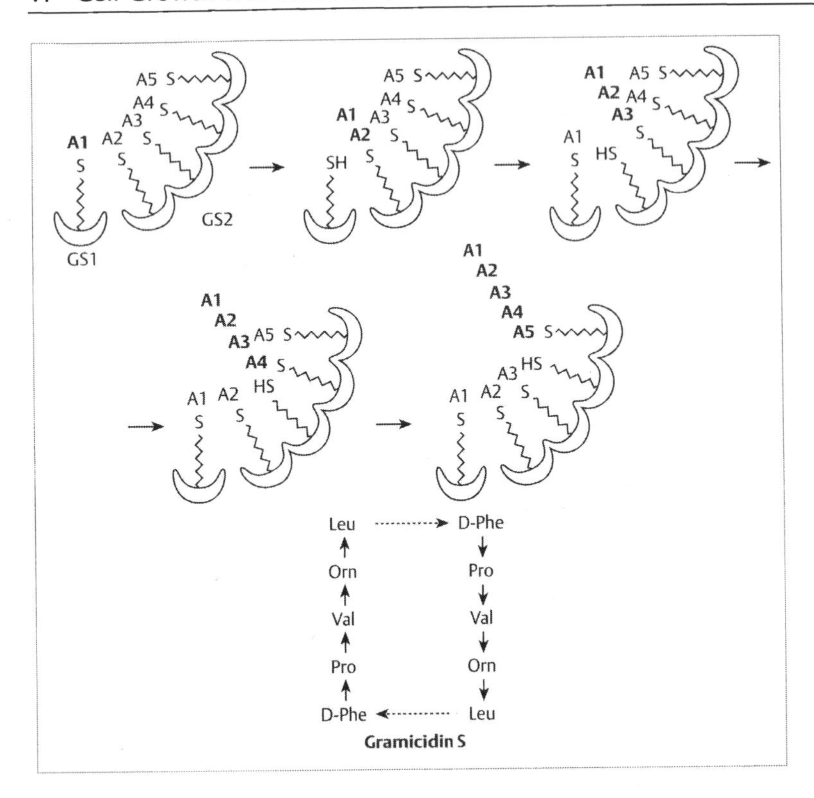

Fig. 27.12 **Thiotemplate mechanism of peptide antibiotic synthesis.** Biosynthesis of gramicidin S in *Bacillus brevis* is shown. The pictorial representation shows that enzyme GS1 has one synthase site carrying D-phenylalanine, represented as A1. Enzyme GS2 has four sites carrying proline (A2), valine (A3), ornithine (A4), and leucine (A5). The pantetheine arms, which transfer the growing chain from one site of the enzyme to the next is represented by ∿∿∿SH (see also Fig. 7.33)

synthase, which has been purified from fungal strains and prokaryotes, including *Streptomyces clavuligerus*. The enzyme activates and binds the amino acids as thioesters (α-aminoadipic acid is activated and bound at its δ-carboxyl group), performs the inversion from L-valine to D-valine, and carries out the condensation steps. It is interesting to note that studies on the sequence of the gene coding for ACV synthetase have revealed the presence of three highly repetitive regions, each containing sequences corresponding to enzyme domains that activate one amino acid.

While a few peptide antibiotics are synthesized on ribosomes, the vast majority are synthesized by a **thiotemplate mechanism**. Activated amino acids are condensed through their carboxyl groups to thiol groups of the multifunctional enzyme, and their sequence in the peptide is determined by the order in which they bind to the enzyme. By this mechanism, for example, the tripeptide that is converted by cyclization into isopenicillin N, is made from which are generated the penicillins and the cephamycins/cephalosporins.

The next reaction is a cyclization, performed by the enzyme isopenicillin N synthase, which leads to the formation of isopenicillin N, the last common intermediate of penicillin and cephalosporin biosynthesis. As illustrated in Figure 27.**13**, the cyclization involves two desaturation steps, in which oxygen is the hydrogen acceptor and Fe^{2+} is a cofactor. First the β-lactam ring is formed, followed by closure of the thiazolidine ring.

In fungi, isopenicillin N may be converted into penicillin G and other hydrophobic penicillins or isomerized to penicillin N, which is then subjected to further transformations. In prokaryotes, the second case has only been observed: the five-membered ring of penicillin N is expanded to a six-membered ring by the enzyme deacetoxycephalosporin C synthase (also denoted as "expandase"). It is noteworthy that the mechanism of this reaction is very similar to that of isopenicillin N formation, which also requires oxygen as co-substrate and Fe^{2+} as cofactor. The next step is again an oxidation, the hydroxylation of the 3-methyl group. The resulting deacetylcephalosporin C is the last common intermediate in the biosynthesis of cephalosporins and cephamycins. In Figure 27.**14**, the reactions that lead from isopenicillin N to either cephalosporin C or cephamycin C are indicated.

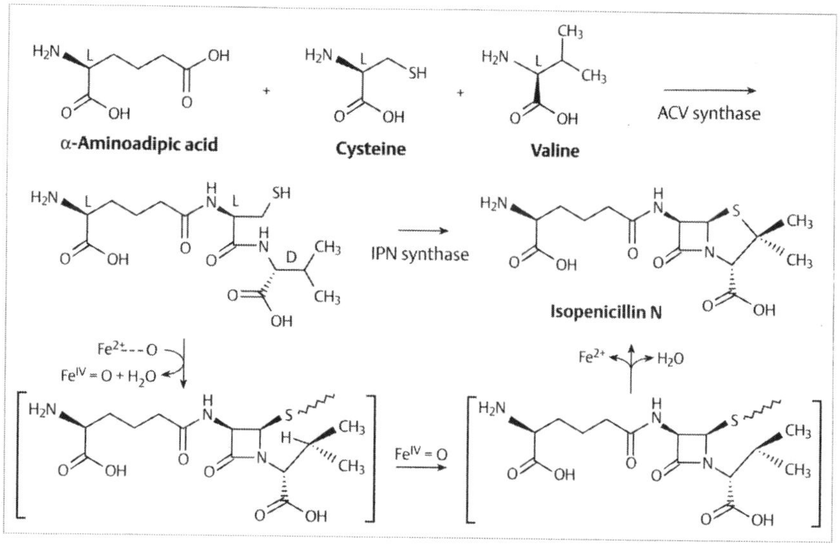

Fig. 27.13 **Biosynthesis of penicillin N.** Penicillin N is the common precursor of penicillins, cephalosporins, and cephamycins (ACV, L-δ-aminodipyl-L-cysteinyl-D-valine; IPN, iso-penicillin N). For further explanations, see text

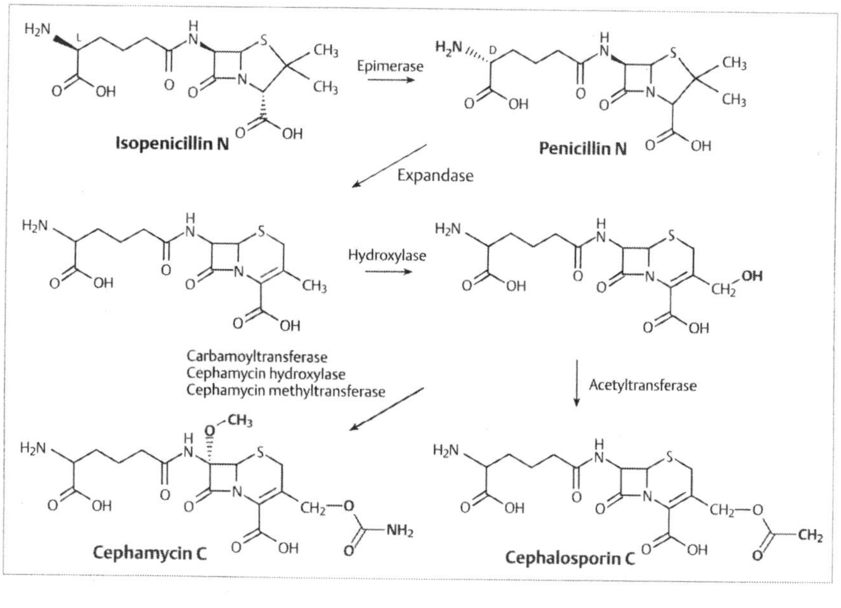

Fig. 27.14 **Biosynthetic pathway of β-lactam antibiotics in streptomycetes.** Producer organisms are *S. clavuligerus*, *S. lipmanii* and *Nocardia lactamdurans*

27.5 Aminoglycoside Antibiotics Produced by Actinomycetes Are Derived from Carbohydrate Oligomerization

The available evidence indicates that the antibiotics having an oligosaccharide structure are formed by the assembly of monomers activated as nucleoside diphosphates at the anomeric carbon. In this respect, their biosynthesis is not different from that of the normal polysaccharides present in bacterial cell walls or surface (see Chapters 7.9 and 23).

The components of the oligosaccharide antibiotics are most often unusual carbohydrates. Two patterns can be distinguished with regard to their formation: (1)

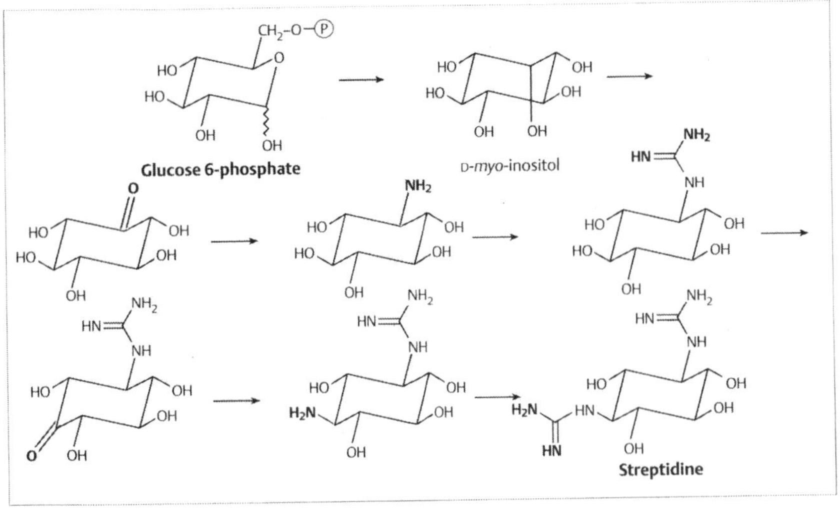

Fig. 27.15 **Biosynthesis of streptidine.** Streptidine is the aminocyclitol moiety of streptomycin. For simplicity, the phosphorylation and dephosphorylation steps have been omitted. For further explanations, see text

carbohydrates normally found in primary metabolism are assembled and then modified, and (2) common sugars, such as glucose or glucosamine, are converted into antibiotic precursors and then condensed step-wise to give the final molecule. In the latter case, the transformation reactions also occur, as a rule, on sugars activated as nucleoside diphosphates at the anomeric carbon. However, there is no rigid distinction between the two cases; both patterns may be present in a single biosynthetic pathway.

27.5.1 Two Different Pathways Give Rise to the Amino-cyclitols Streptidine and 2-Deoxystreptamine

All the important oligosaccharide antibiotics contain an aminocyclitol moiety, which, in the case of streptomycin, is streptidine, and in most of the other cases, is 2-deoxystreptamine. In spite of the similarities of their structures, the biosyntheses of these molecules proceed through different pathways.

The biosynthesis of **streptidine** initiates from D-*myo*-inositol, a common cell metabolite derived from the cyclization of glucose. The pathway consists of two identical series of reactions (Fig. 27.**15**): *myo*-inositol is oxidized at C-2 to *scyllo*-inosose, which is converted to *scyllo*-inosamine by transamination. The latter intermediate is phosphorylated, and the amine is converted into a guanidino function by transfer from arginine of the amidino group [–C(=NH)–NH_2]. After dephosphorylation, the same series of reactions are repeated, starting from the oxidation of the β-hydroxyl group to the amidino group.

In the biosynthesis of **2-deoxystreptamine**, glucose (probably activated as its 6-phosphate) is converted into deoxy-*scyllo*-inosose by a reaction that is similar to the formation of dehydroquinic acid in the aromatic amino acid biosynthetic pathway (Fig. 27.**16**). The similarity is supported by the finding that the reaction involves, as intermediate steps, the oxidation and the reduction of the hydroxyl group at C-4. The keto group is converted into an amino group by transamination. The hydroxyl group adjacent to the methylene is then oxidized to a keto group, and another transamination produces the final molecule.

27.5.2 Examples of Oligosaccharide Antibiotic Biosynthesis

Streptomycin is composed of three moieties, the carbohydrates streptose (a 3-aldehydopentose), *N*-methyl-L-glucosamine, and the aminocyclitol streptidine. How streptidine is synthesized from D-*myo*-inositol was described above. In the biosynthesis of streptose, glucose is activated as deoxythymidine bisphosphate (dTDP-glucose) and isomerized to dTDP-4-keto-L-rhamnose (Fig. 27.**17**). This is then rearranged to

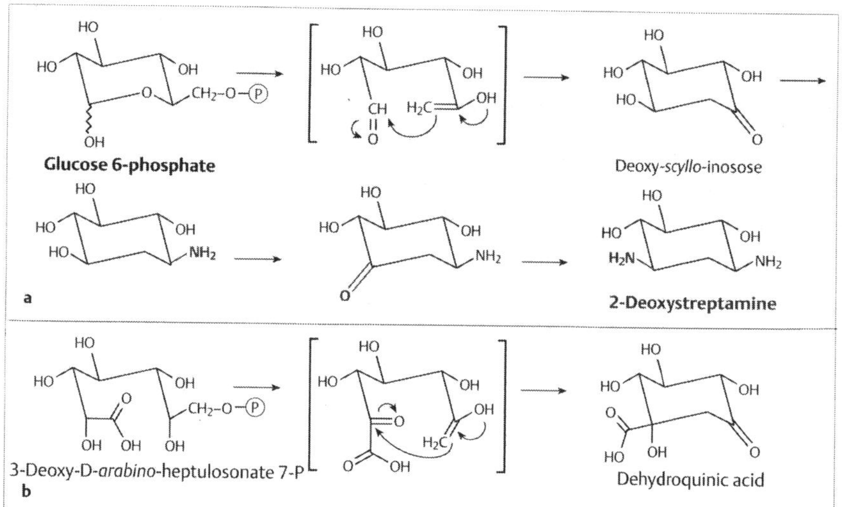

Fig. 27.**16** **Biosynthesis of 2-deoxystreptamine.** This intermediate is the aminocyclitol moiety of several aminoglycoside antibiotics. The mechanism of ring closure (**a**) is similar to that of dehydroquinic acid formation (**b**), which is shown for comparison. For further explanations, see text and Fig. 7.**18**

give dTDP-dihydrostreptose, which is the molecule intervening in the assembly of dihydrostreptomycin. The oxidation of the dihydrostreptose moiety to streptose occurs after the oligomerization process. *N*-Methyl-L-glucosamine originates from D-glucosamine. The biosynthetic steps that formally comprise the inversion of four chiral centers are not understood. There is evidence that D-glucosamine is activated as the UDP ester and that UDP-*N*-methyl-D-glucosamine is an intermediate of the process.

The assembly of the streptomycin molecule proceeds through the addition of dihydrostreptose to streptidine phosphate, followed by the addition of methyl-L-glucosamine to the dihydrostreptose moiety, with formation of dihydrostreptomycin phosphate. This is then converted to streptomycin by an oxidation and a dephosphorylation reaction (Fig. 27.**18**).

An important group of antibiotics, particularly active on *Pseudomonas* and other Gram-negative bacteria, is the **gentamicin sisomicin sagamicin** family. These include not less then 20 members, mostly produced by *Micromonospora* species. In contrast to the biosynthesis of streptomycin, the transformations of the carbohydrate components occur after the basic structure of the molecule has been assembled. The basic structure is formed by the step-wise addition to 2-deoxystreptamine of D-glucosamine and D-xylose (Fig. 27.**19**). Conversion of the 3′-hydroxyl group of xylose into a methylamino group followed by methylation at C-4′ (with inversion of the hydroxyl group) gives gentamicin X_2, the last common intermediate from which all members of the family derive. The subsequent reactions are modifications of the glucosamine moiety. The 6′-

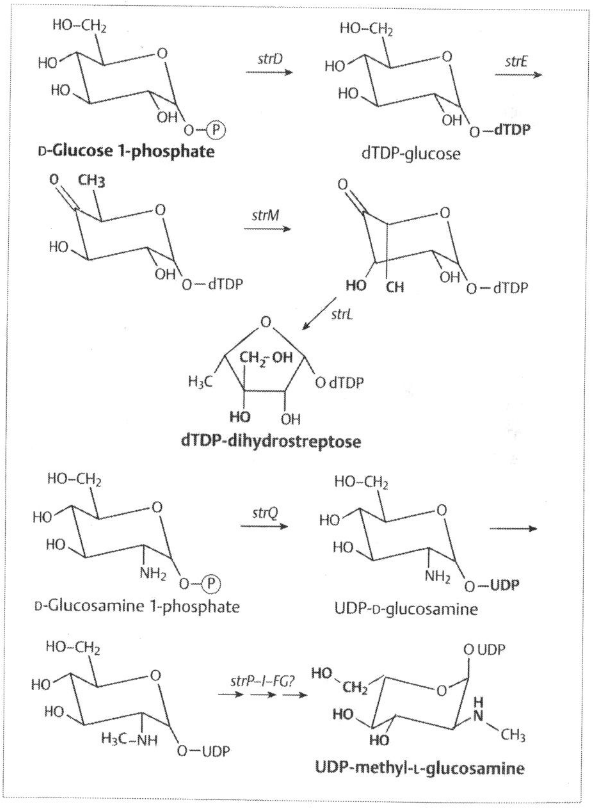

Fig. 27.**17** **Origin of dihydrostreptose and N-methyl-L-glucosamine, the carbohydrate moieties of streptomycin.** The first three steps of dTDP-dihydrostreptose biosynthesis are identical to those of dTDP-rhamnose and other sugars in primary metabolism in Gram-negative bacteria. Nevertheless, the genes governing the reactions, indicated in the figure, are part of the streptomycin (*str*) gene cluster and therefore are part of secondary metabolism

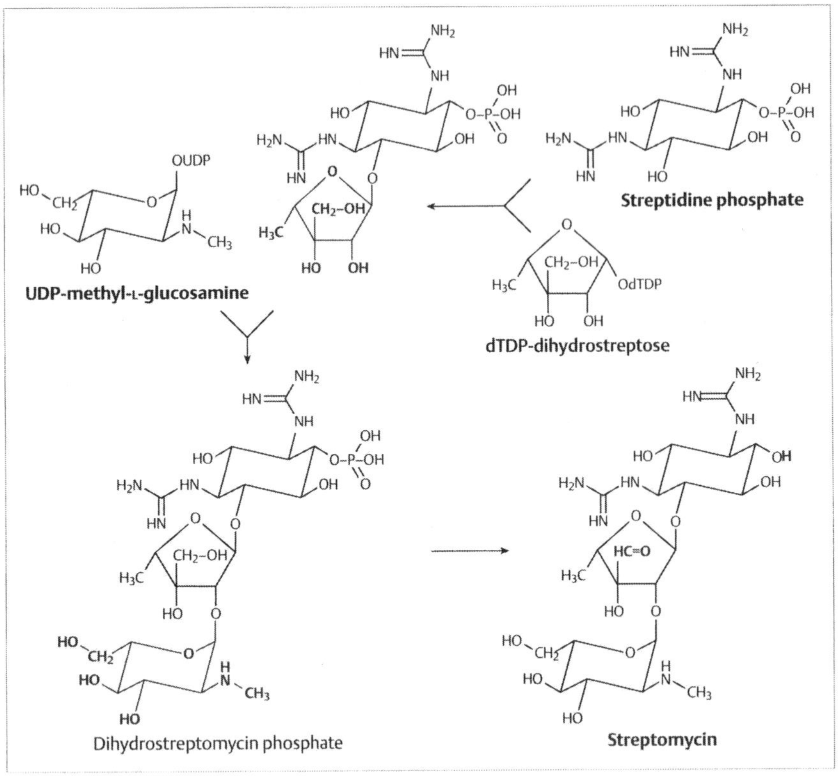

UDP-methyl-L-glucosamine

Streptidine phosphate

dTDP-dihydrostreptose

Dihydrostreptomycin phosphate

Streptomycin

Fig. 27.**18** **Assembly of the strep tomycin molecule.** As an exampl of aminoglycoside biosynthesis th assembly of streptomycin from thre precursors in *Streptomyces griseus* i shown

hydroxyl group is converted into an amino group; dehydration at 4′ gives sisomicin, which, depending on the producing strain, can be excreted as such or reduced to gentamicin C_{1a}, the commercially most important product. In *Micromonospora sagamiensis* fermentations, gentamicin C_{1a} is methylated to give sagamicin, the final product of the biosynthetic process.

Some of the most important commercial antibiotics, the **aminoglycosides**, derive from carbohydrate oligomerization. Again, normal carbohydrates (often in the form of nucleoside diphosphates) are modified after assembly or converted into unusual precursors before assembly. They contain, as a general rule, the amino-cyclitols streptidine or 2-deoxystreptamine, derivatives of glucosamine and complex carbohydrates (e.g., the unusual streptose). Representatives are streptomycin, gentamycins, and sisomycin.

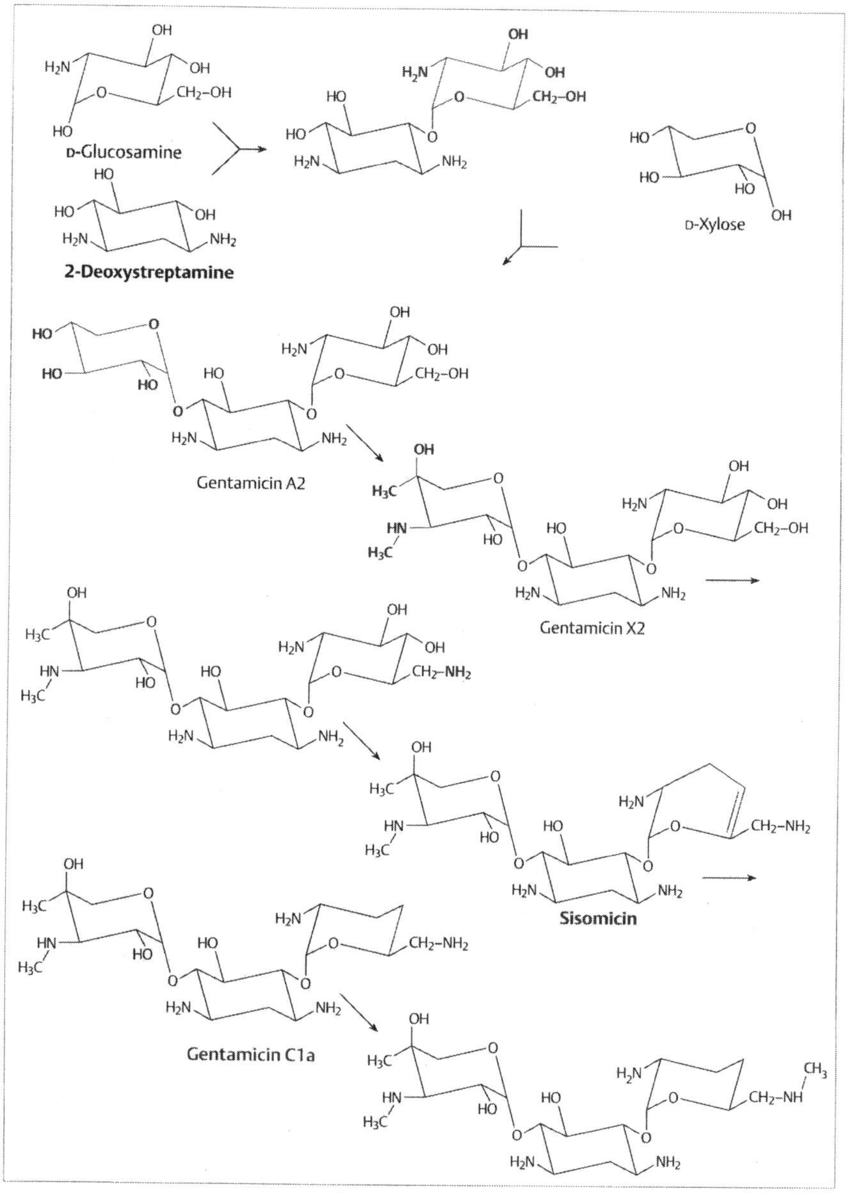

D-Glucosamine

2-Deoxystreptamine

D-Xylose

Gentamicin A2

Gentamicin X2

Sisomicin

Gentamicin C1a

Fig. 27.19 **Biosynthetic pathway to some aminoglycoside antibiotics produced by *Micromonospora* strains.** Branches from the same pathway lead to the production of other antibiotics of the gentamicin family. For further explanation, see text

27.6 Similar to Other Differentiation Processes, Secondary Metabolism Is Highly Regulated

Secondary metabolites are often produced after the exponential growth phase or at submaximal growth rates. The distinction between the growth phase ("**trophophase**") and production phase ("**idiophase**") is sometimes very clear, but in many cases, the idiophase overlaps the trophophase. The timing between the two phases can be manipulated, i.e., the two phases are often distinctly separated in a medium favoring rapid growth, but overlap partially or even completely in a medium supporting slower growth. A secondary metabolite is called "secondary" merely because it has no apparent involvement in the vegetative growth of the producing culture, not because it is produced after growth. Thus, elimination of production of a secondary metabolite by mutation will not stop or slow down growth; indeed, it may increase the growth rate (see Chapter 35.6).

The factors controlling the onset of secondary metabolism are complex and not well understood. They should, however, resemble the factors involved in other global control mechanisms, such as catabolite repression, stringent response, carbon starvation response, in systems using alternative sigma factors and in systems showing temporal control, (e.g., *Caulobacter* cell division and *Bacillus* sporulation). Growth rate and deficiencies in certain nutritional factors are important, but the actual mechanisms involved are not known. The temporal nature of secondary metabolism is certainly genetic in nature, but expression can be influenced greatly by environmental manipulations.

The delay often seen before the onset of secondary metabolism was probably established by evolutionary pressures. Many secondary metabolites have antibiotic activity and could kill the producing cells if made too early. The resistance of antibiotic producers to their own metabolites is well-known (see Chapter 33.5.3). Antibiotic-producing species possess suicide-preventing mechanisms, such as (1) enzymatic detoxification of the antibiotic, (2) alteration of the antibiotic's normal target in the cell, and (3) modification in permeability to allow the antibiotic to be pumped out of the cell and to restrict its re-entry. Such mechanisms are often inducible, but in some cases are constitutive. In the case of inducible resistance, death could result when the antibiotic is produced too early and induction is slow. Delay in secondary metabolite production until the starvation phase makes sense if the product is used as a competitive "weapon" in nature or endogenously as an effector of differentiation. In nutritionally rich habitats, such as the intestines of mammals, where enteric bacteria thrive, secondary metabolite production is not as important as in soil and water, where nutrients limit microbial growth. Thus, secondary metabolites tend not to be produced by enteric bacteria, such as *Escherichia coli*, but are produced by soil and water inhabitants such as bacilli, actinomycetes, and fungi. Nutrient deficiency in nature often induces morphological and chemical differentiation (i.e., sporulation and secondary metabolism, respectively); both are beneficial for survival in the wild. Thus, the regulation of the two types of differentiation is often related.

Whether chromosomal or plasmid-borne, the secondary metabolism genes are usually clustered in prokaryotes, not necessarily as single operons, but in the form of regulons and modulons. Expression of these genes is under strong individual and global control by nutrients, inducers, products, metals, and growth rate. In most cases, regulation is at the level of transcription, translational control may be less common.

Bacterial cell growth is normally coupled to cell division. Since under these conditions all descendants of a cell are (except for mutation) genetically identical (form a **clone**), this growth is called **vegetative**. It is characteristic for cells in the exponential growth phase (also called "**trophophase**"). Upon a switch from the exponential to the stationary growth phase or during slow (linear) growth, global regulatory networks alter cellular metabolism from growing to surviving strategies. These correspond at the morphological level to survival structures (e.g., spores, cell surface rearrangements) and at the metabolic level to "secondary metabolism". Metabolites produced during this "**idiophase**" thus have essential roles in cell survival and competition (see also Chapter 28).

27.6.1 Regulation by Carbon Sources (Carbon Catabolite Repression)

Glucose, which is for many bacteria an excellent carbon source, interferes with the formation of many secondary metabolites. Polysaccharides (e.g., starch), oligosaccharides (e.g., lactose), oils (e.g., soybean oil, methyloleate), and other slowly used carbon sources are often preferable for fermentations yielding secondary metabolites. In media containing a mixture of a rapidly used carbon source and a slowly used carbon source, the former is used first ("diauxie", see Chapter 20.2) to produce cells, but little or no secondary metabolite is formed mainly due to carbon catabolite repression. After the rapidly assimilated compound is depleted, the "second-best" carbon source is used for the idiophase, and repression is relieved. It should be noted that in certain cases (e.g., bacilysin production by *Bacillus subtilis*), glucose is not an interfering carbon source, but other carbohydrates are.

As in other Gram-positive bacteria, cAMP (3,5-cyclic adenosine monophosphate) does not appear to be involved in carbon catabolite repression of secondary metabolism (see Chapter 20.2.4). Repression by glucose is usually exerted at the level of transcription. For example, in the case of phenoxazinone synthase involved in actinomycin biosynthesis, the corresponding mRNA is low in the trophophase, high in the idiophase, and much lower after growth with glucose than with galactose, which is not repressive.

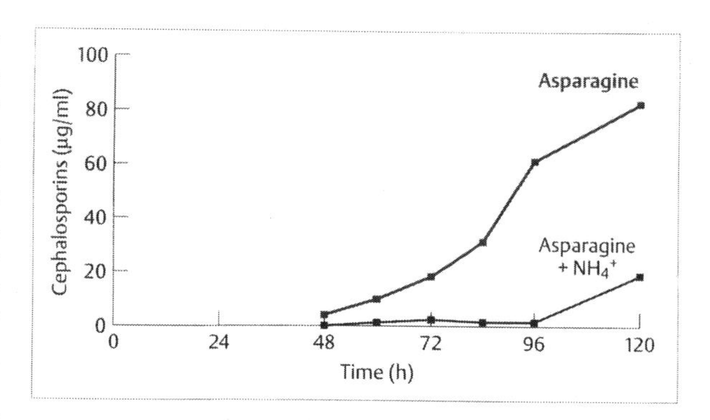

Fig. 27.**20** **Effect of ammonium ion on volumetric cephalosporin formation.** The production in a chemically defined medium by *Streptomyces clavuligerus* NRRL 3585 is shown. □, 15 mM L-asparagine as nitrogen source; ■, 15 mM L-asparagine plus 120 mM ammonium chloride as nitrogen source [after 1]

27.6.2 Regulation by the Nitrogen Source

Many secondary metabolic pathways are negatively affected by nitrogen sources favorable for growth, e.g., ammonium salts (Fig. 27.**20**). As a result, complex fermentation media often include a protein source (such as soybean meal) and defined media contain a slowly assimilated amino acid (such as proline) as the nitrogen source to encourage high production of secondary metabolites. Little information is available concerning the mechanisms underlying the negative effects of NH_4^+ and certain amino acids. Whether or not the regulatory system is similar to the Ntr system of the enteric bacteria (see Chapter 20.4) is under investigation. In the production of tylosin by *Streptomyces fradiae*, the sensitive enzyme appears to be valine dehydrogenase, which is repressed and inhibited by NH_4^+. Valine is the source of carbon atoms for the macrolide ring system.

27.6.3 Regulation by the Phosphorus Source

Regulation by phosphorus sources includes both specific and global controls. A rather specific negative effect of inorganic phosphate arises from its ability to inhibit and/or repress phosphatases. Because biosynthetic intermediates of certain idiolite pathways (e.g., aminoglycoside antibiotics) are phosphorylated, whereas the ultimate product is not, phosphatases are required in biosynthesis. For example, streptomycin biosynthesis by *Streptomyces griseus* includes at least three phosphate-cleavage steps, and the process is very sensitive to the phosphate concentration. The final enzyme in the pathway, which removes phosphate from dihydrostreptomycin 6-phosphate, is inhibited by inorganic phosphate.

Phosphate also has a more general effect than the inhibition or repression of biosynthetic phosphatases; it appears to interfere in many secondary metabolic pathways not known to have phosphorylated intermediates. Such fermentations have to be conducted at levels of free phosphate (usually below 10 mM) that are suboptimal for growth. A number of secondary metabolic enzymes are repressed by phosphate, e.g., those involved in candicidin production in *S. griseus*. Whether the intracellular effector of phosphate regulation is inorganic phosphate itself is unknown. Similarly, the mechanism of general phosphate control of secondary metabolism and a relationship to, for example, the Pho system in *Escherichia coli*, remains to be determined. The possibility that phosphate regulation works by affecting enzyme activity posttranslationally via protein kinases and phosphoprotein phosphatases (see Chapter 19.4) also exists.

27.6.4 Control of Secondary Metabolism Probably Involves All Known Regulatory Mechanisms

Global control by A-factor and related compounds. In many secondary metabolite pathways, primary metabolites induce synthases and thus increase production of the final product. Of greater importance in actinomycete fermentations, however, is the effect of specific regulatory metabolites, e.g., **"A-factor"** (2-*S*-isocapryoyl-3-*R*-hydroxymethyl-γ-butyrolactone) and related γ-butyrolactones (Fig. 27.**21**). They correspond to alarmones or auto-inducers with a role in quorum-sensing (see Chapter 20.1.4). A-factor is an effector that exerts global control on secondary metabolism. It induces both morphological and chemical differentiation in *Streptomyces griseus* and *Streptomyces bikiniensis*, bringing on formation of aerial mycelia, conidia, streptomycin synthases, and streptomycin. A-factor controls at least ten proteins in *Streptomyces griseus* at the transcriptional level. One of these is streptomycin 6-phosphotransferase, an enzyme that functions both in streptomycin biosynthesis and resistance. Antibiotic production and morphological differentiation are related to these γ-butyrolactones (for details see Chapter 25.2) in many or all actinomycetes, but probably in a very complex way (see Fig. 27.**22**).

Control by specific regulatory mechanisms. The role of feedback regulation in controlling secondary metabolism is well-known (Chapter 19). Many secondary metabolites inhibit or repress their own biosynthetic enzymes. For example, kanamycin represses an acetyltransferase involved in its synthesis, and chloramphenicol represses a biosynthetic enzyme, arylamine synthetase. Inhibition of an antibiotic synthase is known in the pathways to bacitracin, gramicidin S, erythromycin, tylosin, and tetracycline. Growth rate control appears to be important in secondary metabolism as stated before. It may be the overriding factor in the cases where nutrient limitation is needed for production of secondary metabolites (e.g., bacitracin). In contrast, both a low growth rate and a particular type of nutrient deficiency are needed to support secondary metabolite biosynthesis in other cases (e.g., thienamycin). Growth rate-dependent biosynthesis of secondary metabolites could involve stability of mRNA, antisense RNA, or regulatory proteins ("sigma factors"), but this is not yet known. Metals control certain pathways of secondary metabolism, e.g., zinc and manganese. Iron stimulates the production of certain secondary metabolites and suppresses production of others. Often, the suppressible secondary metabolites are siderophores (see Chapter 27.7.2), which protect the cells from iron starvation when the concentration of the metal drops below 1 μM. In *Streptomyces pilosus*, iron represses the first enzyme of desferrioxamine biosynthesis, lysine decarboxylase, at the level of transcription.

Production of secondary metabolites eventually stops due to feedback regulation (see above) and inactivation of the synthase(s). For example, most of the synthases involved in production of peptide antibiotics by bacilli disappear a few hours after they are formed. In the producer of gramicidin S, *Bacillus brevis*, the inactivation of the gramicidin S synthetases is oxygen-dependent and independent of protease action. Decay is slowed by the presence of thiols, amino acid substrates, and a utilizable energy source. The sites of inactivation appear to be the SH groups involved in binding the amino acids to the enzymes.

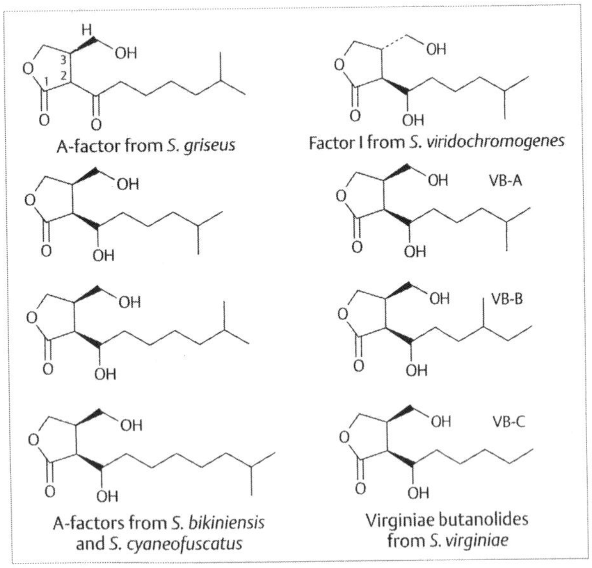

A-factor from *S. griseus*

Factor I from *S. viridochromogenes*

VB-A

VB-B

VB-C

A-factors from *S. bikiniensis* and *S. cyaneofuscatus*

Virginiae butanolides from *S. virginiae*

Fig. 27.**21** **Chemical structures of auto-regulatory factors (A-factors) having a γ-butyrolactone ring from *Streptomyces* spec** [after 2]

> Considering what is known about global regulatory networks in other differentiation systems, it is amazing how little is known about the biochemical pathways and their specific and global control in secondary metabolism. This is even more surprising in view of the central role this metabolism has in biotechnology. The universality of mechanisms in biology, however, suggests the presence of quorum sensing mechanisms (A-factors), sensor/response regulator systems, global regulatory proteins, and alternative sigma factors (see Fig. 27.**22**) at the base of the switching between trophophase and idiophase and in control of secondary metabolism.

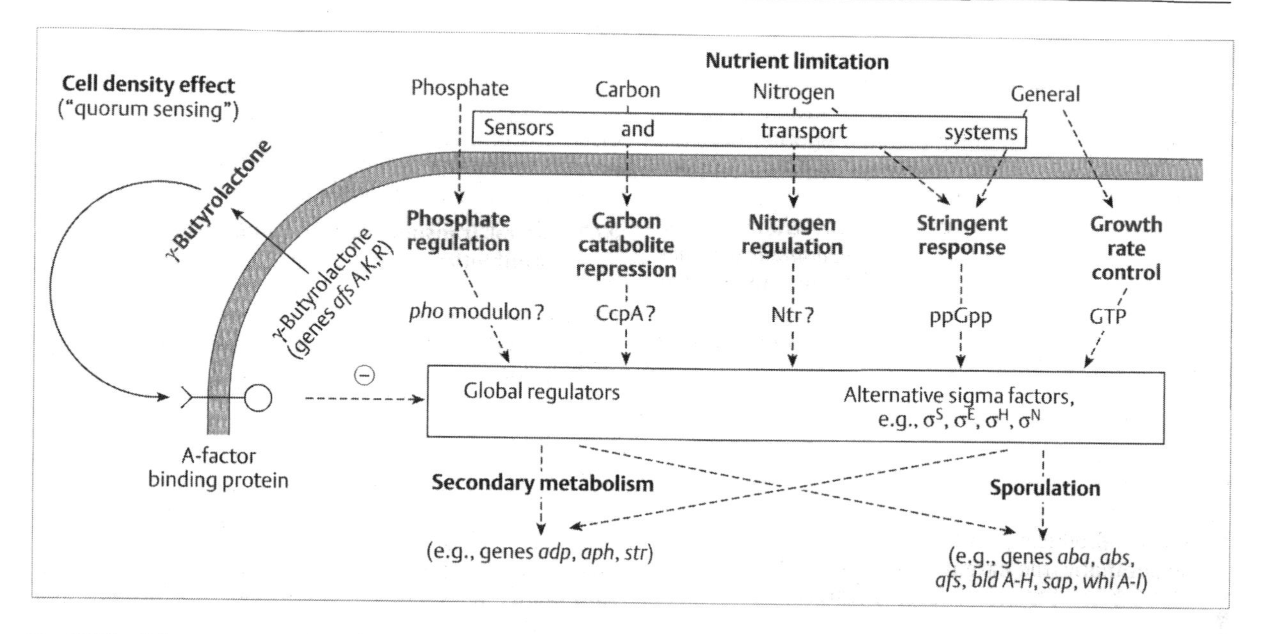

Fig. 27.**22** **Scheme summarizing known stimuli involved in triggering the transition from trophophase to idiophase in *Streptomycetes.*** While the effects of nutrient limitation and of γ-butyrolactones are well established, no regulatory mechanism has been identified. In analogy to other systems of global regulatory networks in Gram-positive bacteria, an essential role for a *pho* modulon, CcpA and *cre* promoters, the nitrogen regulator Ntr, the stringent response, and various alternative sigma factors, e.g., those involved in carbon starvation and in stress control, is postulated. Growth rate control appears to involve a GTP-dependent mechanism, while an A-factor binding protein seems to be involved in quorum sensing. Regulatory connections are indicated by broken arrows and identified genes are described in the text

27.7 Secondary Metabolites Have Essential Functions in Nature

Antibiotics and other secondary metabolites are produced under natural growth conditions and have essential physiological roles. Over 40% of the actinomycetes produce antibiotics when they are freshly isolated from nature and their products are found in soil, straw and agricultural products. For example, siderophores have been found in soil, and microcins, enterobacterial antibiotics, have been found in human fecal extracts. The microcins are thought to be important in bacterial colonization of the human intestinal tract early in life. A further indication of natural antibiotic production is the possession of antibiotic resistance plasmids by most soil bacteria.

The widespread nature of secondary metabolite production, their multiple functions and the preservation of the multigenic biosynthetic pathways in nature indicate that secondary metabolites serve survival functions in organisms that produce them. This is self-evident for pheromones (sexual hormones) (see Chapter 16.2.4).

27.7.1 Many Secondary Metabolites Are Used to Compete With Other Organisms

Bacterial antagonism. Bacterial–bacterial competition works via antibiotics and via bacteriocins ("bacteria-killers"). Agrocin 84, a plasmid-encoded antibiotic of *Agrobacterium rhizogenes* is an adenine derivative that kills strains of plant pathogenic agrobacteria. It is used commercially in the prevention of crown gall disease of plants.

An interesting relationship exists between myxobacteria and their bacterial "diet." Myxobacteria live on other bacteria, and to grow on these bacteria, they require a high myxobacterial cell density. This "population effect" (see environmental or quorum sensing, Chapter 20.1.4) is primarily due to the need for a high concentration of lytic enzymes and antibiotics in the local environment. Thus, *Myxococcus xanthus* fails to grow on *Escherichia coli* unless more than 10^7 myxobacteria per ml are present. At these high cell concentrations, the parent *Myxococcus xanthus* grows, but a mutant that cannot produce antibiotic TA fails to grow. This indicates that the antibiotic is involved in the killing and subsequent nutritional use of other bacteria. From 60 to 80% of myxobacterial isolates produce antibiotics.

Competition also occurs between strains of one species. Phenazine production by *Pseudomonas phenazinium*, to which it is resistant, results in smaller colonies and lower maximum cell densities (but not lower growth rates) than those of nonproducing mutants. Furthermore, the viability of nonproducing mutants in various nutrient-limited media is higher than that of the producing parent. Despite these apparent deficiencies, the producer strain wins out in mixed culture when placed in the above media. The parental strain is able to use its phenazine antibiotic to kill the nonproducing cells, and owing to its resistance to the antibiotic, the parent survives.

Erwinia carotovora subsp. *betavasculorum* is a wound pathogen that causes, for example, vascular necrosis and root rot of sugarbeet. It produces a broad-spectrum antibiotic of unknown structure that is the principal determinant that allows it to compete successfully in the potato against the antibiotic-sensitive *Erwinia carotovora* subsp. *carotovora* strains. Complete correlation exists between antibiotic production in vitro and inhibition of *Erwinia carotovora* subsp. *carotovora* strains in the plant.

Bacteria versus amoebae. Since protozoa use bacteria as food and use these prokaryotes to concentrate nutrients for them, it is not surprising that mechanisms have evolved to protect the bacteria against protozoans such as amoebae. Antibiotically active pigments from *Serratia marcescens* and *Chromobacterium violaceum* (i.e., prodigiosin and violacein, respectively) protect these species from being eaten by amoebae; in the presence of the pigment, the protozoa either encyst or die. Whereas nonpigmented *Serratia marcescens* cells are consumed by amoebae, pigmented cells are not. These results are similar to those obtained with other bacteria such as *Pseudomonas pyocyanea* and *Pseudomonas aeruginosa* and with other microbial products such as pyocyanine and phenazines. Hence, antagonism between amoebae and bacteria in nature is crucially affected by the ability of the latter to produce antibiotics.

Microorganisms versus higher plants. Microorganisms produce more than 150 compounds, called **phytotoxins** or **phytoaggressins**, which are active against green plants; the structures of over 40 are known. Many such compounds also show typical antibiotic activity against microorganisms and could justifiably be classified as either antibiotics or phyto toxins. Examples of bacterial phytotoxins and th defense of the plants through phytoalexins will b discussed in Chapter 34.3; examples of bacterial plan growth stimulants will be discussed in Chapter 34.4.

27.7.2 Metal Transport Agents Can Be Used as Antibiotics

Certain secondary metabolites act as metal transpor agents. One group is the **siderophores** (also known a **sideramines**), which function in uptake, transport, an solubilization of iron (see Chapters 8.8.1 and 33.3.5 Another group is the ionophoric antibiotics, whic function in the transport of certain alkali-metal ions these include the macrotetrolide antibiotics, whic enhance potassium permeability of membranes. Iron transport factors in many cases resemble antibiotic They are on the borderline between primary an secondary metabolites since they are usually no required for growth, but do stimulate growth unde iron-deficient conditions. Over 100 siderophores hav been described. Indeed, all strains of *Streptomyce* *Nocardia*, and *Micromonospora* examined produce suc compounds. Antibiotic activity is due to the ability o these compounds to starve other species for iron whe those species lack the ability to take up the Fe sideramine complex. Such antibiotics include nocarda min and desferritriacetylfusigen. Compounds that spe cifically bind zinc and copper are also known to b produced by microorganisms.

Most living cells have a high intracellular K^+ concentration and a low Na^+ concentration, wherea extracellular fluids contain high Na^+ and low K^+ Production of an ionophore (e.g., a macrotetrolid antibiotic) can serve a survival function as shown b comparison of a *Streptomyces griseus* strain that pro duces a macrotetrolide with its nonproducing mutant In the presence of Na^+, the mutant does not take up K^+ Also, when the strains are grown in high K^+ concentra tions and transferred to a high-Na^+, low-K^+ resuspen sion medium, the parent takes up K^+, but the mutan takes up Na^+ and loses K^+. As a result of thes differences, mutant growth is inhibited by a high-Na^+ low-K^+ environment, but the antibiotic-producin parent grows well.

27.7.3 Symbiosis Between Bacteria and Invertebrates

Antibiotics play a role in the symbiosis between th bacteria of the genus *Xenorhabdus* and nematode parasitic to insects. Each nematode species, member of the Heterorhabditidae and Steinernematidae, is associated with a single bacterial species of *Xenorhabdus*. The bacteria live in the gut of the nematode. When the nematode finds an insect host, it enters, and when in the insect gut, it releases bacteria that kill the insect, allowing the nematode to complete its life cycle. Without the bacteria, no killing of the insect occurs. The bacteria produce antibiotics and keep the insect from being attacked by putrefying bacteria. Two groups of antibiotics have been isolated from two of the bacteria. One group is represented by tryptophan derivatives and the other by 4-ethyl- and 4-isopropyl-3,5-dihydroxy-*trans*-stilbenes.

Symbiosis between intracellular microorganisms and insects similarly involves antibiotics. The brown planthopper, *Nilaparavata lugens*, contains at least two microbial symbionts and lives on the rice plant. One intracellular bacterium is *Bacillus* sp., which produces polymyxin M. Another is *Enterobacter* sp., which produces a peptide antibiotic selective against *Xanthomonas campestris* var. *oryzae*, the "white blight" pathogen of rice. The bacteria act via antibiotic production to protect the insect from invasion by microorganisms and to control competition by bacterial rice pathogens.

Many secondary metabolites are used to compete with other organisms, both prokaryotic and eukaryotic. Secondary metabolites are also used to establish and maintain symbiosis between bacteria and bacteria, bacteria and plants, or bacteria and lower invertebrates.

27.7.4 Secondary Metabolites: Effectors of Differentiation or Products of Differentiation?

From recent work, it has become clear that secondary metabolism is a form of chemical differentiation which parallels morphological differentiation processes, e.g. sporulation in bacilli and in streptomycetes (Fig. 27.**22**), or cell "dormancy" in *Escherichia coli* and other nonsporulators. Both forms of differentiation are triggered through nutrient starvation and require chemosensors (e.g., two-component systems) that signal "starvation" to global regulators (e.g., the Crp system in enteric bacteria, the *cre*-CcpA system in *Bacillus*) and activate new pathways, often through alternative sigma factors. The altered metabolic fluxes provide precursors for pathways of synthesis of antibiotics and other secondary metabolites. These, in turn, are used in defense, for competition, to protect spores, or to function in symbiosis with higher forms of life as already described. They may, however, also function as effectors in morphological and chemical differentiation, for example, as chemical stimuli that trigger signal transduction pathways.

Sporulation initiation (see Chapter 25). Of the various functions postulated for secondary metabolites, the one which has received the most attention is the simultaneous appearance of these compounds, especially antibiotics, with the transition from vegetative cells to spores, i.e. the start of sporulation:

1. Production of peptide antibiotics usually begins at the late-exponential phase of growth and continues during the early stages of the sporulation process in bacilli.
2. Sporulation and antibiotic synthesis are induced by depletion of some essential nutrient.
3. There are genetic links between the synthesis of antibiotics and the formation of spores. Revertants, transductants, and transformants of stage 0 asporogenous mutants of *B. subtilis*, restored in their ability to sporulate, also regain the ability to synthesize antibiotic.
4. Physiological correlations also favor a relationship between the production of antibiotics and spores. As an example, several inhibitors of sporulation inhibit antibiotic synthesis. Furthermore, both processes are repressed by nutrients, including glucose. A high concentrations of manganese is needed for both sporulation and antibiotic synthesis by certain species of *Bacillus*.

Despite the apparent connections between formation of antibiotics and spores, it has become clear that antibiotic production is not obligatory for spore formation. The evidence is the existence of mutants that form no antibiotic, but still sporulate. Such mutants have been found in the case of bacitracin (*Bacillus licheniformis*), mycobacillin (*Bacillus subtilis*), linear gramicidin (*Bacillus brevis*), streptomycin (*Streptomyces griseus*), and methylenomycin A (*Streptomyces coelicolor*). In addition, it has been observed that mutant strains selected for their high antibiotic production ability are often unable to sporulate. This uncoupling seems to imply that the formation of both antibiotics and spores is a consequence (rather than a cause) of the switch from vegetative growth to the differentiation phase, which includes both sporulation and secondary metabolism, and that therefore antibiotics are normally not essential for sporulation.

Although antibiotic production is not obligatory for sporulation, some antibiotics may stimulate the sporulation process. Transfer of exponential-phase populations of *Bacillus brevis* ATCC 8185 (the tyrothricin producer) into a nitrogen-free medium stops growth and restricts sporulation. Supplementation of the medium with tyrocidine induces sporulation. This suggests that the tyrocidine component of tyrothricin is an inducer of sporulation.

Sporulation enhancers are also known in the actinomycetes. One such compound is the antibiotic pamamycin, produced by *Streptomyces alboniger*. In the producing culture, the antibiotic stimulates the formation of aerial mycelia and, thus, that of conidia (see Chapter 25.2.1).

Germination of spores. The close relationship between sporulation and antibiotic formation suggests that certain secondary metabolites involved in germination might be produced during sporulation and that the formation of these compounds and spores could be regulated by a common mechanism or by similar mechanisms.

Germination inhibitors have been found in actinomycetes. The one produced by *Streptomyces viridochromogenes* is a specific inhibitor of ATP synthase. It uncouples respiration from ATP production in the spores, until it is excreted during germination. The antibiotic appears thus to be responsible for maintaining dormancy of the spores. Upon addition of the germinating stimulus Ca^{2+}, the inhibitor is excreted from the spore, the ATP synthase is activated by the Ca^{2+}, ATP is synthesized as glucose is oxidized, and germination ensues.

Gramicidin S (GS) is an inhibitor of the phase of spore germination known as "outgrowth" in *Bacillus brevis*.

1. Exponential growth and initiation of germination (i.e., the darkening of spores) are not inhibited by the antibiotic.
2. Spores of gramicidin-S-negative mutant strains outgrow in 1–2 h, whereas parental strain spores require 6–10 h. Addition of gramicidin S to mutant spores delays their outgrowth. Conversely, removal or inactivation of the antibiotic in the parental spores results in fast outgrowth. The same effect is observed when the parental strain is grown in media that poorly support antibiotic production.
3. The extent of outgrowth delay in mutant spores depends on the antibiotic concentration. The delay in the parent is dependent on the concentration of spores and hence the concentration of gramicidin S. A mixture of parental spores and mutant spores shows parental behavior, i.e., the mixed culture outgrowth is delayed. This indicates that some of the antibiotic externally bound to parental spores is released into the medium. This release could act as a method of communication by which a spore detects crowded conditions and prevents vegetative growth until there is a lower density of *Bacillus brevis* spores. However, proof of such a hypothesis will require experimentation at an ecological level. Alternative hypotheses might be that gramicidin S in and on the dormant and initiated spores protects them from consumption by amoebae, or that antibiotic excretion during germination initiation and outgrowth eliminates microbial competitors in the environment and that the delay in outgrowth is merely "the price the strain must pay" for such protection.

Most secondary metabolites appear at the onset and during the stationary and sporulating growth phase. Rather than being the cause, they are more likely a consequence of this dramatic shift in cell differentiation. A few, however, appear to have an essential role during sporulation initiation and during the germination of the spores.

Further Reading

Baltz, R. H., Hegeman, G. D., and Skatrud, P. L. (1993) Industrial microorganisms: basic and applied molecular genetics. Washington DC: ASM Press

Bennet, J. W., and Bentley, R. (1989) What is a name? Microbial secondary metabolism. Adv Appl Microbiol 34: 1–28

Bibb, M. (1996) The regulation of antibiotic production in *Streptomyces coelicolor* A3(2). Microbiology 142:1335–1344

Corcoran, J. W. (1981) Antibiotics IV—Biosynthesis. Berlin Heidelberg New York: Springer

Cundliffe, E. (1989) How antibiotic-producing organisms avoid suicide. Annu Rev Microbiol 43: 207–233

Demain, A. L. (1989) Function of secondary metabolites, In: Hershberger, C. L., Queener, S. W., and Hegeman, G. (eds.) Genetics and molecular biology of industrial microorganisms. Washington, DC: ASM Press; 1–11

Demain, A. L., and Wolfe, S. (1987) Biosynthesis of cephalosporins. Dev Ind Microbiol 27: 175

Fisher, H.-P., and Bellus, D. (1983) Phytotoxicants from microorganisms and related compounds. Pest Sci 14: 334–346

Hershberger, C. L., Queener, S. W., and Hegeman, G. (1989) Genetics and molecular biology of industrial microorganisms. Washington, DC: ASM Press

Horinouchi, S., and Beppu, T. (1990) Autoregulatory factors of secondary metabolism and morphogenesis in actinomycetes. Crit Rev Biotechnol 10: 191–204

Isono, K. (1988) Nucleoside antibiotics: structure, antibiotic activity and biosynthesis, J Antibiotics 41: 1711–1739

Katz, L., and Donadio, S. (1993) Polyketide synthesis: prospects for hybrid antibiotics, Annu Rev Microbiol 47: 875–912

Kendrick, B. (1986) Biology of toxigenic anamorphs. Pure Appl Chem 58: 211–218

Kleinkauf, H., and von Döhren, H. eds. (1997) Products of secondary metabolism. Biotechnology, vol. 7. Weinheim: VCH

Lancini, G. C., and Lorenzetti, R. (1993) Biotechnology of antibiotics and other bioactive microbial metabolites. New York: Plenum Press

Lancini, G. C., Parenti, F., and Gallo, G. G. (1995) Antibiotics–A multidisciplinary approach. New York: Plenum Press
Neilands, J. B. (1984) Siderophores of bacteria and fungi. Microbiol Sci 1: 9–14
Piggot, P. J., Moran, C. P., and Youngman, P., eds. (1994) Regulation of bacterial differentiation. Washington, DC: ASM Press
Rosenberg, E., and Varon, M. (1984) Antibiotics and lytic enzymes. In: Rosenberg, E. (ed.) Myxobacteria: development and cell interactions. Berlin Heidelberg New York: Springer; 110–125
Singh, B. N. (1945) The selection of bacterial food by soil amoebae and the toxic effects of bacterial pigments and other products on soil protozoa. Br J Exp Pathol 26: 316–325
Strobel, G. A. (1977) Bacterial phytotoxins. Annu Rev Microbiol 31: 205–224
Strohl, W. R. (1997) Biotechnology of antbiotics. New York: Dekker
Swain, T. (1977) Secondary compounds as protective agents. Annu Rev Plant Pathol 28: 479–501
Vining, L. C., and Stuttard, C. (1995) Genetics and biochemistry of antibiotic production. Boston: Butterworth-Heinemann

Sources of Figures

1 Zhang, J., Wolfe, S., and Demain, A. L. (1989) Can J Microbiol 35: 399–402
2 Horinouchi, S., and Beppu, T. (1990) Crit Rev Biotechnol 10: 191–204

28 Adaptation to Extreme Environments

Adaptation to extreme conditions can be viewed as taking place over two quite different time scales. The majority of bacteria have the capacity to withstand a change in an environmental parameter in the direction of one extreme or another and adapt over the time scale of minutes, hours, or days. These organisms rarely express their full growth potential at these extremes, and their adaptive strategies are directed toward survival rather than growth. However, in truly extreme environments, one encounters a group of organisms that have adapted over evolutionary time to survive and to display their greatest growth potential under these conditions. Such organisms are usually referred to as **obligate extremophiles**. This group of organisms have evolved a lifestyle that counters the principal stresses encountered in their preferred growth regime. This chapter deals primarily with short-term adaptation to stress, but places this in the context of adaptation over the evolutionary time scale.

Box 28.1 Stress is exaggerated normality. Growing cells encounter a variety of stresses that influence their biochemical activity. Generally, any change in the environment that provokes a significant change in cell physiology is considered as a "stress state". One of the greatest difficulties is to define the "unstressed state". Rather, all growth conditions should be considered to pose some stress for the organism and, therefore, it should be recognized that adaptation is not an all-or-none phenomenon. Instead, the cell exists in a series of adapted states. As the stress imposed on an organism increases, the changes in the activity of the cells will be qualitative as new activities are expressed, and also quantitative as the balance of existing systems is altered. Many stress responses involve amplification of the activity of enzyme systems that operate under "normal" conditions. What is "normal" varies from one organism to another.

28.1 Adaptation to Stress Requires Physiological Changes and Changes in Gene Expression

28.1.1 Physiological Changes Are Provoked by Stress

The initial response to any imposed stress is an attempt by the cell to correct the perturbation in order to increase the probability of survival. In nearly all cases, this initial response involves existing biochemical activities that may be either non-inducible (i.e., constitutive) or inducible, but expressed at a low level even when the cell is apparently not subjected to a recognizable stress. Changes in gene expression may then take place to establish either new systems or increased levels of the same systems, which enhance survival and which may restore rapid growth. When the cell is unable to cope with the stress, signals may be generated that lead to a global change in the pattern of cell metabolism. Consequently, it is not surprising that the changes in gene expression elicited by one stress are also effected when another apparently unrelated stress is imposed on the cells.

Stress responses show considerable overlap in the activities that are induced. This reflects two facets of stress adaptation; first, there are a limited number of basic systems for eliciting gene expression and therefore, cross-induction of systems that respond to the same signals will occur. Secondly, severe stress of any form will affect the growth rate, and this activates a number of global responses that are the components of the stress **stimulon**.

This integrated response is not unique to extreme stresses, but is a simple extension of conventional biochemical pathways that provide the monomers for growth. When cells grow in a complex medium, they are supplied with the end products of their biosynthetic pathways, and the level of the biosynthetic enzymes is reduced more than tenfold by repression and attenuation of the transcription of their structural genes (see

Chapter 18.4). Carbon flux through the residual enzymes of the pathway is prevented, predominantly by allosteric modulation of the activity (**feedback inhibition**) of key enzymes (see Chapter 19.2). When the cell is subjected to a sudden reduction in the supply of any nutrient, there is an immediate release of feedback inhibition on the existing biosynthetic enzymes, and consequently, the pool of the nutrient may be restored to adequate levels by this enzymatic activity. If demand for this nutrient is very great, the existing biosynthetic activity may not be sufficient to restore the pool, and then further enzyme synthesis may occur. When this is insufficient to meet the needs of the cell, a global response, the **stringent response** (see Chapter 20.3), is initiated, which elicits changes in the expression and activity of many enzymes such that the cell is placed into a survival mode.

The optimum pattern of events for any homeostatic system is shown in Figure 28.**1**. A homeostatic system is any group of enzymes or gene products that, when activated by perturbation of the cell, attempt to restore conditions in the cell such that the functions of the cell are operating either close to the optimum for growth

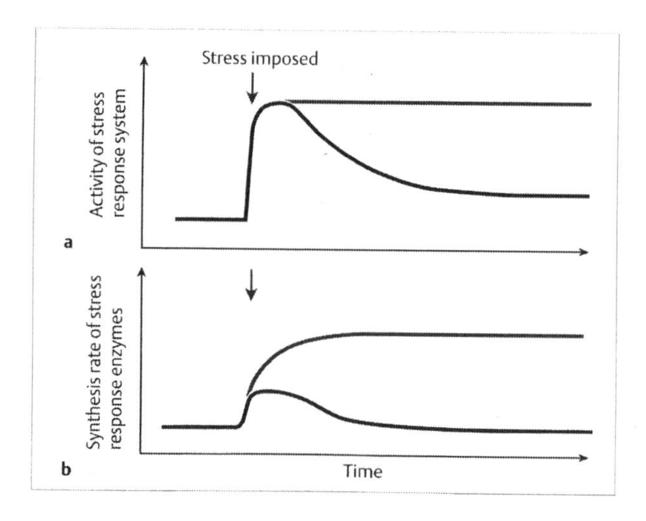

Fig. 28.**1a, b** **Typical pattern for stress response systems.** Prior to the change in the degree of stress, the system has a low activity, usually greater than zero. An increase in stress (red arrow) elicits a rapid change in the activity of the stress response system (**a**), with resultant changes in the rates of synthesis of stress response enzymes (**b**). If the increased activity of the stress response enzymes is sufficient to restore homeostasis, the change in activity is transient (black lines); otherwise, the activity of the stress response system will remain high with high steady-state rates of synthesis of appropriate enzymes (and transport systems) (red lines)

or at rates consistent with survival. Initially, the cell possesses relatively low activity of the "stress response system," either because of repression of enzyme systems or because the existing enzymes are in a quiescent state. Imposition of stress leads to immediate activation of the response system and possibly enhances the transcription of genes whose products further enhance the capacity to adapt to the stress condition. If the activated enzyme systems are able to re-establish the preferred cytoplasmic conditions (i.e., those conducive to rapid growth) the activity of the "stress response system" diminishes in parallel with the stress experienced by the cell. The new steady-state level of activity of the response system will usually be quite different than that of the initial condition, and will also be lower than the peak activity. Under such conditions, novel gene expression may only be transient since the signal for induction is short-lived (Fig. 28.**1**). However, when homeostasis cannot be re-established, novel gene expression may continue and further genes may be expressed. Generally, as a stress persists, the genes expressed are for pathways that are increasingly unattractive solutions for the cell. In stress stimulons, a hierarchy is generally found in which genes encoding the most beneficial solution are expressed when the stress is mild, but the genetic systems for more drastic remedies are activated as the severity of the stress increases.

Hierarchy in stress response. When many different genes or operons are regulated by the same stress signal, the operator sequence for each gene usually displays a different affinity for the regulatory protein. The position of each gene within the hierarchy is determined by the affinity of the operator sequence for the regulatory protein. As the concentration of the stress signal increases, the pattern of gene expression will change as the different threshold level required for the activation of each system is reached. Since stress signals will increase as long as the stress persists, the pattern of gene expression becomes time-dependent. Gene systems that are activated early are usually those that most readily correct the stress. Those expressed later possess a low affinity for the regulatory protein and may represent desperate measures to maintain viability. Such a hierarchy in the response to stress enables the cell to make subtle changes to cope with mild stress, while retaining other measures that potentiate survival under extreme stress. Although hierarchical systems have primarily been seen at the level of gene expression, there is no obvious reason why enzymes involved in stress responses should not be similarly controlled through differential sensitivity to allosteric effectors.

Box 28.2 There is a hierarchy among transport mechanisms for dealing with high turgor. Excessive turgor can be dealt with by cells by efflux of specific solutes and by non-specific release of solutes through stretch-activated channels. When the change in turgor is relatively slow (taking minutes rather than seconds), the solute-specific efflux systems are the primary mechanism for reducing turgor. In contrast, these mechanisms are inadequate when the turgor rise is very sudden and the stretch-activated channels, which by their lack of specificity are the less desirable solution, are employed. The hierarchy in adjusting to excessive turgor is first the inhibition of uptake systems, then activation of solute-specific efflux systems, and if these are inadequate, the use of non-specific stretch-activated channels. Each of these systems must have a different turgor threshold controlling their activity.

There are many different specific mechanisms for sensing stress, but ultimately the only underlying principle is allosteric modulation of protein activity. It may be that the specific effector of a response to stress is actually a phosphorylated protein of, for example, a two-component system (see Chapter 20.4), but the kinase or phosphatase will be allosterically modulated. Not all allosterically modulated proteins will bind low-molecular-weight compounds; some may undergo reversible conformational changes in response to physical parameters, such as water activity, pH, and temperature. The important factor is that the protein has evolved to contain a "sensor" element that responds to stress by a conformational change that elicits a response from the cell.

Alarmones are small molecules that are synthesized in response to a change in environmental conditions. The best examples of such molecules are cAMP, ppGpp, AppppA, and related nucleotides, and, in the case of osmotic stress, the cytoplasmic accumulation of potassium glutamate (see Chapter 20.1). There is a requirement for transmission of a signal to the cytoplasm to elicit alarmone synthesis. Exceptions to this may be: (1) heat shock, where the stress is physical and consequently can act directly on ribosomes and other cytoplasmic proteins, and (2) agents that are freely permeable through the cytoplasmic membrane, such as superoxide ions and weak organic acids. Increases in the cytoplasmic pool of the alarmones can be very rapid after the onset of the stress and are not dependent upon new protein synthesis.

28.1.2 Ion Channels May Play Major Roles in Stress Responses in Bacteria

Over the last few years, it has been recognized that ion channels may play important roles in sensing stress and as components of the homeostatic systems (Tab. 28.**1**). **Ion channels** differ from other transport components in that there is no energy coupling; when the channels are in the open state, the flow of ions and other solutes is driven by their electrochemical gradient. In addition, channels differ from transport proteins in that for each channel opening, many ions (often more than 10^4) will traverse the membrane, whereas for a transport system,

Table 28.**1** **Ion channels in the cytoplasmic membrane of bacterial cells.** Channels are membrane proteins that facilitate the large-scale movement of ions or solutes ($>10^4$/opening). The transition from the open to the closed state can be controlled by chemical (glutathione-gated channels) or physical stimuli (e.g., high turgor for stretch-activated channels and alkaline cytoplasmic pH for sodium ion channels)

Channel name	Function	Relationship to stress
Aquaporin	Water transport	Osmotic stress
Stretch-activated channel	Non-specific solute efflux	Turgor reduction at low osmolarity
Inwardly-rectified channel[1]	Potassium uptake	Not known
Glutathione-gated	Potassium efflux	Protection against electrophiles
Shaker-like channels[2]	Potassium transport	Not known
Sodium channels[3]	Sodium ion influx	pH regulation in alkalophiles

[1] Inward-rectified-channels move K^+ ions into the cell, but are limited in their ability to facilitate K^+ efflux.
[2] Shaker-like channels are structurally similar to channels found in the fruit fly *Drosophila*.
[3] Sodium channels are proposed to mediate the inward movement of Na^+ ions into the alkaliphile, *Exiguobacterium aurantiacum*; their existence is hypothetical.

the number of ions or molecules transported per cycle of the transporter is limited to a small number (frequently one and rarely more than three). All channels are, in essence, allosterically modulated proteins that alter their state of activity either upon the binding of regulatory molecules or upon changes in physical parameters (e.g., pressure, membrane potential). Ion channels have three principal states: open, closed, and inactive. The meaning of open and closed is self-evident, but once channels have been open, they may enter an intermediate state, the inactive state, when they are no longer sensitive to the stimulatory molecules or to changes in physical factors that control their activity (see Chapter 5.6).

A number of channel activities have either been described or hypothesized (see below) (Table 28.**1**). Two major classes of channels that appear to be ubiquitous in bacteria are stretch-activated channels and aquaporins (Table 28.**1**). Both play a role in regulating the turgor pressure. **Aquaporins** facilitate the flow of water through the cytoplasmic membrane and may be in the open state at all times. **Stretch-activated channels** open when the turgor pressure across the membrane is very high. These channels form large pores in the membrane and lead to rapid loss of low-molecular-weight cytoplasmic molecules (e.g., amino acids, sugars, nucleotides), but not larger molecules (such as high-molecular-weight proteins). Activation of these channels also perturbs the cytoplasmic pH and other ion gradients. Regulation of these channels is, therefore, critical to their role in the adaptation to changes in osmotic pressure since unregulated activity would cause severe perturbation of the cell.

28.1.3 Changes in Gene Expression Are Important Aspects of Adaptation to Stress

Over the last decade, three major mechanisms by which bacteria control the expression of stress-related gene systems have been identified: alterations in DNA topology, activation of "two-component systems" (see Chapter 20.4), and alteration of the promoter specificity of RNA polymerase. Changes in DNA topology and in RNA polymerase specificity and the activation of other regulatory proteins usually act in concert to bring about adaptation to stress.

DNA topology affects gene expression. The circular bacterial chromosome is thought to be organized into approximately 50 supercoiled domains (see Chapter 14.2). When a protein seeks its recognition sequence in the DNA major groove, it scans for a particular pattern of groups attached to the bases that are capable of forming bonds with the amino acid side chains of the recognition helix of the protein. In negatively supercoiled DNA, the number of bases per turn of the helix has been altered. Consequently, the affinity of the DNA binding protein for its target sequence will be altered due to the change in strength of the bonds between the amino acids of the recognition helix and the side chains of the bases exposed in the major groove (see Chapter 18.15). This may be one of the mechanisms by which DNA supercoiling changes can affect the expression of stress-regulated genes. Transcription of genes can be enhanced or reduced by changes in negative supercoiling. Increased negative supercoiling can promote DNA strand separation and thus the initiation of transcription. Promoter strength, however, is determined by many different elements of the sequence surrounding the transcription start site, and consequently each promoter has a unique response to changes in DNA supercoiling. For example, the *proU* gene of *Salmonella typhimurium* is one of the best-characterized osmotically regulated genes that appears to be primarily controlled by changes in DNA topology. However, the topology-sensitive sequence (the downstream regulatory element, DRE) does not lie within the promoter, but rather lies at a sequence downstream of the promoter in the first structural gene.

Table 28.**2** **Response of genes to stress and to antibiotics that reduce negative supercoiling of the DNA in enteric bacteria.** Downward arrows: repression; upward arrows: induction of the expression of the indicated genes. Novobiocin inhibits DNA gyrase and thus prevents the introduction of negative supercoils into DNA. See text for the explanation of gene symbols

Condition:	Novobiocin	High osmolarity	Alkaline pH	Anaerobiosis
Negative supercoiling	decreased	increased	increased	increased
Gene: *proU*	↓	↑	↑	↑
tppB	↓	↑		↑
ompC	↓	↑		↑
gyrB	↑			
tonB	↑			↓

Alterations in DNA supercoiling is an important mechanism for effecting bacterial adaptation (see Chapter 18.8) since (1) environmental changes modulate the level of negative DNA supercoiling, and (2) promoters regulated by such changes in the environment respond to changes in DNA supercoiling (Tab. 28.**2**). In *Escherichia coli*, anaerobiosis, alkaline pH, growth temperature, growth phase, and high osmolarity increase the degree of negative supercoiling of DNA. Not surprisingly, many of the genes that are induced by one stress that provokes changes in DNA supercoiling are induced by other conditions that provoke similar changes in DNA topology. The *tppB* (tripeptide permease) locus is induced by anaerobiosis and by high osmolarity, as is the *ompC* locus (outer membrane porin C). Similarly, osmotic induction of the *proU* locus is enhanced by slightly anoxic conditions and by mildly alkaline pH.

Changes in the environment elicit changes in **DNA topology** that may alter the pattern of gene expression. Antibiotics that alter the activity of DNA gyrase can mimic the effects of some environmental changes on gene expression, and consequently, changes in DNA supercoiling are thought to be important in determining the transcription of some environmentally regulated genes. Closely allied with the changes in DNA supercoiling are alterations in DNA topology mediated by "histone-like" proteins that may be potentiators of environmental stress-regulated gene expression.

In recent years, a strong link has been observed between DNA topology and a number of DNA-binding proteins that affect the structure of the DNA (Tab. 14.**2**). Prominent among these proteins is the histone-like protein H-NS of the enteric bacteria, which binds to curved regions of DNA and acts as a transcription silencer. The inactivation of H-NS by mutation renders some stress-regulated gene systems partially constitutive, such as those that are induced by osmotic stress and those that are induced by a higher temperature (off at 28°C, on at 37°C). Mutations affecting the H-NS protein have quite pleiotropic effects. For example, such mutations increase the frequency of gene switching due to site-specific inversion events and affect gene expression (see Chapter 18.8).

Two-component regulator systems are important sensory transducers. The receptor-transmitter/response regulator ("two-component systems") superfamily of proteins are sensory transducers, some of which span the cytoplasmic membrane and are capable of responding to both external and internal signals (see Fig. 20.**17**). The balance of the kinase, phosphotransferase and phosphatase activites determines the concentration of the phosphorylated form of the response regulator protein. The sensor-transmitter is an allosterically-modulated protein that senses stress either directly through a physical stimulus or indirectly through the binding of a small molecule. The response regulator protein is frequently found to possess a low affinity for DNA when non-phosphorylated and the binding affinity is markedly enhanced by phosphorylation of the protein.

Box 28.3 Specific **sigma subunits** regulate separate modulons in different organisms. Although individual sigma subunits are allocated specific regulatory roles within any one organism, in a different bacterium they may be responsible for the transcription of other genes. For example, homologues of σ^{54}, which is involved in transcription of nitrogen-regulated genes in *E. coli*, are active in transcription of some anaerobically regulated genes in *E. coli*, of genes for the synthesis of cell surface structures in *Caulobacter*, *Pseudomonas*, and *Neisseria*, of genes for hydrogen metabolism in diverse genera, and of genes for dicarboxylic acid transport in rhizobia. All of the functions under the control of σ^{54} are specific responses to individual stresses, but they are not all associated with nitrogen starvation.

Alterations in RNA polymerase structure and activity are important determinants in controlling gene expression. The major changes in RNA polymerase activity in stressed cells are at the level of promoter

Table 28.3 **Examples of different sigma subunits for the RNA polymerase from *Escherichia coli***

Core enzyme	Sigma subunit	Gene	Holoenzyme	Genes transcribed
$\alpha_2\beta\beta' = E$	σ^{70}	*rpoD*	$E\sigma^{70}$	Normal metabolic genes
	σ^{32}	*rpoH*	$E\sigma^{32}$	Heat shock genes
	σ^{24}	*rpoE*	$E\sigma^{24}$	Extreme heat shock genes
	σ^{54}	*rpoN*	$E\sigma^{54}$	Nitrogen-regulated genes
	σ^{28}	*rpoF*	$E\sigma^{28}$	Flagella and chemotaxis genes
	σ^{38}	*rpoS*	$E\sigma^{38}$	Growth-phase-regulated genes

selection. RNA polymerase promoter selectivity can be modulated in two different ways. Binding of an allosteric effector to the holoenzyme via one of the core subunits can alter the promoter preference of the polymerase. For example, ppGpp shifts the specificity of RNA polymerase away from transcription of rRNA and tRNA genes in favor of mRNA (see Chapter 20.3). A second widespread mechanism alters promoter specificity of the RNA polymerase by displacement of the normal sigma subunit, σ^{70}, and its replacement by a specific sigma subunit that initiates transcription at a subset of genes with unusual promoter structure (Table 28.**3**).

The synthesis and the activity of alternative sigma subunits is regulated by specific cellular signals that may be generated in response to different stresses. The control of the concentration of the sigma subunit is a major mechanism of determining how much of the alternative form of the holoenzyme of RNA polymerase is formed. The specific accumulation of an alternative sigma subunit ensures that a subset of genes, required to enable the organism to survive a particular stress, is transcribed at the appropriate time.

In organisms that undergo programmed differentiation (e.g., bacilli, clostridia, and actinomycetes), sequential expression of different sigma subunits may be the major mechanism controlling the pattern of development (see Chapters 24 and 25). For other organisms, such as *E. coli*, in which the differentiation pattern associated with starvation is less extreme than in spore-forming organisms, the expression of survival genes is similarly dependent upon a specific sigma subunit, σ^{38}. It is worth noting that motility and chemotaxis genes of the enteric bacteria are also regulated by a specific sigma subunit; this may indicate that these genes are specifically required by the cell to depart from an environment that has changed from beneficial to hostile. In this context, chemotaxis would constitute a stress response, and indeed, the enteric bacteria have been shown to swim away from low pH and high osmolarity environments. Furthermore, the pattern of swimming seen in enteric bacteria in response to chemical stimuli fits the description of a homeostatic response, illustrated in Figure 28.**1**.

Alternative sigma subunits are a widespread mechanism for regulating a large number of genes that are related by their ability to enable cells to adapt to changed circumstances. The major sigma subunit, σ^{70} or its equivalent, controls the expression of proteins that are primarily associated with cell growth. In most organisms, a diverse array of alternative sigma subunits that regulate specific subsets of genes control expression of genes required for adaptation or survival. The accumulation of these sigma subunits usually takes place in response to specific signals and may involve de novo expression of the gene or alterations in the stability of the sigma subunit toward proteases.

28.2 Adaptation to Extreme Temperature

Large-scale shifts in temperature (20°C or more) provoke transient periods of either growth stasis or slow growth, after which normal exponential growth typical of that growth temperature is achieved. Severe temperature downshifts can cause immediate loss of viability, possibly due to disruption of membrane function, leading to loss of solutes. Membrane permeability may also be compromised after sudden shifts to high temperature. Organisms that have evolved to live at high temperatures have evolved proteins and membrane systems that enable them to function under these conditions. For some **thermophilic organisms**, it has been suggested that at very high temperatures (>80°C), the proton circuits can no longer be maintained due to high proton permeability of the membrane. These organisms switch to energy coupling via Na^+ since the permeability of the membrane to this ion is less affected by high temperature (see Chapter 5.6.2). In addition to their effects on membrane structure, high temperatures also lead to protein denaturation, which directly affects metabolic activity.

28.2.1 Physiological Adaptations to Changes in Temperature

In general, there are two important responses of biological systems to an increase in temperature. First, the **fluidity of the lipid bilayer** increases as the temperature rises and may ultimately become leaky to ions, leading to a reduced efficiency of ionic homeostasis and energy transduction. In most bacteria, the lipid composition is changed to maintain a relatively constant viscosity at different temperatures. At low temperature, short-chain fatty acids and unsaturated fatty acids are used to maintain membrane fluidity, with more saturated long-chain fatty acids being inserted as the temperature is raised. Branched-chain fatty acids are sometimes substituted for unsaturated fatty acids as both have the property of preventing close packing of the lipid chains, which would reduce membrane fluidity. The change in lipid composition can usually be achieved without new enzyme synthesis since the activity of the phospholipid biosynthetic enzymes is regulated by temperature. Thermophilic Archaea have solved the problem of membrane stability by the evolution of ether lipids, which are much more stable at high temperature (see Chapter 7.10.5).

Secondly, the **activity of enzyme systems** increases as the temperature is raised, leading to a twofold change in activity for every 10°C. The upper limit for this effect is set by the intrinsic stability of each enzyme, and eventually the thermal motion induced by the elevated temperature leads to protein denaturation and loss of enzyme activity. Many thermophilic bacteria are believed to have evolved enzymes that possess intrinsic heat stability. Relatively few extra bonds are required to generate a thermostable enzyme, and no general pattern has emerged from the study of stable enzymes. Each solution draws upon the different combinations of bonding patterns. Obviously, protein stabilization is achieved by many different routes.

28.2.2 Patterns of Gene Expression Change in Response to Temperature Shifts

Adaptation to temperature change involves the synthesis of many new proteins, with specific subsets of genes associated with cold shock and with heat shock. Transfer of *E. coli* cells from 37°C to 10°C leads to the induction of a number of gene products, some of which may accumulate to be a high percentage of total cell protein. Similar observations have been made with other bacteria of both Gram-positive and Gram-negative genera. Many of the **cold shock genes** can also be induced by the addition of antibiotics that inhibit translation, and the ribosome itself may act as the sensor of cold shock. Both cold shock and inhibition of translation by the antibiotics lower the cytoplasmic pool of ppGpp in *E. coli* cells, and the regulation of transcription of the cold shock genes may be in part controlled by the promoter specificity of RNA polymerase–ppGpp complexes (see Chapter 20.3).

As the environment is progressively cooled, ice begins to form, and it is this that kills bacteria. Ice crystals can damage proteins and membranes, leading to a loss of integrity of the cell. As ice begins to form, the salts surrounding the cell are concentrated, leading to a local increase in osmotic pressure that initiates adaptation to osmotic stress. Consequently, at very low temperatures, the accumulation of compatible solutes (see below) takes place in some organisms, and this may protect some proteins against the damage that ice can inflict. Some bacteria have evolved the capacity to regulate ice formation via ice-nucleation proteins in the outer membrane (e.g., the ice nucleation protein of *Pseudomonas syringae*, see Chapter 34.4.3). The advantage to the cell may be the regularity of the ice structures formed after nucleation by these proteins, which may be intrinsically less damaging to cell structure.

The **heat shock response** refers to transient shifts of 10- to 50-fold in the expression of individual proteins that occur when the organism is shifted to temperatures at the upper extremity of the growth range. In *E. coli* and related organisms, all of these genes are transcribed by RNA polymerase only when it carries a specific sigma subunit, σ^{32} (*rpoH* gene product), in place of the σ^{70} that is utilized for vegetative gene expression. A similar response is seen in other bacteria and in most eukaryotes, and it is now clear that some of the proteins made in the two sets of organisms are analogues that are likely to serve the same function. A second subset of genes induced by very high temperatures has been observed in *E. coli* and is under the control of a separate sigma factor, σ^{24} (*rpoE* gene product). In *E. coli*, these functions include protein degradation, protein assembly and refolding (the chaperone concept), and DNA synthesis (see Chapter 19.4). In addition, there is considerable overlap with other stress responses. For example, the synthesis of catalase, an oxidative stress protein, potentiates survival at high temperatures, possibly by removing peroxides that form more readily at high temperature.

Many of the heat shock proteins are also induced by the addition of high concentrations of ethanol (4–10%) to cells and by exposure of the cells to UV or to nalidixic acid, an inhibitor of DNA gyrase. In each case, overlapping groups of proteins are made in what appears to be a general response ("stimulon") of the cell to stresses that directly affect either protein structure or DNA integrity. Other physiological stresses (pH, osmotic pressure, anaerobiosis) do not appear to induce the heat shock proteins.

The heat shock proteins form part of the normal complement of cell functions. Increased concentrations of these proteins are required for survival under more extreme conditions. Thus, the σ^{32} protein (*rpoH* gene product) is essential for growth across the temperature range of 16–44°C in *E. coli*. This suggests that the heat shock genes, although first detected by their enhanced rates of synthesis after temperature shifts, are an essential part of the growing cell. Induction of the heat shock proteins during pre-adaptation at moderate temperatures increases survival of the organism at higher temperatures. A parallel finding is that low levels of oxidation stress (e.g., by low concentrations of hydrogen peroxide; see below) lead to the induction of catalase and thus poise the cell for surviving exposure to higher concentrations of peroxide. The common link between these stresses may be the synthesis of the **alarmone ApppppA** and related nucleotides, as **stress indicators**. ApppppA and related molecules are made by cells when exposed to UV, ethanol, heat, and a wide variety of oxidants.

28.2.3 Heat Shock Proteins Play Major Roles in Cell Physiology

In most organisms, two major classes of heat shock proteins have been identified: the Hsp70 family and the Hsp60 family. In *E. coli*, the DnaK protein, the product of the *dnaK* gene, is the major member of the **Hsp70 family**, and GroEL and GroES are the representatives of the **Hsp60 family**. The DnaK protein functions with the GrpE and DnaJ proteins to prevent the aggregation of newly synthesized proteins and of unfolded proteins that arise as a consequence of cell stress. The GroEL protein is believed to be a true **chaperone** since it has the capacity to organize folding of newly synthesized and unfolded proteins, whereas the primary function of DnaK appears to be in stabilizing unfolded proteins. Thus, DnaK and GroEL and GroES act in sequence (Fig. 28.**2**).

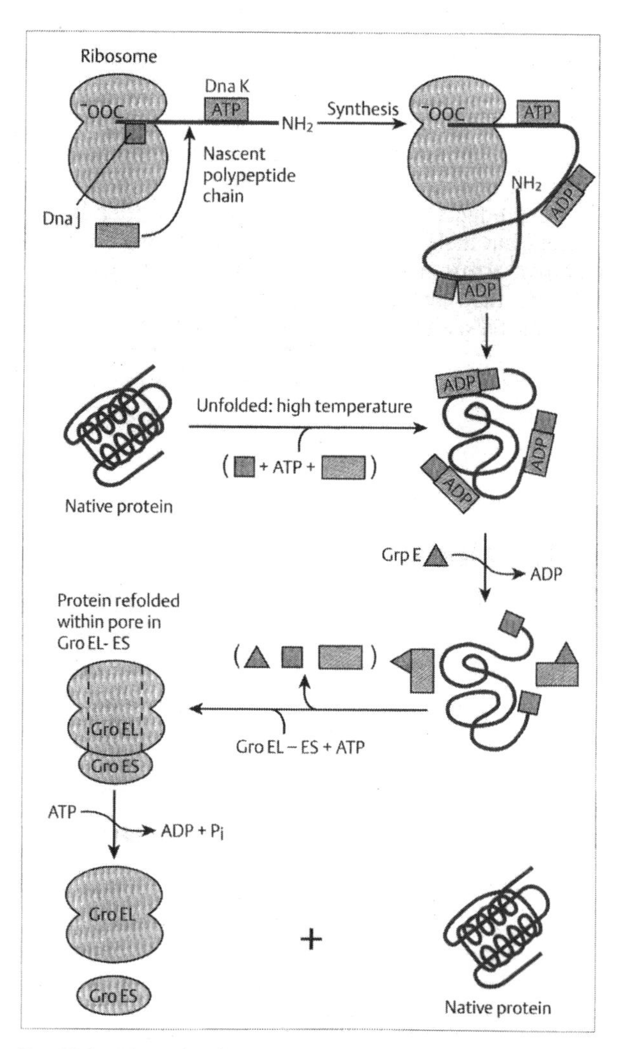

Fig. 28.2 **The role of DnaK, DnaJ, GrpE, GroEL, and GroES in folding of newly synthesized and denatured proteins** [after 1]

Protein folding is a co-operative phenomenon. A protein may consist of a single folded entity or of separately folded domains. Formation of the correct structure is dependent on the sequence of the entire protein, or of that domain. As the protein is being synthesized on the ribosome, the nascent polypeptide must be protected against forming stable, but incorrect, associations with other elements of the polypeptide chain. The greatest probability of formation of incorrect associations lies with the hydrophobic regions, which are normally buried within the protein. It is believed that the Hsp70 family of proteins recognizes hydrophobic sequences 7–8 amino acids in length in the newly synthesized proteins. Hsp70 is also believed to have a role in organizing proteins for export across the cytoplasmic membrane. When the polypeptide is being transported across a membrane, it is usually in an extended conformation, thus exposing short stretches of hydrophobic amino acids that must be bound by Hsp70 proteins to prevent aggregation. Members of the Hsp70 family have also been implicated in the assembly of multi-protein complexes (see Chapters 24 and 25). Thus, the Hsp70 family plays a major role in cell physiology in addition to their more prominent role under conditions where proteins may unfold as a consequence of imposed stress.

Characteristically, the Hsp70 proteins can be co-purified with polysomes, where they can interact with protein chains as small as 55 residues. In contrast, the Hsp60 proteins are free in the cytoplasm and do not associate with ribosomes. The DnaJ and GrpE proteins act with DnaK to effect changes in protein structure. The DnaJ protein is a 41-kDa chaperone, and in free solution, it is the first protein to interact with exposed sequences of an unfolded or denatured protein. This complex is the target for binding by the carboxy-terminal domain of DnaK. Analysis of the amino acid sequences of the Hsp70 homologues has shown that the amino-terminal region is highly conserved and is involved in ATP-binding and hydrolysis. DnaK, DnaJ, and the unfolded protein form a ternary complex in the presence of ATP.

ATP hydrolysis leads to an ADP–DnaK complex with a high affinity for the unfolded protein. Release of the unfolded protein requires GrpE, which induces dissociation of ADP from the complex and facilitates binding of ATP. The ATP–DnaK complex has a low affinity for the unfolded protein, which is then released. Several alternative fates await the released protein: correct folding with the aid of the GroEL and GroES complex (Fig. 28.**2**), incorrect folding and aggregation, or recycling through the DnaK–DnaJ complex. Fourteen GroEL subunits form a complex with a central "hole". A protein to be folded becomes lodged into this "hole" where it is converted from the "molten globule" state and a form which lacks the correct tertiary structure to the correctly folded and mature protein, with the help of the GroES protein.

28.2.4 Heat Shock Proteins Regulate Their own Expression

Following a temperature upshift, the cellular concentration of σ^{32} increases, leading to enhanced rates of transcription of the **heat shock genes**. It is postulated that two mechanisms are used to regulate the level of σ^{32}: (1) the protein is very unstable, with a half-life of approximately 1 min, and (2) the translation of the mRNA for σ^{32} is believed to be regulated. On temperature downshift, σ^{32} appears to be inactivated very rapidly, leading to reduced transcription of the heat shock genes. In vivo, DnaK and other Hsps may act to regulate the accumulation of σ^{32}. Mutants that lack a functional *dnaK* gene express heat shock proteins at high levels even at low temperature, and as a consequence, such mutants are very thermotolerant. Given the instability of σ^{32}, it seems likely that upon heat shock the existing pools of Hsps may alter the balance between incorrectly folded, inactive σ^{32} and the active form of the protein, leading to further expression of the genes encoding the Hsps.

The **heat shock response** typifies a stress response in which the induced genes represent activities that are essential for "normal" growth, but which potentiate survival of environmental extreme when expressed at high levels. Upon heat shock, the existing pools of Hsps may alter the balance between incorrectly folded inactive σ^{32} and the active form of the protein, leading to further expression of the genes encoding the Hsps. Overexpression of these gene products potentiates survival of further heat stress, but their expression is required during normal growth across the temperature range of 16–44 °C to ensure that individual proteins and multi-protein complexes assemble correctly. Thus, chaperones are not simply components of the heat shock response, but reflect the continual need of the cell to possess mechanisms of ensuring a high frequency of correct folding of newly synthesized polypeptides.

28.3 Bacteria Can Adapt to Extreme pH Values

Bacteria are generally classified into three groups with regard to their pH range for growth: **acidophiles** (optimum pH < 4), **neutrophiles** (optimum pH 6–7), and **alkaliphiles** (optimum pH 8–9). Such specialization is an example of a specific way in which organisms have adapted over evolutionary time to the stresses posed by the extremes of environmental pH. The major problems faced by the organisms that grow at the two extremes of the pH range are regulation of the **cytoplasmic pH (pH_i)** at a value close to neutrality and maintenance of the activity of excreted and surface-located proteins. Extremes of external pH can cause irreversible damage to surface-located proteins, such as those involved in cell wall synthesis and nutrient acquisition. This problem has been solved over evolutionary time by the development of proteins that display an intrinsic stability at the extremes of pH. The alkaliphiles, in particular, have been a major source of stable enzymes for use in industrial processes. Regulation of cytoplasmic pH must be achieved over relatively short time spans (minutes rather than hours) to prevent inhibition of cell growth or even cell death. Alkaliphilic bacteria have evolved potent mechanisms of pH homeostasis that allow them to grow at very alkaline external pH, while the cytoplasmic pH remains close to pH 8.

28.3.1 Bacteria Regulate Their Cytoplasmic pH

The ability of an organism to maintain its cytoplasmic pH (pH_i) to within a narrow range despite fluctuations in the external pH is called **pH homeostasis**. For example, in *E. coli*, pH_i changes by less than 0.1 unit per pH unit change in the external pH in the range of external pH 4.5–7.9. When pH_i is perturbed either by potassium limitation or by addition of mineral acids, it returns to the normal range in a potassium-dependent manner (Fig. 28.**3**). Bacteria can sense quite small

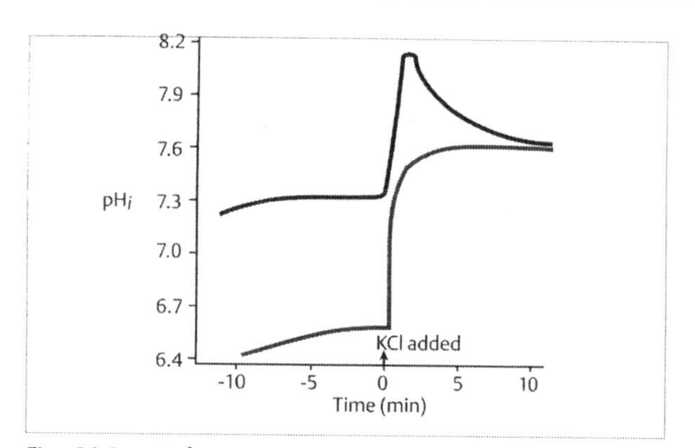

Fig. 28.3 **pH homeostasis in *Escherichia coli*.** Potassium-depleted cells were incubated at either pH 5.3 (red line) or pH 7.1 (black line), and the internal pH was determined. At time zero (arrow), 1 mM KCl was added. Both sets of cells accumulated the same level of potassium [after 2]

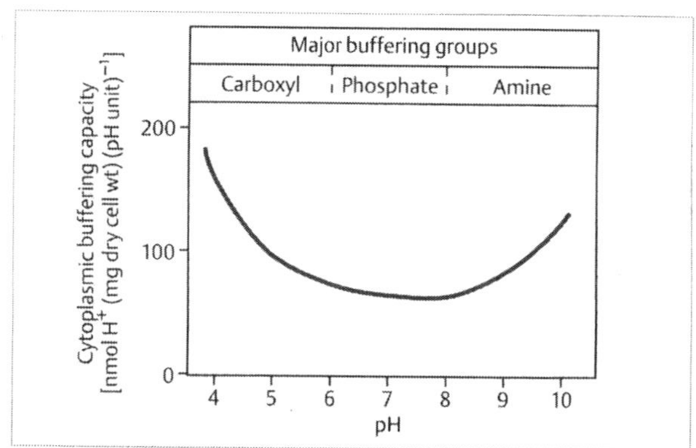

Fig. 28.4 **Cytoplasmic buffering capacity as a function pH.** Buffering capacity rises at the extremes of pH. In the range pH 6–8, where most bacteria hold their pH$_i$, the major buffering capacity is provided by the phosphate backbone of RNA and DNA. However, at extremes of pH, the carboxyl and amine groups of acidic and basic amino acids are more important. Buffering capacity at acid pH may be greater than at alkaline pH because of the contribution of glutamate accumulated by cells as a counterion for potassium

changes in pH$_i$, and alterations in cell activity can result. For example, *E. coli* tumbles more frequently when exposed to either acid pH or weak acids, and thus movement away from an acidic environment is potentiated. Similarly small changes in pH$_i$ can elicit altered patterns of gene expression that potentiate adaptation. Homeostasis is achieved in bacterial cells by a mixture of passive and active mechanisms.

Passive homeostasis. The very low permeability of the membrane to protons and other ions is a major factor preventing large changes in pH$_i$ as the pH of the environment is varied. However, the cytoplasmic membrane is the major site of energy transduction in bacteria and there is very active cycling of protons across the membrane. This circuit contributes little to the problem of pH homeostasis since the protons enter and leave the cell at the same rate. The primary proton pumps do, however, provide the major mechanism by which protons can be extruded from the cell when the pH is required to be at a more alkaline value (see Chapter 5.6.3).

The other factor preventing substantial perturbation of pH$_i$ is the high buffering capacity of the cell, which is derived from the nucleic acid and protein content of the cytoplasm and from cytoplasmic pools of glutamate and polyamines (Fig. 28.**4**). The major sources of buffering capacity are the phosphate groups of RNA and DNA (in the neutral range) and the basic and acidic amino acid side chains of proteins at the extremes of pH; these side chains are most frequently located at the surfaces of proteins. Glutamate and polyamines have pK values at the extremes of the pH range, and only glutamate is accumulated to a level sufficiently high enough to provide significant protection against acid pH$_i$ shifts. Glutamate can be accumulated to 400–500 mM in Gram-negative bacteria and, even though its upper pK value is relatively acidic (pK 4.07), the high concentrations in the cytoplasm could provide significant buffering capacity at low pH$_i$ values. Maximum buffering capacity is found at the extremes of the pH range, where it is approximately twice that found between pH$_i$ 6 and pH$_i$ 8 (Fig. 28.**4**).

Active pH homeostasis is obtained by ion circuits. pH homeostasis can be most simply understood as the interplay between three transmembrane ion circuits; the **proton circuit**, the **potassium ion circuit**, and the **sodium ion circuit** (Fig. 28.5). These three circuits do not play equal roles in all organisms. The Na^+ circuit is very important in the alkaliphiles, but plays a more limited role in acidophilic and neutrophilic bacteria. The potassium ion and proton circuits are important for all bacterial species. The translocation of protons across the membrane generates a membrane potential ($\Delta\psi$) that limits further proton extrusion. Continued proton extrusion can only take place if protons re-enter the cell through one of the energy-transducing protein complexes (e.g., ATP synthase, transport proteins). Very few protons are required to move across the membrane to generate a membrane potential, and this scale of proton movement has no significant effect on cytoplasmic pH. Bulk proton movement can only take place if the membrane potential is transiently dissipated through the movement of cations into the cell. Potassium ion entry fulfills this purpose and leads to the generation of a transmembrane pH gradient (Fig. 28.**3**). Mutants of *E. coli* and *Enterococcus faecalis* that lack potassium uptake systems display reduced pH$_i$ and poor

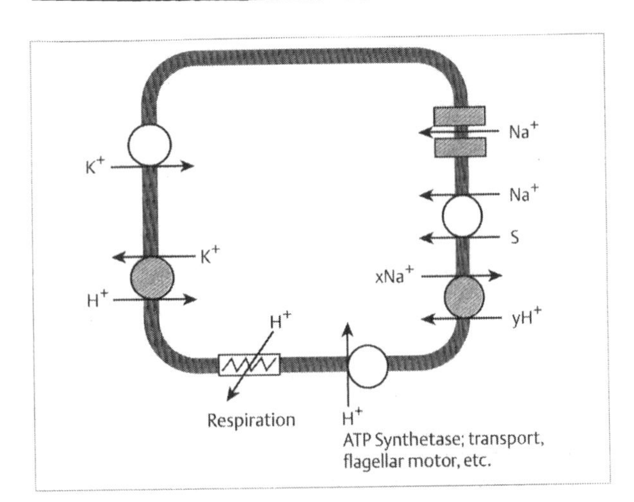

Fig. 28.5 **Ion circuits involved in pH homeostasis in cells.** Three circuits are of major importance: the proton circuit, consisting of the respiratory proton pump (export) and the various entry routes (e.g., ATP synthetase, transport systems, the flagellar motor); the K^+ circuit, consisting of uptake and efflux systems; and the Na^+ circuit, consisting of uptake systems, a putative Na^+ channel, and the antiport. Note that for the Na^+/H^+ antiport, the number of Na^+ ions expelled (x), is less than the number of H^+ imported (y) to make this system electrogenic. Systems that are believed to be regulated by pH_i are shown in red

growth at pH 6, which confirms the importance of these transport systems for alkalinization of the cytoplasm. The steady-state pH_i in *E. coli* is set by the activity of the potassium transport circuit. There is no specific potassium uptake system that is required for pH regulation, merely a requirement for potassium entry to facilitate net proton extrusion by the major proton pumps.

Alkaliphiles face the problem that they must maintain the cytoplasmic pH at a more acidic value than that of their growth environment. To achieve this, alkaliphiles have evolved a **sodium ion circuit** consisting of at least three components (Fig. 28.**5**). The Na^+/H^+ antiport system is the central component of the Na^+ circuit since it facilitates proton entry. The Na^+/H^+ antiport activity is regulated by pH and has its maximum activity at pH values more alkaline than the preferred cytoplasmic pH of the organism. Therefore, it is optimally regulated to effect pH control when the cell experiences a pH value that is too alkaline. Two other components are required to complete the Na^+ circuit. First, Na^+-solute symporters may provide a mechanism for the entry of Na^+, which is an essential substrate for the antiport and must be present at the cytoplasmic face of the membrane for antiport activity to effect acidification of the cytoplasm. However, these systems are primarily involved in the acquisition of nutrients and must, perforce, be active even when Na^+ ions are not required for pH homeostasis. A pH-regulated **Na^+ channel**, which has been proposed as a mechanism by which Na^+ entry could be regulated by the needs of pH homeostasis, would solve this problem. Such a channel would be open when the cytoplasm is too alkaline and would allow Na^+ to enter. The action of the antiport would then expel the Na^+ in exchange for protons and lower the cytoplasmic pH to a more acidic value.

Box 28.4 Multiple ion circuits are required for **pH homeostasis** in alkaliphiles. The importance of the Na^+ circuit for the alkaliphiles is readily exemplified by Fig. 28.**6**. In the absence of sodium ions, *Exiguobacterium aurantiacum* fails to regulate pH_i when incubated at alkaline pH. Addition of sodium ions brings about a rapid restoration of pH_i to a value similar to that seen in growing cells. At high external concentrations, the Na^+ influx exceeds the requirement for pH homeostasis, and pH_i overshoots to more acidic values and is then raised to a more alkaline value by other ion circuits (Fig. 28.**6**). The most probable mechanism for this is the potassium circuit. The alkaliphiles use the maximum interplay of the three ion circuits to regulate cytoplasmic pH.

The **regulation of cytoplasmic pH** in bacterial cells is facilitated by both passive and active mechanisms. The cytoplasmic membrane acts as a barrier to proton entry, and the buffering capacity of the cytoplasm can limit the effects of any protons that do enter the cell. Ion transport cycles are the major active components of pH homeostasis. The proton pumps and the potassium transport systems are the major mechanisms for active pH regulation in bacteria living in the acidic and neutral pH range and act in concert to raise the cytoplasmic pH. In alkaliphiles, Na^+ circuits are of major importance in establishing a pH_i value below the external pH.

28.3.2 Each Organism Has a Unique Niche

Acidophilic bacteria (acidophiles) have evolved to grow with an optimum pH around pH 2–3 (Table 28.**4**). The low permeability of the cytoplasmic membrane to protons and to other cations means that, despite the extremely acidic environment, these organisms do not have major problems in maintaining their cytoplasmic pH near pH 6.5–7.0. Generally the surface-located

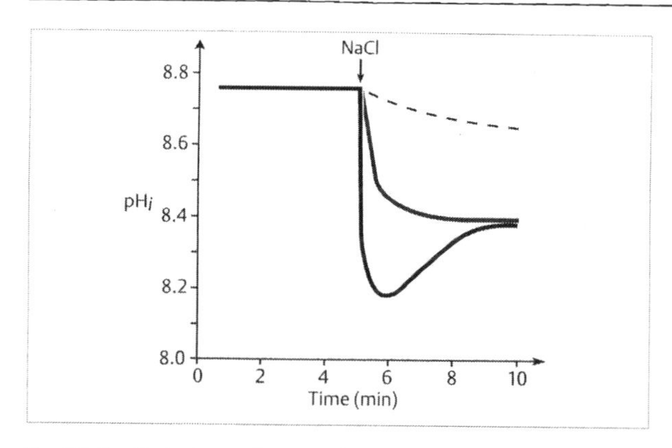

Fig. 28.6 **Na^+-dependent pH homeostasis in *Exiguobacterium aurantiacum*.** Cells were incubated in buffer at pH 9.65 in the presence of K^+, but not Na^+, and the pH_i was measured. After 5 min, NaCl was added. Dashed line, 0.1 mM Na^+; solid black line, 10 mM Na^+; red line, 0.5 mM Na^+ [after 3]

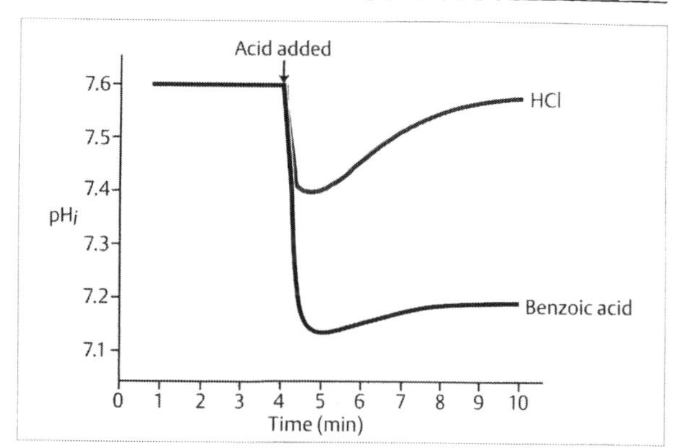

Fig. 28.7 **Changes in pH_i consequent upon the addition of acid to *Escherichia coli* cells.** Cells were incubated in K^+-containing buffer at pH 7.6 (red line) or at pH 6.0 (black line) and either HCl was added to lower the external pH to pH 6 (red line) or 2 mM benzoic acid was added (black line)

proteins of these organisms have evolved to be very resistant to low pH and are often unstable at neutral pH.

Neutrophilic bacteria have pH optima for growth in the neutral range and maintain pH_i near pH 7.6–8.0 (Tab. 28.**4**). Perturbation of the cytoplasmic pH by as little as 0.5 units can cause significant growth inhibition. Similarly, exposure to very low eternal pH can lead to cell death, and most bacteria have evolved adaptive regimes for acquiring tolerance of acid pH (see below). A major source of acid-related stress is the presence of lipid-permeant weak acids in the environment. Although the membrane is intrinsically impermeant to protons, weak acids pass through the membrane with relative ease and liberate protons in the cytoplasm (see Tab. 5.**1**). Consequently, in the presence of lipophilic weak acids (e.g., benzoate, butyrate, or acetate) growth is more severely perturbed than in the presence of mineral acids (e.g., HCl). A sudden change in the external pH of over one unit, induced by the addition of mineral acids, causes a small, but rapid fall in pH_i (Fig. 28.**7**). Recovery of pH_i to a more alkaline value takes place over several minutes and is dependent upon potassium uptake systems. The impermeability of the membrane to protons is the major factor preventing changes in pH_i. In contrast, addition of a lipophilic weak acid, such as benzoic acid (Fig. 28.**7**), results in a permanent change in pH_i due to the ability of benzoate ion to transfer protons across the membrane as the undissociated acid.

There is a synergy between the weak acid and the external pH that arises from the tendency of the bacterial cell to try to maintain the pH_i close to 7.6–8.0. A weak acid will move to its chemical equilibrium determined by its pK and the magnitude of the transmembrane pH gradient. At pH 5, the pH gradient will be 2 units or more, and the theoretical internal concentration of the anion, A^-, of the weak acid will be more than 100-fold greater than the external concentration. If the external concentration of the anion is 1 mM, the cell strives to achieve an internal concentration of 100 mM A^-, which is equivalent to entry of approximately 100 mM H^+ ions into the cytoplasm. This far outstrips the buffering capacity of the cell (Fig. 28.**4**), and as a consequence, the pH_i will decrease, leading to severe growth inhibition and possibly cell death.

Table 28.**4** **Examples of bacteria with different pH optima for growth**

Organism	External pH	Internal pH	Environment
Thiobacillus acidophilum	1–2	6.0–7.0	Coal refuse
Bacillus acidocaldarius	2–6	6.0–7.0	Hot sulfur springs
Escherichia coli	5–8	7.6–8.0	Human gut
Staphylococcus aureus	5–8	7.5–8.0	Skin, food
Bacillus pasteurii	9–11	8.0–8.6	Soil, sewage
Bacillus alcalophilus	8–11.5	8.0–8.6	Widespread in soil
Exiguobacterium aurantiacum	7.5–11	8.2–8.8	Alkaline potato waste

Not all changes in pH_i are detrimental. Many Gram-negative bacteria possess specific potassium export systems that systematically lower pH_i when the cells encounter toxic compounds, such as methylglyoxal. Methylglyoxal is a potent growth inhibitor that damages proteins, nucleic acids, and lipids. Methylglyoxal is produced as a by-product of glycolysis, and its synthesis in large quantities is particularly associated with conditions of carbon source excess when other nutrients are limiting. Exposure of cells to high concentrations of methylglyoxal leads to cell death. The activation of specific glutathione-gated potassium efflux systems in *E. coli* (KefB and KefC) leads to transient acidification of the cytoplasm, which enhances the survival of the cells.

Alkaliphilic bacteria (alkaliphiles) grow optimally between pH 8 and 11 (Tab. 28.**4**) and face three major problems. The first problem is pH_i maintenance, which has been discussed above. Secondly, the enzymes that are on the surface of the cell or that are secreted into the environment must be very resistant to the denaturing effects of extreme pH. There is good evidence that the alkaliphiles have evolved pH-stable enzymes; these are often used for the manufacture of detergents. Finally, the generation of a pH gradient, inside acid, counteracts the energy available from the membrane potential and consequently, the proton motive force, which drives ATP synthesis during respiratory metabolism, may be diminished. In acidophilic and neutrophilic bacteria, the pH gradient is alkaline inside and in thermodynamic terms adds to the driving force for proton entry, the proton motive force. There are two principal solutions to this problem: (1) the membrane potential sustained by these organisms is unusually high, and this compensates for the reversal of the pH gradient, leading to a proton motive force similar in size to that in other bacteria, and (2) some facultatively alkaliphilic bacteria have inducible respiratory chains that couple Na^+ expulsion to electron transport and, using Na^+-based energy transduction, can avoid the consequences of the reversed pH gradient.

Alkaliphilic bacteria can compensate for the reversal of the pH gradient by having a high membrane potential or by coupling Na^+ expulsion to electron transport for pH homeostasis and energy transduction.

28.3.3 Lowering of Cytoplasmic pH Affects Gene Expression

Two classes of pH-induced gene expression events can be defined: (1) expression of genes to correct metabolic deficiencies provoked by low pH, and (2) expression of genes that potentiate survival of the deleterious effects of low pH.

The earliest reports of acid-induced gene expression were the derepression of amino acid decarboxylases at low pH. The regulation of these enzymes is very complex: specific regulatory proteins, changes in DNA topology, and histone-like proteins have been proposed to play significant roles in regulation. The principal physiological advantages of induction of these enzymes is that the products of the enzyme activity are carbon dioxide and polyamines. Carbonic acid (pK 6.2) will be present at only low concentrations at low pH, but is essential for growth. Consequently, the amino acid decarboxylases may be increased to satisfy this metabolic requirement. Simultaneously, the secreted polyamine will cause the pH of the environment to rise towards neutrality. At alkaline pH, deaminases are expressed, possibly reflecting a need for free ammonia, and the weak acid by-products are secreted, lowering the external pH. The pattern of regulation of these two classes of enzymes explains an earlier observation that when broth cultures of *E. coli* cells are adjusted to alkaline or acidic pH, the pH value of the broth gradually adjusts to neutrality.

Survival of bacteria at low pH after pre-adaptation to mildly acidic pH. Bacteria that have been incubated at pH 4.3–6.0 for short periods show much greater survival at pH 3.0 than the same organisms grown at neutral or slightly alkaline pH (Fig. 28.**8**). This adaptation is dependent upon protein synthesis, but the functions of the newly synthesized proteins are un-

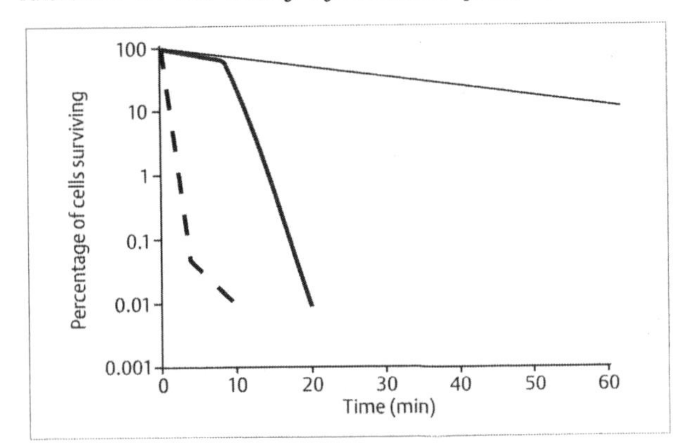

Fig. 28.**8** **Acid tolerance and stationary phase gene expression.** *Escherichia coli* wild-type cells were grown into the early exponential phase at either pH 7.0 (dashed black line) or pH 6.0 (solid black line) and then diluted into identical medium buffered at pH 3; aliquots were taken to determine the number of surviving organisms. An isogenic *rpoS* mutant was pregrown at pH 6.0 and then grown in medium at pH 3 (red line)

known. This phenomenon is strongly augmented by gene expression under the control of σ^{38} (RpoS; see below) (Fig. 28.**8**). The precise signal that regulates the genes of the acid tolerance response is not known. Rapid shifts in external pH lower the cytoplasmic pH transiently, and this change may be sufficient to alter the rate of synthesis of acid-inducible gene products. Regulation of gene expression by pH_i is now well documented (e.g., enhanced transcription of the *rpoS* gene encoding σ^{38}). Cells at neutral pH that have a reduced growth rate or that have entered the stationary phase are also more resistant to the effects of acidic pH due to the expression of the starvation sigma factor σ^{38} (RpoS) (Fig. 28.**8**). However, not all growth-phase-regulated phenomena or pH-induced changes in cell resistance to stress can be attributed to σ^{38} activity, and other mechanisms are still to be identified. Variations in cytoplasmic pH have been linked to increased capacity for DNA repair (increased expressions of the SOS DNA repair modulon at alkaline pH) and to resistance to mutagens and electrophilic reagents at acid pH.

Bacterial cells regulate the cytoplasmic pH within relatively narrow limits. The cytoplasmic pH may vary as little as 0.1 pH unit per external pH unit change, but cells can sense such small changes in pH and initiate adaptive gene expression. Changes in cytoplasmic pH can evoke a wide pattern of changes in cell behavior. Motile bacteria may swim away from conditions that might perturb pH homeostasis (acid environments, high concentrations of weak acids) via pH sensing allied to chemotactic movement. At mildly acidic pH, changes in gene expression take place, partially under the control of σ^{38}, that potentiate the survival of many different types of stress, including exposure to pH extremes.

28.4 Bacteria Adopt a Common Strategy for Surviving Osmotic Stress

Osmotic stress has been studied for many years in a wide range of microorganisms. The problems facing the bacteria are quite straightforward. Enzyme function requires concentrations of solutes in the cytoplasm that are far higher than the organism will experience in the environment and, thus, solutes are concentrated in the cytoplasm. Consequently, there is a tendency for water to flow into the cell, creating a turgor pressure that would burst the membrane if it were not protected by the bacterial cell wall. However, the bacteria have taken this problem one stage further since almost all species accumulate solutes, and ions in particular, to levels that are much higher than are required for biosynthetic processes.

Box 28.5 A. Koch formulated the **surface-stress theory** to account for the evolution of moderately high turgor pressure in bacterial cells. In essence, he proposed that bacteria accumulate substantial solute pools to create a high turgor pressure, enabling them to stretch the cell wall and thereby permitting the elongation of the peptidoglycan chains. Thus, turgor, which arose initially for driving reactions in the cytoplasm, may have evolved to become the force driving reactions at the cell surface. At very high osmolarity, bacteria may cope with reduced turgor by lowering the degree of cross-linking of the cell wall. As a consequence, the cell is enlarged due to a new balance between turgor and the resistance offered by the cell wall.

28.4.1 Compatible Solutes Are Accumulated by Bacteria to Alleviate Osmotic Stress

The essential role for turgor in modelling of the cell wall has created a problem for the cell. Turgor pressure is almost essential to growth and consequently must be maintained despite variations in the external osmolarity. As the external osmolarity is raised, the osmolarity of the cytoplasm must also be increased to maintain turgor. Many enzymes start to lose activity at high osmolarity. Adaptive strategies are based upon maintaining constant turgor by accumulating solutes that are compatible with enzyme function.

Compatible solutes are substances that have the following properties:

1. They are soluble to high concentrations and solutions of up to 1 M in the cytoplasm are not unusual.
2. They are usually either neutral or zwitterionic molecules.
3. The cell membrane exhibits controlled permeability to them, allowing the cytoplasmic pool to be determined by the external osmotic pressure.
4. They do not interfere with enzyme activity and may protect enzymes from denaturation by salts.
5. The best compatible solutes, **betaine**, **ectoine**, **trehalose**, and **proline** (Fig. 28.**9**), show unusual osmotic properties; consequently, lower cytoplasmic concentrations are sufficient to exert the same osmotic pressure as lesser compatible solutes.
6. They are usually end-point metabolites rather than intermediates in a biosynthetic pathway.

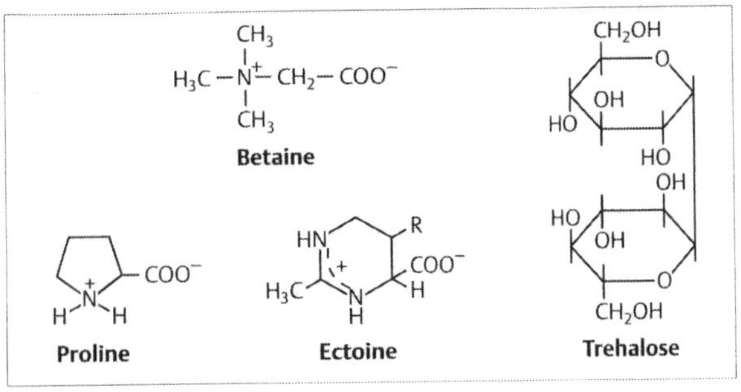

Fig. 28.9 **The structure of some common compatible solutes**

Compatible solutes are generally freely available in the environment because of their de novo synthesis by a number of genera of bacteria, plants, and animals. In addition, some common cell components (e.g., proteins and lipids) can be hydrolyzed to give precursors for compatible solutes, such as proline and betaine. Thus, bacteria in foods or in eutrophic (nutrient-rich) ecosystems are well supplied with sources of compatible solutes.

Gram-positive organisms face a special problem because they have evolved to maintain very large cytoplasmic pools of solutes, leading to turgor pressures of approximately 20 atm (2 MPa), compared with 5–6 atm (0.5–0.6 MPa) for Gram-negative organisms. As a consequence, these organisms accumulate compatible solutes even at low external osmolarities to avoid the stress that would arise from the accumulation of salts to maintain turgor. It has been suggested that Gram-positive organisms evolved in high osmolarity environments, which led to their acquisition of the ability to accumulate solutes to very high levels. Subsequent evolution of a thick cell wall may have enabled their descendents to colonize low osmolarity environments.

The cell must be able to control the accumulation of compatible solutes very tightly so that the internal osmotic pressure can be regulated. Thus, as the osmotic pressure is increased, the pool size of the compatible solute becomes progressively elevated (Fig. 28.**10**). Various microorganisms accumulate a great variety of solutes: betaine, trehalose, glycerol, sucrose, L-proline, D-mannitol, D-glucitol, L-taurine, ectoine, and small peptides. There is no obvious correlation between the identity of the solute accumulated and the ability of the organism to withstand high osmotic pressure. Tolerance of high osmotic pressures is multi-factorial. Enzymes that are located in the periplasm and exposed at the outer surface of the cytoplasmic membrane cannot be protected by compatible solutes. Salt-tolerant organisms possess surface enzymes and structures that are more tolerant of the low water activity than those found in less osmotolerant organisms. In the halobacteria, even cytoplasmic enzymes show specific changes

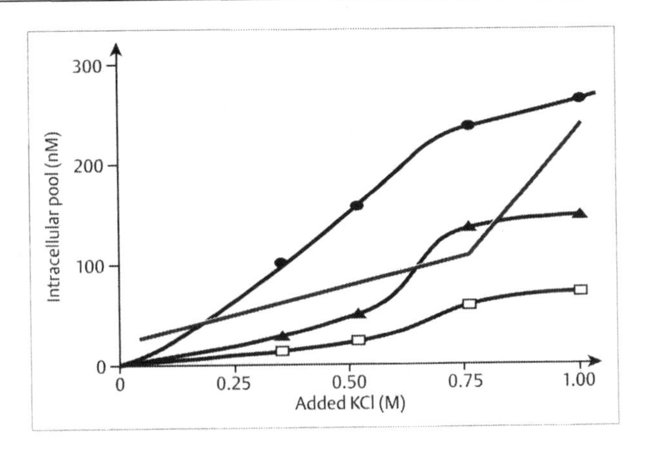

Fig. 28.10 **Intracellular pools of *Streptomyces griseus* as a function of external osmolarity.** The red line indicates the increase in intracellular potassium salts. ● Proline, ▲ glutamine, □ alanine [after 4]

relative to those isolated from organisms growing at low osmolarity. The halobacteria grow optimally in solutions of 3 M NaCl and accumulate high potassium chloride levels (4 M), compared with moderate levels of potassium glutamate (up to 0.8 M) in non-halophilic organisms. Protein structures show enhanced levels of acidic amino acids and a reduced content of hydrophobic amino acids. These proteins bind water two- to fourfold more effectively than equivalent proteins from organisms evolved to occupy low osmolarity niches.

28.4.2 Osmotic Stress Elicits Turgor Regulation and Accumulation of Compatible Solutes

As a result of the changes in gene expression and in protein activity, the physiology of cells growing in media of low osmolarity is markedly different than that of cells adapted to high osmolarity (Tab. 28.**5**). The primary event in **turgor regulation** is the controlled accumulation of potassium and its counterion glutamate. When *E. coli* cells are subjected to a sudden increase in external osmolarity, there is a net outflow of water from the cell via the aquaporin, leading to a loss of turgor. This activates the TrkAEH potassium uptake system, causing further potassium uptake until the turgor is restored (Fig. 28.**11**). If potassium is not available, then the turgor stress persists and leads to the expression of the Kdp transport system, which has a high affinity for potassium. Expression of the *kdp* genes is transient when potassium is freely available and permanent when potassium is limiting. Glutamate is synthesized (or transported from the environment) to a level almost equivalent to that of the increase in potassium (Fig. 28.**11**). At high external osmolarities,

Table 28.5 **Activities regulated by high osmotic pressure in the enteric bacteria**

Function	Genes	Activation	Genetic control
Outer membrane permeability	*ompF*		Repressed
	ompC		Induced
Synthesis of membrane-derived oligosaccharide	*mdoG, H*	Inhibited	Repressed
Potassium uptake	*trkA, E, H*	Activated	Constitutive
	kdpA, B, C		Induced
Betaine uptake	*proP*	Activated	Mild induction
	proU	Activated	Strongly induced
Betaine synthesis	*betAB*		Induced
Choline uptake	*betT*	Activated	Induced
Trehalose synthesis	*otsA, B*	Activated	Induction
Trehalose breakdown	*treA*		Induced
σ^{38}, starvation sigma factor	*rpoS*	Protein stabilized	
Stretch-activated channels	*mscL*	Inhibited[1]	Constitutive
KefA–K^+ efflux	*kefA*	Inhibited at low turgor[2]	Constitutive
BetX–betaine efflux	Not known	Inhibited[3]	Not known

[1] Stretch-activated channels are active when the cell is subjected to rapid lowering of the external osmolarity.
[2] The KefA potassium efflux system plays a major role in K^+ efflux in response to the accumulation of compatible solutes, such as betaine and trehalose.
[3] The BetX betaine efflux system is known to exist, but the regulation of its activity has not been analyzed in detail.

the cytoplasmic levels of potassium glutamate can transiently reach 0.7–0.8 M, sufficient to impair enzyme function. Rapid growth, therefore, is dependent upon secondary responses, principally the accumulation of the **compatible solutes** betaine (*N*-trimethylglycine), proline, and trehalose. The accumulation of these solutes is achieved by the controlled activity of transport systems and enzymes in response to changes in external osmotic pressure. Compatible solutes may be either transported from the environment or synthesized de novo in the cytoplasm. Their accumulation in the cytoplasm is associated with the controlled release of potassium and glutamate (Fig. 28.**11**) and thus reduces the salt concentration in the cytoplasm.

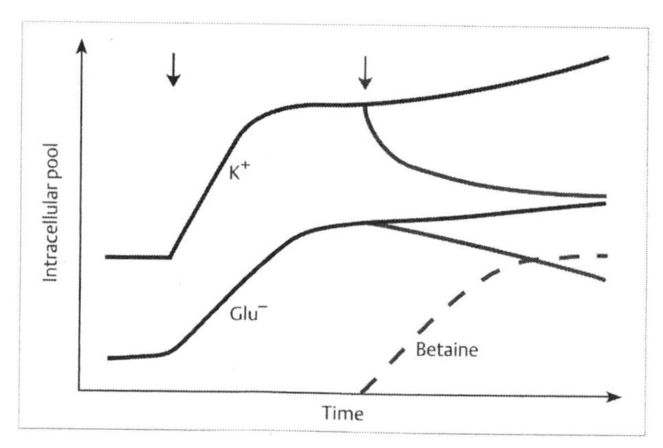

Fig. 28.**11** **Changes in cytoplasmic pools in osmoregulating *Escherichia coli* cells.** The changes in the intracellular pools of betaine, potassium, and glutamate are shown. The osmotic pressure of the environments was raised by the addition of 1 M glucose (black arrow). Betaine was added to the incubation mixture (red arrow) some minutes later. Red lines: pools in the presence of betaine; black lines: pools in the absence of betaine. Note that in most *E. coli* cells, trehalose synthesis would have ensued a few minutes after the increase in external osmolarity, which would also cause loss of potassium glutamate. This has been omitted for clarity

For the majority of organisms, growth potential at high osmotic pressure is determined by the ability of the organism to maintain **turgor** and to establish a balance of solutes in the cytoplasm that is compatible with enzyme function. The accumulation level of **compatible solutes** in the cytoplasm is always determined by the external osmolarity, regardless of whether the solute is transported into the cell or synthesized in the cytoplasm. This implies that there are sensors of osmotic stress that transduce the external signal to regulate enzyme and transport activity. Transport systems are exposed to both the external environment and the cytoplasm, and their activity is regulated by osmotic stress. These systems exert a major controlling influence over solute pools, and consequently, these transport systems may be both sensor and transducer.

28.4.3 The Role of Novel Gene Products in Survival of Osmotic Stress

Bacterial cells undergoing adaptation to osmotic stress show altered patterns of gene expression. Osmotic stress provokes many changes in the cytoplasm, each of which could act as a signal for adaptive gene expression. Two major effectors of the altered pattern of gene expression in the enteric bacteria are altered DNA topology and higher expression of σ^{38}, the sigma factor responsible for growth-phase-dependent gene expression (see below). In addition, transient changes in pH_i, and longer-lasting changes in the levels of solutes, such as potassium glutamate, betaine, polyamines, and ppGpp, may provide both primary and secondary elicitors of gene expression. During osmotically induced growth inhibition, short-term perturbations of metabolism may cause the transient buildup of toxic by-products that elicit induction of other regulatory responses, such as those for DNA repair or destruction of peroxides. For many of the genes that are induced by changes in osmolarity, there is an obvious adaptive function (e.g., the biosynthetic and transport systems for compatible solutes). Other genes that are activatec under conditions of osmotic stress play a less obviou: function. The mixture of signals arising during osmoti stress, therefore, leads to the expression of a range o genes ("stimulon"), some of which are not essential fo osmoadaptation, but which pre-dispose the cell to cop more effectively with other stresses.

As with pH stress, the original signals generated by **osmotic stress** cannot be seen isolated from the overall response of the cell. Altered function of some central activities may generate new signals and stresses, each of which requires gene expression to counter the new stress induced during adaptation. Osmotic stress causes changes in many aspects of cell metabolism, is accompanied by transient changes in pH_i, and also affects the pools of many small molecules. Any of these changes may be sufficient to elicit a specific response. In addition, osmotic stress induces a generalized stress tolerance response in *E. coli* and *Bacillus subtilis*, which is controlled by σ^{38} and σ^{B}, respectively.

28.5 Changes in Growth Rate Elicit an Adaptive Response

Most bacteria are capable of growing at various rates determined by the nutritional complexity of the environment and by physical parameters, such as temperature, pH, and osmotic pressure. Associated with the changing growth rate are a number of adaptive phenomena. The expression of σ^{38} is modulated by the growth rate. This sigma subunit regulates the expression of a large number of genes that are involved with surviving more extreme stresses. Consequently, the growth rate of a cell immediately prior to exposure to stress is a major determinant of its subsequent survival.

In the natural environment, growth rates are limited by the availability of specific nutrients, and consequently, starvation is one of the most prevalent stresses encountered by organisms. The dispersal of many microorganisms depends upon survival outside the host for long periods, and this survival frequently takes place under conditions of adverse pH, osmolarity, and temperature (with low temperature being particularly important). When bacteria are starved for nutrients, a large number of physiological changes take place. In the short term, there is immediate imposition of the stringent response (see Chapter 20.3), and longer starvation may place the organism in a "**resting state**" (sometimes termed "viable but non-culturable", see Chapter 33.5.1). Entry into the starved state is associated with increased rates of protein turnover and the synthesis of novel proteins, many o which are associated with other stress responses. Cell become generally resistant to a number of stresses, sucl as oxidation stress, osmotic stress, acid stress, heat shock and electrophilic stress, and in some bacteria, pathoge nicity is only expressed after a brief period of starvation.

28.5.1 Important Physiological Changes Occur in Starving Cells

The major physiological changes seen in starving cell: are associated with changes in either the energy statu: or the nutrient status, depending upon which specifi growth requirement is limiting. For each nutrien required for growth, there is a specific response, whicl usually leads to improved nutrient acquisition. However, with continued starvation, which leads to sacrificial metabolism, certain specific polymers may b turned over to allow others to be synthesized. As th duration of the starvation phase is extended, howeve cell shrinkage and rounding-up of rod-shaped cell: takes place. These changes will increase the surface-to volume ratio of the cell and will potentiate nutrien acquisition.

In some cases, **nutrient limitation** is perceived as a specific signal of location. For example, iron limitation often leads to the expression of pathogenicity genes as well as systems for iron acquisition, reflecting the host defense response of iron limitation during bacterial invasion. **Carbon limitation** will cause an immediate reduction in the capacity for ATP synthesis in most heterotrophs, but under conditions of carbon excess during other nutrient limitations, altered patterns of metabolism arise. This "**overflow metabolism**" may serve the purpose of reducing energy generation since highly oxidized carbon skeletons are found in the medium (see Chapter 6). The accumulation of by-products in the medium may itself act as a signal for adaptation. For example, induction of synthesis of the antibiotic methylenomycin by streptomycetes occurs after nitrogen limitation has been imposed. Nitrogen limitation leads to excretion of acids into the medium and causes the external pH to decrease. Nitrogen limitation can be mimicked under conditions of nitrogen excess by simply acidifying the medium. Consequently, the metabolism of the cell is profoundly altered by limitation of growth by any nutrient, and this may initiate specific adaptive responses.

In the short term, the major changes seen in starving cells are almost identical to those induced by the stringent response (see Chapter 20.3). Bacterial cells enter the stringent response whenever the supply of either energy or amino acids is limited ("**shift-down**"). Any reduction in growth rate (e.g., that experienced by cells adjusting to an increase in the external osmolarity) will lead to imposition of a stringent response. Conversely, any nutritional "**shift-up**" (e.g., anaerobic to aerobic growth, addition of amino acids to minimal medium) will generally alleviate the degree of stringency associated with a specific value of the growth rate. Major changes in RNA synthesis take place within seconds of the imposition of stringency, resulting in reduced transcription of rRNA and tRNA. The consequence of this is that the bulk of the RNA synthesized is mRNA, and the cell has moved from a state poised for growth (synthesis of rRNA and tRNA needed for protein synthesis) to an adaptation state. Stringency can be either transient or "permanent" and thus conforms to the same pattern as any other stress regime described above. Transient stringency ensures that the rate of synthesis of rRNA, and consequently, of ribosomes, matches the growth potential of the environment. As with many other regulatory systems, the stringent response is set by the cell at a range of "on" values rather than either "on or off." In this context, "permanent" means the time required to acquire a new nutrient source, which in a natural environment may be many weeks.

28.5.2 The Starvation Stimulon Is Related to the Stringent Response

The expression of the "**starvation stimulon**" underpins many of the other stress responses discussed so far. Cells of *E. coli* that possess significant levels of the RpoS protein, σ^{38}, are resistant to high osmolarity, low pH, high temperature, and electrophiles (Fig. 28.**12**). RpoS protein accumulates whenever the growth rate is lowered and not just when growth ceases, and consequently, exponentially growing cells can express the "starvation stimulon" at significant levels. Not all starvation-associated genes are controlled by RpoS; regulatory systems specific to the major nutrient that is in short supply also change the pattern of gene expression. For example, when *E. coli* cells are starved of carbon in a minimal medium, over 50 proteins are synthesized in a relatively specific manner over a period of several hours. Of these proteins, 32 are regulated directly by σ^{38}, and the others respond to the elevation of the concentration of cAMP that occurs after carbon limitation (see Chapter 20.2).

The starvation sigma subunit, σ^{38}, has been recognized as a major factor in enteric bacteria that determines survival of brief periods of starvation and is essential for generation of the generally resistant state. Similar proteins may be present in other bacteria as inferred from the patterns of protein synthesis observed during starvation and from the generality of the phenomenon of multiple sigma subunits controlling blocks of genes associated with adaptation. Expression of σ^{38} is required for the acquisition of tolerance of a wide range of stress conditions. Consequently, this

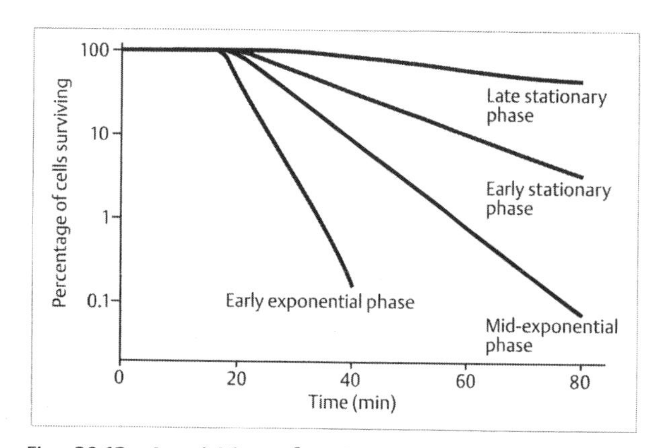

Fig. 28.12 **Acquisition of resistance to electrophiles as growth rate decreases.** *Escherichia coli* cells were grown in a defined medium. The ability of the cells to survive exposure to 100 μM *N*-ethylmaleimide, a potent electrophile, was determined at intervals

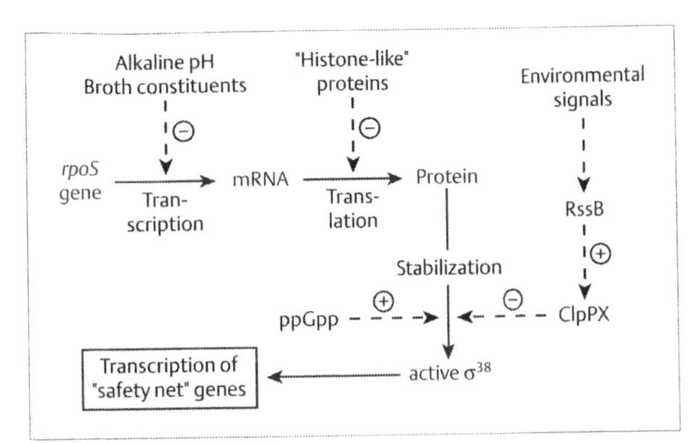

Fig. 28.13 **Factors leading to elevation of the concentration of the RpoS protein in *Escherichia coli* cells.** Transcription of *rpoS* is slightly reduced when the internal pH is maintained above the pH_i (pH 7.5) and by constituents of broth. Translation is inhibited by "histone-like" proteins. ppGpp is believed to be required for stabilization of the RpoS protein. The major determinant of RpoS protein concentration is proteolysis by the ClpPX protease, which is regulated by the RssB response regulator. It is assumed that RssB senses environmental signals relating to growth rate, but it is not known what these signals are

sigma subunit can be considered to lead to the expression of genes that are members of a starvation stimulon and form a "safety net" for the organism in a hostile environment.

Transcription of the *rpoS* gene, which encodes σ^{38}, occurs throughout growth, but is specifically induced by weak acids and by entry into the stationary phase. Much more marked, however, is that the *rpoS* gene is subject to posttranscriptional control such that the σ^{38} protein only accumulates during sub-optimal growth and in the stationary phase (Fig. 28.**13**). During osmotic stress, the growth rate is reduced and accumulation of σ^{38} takes place. Both phenomena can be reversed by compatible solutes, such as betaine. The regulation of accumulation of σ^{38} is very complex (Fig. 28.**13**) and involves control by proteolysis in a manner similar to that for the heat shock sigma factor (see above). Proteolysis of σ^{38} is mediated by the ClpPX protease (see Chapter 19.3.4) that is controlled by the "two-component" system RssB. What environmental signal RssB senses is not clear. In *E. coli*, it has been shown that the synthesis of the σ^{38} is controlled positively by ppGpp whose concentrati pool corresponds inversely with the growth rate und most conditions (see Chapter 20.3). Whether t induction by weak acids or the posttranscriptior control is determined directly or indirectly by t stringent response is not known.

Box 28.6 Virulence factors can be considered as stress responses. Stress responses play two quite different roles in pathogenesis. First, the responses described above enable the cell to thwart the defense mechanisms utilized by higher organisms. Secondly, the sensing of environmental stimuli also enables the bacterial cell to express specific proteins that facilitate invasion. During invasion of the host, several virulence factors are expressed (see Chapter 33.3). The synthesis of these proteins is environmentally regulated via pathways common to regulation of stress-responsive genes. Similarly, survival in the phagosome is dependent upon resistance to acid pH, to toxic superoxide and peroxide ions, and to enzymes and cationic peptides. Each of these properties can be acquired by cells as a result of the "starvation stimulon" under the control of σ^{38}. Therefore, the growth rate has a central role in potentiating invasion of the host.

In growing cells, the major role for the **stringent response** is to balance the synthesis of rRNA and tRNA with the growth rate. The stringent response operates at all growth rates, while the accumulation of σ^{38} occurs as the growth rate decreases. It is important to recognize that σ^{38} is found and is active in exponentially growing cells and is not limited to starving or stationary-phase cells. Expression of a wide range of genes that confer upon the cell the capacity to withstand a plethora of stresses is under the control of σ^{38}. A sudden change in the growth rate, whether caused by nutrient limitation and ppGpp or by imposition of extreme stress, enhances the expression of the "**starvation stimulon**", which is expressed at low rates at all sub-optimal growth rates.

28.6 Conclusion: The Stress Response Is an Exaggerated Normal Response

For many years, adaptations to various stresses have been thought of as specialist functions of the cell by which the organism survives various insults. However, it is more realistic to see these responses within the context of normal growth and metabolism. Adaptation to stress generally involves the amplification of mechanisms that are normally expressed during growth, i.e., a re-balancing of activities. For some genetic systems, there is clearly a major change in activity, but for other systems, the changes are relatively small. The reasons for this are self-evident. Our vision of stress is very much from a human perspective. Conditions that reduce the performance of the organism or that provoke an adaptive response are often defined with care, but less care is often taken to understand what corresponds to a "lack of stress" for the organism. It is primarily for this reason that stress responses are seen as special events rather than as expressions of the integration of growth and survival, both of which are essential traits for a successful microorganism.

Further Reading

Adams, M. W. W. (1994) Enzymes and proteins from organisms that grow near and above 100°C. Annu Rev Microbiol 47: 627–658

Booth, I. R., Douglas, R. M., Ferguson, G. P., Lamb, A. W., and Ritchie, G. Y. (1993) In: Bakker, E. P. (ed.) Alkali cation transport systems in prokaryotes. Boca Raton, Fla.: CRC Press; 291–308

Booth, I. R., Jones, M. A., McLaggan, D., Nikolaev, Y., Ness, L., Wood, C. M., Miller, S., Tötemeyer, S., and Ferguson, G. P. (1996) Bacterial ion channels. In: Konings, W. N., Kaback, H. R., and Lolkema, J. S. (eds.) Handbook of biological physics, vol. 2. Transport processes in eukaryotic and prokaryotic organisms. Amsterdam: Elsevier; 693–729

Csonka, L. N., and Epstein, W. (1996) Osmoregulation. In: Neidhardt, F. C., Curtiss, R., Ingraham, J. L., Lin, E. C. C., Low, K. B., Magasanik, B., Reznikoff, W., Riley, M., Schaechter, M., and Umbarger, H. E. (eds.) *Escherichia coli and Salmonella*: cellular and molecular biology. 2nd edn. Washington, DC: ASM Press; 1210–1224

Demple, B. (1991) Regulation of bacterial oxidative stress genes. Annu Rev Genet 25: 315–339

Dorman, C. J., and NiBhriain, N. (1993) DNA topology and bacterial virulence gene regulation. Trends Microbiol 1: 92–99

Galinski, E. A., and Trüper, H. G. (1994) Microbial behaviour in salt-stressed ecosystems. FEMS Microbiol Rev 15: 95–108

Gross, C. A. (1996) Function and regulation of the heat shock proteins. In: Neidhardt, F. C., Curtiss, R., Ingraham, J. L., Lin, E. C. C., Low, K. B., Magasanik, B., Reznikoff, W., Riley, M., Schaechter, M., and Umbarger, H. E. (eds.) *Escherichia coli and Salmonella*: cellular and molecular biology. 2nd edn. Washington, DC: ASM Press; 1382–1399

Hartl, F.-U., Hlodan, R., and Langer, T. (1994) Molecular chaperones in protein folding: the art of avoiding sticky situations. Trends Biochem Sci 19: 20–25

Hecker, M., Schumann, W., and Völker, U. (1996) Heat-shock and general stress response in *Bacillus subtilis*. Mol Microbiol 19: 417–428

Kjellberg, S., ed. (1993) Starvation in bacteria. New York London: Plenum Press

Koch, A. (1983) The surface stress theory of microbial morphogenesis. Adv Microbial Physiol 24: 301–366

Konings, W. N., Kaback, H. R., and Lolkema, J. S., eds. (1996) Handbook of biological physics, vol. 2. Transport processes in eukaryotic and prokaryotic organisms. Amsterdam: Elsevier

Lin, E. C. C., and Lynch, A. S., eds. (1995) Regulation of gene expression in *Escherichia coli*. Austin, Tx.: Landes

Loewen, P. C., and Hengge-Aronis, R. (1994) The role of the sigma factor σ^S (KatF) in bacterial global regulation. Annu Rev Microbiol 48: 53–80

Miller, K. J., and Wood, J. M. (1996) Osmoadaptation by rhizosphere bacteria. Annu Rev Microbiol 50: 101–136

Padan, E., and Schuldiner, S. (1994) Molecular physiology of Na^+/H^+ antiporters, key transporters in circulation of Na^+ and H^+ in cells. Biochim Biophys Acta 1185: 129–151

Schweder, T., Lee, K.-H., Lomovskaya, O., and Matin, A. (1996) Regulation of *Escherichia coli* starvation sigma factor (σ^S) by ClpXP protease. J Bacteriol 179: 470–476

Slonczewski, J. L., and Foster, J. W. (1996) pH-regulated genes and survival at extreme pH. In: Neidhardt, F. C., Curtiss, R., Ingraham, J. L., Lin, E. C. C., Low, K. B., Magasanik, B., Reznikoff, W., Riley, M., Schaechter, M., and Umbarger, H. E. (eds.) *Escherichia coli and Salmonella*: cellular and molecular biology, 2nd edn. Washington, DC: ASM Press; 1539–1549

Stumpe, S., Schlösser, A., Schleyer, M., and Bakker, E. P. (1996) K^+ circulation across the prokaryotic cell membrane: K^+ uptake systems. In: Konings, W. N., Kaback, H. R., and Lolkema, J. S. (eds.) Handbook of biological physics, vol. 2. Transport processes in eukaryotic and prokaryotic organisms. Amsterdam: Elsevier, 473–500

Yura, T., Nagai, H., and Mori, H. (1994) Regulation of the heat-shock response in bacteria. Annu Rev Microbiol 47: 321–350

Sources of Figures

1 Hartl, F.-U., Hlodan, R., and Langer, T. (1994) Trends Biochem Sci 19: 20–25

2 Kroll, R. G., and Booth, J. R., (1981) Biochem J 198: 691–698

3 McLaggan, D., Selwyn, M. J., and Dawson, A. P. (1984) FEBS Lett 165: 254–258

4 Killham, K., and Firestone, M. K. (1984) Appl Environ Microbiol 47: 301–306

Section VII
Diversity and Systematics

29 Prokaryotic Diversity and Systematics

Prokaryotes are ancient and ubiquitous, but are largely unknown. As compared to the total number of described species of animals, plants, and lower eukaryotes, the number of described prokaryotic species is astonishingly small. This small fraction of eubacteria and archaebacteria (or, as named today, Bacteria and Archaea) of only 0.2% of all species seems unrealistic considering that prokaryotic species probably evolved more than 3.5 billion years ago and occupy all niches that have been investigated for the presence of prokaryotic organisms. Prokaryotes have been found in extreme environments not believed to support life: below salt crusts of dried lakes in the East African valley, in association with deep-sea hydrothermal vents, in fumaroles, geysers, and solfataras, in alkaline lakes, inside Antarctic rocks, and inside invertebrates. Insects, on the other hand, which comprise more than one million species, evolved during the Cambrium period less than 600 million years ago.

The question remains why the number of prokaryotic species is so small. Two main reasons have been identified: (1) The species definition in bacteriology cannot be compared to that used in the classification of higher eukaryotes. The "biological species concept" cannot be applied to prokaryotes that lack gametes and meiosis, and genetic material is frequently exchanged even between remotely related prokaryotic species. Synapomorphic (shared derived) characters, upon which classification of higher eukaryotes is based, are more difficult to detect among prokaryotes, which lack embryogenesis, anatomy, and a multi-structured morphology. The species definition in bacteriology is artificial; it is pragmatic and highly demanding as it requires investigation of phylogenetic distances, chemical analysis of cell constituents, and investigation of physiological and morphological properties. (2) Taxonomists are still unable to culture most prokaryotes. The fraction of species known to exist in the environment that have not yet been cultured is estimated to be as high as 95–99.9%.

Although there appears to be no place on this planet that is not occupied by microorganisms, biologists are still not able to assess the full range of prokaryotic diversity. Prokaryotes were apparently the only life forms on the planet Earth for more than two billion years. Moreover, the genetic diversity of prokaryotes allows them to express a significantly wider range of physiological and biochemical activities than found in higher life forms. The mitochondria and chloroplasts of eukaryotic microorganisms and macroorganisms originated from bacteria. Hence, the eukaryotic cell represents a chimeric structure composed of many genes that are of prokaryotic origin.

Due to their recycling potential, microorganisms play an extremely important role in ecological processes. The following chapters (see Chapters 30–32) describe how essential bacteria are for the maintenance of the geochemical cycles of nitrogen, sulfur, carbon, and metals and they improve, for example, the accessibility of nitrogen for numerous plants and the function of digestive tracts by splitting macromolecules that are indigestible for humans and animals. Consequently, microorganisms play a key role in sustaining life on Earth.

Collections of microorganisms contain strains that can be grown on synthetic growth media. Microorganisms known today must be considered as only a glimpse of the naturally occurring biodiversity hitherto investigated by the application of traditional classification methods. Radioactive tracers, stable isotopes, and microelectrodes have been applied to measure the activity of prokaryotes directly in the environment, and certain natural samples have been characterized reasonably well with respect to the diversity of biochemical or geochemical activity. The inventory of a particular site for prokaryotes and fungi is problematic due to, for example, the minute size of cells, seasonality, changes in morphology, substrate or host dependence, and the ability of many prokaryotes to change into life stages from which they cannot be recovered. It is now generally accepted that microorganisms that grow on isolation plates are the ones best adapted to the artificial growth conditions and are not necessarily those that are metabolically active or abundant in the environment.

As long as information on the identity of the majority of microbial species, their presence and distribution in natural samples, and consequently their role in the global foodweb is lacking, the diversity of microorganisms cannot be fully assessed. Consequently any classification scheme, restricted to strains that are housed in collections, is a flexible construct that must allow changes at any taxonomic level, whenever new information on novel isolates becomes available. In addition, questions whether free-living microorganisms are universally distributed, and whether it is necessary to conserve microbial species in situ for subsequent restoring of the function of an environment, cannot be answered unless knowledge about taxonomy, physiology, and ecological interactions of microorganisms increases.

The environment is an untapped source of novel diversity. Attempts to determine the number of prokaryotic species in a defined environmental sample (usually one to several grams of soil or hundreds of liters of water) have in the past concentrated on certain physiologically and morphologically distinct groups. The recovery of all species from a given environment might not be achieved in the near future considering that the number of species and the range of diversity (genetic, physiological, biochemical) are still unknown. New cultivation methods for the detection and subsequent description of novel organisms are urgently required.

Our inability to provide appropriate growth conditions, which is mandatory to obtain pure cultures for subsequent characterization is not the only reason for the inability to assess microbial diversity. The main reason is the failure to revive so-called "uncultured" (or "resting," "viable, but not culturable") forms. Only recently, through application of molecular methods, such as direct extraction of nucleic acids from the environment, application of polymerase chain reaction (PCR) technologies, and sequence analysis, have microbiologists started to assess the diversity of prokaryotes in environmental samples. The identity of a microbial species can today also be assessed by analysis of its genetic material, preferably by sequence analysis of conserved macromolecules. As sequences normally provide a highly species-specific pattern, a database of sequences is generated against which each new sequence can be compared, and conclusions about the identity of novel isolates can be made. Although sequence comparison does not yet allow a qualitative or quantitative measurement of biodiversity, it can be considered a major breakthrough in microbial ecology. Comparative sequence analysis, followed by identification of sequence dissimilarities, allows the detection of novel groups of prokaryotes, but it alone cannot assess the full range of species richness, species abundance and, above all, the physiological properties and ecological role of an organism.

The vast majority of naturally occurring prokaryotic organisms cannot yet be cultured. Culture collections thus house a minute fraction of extant organisms only, indicating that the full diversity of physiological and biochemical reactions, the biotechnological potential of prokaryotes and their ecological role are still unkown. Comparison of environmental samples, i.e., those of similar physicochemical composition from different parts of the world (marine environment, soil), reveal some similarities in species composition, but also significant differences. Complex environments, such as soil, reveal a higher degree of diversity than environments of extreme conditions caused by pH, salinity, or temperature, such as ore-leaching systems and hot springs. This includes eukaryotic hosts, each of which may contain at least one new prokaryotic species.

The term "microorganism", used in this chapter, will deal with prokaryotic life forms, but a significant portion of taxonomic principles and classification strategies exemplified with bacteria and archaea also hold true for eukaryotic microorganisms. Thus the methods used to classify bacteria are today also being applied to fungi, yeasts, protozoa, and algae.

29.1 Bacterial Systematics Is the Cradle of Comparative Biology

As emphasized by T. Dobzansky, "Nothing in biology makes sense except in the light of evolution." One of the unfortunate side effects of developments in microbial systematics is that the average student or researcher usually only comes in contact with its nomenclature. Hence, far too much attention is paid to changing names or the creation of new taxa, rather than to the underlying theory and philosophy, the improvements in methodology, or an appreciation of the evolution and diversity of microorganisms in their many facets. In order to appreciate these aspects, the scope of systematics and the importance of correct definitions must first be understood.

29.1.1 Systematics Has its Origins in the Work of the Early Natural Historians

In his "Systema naturae", C. von Linné (1707–1778) presented a natural system of classification strongly influenced by the concepts of creationism rather than with the Darwinian concepts of evolution, which were to follow a century later. It is not surprising that systematics had to be refined in a modern context. Zoologists such as G. G. Simpson observed that systematics is the most elementary part of zoology because animals cannot be discussed or treated scientifically until some taxonomy has been established, and the most inclusive part of zoology because systematics utilizes, summarizes, and implements everything that is known about animals, whether morphological, physiological, psychological, or ecological. E. Mayr summarized that systematics is the science of the diversity or organisms, and in a more provocative context, he stated: "in actuality systematics is one of the major subdivisions of biology, broader in base than genetics or biochemistry. It includes not only the service functions of identification and classifying but the comparative study of all aspects of organisms as well as interpretation of the role of lower and higher taxa in the economy of nature and in evolutionary history. It is a synthesis of many kinds of knowledge, theory, and method, applied to all aspects of classification. The ultimate task of the systematist is not only to describe the diversity of the living world but also to contribute to its understanding." In agreement with this concept, systematics has become increasingly complex and is a late chapter in this book. It is clear from such definitions and from the sections that follow that systematics encompasses more than the naming of organisms. Systematics and taxonomy are regarded by some as being virtually synonymous, whereas others clearly distinguish taxonomy as the theory and practice of classifying organisms. **Modern taxonomy** comprises the following features (Fig. 29.**1**): (1) **characterization** (obtaining data on the properties of organisms), (2) **classification** (the theory and process of ordering the organisms), and (3) **nomenclatur** (giving names of appropriate taxonomic rank to th classified organisms). Identification should not b considered part of taxonomy; it shares with taxonom aspects of characterization and uses known taxa as it basis, but is limited by its scope and objectives.

29.1.2 Prokaryotes Must Be Characterized by Their Genotype and Phenotype

The characterization of an organism is dependent on th methods used. In principle, **phenotypic** and **genotypi** data can be distinguished. The **genome** comprises th sum of the information on the genes (**genetic informa tion**), and all observable features of an organisr (whether derived from the phenotype or genotype constitute the **phenome** (**phenetic information**). It i also appropriate to distinguish between informatio carried directly in the genes (genetic) and that carried i the products of the genes (epigenetic).

29.1.3 Classification of Prokaryotes: Subjective Treatment of Objective Data

No classification can be described as the sole classifica tion, and classifications often set out to do differer things. The present trend, for example, to grou microorganisms into risk or hazard groups, recognize four categories that represent increasing degrees c pathogenicity or risk potential to plants, animal: humans, and the environment, and does not serve an other purpose. Another trend takes evolution int consideration and provides a theory of classifyin

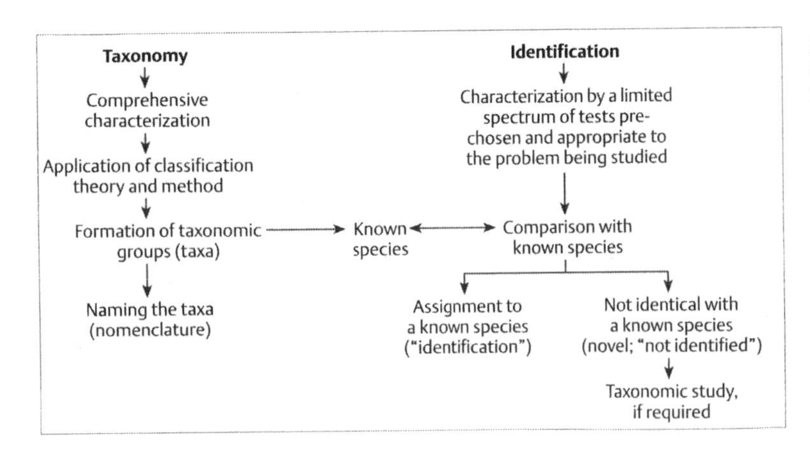

Fig. 29.**1** **Interrelationship between cha acterization, classification, and nomencl ture in the taxonomy of prokaryotes**

Box 29.1 There are various theories of classification, and some are used in parallel:

Essentialism is an ancient theory that classified organisms in the absence of an evolutionary theory according to their essential properties. It originated from creationist thinking, and no consideration is given to the problem of parallelism or convergent evolution.

The proponents of **nominalism** assume that only individuals exist, while all taxonomic categories, such as species, genera, and families, are artifacts of the human mind. The fact that evolution has given rise to groups of organisms sharing common ancestry with differing degrees of biological similarity can be taken as evidence that organisms form groups that are products of biological evolution and not human logic. Nevertheless, the choice of criteria delineating one group from another is thought to be man-made. The basic philosophy is not influenced by the degree to which the perceived groupings are considered as artificial or true reflections of evolution.

Empiricism assumes that neither a theory of evolution nor any other theory of classification is required to create a natural system. The groupings emerge automatically when enough parameters are obtained and intelligently evaluated.

Phenetics is often called "numerical taxonomy" or "numerical phenetics." The rationale and importance of numerical taxonomic analysis is explained in detail in Chapter 29.2.

Cladistics refers to a branch (Greek: klados; branch) of an evolutionary lineage and is probably one of the most controversial areas of systematics. Its adherents range from those who accept the taxonomic principles of G. G. Simpson, to "transformed cladistics" or "pattern cladistics", whose followers do not consider that the course of evolution is important in determining the relationship between taxa. S. T. Cowan stated quite simply that "as we do not know anything about phyletic lines, or convergence, among microbes the term cladistic is inapplicable to microbial taxonomy."

Artificial classifications are usually designed to fulfull certain functions and hence are typical of the applied biology area. Examples are distinguishing antibiotic and non-antibiotic producers, placing members of physiological groups together (e.g., methylotrophs, autotrophs, nitrifiers, methanogens). Artificial classifications should not be considered inferior or inadequate provided they perform the task for which they were developed. This classification could fail to reflect the true diversity of the artificially defined groups. Organisms can be assigned to taxonomic ranks, such as species, genus, and family (see Bergey's Manual of Determinative Bacteriology, editions 1–8); however, no evolutionary implications should be attached to these ranks.

Evolutionary systematics, according to E. Mayr, stems from the concepts of C. Darwin. It considers a number of aspects, including the phenetic concept of "overall similarity," evolutionary divergence of the taxa in question, and their subsequent evolutionary development.

Phylogenetic classification. "Phylogenetic" is derived from phylogeny, which stems from the Greek phylos (tribe or race) and genesis (origin or development). First coined by E. Haeckel (1834–1919), one of Darwin's (1809–1882) closest adherents, the term "Phylogenetik" was a simple transliteration of the older term "Systematik," Phylogeny means different things to different authors, and is often replaced by terms such as "natural relationship" or "genealogy." Phylogeny encompasses the course (over geological time) and the rate and mode of evolution of a lineage. Ideally, a phylogenic classification should take into account the ancestoral taxa and the rate with which they have evolved to give new taxa.

Biological classifications are based on the principle of to what degree the organisms are the same. Thus, various groupings, whether or not they are formally designated taxa (genus, families, etc.), generally bring together those organisms that are to various degrees "the same" in some or all of their properties. Although there are several ways to group microbial strains, the phylogenetic, genealogical relatedness offers the greatest potential since it explains the wide range of genetic and biochemical properties.

organisms that replaces an older theory that did not consider evolution.

29.1.4 The Fundamental Concept of Comparative Biology Is Based on Homologies

Various terms have been introduced to describe the degree of being "the same," and include similarity, affinity, relationship, and homology (Box 29.2). While each of these terms can be applied in any system of classification, there can be subtle changes in definition or interpretation depending upon the theory of classification. In an evolutionary context, relationship (relatedness) and homology infer common past ancestory. Despite the simplicity of the concept and the inability to observe the course of evolution (past and present), the concepts of relationship and homology have received the most attention since the publication of Darwin's "Origin of the Species" in 1859.

Box 29.2 In classical terms, homology can be divided into "true homology" and "false homology." The term homology is restricted to "true homology," while analogy and homoplasy are used for "false homology."

Homology is the sharing by two taxa of a property that is derived from the same or equivalent property of the nearest common ancestor.

Analogy is a similarity not due to common descent, but due to similarity of functions (e.g., wings in insects, birds, and bats).

Homoplasy can arise through parallelism (parallel evolution), convergence (convergent evolution), or reversal (reversal to a less complex or earlier state).

In addition, the following terms are used in the scientific literature:

Isology means similar to a high degree without knowing whether the properties are orthologous, paralogous, or xenologous.

Orthology is equivalent to "true homology."

Paralogy is parallelism usually resulting from gene duplication (hence originally treated as "true homology"). It may be present singly or in combination in the same organism (e.g., the PTS families, many isoenzymes, DNA polymerases, DNA topoisomerases).

Xenology is the acquisition of properties clearly "foreign" to the organism, usually due to horizontal gene transfer or fusion of lineages (e.g., antibiotic sensitivity or the chimeric properties of cryptomonads).

In comparative biology, it is thus essential to know whether a property is homologous (orthologous) or due to homoplasy (parallelism, convergence). The difficulties of determining homology occur, as will be discussed next, in every aspect of biological classification, irrespective of whether it is at the phenotypic (epigenetic) or genetic level.

29.1.5 True Homologies Are Difficult To Determine

The evolution of homologous elements, such as a molecule, organelle and other cellular constituents, requires a common genetic ancestry, and homology can in principle be used to determine phylogenetic, i.e., natural, relationships. Although the temporal scale used today to calculate the evolution of molecules is not precise, the order in which organisms evolved allows the construction of a phylogenetic classification system in which organisms are grouped according to their family history (genealogy) and not necessarily according to their present attributes.

Prokaryotes do not offer easy access to their genealogies for the following reasons: (1) lack of a substantial fossil record and the inability to draw conclusions on genomic and phenotypic properties from fossils prevents the inductive derivation of genealogical lines of descent, and (2) morphological complexity and comparative anatomy, other useful properties of eukaryotes for determining homologies, are lacking in the prokaryotes. Phylogenetic systematics demands congruency between the lines of descent evolved in time and those supraspecific taxa described by taxonomists. A prerequisite for the description of such a taxon of any rank is the recognition that all members to be included originated from one ancestral form, and that properties that evolved in the ancestral form are only found in their descendants.

The main problem taxonomists face is the question whether the occurrence of **apomorphic characters** in different organisms evolved only once (**synapomorphy**) or several times, independently from each other (**convergence**). On the basis of the occurrence of morphological and physiological characters alone, the biologist is unable to decide whether features are homologous or due to parallel or convergent evolution. This is true for all characters used in the past to establish a phylogenetic classification system, such as morphological structures, metabolic pathways, photosynthetic apparatuses, mechanisms of movements, or chemotaxonomic properties.

29.1.6 Nomenclature Is Governed by International Codes

Almost all organisms are given a genus and species binomial name derived from Latin or Greek words. This nomenclature system has its origins in the work of C. Linné and is the only aspect of plant, animal, and microbial taxonomy that is governed by International Codes of Nomenclature. When the Botanical Code was considered unacceptable for prokaryotes, the present International Code of Nomenclature of Bacteria (ICNB) developed from the Botanical Code over several decades. The ICNB differs from the Botanical Code in that only living organisms are allowed as types and taxonomic categories between subgenus and species and names of infrasubspecific forms are not regulated. The cyanobacteria (originally "blue-green algae") are considered "ambiregnal" organisms that can be described by either the ICNB or the Botanical Code. The ICNB recognizes January 1, 1980 as the starting date for the naming of prokaryotic taxa. Only those names that were considered to be properly described were included in the "Approved Lists of Bacterial Names." Names not included lost their standing in nomenclature.

The introduction of the "Approved Lists of Bacterial Names" eliminated numerous confusing synonyms that came to light with improved methods of characterization. As in other Codes, taxonomic names should be made known by publication of an official journal. Although any journal can accept the description of a new bacterial taxon, the taxon can only validly be published in the "International Journal of Systematic Bacteriology" as an original article or in the "Validation Lists," a periodic compilation of names that were published outside the journal. Taxa published as described above should appear without quotation marks (e.g., *Pseudomonas putida*), whereas all other names should appear in quotation marks (e.g., "*Haloarcula californiae*"), denoting that they have not been validated. The main purposes of the code are to prevent a confusing proliferation of names and to lay down rules for naming and for assigning appropriate type strains. However, the code does not govern what methods should be used to characterize an organism, and it does not deal with the theory of classification. An unpopular aspect of the current revisions in microbial taxonomy is the numerous changing of the species or genus names of often familiar organisms (see Tab. 29.**7**).

29.1.7 Idenification Is a Test of the Usefulness of Taxonomy

Identification comprizes the characterization of a strain and the comparison of the data on that strain with data previously classified and named strains. Hence, an organism can only be identified (i.e., shown to be identical with a known taxon) if that taxon is already known. Organisms that have not previously been isolated cannot be identified, must first be recognized as novel, and are then classified within the framework of the existing taxonomy. The purpose and efficiency of a taxonomy is tested by its use in identifying organisms. A key aspect of the existing classification is to what degree it allows the prediction of the position of a new organism within the classification system.

29.1.8 The Modern Classification of Prokaryotes Is Based on Genotypic Information

The potential of unravelling the phylogeny of life by comparing sequences was first recognized by E. Zuckerkandl and L. Pauling (1965). They recognized that biological molecules fall into three categories, based on their information contents: (1) **semantides** [**DNA** (**primary semantides**), **RNA** (**secondary semantides**), and **proteins** (**tertiary semantides**)] which carry genotypic information, i.e., the sequences of these molecules are the historical record of evolution, and the determination of the primary structure provided a powerful approach to measure evolutionary relationships. (2) **episemantic molecules**, which are synthesized under the control of tertiary semantides, such as ATP, carotenoids, and chemotaxonomic markers, and (3) **asemantic molecles**, which are not produced by the organisms themselves and do not express any of the information that this organism contains, such as exogenously supplied vitamins, phosphate ions, oxygen, and viruses.

Zuckerkandl and Pauling stated that "at any level of integration, the amount of history preserved will be greater, the greater the complexity of the elements at that level and the smaller the parts of the elements that have to be affected to bring about a significant change. Under favorable conditions of this kind, a recognition of many differences between two elements does not preclude the recognition of their similarity." This hypothesis was validated by the impressive phylogenetic trees of DNA and proteins (see Chapter 29.4.1). Episemantic molecules were not considered useful for deriving evolutionary conclusions because enzymes with different primary structures can lead to the synthesis of identical episemantic or similar molecules in different organisms as long as the active enzymatic sites are similar.

29.2 Numerical Taxonomy Is an Approach to Cluster Strains on the Basis of Large Sets of Unweighted Phenetic Data

The beginning of the new era in bacterial systematics can be traced back to the introduction of numerical taxonomy. Computer-assisted classification or **numerical taxonomy**, the grouping by numerical methods of taxonomic units or taxa on the basis of their character states, was first applied to bacterial classification by P. H. A. Sneath in 1957. The most widely used approach to numerical taxonomy in bacteriology is based upon the five Adansonian principles outlined in the 18th century by the botanist M. Adanson: (1) the ideal "natural" taxonomy is that in which the taxa have the greatest information content, i.e., they are based on as many characters as possible, (2) every character is given equal weight when constructing "natural" taxa, (3) overall similarity (affinity) is a function of the proportion of characters held in common, (4) distinct taxa are based on correlated features, and (5) affinity is treated independently of phylogeny.

When introduced, the numerical taxonomic procedure was in striking contrast to the prevailing orthodoxy of delineating and recognizing taxa by a few subjectively chosen behavioral, morphological, and staining properties. Taxonomies based on single characters or a series of single characters (monothetic) are notoriously unreliable since they have a low information content and cannot accommodate strain variation ("mutants") or test errors. In contrast, numerical taxonomies have a large information content and can encompass a degree of strain variation. Furthermore, numerically circumscribed groups are polythetic, i.e., no single character is essential to group membership or is sufficient to make an organism a member of a group. The application of **Adansonian taxonomy** led to significant advances in bacterial classification. The main contribution has been the determination of homogeneous groups that can be equated with taxospecies, namely groups of bacteria that share a high proportion of common characters. The taxonomy of many bacterial genera has benefited from a revision of their pre-1960 classification by the application of numerical taxonomic methods; good examples include the genera *Bacillus*, *Mycobacterium*, and *Vibrio*. Numerical taxonomic databases are also valuable information storage and retrieval systems that are replete with information on the phenotypic properties of the constituent taxa and are thereby of interest to scientists with diverse interests and requirements. The steps involved in a typical numerical taxonomic study are shown in Figure 29.**2**.

29.2.1 Choice of Strains and Tests, and Coding of Data May Affect the Classification of Bacteria in Numerical Taxonomic Studies

General considerations. The strength of the numerical taxonomic method depends on its capacity to process large amounts of data on many strains. All of the phenetic characters used reflect a portion, however small, of the genome (chromosomal and plasmid-encoded genes) of the bacterium. In theory, the more phenetic characters that are examined, the better will be the measure of the phenetic relatedness, and hence partially the genetic relatedness, between strains. It is imperative to choose strains, characters, and computer programs judiciously, and always in keeping with the

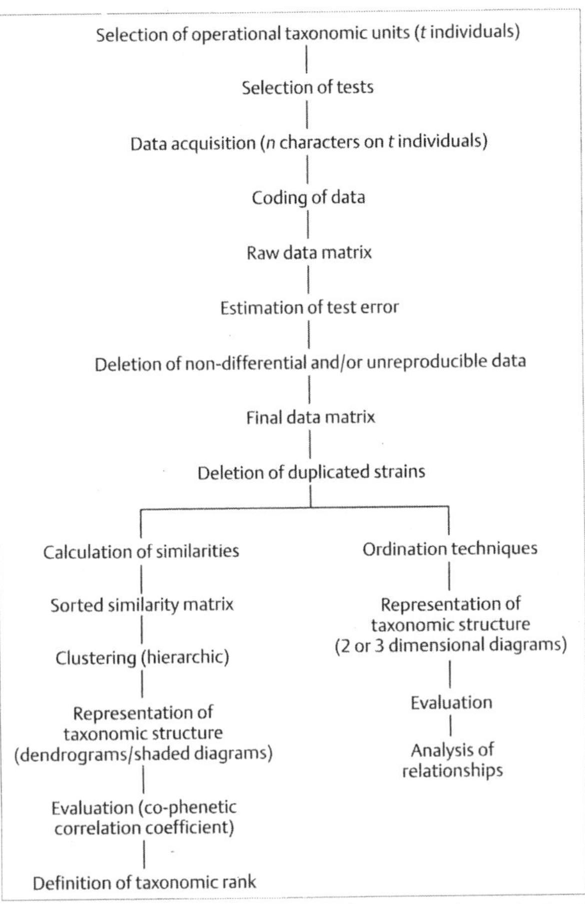

Fig. 29.**2** **Stages in the numerical taxonomic analysis of bacteria**

objective of the intended classification. As the basis of the grouping of strains is similarities among the individuals analyzed, the composition of the groups and overall structure of the classification will be affected by the choice of methods used for calculating the similarities and sorting the organisms into taxa based on overall similarity. Various factors are known or suspected to affect the classification of bacteria in numerical taxonomic studies. These include the choice and number of strains and tests, test of reproducibility, and the selection of data-handling procedures.

Choice of strains. The objects to be classified, usually strains, are referred to collectively as **operational taxonomic units (OTUs)**. Where possible, these should include nomenclatural type strains, well-studied authenticated cultures, and marker strains outside the area of study. Moreover, fresh isolates should be examined since strains that have been repeatedly subcultured may not be good representatives of established taxa. About 10% of the strains should be examined in duplicate and treated as separate OTUs to provide an internal check on test reproducibility. In theory, about 25 strains are needed to define the center and radius of a taxospecies accurately. It is now feasible to examine several hundred OTUs in a single analysis, especially if automated data acquisition procedures are used.

Choice of tests. The characters used in computer-assisted classification are based on the results of tests carried out on the strains under study. It is important to use characters that are genetically stable and not overtly sensitive to experimental or observational uncertainties. The usual procedure is to take a selection of biochemical, cultural, morphological, nutritional, and physiological characters to represent the entire phenome, i.e., the genotype and the phenotype. It is important to have sufficient information to discriminate between taxa. At least 50 and preferable several hundred **characters** are needed, although with high numbers of features, any gain in information decreases disproportionately to the effort involved in securing the data. However, abundance of characters is not sufficient; they should also be of high quality. To achieve this, tests need to be carried out under rigorously standardized conditions.

Rapid data acquisition. The development and application of automated procedures facilitate the rapid generation of databases. **Automated systems** have been used to detect specific enzymes in small amounts of inocula using conjugated enzyme substrates based on the fluoriphores 7-amino-4-methylcoumarin and 4-methylcoumarin. **Conjugated derivatives** of the molecules, when cleaved by the appropriate enzymes, release parent molecules, which are intensely fluorescent in the visible region of the electromagnetic spectrum; the conjugated substrates only exhibit weak fluorescence. In a typical automated system, the conjugated substrates in 96-well microtiter plates are inoculated automatically and incubated for a few hours, and the results are read on a fluorimetric plate reader attached to a computer to allow instant data acquisition. Appropriate cut-off values are set to distinguish positive from negative or weakly positive strains. A similar system based on **microtiter trays** containing 96 substrates and a tetrazolium dye can be used to determine the ability of test strains to use a diverse range of sole carbon compounds for energy and growth.

Coding of data. Operational taxonomic units are coded for a large number of properties or bits of information, each of which is termed a **unit character**. The methods used to code unit characters depend upon the nature of the tests and on the requirements of the computer program(s). Most unit characters tend to be binary or two-state and are coded 1 (positive) or 0 (negative). When more than two grades of reaction are recognized (e.g., weakly and strongly proteolytic), additive coding is usually applied since it preserves information about the magnitude of the character. Thus, a test with three characters states can be converted into binary (i.e., two-state) characters as follows: negative, 0 0 0; weak positive, 1 0 0; moderate positive, 1 1 0; strong positive, 1 1 1.

Qualitative or disordered multi-state characters, such as colony color, may also occur. Such characters are converted into two-state or binary form using the mutually exclusive method of coding where an OTU with a particular character state or property is coded positive (1) for that character state and negative (0) for all the remaining character states. An important character state that can be applied to all types is "missing" or NC (no comparison). Such NC characters arise, for example when a test was not done or was carried out incorrectly. NC characters also arise for logical reasons; an example is spore shape (cylindrical, oval, round) when sporulation is not found. The presence of a high proportion of NC values can give misleading results, but a scattering of NC entries is usually acceptable.

29.2.2 Evaluation of Test Error Is a Crucial Step in Testing the Reliability of Numerical Taxonomic Data

Estimation of test error. By examining cultures in duplicate, initially in a code, and treating them as individual OTUs, the average probability of test error can be estimated from an analysis of test variance. The variances of individual tests between replicate organisms (s_i^2) can be calculated from Eqn. 29.1 given in Box 29.3.

Test error has the general effect of lowering similarities between strains, and when high, of eroding the taxonomic structure. There is a rapid erosion of taxonomic structure when $p > 0.1$. Therefore, information on individual tests that show a probability of test error of over 10% should be excluded from raw $n \times t$ matrices. It is important to balance the detrimental effect on taxonomic structure of individual tests with a high error against information loss of test results deleted from data matrices. Sources of test error include variation in the organisms, test variability, and operator error.

Box 29.3 Test error can be estimated from an analysis of test variance, $s_i^2 = n/2t$, where n is the number of OTUs with discrepancies in the test and t is the total number of duplicated strains. The probability of an erroneous result for an individual test (p_i) is:

$$p_i = 1/2\left(1 - \left[\sqrt{1 - 4s_i^2}\right]\right) \cdot 100 \qquad (29.1)$$

where i is the character and s is the test variance. Individual test variances can be averaged to determine the pooled variance (s^2):

$$s^2 = (1/n)(s_1 + s_2 + \ldots\ldots s_n) \qquad (29.2)$$

where n equals the total number of tests, and s_1, s_2, and s_n correspond to individual test variances for tests 1, 2, and so on to n. The pooled variance may then be used to determine the average probability of an erroneous results (p):

$$p = 1/2\left(1 - \left[\sqrt{1 - 4s^2}\right]\right) \cdot 100 \qquad (29.3)$$

29.2.3 Several Algorithms Are Available for Clustering Strains on the Basis of Similarity Values Derived From the Use of Resemblance Coefficients

Computation of resemblance. The similarities and dissimilarities between test strains can be estimated once an $n \times t$ matrix has been completed and data are entered into the computer. Many resemblance coefficients are known, although few have found favor in bacterial taxonomy. The two most commonly used are the **simple matching** (S_{SM}) and **Jaccard** (S_J) **coefficients**, which measure similarities between OTUs based on binary data. It is good practice to use both coefficients. The S_J coefficient is often applied to ensure that relationships detected using the S_{SM} coefficient are not based on negative correlation, i.e., the OTUs are considered to be similar because they have a high number of negative characters in common. Negative measures are not necessarily a measure of similarity as strains may fail to give positive results for entirely different reasons. For example, some strains may be genetically unable to give a positive response, whereas other organisms may simply be unable to do so under the test conditions. The S_J coefficient is especially useful in studies that involve both fast- and slow-growing organisms.

Hierarchical clustering. OTUs are usually ordered into groups of high overall phenetic similarity by means of one or more commonly used agglomerative clustering methods (Fig. 29.**3**). In general, these begin by searching the similarity matrix for the highest value between any pairs of strains. This pair then forms a group or cluster. The similarities between this group and each of the remaining strains are computed to find the next highest similarity. This may be between two other OTUs or between the previous pair and another OTU. This process continues and at each cycle, strains are added to clusters or clusters join until all of the OTUs are included in a single cluster ($t - 1$ cycles for t OTUs).

Clustering techniques vary in the definition of the similarity between an OTU and a group and, more generally, between two groups. The single linkage (nearest neighbor) technique defines the similarity between two groups as the similarity of the two most similar OTUs, one in each group (Fig. 29.**3**). Average linkage takes the average of all the similarities across the groups. The most popular variant of the average linkage method, the **unweighted pair group method with averages (UPGMA)** takes the simple arithmetic average of the similarities across two groups, with each similarity having equal weight. The similarity levels at which clusters are defined are influenced by the clustering algorithms used.

The most common measure of hierarchicalness is the **co-phenetic correlation coefficient (r)**, which is the correlation coefficient between the two data sets mentioned above and exemplified in Figure 29.**3**. Typical co-phenetic correlation values are in the range of 0.6–0.95. Above 0.8 is usually reasonably good; values below 0.7 suggest that only limited confidence can be given to the relationships expressed in the dendrogram. Complete agreement between dendrograms and similarity matrices cannot be achieved given the taxonomic distortion introduced when representing multidimensional data in two dimensional form. UPGMA clustering always gives the highest co-phenetic correlation of all hierarchical clustering methods. This in itself is a good

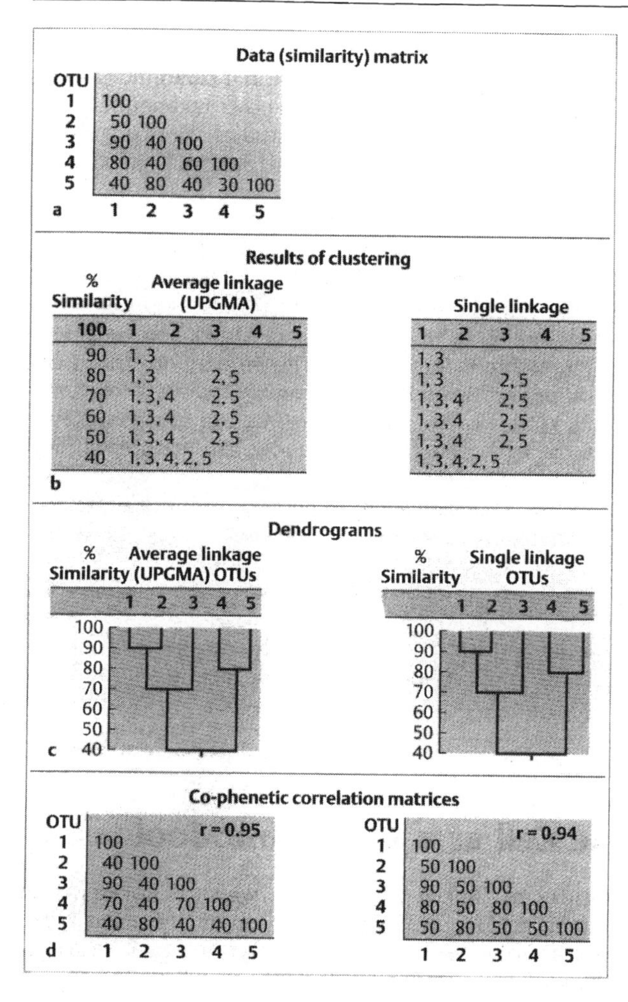

Fig. 29.**3a–d** **Major steps involved in the numerical taxonomic procedure.**
a Test strains are examined for a large number of phenotypic characters, and overall similarities between the strains are expressed as percentages in an unsorted similarity matrix,
b strains that resemble one another are grouped together using clustering algorithms,
c the products of the cluster analyses are expressed as dendrograms, and
d cophenetic correlation values (r) above 0.8 indicate that data are suitable for hierarchical analyses (see text for details)

reason for using this algorithm. The results of clustering are usually presented in the form of a dendrogram where the tips of the branches represent the OTUs and the axis at right angles to the tips is the similarity axis, which shows the similarity values at which the groups form (Fig. 29.**3**).

Non-hierarchic methods are valuable in their own right, but are especially useful when data are not suitable for hierarchical analysis. The non-hierarchical methods most commonly used in bacterial taxonomy are principal components analysis, principal co-ordinates analysis, and non-metric multidimensional scaling. All of these are ordination methods that essentially reduce the number of dimensions (one for each character) in the data so that the relationships given by the entire character set can be summarized by a much smaller number of dimensions, usually two or three, for viewing as maps or models. The co-ordinate axes are chosen so that the first axis expresses the greatest spread or scatter of the OTUs, the second accounts for the next greatest scatter, and so on. Ordination methods are considered satisfactory if the

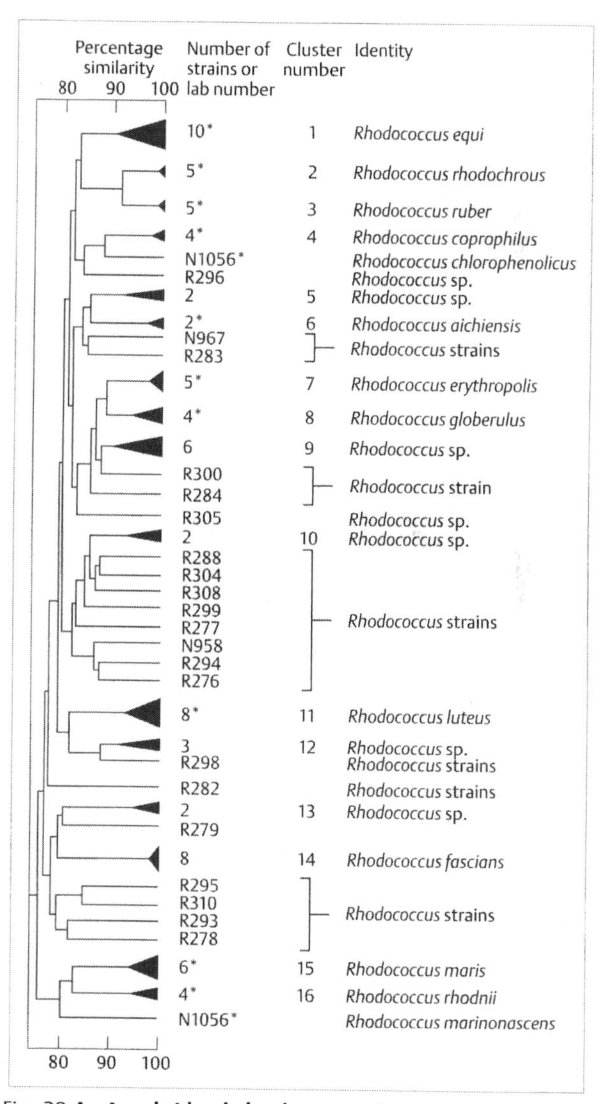

Fig. 29.**4** **An abridged dendrogram showing the relatedness between Rhodococci.** Clusters are defined at or above the 89% similarity level in an S_{SM}, UPGMA analysis. Asterisks denote type strains or clusters containing type strains

variation in the plotted principal coordinates is 40% or more of the total original variation.

29.2.4 The Hierarchic Clustering of Strains Can Be Used to Help Define Their Taxonomic Rank

Definition of taxonomic rank. Results are usually expressed as shaded diagrams or dendrograms (hierarchic methods) or as ordination plots (non-hierarchic methods). Dendrograms are usually simplified by grouping together strains that fuse at pre-determined levels of similarity. These groups or clusters are presented as shaded triangles where the length of the bottom edge of the wedge is proportional to the number of OTUs (Fig. 29.**4**). It is evident from Figure 29.**4** that rhodococci fall into several well-circumscribed taxospecies, not all of which have been validly described.

The results of hierarchical clustering can be used to help decide the taxonomic rank and eventually to assign names to bacterial groups. The resemblance level at which species (and genera) should be separated depends on several factors, notably the strain collection, character set, and statistics employed. It is not always straightforward to define ranks by drawing lines across dendrograms or circles on ordination plots. However, despite such problems, there is usually good congruence between the results of numerical taxonomic surveys and classifications derived from other taxonomic methods, notably DNA relatedness studies. The congruence between taxospecies and genomic species is particularly well-exemplified by rhodococci.

The numerical taxonomic method requires the conversion of information about taxonomic entities into numerical quantities. The primary objectives of numerical taxonomy are to assign individual bacterial strains to homogeneous groups or clusters using large sets of phenetic data and to use features of the numerically circumscribed groups to generate improved identification schemes. Numerical taxonomies are derived from phenetic data, i.e., from similarities between organisms based on observable and recorded characters without regard to ancestry. This means that the relationships between organisms and any hierarchies based on these relationships are phenetic and not phylogenetic.

29.3 Chemical Characterization of the Cell as a Taxonomic Tool

The use of the chemical composition of the cell in classifying the strains, often referred to as **chemotaxonomy** or chemosystematics, is an aspect of taxonomy that uses physico-chemical methods such as gas chromatography (GC), thin layer chromatography (TLC), high performance liquid chromatography (HPLC), and various forms of spectroscopy. In some cases, the chemical composition of the cell allows the rapid characterization or delineation of a taxon, but this is usually supported by other data.

29.3.1 Components of the Cell Wall are Taxonomic Markers

Peptidoglycan, a macromolecule with subtle differences. Initial studies indicated that peptidoglycan occurs universally in the cell wall of prokaryotes; however, this has been shown to be incorrect. In many Archaea (methanogens, non-methanogenic thermophilic Archaea, and the halobacteria) and certain members of the Bacteria (e.g., planctomycetes, which constitute a distinct evolutionary group, and the mycoplasmas), this polymer is absent. The basic structure of peptidoglycan (see Chapter 7.9.5) consists of two glycan chains (comprising an alternating amino sugar and a uronic acid) cross-linked by a peptide, which is linked to the uronic acid (Fig. 23.**1**). The most commonly encountered structure is the murein-type structure found in members of the Bacteria (eubacteria) in which the uronic acid is **muramic acid** and the amino sugar is ***N*-acetyl-D-glucosamine**. Uncommon modifications include phosphorylation or acetylation of the hydroxyl group of the C-6 carbon atom of muramic acid, lack of acetylation of the amino group, or oxidation of the *N*-acetyl residue to give an *N*-glycolyl residue. Whereas little or no variation in the peptidoglycan has been reported within the "classical Gram-negative" organisms, Gram-positive bacteria produce a diverse range of structural variants, in which the amino acid composition of the peptides cross-linking the glycan strands (**interpeptide bridge**) varies (see Chapter 23.1). A complex nomenclature has been developed to cope with the large number of different variants. The one commonly used was developed by O. Kandler and colleagues; two basic structures are recognized: the type A (Fig. 7.**31c**) and type B peptidoglycans (Fig. 29.**5**).

Gram-positive methanogens, which belong within the Archaea, have glucosamine or galactosamine as the amino sugars and *N*-acetyltalosaminuronic acid in their peptidoglycan, which is termed pseudomurein. The

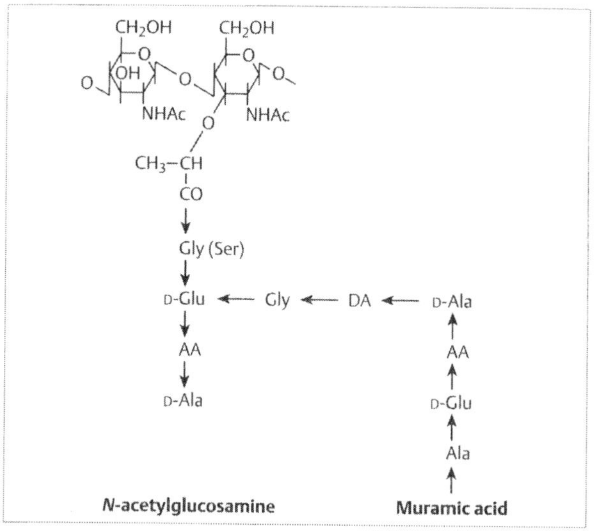

Fig. 29.5 **Generalized structure of B-type peptidoglycan.** DA, Diamino acid (e.g., diaminopimelic acid or lysine); AA, amino acid

most important aspect in the structural variation of these polymers, however, is the variation in the peptide that invariably cross-links the glycan strands (interpeptide bridge). An interesting feature of the interpeptide bridge is the predominance of D-amino acids and the absence of aromatic amino acids in the linking of the peptidoglycan strands, making the polymer resistant to L-proteases such as trypsin or chymotrypsin.

Components other than peptidoglycan. Much less attention has been given to other components in the cell wall of prokaryotes. Those components considered as being additional constituents of the cell wall include teichoic and teichuronic acids (usually in Gram-positive organisms), lipopolysaccharides, and proteins or glycoproteins. In some cases, an outer proteinaceous layer is termed an S-layer; however, in many members of the Archaea in which peptidoglycan is absent, such layers obviously contribute to the maintenance of cell shape and can be regarded as cell wall components. Within the Archaea, there are a variety of non-peptidoglycan cell wall types, ranging from purely proteinaceous, via glycoprotein, to the glycan strands of *Methanosarcina* and *Halococcus* species.

Heteropolysaccharides may have structural analogues in eukaryotes. The two best-documented examples of cell walls in which a heteropolysaccharide structure plays an important role are those found in *Methanosarcina barkeri* and *Halococcus morrhuae*. The structure of the heteropolysaccharide in *Methanosarcina barkeri* (and probably in other *Methanosarcina* species) is essentially a repeating glycan strand. Because of its similarities to chondroitin (a compound produced by eukaryotes), it has been called **methanochondroitin**. The heteropolysaccharide found in *Halococcus morrhuae* is considerably more complex, comprising at least three regions of repeating monomers, each with different degrees of complexity. The polymer is heavily sulfated, a feature also found in the glycoprotein of *Halobacterium salinarum*.

The bacterial cell wall is responsible for maintaining the shape of the cell. Although early studies indicated that peptidoglycans were universally distributed, many exceptions have been found. During the course of evolution, Bacteria and Archaea have developed a range of cell wall components, and the structural variation within the components is important in a modern taxonomic system

29.3.2 Components of the Outer Membrane Also Provide a Rich Spectrum of Taxonomic Markers

Many Gram-negative bacteria, particularly members of the class Proteobacteria, have only a thin peptidoglycan layer and an additional membrane, the **outer membrane**, which lies outside the murein sacculus (structure, see Chapter 2; biosynthesis, see Chapter 23).

The most conservative part of lipopolysaccharides (LPS) is the lipid A region, a reflection of its role in anchoring the molecule in the outer membrane. In members of the genus *Salmonella* (and many other genera), lipid A contains two glucosamine residues linked via a $1 \rightarrow 6$ glycosidic bond (Figs. 23.**3** and 29.**6a**). In *Rhodopseudomonas viridis, Rhodopseudomonas palustris*, members of the genera *Brevundimonas, Bradyrhizobium, Afipia, Nitrobacter, Brucella, Agrobacterium*, and *Rhizobium*, this dimer is replaced by 2,3-diamino-2,3-dideoxy-D-glucose, either as a monomer or as a dimer (Fig. 29.**6b, c**). While this region of lipid A is relatively constant, other variations can be found in the fatty acids attached to the amino sugar backbone. Whereas fatty acids linked to the glucosamine residues are amide-linked to the amino group of glucosamine and esterified to the hydroxyl group of the third carbon atom (Fig. 29.**6b, c**), the fatty acids in the 2,3-diamino-2,3-dideoxy-D-glucose monomer are exclusively amide-linked to the two amino groups. In the dimer found in *Brevundimonas diminuta*, *N*-acyl chains predominate, but there are some *O*-acyl chains. Certain organisms synthesize "mixed" lipid A, in which both glucosamine and 2-diamino-2-dideoxyglucose are present, varying from almost equal proportions to one predominating over the other. In lipid A types, the *N*- or *O*-acyl chains are usually β-hydroxylated fatty acids (3-OH fatty

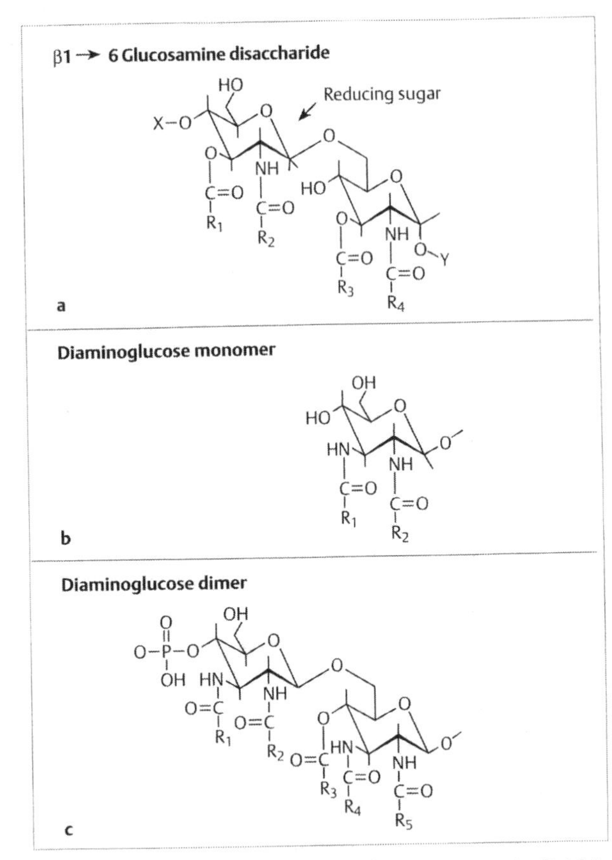

Fig. 29.**6a–c Examples of structural variations in the lipid A region of lipopolysaccharides.**
a $\beta 1 \rightarrow 6$ glucosamine disaccharide,
b diaminoglucose monomer,
c diaminoglucose dimer

acids), although 3-oxo fatty acids have also been reported, and other fatty acids can in turn be attached to them via esterification to the free hydroxyl groups. In some instances, not all of the sugar-backbone-linked 3-OH fatty acids are further esterified, and either only the hydroxylated fatty acids on one glucosamine are further esterified (typical of members of the *Enterobacteriaceae*), or only the amide-linked fatty acids are further esterified (typical of certain members of the β-subclass of the Proteobacteria).

The R-core region is more variable than the lipid A region and can show a degree of variation within members of one species, whereas in other species there is little or no variation. Some variations of interest are the presence or absence of the heptose sugars or of 2-keto-3-deoxy-D-mannose.

The O-chain shows the greatest degree of variation; in some cases it is absent, while the length of the polysaccharide repeating units also varies between taxonomic groups. When the O-chain is absent, this can be a feature of all strains within a species or it can be due to a strain-specific feature. In members of the *Enterobacteriaceae* the absence of an O-chain is responsible for the formation of rough mutants. The serotyping based on the variation of O-chain typing was extensively used for taxonomy of *Salmonella* and other enteric bacteria.

The presence of an outer membrane has been considered to be a feature characteristic of Gram-negative bacteria and appears to be limited to members of the Proteobacteria. The presence of lipopolysaccharides is not a ubiquitous feature of the outer membrane. In organisms not belonging to Proteobacteria, other compounds play an analogous role to lipopolysaccharides in mediating cellular responses. The diversity of compounds detected and the structural variation found within them make such compounds useful taxonomic markers.

29.3.3 Some Taxa Have Additional Taxonomic Signature Markers

Mycolic acids are restricted to a certain group of organisms within the high G + C Gram-positive bacteria, particularly members of the CNM (*Corynebacterium*, *Nocardia, Mycobacterium,* and related taxa) branch of the Actinomycetes. These compounds are unusual covalently bound, long-chain alcohols and fatty acids located on the external surface of the cell. They are generally esterified to an arabinoglycan, which is attached to the peptidoglycan. When present, mycolic acids provide useful additional chemical data for the delineation of taxa since they vary considerably in their total chain length (20–80 carbon atoms), degree of unsaturation, nature of additional substitutions (keto

α-Mycolate

$$CH_3{\cdot}(CH_2)_l{\cdot}X{\cdot}(CH_2)_m{\cdot}Y{\cdot}(CH_2)_n{\cdot}\underset{}{CH(OH)}{\cdot}CH(COOH){\cdot}(CH_2)_p{\cdot}CH_3$$

X = *cis*–CH=CH–, or *cis*–CH–CH– (bridged by CH_2)

a Y = *cis*–CH=CH–, or *trans*–CH=CH–CH(CH_3)–CH_2–, or *cis* –CH=CH– (bridged by CH_2)

α′-Mycolate

$$CH_3{-}(CH_2)_l{-}CH{=}CH{-}(CH_2)_m{-}CH(OH){-}CH(COOH){-}(CH_2)_n{-}CH_3$$

b

Fig. 29.**7 a, b Generalized structures of mycolic acids** Mycolic acids found in members of the genera *Corynebacterium, Nocardia, Mycobacterium,* and related taxa are shown
a α-mycolates,
b α'-mycolates

methoxy, or epoxy groups), and relative lengths of the acyl and alcohol chain of the molecule (Fig. 29.**7**).

29.3.4 The Cytoplasmic Membrane Is Universally Present and Chemically Diverse

Lipids are essential for membrane integrity. Details of the ultrastructure of the cytoplasmic membrane are given in Chapter 2. A wide diversity of lipophilic compounds have been isolated from prokaryotes, many of which are associated with the cytoplasmic membrane (see Box 29.4). These range from non-polar compounds, such as squalenes, to more polar compounds, such as glycolipids that contain four or five sugars. Essentially, membrane lipids can be divided into those containing acyl hydrophilic groups (i.e., fatty acids) or isoprenoid groups (squalene, hopanoids, or polyene carotenoids). Each of these chemical classes ranges in polarity from non-polar (e.g., acyl group: free fatty acids or diglycerides; isoprenoid group: squalene, β-carotene) to polar (acyl group: diacyl polar lipids; isoprenoid group: hopanoids, bacterioruberins).

Pigments are the most conspicuous lipids found in prokaryotes because they impart a color to accumulations of microorganisms. The diversity of pigments includes **carotenoids** (isoprenoid compounds) and non-carotenoids (Figs. 13.**2** and 29.**8**; e.g., xanthomonadin).

Carotenoids have played an important role in the taxonomy of anoxygenic phototrophic bacteria, particularly the purple sulfur and non-sulfur bacteria. For example, those organisms that are brownish yellow when grown under anoxic conditions, but are pink when shaken with air (due to the formation of spheroidinone) all have a high degree of similarity at the 16S rDNA level (i.e., they are members of the genera *Rhodobacter, Rhodovulum, Roseobacter,* and of *Rhodopseudomonas blastica*). (See Chapter 13 for the role of such carotenoids in photosynthesis and Chapter 7.10.4 for their biosynthesis.) The continued importance of carotenoid distribution in prokaryotic taxonomy should not be overlooked, although the same carotenoids may be distributed within distantly related genera.

Polar lipids. Although phospholipids are the most widely known polar lipids, the cytoplasmic membrane may contain phospholipids, glycolipids, aminolipids, polar isoprenoids, and, in certain cases, hopanoids (Fig. 7.**37**). The most commonly encountered lipids comprise a glycerol backbone to which is linked either acyl groups (esterified) or alkyl groups in an ether linkage (Fig. 29.**9**). Occasionally, mixed ester and ether linkages are found, and in the Archaea the side chains are isoprenoid ethers.

Phospholipids are generally named according to the nature of the polar head group (Fig. 29.**10**). Although widely distributed, the only common element among the diverse phospholipids in prokaryotes is the presence of the glycerophosphate group.

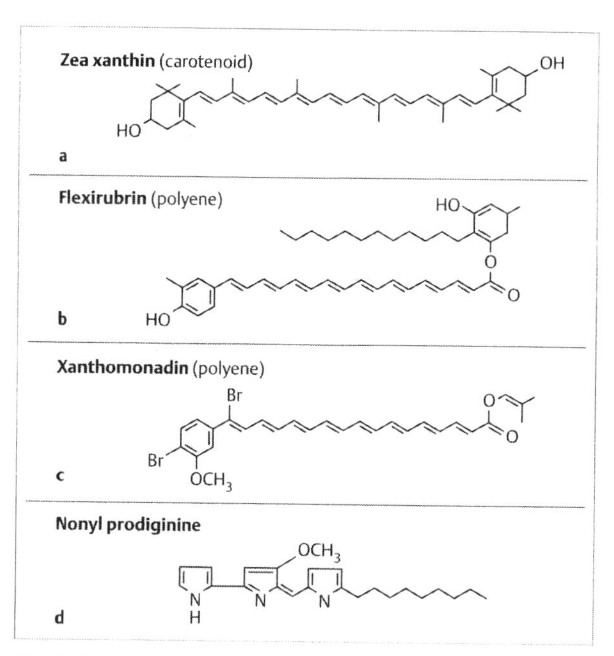

Fig. 29.**8a–d** **Examples of prokaryotic pigments.** Carotenoid and non-carotenoid pigments are shown.
a Zeaxanthin (carotenoid),
b Flexirubin (polyene),
c Xanthomonadin (polyene),
d Nonylprodiginine

Glycolipids in members of the Archaea and many Gram-positive taxa have received particular attention, although they are by no means restricted to these groups. Glycolipids can be divided into mono-, di-, tri-, and tetraglycosyl glycolipids according to the number of sugar groups present. Apart from the number and type of sugars present, another variable feature of structural and potential taxonomic importance is the linkages between the different sugars. Unusual glycolipids include the uronic-acid-containing glycolipids of members of the genus *Brevundimonas*.

Amino lipids. Amino-acid-containing lipids are of significance in a wide range of taxa, although there appears to be no evidence for their universal distribution. The most common types of amino lipids described are those derived from the amino acids serine, glycine, ornithine, lysine, and an ornithine-taurine peptide (Fig. 29.**11**). In many of the chemically characterized amino lipids, a hydroxy fatty acid is linked to the amino group via a peptide bond, and a second fatty acid is linked to the first by esterification of the hydroxy group.

Hydrophobic compounds from the core of membrane bilayers. Fatty acids are the most common

```
                              H
                              |
                         HO-C-H  O
                              |  ||
Diacyl ester              H-C-O-C-R1
                              |
                          H-C-O-C-R2
                              |  ||
a                             H  O

                              H
                              |
                         HO-C-H  O
                              |  ||
Monoacyl monoalkyl        H-C-O-C-R1
-1-enyl ether                 |
                          H-C-O-CH=CH-R2
                              |
b                             H

                              H
                              |
                         HO-C-H  O
                              |  ||
Monoacyl                  H-C-O-C-R1
monoalkyl ether               |
                          H-C-O-CH2-R2
                              |
c                             H

                              H
                              |
                         HO-C-H
                              |
Dialkyl diether           H-C-O-CH2-R1
                              |
                          H-C-O-CH2-R2
                              |
d                             H

                              H
                              |
                         HO-C-H                      H
                              |                      |
Dialkyl tetraether        H-C-O-CH2-R1-CH2-O-C-H
                              |                      |
                          H-C-O-CH2-R2-CH2-O-C-H
                              |                      |
                              H                   H-C-OH
                                                     |
e                                                    H

                              H
                              |
                         HO-C-H  O
                              |  ||
Monoacyl alkyl            H-C-O-C-R1
diol ester                    |
                          H-C-CH2-CH2-R2
                              |
f                             H
```

Fig. 29.**9a–f** **Generalized structure of lipids from the central core of membrane lipids.**
a Diacyl ester,
b Monacyl monoalkyl-1-enyl ether,
c Monoacyl monoalkyl ether,
d Dialkyl diether,
e Dialkyl tetraether,
f Monoacyl alkyl diol ester

components of the hydrophobic region of cell membranes. The hydrophobic region of the polar lipids often, but not always, contains fatty acids. There are over 200 naturally occurring fatty acids (see Chapter 7.10.2). Unusual fatty acids are also known, such as the long-chain dicarboxylic acids (diabolic acids), the hexa- or heptacyclic fatty acids, and the monocarboxylic, mono-alcohol, long-chain alkanes found in certain organisms (Fig. 29.**12**).

Hydroxy fatty acids are common in a wide range of prokaryotes. The most common types of hydroxy fatty

```
    O     H
    ||    |
X-O-P-O-C-H  O
    |     |  ||
    O   H-C-O-C-R1
          |
        H-C-O-C-R2
          |  ||
          H  O
```

R_1 and R_2 are fatty acid residues;
X is an additional functional group: containing phosphate

Phospholipid	X
Phosphatidic acid	H
Phosphatidyl glycerol	$CH_2OH{\cdot}CHOH{\cdot}CH_2OH$
Phosphatidyl ethanolamine	$CH_2OH{\cdot}CH_2NH_2$
Phosphatidyl methyl ethanolamine	$CH_2OH{\cdot}CH_2NHCH_3$
Phosphatidyl dimethyl ethanolamine	$CH_2OH{\cdot}CH_2N(CH_3)_2$
Phosphatidyl choline	$CH_2OH{\cdot}CH_2N^+(CH_3)_3$
Phosphatidyl inositol	$C_6H_{12}O_6$
Phosphatidyl serine	$CH_2OH{\cdot}CHNH_2{\cdot}COOH$

Fig. 29.**10** **Generalized structure of frequently encountere diacyl phospholipids**

acids are 2-OH- (α-) and 3-OH (β-) fatty acids, althougl *iso*-OH (ω 2-OH-) fatty acids have been reported (Fig 29.**13**). Dihydroxy fatty acids of the 2,3-dihydroxy typ have been reported in certain members of the genu *Legionella*. The distribution of hydroxylated fatty acid cannot be generalized since their distribution gives cannot be generalized since their distribution gives complex pattern. They are found in diverse cellula components, including lipopolysaccharides, phospholi

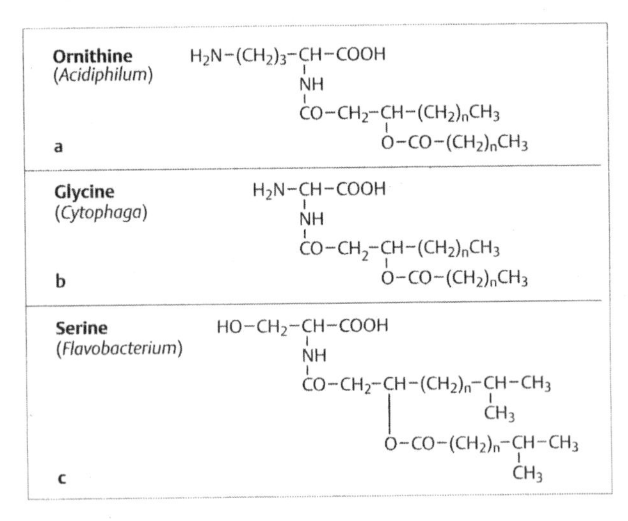

Fig. 29.**11a–c** **Generalized structure of some lipids con taining amino acids and fatty acids.**
a Ornithine (from *Acidiphilum*),
b Glycine (from *Cytophaga*),
c Serine (from *Flavobacterium*)

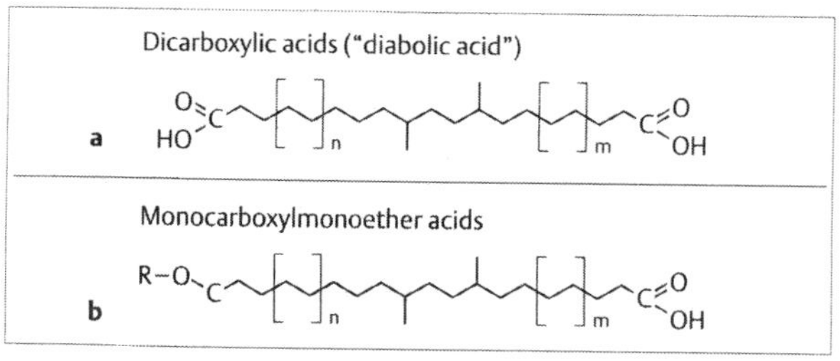

Fig. 29.**12** **Structure of dicarboxylic acids and monocarboxyl-monoether acids**

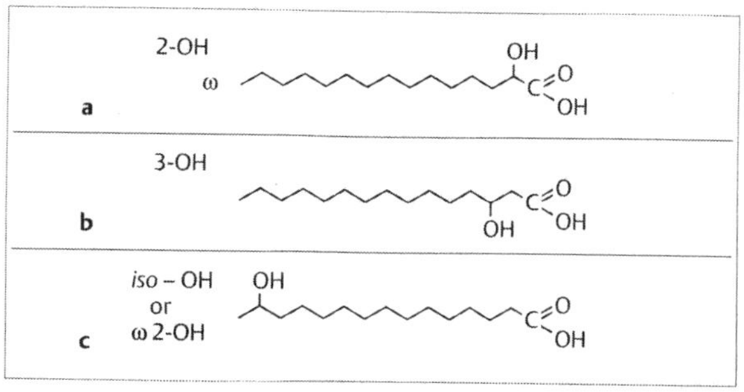

Fig. 29.**13a–c** **Structures of some hydroxylated fatty acids a** 2-OH, **b** 3-OH, **c** *iso*-OH or ω 2-OH

pids, and aminolipids.. 2-Hydroxy fatty acids have been reported in phospholipids of members of the genera *Burkholderia, Nocardioides, Microtetraspora,* and *Saccharothrix.* The presence of 2-hydroxy fatty acids in **sphingolipids** of members of the *Sphingomonas* group, in an amide linkage with the sphingosine residue, provides an interesting diversification of the distribution of hydroxy fatty acids. The unusual *iso*-OH long-chain fatty acids (typically C27–C32) are apparently associated with the lipopolysaccharides of certain members of the α-subclass of the Proteobacteria.

Ether lipids, an unexpected characteristic of the Archaea. The ether lipids found in prokaryotes can be divided into three categories. (Chapter 7.10.5; Fig. 7.**39**): (1) the isoprenoid type, (2) the straight and branched-chain type, and (3) the alk-1-enyl type. The isoprenoid type of ethers were first described in the cell membranes of halobacteria. At first attributed to an adaptation to life at high salinity, the discovery of similar structures in *Thermoplasma acidophila* and members of the genera *Sulfolobus* and *Acidianus* was also taken as an adaptation to extreme growth conditions. Screening a large number of methanogens showed conclusively that they also produced ether lipids with isoprenoid side chains. The presence of these unusual membrane lipids is characteristic of the Archaea, and remains one of the few parameters that quickly allows the recognition of a member of this group,

The ether lipids of these organisms (Figs. 7.**40**, 7.**41**) are unusual in a number of respects. Apart from containing side chains linked in an *sn*-2,3 configuration to glycerol (other non-archaea contain side chains linked in the *sn*-1,2 configuration), these lipids are unique in containing exclusively isoprenoid side chains. The majority of archaeal ether lipids are either glycerol diethers or diglycerol tetraethers in which the C_{40} isoprenoid chains comprise two C_{20} chains linked head to head. One of the interesting consequences of the presence of tetraethers is the availability of the hydroxyl groups at each end of the molecule; these groups can be substituted by polar head groups.

Box 29.4 Some organisms contain atypical membrane components:

Long chain diols can substitute for glycerol. The discovery of alkyl diols in *T. roseum* provided a unique insight into another novel mechanism of cell membrane formation. In the alkyl diols of *T. roseum*, the alkyl side chain functions like a fatty acid, while a single fatty acid is linked to the second hydroxyl group and the first hydroxyl group is substituted with phosphate-containing compounds or undefined glycosyl residues.

Alkylamines are unusual components found in deinococci. In the genus *Deinococcus* the lipid is a complex glycophospholipid in which a phosphatidyl group is linked via a phosphate ester to a glyceric acid residue, which is in turn glycosidically linked to glucose and peptide-linked to a long-chain alkylamine.

Sphingolipids are found in both Eukarya and Bacteria. Sphingolipids are often regarded as typical components of eukaryotic cell membranes. However, both phospho- and glycosphingolipids are found in representatives of certain taxa, including members of the genera *Sphingomonas, Erythrobacter, Sphingobacterium,* and *Bacteroides.* Limited studies on the synthesis of sphingolipids suggest that they are synthesized by the condensation of serine with a long-chain acyl-CoA, followed by decarboxylation and reduction of the 3-keto-derivative.

Capnines may be involved in gliding motility. Capnines are long-chain 1-sulfono-2-hydroxy-substituted alkyl compounds (Fig. 29.**14a**). To date, capnines have only been found in organisms belonging to the *Flavobacterium/Cytophaga/Bacteroides* group. In particular, certain *Cytophaga* and

Capnocytophaga species synthesize these compounds. Free capnines and 2-acyl derivatives have been described (Fig. 29.**14b**). The exact function of these compounds is unclear. Capnines appear to be synthesized by condensation of cysteic acid with a long-chain acyl-CoA and are further modified in a manner analogous to that in sphingolipid synthesis.

Hopanoids are molecular fossils in living organisms. Hopanoids occur in ferns and lichens and in certain fossilized hydrocarbon fractions. These compounds are somewhat similar in structure to steroids, and these two classes of cyclic compounds may have analogous functions in the cell membrane (see Chapter 7). Prokaryotes are not capable of de novo synthesis of steroids. Unlike steroids, which are synthesized via a squalene 2,3-oxide cyclase, which requires oxygen, hopanoids are generally synthesized by squalene cyclase, which does not require oxygen. A variety of hopanoids have been found in prokaryotes ranging from those with only few functional groups to those with tetrol, pentol, or mixed-amino- and hydroxyl-substituted side chains, or to those further derivatized with amino acids, amino sugars, or adenosyl groups. Hopanoids are found predominantly in members of the α-, β-, and γ-subclasses of the Proteobacteria; however, some members of the genus *Alicyclobacillus* also contain hopanoids. None of the members of the Archaea investigated to date produce hopanoids.

The cytoplasmic membrane, present in all prokaryotes, has a similar structural and functional role in all forms of life. Despite this, a diverse range of lipid compounds contribute to the structure of the cell membrane. The diversity of the compounds, their distribution, and their linkage to the underlying synthetic pathways make cytoplasmic membrane components suited for chemotaxonomic purposes.

29.3.5 Compounds Involved in Electron Transport Can Be Used for Taxonomic and Phylogenetic Studies

Cytochromes are constituents of the electron transport system of a wide range of eukaryotic and prokaryotic organism. Detailed studies on the amino acid sequences of cytochromes in the 1960s and early 1970s were important forerunners of sequence analysis in phylo genetic reconstruction (see Chapter 29.1.8).

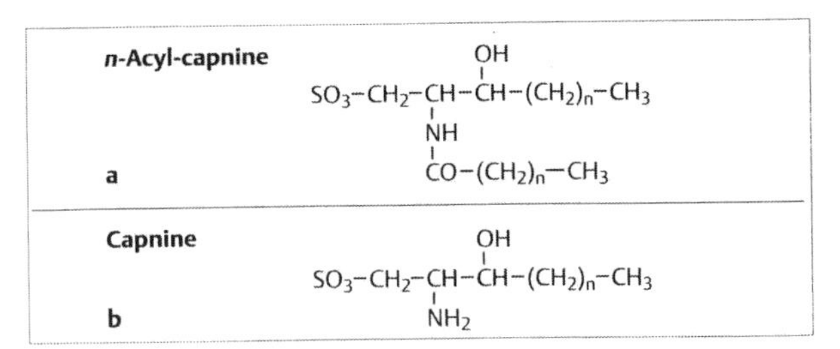

Fig. 29.**14a, b** **Examples of capnine lipids found in member of the genus *Cytophaga*.**
a *n*-acyl capnine,
b capnine

Respiratory quinones are of taxonomic signifi cance because of their structural variation (see Chapter 4 and 11), especially in modification of the ring system or in the isoprenoid side chain.

Ubiquinones (Fig. 4.**8**) are of restricted distribution The predominant variation in ubiquinones is in the length of the isoprenoid side chain, which can vary from Q-7 to Q-14 (based on the major component). Occa sionally, demethoxy derivatives have been found although the significance of their presence is unclear In addition, some obligate methanotrophs synthesize ubiquinones with methylene or with methylene and additional methyl group substitutions in the isoprenoid side chain.

The distribution of **rhodoquinones** appears to be restricted, like that of ubiquinones, to members of the α-, β-, and γ-subclasses of the Proteobacteria. Mod ifications are restricted to the isoprenoid chain length, which varies from RQ-8 to RQ-10.

Plastoquinones appear to be restricted to the cyanobacteria within the prokaryotes. The presence o plastoquinones in higher plants, and in brown, green and red algae is supportive of the ancient-symbion hypothesis, according to which chloroplasts have thei origin in cyanobacteria. This concept is also supported by limited 16S rRNA/rDNA sequencing of chloroplasts.

Despite the importance of ubiquinones in bacteria extensive studies on prokaryotes have shown tha menaquinones are far more widely distributed than ubiquinones. **Menaquinones** (2-methylnaphthoquin ones) are the most common type of respiratory quinones. Some taxa also synthesize demethylmena quinones (naphthoquinones), while others produce monomethylmenaquinones [5 (or 8), 2-dimethyl naphthoquinones] and rarely dimethylmenaquinones [5 (or 8), 6 (or 7), 2-trimethylnaphthoquinoes] (Fig 29.**15**). Methylthionaphthoquinone, which also contains a methylthio group, has also been found. The length o the isoprenoid chain in the menaquinones ranges from

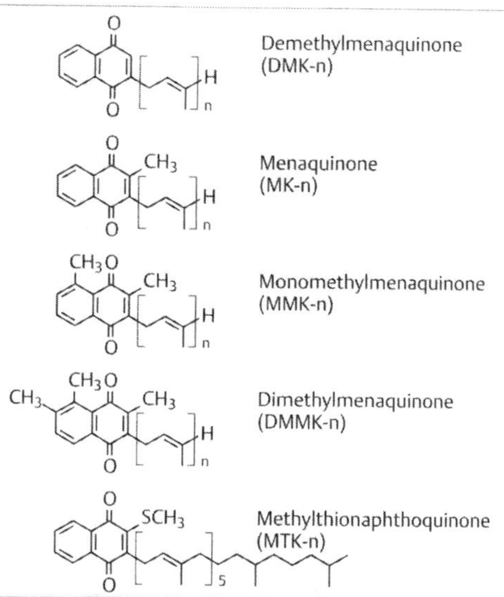

Demethylmenaquinone (DMK-n)

Menaquinone (MK-n)

Monomethylmenaquinone (MMK-n)

Dimethylmenaquinone (DMMK-n)

Methylthionaphthoquinone (MTK-n)

Fig. 29.**15** **Examples of structural variations in the ring nucleus of naphthoquinones**

MK-6 to MK-15. Although many organisms only produce fully unsaturated naphthoquinones, certain organisms (Gram-negative sulfate reducers and members of the high G+C branch of the Gram-positive bacteria) produce partially saturated side chains. The degree of saturation varies from one to four, and each of the positions of unsaturation is highly specific. Each of these structures is evidently the product of enzymes with a high degree of steriospecificity, a feature which is taxonomically useful. An unusual series of menaquinones with a hexacyclic ring in the terminal position of the isoprenoid side chain has been found in all authentic species within the genus *Nocardia*. Certain members of the Archaea contain menaquinones. In members of the genus *Thermoplasma* and members of the family *Halobacteriaceae*, the isoprenoid side chains are either fully unsaturated or dihydrogenated. In members of the genera *Archaeoglobus*, *Thermoproteus*, and *Pyrobaculum*, the side chain of the menaquinones is fully saturated. The presence of a pentacyclic ring in the fully saturated menaquinone from *Pyrobaculum organotrophum* appears to be an unexpected precedent.

Benzothiophenquinones (Fig. 29.**16**) sulfur-containing compounds from sulfur-oxidizing Archaea, are found in members of the order Sulfolobales. Initially described in "*Calderiella acidophila*" (now recognized as *Sulfolobus solfataricus*), calderiella quinone is found in all members of this order examined. Sulfolobus quinone was first indirectly detected in *Sulfolobus brierleyi* (now transferred to the genus *Acidianus*) and was identified in "*Sulfolobus ambivalens*" (a member of the genus *Desulfurolobus* recently transferred to the genus *Acidianus*).

29.3.6 Bacteriochlorophyll and Other Porphyrin Structures Can Be Found in Many Prokaryotes

A diverse range of Mg-tetrapyrrols have been described, ranging from chlorophyll *a*, in cyanobacteria, to bacteriochlorophylls *a*–*e* and *g* in the anoxygenic phototrophic bacteria (Fig. 13.**1** and Tab. 13.**1**). A number of aerobic members of the α-subclass of the Proteobacteria also contain bacteriochlorophyll *a* including members of the genera *Roseobacter*, *Acidiphilum*, and *Methylobacterium*. Many methanogenic Archaea produce a nickel porphyrin compound, Factor F430, which is involved in methanogenesis (Fig. 12.**10**).

Various components found in the respiratory chain show varying degrees of structural variation, which do not significantly alter their functional role. Many of these compounds have a combination of conserved regions (the active site of cytochromes, or the benzoquinone ring structure of ubiquinones) and variable regions (structurally flexible regions within cytochromes, or variations in the isoprenoid chain length and degree of saturation of menaquinones). The combination of conserved and variable regions within the compounds under study enable the construction of taxonomic groups with progressively lower rank.

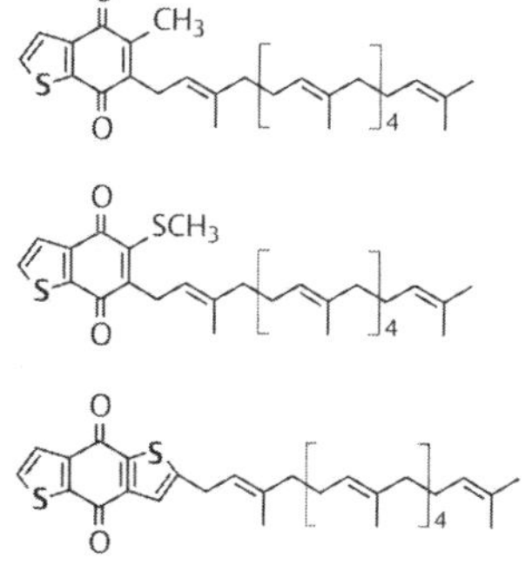

Fig. 29.**16** **Examples of structural variations in benzothiophenquinones**

29.3.7 Polyamines Have a Diverse Range of Functions

Polyamines are long-chain, either straight-chain or branched compounds, with a repeating alkane structure, interspersed with amino groups (see Chapter 7). These compounds do not appear to be universally distributed in microorganisms, and where present, they appear to be involved in a range of different functions. Polyamines have been found rarely covalently bound to peptidoglycan, they may be involved in nucleic acid synthesis and they have been implicated in siderophore functions. Polyamines, where present may be useful markers for the chemical delineation of taxa.

29.3.8 Fingerprints Are a Measure of the Overall Degree of Similarity

Chemical fingerprints It is possible to correlate the presence or absence of certain specific compounds with the taxonomic status of a particular strain. Methods have been adopted that attempt to investigate a large section of the epigenetic information available in whole cells without necessarily attempting to identify the individual components. These methods may be seen as the "epigenetic" versions of methods such as restriction fragment length polymorphism (RFLP), random amplified DNA (RAPD), and ribotyping at the genetic level (see Chapter 29.4). The most commonly used fingerprinting methods are protein profiles, pyrolysis mass spectroscopy, and the recently developed FT-IR methods (Box 29.5). In general, these methods are used where information is desired in a relatively short time on a large number of strains. Their advantages are that they allow the rapid grouping together of strains based on the similarity of the pattern obtained (the phenetic concept of overall similarity plays an important role, see Chapter 29.2). Such methods have their strengths in allowing the assignment of large numbers of strains to known reference strains or species, but they are less suitable for assigning an unknown species to a known higher taxon, particularly with increasing dissimilarity between the known taxon and the new species.

Box 29.5 Some chemical fingerprints are widely used epigenetic taxonomic markers:

Protein profiles. Protein profiling methods generally tend to restrict themselves to the soluble protein fraction obtained from cells grown under controlled conditions. The proteins can then be analyzed by slab gel electrophoresis, 2-D-electrophoresis (See Fig. 20.7), or by capillary electrophoresis. A variation of the method uses highly specific substrates (usually coupled to a redox dye) to detect different isoenzyme activities directly in the developed gels. Protein patterns have been employed to sample a wide range of the phenotype properties of the cell and to provide resolution below the species level when large numbers of strains must be analyzed.

Spectrophotometric methods. Traditionally, the visible spectra of pigmented strains has been used as a taxonomic criterion, such as in anoxygenic phototrophic bacteria, where differences in the carotenoid and bacteriochlorophyll composition give rise to different absorption maxima. Detailed studies on the purified bacteriochlorophylls and carotenoids have provided a sound chemical background. In pigmented prokaryotes, the method can still be used effectively for subgrouping organisms with similar pigmentation (e.g., the group of organisms previously called the "yellow pigmented rods"). The modern spectrophotometric method FT-IR (Fourier transform-infrared) spectroscopy is based on the characteristic absorption maxima of many biological compounds in the infrared region. These properties can be generally attributed to the resonances of the chemical bonds involved. Many compounds containing the same functional group(s) (e.g., peptide bonds, phosphate esters, and carboxylic acid esters) do not give idential spectra because the exact chemical environment causes slight differences in details of the spectrum. Because of the complex chemistry of whole cells used in this technique, it is not easy to predict what factors will affect the nature of the spectrum obtained; however, features such as treatment with antibiotics, differences in the composition of lipopolysaccharides, the onset of sporulation, and different growth media have a clear influence on the fine details of the infrared spectrum obtained.

Pyrolysis mass spectroscopy. Pyrolysis is the fragmentation of a molecule in an inert atmosphere under an appropriate energy influx, such as, heat or laser. In conjunction with gas chromatography, for example, mycolic acid has been analyzed by taking advantage of the cleavage of the molecule between the neighboring hydroxyl and carboxyl groups. Mycolic acid could not otherwise be subjected to gas chromatographic analysis. More recent applications use either pyrolysis, gas chromatography and mass spectrometry, or pyrolysis and mass spectroscopy. The former method gives more detailed mass spectra of individual compounds separated by gas chromatography, whereas the latter method obtains an overall picture of the mass spectrum of all components produced by pyrolysis that can be ionized in the gas phase. The evaluation of the data is generally carried out by suitable computer programs and coupling of the evaluation to neural networks. The method is easy to perform, as is FT-IR, and both require the minimum of sample preparation time and can make use of automatic sampling facilities.

29.3.9 Understanding the Evaluation of the Data Is Critical to Appreciating the Role of Chemotaxonomy

Chemical data have been used to study taxonomic relationships of a variety of groups of microorganisms. Although there is a wealth of data on some groups of microorganisms, our understanding of the chemical diversity of prokaryotes is still incomplete. Further difficulties in the appreciation of the value of chemical data are partly historical and partly due to the lack of concrete theoretical and methodological aspects of data evaluation. Historical problems largely center on the fact that higher taxa defined before the 1970s were often shown to be chemically heterogeneous, which caused taxonomists to have doubts about the usefulness of chemical data in many taxa. However, with our increasing appreciation of the taxonomic heterogeneity of these taxa, the importance of chemical data is being seen in a new light. Although much has been published on the chemical composition of a large number of taxa, little concerning the theory and methods of data evaluation is found in the literature. In the absence of such information, examples of how chemical data are being used (in conjunction with other data) in a modern taxonomic framework is best illustrated using selected examples.

The family Halobacteriaceae, a taxonomic concept based on chemotaxonomy. The various genera within the archaeal family Halobacteriaceae were initially defined by physiological and biochemical parameters, but the chemical composition of the cell membrane has played an increasing role in the past decade. Only recently have 16S rDNA data been shown to give groupings that coincide with those found by chemical analysis (Tab. 29.**1**).

The enterobacteria, a traditional group confirmed in evolutionary studies. The family Enterobacteriaceae was established around a group of organisms of Gram-negative, facultatively anaerobic rods associated with the digestive tract and human diseases. This group of organisms can be regarded as being systematically over-differentiated because of its medical importance and the tendency to create new genera for organisms that cause different medical symptoms (e.g., the genus *Shigella* can be regarded as being a part of the genus *Escherichia*). The closeness of all organisms within this group is illustrated by similarities in the phenotype, chemical composition, and high degree of 16S rDNA sequence similarity. An overview of the chemical composition of the genera within this family is given in Table 29.**2**.

The genus *Pseudomonas*, chemically heterogeneous, or taxonomically unjustified? The genus *Pseudomonas* was originally created in 1894. In the years that followed, an ever increasing number of species were placed in this genus of Gram-negative, aerobic, motile rods. Studies on their chemistry in the late 1960s indicated that many of the species should be removed from this genus. Extensive work on RNA/DNA

Table 29.1 **Chemical composition of members of the family Halobacteriaceae.** PG, phosphatidyl glycerol; PGP, phosphatidyl glycerophosphate; PGS, phosphatidyl glycerosulfate; S-DGD, sulfated-diglycosyl diether; TGD, triglycosyl diether; S-TGD, sulfated triglycosyl diether; TeGD, tetraglycosyl diether; MK, menaquinones

Taxonomic rank	Taxon	Ether lipids	Quinones	Polar lipids
Family	Halobacteriaceae	C20; C20, C25; C20	MK8, MK8 (VIII-H_2)	PG, PGP
Genus	*Haloarcula*	C20; C20,	MK8, MK8 (VIII-H_2)	PG, PGP, TGD-1
	Halobaculum	C20; C20,	MK8, MK8 (VIII-H_2)	PG, PGP, DGD-2
	Halobacterium	C20; C20,	MK8, MK8 (VIII-H_2)	PG, PGP, PGS, TGD-2, S-TGD-2, TeGD
	Halococcus	C20; C20, C25; C20	MK8, MK8 (VIII-H_2)	PG, PGP, TGD-2
	Haloferax	C20; C20,	MK8, MK8 (VIII-H_2)	PG, PGP, DGD-1
	Halorubrum	C20; C20,	MK8, MK8 (VIII-H_2)	PG, PGP
	Natronobacterium	C20; C20, C25; C20	MK8, MK8 (VIII-H_2)	PG, PGP
	Natronococcus	C20; C20, C25; C20	MK8, MK8 (VIII-H_2)	PG, PGP, cyclic PGP

Table 29.2 **Chemical composition of members of the Enterobacteria.** PG, phosphatidyl glycerol; PE, phosphatidyl ethanolamine; DPG, diphosphatidyl glycerol; Qn, ubiquinone n (with n isoprene units); Mkn, menaquinone (with n isoprene units); DMK, dimethymenaquinone (with n isoprene units)

Taxonomic rank	Taxon	Major fatty acids	Quinones	Polar lipids	Hydroxy fatty acids
Family	Enterobacteriaceae	16:0, 16:1, 18:1	Q8	PG, PE, DPG	3OH-14:0
Genus	*Escherichia*	16:0, 16:1, 18:1	Q8, MK8, DMK8	PG, PE, DPG	3OH-14:0
	Salmonella	16:0, 16:1, 18:1	Q8, MK8	PG, PE, DPG	3OH-14:0, 2OH-14:0
	Proteus	16:0, 16:1, 18:1	Q8, MK8, DMK8	PG, PE, DPG	3OH-14:0
	Serratia	16:0, 16:1, 18:1	Q8, MK8	PG, PE, DPG	3OH-14:0, 2OH-12:0
	Citrobacter	16:0, 16:1, 18:1	Q8, MK8	PG, PE, DPG	3OH-14:0
	Yersinia	16:0, 16:1, 18:1	Q8	PG, PE, DPG	3OH-14:0, 3OH-13:0
	Erwinia	16:0, 16:1, 18:1	Q8, MK8, DMK8	PG, PE, DPG	3OH-14:0

hybridization, 16S rRNA cataloguing, and direct sequencing of 16S rRNA or its encoding gene provided evidence for the splitting of the genus *Pseudomonas* into several genera. Many of these genera are only remotely related to the true members of the genus *Pseudomonas*, which centers around the type species of the genus, *Pseudomonas aeruginosa* (see also Table 29.**7**). Comprehensive investigations of the chemical composition of members of the new genera (Tab. 29.**3**) show that each of these can be easily and clearly separated from the species considered to constitute the genus *Pseudomonas*. sensu stricto (in the strict sense).

The microbial cell is a large collection of various organic compounds. The chemical analysis of the microbial cell attempts to analyze individual components as discrete entities (e.g., fatty acids) or to provide an overall picture of the chemical complexity of the cell or parts of its (e.g., by IR spectroscopy).

One of the important aspects in the development of the chemical analysis of microbial cells has been the appreciation that the determinations of certain parameters (such as the isoprenyl ether lipids of members of the Archaea) are powerful and simple routine methods for delineating taxonomic levels. There has been increasing awareness of the significance of chemical data in the half century since the first systematic chemical studies were undertaken, in particular in (1) the methods of evaluating the data, and (2) the biological and potential taxonomic significance of the data in the framework of a modern system.

Table 29.3 **Chemical composition of members of genera formerly in the genus *Pseudomonas*.** PG, phosphatidyl glycerol; PE, phosphatidyl ethanolamine; DPG, diphosphatidyl glycerol; PGL, phosphoglycolipid; UL, uronic-acid-containing lipid; OL, ornithine lipid; Qn, ubiquinone n (with n isoprene units)

Taxonomic rank	Taxon, subclass	Major fatty acids	Quinones	Polar lipids	Formerly
Genus	*Pseudomonas*, γ	16:0, 16:1	Q9	PG, PE, DPG	(Currently) *P. aeruginosa*, type species
	Stenotrophomonas, γ	16:0, 16:1, i-15:0	Q8	PG, PE	*P. maltophilia*
	Methylobacterium, α	16:0, 16:1,18:1	Q10	PG, PE, DPG, PC	*P. radiora*
	Brevundimonas, α	16:0, 16:1,18:1	Q10	PG, PGL, UL	*P. diminuta*
	Sphingomonas, α	16:0, 16:1,18:1	Q10	PG, PE, sphingolipid	*P. capsulata*
	Burkholderia, β	16:0, 16:1,18:1	Q8	PG, PE, OL	*P. cepacia*
	Comamonas, β	16:0, 16:1	Q8	PG, PE	*P. acidovorans*

29.4 Genomic Characterization of Strains and Species

29.4.1 Strains and Species Can Be Identified by Analysis of Protein and Nucleic Acid Patterns

One of the most demanding tasks in microbial systematics is the decision whether a bacterial isolate belongs to a known species or whether it can be considered a representative of a novel species. While appropriately equipped laboratories can handle a small number of organisms in the fields of chemotaxonomy, molecular biology, and physiology, the situation is different when large numbers of strains are involved (e.g., in the pharmaceutical industry and in clinical biology). Here identification must be reliable and rapid and should not require a large spectrum of sophisticated methods. Besides using computer-assisted techniques (Chapter 29.2) and chemotaxonomic techniques (Chapter 29.3), rapid identification can also be achieved by generating patterns from primary semantides, i.e., DNA, RNA and proteins (Tab. 29.**4**); these patterns represent strain-specific fingerprints. The complexity of the patterns depends upon the tools used to cleave, amplify, hybridize, and separate the macromolecules of choice. Separation of DNA fragments by pulsed-field gel electrophoresis, probing of defined genes with labeled probes, and visualization of bands via computerized, laser-analyzed densitometer scanning has improved the resolution and the monitoring of the analyses significantly (see Chapter 17). While protein patterns vary to some extent, as stated previously, the nucleic acid pattern usually is a stable genetic fingerprint. Algorithms are available that transform the banding patterns into dendrograms of similarities.

The decision about which tool to use for strain discrimination needs to be determined experimentally. Because of mutation, even cells within a single colony can differ from one another at the level of the primary structure of DNA (Chapter 15). Hence, the goal of the study decides which banding pattern is useful for answering specific taxonomic questions. Once the optimal set of restriction enzymes, oligonucleotide probes, or oligonucleotide primers has been determined, they can be used in future studies on the same (but not other) group(s) of organisms. Pattern identifications rely on the generation of one-dimensional patterns of semantic molecules or fragments from isolates and from the reference strains to which the isolates are considered to be related. Patterns should neither be too complex to prevent analysis nor too simple to obscure actual genomic dissimilarities. Providing the potential for discrimination, isolates and reference strains that exhibit a high degree of pattern similarity can be considered related. These techniques complement traditional typing methods used mainly in the clinical environment, such as serotyping, biotyping, and phage typing. The decision whether a strain with a unique pattern actually belongs to a described species or should be described as a new species requires more quantitative methods at the genomic level that allows the measurement of the degree of relatedness. The various methods for the characterization of primary semantides listed in Table 29.**4** will be discussed below.

In the **BRENDA** approach, chromosomal DNA is digested individually with different restriction endonucleases and the fragments are electrophoretically separated on a high resolution gel (Fig. 29.**17a**). Depending upon the genome size and enzyme used, the frequencies of restriction sites in DNA vary significantly, leading to profiles of different complexity. The phylogenetic coherency of strains of a species cannot be predicted, and, when the patterns obtained for different strains vary significantly, results of a study might not give immediate answers about the relationships of the unknown isolates. Figure 29.**17a** reveals the clear differences between strains of *E. coli* B and *E. coli* K-12 and demonstrates the genomic heterogeneity of the *E. coli* K-12 strains.

Following transfer of the banding pattern obtained by BRENDA to a membrane, the number of detectable bands is reduced in the **RFLP** approach by hybridization with labeled, cloned random or specific chromosomal DNA fragments having a known or unknown sequence or with labeled synthetic and specific oligonucleotide probes. The presence or absence of the restriction endonuclease sites in the two strains will cause differences in the length of the fragments that contain the targeted gene, and hence a change in the position of the fragments. The complexity of the hybridization

Table 29.**4** **Rapid methods for the characterization of primary semantides**

Bacterial restriction endonuclease nucleic acid digest analysis (BRENDA)
Restriction fragment length polymorphism (RFLP)
Ribotyping
Low-molecular-weight-RNA profiles
Amplified ribosomal DNA restriction analysis (ARDRA)
Random amplified polymorphic DNA (RAPD)
Oligonucleotide probes
Plasmid profiles and plasmid fingerprinting
One-dimensional and two-dimensional protein patterns

patter depends upon the number of target sequences and position of restriction sites.

Analysis of **low-molecular-weight-RNA profiles** takes advantage of sequence differences found in the 5S rRNA and the approximately 60 tRNA species per cell. High-resolution gel electrophoresis leads to taxon-specific separation patterns. Although the majority of class 1 tRNA species are not well-resolved, differences in the length of 5S rRNA and of several class 2 tRNAs give distinct patterns that appear to be species-specific.

Ribotyping is a variation of the RFLP technology in which labeled rRNA-, rDNA-, or gene-specific oligonucleotides serve as a probe. Usually the pattern is more complex than observed after RFLP since rRNA genes are among the few genes in prokaryotes that are present in multiple copies per genome (from 1 to about 12, depending on the taxon). The resolution power of rRNA gene restriction patterns varies with the species studied and the restriction enzyme chosen.

In the **ARDRA** technology PCR-amplified rRNA genes are digested with restriction enzymes and the resulting fragments are separated electrophoretically. Depending on the position of restriction sites, bands of different number and sizes appear, which can be used to screen large numbers of isolates rapidly. Comparison of patterns to those obtained from a database allow assignment of isolates to species or species clusters in those cases where the banding patterns are highly similar.

In the **RAPD** technology, a short oligonucleotide (about 10 nucleotides), usually with random sequence, is used as a primer for PCR (Chapter 17). Depending on the number of complementary or partially complementary target sequences in both DNA strands, DNA fragments are generated whose number and position give an indication about the degree of similarity between strains (Fig. 29.**17b**). For example, using two different random primers, the results of BRENDA analysis was confirmed in that the banding pattern of the *E. coli* B strains was distinct from those of different *E. coli* K-12 strains, which are rather uniform.

Oligonucleotide probing has developed into a powerful identification tool. Short oligonucleotides (up to 50 nucleotides) are extremely useful. These oligomers hybridize to DNA or RNA targets of complementary nucleotide composition in lysed colonies, isolated nucleic acids, or even whole cells (Fig. 17.**11**). Probes are available that target individual genes encoding distinct taxon-specific proteins, plasmids, non-coding DNA (spacers), rRNA genes, or rRNA. The high copy number of rRNA (up to 10^4 per cell) makes it the most favorable target. Ribosomal RNA and rDNA contain sequence motifs that allow the development of taxon-specific probes suitable for the detection of phylogenetically coherent groups above the strain level. Probes used for the detection of almost all medically important bacteria have been described, and many probes are commercially available as part of detection kits. Detection of hybridization has reached a high level of accuracy, and nonradioactive labeling, such as biotinylation, sulfonation, and linkage of fluorescent dyes and digoxigenin-labeled compounds, is replacing radioactive labeling. Techniques that combine non-radioactive labeling with the detection of single cells have made it feasible to identify uncultured endosymbionts and

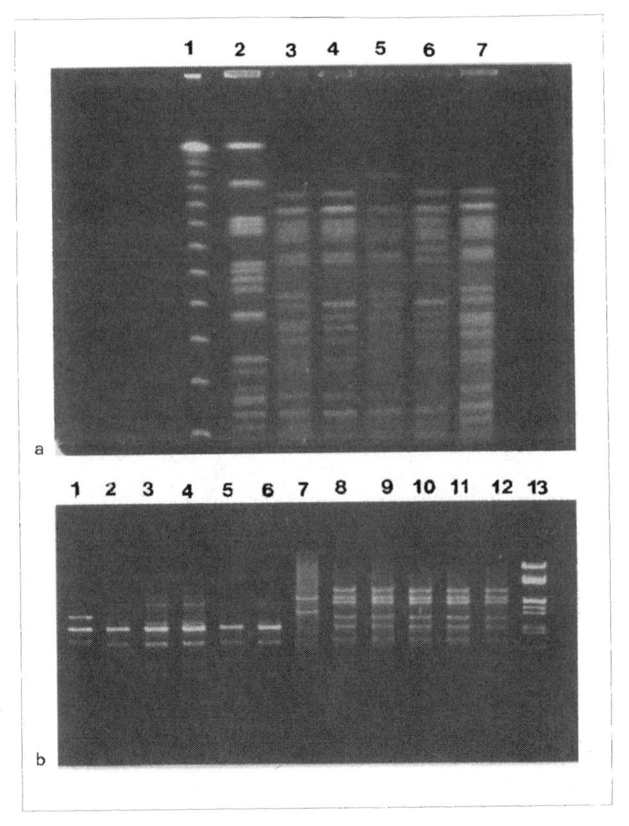

Fig. 29.**17** **Examples of BRENDA and RAPD analysis of genomic DNA.**
a Bacterial restriction enzyme nucleic acid digest analysis (BRENDA) of DNA using enzyme *Sfi*I. Lanes: 1, phage λ DNA ladder; 2, *Escherichia coli* B, wild-type; 3–7, different wild-type strains of *E. coli* K-12 (3–5) and mutants (6, 7). The DNA fragments were separated by pulsed-field gel electrophoresis on a 10% agarose gel.
b Random amplified polymorphic DNA (RAPD) analysis of DNA from strains used in the experiments shown in **a** lanes 2–7, and applied in the same order in lanes 1–6 and 7–12. Lanes: 1–6, primer ARP-2, lanes 7–12, primer ARP-6; lane 13, molecular weight markers. Courtesy of N. Ward-Rainey

parasites, and also microorganisms in their natural habitat.

Plasmid profiles and **plasmid fingerprints**, generated from crude cell lysate by agarose electrophoresis, have been of epidemiological importance for tracing sources of infection by a variety of pathogenic and food-spoiling organisms (Chapter 33). For certain species, plasmid profiling may be the only method to distinguish between strains. Plasmid fingerprinting includes the digestion of isolated plasmid DNA with restriction enzymes, which results in a higher resolution power of the fragments. These fragments can also be analyzed via the RFLP approach.

Electrophoretic analysis of whole cell proteins by **one-dimensional** and **two-dimensional protein patterns** provides a rough measure of the number and physicochemical properties of gene products. Highly standardized conditions and computerized comparison of the electropherograms are necessary to guaranty reproducibility of results obtained from different laboratories (see Chapter 29.3).

Application of molecular methods has significantly improved the circumscription of taxa at all levels of relationships. Knowledge about relatedness among organisms at the level above the genus is informative from an evolutionary and phylogenetic point of view. However, taxonomists working in medical microbiology and in the pharmaceutical industry depend on the availability of pragmatic methods for rapid and reliable identification of strains and species. Clinical and environmental isolates, organisms isolated as food-spoilers and contaminating strains in biotechnological processes need to be identified rapidly to determine whether a strain belongs to a described species or whether it represents a new species, hence possibly having novel properties.

29.4.2 Quantitative Characterization of DNA Is Important in the Determination of Relatedness at the Interspecies and Intraspecies Level

The genome constitutes the total genetic information, mostly present in the chromosome of a prokaryotic cell that is suitable for taxonomic investigation (Fig. 29.**18**) (Chapters 14 and 15). The chromosome size ranges from 80 kb to 120 Mb. Obligately parasitic prokaryotes usually have smaller chromosomes than free-living strains. The chromosome size is a stable character and hence a useful taxonomic feature for strain characterization. The genome (chromosome plus plasmid DNA) size, however, can vary significantly between strains of a species, and this finding, together with its labor-intensive determination, e.g., by CsCl density gradient centrifugation, restricts its taxonomic significance.

Box 29.6 The most widely used method for determining the base composition of DNA depends on the physical property of DNA to denature ("melt") into single-stranded (ss) DNA under the influence of thermal energy. Strand separation is accompanied by an increase in absorbance at 260 nm (hyperchromic shift) of about 30% as compared to that of double-stranded (ds) native DNA; this phenomenon can be monitored spectrophotometrically. The midpoint of absorbance between the double- and single-stranded forms is the **"melting point"**, $\mathbf{T_m}$. The melting temperature at the T_m is a measure of the G + C content. The higher the G + C content, the more thermal energy is required to separate the two DNA strands. Several formula have been described to
calculate the DNA G + C content, the most widely used being

$$\%GC_x = \%GC_{Std} + 2.44[T_{m(x)} - T_{m(Std)}] \quad (29.4)$$

where Std is the standard strain and x is the unknown strain. Alternatively, the second derivative plot of the temperature of inflection (T_i) of the hyperchromic shift can be used to estimate the mol% G + C and requires fewer calculations. Older lit erature also lists DNA G + C values determined by buoyant density, while the HPLC method, developed recently for the determination of the nucleotide concentration of digested DNA is the current method of choice.

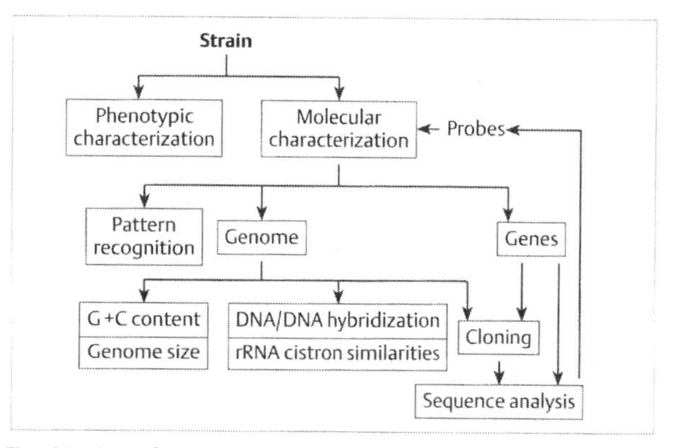

Fig. 29.**18** **Scheme of approaches to measure similarities between DNA and rRNA quantitatively**

The **base composition of DNA**, expressed in mol% G + C (Box 29.6), varies widely with values ranging from 24 (e.g., certain clostridia and mycoplasmas) to 76 mol% (e.g., certain actinomycetes). While the G + C contents of strains of a phylogenetically coherent species usually do not differ by more than a few percent, values between species of a genus may differ more. A range of more than 15 mol% between species of a genus is usually taken as an indication of phylogenetic heterogeneity of the genus (e.g., in *Bacillus, Clostridium*, and *Peptococcus*). Neither genome size nor DNA G+C content can be used to measure phylogenetic distances. This is because the mol% G + C of DNA does not consider the base sequence, which may be significantly different even if the base compositions are identical. The DNA G + C content is an exclusive determinant in that strains that are supposed to be closely related should have a very similar DNA base composition, while strains that show similar DNA G+C values do not necessarily need to be closely related. For example, species of the Gram-positive cocci *Staphylococcus* and *Micrococcus* can be clearly differentiated by their DNA G + C contents (35 and 70 mol%, respectively). Organisms with similar base composition and genome size might not be related; for instance, a G + C content of 60 mol% characterizes members of the unrelated genera *Arthrobacter, Flavobacterium, Paracoccus, Pseudomonas*, and *Spirochaeta*.

29.4.3 Modern Taxonomy Began With the Development of Nucleic Acid Hybridization Techniques

Biologists have witnessed dramatic progress in the elucidation of the nucleic acid sequence of complete genomes. Nevertheless, it is still impossible to obtain sequences of large portions of the genome routinely for measuring the extent of relatedness between taxa. A gross measure of sequence homology can be obtained by DNA/DNA hybridization (Box 29.7). Since this approach gives an average measure of nucleotide similarity of the entire genome, it has advantages over those techniques restricted to the comparison of only individual genes or gene products. The determination of DNA relatedness via hybridization is the standard used to demonstrate the phylogenetic homogeneity at the intraspecific level as well as the distinctness of a species (Fig. 29.**19**). Various hybridization techniques in which hybridization occurs either in free solution or in which one of the DNA samples is immobilized on a membrane filter, are in use. Correlation between similarity values above 30% obtained by different hybridization methods is considered to be good.

Box 29.7 Despite obvious advantages, DNA/DNA hybridization studies have restrictions that must be known prior to interpreting data:

1. Information on which regions of the ss DNA molecules from two organisms are actually participating in the formation of ds DNA hybrids is lacking. For this reason, the term "similarity" is more appropriate than the term "homology", used in older literature (see also Box 29.2).
2. The **discrimination power** of the hybridization approach is very low, and is restricted to the determination of relationships between strains of a species and closely related species. DNAs of strains with a difference in mol% G + C of 10% or more will not hybridize to a measurable extent.
3. Under optimal hybridization conditions (25 °C below T_m of the DNA), two DNA strands must exhibit at least 80–85% sequence complementarity in order to hybridize. Depending in the **sequence similarity** of the reassociating single strands, the hybridization values range between 0% (no hybridization signal) and 100% (as defined by the signal obtained with the identical DNA strands). It is therefore obvious that a given DNA similarity value does not reflect the actual degree of sequence similarity at the level of the primary structure.
4. The **degree of hybridization** is influenced by DNA concentration, fragment size, temperature, salt concentration, denaturing agents, and incubation times, and these conditions must be accurately controlled for reproducible results. In order to increase the relevance of hybridization experiments, a second parameter, the thermal stability of the hybrids formed, is often determined. The presence of 1% unpaired bases within a heteroduplex lowers its T_m value by 1–1.5%. The melting curve of each hybrid is determined individually at the end of the hybridization experiment, and the $T_{m(e)}$ value (e for eluted) of a heterologous hybrid is a measure of the degree of mispaired bases in the hybrid formed during reassociation as compared to the value determined for the homologous hybrid.
5. Unlike sequences, DNA similarity values are not cumulative, and in each experiment, type strains of the group under investigation need to be included as reference strains.

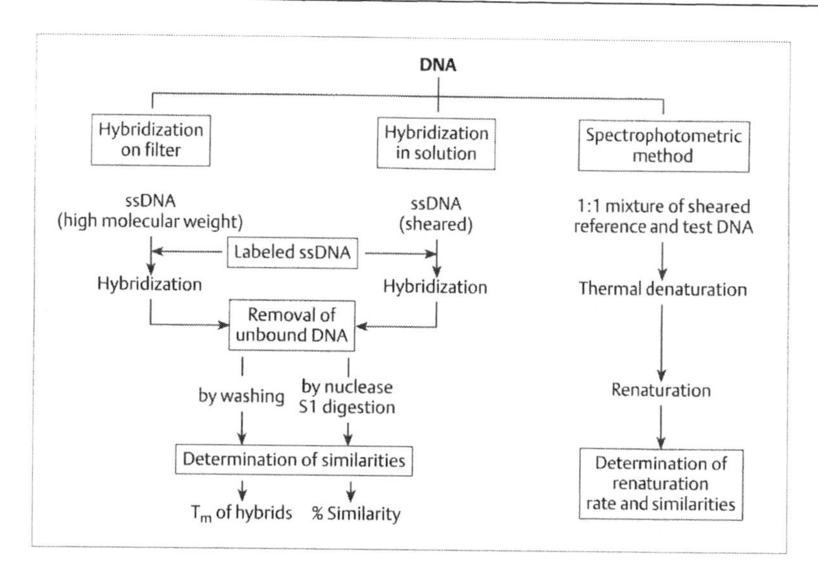

Fig. 29.**19** **Approaches developed for measuring DNA/DNA similarities**

29.4.4 Determination of rRNA Cistron Similarities Led to Insight Into Relatedness at the Interfamily and Intrafamily Level

The nucleotide sequence of genes that code for the large ribosomal RNA (16S and 23S rRNA) species is more conserved than that of the average sequence of the genome. Consequently, analysis of the primary structure of these genes, either directly by sequence analysis (see Chapter 29.5.3) or indirectly by hybridization between rRNA and rDNA cistrons, unravels the degree of remote relationships. The basic mechanisms that cause formation of rRNA and DNA duplexes via reassociation are the same as in DNA/DNA hybridization. Results are presented in two ways: (1) as two-dimensional similarity maps, blotting $T_{m(e)}$ values of DNA/rRNA duplexes against percent rRNA/DNA binding, and (2) as dendrograms based on average linkage clustering of $T_{m(e)}$ values. Highly related strains of a species show $T_{m(e)}$ values of about 80 °C. Depending on the degree of intrafamily and interfamily relationships, good resolution of species is obtained between $T_{m(e)}$ of 80 and 65 °C. The limitation of this method is reached at $T_{m(e)}$ values below 60 °C, where taxa cannot be further discriminated.

All major changes in bacterial systematics witnessed over the last 30 years are mainly based on quantitative measurements of relatedness between DNA and/or ribosomal RNA.

29.4.5 Bacteriologists Work With a Taxonomic Species Definition Rather Than With a Biological Species Concept

The **biological species concept** is associated primarily with the work of zoologists. E. Mayr defined the species as "groups of interbreeding or potentially interbreeding natural populations that are reproductively isolated from other such groups." The problem with this definition in microbiology is that prokaryotes and many eukaryotic microorganisms lack the traditional form of sexuality, and reproductive isolation probably cannot be applied. Exchange of genetic material occurs frequently between prokaryotic organisms, even over large phylogenetic distances, but the extent to which the foreign DNA recombines through homologous recombination with the host DNA differs from taxon to taxon. It has been proposed that the differences in the degree of recombination can be used as a genetic criterion to define a bacterial species. The argument is based on the premise that where recombination between individuals is common (e.g., between closely related strains), phylogenetic trees constructed using different genes should be statistically different. On the other hand, where recombination is rare, (e.g., between strains of less closely related taxa), chromosomal genes of an individual will share a common history, and phylogenetic trees of individuals constructed using different genes should be congruent. Attempts to use this approach to define a species in bacteriology requires gene sequences of many genes from several representatives of a species. Even today, with genome sequencing projects, the task is obviously demanding.

Taxonomists have advanced far beyond the idea formulated approximately 25 years ago that a species is what a competent systematist says it is. At present, it is recommended that prokaryotic strains be allocated to species when they share at least approximately 70% identical base pair sequence (**DNA similarity**) and a difference in the melting point [$T_{m(e)}$] of DNA/DNA homoduplexes and heteroduplexes of less than 5 °C. The borderline value of 70% similarity was chosen because strains shown to be related above this level also show phenotypic similarities in those characters considered to be helpful in species ident ification. The 70% level, however, leaves space for significant genomic, phylogenetic, and phenotypic differences. However, it is recommended to refrain from the description of species as long as differentiating diagnostic characters are lacking.

Even with the generally accepted "species" definition in bacteriology, there is currently no unified concept of a bacterial species with respect to strain diversity or interspecies relationship. Different criteria have been applied historically and a premature nomenclatural change may cause confusion. For example, strains of *Escherichia coli* and *Shigella dysenteriae* are extremely closely related and exhibit DNA hybridization values as high as 89%. Nevertheless, for epidemiological (i.e., practical) purposes, the two taxa are not considered to be strains of the same species, and they are allocated to two different genera. As outlined above, taxonomy is a dynamic discipline with the primary goal being the proper assignment of isolates to a species. New insights into the genotypic and phenotypic properties of a cell may lead to a changing definition. Since the percent DNA similarity level as defined today is arbitrary, taxonomists must be prepared to alter the level and also the concept of using DNA similarity values for species definition when results on novel discriminating properties become available.

The species in bacteriology cannot at present be defined according to the usual biological species concept. A species can be considered as a group of strains characterized by a high degree of similarity in DNA sequence and phenotypic properties. The sum of conserved properties and the extent of DNA reassociation, as measured by DNA/DNA hybridization, plays an important role in the delineation of genomically neighboring species.

29.5 Phylogenetic Trees and Their Interpretation

29.5.1 Only a few Molecules Are Reliable Phylogenetic Markers

Contemporary organisms are the products of evolution, and their structures at all levels of information as defined in Chapter 29.1 (primary, secondary, and tertiary semantides) reflect their evolutionary history. For microorganisms, however, the only historical document is the primary structure of homologous informative molecules. The number and nature of sequence differences among proteins and among genes coding for rRNA and proteins reflect their phylogenies and consequently allow the recognition of pairs or groups derived from a common ancestor. Consequently, contemporary organisms can be ranked according to their evolutionary history.

If changes are selectively neutral, i.e., independent of the overlying phenotype and the selection pressure dictated by its function, it might even be possible to measure the time elapsed since the divergence of the most recent ancestor. The most reliable, "linear" molecular chronometers are randomly evolving "functionless" parts of the genome. In reality, however, all sequenced genes and rRNA species constitute a chimeric structure of portions with varying degrees of conservation depending on their importance for the function of the molecule or the corresponding gene product. Multiple changes accumulated at variable regions will obscure the history of the molecules by stimulating "false" identities and masking the true number of evolutionary events. Given the presence of functionally important and functionless regions, a direct correlation between the degree of sequence dissimilarity and the elapsed time cannot be postulated with certainty. Regions or positions within one molecule with varying degree of sequence conservation are informative for analyzing different phylogenetic levels. While the presence of variable positions is most useful in the determination of close relationships, the more conserved regions report on earlier events in evolution and hence on remote and ancient relationships.

29.5.2 A Reliable Phylogenetic Marker Must Fulfill Certain Prerequisites

The phylogenetic position of species and higher taxa can be determined reliably by analysis of molecular chronometers that meet certain criteria. The most useful molecular chronometers for a given phylogenetic level are molecules with the following characteristics: (1) they have to be universally present among all representatives of a given group of phylogenetically related organisms, (2) they have to be homologous (derived from a common ancestor), as indicated by sequence similarity in combination with constraint in function; given that function is determined by sequence and inversely exacts selection pressure on the sequence, a comparable mode of evolution can only be expected for functionally equivalent molecules, and (3) they have to be "housekeeping" molecules that are genetically stable. Analysis of molecules involved in frequent lateral gene transfer would disturb any phylogenetic conclusions. Results of horizontal gene transfer can be tested by searching for chimeric characters of the molecules potentially resulting from recombination and by comparing the phylogenies of alternative and preferably functionally unrelated markers. An important factor regarding the usefulness of molecules for the reconstruction of major phylogenies is size. Since every individual sequence position can only carry the information on a rather narrow range of evolutionary time, an increasing number of independently evolving positions or regions raises the number of phylogenetic levels that can be detected. The primary structures of phylogenetic marker molecules should contain independently evolving regions (domains), which allow the relevance of local sequence identities or differences for phylogenetic conclusions to be tested.

Reliable phylogenetic markers (molecular chronometers) are ubiquitously distributed, functionally equivalent, and homologous "housekeeping," i.e., vital molecules.

29.5.3 The Large rRNA Molecules Are the Most Frequently Used Phylogenetic Markers

Presently, the most extensively used phylogenetic marker molecules are the larger (16S and 23S) ribosomal RNAs, especially 16S rRNAs (Box 29.8). They provide all prerequisites of a phylogenetic marker molecule, as defined above. Their primary structures are alternating sequences of invariant, more-or-less-conserved to highly variable regions. The frequencies of compositional changes at different positions in the molecules vary greatly. This is illustrated in the schematic secondary structure models of 16S rRNA (Fig. 29.**20a**) and 23S rRNA (Fig. 29.**20b**), which are based on the primary structure of these molecules from *Escherichia coli*. The conservation profiles have been derived from approximately 1800 16S rRNA and 100 23S rRNA primary structures from representatives of all known major phylogenetic groups of the domain Bacteria. Besides their use in unraveling the relatedness among the entire range of taxa, invariant and highly conserved positions are essential for the recognition and alignment of homologous sequences.

Box 29.8 Comparative sequence analysis of 16S rRNAs as a tool for the elucidation of microbial phylogenies was introduced by C. R. Woese approximately 20 years ago. Since then, sequencing methods have improved substantially. Amplification of rRNA genes by the polymerase chain reaction (PCR) provides easy access to material for sequencing. Conserved regions scattered over the rRNA molecules or genes (rDNA) facilitate sequencing because they can be used as target sites of oligonucleotide primers (usually 14–20 bases in length) needed for amplification and sequencing. Thus, a set of not more than 10 primers is sufficient to analyze a wide spectrum of phylogenetically diverse organisms. Usually, the almost complete 16S rRNA genes can be amplified in vitro, using conserved target sites close to the termini of the rDNA. It is also possible to amplify almost complete 23S rRNA genes, preferably by overlapping a few long fragments. Using primers directed against the 3′-terminus of 16S rDNA and the 5′-terminus of 23S rDNA, the intergenic spacer region also becomes accessible for amplification and sequence analysis (Fig. 29.**21**). Purified nucleic acid preparations or crude extracts of bacterial cells can be used for in vitro amplification of rDNA. In principle, a few bacterial cells are sufficient to perform the analysis. Thus, even uncultured bacteria or microbial communities from natural samples are accessible to phylogenetic analysis. In the latter case, the resulting mixture of rDNA fragments with different primary structures can be separated by cloning.

29.5.4 Higher Order Structure of rRNA Molecules Facilitate Sequence Alignment

A prominent feature of rRNA molecules is their capacity to form higher order structures (e.g., helices) by short- and long-distance intramolecular interactions of complementary sequence stretches (Figs. 29.**20** and 29.**22**). Such structures can be predicted on the basis of coordinated nucleotide exchanges at homologous positions in sequences from phylogenetically diverse origins. The under-

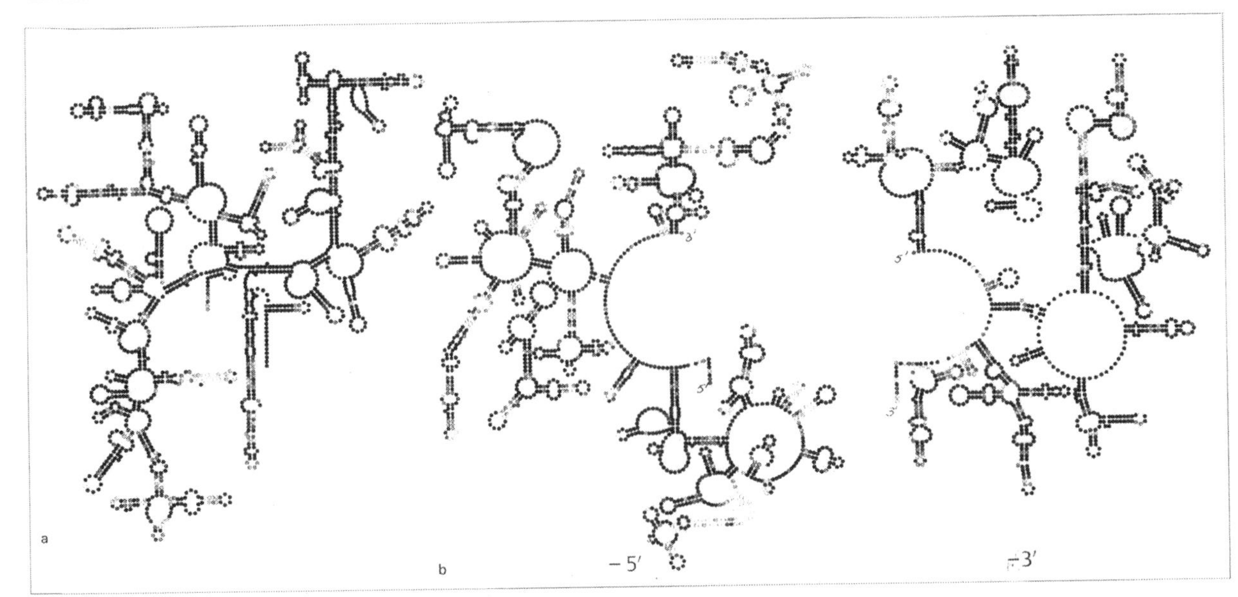

Fig. 29.**20** **Secondary structure of small and large subunit RNAs.** The degree of conservation of individual nucleotide positions is indicated by the various shades of gray, ranging from black (invariant) to white (variable). The determination of the degree of variation is based on the percentage of all bacterial sequences that have the most frequently occurring nucelotide at any given position in common.
a 16S rRNA;
b 23S rRNA; the molecule is divided into the 5′-half and the 3′-half to facilitate the graphic design

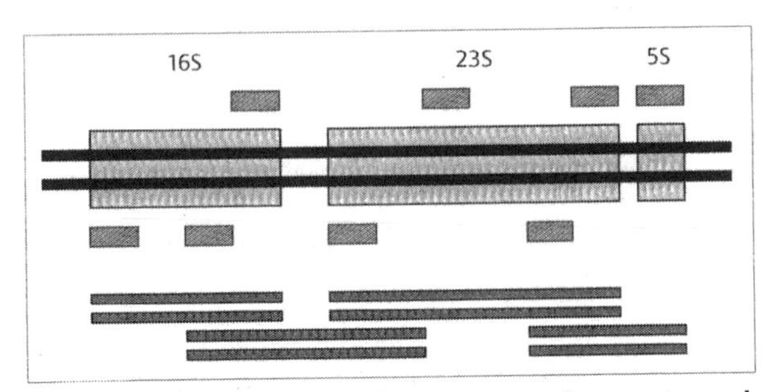

Fig. 29.**21** **Amplification of rRNA gene fragments and intergenic spacers.** The primers and resulting amplification products are indicated. Symbols: double black lines, genomic DNA; hatched double lines, amplified DNA; boxes, rRNA gene; red bars, primer

lying assumption is that conserved function is reflected by structure similarity. Higher order structures are regarded as proven if the same or highly similar structures can be formed in rRNA molecules from phylogenetically moderately related organisms, whose corresponding primary structure stretches differ.

23S rRNA molecules contain about 100 higher order structures and there are about 50 within 16S rRNA molecules. 5′ and 3′ helix halves of double-stranded regions and internal or terminal single-stranded loops can be regarded as homologous elements that can be arranged in the alignment even if their primary structures differ.

Sequence analysis of rRNA genes has become a rapid standard technique. However, one should be aware that in comparison to the billions of years that have passed since the origin of the major lines of descent, the number of informative positions within the molecular sequences (even within large rRNAs) is limited. A reasonable percentage of the positions must remain invariant or highly conserved to manifest the function of the marker molecules. Many of the remaining characters cannot be changed independently and are informative for identical phylogenetic levels. The majority of evolutionary events, therefore, remain undiscovered. The comparative analysis of contemporary sequences allows only a spot check of the course of evolution, while a correct resolution of the roots of phylogenetic groups is often impossible.

29.5.5 Sequence Alignment Should Be Based on True Homology

The quality of phylogenetic trees derived from sequence data depends significantly on the quality of the sequence alignment. A correct alignment ensures that

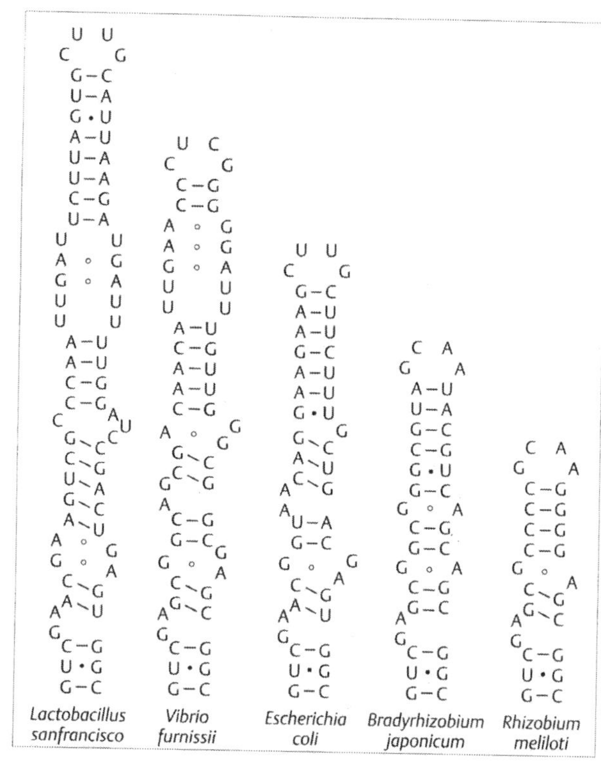

Fig. 29.22 **Higher order structure of rRNA sequences.** Alignments of five homologous sequences of different length (positions 51–108 of the *Escherichia coli* rRNA sequence) are shown

only homologous nucleotides (i.e., residues derived from a common position within the ancestral sequence) are arranged in columns, and consequently are recognized as being identical or different. Given the high content of invariant and conserved positions or regions along the sequence of rRNAs, alignment of these regions is a straightforward procedure (Fig. 29.**23**). However, large deletions and insertions are frequently found. The most prominent example is a stable insertion of about 100 bases in the 23S rRNA of actinomycetes; certain Proteobacteria lack a stretch of approximately 80 bases. Within the variable and highly variable regions that show the highest degree of length variation, it is often difficult or impossible to recognize similarity. The alignment of these regions can be improved in many cases by taking into account the predicted higher-order structure. Nevertheless, these regions are usually omitted from the analysis of taxa above the genus rank.

29.5.6 Several Algorithms Are Available to Evaluate Phylogenetic Trees

The phylogenetic relationships of organisms based on comparative sequence analyses can be graphically visualized. Trees consist of internal and terminal points (nodes) connected by edges (branches). There is a path (series of points connected by edges) for every pair of nodes. The terminal nodes represent the operational taxonomic units (OTU, see Chapter 29.2), which are in most studies the analyzed organisms. In unrooted trees, the interrelationships of the organisms are visualized, whereas in rooted trees, the position of the common ancestor is indicated by an additional point. Two formats of graphic representation are generally used. Radial trees, suitable for the depiction of relationships of a few organisms, resemble living trees. The phylogenetic distance between two nodes (organisms) is measured by adding the distances between the nodes (Fig. 29.**24a**). Dendrograms arrange the organisms in a fork-like fashion, and there are no limitations on the number of organisms in graphically representing phylogenies. Only the horizontal components of connecting lines are summed to read the distances (Fig. 29.**24b**).

If all (terminal and internal) nodes in a tree are well separated by edges of distinct lengths (which is the case when the phylogenetic levels represented by the data set are different enough), it is unlikely that the topology of the tree is affected by the choice of the treeing method or

Vibrio fluvialis	GGGCUACACACGUGCUACAAUGGCGCAUACAGAGGGCGGCCAACUUGCGAAAGUGAGCGAAUCCC
Vibrio vulnificus	GGGCUACACACGUGCUACAAUGGCGCAUACAGAGGGCGGCCAACUUGCGAAAGUGAGCGAAUCCC
Photobacterium phosphoreum	GGGCUACACACGUGCUACAAUGGCGUAUACAGAGGGCUGCAAGCUAGCGAUAGUGAGCGAAUCCC
Escherichia coli	GGGCUACACACGUGCUACAAUGGCGCAUACAAAGAGAAGCGACCUCGCGAGAGCAAGCGGACCUC
Thermotoga maritima	GGGCGACACACGCGCUACAAUGGGCGGUACAAUGGGUUGCGACCCCGCGAGGGGGAGCCAAUCCC
Consensus	****4*******4**********4434****34*4*32**2*2*42****24*34***44*4*4*

Fig. 29.**23** **Alignment of five homologous 16S rDNA sequences** (positions 1220–1265 of the *Escherichia coli* sequence). *, invariant position (in the current data set); numbers 2, 3, and 4 indicate the most abundant base invariant in 2, 3, and 4 sequences

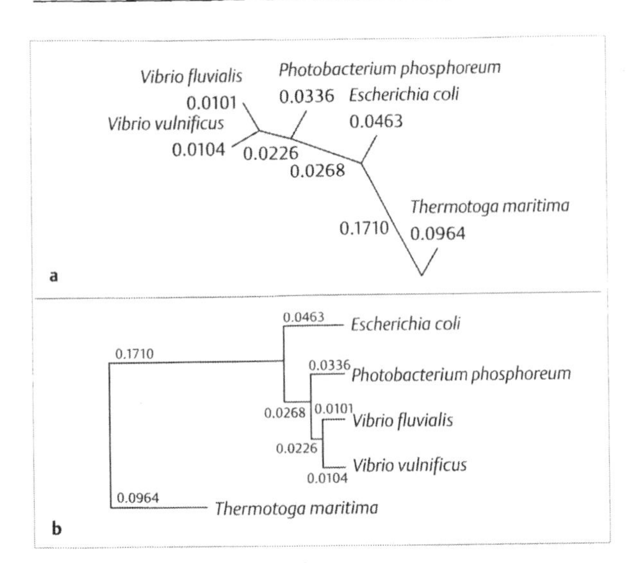

Fig. 29.**24a, b** **Radial tree and dendrogram of 16S rDNA relationships.**
a Radial tree of five organisms. Numbers refer to the phylogenetic distance of segments placed between the nodes and at the end points. The phylogenetic distance between two organisms is obtained by adding the values of the segments that separate them (e.g., the distance between *Vibrio vulnificus* and *Thermotoga maritima* is 0.3272 (0.0104 + 0.0226 + 0.0268 + 0.1710 + 0.0964).
b Dendrogram of the same organisms shown in **a**. The phylogenetic distances are obtained by adding only the values of the horizontal components

other parameters discussed below. However, with the rapidly growing sequence data from sequences of cultured organisms or from environmental 16S rDNA clone sequences, gaps presently observed in the tree will most likely disappear. Thus, in comprehensive trees, many branches will be separated from each other within a narrow range, such as those parts that are already today composed of many closely related species. Examples are seen in those genera that contain a high number of closely related species, such as the *Bacillus, Streptomyces, Micromonospora, Legionella,* and *Agrobacterium/Rhizobium* clusters. Consequently, tree topologies can change locally with the treeing method and data selection applied and their evaluation is a crucial step in phylogenetic analyses.

29.5.7 The Validity of Tree Topologies Can Be Tested by Comparing the Evolution of Different Molecules

The evolution of an organism can be considered as the sum or mean of the evolution of its genes. Thus, the restriction of the phylogenetic analysis to a single molecule, such as 16S rDNA, depicts the evolution of the molecule, but not necessarily the evolution of the organisms from which the molecule was isolated for analysis. The situation improves when the phylogenetic analysis of several functionally independent genes or gene products comes to the same conclusion. Thus, an increasing number of data sets of homologous sequences derived from alternative genealogical markers should provide more insight into the evolution and phylogeny of the organism. These data can also be useful for enhancing the resolution since different genes may have accumulated mutations during different periods of evolutionary time or may have preserved ancestral states of sequence positions for different periods. It is important, however, that care is taken to use only genes that are homologous with reference to origin and function, that are informative within a comparable range of phylogenetic levels, and that are unlikely to be subject to lateral gene transfer. Consequently, if the results of comprehensive phylogenetic analyses based on rRNA data are tested, the information content of alternative markers should cover a comparably broad spectrum of phylogenies.

It is difficult to judge the quality of phylogenetic trees. Numerous factors influence the topology of a branching pattern, which is a dynamic construct and which will change with any new sequence included. Three major types of tree-inferring approaches are commonly used: pairwise distance, maximum parsimony, and maximum likelihood methods. Results of comparative analyses of other conservative molecules responsible for central functions, such as elongation factors, subunits of ATP synthases, and RNA polymerases in general support the rRNA data. Thus, trees based on ribosomal RNAs reflect the evolution of these molecules and most likely the evolution of a major portion of the genome (containing for example, "housekeeping" genes).

The results of comparative analyses of other conservative molecules responsible for central functions, such as elongation factors, subunits of ATP synthases, and RNA polymerases, agree well with the rRNA data. There are some differences in local topologies of published phylogenetic trees derived from the different marker molecules, but a careful re-evaluation of the data shows that there is less phylogenetic information within protein molecules than within rRNAs. Minor discrepancies between phylogenetic trees derived from different marker structures may, therefore, reflect a different resolving power at a given phylogenetic level. Since similar or at least not really conflicting results have been obtained from the analysis of essential, functionally independent molecules, it is unlikely that branching patterns obtained from either of these molecules are artifacts that result from lateral transfer of the underlying genes. This does not hold for nonessential, usually less-conserved genes and gene products. Nevertheless, at lower phylogenetic levels, these molecules may be phylogenetically informative as long as they are ubiquitous and functionally conserved among the particular group of organisms, and lateral gene transfer is unlikely.

29.6 From Early Life Forms to Extant Species of Prokaryotes

29.6.1 Prokaryotes Played and Play a Significantly Greater Role in the Evolution of Life Than Previously Believed

Sequences of macromolecules of proven phylogenetic significance provide biologists with insights into the relationships among modern organisms. By looking down the tree of life, the nature of the common ancestor that once gave rise to many lines of descent and the time frame at which individual groups emerged in the history of the planet Earth can be speculated upon. Conclusions drawn by this approach can be compared to conclusions about the course of evolution as deduced from biochemical properties of prokaryotic organisms.

Until 1977, early life forms and the modern prokaryotic descendants were considered to represent an uninterrupted monophyletic line in which all forms were similar with respect to fundamental aspects of genetic and cellular organization and gene expression. According to this view, it was speculated that the eukaryotic cell evolved from a prokaryotic cyanobacterium (blue-green bacterium) rather recently, about 1 Gy (10^9 years) ago (Fig. 29.**25**). This ancestral eukaryote, considered to be a unicellular alga, would have then given rise to plants, and, after loss of the photosynthetic apparatus, to protozoa, fungi, and higher animals. Textbooks stated that the difference between eukaryotic and prokaryotic cellular organisms is the largest most profound single evolutionary discontinuity. Prokaryotes were originally defined purely by negative characters, i.e., by lacking those properties that are present in eukaryotes (see Tab. 29.**5**), and it was not until the early 1960s that certain traits, such as the presence of 70S ribosomes or peptidoglycan, were used to define prokaryotes in a positive way. Although prokaryotes were granted a kingdom level, the Monera, the fundamental differences at the cellular and molecular level that exist between prokaryotic and eukaryotic organisms were considered less significant than those that exist between members of the eukaryotic kingdoms Metaphyta, Metazoa, Protists, and Fungi.

The endosymbiosis hypothesis, revived in the late 1960s by L. Margulis from an earlier hypothesis from around the turn of the century, postulated that the eukaryotic cell is in fact a chimeric structure, originated by fusion of several prokaryotic lineages. Sequence analysis of semantophoric molecules, first proteins and later ribosomal RNAs, supported Margulis's version of the origin of organelles. The origin of mitochondria and chloroplasts can today be traced back to respiring Proteobacteria and cyanobacteria, respectively.

Prior to the recognition of semantophoric molecules for unraveling the evolution of prokaryotes, genealogical patterns were deduced from analyses of biochemical pathways and metabolic properties of contemporary species. Considering the lack of alternative hypotheses for the origin of life, it is not surprising that an anaerobic and strictly fermentative bacterium was considered to be the ancestor of life in the two main alternative proposals for the evolution of prokaryotes: the "**conversion hypothesis**" of P. Broda and the "segregation hypothesis" of L. Margulis. According to the "**segregation hypothesis**," the evolution of electron transport in anaerobic respirers preceded the evolution of the cyclic electron transport in anaerobic phototrophic organisms and aerobic respirers. Each of these

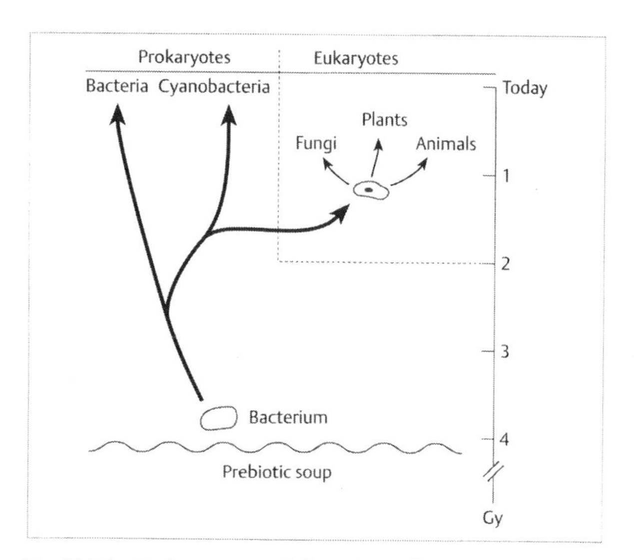

Fig. 29.**25** **Early concept of the origin of eukaryotes from a prokaryotic ancestor**

biochemical properties was thought to have evolved only once during evolution; therefore, clusters of all recent fermentative, anaerobic respirers, and phototrophic, chemoautotrophic, and aerobic types were placed in separate lines of descent. According to the "**conversion hypothesis**," phototrophic organisms evolved directly from the most ancient, fermentative organisms. In a series of events, different types of phototrophs evolved in several lines of descent, some of which became extinct, while others still exist. Reacting to changes in the oxygen level during time, each of the early phototrophic organisms gave rise to individual lines of descent in which metabolically different types evolved (e.g., organotrophs and lithotrophs). Consequently, anaerobes, phototrophs, anaerobic respirers, and aerobic forms evolved independently from each other several times during evolution. The fundamental differences between the two hypotheses thus concern the origin of anoxic photosynthesis and the evolution of respiration. Since Archaea were not yet described when these hypotheses were formulated, they were considered to be special metabolic types of bacteria, and consequently, were included as parts of the general evolution of bacteria.

Results of phylogenetic analyses by and large support the "conversion hypothesis," As indicated above, phototrophic bacteria are found in several lines of descent within Bacteria. Considering the complexity of the photosynthetic apparatus and the diversity of morphological and functional types and the uniformity of the organization and functional principle, it is reasonable to assume that the photosynthetic apparatus evolved monophyletically and the variations developed independently under selective pressure in certain environments. As judged from the position of the most deeply branching phototrophic organism in the 16S rDNA tree, *Chloroflexus*, the photosynthetic apparatus evolved early in the evolution of bacteria. Loss of the

Table 29.5 **Major molecular differences between eukarya and the two domains of prokaryotes (Bacteria and Archaea)**

Trait	Bacteria	Archaea	Eukarya
Organization	Prokaryotic	Prokaryotic	Eukaryotic
Chromosome	Single (or multiple) circular (or linear) chromosome(s)	Single circular chromosome	Several linear chromosomes
Operon	Yes	Yes	No
Promoter	−10 (TATAAT) and variable boxes for various alternative sigma subunits	−25 region, AT-rich region upstream	"TATA" boxes for RNA polymerase II, other for I and III
rRNA genes	16S, 23S, 5S transcriptionally linked; some exceptions	Like bacteria, with variations	18S, 5.8S, 28S transcriptionally linked, 5S unlinked
rRNA size	16S, 23S, 5S	16S, 23S, 5S	18S, 28S, 5.8S, 5S
RNA polymerases	One, but multiple sigma subunits for different promoters	One	Three for three gene classes
Capped mRNAs	No	Probably yes, (some species)	Yes
5′ mRNA leaders	Present and ribosome binding sites in front of each open reading frame	Maybe short or absent	Usually present
3′ poly A tails	Absent or unstable	Probably absent	Usually present, and stable
Ribosome size	70S	70S	80S
Ribosome antibiotic sensitivity	Chloramphenicol and many others	Generally insensitive	Cycloheximide and others
Diphtheria toxin sensitivity	No	Yes	Yes
Initiator tRNA	Usually formylmethionine	Methionine	Methionine

photosynthetic apparatus independently in certain members of all but two lines of descent (green sulfur bacteria, cyanobacteria) and changes in the function of the cyclic electron transport chain to cope with the new nutritional condition led to the evolution of novel metabolic types. The descendant of an anoxic photosynthetic bacterium, predicted to be the ancestor of Gram-positive bacteria, has recently been discovered. *Heliobacterium chlorum*, representing a novel photosynthetic type, is phylogenetically distantly related to clostridia and bacilli. Also, in accord with the "conversion hypothesis" is the finding that several metabolic types, such as fermentation, anaerobic respiration, chemoautotrophy, and aerobic respiration, did not evolve monophyletically, but originated from different ancestors in different main lines of descent. Phototrophic bacteria often share close relationships with chemoautotrophic and aerobic bacteria. Especially the relatedness between *Rhodopseudomonas palustris* and the nitrite-oxidizing, strictly aerobic species *Nitrobacter winogradskii* shows most impressively how the respiration chain evolved from the photosynthetic electron transport chain. It should finally be mentioned that the "conversion hypothesis" is by and large in accord with the branching pattern of sequences of evolutionary conserved macromolecules, while significant differences can be found in the details.

29.6.2 Prokaryotes Constitute Two of the Three Main Lines of Descent, the Domain Archaea and the Domain Bacteria

In 1977, C. R. Woese and G. E. Fox demonstrated by analysis of partial 16S and 18S rRNA sequences that prokaryotes did not constitute a coherent phylogenetic group of organisms. Two main lines, classified as primary kingdoms (Box 29.9), emerged from the analyses of 16S rRNA. The 16S rRNA of each of the main lines was as unrelated to 18S rRNA of the eukaryotes, as they were unrelated to each other. The eubacterial kingdom contained the "typical" bacteria of the *Escherichia coli* type, while members of the archaebacterial kingdom contained several phenotypes not represented in the eubacteria (Tab. 29.**5**). Subsequent analyses of the small subunit rDNA of protozoa revealed that the most ancient eukaryotic organisms were members of *Giardia*, *Microsporidia*, and their relatives. Like prokaryotes, these protozoa possess 16S-rRNA, and they lack mitochondria. The distinctness of the three major rRNA lines of descent was later confirmed by the analyses of the genes coding for some proteins involved in the regulation of translation and in energy-yielding processes.

Box 29.9 The primary kingdom concept was proposed to highlight the cellular distinctness of the three phylogenetically defined groups. Distance-matrix analysis confirmed that all kingdoms were monophyletic, but significant distances separated the kingdoms from each other. Support for this topography came from 5S and 23S rRNA analysis and, although based on a few organisms only, by sequence analysis of the elongation factors EF-Tu and EF-1α and on the β-subunit of ATP synthase. On the other hand, parsimony-based analysis of the same data used by C. R. Woese and colleagues and the presence of apparently distinct differences in the size and topology of ribosomes led to the hypothesis of J. A. Lake, who proposed that archaebacteria did not constitute a phylogenetically coherent group and should be split into three lines: (1) halophiles, believed to belong to the eu(true)-bacteria (presence of a "photosynthetic" apparatus), (2) thermophilic sulfur-metabolizing archaebacteria, named eocytes, which formed a novel and deeply branching line of descent sharing a common ancestry with the eukaryotes (indeed, a substantial number of "eocytic" properties are shared with eukaryotes, but their monophylic origin has rarely been shown), and (3) methanogens, which constitute the genuine archaebacteria.

At present, the arguments are in favor of the concept of C. R. Woese since it is more convincing and takes into consideration the significant coherency of the archaebacteria at the genetic and epigenetic level. Both views are in contrast to the "five kingdom" classification of L. Margulis, which is widely accepted by zoologists and botanists. Here, the prokaryotes constitute a single kingdom (Monera), while the eukaryotes are split into four kingdoms (Plants, Animals, Fungi, and Protists). However, according to molecular and cellular organization, these four kingdoms are virtually identical, and the "five kingdom" concept does not recognize the fundamental differences that distinguish the two groups of prokaryotes enclosed in the Monera.

Reports on the presence of phenotypes in the methanogenic and in the halophilic archaeabacterial, that were peculiar to prokaryotes such as the unusual chemical composition of peptidoglycan and fatty acids, have been known since the early 1970s (see Chapter 29.3). The significance of these findings, the indication that a group of organisms represented a novel prokaryotic group, however, could not be concluded from analyses of these episemantic markers. Following the first reports of the concept of the archaebacteria as possible descendants of evolutionary ancient organisms, the archaebacteria became favorite subjects in investigations on evolution, ultrastructure, genetics, biochemistry, and biophysics. Knowing that archaebacteria predominantly thrive under extreme conditions, which prior to 1977 were not considered to support life, the search for and the isolation of novel archaeal types from hot springs, sulfataras, deep-sea vents, and alkaline lakes resulted in the description of numerous organisms with new physiological and genetic properties.

The three-kingdom classification of Woese was previously replaced by the **Domain concept**. The taxon domain was created above the kingdom level to highlight the importance of the tripartite division of the living world. In order to avoid the impression that the suffix "bacteria" in the names of two of the three kingdoms reflects exclusive evolutionary relationship, the terms **Archaea** and **Bacteria** were proposed to replace the traditional names **eubacteria** and **archaebacteria** (or archaeobacteria). The term **Eukarya** was proposed to replace the term **eukaryotes**. In this concept, kingdoms have been described for the two main evolutionary lineages of Archaea: **Crenarchaeota** for the thermoacidophiles and the strictly anaerobic thermophiles, and **Euryarchaeota** for the halophiles, alkaliphiles, methanogens, some strictly anaerobic thermophiles, and *Thermoplasma* (Fig. 29.**26**). To date no kingdoms have been described for the main lineages of Bacteria for which the terms phyla or divisions are used. The reason is the complexity of the phylogenetic structure of the domain Bacteria. More than 15 main lines of descent have been identified so far, some of which are only separated by small internode distances or contain only a few representatives which are not shown in the overview of the domain Bacteria (Fig. 29.**27**). The relationships within the individual lines are often unexpected, and support for these groupings from the sharing of properties other than similarities in conserved genes is rare.

As the Domain concept is still under discussion, the terms eubacteria and archaebacteria are used side by side with the terms Archaea and Bacteria. It is one of the goals in taxonomy to provide stability in the nomenclature of validly described taxa, and approved names cannot be changed because of changes in ranks of taxa or the introduction of new hierarchic levels. However, names of ranks above the order do not need to be approved, and time will show which names continue to be used throughout the scientific community. In this chapter, the terms Archaea and archaeal and Bacteria and bacterial have been applied since these terms are widely in use today. Despite the abolishment of the suffix "bacteria" from the highest taxonomic rank (Archaea vs. archaebacteria), the stability of validly

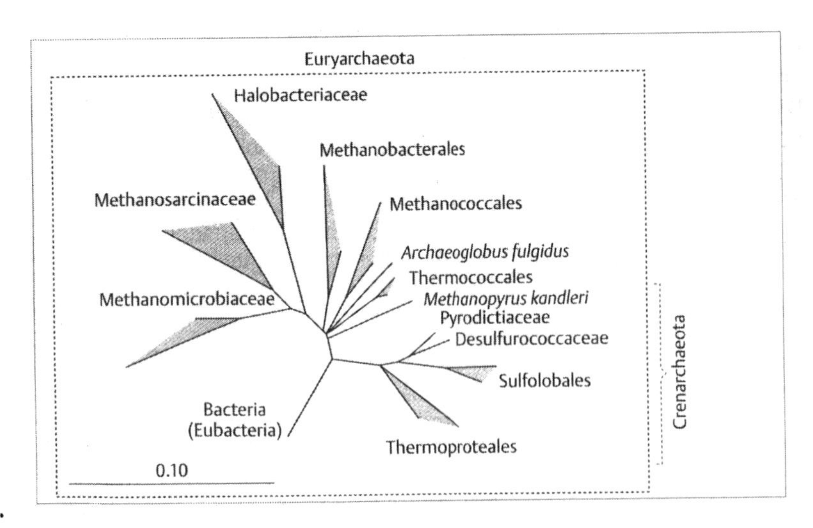

Fig. 29.**26** **Phylogenetic structure of the domain Archaea.** The branching pattern of the main lines of descent as derived from distance matrix analysis is shown. Hatched triangles indicate the phylogenetic depth of kingdoms and phyla. The bar indicates 10% sequence divergence

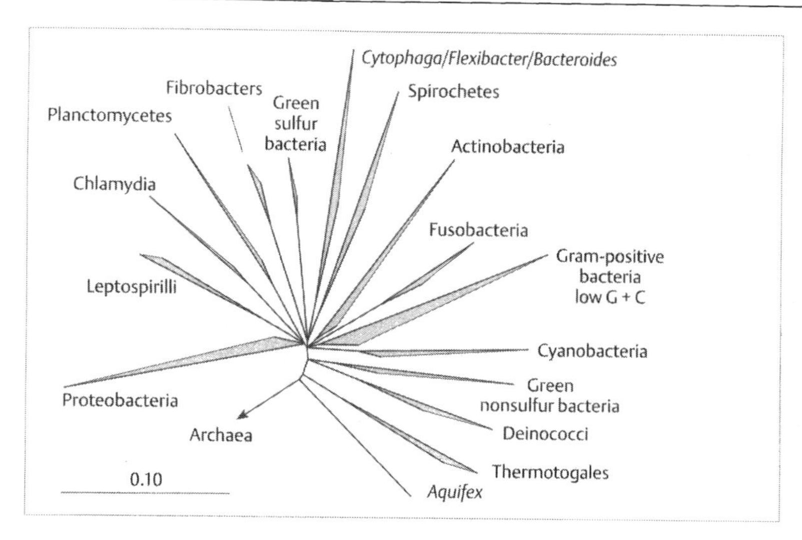

Fig. 29.27 **Phylogenetic structure of the domain Bacteria.** The branching pattern of the main lines of descent as derived from distance matrix analysis is shown. Hatched triangles indicate the phylogenetic depth of kingdoms and phyla. The bar indicates 10% sequence divergence

published names demands that the names of a few archaeal genera, namely *Methanobacterium*, *Natronobacterium*, and *Halobacterium*, will continue to be used as witnesses of their pre-domain history.

29.6.3 The Progenote Is the Hypothetical Ancestor of the Modern Cell Types

The last **common ancestor** of the three major types of extant life has been called the **progenote**, a simple entity more primitive than organisms recognized today. This idea developed shortly after the recognition that the main radiation of the three major lines of descent occurred within an evolutionarily relatively short period and not in more recent history [$\sim$1.5 Gy ($1.5 \cdot 10^9$ years) ago], and that inter-domain differences in molecular structure and processes are more significant than those found within a domain. If it is true that the different types of organisms evolved from a common ancestor within a short time span, this ancestor could not have been a cell in which genetic processes, gene expression, and gene regulation were already fine-tuned. The hypothetical progenote must be viewed as an organism in which higher error rates of DNA and RNA replicating enzymes allowed gene expression to evolve with a significant range of variation and structure. Descendants of the progenote would have maintained central, vital functions such as transcription and translation, but within the individual lines of descent, processes of other homologous functions evolved at different rates, with different modes and solutions, as seen in membrane building blocks, cell envelopes, gene structures, and the presence of domain-specific signature sequences. The **number of main lineages**, however, is an open question. If more lines exist in nature, it is likely that modern approaches of ecology and identification will eventually discover them.

The basic branching pattern of organisms contain three lineages with no information whether any two of these three groups are specifically related to each other. In order to root a tree, an outgroup is needed. Since, obviously, none of the molecules used for genealogical studies offer this opportunity, the primary structure of ancestral genes that were duplicated during an early stage in evolution, was analyzed. This duplication event led to the formation of pairs of homologous genes and proteins, such as those found for the ferredoxin gene and for certain elongation factors. For example, as all organisms contain pairs of genes coding for the elongation factors EF-TU/1a and EF-G/2, the duplication that led to the existence of this gene pair must have already been present in their last common ancestor. Thus, the tree relating all EF-TU sequences can be rooted by using the sequences of EF-G as an outgroup because EF-G is the functional equivalent of EF-TU. The same can be done the other way around. This kind of analysis showed that Archaea and Eukarya share a common ancestry, and hence they are sister taxa, while Bacteria represent an individual line of descent that diverged more closely to the root. However, the question about the unambiguous branching order of the three main lines of descent is still not yet answered; this would require the analysis of more than just one or two of such evolutionary conserved pairs of genes.

29.6.4 Domains Are Defined by Phylogenetic Coherency and a Few Exclusive Phenotypic Properties

Sequence comparisons are genotypic measurements that can reveal differences in phenotype within a phylogenetically coherent group. If, for example, molecular analysis of a few conserved macromolecules point towards the existance of major evolutionary groups, one expects to find additional profound differences at the molecular level that are expressed at the phenotypic level. In many cases, the most rapid approach for the allocation of an unknown organism into a higher taxon is the analysis of rDNA, but the availability of taxon-specific properties is highly desirable. While those characters shared between members of any two of the three kingdoms are not useful for placing strains in the phylogenetically correct kingdom, certain characters have been identified that are exclusive for a particular domain, and hence are of diagnostic value. The optimal "signature" character would be the one that is found exclusively among members of only one domain, and is absent in the two remaining domains. However, the exclusive presence of characters in two domains is valuable for explaining the course of their evolution. Non-homologous phenotypes can occur among members of different domains that mirror common evolutionary origin. On the other hand, homologous properties can actually be found exclusively among members of two domains because of horizontal gene transfer. Optimally, the validity of a character to serve as a reliable "signature" should be tested for each new organism classified as a member of a higher taxon. A summary of major molecular biological differentiating features (Tab. 29.**5**) between members of the domains and characters believed to be indicative for a particular domain must take into account that this knowledge is derived only from a few percent (or a fraction thereof) of known strains.

The isolated position of the Archaea, considered to be an entity as unrelated to Eukarya and Bacteria as the latter two domains are to themselves, was strengthened by the finding that several epigenetic properties were unique to the Archaea (Box 29.10). In addition, certain archaeal taxa exhibit unique properties, such as the presence of coenzymes involved in methanogenesis, the energy-generating purple-membrane-containing bacteriorhodopsin, halorhodopsin or other sensory rhodopsins of halophiles, survival under hyperthermophilic conditions, and certain physiologically unique features (e.g., a modified Entner-Doudoroff pathway) (Box 29.10).

Box 29.10 The following epigenetic properties are unique to all archaeal species:

- Primary structure of rRNA species and some other semantophoric molecules, such as ATP synthase and elongation factor EF-TU.
- Secondary structure of ribosomal RNAs.
- The sacculus polymer is not a peptidoglycan: it is either an outer membrane or an S-layer composed of hexa- or tetragonally arranged proteins or glycoproteins subunits (all Crenarchaeota, thermophilic Euryarchaeota and *Methanococcus*) a "pseudomurein" (*Methanobacterium*), or heteropoly-saccharides (halophiles).
- Resistance to penicillin and D-cycloserine since D-amino acids are absent in the sacculus.
- The single DNA-dependent RNA polymerase (as in bacteria) is insensitive to antibiotics that inhibit the bacterial enzyme.
- Membrane lipid type and linkage. Dominant lipid structure in polar and non-polar lipids is based on the branched isoprene unit. The hydrocarbon chains are linked to glycerol by ether bonds, rather than ester bonds.
- Modification pattern of transfer RNA.

The following properties of archaeal taxa also occur in some or all bacteria and/or eukaryal taxa:

- The DNA polymerase of some methanogens and halophiles is inhibited by aphidicolin, an inhibitor of eukaryal-type DNA polymerases; DNA polymerases from *Methanobacterium thermoautotrophicum* and *Sulfolobus acidocaldarius* resemble those found in bacteria.
- Promoter sequences share features found in Bacteria (polymerase binds to promoter-like structures) and in Eukarya (AT-rich region at position −25 and requirement of associated proteins for initiation of transcription).
- The amino acid sequence of a histone-like protein of *Thermoplasma acidophilum* resembles that of eukaryal histones and that of *E. coli*. A DNA binding protein of *Methanothermus fervidus* resembles that of eukaryal histones.
- The DNA-dependent RNA polymerase is related to Pol II and Pol III of Eukarya.

- Introns are present in tRNAs (some species of *Halobacterium*, *Sulfolobus*, and *Thermoproteus*) and rRNA (*Desulfurococcus*).
- The ribosomal operon structure of 16S-23S-5S is split (thermophiles) or unlinked (*Thermoplasma*) (as in certain bacterial genera, e.g., *Thermus*, *Planctomyces*, *Leptospira*).
- Ferredoxin of the [2Fe-2S]-type of Halobacteria resembles that of chloroplasts and Cyanobacteria (Bacteria). The ferredoxin of the [4Fe-4S]-type of Crenarchaeota is similar to those of bacterial taxa.
- The ribosomes are insensitive to chloramphenicol and streptomycin (as in eukarya), and they bind anisomycin, but not cycloheximide (like eukaryal ribosomes).
- The ribosomes are prokaryotic in size, but halophiles and some methanogens have acidic ribosomal proteins. The protein mass of ribosomes varies greatly, increasing in mass from halophiles and *Methanomicobiales* (Bacteria-like) to thermophilic Euryarchaeota and Crenarchaeota (*Aquifex*-like).

Because Euryarchaeota share some genetic and epigenetic features with members of Bacteria, and because the Crenarchaeota tend to be linked to the Eukarya in other properties, the Archaea themselves do not seem to be phylogenetically coherent (see Tab. 29.7). Due to intrinsic features of the rRNA of slowly evolving organisms, this domain might constitute an artificial higher taxon that is actually composed of two or three evolutionary distinct groupings that lack a common ancestry.

29.6.5 More Than 15 Main Lines of Descent Have Been Identified in the Domain Bacteria

Although the number of recognized archaeal genera has increased from 5 to about 40 since 1977, the number of bacterial genera has remained nearly constant. With the inclusion of cyanobacterial genera that can be described according to the Bacterial Code and to the Botanical Code, the number is about 400. In contrast to the Archaea, where virtually the phylogenetic position of each type strain has been determined, the nearest phylogenetic neighbor of more than 80 bacterial genera to date have not been determined. About 20 distinctly separated, individual lines of descent have been recognized; however, two of them, the Gram-positive bacteria and the Proteobacteria, contain the majority of species. Table 29.**6** is a compilation of higher taxa of the domain Bacteria; the approximate number of genera are indicated. It is obviously more important for students to learn about the rationale for the revolutionary developments in prokaryotic systematics rather than to be confronted with fine details of relationships. Thus, phylogenetic relationships at the genus and species level will not be depicted. Many genera still represent phylogenetically heterogeneous taxa (e.g., *Rhodopseudomonas*, *Pseudomonas*, *Bastobacter*, and *Thiobacillus* among the Proteobacteria, and *Bacillus*, *Clostridium*, *Desulfotomaculum*, and *Peptococcus* among the Gram-positive bacteria. This situation is exemplified with some *Clostridium* species, which contain representatives of genera that are aerobic and spherical non-endosporeformers (Fig. 29.**28**). Future taxonomic rearrangements will hopefully clarify the taxonomic status of these misclassified species. An example of taxonomic progress is the order *Actinomycetales*, which today constitutes a phylogenetically coherent taxon. Sequences are available from almost every type strain of the type species, and the branching pattern of phylo-

Table 29.**6** **Main lines of descent (phyla) of the domain Bacteria** as recognized by analysis of 16S rDNA. The number of genera assigned to each phylum is indicated in parentheses (excluding cyanobacteria)

Phylum *Aquifex* (3)
Thermophilic oxygen reducers (6)
Phylum *Deinococcus* (3)
Phylum *Thermodesulfobacterium* (1)
Phylum *Bacteroides/Cytophaga* (24)
Phylum *Chloroflexus* (6)
Phylum Spirochaeta (10)
Phylum *Leptospirillum* (3)
Phylum Fusobacterium (4)
Phylum Chlamydiae (1)
Phylum Green sulfur bacteria (5)
Phylum Planctomycetales (4)
Phylum Verrucomicrobiales (1)
Phylum *Synergistes* (1)
Phylum *Acidobacterium* (1)
Phylum *Fibrobacter* (1)
Phylum Proteobacteria ($\sim$270)
Phylum Gram-positive bacteria ($\sim$172)
Phylum Cyanobacteria

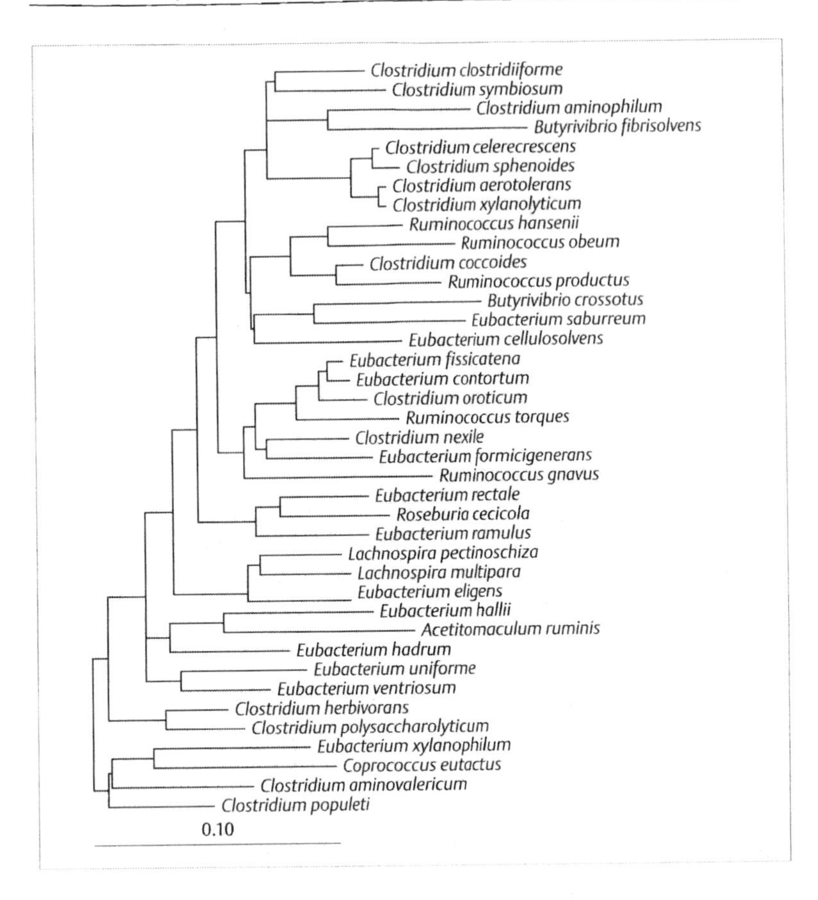

Fig. 29.**28** **Phylogenetic relationships between some *Clostridium* species and taxa.** Strains that display different morphologies, and Gram-staining behavior, and that might lack spore formation are shown. The bar indicates 10% sequence divergence [after 1]

genetic trees of this order can be considered stable (Fig. 29.**29**).

Four bacterial phyla, i.e., those embracing Gram-positive bacteria, the Proteobacteria, the *Bacteroides/Cytophaga/Flavobacterium* taxon, and the Cyanobacteria, have a complex phylogenetic structure. The others, (e.g., those containing *Thermotoga, Chloroflexus, Thermomicrobium* and *Herpetosiphon, Deinococcus* and *Thermus, Chlorobium, Chloroherpeton* and *Clathrochloris*, as well as Spirochaetales, Planctomycetales and some other phyla (Tab. 29.**6**), are defined only by a few or a single species. Broad phylogenetic conclusions covering relationships between the ranks of domains and families have mostly only been inferred from 16S rRNA analysis. Only the relationships of members of the Gram-positive bacteria, the Proteobacteria, and the *Cytophaga* phylum have been analyzed by several different methods, such as DNA/DNA hybridization, determination of rRNA cistron similarities, and sequence analysis of 23S rDNA. As a result, a few of the traditionally defined taxa also are found to constitute phylogenetically coherent groupings, for example, the actinomycetes (see Fig. 29.**29**), the spirochetes (on the basis of morphology), the myxobacteria, a cluster within the δ-subclass of Proteobacteria (on the basis of social behavior), and the cyanobacteria (on the basis of the evolution of oxygen). Thermophily, phototrophy, mechanism of motility, and complex morphology (helical, budding, prosthecate, mycelium) can generally be considered as poor taxonomic markers for defining higher ranks as each of these features is found to occur independently in members that occupy different positions in the phylogenetic tree.

The phylum of **Gram-positive bacteria** contains two major subphyla, each of which contains deeply rooting organisms that stain Gram negative. The Gram-negative *Fusobacterium* and their relatives (Fig. 29.**27**) might be linked to the Gram-positive organisms, but the question whether or not Gram-positive bacteria constitute a phylogenetically coherent group is still a moot

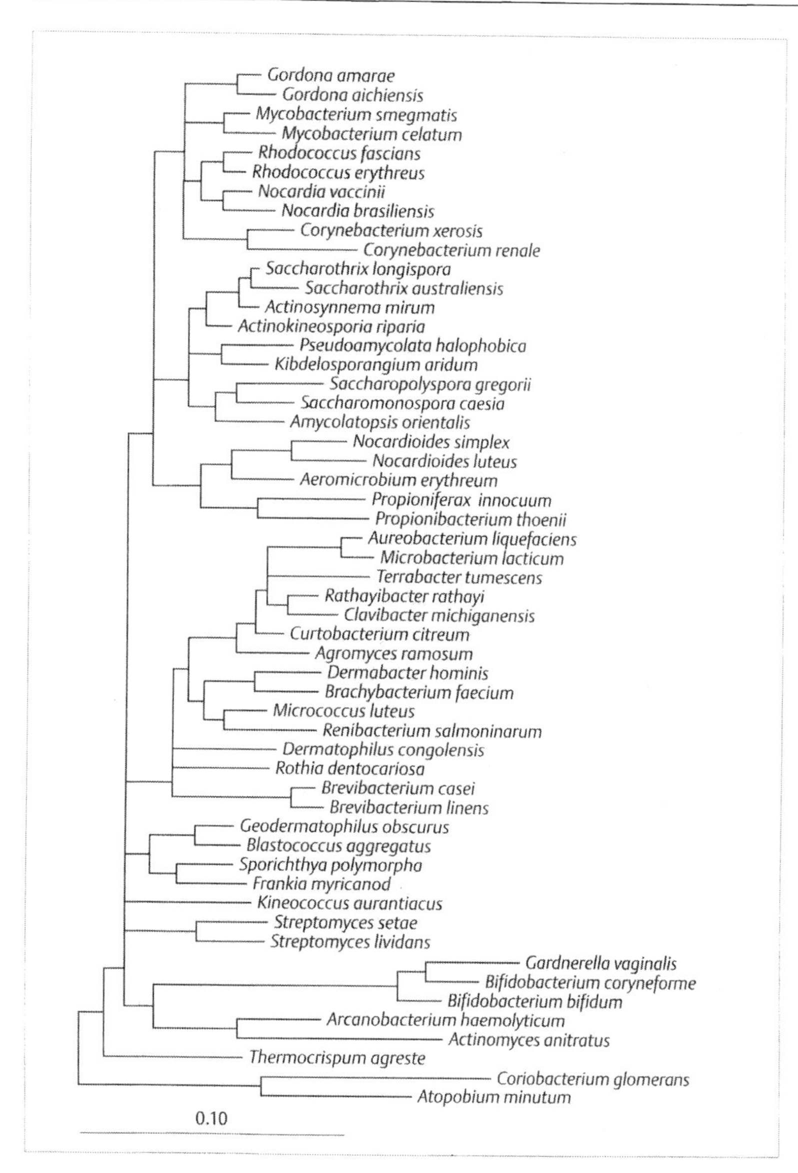

Fig. 29.**29** **Phylogenetic structure of the order Actinomycetales.** The interfamily and intrafamily relationships as derived from distance matrix analysis are shown. The bar indicates 10% sequence divergence

point. The presence of the two major subdivisions correlates with the distribution of the DNA mol% G + C of their members (i.e., the low G + C "clostridial" subline and the high G + C "actinomycetes" subline), although some exceptions have been detected in both subphyla. The 16S-rDNA-based phylogenetic branching pattern of taxa of Gram-positive bacteria is extremely helpful for drawing taxonomic conclusions and for revisions since, in most cases, groups of naturally related species can be detected by superimposing chemotaxonomic and other phenetic characters. The chemistry of cell walls, lipids, fatty acids, polyamines, and isoprenoid quinones, the DNA base composition, relationship to oxygen, and other features are highly evolved in Gram-positive bacteria (see Chapter 29.3), which allows a clear delineation of taxa by combinations of chemical, physiological, and morphological characters. Properties traditionally used as the sole criterion for the description of higher taxa, such as the formation of spores and morphological forms, are losing their significance in a phylogeny-based taxonomy because of their polyphyletic origin or unpredictable loss during evolution.

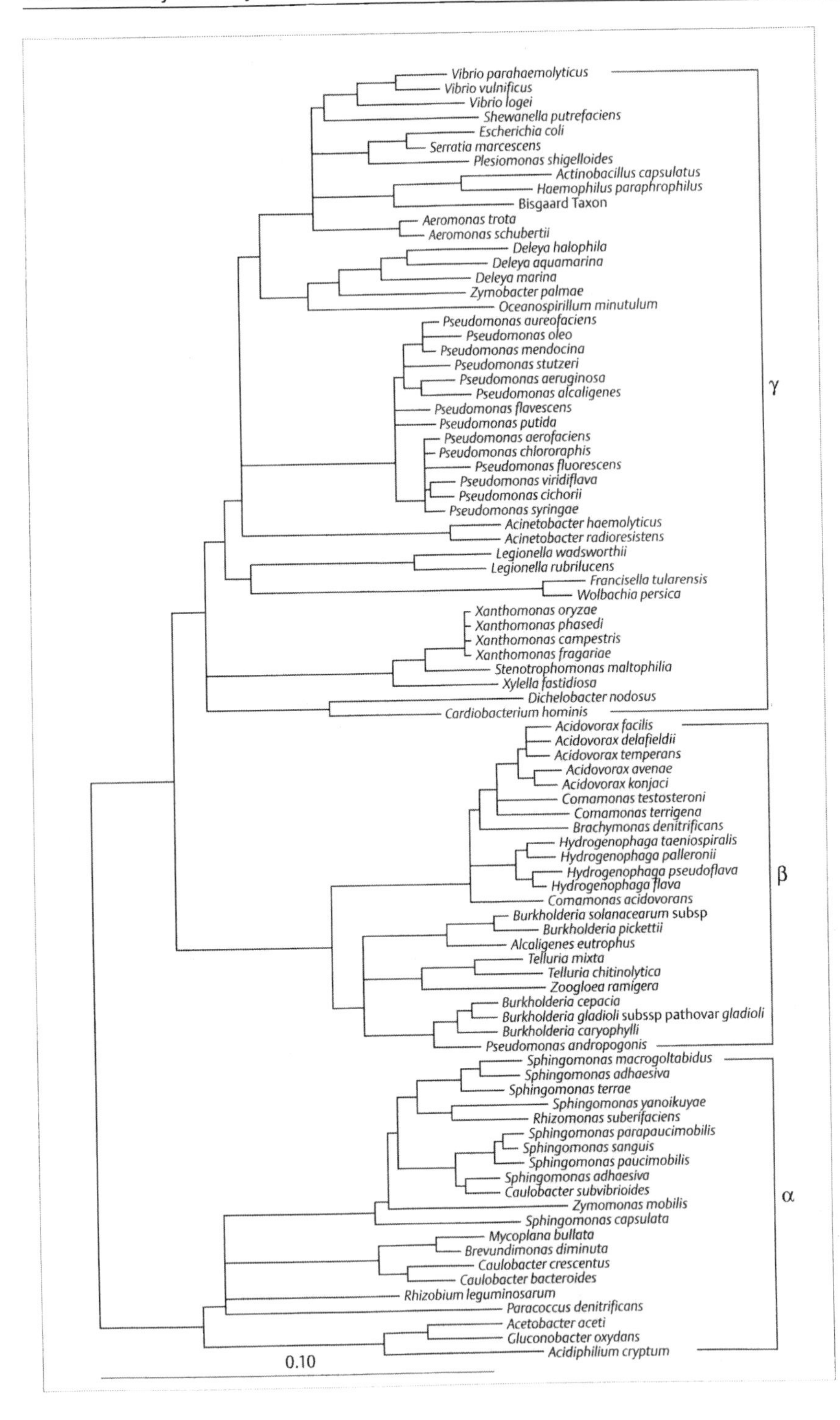

Fig. 29.**30** **Phylogenetic position of some reference genera of the α-, β-, and γ-subclass of the class Proteobacteria.** The bar indicates 10% sequence divergence

The situation is similar within the class Proteobacteria, but is significantly more complex. The subclasses, identified by phylogenetic analyses, are named by Greek letters, from α to ε. More than 220 genera have been identified in this class, most of which are contained in the α-subclass (>70 genera) and the γ-subclass (>90 genera). Hardly any of the classical taxonomic properties believed to express natural relationship, such as phototrophy, gliding motility, budding reproduction, autotrophy, lithotrophy, or habitat, are found to define a major subline exclusively (Fig. 29.**30**). Within short phylogenetic distances, quite different phenotypes can be detected (phototrophs and CO-oxidizers, phototrophs and lithotrophs, nitrogen fixers and plant pathogens), and organisms associated with eukaryotic cells are more closely related than originally believed (e.g., *Agrobacterium, Rhizobium, Brucella*, and *Rochalimaea* within the α-subclass). At a higher level, the composition of the δ-subclass is surprising and cannot be explained by any other molecular feature since there are the highly complex myxobacteria, the bacteriovorous *Bdellovibrio*, and the anaerobic sulfate- and sulfur-reducing Gram-negative bacteria, such as *Desulfovibrio* and *Desulfuromonas*.

Unexpected specific phylogenetic relationships between phenotypically diverse taxa are those between the Gram-positive radiation-resistant deinobacteria and the Gram-negative *Thermus* and between the strictly anaerobic *Bacteroides* and the aerobic *Cytophaga* and *Flavobacteria* lines of descent. Since these members of the two phenotypically very different pairs of organisms were never considered to be phylogenetic neighbors, they were previously never included in joint phenotypic analyses. Support for these groupings is still poor, but it is encouraging to notice that deinobacteria and members of *Thermus* possess lysine in their peptidoglycan, while all members of the *Bacteroides–Cytophaga* phylum are defined by the presence of sphingolipids.

Examples in which organisms are phylogenetically unrelated despite the presence of a common feature of obvious phylogenetic significance are rare. The most prominent example is the lack of relatedness between *Chloroflexus* and *Chlorobium*. Both are taxa characterized by the presence of chlorosomes with similar structure and function, but with different types of reaction centers (see Chapter 13.2.3). Whether such a complex structure evolved independently, whether the genes encoding the proteins involved in the synthesis of chlorosomes were subjected to horizontal gene transfer, or whether the 16S rDNA analysis does not provide us with the approximate course of evolution must await further investigations. The finding that relatives of *Chloroflexus*, such as *Thermomicrobium* and *Herpetosiphon*, lack chlorosomes, while relatives of *Chlorobium*, such as *Chloroherpethon* and *Clathrochloris* contain chlorosomes, provides no clue to the solution.

The availability of the phylogenetic branching pattern of 16S rDNA, available now for more than 2 500 species from more than 80% of all described genera from both domains is a tremendous scientific effort. Its implications on taxonomy and the recognition of early evolutionary events are most obvious. The extent of diversity among the culturable organisms can only be explored after all type strains of described species have been investigated by molecular phylogenetic and phenotype analyses. The determination of the phylogenetic position of an isolate is among the first required steps in its description and classification.

29.6.6 Some Correlation Exists Between the Paleochemical and Geological Records and the Branching Pattern of the 16S rRNA

The fossil record in sedimentary rocks of the Precambrium is sparse. The oldest fossils known are in the 3.8-Gy-old (giga years 10^9 years before present) Isua metasediments (Greenland), the 3.5-Gy-old Waarawoona group (Australia) and the Swartkoppie formation (South Africa). Although of very limited potential for any proof of ancient biochemical properties, the fossil record indicate that life existed as early as 3.2 Gy ago.

The paleochemical record. More informative than the fossil record is the chemical analysis of inorganic molecules of geologically ancient sedimentary rocks (Fig. 29.**31**). The increased ratio of $[^{12}C]_{org}$ over $[^{13}C]_{org}$ in organic matter from sedimentary rocks is commonly considered to indicate the carbon-fixing reaction of photosynthesis or other autotrophic carbon-fixing mechanisms. In biological processes, the slightly lighter $[^{12}C]_{org}$ isotope is enriched as compared to the $[^{13}C]_{org}$ isotope. The discrimination factor δ for $[^{13}C]_{org}$ is $-25‰$. This value, which has also been determined in material from the 3.8 Gy-old rocks of the Isua metasediments, Greenland, has remained relatively constant over billions of years and indicates that the rate of biomass production has virtually been constant during the biological evolution. The distribution of CO_2-fixing mechanisms in members of both domains leads to the assumption that, as pointed out in Chapter 29.6.3 the evolution of the hypothetical ancestor, the progenote,

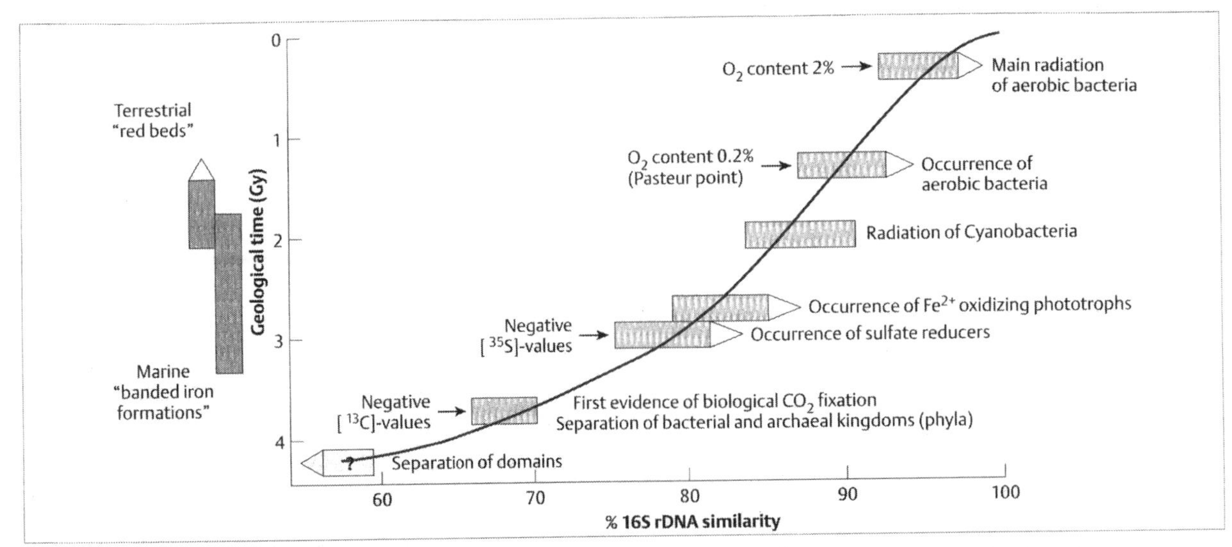

Fig. 29.**31** **Schematic illustration of the occurrence of major phylogenetic groups of bacteria during evolution** Gy (giga years, 10^9 years before present) [after 2]

into the two prokaryotic lines of descent occurred within a rather "short" period (within the first 700 million years) after the formation of the planet Earth. It cannot be excluded that the carbon isotope discrimination already occurred even before 3.8 Gy and the carbon isotopes were not trapped in sedimentary rocks.

Pyrite formation has been considered as the first source of energy and reducing power for CO_2 fixation. In this scenario, developed by G. Wächtershäuser, H_2S plays an important role as a precursor for molecules with catalytic sulfhydryl groups, while pyrites are considered to be precursors for iron-sulfur clusters and ferredoxin. Although only the first reaction below (Eqn. 29.1) has been experimentally verified, it is energetically possible for organisms to develop in this "iron-sulfur" world according to the following reactions:

$$FeS + H_2S\ (\text{aqueous}) \rightarrow FeS_2 + H_2 \qquad \Delta G^{\circ\prime} = -41.9\ \text{kJ/mol} \tag{29.1}$$

$$CO_2\ (\text{aqueous}) + H_2 \rightarrow HCOOH\ (\text{aqueous}) \qquad \Delta G^{\circ\prime} = +30.2\ \text{kJ/mol} \tag{29.2}$$

$$FeS + H_2S\ (\text{aqueous}) + CO_2\ (\text{aqueous}) \rightarrow FeS_2 + HCOOH \qquad \Delta G^{\circ\prime} = -11.7\ \text{kJ/mol} \tag{29.3}$$

Considering their biochemical potential, the thermophilic and chemolithoautotrophic Crenarchaeota may indeed represent ancestral biochemical types.

A second valuable paleochemical analysis is th determination of the ratio at which [^{32}S] is enriched a compared with the heavier [^{34}S] during the biologica reduction of sulfate. No discrimination was detected i the 3.8-Gy-old Isua sedimentary rocks. The earlies significant change in the ratio was found in rocks tha are 3.2–2.8 Gy old. Sulfate respiration, as done b *Desulfotomaculum* (Gram-positive) and *Desulfovibri* and relatives (Proteobacteria), might therefore be more recent adaptation than photosynthesis an methanogenesis. This is in accord with results of 16 rRNA analysis, which separates the sublines embracin these sulfate reducers at approximately 78% similaritie The main radiation of the Gram-negative sulfat reducers occurred even later in evolution (around 82 similarity).

Likewise, the history of atmospheric oxygen contains valuable events that allow a crude calibration c more recent time periods using sequence similaritie There is little doubt that the free oxygen content wa minute before the emergence of plant photosynthesi The source of oxygen in the early Precambrium i unknown, although it is unlikely to have come fron photolysis of water vapor in the high atmosphere. Th existence of the marine "banded iron formation" (abou 3.2–1.9 Gy old), which contains magnetite and hematit can be explained by the activities of oxygenic prokaryotes. *Cyanobacteria*-type organisms were present i 2.7-Gy-old stromatolites, but their identity has not ye been confirmed, and the main radiation of thes organisms occurred more recently than 3.2–2.7 Gy ag

If, however, the formation of marine iron bandings was caused by biological activities, either oxygen must have been produced by organisms for which no descendants have been described, or oxidation of Fe^{2+} to Fe^{3+} occurred non-oxygenically by anaerobic phototrophs. Novel types of phototrophs related to modern phototrophs have recently been shown to exhibit this metabolic activity.

After oxidation of all oxidizable marine compounds, the terrestrial "red beds" (2.0–1.8 Gy ago) were formed. Their existence can be explained by the activity of Cyanobacteria because a rich variety of Cyanobacteria-like microfossils have been found in the stromatolites of the gunflint formation, which is 2.0 Gy old.

It is interesting to note the correlation between the increased oxygen level 1.5–1.9 Gy ago, and the emergence of microaerophilic organisms from anaerobic phototrophic and heterotrophic ancestors in various sublines of the bacterial domain (*Lactobacillus, Streptococcus, Actinomyces*, spirochetes, certain Proteobacteria). Only after the Pasteur point (0.2% O_2) was reached (0.2%, 1.2–1.4 Gy ago) did it become advantageous for organisms to switch from fermentation to aerobic respiration (Fig. 29.**31**). The evolution of aerobic species from microaerophilic fermentative ancestors correlates with rRNA similarities of approximately 87%, while the main radiation of strictly aerobic organisms occurred at similarity values above 92%. This event could have happened at an oxygen partial pressure of 2% in the lower Silurian, about 440 million years ago.

29.7 The Polyphasic Approach to Bacterial Systematics

The ultimate goal of microbial systematics is to relate microbial taxonomy to evolution, thus placing microbiology on a level footing with botany and zoology. Taxonomic units of prokaryotes, arbitrarily defined by humans to facilitate communication, may not correlate with the units that evolved in nature. Consequently, the construction of a hierarchical system is necessarily a compromise that takes into account the inability to understand fully the mode and tempo at which individual bacterial cells evolved, the restriction of basing taxonomic conclusions on a minority of existing organisms, hence on a small fraction of biodiversity only, and the subjectiveness with which taxonomists consider characters to be "important" or less significant.

Eventually, it might be possible to define taxa at all levels exclusively on the basis of genomic information, such as the presence of taxon-specific sequences or nucleotides. However, sequence signatures defined today could become blurred as more sequence information from a larger variety of organisms is included. Even the lowest rank in the hierarchical structure, the strain, is not a coherent collection of genetically identical strains, but is composed of different clones that differ slightly from each other in genotype and phenotype. The error rate of the cellular DNA replication (approx. 10^{-8} per gene per generation) already generates within a single colony mutants that differ (in a few nucleotides) from their ancestral cell. A few generations later, individual clones may have evolved that constitute the beginning of a distinct evolutionary line within the species.

29.7.1 A Fundamental Concept of Taxonomy Is That a Practical System That Serves the User Should Be Chosen

Differences in the rate of evolution within individual lines of descent make it impossible to define a single cut-off point for a taxon at any level, especially because differences at the genomic level do not need to be accompanied by differences at the phenotypic level. However, even if the genetic level provides us with the possibility of creating a classification system based exclusively on sequence idiosyncrasies, the guidelines and standards should, for practical reasons, not be placed so high that they are outside the reach of the majority of taxonomists.

Application of sequence analyses of evolutionarily conserved macromolecules in microbial systematics has provided the framework in which organisms are placed according to their phylogenetic relationships. Nevertheless, modern classification strategies include a large amount of data from the genetic and epigenetic levels. The approach followed in the circumscription of taxa, from species to domain, is termed "polyphasic."

In order to respond to the modern developments in microbial taxonomy, the concept of a "polyphasic" taxonomy has been developed. The major emphasis of the polyphasic system is to establish a framework of phylogenetic groupings and to delineate subsequently these groupings based on suitable phenotypic characteristics. From a taxonomic point of view, the restriction of the wealth of information contained in and expressed by the genome to the mere analysis of the primary structure of macromolecules or to a single property of unknown significance would not take advantage of the evolutionary process at all levels of information.

In order to allow reliable and manageable identification and classification, decisions about the rank of an organism in a hierarchical system is made on the basis of contributions of genotypic and phenotypic properties that distinguish it from neighboring groups. The significance of properties in defining ranks has changed often and will continue to change when new insights into the natural relationships are made available. For example, the use of "phototrophy," or "gliding motility" to describe orders and families has recently been changed to the use of characters suitable for describing genera or species. An important prerequisite of a hierarchical system is that it remains flexible and open to changes in reflecting the course of evolution of properties as accurately as possible, and also to facilitate classification in accordance to the natural relationships of organisms.

29.7.2 From Phylogenetic Classification to Taxonomic Conclusions

How can a meaningful polyphasic classification system be obtained? The most important difference of the new classification system as compared to all bacterial classification systems developed in the past is that it should provide a high degree of stability without abolishing flexibility. The basis for the polyphasic system is the well-founded phylogenetic branching pattern derived by hybridization and sequence analyses. The interfacing classification system successfully started several years ago after the detection of the high degree of correlation between genomic relationships, as revealed by DNA reassociation studies and the distribution of phenotypic traits. The strategy was then extended to lower levels of relationships and included the results of the determination of rDNA cistron similarities. This led to the rearrangement and new description of mainly Gram-negative genera, which are now described by characters of taxonomic significance (Tab. 29.1–29.**3**). With the ability to construct a more complete hierarchical structure, this strategy today is extended to all levels of relationships. The genus *Pseudomonas* is a prime example of misclassification, with members found in the α-, β-, and γ-subclass of *Proteobacteria*. Only those species related to the type species, *P. aeruginosa*, γ-subclass can be considered as authentic pseudomonads. During the last 15 years, several genera have been described that contain strains previously misclassified as *Pseudomonas* species (Tab. 29.**7**).

Knowing the evolutionary line of descent into which a hitherto unclassified organism belongs, allows the determination of its phylogenetic neighbors. If the branching point of the organism is within the radiation of members of a phylogenetically coherent genus, this organism will most likely exhibit the genus-specific characters. Its taxonomic status, i.e., whether it is a strain of a described species or a novel species, depends on the phylogenetic distance to its neighbors and the phenotypic properties. DNA hybridization values above about 70% and high thermal stability of DNA/DNA duplices found between the strain under investigation and the type strain indicate a strain of this species, while lower levels rather indicate a new species. If new discriminating phenotypes cannot be found, it is recommended to refrain from a description until differentiating characters are available. Microbiologists must be aware that the revolutionizing phylogenetic branching pattern might never reflect the actual course of evolution. Phylogenetic reconstructions are based on inferred homologies, but, unless witnessed by the evolutionary history of taxa (i.e., by fossil data) can be considered at best as good approximations. As in previous decades, the one system (or parts thereof) with the highest practicability will succeed against competing systems with less persuasive arguments. The ultimate goal is to establish a hierarchical system in which all taxa show phylogenetic coherency and, at least for ranks below the family level, also a great deal of phenotypic coherency with sufficient differences to distinguish taxa from each other by stable and easily determined characters.

29.7.3 Conclusions: 16 Major Achievements of Phylogenetic Studies and Trends in Modern Taxonomy

Phylogenetic studies had and have a strong impact on changing traditional concepts about the role prokayotes have played in the evolution of life. In summary, these are:

- Based on paleochemical data, life on Earth evolved more than 3.8 Gy ago. The fossil record dates perhaps back to about 3.7 Gy.
- Branching patterns obtained from the analysis of several evolutionarily conserved macromolecules are

Table 29.7 **Recent changes in the classification of prokaryotic species** as a result of new insights into the phylogenetic relatedness at the intergenic and intragenic level, exemplified by *Pseudomonas* species (see also Tab. 29.3)

Former name	Name following reclassification	Membership to subclass of Proteobacteria
Pseudomonas aeruginosa	(Type species of the genus)	γ
Pseudomonas acidovorans	*Comamonas acidovorans*	β
Pseudomonas aminovorans	*Aminobacter aminovorans*	α
Pseudomonas avenae	*Acidovorax avenae*	β
Pseudomonas cepacia	*Burkholderia cepacia*	β
Pseudomonas diminuta	*Brevundimonas diminuta*	α
Pseudomonas flava	*Hydrogenophaga flava*	β
Pseudomonas luteola	*Chryseomonas luteola*	Undetermined
Pseudomonas maltophilia	*Stenotrophomonas maltophilia*	γ
Pseudomonas marina	*Deleya marina*	γ
Pseudomonas mesophilica	*Methylobacterium mesophilicum*	α
Pseudomonas mixta	*Telluria mixta*	β
Pseudomonas oryzihabitans	*Flavimonas oryzihabitans*	Undetermined
Pseudomonas paucimobilis	*Sphingomonas paucimobilis*	α

by and large identical. This excludes the possibility that any one of them has been involved in lateral gene transfer. Hence, the branching pattern is a reflection of the evolution of the molecule and of the phylogeny of the organism.

- Modern prokaryotes are the descendants of ancestral life-forms that evolved from a common ancestor, designated the "progenote".
- The progenote, a hypothetical common ancestor, gave rise to prokaryotes that do not form a phylogenetically coherent group, but constitute two separate life forms: Archaea and Bacteria.
- As detected so far, all modern organisms evolved from one of three major lines of descent, the domains, for which the names Archaea (archaebacteria), Bacteria (eubacteria), and Eukarya (eukaryotes) have been proposed. Since ranks and names above the order are not governed by the International Code of Nomenclature of Bacteria, the use of the rank domain and the names proposed for them, although stimulating from a scientific point of view, are not binding.
- Both prokaryotic domains contain several lines of descent. Kingdoms have been described for the domain Archaea, i.e., Crenarchaeota, Korarchaeota, and Euryarchaeota. The main lines of descent are of similar rank.
- The organization of the genome into introns and exons was likely a feature of the progenote that was subsequently lost in most prokaryotic lines of descent.
- Eukaryotes evolved from a prokaryotic ancestor that was likely more closely related to Archaea than to Bacteria.
- The eukaryotic cell represents a chimeric structure containing genes from both prokaryotic lines of descent. This finding allows the suggestion that gene transfer between representatives of the three major groups of organisms was common in early evolution.
- The extent to which genes within eukaryotic genomes are of prokaryotic origin, from both Bacteria and Archaea is surprisingly high. Representatives of the domain Bacteria are the ancestors of chloroplasts and mitochondria. Archaea and Bacteria still play important roles as symbionts and parasites in all taxa of eukarya.
- The phylogenetic neighbors of the archaea in the eukarya line of descent are protozoa, such as *Giardia* and *Microsporidia*.
- Higher taxa, derived from similarities in phenotype, rarely only reflect the phylogeny of the organism. Apparently, properties used in the past for defining ranks, such as photosynthesis, gliding motilitiy, budding reproduction, mycelium formation, and spore formation, have no exclusive significance in defining families, orders, and classes.
- The classification system presently established for prokaryotes is based primarily on sequence formation of semantophoric molecules (DNAs and proteins) in concert with selected phenotypic characters in order to delineate taxa.
- Sequence information will play an increasingly important role in the design of oligonucleotide probes and PCR primers used in molecular paleontology, ecology, and the diagnosis of microorganisms of medical, environmental, and biotechnical importance.

- Taxonomic ranks are above all subjective units that facilitate communication and allow the formulation of a practical system that serves the user.
- The ultimate goal of microbial systematics is to relate taxonomy to evolution by establishing true homologies at the genomic level. Until this ambitious goal is reached, however, guidelines and standards should not be placed so high that they are outside the reach of daily practicability as needed in the applied microbiology of, for example, medical institutions, environmental agencies, and biotechnical industries.

Further Reading

Balows, A., Trüper, H. G., Dworkin, M., Harder, W., and Schleifer, K.-H. (1991) The prokaryotes. A handbook on the biology of bacteria: ecophysiology, isolation, identification, applications. Berlin, Heidelberg, New York: Springer

Danson, M. J., Hough, D. W., and Lunt, G. G. (1992) The archaebacteria: biochemistry and biotechnology. Biochemical Society Symposium 58. London: Portland Press

Goodfellow, M., and O'Donnell, A. G. (1993) A handbook of new bacterial systematics. New York, London: Academic Press

Goodfellow, M., and O'Donnell, A. G. (1994) Chemical methods in prokaryotic systematics. New York: Wiley

Holt, J. G., Krieg, N. R., Sneath, F. H. A., Staley, J. T., and Williams, S. T., eds. (1994) Bergey's manual of determinative bacteriology, 9th edn. Baltimore: Williams & Wilkins

Li, W.-H., and Graur, D. (1991) Fundamentals of molecular evolution. Sunderland, Mass.: Sinauer

Neidhardt, F. C., Curtiss, R., Ingraham, J. L., Lin, E. C. C., Low, K. B., Magasanik, B., Reznikoff, W., Riley, M., Schaechter, M., and Umbarger, H. E. (eds.) (1996) *Escherichia coli* and *Salmonella*: cellular and molecular biology. 2nd edn. Washington, DC: ASM Press

Pankhurst, R. J. (1991) Practical taxonomic computing. Cambridge: Cambridge University Press

Ratledge, C., and Wilkinson, S. G. (1989) Microbial lipids, vol 1. New York, London: Academic Press

Sackin, M. J., and Jones, D. (1993) Computer-assisted classification. In: Goodfellow, M., and O'Donnell, A. G. (eds.) Handbook of new bacterial systematics. New York, London: Academic Press; 282–313

Schleifer, K.-H. and Ludwig, W. (1989) Phylogenetic relationships among bacteria In: Fernholm, B., Bremer, K., and Jörnwall, H. (eds.) The hierarchy of life. Amsterdam: Elsevier; 103–117

Stackebrandt, E. (1992) Unifying phylogeny and phenotypic properties. In: Balows, A., Trüper, H. G., Dworkin, M., Harder, W., and Schleifer, K.-H. (eds.) The prokaryotes, 2nd edn. Berlin, Heidelberg, New York: Springer; 19–47

Trüper, H. G. (1992) Prokaryotes: an overview with respect to biodiversity and environmental importance. Biodiversity Conservation 1: 227–236

Woese, C. R. (1987) Bacterial evolution. Microbiol Rev 51: 221–271

Woese, C. R., Gutell, R., Gupta, R., and Noller, H. (1983) Detailed analysis of the higher order structure of 16S-like ribosomal ribonucleic acids. Microbiol Rev 47: 621–669

Zuckerkandl, E., and Pauling, L. (1965) Molecules as documents of evolutionary history. J Theor Biol 8: 357–366

Sources of Figures

1 Stackebrandt, E., and Rainey, F. A. (1997) Phylogenetic relationships. In: Rood, J. (ed.) Molecular biology and pathogenesis of the Clostridia. New York London: Academic Press

2 Stackebrandt, E. (1995) Origin and evolution of prokaryotes. In: Gibbs, A., Calisher, C. H., and Garcia-Arenal, F. (eds.) Molecular basis of virus evolution. Cambridge: Cambridge University Press; 224–242

Section VIII
Prokaryotes in the Biosphere

In the preceding chapters, microorganisms were usually treated as pure or defined cultures studied in the laboratory. Isolation of pure cultures marked the beginning of scientific microbiology, and studies with pure cultures provided the basis of our knowledge about microorganisms. Nonetheless, microorganisms are taken from nature into the laboratory, and an understanding of their properties and activities in the laboratory requires the understanding of the natural life conditions under which the respective capacities have evolved.

Understanding life implies the understanding of how life evolved; understanding evolution requires the understanding of the ecological conditions under which evolution occurs.

The following chapters will concentrate on the conditions under which microorganisms live in natural environments, and how microorganisms contribute to the physical, chemical, and biological properties of their environment, as they find it, and how they shape the environment for themselves and for other organisms. The natural environment of a microorganism has some basic differences from the usual laboratory conditions under which the microorganism is grown:

1. In their natural environment, microorganisms are usually exposed to conditions of substrate limitation rather than substrate excess.
2. In nature, a microorganism encounters more than one type of substrate.
3. In nature, a single type of microorganism is seldom alone, it has to compete against or cooperate with others.
4. Natural microbial habitats are spatially heterogeneous, even at rather short distances.
5. Conditions in natural habitats are rarely stable; they change with time, fluctuating in daily or annual cycles.

The aim of ecological research is to understand the single organism or a certain type of organism in its interaction with the abiotic and biotic conditions that shape its environment. **Ecology** is "the entire science of the relations of an organism to its environment to which we can count in a broader sense all conditions of existence" (E. Haeckel, 1866).

Because of their small size, the importance of microorganisms in the maintenance of the biosphere has been underestimated for a long time. It is assumed today that more than 50% of all living cytoplasm is prokaryotic cytoplasm, and this enormous amount of living cell material is actively involved in transformations of organic and inorganic compounds in the environment. Due to their high surface-to-volume ratio, prokaryotic cells are metabolically more active than eukaryotic cells, and this fact increases the relative importance of prokaryotes even further. The existence of microorganisms is noticed much more by their activity than by their optical appearance. Wherever there is a substantial amount of a substrate available in a combination that allows energy exploitation by a living system, a prokaryotic cell that handles this process will be found at a high probability and will multiply to exploit this energy source with its progeny.

To understand the ecology of a certain organism is easy as long as this organism can be studied by direct observation. This is especially difficult with microorganisms. Moreover, the lack of diversity in shape renders it impossible to recognize bacteria by simple optical methods. Even worse, the metabolic capacity and with that the function of a certain organism in its habitat cannot be assessed by such approaches. It is for these reasons that microbial ecology developed only recently, far behind the ecology of animals and plants.

A few definitions are necessary. The basic unit of ecology is the **community** or **biocoenosis**. It makes up the living part of an ecosystem. The term **biota** is synonymous with community and includes the **fauna** (animals), **flora** (plants), and **microflora** (microorganisms, i.e., fungi, protozoa, prokaryotes, and viruses). Another term for biocoenosis is **biome**, which characterizes the communities of certain types of ecosystems (e.g., a fir forest, a desert, an ocean bight). The community has to be considered in relation to the physical and chemical characteristics of the site. The biotic components and the abiotic, physicochemical components make up the ecosystem. The abiotic components of the ecosystems are frequently referred to as the **environment**, although the environment quite often implies biotic components as well. An **ecosystem** is a self-sufficient functioning unit. A complete ecosystem includes primary producers, consumers, and mineralizing organisms. Ecosystems can be as large as an ocean, a pond, a lake, a river, or a forest. An ecosystem is characterized by its structure and function, both with respect to its physicochemical and its biotic components.

Within an ecosystem, a **habitat** can be defined for a single species. The habitat is the location or the dwelling place of a particular organism. A habitat can be defined for a certain type of microorganism as the place where it can be found and perhaps be isolated by suitable techniques. It is the "address" under which a certain organism can be found.

Quite different from the habitat, the **ecological niche** has nothing to do with a spatial assignment, but refers only to the function or the "profession" of the organism. The habitat and ecological niche of organisms need to be known to understand the structure and function of the community within a certain ecosystem. Ecological niches may be broad or very narrow; some organisms are highly specialized in their metabolic capacities or environmental activities (**specialists**) others are very broad (**generalists**).

Among the inhabitants of an ecosystem, the biota, a differentiation is possible between organisms typically encountered there and others found there only accidentally. S. N. Winogradsky defined two different types of microorganisms that occur in soil. **Autochthonous** or **indigenous microorganisms** are typically found in the respective ecosystem and are present at a fair, but stable population size with little variation over time. In contrast, **allochthonous**, **zymogenous**, or **non-indigenous microorganisms** are not typical inhabitants of the respective ecosystem. Their population size varies dramatically with time over several orders of magnitude, depending on available food sources or other life-limiting factors. They grow fast under suitable conditions, but die fast when conditions change, and survive at only very low numbers. This concept, originally developed for the microbiology of soil, can also be extended to other ecosystems. The autochthonous microbial population of a certain ecosystem may not always be easy to define; in many cases, the zymogenous bacteria are much better known because they are much easier to cultivate. Present-day microbial ecology has to deal preferably with the autochthonous rather than the zymogenous microorganisms.

30 Ecophysiology and Ecological Niches of Prokaryotes

A microbial strain isolated in pure culture can be studied in the laboratory with respect to, for example, its metabolic capacity, the substrates it uses and the products it forms, its dependence on environmental parameters, such as temperature, pH, ionic strength, humidity, and light irradiance; its reactions to changes in any of these parameters in space and time; its capacity to move actively; and its capacity to adapt its metabolism to changing conditions. Studies of this kind provide information on the **autecology** of this organism. Such information concerns the potential capacities of an organism and is helpful to understand how an organism might possibly behave in its natural environment. Such studies do not provide any information on how the organism will really react in situ, under the influence of complex conditions that are quite different from those in the laboratory. Nevertheless, the latter, **synecological**, approach is needed to understand how the community of organisms cooperates and how the members of the community depend on each other in a complex natural ecosystem. It is the enormous complexity of many factors acting at once that makes synecological research much more difficult than autecological studies. A synecological view of microbiology also has to include the ecology of higher organisms, unless very special, so-called **extreme environments** are being studied (see Chapter 31.4).

30.1 Substrates in Nature Are Typically Limiting

Laboratory work aims at using fast-growing organisms that are easy to handle and do not clump or stick to surfaces, and to achieve that, bacteria in the laboratory are usually supplied with easily degradable substrates at concentrations that ensure fast growth to high cell densities. The situation in the natural environment of a microorganism is quite different. Although a microorganism in nature can be exposed accidently to high concentrations of substrates (e.g., a spoiling fruit or a decomposing animal cadaver), even there the situation is characterized by competition with other organisms, and the substrate available to the single organism very soon becomes limiting. The typical life condition of a microorganism in nature is a limiting food supply, and even more typically famine, and this situation is created by the activity of the microbes themselves.

30.1.1 Light Is an Energy Source, but Is Also Harmful

Light is the primary source of energy in nearly all ecosystems. Considerable parts of the solar light spectrum that reach the outer atmosphere are absorbed by ozone, dioxygen, and water in the atmosphere, as shown in Figure 30.**1**. Less than 50% of the light energy reaching the Earth's surface is visible light; a small fraction is ultraviolet light and a major part infrared irradiation (heat).

Terrestrial organisms have found ways to protect themselves from the detrimental effects of ultraviolet light, e.g., by the synthesis of suitable protective pigments (carotenoids, melanines), and plants have developed highly refined structures and means to harvest most efficiently those fractions of the visible light that are essential to drive photosynthesis. Aquatic plants, especially algae, and microorganisms do not need strong protection against light irradiation because water itself acts as a protectant.

The penetration of light into a body of water depends on the light wavelength. Infrared light is absorbed strongly by the water molecules, and red light intensity decreases within 2.5-m water depth to 1% of its original intensity. The intensity of ultraviolet light is diminished less by absorption (Fig. 30.**2**), but decreases with depth mainly by scattering and backscattering; the backscattering blue light causes the blue color ("desert colour") of a body of pure water, such as an alpine lake. As a result of absorption and scattering, the light quality changes during its passage through pure water, and the light with wavelengths around 540–560 nm penetrates

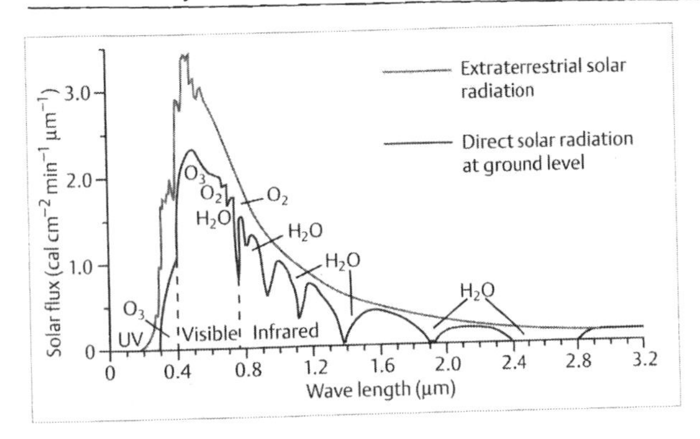

Fig. 30.1 **Comparison of extraterrestrial solar flux and irradiation at the Earth's surface.** The major absorption bands of atmospheric O_2, O_3, and water vapor are indicated [after 1]

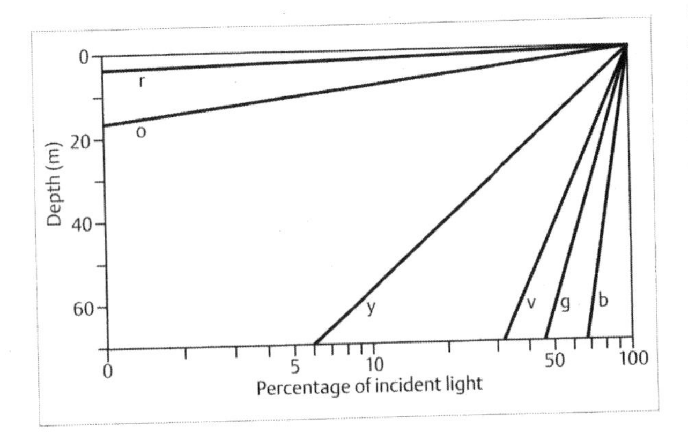

Fig. 30.2 **Transmission of light by distilled water at six different wavelengths.** (r = 720 nm, o = 620 nm, y = 560 nm, g = 510 nm, b = 460 nm, v = 390 nm). The percentage of incident light remaining at various depths is drawn on a logarithmic scale [after 2]

to the greatest depth. This change of light quality with depth is further affected through absorption by dissolved compounds, such as fulvic acids (see Chapter 30.1.2), which absorb primarily blue light, through scattering by suspended particles in the water, as well as through absorption of light in specific wavelength regions by algae and other photosynthetic organisms; the wavelength regions depend on the pigments contained in the organisms. Thus, the light intensity and quality in bodies of water differ individually to a great extent, and depend on the water quality and the amount and types of photosynthetic organisms present.

30.1.2 Organic Substrates Are not Always Easily Available

Organic matter is synthesized in nature mainly by the activity of green plants, especially higher plants, and of algae and cyanobacteria. Plant material can serve as substrate for animals that grow and build their own biomass. Dying plants and animals, leftovers of animal digestion, and plant excretions, for example, make up the food for heterotrophic microorganisms, especially bacteria and fungi, which will finally oxidize these materials to CO_2, water, and inorganic salts, which in turn can serve as substrates for primary production.

Chemically speaking, the substrates of microbial degradation are typically polymers. Dry plant biomass consists of approximately 50% cellulose, 10–25% lignin, 10–20% hemicelluloses and pectin, 5–10% protein, 2–5% lipids, and up to 2% nucleic acids. Animal tissues, on the other hand, are rich in protein (50–80%), including structural proteins such as cartilage, ligaments, hairs and nails, and also consist of 10–15% lipids, 5–10% polysaccharides, and 5% nucleic acids. Dissolved monomers make up less than 5% of the compounds. Monomeric substrates are degraded fast and easily by a wide variety of many organisms. Degradation of polymeric substrates, on the other hand, is a matter for specialists (Chapter 9). Depending on the chemistry of the respective polymers, degradation can be easy and fast, as with starch or globular proteins (Tab. 30.1). Structural polymers such as lignin, keratin, or cellulose have been made to resist microbial attack as long as possible, and can be stable for several hundred or thousand years, depending on the incubation conditions (e.g., oxygen availability, proton concentration, and presence of inhibitory substances, such as tannins). Wooden boards of Roman warships have been preserved in anoxic marine sediments for nearly 2 000 years. Thus lignocellulose is rather stable under such conditions because lignin biodegradation requires molecular oxygen for efficient breakdown. However, the fact that such

Table 30.1 **Estimated half-life times of biological polymers in nature**

Polymer	Half-life
Lignin	20–2000 years
Hair, wool (keratin)	1–2000 years
Humic compounds	2–200 years
Cellulose	0.01–2 months
Starch	1–10 days
Globular proteins	0.1–2 days

wood is corroded to some extent indicates that under such conditions decomposition is possible to a certain degree.

Linkages with other polymers or mineral substrata (surfaces) can further impair degradability. Cellulose, although rather easily degradable as a pure polymer, becomes more recalcitrant when associated with lignin in wood. Adsorption of proteins or nucleic acids to silica or clay surfaces make them nearly inaccessible for depolymerizing enzymes, and their half-life increases by orders of magnitude.

Microbial degradation of organic matter does not always proceed directly to complete mineralization. In the presence of oxygen, specific or unspecific oxygenase reactions transform aromatic and aliphatic substrates to phenolic or alcoholic derivatives, which tend to form radicals and polymerize to large aggregates. The conversion of wood to **humus** in soil is a well-known process; other substrates, such as amino acids, sugars, and oligomers of various kinds, can be bound into this polymeric material, with the consequence that their degradation rate decreases considerably. The molecular mass of such **humic material** can increase to several million Daltons. Terrestrial humic material contains, together with aromatic lignin derivatives, a high proportion of aliphatic constituents derived from sugars and proteins. In lake waters, **aquatic humic acids** represent the counterpart of terrestrial humic material, with molecular masses of 1.5 to approximately 150 kDa. They are poorer in aromatic constituents than terrestrial humic acids and their composition depends on the possible import of allochthonous ligninaceous precursors. Humic materials are turned over slowly by microbial, chemical, and photochemical reactions, and also their constituents change in their chemical properties gradually with every transformation.

Thus, polymeric biomass can be abundant. Organic polymers can make up 70% of dry matter of, for example, a forest surface soil, but their degradation is slow (Tab. 30.**1**). Degradation of these polymers is greatly enhanced by the activity of animals, such as beetles, earthworms, snails, springtails, termites, and ruminants, which decrease the particle size mechanically and allow easier access of microbes to the substrates, either inside their intestines or outside after excretion (Chapter 31.5). Protozoa take up small substrate particles by phagocytosis, whereas prokaryotes have to depolymerize polymers outside the cell by exoenzymes and take up only low-molecular-weight substrates through transport systems in the cytoplasmic membrane.

Thus, biomass degradation by microorganisms has to deal primarily with polymeric substrates. Since degradation of the polymers is usually the rate-limiting step in biomass decomposition, the concentration of free monomers is typically low because they are taken up fast and efficiently by a broad microbial community.

Concentrations of free sugars measured in various habitats vary between a few micromolar in a comparably rich environment, such as the rhizosphere of a plant root, down to picomolar in ocean waters. Free amino acids are typically found in nanomolar concentrations in aquatic environments and in soils. It is obvious, therefore, that microorganisms that grow on sugars or amino acids, for example, have to take up such substrates at very low concentrations. Under such conditions, their uptake activity at low concentrations of substrate, i.e., their **substrate affinity**, is of primary importance for efficient substrate utilization (see Chapters 5 and 30.4). On the other hand, the low concentrations of monomeric organic compounds found in stable natural environments are the consequence of microbial degradative activities and their high substrate affinities. This is true for the organic substrates mentioned here, but in a similar manner also for all other types of substrates.

The substrates for microbial decomposition are mostly polymers of which especially lignocellulose is greatly resistant to degradation. Products of polymer degradation can form radicals and polymerize to humines and fulvines. Monomeric degradation products such as sugars and amino acids are available only in the micromolar to nanomolar concentration range.

30.1.3 Oxygen Is Essential for Obligately Aerobic Microorganisms, but Can Also Be Toxic

Molecular oxygen makes up 21% (about 9 mmol l^{-1}) of air in the atmosphere. Air-exposed microorganisms have to cope with this concentration of a potentially toxic gas, which can be exploited for energy conservation in respiration (see Chapters 4 and 11). In aquatic environments at 4–20 °C, the oxygen in equilibrium with air is in the range of 300–400 μM (Tab. 30.**2**). Although dissolution causes a relative enrichment of oxygen over the other atmospheric gases (N_2, Ar) due to its higher polarity, the small oxygen reservoir dissolved in water can be consumed quite fast by microbial activities. A concentration of organic substrates (calculated as sugar equivalents) in the range of 10 mg l^{-1} is sufficient to consume the total amount of dissolved oxygen in water at 15 °C. Therefore, oxygen-dependent degradation processes in aquatic environments depend on intense replenishment of dissolved oxygen by transport processes, especially by eddy convection (see

Table 30.2 **Solubility of oxygen in pure water in relation to temperature in equilibrium with air at standard pressure**

T (°C)	O_2 (mg l^{-1})	Concentration (μM)
0	14.62	457
4	13.10	409
10	11.29	353
20	9.09	284
30	7.56	236
40	6.41	200
50	6.25	195
100	5.08	159

Chapter 30.6). Oxygen solubility is further decreased by dissolved salts. Seawater (3.5% w/v salinity) contains at a given temperature 20% less oxygen than freshwater. Details will be discussed with the respective habitats in Chapter 31.1. Oxygen in the presence of light can be detrimental because light reactions with cell components can form oxygen radicals, which destroy essential cell constituents such as nucleic acids.

30.1.4 Nitrogen Is the Second Most Important Bioelement

Nitrogen makes up about 16% of biomass and is, after carbon, the most important element for living cells. Inorganic nitrogen as a source for assimilatory metabolism is available in natural habitats in the form of nitrate, nitrite, ammonia, and molecular nitrogen. Concentrations of total bound inorganic nitrogen in natural ecosystems are in the range of 1–500 μM. In cell material, nitrogen is usually found as reduced nitrogen, and **ammonia** is the form in which nitrogen is assimilated or liberated during microbial decay (Chapter 8.4). In habitats rich in molecular oxygen, bound nitrogen occurs mainly as **nitrate**, which for assimilation purposes has first to be reduced to ammonia through assimilatory nitrate reduction. Assimilation of **molecular nitrogen, N_2**, is a metabolic capacity rather widespread among prokaryotic organisms that requires a complex enzyme system, the nitrogenase (Chapter 8.5). The process requires much energy in the form of ATP, and is strictly regulated to avoid energy wastes. **Nitrite** can be detected everywhere where either nitrate is reduced or ammonia is oxidized. Measurable in situ concentrations usually do not exceed a few micromolar; due to its mutagenic potential, higher concentrations of nitrite would be harmful to the biota. For more details on the biochemistry of nitrogen transformations, see Chapters 7.3.1, 8.4, 8.5, and 12.2.2.

30.1.5 Inorganic Sulfur Compounds Are Essential Nutrients and Catabolic Substrates

Sulfur makes up about 0.2% of biomass; it is therefore an essential nutrient for biomass synthesis. In the living cell, sulfur is usually found in its reduced form and assimilated as **hydrogen sulfide** (Chapter 8.7). This form is stable only in strictly anoxic habitats. In the presence of oxygen, H_2S reacts chemically with oxygen to form **thiosulfate** ($S_2O_3^{2-}$), **sulfite** (HSO_3^-), or **elemental sulfur** (S^0), and this reaction is exploited for removal of trace amounts of oxygen with sulfide as reductant for cultivation of strict anaerobes in the laboratory. The spontaneous oxidation of H_2S/HS^- with oxygen is comparably fast at pH 7.5–8.5. The reaction is slow at higher proton activity; another maximum of reactivity occurs at approximately pH 11. In the presence of catalytic amounts of heavy metals or of sulfur-oxidizing microorganisms, sulfide can also be oxidized completely to **sulfate**, which is the predominant form in which sulfur occurs in oxygen-rich environments. Utilization of sulfate for assimilation into cell material requires the presence of an assimilatory sulfate reductase system (Chapter 8.7). Oxic freshwaters contain sulfate in the range of 50–300 μM. Exceptions are waters originating from areas rich in **gypsum** ($CaSO_4$); gypsum-saturated water at 25 °C contains 14 mM sulfate. Seawater is very rich in sulfate, with a sulfate concentration of 28 mM. Evolving H_2S reacts with heavy metal ions, especially iron, according to Reaction 30.1:

$$Fe^{2+} + H_2S \rightarrow FeS + 2\,H^+ \qquad (30.1)$$

Since oxides and carbonates of iron and other heavy metals are usually available in excess, the concentration of free H_2S and HS^- in sediments is typically lower than 100 μM, even in marine sediments. Iron (II) sulfide (**FeS**) precipitates in the form of microcrystals, which absorb light in the entire visible range. Therefore, iron sulfide precipitates look black and are responsible for the dark-gray to black appearance of many anoxic sediments.

FeS can undergo a further reaction with H_2S to form **pyrite** (FeS_2) in which sulfur exists in the oxidation state −1:

$$FeS + H_2S \rightarrow FeS_2 + H_2 \qquad (30.2)$$

Although this reaction releases some free energy ($\Delta G^{\circ\prime} = -38.4$ kJ mol^{-1}), there is no indication to date that it is catalyzed by living organisms. An alternative formation of pyrite from FeS and S^0 is discussed in the literature, but is probably only of limited importance in nature. The pyrite formed is stable in the absence of strong oxidants and can make up the majority of all sulfur components in deeper limnic or marine sediments.

Elemental sulfur (S^0) is nearly insoluble in water; up to 5 mg sulfur dissolves in 1 l of pure water at 25 °C (=0.16 μM). Sulfur is found therefore not in the free water, but mainly in sediments as a minor sulfur fraction, next to pyrite and sulfides, or inside sulfur-metabolizing microbes (sulfur "globules" or "droplets"). Larger deposits of sulfur accumulate around hot springs that release H_2S, which precipitates as elemental sulfur after contact with oxygen in the air.

30.1.6 Phosphorus Does not Undergo Dissimilatory Redox Reactions

Phosphorus occurs in nature nearly exclusively as **phosphate**, meaning in the oxidation state +5. It does not undergo redox reactions in biological energy transformations (Chapter 8.6). Few organophosphorus compounds are known in which also C–P or C–P–C linkages occur (see Chapter 30.2.3).

Phosphate minerals are typically poorly soluble, as in the case of **apatites** [$Ca_5(F, OH, Cl)(PO_4)_3$] or **iron (III) phosphate** ($FePO_4$). The free phosphate concentration in natural waters is therefore low, in the range of a few hundred nanomolar to few micromolar. Phosphate is an essential constituent of biomass, of which it constitutes about 0.5% of dry matter. The so-called "Redfield ratio" gives an average elemental composition of algal biomass in lakes of 106 C:16 N:1 P. Due to its low solubility, phosphorus typically limits primary production, and minor increases in phosphorus supply can dramatically increase total productivity. Phosphate is therefore the key effector in regulating algal productivity in freshwater lake ecosystems. Recent work has demonstrated that heterotrophic bacteria compete successfully with algae and cyanobacteria for the available pools of dissolved inorganic phosphate in lakes.

30.1.7 Redox Processes of Iron and Manganese Are Highly pH-Dependent

Iron is by mass the fourth most important element in the Earth's crust, but makes up less than 0.01% of living cells (Chapter 8.8). When in contact with water, it forms hydroxides and oxides or oxide hydrates. Of special biological importance is the transition of Fe (II) to Fe (III) and the reverse transition which is essential in the action of, for example, cytochromes and iron-sulfur proteins. In anoxic ecosystems, Fe^{2+} is soluble in water up to 0.1–1 μM, depending on the carbonate concentration, and is therefore sufficiently available for living cells. With excess sulfide present, FeS precipitates and the free Fe^{2+} concentration drops to less than nanomolar concentrations. In the presence of oxygen at neutral pH, Fe^{3+} precipitates as $Fe(OH)_3$ and FeO(OH), and the concentration of free Fe^{3+} in pure oxic water is near 10^{-18} M. Thus, free Fe^{3+} is not available to living cells under such conditions, and aerobic bacteria have therefore developed specific uptake systems for iron ions that require specifically synthesized complexing agents, the **siderophores**. In nature, also humic and fulvic acids can act as complexing agents to maintain ferric iron at low concentrations in a dissolved state (see Chapters 8.8 and 31.3.1).

The redox potential of the transition of Fe (II) to Fe (III) depends on the prevailing pH. At a pH < 2.4, the standard redox potential at +770 mV is relevant

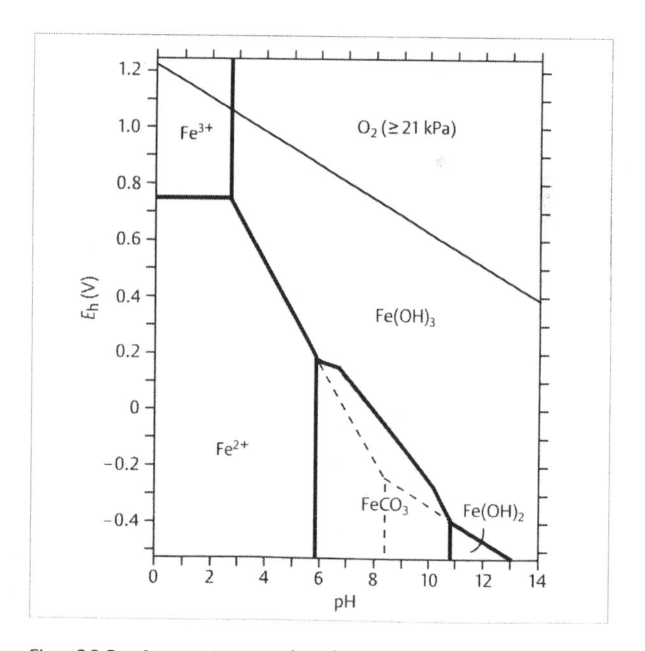

Fig. 30.3 **Approximate distribution of iron species in relation to pH and the prevailing redox potential.** The reactants are assumed to be at 10 mM concentration. Dotted lines represent the situation in the absence of carbonate. The thin line on the upper right refers to reduction of oxygen to water [after 3]

because both forms, Fe^{2+} and Fe^{3+}, are highly soluble and available at equal concentrations. With increasing pH, Fe^{3+} precipitates with hydroxide anions to form $Fe(OH)_3$ and FeO(OH), and the free Fe^{2+} dominates, shifting the effective redox potential to values around +100 mV at neutral pH (Fig. 30.**3**). Thus, the energetics of iron reduction and iron oxidation vary substantially with the prevailing pH.

A similar situation emerges with **manganese**, which oscillates between the redox states Mn (II) and Mn (IV). Again, the solubilities in water of the two ion forms differ considerably, and depend on the prevailing proton activity. Also mixed oxides exist ("Mn_3O_4") that are comparably stable and may trap oxidized and reduced forms in a state that is only poorly bioavailable. The actual redox potentials are influenced by the proton activity at pH > 6.0, and these potentials are in general approximately 500 mV more positive than those of the iron forms.

30.2 Bacteria Are Active in Redox Transformations

Among the various types of redox reactions catalyzed by microorganisms, there is a hierarchy that prefers certain processes over others. Wherever organic material accumulates, a sequence of microbially catalyzed redox events can be observed that is paramount to all microbial ecosystems studied, regardless of whether a freshwater body, a marine sediment, a water-logged soil, a salt pond, sewage sludge, or a microbial mat in a hot spring is considered.

30.2.1 Terminal Acceptors in Oxic and Anoxic Environments

As long as molecular **oxygen** is available, it is the preferred acceptor for electrons derived in the oxidation of organic matter by microbial communities. Oxygen serves as **electron acceptor** at the end of the respiratory chain (see Chapters 4 and 11), but it also acts as cosubstrate in **oxygenase** reactions that are employed in the activation of comparably inert substrates.

Oxygenase reactions depend on the unique biradical nature of the oxygen molecule; there is no parallel to oxygenase reactions in the absence of molecular oxygen.

Due to the low solubility of oxygen in water and its slow diffusion through a body of water, oxygen can soon be used up in an aquatic habitat unless it is replenished by convection processes ("eddy diffusion"). If oxygen is used up, **nitrate** takes over as an alternative electron acceptor and is reduced via nitrite to nitrogen gas, dinitrogen oxide, or ammonia (see Chapter 12.1 and Fig. 30.**4**). Different from oxygen, nitrate or its reduced derivatives can act only as electron acceptors and not as cosubstrates in substrate activation reactions. Nitrate reduction to molecular nitrogen is an activity widespread among many aerobic bacteria that can quickly switch over from aerobic respiration to nitrate-dependent metabolism and back again. Availability of oxygen does not always repress nitrate reduction entirely. A lack of complete repression was observed in the past with some laboratory strains of bacteria studied in detail. Several *Pseudomonas*, *Aeromonas*, *Moraxella*, and *Arthrobacter* strains have been shown recently to be able to reduce nitrate in the presence of oxygen at up to 80% air saturation, i.e., to carry out **aerobic nitrate reduction**. Several fermentative bacteria reduce nitrate only to nitrite and not further; they are responsible to a large extent for the accumulation of nitrite in gastrointestinal tracts upon nitrate loads, e.g., with nitrate-rich drinking water.

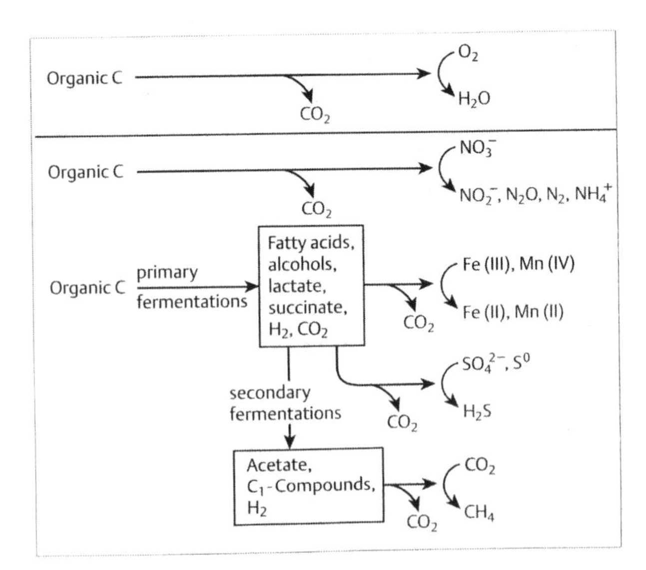

Fig. 30.**4** **Sequence of redox processes coupled to mineralization of organic matter**

The importance of nitrate reduction to ammonia has been disregarded for a long time. Several anaerobic bacteria including sulfate reducers produce ammonia from nitrate and in this way can deliver more electrons to the nitrate molecule than in denitrification. **Nitrate ammonification** is therefore the preferred process in environments rich in organic matter and low in nitrate supply.

Reduction of oxygen and nitrate can be coupled to the complete oxidation of the organic substrate provided, usually including even polymeric substrates. After exhaustion of nitrate, degradation of organic substrates becomes a process shared by different cooperating groups of microorganisms. **Fermentative bacteria** convert polymeric or monomeric substrates to classical fermentation products, including fatty acids, alcohols, lactate, succinate, hydrogen gas, and carbon dioxide (primary fermentations; Chapter 12). The primary fermentation products are oxidized with **iron (III)** and **manganese (IV)** as electron acceptors (Tabs. 10.**2** and 30.**3**). A great variety of iron-reducing bacteria are known today, such as *Geobacter metallireducens* and *Shewanella* spec., which can oxidize a broad spectrum of organic substrates, including sugars, amino acids, and aromatic compounds. Electron transport to iron oxide is obviously coupled with energy conservation. Little is known so far about the path of transfer of the electrons to the practically insoluble iron (hydr)oxide, whether direct cell contact with the iron compound is necessary or whether complexing agents or organic electron carriers such as humic compounds are involved.

Manganese reduction is of minor importance in most freshwater environments, but can play a significant role at places where manganese accumulates, e.g., in certain marine sediments, the Baltic Sea, or in certain rock formations. Whether specialized manganese reducers obtain metabolic energy from this redox process is still unknown.

Table 30.**3** **Redox potentials of major electron carrier systems**, calculated for pH 7.0

Electron carrier system	E_0' (mV)
O_2/H_2O	+810
NO_3^-/N_2	+751
NO_3^-/NO_2^-	+430
NO_3^-/NH_4^+	+363
MnO_2/Mn^{2+}	+390
$FeOOH/Fe^{2+}$	+150
SO_4^{2-}/H_2S	−218
S^0/H_2S	−240
CO_2/CH_4	−244
$2\,H^+/H_2$	−414
$CO_2/(CH_2O)$	−434

In the presence of sulfate, the primary fermentation products are oxidized completely by **sulfate-reducing bacteria** to carbon dioxide, with concomitant reduction of sulfate to sulfide (Chapters 12.1.3 and 12.1.8). No sulfate-reducing bacteria are known that can attack complex polymeric substrates on their own; they depend on the cooperation with primary fermentative bacteria. Thus, the sulfate-dependent oxidation of complex organic matter is basically a two-step process, and the metabolic energy available in the total conversion is shared by both partners.

Due to the high amount of sulfate dissolved in seawater, sulfate reduction is, with respect to the total electron flow, far more important in a marine than in a freshwater habitat. However, also the small amounts of hydrogen sulfide produced in a freshwater sediment serve an important function because hydrogen sulfide reacts with the last traces of oxygen, lowers the redox potential dramatically, and thus prepares the habitat for the development of methanogenic bacteria and their partners.

Sulfur plays a role as an alternative electron acceptor only to a minor degree, e.g., in lake sediments after the turnover phase (see Chapter 31.1.2). Sulfur is reduced by several specialized sulfur reducers, and also by fermentative or sulfate-reducing bacteria.

After complete reduction of sulfate and sulfur to sulfide, **methanogenesis** follows as the last anaerobic degradation process (see Chapter 12.1.7). Due to the comparably small amounts of sulfate available in freshwater habitats, methanogenesis takes over quickly and becomes the dominant terminal conversion process in these environments.

Methanogenic bacteria have a very narrow substrate range, including only one-carbon compounds and acetate. The gap between the diverse mixture of classical fermentation products released in the primary fermentations and the few substrates utilized by methanogens is bridged by the **secondary fermentative bacteria** which are also called **syntrophic fermenters**

The total conversion of complex organic matter to methane and CO_2, the complete dismutation of organic carbon to its most reduced and its most oxidized state, is a three-step process catalyzed by primary fermentative bacteria, secondary fermenters, and methanogens.

Methanogenesis is a dismutation process of organic matter and does not depend on external electron acceptors.

or **obligate proton reducers** (see Chapter 12.2.4). They cooperate closely with methanogenic bacteria (see Chapter 30.7) and convert propionate, butyrate, long-chain fatty acids, and alcohols to acetate, hydrogen, and one-carbon compounds, which are consumed by the methanogens.

The sequence of alternative substrate oxidation processes mentioned here can be looked at as a process in time, describing the temporal sequence of events occurring at a site where organic matter has accumulated in a natural ecosystem. It can also be translated into a **spatial order**, such as that within a lake sediment. The sediment surface may be supplied with oxygen by eddy convection through the water column. A nitrate reduction zone is found underneath the surface, followed by a zone governed by iron and manganese reduction, and in the lower layers, sulfide formation and methanogenesis are found. This spatial arrangement again reflects the order of preference described; the alternative oxidants penetrate by molecular diffusion into the sediment from the water column at similar rates, and are consumed within the sediments in the order mentioned.

Thus, the reaction kinetics translate via diffusion kinetics into a spatial arrangement, which is maintained at a dynamic equilibrium by the metabolic activity of bacteria.

The dimensions of these arrangements can differ substantially. They can span a few millimeters in a microbial mat with high productivity and fast turnover of the intermediates, or they can stretch over several meters in stratified lakes or marine sediments. These dimensions are determined by the diffusion kinetics, the supply of organic matter to be decomposed, and the availability of the alternative oxidants. Details of the respective environments will be discussed in Chapter 31.

The described sequence of biologically catalyzed reduction processes (Fig. 30.**4**) follows the free energy amounts available in the respective redox reactions (see Chapter 12.2). It appears that this sequence is governed by thermodynamics in the sense that the process yielding most energy is preferred over that yielding less energy and so on. The free energy changes of these redox reactions are nothing else but a mathematical transformation of the midpoint redox potentials of the various electron acceptors that are served with electrons from biomass oxidation that are released, on average, at -434 mV (Tab. 30.**3**). The argument that the energetically more favorable process will win over the process releasing less energy is convincing at first sight, but not entirely. Free energy changes of reactions do not determine reaction kinetics. Winning in competition for a limiting substrate, e.g., for an electron donor in the redox processes discussed here, is in the first place a function of **substrate affinity** and efficient substrate uptake at low concentrations (Chapter 30.4). Although it has been observed repeatedly that those organisms that have more energy available in a certain degradation process also have the higher affinity for the respective substrate, this is by no means determined by the reaction energetics per se. It is still an open question how these different aspects of a transformation process are coupled within a living cell. Of course, a bacterium that has more energy available in a transformation process than another bacterium may be able to invest more energy into a costly high-affinity uptake system, but this is a biochemical problem solved individually by different organisms and is not a simple consequence of reaction energetics.

The sequence of the various redox processes is determined primarily by the substrate uptake kinetics in the various metabolic niches.

30.2.2 The Reduced Products of Anaerobic Oxidations Can Be Reoxidized by Lithotrophs

The reduced products formed by the anaerobic oxidation processes mentioned may accumulate in the respective habitats to a certain degree, or may undergo subsequent chemical reaction as, for example, H_2S does (see Chapter 30.1). The reduced products can also be reoxidized when they diffuse into regimes governed by redox systems of a more positive redox potential. These reoxidation processes are typically catalyzed by lithotrophic bacteria (see Chapter 10.6), although some of these oxidations proceed also purely chemically, without biological catalysis.

Hydrogen sulfide can react with molecular oxygen to form mainly thiosulfate, sulfur, and sulfite. This reaction is very slow at low pH; it becomes significant at pH 6.5 and higher. Therefore, acidophilic sulfide oxidizers such as many *Thiobacillus* spec. do not need to compete with chemical sulfide oxidation reactions and can be grown easily in the laboratory with hydrogen sulfide and oxygen provided concomitantly. The situation is more difficult at neutral pH. The neutrophilic *Beggiatoa* spec. has to compete with the chemical oxidation, and cultivation in the laboratory with sulfide and oxygen is possible only in a gradient culture system that separates the two reactants spatially, thus mimicking the situation prevailing in a freshwater sediment (see Chapter 30.6.2). Sulfide can also be oxidized with nitrate by, for example, *Thiobacillus denitrificans*; this combination does not need to compete against a chemical oxidation process and is easy to realize in a laboratory culture. Another alternative is **phototrophic sulfide oxidation** by green or purple sulfide-oxidizing bacteria which occurs in anoxic layers of waters and sediments and combines sulfide oxidation with reduction of CO_2 to cell matter in a light-dependent process.

Also **ferrous iron** oxidizes spontaneously upon exposure to oxygen at neutral pH, but the reaction is slow under strongly acidic conditions. Microbial oxidation of Fe^{2+} with oxygen at low pH, as catalyzed by *Thiobacillus ferrooxidans*, is easy to follow and has been studied in detail. On the other hand, very little is known about iron oxidation by *Gallionella* spec., which oxidizes Fe^{2+} with oxygen at neutral pH and has to compete with the chemical oxidation process. It has been argued earlier that iron oxidation at neutral pH yields too little energy for ATP synthesis. This argument does not apply if the actual redox potentials as determined by the available reactants at these conditions are taken into account (Fig. 30.**3**). Indeed, iron oxidation at neutral pH yields even more energy than under acidic conditions. It is not known how *Gallionella* competes with chemical iron oxidation; again, the biological process is favored by a gradient distribution of the reactants that separates the reduced iron largely from oxygen advection (see Chapter 30.6). Ferrous iron can also be oxidized by several types of nitrate-reducing bacteria or by phototrophic bacteria such as *Rhodomicrobium vannielii*.

Oxidation of **ammonia** requires molecular oxygen as cosubstrate; the initial reaction is catalyzed by a monooxygenase enzyme (Chapter 10.7). Ammonia is oxidized via nitrite to nitrate; thus, nitrite is formed both in nitrate reduction and in ammonia oxidation. Detection of nitrite in an ecosystem indicates that nitrogen compounds are undergoing redox reactions at this site; however, it does not tell us whether the reaction goes towards reduction or oxidation. Evidence for **anaerobic ammonia oxidation** has been provided recently. The process is rather slow, but of major interest for wastewater treatment. It may also be of global importance in marine sediments, but nothing is known so far about its biochemistry.

Oxidation of **methane** by methylotrophic bacteria is again an **aerobic process** that requires molecular oxygen in the primary monooxygenase reaction (Chapter 9.14.4). From the point of reaction energetics, methane oxidation could be possible as well with sulfate, ferric iron, or nitrate, but no one has successfully cultivated bacteria that catalyze such a process. There

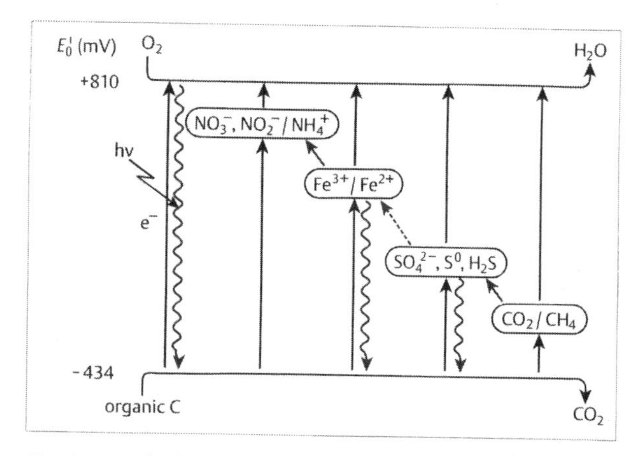

Fig. 30.**5** **Idealized electron flow scheme from organic matter to oxygen via various intermediate electron carrier systems.** (Straight arrows, dissimilatory dark reactions; zig-zag arrows, light-dependent assimilatory reactions; dashed arrow, non-biological redox process)

are reliable indications of **anaerobic methane oxidation** with sulfate as electron acceptor, especially in marine sediments (Chapter 31.2), but nothing is known about the bacteria involved or the biochemistry of the substrate activation reaction.

The various reduction and oxidation processes discussed here are summarized in Figure 30.**5**. The scheme shows the different alternative paths of electron flow from organic matter to the various electron carriers and their further path to oxygen through the lithotrophic reoxidation processes. It illustrates that the alternative carriers are only transient electron storage pools, and that the electrons may finally find their way to oxygen as their terminal acceptor in all cases, either directly or through other intermediate carrier systems. The scheme also includes the phototrophic reactions oxidizing water, ferrous iron, or sulfide. Sulfide oxidation by ferric iron is marked as a dashed arrow because it may be only a chemical process not involving biological activities.

30.2.3 Microbes Catalyze Also Other Redox Processes in Nature

In addition to the redox systems mentioned above, other carriers can play a role as electron acceptors or donors (Tab. 30.**4**; see also Chapter 12). **Reductive dehalogenation** of, for example, chlorinated alkyl or aryl compounds, provides a means of electron release that is exploited by several fermentative, nitrate-reducing or sulfate-reducing bacteria (Chapters 9.17 and 12.19).

Table 30.4 **Redox potentials of electron carrier systems of minor importance** calculated for pH 7.0. (DMSO, dimethyl sulfoxide; DMS, dimethylsulfide; TMAO, trimethylamine *N*-oxide; TMA, trimethylamine)

Electron carrier system	E_0' (mV)
ClO_3^-/Cl^-	+1031
Fe^{3+}/Fe^{2+}	+770 (pH < 2)
Alkyl-Cl/Alkyl+Cl^-	+250–+580
Aryl-Cl/Aryl+Cl^-	+310–+480
Nitrobenzene/Anilin	+416
SeO_4^{2-}/SeO_3^{2-}	+464
SeO_3^{2-}/Se	+256
$HCrO_4^-/Cr^{3+}$	+380
$HAsO_4^{2-}/AsO_2^-$	+154
DMSO/DMS	+160
TMAO/TMA	+130
CO_2/CO	−520
H_2PO_4/PH_3	−680

These reactions do not really degrade the respective compound, but only modify it. Its further mineralization may be a task for entirely different bacteria. Evidence has been provided that reductive dehalogenation can at least partly be catalyzed also non-biologically by free corrinoids or other tetrapyrrols such as cytochromes or factor F_{430}. These free coenzymes are reoxidized by biological activities.

Reduction of **nitroaromatic compounds** to the corresponding amino compounds is another process observed, for example, in polluted groundwaters. This process is not always catalyzed directly by microorganisms, but ferrous iron on the surface of ferric iron minerals can act as the primary electron donor in this system and is replenished by microbial reduction of ferric iron.

Reduction of **chlorate, chromate, selenate, selenite**, and **arsenate** are other examples of reduction reactions that can be employed in remediation of polluted soils or groundwaters. The comparably high standard redox potentials of these compounds make them very likely electron acceptors. However, the high redox potential of, for example, chlorate may also be detrimental to sensitive strict anaerobes.

Trimethylamine-*N*-oxide (TMAO) is a cryoprotectant present in fish tissue, especially in marine fish. It can be reduced by a wide variety of fermentative bacteria, such as enterobacteria, that multiply excessively on the skin of dead fish. The product trimethylamine is volatile and responsible for the typical "fish smell" that people try to avoid by binding the alkaline amine with an acid such as lemon juice (Chapter 12.15).

Dimethylsulfoxide (DMSO) is formed from dimethylsulfide by the action of, for example, phototrophic or chemotrophic sulfur-oxidizing bacteria. D methylsulfide is a product of the partial degrac ation of dimethylsulfoniopropionate, which acts as a osmoregulator in numerous green algae includin macroalgae, and also in the seaweeds *Spartina* an *Zostera* spec., where it accumulates to molar concentra tions. Reduction of dimethylsulfoxide to dimethy sulfide is catalyzed by phototrophic purple bacteria i the dark, as well as by chemotrophic aerobes and nitrat reducers (Chapter 12.1.5).

Reduction of CO_2 to **CO** is the biological redc process with the lowest standard redox potential. It is c major importance in anaerobic bacteria, especially i homoacetogens where CO is formed as an intermediat by CO dehydrogenase (acetyl-CoA synthase) in acetat synthesis. CO can be released in traces by anaerob bacteria containing CO dehydrogenase, but is never major metabolic product (Chapter 12.1.6).

As outlined in Chapter 30.1.6, **phosphorus** remair in biological systems only at its redox state +5, a phosphate. Nevertheless, rumors of microbial reductio of phosphate to phosphane (phosphine, PH_3) persist i the literature, and small (nanomolar) phosphane cor centrations have actually been detected in anox sediments, human feces, and swine manure. The origi of these reduced phosphorus compounds is still enig matic, they may result from hydrolysis of meta phosphides. In any case, phosphane cannot be forme by respirative phosphate reduction. The redox poter tials of the transitions from phosphate via phosphit and phosphorus to phosphane (E_h at −510, −1020, an −456 mV, respectively, or −680 mV for the complet reduction of phosphate to phosphane) are very low an quite difficult to couple to biomass oxidatio (E_0' = −434 mV). Phosphate reduction to a phosphit derivative is catalyzed by, for example, some fungi i the synthesis of carbon phosphorus linkages in seconc ary metabolites (e.g., phosphinothricine), but thes reductions require substantial investment of ATP. Thes partially reduced carbon phosphorus compounds ma also be precursors of the phosphane traces detected i the anoxic ecosystems mentioned.

30.2.4 Microorganisms Act as Chemical Catalysts

From the microbial activities listed above, it appeai that microorganisms act in nature as catalysts t enhance chemical reactions that are thermodynamicall possible, in a manner similar to chemical catalyst Indeed, microorganisms do so in several cases of simpl redox reactions, and they even have to compete wit purely chemical reactions in certain cases, such as i oxygen-dependent oxidation of sulfide or ferrous iron a neutral pH. The same applies, as mentioned, to certain cases of reductive dehalogenation or reduction of nitroaromatic compounds in which free microbial coenzymes (corrinoids, other tetrapyrrols) act as catalysts, or microbial metabolic products (ferrous iron) act as cosubstrates. Actually, it is not always easy to discriminate between microbial activities and purely chemical processes in complex natural environments, especially when chemically unstable reactants are involved or when conditions favor chemical processes, such as at enhanced temperatures in thermal locations. Nevertheless, there are some basic features by which microbes differ from chemical catalysts: 1) microbes can catalyze reactions that would otherwise be extremely slow (aerobic alkane oxidation) or would proceed in a different way (sulfide oxidation to sulfate instead of sulfite, thiosulfate, sulfur, etc.), 2) microbes are typically highly specific for their substrates, and shift the reaction in a defined direction, 3) microbes can combine exergonic with endergonic reactions (e.g., endergonic reduction of sulfate to sulfite with exergonic reduction of sulfite to sulfide) or dissimilatory with assimilatory activities, 4) microbial reaction rates show optima with respect to temperature, pH, osmotic strength, oxygen supply, and redox potential, and show optimal activity within rather narrow limits (approx. 40 °C in temperature and 4 units in pH), 5) the microbial reaction rate can increase with time due to autocatalytic increase of the amount of catalyst (cell growth), but does not always do so (cometabolic activities; see Chapter 30.4.4), 6) microbes depend on cosubstrates (C, N, S, and P sources, etc.) and also interfere with the turnover of such substrates, 7) microbes are sensitive to certain toxic compounds present (phenols, heavy metals, etc.), and 8) microbes are particulate in nature and may produce, besides daughter cells, also extracellular polymers that contribute to a structural organization of their habitat; these structural properties have major influence on transformation efficiency and limitation with respect to space and time (diffusion limitation, transport limitation).

These aspects have to be considered for an understanding of a potentially new, so far unexplained transformation process in a natural environment, or if microbial activities are to be applied for "bioremediation" of, for example, a polluted soil compartment, instead of a chemical or physical (heating, combustion) treatment process (see Chapter 36).

30.3 Growth Kinetics in a Natural Ecosystem

Growth kinetics have been treated in detail in Chapter 6. Here only some aspects relevant to growth and substrate turnover in a natural ecosystem will be considered.

30.3.1 Growth in a Batch Culture Is Unlimited for Only a Short While

The classical growth curve consisting of a lag phase, an exponential growth phase, a stationary phase, and a phase of cell death represents growth in a static (batch) laboratory culture that provides optimal growth conditions for a limited amount of time. However, nonlimiting substrate supply can be granted only for a short time, and in nature such situations are rare exceptions rather than the rule. Microbial growth in nature proceeds under conditions that can be compared with the end of the exponential growth phase, when substrate supply becomes limiting and is insufficient to sustain growth at all. Growth under such conditions is limited by the substrate supply and not by the amount of the (autocatalytically multiplying) cell mass, and the exponential growth curve has to be amended by a capacity term, leading to a "logistic" growth curve:

$$\frac{dN}{dt} = \mu \cdot N \cdot \frac{C - N}{C} \quad (30.3)$$

where N is the number of organisms present, μ is the growth rate, and C is the capacity to which growth of N is possible, as determined by, for example, a growth-limiting substrate. As long as N is small in comparison to C, the last fraction term is close to 1 and is not relevant. However, when N approaches C, the exponential growth is corrected by the capacity term, and the mass increase will go to 0 (leading to a stable final population density).

The capacity term C can refer to a limiting substrate, i.e., the carbon source, the electron donor or acceptor, the nitrogen, sulfur, or phosphorus source, or others. According to the **Liebig principle**, growth will be limited by the substrate that expires first, when all others may still be available in excess. This principle does not apply strictly to the constituents of microbial cells because microbes can adapt the composition of their cell mass to some extent to changing supply conditions. The nitrogen, sulfur, or phosphorus content of cell mass can vary within a factor of two to ten in relation to substrate supply.

The efficiency with which the final capacity C can be reached is determined by the **substrate affinity**, K_S, of the respective organism. The substrate affinity is defined as the substrate concentration that allows growth at half-maximal rate (Chapter 6.8). If growth and substrate turnover are strictly coupled, K_S should also define the substrate concentration at which the substrate turnover rate reaches half of its maximum. K_S values can be determined either directly as substrate uptake kinetics with, for example, a cell suspension, a sediment or soil sample, or in a substrate-limited chemostat culture running at a dilution rate corresponding to the half-maximal growth rate.

30.3.2 The Continuous Culture Is a Good Model for Growth in Natural Systems

In microbial ecology, the chemostat (see Chapter 6.8) can be used as a model system to explore the behavior of microorganisms under conditions of suboptimal substrate supply, regardless what kind of substrate is studied. The chemostat mimics the situation of the natural environment in the sense that cells are maintained under conditions of low but continuous substrate supply over a certain period of time. However, it should be stressed that the chemostat can only mimic just this situation and not life in natural environments. Substrate and cell concentrations at low dilution rates, a situation assumed to be typical of microbial growth in nature, must be examined.

In such a chemostat-like natural system, the substrate concentration is determined exclusively by the half-saturation constant K_S and the relationship between the dilution rate and the maximal growth rate of the organism studied (Chapter 6.8). Extrapolated for the natural environment, this means that in a flowing system at equilibrium, the measurable substrate concentration does not give any information on the amount of substrate turned over, but reflects the substrate affinity of the substrate consumers present and its relationship to the substrate supply rate. Substrate turnover is not exclusively bound to cell mass synthesis; it serves also maintenance purposes which are addressed by the maintenance coefficient m_S (see Chapter 6.8). Maintenance coefficients have been determined with various laboratory strains of bacteria and vary between 0.3 and 10 mmol ATP g^{-1} h^{-1}, depending on the growth conditions (Tab. **6.4**). It is expected that bacteria adapted to low nutrient supply and starvation in nature will have lower maintenance energy requirements (see Chapter 30.3.3).

The **chemostat** mimics the situation in the natural environment in the sense that cells are maintained under conditions of low, but continuous substrate supply. The measurable substrate concentration does not give any information on the amount of substrate turned over, it reflects the consumer's substrate affinity and its relationship to the substrate supply rate. Growth and substrate turnover are not necessarily linked to each other in a stable proportion, but substrate turnover may increasingly uncouple from growth the lower the substrate supply rate becomes.

30.3.3 Bacteria Have Strategies To Cope With Starvation and To Survive Under Stress

Bacteria in nature typically have to face insufficient substrate supply or starvation. Under such conditions, strategies to survive hunger phases without substantial losses in population density are much more important than fast growth in situations of substrate excess. As outlined earlier, S. N. Winogradsky touched on this problem with his concept of autochthonous versus zymogenous bacterial populations in soil. Whereas the fast-growing casual laborers, the zymogenous bacteria, show little ability to endure starvation over prolonged periods of time and die fast, the autochthonous bacteria are better prepared for such situations. Specialists for growth under low substrate supply conditions have been referred to as **oligotrophic bacteria**, as opposed to the well-fed typical laboratory bacteria, the **copiotrophic bacteria**. Unfortunately, the known oligotrophic bacteria are not easy to handle in the laboratory, and little is known about their physiological and regulatory capacities; most research on starvation has been done with laboratory strains of bacteria, such as *Escherichia coli*, *Klebsiella*, or *Paracoccus* (Chapter 28.5), and not with the real "experts" in the field.

It is of major importance for an understanding of the processes encountered to differentiate between different types of starvation. Starvation for a carbon source (meaning usually also the energy source) is for the cell a basically different situation than starvation for a nitrogen, sulfur, or phosphate source, which only prevents cell mass increase, but may still allow energy maintenance at a sufficient rate. Recent research has shown that growing bacterial cell suspensions entering the stationary phase due to limitation of the carbon and energy supply induces the synthesis of a whole series of new proteins that are not expressed in exponentially growing cells. Some of these proteins may be involved in switching on uptake systems for alternative substrates; for most of these gene products, no function has been defined. Starvation is a complex stress situation for the cell that is now being studied by modern molecular techniques.

Strategies for survival under a limitation of the energy substrate have basically two routes: 1) to increase the substrate uptake, even at low supply, or 2) to decrease the cell-internal energy needs. Both concepts occur in nature, and may even be combined. For enhancement of substrate affinity, either the number of uptake systems (permeases, etc.) can be increased to increase the uptake efficiency, or new, high-affinity uptake systems can be induced that ensure substrate supply even at low substrate concentration, e.g., by investment of the proton motive force in proton-driven symport systems (see Chapter 5). Examples of specific high-affinity uptake systems in many well-studied bacteria are known for the uptake of glycerol, phosphate, sugars, and amino acids. Molecular mechanisms of adaptation to starvation are described in Chapter 28.

Another strategy is to increase the cell surface, e.g., by the formation of **prosthecae** or **stalks**. An entire group of bacteria, the budding and appendaged bacteria, including genera such as *Stella*, *Prosthecochloris*, and *Caulobacter* are specialized for a life with low substrate supply, and form stalks and prosthecae that grow longer in laboratory experiments with decreasing substrate concentration. Although these cell extrusions are also involved in cell multiplication and although it has not been proven unequivocally yet whether they are really lined with uptake proteins, the increase in cell surface achieved this way is obviously advantageous under starvation conditions. A disadvantage of this strategy could be enhanced energy losses by proton leakage across the increased cell surface.

A strategy for long-term survival during starvation by a decrease of energy needs is the formation of **spores** (see Chapter 25). Spores represent a very efficient way of dormant life, independent of food supply, and also protected from desiccation and heat stress, for example, in surface soils. Other persistant survival structures found in specific groups of microorganisms are **cysts**, **myxospores**, and **akinetes** (see Chapters 24 and 25).

Many Gram-negative bacteria do not form spores, but instead decrease their internal cell energy needs by decreasing their cell size. With this strategy, also the expenses of maintenance of the membrane potential can be reduced. Formation of such dwarf forms ("**dwarfing**") is a rather widespread ability among bacteria in clean freshwater lakes and in the ocean. The average size of bacterial cells in such nutrient-poor habitats is by a factor of up to 30 smaller than that of non-starving cells, as for marine bacterial isolates studied under conditions of different levels of substrate supply.

The process of dwarfing has been studied in detail with *Vibrio* spec. strain S14. The key steps in this development are similar to those observed in spore formation (Fig. 30.**6**). During the first phase of starvation (0–0.5 h), intracellular protein degradation increases, respiration activity decreases, and new proteins (**Star**vation-**i**nduced, Sti proteins) are synthesized. The second phase, extending from 0.5 to 6 h after growth, is characterized by a series of organizational events including degradation of storage material such as poly-β-hydroxybutyric acid (PHB), decline of protein degradation activity, a shift in membrane fatty acid composition, induction of high-affinity substrate uptake systems, and synthesis of other Sti proteins. The third phase appears as a gradual decline in metabolic activity, a decrease in cell size, and induction of other high-affinity, periplasmic, substrate uptake systems. Also extracellular hydrolytic enzymes may be produced during this phase. At the end, the dwarf cell represents

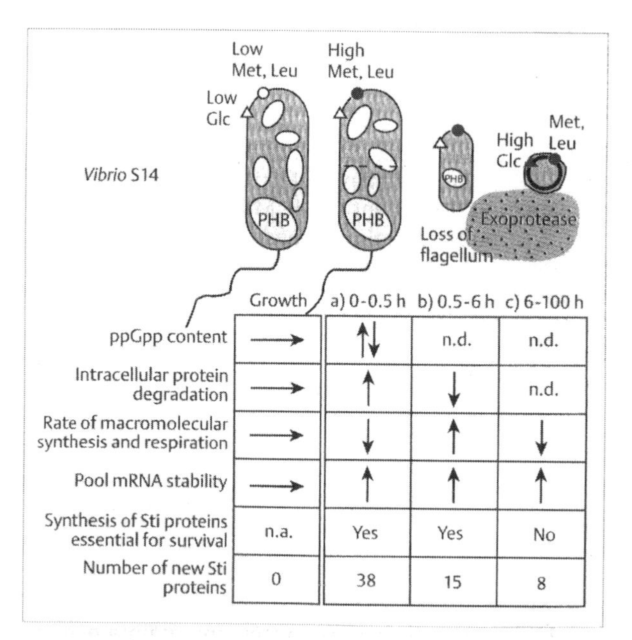

	Growth	a) 0-0.5 h	b) 0.5-6 h	c) 6-100 h
ppGpp content	→	↑↓	n.d.	n.d.
Intracellular protein degradation	→	↑	↓	n.d.
Rate of macromolecular synthesis and respiration	→	↓	↑	↓
Pool mRNA stability	→	↑	↑	↑
Synthesis of Sti proteins essential for survival	n.a.	Yes	Yes	No
Number of new Sti proteins	0	38	15	8

Fig. 30.**6** **Physiological and morphological adaptation to multiple nutrient starvation in *Vibrio* S14.** The 100-h starvation period is divided into three phases (a–c). Arrows indicate an increase (↑) or decrease (↓) in individual responses as compared to exponentially growing (→) cells. "Low" and "high" refer to affinity of uptake systems for glucose and leucine. n.d., not determined; Sti, starvation induced; n.a., not applicable [from 4]

a stable, endurant form that, due to its small size, may be better able to escape capture by sieving grazers, and appears even to resist lysis in digestion vacuoles of grazing protozoa. As a consequence of these modifications, the dwarf cell has a protein pattern entirely different from that of the non-starving cell.

Slime production is also often encountered with starving bacteria. Extracellular slimes may help to store protons, to maintain proton motive force, and to minimize uncontrolled potential losses by proton leakage. More important functions of slime formation, however, are the storage of water during desiccation, and its involvement in **attachment to surfaces**, a strategy that appears to be of major importance to ensure substrate provision under low supply conditions (see Chapter 30.5).

Mechanisms to resist starvation are: increase of high-affinity substrate uptake systems, increase of cell surface, formation of spores, reduction of cell size, modification of cell metabolism by starvation-induced proteins, and slime production.

30.3.4 The Cell Number Is Reduced by Grazing and by Bacteriophages

The production of bacterial cells by growth is counteracted in nature by grazing, infection by bacteriophages, or starvation-dependent death. Many protozoa and also rotatoria and several copepods graze on bacterial populations in aquatic environments. Whereas **ciliates**, **flagellates**, and **copepods** concentrate on suspended bacterial cells that are sieved out of the free water by suitable feeding devices, **amoebae**, **rotatoria**, and **nematodes** feed preferentially on accessible surface-attached bacterial cells. Surface attachment is one of the strategies used efficiently to escape grazing by bacterivorous animals (see Chapter 30.5). On the other hand, larger aggregates of bacteria with decomposing detritus may be ingested by larger animals, including planctivorous fish, which lack suitable devices for sieving of single suspended bacterial cells. In the water column of freshwater lakes, average densities in the range of 10^4–10^6 bacterial cells per ml water are maintained by grazing. The grazing animal community obviously finds its limits of grazing efficiency in this density range. It has been calculated that in a lake water column during the summer season, depending on the trophic status of the lake, 2–12% of the standing bacterial population is grazed per hour by ciliates, flagellates, and daphnids. To balance these losses, doubling times of bacterial growth in the range of 6 h to 2 days are required.

Grazing decreases the size of a certain microbial population and appears to be detrimental only at first sight. However, the population decrease will also decrease the competition for food and thereby increase the food supply, substrate, turnover, and growth rate for the bacterium under predation, maintaining an active community. Thus, the size of a bacterial population is controlled both by nutrient supply and by grazing pressure ("bottom up" and "top down" control), and both effectors influence each other through the bacterial population.

Bacterial biomass is converted by feeding animals to animal biomass. The **feeding efficiency** of this aerobic biomass conversion in laboratory experiments is, due to the high energy yield of aerobic metabolism, near 50%, meaning that half of the food biomass is mineralized to provide the energy for assimilation of the other half into the feeder's biomass. In studies on biomass conversion in lakes, considerably lower efficiency ratios in the range of 20–30% are observed, indicating that limiting food supply and lower growth rates decrease this ratio, similar to the situation of bacteria growing under conditions of limited nutrient supply (Chapter 30.3.2).

The grazers mentioned so far are aerobic animals active in oxic habitats, and little is known about the importance of grazing in anoxic environments. The discovery of facultatively and strictly **anaerobic ciliates**, **flagellates**, and **fungi** in anoxic sediments, the rumen, and other habitats (see also Chapter 30.7) has raised the question of their importance in controlling anaerobic bacterial communities. Feeding experiments with anaerobic ciliates from the anoxic waters of a highly eutrophic lake have shown that the feeding efficiency under such conditions is considerably lower (near 5%) than with aerobic ciliates, namely due to the low energy yields obtainable. Therefore, only small populations of grazing protozoa are maintained in such environments. Attachment to surfaces, especially in sediments, will further limit the food supply for anaerobic grazers, and with that their population sizes.

Bacteriophages also help to control bacterial population sizes in nature. New phages are discovered at a high frequency, and there is hardly any bacterial species known today that is entirely "immune" against phage attack. Recent studies have discovered that phage-like particles are abundant in natural environments. Free viruses have been found at 10^7–10^8 per ml in freshwater lakes, and up to 2% of the bacteria in such waters contain phage-like particles. Nonetheless, the importance of phages in controlling bacterial populations in nature is probably limited to a few situations of high densities of a genetically homogeneous host population, e.g., in blooms of marine cyanobacteria where up to 50% of the total mortality has been attributed to phages. Populations of, for example, heterotrophic bacteria in a water column are too low and too heterogeneous to allow the maintenance of a phage population high enough to become significant in the control of these host bacteria; first estimates attribute about 5–10% of total mortality to phage-induced lysis. No information exists yet on the importance of phages in soils or sediments because microscopic analysis of these habitats is extremely difficult.

30.3.5 Techniques To Quantify and Analyze Microbial Populations in Nature

Quantification of microbial communities in nature requires special techniques different from those used in the laboratory. **Direct microscopic counting** may appear easy at first glance, but is applicable only with water samples. However, even in such material, most bacteria are attached to detritus particles and hardly recognizable by direct microscopy. To visualize bacterial cells in water samples, it is mandatory to stain them specifically with **fluorescent dyes**, such as acridine orange or 4′,6-diamidino-2-phenylindole (DAPI). These dyes penetrate the cytoplasmic membrane and bind either to cellular proteins (acridine orange) or to the double-stranded DNA (DAPI), and allow specific staining of bacterial cells. However, contrary to earlier claims, these dyes do not discriminate between living and non-living cells, and counts obtained this way may include a considerable proportion of dead cells. New dyes that form fluorescent derivatives only after reduction by respiratory enzymes (e.g., 5-cyano-2,3-ditolyl tetrazolium chloride) could allow the specific visualization of actively respiring bacteria. Direct counting with fluorescent dyes cannot be applied with sediment or soil samples because the background of naturally fluorescing compounds in these materials is too high.

Other methods that allow **identification of metabolically active cells** are based on active uptake of radioactive markers and their detection by **autoradiography**. ^{14}C-labeled leucine has been applied as a marker of protein synthesis (and cell growth), and ^{3}H-thymidine as a marker of DNA synthesis (and cell multiplication). both techniques have been used mostly with water and sediment samples in the past, but the information to be gained from these studies bears uncertainties due to partial microbial degradation of these precursors, and is small compared to the experimental effort.

Quantification of bacterial populations by **plating and cultivation**, even on complex and comparably unspecific media, gives very disappointing results. The "total plate counts" obtained, for example, with casein peptone starch agar plates, which have been applied as a standard medium for freshwater bacteria, are 1–2 orders of magnitude lower than those obtained by direct microscopic counting. This means that only 1–10% of the total microbial community can be cultivated with this technique, and other media are by no means better.

Every growth medium and the incubation conditions select for only a small fraction of the total bacterial population present, and all efforts to design less specific, more habitat-adjusted growth media that should give better counting efficiencies have failed so far.

A modification of the plate cultivation technique is the quantification of microbial populations in the **"Most Probable Number (MPN)"** assay. Sampling material is diluted in tubes with liquid growth medium in 1:10 steps in at least three parallel series, and the total number of bacteria originally present can be calculated from the number of tubes containing growth (positive) after sufficient incubation, assuming that one single bacterial cell in the final positive tube will cause visible growth after a sufficient amount of time. The advantage of this method is that also rather delicate bacteria can be grown in such liquid media, avoiding the stress of exposure to the open air on an agar plate, and that metabolic products, for example, can be assessed directly in the dilution tubes. The disadvantage is a considerable statistical uncertainty based on simple "yes" or "no" answers in three independent dilution runs.

Cultivation is more successful if it is directed specifically toward the quantification of defined metabolic groups of bacteria. Of the total population of hydrogen/formate-utilizing methanogenic bacteria that can be quantified directly by their natural fluorescence, 20–50% can be recovered by plate cultivation. Sulfate-reducing bacteria have also been quantified by cultivation rather efficiently when suitable growth media were applied.

All these enumerations, by direct microscopy or by cultivation, imply that microbial cells are dispersed as single units and are optically or experimentally separable as such. In nature, however, bacteria tend to stick to each other or to other surfaces, including plant or animal detritus, and sediment or soil particles, and enumeration by any of the techniques mentioned requires that the cells be resolved from their substratum (surface) without impairing their viability. Shaking, homogenizing, or gentle sonication in the presence of pyrophosphate, complexing agents, or detergents can help in this respect, but it is obvious that all these treatments will also kill some bacteria and thus represent only compromise solutions.

An entirely different strategy for population analysis uses cell components as chemical markers. **Total living biomass** (which is actually mostly microbial biomass in nearly every ecosystem) can be quantified indirectly by determining the **total ATP content**, for example, in a soil sample. In a similar manner, also the **total DNA content** can be taken as a measure of biomass, provided that the extraction procedure differentiates sufficiently between DNA that is still active in living cells and DNA remaining from dead precursors that is preserved by attachment to soil particles. In soil microbiology, the **fumigation method** is rather popular. The majority of biomass present is killed by treatment with chloroform vapor, and the dead biomass is later oxidized by the surviving biomass leading to CO_2 production, which is measured over a 10-day period and taken as a measure of initial biomass content. Assay of **total dehydrogenase activity** by reduction of tetrazolium salts or dimethylsulfoxide is applied as another parameter of total microbial biomass.

More specific information can be gained from analysis of specific cell components. The **fatty acid patterns** in the membrane lipids differ specifically between different taxonomic and metabolic groups of prokaryotes (see Chapter 29.3.4). Unfortunately, only few metabolic groups of prokaryotes contain fatty acid patterns as sufficiently unique as those of sulfate reducers to promise significantly new information from this approach, and it should be mentioned that the fatty acid pattern in membrane lipids is also subject to regulatory changes due to environmental conditions. Moreover, this method has to be calibrated with pure cultures of cultivated bacteria which, as stressed above, make up only a small and not representative fraction of the total population present in the respective sample.

Occurrence of **ergosterol** is used as an indicator of the presence of fungi in a complex microbial community.

Fluorescent antibodies have been applied to identify and recognize certain types of bacteria in environmental samples. Since antibodies, depending on the respective antigens used for their preparation, can be directed to haptenes that are specific for a single strain or specific for a species or genus, the specificity of such antibody binding studies varies considerably. The method is confined again to recognition of those bacteria that can be cultivated and used for antibody preparation; moreover, the surface properties of bacteria in nature and of those cultivated in the laboratory may differ dramatically with respect to slime formation, surface attachment, and other characteristics. Thus, this approach can be applied in certain well-defined situations, but its applicability is rather limited. It has been applied successfully for characterization of methanogens in anoxic sewage digestors.

Quantification of microorganisms in environmental samples can be performed by direct microscopic counting of fluorescence-labeled cells, autoradiography of ^{14}C- or ^{3}H-labeled cells, cultivation on solid or in liquid media combined with dilution techniques, quantification of total ATP or DNA content, quantification of more specific cell components such as fatty acid or chlorophyll, or by labeling with antibodies. All these techniques have their shortcomings.

Molecular biology has provided new tools for population analysis that hold great promise for application in microbial ecology. The 16S rRNA of the vast majority of cultivated strains of bacteria has been sequenced, and these data have been used to derive on the basis of sequence similarities a system of apparent phylogenetic relatedness that is discussed in detail in Chapter 29.4. Oligonucleotides can be synthesized that hybridize specifically to certain regions of rRNA, which, on the basis of comparison with all the sequence data available, can be defined as highly or less variable and as specific for a domain, a phylum, a subphylum, or a smaller similarity cluster within the total system, in some cases down to the species level. Such **rRNA probes** composed of 18 nucleotides or more carry fluorescent dye molecules that stain the respective target bacterium after binding, and make it glow under ultraviolet illumination. Since the living bacterial cell contains several thousand copies of rRNA molecules, sufficient amounts of fluorescent dye can be linked through the probe to the target cell to allow reliable visualization in the fluorescence microscope.

In a first approach, this method can be applied with a specific probe to recognize a certain bacterial strain in a complex community. Less-specific probes allow the assignment of unknown bacteria in such a community to certain phylogenetic groupings. This way, up to 90% of the total microbial population in lake water samples can be attributed to the two major domains of prokaryotes. Another example is given in Figure 30.**7** with bacteria in

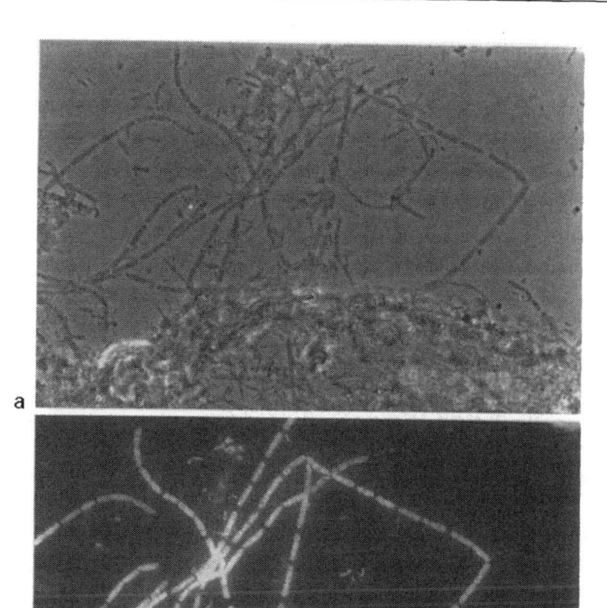

Fig. 30.**7a,b** **Detection of Proteobacteria of the β-subclass in activated sludge by in situ hybridization with a group-specific rRNA-targeted nucleotide probe.**
a Phase-contrast photomicrograph.
b The same picture in UV-illuminated epifluorescence optics. (Courtesy of R. Amann)

an activated sludge sample. With more specific probes, non-cultured prokaryotes living, for example, inside a protozoan host, have been assigned down to the genus level. Molecular probes specific for various groupings within the taxonomic system of the prokaryotes can be combined with different fluorescent dyes that give different color responses upon illumination. Thus, it is possible to resolve within one single preparation of environmental material the affiliation of the microbes present to clusters within the system at various phylogenetic levels. This method has been combined impressively with **confocal laser microscopy** (Chapter 2), thus allowing the visualization of different groups of bacteria at various levels of systematic resolution in a three-dimensionally resolved picture, and to check, for example, for possible spatial interrelationships between them.

Application of the polymerase chain reaction (PCR; Chapter 17.4.3) allows the multiplication of also single molecules of 16S rRNA (in the form of DNA) in extracts from environmental samples, and the comparison of the nucleotide sequence of this DNA with those of known prokaryotes. Thus, also non-cultivated bacteria can be analyzed in detail for their phylogenetic relationship to other prokaryotes (Chapters 29.1 and 29.4). Amplified DNA can be used to design probes specific for hybridization with nucleic acids of unknown microorganisms and thus to obtain information on the diversity of the microbial community present, independent of their culturability. The same technique can be applied to follow the fate of a certain type of non-cultured microorganism through a process of population manipulation. One can control an enrichment process for a new, difficult organism for which the selection criteria are not known, or one can monitor changes in the population composition among non-cultured bacteria, such as in activated sludge under various environmental stress situations.

The amplification of parts of a few rRNA molecules by PCR can also be used for analysis of complex microbial communities. Fragments of genes coding for 16S rRNA are amplified by PCR and the resulting oligonucleotides are separated by **d**enaturing **g**radient **g**el **e**lectrophoresis (DGGE). The band patterns obtained are representative of certain bacterial components within the community studied. This technique allows several microbial components within a complex community to be followed on the basis of their oligonucleotide patterns (Chapter 29.4).

As promising as these molecular probing techniques are for application in microbial ecology, the limitations of these techniques must be pointed out. They may provide excellent means to show where a certain type of microorganism is (its habitat) and to which major branches it belongs within the "phylogenetic" system established by sequence similarities. However, since this system is not homologous with metabolic groupings, not even at the level of the smallest ramifications, it is possible only in very few exceptional cases to attribute a certain metabolic

Single cells or populations of microorganisms can be identified by hybridization with synthesized rRNA probes or hybridization with rRNA probes of PCR-amplified DNA. These techniques can be combined with optical resolution techniques that allow precise localization of even uncultured microorganisms in environmental samples.

function (its ecological niche) to such a non-cultured organism. Furthermore, not all prokaryotes are equally accessible by molecular probing. Although the techniques are being improved continuously, certain target bacteria, especially those with thick cell walls or those surrounded by capsular material and slimes, largely resist DNA extraction or hybridization to RNA with probes. The probing technique is also selective and, therefore, certain bacteria, perhaps important ones, may still be overlooked.

30.4 Substrate Turnover Is Determined by Depolymerization and Substrate Uptake Kinetics

Wherever a substrate is available for microbial degradation, there will be more than one organism able to degrade it, and the number of competitors increases the easier a substrate is to degrade. Such easily degradable substrates are released, e.g., through hydrolysis of a polymer, at a low rate, and this depolymerization is typically the rate-limiting step in the total process of polymer degradation. Competition for the released monomers will be won by the organism that takes up the substrate at low concentration with high efficiency. The chemostat culture has been discussed as a model system that mimics this low but comparably constant substrate supply because this is assumed to be typical of many microbial environments. Competition for spatial advantages, e.g., competition for attachment sites on a structured substrate or for a certain temperature regime in an outflow creek of a hot spring, can also occur, but fighting for substrate is undoubtedly the most widespread and important type of microbial competition observed in nature.

30.4.1 Substrate Affinity Is the Key Factor in Competition for a Substrate

The half-saturation constant K_S (Chapters 6 and 30.3.2) and the maximal growth rate μ_{max} both determine the efficiency of substrate uptake at limiting substrate supply. In Figure 30.**8a**, such substrate saturation curves are compared for three different types of organisms. Organism II has a high substrate affinity, i.e., a low K_S value, and a low maximal growth rate. Organism I has a high K_S and a high maximal growth rate. At high substrate supply, organism I will grow much faster than organism II, and will probably overgrow organism II. At low substrate supply, below the intersection point of both saturation curves, organism II has a higher growth rate than organism I and will outcompete it for the substrate. Organism II is better adapted to a low but constant substrate supply, whereas organism I preferentially takes advantage of sudden substrate pulses. Obviously, organism II represents the autochthonous type of microbes that grow slowly but steadily, and exploit a small but continuous substrate supply; organism I represents the zymogenous type of microbes that grow fast occasionally but are unable to compete successfully at low substrate supply. These two types of organisms can also be mirrored in the common differentiation between *K* and *r* strategists in general

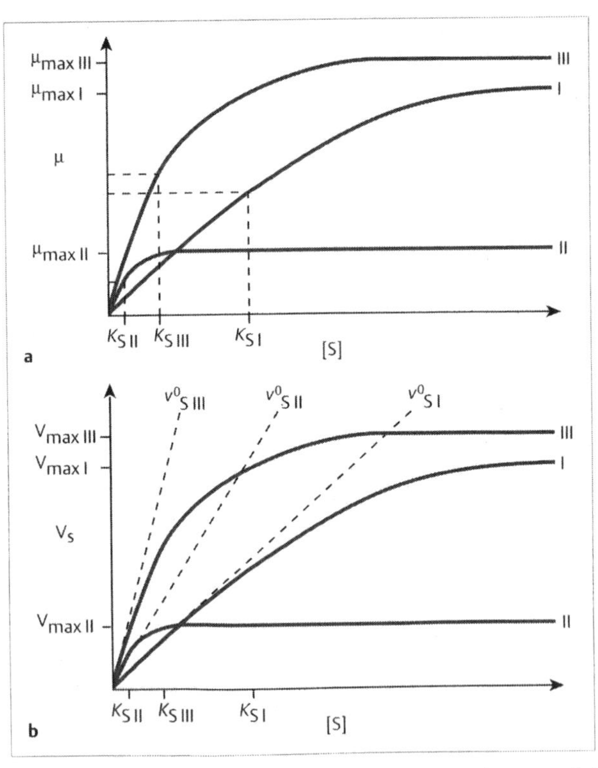

Fig. 30.8 **Influence of substrate concentration on (a) growth rates and (b) substrate uptake kinetics** by three different microorganisms (Organism I, II, and III) with different μ_{max}, K_S, and v^0_S characteristics

ecology, the ones that grow slowly but steadily at maximal exploitation of the system's capacity (***K* strategist**), and those that grow at high rates but with high losses, depending on fluctuations in the system (***r* strategist**). Organism I would be selected in a static culture with high initial substrate concentration, organism II in a chemostat culture at low dilution rate.

The third saturation curve representing organism III illustrates that at limiting substrate supply, not only K_S but also μ_{max} determines the outcome of competition. Although organism III has a higher K_S than organism II, organism III will always outcompete organism II, even at low substrate supply. Therefore, K_S is not sufficient to compare the actual substrate affinity of different organisms. Moreover, K_S is not really a constant, specific for a single organism. It can change, depending on the respective substrate supply conditions. K_S is defined only as an empirical kinetic constant, but it is not really known to what it refers. It can be a function of the saturation constant of the primary uptake system (permease) for the respective substrate, or of the first enzyme metabolizing the substrate. As mentioned in Chapter 30.3.3, new uptake systems of higher affinity may be induced under substrate limitation conditions, leading to a lower K_S of the organism for the substrate. Alternatively, the number of permease molecules can be increased under limitation conditions, leading to a higher substrate uptake rate at limiting substrate supply, although the K_S does not change.

For comparison of competition by different organisms for a limiting substrate, it is more feasible to compare the substrate uptake kinetics rather than growth kinetics. Growth is coupled to substrate turnover via the substrate-dependent growth yield Y_S, according to:

$$\frac{dS}{dt} = \frac{\mu \cdot X}{Y_S} \quad (30.4)$$

and this yield varies individually with the organism concerned.

Based on the Michaelis-Menten approach, the substrate turnover rate v_S can be derived as a function of K_S, the substrate concentration S, and the maximum uptake rate v_{max}, according to

$$v_S = \frac{dS}{dt} = \frac{v_{max} \cdot S}{K_S + S} \quad (30.5)$$

Substrate uptake kinetics (Fig. 30.**8b**) will follow saturation curves similar to those depicted in Figure 30.**8a** if it is assumed for simplicity that all three organisms have the same substrate-dependent growth yields. The **affinity constant** v_S^0 can be defined as the initial substrate uptake rate extrapolated for substrate concentration S = 0, which equals the slope of the saturation curve at S = 0, according to:

$$v_S^0 = \frac{v_{max}}{K_S} \quad (30.6)$$

This term is suitable for comparison of uptake kinetics between different organisms because it describes the actual substrate affinity of an organism, independent of growth parameters. It is obvious from Figure 30.**8b** that the v_S^0 values obtained for the three different organisms can predict the outcome of competition between them more adequately than the half-saturation constants K_S alone could do. v_S^0 has the dimension of time^{-1} and is directly comparable to a first-order reaction rate constant applied in chemical kinetics. It should be emphasized here that this affinity term can be defined for every kind of substrate, not necessarily only for carbon and electron sources, but also for nitrogen, sulfur, phosphorus, magnesium, etc. sources, and the outcome of the competition can be predicted from the values of K_S, v_{max}, and v_S^0, if the substrate supply situation can be sufficiently described.

As an example, take the competition between sulfate-reducing and methanogenic bacteria, e.g., in a freshwater sediment. Both compete for either hydrogen or acetate as their most important substrates, and the sulfate reducers usually outcompete the methanogens as long as sulfate is available. The data compiled in Table 30.**5** illustrate that sulfate reducers and methanogens have similar v_{max} values for hydrogen uptake, but the K_S values of sulfate reducers for hydrogen or for acetate are in general distinctly lower than those of methanogens. The affinity constants calculated for hydrogen uptake differ significantly and thus explain the typical outcome of these competitions. It appears that in both cases the energetically more favorable reaction outcompetes the less favorable reaction because it has the higher substrate affinity, and this is the rule also in the electrochemical series of redox events under oxygen limitation (Chapter 30.2.1). However, as mentioned before, this is not a direct consequence of reaction energetics, and there is no convincing general explanation for this phenomenon.

Table 30.**5** also shows that the K_S values of various freshwater sediments for hydrogen are in the same range as those of pure cultures of sulfate-reducing or methanogenic bacteria. Thus it appears that such bacteria may really determine the hydrogen uptake kinetics of these sediments. This is not at all trivial; for example, hydrogen uptake by soil samples differs significantly from those of defined cultures of hydrogen-oxidizing bacteria, indicating that these bacteria are

Table 30.5 **Growth and substrate uptake kinetics relevant in competition between sulfate-reducing and methanogenic bacteria**

	Sulfate reducers	Methanogens	Freshwater sediment
Hydrogen oxidation			
Free energy change $\Delta G^{\circ\prime}$ (kJ per 8 mol electrons)	−151	−131	
K_S (μM)	0.7–1.9	1.0–6.6	1.4–8.5
V_{max} [nmol min^{-1} (mg protein)$^{-1}$]	110–170	13–220	n.d.
v_S^0	107.8	30.3	n.a.
Acetate degradation			
Free energy change $\Delta G^{\circ\prime}$ (kJ per 8 mol electrons)	−63	−35	
K_S (μM)	67–200	300–500	n.d.

[1]Calculated as median within variation limits (n.d., not determined; n.a., not applicable)

probably not responsible for the observed hydrogen uptake in such soils (see Chapter 32.1.2).

However, such comparisons of affinities between laboratory cultures and environmental samples are dangerous and should be made with great care. It has been mentioned above that microorganisms can exhibit different K_S values depending on growth conditions, whether grown with excess or limiting substrate supply. Also the incubation temperature, possible attachment to surfaces (see Chapter 30.5.3), and the availability of additional substrates (see Chapter 30.4.2) influence the apparent substrate affinity. In general, microorganisms in the complex natural habitat may exhibit half-saturation constants and affinity constants rather different from those determined with defined laboratory cultures.

Turnover of dissolved substrates in a natural ecosystem is determined by the affinity constant v_S^0 of the metabolizing microorganisms. This constant represents the slope of the substrate uptake saturation curve at zero substrate concentration and is directly comparable to the rate constant of a first-order reaction. This affinity constant is not a "constant" in strict terms, but can vary individually with the physiological state of the microorganisms.

30.4.2 Multiple Substrate Supply Changes Uptake Kinetics

Except for a few specialists confined to only one substrate, microorganisms usually are able to use more than one substrate to cover their nutritional needs. With bacteria in the laboratory, it has been observed that in some cases one substrate prevents utilization of another substrate, and this phenomenon is called **diauxie** (Chapter 20.2). However, diauxie is a rather unusual situation provoked in the laboratory by unusually high concentrations of more than one substrate. In chemostat experiments with *Escherichia coli* using low concentrations of glucose and galactose, the classical repression of galactose utilization by glucose is not observed; instead, both are consumed simultaneously. Moreover, the remnant substrate concentration in the culture fluid for every sugar supplied in the mixture is lower than in cultures growing with either substrate alone. In experiments with up to six substrates supplied simultaneously, the remnant substrate concentration becomes a function of the relative contribution of either substrate to the cell's total substrate supply. In general, supply of an additional substrate "B" increases the apparent affinity for substrate "A."

The advantage of simultaneous utilization of substrate mixtures is observed also in experiments with mixed cultures of different bacteria competing in chemostat culture for mixtures of simultaneously supplied substrates. One of these experiments included the obligately lithotrophic sulfur-oxidizing *Thiobacillus neapolitanus*, the facultatively lithotrophic sulfur oxidizer *Thiobacillus* strain A2, and the strictly heterotrophic *Spirillum* spec. in a mineral growth medium containing thiosulfate and/or acetate as energy sources (Fig. 30.**9**). As long as only thiosulfate or only acetate is supplied, the respective specialist wins the competition over the facultative lithotroph able to use either substrate. Obviously, the specialists have the higher affinity for their respective substrates. When both substrates are supplied simultaneously at similar rates, the facultative litho/organotroph dominates the situation and outcompetes the specialists for their respective food

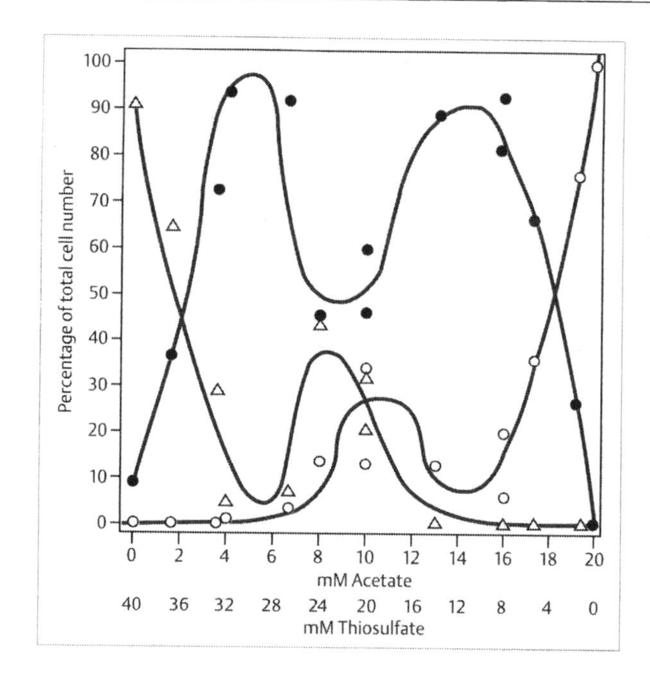

Fig. 30.**9** **Competition between *Thiobacillus neapolitanus* (Δ), *Thiobacillus* A2 (●), and *Spirillum* G7 (○) for thiosulfate and acetate as growth limiting substrates in the chemostat at a constant dilution rate (0.075 h^{-1}).** Relative contributions of the individual species to the total cell density were quantified by microscopic counting after steady states had been reached with every substrate mixture [from 5]

sources because its affinity for these substrates increases by supplying the additional substrate. Another example is that of homoacetogenic bacteria that do not compete successfully against methanogens or sulfate reducers for hydrogen, or against classical fermentative bacteria for sugars, but maintain their population by simultaneous utilization of several different types of substrates due to their higher metabolic versatility.

Generalists do not compete successfully against specialists in their respective field, but maintain themselves in competition by simultaneous utilization of multiple substrates.

30.4.3 Thresholds May Limit Substrate Degradation

Substrate degradation by microbial activities never reaches a "zero" concentration. With all microbial cultures, it is simply a matter of analytical precision to detect remnant substrate concentrations in growth media after growth has ceased, and the same applies to natural environments.

Even extremely "pure" offshore ocean water contains all organic molecules produced by biological activities at low concentrations. These low concentrations of organic compounds illustrate the efficiency of the microbial community in the respective habitat to degrade these compounds, but they also show the limits of these degradative capacities.

These small leftover concentrations of substrates remaining after microbial degradation has ceased are called **threshold concentrations**. They have become detectable only recently by modern trace analytics, especially capillary gas chromatography and high-performance liquid chromatography (HPLC).

It is not really understood why such threshold concentrations exist; one could postulate that every single substrate molecule should be metabolized once it reaches the respective enzymatic apparatus; however, as Figure 30.**10** illustrates, this is not the case. Thresholds of substrate utilization are in the range of nanomolar to picomolar concentrations, and depend on the respective bacteria studied. Due to reasons of analytical sensitivity, most of our knowledge on such threshold concentrations has been obtained with gaseous substrates. Experiments with different metabolic types of hydrogen-oxidizing bacteria have revealed that the threshold concentrations vary with the free energy change of the respective hydrogen oxidation reaction (Tab. 30.**6**), meaning that thresholds are caused by the respective reaction energetics rather than kinetics.

Threshold concentrations for a single organism can be lowered by supply of a cosubstrate, indicating again that the overall energetic situation of the respective organism is concerned. Therefore, thresholds measured

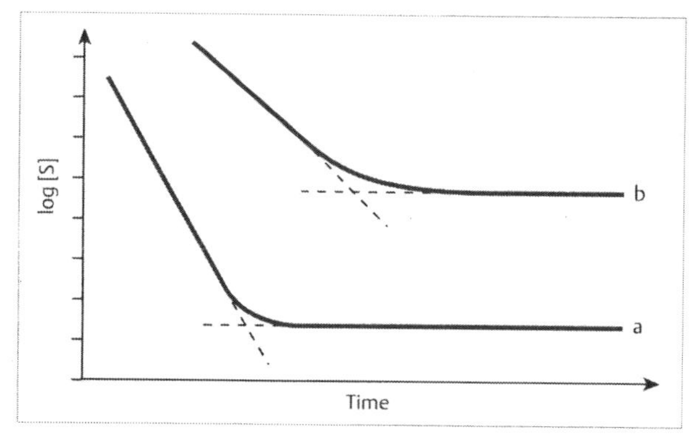

Fig. 30.**10** **Kinetics of substrate uptake by a microbial community** of (a) a high maximal uptake rate, high affinity, low threshold, and (b) of a low uptake rate, low affinity, high threshold

Table 30.6 **Dependence of threshold concentrations of hydrogen oxidation on the redox potential of the electron acceptor system**

Metabolic type	Example	Electron acceptor system	E_0'(mV)	Threshold concentration (Pa)
Homoacetogenic	*Acetobacterium* spec.	CO_2/Acetate	−290	52–95
Methanogenic	*Methanospirillum* spec.	CO_2/CH_4	−244	2.5–10
Sulfur-reducing	*Desulfovibrio* spec.	S^0/H_2S	−240	0.5–2.4
Sulfate-reducing	*Desulfovibrio* spec.	SO_4^{2-}/H_2S	−218	0.8–1.9
Fumarate-reducing	*Wolinella* spec.	Fumarate/Succinate	+30	0.002–0.09
Nitrate-reducing	*Desulfovibrio* spec.	NO_3^-/NH_4^+	+363	<0.002

in the laboratory with pure cultures that are supplied with only one substrate may turn out to be considerably higher than those detectable in natural habitats as the result of microbial multi-substrate utilization. Thresholds are of special concern with respect to degradation of xenobiotic compounds. Especially with potentially toxic compounds, it is not sufficient to ensure that such a xenobiotic is basically degradable; it is desirable to ensure that degradation proceeds to concentrations below a critical level where no harmful consequences with respect to direct toxicity or bioaccumulation are to be expected.

30.4.4 Co-metabolism May Be Detrimental to the Acting Organism

Degradation of organic compounds usually provides energy to the bacterium catalyzing it, and as a result the cell mass of the respective bacterium increases, thus also increasing the rate of substrate degradation. However, some substrates are degraded in a "non-productive" manner, meaning that the respective degradative activity does not provide any advantage to the respective bacterium, which therefore does not increase in cell mass. Such a metabolism was originally termed co-oxidation, and later, more generally, **co-metabolism**. The classical example is ethane oxidation by methane-oxidizing bacteria: the ethanol produced by the initial rather unspecific monooxygenase reaction cannot be metabolized further by the methylotroph and can feed only other bacteria in the environment:

$$CH_4 + O_2 + 2\,[H] \rightarrow CH_3OH + H_2O$$

$$CH_3OH + H_2O \rightarrow CO_2 + 6\,[H]$$

$$CH_3CH_3 + O_2 + 2\,[H] \rightarrow CH_3CH_2OH + H_2O$$

$$CH_3CH_2OH \rightarrow \text{no further oxidation, no electron recovery.}$$

As a result, the ethane-oxidizing activity of a methane-oxidizing population does not increase unless methane is provided as an additional substrate. Actually, ethane oxidation is even detrimental to the methane oxidizer because it invests electrons into the monooxygenase reaction without reclaiming them from the product. Therefore, even the maintenance of a certain level of co-metabolic oxidation activity may require supply of a degradable co-substrate. Co-metabolic activities are of major interest in the degradation of xenobiotic compounds because quite often such compounds are initially attacked by unspecific side activities of degradative enzymes, but do not lead to intermediates that support the respective microorganism any further. Supply of "productive" co-substrates often helps to increase the degradative capacity of microbial communities in waste treatment reactors or in soils. Since such co-metabolic activities do not directly interfere with the energetics of the respective organism, the consequences for substrate conversion kinetics, thresholds, etc. are of major interest, but have not yet been studied in reliable detail.

Co-metabolism does not supply energy or growth substrates to the catalyzing microorganism, and may even be detrimental for the organism. Co-metabolic activities are especially important in the initial attack on xenobiotic compounds.

30.4.5 How Can Substrate Turnover Kinetics in Nature Be Measured?

In the discussion on substrate turnover kinetics in chemostats (Chapter 30.3.2), it was stated that the measurable substrate concentrations remain constant at flow-equilibrium. In this case, the flow rate is known and the turnover kinetics can be determined. In a natural or semi-natural system, a soil or sediment sample, such determinations need different techniques.

The easiest assay may be the analysis of **bulk activities**, such as respiratory activity, which is measured either as oxygen uptake or as CO_2 release by a sample taken from the environment and incubated under controlled conditions similar to those in situ. Degradation of a certain substrate can be followed by analysis of its time-dependent disappearance. This requires addition of this substrate at concentrations significantly higher than those prevailing in the undisturbed system; as a consequence, the measured degradation rate will probably exceed that exhibited in situ. This problem can be avoided by using radioactively labeled substrates, which can be added at very low concentrations, but allow accurate measurements of degradation (e.g., formation of $^{14}CO_2$ from ^{14}C-glucose).

Most often, a substrate is formed and degraded in the same microbial habitat to be studied, i.e., the **turnover of a reaction intermediate** is being dealt with. A simple example of this kind is given in Figure 30.**11** for hydrogen turnover in a soil sample; in this case, the availability of a sensitive analysis allows reliable direct substrate determination. Lowering or increasing the hydrogen concentration in the system under study allows hydrogen formation or hydrogen utilization kinetics to be followed on the way to re-establishment of the equilibrium situation. To make sensitive turnover assays possible also with other substrates, **radiotracer analysis** has proven most suitable. A tracer of high specific activity is injected into the sample, thus changing the in situ concentration only insignificantly, and the disappearance of the tracer with time under in situ incubation conditions is recorded. For analytical reasons, it is often easier to follow the formation of labeled products (Fig. 30.**12**). From the plot of substrate utilization or product

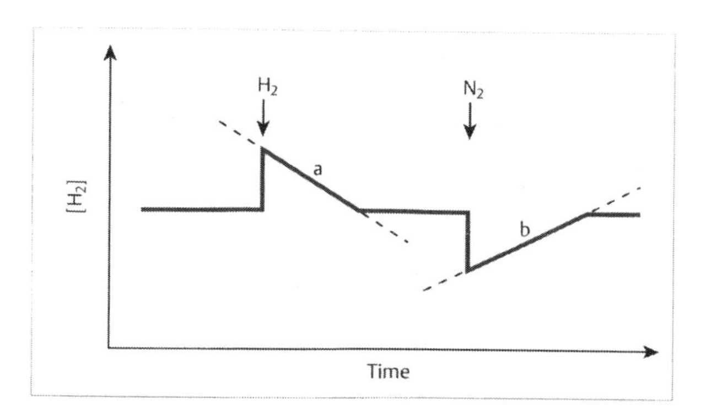

Fig. 30.**11** **Determination of hydrogen turnover in a soil sample.** H_2, addition of hydrogen; N_2 flushing, with nitrogen gas; slope a, rate of hydrogen oxidation; slope b, rate of hydrogen production

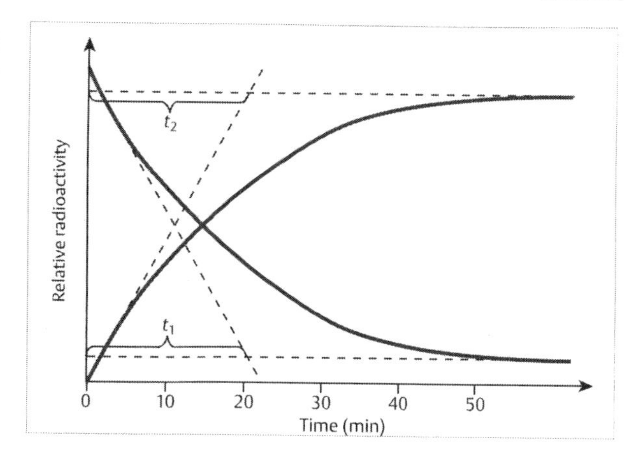

Fig. 30.**12** **Measurement of turnover times with a radioactive tracer.** (t_1 turnover time determined from substrate disappearance; t_2, turnover time determined from product formation). Note that the substrate does not always need to go to zero concentration, and that t_1 and t_2 do not need to be identical

formation over time, a **turnover time** can be determined; this gives the time needed for a full turnover of the pool of the respective intermediate. The reversal of the turnover time is the turnover rate constant r (given as a reciprocal time constant, e.g., h^{-1}) which, multiplied with the in situ concentration, the **pool size**, gives the **turnover rate**, e.g., as nmol substrate per g sampling material per hour:

$$v_s = \frac{dS}{dt} = r \cdot [S] \qquad (30.7)$$

Pool sizes and turnover times are independent figures that provide information on turnover rates only in combination. Compounds present in small pools may turn over rapidly and may be far more important in a transformation process than a compound present in a large pool that changes only very slowly.

A classical example is the case of hydrogen in methanogenic degradation. Hydrogen was first considered irrelevant because it was not detectable until sensitive detectors became available; today it is known that hydrogen is one of the most important intermediates in this process, with turnover times in the range of seconds. In the chemostat as a model system, the pool size is equivalent to the metabolite's concentration in the reactor, and the turnover time is equivalent to the residence time of the medium. Because residence times of all substrates and intermediates are identical in a steady-state situation in the fermenter, the chemostat is a rather artificial system that allows comparison with the natural situation only to a limited extent. The techniques mentioned above are well established and easy to apply; the main problem is in most cases to supply the respective tracer homogeneously without disturbing the internal structure of the sample material. Again, much more reliable information about gas metabolism in natural systems is available than about any other substrates because gases are easy to supply and to analyze. The difficulty increases dramatically the more complex the internal structure of the system under study becomes, e.g., in fine-structured sediment or microbial mat layers in which the spatial organization directly influences the turnover processes (see also Chapter 30.6).

Another problem in these assays is that usually the kinetics of substrate turnover are not as trivial as assumed above. Formation of an intermediate most often follows zero-order kinetics, i.e., the formation rate is independent of the intermediate's concentration if the turnover is not inhibited by accumulation of the intermediate. Utilization may follow zero-order or first-order kinetics, depending on whether the utilizing system is saturated with substrate or not. It may even follow second- or third-order kinetics if additional co-substrates are involved.

Other microbial activities (e.g., hydrolysis of polymers) can be assayed for by incubation experiments with radioactive substrates or with substrate analogues that release fluorescent dyes upon hydrolysis. Such assays usually give only semi-quantitative information on a degradative potential rather than in situ transformation rates because the supramolecular structure of natural polymers may differ substantially from that of the tracer compound. Also the incorporation techniques with labeled thymidine or leucine or the assay of ATP content (Chapter 30.3.5) provide semi-quantitative information on microbial activity in a more general sense.

30.5 Many Bacteria Live Attached to Surfaces

Laboratory microbiology concentrates in most cases on bacterial cells that are suspended free in the liquid growth medium and show little or no tendency to attach to the surfaces of the growth vessel. In fact, attachment to surfaces is highly disliked in laboratory work, and cultures not growing properly in homogeneous suspensions are usually discarded.

A microbial cell in a natural ecosystem, on the other hand, has to cope with surfaces wherever it exists. This is obvious with bacteria living in sediments or soil, or associated with animal or plant tissues, but also the free water of a lake provides ample amounts of surfaces at the sediment/water interface, at the side slopes of the water body, and on small particles such as dead or living phytoplankton and zooplankton and other bacterial cells. Also the boundary facing the air atmosphere can be densely populated by microorganisms. Surfaces are present everywhere, therefore, and they provide habitats for microbial cells that are quite different from the habitat in the freely suspended state.

It is obvious that bacteria degrading polymeric substrates such as cellulose will interact intensely with the substrate surface. However, also apparently "inert" surfaces such as those of mineral particles or poorly degradable organic material represent boundaries of an aqueous system that differ significantly from the free water medium.

Box 30.1 All **natural surfaces** found in nature carry electrical charges, mostly negative charges, due to partial deprotonation of acidic functional groups, e.g., silicic acid, carbonic acid, and humic acid residues. Non-charged surfaces are provided with apolar substrates such as mineral oils, but also the air–water interface is an apolar surface for a suspended microorganism.

Electrostatic forces interact between charged units (ions) and decrease with the square of distance. Opposing charges attract; identical charges repel each other. Divalent counterions can bridge between groups of identical charge. Van der Waals, Debye, and Keesom forces are in general lower than electrostatic forces, always attract and increase with the reciprocal 4th to 6th power of distance. In pure water of low ionic strength, a negatively charged particle will be repelled by a negatively charged surface, but van der Waals, Debye, and Keesom forces counteract them and may lead to bulk attraction at a very short distance. A positively charged particle will be attracted to a negatively charged surface by all types of forces mentioned, and the same applies to a negatively charged particle in the presence of counterions, as this is typically the case in a natural environment. Non-charged surfaces will attract each other exclusively through non-electrostatic forces.

Therefore, any kind of surface exposed to an aqueous phase containing non-charged or charged molecules (ions) will enrich these solutes in a boundary space close to the surface. A bacterial cell is negatively charged on its surface, with Gram-negative bacteria, by acidic lipopolysaccharides and proteins in the outer membrane, and with Gram-positive bacteria, by teichoic acids.

Molecules attracted to a surface can move easily within the boundary layer, but cannot easily leave it, depending on the overall energy of the system. The surface layer can be considered to some extent as a "two-dimensional solution," similar to the situation of lipophilic molecules within biological lipid membranes.

30.5.1 Microbial Attachment to a Surface Is a Multistep Process

Microbial cells underly the aforementioned purely physicochemical interaction with a surface, and this can be the first step in a process that may lead to permanent attachment.

1. A microbial cell is attracted to a surface ("**substratum**") by purely physicochemical interactions, either by electrostatic or by non-electrostatic forces, and thus makes first contact with the respective surface. This attraction is reversible and entirely non-biological; also dead cells show the same behavior. Attraction is counteracted by shearing forces that become more important the more the cell extends into space. The substratum may be already covered by a layer of polymers such as proteins or polysaccharides; in fact, there is hardly a "clean" surface in natural habitats that is not covered with such polymers.

2. The microbial cell enlarges the contact area with the substratum by modifying its shape and by reorganization of charged/non-charged groups on the contact surface, thereby improving interactions with the substratum properties. Also this process is reversible, but only viable cells show this modification. It does not appear to involve protein synthesis because it is not prevented by the addition of chloramphenicol.

3. The cell produces specific adhesion proteins (**adhesins**), organelles (fimbriae), or polysaccharides (**glycocalyx**) that specifically interact with the substratum and bridge the gap of repulsion that exists between similar charges of both units, over distances of 100–300 nm. This step is definitively an induced biological activity governed and controlled by specific genes. Adhesins and polysaccharides are typically produced upon attachment to a suitable surface. Some bacteria produce adhesins constitutively, especially pathogenic bacteria for which adhesin production is a prerequisite for successful host colonization (see Chapter 33.3).

Attachment can also be initiated by hydrophobic interactions. Especially attachment of microbial cells to each other appears to be governed primarily by apolar interactions, and the hydrophobicity of the cell surface is an important factor in the formation of bacterial flocs in, for example, sewage digestors.

Attachment of a bacterial cell to a surface is a process of at least three different steps that are partially physicochemically and partially biologically governed. Physicochemical attraction leads to an enrichment of dissolved and suspended compounds in the surface boundary layer; this is in most cases of advantage for an attached microorganism.

30.5.2 Surface Attachment Provides Several Advantages

Since all molecules, polar or not, are physicochemically enriched in the surface boundary layer, an attached bacterium will find substrate supply higher there than in the free liquid. Surface attachment is a behavior typically observed with bacteria under conditions of low substrate supply; freely suspended bacteria tend to change to an attached lifestyle under substrate limitation, and many typical starvation-adapted bacteria ("oligotrophic bacteria") exhibit attached states as key parts of their life cycle. Because dissolved molecules move easily within the surface boundary layer, the attachment substratum collects substrates for the attached bacterium and transports them within the boundary layer to the microbial cell, which acts as a substrate sink. However, this advantage for an attached cell applies only as long as the number of attached cells is small and does not cover the substratum entirely.

In a running water, a creek or river, cells attached to rock and other surfaces can exploit the substrate supply of large water volumes without spending energy in cell movement.

Detrimental effects of toxic compounds including oxygen can be alleviated because the toxic compound diffuses only from one direction and may be neutralized by a total microbial community instead of a single cell. This aspect is of major concern with respect to the efficiency of disinfectants, e.g., in drinking water treatment or clinical applications. Figure 30.**13** shows

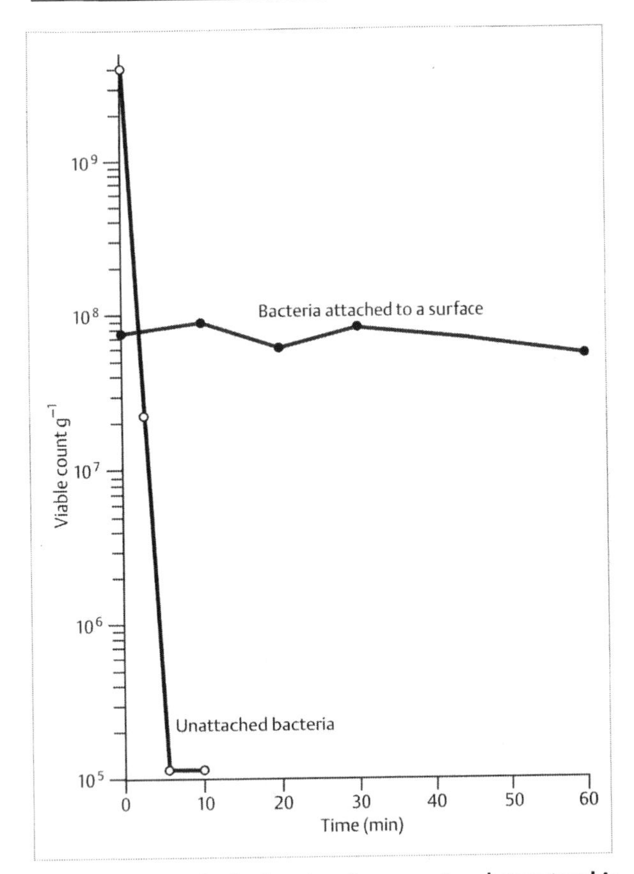

Fig. 30.13 **Survival of naturally occurring heterotrophic bacteria exposed to 2.0 mg chlorine (l water)$^{-1}$ for 1 h as quantified by plate counting.** The chlorine concentration remained nearly constant over the exposure time. The survival of bacteria attached to a surface is compared with the survival of unattached bacteria

that attached bacteria survive chlorine treatment better than unattached bacteria do.

Attached cells escape grazing by ciliates or rotifers much better than freely suspended cells. This effect selects, for example, in activated sludge for the formation of major aggregates of microbes that will sediment easily and maintain the system's operational stability (see Chapter 36).

Attached bacteria can develop cooperative structures with other bacteria much easier and in a more stable fashion than freely floating bacteria. This is of importance especially in the establishment of syntrophic cooperations (see Chapter 30.7). Attached bacterial communities will also find better chances for exchange of plasmids by conjugation or transduction.

Cells attached to surfaces in greater amounts form films, which store water and thus protect the constituent cells from desiccation, e.g., in surface soils.

Different types of bacteria tend toward surface attachment to a greater or lesser extent. Specialists such as *Asticcacaulis*, *Caulobacter* and *Hyphomicrobium* attach constantly with specific organelles or surface parts, or may go obligately through an attached state during their life cycle. Many other bacteria may attach only under environmental conditions favoring attachment, such as low nutrient supply.

In natural environments where substrate supply is nearly always limiting, an overwhelming majority of microbial cells attached to surfaces or combined in aggregates that may contain mineral particles or that may be only aggregates of microbial cells are found. This applies even to environments that may not appear to be nutrient-poor at first sight. The huge freight of particulate material, e.g., in many rivers, provides ample surfaces for attachment for most bacteria.

30.5.3 A Biofilm Is Heterogeneous in its Structure and Dynamics

Due to the natural tendency of microbes to attach to surfaces, also technical appliances, pipes for drinking water supply, waste water discharge, and transport tubes for aqueous solutions in food or paper industry are covered with microorganisms unless such systems are kept strictly sterile. Depending on the supply of degradable substrates, such **biofilms** may develop into slimy layers of considerable dimensions, and the thickness they can reach is basically limited by the substrate supply and the shearing forces to which they are exposed. The transport capacity of pipes and tubings in industrial devices or water supply systems can be severely impaired by biofilms that develop on their inner surfaces, primarily by eddy formation, but also by net decrease of the acting diameter of such tubes.

The establishment of a biofilm on an inert surface goes through a phase of initial cell deposition, which is followed by a growth phase that receives biomass increase both from bacterial growth at the surface and from further cell deposition, and may reach a plateau that is determined by a dynamic equilibrium between growth and sloughing off, which depends on substrate and cell supply and shearing forces (Fig. 30.**14**). With increasing thickness, a biofilm develops a certain degree of heterogeneity with respect to substrate supply and life conditions in general. Surface-associated cells enjoy a comparably rich substrate supply from the liquid, whereas in lower layers, cells may starve and die, leading to structural instability of the film. This heterogeneity is caused by the metabolic activity of the microbes involved and may lead to the establish-

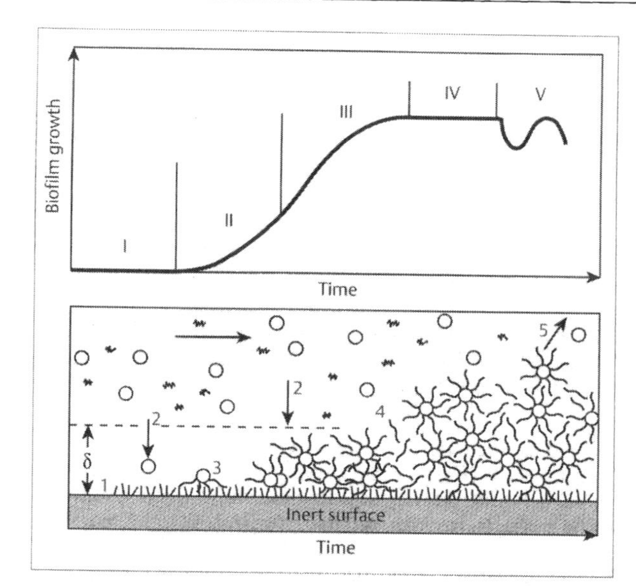

Fig. 30.**14** **Development of a biofilm on an inert surface.** I, lag phase; II, exponential growth; III, limited growth; IV, plateau; V, sloughing; δ, diffusive boundary layer not exposed to eddy shearing forces; 1, polymer deposition; 2, cell deposition; 3, cell attachment; 4, cell multiplication; 5, cell release by sloughing [after 6]

ment of entirely different microbial communities in the upper and lower layers of such films. Since especially oxygen supply can become limiting for the lower cell layers, the dominant type of metabolism in these layers may be anaerobic (e.g., fermentative, nitrate-reducing, or sulfate-reducing) whereas the fermentation products or reduced nitrogen or sulfur compounds diffuse to the surface and are oxidized there by aerobic bacteria.

Recent studies on biofilm microstructure have revealed that they are not necessarily solid homogeneous layers, but may contain channels and open passes for exchange with the overlying aqueous phase. Thus, the biofilm may be free for passage of dissolved compounds and even small particles.

The **biofilm community** consists of a spatially structured, heterogeneous mixture of different organisms and non-biological deposits with specific dynamics of growth and degradation that mirrors on a small scale the types of cooperation and antagonism known for structured oxygen-limited ecosystems in general.

30.5.4 Bacterial Colonies Are Structured With Regard to Oxygen Tension, Substrate Concentration, and the Metabolic State of Cells

A special kind of biofilm is represented by a bacterial colony on an agar plate. At least in a bigger colony, the supply of oxygen from the gas phase and of dissolved growth substrate from the agar phase may differ substantially in the upper and the lower cell layers. In Figure 11.**3**, the distribution of oxygen in a colony of *Bacillus cereus* is illustrated, as determined by microelectrode studies. The oxygen concentration decreases sharply from the colony surface, and there is no oxygen measurable deeper than 20 μm inside the colony. Thus, about 20 layers of bacterial cells are sufficient to consume the atmospheric oxygen entirely, if sufficient substrate is provided from the agar medium. Enzyme assays in cells of *Enterobacter cloacae* that were recovered from thin sections of frozen plate colonies have shown that in lower cell layers, the key enzymes of the citric acid cycle are repressed, but are active in the upper layers. The supply of dissolved substrate from the agar to the surface layers, on the other hand, follows a distribution opposite to that of oxygen. Actually, every cell layer inside such a colony faces life conditions different from those of the layers immediately above or below the layer. Again, this structural heterogeneity is caused and maintained by the bacterial metabolic activity; as soon as the cells are killed, the distribution of oxygen and substrate will equilibrate over the colony and the agar with time.

In a **colony** of living bacteria, steep and opposing gradients of oxygen tension and substrate concentrations are formed. These gradients cause death or change of metabolic activities in the depth of the colony, compared with cells at the surface.

30.5.5 The Formation of Tooth Plaque Is Influenced by the Host Organism and Several Types of Bacteria

Tooth plaques are another special example of a biofilm that, in this case, is constructed by various types of bacteria through specific interactions. Teeth are part of the mouth environment, which, in the case of a healthy human being, houses a microbial community of about 300 different taxa. The environment is characterized by a steady flow of saliva and a broad supply of easily degradable substrates, which altogether make the mouth a continuous culture system of high complexity.

A clean, sterile tooth surface introduced into the human mouth is first covered with saliva, which provides an initial film of organic polymers, i.e., proteins, glycoproteins, and polysaccharides. This organic polymer layer is the substratum for specific adsorption of *Streptococcus* spec., typically *S. oralis*, *S. mutans*, *S. mitis*, or *S. sanguis* as primary colonizers. The *Streptococcus* cells expose adhesin molecules on their surface that specifically recognize surface sugars of other bacteria and thus enhance their deposition specifically. The presence and binding specificity of these adhesins can be assessed with specific antibodies, and the host organism can influence the deposition of additional bacteria through antibodies released into the saliva. Among the secondary colonizers, *Actinomyces naeslundii* dominates, but also strict anaerobes such as *Veillonella atypica* and *Prevotella loescheii* can dominate, the latter organism can be the basis for attachment of other anaerobes such as *Actinomyces israeli* and *Capnocytophaga gingivalis*. In thick plaque layers, other strictly anaerobic bacteria such as *Fusobacterium nucleatum* can establish; their presence already indicates a pathogenic situation of advanced tooth decline that is brought about mainly by lactic acid formation from excess sucrose supply. The tooth plaque environment is in contact with the mouth mucosae, which harbor, among others, spirochetes (e.g., *Treponema* spec.), which can cover and penetrate slimy surfaces and establish there. If plaque layers continue to grow without significant disturbance (brushing), even sulfate-reducing and methanogenic bacteria can establish; this situation clearly indicates that the teeth are severely ill.

30.5.6 Microbes Mediate and Intensify Corrosion Processes

Metal surfaces can provide excellent attachment substrata for bacterial cells, and the colonization of inner surfaces of metal tubes has already been mentioned as a major problem, for example, in freshwater supply systems. However, bacteria can use metals also as electron sources for their energy metabolism, thus corroding and destroying especially steel and iron tubes.

Metallic iron in contact with water becomes polarized and releases ferrous iron ions. The corresponding electrons move within the metal and reduce protons at a different site of the metal surface to form a

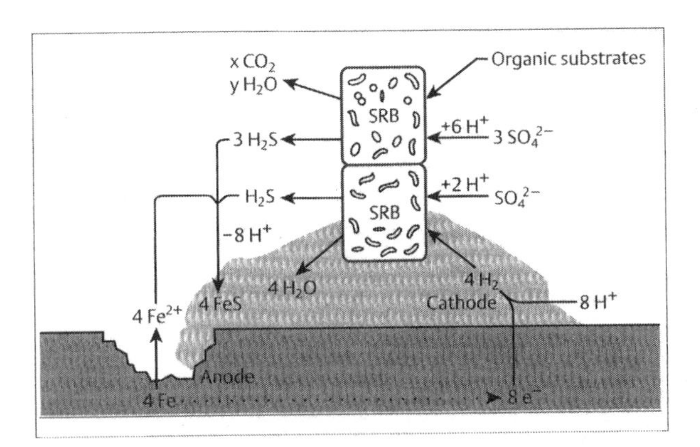

Fig. 30.**15** **Reactions during anaerobic iron corrosion as suggested by the depolarization theory of C. A. H. von Wolzogen-Kühr and I. S. van der Vlugt, with additional sulfate reduction with organic substrates.** Light gray area indicates attached ferrous sulfate. The organotrophically growing sulfate-reducing bacteria (SRB) may or may not be the same as the depolarizing hydrogen-consuming bacteria [after 7]

thin hydrogen gas layer that protects the iron surface to some extent from further oxidation. Sulfate-reducing bacteria attached to such a metal surface take up the hydrogen for reduction of sulfate to sulfide, thus removing the protective hydrogen film and accelerating the corrosion. Supply of additional organic substrates increases the metabolic activity of the sulfate reducers and supports the overall process (Fig. 30.**15**). The accumulating iron sulfide may act as an additional cathode, like a noble metal, and thus increase the corrosive activity of the sulfate-reducing bacteria.

Also limestone and concrete buildings are colonized by bacteria. These bacteria feed on volatile inorganic sulfur and nitrogen compounds, which are supplied through the air as a consequence of excessive fertilization in agriculture, combustion of sulfur-containing coal, and partial nitrogen oxidation (formation of NO_X) in car engines. Lithotrophic bacteria oxidize these compounds to sulfuric and nitric acid, which dissolve the carbonate "glue" of limestone and concrete. The resulting corrosion causes enormous damage, especially with buildings and monuments of historical value.

30.6 Transport Is Accomplished by Convection and Diffusion Along Gradients

In laboratory experiments, bacteria are usually cultivated in homogeneously mixed liquid media. Substrate transport under such conditions is accomplished through physical water mass transport called **advection** or **convection**: the dissolved compound is transported together with its solvent. The same transport system is used in nature for transport of dissolved compounds through major water bodies over long distances, either by horizontal currents or, in stagnant water, in the vertical direction by so-called **"eddy diffusion."** This "diffusion" is actually a turbulent convection process. Horizontal transport processes (waves) induce the formation of eddies, which through whirling motion cause a net transport down into the water body and up again. The diameter of these eddies goes down to about 0.5 mm, depending on the prevailing temperature. Also this process requires mass transport of water bodies on a small scale, therefore, and stops at a distance of about 0.5 mm from a solid surface. On a distance scale shorter than 0.5 mm or in ill-mixed, inhomogeneous or semi-solid media such as sediments, microbial mats, or tissues where such transport is mechanically impeded, transport of dissolved compounds is possible only by **diffusion**.

Box 30.2 Gases in space or solutes within a stagnant solvent move by diffusion that can be described by Fick's first law:

$$J = -D \cdot \frac{dc}{dx} \qquad (30.8)$$

where J is the flux, i.e., the amount of compound moving across a unit area per unit of time, D is the diffusion coefficient which is temperature-dependent and specific for the respective compound, and $\frac{dc}{dx}$ is the concentration change over the distance x. The time needed by a single molecule for diffusive transport along one axis over the distance l is described by the formula

$$t = \frac{l^2}{2D} \qquad (30.9)$$

Since D is a constant, the diffusion time increases with the square of distance. Therefore, transport by diffusion is efficient only at short distances. On the small scale relevant for bacteria and the transport processes in their habitat, diffusive transport takes only milliseconds and is the only relevant transport process. Larger organisms depend to a much higher extent on convective transport inside their bodies and also in the space in which they live.

For natural systems, also characteristics of the matrix in which diffusion occurs need to be considered, such as the porosity and the tortuosity that corrects for the longer random walk that a dissolved molecule has to take due to repeated collisions with particles:

$$J = -\phi \cdot D_S \cdot \frac{dc}{dx} \qquad (30.10)$$

where ϕ is the porosity index (in the range of 0.6–0.99), and D_S represents the apparent diffusion coefficient including the tortuosity effect. This modified version of Fick's first law can be used to understand the net flux of dissolved compounds along concentration gradients from a source to a sink, e.g., from a producer to a consumer. Figure 30.**16** shows a profile of oxygen distribution above and inside a marine sediment. The oxygen concentration is equivalent to air saturation ($\sim$300 μM) down to about 0.5 mm above the surface, where convective eddy transport ceases and only diffusion is possible ("diffusive boundary layer"). Oxygen is consumed in the sediment at all layers down to about 2.5 mm depth. The total rate of oxygen consumption by the sediment can be calculated using Equation 30.10 from the slope of the linear part of the oxygen curve in the diffusive boundary layer. Consumption rates further down in the sediment can be calculated the same way from the slopes at every single depth layer.

Each gradient in a structured system at dynamic equilibrium means a flux; each curvature in such a gradient means a process!

Measurements of diffusion gradients as a means to understand microbial activities in natural habitats has become quite attractive since microelectrodes with tip diameters of 5–20 μm have been constructed. These microelectrodes allow reliable quantitative analysis of redox potentials and of dissolved compounds such as protons (pH), oxygen, hydrogen, hydrogen sulfide, ammonia, nitrate, and nitrite, and similar micro-probes have been developed also for the assay of methane. Thus, the spatial resolution of metabolite distributions has reached dimensions in the range of those of bacterial microcolonies.

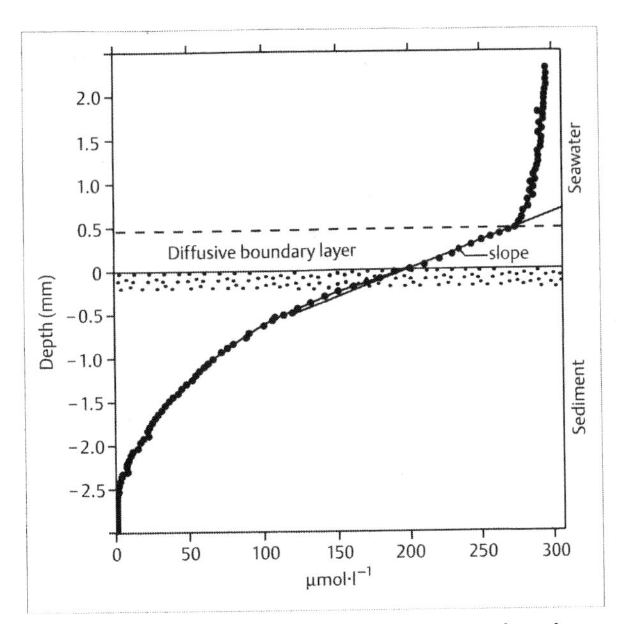

Fig. 30.**16** **Distribution of oxygen over and inside a deep-sea marine sediment.** The oxygen concentration decreases from air saturation through the diffusive boundary layer further into the sediment, and reaches zero at 2.5 mm depth. Note that the slope is linear in the diffusive boundary layer [after 8]

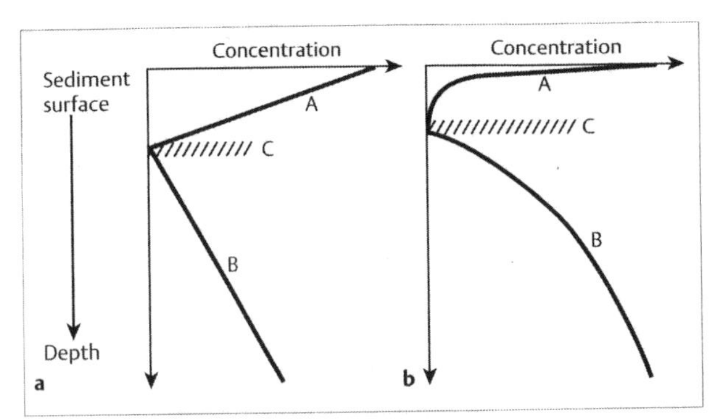

Fig. 30.**17a,b** **Types of metabolite distribution in a structured, stable matrix (agar, sediment) at dynamic equilibrium.**
a Idealized situation with only one source and one sink for metabolites A and B. C, area of disappearance of both metabolites.
b Realistic situation as occurs in natural sediments with actively metabolized molecules

30.6.1 Concentration Gradients Document Dynamics of Fluxes and Activities

Concentration gradients may give information not only about the distribution of dissolved compounds, but also about the fluxes between then and the activities of sources and sinks. For a reliable interpretation of such diffusion gradients, exact data on temperature and other parameters that influence the activities have to be provided. It is also important to known whether the system under study is at dynamic equilibrium (e.g., sulfide distribution in a deep-sea sediment) or whether the system did not reach equilibrium yet as a consequence of, for example, diurnal changes in light or oxygen supply.

Equilibrium situations can be illustrated with two additional examples. If in a sediment the concentration of a compound A decreases linearly from the top to the bottom, and that of compound B from the bottom to the top, this indicates that there is only one source and one sink for these compounds (Fig. 30.**17a**). The simultaneous disappearance of both compounds at the same depth layer C may suggest that both react with each other or are consumed by biological activities in that layer. Calculation of the corresponding fluxes on the basis of the specific diffusion coefficients and the gradient slopes may allow the derivation of a reaction stoichiometry and with this an understanding of the underlying process, regardless of whether the process is catalyzed by microbes or not. More often, distribution gradients of compounds are curved rather than linear, as indicated in Figure 30.**17b**. Compound A, e.g., oxygen, diffuses from the top into the sediment, and is consumed not only in layer C, but already on its way downwards. An opposite curvature, as shown with compound B in Figure 30.**17b**, indicates that this compound, e.g., methane, is produced not only in the deep sediment layers, but also further in the overlying sediment, and is consumed in layer C. Exact resolution of the gradient slopes in every layer of such a structured habitat may provide quantitative information on metabolic processes involved in their transformation (see Chapter 31.2).

30.6.2 "Gradient Bacteria" Live in Gradients of Spatially Separated Substrates

Structured environments such as sediments provide conditions that are required by certain metabolic types of bacteria for optimal activity and survival. This is obvious especially with those bacteria that have to compete for energy metabolism against a chemical process, e.g., the aerobic sulfide-oxidizing bacteria at neutral pH, such as *Beggiatoa* spec. These bacteria cannot be cultivated with hydrogen sulfide and oxygen in a homogeneously mixed culture because under such conditions, the two substrates would react to a large extent with each other, leaving hardly anything for the bacteria. These difficulties in growing aerobic sulfide oxidizers at neutral pH have been solved only recently by an experimental approach that mimics the growth situation of such bacteria in their natural habitat, i.e., by cultivation in **"gradient cultures."**

Such gradient cultures are established in test tubes containing a semi-solid mineral medium with 0.2% agar, a hydrogen sulfide source at the bottom, and air in the headspace. The bacterial inoculum is distributed over the entire medium column. After 4 days of incubation, the *Beggiatoa* cells accumulate in a sharp layer (Fig. 30.**18b**). Analysis of oxygen and sulfide distribution in the tube and in a non-inoculated control tube reveal that the bacteria accumulate just in the layer where oxygen and sulfide meet at a concentration so low that chemical sulfide oxidation is entirely outcompeted. In the control tube, sulfide and oxygen meet at comparably higher concentrations, and the gradient curvatures indicate that both are partly consumed by the chemical reaction over rather long diffusion distances (Fig. 30.**18a**). The gradient slopes in the inoculated tube are much steeper than in the non-inoculated tube, which indicates that the sulfide oxidation activity in this tube is significantly higher.

Gradient systems are a prerequisite for efficient growth of "gradient bacteria" because these systems allow spatial separation of reactive substrates. The gradient systems also illustrate that the biological process, similar to all catalyzed reactions, is far more efficient than the non-catalyzed chemical reaction because the reaction rate at low substrate concentration is increased significantly. With this, the bacteria help to maintain the separation of the reactants, and thus stabilize the chemical structure of their habitat to their own advantage.

Beggiatoa spec. depend for spatial separation of their substrates on sediment or agar as a semi-solid structuring matrix in which the bacteria can move chemotactically, using a creeping motility. Other bacteria that oxidize sulfide at neutral conditions, such as the marine *Thiovulum* spec., do the same in liquid water and have to provide the structuring matrix themselves. *Thiovulum* cells have never been obtained in pure culture, but can be enriched with a piece of easily degradable organic matter, such as a chunk of fish or cheese, as an electron source, wrapped in some gauze and deposited in an aquarium with flowing sea water.

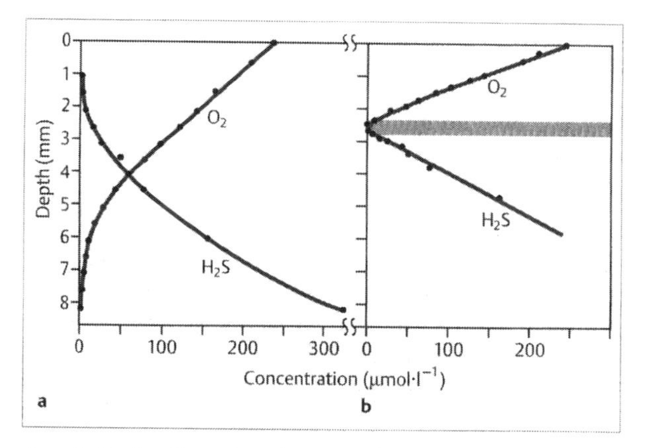

Fig. 30.**18a,b** **Growth of *Beggiatoa* spec. in opposing hydrogen sulfide and oxygen gradients established in a semi-solid agar medium.**
a Sterile control after 4 days of incubation:
b Tube incubated with *Beggiatoa* cells for 3 days. *Beggiatoa* cells accumulate in the gray zone [after 9]

Oxygen-limited microbial degradation of the organic matter soon releases hydrogen sulfide by sulfate reduction at a rather constant rate. After a few days, the substrate chunk is surrounded by a veil of bacterial cells. Microelectrode studies have shown that this cell layer represents a continuous boundary between the inner, reduced, sulfide-rich phase and the outer oxidized environment (Fig. 30.**19**), and that oxygen (and hydrogen sulfide) are consumed at a high rate inside the cell layer. Thus, the cells themselves separate the two different compartments from each other and maintain this separation by active aerobic sulfide oxidation.

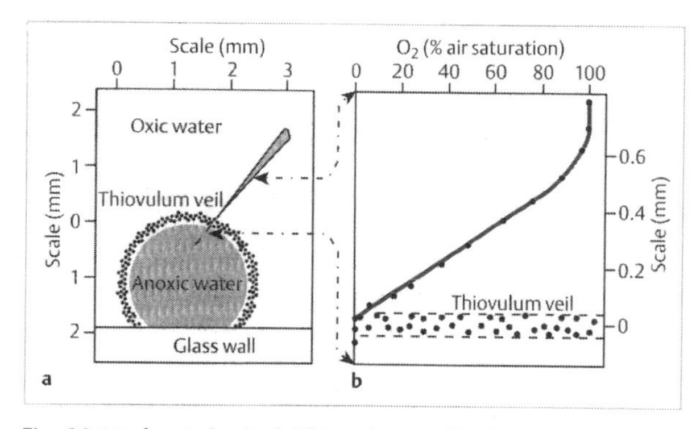

Fig. 30.**19a,b** **Spherical *Thiovulum* veil which encloses an anoxic, sulfide-rich pocket in oxic sea water.** Bacterial oxygen uptake creates a steep diffusion gradient in the 0.6-mm thick diffusive boundary layer surrounding the veil.
a Experimental setup.
b Oxygen gradient outside the *Thiovulum* veil [after 10]

A similar problem as with aerobic sulfide oxidation is posed with aerobic iron oxidation. At neutral conditions, ferrous iron is chemically oxidized with molecular oxygen, and aerobic iron-oxidizing bacteria have to compete against the chemical oxidation process. Iron oxidizers such as *Gallionella* spec. are microaerobic bacteria that develop their activity in the diffusive boundary layer close to ferric iron oxide precipitates, which prevent water convection (eddy convection). Therefore, oxygen can reach these zones only by diffusion, and the diffusive oxygen flux is governed by the microbial consumption activity.

Also **magnetic bacteria** are microaerobic and inhabit the transition zone between reduced and oxidized horizons in freshwater and in organically rich marine sediments. These bacteria are motile and collect in aqueous samples on microscope slides either at the northern or the southern edge. With a small laboratory stirring bar, they can be guided in various directions within such a liquid sample. These bacteria were originally called "magnetotactic bacteria" because of their orientation in the magnetic field. The cells contain chains of small elementary magnets, 35–120 nm in diameter (Fig. 2.**18d**), which are crystals of either magnetite (Fe_3O_4) or greigite (Fe_3S_4), and orient the cell physically in the magnetic field. Some cells contain only one such chain, others contain several chains arranged in parallel. Since even dead cells orient in the magnetic field, this phenomenon is not really a "tactic" behavior, which would include a reaction chain of sensor and effector, and the bacteria were later termed more correctly magnetic bacteria. So far, only few pure cultures of such bacteria have been isolated, and several other strains have been characterized phylogenetically in enrichment cultures by PCR typing. The magnetic bacteria do not form a coherent phylogenetic cluster, but appear in at least two different groups among the proteobacteria. The pure cultures described are microaerobic bacteria; some can also reduce nitrate. They oxidize organic substrates such as fatty acids or succinate; a direct involvement of iron in their metabolism has not been proven. The deposition of large amounts of iron in the magnetic mixed ferrous/ferric form is at least rendered easy in the redox transition zone, where iron can be taken up in the reduced form and oxidized inside.

The ecological function of magnetic cell orientation has been a matter of speculation. It is assumed that the magnetic orientation helps the bacteria to find the redox transition zone again after mechanical disturbance of their habitat. Without magnetic guidance, the bacteria would need to find their home again by trial and error, similar to the usual swimming/tumbling behavior observed with other bacteria (Chapter 20.5.3). The geomagnetic field that enters the ground at slopes that increase toward the North and the South Poles allows the bacteria to swim only in one dimension and provides guiding lines that allow the magnetic bacteria to move efficiently downwards, back to anoxic sediment layers. This interpretation is supported by the finding that magnetic bacteria enriched in the Northern Hemisphere are north-seeking bacteria, those from the Southern Hemisphere are south-seeking bacteria. The polarization can be changed physically by strong magnetic fields; it is not genetically determined. Nonetheless, every enrichment culture also contains a few oppositely oriented cells, which may have little survival chances in natural habitats.

30.7 Microorganisms Cooperate in Various Ways

Whereas competition is quite common in microbial life, cooperation between different types of organisms is less wide-spread among microorganisms. Most aerobic bacteria can perform their energy metabolism, e.g., the degradation of a complex type of organic substrate, without any substantial cooperation with other microbes. Several bacteria, however, degrade a substrate only partially and excrete a degradation intermediate; a different bacterium may take over to complete the oxidation process. Such cooperations are often only facultative in nature, meaning that the two partners do not obligately depend on each other. In other cases, cooperations are based on the transfer of micronutrients, e.g., vitamins or complexing agents for heavy metals. In the laboratory, such dependencies can be overcome by supplying the required micronutrients as medium additives.

More pronounced types of cooperation are typically found in anaerobic microbial communities. The complete conversion of complex organic matter such as cellulose, to methane and carbon dioxide in a natural habitat is possible only by the concerted action of at least four different groups of bacteria, including primary fermentative bacteria, secondary fermentative bacteria, and two types of methanogens (Fig. 30.**20**).

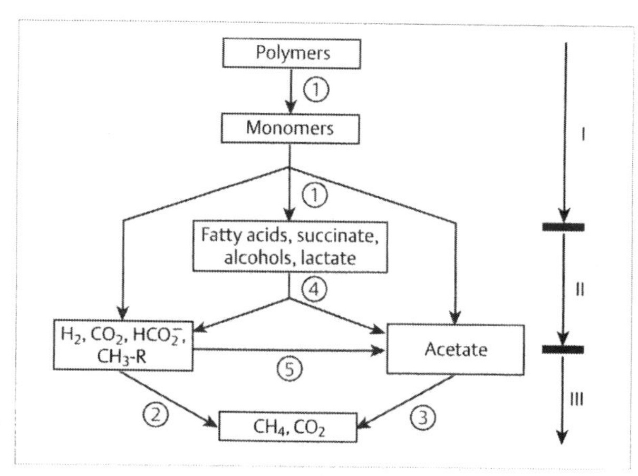

Fig. 30.**20** **Carbon and electron flow in the methanogenic degradation of complex organic matter.** Groups of bacteria involved: 1, Primary fermentative bacteria; 2, hydrogen-oxidizing methanogenic bacteria; 3, acetate-cleaving methanogenic bacteria; 4, secondary fermentative bacteria; 5, homoacetogenic bacteria; I, II, III, steps in degradation

Relationships between different groups of microorganisms differ in their extent. The term "**commensalism**" is applied to case of minimal cooperation between two partners. For example, aerobic and anaerobic bacteria live in the same habitat, and the aerobic bacterium creates by oxygen consumption the conditions under which the anaerobe thrives. In such a case, the anaerobe profits from the activities of the aerobe, but the aerobe obtains no significant advantage or disadvantage.

If such a commensalistic cooperation occurs in a food chain, it is termed "**metabiosis.**" For instance, the methanogenic degradation of fructose can be carried out in an artificial mixed culture. *Acetobacterium woodii* ferments the sugar to three molecules of acetate, and the acetate is then converted to methane and carbon dioxide by *Methanosarcina barkeri*. In this case, the latter member in the food chain profits from the former one, but there is no substantial enhancement of the former chain member, although acetate conversion to neutral products helps to stabilize the pH regime of the system.

The term "**syntrophism**" should be restricted to those cooperations in which both partners depend entirely on each other to perform the metabolic activity observed, and in which the mutual dependence on each other cannot be overcome by simply adding a co-substrate or any type of nutrient. A classical example of this kind is the "*Methanobacillus omelianskii*" culture, which was later shown to be co-culture of two partner organisms, the S-strain and strain M.o.H. Both strains cooperate in the conversion of ethanol to acetate and methane by interspecies hydrogen transfer:

S-strain :

$$2\ CH_3CH_2OH + 2\ H_2O \rightarrow 2\ CH_3COO^- + 2\ H^+ + 4\ H_2$$
$$\Delta G^{\circ\prime} = +19 \text{ kJ per 2 mol ethanol} \quad (30.11)$$

Strain M.o.H. :

$$4\ H_2 + CO_2 \rightarrow CH_4 + 2\ H_2O$$
$$\Delta G^{\circ\prime} = -131 \text{ kJ per mol methane} \quad (30.12)$$

Co-culture :

$$2\ CH_3CH_2OH + CO_2 \rightarrow 2\ CH_3COO^- + 2\ H^+ + CH_4$$
$$\Delta G^{\circ\prime} = -112 \text{ kJ per mol methane} \quad (30.13)$$

The fermentative bacterium cannot be grown in the absence of the hydrogen-scavenging partner organism because the fermentative bacterium carries out a reaction that is endergonic under standard conditions. The first reaction can occur and provide energy for the S-strain only if the hydrogen partial pressure is kept low enough by the methanogen. Therefore, neither partner can grow with ethanol alone, and the degradation of ethanol depends on the cooperation of both.

The term "**consortium**" should be reserved for the description of a symbiotic association of two or more organisms in a structured organization (e.g., the phototrophic consortia "*Pelochromatium*" and "*Chlorochromatium*"). So far, no example of the formation of such specific structures is known among the fermentative or other syntrophic associations.

30.7.1 The Anaerobic Feeding Chain Leading to Methane, Sulfide, and Carbon Dioxide

Due to the above-mentioned metabolic restriction of most strictly anaerobic bacteria, the conversion of complex organic matter to, for example, methane and carbon dioxide, depends on the cooperation of various trophic groups (Fig. 30.**20**). Polymers (polysaccharides, proteins, nucleic acids, also lipids) are first converted to oligomers and monomers (sugars, amino acids, purines, pyrimidines, fatty acids, glycerol), mostly by extracellular hydrolytic enzymes. These depolymerizations are carried out by the **primary fermentative bacteria**, which ferment the resulting monomers further to fatty acids, succinate, lactate, alcohols, and other compounds (Fig. 30.**20**, group 1). Acetate, H_2 and CO_2 and other one-carbon compounds can be used directly by

methanogenic bacteria to convert them to methane and carbon dioxide (Fig. 30.**20**, groups 2 and 3). For degradation of fatty acids longer than two carbon atoms, alcohols longer than one carbon atom, branched-chain and aromatic fatty acids, another group of fermentative bacteria, the so-called **secondary fermenters** or **obligate proton reducers** (Fig. 30.**20**, group 4) is needed. These bacteria convert their substrates to acetate and C-1 compounds, which are subsequently used by the methanogens. Since these secondary fermentative bacteria catalyze reactions that are endergonic under standard conditions (see also Chapter 12.2), they depend on close cooperation with the subsequent partner bacteria, as illustrated above for the "*Methanobacillus omelianskii*" co-culture.

Also in sulfate-rich anoxic habitats such as marine sediments, the primary processes of polymer degradation are carried out by classical fermentative bacteria, which form the fermentation products mentioned above. Many sulfate-reducing bacteria are metabolically much more versatile than methanogenic bacteria and can use all classical fermentation products and oxidize them to carbon dioxide, simultaneously reducing sulfate to sulfide. Although a few sulfate-reducing bacteria that can also use sugars or amino acids have been isolated recently, these bacteria do not compete successfully with classical fermentative bacteria on the same substrates.

Methanogenic degradation of complex organic matter is a three-step process catalyzed by primary and secondary fermentative bacteria, and methanogenic archaea that feed on acetate, CO_2, and H_2. Sulfate-dependent oxidation of organic matter proceeds in a two-step process catalyzed by primary fermentative bacteria and sulfate reducers. Other bacteria may have marginal functions in total substrate turnover.

In methanogenic and sulfate-rich environments, the primary fermentative bacteria (group 1) profit from the activities of the hydrogen-oxidizing partners at the end of the degradation chain. A low hydrogen partial pressure ($<10^{-4}$ bar; <10 Pa) also allows electrons at the redox potential of NADH (-320 mV) to be released as molecular hydrogen, and fermentation patterns can shift to more acetate, CO_2, and hydrogen rather than to ethanol or butyrate formation, and more ATP can be synthesized. *Clostridium butyricum* ferments hexose in pure culture to a mixture of acetate, butyrate, CO_2, and H_2, with an energy yield of 3.3 mol ATP per mol glucose. At a H_2 pressure of 10^{-4} bar (10 Pa), *C. butyricum* forms only acetate, CO_2, and H_2, with an energy yield of 4 mol ATP per mol glucose.

In a well-balanced anoxic sediment or slud reactor in which an active hydrogen-utilizing comm nity maintains a low hydrogen partial pressure, the flu of carbon and electrons goes nearly exclusive throu the "outer" parts of the flow scheme (Fig. 30.**20**), ar reduced fermentation intermediates play only a min role. Nevertheless, the flux through the "central" inte mediates will never be zero because fatty acids a concomitantly formed from fermentation of lipids ar amino acids. The central intermediates become mo important if the hydrogen pool increases for any reaso such as when there is a supply of excess fermentab substrate or when the activity of hydrogenotroph methanogens decreases due to a drop in pH (<6.0) or toxic compounds. Under such conditions, the pools fatty acids will increase and might even shift the p further downwards, with the consequence that hydr genotrophic methanogens will be entirely inactive ar the whole system "turns over." This phenomenon frequently encountered with poorly balanced sewa digesters (see Chapter 36).

The **hydrogen/formate-utilizing methanogens** act as regulators in the total methanogenic conversion process, although they are among the last actors in the play, they conduct the whole process to a maximum of energetic efficiency.

The function of homoacetogenic bacteria (Fi 30.**20**, group 5) in the overall process is less we understood. They connect the pool of one-carbc compounds and hydrogen with that of acetate. Due their metabolic versatility, they can also participate sugar fermentation and in the degradation of speci substrates such as *N*-methyl compounds or methox lated phenols. In certain environments, e.g., at low pH low temperature, they may even successfully compe with hydrogenotrophic methanogens and take ov their function to a varying extent (see Chapter 31.2).

30.7.2 Interspecies Hydrogen Transfer Connects Metabolically Different Bacteria in Syntrophic Associations

Other than "*Methanobacillus omelianskii*," other bacter such as *Thermoanaerobium brockii*, some *Pelobact* species, and ethanol-oxidizing sulfate reducers ca oxidize ethanol in the absence of sulfate by hydroge transfer to a hydrogen-oxidizing methanogenic partn bacterium. Some methanogens can also oxidize ethan directly, without cooperation with syntrophic partner

Table 30.7 **Changes in Gibbs free energies under standard conditions in hydrogen-releasing and hydrogen-consuming reactions.** Representative species of syntrophically fermenting bacteria are mentioned with the respective reactions. All calculations are based on published tables. For H_2S and CO_2, values for the gaseous state were used

	$\Delta G^{\circ\prime}$ (kJ mol^{-1})	Representative bacterial species
Hydrogen-releasing reactions		
Primary alcohols		
$CH_3CH_2OH + H_2O \rightarrow CH_3COO^- + H^+ + 2\ H_2$	+9.6	S-strain, *Desulfovibrio vulgaris* *Pelobacter acetylenicus*
Fatty acids		
$CH_3CH_2CH_2COO^- + 2\ H_2O \rightarrow 2\ CH_3COO^- + H^+ + 2\ H_2$	+48.2	*Syntrophomonas wolfei, S. sapovorans*
$CH_3CH_2COO^- + 2\ H_2O \rightarrow CH_3COO^- + CO_2 + 3\ H_2$	+76.0	*Syntrophobacter wolinii*
$CH_3COO^- + H^+ + 2\ H_2O \rightarrow 2\ CO_2 + 4\ H_2$	+94.9	Strain AOR
Aromatic compounds		
$C_6H_5COO^- + 6\ H_2O \rightarrow 3CH_3COO^- + 2\ H^+ + CO_2 + 3\ H_2$	+49.5	*Syntrophus buswellii*
Amino acids		
$CH_3CH(NH_3^+)COO^- + 2\ H_2O \rightarrow CH_3COO^- + NH_4^+ + CO_2 + 2\ H_2$	+2.7	*Eubacterium acidaminophilum*
Hydrogen-consuming reactions		
$4\ H_2 + 2\ CO_2 \rightarrow CH_3COO^- + H^+ + 2\ H_2O$	−94.9	*Acetobacterium woodii*
$4\ H_2 + CO_2 \rightarrow CH_4 + 2\ H_2O$	−131.0	*Methanospirillum hungatei*
$H_2 + S^0 \rightarrow H_2S$	−33.9	*Wolinella succinogenes*
$4\ H_2 + SO_4^{2-} + H^+ \rightarrow HS + 4\ H_2O$	−151.0	*Desulfovibrio vulgaris*
$H_2C(NH_3^+)COO^- + H_2 \rightarrow CH_3COO^- + NH_4^+$	−78.0	*Eubacterium acidaminophilum*
$Fumarate^{2-} + H_2 \rightarrow Succinate^{2-}$	−86.0	*Wolinella succinogenes*

but they do so very slowly and apparently do not compete successfully against syntrophic associations.

Similar cooperations have been described with syntrophic cultures that degrade fatty acids. An overview of the hydrogen-releasing reactions catalyzed is presented in Table 30.**7**, together with names of syntrophic fermentative bacteria. In general, degradation of fatty acids to acetate and hydrogen or, in the case of propionate, to acetate, hydrogen, and CO_2, are reactions far more endergonic under standard conditions than ethanol oxidation, and the hydrogen partial pressure has to be lowered to substantially lower values ($< 10^{-4}$ bar; 10 Pa) to allow substrate degradation and energy conservation. Syntrophic conversion of acetate to 2 CO_2 and 4 H_2 is catalyzed by a moderately thermophilic (58 °C) homoacetogenic bacterium, strain AOR, which can run acetate oxidation or acetate synthesis in both directions, depending on the external hydrogen concentration. This example illustrates how close to the thermodynamic equilibrium an anaerobic energy metabolism can operate. Also aromatic compounds and amino acids can be oxidatively converted to acetate and CO_2 (and NH_4^+) with concomitant interspecies hydrogen transfer to methanogenic partner bacteria.

In hydrogen-consuming reactions, the function of methanogens can be taken over by homoacetogenic, sulfur-reducing, sulfate-reducing, glycine-reducing, or fumarate-reducing bacteria (Tab. 30.**7**). Thus, also the Stickland fermentation of pairs of amino acids can be uncoupled and be carried out by two partner bacteria cooperating in interspecies hydrogen transfer. One partner oxidizes, for example, alanine to acetate, CO_2, NH_4^+, and hydrogen, and the other one uses hydrogen, for example, for glycine reduction to acetate (see Fig. 12.**32**).

30.7.3 Syntrophic Fatty Acid Oxidation Requires Sharing of ATP Units

The energetic situation of the partner bacteria involved in butyrate conversion to methane and CO_2 is illustrated in Figure 30.**21**. The overall reaction

$$2\ CH_3CH_2CH_2COO^- + 2\ H^+ + 2\ H_2O \rightarrow 5\ CH_4 + 3\ CO_2 \qquad (30.14)$$

yields under standard conditions a $\Delta G^{\circ\prime}$ of − 177 kJ per 2 mol butyrate. With concentrations more similar to

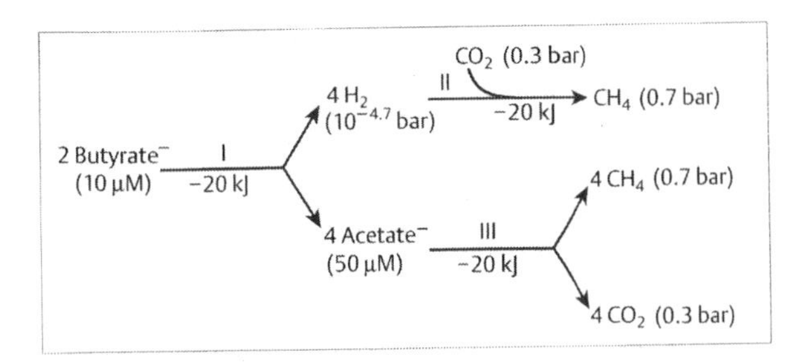

Fig. 30.**21** **Scheme of energy sharing among the three metabolic groups of bacteria** (I, II, III) cooperating in syntrophic conversion of butyrate to methane and carbon dioxide

those in a natural habitat, e.g., a freshwater sediment or a sewage sludge digestor [butyrate, 10 μM; CH_4, 0.7 bar (70 kPa); CO_2, 0.3 bar (30 kPa)], the free energy of this process is −140 kJ per 2 mol butyrate. If the reaction steps involved share the available energy in equal parts, the free energy change is about −20 kJ for every mol partial reaction. With these free energy values, the corresponding concentrations of the fermentation intermediates can be calculated and yields $10^{-4.7}$ bar for hydrogen and 50 μM for acetate. These values agree rather well with those measured in digesting sludge. A similar model can be established also for syntrophic propionate degradation, and again the amount of free energy available to the partner bacteria is in the range of 20–22 kJ per mol partial reaction.

The change in free energy of the total conversion of fatty acids to methane and CO_2 is shared by all partial reactions in approximately equal amounts. This amount is in the range of the minimum energy quantum (about −20 kJ per mol reaction, corresponding to 1/3 ATP equivalent; see Chapter 12.2), which can be exploited for ATP formation.

30.7.4 Interspecies Transfer of Formate or Acetate

Although hydrogen appears to be an ideal carrier for electrons between bacteria of different metabolic types, sometimes also **formate** acts as a carrier. The standard redox potential of the CO_2/formate couple is nearly identical to that of H^+/H_2 (−420 vs −414 mV), and therefore the energetic problems are about the same with both. In the laboratory, syntrophic mixed cultures can be composed that can use either only the formate or only the hydrogen transfer system. In the natural system, both hydrogen and formate may be used as carriers simultaneously.

Also **acetate** is excreted by syntrophic fermentative bacteria and is further metabolized by methanogens. The model introduced above shows that also acetate removal has a profound influence on the total energetics of, for example, butyrate fermentation. In syntrophic conversion of acetone to methane and CO_2, acetate is the only intermediate between both partners that needs to be consumed by the methanogenic partner to avoid inhibition of the primary fermentative bacterium.

Syntrophic associations of two to three microorganisms cooperate in methanogenic degradation of complex organic matter by transfer of hydrogen, formate, or acetate at low concentrations. This allows every partner to gain energy of at least −20 kJ per mol reaction run, which is equivalent to 1/3 ATP unit. All three intermediates may be transferred simultaneously in a certain degradation process, or any one of them may dominate, depending on the type of substrate degraded.

30.7.5 Methane Is Formed in Different Environments

The most important sites of methanogenesis are freshwater sediments and wetlands, anaerobic digestors for technical sewage treatment, and the rumen of ruminants. These environments differ in substrate supply rates, operating temperatures, and turnover times, as summarized in Table 30.**8**. Sediments (see Chapter 31.2) have slow substrate input and operate at low temperature, but have basically unlimited amounts of time available for total conversion of biomass to methane and carbon dioxide. Sewage digestors (Chapter 36) receive high supplies of predigested substrates, which are converted incompletely to methane and carbon dioxide within a limited amount of time (usually 2–4 weeks). Consequently, the concentrations of hydrogen, acetate, and longer fatty acids are slightly higher than in sediments, but still low enough to allow syntrophic degradation of fatty acids. The rumen (Chapter 31.5.1) receives periodically large amounts of easily degradable food, which undergoes fast fermentation to fatty acids and hydrogen at elevated temperature. The average detention time of rumen contents is in the range of 0.5–2 days, which is too short to establish acetate-cleaving methanogens or syntrophic fatty-acid-degrading fermentative communities. Therefore, methane is formed exclusively from one-carbon compounds and hydrogen. Acetate, propionate, and butyrate accumulate to high concentrations and are taken up by the rumen mucosa of the host organism. This system is therefore not optimized towards the most efficient conversion of

Table 30.8 **Characteristics of various methanogenic environments**

	Average temperature (°C)	Average detention time (d)	Concentration of			
			Acetate (μM);	Propionate (μM)	Butyrate (μM)	H_2 (Pa)
Eutrophic freshwater sediment	4–10	"∞"	0.5–200	0.1–20	0.1–10	0.5–5
Sewage digestor	30–35	15–30	5–6000	1–500	1–500	1–10
Rumen	37–39	0.5–2	60 000	20 000	10 000	20–5000

organic matter to methane and carbon dioxide, but towards an efficient fermentation of carbohydrates to fatty acids, in order to ensure optimal feeding of the animal host.

30.7.6 Methanogens Can Cooperate Also With Protozoa

The function of the primary fermentative bacteria (group 1 in Fig. 30.**20**) in conversion of complex organic matter to methane and CO_2 may be taken over also by strictly anaerobic protozoa. Anaerobic fungi, ciliates, and flagellates are known that thrive in entirely anoxic environments under reducing conditions, and some of them are extremely oxygen-sensitive. Since aerobic respiration is not possible in such habitats, anaerobic protozoa do not contain mitochondria. Instead, intracellular organelles called **hydrogenosomes** are present that release hydrogen. The metabolism of these protozoa is fermentative; particles, especially bacterial cells, are ingested into food vacuoles and digested by hydrolysis and fermentation.

Anaerobic protozoa can be associated with symbiotic methanogens, either extracellularly or intracellularly. Rumen ciliates usually carry extracellular hydrogenotrophic methanogens, which can be recognized optically due to their autofluorescence (factor F_{420}). These methanogens appear to be tightly associated with the cell surface at times of low substrate supply and are released after feeding. Ciliates living in strictly anoxic, eutrophic sediments carry methanogenic partner bacteria even inside the cell, quite often closely associated with the hydrogenosomes. Removal of hydrogen and maintenance of a low hydrogen/formate concentration in the cell allows fermentation of complex organic matter mainly to acetate and CO_2. Thus, waste of organic precursors into reduced end products such as ethanol and fatty acids can be avoided, and the fermentative protozoa obtains a maximum ATP yield by fermentation (see Chapter 30.7.1).

The symbiotic methanogen takes over part of the function that mitochondria have in aerobic protozoa. Reducing equivalents are removed by the symbiotic partner, and the eukaryotic host cell runs a fermentative metabolism with a maximum ATP yield. Feeding experiments have documented that methanogen-containing ciliates grow at rates and yields up to 35% higher than those of partner-free control cultures. The methanogenic bacteria can make up as much as 10% of the total biovolume of the ciliate cell.

It is assumed that the hydrogen released by hydrogenosomes stems mainly from pyruvate oxidation to acetyl CoA (pyruvate synthase reaction). Hydrogenosomes contain this enzyme, as well as ferredoxin and hydrogenase (Fig. 30.**22**). Among these strictly anaerobic ciliates, mainly *Trimyema compressum*, *Metopus striatus*, *M. palaeformis*, *Plagiopyla nasuta*, and *P. frontata* have been studied in detail. *Methanobacterium formicicum* and *Methanoplanus endosymbiosus* have, among others, been identified as their intracellular

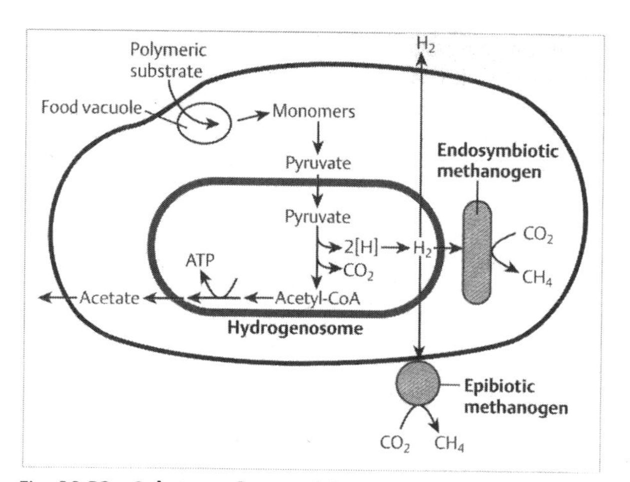

Fig. 30.**22** **Substrate flow and function of the hydrogenosome in a strictly anaerobic ciliate cell** cooperating with either endosymbiotic or epibiotic methanogenic partner bacteria

endosymbionts. Hydrogenosome and methanogenic endosymbiont together form a function entity that replaces mitochondria to some extent in an oxygen-free habitat. In some cases, especially with the comparably large ciliates, such as *Plagiopyla frontata*, hydrogenosomes and methanogens are organized in an alternating sandwich arrangement that allows optimal hydrogen transfer. It has been speculated that hydrogenosomes of strictly anaerobic protozoa have evolved from mitochondria of their aerobic predecessors; other speculations assume a relationship of hydrogenosomes to clostridia. Detailed studies on the cooperation of methanogenic endosymbionts with their protozoan hosts have been hampered by extreme difficulties in handling defined cultures of these protozoa. The protozoa usually depend on living bacteria as food, and only one defined co-culture (axenic culture) of *Trimyema compressum* has been described so far. The methanogenic partner bacteria appear in most cases to be freely suspended within the protozoan cytoplasm; they are typically not surrounded by membranes. Every species of protozoa contains only one type of fluorescent methanogenic endosymbiont. Obviously, the endosymbionts do not derive from methanogens taken up accidently with the food. In some cases, phototrophic or sulfate-reducing bacteria have also been identified inside the cytoplasm of protozoa. It should be mentioned that many other protozoa also contain bacterial endosymbionts, the function of which is entirely unknown.

Many strictly anaerobic protozoa carry **symbiotic methanogens**, either on their surface or inside associated with hydrogenosomes. The symbiotic methanogen takes over part of the function that mitochondria have in aerobic protozoa. Reducing equivalents are removed by the symbiotic partner, and the eukaryotic host cell runs a fermentative metabolism with a maximum ATP yield.

30.7.7 Syntrophy With Sulfur as an Intermediate Between Chemotrophs and Phototrophs

Green sulfur bacteria of the family Chlorobiaceae are known to oxidize H_2S to sulfate, with intermediate accumulation of elemental sulfur, which is stored outside the cell. Enrichment cultures aiming at the isolation of green sulfur bacteria using organic substrates have led to the discovery of a sulfur-coupled phototrophic cooperation ("*Chloropseudomonas ethylica*") in which a sulfur-reducing heterotrophic bacterium, *Desulfuromonas acetoxidans*, oxidizes an organic substrate (e.g., acetate) and reduces sulfur to H_2S, thus restoring the electron needs of the phototroph (Fig. 30.**23**). Thus, sulfur acts as electron carrier between both partners, and is recycled at a high rate. The cooperation becomes closer the smaller the sulfur pool is. Feeding experiments have shown that the total cell yield obtained by the mixed culture exceeds those of the two pure cultures dramatically. The example also shows that storage of sulfur outside the cell (as opposed to the red phototrophic Chromatiaceae, which store sulfur inside the cell) offers the advantage of cooperation with partner bacteria.

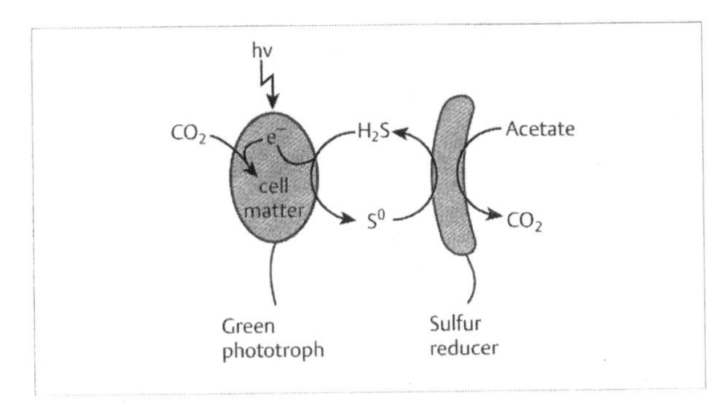

Fig. 30.**23** **Sulfur-based syntrophic cooperation between a green phototrophic bacterium and a chemotrophic sulfur-reducing bacterium**

This cooperation between phototrophic and chemotrophic bacteria has even developed into structured units called **consortia**. Careful microscopic examination of water samples from the redox transition zone of shallow eutrophic lakes and ditches quite often reveals small, irregularly shaped green organisms, which only after mechanical disintegration (e.g., by squeezing) disclose that they are composed of at least two different types of bacteria. A central colorless, rod-shaped bacterium is surrounded by several cells of green or brown phototrophic bacteria attached to the central bacterium's surface. Although the partner bacteria in these consortia have never been brought into defined cultures, they carry names from their original description, such as *Chlorochromatium aggregatum* or *Pelochromatium* spec. (Fig. 30.**24**). It is assumed that the central bacterium is a chemotrophic bacterium that reduces sulfate or sulfur to H_2S, and thus supplies the phototrophic epibionts with their electron source. The immotile green epibionts take advantage of the motility of the central partner, and the whole aggregate even shows a phototactic response, which suggests that some kind of signaling occurs between the two types of bacterial cells.

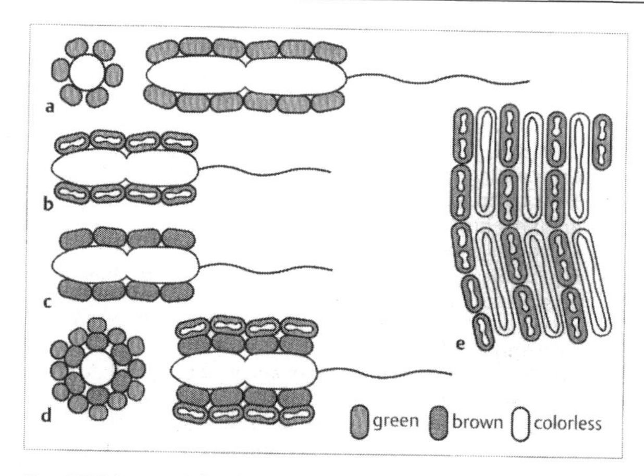

Fig. 30.**24a–e** **Schematic representation of different symbiotic associations of phototrophic green or brown sulfur bacteria with unknown chemotrophic bacteria (colorless).**
a *Chlorochromatium aggregatum*;
b *Chlorochromatium glebulum*,
c *Pelochromatium roseum*;
d *Pelochromatium roseo-viride*;
e *Chloroplana vacuolata*. **a–d** show longitudinal and cross sections; **e** top view [after 11]

30.7.8 There Are Many Other Forms of Cooperation

Enrichment cultures for methane-oxidizing aerobic bacteria usually also contain methanol-oxidizing bacteria, and separation of the two is quite difficult, especially when high substrate concentrations are used. Methane-oxidizing aerobes such as *Methylosinus trichosporium* and *Methylococcus* species are often associated with *Hyphomicrobium* species, an efficient methanol oxidizer. The coculture of both types grows much better on methane than the pure culture of the methane oxidizer does. Obviously, the methanol-oxidizing *Hyphomicrobium* removes toxic amounts of methanol (and/or formaldehyde) that are excreted by the methane oxidizer upon exposure to excessive substrate concentrations.

Filamentous cyanobacteria growing under limitation of bound nitrogen develop around their heterocysts substantial associations of chemotrophic bacteria. Among these are heterotrophic as well as autotrophic hydrogen-oxidizing bacteria, which thrive on organic excretions of the heterocyst and H_2, and protect the oxygen-sensitive nitrogenase system of the heterocyst by extensive respiratory activity.

Many more examples of this kind could be given, and in natural environments cooperation between different metabolic types of microorganisms may be much more the rule than the exception. The study of separate single organisms has quite often obscured such cooperative activities.

Further Reading

Akkermans, A. D. L., van Elsas, J. D., and de Bruijn, F. J. (1995) Molecular microbial ecology manual. Dordrecht: Kluwer

Amann, R., Ludwig, W., and Schleifer, K. H. (1995) Phylogenetic identification and in situ detection of individual microbial cells without cultivation. Microbiol Rev 59: 143–169

Atlas, R. M., and Bartha, R. (1993) Microbial ecology, fundamentals and applications, 3rd edn. Menlo Park, Calif: Benjamin/Cummings

Brock, T. D. (1966) Principles of microbial ecology. Englewood Cliffs, J: Prentice-Hall

Costerton, J. W., Lewandowski, Z., Caldwell, D. E., Korber, D. R., and Lappin-Scott, H. M. (1995) Microbial biofilms. Annu Rev Microbiol 49: 711–745

Egli, T. (1995) The ecological and physiological significance of the growth of heterotrophic microorganisms with mixtures of substrates. In: Jones, J. G. (ed.) Advances in microbial ecology, vol. 14. New York: Plenum Press; 305–386

Fenchel, T., and Blackburn, T. H. (1979) Bacteria and mineral cycling. London New York: Academic Press

Ferry, J. G. (1993) Methanogenesis. Ecology, physiology, biochemistry and genetics. New York: Chapman & Hall

Finlay, B. J. and Fenchel, T. (1992) Methanogens and other bacteria as symbionts of free-living anaerobic ciliates. Symbiosis 14: 375–390

Fletcher, M., and Marshall, K. C. (1992) Are solid surfaces of ecological significance to aquatic bacteria? In: Marshall, K. C. (ed.) Advances in microbial ecology, vol. 6. New York: Plenum Press; 199–236

Geesey, G. G., Lewandowski, Z., and Flemming, H.-C. (1994) Biofouling and biocorrosion in industrial water systems. Boca Raton Ann Arbor London Tokyo: Lewis

Ghiorse, W. C. (1984) Biology of iron-and manganese-depositing bacteria. Annu Rev Microbiol 38: 515–50

Gottschal, J. C. (1993) Growth kinetics and competition: some contemporary comments. Antonie van Leeuwenhoek 63: 299–313

Grigorova, R., and Norris, J. R., eds (1990) Methods in microbiology, vol. 22: Techniques in microbial ecology. London New York: Academic Press

Kjelleberg, S. (1993) Starvation in bacteria. New York London: Plenum Press

Lawrence, J. R., Korber, D. R., Wolfaardt, G. M., and Caldwell, D. E. (1995) Behavioral strategies of surface-colonizing bacteria. In: Jones, J. G. (ed.) Advances in microbial ecology, vol 14. New York London: Plenum Press; 1–75

Leadbetter, E. R., and Poindexter, J. S., eds. (1985) Bacteria in nature, vols. 1–3. New York London: Plenum Press

Lovley, D. R. (1993) Dissimilatory metal reduction. Annu Rev Microbiol 47: 263–290

Lynch, J. M., and Hobbie, J. E. (1988) Microorganisms in action: concepts and applications in microbial ecology. Oxford London: Blackwell

Mitchell, R., ed. (1992) Environmental microbiology. New York: Wiley

Müller, M. (1988) Energy metabolism of protozoa without mitochondria. Annu Rev Microbiol 42: 465–488

Neu, T. R. (1996) Significance of bacterial surface-active compounds in interaction of bacteria with interfaces. Microbiol Rev 60: 151–166

Poindexter, J. S. (1981) Oligotrophy. Fast and famine existence. Adv Microbial Ecol 5: 63–89

Schink, B. (1991) Syntrophism among prokaryotes. In: Balows, A., Trüper, H. G., Dworkin, M., and Schleifer, K. H. (eds) The prokaryotes, 2nd edn. Berlin Heidelberg New York: Springer; 276–299

Schwarzenbach, R. P., Gschwend, P. M., and Imboden, D. M. (1993) Environmental organic chemistry. New York: Wiley

Schwertmann, U., and Cornell, R. M. (1991) Iron oxides in the laboratory. Weinheim: VCH

Stumm, W., and Morgan, J. J. (1981) Aquatic chemistry, 2nd edn. New York: Wiley

Thauer, R. K., Jungermann, K., and Decker, K. (1977) Energy conservation in chemotrophic anaerobic bacteria. Bacteriol Rev 41: 100–180

Wetzel, R. G. (1983) Limnology, 2nd edn. Philadelphia: Saunders

Widdel, F. (1988) Microbiology and ecology of sulfate- and sulfur-reducing bacteria. In: Zehnder, A. J. B. (ed.) Biology of anaerobic microorganisms. New York: Wiley; 469–585

Wimpenny, J. W. T. (1993) Microbial systems: patterns in time and space. Adv Microbial Ecol 12: 469–522

Sources of Figures

1 Wetzel, R. G. (1983) Limnology, 2nd edn. Philadelphia: Saunders; 46

2 Wetzel, R. G. (1983) Limnology, 2nd edn. Philadelphia: Saunders; 56

3 Widdel, F., Schnell, S., Heising, S., Ehenreich, A., Assmus, B., and Schink, B. (1993) Ferrous iron oxidation by anoxygenic phototrophic bacteria. Nature 362: 834–836

4 Östling, J., Homquist, L., Flärdh, K., Svenblad, B., Jouper-Jan, A., and Kjelleberg, S. (1993) Starvation and recovery of *Vibrio*. In: Kjelleberg, S. (ed.) Starvation in bacteria. New York London: Plenum Press; 103–127

5 Gottschal, J. C., de Vries, S., and Kuenen, J. G. (1979) Competition between the facultatively chemolithotrophic *Thiobacillus* A2, an obligately chemolithotrophic *Thiobacillus* and a heterotrophic *Spirillum* for inorganic and organic substrates. Arch Microbiol 121: 241–249

6 Schink, B. (1988) Principles and limits of anaerobic degradation: environmental and technological aspects. In: Zehnder A. J. B. (ed.) Biology of anaerobic microorganisms. New York: Wiley; 771–846

7 Widdel, F. (1988) Microbiology and ecology of sulfate- and sulfur-reducing bacteria. In Zehnder, A. J. B. (ed.) Biology of anaerobic microorganisms. New York: Wiley; 469–585

8 Revsbech, N P., Jørgensen, B. B., and Blackburn, T. H. (1980) Oxygen in the sea bottom measured with a microelectrode. Science 207: 1355–1356

9 Nelson, D. C., Jørgensen, B. B., and Revsbech, N. P. (1986) Growth pattern and yield of a chemoautotrophic *Beggiatoa* sp. in oxygen-sulfide microgradients. Appl Environ Microbiol 52: 225–233

10 Jørgensen, B. B., and Revsbech, N. P. (1983) Colorless sulfur bacteria, *Beggiatoa* spp. and *Thiovulum* spp., in O_2 and H_2S microgradients. Appl Environ Microbiol 45: 1261–1270

11 Pfennig, N. (1980) Syntrophic mixed cultures and symbiotic consortia with phototrophic bacteria: a review. In: Gottschalk, G., Pfennig, N., and Werner, H. (eds.) Anaerobes and anaerobic infections. Stuttgart New York: Fischer, 127–131

31 Habitats of Prokaryotes

In the previous chapter, an attempt was made to understand the behaviour of microorganisms in natural ecosystems in general from the organisms' point of view—how they interact with the abiotic and biotic conditions they find. In this chapter, the habitats themselves, the conditions they offer, the way they are shaped by microbial activities, and how microbial and macrobial communities interact to maintain the respective ecosystem's stability will be discussed.

31.1 Water Is a Perfect Basis for Microbial Life

Water is the most important chemical compound for life on earth, and life originally evolved in an aquatic environment. Organisms living in terrestrial environments have to protect themselves from dessication, especially in extremely dry environments. Organisms living inside major bodies of water are protected not only from dessication, but also from abrupt changes in temperature, due to the high thermal capacity of water. Temperatures in aquatic environments generally change by at most 20–30 °C over the year, whereas a surface soil in a temperate climate region can easily change from −20 to +60 °C. Moreover, water acts as a good solvent for most biologically important compounds, including O_2, CO_2, inorganic salts, and organic substrates. As a consequence, aquatic environments are spatially far less heterogeneous than soils.

The majority of water on Earth is saltwater. **Freshwater** makes up only a small fraction, and a minor proportion undergoes exchange throught the atmosphere (Tab. 31.**1**). Most of the freshwater on the Earth's surface is bound in the polar icecaps, which exchange only very slowly with the atmospheric water budget. The pools of active groundwater, of inland lakes (of which freshwater lakes make up about 50% of all lakes), soil moisture, atmospheric water, and rivers are biologically relevant. Table 31.**1** shows the pool sizes of these various water fractions and their respective renewal (turnover) times. Whereas the small pools of atmospheric water and rivers have average renewal times in the range of 10–11 days, the renewal times of the other pools increase dramatically with their size.

Although freshwater lakes are only of minor importance for the total global water budget, they are among the best-studied ecosystems. Limnology, the science dealing with the study of surface and subsurface freshwaters, developed at the end of the 19th Century and represents perhaps the most advanced part of habitat-oriented ecology. Lakes are comparatively

Table 31.**1** **Distribution of water on Earth**

	Volume (10^3 km^3)	% of total water	% of freshwater	Renewal time (years)
Total	1 458 703	100	–	–
Oceans	1 370 373	94	–	3 000
Deep groundwater	60 000	4	–	5 000
Total accessible freshwater	28 329	1.94	100	–
Polar ice	24 000	1.64	84.5	8 000
Active groundwater	4 000	0.27	14	330
Freshwater lakes	125	0.0086	0.4	1–100
Saline lakes	104	0.0071	0.36	10–1 000
Soil moisture	85	0.0058	0.3	1
Atmospheric water vapor	14	0.00096	0.05	0.027 (10 days)
Rivers	1.2	0.00008	0.004	0.031 (11 days)

well-ordered research objects that are easy to sample and allow physicochemical characterizations with a reasonable amount of effort. The study of oceans requires far more refined techniques and higher expenditure due to their enormous size.

Nevertheless, classical limnological research has entirely overlooked the function of bacteria within the lake ecosystem. Limnologists always used microscopes that did not allow the visualization of bacterial cells because the microscopes had insufficient optical resolution and contrast (bright-field illumination) and also because aquatic bacteria tend to attach to surfaces and are difficult to recognize (see Chapter 30.5). Aquatic microbiology therefore concentrated for a long time on counting bacteria by plate cultivation with media that allowed growth of only a small fraction of the total population present. Only recently have new techniques for specific visualization of microbial cells (see Chapter 30.3.5) combined with sensitive techniques for activity assays in natural samples (see Chapter 30.4.5) shown that microorganisms, especially bacteria, are of eminent importance for nutrient cycling and maintenance of food resources and energy in a lake. Microorganisms make up the largest fraction of biomass in the water column and can no longer be overlooked in their function in the entire ecosystem.

31.1.1 Water Has Unusual Physical and Chemical Properties

The angular shape of the water molecule and the resulting dipole character causes several unusual properties of water. Water is by far the smallest chemical compound that is found at ambient temperature in a liquid state, which occurs because the single molecules interact strongly with each other through hydrogen bridges. In the liquid state, molecules form clusters of 12–70 units, depending on temperature, the size of these aggregates increases with decreasing temperature. Formation and segregation of these clusters releases and binds thermal energy, which causes the unusually high thermal capacity of water. The thermal capacity of pure water is 4.19 kJ kg^{-1} degree^{-1} at 15 °C; solid ice has a thermal capacity of 2.04 kJ kg^{-1} degree^{-1} (see Box 31.1).

In the solid state, water molecules arrange themselves in a crystalline structure based on a tetrahedral arrangement, which leaves ample empty space between the molecules. The density of ice (0.9168 kg l^{-1}) is therefore considerably lower than that of liquid water at 0 °C (0.99987 kg l^{-1}). During the melting process, the crystal structure changes gradually to the cluster structure and reaches a density maximum at 3.94 °C. This **density anomaly** of water is the reason why lakes in their deeper parts usually do not cool down lower than 4 °C, and is of eminent importance for life in lakes in colder climates because it protects the lower water layers from freezing. The **thermal conductivity** of liquid water is very low (0.0057 J cm^{-1} s^{-1} degree^{-1}; transport of heat through a stagnant water body therefore depends nearly exclusively on convection processes.

The high polarity of water makes it an excellent solvent for polar compounds such as salts and CO_2, whereas non-polar gases such as N_2, CH_4, and O_2 are only poorly soluble (see Chapter 30.1.3). CO_2 is formed in the water by mineralization of organic carbon. CO_2 reacts with water to form HCO_3^- and H^+; the pK of this dissociation is 6.3 at 20 °C. The CO_2/HCO_3^- system (bicarbonate system) buffers the free water column and the sediment. The solution capacity for bicarbonate increases with the content of calcium, magnesium, and sodium ions in the water, calcium-rich, so-called **hard-water lakes** have a significantly higher buffering capacity (alkalinity) than **soft-water lakes**. Hard-water lakes are supplied with divalent cations mainly from calcareous rocks in their catchment area; soft-water lakes supply their water in areas dominated by igneous rocks, e.g., in Scandinavia, the Black Forest in Germany, and large parts of North America. The concentration of CO_2 dissolved in pure water in equilibrium with the atmospheric CO_2 content (30 Pa) is near15 μM. Therefore, at the surface of freshwater lakes, CO_2 is released into the atmosphere unless it is used up by photosynthesis (only during the day) or withheld in the water by alkalinity. Because of its high alkalinity, seawater maintains dissolved CO_2 concentrations of about 2 mM at its surface. The **electrical conductivity** of pure water is very low and increases proportionally with the concentration of dissolved ions. Electrical conductivity is therefore used as a measure of dissolved ions in the water, especially of dissolved bicarbonate and its counterions in freshwater lakes (alkalinity) and of salinity in oceans and salt lakes (Box 31.1).

Water is characterized by a high thermal capacity and stability, a density maximum at 3.94 °C, and a low thermal conductivity. Water is an excellent solvent for polar compounds

31.1.2 Heat Uptake Causes Water Stratification

In temperate climates, smaller bodies of water are mixed in the spring from the top to the bottom due to eddy convection, which is originally caused by wind activity. Temperature and dissolved compounds in the water are distributed evenly over the total water column. During the summer, the surface waters warm up, and a warmer surface water body (**epilimnion**) separates from a lower, colder water body (**hypolimnion**) due to thermal stratification. The transition zone between the two layers is called the **metalimnion** (Fig. 31.**1**). During the summer stratification period, the epilimnion and the hypolimnion exchange very little with each other. Heat release in the fall, intensified by wind-induced turbulence, leads again to a turnover that may mix the entire water body down to the sediment surface. In winter, the surface water may cool down and develop an ice cover, which protects and stabilizes an inverted winter stratification with cold water of 0–4 °C on top of a denser water column at 4 °C (Fig. 31.**1**). This annual pattern with complete turnovers in spring and fall is typical for small to medium-sized lakes in temperate climates, so-called **dimictic lakes**. In colder climates, summer stratification may not develop into a stable state, and the winter stratification is interrupted by only one mixing situation (**cold monomictic lake**). In polar regions, lakes never mix because the ice cover never melts (**amictic lakes**). In warmer climates, the winter stratification may not develop (**warm monomictic lakes**). Lake Constance, Southern Germany, is the northernmost representative of this type. In subtropic and tropic regions, there may be no annual mixing pattern but daily temperature changes may cause daily stratification and irregular (oligomictic lakes) or daily (polymictic lakes) and mixing at night, especially in shallow waters.

Mixing does not always impact the whole water column. Depending on the shape of the water body, especially in deep lakes with small surface areas, the mixing events may be incomplete and the lower part of the hypolimnion (**monimolimnion**) may never exchange with surface waters. Such lakes are called **meromictic**. The monimolimnion can accumulate

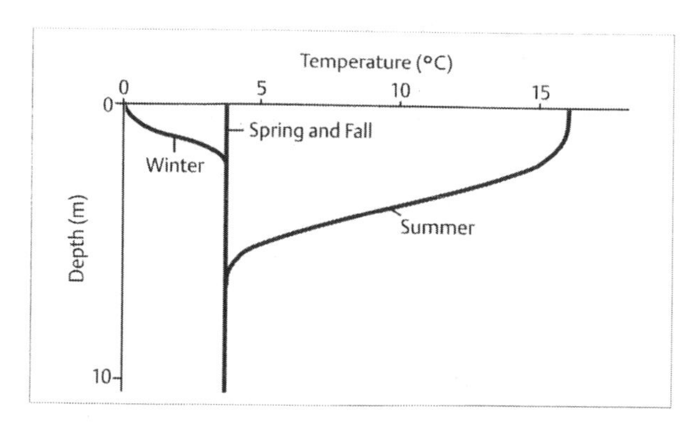

Fig. 31.1 **Temperature profiles in a medium-sized dimictic freshwater lake in a temperate region.** The depth scale varies with the total depth of the lake and its wind exposure

microbially produced gases (e.g., CO_2, CH_4, traces of H_2S) that are released to the atmosphere at irregular intervals, caused, for example, by seismic events. During such an event on Lake Nios, East Africa, in 1986, a total amount of $3 \cdot 10^7$ m^3 gas was released, which killed about 2 000 humans by suffocation and H_2S intoxication. Mixing of the water body is impaired especially by increased salt concentrations in the hypolimnion due to sub-surface uptake of brines or seawater as, for example, in fjords or the Gotland Basin in the Baltic Sea.

Box 31.1 Stagnant water bodies can be characterized physicochemically by parameters that are recorded largely automatically by multiple probing devices. **Temperature** profiles are taken as indicators of possible stratification situations. Highly sensitive temperature sensors can also provide information on horizontal mixing processes in a larger body of water. **Oxygen** is recorded by a Clarke-type electrode. **Electrical conductivity** gives a measure of total dissolved salts, especially bicarbonate, in freshwater lakes. Freshwaters exhibit conductivities in the range of 200–400 $\mu S\,cm^{-1}$, seawater around 50 000 $\mu S\,cm^{-1}$. **pH** measurements, together with electrical conductivity, provide measures of the alkalinity or buffer capacity of the body of water. Surface waters of eutrophic lakes exhibit pH regimes up to pH 9 due to photosynthetic CO_2 depletion. The pH in hypolimnetic waters is neutral to slightly acidic, down to pH 6.7 in hard-water lakes, but considerably lower in soft-water lakes with a high uptake of acidic compounds. **Light penetration** into the water body can be assessed by the Secchi disk. This is a reflecting round metal plate (30 cm in diam.) which is lowered into the water until it cannot be recognized anymore from the surface. The Secchi depth is an indication of the turbidity caused by microorganisms, especially algae, or other suspended organic or inorganic matter in surface waters. Also entirely pure water absorbs and refracts incident light to varying degree, depending on the wavelength (see Chapter 30.1.1); at 80-m depth, there is essentially no light anymore, even in an extremely clean body of water. Quantitative and qualitative changes in light climate can be assessed in the water column by specific light sensors. Various devices have been designed to take water samples free of external contamination from defined depths.

31.1.3 Primary Productivity Is Limited by Various Nutrients

The epilimnion is the site of photosynthetic primary production in lakes. In early spring when all basic nutrients such as silica, bound nitrogen, and phosphate are available, fast-growing diatoms dominate in the open water (**pelagial**) as well as on littoral rock and sediment surfaces. Diatoms and Chrysophyceae depend on silica as one essential nutrient. Once the silica resources have been used up, the population shifts to green algae as the dominant primary producers. The algal biomass produced contains the elements carbon, nitrogen, and phosphorus at a ratio of 106 : 16 : 1 (so-called "Redfield ratio"). The productivity of a body of water is expressed by the terms oligotrophic, mesotrophic, eutrophic, and hypereutrophic, with the terms, in the order written, signifying increasing amounts of nutrient supply. If a surface water is supplied with ample amounts of phosphate, as is typical of **eutrophic** or **hypereutrophic** lakes, bound nitrogen sources will soon become limiting and the conditions select for cyanobacteria such as *Anabaena, Aphanizomenon*, and *Microcystis* species, which can supply their own needs for bound nitrogen by nitrogenase-dependent nitrogen fixation. A mass development of cyanobacteria in surface waters is typically an indicator of a highly eutrophic situation. In freshwater lakes, the phosphate supply governs the extent of productivity (Tab. 31.**2**) beyond other factors such as climate exposure, light exposure, and water chemistry. Slightly acidic soft waters rich in humic material (fulvic acids), so-called **dystrophic lakes**, are significantly less productive than hard-water lakes. The productivity of off-shore oceans is in the same range as that of oligotrophic freshwater lakes, but depends strongly on nutrient supply through water currents. The productivity of coastal seawaters and off-shore upwelling areas supplied with nutrient-rich deep-sea water exhibit productivities in the range of that of mesotrophic to eutrophic lakes (Tab. 31.**2**). Littoral zones, which receive sunlight down to the sediment surface, are significantly more productive than off-shore pelagic bodies of water because nutrients stored in littoral sediments can be reutilized in short cycles. Nutrients buried in sediments of the deep parts of a lake (profundal) contribute to primary productivity only to a smaller degree, due to the thermal stratification and resulting transport impediment.

Table 31.**2** **Productivities of lakes and oceans**

	Total P ($\mu g\, l^{-1}$)	Productivity ($mg\, C\, m^{-2}\, year^{-1}$)
Oligotrophic lakes	< 10	10–200
Mesotrophic lakes	10–50	80–400
Eutrophic lakes	35–100	300–900
Offshore oceans	2.5	10–150
Coastal seawaters	> 10	> 300

Box 31.2 The **activity of primary producers** can be assayed either via CO_2 fixation or O_2 release. E. Steemann-Nielsen introduced in 1952 a technique that measures $^{14}CO_2$ incorporation into cellular biomass. This technique is still the method of choice today due to its high sensitivity. Glass bottles with lake water including plankton are supplied with a small amount of $NaH^{14}CO_3$ and incubated in situ in order to ensure the original temperature and radiation conditions. Dark controls for heterotrophic and chemolithotrophic CO_2 fixation are run in light-shielded bottles. After incubation, the water samples are filtered and the radioactivity in the particulate fraction on the filters is determined as a measure of photosynthetic activity. It should be emphasized that this method measures only the incorporation into particulate material. The dissolved organic excretions of algae (5–30% of total primary production) were overlooked for a long time. This experimental shortcoming explains also why limnologists call the activity of heterotrophic bacteria **secondary production**; dissolved carbon is converted to particulate carbon and becomes measurable as such. Nevertheless, this is a mineralization rather than a production process, and more than 50% of the organic matter is oxidized to CO_2.

31.1.4 Mineralization Activities by Chemotrophic Bacteria

The organic matter synthesized by photosynthetic phytoplankton acts as substrate for herbivorous zooplankton, which are consumed by predatory plankton, which are the food source of fish. This classical food chain (Fig. 31.**2**) may be much more complex, with several more levels of feeding guilds involved. Lysis of phytoplankton cells and release of food particles during "sloppy feeding" on all feeding levels form a pool of particulate and dissolved organic matter called **detritus**. Detritus is consumed by the pelagic heterotrophic bacteria, either directly or after primary depolymerization by extracellular depolymerases. This so-called **microbial loop** also receives support from dissolved

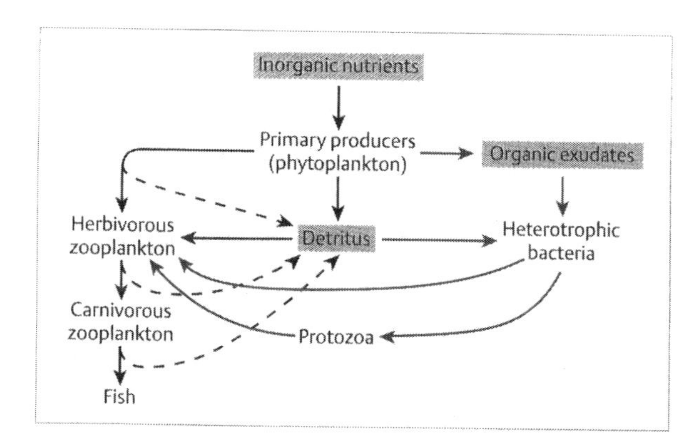

Fig. 31.2 **Transformations of inorganic and organic matter in a freshwater lake ecosystem.** Solid black lines represent the classical food chain from primary producers through secondary producers to fish. Dashed lines refer to the detritus food chain, and solid red lines represent the "microbial loop." Organic exudates and detritus are the main food source of the heterotrophic bacteria

organic exudates (glycolate, glycerol, etc.; 5–35% of total photosynthetic production), which are transformed into particulate bacterial biomass. Bacteria are grazed by herbivorous zooplankton or by protozoa (**detritus food chain**), and their cell material enters the classical food chain at the zooplankton level. The concept of the microbial loop and detritus food chain now attribute far more importance to the function of heterotrophic bacteria than in older views. Bacteria and protozoa are the most important actors in the food chain of a eutrophic lake, with respect to their metabolic activity and their total biomass. The relative importance of the various groups of organisms in a shallow eutrophic lake is comparable to that calculated for a coastal seawater environment (Tab. 31.**3**) in which bacteria and protozoa exhibit by far the highest energy turnover. Since the energy input into the system is a direct function of the surface area, such energy conversions are normalized on surface units, meaning that the activities are integrated over the total water and sediment column. The detritus food chain becomes important especially when only a minor part of organic production is oxidized in the water column and degradation in the sediment takes over a major part of the total mineralization.

In the trophogenic zone, heterotrophic bacteria compete successfully against photosynthetic primary producers for inorganic nutrients such as phosphate. Heterotrophic bacteria gain 50% and more of the total phosphate turnover under phosphate limitation, i.e., at free phosphate concentrations lower than 30 nM. Cyanobacteria and eukaryotic phytoplankton exhibit lower phosphate affinities and profit especially from phosphate pulses provided by lysing cells.

Temporal alterations in nutrient supply rates are among the reasons for the observed diversity of all trophic groups in natural ecosystems. Unlike a chemostat, which maintains stable supply conditions over unlimited times, even a comparably homogeneous ecosystem like a freshwater lake exhibits spatial inhomogeneities and temporal changes that periodically provide advantages for different species within the same feeding guild.

Dissolved organic matter in epilimnetic waters makes up 1–50 mg l^{-1}, particulate organic matter 0.1–40 mg l^{-1}, and both depend on the trophic state. Most of the dissolved detritus is polymeric in nature (fulvic acids, aquatic humic compounds), with turnover times in the range of years or decades. Concentrations of free amino acids in a mesotrophic freshwater lake epilimnion are in the range of 0.1–4 μM, with turnover times in the range of 10–1 000 h during the summer season. Similar values have been obtained for free sugars in the water, with turnover times of 1 d at 3 m depth and 8–25 days at 50 m depth. Cultivation of planktic bacteria (free-floating bacteria) from freshwaters by conven-

Table 31.**3** **Biomass and activities of trophic groups in a coastal seawater environment** [after Fenchel, T. M., and Jørgensen, B. B. (1977) Detritus food chains of aquatic ecosystems: the role of bacteria. In: Alexander, M. (ed.) Advances in microbial ecology, vol. 1, New York: Plenum Press; 1–58]

	Bacteria	Protozoa	Invertebrates	Fish
Total biomass ($g\,m^{-2}$)	2.5	5	0.5	0.5
Number of individuals (m^{-2})	$5 \cdot 10^{12}$	$5 \cdot 10^{7}$	$5 \cdot 10^{2}$	$5 \cdot 10^{-2}$
Individual biomass (g)	$5 \cdot 10^{-13}$	$1 \cdot 10^{-7}$	$1 \cdot 10^{-3}$	10
Metabolic rate ($kJ\,m^{-2}\,d^{-1}$)	21	2.1	0.21	0.021
Metabolic rate ($kJ\,g^{-1}\,d^{-1}$)	8.4	4.1	4.1	0.41

tional plating techniques (see Chapter 30.3.5) have led to isolates that could be attributed to the genera *Pseudomonas, Alcaligenes, Flavobacterium, Cytophaga,* and many others. As outlined in Chapter 30.3.5, these bacteria represent only a small fraction of the total prokaryotic community present. Cell densities assayed by direct counting after fluorescence staining range from 10^4 to 10^6 cells ml^{-1}. These numbers vary by one order of magnitude through the annual seasons and do not differ significantly between oligotrophic and eutrophic lakes. Obviously, the numbers are determined by the grazing efficiency of the zooplankton and protozoa ("**top-down control**") rather than by available food sources ("**bottom-up control**"). The ratio of freely suspended versus particle-associated bacteria varies between 0.3 and 3, with a higher tendency for aggregate formation in nutrient-poor lakes.

Particulate detritus, including remnants (e.g., fecal pellets) of planktic digestion (i.e., digestion by free-floating organisms), sediments slowly to the lake bottom and is degraded by microbial activities during this passage. This particulate organic matter tends to form fragile aggregates up to a few centimeters in diameter which move downwards slowly like snow flakes ("marine snow", "lake snow"). These aggregates exhibit enhanced microbial activity compared to that of the free water. Extracellular hydrolases for degradation of chitin, cellulose, hemicelluloses, proteins, etc. are associated with these particles, together with phosphate and bound nitrogen compounds at enhanced concentrations. The microbial communities attached to these aggregates profit significantly from nutrient enrichment around these snowflakes. Studies with fluorescently labeled nucleotide-specific probes have shown that Proteobacteria of the β- and γ-subclass dominate among the bacterial communities in lake snow in Lake Constance. β-subclass Proteobacteria also dominate in activated sludge, indicating that the strategy of surface attachment may lead to the selection of similar organisms in both cases. However, these flakes also enrich for bacteriophages. Whereas the ratio of free viruses to bacterial cells in free water is about 6:1, this ratio increases to 20–40:1 in lake snow particles, and the relatively high growth rates on these surfaces may be balanced to a significant part by bacteriophage-induced lysis.

Detritus particles, living zooplankton, and bacteria, either free or attached, settle to the bottom at various speeds through gravitation. The metalimnetic water density increase causes a transient particle accumulation, associated with increased degradative activities, and a similar effect is observed at the sediment surface. The sedimentation distance from the productive (**trophogenic**) epilimnetic water layers through the dark **tropholytic,** (dominating consumption) water layers to the sediment determines how far microbial degradation proceeds and how much organic matter reaches the sediment surface; the fraction of primary production reaching the sediment varies between 5 and 80%, depending on lake depth. Oxidation of organic matter causes oxygen consumption, especially in highly productive lakes, and can lead to complete oxygen depletion in hypolimnetic waters (Fig. 31.**3a**). Reduction of alternative electron acceptors such as nitrate, Fe (III), Mn (IV), and sulfate leads to accumulation of their reduced counterparts and, finally, methane is formed. The reduced reaction products are transported upwards through eddy convection and may be reoxidized either phototrophically (H_2S, Fe^{2+}) or by chemotrophic aerobic activities (CH_4 and NH_4^+, but also H_2S and Fe^{2+} in the dark), leading to characteristic distribution profiles along the water column during the summer stratification period (Fig. 31.**3a**). Oligotrophic lakes, on the other hand, show only few chemical changes along the water column (Fig. 31.**3b**). In such lakes, the sediment surface stays oxidized all through the year, and the redox processes discussed occur inside the sediment (see Chapter 31.2).

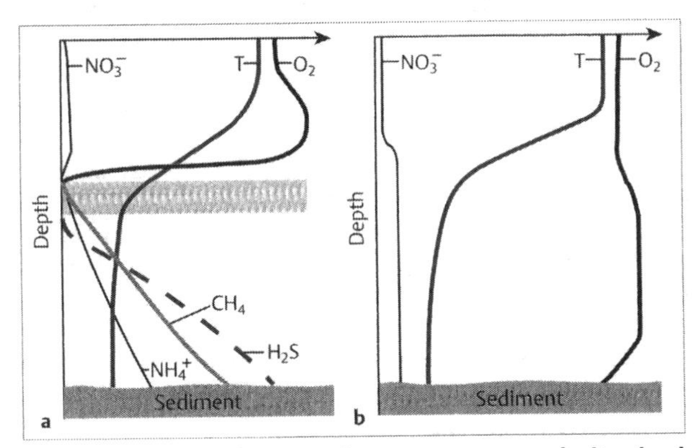

Fig. 31.**3a, b** **Distribution of temperature and dissolved compounds in a eutrophic and an oligotrophic lake during summer stratification in a temperate climate.** **a** In the eutrophic lake, oxygen shows an oversaturation below the water surface and disappears in the metalimnion. Gradients of CH_4 and NH_3 just reach the zone of oxygen disappearance. H_2S may be used up below the oxycline by phototrophic anoxygenic bacteria (light gray zone). NO_3^- appears only in the epilimnic water layers. **b** In the oligotrophic lake, the oxygen content increases in deeper water layers due to higher solubility at low temperature and may decline immediately above the sediment surface. Nitrate is the only dissolved nitrogen species and is consumed in the upper water layers by phytoplankton. Ammonia, methane, and H_2S are not found at significant concentrations. T, temperature; abscissa, concentrations

Aerobic chemotrophic oxidation of methane shows maximum activity in the metalimnetic transition zone where methane is available in the presence of limiting amounts of oxygen. Maximum activity is actually found in those water layers in which oxygen is barely measurable, indicating that these bacteria are really microaerophilic. The same applies to chemolithotrophic reoxidation of ammonia; actually, ammonia oxidation and methane oxidation may interfere with each other because the primary oxygenase enzymes in both cases can hydroxylate either substrate. Reoxidation of methane, ammonia, and other reduced products consumes oxygen in the metalimnetic boundary layer and helps to stabilize the oxygen distribution pattern in this transition zone.

Bacteria and protozoa are the predominant mineralizers of organic matter in lakes. The **microbial loop** and the **detritus food chain** gain importance especially in highly productive waters.

The metalimnetic water layer of a eutrophic lake represents a transition zone which is characterized by steep opposing gradients of reduced and oxidized inorganic and organic metabolites and by accumulation of particulate organic matter due to water density increase. It is a zone of enhanced dissimilatory microbial activity as visualized by increased carbon dioxide concentrations and decreased pH values. Reductive and oxidative microbial activities face each other at a short distance, and redox-active intermediates (nitrogen and sulfur compounds) may turn over in this layer at a high rate.

31.1.5 Anoxygenic Phototrophic Bacteria Reoxidize Reduced Sulfur Compounds

If the oxic/anoxic interface in eutrophic lakes is located close enough to the surface to receive traces of sunlight, sulfide from the hypolimnetic water layers will be mainly oxidized by anoxygenic phototrophic bacteria. These bacteria are adapted to low light intensities and are sensitive to high irradiation. They are active down to water depths of 10–15 m. In an extreme case, phototrophic sulfide oxidation was observed at the chemocline of the Black Sea at 80-m depth. Optimal development of phototrophic bacteria is observed at light intensities of 0.1–5% of the light intensity at the water surface. The process they catalyze is termed **secondary primary production** because they carry out a net photosynthesis from CO_2, but use electron sources that have been derived indirectly from activities of true primary producers (green algae, cyanobacteria, etc.).

Purple and green phototrophic sulfide oxidizers absorb light in those wavelength ranges that penetrate deepest into the water (Fig. 30.**2**) and that are not absorbed significantly by chlorophyll-*a*-containing eukaryotic algae or cyanobacteria. In Figure 31.**4**, absorption spectra of green algae, purple phototrophic bacteria and green phototrophs are compared. Since short-wavelength light shorter than 400 nm and long-wavelength light longer than 700 nm does not penetrate into water to greater depths, only the open "window"

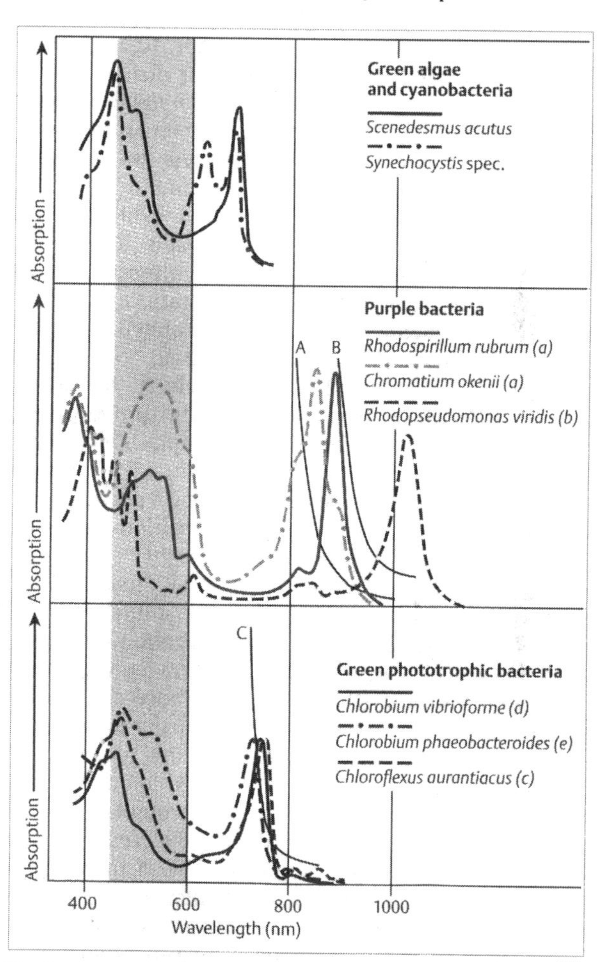

Fig. 31.**4** **Spectral absorption curves of green algae, cyanobacteria, purple bacteria, and green phototrophic bacteria.** The spectra are organized in three different layers as they are typically organized in structured aquatic ecosystems (see text). The types of characteristic major bacteriochlorophylls are indicated in parentheses (*a*, *b*, *c*, *d*, *e*). The absorption properties of infrared filters for selective enrichment of purple bacteria (A), purple bacteria with bacteriochlorophyll *b* (B), and purple and green bacteria (C) have been added for comparison [after 1]. See Figs. 13.**1**–13.**5**. The zone of maximal light absorption in a water layer is shown in light red

between 450 and 600 nm wavelength can be exploited by anoxygenic phototrophs under a layer of oxygenic primary producers. Both purple and green phototrophs exploit this wavelength range through their intense absorption by carotenoids. Mass development of red-pigmented representatives of the Chromatiaceae can stain the water layers below the metalimnetic interface intensely pink. Depending on the lake water chemistry, representatives of the species *Thiopedia, Amoebobacter*, or *Chromatium* dominate in such layers, often associated with syntrophic consortia of the "Pelochromatium" or "Chlorochromatium" type (see Chapter 30.7.7). Green phototrophic sulfur oxidizers find their niche in water layers below the purple bacteria; *Chlorobium limicola* and *Pelodictyon clathratiforme* may dominate. If both purple and green sulfur oxidizers occur in the same habitat, the green bacteria are always found below the layer of purple bacteria. Green bacteria exhibit higher tolerance to sulfide (< 0.5–4 mM) and are better adapted to exploitation of low light intensities than the purple bacteria due to a higher ratio of antenna bacteriochlorophyll to bacteriochlorophyll in the reaction centers (1 000 : 1 as opposed to 100 : 1; see Chapters 13 and 24). An extremely low-light-adapted phototroph from the Black Sea turned out to be a green sulfur-oxidizing phototroph (*Chlorobium phaeobacteroides*).

Development of sulfide-oxidizing phototrophic bacterial layers at the redox transition zone is also influenced by the water (and light) quality. In dystrophic lakes rich in humic material, green phototrophic bacteria dominate over purple sulfur oxidizers because they are more efficient in utilization of the red light that reaches the sulfidic water layers in these lakes.

Purple sulfur bacteria can either move actively by flagella or float by means of gas vacuoles in order to reach optimal positions in the opposing gradients of light and sulfide. Flotation in the water column by gas vesicles (see Fig. 2.**18a–c**) underlies a complex regulation that involves size control of the vesicles and aggregation and disaggregation of cells to increase or minimize the sedimentation velocity. Green sulfur bacteria are non-motile; some contain gas vacuoles. Therefore, the composed consortia "Chlorochromatium" and "Pelochromatium" (see Chapter 30.7.7) have a special advantage because this cooperation not only intensifies the substrate turnover, but also attributes motility and chemotactic orientation to the entire phototrophic aggregate.

Sulfide-oxidizing phototrophs accumulate below the thermally stabilized sulfide/oxygen interface to visible densities because only under these conditions is sulfide protected from chemical oxidation and light is provided in sufficient amounts. The purple non-sulfur phototrophs, on the other hand, do not exhibit preference for a certain location in the water column of a eutrophic lake. They develop to major numbers typically in small ditches, associated with decomposing particulate organic material at the sediment surface. In bigger lakes, they are found at low numbers all along the water column, even in the dark water layers. They do not require association with a certain water layer because the organic substrates they use are stable in the presence of oxygen. They compete successfully against anaerobic or aerobic chemotrophic bacteria in the light because phototrophy provides them with additional energy for more efficient substrate exploitation than the chemotrophic competitors. In the dark, they have no advantage over purely chemotrophic aerobes.

Mass development of sulfide-oxidizing **anoxygenic phototrophic bacteria** is determined by opposing gradients of light intensity and sulfide concentration. Green sulfide oxidizers (Chlorobiaceae) are always found below the purple bacteria (Chromatiaceae) due to their higher light exploitation efficiency and higher tolerance towards sulfide. The non-sulfur purple bacteria are largely independent of water stratification.

31.1.6 Oceans Are Huge Stable Deserts

Oceans cover about 70% of the Earth's surface and represent the largest part of the biosphere; they are ecosystems of enormous stability with respect to temperature and osmotic value. Of the total ocean-covered area, about 8% are continental shelfs with depths down to 200 m, 15% are continental slopes with depths ranging from 200 to 3 000 m. The majority of the oceans (77%) are deep-sea basins with depths of 3 000–6 000 m, and only a small fraction (1.3%) is deeper than 6 000 m (deep-sea trenches, down to 11 000-m depth). Surface waters have temperatures of 8 °C on average, with considerable difference between tropic and polar coastal waters; the deep-sea below 200 m remains at a temperature of 2–4 °C worldwide.

Due to the low ratio of trophogenic to tropholytic bodies of water and intense mixing due to wind influence and water currents, the ocean water is usually saturated with oxygen down to the bottom. Exceptions are found in protected shallow coastal areas and in highly productive water basins such as the Gulf of Mexico, the Black Sea, and parts of the Baltic Sea.

The productive zone of oceans extends to 50–100 m depth at maximum. Primary production is accomplished by green, red, and brown algae, and by cyanobacteria among which unicellular rods and cocci dominate. The primary productivity of shelf regions and of nutrient-rich upwelling zones is 10–20-fold higher than that of off-shore ocean areas (see Tab. 31.**2**), where, according to recent findings, primary productivity is limited by iron availability.

In the open ocean, the organic constituents of sedimenting detritus are nearly completely oxidized on their way down to the bottom, most of it within the first 1 000 m of sedimentation. Concentrations of dissolved and particulate organic carbon in deep-sea water are 1–2 orders of magnitude lower than in freshwater lakes. Deep-sea ocean sediments receive very little organic input (less than 1% of primary production; see Chapter 31.2). Turnover times of dissolved organic matter in deep-sea waters have been calculated to be 2 000–6 000 years.

The numbers of bacteria in surface waters in the ocean according to direct counts are about the same as in freshwater lakes, i.e., 10^4–10^6 cells per ml, indicating again that grazing efficiency rather than food supply controls these populations. However, only a very small fraction of these bacteria ($< 10^2$ per ml) is accessible to cultivation by standard plating methods. Success in cultivating bacteria that are more representative of the dominant populations has been obtained only recently with very dilute liquid media. Bacteria isolated from marine sources by conventional techniques do not form a separate taxonomic cluster as assumed in earlier times, but are dispersed throughout the entire taxonomic system and are closely related on all taxonomic levels to bacteria from other sources. Recent studies with specific rRNA probes have provided evidence that possibly new types of archaea may make up a significant fraction of microbial populations in Antarctic surface waters and other cold ocean regions. Microbial activity in the deep sea is to a large extent restricted to sedimenting particles ("marine snow") which consist of inorganic carbonates and silicates, together with particulate detritus.

The degradative capacity of heterotrophic bacteria in the deep sea is obviously far lower than in surface waters because high pressure inhibits metabolic activity. Actually, the deep sea can be regarded as an extreme environment (see Chapter 31.4) in the sense that only a small number of specialized organisms can proliferate there. Incubation experiments have shown that the degradative activity of the microbial flora deeper than 1 000-m depth decreases to about 1% of that of surface ocean waters. The influence of pressure on metabolic activity is higher at low temperature (2 °C) than at surface temperature (8–10 °C). Increased pressure inhibits activity of nearly all bacteria, independent of the substrates provided and independent of the inoculum size. There are now reliable indications that the deep sea at >3 000-m depth contains bacteria that are especially adapted to high pressure conditions (**obligately barophilic**). They cannot be cultivated at ambient pressure and have been maintained from the deep-sea sampling source at in situ pressure all the way to the pure culture stage in a special sampling and cultivation device, without decompression. Other bacteria exhibit a certain degree of tolerance to enhanced pressure, but their activity also decreases with increasing pressure. The biochemical background of pressure influence on metabolic activities is still enigmatic and awaits elucidation. There are indications that the composition of membrane lipids and cellular proteins is directly influenced by pressure. In any case, the obvious inhibition of microbial degradation at enhanced pressure in deeper waters precludes the abuse of the deep ocean as a dump site for sewage and other human wastes.

Coastal regions of the oceans are ecosystems of high productivity and high microbial activity. In **estuaries**, freshwater from a river meets seawater, and both mix over a certain flow distance that shifts back and forth, depending on the tidal changes. Accordingly, the prevailing salt concentrations vary periodically for sessile organisms, and require refined systems for osmotic adaptation and its regulation. Several anaerobic bacteria have been shown to contain sodium/proton antiporter systems that may be of special value in such environments (see Chapter 28.4).

Marshes are flat coastal areas that are periodically flooded and become dry again with the rise and fall of the tides. Due to repeated supply with nutrients, marshes are very productive habitats in which usually nitrogen-fixing cyanobacteria and diatoms act as primary producers. Organic excretions and lysing cells sustain heterotrophic bacteria, which consume oxygen in the lower layers, leading to complete anoxia and intense sulfate reduction. The resulting sulfide is oxidized predominantly by phototrophic purple and green sulfur bacteria and by aerobic chemolithotrophs. The typical layering of dark-green cyanobacteria over a pink layer of purple phototrophs, followed by a thin layer of green phototrophs and black sulfidic sediment underneath can be observed quite often on beaches ("Farbstreifensandwatt") if the sandy layers are removed stepwise in thin flat slices. Sometimes the purple sulfur-oxidizing phototrophs may develop large, intensely pink areas under thin water films. Most often, however, they remain covered with cyanobacteria and diatoms which develop with time into thick, sturdy

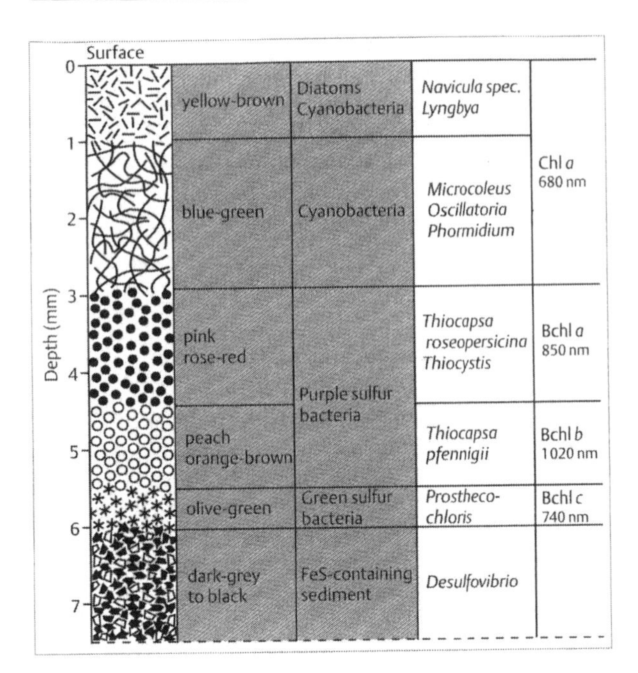

Fig. 31.5 **Schematic cross-section through the laminated microbial mat at Great Sippewisset Salt Marsh.** An impression of the structural organization and the color of the various layers is given on the left. On the right, the representative types of microorganisms and their chlorophylls (Chl) and bacteriochlorophylls (BChl) are listed [after 2]

layers in which *Microcoleus chthonoplastes* dominates as the structuring "backbone." Figure 31.5 shows a section through such a developed **microbial mat**. The layering of these communities is determined by several factors, including an upwards decreasing gradient of sulfide and downwards decreasing gradients of oxygen and light intensity. The latter is further modified by changes in light quality due to the absorption properties of the photosynthetic pigments. Different from the situation deep in a lake, in mats also the long-wavelength range of light can be exploited for energy metabolism, and the absorption bands in that range provide selective advantages to the microbes. These microbial mats therefore represent systems of rather high internal organization. Oxygen production by cyanobacteria in the upper mat layers may also support chemotrophic sulfide oxidizers such as *Beggiatoa* spec., which migrate in the mat following the oxygen distribution pattern in daily cycles. The various interrelationships of the microorganisms have been elucidated only recently by high-resolution microelectrodes, which also allow the assessment of light intensity and light quality on a small scale.

Because such microbial mats in salt marshes release large amounts of sulfide, they are grazed upon only slowly by crabs and other small arthropods and remain intact for many years. Since the metabolism of the various components of the microbial community in such a mat is based substantially on an intense turnover of sulfur compounds, this ecosystem is called a **sulfuretum**.

Due to intense primary production under the influence of periodically repeated fertilization with nutrient-rich coastal seawater, these sulfureta are among the most productive ecosystems known. Intensive carbon dioxide fixation in the upper layers and concomitant alkalinization causes precipitation of carbonates and silicates inside the mat and may convert the mat slowly into a solid, rock-hard material in which the biological components are preserved in a lithified state. It is assumed that **stromatolites** developed this way from microbial mats through biologically induced lithification.

In high-production coastal regions under water layers of more than 100-m depth, sulfide emerging from anoxic sediments is oxidized by chemolithotrophic rather than phototrophic bacteria. Mass developments of marine *Beggiatoa* spec. cells that cover the sediment with 0.5–1-mm-thick white layers have been detected. The marine sulfide oxidizer *Thioploca* spec. develops structured colonies several centimeters in depth (Fig. 31.**6a**), which have been observed recently off the coast of Chile. Some *Thioploca* strains can also oxidize sulfide with nitrate as electron acceptor, and can accumulate nitrate in a specific vacuole inside the cell to more than 1000 times the ambient nitrate concentration, thus alleviating the problem of nitrate and sulfide supply in thick organized layer arrangements. Figure 31.**6b** shows *Thioploca* cells inside a layer with and without internal vacuoles.

Oceans are by far the largest and most stable environments for microbial life, but nutrient supply is usually low. Higher activities of production, consumption, and mineralization are found in the coastal regions, especially in salt marshes where seawater supplies inorganic nutrients at a high rate and an intense turnover of organic matter is maintained by a complex structured community (sulfuretum).

31.1.7 Thermal Vents on the Sea Floor Are Oases in the Desert

Since deep ocean sediments receive very little organic input, the sediments are extremely poor in substrate supply and higher life is restricted to few, dispersed mussels and starfish at a very low density. It was a great

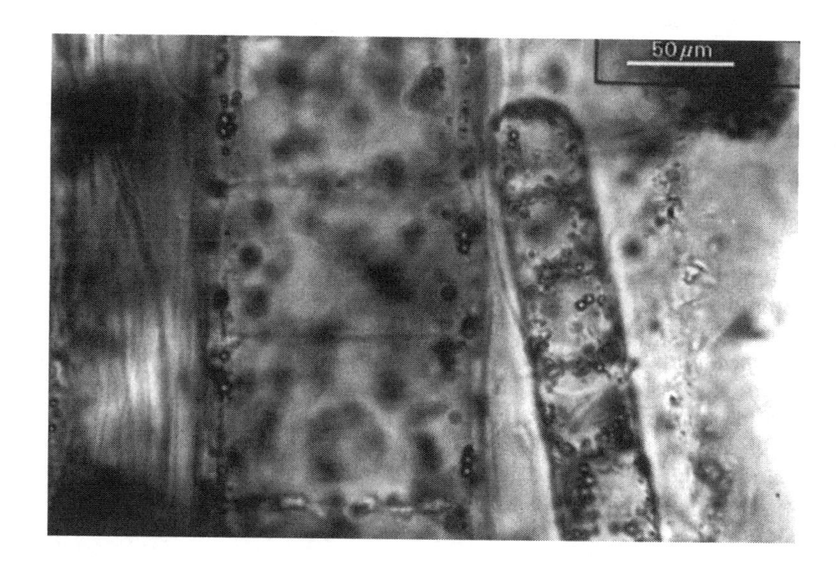

Fig. 31.6 ***Thioploca* spec. in sediments off the coast of Chile.** Two types of *Thioploca* filaments within one sheath. Central vacuoles (invisible) in the bigger filament push sulfur droplets to the cell walls. Bar equals 50 μm. (Courtesy of J. Detmers and J. Küver)

surprise, therefore, when in 1979 rich areas of very high productivity with dense communities of rather unusual forms of animals were found associated with continental fracture zones. These fracture zones are the separation lines on the sea floor where the tectonic plates meet and interact by subduction or separation. At these zones, extending for about 40 000 km through the Atlantic, Pacific, and Indian Oceans, freshly extruded lava contracts upon cooling and allows seawater to penetrate several kilometers downwards into the newly formed crust. At temperatues exceeding 350 °C and high pressures, the seawater reacts with basaltic rock and is transformed to an acidic and highly reduced "hydrothermal fluid" rich in metals, hydrogen sulfide, and molecular hydrogen (Fig. 31.**7**). This fluid can either mix with cold seawater and emanate at low speed and mild temperatures (3–40 °C) into the overlying seawater, or be ejected directly into the cold sea without mixing. In the latter case, iron and manganese salts precipitate as black particle clouds reminiscent of smoke ("black smokers". Fig. 31.**7**). The temperature gradients at the black smokers can be very steep, from 350 to 2 °C over a few decimeters, and mesophilic and extremely thermophilic microorganisms have been isolated from these sites.

It was obvious that the rich fauna existing around these thermal vents in absolute darkness was established on chemosynthetic rather than photosynthetic activity. Sulfide, thiosulfate, methane, hydrogen and reduced iron could act as an electron source for an aerobic, lithotrophic type of metabolism that formed

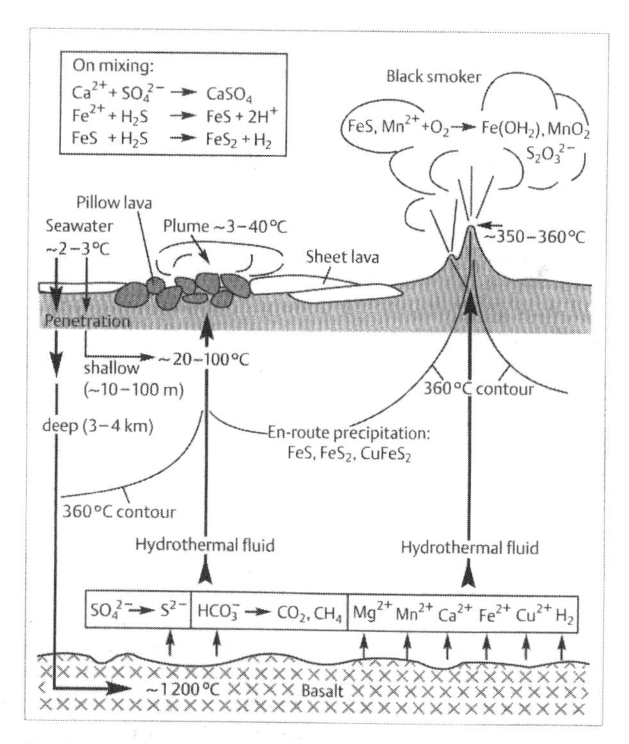

Fig. 31.7 **Scheme of seawater circulation through the oceanic crust at tectonic drift zones.** The flux in warm vents is shown on the left, in hot vents on the right [after 3]

organic matter by autotrophic carbon dioxide fixation. The seawater surrounding the vents is slightly turbid from high densities of bacteria, among which sulfide-oxidizing *Thiomicrospira* species dominate. Other bacteria oxidize reduced iron or manganese, methane, other methyl compounds, or ammonia, and also hydrogen-utilizing aerobic, sulfate-reducing or methanogenic prokaryotes have been isolated. All these organisms appear to grow at 2000–3000-m depth at a relatively high rate, favored by the enhanced in situ temperatures. These free-living bacteria are grazed by amphipods and mussels. However, the surprising size of the dominant types of mussels, e.g., *Calyptogena magnifica* with a body length of 30 cm and weights of up to 800 g, can hardly be explained by exclusive feeding on suspended bacterial cells. These mussels also contain endosymbiotic sulfide-oxidizing bacteria that grow inside the gill cells. Other mussels at hydrothermal vents rich in methane developed similar associations with methane-oxidizing aerobic bacteria. The tube worms *Riftia pachyptila*, which grow to sizes of 2–3 m in length, are among the various unusual types of animals specialized for life on the basis of symbiotic associations with bacterial cells. They contain a special organ, the trophosome, inside the tube that is supplied with O_2, CO_2, and H_2S from the surrounding seawater by the blood circulation system (Fig. 31.**8**). The bacterial symbionts in the trophosome produce organic matter and grow. It is still open to question how the bacteria feed the host animal; physiological experiments have been hampered because the worms do not survive decompression. The bacterial cells could either feed the host by excretion of dissolved organic substrates, or they could be slowly digested by the host, and the digestion could be compensated by bacterial growth.

The thermal vent communities represent exciting ecosystems that maintain themselves in the complete darkness of the deep ocean on the basis of a bacterial chemosynthetic activity and apparently independent of photosynthesis. However, also these communities depend on photosynthetic activity in the euphotic zone; this activity provides them with oxygen as the electron acceptor for oxidation of the reduced constituents of the hydrothermal fluid.

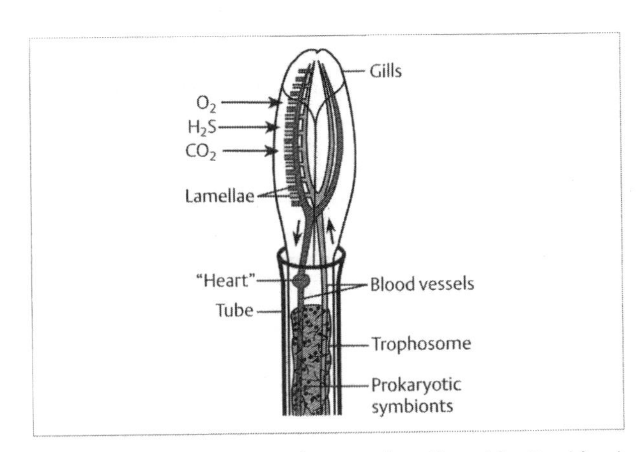

Fig. 31.**8** **Anatomy of *Riftia pachyptila*** with its blood circulation system and the trophosome, which contains the sulfur-oxidizing bacteria

Thermal vents in the deep ocean maintain a complex community of prokaryotes and animals in entire darkness on the basis of bacterial chemosynthesis. Lithotrophic bacteria that oxidize H_2S, H_2, and Fe (II), or methanotrophic bacteria live either free or inside specialized organs of animals and produce bound organic carbon, which supplies a wide variety of animals with carbon and energy.

31.2 Sediments Resemble Digesting Organs and Are History Notebooks

In sediment the inorganic and organic particles settle from the water column and are digested further, mainly by microbial activities. Particulate material settling from the water column accumulates, extracellular enzymes can operate in the condensed material at higher efficiency, organic low-molecular-mass substrates and inorganic nutrients are released, and the overall microbial activity is greatly stimulated in comparison to that in the overlying water column. Sediments are actually the "digesting organs" of bodies of water, especially of shallow lakes. The activity in the sediment depends on the substrate supply; sedimentation rates in eutrophic lakes are considerably higher than in oligotrophic lakes or in oceans. Primary productivity and sedimentation distance determine the rates of sediment deposition at the water bottom and also, together with the water chemistry, its content in organic matter (Tab. 31.**4**).

The material sedimenting from the water column not only serves as the food basis for the sediment-dwelling microbial and macrobial communities, but their inorganic and organic remnants after digestion record the chemical and biological history of the system. Shells of diatoms and foraminifers in marine sediments provide information on ocean temperatures, salinity, and climatic changes; remnants of photosynthetic pigments (chlorophylls, carotenoids) document which primary and secondary producers dominated thousands of years ago. The oldest sedimentary rocks on Earth (about $3.8 \cdot 10^9$ years old) document the presence of liquid water on the planet at that time, and slightly more recent layers contain organic carbon remnants as witnesses of the first autotrophic biotic activities.

31.2.1 Sediment Types Depend on the Chemistry and Biology of the Water Source

The chemistry of sediments is largely determined by the water chemistry and the biological activities in the water column. Particulate organic detritus is usually associated with varying amounts of inorganic material, including silica shells of diatoms and Chrysophyceae, and carbonates. The latter are formed in hard-water lakes in the surface water by CO_2 utilization through photosynthesis and corresponding alkalinization. Small crystals of $CaCO_3$ form, and thin coatings of $CaCO_3$ precipitates are deposited on algal cell surfaces. These carbonates can redissolve in CO_2-oversaturated water layers or can reach the ground to form lime deposits in the sediment (in German: "Seekreide"). Soft-water sediments contain, together with the organic constituents, mainly silicaceous inorganic contents (e.g., clay). Sand grains and bigger mineral particles indicate allochtonous depositions from rivers, etc. In dystrophic, slightly acidic lakes and bogs, the sediment material can be extremely fluffy and consist only of incompletely degraded woody material that is compacted with time to form peat and later brown coal and coal.

The precipitation of carbonates in hard water lakes serves an important function in transport of organic matter into the sediment. Polymeric fulvic acids and monomeric amino acids, for example, are adsorbed to carbonates and transported to the ground. Therefore, hard-water lakes are usually clear, free of the brownish cloudy appearance of soft waters. The binding of organic material to inorganic particles is of major importance for its microbial digestion. Dissolved substrates are transported this way into the sediment, increasing the importance of sedimentary microbial communities for total cycling of organic matter.

Marine sediments of coastal areas are dominated by sandy material in the wave-breaking zones, and organically rich muddy material in protected areas behind pre-coastal islands. Productive off-shore sediments (e.g., in upwelling areas off the west coasts of South America or South Africa) are very fine-grained and comparably rich in organic material. The sediments of the deep sea are extremely poor in organic contents and consist either of carbonates formed in the productive surface waters, or of dust particles that are transported from deserts through the atmosphere and are washed out by rain.

The differences in sedimentation rates (Tab. 31.**4**) are also reflected in the age scales of such sediments. A sediment layer at 20-cm depth in a eutrophic lake may be 100 years old, a sample from a deep sea sediment at the same depth may be nearly 100 000 years old. It is obvious that the substrate availability in each sample is basically different.

Unfortunately, sediments are not always preserved sufficiently to maintain in their layers an undisturbed history notebook. Various mussels and worms dig into the sediment down to 50-cm depths, especially in shallow waters. The activity of digging animals transports oxygen into deeper sediment layers and increases the oxygen-exchanging surface area. The inner surfaces

Table 31.**4** **Sedimentation rates in aquatic ecosystems**

	Annual deposition (mm)	Organic carbon content (% w/w)
Oligotrophic lakes	0.1–2	1–6
Eutrophic lakes	1–5	2–30
Marine upwelling areas of high productivity	0.05–0.3	1–4
Deep sea	0.001–0.02	0.3–0.5

of dwellings of digging worms can exhibit steep gradients of oxidized versus reduced conditions and represent outstanding examples of gradient habitats. The colonization of the sediment by invertebrates can also be used as an indicator of the trophic status of the aquatic system. Whereas oligotrophic sediments are dominated by mussels, echinoderms, and deep-digging arthropods, increasing availability of organic material and subsequent increasing reduction of the sediment layers from below change this diverse community gradually to a dominance of small polychaetes. Finally, a sediment can be entirely reduced or can be dominated at the surface by sulfide-oxidizing *Beggiatoa* or *Thioploca* species (see Chapter 31.1.6), and higher animals do not proliferate.

Box 31.3 Techniques for sampling and characterization of sediments differ extremely with the water depth under which they are located. Whereas coastal and littoral sediments can be sampled easily, deep-lying sediments are accessible only by sampling and measuring devices on ropes or by submersibles. Sediment surface material can be sampled with a grabbing device, such as an Eckman dredge. If the material needs to be maintained in its original layered structure, open tubes (corers) of various sizes are pushed into the sediment either manually or with the help of heavy weights. Multicorers take a series of 4–24 cores simultaneously at one sampling site. Replicate samples obtained this way can indicate how patchy the ground is, i.e., how chemical and biological parameters differ in their horizonal distribution. Deep cores 5–10 m in length are used for the analysis of deep-lying marine sediments and are run into the ground with help of several tons of ballast.

Distribution patterns of metabolites (O_2, NO_3^-, SO_4^{2-}, H_2S, CH_4, NH_4^+, etc.) can be analyzed by classical techniques in thin slices cut from the sampling core. A more elegant technique for high resolution, microelectrodes, is available today (see Chapter 30.6.1). Since distribution patterns of metabolites can change when samples from deep-sea sediments are brought up to the surface, special devices have been developed that are mounted on independent units that land on the sea floors and measure profiles automatically in situ without decompression. Comparisons have shown that, for example, oxygen respiration activities measured in situ are considerably lower than at ambient pressure (see Chapter 31.1.6).

31.2.2 Sediment Chemistry Is Governed by Microbial Activities

Numbers of bacterial cells in sediments exceed those of the overlying water layers by 3–5 orders of magnitude. Direct counts of fluorescently stained cells have given numbers of 10^7–10^{10} bacterial cells per gram, corresponding to biomasses of 1–1000 μg carbon per gram sediment, depending on the trophic status. Numbers are maximal at the sediment surface and decrease by 3–4 orders of magnitude within the upper 50 cm. The deeper material, which is also quite poor in microbially degradable substrates, shows still measurable metabolic activity (e.g., esterase activity or oxygen uptake). Viable microbial cells have been recovered at fair numbers at sediment depths down to several meters, where the sediment material may be more than 100 000 years old. Although most of those bacterial cells were recovered as spores, vegetative cells were also present at such depths.

The chemical conditions in sediments are largely determined by microbial mineralization activities. Since oxygen supply in a structured matrix is restricted to diffusive transport, oxygen is consumed within the uppermost sediment layers, and other electron acceptor systems take over, according to the sequence outlined in Chapter 30.2.1. Whereas the basic pattern of redox processes in oxygen-limited microbial habitats is always the same, the dimensions differ dramatically. The relative availability of the various alternative electron acceptor systems and the deposition rate of organic material on the sediment surface determine the depth scale within which the various processes dominate, and also determine their relative importance in total carbon turnover. Oxygen penetrates into the sediment of a eutrophic lake less than 1 mm deep, into that of oligotrophic lakes or a productive marine sediment by a few centimeters, into an off-shore deep-sea basin sediment by 0.5–1 m. The scales of the other redox processes vary accordingly.

Whereas oxygen and nitrate may be reduced very quickly in the uppermost layers (Fig. 31.**9a**), iron and manganese bear considerable oxidation potential, depending on the local availability of their hydroxides. Freshwaters and seawaters differ especially with respect to their sulfate content. Sulfate reduction can be of marginal importance in freshwater sediments; however, it shifts the redox potential dramatically to low values to enable methanogenesis, which is the dominant anaerobic degradation process in a freshwater sediment. In marine sediments, sulfate is usually available down to considerable depth and determines the redox processes over a large depth scale (Fig. 31.**9b**).

Tab. 31.**5** gives a rough comparison of the electron-accepting capacities of the various alternative oxidation processes in freshwater and marine sediments, assuming that these oxidants are not replenished from the water column. If the sediment is covered with oxic water, oxygen, nitrate, and other acceptors can diffuse into the sediment due to their consumption there, and the resulting fluxes can be calculated from the distribution gradients. The mesotrophic Lake Constance, Germany, which has an annual productivity of about 25 mmol carbon per m^2 per year, is taken as an example. About 20% of this production reaches the sediment (at 100-m average depth), and its oxidation in the sediment is associated with a flux of 4.4 mol O_2 and 0.4 mol NO_3^- per m^2 per year. Sulfate diffuses into the sediment only in late summer and is released to the water column in winter, resulting in no significant overall consumption over the year. Methane does not escape into the water column at significant rates. However, sulfate reduction and methanogenesis are active in the deeper sediment, and sulfide and methane are reoxidized again within the upper sediment layers. Thus, oxygen will always be the dominant electron acceptor in such sediments, either directly or indirectly through other reduced intermediates (see also Fig. 30.**7**). In contrast, the littoral sediment of the same lake releases substantial amounts of methane (about 10 mmol CH_4 per m^2 per year), due to higher productivity, little mineralization in the water column, and insufficient reoxidation. Thus, the relative importance of the various electron acceptors in biomass mineralization varies throughout the year and also between different zones in the same lake, and the net uptake of oxidants may give an insufficient picture of the redox processes active in the sediment.

The major inorganic nutrients can also be regenerated in the sediment. Ammonia is released during biomass digestion in the sediment and diffuses upwards where it is oxidized to nitrate by nitrifying bacteria in the layers where oxygen becomes available. A maximum of nitrate concentration below the sediment surface indicates the zone of nitrate production (Fig. 31.**9b**) from where it diffuses into the overlying water column. Part of nitrate also diffuses downwards and is reduced in mineralization processes, mainly to N_2. Thus, only part of the bound nitrogen transferred into the sediment returns to the water column as bound nitrogen, and another part is lost as N_2.

Phosphate is bound in the water column mainly to calcium ions and Fe (III) hydroxides, and settles to the sediment as $Ca_3(PO_4)_2$ and $FePO_4$. Under reducing conditions, iron is reduced to the ferrous form, and H_2S precipitates with Fe^{2+}:

$$FePO_4 + e^- + H_2S \rightarrow FeS + H_2PO_4^- \qquad (31.1)$$

Thus, phosphate is redissolved and can diffuse back into the water column. This process leads to a recycling of

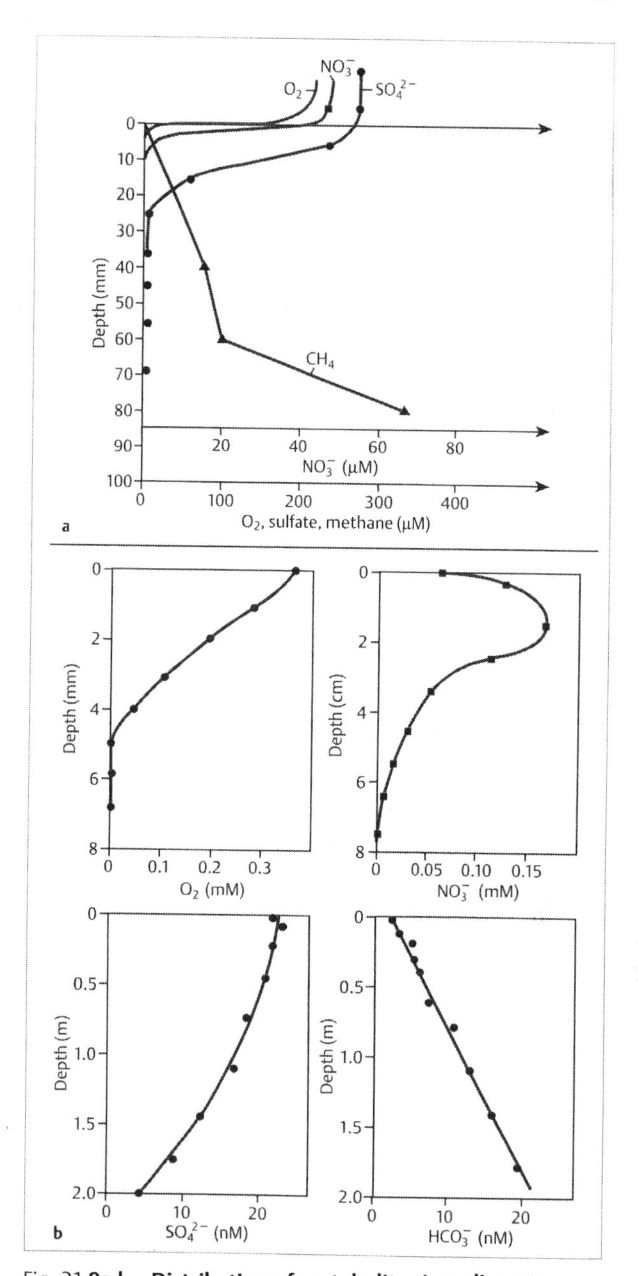

Fig. 31.**9a,b** **Distribution of metabolites in sediments.**
a Freshwater lake sediment (Lake Constance, Germany)
b Coastal sediment off the coast of Denmark. Note the different depth scales

Table 31.5 **Electron-accepting capacities in sediments.** For dissolved compounds, the listed concentrations refer to average values in the oxic water column. The values for iron refer to total iron content (including Fe (II) oxides, Fe (III) oxides, FeS, and FeS_2) in sediments. Manganese is not listed because the availability varies greatly with the location

		Freshwater		Seawater	
Electron acceptor system	Electrons accepted	Concentration (μM)	Electron uptake (μeq/l)	Concentration (μM)	Electron uptake (μeq/l)
O_2	4	400	1600	320	1280
$NO_3^- \rightarrow N_2$	5	200	1000	5	25
$NO_3^- \rightarrow NH_4^+$	8	200	1600	5	40
Fe (III)	1	20000	20000	20000	20000
SO_4^{2-}	8	300	2400	28000	224000
HCO_3^-/CO_2	8	30000	240000 (unlimited)	30000	240000 (unlimited)

phosphate in eutrophic lakes ("phosphate remobilization") if the sediment and the overlying water column are completely reduced. Otherwise, microbial or chemical iron reoxidation at the sediment surface may trap the phosphate and retain it in the deeper sediment.

In marine sediments, the distribution patterns of microbial metabolites are easy to resolve because they extend over a depth of several meters. Figure 31.**10** shows such a profile of microbial metabolites in the pore water of a marine sediment from a high-production area off the coast of South-west Africa. Sulfate reduction exhibits a maximum at a depth of approximately 6 m, and sulfide diffuses upwards, corresponding to a downwards diffusion of sulfate. Between the horizons of nitrate reduction (at 0.5 m) and that of sulfate reduction, there are no redox reactions of dissolved electron carriers that would be detectable in the pore water. Obviously, redox processes in these layers are governed by poorly soluble manganese and iron oxides, and these may be responsible for sulfide reoxidation at 4-m depth. Two other unusual transformations can be observed in Figure 31.10, namely, the disappearance of ammonia at 4-m depth and the disappearance of methane in the sulfate-reducing horizon. Anaerobic oxidation of ammonia and methane are processes that still await elucidation with respect to their microbiology and biochemistry. Thus, the thorough study of metabolite distributions in such sediments can provide evidence of new metabolic activities.

The transformations of organic matter in sediments is called **diagenesis** and over long periods of time leads to organic derivatives that become more and more recalcitrant and differ substantially from their ancestral precursors. Organically rich, deep sediment can develop by microbial and chemical transformation into a paste-like material of slightly oily character called **kerogen**. Whereas all carbohydrate-like structures, functional groups of amino acids, and other energy-rich compounds disappear during diagenesis, the fraction of polyanellated and aliphatic phenols and hydrocarbons increases. Other constituents of biomass such as branched fatty acids and isoprene derivatives, including carotenoids and membrane lipids (steroids, hopanoids), survive largely unchanged and may serve as tracers for a historical analysis of the respective body of water. In some cases, especially the analysis of carotenoids in freshwater lakes has provided detailed information on the lake's productivity, changes in trophic status, and changes in the geological situation of the catchment area. Moreover, the chemical structures preserved in kerogen material document where anaerobic microbial degradative capacities in concert with geochemical transformation activities find their limits.

Sediments are governed largely by anaerobic microbial mineralization processes that may lead to accumulation of reduced electron acceptors. Depending on the oxidation state of the sediment surface, reduced compounds can either be reoxidized within the sediment or be released to the overlying body of water. The further digestion of organic material in the sediment leads to a poorly defined material called kerogen, which contains recalcitrant compounds of biological origin and other compounds that resemble components of mineral oil.

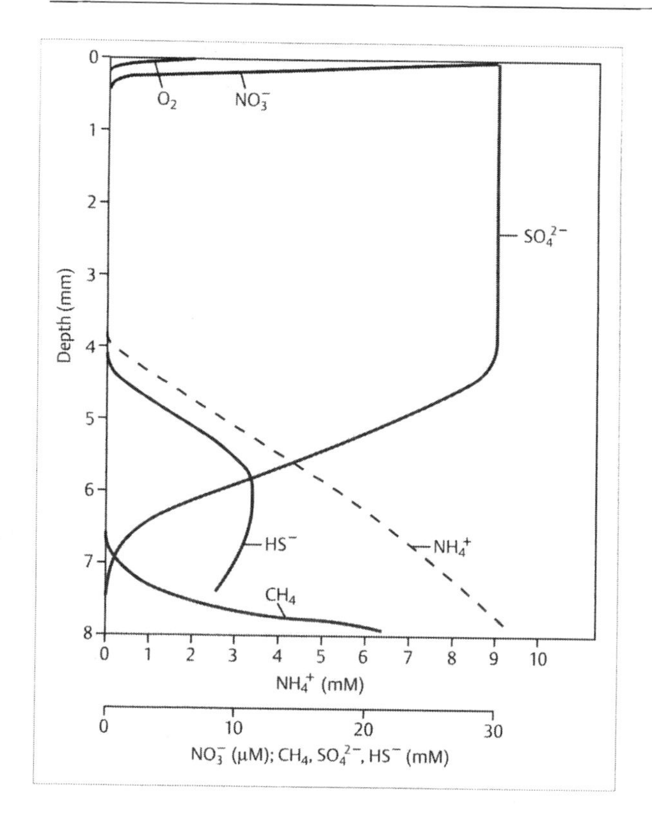

Fig. 31.**10** **Distribution of dissolved metabolites in a marine sediment under a high-production upwelling zone** (Namibia, South-west Africa) at a depth of 1 000 m

31.3 Soil Is Mostly Dry and Heterogeneous

After the oceans, soil is the second most important ecosystem for microbial life; however, basic differences between the ecosystems are obvious. Whereas water in general and especially oceans are rather homogeneous in space, both on a small and on a large scale, and conditions do not change significantly with time, soil is an extremely heterogeneous environment. The degree of heterogeneity and structural complexity actually increases from free water via sediments to soil. At a single sampling site, the properties of soil change from the top to a depth of 30 cm or less. The same applies to soil samples taken at various sites in dimensions of meters, kilometers, and hundreds of kilometers, and again within a single sample on the millimeter and micrometer scale. Moreover, soils undergo extreme variation during a year's cycle with respect to temperature, moisture, osmolarity, acidity, oxygen availability, etc. Soils are the most complex and heterogeneous microbial environments known, and it is not surprising that so little about the microbial ecology in soils is known, despite their enormous importance, for example, for plant growth in general, human nutrition, purification of surface water and groundwater, detoxification of xenobiotics, and stabilization of trace gas concentrations in the atmosphere.

The complexity and heterogeneity of soil as a microbial habitat is also reflected by its genomic diversity. Whereas about 80 different microbial genomes have been detected in a water sample from a fish pond, about 1 000 different genomes were found in coastal marine sediment, and about 10 000 different genomes in a soil sample. It is obvious that microbial ecology in the future will have to deal with soil as its most important subject, for general scientific reasons as well as for economical and agricultural needs.

31.3.1 Chemical Composition and Structure of Soils

First, one has to differentiate between soil and dirt. Whereas nearly every soil fraction can be homogenized, sieved, or artificially mixed, forming different types of dirt of whatever kind, soil is, despite its heterogeneous internal structure, an organized system with a spatial order that arises from a specific history of development. Soil is made up of various components. **Mineral constituents** are usually derived from rocks that decompose by weathering and biological activities, such as penetration of fungal hyphae and plant roots and acid excretions. The chemistry of the rock material determines the chemical reactivity; limestone-based soils are neutral to slightly alkaline and silica-based soils acidify more easily under the influence of, for example, microbial metabolism. Acidification of soils due to excess uptake of nitrogen and sulfur compounds and the microbial oxidation leads (at $pH < 5$) to mobilization of aluminum ions, which may be toxic for plants. **Organic constituents** are provided by plant litter and remnants of animals and microorganisms. These remnants can still be structured (e.g., wooden twigs, chitin carapaces of arthropodes) and are slowly converted into a dark-brown, comparably homogeneous material called **humus** (see also Chapter 30.1.2). Humus is produced primarily by microbial activity which activates aromatic and aliphatic biomass constituents to form radicals that polymerize and produce high-molecular-mass polymers. Worms, insects, and protozoa contribute to disintegration and degradation of organic material. Formerly, humus was considered to consist only of lignin derivatives; today it is known that also sugar and amino acid residues are covalently linked with aromatic residues and can make up a substantial part of humus. Humic material is extracted from soil in NaOH or $Na_4P_2O_7$ solution. The humic material remaining in soil after this alkaline extraction is called **humin** and consists of molecules with masses greater than 100 000. **Humic acids** are precipitated from the alkaline extract by acidification to pH 2. They comprise compounds in the mass range of 10 000–100 000 Da. The material extracted with NaOH that remains soluble at pH 2 are low-molecular-mass (1 000–30 000 Da) compounds called **fulvic acids**. Humin, humic acids, and fulvic acids are the main constituents (> 90%, w/w) of organic matter in soil, which constitutes 10–15% (w/w) of forest surface soil, 0.5–3% of agricultural soil, less than 0.1% of desert soils, and more than 30% in bogs.

Because the formation of humic material proceeds after enzymatic activation largely through uncontrolled radical reactions, the structure of the resulting condensates is not predictable, and no two humic molecules are alike. Carboxylic groups act as ion exchangers and interact with anionic functions of mineral via positively charged metal ions (Fe, Al, Ca, Mn ions; Fig. 31.**11**). Phenolic residues and carboxylic groups form complexes with metal ions; few nitrogen atoms interact with carboxylic groups. Humic material is continuously transformed by extracellular microbial enzymes (lignin peroxidase, laccase, other oxygenases; see Chapter 9.6) with concomitant partial mineralization and further condensation. The organic material also contains the living microbial biomass, largely fungi and bacteria (see Chapter 31.3.2). The predominance of anionic over cationic functional groups in humic material make it

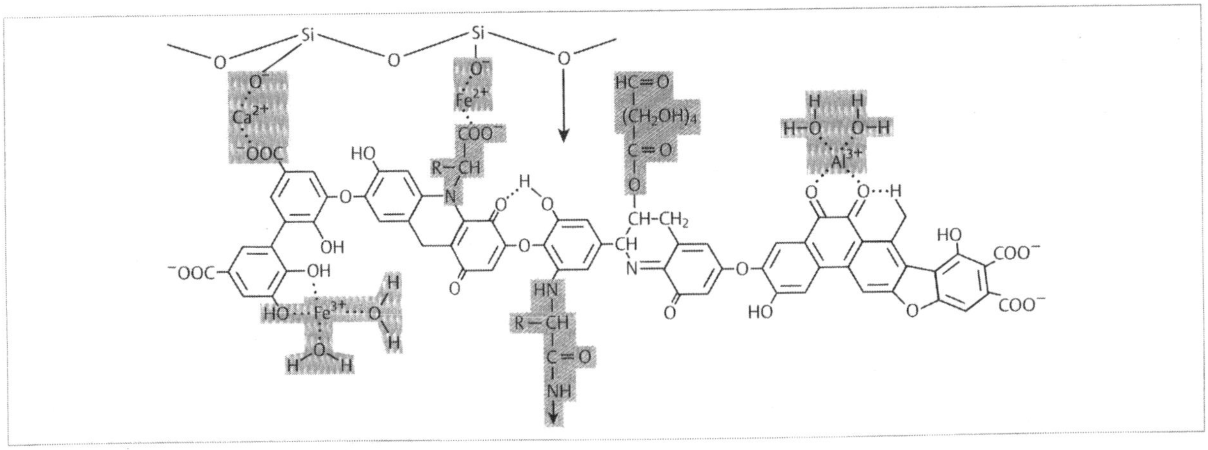

Fig. 31.**11** **Schematic representation of a humic acid molecule** with inclusions of remnants of amino acids and sugars, and its possible association with inorganic soil constituents

an excellent cation exchanger; cations such as ammonium are efficiently bound, whereas anions such as nitrate are washed out. Apolar aliphatic or aromatic compounds may also be bound to the humic constituents through absorption (van der Waals forces).

Box 31.4 A profile of a forest soil in a temperate region is shown in Figure 31.**12**. The uppermost layer (Fig. 31.**12**, **O horizon**) is rich in organic material at various degrees of mineralization, and is usually missing in agricultural soils. The following **A horizon** contains organic material down to a certain depth (**A_h horizon**), but is mainly determined by mineral constituents. The **A_p horizon** (in agriculturally used soils) is that part of soil that is mixed at regular intervals by ploughing. Percolating water leaches dissolved and particulate contents from the A horizon into the **B horizon**, which is often dark brown due to humic inclusions and is rich in accumulated clay. The **C horizon** consists of mineral soil material derived exclusively from the underlying bedrock (R) by weathering. The bedrock or clay layers can seal off the soil column downwards to allow establishment of a groundwater table. Groundwater may saturate the lower soil layers (saturated zone) up to the B horizon, whereas the upper soil parts are typically aerated (unsaturated zone) and have a varying water content that depends on rainfall and climatic exposure.

The **soil texture** determines its density and exchange efficiency with the atmosphere. Depending on the relative particle size, soils are classified as sand, silt, or clay soils and various mixtures of them, depending on the relative particle size. Fungal hyphae, plant roots, and earthworm holes contribute significantly to the formation and stabilization of a porous soil texture that ensures aeration. Soil typing considers chemical, physical, and biological properties. Unfortunately, there is no consistent soil typing system that is internationally accepted; the terminology varies among Russian, English, German, and other systems, and the systems are not directly comparable.

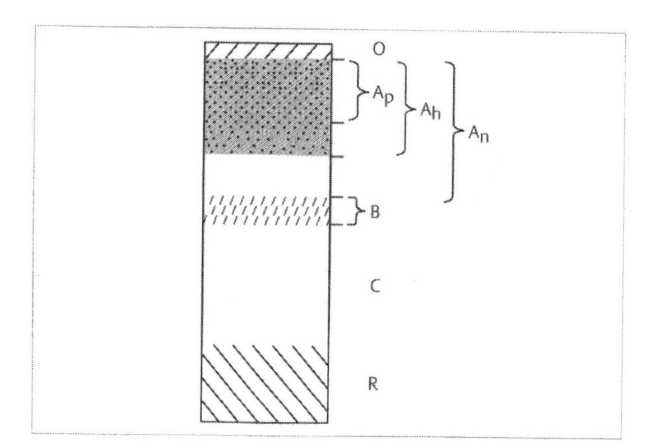

Fig. 31.**12** **Horizons in a woodland soil of temperate regions.** Letters refer to the various soil horizons explained in Box 31.4

The **gas phase** in soil ("soil atmosphere") is usually in equilibrium with atmospheric air. Microbial mineralization processes decrease the oxygen and increase the CO_2 content, especially in the microbially active upper soil layer, thus shifting the pH of the percolating water to slightly acidic values. The air content and with that the oxygen supply in deeper soil layers is largely determined by its porosity (soil texture) and its water content. Microbial oxygen consumption, in concert with inhomogeneities of substrate availability, air supply, and water distribution, promotes the formation of anoxic microenvironments (microniches) in and around soil particles in which alternative anaerobic respiration and fermentation processes take over (see Chapter 30.2.1), concomitant with a decrease in redox potential. Waterlogged soils of dense texture (clays, loams) can easily become entirely anoxic.

The **soil solution** is determined by the solubility of the various organic and inorganic soil constituents, the possible intake from the surface water and the ion-binding capacity of the soil. Rainwater percolating from the top is first saturated with humic residues in the O-horizon (up to 30 mg organic carbon per liter) to form a brown solution, and later with iron, aluminum, calcium, and manganese ions and their complexes until it reaches the groundwater as a colorless mineral-enriched fluid. The concentration of salts in soil solution varies dramatically with the water supply and the ion-binding capacity of the soil. In dry seasons, the leftover water is oversaturated with salts, sometimes similar to that of a salt brine (lateritic soils).

31.3.2 Microorganisms Are Important Constituents of Soil

In a forest soil, fungi make up 86% and bacteria 7% of the total biomass (Tab. 31.**6**). The remaining 7% is shared by the various soil animals. Among the prokaryotic soil microflora, *Actinomycetes* are only of minor importance. In counting experiments after cultivation on agar plates, spore-forming species of the genus *Bacillus* and the non-spore-forming *Arthrobacter* species, which is considered a typical soil bacterium, dominate. In general, Gram-

Table 31.6 **Biomass estimates and annual litter production in a forest soil in a temperate climate**

Group	Dry matter biomass (kg ha^{-1})
Bacteria	36.9
Actinomycetes	0.2
Fungi	453.0
Protozoa	1.0
Nematodes	2.0
Earthworms	12.0
Enchytraeidae	4.0
Mollusks	5.0
Acari	1.0
Collembola	2.0
Diptera	3.0
Other arthropods	6.0
Total microflora	492.1
Total microfauna	36.0
Total biomass	528.1
Annual litter production	7640.0

positive bacteria appear to dominate over Gram-negative bacteria, as far as plate count assays indicate. Unfortunately, rRNA probing techniques have been applied to soils only with limited success, due to the high background autofluorescence. Among the fungal soil inhabitants in neutral soils, yeasts make up only a small fraction (> 2%), but their contribution increases to 40% of the total fungi in acidic soils. The vast majority of soil microorganisms live attached to mineral or humic surfaces in a thin film layer that retains a minimum of water even through periods of drought.

Fungi also play the predominant role in the degradation of organic litter. 70–90% of oxygen consumption in soil is attributed to fungi and 10–20% to bacteria, independent of pH conditions. The microbial biomass of soils is directly correlated with the content of organic matter in various soils. Accordingly, microbial activity is maximal in surface soil and decreases in the soil profile gradually with depth (see Chapter 31.2.5). The data in Table 31.**6** also show that the total biomass present metabolizes 15 times its own mass in litter per year. Assuming a cell yield of 50% (w/w) as typical of aerobic conversions (see Chapter 6), decreased by 80% due to substrate humification (see above), it can be calculated that the soil biota multiplies only by a factor of 2.5 per year, i.e., all organisms on average can double once or twice per year. Although such an estimate is rather rough, it gives a realistic picture of growth rates in a natural ecosystem. Fungi and bacteria cooperate in the mineralization and transformation of plant and animal litter, and the soil microfauna contributes to this process by mechanical disruption, grinding, and homogenization. The further microbial transformation to CO_2 and humic material occurs mostly in the free soil, but to a significant part also inside the intestinal tracts of soil animals and their faeces. Fungi, bacteria, and the soil microfauna are therefore of eminent importance for the formation of humus, which acts as a stabilizing factor and as ion exchanger that retains important nutrients. Also free enzymes, so-called soil enzymes, exist in soil; most biological polymers are degraded outside microbial cells by extracellular hydrolases. Other enzymes may be released from lysing cells after cell death, and little is known about whether these enzymes change their kinetic properties upon release.

Dissolved low-molecular-weight compounds that enter the soil with percolating surface water are also partly degraded and partly integrated into the humus matrix by covalent linking after initial activation by oxygenase reactions. Experiments with radiolabeled aromatic substrates have shown that both fungi and bacteria contribute to this mineralization and immobilization. This applies also to synthetic chemicals, such as plant-protecting agents (e.g., herbicides, insecticides, and fungicides). The majority of these agrochemicals are bound through microbial activities into the polymeric humic soil matrix or inside clay layers, and only a minor fraction (10–30%) is mineralized (or evaporated) within one vegetation period (Fig. 31.**13**). The remnant **bound residues** have turnover times of decades and constitute

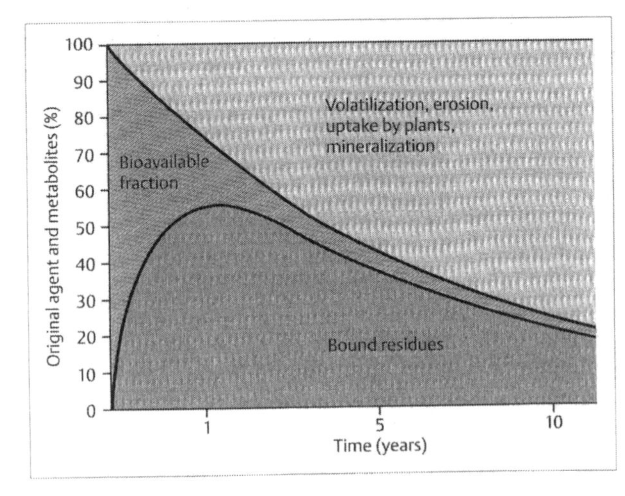

Fig. 31.**13** **Fate of a xenobiotic compound in soil over time.** Such compounds include plant-protecting agents, such as herbicides, insecticides, and fungicides. The time axis is not drawn to scale

an immanent problem in agriculturally utilized soils, especially low-lying soils with a short passage to the groundwater.

The transformation of nitrogen compounds in soil is nearly exclusively a domain of bacteria. **Nitrogen** (N_2) **fixation** is catalyzed by bacteria only (Chapter 8.5), either by free-living bacteria (*Azotobacter, Xanthobacter, Alcaligenes, Bacillus, Clostridium* species) or by symbiotically nitrogen-fixing bacteria, e.g. *Rhizobium*.

The availability of ammonium and nitrate salts as mineral fertilizers has led to an overfertilization of agricultural soils in the past 30 years. Ammonia binds to soil humus quite efficiently (see Chapter 31.2.1), but is converted in oxic soils to nitrate through **nitrification** (Chapter 10.7). Nitrate as an anion is washed out of the productive soil horizon and appears occasionally in groundwater. Ammonia oxidation can be inhibited by addition of nitrapyrin [2-chloro-6-(trichloromethyl)-pyridine; so-called "N-serve"]. In anoxic microniches and especially in water-logged soils, nitrate is reduced, preferentially to N_2, through **denitrification** (Chapter 12.1). This process releases NO and N_2O as side products, which escape into the atmosphere; highly fertilized water-logged soils can release up to 6% of their total nitrogen intake as gaseous nitrogen oxides. NO and N_2O act as greenhouse gases in the atmosphere (see Chapters 32 and 34). 70% of N_2O released into the atmosphere comes from agricultural soils and also from fertilized forest soils.

Soils also act as sources or sinks of other **atmospheric trace gases**. Whereas lowland soils are sources of N_2O, NO, and methane, upland soils act as sinks for hydrogen, methane, and carbon monoxide (Tab. 31.**7**). Since many of these gases are greenhouse gases or interact with tropospheric chemistry, microbial activities in soil fulfill an important function in the stabilization of atmospheric gas budgets. Unfortunately, the biochemical basis of trace gas consumption in soil is only rarely understood. Aerobic methanotrophic, CO-oxidizing, or hydrogen-oxidizing bacteria exhibit uptake kinetics that are basically different from those observed in soil samples. The saturation curve of methane uptake by soil indicates that two different activities coexist in soil: one of high affinity and a low maximum uptake rate and another with low affinity, a high maximum rate, and a high threshold. Thus, the organisms responsible for methane uptake in soil at the low ambient concentrations (1.7 ppmv) are unkown. The situation is similar with CO and hydrogen uptake.

31.3.3 Wetland Soils Are Globally Important Habitats

Wetland soils make up a significant part of the terrestrial Earth surface, especially in the **tundra** areas of the Northern Hemisphere. Primary production on these wetlands is only insufficiently balanced by aerobic mineralization because oxygen has little access to the deeper soil. Therefore, decomposition of organic matter leads primarily to methane or CO_2, and wetlands contribute, with a total amount of 110 Tg methane per year, to 20% of the total methane formation on earth. This methane is partly oxidized at the soil surface by methane-oxidizing bacteria, but the major part escapes into the atmosphere. Tundra soils are one of the sources of atmospheric methane (see Chapter 32).

Rice paddies are agricultural soils that are flooded periodically with freshwater for several months per year. The high plant litter production feeds intensive

Table 31.7 **Contribution of soil to the global cycles of atmospheric trace gases** [from Conrad, R. (1995) Soil microbial processes involved in production and consumption of atmospheric trace gases. In: Jones, J. G. (ed.) Advances in microbial ecology. New York: Academic Press; 207–250]

Trace gas	Lifetime (days)	Mixing ratio (ppbv)[1]	Total budget ($Tg\ yr^{-1}$)	Annual increase (%)	Contribution (%) of soils as source	sink	Impact
N_2O	60 000	310	15	0.2–0.3	70	?	Stratospheric chemistry; greenhouse effect
CH_4	4 000	1 700	540	< 0.8	60	5	Greenhouse effect; tropospheric and stratospheric chemistry
H_2	1 000	550	90	0.6	5	95	Insignificant
OCS	> 350	0.5	12	?	25	?	Aerosol formation
CO	100	100	2 600	1.0	1	15	Tropospheric chemistry
NO	1	< 0.1	60	?	20	?	Tropospheric chemistry

[1] parts per billion volume. 1 ppbv = 10^{-12} molecules of total gas molecules; $Tg = 10^{12} g = 10^6 t$

anaerobic degradation processes during the flooding period. A rice paddy can be considered an intermediate between a freshwater sediment and an agricultural soil, alternating between the two conditions in a yearly cycle. During the flooding period, degradation of organic litter is coupled to sulfate reduction and intensive methanogenesis. Part of the methane formed is reoxidized at the paddy surface where oxygen has access, and another part undergoes reoxidation in the soil microlayer that covers the surface of the rice roots. Like most plants living in water-saturated environments, rice plants exhibit active ventilation of the root system to supply the root cells with oxygen. Thus, methane is transported into the atmosphere and escapes microbial reoxidation, making rice paddies one of the major sources of atmospheric methane. Due to the alternating cycle of rice paddies between dry and flooded situations, the microbial constituents of the paddy soil also have to cope with these changes. Methanogenic bacteria survive the dry period in an oxygen-saturated soil and develop full activity within a few days after flooding. Also other anaerobic bacteria (sulfate reducers, fermentative bacteria) survive the dry period in sufficient numbers, and the same is true for strict aerobes through the anoxic flooded period.

Bogs and **swamps** are wetlands often of low pH ($\geq$2.5) in which degradation of organic matter is impaired by high proton concentrations and by accumulating phenolic compounds. Nevertheless, bogs and peats are by no means sterile environments. They contain considerable amounts of cultivable bacteria (10^4–10^6 cells^{-1}), which are probably active in slow peat transformation.

31.3.4 Microbial Activity Is Still Found Deep in the Ground

In surface soils, microbial activity and microbial biomass are directly correlated with the available amount of degradable substrates (Chapter 31.2.2) and, therefore, both parameters decrease in the soil horizons with decreasing organic content. One would expect that at sufficient depth in the soil, there should be no microbial activity left at all. However, deep sedimentary soils at depths of 200–400 m still exhibit total viable counts of heterotrophic bacteria in the range of 10^4–10^6 (g sediment)$^{-1}$ and also considerable activity of ^{14}C-acetate oxidation, even though there is hardly any degradable substrate available. There are indications that even in solid rock considerable numbers of active bacteria are present that might feed on traces of organic and inorganic substrates that are transported along with groundwater. These indigenous bacteria living in extremely poor environments may fulfill important functions in the removal of trace contaminants in groundwaters (oligotrophic bacteria), but extreme difficulties in cultivation of these organisms has precluded detailed studies. It cannot be ruled out that in many cases the microorganisms found in situ may have been transported there by advection with groundwater streams. Bacteria can be transported through a porous groundwater layer over distances of several hundred meters, a fact that is of special importance for the allocation of freshwater sources free of microbial contaminations e.g, from landfill sites.

Soils are extremely heterogeneous microbial habitats both in space and in time, with respect to temperature, water activity, pH, and supply of inorganic and organic nutrients. Nonetheless, soil is the most important microbial habitat with respect to the maintenance of vegetation, to human nutrition, and to the chemical stability of our atmosphere. Our knowledge on microbial activities in soil is very limited due to the heterogeneity of the soil and the lack of suitable techniques for assessment.

31.4 Extreme Environments Are Habitats for Specialists

Extreme environments have always fascinated mankind, and microbiologists became especially interested in these environments as soon as the first indications of microbial life in boiling hot springs, salt lakes, and highly acidic effluents from coal mine waste piles appeared. However, the definition of what is considered as "extreme" changed with time and always kept a slightly anthropocentric flavor. Temperatures close to the boiling point are undoubtedly "extreme" conditions, and the same is true for temperatures close to the freezing point. On the other hand, a strictly anaerobic bacterium would definitely consider the living conditions of man, in an atmosphere with 21% of a toxic gas, an "extreme" situation.

In a more general manner, extreme environments can be defined as such environments that allow life and survival only to a few taxonomic groups of specialists, meaning that whole taxonomic groups are unable to thrive there and are therefore absent. This concept can be illustrated with the example of temperature dependence. Whereas representatives of all major taxonomic groups of life can thrive and multiply at temperatures of 20–30 °C, vertebrates find their upper limit already at 40 °C, vascular plants and other structurally highly developed life forms in the range of 40–50 °C, and beyond 50 °C, only protozoa, algae, fungi, and prokaryotes can survive. Beyond 60 °C, only prokaryotic life is possible, and beyond 95 °C, a thermal environment appears to contain only archaea as living inhabitants. Thus, the diversity of living organisms decreases with increasing temperature, and a thermal environment, e.g., a hot spring, is inhabited only by comparatively few species (which may still be quite a lot!), and many major taxonomic entities are entirely absent. Similarly, extremely cold places, saturated salt brines, and highly acidic or alkaline biotopes can also be defined as extreme environments because major taxonomic groups are absent, and in most cases, prokaryotes are the only ones to dominate such environments.

The relationship of organisms to specific life conditions are characterized by the epithets "**-philic**" or "**-tolerant**". A **thermophilic** bacterium likes enhanced temperatures and has its optimum of growth at an enhanced temperature. A **thermotolerant** organism can survive and perhaps even thrive at enhanced temperature conditions, but has its optimum temperature in the moderate temperature range. In the same manner, organisms preferring low temperatures (**psychrophilic** or **cryophilic**, **psychrotolerant**), high salt concentrations (**halophilic**, **halotolerant**), high proton activity (**acidophilic**, **acidotolerant**), low proton activity (**alkaliphilic**, **alkalitolerant**), and high pressure conditions (**barophilic** or **piezophilic**, **barotolerant**) are characterized by the respective specific terms.

Scientific interest in extreme environments, beyond a general curiosity about unusual features, focuses on questions regarding adaptation of life to such unusual conditions, and the evolution of organisms able to thrive there. Extreme environments have also been studied as model objects for an understanding of microbial ecology in general. The comparatively small diversity of life forms in such environments should allow an understanding of the cooperation of microorganisms more easily than in the much more complex situation of moderate life conditions, which is characterized by extremely high species diversity and structural complexity. Last, but not least, extreme environments have proven as sources of microorganisms of biotechnological interest. Thermophilic bacteria produce thermostable and thermoactive enzymes used for technical degradation of polymers for the preparation of sugars, under conditions that do not need to be protected from contamination by ambient microorganisms. In the same way, proteases from alkaliphilic bacteria are applied to leather tanning, brewing and in specific detergent preparations, and the same applies to thermophilic lipases used for the saponification of fats in household detergents. Modern molecular biology depends to a large extent on the application of Taq DNA polymerase for controlled synthesis of DNA from DNA templates. The enzyme was isolated from *Thermus aquaticus*, shows optimal activity at 75°C, and can be switched on and off in thermally controlled cycles (PCR, chapter 17.4.3).

31.4.1 Thermophiles Like it Hot

The definition of thermophilic organisms has changed with time. In the 1940s, bacteria growing at temperatures of 40 °C were called thermophiles. From self-heating hay, *Bacillus* and *Clostridium* strains were isolated that exhibited temperature optima in the range of 60–70 °C. Through the 1970s, T. Brock guided a research program on the hot springs in Yellowstone National Park, USA. Numerous thermophilic bacteria were isolated, with temperature optima at 65–75 °C. Since Yellowstone Park is situated at elevations of 1 600–2 200 m above sea level, water boils there at temperatures of 92–94 °C, and higher water temperatures cannot be reached. A new era of research on thermophiles began with the studies by K. O. Stetter, W. Zillig, and H. Jannasch on hot springs at sea level and submarine volcanic emanations. After 1980, numerous extremely thermophilic and hyperthermophilic prokaryotes were isolated, that have temperature maxima up to 110 °C. Such organisms were also called "caldoactive". Today, **thermophilic** bacteria are considered to grow in a temperature range of 40–60 °C, **extremely thermophilic** bacteria at 55–80 °C, and **hyperthermophiles** at temperatures higher than 75 °C and with optima at 80 °C and higher.

Most hyperthermophilic microorganisms known today are Archaea, and only few hyperthermophiles belong to the kingdom of Bacteria. Figure 31.**14** illustrates that hyperthermophilic Archaea and Bacteria are situated on low short branches of the phylogenetic tree of living organisms constructed from 16S RNA sequence similarities; this suggests that they represent early forms of life and that life may have developed originally in thermal environments resembling present-day hot springs or thermal vents. These organisms are considered primary thermophiles, different from the less extreme so-called secondary thermophiles, such as *Bacillus stearothermophilus* and *Thermoactinomyces vulgaris*, which branch off from "higher" twigs of the phylogenetic tree, are closely related to numerous non-

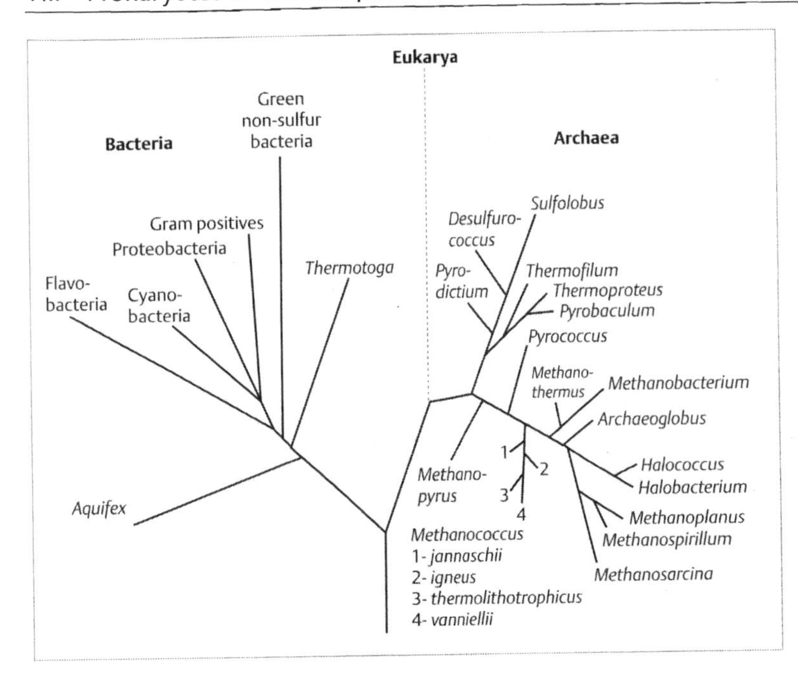

Fig. 31.**14** **Hyperthermophilic representatives in the phylogenetic tree of organisms.** Red lines refer to thermophilic organisms, black lines to non-thermophilic groups [after 4]

thermophilic species, and probably represent secondary developments to thermotolerance and thermophily.

Habitats of thermophilic prokaryotes are not necessarily exotic places. The upper surface layer of soil can heat up on warm summer days to temperatures of 50–60 °C and thus may periodically present suitable growth conditions for thermophiles, especially thermophilic *Bacillus* strains. In compost heaps, organic matter decomposes by the activity of fungi and bacteria, and reaches temperatures of up to 80–85 °C (see Chapter 36). The heat development is partly a side product of the metabolic activity of the microorganisms. The self-heating of insufficiently dried hay is a process analogous to composting; escaping hydrogen has sometimes caught fire and destroyed barns and farm houses. In all these cases, heterotrophic secondary thermophiles predominate that degrade excess amounts of organic matter; hydrogen gas produced as a fermentation side product may serve as an electron source for lithotrophic aerobes and anaerobes.

Extremely thermophilic and hyperthermophilic prokaryotes inhabit marine thermal vents (see Chapter 31.1.7) and terrestrial hot springs, which are found in volcanic areas, such as in Italy, Iceland, and Yellowstone National Park. Such springs are fueled with hot fluids of volcanic origin and are often superheated when they reach the crust surface. Fumaroles release steam at temperatures of 150–450 °C. These hot water emanations often contain H_2S, CO_2, H_2, CO, and other trace gases (see Chapter 31.1.7) together with reduced iron and manganese species; H_2S precipitates upon contact with oxygen in the air as elemental sulfur, which may form yellow layers, especially around fumaroles. It is not surprising that these compounds and elements play a dominant role in the nutrition of hyperthermophiles isolated from such environments.

Neutral or slightly alkaline hot springs can receive some organic matter from exogenous sources, e.g., leaves from surrounding trees. At temperatures higher than 73 °C, autochthonous organic matter has to be synthesized exclusively by chemosynthesis, e.g., through hydrogen oxidation by methanogens or sulfur-, sulfate-, nitrate-, or oxygen-reducing prokaryotes. Sulfide and elemental sulfur and perhaps also carbon monoxide and reduced iron and manganese species are other electron sources for such lithoautotrophic activities (Tab. 31.**8**). Fermentative and anaerobically or aerobically respiring prokaryotes act as consumers; at temperatures lower than 80 °C, also secondary thermophiles (*Bacillus*, *Clostridium*, *Thermoanaerobacter* spec.) become important.

Table 31.8 **Representatives of metabolic groups of hyperthermophiles and extreme thermophiles in terrestrial hot springs**

	T_{max} (°C)	Metabolism
Chemotrophic primary producers		
Pyrobaculum aerophilum	103	$H_2 + NO_3^- \rightarrow NO_2^- + H_2O$
Methanothermus sociabilis	97	$4\ H_2 + CO_2 \rightarrow CH_4 + 2\ H_2O$
Acidianus infernus	95	$H_2 + S^0 \rightarrow H_2S$
Desulfovibrio thermophilus	75	$4\ H_2 + SO_4^{2-} + H^+ \rightarrow HS^- + 4\ H_2O$
Sulfolobus acidocaldarius	85	$2\ S^0 + 3\ O_2 + 2\ H_2O \rightarrow 2\ SO_4^{2-} + 4\ H^+$
Phototrophic primary producers		
Synechococcus lividus	73	Oxygenic photosynthesis
Chloroflexus aurantiacus	70	Anoxygenic photosynthesis
Heterotrophic consumers		
Pyrobaculum islandicum	103	Oxidation of organic compounds with S^0
Thermotoga thermarum	84	Fermentation of organic compounds
Bacillus spec.	85	Oxidation of organic compounds with O_2
Thermus aquaticus	80	Oxidation of organic compounds with NO_3^-, O_2
Thermoanaerobacter brockii	78	Fermentation of organic compounds

Phototrophic primary production is possible only at temperatures lower than 73 °C, where *Synechococcus lividus* can thrive. *S. lividus* forms conspicuous layers in the downflow channel from neutral hot springs (e.g., Octopus Hot Spring in Yellowstone National Park), where the water cools down gradually from more than 90 °C to ambient temperature. As a consequence of flow dynamics, greenish layers develop from the edges to the center of such channels, indicating exactly the temperature limit of this oxygenic photoautotrophic cyanobacterium. After a short time, orange additions are observed in these layers, caused by *Chloroflexus aurantiacus*, an anoxygenic green phototroph that lives on organic excretions of the cyanobacteria. These layers of phototrophic prokaryotes can develop into mats several centimeters thick because above 40–45 °C they are not destroyed by animals. The water chemistry of terrestrial hot springs varies greatly and depends on the chemical properties of the source rock material, and every hot spring can develop an individual microbial community in the temperature gradient of its outflow channel. The oligotrophic heterotroph *Thermus aquaticus* has also been repeatedly isolated from technical water heating systems.

In Table 31.**9**, hyperthermophilic and extremely thermophilic prokaryotes are listed that have been isolated around marine thermal vents (see also Chapter 31.1.7). Due to the extremely steep gradients of temperature around such vents (from > 250 °C to 2 °C within a few meters) an experimental problems, especially in the deep sea, an exact correlation of metabolic functions within such systems is quite difficult to assess.

How can microorganisms survive at enhanced temperatures? First, their enzymes must be adapted to these temperatures, i.e., the enzymes should not coagulate and precipitate as usual proteins do. It was surprising to realize that the primary structure of enzymes of extreme thermophiles is not substantially different from that of isofunctional enzymes from mesophilic organisms. With the few studied enzymes so far, only a few amino acids differ between isofunctional enzymes from thermophiles and mesophiles, but these minor changes in enzymes from thermophiles stabilize the tertiary structure of the enzymes in a way that they stay active into an enhanced temperature range. Extreme thermophiles contain specific chaperones that are organized in so-called thermosomes and help to shape the tertiary structure of proteins specifically for the prevailing high-temperature conditions. This may help to explain why even complete amino acid sequences of hyperthermophilic microbial enzymes show only minor differences in composition to isofunctional enzymes active in the mesophilic range. Further effectors in protein stabilization at higher temperatures are cation bridges and condensed structures due to hydrophobic interactions within the enzyme protein (see Chapter 28.2).

DNA with high G+C content is thermally more stable than DNA of low G+C content (higher "melting points"). However, the DNA of hyperthermophiles does

Table 31.9 **Representatives of metabolic groups of hyperthermophiles and extreme thermophiles found at marine thermal vents**

	T_{max} (°C)	Metabolism
Chemotrophic primary producers		
Methanopyrus kandleri	110	$4\ H_2 + CO_2 \rightarrow CH_4 + 2\ H_2O$
Methanococcus igneus	91	$4\ H_2 + CO_2 \rightarrow CH_4 + 2\ H_2O$
Methanococcus jannaschii	86	$4\ H_2 + CO_2 \rightarrow CH_4 + 2\ H_2O$
Pyrodictium occultum	110	$H_2 + S^0 \rightarrow HS^- + H^+$
Archaeoglobus fulgidus	95	$4\ H_2 + SO_4^{2-} + H^+ \rightarrow HS^- + 4\ H_2O$
Aquifex pyrophilus	95	$2\ H_2 + O_2 \rightarrow 2\ H_2O$
		$2\ S^0 + 3\ O_2 + 2\ H_2O \rightarrow 2\ SO_4^{2-} + 4\ H^+$
Heterotrophic consumers		
Thermotoga maritima	90	Fermentation of organic compounds
Archaeoglobus profundus	90	Oxidation of organic compounds with SO_4^{2-}

not exhibit unusually high G+C contents; they range between 31 and 60 mol%, which is not higher than that of mesophilic bacteria. The DNA may gain additional stability by positive supercoils formed by action of a reverse gyrase, which appears to be present only in extreme thermophiles. In some thermophiles, polyamines (spermidine, thermine) have been detected; these may help to stabilize DNA at enhanced temperatures, but such compounds are also found in mesophiles. Basic proteins (histones) that shift the "melting point" of DNA by up to 30 °C or more, may be more important for stabilizing the DNA at higher temperatures. Obviously, there is no single mechanism for DNA protection in extremely thermophilic and hyperthermophilic microorganisms, rather, there are several stabilizing factors that each act in a different way.

The cytoplasmic membranes of thermophilic bacteria are rich in longer-chain fatty acids and contain less unsaturated fatty acids than those of mesophiles. This increases the "melting temperature" of such lipids and secures optimal stability and fluidity at enhanced temperatures. Membranes of archaea consist of lipids of the glycerol polyprenyl ether type (Chapter 7.10.5), which provide much higher rigidity than ester lipids can, especially if enhanced proton concentrations (pH < 4) act as a further stress factor (see Chapter 31.4.3). Lipids of the octoprenyl diglycerol tetraether type, which span the entire cell membrane, are especially helpful in stabilizing the membrane at high temperatures.

Isolation of prokaryotes at increasing temperatures has raised the question whether there is an upper temperature limit to life and, if so, what it is. Since life needs liquid water, hyperthermophiles growing above 100 °C are found preferentially in environments with enhanced pressure, e.g., at deep-sea vents (see Chapter 31.1.7) where the boiling point of water is higher tha 100 °C. Nonetheless, there are physicochemical limita tions to biochemistry at high temperatures that ar difficult to overcome. At 130 °C, essential cell constituent such as ATP and DNA decompose by thermal hydrolysi Their half-lives shorten to a few minutes or even second and biochemical activities in conventional terms woul hardly be possible under such conditions. Thus, the uppe temperature limit of life has to be assumed to b somewhere between 113 and 130 °C; claims of discover of microbial activities at higher temperatures (250 °C have been disproven later.

Also the pathways of energy and carbon meta bolism in hyperthermophilic prokaryotes may diffe from those known in mesophilic organisms. I hyperthermophilic fermentative archaea, ATP is forme from acetyl-CoA, not through a transacetylation t acetyl phosphate, but directly via an (ADP-forming acetyl-CoA synthetase reaction, perhaps becaus acid anhydride linkages are less stable at enhance temperatures. Also the early steps of hexose degra dation proceed through modified versions of th Embden-Meyerhof-Parnas and the Entner-Doudoro pathway, with less formation of phosphorylate intermediates, but also with smaller ATP gains tha in the corresponding mesophilic pathways (se Chapter 12.2.1). The hyperthermophilic sulfate reduce *Archaeoglobus fulgidus* uses in its one-carbon metabo lism coenzymes that were originally found exclusivel in methanogenic archaea (methanofuran, tetrahydro methanopterin, redox factor F_{420}). Autotrophic carbo dioxide fixation proceeds only through the reductiv tricarboxylic acid cycle and the carbon monoxid dehydrogenase pathway (Chapter 8.1.2); the Calvi cycle is used among the thermophiles only by th cyanobacterium *Synechococcus lividus*.

The molecular mechanisms of adaptation to heat shock are described in Chapter 28.2.

31.4.2 Some Like it Cool

Most of the biosphere of Earth is cold. The deep ocean has a stable low temperature at 2–3 °C, and the polar surface regions are continuously at temperatures in the same range or much lower. Soil surfaces in temperate climates freeze only periodically, depending on their climatic exposure. Specialists for life in such cold environments are called **psychrophilic** organisms, which exhibit maxima of growth rates at 15 °C or lower, and their minimum temperature for growth may even be below 0 °C. Interestingly, eukaryotic algae appear to be important primary producers in such environments. The snow alga *Chlamydomonas nivalis* grows as green vegetative cells in the snow, but develops red-pigmented spores that form pinkish layers on snow and glacier surfaces. Diatoms and green algae have also been detected in permanently frozen seawater in Antarctica.

Psychrotolerant microorganisms are present in soils and waters of temperate regions, and show metabolic activity up to 20–30 °C. Psychrophilic and psychrotolerant microorganisms are of special interest in modern food technology. The preservation of food by cooling and freezing does not necessarily stop microbial activity entirely; instead it selects for psychrotolerant strains able to multiply at low temperature at a low, but significant rate. Milk that has passed pasteurization at the dairy plant is handled only at low temperature. When this milk spoils, this is no longer an activity of classical lactic acid bacteria, but of psychrotolerant *Pseudomonas* strains that turn the milk into a bitter, ill-smelling, and even toxic (primary amines) liquid.

Psychrophilic and psychrotolerant bacteria grow only slowly at temperatures close to the freezing point. They adapt to low temperature through an enrichment of polyunsaturated fatty acids in their membrane lipids, which secures sufficient membrane fluidity and transport activity at low temperature. Enzymes of psychrophilic bacteria are quite sensitive to thermal denaturation and lose activity at slightly enhanced temperatures.

Nonetheless, there is a lower temperature limit to microbial activity at a few degrees below 0 °C, when intracellular water starts to freeze. However, microorganisms survive freezing, especially if induced under protecting conditions, e.g., in the presence of glycerol, sugars, or dimethylsulfoxide, which help to avoid the formation of ice crystals that would destroy cellular substructures, especially membranes. Deep-freezing preserves microbial cultures over several years without detectable damage; however, not all types of prokaryotes are equally suitable to preservation in deep-frozen stocks.

31.4.3 Acidic and Alkaline Environments as Microbial Habitats

Most known microorganisms grow in a narrow pH range around neutrality, between pH 5.5 and 8.5. Some produce organic acids (e.g., lactic acid, acetic acid, propionic acid, and butyric acid) that lower the pH in the surrounding medium to values around pH 3–4. Experts in this process are *Acetobacter* species, which accumulate acetic acid (pK = 4.75) up to 0.8 M concentration, and lactic acid bacteria, which produce lactic acid (pK = 3.8) to 0.2 M concentration. Such bacteria are considered **acidotolerant** because they are able to withstand and survive rather acidic conditions, but are able to grow and multiply also at neutral pH. Slightly acidic conditions also develop from the degradation of organic matter in silage formation, where mainly lactic acid bacteria, but also acetic and butyric acid producers develop moderately acidic conditions that prevent further decomposition of organic matter by the majority of less acidotolerant organisms. This example also illustrates that fluctuations of pH, especially to the acidic side, are not unusual in natural environments rich in organic matter. It should be mentioned that fungi in general are more acid tolerant than bacteria are, and that especially yeasts are selected for in slightly acidic environments, such as fruit juices (pH 2–4), together with lactic acid bacteria.

Conditions of higher acidity are caused by inorganic acids such as sulfuric acid and nitric acid, both of which are produced by microbial activity. Reduced sulfur compounds (H_2S, S^0), which accumulate in substantial amounts around volcanic emanations (hot springs, fumaroles; see Chapter 32.4.1), are oxidized by sulfur-oxidizing bacteria to sulfuric acid,

$$S^0 + 1.5\,O_2 + H_2O \rightarrow SO_4^{2-} + 2H^+ \qquad (31.2)$$

a strong acid (pK = 1.9) that accumulates to several millimolar concentrations, with pH values between 0.5 and 3.0 prevailing in the habitat. Often such places are not only strongly acidic, but also hot (up to 90 °C), creating conditions quite inhospitable to microbial life. Most hot springs exhibit pH values around 2.5; sulfur-free springs are usually slightly alkaline (pH around 8.5).

Reduced sulfur is also a major constituent of sulfidic minerals such as pyrite (FeS_2) and copper sulfides. Although the water solubility of pyrite is very low, it is attacked by acidophilic sulfur- and iron-oxidizing lithotrophs such as *Thiobacillus ferrooxidans*, which oxidizes FeS_2 completely to Fe^{3+} and sulfuric acid. The reaction proceeds in three separate steps: First, FeS_2 is attacked by Fe^{3+}, which is in turn re-reduced; Fe^{2+} and Fe^{3+} are both well soluble at pH < 2.0, and the Fe^{2+}/Fe^{3+} shuttle acts as a redox carrier system between the bacterial cell and the poorly soluble mineral. Thiosulfate is released into solution and oxidized by the same bacteria:

$$FeS_2 + 6\ Fe^{3+} + 3\ H_2O \rightarrow 7\ Fe^{2+} + S_2O_3^{2-} + 6H^+ \quad (31.3)$$

$$7\ Fe^{2+} + 1.75\ O_2 \rightarrow 7\ Fe^{3+} + 3.5\ H_2O \quad (31.4)$$

$$S_2O_3^{2-} + 2\ O_2 + H_2O \rightarrow 2\ SO_4^{2-} + 2\ H^+ \quad (31.5)$$

Reaction 31.3 is fast and purely chemical, and Reaction 31.4 is the rate-limiting step, which is dramatically enhanced by the iron-oxidizing bacterium. The overall reaction releases only one proton per pyrite oxidized:

$$FeS_2 + 3.75\ O_2 + 0.5\ H_2O \rightarrow Fe^{3+} + 2\ SO_4^{2-} + H^+ \quad (31.6)$$

The oxidation of sulfidic minerals and dissolution of metal ions in strongly acidic solutions is a natural process quite common in sulfide-rich mineral deposits. It is applied for economic exploitation of low-grade metal ores, such as in coal refuse piles (see also Chapters 10.6 and 36.7).

Primary production in such strongly acidic environments is typically a chemolithotrophic metabolism, where strongly acidophilic *Thiobacillus* strains (*Thiobacillus thiooxidans*, *T. ferrooxidans*) and others are active down to pH values of 1.0. Acidic hot springs allow proliferation of *Sulfolobus acidocaldarius* (optimal conditions at 70–80 °C and pH 2–3) and *Thermoplasma acidophilum* (45–65 °C and pH 2), or *Acidianus infernus* (up to 95 °C and pH 2–3). The eukaryotic alga *Cyanidium caldarium* catalyzes phototrophic primary production at moderate temperatures (up to 50 °C) and proton activities down to pH 1. Few higher plants have their limits near pH 3.0. Prokaryotic phototrophs are never found at pH values lower than pH 4–5.

Heterotrophic inhabitants of acidic environments are *Bacillus acidocaldarius* (optimal growth at 60 °C and pH 3), and *Thermoplasma acidophilum* and *Acidiphilium cryptum* at moderate temperatures. Some invertebrates and protozoa might even reach areas with pH values around 2, and few vertebrates are found at sites of pH 3.0.

Oxidation of reduced nitrogen compounds by nitrifying bacteria also leads to formation of a strong acid, nitric acid (pK = – 1.3). However, since reduced nitrogen is not available in nature in massive amounts (except, for example, in manure pits), the developing acid does not accumulate to conspicuous degrees and is buffered mostly by carbonate minerals, which, however, undergo acid hydrolysis by this process. The corrosion of buildings by oxidation of partly oxidized nitrogen compounds of the atmosphere has been mentioned earlier (see Chapter 30.5.6). Nitrifying bacteria are not considered as acidophilic organisms; they like even slightly alkaline conditions, as provided by carbonatic rocks and concrete.

The cytoplasm of acidophilic and acidotolerant prokaryotes is usually neutral to slightly acidic (> pH 5.5), while the outer medium can exhibit proton activities 4–5 orders of magnitude higher. This causes a rather high ΔpH across the cytoplasmic membrane, which is balanced in extreme cases by an inverted electric potential to keep the overall proton motive force in the usual range of 180–200 mV and to avoid short-circuits due to excessive voltage across the membrane. The mechanisms of pH homeostasis is described in Chapter 28.3.

Also **alkaline conditions** select for specific organisms specialized for such environments. Not much is known about strictly alkaliphilic organisms. Cultivation of such organisms has to face basic problems with precipitation of common mineral components of growth media. Little is known also about alkaline environments, such as alkaline springs and lakes rich in carbonates (soda lakes in Northern Egypt and Israel, with pH values at 9–10.5). Primary production in such lakes is accomplished by cyanobacteria of the genera *Gloeothece*, *Plectonema*, *Spirulina*, *Anabaenopsis* and others, all of which can grow at pH 9–11. Some green algae of the genus *Chlorella* also can proliferate at pH values up to 12–13. Anoxygenic photosynthesis is carried out by species of the genera *Ectothiorhodospira*, up to a maximum of pH 10 (see also Chapter 31.4.4). Among the known alkaliphilic heterotrophic bacteria are *Bacillus*-like isolates, which grow best in a pH range of 9–11.5. Other heterotrophs of the genera *Flavobacterium*, *Pseudomonas*, *Vibrio*, *Corynebacterium*, and *Arthrobacter* are only alkalitolerant; they can withstand pH values up to 11, but proliferate much better at neutral conditions. The pH homeostasis of alkaliphiles is described in Chapter 28.3.

31.4.4 Saline Environments Are Places of Low Water Activity

Dissolved salt ions are surrounded by envelopes of water molecules and therefore compete with cellular constituents, especially proteins and nucleic acids, for available free water molecules. Dissolved salts therefore cause a similar problem to life as does extreme dryness, in the sense that free water for hydration and maintenance of structural integrity of functional polymers becomes limiting. The availability of free water molecules is easily expressed as the water activity (a_w): it is the ratio of the vapor pressure in the gas phase over the system under consideration at equilibrium (salt solution, soil, etc.), divided by the vapor pressure over pure water. The a_w value therefore changes between 0 and 1. $a_w = 1.00$ refers to pure water, seawater has an a_w value of 0.98, saturated salt brines (salt lakes) have a_w values around 0.75, and dried food and candied fruits may reach a_w values down to 0.70.

Soil is a microbial habitat in which, among other things, also the water availability changes dramatically with time, from $a_w = 1.00$ at water saturation down to $a_w = 0.90$ in drought situations. Typical soil bacteria such as *Bacillus*, *Agrobacterium*, and *Rhizobium* spec. can adapt to these changes. Low water availability acts as a stress factor which triggers first uptake of potassium ions and later the synthesis of specific organic compatible solutes (see Chapter 28.4.1).

Water activities of 0.98 (seawater) do not cause unusual problems to microorganisms in such environments. Sodium ions are selectively transported out of the cell by antiport against protons or potassium ions: the latter require far less free water for hydration. As discussed earlier (Chapter 31.1.6), many bacteria can live alternatively in freshwater or seawater, marine and freshwater bacteria are directly related to each other, and the limits between the two are not clear-cut. Some marine bacteria have a special requirement for sodium chloride and are therefore termed **halophilic**.

Salt lakes with enhanced salt concentrations can form basically in two ways, either by condensation of seawater (so-called **thalassohaline lakes**) or by evaporation of inland surface waters. The salt composition of thalassohaline lakes is qualitatively similar to that of seawater, but the salt concentration is higher due to evaporation. A prominent example is the Great Salt Lake in Utah, USA, which has condensed to about one-tenth of its original water volume. Artificial thalassohaline lakes are seawater evaporation ponds used for production of solar salt. In these series of evaporation basins, seawater is condensed by the energy of sunlight, and dissolved salts precipitate depending on their respective solubility. Sodium chloride, as the most soluble salt, precipitates at rather high purity in the last basins of highest salt concentration.

In most temperate and humic climates, salts dissolved from weathering rocks by rainwater and groundwater are transported through rivers into the sea (exorheic regions). In arid climates, rivers do not always reach the open sea, and may end in terrestrial valleys (endorheic regions) where the water evaporates and the salts concentrate, e.g., in salt pans or salt lakes. Such **athalassohaline lakes** can differ basically in their salt chemistry from seawater-derived lakes, depending on the geochemistry of their catchment area, especially in the amounts of calcium, carbonates, and sulfates. Carbonates precipitate early in the presence of excess calcium ions as $CaCO_3$ (calcite); the saturation concentration of calcite at 25 °C is 133 mM. Precipitation of carbonate leads to a relative enrichment of sulfate and chloride in solution. Such a lake is called a sulfate lake; an example is the Dead Sea, Israel. With more calcium ions, sulfate precipitates as $CaSO_4$ (gypsum; saturation concentration 14 mM), which leads to a relative enrichment of chlorides, together with magnesium, sodium, and potassium ions. In the absence of sufficient calcium ions, carbonate remains in solution and causes a strictly alkaline water reaction (soda lakes, e.g., Mono Lake and Owens Lake, California, USA, and lakes in Wadi Natrun, Egypt). Thus, the evaporation of water and the availability of calcium ions largely determine the development of salt lakes and their water chemistry.

The community structure in extremely saline environments is comparatively simple. In the thalassohaline Great Salt Lake, the southern part reaches in its upper water layers a salinity of about 12% (w/v). The water can turn pinkish-red due to mass developments of the red-pigmented green alga *Dunaliella salina*, which acts as main primary producer. It is grazed upon by the brine shrimp *Artemia salina*. In the microbial community, heterotrophic species of many genera, including *Pseudomonas, Marinomonas, Vibrio, Deleya, Halomonas, Brevibacterium, Micrococcus, Bacillus*, and *Sporosarcina* dominate. In the northern arm of the lake, which is cleaved off from the main part through a causeway and does not receive freshwater in substantial amounts, salinity increases up to 25% (w/v) and leads to precipitations of sodium chloride (saturated brine). In these waters, no eukaryotic organisms survive, and the microbial community consists mainly of representatives of the genera *Halobacterium, Haloferax, Halomonas*, and other extreme halophiles, together with strictly anaerobic species of the genera *Desulfovibrio, Desulfohalobium, Haloanaerobium*, and *Haloanaerobacter*. The water and sediment surface can again be pinkish red, mainly due to the pigments in the purple membranes of the extremely halophilic archaea (see Chapter 13.4). Primary production can be accomplished to a certain extent by extremely halophilic cyanobacteria, but there is also input of organic matter from external sources. The population development in thalassohaline lakes is similar to that in seawater evaporation ponds.

Athalassohaline lakes are also dominated by microbial communities in which halophilic cyanobacteria act as primary producers. Depending on the water alkalinity, also haloalkaliphilic sulfide-oxidizing phototrophs of the genus *Ectothiorhodospira* can develop into large, red-pigmented accumulations. Among the halophilic cyanobacteria, *Aphanothece halophytica* is a real specialist for high salt concentrations, whereas *Microcoleus chthonoplastes* and *Phormidium hendersonii* are halotolerant representatives that prefer lower salt concentrations. Extremely dry environments also select for specialists that can establish and survive under such conditions. Little is known about hot and dry deserts and their microbial inhabitants; typical **xerophilic** organisms are fungi (*Xeromyces bisporus*) and spore-forming bacteria that simply survive in such environments and enjoy short vegetation phases after rain showers. Another type of extremely dry environments are the dry valleys in Antarctica that exhibit low temperatures the year round (−10 °C) and extreme dryness. Some green algae thrive inside the surface layers of rocks where they take advantage of small amounts of dew that collects throughout the night; it is speculated that they even profit from the crystal water of the rock material they inhabit, especially throughout the Antarctic summer. They build small microbial communities with heterotrophic bacteria and yeasts. Lichens are also specialists for such extremely inhospitable life conditions.

Various strategies have been developed to cope with the problem of low water availability. Extreme dryness and high salinity both cause water losses of the cytoplasm to the outer environment through osmosis. To counteract this physicochemical process, the cells accumulate specific compounds inside the cell, so-called **compatible solutes**, which keep a necessary minimum of water inside the cytoplasmic membrane (see Chapter 28.4.1).

31.4.5 High Pressure Environments

The deep sea is a huge environment exposed to high pressure (> 300 bar; 30 MPa). The problem of barotolerance and barophily have been mentioned before (Chapter 31.1.6). Since the deep sea is not really an extreme environment by the definition given above (i.e., because nearly all the major taxonomic groups are present), it is not treated here in more detail.

Extreme environments represent special challenges to their inhabitants because the inhabitants have to cope with extremes of temperature, acidity or alkalinity, low water availability, or sometimes combinations of several such factors. In all cases, prokaryotic cells have proven to be superior to eukaryotes, and environments of really extreme properties are typically inhabited only by prokaryotes. The small size gives the prokaryotic cell high mechanical rigidity; there are usually no internal membranes that could suffer from thermal or other stress; the DNA is comparatively stable due to its circular organization, sometimes even further enhanced by supercoiling; and the cell walls can withstand mechanical and osmotic stresses better than those of higher cells. Ether lipids withstand high temperatures, especially in combination with high ionic activity, much better than ester lipids do, and this may explain why especially thermoacidophilic organisms are typically archaea rather than classical bacteria.

31.5 Associations With Animal Digestive Tracts

The outer surfaces of animals are colonized by microbes to a varying degree, depending on the availability of easily degradable substrates. Also the inner surfaces of animals are inhabited by various microorganisms unless the animal host strictly prevents access of microbes. Especially the digestive tract is a preferred habitat for many microorganisms, including bacteria, fungi, yeasts, and protozoa because it provides an excess supply of degradable food partly pretreated by host enzymes, controlled water activity, and rather stable pH and temperature conditions. Microbes, with their high growth rates, would entirely outcompete the host for its food if the host would not take sufficient means to control such unwelcome guests. The evolution of higher animals implied an optimization of their digestive system including its microbial inhabitants, and various types of competition or cooperation or combinations of both have been developed in which the microbes found their existential basis and function. One should realize that adult humans carry more than ten times as many prokaryotic cells in the large intestine as their own body cells!

The function of microbes in digestion varies to a large degree with the feeding type of the animal. Some animals feed on other animals (**carnivores**), others exclusively on plant materia (**herbivores**), and others are less specialized and feed on both meat and plant diets (**omnivores**). Another group utilizes remnants of biomass that have been predigested by other animals and cannot be attributed to a specific source anymore (**detritivores**). The importance of microorganisms in the digestion of an animal can be estimated from comparisons with germ-free control animals, which can be obtained by cesarean section and subsequent germ-free handling and feeding, or, at least for short-term experiments, by giving heavy doses of antibiotics. In many cases, such germ-free animals exhibit severe signs of malnutrition or die, especially among the herbivores; omnivores, such as mice and rats, can be kept under germ-free conditions for long periods.

The various feeding concepts of animals can be compared with a general scheme of an alimentary tract as shown in Figure 31.**15**. Food enters the mouth where it may be mechanically disrupted to a certain extent and mixed with enzymes (e.g., amylases) and passes the esophagus to the stomach, which is mostly strongly acidic. After passage through the small intestine where low-molecular-mass nutrients are resorbed, the food mash (chyme) moves into the large intestine where it is dehydrated and where remnant substrates are fermented by bacterial inhabitants. The rectum finally releases the feces. This basic organization would be sufficient for a carnivorous animal because a meat diet is relatively easy to digest and does not require more refined arrangements. Animal host and microbial inhabitants compete for the incoming food (competition model). Microbes entering the alimentary tract with the food material are largely killed in the strongly acidic stomach by hydrochloric acid, and fermentative bacteria and protozoa find their chance only in the large intestine where they feed on the leftovers that have not been resorbed by the host in the small intestine.

Animals feeding primarily on plant material depend much more on the help of microorganisms and involve them actively in the digestion process (cooperation model). To allow for sufficient activity of these microbes, the alimentary tract is augmented by fermenting chambers which can be situated either before the stomach (pregastric fermentation chamber, **rumen**) or between the small and large intestine (**cecum**), or in the large intestine itself (postgastric fermentation). In the first case, the cooperating microorganisms are exposed to fresh food and can actively predigest it before it enters the stomach and the small intestine where both the original food and the microbial biomass are digested; in the second case, the fermentative microorganisms see only the leftovers that the animal could not utilize and make those accessible for utilization in the large intestine (combined competition/cooperation model). In the following, a few examples will be described in more detail: the cow rumen as an example of the cooperation model, and the human and the wood-feeding termite as examples of the combined competition/cooperation model.

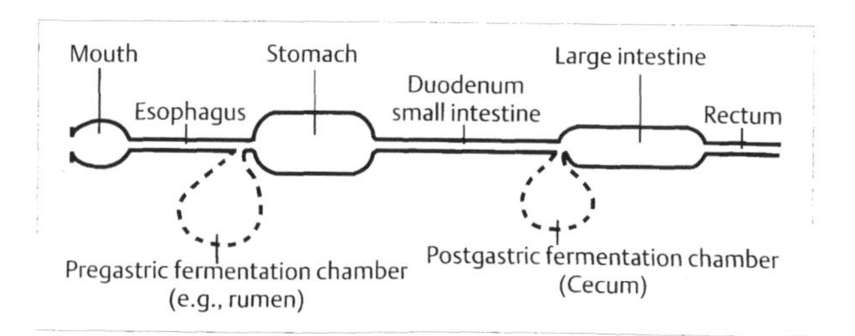

Fig. 31.**15** **Schematic drawing of the digestive system of vertebrates.** Dashed lines refer to fermentation chambers, which are developed only by certain specialists

31.5.1 The Rumen Is a System of Intense Cooperation Between Host and Microbe

The digestive system of ruminants is specifically adapted to the efficient utilization of grass, hay, and straw, which consist mainly of polysaccharides. About 50% of dry grass are hemicelluloses of the polyfructosane and polyxylane type, another 40–50% are cellulose, 2–3% are ligninaceous polymers, and the rest are proteins and lipids. Since the animal has no cellulose-degrading or hemicellulose-degrading enzymes and cellulose is intrinsically intertwined with major part of the hemicelluloses, the majority of the ruminant's feed would be indigestible if it had not developed a refined cooperation with cellulose-degrading anaerobic bacteria.

The rumen of the cow is a dilation of the esophagus, a large fermentation chamber of 100–250-l volume (Fig. 31.**16**). The feed is mixed intensely with saliva in the cow's mouth; the cow secrets 60–120 l saliva per day which is well buffered with bicarbonate. The feed mash is introduced into the rumen where it is mixed and moved around by the rumen wall muscles. Fibrous contents of the diet are sieved out in the reticulum and condensed to small clumps called cuds, which are regurgitated into the mouth and chewed again. Thus, the detention times of rumen contents differ for dissolved and particulate material; the latter is repeatedly regurgitated and swallowed (rumination) to allow optimal degradation.

The rumen is basically an entirely anoxic fermenter that is fed in a semicontinuous manner (Tab. 31.**10**). Oxygen can be introduced in traces with the food and is immediately consumed. After feeding, hydrogen can also accumulate to a few percent concentration in the gas phase, but is soon converted to methane. A wide variety of microorganisms degrade the mainly polymeric diet and ferment it to a mixture of fatty acids,

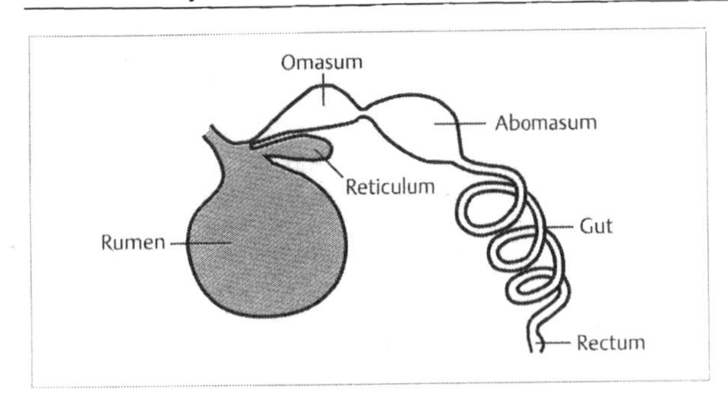

Fig. 31.16 **The digestive system of a ruminant**

roughly according to Reaction 31.7 (glucose stands for hexose equivalents in general):

$$100 \text{ glucose} \rightarrow 113 \text{ acetate} + 35 \text{ propionate} + 26 \text{ butyrate} + 104\ CO_2 + 61\ CH_4 + 43\ H_2O \quad (31.7)$$

As mentioned in Chapter 30.7.2, the methane produced is derived exclusively from the fermentation of one-carbon compounds, e.g., CO_2, formate, and H_2, and is discharged by belching. Acetate-degrading methanogens and secondary fermenters that degrade higher fatty acids do not establish in the rumen because the retention time is too short (0.4–2 days). The fatty acids are resorbed by the rumen mucosa and enter the bloodstream to be further oxidized by the host. The fatty acids amount to 3.7 kg acetate, 1.1 kg propionate, and 0.7 kg butyrate per day; the cow feeds on these fatty acids and not really on plant polysaccharides.

The **microbial flora** of the rumen consists of strictly anaerobic bacteria (10^{10}–10^{11} per g), together with anaerobic ciliates (10^4–10^6 per g). Anaerobic fungi (10^2–10^4 zoospores per g) are partly involved in polymer degradation. An overview of the dominant prokaryotic representatives among more than 200 identified species in the rumen is given in Table 31.**11**. The overall composition of the rumen microflora is rather stable and diverse; no single species makes up more than 3% of the total community.

Rumen protozoa are mainly ciliates that feed on bacterial cells and are also involved in cellulose and starch fermentation. They are not really essential for the functioning of the rumen system, but contribute to the stability of its performance. Some rumen ciliates cooperate with methanogenic partner bacteria associated with their surfaces (see Chapter 30.7.3). Rumen fungi may also help in polymer degradation. Involvement in lignin breakdown has been suggested; however, lignin degradation is hardly possible in the absence of oxygen.

The rumen microflora establishes in new-born calves within a few months; the calves receive the

Table 31.**10** **Physical, chemical, and microbiological characteristics of the rumen**

Characteristic	Property
Physical	
pH	5.5–6.9 (mean 6.4)
Redox potential	−350 to −400 mV
Temperature	37–42 °C
Osmolarity	250–350 meq/l
Dry matter	10–18%
Chemical	
Gas phase (vol. %)	CO_2 (65), CH_4 (27), N_2 (7), O_2 (0.6), H_2 (0.2)
Volatile fatty acids (mM)	Acetic (68), propionic (20), butyric (10), longer chain (2)
Ammonia	2–12 mM
Amino acids	< 1 mM
Soluble carbohydrates	< 1 mM present 3 h postfeeding
Minerals	High [Na^+]; minerals non-limiting
Trace elements/vitamins	Always present; good supply of B vitamins
Growth factors	Branched-chain and aromatic fatty acids, purines, pyrimidines, etc.
Microbiological	
Bacteria	10^{10}–10^{11} g^{-1} (> 200 species)
Ciliate protozoa	10^4–10^6 g^{-1} (predators)
Anaerobic fungi	10^2–10^4 zoospores

Table 31.11 **Dominant prokaryotic species of the rumen microflora**

Cellulose-degrading bacteria
Ruminococcus albus
Butyrivibrio fibrisolvens
Fibrobacter succinogenes
Clostridium locheadii
Hemicellulose-degrading bacteria
Fibrobacter succinogenes
Butyrivibrio fibrisolvens
Ruminococcus albus
Lachnospira multiparus
Starch- and sugar-degrading bacteria
Selenomonas ruminantium
Succinimonas amylolytica
Bacteroides ruminicola
Streptococcus bovis
Lactate-degrading bacteria
Selenomonas lactilytica
Megasphaera elsdenii
Veillonella spec.
Succinate-decarboxylating bacteria
Selenomonas ruminantium
Veillonella parvula
Methanogenic archaea
Methanobrevibacter ruminantium
Methanomicrobium mobile

inocula from their mothers through licking. Calves raised separate from adult companion cattle need much longer to obtain a stable microflora; axenic calves exhibit strong nutritional deficiencies and cannot survive on a grass diet.

Of special importance is the nitrogen supply of the rumen microbiota because the cow's food is rather low in bound nitrogen. The metabolic activity of the microbial community keeps the pool of free amino acids low. Amino groups released in the digestion of the meager protein supply are partly taken up as ammonia by the host mucosa, together with the fatty acid mix produced by the fermentation processes. To secure protein synthesis of the rumen microbiota, part of the urea synthesized in liver does not leave the body with the urine, but instead enters the alimentary tract and the rumen again with the saliva (ureohepatic cycle) and is hydrolyzed there, thus recycling part of the bound nitrogen derived from degradation processes in the host's metabolism. There is no indication of substantial nitrogen fixation in the rumen.

The rumen fluid contains, in addition to the major fatty acids and ammonia, also a mix of partly degraded amino acids, such a branched-chain fatty acids and aromatic fatty acids, which can be taken up by other members of the microbial community and reductively carboxylated and aminated to the corresponding amino acids. In this way many inhabitants of the rumen save efforts for costly de novo syntheses of amino acids. Several rumen microorganisms depend on the provision of such amino acid precursors; this need can be covered in culture by addition of rumen fluid in growth media, in some cases also by defined fatty acid mixtures.

The rumen contents pass into the omasum which acts mainly as a further resorption system in which fatty acids are transferred to the host. The abomasum is an acidic stomach similar to that of non-ruminants, where mainly microbial cell components (400 g bacterial and 300 g protozoal dry cell mass per day) are hydrolyzed by peptic enzymes.The abomasum also secretes **lysozyme**, which helps to digest the bacterial biomass produced in the rumen. A small intestine follows in which parts of the nutrients can be resorbed by the animal host.

Other ruminants are goats and sheep; the increase in sheep keeping, especially in the Third World, is thought to be one of the factors responsible for the global increase in the atmospheric methane concentrations (see Chapter 32). A special case of a "ruminant" is the hoatzin, a non-flying tropical bird that feeds exclusively on leaves. It has developed its crop (which is normally only an organ for efficient mechanical disruption of the food) into a huge fermentation chamber similar to the rumen, and thus ensures sufficient exploitation of its diet by pregastric fermentation.

The **rumen** is a highly developed fermentation chamber that allows an animal lacking cellulases and hemicellulases to degrade and exploit plant material efficiently. The enzymes are excreted by strictly anaerobic, symbiotic bacteria, and a complex microbial community ferments the plant biomass to a mixture of acetate, propionate, and butyrate, which is resorbed by the host animal. Methane is formed as a side product and leaves the rumen through the mouth. Bound nitrogen is partly recycled through the ureohepatic cycle and from digestion of produced microbial cell material to secure a supply sufficient for protein synthesis.

31.5.2 Human Digestion Depends on Competition and Cooperation

Humans are omnivores, meaning that plant material makes up only part of their total diet and they do not exclusively depend on it. The human digestive system has not been optimized for the exploitation of plant fibers. There are no specific dilations to allow extensive fermentative activities, except for the large intestine, which, however, is substantially larger than in a pure carnivore (e.g., a dog). Since a meat-rich diet would easily spoil and perhaps produce toxic derivatives (primary amines, ptoamines) under the influence of fermentative bacteria, the introduced food is first sterilized to avoid uncontrolled fermentation processes. In the stomach, hydrochloric acid is secreted to produce a pH at 0.8–1.5, which is sufficient to kill the majority of introduced microbes. Amylolytic and peptic enzymes are secreted. These enzmes operate under strongly acidic conditions and transform polymeric substrates to sugars, peptides, and amino acids that resorbed in the duodenum. Survival of increased numbers ($>10^2$ per ml) of bacteria in the stomach indicates an irregular, even pathogenic situation, which is often due to insufficient acid production. The acid-tolerant *Helicobactor pylori* can establish in certain areas of moderate acidity in the pylorus region, where it may be one of the causative agents for the development of ulcers.

The duodenum and the anterior part of the small intestine are typically nearly devoid of microorganisms, with overall numbers lower than 10^2–10^3 cells per ml, among which acid-tolerant strains of *Streptococcus, Lactobacillus*, and *Enterococcus* dominate. The pH rises slowly in the anterior small intestine to become slightly alkaline, and becomes neutral with entrance into the large intestine. The number of bacterial inhabitants increases continuously through the posterior part of the small intestine. At the transition zone to the large intestine, the number increases dramatically to 10^{10}–10^{12} cells per ml. Strictly anaerobic bacteria dominate in this community (Tab. 31.**12**) and amount to nearly 50% of the feces material released through the rectum. Aerobic and facultatively aerobic bacteria make up only a minor proportion, $<10^8$ cells per ml. *Escherichia coli* accounts for less than 0.1% of the fecal bacterial community.

The microbial community of the large intestine ferments undigested fibrous food constituents (cellulosic and hemicellulosic material) to a mixture of volatile fatty acids (acetate, propionate, butyrate), together with hydrogen and CO_2 or CH_4. The volatile fatty acids are resorbed by the host into the bloodstream; it is estimated that 10–20% of our nutrition comes from these fatty acids that are resorbed in the large intestine. The hydrogen produced in classical fermentations is either released with the colon gas or consumed by homoacetogenic or methanogenic bacteria. Only one-third of the Western human population maintains stable populations of methanogenic bacteria; "non-methanogenic" individuals are obviously unable to establish methanogenic populations. The ability to form methane appears to be inherited; whole families are either methane-forming or hydrogen forming. The colonic gas is released as flatuses; also here individuals differ in their productivity, from less than 1 ml to 3 l per day. The gas also dissolves in the bloodstream and is released through the lungs. It is therefore easy to assay for methanogenic or non-methanogenic individuals simply by analysis of exhaled air with a sensitive gas chromatograph.

Table 31.**12** **Predominant species isolated from human colon contents**

Species	Number per gram dry matter
Bacteroides thetaiotaomicron	10^{11}
Bacteroides vulgatus	10^{11}
Bacteroides fragilis	10^{10}
Bacteroides distasonis	10^{11}
Bacteroides ovatus	10^{10}
Streptococcus intestinalis	10^{10}
Bifidobacterium adolescentis	10^{10}
Peptococcus prevotii	10^{10}
Peptostreptococcus productus	10^{10}
Strain J 52 (homoacetogenic)	10^{11}
Strain CS7H (homoacetogenic)	10^{9}

The activity of homoacetogenic bacteria in human digestion has long been overlooked because the real dominant species of this metabolic group require cultivation on substrate mixtures. They appear to be numerous (see Tab. 31.**12**) and could act as alternative hydrogen sinks in "non-methanogenic" humans. They can increase the efficiency of food utilization, in the sense that reducing equivalents can produce more acetate to be resorbed by the host. Labeling experiments indicate that one-third to one-fourth of the acetate produced in the colon comes from homoacetogenic CO_2 reduction.

The microbial community in the colon establishes during the first year after birth and remains rather stable throughout an individual's lifetime. Even after serious diarrheas, the microbial community re-establishes quickly to basically the same composition as before. This stability is achieved by surface attachment of the major gut bacteria to the colon epithelium, which secures a stable and continuous seeding. The composition of human colon microflora is rather similar among individuals in the Western hemisphere and is largely independent of their feeding behavior. There is barely a substantial difference in the composition of the microflora between for example, a strict vegetarian and an omnivoric person. Dramatic changes in the composition of the colonic microbial communities can be caused by intensive treatment with antibiotics.

The intensity of microbial fermentations in the large intestine is largely determined by the amount of fermentable substrates supplied. Insufficient resorption of certain food constituents (e.g., of lactose by persons unable to resorb this sugar) can produce increased fermentation activities and, with that, also increased gas production. The same happens when large amounts of plant saccharides are supplied that are not degraded in the duodenum by the host's enzymes. Among these are galactosides (raffinose, stachyose, verbascose), which constitute up to 15% of the dry mass of beans and other leguminous plants.

The strictly reducing conditions in the large intestine and the enormous metabolic versatility of strictly anaerobic bacteria are important with respect to possible side effects of nutrients and food additives. Food colorants are reduced, e.g, by reduction and cleavage of azo-bridges, to amino derivatives; nitro groups are reduced to nitroso or amino groups; phenolics are demethylated or deglycosylated; ketones reduced to secondary alcohols; and halogenated aromatics dehalogenated. Therefore any synthetic compound introduced into human nutrition has to be checked for possible anaerobic transformations to potentially toxic or otherwise adverse products.

The digestion of fibrous plant material takes place exclusively in the large intestine. This appears sufficient because humans are omnivores and depend on such food components only to a limited degree. Our cecum is not developed into a separate fermentation chamber and appears not to serve a special function in our nutrition. In contrast, the horse as a strict vegetarian has a dilated cecum, and its entire nutrition depends on cecal and colonic fermentations of ingested plant matter; the cellulose- and hemicellulose-fermenting *Fibrobacter succinogenes* makes up about 13% of the total microbial biomass in the horse cecum. There is no specific system for nitrogen resorption, and bound nitrogen leaves the horse as undigested microbial biomass and urea with feces and urine. Horse manure is therefore of much higher value as a nitrogen fertiliser than is cow manure.

Certain plant-feeding rodents, such as rabbits, have developed an extended cecum as a fermentation chamber for the digestion of fibrous plant material. The nutrients released are mainly taken up in the large intestine. Rabbits also consume their own feces (**coprophagy**); obviously, this helps the animal to recover microbial cell protein, vitamins, and other micronutrients that are not sufficiently resorbed in the colon.

31.5.3 Termites Have Specialized To Form Many, Different Feeding Types

Among the myriads of insects, there are innumerable types of highly specialized nutrition, and in most cases close associations with symbiotic microorganisms can be expected and await microbiological elucidation. Termites are key mineralizers of litter in tropical and subtropical regions, between 40° Southern and 50° Northern latitude. Due to their global importance and often meager diet, their alimentary system has received special attention. One has to realize that there are about 3 000 termite species with many different feeding behaviors, and generalizations should be made with great caution. The case of a wood-feeding lower termite will be discussed here because this feeding type has been studied in most detail.

A schematic diagram of the organization of the digestive system of the lower termite *Reticulitermes flavipes* is depicted in Figure 31.**17a**. The entire gut system is about 12 mm long and has a total volume of a few microlitres. Measurements of redox potentials indicated that at least the dilated hindgut is a strictly anoxic habitat, with E_0 values in the range of −200 to −300 mV. It was argued on this basis that the termite hindgut represents a small replica of the cow's rumen, and that digestion in this system should follow similar rules. However, the intestinal tract of the termite in vivo is closely associated with air-filled tracheae, and the thin gut epithelium cannot insulate the gut contents from access of oxygen from air. Actually, the low redox potential inside the gut and the obvious absence of measurable amounts of oxygen are features that can only be maintained by active oxygen consumption inside the gut. Studies with microelectrodes have revealed steep concentration gradients of oxygen across the gut epithelium, and allow the calculation of a substantial oxygen flux into the gut. The small size and the high surface-to-volume ratio obviously make the termite gut a microbial habitat entirely different from a cow rumen.

The feeding efficiency of wood-feeding lower termites remains a remarkable example of a cooperative digestion system. *Reticulitermes flavipes* does not produce its own cellulases. The wood material is chewed to small particles by the mandibles and the gizzard of the insect. These particles are digested by a mixed microbial community that fills about 65% of the hindgut volume. The hindgut contains large numbers of protozoa that ferment the cellulosic part of the wood particles to mainly acetate, hydrogen, and CO_2. H_2 and CO_2 are converted to more acetate by homoacetogenic bacteria. Monomeric sugars, disaccharides, and trisaccharides

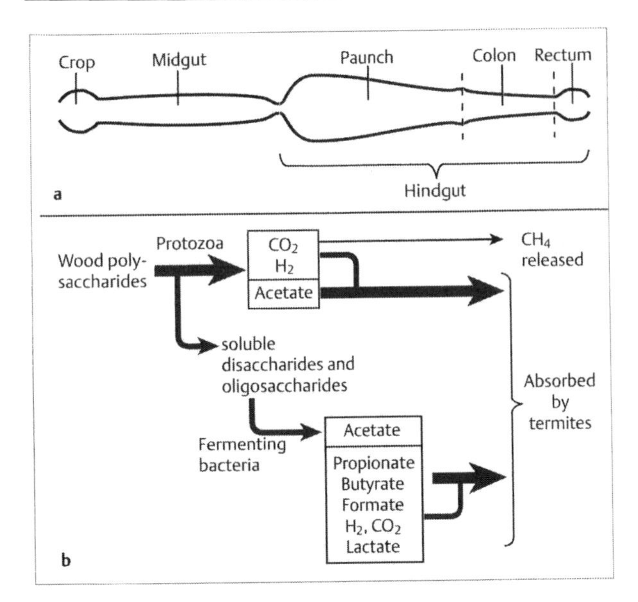

Fig. 31.**17a,b Digestion in a wood-feeding lower termite (*Reticulitermes flavipes*). a** Structural organization of the digestive tract. The overall length of the gut is 10–12 mm. **b** Carbon and electron flow in the intestinal tract of *R. flavipes*. The width of the arrows indicates the relative importance of the respective paths

released by the protozoa are fermented by hindgut bacteria to mainly acetate and small amounts of propionate and butyrate. Although the microbial community contains high numbers of lactic acid bacteria, lactate is barely measurable in the hindgut fluid. There are good indications that the lactic acid bacteria form mainly acetate and transfer part of their electrons to the oxygen diffusing into the gut, thus helping to maintain anoxic conditions. The overall metabolic flux leads from wood cellulose to acetate as the dominant fermentation product, which is resorbed by the insect host (Fig. 31.**17b**). There are also methanogenic bacteria present in high numbers in the *R. flavipes* hindgut; however, no significant amounts of methane are released. If methane is formed in these guts, it is probably immediately reoxidized in the gut's surface layers. Other termites produce substantial amounts of methane and contribute significantly to the methane supply of the atmosphere (see Chapter 32).

It is still unclear how the protozoa that digest the wood particles separate the cellulose moiety of wood from the lignin residues so efficiently. Early reports have claimed that lignin is efficiently degraded in the termite gut, even in the apparent absence of molecular oxygen. Today it is known that oxygen penetrates into the gut lumen and may be involved in partial depolymerization of lignin, but most of the introduced lignin is excreted with the feces. The hindgut microflora also contains large amounts of spirochetes whose metabolic function is entirely unknown because they have never been cultivated.

Since wood is an extremely nitrogen-poor diet, the nitrogen metabolism of wood-feeding termites deserves special attention. The termite converts waste ammonia into uric acid as all insects do, and stores it in the fat body tissue. The uric acid is not excreted, but is released into the hindgut where it is immediately hydrolyzed by a large population (10^8 cells per ml gut volume) of fermentative bacteria. Thus, the nitrogen can be entirely recycled within the insect host and very little bound nitrogen ($< 0.05\%$ of dry mass) is lost with the feces.

Several termites are able to fix nitrogen by the nitrogenase activity of their gut microbiota. In some cases, the measured activities are sufficient to support growth of the whole insect colony with bound nitrogen from N_2; in other cases, the detected nitrogenase activity is far too low to contribute substantially to the overall nitrogen budget.

The digestive system of higher termites is much more complex and consists of several different compartments. The dilated hindgut is a fermenting chamber to which oxygen has limited access. Cellulolytic protozoa are not present in the hindguts of higher termites, regardless of what their feeding basis is. There are wood-feeding higher termites that produce their own cellulases and do not depend on microbial help in this matter. Soil-feeding higher termites live on soil organic matter that they dissolve from the ingested soil; with that they fulfill a function in the tropics similar to that of the earthworms in temperate zones. The physiology and biochemistry of this feeding type still needs to be elucidated; it is known that the ingested soil passes through an alkaline pretreatment before it enters the posterior hindgut for microbial exploitation.

31.5.4 Microbes Also Help in Other Feeding Specializations

Other insects have other types of highly specialized associations with intestinal or external microorganisms for digestion of poorly degradable food sources. The fungus-cultivating termite *Macrotermes* grows fungus gardens (*Termitomyces* species) that produce cellulases to degrade plant fragments, and both the plant litter and the fungal mycelia are consumed by the termite. The wood wasp *Sirex cyaneus* lays eggs into mature wood and cultivates a fungus, *Amylosterium* spec. that provides cellulases and xylanases, and the fungal biomass supports the feeding need of the growing offspring.

An interesting type of symbiotic association was discovered in so-called shipworms. These are actually mussels (Bivalvia: Teredinidae) that bore holes into wooden boards and live inside the board, consuming most of its mass without showing indications of their activity outside. The fragile outer tegument of the board eventually collapses, with disastrous consequences to wooden ships. From the esophagus of these animals, a paired organ branches off. This organ was originally described as the "gland of Deshayes" and recently was shown to be filled with a practically pure culture of an unknown microaerobic bacterium. These bacteria produce cellulase enzymes, thus helping the animal to digest the cellulose moiety of wood. The bacteria can also fix molecular nitrogen and thus supply a source of bound nitrogen to go with the host's extremely meager diet.

Worms and mussels feeding in cooperation, sometimes even in intracorporal arrangements with symbiotic sulfur- or methane-oxidizing bacteria have been mentioned already in the context of the submarine thermal vents (Chapter 31.1.7). Many examples of this kind are known today that live also in shallow coastal sediments, right at the interface between the reduced and oxidized sediment layers. The animal host provides an organized environment with optimal simultaneous supply of oxygen and sulfide or methane, either by water circulation through its burrough or by the blood stream, and feeds either on organic excretions of the symbiont or by digestion of its cells.

Certain nematodes, such as the Stilbonematinae that inhabit oxic/sulfidic interfaces in shallow marine sediments, are covered with sulfide-oxidizing bacteria in amazingly regular surface arrangements (Fig. 31.**18**). The nematode migrates upwards and downwards through the oxic and anoxic sediment layers and exposes the bacteria to oxygen and sulfide in alternating cycles. The bacteria protect the nematode from excess amounts of toxic sulfide and are also selectively grazed upon by the nematode, perhaps through selective shedding. Analysis of the δ ^{13}C ratio of the bacterial and the host biomass indicates that the host feeds nearly exclusively on the autotrophic sulfur oxidizers. The attached bacteria have not yet been cultivated; studies with 16S rRNA probes indicate that they are phylogenetically related to the genus *Thiomicrospira*.

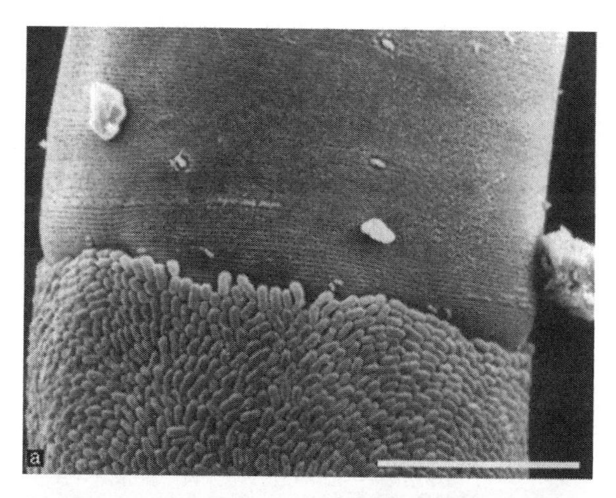

Fig. 31.**18a,b** **Association of sulfide-oxidizing lithotrophic bacteria with the surface of nematodes** (Stilbonematinae). Note the highly organized spatial arrangement. Bar equals 10 µm in each panel.

a Short rods on the surface of *Catanema* species, behind a dilated, non-covered collar, which protects the bacterial lawn from shedding.

b Bacteria attached to the cuticula of *Eubostrichus parasitiferus*. Cells are fixed to the surface with both ends and develop from a length of about 5 µm to filaments of 30 µm in length. (Courtesy of J. Ott and M. Polz)

31.6 Other Associations With Animals

31.6.1 Bacterial Luminescence

Bacterial luminescence is an activity of microorganisms grouped in the genera *Photobacterium* and *Vibrio*. They comprise Gram-negative, facultatively anaerobic bacteria belonging to the Enterobacteriaceae and carry out under anoxic conditions a mixed acid fermentation with acetate, lactate, succinate, formate, ethanol, acetoin, and carbon dioxide as products. They are all typical marine bacteria requiring at least a salinity of 1–2% (w/v) salt in the growth media. The biochemistry of the light emission process is described in Chapter 11.6. The ecological function of the light emission is still a matter of discussion. Luminescent bacteria can be isolated from seawater where they are present at low, but relatively constant numbers (1–100 ml^{-1}), at least over the upper 500 m of the water column. They are easily enriched on fish or shellfish exposed under seawater to low temperature (15 °C). Free-floating single cells do not emit light. The cells excrete a low-molecular-mass lactone compound (e.g., *N-β*-ketocaproylhomoserine lactone) called the autoinducer, which acts as a stimulus to other luminescent bacteria to signal cell density ("quorum sensing") as a function of autoinducer concentration. Once the autoinducer reaches a critical concentration (see Chapters 25.2, 25.3, 27.7.3 and Fig. 20.**17**), the luminescence apparatus is switched on. This regulatory system ensures that the light emission does not waste energy, and that it is turned on only when a sufficient cell density is present to produce a visible light flash. This is the case either in specific light organs of certain fish species, where apparently pure cultures of, for example, *Photobacterium phosphoreum*, *P leiognathi*, or *Vibrio fischeri* are maintained at high cell density. Since fish eggs are sterile upon delivery, it is unclear how the light organ is selectively inoculated, and how culture purity in the light organ is maintained throughout the life of the fish. Antibiotic-like compounds are excreted by such bacteria and may help to control these microbial aggregations.

Other types of luminescent bacteria are inhabitants of the digestive tract of fish and accumulate in the midgut to cell densities that permit light emission. In this case, light may be focused downwards by the swim bladder to reduce visibility of the fish against the skylight and thus decrease its exposure to predation. Enteric luminescent bacteria are excreted with the feces and stay luminescent when at a high cell density. Such light-emitting fecal pellets may have a better chance to be swallowed by other fish, thus reintroducing the bacteria into their preferred habitat. It appears that free-floating single cells of luminescent bacteria in the sea represent only a transition state between assocations with fish where they find their real selective advantage.

The genera *Xenorhabdus* and *Photorhabdus* are terrestrial Gram-negative facultatively anaerobic bacteria belonging to the family Enterobacteriaceae. Both are associated with nematodes, the genus *Xenorhabdus* with the nematode *Steinernema*, the genus *Photorhabdus* with the nematode *Heterorhabditis*; *Photorhabdus* is also luminescent and thereby resembles the above-mentioned marine luminescent bacteria. *Xenorhabdus* and *Photorhabdus* live inside the intestinal tract of juvenile nematodes that attack insect larvae. The bacterial inhabitants are released into the hemocoel of the larvae, proliferate, and kill the insect host by a specific toxin, thus providing permissive conditions for reproduction of the nematodes. After 1–2 weeks, several hundred juvenile nematodes each carrying the bacterial inhabitant in their intestine leave the insect and search for a new host. Nematodes without bacterial inhabitants do not kill the larvae and cannot multiply efficiently. Insect carcasses infected with *Heterorhabditis* and *Photorhabdus* turn red in color and emit light due to the bioluminescence of the bacteria. The *Xenorhabdus/Photorhabdus*-nematode system holdspromise for the control of insect pests through a biological system.

Xenorhabdus and *Photorhabdus* both produce in their infective stages antibiotically active secondary metabolites and two types of protein crystals whose function is still unknown. These crystals are not involved in killing the insect host. This differs from *Bacillus thuringiensis* which contains protein crystals that are the precursors to an insecticidal toxin (see Chapter 34.4.2).

31.6.2 The Human Skin

Human skin is not a homogeneous microbial habitat. The majority of the skin surface is dry and covered with a sturdy layer of dead epidermis cells that do not permit significant bacterial growth. Only in certain protected body areas such as under the arms, between the toes, or in the genital region, is the surface humidity high enough to allow proliferation of bacteria on the skin surface. Most microbial activity is associated with hair follicles and sweat ducts where sufficient substrates and and also protection from mechanical shearing are provided. These capillary spaces are usually free of microbial inhabitants. The entrance ports are protected by a variety of lactic acid bacteria, among which *Staphylococcus* species dominate. Other representatives are *Micrococcus, Corynebacterium, Acinetobacter, Alcaligenes*, and *Propionibacterium* species. The lactic acid bacteria produce acid by excretion an acidic environment that prevents multiplication of other, non-skin-specific microorganisms. Excessive washing destroys this normal skin flora and supports the establishment of possible skin pathogens such as the yeast *Candida albicans*. Increasing humidity and limited access of oxygen improves the living conditions for anaerobic fermentative bacteria such as *Propionibacterium* species and Clostridia with their characteristic fermentation products. *Propionibacterium acnes* is usually a harmless resident on the skin, but may cause severe inflammation of sebaceous glands and hair follicles, a situation known as skin acne.

31.7 Microbes and Plants

Microbes can associate in various ways with higher plants,either with the shoot, the blooms, or the roots. As with animals, the associations of microbes with plants can be either advantageous for the host or disadvantageous, causing plant diseases, wilt, or death. The latter cases are the area of phytopathology, and fungi are of major importance, together with plant pathogenic bacteria of the genera *Erwinia, Xanthomonas*, and others (Chapter 34.3). The non-pathogenic associations of microbes with higher plants will be discussed here.

The shoot of higher plants is covered with a mixture of transient and resident microbial inhabitants, especially on the lower surfaces of leaves where the stomata are localized. These are the sites of increased humidity where also dissolved sugars may be excreted to feed preferentially fermentative bacteria of the genera *Lactobacillus* and *Leuconostoc*, together with *Propionibacterium* species, *Xanthomonas*, and *Pseudomonas* representatives. These acid-producing bacteria act as protectors of the stomata against plant pathogens in a manner similar to that on the animal skin (Chapter 34.4). The microbial communities on leaf surfaces change with the seasons. In spring, bacterial inhabitants dominate, whereas in fall, with decreasing leaf humidity, fungi gain higher importance. The inner lumina of the leaf tissue are considered to be free of living microorganisms, although exceptions have been reported.

The calyces of blooms contain nectar, a sugar-rich liquid that selects for osmophilic microorganisms, especially yeasts. *Candida reukaufii* has been reported to be present in pure culture in such nectar samples.

31.7.1 The Rhizosphere, a Special Microbial Habitat in Soil

The **rhizosphere** is the soil space in close contact with root surfaces (Fig. 34.1). It may extend from a few millimeters up to several centimeters into the soil space, depending on the soil conditions and the type of plant. The rhizosphere is characterized by excretion of dissolved organic substrates (malate, succinate, acetate, glycolate, also sugars) into the surrounding soil. These substrates attract and feed a microbial community that shows a rather high degree of specialization, depending on the plant species. Release of root cells from mucilaginous root surface layers and from the root cap add particulate substrates to support a complex microbiota. Plants excrete up to 20% of their assimilated organic carbon into the root region and shift with that the nitrogen/carbon balance to a relative nitrogen deficiency. Preferentially nitrogen-fixing bacteria are attracted. These thrive in this spatially highly organized ecosystem and contribute bound nitrogen to the overall nutrient balance. Bacteria in the rhizosphere may also support the plant host by the excretion of complexing agents (siderophores) that mobilize heavy metals such as iron and make them also available to the plant. The plant root controls the microbial community to a certain extent by surface-attached glycoproteins, so-called lectins, which differentiate between growth-supporting and possibly pathogenic microorganisms.

Nitrogen-fixing bacteria have developed cooperations with higher plants to varying degrees. These cooperations are described in Chapter 34.2. Monocotyledonous plants (rye grass on sandy beaches, *Calla* grass on haline lateritic soils) specifically enrich *Azospirillum* species in loose association with their root surfaces. These highly motile spirilla stay in a rather open association with the root surface and do not permanently attach or form specific organs with the plant, but enjoy the supply of organic substrates, enrich the rhizosphere with bound nitrogen, and show a rather high degree of specificity. Sugar cane carries symbiotically nitrogen-fixing *Acetobacter diazotrophicus* cells in intercellular spaces in its stems.

Another well-known type of cooperation between plants and microbes are the associations of tree roots with fungi, called the **mycorrhiza**. Many Basidiomycetes and Ascomycetes develop such associations with the roots of *Pinus*, *Laryx*, *Picea*, and also broad-leaf trees such as oak. Mycelia invade the root hairs and develop complex fine-structured networks of root and fungal mycelium, which together act as a highly efficient resorption system for minerals and water.

The roots of plants living in hydromorphic soils are a special microbial habitat. Rice plants and reed live on wet grounds that are at least periodically flooded with water, which prevents access of oxygen. The soil is usually rich in organic substrates and becomes entirely reduced through the activity of fermentative, iron-reducing or sulfate-reducing bacteria, and methanogens. The root tissue may survive such anoxic incubation periods for short periods by a fermentative metabolism (ethanolic fermentation). In the long run, the root tissue is supplied with oxygen from shoot through a gas exchange system, which in turn transports methane through the shoot into the atmosphere. The transition zone between reduced soil and oxygen-supplied root tissue with steep distribution gradients of sulfide, reduced iron, and methane, is a perfect habitat for lithotrophic and methanotrophic bacteria, which take advantage of the close proximity of their substrates in opposing gradients over distances of only few micrometers. The transfer of air oxygen by ventilation in reeds is applied for water purification in specific reed beds (see Chapter 36). Reed areas are among the most productive ecosystems known, with productivities of more than 250 g carbon $year^{-1}$ m^2. Their productivity is enhanced by nitrogen-fixing Enterobacteriaceae (*Klebsiella*, *Enterobacter*), which live in close association with reed plants such as *Carex elata*.

Rice plants are supported by similar enterobacterial root associations. In rice paddies, cyanobacteria also carry out nitrogen fixation. A special case of microbial cooperation is the association of cyanobacteria (*Anabaena azollae*) with the water fern *Azolla* spec. (Fig. 34.**12**).

31.7.2 The Wetwood Syndrome

The dead wood inside larger stems of trees may be colonized by wood-rotting fungi long before the tree shows visible signs of disease. A special situation is the so-called wetwood syndrome, which is caused by the infiltration of aerobic and fermentative bacteria into the inner part of the wooden stems through branch stubs, frost cracks, or root lesions. The mixed microbial community inside the stem also extends into the living tissue, with concomitant degradation of pectin in middle lamellae and pit membranes. The tree loses control over the water supply and the tissue soaks with water from the ground through capillary forces ("wetwood"). The fermentation may develop into an acid fermentation with accumulation of fatty acids (as with oaks, pine, and fir). In softwood trees on low grounds, such as poplars and willows, fermentation may go on to the methanogenic stage, and methane accumulates inside the tree to several bar pressure. These trees are little affected by the bacterial infestation and survive with their inhabitants for several decades. The wetwood syndrome is often associated with trees under stress caused by other environmental factors (soil acidification, excessive nutrient supply, dryness), and deserves economic interest mainly because it decreases the wood quality dramatically.

Further Reading

Austin, B. (1988) Marine microbiology. Cambridge: Cambridge University Press

Bartlett, D. H. (1992) Microbial life at high pressures. Sci Prog 76: 479–496

Breznak, J. A. (1994) Acetogenesis from carbon dioxide in termite guts. In: Drake, H. L. (ed.) Acetogenesis. New York: Chapman & Hall; 303–330

Breznak, J. A., and Brune, A. (1994) Role of microorganisms in the digestion of lignocellulose by termites. Annu Rev Entomol 39: 453–487

Brock, T. D., ed. (1986) Thermophiles. General molecular and applied biology. New York: Wiley-Interscience

Burns, R. G. (1983) Extracellular enzyme-substrate interactions in soil. In: Slater, J. H., Whittenberry, R., and Wimpenny, J. W. T. (eds.) Microbes in their natural environments. Cambridge: Cambridge University Press; 249–298

Conrad, R. (1995) Soil microbial processes involved in production and consumption of atmospheric trace gases. In: Jones, J. G. (ed.) Advances in microbial ecology, vol. 14. New York: Plenum Press; 207–250

Des Marais, D. J. (1995) The biogeochemistry of hypersaline microbial mats. In: Jones, J. G. (ed.) Advances in microbial ecology, vol. 14. New York: Plenum Press; 251–274

Edwards, C., ed. (1990) Microbiology of extreme environments. New York: McGraw-Hill

Hastings, J. W., and Nealson, K. H. (1992) The luminous bacteria. In: Balows, A., Trüper, H. G., Dubikin, M., Harder, W., and Schleifer, K. H. (eds.) The prokaryotes, 2nd edn. Berlin Heidelberg; New York: Springer; 625–639

Hungate, R. E. (1985) Anaerobic biotransformations of organic matter, In: Leadbetter, E. R., and Pointdexter, J. S. (eds.) Bacteria in nature, vol. 1. New York: Plenum Press; 39–95

Jannasch, H. W. (1989) Litho-autotrophically sustained ecosystems in the deep sea. In: Schlegel, H. G., and Bowien, B. (eds.) Biology of autotrophic bacteria. Madison, Wisc.: Science Tech Publishers; 147–166

Jørgensen, B. B. (1980) Mineralization and the bacterial cycling of carbon, nitrogen and sulfur in marine sediment. In: Ellwood, D. C., Hedger, J. N., Latham, M. J., Lynch, J. M., and Slater, J. H. (eds.) Contemporary microbial ecology. New York London: Academic Press; 239–251

Killham, K. (1994) Soil ecology. Cambridge: Cambridge University Press

Lynch, J. M., and Hobbie, J. E. (1988) Micro-organisms in action: concepts and applications in microbial ecology. Oxford London: Blackwell

Nealson, K. H., Schmidt, T. M., and Bleakley, B. (1990) Physiology and biochemistry of ***Xenorhabdus***. In: Gaugler, R., and Kaya, H. K. (eds.) Entomopathogenic nematodes in biological control. Boca Raton, Flor.: CRC Press

Paul, E. A., and Clark, F. E. (1989) Soil microbiology and biochemistry. New York London: Academic Press

Polz, M., Felbeck, H., Novak, R., Nebelsick, M., and Ott, J. (1992) Chemoautotrophic, sulfur-oxidizing symbiotic bacteria on marine nematodes: morphological and biochemical characterization. Microbial Ecol 24: 313–329

Revsbech, N. P., and Sørensen, J. (1991) Denitrification in soil and sediment. New York: Plenum Press

Stumm, W., and Morgan, J. J. (1981) Aquatic chemistry, 2nd edn. New York: Wiley

Tannock, G. W. (1995) Normal microflora: an introduction to microbes inhabiting the human body. London: Chapman & Hall

Tate, R. L. (1995) Soil microbiology. New York: Wiley

Watanabe, I., and Furusaka, C. (1980) Microbial ecology of flooded rice soils. Adv Microbial Ecol 4: 125–168

Wetzel, R. G. (1983) Limnology, 2nd edn. Philadelphia: Saunders

Sources of Figures

1 Pfennig, N. (1967) Photosynthetic bacteria. Annu Rev Microbiol 21: 285–324

2 Nicholson, J. A. M., Stolz, J. F., and Pierson, B. K. (1987) Structure of a microbial mat at Great Sippewissett Marsh, Cape Cod, Massachusetts. FEMS Microbiol Ecol 45: 343–364

3 Jannasch, H. W., and Taylor, C. D. (1984) Deep-sea microbiology. Annu Rev Microbiol 38: 487–514

4 Blöchl, E., Burggraf, S., Fiala, G., Lauerer, G., Huber, R., Rachel, R., Segerer, A., Stetter, K. O., and Völkl, P. (1995) Isolation, taxonomy and phylogeny of hyperthermophilic microorganisms. World J Microbiol Biotechnol 11: 9–16

32 Global Biogeochemical Cycles

The transformation of chemicals in the environment is largely determined by biological processes. Green plants, down to single-celled algae, and cyanobacteria synthesize organic matter from CO_2 and other inorganic sources by using water as electron source and releasing molecular oxygen as oxidized product (**oxygenic photosynthesis, primary production**). The organic matter thus produced is consumed partly by the plants themselves in respiration and partly by animals and man, which oxidize such organic matter to CO_2 (**consumption**). The remains of non-digested plant material and the excrements from animal metabolism are finally degraded by microorganisms, especially bacteria and fungi, which convert organic remnants oxidatively to CO_2 or, under anoxic conditions, to CO_2 and methane (**destruction, mineralization**). The other constituent elements of biomass (nitrogen, sulfur, phosphorus) are released primarily as ammonia, hydrogen sulfide, and phosphate, and may undergo subsequent further transformations (see also Chapter 30.1). In this scheme, microorganisms have found their main niche in the mineralization of organic matter; after all, most prokaryotes we know are heterotrophs. Autotrophic prokaryotes can be active in primary production (cyanobacteria) or in secondary primary production as is true for most anoxygenic phototrophic bacteria (see Chapter 31.1.5). They carry out a net photosynthesis of organic matter but for this process depend on reduced electron sources (H_2S, H_2, Fe^{2+}, organic electron donors) that are usually derived from organic precursors originating in primary production. The same applies to chemolithotrophic prokaryotes that oxidize reduced electron carriers (e.g., H_2, ammonia, H_2S, Fe^{2+}, and CO), which in most cases owe their reduced state to the oxidation of organic matter originally synthesized by primary production; these organisms depend on reduced electron sources that act as intermediate electron carriers in degradation of organic matter (see Figure 30.7). It has been argued that the deep-sea hydrothermal vents are environments based entirely on chemosynthetic primary production and therefore independent of photosynthesis. However, this view is not correct because these systems use as electron acceptor molecular oxygen that is provided by photosynthesis in the upper seawater layers.

Cycles of element transformation can be completed within a certain ecosystem, meaning that there is no net exchange between cycle intermediates and the surrounding environment. Most ecosystems, however, are open systems in the sense that their chemical constituents interchange with those of the outer world, i.e., such an ecosystem (a lake, the ocean, a soil compartment) acts as a source or a sink for a chemical compound, which is usually transported by air or by flowing water. Estimations of global fluxes of chemical compounds are based on measurements of net fluxes between various environmental compartments over time and require numerous single measurements before generalizations on a global scale can be made.

32.1 The Carbon Cycle Is Governed by Photosynthesis

Carbon is present on Earth in six major pools i.e., (1) inorganic carbonates in sedimentary rocks and sediments, (2) organic carbon of biogenic origin deposited as fossil reduced-carbon resources in coal, natural gas, mineral oil, and sediments, (3) carbonate, bicarbonate and dissolved CO_2 in seawater, (4) organic carbon in soil humus, (5) CO_2 in the atmosphere, and (6) organic carbon in living and dead biomass. The latter four pools are mainly involved in the cycling of carbon that is actually a carbon dioxide cycle governed by the photosynthetic activity of terrestrial green plants (Fig. 32.1). These plants produce bound organic carbon in daylight but oxidize about one third of the bound carbon by respiration, mainly at night. The net difference between both processes, in the range of 60 Pg carbon, feeds the consumers and mineralizers, which finally return the carbon in oxidized form into the atmosphere, thus balancing the overall budget of photosynthesis and mineralization and maintaining a CO_2 content in the atmospheric air of approximately

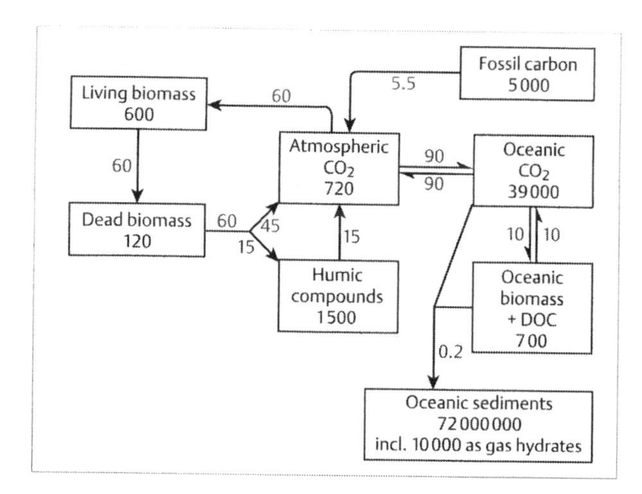

Fig. 32.1 **The global cycle of carbon.** Reservoirs and fluxes are given in Pg carbon. 1 Pg = 10^{15}g = 10^9 metric tons. Reservoirs are depicted in boxes; fluxes are denoted by arrows. DOC = Dissolved organic carbon

0.035% (v/v), equivalent to 350 ppmv (parts per million volumes). These transformations are carried out by a standing crop of biologically active biomass (in the range of 600 Pg organic carbon), which has an average lifetime of about 25 years. It consists nearly exclusively of plant material; consumers and mineralizers make up only a small proportion, less than 2%, of the overall standing crop.

An additional flux of CO_2 release into the atmosphere, in the range of 5.5 Pg, stems from fossil organic material (coal, mineral oil, natural gas) that is burned by human activities. These sources of fossil-bound carbon accumulated in earlier geological times as a result of photosynthetic activity and have been deposited in the earth's crust, mainly in marine sediments, amounting to a total mass of more than $7 \cdot 10^7$ Pg carbon, including biogenic methane bound in deep sediments as gas hydrates. This fossil carbon is balanced in part by approximately 10^7 Pg oxygen in the Earth's atmosphere, which has accumulated there owing to oxygenic photosynthesis.

Burning of fossil fuels is mainly responsible for a slow but continuous increase of the carbon dioxide content of the atmosphere, which has risen from 300 ppmv (parts per million volumes) near the turn of the last century to 350 ppmv at present. This increase to some extent may also be attributed to the large-scale destruction of tropical rain forests, which not only decreases the global photosynthetic activity and the standing crop of plant biomass, but also accelerates soil degradation and desertification with concomitant oxidation of organic carbon stored in the upper soil as humic material.

In the oceans, photosynthesis and mineralization are balanced, thus keeping the CO_2 content of the oceans rather stable. The total CO_2 content of the oceans represents a huge buffer for the atmospheric CO_2 content, and the exchange between atmospheric and ocean-dissolved CO_2 is sufficiently rapid.

The rather simple overview in Figure 32.**1** summarizes mineralization, without stipulating the relative contributions of direct aerobic oxidation and anaerobic processes. As shown below (Table 32.**1**), biogenic methane emission into the atmosphere amounts to an overall flux of approximately 0.3 Pg carbon per year, corresponding to approximately 1% of the total carbon mineralization activity. One could judge from this figure that anaerobic processes are only of minor importance in biomass mineralization. However, this figure only refers to the amount of methane that is released into the atmosphere, without considering how much methane already had been reoxidized before reaching the atmosphere. Moreover, methanogenesis is only the last step in anaerobic biodegradation when all other anaerobic alternatives have been sufficiently exploited. Keeping in mind that nearly all animals base their total metabolism

Table 32.**1** **Rates of annual CH_4 release into the atmosphere**

	CH_4 emission ($Tg\ C \cdot a^{-1}$)	
Source	Annual release	Range
Biogenic		
Natural wetlands	90	75–150
Paddy fields	75	45–130
Ruminants	60	50–75
Termites	40	20–100
Landfills	30	20–60
Oceans	8	4–15
Freshwaters	4	1–20
Total biogenic	**307**	**≈77%**
Abiogenic		
Biomass burning	45	40–80
Gas drilling, venting, pipeline leaks	30	20–40
Coal mining	10	8–30
Methane hydrates	5?	increasing
Volcanoes	0.5	
Automobiles	0.5	
Total abiogenic	**91**	**≈23%**
Total	**398**	

Data calculated for carbon masses. 1 Tg = 10^{12} g = 10^6 metric tons [after R. J. Cicerone, and R. S. Oremland (1988) Global Biogeochem Cycles 2: 299–327]

on a close cooperation with anaerobic fermentative bacteria in their guts (Chapter 31.5), it appears justified to assume that between 20% and 50% of all bound carbon is primarily degraded in the absence of oxygen and that the products of primary anaerobic degradation processes finally are oxidized with molecular oxygen (see Figure 30.**4**).

32.1.1 Methane Is an Important Greenhouse Gas

Methane is the second most important carbon compound in the atmosphere. The methane content in the atmosphere has increased from 0.7 ppmv (parts per million volumes) in the late eighteenth century to 1.7 ppmv today, and the most dramatic increases have taken place within the last 70 years. Methane is formed primarily by anaerobic degradation of organic compounds through the concerted action of fermenting and methanogenic bacteria (see Chapters 12.1.7, and 30.7.1). Their ecological niches are found in all environments where major amounts of organic matter are degraded in the apparent absence of sufficient oxygen (Table 32.**1**). Natural wetlands, including swamps and low-lying soil areas, such as tundra soils, are a major source of atmospheric methane, together with man-made methanogenic environments such as paddy fields and landfills. Higher animals, especially ruminants and termites, contribute significantly to the atmospheric methane supply by intestinal fermentations (see Chapter 31.5). Oceans, although covering more than two thirds of Earth's surface, contribute only a minor part to atmospheric methane. As outlined in Chapters 30.2 and 31.2, methanogenesis is only of minor importance in marine sediments because sulfate as the preferred electron acceptor is available in excess. A major source of methane in marine sediments are methylated amines, which are produced by fish as an antifreeze and apparently are not oxidized by sulfate-reducing bacteria. The small amount of methane released by oceans stems mainly from methanogenesis in rich coastal sediments or from enteric fermentations in marine animals in open ocean waters.

Freshwater lakes contribute only a minor part to atmospheric methane because they comprise only approximately 2% of the global surface. Moreover, methane-producing freshwater lakes are typically stratified during the main production season; methane produced in the sediment is quite efficiently reoxidized at the chemocline (Chapter 31.1.4), and only little methane escapes to the atmosphere during fall mixing. Thus, a well-structured environment releases only a small fraction of the methane originally formed because a major part may have already been reoxidized efficiently within the system.

The same applies also to the other methanogenic habitats. Termite nests harbor considerable numbers of methane-oxidizing bacteria, and the same applies to the rhizosphere of rice plants, thus decreasing the net methane release of these environments. If the water table is lowered in wetlands or tundra soils by periodic climate changes, these environments may turn into methane sinks rather than methane sources owing to the increasing activity of methane-oxidizing bacteria.

Compared to the biological methane sources mentioned, the abiotic sources of atmospheric methane are only of minor importance, contributing altogether less than one quarter of the atmospheric methane budget (Table 32.**1**). Although the ranges of estimated methane fluxes from the various sources are rather wide and the methane emission of some non-biological sources (i.e., the release of methane from underground methane hydrates) may be increasing, it is generally agreed that the observed increase of methane content in the atmosphere is a consequence of the increase of rice cultivation areas and of the number of ruminants, especially sheep, in third-world countries.

Atmospheric methane is degraded to approximately 80% by chemical reactions in the upper atmosphere, especially by oxidation with hydroxyl radicals. About 20% of atmospheric methane is oxidized in soil. As outlined in Chapter 31.3.2, the kinetic properties of methane consumption in soil cannot be easily attributed to classical methane-oxidizing bacteria because the thresholds and half-saturation constants of these bacteria are considerably higher than the concentration of atmospheric methane. There is no doubt that oxidation of atmospheric methane in soil is a biological process; however, the organisms responsible for this activity are so far unknown.

32.1.2 Carbon Monoxide and Hydrogen Are Trace Gases in the Atmosphere

Carbon monoxide (CO) is present in the atmosphere at an average concentration of 100 parts per billion volumes (ppbv), with major differences between the northern and the southern hemispheres and between land and ocean. It is formed in nature mainly by burning of biomass, technical oxidation of non-methane hydrocarbons, other industrial activities, and traffic, in addition to oxidation of atmospheric methane (Table 32.**2**). Biological sources are plants that liberate CO during photorespiration or photooxidation of plant constituents. Degradation of porphyrins, flavonoids,

Table 32.2 **Global cycling of carbon monoxide**

	Flux (Tg C · a^{-1})
Sources	
Biomass burning	430
Vegetation	32
Soil	7.5
Oxidation of non-methane hydrocarbons	385
Industry, household, traffic	274
Oceans	43
Atmospheric CH_4 oxidation	257
Total	**1428**
Sinks	
Release to stratosphere	47
Oxidation by soil	167 = ~12%
Tropospheric oxidation	balance

Data calculated for carbon units. 1 Tg = 10^{12} g = 10^6 metric tons [after R. Conrad (1988) In: Marshall, K. C., (ed.) Advances in microbial ecology, vol 10. New York: Plenum Press; 231–283]

quercetin, and rutin by animals, plants, fungi, and bacteria releases CO from methine bridges. The major part of CO is oxidized with hydroxyl radicals within the troposphere; a small amount escapes into the stratosphere. Only approximately 12% of atmospheric CO is oxidized by biological activities in soil. As with methane oxidation in soil, the microorganisms responsible for this activity are unknown. The known aerobic carboxydotrophic and anaerobic CO-oxidizing bacteria exhibit substrate affinities insufficient for uptake of CO at atmospheric concentration (Chapter 10.9). Again, it has to be assumed that microorganisms differing from the ones known carry out these oxidations in soil; however, these organisms still await isolation and description.

The situation is similar again with atmospheric molecular **hydrogen**. Hydrogen is released mainly by microbial fermentations and as a by-product of nitrogen fixation. Classical aerobic, hydrogen oxidizing (Knallgas) bacteria and anaerobic hydrogen consumers are unable to take up hydrogen at the low concentrations present in our atmosphere. Nonetheless, hydrogen is consumed by soils at these low concentrations by a biological activity that is sensitive to autoclaving but insensitive to antibiotics and other biological inhibitors. The important function of soil as a scavenger of these various trace gases, including also nitrogen and sulfur derivatives, is summarized in Table 31.7.

Prokaryotes and other microbes play a key role in the mineralization of organic carbon and its conversion to carbon dioxide by aerobic and anaerobic activities. The importance of anaerobic mineralization processes has long been underestimated. The major activity of carbon assimilation is oxygenic photosynthesis. Anoxygenic phototrophic and chemotrophic bacteria contribute to CO_2 fixation only to a minor part. Methane is released into the atmosphere mainly by microbial activities, either free or in cooperation with higher animals; methane oxidation is largely a non-biological process. Carbon monoxide is present in the atmosphere only in traces; its formation and its oxidation are only to a small extent catalyzed by microbial activities. Atmospheric methane, carbon monoxide, and hydrogen are oxidized in soil by so far unidentified biotic activities. The actual increase of carbon dioxide and methane concentrations in the atmosphere are both largely due to human activities.

32.2 Prokaryotes Contribute to all Steps of the Nitrogen Cycle

Nitrogen is the second most important bioelement. The element nitrogen changes in the cycle of nitrogen species (Figure 32.**2**) between the redox states −III and +V, depending on the prevailing redox conditions. N_2 is used as a nitrogen source exclusively by nitrogen-fixing prokaryotes that harbor the **nitrogenase** enzyme complex; such organisms act either as free-living individuals or in symbiotic cooperation with plants (see Chapters 8.5, 34.2). The reduced nitrogen species ammonia (or ammonium, NH_4^+, at pH 7.0) is assimilated into cell material by microbes (Chapters 7.3, 8.4) and plants and released again as ammonium during mineralization. Animals can either excrete ammonia directly (fish, other water animals) or synthesize either urea (mammals, most land animals) or uric acid (birds, insects); the latter compounds are excreted as less hygroscopic nitrogen species and are hydrolyzed again to ammonia during their mainly microbial degradation. Plants and aerobic microorganisms can also fulfil their nitrogen needs from nitrate if they have assimilatory nitrate reductase activity (Chapter 8.4). Excess ammonia can be oxidized in the presence of oxygen by nitrifying

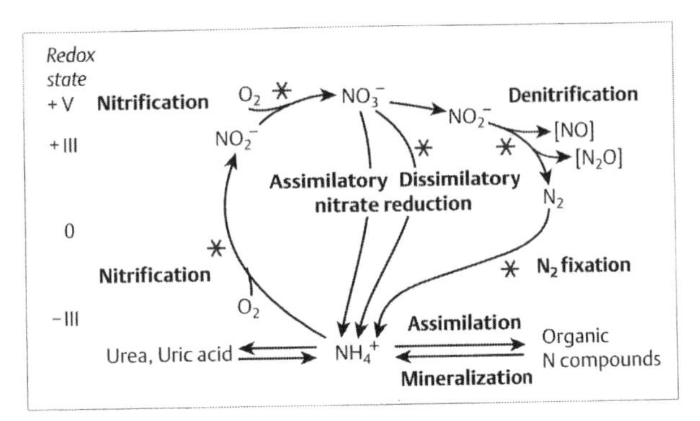

Fig. 32.2 **Key reactions in the cycling of nitrogen species.** The redox state of nitrogen is added on the left. Reactions catalyzed exclusively by prokaryotes are marked with an asterisk

bacteria via nitrite to nitrate (Chapter 10.5); this process depends on the presence of molecular oxygen as cosubstrate for the initial ammonia monooxygenase reaction. (Oxygen-independent oxidation of ammonia with nitrate is a new microbial activity that still awaits biochemical elucidation). Nitrate, besides its assimilatory uptake into plants and microbes, can serve as electron acceptor under anoxic or oxygen-limited conditions (Chapter 12.1). Dissimilatory reduction can lead either to ammonia (**nitrate ammonification**) or, via nitrite, NO, and N_2O to N_2 (**denitrification**), thus balancing the overall flux of N_2 to bound nitrogen. Minor amounts of NO and N_2O can be released into the atmosphere as by-products of denitrification and, after oxidation in the troposphere and stratosphere, are washed out by rain together with other nitrous oxides (NO_x) formed, for example, by lightning, industrial combustion, and oxidation by car engines. The global pools and fluxes of nitrogen species are summarized in Table 32.3, which shows that N_2-binding and N_2-releasing activities are rather well balanced over the ocean but N_2 binding over land exceeds the denitrification rate; the cycle is balanced by release of ammonia into the atmosphere, where it is oxidatively converted to NO_x-species that return to Earth's surface with rainfall. Due to chemical ammonia synthesis (Haber–Bosch process) and concentrated cattle industry, the amount of bound nitrogen in the air and on solid surfaces can reach concentrations that fertilize much of the surrounding area; ammonia in the range of 50–500 μM and nitrate contents of 5–40 μM have been measured in rainwater samples in Germany and the Netherlands. Oxidation of airborne ammonia to nitric acid is one of the key factors in corrosion of buildings (see Chapter 30.5.6).

Table 32.3 **Global pools and fluxes of nitrogen species**

	Land	Oceans
	(Pg N)	
Pools		
Plants	12	0.3
Animals	0.2	0.17
Dead organic matter	300	550
Dissolved inorganic N	160	577
N_2 in overlying atmosphere	1 000 000	2 700 000
	Land	**Oceans**
	(Pg N·a^{-1})	
Fluxes		
Biological N_2 fixation	0.14	0.10
Industrial N_2 fixation	0.06	–
N_2 oxidation by lightning, combustion	0.004	0.004
Renitrification	0.12	0.09
NH_3 release into the atmosphere	0.075	–

Data calculated for nitrogen masses. 1 Pg = 10^{15} g = 10^9 metric tons [after SCOPE Report 21: Bolin, B., and Cook, R. B., eds. (1983) The major biochemical cycles and their interaction. Paris: International Council of Scientific Unions]

Most of the redox processes involving nitrogen compounds are catalyzed exclusively by prokaryotic microoganisms. Binding of molecular nitrogen from the atmosphere is counterbalanced by denitrification; excess release of ammonia due to human activities leads to increased nitrogen accumulation in rainwater and soil. Chemolithotrophic bacteria oxidize ammonia to nitrite and nitrate, nitrate acts as electron acceptor under conditions of oxygen limitation.

32.3 Prokaryotes Participate in all Steps of the Sulfur Cycle

Sulfur, as a further important bioelement, is present in Earth's crust mainly in the lithosphere and in ocean sediments as pyrite and gypsum, also as FeS (Table 32.4). The formation of pyrite from FeS and H_2S in

Table 32.4 **Reservoirs of sulfur species on Earth**

Reservoir	Major form(s)	Total mass (Tg S)
Lithosphere	FeS_2, $CaSO_4$	2.4×10^{10}
Ocean sediments	$CaSO_4$, FeS_2, FeS	2.5×10^9
Seawater	SO_4^{2-}	1.3×10^9
Marine biota	Reduced	30
Soils and land biota	Reduced	2.4×10^5–2.4×10^7
Lakes and rivers	SO_4^{2-}	300
Atmosphere	OCS, SO_4^{2-}, SO_2, DMS, H_2S	4.8

Data normalized for sulfur masses. 1 Tg = 10^{12} g = 10^6 metric tons [after SCOPE Report 21 (see Tab. 32.3) and 48: Howarth, R. W., Stewart, J. W. B., and Ivanov, M. V., eds. (1992) Sulphur cycling on the continents. Wetlands, terrestrial ecosystems, and associated water bodies. Paris: International Council of Scientific Unions] OCS, carbonyl sulfide; DMS, dimethyl sulfide

sediments has been described previously (Chapters 30.1.5, 31.2.2). A further major pool of sulfur is seawater, where sulfate is present at a concentration of 28 mM. Continental resources are considerably smaller, with soils and land biota as the most important sulfur pool. Lakes and rivers are only of minor importance as a sulfur reservoir. The atmosphere contains sulfur compounds, mainly carbonyl sulfide (OCS), sulfate (mainly in aerosols, sea spray), sulfur dioxide, dimethyl sulfide, and hydrogen sulfide.

The biological involvement in the cycling of sulfur compounds is summarized in Figure 32.**3**. Sulfate, the most abundant form of biologically utilized sulfur, is reduced to the H_2S state either by dissimilatory sulfate reduction (Chapter 12.3; *Desulfovibrio, Archaeoglobus* and many others) or by assimilatory sulfate reduction by plants and microorganisms (Chapter 8.3). Organic sulfur-containing cell constituents (cysteine, methionine, coenzyme A, coenzyme M, biotin, thiamine, lipoic acid) are synthesized from sulfur in its reduced state; mineralization of biomass serves to release sulfur as well in the reduced form. In anoxic environments, H_2S can be oxidized by phototrophic bacteria (green and purple phototrophs) to elemental sulfur, which is deposited inside or outside the cell, and further to sulfate (Chapter 31.1.5). Elemental sulfur also can be converted reductively to H_2S by dissimilatory sulfur-reducing bacteria (*Desulfuromonas* spec., *Wolinella* spec., *Pyrodictium* spec., *Thermoproteus* spec.). Aerobic sulfide oxidizers (*Thiobacillus* spec., *Beggiatoa* spec.) oxidize H_sS to sulfate with oxygen as electron acceptor (Chapter 10.6); some representatives (*Thiobacillus denitrificans, Thio-ploca* spec.) can couple sulfide oxidation anaerobically to nitrate reduction. Aerobic oxidation of elemental sulfur to sulfate had been reported for several *Thiobacillus* strains. Nearly all these reactions, except for assimilatory sulfate reduction, are exclusively catalyzed by prokaryotic cells (Figure 32.**3**). Eukaryotes are not directly involved in these sulfur transformations; some animals employ aerobic sulfide-oxidizing bacteria for symbiotic exploitation of the free energy of aerobic sulfide oxidation (see Chapters 31.1.7, 31.5.4).

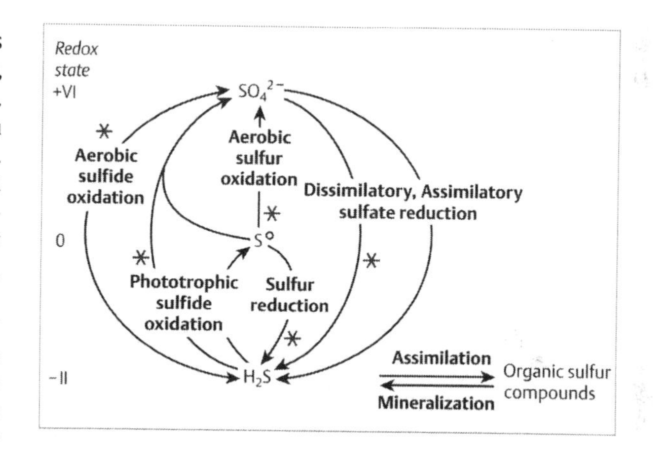

Fig. 32.**3** **Key reactions in the cycling of sulfur species.** The redox state of sulfur is added on the left. Reactions catalyzed exclusively by prokaryotes are marked with an asterisk

Whereas the key reactions of sulfur transformation in nature (Fig. 32.**3**) basically have been known for a long time, some aspects have gained interest only recently. Oxidation of hydrogen sulfide in sediments leads primarily to thiosulfate and traces of sulfite rather

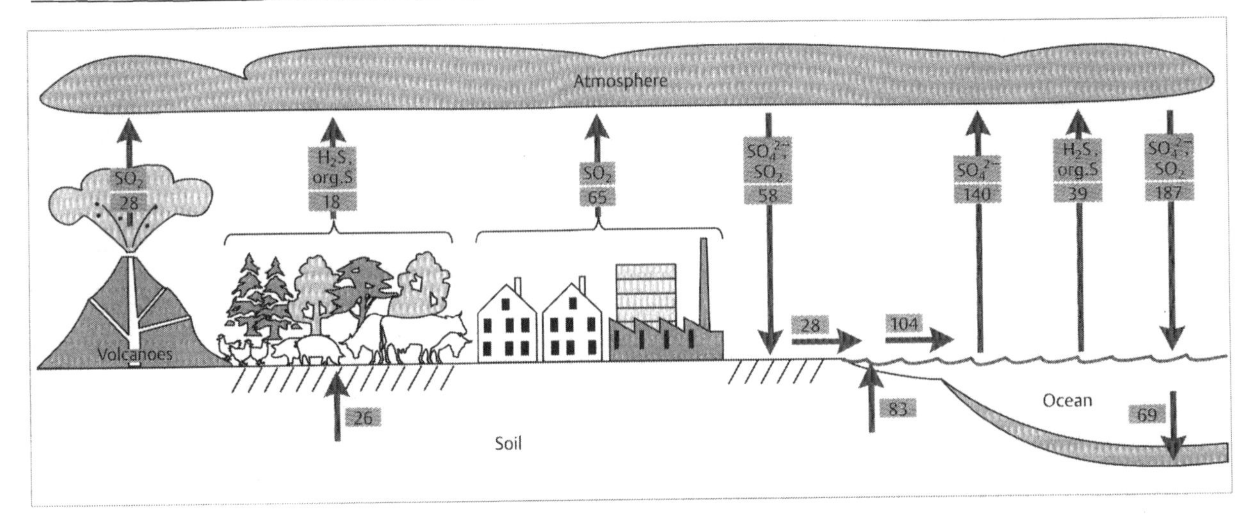

Fig. 32.4 **The global cycle of sulfur.** Numbers denote fluxes in Tg sulfur. 1 Tg = 10^{12} g = 10^6 metric tons. Arrows in the upper part indicate (from left to right): sulfur release by volcanoes, by land biota, by human activities; sulfur deposition to soil; sulfur run-off from land to rivers and lakes; sulfur run-off from rivers and lakes to oceans; release of sulfate from oceans as sea spray; release from oceans as gaseous sulfur compounds (mainly H_2S and dimethylsulfide); sulfur deposition over oceans. In the lower part, weathering of bedrock sulfur to land biota and to rivers and deposition of sulfur in marine sediments are indicated by arrows

than to sulfate; thiosulfate and sulfite may undergo disproportionation to sulfide and sulfate by certain sulfate-reducing bacteria.

Among the gaseous sulfur compounds of importance to the atmosphere, dimethylsulfide turned out to be a rather stable component that is formed mainly in marine habitats by microbial cleavage of the osmolyte dimethylsulfoniopropionate to dimethylsulfide and acrylate, according to

$$(CH_3)_2S^+CH_2CH_2COO^- \rightarrow (CH_3)_2S + CH_2{=}CHCOO^- + H^+$$

This reaction can occur either purely chemically under alkaline conditions, e.g., in microbial mats at CO_2 depletion, or, far more efficiently, by microbial catalysis at neutral pH. It is assumed that dimethylsulfide in the atmosphere is formed to 95% from dimethylsulfoniopropionate cleavage; recently, a further source of dimethylsulfide formation was discovered in which homoacetogenic bacteria transfer methyl groups from methylated phenols (for example, lignin constituents) to sulfide to form methylmercaptan and dimethylsulfide. The geochemical relevance of this new reaction is still unknown; it may play a major role only in terrestrial environments. Dimethylsulfide and methylmercaptan undergo in the atmosphere a chemical oxidation with hydroxyl radicals to methylsulfonate ($CH_3SO_3^-$). Aerobic and phototrophic bacteria can oxidize dimethylsulfide to dimethylsulfoxide [$(CH_3)_2S{=}O$], which can be reduced again by many fermentative and also by aerobic bacteria (see Chapter 30.2.3). Aerobic methylotrophic and anaerobic methanogenic or homoacetogenic bacteria can demethylate dimethylsulfide or methylmercaptan to hydrogen sulfide. The forms of oxidized sulfur in the atmosphere are mostly polar and are washed out by rainwater.

Except for assimilatory sulfate reduction, nearly all steps of the sulfur cycle are dependent on prokaryotic activities. Sulfate is reduced to sulfide in anoxic environments. Sulfide is oxidized in the absence of oxygen by phototrophic or nitrate-reducing bacteria and in the presence of oxygen by aerobic sulfide oxidizers, which also oxidize elemental sulfur. Elemental sulfur is formed by phototrophs and can be either reduced again to sulfide or oxidized to sulfate. Methylated sulfides (e.g., methylmercaptan, dimethylsulfide) are volatile; they are formed mainly by algal and microbial activities and interact in various ways with climatic effectors.

Dimethylsulfide forms mainly over the oceans, in accordance with the above-described mechanisms of formation from a marine precursor. Methylsulfonate and sulfate act as condensation nuclei for cloud formation. Carbonyl sulfide and carbon disulfide are by-products of biochemical and chemical transformation of cysteine, homocysteine, and other organosulfur compounds.

The relative importance of the various fluxes of sulfur compounds between the terrestrial and the marine environment and the atmosphere above both environments are summarized in Figure 32.**4**. Man contributes to the global sulfur cycle significantly by burning of sulfur-containing coal and mineral oil and by thermal oxidation of sulfidic metal ores (for example FeS_2, PbS, ZnS, and HgS), which all release sulfur into the atmosphere as SO_2. More than 50% of the total sulfur emission over land is due to human activities.

32.4 Few Other Elements Undergo Transformations in Cycles

Phosphorus as another important element in biotic processes occurs in biological systems nearly exclusively in the oxidation state +V, as phosphate (Chapters 7.3, 8.6, 30.1.6). Its availability in nature is mainly limited by the low solubility of its calcium and magnesium salts and by its high affinity adsorption to inorganic and organic polymers. In aquatic ecosystems, phosphate is mobilized from inorganic polyphosphates and from organic phosphates (nucleic acid derivatives, nucleotides) by phosphatases that are produced by prokaryotes and algae. Similar activities have been observed also with fungi in soil.

The redox processes involving **iron** and **manganese** species have been discussed previously (Chapter 30.1.7).

Further Reading

Butcher, S. S., Charlson, R. J., Orians, G. H., and Wolfe, G. V., eds. (1992) Global biogeochemical cycles. New York London: Academic Press

Conrad, R. (1995) Soil microbial processes involved in production and consumption of atmospheric trace gases. In: Jones, J. G., (ed.) Advances in microbial ecology, vol 14. New York: Plenum Press; 207–250

Kelly, D. P., and Smith, N. A. (1990) Organic sulfur compounds in the environment: biogeochemistry, microbiology, and ecological aspects. In: Marshall, K. C., (ed.) Advances in microbial ecology, vol 11. New York: Plenum Press; 345–385

King, G. M. (1992) Ecological aspects of methane oxidation, a key determinant of global methane dynamics. In: Marshall, K. C., (ed.) Advances in microbial ecology, vol 12. New York: Plenum Press; 431–468

Taylor, B. F. (1993) Bacterial transformations of organic sulfur compounds in marine environments. In: Oremland, R. S., (ed.) Biogeochemistry of global change. London: Chapman & Hall; 745–781

Section IX
Applied Microbiology

This book has so far described the biology of prokaryotes from different points of view. In order to distinguish between eukaryotes and prokaryotes, the bacterial cell architecture was first presented. This section was followed by the physiological view of prokaryotes. The different metabolic pathways were described in detail, and then the genetic fundamentals of prokaryotes were discussed. A deeper insight into the behavior of prokaryotes became possible as soon as regulatory mechanisms were elucidated. The study of cell growth and differentiation is now one of the most interesting research areas. Conversely, the diversity of prokaryotes and their role in the biosphere still need intensive research efforts in order to understand the principles involved. To summarize, the basic knowledge concerning the biology of prokaryotes is already very extensive. This knowledge can be of use in solving problems affecting human life. Currently, the application of biological knowledge is encompassed within the discipline of biotechnology. Applied microbiology is a subsection which is concerned only with microorganisms. In the following chapters, the aspects of prokaryotes in medicine, agriculture, industrial production, and environmental processes are outlined.

It is worthwhile to study the historical development of the discipline of applied microbiology. Beer, wine, vinegar, cheese, and bread have been produced for thousands of years without the knowledge that these processes are based on the metabolic action of microorganisms. The underlying processes have since been revealed. For example, the production of acetic acid and lactic acid are now well understood. Today, the use of pure cultures allows the controlled fermentation of defined substances, as demonstrated in the field of antibiotics. Due to intensive research efforts in physiology, biochemistry, and genetics, new insights in metabolic networks have been obtained. This research has led to a breakthrough in the application of prokaryotes in industrial, medical, agricultural, and environmental processes. The following chapters summarize the breakthroughs in applied microbiology and outline the developments to be expected in the near future.

Chapter 33 deals with the role of applied microbiology in medicine. One reason for the importance of medical microbiology is the global health problem caused by infectious diseases. Furthermore, new variants of pathogens are still emerging. This chapter on medical microbiology describes the attempts to analyze and understand the interactions between pathogens and their host organisms, including the identification and characterization of virulence factors, host defense mechanisms, and pathogen strategies to overcome defense mechanisms. New techniques allowing the molecular analysis of microbial pathogenicity are also outlined. In addition, diagnosis, therapy, and prophylaxis of infections are discussed.

The application of prokaryotes in agriculture is presented in Chapter 34. The aim is to understand and influence complex interactions between microorganisms and plants. These interactions can be either beneficial, symbiotic, or pathogenic. Although the analysis of symbiosis or pathogenicity is still in the early stages, bacteria such as fluorescent pseudomonads are already used for biocontrol, and rhizobia for symbiotic nitrogen fixation in legumes. The applications of useful soil bacteria in agriculture are outlined. The molecular methods available for the identification and monitoring of microorganisms, no matter whether they are wild-type or genetically modified inocula, or indigenous populations, are also discussed.

Bacteria are able to grow on simple media and to produce compounds by complex metabolic processes not found in other organisms. They therefore represent an ideal source for the production of specific compounds such as food and feed additives in nutrition, or antibiotics in medicine. In addition, bacteria are often used for the production of enzymes which play a role in industry, analytics, and therapy. Chapter 35 describes how bacterial strains can be optimized through classical screening and genetic engineering methods, leading to higher productivity and better utilization of resources. Furthermore, recent developments in bioprocess engineering are presented.

In addition to their ability to synthesize interesting compounds, a large number of bacteria are able to catabolize organic matter. Therefore, bacteria play an essential role in the global cycles of biocompounds. Besides degradable material, there are also complex organic substances that are very stable and slow to decompose. Chapter 36 deals with the use of bacteria in environmental processes for bioremediation of soil, water, and off-gas. This ability is of special interest since the management of environmental pollution is becoming increasingly important in our industrial society. In addition to bioremediation, special characteristics of bacterial enzymes are being utilized, for example, their ability to bind specific compounds is used in the development of biosensors. Finally, the role of bacteria in bioleaching of ores is discussed.

As a result of the development of genetic engineering techniques, applied microbiology has experienced enormous changes. Chapter 37 delineates the expected development from current research to future aspects of prokaryotes in biotechnology. One important aspect is genomic sequencing, which will provide completely new insights into the properties and capacities of bacteria. This will surely revolutionize our understanding of bacteria in agriculture, medicine, and chemical industry, the implications of this new technique are discussed. In addition, the roles of molecular techniques in basic research to understand the environmental processes, biodiversity, and population structures are presented. It is expected that natural bacterial communities which are still mainly unknown due to their enormous diversity and their non-culturable status, will be better understood. The development of regulations initiated by public discussion on the risks and chances of genetic engineering in commercial production is outlined in this chapter.

33 Prokaryotes in Medicine

This chapter deals with a major aspect of applied microbiology the role of microorganisms in infectious diseases. Infectious diseases are caused by pathogenic viruses and bacteria as well as by eukaryotes, such as fungi, protozoa, and multicellular organisms (worms) (see Tab. 33.**1**, 33.**2**). Here the focus will be mainly on pathogenic bacteria. One major aim of medical and veterinary microbiology is to analyze the interaction between pathogenic microorganisms and their host. This includes the identification and characterization of virulence factors and their corresponding genes. The new techniques of molecular biology allow the analysis of regulatory events that influence the expression of virulence genes. In addition to the determination of basic mechanisms of microbial pathogenicity, the new techniques also influence diagnosis, therapy, and prophylaxis of infections.

33.1 Significance of Infectious Diseases

33.1.1 Infectious Diseases Are a Global Health Problem

Worldwide, 1.5–2 billion people suffer from severe cases of infectious diseases. In some countries of Africa, Asia and Middle and South America, infectious diseases are, with 30–50%, the most frequent cause of death among the human population. As indicated in Table 33.**1**, about one billion people suffer from diarrheal diseases every year. About half of the affected people carry bacteria (e.g., intestinal variants of *Escherichia coli*, *Vibrio cholerae*) or viruses as causative agents and the other half suffers from eukaryotic organisms, such as *Entamoeba histolytica*. In addition, nearly 500 million people suffer from "classical" tropical diseases caused by parasites, such as *Plasmodium falciparum*, the causative agent of malaria, and other organisms.

Until the end of the last century, a high percentage of the population in Europe and North America died of infectious diseases caused by "**classical bacterial pathogens**," such as *V. cholerae*, *Salmonella enterica* serovar (sv.) *Typhi*, the causative agent of typhoid fever, and *Mycobacterium tuberculosis*, the causative agent of tuberculosis (Tab. 33.**2**). Such bacteria are specialized to infect higher organisms. They have been designated as

Table 33.**1** **Major infectious diseases in Africa, Asia, and South and Middle America**

Infectious disease	Number of cases in millions[1]	Infectious agent	Type of organism
Diarrheal diseases	1000 per year	*Vibrio cholerae*	Bacterium
		Escherichia coli (intestinal variants)	
		Entamoeba histolytica	Protozoan
Other tropical diseases			
Malaria	267	*Plasmodium falciparum*	Protozoan
Bilharziosis	200	*Schistosoma mansoni*	Multicellular organism
"River blindness"	100	*Onchocerca volvulus*	Multicellular organism
Chargas disease	16–18	*Trypanosoma cruzi*	Protozoan
Leishmaniasis	12	*Leishmania major*	Protozoan
		Leishmania donovanii	Protozoan
Leprosy	11	*Mycobacterium leprae*	Bacterium

[1] Numbers from reports of the World Health Organization (WHO)

Table 33.2 **Major infectious diseases in Europe and North America**

Infectious disease	Infectious agent	Organism
Before the 20th century		
Diarrheal disease	*Vibrio cholerae*	Bacterium
Typhoid fever	*Salmonella typhi*	Bacterium
Diphtheria	*Corynebacterium diphtheriae*	Bacterium
Tuberculosis	*Mycobacterium tuberculosis*	Bacterium
Plague	*Yersinia pestis*	Bacterium
Smallpox	Smallpox virus	Virus
20th century		
Acquired immune deficiency syndrome (AIDS)	Human immune deficiency virus (HIV)	Virus
Respiratory infections	Influenza virus	Virus
	Bordetella pertussis	Bacterium
Urinary tract infections	*Escherichia coli* (uropathogenic variants)	Bacterium
Diarrheal diseases	*Salmonella enterica*	Bacterium
	Escherichia coli (intestinal variants)	Bacterium
	Rota virus	Virus
Nosocomial infections	*Staphylococcus aureus*	Bacterium
	Staphylococcus epidermidis	Bacterium
	Pseudomonas aeruginosa	Bacterium
	Candida albicans	Yeast

obligate pathogens because they usually cause an infectious disease following entry into the human body. By increasing the hygienic standards of the population, establishing broad vaccination programs, and developing strategies for chemotherapy, diseases caused by these "classical" pathogens have been decreased in the developed countries. Presently, only 2–4% of the human population in Europe and North America die from infections. The most numerous infectious diseases occurring in industrialized countries are respiratory infections, mainly caused by viruses, such as the influenza virus, and urinary tract infections, caused by bacteria, such as uropathogenic variants of *E. coli* (Tab. 33.**2**).

A group of microorganisms termed **facultative pathogens** or **opportunists** are becoming increasingly important. In contrast to the "classical" pathogens, opportunistic microbes belong to the normal human microbial flora or are present in the environment. Under certain circumstances (impaired immune response, pregnancy, disturbances of the normal microbial flora), some strains of opportunistic bacteria may cause serious diseases. Strains of the bacteria *Pseudomonas aeruginosa* and *Staphylococcus epidermidis* or isolates of the fungus *Candida albicans*, which are part of the normal body flora, and the environmental bacterium *Legionella pneumophila*, belong to the group of opportunistic microorganisms. Facultative pathogenic organisms play an important role in infections of **immunocompromised patients**. The immunocompetence of these persons is reduced due to either special treatments (e.g., with anticancerogenic drugs or immunosuppressives after organ transplantation) or infections [e.g., by the human immune deficiency virus (HIV), which causes the "acquired immune deficiency syndrome" (AIDS)]. Immunocompromised persons are often hospitalized. Infections acquired at a hospital are designated as **nosocomial infections**. In Germany, more than 1 million cases of nosocomial infections occur yearly.

Especially in developing countries, infectious diseases are a major cause of death. In the industrial countries, infections due to facultative pathogens are becoming increasingly important. Infections acquired in the hospital are termed **nosocomial infections.**

33.1.2 Koch's Postulates Define the Basis for Medical Microbiology

The concept that microorganisms are the causative agents of infectious diseases was developed at the end of the 19th century, especially by Louis Pasteur and Robert Koch. Based on experiments investigating the nature of Anthrax disease (caused by *Bacillus anthracis*) and tuberculosis, R. Koch formulated the following criteria, later called **Koch's postulates**, to classify microorganisms as infectious agents:

1. The presence of microorganisms should be correlated with the occurrence of a specific disease.
2. Microorganisms must be cultivated in pure culture from material of the body of the diseased individual.
3. The isolated species must induce symptoms in experimental animals similar to the symptoms of the disease investigated. The microorganism should be isolated again from the experimentally infected animal. The latter postulate cannot be fulfilled with host-specific pathogens for which no other hosts are available.

Box 33.1 Koch's work also presented the basis for the development of a clear terminology in this area of microbiology.

Symbiosis Interactions between organisms of different species, usually specific and adapted to each other, beneficial for each of them.

Commensalism Interactions between organisms of different species in which only one of the partners takes advantage of the interaction without causing beneficial or harmful effects to its host.

Parasitism Interactions between organisms of different species in which one organism (the parasite) lives on or in a second organism at the expense of this host.

Host Organism in which a parasite lives.

Infection Growth of microorganisms within a host after invasion.

Obligate/facultative pathogen Harmful (parasitic) organisms that cause an infection (and disease).

Pathogenicity Ability of an organism to cause an infection and a disease.

Virulence Degree of pathogenicity of a microorganism.

Virulence or **pathogenicity factors** Properties or products that cause or contribute to pathogenicity (virulence).

In contrast to symbiosis and commensalism, both of which define harmless interactions between organisms of different species, parasitic interactions are harmful for the host organism. **Parasites**, a synonym for the term **pathogens**, can only be defined in relation to their **hosts**. Thus, some parasites, e.g., *Salmonella enterica* sv. *Typhi*, are pathogenic only for humans, while other pathogens are pathogenic only for certain animals, but not for humans. **Pathogenicity**, the ability of microorganisms to cause a harmful infection, is a term always related to a particular bacterial species. It indicates that the species is able to cause an infection in a suitable host. In contrast, **virulence** is a quantitative term used to describe the degree of pathogenicity of a certain strain. For instance, as *Salmonella enterica* sv. *typhimurium* is, in general, able to infect mice, the species is regarded as pathogenic. Different strains of this species, however, show differences in the dosis letalis (LD_{50}) to mice (i.e., 10^3 bacteria of some *Salmonella* strains are enough to kill 50% of mice infected, while 10^8 bacteria of other *S. typhimurium* strains are necessary to lead to the same lethal effect).

Properties or products of pathogenic microorganisms that contribute to microbial pathogenicity (virulence) are termed **virulence** or **pathogenicity factors**. Such factors include adherence factors, necessary for attachment of microorganisms to host tissues; invasins, facilitating the penetration of the pathogen into host cells; toxins, which are able to damage host cells; and factors, such as capsules, lipopolysaccharides, or specific enzymes, which protect the invading microorganisms from the action of the immune system. In addition, iron uptake systems contribute to the pathogenicity of microorganisms. The molecular features of bacterial virulence factors are broadly described in Chapter 33.3.

Infectious organisms should fulfill the criteria of Koch's postulates. Harmful parasites, termed pathogens, have to be described in relation to their host organisms.

33.1.3 Pathogenic Bacteria Infect Different Sites of the Host Body

Infectious organisms are able to spread either from one person to another or from the environment to certain individuals. Figure 33.**1** presents an overview of the routes of dissemination of pathogens that have an impact on the sites of the host body affected. These sites and important pathogens are indicated in Figure 33.**2**. Organisms that are spread via the oral route either directly or by aerosols often cause infections of the upper and lower respiratory tract. Interestingly, *Streptococcus pyogenes* and *Corynebacterium diphtheriae*, which cause the upper respiratory tract diseases scarlet fever and diphtheria, respectively, produce toxins that can disseminate and may act on other sites of the body. Some of the pathogens that cause pneumonia, such as

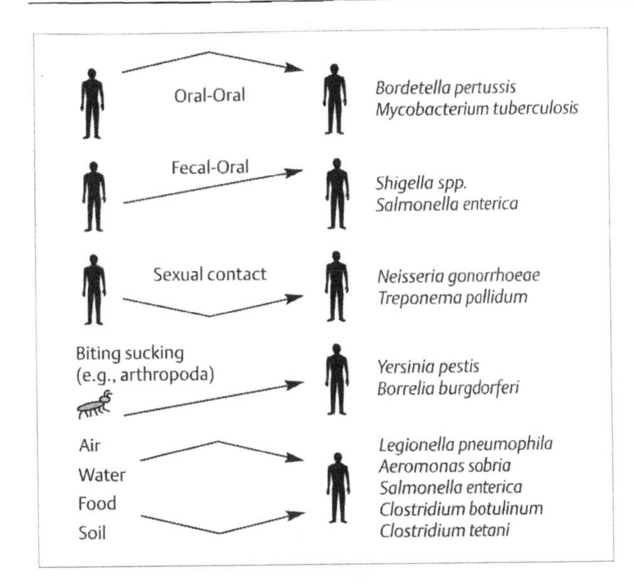

Fig. 33.1 **Transmission of pathogenic microorganisms.** Examples of bacterial pathogens are given on the right

Mycobacterium tuberculosis, are able to multiply inside alveolar macrophages.

Pathogens of the intestinal tract are typical examples of organisms spread from one person to another by the fecal–oral route. They may encounter their hosts through contaminated food or water. Most intestinal pathogens, such as *V. cholerae*, *Shigella* spec., or *S. enterica*, exclusively infect the gastrointestinal tract. *Helicobacter pylori* is a specialist that is able to grow in the stomach of humans. Its colonization is associated with the occurrence of ulcers. In contrast, urinary pathogens, especially uropathogenic *E. coli*, may originate from the patient's own gut flora from where they are transferred to the bladder or the kidney. Pathogenic organisms responsible for sexually transmitted diseases (STD) are transferred from one person to another exclusively by sexual contacts. *Neisseria gonorrhoeae* and *Treponema pallidum*, the causative agents of gonorrhea and syphilis, respectively belong to this group of pathogens.

In addition, microorganisms are transferred by **arthropod bites**. The causative agents of the plague (*Yersinia pestis*) and Lyme disease (*Borrelia burgdorferi*) are disseminated by arthropods. These diseases are termed **zoonoses**. Some of the organisms transferred by arthropods (e.g., *Y. pestis*) are able to multiply within the blood and may cause systemic infections. Other bacteria cause infections that also affect the central nervous system (CNS) (e.g., *Neisseria meningitidis*, *Haemophilus influenzae*). Furthermore, **technical vectors**, such as air conditioners or artificial water systems, can transfer pathogens. *Legionella pneumophila*, the causative agent of Legionnaires' disease, is present in most natural water systems, and only dissemination via aerosols leads to infections. In addition, microorganisms of the soil, such as *Clostridium tetani*, are able to colonize wounds. Some pathogens, such as *Staphylococcus aureus* or *Pseudomonas aeruginosa*, are able to cause various infections at different sites of the body, such as the skin and the respiratory tract (Fig. 33.**2**).

> Microorganisms are able to colonize and infect different regions of the human body. They can be disseminated by various routes and are able to induce different types of infections.

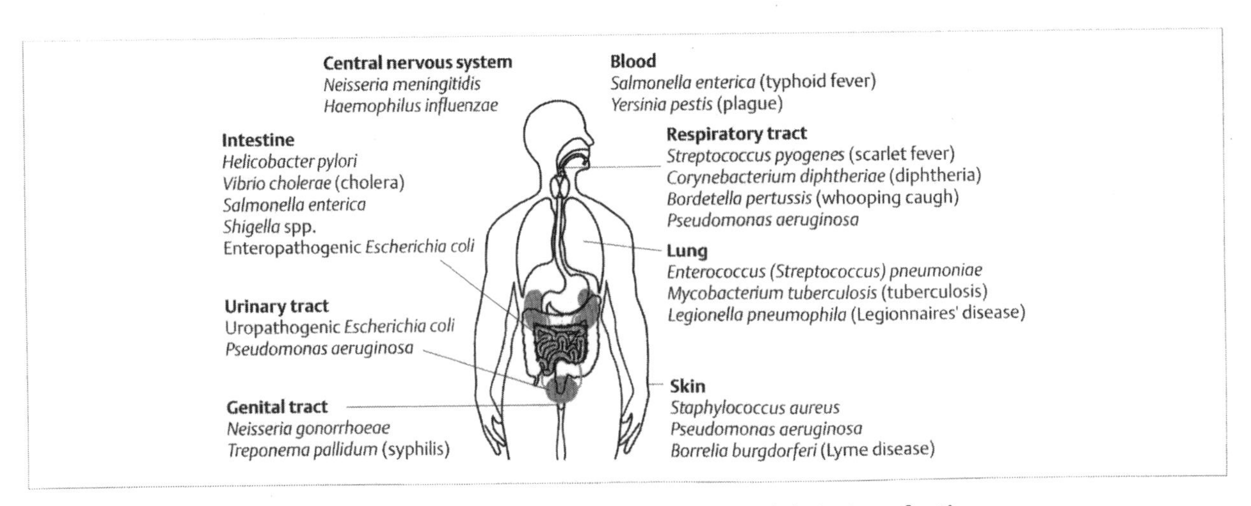

Fig. 33.2 **Infectious diseases caused by important pathogenic bacteria, and their sites of action**

33.1.4 New Variants of Pathogenic Bacteria Are Still Emerging

Among the great variety of prokaryotic microorganisms, about 200 species are considered as pathogenic. The majority of these pathogenic bacteria have been known for a long time, because of which they are considered as "old" or "traditional" pathogens. In addition to these, "new" infectious agents are arising, some of which belong to the group of "new emerging pathogens". During the last 20 years, approximately 30 species have been described as "new" pathogens.

While the properties of some of the "old" pathogens are rather constant, an alteration of important characteristics occurs in many variants. Genes for virulence factors are often located on plasmids or even transposons. Others, especially those coding for toxins, are located on bacteriophages or on special unstable regions of the bacterial chromosomes, termed "pathogenicity islands". All these genes may be transferred to strains of other species or even genera by horizontal gene transfer. Upon acquisition of new virulence genes, new variants of pathogens can emerge.

The occurrence of enterohemorrhagic *Escherichia coli* (EHEC), which causes hemorrhagic colitis and the hemolytic uremic syndrome, is a good example of the development of a "traditional" pathogen with new properties. EHEC strains were first discovered in 1982. Such *E. coli* variants carry a Shiga-like toxin (Slt)-converting prophage, a plasmid carrying the genes for the EHEC hemolysin (e-*hly*), which is another toxin, and an adherence factor (see also Chapters 33.3.1 and 33.3.3). One particular pathogenicity island, which encodes another adhesin as well as proteins involved in invasion, is located on the chromosome (Fig. 33.**3**).

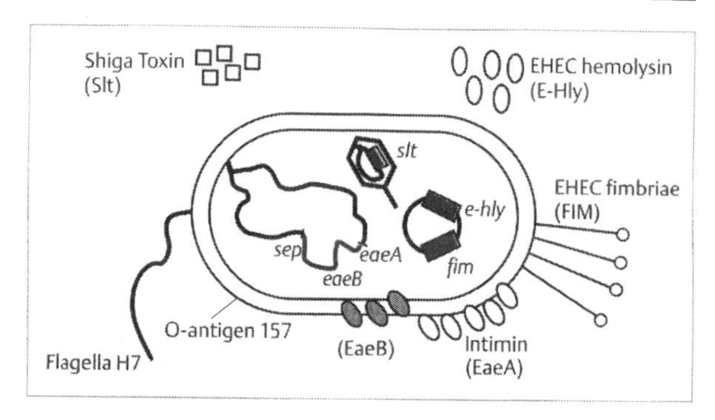

Fig. 33.**3** **Diagram of an enterohemorrhagic *Escherichia coli* (EHEC) bacterium.** Important virulence factors and their corresponding genes, located either on a phage genome, a plasmid, or on a chromosomal pathogenicity island, are indicated. Genes for fimbriae (*fim*), EHEC hemolysin (e-*hly*), attaching and effacing (*eaeA*, *eaeB*), secretion of protein (*sep*), and Shiga-like toxin (*slt*) are also indicated

The occurrence and combination of these genes resulted in the enterohemorrhagic *E. coli* as a new pathogenic *E. coli* variant that had not been observed in the past.

Other "old" pathogens have also changed their properties and are now able to cause infectious diseases at a frequency higher than before (Tab. 33.**3**). Examples are ***Staphylococcus aureus*** strains, which are able to produce "toxic shock syndrome toxins," ***Salmonella enterica*** sv. ***Enteritidis*** isolates of phage type (PT) 4, and ***Vibrio cholerae*** strains of the O serotype O139 that have acquired new *rfb* genes. Another important problem is the spread of drug resistance. Increasing numbers of strains causing tuberculosis and pneumonia (e.g., ***Enterococcus pneumoniae***) are becoming resistant to certain antibiotics. This is due to either point

Table 33.**3** **Bacterial pathogens with "new" properties**

Pathogen	Gene	Property encoded	Underlying mechanism	Disease caused
Enterohemorrhagic *Escherichia coli* (EHEC)	*slt*	Shiga-like toxins	Phage transfer	Enterohemorrhagic colitis
	e-*hly*	EHEC hemolysin	Plasmid transfer	Hemolytic uremic syndrome
Salmonella enterica s.v. *Enteritis* phage type 4	*aer*	Aerobactin	Plasmid transfer	Diarrhea
Vibrio cholerae O139	*rfb*	O-antigen O139	Transfer of chromosomal genes	Cholera
Staphylococcus aureus	*tss*	Toxic shock toxins	Phage transfer	Toxic shock syndromes (TSS)
	mecA	Methicillin resistance	Transfer of chromosomal genes	Nosocomial infections
Streptococcus pyogenes	*speA*	Superantigen	Phage transfer	Toxic-shock-like syndromes (TSLS)
Mycobacterium tuberculosis	*rpoB*	Rifampicin resistance	Point mutations in RNA polymerase gene	Tuberculosis
Enterococcus pneumoniae	*pbp2b*	Penicillin binding protein, penicillin resistance	Transfer of chromosomal genes	Pneumonia

mutations in target genes or to the transfer of genes encoding products that confer resistance to certain antibiotics (Chapter 33.5.3.).

Changes in sanitary standards, behavior, and living conditions of the human population lead to the development of "new" pathogens (Tab. 33.**4**). The increasing standard of diagnostic tools may also be responsible for the detection of "new" pathogens that have been the causative agents of infections for a long time, but were never detected before. Pathogens, such as *Helicobacter pylori*, *Chlamydia pneumoniae*, and *Borrelia burgdorferi*, all of which were discovered during the last 10–15 years, belong to this group of "new emerging pathogens". Furthermore, the increasing rate of HIV infections led to a group of infections, termed "AIDS-related infections". Infectious agents that cause secondary infections in HIV-positive individuals have not been observed at similar frequencies in HIV-negative persons.

One has to distinguish between pathogens that have arisen from "old" pathogens by acquiring certain pathogenicity traits and "new" pathogens, which gained their pathogenic potential by changing environmental conditions or have been discovered only recently.

33.1.5 Molecular Approaches Are Used to Study Pathogenic Bacteria

During the last few years, many genes coding for proteins of pathogenic relevance from different species have been cloned and characterized at a molecular level. The cloned genes have been used as **probes** in DNA–DNA or DNA–RNA hybridization studies to characterize strains with unknown genetic background and/or to analyze the expression and the transcriptional control of virulence genes, such as the toxin genes of *Staphylococcus aureus* or the genes involved in alginate production of *Pseudomonas aeruginosa*, which are regulated by environmental signals and are under the control of a two-component regulatory system (Chapter 33.4.3). In addition, the techniques of **ultrastructural analysis** and **immunological approaches** have been useful methods to describe pathogen-induced events occurring in the host cell in molecular terms, as shown for intracellular pathogens such as *Shigella flexneri* and *Listeria monocytogenes*, the causative agents of shigellosis and listeriosis, respectively.

Table 33.**4** **Recently discovered "new emerging bacterial pathogens"**

Pathogens	Diseases caused
Legionella pneumophila	Legionnaires' disease (pneumonia)
Helicobacter pylori	Inflammation of stomach, gastric ulcer
Chlamydia pneumoniae	Pneumonia
Borrelia burgdorferi	Lyme disease
Non-typical Mycobacteria	AIDS-related infection of respiratory tract
Ehrlichia chaffeensis	Human ehrlichiosis
Bartonella henselae	Cat-scratch disease, bacillary angiomatosis

The study of pathogenic organisms stimulated research in a number of fields of general interest. Thus, toxins have become useful tools to study phenomena of cell biology. For example, actin polymerization was shown to be influenced by toxins of the intestinal pathogen *Clostridium difficile*, and adhesins, such as the M proteins of *Streptococcus pyogenes*, were demonstrated to interact with extracellular matrix proteins. Moreover, the interaction of bacterial adhesins (e.g., the G adhesin of *E. coli* P fimbriae) or toxins (e.g., the B subunit of the cholera toxin) with eukaryotic carbohydrate receptors have been studied in detail. The analysis of protein secretion pathways, such as pathway III, observed in the enteric pathogens *Salmonella enterica*, *Shigella flexneri* and *Yersinia enterocolitica*, as well as that of the α-hemolysin of uropathogenic *E. coli*, is of general interest for molecular biologists. Studies on the structure–function relationship of molecules have employed three-dimensional models of toxins or adhesins and their receptor analogues. In addition, the coordinate regulation of genetic networks, including virulence-associated genes, has been studied (e.g., in *Vibrio cholerae* and *Bordetella pertussis*). Presently, many laboratories are involved in the analysis of in vivo gene expression. The newly described "signature tagging mutagenesis" (Chapter 33.6) seems to be an important approach to identify genes of *Salmonella typhimurium* and other pathogens that are exclusively expressed following interaction of the bacteria with host cells.

Molecular approaches are useful to study the phenomena of microbial pathogenicity. Pathogenicity factors are also used to address issues of general interest in basic research.

33.2 Host Defense Mechanisms

33.2.1 Non-specific Mechanisms May Prevent the Entry of Pathogenic Microorganisms Into the Host

The ability of microbes to infect host organisms depends on the host's **susceptibility** or **resistance** to the pathogen. Pathogens have to compete with the normal flora of the host, which may prevent their growth. Susceptibility of the host organism is also dependent on the nutritional status of the organism, age, and stress factors. **Physical barriers**, such as cilia on the mucosal surfaces, are also of importance. Moreover, the pathogens have to overcome non-specific **chemical barriers**. Low pH of the skin and of certain entry sites, such as the vagina or stomach, may decrease the number of pathogens, as do chemical substances that are part of body fluids (e.g., lysozyme, which hydrolyzes the murein layer of microorganism). Bile salts, anoxic conditions, and the presence of proteinases and other degradative enzymes act selectively on the growth of bacteria in the intestinal tract.

Mucosal surfaces that face certain regions of the human body (e.g., respiratory, gastrointestinal, or urinary tract) contain multiple cell types that may interact with entering microbes. The cells are covered with **mucus**, containing glycoproteins, which is able to bind and eliminate microorganisms. In addition to epithelial cells, **M cells** (microfold cells) are part of the mucosal surface. M cells transport incoming microbes to associated macrophages, which are able to present microbe-specific antigens on their surface. Consequently, a specific T and B cell response is induced, which leads to the production of secretory immunoglobulins A (sIgA) (Chapter 33.2.2). A model of mucosal surfaces is presented in Figure 33.**4**.

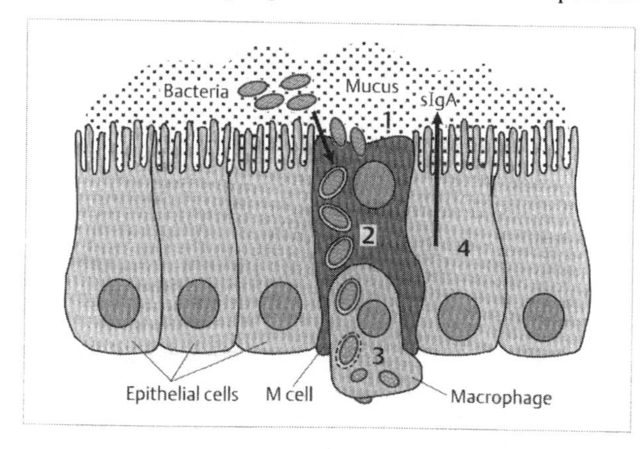

Fig. 33.**4** **Cells of mucosal surfaces.** Epithelial cells, an M cell, a macrophage, and entering bacteria are indicated. (1) Bacteria bind to the M cell. (2) Bacteria are transported to the macrophage via the M cell. (3) Bacteria are digested within the macrophage, and bacterial antigens are presented on the macrophage surface. (4) T and B cells are activated, and secretory immunoglobulin A (sIgA) molecules are secreted

The first active defense line of the host consists of different cellular and humoral factors that attack entering microorganisms. **Professional phagocytic cells** represent **non-specific cellular factors** of the defense system, digest pathogenic microbes, and clear the body of foreign particles. Three phagocytic cells develop from bone marrow stem cells: **Polymorphonuclear neutral** (**PMN**) **leukocytes** or **granulocytes**, **monocytes**, and **tissue macrophages** (Fig. 33.**5a**). Together, these cells form the so-called **reticulo-endothelial system** (**RES**).

Phagocytic cells are able to eliminate microbes by **phagocytosis** (Fig. 33.**5b**). Following attachment of microorganisms to the cell wall of phagocytic cells, the pathogens are ingested. Receptors on the phagocytic cell surface play a crucial role in the uptake of bacterial pathogens and in the interaction of the different parts of the host defense system. Thus, Fc receptors are able to bind bacteria–antibody complexes. C3b receptors have the capacity to bind bacteria that form complexes with the complement factor C3b (see below). This process of uptake of microbes following bacterial binding of opsonins (e.g., antibodies and C3b complement proteins) is termed **opsonization**. Once phagocytized, the microbes become located in a **vacuole** or **phagosome**. Following fusion of the vacuole with special granula or **lysosomes** to a phagolysosome, the microorganisms are attacked by oxygen-dependent and -independent mechanisms. In the phago-lysosome, toxic forms of oxygen, such as superoxide anion ($O_2^{\cdot -}$), hydrogen peroxide (H_2O_2), singlet oxygen ($^1O_2$), and hydroxyl radical ($\cdot OH$) are generated and contribute in addition to other antimicrobial substances and nitrogen intermediates (e.g., NO_2^-, NO_3^-) to the intracellular killing of microbes. Intracellularly living bacteria are able to overcome the bacteriocidal mechanisms of phagocytes and replicate within eukaryotic cells (Chapter 33.3.2).

The **entry of pathogenic microorganisms** into hosts is prevented by non-specific physical and chemical barriers. Phagocytic cells are able to attack entering microorganisms. Invading microorganisms are taken up by phagocytosis, digested, and killed intracellularly. Opsonization, i.e., fixation of microbes by antibodies or complement, stimulates phagocytosis.

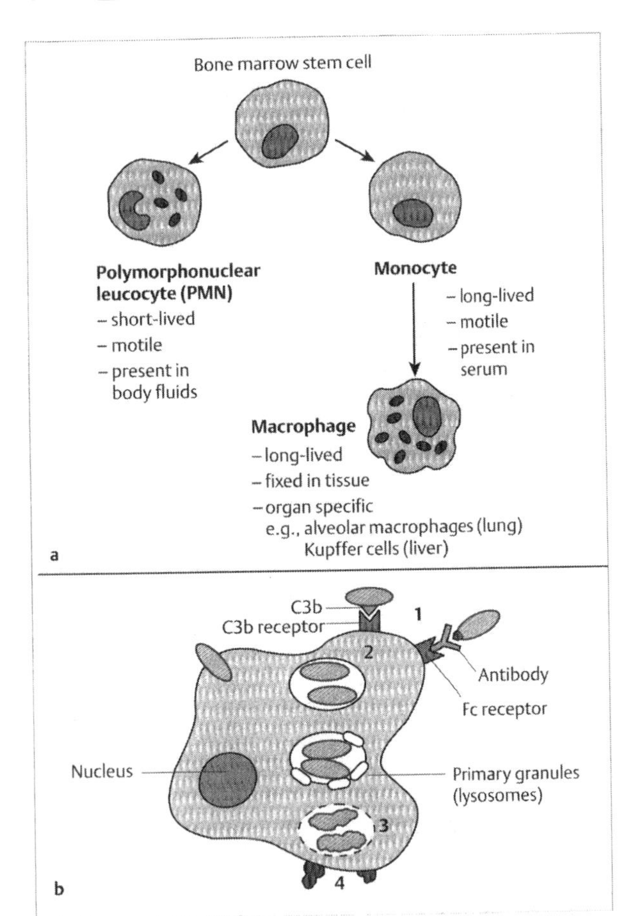

Fig. 33.5a,b **Generation of phagocytic cells and phagocytosis of bacteria. a** The various cell populations are described. **b** (1) Bacteria (in red) bind either via C3b and the C3b receptor or via Fc receptor and antibodies (opsonization) to the phagocytic cell. (2) Bacteria are located in the vacuole (phagosome). (3) Following fusion of vacuole and lysosomes (phagolysosome), bacteria are digested. (4) Bacterial antigens are presented on the surface of the phagocytic cell

The complement system contributes to the **non-specific, humoral host defense**. The complement is a complex system consisting of 26 proteins, most of which are designated C1, C2, C3, etc. The complement molecules are serum proteins and act in a cascade fashion, i.e., the activation of one component results in the activation of another (Fig. 33.**6**). Activation of the complement system occurs via the **classical** or the **alternative way**. Activation of the complement system has various biological consequences, such as:

1. **C3a** and **C5a** cleavage products of C3 and C5 are able to **attract leukocytes**.
2. **C3b** acts as an **opsonin** to enhance phagocytosis.

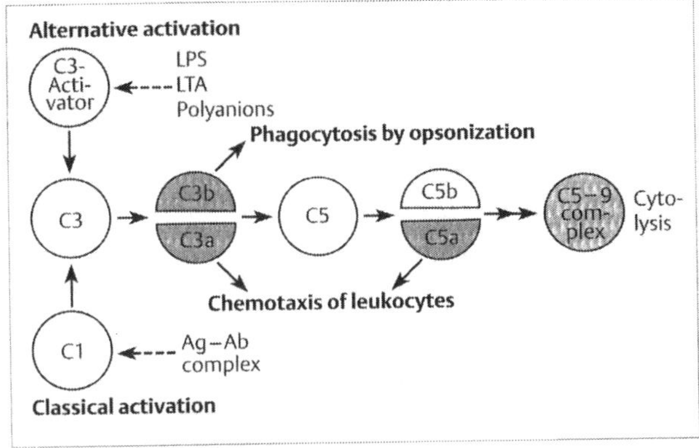

Fig. 33.**6** **Diagram of the complement system.** The complement cascade is activated directly via the alternative way by microbial products, such as lipopolysaccharides (LPS), lipoteichoic acid (LTA) and polyanions or via the classical way by antigen–antibody (Ag–Ab) complexes. Complement molecules (C1, C3, C3a, C3b, C5, C5a, C5, C9) act in cascade and may cause cell lysis of pathogens

3. The **terminal complex C5–9** acts as a **membrane attack complex**. The complex elicits the formation of holes within the bacterial membrane, which cause uncontrolled diffusion of solutes inwards and outwards. Serum-resistant pathogens are able to survive in the presence of complement (Chapter 33.3.8).

Complement is a system of serum proteins that can be activated by microbes. Activated complement leads to several biological phenomena that kill bacteria, and damage membranes.

33.2.2 The Immune Response Against Microbes Is Mediated by B and T Lymphocytes

Higher developed organisms, including humans, have the capacity to react specifically to foreign molecules and cells. This phenomenon is termed the **immune response**. The immune system consists of **B** and **T lymphocytes**, cells that derive from stem cells in the bone marrow. B cells can be activated and converted into plasma cells, which are able to produce antibodies. Antibodies represent the **humoral** and **specific** part of the host defense. During differentiation of B lymphocytes into plasma cells, some B lymphocytes stop differentiation at some point and reside as **memory cells** in the body. Following further antigen stimulation of these memory cells at a later time, they continue their differentiation into plasma cells and rapidly produce antibodies of identical specificity.

Box 33.2 Some important definitions used in relation to the B-cell response are:

Antigen Molecules that react with antibodies and also with activated T cells. Antigens can be molecules of microbial pathogens, such as proteins, lipoproteins, or polysaccharides.

Antigenic determinant (**epitope**) The part of an antigen that reacts with a single antibody. One determinant normally consists of five to ten amino acids or four to five carbohydrate molecules.

Antibody (**immunoglobulin**) Protein molecule directed against antigens or parts of it. Antibodies are secreted by activated B cells (see Fig. 33.7).

Antiserum Serum containing antibodies. A serum containing antibodies of different specificity is termed a **polyclonal antiserum**.

Monoclonal antibody (**Mab**) An Mab is specific for one antigenic determinant (epitope). Monoclonal antibodies are important tools in research and also in the diagnosis of microbial infections (see Chapter 33.5.1).

B lymphocytes are able to secrete antibodies (**immunoglobulins**, **Ig**). Immunoglobulins fall into five major classes: **IgG**, **IgM**, **IgA**, **IgE**, and **IgD**. The **IgG** type is the most common circulating antibody and the prototype of all Ig molecules. IgG molecules are composed of four polypeptides, two heavy chains of 450 amino acids and two light chains of 212 amino acids (Fig. 33.7). The parts of the antibody that show antigenic specificity are termed **variable regions** or **Fab fragments**. The **constant regions** or **Fc fragments**, however, are very similar among antibodies of different specificity.

IgM molecules are large aggregates composed of five IgG-like molecules. **IgA** molecules form dimers in body fluids. Secretory IgA is important for the elimination of extracellular pathogens that colonize mucosal surfaces, such as those of the gastrointestinal tract. **IgE** molecules are involved in allergic reactions, while antibodies of the **IgD** type are like **IgM** molecules present on the surfaces of B cells, where they function as antigen receptors. The interaction of antigens with immunoglobulins on B cells is a prerequisite for the stimulation of B cells and antibody production. Antibodies play an important role in the elimination of extracellular pathogens. In addition, **neutralizing antibodies** are able to inactivate toxins (e.g., diphtheria toxin, pertussis toxin) and certain viruses.

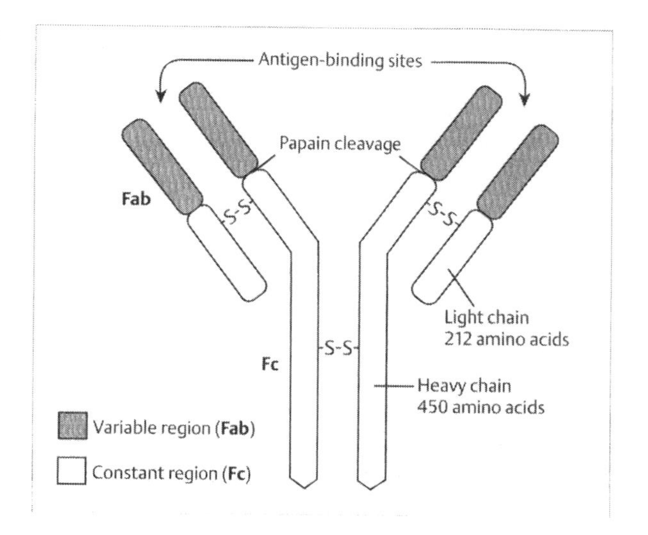

Fig. 33.7 **Diagram of an immunoglobulin G (IgG) molecule.** Protein fragments Fab and Fc formed following papain cleavage are given. S–S, disulfide bridge

During the B-cell response, B lymphocytes are stimulated by specific antigens to secrete antibodies (**immunoglobulins**). Among the five different classes of immunoglobulins, antibodies of the IgG class show the highest concentration in serum.

T lymphocytes play a major role in the **specific cellular immune response** against entering microbes. Two major subsets of T cells are known. **Cytotoxic T cells** (**T_C**) are able to kill infected or transformed host cells. **T helper cells** (**T_H**) produce and secrete **cytokines** (Chapter 33.2.3), which are able to stimulate other cells of the defense system, such as cytotoxic T cells, B cells, and macrophages. Different T cell populations carry **differentiation antigens**, such as CD3, CD4 and CD8, on their surface. While CD3 antigens are present on all T cells, CD4 and CD8 antigens are specific for T_H and T_C cells, respectively (Fig. 33.8). T cells are able to recognize specific microbial antigens in association with **MHC** (**major histocompatibility complex**), molecules (see below). In parallel to B cells, T cells carry specific antigen receptors on their surfaces, called **T-cell receptors** (**TCR**). The T cell receptors exhibit an immunoglobulin-like structure formed by disulfide-linked polypeptides.

The T cell receptor is not able to recognize antigens as free molecules; it recognizes antigens only in a complex with molecules of the major histocompatibility complex "presented" by other host cells, such as macrophages. Macrophages act hereby as "**antigen presenting cells**" (**APCs**). Molecules of the MHC complex are present on the surface of all eukaryotic cell types and are specific for

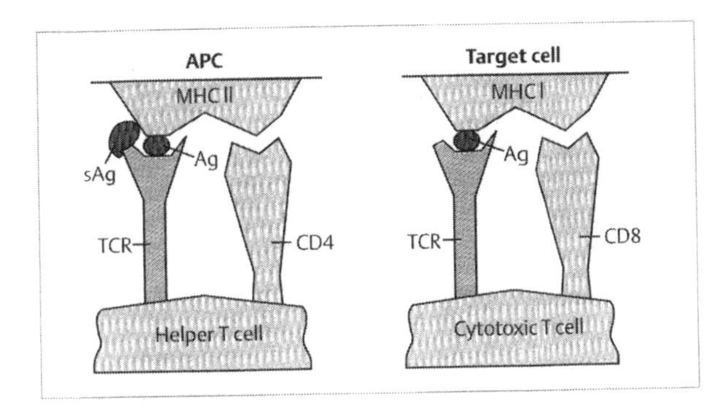

Fig. 33.**8** **Interactions between T cells, an antigen presenting cell (APC), and a target cell.** The major histocompatibility complexes (MHC) I and II, the T cell receptors (TCR), and the cluster of differentiation molecules CD4 and CD8 are indicated. Ag, antigen; sAg, superantigen

certain species, strains, and individuals. Every eukaryotic organism possesses two kinds of MHC, proteins. While the MHC I complex is located on the surface of all nucleated cells, MHC II is only part of cells of the immune system, including macrophages.

The MHC complexes are associated with antigens following intracellular processing. Generally, T helper cells are activated by antigens presented in complex with MHC II molecules, while cytotoxic T cells are stimulated following contact with antigens associated with MHC I complexes. In contrast to "normal" antigens, so-called "**superantigens**" (**sAg**) are able to bind concomitantly to constant regions of the T cell receptor and to MHC II molecules, and thereby initiate a non-specific activation of T helper cells. Certain toxins, such as enterotoxins of *Staphylococcus aureus* and the erythrogenic toxins of *Streptococcus pyogenes*, may act as superantigens. The T cell response plays a major role in protecting the host against viruses and intracellular parasites, such as *Mycobacterium tuberculosis* and *Listeria monocytogenes*.

T lymphocytes play a major role in the cellular immune response, in particular against viruses and intracellular parasites. Cytotoxic T cells are able to eliminate infected cells, while T helper cells are able to activate other cells of the defense systems. T cells are stimulated by antigens when the antigens are presented by macrophages in association with molecules of the major histocompatibility complex (MHC).

33.2.3 Different Parts of the Host Defense System Interact With Each Other

The components of the host defense system do not act independently, but rather show a strong relationship and interdependency. **Cytokines** play a major role in

Table 33.**5** **Molecules of the cytokine network**

Designation	Produced by	Function
Interleukin (IL)-1	Macrophages	Activation of B cells, T cells
Interleukin (IL)-2	T cells	Activation of T cells, B cells
Interleukin (IL)-4	T cells	Activation of B cells, T cells, macrophages
Interleukin (IL)-6	Various cells	B cell differentiation
Interferon (IFN)-α	Macrophages, PMN leukocytes	Antiviral
Interferon (IFN)-γ	T cells	Macrophage activation
Tumor necrosis factor (TNF) α	Macrophages Monocytes	Cachectin, cell death Activation of granulocytes
Tumor necrosis factor (TNF) β	T cells	Lymphotoxin, cell death
Granulocyte/monocyte colony stimulating factor (GM-CSF)	T cells	Stimulation of granulocytes, monocytes
Granulocyte colony stimulating factor (G-CSF)	Macrophages	Stimulation of granulocytes
Monocyte colony stimulating factor (M-CSF)	Macrophages	Stimulation of monocytes

this cooperation. Cytokines are hormone-like glycoproteins secreted by eukaryotic cells. They contribute to stimulation, differentiation, and proliferation of cells of the defense system. Cytokines produced by lymphocytes are also termed **lymphokines**. **Interleukins** are cytokines that preferentially interact with cells of the immune system.

To date, 15 interleukins and several other cytokines have been described. Some important cytokines are indicated in Table 33.**5**. These cytokines are preferentially produced by T cells, monocytes and macrophages. An overview on the network controlled by cytokines is given in Figure 33.**9**.

The different parts of the defense system cooperate to protect the host from foreign particles, including pathogenic microbes. Cytokines are important signal substances involved in the cooperation of immune cells.

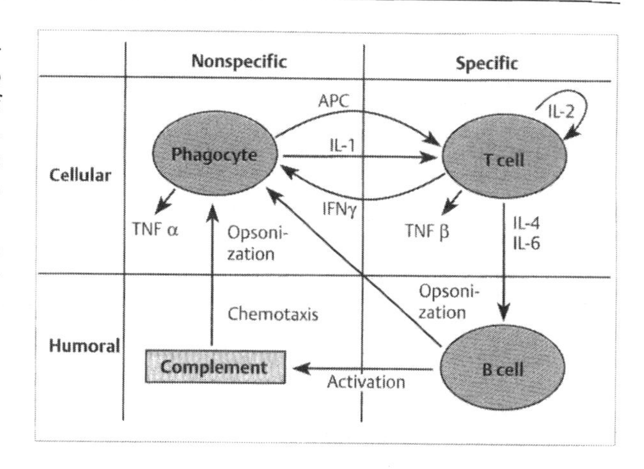

Fig. 33.**9** **Schematic representation of the interactions between different parts of the host defense system.** Important cells types (phagocyte, T cell, B cell), molecules (complement proteins, cytokines) and interactions are given. APC, antigen presenting cell; IL, interleukin; IFNγ, interferon γ; TNF, tumor necrosis factor

33.3 Bacterial Virulence Factors

Microbial pathogens are able to produce virulence or pathogenicity factors. These factors have the capacity to facilitate colonization of microbial pathogens within the host, and to damage and destroy host cells. They also help the invading pathogenic organisms to overcome the host defense system. The following paragraphs describe the architecture and particular aspects of the structure–function relationship between bacterial virulence factors and host cells. Special emphasis is given to the organization of genes involved in virulence and their corresponding gene products.

33.3.1 Adhesins Facilitate Bacterial Colonization of Host Tissues

Adhesins enable bacteria to attach to eukaryotic cells; this is a prerequisite for bacterial pathogens to colonize host tissues successfully (see also Chapter 23.8). Bacterial colonization often starts with a **non-specific aggregation** of bacteria on the basis of hydrophobic interactions between bacterial and host structures. **Adhesion**, however, is a highly specific phenomenon. Certain bacterial adhesins show host as well as tissue specificity, directing the pathogenic organism at a specific site for colonization. Uropathogenic *Escherichia coli* are able to bind to cells of the urogenital tract. *Vibrio cholerae* binds to cells of the gastrointestinal tract and is responsible for diarrheal diseases.

The majority of bacterial adhesins are proteins that recognize **carbohydrate receptors** on eukaryotic cells. As indicated in Figure 33.**10**, bacterial adhesins recognize receptors on eukaryotic cells, such as on epithelial cells, phagocytic cells, or erythrocytes. In addition, glycoproteins of the **extracellular matrix**, such as fibronectin and laminin or integrins, can be used as target structures for bacterial adhesins. The **hemagglutination** assay is a simple assay system to analyze the adherence capacity of bacterial strains. Bacterial adherence factors can be divided into **fimbrial adhesins** and "**non-fimbrial adhesins**." **Fibrillae** represent a thin and flexible form of fimbrial adhesins. **Fimbriae** are structures of the cell envelope consisting of "major" and "minor" subunit proteins (see Figs. 2.**20**, 33.**10**, 33.**11**, and Chapter 23.8).

Genetic and biochemical evidence exists for the carbohydrate binding protein being identical to the major subunit protein ("**major subunit adhesins**") in some fimbriae. This has especially been found for

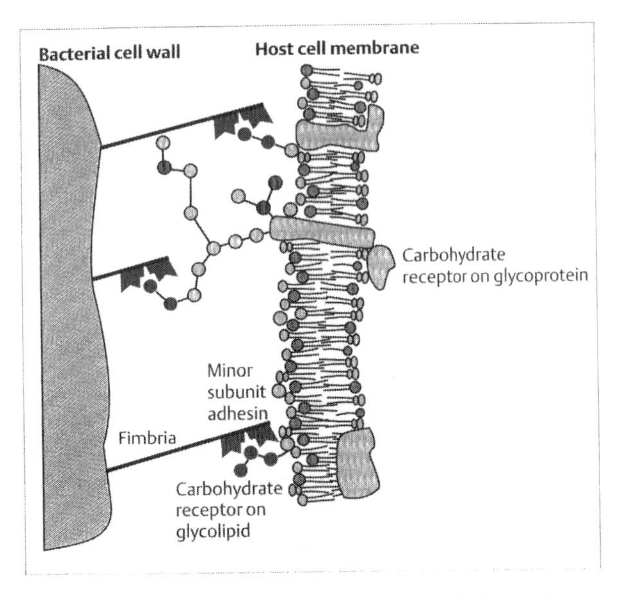

Fig. 33.**10** **Recognition process between a fimbriated bacterium and host cell receptors.** A fimbria with a "minor subunit adhesin" at the tip is indicated on the left. Carbohydrate receptors, which are part of glycolipids and glycoproteins, are shown on the right

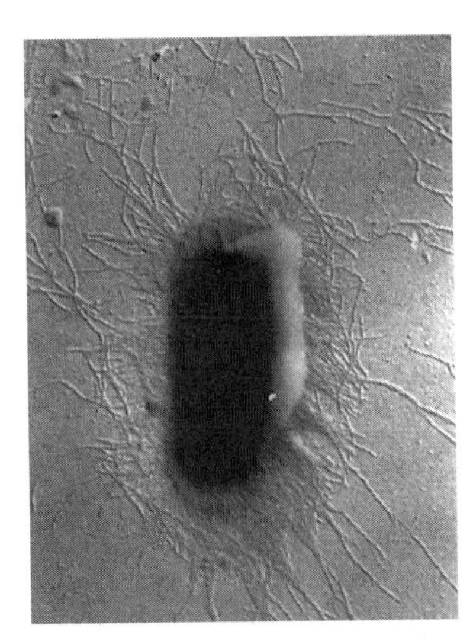

Fig. 33.**11** **Electron micrograph of a pathogenic *Escherichia coli* bacterium with fimbriae**

fimbriae produced by enterotoxigenic *Escherichia coli* (strains K88, K99; Tab. 33.**6**). Other fimbrial adhesins, such as those from type I, P, and S fimbriae of intestinal *E. coli*, show a carbohydrate binding of minor subunits ("**minor subunit adhesins**"). Non-fimbrial adhesins are produced by various microorganisms. The lipoteichoic acid portion of the cell envelope of *Streptococcus pyogenes* is also able to act as an adhesin. In addition, exolipopolysaccharides of various organisms, such as *Staphylococcus epidermidis* and *Pseudomonas aeruginosa*, show adhesive properties.

Bacterial **adhesins** are able to recognize receptor molecules (e.g., carbohydrate moieties) on host cells. Major and minor subunit proteins may act as carbohydrate binding proteins. In addition, fibrillae, non-fimbrial adhesins, lipoteichoic acid, and exolipopolysaccharides show adhesive properties.

Table 33.**6** **Adhesins of pathogenic bacteria**

Adhesin type	Abbreviation	Bacterium	Cellular structure
Type I	Fim	*Escherichia coli*, other Enterobacteria	Fimbriae
P fimbriae	Pap	Extraintestinal *Escherichia coli*	Fimbriae
S fimbriae	Sfa	Extraintestinal *Escherichia coli*	Fimbriae
K88	Fae	Intestinal *Escherichia coli*	Fimbriae
K99	Fan	Intestinal *Escherichia coli*	Fimbriae
Type III	Mrk	*Klebsiella pneumoniae*	Fimbriae
Type IV	Pil	*Neisseria gonorrhoeae*	Fimbriae
	Tcp	*Vibrio cholerae*	Fimbriae
	Pil	*Pseudomonas aeruginosa*	Fimbriae
M Protein	Emm	*Streptococcus pyogenes*	Thin fibrillae
Filamentous **h**em**a**gglutinin	Fha	*Bordetella pertussis*	Non-fimbrial adhesin
Lipo**t**eichoic **a**cid	LTA	*Streptococcus pyogenes*	Bacterial envelope

The fimbrial adhesin gene clusters of pathogenic *Escherichia coli* and the genes coding for type IV fimbriae of various other organisms, including *Neisseria gonorrhoeae*, *Vibrio cholerae* or *Pseudomonas aeruginosa* (Tab. 33.**6**), have been studied in detail. In the case of the *E. coli* adhesins, large operons of about 10 kbp are necessary to encode fimbrial adhesins. The corresponding operon consists of structural genes encoding major and minor adhesin subunit proteins and genes encoding proteins involved in the transport of the fimbrial subunits through the periplasmic space and the outer membrane (see also Chapters 19.4 and 23.8).

The architecture of type IV fimbriae produced by various Gram-negative bacteria is similar to that of fimbriae produced by pathogenic *E. coli*. However, the genes responsible for adhesin production are often not arranged as clusters, but are rather spread over the chromosome. Interestingly, fimbrial adhesins undergo **phase variation** and **antigenic variation**, i.e., one bacterial strain is able to shut the expression of adherence molecules off or on, or to produce adhesins with different antigenic properties. The molecular mechanisms leading to variation in adhesin expression are described in Chapters 33.4.1. and 33.4.2. The M protein of *Streptococcus pyogenes*, one of the best-studied non-fimbrial adhesins, is able to bind to different receptor structures, such as fibronectin. Moreover, it can interact with factor H, an inhibitor of complement activation, thereby leading to serum resistance (Chapter 33.3.8). M proteins also show length variation. To date, more than 60 different M proteins of streptococci have been described.

Table 33.**7** **Intracellular bacteria**

Bacterium	Disease
Facultative intracellular	
Salmonella typhimurium	Diarrheal disease
Yersinia enterocolitica	Diarrheal disease
Enteropathogenic *Escherichia coli*	Dysenteria
Shigella spp.	Shigellosis
Salmonella ***enterica***	Typhoid fever
Listeria monocytogenes	Listeriosis
Mycobacterium tuberculosis	Tuberculosis
Legionella pneumophila	Legionnaires' disease
Yersinia pestis	Bubonic plague
Obligate intracellular	
Rickettsia rickettsii	Rocky Mountain spotted fever
Rickettsia prowazekii	Typhus
Coxiella burnetii	Q fever
Chlamydia trachomatis	Trachoma
Mycobacterium leprae	Leprosy

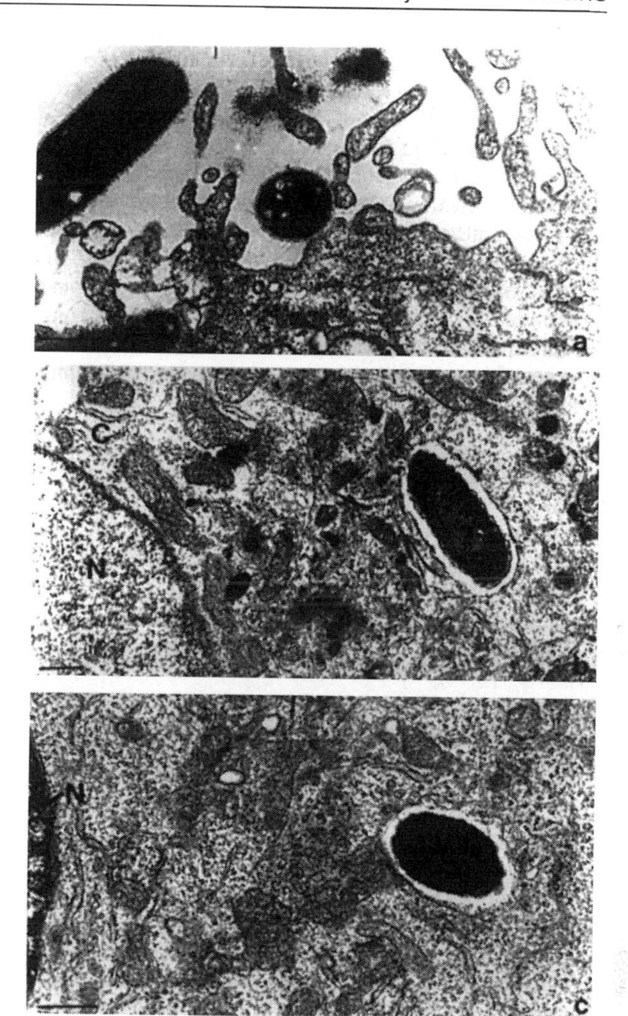

Fig. 33.**12** **Electron micrograph of Gram-negative bacteria before (a) and after (b, c) invasion into epithelial cells.** N, nucleus of epithelial cell; C, cytoplasm of epithelial cell. Bars, 0.5 μm in **a–c**

Fimbrial adhesins are encoded by large gene clusters. Adhesins often show a strong variation in expression.

33.3.2 Invasins Contribute to the Entry of Pathogens Into Host Cells

A number of microbial pathogens do not multiply extracellularly, but rather invade cells of the host organisms. Once in the cell, these intracellular pathogens are able to survive and even multiply within the invaded cells. The term "**invasion**" describes processes triggered by intracellular pathogens that lead to their uptake and intracellular survival within host cells. Virulence factors involved in entry are designated as **invasins**.

A number of pathogenic bacteria are able to grow both outside and inside eukaryotic host cells. Bacteria that preferentially grow within eukaryotic cells are termed **facultative intracellular pathogens** (Tab. 33.**7** and Fig. 33.**12**). Some of these facultative intracellular pathogens invade **non-professional phagocytic cells**, such as epithelial cells (e.g., *Shigella flexneri*). Other pathogens invade **professional phagocytic cells**, such as macrophages (*Mycobacterium tuberculosis*, *Legionella pneumophila*). In contrast, organisms that are strongly adapted to an intracellular life are termed **obligate intracellular pathogens** (e.g., *Chlamydia trachomatis* and *Rickettsia* spec.).

Facultative **intracellular parasites** are able to survive and multiply outside as well as inside host cells. Obligate intracellular parasites are adapted to the intracellular milieu where they commonly multiply.

Box 33.3 Invasion involves the following steps (Fig. 33.**13**):

1. **Adherence to eukaryotic receptors**. Adherence described in Chapter 33.3.1 is a prerequisite for invasion. However, only a limited number of attachment processes are followed by entry of the microbial pathogens into eukaryotic cells. Intracellular pathogens are often able to bind to host cell receptors that are involved in **signal transduction**, such as **integrins**. Integrins are constituents of a large family of transmembrane receptors, which are expressed by a variety of eukaryotic cells. They are normally involved in cell–cell interaction and bind to proteins of the extracellular matrix, such as fibronectin or laminin. Fibronectins bind to integrins via the amino acid motif **Arg–Gly–Asp**. This RGD motif is also part of the filamentous hemagglutinin (Fha), which is involved in binding and uptake of *Bordetella pertussis*, and of an adhesive protein of *Enterococcus faecalis*. Other pathogens use "bridging molecules", such as the complement protein C3b (*Legionella pneumophila, Mycobacterium tuberculosis*) and fibro-nectin (*Treponema pallidum*), which concomitantly bind to adhesins of the pathogen and to integrin receptors.

2. **Signaling and entry**. Microbial attachment to host cells can induce signal transfer processes that may either induce the uptake of pathogens by non-professional phagocytes or enhance their phagocytosis by professional phagocytes. Interestingly, intracellular pathogens are able to express proteins that influence **signal transduction of eukaryotic cells** (Tab. 33.**8**). Yersiniae are able to produce the YopH protein, which has tyrosine phosphatase activity that is also found in eukaryotes. The InvA protein of *Salmonella typhimurium*, capable binding to the epidermal growth factor receptor (EGFR), also affects signal transduction in host cells, leading to activation of the microtubule-associated protein (MAP) kinase by EGFR and subsequent activation of phospholipase A2, which, as a consequence, leads to opening of calcium channels, membrane ruffling, and uptake of bacteria. Particular invasins, identified in intracellular enterobacteria, do not contain a *sec*-dependent signal sequence, but are transferred through the bacterial membrane by a transport mechanism, termed type III secretion system. Type III secretion proteins of *Shigella flexneri*, *Salmonella typhimurium*, *Yersinia enterocolitica*, and enteropathogenic *Escherichia coli* show strong similarity to each other and to proteins involved in the secretion of flagella proteins (see Chapter 19.4).

3. **Survival and persistance within host cells.** Following the invasion of host cells, intracellular pathogens have to overcome the antimicrobial response of eukaryotic cells, especially that of activated macrophages (see Chapter 33.2.1). Intracellular pathogens are able to use different strategies to survive within host cells. Some of the pathogens, e.g., *Legionella pneumophila*, *Chlamydia trachomatis*, and *Mycobacterium tuberculosis*, reside within the vacuole (see Fig. 33.**13**). They are able to avoid fusion of the vacuole with lysosomes, a consequence of which is that microcidal enzymes are not released and the oxidative burst is reduced, enabling the pathogens to survive and multiply within the phagosome. It seems that *Mycobacterium leprae* is able to persist and multiply within the fused vacuole (phagolysosome) without suffering any damage. Other pathogens are able to leave the vacuole and to survive and multiply within the cytoplasm of the host cell. The processes leading to survival of the microorganisms within the host cell cytoplasm have been extensively studied for *Listeria monocytogenes* and *Shigella flexneri*. Both organisms are able to lyse the vacuole membrane by the action of toxins (listeriolysin, IpaB-hemolysin). Once in the cytoplasm, the bacteria are able to induce the polymerization of F actin. The actin microfilaments are attached to one side of the bacterial cell and mediate the intracellular and intercellular migration of the bacteria.

Table 33.**8** **Virulence factors involved in invasion**

Virulence factors	Organisms	Properties
InvA	*Yersinia pseudotuberculosis* *Yersinia enterocolitica*	Integrin $\beta 1$ binding
YopH	*Yersinia pseudotuberculosis* *Yersinia enterocolitica*	Tyrosine phosphatase
Mip	*Legionella pneumophila* *Chlamydia trachomatis*	FK506-binding protein
InvA	*Salmonella typhimurium*	Binding to epidermal growth factor receptor (EGFR)
SpaL–T	*Salmonella typhimurium*	Type III secretion pathway
ActA	*Listeria monocytogenes*	Actin polymerization
IcsA	*Shigella flexneri*	Actin polymerization
MxiA, C	*Shigella flexneri*	Type III secretion pathway
SpaM–S	*Shigella flexneri*	Type III secretion pathway
EaeB	Enteropathogenic *Escherichia coli*	Tyrosine phosphorylation of Hp 90
SepA–C	Enteropathogenic *Escherichia coli*	Type III secretion pathway

Intracellular pathogens have developed different strategies to invade host cells. Following binding and signal transfer, a group of pathogens are able to avoid fusion of the phagosome with the lysosome and to survive within the phagolysosome. Other pathogens survive within the cytoplasm and are able to spread to other cells.

33.3.3 Exotoxins Damage the Host

Toxins are soluble substances that alter the normal metabolism of host cells. They may cause death of the host. A distinction is made between exotoxins and endotoxins. **Exotoxins** are proteins that are usually secreted into the surrounding medium. **Endotoxins** are the lipopolysaccharides of the outer membrane of Gram-negative bacteria, released following lysis of the bacterial cells (see Chapters 23.2 and 33.3.6).

Some pathogenic bacteria, such as *Staphylococcus aureus*, *Streptococcus pyogenes*, or *Pseudomonas aeruginosa*, are able to produce a large variety of toxins, and it is difficult to assess the contribution of the different substances to the infectious process. Other pathogens produce a single or a major toxin that accounts for symptoms of the disease. Examples of such toxins include cholera toxin (*Vibrio cholerae*), diphtheria toxin (*Corynebacterium diphtheriae*), tetanus toxin (*Clostridium tetani*), and botulinum toxin (*Clostridium botulinum*). Protein toxins are the most active substances of biological relevance. For example, 1 g of botulinum toxin would be able to kill about 10 million people.

Genes encoding exotoxins are often located on phages or plasmids. The genes encoding cholera toxin, diphtheria toxin, shiga-like toxins, botulinum toxin, and the erythrogenic toxin of *Streptococcus pyogenes* are part of converting bacteriophages. *Escherichia coli* enterotoxins, tetanus toxin, the toxin involved in scarlet skin syndrome, and others are plasmid-encoded (Tab. 33.**9**). The ST enterotoxin genes of *E. coli* are even part of transposons. It can therefore be concluded that toxin-specific genes have the capacity to spread to non-toxigenic strains.

Exotoxins can be divided into different groups. One criterion for the grouping is the site of action. **Enterotoxins** are defined as a group of exotoxins that act on the intestine and thereby cause massive secretion of fluid into the intestinal lumen. Enterotoxins are produced by a number of bacteria causing diarrheal diseases. Examples are *Staphylococcus aureus*, *Escherichia coli*, *Vibrio cholerae*, *Salmonella* **enterica**, sv. **Enteritidis**, and *Bacillus cereus*. In contrast, **neurotoxins** act on cells of the nervous system. *Clostridium tetani* and *Clostridium botulinum* produce such potent exotoxins.

Exotoxins are protein substances secreted by pathogenic microbes that are able to damage or even destroy host cells. Exotoxins can be divided into different types according to their sites and mechanisms of action.

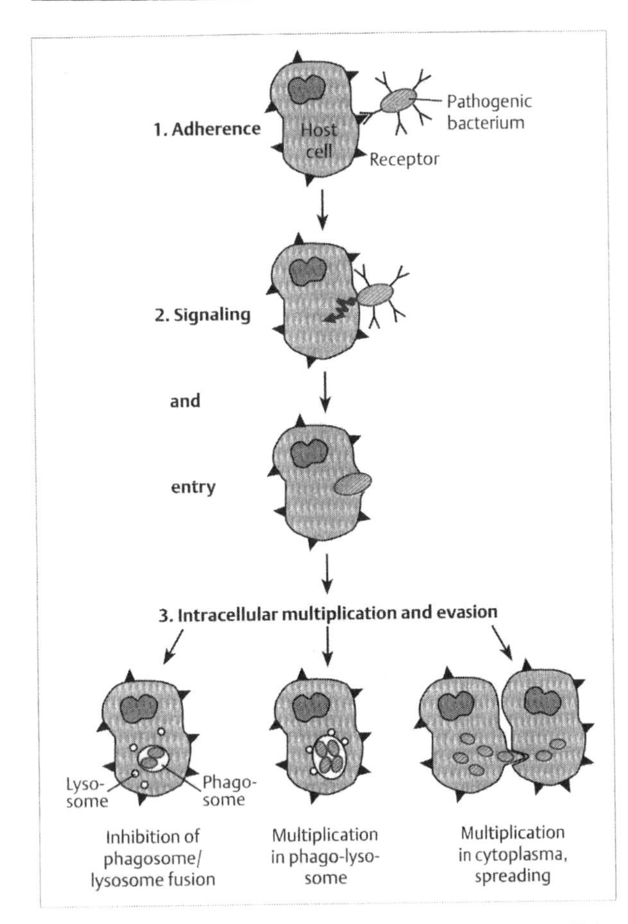

Fig. 33.**13** **Bacterial strategies of invasion and intracellular multiplication.** The following steps are indicated: (1) adherence of a pathogenic bacterium to receptors on the host cell, (2) signal transduction from the bacterium to the host cell and entry of the bacterium into the host cell, and (3) strategies of bacteria to survive and multiply within the host cell

Superantigens (sAgs), such as enterotoxins of *Staphylococcus aureus* and erythrogenic toxin of *Streptococcus pyogenes* form a bridge between the MHC II complex of macrophages and T cell receptors (TCR of T helper cells), which results in a nonspecific stimulation of the immune system. Superantigens have already been described in Chapter 33.2.2 and Figure 33.**8**.

Pore-forming toxins. Particular cytolysins are also termed "**hemolysins**" because they are able to cause lysis of red blood cells. Such toxins are also active on other cells, such as leukocytes and epithelial cells. Depending on their mode of action, they are called "**pore-forming toxins**" (Fig. 33.**14a**). They are able to insert into hydrophobic regions of the eukaryotic cell membrane, which results in disruption of the membrane integrity. The repeat toxins (Rtx) form an important class of pore-forming toxins. Rtx molecules are produced by Gram-negative bacteria, such as *Escherichia coli* (α-hemolysin), *Pasteurella haemolytica* (leukotoxin), and *Proteus* spec. Rtx toxins share a common motif of repeated amino acids. The toxins are transported through the bacterial membrane by secretion pathway I, which includes transport molecules related to the **m**ultiple **d**rug **r**esistance (Mdr) proteins of eukaryotes. The cyclase/hemolysin of *Bordetella pertussis* represents a unique bifunctional toxin, consisting of an adenylate cyclase and a pore-forming toxin. The α-toxin of *Staphylococcus aureus*, the pneumolysin of *Streptococcus pneumoniae*, and streptolysin O of *Streptococcus pyogenes*, which belongs to the group of thiol-activated toxins, are other pore-forming toxins of major import-ance. Electron micrographs of pore-forming toxins are given in Figure 33.**15**.

Hemolysins or cytolysins with enzymatic functions also exist (Fig. 33.**14b**). These enzymes attack the phospholipid layer of the host cell. Since the phospholipid lecithin (phosphatidylcholine) is often utilized as a substrate, these enzymes are called **lecithinases** or **phospholipases**. The phospholipase C molecules, produced by several pathogenic species, such as *Clostridium perfringens*, *Listeria monocytogenes*, *Pseudomonas aeruginosa*, and *Staphylococcus aureus*, play a role in tissue damage.

Particular **cytolysins** (hemolysins) are able to form pores within the eukaryotic membrane, which often results in killing of the host cell. Enzymatic cytolysins are able to hydrolyze membrane phospholipids with the consequence of severe tissue damage.

A–B toxins. While cytolysins act outside the eukaryotic cells where they destroy the membranes of host cells, "**A–B toxins**" act inside the host cell, where they interfere with important cell functions. In most cases, these toxins consist of two parts. The A subunit ("active") exhibits enzymatic activity and is responsible for the pathogenic processes within the host cell. The B subunit ("binding") is able to bind specifically to receptor molecules on the surface of host cells, thereby triggering the internalization of the toxin A subunit into the host cell (Fig. 33.**14c**). A number of toxins of the A–B type have the capacity for **ADP-ribosylation** of various substrates. Depending on the substrates, these toxins interfere with protein biosynthesis, facilitate deregulation of cells, or inhibit actin polymerization. Diphtheria toxin (DT) and cholera toxin (CT) are important members of the group of ADP-ribosylating toxins. While fragment B of DT binds to the eukaryotic receptor,

Table 33.**9** **Important bacterial toxins**

Toxin	Organism	Disease	Mode of action	Location of gene
Toxic shock toxin 1	*Staphylococcus aureus*	Wound infections, skin infections	Superantigen	Bacteriophage
Erythrogenic toxin A	*Streptococcus pyogenes*	Septicemia	Superantigen	Bacteriophage
α-Hemolysin	*Escherichia coli*	Urinary tract infection	Pore formation	Chromosome/ plasmid
α-toxin	*Staphylococcus aureus*	Wound infections, skin infections	Pore formation	Chromosome
Streptolysin O	*Streptococcus pyogenes*	Scarlet fever, pharyngitis	Pore formation	Chromosome
Pneumolysin	*Streptococcus pneumoniae*	Pneumonia	Pore formation	Chromosome
Lecitinase	*Clostridium perfringens*	Gas gangrene	Phospholipase	Chromosome
Phospholipase C	*Pseudomonas aeruginosa*	Skin infections, respiratory tract infections	Phospholipase	Chromosome
Exotoxin A	*Pseudomonas aeruginosa*	Skin infections, respiratory tract infections	ADP-ribosylation	Chromosome
Diphtheria toxin	*Corynebacterium diphtheriae*	Diphtheria	ADP-ribosylation	Bacteriophage
Cholera toxin	*Vibrio cholerae*	Cholera	ADP-ribosylation	Bacteriophage
LT-enterotoxin	*Escherichia coli*	Diarrhea	ADP-ribosylation	Plasmid/ chromosome
Pertussis toxin	*Bordetella pertussis*	Whooping cough	ADP-ribosylation	Chromosome
Shiga toxin	*Shigella dysenteriae*	Dysenteria	28S RNase	Chromosome
Shiga-like toxin	*Escherichia coli*	Diarrheal diseases	28S RNase	Bacteriophage
Tetanus toxin	*Clostridium tetani*	Tetanus	Protease	Plasmid
Botulinus toxin	*Clostridium botulinum*	Botulism	Protease	Bacteriophage

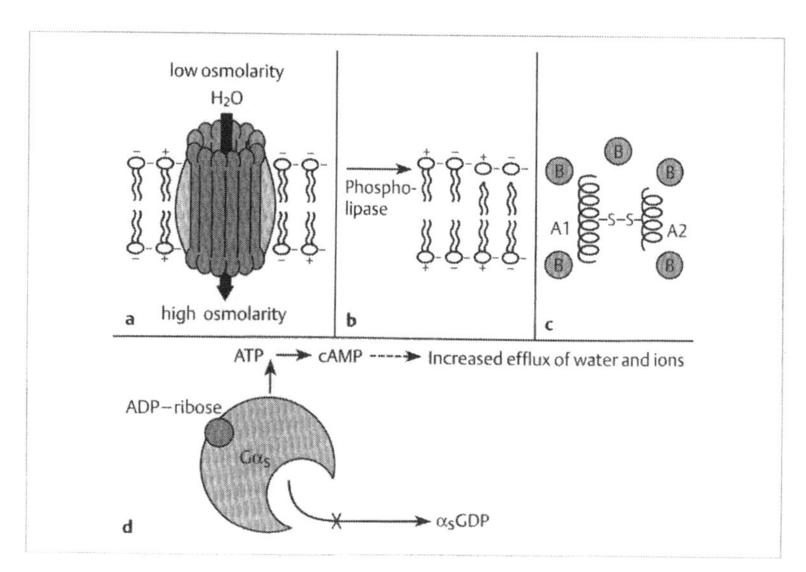

Fig. 33.**14a–d** **Structure and mode of action of toxins.** Various types of toxins are illustrated.
a Pore-forming toxin. The channel formed by these toxins is inserted into the host cell membrane, which leads to swelling and cell lysis.
b Cytolysin with enzymatic function. The polar head group of phospho-lipids is removed by the phospholipase, which leads to destabilization of the host membrane and cell lysis.
c Structure of A–B toxins, here the cholera toxin (CT). The enzymatically active A subunit, consisting of A1 and A2, and the five B subunits, that bind to the host cell membrane, are indicated. S–S, disulfide bridge.
d Action of the cholera toxin CT. The α_S unit of the cellular guanidine nucleotide binding (G) protein is altered by ADP-ribosylation. The GTPase function is inactivated, which leads to increased adenylate cyclase activity. Increased cAMP levels stimulate efflux of electrolytes and water

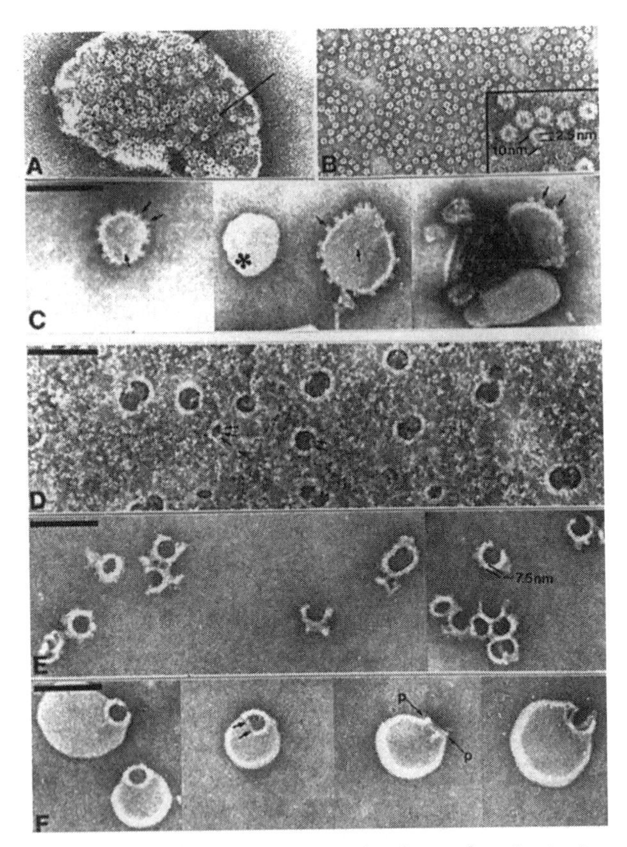

Fig. 33.**15a–f** **Electron micrographs of pore-forming toxins.**
A Fragment of rabbit erythrocyte lysed with staphylococcal α-toxin. Arrows point to the 10-nm ring-shaped structures formed by α-toxins.
B Isolated toxin hexamers in detergent solution.
C Lecithin liposomes carrying reincorporated α-toxin hexamers. Arrows point to α-toxin hexamers incorporated into lecithin liposomes. The star shows a liposome unaffected by α-toxin.
D Negatively stained erythrocyte membrane lysed by streptolysin O. Arrows point to streptolysin-O (SLO)-specific pores incorporated into the erythrocyte membrane.
E Negatively stained isolated oligomers, showing numerous curved rod structures identical to those found in toxin-treated membranes.
F Purified complexes reincorporated into cholesterol-free lecithin liposomes. Arrows point to purified SLO complexes reincorporated into cholesterol-free lecithin liposomes; (p, a lesion seen in profile). Bars 100 nm in all frames [after 1]

fragment A catalyzes the binding of NAD to elongation factor 2 (EF2) of eukaryotic ribosomes. Following ADP-ribosylation, the activity of EF-2 drops dramatically, leading to the inhibition of protein synthesis. Exotoxin A of *Pseudomonas aeruginosa* resembles the diphtheria toxin and works the same way.

The target tissue for CT is the epithelium of the small intestine (Fig. 33.**14d**). The B subunit of CT binds to the G_{M1}, gangleoside of the host cell, the guanidine nucleotide binding (G) protein of the membrane bound complex is inactivated by ADP-ribosylation, and the adenylate cyclase activity is increased. The increased cAMP concentration mediates the active secretion of electrolytes and water into the lumen of the intestine. The change in ionic balance leads to the secretion of large amounts of water into the lumen. The heat labile (LT) enterotoxin of *Escherichia coli* displays the same mechanism of action. **Other A–B toxins** have mechanisms other than ADP-ribosylation. Important examples are Shiga and Shiga-like toxins as well as neurotoxins. Shiga toxin Type I, (ShT) produced by *Shigella dysenteriae* and Shiga-like toxin I (SLT I) are 99% identical, while SLT II and ShT are only 55–57% identical. SLTs are produced by enterohemorrhagic (EHEC) and some enteropathogenic (EPEC) *Escherichia coli* strains, (see Chapter 33.1.4) and other enterobacteria. The toxins have no effect on invasion and intracellular growth of bacteria, but contribute to the severity of the caused diseases. The A subunit is able to cleave a single adenine residue from the 28S RNA of eukaryotic ribosomes by *N*-glycosidase activity, leading to inhibition of peptide elongation.

Tetanus bacilli are able to grow anaerobically within wounds, where they produce the tetanus toxin. The toxin is transported to the central nervous system. Upon entry into the central nervous system, the toxin attaches to nerve synapses. Tetanus toxin acts as a zinc protease and blocks the release of neurotransmitters (e.g., γ-aminobutyric acid, glycine), which leads to inhibition of synapses and thus permits unregulated excitatory synaptic activity (spastic paralysis). The botulinum toxin is often found in contaminated food, where *Clostridium botulinum* grows anaerobically. Botulism is therefore a real intoxication. Botulinum toxin also acts as a zinc protease and binds to presynaptic terminal membranes at the nerve–muscle junction, where it interrupts the signal transmission by interfering with the release of the neurotransmitter acetylcholine (ACH) from synaptic vesicles, the result of which is flaccid paralysis.

Toxins that consist of two fragments (A–B), are able to act within the eukaryotic cell. Some of these toxins exhibit the enzymatic function of ADP-ribosylation.

33.3.4 "Spreading Factors" Contribute to the Distribution of Pathogenic Bacteria

Pathogenic bacteria are able to produce numerous molecules that are not considered as toxins per se because they do not target any particular type of cells, but nevertheless contribute to disease. These proteins can function as "**spreading factors**" by facilitating the distribution of infecting organisms. Several factors are involved in the destruction of extracellular matrix (ECM) molecules of the connective tissues or the fibrin network. Thus, the enzyme **staphylokinase**, produced by *Staphylococcus aureus*, is able to dissolve fibrin clots. **Streptokinase** of *Streptococcus pyogenes* cleaves plasminogen to plasmin, which in turn attracts fibrin. *Bacillus fragilis* and *Clostridium perfringens* produce a **collagenase** (termed κ toxin of *C. perfringens*), which directly destroys collagen, a major constituent of the ECM. **Hyaluronidase**, produced by *S. aureus* and *C. perfringens*, hydrolyzes hyaluronic acid, which is present in the connective tissues.

A number of pathogenic microorganisms, such as *Streptococcus pneumoniae*, *Clostridium perfringens*, *Bacillus fragilis*, *Vibrio cholerae*, and *Salmonella typhimurium*, are able to produce a **neuraminidase** that attacks glycoproteins possessing sialic acid residues. These enzymes are able to lyse serum proteins involved in defense reactions. In addition, highly conserved **zinc-metalloproteases** are produced by Gram-positive (*Listeria monocytogenes*, *Bacillus cereus*) and by Gram-negative pathogens (*Legionella pneumophila*, *Pseudomonas aeruginosa*). The *P. aeruginosa* enzyme is involved in the degradation of elastin and is therefore termed elastase.

Pathogenic microorganisms of the stomach (*Helicobacter pylori*) and of the urinary tract (*Proteus mirabilis*, *Staphylococcus saprophyticus*) produce the enzyme **urease**, which hydrolyzes urea to ammonia and carbon dioxide. Ammonia elevates the pH in the urinary tract, which leads to the precipitation of stones. In the stomach, an elevated pH protects *H. pylori* from gastric acidity.

Microorganisms are able to produce enzymes that support the "spreading" of pathogens in their host organisms.

33.3.5 Iron Acquisition Systems Are Bacterial Growth Factors

Iron is essential for bacterial growth and also for bacterial pathogenicity because bacteria require iron for the synthesis of several compounds, such as cytochromes and iron-sulfur problems. The concentration of free iron is very low in the human body (e.g., $1 \cdot 10^{-18}$ M in human serum) because **eukaryotic iron binding proteins**, such as **lactoferrin** (in the secretory fluid), **transferrin** (in the serum), **ferritin**, and **hemin**, have the ability to bind iron tightly. Bacteria, however, normally need an iron concentration of 0.4–1 µM to grow properly. Therefore, pathogenic bacteria, especially those that multiply extracellularly, have to compete with the host for iron and have, therefore, developed several iron acquisition mechanisms (see Chapter 8.8).

Iron is essential for bacterial growth and for the ability of microorganisms to cause diseases. Bacteria have developed several strategies to acquire iron.

Siderophores are the best-studied iron uptake systems. They are low-molecular-weight compounds (500–1500) that chelate iron with a very high affinity. Two main classes of siderophores can be distinguished: the **catechol type** and the **hydroxamate type**. Enterobactin, which is produced by the majority of pathogenic and non-pathogenic enterobacteria, is the prototype for the catechol type, while aerobactin is a representative of the hydroxamate type. Aerobactin is frequently produced by pathogenic *Escherichia coli* strains, preferentially by blood culture isolates. Interestingly, the aerobactin genes (*aer*) can be located on plasmids or on the bacterial chromosome. The *aer* gene cluster is flanked by insertional elements (IS) 1, and it is speculated that the IS1 sequences may facilitate a transposon-like transfer of *aer*.

Following excretion of siderophores into the environment, the molecules complex Fe^{3+}. The uptake of the complexed iron is described in Chapter 8.8. Some bacteria can utilize siderophores produced by other species.

In addition to the production of siderophores, other iron acquisition systems exist. Some pathogenic bacteria (e.g., *Neisseria gonorrhoeae* and *Haemophilus influenzae*) are able to use eukaryotic compounds, such as transferrin, lactoferrin, ferritin, or hemin, as an iron source. They bind eukaryotic iron-containing molecules on their surface. Other pathogens, such as *Listeria monocytogenes* and *Legionella pneumophila*, secrete an **iron reductase** capable of reducing extracellular Fe^{3+} to Fe^{2+}, which can be taken up by the bacteria. In addition, **exotoxins** are involved in iron aquisition (Chapter 33.3.3). As mentioned above, some exotoxins cause eukaryotic cell death, a consequence of which is that iron is released and can be used by the infecting microorganisms. Interestingly, genes encoding some of the toxins or siderophores are regulated by iron. A repressor molecule, termed **Fur** (**ferric uptake regulator**) can repress genes only following binding of Fe^{2+} (Chapter 33.4.2). The Fur–Fe^{2+} complex is able to repress several genes, including those encoding shiga toxin, shiga-like toxin I, *Serratia* hemolysin, or aerobactin.

Siderophores are small iron-binding molecules produced by several bacteria. In addition, bacteria can use eukaryotic iron-storage proteins for iron aquisition. Eukaryotic cells destroyed by bacterial exotoxins are also used as an iron source.

33.3.6 Endotoxins Are Involved in Many Pathogenic Processes

The endotoxin is part of the **lipopolysaccharide** (**LPS**), which is a major component of the outer membrane of Gram-negative bacteria (for structure and synthesis of LPS, see Chapter 23.2). LPS consists of three components: the lipid A, which is the toxic part, the core polysaccharide, and the O repeating units, encoded by the *rfa* and *rfb* genes, respectively. Unlike exotoxins, endotoxins are not synthesized specially to damage other cells, but to protect the bacterial cell.

LPS is able to exhibit several biological activities, both in the cell-bound form and following release from the bacterial cell. As a major constituent of the bacterial cell wall, LPS can contribute to **bacterial resistance to serum**, i.e., the ability of pathogenic bacteria to overcome the attack of the complement system (Chapter 3.3.8). Certain types of LPS are able to prevent attachment of the complement molecules C3b or C5b so that the membrane attack complex cannot be formed (Chapter 33.2.1). Other LPS types carry long repeating units that form a "steric hindrance" for the binding of the terminal complex on the bacterial surface. In addition, the inhibition of C3 convertase formation by sialic-acid-containing types of LPS contributes to serum resistance.

Endotoxins are released when the bacterial cells die. Low amounts of LPS may cause pyrogenic effects in the host cell, while high doses induce a crisis of the host organisms, termed **septic shock**. Septic shock is often associated with septicemic diseases caused by Gram-negative bacteria. It is a consequence of the release of cytokines, such as IL-1 and TNF, by macrophage cells. In addition, high concentrations of other cytokines, such as IL-6 and IL-8, can be detected (see Chapter 33.2.3). The occurrence of these cytokines and other compounds of the immune system lead to the production of mediators of inflammation (e.g., prostaglandins and leukotrienes), as well as to the activation of the complement and coagulation cascades.

LPS consists of three parts: the lipid A, which is toxic, the core, and the O repeating units. Cell-bound lipopolysaccharide contributes to serum resistance of Gram-negative bacteria. Released endotoxin is a major cause of several pathogenic processes in the host, including septic shock.

33.3.7 Capsules Protect Pathogenic Bacteria

Capsules are networks of polymeric structures located on the surface of microorganisms. Most of the capsules are composed of polysaccharides (see Chapter 23.4). Interestingly, capsules are produced especially by microorganisms that are able to cause invasive diseases, such as meningitis or sepsis. This group of organisms includes *Neisseria meningitidis*, *Haemophilus influenzae*, and *Escherichia coli*, which produce the K1 capsule. Moreover, bacteria that can cause pneumonia, such as *Streptococcus pneumoniae* and *Klebsiella pneumoniae*, also produce capsules. A particular type of *Pseudomonas aeruginosa*, which has been identified as the causative agent of respiratory tract infections in patients suffering from cystic fibrosis (CF), also produces a capsule-like exopolysaccharide, termed alginate. In addition, uropathogenic *E. coli* produce capsules.

Several capsules consist only of polysialic acid. The K1 capsule of *E. coli* contains *N*-acetylneuramic acid in α-2,8 linkage and is identical with the capsule polysaccharide of *N. meningitidis*. Cross-reactivity has also been found with other capsules of different organisms. Thus, the K18 and K100 capsules of *E. coli* resemble those of *H. influenzae* b. Capsules help in the establishment of **evasion strategies** of pathogenic microorganisms by:

1. **Interference with the activation of the complement cascade** (see Chapter 33.2.1). Capsules aid in avoiding the activation of the first step of the complement cascade by hindering the C3bBb complex to assemble on the bacterial cell surface. Consequently, complement-mediated lysis cannot occur.
2. **Presence of structures that resemble host structures.** Polysialic acid of the capsules of *E. coli* K1 and *N. meningitidis* resembles the carbohydrate terminus of the embryonic "**n**eural **c**ell **a**dhesin **m**olecule" (n-CAM; Tab. 33.**10**). These types of capsules are therefore not immunogenic, and the host does not produce antibodies that opsonize the bacterial surface.

Table 33.**10** **Bacterial capsules exhibiting similarity to eukaryotic structures.** n-Cam, **N**eural **c**ell **a**dhesin **m**olecule; 8-α Neu NAc-2, *N*-acetylneuraminic acid; Fru, fructose; Glc, glucose; Gal NAc, *N*-acetyl galactosamine; Glc NAc, *N*-acetylglucosamine

Species	K-Type	Composition	Similarity to
Escherichia coli	K1	8-α Neu NAc-2	n-Cam
	K4	Fru–Glc–Gal NAc	Chondroitin
	K5	Glc–Glc NAc	Heparin
Neisseria meningitidis	b	8-α Neu NAc-2	n-CAM

The majority of bacterial capsules consist of polymerized polysaccharides, which interfere with nonspecific host defense mechanisms, such as complement activation and opsonization.

33.3.8 Pathogenic Bacteria Exhibit Different Defense Mechanisms

Pathogenic microorganisms have evolved strategies to compete with antibacterial host factors. Bacterial molecules, some of which are termed "**evasins**," may prevent recognition of pathogens by phagocytic cells and proteins of the complement cascade. Other mechanisms protect bacteria from immunoglobulins and contribute also to the establishment of infectious agents in the host.

Antiphagocytic substances. Substances of the bacterial cell surface, such as the **M protein** of *Streptococcus pyogenes* (see Chapter 33.3.1), may block the deposition of the C3b opsonin. The M proteins limit the interaction of C3b with receptors on phagocytic cells, such as PMNs, to avoid opsonization (see Chapter 33.2.1). **S-layers**, i.e., proteins that form paracrystallic structures on the bacterial cell surface (Figs. 2.**6c–e**), prevent C3b binding of *Campylobacter fetus* and are involved in virulence of the fish pathogen *Aeromonas salmonicida*.

Superoxide dismutases (SOD), catalases. Bacterial catalases and SODs contribute to detoxification of biologically hazardous reactive oxygen products following phagocytosis (see Chapters 33.2.1 and 11.6). The enzymes are important for the pathogenesis of various bacteria, such as *Staphylococcus aureus*, *Listeria monocytogenes*, and *Shigella flexneri*.

Factors contributing to serum resistance. Especially during extracellular infections, bacteria have to defend themselves against the action of the complement system (see Chapter 33.2.1.). The ability to survive in human serum (**serum resistance**) is a virulence parameter for many pathogens, including uropathogenic bacteria. In addition to LPS and capsules, proteins of the outer membrane of Gram-negative bacteria contribute to serum resistance. The Ail ("**a**ttachment **i**nvasion **l**ocus") protein of *Yersinia enterocolitica*, the plasmid-encoded Rck ("**r**esistance to **c**omplement **k**illing") protein of *Salmonella enterica* sv. *typhimurium*, plasmid-encoded proteins of *E. coli* (TraT, Iss), and proteins encoded by λ phage (Bor) may contribute to serum resistance under special circumstances. The Ail and Rck proteins as well as the Lom (λ phage), OmpX (*Enterobacter cloacae*), and PagC (*Salmonella enterica*, see Chapter 33.3.2) proteins belong to a group of 16–18 kDa proteins produced by Gram-negative bacteria that contribute to virulence by increasing the serum resistance.

Mimicry. Capsules can consist of carbohydrate molecules that resemble host structures and therefore elicit only a weak immune response. The mechanism is termed "**mimicry**." In addition, other bacterial products (e.g., M proteins of *Streptococcus pyogenes*) exhibit epitopes similar to host structures. Some bacteria, such as *Staphylococcus aureus* and *S. pyogenes*, produce **fibronectin binding proteins**, which have the capacity to bind soluble fibronectin, a host glycoprotein, a consequence of which is that a coat of host structures is formed around the bacterial surface, which is only poorly immunogenic.

Complement degradation. *S. pyogenes* is able to produce a C5a protease that especially degrades the complement protein C5a. This enzyme leads to an antichemotactic effect and avoids complement-mediated opsonization of the infecting bacteria.

Antibody binding proteins. Proteins of various microorganisms may act as "Fc receptors", i.e., they have the capacity to bind immunoglobulins specifically. The best-studied example of this particular class of virulence factors is the **protein A** of *S. aureus*. In addition, so-called "M-like proteins" of *S. pyogenes* can also bind IgA as well as IgG proteins. Such antibody binding proteins prevent the specific recognition of pathogens by antibodies and therefore protect the microorganisms from opsonization.

Secretory IgA proteases. Particular pathogenic microorganisms colonize on mucosal surfaces where secretory immunoglobulins of the IgA type (sIgA) are present. Some of these bacteria secrete proteases that specifically cleave sIgA molecules of type 1. The sIgA1 proteases are important virulence factors that allow the bacteria to interfere with the protective host mechanism of IgA1 production on mucosal membranes. Microorganisms causing meningitis or pneumoniae, such as *N. meningitidis*, *H. influenzae*, and *E. pneumoniae*, as well as *N. gonorrhoeae*, the causative agent of gonorrhea, produce IgA proteases.

Interference of pathogenic microorganisms with host factors involves several important bacterial defense mechanisms. Virulence factors protect the pathogens against the phagocytic activity of the host cells and opsonization by complement or immunoglobulins. These factors include binding proteins as well as proteases that specifically cleave complement proteins or IgA1.

33.4 Variation and Regulation of Genes Involved in Virulence

Virulence-associated genes are strictly regulated. Variation of expression can occur independently of the environment, or in response to environmental stimuli (see Fig. 33.**16**).

33.4.1 Mutations and Genomic Rearrangements Lead to Virulence Modulation, Phase Variation, and Antigenic Variation

Pathogenic microorganisms have the capacity to alter properties of the bacterial surface. Such "switch" mechanisms can help microorganisms to escape from the host immune system. One type of variation of bacterial properties is termed "**phase variation**". Phase variation, which leads to the presence or absence of certain surface properties seems to occur randomly. In contrast, "**antigenic variation**" leads to different variants of one surface marker, such as an adhesin or a membrane protein (Fig. 33.**16**). While phase variation and antigenic variation are in general reversible processes, genetic processes leading to **virulence modulation** are irreversible. All these reversible and irreversible processes are consequences of mutations or genomic rearrangements, such as gene duplications, amplification, deletions, or translocations (see Chapter 18.11).

Amplification and deletions of virulence genes. Certain virulence genes, e.g., toxin genes, tend to increase their copy number before or during infections (Fig. 14.**15a**). Cholera toxin genes (*ctx*) or shiga-like toxin (*slt*) genes of enterobacteria produce increasing amounts of toxins as a result of **gene amplification** (Tab. 33.**11**). Amplification of the *slt* and *ctx* genes occurs following induction of converting lysogenic bacteriophages. In addition, small deletions in particular genes or large deletions of chromosomal regions can lead to changes in the virulence pattern of pathogenic organisms. Genes coding for M proteins of *Streptococcus pyogenes* carry DNA structures that contain sequence repeats that tend to undergo **intragenic recombination** processes. Consequently, gene products with altered immunological properties are produced. A 1.5-kb DNA region of the capsule operon of *Haemophilus influenzae* is able to be deleted from the bacterial chromosome, thereby leading to non-capsulated variants.

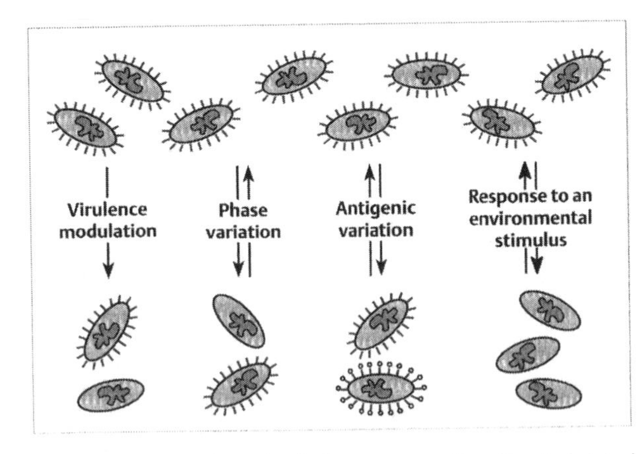

Fig. 33.**16** **Variations in virulence properties.** The fimbriated phenotype (upper part) of a bacterial population changes to a fimbria-negative phenotype (lower part) due to genomic rearrangements following irreversible virulence modulation or phase variation or due to a response to an environmental stimulus. The process of antigenic variation, such as the reversible change from one type of fimbriae to a serologically distinct fimbrial variant, is also shown

In addition, "**pathogenicity islands**" can be deleted from the chromosomes of uropathogenic *Escherichia coli*, *Yersinia pestis*, and *Yersinia enterocolitica*. Variants with altered virulence properties are formed, which may have lost the capacity to produce toxins, adhesins (in the case of *E. coli*), pigmentation, and iron uptake systems (in case of *Y. pestis* and *Y. enterocolitica*). The pathogenicity islands of *Y. pestis*, which comprise 102 kb, can be deleted as a consequence of a recombination of their flanking IS100 elements. In uropathogenic *E. coli*, the islands are located in tRNA genes and are flanked by two short direct repeats of 16–18 bp, one of which disappears following recombination and subsequent deletion. The processes of irreversible virulence modulation may represent adaptation mechanisms of certain species to form variants with increased (following amplification) or decreased (following deletions) virulence.

Phase variation by DNA inversions. Type I fimbriae of enterobacteria (e.g., *E. coli*), undergo phase variation (see Chapter 33.3.1). The main promoter region of the *fim* operon, encoding type I fimbriae, is part of a 314-bp DNA fragment located in front of the structural *fimA* gene (Fig. 33.**17a**). The fragment has the capacity to excise and re-integrate. In the "on" situation, the promoter initiates transcription of the structural *fim* genes, and consequently, fimbriae are produced. If the fragment is locked in the "off" orientation, type I fimbriae are no longer produced. FimB and FimE, two gene products encoded by the *fim* operon, act as site-specific recombinases. While FimB directs the invertable element preferentially into the "on" position, FimE converts the element into the "off" state. Similar processes have been described for flagella antigens

Table 33.**11** **Amplification or deletions of genes leading to virulence modulation**

Bacterium	Virulence factor	Gene(s)	Genetic mechanism involved
Enterohemorrhagic *Escherichia coli* (EHEC)	Shiga-like toxin	*slt*	Amplification
Vibrio cholerae	Cholera toxin	*ctx*	Amplification
Streptococcus pyogenes	M protein	*emm*	Intragenic deletion
Haemophilus influenzae b	Capsule	*cop*	Duplication
		bexA	Deletion
Uropathogenic *Escherichia coli*	Hemolysin	*hly*	Deletion
	P fimbriae	*prf*	
	Cytotoxic necrotizing factor	*cnf*	
Yersinia pestis	Hemin storage	*hms*	Deletion
	Iron uptake	*fyuA, irp*	
Yersinia enterocolitica	Iron uptake	*fyuA, irp*	Deletion

(Hag) of salmonellae (see Chapter 18.11) and for fimbriae of the veterinary pathogen *Moraxella bovis* (the invertable fragment carries structural fimbrial genes).

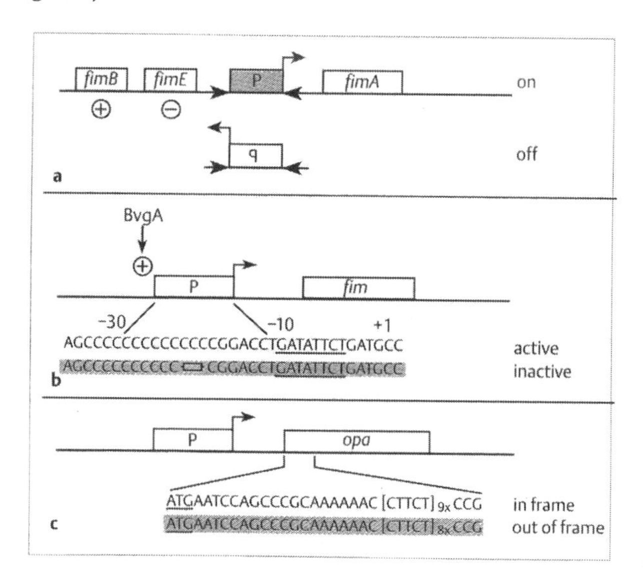

Fig. 33.**17a–c** **Molecular mechanisms of phase variation and antigenic variation.**
a Phase variation of type I fimbriae (Fim) of *Escherichia coli* due to inversion of a DNA fragment carrying the promoter. The gene *fimA* encodes the major subunit protein; *fimB* and *fimE* encode recombinases that influence the insertion of the promoter element.
b Phase variation of fimbriae (*fim*) of *Bordetella pertussis* due to small deletions in the promoter region of the *fim* gene. BvgA acts as an activator.
c Antigenic variation of the Opa proteins of *Neisseria gonorrhoeae*. In the 5′-region of the *opa* genes, small deletions of pentanucleotides (CTTCT) occur, which bring the open reading frame "in" or "out of frame"

Small deletions, insertions, and frameshift mutations. Phase and antigenic variation processes can also be due to small deletions or insertions of one or few base pairs, created by "slipped strand mispair", during replication. Thus, the promoter region of the *fim* genes of *Bordetella pertussis*, coding for fimbriae, carries a stretch of poly C (Fig. 33.**17b**). Deletion of a few base pairs alters the spacing between the – 10 region and a region responsible for the binding of the activator protein BvgA (see Chapter 33.4.3; Fig. 33.**17**) of the *fim* promoter region. This results in the switching from the "on" to the "off" status. Similar processes are involved in phase variation of fimbriae of *Haemophilus influenzae* and antigenic variation of **v**ariable **l**ipo**p**roteins (Vlp) of the veterinary pathogen *Mycoplasma hyorhinis*.

Moreover, the number of repeat motifs located in the 5′-region of some structural genes can be altered due to deletions or insertions that involve a translational frame shift. Thus, the antigenic variation of surface proteins of *Neisseria gonorrhoeae*, termed Opa, is due to the alteration of the number of the CTTCT motifs located in the 5′-coding region of the *opa* gene (Fig. 33.**17c**). The antigenic variation of the Opa proteins is a consequence of the number of CTTCT sequences within the corresponding genes. If nine CTTCT elements are present in the 5′-end of the gene, the leader peptide of Opa is in frame with the rest of the protein. Eight CTTCT elements place the leader out of frame with the remainder of the coding sequence, resulting in the failure to express the protein. From the 8–12 *opa* genes of *N. gonorrhoeae*, only one or two are active at a certain time period. Similar mechanisms of phase and antigenic variation have also been found for a number of virulence factors of various organisms, such as the PilC adhesin of *N. gonorrhoeae*; the *lic* genes, encoding LPS, of *H. influenzae*; and the *sap* genes encoding S-layers of *Campylobacter fetus*.

Genomic rearrangements in virulence genes. Pili of *N. gonorrhoeae* undergo antigenic variation due to rearrangement of the structural *pil* genes. As indicated in Figure 33.**18**, different variants of structural *pil* genes encoding major subunit proteins of the *N. gonorrhoeae* pili (or fimbriae), are located on the chromosome. While one particular copy, *pilE*, is located in an expression site where it is transcriptionally active, other variants, such as *pilS*, are silent copies that do not carry promoter regions and 5′-sequences. The structural *pil* genes consist of variable and constant DNA sequences. New variants of pilin proteins are produced through crossing-over processes that occur between the constant sequences of *pilE* and *pilS* copies. Following intrastrain recombination, *pil* genes located on the chromosome of one particular isolate undergo recombination. *N. gonorrhoeae* is a naturally transformable microorganism; therefore, DNA from lysed cells can be taken up. This can lead to interstrain recombinations between *pilE* and *pilS* sequences originating from different isolates. Phase variation of the Vmp (**v**ariable **m**ajor **p**rotein) of *Borrelia hermsii*, the causative agent of relapsing fever, and of Vlp (**v**ariable **lipo**protein) of *Mycoplasma hyorhinis*, also undergo antigenic variations due to DNA rearrangements.

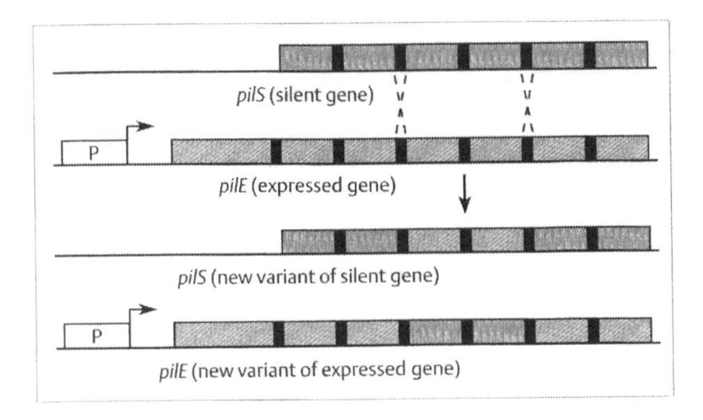

Fig. 33.**18** **Phase variation of pili** (*pil* of *Neisseria gonorrhoeae*. A silent copy (*pilS*) of the major subunit gene and an expression copy (*pilE*) undergo a double crossover, leading to a new variant of the pilus major subunit. Constant regions of the *pil* genes are shaded black; variable regions are given as light red and gray boxes

Pathogenic microorganisms have the capacity to change **surface structures** periodically. These processes, termed virulence modulation, phase variation, and antigenic variation, can help the pathogens to escape from the host immune system. Mutations and genomic rearrangements are involved in the variation mechanisms that lead to the alteration of virulence factors.

33.4.2 Environmental Stimuli Influence the Transcription of Virulence Genes

Many bacterial pathogens (e.g., *Vibrio cholerae, Yersinia enterocolitica*) survive within a natural environment outside the host. However, during infection, the organisms have to multiply inside host organisms and even inside eukaryotic cells. Other pathogens (e.g., *Yersinia pestis, Borrelia burgdorferi*) have two host organisms (e.g., arthropods and vertebrates). Therefore, pathogenic microorganisms are under strong adaptional pressure. Table 33.**12** presents an overview of the stimuli known to influence the activity of virulence genes. However, only limited information is available on stimuli that influence the expression of virulence genes in human or animal host organisms.

Very often, sets of virulence-associated genes are regulated as a consequence of particular environmental **stimuli**. The stimuli are transformed to signals, which are transferred to certain **regulators**, some of which are listed in Table 33.**12**. Regulatory proteins modulate the transcriptional activity of virulence genes, which may be organized in **operons** and regulons. One particular global regulator may be able to induce or repress another regulator gene, leading to cascades of regulation within "**pathogenic networks**". The principles of regulation are described in Chapter 20.

Box 33.4 Some **virulence genes** or operons are regulated by activator and/or repressor molecules encoded by the corresponding operons themselves. An example is given in Figure 33.**19**, which illustrates the regulation of the *pap* operon coding for P fimbriae of uropathogenic *Escherichia coli*. Expression of the *pap* operon depends on various stimuli, such as temperature, carbon source, or growth on solid media. In addition, *pap* expression undergoes phase variation. While the major subunit protein is encoded by the structural gene *papA*, the *papB* and *papI* genes encode two activator proteins, one of which, PapB, is able to bind to the *pap* promoter region. As also shown for other virulence genes, the operon-specific regulators act in concert with other global regulators of the cell. The "catabolite repressor protein" (Crp) is one of those global regulators involved in *pap* expression.

In addition to PapB, PapI, and Crp, the "**l**eucin **r**esponsive regulatory **p**rotein" Lrp is important for *pap* expression. Lrp activates *pap* genes as a consequence of environmental stimuli and is also involved in phase variation of P fimbriae. The Lrp binding sites in the *pap* regulatory region overlap with two GATC sites, which act as recognition sites for the deoxyadenosine methylase (Dam). If Lrp binds to sites 1, 2, and 3, the GATC site II is protected from methylation, while GATC site I is methylated, a consequence of which is that the *pap* operon is shut "off". If Lrp binds to the regulator PapI, the affinity to sites 1, 2, and 3 is reduced, resulting in the methylation of GATC site II, which in turn is a prerequisite for the transcription of the *pap* operon. In addition to PapB, PapI, Crp, Lrp, and Dam, the histone-like protein H-NS is involved in the temperature-dependent regulation of the *pap* operon.

Global regulators of the AraC, LysR, and Fur families also influence the transcriptional activity of virulence genes in many organisms. The "ferric uptake regulator" (Fur) is able to bind iron ions, which act as co-repressors. The Fur–Fe^{2+} complex represses genes encoding siderophores, toxins, superoxide dismutase, and others (see Chapter 33.3.5). Iron is also a co-regulator in the regulation of diphtheria toxin produced by the Gram-positive organism *Corynebacterium diphtheriae*. Calcium is another environmental stimulus that influences gene expression of *Yersinia enterocolitica*.

Moreover, alternative sigma (σ) factors (see Chapters 15 and 28) influence the expression of virulence genes. Most of the virulence genes are transcribed by the "normal" σ-factor σ^{70} (RpoD). Particular virulence genes are expressed by making use of alternative σ-factors. Some of these respond to particular environmental stimuli, such as heat shock (RpoH), nitrogen (RpoN), and temperature (RpoF). Interestingly, the expression of some virulence genes depends on the σ-factor RpoS, which is preferentially used during stationary phase. The influence of σ-factors on regulation of virulence genes illustrates that the changes of bacterial properties during infection resemble processes observed during bacterial differentiation, such as sporulation of bacilli or developmental changes of myxobacteria, which also employ alternative σ-factors and other regulators (Chapter 25).

Table 33.12 **Global regulator proteins controlling expression of virulence genes**

Global regulator	Bacterium	Gene symbol	Environmental stimulus	Virulence factor controlled
Crp	Pathogenic *Escherichia coli*	*crp*	Carbon source	P fimbriae
Crp	*Pseudomonas aeruginosa*	*crp*	Carbon source	Alginate
Lrp	Uropathogenic *Escherichia coli*	*lrp*	Temperature	P fimbriae S fimbriae
Lrp	Enterotoxigenic *Escherichia coli*	*lrp*	Temperature	K99 adhesins
AraC-like	*Yersinia enterocolitica*	*virF*	Temperature	Yop proteins
AraC-like	*Shigella flexneri*	*virF*	Temperature	Ipa proteins VirB protein
AraC-like	Enterotoxigenic *Escherichia coli*	*cfaD* *rns*	Temperature Temperature	CFA I CFA I
LysR-like	*Salmonella enterica* sv. Dublin	*spvR*	Stationary phase	Spv proteins
H-NS	*Shigella flexneri*	*virR*	Temperature	Ipa proteins
H-NS	Uropathogenic *Escherichia coli*	*drdX*	Temperature	P fimbriae S fimbriae
H-NS	*Yersinia enterocolitica*	*ymoA*	Temperature	Yop proteins
H-NS	Enterotoxigenic *Escherichia coli*	*hns*	Temperature	CFA I adhesins
Fur	Pathogenic *Escherichia coli*	*fur*	Iron	Aerobactin Enterobactin Shiga-like toxin I Superoxide dismutase
?	*Yersinia enterocolitica*		Calcium	Yop proteins

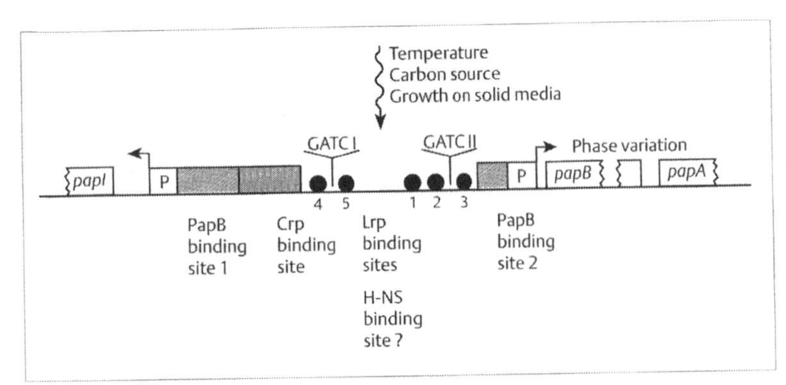

Fig. 33.19 **Regulatory region of the *Escherichia coli pap* operon, which encodes P fimbriae that undergo phase variation.** The genes *papA* (coding for the major subunit protein), *papB* and *papI* (coding for regulators) are given. Binding sites for Lrp (black dots), Crp (shaded boxes), and PapB (open boxes) are indicated. The binding sites for H-NS have not been identified yet. GATC sites act as targets for the Dam methylase. Various stimuli are listed. See Box 33.4 for details

Virulence-associated genes are regulated by activators and/or repressors in response to environmental stimuli. The genes are organized in operons, regulons or modulons which can form "pathogenic networks". A variety of operon-associated regulators and global regulatory factors influence the expression of virulence genes. Often, many regulators and alternative σ-factors act in concert (Chapter 20).

33.4.3 Two-Component Systems Are Involved in the Control of Virulence Gene Expression

The regulation of bacterial virulence factors often involves two-component systems (see Chapter 20.4) These systems are widespread among prokaryotes and even eukaryotes (see Tab. 20.**1**). Various two-component systems of pathogenic bacteria are indicated in Table 33.**13** and Figure 33.**20**. The BvgS protein of *Bordetella pertussis* senses various stimuli and activates the response regulator BvgA by phosphorylation. BvgA is a DNA binding protein that directly activates some of the virulence genes. In addition to the proteins BvgS and BvgA, other regulators are necessary for a correct signal transfer in *B. pertussis*.

Box 33.5 Another example of a two-component system acting in concert with other regulators is the AlgQ/AlgR system of alginate production in *Pseudomonas aeruginosa*. Infections due to *P. aeruginosa* are of major importance in patients suffering from cystic fibrosis (CF), characterized by alterations in the composition of the mucoid slime (see Chapter 33.3.7). *P. aeruginosa* strains isolated from CF patients produce an alginate capsule that protects the bacteria from phagocytosis. High osmolarity activates the *alg* operon, where the sensor AlgQ and the regulator AlgR are of major importance. Other factors, such as the alternative sigma factor AlgU (homologue of RpoE), are also involved in the regulation of alginate production.

Two-component regulatory systems are involved in the environmental response of virulence genes of many pathogenic organisms.

33.4.4 Posttranscriptional Regulation of Virulence Genes Occurs

A wide variety of virulence genes are also regulated at the posttranscriptional level (see Chapters 18.13 and 19). Typical examples involve the forms of regulation listed below.

Translational efficiency. The genes *ctxA* and *ctxB*, encoding the A and B subunits of the cholera toxin are encoded by a single operon that is controlled by a single promoter. The ribosome binding site (Rbs) of the mRNA encoding the B subunit is more efficient than the Rbs of *ctxA*, thus resulting in the production of different amounts of the two toxin subunits.

Translational pausing. The genes *argU* and *leuX*, encoding the rarely used arginine- and leucine-specific tRNAs, are involved in the expression of type I fimbriae of *Escherichia coli* and *Salmonella typhimurium*, probably by causing translational pausing.

Antisense RNA. The outer membrane proteins OmpC and OmpF play a role in the virulence of enteric bacteria such as *S. typhimurium*. The expression of the corresponding genes is regulated by the two-component system EnvZ/OmpR (see Fig. 20.**18**). A small RNA, MicF, is transcribed on the strand opposite to *ompC* and has a promoter that is coregulated with the *ompC* promoter. The antisense RNA prevents translational initiation at the Rbs of the *ompF mRNA*.

Posttranslational processing of proteins involved in virulence. Examples of posttranslational processing

Table 33.**13** **Two-component systems controlling expression of virulence-associated genes**

Bacterium	Two-component system (Sensor/Regulator)	Environmental stimulus	Virulence factors controlled
Bordetella pertussis	BvgS/BvgA	Temperature, nicotinic acid, $MgSO_4$	Filamentous hemagglutinin, pertussis toxin, adenylate cyclase hemolysin, fimbriae,
Pseudomonas aeruginosa	AlgQ/AlgR	Osmolarity, nitrate	Alginate capsule
Vibrio cholerae	ToxR, ToxS, ToxT	pH, temperature, osmolarity	Cholera toxin, Tcp pilus, accessory colonization factor
Salmonella enterica sv. *Typhimurium*	PhoQ/PhoP	Phosphate, magnesium	PagC protein, other *pag-* and *prg*-encoded proteins
Escherichia coli, other enteric bacteria	EnvZ/OmpR	Osmolarity	Other membrane proteins, OmpC, OmpF
Staphylococcus aureus	AgrB/AgrA	pH, growth phase	α-Toxin, coagulase, protein A, fibronectin-binding protein
Streptococcus pyogenes	?/VirR (Mry)	Temperature, anaerobiosis	M protein, C5a peptidase

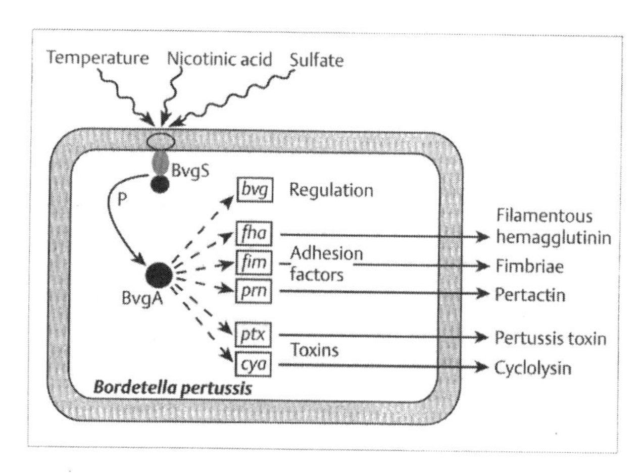

Fig. 33.**20** **The two-component system Bvg (*Bordetella* virulence genes) of *B. pertussis*.** Various stimuli affect via the sensor kinase BvgS, and the response regulator BvgA the transcription of several operons for synthesis of virulence factors. *fha*, filamentous hemagglutinin; *fim*, fimbriae; *prn*, pertactin; *ptx*, pertussis toxin; *cya*, cyclolysin; P, phosphorylation

of proteins (see Chapter 19.3) are the N-methylation of the major subunits of type IV fimbriae of *Vibrio cholerae*, *Pseudomonas aeruginosa*, and other pathogens, activation of A–B toxins from various species by proteolytic nicking, and acylation of the α-hemolysin of uropathogenic *E. coli*.

The synthesis of many proteins involved in virulence is influenced by different mechanisms of posttranscriptional regulation and posttranslational protein processing.

33.5 Diagnosis, Therapy, Prophylaxis: Problems and New Approaches

There is increasing awareness of the importance of precise and fast identification and characterization of pathogens for proper treatment of infectious diseases. This section deals with practical implications of medical microbiology.

33.5.1 Diagnosis of Bacterial Is Improved by New Tools

For clinical microbiologists, it is important to identify pathogenic microorganisms in specimens, such as blood, urine, feces, and tissue samples. The aim of diagnostics is to determine quickly whether specimens belonging to a pathogenic species are present in certain samples. The classical methods to identify microorganisms are staining and microscopic evaluation as well as growth of microorganisms on selective media and differentiation by biochemical tests (see Chapter 29.3). Thus, specific staining procedures exist (e.g., Ziehl-Neelsen staining for the identification of mycobacteria). Particular media [e.g., MacConkey agar, eosine-methylene blue agar, and ENB ("enriched nutrient broth")] are useful for selective growth of pathogens. Anaerobes must be cultivated in media depleted of oxygen. Biochemical growth tests (up to 80 markers in specific test kits, such as glucose fermentation, hydrogen sulfide formation, and citrate utilization) help to characterize microorganisms.

Furthermore, **immunological tests** have been developed to identify pathogenic microorganisms. Polyclonal and monoclonal antibodies (see Chapter 33.2.2) directed against cell surface structures (e.g., outer membrane proteins, flagella, fimbriae) are linked to fluorescent dyes (Fig. 33.**21**). Fluorescence microscopy can help to identify especially those microorganisms that are difficult to cultivate (e.g., *Legionella pneumophila*, *Chlamydia trachomatis*, and *Bacillus anthracis*). Furthermore, microbial antigens (including recombinant proteins) are used to determine the amount of pathogen-specific antibodies in body fluids, such as serum. Western blots, ELISA (**e**nzyme-**l**inked **i**mmuno**s**orbant **a**ssays) or **radioi**mmuno **a**ssays (RIAs) are employed. Using phenotypical tests (e.g., hemagglutination, hemolysis), the **pathogenic potential** of bacteria can be assessed. It is important to characterize the production of virulence factors of facultative bacterial pathogens and to distinguish them rapidly from harmless variants of the same species.

In addition, methods based on recombinant nucleic acid techniques are used to identify specimens of pathogenic origin. For these purposes, **rRNA-specific sequences** are often used to identify bacteria. Oligonucleotide probes can directly be linked to fluorescence dyes and used for in situ hybridizations with rRNA. In addition, the polymerase chain reaction (PCR; see Fig. 17.**18**), using oligonucleic acid primers directed against species-specific rRNA regions or against other species-specific genes, is used to identify pathogens in patients samples. Examples are the identification of non-cultivable microorganisms such as non-typical mycobacteria by in situ hybridization, and the identification of pathogenic bacteria in the environment. The "**viable but not culturable**" (**VBNC**) state represents a survival strategy of Gram-negative bacteria that are exposed to a low-nutrient environment (see Chapter 28). Bacterial pathogens, such as *Vibrio cholerae*, *Shigella dysenteriae*, and *Salmonella enterica*, survive in a VBNC state where they can be identified only by PCR, in situ hybridization, or immunological assays.

Several **methods**, including staining and microscopy, and cultivation and immunological tests, are used **to identify pathogenic microorganisms** in patient samples or to detect virulence factors. Novel nucleic-acid-based methods can be used to characterize the pathogenic potency of bacteria and to identify microorganisms that cannot be cultivated (see Chapters 29.4 and 29.5).

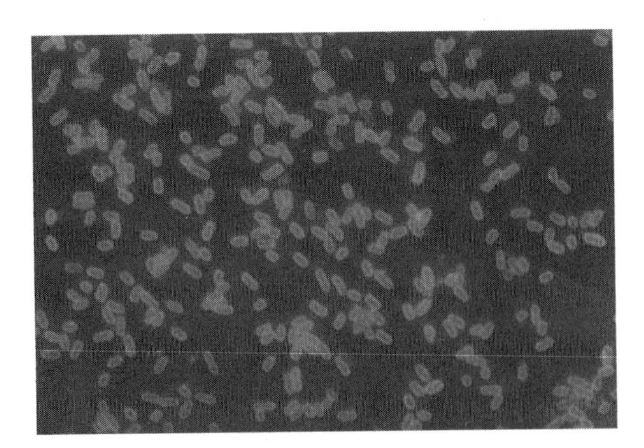

Fig. 33.**21** **Fluorescence-staining of pathogenic *Escherichia coli*.** *E. coli* strain IHE3034, isolated from a case of newborn meningitis (NBM) was labelled with a fluorescein isothiocyanate (FITC) coated antibody directed against the K1 capsule

Other important problems are the identification of the reservoir of an infection and the route of distribution of pathogens. This is especially important for the analysis of nosocomial and "environmental" infections (e.g., Legionnaires' disease). Epidemiology deals with these problems. Classical methods, such as sero-typing, lyso-typing, and bacteriocin-typing, as well as the determination of outer membrane pattern and electro-typing of enzymes are performed to determine the causative agents of outbreaks. Recently, **molecular epidemiology** based on genome analysis has been developed, and represents a more sophisticated approach to distinguish between individual bacterial strains. Plasmid fingerprinting is one of the methods used. In addition, the evaluation of restriction fragment length polymorphism (RFLP), also in combination with Southern hybridizations with defined gene probes, is used (see Chapter 29.4). The analysis of genomic pattern of strains following pulsed-field gel electrophoresis (PFGE) plays a major role in epidemiological studies (see Chapter 29.4). This method uses restriction endonucleases that do not cut the DNA often to produce a small number of large DNA fragments, which can then be separated by PFGE. In addition, different PCR protocols using conserved oligonucleotide primers can be used as fast methods to distinguish between different strains. Using these techniques, "environmental infections" (e.g., caused by *Legionella pneumophila*), food contaminations (e.g., *Salmonella enterica* sv. *Enteritidis*), nosocomial infections (e.g., caused by *Staphylococcus epidermidis*), and other types of infectious diseases can be analyzed.

Genomic mapping techniques, including pulsed-field gel electrophoresis, are used to study the spread of infectious agents in the environment and in hospitals.

33.5.2 Antibiotics Are Used as Drugs Against Bacteria

Antibiotics are low-molecular-weight compounds produced by organisms that kill or inhibit the growth of other organisms, e.g., bacteria. Most antibiotics used in medicine are produced either by bacteria, especially by *Streptomyces* spec., or by fungi (see Chapter 27). Certain antibiotics kill bacteria (**bacteriocidal substances**), whereas others only inhibit growth (**bacteriostatic compounds**). An effective antibiotic should cover a wide spectrum of microorganisms and should exhibit only minor toxicity in humans, that is, its targets should preferentially be molecules not found in eukaryotes. Most of the antibiotics used in therapy (Tab. 33.**14**) are directed against particular bacterial **targets** (see Fig. 33.**22a** and chapter 15.3.4)

Cell wall synthesis. One preferred target is the bacterial cell wall, unique to prokaryotes, or its synthesis. Many antibiotics interfere with the synthesis of the peptidoglycans. The β-lactam antibiotics (e.g., penicillins and cephalosporins) inhibit the transpeptidation reaction involved in the cross-linking step of peptido-

Table 33.**14** **Popular antibiotics used in therapy and targets**

Target/Name	Substance	Effective against
Cell wall synthesis		
Ampicillin	Penicillin (β-lactam)	Gram-negative bacteria (e.g., *Escherichia coli*)
Methicillin	Penicillin (β-lactam)	Gram-positive bacteria (e.g., staphylococci)
Oxacillin	Penicillin (β-lactam)	Gram-positive bacteria (e.g., staphylococci)
Penicillin G	Penicillin (β-lactam)	*Enterococcus pneumoniae*
Cefotaxime	Cephalosporin (β-lactam)	Gram-negative bacteria
Vancomycin	Cyclic glycopeptide	Gram-positive bacteria (e.g., *Staphylococcus aureus*, *Clostridium difficile*)
Teicoplanin	Cyclic glycopeptide	Gram-positive bacteria (e.g., *Staphylococcus aureus*, *Clostridium difficile*)
Protein synthesis		
Gentamycin	Aminoglycoside	Gram-negative bacteria
Kanamycin	Aminoglycoside	Gram-negative bacteria
Streptomycin	Aminoglycoside	Gram-negative bacteria
Erythromycin	Macrolide	Gram-positive bacteria, *Legionella* spec.
Clarythromycin	Macrolide	Gram-positive bacteria
Chloramphenicol	Aromatic biosynthesis	Gram-negative bacteria
Doxycyclin	Tetracycline	Mycoplasma
Nucleic acid synthesis		
Ofloxacin	Quinolone	Gram-negative bacteria
Ciprofloxacin	Quinolone	Gram-negative bacteria
Rifampicin	complex polyketide	Gram-positive bacteria (e.g., *Staphylococcus epidermidis*, mycobacteria)
Metabolism		
Trimethoprim	2,4-diamino-5-(3,4,5-trimethoxybenzyl) pyrimidine	Uropathogenic bacteria

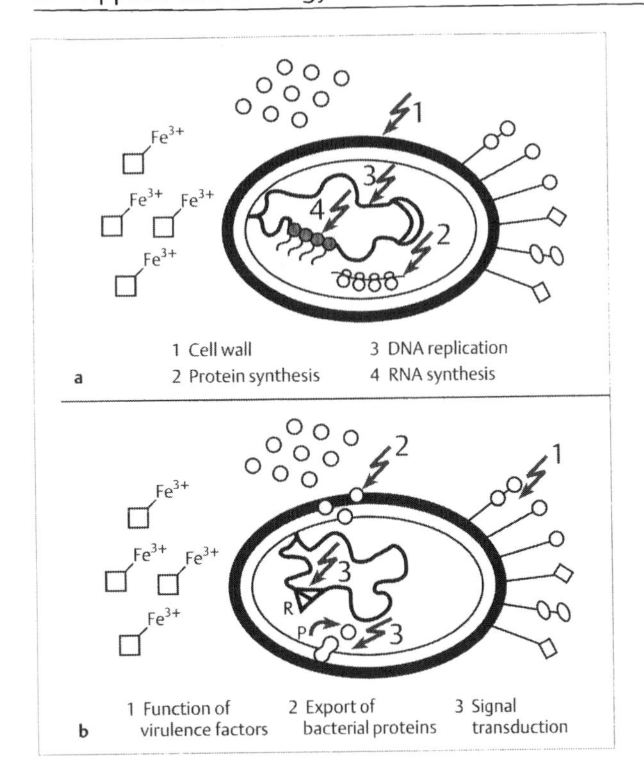

Fig. 33.**22a,b** **Diagram of pathogenic bacteria illustrating real and potential targets for antimicrobial treatment.**
a Targets for antibiotic substances are given.
b Potential targets are indicated. The arrows indicate the targets

glycan biosynthesis (see Chapter 23.1). The glycopeptide antibiotics, exemplified by vancomycin and teicoplanin, inhibit other steps of peptidoglycan synthesis (see Chapter 23.1). These antibiotics have become particular important for the treatment of infections caused by strains of *Staphylococcus aureus* and other Gram-positive pathogens.

Protein synthesis. Another preferred target for antibiotics is the bacterial 70S ribosome and concomitantly protein synthesis. Such antibiotics are specific for the 70S ribosome, which differs extensively from 80S ribosomes of mammalian cells. These antibiotics (e.g., streptomycin and other aminoglycosides, tetracyclines, chloramphenicol, and macrolides) have been discussed in Chapter 15.3.4.

Bacterial nucleic acid synthesis can also be inhibited by antibiotics. Quinolones, used for treatment of infections caused by Gram-negative bacteria, bind to the B subunit of DNA gyrase, thereby inhibiting DNA replication (see Chapter 14.1). Quinolones penetrate macrophages and granulocytes better than most other antibiotics and are consequently useful tools for curing infections caused by bacteria that survive in phagocytes. Rifampicin, which is one of the few antibiotics that can be used to treat tuberculosis, acts by inhibiting bacterial DNA-dependent RNA polymerases.

Metabolism. Other antibiotics are **inhibitors** of enzymes involved in **bacterial metabolism**. For example, trimethoprim inhibits the formation of tetrahydrofolic acid. The synthetic antibacterial compounds of the sulfonamide family act in the same way.

Antibiotics are low-molecular-weight substances that kill bacteria or inhibit bacterial growth. They are produced by bacteria or fungi. Antibiotics have specific targets, such as, the cell wall, the ribosomes, or nucleic acid metabolism.

33.5.3 Occurrence of Antibiotic-resistant Pathogens Enforces the Development of New Drugs

The determination of the sensitivity of pathogens to antimicrobial agents is essential for successful therapy. Filter paper discs, containing known concentrations of different antibiotics, can be placed on a plate with microorganisms. In order to determine the exact **minimal inhibitory concentration** (**MIC**), serial dilutions of the antibiotics are used. Low concentrations of drugs allow the microorganisms to grow, while higher concentrations inhibit their cultivation.

Many pathogens responsible for severe infectious diseases can develop resistance to antimicrobial agents. Thus, Gram-negative bacteria, such as *Escherichia coli* or *Pseudomonas aeruginosa*, may exhibit resistance to one or several antibiotics (**m**ulti**r**esistance, MR). Recently, an increasing number of Gram-positive pathogens showing resistance to methicillin (e.g., *Staphylococcus aureus*) or vancomycin (e.g., *Enterococcus faecalis*), have been isolated. Another problem is the increase of penicillin-resistant pneumococci. A relatively high percentage of mycobacteria show resistance to rifampicin or isoniazol. Resistance mechanisms differ with respect to the antimicrobial drugs and the pathogenic organisms.

Enzymatic inactivation of antibiotics. One of the major resistance mechanisms of Gram-negative bacteria to β-lactam antibiotics is the occurrence of β-lactamases, which cleave the β-lactam ring of the drug and thereby abolish its antibiotic activity. The genes encoding β-lactamases (*bla*) can be located on resistance (R) plasmids or on the chromosome, where they can undergo gene amplification, resulting in an increased level of β-lactamase. Aminoglycosides can be inactivated by modifying enzymes (e.g., **c**hloramphenic **a**cetyl **t**ransferase, CAT), which add specific groups (phosphoryl, acetyl, and others) to the drugs.

Modification of the antibiotic target. β-lactam antibiotics bind to so-called penicillin binding proteins (PBPs) in the bacterial cell wall. Thus, resistance of staphylococci to methicillin is due to a modified PBP2′ (encoded by the *mecA* gene) that is no longer able to bind the drug. In addition, vancomycin resistance is caused by one of the two operons *vanA* and *vanB*, which are regulated by a two-component system in enterococci. The incorporation of **A**la-D-hydroxybutyrate instead of D-**A**la-D-**A**la in the peptidoglycan layer results in vancomycin-resistant strains. The alteration of the 23S rRNA in the ribosome and the alteration of ribosomal proteins (e.g., of L11 in streptomycin-resistant mutants of *Escherichia coli*) lead to resistance of the pathogens to macrolides and tetracycline. In addition, point mutations in the DNA gyrase B subunit (in the case of quinolones) and in the B subunit of RNA polymerase (in the case of rifampicin) result in bacterial resistance to these antibiotics.

Active efflux of antibiotics. Tetracycline resistance can result from an efflux mechanism, by which the drug is pumped out of the cytoplasm. A similar mechanism has been observed for resistance of staphylococci to macrolides.

Point mutations. One way by which bacteria can become resistant to a particular antibiotic, is the occurrence of point mutations in individual strains. For example, *Escherichia coli* strains that become resistant to quinolones can spread among a bacterial population in a hospital. An important problem in clinical microbiology is the occurrence of multiresistant strains.

Horizontal gene transfer, mediated by transformation, conjugative plasmids, or phages, can be a mechanism for the development of multiresistant strains (see Chapters 14.3.2 and 14.4.2). Increased penicillin resistance in clinical isolates of *Enterococcus pneumoniae* is due to the spread of chromosomal genes encoding modified penicillin binding protein by natural transformation.

Conjugative transposons and plasmids. Resistance genes (e.g., *tetM*, encoding resistance to tetracycline, or the *vanB* operon, responsible for vancomycin resistance) may be part of **conjugative transposons**. Such transposon, which are 65–160 kb in size, are located on the chromosomes of Gram-positive bacteria or strains of the Gram-negative genus *Bacteroides*. Following site-specific excision and formation of a "**c**ovalently **c**losed **c**ircle" (ccc), the conjugative transposons are transferred to other strains where they can integrate into the chromosome in Gram-negative strains. **Resistance (R) plasmids** are often spread by conjugation. Very often, these R plasmids have a complex composition, as indicated in Figure 33.**23**. Complex transposons, such as the class I transposon Tn*10*, consisting of a tetracycline-resistance gene and two IS10 elements, or the class II transposons, such as Tn*21*, may be part of such plasmids. In particular strains, resistance genes are part of **integrons**, gene cassettes that carry inserted resistance and integrase genes. The integrase genes allow the mobilization of resistance genes by making use of a site-specific recombination. It is speculated that some of the antibiotic-resistance genes may have originated from antibiotic-producing bacteria, such as *Streptomyces* or *Bacillus*. For example, the enzyme 6′-**a**minoglycoside-**ph**osphotransferase [APH (6′)] which confers resistance to aminoglycoside antibiotics in *Escherichia coli*, shows a high degree of similarity to the same enzyme of *Streptomyces fradiae*.

Horizontal gene transfer accelerates the distribution of **antibiotic resistance genes** among pathogenic bacteria drastically. Especially multiresistant strains are a serious problem in medicine.

Only a small percentage of all globally existing bacteria are resistant to antibiotics. The number of resistant isolates in hospitals, however, is significant due to strong selection. One solution to the resistance

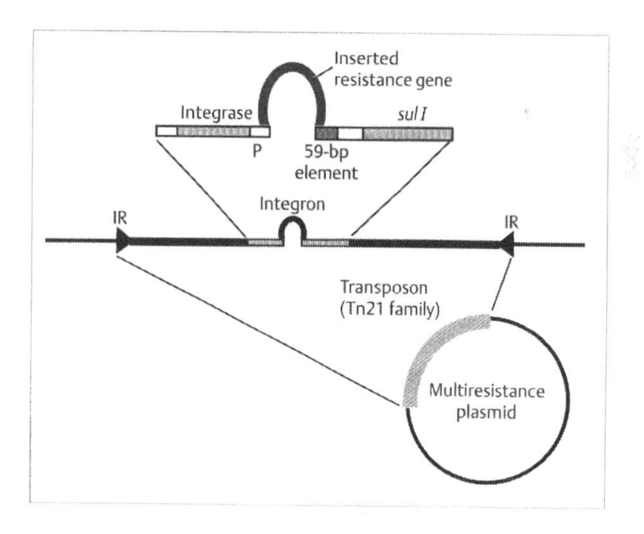

Fig. 33.**23** **Model of a multiresistance (MR) plasmid.** MR plasmids carry collections of transposons. Some transposons, such as the Tn*21* group, bear integrons that allow the integration of gene cassettes harboring resistance genes. These cassettes can be exchanged between different DNA molecules. *P*, promoter, *sul I*, sulfonamide resistance gene; IR, inverted repeat [after 2]

problem may be to increase the hygienic standards in hospitals, but the use of combinations of different drugs in therapy and the avoidance of unnecessary use, such as in virus infection treatment, are especially important. The use of antibiotics in animal feed should be greatly reduced to avoid further selection of resistance strains.

Screening programs have been initiated that aim at the discovery of new antimicrobial substances. Compounds originating from soil samples, marine waters, insects, tropical plants, and other "exotic" places are being screened for antimicrobial activities. In addition, chemical libraries (e.g., "**s**ynthetic **c**ombinatorial **l**ibraries," SCLs) are used to identify new antibacterial compounds. Furthermore, attempts have been undertaken to develop drugs that attack new microbial targets, such as vital enzymes of the cell metabolisms (e.g., ATPsynthases) or structural proteins of the bacterial cell surface. In addition to substances that show bactericidal or bacteriostatic effects, new drugs are under consideration that may **interrupt pathogenesis**. As indicated in Figure 33.**22b**, such drugs can directly interfere with the action of virulence factors (e.g., blocking of adhesin-mediated colonization, inhibition of enzymatic activities of toxins; see Chapters 33.2.1 and 33.2.3) or their secretion (e.g., type III secretion pathway of invasion factors; see Chapter 33.2.2). In addition, transcriptional regulators and signal transduction pathways (e.g., histidine kinases or phosphatases of two-component regulatory pathways and phosphotransferase systems) have been tested as potential targets.

Different and new approaches to screen for new antimicrobial substances and to identify alternative targets may lead to new therapeutic agents against bacterial pathogens.

33.5.4 Vaccines Protect Individuals Against Infectious Agents

Vaccination represents an effective approach to protect individuals from infectious agents. Since the experiments performed by E. von Behring, who provided evidence for the role of "antitoxins" in protection against diphtheria, many vaccines have been developed. Vaccination programs have since contributed enormously to the progress in human health care. Nevertheless, there is an urgent need for novel vaccines against major infectious diseases. This section will concentrate on the development of vaccines against pathogenic bacteria. Immunity to infectious agents can be induced by two ways:

1. **Passive immunization.** An individual receives injections of antisera against an infectious agent or a potent virulence factor, such as antiserum raised against tetanus toxins. Passive immunization is therapeutic.
2. **Active immunization.** An individual receives an injection of an antigen that leads to the development of protective antibodies and/or T cells. An example is detoxified tetanus toxin (termed toxoid). Active vaccination leads to immunity and is prophylactic.

The material used to induce **active immunity** is termed **vaccine**. An ideal vaccine should be safe, effective, and simple to administer. Moreover, it should have the capacity to be used in combination with other vaccines. Multiple strategies are available to design a vaccine. Table 33.**15** provides a summary on the available vaccines used for the immunization of humans against bacterial diseases.

Purified components, such as cell wall extracts, chemically **detoxified toxins** (termed toxoids), or **inactivated microorganisms** are used as vaccines. Such "dead vaccines" are used in vaccination programs

Table 33.**15** **Examples of available vaccines against bacterial infectious diseases**

Causative agent	Disease	Type of vaccine(s)
Corynebacterium diphtheriae	Diphtheria	Toxoid
Clostridium tetani	Tetanus	Toxoid
Bordetella pertussis	Whooping cough	Killed bacteria
Salmonella typhi	Typhoid fever	Killed bacteria, attenuated live bacteria
Mycobacterium tuberculosis	Tuberculosis	Attenuated live bacteria (BCG strain of *Mycobacterium bovis*)
Haemophilus influenzae b	Meningitis	Polysaccharide–protein conjugate
Neisseria meningitidis	Meningitis	Capsule–protein conjugate
Enteroococcus pneumoniae	Pneumonia	Capsule–protein conjugate

Table 33.**16** **Examples of vaccines that are needed against bacterial infectious diseases**

Disease	Causative agent
Diarrheal diseases	*Vibrio cholerae* *Shigella* spec. Pathogenic *Escherichia coli* *Salmonella enterica*
Leprosy	*Mycobacterium leprae*
Gastric ulcer	*Helicobacter pylori*
Lyme disease	*Borrelia burgdorferi*
Meningitis	*Neisseria meningitidis* (improved vaccine)
Pneumonia	*Streptococcus pneumoniae* (improved vaccine)

against various bacterial infections, such as diphtheria, tetanus, and pneumonia. **Live strains** that carry mutations or that are related to the infectious agents are also used such as vaccines developed against *Mycobacterium tuberculosis* on the basis of the *Mycobacterium bovis* strain BCG and the *Salmonella typhi* Ty21a typhoid fever vaccine. Attenuated *Salmonella enterica* strain carrying mutations in genes coding for metabolic enzymes (e.g., *aroA*, *purE*) or in regulatory genes (*phoP*, *crp*) have been used as carrier strains for the expression of foreign antigens.

As indicated in Table 33.**16**, much effort has been invested to develop novel vaccines, such as against infectious agents of the digestive tract (e.g., *Vibrio cholerae*, *Helicobacter pylori*). In order to develop new vaccines, molecular techniques are being applied in rapidly increasing numbers. Thus, **synthetic peptides**, which represent epitopes of certain antigens, or **recombinant antigens**, are used. **Attenuated organisms**, which carry site-specifically introduced mutations, or certain **live vectors** are used for vaccine development. In addition, the development of "**nucleic acid vaccines**" is in progress. Such vaccines consist of DNA molecules containing cloned genes that encode strong antigens. The recombinant DNA molecules are introduced into the host, and the antigens are expressed by antigen-presenting cells, thereby inducing protection.

Vaccination programs are used to induce active immune protection against infectious agents. Passive immunity is induced by injection of protective antibodies. Dead and live vaccines are used for active immunization against various pathogenic organisms.

33.6 Prokaryotes in Medicine: Future Trends and Developments

The techniques of molecular biology have enormously increased our knowledge about infectious agents and their virulence strategies. Especially the concept of "virulence factors" (Chapter 33.3) and the determination of regulation of virulence genes (Chapter 33.4) have been very successful. Techniques of cell biology will stimulate research on pathogen-induced signal transduction mechanisms in host cells. On the other hand, more basic research will be necessary to increase our knowledge about the expression of virulence genes in vivo and about the host stimuli that influence the signalling and regulation of virulence-associated genes. Living host organisms are used to select for genes involved in in vivo virulence.

The **IVET** ("**i**n **v**ivo **e**xpression **t**echnology") system is based on the idea of fusing the *purA* gene of *Salmonella typhimurium*, which is an essential housekeeping gene involved in nucleotide metabolism, to promoters of genes of unknown function (see Chapter 20.1.3). Following their injection into mice, only those bacteria should survive in which the *purA* genes are transcribed from a promoter that is active in the host organism. A number of genes were isolated with this technique; however, as expected, the majority encoded proteins with metabolic functions. The technology of "**s**ignature-**t**agged **m**utagenesis" (**STM**; Fig. 33.**24**) selects mutants of *S. typhimurium* that are not able to survive in mice following transposon mutagenesis. Each mutant is tagged by a specific sequence that allows the reidentification of particular transposon mutant strains following their injection into mice. A number of novel genes of *S. typhimurium* were selected that seem to play a major role in in vivo virulence and to open new therapeutic targets.

Genome sequencing projects for pathogenic bacteria represent another novel technology with high relevance for microbial pathogenicity. The first entirely sequenced genome was that from the human pathogen *Haemophilus influenzae*. Genomes of a number of other pathogenic species have also been sequenced, such as that of *Mycoplasma genitalium*, *Helicobacter pylori*, *Escherichia coli*, and *Staphylococcus aureus*, or are intended to be sequenced, such as that of *Enterococcus pneumoniae*, *Treponema pallidum*, and *Mycobacterium*

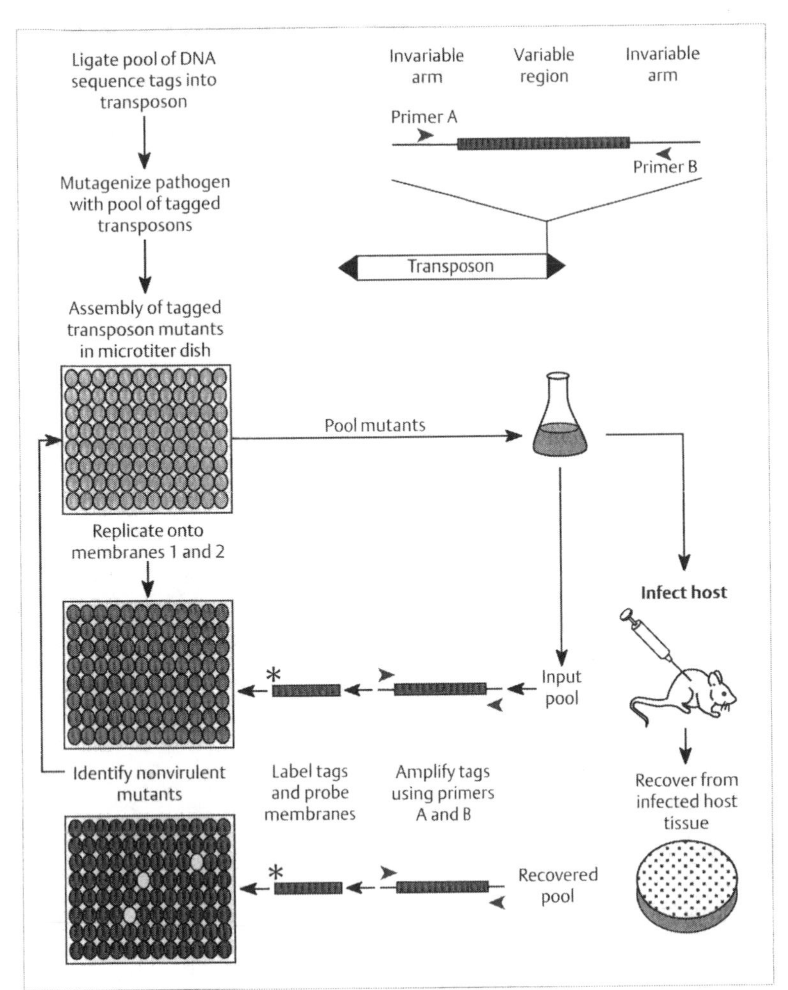

Fig. 33.**24** **Diagram of signature-tagged mutagenesis (STM).** DNA sequence tags are generated by oligonucleotide synthesis and the polymerase chain reaction (PCR). Each tag contains a different central sequence flanked by arms that are common to all the tags. Primers A and B allow the central regions to be amplified and labeled for use as probes. Following ligation of tags into a transposon and transposon mutagenesis of the bacterial pathogen, transposon mutants are assembled in microtiter dishes, pooled, and inoculated into the host. Colony blot hybridization analysis using labeled tags from the inoculum (input pool) and bacteria recovered from the host (recovered pool) allows mutants with attenuated virulence to be identified [after 3]

tuberculosis. These genomic sequences will provide scientists with new information on genes involved in synthesis, transport, and regulation of virulence factors.

Several pathogens are transferred from the environment to host organisms, either directly by means of "technical vectors" (e.g., air conditioning in the case of *Legionella pneumophila*), water or food contamination (e.g., *Vibrio cholerae*, salmonellae), or by vector organisms (e.g., *Borrelia burgdorferi*). In addition, strains causing nosocomial infections can also persist in certain ecological niches. Nevertheless, our basic knowledge about the **ecology of pathogens** is small and needs to be broadened. Thus, questions concerning the natural ecosystems of pathogens (e.g., shigellae), interactions of pathogens and non-pathogenic microorganisms (e.g., amoeba) in the environment, and these interactions as factors that contribute to the survival of microorganisms in habitats outside the host have to be addressed. By increasing our knowledge about the ecology of pathogens, new concepts of treatment and prevention (e.g., by influencing the reservoirs of pathogenic microbes) can be expected.

New techniques will increase our knowledge about host-parasite relationships and will stimulate the development of new drugs, vaccines, and diagnostic tools.

Further Reading

Aktories, K., ed. (1997) Bacterial toxins. London: Chapman & Hall

Amabile-Cuevas, C. F., and Chicurel, M. E. (1992) Bacterial plasmids and gene flux. Cell 70: 189–199

Amann, R., Ludwig, W., and Schleifer, K.-H. (1995) Phylogenetic identification and in situ detection of individual microbial cells without cultivation. Microbiol Rev 59: 143–169

Bhakdi, S., and Tranum-Jensen, J. (1991) Alpha-toxin of *Staphylococcus aureus*. Microbiol Rev 55: 733–751

Brubaker, R. R. (1985) Mechanisms of bacterial virulence. Annu Rev Microbial 39: 21–50

Dorman, C. J. (1994) Genetics of bacterial virulence. Oxford, London: Blackwell

Dougan, G. (1994) The molecular basis for the virulence of bacterial pathogens: implications for oral vaccine development. Microbiology 140: 215–24

Falkow, S. (1988) Molecular Koch's postulates applied to microbial pathogenicity. Rev Infect Dis 10 (Suppl. 2): 274–276

Falkow, S., Isberg, R. R., and Portnoy, D. A. (1992) The interaction of bacteria with mammalian cells. Annu Rev Cell Biol 8: 333–63

Finley, B. B., and Falkow, S. (1989) Common themes in microbial pathogenicity. Microbiol Rev 53: 210–30

Hensel, M., and Holden, D. W. (1996) Molecular genetic approaches for the study of virulence in both pathogenic bacteria and fungi. Microbiology 142: 1040–1058

Isenberg, H. D. (1988) Pathogenicity and virulence: another view. Clin Microbiol Rev 1: 40–53

Miller, V., Kaper, J., Portnoy, D. A., and Isberg, R. R. (1994) Molecular genetics of bacterial pathogenesis. A tribute to Stanley Falkow. Washington DC: ASM Press

Mims, C. A., Dimmock, N., Nash, A., and Stephen, J. (1995) Pathogenesis of infectious diseases, 4th edn. New York London: Academic Press

Mühldorfer, I., and Hacker, J. (1994) Genetic aspects of *Escherichia coli* virulence. Microb Pathog 16: 171–81

Roth, J. A., Bolin, C. A., Brogden, K. A., Minion, F. C., and Wannemuehler, M. J. (1995) Virulence mechanisms of bacterial pathogens. Washington DC: ASM Press

Salyers, A. A., and Whitt, D. D. (1994) Bacterial pathogenesis. A molecular approach. Washington DC: ASM Press

Schaechter, M., Medoff, G., and Schlessinger, D. (1989) Mechanisms of microbial disease. Baltimore: Williams & Wilkens

Vogt, P. K., and Mahan, M. J. (eds.) (1998) Bacterial infections; Close encounters at the host pathogen interface. Curr Topics Microbiol Immunol Vol. 225. Berlin Heidelberg New York: Springer

Sources of Figures

1 Bhakdi, S., and Tranum-Jensen, J. (1991) Microbiol Rev 59: 143–169

2 Amabile-Cuevas, C. F., and Chicurel, M. E. (1992) Cell 70: 189–199

3 Hensel, M., and Holden, D. W. (1996) Microbiology 142: 1049–1058

34 Prokaryotes in Agriculture

Prokaryotes play a significant role in agriculture by participating in the dynamics of biogeochemical cycles (C, N, S, and to a minor extent P, Fe, and Mn), maintaining soil fertility and soil structure, and interacting directly or indirectly with plants (see Chapters 30 and 31). Soil prokaryotes represent the most abundant fraction of the soil microbial biomass, which ranges from 450 to 7000kg/ha (one hectare of soil, at plough depth, corresponds to 2400–2700tons). However, bacteria are not equally distributed in soil. They adhere to mineral and organic matter by means of extracellular polymeric substances, forming microcolonies rather than living as single cells. There are even soil environments where densities as high as 10^8 colony forming units (cfu) per gram of soil dry weight can be found (e.g., the rhizosphere). The term **rhizosphere**, introduced in 1904 by L. Hiltner, indicates the narrow zone of soil surrounding the root (in the range of a few millimeters) that is influenced by the plant root system. Also the direct surface of the root, the rhizoplane, is often heavily colonized by bacteria (Fig. 34.**1**).

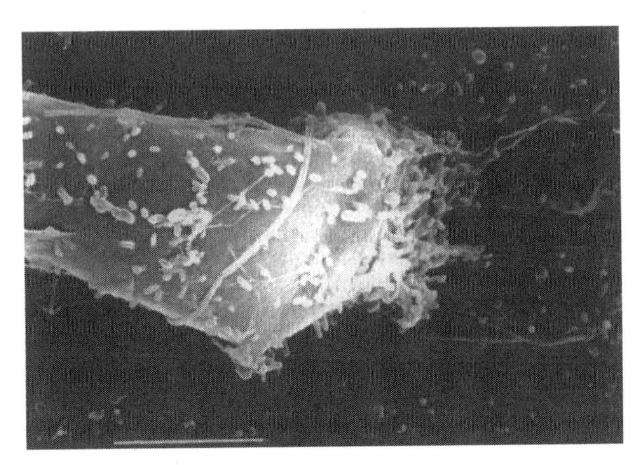

Fig. 34.**1** **Colonization of plant roots by bacteria.** A special area of the rhizosphere is the rhizoplane, i.e., the root surface, at which bacterial cells remain firmly attached after repeated washings. Shown here are rhizoplane bacteria adhering to a wheat root (bar 10μm)

The use of live, pure microbial cultures in agriculture (in contrast to mixed inoculants in the form of "manures") started at the turn of the 20th century in order to improve plant growth and soil fertility. Nowadays, the use of bacterial inoculants extends annually over a considerable acreage (Table 34.**1**) and is expected to increase in the framework of land management practices, such as sustainable agriculture, biological farming, pollution control, and land reclamation. Since more and more genetically modified strains will be generated as potential microbial inoculants, the first section of this chapter will be dedicated to methods available to monitor prokaryotes in soil.

Table 34.**1** **Relevant contribution by inoculants to food production and soil management**

Microorganism	Function	Plant inoculated	Acreage inoculated (ha)
Anabaena	Biofertilizer	Rice	$2 \cdot 10^6$
Azospirillum	Biofertilizer	Cereals (mainly *Triticum durum*, *Zea mays*) and grasses	$1.5–2 \cdot 10^5$
Rhizobium *Bradyrhizobium*	Biofertilizer	Forage and grain legumes	$2–2.4 \cdot 10^7$
Frankia	Biofertilizer	Non-leguminous trees (*Alnus*, *Casuarina*, *Myrica*)	$0.5–1 \cdot 10^3$
Bacillus subtilis *Bacillus thuringiensis* *Pseudomonas*	Biological control agents	Crop plants and trees	$0.5–1 \cdot 10^5$

The acreage inoculated is the estimated average acreage inoculated on an annual basis

34.1 Soil Prokaryotes Can Be Monitored by Various Methods

The potential of recombinant DNA techniques for the production of genetically modified inocula has already been demonstrated for several microorganisms, including *Agrobacterium*, *Bradyrhizobium*, *Clavibacter*, *Pseudomonas*, and *Rhizobium* (Table 34.**2**). When considering the release of wild-type and genetically modified microorganisms (**GMMs**) into the environment, a safe, stable, and effective tracking system is needed to monitor the behavior of an introduced microorganism and to assess the environmental impact and potential risks of such a release (Box 34.**1**).

34.1.1 Plating Techniques Are Used to Select and Quantify Soil Bacteria

Dilution plate counting methods are widely used in the detection, monitoring, and enumeration of indigenous and introduced microbes from environmental samples.

Enrichment techniques. To enrich and quantify indigenous soil bacteria, environmental samples are cultured onto selective media containing specific substances that allow the growth of only certain bacteria or that suppress growth of undesired microorganisms. Many of these selective media rely on intrinsic antibiotic resistances. Nitrite-oxidizing bacteria can be enriched on plates containing KNO_2 as the sole nitrogen source; free-living nitrogen-fixing bacteria, such as *Azotobacter*, can be selected on media containing D-mannitol and no source of combined nitrogen. Since various *Rhizobium* species are resistant to tellurium, selenium, or nalidixic acid, various combinations of metal ions and antibiotics can be used to formulate media selective for *Rhizobium meliloti* or *Rhizobium leguminosarum*. Usually, these media also contain antibiotics that inhibit Gram-positive bacteria, pentachloronitrobenzene to inhibit actinomycetes, and fungicides.

> **Box 34.1 Marker genes** commonly used to tag bacteria are the *lacZ* and the *gusA* genes from *Escherichia coli*, which enable the organism to utilize lactose and β-galactosides, or glucuronides (Gur) respectively. In addition, the corresponding gene products (β-galactosidase and β-glucuronidase, respectively) cleave chromogenic dye substrates (X-Gal and X-Gur, respectively), which results in the formation of a blue color. Consequently, tagged bacteria are identified by their ability to form blue colonies on plates containing X-Gal and X-Gur, respectively. For *Pseudomonas aureofaciens* marked with *lacZ*, a detection limit of 10 cells/g soil has been reported. Similarly, the *xylE* gene from *Pseudomonas* can be used as a reporter gene to tag microorganisms. This gene encodes catechol-2,3-dioxygenase, the activity of which mediates a yellow color reaction. Another convenient reporter system are the *lux* genes, which code for enzymes involved in a light-emitting reaction (bioluminescence, see Chapter 11.6). The photons emitted by bacteria carrying these genes can be detected and enumerated using a scintillation counter or by autoradiography.
>
> Enumeration and examination of bacteria, both in culture and in situ, is also possible using the *inaZ* gene, which encodes a bacterial outer membrane protein that leads to ice nucleation (see Chapter 34.4.3). The detection of populations in the range of 10^4 cells per root has been reported.

Enumeration. To enumerate microbes introduced to the soil, the bacteria are tagged with a marker that allows the introduced microbes to be distinguished from the indigenous microflora. The most popular markers are

Table 34.**2** **Bacterial strains for agricultural use as microbial inoculants, whose traits have been improved through genetic modification**

Microorganism	Altered trait	Purpose of genetic modification
Agrobacterium radiobacter	Deletion of *tra* genes of pAgK84	Biosafety use in bio-control of crown gall
Clavibacter and *Pseudomonas* species	Endotoxin gene from *Bacillus thuringiensis*	Prevention of insect damage in crops
Bradyrhizobium japonicum	Additional copies of *nif*	Increased N_2 fixation
Pseudomonas syringae	Deletion of *inaZ* (ice nucleation gene)	Control of frost damage to plants
Rhizobium meliloti	Additional copies of *nif* or enhanced dicarboxylate transport (*dct* genes)	Increased N_2 fixation

antibiotic resistance genes. As the probability of spontaneously occurring drug resistance is relatively high and in order to avoid problems with the potential transfer of resistance genes, usually more than one drug resistance marker is used. Other frequently used markers are reporter genes, which allow the identification of tagged bacteria by their enzymatic activities, preferably by color reactions or light emission. Examples are given in Box 34.1.

The main advantages of selective dilution plating techniques are the ease of use, the relative inexpensiveness, and ability to statistically analyze the collected data. Also, the sensitivity of detection (10^2 organisms per g soil or tissue) is considered adequate for many purposes. The basic disadvantages of dilution plate counts are that the target organism must be in a culturable state and the marker gene must be expressed. Bacteria frequently enter a physiological state in which they are viable, but are unable to grow on artificial media ("viable, but not culturable"). It is believed that viable cell counts allow recovery of only 1–10% of the total population, with the remainder being in a "dormant" state (see Chapter 28.5). In addition, counting accuracy can be affected by cell clumping and other practical inaccuracies in preparing dilutions.

34.1.2 Immunological and Biochemical Methods Can Also Be Used for Direct Counts

In principle, any molecule that is characteristic for a taxonomic group can be used as a potential marker (see Chapter 29.3). Antibodies (polyclonal or monoclonal antisera) can provide a powerful and reliable way to detect GMMs in the environment. The use of conjugated antibodies has increased the sensitivity of such techniques considerably. The principle of these techniques is outlined in Figure 34.**2**. In the case of the **e**nzyme-**l**inked **i**mmuno**s**orbent **a**ssay (**ELISA**), the antibodies are conjugated with an enzyme, e.g., horseradish peroxidase and alkaline phosphatase, which gives a specific color reaction upon addition of a substrate. Thus, positive cells can be detected by a colorimetric assay and hundreds of colonies can be screened at the same time. Similar techniques, based on antibodies conjugated with radioactive substances (**r**adio**i**mmuno**a**ssays, **RIA**) are equally efficient. However, both techniques have their limitations with respect to direct quantifications in the environment. In contrast, the use of **f**luorescent **a**ntibody (**FA**) techniques has some potential for the study of microorganisms in their natural habitats. In this case, the specific antibody is conjugated to a fluorescent dye, such as fluorescein isothiocyanate, and the antigen–antibody complexes can be visualized by fluorescence microscopy. The FA-immunofluorescence approach has been successfully applied to enumerate *Rhizobium*, *Actinomycetes*, *Nitrobacter*, and other species in the rhizosphere and other natural environments.

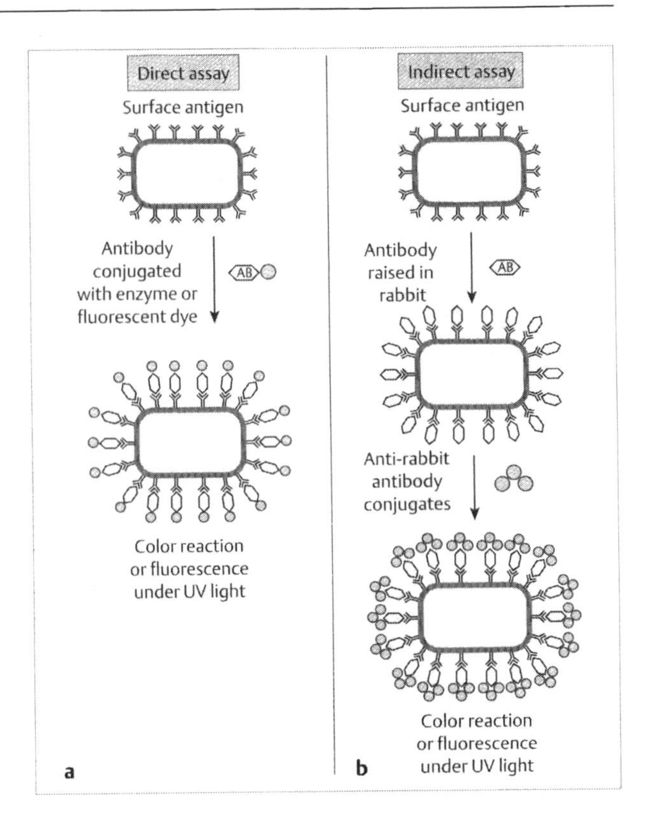

Fig. 34.**2a,b** **Principles of direct and indirect immunological assays using conjugated antibodies.**
a Direct assay Antibodies (AB) against a bacterial antigen (here a surface antigen) are raised and labeled with either an enzyme or a fluorescent dye. After binding of the antibody-conjugates to the antigen, the reaction is detected by a colorimetric assay or by the occurrence of fluorescence under UV light.
b Indirect assay. The primary antibodies, raised, for example, in rabbit, are not labeled. A secondary antibody against the rabbit antibody is raised, for example, in goat, and is labeled with the conjugate. The labeled anti-rabbit antibodies bind to the primary antibody and increase the sensitivity of the reaction detected as shown in **a**

34.1.3 Genomic Characterization of Taxa Is Used Increasingly

Identification methods based on specific and conserved DNA sequences are extensively used and are being more and more improved to increase sensitivity and the range of potential application. The use of molecular genetic markers by **r**estriction **f**ragment **l**ength **p**olymorphism (**RFLP**) analysis has revolutionized the analysis of genetic relationships and genetic diversity (see Chapter 29.4 and Fig. 29.**17**). Originally, the detection of RFLPs relied on gel electrophoresis and Southern hybridization, but recently, the use of PCR for amplification of polymorphic regions (see Chapter 17.4.3) has greatly simplified the procedure. However, potential applications are frequently hampered by the requirement for prior knowledge of DNA sequence information for the design of specific primers.

A modification of the basic PCR system that does not require prior sequence information has been developed. **A**rbitrarily **p**rimed **PCR** (**AP-PCR**) is directed by only one oligonucleotide primer of an arbitrary sequence and produces a characteristic pattern of amplified DNA fragments, with the potential of detecting polymorphisms between strains. Several modifications of this technique, mainly concerning the sequence and length of the random primer used, are available and include **RAPD** (**r**andom **a**mplified **p**olymorphic **D**NA) and **DAF** (**D**NA **a**mplification **f**ingerprinting). For a detailed discussion of these techniques, see Chapter 29.4. AP-PCR techniques are much faster and less expensive than RFLP analyses because the PCR amplification can be performed with bulk DNA isolated directly from the soil. Thus, this method can also be applied to type bacteria that can not be cultivated.

As prokaryotes cannot be numbered and identified by the naked eye, sophisticated methods **to monitor bacteria** had to be developed within reasonable limits of time and money. The classical method of dilution plate counting is still in use, but is being increasingly replaced by refined immunological, biochemical, and molecular techniques. These techniques are extremely helpful in determining the number of bacteria and in identifying specific strains, cultivars, specifically tagged strains, and genetically modified microorganisms (GMMs).

34.2 Bacteria Can Be Used as Biofertilizers

Biofertilizers are microbial inoculants that replace chemical fertilizer. Biofertilizers include free-living, associative, and symbiotic nitrogen-fixing prokaryotes as well as non-prokaryotic organisms, such as mycorrhizal fungi. While the fungi primarily contribute to phosphate availability for the plant, the N_2-fixing organisms supplement the nitrogen requirement and thus make the host plant at least partially independent of combined nitrogen.

The ability to reduce dinitrogen gas (N_2) to ammonia (NH_3) is a property confined to the prokaryotes and is widespread among both eubacteria and archaebacteria (see Chapter 8.5). However, only few bacteria have evolved the ability to form nitrogen-fixing symbioses with higher plants.

34.2.1 Rhizobia Fix Nitrogen in Intracellular Symbioses With Legumes (Fabaceae, Formerly Leguminosae)

Plant nodules. Symbiosis between Rhizobiaceae and leguminous plants results in the formation of a specialized organ (usually on the plant root), called a **nodule** (Fig. 34.3) in which the microsymbiont (*Rhizobium*) converts atmospheric nitrogen to ammonia, the majority of which is translocated to the plant. The symbiotic process involves a complex series of signals between the partners that are highly specific and ensure that only certain rhizobial species can interact with certain leguminous plants (see Fig. 34.**4** and Table 34.**3**).

Most of the 100 agriculturally important legumes belong to a few tribes of the family Papilionaceae (Fabaceae) and include important grain legumes (peas, beans, soybeans) and forage legumes (alfalfa, clover). The legumes fix nitrogen at rates approximating $100\,kg\,ha^{-1}\,year^{-1}$; under intensive agricultural management, even higher rates of fixation have been achieved ($300–600\,kg\,ha^{-1}\,year^{-1}$). Therefore, the *Rhi-*

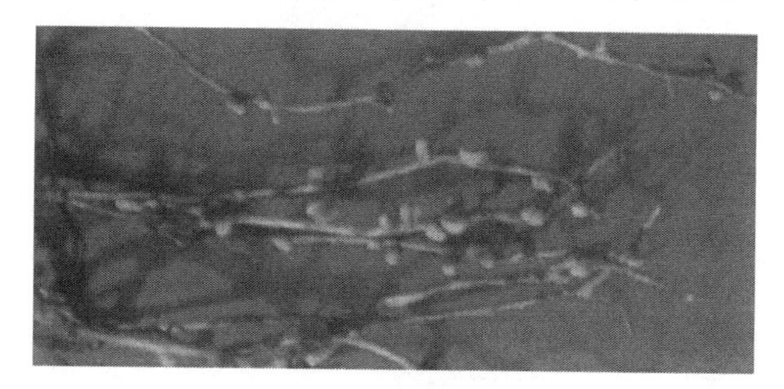

Fig. 34.**3** **The Rhizobium/legume symbiosis.** The root system of a *Vicia faba* plant (faba bean) nodulated by the corresponding microsymbiont *Rhizobium leguminosarum* biovar *viciae* is shown. A biovar is distinguished by physiological properties

Table 34.3 **Species of rhizobia and their host plants.** The taxonomic classification within the Rhizobiaceae is constantly being discussed. For example, some rhizobia within the bean group, which nodulate *Phaseolus vulgaris* and *Leucaena* spec., have been grouped into a new species, *Rhizobium tropici*. A new genus, *Photorhizobium*, has been proposed for photosynthetically active rhizobia isolated from stem nodules of *Aeschenomene*. The list of rhizobial species and host plants is not complete

Genus/species	Host nodulated	Common group name
Rhizobium leguminosarum		
bv. *viciae*	*Pisum, Lens, Vicia*	Pea, Vetch
bv. *phaseoli*	*Phaseolus*	French bean
bv. *trifolii*	*Trifolium*	Clover
Rhizobium meliloti (= *Sinorhizobium meliloti*)	*Medicago, Melilotus, Trigonella*	Alfalfa
Rhizobium loti	*Lotus*	Trefoil
Rhizobium galegae	*Galega*	
Rhizobium spec.	*Leucaena, Acacia, Robinia*	
Bradyrhizobium japonicum	*Glycine*	Soybean
Bradyrhizobium spec.	*Cajanus, Vigna, Arachis, Macroptilium*	Cowpea
Azorhizobium caulinodans	*Sesbania* stem and root nodules	Sesbania
Sinorhizobium fredii	*Glycine*	Soybean

zobium/legume symbiosis is agronomically the most important biological mechanism for adding nitrogen to the soil/plant system.

Bacterioids. A sequential exchange of molecular signals between *Rhizobium* and its host plant (Fig. 34.**4**), involving inducing plant compounds and bacterial nodulation factors, leads to the formation of the root nodule, which is a highly differentiated plant organ. The central tissue is infected by bacteria through infection threads, from which the bacteria are released into the plant cytoplasm, where they divide and differentiate into nitrogen-fixing **bacterioids**. The bacterioids are separated from the plant cytosol by a plant-derived membrane, the peribacteroid membrane. In a fully developed nodule, the plant cells are completely filled with bacterioids (Fig. 34.**5**). During the symbiotic association, the rhizobia inside the nodule receive nutrients from the plant that are derived from photosynthetic activity in the leaves; the plant receives fixed nitrogen in the form of ammonia from the nitrogen-

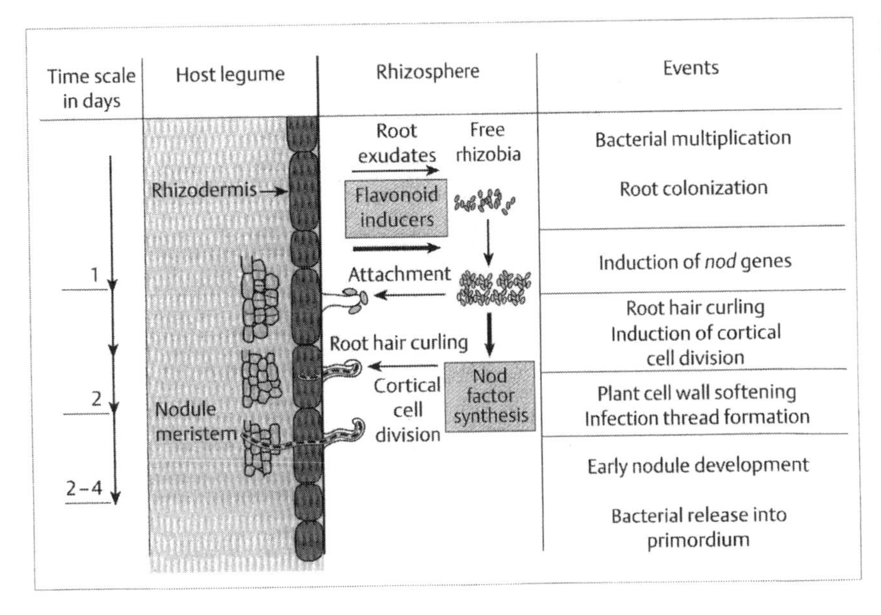

Fig. 34.**4** **Schematic drawing of the early steps of interaction between rhizobia and the legume host plant.** Bacteria are attracted chemotactically by root exudates, and multiply and colonize the root surface. Attachment and binding is thought to involve the action of plant-derived lectins. Specific flavonoid inducers secreted by the plant promote the expression of the rhizobial nodulation (*nod*) genes, which, in response, produce a specific bacterial nodulation signal, the Nod factor. This factor causes defined reactions in the host plant, such as root hair curling (in red) and the induction of mitotic cell divisions in the inner cortex, which leads to a nodule meristem. Bacteria penetrate the root hair and infection threads (in red) are produced through which bacteria infect plant cells, where they are released into the cell cytoplasm

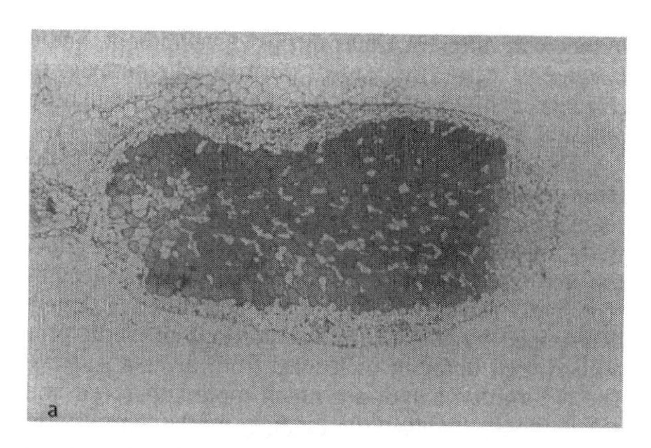

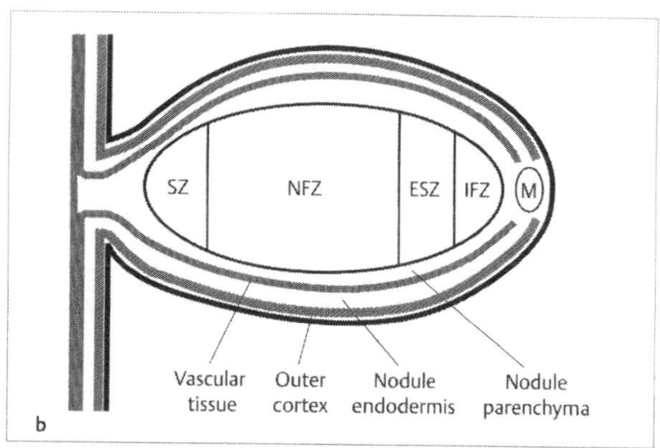

Fig. 34.**5a,b** **The interior of a nodule.** **a** A longitudinal section of a 21-day-old nodule induced by *Rhizobium leguminosarum* biovar *viciae* on a *Vicia hirsuta* plant is shown. The infected plant cells (dark) are completely filled with bacteria (stained with toluidin blue); few of the plant cells (white) are not infected. Within the infected tissue, the zones, schematically outlined in **b** can be detected. M, meristematic zone; IFZ, infection zone containing infection threads; ESZ, early symbiotic zone in which bacteria differentiate into bacteroids; NFZ, nitrogen-fixing zone in which bacteria fix nitrogen; SZ, senescence zone

fixing bacterioids. With time, the nodules age and the bacterioids and the host plant cells undergo senescence.

The anatomy and physiology of the nodule dictates the development of a **microoxic environment**. There is good evidence for the presence of a physical oxygen diffusion barrier located in the outer layers of the nodule, which, together with the very active bacterial respiration, results in an internal oxygen concentration of 30 nM (as compared to 250 mM outside the nodule). In order to provide enough oxygen for the bacterioids to respire, a specific plant protein with high affinity for oxygen, **leghemoglobin**, is synthesized immediately upon the onset of nitrogen fixation.

Nodulation (*nod*) genes. The rhizobial genes responsible for the production of the specific nodulation factor are called *nod* genes. The corresponding operons are often intermingled with other genes, such as *nif* and *fix* genes. The *nod* regulon of *Rhizobium meliloti* is shown in Figure 34.**6**. The so-called **common** *nod* genes (*nodABCIJ*) have been detected in all rhizobia and are functionally interchangeable. Their gene products are involved in the formation of the basic structures of the nodulation factors, which are lipooligosaccharides that contain a chitooligosaccharide chain with three to five β-1,4-linked *N*-acetylglucosamine residues (GlcNAc) (Fig. 34.**7**). An *N*-acetylglucosaminyltransferase activity was ascribed to the NodC protein, suggesting that NodC catalyzes the formation of the β-1,4-linked GlcNAc oligosaccharide chain. The gene product of *nodA* possesses *N*-acyltransferase activity, and the *nodB* gene product is an *N*-deacetylase. It was proposed that these two enzymes are involved in replacing the *N*-acetyl group with an *N*-fatty acyl group at the non-reducing GlcNAc sugar. NodI and NodJ are reported to be involved in the excretion of the nodulation factor.

Box 34.2 The **nodulation factor** synthesized by *R. meliloti* (NodRm-1) is a sulfated β-1,4-tetra-D-glucosamine with three acetylated amino groups and a C16 unsaturated fatty acid at the non-reducing end. The gene products of the *nodH* (sulfotransferase) and *nodPQ* (ATP sulfurylase) genes, which are only present in *R. meliloti*, are involved in sulfatization at the reducing end. The sulfate group is lacking in other nodulation factors and appears to be critical for recognition by alfalfa plants. The Nod factors produced by *R. leguminosarum* biovar *viciae* are characterized by a 6-*O*-acetyl group at the non-reducing sugar; it is proposed to be added by the *nodL* gene product. The *nodFE* genes are found in various *Rhizobium* species, but they are not functionally interchangeable. They are involved in the synthesis of the fatty acid side chain, which differs in the respective nodulation factors with regard to length and extent of unsaturation. NodRm-1, for example, has a C16 fatty acid with two double bonds, while the *R. leguminosarum* bv. *viciae* Nod factor has a C18 fatty acid tail with four double bonds (NodRlv).

Host specificity of the nodulation factor is mediated by specific modifications, which are the product of nodulation genes not conserved between different *Rhizobium* species. These nodulation genes are therefore called ***hsn*** (**h**ost-**s**pecific **n**odulation) **genes**. Examples of various modifications are given in Figure 34.**7** and Box 34.**2**.

Nod factors at picomolar to nanomolar concentrations can elicit responses in the host legumes. Nod factors are active on epidermal cells and root hairs of *M. sativa* and *V. hirsuta*, respectively, and induce root hair deformation and branching, and infection thread formation. However, it should be noted that the same factors can trigger cortical cell division and formation of nodule primordia. The corresponding plant receptors and the receptor genes remain to be elucidated.

Regulation of rhizobial ***nod*** **genes.** Rhizobial *nod* genes are not constitutively expressed, but are induced by plant compounds released by legume roots. Diverse groups of **flavonoids** and **isoflavonoids** (Fig. 34.**8**) act as *nod* gene inducers in *Rhizobium* and *Bradyrhizobium*, respectively. Betaines, such as trigonelline and stachydrine (*N,N*-dimethylproline) also have *nod*-gene-inducing activity in alfalfa/*R. meliloti*. It is interesting to note that different legumes secrete different flavonoid inducers, and that a single plant can synthesize different inducers at different times during development. Some flavonoids can also cause positive chemotaxis in rhizobia and growth enhancement, with regulation potential for the dynamics of the soil microbiota.

The activation of *nod* genes mediated by plant inducers requires NodD, which is a transcriptional regulatory protein. In some species (e.g., *R. meliloti*), there are multiple alleles of *nodD*, which might allow the sensing of different host-specific inducer molecules (Fig. 34.**6**). Flavonoid binding and specificity is largely determined by the C-terminal portions of NodD proteins, which differ in molecules from diverse sources. The N-terminal halves are much more conserved and are probably involved in binding of the proteins to highly conserved sequences (*nod* boxes) upstream of the inducible *nod* genes. Control of *nodD* expression is diverse: in some organisms, *nodD* is not regulated, in others it is subjected to positive (inducible by plant flavonoids) or negative (e.g., auto-repressed) regulation.

Nitrogen fixation by *Rhizobium* bacterioids depends on the low oxygen concentration present in the central part of the nodule, which is believed to function as a physiological signal for triggering gene expression. The main events center around the derepression of

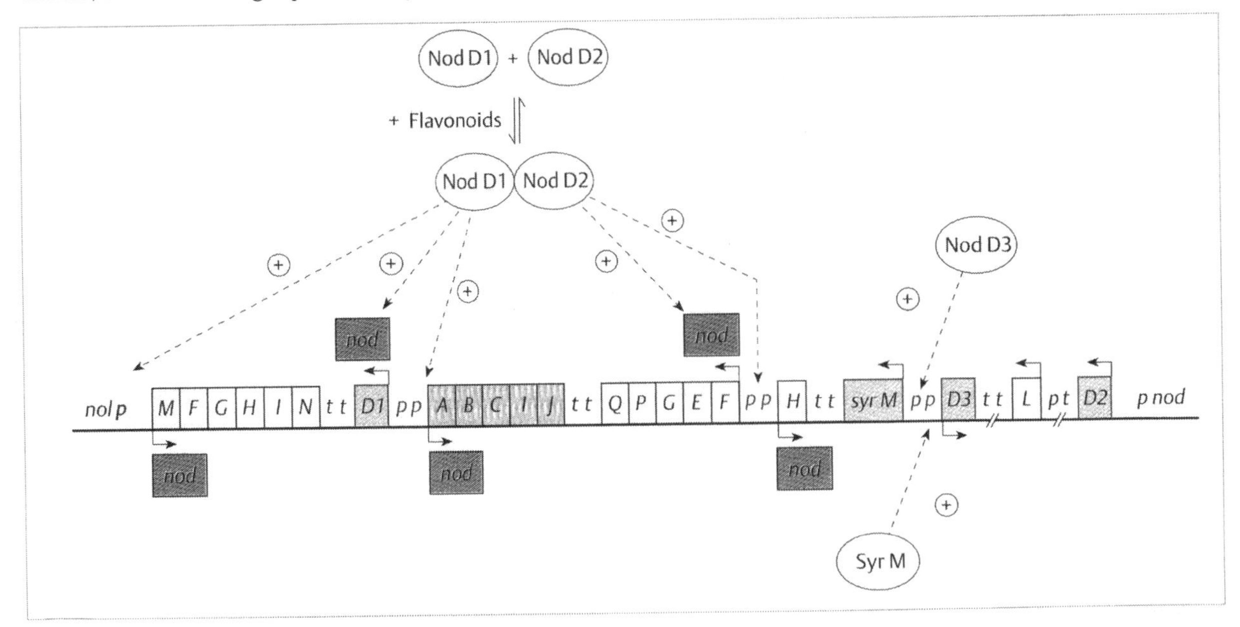

Fig. 34.**6** **The organization and regulation of nodulation genes (*nod* and *nol*) in *Rhizobium meliloti*.** The letters indicate *nod* genes, with the exception of *nol* (mnemonic for nodulation genes beyond the letter "z") and *syr M* (**sy**mbiotic **r**egulator). Common *nod* genes are drawn in light gray, regulatory *nodD* genes in pink, and host-specific nodulation genes in white. Arrow heads in front of genes and operons represent (where indicated, NodD-activated) promoters (*p*) and indicate the direction of transcription to the terminator (*t*). The genes are not drawn to scale. Broken arrows indicate activation (⊕) by activators *Syr M*, Nod D3, and Nod D1 + Nod D2. The inducers for the latter are various flavonoids (see Fig. 34.**8**), and their targets are the so-called *nod*-boxes (red)

C16:2

C18:4

	n	R1	R2	R3	Designation
Rhizobium meliloti	2	SO_3^-	H	C16:2	NodRm1 NodRm-IV(S)
	2	SO_3^-	Ac	C16:2	NodRm-IV(Ac, S)
Rhizobium leguminosarum biovar *viciae*	2	H	Ac	C18:4	NodRlv-IV(Ac)
	3	H	Ac	C18:4	NodRlv-V(Ac)

Fig. 34.7**a,b** **Structural characteristics of rhizobial nodulation factors.**
a The basic structure of the chitooligosaccharide chain of β-1,4-linked *N*-acetylglucosamine residues is shown.
a, b Host specificity of this molecule is achieved by specific modifications of the residues R1 (SO_3^- or H), R2 (H or Ac), and in the fatty acid side chain (R3), which can differ with regard to length and extent of unsaturation. Since different Nod factors have been isolated from a given species or biovar, a uniform nomenclature has been proposed. According to this, NodRm1 is now called NodRm-IV(S): Rm signifies *R. meliloti*, IV indicates four glucosamine residues, and S represents the sulfate at the reducing end of the molecule. From *R. leguminosarum* biovar *viciae* (Rlv), either tetra-(IV) or penta-(V)-glucosamineoligosaccharides have been isolated. These are acetylated (Ac) at the non-reducing end and lack the sulfate group

Flavones

Luteolin: 3′, 4′, 5, 7-Tetrahydroxyflavone

Apigenin: 4′, 5, 7-Trihydroxyflavone

Flavanones

Naringenin: 4′, 5, 7-Trihydroxyflavanone

Hesperitin: 3′, 5, 7-Trihydroxy-4′-methoxyflavanone

Isoflavones

Daidzein: 4′, 7-Dihydroxyisoflavone

Genestein: 4′, 5, 7-Trihydroxyisoflavone

Fig. 34.**8** **Flavonoids and isoflavonoids exuded from legume roots that activate transcription of *nod* genes.** The most effective inducers have hydroxyl substitutions at the 3′ or 4′ position of the B ring and a hydroxyl or glucoside linkage at position 7 of the A ring. While flavones and flavanones are most effective in the induction of *nod* genes in *Rhizobium*, isoflavonoids induce *nod* gene expression in *Bradyrhizobium* and act as antagonists of *nod* gene expression in *Rhizobium meliloti* and *Rhizobium leguminosarum*

nitrogen fixation genes (*nif* and *fix*) to allow the production of the nitrogen-fixing apparatus. As an example, the *nif* and *fix* genes known in *R. meliloti* and their regulation is shown in Figure 34.**9**.

The actual reduction of dinitrogen gas (N_2) in the air to ammonia (NH_3), which is the fixed nitrogen that can be assimilated into amino acids by biological systems, is performed according to the following equation:

$$N_2 + 16\,ATP + 8\,e^- + 8\,H^+ \rightarrow 2\,NH_3 + H_2 + 16\,ADP + 16\,P_i \quad (34.1)$$

(For details, see Chapter 8.5). The cardinal enzyme complex involved in this reaction, the nitrogenase enzyme, is physically and functionally conserved in diverse N_2-fixing organisms. It is composed of two components, the iron-molybdenum protein (MoFe protein) and the iron protein (Fe protein). The MoFe protein (also called dinitrogenase) is an $\alpha_2\beta_2$ tetramer: the α-subunits are encoded by *nifD*, the β-subunits by *nifK*. The tetrameric protein contains a molybdenum-iron cofactor (FeMoCo), which comprises the catalytic site for N_2 reduction. The Fe protein (also called dinitrogenase reductase) is a homodimer, the subunits of which are encoded by *nifH*. Both components are highly oxygen sensitive and are irreversibly inactivated in the presence of oxygen (FeMo protein: $t_{1/2} = 8\,min$; Fe protein: $t_{1/2} = 30\,s$). Additional *nif* and *fix* genes are involved in providing energy (ATP) and reducing equivalents or are involved in the biosynthesis of FeMoCo (NifB, NifE, NifN). The gene products of the *fixABCX* operon could be involved in electron transport to the nitrogenase complex. The proteins encoded by the *fixNOQP* operon, however, could form a membrane-bound cytochrome oxidase, which may be part of a bacteroid-specific respiratory chain, and allow bacteroid respiration under low-oxygen conditions.

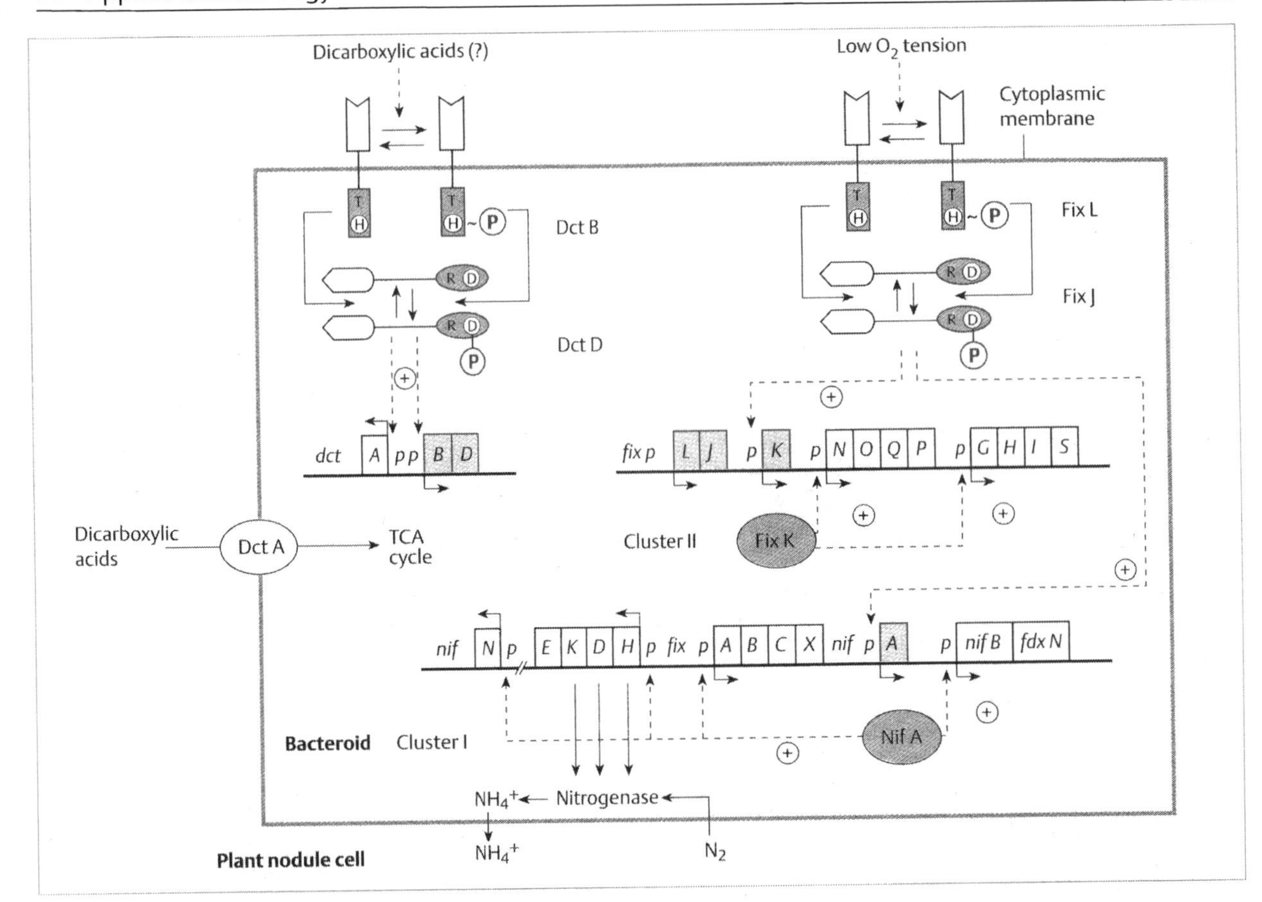

Fig. 34.**9** **Organization and regulation of *nif*, *fix* and *dct* genes in *Rhizobium meliloti*.** In *R. meliloti*, two clusters of *nif* and *fix* genes have been identified, both of which are localized on a very large plasmid (megaplasmid, sym plasmid) of about 1400 kbp. *nifE* is separated from *nifN* by a region of about 30 kb, which carries *nod* genes The gene products of the regulatory genes and the activation (⊕) of their corresponding target genes are indicated (broken arrows). The two-component sensor FixL senses the oxygen concentration. Under low oxygen, it phosphorylates the response-regulator FixJ which activates *fixK* and *nifA* by binding to specific promoter sequences, the so-called "anaeroboxes" for FixK. NifA is the transcriptional activator of the genes in cluster I. The two-component sensor DctB probably senses dicarboxylic acids. It activates the response-regulator DctD, which controls expression of genes *dctB* and *D* and of *dctA*, the structural gene for a dicarboxylic acid transporter. Regulatory problems in pink

Coordinately with the derepression of the *nif* and *fix* genes (Box 34.3), expression of *dctA* (**dic**arboxylate **t**ransport) occurs. *dctA* encodes a permease that is specifically involved in transporting dicarboxylic acids into the nitrogen-fixing bacteroid. During expression of the *nif*, *fix*, and *dct* genes, ammonia assimilation is repressed, thereby ensuring that the majority of the ammonia formed during nitrogen fixation in the bacteroid is partitioned to the plant cell.

Regulation of *nif* and *fix* gene expression. Key regulatory proteins that respond to the oxygen status of the bacteroid function in a cascade system to control the nitrogen fixation process. The system corresponds to a global regulatory network (see Chapter 20). The *nifA* gene product is a key regulatory protein that activates transcription of *nif*/*fix* genes in conjunction with an RNA polymerase containing the alternative sigma factor σ^{54} (RpoN), which has been identified in all nitrogen-fixing organisms. The organization of structural and functional domains of rhizobial NifA proteins is shown in Figure 34.**10**. The activity of the *Rhizobium* NifA proteins is controlled by the oxygen concentration. The mechanism of this activation is not yet clearly understood. Since oxygen sensitivity correlates with the presence of a typical cysteine motif, which could constitute a metal (Fe) binding site, it has been proposed that the redox state of the metal bound to the cysteine residues may influence the conformation and thus the activity of the

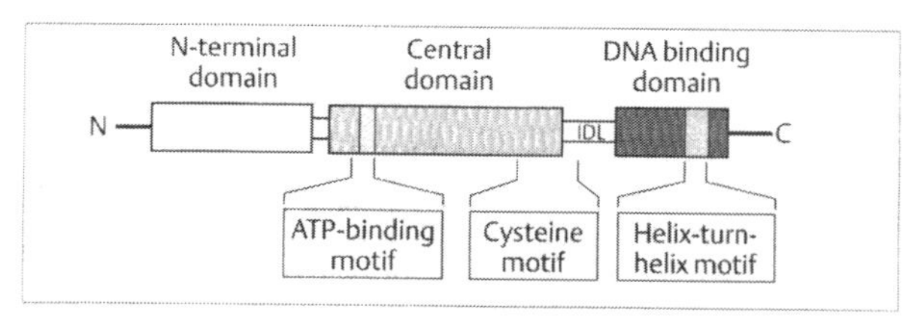

Fig. 34.**10** **Domain structure of rhizobial NifA proteins.** The N-terminal domains of variable lengths have an unknown function. The central domain of 240 amino acids is probably involved in the interaction with RNA polymerase containing σ^{54}. Binding at $-24/-12$ promoter sequences and hydrolysis of ATP is necessary for transcriptional activation. The C-terminus contains a helix-turn-helix motif that is involved in NifA binding to specific upstream activator sequences (UAS), located approximately 100 bp upstream of the transcriptional start sites. All rhizobial NifA proteins are oxygen sensitive and are characterized by an interdomain linker (IDL). A typical cysteine motif (Cys-X_{11}-Cys-X_{19}-Cys-X_4-Cys) is suggested to be involved in oxygen sensitivity of the protein

protein. In addition to this oxygen regulation at the protein level, in some but not all rhizobia, low oxygen concentration is also responsible for the regulation of *nifA* expression at the transcription level. Consequently, oxygen concentration is a regulatory effector that controls *nifA* expression and/or activity and regulates all NifA-dependent *nif* and *fix* genes.

Box 34.3 ***nif*** **and** ***fix*** **genes** identical to those described for *Rhizobium meliloti* have been identified in *R. leguminosarum*, *Bradyrhizobium japonicum*, and *Azorhizobium caulinodans*, but their organization and location are different. In *R. leguminosarum*, *nif* and *fix* genes are clustered on plasmids and interrupted with *nod* genes. In *B. japonicum* and apparently also in *A. caulinodans*, the *nif* and *fix* genes (and the *nod* genes) are located on the chromosome. FixLJ systems have also been identified in *Bradyrhizobium* and *Azorhizobium*. However, in these cases, the FixLJ systems appear not to be directly involved in the oxygen-dependent activation of *nifA*.

The cascade nature of the regulation of *nif* and *fix* gene expression in *R. meliloti* is evident from the dependence of expression of the regulatory genes *nifA* and *fixK* (the transcriptional activator of the *fix* cluster II, such as *fixNOQP*; see Fig. 34.**9**) on a pair of regulatory proteins, FixL and FixJ, which activate their target genes in response to microoxic conditions. FixL and FixJ belong to the family of two-component regulatory proteins (see Chapter 20.4). FixL, a hemoprotein, is an oxygen sensor that is autophosphorylated under low-oxygen conditions. FixJ is an effector protein that can be phosphorylated by FixL and acts as a transcriptional activator of *nifA* and *fixK*.

Members of Rhizobiaceae form intracellular species-specific symbioses with *Fabaceae*. Rhizobia have adapted during evolution to infect root hairs and to induce the formation of root cortical plant meristems, which generate highly organized structures, the root nodules. Low oxygen tension in the nodules and the lack of combined nitrogen result in the induction of the nitrogen-fixing apparatus. After a phase of effective dinitrogen fixation, bacteria and host cells degenerate. The development of the nitrogen-fixing symbiosis depends on an extensive exchange of chemical signals between the symbiotic partners, and is regulated by global regulatory networks of *nod*, *fix*, and *nif* genes.

34.2.2 The Actinomycete *Frankia* Forms Intracellular Nitrogen-Fixing Symbioses With Many Angiosperms

Over 200 angiosperm species out of 8 families are capable of forming intracellular, nitrogen-fixing root nodule symbioses with the actinomycete *Frankia*, a Gram-positive filamentous prokaryote, as an endosymbiont. Among them are important perennial woody shrubs and trees of world-wide distribution.

The best-known association of *Frankia* is with *Alnus* (Fig. 34.**11**). The nodules are perennial and characterized by a coralloid-like structure, called **actinorhiza**, of sometimes considerable size with dichotomous branching. Unlike legume nodules, actinorhiza are modified lateral roots, specified by a central vascular bundle (Fig. 34.**11b**).

Frankia cells enter symbiosis with *Alnus* by root hair infection, and induce cell division in the hypodermis and cortex to form a prenodule. The nodule itself develops in the same manner as a lateral root. Vegetative *Frankia* hyphae with a diameter of 0.5–1.5 μm proliferate and penetrate cells of the developing nodule tissue. Under conditions of nitrogen fixation, specialized structures, the vesicles (4–6 μm in diameter), arise, usually as terminal swellings of hyphal side branches. Mature nitrogen-fixing vesicles are characterized by a highly structured laminated envelope, which separates the endosymbiont from the host plant cytoplasm. The laminate contains unusual hopanoid lipid molecules (see Chapter 7.10.4 and Box 29.**4**), which sometimes form numerous lipid layers around the vesicle. The number of layers correlates with the oxygen concentration in the surrounding environment. These

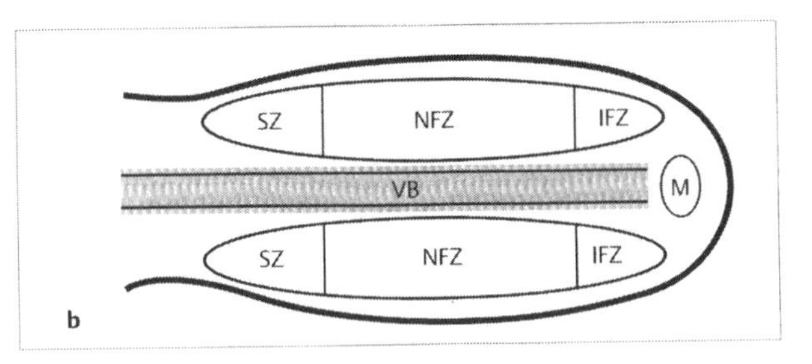

Fig. 34.11a,b **Nodules formed on the root of *Alnus* by the actinomycete *Frankia*.**
a Typical coralloid-like nodule structures or actinorhiza.
b Internal structure of one of these nodule lobes. The zones from the apical to the basal part are shown: M, meristem; IFZ, infection zone with hyphae; NFZ, nitrogen-fixing zone with active vesicles; and SZ, senescence zone. VB, central vascular bundle

lipid envelopes probably function as an oxygen diffusion barrier to provide an internal oxygen concentration low enough for the nitrogenase to be synthesized.

Intracellular nitrogen-fixing root nodule symbioses are formed between the actinomycete *Frankia* and a wide range of angiosperms. In contrast to the nodules formed in the *Rhizobium*/legume symbiosis, the nodules formed in the *Frankia*/*Alnus* interaction, the so-called **actinorhiza**, are modified lateral roots. World-wide, actinorhizas are major contributors to soil nitrogen.

The *Frankia* nitrogenase enzyme complex is identical to that of *Rhizobium* and accordingly, the nitrogenase structural genes, *nifHDK*, the accessory genes (such as *nifEN*, *nifB*), and the regulatory gene *nifA* have also been identified in *Frankia*. No definitive demonstration of the existence of *nod* genes has been provided. Nod factors have been postulated, and flavonoid-like compounds have been purified from *Alnus* seed eluates.

The rate of nitrogen fixation is in the range of many nodulated legumes, i.e., 40–350 $kg\,ha^{-1}year^{-1}$. Bearing in mind the broad range and the world-wide distribution of the host plants (ranging from arctic to tropical habitats), actinorhizas are certainly among the major contributors to the accumulation of soil nitrogen and geochemical nitrogen cycling.

34.2.3 Cyanobacteria Establish Mainly Extracellular Nitrogen-fixing Relationships

Diazotrophic cyanobacteria can fix nitrogen in a free-living form, but also in symbioses. They can establish symbiotic relationships with various eukaryotic organisms, including higher plants such as gymnosperms, angiosperms, ferns, liverworts, and hornworts. Among the angiosperms, only representatives of the genus *Gunnera* can enter N_2-fixing relationships with cyanobacteria (*Nostoc*), which infect stems through special glands and live intracellularly. In contrast to the *Gunnera*/*Nostoc* system, all other cyanobacterial symbioses are extracellular. In gymnosperms (*Cycadaceae*), the microsymbiont infects the coralloid roots; clusters of negatively geotropic roots arise from the hypocotyl, but remain intercellular. In the genus *Azolla*, the nitrogen-fixing filaments occupy specialized leaf cavities and, similarly, in liverwort and hornwort, they are found in cavities in the gametophytic thallus.

Species of the waterfern *Azolla* have extracellular relationships with *Anabaena azollae*. The sporophyte of *Azolla* (Fig. 34.**12a**), floating on the surface of quiet waters, consists of branches bearing alternate, bi-lobed leaves, each comprising a dorsal aerial (containing chlorophyll), and a ventral submerged lobe. Within the upper lobes, there are specialized cavities containing the cyanobacterial filaments, which develop with age into nitrogen-fixing microsymbionts, the **cyanobionts** (Fig. 34.**12b**). Like in the free-living state, differentiation into nitrogen-fixing cells is correlated with heterocyst formation and genetic rearrangement of *nif* and *fix* genes (see Chapter 24.3). However, there are also indications that symbiotic cyanobacterial gene expression is different from that in the free-living state. The percentage of heterocysts in cyanobionts (up to 50%) is higher than in the free-living state (10%), and symbiotic induction of heterocysts may be independent of strict nitrogen control. Thus, it appears that differentiation in symbiotic cyanobacteria is induced at least partly by plant-mediated signals rather than simply by environmental factors.

In leaf cavities, the microsymbiont is usually associated with a population of eubacteria surrounded by an envelope. When the algal pockets are broken, both endophyte and associated bacteria are released in a mucilaginous matrix, rich in amino acids, NH_4^+, and polysaccharides. Among the associated bacteria, different *Arthrobacter* species have been invariably isolated from the leaf cavities and sporocarps of several *Azolla*. Therefore, it appears that this symbiosis is tripartite and more complex than initially thought. The role of associated bacteria has been postulated to be linked to their high respiratory activity, which would favor the establishment of microoxic niches. *Arthrobacter* might also affect, through auxin formation, the hormonal balance during the onset of the association.

Azolla is distributed world wide, but is most common in warm tropical and subtropical waters. Its

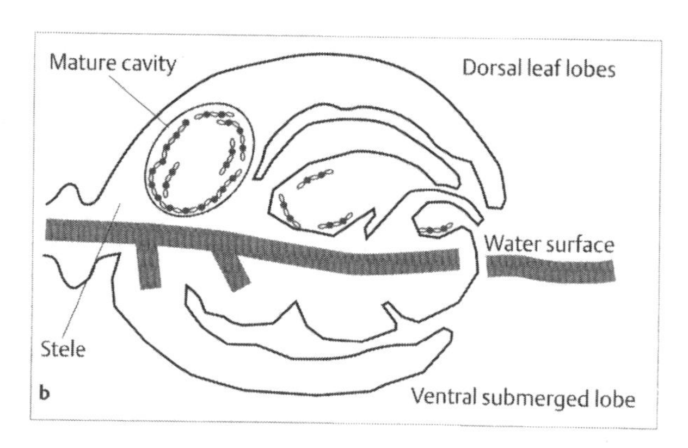

Fig. 34.**12a,b** **The symbiotic association between *Anabaena* and the waterfern *Azolla*.**
a The sporophyt of *Azolla* floats on water surfaces and enters nitrogen-fixing symbiosis with the cyanobacterium *Anabaena azollae* (courtesy of M. Grilli-Caiola).
b Schematic drawing of one leaf. The dorsal and ventral leaf lobes and the development of the symbiosis are indicated. *Anabaena* filaments infect open pores and vegetative cells differentiate into heterocysts. In mature cavities, up to 50% of the cyanobacterial cells are differentiated to heterocysts and fix nitrogen

agricultural use as a forage and green manure crop (especially for rice cultivation) relies upon the rapid invasive ability to colonize water environments, notably calm, poorly aerated waters, where it forms dense surface mats, thus controlling weed growth and insect proliferation. The N_2-fixing ability of the *Azolla*/*Anabaena* symbiosis varies with seasonal and environmental conditions, but is on the average in the order of 50–100 kg ha^{-1} year^{-1}.

The cyanobacterium *Anabaena* is able to differentiate into heterocysts (see Chapter 24.3) and to fix nitrogen in symbiosis with the water fern *Azolla*. In contrast to the *Rhizobium*/legume and *Frankia*/*Alnus* symbioses, the **Anabaena/Azolla association** is extracellular rather than intracellular.

34.3 Bacteria Can Promote Plant Diseases

Although of lower economic importance than diseases caused by fungi and viruses, diseases caused by pathogenic bacteria can lead to considerable losses in crop production and storage. Examples include bacterial wilt of potato, tomato, and other crops; serious damage of staple food crops due to bacterial infection has been reported for rice, wheat, soybean, and potato. Important plant diseases and the responsible bacterial pathogens are listed in Table 34.**4**.

34.3.1 Bacterial Pathogens Have Developed a Broad Range of Pathogenicity and Virulence Mechanisms That Are Tightly Controlled

Successful infection of a plant by microbial pathogens requires mechanisms of attachment, invasion, and inactivation of plant defense. Overall, more than 100 genes may be needed for bacterial pathogenicity (see also Chapter 33 for definition of virulence and patho-

Table 34.**4** **Important plant diseases caused by bacteria**

Symptoms	Examples	Pathogen
Spots and blights	Wildfire (tobacco)	*Pseudomonas syringae* pv.[×] *tabaci*
	Haloblight (bean)	*Pseudomonas syringae* pv. *phaseolica*
	Citrus blast	*Pseudomonas syringae* pv. *syringae*
	Leaf spot (bean)	*Pseudomonas syringae* pv. *syringae*
	Blight (rice)	*Xanthomonas campestris* pv. *oryzae*
	Blight (cereals)	*Xanthomonas campestris* pv. *translucens*
	Spot (tomato, pepper)	*Xanthomonas campestris* pv. *vesicatoria*
Vascular wilts	Ring rot (potato)	*Clavibacter michiganensis* pv. *sepedonicum*
	Wilt (tomato)	*Clavibacter michiganensis* pv. *michiganensis*
	Stewart's wilt (corn)	*Erwinia stewartii*
	Fire blight (pome fruit)	*Erwinia amylovora*
	Moko disease (banana)	*Pseudomonas solanacearum*
	Black rot (crucifers)	*Xanthomonas campestris* pv. *campestris*
Soft rots	Soft rots (numerous)	*Erwinia carotovora* pv. *carotovora*
	Black leg (potato)	*Erwinia carotovora* pv. *atroseptica*
	Pink eye (potato)	*Pseudomonas marginalis*
	Sour skin (onion)	*Pseudomonas cepacia*
Canker	Canker (stone fruit)	*Pseudomonas syringae* pv. *syringae*
	Canker (citrus)	*Xanthomonas campestris* pv. *citri*
Galls	Crown galls (numerous)	*Agrobacterium tumefaciens*
	Hairy root	*Agrobacterium rhizogenes*
	Olive knot	*Pseudomonas syringae* pv. *savastanoi*

× p.v., pathover, a variety of microorganisms with phytopathogenic properties

genicity). In recent years, the application of molecular genetic approaches to bacterial plant pathogens has led to the identification of some of these genes. The genes can be classified into three groups: pathogenicity genes, virulence genes, and host range genes.

Pathogenicity genes are bacterial genes needed for growth on or in plants. A special class thereof are the *hrp* (**h**ypersensitive **r**esponse and **p**athogenicity) genes, which are large clusters of up to 24 genes. Apart from encoding basic pathogenicity functions, such as colonization and growth, *hrp* genes are also involved in the production of proteinaceous signal molecules, which elicit defense responses associated with hypersensitivity in non-host plants. These so-called **harpins** are secreted by a pathway involving other *hrp* gene products (HrpH, HrpI).

The interaction between phytopathogenic bacteria and host plants can be classified into two general categories. **Compatible interaction** occurs between virulent pathogens and susceptible host plants and results in full development of disease symptoms. **Incompatible interaction** results in little or no disease symptoms since the bacterium induces a defense response in the plant. In this interaction, the plant is resistant and the pathogen is avirulent. Incompatible reactions are characterized by a **hypersensitive response** (HR) of the infected plant, which refers to localized cell death that occurs rapidly (12–24h) at the site of pathogen invasion. HR is correlated with the expression of plant defense mechanisms, which inactivate the pathogen or limit its growth and spread inside the plant.

Virulence genes are genes that contribute to the aggressiveness of the pathogen and are necessary for symptom production. General virulence factors include various toxins, **extracellular polysaccharides** (**EPS**), plant growth hormones, and enzymes, such as proteases, cellulases, pectic enzymes, and other cell-wall-degrading enzymes. The regulation of these virulence factors is usually tightly controlled at the level of transcription by global regulatory circuits that respond to specific environmental signals (Box 34.4 and Fig. 34.**13**).

Host range genes determine the host range (i.e., the plant species and cultivars that can be infected) by factors that act positively (specific virulence factors) or negatively (avirulence genes). Unlike the general virulence factors, which affect a wide range of plant species, **h**ost-**s**pecific **v**irulence genes (*hsv* genes) are necessary for pathogenicity on only some of the host plants. **Cullivar** is a variety of a plant species reacting specifically to a pathover. **Avir**ulence genes (*avr* genes) match resistance genes in the host and restrict the host range of a certain bacterial race to certain plant cultivars. The *avr* gene products may be secreted and delivered into the plant cell in an Hrp-mediated way.

Box 34.4 The expression of **virulence factors** is often coordinately controlled by global regulators. For example, in *Pseudomonas solanacearum*, both EPS and cellulase synthesis are regulated by PhcA. Another global regulator, termed PheN, has been identified in *P. tolaasii*. This protein is required for regulation of toxin and protease synthesis, motility, chemotaxis, and siderophore production. PheN in turn is negatively regulated at the transcriptional level at high cell densities, which suggests control by a quorum-sensing system (see Chapter 20). In *Erwinia carotovora*, the extracellular enzymes endopectate lyase, cellulase, polygalacturonase, and protease are coordinately regulated by two transcriptional activators, AepA and AepB (**a**ctivate **e**xtracellular **p**rotein production) in response to plant compounds. In *Erwinia chrysanthemi*, synthesis of the enzymes involved in degradation of pectin is under the control of the transcriptional repressor KdgR (Fig. 34.**13**). There is also a regulatory connection of *hrp* with some virulence and *avr* genes through the regulatory protein HrpS.

34.3.2 Important Bacterial Plant Pathogens Belong to few Genera

Bacteria causing plant diseases occur mainly within the genera of *Agrobacterium*, *Pseudomonas*, *Xanthomonas*, *Erwinia*, *Corynebacterium*, *Mycoplasma*, and *Spiroplasma*. Some of these microbes are obligate pathogens and are unable to survive in soil in the absence of the host, but most of them are competent soil saprophytes. Representative plant pathogens and the diseases caused (see Table 34.**4**) are described in some detail in the following.

***Erwinia* pathogens.** The genus *Erwinia* includes pathogens diverse in character. The soft rot erwinias, *E. carotovora* and *E. chrysanthemi*, produce diverse pectic enzymes (Fig. 34.**13**) and also other extracellular degradative enzymes, such as cellulases, proteases, and xylanases, which cause host cell separation (maceration) that leads to collapsed, soft, and moist tissue (soft rot). Others are necrogens, i.e., they induce local areas of cell death (necrosis), which are visible as spots, blights, or lesions (canker). *E. amylovora* causes most

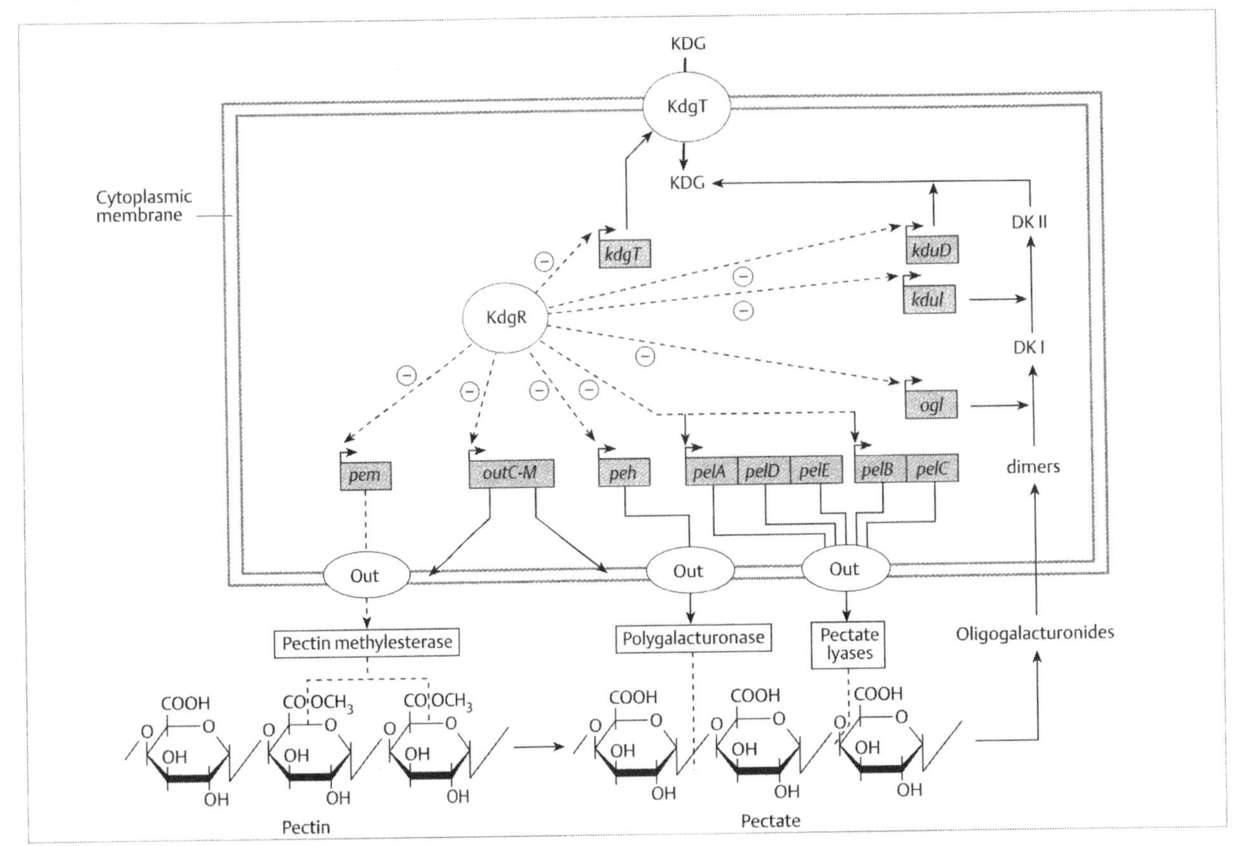

Fig. 34.**13** **The *kdgR* regulon of *Erwinia chrysanthemi*.** *E. chrysanthemi* produces diverse pectic enzymes, three of which are extracellular and are translocated from the periplasm to the external milieu by the Out secretion complex encoded by the *out* operon. These enzymes, pectin methylesterase (encoded by *pem*), the pectate lyase isozymes (encoded by the *pelABCDE* genes), and galacturonase (encoded by *peh*), act in concert to degrade insoluble pectic polymers to assimilable oligogalacturonides. Dimeric molecules are further transformed by the cytoplasmic oligogalacturonide lyase enzyme (*ogl* gene product) to 5-keto-4-deoxyuronate (DKI), which forms 2,5-diketo-3-deoxygluconate (DKII), mediated by the KduI protein. Through the action of KduD, DKII is converted to 2-keto-3-deoxygluconate (KDG), which acts as an inducer for further pectic enzyme synthesis. KDG is eventually degraded to pyruvate and 3-phosphoglyceraldehyde (not shown). The gene *kdgT* encodes a permease for the uptake of external KDG. All the genes boxed are negatively controlled by KdgR, which, in the absence of the inducer, probably binds to conserved KdgR boxes of operators in the promoter regions (arrows) of these genes [after 1]

destructive diseases of pear and other fruit world wide. In this case, the only known pathogenicity factors are exopolysaccharide and a necrosis factor (Box 34.5).

Pseudomonad pathogens. Pathogens belonging to the genus *Pseudomonas* usually show a strict host specificity, but affect a wide range of plants. The majority belong to the species *P. syringae* and is further subdivided into pathovars. The pseudomonad pathogens are primarily necrogenic and cause water-soaked lesions surrounded by chlorotic halos in leaves and fruit of their host plants (spots, blights, canker). The pseudomonads produce a variety of pathogenicity and virulence factors, including EPS, necrosis factors, and various toxins (syringomycin, phaseolotoxin). *P. syringae* pv. *savastanoi* produces indolylacetic acid, which causes plant cell proliferation on olive and oleander stems (olive knot disease). Among the non-fluorescent *Pseudomonas* spec., *P. solanacearum*, which causes bacterial wilt of solanaceous plants, is most important in economic terms, especially in the tropics and in warmer climates throughout the world. Again, the production of EPS was identified as a major virulence factor. *P. tolaasii* is the causative agent of brown blotch disease of mushrooms. The main virulence factor is tolaasin toxin, an extracellular toxin with phytotoxic activity. This toxin is thought to act as a pore former on target cells. It also acts as a biosurfactant, which may aid the spread of bacteria through the host.

Box 34.5 Screening for mutants with **r**educed **vi**rulence (Rvi) in plant pathology assays indicates that there are three main classes of mutants: (1) mutants with defects in the secretion of extracellular enzymes (Out), (2) mutants with defects in motility (Mop), and (3) mutants with defects in the synthesis of exoenzymes.

In *E. carotovora*, quorum sensing (see Chapter 20.1.4) has been implicated in the production of all exoenzymes including an antibiotic called carbapenem. Addition of *N*-(3-oxohexanoyl)homoserine lactone (HSL) to *E. carotovora* cultures results in the production of these extracellular enzymes and antibiotic. A homologue of *Vibrio fischeri* LuxI has been identified in *E. carotovora* (ExpI or RexI) and an ExpI$^-$ mutant is avirulent on potato tubers and is also unable to colonize potato plants. Addition of exogenous HSL to Exp$^-$ mutants reversed this phenotype, i.e., the mutants were pathogenic and able to colonize. The implication of the role of quorum sensing with regard to exoenzyme production and disease is that production of exoenzymes is dependent on the level of HSL present, which is directly related to the population density of the pathogen. Only when sufficient numbers of the pathogen have accumulated will exoenzyme production proceed, leading to disease. This regulatory system appears to be widespread among bacteria and may also be responsible for the regulation of antifungal metabolites and exoenzyme production in *Pseudomonas*, and for the control of conjugation in *Agrobacterium*.

In addition, there appears to be a regulation of exoenzyme synthesis by specific plant compounds. Cultures of *E. carotovora* grown in a minimal salts medium do not synthesize detectable levels of exoenzymes. Addition of celery extract to these cultures results in a dramatic increase in the levels of these exoenzymes. This response is dependent on the presence of the genes *aepA* and *aepB*. This dependence on substances present in plant extracts may account for the host range observed with this pathogen. If a given plant does not contain a certain compound, exoenzyme synthesis does not occur and disease does not progress.

Pathogenic Xanthomonads. Representatives of the genus *Xanthomonas* are also important pathogens and affect a wide range of plants. All species within this group are plant pathogenic. Like *P. syringae*, *Xanthomonas campestris* causes a number of bacterial blights and spots, and can also generate vascular wilts by infection of the vascular system. The production of symptoms could be correlated with the synthesis of EPS and with the presence of necrosis factors and partially also with pectic enzymes.

Among the Gram-positive bacteria, mainly coryneform bacteria (*Corynebacterium*, *Clavibacter*) and some *Streptomyces* are important pathogens. Pathovars of *Clavibacter michiganense* represent major vascular wilt pathogens that colonize xylem vessels and produce extracellular polysaccharides. *Streptomyces* cause common scab of potato and of other below-ground crops.

Several strains of the genera *Pseudomonas*, *Xanthomonas*, *Erwinia*, *Corynebacterium*, and *Streptomyces* are able to infect and colonize plant tissues and to damage their host. A broad array of genes are involved in pathogenicity and virulence. These genes are usually coordinately regulated by global regulatory circuits. There is a fine-tuned interplay between bacterial avirulence and plant resistance genes, through which the host range is determined in a gene-for-gene action. **Gene-for-gene (inter)action** refers to the gene-for-gene hypothesis of H. H. Flor, which suggests that for each gene that confers resistance in the host plant, there is a corresponding gene in the pathogen that confers virulence to the pathogen.

34.3.3 The *Agrobacterium* Grown Gall Disease Is the Best-understood Bacterial Plant Pathogen System

The pathogen *Agrobacterium tumefaciens* induces a neoplastic transformation of the host plant that is physically manifested by gall or tumor formation (crown galls). Crown galls are disorganized tissues. The cells can grow in tissue cultures without growth hormones. The crown galls are important diseases of fruit and nut trees in Australia and California. Transformation of the host plant by the pathogen is achieved by transferring a piece of DNA (tumor DNA, T-DNA) to the plant cell; this T-DNA is integrated into the plant genome (Fig. 34.**14**). The T-DNA directs the synthesis of plant hormones and thus induces tumor formation. The T-DNA also encodes genes for the synthesis of unusual compounds, opines, which are used by *A. tumefaciens* as a carbon and nitrogen source by virtue of opine uptake and catabolism genes.

A. tumefaciens attaches to plant cells via the gene products of *chvA*, *chvB*, *pscA* (or *exoC*), and *att*. Mutants

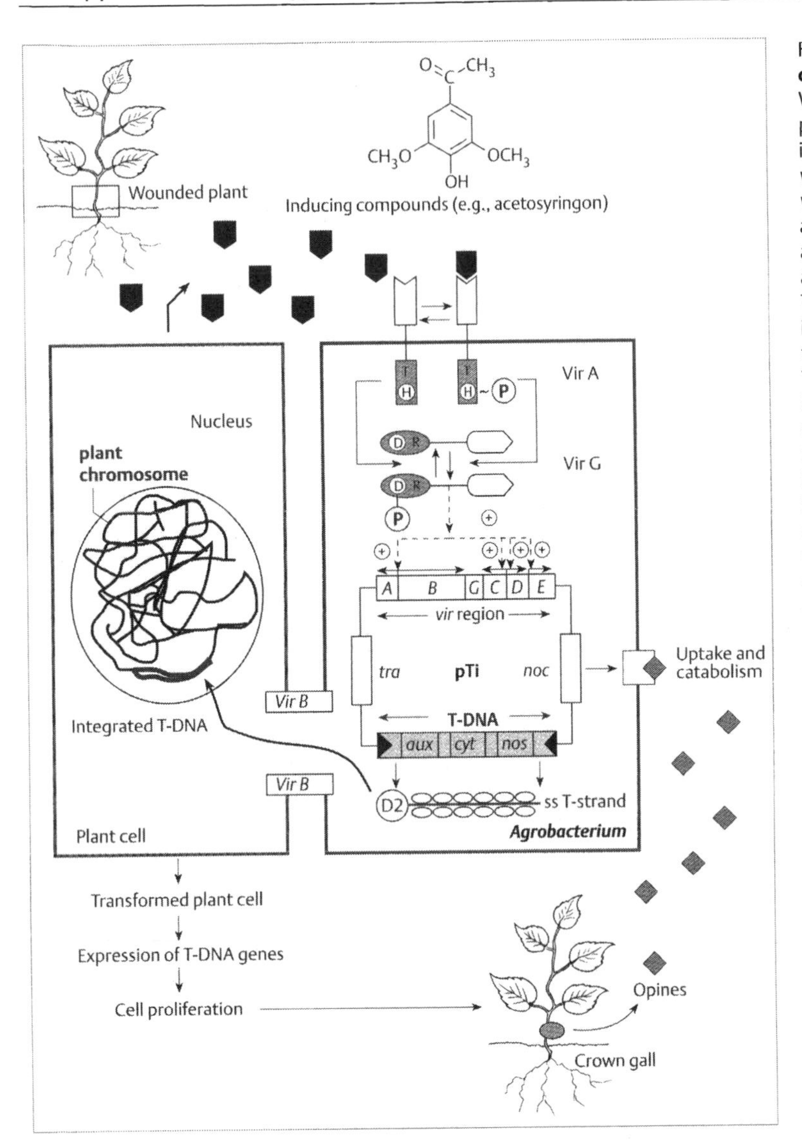

Fig. 34.**14** **Outline of the pathogenic life cycle of *Agrobacterium tumefaciens.*** Wounded plant cells excrete various compounds, e.g., acetosyringon, which act as inducers for the *A. tumefaciens vir* genes. The *vir* region contains six operons. The *virA* and *virG* gene products constitute components of a two-component regulatory system (interacting with the plant-derived inducers) and activate transcription of the other *vir* genes. The products of *virB, virC, virD,* and *virE* are involved in efficient excision and transfer of the T-DNA (for details see Box 34.6). The T-DNA codes for the synthesis of the plant hormones cytokinin (Cyt) and indolylacetic acid (IAA) and carries the region necessary for the production of opines (here *nos* = nopaline synthase). The T-DNA is flanked by inverted repeats (left and right border sequences, indicated by black triangles). The region necessary for opine uptake and catabolism (here *noc* = nopaline catabolism) and the area encoding functions for the conjugal transfer (Tra) are indicated. The genes are not drawn to scale. The entire Ti plasmid (pTi) is approx. 200 kbp, the T-DNA can be up to 20 kbp (depending on the plasmid type), and the *vir* region is approx. 35 kbp. For clarity, the *A. tumefaciens* chromosome is not drawn. Upon transfer of the single-stranded (ss) T-DNA and integration into the plant genome, the transformed plant cells proliferate to form a tumor, the crown gall, which excretes the T-DNA-encoded opines

in these loci are non-pathogenic. ChvA is an inner membrane protein that resembles other export proteins. In *chvA* mutants, the polysaccharide β-1,2-glucan is produced, but is not exported. Consequently, it is thought that ChvA is involved in transporting β-1,2-glucan to the periplasm. In contrast to *chvA* mutants, *chvB* mutants are pleiotropic, showing reduced motility, a failure to bind to plant cells, and avirulence. ChvB is an inner membrane protein involved in β-1,2-glucan synthesis, and *chvB* mutants do not produce the polysaccharide. Other loci of importance in attachment include *exoC* and *att*. ExoC$^-$ mutants lack glucose-phosphate-isomerase and are defective in the production of extracellular polysaccharides. Mutants altered in the *att* locus lack one or more of three minor outer membrane proteins involved in binding to the plant cell.

Virulence genes (*vir*) and T-DNA transfer. The *vir* genes are clustered on a plasmid (**t**umor-**i**nducing plasmid, **Ti plasmid**) that also carries the T-DNA and the genes for the uptake and catabolism of the opines (Fig. 34.**14**). The expression of *vir* genes requires certain phenolic compounds, which are components of plant sap and are excreted by the plant upon wounding. Acetosyringon is an efficient *vir* gene inducer that is produced by tobacco plants; other inducers include coniferyl alcohol, sinapyl alcohol, ferulic acid, and sinapinic acid. *vir* gene induction is also stimulated by the presence of certain sugars, such as glucose, arabinose, and galactose. Optimal conditions can vary between plant host and strain, and this might explain the variation in host range for tumor formation. Induction of the *vir* genes by the plant-derived inducers is mediated by a two-component regulatory system (see Chapters 20.4 and 34.9) which comprises VirA as the sensor component and VirG as the response-regulator. VirG stimulates transcription of the *vir* genes by binding to a 12-bp DNA sequence upstream of the respective promoters (*vir* boxes). There are also chromosomal genes that affect *vir* induction (e.g., *ros*, which is a negative regulator of *virC* and *virD* expression).

The function of the *vir* genes induced by the VirA/VirG system is the excision of the T-DNA strand, its transfer to the plant cell in a conjugation-like process (see Chapter 16.2.5) and integration of the T-DNA into the plant genome. Integration is random, but appears to have a preference for transcriptionally active regions.

Box 34.6 The proteins VirD1 and VirD2 are involved in site-specific excision of T-DNA. VirD1 is a topoisomerase, and together with VirD2, it binds to the borders of the T-DNA, which are specified by inverted repeats. After an endonucleolytical cut at the right T-DNA border, one strand of the T-DNA is displaced and excised. Excision of the T-DNA can be stimulated by binding of the VirC1 protein to a sequence located to the right T-DNA border (over-drive). After nicking of the border repeats of the T-DNA, a linear ss-DNA molecule (T-strand) is generated. VirE2 molecules bind to this ss-DNA strand and may function in protecting the T-strand from cleavage enzymes during transfer to the plant nucleus. The *virB* operon is the largest (9.5 kbp) of the *vir* region and contains essential genes involved in transport of the T-DNA out of the bacterial cell. Most of the *virB* gene products are membrane or periplasmic proteins, and it is proposed that they form a pore through which the T-DNA is excreted. The process by which the T-DNA is taken up into the plant cell, transported into the nucleus, and integrated into the genome is not fully understood. The presence of some of the Vir proteins in the plant cell has been documented and shown to be required for tumorigenesis. VirD2 and VirE2 remain bound to the T-strand during its transport; they contain nuclear localization signals and may be involved in directing the T-DNA to the plant nucleus (see Fig. 16.**15**).

The production of plant hormones and opines. Two genes (*iaaM*, *iaaH*) encode enzymes for the biosynthesis of indolylacetic acid, and one gene (*iptZ*) is involved in the production of cytokinin. As these genes contain plant-specific promoter sequences, they cannot be expressed in *A. tumefaciens*. The uncontrolled production of these hormones in the transformed plant cells gives rise to a tumor-like phenotype where the cells can undergo uncontrolled division even in the absence of exogenous phytohormones. The T-DNA also encodes opine synthases that catalyze the production of opines (condensation products of amino acids and a keto acid or a sugar; Fig. 34.**15**) from plant products in the tumor. Depending on the type of Ti plasmid from which the T-DNA originates, the tumor cells can produce different types of opines (e.g., octopine, nopaline, leucinopine, agrocinopine). The opines formed in the tumor tissue can be metabolized as an energy and nitrogen source solely by *A. tumefaciens* living in the tumor environment. The bacterial genes necessary for opine uptake and catabolism are located on the Ti-plasmid, and thus they can only be used by the *Agrobacterium* strain carrying the respective Ti plasmid (nopaline-type, octopine-type plasmid) and not by other soil bacteria. Thus, *Agrobacterium* can create a favorable ecological niche by manipulating plant cells to divide and synthesize C and N sources for the proliferation and survival of the bacteria (molecular farming).

Hairy root disease. The hairy root disease caused by *Agrobacterium rhizogenes* results from a genetic modification of the plant genome. *Agrobacterium rhizogenes* contains an Ri plasmid (**r**oot-**i**nducing plasmid),

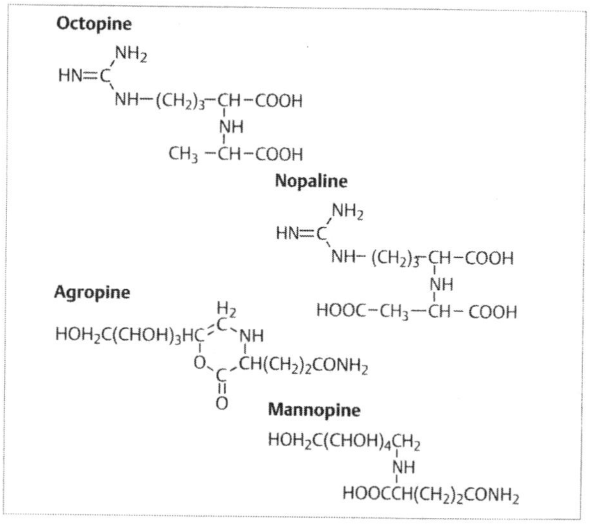

Octopine
$HN{=}C(NH_2)$–$NH{-}(CH_2)_3{-}CH{-}COOH$
NH
$CH_3{-}CH{-}COOH$

Nopaline
$HN{=}C(NH_2)$–$NH{-}(CH_2)_3{-}CH{-}COOH$
NH
$HOOC{-}CH_2{-}CH_2{-}CH{-}COOH$

Agropine
$HOH_2C(CHOH)_3HC{-}CH_2{-}NH$
$O{-}C({=}O){-}CH(CH_2)_2CONH_2$

Mannopine
$HOH_2C(CHOH)_4CH_2$
NH
$HOOCCH(CH_2)_2CONH_2$

Fig. 34.**15** **Structures of some important opines.** Opines are amino acid derivatives that are substituted at their α-N residue, either with pyruvate (e.g., octopine) or α-ketoglutarate (e.g., nopaline). Agropine and mannopine are typical for *Agrobacterium rhizogenes* which carries T-DNA and the *vir* operons. The molecular basis of plant cell transformation by *A. rhizogenes* is similar to that described for *A. tumefaciens*. The *vir* genes are interchangeable between the two species. However, the T-DNA of the Ri plasmid differs from that present on the Ti plasmid; the Ri plasmid T-DNA contains genes, designated *rol* genes, proposed to be involved in the activation of cytokinin conjugates.

The T-DNA represents a natural gene transfer system between prokaryotes and higher plants. This has been exploited to develop gene vectors for introducing recombinant DNA into plants. These systems replace the T-DNA genes with a selectable marker and reduce the size of the original Ti vector plasmid considerably (see Chapters 16.2.5 and 17.7).

Agrobacterium is a unique bacterial pathogen because it genetically manipulates the host plant. By transferring a piece of DNA (T-DNA) into the plant genome, *Agrobacterium* causes the plant cell to proliferate and to produce specific compounds (opines) that can be metabolized solely by the parasitic bacterium. The T-DNA systems of *Agrobacterium tumefaciens* and *Agrobacterium rhizogenes* are natural gene transfer systems between prokaryotes and higher plants and are being used to construct transgenic plants.

34.4 Bacteria Can Enhance Plant Growth and Resistance

A number of soil microorganisms enhance plant growth and plant health and thus improve crop yields using several mechanisms. One group of bacteria can influence plant growth by producing one or more substances that act directly as growth stimulators. In the literature, these "phytostimulators" are often referred to as plant-growth-promoting bacteria. The beneficial effect on the plant of bacteria is related to the bacteria being inhibitory to the growth of plant pathogenic organisms (biological control agents).

34.4.1 Phytostimulators Produce Plant Growth Hormones

In this heterogeneous group are included soil microbes whose main, but not necessarily unique, agrophysiological trait is the production of one or more substances that enhance plant growth directly. Several rhizocenoses, i.e., biocenotic relationships between microbes and roots, have been identified in grasses, e.g., *Azotobacter* with the perennial grass *Paspalum*, and *Campylobacter* with *Spartina*, a grass common in marshes, and agriculturally relevant crops, e.g., *Acetobacter* with sugar cane or sweet potato, *Achromobacter* with rice, and *Azospirillum* with several Graminaeae (barley, rice, sorghum, wheat) or members of Cactaceae, Compositae, Papilionaceae, and Solanaceae. *Herbaspirillum* has been described to form a rhizocenosis with Gramineae (corn, rice, sorghum, wheat) in warm climates, and *Klebsiella* enters rhizocenosis with rice and sweet potato.

The most thoroughly investigated biological system is represented by the *Azospirillum* group and includes the five species *A. amazonense*, *A. brasilense*, *A. halopraefaerens*, *A. irakense*, and *A. lipoferum*. These bacteria, depending on the pedoclimatic situation, can establish themselves in the rhizosphere of plants or enter the root, remaining in the intracellular spaces (endorhizosphere). The bacterial density can be 10^5–10^8 colony forming units per gram of root dry weight. Multiple effects on root morphology and plant physiology due to the presence of azospirilla internal to the root system or around it have been described (Table 34.**5**). Although in tropical and sub-tropical areas, *Azospirillum* is able to fix dinitrogen, thus providing the plant with fixed nitrogen compounds, it is now generally accepted that the beneficial effects to the plant are at least also partially due to phytostimulatory substances produced by the bacterium, i.e., indolylacetic acid (**IAA**). In mutants with an impaired, or quasi-zero IAA biosynthetic ability, the effects on the plant are greatly reduced. The phytostimulatory effects of azospirilla have been exploited agronomically, and microbial inoculants have been industrially developed. The crop yield can be affected, with reported increases of 10–30%, when *A. brasilense*, *A. lipoferum*, or a mixture of both is used as seed inoculants for *Triticum aestivum*, *T. durum*, *T. turgidum*, *Oryza sativa*, *Sorghum bicolor*, or *Zea mays*. In many cases, yields have been reported to remain unaffected in inoculated crops where the nitrogen fertilizers are delivered at 50–70% of the usual dosage.

Table 34.5 **Effects reported following inoculation of plants with *Azospirillum***

Effect on roots	Plant
Increased root diameter and length and density of root hairs	Tomato
Increased number and length of root hairs	Durum wheat
Increased overall root surface during early growth phases	Corn
Increased nutrient assimilation, dry matter formation, and Fe content	Sorghum, durum wheat
Increased respiration and enzymatic activities	Corn

34.4.2 Biocontrol Agents Suppress Plant Pathogenic Organisms

The ability of some natural soils to suppress plant disease has been attributed to the indigenous beneficial rhizosphere microflora. Several bacteria capable of providing substantial disease control have been reported (e.g., *Streptomyces, Agrobacterium, Enterobacter, Erwinia, Bacillus, Serratia,* and fluorescent *Pseudomonas* strains. The nature of biological control by bacterial strains is difficult to determine. One mode of action may be direct killing or suppression of the pathogenic organism; the other type of interaction may be indirect, i.e., competition for nutrients or inhibiting the establishment of the pathogen by increasing the competence and ecological "fitness" of the biocontrol agent.

Bacterial biopesticides. Soil bacteria produce an array of low-molecular-weight compounds, some of which are potential biopesticides that act against fungi, bacteria, or insects (Table 34.**6**). These compounds can be considered as secondary metabolites that preferentially accumulate late in the growth phase. This is consistent with the notion that antimicrobial compounds may have evolved as microbial defense mechanisms to inhibit competitors under stress situations, such as nutrient depletion (see Chapter 27.7). Disease suppression by fluorescent pseudomonads, for example, can be correlated to microbial compounds such as iron-chelating siderophores and metabolites with antimicrobial properties. Pseudomonads produce a range of iron-chelating compounds, including salicylic acid, pyochelins, and fluorescent pseudobactins and pyoverdines (see Chapter 8.8). Fluorescent siderophores have been isolated from soil, and there is considerable genetic and biochemical evidence demonstrating their role in plant growth promotion and biocontrol. Salicylic acid, a precursor of pyochelin and a siderophore in its own right, is also a plant hormone. This metabolite is implicated in the induction of systemic acquired resistance, i.e., plant defense mechanisms that lead to systemic resistance against a number of pathogens.

2,4-Diacetylphloroglucinol and pyoluteorin (Fig. 34.**16**) are metabolites produced by pseudomonads that are largely responsible for the prevention of "damping-off" of seedlings of sugar beet and cotton, caused by *Pythium ultimum* (see also Table 34.**6**).

Volatile compounds such as ammonia and HCN, produced by many rhizosphere strains, are implicated as potentially important metabolites in biocontrol. *Pseudomonas* can produce levels of HCN in vitro that are toxic to certain pathogenic fungi (e.g., *Thielaviopsis basicola*, thereby preventing black root rot of tobacco).

Table 34.**6** **Some examples of specific microbial metabolites implicated in the control of crop diseases**

Disease	Pathogen	Effective metabolite
Take-all of wheat	*Gaeumannomyces graminis* var. *tritici*	Phenazines 2,4.-Diacetylphloroglucinol
Tan spot of wheat	*Pyrenophora tritici*	Pyrrolnitrin
Damping-off of sugar beet, cotton, etc.	*Pythium* sp.	Ammonia Pyoluteorin 2,4-Diacetylphloroglucinol
Black root-rot of tobacco	*Thielaviopsis basicola*	HCN 2,4-Diacetylphloroglucinol
Crown gall of fruit trees	*Agrobacterium tumefaciens*	Agrocin 84
Flax wilt	*Fusarium oxysporum*	Pseudobactin B10

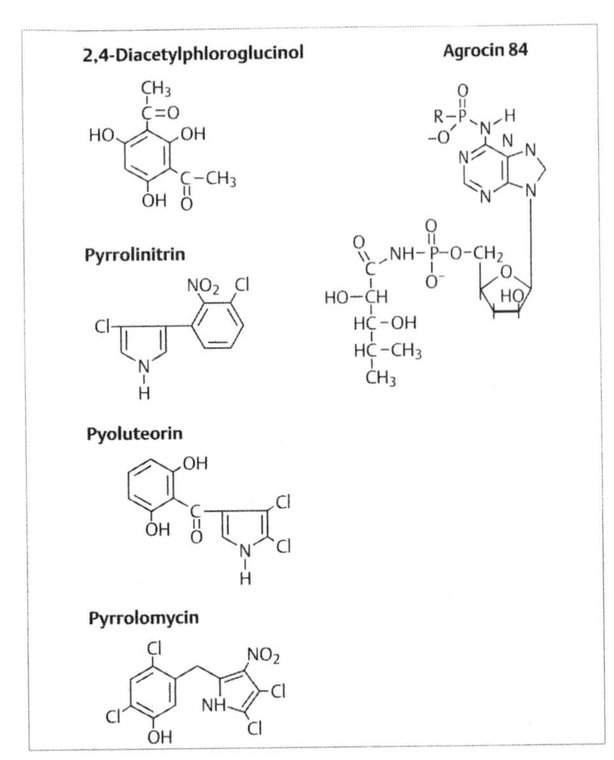

Fig. 34.**16** **Structures of bacterial metabolites active in biocontrol.** 2,4-Diacetylphloroglucinol, pyrrolnitrin, and pyoluteorin are metabolites produced by fluorescent Pseudomonads with biocontrol activity against some phytopathogenic fungi. Agrocin 84, synthesized by *Agrobacterium radiobacter*, inhibits specifically the plant pathogenic bacterium *Agrobacterium tumefaciens*. Pyrrolomycin is an insecticide produced by *Streptomyces fumanus*

On the other hand, high concentrations of HCN are toxic to some plants, and it has been suggested that fluorescent pseudomonads that produce HCN are responsible for a reduction in the yield of certain crops (e.g., potato).

Insecticidal toxins. *Bacillus thuringiensis* produces several substances toxic to insects, including exotoxins and endotoxins. Endotoxins are largely responsible for the insecticidal properties. These proteins (**BT toxins**) are members of a conserved family of polypeptides (130–140 kDa) with conserved and variable domains that determine activity against specific insect species. These protoxins, usually located outside the exospore, are taken up by insect larvae, and the active toxin (30–80 kDa) is released in the alkaline midgut of susceptible larvae.

B. thuringiensis has been used as a biocontrol agent of lepidopteran pests for 30 years. Commercial sales of this biocontrol agent currently represent the major share (90%) of the world biopesticide market. Since its introduction in agriculture and forestry, many improvements have been made to the *B. thuringiensis* (BT) product. These include improved formulation (concentration and the way the agent is applied) and new isolates active against other insect groups (Diptera and Coleoptera) and also nematodes, liver flukes (Trematoda), and mites (Acari). An understanding of the genetics of *B. thuringiensis* has allowed classical genetic approaches (e.g., plasmid curing to remove low toxicity endotoxin genes and plasmid transfer to introduce higher toxicity and broader specificity genes) to generate new and improved strains. More recently, genetic engineering techniques have been used to introduce genes encoding endotoxin into plants and Gram-negative rhizosphere bacteria. Greenhouse experiments with rhizosphere pseudomonads engineered with BT toxin indicate good control of target pests with no side effects on non-target species (e.g., bees). However, of some concern is the demonstration that insect resistance to BT toxins can emerge.

Other Gram-positive organisms also have considerable potential as biological control agents for the control of fungal diseases of crops. For example, *B. subtilis* strains produce a range of toxic metabolites and have the advantage, like *B. thuringiensis*, of being amenable to large-scale fermentation and formulation as a mixture of spores and metabolites with a good shelf-life. *B. subtilis* has activity against a range of fungal pathogens, including *Rhizoctonia solani* and *Botrytis cinerea*. However, the mechanisms involved in the control of fungal diseases is not well understood.

Agrobacterium radiobacter produces a low-molecular-weight bacteriocin, called agrocin 84, which has been successfully used for some years as a biocontrol agent of crown gall since it is toxic to virulent strains of *Agrobacterium tumefaciens*. At the molecular level, the mechanism of action of agrocin 84, a di-substituted nucleotide (Fig. 34.**16**), is thought to be by chain termination of DNA synthesis. Agrocin 84 gains access to the target cell through the opine permease encoded by the Ti virulence plasmid of *A. tumefaciens*, i.e., only strains of *A. tumefaciens* carrying a Ti plasmid are able to take up agrocin 84. Synthesis and resistance to agrocin 84 is also plasmid-encoded in *A. radiobacter* (pAgK84). In the laboratory, this plasmid can be transferred to virulent *A. tumefaciens* at a high frequency in the presence of opines. *A. tumefaciens* transconjugants harboring the pAgK84 plasmid, or derivatives cured of their Ti plasmid, are resistant to agrocin 84 and can no longer be controlled by the bacteriocin. Hence, the conjugal transfer genes of pAgK84 were deleted in a strategy similar to that used to generate Ice^- bacteria (see Chapter 34.4.3). The mutant strain has been successfully introduced as a genetically engineered microbial inoculant for crown gall control.

34.4.3 Mutants of Epiphytic Bacteria Can Prevent Ice Nucleation

Certain epiphytic bacteria are capable of inducing frost damage on crop plants at relatively high temperatures (approximately −2 °C). Plants free of these bacteria show no damage at temperatures as low as approximately −6 °C. It has been estimated that frost damage to crops in the USA alone causes more than 1 billion dollars in losses per year. Various bacterial taxa have been implicated in promoting frost damage, including *Pseudomonas syringae*, *P. fluorescens*, *Erwinia herbicola*, and *Xanthomonas campestris* pv. *translucens*. Strains that induce frost damage can promote enhanced ice nucleation in the laboratory (phenotype Ice^+). The genes responsible for **i**ce **n**ucleation **a**ctivity (***ina***) have been cloned and sequenced. The protein product of the *inaZ* gene is responsible for the ice-nucleation phenotype. The protein (1200 amino acids, molecular weight 118 kDa) consists of an imperfect reiteration of the octapeptide Ala-Gly-Tyr-Gly-Ser-Thr-Leu-Thr. This octapeptide is repeated 122 times throughout the InaZ protein. The periodicity of the reiteration suggests that it binds water in an ordered array, thereby promoting ice formation. An Ice^- deletion mutant of *P. fluorescens* produced by genetic engineering techniques was the subject of considerable media publicity when it was field tested. The Ice^- strain was subsequently used as an inoculant to protect susceptible plants against frost damage. The rationale was that the newly introduced Ice^- bacteria would compete as well as the indigenous population for the limited colonization sites on the leaf surface (phyllosphere). By increasing the inoculum relative to the indigenous population, the severity of frost damage was significantly reduced. This work is an excellent demonstration of the concept of competitive niche exclusion as a powerful tool in the beneficial application of bacterial inoculants in agriculture.

> Bacteria are important plant pathogens, but they can also contribute to plant health and increase yields. The mechanisms by which these beneficial effects for plants are achieved are diverse. Bacteria can directly influence plant growth by producing hormones, or they can indirectly contribute to plant health by controlling plant pathogenic fungi, bacteria, and insects. The BT toxin produced by *Bacillus thuringiensis* is applied as a biocontrol agent of lepidopteran pests, and the bacteriocin agrocin 84, synthesized by *Agrobacterium radiobacter* is used successfully to control crown gall disease.

34.5 Bacteria Can Contribute to Biodegradation of Pollutants in Soil

Highly intensive agriculture requires considerable resources to control pests, such as weeds, insects, nematodes, and plant diseases. Chemical pesticides (herbicides, insecticides, fungicides) are therefore currently an integral part of agricultural practice. Synthetic compounds applied in industrial production are now recognized to be present in great abundance in the soil ecosystem. There is increasing public concern about the ultimate fate of these substances and their potential side effects on soil fertility and human and animal health. The term **bioremediation** refers to the technological use of biological agents (bacteria and fungi) to detoxify chemical pollutants or to reduce their concentration in the environment.

34.5.1 Pollutants Vary From Being Easily Degradable to Being Recalcitrant

The structures of most synthetic pesticides are based on relatively simple hydrocarbon skeletons (aromatic/aliphatic) that carry a variety of substituent groups, for example, halogen, phosphate, and nitro groups (Fig. 34.**17**). Some of these "man-made" substances are similar to natural compounds and can be enzymatically attacked and degraded by a number of microorganisms. Others, however, have structures that are not known in nature (**xenobiotics**), and many of these molecules are resistant to biological degradation, i.e., they are recalcitrant. Therefore, one has to distinguish between two types of xenobiotic compounds:

1. Those (most used now) that are subject to rapid degradation usually within one cropping season. For example, complete degradation of 2,4-D occurs within 4–6 weeks.
2. Recalcitrant pesticides that persist in the environment for long periods of time (months or years). For example, the insecticides parathion and DDT, which was used extensively from the 1930s until its ban in 1979, persist in soil for more than 15 years. Stable metabolites of DDT have been detected in soil and groundwater and even in humans.

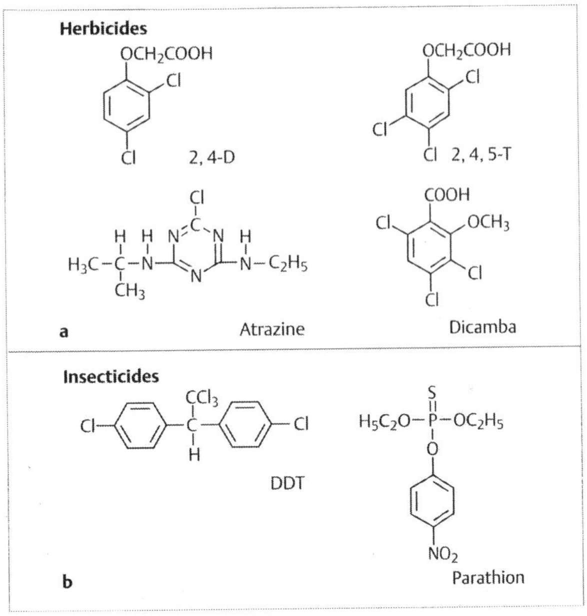

Fig. 34.**17** **Chemical structures of some important (a)** herbicides and **(b)** insecticides. Examples of halogenated compounds include **(a)** the herbicides 2,4-D (2,4-dichlorophenoxy acetic acid), 2,4,5-T (2,4,5-trichlorophenoxy acetic acid), atrazine (2-chloro-4-ethylamino-6-isopropylamino-triazine), and dicamba (3,6-dichloro-2-methoxy-benzoic acid) and **(b)** the insecticide DDT (dichlorodiphenyltrichloroethane). The insecticide parathion (*O*,*O*-diethyl *O*-*p*-nitrophenyl phosphorothioate) (**b**) is an example of an organophosphorous pesticide

Recalcitrance of a molecule is mainly due to unusual chemical bonds or substitutions, which block oxygenation reactions. After removal of these electron stabilizing substituents, the carbon skeleton is then easily mineralized by many soil microorganisms. On the other hand, simple structural changes, such as the addition of one chlorine, can convert a rapidly degradable substrate (e.g., 2,4-D in 4–6 weeks) into a more persistant compound (e.g., 2,4,5-T; 20 weeks).

34.5.2 Bacteria Can Degrade a Range of Xenobiotics

A variety of bacterial groups, among them Gram-negative aerobes (predominantly *Pseudomonas* species), facultative anaerobes (e.g., Enterobacteriaceae), and Gram-positive bacteria (e.g., *Bacillus* and Corynebacteriaceae) have been shown to decompose pesticides and other xenobiotic compounds.

Bacteria that can degrade chloroaromatic pesticides are typical soil inhabitants. 2,4-D, for example, can be rapidly degraded by a variety of bacteria, including *Achromobacter*, *Arthrobacter*, *Corynebacterium*, *Flavobacterium*, *Pseudomonas*, and *Alcaligenes*. 2,4,5-T is completely mineralized to CO_2 by *Pseudomonas cepacia* through a series of enzymatic steps. *Pseudomonas* spec. are also effective in the biodegradation of atrazine and of dicamba, a chemically stable herbicide used to control broad-leaf weeds and several grassy weeds, but which unfortunately also affects growth of certain non-target plants (e.g., soybean). In field studies, strains of *Pseudomonas* were able to mineralize high concentrations of dicamba completely in different natural soils and to protect soybeans from dicamba injuries. Bacteria that convert polychlorinated biphenyls (PCBs), which were introduced into the environment through industrial applications and which now represent a serious environmental pollutant, are ubiquitous. The metabolic pathway elucidated in various pseudomonads is outlined in Fig. 34.**18** and Chapter 9.17.

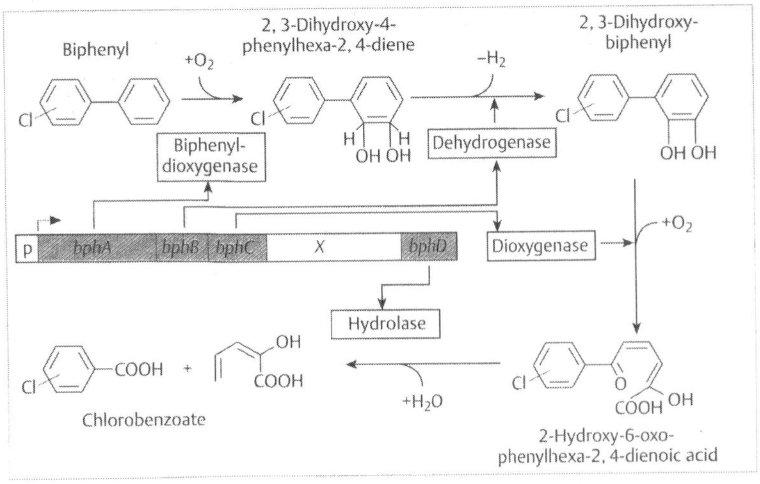

Fig. 34.**18** **Biodegradation of chlorinated biphenyls.** The degradation of chlorobiphenyls as found in a number of bacteria including *Pseudomonas pseudoalcaligenes* is shown. The total size of the *bph* operon in *P. pseudoalcaligenes* KF707 is approx. 10 kb. The genes are not drawn to scale [after 2 and 3]

Organophosphate compounds are active ingredients in many of the insecticides and some of the herbicides used currently. For example, parathion (Fig. 34.**17**) has been extensively used to control insects, but is now banned or is at least used with restrictions. Strains of *Arthrobacter*, *Flavobacterium*, and *Pseudomonas* were isolated that are able to use parathion as a carbon source and to degrade it to *p*-nitrophenol within a few weeks. In most of these cases, the responsible *opd* (**o**rganophosphate **d**egradation) genes are encoded on plasmids.

34.5.3 Co-metabolism and Genetic Engineering May Improve the Biodegradative Potential of Bacteria

Many man-made organic compounds can be completely mineralized to CO_2. Others cannot be used as carbon and energy sources, but they may be biodegraded in the presence of co-metabolites that sustain bacterial growth. The partial or total degradation of non-growth xenobiotic substrates by bacteria growing on other organic material is called **co-metabolism** (see Chapter 30.7). In general, co-metabolic transformations occur relatively slowly, but they are of enormous ecological importance.

However, the end products of partial degradation may be as hazardous as the original xenobiotic. Although some of these products may be further transformed by other bacteria, they also may be stabilized, e.g., as part of the humus. Therefore, microbial consortia or recombinant strains that are capable of complete degradation are desired.

Some of the man-made chemical substances introduced into the soil ecosystem through agricultural and industrial applications can be partially or completely degraded by microorganisms. Biodegradation of xenobiotic compounds, including pesticides, can be ascribed to a few bacterial genera, including *Achromobacter, Alcaligenes, Bacillus, Flavobacterium*, and *Pseudomonas*. Knowledge of the genetics and enzymology of biodegradative processes may help to design recombinant strains and alternative pathways to allow also the mineralization of stable, recalcitrant xenobiotic compounds and to provide bioremediation agents for the decontamination of environments polluted with hazardous chemical mixtures.

The biodegradative potential of natural isolates is generally restricted to specific substances. This is mainly because of the stringency of the first metabolic enzyme, which attacks only a very specific substrate (e.g., 3-chlorobenzoate, but not 4-chlorobenzoate). Therefore, strategies in the construction of new catabolic traits are to expand the substrate profile of the first enzyme by mutation or to introduce genes from another pathway. For example, by introducing a gene encoding 1,2-dioxygenase activity, a recombinant *Pseudomonas* strain was constructed that could oxidize both 3-chlorobenzoate and 4-chlorobenzoate. Another strategy is to clone and combine structural and regulatory genes from various pathways. The construction of new degradative pathways holds considerable potential for the acquisition of desirable catabolic traits in microbes to be used as bioremediation agents.

Further Reading

Atlas, R. M., and Bartha, R. (1993) Microbial ecology: fundamentals and applications. Redwood City, Calif.: Benjamin/Cummings

Barras, F., van Gijsegem, F., and Chatterjee, A. K. (1994) Extracellular enzymes and pathogenicity of soft-rot *Erwinia*. Annu Rev Phytopathol 32: 201–234

Chaudhry, G. R., ed. (1994) Biological degradation and bioremediation of toxic chemicals. London: Chapman & Hall

Cork, D. J., and Krueger, J. P. (1991) Microbial transformations of herbicides and pesticides. In: Neidleman, S. L., and Laskin, A. I. (eds.) Advances in applied microbiology, vol 36. New York: Academic Press; 1–63

Dangl, J. L., ed. (1994) Bacterial pathogenesis of plants and animals. Molecular and cellular mechanisms. Berlin Heidelberg New York: Springer

Denny, T. P. (1995) Involvement of bacterial polysaccharides in plant pathogenesis. Annu Rev Phytopath 33: 173–198

Gresshoff, P. M., ed. (1990) Molecular biology of symbiotic nitrogen fixation. Boca Raton, Flor.: CRC Press

Handelsman J., and Stabb, E.V. (1996) Biocontrol of soilborne plant pathogens. Plant Cell 8: 1855–1869

Head, I.M. (1998) Bioremediation. Microbiol 144: 599–688

Long, S. R., and Staskiawicz, B. J. (1993) Prokaryotic plant parasites. Cell 73: 921–935

Lynch, J. M., ed. (1990) The rhizosphere. New York: Wiley

O'Gara, F., Dowling, D. N., and Boesten, B., eds. (1994) Molecular ecology of rhizosphere microorganisms. Weinheim: VCH

Pueppke, S. G. (1996) The genetic and biochemical basis for nodulation of legumés by rhizobia. Annu Rec Biotechnol 16:1–51

Sources of Figures

1 Barras, F., van Gijsegem, F., and Chatterjee, A. K. (1994) Annu Rev Phytopathol 32: 201–234

2 Furukawa, K. (1994) Genetic systems in soil bacteria for the degradation of polychlorinated biphenyls. In: Chaudhry, G. R. (ed.) Biological degradation and bioremediation of toxic chemicals. London: Chapman & Hall

3 Silvestre, M., and Sondossi, M. (1994) Selection of enhanced polychlorinated biphenyl-degrading bacterial strains for bioremediation: consideration of branching pathways. In: Chaudhry, G. R. (ed.) Biological degradation and bioremediation of toxic chemicals. London: Chapman & Hall

35 Prokaryotes in Industrial Production

Although ignorant of the processes involved and, therefore, practicing an art rather than a science, for thousands of years our ancestors exploited microbial fermentations for food preservation (e.g., cheese, vinegar), flavor enhancement (e.g., bread, soy sauce), and alcoholic beverage production. The chronological development of the fermentation industry may be represented as five overlapping stages, as illustrated in Table 35.**1**.

During the pre-Pasteur era (before 1865), biology had little scientific basis; biological applications were based on tradition in the absence of any fundamental knowledge of the biological principles involved. Louis Pasteur's proof in 1863 that living microbes are the active agents of fermentation was the first step on the long route from descriptive biology toward a real understanding of the microbial processes. His work laid the foundations for the subsequent development of industrial processes for fermentative production of butanol, acetone, butanediol, isopropanol, and other chemicals by various species. The techniques developed for the production of these organic solvents were a major advance in fermentation technology and paved the way for the successful introduction of aseptic aerobic processes in the 1940s. Several different disciplines (biochemistry, microbiology, process engineering) contributed to the newly developing fermentation industry, especially to antibiotic production under sterile conditions. Once the clinical use of penicillin was established, pharmaceutical companies began to consider antibiotics seriously. The use of microorganisms to carry out highly specific and selective enzymatic transformation reactions of pharmaceuticals represented a major breakthrough in the development of steroid hormones for medical use.

The increasing knowledge of microbial metabolism gave rise to the systematic exploitation of the capabilities of bacteria to produce a variety of metabolites and

Table 35.**1** **The stages in the chronological development of biotechnology (examples of products)**

Pre-Pasteur era (before 1865)
- Alcoholic beverages (beers, wines)
- Dairy products (cheeses, yoghurt)
- Other fermented foods (vinegar)

Pasteur era (1865–1940)
- Industrial fermentation (ethanol, butanol, acetone, glycerol)
- Production of organic acids (acetic acid, citric acid, lactic acid)

Antibiotic era (1940–1960)
- Large-scale production of antibiotics (penicillin, streptomycin, chlortetracycline, etc.)
- Microbial steroid transformations (cortisone, testosterone, estrogen)

Post-antibiotic era (1960–1975)
- Microbial production of amino acids (L-glutamate, L-lysine)
- Development of techniques for single-cell-protein production (SCP)
- Production of industrial enzymes (proteases, amylases, glucose isomerases)
- Industrial use of immobilized enzymes and cell technology (glucose isomerases)
- Production of bacterial polysaccharides (xanthan)

Era of synthetic biotechnology (1975–present)
- Recombinat DNA technology (1974)
- First products on the market in 1982 (animal diarrhea vaccines, human insulin)

enzymes. Another breakthrough was in the mass production of enzymes for use, such as in detergents and in the large-scale conversion of starch into glucose and fructose. The very rapid growth of the oil-based industries in the 1960s led to the use of abundantly available mineral oil fractions (e.g., alkanes, gas oil, methanol) in processes for the mass cultivation of microorganisms for use as animal feed (single-cell protein).

In the 1980s, applied microbiology became a major growth area. This change came about primarily through recombinant DNA technology. Whereas mutation and selection had been used to increase the level of pre-existing activity in a microbial cell, recombinant DNA technology could now be used to confer on cells entirely new synthetic capabilities, such as the synthesis of human hormones by *Escherichia coli*. Early work on the commercial applications of recombinant DNA technology centered on the production of proteins. As geneticists have become more skilled, there has been a trend toward cloning all the genes associated with a biosynthetic pathway such that ultimately the range of different bacteria used in commercial processes has diminished. Biological sciences and technologies have changed during this century from largely empirical to quantitative disciplines, thus opening new perspectives for applied microbiology.

35.1 A Limited Number of Strains Have Been Selected and Improved for Industrial Production

Of the large number of known bacterial species, relatively few are currently exploited in industry. Some important products and producer microorganisms are mentioned in Table 35.**2**. While this list is by no means exhaustive, it emphasizes the scope of biotechnology. The principal producer bacteria involved are all chemoorganotrophs and derive their carbon and energy supply from the metabolism of organic compounds. Of the Gram-positive organisms, aerobic endospore-forming bacteria of the genus *Bacillus*, some coryneform bacteria (*Corynebacterium*), and the filamentous bacteria, particularly of the genus *Streptomyces*, are well represented. Gram-negative bacteria include acetic acid bacteria and *Xanthomonas* species. The following attributes are the prerequisites of a microorganism suitable for industrial use:

1. The strain must be available in pure culture (also free of phages) and must be genetically stable.
2. The strain should be nonpathogenic and free of toxins.
3. The strain should grow rapidly also in large-scale culture and should produce the desired product at high yields within a short period of time (3 days or less).
4. If possible, the strain should be able to protect itself against contamination (e.g., through lowering the pH, ability to grow at high temperature, or synthesis of an antibiotic).

With the exception of the food industry, only a few industrial processes directly use strains isolated from nature. Strain improvement is an essential part of process development for microbial products. It helps to reduce costs by developing strains with increased productivity and yield, with the ability to use cheaper raw materials, or with more specialized desirable characteristics, such as improved tolerance to high substrate and/or product concentrations.

Currently, microorganisms that overproduce primary and secondary metabolites are obtained by mutagenesis and selection. In the last few years, automated procedures have been developed that use robotics and microprocessors to increase the numbers of isolated mutants that can be tested per unit time. Thus, the development of many of the highly productive industrial strains has been largely an empirical process ("trial-and-error"). Success in attempts to increase further the productivities and yields of already highly productive strains will largely depend on the availability of detailed information on the metabolic pathways and their regulation.

The introduction of genes into organisms via recombinant DNA techniques is currently a most powerful method for the construction of strains with desired genotypes. The opportunity to introduce heterologous genes and regulatory elements permits construction of metabolic configurations with novel and beneficial characteristics. The improvement of cellular activities by manipulation of enzymatic, transport, and regulatory functions of the cell with the application of recombinant DNA technology is called **metabolic engineering** since in this way changes in metabolism can be targeted.

An early successful use of recombinant bacteria involved altering the nitrogen metabolism of the parent

Table 35.2 **Industrial products from prokaryotes**

Products	Microorganisms
Foods	
Sauerkraut and pickles	*Leuconostoc, Pediococcus,* and *Lactobacillus*
Yoghurt and fermented milk	*Lactobacillus* and *Streptococcus*
Vinegar	*Acetobacter aceti*
Food and feed additives	
Glutamic acid, lysine, and other amino acids	*Corynebacterium glutamicum, Brevibacterium flavum*
Inosinic acid and ribonucleotides	*Corynebacterium glutamicum*
Vitamins	Various bacteria
Enzymes	
Proteases	*Bacillus licheniformis*
α-Amylases	*Bacillus amyloliquefaciens, Bacillus licheniformis*
Glucose isomerase	*Actinoplanes missouriensis, Streptomyces* spec.
Penicillin acylase	*Escherichia coli*
Industrial chemicals	
Ethanol	*Zymomonas mobilis*
n-Butanol, Acetone	*Clostridium* spec.
Lactic acid	*Lactobacillus* spec.
Polysaccharides	
Xanthan	*Xanthomonas campestris*
Dextran	*Leuconostoc mesenteroides*
Alginates	*Azotobacter vinelandii*
Medical products	
Steroids	Mycobacteria and related bacteria
Antibiotics	Actinomycetes and *Bacillus* spec.

culture to obtain a higher yield of **single-cell protein** (SCP). The process modification was developed by ICI (Great Britain) for making SCP from methanol. The pathway used by the methylotrophic bacterium (*Methylophilus methylotrophus* for assimilating ammonia involves an aminotransferase and a glutamine synthase; it consumes one mol of ATP for every mol of ammonia assimilated. In *Escherichia coli*, for example, an alternative pathway uses glutamate dehydrogenase (Gdh) for ammonia assimilation and does not consume any ATP. Therefore, the *gdh* gene from *E. coli* was cloned and expressed in *M. methylotrophus*. The expression of this gene in a glutamate-synthase-negative mutant gave a cell yield 5% higher than that of the parent strain because of the saving in ATP.

Cloning and expression of heterologous genes can be used for extending the catabolic pathway in a bacterium; in this manner, the substrate spectrum can be enlarged. For example, the *E. coli lac* operon has been used previously for the construction of lactose-utilizing strains of *Alcaligenes eutrophus, Corynebacterium glutamicum,* and *Xanthomonas campestris.* Furthermore, application of recombinant DNA techniques to restructure metabolic networks can improve production of metabolites by redirecting metabolite flows; some examples will be presented in this chapter.

Mutagenic procedures and selection have been the main method for improvement of industrial microorganisms. However, in recent years, the application of recombinant DNA technology has become an additional technique to change the metabolism in a purposeful manner (**metabolic engineering**).

35.2 Large-Scale Cultivation of Microorganisms Depends on the Type of the Process

With the recent advances in molecular biology, the scale of an industrial bioprocess has become strongly dependent on the nature of the process. A sizable share of the global demand for a high-value product such as the human growth hormone can be produced in a 200-l bioreactor, whereas traditional products of microorganisms (e.g., amino acids and antibiotics) are produced in 200–400-m^3 bioreactors. The **bioreactor** (traditionally the term fermentor is often used) is the containment system for the biological reactions of a biotechnological process. It should provide the environment for optimization of organism growth and metabolic activity, and it should prevent contamination of the production culture from the environment.

35.2.1 Principles of Bioreactor Design and Culture Media

Sterility. The first large-scale microbial process in the pharmaceutical industry was developed in a stirred tank bioreactor, and although there are numerous designs for industrial bioreactors, this type is still the most widely used. As shown in Figure 35.**1**, a stirred tank bioreactor is typically cylindrical with a slightly curved or almost flat bottom. The material for construction usually is stainless steel, and this must be of high grade if it is not to corrode or leak toxic metal compounds into the growth medium. A key requirement of a successful fermentation is aseptic operation; no contamination of the culture should occur. Thus, the entire bioreactor and ancillary equipment, as well as the growth medium, must be sterilized before inoculation. In addition, the air supplied during the fermentation must be sterile, and there must be no mechanical breaks in the bioreactor that will allow the ingress of contaminating microorganisms. Normally, the growth medium is heat-sterilized in the bioreactor itself by passing steam through the cooling coils and jacket. The air supplied to the microbial culture is sterilized by filtration.

Aeration. As most of the industrial processes are run under oxic conditions, a very important part of the bioreactor is the aeration system. A rapidly growing culture has a very high demand for oxygen. For example, during growth on glucose, about 500 ml O_2 are required for the synthesis of 1 g dry biomass. The fundamental problem in supplying sufficient oxygen is that this gas has a low solubility in aqueous systems. Only 0.3 mM O_2, equivalent to 9 mg/l, dissolves in 1 l of water at 20°C in an air/water mixture at 1 atm (101 kPa). The solubility of oxygen decreases as the temperature and concentration of dissolved solutes increase. For large-scale cultures with high cell densities, the oxygen demand of the bacteria can be met only by forced aeration. In practice, this is achieved by blowing sterile air through the culture. The efficiency with which oxygen is transferred from air bubbles to the liquid phase principally depends on the surface-area-to-volume ratio of the air bubbles and the residence time of the bubbles in the liquid. The smaller the bubbles, the greater will be the surface-area-to-volume ratio and the greater will be the oxygen transfer rate. Similarly, the longer the bubbles remain in the liquid, the greater will be the amount of oxygen that will diffuse into the liquid. One way of decreasing bubble size is to introduce the air through a sparger with multiple small orifices rather than through a single large-bore tube. A second way for increasing the oxygen supply is to agitate the culture broth vigorously with a stirrer. In order to ensure most effective mixing by the impeller, baffles are installed vertically along the inside diameter of the bioreactor. Agitation and aeration of the culture medium can result in excessive foaming; therefore, antifoam control sys-

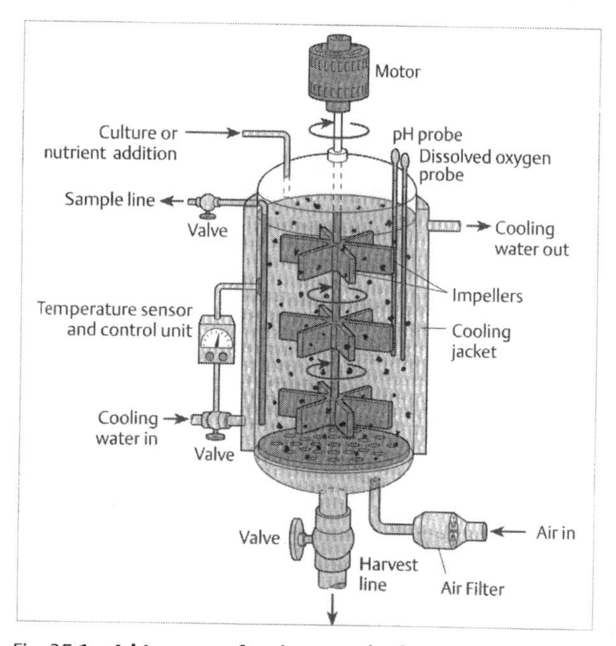

Fig. 35.**1** **A bioreactor for the growth of microorganisms in industrial processes.** Specially designed stirred bioreactors that can be run under oxic or anoxic conditions are often used. Nutrient additions, sampling, and process monitoring can be carried out aseptically

tems are important. Since agitation and aeration used up a considerable part of the input of energy and costs, other systems (e.g., free of stirrers) such as the air-lift bioreactor, have been developed.

Carbon sources. Media used for the cultivation of microorganisms must contain all elements necessary for the synthesis of all material and for the production of metabolic compounds (see Chapters 6 and 8). As 25–70% of the total cost of the fermentation may be due to the carbon and energy source, in industrial processes almost invariably undefined substrate mixtures are used. For example, molasses, a by-product of sugar production, is one of the least expensive sources of carbohydrates used in industrial fermentations. In addition to a large amount of sucrose and raffinose, molasses also contains nitrogen sources, various vitamins, and trace elements. However, a great disadvantage using this substrate is that there is a considerable variation in the composition of molasses depending on the raw material (sugar beet, sugar cane) used for sugar production. Malt extract is an excellent carbon and energy source for many fungi, yeasts, and actinomycetes. In addition, glucose syrup produced by the enzymatic hydrolysis of starch is also frequently used as a fermentation substrate.

Nitrogen sources. As a nitrogen source, "corn steep liquor" is frequently used, which is formed during starch production from corn (maize). The concentrated extract contains about 4% nitrogen, mainly as amino acids, such as alanine, arginine, glutamic acid, isoleucine, threonine, valine, and phenylalanine. Soy meal, the residue from soybeans after the extraction of oil, is often used in antibiotic fermentation. This complex substrate, containing about 50% protein and 30% carbohydrate, is metabolized slowly; thus, carbon catabolite regulation does not occur.

Large-scale industrial bioreactors have to provide optimal culture conditions, such as sufficient oxygen supply, optima of pH, temperature, and substrate supply. Contamination of the cultures must be avoided.

35.2.2 In Modern Industrial Production Processes, at Least pH, Oxygen Tension, and Temperature Are Controlled

Bacteria in nature are subject to a wide range of extracellular factors, such as O_2 tension, temperature, pH, carbon/nitrogen source and micro-element availability (see Chapter 30). The influence of each of these factors has been derived from traditional physiological and biochemical investigations. In all industrial processes, it is essential to optimize total productivity. This can only be achieved by identifying and controlling the many factors that are known to regulate the activity of organisms. Thus, instrumentation of bioreactors has become increasingly important for measuring specific parameters, recording them, and then using this information to improve and optimize the process. The physical and chemical parameters listed in Table 35.**3** can either be measured directly ("on-line") at the bioreactors or can be measured "off-line" in the laboratory. The rate of oxygen uptake within the bioreactor can be measured as the oxygen concentrations in the inlet and outlet gas streams using a paramagnetic oxygen analyzer or a mass spectrometer; also oxygen electrodes are used (Fig. 11.**1**), but the durability varies. Furthermore, electrodes are now available for the determination of dissolved CO_2 levels in the culture broth. Gaseous CO_2 can be measured on-line with spectrophotometric, gas chromatographic, or mass spectrometric methods. There are many types of sensors available to monitor temperature in bioreactors. The energy input from the stirring, aeration, and metabolic oxidation processes (about 50–80 kJ for the synthesis of 1 kg biomass per h) must be removed by a cooling system. The metabolism of most microorganisms results in a change in the pH of the culture medium. Usually, the pH value of the fermentation broth is monitored continuously with a pH electrode, and a fixed pH is maintained by the addition of acid or alkali. Automated sampling devices now allow for on-line analysis of product concentrations and other compounds that can be analyzed by gas chromatography, pressure liquid chromatography, or flow injection analysis. The biological parameters listed in Table 35.**3** must be measured outside of the bioreactor with the exception of the $NADH_2$ measurement, which can be done on-line using a fluorescent method.

Computers serve a variety of functions in fermentation process analysis and control. They can analyze or process the data, present the analysis on display devices, and store it or use it for process control by signaling activation switches, valves, and pumps. Data processing operations include, for example, calculation of rates, yields, productivity, and respiratory quotients. A fully computerized integrated fermentation system requires detailed process models that can detect and respond to changes in culture conditions that may influence cell physiology and productivity. Furthermore, a thorough understanding of the elements and mechanisms controlling the biosynthesis and transport of a

Table 35.3 **Parameters measured in biotechnological processes**

Physical parameters	Chemical parameters	Biological parameters
Temperature	pH	Metabolites
Pressure	Dissolved O_2	Enzyme activity
Power consumption	O_2 and CO_2 in waste gas	DNA and RNA content
Viscosity	Redox potential	$NADH_2$ and ATP content
Flow rates (air and liquid)	Substrate concentration	Protein content
Turbidity	Product concentration	
Weight of fermenter	Ionic strength	

metabolite should make it possible to influence its rate of overproduction predictably. Techniques are currently being developed for the in vivo quantification of carbon fluxes and their control in microbial cell.

Instrumentation of bioreactors is very important for measuring and controlling the many factors that influence the activity of microorganisms; computers now serve a variety of functions in fermentation process analysis and control. In the field of metabolic engineering, it is essential to combine increased knowledge about substrate uptake, metabolic networks, and product secretion with improved biochemical engineering and mathematical modeling.

35.3 Food Industry Is an Important Branch of Biotechnology

Microorganisms have been used for centuries to modify foodstuffs. Fermented foods and beverages constitute a major and important sector of the food industry. This chapter will describe some of the applications of fermentation processes in obtaining dairy products, processing meat, and manufacturing vinegar. Alcoholic beverages are not discussed here as they are produced exclusively with yeasts.

35.3.1 Dairy Products Come From Lactic Acid Fermentation

The manufacture of cultured dairy products is considered to rank second only to the production of alcoholic beverages among the industries that rely on microbiological processes. **Milk fermentation**, usually carried out by various species of *Streptococcus* and *Lactobacillus*, generally causes the dissimilation of lactose to lactic acid. Other reactions that may occur, either during the main fermentation or post-fermentation reactions, produce distinctive milk products. While hard and soft cheeses represent the most important cultured dairy products, other significant product types include yoghurts, sour cream, buttermilk, and kefir. The final product depends on, in addition to the organisms, the character and intensity of the fermentation reactions. There are several major fermentation products that may occur in milk: lactic acid, propionic acid, citric acid, and alcoholic and butyric acid. Of these, lactic acid is the most important and occurs in all milk fermentations. Lactose present in milk is hydrolyzed to glucose and galactose and then fermented to L-, D-, or DL-lactic acid. The formation of a curd by casein at its isoelectric point (pH 4.6) by lactic acid is important in cheese production. Furthermore, by its low pH, soured milk is protected from proteolytic bacterial spoilage. The propionic acid fermentation is important in swiss cheese production as the propionic acid and carbon dioxide formed lead to the typical cheese flavor and hole formation. The particular flavor associated with buttermilk and sour cream is due to a citric acid fermentation. The flavor results from a balance of diacetyl, propionic, and acetic acids, and other related compounds.

Milk fermentations, used for thousands of years, are performed by the indigenous bacteria present in milk. However, today starter cultures (i.e., selected strains) are used to produce more predictable qualities and characteristics in the various milk products. Furthermore, problems related to bacteriophage infection (induction) can be overcome by using bacteriophage-

Table 35.4 **Examples of the main lactic acid bacteria used in milk fermentations**

Product	Organisms
Yoghurt	*Lactobacillus delbrueckii,* sspec. *bulgaricus,* *Streptococcus thermophilus*
Emmenthaler (Swiss) cheese	*Lactobacillus helveticus*
Sour cream, milk	*Lactococcus lactis* sspec. *lactis,* *Lactococcus lactis* sspec. *lactis* biovar *diacetylactis,* *Lactococcus lactis* sspec. *cremoris, Leuconostoc lactis,* *Leuconostoc mesenteroides* sspec. *cremoris*
Gouda and Edam cheese	Same as for sour cream
Kefir	*Lactobacillus kefir,* *Lactobacillus kefiranofaciens,* *Lactococcus lactis*

insensitive (or bacteriophage-free) strains. Commercial starter cultures for the dairy industry consist of various lactic acid bacteria (Tab. 35.**4**).

Cheese production is one of the oldest fermentations known. It is a method of preserving the nutritional value of milk, and has been described by Greek and Roman writers. Although cheese properties vary quite considerably, a number of production steps are common to all. The first is the pasteurization of milk and then the inoculation with starter cultures. This is followed by curdling of the milk, which is generally achieved by the combined processes of acidification by lactic acid formation and of milk-clotting enzymes, such as calf rennet. The milk-coagulating enzymes are characterized by their ability to hydrolyze the κ-casein fraction of milk specifically without attacking the other major casein fractions. κ-Casein in milk stabilizes the milk casein micelles in the presence of calcium in a colloidal suspension. After coagulation, the curd is separated from the whey, and is salted and pressed into forms. Special applications are used for individual cheese types; these include inoculation with various mold spores or bacteria. Ripening allows the microorganisms and enzymes in the curd to hydrolyze fat, protein, and other compounds present. The breakdown of these materials produces the characteristic flavors of the cheese. Cheese ripening can be accelerated by the addition of enzymes (proteinases, lipases) or by using bacterial mutants that produce more of these enzymes.

Yoghurt is prepared from heat-treated milk by inoculation with *Streptococcus thermophilus* and *Lactobacillus bulgaricus* in about equal numbers in the culture to obtain a desirable consistency, flavor and odor. Fermentation at 42–45°C for 4h generates sufficient acidity (pH approx. 4.0) for a satisfactory shelf-life of natural yoghurt. Fermentation and associated equipment in modern yoghurt production is made of stainless steel, and has a capacity up to 10,000l in the largest plants.

The most important **dairy products**, such as hard and soft cheeses, yoghurts, and sour cream, are produced today by the use of commercial starter cultures consisting of a mixture of lactic acid bacteria. *Streptococcus* and *Lactobacillus* species ferment lactose to lactic acid and flavor compounds.

35.3.2 Sauerkraut Is Made From Cabbage

Sauerkraut ("acid cabbage") fermentation is a good example of a natural mixed-culture fermentation in a solid-state matrix. The process starts with shredded cabbage supplemented with 2.5% NaCl placed into a container with a small opening, such that anaerobiosis can be achieved easily by sealing the opening. The microorganisms in the mixture consist mainly of enteric and to a lesser extent of lactic acid bacteria. Complete anoxic conditions are achieved by respiration of the plant materials themselves and consumption of oxygen by aerobic microorganisms. The microbial flora changes to enteric bacteria, i.e., facultatively anaerobic organisms. Since cabbage material has poor buffering capacity, the decrease in pH due to fermentation of lactic acid bacteria inactivates the enteric bacteria and stimulates the further growth of lactic acid bacteria, such as *Leuconostoc mesenteroides*. As this heterofermentative organism (see Chapter 12.2.2) grows along with other microbes, lactic acid, acetic acid, ethanol, D-mannitol, dextran, esters, CO_2, and other compounds, are formed. These compounds stimulate the growth of *Lactobacillus plantarum* and inhibit the growth of undesirable organisms, such as yeasts, in the developing sauerkraut. The last stage of sauerkraut fermentation invariably involves the growth of *L. plantarum*, a homofermentative lactic acid bacterium (see Chapter 12.2.2) that produces the desired final acidity of 1.7% lactic acid. *L. plantarum* also utilizes D-mannitol and thus removes its bitter flavor from the sauerkraut. Sauerkraut fermentation is an example of a microbial succession in naturally fermented food production that utilizes a solid-state fermentation system. A starter culture is not necessary for this fermentation because the predominant bacterium at the end of the fermentation is always *L. plantarum*, regardless of the presence or absence of a starter culture.

Sauerkraut production is an example of a natural mixed-culture fermentation in which enteric and lactic acid bacteria are involved; however, in the last stage of the process, the homofermentative lactic acid bacterium *L. plantarum* is the predominant organism.

35.3.3 Meat Fermentation Results in Flavored and Preserved Products

The majority of fermented meat products consist of dry or semi-dry sausages. Starter cultures in meat fermentation provide lactic acid bacteria to produce the lactic acid required for flavor and low pH. The lower pH increases the juiciness of the product and denatures the meat protein, contributing to the characteristic firm texture. It also antagonizes some pathogens, such as *Salmonella* and *Staphylococcus aureus*. Furthermore, starter culture addition lowers histamine levels and extends the shelf-life of meat. Commercial inoculants, such as *Pediococcus* species, which have good lactic-acid-producing ability, are frequently used. *Staphylococcus carnosus*, a harmless coagulase-negative species, is routinely used in the production of dry sausage. *Micrococcus* species have the ability to reduce nitrates to nitrites in a controlled manner and contribute to *Clostridium botulinum* control. Thus, starter cultures may be used in bacon processing to dissipate any residual nitrite present, thereby lowering or eliminating carcinogenic nitrosamine formation during frying. Meat fermentation technology is being applied to traditionally non-fermented products to enhance flavor development, extend shelf-life, and provide control over food pathogens and toxic chemicals. Research is going on to select rapid acid-producing strains of *Pediococcus* and *Lactobacillus* to shorten fermentation schedules at a wide range of temperatures (20–45°C).

In the production of dry sausages, lactic acid bacteria are important as lactic acid is required for flavor and inhibition of pathogens.

35.3.4 Vinegar Is Produced From Ethanol-containing Liquids

Vinegar, an aqueous solution of acetic acid, is produced by bacterial oxidation of a dilute ethanol solution. The Romans and Greeks, who used diluted vinegar as a refreshing drink, produced it by leaving wine open to air. Vinegar can be produced from any ethanol-containing substance. Although the usual starting materials are wine or cider, it can also be produced from diluted purified ethanol. Vinegar is used as a flavoring ingredient in salads and other foods. Meat and vegetables pickled in vinegar can be stored for years. The annual production of vinegar worldwide is in the range of $1.6 \cdot 10^9$ liters.

The obligate aerobic acetic acid bacteria, which oxidize ethanol to acetic acid, can exist at low pH values; they belong to the closely-related genera *Acetobacter* and *Gluconobacter*. Industrial cultures that tolerate high acidity (13–14% acetic acid) and yield high acetate production rates are selected. Acetic acid formation in these species (see Chapter 9.18) is an incomplete oxidation; oxygen is used as electron acceptor. As shown in Figure 35.**2**, the metabolic process involves conversion of ethanol to acetaldehyde by alcohol dehydrogenase (Adh) and of hydrated acetaldehyde to acetic acid by acetaldehyde dehydrogenase (Ald). Thus, 1 mol of acetic acid is produced from 1 mol of ethanol; from 1 l of 12% (v/v) ethanol, 1 l of 12.4% (v/v) acetic acid is produced. For optimal production, sufficient oxygen is required for the oxidation of $NADH_2$ via the respiratory chain.

There are three different biotechnological processes for the production of vinegar. The **open-vat** or **Orleans method** was the original process and is still used in France. Wine is placed in shallow vats with considerable exposure to air, and the acetic acid bacteria develop as a slimy layer on the top of the liquid. This process is rather

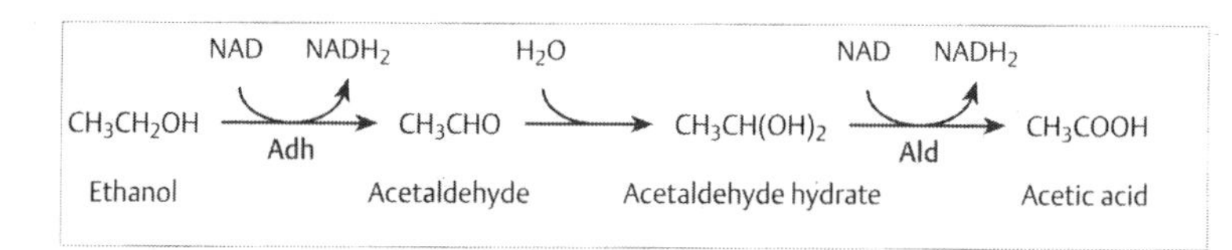

Fig. 35.2 **Oxidation of ethanol to acetic acid by *Acetobacter aceti*.** Adh, alcohol dehydrogenase; Ald, acetaldehyde dehydrogenase

slow since the bacteria come in contact with both the air and the substrate only at the surface. The second system is the **trickling generator process** in which the contact between the bacteria, air, and ethanol is increased by trickling the alcoholic liquid over beechwood shavings (Fig. 35.**3**). The bacteria grow upon the surface of the wood shavings, while the air enters the trickling generator at the bottom and passes upward. Of the ethanol added, 90% is converted to acetic acid at 30°C. The time needed to produce 12% acetic acid in this process is about three days. The third method is a submerged process, which uses the so-called **Frings acetator**. It is widely used for commercial vinegar production by means of a baffled bioreactor containing a bottom-driven turbine for intensive aeration. Typical commercial processes, involving production of 12–15% acetic acid, are carried out semi-continuously. Acetic acid and ethanol concentrations at the start of the cycle are 7–10% and 5%, respectively. Fermentation proceeds at 27–32°C until the alcohol concentration drops to 0.1–0.3%, at which point about one-third of the vinegar is discharged and the vessel is filled with new mash containing 0–2% acetic acid and 12–15% ethanol, and the cycle is repeated. Both acetic acid and ethanol must be present for optimal growth of *Acetobacter*. The ethanol supply is critical, and with less than 0.2% ethanol, the death rate increases; however, the maximal ethanol content should not exceed 5%. The efficiency of the process is high: 90–98% of the alcohol is converted to acid.

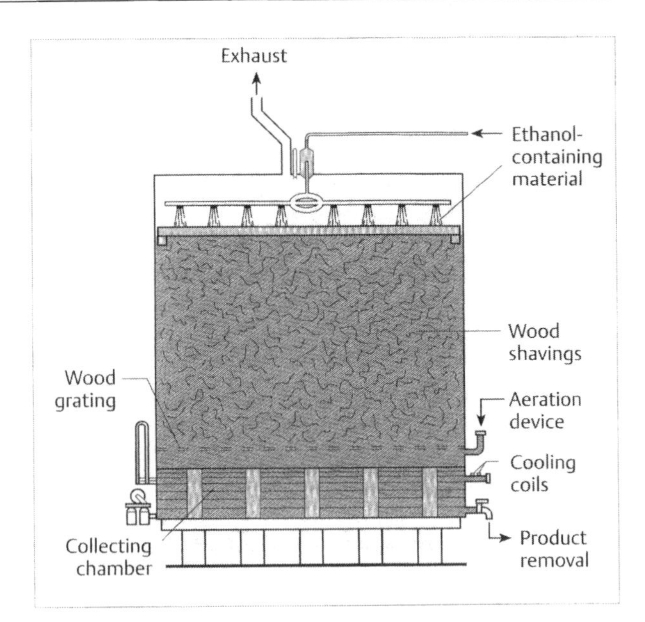

Fig. 35.3 **Diagram of a bioreactor for vinegar production.** See text for details

For the production of vinegar, ethanol is oxidized to acetic acid by the aerobic bacteria *Acetobacter* and *Gluconobacter*, which tolerate high acidity and require an adequate oxygen supply.

35.4 Production of L-Amino Acids and Vitamins Is not a Classical Fermentation Process

In contrast to the fermentation products described before, the production of L-amino acids and vitamins has been developed on the basis of careful studies of metabolic pathways and selection of strains.

35.4.1 Overproduction and Secretion of L-Amino Acids by *Corynebacterium glutamicum* Under Selective Conditions

Demands for amino acids for use in the areas of food and feed additives and drug manufacturing have increased. In medicine, amino acids are used for infusions and as therapeutic agents. Amino acid derivatives are also used in the chemical industry, such as in cosmetics, synthetic leathers, surface-active agents, fungicides, and pesticides. The production methods developed so far may be summarized as follows: (1) protein hydrolysis, (2) chemical synthesis, (3) microbiological production, for example, from carbon sources such as glucose or from chemically synthesized precursors, and (4) enzymatic synthesis. Whereas chemical synthesis produces a racemic mixture, which may require additional resolution, the latter procedures give rise to optically pure amino acids. Because of their importance, the microbial production of L-glutamic acid and L-lysine will be described in this chapter.

Many bacteria are capable of growing on a simple mineral salt medium with glucose as the sole carbon and energy source, ammonium, and phosphate. These bacteria are able to synthesize all the compounds necessary for the living cell from these simple nutrient components. Numerous analyses have indicated that the dry matter of a bacterial cell consists of about 60%

protein, 20% nucleic acids, 10% carbohydrates, and 10% fat. Since the bacterial cell contains very large quantities of protein, it must be able to synthesize amino acids rapidly and efficiently. However, as a rule, only as much of the various amino acids are required for growth are formed in the bacterial cell, i.e., normally, the bacteria do not overproduce and excrete these amino acids into the culture medium. As has been shown by biochemical and molecular biological studies, bacteria have regulatory mechanisms (repression and feedback inhibition through end products) that control the production and excretion of metabolites economically (see Chapters 7.6, 8.6, 9.2).

L-Glutamic acid. Following the increasing demand for monosodium glutamate as a flavoring agent in the mid-1950s, a bacterium was isolated in Japan that excreted large quantities of the amino acid L-glutamic acid into the culture medium. This bacterium, *Corynebacterium glutamicum*, is a short, aerobic, Gram-positive rod capable of growing on a simple mineral salt medium with glucose, provided that biotin is also added. The production of L-glutamic acid is maximal at a critical biotin concentration of 0.5 μg/g of dry cells, which is suboptimal for growth. Excess biotin, which supports abundant growth, decreases the L-glutamic acid accumulation. Biotin is a prosthetic group for the enzyme acetyl-CoA carboxylase, an enzyme also involved in the biosynthesis of fatty acids (Fig. 35.**4**). Thus, limited amounts of biotin cause changes of the fatty acid composition of the cell membrane. The total amount of fatty acids as well as the phospholipid content in glutamate-producing cells are only about half of that of nonproducers grown in a medium with excess biotin. Consequently, the lipid content of the cell membrane was thought to be involved in the regulation of the secretion of L-glutamic acid. Recently, it was demonstrated that L-glutamic acid is not secreted via passive diffusion, but is secreted via a specific active transporter (Fig. 35.**4**). The production rate of L-glutamic acid by *C. glutamicum* is correlated to the high secretion of this amino acid induced by limiting the supply of biotin.

In *C. glutamicum*, glucose is mainly metabolized via the glycolytic pathway into C_3 and C_2 fragments. The key precursor of L-glutamic acid is α-oxoglutarate (α-ketoglutarate), which is formed in the citric acid cycle and then converted to L-glutamic acid by a reductive amination. The enzyme catalyzing this conversion is the NADP-dependent glutamate dehydrogenase. Strains used commercially for L-glutamic acid production have a very low α-ketoglutarate dehydrogenase activity; thus,

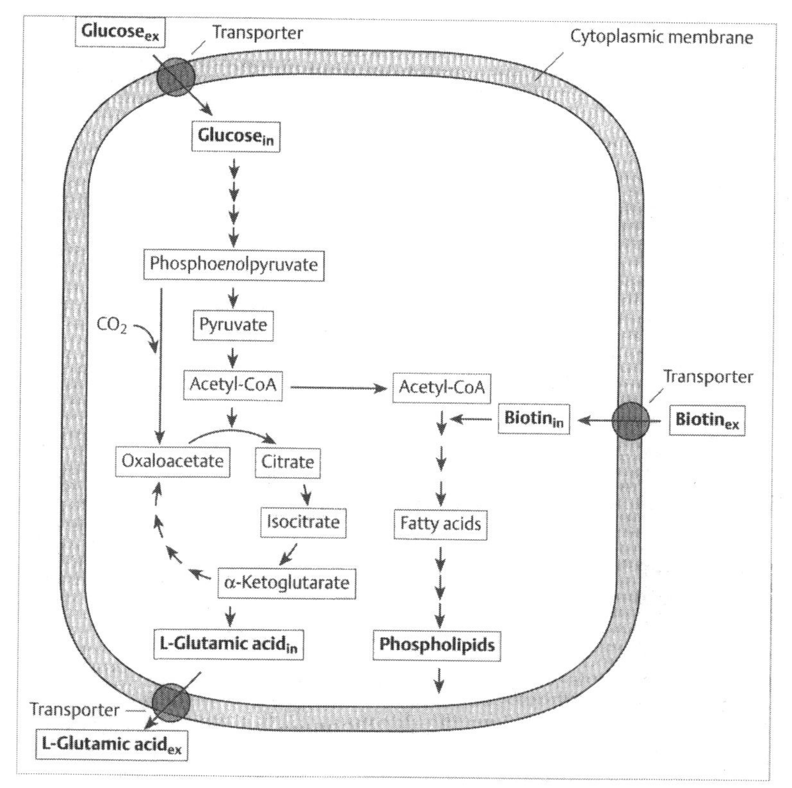

Fig. 35.**4** **Influence of biotin on L-glutamate excretion in *Corynebacterium glutamicum*.** Limiting the supply of biotin causes a reduced phospholipid content of the cytoplasmic membrane and an increased production of L-glutamate (see Figs. 7.**2** and 7.**13**)

the intermediate α-ketoglutarate is only partially further metabolized in the citric acid cycle, and is mainly converted into L-glutamic acid. So far, very little is known about the regulation of these two enzymes in *C. glutamicum*. Oxaloacetate is formed via the phosphoenolpyruvate carboxylase reaction. Mutants with an increased activity of this enzyme showed improved productivity of L-glutamic acid. The overall reaction for L-glutamic acid production from D-glucose is:

$$C_6H_{12}O_6 + NH_3 + 1.5\,O_2 \rightarrow C_5H_9O_4N + CO_2 + 3\,H_2O \qquad (35.1)$$

Thus, the theoretical maximal yield is one mol of L-glutamic acid per mol of glucose metabolized. This represents a 100% molar conversion or 81.7% weight conversion of D-glucose to L-glutamic acid.

The L-glutamic acid production is carried out in stirred baffled bioreactors up to a size of $150\,m^3$. Provision for cooling, dissolved oxygen measurement, and pH measurement and control (usually with ammonium) are required. A temperature between 30 and 35°C and a pH between 7.0 and 8.0 are optimal. The oxygen transfer rate is fairly critical: a deficiency leads to poor glutamate yields, with lactic and succinic acids being formed instead, while an excess causes accumulation of α-ketoglutaric acid. The yield of L-glutamic acid obtained after 2–3 days of incubation is in the order of 50–60% (by weight) of the sugar supplied, and the final concentration is approximately 100 g/l. The annual production exceeds 800000 tons, and the chief use of this amino acid in the form of its monosodium salt is as a flavor enhancer in the food industry.

Bacteria normally do not excrete amino acids in significant amounts into the culture medium since regulatory mechanisms control the synthesis and excretion economically. In *C. glutamicum*, a high **production of L-glutamic** acid can be induced by limiting the supply of biotin required by this bacterium for growth. The citric acid cycle is limited at α-ketoglutarate dehydrogenase, causing a high L-glutamic acid production.

L-Lysine. L-Lysine, an amino acid essential for human and animal nutrition, is mainly used as a feed supplement. Furthermore, it finds pharmaceutical applications in the formulation of diets with balanced amino acid compositions and in amino acid infusions. At present, approximately 300000 tons per year of L-lysine are produced using strains of *C. glutamicum* or subspecies. The wild-type strains of these bacteria do not secrete L-lysine into the culture medium. High-yiel strains were developed through mutation to auxotroph and to antimetabolite resistance.

The pathway for the biosynthesis of L-lysine in (*glutamicum* is illustrated in Figure 35.**5** (see also Fi 7.**15** and Chapter 19.2). The first enzyme, aspartokinas is regulated by concerted feedback inhibition by l threonine and L-lysine. L-Threonine causes feedbac inhibition of homoserine dehydrogenase, while l methionine represses synthesis of this enzyme. Henc a homoserine auxotroph or a threonine and methionin double auxotroph of *C. glutamicum* diminishes th intracellular pool of threonine, reduces its marke feedback inhibitory effect on aspartokinase, and pro motes lysine overproduction (15–30 g/l). Another effec tive technique for obtaining L-lysine-producing strair is the selection of regulatory mutants. Growth of (*glutamicum* is inhibited by an analog of L-lysine, S-(2 aminoethyl)-L-cysteine (AEC). This inhibition is marl

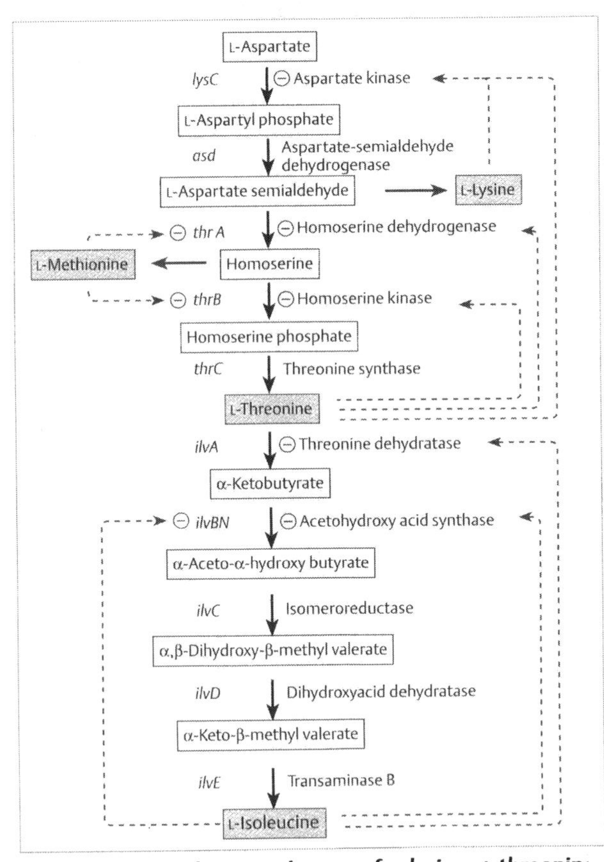

Fig. 35.5 **Biosynthetic pathways of L-lysine, L-threonin and L-isoleucine and their regulation in *Corynebacteriu glutamicum*.** The enzyme activity is regulated by feedbac inhibition of products and repression of respective genes. Se text for details

edly enhanced by L-threonine, but reversed by L-lysine. This implies that AEC behaves as a false feedback inhibitor of aspartokinase. Some mutants, which are capable of growing in the presence of both AEC and L-threonine, contain an aspartokinase that is insensitive to the concerted feedback inhibition; therefore, L-lysine is overproduced (30–35 g/l). L-Aspartate used for L-lysine formation is formed from oxaloacetate by the anaplerotic reaction of phospho*enol*pyruvate carboxylation. A typical time course of L-lysine production with such a mutant of *C. glutamicum* is shown in Figure 35.**6**. The L-lysine concentration is approximately 44 g/l, and the conversion rate relative to the sugar used is between 30 and 40%. By combined overexpression of aspartokinase and dihydrodipicolinate synthase, L-lysine production can be increased by 10–20%.

In addition to all steps considered so far, the secretion of L-lysine into the culture medium must also be noted. Secretion of L-lysine is not the consequence of nonspecific permeability of the cytoplasmic membrane; secretion is mediated by an excretion transporter that is specific for L-lysine. In *C. glutamicum*, L-lysine is excreted in symport with two OH^- (Fig. 35.**7**). The substrate-loaded and the unloaded transporter carry different charges. The velocity of L-lysine excretion is thus influenced by several forces at different individual steps of the translocation cycle, i.e., by the membrane potential, the pH gradient, and the L-lysine gradient. This transporter is a system well-designed for excretion purposes:

1. It has a high K_m value for L-lysine (20 mM) at the internal (cytoplasmic) side, thus preventing unwanted efflux under low internal lysine concentration.

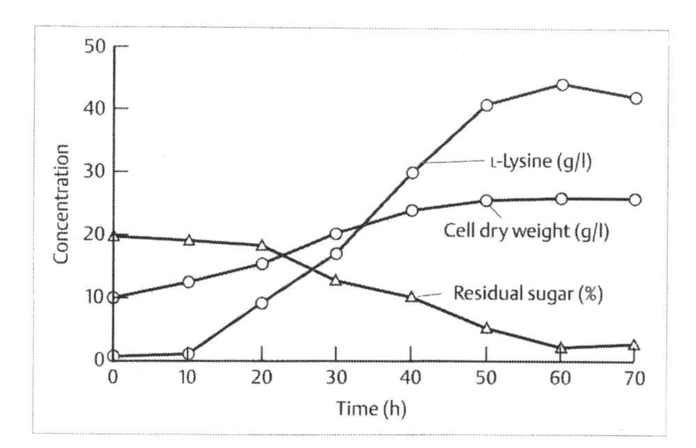

Fig. 35.**6** **Time course of L-lysine fermentation with a mutant of *Corynebacterium glutamicum*.** ○ Lysine production (g/l), ○ cell dry weight (g/l), △ glucose consumption (%)

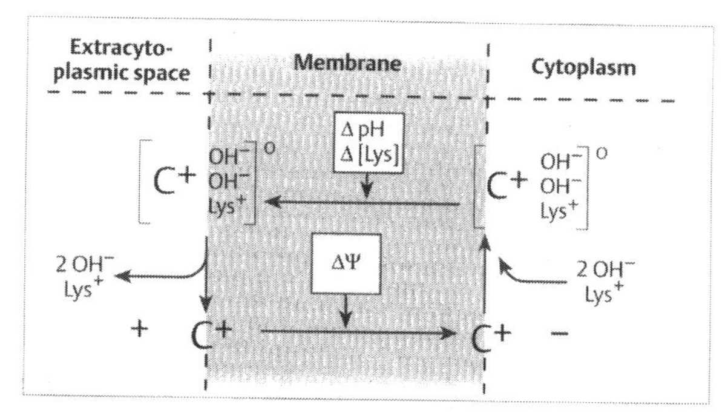

Fig. 35.**7** **Putative mechanisms of L-lysine excretion in *Corynebacterium glutamicum*.** The transporter (C) accepts the substrates L-lysine (Lys^+) and OH^- at the cytoplasmic side and, after re-orientation of the binding site (conformational change), releases the products to the external medium. The transport cycle is modulated by the membrane potential ($\Delta\Psi$), the pH gradient (ΔpH), and the L-lysine gradient (ΔLys) at different steps

2. It is coupled to H^+ and OH^- in a direction opposite to that of uptake systems (see Table 5.**2**).
3. The unloaded carrier is positively charged; thus, the membrane potential is able to drive excretion of L-lysine.

For **L-lysine production**, strains of *C. glutamicum* have been isolated which are inactive in homoserine dehydrogenase or deregulated in aspartokinase. Secretion of L-lysine is mediated by a specific excretion transporter.

An example of strain improvement using recombinant DNA techniques is the amplification of the L-threonine biosynthetic genes in *C. glutamicum*. Amplification of the feedback-inhibition insensitive homoserine dehydrogenase and homoserine kinase in an L-lysine-overproducing strain permits channeling of the carbon flow from the intermediate aspartate semialdehyde toward homoserine, resulting in a high accumulation of L-threonine. The final L-lysine concentration is decreased from 65 g/l to 4 g/l, while the final L-threonine concentration is increased from 0 g/l to 52 g/l (Fig. 35.**8**).

35.4.2 Production of Vitamin B_{12} and Vitamin C Are of Economic Significance

Vitamins are growth cofactors that are increasingly being introduced as food or feed additives, as medical or therapeutic agents, and as health aids. Although a wide variety of vitamins can be produced by microorganisms,

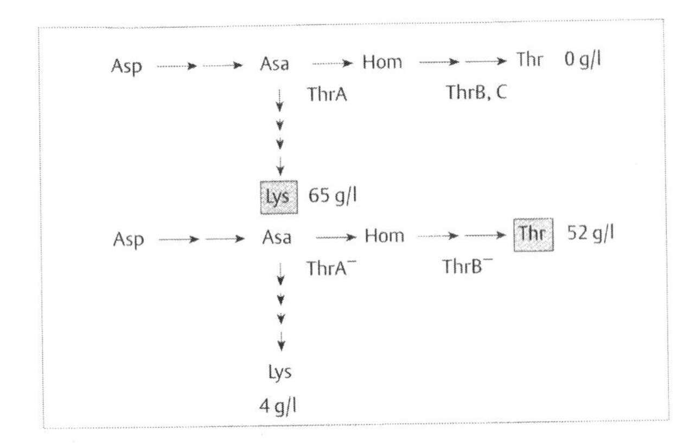

Fig. 35.8 **Diversion of carbon flow by amplification of the *thr* operon for L-threonine biosynthesis into an L-lysine-producing strain of *Corynebacterium glutamicum*.** Asa, aspartate semialdehyde; Hom, homoserine; ThrA, homoserine dehydrogenase; ThrB, homoserine kinase; $ThrA^-$, feedback-inhibition-insensitive homoserine dehydrogenase; ThrC threonine synthase; $ThrB^-$, homoserine kinase feedback in-sensitive

only the bacterial production of vitamin B_{12} and ascorbic acid (vitamin C) are of economic significance so far. **Vitamin B_{12}** (5′-deoxyadenosylcobalamin) is synthesized in nature exclusively by bacteria (Fig. 35.**9**). The requirement of animals for this vitamin is covered by feed intake or by absorption of vitamin B_{12} produced by intestinal bacteria. However, humans obtain vitamin B_{12} only from food since vitamin B_{12} synthesized by microorganisms in the lower intestinal tract cannot be absorbed. Vitamin B_{12} used for human therapy ("pernicious anemia") and as a food or feed supplement was first obtained as a by-product of antibiotic production with *Streptomyces* strains. As the yield was only about 1 mg/l and the demand for vitamin B_{12} increased, bacterial strains with higher yields were selected. For industrial purposes, *Propionibacterium freudenreichii*, *P. shermanii*, and *Pseudomonas denitrificans* are used.

The microaerophilic *Propionibacterium* species produce cobalt corrinoids in conventional media (e.g., in molasses or in corn steep liquor) supplemented with cobalt in the absence of aeration. As these bacteria can synthesize 5,6-dimethylbenzimidazole under oxic culture conditions, a two-stage process was developed. In the first anaerobic stage (2–4 days), the bacteria grow and produce 5-deoxyadenosylcobinamide (Fig. 35.**9**). Subsequently a shift to the aerobic phase (3–4 days) promotes the biosynthesis of 5,6-dimethylbenzimidazole so that 5-deoxyadenosylcobalamin (coenzyme B_{12}) can be produced (40 mg/l). *P. denitrificans* is used in a one-stage process; bacterial growth parallels vitamin B_{12} synthesis under oxic conditions when cobalt and 5,6-demethylbenzimidazole are added as supplements. By mutation and selection, strains have been obtained that produce about 150 mg vitamin B_{12}/l. The current

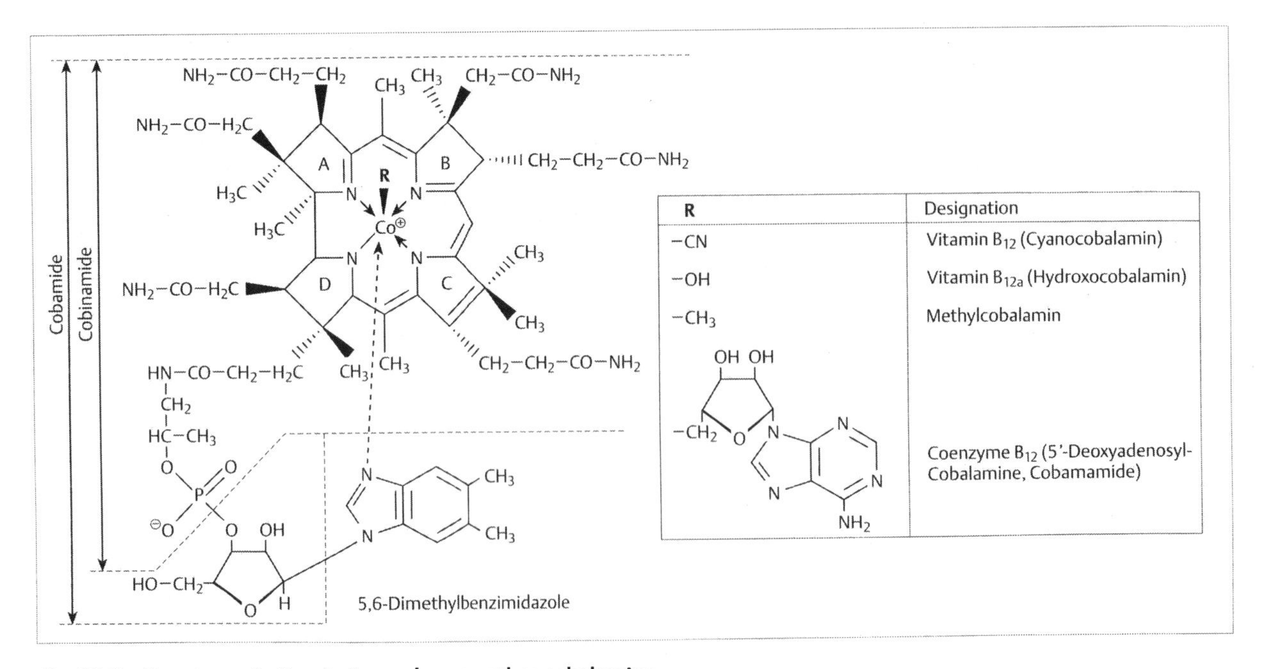

Fig. 35.9 **Structure of vitamin B_{12} and some other cobalamins**

annual world production of this vitamin is estimated at about 12000kg.

Vitamin C (L-ascorbic acid) is important in human and animal nutrition, in medicine, and it is used as an antioxidant in the food industry. Total world production of vitamin C is estimated at 70000 tons per year. It is currently produced by the well-established Reichstein-Grüssner synthesis. This process consists of several chemical steps and one microbial reaction. In the first step, D-glucose is converted into D-glucitol (sorbitol) by chemical hydrogenation. The oxidation of D-glucitol to L-sorbose is carried out by *Gluconobacter oxidans* in a submerged process with vigorous stirring and aeration for sufficient oxygen supply. A quantitative conversion of about 200g D-glucitol/l is completed after about 24h. The L-sorbose is then condensed with acetone to form sorbose diacetone, which is oxidized to 2-keto-L-gulonic acid, which is then converted into L-ascorbic acid. Other industrial microbial processes use a controlled axenic mixed system (*Bacillus megaterium* plus *Gluconobacter* spec.) to convert L-sorbose further, directly into 2-keto-L-gulonic acid.

In addition to the Reichstein-Grüssner synthesis, a two-step microbial process has been developed (Fig. 35.**10**). In the first step, glucose is oxidized to 2,5-diketo-D-gluconate by an *Erwinia* strain; the intermediate products are D-gluconate and 2-keto-D-gluconate. In the second fermentation, a species of *Corynebacterium* converts 2,5-diketo-D-gluconate to 2-keto-L-gulonate. This stereospecific reduction at the C-5 position is catalyzed by an NADPH-requiring 2,5-diketo-D-gluconate reductase. In order to develop a one-step microbial bioconversion of D-glucose into 2-keto-L-gulonate, the pathway of *Erwinia* was enhanced using gene cloning techniques. The gene encoding the 2,5-diketo-D-gluconate reductase was cloned from *Corynebacterium* and expressed in *Erwinia*. With optimized culture conditions, these recombinant strains of *Erwinia produce about 120g 2-keto-*L-gulonate/l within 120h. The molar yield from glucose is approximately 60%. As this process is much simpler than either the current multi-step manufacturing process or the two-stage fermentation method, the conversion of glucose to 2-keto-L-gulonate by a recombinant strain of *Erwinia* may lead to an economical process for vitamin C production.

Vitamin B_{12} (cobalamin) is synthesized in nature exclusively by bacteria. For industrial production, *Propionibacterium freundenreichii, P. shermanii,* and *Pseudomonas denitrificans* are used. **Vitamin C** (ascorbic acid) production is a combination of several chemical and biological transformations. The oxidation of D-glucitol (sorbitol) is carried out by *Gluconobacter oxydans*. Recently, a recombinant strain of *Erwinia* has been constructed that converts glucose to 2-keto-gulonate.

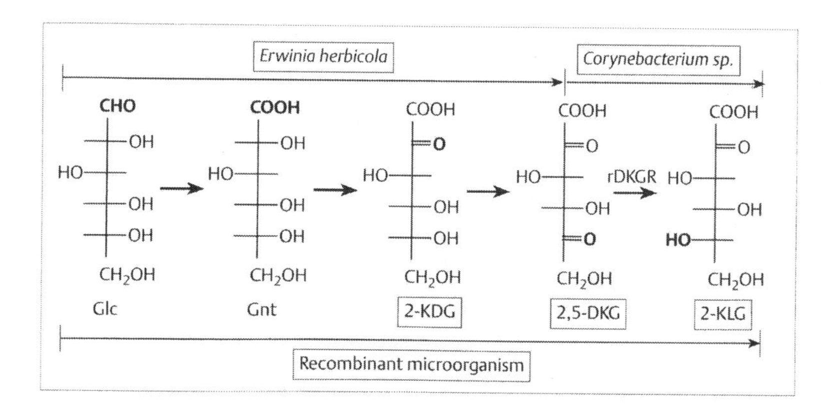

Fig. 35.**10** **Carbohydrate metabolites involved in the bioconversion of D-glucose to 2-keto-L-gulonic acid.** Top: two-step tandem fermentation process; bottom: one-step recombinant process. Glc, D-glucose; Gnt, D-gluconic acid; 2-KDG, 2-keto-D-gluconic acid; 2,5-DKG, 2,5-diketo-D-2-KLG, 2-keto-L-gulonic acid; rDKGR, diekto-Dgluconate reductase, cloned from *Corynebacerium* spec. and expressed in *Erwinia herbicola* for the one-step process

35.5 Steroids and Sterols Can Be Stereospecifically Biotransformed Into the Desirable Hormones

Biotransformations are processes in which microorganisms convert a compound to a structurally related product. They usually comprise only one or a small number of enzymatic reactions, which include hydroxylation, dehydroxylation, epoxidation, oxidation, reduction, hydrogenation, dehydrogenation, esterification, ester hydrolysis, and isomerization. The use of microorganisms for this purpose is preferred to chemical processes when high specificity is required to attack a specific site on the substrate and to prepare a single isomer of a product. This accounts for the high yields typical of biological conversions, which often exceed 90% with microbial cells. Biotransformations proceed at ambient temperatures (20–40°C), an advantage over chemical processes that often call for significant input of energy. Furthermore, in these biological processes, no harmful chemicals are used, and the waste products are biodegradable.

The crucial role of microbes in steroid synthesis has been in connection with the synthesis of the adrenocortical hormones, corticosterone, cortisone, and hydrocortisone, and their therapeutically superior derivatives, such as prednisone, prednisolone, and triamcinolone. Biotransformation centers on the introduction of an oxygen atom at C-11 of the steroid molecules. Following the isolation of cortisone in the 1930s, it was announced in 1949 that the application of cortisone could relieve the pain of patients with rheumatic arthritis. This created a tremendous incentive to provide cortisone by a synthetic route more efficient than extraction from ox adrenal glands. A chemical synthesis from deoxycholic acid was developed, but this route comprised 37 steps and the end product cost $200/g. In 1952, the 11α-hydroxylation of progesterone by a single microbial step was detected using the fungus *Rhizopus arrhizus*. The introduction of this microbial transformation reduced

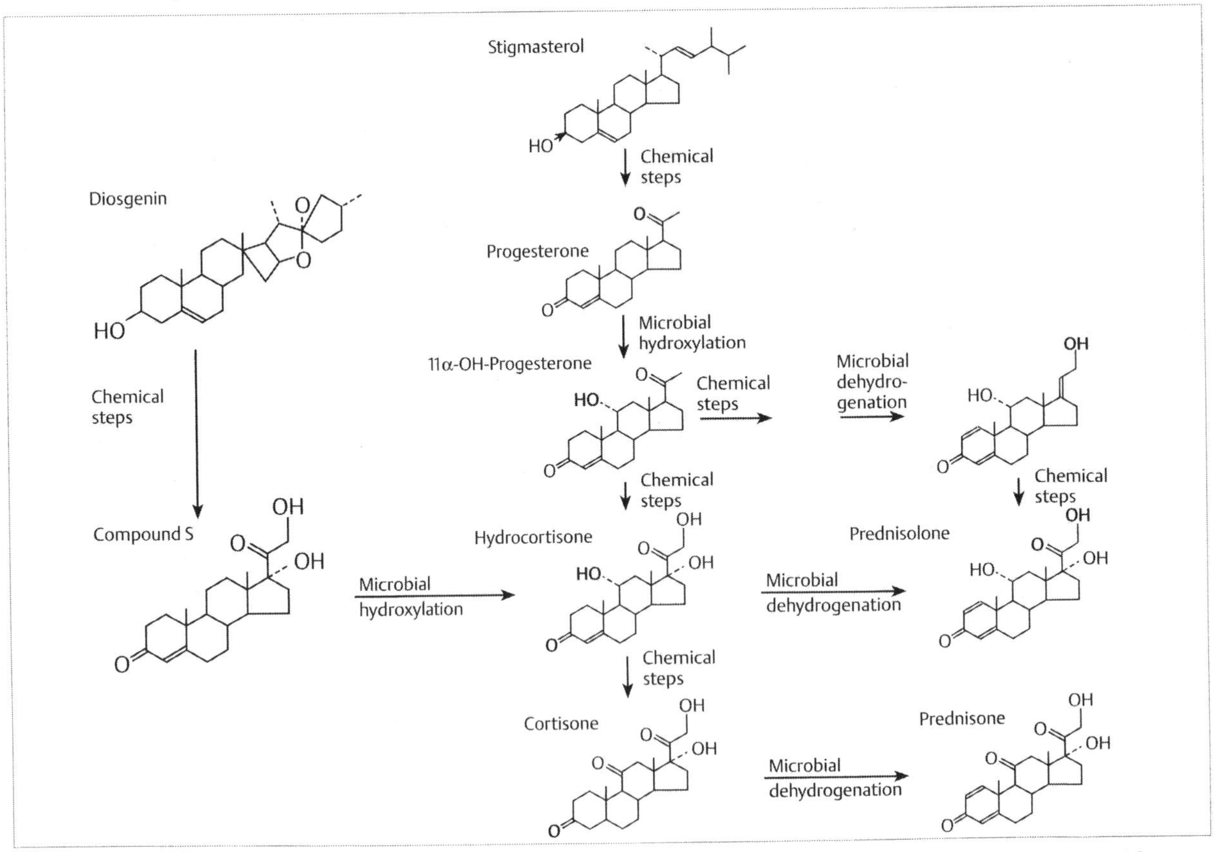

Fig. 35.**11** **Microbial transformations of progesterone.** Diosgenin is extracted from the roots of *Dioscorea composita*. Stigmasterol is extracted from *Glycine max*. The modifications are shown in red

the steps in the synthesis of cortisone to 11, and the cost dropped to \$6/g. Thus, this reaction was decisive for the economic synthesis of adrenocortical hormones and afforded vast possibilities for the preparation of derivatives.

Four biotransformations presently used on an industrial scale are illustrated in Fig. 35.**11**. The 11α-hydroxylation of progesterone, referred to above, is now performed with *Rhizopus nigricans* and yields more than 85% of the desired 11α-hydroxyprogesterone. The 16α-hydroxylation of steroids is effected by *Streptomyces argenteolus*. This reaction is mainly useful for the production of triamcinolone, a 9α-fluorocortisol derivative with an anti-inflammatory activity. The 11β-hydroxylation reaction was first described in the fungi *Cunninghamella blakesleeana* and *Curvularia lunata*; compound S can be 11β-hydroxylated into hydrocortisone with 60–70% yield. Progesterone is by no means the unique substrate for these hydroxylations; Reichstein's compound S and a large variety of steroids are also hydroxylated. The broad substrate specificity of the hydroxylase systems is a great advantage for these processes, allowing selection of the optimal intermediate for hydroxylation in a sequence of chemical and microbial steps to a desired end product. Microbial hydroxylations all involve direct replacement of the hydrogen atom on a given carbon. The hydroxylases are inducible and are both NADPH- and O_2-dependent. The introduction of a double bond in ring A (C-1 dehydrogenation of steroids) can be achieved in high yields using *Streptomyxa affinis* or *Arthrobacter simplex*. This reaction is very important in the production of prednisolone and prednisone (Fig. 35.**12**).

The growing demand for steroids caused a shortage of steroid precursors for bioconversion, such as the compound diosgenin, which is obtained from *Dioscorea composita*. Intensive studies were conducted on the use of low-cost sterols, such as cholesterol or β-sitosterol of animal origin, and stigmasterol from *Glycine max*. Mutants of *Mycobacterium* spec. with inactive C-1-dehydrogenase and/or 9α-hydroxylase are able to transform these sterols into androstendione and androstadiendione, an important intermediate in the synthesis of estrogens. The contribution made by microbial transformations to the overall preparation of clinically important corticoids using the reactions described above is illustrated in Fig. 35.**12**. Progesterone, obtained by the chemical conversion of stigmasterol, is transformed by *Rhizopus nigricans* into 11α-hydroxyprogesterone, a key intermediate. 11α-hydroxyprogesterone is chemically modified to hydrocortisone and cortisone, or dehydrogenated at C-1 by *Streptomyxa affinis* to prednisolone and prednisone, which have high anti-inflammatory activity. Alternatively, diosgenin is chemically converted to Reichstein's compound S, which is 11β-hydroxylated by *Curvularia lunata* into hydrocortisone and further converted into prednisolone by *Arthrobacter simplex*.

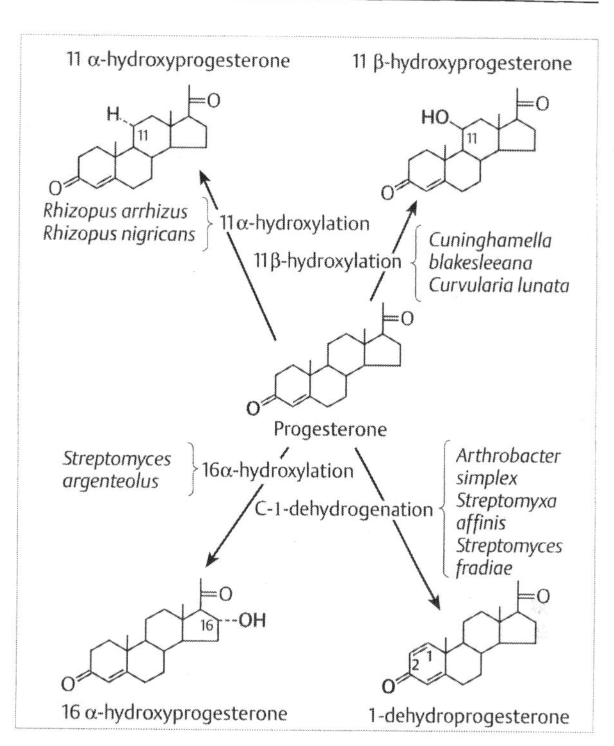

Fig. 35.**12** **Microorganisms involved in the production of therapeutically useful steroids**

Today, the steroid industry consumes 2000 tons of diosgenin, which is used to make, by combined chemical and microbiological processes, products with a market value of well over \$1 billion. Estrogens, progesterone, and androgens are used therapeutically; derivatives of progesterone and estrogens are also used as contraceptives ("the pill"). In addition, steroids are used as sedatives, in antitumor therapy, and as veterinary products. The glucocorticoids are valuable compounds with many therapeutic uses. Steroid biotransformation is second only to antibiotic production as a vital contribution of microbes in the production of pharmaceuticals.

Microbial transformation is decisive for the economic production of therapeutically useful steroids. The stereospecific hydroxylation of a large variety of steroids is catalyzed by microbial hydroxylase systems, which are NADPH- and O_2-dependent.

35.6 Antibiotics Are Products of Microorganisms That Inhibit Metabolic Processes of Other Organisms

While primary metabolites are essential for life and growth of cells, and primary ("housekeeping or vegetative") metabolism functions similarly in all microorganisms, the so-called "secondary" metabolites are seemingly not essential for the vegetative growth of the organisms. Each secondary metabolite is produced by only a few microorganisms, and the synthesis is dependent on environmental conditions. Therefore, recently the terms "individual metabolism" and "individualities" were introduced to replace the terms "secondary metabolism" and "secondary metabolites". The enzymes involved in the production of the individualities are regulated together with differentiation, but separately from the enzymes of primary metabolism (see Chapter 27). Antibiotics are metabolites that inhibit different metabolic processes of other organisms. The observation of Alexander Fleming in 1929 that staphylococcal growth was inhibited by *Penicillium notatum* led to the detection of penicillin and the development of its industrial production. This started the antibiotic era. Currently, more than 9000 antibiotic compounds are known, and several hundred antibiotics are discovered annually.

Antibiotics are used primarily as antimicrobial agents in human disease infection therapy. Other applications include their use as cytotoxic agents against certain tumor types, as disease control agents in veterinary medicine and plant pathology, as food preservatives, and as animal growth promoters. Commercially useful antibiotics are produced primarily by filamentous fungi (Aspergillaceae and Moniliales) and by bacteria, especially of the genera *Streptomyces* and *Bacillus*. About 120 antibiotic types are produced by industrial fermentation, while more than 50 semisynthetic compounds also have clinical applications as antibiotics. Annual worldwide antibiotic production exceeds 100000 tons, with an estimated market value of $5 billion. Selected examples of important antibiotics produced by fermentation for pharmaceutical use are given in Tables 35.**5** and in Chapter 27.

35.6.1 Antibiotics Are Detected by Screening for Inhibitory Features

Most new antibiotics have been discovered by screening. A large number of isolates of possible antibiotic-producing microorganisms from nature are obtained in pure culture, and these isolates are then tested for antibiotic production. The classical method for testing antibiotic activity is the plate-diffusion test. In this test, sensitive test bacteria are seeded into the agar medium to give a fairly dense uniform growth (a lawn). Then culture media from microorganisms that may produce an antibiotic are added, usually on filter paper discs impregnated with the culture filtrates. After 1–2 days of incubation, the inhibition of growth can be detected. Under standardized conditions, the diameter of the zone of inhibition is proportional to the logarithm of the concentration of a given antibiotic; thus, this test can also be used for the quantitative measurement of the antibiotic concentration. Those isolates that show evidence of antibiotic production are then studied further to determine if the antibiotics they produce are new.

The success of a screening procedure is dependent on the development of "intelligent" tests with which known or undesirable antibiotics can be eliminated and compounds with the required properties can be recognized. For example, one mode of resistance to penicillin is through the production of β-lactamase, an enzyme

Table 35.**5** **Examples of important antibiotic types produced by fermentation for pharmaceutical use**

Antibiotic group type	Example	Producing organism	Application
Peptide	Bacitracin	*Bacillus licheniformis*	Use confined to local application because of toxicity
Aminoglycoside	Streptomycin	*Streptomyces griseus*	Mainly used to treat tuberculosis
Macrolide	Erythromycin	*Streptomyces griseus*	Particularly effective against *Staphylococcus* and diphtheroids; low toxicity
Polyene macrolide	Candidin	*Streptomyces viridoflavous*	Widely used for topical anti-fungal application
Tetracycline	Chlortetracycline	*Streptomyces aureofaciens*	Inhibition of almost all Gram-positive and Gram negative bacteria

that splits the β-lactam ring. Inhibitors of β-lactamase might thus prove useful in permitting penicillin therapy against these resistant organisms. For the screening of microbial β-lactamase inhibitors, supernatants of the cultures are placed on agar plates containing penicillin and a β-lactamase-producing bacterial strain. Thus, β-lactamase inhibitors can be detected by growth inhibition of this strain.

Since many pathogenic bacteria have become resistant to most of the common antibiotics, new derivatives of known antibiotics or new antibiotics have to be detected. The testing of toxicity, malignicity, and side effects (e.g., allergic reactions) increases the costs of searching for new antibiotics; therefore, the derivatization of known products is often preferred.

Antibiotics are individualities (secondary metabolites) that inhibit the growth of other organisms. Currently, more than 9000 antibiotic compounds are known, and by using various screening procedures, several hundred antibiotics are discovered annually.

35.6.2 *Streptomycin* Was the First Member of the Aminoglycoside Antibiotics Discovered

The discovery of streptomycin by Waksmann in 1944 was a major medical advance as this was the first antibiotic used in the treatment of tuberculosis. Streptomycin is an aminoglycoside that consists of an aminocyclohexanol compound bonded by a glycosidic linkage to other amino sugars (see Fig. 27.**18**). Over 100 aminoglycosides produced by *Streptomyces* strains are known; in addition to streptomycin, also kanamycin, gentamycin, and neomycin are used clinically primarily against Gram-negative bacteria. The mode of action of these antibiotics is on protein synthesis; streptomycin binds to protein S12 (gene *rpsL*) of the 30S subunit of ribosomes and causes enhanced misreading.

Streptomyces strains produce many aminoglycoside antibiotics that contain amino sugars bonded by glycosidic linkage to other amino sugars. The mode of action of these antibiotics is on protein synthesis.

35.6.3 Tetracyclines Are Important Antibiotics of Widespread Medical Use

Chlortetracycline, the first tetracycline discovered, was isolated from cultures of *Streptomyces aureofaciens* in 1945. The basic structure of the tetracyclines consists of a naphthacene ring system (see Chapter 27.3.1). Various constituents are added to this ring. Chlortetracycline, for instance, has a chlorine atom, whereas oxytetracycline has an additional hydroxy-group and no chlorine. A mutant *Streptomyces* strain blocked in the chlorination reaction excretes tetracycline as the major product. Tetracyclines are broad-spectrum antibiotics that inhibit almost all Gram-positive and Gram-negative bacteria. The mode of action is at the 30S subunit of ribosomes, where binding of aminoacyl-t-RNA to the ribosomal A-site is inhibited (see Chapter 15.3.4).

In the biosynthesis of chlortetracycline, 72 intermediates are involved. Initial stages involve formation of malonamoyl-CoA bound to the enzyme complex anthracene synthase. Malonamoyl-CoA condenses with eight molecules of malonyl-CoA, and cyclization occurs with eventual formation of chlortetracycline. High-yielding tetracycline strains are characterized by a lower rate of glycolysis, and chlortetracycline production may be enhanced by use of the glycolysis inhibitor benzylthiocyanate. Under these conditions, the activity of the pentose phosphate cycle increases. The rate-limiting enzyme in chlortetracycline biosynthesis may be anhydrotetracycline oxygenase, the second to last enzyme in the biosynthetic pathway. Its activity appears to be proportional to the rate of antibiotic synthesis. Synthesis of this enzyme is repressed by phosphate and stimulated by benzylthiocyanate. There is also an inverse relationship between the level of adenylates and the activity of this enzyme. The ATP level or the total adenylate level appears to act as the metabolic effector in catabolite regulation of tetracycline biosynthesis.

Current concentrations of tetracyclines produced industrially, are around 20g/l. Because of the complexity of the biosynthetic pathway, strain yield improvement has depended solely on mutation/selection techniques. Selection of strains resistant to the produced antibiotic is another method that has been applied to improve production capacity. Typical fermentation production media contain sucrose, corn steep, ammonium phosphate, and salts, with pH and temperature maintained at 5.8–6.0 and 28°C, respectively. High aeration rates are necessary, particularly in the biomass growth stages. If glucose is used, continuous feeding is necessary. Because of phosphate repression, tetracycline fermentations are run under phosphate-limited conditions. Production of chlortetracycline in submerged culture may be subdivided into three phases. The first phase is characterized by a rapid increase in biomass and rapid consumption of nutrients. During this phase, the mycelium is characterized by the presence of thick basophilic hyphae with a high RNA content. In the second phase, the growth rate decreases and sometimes ceases, maximum rates of antibiotic synthesis are observed, and the organism begins to differentiate. Hyphal filaments appear thin and contain a low RNA content. In the final phase, lower rates of antibiotic production are observed, and mycelium fragmentation and lysis occur.

> **Chlortetracycline synthesis** is a complex metabolic pathway involving 72 intermediates and a large number of genes. Because of phosphate repression, the culture medium used in commercial production must have limited phosphate concentrations.

35.6.4 Redesign of Antibiotic Pathways by Recombinant DNA Techniques

Because of the large number of genes involved in the biosynthesis of an antibiotic, the genetic research aimed at antibiotic strain development is complex. A major objective of applying modern recombinant DNA techniques to antibiotic strain development is to increase the yield and rate of synthesis and for the production of hybrid or even novel antibiotics. Genes for biosynthetic steps from different organisms can be combined in the some hybrid organism, thus leading to the production of novel metabolites. D. A. Hopwood and colleagues in 1985 described the first experiment testing this idea. Part of the cloned pathway for actinorhodin from *Streptomyces coelicolor* was transformed into a *Streptomyces* strain that produces the compound medermycin (Fig. 35.**13**). The recombinant strain produced an additional antibiotic, identified as mederrhodin. The recombinant plasmid used contained the gene coding for the enzyme that catalyzes the β-hydroxylation of actinorhodin. Thus, in the recombinant, the broad substrate specificity of the enzyme also allowed the hydroxylation of medermycin at the analogous position, producing mederrhodin. With a similar strategy, a mutant of *Saccharopolyspora erythraea*, which was blocked in an early step of erythromycin biosynthesis, was transformed with a DNA library from the oleandomycin producer *Streptomyces antibioticus*. One recombinant strain formed an antibiotic-active compound, which was identified as 2-norerythromycin. In this case, a modified structure lacking a functional group ($-CH_3$) was generated.

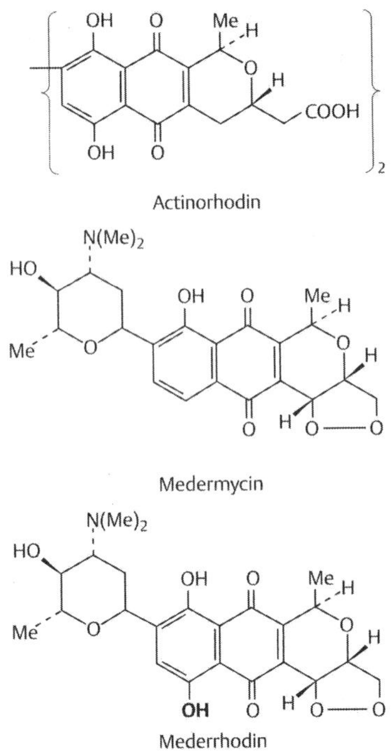

Actinorhodin

Medermycin

Mederrhodin

Fig. 35.**13** **Structures of actinorhodin, medermycin, and mederrhodin**

In each of these examples described, modified structures with a substitution at one carbon atom were made. A greater challenge for obtaining novel antibiotics is to alter the backbone structure of a metabolite. *Streptomyces galilaeus* normally produces aclacinomycin A and B. After transformation with the genes for polyketide synthase, which is involved in the synthesis of actinorhodin, clones were obtained that produce an anthraquinone. This production of an antibiotic with a novel structure is a very promising result. In the near future, we may be able to change structures of antibiotics in a rational way. To reach this goal, it is also important to change the specificity of the biosynthetic enzymes by site-directed mutagenesis and/or genetic engineering.

> Recombinant DNA techniques can be used either to increase the yield and rate of antibiotic synthesis or to obtain novel antibiotics.

35.7 Commerical Exploitation of Bacterial Enzymes

In an empirical manner, enzymes have been applied to practical uses for thousands of years, in the form of crude animal or plant preparations or as a consequence of microbial development, for example, in processes such as cheese making or leather manufacturing. At the turn of this century, it was demonstrated that enzymes retain their activity in cell-free extracts. This observation led to the start of modern applications of enzymes in industry, analytics, and therapy. For example, some microorganisms are used that are able to excrete enzymes into the surrounding medium to break down large organic molecules (proteins, fats, polysaccharides), which otherwise could not be taken up by the organisms.

The first enzyme produced industrially was a fungal amylase, employed as a pharmaceutical agent (for digestive disorders) in the United States as early as 1894. Otto Roehm's patented "laundry process for any and all clothing via tryptic enzyme additives" was announced in 1915. With the progress made in sterile technology, along with the development of antibiotic fermentations, mass cultivation of single organisms became possible. This laid the foundation for the technical production of microbial enzymes, which developed about 25–30 years ago. As shown in Table 35.**6**, these enzymes degrade various biopolymers and find applications for numerous purposes. Their specificity and high reaction rates under mild reaction conditions are favored over competing chemical treatments. These enzymes are mainly produced by *Bacillus* and *Aspergillus* strains. Since these enzymes are excreted in the culture medium, their isolation is rather simple (see Chapter 19.4). The typical enzyme yield from a *Bacillus* fermentation process is estimated to be around 20g excreted material/l in a relatively short time with low-cost carbon and nitrogen sources. This demonstrates the ability of *Bacillus* strains to produce large quantities of enzymes at competitive costs. With the advent of mechanical techniques for release of proteins from microorganisms on a large scale, intracellular enzymes have found wider application in the food, pharmaceutical, and chemical industries, in particular glucose isomerase for production of high-fructose syrups and penicillin acylase for removal of the side chain of penicillins, allowing the subsequent manufacture of semisynthetic penicillins. Other enzymes, such as glucose oxidase and cholesterol oxidase, are widely used for clinical analysis. At present, the market of industrial enzymes amounts to more than $1 billion.

Enzymes have found a broad spectrum of applications in industry as their specificity and high reaction rates under mild conditions are favored over competing chemical treatments. *Bacillus* strains can secrete large quantities of hydrolytic enzymes into the culture medium

35.7.1 For Industrial Purposes, Soluble Enzymes Are Immobilized

Immobilized enzymes offer the advantages of: (1) recovery and re-use of the enzymes in batch processes, or (2) the development of continuously operated enzyme reactors similar to continuous fermentation systems used for microorganisms, and (3) the possibility of multi-enzyme systems. The enzyme may be stabilized by immobilization, but it may also lose activity. There are three approaches to enzyme immobilization (Fig. 35.**14**):

Table 35.**6** **Application of bulk enzymes**

Enzymes	Mode of action	Application
Proteases	Nonspecific hydrolysis of proteins Specific hydrolysis of casein	Detergent additives, leather production, baking Cheese production
α-Amylases	Degradation of starch to oligosaccharides	Starch processing, baking, starch syrup production, sizing of textiles, paper and cardboard production
Glucose isomerase	Isomerization of glucose to fructose	Production of high-fructose syrup
Pectinases	Hydrolysis of pectins	Processing of fruits and vegetables
Cellulases	Hydrolysis of cellulose	Maceration and drying of plant-derived raw materials
Lipases	Splitting of triacylglycerols	Detergent additives

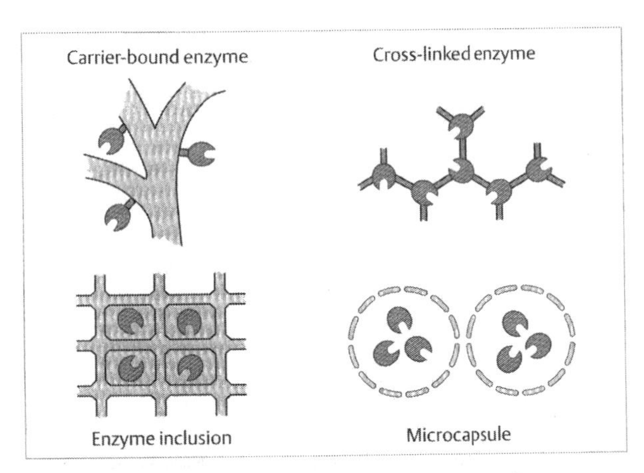

Fig. 35.**14** **Procedures for the immobilization of enzymes**

1. **Bonding of the enzymes to solid supports.** Many supports have been used, including porous glass or ceramic beads, aluminum oxide, synthetic polymers, and cellulose. Methods derived from peptide and protein chemistry are used to attach the enzymes to the support. Formation of covalent bonds has the advantage that these bonds are not reversed by pH, ionic strength, or substrate. However, it is possible that the enzyme is rendered inactive in part or entirely through covalent binding.
2. **Cross-linking with multifunctional agents.** Enzyme molecules are usually linked with each other by a chemical reaction with a bifunctional cross-linking agent such as glutaraldehyde. This compound reacts with amino groups of the enzymes.
3. **Entrapment of enzymes.** Enzymes can be enclosed in microcapsules, gels, or fibrous polymers, which must have pores that are small enough so that the enzyme molecules cannot be washed out, yet large enough to permit the diffusion of low-molecular-weight substrates and products.

When enzymes are used in large-scale processes, it is very often desirable to immobilize them by binding them to solid supports or by entrapping them into polymers.

In the following, some enzymic processes used in industry are described.

35.7.2 Glucanhydrolases Liquify Starch to Sugar

The most important enzymes in the starch saccharification process are 1,4-α-glucan-glucanohydrolases (EC 3.2.11, α-amylases) and α-glucan-glucanohydrolases (E 3.2.13, glucoamylases). α-Amylases are endoenzyme that cleave α-1,4-glucosidic bonds of amylose an amylopectin to yield oligosaccharides of varying chai length. Thermostable α-amylases are used in high temperature liquefaction. The enzyme from *Bacillu amyloliquefaciens* has a temperature optimum of 70°C compared to 92°C for the enzyme isolated from *Bacillu licheniformis*. The latter can be used at temperatures a high as 110°C for short periods of time, and calcium ion and high concentrations of substrate stabilize thi activity. After the dispersion of the starch into aqueou solution (40%), a partial enzymatic hydrolysis occurs a 105°C for 5 min (to reduce viscosity), followed by 1 h a 95°C.

After this first liquifying step, the resulting dextrin are treated with glucoamylase, which accomplishes th further hydrolysis to glucose. Commercially availabl glucoamylases are produced by *Aspergillus niger* o *Rhizopus niveus*. These enzymes possess a low degre of specificity, hydrolyzing α-1,6 bonds at a lower rat than α-1,4 bonds, are stable over a wide range of pH, an exhibit maximum activity at 75°C, although they ar used normally at 65°C. The most important applicatio of glucoamylases is in the production of high-glucos syrups (90–97% glucose), which are used in the pro duction of crystalline glucose or high-fructose syrup Figure 35.**15** summarizes the enzymatic conversion c starch to glucose and fructose.

The isomerization of glucose by immobilize glucose isomerase to produce a high-fructose syrup i one of the outstanding successes of the past 15 year Sweet glucose is converted to the sweeter fructose b glucose isomerase from *Bacillus coagulans, Streptomyce rubiginosus, Actinoplanes missouriensis*, or *Flavobacte ium arborescens*. The enzyme operates at 60–65°C t produce an equilibrium mixture of glucose and fructos These fructose syrups have replaced cane or beet suga in many important uses because fructose tastes eve sweeter than sucrose. More than 2 million tons of high fructose syrup are produced alone in the United State annually.

The enzymatic hydrolysis of starch into glucose and the final conversion into high-fructose syrup have found broad applications in food industry.

35.7.3 Proteases Are Used as Detergent Additives

After amylases, the second most important industri enzymes currently are the proteases. About 500 tons c

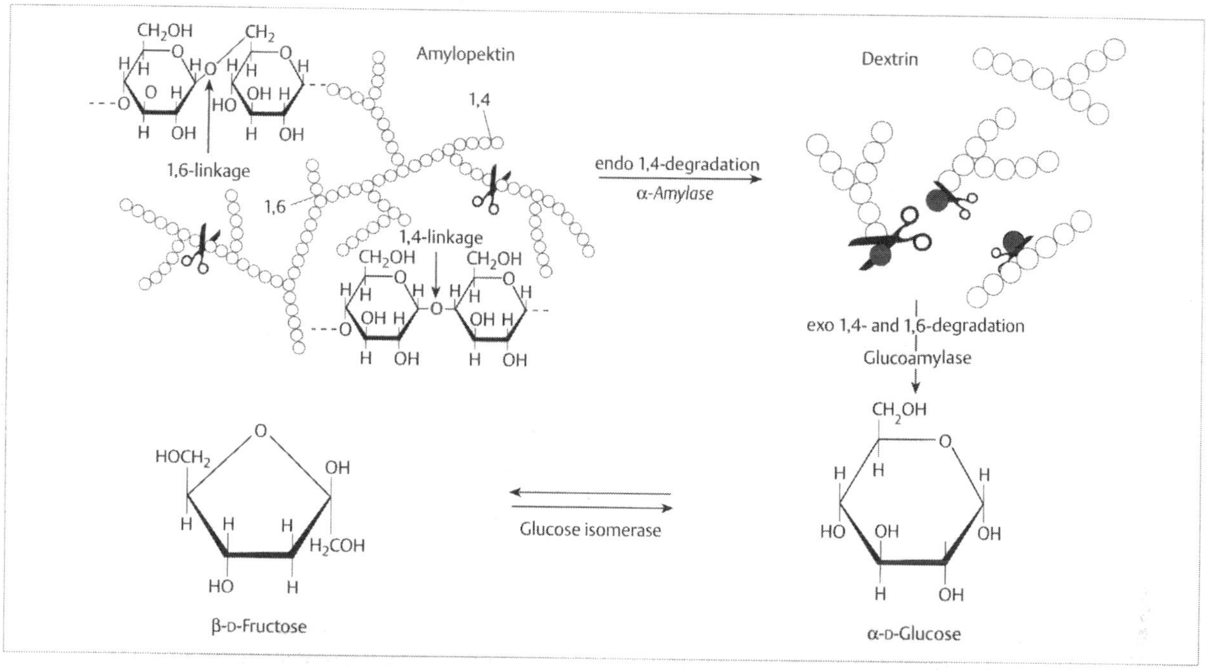

Fig. 35.15 **Enzymatic starch hydrolysis and conversion of glucose into fructose.** Amylopectin is degraded by α-amylase into dextrin, which is further hydrolyzed to glucose by glucoamylase. The isomerization of glucose to fructose is catalyzed by glucose isomerase

these enzymes, which are used primarily in the detergent industry, are produced per year. The tremendous commercial breakthrough came with the production of proteases derived from *Bacillus licheniformis*, in particular subtilisin. By the end of the 1960s, approximately 50% of all detergents manufactured in Europe and the United States contained proteases. Improvements in enzyme activity are continually being sought, particularly with regard to stain-removing ability and increased stability in washing suds. The production of commercial proteases requires strains that produce high yields of extracellular proteases. Since the enzyme yields of wild-type strains are insufficient for industrial utilization, extensive studies have been carried out to increase the yield. The genes of several proteases have been cloned, and protein engineering has been used to develop modified *Bacillus* serine proteases. For example, substitution of the amino acid Met-222 by non-oxidizable amino acids led to proteases with significantly increased stability toward H_2O_2. Production of extracellular proteases is chiefly regulated by the medium composition (see Chapters 7.3 and 19.4). A fed-batch process (nutrients are permanently fed to the bioreactor) is generally used in order to keep the concentration of ammonium ions and amino acids low since these nitrogenous materials repress protease production. Protease production starts toward the end of the exponential phase of growth. Serine proteases do not hydrolyze proteins completely to amino acids. The level of application of proteases in detergents is 0.5% of a preparation that contains 3% active enzyme. Although this results in a relatively low total enzyme concentration, it is sufficient because of the substrate affinity of the enzyme.

Proteases are produced in large amounts by various *Bacillus* strains; these enzymes are used as additives in laundry detergents.

35.7.4 Enzymes That Are Used as Therapeutic Agents

One of the most successful examples of the therapeutic use of enzymes involves the treatment of certain neoplasmas, including "acute lymphocytic leukemia" with asparaginase. These tumor cells lack the capacity to synthesize asparagine and, by maintaining low levels of the amino acid in the body using asparaginase, the neoplastic cells are unable to grow, while normal cells are unaffected. The enzyme is produced by mutants of

Escherichia coli, *Serratia marcescens*, or *Erwinia carotovora*.

Furthermore, an important application of therapeutic enzymes is parenteral administration of firbrinolytic enzymes. Cardiovascular diseases such as heart attack and stroke usually arise from the obstruction of a blood vessel by a clot. The natural formation and dissolution of fibrin, which constitutes the clot, is carefully balanced in the blood stream and tissues. The process of fibrinolysis involves the activation of plasminogen into the proteolytic enzyme plasmin, which acts on fibrin. Treatment, therefore, involves the preparation and administration of an activator of plasminogen. A typical plasminogen activator is streptokinase, which is produced by β-hemolytic streptococci.

35.8 Heterologous Proteins Are Produced by Recombinant DNA Technology

A rapidly increasing number of compounds are being made by the use of genetically engineered organisms, in particular the production of human proteins with therapeutic potential (Tab. 35.**7**). Many of these proteins are synthesized in trace amounts in the body, and before the advent of recombinant DNA technology, it was very difficult to isolate them. Low molecular-weight proteins with few disulfide bridges that do not require glycosylation for activity can be cloned and expressed in bacteria. To produce complex proteins that are properly glycosylated, tissue cultures are most suitable.

Because of its well-known biology and genetics, *E. coli* is so far the preferred bacterium for the production of heterologous proteins. During the last 15 years, a large number of genes have been cloned and expressed in this organism. Under favorable conditions, the total cell protein can comprise up to 30% of the heterologous protein. Often, however, foreign proteins in *E. coli*, especially when overproduced, are subject to proteolysis or they may form insoluble aggregates ("inclusion bodies"). Extensive knowledge of the molecular biology of *Bacillus subtilis* has also led to the recent development of this bacterium as a host for the production of heterologous proteins. An advantage of the *Bacillus* strains is that the foreign proteins can be excreted into the culture medium. The number of heterologous proteins being produced by bacteria are growing rapidly; a few of the major products are described here in more detail.

35.8.1 Human Insulin Was the First Therapeutic Agent Produced by Recombinant DNA Technology

Insulin is a protein produced in the pancreas that is vital for the regulation of carbohydrate metabolism in the body. Diabetes, a disease characterized by insulin deficiency, afflicts, about 1–2% of the population in Europe and the USA. Since the first clinical use of insulin for the treatment of diabetes in 1922, the hormone has been extracted and purified from animal pancreas (beef or pork). Both bovine and porcine insulins differ slightly in amino acid sequence from human insulin, and, therefore, patients treated with these insulins develop circulating anti-insulin antibodies.

Insulin in its active form consists of two polypeptides (A: 21 amino acids, and B: 30 amino acids) connected by two disulfide bridges. These two polypeptides are coded by separate parts of a single insulin

Table 35.**7** **Some therapeutic proteins produced by recombinant DNA technology with bacteria**

Product	Comments	Producers
Human insulin ("Humulin")	Commercially available for treatment of diabetes	Eli Lilly, Genentech, Novo
Human growth hormone	Commercially available to treat pituitary dwarfism	Eli Lilly, Genentech, KABI
Colony stimulating factor (CSF)	Stimulates granulocyte (white cells) differentiation; CSF could be useful in treating leukemia to restore immune competence as an adjunct to chemotherapy	Amgen, Immunex, Genetics Institute, Cetus
Interleukin-2	Useful for cancer therapy in combination with LAK cells or tumor-infiltrating lymphocytes	Amgen, Cetus, Immunex
Gamma interferon	Anti-viral and anti-cancer applications	Amgen, Biogen, Genentech

gene. The insulin gene codes for preproinsulin, a polypeptide containing a signal sequence for excretion of the protein, the A and B polypeptides of the active insulin, and a connecting polypeptide that is absent from mature insulin. Insulin is processed from preproinsulin via proinsulin by enzymatic cleavage of the connecting polypeptide from the A and B chains.

A process for the production of human insulin in bacteria developed by Eli Lilly in collaboration with Genentech consists of initially designing a DNA sequence from the known amino acid sequence of insulin and then chemically synthesizing separate artificial insulin A chain genes and B chain genes (Fig. 35.**16**). Each gene, containing a methionine codon at the 5′-end and stop sequences at the 3′-ends, is individually inserted into the *lacZ* gene (encoding β-galactosidase forming a protein fusion) of two pBR322 plasmids, each of which is transformed into *E. coli* strains. The transformed bacteria hence produce the fusion proteins of the A chain and B chain separately. After lysis of the bacteria, cyanogen bromide treatment, which cleaves proteins only at methionine, allows the separation of the A and B chains from the β-galactosidase part (insulin contains no methionine). The chains are purified and recombined to produce the native two-chain insulin. This product is free from *E. coli* proteins, from endotoxins and pyrogens, is chemically and physically equivalent to human insulin of pancreatic origin, and has full biological activity; it has been marketed by Eli Lilly since 1982.

More recently, an alternative synthetic approach has been adopted, where the gene for proinsulin is cloned in *E. coli*. After synthesis and purification of the proinsulin, native insulin is derived by chemical treatment and by trypsin and carboxypeptidase β digestion; these proteases have no effect on insulin itself. In 1992, the sales of human insulin was approximately $625 million.

35.8.2 Interferons Are Potential Antiviral and Antitumor Agents

One of the most exciting developments arising from use of the recombinant DNA technique has been the synthesis of polypeptides with human interferon activity in *E. coli*. Interferons are proteins produced by cells of most vertebrates in response to invasion by viruses. They induce a virus-resistant state in the infected or induced cell, which is accompanied by the de novo synthesis of a number of proteins. Furthermore, interferons have two other important biological effects: inhibiton of cellular proliferation and modulation of the immune system.

The different classes of interferons are encoded by different genes; they differ immunologically and in their target cell specificity. All the interferons are produced in minute quantities, and this presented again a major obstacle to their purification and their therapeutic use. Therefore, a human α-interferon gene (514 base pairs long) was synthesized, incorporated into a plasmid, and successfully transformed into *E. coli*. Also the human β- and γ-interferon genes were expressed in *E. coli*, yielding products with antiviral activities. This biological potency is very interesting since natural interferons are glycosylated proteins, whereas those produced by recombinant DNA technology in bacteria are not. Some pharmaceutical companies are now producing α- and γ-interferon using recombinant *E. coli* strains, and the products are used as potential antiviral and antitumor agents. The interferons had worldwide sales of $600 million in 1992.

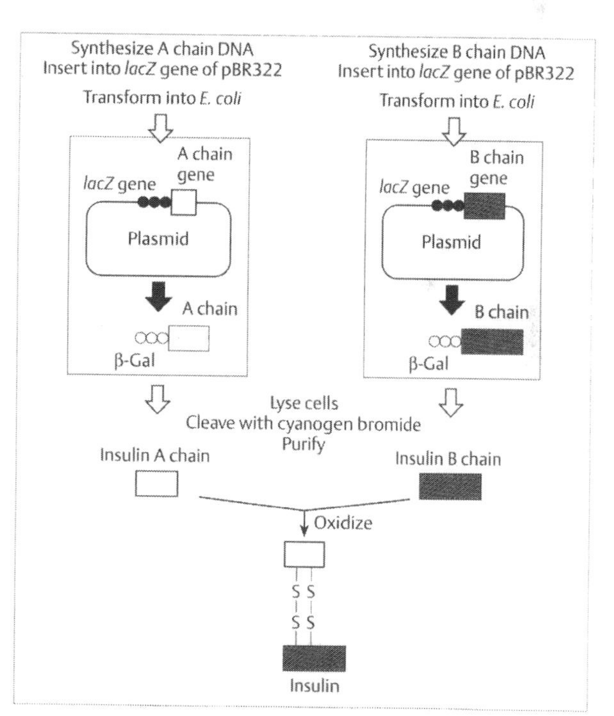

Fig. 35.**16** **Principle of human insulin production using recombinant DNA technology**. The chemically synthesized genes for insulin A chain and insulin B chain are separately inserted into the *lacZ* gene (encoding β-galactosidase) of two pBR322 plasmids. After these plasmids are transformed into *E. coli*, the fusion proteins of β-galactosidase–chain A and β-galactosidase–chain B are produced. The peptides are purified and recombined to form the native two-chain insulin

35.8.3 Viral Antigens Produced by Recombinant Bacteria Are Used as Vaccines

Frequently, inactivated virus preparations are used as vaccines, but there is always a potential danger to the patient if the virus has not been completely inactivated. The antigen in the virus vaccine is the protein coat. Through genetic engineering, the genes for various viral coat proteins have been cloned and expressed in bacteria, hence giving safe and convenient vaccines.

Human virus genes have been expressed in *E. coli*. An example is the genes of the Hepatitis B virus, the causative agent of serum hepatitis, which constitutes a serious and worldwide problem in public health. Fundamental studies of this virus have been seriously hampered because it cannot be obtained in large quantities and cannot be grown in cells in tissue culture. To obtain expression of the viral genes in *E. coli*, fragments of DNA were cloned in plasmid pBR322. Cells carrying the recombinant plasmids were screened for synthesis of the viral core and surface antigens by the radioimmunoassay method using [^{125}I]-labeled antibodies. When injected into rabbits, extracts from the positive clones elicited the formation of antibodies. By suitable subcloning, the yield of the core antigen in *E. coli* was increased. This provides a safe source of the antigen for use (e.g., in diagnosis).

Many viruses have RNA genomes, and in some picorna viruses, such as poliomyelitis virus or "foot-and-mouth-disease" virus, the RNA is translated to give a very large polypeptide that is subsequently processed to yield the various viral capsid proteins. One of these (VP#1) is of particular interest since it stimulates the synthesis of neutralizing antibodies and may thus contribute to the high degree of antigenic variation encountered with this virus. Cloning and analysis of the corresponding segment of the viral genome hence gave an alternative source of vaccine. As shown in Figure 35.**17**, double-stranded cDNA preparations were made from single-stranded foot-and-mouth-disease viral genome by reverse transcriptase and cloned in a plasmid for propagation in *E. coli*. The gene product (VP#1 protein) was purified and shown to be immunogenic when inoculated into animals.

These two examples demonstrate that recombinant DNA technology can be used to produce sensitive diagnostic reagents and viral vaccines in bacteria. However, certain viral vaccines prepared by expression of the viral genes in *E. coli* are poorly immunogenic. Many viral coat proteins are post-translationally processed, generally modified by glycosylation when the virus replicates in the host. The recombinant proteins produced by *E. coli* are unglycosylated, and apparently glycosylation is necessary for these proteins to be fully

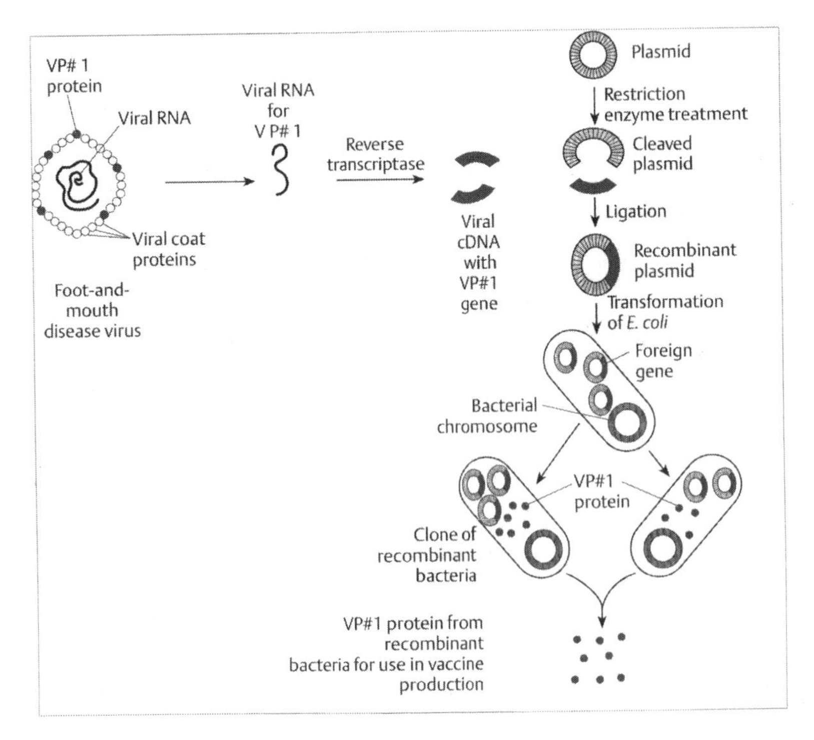

Fig. 35.**17** **Vaccine production using recombinant DNA technology.** Using recombinant DNA techniques, a "foot-and-mouth-disease" vaccine is produced by cloning the viral genes into *Escherichia coli*

immunologically active. Nevertheless, genetically engineered vaccines are likely to become increasingly common because they are safer than attenuated or inactivated virus vaccines, and the production is more reproducible because their genetic makeup can be carefully monitored.

Numerous mammalian proteins with therapeutic potential can be produced in bacteria. Human insulin formed in *E. coli* was the first example of the value of genetic engineering. Furthermore, α- and γ-interferons as well as antiviral vaccines are produced by recombinant *E. coli* strains.

Further Reading

Bailey, J. E. (1991) Toward a science of metabolic engineering. Science 252:1668

Banward, G. J. (1989) Basic food microbiology, 2nd edn. New York: Van Nostrand Reinhold

Crueger, W., and Crueger, A. (1990) Biotechnology: a textbook of industrial microbiology, 2nd edn. Brock, R. D. (ed) Sunderland, Mass: Sinauer Associates

Demain, A. L., and Solomon, N. A., eds. (1986) Manual of industrial microbiology and biotechnology. Washington, DC: ASM Press

Eikmanns, B., Eggeling, L., and Sahm, H. (1993) Molecular aspects of lysine, threonine, and isoleucine biosynthesis in *Corynebacterium glutamicum*. Antonie Van Leeuwenhoek 64:145–163

Enei, H., Yokozeki, K., and Akashi, K. (1989) Recent progress in microbial production of amino acids. Amsterdam: Gordon and Breach

Frazer, W. C., and Westhoff, D. C. (1988) Food microbiology, 4th edn. New York: McGraw-Hill

Glazer, A. N., and Nikaido, H. (1995) Microbial biotechnology: fundamentals of applied microbiology. New York: Freeman

Glick, B. R., and Pasternak, J. J. (1994) Molecular biotechnology: principles and applications of recombinant DNA. Washington, DC: ASM Press

Hollenberg, C. P., and Sahm, H., eds. (1987) Microbial genetic engineering and enzyme technology, Biotec 1. Stuttgart, New York: Gustav Fischer

Hugo, W. B., and Russel, A. D. (1992) Pharmaceutical microbiology, 5th edn. Oxford, England: Blackwell

Krämer, R. (1994) Systems and mechanisms of amino acid uptake and secretion in prokaryotes. Arch Microbiol 162: 1–13

Leatham, G. F., and Himmel, M. E., eds. (1991) Enzymes in biomass conversion. Washington, DC: American Chemical Society

McDaniel, R., Ebert-Khosla, S., Hopwood, D. A., and Khosla, C. (1995) Rational design of aromatic polyketide natural products by recombinant assembly of enzymatic subunits. Nature 375:549–554

Meyers, R. A., ed. (1995) Molecular biology and biotechnology: a comprehensive desk reference. Weinheim: VCH

Omura, S., ed. (1992) The search for bioactive compounds from microorganisms. Berlin, Heidelberg, New York: Springer

Primrose, S. B. (1991) Molecular biotechnology, 2nd edn. Oxford, England: Blackwell

Sahm, H., ed. (1993) Biological fundamentals. In: Biotechnology, vol 1. Rehm, H.-J., and Reed, G. (eds.) Weinheim: VCH

Sahm, H., Eggeling, L., Eikmanns, B., and Krämer, R. (1995) Metabolic design in amino acid producing bacterium *Corynebacterium glutamicum*. FEMS Microbiol Rev 16: 243–252

Sikyta, B. (1995) Techniques in applied microbiology, vol. 31. Progress in industrial microbiology. Amsterdam: Elsevier

Tombs, M. P. (1990) Biotechnology in the food industry. Buckingham: Open University Press

Ward, O. P. (1989) Fermentation biotechnology: principles, processes and products. Chichester: Wiley

36 Prokaryotes in Environmental Processes

Microorganisms, especially prokaryotes and lower fungi, are the most important mineralizers of organic matter, and they play key roles in the global cycling of all bioelements (see Chapters 30, 32.2, and 34). Production, consumption, and mineralization of organic matter have maintained a comparably stable equilibrium on a global scale; only the accumulation of organic carbon in the form of coal, mineral oils, and shale oils represents an exception to this generalization. The degradative activity of the microflora is a self-regulating system: its capacity increases when excess substrate supply is available (microbial growth) and decreases when substrates expire. This adaptative capacity of the microbiota has kept pace also with an increasing human population on Earth; problems have arisen only in densely populated areas where the accumulation of wastes from human activities has exceeded the microbial mineralization capacities in surrounding soils and waters. In ancient Rome, wastes and excreta were removed by a highly refined canalization system and transported via the cloaca maxima to the Tiber River and finally into the Mediterranean Sea. Channel systems of this type have also been reported for Athens (Greece) and cities in Egypt. In the Middle Ages, these technical developments were mostly forgotten, and wastes, including excreta of humans and animals, accumulated in narrow streets, from which they were removed occasionally by the grave diggers, for example. Contamination of drinking water resources was widespread, as were typical water-borne diseases, such as cholera, typhoid fever, and amoeboid dysentery. With the invention of the water closet in 1807 in England, water became the preferred carrier for fecal excrements, and the contaminated water was trickled onto fields of sewage farms outside the larger towns. The growth of towns and whole industrial areas (e.g., the Ruhr district in Germany) in the second half of the nineteenth century led to repeated epidemics of water-borne diseases and the need for a consequent treatment of wastewaters, entirely separated and removed from freshwater resources. Wastewater was first collected in oxygen-limited detention tanks ("Emscher-Brunnen"). Intensive treatment systems (trickling fields, trickling filters, aerobic sewage treatment plants) were developed through the first decades of the twentieth century to the state of the art still in use today. Interest in water purification was first oriented toward the prevention of contamination of drinking water by pathogens. The maintenance of the water of rivers and lakes in a comparably pure state, initially only of secondary interest, was ultimately prompted by the need to maintain sources of drinking water of high quality. The development of sensitive analytical techniques for the assay of contaminants and of an increasing awareness of environmental pollution also have led to activities aimed at the purification of air and of polluted soil compartments, both in which microbially mediated processes are utilized. Not only microbial cells and enzymes have been applied in the development of sensors for the assay of pollutants, but also of reaction metabolites in technical microbiology. Last, the activities of mainly chemolithotrophic prokaryotes already have been applied for a long time in the successful exploitation of low-grade ores in the mining industry.

Box 36.1 There are a number of important terms in wastewater technology. The **biochemical oxygen demand (BOD)** is a measure of the total amount of organic and inorganic pollutants in a given wastewater that can be oxidized by microbial activities. Since it is usually impossible to analyze the organic fraction of wastewater in detail, the BOD gives a measure of the total amount of oxygen that would be consumed if the wastewater were to be discharged untreated into a receiving water, e.g., a river or a lake. The BOD value is given in mg O_2 per liter or per kg wastewater; it is determined by incubation of a wastewater sample with a mixed community of microorganisms, e.g., some activated sludge, and oxygen uptake is followed over 5 days (BOD_5) at a constant temperature. The determination of BOD over 5 days comprises mainly the oxidation of organic contents; ammonia is oxidized later and is not included in this value. There are automatic systems on the market (e.g., "Sapromat") for routine assays. If the composition of the wastewater is known, the oxygen demand can be calculated on the basis of a stoichiometric conversion reaction. If the organic freight is mainly composed of carbohydrates and proteins, the BOD value is roughly equivalent to the dry-matter content of dissolved and suspended pollutants. With more reduced waste constituents (e.g., fats, lipids and hydrocarbons), the BOD value increases correspondingly. The BOD value is also the basis for the taxation of industrial enterprises for their discharge of wastewater: It corresponds to the treatment required before the wastewater is of sufficient quality to be released without impairment of receiving surface waters. Typical BOD values for various types of wastewater are listed in Table 36.**1**.

The **chemical oxygen demand (COD)** gives the value of oxygen equivalents (also in mg O_2 per liter or per kg wastewater) needed for complete chemical oxidation of the wastewater constituents to CO_2, e.g., by oxidation with potassium dichromate in concentrated sulfuric acid at 160°C. The COD refers to complete chemical oxidation, whereas the BOD only includes the constituents that are biologically accessible within a given incubation time. The difference between both values corresponds to that part of the organic and inorganic freight that cannot be efficiently oxidized by microbial activities.

The term **volatile suspended solids (VVS)** refers to that part of the total suspended material in wastewater or in digestor contents that can be volatilized by ashing at 600–650°C. The VSS comprises the particulate suspended material in the wastewater freight and also the living biomass. The VSS plus the ash mass (mineral salts) make up the **total suspended solids (TSS)**. The VSS value is usually the value to which the metabolic activities of digestor contents (e.g., oxygen consumption rate, methane production rate) refer.

Table 36.**1** **Typical BOD loads of various types of wastewater**

Source of wastewater	Biochemical oxygen demand (BOD) (mg $O_2 \cdot l^{-1}$)
Municipal wastewater	300–550
Chemical industry	1000–50000
Breweries, fruit industry	10000–20000
Sugar-beet industry, distilleries	50000–100000
Paper and fiber industry	80000–100000
Slaughterhouse offalls	80000–150000

36.1 Wastewater Treatment Intensifies Natural Biological Degradation Processes

If a creek with relatively clean water is contaminated by wastewater, a typical sequence of events will be observed (Fig. 36.**1**). Due to the high import of degradable organic matter, the oxygen content of the water decreases drastically, and the number of bacteria increases. The decomposition of the complex organic matter releases ammonia, which is further oxidized to nitrate only when the organic matter is nearly depleted and the oxygen content in the water has increased again. The bacterial community is grazed upon by bacterivorous protozoa, which decrease in number when their bacterial food supply slowly disappears. The contents of bound nitrogen species (and of phosphate) increase and give rise to a mass development of green plants and algae; this process serves to increase the oxygen content of the water again.

A **wastewater purification plant** acts in the same manner as the microbial community in the creek does. The plant intensifies the degradative activity of the aerobic microbes by intense aeration and maintains a dense and diverse microbial community for exhaustive degradation of the organic load. Modern wastewater treatment plants not only oxidize organic carbon compounds but also remove bound nitrogen and phosphate from the effluent water.

The composition of typical household wastewater as it reaches a municipal sewage plant is given in Table 36.**2**. The organic load consists mainly of carbohydrates,

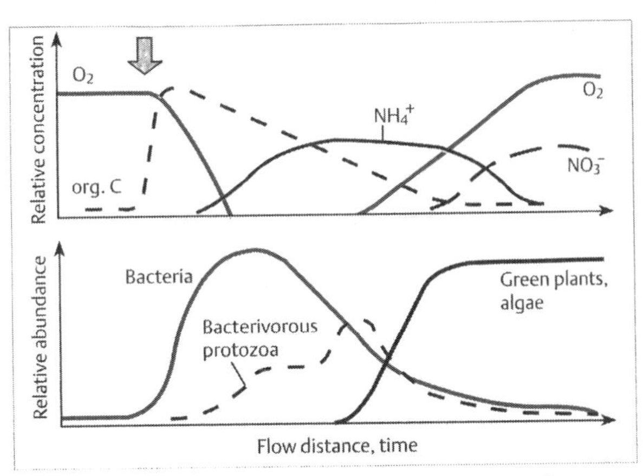

Fig. 36.1 **Contamination of clean creekwater with wastewater.** Arrow denotes point in time of contamination. Relative concentrations of dissolved organic carbon, ammonia, oxygen, and nitrate are documented in the upper diagram; relative abundances of bacteria, bacterivorous protozoa, green plants, and algae are shown in the lower diagram. The flow distance may extend over several hundred meters; the time frame is 10–60 minutes

Table 36.**2** Characteristics of household wastewater and requirements for treated water release •

Organic load	300–550 mg BOD·l^{-1}
50% Carbohydrates	
40% Proteins and urea	
10% Fats and detergents	
30%, approximately, of the organic matter is suspended particulate material	
Inorganic freight	
Phosphate	20 mg P·l^{-1}
Ammonia and nitrates	80 mg N·l^{-1}
Microbial constituents	
10^6–10^8·ml^{-1} bacterial cells, including	
10^4–10^6·ml^{-1} cells of *Escherichia coli*	
≦ 10·ml^{-1} cells of pathogenic bacteria	
10^3–10^5·ml^{-1} lower fungi, yeast cells	
Requirements for treated water release •	
≦ 20 mg BOD·l^{-1}	
≦ 18 mg total N·l^{-1}	
≦ 30 mg suspended solids·l^{-1}	
≦ 10^5 bacterial cells·l^{-1}	
≦ 10 *Escherchia coli* cells·l^{-1}	

• vary with local regulations, BOD, biochemical oxygen demand; ≦, less than

proteins, urea, and fat-like constituents, mainly detergents from body soaps and laundry and cleansing agents. The inorganic constituents derive partly from food preparation and surface runoffs. Among the microbial constituents, nonpathogenic bacteria dominate, mainly *Pseudomonas fluorescens*, *Pseudomonas aeruginosa*, *Proteus vulgaris*, *Bacillus cereus*, *Bacillus subtilis*, *Klebsiella pneumoniae*, *Enterobacter cloacae*, and *Zoogloea ramigera*. The indicator of fecal pollution, *Escherichia coli*, makes up less than 1% of the total bacterial community. Only a very low number of pathogenic bacteria are present.

36.1.1 The Activated Sludge Process, a Continuous Culture With Specific Amendments

The BOD value of domestic wastewater (see Table 36.**1**) necessitates intense aeration to achieve complete oxidation by microbes because the oxygen content of pure water under air is only 10–12 mg $O_2 \cdot l^{-1}$ (see Table 30.**2**). Present-day sewage treatment usually involves three different steps (Figure 36.**2**): In an initial, solely **mechanical treatment** (step 1), major aggregates are sieved out, and suspended material either settles to the bottom (sedimentation) or floats to the top (flotation of fats and oils). This mechanical treatment removes up to 30% of the total waste freight.

The central step for efficient wastewater treatment is the **biological step** (step 2), which is realized in most treatment plants today by the so-called **activated sludge process**. The heart of this process is an intensely aerated basin, the aeration tank, the contents of which are mixed by the forced bubbling of air and, possibly, by additional stirring. The air supply is controlled to maintain in the sludge an oxygen content of 10–40% air saturation. The dissolved and suspended freight of the wastewater is oxidized by a complex microbial community, which removes up to 99% of the total organic freight. Treated wastewater and activated sludge, which mainly consists of microbial biomass, are separated in the settling tank from which the treated water either is discharged directly into the receiving water or undergoes further **chemical treatment** (step 3) for phosphate removal. The microbial biomass settles rather efficiently, and 70–95% of it is recycled back into the aeration tank. This sludge recycling serves three important functions and makes the activated sludge process basically different from a laboratory-scale, continuous-culture system:

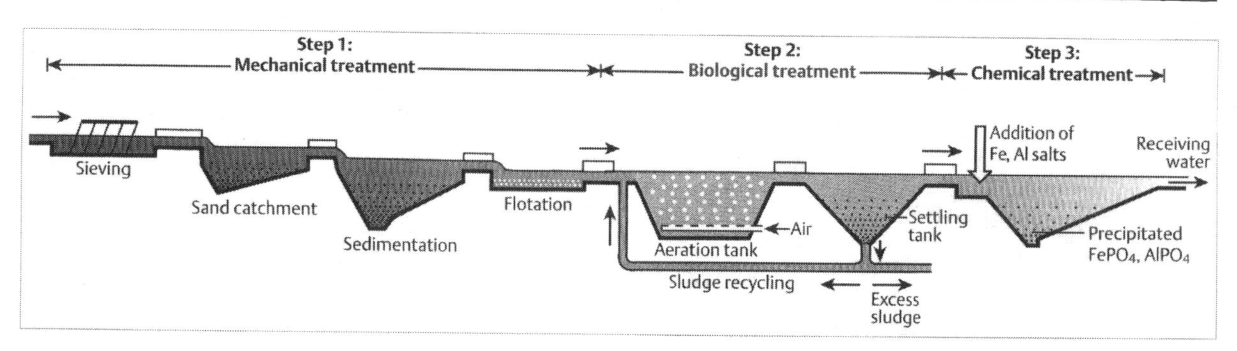

Fig. 36.2 **Operation of a middle-sized sewage treatment plant** with the activated sludge process as central biological step. Step 1: mechanical treatment; step 2: biological treatment; step 3: chemical treatment. Explanations in the text

1. Sludge recycling uncouples the retention time of the sludge from that of the water. Whereas the average retention time of the water is 4–12 h, the sludge has an average retention time of several days to weeks, which also allows the maintenance of slow-growing organisms in the system.
2. Sludge recycling increases the amount of biomass active in the aeration tank far beyond that amount corresponding to the inflowing substrate concentration (compare Chapter 30.3.2). This serves to increase the oxidation rate, to keep the microbial community voracious, and to increase the apparent affinity of the sludge for dissolved substrates ($v_0 = v_{max}/K_S$, see Chapter 30.4.1).
3. Sludge recycling maintains the settling efficiency of the sludge because only those organisms that settle readily are retained in the system whereas microbes that remain suspended in the treated water leave the system. The activated sludge process depends on the sludge-settling efficiency, which is one of the key points at which the system can be controlled.

The system's efficiency is maintained by controlling the delicate balance between sludge recycling and removal of excess sludge. Insufficient recycling decreases the diversity of the microbial community in the system, selecting only for fast-growing organisms and endangering the settling efficiency. Excessive recycling leads to overaging of the sludge and insufficient metabolic activity. A microscopic picture of activated sludge (Figure 36.**3a**, **b**) shows aggregates of bacterial cells, including a few filamentous bacterial cells, some protozoa, and relatively few free-floating bacterial cells. The protozoa (e.g., ciliates of the genera *Paramaecium* and *Vorticella*, and also amoebae) feed mostly on free-floating bacterial cells, as do rotatoria, nematodes, and, sometimes, oligochaeta. They therefore select for bacteria that tend to form aggregates and keep the number of free-floating bacteria low in the treated effluent. Thus, the grazers help to maintain the settling efficiency of the activated sludge. They also fulfill an important function in hygienization of the treated wastewater by removing possible pathogens and indicators of fecal pollution, such as *Escherichia coli*. There are speculations that protozoa not only select for aggregating bacteria by selectively feeding on free-floating cells but also even excrete signalling compounds that induce enhanced surface attachment in bacterial cells.

The bacterial community in sewage sludge is dominated by the aggregate-forming *Zoogloea ramigera* (and related species) and by the filamentous *Leucothrix* and *Thiothrix*. However, the total community is far more complex, and most bacterial constituents neither have been cultivated nor identified. Recent analyses with specific rRNA probes indicate that, beyond the known typical sewage sludge bacteria, representatives of the genera *Paracoccus*, *Caulobacter*, *Hyphomicrobium*, *Nitrobacter*, *Acinetobacter*, *Sphaerotilus*, *Aeromonas*, *Pseudomonas*, *Cytophaga*, *Flavobacterium*, *Flexibacter*, *Haliscomenobacter*, *Arthrobacter*, *Corynebacterium*, *Microthrix*, *Nocardia*, and *Rhodococcus* also occur, together with some cells of *Bacillus*, *Clostridium*, *Lactobacillus*, and *Staphylococcus*. The complex structure of the sewage sludge community is presently being unraveled by analysis with group-specific rRNA-targeted oligonucleotide probes (see Chapter 29.4).

The activated sludge does not always settle efficiently. Bulking sludge is a quite undesired situation in which the biomass floats to the water surface and is discharged into the receiving water. This effect is due to an excess of filamentous bacteria, which form a continuous network between the bacterial aggregates

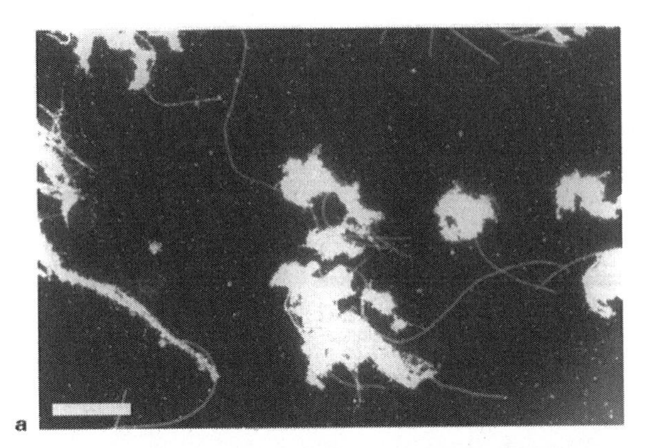

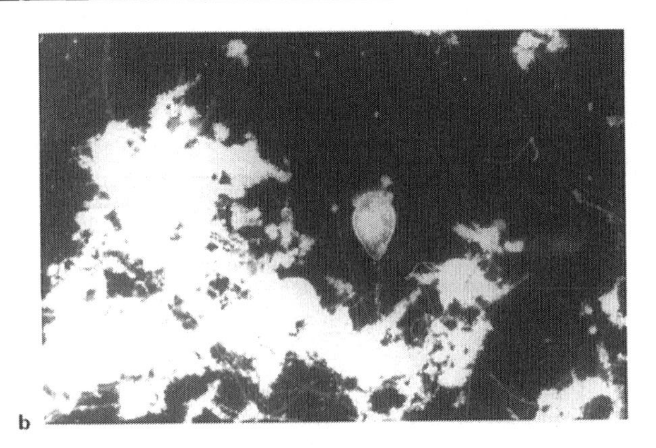

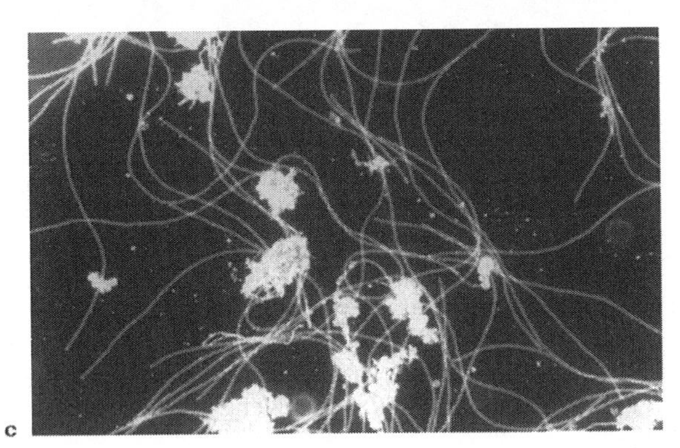

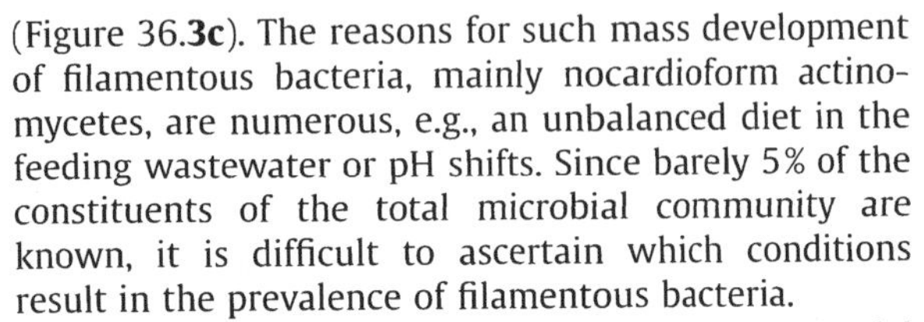

Fig. 36.**3a,b,c** **Dark-field photomicrographs of activated sludge samples.**
a and **b** A well operating activated sludge with a balanced ratio of aggregates and filaments. Note the *Vorticella* spec. in **b**, which selectively feeds on free-floating bacterial cells.
c Bulking sludge. Note preponderance of filamentous bacteria. Bar in **a** represents 50 μm for all three panels

(Figure 36.**3c**). The reasons for such mass development of filamentous bacteria, mainly nocardioform actinomycetes, are numerous, e.g., an unbalanced diet in the feeding wastewater or pH shifts. Since barely 5% of the constituents of the total microbial community are known, it is difficult to ascertain which conditions result in the prevalence of filamentous bacteria.

The excess sewage sludge, together with the material collected in the second sedimentation tank, is usually subjected to **anaerobic post-treatment** unless the sludge can be spread directly onto agricultural fields. In the anoxic sludge reactor, the sludge biomass and other degradable constituents are converted to methane and carbon dioxide, with an average retention time of 3–5 weeks at 30–35°C. Thermophilic treatment (at 50–65°C) already allows sludge discharge after 2–3 weeks due to the enhanced metabolic activity of thermophiles over mesophiles; however, this treatment requires substantial investment in insulation and heating. The gas mixture released, the so-called **biogas**, contains 65–70% CH_4 and 25–30% CO_2, together with traces of N_2 and H_2S. Part of the gas is used to maintain the temperature of the reactor, and the rest may be flared off or used to run gas engines for energy recovery. In most middle-sized sewage-treatment plants in central Europe, efficient gas conversion can provide 30–100% of the energy required for air compression in the aeration tanks.

The **activated sludge process** maintains in a continuous culture system a complex microbial community in an active state at high substrate affinity. Biomass recycling enhances substrate degradation efficiency and serves to maintain also the slow-growing microorganisms in the system.

36.1.2 Alternative Strategies and Post-Treatment

For smaller towns, the activated sludge technique may not be applicable. Alternatives are trickling filters in which the pretreated wastewater flows over a bed of crushed porous rock or slag. A complex biofilm of bacteria and higher organisms, including *Penicillium*, *Aspergillus*, and *Leptomitus*, grows on the carrier surfaces and degrades the dissolved organic matter; in this case, efficient degradation is carried out only by sessile microorganisms. The developing biofilms may be quite heterogeneous, and the substrates may be degraded under oxic, oxygen-limited, and anoxic conditions (see Chapter 30.5.3). The most active, upper layer of these films is only approximately 200 μm thick. Another method involves open oxidation channels in which the waste-water is circulated by the rotation of disks covered with active biofilms. A further alternative method of wastewater treatment, especially in agricultural areas, are reed farms where the wastewater moves slowly through a large reed bed. The reed plant provides aeration through internal gas ventilation (see Chapter 31.7.1) and thus provides oxygen for a complex microbial community in its rhizosphere. The organic freight is degraded in part aerobically and in part anaerobically, and some of the methane formed is released into the atmosphere through the plants.

36.1.3 Nitrogen Is Eliminated by Oxidation and Reduction

The primary goal of wastewater treatment is the removal of organic carbon. Oxidation of the nitrogen freight and removal of dissolved, bound nitrogen in the effluent water was a later development to protect the receiving waters from oxygen depletion. Ammonia, released during protein degradation and urea hydrolysis in the aeration tank, is oxidized by nitrifying bacteria via nitrite to nitrate (nitrification, see Chapter 10.7). Since these bacteria are usually inhibited by dissolved organic matter, complete nitrification can be achieved in the aeration tank only at low levels of organic matter or when a separate aeration system for nitrification follows the settling-tank stage. The nitrate thus formed may be converted to N_2 by **denitrification** (see Chapter 12.1). This requires anoxic conditions and inorganic or organic electron donors, which are supplied either as defined chemicals (hydrogen, methanol, ethanol) or by channeling part of the influent wastewater into the system at this step under strict control. The feasibility of denitrification for water protection is considered doubtful by limnologists because surface waters are usually phosphate-limited, and nitrate also can serve as an alternative electron acceptor in lake sediments. Recent studies have shown that nitrification and denitrification need not be separated entirely from each other; both may occur simultaneously under oxygen limitation when the contents of a treatment basin are insufficiently mixed, and even phosphate accumulation in the biomass (see below) can be achieved at this step to some extent.

Bound nitrogen is eliminated from wastewater through oxidation of ammonia to nitrate and reduction of nitrate to nitrogen gas. Oxidation and reduction can be achieved in the same treatment basin under oxygen limitation; this is typically accompanied by the enrichment of phosphate-accumulating bacteria.

36.1.4 Phosphate Can Be Removed After Biological Accumulation

Removal of dissolved phosphate is advisable, especially when the receiving water is stagnant and traces of phosphates may cause excess eutrophication. Phosphate removal from treated wastewater is no longer of considerable interest because today most laundry detergents are free of polyphosphates. In conventional wastewater treatment, phosphate is precipitated with iron or aluminum salts as $FePO_4$ or $AlPO_4$, respectively, as the final treatment step 3 in Fig. 36.**2**. The chemical treatment follows the mass action law of chemistry: to remove small amounts of phosphate (in the range of a few milligrams per liter), a large excess of precipitating agent is needed. The efficiency of chemical phosphate removal can be enhanced significantly by a biological phosphate-accumulation process. Aerobic bacteria such as *Acinetobacter* spec. and Gram-positive bacteria with DNA of a high G + C (guanine + cytosine) content accumulate phosphate in the form of intracellular polyphosphate granules as an energy storage system (see Chapter 9.4). Under oxygen limitation, when the respiratory electron-transport phosphorylation comes to a halt, polyphosphate is remobilized to form ATP. Such bacteria accumulate inorganic phosphate from the surrounding medium under conditions of sufficient energy supply, and release it again under anoxic conditions (Fig. 36.**4**). Cycling of the activated sludge between oxic and anoxic incubation steps selects for such phosphate-accumulating bacteria and allows a highly efficient phosphate removal either with the sludge biomass or by precipitation in the anoxic incubation tank, in which the phosphate freight has been significantly concentrated (Fig. 36.**4**).

Phosphate cannot be biologically "degraded," but bacterial cells may accumulate phosphate as an energy storage system. This biological phosphate accumulation can be exploited to allow efficient chemical phosphate elimination from wastewater.

36.1.5 Primary Anaerobic Treatment Is Feasible Especially for High Organic Loads

Primary oxic treatment of wastewater is a well-established technology; however, it has its disadvantages. Aerobic oxidation requires a plentiful supply of oxygen, which results in excessive operation costs for compression, for system control, and for the disposal of large amounts of sewage sludge due to the high energy yield of aerobic degradation, in which 30–50% of the organic freight is converted into cell material. This

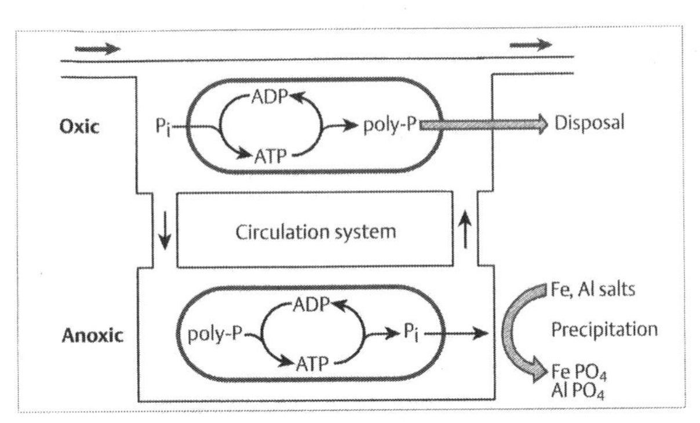

Fig. 36.**4** **Flow scheme of phosphate accumulation in biological phosphate elimination process.** In the presence of oxygen, *Acinetobacter* spec. and other bacteria accumulate polyphosphate (poly-P) as an energy store, and they release inorganic phosphate under anoxic conditions. The excess phosphate can either be disposed of with the sludge biomass or be precipitated chemically in the anoxic incubation basin

sludge production has increased dramatically since sewage-treatment systems have become widespread, and sludge disposal, even after anoxic posttreatment, has become a striking problem especially in densely populated areas. As an alternative, primary anoxic treatment has been reintroduced, especially for high-load wastewaters (see Table 36.**1**); in such cases, its advantages are quite obvious: costs for aeration and sludge disposal can be minimized, and the resulting methane production is of interest for energy exploitation. Primary anoxic treatment converts the bulk of the organic freight to CH_4 and CO_2, and oxic posttreatment only has to deal with the oxidation of ammonia and of sulfide. Nonetheless, anoxic treatment has necessitated specific process developments because anaerobic communities grow far more slowly than aerobic ones, and efficient utilization of anaerobic degradative capabilities requires technical devices to maximize biomass retention in the system, thus largely uncoupling water turnover from biomass turnover. A simple modification of the activated sludge process (Fig. 36.**2**, step 2) to anoxic conditions is not possible because anaerobic protozoa do not graze sufficiently to ensure sludge settling. The methanogenic treatment of wastewater has been made effective by the employment of surface-attachment techniques (fixed-bed and fluidized-bed reactors, Figure 36.**5**), in which the microbial biomass becomes attached to a supporting matrix, such as gravel, plastic or clay structures, or grains of sand.

The **fixed-bed reactor** is simply an anoxic trickling filter since with time it tends to become clogged; it needs to be cleared periodically by back-rinsing. This disadvantage can be overcome in the **fluidized-bed system**, in which the produced biomass is sheared off continuously, maintaining an active young biomass attached to the floating bed particles. The most advanced system for anoxic treatment is the **upflow anaerobic sludge blanket (UASB) reactor**, in which the bacterial cells attach to each other and form solid aggregates of several millimeters diameter. Together with the gas bubbles, the aggregates float upward to screens in the upper part of the reactor and settle again within the reactor vessel due to an increased density caused by internal calcium carbonate precipitates. The development of such aggregates is critical in this system because its efficiency depends on the formation of easily

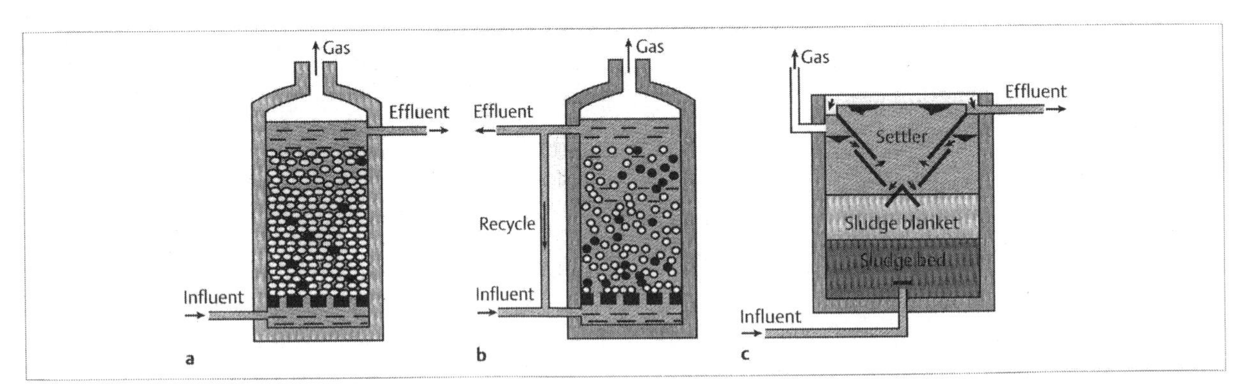

Fig. 36.**5a,b,c** **Modern reactor types for anaerobic treatment of wastewater and sewage sludge.**
a Anaerobic filter system. The biomass is attached to a carrier matrix, which either is fixed in the system (fixed-bed) or floats in the liquid phase (fluidized bed).
b Fluidized-bed reactor with recycling channel for treatment of acidic or otherwise toxic waste materials.
c **U**pflow **a**naerobic **s**ludge **b**lanket (UASB) reactor. The settler screens in the upper part serve to separate gas bubbles and sludge flocs from the
treated effluent

sedimenting aggregates. Once such sludge particles have developed, they are maintained in the system and selected for further propagation. The UASB reactor is applied today worldwide, together with fixed-bed and fluidized-bed reactors, preferentially for treatment of high-load wastewaters in which methane recovery is of economic interest and the low biomass yield of the anaerobic community (5–10% of the organic freight) causes fewer disposal problems. Various modification of these reactor types are made possible by internal recycling devices to cope with, for example, acid/base balance problems. This anoxic technology also is being developed for treatment of low-freight waters such as municipal wastewater, and perhaps the next generation of treatment plants will use a primary anoxic treatment followed by an oxic polishing step. Although the efficiency of methanogenic biomass degradation depends essentially on the close cooperation between the various trophic groups of microorganisms involved (see Chapters 12.1.7 and 30.7), it is sometimes preferable to carry out the process in two separate steps: (1) the formation of mainly fatty acids and lactate (acidification) and (2) the conversion of these intermediates to methane and CO_2 (methanization). This strategy is selected if the substrate load to be treated is supplied in pulses (campaign operation, e.g., in the sugar-beet industry) and if acidification of the substrate occurs readily, e.g., due to a high content of easily fermentable sugars. In other cases, operation in a homogenous, one-step system is more efficient because fatty acids—the degradation of which represents the rate-limiting step in the total mechanization process—are formed only to a minor extent.

Primary anoxic sewage treatment also is advised for certain industrial wastewaters, e.g., those rich in halogenated organic compounds. The higher such compounds are halogenated the more readily they can serve as electron acceptors for anaerobic bacteria. Pentachlorophenol, chlorohexane, chlorinated biphenyls, trichloroethylene, tetrachloroethylene, and other compounds are preferentially reductively dehalogenated under reducing conditions, and may be oxidatively degraded finally in an aerobic polishing process.

Anoxic treatment is especially feasible with wastewaters that, due to high organic loads, promise a high recovery of methane gas without expensive system control. The slow growth of many anaerobic bacteria makes the total process slow; however, efficient biomass retention systems have rendered anoxic treatment of waste materials in many cases a process competitive with or even superior to oxic treatment systems.

36.2 Composting Becomes Fashionable Again

Composting, an aerobic process of biomass oxidation that is accompanied by excessive heat accumulation, is applied today on a large scale for treatment of solid and semisolid household wastes. In compost heaps or technical composting plants, organic matter is decomposed by the activity of fungi and bacteria and reaches temperatures up to 65–70°C. In the first stage, fungi may play an important role together with mesophilic bacteria. With increasing transformation rates and increasing temperature (40–65°C), *Thermoactinomyces*, *Micropolyspora*, and *Thermomonospora* species take over. At higher temperatures (65–75°C), *Bacillus stearothermophilus*, *Bacillus subtilis*, *Bacillus licheniformis*, *Clostridium thermocellum*, and *Thermus* species become dominant. The heat development is a by-product of the metabolic activity of the microorganisms and plays an important role in the **hygienization** of the product, especially in the inactivation of potentially pathogenic bacteria, protozoal parasites, fungi, and worm eggs, for example. Of special interest is the avoidance of *Aspergillus fumigatus*, a fungus that develops in the early heating stage at 35–50°C. Its spores are highly allergenic and can cause severe problems for sensitized individuals or those with lung deficiencies.

Composting is an aerobic degradation of organic material that releases considerable amounts of heat, which serves to hygienize the product. The remaining organic material is rich in humic compounds and can be used for the improvement of garden soil, for example. Anaerobic alternatives to the composting process are being developed.

The **self-ignition** of insufficiently dried hay is a process analogous to composting; escaping hydrogen sometimes catches fire and destroys barns and farmhouses. In all these cases, heterotrophic secondary thermophiles (see Chapter 31.4.1) predominate in the process.

Anaerobic treatment of semisolid household wastes is being developed as an alternative to composting. The advantages are the recovery of methane ga from such waste materials and improved hygienization If the dry-matter content of the material to be dispose of is higher than 50%, usually **incineration** withou biological pretreatment is advised.

36.3 Drinking Water Also May Require Microbiological Pretreatment

The drinking water needs in industrialized countries are to a large extent covered by groundwater or surface waters from rivers and lakes, which are supplied to the distribution network after primary flotation of humic compounds and filtration through sand beds, in which organic trace contaminants are removed by a broad variety of sessile, oligotrophic, aerobic bacteria and protozoa. In some rare cases, the drinking water has to undergo a specific microbial pretreatment, e.g., to remove excess nitrate loads in the presence of added organic substrates. Before the purified water enters the distribution network, it has to undergo sanitation eithe by chlorination with NaClO or, more recently, b ozonation, both of which kill the majority of microbia drinking-water contaminants. However, such oxidation may also activate organic trace contaminants (aquati humic compounds, fulvic acids), which may suppor microbial growth within the distribution system an may even produce chlorinated aromatic compounds a new and even cancerogenic pollutants. Chlorination an ozonation should therefore only be applied when th water is nearly free of organic contaminants.

36.4 Exhaust-gas Treatment, a Further Working Area for Aerobic Bacteria

Treatment of exhaust gases has become mandatory in industrial processes that release toxic or bioactive volatile compounds into the atmosphere. For example, organic solvent vapors released in the painting industry, volatile chlorinated hydrocarbons from dry-cleaning plants and degreasing processes in industry, and ethylene produced by ripening fruit in storehouses all need to be removed if the exhaust gases of such industrial units are to be released into the atmosphere. Charcoal or soil beds originally devised as sorbents for such contaminants support the growth of biofilms of aerobic bacteria that actively degrade these contaminants. Today, beds of synthetic carriers, wood or barl chips, peat, or soil are used through which th contaminated air is passed after primary humidifica tion; the growing biofilm is controlled by highe organisms, thus rendering the cleaning and flushing o these bed systems largely superfluous. In devices for th washing of industrial exhaust gases, which ofte contain toxins, the exhaust gas is run through washin chambers that allow an intensive gas–water exchange The washwater then may be treated separately b microbial oxidation.

36.5 Bioremediation of Soil Is a Growing Market

After water and air, also soil was recognized as an important compartment of the biosphere that may not be contaminated to a large extent without impairing organisms (see also Chapter 34.5). However, the structural inhomogeneity of soil causes substantial problems in the reliable assessment of contamination, and cleanup efforts have to face even worse obstacles. As a consequence of carelessness within the past 50-100 years, old industrial sites, uncontrolled landfill dump sites, gasoline stations, and sites of accidenta chemical spills represent possible hazards to ground water and require consequent treatment in the nea future. More than 50% of the residual pollution world wide is due to mineral-oil contamination. Usually, th contamination begins at the soil surface; the contaminant penetrates the unsaturated soil zone and reaches the boundary layer in contact with the groundwater table. Depending on the density of the contaminating material, some of it may float on the water surface and thus be transported in the upper layer of the groundwater, other portions may dissolve to a certain extent in the groundwater, and any tar-like oil components may sink to the bottom, where they will be transported very slowly. Thus, the contaminants are distributed into various compartments, each of which requires a different method of treatment.

There are basically **five different strategies** for cleaning up a contaminated soil site:

1. The soil remains on the site, and contaminants dissolved in and transported with the groundwater are extracted as such and treated externally. This may leave contaminants of low solubility behind.
2. The on-site treatment is intensified by the addition of oxidants (O_2, NO_3^-), cosubstrates, or emulsifiers to enhance either the transport or the bioavailability of lipophilic contaminants. The additives may be supplied through gassing tubes or by irrigation, and the process may be enhanced further by surface ploughing. This treatment is comparably inexpensive but it may require years or even decades to be completed.
3. The soil is excavated and placed in piles or stacks in which aeration tubes have been inserted to intensify aerobic degradation. This treatment requires several years, depending on the type of contaminant.
4. The soil is excavated and washed, in some cases with tenside additives, and the extracted contaminants subsequently are degraded during biological treatment of the washwater.
5. The entire soil is excavated and incinerated in special furnaces. This is the most expensive way of treatment, which leaves a mineral soil entirely devoid of life. On the other hand, this process is in most cases the fastest method of treatment.

Which of the above-mentioned treatment procedures is applied in each case depends largely on the contaminant, its chemical composition, and the area and depth of its distribution. The soil texture determines whether washing procedures can be applied; an excessive content of organic matter or of fine particles (e.g., clay) may entirely nullify such efforts. If the geological conditions of the contaminated site have prevented groundwater contamination and the surface soil will not be used immediately, e.g., as a building site, long-term, low-cost, biological treatments are preferable to far more expensive physicochemical treatments, which are faster but result in a largely destroyed soil.

36.5.1 Microbiological Aspects of Biological Soil Remediation Processes

Since mineral oil is the most prevalent type of soil contaminant and hydrocarbon oxidation proceeds much more effectively in the presence of oxygen, usually the in situ degradation is limited by the supply of oxygen available to the indigenous microflora of the contaminated soil. Oxygen can be forced into deeper soil layers by compression through a system of metal pipes; since volatile hydrocarbons (e.g., benzene, toluene, xylene) are released into the air by this procedure, usually the reverse procedure is chosen: the air in the soil is removed by suction, and the effluent gas is then treated. Oxygen supply and aqueous extraction can be carried out in a recirculating system that combines intensified in situ oxidation with extraction and ex situ treatment, as Figure 36.**6** shows. The geological and hydrological conditions and the nature and usage of the surrounding area determine to a large extent which type of treatment is feasible.

At sites with highly oxidized contaminants, such as pentachlorophenol (wood preservation), nitroaromatic compounds (explosives production), or highly chlorinated alkanes and alkenes (chloroform, tetrachlormethane, trichlorethylene, or tetrachlorethylene for dry-cleaning or industrial-scale degreasing), the reductive transformation may require a supply of electron donors to intensify the process; this is not the case with

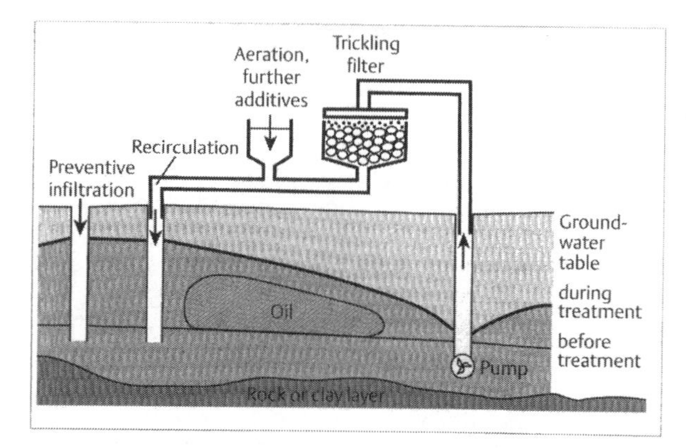

Fig. 36.6 **On-site treatment of oil-contaminated soil by water recirculation and ex situ water treatment.** The groundwater table has been raised to allow efficient extraction; a preventive infiltration of noncontaminated groundwater serves to protect the noncontaminated soil to the left

hydrocarbon contaminants. Inexpensive electron sources (methanol, ethanol, molasses) may be supplied with the circulation water, then the partly reduced contaminants can be treated ex situ by conventional techniques.

Polyannellated aromatic hydrocarbons present a special problem due to their low solubility in water. Biological treatment reaches its limit here, not so much because of lacking degradative capacity but due to the limited capillarity of the contaminants, which reduces their availability to the degradative organisms in the soil. With all water-insoluble contaminants, the transition into a dissolved or otherwise diffusible state is usually the rate-limiting step in degradation; this also applies to liquid hydrocarbons. Adsorption of organic contaminants to mineral surfaces, especially to organic soil constituents, further complicates the situation. The more complex the substrates are, the less likely they will be degraded completely; in most cases, activated (i.e., hydroxylated) derivatives are formed and are covalently linked to the organic soil matrix, especially to humic compounds.

Why do major sites of soil contamination persist at all, although there are microbes that are able to efficiently degrade nearly all types of contaminants? In most cases, the situation is similar to that of a major oil spill in the ocean: although a sufficient number of bacteria are present to degrade the mineral oil, this process is slow because it operates only at the oil-water interface of such a spill, and cosubstrates for degradation (O_2) and for growth (bound nitrogen or phosphorus compounds) may be limiting. This is the case especially with insufficiently water-soluble contaminants. In other cases (e.g., phenols), toxicity may prevent microbial degradation because the in situ concentrations are too high. In any case, dispersal, dissolution, and provision of cosubstrates are the key strategies to be followed to enhance the natural degradation process in soil; these have proven successful also in the cleanup of mineral-oil spills in the ocean. Seeding of the contaminated site with bacteria specalized for degradation of the respective contaminant helps only in very few exceptional cases; usually, the availability of the substrate in situ is the limiting factor for microbial degradation, and organisms provided externally are prone to the same limitations.

The **purification of wastewaters** and air, the disposal of organic wastes, and the cleanup of contaminated soil sites are all processes in which microorganisms, mainly bacteria, act as primary degraders of the contaminants. In all cases, the availability of the substrate or of required cosubstrates (e.g., oxygen, mineral salts) is the factor that limits the degradation rate. This limitation is easily overcome by technical devices for water and air treatment but represents the key obstacle in soil remediation. In many cases, oxygen-independent (anaerobic) processes have proven to be competitive alternatives to the conventional aerobic processes of treatment.

36.6 Biosensors Exploit the Specificity and Diversity of Biochemical Reactions

Biosensors are being applied increasingly to the discontinuous or continuous assay of processes in defined reactor systems or in the environment. Contrary to chemical sensor systems, which are usually based on a chemical reaction for qualitative and quantitative assay of a dissolved reactant, biosensors are based on a primarily biological or biochemical reaction that transforms a certain substrate concentration or change in concentration into a signal, which can be converted into a chemical, electrical, or physical signal that is subsequently enhanced and electronically processed. The elements of a biosensor system are depicted in Figure 36.7. A simple type of biosensor is an electrode by which the concentration of glucose is assessed in a bioreactor. In this case, the biological component is immobilized glucose oxidase, which catalyzes the reaction

$$\text{D-glucose} + O_2 \rightarrow \text{D-glucono-1,4-lactone} + H_2O_2$$

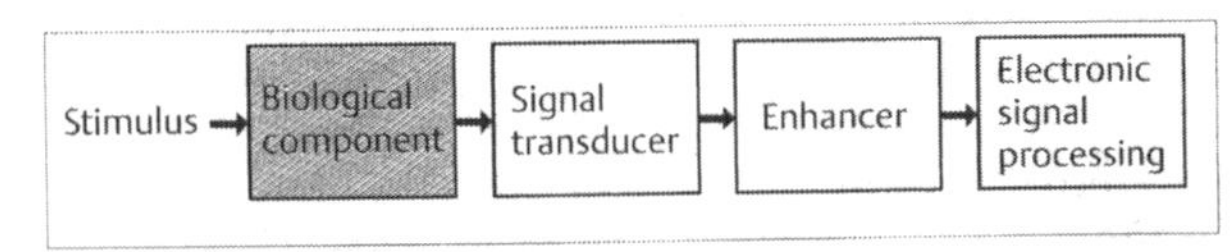

Fig. 36.7 **Operation of a biosensor.** A stimulus (e.g., concentration of a reactant) is monitored by a biological system (enzyme, bacterial cells) connected to a transducer (e.g., oxygen electrode). The transducer releases a signal, which is electronically enhanced before it reaches a processing device

Table 36.3 **Types of biosensors**

Biological component	Transducer	Detection
Immobilized enzymes	Potentiometric electrodes	NH_4^+, H^+, CO_2
	Amperometric electrodes	O_2, H_2
Cells	Field-effect transistors	H^+, NH_3
Antibodies	Photomultipliers in connection with fiber optics	Light emission, luminescence
Cell organelles	Photodiodes with light-emitting diodes	Light absorption

The lactone hydrolyzes rapidly to gluconic acid, and the signal transferred to the transducer is the oxygen concentration. Oxygen diffusion determines the rate of depolarization of an amperometric electrode, which releases an electrical signal. Instead of a single enzyme, several enzymes of a reaction chain also can be co-immobilized in a biosensor to give finally a physically or chemically assessable signal. In other cases, whole bacterial cells, antibodies, or cell organelles are immobilized in a biosensor (see Table 36.**3**). The signals can be coupled to many different types of transducers, which all transform the primary signal into an electronically detectable type of information that can be processed further. The advantage of biosensors over chemical sensing systems is the high specificity and diversity of biological reactions. Problems in the development of biosensors mainly involve the stabilization of the biological components, their integration into a mechanically stabilized matrix (e.g., gels and membranes), their coupling to electronic devices, and their protection from harmful environmental conditions (toxicity), including the possibility of biodegradation (insterility). In general, biosensors have only a limited lifespan, which ranges from days to several months. Modern types of electronic data processing and physical data transformation (e.g., optoelectronics) and recent developments with chromogenic chemical sensors as transducers have added entirely new dimensions to the application fields for biosensors. Today, biosensors are applied widely to process control of chemical and biological reactor systems and to environmental monitoring of specific metabolites, hormones, or toxins.

Biosensors can also be used to assess the freshness of fish by analysis of the ATP content via the luciferin–luciferase reaction. Other applications are in the field of medical diagnostics: continuous control of glucose concentration, the detection of drugs in trace concentrations, and the assay of drug toxicity and mutagenicity, the latter with immobilized defect mutants that upon reversion can be recognized by a chemically detectable activity.

Biosensors are used to monitor chemical substances and their time-dependent changes in concentration fluctuations in the technical industry, medicine, food technology, and environmental science. Highly specific biochemical reactions are coupled with chemical or physical signals that in the end are transformed into electronic information. The high diversity of biological reaction systems and the new developments in chemical and optical monitoring systems make this a quite promising field in applied microbiology.

36.7 Lithotrophic Bacteria Are Active in Metal Mining

The oxidation of elemental sulfur and H_2S leads to the formation of sulfuric acid, which strongly acidifies environments in which reduced sulfur compounds come into contact with atmospheric oxygen, e.g., in volcanic hot springs and fumaroles (see Chapters 31.4.1, 31.4.3). The acid thus produced increases the solubility of many metals, such as iron, copper, uranium, zinc, lead, and aluminum. The acidic effluents from rock materials rich in sulfur and metals are toxic to many life forms due to the acidity and to the high content of dissolved heavy metals. Coal contains not only sulfur linked to organic carbon but also significant amounts of metal sulfides, especially FeS_2 (pyrite), which is easily recognized by its brassy yellow crystals ("fool's gold"). Oxidation of pyrite by iron- and sulfur-oxidizing bacteria, e.g., *Thiobacillus ferrooxidans*, leads to formation of Fe^{3+} and sulfuric acid (see Chapter 31.4.3), an undesirable phenomenon in coal mines owing to the difficult disposal of the toxic and acidic wastewaters (acid mine drainage).

The microbial oxidation of metal sulfides to sulfuric acid and dissolved metal ions also can be exploited as an inexpensive method of leaching low-grade metal ores. The process has especially been developed for the recovery of copper from ores that contain only a few mass per cent copper. Copper sulfide is oxidized aerobically in two steps, according to

$$2\,Cu_2S + O_2 + 4\,H^+ \rightarrow 2\,CuS + 2\,Cu^{2+} + 2\,H_2O$$

$$2\,CuS + 4\,O_2 \rightarrow 2\,Cu^{2+} + 2\,SO_4{}^{2-} \qquad (36.1)$$

The first step, the oxidation of Cu^+ to Cu^{2+}, is a solely microbial oxidation; the second step, the oxidation of sulfide to sulfate, is partly microbial and partly chemical. Oxidation of copper sulfide also can be coupled to the reduction of Fe^{3+} to Fe^{2+}, according to

$$CuS + 8\,Fe^{3+} + 4\,H_2O \rightarrow Cu^{2+} + 8\,Fe^{2+} + SO_4{}^{2-} + 8\,H^+. \qquad (36.2)$$

Since iron is used in the further processing, the oxygen- and the iron-dependent processes take place simultaneously, and Fe^{2+} is reoxidized separately with oxygen by *Thiobacillus ferrooxidans*.

Cu^{2+} can be precipitated as copper metal upon contact with scrap iron, according to

$$Fe^0 + Cu^{2+} \rightarrow Cu^0 + Fe^{2+} \qquad (36.3)$$

This way, scrap iron can be used to recover high-quality copper from low-grade copper ores, and the acidic leaching liquor is recycled within the system whenever feasible. H_2SO_4 has to be added to the process in order to accelerate the proton-requiring first step in the above-mentioned reaction chain.

Since the biological ore-leaching technology requires an intensive oxygen supply, it is most successfully applied in dumps of low-grade ore because the piles of ore usually contain sufficient channels and crevices for aeration. Similarly, zinc and lead also can be extracted from their respective sulfidic ores, and very little sulfuric acid needs to be added to maintain a higher extraction rate in a circulating percolation system. Other metals, such as uranium, can be biologically oxidized from uranium(IV) to uranium(VI) in the presence of ferric sulfate. *Thiobacillus ferrooxidans* appears to act as an oxidizing cocatalyst, according to the equation

$$2\,UO_2 + O_2 + 2\,SO_4{}^{2-} + 4\,H^+ \rightarrow 2\,UO_2SO_4 + 2\,H_2O \qquad (36.4)$$

It is assumed that *Thiobacillus* actually oxidizes ferrous iron to ferric iron, and Fe^{3+} oxidizes UO_2 in a process similar to the leaching of pyrite (see Chapter 31.4.3).

For bioleaching of manganese, reductive conversion of the insoluble MnO_2 to water-soluble Mn^{2+} is required; this conversion can be achieved in the presence of organic waste materials. In this case, the leaching process is a reductive rather than an oxidative one. Finally, gold and silver can be enriched biologically either by oxidative removal of iron salts from pyrite and arsenopyrites that contain auriferrous ores or by reductive removal of manganese from manganese-containing silver ores. In both cases, the desired noble metal accumulates when the less precious contaminant metal oxide is removed.

Further Reading

Agate, A. D. (1996) Recent advances in microbial mining. World J Microbial Biotechnol 12: 487–495

Ahring, B. K. (1995) Methanogenesis in thermophilic biogas reactors. Antonie v Leeuwenhoek 67: 91–102

Aston, W. J. and Furner, A. E. F. (1984) Biosensors and biofuel cells. Biotechnol Gen Engineering Rev. 1: 89–120

Eikelboom, D. H. (1975) Filamentous organisms observed in activated sludge. Water Res 9: 365–388

Kortstee, G. J. J., Appeldoorn, K. J., Bonting, C. F. C., van Niel, E. W. J., and van Veen, H. W. (1994) Biology of polyphosphate-accumulating bacteria involved in enhanced biological phosphorus removal. FEMS Microbiol Rev 15: 137–153

Lettinga, G. (1995) Anaerobic digestion and wastewater treatment systems. Antonie Van Leeuwenhoek 67: 3–28

Swanell, R. P. J., Lee, K., and McDonagh, M. (1996) Field evaluation of marine oil spill bioremediation. Microbiol Rev 60: 342–365

Toerien, D. F., Gerber, A., Lötter, L. H., and Cloete, T. E. (1990) Enhanced biological phosphorus removal in activated sludge systems. In: Marshall, K. C., ed. Advances in Microbial Ecology, vol. 11. New York: Plenum Press; 173–229

Verstraete, W., de Beer, D., Pena, M., Lettinga, G., and Lens, P. (1996) Anaerobic bioprocessing of organic wastes. World J Microbiol Biotechnol 12: 221–238

Zehnder, A. J. B., Ingvorsen, K., and Marti, T. (1982) Microbiology of methane bacteria. In: Hughes, D. E., Stafford, D. A., Wheatley, B. I., Baader, W., Lettinga, G., Nyns, E. J., and Verstraete, W. (eds.) Anaerobic digestion. Amsterdam: Elsevier Biomedical; 45–68

37 Prokaryotes and Man: Chances, Promises, and Risks

Microbes have been useful to humans since the beginning. They have been in use for several thousand years in basic processes of food production, e.g., the making of bread, dairy products, and alcoholic beverages. In the second half of the nineteenth century, the pioneering work of L. Pasteur, R. Koch, and other microbiologists provided a deeper understanding of the underlying processes. In the following decades, the principles of the pure microbial culture led to significant improvements in large-scale production processes (e.g., for lactic acid, citric acid, ethanol, and glycerol) because, up to this time, the growth of undesired microorganisms was hindered solely by conditions selective for the producing organism. The fortuitous discovery of the first antibiotic, penicillin, in 1928 opened a new age in the development of microbial production processes. In the following years, a large variety of antibiotics and products synthesized by microorganisms were discovered, such as amino acids, vitamins, and antibiotics. Despite impressive progress, however, microbiology remained for many years a purely descriptive experimental science; applied microbiology had to rely basically on a "trial and error" approach rather than a logical one based on a sound knowledge of the mechanisms involved (see Table 35.**1**).

37.1 From the Past to the Future: Sequence Analysis of Total Bacterial Genomes Will Revolutionize Applied Microbiology

A new age of biotechnological processes performed by microorganisms began with the discovery of DNA as the genetic material and with the identification of the first tools for the genetic engineering of *Escherichia coli*. These tools led to important new discoveries in physiological processes and to novel strategies in the optimization of production strains. Up to this time, strain improvement had been based on mutation and selection techniques. This process was time-consuming and costly. Recombinant-DNA technologies, however, provided a rapid, efficient, and powerful means for the creation of microorganisms with novel or improved traits. Today, a large number of microorganisms can be engineered at the genetic level to function as biological "factories" for the production of proteins and other components.

Since 1976, it has been possible to determine the sequence of DNA; this provides a broad basis for the understanding of gene function and genetic engineering. DNA sequencing was at first a cumbersome process, but it has improved greatly in terms of throughput and cost. Today, there are highly automated machines and processes to carry out large-scale DNA sequencing, for example, the sequencing of total bacterial genomes in less than one year. The currently most ambitious project is the complete sequencing of the three billion base pairs of the human genome. The information obtained within the scope of these projects not only sheds new light on microbial evolution, but also constitutes a treasure trove of information for any physiological and ecological investigation to be performed on a given microbe. In particular, complex regulatory networks common to different microorganisms can be identified by comparing the sequences of well-characterized regulatory networks to the sequence of an organism under investigation. Since the initial attempt to sequence the *Haemophilus influenzae* genome, several microbial genomes have been determined (Table 37.**1**). These organisms were chosen because an important ecological, medical, or industrial impact was certain or to be expected. Due to the huge amount of DNA sequence data that will then be available, it will be possible to elucidate more rapidly the regulatory and structural functions of newly identified genes. One example is the identification of *Pseudomonas* and *Xanthomonas* genes that are involved in plant pathogenicity, due to their sequence similarity o already well-known genes of the human pathogen *Yersinia*. Based on this knowledge, further experiments can be planned in a straightforward way. Thus, it can be anticipated that, in the future, an analysis of a certain microorganism will only be commenced after its complete sequence is known.

Table 37.1 **Overview of bacterial and archaeal genome sequencing projects**

Microorganism	Size (Mbp)	Important features	Interest
Completed			
Haemophilus influenzae	1.83	Pathogen, invasive childhood infections	Medical
Helicobacter pylori	1.7	Important pathogen, peptic ulcers	Medical
Mycoplasma genitalium	0.58	Smallest genome, rapid evolution	Medical
Mycoplasma pneumoniae	0.8	Pathogen, pneumoniae	Medical
Staphylococcus aureus	2.8	Important pathogen, wound infections	Medical
Synechocystis sp.	3.6	Model organism, oxygenic photosynthesis	Scientific
Methanococcus jannaschii	1.7	Strictly anaerobic, autotrophic Archaeon, hyperthermophilic, barophilic, methanogen	Industrial, scientific
Bacillus subtilis	4.2	Model organism, industrial producer of enzymes	Industrial, scientific
Escherichia coli	4.7	Model organism, producer of fine chemicals and proteins, potential pathogen	Scientific, industrial, medical
In progress			
Aquifex thermophilus	1.6	Thermophilic, deep branching eubacterial phylum, strictly chemolithoautotrophic	Industrial, scientific
Archeoglobus fulgidus	1.7	Hyperthermophilic, barophilic, sulfate-reducing Archaeon, oldest archaeal lineage	Industrial, scientific
Borrelia burgdorferi	0.95	Pathogen, Lyme disease	Medical
Clostridium beijerinckii	6.7	Pathogen, production of organic solvents	Medical, industrial
Deinococcus radiodurans	3.6	Extreme resistance to DNA damage by radiation or chemicals	Scientific
Methanobacterium thermoautotrophicum	1.7	Marine Archaeon from sewage sludge, bioconversion of CO_2 to CH_4	Industrial, environmental
Mycobacterium leprae	2.4	Pathogen, leprosy	Medical
Mycobacterium tuberculosis	4.4	Pathogen, tuberculosis	Medical
Pyrobaculum sp.	1.8	Hyperthermophilic Archaeon, related to the common ancestor of prokaryotes and eukaryotes	Industrial, scientific
Pyrococcus furiosus	2.0	Hyperthermophilic, heterotrophic Archaeon, model organism	Industrial
Rickettsia prowazekii	1.1	Obligate cellular parasite, pathogen, typhus	Medical
Sulfolobus solfataricus	3.1	Thermophilic, sulfur-oxidizing	Industrial
Synechococcus sp.	2.7	Thermophilic marine cyanobacterium, photosynthesis, producer of 5–15% of world's oxygen	Industrial, environmental
Thermotoga maritima	2.0	Thermophilic eubacterium, remote phylum	Industrial, scientific
Treponema pallidum	1.05	Pathogen, syphilis	Medical

37.2 Future Aspects of Prokaryotes in Medicine: New Drugs, Vaccines, and Diagnostic Tools Will Be Based on Knowledge of Molecular Mechanisms

Ever since L. Pasteur and R. Koch postulated that infectious diseases are caused by microorganisms (Chapter 33), mechanisms of pathogenicity have been analyzed in detail for a number of pathogenic microorganisms by using classical microbiological and genetic research techniques. These analyses in general revealed that complex interactions and signal exchanges between pathogenic prokaryotes and animal or human cells determine the course and the outcome of a pathogen attack (see chapter 33).

The key goal for any future research on prokaryotes in medicine will be to understand the pathogenicity process in even more detail and to develop new drugs, vaccines, and diagnostic tools on the basis of this increased knowledge. Up to now, laborious microbiological and molecular-genetic research had to be performed for each pathogen in order to identify its particular mechanism of pathogenicity. This approach might be drastically simplified in the near future by the increasing number of large-scale sequencing projects (Table 37.**1**). Examples for microorganisms whose genomic sequences have been established are *Staphylococcus aureus*, which causes wound infections; *Haemophilus influenzae*, which causes invasive childhood infections; *Helicobacter pylori*, a bacterium involved in peptic ulcers; and *Methanococcus jannaschii*, the first Archaeon. In addition, the genomes of *Mycobacterium leprae* and *Mycobacterium tuberculosis*, which cause leprosy and tuberculosis, respectively, are currently being determined.

It is important to realize that an elucidation of the complete sequence of these genomes is not of pure academic interest. The sequence information alone already can give significant insights into the mechanism of pathogenicity of a particular microorganism by comparison with other microorganisms as stated above. A putative mechanism of pathogenicity for the microorganism under investigation can be proposed and then analyzed functionally in suitable model organisms in order to develop or improve medical treatments. In addition, the complete sequence can help to elucidate regulatory networks involved in pathogenicity and can thus lead to the identification of novel targets for yet-to-be-designed drugs or vaccines. This is particularly envisaged for the analysis of genomes from *Mycobacterium* species, which are responsible for severe (nearly incurable) human diseases (Table 37.**1**).

37.3 Future Aspects of Prokaryotes in Agriculture: Phytopathogenic and Beneficial Bacteria Require More Fundamental Research

Prokaryotes of agricultural relevance can be considered as either phytopathogenic or beneficial for plant growth. Beneficial bacteria are being used increasingly as biofertilizers, phytostimulators, or biocontrol agents in agriculture and forestry (see Chapter 34). Therefore, research focuses on the elucidation of the complex interaction between microbes and plants, the analysis of bacterial communities, and bacterial population genetics. The improvement of molecular methods, such as PCR-based techniques and the application of marker genes for monitoring the bacteria in the environment, will help to gain a deeper insight into the complex interaction of bacteria and plants (Fig. 37.**1** and chapter 34).

In many cases, the pathways of signal exchange that result in either a beneficial or a pathogenic interaction are poorly understood. Strain selection and improvement represent an ongoing effort in the biotechnological industry, and this includes recombinant-DNA techniques. On this basis, efforts will be undertaken to modify bacterial functions in order to improve the bacterial benefits for agriculture. On the other hand, an understanding of pathogenic interactions might lead to improved methods for the control of phytopathogenic bacteria.

37.4 Future Aspects of Prokaryotes in Environmental Processes: Basic Questions Concerning Biodiversity, Community Structures, and Culture Conditions Have not yet Been Answered

Environmental problems caused by man are considered to pose a major threat to the future of mankind. Thus, there is a need to tackle these problems with solutions compatible to the needs of the environment. Hence, the exploitation of natural processes in environmental biotechnology (e.g., the use of prokaryotic organisms for the biodegradation of organic wastes or for the bioremediation of xenobiotic compounds) is of increasing importance. Based on microbial diversity, new metabolic activities of benefit to biotechnology are expected to be detected. Up to now, only 0.1 % of the microorganisms present in terrestrial ecosystems can be cultivated, and only a small proportion of these can be assigned to a known species (see chapters 30, 31, 36).

Fig. 37.1 **Bacterial release experiment** (Agricultural Research Station Braunschweig, Germany, 1994). *Rhizobium meliloti* strains were genetically tagged by a chromosomal insertion of the luciferase gene of the firefly *Photinus pyralis*. These bacteria were released in small soil columns that were closed at the bottom; this made the collection of seepage water possible. Prior to the release, the peat-based rhizobia culture was mixed in the laboratory with soil aliquots removed from the columns. This mixture was manually released into the upper layer of the columns. In order to study the interaction with plants, the host legume (alfalfa) was grown in the columns

Independent of microbial culture methods, the polymerase-chain-reaction (PCR) technology provides an approach to assess the structure of natural microbial communities. By using specific PCR primers, taxonomically relevant ribosomal DNA sequences can be amplified, even if present in minute amounts in environmental samples, and thus made accessible to detailed analyses (see Chapter 29). A method introduced in recent years that might prove useful to describe the diversity of complex natural populations is the **ARDRA** (**a**mplified **r**ibosomal **DNA** **r**estriction **a**nalysis) technique. This technique, which is based on the generation of species-specific fingerprints of microbes, might become a tool for the fast and reliable analysis of community structures (Figure 37.**2**). Based on the DNA sequence information of rDNA, specific, fluorescently-labeled PCR primers can be designed and used for in situ hybridization purposes to unravel the spatial distribution of a target microorganism. In conjunction with the taxonomic position of a given organism, this information could prove useful to define conditions under which presently non-culturable organisms might become culturable. Thus, microbial biodiversity might be exploited in the future for the discovery and development of new products of biotechnological significance.

37.5 Future Aspects of Prokaryotes in Industrial Production: Appropriate Production Strains Will Be Genetically Engineered for an Optimal Metabolic and Process Design

The primary goal in the construction of microbial strains for industrial applications is growth on simple and inexpensive media and the production of a certain compound at the most efficient rate.

Strain construction in former times relied on random mutagenesis and subsequent screening to improve production capacities ("trial and error"). These empirical methods are neither efficient as to time and costs nor are they applicable to any given desired trait. In addition, excessive mutagenesis sometimes leads to strains that are unstable because of lethal metabolic imbalances. The advantage of this method, and hence the major reason for its popularity, is that virtually no knowledge of the metabolism and the genetics of the producing organism and of the desired product is required.

Targeted improvements concerned with the regulated overexpression of production genes in order to bypass metabolic bottlenecks are made possible by genetic engineering. For example, selected genes can be mutagenized in vitro and reintroduced into the producing strain to remove binding sites that mediate feedback inhibition of product formation. Novel metabolic reactions introduced by the insertion of foreign biosyn-

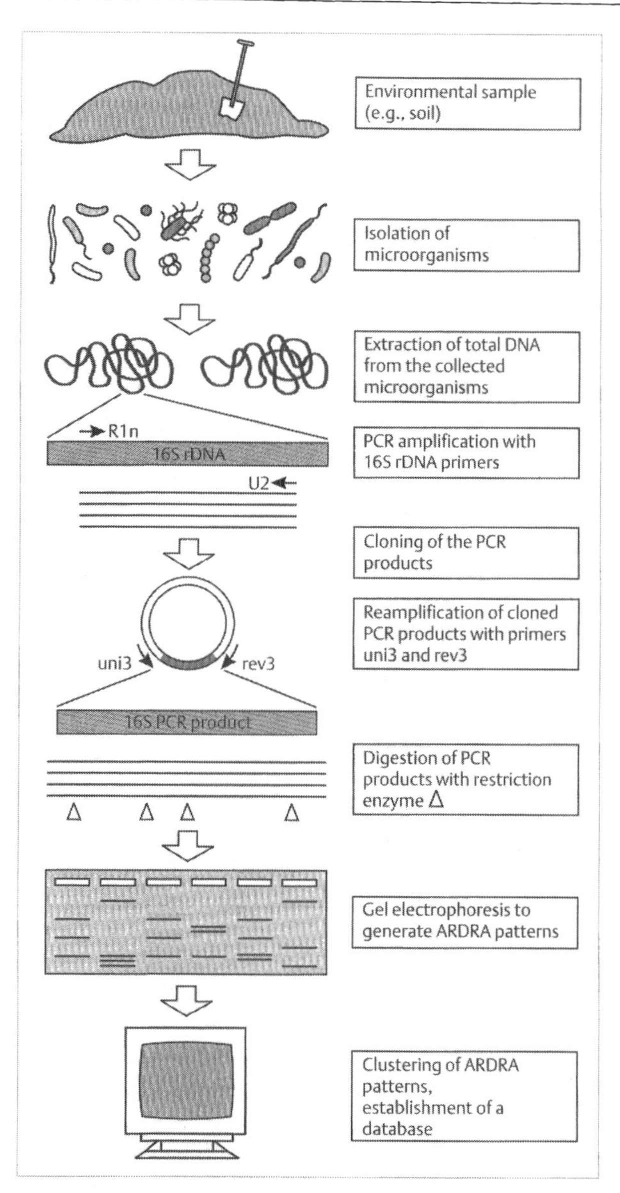

Fig. 37.2 **Flowchart of the procedure to assess bacterial communities using the ARDRA technique.** Bacterial cells are separated from environmental samples (e.g., soil) and, after concentration on filters, total DNA is extracted. By using PCR (polymerase chain reaction) primers complementary to conserved regions of bacterial 16S rDNA genes, the taxonomically relevant DNA regions are amplified and cloned. Digestion of 16S rDNA–PCR products with restriction enzymes is followed by gel electrophoresis of restriction fragments to generate species-specific DNA fingerprints of the organisms present in a sample. The computer-aided processing of ARDRA patterns yields an ARDRA database. Thus, new entries of ARDRA patterns can be identified by comparison with ARDRA patterns already stored in the database. ARDRA, amplified ribosomal DNA restriction analysis

thetic genes ("**transgenic organisms**") allow the production strain to use cheaper substrates or to produce novel products. Thereby, strains can be constructed with respect to metabolism (biochemical or metabolic engineering) or to specific demands of the production process (**process engineering**).

As already mentioned, the construction of successful production strains by genetic engineering requires a deeper knowledge of the genetics of the strain used. It is now possible to determine the genomic sequence of an industrially interesting species of Bacteria or of Archaea in a matter of months (Table 37.**1**). Among the eubacteria under investigation, *Escherichia coli* is the Gram-negative model organism for metabolic flux control and the industrial workhorse in the production of recombinant proteins and fine chemicals (e.g., vitamins and amino acids), and *Bacillus subtilis* is a Gram-positive model organism and a producer of a variety of extracellular proteins used in detergents. *Corynebacterium glutamicum*, the industrial producer of amino acids and nucleotides, is at an earlier stage of investigation (Fig. 37.**3**).

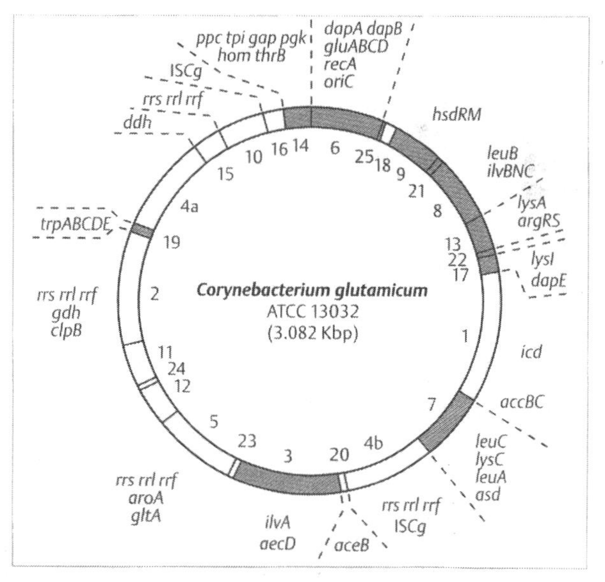

Fig. 37.3 **Physical map of the chromosome of the amino acid–producing bacterium** *Corynebacterium glutamicum*. The circular map is based on the restriction endonuclease *Swa* I. The restriction fragments are numbered according to their size. Apart from repetitive sequences that encode ribosomal RNA (*rrf, rrl, rrs*) or insertion elements (*ISCg*), mostly genes involved in amino acid biosynthesis are shown (in red)

Many of the bacterial genomes that are currently sequenced belong to species that grow under abnormal conditions, for example, at temperatures above the boiling point of water, under high pressures, or at high salinity. From these organisms, novel enzymes of extreme stability are expected to be put to use for a wide range of industrial applications. *Archaeglobus fulgidus* (Table 37.**1**), an example for organisms of this category, is a microorganism that thrives at high temperatures and high pressures associated with deep oil wells. Enzymes produced by this organism might serve to decontaminate industrial or military sites in extreme environments or might prove useful for other processes that convert wastes to commodity or fine chemicals. The information obtained within bacterial genome projects consequently will also lead to the next quantum leap in industrial microbiology.

37.6 Potential Risks and Risk Management of Genetically Engineered Bacteria: Regulations and Laws Have Been Established Worldwide

For microbiology, recombinant-DNA technology has become an indispensable tool at all levels. For applied microbiology, this technology offers a great potential for improving bacteria for use in medicine, agriculture, and industry. Despite many promising aspects, concerns about potential risks caused by the new technology were voiced right from the beginning by responsible geneticists. As a consequence, the Asilomar Conference on recombinant-DNA molecules was held in 1975 (Tab. 37.**2**). A self-imposed moratorium on specific experiments was recommended until the risks could be adequately assessed. In particular, the (accidental or deliberate) release of **genetically engineered organisms (GEOs)**, which through uncontrolled growth could become, for example, "killer bacteria", was forbidden. The capacity to thrive after their release distinguishes GEOs from other noxious agents (e.g., physical ones such as radiation, and chemical ones such as poisons). This capacity and the possible transfer and expression of genes from even distantly related species into any other

Table 37.2 **Landmarks in the historical development of regulations for recombinant-DNA technology in the USA, Japan, and Europe**

Country	Advent, directives
USA	
1972	First experiments involving recombinant DNA
1973	Gordon Research Conference on Nucleic Acids: First concerns about the risks of genetic engineering were discussed
1974	Paul Berg published a letter suggesting a moratorium on certain experiments
1975	Asilomar Conference: Self-imposed moratorium on certain experiments of recombinant-DNA technology
1976	First guidelines for recombinant-DNA research published by the National Institute of Health
Japan	
1979	"Guideline for recombinant-DNA experiments" by the Science and Technology Agency and by the Ministry of Education
1986	"Guideline for industrial application of recombinant-DNA technology" by the Ministry of International Trade and Industry and by the Ministry of Health and Welfare
1989	"Guideline for application of recombinant-DNA organisms in agriculture, forestry, fisheries, and other related industries" by the Ministry of Agriculture, Forestry, and Fisheries
1991	"Guideline for safety assessment of foods and food additives produced by recombinant-DNA techniques" by the Ministry of Health and Welfare
Europe	
1978	"Guidelines for protection against hazards by in vitro recombinant DNA" by the German Federal Government
1986	"Environment and Gene Technology Act" in Denmark
1989	"Genetic Manipulation Regulation" in the United Kingdom
1990	European Community directives concerning the contained use of genetically modified microorganisms (90/219 EEC) and concerning the deliberate release of genetically modified organisms into the environment (90/220/EEC
1990	Passing of the German "Genetic Engineering Act"
1993	First amendment to the German "Genetic Engineering Act"

organism, including man, by means of gene technology (i.e., the construction of **transgenic organisms**), were and are considered the two genuine biological risks related to gene technology.

Consequently, the Asilomar Conference report outlined physical and biological containment standards for recombinant-DNA experiments. In 1976, this report led to the definition of the first set of guidelines for recombinant-DNA research in the U.S.A. by the newly created National Institute of Health (NIH), Recombinant DNA Advisory Committee (Tab. 37.**2**). The guidelines defined a set of graded security devices that involve both physical and biological containments. These devices relied basically on decades of experience gained in handling highly pathogenic microorganisms and viruses in microbial laboratories and in isolation wards of hospitals. The guidelines have been revised repeatedly, primarily due to relaxation of restrictions no longer considered imperative, the results of experience and of control experiments.

In the following years, other countries also established reviewing systems and guidelines. In Japan, experiments involving recombinant-DNA technology are regulated by a variety of guidelines that emphasize the properties of genetically modified organisms. The Japanese guidelines closely follow OECD recommendations and NIH guidelines. The Japanese guidelines are considered to be both easy to follow and highly flexible. They have been simplified further during the past years. In Europe, by contrast, a technology-specific (rather than product-oriented) approach for the regulation of recombinant-DNA technology has been established. Guidelines and regulations have been set up in several European countries. First were the German guidelines of 1978; the first gene technology act was passed in Denmark in 1986. In 1990, the German Genetic Engineering Act followed. As a consequence of the realization of the European Union, the European Community directives of 1990 were implemented into national legislation.

The European Community and several national governments established research programs with the objective of safety assessment of recombinant-DNA technology. Due to increasing experience and to the knowledge gained about the properties of recombinant DNA within living cells, the regulatory procedures that in the beginning were rather restrictive and time-consuming, even for organisms classified as harmless, have been changed several times. The relaxation of regulations has accelerated the use of recombinant-DNA techniques. The security debate, however, which often seems to move from one hot topic to the next, currently is concentrated on two major questions: how should foodstuffs that contain GEOs or their products be handled, and how should the (deliberate) release of GEOs, especially of **genetically engineered microorganisms (GEMs)**, be controlled? The former problem appears to be similar to that caused by any new foodstuff ("Is it a strong allergen?"). The latter problem is that of cultivation, whether the foodstuff is generated by conventional breeding or by recombinant-DNA technology.

Since the beginning of recombinant-DNA technology, genetically debilitated strains have been used as a precautionary measure in biotechnological processes. The biological containment of bacterial strains was achieved by the use of auxotrophic mutants unable to grow in the absence of nutrients rarely found in natural environments or of temperature-sensitive mutants unable to grow in warm-blooded hosts. Biological containment should thus help to control GEMs that accidentally have escaped from physical containment.

Today, GEMs could perform many tasks beneficial to the environment. These include bioremediation of environmental pollutants and the use of GEMs in agriculture to improve crop yields either by inoculation of plants with N_2-fixing organisms or by the use of microorganisms displaying biocontrol ability. The deliberate release of such GEMs, however, raises questions as to their potential impact on ecosystems. In order to minimize such ecological risks, the use of genetically engineered microorganisms that would die after having fulfilled their task in the environment would be desirable. For this purpose, additional biological containment systems have been developed in recent years. These do not simply rely on genetically debilitated strains but rather on strains that carry suicide genes, which are expressed stochastically or which actively kill the GEMs under controlled conditions. While the former strategy is only applicable under certain conditions because mutations that debilitate a strain interfere in most cases with its performance in the field, a major drawback of the latter strategy is that a certain fraction of the cells inevitably escapes the control circuit by spontaneous mutation. Safety assessment experiments have proven thus far that, in principle, the biological containment of GEMs is feasible. A major challenge for the future use of GEMs in environmental biotechnology, however, is the development of strains with a predictable behavior in the field. Since the natural environment is a highly complex system with a wide range of physicochemical and biotic parameters that influence the biodiversity of bacterial populations, release of nonindigenous or genetically modified microorganisms has to take into account that bacteria cannot be recollected. Therefore, **release experiments** should fulfill two prerequisites: (1) The experiments should be performed in a step-by-step procedure, with laboratory and greenhouse experiments that allow the collection of data prior to a deliberate release. (2) The second requirement is the obligatory application of biological containment systems as outlined before.

In the public debate, **horizontal gene transfer**, especially among members of different bacterial species, is still regarded as a major risk of recombinant bacteria. There can be little doubt that horizontal gene transfer plays a large and even essential role in natural bacterial populations. It is even central to the concept of **collective bacterial genomes** (see Chapters 15 and 29), i.e., the sum of all genes available to all cells of a bacterial species. There is direct and indirect evidence from "bacterial archeology" for ample gene transfer in nature (at an evolutionary time scale). Plasmid transfer and mobilization of introduced DNA has been observed in soil, aquatic systems, biofilms, and waste-water systems. If, however, natural gene exchange is relatively high among the prokaryotes, the potential risks of gene transfer among GEMs and other bacteria should not be overemphasized because obviously the nature of the transferred genes and traits is of decisive importance, not gene transfer itself. Nevertheless, to reduce the initial rate of dissemination of a gene in the environment, it is generally agreed that, if possible, the respective gene should be inserted stably in the cellular chromosome rather than on a plasmid or a transposable element. In general, an evaluation of each case is necessary in order to minimize unnecessary risks in experiments with deliberate GEM release.

Especially in the beginning, revolutionary new technologies normally lend themselves to controversial discussions. Some people are afraid that these technologies may constitute uncontrollable dangers or may threaten traditional values and techniques. Others argue that because these methods are revolutionary and new, they are extremely promising and hence must be given any opportunity to be developed. Most geneticists and biologists who use recombinant-DNA techniques constitute a third, more neutral group for which gene technology is but a logical continuation of the previous developments in genetics as initiated by scientists such as C. Darwin, G. Mendel, and B. McClintock. They are convinced that the biological risks outlined above are not radically new and hence can be handled by using reasonable precautions. In their view, gene technology has shown its outstanding value for basic research in an amazingly short time and will also prove its practical value within reasonable expectations.

Finally, they are aware that gene technology will provoke essential ethical, legal, economic, and social questions due to its potential effects on living organisms, including man. At a closer look, however, most if not all of these will be recognized as millennium-old questions, which probably still will be asked in millenniums because perhaps they never can be answered definitively and will have to be asked by each new generation as long as there are human beings.

Further Reading

Ciba Foundation (1990) Science, law and ethics. Human geneti[c] information. Ciba Foundation Symposium 149 Chichester Wiley

Davis, B. D., ed. (1991) The genetic revolution: scientifi[c] prospects and public perceptions. Baltimore London: John Hopkins University Press

Glick, B. R., and Pasternak, J. J. (1994) Molecular biotechnology principles and applications of recombinant DNA. Washington, DC: ASM Press

Lee, T. F. (1993) Gene future: the promise and perils of the new biology. New York: Plenum Press

Old, R. W., and Primrose, S. B. (1994) Principles of gene manipulation: an introduction to genetic engineering, 5th edn. Oxford London: Blackwell

Tudge, C. (1993) The engineer in the garden. Genes and genetics: from the idea of heredity to the creation of life London: Cape

Index

Note: page numbers in *italics* refer to figures and tables

A

C

D

E

G

I

J

K

L

M

Q

S

U

V

Printed in the USA/Agawam, MA
December 30, 2011
563243.007